RIVERS
OF
NORTH AMERICA

EDITED BY

ARTHUR C. BENKE

Department of Biological Sciences
University of Alabama
Tuscaloosa, Alabama

COLBERT E. CUSHING

Streamside Programs
Estes Park, Colorado

ELSEVIER
ACADEMIC
PRESS

AMSTERDAM • BOSTON • HEIDELBERG • LONDON • NEW YORK • OXFORD
PARIS • SAN DIEGO • SAN FRANCISCO • SINGAPORE • SYDNEY • TOKYO

Associate Acquisitions Editor: Kelly Sonnack
Senior Acquisitions Editor: Nancy Maragioglio
Senior Marketing Manager: Linda Beattie
Project Manager: Kristin Macek
Interior Design: Julio Esperas
Composition: SNP Best-set Typesetter Ltd., Hong Kong
Interior Printer: Transcontinental Printing

Elsevier Academic Press
30 Corporate Drive, Suite 400, Burlington, MA 01803, USA
525 B Street, Suite 1900, San Diego, California 92101-4495, USA
84 Theobald's Road, London WC1X 8RR, UK

This book is printed on acid-free paper. ∞

Library of Congress Cataloging-in-Publication Data
Rivers of North America / edited by Arthur C. Benke, Colbert E. Cushing.
    p. cm.
    Includes bibliographical references.
    ISBN 0-12-088253-1
    1. Rivers–North America.   2. Stream ecology–North America.   I. Benke, Arthur C.
II. Cushing, C. E. (Colbert E.)
    QH102.R58 2005
    551.48′3–dc22

                                                                    2005008909

British Library Cataloguing in Publication Data
A catalogue record for this book is available from the British Library

ISBN: 0-12-088253-1

For all information on all Elsevier Academic Press Publications
visit our Web site at www.books.elsevier.com

Printed in Canada
05 06 07 08 09   10 9 8 7 6 5 4 3 2 1

Working together to grow
libraries in developing countries

www.elsevier.com | www.bookaid.org | www.sabre.org

ELSEVIER   BOOK AID
             International   Sabre Foundation

RIVERS

OF

NORTH AMERICA

14595377

fco

CONTENTS

Contributors   xiii
Foreword   xix
Preface   xxi
Acknowledgments   xxiii

1 BACKGROUND AND APPROACH

ARTHUR C. BENKE AND COLBERT E. CUSHING

INTRODUCTION   1
BASIC APPROACH   3
CHAPTER CONTENTS AND BACKGROUND   4
CONCLUDING COMMENTS   16
LITERATURE CITED   16

2 ATLANTIC COAST RIVERS OF THE NORTHEASTERN UNITED STATES

JOHN K. JACKSON, ALEXANDER D. HURYN, DAVID L. STRAYER, DAVID L. COURTEMANCH, AND BERNARD W. SWEENEY

INTRODUCTION   21
PENOBSCOT RIVER   25
CONNECTICUT RIVER   30
HUDSON RIVER   35
DELAWARE RIVER   44
SUSQUEHANNA RIVER   49
ADDITIONAL RIVERS   54
ACKNOWLEDGMENTS   57
LITERATURE CITED   57

3 ATLANTIC COAST RIVERS OF THE SOUTHEASTERN UNITED STATES

LEONARD A. SMOCK, ANNE B. WRIGHT, AND ARTHUR C. BENKE

INTRODUCTION   73
JAMES RIVER   77
CAPE FEAR RIVER   83
SAVANNAH RIVER   88

Contents

OGEECHEE RIVER   93

ST. JOHNS RIVER   99

ADDITIONAL RIVERS   104

ACKNOWLEDGMENTS   108

LITERATURE CITED   108

4 GULF COAST RIVERS OF THE SOUTHEASTERN UNITED STATES

G. MILTON WARD, PHILLIP M. HARRIS, AND AMELIA K. WARD

INTRODUCTION   125

MOBILE RIVER   129

CAHABA RIVER   135

APALACHICOLA–CHATTAHOOCHEE–FLINT RIVER SYSTEM   140

PEARL RIVER   147

SUWANNEE RIVER   152

ADDITIONAL RIVERS   157

ACKNOWLEDGMENTS   161

LITERATURE CITED   162

5 GULF COAST RIVERS OF THE SOUTHWESTERN UNITED STATES

CLIFFORD N. DAHM, ROBERT J. EDWARDS, AND FRANCES P. GELWICK

INTRODUCTION   181

RIO GRANDE   186

SAN ANTONIO AND GUADALUPE RIVERS   192

COLORADO RIVER   198

BRAZOS RIVER   203

SABINE RIVER   208

ADDITIONAL RIVERS   213

ACKNOWLEDGMENTS   215

LITERATURE CITED   215

6 LOWER MISSISSIPPI RIVER AND ITS TRIBUTARIES

ARTHUR V. BROWN, KRISTINE B. BROWN, DONALD C. JACKSON, AND W. KEVIN PIERSON

INTRODUCTION   231

LOWER MISSISSIPPI RIVER   237

WHITE RIVER   246

BUFFALO NATIONAL RIVER   251

BIG BLACK RIVER   256

YAZOO RIVER 259

ADDITIONAL RIVERS 264

ACKNOWLEDGMENTS 266

LITERATURE CITED 266

7 SOUTHERN PLAINS RIVERS

WILLIAM J. MATTHEWS, CARYN C. VAUGHN, KEITH B. GIDO, AND EDIE MARSH-MATTHEWS

INTRODUCTION 283

ARKANSAS RIVER 290

CANADIAN RIVER 294

RED RIVER 299

LITTLE RIVER 304

ADDITIONAL RIVERS 308

ACKNOWLEDGMENTS 310

LITERATURE CITED 311

8 UPPER MISSISSIPPI RIVER BASIN

MICHAEL D. DELONG

INTRODUCTION 327

UPPER MISSISSIPPI RIVER MAIN STEM 331

MINNESOTA RIVER 338

ST. CROIX RIVER 343

WISCONSIN RIVER 347

ILLINOIS RIVER 352

ADDITIONAL RIVERS 358

ACKNOWLEDGMENTS 360

LITERATURE CITED 361

9 OHIO RIVER BASIN

DAVID WHITE, KARLA JOHNSTON, AND MICHAEL MILLER

INTRODUCTION 375

OHIO RIVER MAIN STEM 378

TENNESSEE RIVER 384

CUMBERLAND RIVER 390

WABASH RIVER 396

KANAWHA RIVER 401

ADDITIONAL RIVERS 405

ACKNOWLEDGMENTS 409

LITERATURE CITED 409

10 MISSOURI RIVER BASIN

DAVID L. GALAT, CHARLES R. BERRY JR., EDWARD J. PETERS, AND ROBERT G. WHITE

INTRODUCTION 427

MISSOURI RIVER MAIN STEM 431

YELLOWSTONE RIVER 440

WHITE RIVER 445

PLATTE RIVER 449

GASCONADE RIVER 454

ADDITIONAL RIVERS 459

ACKNOWLEDGMENTS 464

LITERATURE CITED 464

11 COLORADO RIVER BASIN

DEAN W. BLINN AND N. LEROY POFF

INTRODUCTION 483

COLORADO RIVER MAIN STEM 488

GREEN RIVER 496

YAMPA RIVER 502

LITTLE COLORADO RIVER 506

GILA RIVER 511

ADDITIONAL RIVERS 517

ACKNOWLEDGMENTS 521

LITERATURE CITED 522

12 PACIFIC COAST RIVERS OF THE COTERMINOUS UNITED STATES

JAMES L. CARTER AND VINCENT H. RESH

INTRODUCTION 541

SACRAMENTO RIVER 547

SAN JOAQUIN RIVER 552

SALINAS RIVER 558

KLAMATH RIVER 563

ROGUE RIVER 568

ADDITIONAL RIVERS 573

ACKNOWLEDGMENTS 576

LITERATURE CITED 577

13 COLUMBIA RIVER BASIN

JACK A. STANFORD, F. RICHARD HAUER, STANLEY V. GREGORY, AND ERIC B. SNYDER

INTRODUCTION 591

Contents

FLATHEAD RIVER 597

SNAKE/SALMON RIVER 603

YAKIMA RIVER 610

WILLAMETTE RIVER 615

COLUMBIA RIVER MAIN STEM 622

ADDITIONAL RIVERS 632

ACKNOWLEDGMENTS 635

LITERATURE CITED 635

14 GREAT BASIN RIVERS
DENNIS K. SHIOZAWA AND RUSSELL B. RADER

INTRODUCTION 655

BEAR RIVER 661

SEVIER RIVER 667

HUMBOLDT RIVER 673

TRUCKEE RIVER 678

ADDITIONAL RIVERS 683

ACKNOWLEDGMENTS 684

LITERATURE CITED 685

15 FRASER RIVER BASIN
TREFOR B. REYNOLDSON, JOSEPH CULP, RICK LOWELL, AND JOHN S. RICHARDSON

INTRODUCTION 697

FRASER RIVER MAIN STEM 701

THOMPSON RIVER 707

NECHAKO RIVER 712

STUART RIVER 717

ADDITIONAL RIVERS 720

ACKNOWLEDGMENTS 721

LITERATURE CITED 721

16 PACIFIC COAST RIVERS OF CANADA AND ALASKA
JOHN S. RICHARDSON AND ALEXANDER M. MILNER

INTRODUCTION 735

KUSKOKWIM RIVER 741

SUSITNA RIVER 743

KENAI RIVER 746

STIKINE RIVER 751

Contents

SKEENA RIVER   755
ADDITIONAL RIVERS   758
ACKNOWLEDGMENTS   761
LITERATURE CITED   761

17   YUKON RIVER BASIN

ROBERT C. BAILEY

INTRODUCTION   775
YUKON RIVER MAIN STEM   780
TANANA RIVER   786
KOYUKUK RIVER   789
ADDITIONAL RIVERS   791
ACKNOWLEDGMENTS   793
LITERATURE CITED   793

18   MACKENZIE RIVER BASIN

JOSEPH M. CULP, TERRY D. PROWSE, AND ERIC A. LUIKER

INTRODUCTION   805
MACKENZIE RIVER MAIN STEM   809
LIARD RIVER   815
SLAVE RIVER   819
PEACE RIVER   824
ATHABASCA RIVER   829
ADDITIONAL RIVERS   834
ACKNOWLEDGMENTS   837
LITERATURE CITED   837

19   NELSON AND CHURCHILL RIVER BASINS

DAVID M. ROSENBERG, PATRICIA A. CHAMBERS, JOSEPH M. CULP, WILLIAM G. FRANZIN, PATRICK A. NELSON, ALEX G. SALKI, MICHAEL P. STAINTON, R. A. BODALY, AND ROBERT W. NEWBURY

INTRODUCTION   853
SASKATCHEWAN RIVER   861
RED RIVER OF THE NORTH–ASSINIBOINE RIVER   867
WINNIPEG RIVER   873
NELSON RIVER MAIN STEM   879
ADDITIONAL RIVERS   888
ACKNOWLEDGMENTS   888
LITERATURE CITED   888

Contents

20 RIVERS OF ARCTIC NORTH AMERICA

ALEXANDER M. MILNER, MARK W. OSWOOD, AND KELLY R. MUNKITTRICK

INTRODUCTION   903
NOATAK RIVER   908
KUPARUK RIVER   913
SAGAVANIRKTOK RIVER   918
MOOSE RIVER   921
ADDITIONAL RIVERS   926
ACKNOWLEDGMENTS   927
LITERATURE CITED   927

21 ATLANTIC COAST RIVERS OF CANADA

RICHARD A. CUNJAK AND ROBERT W. NEWBURY

INTRODUCTION   939
EXPLOITS RIVER   943
MIRAMICHI RIVER   947
ST. JOHN RIVER   953
MOISIE RIVER   958
ADDITIONAL RIVERS   963
ACKNOWLEDGMENTS   965
LITERATURE CITED   965

22 ST. LAWRENCE RIVER BASIN

JAMES H. THORP, GARY A. LAMBERTI, AND ANDREW F. CASPER

INTRODUCTION   983
ST. LAWRENCE RIVER MAIN STEM   986
OTTAWA RIVER   996
SAGUENAY RIVER   1001
ST. JOSEPH RIVER   1005
ADDITIONAL RIVERS   1011
ACKNOWLEDGMENTS   1013
LITERATURE CITED   1014

23 RIVERS OF MEXICO

PAUL F. HUDSON, DEAN A. HENDRICKSON, ARTHUR C. BENKE, ALEJANDRO VARELA-ROMERO, ROCIO RODILES-HERNÁNDEZ, AND WENDELL L. MINCKLEY

INTRODUCTION   1031
RÍO PÁNUCO   1037
RÍOS USUMACINTA–GRIJALVA   1043
RÍO CANDELARIA (YUCATÁN)   1050

RÍO YAQUI   1054

RÍO CONCHOS   1058

ADDITIONAL RIVERS   1065

ACKNOWLEDGMENTS   1067

LITERATURE CITED   1067

24   OVERVIEW AND PROSPECTS

J. DAVID ALLAN AND ARTHUR C. BENKE

INTRODUCTION   1087

THE VARIETY OF RIVERS   1087

FEW RIVERS ARE PRISTINE   1093

NORTH AMERICA'S RIVERS IN THE TWENTY-FIRST CENTURY   1097

LITERATURE CITED   1101

Appendix: Common and Scientific Names for Plants, Vertebrates,
  and Selected Invertebrates   1105
Glossary   1135
Index of Rivers   1139

CONTRIBUTORS

J. David Allan
School of Natural Resources
University of Michigan
Ann Arbor, Michigan
Overview and Prospects

Robert C. Bailey
Department of Zoology
University of Western Ontario
London, Ontario
Canada
Yukon River Basin

Arthur C. Benke
Aquatic Biology Program
Department of Biological Sciences
University of Alabama
Tuscaloosa, Alabama
*Background and Approach; Atlantic Coast
  Rivers of the Southeastern United States; Rivers
  of Mexico; Overview and Prospects*

Charles R. Berry, Jr.
South Dakota Cooperative Fish and Wildlife
  Research Unit
South Dakota State University
Brookings, South Dakota
Missouri River Basin

Dean W. Blinn
Department of Biological Sciences
Northern Arizona University
Flagstaff, Arizona
Colorado River Basin

R.A. Bodaly
Freshwater Institute
Winnipeg, Manitoba
Canada
Nelson and Churchill River Basins

Arthur V. Brown
Department of Biological Sciences
University of Arkansas
Fayetteville, Arkansas
*Lower Mississippi River and Its
  Tributaries*

Kristine B. Brown
Department of Biological Sciences
University of Arkansas
Fayetteville, Arkansas
*Lower Mississippi River and Its
  Tributaries*

James L. Carter
U.S. Geological Survey
Menlo Park, California
*Pacific Coast Rivers of the Coterminous
  United States*

Andrew F. Casper
GIROQ, Departement de Biologie
Universite Laval
Ste-Foy, Quebec
Canada
St. Lawrence River Basin

Patricia A. Chambers
National Water Research Institute
Environment Canada
Burlington, Ontario
Canada
Nelson and Churchill River Basins

David L. Courtemanch
Maine Department of Environmental
  Protection
Augusta, Maine
*Atlantic Coast Rivers of the Northeastern
  United States*

Joseph M. Culp
National Water Research Institute
Environment Canada
Department of Biology
University of New Brunswick
Fredericton, New Brunswick
Canada
*Fraser River Basin; Mackenzie River Basin;
    Nelson and Churchill River Basins*

Richard A. Cunjak
Department of Biology
University of New Brunswick
Fredericton, New Brunswick
Canada
Atlantic Coast Rivers of Canada

Colbert E. Cushing
Streamside Programs
Estes Park, Colorado
Background and Approach

Clifford N. Dahm
Department of Biology
University of New Mexico
Albuquerque, New Mexico
*Gulf Coast Rivers of the Southwestern
    United States*

Michael D. Delong
Large River Studies Center
Biology Department
Winona State University
Winona, Minnesota
Upper Mississippi River Basin

Robert J. Edwards
Department of Biology
University of Texas-Pan American
Edinburg, Texas
*Gulf Coast Rivers of the Southwestern
    United States*

William G. Franzin
Freshwater Institute
Winnipeg, Manitoba
Canada
Nelson and Churchill River Basins

David L. Galat
Missouri Cooperative Fish & Wildlife Unit
University of Missouri
Columbia, Missouri
Missouri River Basin

Frances P. Gelwick
Ecology and Fisheries Management
Texas A & M University
College Station, Texas
*Gulf Coast Rivers of the Southwestern
    United States*

Keith B. Gido
Division of Biology
Kansas State University
Manhattan, Kansas
Southern Plains Rivers

Stanley V. Gregory
Department of Fisheries and Wildlife
Oregon State University
Corvallis, Oregon
Columbia River Basin

Phillip M. Harris
Aquatic Biology Program
Department of Biological Sciences
University of Alabama
Tuscaloosa, Alabama
*Gulf Coast Rivers of the Southeastern
    United States*

F. Richard Hauer
Flathead Lake Biological Station
University of Montana
Polson, Montana
Columbia River Basin

Dean A. Hendrickson
Texas Memorial Museum
University of Texas
Austin, Texas
Rivers of Mexico

Paul F. Hudson
Department of Geography and the
    Environment
University of Texas
Austin, Texas
Rivers of Mexico

Alexander D. Huryn
Aquatic Biology Program
Department of Biological Sciences
University of Alabama
Tuscaloosa, Alabama
*Atlantic Coast Rivers of the Northeastern
    United States*

Donald C. Jackson
Department of Wildlife & Fisheries
Mississippi State University
Mississippi State, Mississippi
Lower Mississippi River and Its Tributaries

John K. Jackson
Stroud Water Research Center
Avondale, Pennsylvania
*Atlantic Coast Rivers of the Northeastern
    United States*

Karla Johnston
Hancock Biological Station
Murray, Kentucky
Ohio River Basin

Gary A. Lamberti
Department of Biological Sciences
University of Notre Dame
Notre Dame, Indiana
St. Lawrence River Basin

Rick Lowell
National Water Research Institute
National Environmental Effects Monitoring
    Program
Environment Canada
Saskatoon, Saskatchewan
Canada
Fraser River Basin

Eric A. Luiker
National Water Research Institute
Environment Canada
Department of Biology
University of New Brunswick
Fredericton, New Brunswick
Canada
Mackenzie River Basin

Edie Marsh-Matthews
Sam Noble Oklahoma Museum of
    Natural History
University of Oklahoma
Norman, Oklahoma
Southern Plains Rivers

William J. Matthews
Department of Zoology
University of Oklahoma
Norman, Oklahoma
Southern Plains Rivers

Michael Miller
Department of Biological Sciences
University of Cincinnati
Cincinnati, Ohio
Ohio River Basin

Alexander M. Milner
School of Geography
University of Birmingham
Edgbaston, Birmingham
United Kingdom
*Pacific Coast Rivers of Canada and Alaska;
    Rivers of Arctic North America*

Wendell L. Minckley
Department of Biology
Arizona State University
Tempe, Arizona
Rivers of Mexico

Kelly R. Munkittrick
Department of Biology
University of New Brunswick
Saint John, New Brunswick
Canada
Rivers of Arctic North America

Patrick A. Nelson
Freshwater Institute
Winnipeg, Manitoba
Canada
Nelson and Churchill River Basins

Robert W. Newbury
Newbury Hydraulics, Inc.
Okanagan Centre, British Columbia
Canada
*Nelson and Churchill River Basins;
    Atlantic Coast Rivers of Canada*

Mark W. Oswood
University of Alaska Fairbanks
Wenatchee, Washington
Rivers of Arctic North America

Edward J. Peters
Department of Fish & Wildlife
University of Nebraska
Lincoln, Nebraska
Missouri River Basin

W. Kevin Pierson
Ross and Associates Environmental
    Consulting. LTD
Seattle, Washington
Lower Mississippi River and Its Tributaries

N. Leroy Poff
Department of Biology
Colorado State University
Fort Collins, Colorado
Colorado River Basin

Terry D. Prowse
National Water Research Institute
Environment Canada
Department of Geography
University of Victoria
Victoria, British Columbia
Canada
Mackenzie River Basin

Russell B. Rader
Department of Zoology
Brigham Young University
Provo, Utah
Great Basin Rivers

Vincent H. Resh
Department of Environmental Science,
    Policy, and Management
University of California-Berkeley
Berkeley, California
*Pacific Coast Rivers of the Coterminous
    United States*

Trefor B. Reynoldson
Acadia Center for Estuarine Research
Acadia University
Wolfville, Nova Scotia
Canada
Fraser River Basin

John S. Richardson
Department of Forest Sciences
University of British Columbia
Vancouver, British Columbia
Canada
*Fraser River Basin; Pacific Coast Rivers of
    Canada and Alaska*

Rocío Rodiles-Hernández
Departamento de Ecología y Sistematíca Acuática
ECOSUR
San Cristobal de Las Casas, Chiapas
Mexico
Rivers of Mexico

David M. Rosenberg
Freshwater Institute
Winnipeg, Manitoba
Canada
Nelson and Churchill River Basins

Alex G. Salki
Freshwater Institute
Winnipeg, Manitoba
Canada
Nelson and Churchill River Basins

Dennis K. Shiozawa
Department of Zoology
Brigham Young University
Provo, Utah
Great Basin Rivers

Leonard A. Smock
Department of Biology
Virginia Commonwealth University
Richmond, Virginia
*Atlantic Coast Rivers of the
    Southeastern United States*

Eric B. Snyder
Department of Biology
Grand Valley State University
Allendale, Michigan
Columbia River Basin

Michael P. Stainton
Freshwater Institute
Winnipeg, Manitoba
Canada
Nelson and Churchill River Basins

Jack A. Stanford
Flathead Lake Biological Station
University of Montana
Polson, Montana
Columbia River Basin

David L. Strayer
Institute of Ecosystem Studies
Cary Arboretum
Millbrook, New York
*Atlantic Coast Rivers of the Northeastern
    United States*

Bernard W. Sweeney
Stroud Water Research Center
Avondale, Pennsylvania
*Atlantic Coast Rivers of the Northeastern
    United States*

James H. Thorp
Kansas Biological Survey
University of Kansas
Lawrence, Kansas
St. Lawrence River Basin

Alejandro Varela-Romero
Departamento de Investigaciones Científicas
    y Tecnológicas
Universidad de Sonora
Hermosillo, Sonora
Mexico
Rivers of Mexico

Caryn C. Vaughn
Oklahoma Biological Survey
Norman, Oklahoma
Southern Plains Rivers

Amelia K. Ward
Aquatic Biology Program
Department of Biological Sciences
University of Alabama
Tuscaloosa, Alabama
*Gulf Coast Rivers of the Southeastern
    United States*

G. Milton Ward
Aquatic Biology Program
Department of Biological Sciences
University of Alabama
Tuscaloosa, Alabama
*Gulf Coast Rivers of the Southeastern
    United States*

David White
Hancock Biological Station
Murray, Kentucky
Ohio River Basin

Robert G. White
Montana Cooperative Fishery Research Unit
Montana State University
Bozeman, Montana
Missouri River Basin

Anne B. Wright
Department of Biology
Virginia Commonwealth University
Richmond, Virginia
*Atlantic Coast Rivers of the Southeastern
    United States*

FOREWORD

As its primary source of fresh water, North American rivers are the continent's most important natural resource. They are the nerve system of its ecology and define the geology, biology, culture, and civilization of its watersheds. It's no accident that the earliest civilizations evolved beside major river systems, such as the Tigris and Euphrates and the Nile. Rivers irrigated their crops and regularly replenished their fields. They provided food for their tables and water for drinking, bathing, and waste removal. They were an efficient means of transportation, communication, and trade. They were places for baptism, playgrounds for recreation, and sources for artistic inspiration. Later, as humans learned to harness the power of rivers, they became the foundation of the industrial revolution and the staggering economic development that came in its wake.

Yet, despite our dependence on rivers for so many aspects of our lives, we have diverted, drained, dammed, and degraded them to the point where we have killed, in some instances, every living thing in their waters. Meanwhile, we are learning more and more about how important rivers really are. For example, healthy rivers support an enormous diversity of plant and animal life; this intricate web, scientists have discovered, forms more than an interdependent food chain. It also enables the river to process nutrients and pollutants and in doing so, restore itself and combat, within limits, the mess we have all too often made of it.

Now, more than ever, rivers need our help. With waters that flow in a continuum from the source to the sea, rivers are by their very nature a shared resource, a public asset that cannot readily be reduced to private property. The flip side of that, unfortunately, is that rivers have suffered from the "tragedy of the commons"—a resource that is owned by everyone, cared for by no one, and a tempting target for polluters who can profit by privatizing the commons. The Waterkeeper Alliance, which I head, was established to counteract this occurrence. Our primary goal is to motivate millions of people across thousands of watersheds to take ownership of their streams and rivers and to defend them from those determined to steal them from the public and exploit them for private gain. The return of the commons to their rightful owners is a perpetual battle; our success often depends on empowering our volunteers with good information about their common property.

Until now, there has never been a comprehensive effort to detail the state of America's rivers. Arthur Benke and Colbert Cushing have made that effort. *Rivers of North America* is the first complete reference book on the subject—a detailed encyclopedia of information about the physical, biological, hydrological, and ecological characteristics of more than 200 rivers from southern Mexico to the Arctic. This book is much more than a compendium of facts and figures. The authors of each chapter have published extensively on river science, are leading figures in the region about which they write, and are well-known experts in the field. Together these contributors have woven a tapestry of information that is as accessible to lay people as it is to scientists, and that captures the wonderful diversity of our continent's rivers in photographs, color topographic maps, and data summaries.

Rivers of North America is a useful starting point for understanding both the overall pressures on our rivers and the particular attributes of North America's great waterways. In its pages you will find the few remaining pristine rivers that deserve conservation as benchmark systems. And you will also find those that have major problems and for which radical and immediate CPR is required. This volume is a must for river scientists, conservationists, paddlers, river recreationists, sports fishermen—for anyone and everyone concerned about the future of the continent's most valuable natural resource. It should be the standard reference on North American rivers for a long time to come. And for those, such as the Waterkeepers, who work to defend our waters, *Rivers of North America* gives us a new and critical weapon. Our defense is now stronger because we know better what we are defending.

Robert F. Kennedy Jr.
President, Waterkeeper Alliance

PREFACE

For many river scientists, nature-lovers, canoeists, and photographers, rivers are the most fascinating natural features on earth. As the primary source of fresh water, however, they also happen to be the earth's most *important* natural resource—nothing less than our long-term survival depends on them. Thus, we hope this book will not only be informative, but also will be used to promote the stewardship of these vital resources.

Rivers of North America is first and foremost a reference volume for the researcher and river enthusiast alike and is devoted to presenting a compendium of comparable physical, biological, and ecological information about many rivers. Its genesis can be traced to the suggestions of Donna James of Academic Press in early discussions with Art Benke. Although her original vision of a book covering North American rivers was slanted more toward a pictorial or coffee-table book, subsequent deliberations pushed it toward the detailed reference volume that you now hold.

This book was written by scientists, not only for other scientists, but also for students, river conservationists, and lay persons. As a reference volume, this book contains an enormous amount of information that cannot be found in any other single source about many of the major rivers in North America, the result of considerable research and synthesis by the chapter authors. The book can be used by those seeking specific information about rivers of a particular region, those wishing to compare rivers among regions, or those who are simply curious about rivers. As a reference book, it contains information about river animals and plants, ecology, hydrology, geology, geography, river management, river conservation, and the human history of river basins. This book also describes the severe degradation of many rivers that have been heavily exploited for far too long without regard for the natural benefits they bring to mankind—their ecological services, their biodiversity, their incredible beauty, and their opportunities for recreation and fishing. Few people besides river biologists realize, for example, that many unaltered river systems can support several hundred species of invertebrate animals and plants. Few realize that many contain more than 100 species of fishes, even though fishes have been an essential food source for humans in North America for millennia.

We have attempted to make this book as user-friendly as possible, particularly for the non-scientist. For example, there is a color topographic map for every river and color photographs of most rivers that will give a feel for these natural systems that words and statistics alone cannot convey. We also have used common names rather than scientific names for all species for which common names exist (plants, vertebrate animals, and some invertebrate animals). For those who wish to verify the identity of species names, an appendix of common and scientific names is provided. Chapter 1 presents considerable background information for various scientific concepts and terms used in every chapter. A glossary of common scientific terms is also provided for easy reference.

A unique feature of this book is its basin maps. They were created by the Cartography Laboratory in the Department of Geography at the University of Alabama using the United States Geological Survey's web-based National Atlas for the United States, GeoGratis (Natural Resources of Canada) for Canada, and the Hydrolk Elevation Derivative Database for Mexico. Black and white regional maps of river basins and color topographic maps for each river are provided for each chapter. The color maps are simplified to include only basin boundaries, topographic features, major tributaries, major dams, major cities, and delineation of physiographic provinces. Only dams on main-stem rivers and only major dams on tributaries are shown; otherwise important features of many basins would be obscured by a mass of yellow dots (dams). Similarly, only selected cities are shown as reference points.

Almost 190 color photographs illustrating the diversity of North American rivers are included in the book, making it reminiscent of the coffee-table book originally envisioned by the publishers. In many cases, the authors, their colleagues, or the editors provided the photographs. However, the noted author and photographer Tim Palmer contributed more than one-quarter of the photographs in this book. Tim has written eloquently about the flowing waters of the United States and the importance of their conservation. His words and photographs from *America by Rivers*; *Lifelines: The Case for River Conservation, Endangered Rivers and the Conservation Movement*; *The Wild and Scenic Rivers of America*; *The Snake River: Window to the West*, and several others, have inspired river conservationists and scientists alike. His contributions to *Rivers of North America* have not only contributed to the total number of photographs and added to the attractiveness of the book, but they have filled some very important gaps.

A major dilemma we faced in planning this book was how to include a large number of rivers in a single volume and still be reasonably detailed in describing individual rivers. To include a large number of rivers, we decided that all rivers would have a one-page summary of major physical and biological features, including the individual color maps. We borrowed the one-page concept from the excellent compendium of rivers in *Watersheds of the World: Ecological Value and Vulnerability* by Carmen Revenga and others, but we used a substantially different format. In addition to the one-page summaries, up to five rivers per chapter have received more detailed coverage in the text. One-page summaries were placed at the end of each chapter so that comparisons among rivers can be made more easily. This format means that while reading the text, the reader will need to refer to maps and some graphs in the one-page summary at the end of the chapter.

The authors of each chapter have conducted research on many of the rivers in the regions about which they write. Most of them are river ecolo-gists/biologists with a natural orientation toward freshwater fauna and flora. Such scientists usually use interdisciplinary approaches, viewing rivers as complex ecosystems. They thus recognize the essential interactions of the river's biota with the physical environment that includes hydrology, geology, and water chemistry. They also have an acute appreciation of the importance of the conservation of natural ecosystems. Thus, the focus of this book is to describe rivers as natural ecosystems, rather than treating them as resources for exploitation.

After reading about many rivers in this book, one will realize that humans have already exploited many of the rivers in North America to the point where they are seriously degraded as natural ecosystems. One might therefore wonder if it is too late to do anything about the massive construction projects and pollution that have devastated rivers for decades, such as on the Colorado, Columbia, Rio Grande, Missouri, Nelson, and many others. As the editors of this book, we believe that it is not too late. Ecologists and other scientists have shown that rivers can be rehabilitated, but only if governments have the political will, recognize the economic benefits of free-flowing clean rivers, and have an appreciation of their natural features. Scientists and conservation groups (American Rivers, Waterkeeper Alliance, and many others) must educate the public, and the public must apply pressure to politicians so natural rivers will be protected. We sincerely hope that this book will serve as a new source of knowledge for river conservation, and that it will encourage the scientist, conservationist, river manager, lay person, and politician to want to learn more about their unique characteristics.

Arthur C. Benke and Colbert E. Cushing

ACKNOWLEDGMENTS

First and foremost, we want to acknowledge the contributions of chapter authors. Clearly, this book would not have been possible without their first-hand knowledge of these complex river ecosystems. We sincerely thank them for their considerable efforts and patience with the editors in adhering to the strict format that makes the information between chapters comparable.

We are especially grateful to Kelly Sonnack of Academic Press/Elsevier, who was associated with the coordination of book preparation for over two years and played a major role in seeing the book through several phases. We gratefully acknowledge Donna James for her proposal of a book on rivers of North America. Special thanks also go to Chuck Crumly, formerly of Academic Press/Elsevier, for twisting Art Benke's arm to do the book in the first place, and guiding us through the first three years of planning and execution. Thanks also to Keith Roberts, the production editor, particularly for his attention to details in such a complex and atypical format.

We especially want to thank those associated with the creation of the color topographic maps of all river basins used throughout the book. Craig Remington, Director of the University of Alabama Cartography Laboratory (Department of Geography), supervised the maps project. Angela Brink, a geography graduate student, spent many hours creating the maps, and Jennifer White, another graduate student, was instrumental in making many subsequent revisions after Angela graduated.

We thank the many individuals who contributed photographs that greatly enhanced the beauty and appeal of this volume. We are especially grateful to the professional photographer Tim Palmer for his generosity in sharing so many beautiful photos from his collection. We also want to thank Beth Maynor Young, another professional photographer, who contributed several striking photographs for Chapter 4. Thanks also to the many other individuals (besides chapter authors and editors), agencies, and organizations that contributed one or more photographs: E. Benenati, D. Bicknell, C. Bourgeois, J. Boynton, C. S. Brehme, C. Burton, T. Carter, M. Chapman, D. J. Cooper, D. D. Dauble, W. E. Doolittle, K. Ekelund, J. M. Farrell, S. Fend, M. Gautreau, W. Gibbons, Great Valley Museum, G. Harris, T. Harris, D. M. Merritt, NASA, North Dakota Game and Fish Department, M. O'Malley, B. Oswald, R. Overman, S. Parker, G. Pitchford, Raritan Riverkeeper, J. Rathert, M. Rautio, J. Robinson, K. Schiefer, D. Schnoebelen, J. Schwindt, A. H. Siemens, M. C. T. Smith, C. Spence, Tourism Saskatchewan, Virginia Institute of Marine Science, S. Wade, C. White, A. P. Wiens, K. Wilhelm, and G. Winkler.

Art Benke gratefully acknowledges the support of the University of Alabama and its Department of Biological Sciences (Martha Powell, chairperson) for providing the time, library services, and other support to work on this book for an extended period of time. Both editors also wish to thank the many colleagues, chapter authors, and students who provided encouragement and feedback over the past few years.

Most importantly, we wish to express gratitude and appreciation to our wives, Susan Benke and Jackie Cushing, for their patience, encouragement, sacrifices, proofreading, and support. They had to endure our many hours away from home or stuck in front of our computers over the more than four-year span that was necessary for completing this volume.

1

BACKGROUND AND APPROACH

ARTHUR C. BENKE   COLBERT E. CUSHING

INTRODUCTION

BASIC APPROACH

CHAPTER CONTENTS AND BACKGROUND

CONCLUDING COMMENTS

LITERATURE CITED

INTRODUCTION

Rivers are one of the most dramatic features of a continent. They are the inevitable result of precipitation falling across the land, coalescing into streams, and uniting into ever larger streams and rivers. Over millions of years, these networks of flowing waters have delivered sediments and nutrients to downstream areas, sometimes eroding valleys and at other times depositing sediments, before eventually reaching the sea or an inland lake. This movement of water and material has helped shape terrain, created a diversity of freshwater environments, and allowed the evolution of thousand of species of plants, animals, and microbes. Together, these flowing water environments, with their uniquely adapted species, form the river ecosystems that we see today.

The North American continent contains a tremendous diversity of river sizes and types. Rivers range from the frigid and often frozen Arctic rivers of northern Canada and Alaska to the warm tropi-

cal rivers of southern Mexico. They range from the high-gradient turbulent rivers draining the western mountains to the low-gradient, placid rivers flowing across the southeastern Coastal Plain. River size ranges from the tiny Dunk River of Prince Edward Island to the enormous Mississippi, the 2nd longest river in the world and the 9th largest by discharge (Leopold 1994). Such variations in latitude, topography, and size contribute to the great variation in biodiversity and ecological characteristics that we see among the continent's rivers (e.g., Abell et al. 2000).

Total annual discharge from North American rivers is approximately $8200 \, km^3/yr$ or about 17% of the world total (Shiklomanov 1993). The Mississippi is by far the largest river, yet its mean discharge is only 7% of total continental discharge ($580 \, km^3/yr$ or $18,400 \, m^3/s$) (Shiklomanov 1993, Karr et al. 2000). Among the top 25 rivers by discharge, more than a dozen have annual discharge greater than $2000 \, m^3/s$, with the other largest being the St. Lawrence, Mackenzie, Ohio, Columbia, and Yukon (Table 1.1). Of these six largest rivers, all flow to the sea except the Ohio, which contributes almost half the flow of the Mississippi River. The Nelson and Missouri rivers are among the top five in drainage area, but only rank

FIGURE 1.1 Grand Canyon of the Yellowstone River, downstream of Yellowstone Lake in Yellowstone National Park, Wyoming. The Yellowstone River is the longest free-flowing river in the coterminous states, eventually joining the Missouri River (see Chapter 10) in western North Dakota (PHOTO BY A. C. BENKE).

TABLE 1.1 Largest rivers of North America ranked by virgin discharge. All rivers may be found in this book except the Koksoak and La Grande.

	River name	Discharge (m³/s)	Basin area (km²)
1	Mississippi	18,400	3,270,000
2	St. Lawrence	12,600	1,600,000
3	Mackenzie	9,020	1,743,058
4	Ohio	8,733	529,000
5	Columbia	7,730	724,025
6	Yukon	6,340	839,200
7	Fraser	3,972	234,000
8	Upper Mississippi	3,576	489,510
9	Slave (Mackenzie basin)	3,437	606,000
10	Usumacinta	2,687	112,550
11	Nelson	2,480	1,072,300
12	Liard (Mackenzie basin)	2,446	277,000
13	Koksoak (Quebec)	2,420[1]	133,400[2]
14	Tennessee (Ohio basin)	2,000	105,870
15	Missouri	1,956	1,371,017
16	Ottawa (St. Lawrence basin)	1,948	146,334
17	Mobile	1,914	111,369
18	Kuskokwim	1,900	124,319
19	Churchill (Labrador)	1,861	93,415
20	Copper	1,785	63,196
21	Skeena	1,760	54,400
22	La Grande (Quebec)	1,720[1]	96,866[2]
22	Stikine	1,587	51,592
24	Saguenay (St. Lawrence basin)	1,535	85,500
25	Susitna	1,427	51,800
	Additional large basins		
	Rio Grande	~100	870,000
	Colorado	550	642,000
	Arkansas	1,004	414,910

[1] Dynesius and Nilsson (1994)
[2] Leopold (1994)

11th and 15th, respectively, in discharge because their basins receive only moderate precipitation. Three rivers with exceptionally large drainage basins, but not among the top 25 by discharge, are the Colorado, Rio Grande, and Arkansas (see bottom of Table 1.1). The Colorado River and Rio Grande (Bravo) each drain >600,000 km² (among the top ten by basin area) but are located in arid regions and have substantially lower discharge than many rivers draining much smaller basins. In addition to these extremely large rivers and river basins, there are many rivers of moderate to large size (100 to >1000 m³/s) that each flow for several hundred kilometers to the sea or are tributaries of larger rivers.

Although humans have been attracted to rivers throughout North America for more than 12,000 years, it has not been until the past 100 years that industrialization has caused a radical transformation of most rivers. They have been dammed for flood control, hydropower, and navigation; dewatered for human and agricultural consumption; contaminated with waste products; and invaded by many nonnative species. Such activities have seriously degraded water quality, habitat diversity, biological diversity, and ecosystem integrity of rivers throughout most of the continent. In spite of such extensive alterations, rivers have displayed a remarkable degree of resilience, capable of returning to at least seminatural conditions when human impacts are reduced. Fortunately, there are still some rivers that have escaped major human alterations (Benke 1990, Dynesius and Nilsson 1994), particularly those in the Arctic and Northern Pacific (Chapters 16, 17, and 20). Such pristine or lightly altered rivers retain much of the natural physical and biological properties they have had for millennia, and can serve as benchmarks by which to evaluate impacts and restoration success of altered rivers.

As editors of this volume, we recognize that modern societies inevitably must exploit rivers for necessary human needs and not all rivers can retain pristine features. However, any objective evaluation of North American rivers would reveal that we have gone well past a balance between human needs and the need for natural riverine ecosystems (e.g., Palmer 1986, 1993, 1996; National Research Council 1992; Karr et al. 2000; Postel and Richter 2003). Fortunately, the past 35 years have seen a major shift in society's attitudes toward rivers and the need to conserve these valuable natural resources (Boon, Calow, and Petts 1992, Boon, Davies, and Petts 2000). In spite of progress in our treatment of rivers, however, there have been no efforts in North America to comprehensively evaluate the state of its rivers (but see Stanford and Ward 1979, Benke 1990, Palmer 1996, and the U.S. National Park Service Web site for its Nationwide Rivers Inventory [www.nps.gov/ncrc/programs/rtca/nri]) that is comparable to wetlands evaluations (e.g., Cowardin et al. 1979, National Research Council 1992, Dahl 2000, and the U.S. Fish and Wildlife Service Web site for the National Wetlands Inventory [http://wetlands.fws.gov]). Hopefully, the better understanding of North American rivers revealed in this book will help lead to wiser management, sustainability, and restoration of these essential resources.

FIGURE 1.2 Major river basins and regions used in organization of chapters.

BASIC APPROACH

Our basic goal in planning this book was to provide as much information about as many rivers as possible in a single volume. Therefore, working with chapter authors, we have selected a total of 218 rivers throughout the continent. River descriptions are organized into 22 chapters, some of which are represented by a single major river and its tributaries, such as the Missouri River, and others by region, such as the Atlantic Coast rivers of the northeastern United States (Fig. 1.2). Chapter authors were asked to include a range of rivers representing differences in size, physical diversity, biological diversity, ecosystem function, and degree of human influence. This selection strategy enabled us to include most of the major rivers and much of the diversity of rivers present in North America. Selection of rivers in some regions, particularly in the Arctic, was limited by the amount of information available.

Each chapter follows the same format, enabling readers to make comparisons among various physical and biological characteristics. All chapters include the following:

• A regional map highlighting rivers to be described and indicating major political boundaries.
• One-page summaries of up to 12 rivers, including abbreviated descriptions of physical and biological features; graphs of monthly temperature, precipitation, and runoff; and a color topographic map showing major tributaries, cities, dams, and boundaries of physiographic provinces.
• Expanded text for up to five "focus" rivers per region, or a total of 100 for the book.

CHAPTER CONTENTS AND BACKGROUND

This overview provides some background and rationale for the types of information found in each chapter. Each focus river includes several pages of additional description beyond the one-page summaries. This information is organized into the following categories: (1) an introduction, with some early human history; (2) physiography, climate, and land use; (3) geomorphology, hydrology, and chemistry; (4) river biodiversity and ecology; and (5) human impacts and special features.

Physiography, Climate, and Land Use

The character of a river is primarily determined by features of its *drainage basin*, the land area within which it and its tributaries flow.[1] A drainage basin, usually with well-defined boundaries, can be part of or encompass various other types of land designations that may be less precise in their boundaries, but are useful in characterizing a river. One or more drainage basins may be found within a broad *landscape* that is initially characterized by physical features such as geology and topography. Climate,

including temperature and precipitation patterns, further characterizes the landscape and is itself influenced greatly by latitude, global air circulation patterns, ocean currents, the shape of the continent, and mountain ranges. A landscape's physical features and climate together determine its terrestrial *ecosystems* and their biological *communities*. Thus, *landscape*, as used in the general ecological sense, refers to a large area of land composed of many clusters of interacting ecosystems. As we will see, there are various ways of describing the landscape, such as with physiographic provinces and ecoregions, and these can be useful in describing the drainage basin. The important point is that the drainage basin, however its physical and biological features are described, plays a major role in determining the characteristics of the river itself. As the pioneering running-water ecologist H. B. N. Hynes succinctly stated, "the valley rules the stream" (Hynes 1975). Thus, any description of a river is incomplete without characterizing the landscape through which it flows. On the other hand, it should be acknowledged that large drainage basins often intersect multiple landscape categories.

Physiography and Landscape

Each chapter attempts to characterize the landscape through which its rivers flow by describing the physiography, biome, and terrestrial ecoregions of the basin. Physiography is an early physical classification in which the land is divided into *physiographic provinces*: broad-scale subdivisions of the continent based on topographic features, rock type, and geological structure and history. Fenneman (1931) originally described a three-tier hierarchical structure of division, province, and section, and Hunt (1974) provided an updated version and maps of physiographic provinces for the United States and Canada. A physiographic map of the United States that is consistent with Hunt, but also includes sections, can be found on the U.S. Geological Survey (USGS) *Tapestry* Web site (http://tapestry.usgs.gov/physiogr/physio.html). We have prepared a complete physiographic map of North America showing only the *provinces* (Fig. 1.3), primarily based on Hunt and the USGS map, but using Arbingast et al. (1975) for Mexico. The boundaries of these physiographic provinces are also seen on the individual color topographic maps for every river basin in this book, with province names indicated by an acronym (e.g., CP = Coastal Plain).

An early biological classification of the landscape was identification of *biomes*. Biomes are broad global subdivisions of the earth based on terrestrial plant communities (Whittaker 1975) and are described in

---

[1] Drainage basin is also called *watershed* (primarily in North America) and *catchment* (primarily in Europe). All these terms are used synonymously in this book.

1. Arctic Slope
2. Brooks Range
3. Mackenzie Mts.
4. Seward Pen. and Bering
   Coast Uplands
5. Yukon Basin
6. Alaskan Pen. and Aleutian Islands
7. South Central Alaska
8. Arctic Lowlands
9. Baffin Upland
10. Coast Mountains of BC and SE Alaska
11. Rocky Mountains in Canada
12. Great Plains
13. Thelon Plains and Bear
    River Lowland
14. Bear-Slave-Churchill Uplands
15. Athabasca Plain
16. Hudson Bay Lowland
17. Labrador Highlands
18. Superior Upland
19. Laurentian Highlands
20. St. Lawrence Lowland
21. New England/Maritime
22. Adirondack
23. Coastal Plain
24. Piedmont Plateau
25. Blue Ridge
26. Valley and Ridge
27. Appalachian Plateaus
28. Interior Low Plateaus
29. Central Lowland
30. Ozark Plateaus
31. Ouachita
32. Southern Rocky Mts.
33. Wyoming Basin
34. Colorado Plateaus
35. Middle Rocky Mts.
36. Northern Rocky Mts.
37. Columbia Plateau
38. Basin and Range
39. Cascade-Sierra Mts.
40. Pacific Border
41. Lower California
42. Baja California
43. Buried Ranges
44. Sierra Madre Occidental
45. Sierra Madre Oriental
46. Neovolcanic Plateau
47. Sierra Madre Del Sur System
48. Chiapas-Guatemala Highlands
49. Yucatan
50. Central Mesa

FIGURE 1.3 Physiographic provinces of North America. Provinces for Canada and the United States are based on Hunt (1974) and the USGS Tapestry Web site. Provinces for Mexico are based on Arbingast et al. (1975).

most biology texts. Biome categories and boundaries can vary greatly according to author and degree of subdivision, however. One of the most recent and detailed maps was constructed by G. J. Schmidt and is available on the University of Tennessee botany Web site (http://botany1.bio.utk.edu/botany120lect/Biomes/biomemap.htm). We follow this map in defining the major biome categories of North America: *Desert, Chaparral, Temperate Deciduous Forest, Boreal Forest (Taiga), Tundra, Temperate Grassland, Temperate Mountain Forest* (Rocky Mountain and Pacific Coast evergreen forest), *Tropical Rain Forest,* and *Mexican Montane Forest*. Biomes through which each river flows are indicated on the one-page summary for each river.

A more recent and detailed biological approach to landscape classification is characterization by *ecoregions*. Terrestrial ecoregions are relatively large landscape divisions, but much smaller than biomes, that contain a geographically distinct assemblage of natural communities. However, there are different versions of terrestrial ecoregion classification (e.g., Bailey et al. 1994, Omernik 1995, Ricketts et al. 1999). The version used in this book was developed by the World Wildlife Fund (WWF) and provides a detailed classification within the context of a continentwide conservation assessment (Ricketts et al. 1999). As with physiographic provinces and biomes, we also list the major terrestrial ecoregions (including dominant terrestrial vegetation) through which the rivers flow.

Climate

Climate varies greatly over North America and has a major influence on biomes and ecoregions and their associated rivers. Mean annual air temperatures range from 26.5°C in the Lower Usumacinta basin of tropical Mexico to −12.6°C in Barrow, Alaska. Where temperatures consistently fall well below freezing for extended periods, streams and rivers may freeze, particularly in the Tundra and Boreal Forest biomes. In the southern United States and Mexico, with mean daily air temperatures usually above zero, water temperatures in major rivers rarely fall below 10°C. Precipitation in North America can range from <10 cm/yr in the western deserts to >300 cm/yr in the mountains of British Columbia and southern Mexico. However, rivers in arid regions are often sustained by precipitation falling at higher elevations. Precipitation ranges from 20 to 70 cm/yr in the mid-continental grasslands and is usually >100 cm/yr in the eastern part of the United States and Canada. Precipitation declines to <40 cm/yr in northern Canada

and Alaska. The seasonal distribution of precipitation is very significant as well. It can occur relatively evenly throughout the year, as in much of the U.S. east coast, or it can be very seasonal, as on the U.S. west coast (mostly winter precipitation) and throughout Mexico (mostly summer precipitation). To help understand how climate affects rivers and their basins, all river summaries include graphs showing mean monthly values of air temperature and precipitation. Such information can be obtained from many sources, but one which includes over 16,000 cities throughout the world is www.weatherbase.com.

Land Use

Land use can have a major influence on the quality of rivers. It has been known for many decades that "changes in the valley wrought by man may have large effects" on the stream (Hynes 1975). Human activities modify the landscape within drainage basins, and much of this activity results in river degradation through erosion of sediments, transport of pollutants, destruction of river banks, or acceleration of runoff. The major land use categories that degrade rivers are agriculture, commercial forests, and urbanization. All chapter authors have attempted to provide statistical information on land use in river basins. The greatest land use alteration is agriculture, which approaches or exceeds 90% for many sub-basins in the Upper Mississippi (Chapter 8) and Missouri (Chapter 10) basins. Revenga et al. (1998) provided information on land use, area of river basins modified, and percent of basin protected for the major "Watersheds of the World," including 32 of the largest basins in North America (also see http://www.waterandnature.org). The amount of drainage basin areas protected from harmful land-use practices ranges from zero for many (e.g., Mobile River basin) to 29% in the Yukon basin. Protection in most basins is <10%. Protection of the land in most basins is by designation as national or state/provincial parks or wilderness areas. Separate protection of river corridors is usually through their designation by federal (U.S. National Wild and Scenic Rivers, Canadian Heritage Rivers), or state/provincial governments.

River Geomorphology, Hydrology, and Chemistry

This subsection describes the physical and chemical properties of the river itself. River geomorphology involves the physical structure of the river channel and its floodplain. Hydrology describes the pattern and magnitude of river discharge through time and combines with geomorphology to create the physical habitats to which organisms have adapted. Water chemistry is important from at least two perspectives. Natural variation in chemical variables often determines the types of organisms that can survive in a river, but pollutants can alter water chemistry to the point that few, if any, organisms can survive.

River Geomorphology

The geomorphology of river channels can be characterized in many ways (e.g., Leopold 1994, Rosgen 1994). It can be defined in part by slope or gradient (meters of vertical drop per kilometer of river length), shape of river cross section, and composition of bed substrate. Channel shape and behavior today can be a product of millions of years of physical history, including geological uplift, rise and fall of the sea, ancient floods, volcanic activity, and activity of glaciers. A river's medium-term physical history (hundreds to thousands of years) includes deposition (aggradation) of alluvial sediments onto the valley floor, and subsequent erosion (degradation) of alluvial fill, formation of floodplains, creation of river meanders, braiding, and other channel characteristics (Leopold 1994). A river's geomorphology can be altered in a geological instant, however, by major engineering projects, such as dams, levees, and diversions.

The gradient of a river is primarily dependent on the relief of the landscape through which it flows. Although gradients of smaller streams can be >4% (Rosgen 1994), most major rivers have gradients well below 1% (10 m/km). For example, the gradient of the Madison, a Rocky Mountain river, is about 0.35% (3.5 m/km). Gradients in the southeastern Coastal Plain are usually <0.04% (<40 cm/km), and can be as low as 0.002% (2 cm/km) for the St. Johns River in Florida. Substrate composition of the river bed is often associated with river gradient. Substrate composition is a major habitat feature and is also affected by current velocity, rock type, and geological history. Mineral substrates can be solid bedrock or can range in size of bed materials from boulder to cobble, gravel, sand, and fine clays. An organic substrate that is influenced by gradient and in turn influences hydrology and habitats is wood (Gregory et al. 2003).

Rosgen (1994) developed a channel classification scheme using width-to-depth ratio, slope, size of bed substrate, and ratio of floodplain width to bankfull width (also see Leopold 1994). Channel patterns typically fall into one of three types: meandering, straight, and braided (Leopold 1994). The predominant type is meandering channels that twist and turn, often in a sinuous fashion, eroding on the outside of bends and depositing sediments on inside bends (point bars). Braided channels contain multiple streams separated by bars or islands.

River size can be represented in several ways: basin area, mean annual discharge, river length, and stream order. Stream order is a gross means of assessing river size and complexity based on the number of tributaries and how they are linked (Strahler 1957, Leopold et al. 1964, Allan 1995). The smallest permanent stream (based on a USGS topographic map) is a 1st order stream. If two 1st order streams join, they form a 2nd order stream. If two 2nd order streams join, they create a 3rd order stream, and so forth. If a smaller order stream (e.g., 1st) enters a larger stream (e.g., 2nd), it does not change the order of the larger stream. This approach is not easily applied to some regions, particularly in arid parts of the continent. Rivers that might normally be considered as 4th or 5th order may sometimes go dry (or underground) and are really temporary streams, not even designated as 1st order in other regions. Regardless of this and other problems, stream order designations are still widely used today, and were used in development of the River Continuum Concept (RCC), which will be described more fully later in this chapter. Authors of individual chapters usually have provided estimates of order for their rivers, but different approaches have been necessary. Some designations are from published information, some were determined directly from topographic maps, and some were determined indirectly from regressions of order and basin area. Thus, whereas stream order across chapters is not entirely consistent, it does provide some measure of river size within a region.

Hydrology and the Fate of Water

Probably the most fundamental physical feature of a river is its hydrology: the pattern and magnitude of river flow, or discharge, through time. It is now widely accepted that the biodiversity and ecological integrity of rivers are directly dependent on their natural flow regime (Poff et al. 1997). A river's hydrology is part of the water budget of the basin that concerns the balance of water inflows to the basin with outflows from the basin. When precipitation strikes the ground it has three obvious fates: surface runoff, infiltration into the ground, and evaporation. The average annual precipitation over North America is 67 cm, of which only 29 cm (43%) actually runs off or infiltrates the ground and eventually contributes to the flow of rivers (Hornberger et al.

1998). The remaining 38 cm are returned to the atmosphere by evaporation and transpiration by plants. Because it is difficult to separate evaporation from transpiration, together they are called *evapotranspiration*. Storage and discharge from groundwater sustain streams and rivers at "base flow" during periods when no rain occurs. Some streams may even go underground as they dry out on the surface. Because of large differences in climate, hydrological budgets vary greatly across the continent; the fraction of precipitation running off the landscape can be extremely low (<5%) or extremely high (>70%).

Discharge is the actual measurement of flow, usually presented as cubic meters per second (m^3/s) or cubic feet per second (cfs). An enormous amount of information about river discharge is available for rivers throughout North America. A large network of river gauges has been established by the USGS, beginning in 1895 (Leopold 1994). The USGS has collected streamflow data at about 19,000 sites, of which about 7000 have gauges that are currently operational. Real time and historical data (many with >50 years of data) are available through the USGS Web site (http://water.usgs.gov). Extensive data on river discharge also are available through Water Survey of Canada (WSC), a branch of Environment Canada. Data collection began in 1908 and since 1991 has been available on the national HYDAT CD-ROM. HYDAT contains data from more than 2900 active gauging stations, but many additional or WSC-abandoned gauges are maintained by provincial governments and water management agencies (e.g., hydro companies), which report their data to WSC for inclusion in HYDAT. A limited amount of discharge information (British Columbia and the Yukon) is now available on the Environment Canada Web site. In addition, the University of Wisconsin's Center for Sustainability and the Global Environment and the International Hydrological Programme of UNESCO (I. A. Shiklomanov) have developed useful Web sites that provide historical monthly discharge values for many rivers throughout the world (www.sage.wisc.edu/riverdata, http://webworld.unesco.org/water/ihp/db/shiklomanov/index.shtml).

Runoff is a measure of the height of water (centimeters) that drains from a basin over a unit of time (day, month, year). It is calculated by dividing discharge (m^3/s) by basin surface area (km^2) and multiplying by the appropriate conversion of time units. For many rivers throughout the world, runoff can be obtained directly from the International

Hydrological Programme Web site. Such information has sometimes been presented as annual unit area discharge or water yield (km^3 of water/km^2 of land, or m^3 of water/hectare of land). However, runoff, presented simply as cm/mo, is especially useful when compared to precipitation (cm/mo) throughout the year (e.g., Benke et al. 2000). Doing so provides a rough approximation of the water budget for a particular basin (i.e., fraction of precipitation that discharges to the mouth). On average, the difference between annual precipitation and annual runoff should be roughly equal to evapotranspiration losses, assuming no major movements of water among basins from groundwater exchanges. When considered on a daily or monthly basis, however, the difference between precipitation (PPT) and runoff (RO) is not only due to evapotranspiration (ET), but also to groundwater storage or loss (GW) and snow storage or loss (SN). Human storage in reservoirs or transfers between basins (HU) may affect the balance as well. Thus, the short-term fate of precipitation can be represented by the following equation:

$$PPT = RO + ET + GW + SN + HU$$

Seasonal variation in runoff is largely explained by the difference between PPT and RO:

$$(PPT - RO) = ET + GW + SN + HU$$

Seasonal influences on runoff are often apparent by plotting monthly runoff with monthly precipitation, and these plots are provided on the one-page summaries for all rivers. In most cases the major seasonal influences on runoff appear to be seasonal differences in ET and SN. A positive SN for any given month represents snow storage (accumulation on the land), and negative SN represents snow melt. Similarly, a positive GW represents groundwater storage during seasons of high precipitation, such as during the fall and winter on the west coast, and negative GW represents groundwater losses to runoff during the dry season. For some basins, human extraction can be great on an annual basis (>50%), and several examples are described throughout the book and in Chapter 24.

All rivers described in this book contain a figure showing long-term monthly precipitation and runoff. The default interpretation is that the difference between precipitation and runoff is an approximation of evapotranspiration (annual basis). However, this is based on the assumption that there are reliable long-term estimates of runoff and mean basinwide precipitation. This assumption is usually valid for most rivers that have an excellent long-term record of discharge. On the other hand, it is not always possible to obtain good estimates of basinwide precipitation. This is particularly true when there is only a single climate station within a basin containing a spatially heterogeneous pattern of precipitation, such as in mountainous areas. Clearly, if the precipitation estimates available do not reflect basinwide values, then the difference between precipitation and runoff will not provide a good estimate of evapotranspiration. Similarly, long-term losses or gains of water from the basin through groundwater can affect such approximations of evapotranspiration, but such losses are usually unknown. To summarize, plots of monthly precipitation and runoff often can be very useful in interpreting factors affecting seasonal runoff, but they must sometimes be viewed with extreme caution.

Several examples of rivers from across the continent illustrate the factors affecting seasonal runoff (Fig. 1.4). Alaska's Porcupine River (Chapter 17) has very low annual precipitation, but because evapotranspiration is very low, most of the water reaches the mouth of the river (see Fig. 1.4a). As would be expected in an Arctic environment, most of the Porcupine's frozen precipitation that falls during the winter does not flow until snowmelt in late spring. The Moisie River, Quebec (Chapter 21), has much higher annual precipitation but also experiences a strong seasonal runoff pattern due to snow storage in the winter and snowmelt in the spring (see Fig. 1.4b). In contrast, the Umpqua River, Oregon (Chapter 12), has a strong seasonal runoff pattern, primarily responding to a strong seasonal precipitation pattern (see Fig. 1.4c). The Ogeechee River, Georgia (Chapter 3), on the other hand, receives fairly uniform precipitation throughout the year, yet experiences its highest runoff during winter and spring because evapotranspiration is lower than during warm months (see Fig. 1.4d). The combination of low precipitation and high evapotranspiration in the Virgin River basin, Utah (Chapter 11), results in extremely low runoff, often accompanied by increased flow during spring snowmelt from the mountains (see Fig. 1.4e). Damming and water extraction in many rivers of the southwestern United States and northern Mexico reduce flows to zero for much of the year. Damming and flow regulation in eastern rivers can be more subtle but can have significant effects on seasonal runoff patterns. For example, regulation of the Savannah River,

Georgia–South Carolina (Chapter 3), results in considerably less seasonal variation than the adjacent Ogeechee River (compare Fig. 1.4f to Fig. 1.4d).

Water Chemistry

Information on river water chemistry is provided for each river to the extent possible. The availability and quality of river water chemistry data varies widely among rivers of North America. However, because water chemistry is extremely sensitive to human influences, we have asked authors to provide information that can help readers understand both natural water chemistry and chemical changes resulting from human impacts. Among the measures often reported are alkalinity and nutrients. Alkalinity is a measure of the buffering capacity of water and is commonly presented as milligrams of calcium carbonate per liter (mg/L as $CaCO_3$).[2] Natural alkalinity values can fall below 5 mg/L as $CaCO_3$ in low-pH blackwater rivers such as the Satilla River in the Coastal Plain of Georgia (Chapter 3). Alkalinity values are more commonly greater than 20 mg/L as $CaCO_3$, and in regions draining carbonate rocks, values can easily exceed 100 mg/L as $CaCO_3$.

The most common nutrients reported for rivers are phosphorus and nitrogen, either of which may be important in limiting production of algae and plants. Dissolved inorganic phosphorus (commonly presented as soluble reactive phosphorus [SRP], orthophosphate, or phosphate-phosphorus [PO_4-P]) averages about 10 micrograms of phosphorus per liter (µg P/L) among unpolluted rivers throughout the world (Meybeck 1982, Allan 1995, Wetzel 2001). Total dissolved phosphorus (inorganic + organic) averages about 25 µg/L. A recent study of streams and rivers in the United States, however, described natural background concentrations of total phosphorus as varying from 6 µg/L in the xeric west to more than 80 µg/L in the Great Plains (Smith et al. 2003). Waters polluted by agricultural runoff or municipal sewage commonly increase phosphorus levels to at least twice as high as natural background.

Nitrogen concentrations may be presented as total dissolved inorganic nitrogen (DIN), nitrate nitrogen (NO_3-N), nitrite nitrogen (NO_2-N), and ammonia nitrogen (NH_4-N). Nitrate-nitrogen values are commonly measured in rivers and average about 100 µg/L (Wetzel 2001). Median natural background

---

[2] Alkalinity has commonly been presented in units of mg/L as $CaCO_3$, and these are the units provided in this book. However, alkalinity is increasingly being expressed as milliequivalents per liter (1 meq/L = 50 mg/L as $CaCO_3$) (see Wetzel 2001).

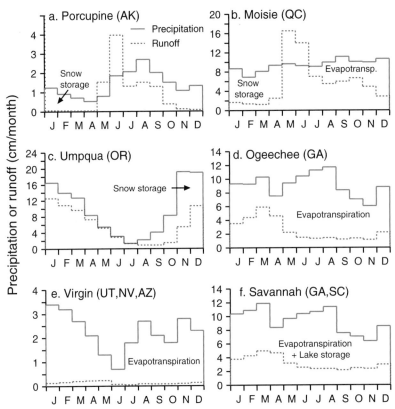

FIGURE 1.4 Patterns of precipitation and runoff for rivers from a diversity of regions in North America. a. Porcupine River (Alaska) showing snowmelt peak of runoff during low precipitation. b. Moisie River (Quebec) showing snowmelt peak of runoff. c. Umpqua River (Oregon) showing runoff peak following seasonal precipitation. d. Ogeechee River (Georgia) showing runoff pattern caused by seasonal changes in evapotranspiration. e. Virgin River (Utah, Nevada, Arizona) showing very low runoff due to low precipitation and high evapotranspiration. f. Savannah River (Georgia, South Carolina) showing flattened runoff pattern due to regulation (compare to Ogeechee).

levels for total nitrogen are about 140 µg/L in the United States but range from less than 20 µg/L in the xeric west to more than 500 µg/L in the Great Plains (Smith et al. 2003). In contrast to the twofold increase in stream phosphorus, actual nitrogen concentrations currently exceed natural background levels by more than sixfold due to both terrestrial (e.g., agriculture) and atmospheric inputs (Smith et al. 2003).

River Biodiversity and Ecology

This subsection attempts to characterize each river as a living system, highlighting the major species of plants and animals and describing ecosystem processes to the extent possible. For more detailed treatments of these subjects, see Allan (1995) and Cushing and Allan (2001). Early descriptions of the biological communities in rivers recognized that

organisms are typically found on or within substrate of the river bed (benthic habitat) or found in the water column. It was soon recognized that river habitats are much more complex than this simplistic picture. The bed substrate can vary greatly in sediment particle size and the water column can vary greatly in current velocity, creating a diversity of physical habitat features that strongly influence the presence of various species. In more recent years the importance of submerged wood (snags), the hyporheic zone, and adjoining floodplain wetlands have all become widely recognized. Wood is introduced to the channel typically by the undercutting of banks lined with trees. The hyporheic zone occurs where porous substrate extends below the bed surface, and through which there is an exchange of water, nutrients, organic matter, and organisms. Floodplain wetlands become directly connected with river channels for short or extended periods as rivers flood and overflow their banks, increasing the width of the river as much as 100-fold. Each of these habitats provides substrate and food for different types of organisms and can make important contributions to ecological processes. The relative occurrence of such habitats varies widely among rivers across the continent.

The great diversity of river types and habitats across North America has resulted in the evolution of a tremendous diversity of biological species that have become adapted to the particular habitats and environmental conditions in which they are found. We provide a brief description of the biodiversity within the major groups of organisms that compose the food chains so vital to the functioning of rivers as living biological systems: (1) algae, (2) plants, (3) invertebrate animals, and (4) vertebrate animals. Common names, rather than scientific names, are used throughout the text for all plants and vertebrates, and for some invertebrates (particularly mollusks). Scientific names of these species may be found in the Appendix.

Other groups of great importance in decomposition of organic matter, in nutrient regeneration, and as food sources for invertebrate animals are the bacteria, protists, and fungi; however, less is known about these groups and they will not be treated further.

We know much more about the continental and regional diversity of some groups than others. For certain groups of animals, their biodiversity, number of extinctions, and risk of extinction are well documented. Of particular concern are freshwater animals, which are highly sensitive to oxygen concentrations, habitat and flow conditions, pollutants, and thermal regimes. The World Wildlife Fund has classified regions of North America into freshwater ecoregions (as they did for terrestrial ecoregions), based primarily on native fish distributions (Abell et al. 2000). All chapters use this classification scheme. At the end of this subsection we will describe the major concepts of rivers ecosystems that address ecological processes.

Algae

Algae and plants are organisms in rivers that produce their own organic matter by photosynthesis; hence, they are *autotrophic* and are frequently referred to as *primary producers*. Algae are often single-celled organisms, but some riverine species occur as multinucleate filamentous forms, visible to the unaided eye as long, green filaments attached to solid substrates. The most abundant algae in streams and rivers are the diatoms, many of which are firmly attached to solid substrates such as rocks, wood, and plant stems. Diatoms and green algae are important components of the *biofilm* (or *periphyton*) that coats these surfaces. Biofilm also includes bacteria, fungi, protists, microinvertebrates, detritus, and other organic matter. In larger rivers (plus lakes and wetlands connected to rivers), algae suspended in the water column (*phytoplankton*) can also be important as a basal energy source for food chains.

Plants

The aquatic plants consist of mosses and flowering plants. The mosses are small, leafy, nonvascular plants that are attached to solid substrates. They form dense, matlike growth forms on stones and wood and may themselves be important habitat for small animals. The vascular flowering plants are primarily found where the current slows and the substrate is conducive to the growth of rooted plants, collectively called *macrophytes*. Huge beds of these plants can be found in certain locations, such as backwaters, pools, and floodplains where conditions are suitable. The macrophytes, much like the mosses, are important as habitat and after they die may be valuable food resources to aquatic animals. This dead organic matter is called *detritus* and forms the base of the detrital food web. In addition to the aquatic macrophytes, a terrestrial plant community often grows along the edges of rivers and streams and is dominated by trees such as willows, alders, cottonwoods, and red maple. This *riparian* vegetation can contribute substantial quantities of dead leaves and wood (detritus) to rivers, which serve as habitat and food sources for many animals. In rivers with broad floodplains, however, extensive macrophytes or floodplain forests can extend for kilometers from the edge of the river channel, and the "riparian" vegetation becomes greatly extended.

Invertebrates

Invertebrates are by far the most abundant and diverse group of animals found in rivers. The term *macroinvertebrate* usually applies to *benthic* (bottom-dwelling) invertebrates that will not pass through a 0.5 mm sieve. Among these are herbivores, which feed directly on primary producers; detritivores, which feed on dead organic matter (detritus); and carnivores, which feed on other animals. Aquatic insects are the most diverse and abundant group of invertebrates in flowing waters. Four orders generally compose the majority of numbers and biomass of insects in most rivers and streams. These are the Diptera (true flies), Ephemeroptera (mayflies), Plecoptera (stoneflies), and Trichoptera (caddisflies). Other important but less abundant insect species are from the orders Coleoptera (aquatic beetles), Hemiptera (true bugs), Odonata (dragonflies and damselflies), and Megaloptera (hellgrammites).

In addition to insects, other common macro-invertebrates in streams and rivers include aquatic crustaceans, mollusks, and oligochaetes (worms). The crustaceans include crayfishes, amphipods, and isopods. The mollusks include bivalves (primarily mussels) and snails. In addition to these macroinvertebrates, there are many smaller but abundant benthic groups that are commonly called *meiofauna*. These nematodes, microcrustaceans, rotifers, and other smaller groups are often the only invertebrates that can live deeper within the interstitial waters of the streambed (hyporheic zone). In addition to the benthic and hyporheic invertebrates, the *zooplankton* of the water column can be important components of the food web. The zooplankton community feeds upon the phytoplankton and typically includes microcrustaceans such as copepods and cladocerans, as well as rotifers.

Biodiversity of invertebrates is best known among the larger animals. There are 338 crayfish species in the United States and Canada, as well as 297 freshwater mussels and 342 freshwater snails (Williams et al. 1992, Lydeard and Mayden 1995, Taylor et al. 1996, Karr et al. 2000, Abell et al. 2000). There are probably in the vicinity of 10,000 aquatic insect species in North America (e.g., see Merritt and Cummins 1996), but we are unaware of separate estimates of stream and rivers species from those found in ponds, lakes, and wetlands. Intensive studies have revealed that there can be several hundred species in a single stream.

Vertebrates

The most widely recognized and important vertebrates occurring in rivers are fishes, simply because they live entirely in the water and are thus more intimately associated with riverine food webs. Fishes have long been an important source of food for native North Americans, and sport and commercial fisheries have been a major industry for over a century. Like the invertebrates, fishes function as herbivores, detritivores, and carnivores. Although representatives of many families of fishes are found in rivers and streams, the Salmonidae (salmon and trout), Cyprinidae (minnows), Catostomidae (suckers), Ictaluridae (bullheads and catfishes), Centrarchidae (sunfishes and bass), Percidae (particularly darters), and Cottidae (sculpins) are certainly among the most common and

important. Depending on the river, however, amphibians, reptiles, birds, and mammals can also be important. Amphibians (salamanders and frogs) and aquatic reptiles (turtles, snakes, and alligators) occur primarily along the edges of streams and rivers, including floodplains. Many birds (e.g., wading birds) and mammals (e.g., beavers, river otters), depend heavily on the riverine and floodplain environment for habitat and food. Totally aquatic mammals such as beluga whales and porpoises can swim well up into the freshwater portions of large rivers.

Fish diversity is the best known of any aquatic group, with more than 1000 species of native freshwater fishes throughout North America (Abell et al. 2000, Nelson et al. 2004). This diversity varies greatly across the continent, with a particularly high concentration of species occurring in the southeastern United States (see Fig. 1.3 for Canada and the United States in Karr et al. 2000). Many rivers of only moderate size (<100,000 km^2 basin) have well over 100 native species (Gido and Brown 1999). A large number of fish species are at risk of extinction in the southeastern United States, where native diversity is high, but a much higher proportion of species are endangered in southwestern states, where native diversity is low (Abell et al. 2000, Karr et al. 2000).

Ecosystem Processes

The foundation for all life and food webs in rivers, as in any ecosystem, is the production of organic matter. Organic matter is produced in rivers and streams by two main processes: *primary production* and *secondary production*. Primary production is the formation of autotrophic biomass (algae and plants) by the process of photosynthesis. Secondary production is the formation of heterotrophic biomass by growth of animals, although the term may also be applied to production of heterotrophic microbes. Secondary producers may obtain their food directly from primary producers (herbivores), after primary producers have died (detritivores), or from feeding on other animals (carnivores).

Two general sources of energy from primary production are often distinguished for running water ecosystems: *autochthonous* (in-stream production) and *allochthonous* (organic matter produced in the terrestrial environment outside the river boundary). Probably the two most important autochthonous energy sources are the diatoms, a major component of periphyton, and the phytoplankton of large rivers. The remaining autochthonous energy sources, filamentous green algae, mosses, and macrophytes, perform important functions (e.g., oxygen production and as habitat), but they are often less important in the overall energy flow of streams and rivers. On the other hand, when macrophytes occur in abundance, particularly in backwaters and floodplain wetlands, they can decompose and become a major autochthonous detrital food source.

Allochthonous sources include any organic matter originating in the terrestrial environment that is transported to the aquatic system via wind, gravity, surface runoff, or some other way. Included here are the more obvious contributions of detritus from riparian or floodplain plants, such as leaves, twigs, and tree trunks (also called *course particulate organic matter* [CPOM]). CPOM (>1 mm diameter) is usually considered the major allochthonous source in small forested streams, but it is also of major importance in floodplain wetlands. CPOM undergoes decomposition by microbes (particularly fungi) and fragmentation by invertebrates, producing fine particulate organic matter (FPOM). FPOM (<1 mm diameter) processed in the low-order tributaries or flushed from floodplains often becomes a dominant form of organic matter and food source in large rivers. FPOM carried by the current is often called *seston*. Dissolved organic matter (DOM) is another component of organic matter that is becoming well recognized as a sizable portion of total organic matter in transport (Webster and Meyer 1997). DOM (<0.5 μm diameter) originates from both in-stream processing (autochthonous), from adjoining floodplain wetlands, and from more distant sources in terrestrial areas (allochthonous). It can be an important energy source for heterotrophic microbes and higher food webs, but much of it is *refractory* and passes through river systems unaltered. Finally, allochthonous *animal* sources of food may include fish carcasses (fishes that die after upstream migration) and terrestrial insect feces (frass), but the latter are less well documented inputs from terrestrial vegetation.

A variety of animals, particularly invertebrates, have evolved a great diversity of feeding strategies to make use of primary and secondary producers in rivers. Animals may be classified into the classic trophic groups: herbivores, carnivores, omnivores, and detritivores. However, many stream and river ecologists have found it more useful to classify benthic invertebrates into *functional feeding groups*: how they obtain their food rather than what they eat (Cummins 1973, Wallace and Webster 1996). The major functional feeding groups are *shredders*, *scrapers* (also called *grazers*), *filtering collectors*, *gathering collectors*, and *predators* (carnivores).

Shredders (including many species of stoneflies, caddisflies, true flies, and crustaceans) obtain their nutrition by consuming microbially conditioned CPOM, producing FPOM in the form of smaller particles and fecal pellets. Scrapers (including many caddisflies, mayflies, and snails), as the name implies, feed by scraping diatoms and periphyton from the surface of solid substrata. Filtering collectors (including some mayflies and many caddisflies, true flies, and mussels) feed on FPOM (seston) from the water column using a variety of nets, leg hairs, modified mouthparts, and other structures to strain the food. In large rivers, filter-feeding zooplankton can be an important part of the food web. Gathering collectors are mainly free-living organisms that feed on FPOM where it has settled on the river bed, and include representatives of all major insect orders. Predators, particularly odonates (dragonflies and damselflies), hellgrammites, true bugs, and several stoneflies, kill and feed on other animals. Other predaceous species can be found in most insect orders.

A common measurement made in streams and rivers to determine the relative dependence on autochthonous or allochthonous energy is the *P/R ratio*. This is defined as the ratio of primary production (which produces oxygen) to community respiration (which consumes oxygen) over a 24-hour period. A value of unity or above denotes an *autotrophic* section of river, where the system is producing sufficient oxygen to meet its needs. A value of <1 denotes a significant *heterotrophic* influence, in which the system is using more oxygen than it produces and dissolved oxygen must be replenished from the atmosphere. Hence, a heterotrophic section of river receives a food subsidy from the terrestrial environment (originally as allochthonous primary production) in the form of CPOM, FPOM, and DOM.

In order to make sense of the variation among streams and rivers in terms of geomorphology, hydrology, habitats, productivity, feeding strategies, and the fate of organic matter produced in rivers basins, various authors have proposed conceptual models. The best known model in use today that describes rivers in a holistic construct is known as the River Continuum Concept (Vannote et al. 1980, Minshall, Petersen et al. 1983, Minshall, Cummins et al. 1985). The basic model, often illustrated as in Fig. 1.5, describes headwater reaches (orders 1 to 3) as heterotrophic, with P/R ratios <1 and an invertebrate community dominated by shredders and collectors. Of more relevance to this book, however, are higher-order rivers. In middle reaches of the continuum, small to medium rivers (typically 4th to 6th order) are

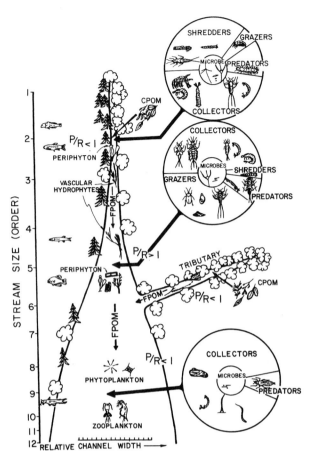

FIGURE 1.5 The River Continuum Concept, showing predicted changes in P/R ratio and functional feeding groups with stream order (from Vannote et al. 1980, with permission).

usually shallow, open to direct sunlight, and have a P/R ratio >1 due to enhanced autochthonous (algal and macrophyte) production. Scrapers are expected to be codominant with filtering and gathering collectors. In the large rivers (7th to 10th order), the channel would widen and deepen and the primary energy source would be excess FPOM produced from the midreaches plus a viable assemblage of phytoplankton. The net result would be P/R ratios <1 and functional feeding groups being dominated by bottom-dwelling filtering and gathering collectors, such as mollusks, oligochaetes, and dipteran larvae, and zooplankton in the water column. Thus, the RCC postulated that the entire river continuum, from headwaters to mouth, is linked in a predictable series of energy changes and functional feeding group adjustments along its course.

Several significant concepts in river ecology have been developed since the RCC, the most important

of which is the linkage of the river with its floodplain. This linkage was described most comprehensively as the "flood pulse concept" by Junk et al. (1989), in which river communities depend on the annual inundation of broad floodplains and the subsequent recession of water levels to the main channel. Junk and colleagues' major deviation from the RCC was to suggest that the lateral exchange of materials with the floodplain will have a much greater influence on the biota and ecosystem function than the organic matter transported from upstream sources. Whereas the authors of the flood pulse concept focused on extremely large rivers such as the Amazon and Mississippi, more recent work has emphasized that the flood pulse can be very important in middle-order rivers as well (e.g., Benke et al. 2000, Tockner et al. 2000). This can be particularly important in the low-gradient rivers of North America, such as in the southeastern Coastal Plain (Benke et al. 2000). The importance of the flood pulse in North American rivers has probably not been as widely recognized because the flows of so many of these rivers are highly regulated by dams.

In addition to the RCC and flood pulse concept, Thorp and Delong (1994) have suggested that a substantial portion of the organic matter consumed by animals in large rivers is either of autochthonous origin (phytoplankton, periphyton, macrophytes) or is produced by the riparian forest, rather than being transported longitudinally from upstream tributaries or laterally from interactions with the floodplain. Furthermore, submerged wood (snags) originating from streamside forests has been of considerable historical importance to many rivers throughout North America, can have a major influence on hydrology and geomorphology (Gregory et al. 2003), and can be of great importance in creating habitat and cover for many animals (Benke et al. 1984, Benke and Wallace 2003). Removal of wood from many rivers in North America for over 100 years has undoubtedly caused a significant shift in their habitat structure, biodiversity, and functional group structure compared to their pristine condition (e.g., Sedell and Froggatt 1984).

In summary, it is difficult to generalize about organic matter budgets and energy sources for food webs in North American rivers. Too few studies have been done that address such questions in rivers having a discharge >25 m^3/s (e.g., see Webster and Meyer 1997), and many authors in this book could only speculate about ecosystem processing. Furthermore, it is difficult to find such a river for study because most rivers are highly regulated. The source of organic matter for a river food web (longitudinal, lateral, or autochthonous derivations) probably depends largely on geographical, geomorphological, and physiographic differences, even in their unaltered state. Human alterations to rivers undoubtedly affect their relative contributions as well. An earlier approach to such questions was examined in Cushing et al. (1995), where authors attempted to put the streams occurring in various broad regions of the world into the context of the RCC.

Human Impacts and Special Features

At the end of each chapter, authors highlight what is special about the river and summarize its major impacts by humans. Special features of each river include such things as the existence of dramatic falls, large floodplain swamps, unique species, or special protection status (e.g., U.S. National Wild and Scenic Rivers, Canadian Heritage Rivers). Impacts to rivers have often been grouped into three major categories: *pollution, flow regulation,* and *water extraction.* A fourth significant impact has come from both intentional and accidental introduction of *nonnative species.* In spite of human alterations, each river is or was unique.

Pollution of rivers in North America has probably been a problem since the 1700s, when the early Europeans developed industrial cities that polluted east-coast rivers. This early pollution was undoubtedly from domestic sewage, a source that remains the second most common pollutant (Karr et al. 2000). Throughout the 1800s and 1900s, the types and amounts of point-source pollution released into rivers created stretches of rivers where little aquatic life could survive for many kilometers downstream of an outfall. In many rivers, persistent toxic chemicals such as mercury, polychlorinated biphenyls (PCBs), and pesticides have accumulated in sediments and magnified through food chains to the point where fishes are unsafe for human consumption (Karr et al. 2000).

Nonpoint sources of pollution also became major problems during the last two centuries, particularly from land use associated with deforestation, agriculture, and urbanization. Deforestation reduces energy inputs that normally are transported to rivers in the form of allochthonous leaf litter and its breakdown products. Urbanization typically accelerates surface runoff patterns and introduces large amounts of both point and nonpoint pollutants. Similarly, agriculture greatly increases silt and nutrient loads to rivers and is the leading pollutant source in the United States

(Karr et al. 2000). Federal laws to reduce pollution in North American rivers have had varied success. Although enforcement of water quality laws in the United States has resulted in significant reductions in point-source pollutants, nonpoint pollution goes largely unchecked (Smith et al. 1987, Karr et al. 2000).

The construction of dams, levees, and channels for flow regulation has been especially devastating to the natural biodiversity and ecology of rivers (Ward and Stanford 1979, National Research Council 1992, World Commission on Dams 2000). Dams have been constructed to control floods, generate electricity, enhance navigation, create recreational areas, and provide water for domestic and industrial use. Unfortunately, the ecological costs associated with dam building have been largely ignored by dam-building agencies and politicians. Dams destroy essential habitats and eliminate river corridors for the passage of materials and migration of biota. Some dams regulate flow to such an extent that downstream reaches may become totally dry or at least prevented from their natural flooding patterns. Hydropower dams required to maintain a minimum flow typically have highly unnatural surges of flow on a daily or weekly basis, creating an enormous stress on biota. Often the water released is devoid of natural sediment loads (left behind in the reservoir), is low in oxygen, and is much colder than natural river temperatures, eliminating many native species for many kilometers below the dam. Dams also may alter seasonal discharge and natural flooding patterns.

The widespread impact of dams on rivers in North America cannot be overstated. The U.S. Army Corps of Engineers (USACE) has built the majority of major dams on rivers of the United States, beginning in the late 1800s. The USACE maintains an inventory of dams that now totals about 76,000 (http://crunch.tec.army.mil/nid/webpages/nid.cfm). This incredible number of dams alone is a strong indication of the major influence that dam-building has had on North American rivers. Benke (1990) estimated that out of over 5.2 million km of rivers in the 48 coterminous United States, there were only 42 relatively small rivers that were reasonably natural and unregulated for at least 200 km of length. Stanford and Ward (1979) reached a similar conclusion in an earlier study. Dynesius and Nilsson (1994) conducted an analysis of large rivers in the northern third of the world, quantifying channel fragmentation and flow regulation. They found that the vast majority of large river systems in the United States and central/eastern Canada were strongly fragmented by dams. Only in

Alaska, British Columbia, and rivers flowing into the Arctic are there many rivers unaffected by dams. Revenga at al. (1998) list the number of large dams (>15 m in height) for many rivers of the world, including 2091 for the Mississippi, 265 for the Colorado, and 184 for the Columbia. Among all the river basins of the world, there are nine with more than five major (>150 m in height) dams, four of which are in North America: Columbia (13), Colorado (12), Mississippi (9), and Nelson (7).

Associated with dam building, and a major reason for the dewatering of major rivers, is water extraction and interbasin transfers. Approximately 8% of freshwater runoff in North America is withdrawn for human use (WRI 2000). In the United States and Canada, the greatest amounts of water are withdrawn for industry (65% and 80%, respectively) (WRI 2000). About 27% of withdrawals are used for agriculture in the United States, but 78% of withdrawals are used for agriculture in Mexico. The greatest problems with extractions are in arid regions. The most noteworthy examples are the Colorado and Rio Grande rivers, where so much water is withdrawn that flow to the sea ceases for much of the year.

The introduction of nonnative species to North American rivers has had a more subtle impact than pollution, flow regulation, and water extraction.[3] Although their effects are difficult to document, nonnative species are believed to have caused the extinction of native species and have resulted in significant changes in the function of river ecosystems (Gido and Brown 1999, Heinz Center 2002). For example, fish management practices have resulted in the widespread introduction of rainbow trout outside of their native habitat. Accidental introductions include the Asiatic clam, which is now distributed throughout much of the United States, and the zebra mussel, which has made its way from the Great Lakes down the major waterways to Louisiana. Sixty percent of 350 watersheds in the coterminous 48 states have 1 to 10 nonnative fish species, 22% have 11 to 20 nonnative species, and 2 watersheds have >40 nonnative species (Heinz Center 2002).

---

[3] We use the term *nonnative species* to indicate species that have intentionally or accidentally been introduced to a river system where they did not previously exist without the influence of humans. Synonyms for nonnative species are *introduced, exotic, alien, nonindiginous,* and *invasive.*

CONCLUDING COMMENTS

The editors and chapter authors of this book have come to realize that our understanding of natural biodiversity and ecology in North America rivers is fair at best. We are well aware of the major factors that threaten river biodiversity and ecosystem integrity such as habitat loss, pollution, and the spread of nonnative species (e.g., Allan and Flecker 1993). However, the biological and ecological data for many rivers are superficial, and sometimes nonexistent, as will become apparent throughout the book. It is difficult to assess the natural ecological characteristics of most rivers now that they are highly modified or polluted. On the other hand, most of the remaining untouched rivers are at such remote locations (e.g., the Arctic) that little research has been conducted in them. There is a great need for major new research initiatives that can lead to better scientific understanding, wider appreciation of their importance, and wiser management. In spite of this large gap in scientific knowledge of rivers, it is imperative that our current knowledge, particularly the extensive information on hydrology, be put to good use in conservation and management if the natural features and biodiversity of rivers are to be retained. Some of the major challenges we face in applying this knowledge to river conservation and management are described in Chapter 24.

LITERATURE CITED

Abell, R. A., D. M. Olson, E. Dinerstein, P. T. Hurley, J. T. Diggs, W. Eichbaum, S. Walters, W. Wettengel, T. Allnutt, C. J. Loucks, and P. Hedao. 2000. *Freshwater ecoregions of North America: A conservation assessment.* Island Press, Washington, D.C.

Allan, J. D. 1995. *Stream ecology: Structure and function of running waters.* Chapman & Hall, London.

Allan, J. D., and A. S. Flecker. 1993. Biodiversity conservation in running waters. *BioScience* 43:32–43.

Arbingast, S. A., C. P. Blair, J. R. Buchanan, C. C. Gill, R. K. Holz, C. A. Marin, R. H. Ryan, M. E. Bonine, and J. P. Weiler. 1975. *Atlas of Mexico.* Bureau of Business Research, 2nd ed. University of Texas at Austin Press, Austin.

Bailey, R. G., P. E. Avers, T. King, and W. H. McNab (eds.). 1994. *Ecoregions and subecoregions of the United States.* USDA Forest Service, Washington, D.C.

Benke, A. C. 1990. A perspective on America's vanishing streams. *Journal of the North American Benthological Society* 9:77–88.

Benke, A. C., I. Chaubey, G. M. Ward, and E. L. Dunn. 2000. Flood pulse dynamics of an unregulated river floodplain in the southeastern U.S. Coastal Plain. *Ecology* 81:2730–2741.

Benke, A. C., T. C. Van Arsdall, D. M. Gillespie, and F. K. Parrish. 1984. Invertebrate productivity in a subtropical blackwater river: The importance of habitat and life history. *Ecological Monographs* 54:25–63.

Benke, A. C., and J. B. Wallace. 2003. Influence of wood on invertebrate communities in streams and river. In S. V. Gregory, K. L. Boyer, and A. M. Gurnell (eds.). *The ecology and management of wood in world rivers*, pp. 149–177. Symposium 37. American Fisheries Society, Bethesda, Maryland.

Boon, P. J., P. Calow, and G. E. Petts. 1992. *River conservation and management.* John Wiley & Sons, Chichester, England.

Boon, P. J., B. R. Davies, and G. E. Petts. 2000. *Global perspectives on river conservation: Science, policy, and practice.* John Wiley & Sons, Chichester, England.

Cowardin, L. M., V. Carter, F. C. Golet, and E. T. LaRoe. 1979. *Classification of wetlands and deepwater habitats of the United States.* Publication FWS/OBS-79/31. U.S. Fish & Wildlife Service, Washington, D.C.

Cummins, K. W. 1973. Trophic relations of aquatic insects. *Annual Review of Entomology* 18:183–206.

Cushing, C. E., and J. D. Allan. 2001. *Streams: Their ecology and life.* Academic Press, San Diego.

Cushing, C. E., K. W. Cummins, and G. W. Minshall. 1995. *Ecosystems of the world: River and stream ecosystems.* Elsevier, New York.

Dahl, T. E. 2000. Status and trends of wetlands in the conterminous United States 1986 to 1997. U.S. Department of the Interior, Fish and Wildlife Service, Washington, D.C.

Dynesius, M., and C. Nilsson. 1994. Fragmentation and flow regulation of river systems in the northern third of the world. *Science* 266:753–762.

Fenneman, N. M. 1931. *Physiography of western United States.* McGraw-Hill, New York.

Gido, K. B., and J. H. Brown. 1999. Invasion of North American drainages by alien fish species. *Freshwater Biology* 42:387–399.

Gregory, S. V., K. L. Boyer, and A. M. Gurnell (eds.). 2003. *The ecology and management of wood in world rivers.* Symposium 37. American Fisheries Society, Bethesda, Maryland.

Heinz Center. 2002. *The state of the nation's ecosystems: Measuring the lands, waters, and living resources of the United States.* Cambridge University Press, Cambridge, England.

Hornberger, G. M., J. P. Raffensperger, P. L. Wiberg, and K. N. Eshleman. 1998. *Elements of physical hydrology.* Johns Hopkins University Press, Baltimore.

Hunt, C. B. 1974. *Natural regions of the United States and Canada.* W. H. Freeman, San Francisco.

Hynes, H. B. N. 1975. The stream and its valley. *Verhandlungen der Internationalen Vereinigung für theoretische und angewandte Limnologie* 19:1–15.

Junk, W. J., P. B. Bayley, and R. E. Sparks. 1989. The flood pulse concept in river-floodplain systems. In D. P. Dodge (ed.). *Proceedings of the international large river symposium (LARS).* Canadian Special Publication of Fisheries and Aquatic Sciences 106, pp. 110–127.

Karr, J. R., J. D. Allan, and A. C. Benke. 2000. River conservation in the United States and Canada. In P. J. Boon, B. R. Davies, and G. E. Petts (eds.). *Global perspectives on river conservation: Science, policy and practice*, pp. 3–39. John Wiley & Sons, Chichester, England.

Leopold, L. B. 1994. *A view of the river.* Harvard University Press, Cambridge, Massachusetts.

Leopold, L. B., M. G. Wolman, and J. P. Miller. 1964. *Fluvial processes in geomorphology.* W. H. Freeman, San Francisco.

Lydeard, C., and R. L. Mayden. 1995. A diverse and endangered aquatic ecosystem of the southeast United States. *Conservation Biology* 9:800–805.

Merritt, R. W., and K. W. Cummins. 1996. *An introduction to the aquatic insects of North America*, 3rd ed. Kendall/Hunt, Dubuque, Iowa.

Meybeck, M. 1982. Carbon, nitrogen, and phosphorus transport by world rivers. *American Journal of Science* 282:401–450.

Minshall, G. W., K. W. Cummins, R. C. Petersen, C. E. Cushing, D. A. Burns, J. R. Sedell, and R. L. Vannote. 1985. Developments in stream ecosystem theory. *Canadian Journal of Fisheries and Aquatic Sciences* 42:1045–1055.

Minshall, G. W., R. C. Petersen, K. W. Cummins, T. L. Bott, J. R. Sedell, C. E. Cushing, and R. L. Vannote. 1983. Interbiome comparisons of stream ecosystem dynamics. *Ecological Monographs* 53:1–25.

National Research Council. 1992. *Restoration of aquatic ecosystems.* National Academy Press, Washington, D.C.

Nelson, J. S., E. J. Crossman, H. Espinosa-Pérez, L. T. Findley, C. R. Gilbert, R. N. Lea, and J. D. Williams. 2004. *Common and scientific names of fishes from the United States, Canada and México.* Bethesda, Maryland: American Fisheries Society, Special Publication 29: 1–386.

Omernik, J. M. 1995. *Level III ecoregions of the continent.* National Health and Environment Effects Research Laboratory, U.S. Environmental Protection Agency, Washington, D.C.

Palmer, T. 1986. *Endangered rivers and the conservation movement.* University of California Press, Berkeley and Los Angeles.

Palmer, T. 1993. *The wild and scenic rivers of America.* Island Press, Washington, D.C.

Palmer, T. 1996. *America by rivers.* Island Press, Washington, D.C.

Poff, N. L., J. D. Allan, M. B. Bain, J. R. Karr, K. L. Prestegaard, B. D. Richter, R. E. Sparks, and J. C. Stromberg. 1997. The natural flow regime: A paradigm for river conservation and restoration. *BioScience* 47:769–784.

Postel, S., and B. Richter. 2003. *Rivers for life: Managing water for people and nature.* Island Press, Washington, D.C.

Revenga, C., S. Murray, J. Abramovitz, and A. Hammond. 1998. *Watersheds of the world: Ecological value and vulnerability.* World Resources Institute, Washington, D.C.

Ricketts, T. H., E. Dinerstein, D. M. Olson, C. J. Loucks, W. Eichbaum, D. DellaSala, K. Kavanagh, P. Hedao, P. T. Hurley, K. M. Carney, R. Abell, and S. Walters. 1999. *Terrestrial ecoregions of North America: A conservation assessment.* Island Press, Washington, D.C.

Rosgen, D. L. 1994. A classification of natural rivers. *Catena* 22:169–199.

Sedell, J. R., and J. L. Froggatt. 1984. Importance of streamside forests to large rivers: The isolation of the Willamette River, Oregon, U.S.A., from its floodplain by snagging and streamside forest removal. *Verhandlungen der Internationalen Vereinigung für theoretische und angewandte Limnologie* 22:1828–1834.

Shiklomanov, I. A. 1993. World fresh water resources. In P. H. Gleick (ed.). *Water in crisis: A guide to the world's fresh water resources*, pp. 13–24. Oxford University Press, New York.

Smith, R. A., R. B. Alexander, and M. G. Wohman. 1987. Water-quality trends in the nation's rivers. *Science* 235:1607–1615.

Smith, R. A., R. B. Alexander, and G. E. Schwarz. 2003. Natural background concentrations of nutrients in streams and rivers of the conterminous United States. *Environmental Science and Technology* 37:3039–3047.

Smock, L. A., E. Gilinsky, and D. L. Stoneburner. 1985. Macroinvertebrate production in a southeastern United States blackwater stream. *Ecology* 66:1491–1503.

Stanford, J. A., and J. V. Ward. 1979. Stream regulation in North America. In J. V. Ward, and J. A. Stanford (eds.). *The ecology of regulated streams*, pp. 215–236. Plenum, New York.

Strahler, A. N. 1957. Quantitative analysis of watershed geomorphology. *American Geophysical Union Transactions* 38:913–920.

Taylor, C. A., M. L. Warren Jr., J. F. Fitzpatrick Jr., H. H. Hobbs III, R. F. Jezerinac, W. L. Pflieger, and H. W. Robison. 1996. Conservation status of crayfishes of the United States and Canada. *Fisheries* 21(4):25–38.

Thorp, J. H., and M. D. Delong. 1994. The riverine productivity model: An heuristic view of carbon sources and organic processing in large river ecosystems. *Oikos* 70:305–308.

Thorp, J. H., M. D. Delong, K. S. Greenwood, and A. F. Casper. 1998. Isotopic analysis of three food web theories in constricted and floodplain regions of a large river. *Oecologia* 117:551–563.

Tockner, K., F. Malard, and J. V. Ward. 2000. An extension of the flood pulse concept. *Hydrological Processes* 14:2861–2883.

Vannote, R. L., G. W. Minshall, K. W. Cummins, J. R. Sedell, and C. E. Cushing. 1980. The river continuum concept. *Canadian Journal of Fisheries and Aquatic Sciences* 37:130–137.

Wallace, J. B., and J. R. Webster. 1996. The role of macroinvertebrates in stream ecosystem function. *Annual Review of Entomology* 41:115–139.

Ward, J. V., and J. A. Stanford (eds.). 1979. *The ecology of regulated streams.* Plenum Press, New York.

Webster, J. R., and J. L. Meyer. 1997. Stream organic matter budgets. *Journal of the North American Benthological Society* 16:3–161.

Wetzel, R. G. 2001. *Limnology: Lake and river ecosystems*, 3rd ed. Academic Press, San Diego.

Whittaker, R. H. 1975. *Communities and ecosystems.* Macmillan, New York.

Williams, J. D., M. L. Warren, Jr., K. S. Cummings, J. L. Harris, and R. J. Neves. 1992. Conservation status of freshwater mussels of the United States and Canada. *Fisheries* 18(9):6–22.

World Commission on Dams. 2000. *Dams and Development: A new framework for decision-making.* Earthscan, Sterling, Virginia.

WRI (World Resources Institute). 2000. *World resources 2000–2001: People and ecosystems, the fraying web of life.* World Resources Institute, Washington, D.C.

2

ATLANTIC COAST RIVERS OF THE NORTHEASTERN UNITED STATES

JOHN K. JACKSON   ALEXANDER D. HURYN   DAVID L. STRAYER
DAVID L. COURTEMANCH   BERNARD W. SWEENEY

INTRODUCTION

PENOBSCOT RIVER

CONNECTICUT RIVER

HUDSON RIVER

DELAWARE RIVER

SUSQUEHANNA RIVER

ADDITIONAL RIVERS

ACKNOWLEDGMENTS

LITERATURE CITED

INTRODUCTION

The Atlantic slope region of the northeastern United States stretches from the Penobscot River in northern Maine (46°N) to the Rappahannock River on the lower Chesapeake Bay (37.5°N). The larger rivers in this region flow south or southeast between the Appalachian Mountains and the Atlantic Ocean (Fig. 2.2). The relatively short distance between the mountains (or other basin divides) and the coast limits both basin area and river length in the region, but the region's abundant precipitation and high runoff result in high average discharge relative to basin area.

The river basins of the northeast have long histories of human occupation, with archaeological evidence at some locations dating back at least 11,000 to 12,000 years. In more recent times, the rivers played important roles in the European colonization of North America and the establishment of the United States. The basins were dominant features in 10 of the original 13 English colonies, and they contributed vast quantities of lumber, minerals, fish, agricultural products, and power for both local consumption and export. The rivers themselves served as economic, transportation, and communication conduits, and early settlers established such cities as Boston, New York, Philadelphia, and Baltimore on sites where there were natural ports and river access. Although urban, suburban, and industrial development have had major impacts on the streams and rivers in the northeast, agriculture, silviculture, mining, dams and diversions, and nonnative species have also significantly affected the quality and quantity of these freshwater systems.

There are six large river basins (>20,000 km²) in the Atlantic U.S.–Northeast region: Penobscot, Connecticut, Hudson, Delaware, Susquehanna, and Potomac. Several smaller river basins (7000 to 15,000 km²) also occur (e.g., Kennebec, Merrimack, Androscoggin, and Rappahannock), as well as numerous coastal and bay streams and rivers

FIGURE 2.1 Merrimack River near Franklin, New Hampshire (PHOTO BY TIM PALMER).

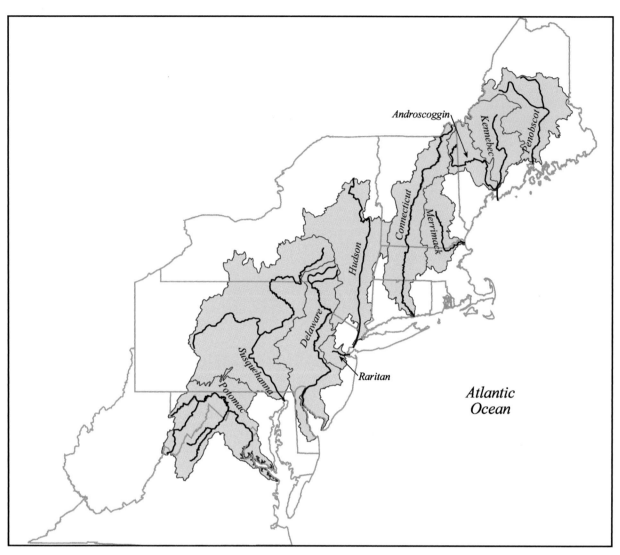

FIGURE 2.2 Atlantic coast rivers of the northeastern United States that are covered in this chapter.

($<5000\,km^2$) (e.g., Saco, St. Croix, Housatonic, Raritan, Mullica, Choptank, and Patuxent). The largest river in the region is the Susquehanna, which is the 22nd largest river in North America that flows to the sea and the 3rd largest in North America that flows into the Atlantic Ocean (after the St. Lawrence and Churchill [Labrador]; Leopold 1994). The Hudson River is one of the most intensively studied rivers in North America, making it one of the best examples to illustrate the physical, chemical, and biological complexity characteristic of large river basins. This chapter addresses in detail five large river basins: the Penobscot, Connecticut, Hudson, Delaware, and Susquehanna. It also provides brief summaries of the physical and biological characteristics of the

Kennebec, Androscoggin, Merrimack, Raritan, and Potomac rivers.

Physiography and Climate

The Atlantic drainage of the northeastern United States is physiographically and climatically diverse. The Appalachian Mountains form the western border of the region and include such well-known ranges as the White Mountains in New Hampshire, the Green Mountains in Vermont, the Catskill Mountains in New York, and the Allegheny Mountains in Pennsylvania and West Virginia. Elevations in the mountains and uplands are frequently 600 to

800 m asl, with peaks reaching 1100 to >1600 m asl. Conditions in the northern mountain ranges of Maine, New Hampshire, and Vermont contrast markedly with those in the flat coastal plain that lies near sea level in Virginia, Maryland, and Delaware. The northern topography is the result of repeated cycles of uplifting and erosion as continental glaciers eroded soils and bedrock and then redeposited them as thin mantles and valley fill. The southern portion did not experience such "smoothing" processes.

The region has seven physiographic provinces (Hunt 1974, Olcott 1995, Trapp and Horn 1997). Three (Appalachian Plateau, New England, and Adirondack) predominate in the northern portion, although limited portions of the Valley and Ridge, Piedmont Plateau, and Coastal Plain also extend up from the south. Four physiographic provinces (Appalachian Plateau, Valley and Ridge, Piedmont, Coastal Plain) dominate the southern portion, although limited portions of the Blue Ridge from the south and the New England province from the north are also found. The northernmost basins originate in the mountains and highlands of the New England province and reach the ocean without entering another physiographic province. In contrast, the southernmost basins originate in the mountains and highlands of the Appalachian Plateau and Valley and Ridge, cross the Piedmont, and end in the Coastal Plain.

The soils and underlying rocks vary greatly across the region (Cuff et al. 1989, Olcott 1995, Trapp and Horn 1997). Sedimentary rocks (e.g., sandstone, shale, carbonate rocks, and in some areas coal) are widespread, especially in the southern portion of the region, whereas metamorphic and igneous rocks (e.g., granite, gneiss, mica schist, slate) are common in the northern portion of the region. There are four common soil orders in the region. Spodosols are mature, acidic soils that are important in the northern basins. Ultisols and alfisols are also mature soils that are often characteristic of the Valley and Ridge, Piedmont Plateau, and Coastal Plain in the southern portion of the region (e.g., Delaware, Susquehanna, and Potomac rivers). Ultisols are thoroughly leached of bases and have low fertility, whereas alfisols have moderate to high levels of bases and are more fertile than spodosols or alfisols. Inceptisols are immature soils (erosion or glaciation have inhibited soil maturation) that do not show horizons and are common in upper portions of the Delaware, Susquehanna, and Potomac rivers.

Mean annual precipitation is generally 100 to 125 cm but can range from 90 to >180 cm (Cuff et al. 1989, Olcott 1995, Trapp and Horn 1997). Although precipitation is abundant throughout the year, its peaks are historically in July to September and are generally associated with tropical storms. Total precipitation tends to be greatest at higher elevations and near the coast. Snowfall occurs throughout the region, but it is far more common in the north and at higher elevations.

This region typically has cold winters and warm summers. Mean annual temperature ranges from 3°C to 13°C. Temperature is colder at higher elevations and further north, and warmer in coastal areas and further south. Coastal areas also tend to experience less variation in temperature than areas further inland.

Basin Landscape and Land Use

There are 10 terrestrial ecoregions in the Atlantic U.S.–Northeast region (Ricketts et al. 1999). The predominant ecoregions are New England/Acadian Forests in the New England province, Northeastern Coastal Forests in the New England province and Piedmont Plateau, Eastern Forest/Boreal Transition in the Adirondack province, Allegheny Highlands Forests in the Appalachian Plateau, and Appalachian/Blue Ridge Forests in the Valley and Ridge. Additional ecoregions include the Eastern Great Lakes Lowland Forests and Appalachian Mixed Mesophytic Forests in the Appalachian Plateau, Southeastern Mixed Forests in the Piedmont Plateau and Coastal Plain, and Atlantic Coastal Pine Barrens and Midatlantic Coastal Forests in the Coastal Plain. Although most of these ecoregions are in the Temperate Deciduous Forest biome, the Eastern Forest/Boreal Transition ecoregion in the Adirondacks and the New England/Acadian Forests ecoregion are partially in the Boreal Forest biome.

Forests were the predominant precolonial land cover in the Atlantic U.S.–Northeast region, although there is evidence that open areas were maintained by fires set by Native Americans and by periodic blowdowns. Forest types vary with latitude and elevation (Braun 1950, Cuff et al. 1989, Ricketts et al. 1999). Appalachian oak forests predominate in the south, northern hardwood forests in the north. Appalachian oak forests commonly contain several oak species, sugar maple, sweet birch, American beech, tulip poplar, and pines. American chestnut, which was common before the blight in the early twentieth century, has since been replaced by hickory. The relative abundance of these species varies among ecoregions.

The transition from Appalachian oak forest to northern hardwood forest occurs in northern Pennsylvania and southern New York (and further north along the coast), and it corresponds in part with the southern limits of glaciation. Characteristic northern hardwood forests include sugar maple, yellow birch, American beech, and eastern hemlock, as well as eastern white pine and red pine. Red spruce and balsam fir become more abundant further north and at higher elevations. White pine and hemlock, which were harvested intensively in the 1800s and early 1900s, were often replaced by hardwood species during reforestation.

The primary forests in the region were harvested 100 to 300 years ago, except in small isolated locations. Some areas have undergone complex evolutions: from forest to farmland or population center, later abandoned and reforested, and finally deforested once again for suburban development. Forests currently represent about 60% of the land cover in most basins in the region, although they are rarely evenly distributed within the basins. For example, forests are often more abundant in areas with steep slopes, whereas agricultural uses and population centers dominate lower gradient areas. Agriculture remains an important feature in the region. Farms tend to be small (0.4 to 0.8 km²) and have a wide range of products compared with farms in other parts of the United States. Dairy, meat, and poultry production and forage crops that support these industries are common, as are vegetable (beans and peas, bell peppers, squash, and tomatoes) and fruit (apples, grapes, peaches, and pears) crops. Whereas forests and agriculture are the predominant land use in the northeastern river basins, industrial, urban, and suburban development continues to expand into agricultural and forested areas. This has connected many of the old colonial cities to form the "northeastern megalopolis," a giant urban zone that extends from Washington, D.C., to Boston and includes 43 million people.

The Rivers

Rivers in the Atlantic U.S.–Northeast region are in either the Chesapeake Bay freshwater ecoregion (which also includes some rivers in the Atlantic U.S.–Southeast region) or the North Atlantic freshwater ecoregion (Abell et al. 2000). The Chesapeake drainages include the Potomac and Susquehanna rivers and numerous smaller drainages on the eastern and western shores of the Chesapeake Bay. The aquatic fauna of the Chesapeake Bay freshwater ecoregion is classified as Continentally Outstanding, with 95 native fish species (two endemic), 14 crayfish species, and 22 unionid mussel species (four endemic). The North Atlantic freshwater ecoregion consists of the drainages from the Delaware Bay north to the Penobscot River in Maine. Its aquatic fauna is classified as Nationally Important, with 98 native fish species (two endemic), 9 crayfish species, and 12 unionid mussel species. Neither of these regions has the degree of biological distinctiveness that characterizes the South Atlantic and Florida ecoregions. This difference probably reflects a number of factors, including extinction associated with glaciation in the northern basins. Although land and marine barriers now separate the coastal basins, millennia of stream migration and capture as well as interbasin connections and dispersal from common refugia during glaciation have contributed to similarities among faunas.

Physicochemical characteristics vary greatly among the rivers. Some are soft, circumneutral, and nutrient poor, and others are hard, alkaline, and nutrient rich. Some are warmwater fisheries, whereas others contain elements of both coldwater and warmwater fisheries. It appears that evapotranspiration is the major factor responsible for the seasonal pattern in runoff. Snowmelt contributes to seasonal variation in discharge in some rivers, whereas others show little or no effect from snowmelt. Variation, reflecting such factors as differences in elevation, river gradient, physiography and geology among branches and tributaries, and proximity to the coast, can also be observed within a basin. Daily tides influence significant portions of the lower reaches of most of these rivers, coming as far as 174 km up the Potomac, 248 km up the Hudson, and 110 km up the Delaware. Much of the tidal portion of the Susquehanna River, however, is "officially" part of the Chesapeake Bay (which is 320 km long), and only 11 km of the Susquehanna is considered tidal. The tidal areas are primarily freshwater, and tidal flows shape channel morphology and determine local flow and related physicochemical conditions. Open access to the sea allows marine animals to enter the rivers, which affects biodiversity and ecosystem function.

Great migratory fish runs, on a par with the better known migrations along the Pacific coast of North America, were once a distinguishing characteristic of the region's rivers. Many millions of anadromous American shad, hickory shad, alewife, blueback herring, striped bass, and Atlantic salmon returned each year to spawn in the rivers and their estuaries. Conversely, catadromous American eels migrated

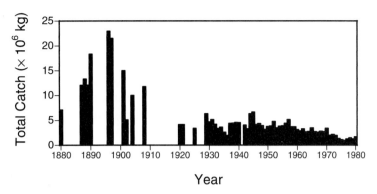

FIGURE 2.3 Total catch of American shad for the U.S. Atlantic coast (i.e., Florida to Maine). Totals should be considered minima. Years with zero catch either had no data or missing data from several states (Atlantic States Marine Fisheries Commission 1985).

from the coastal streams and rivers to spawn in the Sargasso Sea. These large fish migrations were important to the economies of both Native Americans and European settlers. As late as the 1890s, after changes in these basins had presumably begun to affect fish migrations, annual harvests of American shad along the U.S. Atlantic coast were still 12 to 23 million kg or 6 to 13 million fish (Fig. 2.3), and harvests of alewife and blueback herring were similar (13 to 28 million kg or 72 to 148 million fish per year). Unfortunately, dams, pollution, and overfishing decimated most of these annual fish runs long before scientific studies could evaluate their ecological significance, and it has been many years since migratory fish have played a major role in the economy and ecology of this region. With both the public and the rivers themselves disconnected from the once magnificent fish runs, efforts are now being made throughout the region to restore the fisheries and reestablish the connection.

The rivers and their basins in the Atlantic U.S.–Northeast region have been modified significantly over their long histories. Trees have been cut; coal and other minerals mined; land tilled and fertilized; fish harvested; river channels dammed, diverted, and dredged; and waterways polluted with waste from industries, farms, and people. In addition, a wide variety of terrestrial species (e.g., chestnut blight, Dutch elm disease, gypsy moth) and aquatic species (plants, mollusks, and fish) have been introduced to the region. Although the direct and indirect effects of introducing these species are generally unproven, their presence is obvious: Chestnut trees are absent from all basins, nonnative species dominate the region's sport fisheries, and pestiferous zebra mussels have visibly modified the Hudson River.

None of these river basins is considered even close to pristine, but their water quality has improved significantly in the last 50 to 100 years. If current local and national efforts continue this trend, the rivers should be able to provide appropriate water resources for recreation, drinking, fishing, agriculture, manufacturing, and shipping far into the future.

PENOBSCOT RIVER

The Penobscot River arises in western Maine at latitude 46°N (Fig. 2.20). Its drainage (22,253 km²) covers more than a quarter of the state and contains 11,470 stream and river km, 1604 streams (including 188 named streams), and 1224 lakes and ponds (Houtman 1993, PNWRP 2001). Along its southeasterly path, from the border of Quebec to the Gulf of Maine, the Penobscot River cascades through rugged gorges and meanders through tranquil chains of lakes. The western mountains of Maine provide spectacular scenery for this river basin, particularly in the region of the Katahdin massif.

The Penobscot River is the ancestral home of the Penobscot Indian Nation, a tribe once dependent on the river for sustenance and cultural identity (PNWRP 2001). They are Eastern Abenaki— members of the Eastern Algonquian language group (Snow 1978). European exploration of the lower Penobscot River began early in the seventeenth century. Nevertheless, the basin was still largely unpopulated by Europeans as late as 1800. At this time a rapid increase in demand for timber resulted in significant colonization of the Maine forests by Anglo-Americans. By 1840 most of the river valley had a population density of 4 to 19 people/km[2] (Judd et al. 1995), which is comparable to today.

Physiography, Climate, and Land Use

The Penobscot River basin is in the New England (NE) physiographic province (see Fig. 2.20). The upper basin is mountainous with peaks rising above 1500 m asl. The topography of the lower basin is subtler, being covered by a layer of glacial till that is shaped into ribbed moraines, terraces, and spectacular eskers. Glaciomarine clay covers areas lower than 90 m asl. This results in a poorly drained landscape supporting an exceptional richness of wetlands (PNWRP 2001). The topography of the basin began to take shape as glaciers receded from the coast about 13,000 years ago. Since this time the Penobscot River has down cut through till and glaciomarine deposits to produce a channel characterized by extensive bedrock-based waterfalls and rapids that are locally known as "ledge" (Penobscot River Study Team 1972).

The average precipitation for the river basin is 107 cm/yr, including snowfall (240 cm/yr). The maximum monthly precipitation (10 cm) occurs in June and the minimum (6 cm) occurs in February (see Fig. 2.21). Precipitation is greatest in the lower basin. The highest values (~125 cm/yr) have been measured near the Penobscot Bay, whereas the western interior of the basin receives the lowest precipitation (~90 cm/yr; Barrows and Babb 1912). Average air temperature in the lower basin is 9.5°C. The maximum mean temperature (19.6°C) occurs in July and the minimum (−10.1°C) occurs in January.

Despite two centuries of intensive logging (Houtman 1993), the Penobscot River basin remains heavily forested (95% forest and wetland). The upper catchment (New England/Acadian Forests ecoregion) is mantled with spruce–fir (primarily red spruce, white spruce, eastern hemlock, balsam fir). The lower catchment (Northeastern Coastal Forests ecoregion; Ricketts et al. 1999) is a mosaic of spruce–fir, pine (red pine, eastern white pine), and maple–beech–birch stands (silver maple, red maple, American beech, paper birch, gray birch). Acidic bogs and peatlands and mineral-rich swamps and marshes cover almost a third of the basin (Barrows and Babb 1912, Houtman 1993). A small proportion of land cover (5%) consists of farms (hay, potatoes), wood lots, and small urban centers (Houtman 1993). Human density of the basin is sparse, averaging about 8 people/km[2]. Bangor, the largest urban center within the drainage, has a population of 33,000.

River Geomorphology, Hydrology, and Chemistry

The Penobscot River is about 320 km long and begins its flow to the Penobscot Bay at about 500 m asl. Its main channel can be divided into four geomorphic segments: mountainous upland, island/bar, rapids/terrace, and estuary (PNWRP 2001). The mountainous upland segment contains the East and West Branches. Their headwaters are separated from the St. John and Kennebec river drainages by extremely low divides. This has resulted in major diversions of flow to and from the Penobscot drainage in both prehistoric and historic times. Moosehead Lake, which now empties into the Kennebec River drainage, emptied into the West Branch of the Penobscot more than 8000 years ago. Its waters were diverted into the Kennebec River due to a slight tilting of the landscape that resulted from isostatic adjustment following deglaciation. The Telos Canal, constructed around 1840 to facilitate the transportation of timber, diverted water from the St. John drainage into the East Branch and increased the effective Penobscot River basin by 3% (~700 km[2]; Barrows and Babb 1912).

The main stem of the river is formed by the confluence of the West and East Branches and flows for about 158 km. This portion of the river channel contains the island/bar, rapids/terrace, and estuary segments. The island/bar segment is the most extensive, occupying about 80 km of channel (Fig. 2.4). There are more than 119 islands (0.4 ha or more in size), representing a combined area of 1791 ha (Baum 1983). Between the island/bar segment and the estuary is a short (13 km) rapids/terrace segment that is characterized by numerous bedrock-based rapids. The word *Penobscot* translates to "at the place of the descending rocks" and refers specifically to this reach of the river (PNWRP 2001). The rapids/terrace unit terminates near Bangor, where the river becomes tidally influenced. At this point the main channel is greater than 300 m wide. A classic salt-wedge estuary occupies the final 50 km of the main stem before it empties into the Penobscot Bay (Baum 1983, Moore and Platt 1996).

The average annual discharge of the Penobscot River near the point of tidal influence is 402 m[3]/s. The greatest discharge on record (4502 m[3]/s) was measured in April 1987; the lowest (79 m[3]/s) was measured the following September (Houtman 1993). Although precipitation is distributed relatively uniformly over the year, the pattern of discharge is strongly seasonal (see Fig. 2.21) due to winter snow

FIGURE 2.4   Penobscot River at Five Islands Rapids, below Mattawamkeag, Maine (PHOTO BY TIM PALMER).

and ice accumulation and high rates of evapotranspiration during summer (Barrows and Babb 1912). The Penobscot River is usually ice covered from late December to March. Average runoff is about 5 cm/mo (63 cm/yr) (59% of annual precipitation); about 44 cm/yr is lost through evapotranspiration. The highest monthly discharge typically occurs in April, whereas the lowest occurs in September (see Fig. 2.21).

There are currently 116 licensed dams in the Penobscot River basin. These are used for power generation and water-level control (PNWRP 2001). An analysis of discharge records at West Enfield (93 km upstream of the Penobscot Bay) indicates that the natural runoff during April to June and November to December has been reduced by 21% to 27% due to storage. The runoff for January to March and July to October has been increased by 26% to 33% by release of stored water (Penobscot River Study Team 1972).

The Penobscot River is slightly acidic (average pH = 6.6), moderately buffered (mean hardness = 12 mg/L as $CaCO_3$ equivalents), and relatively oligotrophic. Total dissolved nitrogen ranges from 0.15 to 2.60 mg N/L (average = 0.45 mg N/L), total dissolved phosphorus ranges from 0.01 to 0.16 mg P/L (average = 0.03 mg P/L), and total organic carbon ranges from 5.2 to 16.0 mg C/L (average = 9.5 mg C/L). The water temperature ranges from about 0°C in February to 22°C in July, with an annual mean of 9.3°C (http://water.usgs.gov).

River Biodiversity and Ecology

The Penobscot River is in the North Atlantic freshwater ecoregion. Much is known about macroinvertebrate assemblages because of their use in biomonitoring (Davies et al. 1999, PNWRP 2001) and recent concern about the conservation of freshwater mussels (e.g., Nedeau et al. 2000). There is also much known about fisheries biology due to efforts to manage populations of diadromous fishes (Cutting 1963, Baum 1983).

Plants

Red maple, silver maple, and black ash are major trees species, and tussock sedge is a major grass species of seasonally flooded bottomlands. Other common riparian trees are chokecherry, paper birch, gray birch, quaking aspen, tamarack, northern white cedar, and balsam fir. Although American elm is in serious decline, saplings can be numerous. In shallow river reaches with soft sediment, floating-leaf pondweed and bur-reed are important habitat for black fly larvae. The pond-lily and broadleaf arrowhead are abundant in backwaters. Riverweed, listed as threatened in Maine, occurs in a few rocky shoals and rapids.

Invertebrates

Rabeni and Gibbs (1980) conducted the first comprehensive study of benthic invertebrate assemblages of the Penobscot River. They used artificial samplers (rock baskets) and SCUBA to quantify macroinvertebrates from 33 locations along 100 km of the river and two of its major tributaries. Habitats with stony substrata and current velocities >11 cm/s had communities represented by caddisflies (*Brachycentrus*, *Cheumatopsyche*, *Chimarra*, *Helicopsyche*, *Hydropsyche*, *Ithytrichia*, *Macrostemum*, *Neureclipsis*, *Oecetis*), stoneflies (*Agnetina*), mayflies (*Eurylophella*, *Serratella*, *Stenonema*, *Isonychia*, *Baetidae*), and dragonflies (*Neurocordulia*). Macroinvertebrate richness in other habitats was lower. Dragonflies (*Boyeria*) and stoneflies (*Acroneuria*) characterized habitats with stony substrata and current velocities <5 cm/s. Habitats with moderate levels of silt and current velocities of 5–11 cm/s were inhabited by mayflies (*Drunella*, *Stenonema*, *Tricorythodes*), riffle beetles (*Stenelmis*), hydrobiid snails (*Amnicola*), and turbellarians. Habitats with silt substrata and current velocities <5 cm/s were characterized by mayflies (*Eurylophella*, *Stenonema*, *Leptophlebia*, *Paraleptophlebia*), caddisflies (*Nyctiophylax*, *Polycentropus*), aquatic earthworms (Tubificidae), fingernail clams (*Musculium*), leeches (*Helobdella*), isopods (*Asellus*), and amphipods (*Hyallela*). Macroinvertebrate assemblages in highly polluted reaches below pulp mills had high proportions of aquatic earthworms and chironomid midges.

Ten species of freshwater mussels are known in Maine (Nedeau et al. 2000). All occur in the Penobscot River. The most widespread are the eastern elliptio, triangle floater, eastern floater, and eastern lampmussel. These have been recorded from ten or more locations. The creeper, brook floater, and yellow lampmussel occur at six or more locations. The latter three species are of particular conservation concern in Maine and throughout New England (Nedeau et al. 2000).

Vertebrates

Forty-five species of freshwater and 39 species of marine fishes have been recorded from the Penobscot River basin (Baum 1983). Thirty-one freshwater species are permanent residents. Eighteen are either obligatorily or facultatively diadromous, and thus use marine habitats for some part of their life cycle. The assemblage of permanent residents inhabiting the main stem of the river is relatively small and is composed of a handful of warmwater species.

The top predators are the smallmouth bass, a warmwater predator in both riverine and lacustrine habitats, and the chain pickerel, restricted to large pools and impoundments. Other abundant piscivores are the white perch, which is facultatively anadromous, and yellow perch. Abundant invertivores include the creek chub, fallfish, and redbreast and pumpkinseed sunfishes. Omnivores include the brown bullhead, white sucker, and common shiner (FERC 1997). Coldwater species include the piscivorous burbot, restricted to deep pools and impoundments, and the salmonids. Nonmigratory salmonids are scarce in the main stem, however, and resident populations are apparently not sustainable. Brook trout are numerous in smaller tributaries and only occasionally enter the main stem.

The American eel is the only catadromous fish species in the Penobscot River. They are abundant in impoundments and backwaters. Eels are able to travel more than 200 km upstream of the Penobscot Bay. To achieve this they must pass through 11 major hydroelectric projects (FERC 1997). Of 17 anadromous fish species, alewife, American shad, and Atlantic salmon are perhaps best known. Prior to European occupation, millions of alewife migrated as far as 320 km upriver each year (Baum 1983). The current run, however, has been reduced by three orders of magnitude (FERC 1997). American shad were also historically abundant, with 2 million adults running more than 270 km up the river prior to 1800 (Moore and Platt 1996). The current population is probably less than several thousand individuals (PNWRP 2001). Returning Atlantic salmon once penetrated 230 km up the Penobscot River (Moore and Platt 1996). Early in the nineteenth century, 40,000 to 75,000 fish returned each year (Baum 1983). As settlement and development of the basin intensified, populations declined catastrophically. By 1960 Atlantic salmon were extirpated, presumably due to a decline in water quality as the Penobscot basin was industrialized (Cutting 1963, Baum 1983; Fig. 2.5). Atlantic salmon have since been reestab-

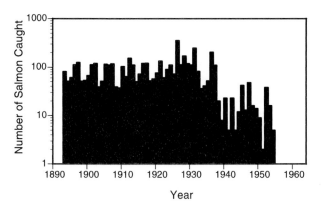

FIGURE 2.5 Atlantic salmon harvest by angling below the Veazie Dam (~43 km upstream of the Penobscot Bay) from 1893 through 1963 (Cutting 1963).

lished (Baum 1983, PNWRP 2001). Since 1990, annual returns of 2000 to 4000 adult fish have occurred regularly, and natural spawning occurs as far as 100 km upstream (Moore and Platt 1996; PNWRP 2001). Nevertheless, 6000 to 10,000 spawning salmon are required for a sustainable population. The survival of this Atlantic salmon population thus depends on a hatchery program (Baum 1983, Moore and Platt 1996).

A diverse assemblage of aquatic and riparian reptiles, birds, and mammals are associated with the Penobscot River. The snapping turtle and the eastern painted turtle are both common. Wood ducks, goldeneyes, and black ducks nest along the main stem. Black ducks, greater scaup, and goldeneyes overwinter in the lower river, and common eiders and surf scoters overwinter in the estuary. Canada geese are common during spring migration (Penobscot Study Group 1972). Bald eagles forage along the main stem during winter and nest in the riparian zone during spring and summer (FERC 1997). Muskrat, beaver, mink, and river otter are common throughout the central and lower drainage. Harbor seals frequent the estuary and pilot whales have appeared as far as 20 km upstream of the Penobscot Bay.

Ecosystem Processes

Like most large rivers, little is known about ecosystem processes in the Penobscot River. Nevertheless, several generalizations can be made. First, the enormous area of wetland moderates fluctuations in river discharge. This results in a highly predictable annual flood pulse. Although numerous dams punctuate the river, its flow is only moderately regulated because these dams tend to be "run-of-the-river." The natural flow regime of the main stem of the

Penobscot River is thus at least nominally intact and the spring flood pulse maintains numerous riparian swamps and wetlands. Second, the acidic, peat-laden soils of the wetlands result in river water that is poor in nutrients but rich in dissolved organic matter. Low nutrient concentrations and low light penetration due to highly stained water indicate that primary production should be low. Production by microbial and animal communities in this river system is thus probably based upon terrestrial carbon. Nevertheless, primary production is probably significant in the shallow rapids and shoals that have not yet been affected by dams.

Special Features and Human Impacts

Baxter State Park protects >800 km² of the upper Penobscot River basin and contains Mt. Katahdin, an important alpine habitat in New England. Henry David Thoreau popularized the Baxter area in *The Maine Woods* (Thoreau 1864). Sunkhaze National Wildlife Refuge, in the lower basin, encloses one of the largest peat bogs in Maine and protects a further 40 km² of land. The Penobscot Indian reservation includes all the islands of the 80 km island/bar segment of the Penobscot River. The Penobscot Indian Nation has continuously occupied these islands for thousands of years.

Following 1800, perhaps the most obvious human impact on the Penobscot River ecosystem is the near extirpation of what were once enormous runs of anadromous fish. This is attributed to dams that obstruct movements of fish, the introduction and establishment of warmwater fish species, and a long legacy of habitat destruction and pollution associated primarily with timber extraction and processing.

The first major dam was constructed near Old Town (~56 km from the Penobscot Bay) during the 1820s; the last was constructed at Mattawamkeag Town (~134 km from the Penobscot Bay) in 1939 (Cutting 1963, Baum 1983). There are currently five dams that span the main stem of the river. Although all of these have functional fishways, this was not always the case. The first fishways were constructed around 1879, almost half a century after the first dam effectively stopped upstream migration of fish 50 to 60 km upstream of the Penobscot Bay. A complex of 21 major dams on the West Branch of the Penobscot, which arguably comprise the largest privately owned hydroelectric development in the United States (Moore and Platt 1996), is notable in that there currently are no fishways. These dams have stopped movement of anadromous fishes into a major portion of the drainage (PNWRP 2001).

The Penobscot River has a rich history of introductions of both salmonids and warmwater gamefish. Some of these were successful, most were not. The first was the chain pickerel, introduced in 1819. The smallmouth bass was introduced in 1869. Both are now abundant and important predators of and competitors with Atlantic salmon. The emerald shiner, a warmwater forage fish, was introduced more recently (Baum 1983), apparently as "bucket bait." From 1874 through 1950, coho salmon, steelhead and rainbow trout, pink salmon, brown trout, and grayling were introduced on one or more occasions. With the exception of the brown trout, which remains as a small population in streams entering the estuary, all introductions of salmonids have failed (Baum 1983).

The Penobscot River was the center of what was arguably the largest basin-specific timber industry in North America. Significant logging had begun by 1815 (Cutting 1963). By 1851 Bangor held the world's record for lumber exports (Cutting 1963). Old-growth timber was virtually exhausted by about 1880, and the timber industry focused on pulp production (Cutting 1963). When the log drives ended in the mid 1960s, over two trillion board feet of lumber (4.8 billion m^3) had been floated down the river (Cayford and Scott 1964). Between 10% and 15% of all logs driven downriver sank ("sinkers") and remain buried in the sediments of impoundments and low gradient river segments. In addition, a century of intensive sawmill operation contributed a layer of bark and sawdust that exceeds 60 cm in places over the bottom of the estuary. This deposit is apparently trapped by the circulating flow of the estuary (Moore and Platt 1996).

Prior to 1960 wastewater from pulp and paper mills, municipalities, and miscellaneous industries resulted in poor water quality throughout most of the main stem. The decomposition of these wastes resulted in extremely low concentrations of dissolved oxygen, and much of this material was contaminated with dioxin and PCBs (Davies et al. 1999). Although severely polluted throughout much of the twentieth century, the water quality of the Penobscot River has shown astonishing improvement since 1975 (Rabeni et al. 1985, Davies et al. 1999). Nevertheless, water-quality issues remain. Major sources of continuing pollution include wastewater from five pulp and paper mills and ten municipal water treatment plants, fish hatcheries, miscellaneous industrial waste discharges, combined sewer overflows, a contaminated

woolen mill, and a host of nonpoint sources dominated by timber harvesting and agriculture (Davies et al. 1999). Tissues from fish sampled during 1999 from locations below pulp and paper mills had dioxin and PCB levels exceeding statutory limits for human consumption (PNWRP 2001).

Perhaps the best integrative measure of the legacy of human impact on the Penobscot River ecosystem is the change in its carrying capacity for Atlantic salmon from first European settlement to the present. The Penobscot River and its tributaries currently provide good water quality and excellent salmonid spawning and nursery habitat (Baum 1983, Davies et al. 1999). Although the total area of this habitat is substantive, when compared with other New England rivers, conservative estimates of the present carrying capacity for Atlantic salmon is only about 6000 to 9000 returning adults (Baum 1983). This is a large decrease compared with estimates of 40,000 to 75,000 returning adults prior to European colonization (Baum 1983). Although these statistics require many assumptions, the apparent order of magnitude decrease in carrying capacity from 1780 to the present indicates the magnitude of structural and functional changes to the Penobscot River ecosystem that have occurred since colonial times. Perhaps the most significant and obvious of these changes was the introduction of warmwater game fish, which act as competitors and predators of Atlantic salmon, and the loss of about 30% of excellent salmon habitat due to channel obstruction and impoundment (Baum 1983).

CONNECTICUT RIVER

The Connecticut River drains the largest watershed in New England and, at 660 km, is its longest river (see Fig. 2.2). Originating as a group of ponds (First to Fourth Connecticut Lakes) in the northern highlands of New Hampshire and Quebec, the river flows southerly through a valley bordered by the two major mountain ranges of New England, the Green and White mountains (Fig. 2.22). Further south, it traverses central Massachusetts and through the coastal plain of Connecticut into Long Island Sound. The drainage encompasses 29,160 km^2. Beginning at an elevation of 568 m asl at Third Connecticut Lake, the river descends rapidly in its first 150 km to an elevation of about 210 m asl at Fifteen Mile Falls near Lancaster, New Hampshire. Thereafter, the descent is more gradual and the natural channel begins to meander (Fig. 2.6). The final 90 km upstream of Long

FIGURE 2.6 Connecticut River near White River Junction, Vermont. White River at upper right (Photo by Tim Palmer).

Island Sound is a tidal estuary that gave the Connecticut its Native American name, "Quinni-tukqut" or "Quoneh-ta-cut," meaning "long tidal river" (Bacon 1906).

There is evidence that the Connecticut River valley was first colonized by Native Americans about 9000 years ago following the retreat of the last glacier. These peoples hunted the megafauna (e.g., mastodon, giant beaver, dire wolf) that inhabited the valley. As those animals disappeared, greater focus was given to the fish and wetland vegetation provided by the river as a food source. The Pequot tribe occupied the lower tidal area of the river, whereas the Mohegan tribe occupied the upper watershed. These peoples also introduced the first agriculture, growing corn, beans, squash, sunflowers, and tobacco (Hauptman and Wherry 1993). In 1614, the Dutch explorer Adriaen Block was the first European to discover the river. Settlement along the river began in 1633 by members of the Plymouth Colony. The rich farmland along the river, coupled with a convenient transportation corridor afforded by the river, led to rapid settlement in the valley. Waterpower provided

by the river allowed early industrialization and by the late 1800s towns along the river had attained prominence in the manufacturing of a variety of goods (Hard 1947).

Physiography, Climate, and Land Use

The Connecticut Valley basin is located within the New England (NE) physiographic province. The basin developed as part of a rift system during the formation of the Atlantic Ocean. The Eastern Border Fault defines the eastern edge of the basin south from Keene, New Hampshire. Beginning over 200 million years ago, rivers deposited gravel to form alluvial fans along the west face of this fault system, and lakes formed repeatedly in the valley as the basin expanded. By the end of the Mesozoic era, erosion and sediment deposition created a large flat plain (Olsen et al. 1989).

Two prominent events in the Cenozoic era shaped the Connecticut Valley: uplift and glaciation. The former plain was uplifted, resulting in extensive erosion and valley development. Continental glaciers

further eroded the valley by excavating bedrock to depths below the present land surface. In low-lying areas, glacial-fluvial, glacial-lacustrine, or glacial-marine sediments were deposited after ice recession. During the Pleistocene, the Connecticut valley was occupied by glacial lakes. The last glacier reached its maximum extent about 20,000 years ago. When the face of the glacier receded to the area of Rocky Hill, Connecticut, deposits blocked the valley, forming a dam that collected meltwater to form a great glacial lake, Lake Hitchcock. At its maximum extent, Glacial Lake Hitchcock extended to an area near St. Johnsbury, Vermont, approximately 320 km. The lake drained about 14,000 years ago. The Connecticut River returned to its valley and began to erode the Hitchcock sediments. The river has left abandoned channels (oxbow lakes) in many areas, floodplains, and where the land was resistant to erosion, narrow valley segments formed with waterfalls and rapids. The lacustrine deposits of Lake Hitchcock created the fertile agricultural soils that later attracted human settlement of the valley.

The river originates in the New England/Acadian Forests ecoregion (Ricketts et al. 1999), a transition area of boreal forest biome in the north to temperate deciduous forest biome in the south, with considerable vertical zonation. The river drains south into the Northeastern Coastal Forests ecoregion. This ecoregion is characterized by a flat gradient. Forests are predominantly hardwood, with red and white oak, sugar maple, beech, and eastern hemlock.

The climate of the Connecticut River watershed is temperate and humid. The mean annual precipitation of about 109 cm is distributed relatively uniformly throughout the year (Fig. 2.23) but ranges from 107 cm at the northern end of the valley to 204 cm in adjacent mountain areas (Garabedian et al. 1998). Mean annual air temperature ranges from 3°C to 10°C in the upper watershed area, with tropical air masses influencing summer weather and polar air masses influencing winter weather. Frozen conditions may persist for four to five months and snow accumulation exceeds 250 cm (Bailey 1980). In the lower watershed, mean annual temperature ranges from 5°C to 13°C, with more humid conditions and oceanic influence. Mean monthly temperatures range from 23.1°C in July to −2.0°C in January (see Fig. 2.23).

Land use varies considerably, from the heavily forested headwaters to highly urbanized areas. Approximately 80% of the watershed is forested and 11% is in agriculture. The rural headwaters support a wood products and tourist-based economy. The

watershed of southern Vermont and New Hampshire and northern Massachusetts is agriculture based, especially dairy, vegetables, and tobacco. The urban/suburban corridor of the Holyoke–Springfield area of Massachusetts to Hartford, Connecticut, supports about 85% of the estimated 2 million people living within the watershed. The lower estuary is characterized by a more rural landscape of small towns. The population of the lower watershed of the river continues to grow as the river traverses the urban corridor connecting Boston and New York City.

River Geomorphology, Hydrology, and Chemistry

The morphology of the river is remarkably symmetric and dendritic in shape, with no singularly large branches or tributaries. The five largest tributaries (Chicopee, Deerfield, Farmington, Westfield, White) are only from 1295 to 1865 km^2 in area, each being less than 7% of total watershed area. After flowing out of First Connecticut Lake, the river flows into the first of many impoundments, Lake Francis. It flows freely for about 150 km until it reaches the next impoundment at Moore Reservoir. Bottom substrate is characterized by coarse eroded material of gravel, cobble, and boulder. Thereupon, the river starts to encounter a series of impoundments formed by the many hydroelectric dams on the river. The contour of the river is a series of steps as water passes from one impoundment to another with short sections of flowing rapids below each dam. Each impoundment provides a gradation of substrate types, from coarser gravel and cobble near the head to predominantly sand and finer organic material in the deeper tail segment. The river reaches a final set of free-flowing rapids at Enfield, Connecticut, above the head of tide.

Average discharge at the river mouth is about 445 m^3/s. Peak discharge occurs in late winter and early spring despite relatively uniform precipitation throughout the year (see Fig. 2.23). Such a strongly seasonal runoff pattern is apparently caused by seasonal variation in evapotranspiration and snow/ice storage. These peak flows (averaging 690 m^3/s) are dampened by storage behind the estimated 125 dams in the drainage. Many of the larger dams are used for power generation and 16 are flood control structures (Merriman and Thorpe 1976, Garabedian et al. 1998). A large portion of the flow of the Chicopee River tributary in Massachusetts is retained in Quabbin Reservoir and diverted as a water supply for the Boston area, approximately 80 km to the east. The Connecticut River contributes about 70% of the total freshwater inflow of Long Island Sound.

The Connecticut is a soft-water river, low in total ion concentration, and circumneutral. Measurements of water chemistry in the lower river, above tidal influence, have a mean specific conductivity of 116 μS/cm, mean calcium and magnesium concentrations of 11.0 mg/L and 1.7 mg/L, respectively, and mean pH of 7.0. The presence of elevated chloride concentrations indicates the chemistry of the lower river may be substantially altered by human activities (Trench 1996). Patrick (1996) notes an N/P ratio for the lower river indicative of organic pollution that may increase the growth of pollution-tolerant algae species.

River Biodiversity and Ecology

The Connecticut River watershed is placed within the North Atlantic Freshwater ecoregion, Long Island Sound subregion (Abell et al. 2000). With the exception of the work of Merriman and Thorpe (1976) in the lower Connecticut River, there have been no comprehensive studies of the river ecosystem. Merriman and Thorpe provide a detailed analysis of the Connecticut River estuary before and after the start-up of the Connecticut Yankee atomic power facility, including analysis of the composition, movement and migration, recruitment, and mortality of the zooplankton, benthic fauna, and fish populations.

Plants

The lower Connecticut River, beginning near its mouth and continuing upstream for a distance of approximately 58 km, contains one of the least disturbed large-river tidal marsh systems in the United States. From a regional standpoint, there are no other areas in the Northeast that support such extensive, high-quality fresh and brackish tidal wetland systems as the Connecticut River estuary. These tidal river waters and marshes provide essential habitat for a variety of fish and an important nursery. Shallow soft bottom areas of the river are populated with wild celery, with fringing populations of broadleaf arrowhead, water plantain, cattail, and pickerel-weed in the fresh intertidal areas and cord grass in the saline intertidal area (Massengill 1976). The river above the estuary does not provide extensive habitat for the development of macrophyte vegetation with the exception of the oxbows and shallow portions of the many impoundments. A number of nonnative species occur that alter the community composition. These

include common reed, purple loosestrife, Eurasian watermilfoil, water chestnut, and flowering rush (Hellquist 2001).

Invertebrates

The upper river above the impounded portions has not been well studied. This flowing habitat will support communities typified by stoneflies (*Acroneuria*), mayflies (*Eurylophella, Serratella, Stenonema*), caddisflies (especially filter-feeding forms such as *Brachycentrus, Chimarra, Hydropsyche, Neureclipsis*), and riffle beetles (*Stenelmis, Psephenus*). The impounded segments support a community typical of lentic habitats. In the impoundment above Holyoke Dam in the 1970s, Patrick (1996) reports from collections made in the 1970s a community dominated by worms (Tubificidae), caddisflies (*Oecetis*), and chironomid midges (*Chironomus, Polypedilum, Microtendipes, Glyptotendipes, Tanytarsus*). Patrick describes a benthic fauna of the river below Holyoke Dam dominated by worms (Tubificidae), snails, pill clams (Pisidiidae), and chironomids. These communities changed following wastewater treatment improvements; however, current data do not exist for this part of the river.

The largest remaining population of the federally listed dwarf wedge mussel in New England is thought to be in the Connecticut River. The Connecticut River has been invaded by the Asiatic clam, but has not been invaded by zebra mussel and may be too soft to support a significant population.

Vertebrates

The most extensive fishery surveys of the river have been made in the lower estuary. Marcy (1976) collected 44 species from 21 families and reported the occurrence of an additional 6 species known to inhabit this area of the river. Ten species were marine and restricted to the brackish portion of the estuary. Ten diadromous species utilize the river for passage, spawning, or nursery habitat. Immature fishes of 28 species were found, indicating the importance of the estuary to fish production.

It is estimated that the nontidal waters of the river support about 64 species (Whitworth 1996, Hartel et al. 2002). Twenty-three species have been introduced, four from outside North America. Like many North American waters, biological composition has been substantially altered by numerous introductions in the watershed. Many fish species that presently dominate the river have been introduced, including brown trout, rainbow trout, common carp, northern pike, bowfin, channel catfish, rock bass, smallmouth bass,

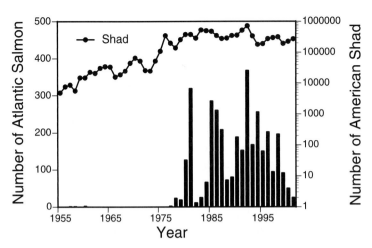

FIGURE 2.7 Number of Atlantic salmon and American shad reported at the Holyoke Dam (Massachusetts) fish lift on the Connecticut River from 1955 through 2001 (Massachusetts Department of Fisheries and Wildlife).

largemouth bass, bluegill, white crappie, black crappie, and walleye. Most of these species are more suited to the altered habitat provided by the slower, deeper impounded waters of the river than many native species. A number of native species adaptable to the altered habitat and increased competition persist, including white sucker; yellow perch; American eel; fallfish; common, golden, and spottail shiners; banded killifish; redbreast and pumpkinseed sunfishes; and brown bullhead. In the upper watershed, a more lotic assemblage of species, including salmonids (native brook trout and introduced brown and rainbow trout), smallmouth bass, and white sucker, occur in greater relative abundance and provide evidence of the shift in species composition that has occurred.

The federally listed shortnose sturgeon is known to occur in the estuary and in several upriver impoundments of the Connecticut River. The shortnose sturgeon occurred in tidal rivers from North Carolina to New Brunswick, but numbers have been decimated in many rivers by dams that have prevented access to spawning areas, pollution, and overexploitation. The eastern silvery minnow is listed as a species of concern in Massachusetts, occurring only in the Connecticut River.

Once abundant, anadromous fish were seriously reduced, predominantly by the construction of dams and, to a lesser extent, water pollution. Atlantic salmon were extirpated over a century ago by dams on the main stem and tributaries that cut off access to spawning habitat. A salmon restoration effort involving hatchery production and stocking and provisions for fish passage at dams has produced a

limited run of fish (Fig. 2.7). A more remarkable restoration occurred for American shad and river herring beginning in the 1950s. The Enfield Dam in Connecticut was breeched, and a fish lift was provided at the Holyoke Dam in Massachusetts and at dams further upriver, providing access for these species to a greater extent of the watershed (see Fig. 2.7). Other diadromous species that have benefited from improved passage and water quality include striped bass, sea lamprey, American eel, alewife, and gizzard shad.

Ecosystem Processes

In its natural state the Connecticut River would be a good model of the River Continuum Concept (Vannote et al. 1980). Arising from a number of small tributaries with steep gradients, the river continually grows in relation to the size of the drainage area, becoming slower and more meandering in the lowland reaches until it reaches a large estuary. Carbon sources of the upper river are linked to its forested watershed. Processing of coarse particulate organic matter in the tributaries contributes smaller fractions to downstream communities. About 300 km of the river have been largely altered by a series of impoundments that change the character and transfer of organic matter. The impoundments are a sink for large particulates, contributing to a larger detritivore population. The impoundments also become the source of a larger fraction of phytoplankton, utilized by planktivores in the impoundments and by filter-feeding organisms in the runs below each dam. Fish populations are dominated by lentic species, better suited to the slow, deep habitats. Primary production in the lower river and estuary is increased by the addition of nutrients that occurs in the lower river from human sources (Patrick 1996).

Human Impacts and Special Features

The Connecticut River was the first large river of the Americas to be significantly modified and controlled for transportation and power. The first dam and canal at Hadley Falls, Massachusetts, was constructed in 1795. Six canals constructed between 1795 and 1829 provided access to the upper reaches of the watershed, and presumably afforded passage for many aquatic species, especially introduced species, to extend their colonization of the river along with the early settlers. The first large power dam was constructed at Holyoke, Massachusetts, in 1848, and completely obstructed passage for upstream migrating fish. It is estimated that upward of 900 dams have been constructed on the river and its tributaries for various purposes to control and store the water, for log-driving in the nineteenth century, and later to generate power and reduce flooding. The Connecticut is one of the most developed rivers in the Northeast, with 16 operating power dams that impound over 300 km of the river, affecting water quality, flow, fisheries, and other river aspects. Federal operating licenses for these dams must be renewed; thus, water management can be expected to change in coming years to accommodate other uses and its natural values.

As a consequence of its long history of human settlement, agriculture, industrialization, and urbanization, the Connecticut River contains a substantial burden of contaminants that affect the river ecosystem and become a significant source of contamination for Long Island Sound. Toxic and nutrient contaminant loads are significant, particularly affecting that portion of the river in Massachusetts and Connecticut. Surveys conducted from 1992 to 1995 found contamination of sediments and tissue with trace metals—chromium, copper, lead, mercury, nickel, and zinc—and organic compounds—chlordane, DDT, PCBs, and various polycyclic aromatic hydrocarbons (PAHs) (Garabedian et al. 1998). Massachusetts and Connecticut have issued consumption advisories for fish and shellfish taken from the river because of high levels of PCBs. In addition, these states have issued consumption advisories for all their freshwaters due to mercury contamination, presumably from atmospheric deposition (MADEP 1998).

The Connecticut River, certain tributaries, the estuary, and Long Island Sound have all been adversely affected by increased nutrient loading, primarily associated with wastewater treatment facilities. Other sources include urban storm water, agriculture, and atmospheric deposition. Overall water quality has improved in recent decades. Trench (1996) and Garabedian et al. (1998) report a decline of total phosphorus since 1980 for the lower Connecticut. This is presumed to be due to improved wastewater treatment and reduction of the use of phosphates in detergents. A similar trend was not found for nitrogen compounds. Further reductions in nutrients are planned for wastewater facilities in the watershed to reduce their effects on Long Island Sound, where algal blooms have caused low oxygen levels.

In recent years, the Connecticut River has earned a number of designations. The rich natural diversity and special qualities of the Connecticut and its watershed have gained both national and international recognition. The tidal wetlands in the lower 58 km of the estuary were designated "wetlands of international importance especially for wildlife" under the Ramsar Convention in 1994. This international treaty identifies wetlands of critical value for wildlife habitat. In 1993, The Nature Conservancy similarly recognized the tidal wetlands of the lower Connecticut River as an area of global ecological importance by naming it among their "Last Great Places" and a target for land conservation and ecological research. In 1991, the entire Connecticut River watershed was designated a National Fish and Wildlife Refuge (NFWR). This unique designation recognizes the national significance of the watershed and the diversity of life that it supports. The Silvio O. Conte NFWR is a model of resource conservation based on a partnership between the U.S. Fish and Wildlife Service, state and local agencies, nonprofit organizations, and landowners. In 1998, the Connecticut River was designated as an American Heritage River. The designation is the result of a valleywide effort led by the Connecticut River Watershed Council and communities and institutions in the four watershed states. As an American Heritage River, the Connecticut receives special attention from government agencies working in partnership with local sponsors to protect the environmental, cultural, and economic values of the river.

HUDSON RIVER

The Hudson is one of North America's most important, best-studied, and fiercely protected rivers. It begins in the Adirondack Mountains and flows south, draining most of eastern New York State (Fig. 2.24). It has been a vital transportation route since the seventeenth century, is part of the Erie Canal system, and is used today for commercial shipping. The Hudson Valley was a center of Native American and early European settlement, and the site of key battles in the Revolutionary War. Its beauty inspired painters of the Hudson River school and helped to motivate the modern environmental movement in the United States. The Hudson's stocks of anadromous fishes have supported commercial fisheries for centuries and still are among the largest remaining populations of striped bass, American shad, and shortnose sturgeon. Today, despite changed land use, altered hydrology, a reshaped channel, invasions of nonnative species, and episodes of serious pollution, including an unresolved problem with PCBs, the Hudson supports diverse biological communities and is used for recreation, drinking water, commercial shipping, and commercial fishing.

The Hudson was an important source of fish and shellfish for the Native Americans of the Algonquin and Iroquois groups who lived in the Hudson Valley before European settlement. Dutch and English settlers drove most of these people from the Hudson basin in the seventeenth and eighteenth centuries, leaving behind Native American names for many places in the basin (e.g., Manhattan, Esopus, Schodack). Despite a long history of European exploration and trade in the Hudson Valley in the seventeenth and eighteenth centuries, the land was not heavily settled or widely cleared for agriculture until after the American Revolution.

Physiography, Climate, and Land Use

The Hudson basin includes parts of several physiographic provinces and ecoregions (Hunt 1974, Isachsen et al. 1991, Ricketts et al. 1999; Fig. 2.24). The northern parts of the basin are in the Adirondack Mountains (AM) province in the Eastern Forest/Boreal Transition ecoregion. The bedrock here is highly metamorphosed plutonic rock, including granitic gneiss, metanorthosite, and olivine metagabbro, and soils are acidic. Much of the central part of the basin is in the Valley and Ridge (VR) province (specifically the Hudson-Mohawk Lowlands and the Taconic Mountains), which includes two ecoregions, the Eastern Great Lakes Lowland Forests to the north and the Northeastern Coastal Forests to the south. This part of the basin is underlain by sedimentary rocks (shale, limestone, sandstone) of Cambrian to Sil-

urian age, which have been metamorphosed in the eastern part of the basin. The Catskill Mountains, part of the Appalachian Plateau (AP) province, form the western edge of the basin. They are composed mainly of shale and sandstone. Finally, the New England (NE) province (Hudson Highlands and Manhattan Prong) and the Piedmont Plateau (PP) province (Newark Lowlands) cover the extreme southern part of the Hudson basin. This area is a complex mix of sedimentary, metamorphic, and igneous rocks.

The entire present-day basin was covered by glaciers, although adjacent parts of the continental shelf were exposed by the low sea levels during full glacial times and were ice-free. Freshwater biota has recolonized the Hudson basin in the 18,000 years since glacial retreat. The Hudson basin has been connected to the Great Lakes and Lake Champlain basins by natural and manmade waterways, allowing many aquatic species to move between these basins.

Climate varies widely across the Hudson basin. Most of the basin is mesic (92 cm annual precipitation at Troy, near the center of the basin; Fig. 2.25), with cold winters and warm summers (mean annual temperature at Troy = 8.9°C). Precipitation, much of which falls as snow, is distributed evenly over the year. The Adirondacks are colder and wetter than Troy, with a mean annual temperature and precipitation of 4.2°C and 99 cm at Indian Lake. The climate at the southern end of the basin is mild and maritime, with a mean annual temperature and precipitation of 12.6°C and 120 cm at New York City.

Before 1700, most of the Hudson basin was covered by mixed hardwood or hardwood–coniferous forests (Braun 1950). Dominant species include eastern hemlock, white pine, sugar maple, red maple, American beech, several species of oak, and formerly American chestnut. Most of the basin was deforested in the late eighteenth and nineteenth centuries, although large tracts of the Adirondacks are uncut. Today, 62% of the basin is forested, 25% is used for agriculture, 8% is urban, and 5% is in other land uses (Wall et al. 1998). Land use varies across the basin (Swaney et al. 1996). Forests cover >95% of the upper basin, agricultural lands cover 30% to 50% of the middle basin, and urban and suburban areas near New York City cover much of the southern part of the basin.

River Geomorphology, Hydrology, and Chemistry

The Hudson can be divided into four sections on the basis of physiography, water chemistry, and human

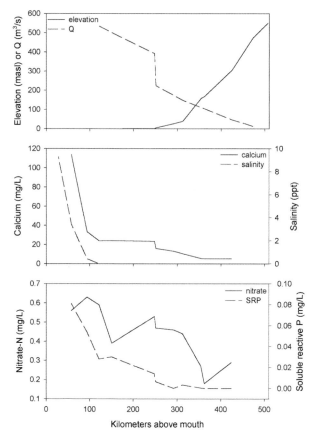

FIGURE 2.8 Longitudinal variation in selected physical and chemical characteristics of the Hudson River (USGS 2002, Lampman et al. 1999).

impacts. From its headwaters at Rkm 507 to the village of Corinth (Rkm 354), the "upper Hudson" flows through the forested Adirondack Mountains. The upper Hudson is a wild high-gradient (2.6 m/km) river largely unaffected by gross human impacts. Much of the river bottom is coarse (gravel, cobbles, boulders, and bedrock), and parts of the river run through dramatic gorges. The water is clear, soft, and low in dissolved phosphorus (Fig. 2.8). Although the flow of the upper Hudson is moderated by many lakes and wetlands in the region, the natural flow is highly variable, with high peaks from snowmelt in March to May. The flow of the upper Hudson is regulated strongly by the outflow from Indian Lake (a reservoir) at Rkm 454 and to a lesser extent by outflows from other lakes.

The "middle Hudson" (Rkm 354 to 248) is a broad, low-gradient, warmwater river flowing through a pastoral landscape. It is distinguished from the upper Hudson by its lower gradient, harder water, and greater human impacts. As the Hudson

enters the Hudson-Mohawk Lowlands near Glens Falls (Rkm 323), its gradient drops to ~0.5 m/km and its calcium content rises (see Fig. 2.8). The sediments are predominantly mud and sand, and beds of rooted aquatic plants are common (Moore 1933, Feldman 2001). Historically, the first serious water pollution entered the Hudson in this region, from industries at Corinth, Glens Falls, Hudson Falls, and Fort Edward. The heavily regulated Sacandaga River enters the Hudson just above Corinth, so the flow of the middle Hudson is significantly moderated. Finally, the Hudson is part of the Champlain Canal between Rkm 248 and Rkm 312, and its channel has been modified by navigational dredging and a series of dams.

Runoff in the middle Hudson shows a strong seasonal pattern (see Fig. 2.25). Although precipitation is lowest from December through March, the melting of accumulated snow and ice result in peak discharge from March through May. The low runoff from July through September is a result of high evapotranspiration during this active growing period.

The "freshwater tidal Hudson" extends 149 km from the last dam at Troy (Rkm 248) to Newburgh (Rkm 99). It is distinguished from the middle Hudson by the entry at Troy of the Hudson's largest tributary, the nutrient-rich Mohawk River, by the dominance of tidal flows in shaping the channel and determining local flow conditions, and by the open passage to the sea, which allows free access to many aquatic animals. The channel of the freshwater tidal Hudson averages 8 m deep and 0.9 km wide, and the gradient is scarcely measurable (~1 cm/km). Most of the bottom is sand or mud, but patches of cobbles and exposed bedrock occur throughout the reach. Large beds of rooted plants are common. Mean discharge at the mouth of the Hudson is almost 600 m³/s. However, the entire freshwater tidal Hudson has tides of 0.8 to 1.6 m, which reverse the direction of water flow every six hours and prevent stratification. Tidal flows are much larger than net freshwater flows. Summer water temperatures often exceed 26°C (Ashizawa and Cole 1994), and much of the freshwater tidal Hudson is ice covered in cold winters. The water in the freshwater tidal Hudson is hard, rich in nutrients, and usually free from oceanic salinity (see Fig. 2.8).

The "brackish Hudson" extends 99 km from Newburgh to New York City (Fig. 2.9). This section of the river physically resembles the freshwater tidal Hudson, with a large channel, beautiful scenery, no measurable slope, soft sediments, and strong tidal flows, but typically contains significant sea salt

FIGURE 2.9 Hudson River above West Point, New York (PHOTO BY TIM PALMER).

during at least part of the year (see Fig. 2.8). The brackish Hudson includes both narrow, deep reaches (up to 66 m deep between Rkm 70 and 90) and broad, shallow bays like Haverstraw Bay, which is 5 km wide. The water in the brackish Hudson is hard and nutrient rich; its composition varies spatially and seasonally according to the balance between freshwater flow and inputs from the sea (the latter carrying effluent from New York City upriver).

River Biodiversity and Ecology

As many of the rivers in this chapter, the Hudson is found within the North Atlantic Freshwater ecoregion. Unlike most of these rivers, however, it has been one of the most intensively studied large rivers in North America.

Algae, Cyanobacteria, and Protists

The phytoplankton of the upper and middle Hudson River has not been described, although the middle Hudson presumably contains a rich community during low flow. The freshwater tidal Hudson supports a rich phytoplankton community. Large diatoms strongly dominate the community, but cyanobacteria and other forms were abundant before the zebra mussel invasion (Smith et al. 1998). In the brackish Hudson, diatoms, green algae, dinoflagellates, and cyanobacteria are abundant (Weinstein 1977). Little is known about attached algae anywhere in the Hudson, although they probably are an important part of the food web.

Plants

Aquatic macrophytes are rare in most of the upper Hudson (Moore 1933), probably as a result of unstable flows, coarse sediments, and shading by streamside vegetation. Aquatic plants are important elsewhere in the Hudson. The middle Hudson contains large beds of aquatic plants (wild celery, pickerel-weed, water star-grass, water chestnut, slender water nymph, waterweed, water bulrush, various pondweed, needle-rush) (Moore 1933, Feldman 2001). Nearly pure beds of wild celery and the nonnative water chestnut cover ~10% and 5%, respectively, of the freshwater tidal Hudson (Findlay et al. 2005). Mixed emergent vegetation (cattail, common reed, yellow water lily, bulrush, pickerel-weed, southern wild rice, broadleaf arrowhead, arrow-arum) lives in marshes along the shore. Beds of submersed plants are rare further downriver, and are dominated by wild celery, the nonnative curly-leaved pondweed, slender water nymph, Eurasian watermilfoil, and water star-grass. Marshes in the brackish Hudson contain common reed, cord grass, cattail, spike grass, arrow-arum, swamp rose-mallow, and pickerelweed (Weinstein 1977).

Invertebrates

The upper Hudson supports a diverse community of "clean-water" insects, including the beetle *Stenelmis concinna*, the mayflies *Isonychia bicolor*, *Pseudocloeon* sp., *Epeorus* sp., and *Leucrocuta* sp., the caddisflies *Helicopsyche borealis*, *Hydropsyche sparna*, *Brachycentrus appalachia*, and *Chimarra* spp., and the chironomid midge *Micropsectra* sp. (Boyle 1979; Bode et al. 1993, 1996). Stoneflies, hell-grammites, and black flies also are common.

The middle Hudson River presumably supports zooplankton, at least during low water, but it has not been studied. Scattered studies of the zoobenthos (Simpson et al. 1972; Feldman 2001; Bode et al. 1993, 1996) have shown that chironomid midges (including *Rheotanytarsus*, *Polypedilum*, *Orthocladius*), caddisflies (*Neureclipsis*, *Chimarra*, *Hydropsyche*, *Cheumatopsyche*), the mayfly *Stenonema modestum*, and oligochaetes are common in the sediments. Plant beds are inhabited by a rich fauna of invertebrates, including chironomids, nematodes, sphaeriid clams, oligochaetes, the amphipod *Hyalella azteca*, the isopod *Caecidotea*, the mayfly *Caenis*, several gastropods (*Ferrissia*, *Amnicola*, *Gyraulus*), several caddisflies, and several odonates.

More than 200 species of invertebrates inhabit the freshwater tidal Hudson (Strayer and Smith 2001, Pace and Lonsdale 2003). The zooplankton is dominated by protozoans and rotifers, the cladocerans *Bosmina freyi* and *Leptodora kindtii*, and the copepods *Diacyclops thomasi* and *Eurytemora affinis*. The zooplankton is small-bodied and sparse compared to that of lakes, presumably because of constant washout downriver (Pace et al. 1992). The zoobenthos is dominated by tubificid oligochaetes (chiefly *Limnodrilus hoffmeisteri*), bivalves (union-ids, sphaeriids, and the zebra mussel), the amphipod *Gammarus tigrinus*, and chironomids. Since 1992, the zebra mussel has dominated the zoobenthos of the freshwater tidal Hudson and has greatly reduced populations of native planktivores (unionid mussels, sphaeriid clams, and *Chaoborus*).

The zooplankton of the brackish Hudson is dominated by the copepods *Eurytemora affinis*, *Acartia tonsa*, and *Acartia hudsonica*. Other copepods, protozoans, ctenophores, rotifers, cladocerans, mysids, and larvae of benthic animals can also be abundant (Pace and Lonsdale 2005). The zoobenthos is dominated by polychaete (*Marenzelleria viridis*, *Streblospio benedicti*, *Eteone heteropoda*, *Nereis succinea*, *Polydora* spp.) and oligochaete worms, amphipods (*Leptocheirus plumulosus*, *Monoculodes edwardsi*), and bivalves (Atlantic rangia, Baltic macoma, soft-shell clam). Zoobenthic densities are low (<5000/m²) in the transition zone between fresh and brackish water (Rkm 30 to 100), but high (>10,000/m²) in the lower Hudson (Rkm 0 to 30) (Strayer 2005).

Vertebrates

The Hudson contains a rich, ecologically important, and economically valuable fish community. More than 200 species of fishes have been recorded from the Hudson basin (Smith and Lake 1990). About 95 of these are marine and essentially restricted to the brackish Hudson. Of the remaining species, 10 are diadromous and occur regularly above Rkm 99, 70 are native freshwater species, and 33 are nonnative freshwater species.

The fish community changes dramatically from the headwaters of the Hudson to its mouth. The cold waters of the upper Hudson are inhabited by just a few species, including brook trout, white sucker, blacknose and longnose dace, cutlips minnow, creek chub, and common shiner (Moore 1933, Smith 1985). Parts of the upper Hudson basin, isolated by waterfalls, probably contained no fish before post-Columbian stocking. The middle Hudson contains a rich community of warmwater fish, including white sucker, common carp and other minnow species, brown bullhead, chain pickerel, American eel, trout-perch, walleye, yellow perch, tessellated darter, and several sunfishes, including smallmouth and large-mouth bass. Below the Troy dam, the mixed warmwater fish community is supplemented by large populations of diadromous and estuarine species, including shortnose sturgeon, American shad, blue-back herring, alewife, American eel, white catfish, fourspine stickleback, striped bass, and white perch. Marine or brackish-water fish like the menhaden, bay anchovy, Atlantic tomcod, bluefish, weakfish, and flounder are increasingly abundant toward the sea (Smith 1985, Smith and Lake 1990).

The shortnose sturgeon, the only species in the river that is protected as rare or endangered, is listed as "endangered" by the U.S. Fish and Wildlife Service and the New York State Department of Environmental Conservation. Shortnose sturgeon live their entire lives within the Hudson estuary, spawning in the upper estuary (Rkm 200 to 248), and migrating throughout the rest of the estuary to feed and overwinter (Dovel et al. 1992). The Hudson's population, at 10^4 to 10^5 fish, is thought to be one of the largest remaining populations of this species (National Marine Fisheries Service 1998).

Dozens of species of other vertebrates live in and along the Hudson and are sometimes important in its ecology. Especially notable are waterfowl, including the nonnative mute swan, which is now common on the freshwater tidal and brackish Hudson, and the bald eagle, which breeds and overwinters along the tidal Hudson. Beavers are abundant in the upper Hudson and in tributaries throughout the basin and are important because of their key role as habitat modifiers. Marine mammals (harbor seals, harbor porpoises, dolphins, and even whales) are seen occasionally in the tidal Hudson River as far north as Troy (Rkm 248).

Ecosystem Processes

Ecosystem processes are best known in the freshwater tidal section of the Hudson. The organic carbon budget of the freshwater tidal Hudson is strongly dominated by allochthonous inputs, advective losses, and respiration by bacteria and zebra mussels (Table 2.1). Autochthonous production is limited by light rather than nutrients (Cole and Caraco 2005). Outputs of nitrogen are much less than inputs, suggesting that burial and denitrification may be important (Lampman et al. 1999). Even though allochthonous inputs dominate organic matter inputs and probably support much of microbial respiration (Findlay et al. 1998, Cole and Caraco 2001), autochthonous production appears to provide crucial support to the upper food web. At the least, populations of zooplankton, zoobenthos, and fish changed radically following the loss of phytoplankton and increase in macrophyte production that accompanied the zebra mussel invasion.

Five interacting factors control the ecosystem in the freshwater tidal Hudson. First, land use in the basin controls the delivery of silt, organic matter, and other materials into the river. Because primary production is light-limited and inputs of silt help to determine water clarity, phytoplankton could be much more productive if silt inputs were cut (Caraco

TABLE 2.1   Organic carbon budget (g C m^{-2} yr^{-1}) for the freshwater tidal Hudson River, before and after the zebra mussel invasion.

	Preinvasion	Postinvasion
Inputs		
Allochthonous	650	650
Phytoplankton GPP	330	80
Macrophyte GPP	41	44
Total	1000	780
Outputs		
Phytoplankton R	280	70
Macrophyte R	37	37
Bacterioplankton R	116	116
Zooplankton R	8	2
Zebra mussel R	0	83
Other zoobenthic R	9	5
Advection	636	482
Total	1086	795

Modified from Cole and Caraco (2005), who summarized data from several sources.
Many data are approximate. Benthic algal activity and benthic bacterial respiration are omitted because they have not been well studied in the Hudson. GPP = gross primary production, R = respiration.

et al. 1997). Land use in the watershed also controls the amount and kind of organic matter that is brought into the river (Howarth et al. 1996, Swaney et al. 1996). Second, inputs of fresh water set the residence time of water in the river. Residence time must be long enough during the growing season for plankton to develop. In wet years, residence time is short and plankton is sparse (Fig. 2.10). Other parts of the ecosystem depend on plankton, so the effects of hydrology probably extend to processes like organic matter processing, nutrient cycling, and fish recruitment (see Fig. 2.10). Third, the morphology of the river critically determines where primary production by phytoplankton and benthic plants can occur. Fourth, primary production by phytoplankton and benthic plants are negatively correlated because phytoplankton populations decrease water clarity and so can modulate production by benthic plants. This interaction is important because it can stabilize the primary production of the entire ecosystem: When phytoplankton populations decline, benthic primary production can increase, and when phytoplankton populations increase, benthic primary production will decrease. This interaction probably reduced the impact of the zebra mussel invasion on the Hudson's ecosystem. Fifth, the waxing and waning of biologi-

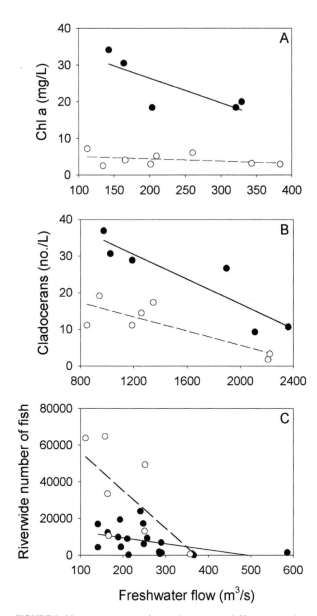

FIGURE 2.10 Sensitivity of populations at different trophic levels in the freshwater tidal Hudson River to freshwater flow, before (black dots) and after (white dots) the zebra mussel invasion. A. Phytoplankton and flow, 15 May to 1 October (Cole and Caraco 2005); B. Planktonic cladocerans and flow, annual means (Pace and Lonsdale 2005); C. Number of young-of-year redbreast sunfish, August to September. Note the interaction between flow and grazing by the zebra mussel (Cole and Caraco 2005, Pace and Lonsdale 2005, Strayer et al. 2005).

cal populations can affect ecosystem function. The best example is the recent invasion of the Hudson by zebra mussels, which changed water clarity, concentrations of dissolved nutrients and oxygen, phytoplankton biomass and composition, bacterio-

plankton, zooplankton, zoobenthos, and fish (Caraco, Cole, Raymond et al. 1997; Caraco, Cole, Findlay et al. 2000; Strayer, Caraco et al. 1999; Strayer, Hattala, and Kahnle 2004; Strayer and Smith 2001). Other species may likewise control ecosystem processes, but have not been well studied in the Hudson. The nonnative aquatic plant water chestnut affects dynamics of dissolved oxygen and probably nutrients (Caraco and Cole 2002), and the large populations of juvenile anadromous fish probably have far-reaching impacts.

Four of these five key factors are strongly affected by humans. Humans control land use in the watershed and thereby alter inputs of silt and organic matter. Humans also changed the hydrology of the Hudson, reducing spring flows and increasing summer flows, possibly reducing plankton populations. The morphology of the Hudson has been vastly altered by attempts to better suit the river to navigation. In particular, large shallow-water areas in Rkm 180 to 248 have been filled or deepened, making this part of the river less hospitable to phytoplankton and benthic plants. Finally, humans have changed biological populations in the river through the introduction of alien species and the exploitation of fish and shellfish.

Human Impacts and Special Features

Although the Hudson is a beautiful river with a rich biota, it is far from being a natural river. Humans have physically altered the river channel, regulated the river's flow, polluted its water, harvested its inhabitants, and introduced nonnative species of plants and animals. The channel alteration that began in the early nineteenth century to make the river more suitable for navigation, afford access to rail lines along both shores, make the shoreline more hospitable to human use, and eliminate "waste" wetlands vastly transformed the tidal river by 1900. Particularly in the upper freshwater tidal Hudson, the channel was deepened and narrowed by dredging, bulkheading, and filling of wetlands and backwaters with dredge spoils (Fig. 2.11). Rail lines between the main channel and coves created new wetlands and restricted access between these marginal shallows and the main channel, and everywhere shorelines were straightened and hardened with riprap, bulkheads, and concrete (Ellsworth 1986). The middle Hudson River was transformed into a series of slowing-moving pools by low-head dams and locks.

The Hudson and its tributaries have been extensively dammed for power production, flood control,

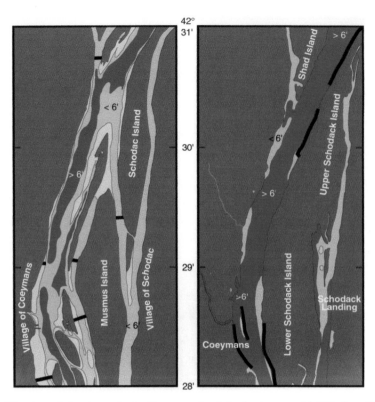

FIGURE 2.11 Modification of the channel of a 5 km stretch of the freshwater tidal Hudson River between 1820 (left) and 1970 (right) showing loss of channel complexity and shallow-water habitats. Key: red = dry land, yellow = intertidal zone, light blue = shallow water (<1.8 m deep), dark blue = deep water (>1.8 m deep), heavy black lines = dikes or bulkheads to constrain the channel (John W. T. Ladd unpublished).

and water supply, but the Hudson has been less altered than many American rivers. The main-stem Hudson is blocked by 14 dams, but they extend over a limited geographic area (Rkm 248 to 354) and do not alter the flow regime greatly. Most significantly, the lower 248 km of the Hudson are undammed, allowing diadromous fish access to a long reach of river (although many tributaries are now inaccessible because of dams near their mouths). Two dams on tributaries (Indian River and Sacandaga River in the Adirondacks) significantly alter the Hudson's flow. Both were built for flood control, and moderate flood peaks and summer low flows. Although the operation of these dams has had devastating ecological impacts on the Indian and Sacandaga rivers, it probably has not had a major impact on the Hudson proper.

Pollution has affected large reaches of the Hudson. Much of the wood cut in the southeastern Adirondacks was processed in the towns along the middle Hudson for pulp, paper, or timber. Consequently, the Hudson between Corinth (Rkm 354) and Hudson Falls (Rkm 312) was badly polluted by pulp

and paper wastes, sawdust, dyes, and sewage for much of the late nineteenth and twentieth centuries. As late as 1972, Simpson et al. (see also Boyle 1979) described "large clumps of paper waste floating on [the water] surface," a bottom of "grayish muck," widespread sewage bacteria (*Sphaerotilus natans*), and low dissolved oxygen in the middle Hudson. This pollution has been largely controlled, and water quality in the middle Hudson has vastly improved (Bode et al. 1993, 1996). Other significant sources of pollution in the middle Hudson are General Electric plants in Fort Edward and Hudson Falls, which released 90,000 to 600,000 kg of PCBs into the Hudson in the mid-twentieth century (Baker et al. 2001). PCBs now contaminate the entire Hudson and its biota downriver from Fort Edward, and have resulted in closures and severe restrictions on commercial and sport fisheries throughout much of the Hudson. Because PCBs are long-lived, the ecosystem will remain contaminated unless action is taken to address "hot spots" of PCB contamination in the middle Hudson. Before the 1970s, many cities released raw or partially treated sewage into the

Hudson. These releases were large enough in the middle Hudson (see previous discussion), the Albany–Troy area, and around New York City to seriously deplete dissolved oxygen and make large parts of the river unsuitable for aquatic life and human recreation. Improvements in water treatment after the Clean Water Act of 1972 greatly reduced sewage pollution in the Hudson. Finally, industrial discharges have contaminated parts of the Hudson with toxins. The most severe example probably is Foundry Cove (Rkm 86), which was contaminated with cadmium and nickel from a battery factory. An estimated 22 tons of cadmium were released into the cove, and cove sediments contained up to 22% cadmium (Wallace et al. 1998). Foundry Cove was designated as a federal Superfund site and remediated in the mid-1990s.

The Hudson has supported commercial fisheries for fishes, blue crabs, and eastern oysters. American shad are by far the most important species (and the only fish now caught commercially), but Atlantic sturgeon, striped bass, American eel, white perch, and other species also were taken. Catches of shad have been highly variable, but exceeded 10^6 kg in some years (Fig. 2.12). The striped bass fishery, closed in 1976 because of PCB contamination, was much smaller but historically important (see Fig. 2.12). The Hudson River contributed to the millions of kilograms of Atlantic sturgeon caught every year along the East Coast in the late nineteenth century, but by the time the fishery was closed in the mid-1990s because of dwindling stocks the commercial catch had fallen to <100,000 kg rangewide, 3700 kg of which came from the Hudson (NYSDEC 1994).

Blue crabs occur as far north as Troy and are caught commercially well into the freshwater tidal Hudson; recent landings in the Hudson have averaged ~40,000 kg/yr (NYSDEC 1993). Although eastern oysters were taken commercially in the brackish Hudson in the nineteenth century (Stanne et al. 1996) and still occur in the river, there is currently no commercial fishery for eastern oysters in the Hudson. These fisheries have provided important economic benefits to the local economy but also have almost certainly affected fish stocks in the river. Despite recent closures and restrictions from PCB contamination, sport fisheries on the Hudson are worth millions of dollars every year (Connelly and Brown 1991).

Finally, humans have introduced hundreds of nonnative species into the Hudson and its tributaries. Mills et al. (1996) catalogued 113 species of macroscopic plants and animals from the freshwater parts of the basin, and microscopic and brackish-water species would surely add >100 species to this total. Many of these aliens are among the most familiar and ecologically important species in the river: The zebra mussel, largemouth and small-mouth basses, common carp, brown trout, purple loosestrife, water chestnut, and Atlantic rangia all are nonnative species that are common in the Hudson. Nearly all aquatic habitats in the basin contain at least one dominant nonnative species. New species of aliens appear to be entering the freshwater parts of the basin at a rate of ~6 species/decade (Mills et al. 1996); including the brackish parts of the river would substantially increase this estimate.

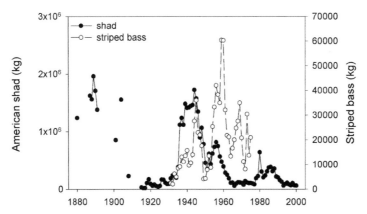

FIGURE 2.12 Reported commercial landings of American shad and striped bass from the Hudson River (shad data compiled by Kathryn Hattala [personal communication] from various sources; striped bass data from McLaren et al. 1988).

DELAWARE RIVER

The Delaware River flows south and forms part of the boundary between New York, Pennsylvania, New Jersey, and Delaware (Fig. 2.26). About 50.3% of the basin lies in Pennsylvania, with the rest in New Jersey (23.3%), New York (18.5%), and Delaware (7.9%). The Delaware basin lies between latitudes 39° 52′ and 42° 27′ N and comprises <0.4% of the land area of the United States (~33,041 km²). It ranks only 42nd in North America in terms of mean annual flow (422 m³/s at Fort Mifflin, Pennsylvania; Leopold 1994). However, it contains the nation's largest freshwater port (Philadelphia) and provides water to about 10% of the nation's people, including 7.3 million living in the basin and another 10 million living outside (including New York City). More than 2411 billion liters of water were withdrawn from the Delaware River basin in 1996. There are over 1250 permitted waste-water discharges in the basin, including 360 municipal sewage plants. This activity makes the Delaware one of the most intensively used rivers in the United States (Majumdar et al. 1988, www.state.nj.us/drbc/thedrb.html 2002).

Even though its watershed had been occupied by humans for about 12,000 years, and Algonkin natives greeted him on his arrival, the Delaware River was "discovered" by Henry Hudson in 1609 while pursuing a northwest passage to China (Custer 1996). In 1620 Captain Samuel Argall named the river for Lord de la Warr, who was then governor of Virginia (Majumdar et al. 1988). Sweden established the first European settlement in 1630 at what is now Wilmington, Delaware, as a home for deported convicts. William Penn founded the basin's largest city, Philadelphia, in 1682. As evidenced by the name Pennsylvania (or "Penn's Woods"), the Delaware watershed encountered by the early Europeans was dominated by forests (Majumdar et al. 1988). Yet by 1700, colonists of Dutch, British, and German ancestry had cleared the forests, transformed the lower Delaware basin into the "granary of America," and begun harvesting its seemingly limitless stands of trees in the upper basin. The river and its tributaries soon became the main arteries for shipping grain, flour, and timber products to other colonies, the West Indies, and elsewhere (Majumdar et al. 1988). Thus, it is not surprising that the crossing of the Delaware River and capture of the Trenton, New Jersey, portion of the watershed by General George Washington on Christmas Eve 1776 proved to be the pivotal victory for the continental army during the American Revolution.

Physiography, Climate, and Land Use

Situated in the Temperate Deciduous Forest biome, the Delaware River passes through three terrestrial ecoregions and, for a medium-size river, an unusually large number (5) of major physiographic provinces (Ricketts et al. 1999; see Fig. 2.26). The sequence from headwaters to the mouth includes (1) Allegheny Highland Forests ecoregion, including the Appalachian Plateau (AP) province underlain by the Catskill formation (sandstone with shale and conglomerates); (2) the Appalachian/Blue Ridge Forests ecoregion, including the mountainous Valley and Ridge (VR) province (sandstone, shale, and limestone); and (3) the Northeastern Coastal Forests (NCF) ecoregion. The NCF ecoregion is large and includes the highly weathered Great Valley province (quartzite), the Piedmont Plateau (PP) province (a mix of igneous and metamorphic rocks [granite and gneiss], metamorphosed sedimentary rocks [schist, phyllite, and quartzite], and slightly metamorphosed limestone and dolomites), the Triassic Lowlands province (arkosic sandstone, shale, siltstone, and conglomerates), and the Coastal Plain (CP) province (sands, clays, and gravels; Majumbar et al. 1988). Topographic relief in the watershed ranges from 698 m asl in the Appalachian Plateau at High Knob, New York, to sea level at the Delaware Bay (Majumdar et al. 1988).

The Delaware basin has a humid continental climate. Average annual air temperature varies from about 8.8°C (upper basin) to about 12.2°C (lower basin), with annual monthly temperatures for the northern/southern end of the watershed ranging from a mean of −5.0/1.6°C in January to 18.3/25°C in July (Fig. 2.27). Annual extreme temperatures at the southern and northern end of the basin are about −36°C to −23.8°C and 34.4°C to 40.5°C, respectively. Relative humidity averages about 65% to 72% for the region (*Climate and Man* 1941, cited in Majumdar et al. 1988).

Mean annual precipitation is fairly uniform throughout the basin (see Fig. 2.27), averaging about 105.1 cm/yr near Philadelphia and 107.9 cm/yr for the entire basin. February and July to August are usually the driest and wettest periods, respectively (www.state.nj.us/drbc/thedrb.html 2002). Extreme precipitation (15 to 20 cm per storm) occurs as hurricanes about once a decade (Majumdar et al. 1988).

Vegetation, past and present, reflects the watershed's topographic, geologic, and climatic features. Pollen analysis shows that the basin was historically dominated by forest, with the following temporal sequence: pine–spruce–fir (13,000 to 10,000 years ago), pine–spruce–oak forest (8500 to 10,000 years ago), hemlock–oak–pine forest (5000 to 8500 years ago), oak–pine–hickory (2500 to 5000 years ago), mixed oak–chestnut (2500 years ago to present; Custer 1996). The present forest falls into two categories: Appalachian oak forest—a deciduous forest in the south/central part of the basin dominated by oak trees (white and northern red)—and, to the north, the Northern Hardwood Forest—a deciduous forest with sugar maple, yellow birch, and American beech predominating in addition to some evergreen trees (pines, hemlock) (Cuff et al. 1989).

Today, the basin has about 7.3 million people (or about 214/km^2; www.state.nj.us/drbc/thedrb.html 2002) and is about 60% forest, 24% agriculture, 9% urban, and 7% surface water and miscellaneous land cover (http://nj.usgs.gov/delr/factsheet.html 2002). However, land cover can vary greatly from region to region (e.g., the adjacent West Branch, East Branch, and Neversink subbasins vary as follows, respectively: row-crop agriculture, 1.5%, 0.4%, 0%; grass-pasture lands, 28%, 12%, 0.8%; forest, 56%, 81%, 98%; and impervious surface, 0.7%, 0.1%, 0.1%; Stroud Water Research Center 2001). The basin also contains about 2270 km^2 of freshwater wetlands, mostly nontidal and palustrine marshes located in New Jersey (43%), New York (27%), Pennsylvania (19%), and Delaware (11%).

River Geomorphology, Hydrology, and Chemistry

The Delaware is a 7th order river (estimated based on basin area) fed by 216 tributaries (www.state.nj.us/drbc/thedrb.html 2002). It begins as the East and West branches in the Catskill Mountains of New York. The branches join in Hancock, New York, and the river then flows southeast for 530 km to the sea. Although dammed in its headwaters region, the middle and lower sections comprise the longest free-flowing reach of river east of the Mississippi River (www.state.nj.us/drbc/thedrb.html 2002). The upstream middle portion near Narrowsburg, New York, is narrow and deep (maximum depth of 34.4 m) and a relatively steep gradient of 1.4 m/km produces Class I and Class II rapids downstream until Port Jervis, New York, where it widens again. From Port Jervis to Easton, Pennsylvania, the river gradient is half as steep at 0.7 m/km, but this reach also contains several sets of Class I and Class II rapids (Fig. 2.13). The river then flattens slightly to 0.5 m/km as it flows from Easton to the Fall Line at Trenton, New Jersey. The navigable portion of the

river begins at Trenton with a gradient of 3.5 cm/km to Marcus Hook (near Philadelphia). The Delaware contains many islands, with more than 20 in the Port Jervis to Stroudsburg, Pennsylvania, reach alone.

Annual discharge of the Delaware River (1913 to 2000) at Trenton averages 333 m^3/s, with mean monthly discharge ranging from 163 m^3/s (September) to 630 m^3/s (April; see Fig. 2.27). Mean annual runoff ranges from 45.7 to 71.1 cm for the basin. Runoff varies throughout the basin due to the many lakes, ponds, and wetlands formed during Pleistocene glaciation, as well as factors such as geology, topography, land use, vegetative cover, and size of the watershed (Page and Shaw 1977, cited in Majumdar et al. 1988). The annual pattern of runoff relative to precipitation seems to reflect a strong seasonal influence of evapotranspiration and, to a lesser degree, snow storage (see Fig. 2.27). The highest flow on record at Trenton was 9316 m^3/s on August 20, 1955, during Hurricane Diane. The lowest historic flow was 35 m^3/s on July 10, 1965 (http://water.usgs.gov/nj/nwis/monthly/site.html 2002).

The natural flow pattern of the Delaware has been altered by reservoirs constructed between 1926 (Toronto Reservoir) and 1979 (Blue Marsh Reservoir). For example, river flow at Trenton for March (a period of peak reservoir storage) averaged about 28.5% higher for the prereservoir period of 1913 to 1922 (716 m^3/s) than for the postreservoir period of 1980 to 2000 (511 m^3/s). This is especially noteworthy because the 1980 to 2000 March period was actually slightly wetter than the years 1913 to 1922 (avg: 113.2 versus 104.7 cm/yr and 9.24 versus 8.63 cm/March). Three large reservoirs (Cannonsville, Pepacton, and Neversink) owned by New York City contribute significantly to the modified flow. The volume and timing of water releases from these reservoirs has been controversial. A 1983 agreement stipulates that conservation releases from all three reservoirs be used to augment flows in the Delaware, especially during the summer (Majumdar et al. 1988).

The chemistry of the Delaware River at Trenton reflects local geology and anthropogenic enrichment and varies seasonally, ranging in 2001 as follows: pH (7.1 to 9.0), dissolved oxygen (8.9 to 15.6 mg/L), turbidity (<2 to 200 NTU), hardness (25 to 80 mg/L), alkalinity (19 to 55 mg/L as CaCO$_3$), Cl (9.2 to 30.3 mg/L), SO$_4$ (8.8 to 21.3 mg/L), NH$_4$-N (<0.03 to 0.07 mg/L), NO$_2$-N + NO$_3$-N (0.56 to 1.38 mg/L), PO$_4$-P (<0.018 to 0.067 mg/L), total P (0.031 to 0.266 mg/L), and dissolved organic carbon (0.4 to 12.0 mg/L).

FIGURE 2.13 Delaware River above Belvidere, New Jersey (PHOTO BY A. C. BENKE).

River Biodiversity and Ecology

The Delaware is the southernmost river of the North Atlantic freshwater ecoregion, which extends up the East Coast into southern Nova Scotia and southern New Brunswick (Abell et al. 2000). The biology of both the upper and lower Delaware is relatively well studied, with several surveys of algae, invertebrates, and fishes.

Algae, Cyanobacteria, and Protists

A recent quantitative study of algae, primarily diatoms, on the middle reaches of the Delaware near Belvidere, New Jersey, revealed an average of 41 species per sample (ANSP 2000). The five most common species were (in rank order) *Cocconeis placentula* (36.1%), *Navicula minima*, *Navicula pusilla*, *Nitzschia amphibia*, and *Amphora veneta* (3.9%). Patrick (1994) provides qualitative data on the historical (1957 to 1959) composition of algae for the 11 km of tidal freshwater river. She reported 99 species of diatoms and 25 species of green algae and cyanobacteria upstream from Philadelphia but only

about 29 species of diatoms and 22 species of green algae and cyanobacteria downstream of Philadelphia. Phytoplankton estimates for the 1977 to 1988 period in the freshwater tidal Delaware River were quite variable, with median chlorophyll concentrations and phytoplankton production values for the period of 8.1 μg/L and 14.6 mmol C m^{-3}d^{-1}, respectively, and peak values occurring in the summer (Frithsen et al. 1991). The summer peak was confirmed by Marshall (1992), who also showed that mean monthly phytoplankton concentrations in the tidal river near Trenton (5.7 to 10^5 cells/L) were only about half the level observed downstream near Philadelphia (10.4 to 10^5 cells/L) or Marcus Hook, Pennsylvania (9.8 to 10^5 cells/L).

Plants

Twelve species of macrophytes have been collected from the tidal river since 1970. Common species are wild celery, Eurasian watermilfoil, ditchmoss, slender water nymph, horned poolmat, curly-leaved pondweed, sago pondweed, common hornwort, and water-starwort. Another 14 species, that were observed earlier have not been collected since 1970 (e.g., white water-crowfoot, common bladderwort, floating-leaf pondweed, water star-grass, common water nymph; Majumdar et al. 1988).

Invertebrates

The headwater and upper main-stem reaches of the Delaware contain a fairly abundant and diverse macroinvertebrate community. In 1983, riffle areas of the East Branch below Pepacton Reservoir averaged about 13,545 macroinvertebrates/m², with chironomid (midge) insects comprising 47% of the community (Stroud Water Research Center, unpublished data). More recently, riffle areas of the river at Belvidere, New Jersey (see Fig. 2.13), averaged about 19,270 macroinvertebrates/m² (range 8,800 to 54,400; ANSP 2000) and contained about 68 taxa. The dominant genera and families of the headwater and upper main-stem reaches of the river are alderflies (*Sialis*), amphipods (*Gammarus*), beetles (*Optioservus, Oulimnious, Psephenus, Promesia, Stenelmis*), caddisflies (*Apatania, Brachycentrus, Ceraclea, Cheumatopsyche, Chimarra, Glossosoma, Helicopsyche, Hydropsyche, Lepidostoma, Macrostemum, Nectopsyche, Neophylax, Neuroclipsis, Oecetis, Rhyacophila*), clams (*Corbicula*), true flies (*Antocha, Chironominae, Hemerodromia*), dragonflies (*Argia, Gomphidae*), mayflies (*Acentrella, Baetis, Epeorus, Eurylophella, Isonychia, Leucrocuta, Serratella, Stenonema, Tricorythodes*), leeches, mites (*Hydracarina*), moths (*Petrophila*), snails (*Ferissia, Hydrobiidae, Physidae*), stoneflies (*Acroneuria*), and worms (*Oligochaeta*).

In the lower tidal Delaware, surveys during 1957 to 1959 revealed 38 insect and 23 noninsect macroinvertebrate taxa for the reach upstream of Philadelphia, but only 6 insect and 20 noninsect macroinvertebrate species for the reach immediately below Philadelphia (Patrick 1994). ECS (1993) reported that mean densities (and biomass) of macroinvertebrates during 1992 and 1993 varied from 2951/m² (35.5 g dry mass/m²; 34.4 g/m² were Asiatic clams) near Trenton to 3901/m² (3.4 g/m²) near Philadelphia to 2621/m² (2.9 g/m²) at the downstream freshwater limit near Chester, Pennsylvania. Although their survey revealed 129 taxa representing nine phyla, the benthic community was dominated by oligochaete worms (range 52 to 75%) and chironomid insects (range 14 to 24%).

Vertebrates

There are currently about 105 species of fish representing 33 families inhabiting the Delaware River basin (Schmidt 1986). The main stem alone has 53 species (15 families). Many species (39) in the basin are nonnative. Recent surveys for the reach near Belvidere, New Jersey, revealed 32 species (10 families) of fish (ANSP 1995, 2000), with the 10 most abundant (in rank order) being margined madtom (11.5% of total catch), shield darter (9.6%), swallowtail shiner (9.1%), American eel (9.0%), redbreast sunfish (8.5%), tessellated darter (8.2%), satinfin shiner (7.9%), spottail shiner (6.1%), fallfish (5.8%), and white sucker (4.2%).

For the lower Delaware, Patrick (1994) lists 27 fish species in the reach between Yardley and Bristol, Pennsylvania, in her surveys from 1957 to 1959. The most common species were golden shiner, spottail shiner, satinfin shiner, pumpkinseed, black crappie, banded killifish, blueback herring, and the Mississippi silvery minnow. The same survey found only 17 species in the freshwater tidal reach below Philadelphia, with only the blueback herring and the mummichog being common. More recent surveys (see Horwitz 1986 for review) indicate that as many as 95 species of fish inhabit the lower Delaware River, including the endangered shortnose sturgeon and threespine stickleback. According to Horwitz's (1986) review, 78 of the 95 species (82%) are native freshwater or estuarine species.

Although scientific details of the seventeenth-century condition of fisheries in the Delaware River are lacking, William Penn complained that the shortnose sturgeon were a danger to small boat safety as they jumped into the air by the dozens near Philadelphia (Wildes 1940, cited in Majumdar et al. 1988). However, pollution brought a swift decline to the Delaware River fishery. The American shad industry is a case in point. Early records near Gloucester, New Jersey, show an average annual catch of 131,229 fish for the years 1818 to 1822, with a subsequent decline to 43,915 for the period from 1870 to 1873, at which point the fishery was declared economically nonviable (Howell and Slack 1871, cited in Majumdar et al. 1988). Near Lambertville, New Jersey, the annual shad catch declined throughout the first half of the twentieth century, reaching zero by 1953 (Majumdar et al. 1988). Apparently, dissolved oxygen levels of <2 mg/L along a 48 km section of river near Philadelphia blocked shad migration during the period (Majumdar et al. 1988). Pollution abatement measures have now restored shad migration and data for juvenile shad indicate a fairly steady population since about 1981 (New Jersey Department of Fish, Game, and Wildlife, personal communication).

Besides fishes, other important vertebrates associated with the main river include the northern water snake, snapping turtle, eastern mud turtle, eastern painted turtle, spotted turtle, red bellied turtle, Kemp's ridley turtle, beaver, and river otter. In addition, some vertebrate species in the basin are on the federal list of endangered species, including the short-nose sturgeon, Kemp's ridley turtle, bald eagle, and peregrine falcon.

Ecosystem Processes

To date, the various ecological data have not been integrated into a comprehensive model describing the dynamics of either the upper or lower main stem of the Delaware River. Regardless, some general patterns are apparent. The headwaters of the ecosystem were highly disturbed in the 1970s following construction of large reservoirs on the East and West branches of the Delaware River. Much of this disturbance has been mitigated by adoption of a minimum flow program, but the seasonal temperature regime in the headwaters remains altered by the bottom-release reservoirs. Nevertheless, the presence of a coldwater fishery and a diverse and abundant (10,000 to 20,000 individuals/m^2) macroinvertebrate community in both the headwaters and free-flowing middle main stem of the Delaware River suggest that the ecosystem has a diverse and abundant food base (both allochthonous and autochthonous organic inputs) that is being processed extensively. In contrast, although the tidal freshwater portion of the Delaware River ecosystem is much healthier than it was 50 years ago, the structure and abundance of its aquatic plant and animal communities remain compromised.

Human Impacts and Special Features

Water quality in the lower Delaware declined steadily following European colonization of the basin in the mid-1600s (see extensive review by Patrick et al. 1992). Water-quality surveys in 1799 indicated significant pollution around Philadelphia. A 1915 sanitary survey reported a mean dissolved oxygen level of 2.9 mg/L near Philadelphia, and 50 years later readings were only 0.2 mg/L in the lower Delaware (Majumdar et al. 1988). Contamination reached its peak during World War II, when the river's odor nauseated dock workers, its water blackened the hulls and decks of vessels, and pollution caused frequent fish kills near Philadelphia. The Interstate Commission on the Delaware River Basin (INCODEL), founded in 1936, established the first water-quality

standards for the river. By summer 1985, dissolved oxygen levels had improved to 4 mg/L (Majumdar et al. 1988). Unfortunately, toxic substances have also impacted the river. A nationwide survey (1975 to 1979) ranked the Delaware River at Trenton eighth—and its largest tributary, the Schuylkill, first—in the number of detections of organochlorine pesticides in stream bed sediments (Gilliom et al. 1985, cited in Majumdar et al. 1988).

Unfortunately, water quality in the upper Delaware has also been impacted by humans. The demand for wood and wood products by Philadelphia and New York led to almost complete deforestation of the basin by the end of the nineteenth century. Philadelphia's enormous economic growth in the early 1800s triggered the need for a transportation network to penetrate into portions of the basin far from the river's main stem. The historic construction of canals provided this penetration and contributed greatly to rapid development of the basin by humans. In 1824, the Schuylkill Navigation Company established one of the nation's first canals. No other river in the nation has more canals serving one river valley as the Delaware (Majumdar et al. 1988). These canals facilitated the large-scale development of coal mining activities and agriculture in the basin, which has left a legacy of acid pollution and stress related to sediment and chemical laden runoff.

Permitted waste-water discharges to the river are presently a big threat to the basin. The West Branch in New York currently receives effluent (hence, nutrients) from nine permitted discharges (about 10,456 m^3/d of effluent) and the East Branch from five (895 m^3/d). In addition, the cold, bottom-water discharge from New York reservoirs greatly modifies the natural temperature pattern of the river for a distance of >50 km (Sweeney et al. 1986).

Water quantity or river flow also became an issue in the 1920s when New York City added the Delaware River basin to its water supply system. In 1931, litigation involving the U.S. Supreme Court (later modified in 1954) granted New York access to the water but required a minimum flow downstream of any reservoirs constructed. Today, the 1954 modified ruling still assures that discharge from the New York reservoirs meets the river's downstream flow requirements.

Despite the historical challenges, the river today is much improved from the river of the 1950s. In fact, it was among the first rivers to be authorized for inclusion in the National Wild and Scenic River System, with 241 of the 454 km main river included in the system (Majumdar et al. 1988). A 64 km reach of this Wild and Scenic River, from Milford, Pennsylvania, to the Delaware Water Gap, is further protected as the Delaware Water Gap National Recreation Area. Ironically, this stretch of river was originally purchased by the federal government to build a large main-stem dam just upstream from the Water Gap at Tocks Island Dam. The plan was eventually abandoned due to public pressure and environmental impact studies. In addition to the main stem, significant portions of tributaries have been designated wild and scenic, including White Clay Creek, which is the last Pennsylvania tributary to the Delaware and the first entire watershed to be added to the National Wild and Scenic River System.

SUSQUEHANNA RIVER

The Susquehanna River is a major landscape feature in Pennsylvania, New York, and Maryland, as well as the principal tributary of the Chesapeake Bay, the largest freshwater estuary in North America, and the second largest estuary in the world (Fig. 2.28). The river begins at Lake Otsego in Cooperstown, New York (43.0°N), and flows 721 km to Havre de Grace, Maryland, on the Chesapeake Bay (39.5°N). Along the way, it crosses high plateaus, rugged mountains and ridges, and fertile valleys. The drainage of the Susquehanna River covers 71,432 km² and includes almost 50% of Pennsylvania and 13% of New York. Within the basin are 50,190 km of named streams and 2293 lakes, reservoirs, and ponds (Stranahan 1993, Edwards 1996). The Susquehanna River is the 22nd largest river in North America that flows to the sea and the 3rd largest (after the St. Lawrence and Churchill [of Labrador]) that flows into the Atlantic Ocean (Leopold 1994). Because the river channel is relatively wide, the Susquehanna has been described as "a mile wide and a foot deep" and has earned the distinction of being the longest commercially nonnavigable river in North America.

The north–south orientation of the Susquehanna River made it an important economic, transportation, and communication artery connecting the Mid-Atlantic region (including the Chesapeake Bay) with the interior of upstate New York. The river and its valley also provided valuable resources to both Native Americans (primarily the Susquehannocks in the 1600s, after whom the river was named) and European colonists. Evidence of human occupation and use of basin resources dates back at least 9600 years, and Native American occupation in the valley is thought to have been continuous from 11,000

years ago through the American Revolution (Custer 1996). Native Americans were few in number and were concentrated in the southern portion of the basin at the time of the American Revolution. Following European colonization, the basin has yielded vast quantities of lumber, coal, fish, agricultural products, and power, and in the process has been modified significantly relative to its prehistoric conditions.

Physiography, Climate, and Land Use

The Susquehanna River basin includes four physiographic provinces: the Appalachian Plateau (AP) in the north, the Valley and Ridge (VR) in the middle, a small portion of the Blue Ridge (BL) and the Piedmont Plateau (PP) in the south (see Fig. 2.28). Most of the basin lies in the Appalachian Plateau and Valley and Ridge. The Appalachian Plateau is relatively flat with deep, steep valleys, for much of the region was glaciated during the Wisconsinan and earlier glacial episodes (Sevon and Fleeger 1999). Some valleys have deep deposits of glacial fill, and outwash sediments from the glaciers extend downstream as far as Conestoga Creek on the main stem. The glacial fill and outwash form terraces up to 10 m high and 1 km wide immediately adjacent to the river. Folded and faulted rocks in the Valley and Ridge form steep mountains and ridges separated by valleys. The upland soils (primarily inceptisols) in the Appalachian Plateau and Valley and Ridge tend to be poorly developed and infertile, whereas valley soils range from poor to excellent (Cuff et al. 1989). The Piedmont's rolling hills and fertile soils (alfisols and ultisols) contain some of the most productive farmland in eastern North America. The quality of the soils reflects their maturity and parent material, with mature soils over metamorphic rock or limestone the most fertile and immature soils in glaciated or steep, mountainous areas the least fertile in the basin. Bituminous coal fields underlie the westernmost portions of the Appalachian Plateau, and distinct beds of anthracite coal are found in the eastern Appalachian Plateau and Valley and Ridge.

There are three predominant terrestrial ecoregions in the Susquehanna River basin: the Allegheny Highlands Forests in the Appalachian Plateau, the Appalachian/Blue Ridge Forests in the Valley and Ridge, and the Northeastern Coastal Forests in the Piedmont Plateau (Ricketts et al. 1999). The Allegheny Highlands Forests are part of the northern hardwood forests that commonly include American beech, sugar maple, eastern hemlock, and yellow and sweet birch, as well as eastern white pine, red pine, white ash, and black cherry. The Appalachian/Blue Ridge and Northeastern Coastal forests are predominantly Appalachian oak forests that contain several oak and hickory species, sugar maple, sweet birch, American beech, tulip popular, and pines.

The climate in this region is humid continental. Average annual temperature is 9.7°C, with warmest conditions in June to August and coldest conditions in January to February (Fig. 2.29). Total precipitation ranges from 79 cm/yr (e.g., near Corning, New York) to 129 cm/yr (along the basin boundary east of Sunbury, Pennsylvania), and averages 98 cm/yr (6 to 10 cm/mo) across the basin. Precipitation is greatest in May to July (9 to 10 cm/mo) and least in January to February (6 to 7 cm/mo) (see Fig. 2.29).

Land use in the Susquehanna River basin ranges from state forests and game lands to modern urban centers (Cuff et al. 1989). Forests cover 63% of the basin, agriculture 26% (19% cropland, 7% pasture), and urban development 9%. Land use patterns vary between the upper and lower basins (i.e., above and below Sunbury). For example, forests are most common in the upper basin, especially in the areas with steep terrain (Edwards 1996). Most of these forests have regrown after intensive timber harvests between 1850 and 1900 had removed most of the trees (Stranahan 1993). Agricultural activities, which are primarily dairy and livestock related, are concentrated in the flat river valleys and plateaus, especially in the lower basin. The lower basin is also the location of the major hydropower facilities, as well as more than half of the basin's 3.97 million people (56 people/km²).

River Geomorphology, Hydrology, and Chemistry

The upper Susquehanna River basin's two major branches originate in the Appalachian Plateau and converge in the Valley and Ridge near Sunbury, Pennsylvania (see Fig. 2.28). The North Branch begins as the outflow from Lake Otsego in New York and drains 29,275 km² as it flows 521 km to its confluence with the West Branch. The West Branch begins about 20 km west of Altoona, Pennsylvania, and drains 18,109 km² (Stranahan 1993, Edwards 1996). The lower Susquehanna includes one major tributary (the Juniata River) and crosses both the Valley and Ridge and Piedmont before ending at Chesapeake Bay (Fig. 2.14).

Very little of the Susquehanna has tidal influences, and the tidal/estuarine characteristics it does have are actually attributes of the Chesapeake Bay rather than the river itself. Because the Susquehanna River is relatively wide and shallow, distinct riffle pool sequences are visible during low and moderate flow levels. Prominent features in the river channel throughout the basin are islands, the most famous of which is Three Mile Island with its nuclear power facility.

Average discharge is 438 m³/s for the North Branch (at Danville, Pennsylvania), 309 m³/s for the West Branch (at Lewisburg, Pennsylvania), and 1153 m³/s at the Conowingo Dam, Maryland (16 km from the mouth). The Susquehanna River contributes almost 60% of all freshwater inputs into Chesapeake Bay (Porse 2000). Because the four major water retention/hydroelectric dams are all within 100 km of the river's mouth, the seasonal flow regime is relatively unmodified throughout most of the basin. Discharge exhibits a strong seasonal pattern, with discharge in March and April several times greater than that in August and September (see Fig. 2.29). Rainfall exceeds evapotranspiration in all months except May to August (Cuff et al. 1989), whereas runoff slightly in excess of precipitation in March and April is presumably due to snowmelt. It appears that evapotranspiration is the major factor responsible for the seasonal pattern in runoff.

The Susquehanna River is renowned for its frequent floods, and major floods have occurred every 14 to 20 years (Stranahan 1993, Porse 2000). Maximum discharge recorded at the Conowingo Dam was 32,002 m³/s following Hurricane Agnes in June 1972. Minimum discharge was 43 m³/s in September 1964 (Porse 2000). Flows in excess of 11,328 m³/s (about 10 times mean annual flow) occur an average of twice per year (Porse 2000). A combination of conditions—intense precipitation from tropical storms, a river topography that produces ice dams, less pervious soils, and even possibly land use such as the deforestation of the late nineteenth century—contribute to this pattern (Stranahan 1993, Porse 2000). The economic and social impact of these floods has been particularly great in areas where the wide terraces adjacent to the river have attracted agricultural and urban development.

Water chemistry varies greatly within the basin in response to natural variations in underlying soils as well as anthropogenic activities. Some streams naturally have high alkalinity and pH and abundant nutrients (e.g., in the Valley and Ridge, with mature soils derived from carbonate bedrock), whereas other streams may have low alkalinity and pH and be nutrient poor (e.g., in glaciated portions of the

FIGURE 2.14 Susquehanna River above Harrisburg, Pennsylvania (PHOTO BY TIM PALMER).

Appalachian Plateau, with shallow, immature soils). In the main stem, pH is 6.9 to 9.0, alkalinity is 40 to 110 mg/L as $CaCO_3$, NO_3-N is 0.29 to 1.49 mg/L, and PO_4-P is <0.02 to 0.1 mg/L (Rowles and Sitlinger 2000). pH and alkalinity are lowest in areas affected by acid mine drainage, whereas nitrate and phosphate are highest in areas where agriculture is most intense.

River Biodiversity and Ecology

The Susquehanna River is in the Chesapeake Bay freshwater ecoregion (Abell et al. 2000) and was connected to other Chesapeake tributaries and nearby Atlantic drainages when sea levels dropped during past glacial periods (Hocutt et al. 1986). Ecological studies of the Susquehanna River have generally been limited to qualitative and quantitative collections associated with environmental monitoring, taxonomic surveys, or fisheries management. These collections occur most often during low summer flow, when more of the river can be waded.

The taxa collected give some indication of organic matter processes and ecosystem function, which otherwise remain unexamined.

Algae, Cyanobacteria, and Protists

The wide, shallow riverbed of the Susquehanna River supports a variety of algae at relatively high densities. Collections by the Academy of Natural Sciences of Philadelphia on the North Branch near Meshoppen Creek found over 100 species, most of which were diatoms (Bacillariophyceae), especially *Cyclotella atomus*, *Cocconeis placentula* var. *lineata*, and *Nitzschia palea*. Common green algae (Chlorophyceae) in these collections were *Cladophora glomerata*, *Hydrodictyon reticulatum*, *Rhizoclonium hieroglyphicum*, *Oedogonium*, and *Spirogyra*, whereas *Schizothrix calcicola* was the common cyanobacteria. Algal biomass during low flow can average 30 to 90 mg chlorophyll a/m^2 in the riffles at these sites (Jackson and Sweeney, unpublished data), with greater accumulations associated with mats of filamentous green algae (e.g., *C. glomerata*, *R. hieroglyphicum*). Reduced light from fine particle transport and accumulation appears to limit algal biomass in pools and some depositional areas.

Plants

Although macrophytes are not common at most sites, they do occur and can be abundant at some sites (e.g., water star-grass, jewel weed, water willow, lizard tail). As with most rivers in the region, several nonnative plants (e.g., watercress, Eurasian water-milfoil, purple loosestrife, curly-leaved pondweed) are now common in some locations.

Invertebrates

Macroinvertebrates are both abundant and diverse, with densities in riffles often averaging 10,000 to 30,000/m^2 (e.g., Jackson, Sweeney et al. 1994; Jackson, Horwitz, and Sweeney 2002). Although the exact number of aquatic invertebrate species in the basin is not known, it is well over 100. For example, 120 aquatic insect species alone (including 27 mayflies, 20 beetles, 17 caddisflies, and 30 dipterans) were recently collected from a North Branch site downstream of Meshoppen Creek (Jackson and Sweeney, unpublished data). In riffles, common macroinvertebrates include those that feed on attached algae (e.g., baetid [*Baetis, Pseudocloeon, Centroptilum*] and heptageniid [*Stenonema, Leucrocuta*] mayflies, numerous chironomid midges), fine particles suspended in the water (e.g., hydropsychid caddisflies [*Hydropsyche, Cheumatopsyche*], isonychiid mayflies, and the chironomid *Rheotanytarsus*), and fine particles on the surface or in sediments (e.g., anthopotamid [*Anthopotamus*] mayflies and many chironomids). In contrast, common macroinvertebrates in depositional areas are predominately those that feed on fine particles on the surface or in sediments (e.g., oligochaete worms, several chironomids [*Tanytarsus, Polypedilum*], elmid [*Stenelmis, Optioservus, Dubiraphia*] beetles, and anthopotamid [*Anthopotamus*] mayflies). There are two crayfish species native to the main stem and branches of the Susquehanna River (Ted Nuttall, personal communication). The spiny-cheek crayfish is found in slower areas, whereas *Orconectes obscurus* is found in faster areas, especially in the north. In addition, the Appalachian brook crayfish may be found entering from smaller tributaries in mountainous sections, and the nonnative rusty crayfish can be common in the lower basin.

Although mollusks have been well inventoried in the Susquehanna basin (Strayer and Fetterman 1999, A. E. Bogan, personal communication), their ecological role is not well studied. The most abundant mollusks are small bivalves (Sphaeriidae), limpids (Ancylidae), and snails (e.g., Physidae, Viviparidae) (e.g., Jackson et al. 1994). There are a total of 13 unionid species in the basin, five of which are widespread and sometimes common: eastern elliptio, triangle floater, elktoe, creeper, and yellow lampmussel. Strayer and Fetterman (1999) found little evidence of large changes in mussel communities in the New York portion of the North Branch between 1955 to 1965 and 1996 to 1997. However, the brook floater clearly declined in range, which may be due to hybridization with the elktoe. Asiatic clams are established in the lower Susquehanna River. Adult zebra mussels were only recently collected in Eaton Brook Reservoir (2000) and Canadarago Lake (2002), and appear established.

Vertebrates

At least 103 fish species are found in the Susquehanna River basin (Hocutt et al. 1986, Argent et al. 1998), and densities can be as high as 6/m^2 (e.g., Jackson et al. 2002). The native fish fauna is similar to nearby Atlantic drainages but includes some species also found in the northern Mississippi basin (Hocutt et al. 1986). Only the Maryland darter is endemic; 14 species spend some portion of their life in marine or brackish waters (Hocutt et al. 1986) and at least 29 species are not native to the basin (Argent et al. 1998). Reductions in species distribution have occurred for 29 species in the last 100 years, including the loss of 4 to 7 species from several watersheds (Argent et al. 1998). The impact of nonnative species, which are now the dominant insectivorous and piscivorous fishes at many locations in the basin (e.g., Jackson et al. 2002), on native fishes and other components of the ecosystem is unknown.

Fishes, especially the migratory American shad, alewife, blueback herring, and American eel, played an important role in the lives of the native Susquehannocks and early European settlers in the Susquehanna basin (Custer 1996). All were abundant, but shad were considered especially valuable. Accounts from the 1800s and early 1900s are largely anecdotal, but the shad harvests clearly included hundreds of thousands and even millions of fish (Anonymous 2000). Harvests plummeted by the early twentieth century in response to changes in the basin. The Susquehanna River still supports an active warmwater sport fishery that is focused primarily on several nonnative fishes, especially smallmouth bass, but also channel catfish, rock bass, walleye, and muskellunge. Popular native species include striped bass and white perch, both of which are most abundant below the Conowingo dam. A variety of native and nonnative sunfishes are also frequently caught in the river, even though they are generally thought of as pond species.

The Susquehanna River basin supports a wide variety of additional aquatic vertebrates, such as beaver, muskrat, turtles, snakes, blue heron, bald eagle, and various ducks and geese.

Ecosystem Processes

Although intensive ecosystem studies have not been conducted on the Susquehanna, observations of the flora and fauna give some indication of the structure and function of the river. First, benthic algae and fine particles (benthic and suspended) are abundant and the major food resources for invertebrate consumers in the river. Second, benthic algae may be a more important primary food resource than in rivers of similar size because light reaches the river bottom in the shallow riffles. Third, abundant food, high macroinvertebrate densities, and warm water temperatures suggest that invertebrate production is relatively high. Finally, high invertebrate production contributes to the high densities of insectivorous and piscivorous fishes. The impact of fish introductions and losses has not been studied to date.

Human Impacts and Special Features

The Susquehanna River basin is known for the broad river channel with shallow, rocky reaches and numerous islands. It is also known for its abundance of natural resources. Although human activities in the basin over the last 300 years have been well chronicled, their impact has not. One reason for this is that residents of the basin have relied on groundwater and tributaries for drinking water, rather than the Susquehanna River, and with the exception of shad fisheries, human activities were generally not dependent on the environmental quality of the river. The three activities that have had the greatest effect on the river are logging in the middle to late 1800s, coal mining from the late 1890s to the middle 1900s, and dams/diversions in the early 1800s and then again in the early 1900s. Portions of the river still appear to be responding to those activities.

Forests (white pine, hemlock, oak, chestnut) were vital in the development of this region, and the Susquehanna River played a key role in transporting and processing cut timber. By 1850 the best stands were exhausted in northern and eastern areas. However, lumbering on the West Branch continued into the early twentieth century. Some cleared lands quickly became productive croplands and pastures, whereas others were abandoned immediately or after agricultural efforts failed. Today hardwood and mixed-hardwood forests cover 63% of the basin, including those areas that were primarily pine and hemlock. Croplands and pastures now account for 26% of land use in the basin. There is little documentation of the impact of forest removal on the river, but it presumably affected flow regime and water quality. Sediment transport between 1890 and 1940 was at least nine times greater than it had been 200 years earlier (Reed 2000). Two-thirds of the increase was due to soil erosion, primarily from forestry and agriculture, and the remaining third was caused by coal mining. Sediment transport has declined significantly, but remains a great concern (along with nutrients) for the Chesapeake Bay.

Coal is plentiful throughout much of the Susquehanna basin, with some of the world's largest deposits of anthracite in the north and east and vast bituminous coal fields in the west. Over the last 200 years almost 30 billion tons of coal has been mined in Pennsylvania, and much of it came from the Susquehanna basin. Coal waste and acid mine drainage decimated great lengths of the river and its tributary streams. Coal silt was so abundant that it spawned the Susquehanna's Hard Coal Navy, a fleet of over 200 vessels that dredged the dust from the river bottom (Stranahan 1993). From 1920 to 1950, 3 million tons of coal were recovered from behind the Holtwood Dam, and from 1951 to 1973, about 10 million tons of coal dust were recovered from behind the Safe Harbor Dam (O'Donel 2000). The coal industry is now relatively small, but its environmental impact (primarily from acid mine drainage) continues in the basin (Edwards 1996).

Dams and Susquehanna River fisheries (especially for American shad) have long been entwined, as residents in the 1700s recognized that even low-head dams blocked fish migrations in tributaries (Anonymous 2000). Shad fisheries flourished until the 1830s, when main-stem dams were constructed to supply water to the 950 km canal system associated with the river. In addition to the role played by the dams, excessive harvests and poor water quality associated with timber and coal development and industrial and municipal effluents may have contributed to the shad harvest decline. Shad migrating up the Chesapeake Bay had access to very little of the Susquehanna River following the completion of the York Haven hydroelectric dam at Rkm 90 (1905), the Holtwood hydroelectric dam at Rkm 40 (1910), and finally the Conowingo Dam at Rkm 16 (1928). Remnants of the fisheries that remained in the lower

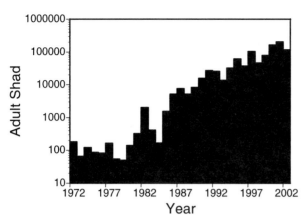

FIGURE 2.15   Number of adult American shad returning to the Susquehanna River and captured at the fish lift on the Conowingo Dam, Maryland. No fish passed the dam between 1928 and 1972.

river and upper Chesapeake Bay declined in the 1970s, and shad fishing has been prohibited in these areas since 1980 (Anonymous 2000).

Although economic development and population growth have produced some grim periods in the environmental history of the Susquehanna Basin, significant improvement has occurred (e.g., Reed 2000, Rowles and Sitlinger 2000) and the future appears bright. River biota have responded positively to improved water and habitat quality. For example, the number of aquatic insects collected at a North Branch site near Mehoopany Creek has almost doubled since 1974 (Jackson and Sweeney, unpublished data). Only 11% of the 28,038 km of streams and rivers recently assessed did not support their designated use (Edwards 1996). Acid mine drainage continues to be the major cause of degradation at these sites, followed by agricultural effluents. Finally, the fishery remains productive, and migratory fish are responding positively to stocking programs and modern fishways (Fig. 2.15).

ADDITIONAL RIVERS

The adjacent Kennebec and Androscoggin river basins are located entirely within the New England physiographic province (Figs. 2.30 and 2.32). Their forested headwaters arise in the rugged mountains of western Maine (Fig. 2.16), and after leaving the mountains, the rivers flow across the gently rolling agricultural landscape of the lower basin (Fig. 2.17). The two rivers join downstream of Lewiston and Augusta, just before flowing into the Atlantic Ocean.

FIGURE 2.16   Androscoggin River near Wilsons Mills, Maine (PHOTO BY TIM PALMER).

There are numerous lakes and wetlands throughout their catchments, which act to dampen changes in flow. The peat-laden soils of these wetlands result in river water that is poor in nutrients but rich in dissolved organic matter. Consequently, aquatic productivity is low to moderate. A few of the lakes —Moosehead Lake (Kennebec River) in particular— are exceptionally large. Like most rivers of New England, they have been dammed for centuries. Their flow is only moderately regulated, however, because most of the dams are "run-of-the-river." The Edwards Dam on the Kennebec River at Augusta was the first major dam on a major river that was ordered to be removed by the federal government. It was breached in 1999.

The Merrimack River is recovering from a history of intensive industrial use. Rising in the largely forested White Mountains of New Hampshire, the Merrimack flows swiftly south and east into the urban and suburban landscape of southern New

FIGURE 2.17   Kennebec River at Solon, Maine (Photo by Tim Palmer).

Hampshire and Massachusetts (Fig. 2.34). The Hubbard Brook Experimental Forest, a site of intensive long-term ecological research, is in the headwaters of the Merrimack. The Merrimack's water is clear and soft. The Merrimack was so heavily impounded (there are more than 500 dams in its basin) that it was called "America's hardest working river." These dams and pollution from paper mills and other industries led to the disappearance of important migratory fish species in the nineteenth century. Dramatic improvements in water quality and installation of fish passages are helping to restore American shad and Atlantic salmon to the Merrimack (Fig. 2.1).

The Raritan River is the longest river found entirely in New Jersey, and it is an excellent example of the urban/suburban conditions that characterize portions of many rivers and streams in this region. The Raritan begins in the highlands of the New England province and crosses the Piedmont Plateau and Coastal Plain before emptying into the Raritan Bay near the mouth of the Hudson River (Fig. 2.36). The basin is highly fragmented; over 250 dams (most are <8 m high) slow and divert river flow and

block fish migrations and extensive agriculture and urban/suburban development (19% and 36% of current land cover, respectively) have resulted in small, isolated forest areas. Impervious cover is presently 11% across the basin, and is between 10% and 25% in most watersheds. A majority of mainstem and tributary sites show evidence of moderate or severe impairment; only 38% were classified as unimpaired. Even so, the Raritan supports a diverse flora and fauna and represents an important aquatic resource in the region. Efforts to maintain or improve water quality and quantity are underway, but they face resistance from continued urban development (Fig. 2.18).

The Potomac River begins its 616 km journey to the Chesapeake Bay in the Allegheny Mountains of West Virginia (Fig. 2.38). Along that route it crosses five physiographic provinces and five terrestrial ecoregions. Although best known as a scenic backdrop for the U.S. capital and Arlington and Alexandria, Virginia, most of the Potomac basin is still undeveloped forests and open agricultural fields and the river is relatively free flowing (Fig. 2.19). Most (80%) Potomac basin residents live in the

FIGURE 2.18 Raritan River at Route 1 bridge between New Brunswick and Edison, New Jersey (Photo by Raritan Riverkeeper).

FIGURE 2.19 Potomac River near Harper's Ferry, West Virginia (Photo by A. C. Benke).

Washington metropolitan area. Biological diversity and productivity are similar to the James, Susquehanna, Delaware, and Hudson Rivers. Likewise, the Potomac River has many of the same environmental problems associated with more than 250 years of intensive use. The most common sources of stream impairment are excessive nutrients and sediments; urban, industrial, and agricultural toxins; and acid mine drainage. Efforts over the last 50 years have significantly improved water and habitat quality in the river.

ACKNOWLEDGMENTS

The staff of the Old Town office of the U.S. Fish and Wildlife Service and Dr. John Moring of the Cooperative Fish and Wildlife Research Unit at the University of Maine provided references and publications regarding the Penobscot River. Vivian Butz Huryn and Michael Chadwick provided helpful comments on an early draft for the Penobscot River. Bob Bode, Kathy Hattala, and colleagues at the Institute of Ecosystem Studies provided helpful advice and data for the Hudson, and John Ladd allowed us to use his beautiful Figure 2.11. The Delaware River Basin Commission, Susquehanna River Basin Commission, and Interstate Commission on the Potomac River Basin provided access to their publications and libraries. Michael Hendricks provided recent shad data for the Susquehanna River, and Richard St. Pierre helped us access historical shad data for the Atlantic region. This is a contribution to the program of the Institute of Ecosystem Studies, and contribution number 2003001 for the Stroud Water Research Center.

LITERATURE CITED

Abell, R. A., D. M. Olson, T. H. Ricketts, E. Dinerstein, P. T. Hurley, J. T. Diggs, W. Eichbaum, S. Walters, W. Wettengel, T. Allnutt, C. J. Loucks, and P. Hedao. 2000. *Freshwater ecoregions of North America: A conservation assessment*. Island Press, Washington, D.C.

Academy of Natural Sciences of Philadelphia (ANSP). 1995. Biological and mixing-zone studies on the Upper Delaware River: Final report. Report no. 95-15F. Academy of Natural Sciences of Philadelphia, Pennsylvania.

Academy of Natural Sciences of Philadelphia (ANSP). 2000. Biological studies on the Upper Delaware River: Draft report. Report No. 00-4F. Academy of Natural Sciences of Philadelphia, Pennsylvania.

Anonymous. 2000. *Migratory fish restoration and passage on the Susquehanna River*. Pennsylvania Fish and Boat Commission, Harrisburg.

Argent, D. G., R. F. Carline, and J. R. Stauffer Jr. 1998. Changes in the distribution of Pennsylvania fishes: The last 100 years. *Journal of the Pennsylvania Academy of Science* 72:32–37.

Ashizawa, D., and J. J. Cole. 1994. Long-term temperature trends of the Hudson River: A study of the historical data. *Estuaries* 17:166–171.

Atlantic States Marine Fisheries Commission. 1985. Fishery management plan for American shad and river herrings. Fisheries Management Report 6. Atlantic States Marine Fisheries Commission, Washington, D.C.

Bacon, E. M. 1906. *The Connecticut River and the valley of the Connecticut*. G.P. Putnam's Sons, New York.

Bailey, R. G. 1980. Ecoregions of the United States. U.S. Forest Service Miscellaneous Publication 1391. U.S. Forest Service, Washington, D.C.

Baker, J. E., W. F. Bohlen, R. Bopp, B. Brownawell, T. K. Collier, K. J. Farley, W. R. Geyer, and R. Nairn. 2001. PCBs in the upper Hudson River: The science behind the dredging controversy. White paper prepared for the Hudson River Foundation, New York.

Barrows H. K., and C. C. Babb 1912. Water resources of the Penobscot River basin, Maine. Water Supply Paper 279, U.S. Geological Survey, Washington, D.C.

Baum, E. T. 1983. *The Penobscot River: An Atlantic salmon river management report*. State of Maine, Atlantic Sea Run Salmon Commission, Bangor.

Bode, R. W., M. A. Novak, and L. E. Abele. 1993. *20 year trends in water quality of rivers and streams in New York State based on macroinvertebrate data 1972–1992*. New York State Department of Environmental Conservation, Albany.

Bode, R. W., M. A. Novak, and L. E. Abele. 1996. Biological stream (macroinvertebrate) data. In J. Myers (program coordinator), *Rotating Intensive Basin Studies Water Quality Assessment Program: The Upper Hudson river drainage basin biennial report 1993–94*, Appendix B. New York State Department of Environmental Conservation, Albany.

Boyle, R. H. 1979. *The Hudson River: A natural and unnatural history*. W. W. Norton, New York.

Braun, E. L. 1950. *Deciduous forests of eastern North America*. The Free Press, New York.

Caraco N. F., and J. J. Cole. 2002. Contrasting impacts of a native and alien macrophyte on dissolved oxygen in a large river. *Ecological Applications* 12:1496–1509.

Caraco, N. F., J. J. Cole, S. E. G. Findlay, D. T. Fischer, G. G. Lampman, M. L. Pace, and D. L. Strayer. 2000. Dissolved oxygen declines in the Hudson River associated with the invasion of the zebra mussel (*Dreissena polymorpha*). *Environmental Science and Technology* 34:1204–1210.

Caraco, N. F., J. J. Cole, P. A. Raymond, D. L. Strayer, M. L. Pace, S. E. G. Findlay, and D. T. Fischer. 1997. Zebra mussel invasion in a large, turbid river: Phytoplankton response to increased grazing. *Ecology* 78:588–602.

Cayford J. E., and R. E. Scott. 1964. *Underwater logging*. Cornell Maritime Press, Cambridge, Maryland.

Climate and Man, Yearbook of Agriculture. 1941. U.S. Department of Agriculture, Washington, D.C.

Cole, J. J., and N. F. Caraco. 2001. Carbon in catchments: Connecting terrestrial carbon losses with aquatic metabolism. *Marine and Freshwater Research* 52: 101–110.

Cole, J. J., and N. F. Caraco. 2005. Primary production and its regulation in the tidal-freshwater Hudson River. In J. S. Levinton, and J. R. Waldman (eds.), *The Hudson River estuary.* Oxford University Press, New York.

Connelly, N. A., and T. L. Brown. 1991. Net economic value of the freshwater recreational fisheries of New York. *Transactions of the American Fisheries Society* 120:770–775.

Cuff, D. J., W. J. Young, E. K. Muller, W. Zelinsky, and R. F. Abler. 1989. *The atlas of Pennsylvania.* Temple University Press, Philadelphia, Pennsylvania.

Custer, J. E. 1996. *Prehistoric cultures of eastern Pennsylvania.* Anthropological series no. 7. Pennsylvania Historical and Museum Commission, Harrisburg.

Cutting R. E. 1963. *Penobscot River salmon restoration.* Maine Atlantic Sea Run Salmon Commission, Bangor.

Davies S. P., L. Tsomides, J. L. DiFranco, and D. L. Courtemanch. 1999. *Biomonitoring retrospective: Fifteen year summary for Maine rivers and streams.* Division of Environmental Assessment, Bureau of Land and Water Quality, Augusta, Maine.

Dovel, W. L., A. W. Petkovitch, and T. J. Berggren. 1992. Biology of the shortnose sturgeon (*Acipenser brevirostrum* Lesueur, 1818) in the Hudson River estuary, New York. In C. L. Smith (ed.), *Estuarine research in the 1980s*, pp. 187–227. State University of New York Press, Albany.

Environmental Consulting Services (ECS). 1993. Survey of benthos: Delaware estuary: From the area of the C&D Canal through Philadelphia to Trenton. Report to the Delaware Estuary Program.

Edwards, R. E. 1996. *The 1996 Susquehanna River basin water quality assessment 305(B) report.* Publication 176. Susquehanna River Basin Commission, Harrisburg, Pennsylvania.

Ellsworth, J. M. 1986. Sources and sinks for fine-grained sediment in the lower Hudson River. *Northeastern Geology* 8:141–155.

Feldman, R. S. 2001. Taxonomic and size structures of phytophilous macroinvertebrate communities in *Vallisneria* and *Trapa* beds of the Hudson River, New York. *Hydrobiologia* 452: 233–245.

Federal Energy Regulatory Commission (FERC). 1997. *Final environmental impact statement: Lower Penobscot River basin, Maine.* FERC/FEIS-0082. Federal Energy Regulatory Commission, Washington, D.C.

Findlay, S., R. L. Sinabaugh, D. T. Fischer, and P. Franchini. 1998. Sources of dissolved organic carbon supporting planktonic bacterial production in the tidal freshwater Hudson River. *Ecosystems* 1:227–239.

Findlay, S., C. Wigand, and W. C. Nieder. 2005. Submersed macrophyte distribution and function in the tidal freshwater Hudson River. In J. S. Levinton and J. R. Waldman (eds.), *The Hudson River estuary.* Cambridge University Press, New York.

Flanagan, S. M., M. G. Nielsen, K. W. Robinson, and J. F. Coles. 1999. *Water-quality assessment of the New England coastal basins in Maine, Massachusetts, New Hampshire, and Rhode Island: Implications for water quality and aquatic biota.* Water-Resources Investigations Report 98-4249. U.S. Geological Survey, Pembroke, New Hampshire.

Flynn, K. C., and W. T. Mason (eds.). 1978. *The freshwater Potomac: Aquatic communities and environmental stresses.* Interstate Commission on the Potomac River Basin, Rockville, Maryland.

Frithsen, J. B., K. Killam, and M. Young. 1991. *An assessment of key biological resources in the Delaware Estuary.* Delaware Estuary Report. Versar, Columbia, Maryland.

Garabedian, S. P., J. F. Coles, S. J. Grady, E. C. T. Trench, and M. J. Zimmerman. 1998. Water-quality assessment of the Connecticut, Housatonic, and Thames River Basins, 1992–95. U.S. Geological Survey Circular 1155, U.S. Geological Survey, Washington, D.C.

Gilliom, R. J., R. B. Alexander, and R. A. Smith. 1985. Pesticides in the nation's rivers, 1975–1980, and implications for future monitoring. Water Supply Paper 2271, U.S. Geological Survey, Washington, D.C.

Grady, S. J., and S. P. Garabedian. 1991. National Water Quality Assessment Program: The Connecticut River and Long Island coastal rivers. Open-File Report 91-159, U.S. Geological Survey, Reston, Virginia.

Hard, W. 1947. *The rivers of America: The Connecticut.* Rinehart, New York.

Hartel, K. E., D. B. Halliwell, and A. E. Launer. 2002. *Inland fishes of Massachusetts.* Natural History of New England no. 3. Massachusetts Audubon Society Press, Lincoln.

Hauptman, L. M., and J. D. Wherry (eds.). 1993. *The Pequots of southern New England: The rise and fall of an American Indian nation.* University of Oklahoma Press, Norman.

Hellquist, C. B. 2001. *A guide to selected non-native aquatic species of Massachusetts.* Massachusetts Department of Environmental Management, Boston.

Hocutt, C. H., R. E. Jenkins, and J. R. Stauffer Jr. 1986. Zoogeography of the fishes of the central Appalachians and central Atlantic coastal plain. In C. H. Hocutt and E. O. Wiley (eds.), *The zoogeography of North American freshwater fishes*, pp. 161–211. John Wiley and Sons, New York.

Hoover, E. E. (ed.). 1938. *Biological survey of the Merrimack watershed.* New Hampshire Fish and Game Department, Concord.

Horwitz, R. J. 1986. Fishes of the Delaware Estuary in Pennsylvania. In S. K. Majumdar, F. J. Brenner, and A. F. Rhoads (eds.), *Endangered and threatened species programs in Pennsylvania and other states: Causes, issues, and management*, pp. 177–201. Pennsylvania Academy of Science, Easton.

Houtman, N. R. 1993. *Natural resources highlights: Penobscot River watershed*. Water Resources Program, University of Maine, Orono.

Howarth, R. W., R. Schneider, and D. Swaney. 1996. Metabolism and organic carbon fluxes in the tidal freshwater Hudson River. *Estuaries* 19:848–865.

Howell, B. P., and J. H. Slack. 1871. First annual report of the commissioners of fisheries of the state of New Jersey. Murphy and Bechtel, Trenton, New Jersey.

Hunt, C. B. 1974. *Natural regions of the United States and Canada*. W. H. Freeman, San Francisco, California.

Isachsen, Y. W., E. Landing, J. M. Lauber, L. V. Rickard, and W. B. Rogers (eds.). 1991. Geology of New York: A simplified account. New York State Museum Educational Leaflet 28.

Jackson, J. K., R. J. Horwitz, and B. W. Sweeney. 2002. Effects of *Bacillus thuringiensis israelensis* on black flies and nontarget macroinvertebrates and fish in a large river. *Transactions of the American Fisheries Society* 131:910–930.

Jackson, J. K., B. W. Sweeney, T. L. Bott, J. D. Newbold, and L. A. Kaplan. 1994. Transport of *B.t.i.* and its effect on drift and benthic densities of nontarget macroinvertebrates in the Susquehanna River, northern Pennsylvania. *Canadian Journal of Fisheries and Aquatic Sciences* 51:295–314.

Judd, R. W., E. A. Churchill, and J. W. Eastman (eds.). 1995. *Maine: The pine tree state from prehistory to the present*. University of Maine Press, Orono.

Kennen, J. G. 1999. Relation of macroinvertebrate community impairment to catchment characteristics in New Jersey streams. *Journal of the American Water Resources Association* 35:939–955.

Lampman, G. G., N. F. Caraco, and J. J. Cole. 1999. Spatial and temporal patterns of nutrient concentration and export in the Hudson River. *Estuaries* 22:285–296.

Leopold L. B. 1994. *A view of the river*. Harvard University Press, Cambridge, Massachusetts.

Levinton, J. S., and J. R. Waldman (eds.). 2005. *The Hudson River estuary*. Cambridge University Press, New York.

Limburg, K. E., M. A. Moran, and W. H. McDowell. 1986. *The Hudson River ecosystem*. Springer-Verlag, New York.

Majumdar, S. K., E. W. Miller, and L. E. Sage. 1988. *The Delaware River*. Pennsylvania Academy of Science, Easton.

Marcy, B. C. 1976. Fishes of the lower Connecticut River and the effects of the Connecticut Yankee plant. In D. Merriman and L. M. Thorpe (eds.), *The Connecticut River ecological study: The impact of a nuclear poser plant*, pp. 61–113. Monograph 1. American Fisheries Society, Washington, D.C.

Marshall, H. G. 1992. Assessment of phytoplankton species in the Delaware River estuary. Delaware River Estuary Program Report. Old Dominion University Research Foundation, Norfolk, Virginia.

Massachusetts Department of Environmental Protection (MADEP). 1998. *Connecticut River basin 1998 water quality assessment report*. Massachusetts Department of Environmental Protection, Boston.

Massengill, R. R. 1976. Benthic fauna: 1965–1967 versus 1968–1972. In D. Merriman and L. M. Thorpe (eds.), *The Connecticut River ecological study: The impact of a nuclear poser plant*, pp. 39–53. Monograph 1. American Fisheries Society, Washington, D.C.

McLaren, J. B., R. J. Klauda, T. B. Hoff, and M. Gardinier. 1988. Commercial fishery for striped bass in the Hudson River, 1931–1980. In C. L. Smith (ed.), *Fisheries research in the Hudson River*, pp. 89–123. State University of New York Press, Albany.

Merriman, D., and L. M. Thorpe (eds.). 1976. *The Connecticut River ecological study: The impact of a nuclear power plant*. Monograph 1. American Fisheries Society, Washington, D.C.

Mills, E. L., D. L. Strayer, M. D. Scheuerell, and J. T. Carlton. 1996. Exotic species in the Hudson River basin: A history of invasions and introductions. *Estuaries* 19:814–823.

Moore, B., and D. D. Platt. 1996. The river and its watershed. In D. D. Platt (ed.), *Penobscot: The forest, river and bay*, pp. 22–31. Island Institute, Rockland, Maine.

Moore, E. (ed.). 1933. *A biological survey of the upper Hudson watershed*. Supplement to the 22nd annual report of the State of New York Conservation Department. L. B. Lyon, Albany, New York.

National Marine Fisheries Service. 1998. Recovery plan for the shortnose sturgeon (*Acipenser brevirostrum*). Prepared by the Shortnose Sturgeon Recovery Team for the National Marine Fisheries Service, Silver Spring, Maryland.

Nedeau, E. J., M. A. McCullough, and B. I. Swartz. 2000. *The freshwater mussels of Maine*. Maine Department of Inland Fisheries and Wildlife, Augusta.

New Jersey Water Supply Authority (NJWSA). 2000. *Setting of the Raritan River basin*. New Jersey Water Supply Authority, Clinton.

New Jersey Water Supply Authority. 2002a. *Landscape of the Raritan River basin*. New Jersey Water Supply Authority, Clinton.

New Jersey Water Supply Authority. 2002b. *Surface water and riparian areas of the Raritan River basin*. New Jersey Water Supply Authority, Clinton.

New York State Department of Environmental Conservation (NYSDEC). 1993. *Hudson River estuary quarterly issues update and state of the Hudson report*. Vol. 2, no. 4. Hudson River Estuary Management Program, New Paltz, New York.

New York State Department of Environmental Conservation. 1994. *Hudson River estuary quarterly issues update and state of the Hudson report*. Vol. 3, no. 4. Hudson River Estuary Management Program, New Paltz, New York.

O'Donel, O. L. 2000. Dams on the lower Susquehanna. In *Proceedings of the sediment symposium*, pp. 32–35. Publication no. 216E. Susquehanna River Basin Commission, Harrisburg, Pennsylvania.

Olcott, P. G. 1995. *Ground water atlas of the United States: Connecticut, Maine, Massachusetts, New Hampshire, New York, Rhode Island, Vermont.* Hydrologic Investigations Atlas HA 730-M. U.S. Geological Survey, Reston, Virginia.

Olsen, P. E., R. W. Schlische, and P. J. W. Gore (eds.). 1989. *Tectonic, depositional, and paleoecological history of early Mesozoic rift basins, eastern North America.* International Geological Congress Field Trip T-351. American Geophysical Union, Washington, D.C.

Pace, M. L., S. E. G. Findlay, and D. Lints. 1992. Zooplankton in advective environments: The Hudson River community and a comparative analysis. *Canadian Journal of Fisheries and Aquatic Sciences* 49: 1060–1069.

Pace, M. L., and D. J. Lonsdale. 2005. Ecology of the Hudson zooplankton community. In J. S. Levinton and J. R. Waldman (eds.), *The Hudson River estuary.* Cambridge University Press, New York.

Page, L. V., and L. C. Shaw. 1977. *Low-flow characteristics of Pennsylvania streams.* Water Resources Bulletin no. 12. Pennsylvania Department of Environmental Resources, Harrisburg.

Patrick, R. 1994. *Rivers of the United States.* Vol. 1, *Estuaries.* John Wiley and Sons, New York.

Patrick, R. 1996. *Rivers of the United States.* Vol. 3, *The eastern and southeastern states.* John Wiley and Sons, New York.

Patrick, R. P., F. Douglas, D. M. Palavage, and P. M. Stewart. 1992. *Surface water quality: Have the laws been successful.* Princeton University Press, Princeton, New Jersey.

Penobscot River Study Team. 1972. *Penobscot River study.* Vol. 1. Technical Report no. 1. University of Maine, Environmental Studies Center, Orono.

Penobscot Nation Water Resources Program (PNWRP). 2001. *The Penobscot Nation and the Penobscot River basin: A watershed analysis and management (WAM) project.* Penobscot Nation Water Resources Program, Department of Natural Resources, Indian Island, Maine.

Porse, N. C. 2000. Hydrology of the Susquehanna River basin. In *Proceedings of the Sediment Symposium*, pp. 16–27. Publication no. 216E. Susquehanna River Basin Commission, Harrisburg, Pennsylvania.

Rabeni C. F., S. P. Davies, and K. E. Gibbs. 1985. Benthic invertebrate response to pollution abatement: Structural changes and functional implications. *Water Resources Bulletin* 21:489–497.

Rabeni C. F., and K. E. Gibbs. 1980. Ordination of deep river invertebrate communities in relation to environmental variables. *Hydrobiologia* 74:67–76.

Reed, L. 2000. Where did all the sediment come from? In *Proceedings of the Sediment Symposium*, pp. 28–

35. Publication no. 216E. Susquehanna River Basin Commission, Harrisburg, Pennsylvania.

Ricketts, T. H., E. Dinerstein, D. M. Olson, C. J. Loucks, W. Eichbaum, D. DellaSala, K. Kavanaugh, P. Hedao, P. T. Hurley, K. M. Carney, R. Abell, and S. Walters. 1999. *Terrestrial ecoregions of North America: A conservation assessment.* Island Press, Washington, D.C.

Rowles, J. L., and D. L. Sitlinger. 2000. *Assessment of the interstate streams in the Susquehanna River basin: Monitoring report no. 13, July 1, 1998–June 30, 1999.* Publication no. 211. Susquehanna River Basin Commission, Harrisburg, Pennsylvania.

Scarola, J. F. 1973. *Freshwater fishes of New Hampshire.* New Hampshire Fish and Game Department, Concord.

Schmidt, R. E. 1986. Zoogeography of the northern Appalachians. In C. H. Hocutt and E. O. Wiley (eds.), *The zoogeography of North American freshwater fishes*, pp. 137–159. John Wiley and Sons, New York.

Sevon, W. D., and G. M. Fleeger. 1999. *Pennsylvania and the Ice Age,* 2nd ed. Pennsylvania Geological Survey, Harrisburg.

Simpson, K. W., T. B. Lyons, and S. P. Allen. 1972. *Macroinvertebrate survey of the upper Hudson River, New York.* New York State Department of Health Environmental Health Report 2, Albany.

Smith, C. L. 1985. *The inland fishes of New York state.* New York State Department of Environmental Conservation, Albany.

Smith, C. L., and T. R. Lake. 1990. Documentation of the Hudson River fish fauna. *American Museum Novitates* 2981.

Smith, T. E., R. J. Stevenson, N. F. Caraco, and J. J. Cole. 1998. Changes in phytoplankton community structure during the zebra mussel (*Dreissena polymorpha*) invasion of the Hudson River (New York). *Journal of Plankton Research* 20:1567–1579.

Snow, D. 1978. Eastern Abenaki. In B. G. Trigger (ed.), *Handbook of North American Indians.* Vol. 15, *Northeast*, pp. 137–147. Smithsonian Institution, Washington, D.C.

Stanne, S. P., R. G. Panetta, and B. E. Forist. 1996. *The Hudson: An illustrated guide to the living river.* Rutgers University Press, New Brunswick, New Jersey.

Stranahan, S. Q. 1993. *Susquehanna, river of dreams.* Johns Hopkins University Press, Baltimore, Maryland.

Strayer, D. L. 2005. The benthic animal communities of the riverine Hudson River estuary. In J. S. Levinton and J. R. Waldman (eds.), *The Hudson River estuary.* Cambridge University Press, New York.

Strayer, D. L., N. F. Caraco, J. J. Cole, S. Findlay, and M. L. Pace. 1999. Transformation of freshwater ecosystems by bivalves: A case study of zebra mussels in the Hudson River. *BioScience* 49:19–27.

Strayer, D. L., and A. R. Fetterman. 1999. Changes in the distribution of freshwater mussels (Unionidae) in the upper Susquehanna River basin, 1955–1965 to 1996–1997. *American Midland Naturalist* 142: 328–339.

Strayer, D. L., K. A. Hattala, and A. W. Kahnle. 2004. Effects of an invasive bivalve (*Dreissena polymorpha*) on fish communities in the Hudson River estuary. *Canadian Journal of Fisheries and Aquatic Sciences* 61:924–941.

Strayer, D. L., and L. C. Smith. 2001. The zoobenthos of the freshwater tidal Hudson River and its response to the zebra mussel (*Dreissena polymorpha*) invasion. *Archiv für Hydrobiologie Supplementband* 139:1–52.

Stroud Water Research Center. 2001. *Water quality monitoring in the source water areas for New York City: An integrative watershed approach.* First year report. Stroud Center Contribution no. 2001007. Strond Water Research Center, Avondale, Pennsylvania.

Swaney, D. P., D. Sherman, and R. W. Howarth. 1996. Modeling water, sediment and organic carbon discharges in the Hudson–Mohawk basin: Coupling to terrestrial sources. *Estuaries* 19:833–847.

Sweeney, B. W., D. H. Funk, and R. L. Vannote. 1986. Population genetic structure of two mayflies (*Ephemerella subvaria*, *Eurylophella verisimilis*) in the Delaware River drainage basin. *Journal of the North American Benthological Society* 5:253–262.

Trapp, H., Jr., and M. A. Horn. 1997. *Ground water atlas of the United States: Delaware, Maryland, New Jersey, North Carolina, Pennsylvania, Virginia, West Virginia.*

Hydrologic Investigations Atlas HA 730-L. U.S. Geological Survey, Reston, Virginia.

Trench, E. C. T. 1996. Trends in surface water quality in Connecticut. 1969–88. Water-Resources Investigations Report 96-4161. U.S. Geological Survey, Hartford, Connecticut.

Thoreau, H. D. 1864. *The Maine woods.* Ticknor and Fields, Boston.

Vannote, R. L., G. W. Minshall, K. W. Cummins, J. R. Sedell, and C. E. Cushing. 1980. The river continuum concept. *Canadian Journal of Fisheries and Aquatic Sciences* 37:130–137.

Wall, G. R., K. Riva-Murray, and P. J. Phillips. 1998. *Water quality in the Hudson River basin, New York and adjacent states, 1992–1995.* U.S. Geological Survey Circular 1165, U.S. Geological Survey, Denver, Colorado.

Wallace, W. G., G. R. Lopez, and J. S. Levinton. 1998. Cadmium resistance in an oligochaete and its effect on cadmium trophic transfer to an omnivorous shrimp. *Marine Ecology Progress Series* 172:225–237.

Weinstein, L. H. (ed.). 1977. *An atlas of the biologic resources of the Hudson estuary.* Boyce Thompson Institute for Plant Research, Yonkers, New York.

Whitworth, W. R. 1996. *Freshwater fishes of Connecticut,* 2nd ed. State Geological and Natural History Survey of Connecticut Bulletin 114, Hartford.

Wildes, H. E. 1940. *The Delaware.* Farrar and Rinehart, New York.

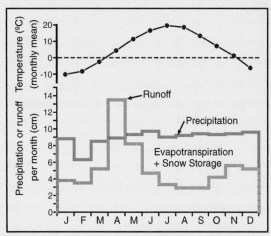

FIGURE 2.21 Mean monthly air temperature, precipitation, and runoff for the Penobscot River basin.

PENOBSCOT RIVER

Relief: 1607 m
Basin area: 22,253 km²
Mean discharge: 402 m³/s
River order: 6 (approximated)
Mean annual precipitation: 107 cm
Mean air temperature: 9.5°C
Mean water temperature: 9.3°C
Physiographic province: New England (NE)
Biome: Temperate Deciduous Forest
Freshwater ecoregion: North Atlantic
Terrestrial ecoregions: New England/Acadian Forests,
    Northeastern Coastal Forests
Number of fish species: 45
Number of endangered species: 1 fish
Major fishes: alewife, American eel, American shad,
    Atlantic salmon, brown bullhead, burbot, chain
    pickerel, common shiner, creek chub, fallfish,
    pumpkinseed, redbreast sunfish, smallmouth bass,
    white sucker, white perch, yellow perch
Major other aquatic vertebrates: eastern painted
    turtle, snapping turtle, muskrat, beaver, mink,
    river otter
Major benthic invertebrates: mayflies (*Centroptilum, Ephemerella, Eurylophella, Heptagenia, Stenonema*), stoneflies
    (*Acroneuria*), caddisflies (*Cheumatopsyche, Hydropsyche, Macrostemum, Neureclipsis, Polycentropus*), hellgrammites
    (*Corydalus*), mollusks (*Amnicola, Elliptio, Musculium*), beetles (*Stenelmis*), odonates (*Calopteryx, Hagenius*), crayfish
    (*Cambarus, Orconectes*)
Nonnative species: chain pickerel, smallmouth bass, brown trout
Major riparian plants: balsam fir, red maple, silver maple, paper birch, black ash, quaking aspen, chokecherry, northern white
    cedar, American elm (declining)
Special features: River islands north of Old Town comprise Penobscot Indian reservation; Baxter State Park and Mt. Katahdin;
    Sunkhaze National Wildlife Refuge
Fragmentation: 5 major dams span main stem; 111 additional licensed dams (many hydroelectric) on tributaries.
Water quality: 4 pulp and paper mills, 9 major municipal discharges. Major pollutants are dioxin, PCBs, mercury, organic
    waste. Total nitrogen = 0.45 mg N/L, total dissolved phosphorous = 0.03 mg P/L, total organic carbon = 9.5 mg/L,
    hardness = 12.0 mg/L as $CaCO_3$, pH = 6.6, specific conductance = 61.3 µS/cm
Land use: 95% forest and wetland, 5% agriculture and urban
Population density: 8 people/km²
Major information sources: Barrows and Babb 1912, Cutting 1963, Rabeni and Gibbs 1980, Baum 1983, Davies et al. 1999,
    PNWRP 2001

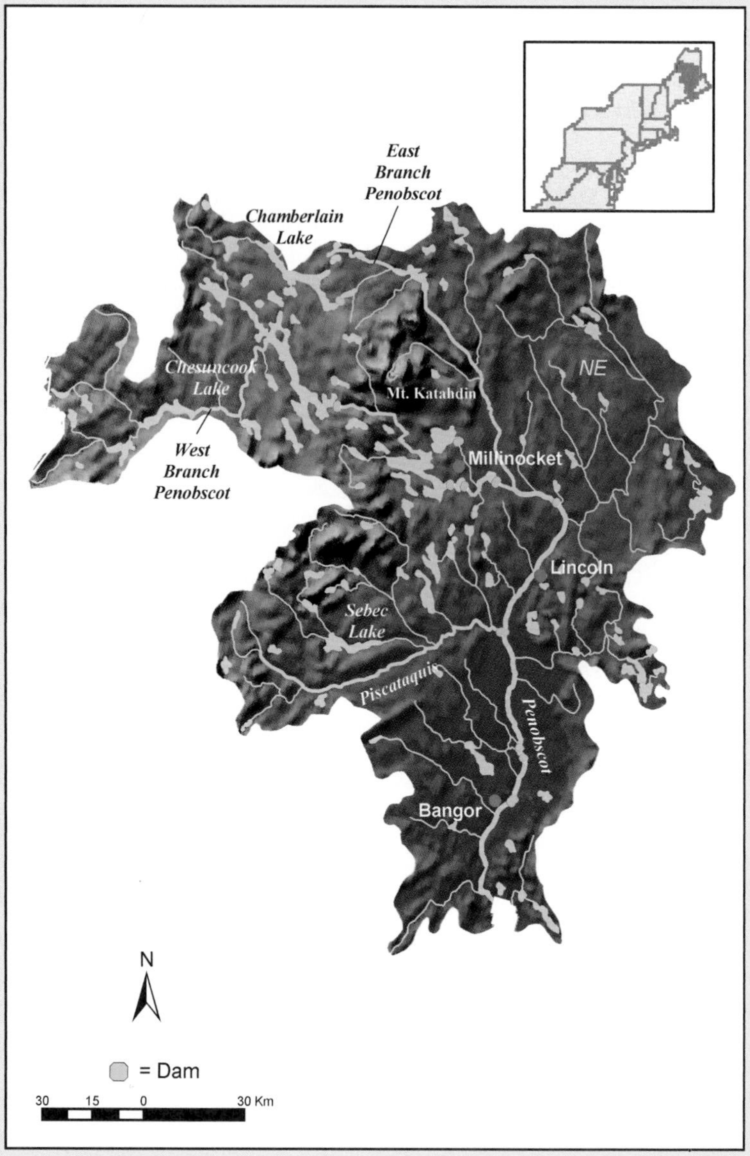

FIGURE 2.20 Map of the Penobscot River basin.

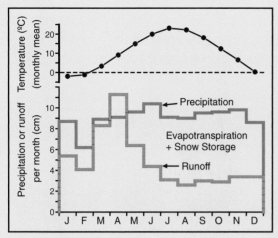

FIGURE 2.23 Mean monthly air temperature, precipitation, and runoff for the Connecticut River basin.

CONNECTICUT RIVER

Relief: 1917 m
Basin area: 29,160 km^2
Mean discharge: 445 m^3/s
River order: 7 (approximated)
Mean annual precipitation: 109 cm
Mean air temperature: 10.5°C
Mean water temperature: NA
Physiographic province: New England (NE)
Biome: Temperate Deciduous Forest
Freshwater ecoregion: North Atlantic
Terrestrial ecoregions: New England/Acadian Forests, Northeastern Coastal Forests
Number of fish species: 64 freshwater, 44 estuarine
Number of endangered species: 1 fish, 1 mussel
Major fishes: anadromous: American shad, blueback herring, sea lamprey, striped bass; freshwater: brook trout, American eel, white sucker, yellow perch, fallfish, common shiner, golden shiner, spottail shiner, banded killifish, redbreast sunfish, pumpkinseed, brown bullhead
Other aquatic vertebrates: beaver, river otter, northern water snake, snapping turtle, bald eagle, bank swallow, common loon, common merganser, belted kingfisher, great blue heron
Major benthic invertebrates: stoneflies (*Acroneuria*), mayflies (*Eurylophella, Serratella, Stenonema*), caddisflies (*Brachycentrus, Chimarra, Hydropsyche, Neureclipsis, Oecetis*), beetles (*Stenelmis, Psephenus*), chironomid midges (*Chironomus, Polypedilum, Microtendipes, Glyptotendipes, Tanytarsus*), mollusks (Pisidiidae), worms (Tubificidae)
Nonnative species: 20 fish species (brown trout, smallmouth bass, largemouth bass, black crappie, white crappie, northern pike, common carp, rainbow trout, bowfin, bluegill), Asiatic clam
Major riparian plants: Estuary: cord grass. Freshwater: wild celery, broadleaf arrowhead, cattail, pickerelweed, purple loosestrife, common reed
Special features: Watershed, including tidal wetlands, has received national and international recognition of its ecological uniqueness and value.
Fragmentation: 16 large hydroelectric dams
Water quality: pH = 6.5 to 7.5, alkalinity = 20 to 47 mg/L as CaCO$_3$, specific conductance = 116 μS/cm, NO$_3$-N + NO$_2$-N = 0.19 to 0.61 mg/L, total phosphorus P = 0.04 to 0.16 mg/L
Land use: 80% forest, 11% agriculture, 9% other
Population density: 69 people/km^2
Major information sources: Merriman and Thorpe 1976, Patrick 1996, Garabedian et al. 1998

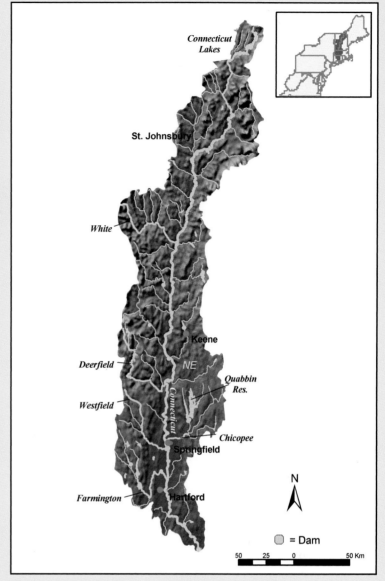

FIGURE 2.22 Map of the Connecticut River basin.

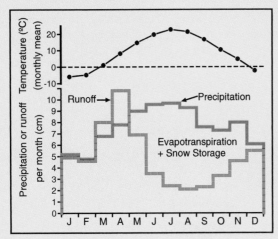

FIGURE 2.25 Mean monthly air temperature, precipitation, and runoff for the Hudson River basin.

HUDSON RIVER

Relief: 1629 m
Basin area: 34,615 km²
Mean discharge: 592 m³/s
River order: 7 (approximated)
Mean annual precipitation: 92 cm
Mean air temperature: 8.9°C
Mean water temperature: 12.4°C
Physiographic provinces: Adirondack Mountains (AD), Valley and Ridge (VR), Appalachian Plateau (AP), New England (NE), Piedmont Plateau (PP)
Biome: Temperate Deciduous Forest
Freshwater ecoregion: North Atlantic
Terrestrial ecoregions: Eastern Forest/Boreal Transition, Eastern Great Lakes Lowland Forests, Northeastern Coastal Forests
Number of fish species: >200 (70 native freshwater, 95 brackish)
Number of endangered species: 1 fish
Major fishes: American eel, American shad, blueback herring, alewife, white catfish, white sucker, common carp, cutlips minnow, blacknose dace, creek chub, fallfish, common shiner, spottail shiner, brook trout, Atlantic tomcod, banded killifish, striped bass, white perch, smallmouth bass, largemouth bass, tessellated darter
Major other aquatic vertebrates: snapping turtle, mute swan, Canada goose, beaver
Major benthic invertebrates: mollusks (zebra mussel, eastern elliptio, *Ferrissia*, *Amnicola*, *Gyraulus*), worms (*Limnodrilus hoffmeisteri*), crustaceans (*Gammarus tigrinus*, *Hyalella*, *Caecidotea*), mayflies (*Stenonema*, *Caenis*), caddisflies (*Neureclipsis*, *Chimarra*, *Hydropsyche*, *Cheumatopsyche*)
Nonnative species: curly-leaved pondweed, water-chestnut, Eurasian watermilfoil, purple loosestrife, mud bithynia, zebra mussel, dark falsemussel, Atlantic rangia, common carp, brown trout, northern pike, rock bass, smallmouth bass, largemouth bass, black crappie, mute swan
Major riparian plants (freshwater tidal Hudson): silver maple, red maple, cottonwood, sycamore, willows, common reed, narrowleaf cattail
Special features: Drains much of Adirondack Mountains; long (248 km) intertidal zone; important anadromous fishery
Fragmentation: 14 dams in middle section; no dams on lower 248 km or in upper river
Water quality: highly variable: soft (<25 mg/L as CaCO₃) and nutrient-poor (soluble reactive phosphorus <10 µg/L) in headwaters; moderately hard (~50 mg/L as CaCO₃) and nutrient-rich (SRP ~20 mg/L) in middle section; brackish (>2 psu) and nutrient-rich (SRP >25 µg/L) in lower section
Land use: 62% forest, 25% agriculture, 8% urban, 5% other
Population density: 135 people/km² (approximated); 0/km² in headwaters to >10,000/km² in Manhattan
Major information sources: Weinstein 1977, Boyle 1979, Limburg et al. 1986, Stanne et al. 1996, Levinton and Waldman 2005

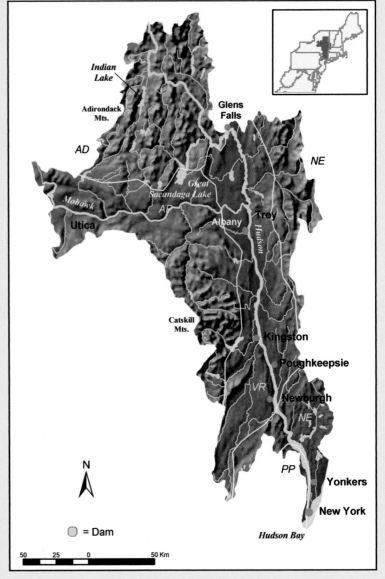

FIGURE 2.24 Map of the Hudson River basin. Physiographic provinces are separated by yellow lines.

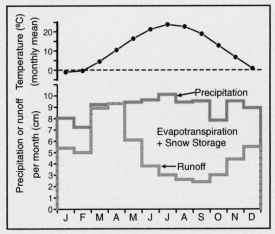

FIGURE 2.27 Mean monthly air temperature, precipitation, and runoff for the Delaware River basin.

DELAWARE RIVER

Relief: 698 m
Basin area: 33,041 km^2
Mean discharge: 422 m^3/s
River order: 7 (approximated)
Mean annual precipitation: 108 cm
Mean air temperature: 11.8°C
Mean water temperature: 13.8°C
Physiographic provinces: Appalachian Plateau (AP), Valley and Ridge (VR), New England (NE), Piedmont Plateau (PP), Coastal Plain (CP)
Biome: Temperate Deciduous Forest
Freshwater ecoregion: North Atlantic
Terrestrial ecoregions: Northeastern Coastal Forests, Allegheny Highlands Forests, Appalachian/Blue Ridge Forests
Number of fish species: 105
Major fishes: American eel, American shad, black crappie, bluegill, blue catfish, brown bullhead, brown trout, carp, channel catfish, fallfish, largemouth bass, river chub, river herring, rainbow trout, redbreast sunfish, rock bass, smallmouth bass, striped bass, walleye, white perch, yellow perch
Endangered species: 1 mussel, 1 fish, 1 turtle, 2 birds
Major other aquatic vertebrates: northern water snake, snapping turtle, eastern mud turtle, painted turtle, spotted turtle, red bellied turtle, beaver, river otter
Major benthic invertebrates: alderflies (*Sialis*), crustaceans (*Gammarus, Cyathura*), beetles (*Optioservus, Psephenus, Stenelmis*), caddisflies (*Brachycentrus, Cheumatopsyche, Chimarra, Hydropsyche, Lepidostoma, Rhyacophila*), mayflies (*Acentrella, Baetis, Epeorus, Eurylophella, Isonychia, Stenonema, Tricorythodes*), stoneflies (*Acroneuria*), mollusks (*Ferissia*)
Nonnative species: Brazilian waterweed, brittle naiad, Carolina fanwort, dotted duckweed, Eurasian watermilfoil, European water clover, hydrilla, parrot feather, pond water starwort, purple loosestrife, sacred lotus, watercress, water lettuce, Asiatic clam, bluegill, brown trout, carp, channel catfish, largemouth bass, smallmouth bass, rock bass, walleye
Major riparian tree species: American beech, American chestnut, American hornbeam, bitternut hickory, sweet birch, black cherry, black locust, black walnut, butternut, eastern hemlock, catalpa, chestnut oak, hackberry, pignut hickory, red elm, red maple, northern red oak, shagbark hickory, silver maple, sour gum, sugar maple, tulip poplar, white ash, white oak
Special features: Longest undammed main-stem river (530 km) in eastern United States; several km of main stem and tributaries designated National Wild and Scenic Rivers; Delaware Water Gap
Fragmentation: 16 major dams on tributaries
Water quality: Port Jervis, New York: pH = 7.40, alkalinity = 11.36 mg/L as CaCO$_3$, NO$_3$-N = 0.23 mg/L, PO$_4$-P = 0.012 mg/L; Trenton, New Jersey: pH = 7.96, alkalinity = 43.54 mg/L as CaCO$_3$, NO$_3$-N = 0.87 mg/L, PO$_4$-P = 0.047 mg/L
Land cover: 60% forest, 24% agriculture, 9% urban, and 7% surface water or other
Population density: 214 people/km^2
Major information sources: Schmidt 1986, Majumdar et al. 1988, Cuff et al. 1989, Patrick et al. 1992, Patrick 1994, DRBC 2002

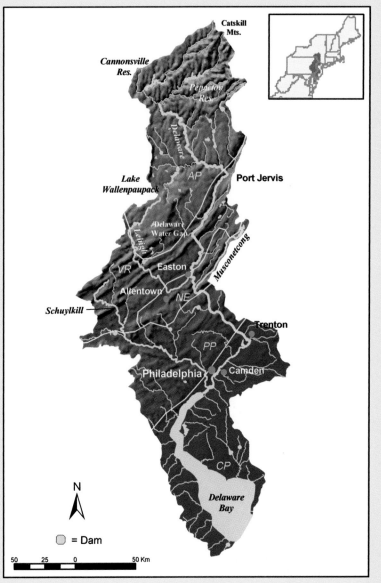

FIGURE 2.26 Map of the Delaware River basin. Physiographic provinces are separated by yellow lines.

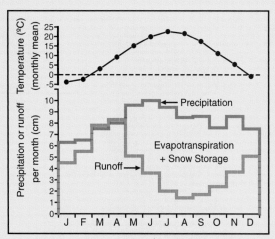

FIGURE 2.29 Mean monthly air temperature, precipitation, and runoff for the Susquehanna River basin.

SUSQUEHANNA RIVER

Relief: 959 m
Basin area: 71,432 km^2
Mean discharge: 1153 m^3/s
River order: 7 (approximated)
Mean annual precipitation: 98 cm
Mean air temperature: 9.7°C
Mean water temperature: 14°C
Physiographic provinces: Appalachian Plateau (AP), Valley and Ridge (VR), Piedmont Plateau (PP), Blue Ridge (BL)
Biome: Temperate Deciduous Forest
Freshwater ecoregion: Chesapeake Bay
Terrestrial ecoregions: Allegheny Highlands Forests, Appalachian/Blue Ridge Forests, Northeastern Coastal Forests, Appalachian Mixed Mesophytic Forests
Number of fish species: 103
Number of endangered species: 2 birds
Major fishes: American eel, American shad, blueback herring, alewife, rock bass, smallmouth bass, channel catfish, walleye, muskellunge, striped bass, spotfin shiner, banded darter, bluntnose minnow, margined madtom, bluegill

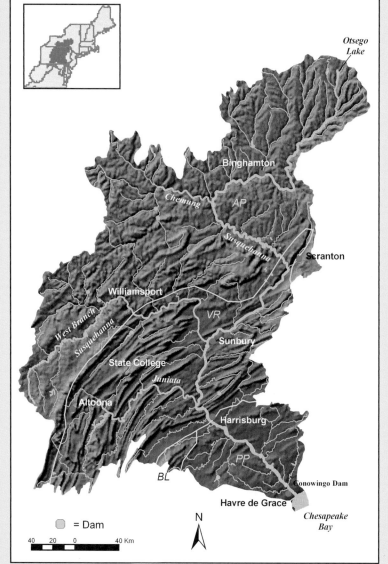

FIGURE 2.28 Map of the Susquehanna River basin. Physiographic provinces are separated by yellow lines.

Major other aquatic vertebrates: beaver, muskrat, common map turtle, eastern painted turtle, snapping turtle, wood turtle, northern water snake, great blue heron, bald eagle, peregrine falcon, mallard duck, wood duck, Canada goose
Major benthic invertebrates: mayflies (*Baetis, Pseudocloeon, Centroptilum, Stenonema, Leucrocuta, Isonychia, Serratella, Anthopotamus*), caddisflies (*Hydropsyche, Cheumatopsyche, Hydroptila*), crustaceans (spinycheek crayfish), mollusks (eastern elliptio, yellow lampmussel), stoneflies (*Agnetina*), beetles (*Stenelmis, Optioservus, Dubiraphia*), chironomid midges (*Rheotanytarsus, Polypedilum, Tvetenia, Chironomus, Dicrotendipes*)
Nonnative species: 27 fishes (smallmouth bass, channel catfish), Asiatic clam, zebra mussel, rainbow mussel, rusty crayfish, watercress, Eurasian watermilfoil, purple loosestrife, curly pondweed, Japanese knotweed, mile-a-minute weed
Major riparian plants: sycamore, tulip poplar, red maple, silver maple, river birch, black willow, black cherry, American beech, American elm, black locust
Special features: Broad river channel, numerous islands, largest commercially non-navigable river in U.S., short tidal/estuarine section for a coastal river.
Fragmentation: >100 dams, first major dam 16 km from mouth
Water quality: pH = 6.9 to 9.0, alkalinity = 40 to 110 mg/L as CaCO$_3$, NO$_3$-N = 0.29 to 1.49 mg/L, PO$_4$-P = <0.02 to 0.1 mg/L; pollution from acid mine drainage and agricultural runoff
Land use: 63% forest, 20% cropland, 9% urban, 7% pasture
Population density: 56 people/km^2
Major information sources: Hocutt et al. 1986, Stranahan 1993, Edwards 1996, Rowles and Sitlinger 2000, Susquehanna River Basin Commission

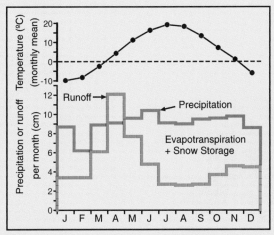

FIGURE 2.31 Mean monthly air temperature, precipitation, and runoff for the Kennebec River basin.

KENNEBEC RIVER

Relief: 1234 m
Basin area: 13,944 km²
Mean discharge: 257 m³/s
River order: 6 (approximated)
Mean annual precipitation: 108 cm
Mean air temperature: 11.9°C
Mean water temperature: 10.2°C
Physiographic province: New England (NE)
Biome: Temperate Deciduous Forest
Freshwater ecoregion: North Atlantic
Terrestrial ecoregions: New England/Acadian Forests, Northeastern Coastal Forests
Number of fish species: 48
Number of endangered species: 1 fish
Major fishes: alewife, blueback herring, Atlantic salmon (including migratory and landlocked populations), American shad, striped bass, rainbow smelt, smallmouth bass, brook trout, brown trout, rainbow trout, white perch, yellow perch, white sucker, fallfish
Major other aquatic vertebrates: eastern painted turtle, snapping turtle, beaver, muskrat, river otter
Major benthic invertebrates: mayflies (*Ameletus Caenis, Epeorus, Ephemerella, Eurylophella, Heptagenia, Stenacron, Stenonema*), caddisflies (*Cheumatopsyche, Chimarra, Macrostemum, Mystacides, Neureclipsis, Polycentropus*), hellgrammites (*Corydalus*), beetles (*Macronychus*), odonates (*Calopteryx*), crustaceans (*Asellus, Cambarus, Hyallela, Orconectes*), mollusks (*Alasmidonta, Amnicola, Anodonta, Campeloma, Elliptio, Lampsilis*)
Nonnative species: purple loosestrife, rusty crayfish, common carp, white catfish, gizzard shad, brown trout, rainbow trout, smallmouth bass, largemouth bass, black crappie, northern pike, European rudd
Major riparian plants: balsam fir, red maple, paper birch, eastern white pine, chokecherry, northern red oak, eastern hemlock
Special features: First major dam (Edwards Dam) on major river breached by U.S. government (1999); Moosehead Lake (311 km²); Merrymeeting Bay—large freshwater tidal estuary
Fragmentation: 8 hydroelectric dams
Land use: 82% forest, 6% agriculture, 2% urban, 10% surface water; upper basin largely forested; lower basin with significant agriculture
Water Quality: total nitrogen = 1.4 mg/L, total phosphorous = 0.03 mg/L, total organic carbon = 7.0 mg/L, hardness = 17.9 mg/L as CaCO₃, pH = 7.1, specific conductance = 66.9 µS/cm. Pollution source from two pulp and paper mills
Population density: 15 people/km²
Major information sources: Patrick 1996, Davies et al. 1999

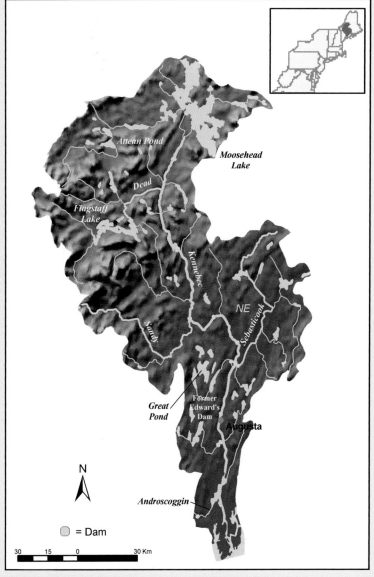

FIGURE 2.30 Map of the Kennebec River basin.

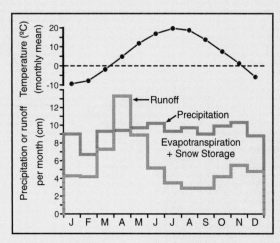

FIGURE 2.33 Mean monthly air temperature, precipitation, and runoff for the Androscoggin River basin.

ANDROSCOGGIN RIVER

Relief: 1234 m
Basin area: 8451 km²
Mean discharge: 175 m³/s
River order: 6 (approximated)
Mean annual precipitation: 111 cm
Mean air temperature: 10.9°C
Mean water temperature: 9.7°C
Physiographic province: New England (NE)
Biome: Temperate Deciduous Forest
Freshwater ecoregion: North Atlantic
Terrestrial ecoregions: New England/Acadian Forests, Northeastern Coastal Forests
Number of fish species: 33 (27 native), 7 estuarine
Number of endangered species: 1 fish
Major fishes: brown trout, rainbow trout, smallmouth bass, white perch, yellow perch, white sucker, fallfish
Major other aquatic vertebrates: beaver, muskrat, river otter

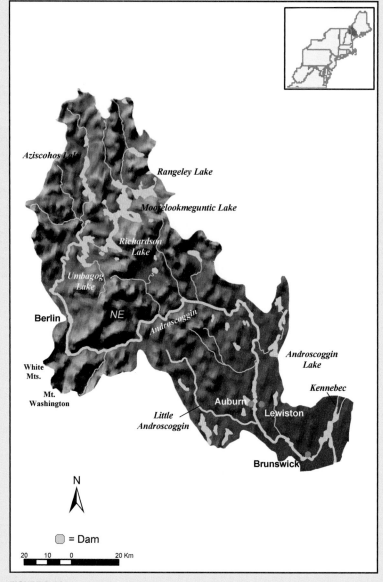

FIGURE 2.32 Map of the Androscoggin River basin.

Major benthic invertebrates: mayflies (*Ameletus, Caenis, Ephemerella, Eurylophella, Heptagenia, Stenacron, Stenonema*), caddisflies (*Cheumatopsyche, Chimarra, Macrostemum, Mystacides, Neureclipsis, Polycentropus*), hellgrammites (*Corydalus*), beetles (*Macronychus*), crustaceans (*Asellus, Hyallela, Orconectes*), mollusks (*Alasmidonta, Amnicola, Anodonta, Campeloma, Elliptio, Strophitus, Lampsilis, Margaritifera*)
Nonnative species: purple loosestrife, rusty crayfish, calico crayfish, brown trout, rainbow trout, smallmouth bass, largemouth bass, northern pike, spottail shiner, common carp, white catfish
Major riparian plants: balsam fir, red maple, paper birch, American witch hazel, eastern white pine, chokecherry, northern red oak, eastern hemlock
Special features: Lake Umbagog National Wildlife Refuge
Fragmentation: 14 hydroelectric dams from source at Umbagog Lake to tidewater
Land use: 86% forest, 5% agriculture, 2% urban, 7% surface water; upper basin largely forested; lower basin with significant agriculture
Water quality: Total nitrogen = 0.77 mg/L, total phosphorous = 0.06 mg/L, total organic carbon = 7.5 mg/L, hardness = 19.1 mg/L as CaCO₃, pH = 6.7, specific conductance = 101.8 µS/cm. Pollution source from three large pulp and paper mills
Population density: 25 people/km²
Major information sources: Davies et al. 1999

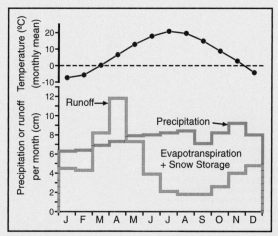

FIGURE 2.35 Mean monthly air temperature, precipitation, and runoff for the Merrimack River basin.

MERRIMACK RIVER

Relief: 1563 m
Basin area: 12,986 km^2
Mean discharge: 235 m^3/s
River order: 6 (approximated)
Mean annual precipitation: 92 cm
Mean air temperature: 7.3°C
Physiographic province: New England (NE)
Biome: Temperate Deciduous Forest
Freshwater ecoregion: North Atlantic
Terrestrial ecoregions: New England/Acadian Forests, Northeastern Coastal Forests
Number of fish species: 50
Number of endangered species: 1 fish
Major fishes: American eel, alewife, brook trout, brown trout, rainbow trout, chain pickerel, fallfish, common shiner, white sucker, white perch, smallmouth bass, largemouth bass, black crappie, bluegill
Major benthic invertebrates: caddisflies (Hydropsychidae, Philopotamidae, Leptoceridae), chironomid midges, crustaceans (Gammaridae), snails, worms
Nonnative species: common carp, brown trout, smallmouth bass, largemouth bass, black crappie
Major riparian plants: NA
Fragmentation: high (>500 dams in basin, several on main stem)
Water quality: formerly very poor, now improved: pH = 6 to 7; alkalinity = 8 to 10 mg/L as CaCO$_3$; NO$_3$-N = 0.1 to 0.4 mg/L; PO$_4$-P = <0.01 to 0.05 mg/L
Land use: 75% forest, 13% urban, 6% agriculture, 5% surface water, 1% other
Population density: 156 people/km^2 (approximated)
Major information sources: Hoover 1938, Scarola 1973, Flanagan et al. 1999

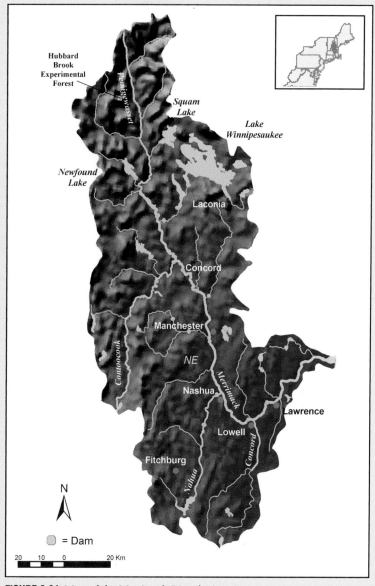

FIGURE 2.34 Map of the Merrimack River basin.

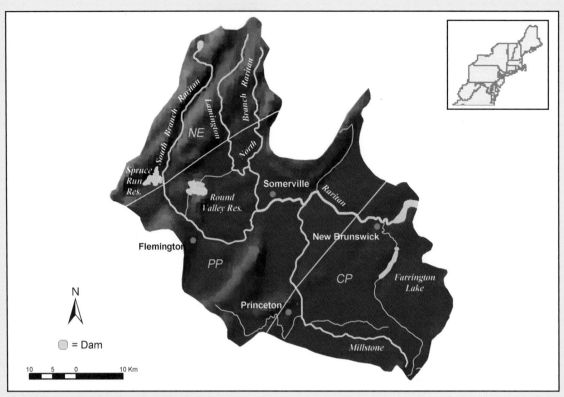

FIGURE 2.36 Map of the Raritan River basin. Physiographic provinces are separated by yellow lines.

RARITAN RIVER

Relief: 430 m
Basin area: 2862 km²
Mean discharge: 34 m³/s
River order: 5 (approximated)
Mean precipitation: 118 cm
Mean air temperature: 10.9°C
Mean water temperature: 13.6°C
Physiographic provinces: New England (NE), Piedmont Plateau (PP), Coastal Plain (CP)
Biome: Temperate Deciduous Forest
Freshwater ecoregion: North Atlantic
Terrestrial ecoregions: Northeastern Coastal Forests
Number of fish species: 88 (including 23 estuarine)
Number of endangered species: 1 mussel, 1 turtle
Major fishes: white sucker, blacknose dace, American eel, tessellated darter, longnose dace, redbreast sunfish, spottail shiner, common shiner, banded killfish, rock bass, bluegill, brown trout, brook trout, smallmouth bass, largemouth bass, American shad, alewife
Major other aquatic vertebrates: beaver, muskrat, river otter, eastern painted turtle, common snapping turtle, northern water snake, Canada goose, mallard duck, wood duck, great blue heron, osprey
Major benthic invertebrates: mayflies (*Anthopotamus, Baetis, Centroptilum, Leucrocuta Pseudocloeon, Stenonema*), caddisflies (*Hydropsyche, Cheumatopsyche*), chironomid midges (*Rheotanytarsus, Tanytarsus, Polypedilum*)
Nonnative species: 25 fishes (smallmouth bass, largemouth bass, channel catfish), Asiatic clam, Eurasian watermilfoil, purple loosestrife, phragmites
Special features: Longest river entirely in New Jersey; heavily urbanized/suburbanized
Fragmentation: High (>250 dams in basin; 57 for flood control; only 5 are >15 m high)
Water quality: pH = 7.3 to 7.9, alkalinity = 36 to 53 mg/L as $CaCO_3$, total N = 0.67 to 2.50 mg/L, PO_4-P = 0.08 to 0.25 mg/L, conductivity = 234 to 762 µS/cm
Land use: 27% forest, 36% developed, 19% agriculture, 15% wetland, 2% water, 1% barren
Population density: 419 people/km²
Major information sources: Kennen 1999; NJWSA 2000, 2002a, 2002b

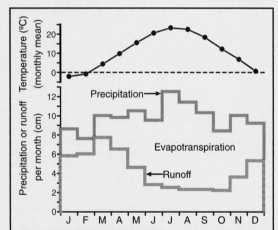

FIGURE 2.37 Mean monthly air temperature, precipitation, and runoff for the Raritan River basin.

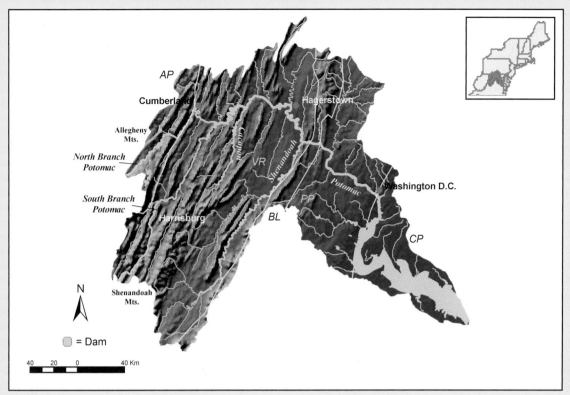

FIGURE 2.38 Map of the Potomac River basin. Physiographic provinces are separated by yellow lines.

POTOMAC RIVER

Relief: 1481 m
Basin area: 37,995 km²
Mean discharge: 320 m³/s
River order: 7 (approximated)
Mean annual precipitation: 99 cm
Mean air temperature: 12.3°C
Mean water temperature: 14.0°C
Physiographic provinces: Appalachian Plateau (AP), Valley and Ridge (VR), Piedmont Plateau (PP), Blue Ridge (BL), Coastal Plain (CP)
Biome: Temperate Deciduous Forest
Freshwater ecoregion: Chesapeake Bay
Terrestrial ecoregions: Appalachian/Blue Ridge Forests, Southeastern Mixed Forests, Appalachian Mixed Mesophytic Forests, Middle Atlantic Coastal Forests, Northeastern Coastal Forests
Number of fish species: 95 species (65 native)
Number of endangered species: 1 fish, 1 mussel

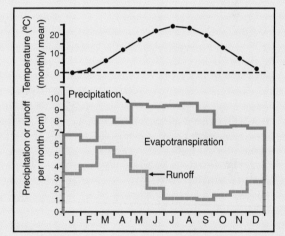

FIGURE 2.39 Mean monthly air temperature, precipitation, and runoff for the Potomac River basin.

Major fishes: smallmouth bass, channel catfish, spottail shiner, spotfin shiner, bluntnose minnow, redhorse sucker, redbreast sunfish, bluegill, tessellated darter
Major other aquatic vertebrates: beaver, muskrat, common snapping turtle, eastern painted turtle, great blue heron, bald eagle, osprey, Canada goose, mallard duck
Major benthic invertebrates: mayflies (*Anthopotamus, Caenis, Ephoron, Serratella, Stenonema, Tricorythodes*), caddisflies (*Hydropsyche, Cheumatopsyche, Hydroptila, Potamyia, Macrostemum*), chironomid midges (*Thienemannimyia, Tanytarsus, Cricotopus, Polypedilum, Cryptochironomus, Synorthocladius*), beetles (*Stenelmis*)
Nonnative species: smallmouth bass, channel catfish, common carp, several sunfishes, northern pike, muskellunge, Asiatic clam, hydrilla, Eurasian watermilfoil, common hornwort, purple loosestrife, phragmites
Special features: second-largest tributary to Chesapeake Bay, 174 km tidal/estuarine section, no native piscivores originally above Great Falls
Fragmentation: High, but only three impoundments >4 km²; >60 blockages in the Anacostia basin being removed or altered for fish passage.
Water quality: pH = 7.3 to 8.0, alkalinity = 36 to 86 mg/L as $CaCO_3$, NO_3-N = 0.58 to 1.48 mg/L, PO_4-P = <0.01 to 0.3 mg/L, conductivity = 194 to 491 µS/cm
Land use: 58% forest, 5% developed, 32% agriculture, 4% water, 1% wetland, 1% barren
Population density: 138 people/km²
Major information sources: Flynn and Mason 1978, Hocutt et al. 1986, Patrick 1996, Interstate Commission on the Potomac River Basin

3

ATLANTIC COAST RIVERS OF THE SOUTHEASTERN UNITED STATES

LEONARD A. SMOCK   ANNE B. WRIGHT   ARTHUR C. BENKE

INTRODUCTION

JAMES RIVER

CAPE FEAR RIVER

SAVANNAH RIVER

OGEECHEE RIVER

ST. JOHNS RIVER

ADDITIONAL RIVERS

ACKNOWLEDGMENTS

LITERATURE CITED

INTRODUCTION

The Atlantic slope region of the southeastern United States encompasses a broad geographic area from 38°N to 26°N latitude, ranging from central Virginia to eastern Florida (Fig. 3.2). The northern part of the region includes the Southern Appalachian Mountains along its western boundary to the flat coastal areas along the Atlantic Ocean. Below 31°N latitude in southern Georgia and Florida, however, there is only Coastal Plain. The region has abundant rainfall, moderate to warm temperatures, historically dense forests, and a landscape mosaic of forests, wetlands, agriculture, and urbanized areas. A series of rivers drain the region, most flowing in a primarily south-easterly direction to the Atlantic coast (Garman and Nielsen 1992, Smock and Gilinsky 1992, Patrick 1996). The diverse geology, physiography, and climate of the region have produced rivers that have variable geomorphic, hydrologic, and biological characteristics. Many of the larger rivers have their headwaters in the rugged Appalachian Mountains, whereas others originate among rolling Piedmont hills or on the flat Coastal Plain.

Flowing here are some of the most historic rivers on the continent. They had a profound influence on Native American civilizations that occupied these basins for over 11,000 years until well into the eighteenth century. The first Europeans to occupy this region were the Spanish missionaries who settled along the coast in the early 1500s and remained for more than a century. The establishment of Jamestown in 1607, along the James River in Virginia, marked the British arrival in North America that would eventually push Native Americans out of the southeastern Atlantic region. British colonists established towns at the mouths of major rivers that would eventually become major cities. They used the rivers as corridors for inland incursions that would help develop the agricultural and commercial strength of the United States.

Sixteen major rivers are found along the south-eastern Atlantic slope, from the York River on the

◄ FIGURE 3.1 Altamaha River, about 4 km upstream of I-95 (Photo by R. Overman).

73

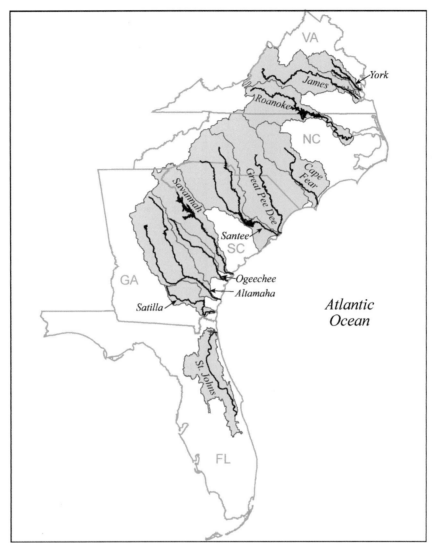

FIGURE 3.2 Atlantic coast rivers of the southeastern United States that are covered in this chapter.

northernmost border to rivers on the eastern coast of Florida. The mouths of major rivers occur about every 100 km along the coast (see Fig. 3.2). We will discuss five major rivers that together cover a considerable range in natural diversity and human impacts: James, Cape Fear, Savannah, Ogeechee, and St. Johns. Abbreviated descriptions of physical and biological information are provided for six additional rivers: York, Roanoke, Great Pee Dee, Santee, Altamaha, and Satilla.

Physiography and Climate

The southeastern Atlantic slope region is an area of varied physiography that includes four physiographic provinces. Along its western edge the region encom-

passes the unglaciated topography of the eastern slopes of the Southern Appalachian Mountains, comprised of the Valley and Ridge and Blue Ridge provinces. The Blue Ridge province tapers into the rolling hills of the Piedmont Plateau, and the Piedmont abruptly drops down to the broad, flat Coastal Plain. Geologically old and highly weathered rock and soil characterize the area.

The Valley and Ridge, Blue Ridge, and Piedmont are areas of heavily dissected and primarily highly metamorphosed rock of Paleozoic age, with occasional areas of igneous and sedimentary rock. The Valley and Ridge consists of parallel valleys and ridges that extend south through Georgia and into Alabama at an elevation typically 300 to 900 m asl. The underlying rock is varied, with bands of limestone, dolomite, shale, sandstone, cherts, and marble. Numerous springs and caves are scattered throughout the area. The Blue Ridge, terminating to the south in northern Georgia, is the easternmost ridge of the Appalachian Highlands, rising to peaks of more than 1700 m asl. It varies from narrow ridges to plateaus and broad mountains formed from primarily metamorphic rock. The Blue Ridge has some of the highest floral and faunal biodiversity in the eastern United States.

The Piedmont Plateau is a transition zone between the mountains to the west and the flat coastal areas to the east. Cutting through the central portions of Virginia, the Carolinas, and north-central Georgia, these rolling uplands of largely metamorphic rock with igneous intrusions generally range in altitude from 150 to 600 m asl, the higher altitudes occurring to the south. At its eastern boundary, the metamorphic rock of the Piedmont dips below the sedimentary deposits of the Coastal Plain. This boundary is termed the Fall Line, where a steep drop in elevation causes rivers to quickly descend through a series of rapids before leveling off in the Coastal Plain.

The Coastal Plain, a former sea bottom, is up to 200 km wide along the southeastern Atlantic and contains Cretaceous, Tertiary, and Quaternary sediments that drop gently from the Piedmont to the Atlantic Ocean. Its topography is that of a broad plain, with low relief and numerous wetlands and only a few inland ridges paralleling the coast. Elevation is usually <150 m asl, and much of it is <30 m asl.

Soils of the region are old and highly weathered, with ultisols predominating over much of the area. These are light-colored, acidic soils with a low cation-exchange capacity and a sandy or loamy surface horizon and a loamy to clay subsurface. Spodosols, with a high organic content and a high water-retention capacity, are common along the Florida coast.

The climate of the region ranges from humid continental in the north to humid subtropical in the south. Temperate summers and cool winters prevail in the north and at higher altitudes, grading to hot wet summers and mild winters in the subtropical southern region. Mean monthly air temperature across the region ranges from highs of 24°C to 28°C to lows of 2°C to 11°C. Precipitation is ample and fairly evenly distributed over the year, autumn being the driest period. Nearly all precipitation occurs as rainfall. The range in mean annual precipitation over the region is about 100 to 140 cm, being highest to the south and along the coast. Tropical storms and hurricanes cause short periods of high precipitation primarily during late summer and autumn.

Basin Landscape and Land Use

The southeastern Atlantic river basins all occur in the Temperate Deciduous Forest biome, which is further divided into four terrestrial ecoregions (Ricketts et al. 1999). These ecoregions generally lie along a northeast to southwest axis. The mountainous Valley and Ridge and Blue Ridge physiographic provinces roughly correspond to the Appalachian/Blue Ridge Forests ecoregion, which originally contained dense oak and American chestnut communities. The Piedmont roughly corresponds to the Southeastern Mixed Forests ecoregion, consisting of oak, hickory, and pine forests. The Coastal Plain is divided into the Middle Atlantic Coastal Forests (Virginia and the Carolinas) and the Southeastern Conifer Forests (Georgia and Florida) ecoregions. These Coastal Plain ecoregions supported extensive pine forests where fire suppressed the growth of oak, gum, beech, southern magnolia, and other deciduous trees. Winding through the coastal uplands are broad floodplain forests, intimately associated with both major and minor river systems. They range up to several kilometers in width along the major rivers and support highly productive bottomland hardwood forests, typically dominated by bald cypress, swamp black gum, and water tupelo, which exist as a broad ecotone between upland areas and river channels.

Although the region originally was covered by deciduous and pine forests, today the landscape is highly fragmented. The region was among the first in North America to be colonized and impacted by European settlers. Widespread forest clearing began in earnest during the eighteenth century and peaked during the nineteenth and early twentieth centuries, resulting in heavy soil erosion and input of sediment to streams and rivers. Today, the landscape includes a mosaic of second-growth forests, farmland, old fields, and timbered and urbanized areas. The Valley and Ridge, Blue Ridge, and Piedmont were highly cultivated, although large portions of those regions have reverted to pine and hardwood forests. Extensive areas of the Coastal Plain presently support regrowth forests, although much of that land consists of managed monoculture pine forests. The southeastern pine forests are one of the two principal lumber-producing regions of the United States. Cropland covers from 15% to nearly 50% of the Coastal Plain sections of the river basins, with higher crop cover occurring to the north, where soils are better drained, than to the south. The major cities in the region are often the oldest cities, which either developed along the coast (Norfolk, Wilmington, Charleston, Savannah, Jacksonville) or near the Fall Line (Richmond, Raleigh, Columbia, Augusta), where rivers could be most easily crossed without bridges. These growing urban areas have put continuous pressures on their rivers over the past two to three centuries.

The recent revegetation of much land originally under cultivation has resulted in a decrease in soil erosion, although all rivers of the region continue to receive significant inputs of soils and nutrients as nonpoint-source pollutants. The ongoing inputs of soil, along with the existing bed load in tributaries accumulated from decades to centuries of soil erosion, continue to provide a high input of sediment to most rivers of the region. As such, suspended and bed load solids are the most widespread and prevalent water pollution problem throughout the region. Most of the rivers also show local effects from municipal and industrial point-source discharges, and most are significantly impacted by impoundments. All of the rivers support heavy recreational use by boaters and fisherman.

The Rivers

There are three freshwater ecoregions within the southeastern Atlantic slope region, divided primarily as a function of latitude and drainage basin, according to the classification of Abell et al. (2000). These include the southern portion of the Chesapeake Bay ecoregion, the South Atlantic ecoregion (southern Virginia to eastern Georgia), and the Florida ecoregion (southern Georgia and Florida).

The rivers covered in this chapter are quite varied in their geomorphology, hydrology, chemistry, and biology. Some of these differences, such as in mean annual temperature, are largely due to the region extending over 10 degrees of latitude. Other differences are caused by the variation in geology, dominant vegetation, and land use within the river basins. Although there are many differences among the rivers, Garman and Nielsen (1992) provide a useful scheme that classifies the rivers into four groups based on similarities in their physicochemical and ecological characteristics. The groups cluster roughly according to latitude and mostly overlap the freshwater ecoregion classification scheme of Abell et al. (2000).

The first group includes the York, James, and Roanoke rivers, the northernmost rivers with the coolest water and widest fluctuations in water temperature (see Fig. 3.2). They originate in the Virginia mountains and Piedmont, have a higher gradient than the other rivers, and have a generally rocky substratum of gravel, cobble, and bedrock until reaching their coastal areas. Their water is relatively clear and slightly alkaline. The York and James are included in the Chesapeake Bay freshwater ecoregion and the Roanoke is in the South Atlantic freshwater ecoregion. The Roanoke, however, has many characteristics similar to the more northern basins, and other characteristics, such as its fish assemblage, that are transitional to the rivers to the south.

The largest group of rivers occurs immediately south of the Roanoke and encompasses most of the major rivers of the Carolinas and parts of Georgia, corresponding to the South Atlantic freshwater ecoregion (see Fig. 3.2). Included in this group are the Tar, Neuse, and Cape Fear of North Carolina, the Great Pee Dee and Santee-Cooper systems flowing across the Carolinas, the Savannah River along the South Carolina–Georgia border, and the Ogeechee and Altamaha rivers of Georgia. These rivers have their headwaters primarily in the mountains or Piedmont, but are of lower gradient than rivers of the first group. Their sediment is mainly sand and silt, with some rocky areas primarily upriver. The rivers are high in suspended solids, especially in their Piedmont section, and have neutral to slightly acidic waters that are stained in their Coastal Plain sections because of high dissolved organic carbon (DOC) concentrations.

The watersheds of the third group of rivers lie entirely within the Coastal Plain and are found within both the South Atlantic and Florida freshwater ecoregions. They include the Edisto of South Carolina, the Satilla of Georgia, and the St. Marys along the Georgia–Florida border. These are low-gradient rivers with sediment of fine sand and silt and no rocky areas. They are acidic, low alkalinity, blackwater systems, and their DOC concentrations often range up to 50 mg/L.

The St. Johns River typifies a fourth type of river that occurs exclusively in the Florida freshwater ecoregion and has characteristics very different from the other rivers covered in this chapter. They are wide and shallow systems with very low gradients. Although all of the rivers covered in this chapter are tidal, with estuaries at their mouths, the Florida rivers have extensive salt and fresh tidal areas extending far upriver. Their highly alkaline water chemistry, fringing salt marshes, and mix of freshwater and estuarine biota reflect their close connection with the ocean. An exception among Florida rivers is the extensively channelized Kissimmee River, which flows in a southerly direction before emptying into Lake Okeechobee.

The large rivers originating in the mountains offer the greatest change in aquatic biota because they each begin as high-gradient systems and end as low-gradient systems, during which time they traverse substantially different geological formations and substrata. At higher elevations their flow is typically turbulent, but once they enter the Coastal Plain their flow becomes laminar as they meander through broad floodplain forests that have their own unique plant communities. In those rivers found entirely within the Coastal Plain, there is only laminar flow within the wide floodplains.

Although precipitation is relatively uniform throughout the region, natural discharge regimes are strongly seasonal. Highest mean discharge occurs during the winter and spring, driven largely by seasonal changes in evapotranspiration. Under the natural regime of low evapotranspiration and high discharge in winter and spring, floodplain forest swamps are inundated for weeks to months, an important factor in the ecology of both the floodplains and main river channels (Benke et al. 2000). Hurricanes also are an integral aspect of the hydrology and ecology of rivers in the southeast, occasionally causing extensive flooding. Though these rivers can go decades without a significant impact from a hurricane, they can just as quickly be struck by a series of hurricanes that can have devastating effects both on the river and the people in the basin.

For most of the southeastern Atlantic slope rivers, natural flooding has been substantially reduced or altered by dams in upstream reaches and by channelization for navigation in the lower reaches. Although a few rivers in the region have no major dams (e.g., the Ogeechee and Satilla), others such as the Roanoke, Santee, and Savannah are highly fragmented by impoundments. The impoundments were constructed for a variety of intended uses, in particular hydroelectric power generation and flood control, but also drinking and industrial water supply, low-flow augmentation, recreation, and fish and wildlife. Besides altering hydrographs, with associated effects on river and floodplain ecology, the dams have affected water chemistry, the downstream transport of sediment, and the structure and functioning of the biological communities of the rivers. The decrease in migratory fish runs resulting from damming the

rivers, for example, has had important effects on the nutrient dynamics and food webs of many rivers of the region (Garman and Macko 1998).

An important natural feature of most Coastal Plain rivers is the occurrence of wood (snags) that is deposited when trees are undercut along the banks. Snags provide an important habitat, particularly for invertebrates, fishes, and reptiles, and thus likely enhance overall biodiversity and faunal productivity (Benke, van Arsdall et al. 1984, Benke, Henry et al. 1985). Unfortunately, wood has been removed from many of these rivers by snagboats for over a century because of the hazard it creates for recreational and commercial navigation.

Rivers of the southeastern Atlantic slope have a high degree of biological distinctiveness. The South Atlantic and Florida ecoregions are considered to be globally outstanding, and the Chesapeake Bay ecoregion is considered continentally outstanding (Abell et al. 2000). The South Atlantic ecoregion is the richest, supporting many endemics among at least 177 species of fishes, 39 species of unionid mussels, and 56 species of crayfishes, although not all of these are riverine species. Unfortunately, the conservation status of both the South Atlantic and Florida ecoregions is currently endangered and anticipated to become critical within the next 20 years (Benz and Collins 1997, Abell et al. 2000). At least 47 species of fishes and mussels are at risk of extinction in the South Atlantic (Master et al. 1998).

The primary perturbations to the rivers of this region include dams, nonpoint-source runoff, point-source discharges, channelization, and snag removal. Nearly all of the rivers have been impounded at or above the Fall Line, altering their geomorphology, hydrology, and ecology. The proliferation of non-native species also is affecting the ecology of many of the rivers. The Asiatic clam is widespread and common throughout the region, although limited by low Ca concentrations in some rivers, such as the Satilla. Many species of nonnative fishes now are widespread and abundant, have substantially altered riverine food webs, and are having considerable negative impact on other species of fishes.

JAMES RIVER

Arising in the western mountains of Virginia at the confluence of the Jackson and Cowpasture Rivers, the James River flows eastward for about 540 km to the southern portion of the Chesapeake Bay (Fig. 3.13). It drains a large portion of central Virginia

FIGURE 3.3   James River at Blue Ridge Parkway, northwest of Lynchburg, Virginia (PHOTO BY TIM PALMER).

(basin area = 26,164 km²), including extensive areas of mountains, rolling Piedmont hills, and flat Coastal Plain. Along its path to the bay the main-stem James changes from a rushing, cool 5th order river to a broad, meandering tidal 7th order river. Much of the river is very scenic, and water quality is generally quite high (Fig. 3.3). Although a number of low-head dams occur along the river, the James has been spared the large impoundments that occur on many of the other major rivers of the southeastern Atlantic slope. Fishing is excellent and includes one of the premier smallmouth bass recreational fisheries on the continent.

The James is one of the most historic rivers in North America. Paleo-Indians using Clovis points were in the basin about 12,000 years ago, and bands using pre-Clovis points may have been present 6000 years earlier. A number of Native American tribes developed throughout the basin, including the Powhatans in coastal areas and the Monacans in the Piedmont. The river was the focal point for many of their activities as well as those of the earliest English colonial settlements, beginning with Jamestown and

Captain John Smith's explorations of the lower river in 1607. The shores of the lower river became lined with historic tobacco plantations, producing the crop that fueled the early economy of the area. The river served both the North and South throughout much of the Civil War, being a key pathway for invasion into the Confederacy during several Northern campaigns. Some of the nation's first ironworks and flour and paper mills were built along the river, helping to make the river a major artery for commerce then as it is now.

Physiography, Climate, and Land Use

The James River basin flows primarily along an east–west axis at 37°N latitude and crosses the Valley and Ridge (VR), Blue Ridge (BR), Piedmont (PP), and Coastal Plain (CP) physiographic provinces (see Fig. 3.13). This area corresponds to the Appalachian/Blue Ridge Forests, Southeastern Mixed Forests, and Middle Atlantic Coastal Forests terrestrial ecoregions (Ricketts et al. 1999). The predominant trees throughout the basin include a variety of oaks, hickories, sweetgum, tuliptree, and loblolly pine. Soils are generally highly weathered ultisols, their characteristics varying across the physiographic provinces.

Annual mean air temperature is about 14°C, with monthly mean temperatures typically ranging from 2°C in January to 26°C in July (see Fig. 3.14). Annual precipitation averages 108 cm, falling primarily as rainfall and being relatively evenly distributed over the year.

The landscape, once dominated by the extensive Temperate Deciduous Forest biome, is still largely rural but is greatly influenced by agriculture and silviculture. About 71% of the basin is forested, 23% is in agriculture, and 6% is urbanized (http://www.dcr.state.va.us/waterways/the_problem/ watersheds_and_you/p_james_river_watershed.htm). Three major urban areas occur along the river: Lynchburg in the upper Piedmont, Richmond at the Fall Line, and Norfolk and other cities at the river's mouth. Population density, at a mean of 96 people/ km^2, varies considerably across the basin, from the highly urbanized areas to large rural expanses.

River Geomorphology, Hydrology, and Chemistry

The river can be separated into five sections, each defined in terms of its physical, hydrological, and chemical characteristics and constituting an area of the river of relatively similar habitat and biotic characteristics (adapted from Garman and Smock 1999). Though the headwaters arise high in the mountains, the main-stem James River arises at an elevation of 328 m asl in its Valley and Ridge section. This part of the river, with a channel gradient of 90 cm/km and a mean width of 75 m, is distinguished by well-developed riffles, runs, and pools. A diversity of depth and velocity regimes exist, providing a wide variety of habitats for riverine biota. The river's sediment here consists primarily of boulders, cobble, and gravel that is well sorted with little embedding. The highest concentrations of benthic organic matter in the river occur here, both as large woody debris and finer particles. The water chemistry of this section also is distinct from downstream areas, with the mean conductivity (271 µS/cm), pH (8.3), alkalinity (81 mg/L as CaCO$_3$), and hardness (111 mg/L as CaCO$_3$) of the water all higher than downstream, reflecting inputs from tributaries that flow through areas of limestone.

The river's physical and biological characteristics change substantially as it flows into an impounded section above the city of Lynchburg, where a series of three dams regulate flow through shallow impoundments. The habitat produced by these dams is intermediate between a free-flowing river and that of the more typical deep and broad reservoirs of the southeast. Although the habitat here is less diverse and conducive to riverine biota than elsewhere along the river, the impoundments do increase overall habitat and biotic diversity when viewed on the scale of the entire river.

The James then winds its way eastward for 220 km across the Piedmont. This is a long stretch where the river widens to a mean of 170 m and the channel gradient, at 45 cm/km, is half that in the Valley and Ridge section. A riffle-run-pool geomorphology is evident, but the riffles are not nearly as well developed or extensive as upriver. A variety of sediment particle sizes occurs, including some extensive areas of bedrock, and many of the riffles and especially the runs are highly embedded with gravel and sand. The water chemistry of this section of the river shows the influence of inputs from tributaries that drain areas with predominately crystalline rock, resulting in lower conductivity (183 µS/cm), pH (7.7), alkalinity (61 mg/L as CaCO$_3$), and hardness (71 mg/L as CaCO$_3$) than upriver.

At Richmond the river begins its descent through the Fall Line to the Coastal Plain. The Fall Line along the James is about 15 km in length, dropping about 2 m/km. Fast-flowing water, extensive outcroppings of bedrock, and riffles of well-sorted boulders and cobble characterize this section of the river, famous for its Class IV white-water rapids. Intermixed among these riffles are extensive depositional areas of shifting sand that embed boulders and overlay bedrock. A number of old low-head dams span the channel, all recently having been breached or having fish passages installed to provide for the upstream migration of anadromous fishes.

The character of the river changes dramatically below the Fall Line. The channel quickly widens as it meanders for about 170 km through the Coastal Plain with a gradient of about 20 cm/km. The river here is tidal, with freshwater tides predominating for about 70 km down to the city of Hopewell. Although numerous oxbows and islands occur in the lower James, the river's channel, with sediment of sand and silt, has been substantially straightened and deepened by dredging to accommodate deepwater shipping up to Richmond's port facilities.

Annual mean discharge at Richmond, just above where the river becomes tidal, is 213 m^3/s, or about four times greater than at the river's origin. The

James River has highest discharge during late winter and spring and lowest flows in late summer and autumn (see Fig. 3.14). With highest precipitation in July and August, it appears that the river's hydrology is primarily influenced by seasonal changes in evapotranspiration. Tropical storms from midsummer through autumn occasionally result in much flooding along the river's length. Flow in the river is stabilized only slightly by the series of low dams at and above Lynchburg. The daily tidal amplitude just below the Fall Line is about 1.1 m.

River Biodiversity and Ecology

Given the river's ecological and economic importance to the region, there have been surprisingly few comprehensive studies of the ecology of the James River, which lies within the Chesapeake Bay freshwater ecoregion (Abell et al. 2000). The studies that have been conducted have focused on documenting the biodiversity of the flora and fauna of the river. Almost no attention has been given to examining the structure and function of the aquatic communities, their ecological interactions, or aspects of ecosystem dynamics.

Algae, Cyanobacteria, and Protists

Patrick (1996) summarized the findings of previous studies on the algae and cyanobacteria of the river, noting that diatoms such as *Navicula, Cymbella, Fragilaria, Nitschia,* and *Melosira* were consistently among the dominant periphyton in the river. Higher richness of algal taxa was found along the channel margins associated with silt and detritus than on rocks in midchannel, but highest densities of periphytic algae occurred on midchannel rocks.

Plants

Submerged macrophytes are uncommon in the swift-flowing upper section of the James River but become common to abundant during the summer and autumn throughout the Piedmont. About one-third of the river bottom in the Piedmont section was covered with submerged macrophytes during midsummer (Sprenkle et al. 2004). By far the most abundant species is grassleaf mudplaintain; other common species are sago pondweed, Illinois pondweed, American eelgrass, southern waternymph, and waterweed. American water willow occurs as an emergent species along the banks of much of the river. Macrophytes also occur in the tidal James River, but submerged species there probably have been reduced by low light penetration because of turbid waters.

A narrow corridor of dense riparian vegetation, primarily of sycamore, swamp black gum, river birch, American elm, red maple, and ash-leaf maple, shades channel margins along much of the nontidal river, providing a buffer between the river and the surrounding agricultural landscape. The riparian areas and floodplain forests along the lower James are composed primarily of sycamore, swamp black gum, red maple, river birch, and stands of bald cypress.

Invertebrates

A mosaic of habitats are available to invertebrates along the length of the James River. On a river-basin scale, these habitats range from the fast-flowing, erosional riffles in the upper reaches of the river to slow-flowing depositional areas in the lower section and impounded areas. Large woody debris and extensive macrophyte beds provide other important habitats, and the tidal fresh- and saltwater sections of the river add an additional environment for invertebrates. These and other aspects of the river's physical and chemical features result in a heterogeneous riverine environment that supports a high biodiversity of invertebrates (Smock and Mitchell 1991, Patrick 1996).

Significant differences in invertebrate species composition and abundance occur among the primary habitats in the river. Densities are highest in riffles, whereas high numbers of large species, including clams, mussels, snails, and crayfish, cause standing crop biomass to be highest in the slower flowing reaches and among macrophytes (Smock and Mitchell 1991). The Valley and Ridge section supports the highest species richness and most distinctive invertebrate community in the river. Stoneflies, baetid and heptageniid mayflies, and riffle beetles typify the fauna of this section, with many species occurring only there or in far lower numbers downriver. The invertebrate fauna of the Piedmont and Fall Line sections also is varied, resulting from the mix of erosional and depositional areas and the abundant macrophyte beds. Overall abundance of invertebrates, however, is higher in the Valley and Ridge section. The predominant species in the Piedmont and Fall Line sections include the nonnative Asiatic clam *Corbicula,* the pea clam *Pisidium* spp., the snails *Elimia* and *Somatogyrus,* the mayflies *Tricorythodes* and *Stenonema,* and the caddisflies *Hydropsyche, Cheumatopsyche,* and *Polycentropus.*

The invertebrate community changes considerably as the river drops into the Coastal Plain, with species richness there about half that of the Piedmont section (Garman and Nielsen 1992). Species of oligochaete worms, chironomid midges, and clams predominate in the sediments of the main channel; the mayfly *Caenis* occurs frequently in backwater areas. Many of these same taxa also are typical of the upstream impounded section of river. In addition, a variety of estuarine species appear in the tidal area.

Filter feeders and scrapers dominate the trophic structure of the river's invertebrate community (Garman and Smock 1999). Filter feeders are abundant throughout the river, suggesting high concentrations of seston in the water column. Among the dominant filter feeders are *Corbicula* and *Pisidium*, occurring primarily in fine sediments. The net-spinning caddisflies *Hydropsyche* and *Cheumatopsyche* are common in riffles, whereas the long trumpet nets of *Neureclipsis* caddisflies are frequently encountered on large rocks in the reaches. The filter-feeding chironomid *Rheotanytarsus* and species of black flies are common throughout the river, especially on macrophytes in fast-flowing areas.

The abundance of periphyton covering the sediments and other substrata along much of the river through the Fall Line results in scrapers also being an important component of the invertebrate community. The snails *Elimia* and *Somatogyrus* are common throughout the river. Several species of mayflies, such as *Stenonema*, *Baetis*, and *Heptagenia*, and elmid beetles also are among the more common scrapers.

A wide variety of predaceous invertebrates occur in the river, the most conspicuous being the hellgrammite *Corydalus cornutus*. Common in riffles with well-sorted cobble sediment, this species in turn is important prey for several species of fishes. Crayfish feed heavily on the abundant snails and macrophytes in the river, though their numbers are limited by the availability of nonembedded cobble substrata that provide a refuge from heavy predation pressure from smallmouth bass (Mitchell and Smock 1991).

Although the James River has not garnered the attention that some other Virginia rivers have in terms of its freshwater mussel populations, numerous species do occur in the river and several are included on lists of protected species. Johnson (1970) reported that 12 species of Unionidae mussels occur in the nontidal portion of the drainage basin. By far the most common mussel in the river is the eastern elliptio, found throughout the drainage basin. The yellow lance, listed as "rare" to "very rare" both in the state and globally, is rare but of regular occurrence in the river. Other mussels, including the brook floater, Atlantic pigtoe, green floater, eastern pondmussel, and James spinymussel, all appearing on federal or state lists of threatened and endangered species, have been reported in the James River (e.g., Johnson 1970, Garman and Smock 1999). All of these species, however, were found in isolated locations at best, and their current distribution and abundance are unknown.

Vertebrates

The fish community of the river is quite diverse, with at least 13 families and 75 species occurring in the nontidal portion of the river (G. C. Garman, personal communication). These species represent three-fourths of all freshwater species occurring throughout the entire James River basin (Jenkins and Burkhead 1994). In addition, many estuarine species occur in the lower tidal river.

The families Cyprinidae, Centrarchidae, Catostomidae, and Percidae comprise over 70% of the species in the nontidal river (Garman and Smock 1999). The Cyprinidae dominate fish diversity. Among the most common cyprinids are the bull chub, satinfin shiner, rosefin shiner, and spottail shiner. The James River ichthyofauna, dominated by cyprinids, more closely resembles fish communities of Atlantic slope rivers to its north than to the south, where sunfishes and suckers typically are most common (Garman and Nielsen 1992).

The majority of fishes in the river are trophic generalists. The numerically dominant cyprinids and centrarchids feed heavily on immature mayflies, caddisflies, and terrestrial arthropods during the summer, the latter suggesting a strong trophic link between riparian areas and the river channel. An increasing prevalence of planktivores such as gizzard and threadfin shad and white mullet at and below the Fall Line suggest a decreasing reliance on benthic and terrestrial food sources as the river descends into the Coastal Plain. Various piscivores, in particular largemouth bass and the nonnative smallmouth bass and flathead catfish, are the primary apex predators in freshwater areas.

The most abundant species in the Valley and Ridge section include two cyprinids, the bull chub and the rosefin shiner, and several sunfishes. In the Piedmont, most of these species continue to be quite common, along with the satinfin shiner, the margined madtom, and the American eel. The common carp, catostomids, including the golden and shorthead redhorse, and several sunfishes are among the dominant species in the impounded areas of the river.

The fishes at the Fall Line and in the Coastal Plain sections are markedly different from those upriver. High numbers of gizzard shad occur along with various minnows, sunfishes, and bullheads. Substantial numbers of anadromous shads and herrings also are present seasonally. In addition, estuarine species such as Atlantic needlefish, hogchokers, summer flounder, and white mullet can be found as far upriver as the Fall Line during the late summer and autumn (Garman and Smock 1999).

Three species of fishes are endemic to the James River basin, but all occur in small tributaries and are not regular inhabitants of the main stem (Jenkins and Burkhead 1994). The James has few endemic species compared to more southern rivers, being more typical of mid-Atlantic rivers to the north.

The lower James provides critical habitat for one of the largest populations of bald eagles along the Atlantic coast. Many other species of birds, including a wide array of piscivorous species, occur along the river. The James and its floodplain also support a variety of amphibians and reptiles, including disjunct populations of cottonmouth at the northern edge of its range. The lower river and some of its larger tributaries have populations of the glossy crayfish snake and the greater siren, both included as very rare or extremely rare on Virginia's Rare Animal List.

Ecosystem Processes

The lack of information on ecosystem-level processes for a river of the size and ecological importance of the James is striking. No quantitative studies have been conducted to determine levels of primary or secondary production throughout the freshwater section of the river, and almost no data except for anecdotal information exist on food web structure and dynamics. Epilithic algae no doubt are an important part of primary production in the river above the Fall Line. The deeper, more turbid water in the Coastal Plain likely limits periphyton production, but phytoplankton production may be locally high in the slower flow and extensive backwaters of the lower river. The abundance of macrophytes in the middle section of the river suggests that they may be an important component of primary production there and that they provide much detritus during their senescence. Inputs of allochthonous detritus occur primarily from tributaries rather than from riparian areas. Many of the fringing wetlands that historically occurred along the lower river have disappeared, removing what may have been important inputs of dissolved and particulate organic matter directly to the river.

The river's food web above the Fall Line likely is based primarily on periphyton production and macrophyte-derived detritus, whereas inputs of allochthonous detritus and phytoplankton production probably are of secondary importance. Much of the material passes to filter feeders and scrapers, which compose the most abundant primary consumers in the river. Their production must be substantial in order to support the high numbers and biomass of both invertebrate and vertebrate secondary consumers throughout the river.

Human Impacts and Special Features

In spite of three centuries of human exploitation, the James River remains highly scenic and supports a diverse and productive biota. Alteration of the river's hydrology by dams also has been minimal, unlike many of the major rivers to the south that are highly fragmented by large impoundments. Although the river receives inputs of both point and nonpoint pollutants along its entire length, water quality in much of the river down to the Fall Line is quite good, as evidenced by both physicochemical and biological indicators.

Trends from water-quality monitoring programs show that water quality has generally increased over the past decade (e.g., Sprague et al. 2000). Several urbanized areas and industries in the upper river cause local water-quality problems, but the most significant problems occur in the lower James. Fecal coliform concentrations increase greatly at the Fall Line, primarily due to the antiquated combined sewer outfalls in Richmond that discharge directly to the river during storms. A high incidence of free-living pathogenic amoebae, including *Acanthamoeba* and *Naegleria*, have been found associated with lesions on a variety of fish species in the lower James River (Webb et al. 2002). Their distribution and abundance may be associated with degraded water quality.

Much of the emphasis on reducing inputs of nutrients and sediment to the river is driven by programs to improve water quality in the Chesapeake Bay. About 38% of the nitrogen and 28% of the phosphorus entering the river originate from point sources (Virginia Department of Conservation and Recreation Web site for the James River Watershed: http://www.dcr.state.va.us/waterways/the_problem/watersheds_and_you/p_james_river_watershed.htm), with the remainder coming from nonpoint sources. The input of sediment from land erosion, however, is the primary pollution problem for much of the upper river. The heavy load of sediment and resulting embedding of benthic habitats has caused substantial loss of critical habitat for periphyton, invertebrates, and fishes.

The lower James River was the site of one of the nation's most publicized incidents of river pollution. A manufacturing plant discharged the chlorinated insecticide Kepone from 1966 to 1975 to the lower river (Cutshall et al. 1981). The widespread contamination of sediments and biota and accumulation of Kepone up the food chain raised concerns of carcinogenic effects for humans through consumption of fishes and shellfish. A ban on commercial fishing in the lower James decimated a flourishing industry (Diaz 1989). Though the ban was lifted in 1989, much Kepone remains buried in the sediments and continues to be monitored today because of fears of its resuspension during storms or dredging operations.

Nonnative fishes have had a major impact on the ecology of the James River. Over one-third of the fish species in the nontidal James River have been introduced over the past 150 years (Garman and Smock 1999). Possibly the most successful is the smallmouth bass. Brought from the Ohio River drainage around 1870 (Snyder et al. 1996), its population has turned the James River into one of the premier smallmouth rivers in North America. Other introduced species, such as the channel and blue catfish, muskellunge, and several species of sunfishes, also have become local favorites of anglers. The historical fish community, however, was dominated numerically by small insectivorous minnows and sunfishes as well as anadromous species. The establishment of thriving populations of large nonnative piscivorous species likely has had a significant effect on the river's trophic structure, given that some of the species feed at least one trophic level higher than do the native piscivores (Garman and Smock 1999).

Another anthropogenic disturbance that has had a major impact on the ecology of the river was the construction of the low- and medium-head dams in the Richmond and Lynchburg areas in the 1800s. These dams became barriers to migratory fishes, halting previously large runs of clupeids, including American and hickory shad, blueback herring, and alewife. Along with water-quality and habitat degradation in the lower river and its tributaries, the dams caused these species to steadily decline to near complete collapse in the late 1970s. Recent stocking programs, breaching of dams, and construction of fish passages have focused on revitalizing stocks and restoring access to the river above the Fall Line. Efforts such as these and the generally successful programs focused on improving water quality are helping to restore migratory fish runs and other aspects of the biological integrity of this river.

CAPE FEAR RIVER

The Cape Fear River basin, located entirely within North Carolina, is the largest basin within the state (basin area = 24,150 km^2). The river's headwaters lie in the Piedmont hills of the north-central part of the state, whereas the main-stem Cape Fear River flows primarily through the Coastal Plain (Fig. 3.15). Two large blackwater tributaries, the Black (called "South River" upstream) and Northeast Cape Fear rivers, join the Cape Fear near its mouth at Wilmington. The basin is highly populated and developed but also supports many rare species and includes some of the ecologically most significant riverine and wetland habitats in the state. Among these are blackwater rivers, broad floodplain swamps, Carolina Bay lakes, and an extensive estuarine system.

The river basin was historically inhabited by various Native American tribes, primarily hunter-gatherers whose subsistence included the bountiful game and fishes of the Cape Fear River and its tributaries and floodplain. Giovanni da Verrazano, sailing under a French flag, was the first European explorer to the region, briefly visiting the mouth of the Cape Fear in 1524. Two years later, a Spanish colony was established, but disease and starvation soon led to failure of the colony. The first permanent English settlers arrived in the 1660s, their numbers steadily increasing though the late 1600s and early 1700s. Encroachment on Native American land led to a series of Indian wars that ended in the early 1700s, followed by rapid expansion by the settlers upriver.

The initial settlers, who first established rice plantations near the mouth of the river, undertook explorations upriver. Accounts from that time attest to the beauty and condition of the river and surrounding area. "The whole Country consists of stately Woods, Groves, Marshes and Meadows; it abounds with variety of as brave Okes as Eye can behold, great Bodies tall and streight from 60 to 80 foot, before there be any Boughs. . . . Here are as brave Rivers as any in the World." (*A Brief Description . . .* 1944). Although the river was deep, the ability to navigate it and its primary tributaries was greatly impeded by the abundance of snags in the channel: "Small Craft [could go far upriver] were it not for a multitude of Logs that have fallen into the Rivers, which are so heavy and solid that they lie at the bottom, and many of them show but little Appearance of Decay. . . . In some Places we saw whole Heaps jambed together, almost from Side to Side, and so firm that they are immovable, being sound, heavy, fast and deep in the Sand, other-wise this would be a fine river." (Meredith 1731).

Even with its abundance of snags, however, the Cape Fear was heavily used by settlers, employing boats made from cypress logs to transport tobacco, timber, produce, and furs. River "improvement projects," begun in 1815 to remove the snags, led to the river becoming easily navigable by larger ships far upriver, with the first steamboats plying its waters in 1818. With the opening of the river came a rapid expansion of settlements upriver; today the Cape Fear River basin is the most densely populated basin in North Carolina.

Physiography, Climate, and Land Use

The Cape Fear River flows in a southeasterly direction between 36°N to 34°N latitude. The basin lies within two physiographic provinces, about one-third of the basin being in the Piedmont (PP) and two-thirds in the Coastal Plain (CP) (see Fig. 3.15). The geology of the Piedmont, where the majority of the river's headwaters flow, is dominated by old crystalline rock with areas of sedimentary rock. The red and yellow soils of the Piedmont are predominately highly weathered ultisols with a sandy-clay or silty-clay texture. The Coastal Plain consists of flat terraces of sand, silt, clay, and limestone, with extensive areas of wetlands, including pocosins and riverine floodplains. The soils of this area are primarily sand and clay, with upland areas having well-drained soils, whereas those of the lowlands are poorly drained. The river basin lies within the Temperate Deciduous Forest biome, historically covered by dense hardwood forests of oak, hickory, and sweetgum, and pine forests of loblolly, slash, and shortleaf pine. The basin's Piedmont section roughly corresponds to the southeastern Mixed Forests ecoregion, whereas the Coastal Plain section generally corresponds to the Middle Atlantic Coastal Forests ecoregion (Ricketts et al. 1999).

The climate over the basin is one of warm, humid summers and mild winters. Annual mean air temperature over the basin is 16°C, with monthly mean temperatures typically ranging from 4°C in January to 26°C in July and August (Fig. 3.16). Annual precipitation averages 119 cm, nearly all of which is rainfall. Precipitation is distributed fairly evenly over the year, with the driest period from October through December and the wettest period during July and August. Precipitation during the summer and autumn is greatly influenced by tropical storms and hurricanes.

The Cape Fear River basin is the most industrialized and populated basin in the state, with significant changes in land cover from historical times. As of 1992, about 56% of the basin was forested, 24% in agriculture, 9% urbanized, and 11% in other land uses, including open water (NCDENR 2000). Urban land cover increased 43% from 1982 to 1992, whereas uncultivated cropland increased by 18% during that time. Cotton and tobacco are the primary crops, and extensive swine, poultry, and timber operations occur within the basin.

The basin supports 27% of the state's population at a density of 69 people/km^2, with the highest density occurring in the basin's Piedmont headwaters. The major urban and industrial areas are the Greensboro–Burlington–High Point area at the head of the basin, Durham–Chapel Hill in the eastern Piedmont, Fayetteville at midbasin, and Wilmington near the mouth of the river.

River Geomorphology, Hydrology, and Chemistry

The main stem of the Cape Fear River arises near the Fall Line at the confluence of the Haw and Deep rivers at an elevation of about 83 m asl, just to the southwest of Raleigh (see Fig. 3.15). Immediately above the confluence is the B. Everett Jordan Reservoir, created in 1981 on the Haw and New Hope rivers. At 5642 ha, it is the largest impoundment in the basin, operated by the U.S. Army Corps of Engineers for flood control, water supply, recreation, fish and wildlife, and augmentation of low flows.

The main-stem Cape Fear flows as a 6th order river for 518 km, first dropping down across the Fall Line at a rate of about 1 m/km and then flattening out as it winds across the Coastal Plain (Fig. 3.4). The river's gradient from its origin to its mouth is about 6 cm/km. Water depth in the channel typically is 1 to 2 m, flowing over a bed primarily of sand with patches of silt along the banks (Patrick 1996). The few riffles that occur are along the Fall Line. The channel often is incised into steep clay banks. Snags occur frequently throughout the river, primarily near the banks, although the number and sizes of snags must be nowhere near what they were historically.

The river's hydrology is affected to some degree by the many impoundments throughout the basin, including the Jordan Reservoir on the river's primary tributary. The Cape Fear main stem, however, is generally free flowing, having only three low-head locks and dams, built in the early 1900s, along its course.

FIGURE 3.4 Cape Fear River near Erwin, North Carolina (PHOTO BY TIM PALMER).

Annual mean discharge is about 217 m³/s. Flow is highest during winter and spring and lowest during autumn, driven by evapotranspiration but also occasionally greatly impacted by hurricanes (Fig. 3.16). The majority of flow along much of the river is from its Piedmont tributaries, as its two primary Coastal Plain tributaries, the Black (or South) River and the Northeast Cape Fear River, enter near the upper estuary. The final 55 km of river from above Wilmington to the ocean is a tidal, estuarine basin.

Because most of the flow in the Cape Fear originates from its Piedmont tributaries, the river's chemistry primarily reflects the characteristics of those waters and is far less influenced by the smaller inputs of Coastal Plain blackwater tributaries than are most other southeastern Atlantic slope rivers. The Cape Fear is a slightly acidic, soft-water river, with conductivity typically around 80 to 110 μS/cm. Water temperature averages 17°C, rarely falling below 5°C or going above 30°C. The river is highly turbid, even during low flow but especially during rain events, as a result of erosion from the large proportion of developed land in the Piedmont portion of the basin. The

river acquires more of the characteristics of a blackwater river system as it flows deeper into the Coastal Plain, receiving water with high color caused by dissolved organic compounds from tributaries and its extensive floodplain. Color in the river is highest in the spring and early summer when the channel is most closely connected with its floodplain.

River Biodiversity and Ecology

The Cape Fear, lying within the South Atlantic Freshwater ecoregion (Abell et al. 2000), has ecological characteristics similar to those of the Coastal Plain sections of other southeastern rivers. It is characterized by a meandering channel, slow current, and primarily sandy sediment. Snags are the only stable substratum in the river except for some rocky shoals along the Fall Line. The river has a close connection with its often broad, seasonally inundated floodplain. These characteristics establish the template that determines the composition and structure of the river's flora and fauna and the ecosystem dynamics of the river.

Even with its high population and extensive industrialization and agricultural activities, the Cape Fear basin supports several areas of outstanding ecological importance. The U.S. Fish and Wildlife Service has designated areas of the Deep, Rocky, and Haw rivers, all major Piedmont tributaries to the Cape Fear, as critical habitat for rare aquatic fishes, mussels, and insects. The South-Black river system is named an "Outstanding Resource Waters" and is one of the best examples of Coastal Plain blackwater rivers. The river supports many rare fishes and mussels and an ancient cypress-gum swamp with trees dating to over 1600 years ago. The Sandhills, occurring in an area with deep sandy soils, longleaf pine forests, streamhead pocosins, and mixed hardwood and Atlantic white cedar swamps, has streams with clean sand sediments and many rare species. The Cape Fear's extensive and highly productive estuary has four areas designated as "Outstanding Resource Waters."

Algae, Cyanobacteria, and Protists

Patrick (1996) summarizes studies by the Academy of Natural Sciences of Philadelphia on the river's algae and cyanobacteria, of which about 200 taxa have been found in the river. An additional 160 taxa of protists also have been identified from the river (Patrick 1996). The periphyton in the channel is dominated by cyanobacteria and diatoms, among the more common genera being *Oscillatoria*, *Melosira*, *Eunotia*, *Gomphonema*, *Navicula*, and *Cyclotella*. Green algae are most commonly encountered along the banks on silty sediment, debris, and snags in slow flowing areas. Phytoplankton occur in the upper river, primarily as a result of their being washed out of Jordan Reservoir, and in the lower river during periods of low flow.

Plants

Macrophytes are not common along much of the channel except in backwater areas. The river's riparian zone and floodplains are largely forested. Sycamore and ash are common along the riparian zone of the upper river and in elevated, drier areas downriver. The broad floodplains support bald cypress, water tupelo, swamp black gum, red maple, sweetgum, oaks, and other hardwoods. Far downriver the hardwood swamps give way to estuarine marshes where cordgrass and rush are the dominant species.

Invertebrates

The species composition and structure of the invertebrate community of the Cape Fear is similar to that of other large Coastal Plain rivers (Patrick 1996, Mallin et al. 2000). Patrick (1996) lists about 150 taxa of macroinvertebrates found in the upper river. Collector-gatherers are numerically dominant in the sediment and collector-filterers are dominant on snags. Oligochaete worms, chironomid midges, sphaeriid clams (*Sphaerium*, *Pisidium*), and the nonnative Asiatic clam are among the most abundant taxa in the sediment. These taxa and the mayfly *Tricorythodes*, freshwater grass shrimp (*Paleomonetes*), dragonflies (*Gomphus*, *Neurocordulia*), damselflies (*Argia*, *Enallagma*), chironomids, and crayfishes are common along the banks and among debris accumulations in the channel.

The most diverse assemblage of macroinvertebrates in the river occurs on snags, where the most frequently encountered taxa are mayflies (*Baetis*, *Stenonema*, *Tricorythodes*), elmid beetles (*Ancyronyx*, *Macronychus*), black flies (*Simulium*), chironomids (*Rheotanytarsus*), dragonflies (*Boyeria*), damselflies (*Calopteryx*), and hydropsychid caddisflies (*Hydropsyche*, *Cheumatopsyche*, *Chimarra*). The caddisflies seemingly are not as abundant in the Cape Fear River as in other rivers along the southeastern Coastal Plain, possibly because the high silt load in the river decreases the efficiency of their filterfeeding activity.

Unionid mussels once were diverse and abundant in the river (Patrick 1996), but the number of species present and their densities have declined over the years, probably in response to decreasing water quality and competition from the Asiatic clam. The North Carolina Nongame and Endangered Wildlife Fund lists 15 species of mussels within the Cape Fear River basin as endangered, threatened, or of special concern.

Vertebrates

At least 95 species of fishes live in the Cape Fear River (NCDENR 2000). Studies from the 1960s show that bluegill, pumpkinseed, longnose gar, spotted sucker, whitefin shiner, largemouth bass, snail bullhead, white catfish, and channel catfish were among the more abundant species (Patrick 1996). More recent studies, however, have shown a change in the relative abundance of fishes in the river linked to the introduction of several nonnative species (Moser and Roberts 1999, Mallin et al. 2000). The nonnative blue catfish now is one of the more abundant species in the river. Flathead catfish were introduced in 1966 and now are abundant and the dominant predator in the river. They have caused the near extirpation of native catfishes and the channel catfish, the latter having been introduced in the early 1900s (Moser and Roberts 1999, Mallin et al. 2000). Among other nonnative species, common carp are ubiquitous and grass carp are becoming more common. White bass and white perch were introduced into Jordan Reservoir, the latter species being highly successful in the reservoir and both species having dispersed downriver. Overall, from about 50% to 100% of the fishes captured in gill nets in a recent study in the lower river were nonnative species (Mallin et al. 2000).

Seven species of migratory fishes are reported from the river. American shad, hickory shad, and blueback herring are important commercial and recreational species, although their numbers are declining. Atlantic and shortnose sturgeon, once plentiful in the river, are federally listed as endangered. The last shortnose sturgeon was captured in the river in 1993 (NCDENR 2000). The striped bass population is far below historical numbers. A moratorium on the taking of stripers in North Carolina was put into effect from 1985 to 1990. Although populations rebounded in most other rivers with a moratorium, stripers have not faired as well in the Cape Fear. Predation by nonnative catfishes and reproductive competition from hybrid striped bass may be hindering their recovery (NCDENR 2000).

A wide variety of vertebrates other than fishes occur in the Cape Fear. Water snakes are abundant and cottonmouths are occasionally encountered along the river. Turtles such as sliders and river cooters are frequently observed along the banks and on snags projecting above the water's surface. Snapping turtles and mud turtles also are common. The federally endangered West Indian manatee is an infrequent visitor to the lower river.

Ecosystem Processes

Information on ecosystem-level processes such as energy flow and nutrient cycling is generally lacking for the Cape Fear. The river no doubt functions much like other large Coastal Plain rivers. Primary production likely is low, dominated by periphyton on snags and other structures in the channel. Periphyton growth overall, however, is limited by low light caused by the high turbidity of the water (Mulholland and Lenat 1992, Mallin et al. 2000). Phytoplankton production is highest below the Jordan Reservoir from algae washing out of the impoundment, and also in backwaters and near the river's mouth, where flushing rates are low. Macrophyte production is low because of the lack of extensive plant beds throughout much of the river. The river likely is a highly heterotrophic system, the primary source of energy being allochthonous organic matter washed in from tributaries and from the fringing floodplain. The latter is a source primarily during the winter–spring period, when the broad floodplain is inundated and water exchanges directly with the channel.

Levels of primary and secondary consumer production are unknown. As in other Coastal Plain rivers, invertebrate production likely is predominately on snags and in backwaters, where organisms can complete their life cycles without being disturbed by shifting sand during high-flow events. Many fishes in the river no doubt are dependent on invertebrate production from snags for food (e.g., Benke et al. 1985).

Human Impacts and Special Features

The Cape Fear River basin supports a variety of ecologically and economically important riverine habitats. The river system is known for its broad floodplains and scenic beauty and the many rare species of flora and fauna that occur within the basin. Ever rising demands are being placed on the river, however, because of increasing population and industrial, commercial, and agricultural development in the basin. The river provides water for businesses, residential users, and irrigation. It serves as a commercial transportation route and receives heavy recreational use for boating and fishing as well as supporting some commercial fisheries in the lower river. Its waters and floodplain provide important habitat for fishes and other wildlife. Even with these many existing demands on the water resources of the basin, it is estimated that water use throughout the basin will increase 95% from the early 1990s to 2020 (NCDENR 2000).

At the same time that there are increasing demands on the river to provide a safe and reliable source of water, it also is heavily used for waste assimilation. There presently are about 280 permitted wastewater discharges in the basin, the majority of them to tributaries in the Piedmont section of the basin (NCDENR 2000). Nonpoint-source inputs to the river and its tributaries also are widespread and significant. Large quantities of nutrients and sediment enter the river from land deforested primarily for agricultural, timbering, and urban land uses. These many point discharges and nonpoint-source inputs have resulted in over 20% of the monitored waters in the basin being rated as impaired.

The river showed a general trend of decreasing water quality into the 1980s, with conductivity, nutrient concentrations, and sediment load increasing as pH and dissolved oxygen concentrations decreased. Improvements in both industrial and municipal wastewater treatment, as well as better land management, have slowed or reversed that trend. The many textile mills that once were a major source of a variety of pollutants to the Haw River, including organic compounds and trace metals, have now largely been abandoned. The result has been an improvement in water quality in the Cape Fear's primary tributary.

The Cape Fear has been spared major impoundments along its main stem. The Jordan Reservoir, however, located just above where the main stem originates, does have major impacts on the hydrology and ecology of the river. In addition, there have been long-standing concerns about nutrient trapping and eutrophication in the impoundment given the significant inputs of municipal wastewater to its tributaries and the low flushing rate of most of the reservoir (Moreau and Challa 1985).

The Cape Fear basin has a long history of being impacted by hurricanes, but the late 1990s brought a succession of unusually severe storms. Although hurricanes are natural events to which these river systems are adapted, human development in the basin, especially in the Cape Fear's floodplain, greatly increased their negative impact on water quality and the biota of the river. A primary problem was the flooding of numerous industrial hog and poultry operations located in and near the floodplain, causing the release of large quantities of at best partially treated animal wastes to the river (Mallin, Posey, Moser et al. 1999, Mallin, Posey, Shank et al. 1999, Mallin 2000). In addition, numerous municipal wastewater treatment facilities were flooded, releasing untreated human waste to the river. Severe dissolved oxygen depletion occurred along with high inputs of sediments, nutrients, and fecal coliform bacteria. The benthic invertebrate community was negatively impacted and fish kills were numerous (Mallin, Posey, Moser et al. 1999, Mallin, Posey, Shank et al. 1999). The fish community now may be experiencing the cumulative impacts of the succession of hurricanes during this period, with both species richness and abundance showing a declining trend (Mallin, Posey, Moser et al. 1999).

SAVANNAH RIVER

The Savannah River is among the most hydrologically, ecologically, and historically important rivers of the southeastern United States, forming the border between South Carolina and Georgia along its entire 476 km length (Fig. 3.17). Its headwaters, arising in the cool, swift-flowing streams of the Blue Ridge Mountains, coalesce in the upper Piedmont to form the main stem of this large warmwater river (basin area = 27,414 km^2). At its point of origin and along much of its route among the rolling hills of the Piedmont, the river is mainly a series of large hydroelectric impoundments (Fig. 3.5). Below the Fall Line, however, the Savannah is a long, free-flowing river, meandering through the Coastal Plain and bordered by broad riverine swamps.

From the tenth to the fifteenth centuries, well before the arrival of Europeans, the Savannah River basin was occupied by prehistoric chiefdoms of the Mississippian culture. For unknown reasons, the Mississipians disappeared from the Savannah basin between about 1450 and 1600, eventually being replaced during the seventeenth century by the Creek Confederacy in the lower river basin and tribes of the Cherokee Nation in the upper basin. Spanish influence came to the region during the early 1500s, notably when the explorer Hernando de Soto crossed the upper basin in 1540 searching for gold, and Jesuit and Franciscan missions became established along the coast, remaining for more than 100 years. It was not until the arrival of the English, led by James Oglethorpe in 1733, that European settlement spread out from the coast. These settlers slowly pushed upriver through the mid 1700s until after the Revolutionary War, when they rapidly spread to the headwaters of the river. From 1777 until 1796, the Georgia state capital alternated between Savannah, located about 35 km above the river's mouth, and Augusta, located at the Fall Line on the river.

Agriculture in the Savannah basin flourished into the mid-nineteenth century, with much of the primary cash crops of cotton, rice, and tobacco being grown on plantations supported by slave labor. During the Civil War, the North maintained a blockade of the river, limiting its use as a supply line for the South. Savannah fell to General Sherman in late 1864 during his "march to the sea." With the end of the Civil War the area fell into an agricultural depression as the plantation-based economy disintegrated. It was not until the early 1900s that an economy based on forestry, shipping, and a restructured agricultural system revitalized the area.

Physiography, Climate, and Land Use

The Savannah River flows in a southeasterly direction between 34°N to 32°N latitude. About 55% of

FIGURE 3.5 Savannah River below Thurmond Dam, upstream of Augusta, Georgia (PHOTO BY A. C. BENKE).

the basin is in Georgia, 43% in South Carolina, and 2% in North Carolina. The basin lies within three physiographic provinces, 10% in the Blue Ridge (BL), 56% in the Piedmont (PP), and 34% in the Coastal Plain (CP) (see Fig. 3.17). The red soils of the Piedmont are predominately highly weathered ultisols with a sandy-clay or silty-clay texture. Soils of the Coastal Plain are well-drained sands in the uplands and poorly drained sands and clays in the lowlands. The basin lies within the Temperate Deciduous Forest biome and encompasses four terrestrial ecoregions (Ricketts et al. 1999): Appalachian/Blue Ridge Forests, Southeastern Mixed Forests, Middle Atlantic Coastal Forests, and the Southeastern Conifer Forests. The predominant upland pine forests range from white pines in the mountains to loblolly and shortleaf pine in the Piedmont and longleaf and slash pine in the Coastal Plain. Hardwoods are more common at higher elevations and include oaks, hickory, poplar, and maple.

The climate over much of the basin is subtropical, with hot, humid summers and mild winters. Annual mean air temperature is 18°C, with monthly mean temperatures typically ranging from 7°C in January to 27°C in July and August (Fig. 3.18). Annual precipitation averages 114 cm, nearly all of which is rainfall. Precipitation is distributed fairly evenly over the year, with the driest months from September to November.

Though historically covered by deciduous and pine trees, today the landscape is a mix of land uses. About 65% of the basin is forested, 22% in crop or pasture, 4% urbanized, and 9% in other land uses. Forest products are a major component of the region's economy. Agriculture is based on a mix of animal operations and crops, especially cotton, peanuts, tobacco, and grain. The major urban and industrial areas are Savannah, with a deep-draft harbor; Augusta, with an inland port at the Fall Line; and Anderson, South Carolina, located near the origination of the river's main stem.

River Geomorphology, Hydrology, and Chemistry

The Savannah arises from the merging of the Seneca and Tugaloo rivers in the upper Piedmont at an elevation of about 190 m asl. It arises not as a free-flowing river but rather as an impoundment (see Fig. 3.17). Hartwell Lake, a 22,663 ha reservoir completed in 1963, forms the head of the main stem. Just 48 km below Hartwell Dam is the Richard B. Russell Lake, a 10,785 ha reservoir completed in 1985. Farther downstream is the J. Strom Thurmond Lake, completed in 1954 and at 28,329 ha is the largest of the river's impoundments (see Fig. 3.5). All three impoundments are operated for hydroelectric power generation, flood control, recreation, and fish and wildlife.

The three impoundments create a chain of reservoirs along most of the approximately 160 km length of the Piedmont section of the Savannah. In the few stretches that are not impounded, especially from the lower lake down to the Fall Line at Augusta, the river is about 110 m wide and has a series of well-developed riffles. River banks typically are 1 to 3 m high, consist of sand-clay, and are mostly lined with trees (Sehlinger and Otey 1980).

The river is impounded again for a short distance at Augusta by several low-head locks and dams, but otherwise drops quickly through a series of riffles into the Coastal Plain. Thereafter the channel broadens from 75 to 100 m wide to nearly 200 m near Savannah (Sehlinger and Otey 1980). Channel gradient across the Coastal Plain is about 20 cm/km, compared to 50 cm/km across the Piedmont. The river's gradient from headwaters to mouth is 3.5 m/km. Dredging in the lower river maintains depth at about 3 m. The Coastal Plain section is free flowing and meanders through floodplain forests that occasionally reach 3 to 4 km in width. Oxbow lakes are common in the floodplain, with 24% of the channel length in oxbows (Schmitt and Hornsby 1985). The sediment consists primarily of sand with some silty areas (Patrick et al. 1967). Snags are common, especially along the banks. Above Savannah, the floodplain swamps give way to tidal marshes, including the Savannah National Wildlife Refuge, where numerous side branches of the river wind through the marshes before the river passes the city and empties into the Atlantic Ocean.

Annual mean discharge 98 km above the river's mouth is 319 m^3/s. The river is tidal over its lower 70 km. Flow in the river, especially in the Piedmont, is highly regulated by the dams, dampening the natural seasonal changes of highest discharge in winter and spring and lowest in late summer and autumn (see Fig. 3.18). Prior to construction of the dams, high levels of evapotranspiration would reduce the dependable minimum flow in the river to 32 m^3/s (U.S. Army Corps of Engineers 1981), or a runoff

of only 0.30 cm/mo. The J. Strom Thurmond Dam, however, now provides a minimum flow of 164 m^3/s, or a runoff of about 1.55 cm/mo.

The Savannah is a neutral to slightly acidic, soft-water river, the majority of its flow being derived from the crystalline geology of the Piedmont. Alkalinity averages about 20 mg/L as $CaCO_3$. Water temperature can exceed 30°C and rarely falls below 8°C. The river carries a high sediment load during storms, although turbidity is ameliorated some by the settling of solids in the impoundments. Within the Coastal Plain, the river becomes a blackwater system, highly stained by dissolved organic compounds derived primarily from the floodplain. Color in the lower river typically is 30 to 80 Pt units, highest in the spring and early summer and lowest in the autumn when connectivity between the channel and floodplain is at its lowest (Patrick 1996). Dissolved organic carbon concentrations in the lower river typically are 4 to 8 mg/L and constitute the majority of the total organic carbon transported by the river (Patrick 1996). Water salinity is about 15 ppt near Savannah and the salt wedge travels 30 km above the city.

River Biodiversity and Ecology

The Savannah River, within the South Atlantic freshwater ecoregion (Abell et al. 2000), provides a variety of habitats that support a diverse flora and fauna. Much is known of the organisms inhabiting the river because of long-term, ongoing studies begun in 1951 about 40 km below the Fall Line. Led by Dr. Ruth Patrick, scientists from the Academy of Natural Sciences of Philadelphia have been studying the river at the U.S. Department of Energy's Savannah River Plant (SRP). They and researchers associated with the SRP and the Savannah River Ecology Laboratory of the University of Georgia have produced a considerable body of information on the physicochemical characteristics, flora and fauna, and ecology of the river, its tributaries, and floodplain.

Algae, Cyanobacteria, and Protists

Over 800 species of algae, cyanobacteria, and protists have been identified in the river (see Patrick 1996 for species lists). Diatoms dominate, in particular *Melosira*, *Gomphonema*, *Fragileria*, *Navicula*, *Eunotia*, and *Achnanthes*. Green algae also are common, including *Chaetophora*, *Closterium*, *Stigeoclonium*, *Oedogonium*, and *Spirogyra*. Less common are cyanobacteria, such as *Calothrix*, *Oscillatoria*, *Microcoleus*, and *Phormidium*, and the red algae *Compsopogon* and *Batrachospermum*. Lentic species of phytoplankton prevail in the impoundments. Periphyton peak during the summer, biomass on artificial substrates being 2.6 to 5.5 g/m², and chlorophyll *a* ranging up to 95 mg/m² from May to August (Specht et al. 1984). The highest densities of phytoplankton in the lower river also occur during the summer (Patrick 1996). Thick algal mats often occur in the sloughs and backwaters.

Plants

Among the most abundant macrophytes in the river are alligatorweed, coon's tail, and waterweed. The moss *Leskea obscura*, and to a lesser extent the liverwort *Porella pinnata*, are common on snags (Cudney and Wallace 1980). Nonnative hydrilla has been established in the J. Strom Thurmond Lake since 1995 and has become a serious problem. Cordgrass and rush are the predominant species in the estuarine marshes along the lower 45 km of the river (Schmitt and Hornsby 1985). The river's riparian zone and floodplains are predominately forested. Among the most common species of trees in the floodplain are bald cypress, water tupelo, swamp black gum, red maple, water ash, water oak, and sweetgum.

Invertebrates

Numerous studies have reported on the over 300 taxa of macroinvertebrates in the Savannah and on their associations with different habitats within the river (summarized by Patrick 1996). Collector-gatherers and collector-filterers together comprise about 75% of the invertebrate biomass in the free-flowing river (Specht et al. 1984). Most of that biomass is filter-feeders, which are abundant on snags throughout the river (Cudney and Wallace 1980) and also on rocks downstream of the impoundments, the discharges from which provide filter-feeders an abundant and nutritious supply of sestonic algae and zooplankton. Predaceous species comprise about 19% of the biomass, shredders 4%, and scrapers 2% (Specht et al. 1984).

The invertebrate fauna of the impoundments is typical of that of large southeastern reservoirs, with many species of zooplankton and benthic invertebrates occurring in the profundal and littoral areas. The benthic invertebrates are numerically dominated by chironomid midges and oligochaete worms. High numbers of phantom midge larvae (*Chaoborus*) also are found in the impoundments, their third and fourth instars occurring in the profundal sediments during the day and migrating into the water column to feed on zooplankton at night.

The riffles in the upper river support a diverse array of species. Among the more common insects are mayflies (*Baetis*, *Stenonema*), caddisflies (*Cheumatopsyche*, *Hydropsyche*, *Macrostemum*, *Brachycentrus*, *Chimarra*), hellgrammites (*Corydalus cornutus*), black flies (*Simulium*), and a variety of chironomids, especially *Rheotanytarsus*. The sand sediment of the lower river supports many chironomids and oligochaetes as well as lesser numbers of other taxa such as beetles, dragonflies, and damselflies. The nonnative Asiatic clam is the most abundant bivalve mollusk and *Physella heterostropha* the most common snail in the river.

Within the Coastal Plain section of the river, snags and backwater areas support the most diverse and productive assemblage of invertebrates. As in other sand-bottomed rivers of the southeast, snags provide the only stable habitat for invertebrates in the main channel and hence are heavily colonized. Among the more frequently encountered taxa on snags are mayflies (*Baetis*, *Heptagenia*, *Stenonema*), dragonflies (*Boyeria*), damselflies (*Calopteryx*), elmid beetles (*Ancyronyx*, *Macronychus*), hellgrammites (*Corydalus*), black flies (*Simulium*), and chironomids (*Rheotanytarsus*). Also abundant on snags are filter-feeding caddisflies of the genera *Hydropsyche*, *Cheumatopsyche*, *Macrostemum*, *Chimarra*, and *Neureclipsis*.

Macroinvertebrates of the bank and backwater areas also are diverse and presumably productive. Many species of benthic copepods are abundant, as are oligochaetes, chironomids, beetles, and sphaeriid clams (*Musculium*, *Pisidium*). Decapods are more abundant in the floodplains than in the main channel (Hobbs et al. 1976). The mayfly *Tortopus* occurs primarily along the upper river, where their burrows are found in the vertical walls of the silt-clay banks.

About 20 species of unionid mussels are reported in the Savannah River basin, with the eastern elliptio and the variable spike being the most common (Johnson 1970, Britton and Fuller 1979, Chris Skelton, Georgia Department of Natural Resources [GDNR], personal communication). Overall abundance of unionids in the river has declined since the 1950s, possibly as an effect of the invasion of Asiatic clams or human alteration of the river's environment (Patrick 1996). The Atlantic pigtoe mussel is federally listed as endangered; seven other mussels designated as being of special concern on South Carolina's list of protected species occur in the river basin, primarily in the river's tributaries. Several species of crayfish in the basin also are on state protected species lists.

Vertebrates

The Savannah River basin supports at least 21 families and 106 species of freshwater fishes (Chris Skelton, GDNR, personal communication). The most species-rich families are the Cyprinidae (28 species), Centrarchidae (20), Catostomidae (12), Percidae (12), and Ictaluridae (10). Among the most abundant species in the main stem along the Upper Coastal Plain are the eastern silvery minnow, spottail shiner, redbreast sunfish, and spotted sucker (McFarlane et al. 1979). Fish biomass, however, is dominated by spotted sucker, silver redhorse, common carp, bowfin, gizzard shad, and channel catfish (McFarlane et al. 1979, Schmitt and Hornsby 1985). Striped mullet are abundant in the Savannah estuary and the lower part of the river.

Among the more important migratory species in the river are the anadromous American shad and blueback herring and the catadromous American eel. Striped bass, normally a migratory species, likely are not migratory in the Savannah. Stocks of this species have been decreasing since the mid 1980s with the loss of their primary spawning grounds near Savannah due to salt water encroachment caused by harbor improvement projects (GDNR 2001b).

The Savannah River supports a thriving sport fishery in both the impoundments and lower river. The primary focus of sport fishermen is on largemouth bass, black crappie, channel catfish, striped bass, American shad, bluegill, redear sunfish, redbreast sunfish, and yellow perch (GDNR 2001b). The river supports only a small commercial fishery, primarily for ictalurids (Schmitt and Hornsby 1985).

Seven species of fishes in the basin are on Georgia or South Carolina lists of protected species. Among these is the federally endangered shortnose sturgeon. The robust redhorse, once thought extinct in the river, was found in 1997 in the shoals below the J. Strom Thurmond Dam down to Augusta (GDNR 2001b). A program stocking this species in the Broad River, a major tributary to the Savannah, is now underway.

The Savannah River supports a high diversity of vertebrates besides fishes. At least 9 species of frogs, 17 salamanders, 9 snakes, 11 turtles, and 1 crocodilian occur in the river or its floodplain (John Jensen, GDNR, personal communication). Water snakes, cottonmouths, snapping turtles, sliders, and river cooters are common in the river, and the American alligator is occasionally seen. The river and its floodplain also are important avian habitats, with the Savannah National Wildlife Refuge near the mouth of the river an important migratory and wintering ground for waterfowl.

Ecosystem Processes

Little information exists on ecosystem-level processes in the river. High phytoplankton cell counts, high periphyton biomass, and locally abundant macrophytes combined with warm water temperatures suggest that the Savannah is a productive system. However, the impoundments along the upper river no doubt have substantially altered the river's energy flow and metabolism. The river's energy base historically would have depended largely on inputs of allochthonous organic matter from tributaries, riparian areas, and the extensive floodplain along the lower river. The impoundments, however, provide a habitat for high phytoplankton production, some portion of which passes to downstream food webs. Thus, the river likely is far more autotrophic now than historically, though it probably still is a heterotrophically based system in the Coastal Plain.

Primary and secondary consumer production likely is high based on the abundance of invertebrates and fishes throughout the river. Invertebrate production is predominately of zooplankton in the impoundments and on snags and in backwaters in the Coastal Plain. Annual production of net-spinning caddisflies on snags was estimated at 12 to 36 g ash-free dry mass/m^2 of snag surface area (Cudney and Wallace 1980); total consumer production on snags probably is considerably higher than this. Fish production, higher in the river's backwaters than in the main channel, must be substantial given estimates of their standing stock biomass (e.g., Patrick 1996).

Human Impacts and Special Features

The Savannah River serves a multitude of uses. Foremost are withdrawals for drinking and industrial use, power generation, navigation, and sport fishing and other recreational activities. Nearly 14 million people use Hartwell Lake annually for recreational purposes, making it one of the three most visited Army Corps of Engineers impoundments in the nation. Another 7 million people use the J. Strom Thurmond Lake and 10 million use the Richard B. Russell Lake (GDNR 2001b). Even with the heavy use of the Savannah, however, water quality in the river overall is quite good and has been improving over the past few decades. The designated use of the river down to the Thurmond Dam is for recreation, and from there to Augusta it is for drinking water. Portions of the lower part of the river are designated for fishing and drinking water (GDNR 2001b). The Savannah has been identified as one of 26 rivers in North America with a mean annual virgin discharge >350 m³/s that are highly fragmented by dams (Dynesius and Nilsson 1994).

Although water quality is generally good, ecologically the Savannah is a highly modified system. The impoundments alter the daily and seasonal hydrograph, reduce downstream sediment load, and negatively impact water quality through hypolimnetic releases of cold, oxygen-depleted water during the late summer and autumn. The seasonal impact of the dams on the river's hydrograph can be seen by comparing runoff in the Savannah with that in the neighboring Ogeechee River, which is unregulated and likely represents the natural flow regime for rivers in the region (compare Figs. 3.18 and 3.20). The relative differences in runoff between dry and wet periods over the year are far smaller in the Savannah than in the Ogeechee. Summer and autumn runoff in the Savannah is augmented by releases from the dams, maintaining higher than natural flow during this time, whereas winter and spring runoff is decreased by the impoundments, in particular affecting the extent of inundation of downstream floodplains.

The river below the Fall Line also is impacted by dredging and channelization, performed to maintain a shipping channel and for flood control. Channelization has reduced the length of the river by 13% (Schmitt and Hornsby 1985). Flow regulation and channelization together have reduced the frequency and magnitude of downstream flooding, allowing development in the floodplain as well as causing a decrease in the size, inundation period, and probably the productivity of the floodplain.

The river is impacted by various industrial and municipal discharges, especially in the vicinity of Augusta and Savannah, as well as by nonpoint sources from timbering, agricultural, and urban land uses that result in the input of considerable quantities of sediment and nutrients. In addition, habitat alteration in the floodplain, heavy recreational fishing pressure, the introduction of nonnative sport fishes, the blocking of anadromous fish runs by dams, and removal of the historically abundant snags from the channel to aid navigation all have altered the ecology of this once prime example of a southeastern river.

OGEECHEE RIVER

The Ogeechee River is a medium-size river (basin area = 13,500 km²) in eastern Georgia that arises from spring-fed streams in the Piedmont physiographic province but flows most of its 400 km length through the Coastal Plain (Fig. 3.19). During most of its length the Ogeechee is a scenic, low-gradient blackwater river meandering within a broad forested swamp, then through vast marshes, before emptying into Ossabaw Sound about 24 km south of Savannah. The Ogeechee is one of the few rivers along the southeastern Atlantic slope that flows its entire length without any major dams, including its largest tributary, the 137 km Canoochee River. Water quality is generally good along most of the river, fishing is popular among local residents, and paddle trips are common (Sehlinger and Otey 1980).

Like many eastern rivers, the Ogeechee has an interesting history of human influence dating to the arrival of the Paleo-Indians 11,500 years ago. For several hundred years before the arrival of Europeans, the Ogeechee basin and much of the Georgia Coastal Plain were occupied by prehistoric Indian chiefdoms of the Mississippian culture. European influence began with the establishment of Spanish missions along the coast from the mid 1500s to late 1800s. By this time, the chiefdoms had disappeared and their descendents across much of middle Georgia became known as the Creeks. In the early eighteenth century, the British took control of the Georgia coast, as James Oglethorpe founded Savannah in 1733. Oglethorpe's treaties with the Creek Indians began the cession of land in the Ogeechee and surrounding basins to Georgia that would continue after the Revolutionary War. By 1790, seven years after the war ended, the entire Ogeechee basin became the central portion of the new state of Georgia as the remainder of the Native Americans was expelled. In 1796, the town of Louisville was established in the upper Ogeechee basin as the third of five state capitals. Little more than 50 years later, Civil War General Sherman's "march to the sea" crossed the basin to capture Fort McAllister, built on the southern shore of the Ogeechee to protect Savannah's southern flank.

Physiography, Climate, and Land use

The Ogeechee flows in a southeasterly direction between 34°N to 32°N latitude (see Fig. 3.19). The upper 5% of the basin is in the Piedmont (PP) physiographic province, but the river soon crosses the Fall Line, where it flows over shallow rocky shoals. The Ogeechee then flows through the somewhat hilly upper Coastal Plain (CP) (57%), and then through the nearly flat lower Coastal Plain (38%). Soils of the

Coastal Plain consist of sedimentary deposits that are sandy, porous, and vary from zero to 150 m in depth. These sediments overlay the extensive Upper Floridan aquifer, containing water from limestone and dolomite rocks. The basin encompasses three terrestrial ecoregions on its way to the coast: Southeastern Mixed Forests, Middle Atlantic Coastal Forests, and Southeastern Conifer Forests (Ricketts et al. 1999). The predominant upland trees are slash, longleaf, and loblolly pines, with hardwoods such as oak, hickory, sweetgum, and poplar scattered throughout the basin.

The Ogeechee basin has a subtropical climate, characterized by hot summers and mild winters. Annual mean air temperature is 18°C, ranging from a monthly mean of 8°C in January to 26°C in July (see Fig. 3.20). However, daily nighttime temperatures commonly fall below 0°C in winter and daytime temperatures exceed 38°C in summer. Annual precipitation is relatively high and uniform over the basin, with averages ranging from 100 to 130 cm. Precipitation falls mostly as rain and is distributed fairly evenly through the year, as illustrated by a high of 12 cm in August and a low of 6 cm in November at Millen, Georgia, in the upper Coastal Plain.

Human population density in the basin is 30 people/km², which is deceptively high because Statesboro (population = 21,000) is the largest city wholly within the basin. Near the coast, however, intensive development around Savannah has occurred, increasing the basinwide average density. In spite of a relatively low density throughout most of the basin, very little of the Ogeechee basin forests remain in a natural state due to agricultural conversion during the late 1700s and early 1800s, particularly with cotton and tobacco in the uplands and rice along the tidal portions of the river where marshes were drained. During the twentieth century, agriculture became more diversified (soybeans, tobacco, cotton, peanuts, corn, cattle, poultry), particularly after World War I, when the boll weevil devastated cotton production. Furthermore, extensive pine forests were planted over former croplands, particularly in the lower Coastal Plain, where pulpwood production became extensive. Currently, about 54% of the Ogeechee basin is covered with forests (mostly managed), 18% with crops or pasture, 1% urban, 17% forested wetlands, and most of the remaining 11% as nonforested wetland (GDNR 2001a). Floodplain (wetland) forests with native species remain intact throughout most of the basin in spite of having been extensively logged in the past; rice plantations near the coast have reverted back to marshlands. The lower section of the Canoochee River passes through the Fort Stewart Military Reservation, which occupies about 1130 km², mostly in managed forest.

River Geomorphology, Hydrology, and Chemistry

The Ogeechee basin begins at an elevation of about 200 m asl on the edge of the Piedmont Plateau, and the average gradient over its entire length is about 50 cm/km (Meyer 1992). Below the Fall Line, its gradient is eventually reduced to only 20 cm/km, with laminar flow. Channel width of the Ogeechee increases from 10 to 15 m just below the Fall Line, to about 60 m when it is joined by the Canoochee River, to >100 m just before emptying into Ossabaw Sound. Once in the Coastal Plain, the major substrate of the river bed is shifting sand, but backwaters along the many bends of the river have substantial organic deposits. Sandbars are commonly observed at low water during summer months. Extensive snags that originate from undercut riparian trees are a major feature of the channel along most of its length. In contrast to many rivers throughout North America in which wood has been removed by snagging operations, particularly during the nineteenth and twentieth centuries, the Ogeechee retains a substantial amount. Such wood has important influences on the geomorphology of the river, but more obviously it provides stable habitat for many aquatic animals, given the unstable nature of the riverbed. Unfortunately, the same wood habitat that perpetuates biological diversity and productivity is not always appreciated by the public because of the boating hazards it creates.

Much of the Ogeechee meanders within heavily forested swamps that can be 1 to 2 km in width, with many horseshoe bends and backwaters (Fig. 3.6). However, scenic sand and clay bluffs also occur intermittently among its length. The Ogeechee's broad forested swamp diminishes just below its confluence with the Canoochee as the floodplain is transformed into a vast marsh. The floodplain provides a highly variable environment with diverse habitat for both aquatic and terrestrial species. Aquatic life abounds during the flooded periods in winter, but when most of the waters recede during the warmer months, terrestrial species use the forest floor. Even during dry periods when the river is no longer flooding, floodplain pools and oxbows persist in many portions of the swamp, replenished by groundwater and rainfall.

Mean discharge for the Ogeechee River at its mouth is about 115 m³/s, but there is usually a strong

FIGURE 3.6 Ogeechee River, upstream of I-16 (PHOTO BY A. C. BENKE).

seasonal pattern that typically varies from only 10 to 20 m³/s in summer to over 400 m³/s in winter and spring. Rainfall is relatively uniform throughout the year, suggesting that the pattern of runoff is due to seasonal variation in evapotranspiration, as illustrated by a site near Interstate Highway 16, about 63 km from the mouth (Benke et al. 2000; see Fig. 3.20). As discharge increases during winter and spring, ever-increasing areas of floodplain become inundated and substantial flooding can continue for weeks or months. Thus, virtually every year floodplain inundation effectively increases the width of the river near I-16 from about 33 m to more than a kilometer. Because the Ogeechee is a free-flowing system with no regulation, the seasonal pattern of flooding can serve as a benchmark for the type of hydrological pattern that should be expected (and sought) in attempts to restore the hydrological regimes of similar rivers that are now regulated for power generation and flood control. The Ogeechee experiences a tidal influence just upstream of the confluence with the Canoochee, approximately 50 km from its mouth. The saltwater wedge also extends nearly this far upstream.

The Ogeechee is a blackwater river, often tea-colored as a result of dissolved organic carbon consisting mostly of fulvic acids that are leached from the terrestrial environment (Meyer 1992, Smock and Gilinsky 1992). DOC ranges from 4 to 34 mg/L, with a mean of 13 mg/L, and color ranges from 35 to 92 Pt units (Meyer 1992). The Ogeechee does not have the low pH and low alkalinity typical of many blackwater rivers because it receives discharge of carbonate-rich waters from the Upper Floridan aquifer, particularly from Magnolia Springs near Millen. Alkalinity ranges from about 10 to 40 mg/L as $CaCO_3$ (mean = 23 mg/L), and pH ranges from 6.6 to 7.2 (mean = 7.0). In spite of extensive agriculture in the basin, NO_3-N concentrations typically are <0.3 mg/L and total phosphorus is usually <1 mg/L. Dissolved oxygen concentrations are usually above 6 mg/L but can fall below 4 mg/L during droughts due to high amounts of organic matter (as will be discussed later). Water temperatures in the main channel at the I-16 study site typically range from a daily low of about 10°C to a daily high of 30°C, with an annual mean of about 19°C.

River Biodiversity and Ecology

The Ogeechee River lies wholly within the South Atlantic freshwater ecoregion (Abell et al. 2000). The ecological characteristics of the Coastal Plain portion of the Ogeechee are well known based on studies conducted during the 1980s at a site near I-16 (e.g., Meyer 1990, 1992; Benke 2001). Most of these studies focused on ecological processes, such as primary and secondary production, organic matter processing, ecosystem metabolism, and the role of microbes in food webs. Only the invertebrate taxa were intensively inventoried, but information was also obtained for protists. However, relatively complete information is available from the GDNR and other publications for fishes, amphibians, reptiles, crayfish, and mussels.

Algae, Cyanobacteria, and Protists

Water-column protists of the Ogeechee River have been studied to understand their taxonomic composition and their role in the microbial food web (Carlough 1989, Carlough and Meyer 1989). Protist densities were very high (typically >2 to 10^6/L). They included small flagellates such as choanoflagellates and *Spermatozopsis*, as well as large flagellates such as cryptomonads (*Cryptomonas*, *Chroomonas*, *Chilomonas*), the dinoflagellate *Glenodinium*, and various euglenoids (*Euglena*, *Distigma*, *Phacus*). Colonies of *Volvox*, *Eudorina*, and *Dinobryon* were also present. Some limited collections from the water column (A. C. Benke, unpublished data) also included diatoms (*Achnanthes*, *Melosira*, naviculoids), green algae (*Scenedesmus*, *Ankistrodesmus*), and Cyanobacteria (*Lyngbya* and *Oscillatoria*). Algae were also found as components of the biofilm on snags (including the filamentous alga *Tribonema*, as well as *Microspora* and *Melosira*) (Couch and Meyer 1992). The complete algal assemblage is probably similar to that found in the adjacent Savannah River (Patrick 1996).

Plants

The dominant plants along the Ogeechee River are those found in the swamp forest, an integral feature of this river–floodplain system. Among the most common tree species in the floodplain near the I-16 field site are bald cypress, swamp black gum, red maple, water oak, and sweetgum, with willows found along sandbars (Meyer 1992). The floodplain forest subsidizes the food webs of the river with its high production of leaves, which partially decompose on the forest floor and a large portion of which are ultimately flushed into the main channel (e.g., Meyer 1990). Aquatic macrophytes are not a common feature of the main channel, except for those found within quiet backwater and floodplain pools, such as the yellow pond-lily. Mosses and liverworts commonly are found growing on snags and tree trunks at the river's edge and in the floodplain forest. As the river enters the tidewater region, the floodplain forests give way to vast marshes of rush and cordgrass.

Invertebrates

Over 270 species of freshwater invertebrates have been collected within the channel and floodplain habitats at the I-16 field site (A. C. Benke, unpublished data). This includes 208 species of aquatic insects, with 31 mayflies, 18 caddisflies, 34 beetles, 13 stoneflies, 59 dipterans, 37 dragonflies and damselflies, and 10 mollusks.

Snags are the most stable habitat in the main channel and are heavily colonized by a high diversity (>108 species) of aquatic invertebrates with a mean density of >97,000/m^2 of wood surface (e.g., Benke 2001, Benke et al. 2001). Snag invertebrates primarily consist of species that feed by filtering particles from the water or gathering food from the wood substrate. Among the more abundant filter-feeders are hydropsychid caddisflies (*Hydropsyche*, *Cheumatopsyche*), Tanytarsini midges (*Rheotanytarsus*), black flies (*Simulium*), and the mayfly *Isonychia*. Among the more abundant gatherers are over 20 mayfly species (*Stenonema*, *Baetis*, *Tricorythodes*, *Ephemerella*), as well as several elmid beetles (*Stenelmis*, *Macronychus glabratus*) and many chironomid midges (*Polypedilum*, *Rheocricotopus*, *Stenochironomus*). There is also a diverse assemblage of predaceous insects, including hellgrammites (*Corydalus cornutus*), perlid stoneflies (*Paragnetina kansensis*, *Perlesta placida*, *Neoperla clymene*), and dragonflies (*Neurocordulia molesta*, *Boyeria vinosa*).

Invertebrates of the sandy bed include oligochaete worms (enchytraeids and tubificids) and a diverse assemblage of chironomids (*Rheosmittia*, *Cricotopus*, *Cladotanytarsus*, *Polypedilum*, *Cryptochironomus*, and *Robackia*) (Stites 1986). Trails of the burrowing dragonfly larvae *Progomphus obscurus* commonly are observed along the edges of sandbars at low water. By far the most common mollusk at the I-16 site is the nonnative Asiatic clam. Other mollusk taxa include the sphaeriids *Sphaerium* and *Musculium* and the gastropods *Amnicola*, *Gyraulus*, *Menetus*, *Physella*, and *Viviparus* (Stites 1986). At least 12 species of unionid mussels have been reported from the Ogeechee River (Johnson 1970; Chris Skelton, GDNR, personal communication), but these were not apparent at the I-16 site. The Atlantic pigtoe mussel is the most endangered mollusk. At least 16 crayfish species, most in the genera *Cambarus* and *Procambarus*, occur in the Ogeechee, none of which are considered to be endangered (Chris Skelton, GDNR, personal communication).

The inundated floodplain forest also supports an extensive assemblage of aquatic invertebrates (Benke et al. 2001), consisting primarily of oligochaete worms, small mollusks, dipterans, and crustaceans. Although the densities and production of floodplain invertebrates are not nearly as high as that found on snags, the large floodplain area results in the greatest overall total abundance and biomass of invertebrates in the river system. At least 34 species of microcrustaceans (zooplankton and bottom dwellers) have been found in the inundated floodplain swamp, including cladocerans (*Alona*, *Camptocercus*, and *Chydorus*), copepods (*Acanthocyclops*, *Diacyclops*, and *Mesocyclops*), and ostracods (*Candona parvula*, *Cypria turneri*) (Anderson 1995).

Vertebrates

The fish assemblage of the Ogeechee is quite diverse, particularly for a river that is primarily in the Coastal Plain. Various sources suggest at least 80 species, including marine species sometimes found in freshwater (Swift et al. 1986, Schmitt 1988, Warren et al. 2000; Chris Skelton, GDNR, personal communication). There are at least 15 species of Cyprinidae (carps and minnows) and 13 species of Centrarchidae (sunfishes). Most of the Centrarchidae and many small species (Cyprinidae and others) are insectivores. Several species of the Centrarchidae are snag-feeding insectivores, particularly redbreast sunfish, warmouth, spotted sunfish, and redear sunfish (Meyer 1990, 1992). Bottom-feeding fishes include spotted sucker and at least six species of Ictaluridae. Among the most abundant piscivorous species are largemouth bass, longnose gar, bowfin, and chain pickerel. These piscivores feed primarily on the smaller forage fishes such as shiners, minnows, and darters. Several species are floodplain spawners, as in other Coastal Plain rivers, including the anadromous American shad (Meyer 1990).

The GDNR (http://georgiawildlife.dnr.state.ga.us) describes the Ogeechee as excellent for redbreast

sunfish fishing, with bluegill, redear, black crappie, and spotted sunfish as other popular species. Largemouth bass are plentiful but represent only 3% of the harvest. The most common catfish in the river is the snail bullhead; the white catfish is found in the lower reaches. The Ogeechee is surprisingly free of nonnative fish species; it is unclear whether the channel catfish is native (Warren et al. 2000). Fisheries managers are much more concerned about the flathead catfish, which has not yet invaded the river. Only the shortnose sturgeon, a diadromous species, and the robust redhorse sucker are considered rare or endangered in the Ogeechee (Warren et al. 2000). The robust redhorse, once thought extinct, has been introduced to the Ogeechee. Although native to adjacent basins (Savannah and Altamaha), it is uncertain whether it is native to the Ogeechee. The abundance of the anadromous American shad decreased dramatically over the latter half of the twentieth century, apparently due to commercial fishing (Schmitt 1988).

The amphibians and reptiles directly associated with the Ogeechee River also are diverse. Seven species of frogs, 13 salamanders, 9 snakes, 12 turtles, and 1 crocodilian are found either in the main stem of the river or in the floodplain much of the year (John Jensen, GDNR, personnel communication). Considerable overlap occurs with species found in the much larger Savannah and Altamaha Rivers, which only have a few additional species. Among the more charismatic amphibians are the large salamanders: the two-toed amphiuma, the greater siren, the lesser siren, and the dwarf waterdog. Only a few amphibians are on Georgia's protected list or are of special concern: the dwarf waterdog, Brimley's chorus frog, and the many-lined salamander. Water snakes are very common in the Ogeechee and are often mistaken for the only poisonous aquatic snake, the cottonmouth. Among the more common turtles is the river cooter, often seen basking on snags projecting from the water. The snapping turtle, two species of softshell turtles, and the American alligator are among the more fascinating but elusive species. The rainbow snake and the spotted turtle are the only species of special concern.

Ecosystem Processes

There is extensive literature on ecosystem processes in the Ogeechee River, only a portion of which will be described here. Plant production in river swamps is typically higher than in upland areas and supports a high plant diversity in contrast to much of the upland that has been transformed into pine forest monoculture. Much of the litter fall partially decomposes on the swamp floor (Cuffney and Wallace 1987) and, along with its bacterial assemblage, it ultimately is flushed as dissolved or fine particulate organic matter to the river. The runoff containing this bacteria-rich organic matter becomes even further enriched with protists, algae, and drifting animals as it flows in the river channel. This seston becomes the major food source on which many species of filtering and gathering invertebrates depend (Meyer 1990, 1992). With the high inputs of organic matter to the river, community respiration rates of the main channel water column are relatively high, with an annual average of $6.7\,g\,O_2\,m^{-2}\,day^{-1}$ at the I-16 field site (Meyer 1992). In contrast, gross primary production in the water column is only moderate, with an annual average of $2.2\,g\,O_2\,m^{-2}\,day^{-1}$. Thus, the Ogeechee River is heterotrophic, with allochthonous organic matter from the floodplain swamp contributing to an annual Production/Respiration (P/R) value of only 0.25. A more complete description of the organic matter dynamics of the Ogeechee can be found in Meyer et al. (1997).

The nutritious seston provides an unlimited food supply for the snag inhabitants that are able to gather this material as it is intercepted by the wood itself or filtered from the water by the silken nets and specialized appendages of many species. This food subsidy from the flowing waters results in high invertebrate production on snags ($>100\,g$ dry mass $m^{-2}\,yr^{-1}$ of snag surface; Benke et al. 2001). Chironomids, caddisflies, and mayflies are especially productive, and they in turn are preyed upon by hellgrammites, dragonflies, and stoneflies. Invertebrate production and diversity are much higher on the snag habitat (a biodiversity hot spot) than in the shifting sand bed of the main channel, a habitat of low stability. The importance of the snag habitat is further shown by high densities of invertebrate drift (>20 animals/m^3 of water), most of which originate from snags, and the heavy use of snag prey as food by many species of fishes (Benke et al. 1985).

Human Impacts and Special Features

The Ogeechee River is one of only 42 reasonably natural rivers of at least 200 km in length that is free flowing in the coterminous United States (Benke 1990). Thus, with no major dams, channelization, or water diversions on the Ogeechee, discharge and inundation patterns are unaltered (Benke et al. 2000). Although the floodplain swamp forest has been exploited for over a century, it remains relatively intact along most of the river (i.e., most areas have

not been drained). Virtually all of the floodplain forest is second growth, and enormous stumps of previously harvested cypress trees are still visible in the swamp. During the 1980s, the Ogeechee was under consideration for the Federal Wild and Scenic River system.

Water quality is very good in the Ogeechee River and has been well summarized by Meyer (1992) and GDNR (2001a). Although nitrate and phosphate are not especially high (see previous text), the combined inputs from municipal wastewater treatment plants and agricultural runoff sometimes cause phytoplankton blooms during low summer discharge. Higher concentrations in the spring at the time of cropland fertilization implicate agriculture. There is relatively little sewage pollution in the Ogeechee, with five-day biochemical oxygen demands low, and reports of high fecal coliform counts are infrequent. Although dissolved oxygen is periodically low, it is what might be expected in an unpolluted blackwater Coastal Plain river rather than being caused by cultural waste inputs. Approximately 3% of the total flow of the Ogeechee River is withdrawn from both groundwater and surface water sources. Attempts to release pollutants into the river and develop landfills in floodplains are strongly opposed by local conservation groups such as the Ogeechee River Valley Association as well as statewide organizations.

The only nonnative aquatic animal in the Ogeechee River that may have caused ecological problems is the Asiatic clam. It apparently arrived in the mid 1970s as it did in many rivers of the southeastern Atlantic slope, with the exception of low-pH rivers such as the Satilla. Although the clam is relatively abundant, its production is not very high and there are often die-offs, suggesting that the low-to-moderate alkalinity in the Ogeechee may be near the clam's tolerance limits. Although other bivalves are not abundant in the vicinity of the I-16 study site, we are unaware of any evidence showing the Asiatic clam having a negative effect on native species or significantly altering the natural ecosystem functioning of such systems.

In spite of extensive agricultural and forestry development in the uplands, the Ogeechee River remains one of the more natural free-flowing rivers in the eastern United States and can serve as a benchmark for restoration of other rivers of the region. The absence of large cities, industries, and dams, and the presence of a wide forested floodplain swamp, are major factors that have allowed the Ogeechee to remain in a relatively natural state. Vigilant conservation groups will probably play a major role in

FIGURE 3.7 St. Johns River, north end of Lake George, Florida (PHOTO BY M. O'MALLEY/ST. JOHNS RIVER WATER MANAGEMENT DISTRICT).

preserving the Ogeechee and its floodplain as development pressures increase. Increased growth in the Savannah metropolitan area, which lies on the edge of the lowest part of the Ogeechee basin, has placed heavy water demands on the Upper Floridan aquifer near the coast. Potential use of the Ogeechee River as a supplementary water source could have severe impacts on the ecological conditions of the lower river reaches.

ST. JOHNS RIVER

The Native Americans of the Timucua tribe called it Welaka, roughly translating as "the Chain of Lakes," for that is how the St. Johns River appears. The longest river in Florida is more a series of interconnected lakes than a swift-flowing river (see Fig. 3.21). It is a broad, shallow, blackwater system that widens into a number of lakes as it drains much of northeast Florida on its way to the Atlantic Ocean (Fig. 3.7). It is a tidal river with an extended estuary,

in which tides influence its hydrology and ecology far upriver into broad freshwater marshes. As such, its geomorphology and hydrology are quite different from the other Atlantic slope rivers described in this chapter.

The earliest Native Americans probably moved into the area around 12,000 years ago, living off the river and bountiful plants and game throughout the basin. The French Huguenots founded the first colony in 1564 along the river near present-day Jacksonville. A Spanish force drove the French from the region the following year, beginning 250 years of Spanish influence in the area. American settlers moved into the river basin in earnest beginning around 1819 when Spain ceded Florida to the United States. Clashes between Native Americans and Europeans began almost with the earliest settlements and became more frequent as more and more of the former's land was lost to plantations. Hostilities finally stopped at the end of the Seminole Wars in 1842. Since that time, agricultural and then commercial development in the basin has been an important component of the growth and vitality of the region.

Physiography, Climate, and Land Use

The St. Johns River basin lies completely within the Coastal Plain (CP) physiographic province at latitude 28°N to 30°N, draining a watershed of rolling hills and lowlands set within the Southeastern Conifer Forests and Florida Sand Pine Scrub ecoregions (Ricketts et al. 1999). Its watershed is located in an area of Pleistocene barrier islands, coastal dunes and ridges, and estuarine marshes and lagoons (Cooke 1945). The basin began to take shape with the formation of barrier islands 16 to 25 km off the Florida peninsula. Those islands and the coastal dunes and ridges on the mainland eventually enclosed a long shallow bay running parallel to the coastline. As ocean levels dropped, the bay turned from an estuarine to a freshwater system. The Oklawaha River, now the largest tributary to the St. Johns, originally drained directly into the bay but then was captured by the newly forming St. Johns. The north–south coastal axis of the basin gives the St. Johns the distinction of being one of the few rivers on the North American continent that flows in a northerly direction.

The river basin today lies on the flat Pamlico terrace, with the highest elevation in the basin at 107 m asl. Rich spodosol and histosol soils occur throughout the region, with vegetation characterized by oaks, maples, magnolias, bays, and palmettos. Lowland areas are poorly drained, resulting in numerous lakes and extensive wetlands throughout the basin, with the marshes underlain by fibrous peat deposits. Portions of the basin occur in a karst topography of limestone and dolomite, particularly along the basin's western boundary, where the watershed sits on Ocala limestone and the Floridan aquifer (Cooke 1945).

The climate of the river basin is humid subtropical. Summer months are warm, with mean air temperature from June through August of 27°C (see Fig. 3.22). Winter months are mild, with a mean temperature of 12°C. The annual mean temperature is 20°C. Air temperatures occasionally drop to freezing during many winters, but most frequently only in the northern portion of the basin. Over half of the basin's annual rainfall occurs from June through September, primarily as thunderstorms. The mean annual rainfall for the basin is about 131 cm.

Land use throughout the river basin is mixed, significant portions of the basin remaining in wetlands and open water (24%) and upland forested areas (45%), but much of the land is converted to crop and livestock agriculture (25%) and urbanized areas (6%). The primary urban areas in the basin include Orlando, Deland, Ocala, and Jacksonville. Population density in the basin is quite high at 78 people/km^2.

River Geomorphology, Hydrology, and Chemistry

The river arises in a broad marsh located north of Lake Okeechobee at 8 m asl. From there it flows 460 km north to Jacksonville, where it takes an abrupt turn to the east and travels an additional 40 km to its mouth at the Atlantic Ocean. The river's overall gradient is only 2 cm/km, making it one of the flattest major rivers in North America. Along its slow meandering route the channel passes through eight broad lakes and has an immediate connection to at least five additional lakes. The river has three relatively distinct sections, based on differences in channel shape, sediment, hydrology, and ecology (Burgess and Franz 1978). All are encompassed by the St. Johns River Water Management District, a state agency charged with preserving and managing the ground and surface water resources throughout the river's basin.

The upper section, which includes the initial 121 km, begins as indistinct channels in dense headwater marshes. Distinct but shallow multiple channels gradually develop and then pass through a series of lakes, some formed from remnant coastal lagoons. The lakes are all small and shallow, with maximum depths typically of 1 to 3 m. They support a dense growth of emergent, floating, and submerged macrophytes, with occasional sections of channel through the lakes free of macrophytes. Sediment in the channel and lakes is a thick peat with some areas of sand. The floodplain historically was quite broad but has been much reduced, largely by the many canals built to drain land for agricultural purposes. Partway through this section the river develops distinct banks and has a firm sand substrate, but then again broadens into a valley of braided channels, palmetto hammocks, and marshes. Just above Lake Harney the river receives inputs from salt springs that raise surface water salinities to 11 ppt (DeMort 1991).

The middle section of river runs from Lake Harney to the confluence with the Oklawaha River, a distance of 161 km, much of which is in the Ocala National Forest. The channel through this section alternates between having elevated, well-defined banks to flowing through broad shallow lakes, including Lake George, the largest of the lakes along the river at over 20 km wide. Sediments are typically sand with areas of peat. Below where the Wekiva River enters the St. Johns north of Orlando, the channel deepens to 3 m and broadens to a mean width of 100 m. Dredging and channelization are employed to maintain a navigation channel throughout this middle section and down to the river's mouth.

The lower section includes that portion of river from just below Lake George, where the Oklawaha enters, to the river's mouth, a distance of about 200 km. Much of the river basin here is mixed forest and urban land. Down to Jacksonville the river widens from 1.5 km to 5 km, but with a depth still generally no greater than 3 m. At Jacksonville, where the river is heavily used for the city's thriving industrial port, the channel deepens to about 9 m, with some scoured and dredged areas that are 18 m deep (DeMort 1991). The river through this section is tidal, with freshwater marshes grading into salt marshes.

The primary sources of water to the river are from rainfall draining from marshes at the head of the basin and from groundwater from karst aquifers, especially the Floridan Aquifer. Several of the main tributaries, such as the Wekiva and Econlochhatchee rivers, have extensive inputs of water from karst formations and shallow aquifers, providing flow even during prolonged dry periods (DeMort 1991). Mean annual discharge in the river at Jacksonville is 222 m^3/s. Mean monthly discharge is remarkably uniform throughout the year in spite of substantially higher rainfall from June through September than during the rest of the year (see Fig. 3.22). The ability of the basin's wetlands to store water during the wettest months, higher evapotranspiration in the summer, and the substantial and relatively constant groundwater inputs to the river system all contribute to maintaining the uniform discharge over the year.

Tides have a major impact on the hydrology and chemistry of much of the river. A salinity wedge regularly reaches 42 km upriver and occasionally 90 km or more (Mason 1998). The lower river near Jacksonville is mesohaline at 15 to 18 ppt, increasing to 32 to 35 ppt at its mouth. Freshwater tides typically occur up to Lake George 160 km inland and occasionally farther upriver, affecting nearly half of the river's length. Tidal amplitude at the river's mouth is about 1.5 m. The low gradient and tidal action of the river make it very sluggish, with the direction of flow often determined by the prevailing winds and tides (Kautz 1981).

The mean annual water temperature of the river is about 22°C, typically ranging from 9°C to 30°C,

but occasionally reaching up to 38°C. High concentrations of dissolved organic carbon leaching from wetland vegetation result in the highly stained, blackwater appearance of the river, with its color often >200 Pt units (Aldridge et al. 1998). The river is neutral to slightly basic and highly buffered (Garman and Nielsen 1992). Conductivity and total dissolved solids are higher, and turbidity lower, than in other northern Florida rivers (Kautz 1981). Upwellings from salt springs cause pockets of higher salinity (>5 ppt) in otherwise freshwater portions of the river. The river's hydrology results in a high retention time for its water and dissolved and suspended constituents. Retention times in the lower river of 3 to 4 months greatly affect the river's water quality and ecology. Water quality is highly variable among the upper, middle, and lower sections of the river.

River Biodiversity and Ecology

The St. Johns River is located entirely within the Florida freshwater ecoregion (Abell et al. 2000). It supports a diverse flora and fauna. Plant and animal diversity is high in the river system because of the variety of aquatic habitats along the river, including riverine, lake, and wetland areas, as well as the strong estuarine influence far upriver.

Algae and Cyanobacteria

Little information exists on the periphyton in the river and its wetland. Phytoplankton species richness is high; DeMort (1991) noted that at least 343 species had been identified from the river. Cyanobacteria typically dominate in the upper river and diatoms in the lower river. Among the more common phytoplankton are *Cylindrospermopsis raciborskii*, *Microcystis incerta*, *Lyngbya contorta*, *Melosira italica*, *Skeletonema costatum*, *Anabaena circinalis*, *Pediastrum simplex*, and *P. duplex* (D. Dobberfuhl, personal communication; DeMort 1991). The marine benthic green algae *Ulva latuca* and *Enteromorpha intestinalis* are common in the estuarine section of the river. Chlorophyll *a* concentrations generally range from 4 to 20 µg/L in the upper sections of the river, but considerably higher concentrations frequently occur in the middle and lower sections. For example, the lower section can reach 174 µg Chl/L and averages around 22 µg Chl/L (D. Dobberfuhl, personal communication).

Plants

A diverse array of species of macrophytes occurs throughout the river system, with over 300 species recorded from the river and its floodplain. American eelgrass is the most abundant species in the channel, and waving beds of this species occur throughout much of the freshwater portion of the river. Other common submerged species in the channel and lakes include coon's tail, pondweed, widgeon grass, and southern waternymph.

The extensive marshes in the headwaters, fringing wetlands, and near-bank areas of the river's channel and lakes are characterized by a variety of emergent and floating-leaved species, with the species assemblage depending primarily on elevation and salinity. Typical species in freshwater areas include maidencane, pickerelweed, arrowhead, swamp saw-grass, cattail, yellow pond-lily, and American white waterlily. Bald cypress occurs along sections of the lower river and its tributaries. Halophytes occur in the extensive estuarine marshes and also in patches in otherwise freshwater areas where salt springs emerge. Needlegrass rush and smooth cordgrass dominate in high and low salt marshes, respectively; sand cordgrass and other halophytes dominate in inland areas with increased salinity. A variety of non-native macrophytes have been introduced to the river, with common water hyacinth and hydrilla the most abundant.

Invertebrates

The slow flow of the river and its many lakes allows far higher numbers of zooplankton to occur in the St. Johns River than in most other rivers. Zooplankton assemblages in the freshwater section of the river are composed predominately of cyclopoid copepods and rotifers. Estuarine species predominant in the lower river, with calanoid copepods and barnacle nauplii, in particular, increasing in number downriver but often occurring far upriver depending on river hydrology and time of year.

The benthic invertebrate assemblage also is a mix of freshwater and estuarine species. Studies suggest high species richness, in part because of the mix of freshwater and estuarine species and also because of the heterogeneous environment provided by the abundant submerged aquatic vegetation. Mason (1998) refers to studies that document the occurrence of nearly 300 macrobenthic taxa in the lower river. Several species of endemic crayfish and at least 14 species of endemic gastropods occur in the drainage basin (Garman and Nielsen 1992). A diverse, though probably declining, unionid mollusk fauna also occurs in the basin (Johnson 1970).

Chironomid midges and tubificid (oligochaete) worms are the most common species in the silt and organic sediments, where low oxygen concentrations, especially in the lakes during the summer and autumn, affect species composition and abundance. Abundance of invertebrates is higher in these sediments than in sand sediment (Mason 1998), where the clams *Corbicula fluminea*, *Rangia cuneata*, and various sphaeriids are among the more obvious species. *Rangia* is the most abundant bivalve in the lower section of the river (DeMort 1991).

Macrobenthic biodiversity is highest in beds of aquatic vegetation. The baetid mayfly *Callibaetis floridanus* is ubiquitous and often abundant in freshwater littoral zones and fringing wetlands. Many hydroptilid caddisflies also are common among the vegetation, including species of *Hydroptila*, *Orthotrichia*, and *Oxythira*. A variety of species of other mayflies, caddisflies, odonates, coleopterans, and hemipterans, as well as many species of dipterans, also occur among the vegetation. Amphipods, including *Hyalella azteca* and various gammarid species, as well as mysid and grass shrimp, pulmonate snails (e.g., *Physella*), and many species of tubificid and naidid worms are the most common noninsects in macrophyte beds.

Freshwater and estuarine species coexist throughout much of the lower river, although the number of freshwater species is low in the last 60 km of the river. Crustaceans replace insects as the dominant taxa as salinity increases downriver. It is not unusual to find high numbers of chironomids and tubificids in mud and peat beds, barnacles and mussels encrusted on nearby submerged objects, and mayflies, caddisflies, and estuarine crabs together inhabiting fringing macrophyte beds. The Harris (or white-fingered) mud crab is widespread, occurring from the river's mouth to far upriver (DeMort 1991). Economically important white shrimp occur over 200 km upriver, and blue crab often are abundant upriver to Lake George and occur over 300 km upriver (DeMort 1991). Salt springs allow the establishment of patches of estuarine species within an otherwise freshwater landscape far upriver.

Vertebrates

The St. Johns River is one of the more species-rich rivers in terms of fishes along the southeastern Atlantic slope, supporting a diverse warmwater and euryhaline fish assemblage. At least 75 freshwater species occur in the drainage basin, representing 23 families (Garman and Nielsen 1992). Of those, 55 species occur regularly in the main stem. Tagatz (1967) noted an additional 115 euryhaline species occurring in the river, often far upriver because of the influence of tides and the salt springs. Numerous estuarine and marine species are found in the lower river. The freshwater assemblage is dominated by invertivores. Of the freshwater species, centrarchids are the most abundant and diverse group, with 15 species occurring in the river. Centrarchids compose the majority of fishes in the lower river, with bluegill and redbreast sunfish alone composing 60% of the individuals (Bass 1991). Other species-rich families in the river include the Cyprinodontidae, Gobiidae, and Clupeidae.

The river serves as an important nursery for many marine species and also supports an important commercial fishery focused on American eel, channel catfish, and white catfish. Freshwater sport fishing concentrates on largemouth bass, black crappie, and bluegill (DeMort 1991). Six anadromous species occur in the river, with American shad the most abundant and economically important. A thriving striped bass population likely is nonmigratory (Garman and Nielsen 1992). Several nonnative species have been introduced, and three endemic subspecies are reported from the river, including the Lake Eustis minnow, Florida largemouth bass, and a subspecies of the pugnose minnow (Tagatz 1967, Burgess and Franz 1978).

The river, its tributaries, and its basin harbor many species of reptiles, amphibians, birds, and mammals. Reptiles and amphibians are abundant throughout the basin, including alligators, turtles, snakes, and frogs. Many aquatic birds, including anhingas, herons, egrets, ibises, limpkins, pelicans, and ducks, flourish along the river and in its marshes and lakes. The river basin harbors a population of the endangered Everglades snail kite that has been making a comeback since the early 1990s. Brown pelicans were listed as endangered in 1970 but were removed from the list in 1985 due to population recovery. Continual loss of habitat through the drainage of wetlands is the likely cause of the decline of the endangered wood stork. Bobcat and river otter occur in the floodplains, as do dwindling numbers of the endangered Florida panther. The federally endangered West Indian manatee is frequently found at many locations along the river.

Ecosystem Processes

The trophic basis of the river's food web is a mix of algae, including periphyton and phytoplankton, particulate detritus from senescing vascular plants both from within the channel and from the fringing wetlands, and DOC, which supports microbial communities, primarily from emergent plants throughout the wetlands. Primary production in the river system appears to be quite high and dominated by vascular plants. The contribution of phytoplankton to primary production can be substantial in the channel and fringing lakes and also increases downriver as the channel deepens, aided by the low flushing rate of the river. Phytoplankton production likely is light limited, both through shading by the abundant macrophytes and by the high concentrations of DOC that cause high light attenuation in the water column. High turbidity, especially in the lower river, and phosphorus limitation because of high Fe concentrations also may affect phytoplankton production in the river. Although phytoplankton production is high today, it is likely that prior to intensive development the river was dominated by allochthonous detrital-based production (D. Dobberfuhl, personal communication). With development and ensuing eutrophication, the system has made a relatively dramatic shift to autochthonous algal production.

Production of primary consumers in the river and its floodplain likely is very high. The year-round high water temperature and high availability of food, particularly periphyton, but also decaying macrophyte tissue, no doubt provide considerable food resources to the abundant invertebrate consumers in the river system. Production of several species of chironomids must be very high, as they occasionally reach nuisance status. The exuviae of the chironomids *Chironomus plumosus*, *C. decorus*, and *Glyptotendipes lobiferus* form sufficiently dense mats that they clog channels in the floodplains (Mason 1998). Secondary consumer production also must be high given the abundance of fishes and other vertebrates in the river system.

Human Impacts and Special Features

The St. Johns River is unique among the major rivers along the Atlantic slope. Its river–lake geomorphology, extremely flat gradient, and extensive estuarine influence greatly affect the biota and ecosystem functioning of the river. The river is strongly coupled to its broad floodplain marshes and swamps in terms of hydrology, geochemistry, and food web dynamics. The variety of aquatic habitats supports a diverse flora and fauna, made even more interesting by the coexistence far inland of both freshwater and marine species. In addition, the St. Johns is one of the few free-flowing large rivers in the southeastern United States, the low gradient of the system having precluded the construction of major impoundments along the river.

The St. Johns river basin, however, has a long history of use by humans that has resulted in a highly modified landscape. The basin continues to absorb a rapid increase in population. Jacksonville is a major maritime business center, and the ever-increasing shipping that moves through its port affects the lower St. Johns through dredging operations and water-quality issues. The river supports a high degree of agricultural, industrial, and recreational uses. The hydrology of the system, and especially inflow to the river, has been greatly changed over the past century. Draining of fringing marshlands began in earnest in the early 1900s. Flow patterns have been altered by canals and levees as well as by diversion of water from the upper drainage basin into neighboring river systems. These modifications have resulted in a flood-plain much reduced in size and function from its natural state. Along the upper St. Johns River, for example, over 80% of the floodplain has been highly modified. Water withdrawal from the river and its tributaries occurs for irrigation as well as for drinking water and industrial use. The result of the many hydromodification projects has been an altered hydroperiod and reduced discharge in the river.

Water quality in the river is affected by point discharges and especially by nonpoint runoff from agricultural land. The impact of these inputs is exacerbated by the high retention time of water, and thus pollutants, because of the low flushing rate of the river. Increasing rates of sedimentation and eutrophication result as decreasing discharge in the river causes even longer retention rates. Pesticide concentrations are elevated from agricultural runoff. Metal concentrations are elevated in urbanized areas, especially in the lower river near Jacksonville, where the highest industrial inputs occur.

The introduction of nonnative plants and animals, including both invertebrates and vertebrates, is a problem in the river basin as it is throughout much of Florida. Two species that have had considerable impact are the macrophytes water hyacinth and hydrilla. Hyacinth, introduced in the late 1800s, forms dense floating mats that can cover extensive areas of the river and lakes. Hydrilla, probably introduced in the 1970s, also has proliferated, especially in the lakes. Both are nuisance species, decreasing biodiversity, reducing fish stocks, increasing organic deposits, and impeding navigation. Extensive hyacinth and hydrilla control programs using herbicides are ongoing in the lower river to control these species and maintain an open navigation channel.

Several restoration and management projects are underway to address the many issues facing the St.

Johns. An effort to restore wetlands and a more natural hydroperiod in the upper basin is being undertaken by the St. Johns River Water Management District and the Army Corps of Engineers. Ecosystem management efforts in the middle basin are directed toward restoring water quality in the riverine lakes, especially Lake George. Programs focused on water-quality issues, including both point and nonpoint inputs, as well as restoration of degraded aquatic habitat, are ongoing in the lower river basin.

ADDITIONAL RIVERS

The York River, located in the Piedmont and Coastal Plain of Virginia, is part of the Chesapeake Bay watershed (see Fig. 3.23). The main-stem river, formed below the Fall Line at the confluence of the Pamunkey and Mattaponi Rivers, is tidal over its entire length (Fig. 3.8). Over two-thirds of the basin is forested, including extensive floodplain forests, resulting in relatively high water quality throughout much of the basin. The largely intact marshes along the upper reaches of the York are included in the Chesapeake Bay Natural Estuarine Research Reserve System. The York and its tributaries are largely free flowing, the only major impoundment being Lake Anna in the upper watershed of the North Anna River, allowing strong runs of anadromous clupeid fishes. Recreational fishing is excellent, with abundant largemouth bass, pickerel, and white perch. Rapidly increasing urban sprawl and timbering, however, threaten the water quality and scenic beauty of this river.

The Roanoke River arises in the Blue Ridge Mountains of Virginia and flows in a southeasterly direction through the Piedmont and Coastal Plain before emptying into Albemarle Sound in North Carolina (see Figs. 3.9, 3.25). Forest predominates across much of the basin. The river is highly fragmented by large impoundments along much of its length. Despite this hydrological regulation, the lower Roanoke River floodplain contains some of the largest intact and least-disturbed bottomland forest floodplains along the Atlantic coast. The ancient bald cypress and water tupelo forests in the lowest section of the basin are up to 8 km wide. The river is an important habitat for anadromous fishes and other wildlife. It harbors a diverse biota, including a variety of endemic species as well as many state and federally listed threatened, rare, and endangered species.

FIGURE 3.8 Mattaponi River, one of two major branches of the York River, just before it joins the York near West Point, Virginia (PHOTO FROM VIRGINIA INSTITUTE OF MARINE SCIENCE).

The Great Pee Dee River has its headwaters in the Blue Ridge Mountains in western North Carolina (see Fig. 3.27). The river drains large portions of the Piedmont and Coastal Plain of the Carolinas (Fig. 3.10). Much of the basin is in agriculture or has been urbanized, with only about half of the landscape forested, resulting in the river carrying a heavy silt load. A series of impoundments occur along the upper half of the river (in the Piedmont), regulating the river's hydrology. The wide river channel winds through broad water tupelo and bald cypress bottomland forests along its Coastal Plain section. The river and its floodplain support a diverse fauna, including various endemic species in its tributaries. A

primary blackwater tributary to the river, the Lumber River, is a National Wild and Scenic River.

Tributaries of the Santee River originate in the Blue Ridge Mountains, but the major drainage occurs in the Piedmont of the Carolinas (see Fig. 3.29). After entering the Coastal Plain, the Congaree and Wateree rivers join to form the Santee as the primary waterway of central South Carolina. Although the Santee flows only within the Coastal Plain, its basin extends across a large portion of the western Carolinas, making it the second largest river basin along the U.S. Atlantic coast. The river's hydrology is highly regulated by dams along its tributaries and main stem. Lake Marion, on the main stem, is a broad

FIGURE 3.9 Roanoke River below Altavista, Virginia (PHOTO BY TIM PALMER).

FIGURE 3.10 Great Pee Dee River, upstream of Route 32 near Brownsville, South Carolina (PHOTO BY A. C. BENKE).

FIGURE 3.11 Congaree River, one of two major branches of Santee River, downstream of Congaree National Park, South Carolina (Photo by A. C. Benke).

impoundment behind an 8-mile-long dam. In addition, the flows of the Santee and the Cooper rivers, the latter originally a short tidal river, have been connected by water-diversion projects. The river basin is well known for its recreational fishing. One of the most significant tracts of virgin bottomland forest in the eastern United States exists in the Congaree National Park, near the point of origin of the Santee (Fig. 3.11).

The Altamaha River lies entirely within Georgia and is one of the largest river basins in the eastern United States (see Fig. 3.31). It begins with the Ocmulgee and Oconee rivers, which originate in the Piedmont province. The Ocmulgee and Oconee cross the Fall Line and eventually form the Altamaha, which meanders within broad forested swamps of the Coastal Plain (Fig. 3.1). As one of the most biologically rich rivers draining into the Atlantic, the Altamaha has been designated a Bioreserve by The Nature Conservancy and is on their list of "last great places" in the world. The river is known for its outstanding largemouth bass and sunfish fishery. The recent illegal introduction of the flathead catfish has resulted in a significant decline of redbreast sunfish, and the flathead catfish itself has become a primary game species. Although there are major dams on the Oconee and Ocmulgee, their lower reaches and the main-stem Altamaha are free flowing. The river faces major threats from development around Atlanta in the upper basin, particularly by newly proposed dams that threaten to alter natural hydrological regimes.

The Satilla River basin is located entirely in the lower Coastal Plain of Georgia, within a forested and agricultural landscape (see Fig. 3.33). It is a true blackwater river, with highly stained water and a pH that typically fluctuates between 4 (at high water) and 6 (low water). The Satilla's almost-white sand bars, blackwater, and broad floodplain swamps are striking (Fig. 3.12). Like the Ogeechee, the Satilla is one of the few relatively natural rivers >200 km in length in the coterminous states that is unimpounded from headwaters to mouth. Its biological diversity and productivity depend heavily on snags in the main channel for habitat and on its floodplain swamps. However, the Satilla's diversity is not as high

FIGURE 3.12   Satilla River, about 16 km downstream of Atkinson, Georgia (PHOTO BY A. C. BENKE).

as the more alkaline rivers that drain the Piedmont and mountains farther inland. Although the basin is mostly forested, it faces threats from timber harvesting within its floodplain swamps.

ACKNOWLEDGMENTS

We thank Chris Skelton and John Jensen, both of the Georgia Department of Natural Resources, for the considerable information they provided on the occurrence of crayfishes, amphibians, reptiles, and fishes in the Georgia rivers included in this chapter. Greg Garman, Steve McIninch, and Charles Blem, all of Virginia Commonwealth University, provided information on the distribution, status, and ecology of amphibians, reptiles, and fishes in the Virginia rivers. Dean Dobberfuhl provided information on the St. Johns River. Thanks to R. Overman and M. O'Malley for photos of the Altamaha and St. Johns, respectively.

LITERATURE CITED

Abell, R. A., D. M. Olson, T. H. Ricketts, E. Dinerstein, P. T. Hurley, J. T. Diggs, W. Eichbaum, S. Walters, W. Wettengel, T. Allnutt, C. J. Loucks, and P. Hedao. 2000. *Freshwater ecoregions of North America: A conservation assessment.* Island Press, Washington, D.C.

Aldridge, F. J., A. D. Chapman, C. L. Schelske, and R. W. Brody. 1998. Interaction of light, nutrients and phytoplankton in a blackwater river, St. Johns River, Florida, USA. *Verhandlungen der Internationalen Vereinigung für Theoretische und Angewandte Limnologie* 26: 1665–1669.

Anderson D. H. 1995. Microcrustacean growth and production in a forested floodplain swamp. Ph.D. diss., University of Alabama, Tuscaloosa.

Bass, D. G., Jr. 1991. Riverine fishes of Florida. In R. J. Livingston (ed.). *The rivers of Florida*, pp. 65–83. Springer-Verlag, New York.

Benke, A. C. 1990. A perspective on America's vanishing streams. *Journal of the North American Benthological Society* 9:77–88.

Benke, A. C. 2001. Importance of flood regime to invertebrate habitat in an unregulated river-floodplain ecosystem. *Journal of the North American Benthological Society* 20:225–240.

Benke, A. C., I. Chaubey, G. M. Ward, and E. L. Dunn. 2000. Flood pulse dynamics of an unregulated river floodplain in the southeastern U.S. Coastal Plain. *Ecology* 81:2730–2741.

Benke, A. C., R. L. Henry III, D. M. Gillespie, and R. J. Hunter. 1985. Importance of snag habitat for animal production in southeastern streams. *Fisheries* 10:8–13.

Benke, A. C., T. C. van Arsdall Jr., D. M. Gillespie, and F. K. Parrish. 1984. Invertebrate productivity in a subtropical blackwater river: The importance of habitat and life history. *Ecological Monographs* 54:25–63.

Benke, A. C., J. B. Wallace, J. W. Harrison, and J. W. Koebel. 2001. Food web quantification using secondary production analysis: Predaceous invertebrates of the snag habitat in a subtropical river. *Freshwater Biology* 46:329–346.

Benz, G. W., and D. E. Collins (eds.). 1997. *Aquatic fauna in peril: The southeastern perspective.* Southeast Aquatic Research Institute special publication 1. Lenz Design and Communications, Decatur, Georgia.

A brief description of the province of Carolina on the coasts of Floreda. 1944. Facsimile of a pamphlet from the late 1600s. University of Virginia Press, Charlottesville, Virginia.

Britton, J. C., and S. L. H. Fuller. 1979. *The freshwater bivalve Mollusca (Unionidae, Sphaeriidae, Corbiculidae) of the Savannah River plant, South Carolina.* Report SRO-NERP. Savannah River Ecology Laboratory, Aiken, South Carolina.

Burgess, G. H., and R. Franz. 1978. Zoogeography of the aquatic fauna of the St. Johns River system with comments on adjacent peninsula faunas. *American Midland Naturalist* 100:160–170.

Carlough. L. A. 1989. Fluctuations in the community composition of water-column protozoa in two southeastern blackwater rivers (Georgia, USA). *Hydrobiologia* 185:55–62.

Carlough, L. A., and J. L. Meyer. 1989. Protozoans in two southeastern blackwater rivers and their importance to trophic transfer. *Limnology and Oceanography* 34:163–177.

Cooke, W. C. 1945. Geology of Florida. *Florida State Geological Survey Bulletin* 29:1–339.

Couch, C. A., and J. L. Meyer. 1992. Development and composition of the epixylic biofilm in a blackwater river. *Freshwater Biology* 27:43–51.

Cudney, M. D., and J. B. Wallace. 1980. Life cycles, microdistribution and production dynamics of six species of netspinning caddisflies in a large southeastern (U.S.A.) river. *Holarctic Ecology* 3:169–182.

Cuffney, T., and J. B. Wallace. 1987. Leaf litter processing in Coastal Plain streams and floodplains of southeastern Georgia, U.S.A. *Archiv für Hydrobiologie Supplement* 76:1–24.

Cutshall, N., I. Larsen, and M. Nichols. 1981. Man-made radionuclides confirm rapid burial of Kepone in James River, Virginia sediments. *Science* 213:440–442.

DeMort, C. L. 1991. The St. Johns River system. In R. J. Livingston (ed.). *The rivers of Florida*, pp. 97–129. Springer-Verlag, New York.

Diaz, R. J. 1989. Pollution and tidal benthic communities of the James River Estuary, Virginia. *Hydrobiologia* 180:195–211.

Dynesius, M., and C. Nilsson. 1994. Fragmentation and flow regulation of river systems in the northern third of the world. *Science* 266:753–762.

Garman, G. C., and L. A. Nielsen. 1992. Medium-sized rivers of the Atlantic Coastal Plain. In C. T. Hackney, S. M. Adams, and W. H. Martin (eds.), *Biodiversity of the southeastern United States: Aquatic communities*, pp. 315–349. John Wiley and Sons, New York.

Garman, G. C., and S. A. Macko. 1998. Contribution of marine-derived organic matter to an Atlantic coast, freshwater, tidal stream by anadromous clupeid fishes. *Journal of the North American Benthological Society* 17:277–285.

Garman, G. C., and L. A. Smock. 1999. *Critical habitat requirements for living resources of the nontidal James River, Virginia.* Virginia Department of Conservation and Recreation, Richmond.

Georgia Department of Natural Resources (GDNR). 2001a. *Ogeechee River basin management plan 2001.* Environmental Protection Division, Georgia Department of Natural Resources, Atlanta.

Georgia Department of Natural Resources. 2001b. *Savannah River basin management plan 2001.* Environmental Protection Division, Georgia Department of Natural Resources, Atlanta.

Hobbs, H. H., III, J. H. Thorp, and G. E. Anderson. 1976. *The freshwater decapods and crustaceans (Paleomonidae, Cambaridae) of the Savannah River Plant, South Carolina.* Savannah River Ecology Laboratory, Aiken, South Carolina.

Hocutt, C. H., R. E. Jenkins, and J. R. Stauffer, Jr. 1986. Zoogeography of the fishes of the Central Appalachians and Central Atlantic Coastal Plain. In C. H. Hocutt and E. O. Wiley (eds.). *The zoogeography of North American freshwater fishes*, pp. 161–265. John Wiley and Sons, New York.

Hottell, H. E., D. R. Holder, and C. E. Coomer Jr. 1983. *A fisheries survey of the Altamaha River.* Georgia Department of Natural Resources, Atlanta.

Jenkins, R. E., and N. M. Burkhead. 1994. *Freshwater fishes of Virginia.* American Fisheries Society, Bethesda, Maryland.

Johnson, R. I. 1970. The systematics and zoogeography of the Unionidae (*Mollusca:Bivalvia*) of the southern Atlantic Slope region. *Bulletin of the Museum of Comparative Zoology at Harvard University* 140:263–449.

Kautz, R. S. 1981. Fish populations and water quality in north Florida rivers. *Proceedings of the Annual Conference of the Southeast Association of Fish and Wildlife Agencies* 35:495–507.

Mallin, M. A. 2000. Impacts of industrial animal production on rivers and estuaries. *American Scientist* 88:26–37.

Mallin, M. A., M. H. Posey, M. L. Moser, M. R. McIver, T. D. Alphin, S. H. Ensign, G. C. Shank, and J. F.

Merritt. 1999. *Environmental assessment of the Lower Cape Fear River system, 1998–1999.* Report Number 99-02. Center for Marine Science Research, University of North Carolina at Wilmington.

Mallin, M. A., M. H. Posey, G. C. Shank, M. R. McIver, S. H. Ensign, and T. D. Alphin. 1999. Hurricane effects on water quality and benthos in the Cape Fear watershed: Natural and anthropogenic impacts. *Ecological Applications* 9:350–362.

Mallin, M. A., M. H. Posey, M. R. McIver, S. H. Ensign, T. D. Alphin, M. S. Williams, M. L. Moser, and J. F. Merritt. 2000. *Environmental assessment of the Lower Cape Fear River system, 1999–2000.* Report Number 00-01. Center for Marine Science Research, University of North Carolina at Wilmington.

Mason, W. T., Jr. 1998. Macrobenthic monitoring in the lower St. Johns River, Florida. *Environmental Monitoring and Assessment* 50:101–130.

Master, L. L., S. R. Flack, and B. A. Stein (eds.). 1998. *Rivers of life: Critical watersheds for protecting freshwater biodiversity.* The Nature Conservancy, Arlington, Virginia.

McFarlane, R. W., R. A. Frietsche, and R. D. Miracle. 1979. Community structure and differential impingement of Savannah River fishes. *Proceedings of the Annual Conference of the Southeast Association of Fish and Wildlife Agencies* 33:628–638.

Meredith, H. 1731. *Pennsylvania Gazette,* April 29 to May 13. Reprinted in E. G. Swem (ed.). 1922. *An account of the Cape Fear country, 1731.* C. F. Heartman, Perth Amboy, New Jersey.

Meyer, J. L. 1990. A blackwater perspective on riverine ecosystems. *BioScience* 40:643–651.

Meyer, J. L. 1992. Seasonal patterns of water quality in blackwater rivers of the Coastal Plain, Southeastern United States. In C. D. Becker and D. A. Neitzel (eds.). *Water quality in North American river systems,* pp. 249–276. Battelle Press, Columbus, Ohio.

Meyer, J. L., A. C. Benke, R. T. Edwards, and J. B. Wallace. 1997. Organic matter dynamics in the Ogeechee River, a blackwater river in Georgia, USA. In J. R. Webster and J. L. Meyer (eds.). Stream organic matter budgets, pp. 82–87. *Journal of the North American Benthological Society* 16:3–161.

Mitchell, D. J., and L. A. Smock. 1991. Distribution, life history and production of crayfish in the James River, Virginia. *American Midland Naturalist* 126:353–363.

Moreau, D. H., and R. Challa. 1985. *An analysis of the water balance of B. Everett Jordan Lake including estimates of flows between segments.* Report no. 221. University of North Carolina Water Resources Research Institute, Raleigh.

Moser, M. L., and S. B. Roberts. 1999. Effects of non-indigenous ictalurid introductions and recreational electrofishing on native ictalurids of the Cape Fear River drainage, North Carolina. In E. R. Irwin, W. A. Hubert, C. F. Rabeni, H. L. Schramm Jr., and T.

Coon (eds.). *Catfish 2000: Proceedings of the international ictalurid symposium,* pp. 479–486. Symposium 24. American Fisheries Society, Bethesda, Maryland.

Mulholland, P. J., and D. R. Lenat. 1992. Streams of the southeastern Piedmont, Atlantic drainage. In C. T. Hackney, S. M. Adams, and W. H. Martin (eds.). *Biodiversity of the southeastern United States: Aquatic communities,* pp. 193–231. John Wiley and Sons, New York.

North Carolina Department of Environment and Natural Resources (NCDENR). 2000. *Cape Fear River basinwide water quality plan.* Division of Water Quality, Raleigh, North Carolina.

North Carolina Department of Environment and Natural Resources. 2001. *Roanoke River basinwide water quality plan.* Division of Water Quality, Raleigh, North Carolina.

Patrick, R. 1996. *Rivers of the United States.* Vol. 3, *The eastern and southeastern states.* John Wiley and Sons, New York.

Patrick, R., J. Cairns Jr., and S. S. Roback. 1967. An ecosystematic study of the fauna and flora of the Savannah River. *Proceedings of the Academy of Natural Sciences of Philadelphia* 118:109–407.

Ricketts, T. H., E. Dinerstein, D. M. Olson, C. J. Loucks, W. Eichbaum, D. DellaSala, K. Kavanaugh, P. Hedao, P. T. Hurley, K. M. Carney, R. Abell, and S. Walters. 1999. *Terrestrial ecoregions of North America: A conservation assessment.* Island Press, Washington, D.C.

Schmitt, D. N. 1988. *A fisheries survey of the Ogeechee River.* Georgia Department of Natural Resources, Game and Fish Division, Atlanta.

Schmitt, D. N., and J. H. Hornsby. 1985. *A fisheries survey of the Savannah River.* Final report, Project F-30-12. Georgia Department of Natural Resources, Game and Fish Division, Atlanta.

Sehlinger, B., and D. Otey. 1980. *A paddler's guide to Southern Georgia,* 2nd ed. Menasha Ridge Press, Birmingham, Alabama.

Smock, L. A., and E. Gilinsky. 1992. Coastal Plain blackwater streams. In C. T. Hackney, S. M. Adams, and W. H. Martin (eds.). *Biodiversity of the southeastern United States: Aquatic communities,* pp. 271–313. John Wiley and Sons, New York.

Smock, L. A., and D. J. Mitchell. 1991. *James River mainstem investigation: Macroinvertebrates.* Federal Aid in Fish Restoration Project F-74-R. Virginia Department of Game and Inland Fisheries, Richmond.

Snyder, J. A., G. C. Garman, and R. W. Chapman. 1996. Mitochondrial DNA variation in native and introduced populations of smallmouth bass, *Micropterus dolomieu. Copeia* 1996:995–998.

Specht, W., H. J. Kania, and W. Painter. 1984. *Annual report on the Savannah River Aquatic Ecology Program, September 1982–August 1983.* Vol. 2. Report ECS-SR-9. Environmental & Chemical Sciences, Inc., Aiken, South Carolina.

Sprague, L. A., M. J. Langland, S. E. Yochum, R. E. Edwards, J. D. Blomquist, S. C. Phillips, G. W. Shenk, and S. D. Preston. 2000. *Factors affecting nutrient trends in major rivers of the Chesapeake Bay watershed.* Water-Resources Investigations Report 00-4218. U.S. Geological Survey, Richmond, Virginia.

Sprenkle, E. S., L. A. Smock, and J. E. Anderson. 2004. Distribution and growth of submerged aquatic vegetation in the Piedmont section of the James River, Virginia. *Southeastern Naturalist* 3:517–530.

Stites, D. L. 1986. Secondary production and productivity in the sediments of blackwater rivers. Ph.D. diss., Emory University, Atlanta.

Swift, C. C., C. R. Gilbert, S. A. Bortone, G. H. Burgess, and R. W. Yerger. 1986. Zoogeography of the freshwater fishes of the southeastern United States: Savannah River to Lake Pontchartrain. In C. H. Hocutt and E. O. Wiley (eds.). *The zoogeography of North American freshwater fishes*, pp. 213–265. John Wiley and Sons, New York.

Tagatz, M. E. 1967. Fishes of the St. Johns River, Florida. *Quarterly Journal of the Florida Academy of Science* 30:25–50.

U.S. Army Corps of Engineers. 1981. *Water resources development in Georgia.* South Atlantic Division, U.S. Army Corps of Engineers, Atlanta, Georgia.

Warren, M. L. Jr., B. M. Burr, S. J. Walsh, H. L. Bart Jr., R. C. Cashner, D. A. Etnier, B. J. Freeman, B. R. Kuhajda, R. L. Mayden, H. W. Robinson, S. T. Ross, and W. C. Starnes. 2000. Diversity, distribution, and conservation status of the native freshwater fishes of the southern United States. *Fisheries* 25(10):7–29.

Webb, S. R., G. C. Garman, S. P. McIninch, and B. L. Brown. 2002. Amoebae associated with ulcerative lesions of fish from tidal freshwater of the James River, Virginia. *Journal of Aquatic Animal Health* 14:68–76.

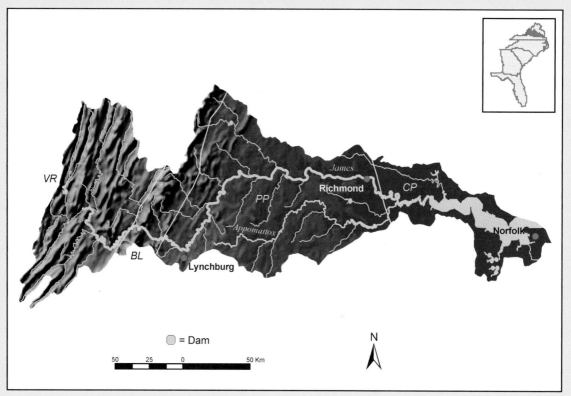

FIGURE 3.13 Map of the James River basin. Physiographic provinces are separated by yellow lines.

JAMES RIVER

Relief: 1250 m
Basin area: 26,164 km²
Mean discharge: 227 m³/s
River order: 7
Mean annual precipitation: 108 cm
Mean air temperature: 14°C
Mean water temperature: 16°C
Physiographic provinces: Valley and Ridge (VR), Blue Ridge (BL), Piedmont Plateau (PP), Coastal Plain (CP)
Biome: Temperate Deciduous Forest
Freshwater ecoregion: Chesapeake Bay
Terrestrial ecoregions: Appalachian/Blue Ridge Forests, Southeastern Mixed Forests, Middle Atlantic Coastal Forests
Number of fish species: 109
Number of endangered species: 3 fishes, 1 amphibian, 1 reptile, 6 mussels

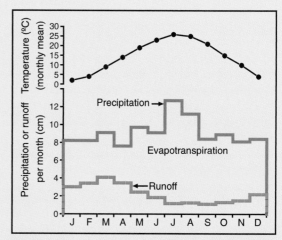

FIGURE 3.14 Mean monthly air temperature, precipitation, and runoff for the James River basin.

Major fishes: American eel, American shad, hickory shad, blueback herring, alewife, gizzard shad, common carp, bull chub, quillback sucker, satinfin shiner, spottail shiner, flathead catfish, blue catfish, white perch, striped bass, redbreast sunfish, bluegill, smallmouth bass, largemouth bass
Major other aquatic vertebrates: cottonmouth, water snakes, painted turtle, musk turtle, river cooter, red-bellied turtle, snapping turtle, muskrat, beaver
Major benthic invertebrates: mayflies (*Tricorythodes*, *Stenonema*, *Baetis*, *Caenis*), caddisflies (*Hydropsyche*, *Cheumatopsyche*, *Polycentropus*), hellgrammites (*Corydalus*), beetles (*Macronychus*), midges (*Rheotanytarsus*), crustaceans (*Orconectes*), bivalves (*Elliptio*, *Corbicula*, *Pisidium*), snails (*Elimia*, *Somatogyrus*)
Nonnative species: Asiatic clam, smallmouth bass, largemouth bass, rock bass, bluegill, flathead catfish, blue catfish, channel catfish, muskellunge, walleye, threadfin shad
Major riparian plants: sycamore, swamp black gum, river birch, American elm, red maple, ash-leaf maple, bald cypress
Special features: drains from four physiographic provinces, no large impoundments, class IV white-water rapids at Fall Line
Fragmentation: 12 low-head dams, some partially breached, that regulate flow in about 10% of nontidal river
Water quality: major pollutants are sediment, combined sewer-stormwater runoff, Kepone; pH = 7.4, alkalinity = 52 mg/L as $CaCO_3$, NO_3-N = 0.25 mg/L, PO_4-P = 0.02 mg/L
Land use: 71% forest, 23% agriculture, 6% urban
Population density: 96 people/km²
Major information sources: Jenkins and Burkhead 1994, Garman and Nielsen 1992, Patrick 1996, G. C. Garman personal communication, www.dcr.state.va.us/waterways/the_problem/watersheds_and_you/p_james_river_watershed.htm

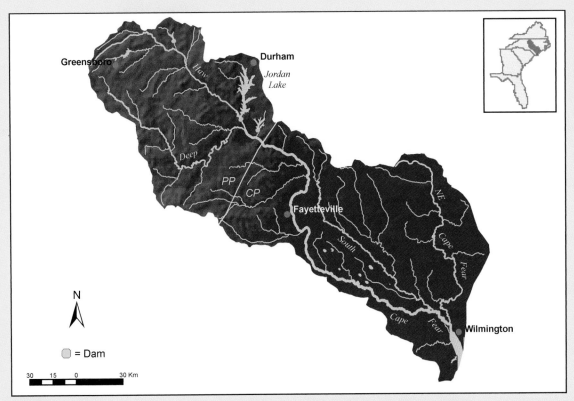

FIGURE 3.15 Map of the Cape Fear River basin. Physiographic provinces are separated by yellow line.

CAPE FEAR RIVER

Relief: 305 m
Basin area: 24,150 km²
Mean discharge: 217 m³/s
River order: 6
Mean annual precipitation: 119 cm
Mean air temperature: 16°C
Mean water temperature: 17°C
Physiographic provinces: Piedmont Plateau (PP), Coastal Plain (CP)
Biome: Temperate Deciduous Forest
Freshwater ecoregion: South Atlantic
Terrestrial ecoregions: Southeastern Mixed Forests, Middle Atlantic
 Coastal Forests
Number of fish species: 95
Number of endangered species: 8 fishes, 1 mammal, 15 mussels
Major fishes: American eel, American shad, hickory shad, blueback
 herring, gizzard shad, common carp, spotted sucker, shiners,
 darters, channel catfish, bluegill, pumpkinseed, largemouth bass, striped bass
Major other aquatic vertebrates: cottonmouth, water snakes, painted turtle, musk turtle, river cooter, slider, mud turtle,
 snapping turtle, muskrat, bull frog, river otter, beaver
Major benthic invertebrates: mayflies (*Tricorythodes, Caenis, Stenonema, Baetis*), caddisflies (*Hydropsyche, Cheumatopsyche*),
 damselflies (*Argia*), hellgrammites (*Corydalus*), beetles (*Stenelmis, Macronychus*), bivalves (*Elliptio, Corbicula, Sphaerium,
 Pisidium*), crustaceans (*Paleomonetes*)
Nonnative species: Asiatic clam, smallmouth bass, white crappie, flathead catfish, channel catfish
Major riparian plants: sycamore, sweetgum, swamp black gum, red maple, bald cypress, water tupelo, ashes, oaks
Special features: lower tributaries (Black River, Northeast Cape Fear River) are classic southeastern blackwater rivers with broad
 hardwood floodplains; Carolina Bay lakes in basin
Fragmentation: 3 locks and dams on main stem; large impoundment on primary tributary (Haw River)
Water quality: pH = 6.9, alkalinity = 29 mg/L as CaCO₃, NO₃-N = 0.83 mg/L, PO₄-P = 0.07 mg/L
Land use: 56% forest, 24% agriculture, 9% urban, 11% other
Population density: 69 people/km²
Major information sources: Hocutt et al. 1986, Garman and Nielsen 1992, Patrick 1996, NCDENR 2000

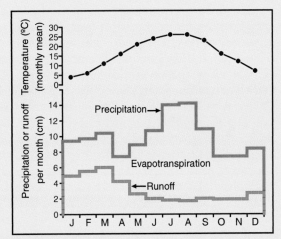

FIGURE 3.16 Mean monthly air temperature,
precipitation, and runoff for the Cape Fear River basin.

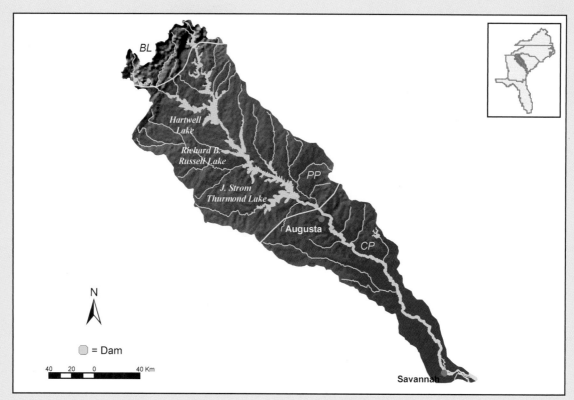

FIGURE 3.17 Map of the Savannah River basin. Physiographic provinces are separated by yellow lines.

SAVANNAH RIVER

Relief: 1743 m
Basin area: 27,414 km^2
Mean discharge: 319 m^3/s
River order: 7
Mean annual precipitation: 114 cm
Mean air temperature: 18°C
Mean water temperature: 17°C
Physiographic provinces: Blue Ridge (BL), Piedmont Plateau (PP), Coastal Plain (CP)
Biome: Temperate Deciduous Forest
Freshwater ecoregion: South Atlantic
Terrestrial ecoregions: Appalachian/Blue Ridge Forests, Southeastern Mixed Forests, Middle Atlantic Coastal Forests, Southeastern Conifer Forests
Number of fish species: 106
Number of endangered species: 7 fishes, 4 amphibians, 2 reptiles, 8 mussels, 3 crayfishes

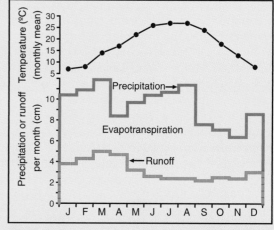

FIGURE 3.18 Mean monthly air temperature, precipitation, and runoff for the Savannah River basin.

Major fishes: bowfin, American eel, American shad, blueback herring, common carp, eastern silvery minnow, shiners, silver redhorse, spotted sucker, channel catfish, striped bass, black crappie, redbreast sunfish, bluegill, largemouth bass
Major other aquatic vertebrates: American alligator, cottonmouth, water snakes, snapping turtle, musk turtle, river cooter, slider, river frog, muskrat, river otter, beaver
Major benthic invertebrates: mayflies (*Baetis, Caenis, Isonychia, Stenonema, Tricorythodes*), caddisflies (*Cheumatopsyche, Chimarra, Hydropsyche, Macrostemum, Neureclipsis, Oecetis*), beetles (*Ancyronyx, Macronychus, Stenelmis*), midges (*Rheotanytarsus*), black flies (*Simulium*), crustaceans (*Hyalella*), bivalves (*Elliptio, Corbicula, Pisidium*), snails (*Physella*)
Nonnative species: hydrilla, waterweed, Asiatic clam, flathead catfish, channel catfish
Major riparian plants: bald cypress, water tupelo, swamp black gum, sweetgum, water hickory, red maple, sycamore, oaks
Special features: broad forested floodplain swamp throughout the Coastal Plain
Fragmentation: several large dams in Piedmont
Water quality: major pollutants are sediments and nutrients from nonpoint inputs; pH = 6.9, alkalinity = 20 mg/L as CaCO$_3$, NO$_3$-N = 0.26 mg/L, total PO$_4$-P = 0.05 mg/L
Land use: 65% forest, 22% agriculture, 4% urban, 9% other
Population density: 35 people/km^2
Major information sources: Swift et al. 1986, Garman and Nielsen 1992, Patrick 1996, GDNR 2001b

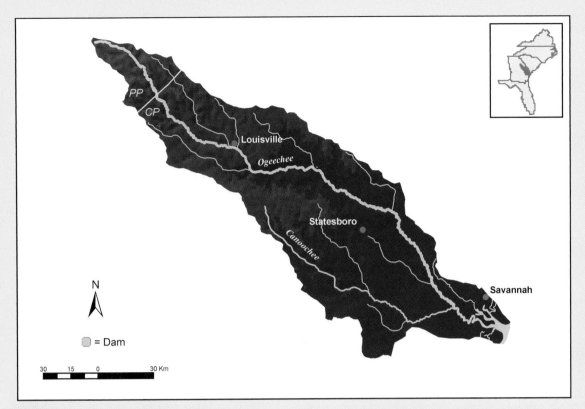

FIGURE 3.19 Map of the Ogeechee River basin. Physiographic provinces are separated by yellow line.

OGEECHEE RIVER

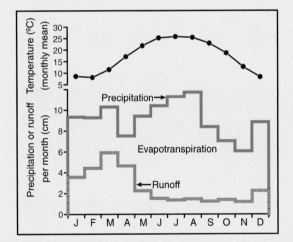

FIGURE 3.20 Mean monthly air temperature, precipitation, and runoff for the Ogeechee River basin.

Relief: 200 m
Basin area: 13,500 km²
Mean discharge: 115 m³/s
River order: 6
Mean annual precipitation: 113 cm
Mean air temperature: 18°C
Mean water temperature: 19°C
Physiographic provinces: Piedmont Plateau (PP), Coastal Plain (CP)
Biome: Temperate Deciduous Forest
Freshwater ecoregion: South Atlantic
Terrestrial ecoregions: Southeastern Mixed Forests, Southeastern Conifer Forests, Middle Atlantic Coastal Forests
Number of fish species: >80
Number of endangered species: 6 fishes, 2 amphibians, 2 reptiles, 1 mussel
Major fishes: American eel, longnose gar, bowfin, snail bullhead, redbreast sunfish, spotted sunfish, bluegill, largemouth bass, chain pickerel, spotted sucker, shiners, American shad, black crappie, warmouth, redear sunfish, white catfish, chubs, darters, silversides
Major other aquatic vertebrates: cottonmouth, water snakes, softshell turtles, river cooter, American alligator, river frog, sirens, treefrogs, dusky salamanders, muskrat, river otter, beaver
Major benthic invertebrates: caddisflies (*Hydropsyche, Cheumatopsyche, Hydroptila, Chimarra*), stoneflies (*Perlesta, Paragnetina, Taeniopteryx, Pteronarcys*), mayflies (*Stenonema, Baetis, Isonychia, Ephemerella, Tricorythodes, Caenis*), dragonflies (*Neurocordulia, Boyeria*), hellgrammites (*Corydalus*), beetles (*Stenelmis, Macronychus*), crustaceans (*Lirceus*)
Nonnative species: Asiatic clam
Major riparian plants: swamp black gum, water tupelo, bald cypress, red maple, water oak, laurel oak, sweetgum, water hickory
Special features: broad forested floodplain swamp (>1 km width); extensive submerged wood habitat; one of few natural free-flowing rivers in coterminous 48 states
Fragmentation: no major dams
Water quality: free of major pollutants; pH = 7.0, alkalinity = 23 mg/L as $CaCO_3$, NO_3-N = 0.10 mg/L, total P = 0.05 mg/L
Land use: 54% forest, 17% forested wetlands, 18% agriculture, 1% urban
Population density: 30 people/km²
Major information sources: Swift et al. 1986, Meyer 1990, 1992, Benke 2001, Benke et al. 2000, GDNR 2001a

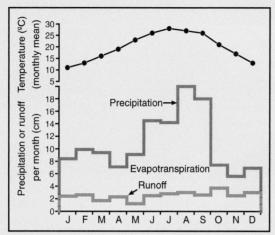

FIGURE 3.22 Mean monthly air temperature, precipitation, and runoff for the St. Johns River basin.

ST. JOHNS RIVER

Relief: 107 m
Basin area: 22,539 km^2
Mean discharge: 222 m^3/s
River order: 5
Mean annual precipitation: 131 cm
Mean air temperature: 20°C
Mean water temperature: 22°C
Physiographic province: Coastal Plain (CP)
Biome: Temperate Deciduous Forest
Freshwater ecoregion: Florida
Terrestrial ecoregions: Southeastern Conifer Forests, Florida Sand Pine Scrub
Number of fish species: >75 freshwater, 115 euryhaline
Number of endangered species: 1 mammal, 4 fishes, 2 reptiles, 2 birds
Major fishes: bluegill, redbreast sunfish, American eel, channel catfish, white catfish, largemouth bass, black crappie, American shad, striped bass
Major other aquatic vertebrates: cottonmouth, water snakes, snapping turtle, mud turtle, musk turtle, Florida cooter, bullfrog, treefrogs, American alligator, river otter, muskrat, beaver, West Indian manatee, brown pelican, herons, egrets, ibis

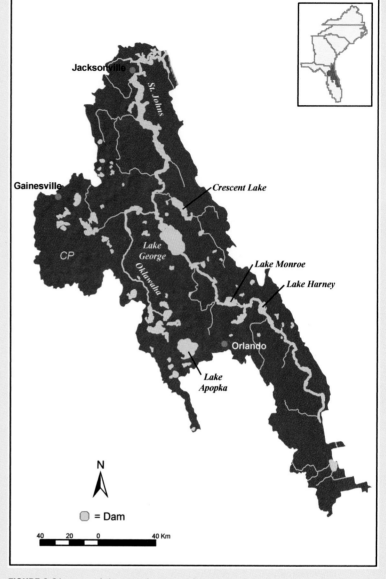

FIGURE 3.21 Map of the St. Johns River basin.

Major benthic invertebrates: mayflies (*Callibaetis floridanus*, *Stenacron floridense*), caddisflies (*Oecetis*, *Hydroptila*, *Orthotrichia*, *Cyrnellus*), midges (*Chironomus*, *Glyptotendipes*), oligochaete worms (*Limnodrilus*), bivalves (*Corbicula*, *Sphaerium*, *Pisidium*), snails (*Physella*), crustaceans (*Hyalella*, *Gammarus*, *Penaeus*, *Callinectes*, *Rangia*, *Rhithropanopeus*)
Nonnative species: water hyacinth, hydrilla, Eurasian watermilfoil, common salvinia, parrot feather, Asiatic clam, nutria, numerous other plant and animal species
Major riparian plants: maidencane, pickerelweed, arrowhead, sawgrass, cattail, rush, cordgrass, coastal plain willow
Special features: channel–lake geomorphology; broad marsh floodplain; tidal influence far upriver; extremely low gradient
Fragmentation: no major dams on main stem, dam on Oklawaha River; altered hydroperiod due to water diversions
Water quality: pH = 7.5; alkalinity = 95 mg/L as CaCO$_3$, NO$_3$-N = 0.25 mg/L, PO$_4$-P = 0.05 mg/L
Land use: 45% forest, 25% agriculture, 24% wetlands and water, 6% urban
Population density: 78 people/km^2
Major information sources: Bass 1991, DeMort 1991, Garman and Nielsen 1992

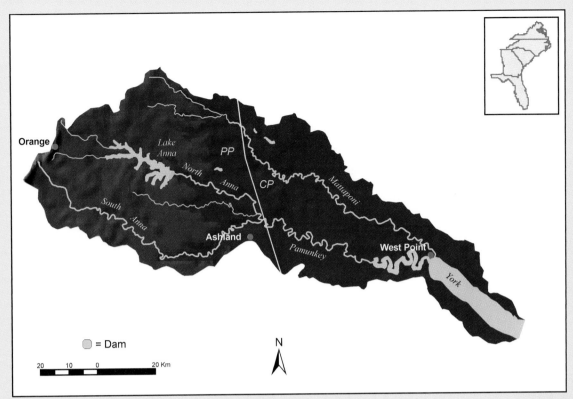

FIGURE 3.23 Map of the York River basin. Physiographic provinces are separated by yellow line.

YORK RIVER

Relief: 365 m
Basin area: 6892 km²
Mean discharge: 45 m³/s
River order: 6
Mean annual precipitation: 108 cm
Mean air temperature: 14°C
Mean water temperature: 15°C
Physiographic provinces: Piedmont Plateau (PP), Coastal Plain (CP)
Biome: Temperate Deciduous Forest
Freshwater ecoregion: Chesapeake Bay
Terrestrial ecoregions: Southeastern Mixed Forests, Middle Atlantic Coastal Forests
Number of fish species: 75
Number of endangered species: 2 fishes, 1 amphibian, 4 mussels
Major fishes: longnose gar, American eel, American shad, blueback herring, pickerels, blue catfish, white perch, striped bass, bluegill, redbreast sunfish, largemouth bass, yellow perch

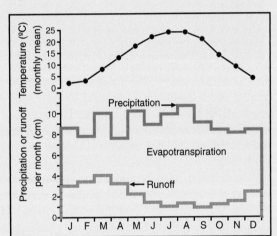

FIGURE 3.24 Mean monthly air temperature, precipitation, and runoff for the York River basin.

Major other aquatic vertebrates: northern water snake, brown water snake, painted turtle, musk turtle, mud turtle, red-bellied turtle, snapping turtle, muskrat, river otter, beaver
Major benthic invertebrates: mayflies (*Baetis, Stenonema, Eurylophella, Caenis*), caddisflies (*Cheumatopsyche, Hydropsyche, Chimarra, Hydroptila*), hellgrammites (*Corydalus*), beetles (*Macronychus*), midges (*Rheotanytarsus*), crustaceans (*Caecidotea*), mussels (*Elliptio, Pisidium, Corbicula*), snails (*Physella*)
Nonnative species: common carp, grass carp, channel catfish, blue catfish, white crappie, black crappie, bluegill, smallmouth bass, largemouth bass, spotted bass, Asiatic clam
Major riparian plants: sycamore, swamp black gum, river birch, American elm, red maple, ash-leaf maple, bald cypress
Special features: strong anadromous fish runs and intact historical anadromous fish spawning grounds
Fragmentation: 1 major dam on primary tributary (North Anna River)
Water quality: pH = 6.9, alkalinity = 18 mg/L as $CaCO_3$, NO_3-N = 0.28 mg/L, PO_4-P = 0.08 mg/L
Land use: 73% forest, 19% agriculture, 8% urban
Population density: 54 people/km²
Major information sources: Jenkins and Burkhead 1994, http://www.dcr.state.va.us/waterways/the_problem/watersheds_and_you/p_york_river.htm

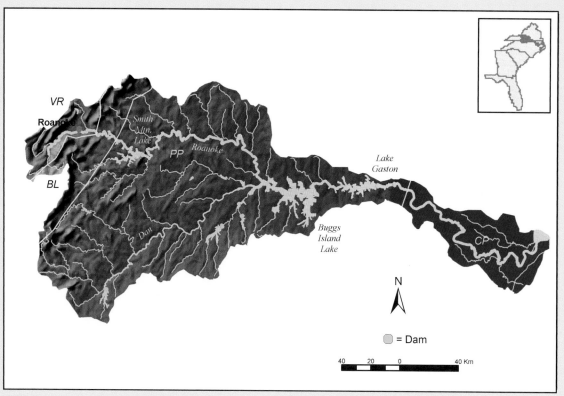

FIGURE 3.25 Map of the Roanoke River basin. Physiographic provinces are separated by yellow lines.

ROANOKE RIVER

Relief: 920 m

Basin area: 25,326 km²

Mean discharge: 232 m³/s

River order: 6

Mean annual precipitation: 108 cm

Mean air temperature: 14°C

Mean water temperature: 16°C

Physiographic provinces: Valley and Ridge (VR), Blue Ridge (BL), Piedmont Plateau (PP), Coastal Plain (CP)

Biome: Temperate Deciduous Forest

Freshwater ecoregion: South Atlantic

Terrestrial ecoregions: Appalachian/Blue Ridge Forests, Southeastern Mixed Forests, Middle Atlantic Coastal Forests

Number of fish species: 119

Number of endangered species: 9 fishes, 2 amphibian, 7 mussels

Major fishes: American eel, American shad, hickory shad, blueback herring, alewife, gizzard shad, redhorses, shiners, darters, striped bass, white perch, redear sunfish, bluegill, smallmouth bass, largemouth bass, Roanoke bass, black crappie, yellow perch

Major other aquatic vertebrates: cottonmouth, water snakes, painted turtle, musk turtle, river cooter, sliders, mud turtle, snapping turtle, bull frog, muskrat, river otter, beaver

Major benthic invertebrates: mayflies (*Stenonema*, *Baetis*), caddisflies (*Hydropsyche*, *Cheumatopsyche*, *Polycentropus*), hellgrammites (*Corydalus*), beetles (*Macronychus*), midges (*Rheotanytarsus*), damselflies (*Argia*), crustaceans (*Orconectes*), bivalves (*Elliptio*, *Corbicula*, *Pisidium*), snails (*Elimia*, *Somatogyrus*)

Nonnative species: Asiatic clam, smallmouth bass, rock bass, bluegill, flathead catfish, channel catfish, threadfin shad, walleye

Major riparian plants: sycamore, swamp black gum, American elm, red maple, ash-leaf maple, bald cypress, water tupelo, green ash, water ash, swamp chestnut oak

Special features: floodplain in the Coastal Plain supports one of the largest tracts of intact and largely undisturbed bottomland hardwood forests on the Atlantic coast; six endemic species of fishes occur in the basin

Fragmentation: strongly fragmented by large dams on main stem

Water quality: pH = 6.8, alkalinity = 32 mg/L as CaCO₃, NO₃-N = 0.16 mg/L, PO₄-P = 0.03 mg/L

Land use: 68% forest, 25% agriculture, 3% urban, 4% other

Population density: 31 people/km²

Major information sources: Jenkins and Burkhead 1994, NCDENR 2001

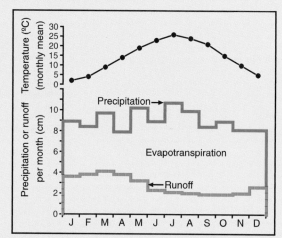

FIGURE 3.26 Mean monthly air temperature, precipitation, and runoff for the Roanoke River basin.

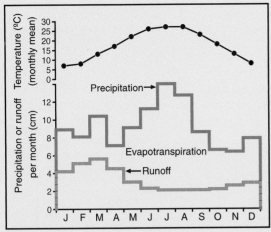

FIGURE 3.28 Mean monthly air temperature, precipitation, and runoff for the Great Pee Dee River basin.

GREAT PEE DEE RIVER

Relief: 1090 m
Basin area: 27,560 km²
Mean discharge: 371 m³/s
River order: 7
Mean annual precipitation: 111 cm
Mean air temperature: 17°C
Mean water temperature: 17°C
Physiographic provinces: Blue Ridge (BL), Piedmont Plateau (PP), Coastal Plain (CP)
Biome: Temperate Deciduous Forest
Freshwater ecoregion: South Atlantic
Terrestrial ecoregions: Appalachian/Blue Ridge Forests, Southeastern Mixed Forests, Middle Atlantic Coastal Forests
Number of fish species: 101
Number of endangered species: 6 fishes, 1 reptile, 8 mussels (North Carolina only)
Major fishes: bowfin, American eel, American shad, blueback herring, gizzard shad, common carp, eastern silvery minnow, shiners, silver redhorse, channel catfish, striped bass, white perch, redbreast sunfish, bluegill, redear sunfish, largemouth bass, black crappie
Major other aquatic vertebrates: cottonmouth, northern water snake, brown water snake, mud turtle, snapping turtle, musk turtle, river cooter, slider, river frog, bull frog, muskrat, river otter, beaver
Major benthic invertebrates: mayflies (*Baetis, Caenis, Stenonema, Tricorythodes*), caddisflies (*Cheumatopsyche, Chimarra, Hydropsyche, Oecetis*), beetles (*Macronychus, Stenelmis*), midges (*Rheotanytarsus*), black flies (*Simulium*), bivalves (*Corbicula, Pisidium*)
Nonnative species: Asiatic clam, flathead catfish, channel catfish
Major riparian plants: bald cypress, water tupelo, swamp black gum, sweetgum, water hickory, red maple, oaks
Special features: broad hardwood floodplain forests in Coastal Plain; Lumber River tributary is National Wild and Scenic River; Lumber and Lynches rivers are among few natural free-flowing rivers in coterminous 48 states
Fragmentation: several dams in Piedmont
Water quality: major pollutants are sediments and nutrients from nonpoint inputs; pH = 6.9, alkalinity = 22 mg/L as $CaCO_3$, NO_3-N = 0.46 mg/L, PO_4-P = 0.04 mg/L
Land use: 58% forest, 28% agriculture, 8% urban, 6% other
Population density: 49 people/km²
Major information source: Hocutt et al. 1986

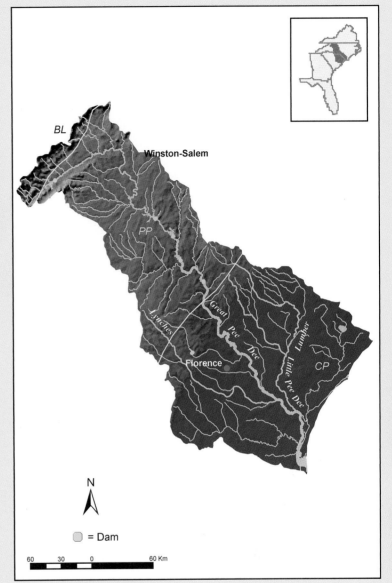

FIGURE 3.27 Map of the Great Pee Dee River basin. Physiographic provinces are separated by yellow lines.

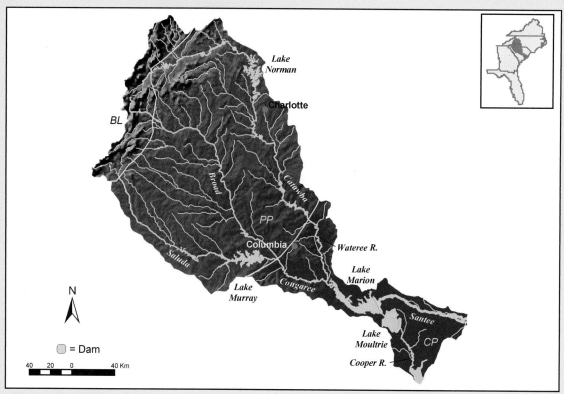

FIGURE 3.29 Map of the Santee River basin. Physiographic provinces are separated by yellow lines.

SANTEE RIVER

Relief: 1789 m

Basin area: 39,500 km^2

Mean discharge: 434 m^3/s

River order: 7

Mean annual precipitation: 125 cm

Mean air temperature: 17°C

Mean water temperature: 18°C

Physiographic provinces: Blue Ridge (BL), Piedmont Plateau (PP), Coastal Plain (CP)

Biome: Temperate Deciduous Forest

Freshwater ecoregion: South Atlantic

Terrestrial ecoregions: Appalachian/Blue Ridge Forests, Southeastern Mixed Forests, Middle Atlantic Coastal Forests

Number of fish species: 125

Number of endangered species: 5 fishes, 2 reptiles

Major fishes: bowfin, American eel, American shad, blueback herring, gizzard shad, spotted sucker, shiners, channel catfish, striped bass, black crappie, redbreast sunfish, redear sunfish, bluegill, largemouth bass

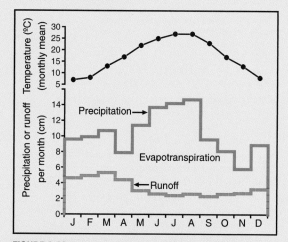

FIGURE 3.30 Mean monthly air temperature, precipitation, and runoff for the Santee River basin.

Major other aquatic vertebrates: cottonmouth, water snakes, snapping turtle, mud turtle, musk turtle, river cooter, slider, river frog, bullfrog, muskrat, river otter, beaver

Major benthic invertebrates: mayflies (*Baetis, Caenis, Stenonema, Tricorythodes*), caddisflies (*Cheumatopsyche, Chimarra, Hydropsyche, Oecetis*), beetles (*Ancyronyx, Macronychus, Stenelmis*), midges (*Rheotanytarsus*), black flies (*Simulium*), bivalves (*Corbicula, Sphaerium*), snails (*Physella*)

Nonnative species: Asiatic clam, flathead catfish, channel catfish, white crappie, hydrilla

Major riparian plants: sycamore, bald cypress, water tupelo, swamp black gum, sweetgum, red maple, hickories, oaks

Special features: second largest river basin on Atlantic coast of United States

Fragmentation: many dams on main stem and major tributaries

Water quality: pH = 6.9, alkalinity = 23 mg/L as CaCO$_3$, NO$_3$-N = 0.2 mg/L, total PO$_4$-P = 0.03 mg/L

Land use: 64% forest, 26% agriculture, 6% urban, 4% other

Population density: 65 people/km^2

Major information sources: Hocutt et al. 1986, Patrick 1996

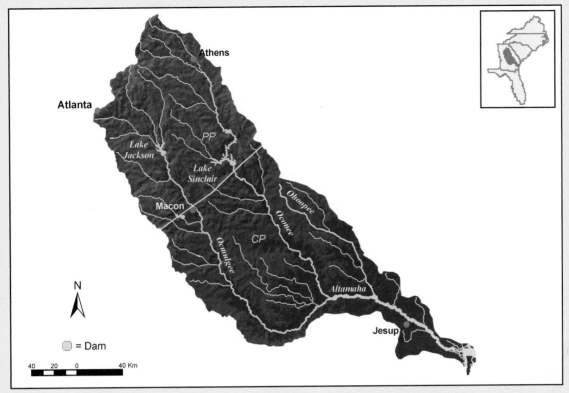

FIGURE 3.31 Map of the Altamaha River basin. Physiographic provinces are separated by yellow line.

ALTAMAHA RIVER

Relief: 372 m

Basin area: 37,600 km^2

Mean discharge: 393 m^3/s

River order: 7

Mean annual precipitation: 130 cm

Mean air temperature: 19°C

Mean water temperature: 20°C

Physiographic provinces: Piedmont Plateau (PP), Coastal Plain (CP)

Biome: Temperate Deciduous Forest

Freshwater ecoregion: South Atlantic

Terrestrial ecoregions: Southeastern Mixed Forests, Southeastern Conifer Forests

Number of fish species: 93

Number of endangered species: 1 mammal, 12 fishes, 2 amphibians, 2 reptiles, 7 mussels, 1 crayfish

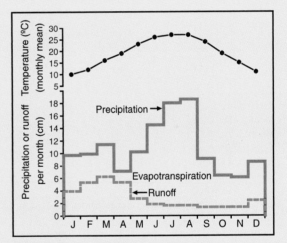

FIGURE 3.32 Mean monthly air temperature, precipitation, and runoff for the Altamaha River basin.

Major fishes: bowfin, American eel, American shad, gizzard shad, chain pickerel, carp, minnows, shiners, silver redhorse, spotted sucker, carpsucker, bullhead catfish, channel catfish, black crappie, bluegill, redear sunfish, warmouth, largemouth bass, hogchoker

Major other aquatic vertebrates: cottonmouth, water snakes, softshell turtles, river cooter, American alligator, river frog, sirens, treefrogs, dusky salamanders, West Indian manatee, muskrat, river otter, beaver

Major benthic invertebrates: caddisflies (*Hydropsyche, Cheumatopsyche, Macrostemum, Chimarra, Neureclipsis*), mayflies (*Isonychia, Stenonema, Baetis, Tricorythodes, Caenis*), stoneflies (*Paragnetina*), dragonflies (*Neurocordulia, Boyeria*), damselflies (*Argia*), hellgrammites (*Corydalus*), beetles (*Stenelmis*), midges (*Polypedilium, Rheotanytarsus*), black flies (*Simulium*), bivalves (*Corbicula, Elliptio*)

Nonnative species: Asiatic clam, flathead catfish

Major riparian plants: swamp black gum, water tupelo, bald cypress, water hickory, red maple, sweetgum, oaks

Special features: broad forested floodplain swamp; designated as a "Bioreserve" by the Nature Conservancy

Fragmentation: dams in Piedmont portions of two primary tributaries (Ocmulgee and Oconee rivers); no main-stem dams

Water quality: pH = 7.2; alkalinity = 30 mg/L as CaCO$_3$, total NO$_3$-N = 0.35 mg/L, total P = 0.09 mg/L

Land use: 64% forest, 26% agriculture, 3% urban, 7% other

Population density: 28 people/km^2

Major information sources: Hottell et al. 1983, Swift et al. 1986

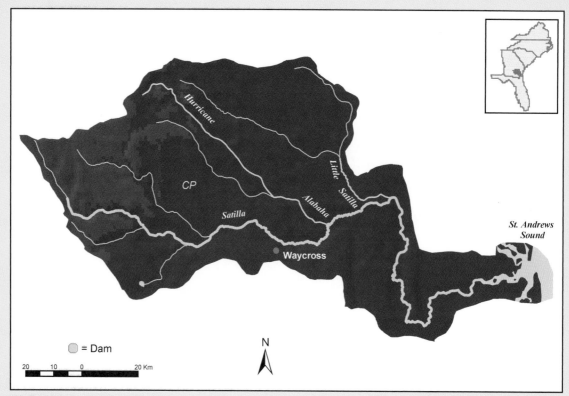

FIGURE 3.33 Map of the Satilla River basin.

SATILLA RIVER

Relief: 107 m
Basin area: 9143 km^2
Mean discharge: 65 m^3/s
River order: 6
Mean annual precipitation: 126 cm
Mean air temperature: 19°C
Mean water temperature: 20°C
Physiographic province: Coastal Plain (CP)
Biome: Temperate Deciduous Forest
Freshwater ecoregion: Florida
Terrestrial ecoregion: Southeastern Conifer Forests
Number of fish species: 52
Number of endangered species: 2 fishes, 1 amphibian, 2 reptiles,
   1 mussel

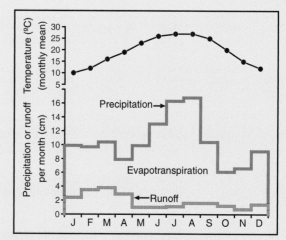

FIGURE 3.34 Mean monthly air temperature, precipitation, and runoff for the Satilla River basin.

Major fishes: bowfin, American eel, chain pickerel, spotted sucker, channel catfish, yellow bullhead, black crappie, bluegill, redbreast sunfish, warmouth, largemouth bass, brook silverside, eastern mosquitofish, topminnow, pirate perch, spotted sunfish, minnows, darters
Major other aquatic vertebrates: cottonmouth, water snakes, softshell turtles, river cooter, loggerhead musk turtle, American alligator, bullfrog, treefrogs, dusky salamanders, river otter, muskrat, beaver, West Indian manatee
Major benthic invertebrates: caddisflies (*Hydropsyche, Cheumatopsyche, Macrostemum, Chimarra*), stoneflies (*Perlesta, Acroneuria*), mayflies (*Stenonema*), dragonflies (*Neurocordulia, Boyeria*), hellgrammites (*Corydalus*), beetles (*Stenelmis, Ancyronyx*), black flies (*Simulium*), midges (*Rheotanytarsus, Polypedilium, Rheosmittia*)
Nonnative species: flathead catfish
Major riparian plants: swamp black gum, water tupelo, bald cypress, water hickory, river birch, black willow, red maple
Special features: broad forested floodplain swamp; one of few natural free-flowing rivers in coterminous 48 states
Fragmentation: no dams on main stem; proposed titanium mining in floodplain near Atkinson
Water quality: pH = 5.8; alkalinity = 4 mg/L as CaCO$_3$, NO$_3$-N = 0.19 mg/L, PO$_4$-P = 0.05 mg/L
Land use: 72% forest, 26% agriculture, 1% urban, 1% other
Population density: 11 people/km^2
Major information sources: Benke, van Arsdall et al. 1984, Benke, Henry et al. 1985, Swift et al. 1986, Meyer 1992

4

GULF COAST RIVERS OF THE SOUTHEASTERN UNITED STATES

G. MILTON WARD   PHILLIP M. HARRIS   AMELIA K. WARD

INTRODUCTION

MOBILE RIVER

CAHABA RIVER

APALACHICOLA–CHATTAHOOCHEE–FLINT
   RIVER SYSTEM

PEARL RIVER

SUWANNEE RIVER

ADDITIONAL RIVERS

ACKNOWLEDGMENTS

LITERATURE CITED

INTRODUCTION

The river basins of the eastern Gulf Coast lie west of the Atlantic slope and east of the Mississippi River (Fenneman 1938). The climate of the region is moderate, with warm summers, mild winters, and abundant rainfall. The region has abundant water resources, including seven major rivers that arise and flow through five physiographic provinces in five states to empty into the Gulf of Mexico (Fig. 4.2). Eastern Gulf Coast rivers encompass a rich variety of aquatic habitats and resources. In addition to the many upland streams and large rivers, there are large swamps, such as the Okefenokee Swamp and the Mobile River Delta, wide floodplain swamps, and oxbow lakes, which occur along all of the major rivers and many of the smaller coastal plain rivers. Limestone springs derived from the Floridan aquifer arise in southwestern Georgia, southeastern Alabama, and northern Florida. Today, the main stems of many major rivers of the eastern Gulf are

severely fragmented by numerous hydroelectric and navigation dams. Their main-stem channels have been deepened, and many riverine fauna have been replaced by lentic (reservoir) fauna (Ward, Ward, and Harris 1992). At the lower end of many basins are moderate to large estuaries, such as the Apalachicola, Mobile, Choctawhatchee, and Pearl, which are now, or were at one time, important centers of commercial and recreational fisheries.

Humans have inhabited these river basins for at least 12,000 years. Although hunter-gatherer societies occupied much of this time, the more sedentary lifestyle of the Mississippian culture dominated the southeastern region from 900 to 1500, reaching a peak between 1200 and 1400 (Jenkins and Krause 1986). By the time of the arrival of the first Europeans, the expansive Mississippian culture was already in decline, largely attributed to population growth that outstripped available resources, intense intertribal competition, and climate change (Scarry 1996). The important tribes in the region were the Creek (Upper and Lower), Choctaw, Chickasaw,

FIGURE 4.1 Tensaw Delta, a distributary of the lower Mobile River, northeast of Mobile, Alabama (PHOTO BY BETH MAYNOR YOUNG).

125

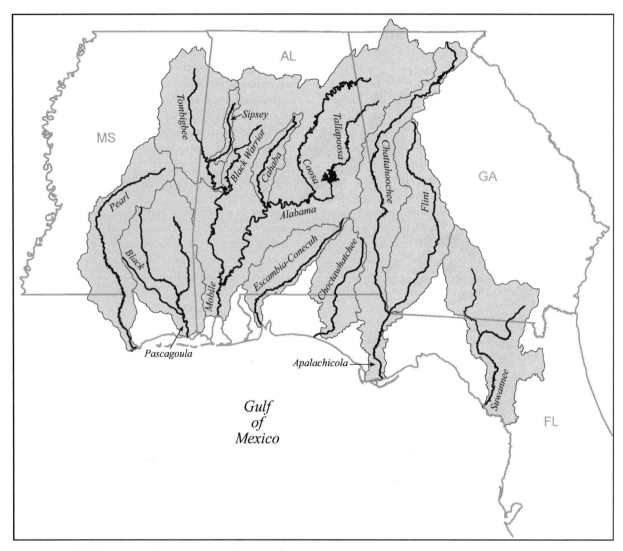

FIGURE 4.2 Gulf Coast rivers of the southeastern United States that are covered in this chapter.

Cherokee, Apalachee, and Timucua. The first European exploration was by the Spanish, and later colonization was by the Spanish, French, and English, who all partnered with Native Americans to vie for control of important lands, especially commercial and military corridors such as rivers and bays. After the war with the French in 1756, the English gained official control after the Treaty of Paris ceded land east of the Mississippi to England in 1763 (Jackson 1995).

Of the seven major rivers that reach the eastern Gulf of Mexico, four will be described in some detail: the Suwannee, Apalachicola, Mobile, and Pearl. A Mobile River tributary, the Cahaba River, also will be covered in detail. Abbreviated descriptions of physical and biological characteristics are given for the remaining three rivers that reach the Gulf of Mexico (Pascagoula, Escambia–Conecuh, and Choctawhatchee), as well as two other tributaries of the Mobile River (Upper Tombigbee and Sipsey) and one of the Apalachicola River (Flint).

Physiography and Climate

The dominant physiographic province through which the rivers flow is the Coastal Plain (Fenneman 1938), more specifically the sections named the Eastern Gulf Coastal Plain and the Floridan. The region includes catchments from the Pearl River basin in the west to the Suwannee River in the east, extending 160 to 320 km inland from the present coastline and the same distance offshore in the Gulf of Mexico as the continental shelf. The Coastal Plain is an elevated sea bottom with relief less than 150 m (Hunt 1967). In general, the formation dips seaward, and progressively younger sediments of Cretaceous, Tertiary, and Quaternary age trend north to south. During the last 60 million years the shores of the Gulf of Mexico have advanced and receded from what is now the northern edge of the Coastal Plain south to its present location (Hunt 1967). The Cretaceous-age shoreline that now separates the Eastern Gulf Coastal Plain from the Appalachian Highlands is a crescent-shaped boundary known as the Fall Line. In the eastern Gulf region it stretches from mid-Georgia through Alabama, intersecting the cities of Macon, Columbus, Montgomery, and Tuscaloosa before it turns northward into northeastern Mississippi. The Fall Line is characterized by an abrupt change in river channel gradient, as either falls or shoals, making further upstream commercial navigation impossible. Now, however, these shoals and falls on the larger rivers are submerged beneath reservoirs.

The Piedmont, Appalachian Plateau, and Valley and Ridge physiographic provinces lie upgradient of the Fall Line. The Appalachian Plateau is an area of deeply dissected terrain, where incised canyons and steep slopes have formed. Surface rocks are mostly sandstone of Pennsylvanian age and are often coal-bearing (Adams et al. 1926). The Piedmont is a gently rolling upland underlain with igneous and metamorphic crystalline rocks of Paleozoic age. Rock types are mainly schist, gneiss, granite, and quartzite. The Valley and Ridge is characterized by folded terrain trending northeast to southwest. Underlying rocks are carbonate overlain by sandstone (Adams et al. 1926). Carbonate is often exposed in valley bottoms where erosion has removed the overlying sandstone caps. The headwaters of the Mobile and Apalachicola rivers arise in the Blue Ridge province. However, these very steep terrains comprise <1% of the basins.

The majority of soils of the eastern Gulf Coast river basins are classified as ultisols. Such soils lie south of the advance of the last glaciation and develop in humid climates with tropical to subtropical temperatures and under forested or forested/grassland conditions. They are characterized as intensely weathered, low in basic cations (therefore acidic), and with clay deposition in the B horizon. Indeed, the red hills, so prominent in outcrops across the south, are exposed B horizons. In an earlier classification system, these soils were identified as Yellow and Red Podzolics, containing high concentrations of iron. Soils vary in depth, but

typically develop over crystalline rock, such as sandstone, shale, limestone, granite, gneiss, and schist. Coastal Plain sediments, in some places up to 1219 m thick, originated principally from recent marine sediments, with sand, sandy limestone, and clays predominating. Alluvial soils border all the larger Coastal Plains river channels.

The climate of the eastern Gulf region is generally warm and humid, with long summers and mild winters. River basins span 4 to 5 degrees of latitude, generating a north–south temperature gradient with coolest average annual temperatures in the mountains of northern Georgia (15°C) and warmest temperatures near the Gulf Coast (21°C). Rainfall is abundant, with greater precipitation along and near the coast. Average annual rainfall is 150 to 165 cm along the Gulf Coast but declines inland to 120 to 135 cm. Drier inland areas occur in central Georgia, whereas wetter areas are westward into Mississippi. Orographic effects in the Blue Ridge Mountains of northern Georgia result in higher precipitation (150 cm) than is the case in the lower elevations of the Piedmont (Couch et al. 1996).

Basin Landscape and Land Use

All basins described in this chapter are in the Temperate Deciduous Forest biome, where abundant rainfall and high humidity support a heavily forested region. Rivers drain four terrestrial ecoregions (Ricketts et al. 1999) that are dominated by Southeastern Mixed Forests and Southeastern Conifer Forests. The Southeastern Conifer Forests extend along the Gulf Coast from southeastern Louisiana to the Florida peninsula. Once composed mostly of longleaf pine that was maintained by understory fires, these forests are now dominated by mixed hardwoods and pine. Fire suppression, agriculture, and timbering have led to this shift in species composition. Northward lies the Southeastern Mixed Forests. Dominated by oak, hickory, and pine, this ecoregion extends in a band across Mississippi, Alabama, and Georgia and then further northeastward to Maryland. The northern area of the Mobile River basin west of the Appalachian Mountains contains vegetation of the Appalachian Mixed Mesophytic Forests. Lower-elevation forests are dominated by oaks, hickories, and magnolia, but forests may contain more than 30 canopy species. The northeastern portions of the Mobile River and the upper Apalachicola basin contain deciduous vegetation of the Appalachian/Blue Ridge Forests. As a result of a great diversity of soils, topography, climate, and long-term geologic stability, these forests contain some of the most species-rich faunal and floral communities of any of the world's temperate forests. Depending on elevation, vegetation is dominated either by mixed oaks or spruce forests.

A common misperception among many who have not traveled in the southeastern United States is that the landscape is dominated by plantations and row crops. Although there are some areas that are heavily agricultural, most of the eastern Gulf Coast rivers flow through forested or wetland landscapes. The major basins vary in forested land cover from 68% in the Mobile drainage to 38% in the Suwannee (G. M. Ward, unpublished data). Timber production is a major industry throughout the south, which often dictates forest composition. The Mobile River basin is dominated by deciduous and mixed deciduous forests, whereas the Pearl River and rivers flowing through Georgia and Florida drain land covered mainly by evergreen and mixed forests. Forested wetlands also constitute an important land-cover type in the Suwannee, Pearl, and Mobile river basins, which contain wetlands dominated by woody vegetation that constitute 22%, 12%, and 10%, respectively, of land cover in the basins. These percentages are mostly driven by the sizes of the basins, particularly the amount of upland landscape present in the catchments.

Most agriculture occurs in the coastal plain region's eastern Gulf Coast river catchments. The deep, fertile soils and abundant rainfall of the area support row crops and hay/pasture cover types. Recently constructed land-use coverage data from the 1990s (Vogelmann et al. 2001) show that, on average, basins in the eastern Gulf Coast region contain ~25% agricultural land. The minimum percentage occurs in the Mobile River basin, which has only 18% agricultural land cover. Row crops are more abundant than hay/pasture lands from the Choctawhatchee River east (16% to 22% versus 8% to 11%), whereas hay/pasture is more prevalent from the Escambia River west to the Pearl River (G. M. Ward, unpublished data).

Urban land cover is not abundant in any basin as a percentage of total area, but the Apalachicola and Mobile rivers contain sizeable urban centers. The city of Atlanta lies near the headwaters of both rivers, and Birmingham lies in the heart of the Mobile River basin. Population pressures on these two rivers are dramatic. In addition to water-quality problems caused by sediment and domestic effluents, both rivers are now part of a legal battle over water-allocation rights. Atlanta has joined other cities in

more arid regions of the world as an area that has outstripped its capacity to provide enough water to support growth. The proposals by Atlanta to remove substantially more water from the Chattahoochee and Coosa rivers than is consistent with current agreements has prompted the threat of litigation by Alabama and Florida. Removal of water from the Chattahoochee River threatens the commercial fisheries industry in the Apalachicola estuary in Florida, whereas the removal of water from the Coosa River (upper Mobile River basin) threatens to stifle future growth in Alabama.

The Rivers

There are five freshwater ecoregions within the eastern Gulf of Mexico region, arranged by longitude. These include portions of the Mississippi Embayment and Florida ecoregions, all of the Mobile Bay and Apalachicola ecoregions, and much of the Florida Gulf ecoregion. Two of these, the Mobile Bay and the Apalachicola, follow watershed boundaries of the Mobile and Apalachicola river basins, whereas the remaining three ecoregions include multiple river basins.

The eleven rivers covered in this chapter vary widely in size, hydrology, geomorphology, and water chemistry, but experience a similar climate and exhibit many biological similarities. All rivers in this region flow from north to south into the Gulf of Mexico, and all are located between 84°W and 91°W longitude and 30°N and 35°N latitude (see Fig. 4.2). The seven rivers that reach the Gulf of Mexico drain over 265,000 km^2, with an approximate combined flow of 4000 m^3/s. These seven basins range in size from 11,000 to 111,000 km^2 and in discharge from 200 to 1914 m^3/s. All 11 basins covered in this chapter lie, either completely or substantially, in Coastal Plain sediments. The two largest river basins in the region extend northward to the Appalachian Highlands and Tennessee River divide, whereas all others arise in the Coastal Plain (see Fig. 4.2).

The Mobile and Apalachicola rivers drain high-gradient mountain and upland areas in Tennessee, Georgia, and Alabama before crossing into the gently rolling topography of the Coastal Plain. These rivers flow through a diversity of geologic material of variable geologic age and variable physiography. The rivers here are ancient. Thus, the upper portions of the two largest basins are much older than the southern parts. The Coastal Plain portions of the eastern Gulf rivers became exposed at approximately the same time period. Therefore, they are of similar ages and possess many physiographic characteristics in common. The rock types in the upland areas were formed hundreds of millions of years ago (Lineback 1973). It is likely that a number of the rivers in the region that traverse these upland areas are much more than 25 million years old and, therefore, older than the oldest extant lake in the world, Lake Baikal (Martens 1997). As a result of climate, age, and geologic and physiographic diversity, the flowing waters of the eastern Gulf are one of the most biologically diverse regions in North America (Ward, Ward, and Harris 1992, Lydeard and Mayden 1995).

The upper portions of the Apalachicola and Mobile basins flow through channels with bedrock, boulder, and cobble substrata, whereas Coastal Plain portions of all rivers are low gradient and slow flowing, with sandy substrata and substantial amounts of woody debris. Several rivers have blackwater streams, for example, the Escatawpa (a tributary of the Pascagoula River) and other small, coastal rivers not further described (e.g., the Blackwater River in Florida).

The signature biological characteristic of all basins is the breadth of biological diversity. This diversity is most evident in the fishes, mollusks, crayfishes, and caddisflies. Explanations note that both the Mobile and Apalachicola basins are relatively large, geologically old, and physiographically and geologically diverse. The region has not been glaciated and may at one time have had hydrological connections to what is now the Tennessee drainage.

All the river systems described here are substantially impacted by human activities and face continuing threats. Throughout the history of the eastern Gulf Coast region, rivers have been used for transportation, food, and hydropower generation (mechanical and electrical), as well as for domestic, agricultural, and industrial water supply. During the 1800s river corridors were a major component of the economic system, primarily for transport of cotton to seaports such as Mobile, Alabama, and Apalachicola, Florida (Willoughby 1999). Upstream travel was halted at the Fall Line (e.g., Falls of the Chattahoochee), the boundary between the Coastal Plain and upland regions, because of the occurrence of shoals and waterfalls. Cities developed near the Fall Line as initial transportation portals to downstream coastal ports.

The Mobile and Apalachicola rivers are highly fragmented, both along their main stems and in their major tributaries, because of the historic emphasis on transportation and use of the rivers for other economic purposes. Since colonial times the energy of the rivers has been harnessed to produce power for industrial and commercial uses. During the antebellum period, wing dams were often constructed to divert water through factories to drive machinery in grist mills and iron foundries. The first dam for hydroelectric power generation in the Mobile River basin was completed in 1914 (Owen 1949) and was soon followed by one on the Apalachicola River (Willoughby 1999). Improvements for navigation in the Mobile River basin, authorized by Congress in 1874, came to fruition in 1895 with completion of a lock and dam on the Black Warrior River (Mettee et al. 1989).

Today, the main stems of virtually all the major rivers of the eastern Gulf are severely fragmented by numerous dams for hydroelectric power and navigation. Their main-stem channels have been deepened, and many riverine fauna have been replaced by lentic (reservoir) fauna (Ward, Ward, and Harris 1992). Coastal Plain rivers are, however, less fragmented, but only the Sipsey, Choctawhatchee, Pascagoula, and Suwannee are free-flowing, and within the Mobile River basin the Cahaba River contains only one small impoundment in the most upgradient headwaters.

MOBILE RIVER

"Opposite this bluff, on the other side of the river, is a district of swamp or lowland, the richest I ever saw, or perhaps anywhere to be seen: as for the trees, I shall forbear to describe them, because it would appear incredible; let it suffice to mention that the Cypress, Ash, Platanus, Liquidambar, and others are by far the tallest, straightest, and every way the most enormous that I have seen or heard of."
William Bartram, describing the Mobile River Delta, north of Mobile Bay, in *Travels through North & South Carolina, Georgia, East and West Florida* (1791; Doren 1928).

The Mobile River system, which drains land in Alabama, northwestern Georgia, southeastern Tennessee, and northeastern Mississippi, is the largest basin in the eastern Gulf region (Fig. 4.12) and one of the largest in North America (see Table 1.1). The headwaters lie in the mountains of northern Georgia, northern Alabama, and the Coastal Plain of northeastern Mississippi. The river drains south through forests and agricultural lands, eventually flowing through a large forested swamp delta north of the city of Mobile, Alabama (Fig. 4.1). The river then flows into Mobile Bay before discharging into the Gulf of Mexico.

The Mobile River system is composed of seven major rivers (see Fig. 4.12). Officially, only the 80 km reach north of Mobile Bay is known as the Mobile River, formed by the convergence of the Alabama and the Tombigbee rivers. The Tombigbee River receives two major tributary rivers, the Black Warrior River and the Sipsey River. The Alabama River has three major tributaries, including the Cahaba River, which enters near Selma, Alabama, as well as the Tallapoosa and Coosa rivers, which meet near Wetumpka, Alabama. Seventy-one percent of the basin lies in Alabama, 14% in Mississippi, 13% in Georgia, and 2% in Tennessee.

European exploration of the Mobile River basin began in the mid-sixteenth century. From 1539 to 1543, Spanish explorer Hernando De Soto led a 5635-km trek that originated in northern Florida and followed many rivers throughout the southeastern region. The journey eventually contributed to the earliest known map that detailed interior regions of North America, including locations of major rivers (Clayton et al. 1993a, 1993b; Hudson 1993; Galloway 1997). At the time, over 30 Native American groups were documented that occupied villages clustered along major waterways in Georgia, Alabama, Mississippi, and Florida (Swanton 1953). Many river names in the region are derived from the languages of these Native American groups (Read 1984).

During the latter part of the seventeenth and eighteenth centuries the French established forts within much of the basin, typically in strategic areas along the Tombigbee and Alabama rivers, in order to secure trading routes. In 1702, Pierre LeMoyne d'Iberville established a settlement on the northwest shoreline of Mobile Bay that he named Fort Louis de la Louisiane, but which was more commonly called La Mobile, "the rowers" (present-day Mobile), after the Native Americans in the area (McWilliams 1981). Conflicts between French traders and later English settlers were commonplace in the latter 1700s and early 1800s as immigrants chose wooded regions and river bottom lands for their homes and farms. However, battles with Native Americans largely ended in March 1814, when Andrew Jackson led a U.S. victory at Horseshoe Bend, so named because of its location on a curve in the Tallapoosa River (Jackson 1995).

Physiography, Climate, and Land Use

The physiography of the Mobile Basin is quite varied and, geologically, the basin is among the most diverse on the continent (see Fig. 4.12). Elevations range from approximately 1250 m asl in the Blue Ridge (BL) physiographic province in the northeastern headwaters to sea level at Mobile, Alabama. Much of the basin (56%) is in the Coastal Plain (CP) province and therefore occurs at elevations <150 m asl. Sixteen percent each lies in the Appalachian Plateau (AP) and Piedmont (PP) provinces, with 12% in the Valley and Ridge (VR). Rocks in the upper basin are of Paleozoic age (Lineback 1973). Sandstones and shales of Pennsylvanian age are present in the Appalachian Plateau, whereas carbonate, shale, and sandstone rocks dominate the Valley and Ridge. Crystalline igneous and metamorphic rocks, such as schists, slate, granites, gneiss, and quartzite, underlie the Piedmont and Blue Ridge (Lacefield 2000). In the upper Mobile River basin, riverine precursors of the present-day upper Black Warrior, Cahaba, upper Coosa, and Tallapoosa rivers may well have flowed since Mesozoic times (65 to 245 million years ago). Over the past several 100 million years the Mobile River basin has been altered by geological events that have reconfigured the land, altered drainage patterns, and changed sea levels. Close phylogenetic linkages between fishes in the Mobile and Tennessee drainages clearly suggest stream capture of Tennessee streams by an eroding Mobile drainage. In the Coastal Plain, unconsolidated sands, gravels, clays, and limestone of Cretaceous to Tertiary age are present. Portions of eastern Gulf drainages have been periodically inundated by oceanic transgressions, and at other times the Coastal Plain has extended into the Gulf of Mexico well past its present-day coastline. The last high-sea period was during the Miocene, 5 to 20 million years ago.

The Mobile River Basin lies in four terrestrial ecoregions. A small portion of the lower basin along the Mobile River floodplain and delta lies in Southeastern Conifer Forests. Although the uplands of this ecoregion were historically dominated by longleaf pine, the vast lowland delta region consisted primarily of cypress swamps. Upgradient, to the northern edge of the Coastal Plain, lie Southeastern Mixed Forests, dominated by oak, hickory, and pines. Mid-elevations in the Piedmont and Appalachian Plateau are covered by Appalachian Mixed Mesophytic Forests, consisting primarily of oaks, hickories, magnolia, elm, and pines. At higher elevation in the northeastern portions of the basin lie Appalachian Blue Ridge Forests, dominated by mixed oaks and other hardwoods.

The climate in the region is warm and humid. Mean annual air temperatures range from 21°C near the coast to 15.5°C in the northern part of the basin. Maximum summer monthly air temperatures occur in July and August, and winters are generally mild, with a minimum in January (see Fig. 4.13). Average annual precipitation ranges from 127 to 152 cm. Snowfall is not an important contribution over the majority of the basin but can be more significant at higher elevations in Georgia and Tennessee. The highest precipitation is concentrated in southern Alabama, near the coast. Only slight seasonal variations in rainfall occur within the Mobile Basin, with January through March somewhat higher than September through October (see Fig. 4.13). Summer precipitation exhibits substantial interannual variability, highly dependent on the extent of convectional rainfall and tropical storm activity.

The basin is mostly forested (68%; G. M. Ward, unpublished data). Forest types include hardwoods and mixed deciduous–coniferous in the northern parts of the basin contrasted with mostly pine forest in the Coastal Plain. Agricultural land that supports row crops (corn, soybeans, hay, and cotton) and confined animal feeding operations (poultry and livestock) constitute 18% of the basin. Wetlands cover approximately 7% of the basin, mostly confined to the Coastal Plain. Urban land use is a small fraction of the total basin area (<2%), but a population of 4.9 million resides in the basin. Major population centers in Alabama include Birmingham, Mobile, Montgomery, and Tuscaloosa. Significant population pressures increasingly occur throughout the basin, particularly from the Birmingham and Atlanta areas.

River Geomorphology, Hydrology, and Chemistry

Drainage network type and substrate composition in Mobile River channels are determined by the parent material through which the river flows. Because the Tombigbee River arises in the sandy, unconsolidated Coastal Plain, the main-stem channel is low gradient and sediments are fine sand, silt, and mud. Headwater and main-stem channels for other major tributaries arise in rocky, bedrock-dominated areas in the Appalachian Plateau, Valley and Ridge, and Piedmont (Ward, Ward, Harlin et al. 1992). Higher-gradient headwater streams (>1 m/km) as well as downstream channels are dominated by bedrock downstream to the Fall Line. At present, only the main channel of the Cahaba River remains unaltered. Below the Fall Line, channel slopes are very low

(12 cm/km) and substrates in all rivers consist of fine sand, silt, and mud.

River channels flowing through Coastal Plain sediments are typically low gradient and meander through floodplain and bottomland forests. The main stem of the Mobile River below the confluence of the Alabama and Tombigbee rivers is flanked by a wide alluvial plain deposited over Holocene time (the last 10,000 years) (Smith 1988). The main channel divides a few kilometers downstream into the Mobile and the Tensaw rivers (see Fig. 4.1) and then divides further into a complex of distributaries before emptying into Mobile Bay. This southernmost part of the Mobile River basin is known as the Mobile River Delta and covers about 100,000 to 140,000 ha. Vegetation within the Delta is composed of 87% forested wetlands, 9% marshes, and 3% submersed grass beds (Stout et al. 1982). Once dominated by cypress, the last two centuries have resulted in a shift to dominance by water tupelo.

Mean annual flow of the Mobile River is approximately 1914 m³/s. Fifty-two percent (995 m³/s) is contributed by the Alabama River and 47% (899 m³/s) is from the Tombigbee River. The principal tributary to the Tombigbee River, the Black Warrior River, drains the Appalachian Plateau and contributes 277 m³/s. Major tributaries to the Alabama River, the Cahaba, Coosa, and Tallapoosa rivers, have a combined flow of 680 m³/s. Peak runoff occurs in March, coinciding with a period of high precipitation (see Fig. 4.13). Summer base flow in the river is typically reached by early July and remains stable until early winter. Variations in summer–fall precipitation appear to have little influence on stream flow, as most of this moisture is either lost to evapotranspiration or remains stored in soil.

Water chemistry in the basin varies greatly, largely as a function of the underlying parent material from which it originates (G. M. Ward, unpublished data). Waters from the Coastal Plain are naturally low in alkalinity (<2 mg/L as $CaCO_3$) and specific conductance (<20 µS/cm) and are approximately circumneutral in pH. Waters originating in the Appalachian Plateau have similar low ionic strength because the sandstone cement is largely silica rather than carbonate. Where carbonate terrain is exposed (e.g., in many areas in the Valley and Ridge physiographic province as well as parts of some other provinces), ionic strength of the water increases and is reflected in alkalinity values above 100 mg/L as $CaCO_3$, specific conductance >100 µS/cm, and pH >7.5. Chemistry of Piedmont water tends to be intermediate in values for pH (~7.0 to 7.5) alkalinity (16 to 17 mg/L as CaCO$_3$), and specific conductance (33 to 39 µS/cm) compared to water in streams and rivers from other provinces in the basin.

River Biodiversity and Ecology

The Mobile River basin lies in the Mobile Bay freshwater ecoregion. Abell et al. (2000) list the species richness of fishes, mussels, and crayfishes in this ecoregion as being globally outstanding. Largely as a result of its size, age, geological and physiographical heterogeneity, and absence of past glaciation, the biodiversity of many groups of freshwater fauna within Alabama and the Mobile River is among the greatest in North America and for some groups the greatest biodiversity worldwide (Lydeard and Mayden 1995, Stein et al. 2000). Some of that biodiversity was lost as a result of habitat modification as the river was harnessed for hydroelectric power and navigation over the past century. Many additional species of fishes and mollusks are on the decline and face extinction. Table 4.1 provides data for species richness for the Mobile River system and for the state of Alabama.

Algae and Cyanobacteria

Few published investigations have focused on the algae in large rivers of the Mobile River system. Diatoms and green algal communities dominate the phytoplankton in the highly regulated Black Warrior River, with much less representation from cyanobacteria, euglenoids, cryptophytes, and dinoflagellates (Ratnasabapathy and Deason 1977, Joo 1990). Generally similar phytoplankton composition was reported for Weiss Reservoir, an impoundment on the Coosa River, where diatoms dominate in winter and spring months, and green algae were more abundant in summer and fall (Bayne, Seesock, Emmerth et al. 1997). Common genera included algae tolerant of organic pollution such as *Ankistrodesmus*, *Chlamydomonas*, and *Melosira*. Periphyton communities in large rivers are typically dominated by diatoms and green algae, although substantial numbers of cyanobacteria can also occur at some sites.

Plants

Riparian plant assemblages along the main-stem channels in the Mobile River differ greatly depending on whether a location is above or below the Fall Line. Above the Fall Line, river channels have very narrow or nonexistent alluvial areas. Riparian vegetation includes river birch, American hornbeam, American sycamore, American holly, American beech, sweetgum, yellow poplar, red maple, and black gum. Below the Fall Line, rivers flow through unconsolidated Cretaceous and Tertiary sediments. Here, channels meander and overbank flows are more common than in upgradient areas. As a result, bottomland hardwood forests and swamp forests are predominant. Common riparian species are bald cypress, eastern cottonwood, water oak, southern red oak, swamp tupelo, water tupelo, and Carolina ash.

Invertebrates

Of the 335 species of North American crayfishes (Abell et al. 2000), 81 species (21%) are known from Alabama (www.natureserve.org/explorer 2002), and 60 (17%) can be found in the Mobile River system (see Table 4.1) (Abell et al. 2000). Common crayfishes inhabiting the basin are *Oronectes (Hespericambarus) perfectus*, *Oronectes (Tridellescens) holti*, and *Procambarus (Pennides) versutus* (McGregor et al. 1999).

Historically, the freshwater gastropod fauna of the Mobile River system was represented by nine families and 118 species, making it one of the most diverse basins for this group worldwide (see Table 4.1) (Bogan and Pierson 1993b). The centers of richness were the Coosa River (82 species) and the Cahaba River (36 species). The fauna included six endemic genera: *Tulotoma* (Viviparidae), *Clappia*,

TABLE 4.1 Summary of faunal species richness in the Mobile River system.

	Crayfishes	Mussels	Snails	Turtles	Fishes
Species in North America	335	283	342	44	792
Species in Alabama (%)	81 (21)	171 (58)	147 (43)	23 (52)	303 (38)
Species in Mobile River basin (recent/historic)	60	75	80/118	20	236
Species endemic to Alabama watersheds, including adjoining states	NA	58	113	5	106
Mobile River species endangered or threatened	NA	19	7	3	12

Modified from Bogan and Pierson 1993a, 1993b; Lydeard and Mayden 1995; and Abell et al. 2000.

Lepyrium (Hydrobiidae), *Gyrotoma* (Pleuroceridae), *Amphigyra*, and *Neoplanorbis* (Planorbidae). The Pleuroceridae was the most diverse family (76 species), with the genera *Pleurocera*, *Leptoxis*, and *Elimia* exhibiting their greatest radiation in the Coosa River drainage. Unfortunately, recent surveys of the aquatic gastropod fauna (Bogan and Pierson 1993b) have documented population declines, decreases in species' ranges, and the loss of a major portion of gastropod diversity, especially in the Coosa River, where 26 species are now presumed extinct. These include all six recognized species of *Gyrotoma* and all but one species of *Leptoxis*. The endemic genus *Tulotoma*, formerly widespread in the main channel of the Alabama and Coosa rivers, was until recently presumed extinct (Hershler et al. 1990).

Freshwater mussels (also known as naiads, unionids, or clams) are worldwide in distribution but reach their greatest diversity in North America. In particular, the southeastern United States and the Mobile River system are "hotspots" of that diversity. Of the 283 species of North American freshwater mussels (Williams et al. 1993), 171, or 58%, are found in rivers of Alabama, with 75 species in the Mobile River system alone. Historically, all the major tributary rivers in the Mobile River system contained large and diverse assemblages of mussels. Many species have distributions limited to one or only a few narrow drainages. Many of these are endemic. There are 11 species of mussels endemic to Alabama and 58 species that are endemic to watersheds within Alabama or adjoining states. These numbers do include, however, species from the Tennessee River, an area not strictly covered by the present description.

As of 1994, of 171 mussel species previously documented in Alabama, 28 were extinct, 51 endangered, 20 threatened, and 37 of special concern (for a total of 131, or 78%). Fourteen of the extinctions (*Pleurobema* spp.) were from the Mobile River system (Neves et al. 1997). The decline in mussel biodiversity has been attributed to habitat degradation from impoundment, channel modifications, sediment deposition, and toxic effluents (Williams et al. 1992). Some common mussels collected frequently in the Mobile River system, specifically the Alabama and Lower Tombigbee rivers, are the southern fatmucket, yellow sand shell, little spectaclecase, Alabama orb, and bleufer (McGregor et al. 1999).

Flow alterations to the main-stem channels in the Mobile River system have greatly simplified the aquatic insect fauna of these channels. In the Tombigbee and Black Warrior rivers the large river mayflies,

such as *Hexagenia*, *Pentagenia vittegera*, and *Tortopus*, were common, as were *Hydropsyche* spp. and *Potamyia flava* among the caddisflies (U.S.A.C.E. 1987). Where unimpounded, river channel faunas are much richer. Frequently encountered are the stoneflies *Acroneuria* and *Isoperla* and the mayflies *Hexagenia*, *Stenacron*, *Stenonema*, and *Isonychia*. The caddisflies of the Mobile River system have been extensively surveyed by Harris et al. (1991). Light trap collections at 63 sites on the major rivers yielded 91 species in 31 genera from 11 families. The most commonly encountered genera were *Hydropsyche*, *Potamyia*, *Cheumatopsyche*, *Hydroptila*, *Ceraclea*, and *Oecetis*.

Vertebrates

The Mobile River basin supports 236 species of fishes, of which approximately 47 species are endemic (Mettee et al. 1996, Boschung and Mayden 2004). The fish community of the Mobile River basin consists of 26 families occurring in the nonestuarine portion of the river (Mettee et al. 1996, Boschung and Mayden 2004). Species from an additional 17 families of marine or brackish-water fishes can be found in the Mobile Delta and in the lower portions of the basin's rivers below the Fall Line. Migratory anadromous and catadromous species found in the basin include Gulf sturgeon, American eel, and Alabama shad. The families Cyprinidae, Percidae, Catostomidae, and Centrarchidae comprise over 56% of primary freshwater fishes found in the basin (Mettee et al. 1996, Boschung and Mayden 2004). Among these families, the Cyprinidae (54 species) and Percidae (48 species) are the dominant families. Two species are more commonly found in rivers above the Fall Line (Alabama shiner, riffle minnow), and seven species are more common in rivers below the Fall Line (pretty shiner, undescribed speckled chub, silver chub, emerald shiner, silverside shiner, fluvial shiner, and bluntnose minnow). Fourteen species of minnows are endemic to the rivers and streams of the basin, including the endangered Cahaba shiner and threatened blue shiner (Mettee et al. 1996, Boschung and Mayden 2004).

Similar to minnows, two species of darters (blackbanded darter and Mobile logperch) are common throughout the rivers of the Mobile River basin. No species of darters are restricted to large rivers above the Fall Line. In contrast, six species (naked sand darter, southern sand darter, crystal darter, rock darter, river darter, and saddleback darter) are found in rivers below the Fall Line. Of the 25 species of darter endemic to this area, 3 species (southern sand darter, rock darter, and Mobile logperch) are primarily riverine species (Mettee et al. 1996, Boschung and Mayden 2004).

Although the Centrarchidae (sunfishes and basses, 17 species) and Catostomidae (suckers, 13 species) are not as diverse as the Cyprinidae and Percidae, these fishes are widespread throughout the basin. All species of sunfishes and basses are the object of recreational fisheries, although most fishing effort is focused on largemouth bass, striped bass, black and white crappie, various sunfishes, and channel and blue catfish.

The Alabama sturgeon is the rarest riverine fish in North America and is endemic to this basin (Mettee et al. 1996, Boschung and Mayden 2004). An extensive commercial fishery existed for the Alabama sturgeon during the late 1800s and early 1900s. However, by the last decade of the twentieth century only nine specimens were collected, indicating an extreme decrease in abundance over the last century (Mayden and Kuhajda 1996). This species received a considerable amount of media and political attention prior to the U.S. Fish and Wildlife Service listing this species as endangered in 2000 (Scharpf 2000).

The Mobile River basin supports many species of reptiles, amphibians, mammals, and birds (see Table 4.1). Over 50% of the North American turtle fauna can be found in Alabama, and 45% can be found in the Mobile River system. Reptiles associated with large rivers in the basin include the American alligator, alligator and common snapping turtles; common, stripeneck, and flattened musk turtles; eastern mud turtle; southern and northern black-knobbed map turtles; Alabama map turtle; Eastern chicken turtle; Southern painted turtle; yellowbelly and red-eared sliders; Alabama redbelly turtle, Gulf Coast smooth softshell turtle; Gulf Coast spiny softshell; cottonmouth, and yellowbelly and diamondback water snakes. The flattened musk turtle (federally listed as threatened), southern and northern black-knobbed map turtles, and the Alabama redbelly turtle (federally listed as endangered) are endemic to the basin. A variety of amphibians, including frogs, salamanders, newts, sirens, and amphiuma, are found throughout the basin. Mudpuppies can be found in many rivers. Mammals commonly found throughout the basin include the American mink, beaver, and river otter. Many federal and state wildlife management areas and wildlife refuges occur in the basin and offer opportunities for observing migratory waterfowl, bald eagles, osprey, brown and American white pelicans, and cormorants.

Ecosystem Processes

The primary research focus in rivers of the Mobile River system has been on faunistic studies, mostly of fishes and mollusks, with relatively little information on ecosystem processes, although limited information is available on several main-stem reservoirs. A variety of chemical and biological indicators show that reservoirs are becoming more eutrophic in response to enhanced nutrient levels (Bayne, Weesock, Reutebuch et al. 1997, Bayne, Seesock, Emmerth et al. 1997). Lake Neely Henry, a shallow reservoir with a large surface area (maximum depth 3.3 m, surface area 4547 hectares) on the Coosa River serves as a drinking-water source and a recreation area for nearby towns (Bayne, Weesock, Reutebuch et al. 1997). Since 1948 it has been receiving industrial and municipal pollution that has caused water-quality problems. Excessive growth of phytoplankton in response to both high phosphorus and nitrogen concentrations has resulted, and phytoplankton primary productivity values ($>1\,\mathrm{g\ m^{-2}\,d^{-1}}$) place it in the eutrophic category. A similar reservoir, Weiss Lake, is just upgradient from Lake Neely Henry and also has problems associated with excess nutrient inputs as well as toxic contaminants (Bayne, Seesock, Emmerth et al. 1997). Both chlorophyll *a* concentrations and phytoplankton primary productivity indicate that this reservoir is also eutrophic.

Human Impacts and Special Features

The Mobile River is one of the largest river basins in North America and stands out among all rivers in North America for its geological and biological diversity. A huge diversity of fishes, mollusks, and crayfishes exists within its headwaters, tributary rivers, and main stem. Although flow in many of the rivers has been altered, two important tributaries, the Sipsey and Cahaba rivers, have remained unaltered. Two areas within the basin now have federal protection, the Little River Canyon in northeast Alabama and the Sipsey Wilderness in northwest Alabama. Much of the swamplands in the Mobile River Delta also remain relatively undeveloped, and recently thousands of hectares have received state protection.

The Mobile River is a highly fragmented system, as illustrated by the history of the Black Warrior River. Before the 1890s, commercial river navigation was possible only as far north as Montgomery (on the Alabama River) and Tuscaloosa (on the Black

Warrior River). Even travel to these cities was restricted to high-flow winter months because of low water levels during other times of the year and the large number of shoals and debris jams (Mettee et al. 1989). Prior to navigational improvement, surveys of the Black Warrior River by the U.S.A.C.E in the 1870s and 1880s documented in detail the numbers of shoals, bars, debris dams, and snags present along the channel of the Black Warrior River (U.S. House 1875, U.S. Senate 1881). They also showed that the river width decreased from approximately 122 m at the Fall Line near Tuscaloosa to only 46 m at Demopolis, 80 km downstream. The first generation of locks and dams on the Lower Tombigbee and Black Warrior rivers was completed in 1895. By 1915, 17 dams had been constructed and linked the port of Mobile to the iron- and steel-producing region near Birmingham.

The upper reaches of two major tributary rivers, the Coosa and Tallapoosa, are also highly fragmented by the presence of hydroelectric power dams. In the early 1900s, the precursor of what is today the Southern Company privately financed and constructed a series of dams. These large impoundments on the main stems of both the Coosa and Tallapoosa rivers and their regulated outflows have completely altered the habitat and natural hydrologic regimes of the rivers.

The Alabama–Coosa–Tallapoosa (ACT) sub-basin in the Mobile River system and the Apalachicola–Chattahoochee–Flint (ACF) basin in neighboring Georgia and Florida are currently the intense focus of bitter interstate water-allocation conflicts among Alabama, Georgia, and Florida. One contentious issue is that water transfer out of the ACT basin and into the ACF basin for use by Atlanta may cause insufficient instream flow within the Mobile basin, which will lead to impairment of downgradient rivers and Mobile Bay. This issue in the context of human impacts on the ACF will be discussed in another section of this chapter.

Historically, streams and rivers in the Mobile River basin existed with little turbidity. However, high concentrations of suspended solids from non-point-source agricultural runoff, urbanization and development, and point-source effluents are now major perturbations. Ryan (1969) estimated sediment yields in the Mobile River of 4.263 billion kg annually from suspended sediments (range of 1.814 billion to 7.257 billion kg). Bedload was as much as 0.453 billion kg, 30% of which was retained in the Mobile Delta and 70% transported into Mobile Bay.

Although yields over the last 50 years may be lower than those in the first half of the twentieth century (due to reductions in agricultural acreage and increased forest cover), recent increased mining activity and construction due to urbanization in the basin have very likely led to higher loads.

Complicating the issue of sediment transport is the potential downgradient impact of the many dams now in place on the major tributaries in the Mobile River system. Changes in types of sediment carried to the Mobile Delta may result from dams because it is unclear whether dams trap both large and fine sediments or only large, heavy sand particles, which would leave small particles in transport. As a result of heavy sediment loading from both dam outflow and other sources, the capacity of the Delta to trap sediment may diminish over time, decreasing its effectiveness as a sediment and nutrient sink. The regulated outflow from many upgradient dams also poses threats to the natural seasonal periodicity of fresh water/salt water ebb and flow in Mobile Bay, which may cause shifts in the zone of saline and fresh water mixing (Delta Project 1998, Mobile Press Register 1998; www.al.com/specialreport/mobileregister/?delta.html).

CAHABA RIVER

The Cahaba River is a 6th order tributary to the Alabama River of the Mobile basin and one of the longest free-flowing rivers in the eastern Gulf region. From its headwaters northeast of Birmingham (Fig. 4.14), the Cahaba River flows in a generally southward direction for 304 km until joining the Alabama River near Selma, Alabama. It is one of the most biologically striking and most threatened rivers in the eastern Gulf region, important because of its lack of impoundments; the large fish, mussel, and gastropod fauna that reside there; and the threats to this fauna and general water quality from population pressures emanating primarily from the city of Birmingham. Active grassroots environmental movements, such as the Alabama Rivers Alliance and the Cahaba River Society, have emerged in the region recently and have become persuasive and effective advocates for preservation, balanced growth, and environmental education. Among the rivers in the Mobile River system, protection of the Cahaba River is a very high priority.

The presettlement history within the Cahaba River basin is similar to that described for the Mobile River basin. In addition to those comments about woodland and Mississippian cultures, the Cahaba River was a natural boundary separating two major tribes inhabiting the region. Choctaws were dominant to the west and Creek Indians to the east. The name "Cahaba" was apparently derived from the Choctaw language, meaning "waters above." This has been variously interpreted as meaning water from the sky, or from mountain springs, or as a gift from above. The most well-known event of de Soto's expedition through the southeastern United States took place within this basin. The Battle of Mabila (1540) was a confrontation between the Spanish and the head of one of the most powerful remaining chiefdoms of the region, the imposing Chief Tascaluza. Because of harsh treatment from the Spanish, Chief Tascaluza and his warriors engaged de Soto's soldiers in a battle, likely near the confluence of the Cahaba and Alabama rivers, that resulted in thousands of Native American deaths and was the bloodiest battle in North America until the Civil War (Jackson 1995). The unequivocal siting of this battle, however, remains a fascinating mystery (Hudson 1993, Jackson 1995).

Physiography, Climate, and Land Use

The Cahaba River arises from seeps and springs on Mt. Cahaba in St. Clair County, northeast of Birmingham. The headwaters of the Cahaba lie in the Valley and Ridge (VR) physiographic province, a series of parallel southeastern to northeastern–oriented ridges formed by massive sandstone and conglomerate beds of the Pottsville and Parkwood formations (Pierson et al. 1989). Downstream the river crosses the Fall Line into the Coastal Plain (CP) province near Centreville, Alabama, approximately 144 km from its headwaters. The river then passes through two Coastal Plain physiographic districts, the Fall Line Hills and the Black Belt. The terrain just downgradient from the Fall Line, the Fall Line Hills, is dissected upland by a few broad or flat divides (Fenneman 1938). Altitudes of the ridges reach 213 m asl but decline to the south and southwest. Much of the area consists of unconsolidated sands of Cretaceous age carved to maturity by valleys 30 to 60 m deep. South of the Fall Line Hills lies the Black Belt, so named for the deep, black soil formed on the underlying Selma chalk formation. Here, ridges occur 20 m above stream channels. The large rivers that traverse this district have cut down through the chalk and eroded deep channels (Fenneman 1938), as evidenced by the high bluffs along the Tombigbee River. Surface-water supplies are rather deficient.

Therefore, small streams are not numerous, and many are intermittent.

The Cahaba River basin lies in two terrestrial ecoregions, the Appalachian Mixed Mesophytic Forests and the Southeastern Mixed Forests. In Alabama, the northern border of the Southeastern Mixed Forests closely corresponds to the Fall Line. Below the Fall Line, the vegetation in the basin is largely oak, hickory, and pines. Above the Fall Line, vegetation in the Appalachian Mixed Mesophytic Forests consists primarily of oaks, hickories, magnolia, elm, and pines.

Climate of the basin is humid and subtropical, with an average annual temperature of 16.7°C and average monthly temperatures ranging from 5°C in January to 26°C in July (Fig. 4.15). Annual rainfall averages 138 cm. There are no strong seasonal trends in precipitation, but lowest rainfall typically occurs in late summer–early fall and highest in winter–early spring (see Fig. 4.15).

Land cover in the basin is mostly forests (77%; G. M. Ward, unpublished data). The upper basin contains rocky terrain and shallow soils. Although agriculture accounts for 11% basinwide, it is much more extensive in the lower half of the basin, which lies in the Coastal Plain. Row crops, such as corn, cotton, hay, and wheat, are the major cover types. Riparian wetlands (6%) flank the river throughout the Coastal Plain. Land use in the upper Cahaba River basin is increasingly urban. Birmingham continues to expand, and although urban areas now comprise only 2% of the land cover, rapidly expanding suburban and other commercial developments along the Cahaba River seriously threaten the drinking-water supply for almost a million residents.

River Geomorphology, Hydrology, and Chemistry

The headwaters of the Cahaba River derive from local precipitation in sandstone-dominated terrain at an altitude of 300 m. The channel falls approximately 200 m from its headwaters until reaching the relatively low-gradient reaches lying in the Coastal Plain (1.3 m/km). Stream channels in the headwaters are narrow and bedrock-boulder dominated, and as a result of faulting, the stream network pattern is rectangular. Further downstream, river widths increase, but high bluffs constrain the channel and prevent floodplain development. Large shoal areas are common, many of which support stands of aquatic plants (Fig. 4.3). Downstream, sandstone bluffs give way to carbonate outcrops where the river has

FIGURE 4.3   Cahaba River at "Lily Shoals" near Marvel, Alabama, southwest of Birmingham, with the shoals spider lily (or Cahaba lily) in bloom (PHOTO BY BETH MAYNOR YOUNG).

eroded through the younger, overlying Pottsville sandstone. As a result of differential weathering, small tributary basins exhibit resistant chert caps on the ridges and carbonate exposures in the streambeds. As the river passes over the Fall Line near Centreville, the bluffs diminish, channel slopes decline to approximately 20 cm/km, and a wide channel with slower flow develops. The unconsolidated alluvial sediments of the Coastal Plain allow the lower Cahaba River to meander across a wide floodplain. The channel of the lower Cahaba is wide (50 m) and deep (4 to 6 m) with shear banks. The many point bars consist of gravel mixed with sand.

The Cahaba River drains 4730 km^2, with mean daily flows of 80 m^3/s. The upper basin is primarily precipitation fed, whereas the lower basin in the Coastal Plain receives substantial contributions of groundwater (Pierson et al. 1989). Maximum flows (see Fig. 4.15) occur from February to April (141 to 177 m^3/s), and low flows occur in September or October (~25 m^3/s), largely the result of lower rainfall and increased evapotranspiration. Of the 138 cm of precipitation, 56 cm (40%) is contained in runoff.

The stream-water chemistry in the Cahaba at any one location is determined by the proportional mixture of the effects of weathering in the surrounding basin and the influences of upstream human activity. The mostly pristine headwaters above Birmingham are slightly alkaline pH (7.2 to 7.7), relatively low specific conductance (30 to 160 μS/cm), low alkalinity (15 to 35 mg/L as $CaCO_3$), low NO_2-N + NO_3-N (0.025 to 0.307 mg/L), and PO_4-P (0.002 to 0.01 mg/L) (Shepard et al. 1994). However, over a distance of 80 km downgradient from the headwaters, 26 large sources of domestic sewage and industrial discharges contribute substantial amounts of nitrogen and phosphorus to the river. By the time the Cahaba River reaches the Coastal Plain, pH increases slightly (7.3 to 8.3), as does specific conductance (100 to 225 μS/cm), alkalinity (50 to 110 mg/L as $CaCO_3$), NO_2-N + NO_3-N (0.25 to 0.30 mg/L), and PO_4-P (0.003 to 0.025 mg/L). Although waters of unperturbed streams in the Coastal Plain typically have very low ionic strength and low inorganic N and P (<0.025 mg/L N and <0.005 mg/L P), water chemistry near the mouth of the Cahaba River reveals substantially increased NO_2-N + NO_3-N (0.28 to 0.64 mg/L), likely from agricultural runoff.

River Biodiversity and Ecology

The Cahaba River lies within the Mobile Bay ecoregion, an area that encompasses the entire Mobile river drainage (Abell et al. 2000). The Cahaba River is located in the approximate center of the ecoregion. Except for the issue of river fragmentation, many of the characteristics of the entire ecoregion are shared with the Cahaba River. Issues associated with land-use change, degraded water quality, and water allocation are challenges facing conservationists within the basin. For a river of its size, the Cahaba River supports an incredible diversity of aquatic life. Information on fishes, mussels, gastropods and caddisflies is the most complete. However, our knowledge of many other groups, which have not been systematically inventoried, is much less complete. Unfortunately, the number of extinct, endangered, or threatened species is also high.

Algae and Cyanobacteria

The main channel of the Cahaba is open and receives full sun. Predominant periphytic algae include diatoms (*Melosira*, *Cymbella*, and *Fragilaria*), green algae (*Cladophora*, *Ulothrix*, *Spirogyra*, *Mougeotia*, *Chaetophora*, and *Stigeoclonium*), and cyanobacteria (*Schizothrix*, *Rivularia*, *Anabaena*, *Cylindrospermum*, and *Microcoleus*). In areas near and downstream of Birmingham, diatom diversity is low and *Cladophora* has become the most dominant

and widespread alga in the Cahaba River channel, largely as a result of overgrowth due to high nutrient concentrations emanating from the surrounding urban and suburban landscape, including wastewater treatment plants (Howard et al. 2002).

Plants

Shallow shoals within the main stem are often occupied by an abundance of aquatic vascular plants. Perhaps most visually striking is the presence of the shoals spiderlily (or Cahaba lily), which blooms with a profusion of large white blossoms in May (see Fig. 4.3). Once more widely distributed than now, the Cahaba lily exists in only a very limited number of shoals and is considered endangered. American water willow is frequently found in association with shallow shoals as well as along the channel margin. In the faster flowing midchannel areas, riverweed can be found on bedrock and large boulders. The moss *Fontinalis* is also frequently found on the streambed, typically in association with *Cladophora*. The channel upstream of the Fall Line is largely constrained by rocky bluffs and has little riparian development. However, downstream of the Fall Line the channel meanders through soft alluvial sediments and is flanked by bottomland hardwood forests containing water tupelo, bald cypress, pine, American hornbeam, American beech, southern red oak, water oak, live oak, yellow poplar, sweetgum, American sycamore, American holly, and red maple, among other species.

Invertebrates

The Cahaba River harbors a wide array of insects, mussels, and gastropods. The earliest survey of mussel diversity in the Cahaba River was conducted by Henry van der Shalie (1938). Forty-two species were identified at that time, but a later resurvey by Baldwin (1973) recorded only 31 species. By 1994, only 27 species were documented (Shepard et al. 1994). At present, 20 species appear to have stable populations, including the Alabama orb, southern fatmucket, yellow sandshell, little spectaclecase, and the nonnative Asiatic clam. Nine species are considered threatened or endangered.

Historically, the Cahaba River supported 36 species of gastropods. Bogan and Pierson (1993a) concluded that by 1993 the gastropod fauna in the Cahaba had declined 33%. They listed 24 species, including 16 species that were candidates for listing as threatened or endangered species. Morales (1990) listed 10 gastropods (*Elimia cahabensis*, *E. clara*, *E. pupoidea*, *E. showalteri*, *Leptoxis*, *Somatogyrus*,

Physella, Ferissia, Micromenetus, and *Fossaria*) from a single riffle habitat in the 5th order Little Cahaba River, a major tributary to the main-stem Cahaba. This study found a trend toward increasing gastropod species richness with increasing stream order.

The high diversity of invertebrates in the Cahaba River extends to aquatic insects as well. Harris et al. (1991) recorded 342 species of caddisflies in the state of Alabama, approximately 25% of the North American fauna. The Cahaba River basin contained 156 species. At 16 sites along the main stem of the Cahaba, Harris et al. recorded 13 families, 37 genera, and 120 species. Species densities at these main-stem sites ranged from 25 to 59 species per site. The most abundant families were Hydropsychidae, Polycentropodidae, and Hydroptilidae. The genera, *Hydropsyche, Cheumatopsyche, Hydroptila, Ceraclea,* and *Cyrnellus* were commonly collected. Faunistic studies of other orders are not as complete, and accurate richness estimates for the Cahaba River are not now available. However, frequently encountered noncaddisfly genera in the main river are the damselfly *Enallagma*; the stoneflies *Acroneuria, Isoperla,* and *Perlesta*; the mayflies *Baetis, Stenacron, Stenonema, Eurylophella, Serratella,* and *Isonychia*; the riffle beetles *Macronychus* and *Stenelmis*; and chironomid midges (e.g., *Ablabesmyia, Polypedilum,* and *Rheotanytarsus*).

Vertebrates

The fish fauna of the Cahaba River is more diverse than any other comparably sized river in North America. There are 135 species of fishes in the Cahaba River basin, predominantly cyprinids (36 species) and percids (23 species) (Mettee et al. 1996; Boschung and Mayden 2004). Commonly encountered minnows include bluehead and river chub, largescale stoneroller, and Alabama and blacktail shiner; common darters include greenbreast, rock, speckled, and blackbanded darter and Mobile logperch. The Cahaba River has several species listed as endangered, threatened, or vulnerable, including the Alabama sturgeon, Cahaba shiner, Coosa madtom, goldline darter, crystal darter, and freckled darter (Warren et al. 2000); the Cahaba shiner was thought to be endemic to the Cahaba River, but another population was recently discovered in the Locust Fork of the Black Warrior River (Boschung and Mayden 2004). The blue shiner and Alabama shad have been extirpated from this system, and the relative abundances of disturbance-sensitive species, such as the coal and greenbreast darters and Cahaba shiner, have declined due to the effects of extensive urbanization

in the upper portion of the river, which flows through the Birmingham metropolitan area (Onorato et al. 2000). In contrast, the relative abundances of sunfishes have increased in the Cahaba River, possibly due to increased reservoir habitat resulting from the formation of Lake Purdy on the "upper" Little Cahaba River.

Although sunfishes and basses (15 species) and catfishes (11 species) are not as diverse as minnows and darters, these fishes are widespread throughout the Cahaba River drainage. All species of sunfishes, basses, and catfishes are the object of recreational fisheries, although most fishing effort is focused on largemouth bass, black and white crappie, various sunfishes, and channel, blue, and flathead catfish.

The Cahaba River, its tributaries, and the basin support many species of reptiles, amphibians, mammals, and birds. Reptiles associated with the Cahaba River proper include alligator and common snapping turtles, common and stripeneck musk turtles, Eastern mud turtle, Alabama map turtle, Northern black-knob sawback map turtle, Eastern chicken turtle, Southern painted turtle, river cooter, yellowbelly and red-eared sliders, Gulf Coast smooth softshell turtle, Gulf Coast spiny softshell turtle, cottonmouth, yellowbelly, and diamondback water snakes. A variety of amphibians, including frogs, salamanders, newts, and sirens, are found throughout the basin; mudpuppies, or waterdogs, can be found in the main-stem Cahaba River. Mammals commonly found in the basin include beaver, mink, raccoon, and river otter. Cahaba River Wildlife Management Area, Cahaba River National Wildlife Refuge, and Talladega National Forest offer opportunities for bird watching along the Cahaba River, especially for migratory waterfowl and bald eagles.

Ecosystem Processes

Relatively little is known regarding most ecological processes within the main stem of the Cahaba River. Above the Fall Line, the main channel of the Cahaba River is constricted but receives full sun and is shallow with mostly bedrock/cobble substrates. Consequently, benthic processes likely predominate. Historically, the river has run clear, but now significant erosion and sedimentation from the Birmingham area increasingly threaten the physical and biological structure of the system. Below the Fall Line, the river is deeper, with unconsolidated sediments. Certainly, water-column processes would be more important here than above the Fall Line. Because of the lack of impoundments, there is high connectivity between the river and the adjacent floodplain, although agricultural use of bottomland areas is common.

Human Impacts and Special Features

The Cahaba River is one of the longest free-flowing rivers in the eastern Gulf region and is noted for its incredible biological diversity, geologic diversity, and scenic beauty. The locale is attractive to humans as a place to live and work, which is currently the main threat to the Cahaba. The river flows through the heart of one of the fastest growing commercial and residential areas in the region, and the vast majority of land along the river is privately held. Consequently, little is contained within preservation areas. In 2000, however, the U.S. Congress established the Cahaba River Wildlife Refuge, and has subsequently appropriated funds to purchase 1415 ha along a 5.6 km reach of critically important river habitat. Protection of habitat for the goldline darter, Cahaba shiner, round rocksnail, and cylindrical lioplax snail are among the management objectives.

Although many reaches of the main-stem Cahaba and tributaries are classified in the Nationwide Rivers Inventory (but not protected), several main-stem reaches and tributaries are included in the state's 303(d) list as severely impaired. Shades Creek and Bucks Creek are two large tributary creeks that receive industrial and domestic sewage effluents from around Birmingham, delivering these effluents to the Cahaba main stem. Recent studies by the EPA (Howard et al. 2002) document these impacts. Many parts of the wastewater transport system in Birmingham have been leaking for decades, and the city is now under a federal court order to upgrade the sewer system and come into compliance with clean water standards.

In addition to excessive nutrient inputs and periphytic growth in the Cahaba, excessive sediment loading from construction sites and suburban development near Birmingham has negatively impacted the river downstream. Large amounts of fine sediments are visible in channel storage and in transport following rain events. Evidence of declines in fish species richness throughout the upper Cahaba from historical levels has been developed by Onorata et al. (2000). Most at risk are crevice spawning fish species that lose oviposition sites, as well as the aquatic insects that use the spaces for habitat.

Unlike many of the other major tributaries in the Mobile River system, fragmentation of the Cahaba River is not a major issue. Only one impoundment exists (Lake Purdy; see Fig. 4.14), located in a headwater tributary. Water is stored here for withdrawal by Birmingham during summer, when Cahaba River surface flow is too low to support withdrawals directly from the main channel. As a result of the use of the Cahaba River as the primary water supply for Birmingham, the quantity and quality of drinking water are now of primary environmental concern in the upper Cahaba Basin. The city has now reached the maximum amount it can withdraw at the current intake site on the river. Thus, other sources are being sought. At the same time, the basin surrounding the water source is undergoing rapid development. The threat of degradation to the city's water supply caused by additional sewage disposal and erosion is great. Currently, efforts are in place to build a coalition of support among several political units (Birmingham, several small cities, Jefferson County) to produce a land-use development plan for the upper basin that can be a guide for a more balanced economic growth pattern and the preservation of water quality in the upper Cahaba River basin.

APALACHICOLA–CHATTAHOOCHEE–FLINT RIVER SYSTEM

The Apalachicola River is formed by the confluence of two large tributaries, the Chattahoochee and Flint rivers, near the Georgia–Florida state line (Fig. 4.16). These rivers meet at what is now Lake Seminole, a 15,200 ha reservoir formed by the Jim Woodruff Lock and Dam. Below the lake the Apalachicola River flows southward for 170 km to Apalachicola Bay and on to the Gulf of Mexico. The basin occupies an area of 50,688 km^2 between the Atlantic Slope and the Mobile River catchment and is the 21st largest river basin in the coterminous United States (Leopold 1994). The entire basin is often called the ACF river system.

The 692 km Chattahoochee River originates in northeastern Georgia, flows southwest past Atlanta, then south to Lake Seminole, forming the border between Alabama and Georgia (see Fig. 4.16). The upper Chattahoochee is a free-flowing, high-gradient river, whereas the middle Chattahoochee is highly fragmented. The headwaters of the Flint River arise just south of Atlanta. The Flint drains substantial amounts of agricultural land in central and southwestern Georgia, flowing 560 km in a south-southwest direction into Lake Seminole. Major tributaries include Ichawaynochaway, Chickasawhatchee, Kinchafoonee, and Muckalee creeks.

Below Lake Seminole, the Apalachicola River flood-plain, the largest in Florida, is 114 km long and 1.6 to 8 km wide. The floodplain covers 453 km² from the outlet at Lake Seminole to the point of tidal influence, approximately 40 km above Apalachicola Bay (Light et al. 1998). Much of the Apalachicola catchment lies in Georgia (73%). The remainder is divided between Alabama (14%) and Florida (13%) (Couch et al. 1996). The Chattahoochee catchment occupies 22,714 km², or 44%, of the basin, and the Flint River occupies 21,911 km², or 43%. The area south of Lake Seminole covers 6734 km², or 13%, of the basin.

Early settlement of the ACF basin by humans occurred at the same time as other river systems in the southeastern region (>12,000 years ago). The native peoples depended primarily on a hunting-gathering lifestyle prior to developing into the agriculturally based Mississippian culture. By the 1700s the Creek Indians (Muskogeans) emerged as the dominant group in the ACF region as well as most of present-day Georgia and Alabama. "Creek" was a term used by the English to describe them because they lived along rivers and streams, whereas "muscogee," a Native American term, denoted "flooded land." The Creek towns were grouped in two areas defined by upper and lower trading routes. (Smith 2004). The Lower Creeks encompassed villages from the lower ACF valleys and were especially powerful in the southeastern region (Worth 2000).

Another distinctive group of this region, the Apalachee Indians, occupied land in the Florida panhandle east of the lower part of the Apalachicola basin (McEwan 2000). The Apalachees established several village centers, which de Soto and his band encountered when they overwintered in the area in 1539–1540. After initial battles with de Soto and other Europeans, the Apalachees eventually established a more cordial relationship with the Spanish before their villages were devastated by attacks from Creek and English combatants by the early 1700s (McEwan 2000).

Physiography, Climate, and Land Use

The path of the ACF river system flows through three physiographic provinces (see Fig. 4.16). These include a small portion of the Blue Ridge (BL) (<1%), the Piedmont Plateau (PP), and the Coastal Plain (CP). The headwaters of the Chattahoochee River arise in the Blue Ridge, a region of rugged mountains and ridges that range from 914 to 1067 m asl. Much

of the upper and middle Chattahoochee basin lies in the Piedmont. Below Columbus, Georgia, the river flows through hilly, Coastal Plain terrain until reaching Lake Seminole. The headwaters of the Flint begin in the upper Piedmont near Atlanta, although the majority of the river lies in the Coastal Plain. In southern Georgia, the river traverses karst topography, the Dougherty Plains, which is characterized by outcrops of Ocala and Suwannee limestone. Sinkholes with associated ponds and marshes are present, as are small, but often intermittent, streams. Below Lake Seminole the Apalachicola River is bounded on the east by steep bluffs through a region named the Tallahassee Hills and on the west by the Marianna Lowlands. The river then reaches the Gulf Coast Lowlands, where the floodplain widens to 5 to 8 km (Couch et al. 1996).

The ACT system lies in three terrestrial ecoregions. The lowest portion of the ACF basin, from above Lake Seminole to the Gulf of Mexico, lies within Southeastern Conifer Forests. This ecoregion is dominated by pine and mixed hardwoods. The broad floodplain of the Apalachicola River occurs here. Upgradient, Southeastern Mixed Forests, dominated by oak, hickory, and pines, covers much of the Flint and Chattahoochee basins. At higher elevations in the northeastern portion of the Chattahoochee River lie Appalachian/Blue Ridge Forests, dominated by mixed oaks and other hardwoods.

Climate in the ACF basin is characterized as warm and humid. Average annual temperatures range from 16°C in the north to 21°C in the south. Winter temperatures range from 4.5°C to 12°C, and summer maxima reach 24°C to 27°C. The mean annual temperature at Columbus, Georgia, is 18.3°C (Fig. 4.17). Precipitation is high (140 cm/yr) in the northern mountains as a result of orographic lifting of moist air from the Gulf of Mexico. High precipitation also occurs in coastal areas, which have an annual average of 152 cm (Couch et al. 1996). Mean annual precipitation at Columbus is 128 cm and is spread relatively evenly throughout the year, although somewhat lower in September and October (see Fig. 4.17). The lowest rainfall occurs in the Flint River basin, where average annual precipitation is 114 cm. Almost all of the precipitation is rain, with significant snowfall occurring in only a small northern portion of the basin.

Three major soil orders occur within the basin. Like much of the eastern Gulf region, ultisols cover the Piedmont and much of the Coastal Plain. However, a significant portion of the Flint River basin contains entisols. These geologically young soils are often infertile and drought prone because they are deep, sandy, and highly erosive (Couch et al. 1996). Poorly drained spodzols occur along the Apalachicola River.

Land use in the ACF basin is 55% forest, 25% agriculture, 10% wetlands, and <3% urban (G. M. Ward, unpublished data). Forestlands consist of second-growth timber (mixed deciduous, coniferous) and planted pine plantations. Approximately 25% of forest cover is silviculture, concentrated in northern Florida, the Piedmont south of Atlanta, and the Coastal Plain near the Fall Line. Agriculture is a mixture of primarily row crops (peanuts, corn, soybeans, and cotton), pasture, orchards, and confined animal feeding operations. Row-crop agricultural acreage has been on the decline in recent decades, but there have been increases in poultry and livestock production. Row crops dominate in the Coastal Plain, whereas pastures and poultry/livestock production dominate in the Piedmont.

River Geomorphology, Hydrology, and Chemistry

The Chattahoochee River arises in steep, mountainous terrain in north Georgia. Northeast-trending ridge lines strongly control the direction of flow of the Upper Chattahoochee to the southwest. The presence of fractures causes the main channel and its tributaries to follow a rectangular drainage pattern. Upper reaches are characterized by high gradients and numerous waterfalls. Above Lake Lanier, a reservoir created by Buford Dam north of Atlanta, Georgia, the river is free-flowing, with an average slope of 6.1 m/km (Fig. 4.4). River elevation drops 30 m below Buford Dam, and the Chattahoochee flows again south-southwest, falling approximately 100 m over 200 km to West Point Lake. West of Atlanta at the Alabama–Georgia state line the Chattahoochee turns south and flows through more rolling topography. The "Falls of the Chatta-

FIGURE 4.4 Chattahoochee River (Apalachicola system) below Buford Dam (Lake Lanier), north of Atlanta. Note tan waterline on cliffs indicates daily/weekly fluctuations in water level from hydropower releases (PHOTO BY A. C. BENKE).

hoochee" near Columbus, Georgia, denote the junction of the Piedmont and the unconsolidated Cretaceous and Tertiary sediments of the Coastal Plain (Couch et al. 1996). These falls are now submerged as a result of an impoundment but were similar to others along major rivers in Alabama where sharp river gradients limited further commercial transportation upstream. Below the W. F. George Reservoir, the Chattahoochee River winds slowly through Coastal Plains sediments until meeting Lake Seminole. Channel gradients here are similar to other Coastal Plain rivers, approximately 35 cm/km.

Geomorphologically, the Flint River traverses three distinct zones (Hicks and Opsahl 2004). Between the headwaters and the Fall Line, the Upper Flint River has broad riparian floodplains and moderate slopes (63 cm/km) throughout much of its length. However, in the lower reaches of the upper Flint the river valley narrows, the gradient steepens (5.2 m/km) into white-water rapids, and rocky shoals provide habitat for aquatic communities derived from both the Piedmont and Coastal Plain. Because the river is underlain by crystalline rocks, bedrock and cobble substrates are common. Below the Fall Line, the river exhibits deeply incised sandy banks and wide floodplains. The Middle Flint River is deep, wide, and slow, with soft, sandy sediments. Channel slopes again lessen (28 cm/km) to that typical of Coastal Plains Rivers. Overbank flows are common, and a broad alluvial floodplain has formed. Below Lake Blackshear, the lower Flint River traverses Ocala limestone and the river sediment shifts from sand to rocky limestone shoals. Here in the lower Flint River, unlike other eastern Gulf coastal basins (Suwannee River excepted), many springs emerge from the underlying Floridan aquifer. The largest, Radium Springs, discharges >3 m^3/s. Downstream at Bainbridge, Georgia, the shoals are submerged, and the river is again wide, slow, and deep before disappearing into Lake Seminole. Below Lake Seminole, the Apalachicola River is a wide, low-gradient floodplain river (7 cm/km) before emptying into Apalachicola Bay, an estuary widely known for its fishery and shellfish production.

The mean daily discharge of the Apalachicola River at Sumatra, Florida, is 759 m^3/s, ranging from a high of 5041 m^3/s to a low of 164 m^3/s (Couch et al. 1996). Eighty percent of the flow of the Apalachicola is contributed by the Chattahoochee and Flint rivers. The Chipola River, a primarily spring-fed tributary, contributes 11%, and the remaining <10% is derived from groundwater and overland flow. Because of higher precipitation, the Chattahoochee basin provides more runoff than the Flint River in most years. This is reversed in extremely dry years, when groundwater inputs to the Flint River result in more discharge than that of the Chattahoochee (Couch et al. 1996). Throughout the ACF basin, highest flows occur from January to April, when precipitation is high, and evapotranspiration is low (see Fig. 4.17). Low-flow periods occur from September through November. River flow in the Chattahoochee and Flint is largely precipitation driven, but substantial groundwater inputs are derived from the Floridan aquifer (karstic Ocala limestone) in southern Georgia, particularly in the Flint River basin. Undoubtedly, river hydrographs throughout the regulated sections of the ACT basin are modified by reservoir operations and surface/groundwater extractions (particularly in the Flint).

Water-quality investigations on the ACF basin have recently been conducted (1992–1995) as part of the USGS NAWQA program (Frick et al. 1998). Upstream of Atlanta, NO$_3$-N concentrations (given as 25th to 75th percentile) in the Chattahoochee River ranged from 0.1 to 0.3 mg/L, whereas downstream of Atlanta concentrations increased to 1 to 2 mg/L. Apalachicola River concentrations ranged from 0.3 to 0.5 mg/L. Total phosphorus concentrations in the Chattahoochee River upstream of Atlanta ranged from 0.1 to 0.2 mg/L, whereas downstream of Atlanta total phosphorus concentrations greatly increased to 0.7 to 1.5 mg/L. Apalachicola River total phosphorus concentrations ranged from 0.2 to 0.4 mg/L.

River Biodiversity and Ecology

The Apalachicola River basin lies within the Apalachicola freshwater ecoregion (Abell et al. 2000). This ecoregion contains all of the Apalachicola basin as well as the nearby Ecofina River and differs from adjacent ecoregions by having its headwaters well above the Fall Line. As a consequence, the Apalachicola ecoregion is distinct from nearby coastal drainages by having a much richer diversity of fishes, mollusks, crayfishes, and herpetofauna.

Algae and Cyanobacteria

Periphyton communities in the Chattahoochee River are typically dominated by diatoms and green algae mixed with cyanobacteria and dinoflagellates at some sites. Abundances generally increase downriver compared to the most upgradient sites (Environmental Protection Division 1974, 1984). Multiple years of periphyton sampling in the Flint River indicate that periphyton communities are typically dominated by diatoms, with fewer numbers of green algae and cyanobacteria (Environmental Protection Division, 1978, 1981, 1984). Cyanobacteria at some sites outnumbered green algae in the periphyton.

Plants

As was the case in the Cahaba River, shoals in the Flint River along the Fall Line support stands of the shoals spider lily. Historically known from South Carolina to Alabama, these now endangered plants are highly specialized for life in rapidly flowing habitats and are noted for large, showy, white flowers that are highly fragrant. Along the Flint River other aquatic macrophytes are also present, such as riverweed, cattails, bur-reed, coontail, and water primrose. Near Lake Seminole along the Flint River are water hyacinth, arrowhead, and water pennywort. The alluvial corridors of the lower Flint and Chattahoochee rivers are dominated by bottomland forests, with cypress, oaks, pines, sycamore, sweetgum, and willow. Bottomland forests along the Apalachicola River contain water tupelo, swamp tupelo, Ogeechee tupelo, bald cypress, Carolina ash, water hickory, sweetgum, overcup oak, green ash, and sugarberry (Leitman et al. 1983). In depressions and low-elevation areas, stands of tupelo and cypress (swamps) exist, some inundated year round, but others may lack standing water during the October–November dry period (Light et al. 1998). Near the mouth of the Apalachicola River lowland forests give way to freshwater and salt marshes around Apalachicola Bay. Here, sawgrass and cattails dominate freshwater marshes and needlegrass rush, smooth cordgrass, and saltgrass dominate the salt marshes.

Invertebrates

With the exception of the mollusks and crayfishes, knowledge of the richness and distribution of invertebrates in the ACF is limited (Couch et al. 1996). Thirty crayfish species were recorded by Hobbs (1942, 1981), 15 in the Apalachicola and 20 in the Chattahoochee/Flint. Six species each are endemic to the Flint and to the Chattahoochee. Hobbs and Hart (1959) reported 21 species of crayfishes from the Apalachicola River portion of the basin. The karst topography of southern Georgia also provides habitat for a diverse subterranean faunal assemblage. Two well-known troglobites (cavedwelling species) from southwestern Georgia are the

blind cave salamander and the Dougherty cave crayfish (Golladay and McIntyre 2004).

Historically, as many as 33 species of mussels and 83 species of freshwater snails have been found in the ACF basin (Box and Williams 2000, Nordlie 1990, Thompson 1984). The Apalachicola, Flint, and Chattahoochee were all rivers with large mussel faunas. Twenty-nine species of mussels are known from the Flint River and 30 from the Chattahoochee (Box and Williams 2000). Recent surveys in the Flint relocated 22 species, but only 5 in the Chattahoochee. Species commonly encountered are eastern elliptio, elephantear, yellow sandshell, and round pearlshell. From the earliest collections of mollusks from the ACF in 1834, taxonomists recognized that the ACF fauna differed substantially from that of the Atlantic slope and that of more western Gulf of Mexico basins. To an extent, there is mixing of species from adjacent basins, but there is additional uniqueness in the species composition beyond simple mixing. Clench and Turner (1956), noting the numerous close phylogenetic relationships of the ACF fauna with faunas in other river systems, concluded that the northwest Florida mollusk fauna was originally derived from areas northwest of the rivers, very likely the Coosa–Alabama and Tennessee rivers.

The free-flowing upper Chattahoochee River has a rich aquatic insect fauna, typical of the region. Stoneflies, such as *Paragnetina*, and mayflies, such as *Stenonema* and *Baetis*, are common, as are net-spinning caddisflies (*Hydropsyche* and *Cheumatopsyche*). Chironomid midges, such as the net-spinning *Rheotanytarsus*, are frequently encountered. Downstream, below river impoundments and near urbanized areas, the aquatic insect fauna is restricted to chironomids and oligochaete worms. Downstream in the Coastal Plain, below impacts of most of the impoundments, the mayflies *Stenonema*, *Tricorythodes*, and *Baetis*, the caddisfly *Cyrnellus*, and many species of chironomids can be found (Environmental Protection Division 1974) associated with fine sediments, woody debris, and aquatic plants.

Vertebrates

The Apalachicola River basin supports more species of freshwater fishes than do adjacent Coastal Plain river basins because its headwaters lie well above the Fall Line (Swift et al. 1986). Sixteen families and 104 species of freshwater and estuarine fishes are found in the Apalachicola River basin (Swift et al. 1986, Page and Burr 1991, Abell et al. 2000). Of the nine species endemic to the Apalachicola River basin, three species (greater jumprock, grayfin redhorse, and shoal bass) are found primarily in medium to large rivers. Several estuarine species, particularly killifishes and gobies, occur in the lower, brackish-water portions of the river. Migratory anadromous and catadromous species found in the Apalachicola River include Gulf sturgeon, American eel, and Alabama shad. These species were once abundant in this system, but construction of the Jim Woodruff Lock and Dam has reduced their migration routes and access to spawning grounds (Livingston 1992).

The families Cyprinidae (26 species), Centrarchidae (14 species), and Ictaluridae (8 species) comprise approximately 46% of primary freshwater fishes found in the Apalachicola River basin. Only six species of darters are found in the Apalachicola and Flint rivers; the blackbanded darter is widespread throughout these rivers. Two introduced minnows, common and grass carp, are common in the river below Jim Woodruff Lock and Dam. All species of sunfishes, basses, and catfishes are the object of recreational fisheries in the Apalachicola River drainage, although most fishing effort is focused on largemouth bass, redeye bass, shoal bass, black and white crappie, various sunfishes, and channel and blue catfish. The Apalachicola River has produced state fishing records for common carp, redeye bass, spotted bass, striped bass, and white bass (Florida Fish and Wildlife Conservation Commission 2003).

The Apalachicola River basin supports the highest species density of terrestrial and aquatic amphibians and reptiles in all of North America north of Mexico (Livingston 1992). Reptiles associated with the Apalachicola River basin include American alligator, redbelly water snake, brown water snake, cottonmouth, snapping turtle, alligator snapping turtle, Barbor's map turtle (endemic to the Apalachicola River basin), river cooter, Florida cooter, Florida redbelly turtle (westernmost population), yellowbelly slider, striped mud turtle, common musk turtle, Florida softshell turtle, and Gulf Coast spiny softshell turtle. A variety of amphibians, including frogs, salamanders, newts, sirens, and amphiuma, are found throughout the basin.

Mammals commonly found in the rivers of the Apalachicola River basin include beaver, mink, raccoon, and river otter; the West Indian manatee is seen only rarely in the Apalachicola River (Lefebvre et al. 1989). Along the Apalachicola River, The Nature Conservancy's Apalachicola Bluffs and Ravines Preserve, Apalachicola National Forest, and Apalachicola Wildlife Management Area offer opportunities for viewing wildlife. These areas provide feeding, resting, and winter habitat for many migratory bird species, including brown and American white pelicans, osprey, and cormorants. Bald eagles, Mississippi kites, and swallow-tailed kites can frequently be seen in these wildlife reserves.

Ecosystem Processes

Within the Apalachicola River floodplain are a wide variety of aquatic habitats that vary in character based on elevation above the main channel and specific geomorphic features, which may be linked to elevation and hydrology. Light et al. (1998) examined the quantity and variety of aquatic habitats within the floodplain in relation to river discharge. They concluded that during annual high water periods (>1400 m³/s), 95% of the floodplain surface is inundated, connected to the main channel, and all habitats are flowing. As flows decrease, less of the floodplain terraces are inundated and higher-elevation streams cease to flow. At median annual flow (464 m³/s), only 10% of the floodplain is connected to aquatic habitat, mostly backwater tupelo–cypress swamps. At median annual low flow <1% of the floodplain is inundated and there is little connectivity to the river. At very low flow, on average once every 20 years, only isolated aquatic habitats are present, again comprised mostly of backwater tupelo–cypress swamps. In an altered flow regime, as would certainly be the case if Atlanta were to withdraw large amounts of water from the Chattahoochee River, substantial disconnections between the floodplain and river would occur. Certainly there would be losses of bottomland forest as mortality and recruitment patterns of forest trees changed. Fish community structure would also likely change.

The Joseph W. Jones Ecological Research Center, an 11,800 ha ecological reserve located along the Flint River in southwestern Georgia, maintains active freshwater research that includes investigations of lime-sink wetlands, seasonally flooded riparian zones, swamps, streams, and rivers in the region. The lower Flint River and its floodplain, for example, have been the sites of studies that have resulted in a better understanding of the roles of intact, forested floodplains in mitigating the effects of large floods (Michener et al. 1998). During major floods, intact forested floodplains reduce sediment and nutrient loading and increase stream biotic integrity through inputs of coarse woody debris. Fauna such as fishes and mussels are extraordinarily resilient to extreme flooding, very likely because of flow mitigation and other effects (e.g., presence of refugia) of the intact Flint River floodplain. Other studies in this region have indicated that reduction of particulate organic matter (POM) in rivers during reduced flow and drought conditions may result in long-term declines in secondary production for those food web components that depend on detrital resources (Golladay 1997, Golladay et al. 2000).

Human Impacts and Special Features

The Apalachicola River is the second largest source of flow into the eastern Gulf of Mexico. The basin is large, extending from the Gulf of Mexico to the Blue Ridge, and supports a large estuarine fishery in Apalachicola Bay, agricultural water needs, and river transportation, as well as a large and growing urban population around Atlanta. It is noted for its richness of fishes, herpetofauna, and mollusks. For years, river water quality has been in decline, but citizen-based efforts are now underway to reverse that trend, and improvements have been seen (Couch et al. 1996).

Recent U.S. Geological Survey studies of the ACF basin have concluded that land use has played an important role in determining the concentrations of nutrients and metals in stream water and sediments (Frick et al. 1998). NO_3-N, NH_4-N, and PO_4-P concentrations are highest in tributaries draining catchments dominated by poultry production and urban and suburban land use. Nutrient signals are strongest during high flow conditions. Similar results were found for the main-stem Chattahoochee River downstream from Atlanta. Low nutrient concentrations predominated in parts of the river draining relatively undisturbed forested land as well as in the Chattahoochee River upstream from Atlanta and in the Apalachicola River near its mouth.

Although urban and suburban land use usually accounts for only a small fraction of basin area, their impacts on river water quality far exceed their proportionality in land area. As the percentage of urban land use increases within a watershed, nutrients, pesticides, trace elements, and organic compounds occur at higher concentrations in streams. Highest concentrations of Hg, Zn, Pb, and Cd in riverbed sediments occur in urban and suburban watersheds draining portions of metropolitan Atlanta and Columbus, Georgia, as well as in main-stem reservoirs downstream from Atlanta. Concentrations increase in direct proportion to the amount of industrial land and transportation corridors that occur in these watersheds. The source of these metals is stormwater runoff from impervious surfaces as well as

local and regional industrial emissions. Analyses of sediment cores from reservoirs downstream from urban areas reveal that metals reached maximum concentrations during the late 1960s and mid-1970s but began to decline after the mid-1970s.

River flow in the Chattahoochee–Flint system is heavily regulated by dams. Thirteen reservoirs occur on the Chattahoochee River, four of which are large (Couch et al. 1996). Two of the three dams on the Flint River have also created large impoundments. Dams are primarily used for hydropower, navigation, and domestic water supplies, particularly for Atlanta. Although annual flow has not been substantially altered, daily flow variations can be great as dams supplement power supplies during periods of peak electrical usage. For example, below the Buford Dam (Lake Lanier) in January 2004 the typical 24-hr fluctuation in gage height was 1.2 m and the 24-hr variation in discharge was about tenfold (approximately 18 to 180 m^3/s) (see Fig. 4.4).

Water use and water allocation are major issues within the basin. Surface-water withdrawals for use in Atlanta and groundwater withdrawals for irrigation, primarily in the Flint River basin, are already substantial, and future demand is expected to increase. In 1990, 7.9 million m^3 of water were removed from the ACF basin daily (Marella et al. 1993), most for use in Georgia (82%). In 2000, approximately 1.75 million m^3 were withdrawn for public water supplies in the Atlanta area, largely from surface water sources (http://gaz.er.usgs.gov/gawater/waterusega.cfm). In the Coastal Plain, particularly in the Flint River basin, groundwater extractions for irrigation have converted many permanent streams and wetlands into intermittent ones (S. Golladay, personal communication). Approximately 20% of surface- and groundwater extractions are not returned to either surface or ground water.

The ACF river basin and the Alabama-Coosa-Tallapoosa (ACT) river subbasin of the Mobile River system are currently linked in water-allocation conflicts that include Alabama, Georgia, and Florida. Rivers in both basins have headwaters in northern Georgia that are used as a water source for Atlanta. Downgradient portions of these rivers in Alabama and Florida discharge into economically and recreationally important estuaries. Water-supply needs to support the explosive growth of Atlanta have driven a request to withdraw significantly more water from the ACT and ACF than previous agreements have allowed. Atlanta's thirst, if satiated, is expected to have far-reaching negative effects on downgradient communities and economic interests. To address this issue, the ACT and ACF compacts were passed by Congress in November 1997, "for the purposes of promoting interstate comity, removing causes of present and future controversies, equitably apportioning the surface waters of the ACT–ACF compact engaging in water planning, and developing and sharing common data bases." Despite repeated efforts to complete negotiations in the years since the compacts were passed, formal agreements among the states were not yet completed in early 2004. In 2000, American Rivers listed the tristate river basins (ACT and ACF) among 13 of the most endangered rivers in the United States with threats from water withdrawals, dams, urban sprawl and nonpoint pollution. The ACT–ACF conflict is the first example of a water-allocation dispute of this scope in a humid, wet region of the world. It reinforces the view of many that maintenance of sufficient amounts of high-quality fresh water for human use is and will remain the most critical issue of the twenty-first century (e.g., Postel 1996, Postel et al. 1996).

PEARL RIVER

The Pearl River basin occupies an area of $21,999\,km^2$ between the Pascagoula and Mississippi rivers. The basin lies entirely in the Coastal Plain (CP), mostly in the state of Mississippi, but with a portion of the southwestern quadrant in Louisiana (Fig. 4.18). The headwaters arise in rolling hills in east-central Mississippi and the river forms at the confluence of Nanawaya and Tallahaga creeks. Other significant tributaries include the Yockanookanay and Strong rivers and Bogue Chitto Creek. Along much of its length the river is bordered by wide bottomland forests, backwaters, and floodplain swamps (Fig. 4.5). The timber industry and manufacturing of wood products dominate the economy of the lower basin, whereas poultry and soybeans are major parts of the upper basin economy.

Prior to the arrival of Europeans, the Choctaw Indians were the dominant Native Americans in the Pearl River basin. Their lands were bounded by Chickasaw Indians to the north, Creek Indians to

FIGURE 4.5 Pearl River near Slidell, Louisiana (PHOTO BY G. M. WARD).

the east, and the Mississippi River on the west. The Choctaw language is linguistically part of the Muskhogean family of tribes, which included Creeks, Chickasaws, Seminoles, and Apalachi. The Muskhogean tribes were thought to number more than 250,000 before contact with Europeans, and the Choctaw were considered the largest fraction. Population estimates for the Choctaws in 1685 were approximately 28,000 (Johnson 2000) but declined to 14,700 by 1790. The first European contact was the Hernando de Soto expedition of 1539 to 1543. After de Soto's travels through Mississippi, there was no European contact with interior southeastern Native Americans for more than 100 years. The second contact with the Choctaw was the French from Louisiana (Woods 1980) and the English from Charles Town (later Charleston, South Carolina) (Johnson 2000). The early English trade interest was for slaves, whereas the French needed a stable commodities trade for the survival of a new colony at Fort Biloxi. Pierre LeMoyne d'Iberville established the Biloxi colony in 1702 to counter the Spanish economic and military presence in Florida (McWilliams 1981, Thigpen 1965). During his early explorations along the Gulf Coast, d'Iberville is credited with naming the Pearl River after the pearls his soldiers found near the mouth of the river.

Recognizing the value of the Pearl River as a transportation corridor into the wilderness, the French explored and mapped the river in 1731, establishing a trading post near present-day Jackson, Mississippi. The wide river made shipping easy. By the 1830s, when cotton and timber industries began to flourish, steamboats plied the river far upstream, and the river was an invaluable economic asset for shipping crops to ports on the Gulf Coast (Thigpen 1965). The virgin hardwood and pine forests were cut and the river was used to raft logs from the interior down to mills on the coast (MDEQ 2000). Economic growth was halted by the Civil War, and the economy remained in ruins for decades afterward. The depressed economy greatly reduced river traffic, and the region soon depopulated. It was not until the mid-1900s that the region began to recover. A rejuvenated timber industry led this growth, but by that time the river was no longer the vital transportation corridor it once had been.

Physiography, Climate, and Land Use

The Pearl River drains Quaternary and Tertiary clastic sediments (primarily sand and clay) of the Eastern Gulf section of the Coastal Plain (CP) phys-

iographic province (Hudson and Moussa 1997). Sediments in much of the northern and southern portions of the basins are very sandy and porous, acting as aquifer recharge areas (Lang 1972). One physiographic district, the Jackson Prairie, is underlain by mostly clay formations, permitting the siting of the basin's major impoundment, Ross Barnett Reservoir.

Two terrestrial ecoregions lie in the basin. Approximately the upper half of the basin is within the Southeastern Mixed Forests. Dominant vegetation is hardwoods, such as oaks, sweetgum, black gum, water tupelo, beech, hickory, and yellow poplar, as well as loblolly, shortleaf, and longleaf pine. The lower half of the basin is covered by the Southern Conifer Forests. Canopy vegetation is primarily loblolly, longleaf, and slash pine, as well as live oak, magnolia, pecan, sweetgum, black gum, and water tupelo.

The climate in central and southern Mississippi is humid subtropical. Annual rainfall is high, averaging 132 cm in the northern portion of the basin and 163 cm near the Gulf Coast (Green 2000). November through April are the wettest months (11 to 15 cm/mo), whereas August through October are the driest (8 to 9 cm/mo) (Fig. 4.19). Mean annual temperature at Jackson, Mississippi, is 17.7°C, with mean monthly temperature exceeding 27°C in midsummer and falling to 7°C in January (see Fig. 4.19).

Much of the Pearl River basin remains forested (58%), with 25% agriculture, 12% wetland, and <2% urban or suburban (G. M. Ward, unpublished data). The main population center is the capital city, Jackson, with a regional population of 202,000. Mississippi is one of the least urbanized states in the United States and, in general, the basin is not densely populated, with approximately 31 people/km^2.

River Geomorphology, Hydrology, and Chemistry

The Pearl River arises in Neshoba County, Mississippi, flowing southwesterly for 209 km to Jackson. The Ross Barnett Dam is located about 5 km northeast of Jackson, creating the only large impoundment in the basin (see Fig. 4.18). Below Jackson, the river turns south and continues for an additional 375 km before dividing into two major distributaries, the East and West Pearl rivers. This final 75 km is a complex of interconnected channels that eventually empty into Lake Borgne, which is connected to the Gulf of Mexico via the Mississippi Sound near Slidell,

Louisiana. The lower 98 km of the river forms a boundary between Mississippi and Louisiana.

The Pearl River, a 6th order river, falls 210 m from its headwaters in east-central Mississippi to the Gulf of Mexico coast, with an average slope of 19 cm/km. The channel varies in width from 30 to 300 m (Wiche et al. 1988) and is characterized by tortuous bends, shoals, heavily vegetated banks, and other obstructions that restrict flow. The river bed is composed primarily of sand and silt. The valley bottom is comprised of multiple interconnecting channels separated by densely vegetated islands. Many channels are stagnant at low water but are active during periods of high stage (Hudson and Moussa 1997).

The mean daily discharge of the Pearl River, 373 m^3/s, makes it the fourth largest runoff reaching the eastern Gulf of Mexico. As with other basins in the region, flows are high from December to May and low through summer and fall (see Fig. 4.19). The highest mean monthly runoff occurs in March and the lowest in September. Although the precipitation pattern is not strongly seasonal, periods of low flow coincide with slightly lower rainfall, maximum temperatures, and higher evapotranspiration.

Because the Pearl River drains Coastal Plain sediments, ionic strength of the water is not high. Specific conductance values ranges from 50 to 150 μS/cm (http://nwis.waterdata.usgs.gov/la/nwis/gwdata 2003). Long-term average data from a USGS water-quality station at Bogalusa, Louisiana, reveal that pH was 6.5, PO_4-P was 0.026 mg/L, and NO_3-N + NO_2-N was 0.213 mg/L. Although the Pearl River was once described as crystal clear, the upper two-thirds is now often turbid (http://www.ms.water.usgs.gov/ms_proj/eric/pearl.html 2003). As with other rivers draining into the Gulf of Mexico, the majority of sediment transport in the Pearl occurs during frequent moderate discharge events (Hudson and Moussa 1997). In the lower portion of the basin, mean monthly water temperature varies between 10°C and 26°C.

River Biodiversity and Ecology

The Pearl River basin is located entirely within the Mississippi Embayment Freshwater ecoregion (Abell et al. 2000). The ecoregion is generally noted for a large richness of fishes and moderate richness in mussels and crayfishes. Although the Mississippi Embayment Freshwater ecoregion extends over an area much larger than just the Pearl River basin, the biological characteristics of the Pearl River coincide closely with those of the larger ecoregion. Fish species richness in the basin is particularly high, with 119 species known to occur. Crayfish endemism is also high.

Plants

Certainly the most visually striking aspect of the Pearl River is the vast riparian bottomland forests and riparian swamp wetlands that flank the river. Because the river lies in Coastal Plain sediment for its entire distance, such forests trace the entire length of the river but enlarge substantially below Jackson. The Mississippi Natural Heritage Program (2002a) has identified 63 nonmarine plant communities statewide, categorized according to the dominant species and habitat characteristics. The three major categories that include riverine-related vegetation include Riverfront Forests/Herblands, Wet Palustrine, and Swamp Forests. Within the Riverfront Forests/Herblands category, the three most abundant communities are dominated by eastern cottonwood, black willow, river birch, sycamore, and silver maple. The Wet Palustrine category is composed of several bottomland hardwood forests communities. Dominant species are laurel oak, willow oak, water oak, sugarberry, American elm, green ash, overcup oak, and water hickory. Swamp Forest is composed of eight communities that have standing water for much of the year and are dominated by various combinations of bald cypress, pond cypress, black gum, water hickory, overcup oak, water tupelo, sweetbay, red maple, and slash pine.

Invertebrates

Historically, the Pearl River contained an abundance of mussels and crayfishes. The Mississippi Natural Heritage Program (2002b) lists 51 species of unionid mollusks and 36 species of crayfishes from the state of Mississippi on its special animals tracking list, which indicates that these species are of special concern because of their rarity or vulnerability to habitat perturbation. The list included aquatic insects as well, with two mayflies, ten dragonflies, two stoneflies, and one caddisfly. Recent preliminary surveys in the upper Pearl and Pascagoula rivers (Haag and Warren 1995) suggest perhaps 18 species of mussels could be present in the Pearl River drainage. Some commonly encountered species would be the pondshell, giant floater, and yellow sandshell. Given the substrate and flow similarities among Coastal Plain rivers along the eastern Gulf of Mexico, invertebrate communities from the Pearl are likely to be similar to those in the Pascagoula River. Sandy sediments, woody debris, and accumulations of fine and particulate organic matter are substrates where one would likely encounter the majority of riverine aquatic insects. Particularly abundant would be the mayflies *Stenonema*, *Baetis*, *Caenis*, *Tricorythodes*, and *Isonychia*. Stoneflies, such as *Paragnetina*, *Neoperla*, and *Acroneuria*, and caddisflies, such as *Hydropsyche*, *Cheumatopsyche*, *Hydroptila*, *Chimarra*, and *Ceraclea*, would also be numerous here.

Vertebrates

The Pearl and Pascagoula rivers of southern Mississippi mark the westernmost extent of many fish species associated with the Mobile River basin and adjoining coastal river basins (Swift et al. 1986, Mettee et al. 1996, Ross 2001). The Pearl River and lower Mississippi River also form a biogeographic divide between several closely related sister species (e.g., blue sucker and southeastern blue sucker). The fish community of the Pearl River consists of 18 families and 119 species that occur in the nonestuarine portion of the river (Ross 2001). Several estuarine species, particularly killifishes and gobies, occur in the lower, brackish-water portions of the river. Migratory anadromous and catadromous species found in the Pearl River include Gulf sturgeon, American eel, and Alabama shad.

The families Cyprinidae, Percidae, Ictaluridae, and Centrarchidae comprise over 66% of primary freshwater fishes found in the Pearl River (Ross 2001). The Cyprinidae (28 species) and Percidae (23 species) are the dominant families found in this system. The most common cyprinids are the blacktail shiner, silverjaw minnow, cypress minnow, speckled chub, silver chub, emerald shiner, and bullhead minnow. Common percids include harlequin darter, brighteye darter, speckled darter, dusky darter, river darter, Gulf logperch, and saddleback darter. Other darters less frequently encountered in the Pearl River drainage, either because of their preference for faster, deeper water or because of population reductions, include the crystal darter, freckled darter, and pearl darter. The Pearl darter is endemic to the Pearl and Pascagoula River drainages; however, populations in the Pearl River may be extirpated because of extensive shoreline development and cultivation (Ross 2001).

Although the Centrarchidae (sunfishes and basses, 15 species) and Ictaluridae (catfishes, 13 species) are not as diverse as the Cyprinidae and Percidae, these fishes are widespread throughout the Pearl River drainage. All species of sunfishes, basses, and catfishes are the object of recreational fisheries in the Pearl River drainage, although most fishing effort is focused on largemouth bass, black and white crappie, various sunfishes, and channel and blue catfish. Ross Barnett Reservoir, on the upper Pearl River, has produced state fishing records for smallmouth buffalo, paddlefish, and bowfin (www.mdwfp.com/fishing_records.asp 2002).

The Pearl River, its tributaries, and the basin support many species of reptiles, amphibians, mammals, and birds. Reptiles associated with the Pearl River proper include American alligator, alligator snapping turtle, common snapping turtle, stripe-neck musk turtle, ringed map turtle (endemic to the Pearl River; federally listed as threatened), Gulf Coast spiny softshell turtle, and cottonmouth. A variety of amphibians, including frogs, salamanders, newts, sirens, and amphiuma, are found throughout the basin; mudpuppies can be found in the main-stem Pearl River. Mammals commonly found in the river include raccoon and river otter. Three Mississippi Wildlife Management Areas (WMA) (Nanih Waiya, Pearl River, and Lower River) and the Bogue Chitto National Wildlife Refuge (NWR) offer opportunities for bird watching throughout the Pearl River basin. Lower River WMA and Bogue Chitto NWR provide feeding, resting, and winter habitat for many migratory bird species, including brown and American white pelicans, osprey, and cormorants. National attention has been given recently to the swamps of the lower Pearl River because of a putative sighting of an ivory-billed woodpecker, which were thought to have been extinct since 1951. The initial sighting remains unconfirmed, and subsequent expeditions into this area to locate ivory-billed woodpeckers have been unsuccessful.

Ecosystem Processes

No ecosystem scale studies have occurred on the Pearl River. However, given the nature of the river, some ecosystem properties can be assumed. South of Jackson, along the middle Pearl River, lies a wide floodplain of bottomland forests and swamps that is still substantially connected with the main-stem river. However, in the lower Pearl, numerous hydrologic-control structures designed for flood protection occur in low-lying areas along the river near Picayune, Mississippi. The bottomland forests have historically been very productive. Descriptions of Native American life in the bottomland forests along the Pearl River refer to thousands of springs and swamp streams teeming with fishes and an abundance of bear, fox, panther, deer, beaver, and wolf, among many others (Thigpen 1965, Bremer n.d.). The river runs deep, and often very turbid, particularly after storms. The streambed substrata consist of sand and fine organic material, but coarse woody debris is present along the river banks. The majority of primary production and any intense invertebrate secondary production would likely occur along the river edges and in the floodplain during periods of inundation. The many snags along the banks and the high floodplain connectivity would provide an abundance of habitat and cover for fishes and invertebrates.

Human Impacts and Special Features

The Pearl River is an example of a large coastal river arising within and flowing through unconsolidated Coastal Plain sediments. Relief in the basin is low, thus river channel gradients are also low. High annual precipitation, a large volume of runoff, and low relief lead to frequent flooding. Wide floodplain forests exist all along the length of the river. Weathering of the largely silicate sediments results in low ionic and nutrient content in river waters under natural conditions. Despite the relatively low density of human population, past human activity in the basin has changed the river water from one noted for its clarity to one noted for the sediment it carries. Despite this fact, the river and its adjacent bottomland forest retain a sense of wildness and allure. Indeed, the lower Pearl River is an incredible maze of branching channels winding through a labyrinth of wild cypress swamps and marshlands.

Although the Pearl River does not exhibit extensive fragmentation as do many other eastern Gulf Coast rivers, significant river modifications have taken place. Near Jackson, the Ross Barnett Reservoir was constructed on the river main stem to provide for recreation, flood control, and water supply. It has a surface area of $135\,km^2$ and a storage capacity of $382\,hm^3$ (Spiers and Dalsin 1972). In the lower basin, periodic flooding on the Pearl River has long been an important regional issue, but only in the past several decades has property loss been significant. Major engineering projects on the lower Pearl River, completed by the U.S. Army Corps of Engineers, effectively rerouted river flow away from economically sensitive areas. However, following completion of the diversion project, the original east river channel dried up completely during low flow periods of the year. The USACE has now installed weirs to protect cities but also maintains year-round flow in the East Pearl River (MDEQ 2000).

In general, water quality is rated as fair in the upper Pearl basin and fair to good in the lower basin (MDEQ 2000). However, like most rivers in North America, water quality in the Pearl River has been substantially impacted by human activities. The main pollutants of concern in the Pearl River basin included siltation, organic enrichment, excess nutrients, pesticides, and pathogens (MDEQ 2000). Other listed causes of significant impairment to streams in the basin were mercury, low pH, and PCBs. Both nonpoint-source inputs, such as agriculture, timber, and urbanization, and point-source inputs, such as industrial, manufacturing and municipal discharges, oil and gas production, and mining, are important components of water-quality impairments in the basin.

Row-crop agriculture and timber industries have long histories in the basin. Improper management of forages, corn, soybeans, and cotton has lead to substantial inputs of sediments, nutrients, and pesticides into streams and rivers. During the past 50 years, the amount of row-crop acreage has decreased, replaced by timber production. The primary nonpoint-source input related to the timber industry in Mississippi is sediment loading to streams from forest roads, skid trails, and other activities that expose mineral soil (MDEQ 2000). More recently, negative water-quality impacts of urbanization and confined animal feeding operations, such as poultry and cattle, have been increasing in prominence. The Pearl River basin remains mostly rural, but growth throughout the basin, particularly around Jackson, has significantly increased the loading of fertilizer, pesticides, oils and greases, and heavy metals. Sediment loading from construction sites is perhaps the most serious urban impact problem (MDEQ 2000). Eroding stream channels further increase sediment loads and result in more river habitat losses.

Municipal and industrial point-source discharges into the Pearl River are also an important component of water-quality degradation within the basin. The majority of outfalls on the river are located south of Jackson (MDEQ 2000). North of Jackson, the river is less impacted because the city draws the majority of its municipal water supply from Ross Barnett Reservoir, which lies on the river main stem. At present, 122 industrial and 5 industrial park dischargers hold NPDES permits within the basin, which include timber products and energy production, chemical, agricultural, and metal manufacturing, oil and gas producers, and sand/gravel mines. Hazardous waste sites also pose potentially serious localized health problems within the basin (MDEQ 2000). There are 230 known sites where hazardous waste releases (both planned and unplanned) have

FIGURE 4.6   Lower Suwannee River (Photo by Tim Palmer).

occurred, and another 637 facilities that generate quantities of hazardous materials that are stored on site. At present, contamination is largely a surface-water issue in the Pearl River basin, but the potential for groundwater contamination remains a threat.

SUWANNEE RIVER

The Suwannee River is a 6th order river that originates as a blackwater stream in the Okefenokee Swamp in southern Georgia (Fig. 4.20). From there the river flows in a generally southward direction for 394 km through north Florida and empties into the Gulf of Mexico. The drainage network is composed of the upper Suwannee plus three major tributaries, the Alapaha, Withlacoochee, and Santa Fe rivers. The Withlacoochee and Alapaha both arise in southern Georgia, flowing south-southeast to their confluence with the upper Suwannee near Ellaville, Florida. Below Ellaville, the middle Suwannee River runs south and east for approximately 158 km to its confluence with the Santa Fe River below the town of Branford, Florida. Below this junction, the lower Suwannee widens and the floodplain broadens before the river reaches the Gulf of Mexico (Fig. 4.6). Below the confluence with the Withlacoochee, the Suwannee is navigable to Ellaville, once the head of commercial steamboat navigation.

Before the arrival of the Spanish, approximately 35 local chiefdoms, referred to as the Timucua, occupied land between the Apalachee tribes on the west and the Atlantic Ocean on the east (Milanich 2000). The Suwannee River cultures are thought to have been established more than a thousand years before the arrival of the Spanish. Linguistic studies suggest that the Timucua were derived from South America, but the archaeological evidence clearly indicates a southeast origin. Prior to the arrival of the Spaniards there were an estimated 200,000 Timucua occupying the northern third of peninsular Florida as well as southeastern Georgia as far as the Altamaha River. They farmed in the summer and in winter utilized the bays and rivers for shellfish and snails. The Timucua were the first tribes in what became known as La Florida to be encountered by the Spanish expeditions of Navarez and de Soto. The Timucua had continual contact with Europeans until their demise in the mid-

1700s, caused primarily by other natives, who sold the Timucua as slaves to the English.

Physiography, Climate, and Land Use

The Suwannee River basin lies within the Coastal Plain (CP) physiographic province, an area with relatively little relief. Upper reaches of the Suwannee drainage lie within the Tifton uplift and Okefenokee basin physiographic units. The Tifton uplift is a plain of low relief with uplands lying 15 to 60 m above relatively narrow valleys. The Okefenokee basin is characterized by very low relief, numerous and extensive swamps, and local sand ridges. Sediments of the upper Suwannee River basin are of marine origin and typical of Coastal Plain basins. They are highly weathered and poorly consolidated, dominated by sands, clays, and gravels of Miocene to Holocene age (Environmental Protection Division 2002). Soil types in the Suwannee River basin vary depending on elevation and underlying geology. Soils in the Alapaha River basin are well-drained, with loamy-sand surface soils and loamy subsoils, whereas those in the Withlacoochee are less well-drained spodzols (sandy soils where a layer of aluminum and organic matter has accumulated due to the poor drainage). Further eastward in the Okefenokee Swamp, soils are highly organic, extremely acid, and saturated or covered with water much of the year.

The hydrological and biological characteristics of the Suwannee River are intimately tied to the physiography and underlying geology of the region. What is now the panhandle and peninsula of Florida is the emergent portion of a larger geologic feature called the Floridan Plateau, which consists of sand, clay, and limestone strata to a depth of a kilometer or more (Rosenau et al. 1977). The upper part of these limestone and dolomite deposits contains an extensive aquifer that covers all of Florida as well as portions of southeast Alabama, southern Georgia, and southwest South Carolina. In Florida it is known as the Floridan aquifer. Principally an artesian aquifer, the Floridan aquifer contains large volumes of solution channels, caverns, and sinkholes through which surface water exits and enters the system. Impermeable surface deposits confine much of the Floridan aquifer, but in the Suwannee and nearby basins these deposits have been eroded to expose large numbers of springs emanating from the underlying aquifer.

The Suwannee basin lies completely within the Southeastern Conifer Forests terrestrial ecoregion. In Georgia, the basin is contiguous with well-known longleaf pine forests of southeastern Alabama and southwestern Georgia. However, little is left of the native stands, and the region now contains many species of pine as well as mixed oak–pine forests. Much of the original conifer and hardwood forest has been replaced by commercial forests, largely converted into commercial pine species. In the upper Suwannee basin in Georgia, 62% of forestland is in commercial forest (Environmental Protection Division 2002). In Florida, the upper Suwannee was historically dominated by north Florida flatwoods, upland hardwood hammocks, longleaf pine–turkey oak hills, and swamp hardwoods/shrub bogs (SRWMD 2001). The middle Suwannee regions contained longleaf pine–turkey oak, mixed hardwoods and pine, upland hardwood hammocks, and swamp hardwoods.

Climate in the Suwannee River basin is characterized by hot summers, mild winters, and abundant rainfall. Air temperature patterns for the Suwannee River basin are similar to that of the other eastern Gulf basins. Mean annual temperature for the basin is 20.2°C and ranges from 18.3°C in upper portions of the basin to 22.2°C in the south. Typical mean daily air temperature for the warmest months, July to August, is 27°C, and for the coolest month, January, is 11°C (see Fig. 4.21). Mean annual precipitation (134 cm) is similar to other eastern Gulf basins and ranges from 114 cm/yr in Georgia to 142 cm/yr in Florida. Precipitation occurs as rainfall, and basinwide is relatively evenly distributed throughout the year, although midsummer to late fall may be somewhat drier (see Fig. 4.21). Interestingly, near the Gulf Coast (Wilcox, Florida), numerous convectional thunderstorms in summer form a pattern whereby maximum precipitation occurs from June through August.

The eastern headwaters of the Suwannee basin are protected within the Okefenokee Wildlife Refuge. South of the refuge are extensive swamps, little agriculture, and little urbanization. In the western headwaters, the Withlacoochee and Alapaha rivers, agriculture is much more abundant, and there are many small cities. Some subbasins are up to 80% agriculture. Although agriculture has had little impact on many main-stem riparian areas, smaller tributaries have had substantial riparian vegetation removed. Much of the Suwannee basin remains forested (38%), although a significant fraction is in managed pine plantations (G. M. Ward, unpublished data). Of the remainder, 30% was in agriculture, 22% was wetland, and <1% urban or suburban. The main population centers are Lake City, Florida, and Valdosta, Georgia.

River Geomorphology, Hydrology, and Chemistry

Knowledge of the underlying geology of the Suwannee River basin is important to understanding its geomorphology, hydrology, and water chemistry. The Suwannee River has a rather complex geomorphology reflecting headwaters that lie in either Coastal Plain uplands, swamps, or spring runs. The headwaters of the upper Suwannee River arise from the Okefenokee Swamp. The main-stem river (upper Suwannee) and two major western tributaries, the Withlacoochee and Alapaha rivers, all flow initially over low-permeability sediments in Georgia that limit groundwater recharge and permit permanent flow. Streambed sediments in the Alapaha and Withlacoochee rivers are composed of fine sand or mud, but where the stream has eroded down to the underlying rock the streambed may be composed of limestone bedrock and cobble. Downstream of the Georgia–Florida boundary the sediments underlying the riverbed in all channels change to a porous, karstic formation called Ocala limestone. The formation contains numerous solution channels and sinkholes within the main channels. At low flow, sinkholes in the Alapaha and Santa Fe rivers absorb all of the flow from their upper segments, leaving the lower portion of these tributaries dry for several months at a time. From the point of groundwater recharge, the stream emerges several kilometers downstream (U.S. Study Commission 1963, SRWMD 2001). The main-stem Suwannee River and many of its tributaries have deeply incised channels where the stream has cut through the shallow overburden and into the underlying limestone. Limestone outcrops line the banks of the Upper Suwannee, and the riverbed may be sandy or limestone bedrock. White sand beaches often occur in the bends of the river. Low falls may occur where limestone shoals have been highly eroded. Channel slopes in the Suwannee are quite low. From the outlet of the Okefenokee Swamp at 28 m asl the river falls 13 m over the first 80 km (16 cm/km) to a point near Ellaville. Over the last 314 km to the Gulf of Mexico channel slope is only 5 cm/km.

The Suwannee River is the second largest river in Florida, with a mean annual flow of 294 m^3/s. High flows occur in mid to late spring and low flows in November and December (see Fig. 4.21). The upper reaches of the Suwannee tend to have more flow variability, whereas the lower reaches are more stable, primarily a result of the hydrologic buffering offered by the many springs that enter the middle and lower

Suwannee (Mattson et al. 1995). At times flow in the lower Suwannee consists entirely of spring inputs. Among the eastern Gulf Coast rivers described in this chapter, the Suwannee River has the lowest annual runoff (36 cm) and the lowest runoff as a percentage of rainfall (27%). At least two explanations are possible. One is a higher rate of evapotranspiration. Given the amount of wetland habitat and the warm, long summers, higher evapotranspiration is certainly a possibility, although studies such as Bidlake et al. (1995) from peninsular Florida would not support a rate higher than other eastern Gulf basins. Another possibility is greater infiltration into the underlying aquifer and subsequent subsurface transfer of water to other basins or the sea. Certainly, surface watershed boundaries in the Suwannee River are not likely to coincide with subsurface water divides.

The chemistry of Suwannee River water varies substantially from north to south along the river's course as a result of changes in the underlying geology. Water entering the Suwannee River from the Okefenokee Swamp is characteristic of the swamp itself, highly stained (200 to 600 Pt/Co units), acidic (pH 3.2 to 7.4), low alkalinity (0.50 to 23 mg/L as CaCO$_3$), and low conductivity (50 to 122 uS/cm) (Mattson et al. 1995). Water from the Withlacoochee, derived from runoff in upland areas of southern Georgia, has a lower color (10 to 250), is circumneutral (pH 6.6 to 7.4), with higher alkalinity (15 to 120) and conductivity (82 to 258). As the river flows downstream, spring inputs contribute substantial quantities of calcareous water containing high concentrations of calcium, magnesium, and bicarbonates, which increase the pH (7.1 to 8.3), alkalinity (110 to 160), and conductivity (138 to 364) in downstream reaches. Historical patterns for nitrogen and phosphorus are more difficult to discern due to the long-term loading from agricultural and industrial sources along the river. From 1989 to 1991, total N in the lower Suwannee ranged from 0.31 to 1.69 mg/L and total P ranged from 0.05 to 0.57 mg/L.

River Ecology and Biodiversity

The Suwannee lies entirely within the Florida freshwater ecoregion (Abell et al. 2000). The northern boundary of this ecoregion follows the northern edge of the Suwannee River basin and includes coastal rivers west to the Apalachicola River and all of peninsular Florida. The Suwannee River is typical of the ecoregion in containing a wide variety of habitats, including upland and lowland streams, swamps, and springs. Unlike other eastern Gulf rivers, the numerous large springs derived from the Floridan aquifer harbor a rich and endemic crayfish fauna. But in contrast to other rivers described earlier, the fish and mollusk fauna of the Suwannee are not particularly rich.

Algae and Cyanobacteria

Large spring-fed streams in north central Florida, such as Ichetucknee Springs, have long been known for their abundance and high productivity of algae and submerged macrophytes. Spring-fed tributaries in the Suwannee drainage were included in the periphyton classification scheme of Whitford (1956), in which the lower Santa Fe River and Ichetucknee Springs were characterized as a *Cocconeis-Stigeoclonium* type. Within the type, diatoms dominated, with the genus *Stigeoclonium* a subdominant. More recent studies by Mattson et al. (1995) in the lower Suwannee and Santa Fe rivers revealed that algal communities attached to glass slides were heavily dominated by diatoms, particularly *Achnanthes*, *Cocconeis*, *Gomphonema*, *Melosira*, *Navicula*, and *Synedra*. Green algae were represented by *Protoderma viride*, *Scenedesmus acuminatus*, and *Stigeoclonium* sp.

Plants

The large spring-fed tributaries in the Suwannee drainage also have large stands of aquatic plants. For example, the Ichetucknee Springs complex contains 12 species of submerged macrophytes (Nordlie 1990). The lower Santa Fe River, into which the Ichetucknee Springs complex drains, is dominated by muskgrass (the multicellular alga *Chara* spp.), loose watermilfoil, American eelgrass, and springtape (Mattson et al. 1995).

Perhaps the most visible characteristic of the riparian vegetation along the Suwannee River is cypress trees. Bald cypress can be found along the main channel and in sloughs, backwaters, and tributaries along the length of the Suwannee, Alapaha, and Santa Fe rivers (Duryea and Hermansen 2000). A recent study concluded that 77 tree, shrub, and woody vine species were present in the lower Suwannee River riparian and bottomland areas, a high species richness relative to other wetland forests (Light et al. 2002). Fourteen forest types were identified, which could be generalized into three categories: riverine high bottomland hardwoods dominated by live oak; riverine low bottomland hardwoods dominated by five species of oak, the most important being laurel oak; and riverine

swamps, occurring in the lowest and wettest areas, dominated by bald cypress.

Invertebrates

Sand, wood, submerged macrophytes, and limestone outcrops constitute the primary substrates available for invertebrates in the Suwannee River and its major tributaries. In the upper tributaries, mayflies, such as *Stenonema*, *Caenis*, *Baetis*, *Tricorythodes*, and *Habrophlebiodes* are commonly found. Stoneflies, such as *Neoperla*, *Acroneuria*, and *Perlinella*, as well as caddisflies, such as *Hydropsyche*, *Cheumatopsyche*, *Macrostemum*, and *Chimarra*, are abundant (Environmental Protection Division 1973). Hester-Dendy samplers from the upper and lower Santa Fe River accumulated 20 to 40 taxa, consisting primarily of chironomid midges, mayflies (Heptageniidae, Tricorythodidae, and Baetidae), and hydropsychid caddisflies (Mattson et al. 1995). Samplers in spring-influenced reaches also included *Elimia*, hydrobiid snails, limpets, and crustaceans, particularly *Hyallela azteca*. Communities on woody substrates were substantially more species rich than softer, sandy substrates. Sandy substrates in the lower Santa Fe River harbored chironomids, oligochaete worms, and mollusks, including *Campeloma* and *Corbicula*, as well as sphaeriid (*Sphaerium*, *Musculum*) and unionid clams (*Elliptio*, *Villosa*). Macrophyte beds were dominated by chironomids, particularly *Dicrotendipes*, mayflies (*Baetidae*, *Tricorythodes albilineatus*), and *H. azteca*. The Suwannee River system also harbors a troglobytic fauna, found in the many springs and sinkholes (Nordlie 1990). Most of the fauna are crustaceans (Amphipoda, Isopoda, and Decapoda) and have highly restricted distributions. Biodiversity in the Suwannee River appears to be greatly influenced by inputs from calcareous streams. Certain taxa groups, particularly mayflies, mollusks, and oligochaetes, exhibit increased species richness and increased abundance in waters with increasing alkalinity.

Vertebrates

The Suwannee River forms a divide between the fish faunas of the panhandle and peninsular Florida; approximately 50% of Florida's native freshwater fish species occur in or west of the Suwannee River drainage (Bass 1991). Species differences between the two areas are due to the loss of certain species, especially minnows, in peninsular Florida (Bass 1991). Seventeen families and 81 species of freshwater and estuarine fishes are found in the Suwannee River drainage (Swift et al. 1986, Page and Burr 1991, Abell et al. 2000). There are no endemic fish species in the Suwannee River; the Suwannee bass, however, is restricted to the Suwannee River and Ochlockonee River drainages (Page and Burr 1991). Several species of euryhaline fishes, particularly topminnows and killifishes, occur in the lower portions of the river. Migratory anadromous and catadromous species found in the Suwannee River include Gulf sturgeon, striped bass, American eel, and Alabama shad; the Suwannee River is the easternmost limit in the distribution of Alabama shad (Page and Burr 1991).

The Cyprinidae (12 species), Centrarchidae (14 species), and Ictaluridae (9 species) comprise approximately 60% of primary freshwater fishes found in the Suwannee River basin; the redbreast sunfish is the most numerically abundant fish in this river (Bass 1991). All species of sunfishes, basses, and catfishes are the focus of sport fisheries. The Suwannee River has produced state fishing records for the Suwannee bass, redbreast sunfish, and spotted sunfish (http://floridafisheries.com/Fishes/sci-name.html 2003).

There are no endemic species of reptiles or amphibians in the Suwannee River; the Suwannee cooter, however, has a restricted distribution and is found primarily in the Suwannee River south to the vicinity of Tampa Bay (Conant and Collins 1991). Reptiles commonly found in the Suwannee River include the American alligator, alligator snapping turtle, Florida softshell turtle, Florida snapping turtle, Suwannee cooter, Florida cooter, Florida redbelly turtle, Florida cottonmouth, Gulf saltmarsh snake, redbelly, banded, brown, and Florida green water snakes, and the North Florida swamp snake. A variety of amphibians, including frogs, salamanders, newts, sirens, and amphiuma, are found throughout the Suwannee River drainage (Conant and Collins, 1991).

Mammals commonly found in the Suwannee River include beaver (found primarily in the upper river basin), mink, raccoon, river otter, and the West Indian manatee. Manatees are especially common during the winter at Manatee Springs State Park, which provides a boardwalk and viewing deck for observation of these animals. Suwannee River State Park lies at the confluence of the Suwannee and Withlacoochee Rivers; beavers are commonly seen in this area.

A number of migratory and resident birds can be found in the Suwannee River drainage, including brown and American white pelicans, osprey, cormorants, anhingas, swallow-tailed kites, and bald eagles.

Ecosystem Processes

A number of studies of primary productivity of spring runs in northern Florida have been published. Although located in a drainage adjacent to the Suwannee River, Silver Springs is perhaps the most notable spring study, because it was one of the first to examine primary production in a large flowing-water system (Odum 1957a). First-magnitude springs here are noted for stable temperatures, nutrient-rich water, and stable hydraulic conditions. It is exactly these conditions that lead to high standing-stock biomass and high productivity of aquatic macrophytes and periphyton. Primary productivity in Silver Springs was calculated to be $>3000\,g\ C\ m^{-2}\,yr^{-1}$, mostly by a single aquatic plant population, the awl-leaf arrowhead, and its attached epiphytes (Odum 1957a). Gross primary production measurements from five first-magnitude springs in northern Florida (including springs within the Suwannee basin) ranged from 2 to $24\,g\ C\ m^{-2}\,d^{-1}$ (Nordlie 1990, recalculated from Odum 1957b). From Manatee Springs, a first-magnitude spring in the lower Suwannee, a value of $7.28\,g\ C\ m^{-2}\,d^{-1}$ was reported. From these and other studies, there is ample evidence that Florida spring ecosystems are among the more productive aquatic ecosystems in the world (e.g., Duarte and Canfield 1990).

Northern Florida freshwater swamps also exhibit high primary productivity. Typically, floodplain swamps exhibit the highest net primary productivity, whereas cypress swamps, supplied primarily by rainfall, are much less productive. Spring-run swamps are intermediate. Regulation of primary productivity appears to be related to nutrient delivery rates through stream flow. Riverine and floodplain swamps receive nutrient subsidies from overbank flows, and spring-run swamps from groundwater (Ewel 1990). Primary productivity (above ground net primary production) in Florida cypress swamps ranges from 700 to $1600\,g\ m^{-2}\,yr^{-1}$ (Mitsch and Gosselink 2000). Hydrological, biochemical, and productivity studies in the Okefenokee Swamp indicate that secondary productivity of bacteria and fishes is also high in this blackwater ecosystem (Blood 1981).

Human Impacts and Special Features

The Suwannee River is distinctive among eastern Gulf Coast rivers in that its headwaters arise in a large swamp ecosystem, the Okefenokee Swamp. The Okefenokee is a $2250\,km^2$ wilderness that has been protected since 1937 and is now part of the National Wilderness Protection System. Although there were attempts to commercialize the Okefenokee Swamp in the late 1800s and early 1900s, all failed. However, remnant canals and other commercial infrastructure of the time can still be seen. Ecological habitats in the Okefenokee include deep pools of highly stained water with cypress trees dominant, floating islands (hammocks) covered with trees and shrubs, open grassland prairies, and mats of floating peat. The Okefenokee is drained by two rivers, the St. Marys and the Suwannee, with most of the water carried by the latter.

The signature characteristic of the Suwannee River is the large number of springs that flow into the river. These springs greatly influence river hydrology, geochemistry, and biota. The upper, middle, and lower Suwannee as well as major tributary rivers are fed by hundreds of large and small springs that flow from the underlying Floridan aquifer. At least 97 have been named, including eight first-magnitude springs (those with a discharge >2.8 m³/s). Of the 27 first-magnitude springs in Florida, 30% are in the Suwannee basin. Some of the more well-known springs are Suwannee Springs, Manatee Springs, and Ichetucknee Springs. Recharge of these aquifers occurs in southern Georgia, the northern Florida panhandle, and central Florida.

Despite the pristine nature of much of the Suwannee River basin, serious environmental challenges exist, particularly nitrate and phosphate contamination of surface and ground water. Phosphate rock is mined in the upper Suwannee basin in Florida, leading to substantial surface-water loadings from the mining operations. Substantial loading of phosphate also occurs in the Withlacoochee River as a result of agricultural runoff. High concentrations of nitrate have been found in surficial aquifers and surface waters of the Suwannee River. Nitrate concentrations in excess of the EPA drinking water standard (10 mg/L) have been found in 33% of groundwater wells tested in agriculturally dominated landscapes in the Suwannee River basin (Berndt et al. 1998). Concentrations were highest in shallow wells in agricultural settings as compared to mixed-use, forested, or urban settings. Wells near row-crop agriculture and confined animal feeding operations were most impacted. Fortunately, these shallow groundwaters lie above the upper Floridan Aquifer, which serves as the drinking-water source for many residents of the basin. It is, however, obvious that the drinking-water aquifer for this region is at serious risk of contamination. Surface-water nitrate concentrations tended to be lower than groundwater, but in agricultural settings, such as the Withlacoochee River, surface-water nitrate concentrations were up to 0.25 mg/L. Nitrate loading to the lower Suwannee River from springs is a major concern. During low-flow periods spring input can contribute 100% of downgradient discharge. Many of these springs contain nitrate concentrations between 1 and 2 mg/L, raising the possibility of surface-water degradation in this portion of the Suwannee, highly prized for its recreational value. Interestingly, at high flows the river can recharge the underlying Floridan aquifer, potentially introducing contaminants to the groundwater directly from the river (Crandall et al. 1999).

In order to protect and conserve the Suwannee River from environmental threats, the Suwannee River Water Management District (SRWMD) has been very active over the past few years, acquiring sizeable amounts of riparian habitat for conservation and flood control. At present, the SRWMD owns 41,100 ha in the basin, including 63% of river frontage in the upper Suwannee. Land-acquisition plans are in place to continue purchase of critical riparian areas to protect the main-stem river and important springs and spring runs.

ADDITIONAL RIVERS

The Choctawhatchee River is an alluvial river that arises in southeast Alabama and flows southward through a rolling, rural, and largely forested landscape into Choctawhatchee Bay in northwest Florida (Fig. 4.22). It is one of the longest free-flowing rivers (273 km) in the southeastern United States. The major tributaries in Alabama derive their flow from local precipitation that is low in ionic content, whereas the principal tributaries in Florida receive flow from calcium-rich springs of the Floridan aquifer. The river corridor is undeveloped and fringed by biologically rich bottomland hardwood forests, marshes, and tupelo–cypress swamps (Fig. 4.7). A general decline in invertebrate and fish biodiversity has occurred over the past decades, with much of the historically known mussel fauna either lost or in decline (Blalock-Herod et al. 2000). Overall, river water quality is good, but sedimentation is a major threat to economically valuable commercial and recreational fisheries and shellfisheries in Choctawhatchee Bay.

The Escambia River originates as the Conecuh River in the gently rolling Coastal Plains hills of south-central Alabama (Fig. 4.24). Four major tributaries join the Conecuh River on its southwestern path before it crosses into Florida and becomes the

FIGURE 4.7 Choctawhatchee River at High Bluff, Alabama, west of Dothan (Photo by Beth Maynor Young).

Escambia River that flows south through the Florida panhandle to Escambia Bay (Fig. 4.8). Water in the main stem is deep and slow moving, with channels flanked by lowland hardwood forests and tupelo–cypress swamps. Two impoundments used for hydroelectric power occur in the Conecuh River. Water quality is good, although dairy and agricultural runoff, silviculture, urbanization, and municipal/industrial wastewater discharges are threats. Escambia Bay once had a thriving commercial and recreational fishery and shellfishery, much of which is now lost as a result of pollution. A sizable population of the threatened Gulf sturgeon (*Acipenser*

oxyrinchus desotoi), once thought lost to this river, is found in the upper reaches, which the U.S. Fish and Wildlife Service (2002) proposes to list as critical habitat.

The Flint River (Fig. 4.9), located in central Georgia, drains parts of the Piedmont and Coastal Plain. The main stem flows from south of Atlanta south to Lake Seminole (Fig. 4.26). Information about the hydrology, geomorphology, and biota of the Flint River system is provided in the section on the Apalachicola–Chattahoochee–Flint river system in this chapter. Water quality within the basin is generally good. However, threats to water quality

FIGURE 4.8 Escambia River in the western panhandle of Florida, upstream from Pensacola (PHOTO BY BETH MAYNOR YOUNG).

include nonpoint-source loadings, such as erosion and sedimentation, agriculturally derived nutrient loading, metals in urban runoff, and fecal coliform bacteria in both urban and rural runoff. Also, large amounts of water withdrawn for irrigation from both surface water and the upper Floridan aquifer threaten water availability. Anticipated future demands from industry, agriculture, and municipalities in the basin, especially Atlanta, may be difficult to satisfy with existing supplies.

The Pascagoula River arises in sandy, low-rolling hills of the Coastal Plain of southeastern Mississippi and empties into the Gulf of Mexico. The main stem is formed by the confluence of several tributaries (Fig. 4.28), and one of them, Black Creek, is a National Wild and Scenic river. Slow flowing and low gradient, the Pascagoula River flows through low, broad floodplains dominated by bottomland hardwood forests and tupelo–cypress swamps (Fig. 4.10). The basin is mostly covered by coniferous and mixed deciduous forests, and silviculture is an important industry. Fragmentation of the river by dams and reservoirs is low, although water extractions occur

for irrigation, livestock production, and industry. Although most streams in the basin are clear water, the Escatawpa River, which joins the main stem near the coast, is a blackwater stream. Water quality is generally good, except near industrial outfalls and population centers. Mercury contamination in bass and catfish has caused consumption advisories since 1996 throughout the main stem (Mississippi Department of Environmental Quality 2001).

The Tombigbee River is one of the two major tributaries of the Mobile River system. Upstream of its confluence with the Black Warrior River, the Tombigbee is referred to as the Upper Tombigbee River (Fig. 4.30). Free flowing until the 1970s and noted for its high diversity of fishes and mollusks, construction of the Tennessee–Tombigbee Waterway has had a major hydrologic and ecologic impact on the river. Dedicated in 1985, the waterway is 377 km of channelized river that traverses parts of west-central Alabama and northeastern Mississippi, connecting the upper Tombigbee River with the Tennessee River, a connection that created a commercial transportation route between Mobile, Alabama, and the

FIGURE 4.9   Flint River near Thomaston, Georgia, west of Macon; wide and shallow section of river upstream from Yellow Jacket Shoals (Photo by Beth Maynor Young).

Mississippi River basin. Hydrologic alterations included 10 locks and dams, river meanders that were cut off (creating many new oxbow lakes), newly dredged channels that completely bypassed small upstream reaches, and a canal across the watershed divide. The decision to construct the waterway and its projected benefits remain controversial. It is lauded as an important transportation route that has also opened lake recreational areas and education centers. However, the actual economic impact is questionable, whereas the environmental consequences have been profound. The channelization has separated the upper Tombigbee River from its flood-plain, dramatically reduced the diversity of riverine fauna, and made the Mobile basin more vulnerable to invasion by nonnative species.

In contrast to the environmental alteration of the upper Tombigbee River, the Sipsey River is a free-flowing tributary winding through the Coastal Plain in west-central Alabama before joining the upper Tombigbee (Fig. 4.30). The main stem is low gradient and flows through extensive marshes, bottom-land hardwood floodplains, and tupelo-cypress wetlands. As with other undeveloped Coastal Plain rivers, the Sipsey River has substantial quantities of snag habitat (Fig. 4.11). Although the main stem is

FIGURE 4.10 Pascagoula River: Oxbow Lake near confluence of Chickasawhay and Leaf rivers (PHOTO BY BETH MAYNOR YOUNG).

without major tributaries, it often consists of multiple parallel channels, many of which dry in summer but are filled and swiftly moving during winter. The river is ecologically significant because of a lack of impoundments and a rich fauna that includes 35 species of mussels and 91 species of fishes. To protect some of this biological diversity, Alabama recently purchased 1214 km² of bottomland forest and swampland bordering the river. Ionic strength of the water is low, and water quality is considered good, although municipal wastewater discharges around several small cities create localized inputs high in organic matter and inorganic nitrogen. Urban

development, logging, agriculture, and mining in the basin have increased sediment load over the years.

ACKNOWLEDGMENTS

The extraordinary photograph of the Cahaba and several other rivers described in this chapter are those of Beth Maynor Young. We greatly appreciate her generosity through the contribution of these images. We also appreciate data provided by Fred Howell, University of Southern Mississippi; Tom Kenncdy, University of Alabama; and Robert A. Mattson, Suwannee River Water Management District. Our knowledge of the region has greatly benefited

FIGURE 4.11   Sipsey River near Elrod, Alabama, west of Tuscaloosa. Photo taken at very low water, exposing extensive submerged wood (snags) (PHOTO BY A. C. BENKE).

from previous research supported by NSF EAR-0083752, NSF BSR-8818810, and resources from the Center for Freshwater Studies at the University of Alabama.

LITERATURE CITED

Abell, R. A., D. M. Olson, T. H. Ricketts, E. Dinerstein, P. T. Hurley, J. T. Diggs, W. Eichbaum, S. Walters, W. Wettengel, T. Allnutt, C. J. Loucks, and P. Hedao. 2000. *Freshwater ecoregions of North America: A conservation assessment.* Island Press, Washington, D.C.

Adams, G. I., C. Butts, L. W. Stephenson, and W. Cooke. 1926. *Geology of Alabama.* Special Report no. 14. Geological Survey of Alabama, Tuscaloosa.

Alabama-Coosa-Tallapoosa river basin compact. 1997. Public law no. 105-105, 111 statute 2233.

Baldwin, C. S. 1973. Changes in the freshwater mussel fauna in the Cahaba River over the past 40 years. Master thesis. Tuskegee Institute, Tuskegee, Alabama.

Bass, D. G. J. 1991. Riverine fishes of Florida. In R. J. Livingston (ed.). *The rivers of Florida*, pp. 65–83. Springer-Verlag, New York.

Bass, G., P. Shafland, and B. Wattendorf. 2003. A checklist of Florida's freshwater fishes, with photos. Florida Fish and Wildlife Conservation Commission. www.floridafisheries.com/fishes/sci-name.html.

Bayne, D. R., W. C. Seesock, P. P. Emmerth, and F. Leslie. 1997. *Weiss Lake, Phase I Diagnostic/Feasibility Study, Final Report.* Alabama Department of Environmental Management, Montgomery.

Bayne, D. R., W. C. Seesock, E. Reutebuch, and D. Watson. 1997. *Lake H. Neely Henry, Phase I, Diagnostic/Feasibility Study, Final Report.* Alabama Department of Environmental Management, Montgomery.

Berndt, M. P., H. H. Hatzell, C. A. Crandall, M. P. Turtora, J. R. Pittman, and E. T. Oaksford. 1998. Water quality in the Georgia–Florida Coastal Plain, Georgia and Florida, 1992–1996. U.S. Geological Survey Circular 1151. Government Printing Office, Washington, D.C.

Bidlake, W. R., W. M. Woodham, and M. A. Lopez. 1995. Evaporation from areas of native vegetation in west-central Florida. U.S. Geological Survey Water-Supply Paper 2430. Government Printing Office, Washington, D.C.

Blalock-Herod, H. N., J. J. Herod, and J. D. Williams. 2000. Interim report: Freshwater mussels of the

Choctawhatchee River drainage in Alabama and Florida. Report to U.S. Fish and Wildlife Service, Panama City, Florida.

Blood, E. C. 1981. Surface water hydrology and biochemistry of the Okefenokee Swamp. Ph.D. diss., University of Georgia, Athens.

Bogan, A. E., and J. M. Pierson. 1993a. *Survey of the aquatic gastropods of the Cahaba River Basin, Alabama: 1992.* Final report submitted to Alabama Natural Heritage Program, Montgomery.

Bogan, A. E., and J. M. Pierson. 1993b. *Survey of the aquatic gastropods of the Coosa River Basin, Alabama: 1992.* Final report submitted to Alabama Natural Heritage Program, Montgomery.

Boschung, H. 1989. Atlas of fishes of the Upper Tombigbee River drainage, Alabama–Mississippi. *Proceedings of the Southeastern Fishes Council* 19:1–104.

Boschung, H. T., Jr., and R. L. Mayden. 2004. *Fishes of Alabama.* Smithsonian Books, Washington, D.C.

Box, J. B., and J. D. Williams. 2000. Unionid mollusks of the Apalachicola basin in Alabama, Florida, and Georgia. *Bulletin of the Alabama Museum of Natural History* 21:1–143.

Bremer, Christopher. n.d. *The Chatta Indians of Pearl River: An outline of their customs and beliefs.* Picayune Job Press, New Orleans (University of Alabama Rare Book Collection, Call Number E99.C8 B7x).

Clayton, L. A., J. V. Knight Jr., and E. C. Moore (eds.). 1993a. *The De Soto Chronicles.* Vol. 1. University of Alabama Press, Tuscaloosa.

Clayton, L. A., J. V. Knight Jr., and E. C. Moore (eds.). 1993b. *The De Soto Chronicles.* Vol. 2. University of Alabama Press, Tuscaloosa.

Clench, W. J., and R. Turner. 1956. Freshwater mollusks of Alabama, Georgia, and Florida from the Escambia to the Suwannee River. *Bulletin of the Florida State Museum. Biological Sciences* 1:97–220.

Conant, R., and J. T. Collins. 1991. *A field guide to reptiles and amphibians: Eastern and central North America.* The Peterson Field Guide Series. Houghton Mifflin, New York.

Couch, C. A., E. H. Hopkins, and P. S. Hardy. 1996. Influences of environmental settings on aquatic ecosystems in the Apalachicola–Chattahoochee–Flint River basin. U.S. Geological Survey Water-Resources Investigations Report 95-4278. U.S. Geological Survey, Atlanta, Georgia.

Crandall, C. A., B. G. Katz, and J. J. Hirten. 1999. Hydrochemical evidence for mixing of river water and groundwater during high-flow conditions, lower Suwannee River basin, Florida, USA. *Hydrogeology Journal* 7:454–467.

Doren, M. V. (ed.). 1928. *The travels of William Bartram.* Macy-Masius, New York.

Duarte, C. M., and D. E. Canfield Jr. 1990. Macrophyte standing crop and primary productivity in some Florida spring-runs. *Water Resources Bulletin* 26:927–934.

Duryea, M. L., and L. A. Hermansen. 2000. Cypress: Florida's majestic and beneficial wetlands tree. Circular 1186. University of Florida Cooperative Extension Service, Institute of Food and Agricultural Sciences, University of Florida, Gainesville.

Environmental Protection Division. 1973. *Water quality investigation of the Suwannee River basin in Georgia.* Georgia Department of Natural Resources, Atlanta.

Environmental Protection Division. 1974. *Water quality monitoring data for Georgia streams, 1973.* Vol. 2. Georgia Department of Natural Resources, Atlanta.

Environmental Protection Division. 1978. *Water quality monitoring data for Georgia streams, 1977.* Georgia Department of Natural Resources, Atlanta.

Environmental Protection Division. 1981. *Water quality monitoring data for Georgia streams, 1980.* Georgia Department of Natural Resources, Atlanta.

Environmental Protection Division. 1984. *Water quality monitoring data for Georgia streams, 1983.* Georgia Department of Natural Resources, Atlanta.

Environmental Protection Division. 2002. *Suwannee River basin management plan 2002.* Georgia Department of Natural Resources, Atlanta.

Ewel, K. C. 1990. Swamps. In R. L. Myers and J. J. Ewel (eds.). *Ecosystems of Florida*, pp. 281–323. University of Central Florida Press, Orlando.

Fenneman, N. M. 1938. *Physiography of the eastern United States.* McGraw-Hill, New York.

Frick, E. A., D. J. Hippe, G. R. Buell, C. A. Couch, E. H. Hopkins, D. J. Wangsness, and J. W. Garrett. 1998. Water quality in the Apalachicola–Chattahoochee–Flint River basin, Georgia, Alabama, Florida, 1992–1995. U.S. Geological Survey Circular 1164. Government Printing Office, Washington, D.C.

Galloway, P. (ed.). 1997. *The Hernando de Soto expedition: History, historiography, and "discovery" in the Southeast.* University of Nebraska Press, Lincoln.

Golladay, S. W. 1997. Suspended particulate organic matter concentration and export in streams. In J. R. Webster and J. L. Meyer (eds.). Stream organic matter budgets, pp. 122–131. *Journal of the North American Benthological Society* 16:3–161.

Golladay, S. W., and K. McIntyre. 2004. Fauna and flora of the Flint River. In *The natural Georgia series: The Flint River.* Sherpa Guides Web site: www.sherpaguides.com/georgia/flint_river/wildnotes/index.html

Golladay, S. W., K. Watt, S. Entrekin, and J. Battle. 2000. Hydrologic and geomorphic controls on suspended and particulate organic matter concentration and transport in Ichawaynochaway Creek, Georgia, USA. Archive für Hydrobiologie 149:655–678.

Green, A. C. 2000. The distribution and speciation of mercury in the Pearl River drainage basin, Mississippi. Master's thesis, University of Alabama, Tuscaloosa.

Haag, W. R., and M. L. Warren Jr. 1995. Current distributional information on freshwater mussels (family Unionidae) in Mississippi national forests. General Technical Report SO-119. U.S. Department of Agriculture Forest Service, Southern Forest Experiment Station, New Orleans.

Harris, S. C., P. E. O'Neill, and P. K. Lago. 1991. *Caddisflies of Alabama.* Bulletin 142. Geological Survey of Alabama, Tuscaloosa.

Hershler, R., J. M. Pierson, and R. S. Krotzer. 1990. Rediscovery of *Tulotoma magnifica* (Conrad) (Gastropoda: Viviparidae). *Proceedings of the Biological Society of Washington* 103:815–824.

Hicks, D. W., and S. P. Opsahl. 2004. The natural history of the Flint River. In *The natural Georgia series: The Flint River.* Sherpa Guides Web site: http://www.sherpaguides.com/georgia/flint_river/natural_history/index.html

Hobbs, H. H. Jr. 1942. The crayfish of Florida. *University of Florida Biological Science Series* 3:1–179.

Hobbs, H. H. 1981. *The crayfish of Georgia.* Smithsonian Contributions to Zoology 318:1–549.

Hobbs, H. H., Jr., and W. C. Hart Jr. 1959. The freshwater decapod crustaceans of the Apalachicola drainage system in Florida, southern Alabama and Georgia. *Bulletin of the Florida State Museum. Biological Sciences* 4:145–191.

Howard, H. S., B. Quinn, M. C. Flexner, and R. L. Raschke. 2002. Cahaba River: Biological water quality studies, Birmingham, Alabama. March/April, September and July 2002. USEPA, Region 4, Science and Ecosystem Support Division, Ecological Support Branch, Atlanta.

Hudson, J. R. 1993. *Looking for De Soto: A search through the South for the Spaniard's trail.* University of Georgia Press, Athens.

Hudson, P. F., and J. Moussa. 1997. Suspended sediment transport effectiveness of three large impounded rivers, U.S. Gulf Coastal Plain. *Environmental Geology* 32:263–273.

Hunt, C. B. 1967. *Physiography of the United States.* W. H. Freeman, San Francisco, California.

Jackson, H. H., III. 1995. *Rivers of history: Life on the Coosa, Tallapoosa, Cahaba, and Alabama.* University of Alabama Press, Tuscaloosa.

Jenkins, N. A., and R. A. Krause. 1986. *The Tombigbee watershed in Southeastern prehistory.* University of Alabama Press, Tuscaloosa.

Johnson, G. C., R. E. Kidd, C. Journey, H. Zappia, and J. B. Atkins. 2002. Environmental setting and water-quality issues of the Mobile River basin, Alabama, Georgia, Mississippi, and Tennessee. U.S. Geological Survey Water-Resources Investigations Report 02-4162. Government Printing Office, Washington, D.C.

Johnson, J. K. 2000. The Chickasaws. In B. G. McEwan (ed.). *Indians of the greater Southeast: Historical archaeology and ethnohistory*, pp. 85–121. University Press of Florida, Gainesville.

Joo, G.-J. 1990. Limnological studies of oxbow lakes in the southeastern United States: Morphometry, physicochemical characteristics and patterns of primary pro-

ductivity. Ph.D. diss., University of Alabama, Tuscaloosa.

Lacefield, J. 2000. *Lost worlds in Alabama rocks: A guide to the state's ancient life and landscapes.* Alabama Geological Society, Tuscaloosa.

Lang, J. 1972. Geohydrologic summary of the Pearl River Basin, Mississippi and Louisiana. U.S. Geological Survey Water-Supply Paper 1899:M1–M44.

Lefebvre, L. W., T. J. O'Shea, G. B. Rathbun, and R. C. Best. 1989. Distribution, status, and biogeography of the West Indian manatee. In C. A. Woods (ed.). *The biogeography of the West Indies: Past, present, and future*, pp. 567–620. Sandhill Crane Press, Gainesville, Florida.

Leitman, H. M., J. E. Sohm, and M. A. Franklin. 1983. Wetland hydrology and the tree distribution of the Apalachicola River floodplain, Florida. U.S. Geological Survey Water-Supply Paper 2196. Government Printing Office, Washington, D.C.

Leopold, L. 1994. *A view of the river.* Harvard University Press, Cambridge, Massachusetts.

Light, H. M., M. R. Darst, and J. W. Grubbs. 1998. Aquatic habitat in relation to river flow in the Apalachicola River floodplain, Florida. U.S. Geological Survey Professional Paper 1594:1–77.

Light, H. M., M. R. Darst, L. J. Lewis, and D. A. Howell. 2002. Hydrology, vegetation, and soils of riverine and tidal floodplain forests of the lower Suwannee River, Florida, and potential impacts of flow reductions. U.S. Geological Survey Professional Paper 1656A:1–124.

Lineback, N. G. 1973. *Atlas of Alabama.* University of Alabama Press, Tuscaloosa.

Livingston, R. J. 1992. Medium sized rivers of the Gulf Coastal Plain. In C. T. Hackney, S. M. Adams, and W. H. Martin (eds.). *Biodiversity of the Southeastern United States*, pp. 351–385. John Wiley and Sons, New York.

Lydeard, C., and R. L. Mayden. 1995. A diverse and endangered aquatic ecosystem of the southeast United States. *Conservation Biology* 9:800–805.

Lydeard, C., B. McLane, and W. Duncan. 2003. *An ecological portrait of the Black Warrior River watershed.* Alabama Rivers Alliance, Birmingham.

Marella, R. L., J. L. Fanning, and W. S. Mooty. 1993. Estimated use of water in the Apalachicola–Chattahoochee–Flint river basin during 1990 with state summaries from 1970 to 1990. U.S. Geological Survey Water-Resources Investigations Report 93-4084. Government Printing Office, Washington, D.C.

Martens, K. 1997. Speciation in ancient lakes. *Trends in Ecology and Evolution* 12:177–182.

Mattson, R. A., J. H. Epler, and M. K. Hein. 1995. Description of benthic communities in karst, spring-fed streams of north central Florida. *Journal of the Kansas Entomological Society* 68:18–41.

Mayden, R. L., and B. R. Kuhajda. 1996. Systematics, taxonomy, and conservation status of the endangered

Alabama sturgeon, *Scaphirhynchus suttkusi* Williams and Clemmer (Actinopterygii, Acipenseridae). *Copeia* 1996:241–273.

McEwan, B. G. 2000. The Apalachee indians of northwest Florida. In B. G. McEwan (ed.). *Indians of the greater Southeast: Historical archaeology and ethnohistory*, pp. 57–84. University Press of Florida, Gainesville.

McGregor, S. W., and P. E. O'Neill. 1992. The biology and water-quality monitoring of the Sipsey River and Lubbub and Bear creeks, Alabama, 1990–1991. Circular 169. Geological Survey of Alabama, Tuscaloosa.

McGregor, S. W., T. E. Shepard, T. D. Richardson, and J. F. Fitzpatrick Jr. 1999. A survey of the primary tributaries of the Alabama and Lower Tombigbee rivers for freshwater mussels, snails, and crayfish. Circular 196. Geological Survey of Alabama, Tuscaloosa.

McWilliams, R. G. 1981. *Iberville's Gulf journals.* University of Alabama Press. Tuscaloosa.

Mettee, M. F., P. E. O'Neil, and J. M. Pierson. 1996. Fishes of Alabama and the Mobile Basin. Oxmoor House, Birmingham, Alabama.

Mettee, M. F., P. E. O'Neill, J. M. Pierson, and R. D. Suttkus. 1989. Fishes of the Black Warrior River system in Alabama. Bulletin 133. Geological Survey of Alabama, Tuscaloosa.

Michener, W. K., E. R. Blood, J. B. Box, C. A. Couch, S. W. Golladay, D. J. Hippe, R. J. Mitchell, and B. J. Palik. 1998. Tropical storm flooding of a coastal plain landscape. *BioScience* 48:696–705.

Milanich, J. T. 2000. The Timucua Indians of northern Florida and southern Georgia. In B. G. McEwan, (ed.). *Indians of the greater Southeast: Historical archaeology and ethnohistory*, pp. 1–25. University Press of Florida, Gainesville.

Mississippi Department of Environmental Quality (MDEQ). 2000. *Pearl River Basin status report 2000.* Mississippi Department of Environmental Quality, Jackson.

Mississippi Department of Environmental Quality. 2001. *Pascagoula River Basin status report 2001.* Mississippi Department of Environmental Quality, Jackson.

Mississippi Natural Heritage Program. 2002a. *List of ecological communities.* Museum of Natural Science, Mississippi Department of Wildlife, Fisheries, and Parks, Jackson.

Mississippi Natural Heritage Program. 2002b. *Special animal tracking list.* Museum of Natural Science, Mississippi Department of Wildlife, Fisheries, and Parks, Jackson.

Mitsch, W. J., and J. G. Gosselink. 2000. *Wetlands.* 3rd ed. John Wiley and Sons, New York.

Morales, J. B. T. 1990. The distribution of freshwater snails among different ordered streams of the Cahaba River drainage, Alabama. Master's thesis, University of Alabama, Tuscaloosa.

Mount, R. H. 1975. *The reptiles and amphibians of Alabama.* University of Alabama Press, Tuscaloosa.

Myers, R. L., and J. J. Ewel (eds.). 1990. *Ecosystems of Florida.* University of Central Florida Press, Orlando.

Neves, R. J., A. E. Bogan, J. D. Williams, S. A. Ahlstedt, and P. W. Hartfield. 1997. Status of aquatic mollusks in the Southeastern United States: A downward spiral of diversity. In G. W. Benz and D. E. Collins (eds.). *Aquatic fauna in peril: The Southeastern perspective*, pp. 43–85. Lenz Design & Communications, Decatur, Georgia.

Nordlie, F. G. 1990. Rivers and springs. In R. L. Myers and J. J. Ewel (eds.). *Ecosystems of Florida*, pp. 392–426. University of Central Florida Press. Orlando.

Odum, H. T. 1957a. Primary production measurements in eleven Florida springs and a marine turtle grass community. *Limnology Oceanography* 2:85–97.

Odum, H. T. 1957b. Trophic structure and productivity of Silver Springs, Florida. *Ecological Monographs* 27:55–112.

Onorato, D. R., A. Angus, and K. R. Marion. 2000. Historical changes in the ichthyofaunal assemblages of the upper Cahaba River in Alabama associated with extensive urban development in the watershed. *Journal of Freshwater Ecology* 15:47–63.

Owen, M. B. 1949. *The story of Alabama: A history of the state.* Vol. 1. Lewis Historical Publishing Company, New York.

Page, L. M., and B. M. Burr. 1991. *A field guide to freshwater fishes: North America north of Mexico.* Houghton Mifflin, New York.

Pierson, J. M., W. M. Howell, R. A. Stiles, M. F. Mettee, P. E. O'Neill, R. D. Suttkus, and J. S. Ramsey. 1989. *Fishes of the Cahaba River system in Alabama.* Bulletin 134. Geological Survey of Alabama, Tuscaloosa.

Postel, S. 1996. *Dividing the waters: Food security, ecosystem health, and the new politics of scarcity.* Worldwatch Institute, Washington, D.C.

Postel, S., G. Daily, and P. Erhlich. 1996. Human appropriation of renewable freshwater. *Science* 271: 785–788.

Ratnasabapathy, M., and T. R. Deason. 1977. Phytoplankton of the Black Warrior River, Alabama. *Phytologia* 37:1–22.

Read, W. A. 1984. *Indian place names in Alabama.* University of Alabama Press, Tuscaloosa.

Ricketts, T. H., E. Dinerstein, D. M. Olson, C. J. Loucks, W. Eichbaum, D. DellaSala, K. Kavanaugh, P. Hedao, P. T. Hurley, K. M. Carney, R. Abell, and S. Walters. 1999. *Terrestrial ecosystems of North America: A conservation assessment.* Island Press, Washington, D.C.

Rosenau, J. C., G. L. Faulkner, C. W. Hendry Jr., and R. W. Hull. 1977. Springs of Florida. Florida Geological Survey Geological Bulletin no. 31, revised. (pdf document accessible at www.flmnh.ufl.edu/springs_of_fl/aaj7320/index.html; March 4, 2003)

Ross, S. T. 2001. *Inland fishes of Mississippi.* University Press of Mississippi, Jackson.

Ryan, J. J. 1969. A sedimentologic study of Mobile Bay, Alabama. Florida State University Department of Geology Contribution no. 30, Tallahassee.

Scarry, J. F. (ed.). 1996. *Political structure and change in the prehistoric Southeastern United States*. University of Florida Press, Gainesville.

Scharpf, C. 2000. Politics, science, and the fate of the Alabama sturgeon. *American Currents* 26:6–14.

Shepard, T. E., P. E. O'Neil, S. W. McGregor, and S. C. Harris. 1994. *Water quality and biomonitoring studies in the upper Cahaba River drainage of Alabama*. Bulletin 160. Geological Survey of Alabama, Tuscaloosa.

Smith, S. M. L. 2004. The Flint River valley: Shaped by Indians, agriculture, war, and industry. In *The natural Georgia series: The Flint River*. Sherpa Guides Web site: http://www.sherpaguides.com/georgia/flint_river/cultural_history/index.html

Smith, W. E. 1988. *Geomorphology of the Mobile Delta*. Bulletin 132. Geological Survey of Alabama, Tuscaloosa.

Spiers, C. A., and G. J. Dalson. 1979. Water for municipal and industrial development in Hinds, Madison, and Rankin Counties, Mississippi. Mississippi Research and Development Center, Jackson.

Stein, B. A., L. S. Kutner, and J. S. Adams. 2000. *Precious heritage: The status of biodiversity in the United States*. Oxford University Press, New York.

Stout, J. P., M. J. Lelong, H. M. Dowling, and M. T. Powers. 1982. Wetland habitats of the Alabama Coastal Zone, III. An inventory of wetland habitats of the Mobile-Tensaw River Delta. Alabama Technical Report no. CAB 81-49A, Daphne, Alabama. 25 p.

Suwannee River Water Management District (SRWMD). 2001. *Florida Forever 2002 workplan*. Suwannee River Water Management District. Live Oak, Florida.

Swanton, J. R. 1953. *The Indian tribes of North America*. Bureau of American Ethnology Bulletin 145. Smithsonian Institution Press, Washington, D.C.

Swift, C. C., C. R. Gilbert, S. A. Bortone, G. H. Burgess, and R. W. Yerger. 1986. Zoogeography of the freshwater fishes of the southeastern United States: Savannah River to Lake Pontchartrain. In C. H. Hocutt and E. O. Wiley (eds.), *The zoogeography of North American freshwater fishes*, pp. 213–266. John Wiley and Sons, New York.

Thigpen, S. G. 1965. *Pearl River: Highway to glory land*. Kingsport Press, Kingsport, Tennessee.

Thompson, F. G. 1984. *Freshwater snails of Florida: A manual for identification*. University of Florida Press, Gainesville.

U.S. Army Corps of Engineers (U.S.A.C.E.). 1987. Final supplement to the final environmental impact statement: Black Warrior and Tombigbee rivers Alabama (maintenance) COESAM/PDEI–87/1c01. Mobile, Alabama.

U.S. Fish and Wildlife Service, Mississippi Ecological Services Field Office. 2002. Endangered and threatened species of Mississippi. http://southeast.fws.gov/jackson/MsCo_TE.html

U.S. House of Representatives. 1875. Examination and survey of Black Warrior River, from Locust Fork to its mouth, Alabama. House Executive Documents 75–76. 43rd Congress, 2nd session (Part 2, Appendix T-7, 16–26). U.S. Government Printing Office, Washington, D.C.

U.S. Senate. 1881. Survey of Black Warrior River, from Tuscaloosa to Forks of Sipsey and Mulberry, Alabama. Senate Executive Documents 42. 46th Congress, 3rd session (Appendix K 19, 1218–1221). U.S. Government Printing Office, Washington, D.C.

U.S. Study Commission. 1963. Plan for development of the land and water resources of the southeast river basins. Appendix 5. Suwannee River. U.S. Study Commission, Southeast River Basins. Atlanta, Georgia.

Van der Schalie, H. 1938. The naiades (freshwater mussels) of the Cahaba River in northern Alabama. Occasional Papers of the Museum of Zoology 392. University of Michigan, Ann Arbor.

Vogelmann, J. E., S. M. Howard, L. Yang, C. R. Larson, B. K. Wylie, and N. Van Driel. 2001. Completion of the 1990s National land cover data set for the conterminous United States from Landsat Thematic Mapper data and ancillary data sources. Photogrammetric Engineering and Remote Sensing 67:650–662.

Ward, A. K., G. M. Ward, J. Harlin, and R. Donahoe. 1992. Geological mediation of stream flow and sediment and solute loading to stream ecosystems due to climate change. In P. Firth and S. G. Fisher (eds.). Global climate change and freshwater ecosystems, pp. 116–142. Springer-Verlag, New York.

Ward, A. K., G. M. Ward, and S. C. Harris. 1992. Water quality and biological communities of the Mobile River drainage, eastern Gulf of Mexico region. In C. D. Becker and D. A. Neitzel (eds.). *Water quality in North American river systems*, pp. 279–304. Battelle Press, Columbus, Ohio.

Warren, M. L., Jr., B. M. Burr, S. J. Walsh, H. L. Bart, R. C. Cashner, D. A. Etnier, B. J. Freeman, B. R. Kuhajda, R. L. Mayden, H. W. Robison, S. T. Ross, and W. C. Starnes. 2000. Diversity, distribution, and conservation status of the native freshwater fishes of the southern United States. *Fisheries* 25:7–31.

Whitford, L. A. 1956. The communities of algae in the springs and spring streams of Florida. *Ecology* 37:433–442.

Wiche, G. J., J. J. Gilbert, D. C. Froelich, and J. K. Lee. 1988. Analysis of alternative modifications for reducing backwater at Interstate Highway 10 crossing of the Pearl River near Slidell, LA. U.S. Geological Survey-Water-Supply Paper 2267. U.S. Geological Survey, Washington, D.C.

Williams, J. D., M. L. Warren, K. S. Cummings, J. L. Harris, and R. J. Neves. 1993. Conservation status of freshwater mussels of the United States and Canada. *Fisheries* 19:6–22.

Williams, J. D., S. L. H. Fuller, and R. Grace. 1992. Effects of impoundments on freshwater mussels (Mollusca: Bivalvia: Unionidae) in the main channel of the Black Warrior and Tombigbee rivers in western Alabama. *Alabama Museum of Natural History Bulletin* 13: 1–10.

Willoughby, L. 1999. *Flowing through time: A history of the lower Chattahoochee River.* University of Alabama Press, Tuscaloosa.

Woods, P. D. 1980. French–Indian relations on the southern frontier, 1699–1762. UMI Press, Ann Arbor, Michigan.

Worth, J. E. 2000. The lower Creeks. In B. G. McEwan (ed.). *Indians of the greater Southeast: Historical archaeology and ethnohistory*, pp. 265–298. University Press of Florida, Gainesville.

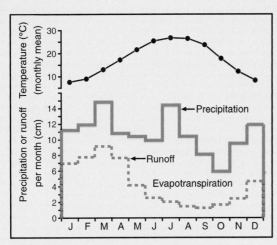

FIGURE 4.13 Mean monthly air temperature, precipitation, and runoff for the Mobile River basin.

MOBILE RIVER

Relief: 1278 m
Basin area: 111,369 km²
Mean discharge: 1914 m³/s
River order: 8
Mean annual precipitation: 128 cm
Mean air temperature: 17.4°C
Mean water temperature: 19.9°C
Physiographic provinces: Coastal Plain (CP), Valley and Ridge (VR), Appalachian Plateau (AP), Piedmont Plateau (PP), Blue Ridge (BL)
Biome: Temperate Deciduous Forest
Freshwater ecoregion: Mobile Bay
Terrestrial ecoregions: Appalachian/Blue Ridge Forests, Appalachian Mixed Mesophytic Forests, Southeastern Mixed Forests, Southeastern Conifer Forests
Number of fish species: 236
Number of endangered species: 12 fishes, 3 reptiles, 19 mussels, 7 snails
Major fishes: paddlefish, Gulf sturgeon, Alabama sturgeon, gars, shads, highfin carpsucker, southeastern blue sucker, spotted sucker, river

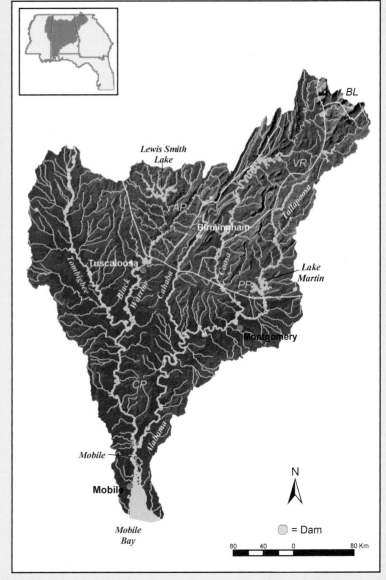

FIGURE 4.12 Map of the Mobile River basin. Physiographic provinces are separated by yellow lines.

redhorse, golden redhorse, blacktail redhorse, channel catfish, flathead catfish, white bass, striped bass, warmouth, green sunfish, bluegill, longear sunfish, spotted bass, largemouth bass, white crappie, black crappie, freshwater drum
Major other aquatic vertebrates: river otter, American alligator, cottonmouth, diamondback water snake, alligator snapping turtle, common snapping turtle, flattened musk turtle, southern black-knobbed map turtle, Alabama map turtle, southern painted turtle, Alabama redbelly turtle, spiny softshell turtle, yellowbelly slider, red-eared slider
Major invertebrates: mollusks (rough fatmucket, southern fatmucket, yellow sandshell, little spectaclecase, Alabama orb, bleufer, *Pleurocera*, *Leptoxis*, *Elimia*), mayflies (*Hexagenia*, *Stenacron*, *Stenonema*, *Isonychia*, *Baetis*), stoneflies (*Acroneuria*, *Perlesta*, *Isoperla*), caddisflies (*Hydropsyche*, *Cheumatopsyche*, *Hydroptila*, *Ceraclea*, *Oecetis*)
Nonnative species: common carp, silver carp, bighead carp, grass carp, red shiner, fathead minnow, muskellunge, rainbow trout, brown trout, brook trout, palmetto bass, smallmouth bass, yellow perch, blue tilapia, white catfish, goldfish
Major riparian plants: bald cypress, eastern cottonwood, swamp cottonwood, mockernut hickory, river birch, American hornbeam, American beech, southern red oak, water oak, live oak, American elm, yellow poplar, sweetgum, American sycamore, American holly, red maple, black gum, water tupelo, swamp tupelo, Carolina ash
Special features: largest flow into eastern Gulf of Mexico and 5th largest of rivers in United States that reach the sea; high diversity of fishes, turtles, mollusks, insects; high levels of endemism and high levels of threatened/endangered species; large pristine forest at mouth
Fragmentation: 36 major dams
Water quality: pH = 7.3, alkalinity = 40.8 mg/L as $CaCO_3$, NO_3-N = 0.27 mg/L, PO_4-P = 0.02 mg/L
Land use: 68% forest, 18% agriculture, 7% wetland, <2% urban, 5% other
Population density: 43.9 people/km²
Major information sources: Boschung and Mayden 2004, Johnson et al. 2002, Harris et al. 1991, Lydeard et al. 2003

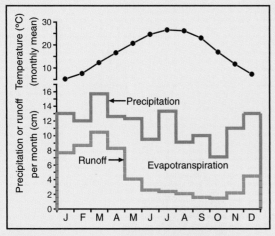

FIGURE 4.15 Mean monthly air temperature, precipitation, and runoff for the Cahaba River basin.

CAHABA RIVER

Relief: 274 m
Basin area: 4730 km²
Mean discharge: 80 m³/s
River order: 6
Mean annual precipitation: 138 cm
Mean air temperature: 16.7°C
Mean water temperature: 18.1°C
Physiographic provinces: Valley and Ridge (VR), Coastal Plain (CP)
Biome: Temperate Deciduous Forest
Freshwater ecoregion: Mobile Bay
Terrestrial ecoregions: Southeastern Mixed Forests, ppalachian Mixed Mesophytic Forests
Number of fish species: 135
Number of endangered species: 3 fishes, 4 mollusks
Major fishes: paddlefish, Alabama sturgeon, gars, shads, blacktail shiner, quillback, highfin carpsucker, southeastern blue sucker, spotted sucker, river redhorse, golden redhorse, blacktail redhorse, blue catfish, channel catfish, flathead catfish, white bass, warmouth, green sunfish, bluegill, longear sunfish, spotted bass, largemouth bass, white crappie, black crappie, freshwater drum

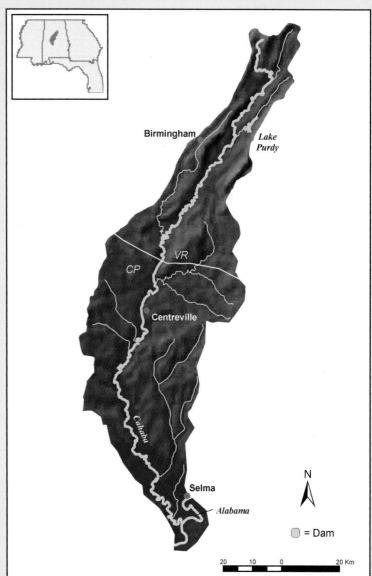

FIGURE 4.14 Map of the Cahaba River basin. Physiographic provinces are separated by yellow line.

Major other aquatic vertebrates: river otter, American alligator, alligator snapping turtle, common snapping turtle, common musk turtle, stripeneck musk turtle, eastern mud turtle, Alabama map turtle, northern black-knob sawback map turtle, Gulf Coast smooth softshell turtle, Gulf Coast spiny softshell turtle, southern painted turtle, river cooter, cottonmouth water snake, yellowbelly water snake, diamondback water snake
Major invertebrates: mollusks (southern fatmucket, Alabama orb, elephant ear, bleufer, three-horned wartyback, *Elimia*, *Pleurocera*, *Leptoxis*, *Somatogyrus*), mayflies (*Stenacron*, *Stenonema*, *Tricorythodes*, *Eurylophella*, *Serratella*), stoneflies (*Acroneuria*, *Perlesta*, *Isoperla*), caddisflies (*Cheumatopsyche*, *Hydropsyche*, *Hydroptila*, *Cyrnellus*, *Ceraclea*)
Nonnative species: grass carp, common carp, fathead minnow, white catfish, palmetto bass, smallmouth bass, Asiatic clam
Major riparian plants: bald cypress, eastern cottonwood, swamp cottonwood, mockernut hickory, river birch, American hornbeam, American beech, southern red oak, water oak, live oak, American elm, yellow poplar, sweetgum, American sycamore, American holly, red maple, black gum, water tupelo, swamp tupelo, Carolina ash
Special features: longest free-flowing river in southeastern Gulf Coast region; most fish species for its size in North America; very high mollusk diversity
Fragmentation: small impoundment in headwater tributary
Water quality: pH = 7.7, alkalinity = 84.0 mg/L as $CaCO_3$, NO_3-N = 0.34 mg/L, PO_4-P = 0.02 mg/L
Land use: 77% forest, 11% agriculture, 6% wetland, 2% urban, 4% other
Population density: 33 people/km²
Major information sources: Pierson et al. 1989, Bogan and Pierson 1993a, Onorato et al. 2000, Shepard et al. 1994

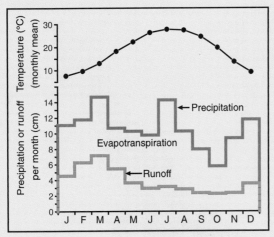

FIGURE 4.17 Mean monthly air temperature, precipitation, and runoff for the Apalachicola River basin.

APALACHICOLA-CHATTAHOOCHEE-FLINT RIVER

Relief: 1066 m

Basin area: 50,688 km²

Mean discharge: 759 m³/s

River order: 8

Mean annual precipitation: 128 cm

Mean air temperature: 18.3°C

Mean water temperature: 20.6°C

Physiographic provinces: Blue Ridge (BL), Piedmont Plateau (PP), Coastal Plain (CP)

Biome: Temperate Deciduous Forest

Freshwater ecoregion: Apalachicola

Terrestrial ecoregions: Appalachian/Blue Ridge Forests, Appalachian Mixed Mesophytic Forests, Southeastern Mixed Forests, Southeastern Conifer Forests

Number of fish species: 104

Number of endangered species: 1 fish, 1 reptile

Major fishes: Gulf sturgeon, longnose gar, American eel, Alabama shad, gizzard shad, Apalachicola redhorse, channel catfish, white catfish, yellow bullhead, brown bullhead, spotted bullhead, striped bass, largemouth bass, shoal bass, black crappie, redbreast sunfish, warmouth, bluegill

Major other aquatic vertebrates: river otter, American alligator, cottonmouth, alligator snapping turtle, Barbor's map turtle, striped mud turtle, common musk turtle, Florida softshell turtle, Gulf Coast spiny softshell turtle, Florida redbelly turtle, Florida cooter, river cooter, yellowbelly slider, Alabama waterdog

Major invertebrates: mollusks (eastern elliptio, variable spike, little spectaclecase, elephantear, *Campeloma*, *Elimia*, *Viviparus*, *Lioplax*, *Planorbella*), mayflies (*Hexagenia*, *Stenonema*, *Baetis*, *Caenis*), stoneflies (*Acroneuria*, *Neoperla*, *Paragnetina*, *Perlesta*, *Clioperla*, *Taeniopteryx*), caddisflies (*Ceraclea*, *Cheumatopsyche*, *Hydropsyche*, *Oecetis*)

Nonnative species: common carp, grass carp, green sunfish, orange spotted sunfish, walking catfish, goldfish, tilapia, Asian swamp eel, Asiatic clam

Major riparian plants: water tupelo, Ogeechee tupelo, bald cypress, Carolina ash, swamp tupelo, sweetgum, overcup oak, planer tree, green ash, water hickory, diamond-leaf oak, American elm, American hornbeam, water oak, red maple, sweetbay

Special features: second largest drainage into eastern Gulf of Mexico, high diversity of channel gradients and geology; historically diverse mollusk, crayfish, and fish fauna

Fragmentation: highly fragmented, with 16 main-stem dams, 13 on Chattahoochee River and 3 on Flint River

Water quality: pH = 7.4, alkalinity = 37.9 mg/L as $CaCO_3$, NO_3-N = 0.33 mg/L, PO_4-P = 0.02 mg/L

Land use: 55% forest, 25% agriculture, 10% wetland, 2% urban, 8% other

Population density: 51.3 people/km²

Major information sources: Box and Williams 2000, Clench and Turner 1956, Conant and Collins 1991, www.gwf.org/protectedanimals.htm 2002a, Leitman et al. 1983, www.nanfa.org/NANFAregions/ga/ga_fw_fishes.htm 2002, Myers and Ewel 1990

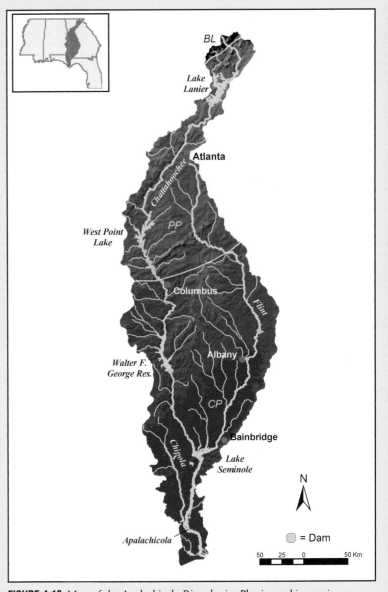

FIGURE 4.16 Map of the Apalachicola River basin. Physiographic provinces are separated by yellow lines.

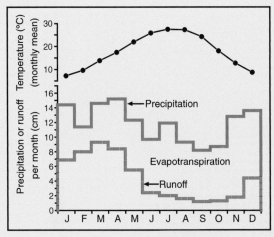

FIGURE 4.19 Mean monthly air temperature, precipitation, and runoff for the Pearl River basin.

PEARL RIVER

Relief: 210 m
Basin area: 21,999 km^2
Mean discharge: 373 m^3/s
River order: 6
Mean annual precipitation: 142 cm
Mean air temperature: 17.8°C
Mean water temperature: 19.2°C
Physiographic province: Coastal Plain (CP)
Biome: Temperate Deciduous Forest
Freshwater ecoregion: Mississippi Embayment
Terrestrial ecoregions: Southeastern Mixed Forests, Southeastern Conifer Forests
Number of fish species: 119
Number of endangered/threatened species: 1 fish, 2 reptiles, 1 mussel
Major fishes: paddlefish, Gulf sturgeon, alligator gar, gizzard shad, highfin carpsucker, southeastern blue sucker, smallmouth buffalo, blacktail redhorse, yellow bullhead, channel catfish, flathead catfish, warmouth, green sunfish, bluegill, longear sunfish, redspotted sunfish, spotted bass, largemouth bass, white crappie, black crappie, drum

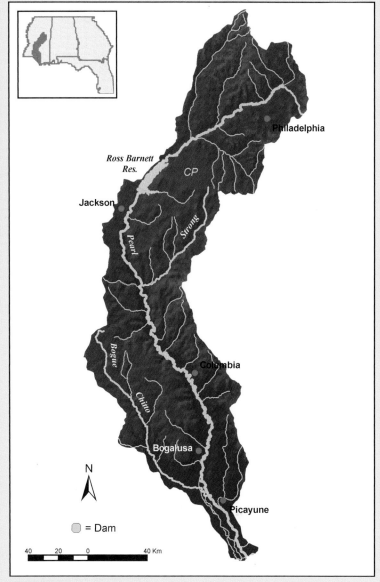

FIGURE 4.18 Map of the Pearl River basin.

Major other aquatic vertebrates: river otter, American alligator, alligator snapping turtle, common snapping turtle, stripeneck musk turtle, Pascagoula map turtle, ringed map turtle, Gulf Coast spiny softshell turtle, cottonmouth
Major invertebrates: mollusks (paper pondshell, giant floater, *Lampsilis teres*, Asiatic clam, *Viviparus*), mayflies (*Stenonema, Baetis, Caenis, Tricorythodes, Isonychia*), stoneflies (*Paragnetina, Acroneuria, Neoperla*), caddisflies (*Hydropsyche, Cheumatophsyche, Hydroptila, Chimarra, Ceraclea*)
Nonnative species: fathead minnow, common carp, goldfish, Asiatic clam
Major riparian plants: eastern cottonwood, black willow, river birch, sycamore, silver maple, laurel oak, willow oak, water oak, sugarberry, American elm, green ash, overcup oak, water hickory, bald cypress, pond cypress, black gum, water hickory oak, water tupelo, sweetbay, red maple, slash pine
Special features: low-gradient Coastal Plain river with swamps and wide floodplain forests; large coastal estuary; high fish species richness
Fragmentation: one large impoundment, Ross Barnett dam, near Jackson, Mississippi
Water quality: pH = 6.4, alkalinity = 11.2 mg/L as CaCO$_3$, NO$_3$-N = 0.21 mg/L, PO$_4$-P = 0.03 mg/L
Land use: 58% forest, 24% agriculture, 12% wetland, <2% urban, 3% other
Population density: 42 people/km^2
Major information sources: Boschung and Mayden 2004, Conant and Collins 1991, Mount 1975, Ross 2001, Swift et al. 1986

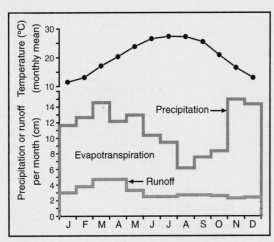

FIGURE 4.21 Mean monthly air temperature, precipitation, and runoff for the Suwannee River basin.

SUWANNEE RIVER

Relief: 140 m
Basin area: 24,967 km²
Mean discharge: 294 m³/s
River order: 7
Mean annual precipitation: 134 cm
Mean air temperature: 20.2°C
Mean water temperature: 19.7°C
Physiographic province: Coastal Plain (CP)
Biome: Temperate Deciduous Forest
Freshwater ecoregion: Florida Gulf
Terrestrial ecoregion: Southeastern Conifer Forests
Number of fish species: 81
Number of endangered species: 1 fish
Major fishes: Gulf sturgeon, longnose gar, Florida gar, gizzard shad, threadfin shad, blacktail shiner, spotted sucker, channel catfish, white catfish, flathead catfish, snail bullhead, yellow bullhead, brown bullhead, spotted bullhead, striped bass, Suwannee bass, largemouth bass, black crappie, warmouth, redbreast sunfish, spotted sunfish, bluegill

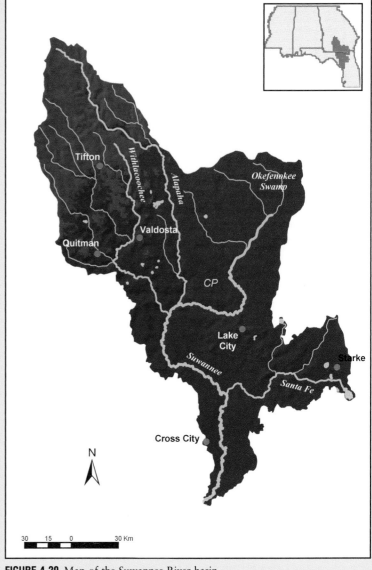

FIGURE 4.20 Map of the Suwannee River basin.

Major other aquatic vertebrates: river otter, West Indian manatee, American alligator, alligator snapping turtle, Florida snapping turtle, Florida softshell turtle, Florida redbelly turtle, Suwannee cooter, Florida cooter, Florida cottonmouth, Gulf saltmarsh swamp snake, North Florida swamp snake, redbelly water snake, banded water snake, Florida green water snake, brown water snake
Major invertebrates: mollusks (*Elliptio, Villosa, Quincucina, Musculium, Campeloma, Elimia, Lioplax, Micromenetus, Notogilla*), mayflies (*Baetis, Pseudocloeon, Stenacron, Stenonema, Tricorythodes*), stoneflies (*Acroneuria, Neoperla, Attaneuria, Paragnetina, Perlesta*), caddisflies (*Cheumatopsyche, Hydropsyche, Hydroptila, Ceraclea, Oecetis, Chimarra*)
Nonnative species: American shad, grass carp, blue catfish, flathead catfish, wiper, Asiatic clam
Major riparian plants: bald cypress, water elm, swamp laurel oak, overcup oak, live oak, sand live oak, sweetgum, river birch, planer tree, cabbage palm, red maple, water tupelo
Special features: large free-flowing Coastal Plain river with headwater swamps (Okefenokee Swamp); swamps and wide floodplain forests in mid and lower reaches; large hydrologic inputs from many springs flowing from underlying Floridan aquifer
Fragmentation: no impoundments
Water quality: pH = 7.0. alkalinity = 116.3 mg/L as CaCO₃, NO₃-N = 0.50 mg/L, PO₄-P = 0.20 mg/L
Land use: 38% forest, 30% agricultural, 22% wetland, 1% urban, 9% other
Population density: 22 people/km²
Major information sources: Bass et al. 2003, http://nas.er.usgs.gov/fishes/index.html 2003, Myers and Ewel 1990, www.flmnh.ufl.edu/natsci/herpetology/FL-GUIDE/Flaherps.htm 2003, Mattson et al. 1995, Light et al. 2002

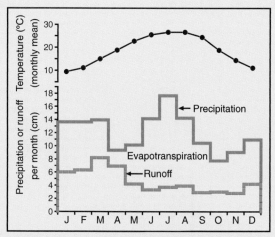

FIGURE 4.23 Mean monthly air temperature, precipitation, and runoff for the Choctawhatchee River basin.

CHOCTAWHATCHEE RIVER

Relief: 179 m
Basin area: 12,033 km^2
Mean discharge: 212 m^3/s
River order: 6
Mean annual precipitation: 144 cm
Mean air temperature: 18.7°C
Mean water temperature: 20.0°C
Physiographic province: Coastal Plain (CP)
Biome: Temperate Deciduous Forest
Freshwater ecoregion: Florida Gulf
Terrestrial ecoregions: Southeastern Conifer Forests, Southeastern Mixed Forests
Number of fish species: 80
Number of endangered species: 1 fish
Major fishes: Gulf sturgeon, spotted gar, longnose gar, American eel, Alabama shad, gizzard shad, threadfin shad, blacktail shiner, quillback, highfin carpsucker, blacktail redhorse, yellow bullhead, brown bullhead, channel catfish, striped bass, warmouth, bluegill, green sunfish, longear sunfish, spotted bass, largemouth bass, black crappie

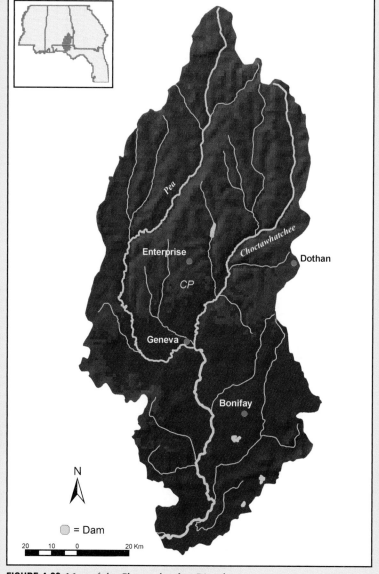

FIGURE 4.22 Map of the Choctawhatchee River basin.

Major other aquatic vertebrates: river otter, American alligator, Alabama waterdog, Florida cottonmouth, eastern cottonmouth, brown water snake, alligator snapping turtle, Florida softshell turtle, Gulf Coast spiny softshell turtle, common snapping turtle, eastern chicken turtle, mud turtle, stripeneck musk turtle, loggerhead musk turtle, yellowbelly turtle, Florida cooter, river cooter, stinkpot
Major invertebrates: mollusks (little spectaclecase, southern rainbow, variable spike, yellow sandshell, *Campeloma*, *Lioplax*, *Pomacea*, *Viviparus*), mayflies (*Isonychia*, *Leptophlebia*, *Eurylophella*), stoneflies (*Neoperla*, *Paragnetina*, *Perlesta*, *Clioperla*, *Isoperla*, *Taeniopteryx*), caddisflies (*Cheumatopsyche*, *Hydropsyche*, *Hydroptila*, *Oecetis*, *Phylocentropus*)
Nonnative species: grass carp, common carp, palmetto bass, yellow perch, unidentified pacu (*Colossoma* or *Piaractus*), Asiatic clam
Major riparian plants: bald cypress, pond cypress, black gum, water tupelo, swamp tupelo, Ogeechee tupelo, sweetgum, red maple, sweetbay, river birch, black titi, red titi, Atlantic white cedar, eastern cottonwood, swamp cottonwood, mockernut hickory, water hickory, American beech, water oak, live oak, overcup oak, American elm, yellow poplar, American sycamore
Special features: low-gradient Coastal Plain river with swamps and wide floodplain forest; one of few free-flowing southeastern Gulf of Mexico rivers; flows through lightly populated landscape into commercially valuable estuary
Fragmentation: no impoundments
Water quality: pH = 7.5, alkalinity = 34.8 mg/L as CaCO$_3$, NO$_3$-N = 0.17 mg/L, PO$_4$-P = 0.02 mg/L
Land use: 57% forest, 25% agriculture, 9% wetland, 1% urban, 8% other
Population density: 17.8 people/km^2
Major information sources: Blalock-Herod et al. 2000, Boschung and Mayden 2004, www.flmnh.ufl.edu/natsci/herpetology/FL-GUIDE/Flaherps.htm 2003, http://nas.er.usgs.gov/fishes/index.html 2003

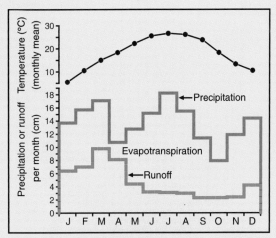

FIGURE 4.25 Mean monthly air temperature, precipitation, and runoff for the Escambia–Conecuh River basin.

ESCAMBIA–CONECUH RIVER

Relief: 180 m
Basin area: 10,963 km²
Mean discharge: 196 m³/s
River order: 6
Mean annual precipitation: 164 cm
Mean air temperature: 18.0°C
Mean water temperature: 20.4°C
Physiographic province: Coastal Plain (CP)
Biome: Temperate Deciduous Forest
Freshwater ecoregion: Florida Gulf
Terrestrial ecoregions: Southeastern Conifer Forests, outheastern Mixed Forests
Number of fish species: 102
Number of endangered species: 0
Major fishes: Gulf sturgeon, spotted gar, longnose gar, Alabama shad, gizzard shad, threadfin shad, blacktail shiner, quillback, highfin carpsucker, river redhorse, blacktail redhorse, yellow bullhead, brown bullhead, channel catfish, flathead catfish, striped bass, warmouth, bluegill, green sunfish, longear sunfish, spotted bass, largemouth bass, black crappie

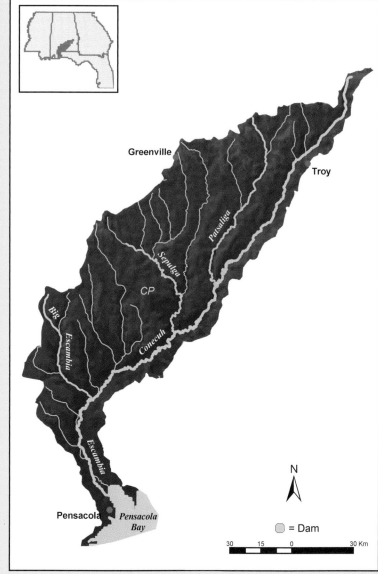

FIGURE 4.24 Map of the Escambia–Conecuh River basin.

Major other aquatic vertebrates: river otter, American alligator, Alabama waterdog, Florida cottonmouth, eastern cottonmouth, brown water snake, alligator snapping turtle, Florida softshell turtle, Gulf Coast spiny softshell turtle, common snapping turtle, eastern chicken turtle, mud turtle, Escambia map turtle, loggerhead musk turtle, stripeneck musk turtle, yellowbelly turtle, Florida cooter, river cooter, stinkpot
Major invertebrates: mollusks (*Somatogyrus, Pomacea*), mayflies (*Baetis, Isonychia, Leptophlebia, Hexagenia, Eurylophella*), stoneflies (*Acroneuria, Neoperla, Attaneuria, Paragnetina, Clioperla, Isoperla*), caddisflies (*Ceraclea, Cheumatopsyche, Hydropsyche, Hydroptila, Oecetis, Oxyethira, Phylocentropus*)
Nonnative species: grass carp, common carp, palmetto bass
Major riparian plants: bald cypress, eastern cottonwood, swamp cottonwood, mockernut hickory, river birch, American hornbeam, American beech, southern red oak, water oak, live oak, American elm, yellow poplar, sweetgum, American sycamore, American holly, red maple, black gum, water tupelo, swamp tupelo, Carolina ash
Special features: low-gradient Coastal Plain river with swamps and wide floodplain forests in lower reaches; empties into a once productive estuary
Fragmentation: largely free flowing, although two moderate-size impoundments on main stem
Water quality: pH = 7.3, alkalinity = 24.3 mg/L as $CaCO_3$, NO_3-N = 0.11 mg/L, PO_4-P = 0.02 mg/L
Land use: 72% forest, 15% agriculture, 7% wetland, <1% urban, 5% other
Population density: 33 people/km²
Major information sources: Boschung and Mayden 2004, Conant and Collins 1991, www.flmnh.ufl.edu/natsci/herpetology/FL-GUIDE/Flaherps/htm 2003, Mount 1975, http://nas.er.usgs.gov/fishes/index.html 2003

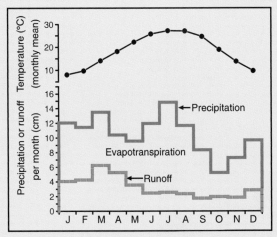

FIGURE 4.27 Mean monthly air temperature, precipitation, and runoff for the Flint River basin.

FLINT RIVER

Relief: 294 m
Basin area: 22,377 km²
Mean discharge: 283 m³/s
River order: 7
Mean annual precipitation: 126 cm
Mean air temperature: 18.2°C
Mean water temperature: 19.8°C
Physiographic provinces: Coastal Plain (CP), Piedmont Plateau (PP)
Biome: Temperate Deciduous Forest
Freshwater ecoregion: Apalachicola
Terrestrial ecoregions: Southeastern Conifer Forests, Southeastern Mixed Forests
Number of fish species: 71
Number of endangered species: 1 fish, 1 reptile, 3 mollusks
Major fishes: Gulf sturgeon, longnose gar, American eel, Alabama shad, gizzard shad, spotted sucker, Apalachicola redhorse, greater jumprock, channel catfish, white catfish, snail bullhead, brown

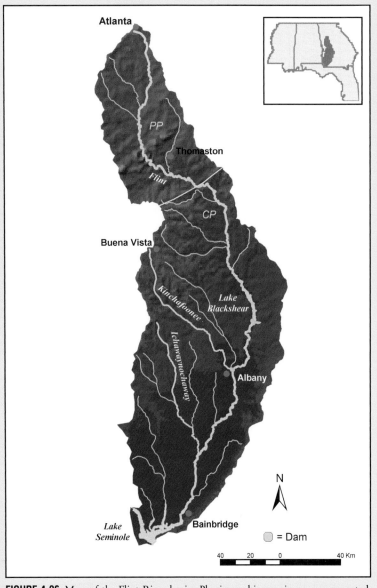

FIGURE 4.26 Map of the Flint River basin. Physiographic provinces are separated by a yellow line.

bullhead, spotted bullhead, yellow bullhead, striped bass, black crappie, shoal bass, largemouth bass, warmouth, redbreast sunfish, bluegill
Major other aquatic vertebrates: river otter, American alligator, Alabama waterdog, redbelly water snake, brown water snake, cottonmouth, snapping turtle, alligator snapping turtle, Barbor's map turtle, Florida redbelly turtle, striped mud turtle, common musk turtle, Florida softshell turtle, Gulf Coast spiny softshell turtle, Florida cooter, river cooter, yellowbelly slider
Major invertebrates: mollusks (*Elliptio, Toxolasma, Uniomerus, Elimia, Lioplax, Villosa*), mayflies (*Stenonema, Tricorythodes, Baetis*), stoneflies (*Paragnetina*), caddisflies (*Hydropsyche, Cheumatopsyche, Cyrnellus*)
Nonnative species: common carp, grass carp, flathead catfish, green sunfish, orange spotted sunfish, spotted bass, striped bass, goldfish, tilapia, walking catfish, Asian swamp eel
Major riparian plants: spruce pine, eastern hemlock, river birch, American hornbeam, American beech, white oak, water oak, laurel oak, American elm, sugarberry, umbrella magnolia, sweetbay, yellow poplar, sweetgum, American sycamore, red maple, box elder, water tupelo, black gum
Special features: second largest tributary to Apalachicola; karst terrain in lower reaches; habitat for Gulf sturgeon
Fragmentation: three dams, two large reservoirs on main stem for hydroelectric production
Water quality: pH = 7.3, alkalinity = 33.0 mg/L as CaCO₃, NO₃-N = 0.474 mg/L, PO₄-P = 0.030 mg/L
Land use: 49% forest, 34% agriculture, 9.5% wetland, 1.5% urban, 5% other
Population density: 26.8 people/km²
Major information sources: Conant and Collins 1991, www.gwf.org/protectedanimals.htm 2002a, www.gwf.org/commonreptiles.htm 2000b, North American Native Fishes Association 2002, http://nas.er.usgs.gov/fishes/index.html 2003

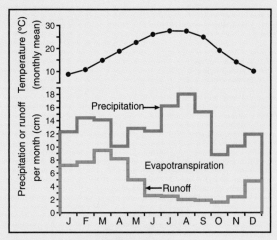

FIGURE 4.29 Mean monthly air temperature, precipitation, and runoff for the Pascagoula River–Black Creek River basin.

PASCAGOULA RIVER–BLACK CREEK

Relief: 198 m

Basin area: 24,599 km²

Mean discharge: 432 m³/s

River order: 6

Mean annual precipitation: 156 cm

Mean air temperature: 18.7°C

Mean water temperature: 19.7°C

Physiographic province: Coastal Plain (CP)

Biome: Temperate Deciduous Forest

Freshwater ecoregion: Mississippi Embayment

Terrestrial ecoregions: Southeastern Mixed Forests, Southeastern Conifer Forests

Number of fish species: 114

Number of endangered species: 1 fish, 2 reptiles

Major fishes: paddlefish, Gulf sturgeon, spotted gar, longnose gar, gizzard shad, highfin carpsucker, southeastern blue sucker, smallmouth buffalo, blacktail redhorse, yellow bullhead, channel catfish, flathead catfish, shadow bass, warmouth, green sunfish, bluegill, redspotted sunfish, spotted bass, largemouth bass, white crappie, black crappie, freshwater drum

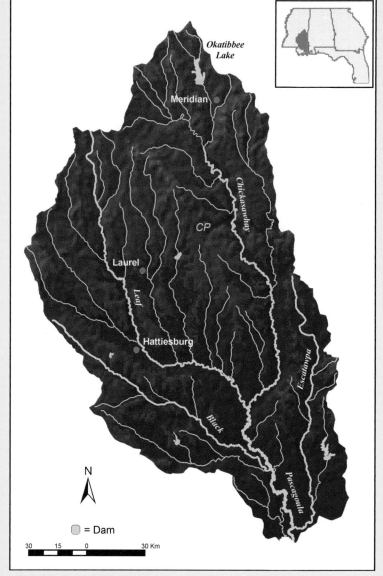

FIGURE 4.28 Map of the Pascagoula River–Black Creek basin.

Major other aquatic vertebrates: beaver, river otter, mudpuppy, cottonmouth, American alligator, alligator snapping turtle, common snapping turtle, stripeneck musk turtle, Pascagoula map turtle, yellow blotched map turtle, Gulf Coast spiny softshell turtle

Major invertebrates: mollusks (threeridge, elephantear, *Lampsilis teres*, ebonyshell, giant floater, paper pondshell, Asiatic clam, *Viviparus*), mayflies (*Stenonema, Baetis, Caenis, Tricorythodes, Isonychia*), stoneflies (*Paragnetina, Neoperla, Acroneuria*), caddisflies (*Hydropsyche, Cheumatopsyche, Hydroptila, Chimarra, Ceraclea*)

Nonnative species: bigheaded carp, fathead minnow, pirapitinga, Tilapia spp.

Major riparian plants: eastern cottonwood, black willow, river birch, sycamore, silver maple, laurel oak, willow oak, water oak, sugarberry, American elm, green ash, overcup oak, water hickory, bald cypress, pond cypress, black gum, water hickory oak, water tupelo, sweetbay, red maple, slash pine

Special features: low-gradient blackwater ecosystem with swamps and wide floodplain forests; free flowing; 34 km reach of Black Creek classified as Wild and Scenic

Fragmentation: no dams on main stem or major tributaries

Water quality: pH = 6.2, alkalinity = 13.1 mg/L as CaCO₃, NO₃-N = 0.09 mg/L, PO₄-P = 0.10 mg/L

Land use: 66% forest, 17% agriculture, 11% wetland, <1% urban, 5% other

Population density: 29 people/km²

Major information sources: Boschung and Mayden 2004, Conant and Collins 1991, www.flmnh.ufl.edu/natsci/herpetology/FL-GUIDE/Flaherps/htm 2003, Mount 1975, http://nas.er.usgs.gov/fishes/index.html 2003

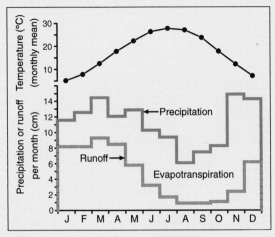

FIGURE 4.31 Mean monthly air temperature, precipitation, and runoff for the Upper Tombigbee River basin.

UPPER TOMBIGBEE RIVER

Relief: 245 m
Basin area: 18,800 km^2
Mean discharge: 336 m^3/s
River order: 6
Mean annual precipitation: 134 cm
Mean air temperature: 18.7°C
Mean water temperature: 18.8°C
Physiographic province: Coastal Plain (CP)
Biome: Temperate Deciduous Forest
Freshwater ecoregion: Mobile Bay
Terrestrial ecoregion: Southeastern Mixed Forests
Number of fish species: 122
Number of endangered species: 1 fish, 5 mussels
Major fishes: paddlefish, Alabama sturgeon, gars, shads, quillback, highfin carpsucker, smallmouth buffalo, southeastern blue, spotted sucker, river redhorse, golden redhorse, blacktail redhorse, blue catfish, channel catfish, flathead catfish, warmouth, green sunfish, bluegill, longear sunfish, redear sunfish, redspotted sunfish, spotted bass, largemouth bass, white crappie, black crappie, freshwater drum

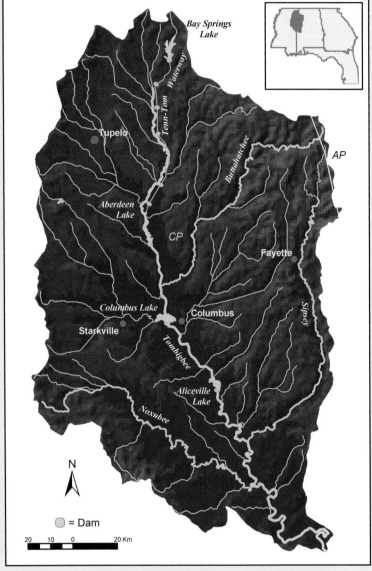

FIGURE 4.30 Map of the Upper Tombigbee River basin. Physiographic provinces are separated by a yellow line.

Major other aquatic vertebrates: river otter, American alligator, alligator snapping turtle, common snapping turtle, common musk turtle, stripeneck musk turtle, eastern mud turtle, northern black-knob sawback map turtle, eastern chicken turtle, southern painted turtle, Gulf Coast smooth softshell turtle, spiny softshell turtle, river cooter, red-eared slider, cottonmouth, yellowbelly water snake, diamondback water snake
Major invertebrates: mollusks (ebonyshell, Alabama orb, fragile papershell, threehorn wartyback, yellow sandshell, Asiatic clam, *Elimia, Somatogyrus, Lioplax*), mayflies (*Hexagenia, Stenonema, Baetis, Heptagenia, Tricorythodes, Isonychia*), stoneflies (*Neoperla*), caddisflies (*Hydropsyche, Cheumatopsyche, Chimarra, Cyrnellus*)
Nonnative species: grass carp, common carp, smallmouth bass, palmetto bass, yellow perch, Asiatic clam
Major riparian plants: bald cypress, eastern cottonwood, swamp cottonwood, mockernut hickory, river birch, American hornbeam, American beech, southern red oak, water oak, live oak, American elm, yellow poplar, sweetgum, American sycamore, American holly, red maple, black gum, water tupelo, swamp tupelo, Carolina ash
Special features: historically noted for diverse fish and mussel fauna; currently Upper Tombigbee is location of Tennessee–Tombigbee waterway connecting Mobile and Tennessee river systems
Fragmentation: numerous navigational locks and dams part of Tennessee–Tombigbee Waterway
Water quality: pH = 7.2, alkalinity = 43.0 mg/L as CaCO$_3$, NO$_3$-N = 0.27 mg/L, PO$_4$-P = 0.06 mg/L
Land use: 64% forest, 20% agriculture, 12% wetland, 2% urban, 2% other
Population density: 18 people/km^2
Major information sources: Boschung and Mayden 2004, Mettee et al. 1996, McGregor et al. 1999

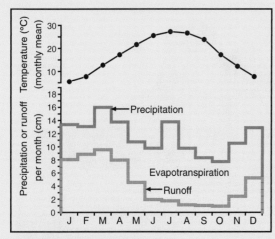

FIGURE 4.33 Mean monthly air temperature, precipitation, and runoff for the Sipsey River basin.

SIPSEY RIVER

Relief: 229 m
Basin area: 2044 km²
Mean discharge: 34 m³/s
River order: 5
Mean annual precipitation: 139 cm
Mean air temperature: 17.0°C
Mean water temperature: 17.1°C
Physiographic provinces: Coastal Plain (CP),
    Appalachian Plateau (AP)
Biome: Temperate Deciduous Forest
Freshwater ecoregion: Mobile Bay
Terrestrial ecoregion: Southeastern Mixed Forests
Number of fish species: 91
Number of endangered species: none
Major fishes: spotted gar, longnose gar, gizzard shad,
    threadfin shad, blacktail shiner, quillback,
    smallmouth buffalo, Alabama hog sucker, spotted
    sucker, river redhorse, golden redhorse, blacktail
    redhorse, channel catfish, warmouth, bluegill,
    longear sunfish, redear sunfish, redspotted sunfish,
    spotted bass, largemouth bass, freshwater drum
Major other aquatic vertebrates: river otter,
    cottonmouth, yellowbelly water snake, diamondback water snake, American alligator, alligator snapping turtle, common
    snapping turtle, common musk turtle, stripeneck musk turtle, northern black-knob sawback map turtle, eastern chicken
    turtle, southern painted turtle, Gulf Coast smooth softshell turtle, spiny softshell turtle, river cooter, red-eared slider
Major invertebrates: mollusks (southern fatmucket, yellow sandshell, little spectaclecase, Alabama orb, bleufer), mayflies
    (*Isonychia, Eurylophella, Serratella, Baetis, Caenis*), stoneflies (*Acroneuria, Perlesta, Isoperla, Neoperla, Taeniopteryx*),
    caddisflies (*Cheumatopsyche, Hydropsyche, Hydroptila, Ceraclea, Oecetis, Chimarra*)
Nonnative species: common carp
Major riparian plants: bald cypress, eastern cottonwood, swamp cottonwood, mockernut hickory, river birch, American
    hornbeam, American beech, southern red oak, water oak, live oak, American elm, yellow poplar, sweetgum, American
    sycamore, American holly, red maple, black gum, water tupelo, swamp tupelo, Carolina ash
Special features: lightly populated low-gradient Coastal Plain river with swamps and wide floodplain forests; high diversity of
    fish and mollusks
Fragmentation: no impoundments
Water quality: pH = 5.9 to 7.7, alkalinity = 4 to 28 mg/L as CaCO₃, NO₃-N = 0.35 mg/L, PO₄-P = 0.03 mg/L
Land use: 75% forest, 10% agriculture, 9% wetland, <1% urban, 5% other
Population density: 11 people/km²
Major information sources: Boschung 1989, Boschung and Mayden 2004, Conant and Collins 1991, McGregor and O'Neill
    1992, Mettee et al. 1996, Mount 1975, Pierson et al. 1989

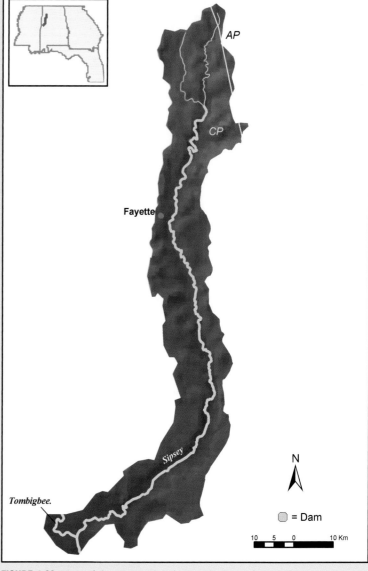

FIGURE 4.32 Map of the Sipsey River basin. Physiographic provinces are separated by a yellow line.

5

GULF COAST RIVERS OF THE SOUTHWESTERN UNITED STATES

CLIFFORD N. DAHM   ROBERT J. EDWARDS   FRANCES P. GELWICK

INTRODUCTION

RIO GRANDE

SAN ANTONIO AND GUADALUPE RIVERS

COLORADO RIVER

BRAZOS RIVER

SABINE RIVER

ADDITIONAL RIVERS

ACKNOWLEDGMENTS

LITERATURE CITED

INTRODUCTION

River catchments flowing into the Texas portion of the Gulf of Mexico encompass a broad geographic area, with latitude ranging from around 38°N in southern Colorado to 25°N in northern Mexico and longitude ranging from about 108°W in western New Mexico to 93°W in western Louisiana (Fig. 5.2). The northwestern part of the region includes the southern Rocky Mountains and the southeastern portion includes the flat coastal areas along the Gulf of Mexico. The region is traversed by a strong decreasing rainfall gradient from east to west and a temperature gradient from north to south that strongly influences vegetation, land use, and river flow. Western rivers originating in Colorado and New Mexico are snowmelt dominated and flow from snow-fed perennial streams. These rivers encounter increasingly arid conditions as they flow southward through grasslands, shrublands, and deserts. North-eastern rivers of the western Gulf flow through prairies, pine forests, and cypress-lined bayous.

Rivers crossing the desert and semiarid western parts of the region produce considerably less discharge to the Gulf of Mexico than the rivers of east Texas. The diverse geology, physiography, and climate of the region have produced rivers that have variable geomorphic, hydrologic, chemical, and biological characteristics. Many of these rivers are of great length, with the Rio Grande, Brazos, Pecos, and Colorado among the hundred longest rivers in the world (World Almanac Books 2003).

Human habitation within catchments that flow into the western Gulf of Mexico dates back to the Clovis culture of Paleo-Indians nearly 12,000 years ago. Clovis culture was first discovered from excavations in the early 1930s near Clovis, New Mexico, in the Pecos River drainage basin. These big-game hunters are the earliest definitively dated human populations in the Americas. The Folsom culture also was an early hunting group of Paleo-Indians in the region and was present until about 2500 years ago. Highly developed Native American civilizations became established 1000 to 2000 years ago in both

◄ FIGURE 5.1 Rio Grande at Big Bend National Park, western Texas. (PHOTO BY A. D. HURYN).

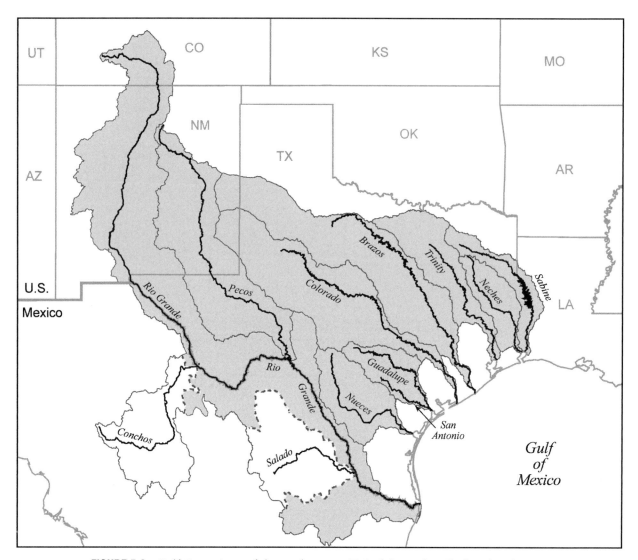

FIGURE 5.2 Gulf Coast rivers of the southwestern United States described in this chapter.

the eastern and western areas of rivers flowing into the western Gulf of Mexico. The Caddoan period of 780 to 1260 years ago in the Sabine basin was the most southwestern example of the Mississippian mound-building culture. The Anasazi culture from 600 to 1700 years ago in the Rio Grande and Colorado basins established thriving cities and cultural centers. When early Spanish explorers visited the region in the sixteenth century, they found (1) Caddoan tribes in east Texas, (2) Tonkawa, Waco, Apache, Karankawa, and Comanche tribes in central and west Texas, (3) Pueblo, Hopi, Zuni, Navajo, and Apache tribes in New Mexico, and (4) Jumano, Pataragueye, and Coahuiltecan tribes in the Rio Grande border region between Texas and Mexico.

Eleven major rivers discharge into the western Gulf of Mexico from the United States, including the Rio Grande, which borders with Mexico. They are located in generally narrow catchments that are from two to five times longer than their average widths and they have generally developed dendritic drainage systems. Extending from southwest to northeast, the rivers are the Rio Grande, Nueces, Guadalupe/San Antonio, Lavaca, Colorado, Brazos, San Jacinto, Trinity, Neches, and Sabine. A subset of five major river systems that together cover a considerable range in natural diversity and human impacts is discussed within this region. These are the Rio Grande, Guadalupe/San Antonio, Colorado, Brazos, and Sabine. In addition, one-page summaries of physical and biological information are provided for four

(Pecos, Nueces, Trinity, and Neches) of the six additional rivers. Two major Mexican tributaries of the Rio Grande (Conchos and Salado) are described in Chapter 23.

Physiography and Climate

The rivers of the western Gulf of Mexico are found in areas of differing geography that include eight physiographic provinces (Hunt 1974). The Rio Grande basin contains seven of the eight physiographic provinces in the region. These provinces include the Southern Rocky Mountains (SR), Colorado Plateau (CO), Basin and Range (BR), Great Plains (GP), Coastal Plain (CP), Sierra Madre Occidental (SC), and the Sierra Madre Oriental (SO). The eighth physiographic province in the region is the Central Lowland (CL), in which parts of the Colorado, Brazos, and Trinity catchments are located. The Sabine and Neches drainage basins are completely within the CP, whereas the San Antonio/Guadalupe and Nueces basins include the CP and GP. The Colorado, Brazos, and Trinity catchments include the CL, GP, and CP, and the Pecos basin includes the SR, BR, and GP.

The Rio Grande in southern Colorado and New Mexico flows through the Rio Grande rift. The rift transects New Mexico from south to north and extends into the headwaters region of the Rio Grande in southern Colorado. The rift consists of a series of grabens (down-dropped blocks). In the northern part of the rift, ancient Precambrian rocks cap the massive uplifts of the Sangre de Cristo and Brazos ranges. Recent Tertiary and Quaternary volcanics are on the west side of the Rio Grande graben. In central New Mexico, uplifted blocks form the Sandia and Manzano mountains with capping Pennsylvanian limestones. In southern New Mexico, the rift is bordered by complex uplifted ranges of Precambrian, Paleozoic, or lower Tertiary rocks. The Rio Grande in New Mexico flows through valleys with Quaternary and young Tertiary sediments filling the rift, sometimes to depths exceeding 10 km.

The surface geology of Texas and eastern New Mexico reflects the more tectonically stable characteristics of this region in recent Earth history. Coastal Plain areas have surface geology of generally Quaternary, Pliocene, Miocene, Oligocene, and Eocene ages. Rich oil and gas deposits from Cretaceous, Permian, Pennsylvanian, Mississippian, Jurassic, and Triassic periods dominate the geology of central Texas, whereas the panhandle region of Texas is mostly of Pliocene, Miocene, Oligocene, and Permian

age. Complex heterogeneous surface geology predominates in west Texas and much of eastern New Mexico. Rivers of the western Gulf commonly flow through multiple geological provinces from headwaters to mouth.

Soils of the region are highly varied, reflecting strong climatic gradients, varying geologic ages and structures, and differential tectonics. Calcareous and/or gypsum-rich soils are common in the more arid parts of the region, and the pH of these soils is usually neutral or slightly basic. Prairie soils are commonly slightly acidic sandy to clayey loams. Forest soils of east Texas are generally light-colored, acidic loams. Soils of the coastal plains range from light-colored to light-brown acid loams of various textures and sands.

The climate of the region ranges markedly, from humid continental in eastern Texas, to humid subtropical in the south in Texas, to alpine in southern Colorado and northern New Mexico, to desert in southern New Mexico and western Texas. Mean monthly air temperatures range from −10°C in winter in the valleys of the headwater region of the Rio Grande to 30°C in the summer months on the Coastal Plain of Texas. Winter mean monthly temperatures range from 15°C near the mouth of the Rio Grande in southern Texas to −15°C in the mountains of southern Colorado where the headwaters of the Rio Grande originate. Annual rainfall ranges from greater than 125 cm in the easternmost basins to less than 20 cm in basins in central New Mexico and western Texas. Precipitation also increases from south to north in the western portion of the region, from the deserts of southern New Mexico and western Texas (<20 cm/yr) to mountain peaks of the Rocky Mountains in southern Colorado (>125 cm/yr). Precipitation occurs as rainfall in the Coastal Plain, with an increasing role for snowfall as elevation increases to the northwest. Tropical storms and hurricanes cause short periods of high precipitation, primarily during late summer and autumn throughout much of the region.

Basin Landscape and Land Use

The rivers of the western Gulf of Mexico pass through 16 highly diverse terrestrial ecoregions, which include forests, grasslands, prairies, savannahs, shrublands, and deserts (Ricketts et al. 1999). The Colorado Rockies Forests ecoregion is a high-elevation montane area in the northwestern portion of the region and includes the headwaters of the Rio Grande and Pecos River. The high-elevation plateau of eastern New Mexico and the panhandle part of Texas are in the Western Short Grasslands ecoregion and include the headwaters of the Colorado and Brazos rivers. Three xeric shrubland or desert ecoregions are traversed by the Rio Grande: the Colorado Plateau Shrublands, Chihuahuan Desert, and Tamaulipan Mezquital. Along the lower Rio Grande is drainage from the Sierra-Madre Oriental oak–pine forests, Sierra-Madre Occidental pine–oak forests (Conchos drainage), Tamaulipan Pastizal, and Tamaulipan Matorral (around Monterrey). Four grassland, savanna, or shrubland ecoregions are located in central Texas. These ecoregions are the Edwards Plateau Savannas, the Central and Southern Mixed Grasslands, the Central Grassland/Forest Transition Zone, and the Texas Blackland Prairies. Gulf Coast terrestrial ecoregions include the East Central Texas Forests, Piney Woods Forests, and Western Gulf Coastal Grasslands.

Geomorphology of the rivers also varies markedly. Snowmelt-fed rivers like the Rio Grande originate in high-gradient montane regions with coarse substrates and high velocities. The Rio Grande also flows through an active tectonic rift valley. The Brazos and Colorado rivers start on or near the high plateau of eastern New Mexico and the Texas panhandle before passing through grasslands, prairies, savannahs, and forests. The San Antonio and Guadalupe rivers emerge as large springs from the karst region of the Edwards plateau. The Sabine River begins at a much lower elevation in mesic east Texas in a forested region with higher precipitation than the rest of the region.

The Rio Grande basin in southern Colorado is largely forested in the uplands, with agriculture in the San Luis Valley. After entering New Mexico, the Rio Grande basin is mountainous with canyonlands and valleys until entering the middle Rio Grande basin of central New Mexico. Agricultural land use in this region is focused along the riverine corridor, with grassland, shrubland, and forest moving away from the riverine corridor. Urban centers also are concentrated along the riverine corridor, with over half the population of New Mexico (almost one million people) living along this segment of the Rio Grande. South of Elephant Butte and Caballo reservoirs, the Rio Grande basin is a mix of agriculture along the river corridor, desert shrubland and grassland in the uplands, and urban centers in southern New Mexico, west Texas, and northern Mexico. Forested mountains, Chihuahuan desert, and riparian floodplain characterize the border region between Texas and Mexico. When the Río Conchos of Mexico

enters the Rio Grande (or Río Bravo del Norte), the river again flows continuously until a series of reservoirs impounds water for agricultural use in the lower Rio Grande basin.

The headwater region of the San Antonio River is largely urbanized, whereas the headwaters of the Guadalupe River begin in relatively undeveloped ranch lands. The spring-fed rivers and their tributaries are major attractions for cities like San Antonio and San Marcos. The springs at San Marcos are the most dependable springs in the region and are renowned for their biological diversity. Impoundments of these rivers provide for water supply, recreation, electrical power, and agriculture. Land use in the lower portions of these basins is largely agriculture, ranching, and wildlife refuges.

The Colorado and Brazos rivers arise near the Texas–New Mexico border and flow to the Gulf of Mexico. European-Americans from the eastern United States originally settled between these rivers on land grants to Stephen F. Austin. Early settlers first had to cross the Trinity River, which now connects Dallas–Fort Worth to Galveston Bay, just east of Houston, which lies at the mouth of the San Jacinto River. These three watersheds now contain more than half of the present population of Texas, and human activities have stressed the natural capacity of these rivers to cleanse themselves, especially in the Trinity and lower San Jacinto watersheds. Land use in these basins includes grazing land, irrigated cropland, and urban land. Urbanization is one of the most rapidly growing land uses in the Colorado, Brazos, and Trinity river basins.

The Sabine and Neches rivers drain the parts of Texas with the highest annual precipitation. The Sabine discharges the most water into the Gulf of Mexico of all the rivers of the western Gulf of Mexico. The Neches River basin contains a wealth of native plants and wildlife set aside as the Big Thicket National Preserve in 1974 and designated an International Biosphere Reserve by the United Nations Educational, Scientific, and Cultural Organization (UNESCO). Although presently fragmented, it forms corridors of old-growth bottomland hardwoods growing in soil enriched by frequent floods. Forestry, recreation, and petrochemicals are major land uses in the basins.

Dominant land use in basins that discharge into the western Gulf of Mexico shifts longitudinally. Eastern basins such as the Sabine and Neches are predominantly forested. The Trinity basin also is predominantly forested but with significant grasslands, croplands, and urban areas. Grasslands or rangelands dominate in the central part of the region in catchments of the Brazos, Colorado, San Antonio/Guadalupe, and Pecos. Shrublands are the predominant land type in the most western and southern basins (Rio Grande and Nueces) of the region. Urbanization is rapidly changing the landscape in the region, with the Trinity and San Antonio/Guadalupe basins the most urbanized basins presently.

The Rivers

There are eight freshwater ecoregions within the western Gulf rivers of this chapter (Abell et al. 2000). These include the Upper Rio Grande in parts of Colorado, New Mexico, Texas, and Mexico; the Pecos in parts of New Mexico and Texas; the Lower Rio Grande in parts of Texas and Mexico; the Rio Conchos in Mexico; the Rio Salado in Mexico; the Rio San Juan in Mexico; the West Texas Gulf in Texas; and the East Texas Gulf in parts of Texas, Louisiana, and New Mexico.

The western rivers of the region (Rio Grande and Pecos) originate in the southern Rocky Mountains of Colorado and New Mexico (see Fig. 5.2). These rivers are high gradient, with hydrology dominated by spring snowmelt. Upper reaches of these rivers are higher gradient than other rivers in the region, with a rocky substratum of gravel, cobble, boulders, and bedrock. The western portions of the Rio Grande and the Pecos are in the Upper Rio Grande and the Pecos freshwater ecoregions, respectively (Abell et al. 2000). The lower Rio Grande, or Río Bravo del Norte as it is called in Mexico, is a separate freshwater ecoregion (Abell et al. 2000). This ecoregion encompasses much of the boundary between the United States and Mexico. The river in this part of the drainage basin is lower gradient, with tributaries from Mexico such as the Rio Conchos providing much of the flow.

Only one main river, the Nueces, is located in the West Texas Gulf freshwater ecoregion (Abell et al. 2000). Arising on the Edwards Plateau, this river passes through canyonlands in semiarid landscapes to the Gulf Coast area. The East Texas Gulf freshwater ecoregion (Abell et al. 2000) includes the San Antonio/Guadalupe, Colorado (Texas), Brazos, Trinity, Neches, and Sabine.

Annual peak flows in rivers like the Rio Grande and Pecos are quite predictable. Rivers originating from karst aquifers from the Edwards Plateau (San Antonio, Guadalupe, and Nueces) are spring fed and sustain strong base flow year-round. Flows in these rivers become less predictable further downstream on the coastal plains. Peak flows in the rivers of eastern Texas are more unpredictable, with high flows possible throughout the year. Tropical storms can produce high flows in the fall, and heavy precipitation in winter and spring commonly results in floodplain forest swamps being inundated for weeks or months. However, natural flooding is substantially reduced or altered by dams and levees in upstream reaches in most rivers of the western Gulf of Mexico, and the lower reaches of these rivers are now channelized for navigation and irrigation.

An important natural feature of most lowland rivers of the Gulf Coast of Texas is the occurrence of wood (snags) as rivers undercut banks and support for trees fails. Large debris dams are documented in the history of several rivers in the region (Sabine, Colorado, Brazos, Trinity, and Neches). Navigation enhancement in the nineteenth century removed most of the major logjams that occurred naturally in the lowland portions of the rivers. Snags provided an important habitat, particularly for invertebrates, fishes, and reptiles, and thus likely enhanced overall biodiversity and faunal productivity. Unfortunately, wood has been removed from these rivers by snagboats for over a century because of hazards woody debris create for recreational and commercial navigation. More recently, the lower reaches of some of these rivers such as the Rio Grande have become choked with aquatic macrophytes, many nonnative to the region, and at times have ceased to flow to the Gulf of Mexico.

Rivers that discharge into the western Gulf of Mexico have considerable variability in their hydrology, chemistry, and biology. The Upper Rio Grande, Pecos, Lower Rio Grande, East Texas Gulf, and West Texas Gulf ecoregions are all considered to be continentally outstanding in their biological attributes (Abell et al. 2000). The current conservation status of the Pecos, West Texas Gulf, and East Texas Gulf ecoregions is vulnerable, whereas the Upper Rio Grande is endangered and the Lower Rio Grande is critical. Abell et al. (2000) anticipate that the Upper Rio Grande will become critical and the Pecos will become endangered within the next 20 years. Rivers of the Texas Gulf Coast contain more than 160 species of freshwater fishes, over 50 species of unionid mussels (Howells et al. 1996), and 38 species of crayfishes, although not all crayfishes are riverine species. At least 42 species of freshwater fishes and nine species of mussels are at risk of extinction. Major threats to rivers of the region are urban development, wastewater effluent, water extraction, nonpoint-source pollution, nonnative species, and impoundments.

RIO GRANDE

The Rio Grande is 2830 km in length and is the 5th longest river in North America and 24th longest river in the world. Its headwaters begin at approximately 3700 m asl in the San Juan Mountains of southern Colorado. After entering New Mexico, the river bisects the state and then flows in a generally southeastern direction as it forms the shared border between Texas and Mexico before it empties into the Gulf of Mexico near Brownsville, Texas (Fig. 5.11). Because of its length, the Rio Grande travels through a wide variety of habitats, ranging from mountain forests to chaparral, high mountain desert, and lowland brush country. The watershed encompasses about 870,000 km^2; however, a large proportion of the river's basin is arid or semiarid, with a number of endorheic (inwardly draining) subbasins such that only about half of the total area, or about 450,000 km^2, actually contributes to the river's overall flow.

The Rio Grande basin was home to the Clovis culture, a group of Paleo-Indians first known through artifacts unearthed near Clovis, New Mexico. The oldest Clovis cultural artifacts have been dated to about 12,000 years ago. Clovis groups were big-game hunters, and the earliest known definitively dated human populations in the Americas. The Rio Grande basin in New Mexico also was home to Anasazi culture from about 1700 to 600 years ago. Some of the descendents of the Anasazi culture that built the great ancient cities at Mesa Verde, Colorado, Chaco Canyon, New Mexico, and Canyon de Chelly, Arizona, are thought to have immigrated to the Rio Grande and Little Colorado during severe droughts in the thirteenth century. Many Indian pueblos that used the river for crop irrigation were present at the time of early Spanish exploration. Numerous thriving settlements were found along the Rio Grande in New Mexico by Spanish conquistadors in 1540. The region along the shared Texas–Mexico border also has had a long history of human habitation. Originally, major Indian groups inhabited the area and left numerous relics and pictographs throughout the region. The major tribes included the Jumanos, a group of primarily hunters and traders in the trans-Pecos region of western Texas, the Pataragueyes, a group of pueblo farmers along the Rio Grande near the El Paso area downstream to about Big Bend National Park, and the Coahuiltecans, who were primarily hunter-gatherers in the south Texas brushlands region. Because of the remote location of this latter group, they had little contact with early Europeans. Other tribes that traveled through the area and traded with indigenous groups included various Apache and, later, Comanche tribes.

Spanish explorers first sighted the mouth of the Rio Grande in 1519 during the Spanish expeditions by Alonso Alvarez de Pineda, who sailed out of Jamaica. Stories of splendid cities and gold along the river sparked exploration by Captain Francisco Vasquez de Coronado in 1540 and 1541. Although his expedition failed to find gold, the expedition described pueblos along the Rio Grande and one of its main tributaries, the Pecos. In 1598 Don Juan de Onate began settlements along the Rio Grande in the El Paso area with 100 families and 300 single men. Missions and settlements developed along the river valley until 1680, when a Pueblo revolt drove the Spanish out of the upper Rio Grande. The Spanish returned in 1692 and brutally reasserted their dominance along the river and within the basin.

Spanish control of the region began to weaken with Anglo-American colonization of Texas in the early 1800s and the Mexican revolt against Spain that led to Mexican freedom in 1821. Political unrest in Mexico and increasing numbers of Anglo-Americans in Texas set the stage for the Texas Revolution in 1835. In 1836, Texas put forth a claim for the Rio Grande from mouth to source as the southwestern and western boundary of the Republic of Texas. The Rio Grande was finally recognized by Mexico as the Texas boundary with Mexico in the Treaty of Guadalupe Hidalgo in 1848 at the end of the Mexican War (Horgan 1984).

Physiography, Climate, and Land Use

The Rio Grande flows through seven physiographic provinces including the Southern Rocky Mountains (SR), Colorado Plateau (CO), Basin and Range (BR), Great Plains (GP), Coastal Plain (CP), Sierra Madre Occidental (SC), and Sierra Madre Oriental (SO) (see Fig. 5.11). Piñons, junipers, and sagebrush in the semiarid north of New Mexico give way to mesquite, creosote bush, cactus, and other drought-tolerant plants in the south of New Mexico and west Texas. The landscape in the lower basin is primarily hilly scrub and brush, gradually less hilly as the river approaches the Gulf of Mexico. Ricketts et al. (1999) show eight terrestrial ecoregions within the Rio Grande basin (Colorado

Rockies Forests, Colorado Plateau Shrublands, Chihuahuan Desert, Tamaulipan Mezquital, Sierra Madre Occidental pine–oak Forests, Sierra Madre Oriental oak–pine Forests, Tamaulipan Matorral, and Tamaulipan Pastizal).

Climate changes markedly from the headwaters to the mouth of the Rio Grande. Climate conditions in the San Luis Valley of southern Colorado are cold and dry below the San Juan Mountains, which hold the winter snows that sustain the headwater streams of the Rio Grande. Average temperatures in Alamosa, Colorado (37.45°N, 105.86°W), at 2297 m asl range from −9.6°C in January to 18.2°C in July. Annual average precipitation is 18.1 cm, with a peak in August and a minimum in February. Average temperatures in Albuquerque, New Mexico (35.03°N, 106.61°W), in the Middle Rio Grande of central New Mexico at 1617 asl range from 1.2°C in January to 25.8°C in July (Fig. 5.12). Average annual precipitation is 21.5 cm, with a peak of 3.9 cm in August and a minimum of 1.0 cm in January (see Fig. 5.12). Average temperatures in El Paso, Texas (31.80°N, 106.40°W), in westernmost Texas at 1194 m asl range from 6.0°C in January to 27.9°C in July. Average annual precipitation is 21.8 cm, with a peak in August and a minimum in April. Average temperatures near the mouth of the Rio Grande at McAllen, Texas (30 m asl) range from 14.7°C in January to 29.8°C in August. Average annual precipitation is 57.7 cm, with peaks in May and September and a minimum in March. Climate in the basin ranges from cold temperate to subtropical along the 2830 km of the Rio Grande.

Land use in the Rio Grande basin includes forest (14%), cropland (5%), shrubland (43%), grassland (31%), and urban (7%). Most of the basin is either desert shrubland or desert grassland. Population density was recently estimated at 16 people/km² (Revenga et al. 1998). Urban areas are growing fast, with border towns between the United States and Mexico, such as El Paso, Juárez, Piedras Negras, Laredo, Nuevo Laredo, Brownsville, and Matamoros, growing particularly rapidly. Irrigated agriculture from the Rio Grande is most prevalent in the San Luis Valley of southern Colorado, the Middle Rio Grande and Mesilla valleys of New Mexico, and the lower Rio Grande Valley of southern Texas. Principal crops vary along the river, with Colorado and northern New Mexico growing potatoes and alfalfa, southern New Mexico and west Texas growing cotton, peppers, onions, and pecans, and the lower Rio Grande Valley growing citrus fruits, vegetables, and cotton.

River Geomorphology, Hydrology, and Chemistry

The headwaters of the Rio Grande begin along the continental divide in the San Juan Mountains in southern Colorado. The Rio Grande begins as a clear spring- and snow-fed mountain stream draining forested slopes. The river next passes through the San Luis Valley of southern Colorado. The Rio Grande then flows east and south into New Mexico through a canyon 113 km long and up to 250 m deep. The Rio Grande exits the canyon country and is impounded north of Albuquerque by Cochiti Dam, completed in 1975. The Rio Grande flows southward from Cochiti Reservoir through the Middle Rio Grande valley of central New Mexico to Elephant Butte Reservoir, completed in 1916 (Fig. 5.11). Elephant Butte is the largest reservoir in New Mexico.

The Rio Grande flows south from Elephant Butte Reservoir to Caballo Reservoir and into southern New Mexico. Approximately 32 km north of El Paso, Texas, the Rio Grande forms the New Mexico–Texas border. The river then trends generally southeasterly to the Gulf of Mexico, comprising the United States–Mexico border beginning in the El Paso/Juárez area. Below Presidio/Ojinaga, the Rio Grande increasingly flows through a more mountainous region, first continuing its southeasterly flow and then cutting northeast through the Big Bend country of Texas and Mexico. The northern portion of this large bend on the U.S. side of the border comprises Big Bend National Park and represents one of the most isolated environments within the U.S. National Park system. The Rio Grande cuts three spectacular sheer-walled canyons between 457 and 518 m in depth across this faulted region and then continues northeasterly for several hundred km until it again turns to the southeast for the remainder of its course (Fig. 5.1). Below Big Bend, Amistad Dam (completed 1969) and Falcon Dam (completed 1953) impound the Rio Grande along the border for irrigation and flood control. The Rio Grande increasingly has not reached the ocean as human consumption of river water increases in both the United States and Mexico. In the lowermost reach, the Rio Grande wanders sluggishly across the coastal plain to its delta on the Gulf of Mexico. The lowermost river delta area is primarily flat, with clay, loam, and sandy soils that have been transformed into a large agricultural area.

River gradient of the Rio Grande is greatest in reaches in northern New Mexico and southern Colorado (>3 m/km). Gravel, cobble, and boulders dominate the substrates in this reach. The Middle Rio

Grande in New Mexico has an average gradient of 91 cm/km from Cochiti Reservoir to Elephant Butte Reservoir. A shifting sand bottom was characteristic historically, but the dams and channelization have increased the extent of gravel and cobble in various portions of this reach. The reach from Elephant Butte to El Paso, Texas, is largely channelized and serves to deliver irrigation and municipal water from Elephant Butte and Caballo reservoirs. The average gradient is 57 cm/km. The reach from El Paso to Laredo, Texas, grows steeper (~70 cm/km) and includes the canyonlands of Big Bend National Park. The lowest gradients are in the Coastal Plains section of the river (~20 cm/km).

Geomorphology of the Rio Grande in New Mexico and Colorado is closely associated with the Rio Grande rift and human depletion of surface waters. Jones and Harper (1998) showed that channel avulsions (abrupt shifts in channel location) were common in segments of the Rio Grande in south-central Colorado before major discharge changes due to irrigated agriculture. Avulsions decreased from 19 (1875 to 1941) to 2 (1941 to present). The active channel of the Middle Rio Grande of central New Mexico decreased in area from 8973 ha in 1935 to 4345 ha in 1989 (Crawford et al. 1993). The historical Rio Grande in the southern Rio Grande rift in the Mesilla basin of southern New Mexico was an incised pebbly sand bedload stream prior to completion of Elephant Butte Dam in 1916 (Mack and Leeder 1998). Sand and silt now predominate. Major avulsions occurred in this reach of the Rio Grande between 1844 and 1912 but ceased after completion of Elephant Butte Dam. Channel geomorphology between Fort Quitman and Presidio, Texas, also has undergone major changes as upstream discharge has declined (Everitt 1993). Channel capacity has decreased and stabilized and the river has shortened, steepened, and migrated away from major tributaries. Nonnative riparian plant communities have invaded a majority of the now stabilized former river channel.

Hydrologically, Rio Grande peak flow historically occurred from April to June in the river above the confluence of the Río Conchos near Presidio, Texas, due to montane snowmelt in the mountains of southern Colorado and northern New Mexico. The Río Conchos is the largest tributary to the Rio Grande, and peak flows downstream of the confluence with the Río Conchos often occur in August, September, or October during the monsoons of northern Mexico. The principal tributaries of the Rio Grande are the Pecos, Devils, Chama, and intermittent Puerco rivers in the United States and the Conchos, Salado, and San Juan rivers in Mexico.

Flow upstream of Cochiti Dam in New Mexico remains snowmelt dominated. Average flow at Otowi gage immediately upstream of Cochiti Reservoir based on a 97-yr record has been 43 m³/s, with peak flows most commonly in May. Discharge downstream of Cochiti Dam through Albuquerque, New Mexico, has averaged 34 m³/s, reflecting both reductions for agriculture and evaporation losses from the reservoir. The channelized Rio Grande flows about 300 km from the outlet of Cochiti Dam to Elephant Butte Reservoir, with additional agricultural diversions along the reach, sometimes causing the river bed to go dry.

Water release from Elephant Butte Dam is about 27 m³/s and is largely diverted for crop irrigation in southern New Mexico. Water reaching Texas and Mexico is diverted to the United States (~11 m³/s) as the main supply canal for El Paso and to Mexico (~2 m³/s) for irrigation of the Juarez Valley. There is a significant reduction in the flow of the Rio Grande following these diversions. Downstream from the El Paso/Juarez area, nearly all of the water remaining in the river stems from wastewater and agricultural return flows (Miyamoto et al. 1995, Schmandt et al. 2000). This limited amount of water provides the majority of the flow of the river for several hundred kilometers downstream to Presidio/Ojinaga, where the Río Conchos enters the Rio Grande. Annual average flow of the Rio Grande immediately above the Río Conchos confluence is only about 3 m³/s.

The Río Conchos historically contributed an average annual flow of about 30 m³/s of water into the Rio Grande at its confluence at the cities of Presidio and Ojinaga (but see Chapter 23), forming the main flow of the Rio Grande in the stretch between the confluence and Amistad Reservoir (Eaton and Hurlburt 1992, IBWC 1990, TNRCC 1994). The Pecos River and the Devils River historically contributed about 9 and 11 m³/s to the flow of the Rio Grande, respectively, although these flows have been reduced substantially. All of these flows are stored at Amistad International Reservoir. Amistad Dam, opened in 1968, averages an annual discharge of approximately 65 m³/s. Nearly half of this is taken into the Maverick Canal located 28 km south of Del Rio, Texas, for power generation and irrigation and most is returned to the river upstream of Eagle Pass, Texas. There are few diversions between Eagle Pass and Laredo, and the river discharges approximately 89 m³/s at Laredo, Texas. The Río Salado enters Falcon Reservoir downstream from Laredo with an annual contribution of approximately $15 \, m^3/s$ for a combined flow of about $95 \, m^3/s$ at Falcon International Reservoir. Below Falcon Reservoir, the Río San Juan (~$14 \, m^3/s$) flows into the Rio Grande from the Mexican side at Camargo. Two large diversions and numerous other pumps divert water between Rio Grande City and Brownsville, Texas, for irrigation and municipal supplies. Although mean virgin discharge of the Rio Grande probably exceeded $100 \, m^3/s$ by the time the river reached Brownsville (near the mouth), the current flow is only ~$37 \, m^3/s$. Much of this comes during flood events, in particular hurricanes and tropical storms (Miyamoto et al. 1995, Schmandt et al. 2000).

Irrigated agriculture is the primary use of the Rio Grande surface flow throughout the basin. Below Elephant Butte Reservoir, 35,200 ha of cropland are irrigated in New Mexico. In the El Paso and Juarez Valleys, approximately 17,000 ha are irrigated, and an additional 2000 ha are irrigated between Fort Quitman and Amistad Reservoir on the Texas side of the border. Between Amistad and Falcon reservoirs an estimated 16,300 ha are irrigated on the Texas side, mostly in Maverick County. Below Falcon Reservoir, irrigated agriculture is especially great; over 310,000 ha are irrigated on the Texas side and approximately 292,000 ha are irrigated on the Mexican side of the river. Evaporation from major reservoirs also is substantial in the region and has been estimated to exceed the quantity of water used for municipal purposes in the basin. Municipal water usage from the Rio Grande averages about $3 \, m^3/s$ on the Texas side and $1.5 \, m^3/s$ on the Mexican side. This usage is approximately 3% to 5% of the agricultural consumption of the river along this reach. Relatively few industrial users occur within the basin. The major industrial user of the Rio Grande is the Laredo Power Plant, which consumes only $0.05 \, m^3/s$. Population in the basin was about 13 million inhabitants as of 1990. The population along the Texas border increased by 27% between 1980 and 1990, and the Mexican population along the border increased 26% in those same years.

Runoff represents about 15% of precipitation in the high-elevation regions in the upper Rio Grande basin as measured in the Albuquerque reach (see Fig. 5.12). Snowmelt hydrology dominates in this portion of the basin, with peak runoff from April through June. Human appropriation of Rio Grande water increases downstream, and the basin is increasingly xeric in southern New Mexico and west Texas. The Rio Grande failed to reach the Gulf of Mexico for multiple months in 2002 and 2003. Evapotranspira-

tion, groundwater recharge, and human appropriation of Rio Grande water has resulted in less than 1% of basin precipitation reaching the Gulf of Mexico in recent years.

Water chemistry and water quality change markedly throughout the basin. Nutrient loading in the upper basin occurs in the Alamosa Valley of Colorado, from the wastewater treatment plant in Albuquerque, New Mexico, in the Mesilla Valley, and from the cities of El Paso, Texas, and Juarez, Mexico. The Upper Rio Grande from Colorado to El Paso, Texas, was a National Water Quality Assessment (NAWQA) site from 1992 to 1995 (Ellis et al. 1993, Carter and Porter 1997, Levings et al. 1998). Major findings include the increasing incidence of pesticides downstream, trace element elevations in the Creede, Colorado, mining district, an environmentally stressed fish community, and significant habitat degradation in many reaches. Total dissolved solids are a concern in the El Paso and Fort Quitman, Texas, reach due to evapotranspiration and naturally occurring sources of salinity derived from deep groundwater sources (Hibbs and Boghici 1999). Fecal coliforms, nutrients, low dissolved oxygen, pesticides, herbicides, metals, and organic contaminants are significant concerns in the borderlands between the United States and Mexico.

River Biodiversity and Ecology

The Rio Grande basin contains six freshwater ecoregions, including the Upper Rio Grande, Lower Rio Grande, Pecos, Rio San Juan, Rio Salado, and Rio Conchos (Abell et al. 2000). The Guzmán ecoregion is also part of the Rio Grande complex (Abell et al. 2000) but is endorheic and thus does not actually drain into the Rio Grande system. For such a large and important river basin, the biodiversity and ecology of the Rio Grande is not particularly well studied.

Algae and Cyanobacteria

Published information on the algal communities of the Rio Grande is sparse. *Cladophora*, *Zygnema*, and periphytic diatoms (*Synedra* and *Cymbella*) were common in the Pecos River (Davis 1980a). Dense growths of filamentous green algae (*Cladophora*, *Vaucheria*, and *Ulothrix*) also were reported on the streambed of the Pecos River in certain areas. Unpublished data from the Rio Grande NAWQA study by the USGS on the Rio Grande in southern Colorado and New Mexico found benthic diatoms (*Nitzschia*, *Navicula*, *Achnanthes*, *Cocconeis*, *Fragilaria*,

Staurosira, Synedra, Amphora, Luticola, Caloneis, and *Tryblionella*), cyanobacteria (*Oscillatoria, Spirulina, Schizothrix,* and *Lyngbya*), and green algae (*Cladophora, Closterium, Cosmerium, Pediastrum,* and *Scenedesmus*) to be common (Scott Anderholm, unpublished data).

Plants

Riparian vegetation changes along the considerable length of the Rio Grande. Lowland riparian areas in the upper Rio Grande of Colorado and central and northern New Mexico were historically dominated by cottonwoods, with willows and a variety of native shrubs (Snyder and Miller 1992, Ellis et al. 2002). Unfortunately, nonnative species increasingly dominate riparian areas in this region (Howe and Knopf 1991, Everitt 1998, Ellis et al. 2002). Saltcedar and Russian olive are particularly successful invaders. This has led to considerable research on the interactions between native and non-native species and methods for control of invasive species (e.g., Shafroth et al. 1995; Taylor and McDaniel 1998; Taylor et al. 1999; Sprenger et al. 2001, 2002). Litter dynamics of native and nonnative vegetation in riparian areas also has been studied in detail (e.g., Molles et al. 1995, Ellis et al. 1999).

Nonnative species also play a major role in the riparian zones of southern New Mexico and western Texas. Saltcedar dominates much of the riparian areas until the confluence of the Rio Grande with the Rio Conchos. Dominant riparian tree species in the lower Rio Grande below the confluence of the Rio Conchos include mesquite, hackberry, cedar elm, anacua, black willow, and retama. Nonnative aquatic macrophytes, including water hyacinth and hydrilla, increasingly choke the channel of the lower Rio Grande (Everitt et al. 1999). Introduced grasses, such as Guinea grass and buffelgrass, also are becoming dominants in many areas of the riparian zone of the lower Rio Grande (Lonard et al. 2000).

Invertebrates

Data on the aquatic invertebrates of the Rio Grande are limited. Ward and Kondratieff (1992) reported 19 species of mayflies, 21 species of stoneflies, and 25 species of caddisflies occurring in the Colorado portion of the Rio Grande drainage. Sampling in the canyon country of the Rio Grande in northern New Mexico, Shaun Springer (New Mexico Environment Department, unpublished data) found the dominant taxa to be hydropsychid caddisflies; the

caddisflies *Brachycentrus, Leucotrichia,* and *Stactobiella*; the mayflies *Baetis tricaudatus* and *Tricorythodes*; and various true flies, such as *Atherix*, black flies, and chironomid midges (e.g., *Cricotopus*). Davis (1980b) sampled the aquatic invertebrates of the Rio Grande at seven sites in Texas from El Paso to Del Rio. Water quality and quantity varied substantively along this reach of the Rio Grande. Poor water–quality sites with lower mean diversity (e.g., El Paso, Zaragosa International Bridge, and Upper Presidio) were dominated by the Asiatic clam; oligochaete worms; the mayfly *Homoeoneuria*; the caddisfly *Cheumatopsyche*; the chironomids *Hydrobaenus, Dicrotendipes,* and *Cricotopus*; and other true flies (*Palpomyia tibialis*, psychodids, and the black fly *Simulium*). Sites with higher water quality and greater mean diversity (e.g., Lower Presidio, Santa Elena Canyon, Foster's Ranch, and Del Rio) were dominated by the caddisflies *Cheumatopsyche, Hydroptila,* and *Protoptila*; the mayflies *Thraulodes, Traverella, Choroterpes mexicanus,* and *Tricorythodes*; the chironomids *Orthocladius* and *Cricotopus trifascia*; the black fly *Simulium*; and the Asiatic clam. Surface-active arthropod communities of the Rio Grande riparian forests also have been described (Ellis et al. 2000, Ellis et al. 2001). Nonnative isopods (*Armadillidium vulgare* and *Porcellio laevis*) were abundant, along with native crickets and carabid beetles.

The Rio Grande once had a much larger group of freshwater mussels (Unionidae) than are presently found. At least 16 species of unionid mussels once occurred within the Rio Grande drainage, but these are among the fastest declining groups in the basin, in part because of their environmental sensitivity and intolerance to degraded conditions. Some mollusks are expanding in the basin, such as the introduced and highly tolerant Asiatic clam, which are found to be locally abundant at sites throughout the Rio Grande (Howells 2001). Other common bivalves in the lower Rio Grande of Texas include paper pondshell, tampico pearly mussel, and yellow sandshell mussel (Neck and Metcalf 1988).

Vertebrates

Approximately 166 species of fishes have been found in the Rio Grande when both freshwater (86) and brackish water (80) species are considered. At least 34 of these species are considered rare or endangered and many appear on the endangered species list of the United States. Endangered species include shovelnose sturgeon, American eel, a number of minnows (Mexican stoneroller, Maravillas red shiner, proserpine shiner, manatial roundnose minnow, Devils River minnow, Rio Grande chub, Chihuahua chub, Rio Grande silvery minnow, Chihuahua shiner, Rio Grande shiner, Pecos bluntnose shiner, Tamaulipas shiner, ornate shiner), two suckers (west Mexican redhorse, blue sucker), three catfishes (headwater catfish, Chihuahua catfish, and a unique form of blue catfish in the Rio Grande in west Texas), a trout (Rio Grande cutthroat trout), and four pupfish (Leon Springs pupfish, Comanche Springs pupfish, Pecos pupfish, Conchos pupfish). In addition, there are four livebearers (Big Bend gambusia, blotched gambusia, Pecos gambusia, and an undescribed species of gambusia from the Del Rio area), a darter (Rio Grande darter), and a number of coastal forms (opossum pipefish, snook, fat snook, river goby). Several species of fishes have gone extinct in the basin, including phantom shiner, bluntnose shiner, Amistad gambusia, and very likely blackfin goby.

The fishes of the Rio Grande basin are dominated by a rich minnow (Cyprinidae) assemblage. Where spring systems exist, unique components of pupfishes and livebearers are found. More than 16 families of fishes also are found in the nontidal portions of the basin (Miller 1977, Hubbs et al. 1991), excluding the mountain headwaters, and most localities also include one or two gars (Lepisosteidae), one or two species of shad, several suckers, several catfishes, sunfishes, a drum, and a single species each of *Astyanax* and *Cichlasoma*, species that are far better represented in more southern drainages. Compared to other western Gulf drainages, the darter fauna is greatly reduced; however, several species are found in limited areas in the basin. Dams along the river have greatly restricted the natural range of the American eel and greatly expanded the range of other species such as the inland silverside. One of the greatest problems for the fishes of the basin is the dewatering of the watercourse. This is especially evident in the region downstream from Albuquerque, New Mexico, and in the lowermost Rio Grande near its mouth, where its characteristic freshwater fauna has largely been replaced by estuarine and marine forms (Edwards and Contreras-Balderas 1991). The closing of the mouth in 2001 from lack of freshwater inflows has temporarily prevented the invasion of marine species into the lower watercourse.

Many bird species in the basin, such as the common yellow-throat, great blue heron, snowy egret, black-crowned night heron, white-faced ibis, belted kingfisher, and green kingfisher, are dependent on the Rio Grande. Aquatic reptiles of the Rio Grande include the plainbelly water snake, bullfrog, Rio Grande leopard frog, snapping turtle, box turtle, and western ribbon snake. Beaver, mink, and nutria are found in the river, and the American alligator occurs in the Coastal Plain of the Rio Grande.

Ecosystem Processes

A major change in the Rio Grande in recent years has been the disconnection of the river from the floodplain (Molles et al. 1998). River–floodplain ecosystems are some of the most productive and diverse ecosystems in the world. Fragmentation of river channels by dams, diversions, and depletions reduces productivity and simplifies the structure of the riparian ecosystem. Elimination of the flood pulse in May and early June in the Rio Grande Valley of central New Mexico has reduced germination and establishment of native cottonwoods and willows, favored invasions by saltcedars and Russian olives, and increased the incidence and destructiveness of fires. Managed flooding as a restoration tool has been recommended to promote native populations and communities of plants and animals and to stimulate decomposition and nutrient cycling (Crawford et al. 1993, Molles et al. 1995, Molles et al. 1996, Molles et al. 1998).

Large numbers of overwintering birds in the Rio Grande basin influence nutrient cycling at the landscape scale. Kitchell et al. (1999) showed that wintering waterfowl were a major source of nutrients to riverine wetlands at Bosque del Apache National Wildlife Refuge in central New Mexico. Geese increased nutrient loading by up to 40% for total nitrogen and 75% for total phosphorus. Stable isotopes of nitrogen showed the importance of alfalfa and corn derived from foraging birds in the food webs of the ponds, with fishes and crayfishes deriving much of their nitrogen from these sources.

Water availability is a critical limiting resource in riverine corridors in arid and semiarid regions of the world. Dahm et al. (2002) examined the role of riparian evapotranspiration (ET) by native and nonnative riparian tree species in surface water losses of a 320 km reach of the Rio Grande in central New Mexico. Tower-based micrometeorological measurements using eddy covariance methodology were employed during the growing season to determine annual ET (74 to 123 cm/yr at two saltcedar and two cottonwood sites). Results were scaled to the 320 km corridor using satellite imagery and leaf area index (LAI) measurements. Riparian ET was estimated to be between 20% and 33% of total surface water losses along this reach of the Rio Grande.

Human Impacts and Special Features

The Rio Grande has been called "Great River" because of its prominent role in the history of the American Southwest (Horgan 1984). The Rio Grande is the fifth longest river in North America after the Mississippi/Missouri, Mackenzie, Yukon, and St. Lawrence. The river basin is large at 870,000 km², and the "Great River" has been a major influence on Native American, Hispanic, and European cultures for centuries. Predictable snowmelt hydrology long provided the water, sediment, and nutrients for human agriculture along the riverine corridor.

The Rio Grande presently is one of the most impacted rivers in the world. Both water-quantity and water-quality issues are major concerns. The central portion of the Rio Grande from Elephant Butte and Caballo dams to the confluence of the Rio Conchos operates largely as a ditch for water delivery for agriculture and rapidly growing municipalities. Amistad and Falcon dams in the lower border reach further regulate flow and divert water for irrigation. The Rio Grande failed to reach the Gulf of Mexico in much of 2002 and 2003. Water-quality problems include elevated salinity, nutrients, bacteria, metals, pesticides, herbicides, and organic solvents. In addition, riparian areas in most parts of the basin are in decline, with nonnative species dominating in many reaches.

The segment of the Rio Grande from the Colorado state line southward for almost 110 km is designated as a National Wild and Scenic River. White-water rafting through gorges up to 250 m deep is a primary recreational opportunity. Bosque del Apache National Wildlife Refuge in central New Mexico is the winter home to tens of thousands of waterfowl, with large populations of snow geese and sandhill cranes. Chamizal National Memorial in El Paso, Texas, was established along the Rio Grande to commemorate peaceful settlement of a century-long boundary dispute between Mexico and the United States caused by migration of the Rio Grande river channel. The Rio Grande is again designated as a Wild and Scenic River for 315 km from the Coahuila/Chihuahua, Mexico, state border below its confluence with the Rio Conchos to the Terrell/Val Verde county line. Approximately 111 km of this segment of Wild and Scenic River makes up the boundary of Big Bend National Park.

SAN ANTONIO AND GUADALUPE RIVERS

The San Antonio River arises from a series of medium to small springs about 6 km north of downtown San Antonio, Texas. Although numbering more than 100, many of these springs flow only during periods of very wet weather. The river flows 288 km southeasterly until it joins the Guadalupe River near Tivoli, Texas, just before it empties into San Antonio Bay on the Gulf of Mexico (Fig. 5.13). The Guadalupe River rises as springs in two forks, the North Fork (30°06′N, 99°39′W) and South Fork (30°04′N, 99°20′W). After the confluence of the two forks near Hunt, Texas, the Guadalupe River proper flows southeast 368 km before reaching its mouth at San Antonio Bay (28°26′N, 96°48′W). The principal tributaries of the Guadalupe are the Comal and San Marcos rivers, which contain the two largest springs in the southwest. The Guadalupe flows across the Edwards Plateau, the Balcones fault line, and the Coastal Plain (Fig. 5.3). Together, the two rivers drain an area of 26,231 km². The spring flow of the San Antonio and the Guadalupe and their principal tributaries keeps the discharge of the rivers steadier than that of most western Gulf Coast rivers.

The San Antonio and Guadalupe basins have supported human habitation for nearly 10,000 years. Early Spanish explorers encountered Tonkawa, Waco, Lipan Apache, and Karankawa Indians. The San Antonio was probably first encountered by Spanish explorers in 1535, when Álvar Núñez Cabeza de Vaca crossed the San Antonio. The river was named for San Antonio de Padua on June 13, 1691. The present headwater areas of the San Antonio impressed Spanish missionaries, and the San Antonio de Valero Mission was established on the east bank of the river on May 1, 1718. Plentiful water for drinking, irrigation, and power made the San Antonio River the center of Spanish activities in the province of Texas. The name Guadalupe has been applied to at least the lower course of the Guadalupe River since 1689, when Alonso De León named the stream. Spanish settlement along the Guadalupe began in the 1720s, when several missions were established. Additional settlements along the river were established by settlers from Spain, Mexico, and the United States up through the beginning of the Texas Revolution in 1835. The first shot of the Texas Revolution occurred at Gonzales, Texas, on the south bank of the Guadalupe on October 2, 1835. The San Antonio River also played a prominent role in the

FIGURE 5.3 Guadalupe River above Rt. 311 (PHOTO BY TIM PALMER).

Texas Revolution. Numerous battles were fought along the San Antonio, including the Alamo, Concepción, and Grass Fight. The siege of Bexar and the Goliad Massacre also occurred in close proximity to the San Antonio River. After the Texas Revolution, settlements were begun further upstream on the Guadalupe River, including Seguin (1839), New Braunfels (1845 by German settlers), and Kerrville (1856). Construction of railroads in the 1880s through the middle and upper Guadalupe valley brought many new residents. Although early attempts at navigation on the Guadalupe were limited due to large snags and small waterfalls, its steady flow made the river attractive for hydropower and numerous dams were built beginning in the 1800s. Larger dams for flood control were built, with Canyon Lake, completed in 1964, a major source of flood control for the lower basin. During the latter 1800s, the San Antonio River was prominent in the development of the city of San Antonio. More than fifty bridges in the city now span the San Antonio and the river flows for 24 km through San Antonio, Texas. The Riverwalk or "Paseo de Río" in downtown San Antonio is one of the principal attractions of the city.

Physiography, Climate, and Land Use

The San Antonio and Guadalupe rivers flow through two physiographic provinces, the Great Plains (GP) and the Coastal Plain (CP). Originating from the Edwards Plateau section of the Great Plains province, the rivers flow through the highly dissected limestone of the Edwards Plateau, the rolling, hilly Blackland Prairies, and the low-relief Coastal Plain (see Fig. 5.13). Beginning in the Edwards Plateau, the rivers pass through four terrestrial ecoregions (Edwards Plateau Savannas, Texas Blackland Prairies, East Central Texas Forests, and Western Coastal Grasslands) as they descend to San Antonio Bay. Common upland vegetation types from the Edwards Plateau to the coast are juniper–oak–mesquite savannah, blackland prairie grasses, oak hickory forest, oak savannah, and coastal prairie grasses. Soils include dark calcareous stony clays and clay loams in the upper basin, light brown to dark gray acid sandy and clayey loams in the middle basin, and dark-colored neutral to slightly acid clay loams and clays in the lower basin.

Average annual temperature at San Antonio, Texas, is 20.3°C, with minimum average monthly temperature of 9.6°C in January and maximum average monthly temperature of 29.4°C in July. Average annual precipitation is 74 cm, with minimum average monthly precipitation of 4.2 cm in January and maximum average monthly precipitation of 10.2 cm in May. Conditions are similar in the Guadalupe River basin. For example, annual average temperature at Cuero, Texas, is 21.0°C, with minimum average monthly temperature of 11.0°C in January and maximum average monthly temperature of 29.0°C in July and August (Fig. 5.14). Average annual precipitation is 81 cm, with minimum average monthly precipitation of 4.0 cm in March and maximum average monthly precipitation of 9.6 cm in May (see Fig. 5.14).

Land use in the San Antonio/Guadalupe basin is a mix of range (60%), agriculture (15%), and urban (25%). Population densities in 2000 ranged from 433 people/km^2 in the San Antonio area to 6 people/km^2 in rangeland-dominated counties. Overall, the basin averages approximately 85 people/km^2. Agricultural products include livestock, cotton, peanuts, vegetables, poultry, rice, and dairy products. Croplands increase from the headwaters to the mouth of the rivers. The cattle industry was important in this region in the early years of settlement, whereas sheep and goats now graze the sparse grasslands on the poor soils of the limestone uplift in the upper portions of the basin. Tourism and recreation are increasingly important aspects of land use throughout the basin. For example, people are rapidly buying up Hill Country (the commonly used name for the Edwards Plateau) property as tourism has begun to edge out agriculture as an economic base in the region. Urbanization with increased emphasis on recreation and tourism is expanding substantially in the San Antonio/Guadalupe drainage basin.

River Geomorphology, Hydrology, and Chemistry

The upper segments of the San Antonio and Guadalupe flow through limestone terrain. Limestone bluffs are interspersed with less confined river reaches. The rivers pass over the Balcones fault line with higher gradient reaches and occasional small waterfalls. After the Balcones Fault Zone, the rivers transition into the Coastal Plain, where they become low-gradient meandering systems with sand bars and wider, less-constrained channels, except where channelized by levees.

Springs from the Edwards Plateau and the Balcones Fault Zone sustain good base flow conditions in both the San Antonio and Guadalupe rivers throughout the year and especially during times of drought. A 38-year U.S. Geological Survey record for the San Antonio near Elmendorf, Texas, has mean monthly flows that range from $11\,m^3/s$ in August to $27\,m^3/s$ in June. Mean values over approximately the same period for the Guadalupe River at Cuero, Texas, range from $32\,m^3/s$ in August to $92\,m^3/s$ in June, although these values were to some degree determined by the operation of Canyon Dam, for which construction began in 1958. Before this time, peak discharge occurred in May, rather than June, based on records from 1935 through 1957 at the Victoria gaging station. Mean discharge (38 years) for the San Antonio is about $21\,m^3/s$ and for the Guadalupe is about $58\,m^3/s$, for an approximate total of $79\,m^3/s$ at their confluence just before emptying into San Antonio Bay. Annual runoff for the Guadalupe basin is about 14 cm, with the lowest monthly runoff (0.7 cm) in August and the highest (1.9 cm) in June (see Fig. 5.14). Runoff for the San Antonio basin is somewhat lower than for the Guadalupe. Base flows in the San Antonio during drought remain above $2\,m^3/s$, with peak monthly mean flows as high as $240\,m^3/s$ during exceptionally wet periods. Flows are somewhat more variable in the Guadalupe River. The Guadalupe River at Cuero had minimum monthly flows of $<3\,m^3/s$ during a strong drought in 1984 and maximum monthly flows of nearly $900\,m^3/s$ during an extremely wet month in October 1998. Flooding from rainfall and hurricanes is characteristic of this area. Long-term averages of monthly discharge show peak discharge in June and minimum discharge in August.

Water quality of the spring sources for the San Antonio and Guadalupe rivers is generally good; however, both rivers experience localized organic and chemical pollutant inputs from urban sources. The Edwards aquifer has had no major water-quality problems to date. The water chemistry reflects the karst nature of the aquifer, with calcium the major cation and bicarbonate the major anion. The chemistry of the Guadalupe and San Antonio rivers also shows a major influence of the large springs emanating from the limestone geologic units. The pH ranges from about 7.0 to 9.0. Specific conductance is normally in the 700 to $1100\,\mu S/cm$ range, except for drops to lower levels during storms. The dominant cations are calcium, sodium, and magnesium, and the dominant anions are bicarbonate, chloride, and sulfate. The water has high alkalinity ($\sim200\,mg/L$ as $CaCO_3$) and hardness ($\sim300\,mg/L$ as $CaCO_3$). Dissolved oxygen normally is between 3 and $14\,mg/L$. The rivers of the San Antonio/Guadalupe basin commonly show a shift from the primary influence of groundwater from springs to a more runoff-

dominated and anthropogenically influenced character as they flow from their sources to the downstream lowlands (Groeger et al. 1997).

River Biodiversity and Ecology

The San Antonio and Guadalupe are the westernmost rivers within the East Texas Gulf freshwater ecoregion, which includes all the remaining Texas rivers to the east (Abell et al. 2000). The Guadalupe River is considered to be particularly important in this ecoregion because of the number of endemic species, mostly associated with the karst bedrock of the Edwards Plateau.

Algae

Limited published information is available on the algae of the Guadalupe and San Antonio rivers. Sherwood and Sheath (1999) studied the macroalgae and epilithic diatoms of the San Marcos River and the Comal River, major tributaries of the Guadalupe. Twelve species of macroalgae were identified. Macroalgae included Cyanophyta (*Lyngbya taylorii*), Chlorophyta (*Cladophora glomerata, Dichotomosiphon tubersosus, Hydrodictyon reticulatum,* and *Oedogonium sp.*), Chrysophyta (*Tribonema regulare*), and Rhodophyta (*Audouinella pygmaea, Batrachospermum globosporum, B. involutum, Hildenbrandia angloensis, Sirodotia huillensis,* and *Thorea violacea*). Sixty-eight species of epilithic diatoms were identified during a 15-month seasonality study. Common genera of epilithic diatoms included *Achnanthes, Achnanthidium, Cocconeis, Denticula, Encyonema, Eunotia, Fragilaria, Gomphonema, Navicula, Nitzschia, Planothidium, Staurosira,* and *Synedra*.

Plants

The mature forests of the San Antonio and Guadalupe rivers are dominated by pecan, Texas sugarberry, and bald cypress. The dominant riparian species all have eastern affinities and the floodplain forests of the Edwards Plateau represent the westernmost extension of their ranges. Other common riparian species are cedar elm, Virginia creeper, Texas persimmon, red mulberry, greenbrier, box elder, cottonwood, gum bumelia, and black walnut (Ford and Van Auken 1982; Bush and Van Auken 1984; Van Auken and Bush 1985, 1988). In secondary successional studies on terraces along the San Antonio River, recolonization occurred within five years of disturbance. Retama, mesquite, desert hackberry, huisache, and Texas sugarberry were early colonizers and these, in addition to cedar elm, box elder, pecan, American elm, and gum bumelia, were important secondary colonizers that became more dominant after 25 years (Van Auken and Bush, 1985). Van Auken and Bush (1988) studied black willow and cottonwood along the edge of the San Antonio River. Highest densities were within 10 m of the river edge, but the largest trees were 15 to 20 m from the river edge. Periodic disturbance is apparently a requirement for maintaining black willow and cottonwood in these riparian forests.

Common native aquatic plants include fanwort, coontails, pennywort, water primrose, Illinois pondweed, grassy arrowhead, and water celery (Owens et al. 2001). In addition, a number of nonnative aquatic plants have flourished in the headspring areas, including hydrilla, elephant ears, East Indian hygrophila, Eurasian watermilfoil, and water lettuce. There is one listed endangered species of aquatic plant in the basin. Texas wild rice is known only from the San Marcos River spring run, associated with the Edwards aquifer, and presently is a listed endangered species. Texas wildrice faces threats from dewatering of the aquifer upon which the San Marcos springs depends, impacts from recreational users, impacts from nonnative organisms (especially from nonnative aquatic plants and mammals), habitat modifications, and problems stemming from inhabiting an urban stream.

Invertebrates

Common aquatic invertebrates of the San Antonio and Guadalupe rivers include caddisflies (*Chimarra, Cheumatopsyche, Oxyethira, Smicridea, Hydroptila, Atopsyche erigia*), mayflies (*Dactylobaetis mexicanus, Leptohyphes vescus, Tricorythodes albilineatus, T. curvatus, Choroterpes mexicanus, Thraulodes gonzalesi, Baetodes alleni*), aquatic beetles (*Microcylloepus pasillus, Hexacylloepus ferruginues, Neoelmis caesa*), and chironomid midges (*Cricotopus, Rheotanytarsus exiguous, Polypedilum convictum, Orthocladius, Pseudochironomus*). Short (1983) described the normally spring-dwelling caddisfly *Atopsyche erigia* from tailwaters of dams on the Guadalupe River. McCafferty and Provonsha (1993) described the new mayfly species *Baetodes alleni*, a large larval mayfly from the Guadalupe River. Other common aquatic invertebrates include true flies (*Hemerodromia*), true bugs (*Ambrysus circumcinctus*), amphipod crustaceans (*Hyallela azteca*), and hellgrammites (*Corydalus cornutus*).

Young and Bayer (1979) carried out a detailed study of dragonfly nymphs (Odonata: Anisoptera) of the Guadalupe River drainage basin. They collected samples from 56 lotic and 14 lentic sites and found 44 species of dragonfly nymphs. When combined with specimens and descriptions from historic collections, a total of 61 species have been described for the Guadalupe River basin.

Fifteen species of mussels have been reported from the San Antonio and Guadalupe river basins. Two species are considered rare (Texas pimpleback and false spike). One species, the golden orb, is a species of concern among the mussels in the basin (Howells et al. 1996). The introduced Asiatic clam occurs widely in the basin. In addition, three endangered species of aquatic invertebrates are associated with threatened springs and cave ponds associated with the Edwards aquifer: Comal Springs dryopid beetle (*Stygoparnus comalensis*), the Comal Springs riffle beetle (*Heterelmis comalensis*), and the Peck's cave amphipod (*Stygobromus pecki*) (www.edwardsaquifer.net/species.html).

Vertebrates

Eighty-eight species of fishes have been found in the Guadalupe and San Antonio rivers and their tributaries, of which 28 are nonnative. One species, the San Marcos gambusia, is extinct, four fish species are endangered, and two fish species are of concern. The endangered fish species are the headwater catfish, the endemic widemouth blindcat, toothless blindcat, and fountain darter, which is also listed as federally endangered. The widemouth blindcat and toothless blindcat are unique in that they inhabit the Edwards aquifer beneath the city of San Antonio. The fish species of concern are the Guadalupe bass and the blue sucker. The Guadalupe bass, the official state fish of Texas, is a central Texas endemic species of bass that has been shown to hybridize with nonnative smallmouth bass (Edwards 1979).

Where the rivers flow over the Cretaceous-age limestone bedrock of the Edwards Plateau there is a mixture of riffles and pools, and fishes tend to segregate within these mesohabitats. Pool habitats contain a variety of sunfishes, including largemouth bass, Guadalupe bass, bluegill, longear sunfish, redear sunfish, and the nonnative redbreast sunfish. Where significant springflows are present, such as the San Marcos River, spotted sunfish also are found abundantly. Other pool inhabitants include minnows, such as blacktail and red shiners, and central stonerollers, gray redhorse, channel catfish, western mosquitofish, and the nonnative Rio Grande cichlid.

Riffles are often inhabited by Texas shiners, young stonerollers, young flathead catfish, several darters (including the dusky, orangethroat, and greenthroat darters), and the Texas logperch. In more downstream reaches of the San Antonio/Guadalupe basin, the Guadalupe bass is replaced by the spotted bass, the greenthroat darter is replaced by the bluntnose darter, and the river darter replaces the dusky darter. In addition, more large river-adapted species such as spotted and alligator gars, gizzard and threadfin shads, bullhead minnows, and a variety of suckers (such as the river carpsucker and the smallmouth buffalo) are more commonly encountered (Hubbs et al. 1953, Young et al. 1973).

Edwards (2001) has shown that nonnative fishes in the upper San Antonio River are having a substantial impact upon native fishes. Seven native species were captured, including the western mosquitofish, three shiners (Texas, red, and mimic shiners), and three sunfishes (largemouth bass, longear sunfish, and spotted sunfish). Eleven nonnative species were taken from the same area. These nonnatives made up 61% of the species, 17% of the individuals, and 62% of the biomass. Some noticeable native species were missing from the assemblage, including a complete absence of the orangethroat and greenthroat darters, the Texas logperch, the native catfish, and the Guadalupe bass. Urban areas in the upper basin are thought to have a major impact upon the conditions leading to this assemblage of fishes in the upper San Antonio.

Aquatic nonfish vertebrates include the cottonmouth, diamondback and plainbelly water snakes, American alligator, Gulf Coast toad, North American bullfrog, Texas river cooter, red-eared slider, an occasional beaver, the even rarer river otter, and the ever-present and widespread nonnative nutria. Other endangered vertebrate species in the basin include the golden-cheeked warbler, whooping cranes near the coast, and the Texas blind salamander. The threatened San Marcos salamander also is endemic in the basin. In addition, the endemic Cagle's map turtle of the Guadalupe River is a candidate for listing.

Ecosystem Processes

Stanley et al. (1990) assessed nutrient limitation of periphyton and phytoplankton in the upper Guadalupe River. Nutrient-diffusing substrates were used to study periphyton responses, and nutrient amendments to glass bottles were used to study the response of phytoplankton. Most of the nutrient-diffusing experiments showed greater periphyton chlorophyll *a* when phosphorus was added to the substrate. More variability was noted in the bottle experiments for phytoplankton response, but the more responsive nutrient was nitrogen. In general, phytoplankton tended to be nitrogen limited and periphyton tended to be phosphorus limited in the upper Guadalupe River.

Epperson and Short (1987) studied annual production of the predaceous stream hellgrammite *Corydalus cornutus* (Megaloptera) in the Guadalupe River. Annual production at five sites decreased from upstream to downstream. Production values decreased from 22.9 to 4.6 g AFDM $m^{-2} yr^{-1}$. These production values are some of the highest reported for a single species of aquatic insect.

Human Impacts and Special Features

The Edwards aquifer is a vast groundwater ecosystem that contains one of the most diverse, unique, and significant biological assemblages in the world and feeds the two largest and most stable spring systems in the western Gulf Coast region. Because of the constancy of the water temperatures and flows of the San Marcos and Comal rivers, a unique and extremely diverse group of endemic organisms have also developed in these ecosystems. Many of the species found in these systems are found nowhere else. These springs and the ecosystems that support them are the main habitats for which the first ecosystem-based endangered species recovery plan was developed. This plan attempts to protect the threatened and endangered species of aquatic plants and animals from loss of habitat due to reduced spring flows resulting from increased pumping and subsequent draw down of the Edwards aquifer, as well as threats from nonnative species, recreation, and modifications such as dams, bank stabilization, and control of aquatic vegetation and factors that decrease water quality (U.S. Fish and Wildlife Service 1996). The aquifer also serves agricultural, industrial, recreational, and domestic needs for about two million people, including the cities of San Antonio, San Marcos, and New Braunfels. Numerous other springs are also present in the San Antonio/Guadalupe basin due to the highly fractured nature of the Cretaceous-age limestone karst topography. In Texas, where right of capture is the main law applied to groundwater use, there is a growing trend for major aquifer authorities to research and manage groundwater reserves and withdrawals. The Guadalupe Delta Wildlife Management Area, Matagorda Island Wildlife Conservation Area, and Aransas National Wildlife Refuge provide habitat for wintering whooping cranes, 19 endangered or threatened species, and 300 species of birds.

The San Antonio and Guadalupe rivers are no longer free flowing. A number of dams for flood control and water storage are located throughout their river courses. Canyon Dam, the largest on the Guadalupe River, has had a profound effect, especially below the dam due to hypolimnetic release flows that have dramatically changed downstream water-temperature regimes. In a comparison of the fishes immediately above and below Canyon Reservoir and from preimpoundment surveys completed 25 years earlier, Edwards (1978) found that species such as Guadalupe bass, channel catfish, speckled chub, mimic shiner, Rio Grande cichlid, longnose gar, gizzard shad, and redear sunfish were either absent or very limited below the dam. Other species, such as orangethroat darters, have become extremely abundant and have extended their normal late winter and spring spawning season to nearly year-round spawning in the tailrace waters. A put-and-take trout fishery has been established in this area below the dam.

Because of the tremendous growth of the region in the past century and the long history of human habitation, water has been increasingly pumped from the Edwards aquifer, placing stress on the biological resources within the aquifer and the river biota as well. In times of severe drought, spring flows from the Edwards aquifer, especially into the San Marcos River, provide the major source of water for the downstream portion of the Guadalupe River and to the estuaries at their terminus. Although the dams minimize water-level fluctuations following storm-related flood events, significant flooding in the San Antonio/Guadalupe basin occurred during severe storms in 1998 and 2000. In addition, as cities using the waters from the San Antonio and Guadalupe rivers have grown, urban pollution has become a more frequent and persistent problem. Similarly, the long-term increase in the number of nonnative species is having a negative impact upon the native biological assemblages in these rivers.

COLORADO RIVER

The basin of the Colorado River of Texas begins in the very arid region of southeastern New Mexico and western Texas, where rainfall is infrequent and streams are intermittent (Fig. 5.15). Bending southeastward, the river flow becomes permanent through increased rainfall and freshwater spring flows from the Edwards Plateau as it winds through the karst topography of Cretaceous limestone known as the "Texas Hill Country." The area is dotted with rural farms, ranches, and fruit orchards, but ironically "ranchettes" and suburbs have grown rapidly as people "escape" from increasing urbanization. The state capital of Texas, Austin, is situated where the Colorado River exits the hill country at the Balcones Escarpment and enters the Coastal Plain on its way to Matagorda Bay. What were once dry-land cotton and grain fields in the lower Colorado River have become irrigated rice fields and row crops, as cotton and sorghum farming have moved upriver where surface water is supplemented by pumping from the Ogallala aquifer. Total river length is approximately 1560 km, with a drainage area of about 103,300 km^2 (roughly the area of Kentucky). The name Colorado, Spanish for red, is evidently a misnomer because the river is clear and appears to have been so at the time of exploration by Cabeza de Vaca in the 1530s. Consensus is that the name Colorado was applied to the Brazos but the names were interchanged during the seventeenth-century period of Spanish exploration and the name was well established before the end of Spanish Texas following the successful Mexican War of Independence from Spain in 1821.

Paleo-American Indians in this area included the Folsom culture that hunted a now-extinct form of giant bison using only spears and spear-throwers. These cultures waned by 2500 years ago and became integrated into the many more localized Archaic hunter and gatherer cultures. Later infused by cultures from Central and South America, these tribes grew into the agricultural and pottery-making cultures of the Neo-American stage. Comanches also lived along the upper Colorado River and were among the first of the Plains Indians to use horses to raid other tribes and white settlers along the river. Lipan Apaches were found along the western basin, and they got along with the immigrating white Texans, in part by fighting against the Comanches. The Tonkawas, who settled along the main river and tributaries of the Colorado River in the Edwards Plateau, also fought against the Apaches with white settlers. Several Karankawa tribes lived along the coast and had little contact with European settlers, apart from Cabeza de Vaca (1528–1536) and La Salle (1685–1689), until 1720, when French and Spanish expeditions fought over this territory. When European diseases took their toll, the remaining Karankawa and Tankawa people were rounded up and given some lands, but they were later expelled from Texas to Indian Territory in present-day Oklahoma. Little Hispanic influence occurred, partly because the early colonists of Texas from the United States and Central Europe settled near the banks and mouth of the Colorado.

Physiography, Climate, and Land Use

The Colorado River mostly drains the Great Plains (GP) physiographic province, with a small amount from the Central Lowlands (CL), and finally enters the Coastal Plain (CP) (see Fig. 5.15). Major biomes include Temperate Grasslands and Temperate Deciduous Forests, and terrestrial ecoregions include the Western Short Grasslands, Central and Southern Mixed Grasslands, Edwards Plateau Savannas, Texas Blackland Prairies, East Central Texas Forests, Western Gulf Coastal Grasslands, and Central Forest/Grassland Transition Zone. Native vegetation from headwaters to mouth includes plains grassland, desert shrub, mesquite savanna, juniper–oak–mesquite savanna, blackland prairie grasses, oak–hickory forest, oak savanna, and coastal prairie grasses.

Climate varies considerably along the Colorado River basin. In the upper basin, annual mean temperature in Ballinger, Texas, is 18.3°C and monthly mean temperature ranges from 7°C to 28°C; annual mean precipitation is 61 cm and monthly mean precipitation ranges from 3 cm during November through March to 9.8 cm in May. Further south in Austin, Texas, annual mean temperature is 20.3°C and monthly mean temperature ranges from 9°C in January to 29°C in July and August; annual mean precipitation is 82 cm and monthly mean precipitation ranges from 4.7 cm in July to 11.8 cm in May (Fig. 5.16). In the lower basin at Matagorda, annual mean temperature is 21.5°C and monthly mean temperature ranges from 12°C to 29°C; annual mean precipitation is 112 cm and monthly mean precipitation ranges from 5.7 cm in March to 14.3 cm in August.

Land use in the basin is 55% range, 30% agriculture, and 15% urban. Major crops include cotton and grains such as sorghum and rice, and livestock (goats, sheep, horses, and cattle), dairy, and poultry are additional important agricultural activities. The upper basin was estimated to be 19% cropland in 1985, with most of the rest of the upper basin consisting of mesquite and rangeland/prairie grasses (Shirinian-Orlando and Uchrin 2000) supporting cattle ranging among oil wells on the Permian Basin–influenced landscape. Major towns along the Colorado include Big Spring, Ballinger, Austin, Wharton, Bay City, and Matagorda. Population densities in 2000 for counties within the basin averaged 35 people/km² and ranged from 3 to 88 people/km².

River Geomorphology, Hydrology, and Chemistry

The Colorado River along its upper reaches does not flow year-round until its confluence with the Concho River at O. H. Ivie Lake (see Fig. 5.15). Downstream of Big Spring, in the upper reach, it passes through rolling prairies with shallow and variable flow in a generally low-gradient channel (50 cm/km). The middle basin includes several more constrained reaches as the river passes through the Hill Country and the Llano Uplift. The river reach just upstream of Lake Buchanan flows through high limestone bluffs over a bed of bedrock and gravel. Within the Pedernales and Llano rivers and smaller tributaries that enter the Colorado River below Lake Buchanan, white-water enthusiasts can experience reaches of white water as well as gently meandering streams. Below Austin the channel is wide and water moves slowly over a bed of sand and gravel (Fig. 5.4). As it becomes a coastal river, it develops steeper banks and large sand bars that dissect the deep alluvial sediments. Baker and Penteado-Orellana (1977) studied the alluvial sediments of the lower Colorado River and showed that Quaternary climate in central Texas alternated between arid and humid phases, with the recent transition to humid conditions accompanied by more uniform stream-flow characteristics.

Hydrology of the Colorado River also changes substantially as the river moves from intermittent headwaters to its mouth. Average annual discharge to the Gulf of Mexico is about 75 m³/s. Interestingly, discharge of the Colorado River of Texas at its mouth now exceeds the average discharge of the more famous Colorado River as it discharges into the Gulf of California (Chapter 11). Specifically, the "big" Colorado at one time had a virgin discharge of 550 m³/s but now discharges only about a mean of 40 m³/s because of extractions and interbasin transfers and often goes dry before reaching the sea. River flow of the Colorado River of Texas is highly variable. For example, mean monthly discharge near the mouth of the Colorado has ranged from 2 m³/s during the drought of the early 1950s to 1200 m³/s during the winter floods of 1992. Average monthly peak discharge at the mouth occurs in June (~127 m³/s), with low flows normally occurring in August (~23 m³/s). These are relatively low discharge values for a basin this size, giving a very low annual runoff of only

FIGURE 5.4   Colorado River at Wharton, Texas (PHOTO BY TIM PALMER).

about 3 cm (see Fig. 5.16). The large difference between precipitation and runoff is probably due to evapotranspiration, but there may be groundwater and irrigation losses as well.

Water chemistry reflects both climatic variation and land-use patterns within the basin. The Upper Colorado basin suffers from salinization of soil, groundwater, and surface water (Slade and Buszka 1994, Shirinian-Orlando and Uchrin 2000). Total dissolved solids for the Upper Colorado River have exceeded 12,000 mg/L, with values above 1000 mg/L commonly reported. Along with natural salt deposits, activities associated with oil and gas exploration and production are another major source of dissolved solids. Water-quality concerns in the middle basin involve periodic excursions of low dissolved oxygen (<3.0 mg/L) and high NO_3-N + NO_2-N (>2.8 mg/L). High chlorophyll *a* (>21.4 µg/L) for reservoirs in the Highland Lakes region due to nonpoint nutrient sources causes periodic decreased visibility. The lower basin below Austin has shown improved water quality in the past decade as a result of improved

wastewater treatment facilities. Contact recreation (e.g., swimming), however, is often discouraged due to high counts of fecal coliform bacteria. Water-quality problems in the Austin area have led to studies on meeting dissolved oxygen standards by using pollution-offset permits to lower biological oxygen demand in the river (Letson 1992). Maintaining adequate downstream flow also is a concern and models of the Lower Colorado have been used to manage drought (Martin 1991) and predict discharge using a linked geographic information system (GIS)/hydrologic model (Rosenthal et al. 1995).

River Biodiversity and Ecology

The Colorado River is located within the East Texas Gulf freshwater ecoregion, an area stretching from the San Antonio/Guadalupe basin to the Sabine (Abell et al. 2000). In general, the biodiversity and ecology of the Colorado are not well studied, although the distribution of vertebrates, fishes in particular, is fairly well known.

Algae

Little information is available on the algae of the Colorado River, but reservoirs on the Colorado have suffered from episodic blooms of the golden alga *Prymnesium parvum*. The algae give the water a yellowish tint and have been responsible for fish kills numbering in the hundreds of thousands. E. V. Spence Reservoir on the upper Colorado (see Fig. 5.15) has been especially affected by this alga.

Plants

In the upper river, the saline waters have influenced riparian vegetation, where nonnative saltcedar now grows extensively. In the middle basin, riparian vegetation includes elm, willow, sycamore, and buttonbush. In backwater habitats, bald cypress and such macrophytes as smartweed, spiderlily, arrowhead, and white waterlily are found. Nonnative macrophytes include hydrilla, Eurasian watermilfoil, and water hyacinth from the middle and lower Colorado River, curly pondweed and water spangles in the middle Colorado, and water lettuce in the lower Colorado. The lower basin, once the site of massive logjams derived from lowland riparian vegetation, is now primarily a mixed hardwood riparian forest.

Invertebrates

Few studies have characterized the invertebrates of the Colorado River in Texas. As part of the ecoregion assessment project for EPA Region VI, macroinvertebrates that characterize least-disturbed streams are listed in Bayer et al. (1992). In most western streams of the Colorado basin, EPT (Ephemeroptera–Plecoptera–Trichoptera) indices are predominantly due to mayflies and caddisflies, but range widely. In small headwaters streams where salinity is high and conditions are harsh, EPT values are near 7 in assemblages dominated by salinity-tolerant clams (*Sphaerium*) and oligochaete worms (*Limnodrilus*). EPT values reach 23 in larger streams with a wider range of habitats and more permanent flow, with predominantly salinity-intolerant species, including the shiny peaclam, caddisflies (*Helicopsyche*, *Cheumatopsyche*), and beetles (*Microcylloepus pusillus*, *Stenelmis cheryl*). Spring-fed streams of the Edwards Plateau region have EPT indices of 15 to 18 and are characterized by salinity-intolerant species of mayflies (*Tricorythodes albilineatus*), caddisflies (*Chimarra*, *Hydroptila*), chironomid midges (*Polypedilum convictum*, *Rheotanytarsus exiguous*), predatory ceratopogonid midges (*Probezzia*), and damselflies (*Argia*). In streams of the Coastal Plain, additional species include mayflies (*Caenis*, *Fallceon quilleri*, *Stenacron*), beetles (*Stenelmis occidentalis*), limpets (two-ridged ramshorn), and tubificid worms (*Nais pardalis*, *N. communis*).

The native ranges of several crayfish species include the Colorado River drainage (*Cambarellus shufeldtii*, *Procambarus clarkii*, *P. acutus acutus*, and *Orconectes palmeri longimanus*). The Cajun dwarf crayfish is a Texas endemic with a small range that includes the Colorado River coastal region. Prehistoric inhabitants of the Colorado River drainage used freshwater mussels for shells, meat, and pearls, as did early Spanish and European settlers, especially along the Concho and Llano rivers. Fifteen species of mussel have been reported from the Colorado River (Howells et al. 1996). Common or widely distributed species include giant floater, paper pondshell, giant washboard, bleufers, southern mapleleaf, Texas lilliput, pistolgrip, and tapered pondhorn. Four species are considered rare (smooth pimpleback, Texas pimpleback, false spike, Texas fawnsfoot), and Texas fatmucket and golden orb are species of concern. The introduced Asiatic clam occurs throughout much of the riverine corridor, and native fingernail clams and peaclams, although widely distributed, are less common.

Vertebrates

There are 98 species of fishes in the Colorado River, of which 26 are nonnative. Common species in more saline and harsh summer environments of the upper Colorado River are red shiner, fathead minnow, western mosquitofish, channel catfish, and green sunfish. In fresher water and spring-fed streams, blacktail shiner, central stoneroller, bullhead minnow, roundnose minnow, Texas shiner, sand shiner, mimic shiner, gray redhorse, orangethroat darter, Texas logperch, bluegill, longear sunfish, redear sunfish, and largemouth bass are more abundant. In lowland streams nearer the coast, dominant fishes include pugnose minnow, yellow bullhead catfish, tadpole madtom, pirate perch, dusky darter, warmouth, white crappie, spotted gar, river carpsucker, and gizzard shad. Nonnative redbreast sunfish and Rio Grande cichlid have been introduced widely. The blue sucker occurs in the larger river, and Guadalupe bass (state fish of Texas) occurs in streams of the Edwards Plateau; both are listed as Texas species of concern.

Common aquatic snakes include the diamondback water snake, plainbelly water snake, and cottonmouth. The endangered Concho water snake has one of the smallest ranges of any North American snake species, approximately 25 km of impounded shoreline and 396 km of streams within the upper Colorado River basin. The endangered, neotenic, Barton Springs salamander lives wholly within Zilker Park in Austin, Texas, but a captive breeding program by the U.S. Fish and Wildlife Service maintains populations in the Dallas, Texas, Aquarium and the Midwest Science Center in Columbia, Missouri. In the warm seasons, Strecker's chorus frog and plains and southern leopard frogs are seen and heard. Turtles frequently seen on woody debris along the riverbank include red-eared slider, common snapper, stinkpot, Texas river cooter, and yellow mud and Texas map turtles. Both smooth and spiny softshell turtles are common along the river, and American alligators are seen closer to the coast.

The Colorado River basin is home to numerous bird species, including many threatened and endangered species associated with the river. The rare green kingfisher and more common belted kingfisher are found here. Wood ducks and shorebirds flock to reservoirs of the high plains during winter. Near the coast, Attwater's Prairie Chicken National Wildlife Refuge and Eagle Lake are home to thousands of ducks and geese during winter. Whooping cranes, white-faced ibis, reddish egrets, piping plovers, brown pelicans, wood storks, and 37 species of shorebirds can be viewed along the Matagorda Peninsula at the mouth of the Colorado River.

Ecosystem Processes

Very little ecosystem research has been reported from the Colorado River basin. However, filter feeders are the most abundant macroinvertebrate guild, followed by gatherers and grazers, then predators, and finally shredders and miners, based on the Texas Clean Rivers surveys (Bayer et al. 1992). This suggests a strong influence of smaller size fractions of particulate organic matter as energy sources. This is consistent with the arid climate across the primarily western headwaters, where meager gallery forests and prairie grasses grow along intermittent and shallow headwater streams. In addition, range and croplands augment nutrient supplies, supporting the growth of stream periphyton in these shallow and well-lighted streams and rivers.

Human Impacts and Special Features

Among the more interesting features of the Colorado River is that it begins its flow among the rolling prairies but then flows through scenic canyons in the Edwards Plateau. The river and its tributaries pass through a large region of metamorphic and igneous rock (the Llano Uplift), with spectacular reddish granite. The state capitol in Austin was built using granite from the area. Below Austin, the river widens and slowly meanders to the coast. All along its length, however, the nature of the Colorado River has been greatly altered, particularly by dams, pollution, and introduction of nonnative species.

Attempts were made to make the Colorado River navigable by removing large logjams that hindered transportation in the mid-1800s. After the Civil War, the Colorado River was no longer used for transportation, and 25 reservoirs now make it the most heavily dammed river in Texas. Three large reservoirs, Lake J. B. Thomas, E. V. Spence Reservoir, and O. H. Ivie Reservoir are located on the upstream portion of the river (see Fig. 5.15). Below O. H. Ivie Reservoir the modified flow regime and reduced sediment load has increased channel downcutting and artificial riffles have been constructed to provide shallower flowing habitat to facilitate foraging by the endangered Concho water snake.

The Highland Lakes region of central Texas consists of seven reservoirs along 137 km of river northwest of Austin (see Fig. 5.15). The seven lakes from upstream to downstream are Lake Buchanan, Inks Lake, Lake LBJ, Lake Marble Falls, Lake Travis, Lake Austin, and Town Lake. Urbanization in the Austin area has increased rapidly, and the number of water surveys reporting excessive coliform bacterial counts due to nonpoint pollution sources also has increased. The total volume of water from the Colorado River is overallocated, but flow management takes advantage of asynchronous demands within the drainage basin. Hydropower and flood-control influence discharge downstream of the Highland Lakes. Despite flood-control dams, flooding still occurs in the uplands following intense thunderstorms, when rain rapidly washes across limestone bedrock and through permeable soils. Flooding also occurs associated with hurricanes in the Coastal Plain. Flows are provided to irrigate cotton, sorghum, and rice fields and to augment municipal and industrial water supplies during dry periods. This is reflected in average monthly runoff in summer, as average monthly runoff increases in summer despite lower rainfall because dam releases satisfy water users downstream.

Several fish species have been introduced either as bait or sport fishes. The genetic integrity of Guadalupe bass (the Texas state fish) has been introgressed by nonnative smallmouth bass that have been stocked as a cool water game species. Other stocked nonnative game fish include walleye, sauger, northern pike, striped bass, and rainbow trout. Nonnative aquatic plants have become established following introductions via ornamental water gardens, aquaria, and transport on boats and trailers. These include water hyacinth, hydrilla, water lettuce, Eurasian watermilfoil, and alligatorweed. Nonnative grass carp has been introduced to control overgrown vegetation in reservoirs, but only sterile certified triploids are permitted. Blue tilapia is commonly found in cooling reservoirs and warm backwaters, having escaped from aquaculture facilities or illegally introduced as food fish or bait. Common carp are widespread, having been introduced from Europe and China as food fish in the 1800s. Many mistakenly believe that common carp are native to the Colorado River.

BRAZOS RIVER

Thirsty Spaniards called the river "Los Brazos de Dios" (the arms of God) and Native Americans called the river Tokonohono. The Brazos River arises at the confluence of the Salt Fork and Double Mountain Fork (33°16′N, 100°01′W) and flows about 1390 km southeasterly across Texas to its mouth on the Gulf of Mexico southwest of Houston (Fig. 5.17). The Brazos has the greatest channel length entirely within Texas and is the third longest of all Texas rivers. Its drainage basin is about 115,600 km², with 94% in Texas. The Brazos heads in New Mexico, and its forks drain much of the southern panhandle and parts of west Texas surrounding Lubbock, Texas, on the Llano Estacado. Both dry, hot summers and sudden icy winds and winter blizzards color and shape the landscape, and long cycles of drought often are devastating to inhabitants.

Although commonly appearing like a sunscorched desert in summer, showers change the land into a rich green carpet of buffalo, grama, and bunch grasses that once supported large numbers of buffalo, pronghorn antelope, and deer. Little is known archaeologically of the Paleo-Indians that would become the nomadic Apache Plainsmen who hunted here before acquiring horses. The Apaches also gardened and traded with agricultural Pueblo Indians to secure their livelihood and to adapt to the erratic and unpredictable movement patterns of buffalo herds. Living in small villages, the Apaches were eventually displaced by mounted Comanches who raided from the north and by the Spanish from the south. Comanches lived in the Brazos catchment in loose bands and

camped along the rivers, but they rarely ate fishes or fowl unless starving.

Brackish water, red clay, and ancient sand and silt washed down from the Rocky Mountains discouraged travelers and settlers along the Salt Fork of the Brazos. Downstream, the freshwater of the White River emerging from Blanco Canyon improved river conditions substantially. This is the area where Coronado likely staged his search for Quivira and the seven cites of Cibola in 1541. The river exits through the palisades of the Staked Plain and drops into the Central Lowlands, topographically flat except for occasional sandstone buttes and towers. A largely Hispanic population now farms croplands and controls the irrigation systems. Longhorn and other cattle, oil, and gas production dominate the rangeland north of Abilene, Texas. After the Clear Fork enters the Brazos west of Fort Worth, Texas, and leaves the Central Lowlands, the Brazos briefly re-enters the Great Plains and then flows more southward to Waco, Texas, and onto the Coastal Plain.

Downstream of Waco, the Brazos meanders through bottomland hardwoods and sandy floodplains renewed by periodic excursions over natural river levees. The native people called Tonkawas (meaning "they all stay together") by the Wacos in this area and called Tickanwatic by themselves (meaning "the most human of people") lived as clans in this region. Part of the Plains prehorse culture, these clans hunted buffalo and small game animals, fished, used dogs as pack animals and food, and had no agriculture, but gathered pecans, acorns, herbs, fruit, and seeds along the river. The timid Tonkawas, reduced in number by disease and raiding Comanches, were forced out of this area in the nineteenth century by European settlers.

Spanish explorers first recorded the mouth of the Brazos River in 1519. Early maps located fords on the Brazos, but Spanish activity was limited within the basin. The Brazos played a prominent role in the history of modern Texas. European settlement began in December 1821, when the first of Stephen F. Austin's settlers arrived at the mouth of the Brazos. Early Anglo-American settlers located their first capital at Washington-on-the-Brazos, where the Navasota River enters the Brazos south of College Station, Texas. Steam navigation reached the Brazos around 1830 and serviced inland cities up to 400 km upriver from the Gulf of Mexico.

Physiography, Climate, and Land Use

The Brazos flows through three physiographic provinces beginning with the Great Plains (GP) of the

Texas panhandle and eastern New Mexico (see Fig. 5.17). The Brazos then passes through the Central Lowlands (CL), emerges back into the Great Plains, and then flows across the Coastal Plain (CP) province into the Gulf of Mexico. The Brazos also passes through six terrestrial ecoregions along its course: Western Short Grasslands, Central and Southern Mixed Grasslands, Central Forest/Grassland Transition Zone, Texas Blackland Prairies, East Central Texas Forests, and Western Gulf Coastal Grasslands (Ricketts et al. 1999). Although much of the basin is modified by ranching and agriculture, natural upland vegetation begins with bunched growth forms, including short grasses adapted to seasonal drought, fire, and grazing, such as grama and buffalo grass, then mixed grasses, such as little bluestem and western wheatgrass, and then taller grasses, such as Indian grass, on the Blackland Prairie and riparian trees, such as cottonwood, hackberry, and elm. When the river enters the East Central Texas Forests ecoregion, it includes a greater dominance of upland trees, such as post and blackjack oaks, and in the bottomlands black and mockernut hickories and pecan. Vegetation in the Western Gulf Coastal Grasslands shifts from upland tallgrass species to those found in more saline soils, such as gulf cordgrass, sedges, and rush, bulrush, and salt grass.

Climate varies greatly from the upper basin in the high plains to the middle basin in the prairies and woodlands to the coastal plains of the lower basin. For example, average annual temperature and rainfall at Lubbock in the upper Brazos Basin are 15.6°C and 47 cm, respectively, with average monthly temperatures ranging from 4°C to 27°C and monthly precipitation ranging from 1.4 cm in January to 7.0 cm in May. Average annual temperature and rainfall at Waco, in the central Brazos Basin, are 19.2°C and 81 cm, respectively. Mean monthly temperatures range from 7°C in January to 30°C in July and August and monthly precipitation ranges from 4.4 cm in January to 12.0 cm in May (Fig. 5.18). The lower Brazos basin has greater precipitation and warmer temperatures. Freeport, at the mouth of the Brazos, has an average annual temperature of 20.9°C and an average annual precipitation of 128 cm. Monthly mean temperatures at Freeport range from 11.4°C in January to 28.8°C in July and August, with peak precipitation of 18.4 cm in September and minimum precipitation of 6.6 cm in March.

Land use in the basin is a mix of grazing, agriculture, and urban development, with remnants of native vegetation throughout the region. The basin is approximately 57% grassland, 24% cropland, 16%

urban and suburban, and 3% forest. Major cities in the Brazos basin are Lubbock, Abilene, Waco, Temple, College Station, and Freeport. Houston abuts the region on the east near the river mouth. Cattle, cotton, grain sorghum, and wheat are grown in the upper basin. The middle basin produces peanuts, dairy products, livestock, poultry, and grains. The lower basin produces rice, cotton, livestock, poultry, and dairy products.

River Geomorphology, Hydrology, and Chemistry

The Brazos River shows varied geomorphic characteristics as it crosses Texas. The Upper Brazos of the high plains is generally a broad, shallow, sandy, spatially intermittent river in short grass country. Canyonlands exist at the breaks of the Llano Estacado and the Caprock escarpment. The Middle Brazos was relatively unspoiled until numerous dams were constructed in this section of the river beginning in the 1940s and continuing through the 1980s (Gillespie and Giardino 1996). Much of the Brazos was entrenched and confined in narrow valleys with steep sides or bluffs. Near Waco the topography changes to gently rolling hills and the river is less constrained. The Lower Brazos of the Coastal Plain becomes a deep, broad river in agricultural lands (Fig. 5.5). The Brazos starts at an elevation of 450 m asl and stream gradients diminish from 66 cm/km to 9 cm/km as the river flows from its headwaters to the mouth.

There have been some detailed geomorphic studies of the Brazos River. Gillespie and Giardino (1996) examined migratory rates of meanders in the Middle and Lower Brazos as the river adjusts to numerous dams built recently on the main stem and tributaries. This research discussed the effects of flow regulation on channel stability and determined an index to assess channel stability where flow alteration is occurring. Gillespie and Giardino (1997) included data for 125 bends over 260 km in the Middle and Lower Brazos from the 1930s to 1988. Migration rate has decreased substantially as regulated flows have diminished peak flows and reduced suspended sediment loads. Ratzlaff (1981) studied the mechanisms of meander development and cutoff in the Brazos. Blackburn et al. (1982) studied the role of saltcedar on sedimentation in the Upper Brazos. Flow regulation through dams and invasion by nonnative plant species into riparian zones are modifying the geomorphology of the Brazos River. Saltcedar infes-

FIGURE 5.5 Brazos River at Rt. 105 near College Station, Texas (PHOTO BY TIM PALMER).

tations in the Upper Brazos are affecting channel structure, sediment deposition, and river hydrology. Channel width has been reduced by about 90 m in a 129-km river reach from Clear Fork to Seymour. Approximately 3 m of deposition has occurred within the saltcedar-infested channel reaches, and higher flood stages now occur for similar flow volumes such that inundation patterns are substantially larger now for comparable flows.

Mean discharge for the Brazos River is 249 m³/s. Given the large size of the basin, however, runoff is quite low, ranging from only 0.2 cm/mo during the winter to 0.8 cm/mo in May (see Fig. 5.18). The annual total runoff of about 4 to 5 cm is only 5% of annual precipitation and indicates high evapotranspiration and other losses. The discharge regime is strongly regulated by the dams upstream of Waco on both a daily and seasonal basis, with daily discharge commonly being reduced to <1 m³/s below Lake Whitney Dam.

Water chemistry reflects the predominant marine clays, limestone and sandstone geology, and agricultural land use. Water is mildly alkaline and salty (pH 7.6, alkalinity 133 mg/L as $CaCO_3$, specific conduc-

tance 733 µS/cm) due to marine salts deposited in the region of the upper basin as an ancient inland sea evaporated. Nutrient loading (dairy farms) in the middle basin causes late summer algal blooms, especially in intermittent reaches. In recent years, toxic blooms of golden alga linked to nutrient loading in some reservoirs have caused multiple fish kills. Water-quality parameters of primary concern in the basin include natural salinity, atrazine, perchlorate, phosphorus, dairy wastes, and dissolved oxygen.

River Biodiversity and Ecology

The Brazos River flows through the East Texas Gulf freshwater ecoregion (Abell et al. 2000). Unfettered by dams downstream of Waco, the rich biodiversity in the lower drainage is contained in habitats maintained by natural fluvial processes.

Algae and Cyanobacteria

Little published information is available on the algae of the Brazos River. A general survey done for the Brazos River Authority in early fall (Winemiller and Gelwick 1999) lists tentative genera of diatoms

(*Nitzschia, Navicula, Cymbella, Gomphonema, Diatoma, Synedra, Navicula, Tabellaria, Cocconema, Cosmarium*), unicellular green algae (*Ankistrodesmus, Characium*), filamentous green algae (*Rhizodonium, Cladophora, Oedogonium, Spirogyra, Tribonema, Mougeoutia, Ulothrix*), and cyanobacteria (*Anabaena, Oscillatoria*). There has been recent concern about a toxic golden alga (*Prymnesium parvum*) in reservoirs of the Brazos Basin. This golden alga is usually found in estuaries, but it now has been found in numerous inland water bodies throughout Texas. Possum Kingdom Lake on the Brazos was especially affected in 2000 and 2001, with thousands of fishes killed.

Plants

Riparian vegetation in the Upper Brazos was dominated by obligate and facultative phreatophytes, such as mesquite, baccharis, cottonwood, willow, elm, hackberry, and sumac (Blackburn et al. 1982). Nonnative saltcedar has increasingly affected riparian areas in the Upper Brazos (Busby and Schuster 1971, Blackburn et al. 1982). Blackburn et al. (1982) estimated that coverage by saltcedar in the Upper Brazos from the confluence of Clear Creek to Seymour (129 km) changed from small patches in the 1930s to 57% of the original river channel by 1979. Early settlers described the Middle Brazos as wooded sections frequently containing great pecan and oak trees interspersed with more open country with broad prairies. The Lower Brazos originally passed through forests, transporting snags and forming debris dams throughout much of its lower reach. Today it is partially channelized by levees, with sand margins and mixed riparian forests.

Macrophytes such as water willow are generally limited to shallow, sluggish, or slow-flowing margins of streams. Other native plants in oxbows, ponds, and reservoirs include spatterdock, water shield, duckweed, arrowhead, eelgrass, water primrose, Illinois pondweed, spikerush, and water stargrass. Many of these are being planted in an attempt to establish native species before introduced nonnative species become further established. Two nonnative macrophytes that occur in both the Middle and Lower Brazos drainage are alligatorweed and hydrilla. In addition, nonnative water hyacinth, water lettuce, dotted duckweed, and giant salvinia have been reported in the lower drainage.

Invertebrates

Kenneth Stewart and colleagues at the University of North Texas have studied the aquatic insects of the Brazos drainage basin in north-central Texas, with particular emphasis on caddisflies. Forty-two species of caddisflies distributed among nine families were reported in the Middle Brazos River and tributaries (Moulton et al. 1993). At a site downstream of Possum Kingdom Lake on the Brazos River, 4 of 22 species were particularly abundant and productive (*Hydropsyche simulans, Cheumatopsyche lasia, Cheumatopsyche campyla,* and *Chimarra obscura*). This study brought the total of known caddisfly species in Texas to 106, with 8 new species described from Texas and the Brazos drainage. Cloud and Stewart (1974) studied caddisfly drifting behavior in the Brazos and found *Cheumatopsyche campyla, C. lasia,* and *Hydropsyche simulans* dominating caddisfly drift.

Surveys for the statewide Clean Rivers Program (Bayer et al. 1992) also provide information on least disturbed streams in the Brazos basin. EPT indices range from 4 to 7 in the uppermost reaches in the Great Plains north and west of Abilene, where harsh, hot summers, high salinity, and cold winters limit populations. Predominant species are chironomid midges (*Polypedilum scalaenum, P. convictum, Rheotanytarsus exiguous, Cryptochironomus fulvus*), biting midges (*Bezzia*), caddisflies (*Cheumatopsyche*), tolerant oligochaete worms (*Limnodrilus hoffmeisteri*), and beetles (*Stenelmis sexlineata, S. occidentalis, Berosus subsignatus*). EPT indices range from 8 to 12 in the more mesic Central Lowlands north of Waco, where streams run through croplands and post oak–juniper forests. Here, assemblages are dominated by salinity-intolerant species of caddisflies (*Chimarra*), chironomids (*Orthocladius, Cricotopus bicinctus*), beetles (*Psephenus texanus*), and mayflies (*Stenonema*). Other common species include caddisflies (*Hydroptila*), chironomids (*Tanytarsus glabrescens, T. guerlus*), mayflies (*Choroterpes mexicanus, Fallceon quilleri*), stoneflies (*Perlesta placida, Neoperla clymene*), and damselflies (*Argia*). EPT indices range from 10 to 18 in the lower reaches of the Central Lowlands north and west of Temple, before the Brazos reenters the Great Plains. Streams run through diversified croplands, ranches, and pastures. Most abundant species include mayflies (*Thraulodes gonzalesi, Tricorythodes albilineatus*), caddisflies (*Chimarra, Cheumatopsyche, Oxyethira*), and chironomids. Additional chironomids in the lower Central Lowlands include *Dicrotendipes neomodestus* and *Pseudochironomus*. EPT indices range from 9 to 14 in the Coastal Plain streams between Waco and College Station, where post oak woods and bottomlands were cleared for crops. Abundant species include caddisflies (*Smicridea, Cheumatopsyche*), mayflies (*Fallceon quilleri*), tipulid flies (*Hexatoma*), chironomids (*Pentaneura, Meropelopia*), and predatory dragonflies (*Brechmorhoga mendax, Erpetogomphus*). In streams heavily impacted by cropland and livestock, EPT indices range from 0 to 5 and are dominated by poor-water-quality tolerant species of oligochaete worms (*Dero digitata, Limnodrilus maumeensis, L. hoffmeisteri, Quistadrilus multisetosus*) and phantom midges (*Chaoborus*). EPT indices range from 8 to 12 in streams west of Houston, where former grasslands are now used for rice and pasture. Common species include caddisflies (*Cheumatopsyche*), mayflies (*Baetis pygmaeus*), beetles (*Stenelmis occidentalis*), and limpets (creeping ancylid).

Crayfish species in the Brazos Basin include *Cambarellus puer, C. texanus, Fallicambarus hedgpethi, Orconectes causeyi, O. palmeri longimanus, O. virilis* (virile crayfish), *Procambarus acutus acutus,* and *P. clarkii* (red swamp crayfish). Nineteen species of mussels have been described in the Brazos drainage, including bleufer, western pimpleback, pistolgrip, southern mapleleaf, Texas lilliput, pondhorn, and tapered pondhorn. Three species are presently rare (smooth pimpleback, false spike, and Texas fawnsfoot), and one species is of concern (golden orb). The nonnative Asian clam, now abundant in the Brazos drainage, was probably introduced in 1972 and 1973 and was widespread and abundant by 1980 (Fontanier 1982).

Vertebrates

There are 93 species of fishes in the freshwaters of the Brazos Basin. Dominant species in the most western streams with highest salinities and intermittent flows are Red River pupfish, plains killifish, plains minnow, and smalleye shiner (now a candidate for listing due to reduced populations downstream of dams). Additional common species eastward into the Central Lowlands where flow is more permanent include red shiner, bullhead minnow, channel catfish, bluegill, green sunfish, orangespotted sunfish, largemouth bass, and mosquitofish. Pool-riffle habitat heterogeneity increases in the Central Lowlands around Abilene, where woody debris and overhanging vegetation contribute additional cover and terrestrial sources of prey. Additional common fishes in the Central Lowlands include the central stoneroller minnow, yellow and black bullhead catfish, blacktail shiner, golden shiner, mimic shiner, sharpnose shiner (a candidate for listing due to populations reduced above dams), redear sunfish, spotted bass, blackstripe

topminnow, and orangethroat darter. Larger streams also contain dusky darter, smallmouth buffalo, spotted sucker, flathead catfish, and longnose gar. As the Brazos flows onto the Coastal Plain below Waco, the river forms numerous oxbows, sloughs, and swamps with abundant vegetation. Species adapted to these habitats and seasonal flooding become more abundant. These fishes include banded pygmy sunfish, pirate perch, slough darter, tadpole madtom, gizzard shad, warmouth, white crappie, and spotted gar.

Twenty-one nonnative fish species have been introduced, including species for sport (northern pike, sauger, walleye, rainbow trout, striped bass, smallmouth bass, Rio Grande cichlid), food (common carp, blue tilapia), bait (rudd), and prey (inland silverside). Two minnow species (smalleye shiner and sharpnose shiner) are candidates for federal listing, and the blue sucker (primarily found in large rivers) is listed as a Texas species of concern.

Common aquatic snakes in the Brazos Basin include cottonmouth, copperhead, water snakes (broad-banded, diamondback, green, and blotched), and glossy crayfish snake. The Harter's water snake is restricted to the upper Brazos River drainage. It is found in about 303 km of stream plus two reservoirs. It and the other subspecies (Concho water snake) are the only endemic Texas snakes (Scott et al. 1989). Common amphibians include cricket and green frogs, spotted and Strecker's chorus frogs, and Rio Grande, southern, and plains leopard frogs. Turtles include Texas and Ouachita map turtles, red-eared slider, and stinkpot. Midland smooth and spiny softshell turtles and American alligators are more common in the Lower Brazos. The Lower Brazos also is a haven for birds. Freeport usually wins the unofficial competition for most bird species seen in Texas (>200), which include the rare masked duck (usually found in the West Indies and Mexico). Tens of thousands of ducks, geese, and shorebirds winter along the southern shores of the Brazos. Beaver "slides" along the river and drifting "beaver sticks" indicate their activities throughout the basin. Nonnative nutria also have become residents in the Lower Brazos.

Ecosystem Processes

Limited ecosystem research has been carried out on the Brazos River. Clean Rivers Surveys indicate numerical dominance by macroinvertebrate guilds of filterers and grazers, followed by gatherers, miners, and finally predators (Bayer et al. 1992). These data suggest the importance of FPOM as an energy source. Winemiller et al. (2000) reported that oxbow lakes in the

Middle–Lower Brazos riverine corridor were eutrophic, with chlorophyll *a* levels of 15 to 640 μg/L. Rates of nutrient recycling measured at the water–sediment interface of two oxbows near College Station were among the highest reported for freshwater ecosystems (J. B. Cotner, University of Minnesota, unpublished data). Oxbows trap sediment as the velocity of floodwaters declines, and nutrient mineralization rates are extremely high (Shormann and Cotner 1997). Total dissolved nitrogen was inversely correlated with species diversity, indicating that the most nutrient-rich systems were also relatively harsh habitats and limiting to many fish species. Research on a Navasota River tributary (Carter Creek) dominated by wastewater effluent from multiple sources indicated that algae were not nutrient limited, but rather were constrained primarily by frequent scouring events. Late successional stages (dominated by cyanobacteria) were reached only during brief periods of low or no rainfall. Moulton et al. (1993) reported very high rates of secondary production by caddisflies in the Middle Brazos River west of Fort Worth. Production rates from 1980 to 1982 averaged 55 g AFDM $m^{-2} yr^{-1}$, with average monthly densities around 16,000 individuals/m^2. These production rates and standing stocks are among the highest measured for stream insects (Benke 1984), suggesting overall production of aquatic invertebrates is high.

Human Impacts and Special Features

Characteristics of the Brazos River change greatly over its length. From the Brazos de Dios across the high Llano Estacado of the Great Plains, where it flows only intermittently in the summer and rarely freezes in the winter because of its high salinity, the Brazos cuts through canyonlands and briefly enters the Central Lowlands before crossing back into the Great Plains and finally onto the flat Coastal Plain.

Upstream, populations of minnows that depend on seasonal high flows for upstream movement and reproduction have been reduced by both harsh environmental conditions and disruption of flows caused by damming and irrigation (Wilde and Ostrand 1999). John Graves (1960), in *Goodbye to a River*, wrote a moving tribute to natural free-flowing rivers and points to the cumulative consequences of building dams on rivers like the Brazos. There are more than a dozen reservoirs within the Brazos basin (see Fig. 5.17). Examples include Possum Kingdom, Granbury, and Whitney, which capture water just after the forks of the Brazos join. Apart from Lake

Brazos (formed by a low-head dam at Waco), the main stem of the Brazos in the lower basin flows freely for over 640 km to the Gulf of Mexico, due in part to the shallow gradient of the Coastal Plain. Wide meanders have formed many oxbow lakes at various distances from the present main stem of the river, which reflects their various histories of inundation and reconnection to the river. The Brazos River, deprived of much of its sediments by many dams, has cut deeper into its channel, and less frequently reconnects with the oxbow lakes. The centennial flood of 1992 had a major impact on these oxbow lakes and their biota, providing a significant source of recruitment to the river fishes assemblage (Winemiller et al. 2000).

Few towns have grown up along the lower river because floods still come often enough that they carry away property and huge sections of riverbank. However, growing cities (Fort Worth and Houston) outside the Brazos drainage basin have either already captured Brazos water or plan to do so in the near future. The addition of an off-channel reservoir for water pumped from the Brazos is presently planned to supply water for Houston. A major concern of this proposed project is that flows will be severely reduced during critical periods in the summer. In addition, sediment from the main channel of the Lower Brazos has been utilized extensively as a source of sand for construction of infrastructure in southeast Texas. Concern has recently arisen over sand removal and effects of this practice on sediment transport to the Gulf of Mexico and ultimately beach erosion on the coast.

SABINE RIVER

The Sabine River draws its name from the Spanish word for cypress. It arises from low hills northeast of Dallas, Texas, in three main branches (Cowleech Fork, Caddo Fork, and South Fork) and flows approximately 890 km to the Gulf of Mexico (Fig. 5.19). The former juncture of these three branches is now inundated by Lake Towakoni, constructed in 1958. A fourth branch, Lake Fork Creek, joins the main river about 64 km downstream of the former juncture of the other three branches. The Sabine initially flows southeasterly through fertile east Texas farmland of the blackland prairies, which supports sorghum, cotton, beef and dairy cattle, sheep, and poultry. Then it flows through the timberlands, where logging is the major industry, and eventually turns south to flow through Toledo Bend Reservoir, where the Sabine becomes part of the Texas and Louisiana border at the 32nd parallel. The total drainage basin is approximately 25,270 km², with about three-quarters in Texas and one-quarter in Louisiana (see Fig. 5.19). As the Sabine descends deeper into the coastal prairie, it forms a maze of wetlands, with massive trees draped in Spanish moss. Running a parallel course, the Neches River to the west joins the Sabine at Sabine Lake, which empties into the Gulf of Mexico through Sabine Pass. Annual discharge from the Sabine River into the Gulf of Mexico makes it the largest river in Texas by volume (Bartlett 1984).

Archaeological excavations have discovered human habitation within the Sabine River basin, beginning with the 12,000-year-old Clovis culture. The Caddo period, from about 780 to 1260 years ago, was a peak of Indian development in the region. Caddo Indians were a loose confederation of more than two dozen tribes allied to the Natchez tribes of the Lower Mississippi Valley who shared a common language and culture that indicated their likely ancestral migration from the Caribbean along the Gulf Coast during prehistoric times. Living in permanent farming villages, they built flat-topped earthen temple mounds and grass houses and achieved the highest level of cultural development of any Texas Indian tribe. Their civilization spread west to the Trinity River and east to the Atlantic Coast. They fished using spears, nets, and trotlines, methods still used today, and wove reed baskets, made various types of pottery, and made bows made of *bois d'arc* (Osage orange). Tribes speaking Atakapan (meaning "cannibal" in the Choctaw language) included the Bidai, Deadose, Patiri, and Akiosa, who lived along the middle and lower Sabine River and Sabine Lake. They tattooed their bodies and faces, used dugout canoes, hunted and fished, and gathered bird eggs and shellfish. Following frequent contact with Europeans from the sixteenth to eighteenth centuries, these civilizations collapsed due to epidemic diseases.

In the early 1800s, displaced Cherokee people began moving into East Texas. Treaties, first by Mexicans, then by Sam Houston and the Texas Republic, had promised them land between the Neches and Angelina rivers. Nevertheless, the Texas federal government voided the treaties and the Cherokees were removed to Indian Territory north of the Red River in Oklahoma. The Alabama and Coushatta Indians migrated to the Neches and Trinity River basins from the southeastern United States after the French and Indian War ended in 1763. Of all tribes who once lived in East Texas, only the migrant Alabama–Coushatta confederation currently holds

territory. Their reservation is on the north edge of the Big Thicket National Preserve (primarily located along the Neches River, west of Lake Steinhagen), itself a remnant of the primitive Big Thicket Region that once stretched from the Sabine River to the Brazos and from Lufkin to Houston and Beaumont.

The first European to view the Sabine was the Spanish explorer and cartographer Alonso de Piñeta in 1519. The Spanish explorer Domingo Ramón gave the river its name in 1716. The importance of the Sabine grew after the United States acquired the Louisiana Territory (1803) from the French. The Adams–Onis Treaty of 1819 passed ownership of Florida to the United States and the Sabine River was the boundary between the United States and Spanish Texas to 32°N latitude. During the Civil War the Battle of Sabine Pass in 1863 was an important Confederate victory that kept the Sabine open for transport through the Union blockade.

Physiography, Climate, and Land Use

The maximum width of the Sabine River drainage basin is only about 72 km, and the basin is fully within the Coastal Plain (CP) physiographic province (see Fig. 5.19). This area corresponds to the Texas Blackland Prairies, East Central Texas Forests, Piney Woods Forests, and Western Gulf Coastal Grassland terrestrial ecoregions of Ricketts et al. (1999). Soils are generally light-colored acid sandy loams, clay loams, and sands with light brown to dark gray loams and dark-colored calcareous clays in the upper basin. The high diversity of vegetation within the lower basin reflects the geological history within the Big Thicket region, formed by changing climate and multiple advances and recessions of ocean shorelines, which mixed and overlaid different soil types. Dominant plant species are longleaf, shortleaf, and loblolly pines, American beech, bald cypress, swamp tupelo, basket oak, American sycamore, river birch, sweetgum, black willow, and water oak trees, along with palmettos (some up to 3 m tall), Spanish moss, eastern grama grasses, Indian grass, and panicum.

Annual mean air temperature in the basin is about 18°C, with monthly mean temperatures typically from 7°C to 28°C annually. For example, Carthage, Texas, averages 7°C in January and 28°C in July (Fig. 5.20). Annual precipitation averages 127 cm as rainfall, with a range of about 100 to 145 cm throughout the basin. Long-term records in Carthage report the greatest precipitation of 13.1 cm on average in May and lowest precipitation of 6.4 cm on average in August, although precipitation is relatively evenly distributed throughout the year.

The landscape is still largely rural but is greatly influenced by agriculture and silviculture. The basin is approximately 67% forest, 15% grassland, 10% agriculture, and 8% urban. Mean population density within the Sabine River basin is 18 people/km^2 but varies considerably across the basin, from large expanses with <6 people/km^2 to urbanized areas of >160 people/km^2. The economic base in the Sabine drainage basin traditionally has been petroleum. Field production is centered around Longview, Texas, whereas petrochemical refining and transportation as well as offshore oil-rig construction dominate the economy at Port Arthur on Sabine Pass (Ridenour 1998), which with Orange and Beaumont form the "Golden Triangle" of oil refining and export in east Texas.

River Geomorphology, Hydrology, and Chemistry

The Sabine River channel is characterized by flat slopes (mean gradient of 22 cm/km) and wide, forested floodplains (Fig. 5.6). The Sabine River arises at an elevation of 198 m asl and is joined by various tributaries before entering Lake Tawakoni behind Iron Bridge Dam at an elevation of about 134 m asl. Below the dam, the Sabine River flows through Texas to Toledo Bend Reservoir on the Texas and Louisiana border at an elevation of about 52 m asl. Toledo Bend Reservoir is a large impoundment covering about 75,480 ha, making the reservoir the largest body of water in Texas and in the South. The Sabine River then flows south to Sabine Lake, which is formed by the confluence of the Neches and Sabine rivers at about 2 m asl. The lake is drained by the Intracoastal Waterway and Sabine Pass into the Gulf of Mexico.

Annual discharge of the Sabine is 238 m^3/s and of the Neches is 179 m^3/s through Sabine Lake into the Gulf of Mexico. Average runoff during the period from 1941 to 1967 was about 2.07 cm/mo. The Sabine River maintains strong flows most of the year (see Fig. 5.20). High rainfall rates in spring produce frequent flooding of low-lying areas, and large floods occur on average with about a five-year return interval. Tropical storms, which come from midsummer through autumn, result in flooding along the lower river. Floods generally rise and fall slowly. The low parts of the Sabine Basin often remain inundated for days or weeks during floods. The two large reservoirs

FIGURE 5.6 Sabine River below Rt. 12 upstream of Orange, Texas (PHOTO BY TIM PALMER).

(Lake Tawakoni at the junction of the South and Cowleech forks and Toledo Bend Reservoir on the Texas and Louisiana border) provide some flow regulation. For example, discharge from Toledo Bend Reservoir is generally highest from January through March, in anticipation of spring rainfall, and lowest from September through November to retain floodwaters produced by tropical storms. However, this only slightly dampens the expected seasonal flow patterns (see Fig. 5.20).

Water chemistry of the Sabine River reflects a lowland river in a high-precipitation zone. Mean annual water temperature is 21°C. Mean pH is 6.7, mean conductivity is 145 µS/cm, and mean alkalinity is 19 mg/L as CaCO$_3$. Water-quality concerns within the basin include periodically low dissolved oxygen levels (<2 mg/L), routinely low dissolved oxygen (<4 mg/L), high chemical oxygen demand (>80 mg/L), high total organic carbon (>40 mg/L), high nutrient levels (NH$_4$-N > 1.6 mg/L, NO$_3$-N + NO$_2$-N > 3.9 mg/L, and PO_4-P > 0.9 mg/L), high fecal coliforms (>3000 colonies/100 mL), and the presence of the herbicide atrazine. Water-quality problems are likely due to municipal and industrial wastewater discharge and stormwater runoff from agricultural, industrial, and urban areas.

River Biodiversity and Ecology

The Sabine is the easternmost large-river basin within the East Texas Gulf freshwater ecoregion of Abell et al. (2000). Aside from biological surveys and water-quality assessments, relatively little information is available on the ecology of the Sabine River.

Plants

Below Toledo Bend Reservoir is an extensive reach of sandy river crossed by only one road and lined by southern bald cypress in the lower 80 km of the river. Other common lowland tree species include sweetgum, water oak, black gum, water tupelo, magnolia, elm, cottonwood, hickory, walnut, maple, American beech, and ash. Submerged and emergent native aquatic plants include arrowhead, smartweed, and buttonbush. Spiderlily, with its narrow, elongated white petals, is a distinctive member of the Amaryllis family, seen commonly along streams and marshy places, where it forms large colonies as bulbs divide. Nonnatives of concern include hydrilla, water hyacinth, water lettuce, Eurasian watermilfoil, alligatorweed, torpedo grass, and giant salvinia. In human-built reservoirs and ponds, conditions are especially ideal for nonnative floating and submerged species because there are few native competitors. Nonnatives are often spread during floods and by hitchhiking on recreational vehicles and equipment.

Invertebrates

Very little published information exists on the aquatic invertebrates of the Sabine River. We examined data from the Texas Clean Rivers Program for surveys of least-disturbed streams above and below Toledo Bend Reservoir (Bayer et al. 1992). EPT indices ranged from 5 to 12. Abundant species in more sluggish silt-bottom areas included chironomid midges (*Rheotanytarsus exiguous, Stenochironomus, Tanytarsus guerlus, T. glabrescens*), oligochaete worms (*Limnodrilus hoffmeisteri*), and isopod crustaceans (*Asellus laticaudatus*). In tannin-stained systems with abundant woody debris, cypress knees, and roots, common species included snails (*Campeloma decisum*), caddisflies (*Hydropsyche* and *Cheumatopsyche*), beetles (*Stenelmis grossa and*

Macronychus glabratus), dobsonflies (*Corydalus cornutus*), and mayflies (*Caenis*).

There are 32 species of mussels in the Sabine River (Howells et al. 1996; Howells 1997, 2001). Common species include threeridge, giant floater, paper pondshell, Louisiana fatmucket, yellow sandshell, fragile papershell, pond mussel, washboard, threehorn wartyback, bankclimber, bleufer, southern mapleleaf, western pimpleback, lilliput, Texas lilliput, tapered pondhorn, and pondhorn. Rock pocket-book also is widely distributed but typically rare to uncommon in collections. Round pearlshell typically occurs in lower reaches of coastal rivers like the Sabine near the freshwater–saltwater interface but is becoming more abundant in irrigation canals and reservoirs. Fawnsfoot and deertoe reach the southeastern limits of their ranges in eastern Texas. Little spectaclecase apparently was once more abundant in many streams of the Sabine basin but has declined in recent years. A number of unionids are extremely rare. Three species of concern due to declining abundance are southern hickorynut, Texas heelsplitter, and Louisiana pigtoe. Declines are likely due to increasing abundance of the highly successful nonnative Asiatic clam and degraded water quality. Considerable concern exists within the basin that the nonnative zebra mussel will soon invade.

Crayfish species in the Sabine River include *Fallicambarus hedgpethi, Faxonella beyeri, Orconectes difficilis* (painted crayfish), *O. lancifer* (stilt crayfish), *O. nais, O. palmeri longimanus, Procambarus acutus acutus, P. dupratzi, P. simulans,* and two commercially important species, *P. clarkii* (red swamp crawfish) and *P. zonangulus* (the white river crawfish). *Procambarus kensleyi* occurs in the Sabine, Neches, Trinity, and San Jacinto rivers. It was classified as a species of special concern by the American Fisheries Society Endangered Species Committee (Taylor et al. 1996). The Kisatchie stream crawfish (*Orconectes maletae*) also is threatened, and the Brazoria crayfish (*Procambarus brazoriensis*) and *Procambarus nigricinctus* are considered endangered in the Sabine River basin.

Vertebrates

The Sabine River is home to 104 species of freshwater fishes. Based on surveys for the Clean Rivers Program (Bayer et al. 1992) upstream of Toledo Bend Reservoir, common or widely distributed native fish species included red shiner, blacktail shiner, ribbon shiner, weed shiner, bullhead minnow, yellow bullhead catfish, tadpole madtom, green sunfish, warmouth, bluegill, longear sunfish, redear sunfish, spotted sunfish, largemouth bass, pirate perch, blackspotted topminnow, bluntnose darter, slough darter, and dusky darter. Downstream of Toledo Bend Reservoir additional species included spotted gar, cypress minnow, Mississippi silvery minnow, golden shiner, mimic shiner, banded pygmy sunfish, dollar sunfish, bantam sunfish, spotted sunfish, black crappie, blackstripe topminnow, western mosquitofish, scaly sand darter, and mud darter.

Sixteen species of fishes are nonnative in the Sabine River basin, including game species (striped bass, smallmouth bass, rainbow trout, redbreast sunfish, sauger, and walleye), and species used as bait (rudd and goldfish), forage (inland silversides), food (common carp), weed control (grass carp), and other nongame species (Mexican tetra). The paddlefish is endangered in Texas, and an active stocking program is in place to try to revitalize the species upstream of Toledo Bend. The blue sucker, creek chubsucker, and western sand darter are additional fish species of concern.

A wide array of bird species (e.g., brown pelican, great blue heron, little blue heron, tricolored heron, great egret, snowy egret, white ibis, wood stork, and spotted sandpiper) occur along the Sabine River. The J. D. Murphree Wildlife Management Area on Taylor Bayou, which enters Sabine Lake from the west, contains alligator, river otter, beaver, nutria, muskrat, white-faced ibis, anhinga, purple gallinule, common snipe, and the largest population of canvasback ducks in Texas. Common turtle species include common snapper, Mississippi mud turtle, and red-eared slider, and less-common species of note include alligator snapping turtle, razorback musk turtle, Sabine map turtle, smooth softshell, and pallid spiny softshell. Nonvenomous water snakes commonly found are yellowbelly, diamondback, green, and broad-banded. The venomous cottonmouth is commonly seen curled up in root wads and on semisubmerged logs or tree branches overhanging the river. The Gulf Coast waterdog, southern and plains leopard frogs, and cricket frog also occur in the Sabine drainage.

Ecosystem Processes

The broad, flat, well-vegetated watershed of the Sabine River floodplain includes a diverse assemblage of clay- and sand-bottom streams and blackwater habitats (Bianchi et al. 1996). Clean Rivers Program surveys of least-disturbed streams in the Sabine system indicate that invertebrate filter feeders and collector-gatherers dominate the macroinvertebrate assemblages, followed by miners, shredders, grazers, and finally predators. This reflects the heavy shade of the riparian habitats and instream substrata that provides little habitat for growth of periphyton. The streams and rivers of the Sabine basin rely largely on allochthonous energy sources. Bianchi et al. (1996) studied some of these habitats concerning dissolved organic carbon and flux of methane to the atmosphere. Plant-mediated transport of methane was important, with large spatial variability in fluxes. Methane emission rates were highest after the longest period of flooding during the sampling period. DOC concentrations were very high, with values between 15 and 50 mg C/L.

Ridenour (1998) modeled biological oxygen demand (BOD) on the Sabine River at two gages downstream from a point source located in Longview. The model also evaluated the amount of remaining BOD entering Toledo Bend Reservoir under different flow regimes. Hydraulic geometry, discharge, and temperature were critical inputs to the model. The fraction of BOD remaining in the river as it entered the reservoir was greatest in the winter and spring, with discharge about twice as important as temperature in accounting for the difference.

Human Impacts and Special Features

The western edge of the drainage of the Sabine River (at about the level of Toledo Bend Reservoir) and south to near Houston and Beaumont contains a rich and diverse biological heritage that developed upon ancient sand dunes and beaches of a fossil sea bottom. The swamps and bogs are remnants of prehistoric lagoons and ponds that were created when dunes trapped retreating seawater between ice ages. However, loss of these freshwater wetlands in the lower reaches of the Sabine is an ongoing ecosystem concern (White and Tremblay 1995). Since the 1950s, more than 5000 ha of vegetated wetlands have been submerged due to subsidence in the Sabine Lake region. Human-induced contributions to subsidence include groundwater withdrawal and oil and gas extraction.

Urbanized areas and industries in the lower river cause the most significant water-quality problems in the basin. Sources of concern are oil refineries, petrochemical plants, and dredged deep-water canals constructed near the end of the nineteenth century for shipping. Intensive agriculture in parts of the basin also has caused periodic water-quality problems in the upper reaches of the Sabine River.

Nonnative species also have had significant impacts on the Sabine drainage basin. The Asiatic clam is abundant throughout the riverine corridor and may be a key species in terms of the effects of its filter feeding on ecosystem functions such as nutrient cycling and trophic dynamics. The Asiatic clam also may be a factor in the decline of many of the indigenous species of unionid mussels. Additional declines of mussels are linked to increased silt loading and floods following clear-cutting of timber on sandy soils or overgrowth of littoral habitats by nonnative aquatic vegetation (hydrilla, water hyacinth, and giant salvinia). The introduction of a number of game fishes, such as walleye, striped bass, and smallmouth bass, is likely to have negatively affected native fish populations in the basin. Ironically, however, consumption advisories exist for game fishes. Game fishes accumulate mercury in their flesh due to anthropogenic changes within the sediments of impounded waters, which influence the toxicity of metals despite their natural occurrence in the basin.

An additional anthropogenic disturbance that has had a major impact on the ecology of the Sabine River is the construction of large hydropower and flood-control reservoirs on the main river (Lake Tawakoni and Toledo Bend), as well as multiple tributary reservoirs. These dams are barriers to migratory fishes, halting previously large runs of anadromous and catadromous fishes (paddlefish and American eel), as well as seasonal movements of striped mullet. Along with water-quality and habitat degradation in the lower river and its tributaries, the dams have caused populations of these species to steadily decline. Recent stocking programs (e.g., paddlefish) and planned construction of fish passages have focused on revitalizing stocks and restoring access to the river. Large reservoirs also can influence upstream tributaries by limiting movement (due to higher populations of piscivores and unfavorable lentic habitats) of more fluvial species that move among headwater streams to maintain populations, or require downstream movement to larger habitats to complete their life cycles. Upstream populations of lentic species can be artificially supplemented from downstream populations that thrive in reservoirs.

ADDITIONAL RIVERS

The Pecos River begins in the Sangre de Cristo Mountains of northern New Mexico and flows through temperate forest, grassland, and desert before joining the Rio Grande on the Mexican border (Figs. 5.21 and 5.7). The Pecos River drops about 3350 m along its course of 1175 km. The river course

FIGURE 5.7 Upper Pecos River, New Mexico (PHOTO BY TIM PALMER).

is over 300 m below the surrounding land in much of the northern reaches. In places, the river disappears into the porous rock layers and emerges downstream. The river is dammed at Alamogordo Reservoir, McMillan Lake, Avalon Lake, and Red Bluff Reservoir. The Pecos flows through a beautiful, steep-sided, twisting canyon in the lower reaches before joining the Rio Grande. Water delivery from New Mexico to Texas from the Pecos has been contentious over the years, as the river is heavily appropriated for agricultural and urban use (Hayter 2002). This large but extremely arid basin (annual precipitation of 28 cm at Pecos, Texas), with high evapotranspiration and heavy water extraction, results in very little discharge into the Rio Grande (2 m³/s), with an annual runoff of <0.1 cm.

The Nueces River arises at an altitude of 730 m on the Edwards Plateau of southwest Texas and flows in a generally southeasterly direction to the Gulf of Mexico (Figs. 5.23 and 5.8). The Nueces is the only

FIGURE 5.8 Nueces River near its headwaters at the north Barksdale Highway 335 (Photo by R. Edwards).

FIGURE 5.9 Trinity River below Livingston Dam, Texas (Photo by Tim Palmer).

major river basin within the West Texas Gulf freshwater ecoregion (Abell et al. 2000). Pecan trees along the river gave rise to the Spanish name. The river course and drainage basin are in a predominantly rural area of Texas. The river carves distinctive canyonlands as it descends the Balcones escarpment. Below the Balcones escarpment, the river flows through the Coastal Plain in cattle-ranching country. A major dam blocks the river, forming Lake Corpus Christi in the lower reach. Below the dam the Nueces is a meandering river in its lower course, with citrus and truck farming. The Nueces River enters the Gulf of Mexico at Nueces Bay, an arm of Corpus Christi Bay. Although not as arid as the Pecos River, annual runoff is still low at 1.6 cm, producing a discharge of only 20 m³/s (see Figs. 5.24).

The Trinity River is entirely within the state of Texas, arising south of Wichita Falls and flowing 815 km to Galveston Bay (Fig. 5.25). The river flows initially southeast to Fort Worth, where the Clear Fork joins the West Fork. The Trinity continues eastward to Dallas and then flows south-southeast to Trinity Bay, the northeastern arm of Galveston Bay (Fig. 5.9). Several large dams dot the Trinity River catchment and many large urban areas occur along the river course. With an annual precipitation of 115 cm, the discharge is quite high at 222 m³/s, even though annual runoff (16 cm) represents only 14% of precipitation (see Fig. 5.26).

The Neches River arises at an elevation of 150 m in the rolling hills of the Eastern Timbers area of Texas and flows 666 km to the Gulf of Mexico (Fig. 5.27). The river initially flows southeast through mixed hardwood and southern pine forests (Fig. 5.10). The river turns south near Jasper and flows into low-lying coastal prairie near Beaumont, Texas. Rice agriculture and oil industry dominate along this portion of the Neches River. The Sabine–Neches deepwater channel links Beaumont with Sabine Lake, which in turn accesses Port Arthur, the Gulf Intracoastal Waterway, and the Gulf of Mexico. Precipitation in the basin is high at 136 cm, distributed fairly evenly throughout the year (see Fig. 5.28). About 16% of precipitation appears as runoff, which is highest in winter and spring. Several large reservoirs have been constructed in the basin, the largest of which is Sam Rayburn Lake, located on the Angelina tributary (see Figs. 5.27).

ACKNOWLEDGMENTS

We wish to thank the many individuals who provided information for this chapter. Jim Thibault was instrumental in putting together much of the data on each river basin and proofread the entire manuscript. Scott Anderholm provided data on water quality and algae in the Rio Grande. Shann Springer provided unpublished data on the aquatic invertebrates of the Rio Grande in New Mexico. David Bauer provided a detailed bibliography on the aquatic invertebrates of Texas. Diana Northup assisted with the literature review for the chapter. In addition, numerous other colleagues helped assemble necessary information for this survey. We sincerely thank all who helped assemble the material in this chapter.

FIGURE 5.10 Neches River above Rt. 59, Texas (PHOTO BY TIM PALMER).

LITERATURE CITED

Abell, R. A., D. M. Olson, E. Dinerstein, P. T. Hurley, J. T. Diggs, W. Eichbaum, S. Walters, W. Wettengel, T. Allnutt, C. J. Loucks, and P. Hedao. 2000. *Freshwater ecoregions of North America: A conservation assessment.* Island Press, Washington, D.C.

Baker, V. R., and M. M. Penteado-Orellana. 1977. Adjustment to Quarternary climatic change by Colorado

River in central Texas. *Journal of Geology* 85: 395–422.

Bartlett, R. A. 1984. *Rolling rivers: An encyclopedia of America's rivers.* McGraw Hill, New York.

Bayer, C. W., J. R. Davis, S. R. Twidwell, R. Kleinsasser, G. Linam, K. Mayes, and E. Hornig. 1992. Texas aquatic ecoregion project: An assessment of least disturbed streams. I. Presentation of results. Draft copy. Texas Natural Resource Conservation Commission (Texas Water Commission), Austin, Texas Parks and Wildlife Department, Austin, U.S. Environmental Protection Agency, Region VI, Dallas.

Benke, A. C. 1984. Secondary production of aquatic insects. In V. H. Resh and D. M. Rosenberg (eds.). *The ecology of aquatic insects*, pp. 289–322. Praeger, New York.

Bianchi, T. S., M. E. Freer, and R. G. Wetzel. 1996. Temporal and spatial variability, and the role of dissolved organic carbon (DOC) in methane fluxes from the Sabine River Floodplain (Southeast Texas, USA). *Archiv für Hydrobiologie* 136:261–287.

Blackburn, W. H., R. W. Knight, and J. L. Schuster. 1982. Saltcedar influence on sedimentation in the Brazos River. *Journal of Soil and Water Conservation* 37:298–301.

Busby, F. E., Jr., and J. L. Schuster. 1971. Woody phreatophyte infestation of the middle Brazos River floodplain. *Journal of Range Management* 23:285–287.

Bush, J. K., and O. W. Van Auken. 1984. Woody-species composition of the upper San Antonio River gallery forest. *Texas Journal of Science* 36:139–148.

Carter, L. F., and S. D. Porter. 1997. Trace-element accumulation by *Hygrohypnum ochraceum* in the upper Rio Grande basin, Colorado and New Mexico, USA. *Environmental Toxicology and Chemistry* 16:2521–2528.

Cloud, T. J., and K. W. Stewart. 1974. Seasonal fluctuations and periodicity in the drift of caddisfly larvae (Trichoptera) in the Brazos River, Texas. *Annals of the Entomological Society of America* 67:805–811.

Crawford, C. S., A. C. Culley, R. Leutheuser, M. S. Sifuentes, L. H. White, and J. P. Wilber. 1993. *Middle Rio Grande ecosystem: Bosque biological management plan.* U.S. Fish and Wildlife Service, District 2, Albuquerque, New Mexico.

Dahm, C. N., J. R. Cleverly, J. E. A. Coonrod, J. R. Thibault, D. E. McDonnell, and D. J. Gilroy. 2002. Evapotranspiration at the land/water interface in a semi-arid drainage basin. *Freshwater Biology* 47:831–843.

Davis, J. R. 1980a. Species composition and diversity of benthic macroinvertebrate populations of the Pecos River, Texas. *Southwestern Naturalist* 25:241–256.

Davis, J. R. 1980b. Species composition and diversity of benthic macroinvertebrates in the upper Rio Grande, Texas. *Southwestern Naturalist* 25:137–150.

Eaton, D. J., and D. Hurlburt. 1992. Challenges in the binational management of water resources in the Rio Grande/Río Bravo. U.S.–Mexican Policy Report no. 2. U.S.–Mexican Policy Studies Program, Lyndon B. Johnson School of Public Affairs, University of Texas at Austin.

Edwards, R. J. 1978. The effect of hypolimnion reservoir releases on fish distribution and species diversity. *Transactions of the American Fisheries Society* 107:71–77.

Edwards, R. J. 1979. A report of Guadalupe bass (*Micropterus treculi*) × smallmouth bass (*M. dolomieui*) hybrids from two localities in the Guadalupe River, Texas. *Texas Journal of Science* 31:231–238.

Edwards, R. J. 2001. New additions and persistence of the introduced fishes of the upper San Antonio River, Bexar County, Texas. *Texas Journal of Science* 53:3–12.

Edwards, R. J., and S. Contreras-Balderas. 1991. Historical changes in the ichthyofauna of the lower Rio Grande (Rio Bravo del Norte), Texas and Mexico. *Southwestern Naturalist* 36:201–212.

Ellis, L. M., C. S. Crawford, and M. C. Molles Jr. 2001. Influence of annual flooding on terrestrial arthropod assemblages of a Rio Grande riparian forest. *Regulated Rivers: Research and Management* 17:1–20.

Ellis, L. M., C. S. Crawford, and M. C. Molles Jr. 2002. The role of the flood pulse in ecosystem-level processes in southwestern riparian forests: A case study from the middle Rio Grande. In B. A. Middleton (ed.). *Flood pulsing in wetlands: Restoring the natural hydrological balance*, pp. 51–108. John Wiley and Sons, New York.

Ellis, L. M., M. C. Molles Jr., and C. S. Crawford. 1999. Influence of experimental flooding on litter dynamics in a Rio Grande riparian forest, New Mexico. *Restoration Ecology* 7:193–204.

Ellis, L. M., M. C. Molles Jr., C. S. Crawford, and F. Heinzelmann. 2000. Surface-active arthropod communities in native and exotic riparian vegetation in the middle Rio Grande Valley, New Mexico. *Southwestern Naturalist* 45:456–471.

Ellis, S. R., G. W. Levings, L. F. Carter, S. F. Richey, and M. J. Radell. 1993. Rio Grande valley, Colorado, New Mexico, and Texas. *Water Resources Bulletin* 29:617–646.

Epperson, C. R., and R. A. Short. 1987. Annual production of *Corydalus cornutus* (Megaloptera) in the Guadalupe River, Texas. *American Midland Naturalist* 118:433–438.

Everitt, B. 1993. Channel response to declining flow on the Rio Grande between Fort Quitman and Presidio, Texas. *Geomorphology* 6:225–242.

Everitt, B. L. 1998. Chronology of the spread of tamarisk in the central Rio Grande. *Wetlands* 18:658–668.

Everitt, J. H., C. Yang, D. E. Escobar, C. F. Webster, R. I. Lonard, and M. R. Davis. 1999. Using remote sensing and spatial information technologies to detect and map two aquatic macrophytes. *Journal of Aquatic Plant Management* 37:71–80.

Fontanier, C. E. 1982. The distribution of Corbicula (Bivalvia, Corbiculidae) in the Brazos River system, Texas, 25 August–12 November 1980. *Texas Journal of Science* 34:5–15.

Ford, A. L., and O. W. Van Auken. 1982. The distribution of woody species in the Guadalupe River floodplain forest in the Edwards Plateau of Texas. *Southwestern Naturalist* 27:383–392.

Gillespie, B. M., and J. R. Giardino. 1996. Determining the migratory activity index for a river: An example from the Brazos River, Texas. *Zeitschrift für Geomorphologie* 40:417–428.

Gillespie, B. M., and J. R. Giardino. 1997. The nature of channel planform change: Brazos river, Texas. *Texas Journal of Science* 49:109–142.

Graves, J. 1960. *Goodbye to a river: A narrative*. Alfred A. Knopf, New York.

Groeger A. W., P. F. Brown, T. E. Tietjen, and T. C. Kelsey. 1997. Water quality of the San Marcos River. *Texas Journal of Science* 49:279–294.

Hayter, D. J. 2002. Pecos River. Handbook of Texas Online Web site: http://www.tsha.utexas.edu/handbook/online/articles/view/PP/rnp2.html.

Hibbs, B. J., and R. Boghici. 1999. On the Rio Grande aquifer: Flow relationships, salinization, and environmental problems from El Paso to Fort Quitman, Texas. *Environmental and Engineering Geoscience* 5:51–59.

Hobbs, H. H., Jr. 1976. *Crayfishes (Astacidae) of North and Middle America*. U. S. Environmental Protection Agency, Water Pollution Control Research Series 18050 ELDO5/72. Washington, D.C.

Horgan, P. 1984. *Great river: The Rio Grande in North American history*, 4th ed. Wesleyan University Press, Hanover, New Hampshire.

Howe, W. H., and F. L. Knopf. 1991. On the imminent decline of Rio Grande cottonwoods in central New Mexico. *Southwestern Naturalist* 36:218–224.

Howells, R. G. 1997. Status of freshwater mussels (Bivalvia: Unionidae) of the Big Thicket Region of Eastern Texas. *Texas Journal of Science Supplement* 49:21–34.

Howells, R. G. 2001. *Status of freshwater mussels of the Rio Grande, with comments on other bivalves*. Inland Fisheries Division Report. Texas Parks and Wildlife Department, Austin.

Howells, R. G., R. W. Neck, and H. D. Murray. 1996. *Freshwater mussels of Texas*. Inland Fisheries Division Report. Texas Parks and Wildlife Department, Austin.

Hubbs, C., R. J. Edwards, and G. P. Garrett. 1991. An annotated checklist of the freshwater fishes of Texas with keys to the identification of species. *Texas Journal of Science Supplement* 43:1–56.

Hubbs, C., R. A. Kuehne, and J. C. Ball. 1953. The fishes of the upper Guadalupe River, Texas. *Texas Journal of Science* 5:216–244.

Hunt, C. B. 1974. *Natural regions of the United States and Canada*. W. H. Freeman, New York.

International Boundary and Water Commission (IBWC). 1990. Flow of the Rio Grande and related data. Water Bulletin no. 61. El Paso, Texas.

Jones, L. S., and J. T. Harper. 1998. Channel avulsions and related processes, and large-scale sedimentation patterns since 1875, Rio Grande, San Luis Valley, Colorado. *Bulletin of the Geological Society of America* 110:411–421.

Kitchell, J. F., D. E. Schindler, B. R. Herwig, D. M. Post, M. H. Olson, and M. Oldham. 1999. Nutrient cycling at the landscape scale: The role of diel foraging migrations by geese at the Bosque del Apache National Wildlife Refuge, New Mexico. *Limnology and Oceanography* 44:828–836.

Letson, D. 1992. Simulation of a two-pollutant, two-season pollution offset system for the Colorado River of Texas below Austin. *Water Resources Research* 28:1311–1318.

Levings, G. W., D. F. Healy, S. F. Richey, and L. F. Carter. 1998. Water quality in the Rio Grande Valley, Colorado, New Mexico, and Texas, 1992–95. Circular 1162. U.S. Geological Survey, Albuquerque, New Mexico.

Linam, G. W., L. J. Kleinsasser, and K. B. Mayes. 2002. Regionalization of the Index of Biotic Integrity for Texas Streams. River Studies Report no. 17. Resource Protection Division, Texas Parks and Wildlife Department, Austin.

Lonard R. I., F. W. Judd, J. H. Everitt, D. E. Escobar, M. R. Davis, M. M. Crawford, and M. D. Desai. 2000. Evaluation of color-infrared photography for distinguishing annual changes in riparian forest vegetation of the lower Rio Grande in Texas. *Forest Ecology and Management* 128:75–81.

Mack, G. H., and M. R. Leeder. 1998. Channel shifting of the Rio Grande, southern Rio Grande rift: Implications for alluvial stratigraphic models. *Sedimentary Geology* 117:207–219.

Martin, Q. 1991. Drought management plan for Lower Colorado River in Texas. *Journal of Water Resources Planning and Management* 117:645–661.

McCafferty, W. P., and A. V. Provonsha. 1993. New species, subspecies, and stage descriptions of Texas Baetidae (Ephemeroptera). *Proceedings of the Entomological Society of Washington* 95:59–69.

Miller, R. R. 1977. Composition and derivation of the native fish fauna of the Chihuahuan Desert region. In R. H. Wauer and D. H. Riskind (eds.). *Transactions of the Symposium on the Biological Resources of the Chihuahuan Desert Region, United States and Mexico*, pp. 365–382. National Park Transaction and Proceedings Series. Department of the Interior, Washington, D.C.

Miyamoto, S., L. B. Fenn, and D. Swietlik. 1995. Flow, salts, and trace elements in the Rio Grande: A review. Technical Report 169. Texas A&M University System, Texas Agricultural Experiment Station, Texas Water Resources Institute, College Station.

Molles, M. C. Jr., C. S. Crawford, and L. M. Ellis. 1995. Effects of an experimental flood on litter dynamics in the middle Rio Grande riparian ecosystem. Regulated Rivers: Research and Management 11:275–281.

Molles, M. C. Jr., C. S. Crawford, L. M. Ellis, and J. R. Thibault. 1996. Influences of flooding on dynamics of coarse woody debris and forest floor organic matter in riparian forests. *Bulletin of the North American Benthological Society* 13:154–155.

Molles, M. C. Jr., C. S. Crawford, L. M. Ellis, H. M. Valett, and C. N. Dahm. 1998. Managed flooding for riparian ecosystem restoration. *BioScience* 48:749–756.

Moulton, S. R., II, D. Petr, and K. W. Stewart. 1993. Caddisflies (Insecta, Trichoptera) of the Brazos River drainage in North-Central Texas. *Southwestern Naturalist* 38:19–23.

Neck, R. W., and A. L. Metcalf. 1988. Freshwater bivalves of the lower Rio Grande, Texas. *Texas Journal of Science* 40:259–268.

Owens, C. S., J. D. Madsen, R. M. Smart, and R. M. Stewart. 2001. Dispersal of native and nonnative aquatic plant species in the San Marcos River, Texas. *Journal of Aquatic Plant Management* 39:75–79.

Ratzlaff, J. R. 1981. Development and cutoff of Big Bend Meander, Brazos River, Texas. *Texas Journal of Science* 33:121–129.

Revenga, C., S. Murray, J. Abramovitz, and A. Hammond. 1998. *Watersheds of the world: Ecological value and vulnerability*. World Resources Institute, Washington, D.C.

Ricketts, T. H., E. Dinerstein, D. M. Olson, C. J. Loucks, W. Eichbaum, D. DellaSala, K. Kavanagh, P. Hedao, P. T. Hurley, K. M. Carney, R. Abell, and S. Walters. 1999. *Terrestrial ecoregions of North America: A conservation assessment*. Island Press, Washington, D.C.

Ridenour, G. S. 1998. The sensitivity of fluvial point-source models to hydraulic geometry, temperature, and discharge: Implications for hypoxia in large water bodies. *Physical Geography* 19:301–317.

Rosenthal, W. D., R. Srinivasan, and J. G. Arnold. 1995. Alternative river management using a linked GIS-hydrology model. *Transactions of the American Society of Agricultural Engineering* 38:783–790.

Schmandt, J., I. Águilar-Barajas, M. Mathis, N. Armstrong, L. Chapa-Alemán, S. Contreras-Balderas, R. Edwards, J. Hazleton, J. Navar-Chaidez, E. Vogel, and G. Ward. 2000. Water and sustainable development in the binational lower Rio Grande/Río Bravo basin. Final Report to EPA/NSF Water and Watersheds grant program (Grant No. R 824799–01–0). Houston Advanced Research Center, Center for Global Studies, The Woodlands, Texas.

Scott, N. J., Jr., T. C. Maxwell, O. W. Thornton Jr., L. A. Fitzgerald, and J. W. Flury. 1989. Distribution, habitat, and future of Harter's water snake, *Nerodia harteri*, in Texas. *Journal of Herpetology* 23:373–389.

Shafroth, P. B., J. M. Friedman, and L. S. Ischinger. 1995. Effects of salinity on establishment of *Populus-fremontii* (cottonwood) and *Tamarix-ramosissima* (saltcedar) in southwestern United States. *Great Basin Naturalist* 55:58–65.

Sherwood, A. R., and R. G. Sheath. 1999. Seasonality of macroalgae and epilithic diatoms in spring-fed streams in Texas, USA. *Hydrobiologia* 390:73–82.

Shirinian-Orlando, A. A., and C. G. Uchrin. 2000. A method for determining salt sources in surface waters. *Journal of the American Water Resources Association* 36:749–757.

Shormann, D. E., and J. B. Cotner. 1997. The effects of benthivorous smallmouth buffalo (*Ictiobus bubalus*) on water quality and nutrient cycling in a shallow floodplain lake. *Journal of Lake and Reservoir Management* 13:270–278.

Short, R. A. 1983. Occurrence of the caddisfly *Atopsyche erigia* in Texas' Guadalupe River below Canyon Reservoir. *Texas Journal of Science* 35:243–244.

Slade, R. M., Jr., and P. M. Buszka. 1994. Characteristics of streams and aquifers and processes affecting the salinity of water in the Upper Colorado River Basin, Texas. Water-Resources Investigations Report 94–4036. U.S. Geological Survey, Austin, Texas.

Snyder, W. D., and G. C. Miller. 1992. Changes in riparian vegetation along the Colorado River and Rio Grande, Colorado. *Great Basin Naturalist* 52:357–363.

Sprenger, M. D., L. M. Smith, and J. P. Taylor. 2001. Testing control of saltcedar seedlings using fall flooding. *Wetlands* 21:437–441.

Sprenger, M. D., L. M. Smith, and J. P. Taylor. 2002. Restoration of riparian habitat using experimental flooding. *Wetlands* 22:49–57.

Stanley, E. H., R. A. Short, J. W. Harrison, R. Hall, and R. C. Wiedenfeld. 1990. Variation in nutrient limitation of lotic and lentic algal communities in a Texas (USA) river. *Hydrobiologia* 206:61–71.

Taylor, C. A., M. L. Warren Jr., J. F. Fitzpatrick, H. H. Hobbs III, R. F. Jezerinac, W. F. Pflieger, and H. W. Robison. 1996. Conservation status of crayfishes of the United States and Canada. *Fisheries* 21:25–38.

Taylor J. P., and K. C. McDaniel. 1998. Restoration of saltcedar (*Tamarix* sp.)-infested floodplains on the Bosque del Apache National Wildlife Refuge. *Weed Technology* 12:345–352.

Taylor, J. P., D. B. Webster, and L. M. Smith. 1999. Soil disturbance, flood management, and riparian woody plant establishment in the Rio Grande floodplain. *Wetlands* 19:372–382.

Texas Natural Resource Conservation Commission (TNRCC). 1994. Regional assessment of water quality in the Rio Grande basin including the Pecos River, the

Devils River, the Arroyo Colorado and the Lower Laguna Madre. Texas Natural Resource Conservation Commission, Austin.

U.S. Fish and Wildlife Service. 1996. *San Marcos and Comal Springs and associated aquatic ecosystems (revised) recovery plan.* U.S. Fish and Wildlife Service, Albuquerque, New Mexico.

Van Auken, O. W., and J. K. Bush. 1985. Secondary succession on terraces of the San Antonio River. *Bulletin of the Torrey Botanical Club* 112:158–166.

Van Auken, O. W., and J. K. Bush. 1988. Dynamics of establishment, growth, and development of black willow and cottonwood in the San Antonio River forest. *Texas Journal of Science* 40:269–277.

Ward, J. V., and B. C. Kondratieff. 1992. *An illustrated guide to the mountain stream insects of Colorado.* University Press of Colorado, Boulder.

White, W. A., and T. A. Tremblay. 1995. Submergence of wetlands as a result of human-induced subsidence and faulting along the upper Texas Gulf-Coast. *Journal of Coastal Research* 11:788–807.

Wilde, G. R., and K. G. Ostrand. 1999. Changes in the fish assemblage of an intermittent prairie stream upstream from a Texas impoundment. *Texas Journal of Science* 51:203–210.

Winemiller, K., and F. Gelwick. 1999. Assessment of ecological integrity of streams in the Brazos-Navasota River watershed based on biotic indicators. Final Project Report to the Brazos River Authority by the Texas Agricultural Experiment Station, College Station, Texas.

Winemiller, K. O., S. Tarim, D. Shormann, and J. B. Cotner. 2000. Fish assemblage structure in relation to environmental variation among Brazos River oxbow lakes. *Transactions of the American Fisheries Society* 129:451–468.

World Almanac Books. 2003. *The world almanac and book of facts.* World Almanac Books, New York.

Young, W. C., and C. W. Bayer. 1979. The dragonfly nymphs (Odonata: Anisoptera) of the Guadalupe River basin, Texas. *Texas Journal of Science* 31:85–97.

Young, W. C., B. G. Whiteside, G. Longley, and N. E. Carter. 1973. The Guadalupe–San Antonio–Nueces River Basin Project: Review of existing biological data. Final Report to Texas Water Development Board, Austin, Texas.

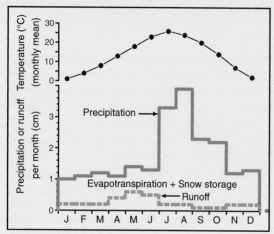

FIGURE 5.12 Mean monthly air temperature, precipitation, and runoff for the Rio Grande basin.

RIO GRANDE

Relief: 4272 m

Basin area: 870,000 km² (~450,000 km² contributing water)

Mean discharge: 37 m³/s (virgin >100 m³/s)

River order: 7

Mean annual precipitation: 21 cm

Mean air temperature: 13°C

Mean water temperature: 14°C

Physiographic provinces: Southern Rocky Mountains (SR), Colorado Plateau (CO), Basin and Range (BR), Great Plains (GP), Coastal Plain (CP), Sierra Madre Occidental (SC), Sierra Madre Oriental (SO)

Biomes: Desert, Temperate Grasslands, Temperate Mountain Forest, Mexican Montane Forest

Freshwater ecoregions: Upper Rio Grande, Lower Rio Grande, Pecos, Rio San Juan, Rio Salado, Rio Conchos

Terrestrial ecoregions: 8 ecoregions (see text)

Number of fish species: ≥86 freshwater, ≥80 estuarine

Number of endangered species: ≥16 fishes, several mollusks, 6 birds

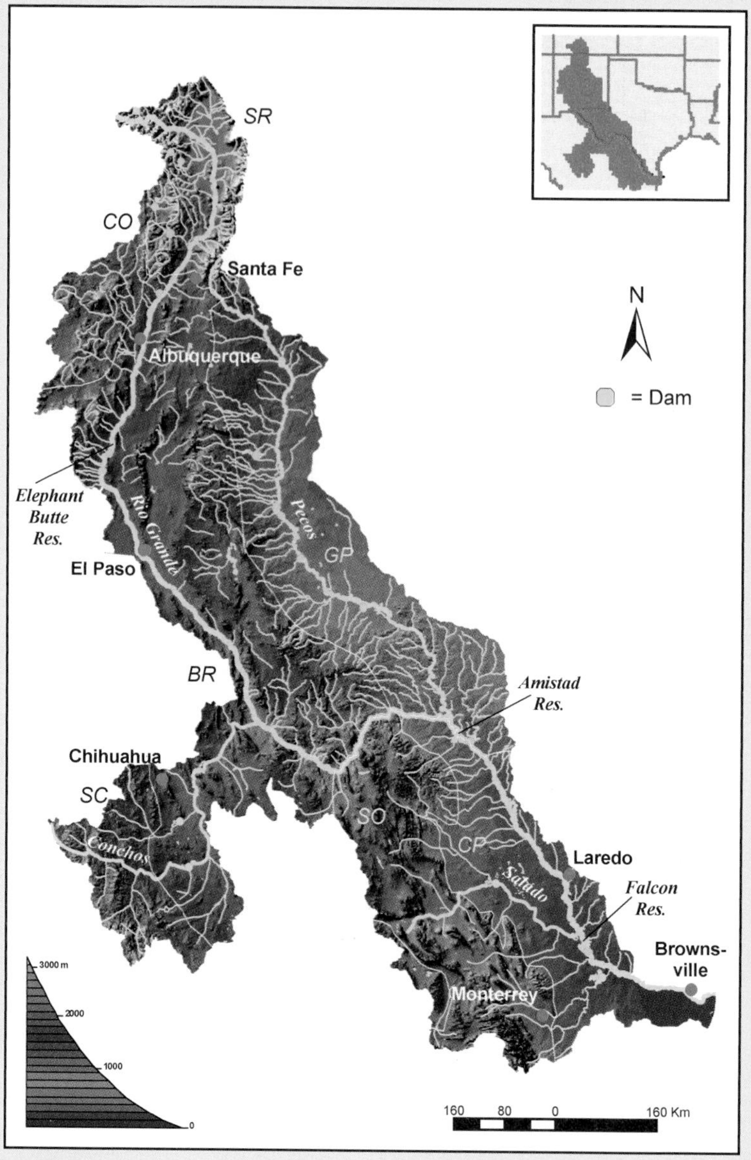

Major fishes: Rio Grande cutthroat trout, red shiner, Rio Grande silvery minnow, fathead minnow, white sucker, blue sucker, river carpsucker, western mosquitofish, largemouth bass, bluegill, longnose gar, threadfin shad, Rio Grande shiner, Tamaulipas shiner, longnose dace, Mexican tetra, sailfin molly, Amazon molly, longear sunfish, Rio Grande cichlid

Major other aquatic vertebrates: common yellow-throat, great blue heron, snowy egret, black-crowned night heron, white-faced ibis, belted kingfisher, green kingfisher, plainbelly water snake, American alligator, beaver, mink, nutria, bullfrog, Rio Grande leopard frog, snapping turtle, painted turtle, box turtle, western ribbon snake

Major benthic invertebrates: caddisflies (*Cheumatopsyche, Brachycentrus, Leucotrichia, Stactobiella, Hydroptila, Protoptila*), mayflies (*Baetis tricaudatus, Tricorythodes, Thraulodes, Traverella, Choroterpes mexicanus*), chironomid midges (*Cricotopus, Orthocladius*), other true flies (*Atherix, Simulium*), Asiatic clam

Nonnative species: ≥23 fishes (common carp, blue tilapia, inland silversides); saltcedar, Russian olive, Siberian elm, white mulberry, Guinea grass, buffelgrass, water hyacinth, hydrilla, water lettuce, alligatorweed, Asiatic clam, nutria

Major riparian plants: cottonwoods, willows, saltcedar, Russian olive, mesquite, hackberry, cedar elm, anacua, black willow, retama, Guinea grass, buffelgrass

Special features: Bosque del Apache National Wildlife Refuge, Chamizal National Memorial, Big Bend National Park, National Wild and Scenic River designation for segments in New Mexico and Texas

Fragmentation: highly fragmented, with five main-stem and numerous tributary dams

Water quality: variable, but generally decreasing downstream; salinity problems in southern New Mexico and western Texas; pH = 8.0, conductivity = 377 µS/cm, mean alkalinity 136 mg/L as CaCO₃ at Albuquerque

Land use: 14% forest, 5% cropland, 7% urban, 43% shrubland, 31% grassland

Population density: 16 people/km²

Major information sources: Hubbs et al. 1991, Revenga et al. 1998

FIGURE 5.11 Map of the Rio Grande basin. Physiographic provinces are separated by yellow lines.

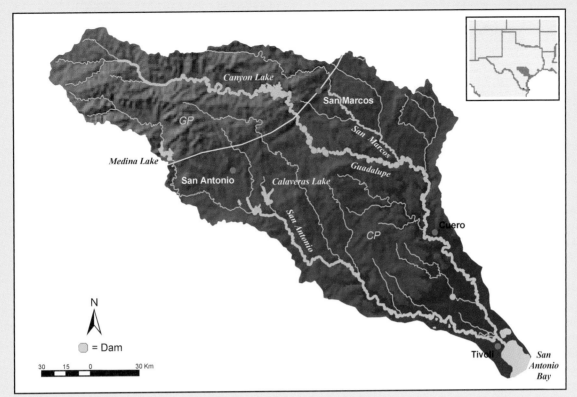

FIGURE 5.13 Map of the Guadalupe–San Antonio River basin. Physiographic provinces are separated by a yellow line.

SAN ANTONIO AND GUADALUPE RIVERS

Relief: 700 m
Basin area: 26,231 km²
Mean discharge: 79 m³/s
River order: 7
Mean precipitation: 81 cm
Mean air temperature: 21°C
Mean water temperature: 23°C
Physiographic provinces: Great Plains (GP), Coastal Plain (CP)
Biomes: Temperate Grasslands, Temperate Deciduous Forest
Freshwater ecoregion: East Texas Gulf
Terrestrial ecoregions: Edwards Plateau Savannas, Texas Blackland Prairies, East Central Texas Forests, Western Gulf Coastal Grasslands
Number of fish species: ≥88 (60 native)
Number of endangered species: ≥7 fishes, several amphibians, 3 spring/cave pool-associated aquatic insects, 1 plant

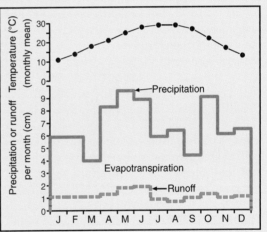

FIGURE 5.14 Mean monthly air temperature, precipitation, and runoff for the Guadalupe–San Antonio River basin.

Major fishes: largemouth bass, Guadalupe bass, bluegill, longear sunfish, redear sunfish, spotted sunfish, blacktail shiner, red shiner, central stonerollers, gray redhorse, channel catfish, western mosquitofish, Texas shiner, Texas logperch, river carpsucker, smallmouth buffalo, spotted bass, dusky darter, orangethroat darter, greenthroat darter, bluntnose darter, river darter
Major other aquatic vertebrates: beaver, northern parula warbler, prothonotary warbler, Louisiana waterthrush, great blue heron, snowy egret, white-faced ibis, belted kingfisher, green kingfisher, American alligator, Texas map turtle, Cagle's map turtle, smooth softshell turtle, spiny softshell turtle, plainbelly water snake, Western cottonmouth, nutria, mink
Major benthic invertebrates: caddisflies (*Chimarra, Cheumatopsyche, Atopsyche, Hydroptila*), mayflies (*Dactylobaetis, Tricorythodes, Choroterpes, Thraulodes*), beetles (*Microcylloepus, Hexacylloepus, Neoelmis*), chironomid midges (*Cricotopus, Rheotanytarsus*), true bugs (*Ambrysus*), amphipod crustaceans (*Hyallela*), hellgrammites (*Corydalus*)
Nonnative species: at least 27 fishes (tilapia, crappie, walleye, smallmouth bass, goldfish, grass carp), nutria, Asiatic clam, ramshorn snail, elephant ears, alligatorweed, water hyacinth, hydrilla, water lettuce
Major riparian plants: pecan, Texas sugarberry, bald cypress, cedar elm, Virginia creeper, Texas persimmon, red mulberry, greenbrier, box elder, black walnut, cottonwood, gum bumelia, black willow, American elm
Special features: large artesian aquifers (associated with Edwards aquifer), San Marcos spring
Fragmentation: major dam on Upper Guadalupe (Canyon); smaller dams on San Antonio
Water quality: generally good; pH = 7.0 to 9.0, alkalinity ~200 mg/L as $CaCO_3$, specific conductance 700 to 1100 µS/cm
Land use: 60% range, 15%, agriculture, 25% urban
Population density: 85 people/km²
Major information sources: Hubbs et al. 1991, Guadalupe-Blanco River Authority (http://www.gbra.org/), San Antonio River Authority (http://www.sara-tx.org)

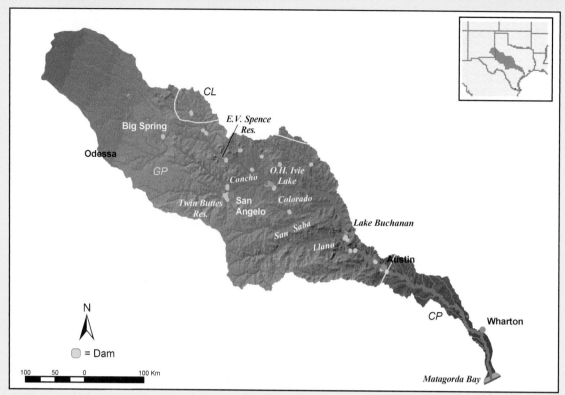

FIGURE 5.15 Map of the Colorado River basin. Physiographic provinces are separated by yellow lines.

COLORADO RIVER

Relief: 1195 m

Basin area: 103,341 km²

Mean discharge: 75 m³/s

River order: 7

Mean annual precipitation: 82 cm

Mean air temperature: 20°C

Mean water temperature: 22°C

Physiographic provinces: Central Lowland (CL), Great Plains (GP), Coastal Plain (CP)

Biomes: Temperate Grasslands, Temperate Deciduous Forest

Freshwater ecoregion: East Texas Gulf

Terrestrial ecoregions: seven ecoregions (see text)

Number of fish species: >98 (72 native)

Number of endangered species: >4 fishes, 2 salamanders, 1 snake, 5 mussels

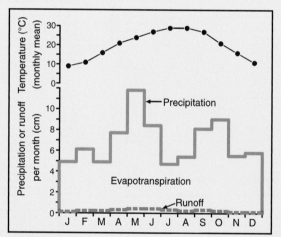

FIGURE 5.16 Mean monthly air temperature, precipitation, and runoff for the Colorado River basin.

Major fishes: spotted gar, longnose gar, red shiner, bullhead minnow, central stoneroller minnow, suckermouth minnow, blue sucker, gray redhorse, yellow bullhead, channel catfish, blue catfish, western mosquitofish, blackstripe topminnow, longear sunfish, spotted bass, Guadalupe bass, orangethroat darter, greenthroat darter, Texas logperch, bigscale logperch

Major other aquatic vertebrates: beaver, great blue heron, snowy egret, white-faced ibis, belted kingfisher, green kingfisher, Texas map turtle, Texas River cooter, red-eared slider, smooth softshell turtle, spiny softshell turtle, American alligator, plainbelly water snake, Concho water snake, Harter's water snake, western cottonmouth, gray treefrog, cricket frog, southern leopard frog

Major benthic invertebrates: caddisflies (*Cheumatopsyche, Helicopsyche, Hydroptila*), mayflies (*Tricorythodes, Caenis*), beetles (*Stenelmis*), damselflies (*Argia*), hellgrammites (*Corydalus*), chironomid midges (*Rheotanytarsus, Tanytarsus*), crayfishes (*Cambarellus, Procambarus*), snails (*Physa*), bivalves (*Sphaerium*, Texas fatmucket, Texas pimpleback, Texas fawnsfoot)

Nonnative species: common carp, grass carp, rudd, walleye, northern pike, nutria, saltcedar, water hyacinth, hydrilla, water lettuce, giant salvinia, Eurasian watermilfoil, alligatorweed

Major riparian plants: live oak, red oak, sugarberry, sycamore, elm, black willow, eastern cottonwood, pecan

Special features: three reaches are Texas Natural Rivers System candidates by U.S. Park Service; 18-m Gorman Falls upstream of Lake Buchanan; class IV rapids at Crabapple Creek; 15-m falls into collapsed grotto at Hamilton Pool Preserve

Fragmentation: most heavily dammed river in Texas (25 hydroelectric and water-supply reservoirs)

Water quality: pH = 7.9, alkalinity = 163 mg/L as CaCO₃, specific conductance = 528 μS/cm, occasional fish kills from storm runoff and low oxygen, salinity problems in upper basin, toxic golden alga in some reservoirs

Land use: 55% range, 30% agriculture, 15% urban

Population density: 35 people/km²

Major information sources: Hubbs et al. 1991, Bayer et al. 1992, Howells et al. 1996, Linam et al. 2002, Lower Colorado Water Authority (http://www.lcra.org), Upper Colorado Water Authority (http://www.ucratx.org)

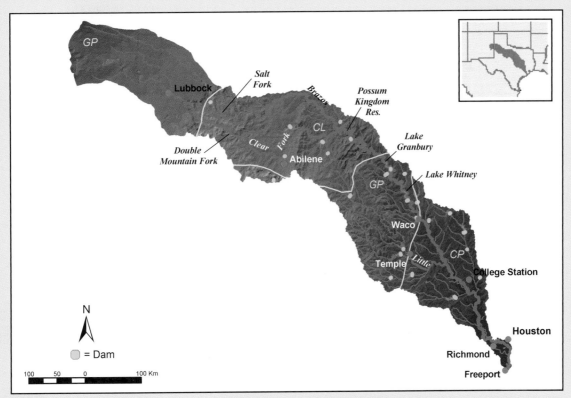

FIGURE 5.17 Map of the Brazos River basin. Physiographic provinces are separated by yellow lines.

BRAZOS RIVER

Relief: 1204 m
Basin area: 115,566 km²
Mean discharge: 249 m³/s
River order: 8
Mean annual precipitation: 81 cm
Mean air temperature: 19°C
Mean water temperature: 21°C
Physiographic provinces: Central Lowland (CL), Great Plains (GP), Coastal Plain (CP)
Biomes: Temperate Grasslands, Temperate Deciduous Forest
Freshwater ecoregion: East Texas Gulf
Terrestrial ecoregions: six ecoregions (see text)
Number of fish species: >93 (≥72 native)
Number of endangered species: >4 fishes, 4 mussels
Major fishes: longnose gar, spotted gar, western mosquitofish, blackstripe topminnow, red shiner, blacktail shiner, ribbon shiner, pugnose minnow, bullhead minnow, bluegill, green sunfish, longear sunfish, white crappie, largemouth bass, spotted bass, slough darter, orangethroat darter, tadpole madtom, channel catfish, blue catfish, flathead catfish, smallmouth buffalo, river carpsucker

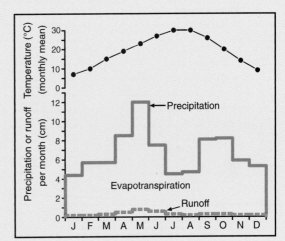

FIGURE 5.18 Mean monthly air temperature, precipitation, and runoff for the Brazos River basin.

Major other aquatic vertebrates: beaver, river otter, Harter's water snake, great blue heron, snowy egret, black-crowned night-heron, white-faced ibis, kingfishers, softshell turtles, American alligator, plainbelly water snake, western cottonmouth
Major benthic invertebrates: caddisflies (*Cheumatopsyche*), mayflies (*Choroterpes, Tricorythodes*), chironomid midges (*Tanytarsus, Rheotanytarsus*), beetles (*Stenelmis*), crayfishes (*Cambarellus puer, Procambarus acutus*), bivalves (bleufer, washboard, Texas lilliput, *Sphaerium transversum*), snails (*Physella virgata*), oligochaete worms (*Limnodrilus, Dero*)
Nonnative species: ≥21 fishes (grass carp, striped bass, rainbow trout, redbreast sunfish, rudd), saltcedar, Asiatic clam, nutria, water hyacinth, hydrilla, giant salvinia, Eurasian watermilfoil, alligatorweed
Major riparian plants: baccharis, cottonwood, willow, elm, hackberry, pecan, sumac, poison ivy, arrowhead, eelgrass, water primrose, Illinois pondweed, spikerush, water star grass
Special features: Llano Estacado; Blanco Canyon; dinosaur tracks in Paluxy River bank; class III rapids at Tonkawa falls west of Waco; floods reconnect river with floodplain oxbow lakes in lower river
Fragmentation: highly fragmented river, with 132 large dams in basin
Water quality: pH = 7.6, alkalinity = 133 mg/L as CaCO₃, specific conductance = 733 µS/cm, natural salinity high in upper basin, nutrient loading (dairy farms) in middle basin, toxic golden alga in some reservoirs
Land use: 3% forest, 24% cropland, 16% urban, 57% grassland
Population density: 20 people/km²
Major information sources: Hubbs et al. 1991, Bayer et al. 1992, Howells et al. 1996, Linam et al. 2002, Revenga et al. 1998, Brazos River Authority (http://www.brazos.org)

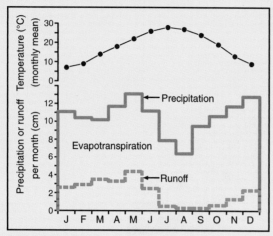

FIGURE 5.20 Mean monthly air temperature, precipitation, and runoff for the Sabine River basin.

SABINE RIVER

Relief: 198 m
Basin area: 25,268 km²
Mean discharge: 238 m³/s
River order: 8
Mean annual precipitation: 127 cm
Mean air temperature: 18°C
Mean water temperature: 21°C
Physiographic province: Coastal Plain (CP)
Biome: Temperate Deciduous Forest
Freshwater ecoregion: East Texas Gulf
Terrestrial ecoregions: Texas Blackland Prairies, East Central Texas Forests, Piney Woods Forests, Western Gulf Coastal Grasslands
Number of fish species: >104 (≥88 native)
Number of endangered species: >4 fishes, 2 crayfishes
Major fishes: bowfin, spotted gar, alligator gar, freshwater drum, river carpsucker, blacktail redhorse, flier, creek chubsucker, grass pickerel, golden topminnow, blackspotted topminnow, brook silverside, white bass, yellow bass, dollar sunfish, redear sunfish, largemouth bass, spotted bass, black crappie, scaly sand darter, goldstripe darter, cypress darter, bigscale logperch

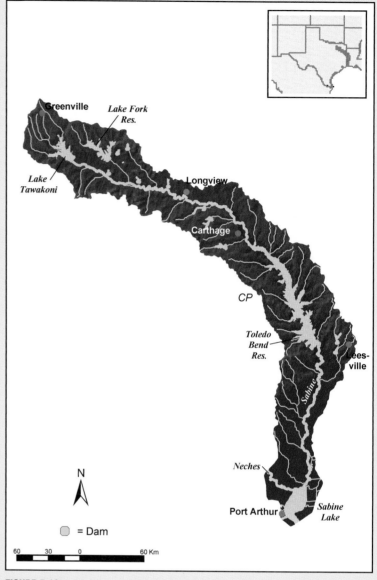

FIGURE 5.19 Map of the Sabine River basin.

Major other aquatic vertebrates: alligator snapping turtle, Sabine map turtle, smooth softshell turtle, spiny softshell turtle, American alligator, Gulf Coast waterdog, pickerel frog, green water snake, western cottonmouth, wood duck, canvasback, bald eagle, white-faced ibis, anhinga, great blue heron, snowy egret, purple gallinule, belted kingfisher, river otter, nutria
Major benthic invertebrates: caddisflies (*Hydropsyche, Cheumatopsyche*), hellgrammites (*Corydalus*), beetles (*Stenelmis*), chironomid midges (*Tanytarsus, Rheotanytarsus, Stenochironomous*), mayflies (*Caenis*), crayfishes (*Orconectes causeyi, Procambarus clarkii*), amphipod crustaceans (*Gammarus*), bivalves (Texas pigtoe, threeridge, little spectaclecase)
Nonnative species: ≥15 fishes (walleye, smallmouth bass, common carp, grass carp, rudd, striped bass, Mexican tetra), Asiatic clam, nutria, water hyacinth, hydrilla, giant salvinia, water spangles, Brazilian waterweed, Eurasian watermilfoil, parrotfeather, duck lettuce, alligatorweed, torpedo grass
Major riparian plants: bald cypress, sweetgum, water oak, black gum, water tupelo, magnolia, elm, cottonwood, hickory, walnut, maple, American beech, ash, palmetto, arrowhead, smartweed, buttonbush, spiderlily
Special features: 32 mussel species, scenic reach for 80 km below Toledo Bend Reservoir, Blue Elbow Swamp near Orange
Fragmentation: highly fragmented due to both low- and medium-head dams; Toledo Bend Reservoir is largest in the south
Water quality: pH = 6.7, alkalinity = 19 mg/L as CaCO₃, specific conductance = 145 µS/cm; problems include low oxygen, high fecal coliforms, nutrient enrichment, atrazine, mercury
Land use: 67% forest, 10% agriculture, 8% urban, 15% grassland
Population density: 18 people/km²
Major information sources: Hubbs et al. 1991, Bayer et al. 1992, Howells et al. 1996, Linam et al. 2002, Sabine Water Authority (http://www.sra.dst.tx.us)

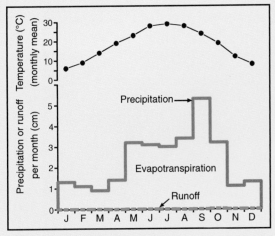

FIGURE 5.22 Mean monthly air temperature, precipitation, and runoff for the Pecos River basin.

PECOS RIVER

Relief: 4012 m
Basin area: 113,960 km²
Mean discharge: 2 m³/s (virgin ≥9 m³/s)
River order: 5
Mean precipitation: 28 cm
Mean air temperature: 18°C
Mean water temperature: 21°C
Physiographic provinces: Southern Rocky Mountains (SR), Basin and Range (BR), Great Plains (GP)
Biomes: Temperate Mountain Forest, Deserts, Temperate Grasslands
Freshwater ecoregion: Pecos
Terrestrial ecoregions: Colorado Rockies Forests, Western Short Grasslands, Chihuahuan Desert, Colorado Plateau Shrublands
Number of fish species: ≥70
Number of endangered species: ≥12 fishes, 3 snails, and 1 amphipod proposed for listing
Major fishes: red shiner, inland silverside, Pecos pupfish, western mosquitofish, rainwater killifish, roundnose minnow, proserpine shiner, channel catfish, Rio Grande cichlid, Mexican tetra, green sunfish, largemouth bass
Major other aquatic vertebrates: great blue heron, snowy egret, black-crowned night-heron, white-faced ibis, belted kingfisher, green kingfisher, plainbelly water snake, beaver, muskrat
Major benthic invertebrates: caddisflies (*Ithytrichia, Cheumatopsyche, Hydroptila*), mayflies (*Choroterpes, Thraulodes, Tricorythodes, Traverella*), beetles (*Berosus*), true bugs (*Cryphocricos*), chironomid midges (*Tanytarsus, Dicrotendipes, Pseudochironomus, Microtendipes, Cricotopus*), amphipod crustaceans (*Hyallela azteca*)
Nonnative species: ≥19 fishes (grass carp, goldfish, common carp, rudd, rainbow trout, white crappie, walleye, smallmouth bass, redear sunfish), saltcedar, Russian olive, water hyacinth, hydrilla, Eurasian watermilfoil, alligatorweed
Major riparian plants: saltcedar, mesquite, cottonwood, four-winged saltbush, Russian olive, willow
Special features: Bitter Lake National Wildlife Refuge with high biodiversity of dragonflies and damselflies; river passes through deep gorges within limestone terrain
Fragmentation: several impoundments, including Red Bluff Reservoir upstream of Pecos, Texas; major water diversions near Pecos and Grandfalls
Water quality: pH = 7.9, alkalinity = 127 mg/L as CaCO₃, specific conductance = 3326 μS/cm, salinity problems in lower basin, toxic golden alga in reservoirs
Land use: 40% shrubland, 45% grassland, 10% agriculture, 3% forest, 2% urban
Population density: 3 people/km²
Major information sources: Davis 1980a, Hubbs et al. 1991, Hayter 2002

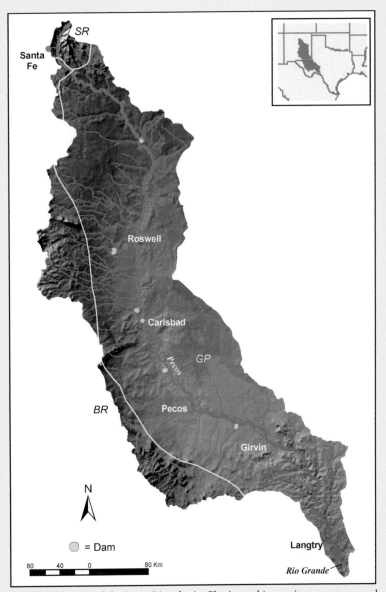

FIGURE 5.21 Map of the Pecos River basin. Physiographic provinces are separated by yellow lines.

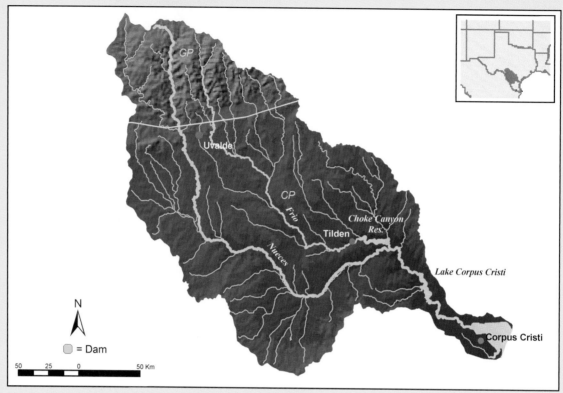

FIGURE 5.23 Map of the Nueces River basin. Physiographic provinces are separated by a yellow line.

NUECES RIVER

Relief: 730 m
Basin area: 43,512 km²
Mean discharge: 20 m³/s
River order: 7
Mean precipitation: 61 cm
Mean air temperature: 21°C
Mean water temperature: 23°C
Physiographic provinces: Great Plains (GP), Coastal Plain (CP)
Biome: Temperate Grasslands
Freshwater ecoregion: West Texas Gulf
Terrestrial ecoregions: Edwards Plateau Savannas, Tamaulipan Mezquital, East Central Texas Forests, Western Gulf Coastal Grasslands
Number of fish species: ≥66
Number of endangered species: ≥3 fishes
Major fishes: longnose gar, spotted gar, Mexican tetra, Nueces roundnose minnow, plateau shiner, Texas shiner, channel catfish, sailfin molly, western mosquitofish, largemouth bass, longear sunfish, bluegill, greenthroat darter, Rio Grande cichlid

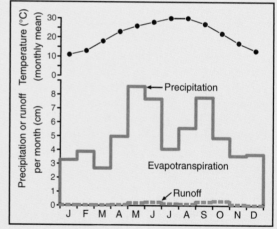

FIGURE 5.24 Mean monthly air temperature, precipitation, and runoff for the Nueces River basin.

Major other aquatic vertebrates: beaver, great blue heron, snowy egret, white-faced ibis, belted kingfisher, green kingfisher
Major benthic invertebrates: caddisflies (*Chimarra, Cheumatopsyche, Hydroptila*), mayflies (*Fallceon, Dactylobaetis, Tricorythodes, Choroterpes, Thraulodes*), beetles (*Microcylloepus, Hexacylloepus, Neoelmis*), chironomid midges (*Cricotopus, Rheotanytarsus*), true flies (*Hemerodromia, Simulium*), amphipod crustacean (*Hyallela*), hellgrammites (*Corydalus*)
Nonnative species: ≥20 fishes (goldfish, grass carp, common carp, golden shiner, rudd, rainbow trout, inland silverside, striped bass, smallmouth bass, Guadalupe bass, walleye, Rio Grande cichlid)
Major riparian plants: pecan, Texas sugarberry, bald cypress, cottonwood, black willow, cedar elm, Texas persimmon, red mulberry, greenbrier, box elder, black walnut, American elm
Special features: arises from springs in Edwards Plateau, tricanyon area of Nueces, Frio, and Sabinal highly scenic
Fragmentation: two major reservoirs in lower basin
Water quality: pH = 7.8, alkalinity = 142 mg/L as CaCO₃, specific conductance = 1211 μS/cm
Land use: 55% shrublands, 25% rangelands, 15% agriculture, 5% urban
Population density: 16 people/km²
Major information sources: Hubbs et al. 1991, Nueces River Authority (http://www.nueces-ra.org)

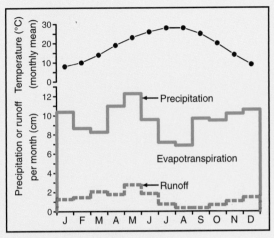

FIGURE 5.26 Mean monthly air temperature, precipitation, and runoff for the Trinity River basin.

TRINITY RIVER

Relief: 362 m
Basin area: 46,540 km^2
Mean discharge: 222 m^3/s
River order: 7
Mean annual precipitation: 115 cm
Mean air temperature: 19°C
Mean water temperature: 21°C
Physiographic provinces: Central Lowland (CL), Great Plains (GP), Coastal Plain (CP)
Biomes: Temperate Grasslands, Temperate Deciduous Forest
Freshwater ecoregion: East Texas Gulf
Terrestrial ecoregions: Central Forest/Grassland Transition Zone, Texas Blackland Prairies, East Central Texas Forests, Piney Woods Forests, Western Gulf Coastal Grasslands
Number of fish species: ≥99
Number of endangered species: 3 fishes, 1 crayfish, 3 mussels
Major fishes: spotted gar, threadfin shad, channel catfish, freckled madtom, pirate perch, river carpsucker, smallmouth buffalo, red shiner, ribbon shiner, weed shiner, bullhead minnow, pugnose minnow, golden topminnow, blackstripe topminnow, western mosquitofish, dollar sunfish, largemouth bass, spotted bass, white crappie, bluntnose darter, slough darter, dusky darter
Major other aquatic vertebrates: beaver, nutria, American alligator, cottonmouth, yellow-bellied water snake, diamondback water snake, red-eared slider, Texas river cooter, common snapping turtle, cricket frog, green treefrog, gray treefrog, southern leopard frog, bald eagle, osprey, anhinga, wood duck, pintail, canvasback, great blue heron, greenback heron
Major benthic invertebrates: caddisflies (*Cheumatopsyche*, *Chimarra*), mayflies (*Caenis*, *Isonychia*, *Heptagenia*), beetles (*Stenelmis*), chironomid midges (*Stictochironomus*), true bugs (*Rhagovelia*), bivalves (Louisiana fatmucket, yellow sandshell, bleufer, western pimpleback, tapered pondhorn)
Nonnative species: 20 fishes (grass carp, common carp, rudd, northern pike, rainbow trout, inland silverside, striped bass, redbreast sunfish, smallmouth bass, walleye, blue tilapia), nutria, Asian clam, hydrilla, water hyacinth, giant salvinia
Major riparian plants: red maple, river birch, water hickory, pecan, black hickory, hackberry, honey locust, water elm, water oak, sumac, black willow, American elm, water willow, water-pennywort, smartweed, bulrush, cattail, sedge, spikerush
Special features: Trinity River National Wildlife Refuge (permanently flooded swamps); Richland Creek and Keechi Creek Wildlife Management Areas (numerous periodically flooded oxbows)
Fragmentation: highly fragmented, with 21 major reservoirs in basin
Water quality: pH = 7.5, alkalinity = 109 mg/L as CaCO$_3$, specific conductance = 594 µS/cm, pollution near cities and agricultural areas due to low oxygen, high fecal coliforms, high total dissolved solids, algal blooms, atrazine, pesticides/ PCBs, metals
Land use: 35% forest, 15% cropland, 30% urban, 20% grassland
Population density: 98 people/km^2
Major information sources: Hubbs et al. 1991, Bayer et al. 1992, Howells et al. 1996, Linam et al. 2002, Trinity River Authority (http://www.trinityra.org/trahp.htm)

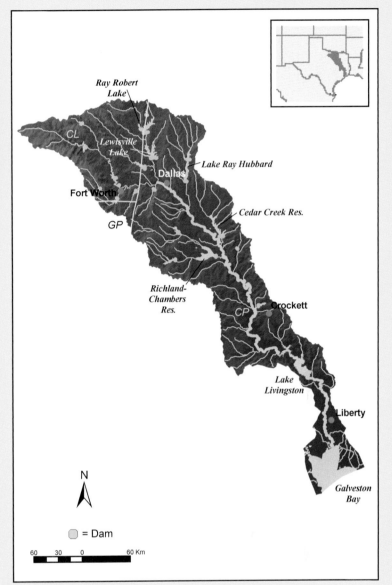

FIGURE 5.25 Map of the Trinity River basin. Physiographic provinces are separated by a yellow line.

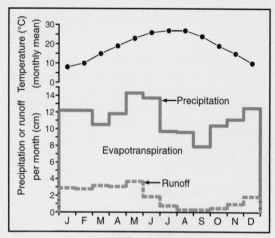

FIGURE 5.28 Mean monthly air temperature, precipitation, and runoff for the Neches River basin.

NECHES RIVER

Relief: 153 m
Basin area: 25,929 km²
Mean discharge: 179 m³/s
River order: 7
Mean precipitation: 136 cm
Mean air temperature: 19°C
Mean water temperature: 21°C
Physiographic province: Coastal Plain (CP)
Biome: Temperate Deciduous Forest
Freshwater ecoregion: East Texas Gulf
Terrestrial ecoregions: East Central Texas Forests, Piney Woods Forests, Western Gulf Coastal Grasslands
Number of fish species: ≥96
Number of endangered species: ≥4 fishes, 1 crayfish
Major fishes: alligator gar, channel catfish, freshwater drum, blacktail redhorse, longear sunfish, spotted sunfish, pirate perch, banded pygmy sunfish, flier, spotted bass, grass pickerel, blackspot shiner, ribbon shiner, Sabine shiner, weed shiner, bullhead

FIGURE 5.27 Map of the Neches River basin.

minnow, blackspotted topminnow, western mosquitofish, brook silverside, cypress darter, scaly sand darter, bluntnose darter
Major other aquatic vertebrates: beaver, river otter, mink, nutria, muskrat, cottonmouth, yellow-belly water snake, diamondback water snake, Graham's crayfish snake, Missouri river cooter, Sabine map turtle, alligator snapping turtle, spiny softshell turtle, green frog, American alligator, bald eagle, pied-billed grebe, anhinga, wood duck, belted kingfisher, great egret
Major benthic invertebrates: mayflies (*Caenis*), caddisflies (*Hydropsyche*, *Cheumatopsyche*, *Hydroptila*, *Chimarra*), damselflies (*Calopteryx*, *Lestes*, *Enallagma*), dragonflies (*Gomphus*, *Libellula*), beetles (*Stenelmis*), bivalves (threeridge, giant floater, paper pondshell, Louisiana fatmucket, yellow sandshell, washboard)
Nonnative species: 15 fishes (threadfin shad, goldfish, grass carp, common carp, rudd, Mexican tetra, rainbow trout, sailfin molly, inland silverside, white bass, striped bass, redbreast sunfish, smallmouth bass, white crappie, walleye), white heelsplitter, flat floater
Major riparian plants: palmetto, bald cypress, black willow, river birch, sycamore, sweetgum, black gum, willow oak, water oak, swamp tupelo, water hickory, southern magnolia, water willow, cedar elm, cottonwood, pecan, black oak, arrowhead, water smartweed, buttonbush
Special features: Big Thicket Reserve (Man and the Biosphere Program)—remnant of large area developed on ancient sand dunes and beaches of a fossil sea; number-one scenic river in east Texas (National Park Service 1995)
Fragmentation: 3 major reservoirs in basin (Palestine, Sam Rayburn, and Steinhagen)
Water quality: pH = 6.7, alkalinity = 25 mg/L as $CaCO_3$, specific conductance = 213 µS/cm; pollution problems in lower reaches due to salt intrusion and siltation during construction of roads and timber cutting
Land use: 65% forest, 15% agriculture, 5% urban, 15% grassland
Population density: 11 people/km²
Major information sources: Hubbs et al. 1991, Bayer et al. 1992, Howells et al. 1996, Linam et al. 2002, Angelina and Neches River Authority (http://www.anra.org)

228

6

LOWER MISSISSIPPI RIVER AND ITS TRIBUTARIES

ARTHUR V. BROWN   KRISTINE B. BROWN
DONALD C. JACKSON   W. KEVIN PIERSON

INTRODUCTION

LOWER MISSISSIPPI RIVER

WHITE RIVER

BUFFALO NATIONAL RIVER

BIG BLACK RIVER

YAZOO RIVER

ADDITIONAL RIVERS

ACKNOWLEDGMENTS

LITERATURE CITED

INTRODUCTION

The Mississippi River basin is the largest in North America (~3.27 million km²; Baker et al. 1991, Karr et al. 2000) and the 3rd largest in the world, exceeded in area only by the Amazon basin in South America and the Congo in Africa (Leopold 1994). It extends from 37°N to 29°N latitude, covers nearly 14% of the North American continent, and drains 41% to 42% of the area of the conterminous United States. The Mississippi River, at 6693 km, is second only to the Nile in length. Its estimated virgin discharge of 18,400 m³/s ranks 9th in the world (Leopold 1994, Dynesius and Nilsson 1994). The Lower Mississippi River (LMR) extends from the confluence with the Ohio River at Cairo, Illinois, at Rkm 1536, where the bed is about 100 m asl, to the Gulf of Mexico (Fig. 6.2). Unlike all other rivers considered in this book, the LMR main stem receives most of its flow from major upstream tributaries that are considered in other chapters: the Missouri River (Chapter 10), the Upper Mississippi River (Chapter 8), and the Ohio River (Chapter 9). The basin area of the LMR and its tributaries is about 880,000 km². The focal

area for this chapter (460,000 km²) excludes the upper and middle reaches of the Arkansas and Red rivers, which enter this area from the west (see Chapter 7), and roughly corresponds to three freshwater ecoregions: Mississippi Embayment (excluding the Pearl River, which enters directly into the Gulf of Mexico east of the Mississippi), Ozark Highlands, and Ouachita Highlands (Abell et al. 2000). Thus, this area includes drainages that cover most of Arkansas, Mississippi, and Louisiana and portions of southern Missouri, western Tennessee, and western Kentucky.

The LMR has a long history of human habitation for this continent: as long as 16,000 years ago as indicated by artifacts in the basin. The last major cultural development by Native Americans initiated in the eastern United States was the Mississippian. The Mississippian culture developed from the Caddoan along the LMR from the eighth century through the middle of the thirteenth century. The new agricultural methods, religious practices, and social concepts resulted in a sedentary society that built earthen mounds with wooden buildings on them, often surrounded by walled fortifications with timber palisades

◄ FIGURE 6.1 Current River above Van Buren, Missouri (PHOTO BY TIM PALMER).

231

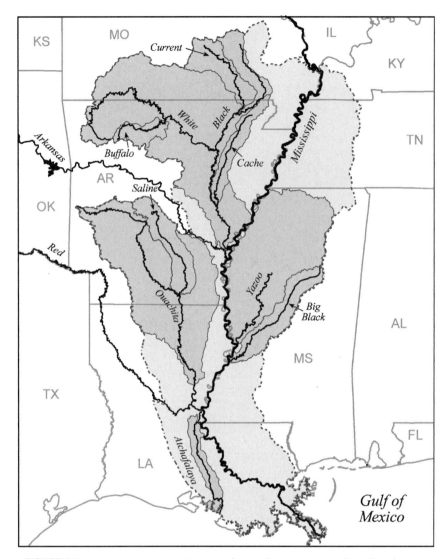

FIGURE 6.2   Lower Mississippi River and its tributaries covered in this chapter.

and bastions. At the time Europeans arrived, clans of Choctaw, Chippewa, Koroa, Taensa, Chickasaw, Tunica, Yazoo, Pascagoula, Natchez, Biloxi, Alibamu, and many others lived in the LMR valley and successfully exploited the abundant natural resources of the rivers and their wetlands. Although first explored by the Spanish beginning with Hernando de Soto, the LMR was first settled by the French. La Salle claimed the river basin for France in 1682. France ceded it to Spain in 1762, but in 1800, Napoleon Bonaparte managed to take it back again. France sold the "Louisiana Purchase" to the United States, doubling its size, for about $20 million in 1803. This land area includes the entire Mississippi River basin west of the main channel, which is a large majority of the basin. Early transportation on the rivers of the region was primarily downstream by canoes, and later by flat-

boats. Keelboats were narrower and more stream-lined, and were paddled, pushed, sailed, and pulled with ropes upstream with considerable expenditure of effort. The first steamboat paddle wheeler (the *New Orleans*, built in Pittsburgh, Pennsylvania) plied the LMR beginning in 1811 and survived the New Madrid earthquakes that occurred the same year. The earthquakes realigned much of the river channel, causing it to flow northward for several days, and also created Reelfoot Lake in northwestern Tennessee, the site of a biological field station from 1931 to 1969. An era when huge cotton plantations were owned by rich caucasian families using African-American slave labor coincided with the history of steam-powered paddle wheelers in the Mississippi River and its tributaries. This era (antebellum South) was ended after about 50 years by the Civil War.

The Acadian people, who came to be known as Cajuns, settled the swamps of southern Louisiana, especially in the Atchafalaya basin, in 1755 during the French and Indian War, when they were forced to leave the Acadia region of Canada by the British. Cajuns have been, and many remain, industrious modern hunter-gatherers of wildlife (shrimp, crayfish, oysters, fishes, waterfowl, nutria) from the productive rivers, estuaries, and bayous, but also grow a variety of crops, including corn, yams, sugar cane, and melons. More recently many Cajuns have become rather wealthy working in the oil fields and petrochemical industries that have developed in the LMR delta. Thus, Cajuns have had a major impact on the LMR.

In this chapter we will focus on the LMR main stem and four of its tributaries: the White, Buffalo, Yazoo, and Big Black rivers. Five additional rivers that are entirely within this region are also addressed but will receive less attention: the Atchafalaya, Cache, Ouachita, Saline, and Current. Of these nine tributaries, seven are west of the Mississippi and two (Yazoo and Big Black) enter from the east.

Physiography and Climate

Besides the regions containing its upstream tributaries, the LMR drains from three physiographic provinces that are very distinct: Coastal Plain (CP), Ouachita province (OP), and Ozark Plateaus (OZ) (Fig. 6.8). The Coastal Plain is a large, flat plain of deep, fine alluvial deposits. It has a large variety of habitats despite lack of much topographic relief or coarse substrates. The Ozark Plateaus are highlands consisting of an area of uplifted plateaus composed of nearly horizontal strata of limestones, sandstones, and shales that have been radially eroded to form steep valleys. The Ozark Plateaus are characterized by karst topography, with numerous caves and springs. The Ouachita province is an area of east–west trending ridges and valleys extending through west-central Arkansas south of the Arkansas River into east-central Oklahoma. The Ouachita Mountains are similar topographically to the Appalachian Mountains but not as tall (152 m to 853 m asl). The region is complex stratigraphically and lithologically, with many folds and faults caused by uplifts through formerly horizontal deposits of sandstones, shales, and novaculites (Croneis 1930). Quartz crystals and flat stones suitable for building are abundant, and some coal is found in the Ouachitas, so there are many small mines in the area to extract these resources.

The climate of this entire region ranges from temperate midcontinental in the northern and western portions to subtropical and very humid in the south. The mean annual temperature is about 16°C, with an annual range from 0°C to 35°C. Rainfall ranges from just over 100 cm/yr in the west to over 160 cm/yr in the warm, moist delta region of southern Louisiana. Heavy rains can occur at any time of year, but winter to early summer is the wettest, and late summer to mid-fall is the driest. As a result of somewhat lower precipitation and higher evapotranspiration during the summers, many headwater streams are ephemeral or intermittent. Hurricanes occasionally lash the coast of Louisiana and Mississippi, and tornados are common throughout the region. These storms seriously impact fisheries in the productive Mississippi delta region and Atchafalaya basin by making the backwaters and swamps anoxic for many weeks afterward (Sabo et al. 1999).

Basin Landscape and Land Use

The LMR basin contains parts of six terrestrial ecoregions as depicted by Ricketts et al. (1999): Central U.S. Hardwood Forests, Ozark Mountain Forests, Piney Woods Forests, Southeastern Mixed Forests, Western Gulf Coastal Grasslands, and Mississippi Lowland Forests. Unlike most of the Upper Mississippi River basin, most of which occurs in grasslands, the LMR basin occupies an area where the climax vegetation of most of the region is forest, although 77% of the forests are presently cleared (Mitsch and Gosselink 1993) for agricultural production and other uses.

Vegetationally, the natural landscape of the LMR region was forested but interspersed with prairies. The vegetation was, and continues to be, a complex, patchy mosaic, with the floral composition of each patch resulting principally from soil type, climate, frequency and duration of inundation, fire history, and the long history of anthropogenic disturbance (Williams 1992, Pyne 1997). More recently, protection from flooding and fire disturbances has changed the vegetational landscape of the LMR region. Presently 50% to 90% of the obvious vegetation consists of diverse crops of food and fiber dominated by cotton, soybeans, rice, maize, and loblolly pine. The deciduous forests in the basin are diverse and vary among ecoregions, but generally they are composed of oaks and hickories, with gums, hackberries, sycamores, birches, and willows becoming more abundant nearer to water. Some of the same genera, but different species, occur in swamps (e.g.,

Nuttall's oak, live oak, and water hickory), with bald cypress and water tupelo replacing them in the wettest areas. Longleaf pines were dominant to subdominant in some areas with sandy, acid soils, but they have been extensively replaced by faster-growing loblolly pines by silviculturalists.

Grasslands in the LMR region were very diverse, but tall grass prairies dominated by big and little bluestem, Indian grass, and switch grass were most common. The prairies often blend into savannas formed by scattered longleaf pines. Further south these grasses are replaced by panic grasses, sea oats, a large variety of sedges including sawgrass, and other species that can withstand wetter, tighter soils and more saline conditions. Palmettos are a common understory species in many areas of the southern Coastal Plain.

Wetlands of a large variety of types occur in this area, primarily along the larger river courses, and are integral components of the riverine ecosystems. The floodplain of the LMR is one of the largest in the world and contains $36,000 \, km^2$ of wetland habitats. The Lower Mississippi main stem has 26 tributary streams and 242 lakes >8 ha in surface area located in its floodplain. Many of these are oxbow lakes and back swamps that form behind natural levees in the floodplains of the larger rivers (Mississippi, Arkansas, White, Yazoo, Big Black, Cache, St. Francis, Tensas, Atchafalaya). Large areas of wetland associated with these rivers are known as bottomland hardwood forests, which are a type of swamp. Nuttall's oak, pin oak, and hackberry trees dominate higher ground in the bottomland forests, with willows, cypress, and water tupelo in the areas that are flooded for longer periods. The Atchafalaya River basin alone has the 3rd largest continuous wetland area in the United States and contains about 30% of all bottomland hardwood forests in the Mississippi Alluvial Plain. The Black Swamp in the Cache River basin is another vast (60,000 ha) bottomland hardwood wetland in the LMR (Kress et al. 1996).

More of the agricultural land is maintained as cleared fields for row-crop production in the rich lowlands of the Coastal Plain region than in the Ozark Plateaus and Ouachita province. In the Upper Gulf Coastal Plain, comprising the northeastern third of the Mississippi Embayment, agriculture makes up 54% of land use (Gonthier 2000). In the remainder, which is the Mississippi Alluvial Plain, 80% is used for row-crop agriculture. Silviculture of loblolly pine in the Ouachita province and various hardwoods (e.g., oaks) in the Ozark Plateaus remains a dominant land use where the hill slopes and soil types are

not suitable for other forms of agricultural production. Thus, about half of these regions remain forested. Native Americans of the Mississippian culture deforested much of the Coastal Plain in the thirteenth century for fuel and building material. Other Native Americans were largely responsible for maintaining extensive prairies in the Ozark and Ouachita highlands by use of fire before Europeans colonized the area and began fire-control practices. Some of the fires were accidental, but others were set to provide fresh grazing to attract American bison (and later to provide grasses for their horses), to provide areas for growing vegetables (primarily beans, corn, and squash), to harass enemies, to control ticks and insects, and to drive animals during hunting expeditions. European diseases introduced by the de Soto expedition and others devastated the Native Americans, which allowed forests to regrow before extensive exploration and colonization by Europeans.

The LMR valley is highly productive land and has been cleared for intensive farming to produce cotton, soybeans, rice, and corn. Row-crop production of these plants dominates the LMR Coastal Plain. The fertilizers, pesticides, and herbicides used to grow these crops, and sediments from the plowed fields, are washed into the river. A typical form of privately owned agricultural operation in the Interior Highlands (Ozark Plateaus and Ouachita province) consists of a confined animal production facility for poultry or swine combined with pastures for beef cattle production. The manure from the confined animals and litter (rice hulls, shavings) used in the buildings are spread on the pastures to fertilize grasses for cattle. Part of the grassland is protected from grazing, cut, and baled as hay to be fed in winter. Other types of farms are present (dairies, orchards, vineyards, etc.), but they collectively comprise less than 5% of land use. The pastures that receive the manure from the extensive confined animal production facilities are often adjacent to streams, and runoff from these fields contributes substantial quantities of nonpoint-source nutrients (organic matter as well as nitrates and phosphates). The cattle that graze the pastures have access to the streams and are a source of physical disturbance as well as enriching nutrients. Silviculture, especially of loblolly pine, also contributes nonpoint-source pollution.

The Rivers

The Mississippi River main stem is obviously the primary river of this region. After the confluence with the Missouri River the Mississippi is a 10th order river and retains this size designation throughout the Lower Mississippi reach. The three largest tributaries (Arkansas, White, and Red rivers) are all 7th order rivers that enter from the west. The Yazoo River entering from Mississippi is a 6th order stream, but with less flow than the others. As the rivers in this region approach the Mississippi River, they all take on the character of large, lowland rivers; they are influenced by crossing the relatively flat plain of fine alluvial deposits called the Mississippi Delta. However, many of them originate in highlands and are much different in their headwaters and midreaches. Those coming from the east originate just beneath the western slopes of the Appalachian Mountains but are all wholly contained within the Coastal Plain physiographic province or Mississippi Embayment freshwater ecoregion. Those from the west have such diverse origins as the Ozark Plateaus (White, St. Francis), southern Rocky Mountains (Arkansas), Ouachita province (Ouachita), and Great Plains (Red). Streams that arise within the lowlands of the Coastal Plain (e.g., Homochitto, Cache) do not have the dramatic changes in character along their lengths that are characteristic of those entering the Mississippi from the west.

Although several classifications of aquatic habitats in the Mississippi River valley exist (Baker et al. 1991), the habitats seem to vary principally in flow rate, depth, frequency of connection to the river, permanence, and riparian influence. The extent and relative abundance of the different habitat types change with river stage. Construction of levees, floodways, cutoffs, dikes, revetments, locks and dams, and tributary basin modifications by the U.S. Army Corps of Engineers (USACE) to control floods and allow shipping and barge traffic has altered the relative percentages of the different types of habitat and formed novel habitat types (Baker et al. 1991). Maintenance of a single, large, relatively straight, deep channel in the Mississippi, Arkansas, White, and Yazoo rivers has likely had significant impacts on the fishes, macroinvertebrates, and other biota, but these impacts are not well documented, principally because of the difficulties involved in studying the biota of large rivers.

The Mississippi River presently has a meandering channel form (although highly artificially regulated), but it and other large North American rivers have alternated between meandering and braided channel forms for about two million years as climatic conditions changed amounts of runoff and sedimentation (Saucier 1968). During periods of high sedimentation

(as from glacial outwash) and diminished runoff, braided channel patterns developed because flows were insufficient to move the bedload and instead divided into multiple, anastomosing, shallow channels across the top of the deposits. Braided channels have a different array of habitat types from meandering channels. The main-stem Lower Mississippi and its larger tributaries (Arkansas, White, Yazoo, etc.) did not have the same type of channel form simultaneously, thus providing a larger variety of habitat types for riverine fauna in this region. The lower Arkansas River presently has a braided channel form. This is especially obvious below the artificial cutoff to the White River that facilitates barge traffic. Above this point, dredging and other control measures by the USACE maintain one large channel resembling a meandering channel form to the inland terminus of the McClellan-Kerr Navigation System near Tulsa, Oklahoma, upstream of which the river channel remains braided. Thus, in the LMR the diverse array of habitats seen along river continua from headwaters to downstream reaches is enhanced by variation in channel form of the large rivers through space and time, as well as by the variety of floodplain habitat types.

Wetlands are an important riverine habitat in the region, although the wetlands in the LMR have been largely isolated from the large rivers by construction of levees and substantially reduced in area by clear cutting, draining, impounding, and channelization. Wetlands associated with the Ozark and Ouachita upland streams are much less extensive, but are important ecological components of the watershed ecosystems. Many of the Ozark streams larger than 3rd order have well-developed floodplains complete with floodplain ponds and small oxbow lakes used as spawning and nursery areas by several species of fishes (e.g., white crappie, common carp, black and yellow bullhead catfish).

The smaller rivers of the LMR region range from clear, cool, mountain streams with bedrock outcroppings, coarse gravel bedloads, and moderate slopes that are influenced by karst limestone topography, like the Buffalo, Current, and Kings rivers of the Ozark Highlands, to those that originate in the comparatively low lands of the Coastal Plain, with fine, alluvial substrate, higher turbidity, and very modest slope (Big Black, Yazoo, Homochitto), with many others of distinctive character somewhere in between (e.g., Ouachita and Saline rivers in the Ouachita province). Streams of the Ouachita province resemble those of the Ozark Plateaus in some respects (cool, clear, gravel bed, intermittent headwaters), but the Ouachita province lacks limestone, so Ouachita streams have less alkalinity (~20 μS/cm specific conductance), and this renders them less productive. The formerly horizontal stratigraphy of the Ouachitas has been uplifted and folded, so Ouachita streams have a less predictable shape than those in the Ozark Plateaus. Also, pine trees occur in the riparian zones of Ouachita streams but not of Ozark streams.

The Atchafalaya River distributary and its associated wetlands are unique. It is a distributary of the LMR that is about 224 km long and is made up largely of several old, interconnected, Mississippi River channels (oxbow lakes). In its natural state, the Atchafalaya was simply the primary watercourse in a complex system of interconnecting bayous in a large swamp. The USACE has extensively modified the Atchafalaya basin to enhance its use as an overflow outlet for the Mississippi River main channel, including water-control structures, levees, dikes, and revetments. The Atchafalaya River is further complicated because it has captured the Red River, which was a tributary to the Mississippi. The basin is very flat, so flow direction in side channels (locally called bayous) of the Atchafalaya basin often depend on comparative inputs from the Red River and the Mississippi River as controlled by USACE flood-control structures.

Reservoirs are a major feature of rivers in the Ozark and Ouachita highlands. They are as aneurisms in the streams and rivers. Many reasons have been given to justify their construction, including flood control, hydroelectric power generation, municipal water supply, recreation, and economic development. Reservoirs are also justified by the need for water for steam and cooling for electricity generating plants powered by fossil fuels and nuclear power. Many of the older reservoirs were built by the Works Progress Administration (WPA) and/or Civilian Conservation Corps (CCC) in the 1930s. Most of the larger reservoirs have since been constructed by the USACE, but the Arkansas Game and Fish Commission has continued to build many smaller ones using federal subsidies provided from taxes on sales of fishing equipment (Pittman-Robertson Act funds). Most of the larger reservoirs have deep-release structures, and the waters from them have altered temperature cycles both annually (dampened) and daily (increased in summer). To mitigate loss of warmwater fish habitat, various species of trout are stocked in these tail waters. Some provide only put-and-take fisheries, but others have reproducing populations of various species of trout. Reservoirs are less numerous in the Coastal Plain,

primarily because sites for them are less numerous in the flatter terrain. The highest elevations in Mississippi and Louisiana are 246 m and 163 m asl, respectively.

As the largest river basin in North America, fish diversity in the Mississippi River is expectedly high. At least 375 freshwater species in 31 families are found in the entire basin (Burr and Mayden 1992, Revenga et al. 1998, Abell et al. 2000). Inclusion of the predominantly marine species that exploit the rich estuaries where the river meets the Gulf of Mexico (e.g., striped mullet, bay anchovy) would add another 50 to 60 species to this count. The LMR section includes three freshwater ecoregions: the Mississippi Embayment, the Ozark Highlands, and the Ouachita Highlands (Abell et al. 2000). The Mississippi Embayment alone has 206 species of freshwater fishes, making it comparable to other freshwater ecoregions with more than 200 species, such as the Teays–Ohio, Tennessee–Cumberland, and Mobile Bay (Abell et al. 2000, Boschung and Mayden 2004). The Ozark Highlands and Ouachita Highlands freshwater ecoregions add to the already rich fish diversity of the Mississippi Embayment. Together, these ecoregions are also extremely rich in mollusks, crayfish, and aquatic herpetofauna. The Ozark Highlands and Ouachita Highlands have a remarkably diverse crayfish fauna, including six species of stygobitic (cave-adapted) forms (Abell et al. 2000). Although the large rivers of this region have experienced dramatic changes through geologic (and thus, evolutionary) time, the diverse endemic aquatic fauna indicates that the rivers were not all simultaneously disturbed. Many fish species have had a continuous presence here since at least the Miocene (Baker et al. 1991). Refugia were present for the organisms somewhere in the basin even during successive embayments; thus, survivors could recolonize the entire region from these refugia after disturbances. The Ozark and Ouachita highlands escaped recent inundations that occurred in the Mississippi Embayment freshwater ecoregion and were south of the maximum extent of glaciation, allowing survival of endemics (Gordon et al. 1979). The LMR basin provides habitat for numerous species of waterfowl and other avifauna that use the rivers and associated wetlands, some seasonally and others year-round (Twedt and Loesch 1999, Stanturf et al. 2000). The region is the most important in North America as a winter refuge for migrating waterfowl and other wetland birds and as nesting habitat for neotropical migratory birds in spring and summer. The production of aquatic invertebrates is important to many

FIGURE 6.3   Mississippi River at New Orleans, Louisiana (Photo by Tim Palmer).

species of these birds, connecting the riverine and ter-restrial (or wetland) ecosystems.

LOWER MISSISSIPPI RIVER

The Lower Mississippi River represents the final reach of the largest river system in North America. The main stem primarily flows in a southerly direction as it traverses the Mississippi Alluvial Plain section of the Coastal Plain physiographic province from Cairo, Illinois (Rkm 1536), to Baton Rouge, Louisiana (Rkm 378) (Fig. 6.8). It then turns in a southeasterly direction as it crosses the Deltaic Plain

to Head of Passes (Rkm 0), and actually continues another 35 km to the Gulf of Mexico (Fig. 6.3). As previously mentioned, most of the flow of the LMR is due to its large tributaries that are not otherwise considered in this chapter: the Missouri, Upper Mississippi, Ohio, Arkansas, and Red rivers. Thus, the origin of the LMR at Cairo (from the Missouri, Upper Mississippi, and Ohio rivers) provides a flow that is about 80% of total flow reaching the Gulf of Mexico.

Historically, the LMR has been an extremely important provider of resources and trade routes for humans for at least 16,000 years. In 1541, Hernando de Soto found a large group of Native Americans that were "well-provisioned and socially disciplined" on the LMR below present-day Memphis. He was met by about 200 canoes and over 7000 warriors who repeatedly attacked his small group (Wells 1994). In contrast, the French explorers Marquette and Joliet saw only two small bands of natives when they explored the river by canoes from the Ohio to the Arkansas River in 1673, 132 years later. Several ideas have been advanced to explain the dramatic decline of Native Americans in this region during that period of limited historical accounts. Some think it was due to climate change and depletion of resources. Others blame it on diseases introduced and anxieties induced by the Spanish. Some ethnographers contend that European diseases resulted in death for up to 95% of the Native Americans (Dobyns 1983). The Native Americans may have feared the return of the Spanish (with their firearms and diseases) in greater numbers to this area, so those who survived the introduced diseases moved away, largely to the Ozark and Ouachita highlands to the west. With the exception of Memphis, Tennessee, and Baton Rouge and New Orleans, Louisiana, much of the region along the LMR has lower human populations now than it did when de Soto first arrived over 460 years ago.

Physiography, Climate, and Land Use

The LMR main stem is located within the Coastal Plain (CP) physiographic province, or more specifically within a geographical subunit known as the Mississippi Embayment, an area that was inundated by the ocean during several warm interglacial periods (see Fig. 6.8). This area is popularly known as "the Delta" and is essentially a broad, rather flat, shallow trough through the center of which flows the Mississippi River. The entire area consists of fine alluvial deposits and may have formed as early as 50 to 60 million years ago when formation of the Rocky Mountains and other geological events changed the formerly westward drainage of the North American continent toward the south (Dott and Batton 1981). The LMR main stem primarily drains the Mississippi River Alluvial Plain section of the Coastal Plain, but also drains smaller areas of the East Gulf Coastal Plain and West Gulf Coastal Plain sections (Hunt 1974). Part of the East Gulf Coastal Plain is also known as the Mississippi Valley Loess Plains (Omernik 1987, Kleiss et al. 2000). The western margin of the Loess Plains, along the eastern side of the Mississippi River, consists of windblown silts that formed the Loess Hills, which are about 100m taller

than the Mississippi River Alluvial Plain, a significant feature in this flat landscape.

Terrestrial ecoregions, as described by Ricketts et al. (1999), surrounding the LMR include the Mississippi Lowland Forests, Central U.S. Hardwood Forests, Ozark Mountain Forests, Piney Woods Forests, Southern Mixed Forests, and Western Gulf Coastal Grasslands. Other authors (e.g., Arkansas Department of Planning 1974, Bailey 1994, Omernik 1995) separate the Ozark Mountain Forests from the Ouachita Mountain Forests, as is done for freshwater ecoregions (Abell et al. 2000). In either case, both the Ozark and Ouachita mountains are included in this area. The Mississippi Lowland Forests extend for the entire length of the LMR main stem (except for the downstream end that passes through the eastern tip of the Western Gulf Coastal Grasslands) and is comprised of bottomland hardwood forests. These bottomland forests were dominated by cypress, gum, hickory, oak, and cedar trees when Europeans arrived in North America but have been extensively logged and cleared for agriculture (Sharitz and Mitsch 1993). Hydroperiod (duration of flooding) is an important environmental condition separating the different plant associations in the Mississippi Lowland Forests. Four major subdivisions of lowland forests are recognized: (1) river swamp forests that are always wet are dominated by cypress and water tupelo; (2) lowland swamp forests are more diverse and include water hickory, red maple, green ash, and river birch; (3) backwaters and flats are even richer in diversity and add sweetgum, American sycamore, laurel oak, and willow oak; and (4) upland transitional forests are an oak–hickory–pine association. Ozark and Ouachita Mountain Forests are similar in that they are both oak–hickory climax associations but are distinguished vegetationally primarily by the lesser amount of pine in the Ozarks. They differ floristically and faunistically in detail (at the species level) largely as a result of their distinct lithology and stratigraphy (i.e., edaphic factors). The Ozarks have horizontal strata of limestones, dolomites, sandstones, and shales that have been uplifted as plateaus and radially eroded. The Ouachitas have much less soluble lithology (very little limestone), with novaculites included, and were folded and uplifted, creating complex angular, often vertical, strata.

The climate of the LMR valley ranges from mild, midcontinental where it begins with the confluence of the Ohio to warm, subtropical in southern Louisiana. Over the region, mean monthly temperatures range from 6°C in January to 27°C in July (Fig. 6.9). Precipitation ranges from 7cm during August to over 14.5 cm in December, totaling 120 to 160 cm/yr. Rainfall is consistently high from November through July (>11.7 cm/mo) before falling below 10 cm/mo from August through October. However, it is humid year-round. In contrast, mean monthly precipitation for the entire Mississippi River basin is only about 81 cm, substantially lower than the LMR. Furthermore, precipitation peaks in May at <10 cm/mo and falls to <5 cm/mo during January and February (see Fig. 6.9).

The combination of rich soils, ample moisture, warmth, and the highest insolation on the planet during the growing season (600 cal cm^{-2} d^{-1}) give this region an extremely high potential for primary producers (Mac et al. 1998). Thus, it is not surprising that >80% of the land has been cleared of forests and is used for row-crop agriculture. Of the remainder, only 16% is forested and 4% is urbanized. Inside the levees, crop production is risky due to flooding, so silviculture is more common. Land and water inside the levees are also leased for recreational hunting and fishing.

River Geomorphology, Hydrology, and Chemistry

At their confluence at Rkm 1536, the Upper Mississippi and Ohio rivers are similar regarding discharge (5923 m^3/s and 8733 m^3/s, respectively) and the resulting river and its enormous volume of water (~14,600 m^3/s) has a substantial chemical and physical (e.g., flow, temperature, turbidity) inertia and is changed only slightly by inputs from the large western tributaries that enter the LMR. The combined White and Arkansas rivers contribute only about 10% of the proportional volume. The largest tributary rivers that enter from the east (Big Black and Yazoo) contribute only about 1% and 3%, respectively, of the volume of the LMR at their confluences. Despite the large flow near the mouth of the LMR (>15,200 m^3/s, not including the Atchafalaya distributary), the cities of Baton Rouge and New Orleans and the numerous petrochemical industries near them degrade the quality of the water by depleting oxygen and lowering pH, as well as by adding toxic compounds.

The LMR traces a very sinuous path across the Mississippi Alluvial Plain to Baton Rouge, complicated by sections where the river has cut off meanders that have become too extensive and has recaptured other abandoned channels. Deep pools occur along the outside margins of bends, with exten-

sive sandbars on the inside of curves, and where the river changes directions between the pools the thalweg (deepest part of the channel) crosses over shallow bars. Huge sand dunes (up to 10 m tall) to smaller ripples occur across the bottom of the river to further complicate the form of the riverbed and flow patterns in the main channel habitat. These supply respite from the fast (up to 5 m/s) flow for invertebrates and fishes. The three-dimensional shape of the river is complex and naturally changing, but it has some elements of predictability. At base flow, the river channel is fairly incised between steep bluffs, especially in the upper portion of this reach and where it encounters the Loess Hills along the eastern bank. Although the main channel of the river certainly varies in habitat, the marginal and floodplain areas add much more variety of aquatic habitats to the total riverine ecosystem. The character of the marginal habitats is strongly influenced by both the flow in the main channel (which may be the result of rainfall and/or snowmelt in the headwaters without contributions from local rainfall) and flow conditions in the tributaries. The stage of the main channel of the Mississippi River is occasionally high enough to impound tributaries and essentially create temporary lakes upstream from the confluence of each of them. Snowmelt runoff from the Upper Mississippi River basin often causes flooding in the LMR and impoundment of its tributaries when runoff in the southern tributaries is not large.

In the LMR, habitats (e.g., pools) expand, contract, and move about with changes in conditions, especially of flow (stage). Some habitats are dry or stagnant (lentic) at low river stages but several meters deep and flowing at higher stages. Mobile (or vagile) biota move with varying conditions to remain in their preferred habitats. Baker et al. (1991) compared five different classification schemes, including the widely adopted Computerized Environmental Resources Data System (CERDS) delineation of Cobb and Williamson (1989), and proposed one of their own based on a multivariate approach that has 12 categories that "feels right" and is intellectually pleasing. Main-stem habitat types include channel, natural steep bank, revetment, lotic sandbar, lentic sandbar, and pool. Floodplain habitats are oxbow lake, borrow pit, seasonally inundated floodplain, floodplain pond, and tributary. Sloughs (contiguous and isolated) are intermediate between main-stem and floodplain habitat.

In an excellent review of aquatic habitats and biota of the LMR, Baker et al. (1991), estimated amounts of each category at low and high river stages

TABLE 6.1   Limnological characteristics of Lower Mississippi River lotic and lentic habitats during low flow conditions.

	Temperature (°C)	pH	Turbidity (NTUs)	Specific Conductance (μmhos/cm)	Suspended Solids (mg/L)	Dissolved Solids (mg/L)	Dissolved Oxygen (mg/L)	Chlorophyll *a* (mg/L)	Nutrients	Zoo-plankton
Lotic	18–29	7–8	10–65	450–500	50–75	250–300	6–7	50–70	High	Moderate
Lentic	20–34	5–9	5–15	250–675	10–35	300–450	4–20*	50–150	Low	High

Baker et al. 1991.

* Dissolved oxygen of small floodplain ponds is commonly reduced to 0.

presently and prior to human modification, then described the physical, chemical, and biotic characteristics by habitat type. The most dramatic changes resulting from modification by the U.S. Army Corps of Engineers has been reduction of all natural floodplain habitats, especially those outside the levees. Borrow pits (for levee construction) and revetments were not present naturally. Main-channel pool habitat has been increased slightly at low stages because dikes form pools, but all other habitat types have been reduced, including main-channel length and width.

Physical, chemical, and, therefore, biological, characteristics of these habitat types differ substantially (see Baker et al. 1991). Obviously, floodplain habitats have closer association with riparian vegetation and are lentic except during high flows. Sloughs are intermediate, whereas other main-channel habitats are largely isolated from riparian influence by extensive exposed sandbars and revetted banks. Most physical and chemical characteristics in flowing habitats are different from those in lentic habitats, including pools and lentic sandbar habitats connected to the main channel during low flow periods (Table 6.1). In general, the flowing habitat types are cooler, more turbid, richer in nutrients, and less variable for most parameters. During high flows all habitats tend to take on characteristics of the main channel; temperatures are lower (by 2°C to 18°C), suspended solids are higher, and chlorophyll *a* concentration is lower.

The current velocity of the LMR main channel in midstream varies from about 0.9 to 2.4 m/s during low to moderate flows but can increase up to 5 m/s during high flows, with considerable variability laterally and with depth. Large eddies (upstream flows) are common and may exceed 250 m in length and extend 150 m out from the bank. At high current velocities the water keeps large quantities of sediment in suspension (100 to 1000 mg/L). About 2×10^8

metric tons of suspended sediments are transported past Vicksburg, Mississippi, and on to the Gulf of Mexico each year (Meade 1995). The average turbidity of the water in the LMR (10 to 65 NTUs) reduces light penetration to typically less than 0.3 m, keeping primary production low even though nutrient concentrations are moderate to high in the main channel habitats. Dissolved oxygen exceeds the 75% saturation level over 90% of the time in this portion of the LMR and is typically 6 to 12 mg/L above Baton Rouge. Below Baton Rouge, dissolved oxygen commonly dips below 4 mg/L and the Atchafalaya distributary often becomes anoxic during floods, especially those induced by hurricanes, due to oxygen demand created by introduced and resuspended dissolved and particulate organic matter.

The LMR at Vicksburg, Mississippi (Rkm 704), has a mean flow of 17,075 m³/s (USGS data from 1932 to 1998) midway between Cairo and Head of Passes (Rkm 0, but actually 35 km from the Gulf). Annual runoff at Vicksburg, which drains a total area of 2.96 million km², is about 18 cm. Runoff peaks with spring snowmelt in April (2.4 cm/mo) and is lowest in September (0.7 cm/mo), apparently the result of higher evapotranspiration during the summer months (see Fig. 6.9). Downstream of Vicksburg, additional flows include contributions from the Big Black, Ouachita, and Red rivers to increase the amount that enters the Gulf of Mexico to an average 18,112 m³/s. The virgin mean annual discharge of the Mississippi River was ~18,400 m³/s (Dynesius and Nilsson 1994). Upstream of Baton Rouge, the flow becomes divided, with the principal distributary being the Atchafalaya via the Old River and associated control structures operated by the USACE. The Atchafalaya receives 19% to 50% (typically 25%) of the flow of the Mississippi River through the control structures at Old River (Rkm 480), but flows only 224 km to the Atchafalaya Bay in the Gulf of Mexico. Because the Atchafalaya River reaches the Gulf of Mexico in less than half the distance as the main-stem Mississippi River (480 km), it is about twice as steep as the main route.

The alluvial fan across the lower delta includes several distributary rivers that were main channels at some time in the past. As the main channel deposits sediments and increases in length, at some point this creates a shorter, steeper route to the Gulf of Mexico and the river changes course. The Atchafalaya River would likely be the main channel for the Mississippi River presently if it were not for the intervention of the USACE. Natural shifts in the main channel have occurred about every 1000 years, and the present channel has been the primary one for about 1000 years (McPhee 1989).

Below Baton Rouge the river has less slope (8 cm/km) and is straighter, narrower, and more uniformly deep than the reach above Baton Rouge, where the slope is 24 cm/km. In part its character is due to natural conditions of the Deltaic Plain (less slope and finer, cohesive sediments that are less prone to erosion), but the geomorphology is also the result of maintenance of a 12 m deep channel for oceangoing vessels by the USACE. Here the habitats for biota are considerably less variable and the floodplains are almost nonexistent. Crossing the Deltaic Plain, the river is more narrowly confined by levees, and revetments line most of the banks, but dikes are not used in this section of the river. The deep, swift channel continues into the Gulf and out to the edge of the continental shelf before most of the suspended sediments are deposited. Shoreline losses in Louisiana are largely due to loss of aggradation by these sediments as the river channel has been made more narrow and swift, but also by land subsidence caused by extraction of oil, gas, and water, and by increasing sea level by glacial melt. Louisiana has no beaches, and none are likely to develop under these conditions.

Because the origin of water in the LMR is primarily from upstream tributaries, water chemistry is greatly influenced by upstream landscapes. This is exacerbated by the large volume of water, which gives it chemical as well as thermal and physical inertia. Rather than being reflective of the Coastal Plain through which it flows, the LMR water is relatively alkaline, with a pH of about 7.8 and an alkalinity ranging from 140 to 182 mg/L as $CaCO_3$. Water chemistry is strongly influenced by runoff from application of chemicals, including nutrients, in agricultural areas throughout the basin. Phosphate is relatively high, commonly ranging from 0.06 to 0.20 mg/L, but NO_3-N is extremely high, typically about 1.5 mg/L in the LMR. Although high, these elevated NO_3-N levels are actually lower than values found in the upper Mississippi River (see Chapter 8). Baton Rouge, Louisiana, and the agrichemical and petrochemical industries nearby further enrich and degrade the water chemistry of the river as a final insult before it enters the Gulf of Mexico.

River Biodiversity and Ecology

Biodiversity of the LMR main stem is higher than would be expected for a river that flows through a single freshwater ecoregion, the Mississippi Embayment (Abell et al. 2000), perhaps because the river is very large. Tributaries within the Mississippi Embayment appear to add substantial biodiversity to that found in the main stem alone. Habitat diversity is due primarily to variations in flow and depth along its course and in its floodplain habitats (oxbows, etc).

Algae and Cyanobacteria

Information on the taxonomic composition and ecological significance of phytoplankton and periphyton in the main-stem LMR is surprisingly scarce, except for suggestions that the river, delta, and estuarine marshes are all detritus-based systems (Day et al. 1973, Mac et al. 1998). The dominant forms of phytoplankton in this general region are cyanobacteria during warmer months, changing to diatoms during winter. Some information on phytoplankton and periphyton from the Atchafalaya distributary has been summarized by Patrick (1998) and is probably similar to what would have been found in the Mississippi delta marshes before human alterations confined the lower river channel between large levees. For example, phytoplankton was dominated by diatoms, including *Melosira*, *Stephanodiscus*, and *Skeletonema*, and also included the green algae *Scenedesmus* and *Crucigenia*.

Plants

Extensive floodplain forests associated with the river are composed largely of hardwood species. Four types of bottomland hardwood forests have been identified by dominant tree associations (Smith 1996). In order of increasing period of annual inundation of the land they are as follows: willow oak and sweetgum; Nuttall's oak and green ash; overcup oak and water hickory; water tupelo and bald cypress. The river delta area that forms the extensive swamps and bayous of southern Louisiana has extensive areas of cypress and tupelo stands, but these give way to marshes dominated by aquatic macrophytes toward the coast. The diverse and dynamic macrophytic associations have attracted several investigators (Conner et al. 1981; Mitsch and Gosselink 1993; Mac et al. 1998). The patterns of vegetation in these lush subtropical marshes are constantly changing in response to water levels, salinity, hurricanes, and invading plant and animal species. Many of the macrophytes are nonnative species that have invaded the wetlands (water hyacinth, alligatorweed, hydrilla, and Eurasian watermilfoil). Nutria, rodents introduced from South America, are very abundant, and they consume large amounts of macrophytes and young trees, including cypress (Conner 1988). As much as 70% of the deltaic marshes consist of floating-island plant associations (O'Neil 1949, Sasser et al. 1996). One of the most abundant associations is dominated by maidencane. The marsh plants that senesce produce a detritus substance, locally called "coffee grounds," that accumulates along wetland tributaries (bayous) and is eventually transported to estuaries and the Gulf by floodwaters. This detritus is refractory, but when suspended by storms and floods heterotrophic bacteria metabolize enough of it to induce hypoxia (Sabo et al. 1999).

Invertebrates

Invertebrates of the main channel include abundant caddisflies (especially *Hydropsyche orris* and *Potamyia flava)* on hard substrates and burrowing mayflies in hard clay (*Pentagenia, Tortopus*) and mud (*Hexagenia*). Channel habitats also have numerous scraping mayflies (*Stenonema*) on hard substrates of dikes and the articulated concrete mattresses (ACMs) installed by the USACE to control bank erosion. Chironomid midges and oligochaete worms are abundant in all habitat types of the LMR. The Asiatic clam has been abundant in sand and gravel habitats of rivers throughout this region since its colonization in the 1970s. More recently the LMR experienced an invasion of zebra mussels. Asiatic clams and zebra mussels are very efficient filter feeders that capitalize on the abundant fine particulate organic matter (FPOM) in the river. They are consumed in large quantities by catfish but are still numerous. Meiofauna have also been reported to be abundant in lotic as well as lentic habitats of the river.

Lentic habitats of abandoned channels (created naturally and by actions of the USACE) and floodplains have diverse and abundant taxa of chironomids (e.g., *Rheotanytarsus, Chironomus, Coelotanypus, Robackia, Glyptotendipes*), oligochaetes (*Limnodrilus, Hyodrilus, Nais*), and fingernail clams (*Sphaerium*). The soft sediments of these nonflowing habitats also have numerous phantom midges (*Chaoborus*). Mayflies (especially *Hexagenia* and other burrowers) are common, but stoneflies are rare anywhere in the LMR.

Vertebrates

Freshwater fish assemblages of the LMR have been studied more than other biota, but most accounts are limited to occurrence of species, leaving ecological relationships poorly understood. Perhaps as many as 150 species of freshwater fishes once inhabited the main-stem LMR (Fremling et al. 1989), but Baker et al. (1991) reported that only 91 fish species maintain reproducing populations there, discounting recently introduced species (e.g., grass carp), strays from small tributaries, and marine species. The number of fish species in the main-stem LMR is therefore less than half that estimated for the Mississippi Embayment ecoregion as a whole (206) (Abell et al. 2000). This ecoregion includes several Mississippi tributaries, such as the lower reaches of the Arkansas and White rivers, as well as the Big Black and Tensas rivers. Some of the fish taxa in the LMR are common to abundant in nearly all habitats (e.g., channel catfish, common carp, river carpsucker, freshwater drum, gizzard shad, threadfin shad, spotted gar, short-nose gar, smallmouth buffalo, bigmouth buffalo, white crappie). Other taxa are found almost exclusively in flowing habitats, like blue catfish, blue suckers, river darters, and shovelnose sturgeon. Many more are common only in lentic habitats of pools, abandoned channels, and floodplains, including grass pickerel, chain pickerel, black bullhead, yellow bullhead, pirate perch, western mosquitofish, four top minnows (Fundulidae), nine cyprinids (Cyprinidae), and seven sunfish (Centrarchidae) species.

Many of these abundant fish species are invertivorous and/or piscivorous as adults. Gizzard and threadfin shad are planktivorous, but also feed on the bottom, as indicated by the presence of sand and detritus in their guts (Baker and Schmitz 1971). Bottom feeding may be a good way to obtain the abundant meiofauna on the surface of benthic substrates. Native phytophagous (other than algae) and detritivorous fishes are almost nonexistent in Nearctic waters. However, the grass carp has been imported from China and widely stocked by the Arkansas Game and Fish Commission to control aquatic macrophytes. Grass carp are thought to have established reproducing populations in the LMR (Robison and Buchanan 1988). The pallid sturgeon is the only LMR main-stem fish listed as endangered by the U.S. Fish and Wildlife Service, although a few other species are listed by state agencies.

Amphibians and reptiles of the LMR are also diverse and abundant, especially toward the southern coastal zone. American alligators are abundant, along with a large variety of turtles (e.g., common snapping turtle, alligator snapping turtle, both smooth and spiny softshell turtles, red-eared turtle, Mississippi mud turtle), snakes (e.g., diamondback water snake, broad-banded water snake, western cottonmouth), and frogs (bullfrog, pig frog, spring peeper, cricket frog). Semiaquatic mammals, in addition to the nutria, include beaver, muskrat, raccoon, mink, and river otter. Wetland forests are also habitat for white-tailed deer, red and gray squirrels, swamp rabbits, black bears (subspecies classified as threatened), and endangered red wolves.

Many of the species of fishes, reptiles, mammals, and birds that occur in the LMR and its wetlands have commercial value and/or are pursued by those who hunt and fish. Populations of these animals have been studied more than others and are of particular interest to state and federal management agencies, but all species have some legal protection.

Ecosystem Processes

Ecosystem processes in the LMR may be almost as diverse as the habitat types in this large floodplain river. However, the main-channel community appears to depend primarily on species that intercept FPOM being transported downstream, in accord with the River Continuum Concept (RCC) of Vannote et al. (1980). Most invertebrates and some fishes in the main channel are collector-filterers (caddisflies, mussels, paddlefish) or collector-gatherers that exploit the rich seston of primarily allochthonous origin. The more lentic habitats are populated by burrowing collector-gatherers (mayflies, midges, oligochaetes), which use the same fine particles as they settle. Communities in the marginal habitats (sloughs, oxbows, etc.) appear to be more dependent on locally derived organic matter, whether autochthonous or allochthonous. Contribution of heterotrophic microbial production to higher trophic levels, especially via meiofauna linkages, may be very significant in the LMR. River interactions with floodplain wetlands, as would be expected in such a large lowland river according to the flood pulse concept of Junk et al. (1989), appear to be substantially reduced by the extensive engineering projects of the USACE.

There is a paucity of data for primary and secondary production, food web structure, and even habitat use by fishes, making any statements about trophic dynamics in the LMR tenuous. Primary production (P) in flowing habitats is probably kept low by turbid, turbulent conditions, and low in lentic habitats by limited nutrient availability and shading by trees. Respiration (R) is probably high in response to temperature. Thus, the P/R ratio in most habitats most of the year is probably less than 1, in agreement with the RCC for high-order rivers. Most of the habitat types are probably allochthonous-supported, heterotrophic systems, especially if the bottomland hardwoods and other swamp vegetation are not considered part of the aquatic producers.

Ecosystem processes are largely governed by characteristics of physical habitat, and the LMR habitats have been significantly altered by the USACE. Effects of modifications to the habitat structure of the LMR by the USACE on community structure and trophic dynamics are impossible to completely assess but have been extensive. Removal of snags for the past 175 years took away much of the natural habitat that provided firm attachment sites for invertebrates, most of which are filter-feeders. This left hardened clay banks that are extensively colonized by burrowing mayflies (*Hexagenia* spp., *Tortopus incertus*, *Pentagenia vittigera*), which are collector-gatherers, and by some filtering hydropsychid caddisflies (Beckett et al. 1983, Beckett and Pennington 1986). Sand substrates in the flowing-channel habitats contain many chironomids, oligochaetes, microturbellarians, and other meiofauna, but few macroinvertebrates (Wright 1982, Baker et al. 1991). Abandoned channels (sloughs, oxbows) have high densities of phantom midges, tubificid worms, and fingernail clams (Sphaeridae) (Mathis et al. 1981). Rocky substrates provided by dikes, ACMs, and other revetment materials (e.g., riprap) have extensively supplanted snags as suppliers of firm substrate in the LMR. These rocky substrates have the most dense macroinvertebrate assemblages of any habitat type, and they are dominated by net-spinning caddisflies (*Hydropsyche orris*, >60%) and tube-building midges (~24%), especially *Rheotanytarsus* (Mathis et al. 1982, Way et al. 1995). Revetments now cover many former hard clay substrates and soft mud habitats used by burrowing mayflies, but dike fields (materials deposited around the dikes by currents) also contain these habitats (Beckett and Pennington 1986). The GIS-based system used by the USACE (CERDS; see Cobb and Willamson 1989) has been used to estimate changes in the proportion of habitats (other than snags) before and after modifications to the LMR (Baker et al. 1991). However, there is still no way to accurately estimate the former densities of snags in the river or of invertebrates on them. It seems possible that the community structure and trophic dynamics of the system may be substantially similar to those in place before extensive alterations, as introduced rock and concrete structures have replaced snags.

It is likely that enrichment of the water with organic matter (dissolved organic matter, FPOM, bacteria) and inorganic nutrients from point and nonpoint sources of pollution have increased microbial metabolism in the main-channel habitats, but limiting contributions of these substances from the floodplain may have offset this. One thing is certain; the amount of wetland habitats in the floodplain has been extensively reduced, and flood pulses of the river are denied access to the former wetlands, many of which are now cleared agricultural cropland. The enormous anthropogenic disturbances may have changed the total amount and relative abundances of taxa in the LMR, just as they have changed the total and relative amounts of the various habitat types, without substantially altering the list of taxa (irrespective of relative abundance) and the dominant ecological processes. Decrease in the efficiency of use of resources by the community may have been the major consequence of the disturbances, and this has likely resulted in the enormous dead zone that has developed in the Gulf of Mexico. Transport of water and material from the river to the Gulf has been enormously increased, and this has substantially reduced the utilization of organic and inorganic nutrients as they enter and pass through the LMR to the Gulf.

Human Impacts and Special Features

The LMR represents the lower reach of the largest river in North America and one of the largest rivers in the world. Flowing primarily through low-relief Coastal Plain, the LMR and its alluvial plain formed one of the great river–floodplain ecosystems of the world. Unfortunately, the ecological integrity of this enormous ecosystem has been strongly challenged at the physical, chemical, and biological levels by extensive physical, chemical, and biological alterations. The major alterations are the result of (1) engineering projects to improve navigation and reduce flooding; (2) chemical contaminants from industrial, urban, and agricultural activities in >40% of the conterminous United States; and (3) the introduction of numerous nonnative species.

The need for navigation projects intensified with the advent of boats powered by steam engines, like the early paddle wheelers. Later, with more powerful

boats with diesel engines, the Mississippi River quickly became an important avenue for bulk transportation of materials from the North American heartland to the ports of Baton Rouge and New Orleans, where they could be transferred to oceanic vessels. This development was an important impetus for government intervention to "improve" the channels for navigation. Huge, devastating floods also increased the need for control of the LMR waterways. The earliest recorded large flood was in 1543 by de Soto's expedition and the most recent was in 1993. Because of the need for cheap commercial transportation of materials and the enormous flooding potential in this region, a series of governmental actions have given extensive authorization to the USACE and other agencies to perform huge projects to enhance navigation and control flooding.

Levees were the earliest form of flood control and were first built to protect New Orleans. By 1735, levees lined the river from 48 km above to 19 km below the city. By 1844, levees were nearly continuous upstream to the confluence with the Arkansas River. Now there are nearly 3000 km of levees 9.25 m high along the LMR and very nearly another 1000 km along its tributaries. The 600,000 ha of floodplain remaining inside the LMR levees is only about 10% of the area inundated by large floods before containment (Baker et al. 1991). Levees are primarily responsible for denying the river access to its rich floodplains and disruption of the natural ecological functioning, as described by Junk et al. (1989) as the flood pulse concept.

Cutoffs are events that normally occur during floods when the river takes a shorter path downstream by either cutting across a point bar or the neck of a meander. These may become permanent features when the flood subsides. They straighten the channel and increase the slope of the reach, causing the water to flow faster. Cutoffs are desirable to the USACE because they shorten, straighten, and deepen navigation pathways, as well as allow floodwaters to subside more quickly. The USACE has artificially constructed cutoffs that have shortened the LMR channel. Natural cutoffs formerly shortened the channel also, but by a lesser extent and were offset by meandering, which is now thwarted by revetments. The exact amount of shortening of the channel by the USACE is difficult to estimate, but it has been close to 500 km, about 25% to 30% of the former length of the LMR (see Baker et al. 1991).

Floodways and control structures have been constructed to divert excess water from the main channel during only the highest flows. The Atchafalaya River distributary has associated floodways that parallel it, within separate levees, to the Gulf of Mexico. The Atchafalaya Floodway is on the west side of the Atchafalaya River and the Morganza Floodway is on the east. Control structures at Old River can be manipulated to control flows down the Atchafalaya River and the two floodways independently of each other. When the Mississippi River discharge reaches 42,613 m³/s at Baton Rouge it can be partially diverted into the Atchafalaya River through the Morganza and West Atchafalaya floodways or into the Gulf of Mexico via the Bonnet Carre Spillway and the Rigolets (Everett 1971). In addition to the Atchafalaya, the Bonnet Carre Spillway above New Orleans can be used to discharge floodwaters from the Mississippi River to Lake Pontchartrain. The control structures have essentially dried for much of the year what were formerly wetlands by not allowing water to flow through them other than during very large floods.

Removal of snags from the Ohio and Mississippi rivers was initiated by the U.S. War Department in 1824 and is still continued in the LMR and its tributaries using special barges for this purpose (Brown and Matthews 1995). Dredging is also used to maintain deep channels at crossings (where the thalweg changes from one side of the river to the other). There are >200 crossings between Cairo, Illinois, and Baton Rouge that require periodic dredging of 32 to 81 million m³ annually to maintain a 4 m deep channel (Keown et al. 1981). Another 60 million m³/yr are dredged below Baton Rouge to maintain a 12 m deep channel to the Gulf of Mexico. Dredging introduces large volumes of fine sediments to the river water and produces a plume of turbidity downstream. It also resuspends contaminants that have become buried or adsorbed to sediment particles.

Revetments have been used extensively on the LMR to keep banks from eroding and to prevent channels from meandering. Logs, fences, gabions, rocks, tires, and automobile bodies have all been used as revetments, but since the mid-1940s ACMs have been used almost exclusively on the Mississippi River. Almost 2000 km of ACMs are in place now, and eventually nearly 50% of the banks along the LMR will be covered with them. Cypress trees several hundred years old are sometimes removed to install ACMs. Such actions diminish the natural interaction between the river and its riparian zone, as described by Thorp and Delong (1994) as the riverine productivity model (RPM).

Dikes are partial dams that also have been used throughout much of the LMR. They are used to control the aggradation and degradation of sediments by controlling the flow of the river, thereby encouraging the river to maintain a single, straight, deep channel. They were originally made of wood from about 1880 until 1960, but now they are constructed of large limestone rocks. Dikes are usually used in parallel groupings extending nearly perpendicular to the flow. They are very large, averaging 630 m long and 10 m wide at the top. Nearly 500 km of dikes occur between Cairo and Baton Rouge.

Reservoir and lock-and-dam construction on tributaries to the Mississippi River is part of the Mississippi River and Tributaries Project plan to control floods and enhance commercial navigation. Extreme high and low flows can be moderated by operation of the dams. Many of the larger tributaries (e.g., Arkansas, White) have also been modified for navigation by barges requiring an 8 ft. (2.4 m) deep channel. Currently the USACE plans to increase the depth of the channel in the Lower White River to 9 ft. (2.7 m) to facilitate passage of larger barges. The plan required construction of a set of locks and dams near the mouth of the river (Montgomery Point), as well as extensive dredging and installation of dike fields. These alterations will disrupt natural ecological functioning of the White and Mississippi rivers, and are probably not sufficiently justified economically (Wright 2000).

In addition to physical alterations, the LMR receives inputs of municipal wastes, industrial effluents, and agricultural runoff from over 40% of the conterminous 48 states. Inputs now move quickly to the Gulf of Mexico through the shorter, steeper, deeper channel, especially during floods, by design of the USACE. This creates large problems downstream in the estuary and the Gulf. The estuary is rapidly receding because it no longer receives sediments or as much water from the river. An expansive summer dead zone of >20,000 km² resulting from oxygen depletion has developed in the Gulf. Primary production stimulated by agricultural nutrients (especially nitrogen) in runoff is thought to be the cause (McIsaac et al. 2001), but a more parsimonious explanation might be found by examining the dissolved and particulate organic carbon loading of the Gulf. The total oxygen demand of the normal LMR flow (>18,000 m³/s) to the Gulf must be immense without additional organic matter via enhanced photosynthesis. The swamps, estuaries, and bays may be predominantly heterotrophic systems with low algal primary production, but they are highly productive of aquatic macrophytes and the swamps receive large amounts of litterfall from trees. During floods, turbidity limits aquatic photosynthesis, but accumulated organic matter in the floodplain wetlands is swept into the Gulf as the floodwaters flow out over the more dense salt water. This tremendous load of organic matter is respired by aerobic heterotrophic bacteria, reducing dissolved oxygen levels, and the refractory portion persists for many weeks, holding down oxygen concentrations. If this hypothesis is correct, even significant reduction of nitrogen will have little effect on reducing the dead zone. The dead zone may result from altered flow dynamics (controlled by the USACE) that seasonally flush enormous amounts of dissolved and particulate organic matter into the Gulf, more than from inorganic nutrient additions.

WHITE RIVER

The 1159 km long White River is a major tributary to the Lower Mississippi River, with a basin area of 72,189 km². It originates in the Boston Mountains of the Ozark Plateaus near Boston, Arkansas (35.8°N, 93.6°W), at 785 m asl and first flows west, then circles north through Missouri and southeast back into Arkansas (Fig. 6.10). The White River is strongly fragmented by dams along its main stem, but has two major tributaries (Buffalo and Current) that are protected and free flowing. Near Newport, Arkansas, the White is joined by the Black River as it leaves the Ozark Highlands and flows south across the Mississippi Alluvial Plain to its confluence with the LMR. The dendritic White River attains 5th order by its confluence with the West Fork of the White River and 7th order when joined by the Black River. The Lower White River near its confluence with the Mississippi was connected to the Arkansas River in 1969 by the USACE as part of the McClellan-Kerr Navigation Project on the Arkansas River that passes through Little Rock and on to Tulsa, Oklahoma. Several locks and dams regulate flow to facilitate navigation in both rivers.

Native Americans have occupied the Ozark Plateaus area for at least 10,000 years and the Lower White River in the Mississippi Alluvial Plain for 16,000 years. No major cultures developed in the Ozark region, but diverse cultures from other areas, including the Cahokian, Caddoan, and Mississippian cultures, were represented here (Gerlach and Wedenoja 1984). The Osage, Quapaw, and lesser numbers of other tribes were in the Ozark region of the Upper White River when Europeans arrived. The tribes mentioned earlier for the Mississippi River were also along the Lower White, and population densities in the sixteenth century were higher than they are now. In the late eighteenth and early nineteenth centuries the U.S. government relocated many eastern tribes to the Ozarks, including the Cherokee, Delaware, Shawnee, and Kickapoo. The first European explorers were Spanish, but the first to settle in the White River basin were French, who were replaced by Scotch-Irish. Early white settlers came from Kentucky and Tennessee to exploit the abundant fishes and wildlife along the Lower White River (Keefe and Morrow 1994), as did the Native Americans who preceded them. The Ozarks provided diverse but not abundant resources to people, and this was reflected in how they lived. The river provided fishes, game, fertile floodplains for crops, and travel corridors.

Physiography, Climate, and Land Use

The headwaters and midreaches of the White River are mostly in the Ozark Plateaus (OZ) physiographic province, with a small amount in the Ouachita province (OP) (see Fig. 6.10). The headwaters of the White are in an area of karst topography where the streams have cut and dissolved deep (to 200 m) valleys through limestone and dolomite deposits. In most places, soils of the Ozarks are too thin for row-crop agricultural production and much has remained forested. The more northern part of this upper basin is primarily in the Central U. S. Mixed Hardwood Forests terrestrial ecoregion (Ricketts et al. 1999). The southern portion of the upper basin drains the more rugged mountains of the Ozark Mountains Forests ecoregion. The native forests support good stands of hardwoods dominated by oaks (red, white, black) and hickories (shagbark, bitternut). Also found are flowering dogwood, eastern red cedar, and shortleaf pine. Floodplains consist of a band of flatter, alluvial deposits whose width is proportional to the size of the streams. Most of the floodplains have been cleared of their diverse riparian vegetation and planted with fescue, a nonnative pasture grass. The fescue has extremely small seeds that are not used by native birds and other wildlife. These floodplains receive applications of large amounts of manure and litter from confined animal agricultural operations.

The White River changes almost abruptly from a mountain stream to a lowland river at Newport, Arkansas, as it moves from the Ozarks Plateaus into the Mississippi River Alluvial Plain section of the

Coastal Plain (CP) physiographic province, crossing comparatively flat, deep, transported (alluvial and windblown loess) deposits (see Fig. 6.10). As described for the Mississippi River main stem, this alluvial plain contains the Mississippi Lowland Forests, a terrestrial ecoregion from which much of the natural forests have been cleared (Ricketts et al. 1999). These forests were dominated by oak, hickory, and pine in the more upland areas and a diverse and extensive bottomland forest in the flood-prone areas.

Precipitation in the White River basin is primarily rainfall and trends from approximately 110 cm/yr in the uplands to 130 cm/yr in the lowlands. Average monthly precipitation is fairly evenly distributed through the year, but is lowest in January (7.4 cm) and highest in May (12.6 cm) (Fig. 6.11). Mean annual temperatures are 14°C (range 1°C to 26°C) in the uplands and 16.4°C (3.7°C to 27.4°C) in the lowlands. The highest mean monthly temperatures are in July (27°C) and the lowest in January (2°C).

Northwestern Arkansas in the headwaters region and Missouri in the midreaches are experiencing rapid urbanization, but most of the basin is sparsely populated (18 people/km^2). Land use in the Ozark Plateaus area is dominated by silviculture and beef cattle production, but row-crop production of soybeans, cotton, and rice is characteristic of the lower basin. Forests cover about 70% in the upper and middle regions, but only about 8% in the lower region. Agriculture (excluding silviculture) is about 28% in the upper and middle regions but increases greatly to 83% in the lower region. Increasing amounts of land are being placed in the federal Crop Reserve Program in both parts of the basin. The last 95.5 km of the river flow through the 457.5 km^2 White River National Wildlife Refuge. This refuge was established primarily for waterfowl, but harbors many other species, including a relict population of native black bears that are genetically distinct from those that have been reintroduced from Wisconsin in the upper portion of the basin.

River Geomorphology, Hydrology, and Chemistry

The White River, like most Ozark streams, originates as an ephemeral debris-regulated channel but becomes an intermittent, gravel-bed stream with distinct pool and riffle structure while still in the 1st order reach. The floodplain also develops rapidly downstream. Within the headwaters reach (orders 1 to 3), the stream changes from a channel dominated by bedrock outcroppings and boulders that the stream cannot move to one with more gravel whose channel shape is determined largely by flow patterns during floods. During most years the White has perennial flow in the 3rd order reach, however, the 3rd order occasionally becomes intermittent during summer (i.e., without surface flow between pools). Overall, the slope of the White River in the headwaters is about 1.5 m/km. Small reservoirs used for municipal water supplies, flood control, and recreation are common on 2nd and 3rd order tributaries, including the main stem. However, the Buffalo and Current rivers have been protected from damming so far. Large reservoirs, primarily for generation of electric power and flood control, are numerous in the White River basin on 4th and 5th order reaches. The largest of these are Beaver Lake, Table Rock Lake, Lake Taneycomo, and Bull Shoals Lake on the main stem, with Norfolk Lake on the North Fork and Greers Ferry Lake on the Little Red River (see Fig. 6.10). Thus, the midreaches of many of the streams in the White River basin are impounded by dams that release hypolimnetic water. As a result, water temperatures are cooler than normal in summer and warmer in winter in most of the Ozark Plateaus portion of the basin. The reservoirs are also a major impact on other physical and chemical characteristics of the river. There are no reservoirs below Newport, Arkansas, because the flat terrain is not as suitable for their construction. The Lower White River below Newport is a large lowland river where the riffle–pool structure has given way to a sinuous sand-bed channel with extensive floodplain development. As it crosses the Mississippi Alluvial Plain it becomes more like the Lower Mississippi River in most respects, although the White is less impacted by humans.

Mean annual discharge for the White River is 979 m^3/s, with an annual runoff of almost 43 cm. Mean monthly runoff shows a seasonal pattern from lowest in September (1.5 cm) and highest in March (5.6 cm) (see Fig. 6.11). The natural runoff patterns are influenced by higher precipitation in winter months and higher evapotranspiration in the summer. However, the pattern of runoff is also influenced by periods of conservation and release of water through dams for generation of hydroelectric power and flood control. Flow in the headwaters of the White is seasonal, with predictable intermittent conditions (isolated pools) developing during each summer and a period of continuous flow from about November to June in most years. The hydrograph of a typical water year for the upper portion of the river is quite spiky because the water rises and falls rapidly, with changes of water level of 1 to 3 m fairly common. Flooding (exceeding bankfull capacity) is not predictable by month or even by season, even though average rainfall is higher during fall and spring. Flow in the Lower White River is protected by instream minimum flow regulations established by the state of Arkansas. Although the regulations are not laws, they are agreements with the management agencies to maintain minimum flows to protect water quality, navigation, and fish and wildlife resources. Proposed increases in width and depth of the barge canal and plans for withdrawal of water for an immense irrigation project seriously threaten the near-natural hydrology of the Lower White River.

Chemical water quality in the White River is of growing concern, in addition to issues regarding geomorphology (channelization, etc.) and water quantity (especially irrigation). Mean annual nutrient levels (PO_4-P = 0.07 mg/L, NO_3-N = 0.39 mg/L) are fairly high and are increasing at some monitoring sites (e.g., below Beaver Dam in northwestern Arkansas). In the Lower White at St. Charles, Arkansas, NO_3-N ranges from 0.1 to 0.5 mg/L and PO_4-P ranges from 0.01 to 0.15 mg/L. Water released from the reservoirs during summer and fall has very low dissolved oxygen concentrations, and this has negative impacts on downstream aquatic communities. Fish kills have been documented as a result of this, and chronic impacts on fishes are suspected. Limestone deposits in the Ozark Plateaus buffer the White River within pH extremes of 6 to 9, with an average pH of 7.65, and contribute to the average alkalinity of 110.5 mg/L as $CaCO_3$. Turbidity is typically less than 10 NTUs in the headwaters and below the large reservoirs, but increases to 20 to 30 NTUs at St. Charles in the Lower White River. During periods when the river has not been flushed out recently by high flows the water has a greenish-white tinge due to fine colloids dissolved from the limestone in the basin. Overall, environmental quality of the headwaters of the White is less than that of adjacent streams in the Boston Mountains (some of which are tributaries to the White) because it has experienced more forest removal and agricultural development (Radwell 2000).

River Biodiversity and Ecology

The White River basin occurs in two freshwater ecoregions, the Ozark Highlands and the Mississippi Embayment (Abell et al. 2000). Understanding of the biological communities and ecological relationships is much more thorough for the Ozark Highlands than

for the lower half that crosses through the Mississippi Embayment. Levels and patterns (spatial and temporal) of primary and secondary production are not well known, but energetic interactions among invertebrates and fishes have received more attention, at least in the Ozark Highlands. The Lower White, being a large river in the Mississippi Embayment, shares many biota and ecological characteristics with others of that ecoregion, including the LMR. Like the LMR, it has been less thoroughly studied, probably because of the difficulty involved in sampling large rivers and its distance from major universities.

Algae and Cyanobacteria

Little is known about the algae of the White River per se, but studies of streams from the same watershed and ecoregion infer that a diverse array of periphytic diatoms are very important, especially in winter. The diatoms are limited by silica, and by early spring they exhaust the supply and other forms become more abundant. These include an extensive list of filamentous, colonial, and unicellular green algae (e.g., *Cladophora*, *Eudorina*, *Cosmarium*) and cyanobacteria (e.g., *Anabaena*, *Microcystis*, *Formidium*). The macroalga *Chara* develops extensive localized patches. Flagellated mixotrophic forms (*Euglena*, etc.) make some summer pools with limited flow emerald green, especially if the pools are frequented by cattle trying to keep cool and escape insects. Plankton are rare in free-flowing portions of the White, but are found in the large (>30 × 600 m) pools in the 3rd to 5th order reaches. Periphyton and plankton have not been studied in the Lower White, but neither are likely to be abundant due to shifting sand substrates, rather turbid water conditions, and fluctuating water levels.

Plants

Water willow, an emergent macrophyte with a lanceolate leaf similar to that of willow trees, is abundant in and along all Ozark and Ouachita streams. It is found in almost all shallow areas that receive ample sunlight and escape the harshest physical disturbances (flood impacts, canoe traffic, cattle crossings). Water willow is becoming more abundant as canopy cover is removed by disturbances in the riparian forests. Smartweed, numerous sedges, and river cane are common along the White River margins and form dense stands in some areas. Numerous vines occur along the riverbanks too, including several species each of poison ivy, grapes, and greenbriers, and cucumber vine in the lower reaches. Gravel bars and stream margins are colonized by a variety of herbaceous plants but also by willows, witchhazel, sycamore, river birch, red maple, buttonbush, and cottonwood. Adjacent floodplain forests are diverse and include many of these same woody plants and a mixture of American elm, green ash, box elder, hackberry, sweetgum, Nuttall's oak, and many others. Bald cypress and water tupelo are abundant in and along the Lower White River, but water willow is rare. Nonnative plants are not a serious problem along the main stem of the White River, although some aquatic macrophytes that are not normally seen along the river are abundant below some of the dams (e.g., elodea). Eurasian watermilfoil and coontail have created problems in shallow oxbow lakes.

Invertebrates

One striking characteristic of the White River invertebrate assemblage is the abundance and diversity of crayfishes, with 15 endemic epigean species and 3 more in the basin that are cave-adapted (Rabeni et al. 1995, Pflieger 1996). One species, *Orconectes longidigitus*, grows to over 15 cm carapace length in the White River headwaters and in Beaver Reservoir. Another unusual characteristic is the abundance and diversity of mollusks, with about 100 known historically and 58 extant in the basin (Gordon 1982, Gordon et al. 1994). The White River was one of the main sources of mussels for the button industry during the first half of the twentieth century, and mussel shells are still harvested and shipped to Japan for seeding cultured pearls. The fat pocketbook mussel is an endangered species that occurs in the Lower White. Some of the most common mussels in the White River include threeridge, Ozark pigtoe, pimpleback, pocketbook, ladyfinger (or spike), rabbitsfoot, mucket, squawfoot, and pistolgrip.

Smaller macroinvertebrate taxa are moderately diverse and usually include large numbers of mayflies (*Stenonema, Hexagenia*), caddisflies (*Cheumatopsyche, Chimarra*), chironomid midges, beetles (*Stenelmis, Psephenus*), dragonflies (*Gomphus*), and damselflies (*Coenagrion, Argia*) (Brown et al. 1983, Rabeni et al. 1995, Doisy et al. 1997). Oligochaete worms and chironomids are locally abundant in organically rich, fine sediments. Stoneflies, especially the winter genera, are present but no longer occur in large numbers as they did 30 years ago. Meiofauna are abundant, especially rotifers and copepods, through all seven orders of the river.

Vertebrates

The entire White River basin contains at least 163 native fish species, 11 of which are endemic to the basin (Killgore and Hoover 2003). Killgore and Hoover (2003) collected 97 species in the Lower White from 1996 to 2000. The most abundant and diverse families of fishes in the White River are Cyprinidae, Percidae (darters), Centrarchidae, Catastomidae, and Ictaluridae. Blacktail shiner, Mississippi silvery minnow, emerald shiner, mimic shiner, bullhead minnow, blue catfish, channel catfish, bluegill, spotted bass, and western sand darter are often the most abundant fishes at sites in the Lower White River. Typically, 29 to 37 fish species are collected per site in this region (Keith 1987, Buchanan 1997). The Ozark Highlands portions of the White River and its tributaries have popular sport fisheries for centrarchid species, especially smallmouth bass. The Lower White River floodplain lakes also have an exceptional centrarchid sport fishery for largemouth bass and white crappie, in addition to a very productive commercial fishery for primarily carp, suckers, gars, drum, and catfish. The pallid sturgeon is an endangered species that occurs in the Lower White but has not been collected there recently. Its congener, the shovelnose sturgeon, is fairly common. The nonnative common carp and western mosquitofish have become abundant throughout the White River system, and many other nonnative species are common there.

Amphibians and reptiles are also abundant in the White River, with numerous species of aquatic and semiaquatic snakes, turtles, frogs, and salamanders. These include the western cottonmouth and several species of water snakes, common and alligator snapping turtles, stinkpot turtles, softshell turtles, and map and false map turtles. Bullfrogs, green frogs, pickerel frogs, and southern leopard frogs are common. American alligators are abundant along the Lower White and in its floodplain waters.

Muskrats, beaver, river otters, raccoons, and mink are mammals commonly seen in or near the White River along its entire length. The White River provides winter habitat for bald eagles and numerous migratory waterbirds (double-crested cormorants, anhingas, grebes, etc.) and waterfowl (ducks and geese). Great blue herons and belted kingfishers are abundant local residents. Nearly $1219\,km^2$ of lands along the Lower White River are protected in three national wildlife refuges and seven state wildlife management areas, primarily for the protection of migratory waterfowl.

Ecosystem Processes

Plant, invertebrate, and vertebrate animal species assemblages of the White River are diverse and abundant, and so are the physical and chemical environments. Therefore, ecological interactions and bioenergetics pathways are complex, but some fascinating attributes are known about them, especially for the Ozark Highlands region. Flashy hydrologic regimes remove most litterfall from rivers to the floodplains in headwaters reaches before aquatic organisms can consume it (Brown and Matthews 1995), but that remaining in the water is rapidly processed (decay rate k = 0.01 to 0.03; Brown et al. 1983). Thus, community P/R ratios are high, even in the headwaters, and shredders are rare (A. V. Brown, unpublished data). Production in shallow habitats (riffles, runs) is about twice that of respiration rates in May, June, and October, but the P/R ratio in pools is ~0.5 (Whitledge and Rabeni 2000). In their study of the Jack's Fork, Missouri, a tributary to the White, Whitledge and Rabeni (2000) found that primary production ranged from 749 mg $O_2 \cdot m^{-2} \cdot d^{-1}$ in a riffle during July to 30 mg $O_2 \cdot m^{-2} \cdot d^{-1}$ in a pool in October. The distinct riffle–pool structure of most of the White River and its tributaries is a physical template that overshadows the longitudinal continuum (Vannote et al. 1980, Brussock and Brown 1991, Doisy and Rabeni 2001). Distribution and movement of biota, interspecific interactions, and trophic dynamics are largely characteristic of and regulated by habitat patches of distinct flow, depth, and substrate composition (Brown and Brussock 1991, Doisy et al. 1997, Peterson and Rabeni 2001). Disturbances that impact the distribution of these physical habitat units, such as gravel mining, significantly alter the ecological functioning of gravel-bed streams like those in the Ozark Plateaus portion of the White River basin (Brown et al. 1998, Zweifel et al. 1999, Rabeni 2000).

Crayfishes are exceptionally important ecological components of the White River basin biota (Whitledge and Rabeni 1997). Half of the total benthic community production and invertebrate consumption of food in the Current and Jacks Fork rivers has been attributed to two crayfish species (golden and spot-handed; Rabeni et al. 1995). The crayfishes appear to have major impacts on algae, detritus, and invertebrates that they consume, and on several species of fishes that prey upon them (Rabeni 1992).

Ecology of the Lower White River, like other large rivers of the LMR basin, is poorly understood. The Lower White appears to be a very productive system, judging from the large harvests of fishes and mollusks that it sustains, although their harvests are much lower than they were 80 years ago (Shirley 2000, Ken Shirley, personal communication, 2002). The Lower White also must have a very rich seston to support the exceptionally abundant filter-feeding invertebrates (hydropsychid caddisflies, Asiatic clams, zebra mussels, 58 species of other native mussels) and fishes (gizzard and threadfin shad, paddlefish, larval fish of many species).

Human Impacts and Special Features

The White is a very special North American river. It is the largest river draining the Ozark Plateaus, and two of its tributaries have been designated as National Rivers (Buffalo and Current/Jacks Fork). The human population density has been low, but it is now growing rapidly. The Lower White basin still experiences a fairly normal hydrologic flooding regime of its extensive floodplain forest despite the large main-stem reservoir system, so the flood pulse ecological rhythm (Junk et al. 1989) remains close to natural. Aquatic species diversity is high. The Lower White is also the most important winter refuge for waterfowl and neotropical migrant birds in North America and harbors a relict population of black bears. Although the White has several outstanding features, it has been substantially altered with major dams on its main stem and is now seriously threatened by additional human impacts, some of which are enormous.

Several large hydroelectric dams in midreaches of the White River have dramatically changed its ecological characteristics, making it a discontinuous system (Ward and Stanford 1983). The dams reset the river continuum so that the tailwater reaches resemble headwater reaches in several respects. On the other hand, some of the reservoirs mimic large rivers (especially Taneycomo) and others are more like lakes (Bull Shoals, Table Rock). The lentic reservoir habitats resemble lakes in some fundamental respects, but they are deepest near the dams rather than the center and have very poorly developed littoral zones. The reservoirs have an upstream–downstream nutrient gradient from eutrophic upstream to almost oligotrophic near the dams and are warm, monomictic systems (i.e., usually do not freeze over in winter). The algal communities change dramatically along this gradient. Zooplankton species assemblages are dominated by rotifers, with *Keratella cochlearis* and *Polyarthra vulgaris* often most abundant. Chironomids (midges) are the dominant benthic invertebrates in the reservoirs, joined by chaoborids (phantom midges) below the depth of the thermocline.

The reservoirs support moderately productive warmwater fisheries for a variety of centrarchids and ictalurids in the river and nonnative fishes like white bass and striped bass stocked in the reservoirs. Most native warmwater fish species (e.g., largemouth bass, spotted bass, crappie, channel catfish) flourish in the reservoirs, especially with supplemental stockings from the Arkansas Game and Fish Commission. Fisheries for walleye and sauger have been developed but are not as successful. Bluegill, redear, and green sunfish are abundant in the reservoirs also. Gizzard and threadfin shad provide additional "forage fishes" for the larger predatory sport fishes. Lake Taneycomo is small and strongly influenced by cold water released from Table Rock Lake, so it is more riverine and has a fishery that includes stocked trout. Tailwaters of the reservoirs have been stocked with several species of trout and some of the species now have largely self-sustaining populations, whereas others are primarily put-and-take fisheries. Below the dams most of the native invertebrates, including mollusks, have been extirpated, and now amphipods and isopods are abundant. Sculpins and red horse suckers are abundant native fish species.

In contrast to the dams on the Upper White River, the Lower White is dredged and snagged to maintain a 2.4 m deep barge canal to Newport. However, it lacks the revetments, dikes, cutoffs, levees, and locks characteristic of the LMR until it reaches the area subject to flooding from the Mississippi. Presently it retains access to extensive areas of its floodplains, although large areas have been cleared and drained for agriculture, and water flows swiftly to the deepened Mississippi River channel instead of flooding all of the riparian lowlands as frequently as it once did. Some of the cleared bottomland hardwood forests are now being replanted with oaks. However, current plans for deepening and widening the barge canal to 2.7 m × 76 m and for a huge irrigation system (Grand Prairie Area Demonstration Project) would lower water levels in the White and seriously diminish the flood pulse aspect of this large-river ecosystem (Sutton 2000). The proposed barge canal and irrigation project would reduce the habitat available for many of the river's aquatic species (Chordas and Harp 1991, Gordon et al. 1994, Buchanan 1997) and devastate the remaining rich floodplain habitats (Sutton 2000, Wright 2000).

BUFFALO NATIONAL RIVER

The Buffalo National River (BNR) originates in the Boston Mountains a few kilometers east of the origin

of the White River, but it is on the other side of a large dome (Salem Plateau) and flows east-northeast to its confluence with the White (see Figs. 6.10 and 6.12). It begins at an elevation of about 701 m asl and drops 579 m during its 238 km course across the Springfield Plateau into the Salem Plateau. Although it is a relatively small river, about 190 km of the Buffalo's length and 392 km² of its 3465 km² watershed were protected by Congress in 1972 by an act declaring it a National River and establishing the long, narrow park as a National Scenic Riverway. Furthermore, the headwaters area and last 32 km of the river are both owned by the U.S. Forest Service and have been set aside as Wilderness Areas. Thus, the entire length of the river is publicly owned and protected, a rare occurrence in the lower 48 states. Approximately 25 named tributaries enter the river and nearly all of them originate on and flow through private property before entering the narrow park along the river (Fig. 6.4).

The presence of terrace village sites and bluff shelters rich in Native American artifacts reveals that archaic tribes lived in the area as much as 10,000 years ago. More recently, tribes of Quapaw and Osage people inhabited this land, which is rich in fishes, shellfish, edible plants, medicinal plants, and wildlife resources. Presently the permanent human population (7 people/km²) in this rough terrain is probably no higher than it was thousands of years ago. However, the park campgrounds and river attract large numbers of campers, hikers, and canoers in spring and summer (Springer et al. 1977).

Physiography, Climate, and Land Use

The Buffalo National River is in the Boston Mountains section of the Ozark Plateaus (OZ) physiographic province (Hunt 1974) (see Fig. 6.12). The BNR is in an area of karst limestone topography. The limestone is soluble in water, so many fissures, collapsed dolines (sink holes), caves, and springs occur along the river course and throughout the watershed. Bluffs along the river are as high as 152 m. Diverse habitats range from xeric south-facing bluffs, open glades on high rock outcroppings with thin soils, mesic hillside forests, and riparian areas on alluvial floodplains, to damp seeps, caves, and springs along margins of the river. The limestones and associated horizontal strata of dolomites, sandstones, shales, and chert intrusions accumulated here primarily during the Ordovician and Mississippian periods under marginal and shallow marine bays, interrupted by several intervals of uplifts and erosion during the Silurian and

FIGURE 6.4   Buffalo National River, Arkansas (Photo by A. C. Benke).

Devonian. Soils are generally shallow and stony and of medium texture, but are very diverse typologically. Soil pH is patchy and variable (4.2 to 8.6) in the watershed and interacts with slope, aspect, elevation, and moisture to influence distribution of vegetational associations.

The BNR is in the Ozark Mountain Forests terrestrial ecoregion (Ricketts et al. 1999). Primary upland trees include red oak, white oak, post oak, black oak, and hickories, with eastern red cedars in areas with shallow soils and on disturbed sites. Cedar glades occur as patches, especially on south- and west-facing slopes, and harbor interesting plant associations, including diverse herbaceous species and grasses. Shortleaf pine occurs with some of the upland hardwoods and as almost pure stands on some drier sites with thin, acid soils.

Climate of the BNR watershed is a little cooler (annual mean temperature 15°C) and less humid than that of other subbasins of the Lower Mississippi River (Fig. 6.13). July is the warmest month (26°C) and January is the coldest (2°C). Precipitation is rather evenly distributed over the year, with a low of 5.6 cm in January and a high of 12.6 cm in May, and averages just over 100 cm/yr. Most precipitation falls as rain, although several centimeters fall as snow each winter.

Land use in the BNR basin is rapidly changing due to increased timber harvests as trees have matured after intense cutting in the 1930s and due to population increases, partly as a result of attention it received as the first National River. The BNR basin may be about 90% forested, with most of the remainder in pasture (Davis and Bell 1998). However, a study by Hofer et al. (1995) reported much less forest cover (73%) and calculated that 14.8% of the watershed that was forested in 1979 had been converted to pasture and "urban/barren" areas by 1992. There are no truly urban areas along the river now, although during a lead-mining and timber-harvesting era about a century ago as many as 5000 persons lived in Rush near the confluence with the White River. Rush is now a ghost town. Other than loss of forested area in the basin, which is the major cause of decline of environmental quality of streams in the region (Radwell 2000), the Buffalo is threatened by excessive traffic by tourists, like many other national parks.

The rough topography along the Buffalo and some of the other streams has protected them from some anthropogenic disturbances. Timber could not be harvested in the steep valleys by "snaking out" the saw logs with draft animals in the 1890 to 1930 period of intense timber harvest, or now with motorized vehicles, although some logs were cut near the rivers and floated downstream to mills during the earlier logging boom. Cable logging is not done here like it is in areas with larger stands of more valuable timber. The fact that these more pristine streams are topographically different from some others in the Ozark region (Mulberry, Illinois) may make them less suitable as good regional reference streams because their differences may be the result of their topography as much as their low level of disturbance.

Geomorphology, Hydrology, and Chemistry

The Buffalo River, although fairly steep (2.4 m/km), has abundant chert (a form of flint) gravel that was embedded in the limestone that has been dissolved away. The substrate is coarse, ranging from exposed bedrock to granule-size gravel (Wentworth 1922), with limited sands and fine sediments. The moderately steep slopes, abundant gravel, and stream power during floods (which are occasionally huge) result in distinct riffle–pool channel structure that largely conforms to the riffle interval (5 to 7 stream widths) described by Leopold et al. (1964). However, many Buffalo River meanders encounter immense limestone bluffs that do not yield to its currents and dictate the stream's geomorphology.

Hydrologically, the BNR is quite flashy, with a mean discharge of approximately 48 m^3/s at its mouth. It has a hydrologic regime typical of streams that are dominated by surface runoff, with little groundwater storage capacity. Actually, rainfall does not travel far before finding passageways to the water table in the karst terrain, but it resurges just as rapidly as springs, often with little filtration through the thin soils of the area. Low flows of around 1.8 m^3/s occur about every two years and typically last for 30 days or longer, and bankfull flows of 1275 m^3/s have a similar return interval (<2.5 yr). Floods are not very predictable but do occur more commonly during January to May, when soils are nearer saturation. However, the runoff pattern is not the result of snowmelt as is typical of streams in more northern climates. Somewhat lower rainfall, rapid runoff during rainstorms, and high evapotranspiration rates during summer and early fall cause the 1st order streams in the basin to dry completely and the river to stand in isolated pools through the 3rd order reaches during this time. Thus, mean monthly runoff is highest from March through May (>6 cm/mo) and lowest during summer (<1 cm/mo) (see Fig. 6.13).

Water chemistry of the BNR is strongly influenced by the lithology of the basin. Concentrations of calcium, magnesium, and sulfate all increase downstream during base flow conditions. The calcium carbonate content of the limestone buffers the pH of the acid rain (~4.4 to 4.5) that enters the Buffalo, so pH is normally between 7 and 8 and rarely below 7 except following rainstorms. Inorganic plant nutrients are generally low (NO_3-N = 0.14 mg/L, PO_4-P = 0.06 mg/L), as are coliform bacteria, with very few exceptions (Babcock and MacDonald 1973). Coliforms and nutrients are higher downstream, which may result from increased agricultural and domestic sewage (septic systems) nonpoint-source inputs. The annual mean percentage saturation of dissolved oxygen actually exceeds 100% due to supersaturation that is frequently more than 110% in large, quiet pools (Meyer and Woomer 1978). Dissolved organic carbon (<1 mg/L) is generally low and the stream is exceptionally clear (<5 NTU).

River Biodiversity and Ecology

The BNR is probably the least disturbed major tributary within the Ozark Highlands freshwater ecoregion (Abell et al. 2000). Since designation of the Buffalo as a National River and establishment of the national park there have been many surveys of its physical, chemical, and biological properties but very few of the results are published in peer-reviewed scientific journals (but contact the Arkansas Water

Resources Center at the University of Arkansas, Fayetteville). Few studies have been performed in the Buffalo that address ecological interactions and processes (trophic dynamics, energetics, nutrient cycling), but these can be inferred from studies of other streams in the region. Recent unpublished studies have focused on impacts of summer intermittent conditions on ecological interactions in the isolated pools and how they affect fish and meiofauna species assemblages.

Algae, Cyanobacteria, and Protists

Over 270 algal taxa are recorded from the BNR (Springer et al. 1977). Diatoms are the major algae, with various species of *Cymbella, Achnanthes, Navicula, Coconeis,* and *Nitschia* most abundant. However, diatoms peak in late winter to early spring as silica levels are depleted, and they are replaced in summer by a broad spectrum of algae, especially greens and cyanobacteria. *Cladophora, Spirogyra,* and the macroalga *Chara* form extensive patches in summer and autumn, with cyanobacteria (*Anabeana, Lyngbya, Oscillatoria, Synechococcus*) becoming principal components of epilithic biofilms (Rippey and Meyer 1975). Flagellates including *Euglena, Dinobryon, Synura, Pandorina,* and *Trachelomonas* are abundant in languid pools, especially those rich with organic matter. Periodic spates reset or interrupt this seasonal succession. Plankton are rare in the Buffalo, but include *Pediastrum, Staurastrum, Ceratium, Stephanodiscus,* and *Pandorina.* Desmids (e.g., *Cosmarium, Closterium*) are also common, as are epiphytes (e.g., *Gomphonema, Coconeis*) of water willow and some of the larger filamentous algae (e.g., *Cladophora*).

Plants

Water willow is an abundant aquatic macrophyte in the Buffalo River. Although it does not appear to be grazed by terrestrial or aquatic herbivores, it is ecologically significant in other ways. For example, it provides surface for attachment of epiphytic algae that are grazed by aquatic snails, including *Elimia potosiensis.* The rhizomes and roots stabilize substrates. The physical structure of the plants, which occur in dense stands, provide preferred habitat for many macroinvertebrate and fish taxa, both predators and prey. The dead shoots and leaves of water willow contribute detritus.

Riparian forests are not as distinct and extensive along the Buffalo as they are along streams with less slope and more extensive floodplain development. The ephemeral to intermittent flow patterns of head-

water reaches indicate the dry conditions that occur along the Buffalo, especially on steep south-facing slopes. Gravel bars are colonized by black willow, Ward's willow, witchhazel, and, nearer the water, smartweed, water willow, and rushes. Stream sides are lined with sycamore, river birch, silver maple, cottonwood, and buttonbush (Dale 1973). Large stands of giant river cane dominate the herb layer in some places. Floodplains may have some of these same plants but also green ash, box elder, and sweetgum. Nonnative plants include Japanese honeysuckle, multiflora rose, loblolly pine, mimosa, and fescue, but none are very abundant near the river.

Invertebrates

Macroinvertebrates are not very abundant in the BNR but are just as diverse as other tributaries of the White River. Mayflies are the most abundant and diverse insect taxon throughout the length of the Buffalo. Common genera of mayflies include *Isonychia, Baetis, Heptagenia, Centroptilum, Stenonema, Caenis, Ephemerella,* and *Ephoron.* Stoneflies (e.g., *Acroneuria, Neoperla, Perlesta, Taeniopteryx*) and caddisflies (especially *Cheumatopsyche,* but also *Hydropsyche, Polycentropus, Chimarra*) are well represented. Beetles, especially genera like *Lutrochus, Psephenus, Stenelmis,* and *Helicus,* are common and occasionally codominant with the mayflies at a site. True flies such as chironomid midges do not appear as abundant in the BNR as in other rivers. Dragonflies (*Hagenius, Macromia, Neurocordulia, Gomphus*) and damselflies (*Enallagma, Hetaerina, Argia, Lestes*) are diverse but not numerous. Hellgrammites (especially *Corydalus* and *Chauliodes*) are common but in low densities.

Crayfish species (*Orconectes* spp.) are an important invertebrate in the Buffalo, as they are in other Ozark streams (Rabeni et al. 1995). In general, Ozark Highlands streams are remarkable for the diversity and uniqueness of their crayfishes (Abell et al. 2000). Crayfishes are opportunistic feeders that generally prefer macroinvertebrates for food. Mollusca are also common in the Buffalo but are not as numerous as in the Lower White (Gordon 1982). Snails (*Elimia, Physella*) are important grazers. The nonnative Asiatic clam has become abundant in the BNR, as in other rivers of the LMR region. Standing crop biomass of all benthic macroinvertebrates is generally low, but to our knowledge no estimates of invertebrate secondary production have been performed in the BNR.

The meiofauna species assemblage of the BNR is composed primarily of copepods, cladocerans, ostracods, and mites (Hydrachnidia). In other streams of the region the dominant meiofauna are rotifers and tiny chironomids, but in the BNR they are minor components. When the headwater pools cease to flow and become isolated, resulting in lentic habitats during summer and fall, two meiofauna species, *Bosmina longirostris* and *Keratella cochlearis*, rapidly proliferate. These two species commonly account for ~90% of the density of planktonic meiofauna in isolated BNR pools.

Vertebrates

Over 66 species of fishes occur in the Buffalo National River (Becker and Kilambi 1975, Kilambi and Becker 1977). The most diverse and abundant are Percidae (mostly darters) and Cyprinidae (minnows), with the darters more numerous in riffles and minnows more abundant in pools. Centrarchids (e.g., longear sunfish, smallmouth bass, Ozark bass) and ictalurids (Ozark madtom and slender madtom) are also common. The Buffalo is an excellent smallmouth bass stream, but fishermen who use small lures or baits must catch far more of the colorful longear sunfish than smallmouth bass because they outnumber them more than 10 to 1. The most abundant cyprinids are the duskystripe shiner, telescope shiner, rosyface shiner, and central and largescale stonerollers. Abundant darters include the yoke darter, rainbow darter, Arkansas saddled darter, and banded darter. Banded sculpin are also fairly abundant.

The Buffalo River has a moderate diversity of amphibians and reptiles, with several snakes (western cottonmouth, midland water snake), turtles (softshell turtle, map turtle, snapping turtle), and frogs, but none are very abundant. Common mammals include beaver, muskrat, raccoon, and mink. River otters are present but not abundant. Great blue herons and belted kingfishers are common year-round, and a few bald eagles can be seen there in winter, as is true of all Ozark rivers.

Ecosystem Processes

Although the physical, chemical, and biological elements of the Buffalo National River have been fairly extensively surveyed and reported in governmental agency reports and publications, almost no studies of ecosystem processes (primary/secondary production, metabolism, food web analyses, trophic dynamics, etc.) have been performed. These can only be inferred from the survey data, studies in the other tributaries of the White River, and studies of other streams in the Ozark Plateaus. Large woody debris is scarce in the Buffalo, as in most Ozark streams. Several factors cause this: removal of snags by canoe rental agencies and government employees, rapid decay of wood in this area, and removal to the floodplains by large floods. Flashy hydrology and lack of retention result in most litterfall being flushed out of the stream channel to the floodplains. Dissolved organic matter is also low (<1 mg/L). Primary production is limited by low nutrient concentrations. Thus, low organic matter resources limit secondary production. From perusal of invertebrate species abundances, very few shredders are present and grazers are most abundant. Grazing fishes (central and largescale stonerollers) are also numerous, but most fishes are invertivorous (darters and small cyprinids) or piscivorous (bass). Thus the river is probably an autotrophic system despite being in a forested area. The clear water and clean gravel of the river are indicative of its trophic status.

Human Impacts and Special Features

The Buffalo National River is one of only 42 natural free-flowing rivers >200 km long in the conterminous 48 United States (Benke 1990). Water quality in the Buffalo has been studied fairly extensively and is exceptionally good. The Buffalo and several of the other streams flow down the sides of the Ozark dome from which they originate (Salem Plateau) in a radial pattern (Big Piney Creek, Mulberry River, Crooked Creek, Illinois Bayou, War Eagle Creek). They are considered to be among the best examples in the United States of high-quality reference streams (Omernik 1995, Radwell 2000). Water quality of rivers in this region is closely tied to the percentage of forested area in the watershed (Radwell 2000). This is of particular concern to managers of the BNR, as most of the watershed is in private ownership despite the fact that the full length of the waterway and riparian zone of the main-stem river is in public ownership. Several studies of land use in the basin have been performed (Scott and Smith 1994, Hofer et al. 1995, Davis and Bell 1998), and although they do not agree on the details it is apparent that the watershed is being converted from forest to pasture. The current outbreak of red oak borers resulting from several very dry summers that stressed the trees is adding to the problem.

Although the BNR is an uncommon river in a scenic and relatively undeveloped watershed ecosystem, development of the basin is occurring fairly rapidly. People who live in the watershed do not want to leave it, yet require income for their families. Thus, many cut trees, clear pastures for livestock, build cabins to rent to tourists, rent canoes, or live and work in the basin as employees of the National Park Service. All of this contributes to declining environmental quality in the basin. Ironically, part of the environmental degradation is the result of actions taken to protect it. However, the Buffalo would have been dammed by the USACE, like most other tributaries to the White River, at Gilbert and/or Lone Rock 40 to 50 years ago except for strong opposition from the National Park Service, Ozark Society, Arkansas Audubon Society, and The Nature Conservancy (Albright 1957–1993). A recent controversy concerns a proposed reservoir on Bear Creek, one of the major tributaries to the BNR. In August 2002 the USACE granted a permit to build a drinking-water reservoir on Bear Creek, one of the major tributaries of the Buffalo, in spite of opposition from the National Park Service, U.S. Fish and Wildlife Service, and many conservation organizations, in apparent violation of federal law.

BIG BLACK RIVER

The Big Black River is the least disturbed river in the Delta region of western Mississippi and is a prime candidate for inclusion in Mississippi's natural and scenic waterways system (Mareska and Jackson 2002). It originates in the hills of north-central Mississippi, flows southwesterly across the state for 434 km, and discharges directly into the Mississippi River 43 km downstream from Vicksburg, draining a watershed area of approximately 8770 km^2 (Fig. 6.14). There are no dams on the main stem of the river, but there are small impoundments on some of the tributary streams. Channel modification for navigation and flood control has been minimal, and there has been virtually no maintenance on these public works projects since 1955 (U.S. Army Corps of Engineers 1964). Recovery processes (Jackson 2000) are advanced for the Big Black River.

During the Civil War, gunboat battles raged on the Big Black River. It is still possible to find artifacts from the period following flood-induced scouring of the stream channel. As a consequence of logjams, fallen trees, and snags (large woody debris), boat navigation now is restricted to small craft, and often even these (excepting canoes) cannot travel along the system unless a path is cleared through debris with a chain saw.

Physiology, Climate, and Land Use

The Big Black River drops 152 m as it courses through the Coastal Plain (CP) physiographic province, beginning in the Eastern Gulf Coastal Plain section and then moving into the Mississippi River Alluvial Plain section near its mouth (see Fig. 6.14). It drains an area representing the westernmost portion of the Southeastern Mixed Forests terrestrial ecoregion, whose natural vegetation is characterized by oak–hickory–pine forests (Ricketts et al. 1999). Higher elevations within the watershed are mixtures of longleaf pine, oaks, hickories, elms, Eastern red cedar, and herbaceous vegetation. Pastureland in the watershed typically is a mixture of broomsedge, nonnative fescue, and Dallas grass. Terrestrial ecotones in the watershed support Osage orange, black locust, honey locust, persimmon, blackberry, honey suckle, and the nonnative privet and kudzu.

Rainfall in the Big Black River watershed averages around 135 cm/yr, with highest rainfall typically 12 to 14 cm/mo during winter/spring and 8 to 10 cm/mo during summer (Fig. 6.15). Mean annual air temperature is approximately 18°C, with temperatures frequently exceeding 30°C during summer and occasionally falling below −5°C during winter. The wet temperate climate coupled with a long growing season encourages rapid regeneration of forests (typically <20 years) along upper sandbar terraces of riparian zones and throughout the watershed.

The watershed of the Big Black River is sparsely populated (<25 people/km^2) and is a mixture of forest (~54%), row-crop agriculture (~35%), and pastureland (~11%) (Insaurralde 1992). Most forested lands are natural mixes of hardwoods and pines, but planted pine forests are common. These landscape characteristics have changed little since the 1960s (Holman et al. 1993).

Geomorphology, Hydrology, and Chemistry

The Big Black River is a low-gradient stream with a slope ranging from approximately 50 cm/km in its upper reaches to approximately 20 cm/km in downstream reaches. Channel width in the upper reaches near Kilmichael, Mississippi, is approximately 30 m, whereas in the lower reaches near Bovina, Mississippi, the width increases to around 80 m (Holman et al. 1993). The river is a mosaic of shallow runs, deep holes, steep cut banks, and expansive sand bars, lined by riparian forest comprised primarily of bottomland hardwood assemblages (Fig. 6.5). Highly erodible soils comprised primarily of sand, silt, and

FIGURE 6.5 Big Black River, Mississippi (Photo by D. Jackson).

clays dominate the Big Black River watershed. Hard rock features are rare or absent, and when present are typically in the form of gravel deposits. Scour and fill processes result in lateral channel movements that locally can exceed several meters a year.

Mean discharge in the Big Black River is 107 m³/s based on a 66-yr record (1936 to 2002) at the USGS gaging site near Bovina, Mississippi. However, extremes in local and regional weather conditions can lead to extensive flooding as well as severely reduced streamflows. For example, maximum and minimum flows recorded for the Bovina station are approximately 2600 m³/s and 1.5 m³/s, respectively (Mississippi Department of Environmental Quality 1992). Holman et al. (1993) reported that highest discharges occur from February through April (>200 m³/s), and that lowest discharges are from August through October (15 to 41 m³/s). Monthly runoff ranges from an average of 0.61 cm/mo in September to 8.07 cm/mo in March (see Fig. 6.15) and is influenced by seasonal differences in evapotranspiration and rainfall patterns.

Water quality (U.S. Geological Survey 2000b, 2001) for the Big Black River is considered good. Suspended sediment varies seasonally but averages around 135 mg/L. Dissolved oxygen concentration typically exceeds 5.0 mg/L regardless of season, and pH is fairly stable, ranging between 6.8 and 7.2. Total alkalinity and hardness both are around 50 mg/L as $CaCO_3$. Nitrate-nitrogen typically is <0.3 mg/L and reflects a watershed primarily devoted to forestry and wildlife management rather than intensive agriculture. Mean concentrations for PO_4-P are somewhat elevated, ranging from 0.18 mg/L during the dry period to 0.31 mg/L during the wet season (Holman et al. 1993).

River Biodiversity and Ecology

The Big Black River is a relatively intact, functional floodplain river ecosystem within the Mississippi Embayment freshwater ecoregion (Abell et al. 2000). Although there has been research on invertebrates and fishes, the river has not been intensively studied, particularly for ecosystem processes.

Plants

As a result of variable flow regimes and fairly high turbidity, aquatic vegetation is not common in littoral zones and lower elevations of sand and mud bars along the Big Black River. In the higher elevations of riparian zones on the inside of meanders there are assemblages of American sycamore, red maple, river birch, black willow, green ash, and cottonwood. Beyond these zones, where the bars make the transition to *terra firma*, and along the outside of channel meanders, where current erodes the stream bank, there are hardwood forests dominated by aggregates of various oaks, bald cypress, and, on floodplain terrace ridges, black walnut.

Invertebrates

Invertebrates in the Big Black River are best known from drift net studies by Insaurralde (1992). Major taxa were the mayflies *Baetis*, *Cinygmula*, and *Tricorythodes*; the dragonfly *Gomphus*; the caddisflies *Hydropsyche*, *Cheumatopsyche*, and *Nectopsyche*; the beetles *Ancyronyx* and *Stenelmis*; the hellgrammites *Corydalus* and *Chauliodes*; various chironomid midges; and the blackfly *Prosimulium*. Collectively these insects contributed nearly 60% by weight to the overall drift net catches. Caddisflies originated mostly from snags, whereas the chironomids originated mostly from stream bottom substrates (Insaurralde 1992). Hartfield and Rummel (1985) documented 34 species of mussels from the Big Black River but in 2002, 38 species were noted (Mississippi Department of Wildlife, Fisheries, and Parks 2002). Some common mussels are the mucket, rock pocketbook, butterfly, southern pocketbook, and southern mapleleaf. The Big Black River has two species of mussels that are on Mississippi's endangered species list: the pyramid pigtoe and the rabbitsfoot. Archaeological remains of sheepnose and western fanshell exist in the river, but currently there are none known to be alive in the system (Mississippi Department of Wildlife, Fisheries, and Parks 2002).

Vertebrates

Ross (2001) provides a comprehensive list of 112 fish species from the Big Black River. Principal larger fishes are flathead catfish, blue catfish, channel catfish, smallmouth buffalo, bigmouth buffalo, black buffalo, freshwater drum, white crappie, black crappie, largemouth bass, gizzard shad, bluegill, longnose gar, spotted gar, blue sucker, and paddlefish (Holman et al. 1993). Flathead catfish is the most important fishery resource in the river, although productive fisheries also exist for blue catfish, channel catfish, and, to a lesser extent, sunfishes (Centrarchidae). Principal smaller fishes are blacktail shiner, emerald shiner, striped shiner, creek chubsucker,

freckled madtom, blackspotted topminnow, central stoneroller, scaly sand darter, slough darter, logperch, and dusky darter (Holman et al. 1993).

Other vertebrates associated with aquatic environments of the Big Black River include mammals (beaver, river otter, raccoon, mink), reptiles (western cottonmouth, common snapping turtle, alligator snapping turtle, slider, Mississippi map turtle), and birds (great blue heron, kingfisher, wood duck, and, seasonally, various species of migrating waterfowl, particularly mallard). Beaver, otter, raccoon, and mink support local trapping, with a major fur buyer and distributor located in nearby Kosciusko, Mississippi. Migratory waterfowl provide the foundation for seasonal hunting in the numerous hunting clubs that are operative along the river. Riparian zones and floodplains harbor high densities of whitetail deer, wild turkey, gray squirrel, fox squirrel, gray fox, coyote, armadillo, cottontail, swamp rabbit, and numerous raptors and neotropical migrant birds.

Ecosystem Processes

Little research has been done on ecosystem processes in the Big Black River, and the description here is based to a large extent on field observations and the literature from other river–floodplain systems, including studies in the Mississippi River basin. Heterotrophic processes are the principal bioenergetic driving forces for floodplain river ecosystems in Mississippi. Allochthonous organic materials can enter the stream directly from riparian zones or can blow or wash in from more remote locations in the watershed. Floods set into motion incorporation of extrachannel allochthonous organic materials (Junk et al. 1989, Bayley 1995, Thorp and Delong 1994, Sparks 1995). Along the transitional front separating the aquatic environment from the terrestrial environment, rapid nutrient exchange occurs (Bayley 1989). These nutrients leach from allochthonous organic materials and organic components of soils and are utilized by microbes (Fisher and Likens 1973, Anderson and Sedell 1979). These processes, in conjunction with the (seasonally) warmer, clearer, shallow water of the floodplain relative to main-stream channels, encourage production of plankton and other microflora and microfauna on the floodplain that are used as forage by benthic macroinvertebrates and early life history stages of fishes. Coarse particulate organic matter (CPOM) remaining after nutrient leaching can serve as invertebrate habitat (Brown and Matthews 1995). As this CPOM is further colonized by microbes and broken into smaller particles, it is consumed by macroinvertebrates (Cummins 1974, Marzolf 1978) that in turn become potential forage items for fishes. Ultimately, the dynamics of this secondary production are expressed most vividly in terms of the river's excellent fisheries, and particularly those targeting catfishes (Ictaluridae) (Jackson and Francis 1993). Low flow periods during summer and autumn concentrate aquatic biota within the stream channel and in backwater seasonal lentic environments (e.g., ephemeral pools, sloughs, oxbow lakes).

Flooding also introduces snags (large woody debris) into the main channel of a floodplain river ecosystem. In streams such as the Big Black River that have fine-grained substrates, snags provide important habitat and refuge for fishes and attachment sites for macroinvertebrates (Gorman and Karr 1978, Benke et al. 1985, Insaurralde 1992, Skains 1992, Brown and Matthews 1995).

The general well-being of the Big Black River as a floodplain river ecosystem fishery is best reflected by the stock structure and population dynamics of flathead catfish, the top predator within the Big Black River fish assemblage (Jackson 1999). Using low-altitude (300 m) aerial photography, Insaurralde (1992) characterized Big Black River riparian zones and estimated the abundance of snags in the stream channel. Then he related these environmental characteristics to flathead catfish biology. He found that the abundance of flathead catfish in the Big Black River was directly related to the proportion of the riparian zone in mature forest ($R^2 = 0.77$), and the proportion of the flathead catfish population >41 cm was related to the number of snags/km ($R^2 = 0.61$).

Human Impacts and Special Features

Like the Buffalo National River, the Big Black River is one of only 42 natural free-flowing rivers >200 km in length in the lower 48 states (Benke 1990). Its significance is further highlighted in this regard in that it is a functioning floodplain river ecosystem in a fairly stable, primarily forested watershed. Most floodplain rivers in the United States have been ravaged by flood-control programs, including channelization, dredging, clearing of riparian zones, removal of large woody debris, damming, or combinations of these impacts. With the exception of the river's lowermost reaches near Vicksburg, Mississippi, the Big Black River has been spared these insults and subsequently deserves recognition as a state and national treasure. It is replete with aesthetic beauty as it meanders through its floodplain and associated bottomland forests. Its natural state

encourages fisheries and wildlife resources, and subsequently human interactions with the river and these resources. As of 2003 there were no fish consumption human health advisories for the Big Black River.

During low flow periods (May to October) the Big Black River is an extremely popular fishing destination. Recreational anglers, particularly those targeting catfishes, invest considerable effort in fishing the system, primarily with multihook longlines baited and deployed in deeper pools and left to fish passively overnight. In addition, during May, June, and July there is a special season for recreational hand fishing. The Big Black River is noted by hand fishers for its exceptionally large flathead catfish (commonly >25 kg). In this fishery, the fisher enters the water, feels under the water with hands or feet for cavities under the stream bank or within logjams or root wads, and upon encountering a fish grabs it by the jaw (often the fish grabs the fisher) and attempts to wrestle it either into a waiting boat or onto the stream bank (Francis 1993).

Wildlife considered game species are subject to hunting by local human populations. This hunting is generally within the constraints of established clubs with exclusive access to private property. Current trends in land purchase and development throughout the watershed, primarily for forestry, wildlife management, and hunting purposes, should maintain the integrity of the Big Black River ecosystem and its various tangible and intangible resources.

YAZOO RIVER

The Yazoo River is a large floodplain river ecosystem that encompasses approximately 35,000 km² in north-central and northwestern Mississippi (Fig. 6.16). It has six principal tributary rivers: the Coldwater, Sunflower, Tallahatchie (including the Little Tallahatchie), Yocona, Skuna, and Yalobusha rivers. All but the Sunflower River originate in the hill country of northern Mississippi and flow generally west and southwest, converging in the Mississippi River Alluvial Plain (or Delta) to form the main stem of the Yazoo River near Greenwood, Mississippi. The Sunflower River originates within the Delta, west of Greenwood, courses primarily north to south, and enters the Yazoo River approximately 50 km from the Yazoo River's confluence with the Mississippi River near Vicksburg.

For the last 3500 years the Yazoo River system has been used by at least four cultures. Mound Builders were the first known inhabitants of the region, arriving around 3500 years ago (Smith 1954). By the late seventeenth century they were decimated by diseases transported into the region by the exploration party of Hernando de Soto in 1541. Choctaw and Chickasaw tribes settled the region during the eighteenth century and engaged in agriculture, hunting, and fishing. Native American tribes were driven out of the region by European-American settlers via several U.S. government treaties during the early nineteenth century. These settlers converted large tracts of land from bottomland hardwood forest to agricultural land conducive for row-crop production of cotton but had to contend with expansive regionwide flooding for nearly half of each year. Some built private levees to protect their lands from flooding, but this tended to generate problems with neighbors downstream and frequently stimulated hostility.

After the great flood of 1927 the U.S. Congress authorized massive flood-control projects throughout the Yazoo River system. Subsequently, during the middle of the twentieth century all of the principal tributaries of the Yazoo River with the exception of the Sunflower River (Coldwater, Little Tallahatchie, Yocona, Yalobusha) were dredged, cleared, snagged, and/or channelized and had upstream flood-control dams constructed, generally near the boundary separating the eastern hill country from the Delta (Jackson and Jackson 1989, Jackson et al. 1993).

Physiography, Climate, and Land Use

The Yazoo River system is contained entirely within the state of Mississippi and is positioned in the Coastal Plain (CP) physiographic province (see Fig. 6.16). The upper reaches of principal tributary streams are in uplands of the Eastern Gulf Coastal Plain section of the province. The lower reaches of the system extend into and across the Mississippi Alluvial Plain section. Upper and eastern tributaries drain portions of the Central U.S. Hardwood Forests (native oak and hickory) and Southeastern Mixed Forests (native oak, hickory, and pine) terrestrial ecoregions, respectively (Ricketts et al. 1999). However, most of the main stem, including the main stem of the Tallahatchie, and all of the western tributaries drain from the Mississippi Lowlands Forests ecoregion. Natural vegetation of the Mississippi lowlands once also included oak–hickory–pine forests and various other bottomland hardwood species, but only remnants of native forest remain, most having been converted to agriculture.

The prevailing climate is one of hot, humid summers and cool, moist winters (Ye 1996). Heaviest rainfall typically occurs during winter and spring, with average precipitation from December through June of approximately 12.5 cm/mo (Fig. 6.17). The driest period is August through October, with average precipitation of approximately 7.6 cm/mo. Normal summer temperatures can exceed 27°C and normal winter temperatures can be below 7°C.

Land use in the upper reaches of tributary streams is primarily forestry (including U.S. National Forests) and diversified agriculture (including livestock grazing). In the lower reaches, row-crop agriculture prevails, although throughout the system there are still large tracts devoted to timber production and wildlife management. Insaurralde (1992) utilized satellite imagery (LANDSAT) and high-altitude aerial photography (NAP) to characterize the landscape of the Tallahatchie River's lower reaches (typical of the Yazoo River basin) and reported that approximately 80% was in row-crop agriculture (primarily cotton and soybeans) and approximately 15% was bottomland hardwood forest. The remainder was pasture and small tracts of pine or mixed pine–hardwood forestland. Overall, approximately 60% of the Yazoo River basin is devoted to agriculture (U.S. Geological Survey 2002).

The entire Yazoo River basin is a major wintering area for migrating waterfowl. The international significance of the region in this regard has resulted in the establishment of numerous state wildlife management areas and federal wildlife refuges throughout the Yazoo River basin, primarily dedicated to waterfowl. Large tracts of private lands also are being converted to these purposes (Zekor and Kaminski 1987, Uihlein 2000).

Geomorphology, Hydrology, and Chemistry

Stream reaches upstream from Yazoo's flood-control reservoirs course through relatively soft substrates comprised of clay, sand, and silt mixed with occasional gravel deposits. Deforestation coupled with poor agriculture practices during the 1800s and 1900s resulted in massive regionwide land erosion that ultimately forced large numbers of people to leave the area. In addition to rendering much of the land uninhabitable and unsuitable for farming, this erosion filled the tributary streams and main channels of the rivers with sediment. Through reforestation programs (including establishment and development of national forests throughout northern Mississippi) and implementation of soil conservation

FIGURE 6.6 Yazoo River, Mississippi (Photo by D. Jackson).

practices these destructive fluvial processes were substantially reduced by the 1980s. In this last regard, the USDA Sedimentation Laboratory (Oxford, Mississippi) has been particularly instrumental through development and implementation of stream rehabilitation programs throughout the region (e.g., Shields et al. 1998). Rehabilitated streams in these upper reaches of the Yazoo River basin have relatively stable stream banks, heterogeneous instream habitat, clear water (relative to most streams in Mississippi), and recovering fish assemblages, as evidenced by development of local recreational fisheries targeting spotted bass. These upland tributaries typically have bankfull widths of <20 m and, at minimum flows, average depths of <2 m.

Jackson et al. (1993) described the riverine environments in the lower reaches of the Yazoo River's principal tributary streams downstream from the flood-control dams. Generally, at bankfull flows (common in winter and spring) the main channels are <50 m wide, with depths of <10 m. Under low flow conditions (summer and autumn) widths and depths can be considerably less. The Yazoo River itself (Fig. 6.6) has somewhat greater bankfull widths (75 to 175 m), with depths occasionally >20 m. Even under low flow regimes, the main channel of the Yazoo River typically can accommodate shallow draft barge navigation as far upstream as Greenwood, Mississippi. Historically, however, steamboats traveled upstream as far as the current Sardis Dam (near Oxford, Mississippi) on the Little Tallahatchie River, and as far as Avalon, Mississippi, on the Yalobusha River.

The rivers in the Yazoo River system are usually very turbid, particularly in their lower reaches, with Secchi transparencies commonly <10 cm. Substrates are comprised of sand, loamy silt, and clay, with occasional patches of small gravel. Large woody debris originating from riparian zones is common as isolated snags or more complex logjams. Most stream reaches are lined by riparian forest, but these forests can be reduced to narrow strips of trees as a result of land clearing for agricultural purposes. Under high flow regimes this practice commonly results in bank collapse, channel meandering, loss of cropland, and introductions of sediment, pesticides, and large woody debris into the river channels.

The design discharge from the region's flood-control projects was established at 566 m³/s for the Yazoo River at Greenwood, Mississippi. The USGS gaging site at Greenwood provides a long-term mean annual discharge of 296 m³/s. In order to protect farmland from flooding, the dams are operated to ensure that the discharge does not exceed 312 m³/s during the crop season (U.S. Army Corps of Engineers 1991). Mean monthly runoff and flooding are highest in winter and spring and are typically minimal during summer and autumn, as would be expected without dam regulation (see Fig. 6.17). However, the dams reduce peak flooding. Discharge from the flood-control dams is determined by rule curves regulating reservoir pool levels to ensure ample flood storage capacity while maintaining secondary socioeconomic benefits associated with recreational uses of the reservoirs (e.g., fishing and boating). Instream flow management for purposes other than flood control have for all practical purposes been ignored, although Ye (1996), Cloutman (1997), Cloutman et al. (1999), and Jackson and Ye (2000) have clearly demonstrated relationships between stream hydrology and principal fishery resources (catfishes and buffalo fishes) and angling in the lower reaches of the Yazoo's principal tributary streams. Additional relationships (impacts) are probable with regard to aquatic fauna throughout the ecosystem, because under current management flood pulses are dampened (but prolonged) during winter and spring and stream flows downstream from the dams during summer and autumn often are greater than during these seasons prior to the dams.

In the extreme lower reaches of the Yazoo River, at its junction with Steele Bayou just upstream from Vicksburg, Mississippi, the USGS (2002) reported mean monthly discharge ranged from 294 to 869 m³/s from 1996 to 2000, with an annual mean discharge of 522 m³/s. During this period discharge from the Yazoo River represented 2.8% of the Mississippi River flow at Vicksburg.

Water-quality data for the Yazoo River system (U.S. Geological Survey 2000b, 2001, 2002) reflect land use patterns (primarily agricultural enterprise) coupled with the dynamics of scour and fill processes through deep alluvial deposits. These deposits are nutrient rich and typically slightly acidic to somewhat basic (pH 6.7 to 7.7). Application of agricultural fertilizer is considered the primary source for elevated levels of NO_3-N (range 0.05 to 1.2 mg/L; mean 0.20 mg/L) and PO_4-P (range 0.11 to 0.94 mg/L; mean 0.29 mg/L) in the rivers. Suspended sediments can be high seasonally (range 52 to 468 mg/L;

mean 168.4 mg/L), and usually can be traced to sheet flow across unprotected agricultural lands. Dissolved oxygen typically is sufficient for warmwater stream fishes (range 4.2 to 11.3 mg/L; mean 7.7 mg/L). Water temperature averages 21.5°C but can range from 0°C during winter to nearly 30°C during summer.

River Biodiversity and Ecology

The Yazoo River is a highly modified floodplain river ecosystem (Jackson and Ye 2000) located within the Mississippi Embayment freshwater ecoregion. Broad-scale ecological studies of the river are sparse, but the fish community is well described, and there have been several studies of the fisheries in the basin.

Plants

Most of the floodplains of the Yazoo River system are converted agricultural lands. Remnant bottomland hardwood forests are dominated by various oaks, bald cypress, black walnut, and pecan. Along recovering riparian corridors, black willow, cottonwood, American sycamore, green ash, red maple, and river birch prevail. Mature riparian zones are dominated by oaks, water tupelo, and bald cypress. Aquatic vegetation within and along main-stream channels is absent or rare as a result of variable flow regimes. However, during drought conditions when low stream flows are maintained for extended periods, exposed mud bars can develop thick stands of herbaceous vegetation. Backwater environments (e.g., sloughs, swamps, oxbow lakes) with low turbidity frequently develop assemblages of aquatic vegetation comprised of various combinations of water primrose, spikerush, alligatorweed, water shield, cattail, American lotus, pondweed, and chara.

Invertebrates

Large woody debris in the Yazoo River system provides important attachment sites for benthic macroinvertebrates, primarily caddisflies and mayflies, whereas other macroinvertebrates, such as chironomid midges, tend to be associated with bottom sediments (Insaurralde 1992). Major taxa were the caddisflies *Hydropsyche*, *Cheumatopsyche*, and *Nectopsyche*; the mayflies *Baetis*, *Caenis*, *Cinygmula*, *Stenonema*, and *Tricorythodes*; the true flies *Chaoborus*, *Prosimulium*, and *Simulium*; and the hellgrammites *Corydalus* and *Chauliodes*.

Collectively, throughout the Yazoo River system there are 44 species of mussels of which 13 species are imperiled (threatened or endangered), primarily as a result of flood-control programs coupled with poor agricultural practices (Mississippi Department of Wildlife, Fisheries, and Parks 2002). The Sunflower River has the only known populations of the sheepnose and the muckett in Mississippi (Mississippi Department of Wildlife, Fisheries, and Parks 2002). Other common mussels in the Yazoo River system are the rabbitsfoot and the pyramid pigtoe.

Vertebrates

The Yazoo River and its principal tributaries are inhabited primarily by fishes best described as fluvial habitat generalists (Cloutman 1997). A comprehensive list of 119 fish species in the Yazoo is provided by Ross (2001). Catfishes (Ictaluridae), suckers (Catostomidae), and gars (Lepisosteidae) are the dominant large species in lotic environments of the Yazoo River system. Sunfishes (Centrarchidae), gars, and suckers dominate the system's backwater lentic environments. Minnows (Cyprinidae), small centrarchid sunfishes, topminnows (Cyprinodontidae), madtoms (Ictaluridae), mosquitofishes (Poeciliidae), and darters (Percidae) dominate the small species assemblages.

Dominant catfishes are channel catfish, flathead catfish, and blue catfish, all of which support important recreational, subsistence, and small-scale commercial fisheries. Smallmouth buffalo, bigmouth buffalo, longnose gar, and spotted gar also contribute to small-scale commercial fisheries. Principal smaller fishes are blacktail shiner, emerald shiner, bluntface shiner, bullhead minnow, freckled madtom, and orangespotted sunfish (Cloutman et al. 1999).

Common aquatic/semiaquatic mammals are beaver, river otter, mink, and raccoon. Principal amphibians and reptiles are bullfrog, leopard frog, western cottonmouth, diamondback water snake, redbellied water snake, slider, common snapping turtle, alligator snapping turtle, and Mississippi map turtle. The American alligator is present but not common. Common resident aquatic birds are great blue heron, little blue heron, green heron, and wood duck. The rivers and associated floodplains (including croplands) seasonally are used by migratory waterfowl, such as snow goose, white-fronted goose, mallard, gadwall, scaup, American widgeon, pintail, and double-crested cormorant.

Ecosystem Processes

There has been little in the way of ecosystem studies of the Yazoo River system, either before or after the dams in the upper reaches or channelization in the lower reaches of the basin. Being a Coastal Plain river, this river historically would have experienced important connections with its naturally broad floodplain during floods that normally occurred in the winter and spring. Although still subject to flood pulses (Junk et al. 1989) that inundate adjacent floodplains (including agricultural lands), the river is disconnected from many sections of its floodplain by an extensive levee system. This restricts incorporation of organic materials into the aquatic component of the ecosystem, secondary production of benthic macroinvertebrates, and access to these floodplain resources by aquatic fauna in the river channels.

Dams produce their own unique environmental processes in the reservoirs and in tailwaters immediately downstream from the impoundments (see review by Jackson and Marmulla 2000). The semilentic environments of the reservoirs are driven primarily by autotrophic processes of various seasonal phytoplankton assemblages but suspended materials (primarily colloidal clay particles) attenuate the euphotic zone, resulting in a restricted trophogenic zone that is most extreme during summer. Thermal and associated chemical stratification are abrupt, with strong gradients that persist until late September or October. Early autumn fronts with brisk winds can create internal seiches that bring deoxygenated, low pH waters into littoral trophogenic zones. This can cause massive kills of lacustrine benthic macroinvertebrates. In addition, cool, cloudy days during late summer and early autumn, coupled with cooling rains, can break down thermoclines and result in complete lake turnovers (mixing of epilimnetic and hypolimnetic waters). These can result in deoxygenated water throughout the reservoirs and corresponding fish kills.

Tailwaters below the flood-control dams typically receive epilimnetic water from their respective reservoirs throughout the year, flushing plankton and other organic seston through the system throughout the entire year. This greatly alters the natural food resources, but they are used by an abundance of filter-feeding benthic macroinvertebrates (e.g., caddisflies, blackflies, bivalve mollusks) below dams that are themselves utilized as the foundation for extremely productive, albeit localized, fisheries. White bass, white crappie, blue catfish, and paddlefish are principal fisheries resources in the tailwaters. Well-developed facilities and access for anglers exploiting these fisheries resources are provided through cooperative arrangements between the U.S. Army Corps of Engineers (Vicksburg District) and the Mississippi Department of Wildlife, Fisheries, and Parks.

Downstream from the tailwaters, stream channels are periodically subject to channel dredging, removal of large woody debris, clearing of riparian forests, and, in some cases, have been channelized and leveed (see review by Hubbard et al. 1993). These impacts can be devastating to floodplain river ecosystems because they result in homogeneous channel environments and remove the principal substrate (i.e., large woody debris) used as attachment sites by most benthic macroinvertebrates. This undoubtedly reduces secondary production, reduces contributions of allochthonous materials from riparian zones, and tends to reduce overbank flooding. The normal incorporation of allochthonous organic materials into biological production processes and fish and invertebrate access to seasonally (winter) warmer, clearer floodplain water is greatly diminished compared to natural processes. However, throughout the Yazoo River system dredging schedules have been neglected. This has allowed the rivers to recover substantially, even to the point where many stream reaches, floodplains, and their associated hydrologic regimes have characteristics not unlike those of unaltered lowland stream reaches elsewhere in the region. Corresponding to this neglect are increased benefits to human society, in consumptive (e.g., fisheries) and nonconsumptive dimensions (e.g., aesthetics and existence values).

Human Impacts and Special Features

As described, the Yazoo River basin is a highly modified ecosystem, primarily from construction of dams on the upstream tributaries and channelization, dredging, floodplain clearing, and other abuses in the lower river. Before these alterations the river was highly dynamic, depending on extensive floodplain forests, highly variable river–floodplain interactions, and accumulations of wood in channels as essential habitats for river animals. The altered hydrology from dams and channel maintenance continues to have negative impacts on aquatic fauna that have evolved in more variable environments.

Although terribly abused during the past century, the Yazoo River ecosystem has shown some degree of recovery thanks to a mild temperate climate, substantial rainfall, deep, nutrient-rich soils, and, most of all, an evolving sense of natural resources stewardship and stream conservation throughout the region. The traditional advocates for river ecosystem destruction are losing voice and political influence throughout the region as a result of changing demographics. Throughout Mississippi there is growing sentiment that people must learn to live with the rivers rather than be pitted against them. However, constant vigilance is required in order to keep those who would abuse the river ecosystems in check (Jackson and Jackson 1989). To this end, river conservation educational efforts have been initiated that emphasize fisheries and sociological/cultural connections to the rivers and their natural resources, and keep the issues alive and active in public forums.

Through public school education programs, civic and religious group involvement, legislative initiatives, and support from nonprofit conservation organizations and professional societies, the public is learning that the flood-control dams throughout the Yazoo River basin have negative impacts from river fishery perspectives, but these have been ameliorated to a certain extent by development of the reservoir and tailwater fisheries (Jackson and Marmulla 2000). In addition, and in spite of flood-control programs, the public is becoming aware that flooding is good because it maintains lateral connectivity between the rivers and their respective floodplains (Junk et al. 1989, Flotemersch et al. 1999) and enhances overall system productivity and floodplain river fisheries (Cloutman 1997, Jackson et al. 1993, Jackson and Ye 2000).

ADDITIONAL RIVERS

The Atchafalaya River is a 224 km long distributary that receives 20% to 50% of the LMR's annual flow and all of the flow from the Red River (Fig. 6.18). Removal of a 30 km long logjam in the Atchafalaya near the former confluence of the Mississippi and Red rivers in 1855 allowed both rivers to begin flowing down the Atchafalaya. Numerous subsequent actions of the USACE (dredging, levee construction, etc.) culminating with construction of the Old River Control Structure in 1963 shaped the Atchafalaya into a deep (24 to 55 m) river with the 6th largest average flow in North America ($5178 m^3/s$). The Atchafalaya Basin Floodway ($2129 km^2$) contains the largest bottomland hardwood forest ($>1500 km^2$) in North America. The floodway, which is confined by levees (see Fig. 6.18), consists of a natural maze of distributary channels, bayous, lakes, and cypress–tupelo gum swamps that is now crisscrossed with canals dug to facilitate oil and gas extraction. During high water the floodway becomes a 24 to 32 km wide sheet of water flowing to the Gulf of Mexico. Physical and chemical conditions (shade, turbidity, low nutrients) do not support high levels of aquatic primary production (Bryan et al. 1975). However, the trees and macrophytes are very productive and provide high levels of particulate and dissolved organic matter to the system.

The Cache River, paralleled by a major tributary, Bayou De View, flows southward 229 km along the western edge of the Coastal Plain in northeastern Arkansas (Fig. 6.20). The long narrow (<29 km) basin covers 5240 km² of nearly flat alluvial deposits of clay, silts, and sand. Natural vegetation in the basin consists of cypress–tupelo gum swamps and seasonally flooded (March to May), diverse bottom-land hardwood forests that include overcup oak, Nuttall's oak, water hickory, green ash, cottonwoods, willows, and buttonbush (Smith 1996). These wetlands constitute one of the largest and least-disturbed tracts of this habitat type (~60,000 ha) remaining in the LMR basin. Hydrology of the system is also fairly natural, with low flow in early to midsummer, elevated in late summer and fall (Fig. 6.21). Some impacts are associated with irrigating rice in the basin (Wilber et al. 1996) and with removal of about 65% of the forest cover for production of cotton, rice, and soybeans (Kress et al. 1996). The Cache River and associated wetlands are an intact flood-pulse ecosystem (Junk et al. 1989), with production of fishes significantly enhanced by their exploitation of the seasonally flooded forests (Killgore and Baker 1996).

The Ouachita River basin significantly enhances the habitat diversity and biotic diversity found in the Lower Mississippi River basin. Approximately the first 400 km of the 974 km long river flows east across the steep (1.64 m/km), rocky terrain of the Ouachita province. Near Hot Springs, Arkansas, the river turns to the south-southeast to cross the flatter (0.17 m/km), fine alluvium of the Coastal Plain to its confluence with the Tensas River, which together become the Black River in northern Louisiana (Fig. 6.22). The Black River soon thereafter joins the Red River just before it empties into the Atchafalaya River distributary of the Mississippi River. The Ouachita is naturally different in these two distinct physiographic provinces, but the differences are increased by three hydroelectric and flood-control reservoirs along the main stem just before it enters the Coastal Plain (see Fig. 6.22). Two other main-channel locks and dams facilitate the small amount of barge traffic up to Camden, Arkansas. The large watershed (64,206 km²) is less impacted by land use, with 84% forested and only 9% cleared for pasture. However, silviculture (Brown et al. 1997, Smith et al. 2001), mining (barite, sand, rocks, gravel), and petroleum extraction/refining have impacted the biota. A special feature is the 263 km² Felsenthal National Wildlife Refuge centered around a 60 km² natural lake in the Ouachita River just north of the Louisiana border.

The Saline River (from its source as the Alum Fork) is a 328 km long gravel-bed stream whose watershed drains the eastern Ouachita province. Near Benton, Arkansas, it enters the Coastal Plain and continues about 186 km to its confluence with the Ouachita River in the Felsenthal National Wildlife Refuge (Fig. 6.24). The headwater streams (Alum Fork, North Fork, Middle Fork, and South Fork) are listed by the Arkansas Department of Environmental Quality as Extraordinary Resource Waters for their exceptional water quality and the presence of an endangered mussel, the Arkansas fatmucket; an endangered fish, the Ouachita madtom; and several other species of special concern (e.g., taillight shiner, peppered shiner, and southern pocketbook mussel). There are numerous small reservoirs in the headwaters tributaries, including the 555 ha Lake Winona on Alum Fork and 167 ha Lake Norrell on North Fork. However, downstream from these the Saline is the only river in the Ouachita province that flows >200 km without being impounded (Fig. 6.7). Water quality remains good to excellent in lower reaches of the Saline and it is an excellent smallmouth bass stream.

The Current River flows southeasterly from its source in Missouri for 215 km to its confluence with the Black River in Arkansas (Fig. 6.26). This river and its chief tributary, the Jack's Fork, are spring fed with substantial flow all year (mean discharge 77 m³/s; Fig. 6.27), unlike most other Ozark streams that experience seasonal low flows. The Current and Jack's Fork were protected from reservoir construction by the Ozark National Scenic Riverways Act in 1964, and the riverway is now managed by the National Park Service. The Current is one of the few remaining natural, free-flowing rivers in the lower 48 conterminous United States (Fig. 6.1). Both streams have gradients of about 1 m/km, have cherty limestone gravel-cobble substrates, and distinct riffle–pool geomorphology. The clean, cool waters of the Current and Jack's Fork rivers and the scenic Ozark Plateaus landscape attract numerous visitors, most of whom traverse the rivers by float tubes, canoes, rafts, john boats, or jet skis during the summer. The macroinvertebrate assemblage is moderately diverse and productive (Rabeni et al. 1995), and crayfishes account for half of invertebrate production. Smallmouth bass, rock bass, and longear sunfish are the most abundant sportfishes among the approximately 117 fish species.

FIGURE 6.7   Saline River near Malvern, Arkansas (PHOTO BY D. JACKSON).

ACKNOWLEDGMENTS

We are indebted to John (Sonny) Hall and his wife Carolyn Hall for several guided tours by numerous means of travel (hiking, boats, ATVs, auto) in the region where the Arkansas and White rivers flow into the Mississippi River. We appreciate Jack Kilgore and Jan Hoover and Ken Shirley for sharing unpublished data with us. Lists of endangered species for the Atchafalaya and Mississippi rivers in Louisiana were supplied by the Louisiana Natural Heritage Program. Assistance with lists of endangered species in Arkansas was provided by the Arkansas Game and Fish Commission. The USGS Nonindigenous Aquatic Species Database was very useful for identifying nonnative species in several of the rivers.

LITERATURE CITED

Abell, R. A., D. M. Olson, E. D. Dinerstein, P. T. Hurley, J. T. Diggs, W. Eichbaum, S. Walters, W. Wettengel, T. Allnut, C. J. Loucks, and P. Hedao. 2000. *Freshwater ecoregions of North America*. Island Press, Washington, D.C.

Albright, G. 1957–1993. *The Buffalo River*. Scrapbooks and publications. University of Arkansas Library, Special Collections, MC 1295, Fayetteville.

Anderson, N. H., and J. R. Sedell. 1979. Detritus processing by macroinvertebrates in stream ecosystems. *Annual Review of Entomology* 24:353–377.

Arkansas Department of Environmental Quality. 2000. Water quality inventory report (pursuant to section 305b of the Federal Water Pollution Control Act). Arkansas Department of Environmental Quality, Little Rock.

Arkansas Department of Planning. 1974. Arkansas Natural Area Plan. AR. 248. Arkansas Department of Planning, Little Rock.

Babcock, R. E., and H. C. MacDonald (eds.). 1973. *Preliminary reconnaissance water quality survey of the Buffalo National River*. Water Resources Research Center Publication no. 19. University of Arkansas, Fayetteville.

Baker, C. D., and E. H. Schmitz. 1971. Food habits of adult gizzard shad and threadfin shad in two Ozark reservoirs. In G. E. Hall (ed.). *Reservoir fisheries and limnology.* American Fisheries Society Special Publication no. 8, pp. 3–11. American Fisheries Society, Washington, D.C.

Baker, J. A., K. J. Killgore, and R. L. Kasul. 1991. Aquatic habitats of fish communities in the Lower Mississippi River. *Aquatic Sciences* 3:313–356.

Bailey, R. G. 1994. *Ecological classification for the United States.* USDA Forest Service, Washington, D.C.

Bayless M., and C. Vitello. 2001. White River watershed inventory and assessment. Missouri Department of Conservation, Springfield. Missouri Department of Conservation Web site: www.conservation.state.mo.us/fish/watershed/whriver/contents/390cotxt.htm

Bayley, P. B. 1989. Aquatic environments in the Amazon Basin, with an analysis of carbon sources, fish production and yield. In D. P. Dodge (ed.). *Proceedings of the International Large River Symposium (LARS).* Canadian Special Publication of Fisheries and Aquatic Sciences 106, pp. 399–408.

Bayley, P. B. 1995. Understanding large river-floodplain ecosystems. *Bioscience* 45:153–158.

Becker, D. A., and R. V. Kilambi. 1975. Ichthyofauna study. In R. E. Babcock and H. C. MacDonald (eds.). *Buffalo National River ecosystems: Part I.* Arkansas Water Resources Research Center Publication no. 34, pp. 139–149. University of Arkansas Press, Fayetteville.

Beckett, D. C., C. R. Bingham, L. G. Sanders, D. B. Mathis, and E. M. McLemore. 1983. Benthic macroinvertebrates of selected aquatic habitats of the Lower Mississippi River. Technical Report E-83-10. U.S. Army Engineer Waterways Experiment Station, Vicksburg, Mississippi.

Beckett, D. C., and C. H. Pennington. 1986. Water quality, macroinvertebrates, larval fishes, and fishes of the Lower Mississippi River: A synthesis. Technical Report E-86-12. U.S. Army Engineer Waterways Experiment Station, Vicksburg, Mississippi.

Benke, A. C. 1990. A perspective on America's vanishing streams. *Journal of the North American Benthological Society* 9:77–88.

Benke, A. C., R. L. Henry III, D. M. Gillispie, and R. J. Hunter. 1985. Importance of snag habitat for animal production in southeastern streams. *Fisheries* 10(5):8–13.

Boschung, H. T., Jr., and R. L. Mayden. 2004. *Fishes of Alabama.* Smithsonian Books, Washington, D.C.

Brown, A. V., and P. P. Brussock. 1991. Comparisons of benthic invertebrates between riffles and pools. *Hydrobiologia* 220:90–108.

Brown, A. V., Y. Aguila, K. B. Brown, and W. P. Fowler. 1997. Responses of benthic macroinvertebrates in extremely small intermittent streams to silvicultural practices. *Hydrobiologia* 347:119–125.

Brown, A. V., M. M. Lyttle, and K. B. Brown. 1998. Impacts of gravel mining on gravel bed streams. *Transactions of the American Fisheries Society* 127:981–997.

Brown, A. V., and W. J. Matthews. 1995. Stream ecosystems of the central United States. In C. E. Cushing, K. W. Cummins, and G. W. Minshall (eds.). *Ecosystems of the world.* Vol. 22: *River and stream ecosystems*, pp. 89–116. Elsevier Science B.V., Amsterdam.

Brown, A. V., L. D. Willis, and P. P. Brussock. 1983. Effects of sewage pollution in the White River, Arkansas. *Proceedings of the Arkansas Academy of Science* 37:13–18.

Brussock, P. P., and A. V. Brown. 1991. Riffle-pool geomorphology disrupts longitudinal patterns of stream benthos. *Hydrobiologia* 220:109–117.

Bryan, C. F., F. M. Truesdale, and D. S. Sabins. 1975. *Limnological studies of the Atchafalaya River basin.* U.S. Department of the Interior, Washington, D.C.

Buchanan, T. M. 1997. The fish community of Indian Bayou, a coastal plain stream of remarkable species richness in the lower White River drainage of Arkansas. *Journal of the Arkansas Academy of Science* 51:55–65.

Burr, B. M., and R. L. Mayden. 1992. Phylogenetics and North American freshwater fishes. In R. L. Mayden (ed.). *Systematics, historical ecology, and North American freshwater fishes*, pp. 18–75. Stanford University Press, Stanford, California.

Chordas, S. W., and G. L. Harp. 1991. *The aquatic macroinvertebrates of the White River National Wildlife Refuge.* Arkansas Water Resources Research Center Publication no. 154. University of Arkansas, Fayetteville.

Cloutman, D. G. 1997. Biological and socio-economic assessment of stocking channel catfish in the Yalobusha River, Mississippi. Ph.D. diss., Mississippi State University.

Cloutman, D. G., G. R. Hand, C. A. Chisam, and D. C. Jackson. 1999. Biological and socio-economic assessment of supplemental stocking of catchable-sized channel catfish in the Yalobusha River, Mississippi. Final Report, Federal Aid Project F-111. Mississippi Department of Wildlife, Fisheries, and Parks, Jackson.

Cobb, S. P., and A. N. Williamson. 1989. The computerized environmental resources data system (CERDS): A geographic information system for environmental data on the leveed floodplain of the Lower Mississippi River. Lower Mississippi Environmental Program Report no. 17. U.S. Army Corps of Engineers, Mississippi River Commission, Vicksburg, Mississippi.

Conner, W. H. 1988. Natural and artificial regeneration of bald cypress (*Taxodium distichum* [L.] Rich.) in the Barataria and Lake Verret basins of Louisiana. Ph.D. diss., Louisiana State University, Baton Rouge.

Conner, W. H., J. G. Gosselink, and R. T. Parrondo. 1981. Comparison of the vegetation of three Louisiana swamp sites with different flooding regimes. *American Journal of Botany* 68:320–331.

Croneis, C. 1930. Geology of the Arkansas Paleozoic area. *Arkansas Geological Survey Bulletin* 3, 457 pp.

Cummins, K. W. 1974. Structure and function of stream ecosystems. *Bioscience* 24:631–641.

Dale, E. E. 1973. Vegetation and site characteristics. In R. E. Babcock and H. C. MacDonald (eds.). *Preliminary reconnaissance water quality survey of the Buffalo National River*. Water Resources Research Center Publication no. 19, pp. 85–98. University of Arkansas, Fayetteville.

Davis, J. V., and R. W. Bell. 1998. Water-quality assessment of the Ozark Plateaus Study Unit, Arkansas, Kansas, Missouri, and Oklahoma: Nutrients, bacteria, organic carbon, and suspended sediment in surface water, 1993–1995. U.S. Geological Survey Water-Resources Investigations Report 98–4164. Little Rock, Arkansas.

Day, J. W., Jr., W. G. Smith, P. Wagner, and W. Stowe. 1973. Community structure and carbon budget in a salt marsh and shallow bay estuarine system in Louisiana. Sea Grant Publication LSU-SG-72-04. Louisiana State University Center for Wetland Resources, Baton Rouge.

Dobyns, H. F. 1983. *Their numbers become thinned: Native American population dynamics in Eastern North America*. University of Tennessee Press, Knoxville.

Doisy, K. E., and C. F. Rabeni. 2001. Flow conditions, benthic food resources, and invertebrate community composition in a low-gradient stream in Missouri. *Journal of the North American Benthological Society* 20:17–32.

Doisy, K. E., C. F. Rabeni, and D. L. Galat. 1997. The benthic insect community of the lower Jacks Fork River. *Transactions of the Missouri Academy of Science* 31:19–36.

Dott, R. H., and R. L. Batten. 1981. *Evolution of the Earth*, 3rd ed. McGraw-Hill, New York.

Dynesius, M., and C. Nilsson. 1994. Fragmentation and flow regulation of river systems in the northern third of the world. *Science* 266:753–762.

Everett, D. E. 1971. Hydrologic and quality characteristics of the Lower Mississippi River. U.S. Geological Survey Technical Report no. 5. Louisiana Department of Public Works, Baton Rouge.

Fisher, S. G., and G. E. Likens. 1973. Energy flow in Bear Brook, New Hampshire: An integrative approach to stream ecosystem metabolism. *Ecological Monographs* 43:421–439.

Flotemersch, J. E., D. C. Jackson, and J. R. Jackson. 1999. Channel catfish movements in relation to river channel–floodplain connections. *Proceedings of the Annual Conference Southeastern Association of Fish and Wildlife Agencies* 51(1997):106–112.

Francis, J. M. 1993. Recreational handgrabbing as a factor influencing flathead catfish stock characteristics in two Mississippi streams. Master's thesis, Mississippi State University.

Fremling, C. R., J. L. Rasmussen, R. E. Sparks, S. P. Cobb, C. F. Bryan, and T. O. Claflin. 1989. Mississippi River fisheries: A case history. In D. P. Dodge (ed.), *Proceedings of the International Large River Symposium (LARS)*. Canadian Special Publication of Fisheries and Aquatic Sciences 106, pp. 309–351.

Gerlach, R. L., and W. Wedenoja. 1984. *The heritage of the Ozarks*. August House, Little Rock, Arkansas.

Gonthier, G. J. 2000. Water quality in the deep tertiary aquifers of the Mississippi Embayment, 1996. U.S. Geological Survey Water Resources Investigations Report 99–4131. U.S. Geological Survey, Pearl, Mississippi.

Gordon, M. E. 1982. Mollusca of the White River, Arkansas and Missouri. *Southwestern Naturalist* 27:347–352.

Gordon, M. E., S. W. Chordas, G. L. Harp, and A. V. Brown. 1994. Aquatic mollusca of the White River National Wildlife Refuge, Arkansas, U.S.A. *Walkerana* 7(17–18):1–9.

Gordon, M. E., L. R. Kraemer, and A. V. Brown. 1980. Unionaceae of Arkansas: Historical review, checklist, and observations on distributional patterns. *Bulletin of the American Malacological Union, Inc.* 1979:31–37.

Gorman, G. T., and J. R. Karr. 1978. Habitat structure and stream fish communities. *Ecology* 59:507–515.

Hartfield, P. D., and R. G. Rummel. 1985. Freshwater mussels (Unionidae) of the Big Black River, Mississippi. *Nautilus* 99(4):116–119.

Hofer, K. R., H. D. Scott, and J. M. McKimmey. 1995. Spatial distribution of the surface geology and 1992 land use of the Buffalo River watershed. Arkansas Water Resources Center, Publication 174. Fayetteville, Arkansas.

Holman, T. H., J. Skains, and D. Riecke. 1993. The Big Black River, Mississippi: A case history. In L. W. Hesse, C. B. Stalnaker, N. G. Benson, and J. R. Zuboy (eds.). *Proceedings of the Symposium on Restoration Planning for the Rivers of the Mississippi River Ecosystem*. Biological Report 19, pp. 266–281. U.S. Department of the Interior, National Biological Survey, Washington, D.C.

Hubbard, W. D., D. C. Jackson, and D. J. Ebert. 1993. Channelization. In C. F. Bryan and D. A. Rutherford (eds.). *Impacts on warmwater streams: Guidelines for evaluation*, pp. 135–156. Southern Division, American Fisheries Society, Bethesda, Maryland.

Hunt, C. B. 1974. *Natural regions of the United States and Canada*. W. H. Freeman and Company, San Francisco.

Insaurralde, M. S. 1992. Environmental characteristics associated with flathead catfish in four Mississippi streams. Ph.D. diss., Mississippi State University.

Jackson, D. C. 1999. Flathead catfish: Biology, fisheries and management. Proceedings of the First International Ictalurid Symposium. *American Fisheries Society Symposium* 24:23–35.

Jackson, D. C. 2000. Distribution patterns of channel catfish (Ictaluridae: *Ictalurus punctatus*) in the Yalobusha River floodplain river ecosystem. Proceedings of the Seventh International Symposium on the Ecology of Fluvial Fishes. *Polish Archives of Hydrobiology* 46(1):63–72.

Jackson, D. C., N. J. Brown-Peterson, and T. D. Rhine. 1993. Perspectives for rivers and their fisheries resources in the upper Yazoo River basin, Mississippi. In L. W. Hesse, C. B. Stalnaker, N. G. Benson and J. R. Zuboy (eds.), *Proceedings of the Symposium on Restoration Planning for Rivers of the Mississippi River Ecosystem.* Biological Report 19, pp. 255–265. U.S. Department of the Interior, National Biological Survey, Washington, D. C.

Jackson, D. C., and J. M. Francis. 1993. Effects of different exploitation rates on riverine populations of flathead catfish (*Pylodictis olivaris*) in Mississippi. Completion Report, Federal Aid to Sportfish Restoration Project F-90-1. Mississippi Department of Wildlife, Fisheries, and Parks, Jackson.

Jackson, D. C., and J. R. Jackson. 1989. A glimmer of hope for stream fisheries in Mississippi. *Fisheries* 14(3):4–9.

Jackson, D. C., and G. Marmulla. 2000. The influence of dams on river fisheries. Thematic Reviews on Environmental Issues for the World Commission on Dams Secretariat, Cape Town, South Africa. United Nations Food and Agriculture Organization, Rome.

Jackson, D. C., and Q. Ye. 2000. Riverine fish stock and regional agronomic responses to hydrological and climatic regimes in the upper Yazoo River basin. In I. G. Cowx (ed.). *Management and ecology of river fisheries*, pp. 242–257. Fishing News Books, Blackwell Science, London.

Junk, W. J., P. B. Bayley, and R. E. Sparks. 1989. The flood pulse concept in river-floodplain systems. In D. P. Dodge (ed.), *Proceedings of the International Large River Symposium (LARS).* Canadian Special Publication of Fisheries and Aquatic Sciences 106, pp. 110–127.

Karr, J. R., J. D. Allan, and A. C. Benke. 2000. River conservation in the United States and Canada. In P. J. Boon, B. R. Davies, and G. E. Petts (eds.). *Global perspectives on river conservation: Science, policy and practice*, pp. 3–40. John Wiley and Sons, Chichester, England.

Keefe, J. F., and L. Morrow. 1994. *The White River chronicles of S. C. Turnbo.* University of Arkansas Press, Fayetteville.

Keith, W. E. 1987. Distribution of fishes in reference streams within Arkansas' ecoregions. *Proceedings of the Arkansas Academy of Science* 41:57–60.

Keown, M. P., E. A. Dardeau, and E. M. Causey. 1981. Characterization of the suspended-sediment regime and bed-material gradation of the Mississippi River Basin. U.S. Army Engineer Waterways Experiment Station Potamology Program (P-1) Report 1. 2 vols. U.S. Army Engineer Waterways Experiment Station, Vicksburg, Mississippi.

Kilambi, R. V., and D. A. Becker. 1977. Population dynamics and species diversity of ichthyo-parasitofauna of the Buffalo National River. Arkansas Water Resources Research Center Publication no. 48. University of Arkansas Press, Fayetteville.

Killgore, K. J., and J. A. Baker. 1996. Patterns of larval fish abundance in a bottomland hardwood wetland. *Wetlands* 16:288–295.

Killgore, J., and J. J. Hoover. 2003. White River navigation to Newport, Arkansas general re-evaluation project: Fishery studies. Prepared for U.S. Army Engineer District, Memphis. Engineer Research and Development Center, Waterways Experiment Station, Vicksburg (unpublished report).

Kleiss, B. A., R. H. Coupe, G. J. Gonthier, and B. G. Justus. 2000. Water quality in the Mississippi Embayment; Mississippi, Louisiana, Arkansas, Missouri, Tennessee, and Kentucky, 1995–98. U.S. Geological Survey Circular 1208. U.S. Geological Survey, Pearl, Mississippi.

Kress, M. R., M. R. Graves, and S. G Bourne. 1996. Loss of bottomland hardwood forests and forested wetlands in the Cache River basin, Arkansas. *Wetlands* 16: 258–263.

Leopold, L. B. 1994. A view of the river. Harvard University Press, Cambridge, Massachusetts.

Leopold, L. B., M. G. Wolman, and J. P. Miller. 1964. *Fluvial processes in geomorphology.* Freeman, San Francisco.

Mac, M. J., P. A. Opler, C. E. P. Haecker, and P. D. Doran (eds.). 1998. *Status and trends of the nation's biological resources.* 2 vols. U.S. Department of the Interior, U.S. Geological Survey, Reston, Virginia.

Mareska, J. F., and D. C. Jackson. 2002. Use of shadow bass stock characteristics to evaluate natural and scenic waterways in Mississippi. *Proceedings of the Annual Conference of Southeastern Association of Fish and Wildlife Agencies* 54(2000):167–178.

Marzolf, G. R. 1978. The potential effects of clearing and snagging on stream ecosystems. FWS/OBS-78-14. U.S. Fish and Wildlife Service, Washington, D.C.

Mathis, D. B., C. R. Bingham, and L. G. Sanders. 1982. Assessment of implanted substrate samplers for macroinvertebrates inhabiting stone dikes of the Lower Mississippi River. Misc. Paper E-82-1. U.S. Army Engineer Waterways Experiment Station, Vicksburg, Mississippi.

Mathis, D. B., S. P. Cobb, L. G. Sanders, A. D. Magoun, and C. R. Bingham. 1981. Aquatic habitat studies on the Lower Mississippi River, river mile 480 to 530. Report 3: Benthic macroinvertebrate studies—Pilot Report. Misc. Paper E-80-1. U.S. Army Engineer Waterways Experiment Station, Vicksburg, Mississippi.

McIsaac, G. F., M. B. David, G. Z. Gertner, and D. A. Goolsby. 2001. Eutrophication: Nitrate flux in the Mississippi River. *Nature* 414:166–167.

McPhee, J. 1989. Atchafalaya. In *The control of nature*, pp. 3–92. Farrar Straus Giroux, New York.

Meade, R. H. (ed.). 1995. Contaminants in the Mississippi River 1987–92. U.S. Geological Survey Circular 1133. Denver, Colorado.

Meyer, R. L., and N. Woomer. 1978. Water quality and phycological studies. In R. E. Babcock (ed.). *Buffalo National River ecosystems.* Part IV. Arkansas Water Resources Research Center Publication no. 58, pp. 1–36. University of Arkansas Press, Fayetteville.

Mississippi Department of Environmental Quality. 1992. *Mississippi water quality report.* Mississippi Department of Environmental Quality, Jackson.

Mississippi Department of Wildlife, Fisheries, and Parks. 2002. Freshwater mussels of Mississippi. Mississippi Department of Wildlife, Fisheries, and Parks (annotated poster). Jackson, Mississippi.

Mitsch, W. J., and J. G. Gosselink. 1993. *Wetlands,* 2nd ed. Van Nostrand Reinhold, New York.

Omernik, J. M. 1987. Ecoregions of the conterminous United States. *Annals of the Association of American Geographers* 77:118–125.

Omernik, J. M. 1995. Ecoregions: A spatial framework for environmental management. In W. S. Davis and T. P. Simon (eds.). *Biological assessment and criteria: Tools for water resource planning and decision making,* pp. 49–62. CRC Press, Boca Raton, Florida.

O'Neil, T. 1949. *The muskrat in the Louisiana coastal marshes.* Louisiana Wildlife and Fisheries Commission, New Orleans.

Patrick, R. 1998. *Rivers of the United States.* Vol. 4, part A: *The Mississippi River and tributaries north of St. Louis.* John Wiley and Sons, New York.

Peterson, J. T., and C. F. Rabeni. 2001. Evaluating the physical characteristics of channel units in an Ozark stream. *Transactions of the American Fisheries Society* 130:898–910.

Pflieger, W. L. 1996. *The crayfishes of Missouri.* Missouri Department of Conservation, Columbia.

Pyne, S. J. 1997. *Fire in America: A cultural history of wildland and rural fire,* 2nd ed. University of Washington Press, Seattle.

Rabeni, C. F. 1992. Trophic linkage between stream centrarchids and their crayfish prey. *Canadian Journal of Fisheries and Aquatic Sciences* 49:1714–1721.

Rabeni, C. F. 2000. Evaluating physical habitat integrity in relation to the biological potential of streams. *Hydrobiologia* 422/423:245–256.

Rabeni, C. F., M. Gossett, and D. D. McClendon. 1995. Contribution of crayfish to benthic invertebrate production and trophic ecology of an Ozark stream. *Freshwater Crayfish* 10:163–173.

Radwell, A. 2000. Ecological integrity assessment of Ozark rivers to determine suitability for protective status. Master's thesis, University of Arkansas, Fayetteville.

Revenga, C., S. Murray, J. Abramovitz, and A. Hammond. 1998. *Watersheds of the world: Ecological value and vulnerability.* World Resources Institute and Worldwatch Institute, Washington, D.C.

Ricketts, T. H., E. Dinerstein, D. M. Olson, C. J. Loucks, W. Eichbaum, D. DellaSala, K. Kavanagh, P. Hedao, P. T. Hurley, K. M. Carney, R. Abell, and S. Walters. 1999. *Terrestrial ecoregions of North America.* Island Press, Washington, D.C.

Rippey, L. L., and R. L. Meyer. 1975. Spatial and temporal distribution of algae and associated parameters. In R. E. Babcock and H. C. MacDonald (eds.). *Buffalo National River ecosystems.* Part I. Arkansas Water

Resources Research Center Publication no. 34, pp. 103–115. University of Arkansas Press, Fayetteville.

Robison, H. W., and T. M. Buchanan. 1988. *Fishes of Arkansas.* University of Arkansas Press, Fayetteville.

Ross, S. T. 2001. *Inland fishes of Mississippi.* Mississippi Department of Wildlife, Fisheries, and Parks, Jackson.

Runner, M. S., D. P. Turnipseed, and R. H. Coupe, 2002. Streamflow and nutrient data for the Yazoo River below Steele Bayou near Long Lake, Mississippi, 1996–2000. U.S. Geological Survey. Water Resources Investigations Report 02-4215. Pearl, Mississippi. 35 pp.

Rutherford, D. A., K. R. Gelwicks, and W. E. Kelso. 2001. Physicochemical effects of the flood pulse on fishes in the Atchafalaya River Basin, Louisiana. *Transactions of the American Fisheries Society* 130:276–288.

Rutherford, D. A., W. E. Kelso, C. F. Bryan, and G. C. Constant. 1995. Influence of physicochemical characteristics on annual growth increments of four fishes from the Lower Mississippi River. *Transactions of the American Fisheries Society* 124:687–697.

Sabo, M. J., C. F. Bryan, W. E. Kelso, and D. A. Rutherford. 1999. Hydrology and aquatic habitat characteristics of a riverine swamp: II. Hydrology and the occurrence of chronic hypoxia. *Regulated Rivers: Research and Management* 15:525–542.

Sasser, C. E., J. G. Gosselink, E. M. Swenson, C. M. Swarzenski, and N. C. Leibowitz. 1996. Vegetation, substrate, and hydrology in floating marshes in the Mississippi River Delta Plain wetlands, U.S.A. *Vegetation* 122:129–142.

Saucier, R. T. 1968. A new chronology for braided stream surface formation in the Lower Mississippi Valley. *Southeastern Geology* 9:65.

Scott, H. D., and P. A. Smith. 1994. The prediction of sediment and nutrient transport in the Buffalo River watershed using a geographic information system. Arkansas Water Resources Center Publication no. 167. University of Arkansas, Fayetteville.

Sharitz, R. R., and W. J. Mitsch. 1993. Southern floodplain forests. In W. H. Martin, S. G. Boyce, and A. C. Echternacht (eds.). *Biodiversity of the southeastern United States: Lowland terrestrial communities,* pp. 311–372. John Wiley and Sons, New York.

Shields, F. D., Jr., S. S. Knight, and C. M. Cooper. 1998. Rehabilitation of aquatic habitats in warmwater streams damaged by channel incision in Mississippi. *Hydrobiologia* 382:63–86.

Shirley, K. E. 2000. The White River: A case history of man's effects on a big river. In J. Harel and K. Steele (eds.). White River Forum II. Arkansas Water Resources Center Publication no. MSC-287. Arkansas Water Resources Center, Fayetteville, Arkansas.

Skains, J. A. 1992. Linear home range and movements of flathead catfish (*Pylodictis olivaris*) in two Mississippi streams. Master's thesis, Mississippi State University.

Smith, F., A. V. Brown, M. Pope, and J. L. Michael. 2001. Benthic meiofauna responses to five forest harvest methods. *Hydrobiologia* 464:9–15.

Smith, F. E. 1954. *The Yazoo River.* Rinehart, New York.

Smith, R. D. 1996. Composition, structure and distribution of woody vegetation on the Cache River floodplain, Arkansas. *Wetlands* 16:264–278.

Sparks, R. E. 1995. Need for ecosystem management of large rivers and their floodplains. *Bioscience* 45:168–182.

Springer, M. D., E. B. Smith, D. G. Parker, R. L. Meyer, E. E. Dale, and R. E. Babcock. 1977. *Buffalo National River ecosystems.* Part III. Water Resources Research Center Publication no. 49-A. University of Arkansas Press, Fayetteville.

Stanturf, J. A., E. S. Gardiner, P. B. Hamel, M. S. Devall, T. D. Leiniger, and M. E Warren. 2000. Restoring bottomland hardwood ecosystems in the lower Mississippi alluvial valley. *Journal of Forestry* 98:10–16.

Sutton, K. 2000. Pondering the future. *Arkansas Wildlife*, November–December, 33.

Thorp, J. H., and M. D. Delong. 1994. The riverine productivity model: An heuristic view of carbon sources and organic processing in large river ecosystems. *Oikos* 70(2):305–308.

Twedt, D. J., and C. R. Loesch. 1999. Forest area and distribution in the Mississippi alluvial valley: Implications for breeding bird conservation. *Journal of Biogeography* 26:1215–1224.

Uihlein, W. B., III. 2000. Extent and distribution of waterfowl habitat managed on private lands in the Mississippi Alluvial Valley. Ph.D. diss., Mississippi State University.

U.S. Army Corps of Engineers. 1964. *Big Black River comprehensive basin study, plan of survey.* U.S. Army Corps of Engineers, Vicksburg, Mississippi.

U.S. Army Corps of Engineers. 1991. *Supplement 1 to the final environmental impact statement on the operation and maintenance (of) Arkabutla Lake, Enid Lake, Grenada Lake, and Sardis Lake, Mississippi.* U.S. Army Corps of Engineers, Vicksburg, Mississippi.

U.S. Geological Survey. 2000a. *Water resources data Arkansas water year 1999.* U.S. Geological Survey, Little Rock, Arkansas.

U.S. Geological Survey. 2000b. *Water resources data Mississippi water year 1999.* U.S. Geological Survey, Pearl, Mississippi.

U.S. Geological Survey. 2001. *Water resources data Mississippi water year 2000.* U.S. Geological Survey, Pearl, Mississippi.

U.S. Geological Survey. 2002. Streamflow and nutrient data for the Yazoo River below Steele Bayou near Long Lake, Mississippi, 1996–2000. Water Resources Investigations Report 02-4215. National Water Quality Assessment Program. Mississippi Embayment Study Unit, Pearl, Mississippi.

Vannote, R. L., G. W. Minshall, K. W. Cummins, J. R. Sedell, and C. E. Cushing. 1980. The river continuum concept. *Canadian Journal of Fisheries and Aquatic Sciences* 37:130–137.

Ward, J. V., and J. A. Stanford. 1983. The intermediate-disturbance hypothesis: An explanation for biotic diversity pattern in lotic ecosystems. In T. D. Fontaine and S. M. Bartell (eds.). *Dynamics of lotic ecosystems*, pp. 347–356. Ann Arbor Science Publishers, Ann Arbor, Michigan.

Way, C. M., A. J. Burkey, C. R. Bingham, and A. C. Miller. 1995. Substrate roughness, velocity refuges, and macroinvertebrate abundance on artificial substrates in the lower Mississippi River. *Journal of the North American Benthological Society* 14:510–518.

Wells, M. A. 1994. *Native land: Jackson, Mississippi, 1540–1798.* University Press of Mississippi, Jackson.

Wentworth, C. K. 1922. A grade scale and class terms for clastic sediments. *Journal of Geology* 30:377–392.

Whitledge, G. W., and C. F. Rabeni. 1997. Energy sources and ecological role of crayfishes in an Ozark stream: Insights from stable isotopes and gut analysis. *Canadian Journal of Fisheries and Aquatic Sciences* 54:2555–2563.

Whitledge, G. W., and C. F. Rabeni. 2000. Benthic community metabolism in three habitats in an Ozark stream. *Hydrobiologia* 437:165–170.

Wilber, D. H., R. E. Tighe, and L. J. O'Neil. 1996. Associations between changes in agriculture and hydrology in the Cache River basin, Arkansas, USA. *Wetlands* 16:366–378.

Wilkerson, T. F., Jr. 2003. Current River watershed inventory and assessment. Missouri Department of Conservation, West Plains, Web site: www.conservation.state.mo.us/fish/watershed/current/contents/080cotxt.htm

Williams, M. 1992. *Americans and their forests.* Cambridge University Press, New York.

Wiseman, J. B. 1982. A study of the composition, successful relationships, and floristics of Mississippi River floodplain forests in parts of Washington, Bolivar, and Sharkey counties, Mississippi. Ph.D. dissertation. Mississippi State University. 248 pp.

Wright, S. 2000. Where river flows through forest. *National Wildlife* 38(4):38–44.

Wright, T. D. 1982. Aquatic habitat studies on the Lower Mississippi River, river mile 480 to 530, Report 8: Summary. Misc. Paper E-80-1. U.S. Army Engineer Waterways Experiment Station, Vicksburg, Mississippi.

Ye, Q. 1996. Riverine fish stock and regional agronomic responses to hydrologic and climatic regimes in the upper Yazoo River basin. Ph.D. diss., Mississippi State University.

Zekor, D. T., and R. M. Kaminski. 1987. Attitudes of Mississippi Delta farmers toward private-land waterfowl management. *Wildlife Society Bulletin* 15:346–354.

Zweifel, R. D., R. S. Hayward, and C. F. Rabeni. 1999. Bioenergetics insight into black bass distribution shifts in Ozark border region streams. *National American Journal of Fisheries Management* 19:192–197.

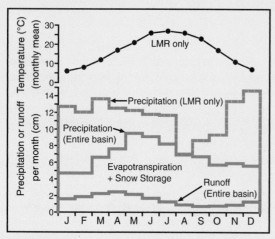

FIGURE 6.9 Mean monthly air temperature, precipitation, and runoff for the Lower Mississippi River basin.

LOWER MISSISSIPPI RIVER

Relief: 826 m (LMR only), 4141 m (entire basin)
Basin area: 3.27×10^6 km^2
Mean discharge: 18,400 m^3/s
River order: 10
Mean annual precipitation: 140 cm (LMR only), 94 cm (entire basin)
Mean air temperature: 17°C (LMR only)
Mean water temperature: 16°C
Physiographic provinces: Coastal Plain (CP), Ouachita Province (OP), Ozark Plateaus (OZ)
Biome: Temperate Deciduous Forest
Freshwater ecoregions: Mississippi Embayment, Ozark Highlands, Ouachita Highlands
Terrestrial ecoregions: Mississippi Lowland Forests, Central U.S. Hardwood Forests, Ozark Mountain Forests, Piney Woods Forests, Southeastern Mixed Forests
Number of fish species: 375 (entire Mississippi basin)
Number of endangered species: 3 fishes, 2 mussels, 1 bird (main stem only)
Major fishes: longnose gar, shortnose gar, shovelnose sturgeon, bowfin, gizzard shad, threadfin shad, central silvery minnow, speckled chub, silver chub, emerald shiner, river shiner, silverband shiner, mimic shiner, river carpsucker, blue sucker, smallmouth buffalo, blue catfish, channel catfish, flathead catfish, inland silverside, white bass, sauger, freshwater drum
Major other aquatic vertebrates: American alligator, western cottonmouth, snapping turtles, softshell turtles, Mississippi mud turtle, red-eared turtle, bullfrog, pigfrog, southern leopard frog, beaver, muskrat, river otter, nutria, great blue heron, cormorant, cattle egret, ibis, belted kingfisher
Major benthic invertebrates: mayflies (*Pentagenia, Tortopus, Stenonema, Baetis*), caddisflies (*Hydropsyche, Potamyia*), true flies (*Chaoborus, Rheotanytarsus*), crustaceans (*Procambarus, amphipods*), mollusks (*Sphaerium*), oligochaete worms (*Limnodrilus, Branchiura, Nais*), hydrozoans (*Cordylophora, Hydra*), turbellarian flatworms (*Dugesia*)
Nonnative species: Asian clam, zebra mussel, common carp, grass carp, silver carp, bighead carp, striped bass, rainbow smelt, rainbow trout, chinook salmon, American shad, greenhouse frog, nutria, alligatorweed, wild taro, water hyacinth, Peruvian water grass, Eurasian watermilfoil, water lettuce, curly pondweed, water spangles, dotted duckweed
Major riparian plants: willow, bald cypress, water tupelo, Nuttall's oak, swamp chestnut oak, overcup oak, sweetgum, ash, river birch, cottonwood, Eastern hophornbeam, water hickory, pecan, buttonbush, drummond maple, dwarf palmetto, palmettos, greenbriers, poison ivy, giant river cane, maidencane, alligatorweed
Special features: third largest river basin in world, with extensive floodplain habitats
Fragmentation: no dams on LMR main stem but highly modified for navigation and flood control
Water quality: pH = 7.7, alkalinity = 161 mg/L as CaCO$_3$, NO$_3$-N = 1.4 mg/L, PO$_4$-P = 0.13 mg/L
Land use: 80% agriculture, 16% forest, 4% urban/other (LMR); 57% agriculture, 28% forest/shrub, 14% urban (entire basin)
Population density: 10 people/km^2 (LMR only)
Major information sources: Baker et al. 1991, Rutherford et al. 1995

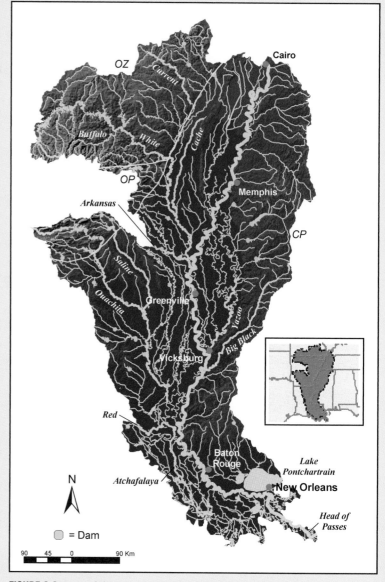

FIGURE 6.8 Map of the Lower Mississippi River basin. Physiographic provinces are separated by yellow lines.

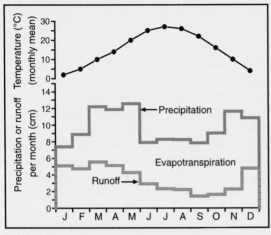

FIGURE 6.11 Mean monthly air temperature, precipitation, and runoff for the White River basin.

WHITE RIVER

Relief: 731 m
Basin area: 72,189 km^2
Mean discharge: 979 m^3/s
River order: 7
Mean annual precipitation: 117 cm
Mean air temperature: 15°C
Mean water temperature: 19°C
Physiographic provinces: Ozark Plateaus (OZ), Ouachita Province (OP), Coastal Plain (CP)
Biome: Temperate Deciduous Forest
Freshwater ecoregions: Ozark Highlands, Mississippi Embayment
Terrestrial ecoregions: Ozark Mountain Forests, Central U.S. Hardwood Forests, Mississippi Lowland Forests
Number of fish species: 163
Number of endangered species: 5 mussels, 10 fishes
Major fishes: longnose gar, gizzard shad, central stoneroller, Mississippi silvery minnow, Ozark minnow, bullhead minnow, emerald shiner, bigeye shiner, duskystripe shiner, blacktail shiner, mimic shiner, northern hogsucker, black redhorse, channel catfish, blue catfish, Ozark bass, longear sunfish, smallmouth bass, white crappie, rainbow darter, orangethroat darter, banded sculpin
Major other aquatic vertebrates: American alligator, snapping turtle, Mississippi mud turtle, map turtle, slider, softshell turtles, cottonmouth, midland water snake, yellow-bellied water snake, cricket frog, bullfrog, green frog, southern leopard frog, pickerel frog, muskrat, beaver, river otter, raccoon, mink, great blue heron, belted kingfisher
Major benthic invertebrates: mayflies (*Tortopus*, *Pentagenia*, *Baetis*, *Stenonema*, *Caenis*), stoneflies (*Acroneuria*), caddisflies (*Chimarra*, *Hydropsyche*, *Agapetus*), hellgrammites (*Corydalus*), beetles (*Stenelmis*, *Psephenus*, *Optioservus*), crustaceans (*Orconectes*), mollusks (threeridge, rabbitsfoot, pimpleback, pocketbook, and pistolgrip mussels; *Elimia*)
Nonnative species: Asian clam, zebra mussel, grass carp, common carp, silver carp, bighead carp, yellow perch, sauger, walleye, redeye bass, striped bass, white bass, American shad, threadfin shad, northern pike, muskellunge, chain pickerel, brown bullhead, cutthroat trout, rainbow trout, brown trout, brook trout, lake trout, Eurasian watermilfoil, water hyacinth, duck lettuce
Major riparian plants: willows, witchhazel, American sycamore, river birch, red maple, buttonbush, cottonwood, American elm, green ash, box elder, sugarberry, sweetgum, Nuttall's oak, bald cypress, water tupelo, hickories, river cane, poison ivy, greenbrier, cucumber vine, smartweed
Special features: karst topography in headwaters, extensive bottomland hardwood forest floodplains; two tributaries are National Rivers
Fragmentation: 4 large dams on main stem, 3 large dams on tributaries
Water quality: pH = 7.65, alkalinity = 110.5 mg/L as CaCO$_3$, NO$_3$-N = 0.39 mg/L, PO$_4$-P = 0.07 mg/L
Land use: 28% agriculture, 70% forest, 1% urban (upper/middle region); 83% agriculture, 8% forest, 1% urban (lower region)
Population density: 18 people/km^2
Major information sources: Arkansas Department of Environmental Quality 2000, Bayless and Vitello 2001, U.S. Geological Survey 2000a, Ken Shirley (personal communication)

FIGURE 6.10 Map of the White River basin. Physiographic provinces are separated by yellow lines.

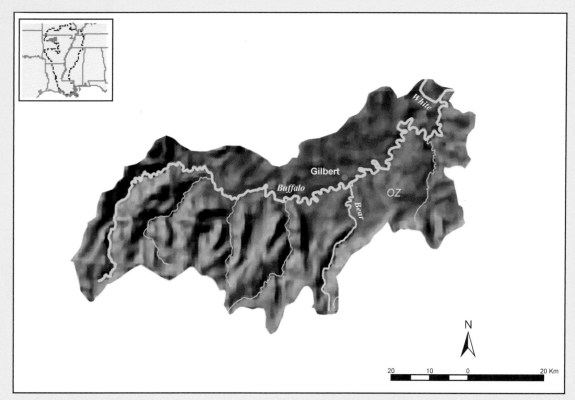

FIGURE 6.12 Map of the Buffalo National River basin.

BUFFALO NATIONAL RIVER

Relief: 666 m
Basin area: 3465 km^2
Mean discharge: 48 m^3/s
River order: 4
Mean annual precipitation: 107 cm
Mean air temperature: 15°C
Mean water temperature: 13°C
Physiographic province: Ozark Plateaus (OZ)
Biome: Temperate Deciduous Forest
Freshwater ecoregion: Ozark Highlands
Terrestrial ecoregion: Ozark Mountain Forests
Number of fish species: >66
Number of endangered species: 9 riparian vascular plants, 1 mussel
Major fishes: longear sunfish, smallmouth bass, largemouth bass, spotted bass, Ozark bass, largescale stoneroller, duskystripe shiner, rosyface shiner, telescope shiner, bigeye shiner, Ozark madtom, slender madtom, channel catfish, banded sculpin, greenside darter, rainbow darter, Arkansas saddled darter, yoke darter, banded darter, gilt darter, Ozark minnow, bigeye chub, northern hogsucker

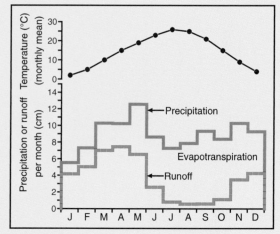

FIGURE 6.13 Mean monthly air temperature, precipitation, and runoff for the Buffalo National River basin.

Major other aquatic vertebrates: beaver, river otter, mink, common snapping turtle, map turtle, slider, midland smooth softshell turtle, western cottonmouth, midland water snake, yellow-bellied water snake, red river waterdog, cricket frog, bullfrog, green frog, southern leopard frog, pickerel frog, great blue heron, belted kingfisher
Major benthic invertebrates: mayflies (*Pseudocloeon, Heptagenia, Stenonema, Ephemerella, Isonychia, Baetis, Ephoron*), stoneflies (*Perlesta*), beetles (*Lutrochus*), caddisflies (*Agapetus, Cheumatopsyche*), crayfishes (*Orconectes*) mollusks (mucket, spike, plain pocketbook, Ozark broken-ray, Ozark pigtoe, and bleedingtooth mussels; *Elimia potosiensis, Pleurocera acuta*)
Nonnative species: freshwater jellyfish, Asian clam, common carp, fathead minnow, western mosquitofish, largemouth bass, smallmouth bass, walleye, rainbow trout
Major riparian plants: American elm, green ash, silver maple, box elder, American sycamore, river birch, black willow, Ward's willow, sandbar willow, cottonwood, sweetgum, witchhazel, buttonbush, giant river cane, sea oats, sedges, water willow
Special features: almost pristine main stem protected as National River; one of few natural free-flowing rivers in conterminous 48 states
Fragmentation: none
Water quality: pH = 7.6, alkalinity = 102.3 mg/L as CaCO$_3$, NO$_3$-N = 0.14 mg/L, PO$_4$-P = 0.06 mg/L
Land use: 9.5% agriculture, 90% forest, 0.5% urban/other
Population density: 7 people/km^2
Major information sources: Arkansas Department of Environmental Quality 2000, U.S. Geological Survey 2000a

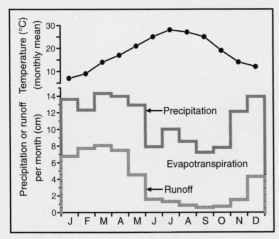

FIGURE 6.15 Mean monthly air temperature, precipitation, and runoff for the Big Black River basin.

BIG BLACK RIVER

Relief: 152 m
Basin area: 8770 km^2
Mean discharge: 107 m^3/s
River order: 4
Mean annual precipitation: 135 cm
Mean air temperature: 17.7°C
Mean water temperature: 17.2°C
Physiographic province: Coastal Plain (CP)
Biome: Temperate Deciduous Forest
Freshwater ecoregion: Mississippi Embayment
Terrestrial ecoregions: Southeastern Mixed Forests, Central U.S. Hardwood Forests, Mississippi Lowland Forests
Number of fish species: 112 (native)
Number of endangered species: 1 fish, 1 reptile, 2 mussels
Major fishes: flathead catfish, blue catfish, channel catfish, smallmouth buffalo, bigmouth buffalo, black buffalo, freshwater drum, white crappie, black crappie, largemouth bass, gizzard shad, bluegill, longnose gar, spotted gar, blue sucker, paddlefish, blacktail shiner, emerald shiner, striped shiner, creek chubsucker, freckled madtom, blackspotted topminnow, central stoneroller, scaly sand darter, slough darter, logperch darter, dusky darter

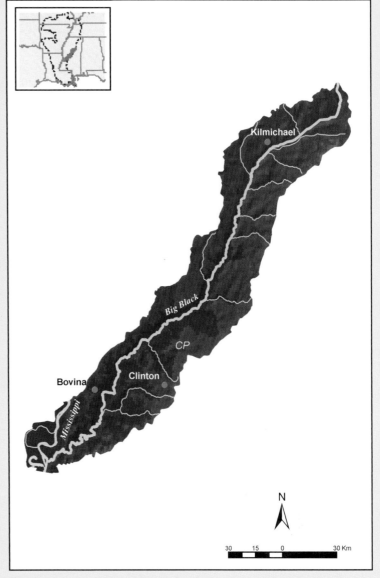

FIGURE 6.14 Map of the Big Black River basin.

Major other aquatic vertebrates: beaver, river otter, cottonmouth, red-bellied water snake, common snapping turtle, alligator snapping turtle, slider, Mississippi map turtle, great blue heron, green heron, little blue heron, wood duck, mallard, kingfisher
Major benthic invertebrates: mayflies (*Baetis, Cinygmula, Tricorythodes*), caddisflies (*Hydropsyche, Cheumatopsyche, Nectopsyche*), hellgrammites (*Corydalus, Chauliodes*), beetles (*Ancyronyx, Stenelmis*), dragonflies (*Gomphus*), black flies (*Prosimulium*), mussels (mucket, rock pocketbook, butterfly, mapleleaf)
Nonnative species: Asian clam, common carp, grass carp, goldfish, striped bass, bluespotted sunfish, dotted duckweed, parrot feather, sacred lotus, water lettuce
Major riparian plants: American sycamore, red maple, river birch, black willow, green ash, cottonwood, water oak, willow oak, water tupelo, bald cypress, honey suckle, American lotus, waterlily, water shield, pondweed, cattail, water primrose, alligatorweed
Special features: only free-flowing river in Mississippi that flows directly into Mississippi River; intact forested floodplain; numerous Civil War relics (sunken gunboats)
Fragmentation: no dams; only portions of lowermost reaches impacted by channel modification
Water quality: suspended sediment varies seasonally, with average of 134.8 mg/L, pH = 7.0, alkalinity = 48.4 mg/L as CaCO$_3$, hardness = 49.8 mg/L as CaCO$_3$, NO$_3$-N = 0.24 mg/L, PO$_4$-P = 0.25 mg/L
Land use: 35% agriculture, 11% pastureland, 54% forest
Population density: 25 people/km^2
Major information sources: Holman et al. 1993, Insaurralde 1992

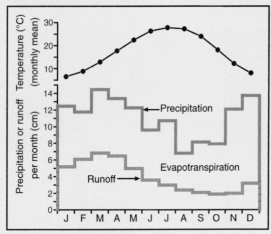

FIGURE 6.17 Mean monthly air temperature, precipitation, and runoff for the Yazoo River basin.

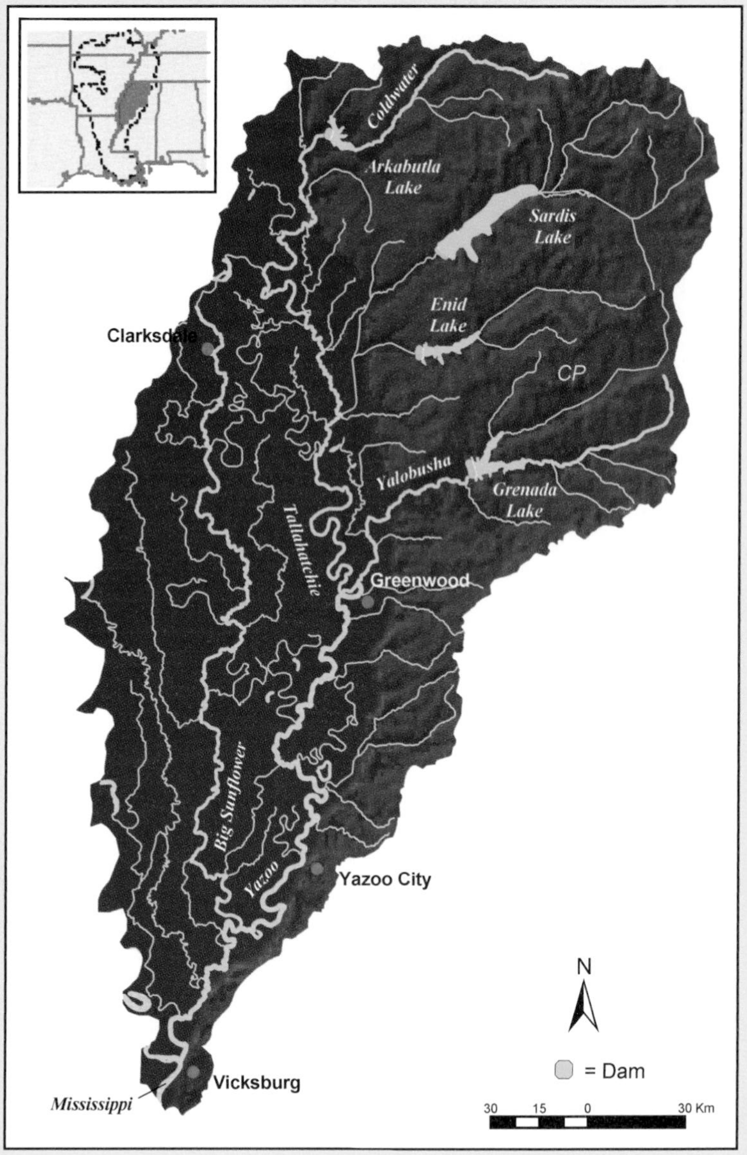

YAZOO RIVER

Relief: 195 m
Basin area: 35,000 km^2
Mean discharge: 523 m^3/s
River order: 6
Mean annual precipitation: 134 cm
Mean air temperature: 17.6°C
Mean water temperature: 21.5°C
Physiographic province: Coastal Plain (CP)
Biome: Temperate Deciduous Forest
Freshwater ecoregion: Mississippi Embayment
Terrestrial ecoregions: Mississippi Lowland Forests
Number of fish species: 119 (native)
Number of endangered species: 1 threatened reptile (federal); 2 endangered fishes (state), 13 endangered mussels (federal)
Major fishes: flathead catfish, blue catfish, channel catfish, smallmouth buffalo, bigmouth buffalo, freshwater drum, white crappie, black crappie, largemouth bass, gizzard shad, bluegill, longnose gar, spotted gar, blue sucker, paddlefish, black buffalo, striped bass
Major other aquatic vertebrates: beaver, river otter, cottonmouth, bullfrog, common snapping turtle, alligator snapping turtle, Mississippi map turtle, slider, great blue heron, wood duck
Major benthic invertebrates: mayflies (*Baetis*, *Caenis*, *Cinygmula*, *Stenonema*, *Potamanthus*), caddisflies (*Cheumatopsyche*, *Hydropsyche*, *Nectopsyche*), hellgrammites (*Chauliodes*, *Corydalus*), beetles (*Peltodytes*, *Berosus*, *Atrichopogon*), true flies (*Chaoborus*, *Simulium*, *Prosimulium*, *Hemerodromia*)
Nonnative species: freshwater jellyfish, Asian clam, common carp, grass carp, bighead carp, goldfish, white bass, striped bass, yellow perch, walleye, fathead minnow, American shad, tilapia, alligatorweed, water hyacinth, dotted duckweed, parrot feather, water lettuce
Major riparian plants: American sycamore, red maple, river birch, black willow, green ash, cottonwood, water oak, willow oak, water tupelo, bald cypress, honey suckle, water primrose, spikerush, alligatorweed, water shield, cattail, American lotus, waterlily, pondweed
Special features: tremendously productive recreational and "artisanal" fisheries for catfish; floodplains internationally important overwintering areas for migratory waterfowl
Fragmentation: major dams on tributaries (Coldwater, Little Tallahatchie, Yocona, Yalobusha–Skuna); extensive channelization
Water quality: suspended sediments = 168.4 mg/L, pH = 7.0, alkalinity = 45 mg/L as CaCO$_3$, NO$_3$-N = 0.20 mg/L, PO$_4$-P = 0.29 mg/L
Land use: 60% to 80 % agriculture, 15% to 30% forest, 5% to 10% other
Population density: 16 people/km^2
Major information sources: www.mdwfp.state.ms.us/museum/html/research, www.msstate.edu/Dept/GeoSciences/climate, Runner et al. 2002, Mississippi Department of Environmental Equality 1992, Wiseman 1982, Jackson et al. 1993, Insaurralde 1992, Jackson and Ye 2000

FIGURE 6.16 Map of the Yazoo River basin.

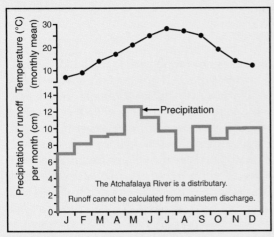

FIGURE 6.19 Mean monthly air temperature, precipitation, and runoff for the Atchafalaya River basin.

ATCHAFALAYA RIVER

Relief: 15 m
Basin area: 8345 km^2
Mean discharge: 5178 m^3/s
River order: not applicable
Mean annual precipitation: 153 cm
Mean air temperature: 20°C
Mean water temperature: 22°C
Physiographic province: Coastal Plain (CP)
Biome: Temperate Deciduous Forest
Freshwater ecoregion: Mississippi Embayment
Terrestrial ecoregion: Mississippi Lowland Forests
Number of fish species: 181
Number of endangered species: 1 mammal, 3 fishes, 1 reptile, 3 birds, 1 riparian plant
Major fishes: redear sunfish, bluegill, smallmouth buffalo, white crappie, black crappie, largemouth bass, warmouth, white bass, spotted gar, alligator gar, channel catfish, blue catfish, gizzard shad, threadfin shad, freshwater drum, bowfin, common carp, emerald shiner, silverband shiner, mimic shiner, golden shiner, mosquitofish, inland silverside

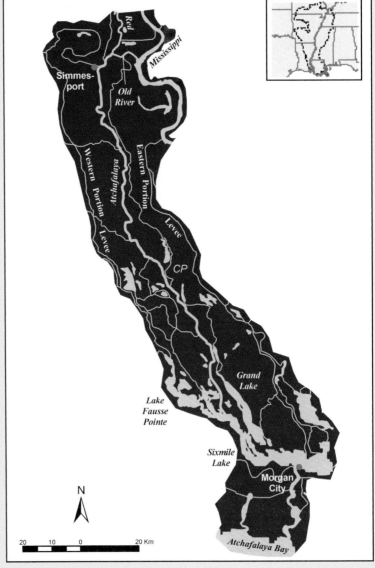

FIGURE 6.18 Map of the Atchafalaya River basin.

Major other aquatic vertebrates: American alligator, western cottonmouth, yellow-bellied water snake, bullfrog, pigfrog, southern leopard frog, beaver, muskrat, river otter, mink, nutria, great blue heron, green heron, cormorant, belted kingfisher, egrets, ibis, anhinga, wood duck
Major benthic invertebrates: mayflies (*Tortopus*, *Pentagenia*), caddisflies (*Hydropsyche*), true flies (*Coelotanypus*, *Polypedilum*, *Chaoborus*), crustaceans (White River crawfish, red swamp crayfish, *Palaemonetes kadiakensis*, *Lirceus*, *Gammarus*, *Hyalella*), mollusks (*Sphaerium*, mapleleaf mussel, *Anodonta*, *Physa*), oligochaete worms (*Limnodrilus*)
Nonnative species: Asian clam, zebra mussel, grass carp, silver carp, bighead carp, common carp, nutria, water hyacinth, Eurasian watermilfoil, hydrilla, alligatorweed, wild taro, horsefly's eye, Brazilian water weed, dotted duck weed, marshweed, Uruguay seedbox, parrot feather, brittle naiad, watercress, rice, duck lettuce, torpedo grass, water lettuce, water spangles
Major riparian plants: bald cypress, black willow, water tupelo, drummond maple, cottonwood, river birch, American sycamore, sweetgum, sugarberry, buttonbush, ash, smartweed, fanwort, coontail, Eurasian watermilfoil, pondweed, duckweed, frogbite, bladderwort, maidencane, cattails, dwarf spikerush, purple ammania, palmetto
Special features: 3rd largest continuous wetland area in United States, with about 30% of all bottomland hardwood forests in Mississippi Alluvial Plain; Mississippi River distributary that captured Red River
Fragmentation: no dams, but extensive levees; discharge determined by control structures (dams and floodgates) linking Atchafalaya to Mississippi and Red rivers
Water quality: pH = 7.4, alkalinity = 105 mg/L as CaCO$_3$, NO$_3$-N = 1.2 mg/L, PO$_4$-P = 0.22 mg/L, becomes anoxic during summer low flow
Land use: 33.5% forest/wetland, 63% agriculture, 3.5% urban/other
Population density: 18 people/km^2
Major information sources: Patrick 1998, Rutherford et al. 2001, Sabo et al. 1999, U.S. Geological Survey 2000a

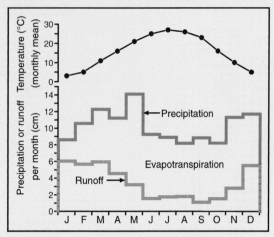

FIGURE 6.21 Mean monthly air temperature, precipitation, and runoff for the Cache River basin.

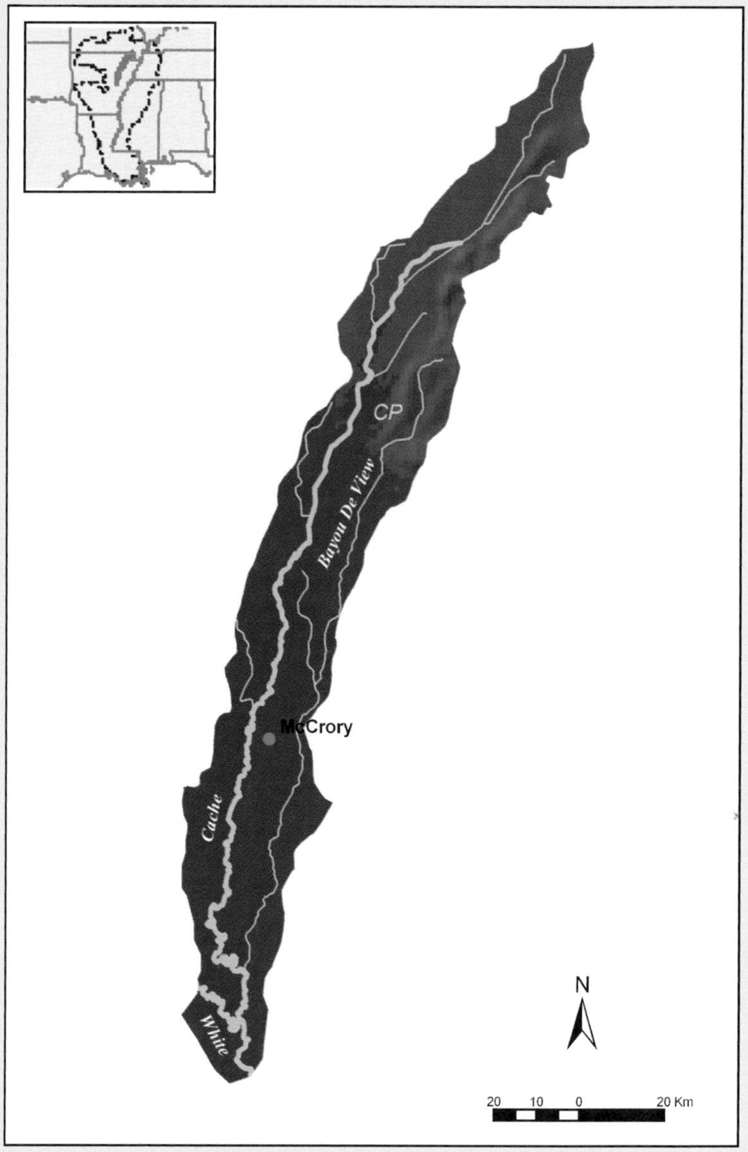

CACHE RIVER

Relief: 50 m
Basin area: 5227 km²
Mean discharge: 68 m³/s
River order: 4
Mean annual precipitation: 126 cm
Mean air temperature: 16°C
Mean water temperature: 17°C
Physiographic province: Coastal Plain (CP)
Biome: Temperate Deciduous Forest
Freshwater ecoregion: Ozark Highlands
Terrestrial ecoregion: Mississippi Lowland Forests
Number of fish species: 32
Number of endangered species: 1 mussel
Major fishes: gizzard shad, pugnose minnow, bullhead minnow, spotted sucker, channel catfish, tadpole madtom, blackspotted topminnow, flier, black crappie, blacktail shiner, mud darter, bluntnose darter, slough darter, cypress darter, speckled darter, river darter, logperch, spotted gar, shortnose gar, smallmouth buffalo, mosquitofish, longear sunfish, orangespotted sunfish, bluegill

FIGURE 6.20 Map of the Cache River basin.

Major other aquatic vertebrates: beaver, muskrat, river otter, mink, great blue heron, alligator snapping turtle, snapping turtle, stinkpot turtle, map turtle, Mississippi map turtle, slider, midland smooth softshell turtle, spiny softshell turtle, western cottonmouth, yellow-bellied water snake, midland water snake, green water snake, broad-banded water snake, bullfrog, green frog, southern leopard frog, marbled salamander
Major benthic invertebrates: mayflies (*Baetis*, *Caenis*, *Stenacron*), caddisflies (*Cheumatopsyche*, *Pycnopsyche*), alderflies (*Sialis*), beetles (*Berosus*, *Macronychus*), midges (*Tribelos*), damselflies (*Argia*), crustaceans (*Orconectes*, *Gammarus*, *Palaemonetes*), mussels (threeridge, bankclimber, pimpleback, mapleleaf), oligochaete worms (*Dero*)
Nonnative species: Asian clam, common carp, grass carp, bighead carp, chain pickerel, grass pickerel, sauger, Eurasian watermilfoil
Major riparian plants: overcup oak, Nuttall's oak, water oak, willow oak, red maple, water hickory, American elm, water elm, persimmon, bald cypress, pumpkin ash, honey locust, sweetgum, stiff dogwood, American hornbeam, swamp privet, buttonbush, Virginia sweetspire
Special features: contains Black Swamp, one of largest contiguous bottomland forests in Lower Mississippi Valley; one of few natural free-flowing rivers in conterminous 48 states
Fragmentation: no dams but extensive channelization in upper and middle reaches
Water quality: pH = 7.5, alkalinity = 81.6 mg/L as CaCO₃, NO₃-N = 0.25 mg/L, PO₃-P = 0.25 mg/L
Land use: 81% agriculture, 17% forest, 1.5% urban/other
Population density: 23 people/km²
Major information sources: Arkansas Department of Environmental Quality 2000, U.S. Geological Survey 2000a

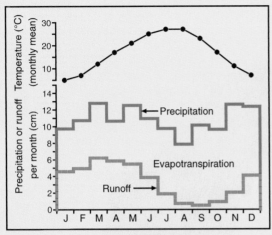

FIGURE 6.23 Mean monthly air temperature, precipitation, and runoff for the Ouachita River basin.

OUACHITA RIVER

Relief: 810 m
Basin area: 64,454 km^2
Mean discharge: 843 m^3/s
River order: 6
Mean annual precipitation: 130 cm
Mean air temperature: 17°C
Mean water temperature: 16°C
Physiographic provinces: Ouachita Province (OP),
    Coastal Plain (CP)
Biome: Temperate Deciduous Forest
Freshwater ecoregions: Ouachita Highlands, Mississippi
    Embayment
Terrestrial ecoregions: Piney Woods Forests, Ozark
    Mountain Forests, Mississippi Lowland Forests
Number of fish species: 80
Number of endangered species: 10 mussels, 2 fishes
Major fishes: spotted gar, longnose gar, northern hog
    sucker, spotted sucker, black redhorse, golden
    redhorse, central stoneroller, bigeye shiner, rosyface
    shiner, redfin shiner, steelcolor shiner, bluntnose
    minnow, blackspotted topminnow, channel catfish,
freckled madtom, brook silverside, green sunfish, spotted bass, orangebelly darter, greenside darter, channel darter
Major other aquatic vertebrates: snapping turtle, stinkpot turtle, Mississippi mud turtle, Ouachita map turtle, southern painted
    turtle, midland smooth softshell turtle, western spiny softshell turtle, western cottonmouth, broad-banded water snake,
    yellow-bellied water snake, diamondback water snake, green water snake, bullfrog, green frog, southern leopard frog, great
    blue heron, belted kingfisher, river otter, mink
Major benthic invertebrates: stoneflies (*Amphinemura*), mayflies (*Isonychia, Caenis, Stenonema*), caddisflies (*Hydroptila,*
    Agapetus, Chimarra), hellgrammites (*Corydalus*), beetles (*Stenelmis*), crustaceans (*Orconectes, Caecidotea, Lirceus*),
    mollusks (Wabash pigtoe, mucket, flutedshell, threeridge, giant floater, and Louisiana fatmucket mussels, *Physa*)
Nonnative species: Asian clam, threadfin shad, grass carp, common carp, silver carp, bighead carp, northern pike, muskellunge,
    chain pickerel, blue catfish, white catfish, striped bass, sauger, walleye, rainbow trout, brown trout, brook trout,
    alligatorweed, wild taro, water hyacinth, yellow iris, parrot feather, water lettuce, hydrilla, Eurasian watermilfoil
Major riparian plants: American sycamore, sweetgum, shortleaf pine, loblolly pine, American hornbeam, eastern hophornbeam,
    red oak, beech, American holly, Ozark witchhazel, water tupelo, poison ivy, greenbrier, smartweed
Special features: geothermal hot springs in basin; natural reservoir formed by Felsenthal basin
Fragmentation: 5 impoundments on main stem; 1 large impoundment on each of two major tributaries
Water quality: pH = 7.1, alkalinity = 22.5 mg/L as CaCO$_3$, NO$_3$-N = 0.19 mg/L, PO$_4$-P = 0.05 mg/L; mercury levels have resulted
    in a fish consumption advisory; thermal impacts of reservoirs
Land use: 13% agriculture, 81% forest, 6% urban/other
Population density: 16 people/km^2
Major information sources: Arkansas Department of Environmental Quality 2000, U.S. Geological Survey 2000a

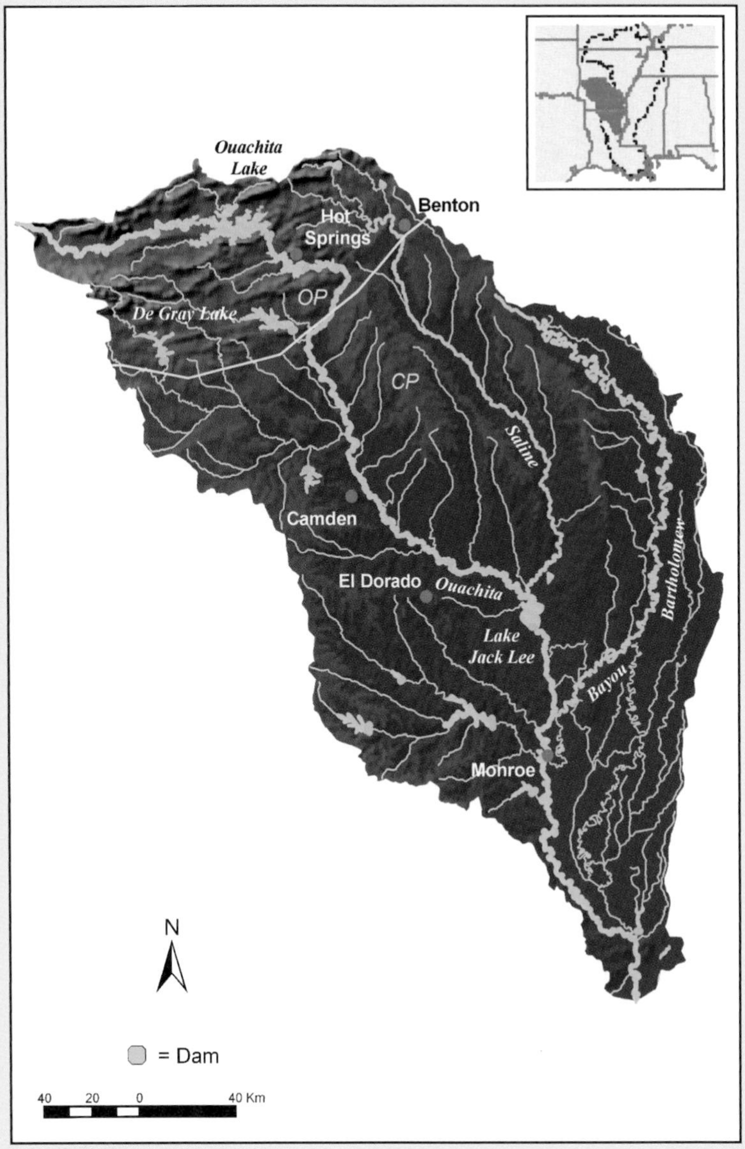

FIGURE 6.22 Map of the Ouachita River basin. Physiographic provinces are separated by a yellow line.

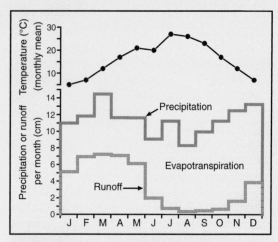

FIGURE 6.25 Mean monthly air temperature, precipitation, and runoff for the Saline River basin.

SALINE RIVER

Relief: 532 m
Basin area: 5465 km^2
Mean discharge: 89 m^3/s
River order: 5
Mean annual precipitation: 130 cm
Mean air temperature: 17°C
Mean water temperature: 17°C
Physiographic provinces: Ouachita Province (OP), Coastal Plain (CP)
Biome: Temperate Deciduous Forest
Freshwater ecoregion: Ouachita Highlands
Terrestrial ecoregion: Piney Woods Forests
Number of fish species: 85
Number of endangered species: 3 mussels, 1 fish
Major fishes: smallmouth bass, largemouth bass, spotted bass, warmouth, shadow bass, longear sunfish, bluegill, green sunfish, banded pygmy sunfish, black crappie, channel catfish, Ouachita madtom, cypress darter, taillight shiner, peppered shiner, redfin shiner, big eye shiner, striped shiner, steelcolor shiner

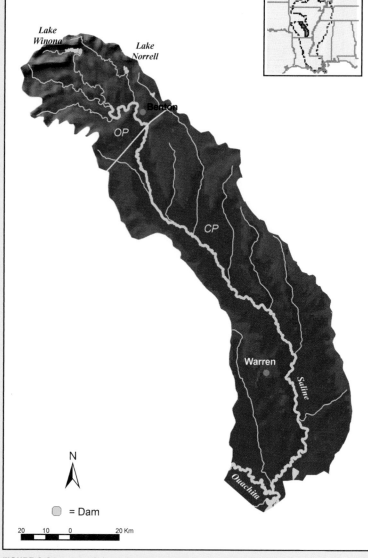

FIGURE 6.24 Map of the Saline River basin. Physiographic provinces are separated by a yellow line.

Major other aquatic vertebrates: snapping turtle, Ouachita map turtle, midland smooth softshell turtle, midland water snake, yellow-bellied water snake, diamondback water snake, green water snake, western cottonmouth, red river waterdog, green frog, bullfrog, southern leopard frog, pickerel frog, belted kingfisher, great blue heron, muskrat, beaver, mink, river otter
Major benthic invertebrates: mayflies (*Stenonema*, *Isonychia*), stoneflies (*Neoperla*, *Amphinemura*), caddisflies (*Chimarra*, *Cheumatopsyche*), hellgrammites (*Corydalus*), beetles (*Psephenus*), crustaceans (*Orconectes*), mussels (ladyfinger, fluted shell, Ouachita kidneyshell, black sandshell, Wabash pigtoe, Louisiana fatmucket, squawfoot)
Nonnative species: freshwater jellyfish, Asian clam, grass carp, common carp, goldfish, walleye, chain pickerel, blue catfish, threadfin shad, fathead minnow
Major riparian plants: water oak, willow oak, red oak, sweetgum, American sycamore, black willow, river birch, buttonbush, smooth alder, eastern hophornbeam, American hornbeam, common winterberry, haws, water willow, smartweed
Special features: one of the few natural free-flowing rivers in conterminous 48 states (last in Ouachita Mountain area); excellent float and fishing river
Fragmentation: no major dams, 2 small ones in headwaters
Water quality: pH = 7.0, alkalinity = 39.9 mg/L as CaCO$_3$, NO$_3$-N = 0.1 mg/L, PO$_4$-P = 0.05 mg/L
Land use: headwaters >90% forest, remainder >50% agricultural
Population density: 17 people/km^2
Major information sources: Arkansas Department of Environmental Quality 2000, Robison and Buchanan 1988, U.S. Geological Survey 2000a

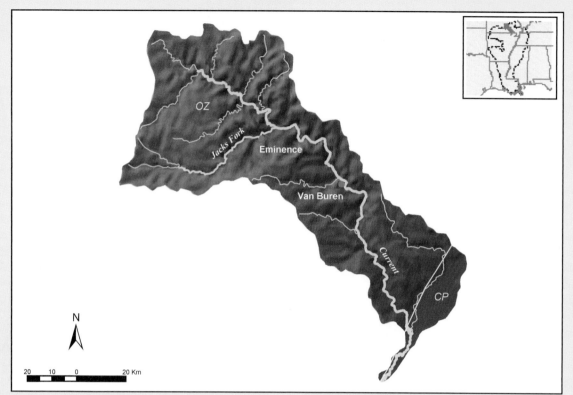

FIGURE 6.26 Map of the Current River basin. Physiographic provinces are separated by a yellow line.

CURRENT RIVER

Relief: 372 m
Basin area: 6776 km²
Mean discharge: 77 m³/s
River order: 6
Mean annual precipitation: 123 cm
Mean air temperature: 15°C
Mean water temperature: 17°C
Physiographic provinces: Ozark Plateaus (OZ), Coastal Plain (CP)
Biome: Temperate Deciduous Forest
Freshwater ecoregion: Ozark Highlands
Terrestrial ecoregion: Central U.S. Hardwood Forests
Number of fish species: 117
Number of endangered species: 5 mussels, 1 snail, 2 fishes, 1 hellbender
Major fishes: smallmouth bass, rock bass, longear sunfish, northern hog sucker, central stoneroller, rosyface shiner, telescope shiner, Arkansas saddled darter, greenside darter, rainbow darter, fantail darter, stargazing darter, mountain madtom, blackspotted topminnow

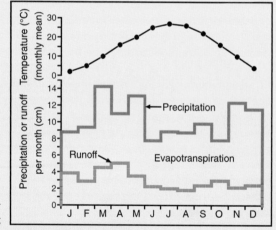

FIGURE 6.27 Mean monthly air temperature, precipitation, and runoff for the Current River basin.

Major other aquatic vertebrates: snapping turtle, softshell turtle, Mississippi map turtle, midland water snake, western cottonmouth, green frog, beaver, river otter, muskrat, raccoon, mink, great blue heron, belted kingfisher, green-backed heron
Major benthic invertebrates: mayflies (*Stenonema, Baetis*), stoneflies (*Neoperla, Leuctra*), caddisflies (*Hydropsyche, Ceratopsyche*), hellgrammites (*Corydalus*), beetles (*Stenelmis*), crustaceans (*Orconectes luteus, O. punctimanus*), mollusks (Ozark pigtoe, Ozark broken-ray, round pigtoe, rainbow mussels; *Amnicola, Physa, Elimia*)
Nonnative species: Asian clam, common carp, chain pickerel, grass pickerel, white bass, walleye, rainbow trout, brown trout
Major riparian plants: American sycamore, box elder, American elm, winged elm, slippery elm, black willow, river birch, hackberry, silver maple, sugar maple, bur oak, green ash, white ash, common witchhazel, spicebush, pawpaw, American hornbeam, flowering dogwood, hawthorns, poison ivy, water willow, smartweed
Special features: 161 km protected as Ozark National Scenic Riverways; numerous large springs provide 60% of base flow; one of few natural free-flowing rivers in conterminous 48 states
Fragmentation: none
Water quality: pH = 7.8, alkalinity = 145 mg/L as $CaCO_3$, NO_3-N = 0.25 mg/L, PO_4-P = 0.05 mg/L
Land use: 17% agriculture, 83% forest, <1% urban
Population density: 6 people/km²
Major information sources: Doisy and Rabeni 2001, Rabeni 1992, Rabeni 2000, Rabeni et al. 1995, U.S. Geological Survey 2000a, Wilkerson 2003

7

SOUTHERN PLAINS RIVERS

WILLIAM J. MATTHEWS   CARYN C. VAUGHN
KEITH B. GIDO   EDIE MARSH-MATTHEWS

INTRODUCTION

ARKANSAS RIVER

CANADIAN RIVER

RED RIVER

LITTLE RIVER

ADDITIONAL RIVERS

ACKNOWLEDGMENTS

LITERATURE CITED

INTRODUCTION

Two large, separate river basins, the Arkansas and the Red, drain the southern Great Plains region of the United States south of the Kansas River and north of the Texas–Gulf coastal drainages (Fig. 7.2). All major rivers in the region drain generally from northwest to southeast and are tributaries of the Mississippi River. The Southern Plains region includes all of Oklahoma, much of western and central Arkansas, and parts of eastern New Mexico, Colorado, Kansas, north Texas, and western and central Louisiana. The region is characterized by shortgrass prairie in the west, mixed or tallgrass prairie in the midsection, and forests in the east. A general description of rivers in the southern Great Plains is in Matthews (1988), Matthews and Zimmerman (1990), and Brown and Matthews (1995).

The largest rivers (Arkansas, Canadian, Red, Washita, Cimarron) all have upper main stems that lack flow at times and mid- and downstream reaches that are wide, shallow, and sand or mud bottomed. They are some of the hottest and harshest aquatic habitats on Earth, with water temperatures reaching near 40°C when exposed to full sun under low flow conditions. Hefley (1937), describing the South Canadian River near Norman, Oklahoma, wrote,

"Probably no more ecologically dynamic region exists: the seasonal, diurnal, and yearly fluctuations of meteorological factors are great and sudden; the course of the river changes with each succeeding rain and the shifting sand . . . is constantly being moved by wind and water." Adding to this the intense summer heat and winter cold of the region, organisms of these rivers are challenged by harsh, rapidly changing environmental conditions (Matthews and Hill 1979, Matthews 1987, Matthews and Zimmerman 1990).

People occupied the Arkansas and Red basins 11,500 to 10,000 years ago as hunters of the last ice age large animals (mammoths, mastodons, and big horned bison), roaming from central Texas, where they apparently spent winters fashioning and refurbishing tools from cherts of the region. By 10,500 years ago, the Ouachita Mountain parts of the Arkansas and Red basins were primarily grasslands, with sparse, riparian woodlands of oak and pine. Hunter-gatherers there and in the Ozarks focused on deer, turkey, small game, and wild plant foods. By 7500 years ago the Ozarks and Ouachitas were denuded of most woodlands, as a hot, dry Altithermal climate prevailed until about 4500 years ago. Recovery from the adverse effects of the Altithermal climatic regime was slow. Substantial numbers of

◄ **FIGURE 7.1** Red River, Oklahoma (PHOTO BY W. J. MATTHEWS). *283*

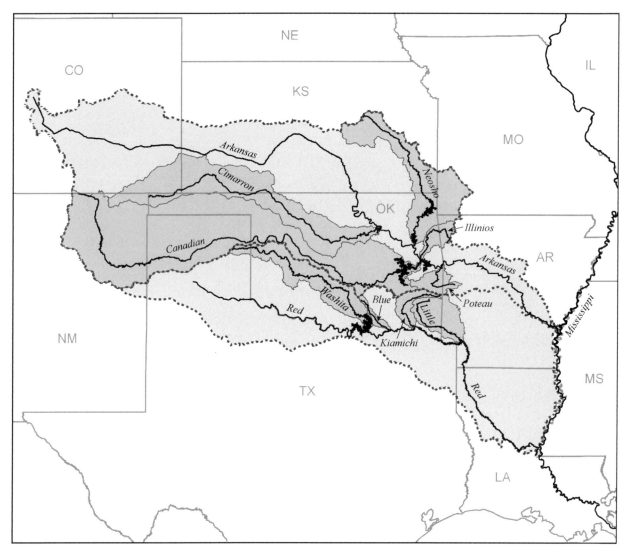

FIGURE 7.2   Southern Plains rivers covered in this chapter.

hunter-gatherers came to occupy the eastern parts of the Arkansas and Red basins, and by 2000 years ago they were adopting the bow and arrow and manufacturing pottery. By 1500 years ago these people were becoming farmers of corn, beans, and squash, and by 900 years ago major populations of these people occupied the fertile valleys along the Grand, Arkansas, Poteau, Little, and Red rivers in eastern Oklahoma and adjacent Arkansas. Meanwhile, to the west, sizeable societies of part-time bison hunters and farmers were spreading along the Washita River in Oklahoma, the South Canadian River in the Texas panhandle, and the Beaver (North Canadian) River in the Oklahoma panhandle (D. Wyckoff, personal communication).

By 500 years ago these people were undertaking major migrations and social change as climatic fluc-

tuations adversely affected farming and other native societies moved into the plains from the Mississippi Valley and the Great Basin. It was during this transitional period that French and Spanish explorers began recording native inhabitants and lifeways in the basins of the Red and Arkansas rivers (D. Wyckoff, personal communication). The Arkansas and Red basins first came under European control under the claims of Spanish explorers like Coronado and de Soto. By the early 1700s many French explorers, trappers, and traders came into the region, making contact with and in many cases marrying native people. Spain was recognized by other European nations as the owner of the region in treaties of 1762–1763 (Morris et al. 1986), but transferred ownership of "Louisiana" to France in negotiations in 1800–1802. The United States purchased

"Louisiana" for $15,000,000 in 1803, by which most of the Red and Arkansas river basins, along with the Missouri and western Mississippi basins, became permanently owned by the United States. Following this purchase, numerous military expeditions throughout the West provided the first records of natural history of the region, and the stage was set for European dominance of the Red and Arkansas river basins.

In spite of similarities among the larger rivers, the streams of the region are so diverse overall that they represent many of the types of rivers in North America, ranging from shallow, unstable sand-bed rivers, to tumultuous montaine headwaters, to small rivers in upland valleys of modest gradient. We selected four focus rivers in the region, including the Arkansas, Canadian, Red, and Little rivers. The Canadian and Little river systems are substantial tributaries of the Arkansas and Red rivers, respectively, but their dominance of the landscape or their unique faunas cause us to give them special attention. Of the many lesser rivers in the southern Great Plains, seven best represent the wide range of diverse physical, hydraulic, and floral–faunal characteristics of streams of the southern Great Plains: the Cimarron, Neosho (Grand), Illinois, and Poteau in the Arkansas River basin and the Washita, Blue, and Kiamichi in the Red River basin. Some of these smaller rivers differ substantially from most other streams in the region (e.g., the calcareous, marl-depositing Blue River), and we included them for their unique features.

Physiography and Climate

Physiographic provinces included in the region are Southern Rocky Mountains, Great Plains, Osage Plains section of the Central Lowland province, Coastal Plain, Ozark Plateaus, and Ouachita Province (Hunt 1974). The latter two provinces are known collectively as the "Interior Highlands" in many treatments of fauna, particularly fishes. The region is quite diverse (Brown and Matthews 1995) and includes some of the highest mountains in the conterminous United States, vast expanses of flat plains, parts of the Gulf Coastal Plain, the highly eroded Ozark Mountains, the ridge and valley structures of the Ouachita Mountains, and numerous smaller but important uplifted areas, including the Flint Hills in Kansas and the Wichita and Arbuckle mountains in Oklahoma. Outside the mountainous areas the landscape of the region appears relatively flat, but the plains drop from an elevation exceeding

1200 m at the base of the Rocky Mountains to 50 m asl or less in the east. A major feature of the western part of the region is the Llano Estacado or "Staked Plain," which rises abruptly along a long north–south line in the Texas and Oklahoma panhandles, forming a high level plain of immense proportions.

Geologically, the region is complex (Hunt 1974). It was not glaciated in the Pleistocene, but glaciation to the north had strong influence on river courses and connectivity in the southern plains (Cross et al. 1986) and likely on distributions of aquatic organisms. The Arkansas and Canadian rivers arise in granitic-volcanic terrain as high-gradient, turbulent streams over boulder-strewn channels, with the upper Arkansas River sufficiently large and high gradient to support a white-water rafting industry.

Thus, at the extreme western edge of the region are mixed volcanic and metamorphic rock and Tertiary or Mesozoic sedimentary deposits, mostly of marine origin. Dominance of marine sedimentary deposits continues east onto the Great Plains, with streambeds also characterized by outwash from the Rocky Mountains (Hunt 1974, Brown and Matthews 1995). Much of the region in eastern Kansas, most of Oklahoma, and western Arkansas includes marine and continental sediments of Mississippian, Pennsylvanian, and Permian age, with sedimentary Cretaceous formations in southeast Oklahoma and southwest Arkansas. The Red River from south-central Oklahoma through Louisiana passes mostly over Quaternary sedimentary deposits of fluvial origins, as does the Arkansas River after it drops off the Fall Line onto the Mississippi Delta of the Coastal Plain (Hunt 1974). As a consequence of geologic diversity between and within the river drainages, the physical structure of rivers in the southern Great Plains ranges from high-gradient riffle–pool headwaters to wide, shallow, sandy river main stems, with substrates varying from boulder and bedrock to fine river sands.

Extensive geological and podological research near the Great Bend of the Arkansas River documents soil development and surface stability about 36,000 years ago, followed by wind and water erosion until some 22,000 years ago. The region was a C_4 dominated grassland, where horses, camels, and mammoths were the prevalent large herbivores. By 21,000 years ago the Wisconsinan glaciation was nearing its maximum. Silt from the glacial front was deposited in the Arkansas basin, and macrofossil finds indicate that at least scattered stands of spruce woodlands were present along the middle Arkansas River during full glacial times. To the east, in the

Grand (Neosho) watershed, spruce and pine woodlands persisted until at least 12,000 years ago. During full glacial times the High Plains and their eroded eastern margins in the Red and Arkansas basins were lush grasslands where mammoth, camel, and big horned bison were common (D. Wyckoff, personal communication).

Total rainfall and its temporal distribution has long been the bane of settlers in the western parts of the region. Thornthwaite (1941) noted about the Great Plains, "Men have been badly fooled by the semiarid regions because they are sometimes humid, sometimes desert, and sometimes a cross between the two." The Llano Estacado of eastern New Mexico and west Texas, through which the Red River, Canadian River, and southern tributaries of the Arkansas River pass, is so lacking in surface water that a patrol of "Buffalo Soldiers" of the 10th Cavalry, somewhere near the headwaters of the Red River in 1877, drank the blood and urine of their horses in an attempt to survive (Leckie 1967).

Climate differs markedly from southeast to northwest. Louisiana and eastern Arkansas are in a region of hot summers (daily average temperature about 28°C in July) and moderate winters. In the western part of these river basins, summers are also hot, but winters can have extended periods of cold and true blizzard conditions. Snow cover is not persistent, however, except at high elevations in the mountainous westernmost parts of the Arkansas and Canadian rivers. Air temperatures in the headwaters of the Canadian and Arkansas rivers in Colorado–New Mexico are cold, influenced by elevation, but even on the lower-altitude western plains in the upper parts of these basins mean air temperature in January is below freezing. Mean annual precipitation decreases from >120 cm in Arkansas or Louisiana to about 20 cm in the west, in the rain shadow of the Rocky Mountains (Brown and Matthews 1995). Because of its central location in the North American continent, air masses from the Pacific Ocean, Gulf of Mexico, and Canada converge in the region, resulting in some of the most violent weather on Earth (e.g., central Oklahoma averages more tornados per year than any other location, and the most powerful tornado ever recorded hit Oklahoma City on May 3, 1999).

Basin Landscape and Land Use

Biomes within the southern Great Plains region include Temperate Mountain Forest, Temperate Grasslands, and Temperate Deciduous Forest. Terrestrial ecoregions in the southern Great Plains include, roughly from west to east, the Colorado Rockies Forests, Western Short Grasslands, Central and Southern Mixed Grasslands, Flint Hills Tall Grasslands, Central Forest/Grassland Transition Zone, Ozark Mountain Forests, Piney Woods Forests, and Mississippi Lowland Forests (Ricketts et al. 1999). Major rivers like the Red, Canadian, and Arkansas flow through and cut across the terrestrially defined ecoregions but have such distinctive faunas that some states (e.g., Oklahoma) have officially recognized an additional "Big Rivers Ecoregion" that represents these unique large rivers and their associated riparian zones. A prominent feature within the Central Forest/Grassland Transition Zone is the "Crosstimbers," a mosaic of forest, woodland, savanna, and prairie vegetation dominated by post oak and blackjack oak (Hoagland et al. 1999) about 100 km wide east to west and extending from southern Kansas to north Texas.

Before European settlement many river main stems in the east flowed through vast forests, and even in the western part of the region riparian forests of cottonwood, willow, or chinaberry existed before they fell to the axes of early settlers in need of timber and firewood. One enormous cottonwood near the Canadian River in present-day north Texas was reported by Capt. Randolph Marcy as being 19.5 feet (almost 6 m) in circumference. Historically, vast forests covered the Interior Highlands and unbroken grasslands extended to the west. Forests of the Ozark Mountains remain relatively like the original oak–hickory forest and probably have had the least cultural modification overall. The Ouachita Mountains in south Arkansas and eastern Oklahoma were dominated in presettlement times by relatively open parkland of large pine trees interspersed with grasses.

Beyond the immediate proximity of the rivers, much of the western part of the region was covered in presettlement times by seemingly endless plains of short grasses. Near the headwaters of the Red and Canadian rivers, Marcy described "the elevated plateau of the Staked Plain, where the eye rests upon no object of relief within the scope of vision," and that "the grass upon the Staked Plain is generally a very short variety of misquite [sic], called buffalograss, from one to two inches in length, and gives the plains the appearance of an interminable meadow that has been recently mown very close to the earth" (Foreman 1937).

Rivers have played major roles in the European human history of the region. In 1806, Capt. Zebulon Pike's expedition followed the Arkansas River from Great Bend west to the Rocky Mountains, but a detachment under Lt. James Wilkerson turned downstream at Great Bend to follow the Arkansas past the present site of Webbers Falls, Oklahoma, thence to New Orleans. These explorations provided some of the earliest accounts of the landscape and natural history of the Arkansas River region. Other pre–Civil War explorations of the Arkansas River region in Kansas–Colorado were by parties under command of or including Maj. Stephen Long, Thomas Say, Jacob Fowler, Jedediah Smith, John C. Fremont, and Washington Irving. Further south, substantial explorations and descriptions of natural history were by Col. George Sibley (Chikaskia River and Salt Fork of the Arkansas), Maj. Stephen Long (Red River, Kiamichi River, Mountain Fork, Poteau River), Thomas Nuttall (botanical descriptions of streams and mountains in southeast Oklahoma, then the Arkansas, Grand, Verdigris, Canadian, North Canadian, Deep Fork, and Cimarron rivers, and native peoples encountered therein), and a variety of other military or trading expeditions (Morris et al. 1986). Highly significant in knowledge of natural history of the region were the "Pacific Railroad Surveys" westward across the region under the command of Whipple, Marcy, and others, completed by the War Department in the 1850s.

In the mid-1800s, military expeditions under Marcy and Whipple followed the major east–west rivers like the Canadian and Red to seek routes for railroads, trade and commerce, or westward settlement. Steamboats navigated the Arkansas and Red rivers upstream to eastern Oklahoma by the early 1800s, and rivers became corridors along which movements of settlers and military were common. Throughout the 1800s, native peoples from elsewhere in the United States were displaced from their homelands and moved to the region that would become Oklahoma. The major original native people in the Red and Arkansas basins prior to the European-mediated displacements of the 1800s were the Kichai and Caddo in Louisiana, eastern Oklahoma, and southwest Arkansas; the Quapaw and Osage in southern Oklahoma, northeast Oklahoma, Arkansas, and Kansas; the Wichita along the Red River; and the Comanche, Kiowa-Apache, and Apache on the high plains to the west. Arapaho and Cheyenne lands were generally north of the Arkansas River basin (Socolofsky and Self 1988), but these nomadic peoples of the high plains no doubt were also present in the Southern Plains region at times.

European settlement of the Arkansas or Red River basin was advanced by establishment of military forts throughout the region from about 1820 onward, by the existence of federal wagon roads after 1849, by stage routes through the region after 1850, by cattle trails from Texas that crisscrossed the region from about 1840 to 1897, by sheep trails up the Arkansas River from the 1870s to 1900, and by establishment of railroads with access to commerce in western Kansas and eastern Colorado after 1865, followed by rail lines throughout the region by the 1870s and 1880s. Statehood promoted settlement by the establishment of permanent governments, with the states in the Southern Plains region granted statehood as follows: Louisiana (1812), Arkansas (1836), Texas (1845), Kansas (1861), Colorado (1876), and Oklahoma (1907).

Europeans dramatically changed the region by plowing the prairies, cutting forests, mining, altering patterns of stream flow, and extracting oil and natural gas. The region is now a mosaic of private timber production, large areas of forest under federal ownership and management, row-crop production, and cattle ranching, with only a few large cities and very limited heavy manufacturing. In recent decades, large-scale swine and poultry production has sharply increased, threatening water quality. Feedlots for cattle also cause local water-quality problems. Plowing of native prairie caused increases in silt and associated losses of some fish species in streams (Cross and Moss 1987). Fire suppression on former prairies allowed encroachment of trees, and the introduced cattle imposed grazing patterns very different from those of the wide-ranging native bison they replaced. The last "virgin forest" in the central United States (in the Ouachita Mountains) was harvested for timber before 1950 (Smith 1986) and replaced by pine monoculture, now cut mostly for pulp, particle board, and similar products.

As the vast southern bison herd was annihilated in the late 1800s, open ranging of cattle began, followed by huge privately owned cattle ranches hundreds of square kilometers in size or larger (e.g., in the upper Red River region). Where water was available, row crops of wheat, corn, sorghum, milo, cotton, and, more recently, peanuts came to dominate the landscape. Center-pivot irrigation allowed extension of row crops farther west than formerly possible, but decimated important aquifers like the Ogallalah, which formerly recharged many prairie streams but now is reduced and imperiled. A dramatic example of irrigation and aquifer depletion is in the Oklahoma Panhandle. There, a sharp increase in numbers of high-capacity irrigation wells (since about 1960) coincided with a dramatic increase in the number of "no-flow" days per year in the Beaver (upper North Canadian) River. No-flow days increased from fewer than 20 before 1960 to almost 100 in many of the years from 1980 to the present (http://nwis.waterdata.usgs.gov/ok/nwis/discharge).

Discovery of oil and gas resulted in further changes to the landscape in the 1900s, with negative impacts on some streams from salt water and other byproducts of drilling. In some areas, mining for zinc, lead, or other minerals contaminated large terrestrial areas and polluted streams like Tar Creek in northeast Oklahoma. Limited coal mining in the east and some gold mining in the west has had local impacts on streams. However, the region largely lacks raw materials, such as iron, copper, or coal, that support heavy manufacturing, so rivers here escaped some of the pollution problems. An increasing lack of water relative to demands, siltation from agriculture, local sewage or agricultural pollution due to large agribusinesses, impacts of impoundments, and generally poor water quality or physical conditions for biota in western parts of the region remain the most serious challenges to streams in the region. Rivers to the east, including the Little, Kiamichi, Neosho (Grand), and Illinois, may be some of the "best" in North America in retaining much of their original presettlement biodiversity and aesthetic quality. However, some, like the Blue River, are under increasing user pressure and erosion of overall quality.

Ultimately, the largest cities in the region grew on or near major rivers, including Little Rock and Fort Smith, Arkansas; Shreveport, Louisiana; Tulsa and Oklahoma City, Oklahoma; and Wichita, Kansas. With construction of the Kerr-McClellan navigation system from 1957 to 1970 widening and modifying the Arkansas and Verdigris rivers, Tulsa, Muskogee, and Little Rock became major barge ports, linking those inland cities to overseas commerce. The construction of numerous locks and dams along that system, in addition to the large dams and reservoirs on most river systems in the region, have changed irreversibly the channel configurations and flow schedules of these systems (although not necessarily altering overall annual discharge for river basins).

The Rivers

Rivers of the region flow through the Southern Plains, Central Prairie, Ozark Highlands, Ouachita Highlands, and Mississippi Embayment "freshwater ecoregions" (Abell et al. 2000). However, these freshwater ecoregions are very large and therefore generalized, and there is great diversity of streams within any one ecoregion. The rivers included in this chapter fall into two rather different groups on the basis of their upland versus lowland characteristics. The group including the Arkansas, Canadian, Red, Washita, and Cimarron rivers represents typical large to medium-size low-gradient prairie main stems with wide, shallow, braided, unstable sand-bed channels, often carrying heavy loads of large wood snags washed in by floods. These form massive brush piles around bridge abutments, requiring regular removal in some rivers (e.g., Washita River in southern Oklahoma). These rivers are highly distinctive as habitats or "ecoregions" within the region, containing animals and plants (e.g., some fishes and riparian vegetation) not found elsewhere. The other rivers, including the Little, Kiamichi, Blue, Illinois, Neosho (Grand), and Poteau, arise in upland areas as stony-bottomed streams and are typified in their upper reaches by relatively clear water, differentiation into swift riffles and deep pools, and relatively diverse and stable structural features. As streams of this latter group leave the uplands they become more turbid, sluggish, and incised but lack the braided sand-bed channels that typify rivers of the other group.

However, even within these groups each river has unique features of topography and biota, hence their inclusion here to provide a cross-section of rivers typical of the southern Great Plains region. The main-stem Red River arises in the Texas Panhandle near Amarillo, Texas, and lacks any montaine snowmelt influence. The Arkansas River and the Canadian River originate in the Rocky Mountains, and thus are influenced by snowmelt in their upper reaches. However, as they emerge from the mountains onto the dry western plains they lose water from evaporation or agricultural withdrawals and become small streams in western Oklahoma and Kansas.

In sharp contrast to low-gradient rivers flowing eastward across the prairies and grasslands, other rivers in the region arise in heavily forested, uplifted areas like the Ozark and Ouachita mountains in Oklahoma–Arkansas, the Arbuckle Mountains of Oklahoma, and the Flint Hills of Kansas. The Neosho (Grand), Illinois, Blue, Kiamichi, and Little rivers are strongly influenced in physiognomy by the uplifted areas where they arise and through which they flow for much of their length. Streams originating in these uplands typically have high-gradient headwaters with strong base flow (in spite of some dramatic drought conditions in recent years), more benign environmental conditions, clear water, and stony bottoms with much more well-defined riffle–pool configurations than for prairie streams.

Large to modest-size springs dominate base flow in the Ozarks and Arbuckles, whereas the Ouachita uplands have fewer large springs and their streams arise from runoff and more gradual influx of subsurface water. Rivers in the eastern mesic part of the region generally have more reliable flow, complex physical structure, and diverse faunas than rivers further west.

Late-summer drying of streams in the region is a function of both evapotranspiration and lowered rainfall. Shallow river main stems in the midportion of the region can enter winter under drought conditions if autumn rains are lacking, and under those conditions icing can be substantial in main channels. To the southeast, icing of river main stems typically does not occur. In most of the region spates can occur in any month of the year, and flow patterns are unpredictable (Resh et al. 1988, Brown and Matthews 1995). The rivers influenced by snowmelt may have more predictable seasonal patterns, but their flow also becomes erratic after entering the plains.

Extreme rainfall events result in annual or more frequent bankfull spates and streambed scouring in many of the smaller tributaries to the main rivers. For example, in October 1981, approximately 43 cm of rain fell in a two-day period in parts of south Oklahoma, resulting in massive flooding of small streams and alteration and complete scouring and reshaping of their physical structure. Such events also cause widespread flooding of main rivers, rearrangement of their sand-bed substrates and braided flow patterns, and extreme water-level fluctuations in main-stem reservoirs. Conversely, severe drought years have occurred in the region in recent decades (e.g., 1977–1979, 1981, 1991, 1998, and 2000), with long reaches of streams dewatered or reduced to small, isolated pools and reservoirs reaching very low levels. Under such conditions temperature and oxygen stress can be extreme, and animals crowded into shrinking habitats may experience density-dependent as well as density-independent constraints on their growth, survival, or population sizes (Matthews et al. 2001). Although the fauna is largely resistant to or adapted for physical stress, some evidence now suggests that recurrent drought may be having impacts on stream fishes (W. J. Matthews, unpublished data) and mussels (D. Spooner and C. C. Vaughn, unpublished data).

The rivers that originate in the Rocky Mountains (Canadian, Arkansas) arise as retentive, debris-regulated channels that become alluvial gravel beds with fluvially formed riffle–pool structure farther downstream and wide, braided sandbed channels after entering the plains (Brown and Matthews 1995). Other streams originating in the region arise as alluvial gravel riffle–pool channels or sand-bed or mud-bottomed streams. In the prairie portions of rivers like the upper Arkansas, Red, Canadian, Cimarron, and Washita, harsh physical conditions owing to solar heating, extreme and extended winter cold, and unpredictable drought or flooding limit development of the biota and may disrupt the transitions in flora and fauna hypothesized for more stable systems. Rivers in the prairie may be characterized by local patchiness of habitat and physicochemical conditions (e.g., local refugia formed at mouths of creeks entering river main stems). There, deeper shaded refugia, compared to the shallow exposed main stems, may provide less harsh thermal, oxygen, or other physical conditions (Matthews and Hill 1979). Sources of energy for streams in the western and central parts of the southern Great Plains likely differ from those in mesic forested regions in that headwaters of many systems lack riparian forest, and energy may come directly from autochthonous production (periphyton, biofilms) or from prairie grasses as allochthonous inputs.

In rivers arising in the mesic uplands to the east (including the Little, Kiamichi, Blue, Illinois, Neosho [Grand], and Poteau rivers) base flow (often spring fed) is more reliable, harsh physical conditions are less common, and a more predictable longitudinal zonation or continuum of biota and ecosystem processes may exist. However, most of these rivers are interrupted by dams, ranging from very large flood-control/hydropower structures to local low-head dams for municipal water supply or other uses, all potentially disrupting natural transitions of organisms and processes from headwaters to lower rivers. Finally, modern distribution of some aquatic organisms may be influenced by a large "Pre-Glacial Plains Stream" (Metcalf 1966, Cross et al. 1986) that hypothetically cut across present-day east–west river systems, draining into the lower Mississippi River or Gulf of Mexico, thus providing an avenue of north–south movement for stream fishes in the Pleistocene.

All large rivers and most small ones in the region have been dammed, resulting in reaches below dams in which physical conditions are markedly altered (e.g., increased water clarity due to trapping of silt in the reservoir; increased stony substrates in some cases). Rivers have been altered by clearing and snagging rivers for boat passage, contamination by salt water from oil production, and interbasin water transfers (Red to Trinity river basins). However, many physical features of main stems in the central part of the region remain similar to presettlement/first explorer reports (Matthews 1988).

ARKANSAS RIVER

The Arkansas River basin is the largest in the lower Great Plains, draining the western Mississippi River basin south of the Missouri River basin (Fig. 7.7). It originates in central Colorado near some of the tallest peaks in North America. Denver may be the "Mile High" City, but the Arkansas River could claim the title of "two mile" river because it drops almost exactly two miles in elevation from its headwaters to its confluence with the Mississippi River. In its course it passes through highly changing terrain, climate, and land use. It has strong flow in the Rocky Mountains, loses water to evaporation and withdrawal in eastern Colorado and western Kansas (Ferrington 1993), then recharges to be a major plains river through the rest of its path to the Mississippi River. In Oklahoma, the main-stem Arkansas River is joined by the Salt Fork, Cimarron, Verdigris, Neosho (Grand), Illinois, and Canadian rivers. The Arkansas becomes a 7th order river at the confluence of the Verdigris, at which it remains to its confluence with the Mississippi River. Some old maps show the large White River of north Arkansas and southern Missouri as joining the Arkansas River before the latter enters the Mississippi River, but at present the White River flows directly into the Mississippi River and there is no direct connection of the White and Arkansas rivers (except perhaps during periods of major flooding).

The lower Arkansas River basin was occupied from 10,000 to 2000 years ago by Paleo-Indians similar to those of known bluff-dwelling sites in the Ozarks and Ouachita mountains, who produced characteristic projectile points and stone tools (Hanson and Moneyhon 1989). In the "formative period" of Paleo-Indian culture in the lower Arkansas basin, permanent villages appeared, and major mound-building cities arose. For example, the Spiro Mound, near the Arkansas River in eastern Oklahoma, produced a rich treasure of detailed and intricate artifacts, making clear that this was a major center of prehistoric culture. By the mid 1500s the Spanish explorer de Soto reported large native populations along the Arkansas River, with substantial fortified villages (Hanson and Moneyhon 1989). In the western Arkansas basin nomadic peoples like the

Kiowa, Comanche, Arapaho, and Cheyenne roamed the plains, with their mobility vastly increased and lifeways highly altered by dispersion of European horse onto the plains in the 1600s to early 1700s (Beck and Haase 1989). The upper Arkansas River basin was first explored with an eye toward permanent settlement beginning in about 1806, after the Louisiana Purchase, and during the next 50 years a substantial number of military and private expeditions used the upper Arkansas River as a conduit to the west.

Physiography, Climate, and Land Use

The Arkansas River basin flows primarily in an east-southeasterly direction, from latitude 39°N to latitude 34°N, passing through six physiographic provinces, including Southern Rocky Mountains (SR), Great Plains (GP), Central Lowland (CL), Ozark Plateaus (OZ), Ouachita Province (OP), and Coastal Plain (CP). This area includes the Western Short Grasslands, Central and Southern Mixed Grasslands, Central Forest Grassland Transition Zone, Ozark Mountain Forests, and Mississippi Lowland Forests terrestrial ecoregions (Ricketts et al. 1999). There is a wide range of vegetation, including coniferous uplands in the Rocky Mountains, short native grasses in the west, mixed to tallgrass prairie in the east, slopes covered with deciduous forests in the Ozarks, and dense Coastal Plain forest downstream in eastern Arkansas.

Climate varies greatly along the Arkansas River from central Colorado to eastern Arkansas. Near its headwaters in Colorado, mean monthly air temperatures typically range from −8°C in January to 13°C in July, with average annual rainfall about 51 cm. Overall, precipitation is greatest in May or June and lowest from December to February (Fig. 7.8). Mean temperatures are hottest in July and August and lowest in January. To the east, the climate is warm-subtropical, with generally hot, humid summers and mild winters and only occasional extreme cold events. At Pine Bluff, Arkansas, near the far eastern end of the basin, mean January temperature is 7°C, mean July temperature is 29°C, and rainfall exceeds 100 cm per year. In much of Colorado and western Kansas, January daily average temperatures are below freezing, but in eastern Oklahoma and Arkansas, January temperatures average well above freezing. From Colorado through western Kansas there is too little rainfall (<50 cm/yr) for farming without irrigation. As the river passes through Kansas and Oklahoma it traverses sharp increases in rainfall, passing the 100 cm/yr isopleth in northeast Oklahoma.

Land use in the Arkansas basin varies with climate. In arid lands to the west, rangelands and grazing of cattle dominate, with row-crop agriculture, particularly wheat, increasing with rainfall to dominate the landscape through much of Kansas. In Oklahoma and Arkansas, row crops and livestock dominate the immediate river valleys, but the surrounding hills and mountains are heavily wooded. The entire region, with the exception of major cities like Pine Bluff, Little Rock, Fort Smith, Muskogee, Tulsa, and Wichita, is predominantly rural; oil and gas production is a major industry. There is little heavy manufacturing in the region; hard-mineral extraction is limited or localized and has no substantial impact on the river. Estimated percentages of land use outside of cities are, from U.S. Department of Agriculture maps, 50% rangeland and 50% cropland in the western parts of the basin, compared to 50% forest, 15% cropland, and 25% pasture further to the east.

River Geomorphology, Hydrology, and Chemistry

The Arkansas River is formed by creeks near Leadville, Colorado, at 3010 m asl, 30 km north of Mt. Elbert, the highest mountain in the state (and in the Arkansas River basin) at 4370 m asl. The Rocky Mountains in central Colorado are volcanic and metamorphic rock interspersed with sedimentary marine deposits; thus, the upper Arkansas River main stem flows through steep and rugged terrain en route to the plains. Near Canyon City, Colorado, it flows through the Royal Gorge, where one of the highest suspension bridges in the world passes over the turbulent main stem 320 meters below.

Downstream, it leaves the Rocky Mountains and flows near Pueblo, Colorado, onto the plains of east-central Colorado, becoming a low-gradient river through grazing and agricultural lands. Crossing Kansas, Oklahoma, and western Arkansas, the river traverses marine and continental sediments of Mississippian, Pennsylvanian, and Permian age, bisecting the formerly continuous Interior Highland (Ozark and Ouachita mountains) and finally reaching Quaternary sediments of the Mississippi Embayment. Before leaving Colorado the Arkansas is joined by the Purgatoire River, which drains very arid lands of southeastern Colorado, and forms John Martin Reservoir, the first large reservoir on the main stem.

Eastward in Kansas, it passes Garden City and Dodge City, losing water to become ephemeral upriver of Great Bend. The most severe dewatering has been in a reach from Syracuse to Great Bend, Kansas, where complete loss of surface flow has occurred in several places for as much as a year (Ferrington 1993). Serious declines in surface flow coincided with increases in groundwater withdrawal in the early 1970s (Ferrington 1993). Near Great Bend it receives water from the Pawnee River and passes near the famous Cheyenne Bottoms Wildlife Area, which is a critical wetland for waterfowl and a wide variety of birds in the Central Flyway. Downstream from Great Bend the Arkansas thus is again a "river," about 20 m to more than 100 m wide in places, with moderate to swift current over coarse gravel and cobble in riffles and sand-bottomed pools to nearly a meter deep. Flowing through Wichita, Kansas, it has a braided, sandy channel, and beyond Wichita it is joined by the sandy Ninnescah River and the turbid Walnut River and is a sizeable stream as it crosses into Oklahoma. Immediately below the Kansas–Oklahoma border it becomes Kaw Lake, gains the Salt Fork of the Arkansas, and flows into Keystone Lake, impounded at the juncture of the Cimarron and Arkansas rivers.

Below Keystone Lake the Arkansas River flows through downtown Tulsa, Oklahoma, where it has a sand-bottomed, braided channel hundreds of meters wide, with flow strongly controlled by releases from the reservoir. Near Muskogee, Oklahoma, it is joined by the channelized Verdigris River and becomes the Kerr-McClellan Navigation System. The river passes through locks and dams and the Robert S. Kerr reservoir, exiting Oklahoma near Ft. Smith, Arkansas. At Ft. Smith the stream is a large, deep river several hundred meters wide, with mostly sand banks, and the channel is controlled by wing dikes and other navigational developments. Near Russellville, Arkansas, the river enters Dardanelle Reservoir, the last large man-made impoundment on the river. In central Arkansas the river gains more barge traffic at the port of Little Rock, then flows southeast through the large forests and rich farmlands of the Mississippi Delta to its confluence with the Mississippi River. From Little Rock downstream the river is low gradient, with broad meanders, backwater bays and sloughs, and flooded riparian forests, but is frequently interrupted by locks and dams of the Kerr-McClellan Navigation System. In its total course from headwaters in the Rocky Mountains to its confluence with the Mississippi River the Arkansas River has a mean overall slope of 185 cm/km.

Virgin discharge (1928 to 1939, before large dams) averaged 1004 m³/s downstream in the basin at Little Rock, Arkansas. For the entire period of record at Little Rock the mean annual discharge "prior to regulation" (1928 to 1969) was 1118 m³/s, whereas after upstream regulation by locks and dams, the mean annual discharge from 1970 to 2000 was 1389 m³/s (ranging from 360 to 2711 m³/s). Thus, overall flow statistics show that in spite of water withdrawal in Colorado and west Kansas and the highly regulated nature of the system, as much water is delivered downstream at Little Rock as during virgin flow conditions. Slightly upstream, near Van Buren, Arkansas, mean annual discharge only decreased from 885 to 826 m³/s after regulation by Lake Eufaula (on the Canadian River, a large tributary), further evidence of only modest changes in discharge relative to virgin flow conditions. Runoff (0.3 to 1.5 cm/mo) is low in the Arkansas River basin because it includes arid lands, with most of the precipitation lost as evapotranspiration before it ever reaches the mouth (see Fig. 7.8).

The headwaters near Leadville, Colorado, are cold even in summer (e.g., 10°C) and low in ion concentrations. Typical values at three USGS headwater sites include conductivity of 85 to 150 μS/cm, pH of 6.3, hardness of 30 to 50 mg/L as $CaCO_3$, and calcium, magnesium, sodium, and chloride all <10 mg/L. Crossing the plains, the river changes chemically so that below Great Bend, Kansas, the main stem is characterized by pH typically >8.0, conductivity of 1000 to 3000 μS/cm, total hardness of 300 to 500 mg/L as $CaCO_3$, alkalinity of 150 to 300 mg/L as $CaCO_3$, and ions dominated by sodium (300 to 500 mg/L) and chloride (200 to 800 mg/L), but with substantial magnesium (20 to 40 mg/L), calcium (100 to 130 mg/L), and sulfate (150 to 250 mg/L). From Ft. Smith downstream the main stem typically has pH of 7.5 to 8.3, conductivity of about 400 to 900 μS/cm, total hardness of 100 to 200 mg/L as $CaCO_3$, alkalinity of 80 to 130 mg/L as $CaCO_3$, sodium of 30 to 80 mg/L, chloride of 40 to 150 mg/L, magnesium of 10 to 15 mg/L, calcium of 30 to 50 mg/L, and sulfate of 40 to 90 mg/L. In the lower river in east Arkansas, where there is much row-crop agriculture, nutrients are high: Total organic nitrogen is typically 0.5 to 2.0 mg/L and sometimes as high as 4.0 mg/L, and total phosphorus is about 0.05 to 0.20 mg/L. Water temperatures of the main river in eastern Arkansas have reached 32°C in some summers and as low as 4°C in winter. Average year-round water temperature for the basin is estimated at about 18°C.

River Biodiversity and Ecology

As mentioned, the Arkansas changes greatly along its course, and in so doing passes through four freshwater ecoregions: Southern Plains, Central Prairie, Ozark Highlands, and Mississippi Embayment (Abell et al. 2000). The Ozark Highlands ecoregion is largely defined by its high-gradient tributaries that drain into the Arkansas River; the main stem remains a low-gradient river that does not itself pass through the Ozark Mountains before entering the Mississippi Embayment. There are no comprehensive, basinwide studies of the flora and fauna of the Arkansas River. Sections of the river are treated in appropriate "state" floral and faunal guides, and there have been substantial studies of fishes in Kansas and Colorado by the University of Kansas, Fort Hays State University, and Colorado State University.

Algae

Relatively little is known about algal communities of the Arkansas River. Wilhm et al. (1978), however, reported that phytoplankton was dominated by diatoms (109 of 128 taxa), with *Cyclotella* and *Melosira* the most abundant genera. Benthic diatoms included *Navicula*, *Surirella*, *Nitzschia*, *Synedra*, *Cocconeis*, *Amphiprora*, and *Gomphonema*.

Plants

Riparian areas are dominated by silver maple, box elder, bur oak, and red oak. Sugarberry is common. Cottonwood–willow woodlands are common on floodplains throughout the watershed. Cattails and American bulrush occur in temporarily flooded sloughs. Within-channel macrophytes are not a noteworthy feature of the main-stem Arkansas River.

Invertebrates

The Arkansas River supports a diverse array of macroinvertebrates. Hard substrates (clear, rocky areas of the upper river and woody snags throughout) support grazing mayflies such as *Stenonema* and *Heptagenia*, filtering caddisflies such as *Cheumatopsyche* and *Hydropsyche*, the stonefly *Neoperla*, and black flies (*Simulium*). Further downstream the mayflies *Caenis* and *Hexagenia* (a burrower) become more common. Chironomid midges associated with hard substrates include *Polypedilum*, *Rheotanytarsus*, and *Glyptotendipes* (Wilhm et al. 1978). Crayfishes include *Orconectes palmeri*,

O. virilis, O. nais, Procambarus simulans, and *P. acutus* (Reimer 1969, E. A. Bergey, personal communication). Goldhammer and Ferrington (1992) demonstrated the importance of "epirheic" zones of capillary water fringe habitats as sources of secondary production of aquatic invertebrates in the Cimarron River, a tributary of the Arkansas River.

Lower reaches of the Arkansas River contain a diverse "big-river" mussel assemblage dominated by the commercially important washboard mussel, threeridge, and mapleleaf. Other common mussels include pink papershell, bleufer, plain pocketbook, fluted shell, fragile papershell, pimpleback, fawnsfoot, and pondhorn (Branson 1982, 1983, 1984, C. C. Vaughn, unpublished data). Two nonnative bivalves have invaded the Arkansas River: the Asian clam and the zebra mussel. Zebra mussels were introduced via barges in the early 1990s and are common in several main-stem impoundments, particularly Kerr Reservoir.

Vertebrates

The Arkansas River basin has 141 known species of native fishes and about 30 nonnative species. The number of native species occurring in the lower, middle, and upper river are 117, 111, and 64, respectively (Cross et al. 1986), with the fewest species in the far western reaches. Some species like longear sunfish occur widely in habitats from main stem to small tributaries, but others, like the river shiner or river darter, are mostly in the main channel. Native trout occur in the Arkansas River headwaters. The Arkansas River shiner, a federally threatened species, occurs in the main channel of the lower Arkansas River. Other main-stem-limited taxa of special interest are small minnows of the speckled chub complex, with the peppered chub (now rare or extinct) known from middle and upper portions of the system, and the shoal chub present in middle and lower parts of the drainage (Eisenhour 1999). The speckled chubs and Arkansas River shiner thus are unique faunal elements of the main stem. Collections made with small-meshed seines produced 7 and 9 species at a single site in the main stem of the upper Arkansas River at Great Bend, Kansas, compared to 16 and 17 species with about the same effort in shallow edges at a middle main-stem site at Webers Falls, Oklahoma. Below Ft. Smith in western Arkansas, T. M. Buchanan (personal communication) typically finds 18 to 20 species in similar seining samples in the main stem. The Arkansas River main stem in the lower parts of the basin is occupied by game fish like large-mouth bass and spotted bass and the introduced striped bass, whereas upriver, where the main stem is smaller and shallower, white bass are more common piscivores. Channel catfish and blue catfish are also common in the main river. Formerly present anadromous species like American eel may persist lower in the basin, but their upstream passage through high dams is unlikely, so they would now be rare in the system upstream. Gars of the genus *Lepisosteus* are common in the lower main channel and associated backwaters or oxbow lakes.

Major aquatic amphibians or reptiles of the upper Arkansas River main stem and associated habitats (Colorado or Kansas) include plains leopard frog, American bullfrog (introduced in Colorado), snapping turtle, yellow mud turtle, common slider, midland smooth softshell turtle, western spiny softshell turtle, and northern water snake. Farther downstream (Oklahoma or Arkansas) are common snapping turtle, false map turtle, red-eared turtles, and several species or subspecies of water snakes. Cottonmouth are found from east Oklahoma downstream through Arkansas. Major riparian birds and mammals include white pelican (which sometimes overwinters), great blue heron, green heron, belted kingfisher, beaver (which are broadly increasing), and muskrat. River otter is an Oklahoma state species of special concern in the Arkansas River system, whose populations may be increasing due to stocking programs. River otter are increasing in Arkansas (Sealander and Heidt 1990), as are the nonnative nutria. Nonvegetated beaches and sandbars of the Arkansas River support breeding populations of the federally endangered interior least tern, including a protected population within the city of Tulsa.

Ecosystem Processes

The structure and function of Arkansas River ecosystems, from montaine headwaters to confluence with the Mississippi River, have not been studied comprehensively. Productivity in the headwaters is probably dominated by availability of coarse rock substrates and relatively clear water and perhaps nutrient limited, whereas far downstream the main river is sufficiently large and turbid that photosynthesis is probably restricted to upper parts of the water column and hard substrate processes are probably of minimal importance. The river downstream in Arkansas and Oklahoma is too deep in midchannel for substantial development of large snag piles, although wood is probably biologically important in backwaters or side channels. Comprehensive studies of such processes for the lower main stem are lacking. Capillary fringe habitats (Goldhammer and Ferrington 1992), which have become scarcer since European settlement, may nonetheless be important sites of secondary production.

Many upland tributaries are stony bottomed, with algal communities forming thick coatings on substrates. In these tributaries, large standing crops of central stonerollers, other algae-eating minnows (Ozark minnow, southern redbelly dace), crayfishes, and snails likely have a strong influence on benthic algae and autochthonous primary productivity. In contrast, tributaries in prairie regions are typically muddy, with shifting mud–sand bottoms and a poorly developed benthic algal flora.

Human Impacts and Special Features

Within the southern Great Plains the Arkansas River is unique as the river most likely influenced by snowmelt from the Rocky Mountains and probably crossing more diverse landscapes than any other in the region. Its headwaters are high in the Rocky Mountains, followed by its passage as substantial rapids through rock-bound canyons, thence out onto the plains, where its loss and gain of water is dramatic. Crossing Kansas, what was a brawling mountain river is reduced to such a small size that one can literally step across it in places; then, to the east it again gains water to become a major river and a watercourse famous in European history, now dotted with large dams and their associated reservoirs.

Military posts were present on the Arkansas River by the early 1800s, and steamboats traveled upriver to Muskogee by that time. Throughout most of the history of European settlement the main Arkansas River remained largely unaltered, but the latter half of the 1900s saw marked changes, with modification by the Kerr-McClellan Navigation System upstream to Tulsa, addition of numerous large dams, and withdrawal of water in the western half of the basin. As previously noted, loss of water from the main channel of the river in western Kansas is extreme, owing to depletion of aquifers and removal of water for irrigation. This devastation of the main river channel must have had tremendous, albeit unknowable, impacts on the animals and plants that depended on the river in that reach of the plains in pre-European times. In addition, the channel has been modified by wing dikes and other navigation-control structures. Major impacts in the middle and lower river are locks and dams for boat passage, conversion of much of the midreach into the Kerr-McClellan Navigation Channel, and construction of major on-channel reservoirs (Lake Dardanelle, Arkansas; Keystone Lake, Oklahoma). Major cities on the main-stem river include Wichita, Kansas, and Tulsa, Oklahoma, where major oil refineries are alongside the river. Zebra mussels have invaded the river via barges in the navigation channel.

CANADIAN RIVER

The Canadian River, with the North Canadian River (Beaver River in the Oklahoma Panhandle), is the longest tributary of the Arkansas River (Fig. 7.9). The main stem of the Canadian River is the "south" Canadian River, which flows eastward between 37°N and 35°N latitude and is joined by the North Canadian River and the Deep Fork River in Lake Eufaula in eastern Oklahoma just before the system joins the Arkansas River. We focus on the "south" Canadian River in this description. The Canadian, like the Arkansas, originates as a high-gradient stream in the southern Rocky Mountains, crosses vast, arid plains, where it courses through earthen-walled canyons or past steep bluffs, and then becomes a wide, sand-bed river through western and central Oklahoma. Through much of its midreach the river is shallow, flowing over mobile sand-bed substrates and exposed to intense solar heating in summer (Fig. 7.3). It is possible that no large river on Earth has a harsher or more rapidly changing thermal environment to challenge the existence of fishes and aquatic invertebrates (Hefley 1937, Matthews and Hill 1979).

Human history in the Canadian River basin is similar in many ways to that of the geographically parallel Arkansas and Red rivers. The same Paleo-Indians that roamed the upper Arkansas and Red rivers no doubt also included the Canadian in their journeys, as did many early European explorers. In recent presettlement history, the Kiowa, Kiowa-Apache, and Comanche dominated the upper Canadian basin. As some of the original native peoples were displaced by tribes from the east, nations like the Sac and Fox, Shawnee, Seminole, Creek, Choctaw, Kickapoo, and Iowa were established in the eastern Canadian basin, and the western Canadian basin became territory of the Wichita, Caddo, Cheyenne, and Arapaho. Important early forts in the Canadian basin were Fort Holmes and Camp Arbuckle on the Canadian River and Fort Reno and Camp Supply on the North Canadian River. Military roads connecting these forts helped promote movement of people and goods. The

FIGURE 7.3 South Canadian River, Oklahoma (Photo by W. J. Matthews).

California Road, from Fort Smith on the Arkansas–Oklahoma border, followed the Canadian River through Oklahoma and westward, retracing the route initially followed by Whipple's expedition as part of the Pacific Railroad Surveys. The western Canadian River basin became permanently settled as open-range cattle herding gave way to farming and fencing of the plains in the late 1800s.

Physiography, Climate, and Land Use

The Canadian River passes through four physiographic provinces (Southern Rocky Mountains [SR], Great Plains [GP], Central Lowland [CL], Ouachita Province [OP]) and three terrestrial ecoregions (Western Short Grasslands, Central and Southern Mixed Grasslands, Central Forest Grassland Transition Zone) (see Fig. 7.9). The Canadian River basin originates in northeastern New Mexico and southern Colorado in a region of mixed Cretaceous sedimentary deposits and rocks of volcanic origin. It crosses a broad belt of Tertiary sedimentary rock in the Oklahoma–Texas panhandle region, Quaternary sedimentary deposits in western Oklahoma, a broad region

of Upper Paleozoic unmetamorphosed sedimentary deposits in central Oklahoma, and then metamorphic rock related to the Ouachita uplift en route to its confluence with the Arkansas River (Hunt 1974). Soils in the western half of the Canadian basin are mostly alkaline red or red-brown prairie soils, whereas the eastern part of the basin flows through a region of neutral prairie soils (Hunt 1974). Original vegetation varied from short grasses in the west to mixed and tallgrass prairie in central Oklahoma, intermixed with the Crosstimbers region, which is dominated by low-growth form oak forests.

Like the upper Arkansas River, the Canadian River initially flows through a very arid region, with rainfall less than 40 cm/yr and high rates of evaporation. Summers are hot, with low humidity, and winters cold, with snow and blizzard conditions common. In eastern Oklahoma the Canadian River flows through more mesic country, with annual rainfall averaging 100 to 120 cm/yr. On average for the entire basin, rainfall is greatest in May, declining through the summer and autumn to winter low precipitation in December and January (Fig. 7.10). The upper Canadian system averages only about 175 to 180 frost-free days per year, whereas in east Oklahoma 210 to 220 days per year are without killing frost. Air temperatures in most of the Canadian River basin average about 2°C in January and 27°C in August (see Fig. 7.10).

Land use in the upper Canadian (New Mexico and Texas) is dominated by grazing and short grasslands, with Amarillo, Texas, the only major city near the river. Its course through Oklahoma is mostly rural, in landscapes dominated by cattle ranching, some oil production, and wooded bluffs or terraces. From central Oklahoma eastward, farms and ranches are common, with grazing land, wheat, cotton, milo, and substantial wooded areas. Land use overall is about 50% rangeland or pasture and 30% cropland but can be up to 55% forested in the eastern part of the basin.

River Geomorphology, Hydrology, and Chemistry

The Canadian River arises at about 1970 m asl near Raton, New Mexico, and is joined within the mountains by the Mora River, which arises south of Taos, New Mexico. The average slope of the entire Canadian River main stem from the headwaters to its confluence is 2.8 m/km. After exiting the mountains it crosses an arid, short-grass region and flows into Conchas Reservoir, a relatively small impoundment. Below Conchas Reservoir the river runs through steep-walled canyons, and Revuelto Creek adds water from the south before the river enters the Texas Panhandle. Crossing the Texas Panhandle north of Amarillo, it flows within about 50 km of the headwaters of the Red River, and early explorers were sometimes confused about which river they were on. Lake Meredith, near Borger, Texas, regulates flow of the river, which becomes a modestly wide, shallow, sand-bed river downstream through Texas and into western Oklahoma. In west Oklahoma the river makes three major loops, passing the Antelope Hills and the Black Kettle national grasslands, then flowing almost directly southeast for about 250 km past Norman, Oklahoma. In west Oklahoma it is a widening, sand-bedded river, with steep earthen-bluff banks that are 20 to 30 meters high in places. In central Oklahoma the Canadian flows through a riverbed several kilometers wide, with an actual sand bed half a kilometer or more wide in places and with flow highly variable and braided.

This is the area described by Helfley (1937) and the site of studies on fish habitat selection and

tolerance as a function of physical stressors (Matthews and Hill 1979, 1980, Matthews 1987). Hefley (1937) described flood formation of terraces of the Canadian River and the physical dynamics of the unstable floodplain. This condition remains in spite of upstream dams built since Hefley's writing. The main-channel Canadian River in central Oklahoma still shrinks in hot summers to be so narrow that a child can step across, but occasional rains convert the Canadian at the same locality into a raging river with standing waves of a meter high or more, running "full" from bank to bank. The same aeolian processes described by Hefley dominate the architecture of the sand bed outside the wetted channel, leaving wide expanses of sand bars sculpted by the wind.

Near Ada, Oklahoma, the Canadian turns to flow slightly northeast to its confluence with the Arkansas River. En route it remains a wide, shallow, sand-bed river, characterized by an unstable and shifting sand bed, with relatively steep, incised bluffs on the outside of bends in the river and wide expanses of sand bars and flat floodplain with willow and small cottonwoods on the inside of bends. It is joined from the north by the Little River, which originates in Norman, Oklahoma (a different "Little River" than the one in southeast Oklahoma), then joins the North Canadian River and Deep Fork River in forming Lake Eufaula, the largest man-made impoundment in Oklahoma. After leaving Lake Eufaula, the Canadian, now carrying water from all its major tributaries, flows a few kilometers across the Arkansas River floodplain and joins that main stem in Robert S. Kerr Lake, near Webbers Falls, Oklahoma.

The system is fed by snowmelt in New Mexico, with flow controlled by Conchas Lake, New Mexico, and Lake Meredith, Texas, and, in its lowest reaches in east Oklahoma, by Lake Eufaula. Flow is highly variable through most of Oklahoma because of the occurrence of extreme rainfall events, at any time of year, that create river flooding, and summer desiccation, which in some years reduces the main stem to not more than a meter wide (W. J. Matthews, personal observation). In its headwaters near Logan, New Mexico, the Canadian River has a recent (since dam construction upriver) 30-year (1970 to 1999) mean discharge of about 1.3 m³/s, with a few high discharge years (four years in the series had average discharge >2.8 m³/s). In sharp contrast, virgin flow in that reach from 1927 to 1947 had annual mean stream flow an order of magnitude higher, averaging 12.1 m³/s (http://waterdata.usgs.gov/nwis/sw). Downstream, near Whitefield, Oklahoma, mean annual discharge of the Canadian River before regulation by Eufaula Lake (1939 to 1963) averaged 174 m³/s, compared to 120 m³/s from 1964 to 1984, after the lake was built. Basinwide runoff in this mostly arid area is extremely low, ranging from only 0.15 cm/mo in August to 1.04 cm/mo in May (see Fig. 7.10), due to high evapotranspiration.

With the exception of its headwaters in the Rocky Mountains, the Canadian River flows mostly over very unstable sand and mud substrates with a braided channel that changes continually. Each stage rise changes the patterns of flow and sandbars in the river bed, and stable habitats are largely nonexistent (Matthews and Hill 1980). The Canadian River has a highly shifting channel in its lower portions, often eroding away or depositing new farmlands. Oklahoma law allows a landowner with "new" land deposited by the river to use it, but the landowner on the other side, whose land is washed away, is merely "out of luck." The main stem in central Oklahoma has a bed several hundred meters wide, but flow over these sandy beds is often reduced to a few meters wide by late summer (Hefley 1937, Matthews and Hill 1979), with flow as little as 1 to 2 m³/s. During extreme drought in the early 1950s, many kilometers of the main stem in central Oklahoma were reported by local residents to be completely dewatered. After rains the channel can run full, with swift and turbulent waters bank to bank over the sand bed, but by late summer the Canadian River near Norman is often reduced to a series of pools connected by scant flow in shallow channels 1 to 2 m wide. Under those conditions, solar heating is extreme, and the biota is impacted by high water temperatures and low oxygen alternating with supersaturation in backwaters. These conditions apparently have strong influence on the distribution of fishes, which sometimes are concentrated in huge numbers in microhabitats that offer slight survival advantages (Matthews and Hill 1979, 1980).

At Logan, New Mexico, the Canadian has occasional high specific conductance (1200 to >9000 μS/cm, depending on flow), pH ranging from 8.0 to 8.3, and is dominated chemically by sodium and chloride (150 to 1900 mg/L and 60 to 2800 mg/L, respectively) but with high concentrations of sulfate (300 to 500 mg/L). Hardness ranges from about 300 to 600 mg/L as $CaCO_3$, and alkalinity is about 200 to 330 mg/L as $CaCO_3$. By Amarillo, Texas, values remain approximately in those ranges. Alkalinity remains high and the river is well-buffered through central Oklahoma, with values at Calvin, Oklahoma, approximating specific conductance of 600 to 1200 μS/cm, pH 8.2 to 8.5, total

hardness about 300 mg/L as $CaCO_3$, alkalinity about 200 mg/L as $CaCO_3$, and ions like calcium, magnesium, sodium, and chloride ranging from 20 to 150 mg/L. At this site nutrients in recent years have ranged from 0.1 to 0.5 mg/L total nitrogen and 0.04 to as high as 0.49 mg/L total phosphorus. In central Oklahoma, water temperature in the main channel regularly reaches or exceeds 36°C on summer afternoons, when direct sunlight strongly heats the shallow waters (less than 1 m deep) of the exposed sand-bed channel. The river in central Oklahoma ices along the shore in some winters, with main-channel water temperatures measured at 0°C to 1°C (W. J. Matthews, unpublished data).

River Biodiversity and Ecology

Unlike the Arkansas River, which traverses four freshwater ecoregions, the Canadian River passes only through the Southern Plains freshwater ecoregion (Abell et al. 2000), and, not surprisingly, its overall biological diversity is substantially less than the main river. Not only is its western portion even harsher than the main-stem Arkansas, but it does not flow as far into the more mesic east. Ecological studies of the Canadian are limited, with the most information available on the fish community.

Plants

The Canadian River basin has a distinct vegetation gradient from east to west. To the east, patches of silver maple and box elder occur on stream banks. An elm–hackberry–ash association is common in lower reaches of the river in Oklahoma. From the central to the western parts of the basin mixed grasses are common along the rivers edge, with cottonwood–willow woodlands common on floodplains, as well as patches of salt-cedar and sandbar willow. Salt-cedar, a major invader upstream, becomes limited downstream. Oak forest occurs on the upper terraces.

Invertebrates

The sand and clay sediments of the Canadian River support a limited invertebrate fauna, and most invertebrates are associated with snag habitats. Oligochaetes worms (*Limnodrilus*) and midges (e.g., *Bezzia*, *Chironomus*, *Cryptochironomus*, *Paratendipes*) are abundant. The most common mayflies are *Tricorythodes* and *Caenis*. The caddisfly fauna is dominated by filter-feeding *Cheumatopsyche* and *Hydropsyche* (Bass and Walker 1992, Wilhm et al. 1978). Crayfishes include *Orconectes nais*, *Procam-*

barus simulans, and *P. acutus* (Reimer 1969, E. A. Bergey, personal communication). Mollusks include *Physa* and the fingernail clams *Sphaerium* and *Pisidium*. At least 11 unionid species are known from the river, including pink papershell, fragile papershell, yellow sandshell, and white heelsplitter (Branson 1982, 1983, 1984). The introduced Asian clam is common.

Vertebrates

Cross et al. (1986) considered about 63 species of fishes to be native to the Canadian River system. Fish distribution in this system changes more gradually from west to east than in the Red River, with its salt gradients. One federally threatened fish species (the Arkansas River shiner) in the South Canadian River is now much reduced from its former range, but it was common in 1978 at least as far upstream as Revulito Creek, New Mexico (W. J. Matthews, personal observations). Marked differences exist in distribution of some fish species between North and "south" Canadian rivers (e.g., no Arkansas River shiners in the North Canadian). Like the Red and Arkansas rivers, there is a main-stem "big-river" fish fauna in the wide, shallow, sandy main-channel Canadian River that differs from that in tributary creeks (Matthews and Hill 1979, 1980). Typical and abundant (at least formerly) main-stem species include red shiners (also common in creeks), Arkansas River shiner, plains minnow, bullhead minnow, and emerald shiner.

Fish habitat use is strongly influenced by high temperatures and other physical stressors in summer, as temperatures in the main channel approach lethal limits for all species (Matthews and Hill 1979, Matthews 1987, Matthews and Zimmerman 1990). Important fish habitat is found at or near creek inflows, where pools typically are deeper and more stable than in the shallow main stem. These edge habitats also support fathead minnows in backwaters, longear and green sunfishes, and a limited number of channel catfish, largemouth bass, and gizzard shad. The Red River pupfish, formerly restricted in Oklahoma to the Red River basin, was introduced to the Canadian River in central Oklahoma. However, numbers of species at any site reflect the lower number of species in this basin compared to other main rivers, with typically only about a dozen species taken at any one seining site.

River otter, an Oklahoma state species of special concern, is found in eastern parts of the Canadian River. Nonvegetated beaches and sandbars of the Canadian River support breeding populations of the federally endangered interior least tern (Byre 2000). Smooth softshell turtles are commonly found burrowing into the sand bed, and snapping turtle, common slider, yellow mud turtle, and stinkpot turtle are found. Water snakes are common. Beavers are common in tributary creeks near their confluence with the main river channel, and beaver cuttings are commonly seen along the main stem as well. Great blue herons (sometimes in large numbers), green herons, and little blue herons occur along the river. Nesting colonies of great blue herons occur in large trees on upper terraces of the river at Norman, Oklahoma.

Ecosystem Processes

There are no comprehensive studies of ecosystem processes in the Canadian River or its large tributaries. However, in that much of this river flows through grasslands or former prairie, we would suspect that ecosystem processes are dominated by inputs of grasses or smaller amounts of tree-derived vegetation, combined with periodic high primary productivity within the streambed. For example, in autumn of some years, major portions of the streambed can become encrusted with a thick layer of algae that appears to be highly productive and is, in fact, directly eaten by some of the common fishes (W. J. Matthews, personal observations). An additional important organic input to the river itself is probably derived from the encroachment of seedlings of cottonwood and other woody plants, which rapidly grow on the dried portions of the streambed at low water. Upon the next flooding, this material is uprooted and washed into the river, where it no doubt provides substantial nutrients by its decomposition (W. J. Matthews, personal observations). However, retention of particulate carbon is probably low, in that stage rises substantially move the easily shifted sand bed.

Human Impacts and Special Features

The Canadian River was described by Hefley (1937) as one of the harshest environments on Earth, and heating of water can be extreme in summer (W. J. Matthews, unpublished data). The Canadian is probably one of the most dynamic and variable river environments anywhere in the world, ranging from a mere trickle in late summer to a boldly flooding river following rains. This natural flow cycle apparently was important as a stimulus to reproduction by some of the common fishes, which are now substantially reduced in number as a result of flow moderation by upstream dams.

Two dams influence flow, and there is a strong influence of agriculture and oil production on the river. Near Norman the riverbed is extensively disturbed by commercial sand mining, but it also is protected in numerous places by private landowners and The Nature Conservancy, with nesting refuges on sandbars for the federally endangered interior least tern. No major cities are on the primary main stem, but the North Canadian River flows through Oklahoma City, where it recently has been dammed in several places to provide water for a new tourist canal system in the refurbished downtown. Lake Overholser is an "off-channel" impoundment on the North Canadian River near Oklahoma City, and an artificially heated power plant lake northeast of Oklahoma City apparently allowed establishment of a population of nonnative blue tilapia (Pigg 1978), which reproduce and invade the North Canadian River in large numbers in some years, as far as 50 to 75 km downstream to near Seminole (Matthews and Gelwick 1990). The Canadian River and its drainage generate controversy between the states of Texas and Oklahoma, with Oklahoma officials alleging that Texas is wrongfully storing water in a reservoir on a Canadian basin tributary in the Texas Panhandle (*Daily Oklahoman*, August 26, 2002).

RED RIVER

The Red River is the 2nd largest river basin in the southern Great Plains (169,890 km^2), arising at about 1050 m asl as mostly dry or ephemeral creeks converging in Palo Duro Canyon in the Texas Panhandle, and joining the Mississippi River (and/or the Atchafalaya River) in eastern Louisiana at about 25 m asl (Fig. 7.11). The Red River drains some of the driest areas of the Southern Plains, consisting of the western portion of the Texas Panhandle, where surface water is scarce, playa lakes appear and disappear, and permanent water in streams is rare. It is in or near this area that troopers of the U.S. 10th Cavalry died of thirst in the 1870s. From Eastern Oklahoma through Arkansas and Louisiana, the Red enters a more mesic climate and is joined by a large tributary, the Ouachita River (see Chapter 6), just before it reaches the Mississippi River.

There has been considerable debate over whether North Fork or Prairie Dog Town Fork in western Oklahoma is the true main stem of the Red River. Prairie Dog Town Fork was followed by Marcy's military expedition of 1852 to find the source of the Red River, and, on the basis of lengthy congressional and judicial findings (Tyson 1981), is considered the main fork of the Red River. On the other hand, North Fork is nearly as long and typically carries more water (9.6 m^3/s for North Fork at Headrick; 3.2 m^3/s for Prairie Dog Town Fork near Childress), but legal decisions have agreed with Marcy's determination that the Prairie Dog Town Fork is "the river."

Native peoples along the Red River basin had sharply differing cultures even before Europeans entered the region. To the east were cultures of the eastern Piney Woods, dominated by tribes of the Caddoan Confederacy and characterized by a sedentary lifestyle in an area of plentiful game and favorable conditions for growing crops. In the middle Red River region, the Wichita and Tonkawa tribes roamed the prairies, living in skin tepees or pole and brush structures, with seasonal hunting and limited farming. In the western Red River basin, the Lipan Apache were the originally dominant native peoples but were displaced by the Comanche from the north in the 1700s, after which the upper Red basin was dominated by the Comanche until Europeans took control of the region in the late 1800s.

During European settlement after the American Civil War the upper Red River area in Oklahoma and the Texas Panhandle was a region of war or lawlessness involving Comanches, the U.S. Calvary "Buffalo Soldiers" of Fort Sill and other frontier forts, infamous outlaws and outlaw towns, cattle barons with huge holdings, and lawmen of legendary proportions. Perhaps no other region in the southern Great Plains so epitomizes the "Wild West" in all its inglorious forms. It has been immortalized in popular historical novels by Larry McMurtry (*Lonesome Dove*) and Elmer Kelton (*The Far Canyon*), in songs ("Panhandle Wind" by Bill Staines), and in films (*Red River*, starring John Wayne). However, despite wishes to the contrary by some residents, the plaintive love song "Red River Valley" was actually written about a cowboy at the "Red River of the North" in Minnesota.

The Red River forms the long-debated boundary between Oklahoma and Texas, with a long history of legislative and judicial arguments, a U.S. Supreme Court judgment to legally identify the main fork, and a recent agreement by the states setting the southern border of Oklahoma at the vegetated edge of the water line on the south bank. The focus of this debate was also on Greer County, Oklahoma, between the two main forks of the Red River in southwest Oklahoma, whose ownership was bitterly contested by Texas versus "the United States" for decades, and which was known as the "Empire of Greer" in some early documents and now is referred to locally as "Old Greer County." As recently as 1931, National Guard units of the two states risked armed hostilities over ownership of the river and its oil-bearing riverbed, culminating in the peacefully ended "Red River Bridge War" (Fugate and Fugate 1991) over the right to establish toll fees for crossing.

Physiography, Climate, and Land Use

The Red River passes through four physiographic provinces (Great Plains [GP], Central Lowland [CL], Ouachita Province [OP], and Coastal Plain [CP]) and four terrestrial ecoregions (Western Short Grasslands, Central and Southern Mixed Grasslands, Central Forest Grassland Transition Zone, Piney Woods Forests) (see Fig. 7.11). The upper Red River originates at about 35°N latitude on the elevated plain known as the Llano Estacado in the Texas Panhandle at the eastern edge of primarily Tertiary deposits of sedimentary and mostly marine origin (Hunt 1974). In the eastern Texas Panhandle and western Oklahoma to about Lake Texoma, the Red River passes mostly through nonmetamorphosed marine and continental sedimentary deposits of Mississippian, Pennsylvanian, and Permian age. In western and central Oklahoma, the basin also drains streams from large granitic uplifted areas in the Wichita Mountains and other smaller and isolated granitic areas such as the locally important Tishomingo Granite. East of Lake Texoma, the Red River drains Cretaceous formations that abound in ammonites and "sea biscuit" echinoderm fossils, but the river main stem soon enters Quaternary sedimentary deposits, through which it flows to its mouth in eastern Louisiana at about 31°N (Hunt 1974).

The Red River basin in Texas and western Oklahoma is characterized by generally red to red-brown alkaline soils (Hunt 1974). In central Oklahoma and north Texas the basin drains generally neutral prairie soils, then drains mostly acidic red or yellow podzol soils in east Oklahoma and much of Louisiana and crosses alluvial soil in extreme eastern Louisiana. Vegetation generally reflects the major terrestrial ecoregions drained by the basin, with short and mixed grasses in the west, a transitional zone of short deciduous oak forest ("Crosstimbers") and taller native grasses, and the heavily forested mixed coniferous and deciduous regions of eastern Oklahoma and Louisiana.

Climatologically, the Red River flows from arid grasslands in the Texas Panhandle, receiving less than 50 cm/yr of rainfall on average with frequent blizzard conditions in winter, to areas in Louisiana that receive more than 140 cm/yr of rain and average about 250 frost-free days per year, with snow a rarity. The region as a whole has generally mild to cold continental winters, but summers are very hot, with July and August monthly means near 30°C (Fig. 7.12). In the upper Red River basin, average daily temperatures in January are about 2°C, in contrast to about 8°C in the lower basin in Louisiana. From April through autumn the middle and eastern Red River basin often has massive rain fronts coming in from the Gulf of Mexico or near daily afternoon thundershowers, some of which reach violent mesocyclone proportions. Greatest average monthly precipitation is in the month of May (see Fig. 7.12).

The Red River now flows mostly through cattle ranching or row-crop farming areas, with Shreveport, Louisiana, and Alexandria, Louisiana, the only large cities directly on the river. Oil and gas production are prominent features of land use near the river, with some large oil fields in southwest Oklahoma. The region is mostly rural, with many small municipalities in counties adjoining the river, plus the moderately sized cities of Wichita Falls and Sherman-Denison, Texas, and Texarkana, Arkansas–Texas. In the west, land use is about 40% to 60% rangeland and 30% cropland, whereas to the east it is about 50% forested, 20% cropland, and 10% pasture.

River Geomorphology, Hydrology, and Chemistry

Originating outside the Rocky Mountains, the Red River has no substantial influence of snowmelt and lacks montaine outwash in its sandy riverbed. The channel form throughout essentially all of the basin is a gently sloping floodplain a kilometer or more wide, bordered in some reaches by moderately sloping, rounded bluffs 20 to 30 m in height (Fig. 7.1). The exception is the steep-walled Palo Duro Canyon, through which the headwaters of the Prairie Dog Town Fork of the river flows. Beyond Palo Duro the river is essentially low gradient for its entire course, lacking waterfalls or any steep drops in elevation.

The main-stem Red River (Prairie Dog Town Fork) originates in Palo Duro Canyon in the Llano Estacado (High Plains) south of Amarillo, Texas, flows eastward through the relatively arid Texas Panhandle to the western border of Oklahoma, then forms the southern border of Oklahoma with Texas. The river flows almost entirely through terrain of relatively low relief, resulting in a mean slope for the entire main stem of only 65 cm/km. In the Oklahoma–Texas reach it is joined by the North Fork and Salt Fork from Oklahoma and the Wichita River from Texas. Flow in even the largest of these forks remains uncertain in dry years, "pooling up" in some summers as far downstream as Burkburnett, Texas. Further downstream, where at least some flow persists in most years, it remains a sandy-muddy river, the crossing of which has vexed Europeans since Marcy's expedition in the 1850s and the first cattle drives. The Red becomes a substantial river from about Wichita Falls, Texas, downstream to Lake Texoma, a major on-channel impoundment at the junction of the Washita and Red rivers (see Fig. 7.11). Only below Lake Texoma, in the mesic region from eastern Oklahoma through Louisiana, does the river take on a more consistent flow in a defined channel, bordered by large deciduous forests and, in some places, oxbow lakes and overflow swamps. Below Lake Texoma the river is joined by the Blue, Boggy, and Kiamichi rivers from the north in Oklahoma, by the Little River in Arkansas, and by the Sulfur River and Cypress Bayou–Caddo Lake systems that originate in Texas. En route east, the main-stem Red River flows from arid grasslands to increasingly mesic forests, but throughout most of its length it has a sand-bed channel with massive bars and navigation hazards. The Cypress Bayou system, a substantial tributary in northeast Texas, has one of the largest remaining cypress forests in the region, with huge old trees in vast, winding river channels and overflow fringing swamps.

The Red River in eastern Oklahoma–north Texas and deep into Louisiana was the location of the incredible "Great Raft" or "Red River Raft" of drift logs and mud that existed before European settlement and was reported by early explorers (1806) to be as much as 240 km long and 30 km wide, blocking the channel and causing the river to spill over into countless oxbow and side-channel shallow lakes. A channel through the Great Raft as far as present-day Shreveport was first opened between 1833 and 1836 by the "snag steamboat" invented by Capt. Henry Shreve, but by 1839 the river had rebuilt and closed the raft. The Great Raft defied all attempts to remove it until intensive efforts were resumed in 1872 employing snag boats, "saw boats," "crane boats," and nitroglycerin. The Great Raft finally was cleared

by about 1900, and the river remained permanently open thereafter (Tyson 1981).

Mean virgin discharge of the Red River (1928 to 1944) was 852 m^3/s, with basinwide runoff varying from very low in late summer (about 0.3 cm/mo) to about 2.5 cm/mo in late spring (see Fig. 7.12). A relatively high fraction of precipitation is lost as evapotranspiration. For the Red River downstream of Lake Texoma, in east Oklahoma, mean annual discharge was about 260 m^3/s before regulation by the dam, but from 1945 to 2000, mean annual discharge was identical, at 260 m^3/s. Drastic variation has, however, been observed in that period, from a low of 77 m^3/s in 1964 to a maximum of 652 m^3/s in 1990.

Upper portions of all forks have unpredictable flow. The main stem as far downstream as Wichita Falls, Texas, is subject to extended periods of "noflow" and pooling up. Downstream of the forks it is a perennial, wide, shallow, sand–mud bed river, with discharge that varies greatly. The most likely months for storm-induced spates in tributaries are April, May, and midautumn. The Red River main stem can rise rapidly after heavy rains in the basin and can produce extreme flooding and turbidity in Lake Texoma, with strong effects on fishes and zooplankton.

Water chemistry reflects the harsh arid country through which the river flows. Marcy's 1852 expedition to the headwaters in Palo Duro Canyon was forced to endure bad water conditions, with the chemically charged Prairie Dog Town Fork causing severe gastrointestinal distress for army troopers who attempted to drink it. In the eastern Texas Panhandle, north Texas west of Wichita Falls, and extreme southwest Oklahoma, this river has (for "freshwater") extremely high salinity, approaching or exceeding that of seawater, because of a dozen or more brine springs from Permian strata. Downstream in Oklahoma, tributaries of the Red River flow through a large gypsum region, resulting in inputs of sulfates to the river.

Typical chemical values in the Prairie Dog Town Fork near Childress, Texas, include pH 7.7 to 8.2, alkalinity 75 to 110 mg/L as $CaCO_3$, total hardness 1800 to 5200 mg/L as $CaCO_3$, specific conductance 16,000 to 72,000 µS/cm, sodium 12,000 to 17,000 mg/L and chloride 17,000 to 28,000 mg/L (greater, overall, than for seawater), and calcium 600 to 1500 mg/L and magnesium 100 to 360 mg/L, suggesting the laxative properties of the water discovered by Marcy's army troops. In contrast, below Lake Texoma the main-stem Red River near DeKalb, Texas, has calcium and magnesium ranging around only about 25 to 70 mg/L and 5 to 25 mg/L, respectively, sodium and chloride values of only 10 to 100 mg/L and 10 to 140 mg/L, respectively, and specific conductance about 200 to 950 μS/cm, depending on flow and discharge. The Red River near DeKalb, Texas, has typical nutrient concentrations ranging 0.1 to 0.3 mg/L total nitrogen, and 0.03 to 0.13 mg/L total phosphorus. Water temperatures in the main channel of the Red River regularly exceed 36°C in late summer and have been recorded as high as 39°C, making it one of the hottest large rivers on Earth, because of intense solar heating of the shallow water in the unshaded channel (Matthews and Zimmerman 1990).

River Biodiversity and Ecology

The Red River flows primarily through three freshwater ecoregions, the Southern Plains, the Ouachita Highlands, and the Mississippi Embayment (Abell et al. 2000). The Red River does not actually pass through the Ouachita Highlands, but several of its tributaries drain the highlands before entering the Coastal Plain and subsequently joining the main stem. Thus, the main stem remains a low-gradient river throughout its length as it enters the Mississippi Embayment ecoregion before joining with the Atchafalaya–Mississippi rivers. The overriding ecological feature of the upper Red River (in the Southern Plains above Lake Texoma) is strong structuring of biota by salinity gradients produced in the Permian Redbeds, with high fish diversity downstream reduced to as few as only two very hardy and salt-tolerant fish species in some of the headwaters (Echelle et al. 1972). Below Lake Texoma the river has relatively consistent flow, essentially never pools up or goes dry, and salinity gradients that would structure biota are lacking.

Plants

The basin as a whole has a gradient from high plains vegetation in the west to southern bottomland hardwood in the east. Cottonwood–willow woodlands are common on floodplains throughout the watershed. Stream banks contain patches of box elder and silver maple. Second growth slippery elm, sweetgum, black oak, and post oak are common. Eastern oxbows support swamp oak. Second growth loblolly pine, both naturally occurring and in plantations, occurs in the lower watershed.

Invertebrates

The shifting clay and sand sediments of the main-stem Red River, coupled with the high conductivities, create a harsh environment for many benthic invertebrates, and there have been few comprehensive studies of the benthic fauna. Oligochaetes, burrowing mayflies (*Hexagenia*), and chironomid midges (*Chironomus*) are common in sediments. Most other invertebrates are associated with snag habitats, including the chironomids *Glyptotendipes*, *Dicrotendipes*, and *Rheotanytarsus* (Sublette 1953, Vaughn 1982). Mayflies include *Caenis* and *Stenonema*, and the caddisflies *Chimarra*, *Cyrnellus*, *Hydropsyche*, *Cheumatopsyche*, *Oecetis*, *Hydroptila*, and *Triaenodes* are common (Resh et al. 1978). Crayfishes include *Orconectes palmeri*, *O. nais*, *O. virilis*, and *Procambarus simulans* (Reimer 1969, E. A. Bergey, personal communication). The Red River contains a large population of the Mississippi grass shrimp (Cheper 1988), an omnivore that thrives in the high-conductivity water. Upper reaches of the Red River are fed by hypersaline (43 ppm) springs that contain grapsoid crabs, *Hemigrapsus estellinensis* (Creel 1964).

The mussel fauna of the main-stem Red River is depauperate for a river of its size, largely because many species do not fare well in the shifting sediments. The fauna is dominated by pink papershell, fragile papershell, and bleufer. Other mussel species known from the main-stem Red River include white heelsplitter, yellow sandshell, threehorn wartyback, giant floater, maple leaf, pimpleback, lilliput, pistolgrip, and paper pondshell (Branson 1982, 1983, 1984; W. J. Matthews et al. unpublished data; Valentine and Stansbery 1971; Vaughn 2000). The nonnative Asian clam occurs throughout the river. In contrast to the main-stem Red River, tributaries to the river contain a diverse mussel fauna that has been well studied (Isely 1924, Valentine and Stansbery 1971, Vaughn 1997). The easternmost tributaries arising in the Ouachita uplands are particularly diverse. For example, 31 species of mussels are known from the Little River (Vaughn and Taylor 1999) and 29 from the Kiamichi River (Vaughn et al. 1996). Two federally endangered mussels, the Ouachita rock pocketbook and the scaleshell mussel, occur in Red River tributaries (Vaughn and Pyron 1995, ONHI 2001).

Vertebrates

A total of 171 fish species, 152 of them native, was reported by Cross et al. (1986), and additional introduced species are now present. A sharp contrast exists between the 133 native fish species in the lower Red River and the 56 in the upper Red River above Lake Texoma (Cross et al. 1986). The upper reaches that dry often or have very high salt loads have environmental conditions so harsh that only two species (Red River pupfish and plains killifish) occur in some headwater reaches. Farther downstream, as salt concentrations are diluted by tributaries, a more speciose fish assemblage develops, with upstream limits to many species set by salinity gradients (Echelle et al. 1972). Below the highest salinity, upper reaches of the Red River are dominated by a recognizable "big-river" fish fauna, notably minnows like red shiner, chub shiner, plains minnow, silver chub, and the endemic Red River shiner. Main-channel members of the speckled chub complex include the prairie chub in the upper river and the shoal chub in the lower river, with the species sympatric near Wichita Falls (D. Eisenhour, personal communication). Nearing Lake Texoma additional minnow species like bullhead minnow and emerald shiner become more common, as do gizzard shad, channel catfish, introduced inland silversides, various sunfish species, white crappie, largemouth bass, white bass, and the introduced striped bass. In addition, gars are common above Lake Texoma, and paddlefish have been reintroduced above the reservoir.

Lake Texoma has a relatively stable fish assemblage (Gido et al. 2000) dominated both by native species like red shiners, blacktail shiners, gizzard shad, blue catfish, channel catfish, black bass, and various sunfishes, and abundant introduced species like striped bass, threadfin shad, and inland silversides. Lake Texoma and Keystone Reservoir on the Arkansas–Cimarron rivers are two of very few artificial impoundments with reproducing populations of the prized striped bass. Here, for unknown reasons likely related to salinity, striped bass exhibit spawning runs, resulting for Lake Texoma in a strong, naturally reproducing striped bass fishery with an economic value to the local economy estimated at $26 million per year.

Downstream from Lake Texoma the river remains dominated by unstable sand substrates and fishes tolerant of those conditions, with shads, catfishes, minnows, sunfishes, gars, and suckers (Catostomidae) common. The majority of the habitat is shifting sand-bottom, which supports numerous minnows (red shiner, blacktail shiner, speckled chub, emerald shiner, ghost shiner, and bullhead minnow), inland silversides, western mosquitofish, longear sunfish, bluegill, largemouth bass, and several darter species like western sand darter, Johnny darter, and bigscale logperch. Larger-bodied fishes that occur in the deeper portions of the main channel include shovelnose sturgeon, blue sucker, river carpsucker, and golden redhorse. In addition, there are low-velocity backwater or deep pool habitats that contain several gar species (shortnose, longnose, spotted), gizzard and threadfin shad, smallmouth and bigmouth buffalo, and common carp. After passing the Arkansas–Louisiana border, the river winds southeast through Louisiana, with a fish fauna typical of lowland streams of the Gulf Coastal Plain but still dominated by many of the same groups. The separation of the river into an upper and lower reach by Cross et al. (1986) not only reflects the general geography of the river, but the features of its native fish communities as well.

Lake Texoma is home to numerous visitors, such as gulls and shorebirds from the Gulf Coast, and bald eagles are very common during winter. Osprey are seen year-round at this impoundment. White pelicans are common in autumn, as they use Lake Texoma as a stopover, and sometimes overwinter. Many great blue herons and green herons, and an increasing number of cormorants, occur at Lake Texoma. Beaver are becoming very abundant around Lake Texoma, and nutria and muskrat are commonly associated with the river in Louisiana. Amphibians and reptiles associated with Red River in Oklahoma and eastward include snapping turtles, occasional alligator snapping turtles (from Lake Texoma eastward), plain-bellied water snake, false map turtle, yellow mud turtle, and cottonmouth from below Lake Texoma and eastward. We have never observed a cottonmouth in Lake Texoma, but they are common in the Blue River, the first substantial tributary east of the reservoir.

Ecosystem Processes

Relatively little is known about ecosystem processes in the main-stem Red River. However, in the Little Washita River (tributary to Washita River, thence to Red River) of south-central Oklahoma, riparian cover upstream was directly related to availability of whole leaves in the stream, and long reach lengths served as sources of detrital input at a given point (Johnson and Covich 1997). Johnson and Covich (1997) also found that organic content of benthic materials declined from headwaters to lower in the Little Washita River, and that floods reduced coarse POM in headwaters and increased the amount of fine organic material (10 to 360 µm) that were in suspension downstream.

Ecosystem factors like amounts and kinds of periphyton, primary productivity, ash-free dry mass (AFDM) and percentage of organic material in the substrate, size fractions of particulate organic matter, bacteria, and invertebrate standing crops, and carbon–nitrogen ratios in Brier Creek, Marshall County, Oklahoma, a tributary of the Red River (in Lake Texoma), are strongly influenced by presence of the algae-grazing stoneroller minnows (Power et al. 1985, Matthews et al. 1987, Gelwick and Matthews 1992). A trophic cascade involving piscivorous black bass, central stoneroller minnows, and algal density and composition has been shown to influence primary productivity in pools (Power et al. 1985). Red shiners from the Washita River enhanced benthic primary productivity in experimental streams (Gido and Matthews 2001). In addition, another dozen fish species from several trophic or functional groups from Red River tributaries have been found to alter benthic primary productivity (C. Hargrave, K. Gido, W. Matthews, unpublished data.). Gido (1999) found in experimental pens in Lake Texoma that large-bodied, benthic-feeding fishes like river carpsucker and gizzard shad changed densities of chironomid larvae in benthic substrates, and that excretion by these fishes can be in sufficient amounts to account for substantial nutrient inputs to the reservoir.

Human Impacts and Special Features

Notable human impacts include the influence of agriculture (wheat, cattle) and of oil production in headwaters, and the construction of Lake Texoma, a 36,000 ha reservoir, at the juncture of the Washita and Red rivers. Interbasin water transfer from Lake Texoma to a reservoir of the upper Trinity River, Texas, is possible via a conveyance system completed in 1993. The largest city on the main stem is Shreveport, Louisiana, but the growing cities of Sherman-Denison and Wichita Falls, Texas, have a growing desire for Red River water for municipal uses. The Dallas–Ft. Worth metropolis also has increasing potential to use Red River water via interbasin transfer.

The mussel fauna of the Red River drainage has been significantly impacted by human activity. Vaughn (2000) resampled 19 sites in the drainage in Oklahoma and Texas that had been sampled in the 1910s and in some cases in the 1960s. Species richness declined at 89% of the sites. Local extinction rates were significantly greater than local colonization rates, indicating that mortality of mussels is significantly exceeding recruitment in the region.

In spite of all the human-induced changes, the Red River still retains many of its characteristics from presettlement times. The upper river, in its wide, shallow, and sandy nature, still reflects conditions first described by Marcy in the 1850s, although many small creeks and tributaries no longer flow freely or as clear as Marcy found them. Downriver, Lake Texoma blocks upstream migration of anadromous fish like American eel and traps huge loads of silt and woody debris borne by the upper river on flood flows. Further downstream, in spite of removal of the Great Raft, which formerly blocked human passage on the river, the river remains relatively unstable, and navigation of the river for commercial purposes is impractical along the Oklahoma–Texas border. From Lake Texoma onward, the river flows almost entirely through rural regions with small populations (with the exception of Shreveport and Alexandria, Louisiana), and it still overflows into oxbows and swamps throughout much of its lower course. Some of the remaining largest cypress forests and swamps in Texas remain along the Cypress Bayou system, which is a major tributary of the Red basin in northeast Texas. The Red River has thus resisted most efforts of humans to tame its unruly nature. Lake Texoma minimizes flood losses downstream when massive hurricane-derived rain fronts stall over the basin, but otherwise much of the river remains much as it was as seen by the first explorers—a relatively "wild" place where human impacts seem rather small. However, the Army Corps of Engineers has begun a massive program to reduce chlorides in the upper basin, which, if ever completed, will threaten the existing, natural salinity gradient that is the template for much of the distribution of flora and fauna in the upper river. In addition, if the waters of the upper Red River were lower in salinity such that they could be directly used for irrigation, water withdrawals would no doubt increase, and hydrological estimates suggest that "no flow" days in the upper basin might be tripled annually.

LITTLE RIVER

The Little River, a major tributary of the Red River, drains $10,720 km^2$ in southeastern Oklahoma and southwestern Arkansas (Fig. 7.13). This system is the antithesis of many of the other rivers in the region, with the upper main stem or larger tributaries representing mostly high-quality habitat flowing from rocky uplands of the southern slopes of the Ouachita Mountains. Outside the uplands the Little River becomes a low-gradient stream, with large gravel riffles interspersed with long, deep pools. Biodiversity of this system is some of the highest in the southern Great Plains, and in some reaches it contains highly diverse local aquatic faunas at the ecotone between upland and lowlands. Highly diverse fish, benthic invertebrate, and mussel faunas exist, and numerous species (e.g., about 15 fish species) reach their westward range limits in the Little River basin. The Little River and some of its tributaries are noteworthy for their remaining native mussel fauna, including some federally endangered species and several fish species of concern or with federal protection are in these streams.

Early human history of the region was dominated in prehistoric times by groups of woodlands cultures, with a mixture of hunting and gathering and planting of crops like corn, squash, beans, pumpkins, and sunflowers. The Kichai, Caddo, and Kadohadacho peoples dominated recent prehistory, when as many as 8000 Caddo, many in permanent settlements, lived in the region. As native peoples from the east were displaced, the Little River basin was eventually dominated by the Choctaw Nation, with establishment of early settlements like Eagletown and a prosperous culture dominated by farming and ownership of plantation-style homes and lifeways. During the 1900s native forest gave way to major pine plantations, and the Little River basin has a later history dominated more by the major land-holding timber companies, as well as national forests maintained by the federal government.

Physiography, Climate, and Land Use

The Little River passes through two physiographic provinces (Ouachita Province [OP], Coastal Plain [CP]) and three terrestrial ecoregions (Ozark Mountain Forests, Central Forest Grassland Transition Zone, Piney Woods Forests), flowing between about 35°N and 34°N (see Fig. 7.13). Geology of the region is dominated by the Ouachita uplift, in which all major tributaries (Glover, Mountain Fork, Rolling Fork, Cossatot, and Saline rivers), as well as the Little River proper, originate. The Ouachita Mountains exhibit "ridge and valley" structure, with long (as much as 50 km), steep, but narrow (only a few kilometers wide in most places) mountain ridges, with rivers flowing through the valley floors between the ridges. Rock is mostly sedimentary, of marine origin, but much is metamorphosed to form shales, quartzites, and similar noncalcareous bedrock. After flowing generally southward out of the Ouachita uplift, the Little River crosses Cretaceous sedimentary deposits and then enters Quaternary sedimentary deposits in the lowlands before flowing into the Red River. Soils of the basin are typically acidic red or yellow podzol (Hunt 1974). Dominant vegetation of the Ouachita Mountain uplands is conifer forest, apparently mixed in presettlement times with tall grasses in relatively open parkland. Downstream in the lowland parts of the basin, Coastal Plain deciduous forests mixed with conifer dominate the landscape.

Climate in the Little River basin is hot and humid in summer, with influence from Gulf Coastal weather patterns, and winters are milder than in most of the Great Plains. In the Little River basin, rainfall averages in excess of 100 cm/yr, and there are about 220 to 240 frost-free days per year. Mean monthly precipitation is highest in May (16 cm), with lows in January and August (7 cm) (Fig. 7.14). Air temperature in the Little River basin averages about 4°C in January and 26°C in August.

Land use in the uplands is almost entirely commercial timber, National Forest, or cattle ranching, and there are extensive cattle ranching, small farms, and some timber in the lowlands. Commercial poultry houses have increased in recent decades, and there is a major processing plant near the Little River south of Broken Bow, Oklahoma, that has locally polluted the Little River at times. There are few row crops, and the region largely lacks impacts of oil production. The lower main-stem Little River in Oklahoma is in the Little River Wildlife Refuge (U.S. Fish and Wildlife Service), but threats to the system continue from plans to transfer water out of state or to channelize some main-stem reaches. There are no large cities in the Little River basin; the region remains almost completely rural or in commercial timber, and the streams are prized by fishermen, hunters, and recreationists. Land use in the Little River basin is about 75% forest, 10% cropland, and 15% pasture, with no large urban areas.

River Geomorphology, Hydrology, and Chemistry

All major tributaries and the main-stem Little River originate in steep uplands of the Ouachita Mountains of Oklahoma or Arkansas, with typically rocky (boulder/bedrock) riffles and large pools with boulder-strewn bottoms. Here, the water is sufficiently clear that underwater observation of fishes is a useful study method. Upon exiting the uplands, the Little River flows onto the Coastal Plain, where it becomes deeply incised into sandy or clay soils, with highly developed floodplain/riparian forests and long, wide riffles separating pools often as much as a kilometer or more long. Overflow ponds or oxbow lakes, as well as fringing swamps, formerly dominated the landscape of the lower Little River, although vast, large swamps are now drowned under Millwood Reservoir (see Fig. 7.13).

The main-stem Little River originates on the southern slopes of the Kiamichi and Winding Stair mountains at about 460 m asl in remote, rugged uplands in southeast Oklahoma near Honobia, then flows southwest and eventually eastward in a wide arc (see Fig. 7.13). Its headwaters are boulder-cobble and gravel, with well-developed pools and swift rapids until the main stem forms Pine Creek Reservoir. Below the reservoir the river retains pool–riffle structure over a mostly gravel bed, with water willow a common structural feature in the shallows. The Little River then turns eastward, widening and deepening, flowing through cypress riparian forest, and entering the Little River Wildlife Refuge near Idabel, Oklahoma. It flows parallel to and north of the Red River to the Arkansas border and gains major tributaries, including (west to east) the Glover, Mountain Fork, Rolling Fork, Cossatot, and Saline rivers (see Fig. 7.13). The entire system flows into and forms Millwood Reservoir northeast of Ashdown, Arkansas, inundating formerly huge swamplands. Below Millwood Reservoir the Little River enters the Red River about 2 km west of Fulton, Arkansas, at an elevation of 70 m. Mean slope for the entire Little River main stem is 2.1 m/km.

Although arising in uplands, headwaters are not fed by any large springs; thus, flows in upper reaches can be tenuous in late summer. In spite of this, the location of these streams in a relatively mesic area makes the Little River a large tributary of the Red River. Mean annual virgin discharge at Idabel, Oklahoma, from 1930 to 1968 (before closure of the upstream Pine Creek Reservoir) was about 45 m³/s. Average annual discharge at Idabel from 1971 to 2000 was 53 m³/s, suggesting no major change of discharge as a result of regulation. The highest annual mean discharge at Idabel was 89 m³/s, whereas the lowest, during the worst drought on record in 2000, was only 18 m³/s. Average annual discharge for the entire basin, measured at Millwood Dam, Arkansas, is 183 m³/s. Monthly runoff for the entire basin ranges from only 0.8 cm/mo in late summer (apparently due to high summer evapotranspiration) to more than 8 cm/mo in spring (see Fig. 7.14).

Water chemistry at one site on the main-stem Little River included pH 7.2, alkalinity 24 mg/L as CaCO₃, and NO₃-N 0.33 mg/L. Total phosphorus can be as high as 0.36 mg/L, but typically ranges from 0.06 to 0.09 mg/L. The lower pH and alkalinity of the Little River or its major tributaries compared to other rivers of the Southern Plains is due to its origin in noncalcareous uplands, with subsequent passage through conifer forest and the Coastal Plain. For example, a large tributary, the Cossatot River, has pH values as low as 5.9, hardness ranging from 9 to 16 mg/L as CaCO₃, and low concentrations of ions like calcium, magnesium, sodium, and chloride, with specific conductance only 48 µS/cm. Measured water temperatures in the main channel near Idabel, Oklahoma, have ranged in recent decades from 0°C to 33°C, so this is clearly a "warmwater" river despite its origins in the Ouachita Mountains.

River Biodiversity and Ecology

The Little River is part of the Ouachita Highlands freshwater ecoregion (Abell et al. 2000). The river is known for its high aquatic biodiversity and was identified by The Nature Conservancy as a critical watershed for protecting freshwater biodiversity based on its diverse and healthy fish and mussel fauna (Master et al. 1998). It is one of the better-known river ecosystems in the Southern Plains, with extensive surveys of fishes, mussels, and other invertebrates by C. C. Vaughn and collaborators.

Plants

Common riparian corridor species include river birch, sycamore, smooth alder, sugar maple, and box elder (Hoagland 2000). Lower portions of the river flow through bottomland hardwood forest characterized by willow oak and blue beech (Hoagland et al. 1996). Sloughs and swamps along the river contain bald cypress. Extensive growths of water willow develop in shallow areas of low flow throughout the river.

Invertebrates

The Little River harbors 31 species of mussels, including regional endemics such as the Ouachita kidneyshell and the Ouachita creekshell and nationally declining species such as the rabbitsfoot and butterfly (Vaughn and Taylor 1999). Comparisons of historical mussel distributions (Isely 1924, Valentine and Stansbery 1971) with current distributions reveal no species extirpations from the river (Vaughn 2000), although populations below impoundments have been severely impacted (see the Human Impacts and Special Features section for this river). A small population of the federally endangered Ouachita rock pocketbook mussel occurs in the river (C. C. Vaughn, unpublished data).

Upper portions of the Little River are rocky, and cobble–gravel riffles occur throughout the river, so snags are not as important a habitat for invertebrates as in sandy-bottomed southern plains rivers. Limpets (*Ferrissia*) and snails (*Physa, Elimia, Helisoma*) are common throughout the river. Crayfishes include *Orconectes palmeri* and *Procambarus acutus* (Reimer 1969, E. A. Bergey, personal communication). The Little River contains an abundant and diverse insect fauna, including beetles (*Stenelmis, Psephenus*), mayflies (*Pseudocleon, Ephemerella, Heptagenia, Leucrocuta, Stenacron, Stenonema, Paraleptophlebia, Isonychia*), damselflies (*Argia*), stoneflies (*Neoperla, Acroneuria*), and caddisflies (*Cheumatopsyche, Helicopsyche, Chimarra,* and *Polycentropus*). (C. C. Vaughn, unpublished data).

Vertebrates

The fish fauna of the Little River is quite diverse for a system of such small size, with many small-bodied insectivore–omnivore species. Jenkins (1956) reported 87 species from the preimpounded basin, and a count of species from the "Fish Atlas" (Lee et al. 1980) and our own collections shows about 110 native species in the drainage. This unusually high number of fish species for such a small drainage reflects its geographic position, with headwaters in high-gradient stony streams of the Ouachita Mountains, its passage onto the low-gradient, soft-bottomed substrates, and eventual entry to the Gulf Coastal Plain. Thus, the headwaters have typical high-gradient upland fish species such as Ouachita Mountain shiner, steelcolor shiner, and orangebelly darter. In the midreach of the Little River, redfin shiner, harlequin darter, and numerous other darters were formerly common. The dominant game fish of the uplands is native smallmouth bass. After entering the Coastal Plain, upland species drop out and a largely different suite of species occurs, including, in the river or its swampy backwaters, taxa such as the rare bluehead shiner, pirate perch, pygmy sunfish, Blair's starheaded topminnow, redhorse suckers of the genus *Moxostoma*, grass pickerel, flier, bantam sunfish, and dollar sunfish. Downstream, dominant predaceous or game fishes include largemouth bass and gars. However, some species, like steelcolor shiner and orangebelly darter, are common throughout most of the system. In some off-channel habitats like overflow ponds, pirate perch and pygmy sunfish dominate and are extremely abundant. The main channel can yield very diverse local assemblages. Eight darter (*Etheostoma, Percina*) species were taken in a single gravel riffle, and as many as 29

species were in some individual seining collections (W. J. Matthews and C. C. Vaughn, personal observations).

One federally threatened fish species (leopard darter) is endemic to this system, as is the Ouachita Mountain shiner. Several other fish species, such as peppered shiner and orangebelly darter, although not endemic, are of very limited ranges geographically, with strongholds in this drainage. The crystal darter, now becoming uncommon throughout its range, has a substantial population in the Little River near Idabel. Other species of particular interest known from the Little River drainage include the southern brook lamprey, taillight shiner, and blackside darter, all state species of special concern in Oklahoma.

Aquatic and riparian areas of the Little River also support alligator snapping turtle and mole salamander, considered species of special concern in Oklahoma. Swamp rabbits are present in the riparian zone. Snapping turtle, common slider, razor-backed musk turtle, diamondback and northern water snakes, and cottonmouth, occur in the river proper. As elsewhere in the southern Great Plains, beaver are common, and American alligator could occur in the easternmost parts of Little River. Herons and kingfisher can be expected. Large vulture roosts are found on gravel bar islands in the river from below Pine Creek Reservoir to the Arkansas state line.

Ecosystem Processes

Ecosystem processes in the main-stem Little River have not been studied. In some tributaries, mussel beds (Spooner 2002, C. C. Vaughn et al. unpublished data) and abundant central stoneroller minnows may alter nutrients, primary productivity, standing crops of invertebrates, or ecosystem processes, but studies of those effects have not been made in this system. Headwaters in this river system are typically clear, and where riparian canopy is open, stones are coated with a rich algal covering, and autochthonous primary productivity appears high. Farther downstream, outside the uplands, the Little River becomes an incised, more turbid stream, with less potential for autochthonous primary productivity.

Human Impacts and Special Features

The Little River system, including the main stem and its array of large tributaries, collectively represent one of the most diverse riverscapes in North America relative to their short length. All major branches of the system arise on steep slopes of the Ouachita uplands, then drain long valley floors until they emerge from the mountains to flow across the low-gradient, historically swampy Coastal Plain. Thus, in their relatively short distances, the branches of the Little River system mimic the headwater to lower river differences that are often seen in river systems over distances of thousands of miles, and also show clearly the biological phenomenon of a sharp ecotone between uplands and lowlands. As a result, these rivers, including the Little River main stem, depict strong longitudinal differentiation of taxa very well. In addition, they harbor rare or at-risk species, including some federally threatened and endangered fishes or invertebrates, and are marked by very high diversity of some groups, like darters.

The greatest human impacts in the region include dams on the main stem and all main tributaries except the Glover River (which remains one of the few free-flowing rivers in the region) and effects of the timber industry. Following final clearing of the last major old-growth forest in the central United States early in the twentieth century, vast tracts of the region were converted under timber company ownership to pine monoculture, replacing the original large pine-savannah and mixed deciduous forest with grow-and-cut forests. In some of the streams where timber is harvested, there has been a noticeable increase in silt on streambed surfaces in the last three decades (W. J. Matthews, personal observations). Large areas also are managed by the USDA Forest Service, and there have been recent land trades between private holders and the Forest Service that now place more of the rivers under federal authority. In some reaches, forest industries create pollutants, such as an outfall from a large mill near Wright City, Oklahoma, that apparently reduced native fish assemblages between the 1970s and 1990s (W. J. Matthews, personal observations). Additional pollutants can come from poultry processing effluents.

The Little River basin is most influenced by the three largest impoundments (see Fig. 7.13). The main stem of the river is impounded by 1644 km² Pine Creek Reservoir, used for flood control, water supply, and recreation. A major tributary of the Little River, the Mountain Fork River, is impounded by 1952 km² Broken Bow Reservoir, used for generation of hydropower, flood control, water supply, and recreation. Outflow from Broken Bow Reservoir enters the Little River via the Mountain Fork River, 64 km downstream of Pine Creek dam. Vaughn and Taylor (1999) examined the distribution and abundance of mussels at 37 sites along a 240 km length of the river. They observed an extinction gradient downstream from the two impoundments, with a gradual, linear

increase in mussel species richness and abundance with increasing distance from the reservoirs. Mussel species distributions were significantly nested, with only sites furthest from the reservoirs containing rare species. The extreme downstream portion of the basin is now impounded by Millwood Reservoir, which drowned vast shallow swamplands of southwest Arkansas. This chapter's senior author vividly recalls as a child crossing those "spooky" swamps late at night on the old highway from Nashville to Ashdown, Arkansas, across what seemed like interminable wooden bridges, while his parents talked in hushed tones in the front seat of the old Nash Rambler station wagon and glanced nervously about, probably hoping the tubes in the tires wouldn't fail.

ADDITIONAL RIVERS

The Cimarron River is a large, shallow, low-gradient prairie tributary to the Arkansas River, interesting for its harsh physical conditions and desiccation of substantial reaches in the upper river in most summers. It arises in the Great Plains and flows through the Central Lowland before terminating in Keystone Lake (Fig. 7.15). It is representative of medium-size southern plains rivers, with an unstable, braided sand bed and highly fluctuating physical conditions. The Cimarron River has low basinwide runoff (<1 cm/mo year-round) due to very high evapotranspiration (Fig. 7.16). There has been widespread introduction of the Red River shiner from outside the drainage, coincident with dramatic loss of the native Arkansas River shiner. Red River pupfish has become abundant in at least one tributary to the Cimarron (D. McNeely, personal communication).

The Neosho (Grand) system also is a large tributary of the Arkansas River, arising in the Flint Hills of Kansas (Fig. 7.17). Endemic fish species and isolated or relict populations of other formerly widespread species exist in the drainage, such as the Topeka shiner, isolated populations of cardinal shiner and southern redbelly dace, and the endemic and federally threatened Neosho madtom. The lower main stem of the Neosho (Grand) system is dominated by three large tailwater-to-headwater reservoirs, with little free-flowing river remaining.

The Illinois River, with its major tributary, the Baron Fork, is a large Ozark river system in the Arkansas River basin (Fig. 7.19). It arises in karst topography, with many spring-fed tributaries, caves, and gravel-bottomed riffle–pool reaches. The Illinois River contains some of the highest-quality stream

FIGURE 7.4   Illinois River, Oklahoma (PHOTO BY C. C. VAUGHN).

habitat in the region and is the only major tributary flowing into the Arkansas River that represents an Ozark uplift, calcareous stream type (Fig. 7.4). Tenkiller Reservoir is a major on-river impoundment on the lower Illinois River. There is a particularly diverse (101 species) fish fauna (Moore and Paden 1950) for such a small basin, with many Ozark-limited fish species like the Ozark minnow, cardinal shiner, banded sculpin, slender madtom, greenside darter, and banded darter as prominent members of the fauna.

The Poteau River is a substantial southern tributary to the Arkansas River in east Oklahoma and west Arkansas. It arises in the steep Ouachita Mountains and feeds, along with other large upland tributaries, into the impounded Lake Wister, after which there is a long portion of low-gradient river and densely wooded riparian habitat before it enters the Arkansas River (Fig. 7.21). Numerous locally restricted fish species exist, like the relatively rare blackside darter and longnose darter. The Poteau River exhibits extremes of seasonal runoff, from an average as low as 0.25 cm/mo in August to more than 6 cm/mo in winter (Fig. 7.22).

The Washita River is the largest low-gradient, western tributary of the Red River that flows into Lake Texoma (Fig. 7.23). An unstable mud–sand riverbed and steeply incised, erosive red earth banks makes this one of the most turbid, silt-laden streams in North America (Fig. 7.5). The Washita River basin is heavily affected by agriculture, with cattle farming and row crops dominating the landscape, along with oil and gas operations. The native fish fauna historically was somewhat limited even before human impacts, with gaps in distribution of some minnow species. However, it has an excellent sports fishery for native flathead catfish and channel catfish.

The Blue River is a relatively short but faunally important tributary of the Red River, fed by large springs in south-central Oklahoma (Fig. 7.25). The upper portion of the river is strongly marl-depositing, and marl dikes form pools and small waterfalls throughout much of the upper main stem. There are no major dams on this system. Two locally unique forms of fishes (orangebelly darter and striped shiner) are distinctive morphologically or genetically from those taxa elsewhere in their range and warrant species-level recognition. Also, the redspot

FIGURE 7.5 Washita River, Oklahoma (Photo by W. J. Matthews).

chub is found only here and in the Ozarks. A reach of several kilometers is owned by the state conservation department for public access, particularly a stocked trout fishery. The latter results in heavy vehicle traffic in and near this part of the river. Mussels (24 species) have been extirpated from 75% of the Blue River over the last 30 years, probably primarily due to siltation from riparian clearing and agriculture (Vaughn 1997).

The Kiamichi River is an upland tributary of the Red River. It arises just east of the Arkansas–Oklahoma border, flows westward into Oklahoma, then south to join the Red River near Antlers, Oklahoma (Fig. 7.27, Fig. 7.6). It flows through a narrow river valley floor, bordered on both sides by steep slopes of long ridge-and-valley mountains. Its flow is altered about halfway down the main stem by the off-channel Sardis Reservoir, which controls inflow from a large tributary creek, and the lower main stem is altered by Hugo Reservoir. The Kiamichi River arises as a clear stream with riffles and long wide pools flowing over stony cobble–boulder substrate or well-developed gravel bars. Water willow is a prominent feature of shallow riffles and pool edges throughout

the upper two-thirds of the river. It possesses an important native fish and mussel fauna, with two federally protected mussel species. The Kiamichi River has been particularly hard hit by drought in recent summers, with long reaches of typically flowing headwaters badly dewatered, which has had impacts on native mussel beds.

ACKNOWLEDGMENTS

We thank Don Wyckoff, associate curator of archaeology at the Sam Noble Oklahoma Museum of Natural History (SNOMNH), for advice on prehistory of native peoples and climates in the region. We also thank David Edds (Emporia State University) for information on the Neosho River, Elizabeth Bergey (Oklahoma Biological Survey) and the Oklahoma Conservation Commission for unpublished data, Oklahoma Natural Heritage Inventory for information on endangered species, Donna Cobb (OU Biological Station) for assistance in searching databases, Jason B. Jackson (assistant curator of ethnology, SNOMNH) for reviewing our sections on contemporary and recent native peoples, Bruce Hoagland (Oklahoma Biological Survey) for information and verification of riparian plant species, and

FIGURE 7.6 Kiamichi River, Oklahoma (Photo by C. C. Vaughn).

the following curators at SNOMNH for advice on taxonomy of vertebrates: Jan Caldwell, Nick Czaplewski, and Gary Schnell. C. Vaughn was supported during preparation of the manuscript by NSF DEB 9870092, and information provided by K. Gido was gathered while supported by the USGS National Gap Analysis Program.

LITERATURE CITED

Abell, R. A., D. M. Olson, E. D. Dinerstein, P. T. Hurley, J. T. Diggs, W. Eichbaum, S. Walters, W. Wettengel, T. Allnut, C. J. Loucks, and P. Hedao. 2000. *Freshwater ecoregions of North America: A conservation assessment*. Island Press, Washington, D.C.

Bass, D., and V. Walker. 1992. A preliminary report of invertebrates from hyporheic sediments of the North Canadian River. *Proceedings of the Oklahoma Academy of Science* 72:3–4.

Beck, W. A., and Y. D. Haase. 1989. *Historical atlas of the American west*. University of Oklahoma Press, Norman.

Branson, B. A. 1982. The mussels (Unionacea: Bivalva) of Oklahoma–Part 1: Ambleminae. *Proceedings of the Oklahoma Academy of Science* 62:38–45.

Branson, B. A. 1983. The mussels (Unionacea: Bivalva) of Oklahoma–Part 2: The Unioninae, Pleurobemini and Anodontini. *Proceedings of the Oklahoma Academy of Science* 63:49–59.

Branson, B. A. 1984. The mussels (Unionacea: Bivalva) of Oklahoma–Part 3: Lampsilinae. *Proceedings of the Oklahoma Academy of Science* 64:20–36.

Brown, A. V., and W. J. Matthews. 1995. Stream ecosystems of the central United States. In C. E. Cushing, K. W. Cummins, and G. W. Minshall (eds.). *River and stream ecosystems of the World*, pp. 89–116. Elsevier Press, Amsterdam.

Byre, V. J. 2000. Productivity, habitat assessment, and management of least terns nesting along the Canadian River in central Oklahoma. *Occasional Papers Sam Noble Oklahoma Museum of Natural History* 8:1–13.

Cheper, N. J. 1988. *Palaemonetes kadiakensis* Rathburn in Oklahoma (Crustacea: Decapoda). *Proceedings of the Oklahoma Academy of Science* 68:77–78.

Covich, A. P., W. Shepard, E. A. Bergey, and C. S. Carpenter. 1978. Effects of fluctuating flow rates and water levels on chironomids: Direct and indirect alterations of habitat stability. In J. H. Thorp and J. W. Gibbons (eds.). *Energy and environmental stress in aquatic systems*, pp. 141–156. ERDA Symposium Series. Oak Ridge Press. Oak Ridge, Tennessee.

Creel, G. C. 1964. *Hemigrapsus estellinensis*: A new grapsoid crab from North Texas. *Southwestern Naturalist* 8:236–241.

Cross, F. B., R. L. Mayden, and J. D. Stewart. 1986. Fishes in the western Mississippi drainage. In C. H. Hocutt and E. O. Wiley (eds.). *The zoogeography of North American freshwater fishes*, pp. 363–412. John Wiley and Sons, New York.

Cross, F. B., and G. A. Moore. 1952. The fishes of the Poteau River, Oklahoma and Arkansas. *American Midland Naturalist* 47:396–412.

Cross, F. B., and R. E. Moss. 1987. Historic changes in fish communities and aquatic habitats in plains streams of Kansas. In W. J. Matthews and D. C. Heins (eds.). *Community and evolutionary ecology of North American stream fishes*, pp. 155–165. University of Oklahoma Press, Norman.

Echelle, A. A., A. F. Echelle, and L. G. Hill. 1972. Interspecific interactions and limiting factors of abundance and distribution in the Red River pupfish, *Cyprinodon rubrofluviatilis*. *American Midland Naturalist* 88:109–130.

Echelle, A. A., and G. D. Schnell. 1976. Factor analysis of species associations among fishes of the Kiamichi River, Oklahoma. *Transactions of the American Fisheries Society* 105:17–31.

Eisenhour, D. J. 1999. Systematics of *Macrhybopsis tetranema* (Cypriniformes: Cyprinidae). *Copeia* 1999:969–980.

Ferrington, L. C., Jr. 1993. Endangered rivers: A case history of the Arkansas River in the Central Plains. *Aquatic Conservation: Marine and Freshwater Ecosystems* 3:305–316.

Foreman, G. 1937. *Adventure on Red River: Report on the exploration of the Red River by Captain Randolph B. Marcy and Captain G. B. McClellan.* University of Oklahoma Press, Norman.

Fugate, F. L., and R. B. Fugate. 1991. *Roadside history of Oklahoma.* Mountain Press, Missoula, Montana.

Gelwick, F. P., and W. J. Matthews. 1992. Effects of an algivorous minnow on temperate stream ecosystem properties. *Ecology* 73:1630–1645.

Gido, K. B. 1999. Ecosystem effects of omnivorous fishes in Lake Texoma (Oklahoma–Texas). Ph.D. diss., University of Oklahoma, Norman.

Gido, K. B., and W. J. Matthews. 2001. Ecosystem effects of water column minnows in experimental streams. *Oecologia* 126:247–253.

Gido, K. B., W. J. Matthews, and W. C. Wolfinbarger. 2000. Long-term variation in fish assemblages of an artificial reservoir: Stability in an unpredictable environment. *Ecological Applications* 10:1517–1529.

Goldhammer, D. S., and L. C. Ferrington Jr. 1992. Emergence of aquatic insects from epirheic zones of capillary fringe habitats in the Cimarron River, Kansas. In J. A. Stanford and J. J. Simons (eds.). Proceedings of the First International Conference on Ground Water Ecology, pp. 155–164. American Water Resources Association, Bethesda, Maryland.

Gordon, M. E., A. V. Brown, and L. R. Kraemer. 1979. Mollusca of the Illinois River, Arkansas. *Arkansas Academy of Science Proceedings* 33:35–37.

Hanson, G. T., and C. H. Moneyhon. 1989. *Historical atlas of Arkansas.* University of Oklahoma Press, Norman.

Hefley, H. M. 1937. Ecological studies on the Canadian River floodplain in Cleveland County, Oklahoma. *Ecological Monographs* 7:345–402.

Hoagland, B. W. 2000. The vegetation of Oklahoma: A classification for landscape mapping and conservation planning. *Southwestern Naturalist* 45:385–420.

Hoagland, B. W., I. H. Butler, F. L. Johnson, and S. Glenn. 1999. The cross timbers. In R. C. Anderson, J. S. Fralish, and J. M. Baskin (eds.). *Savannahs, barrens, and rock outcrop plant communities of North America*, pp. 231–145. Cambridge University Press, Cambridge.

Hoagland, B. W., L. R. Sorrels, and S. M. Glenn. 1996. Woody species composition of floodplain forests of the Little River, McCurtain and LeFlore Counties, Oklahoma. *Proceedings of the Oklahoma Academy of Science* 76:23–29.

Hunt, C. B. 1974. *Natural regions of the United States and Canada.* W. H. Freeman and Company, San Francisco.

Isely, F. B. 1924. The freshwater mussel fauna of eastern Oklahoma. *Proceedings of the Oklahoma Academy of Science* 197:266–270.

Jenkins, R. M. 1956. The fishery resources of the Little River system, McCurtain County, Oklahoma. Report no. 55. Oklahoma Fisheries Research Laboratory, Norman.

Johnson, S. L., and A. P. Covich. 1997. Scales of observation of riparian forests and distributions of suspended detritus in a prairie river. *Freshwater Biology* 37:163–175.

Johnson, S. L., and C. C. Vaughn. 1995. A hierarchical study of macroinvertebrate recolonization of disturbed patches along a longitudinal gradient in a prairie river. *Freshwater Biology* 34:531–540.

Leckie, W. H. 1967. *The Buffalo Soldiers: A narrative of the Negro cavalry in the West.* University of Oklahoma Press, Norman.

Lee, D. S., C. R. Gilbert, C. H. Hocutt, R. E. Jenkins, D. E. McAllister, and J. R. Stauffer Jr. 1980. *Atlas of North American freshwater fishes.* North Carolina State Museum of Natural History, Raleigh.

Lindsay, H. L., J. C. Randolph, and J. Carroll. 1983. Updated survey of the fishes of the Poteau River, Oklahoma and Arkansas. *Proceedings of the Oklahoma Academy of Science* 63:42–48.

Magdych, W. P. 1984. Salinity stresses along a complex river continuum: Effects on mayfly (Ephemeroptera) distributions. *Ecology* 65:1662–1672.

Master, L. L., S. R. Flack, and B. A. Stein. 1998. *Rivers of life: Critical watersheds for protecting freshwater biodiversity.* Nature Conservancy, Arlington, Virginia.

Matthews, W. J. 1987. Physicochemical tolerances and selectivity of stream fishes as related to their geographic ranges and local distributions. In W. J. Matthews and D. C. Heins (eds.). *Community and evolutionary ecology of North American stream fishes*, pp. 111–120. University of Oklahoma Press, Norman.

Matthews, W. J. 1988. North American prairie streams as systems for ecological study. *Journal of the North American Benthological Society* 7:387–409.

Matthews, W. J., K. B. Gido, and E. Marsh-Matthews. 2001. Density-dependent overwinter survival growth of red shiners from a southwestern river. *Transactions of the American Fisheries Society* 130:478–488.

Matthews, W. J., and F. P. Gelwick. 1990. Fishes of Crutcho Creek and the North Canadian River near Oklahoma City: Urbanization and temporal variability. *Southwestern Naturalist* 35:403–410.

Matthews, W. J., and L. G. Hill. 1979. Influence of physicochemical factors on habitat selection by red shiners, *Notropis lutrensis* (Pisces: Cyprinidae). *Copeia* 1979: 70–81.

Matthews, W. J., and L. G. Hill. 1980. Habitat partitioning in the fish community of a southwestern river. *Southwestern Naturalist* 25:51–66.

Matthews, W. J., A. J. Stewart, and M. E. Power. 1987. Grazing fishes as components of North American stream ecosystems: Effects of *Campostoma anomalum*. In W. J. Matthews and D. C. Heins (eds.). *Community and evolutionary ecology of North American stream fishes*, pp. 128–135. University of Oklahoma Press, Norman.

Matthews, W. J., and E. G. Zimmerman. 1990. Potential effects of global warming on native fishes of the southern Great Plains and the Southwest. *Fisheries* 15:26–32.

Metcalf, A. L. 1966. Fishes of the Kansas River system in relation to zoogeography of the Great Plains. *University of Kansas Publications, Museum of Natural History* 17:23–189.

Moore, G. A., and J. M. Paden. 1950. The fishes of the Illinois River in Oklahoma and Arkansas. *American Midland Naturalist* 44:76–95.

Morris, J. W., C. R. Goins, and E. C. McReynolds. 1986. *Historical atlas of Oklahoma*, 3rd ed. University of Oklahoma Press, Norman.

Morris, W. K., and M. O. Madden. 1978. Benthic macroinvertebrate communities aid water quality evaluation of the Washita River. *Proceedings of the Oklahoma Academy of Science* 58:93–97.

Moulton, S. R., and K. W. Stewart. 1996. *Caddisflies (Trichoptera) of the interior highlands of North America.* Memoirs of the American Entomological Institute, 56. American Entomological Institute, Gainesville, Florida.

Nulty, M. L. 1980. Ecology of caddisflies (Trichoptera: Hydropsychidae) in a Neosho River riffle. *Emporia State Research Studies* 18:1–30.

Obermeyer, B. K., D. R. Edds., C. W. Prophet, and E. J. Miller. 1997. Freshwater mussels (Bivalvia: Unionidae) in the Verdigris, Neosho and Spring River basins of Kansas and Missouri, with emphasis on species of concern. *American Malacological Bulletin* 14:41–55.

Oklahoma Natural Heritage Inventory (ONHI). 2001. Database of Oklahoma's endangered, threatened, and rare species. Oklahoma Biological Survey Web site: http://www.biosurvey.ou.edu

Pigg, J. 1978. The tilapia *Sarotherodon aurea* (Steindachner) in the North Canadian River in central Oklahoma. *Proceedings of the Oklahoma Academy of Science* 58:111–112.

Pigg, J., and L. G. Hill. 1974. Fishes of the Kiamichi River. *Proceedings of the Oklahoma Academy of Science* 54:124–130.

Power, M. E., W. J. Matthews, and A. J. Stewart. 1985. Grazing minnows, piscivorous bass and stream algae: Dynamics of a strong interaction. *Ecology* 66:1448–1456.

Reimer, R. D. 1969. A report on the crawfishes (Decapoda: Astacidae) of Oklahoma. *Proceedings of the Oklahoma Academy of Science* 48:49–65.

Resh, V. H., A. V. Brown, A. P. Covich, M. E. Gurtz, H. W. Li, G. W. Minshall, S. R. Reice, A. L. Sheldon, J. B. Wallace, and R. C. Wissmar. 1988. The role of disturbance in stream ecology. *Journal of the North American Benthological Society* 7:433–455.

Resh, V. H., D. S. White, and S. J. White. 1978. Lake Texoma caddisflies (Insecta: Trichoptera): 1. Species present and faunal changes since impoundment. *Southwestern Naturalist* 23:381–388.

Ricketts, T. H., E. Dinerstein, D. M. Olson, C. J. Loucks, W. Eichbaum, D. DellaSala, K. Kavanagh, P. Hedao, P. T. Hurley, K. M. Carney, R. Abell, and S. Walters. 1999. *Terrestrial ecoregions of North America: A conservation assessment.* Island Press, Washington, D.C.

Shackelford, J., and J. Whitaker. 1997. Relative abundance of the northern river otter, *Lutra canadensis*, in three drainage basins of southeastern Oklahoma. *Proceedings of the Oklahoma Academy of Science* 77: 93–98.

Sealander, J. A., and G. A. Heidt. 1990. *Arkansas mammals: Their natural history, classification and distribution.* University of Arkansas Press, Fayetteville.

Smith, K. L. 1986. *Sawmill: The story of cutting the last great virgin forest east of the Rockies.* University of Arkansas Press, Fayetteville.

Socolofsky, H. E., and H. Self. 1988. *Historical atlas of Kansas*. University of Oklahoma Press, Norman.

Solley, W. B., R. R. Pierce, and H. A. Perlman. 1998. Estimated use of water in the United States in 1995. Circular 1200. U.S. Geological Survey, Denver, Colorado.

Spooner, D. E. 2002. A field experiment examining the effects of freshwater mussels (Family: Unionidae) on sediment ecosystem function. Master's thesis, University of Oklahoma, Norman.

Sublette, J. E. 1953. The physico-chemical and biological features of Lake Texoma (Denison Reservoir), Oklahoma and Texas: A preliminary study. *Texas Journal of Science* 7:164–182.

Thornthwaite, C. W. 1941. Climate and settlement in the Great Plains. In G. Hambridge and M. J. Drown (eds.). *Climate and man: Yearbook of agriculture*, pp. 177–187. U.S. Department of Agriculture, Washington, D.C.

Tyson, C. N. 1981. *The Red River in southwestern history*. University of Oklahoma Press, Norman.

U.S. Geological Survey. 1974. Geologic Map of the United States.

Valentine, B. D., and D. H. Stansbery. 1971. An introduction to the naiads of the Lake Texoma region, Oklahoma, with notes on the Red River fauna (Mollusca: Unionidae). *Sterkiana* 42:1–40.

Vaughn, C. C. 1982. Distribution of chironomids in the littoral zone of Lake Texoma, Oklahoma and Texas. *Hydrobiologia* 89:177–188.

Vaughn, C. C. 1997. Catastrophic decline of the mussel fauna of the Blue River, Oklahoma. *Southwestern Naturalist* 42:333–336.

Vaughn, C. C. 1998. Determination of the status and habitat preference of the Neosho Mucket in Oklahoma. Report to the Oklahoma Department of Wildlife Conservation, Oklahoma City, Oklahoma.

Vaughn, C. C. 2000. Changes in the mussel fauna of the middle Red River drainage: 1910–present. In R. A. Tankersley, D. I. Warmolts, G. T. Watters, B. J. Armitage, P. D. Johnson, and R. S. Butler (eds.). *Freshwater mollusk symposium proceedings*, pp. 225–232. Ohio Biological Survey, Columbus.

Vaughn, C. C., C. M. Mather, M. Pyron, P. Mehlhop, and E. K. Miller. 1996. The current and historical mussel fauna of the Kiamichi River, Oklahoma. *Southwestern Naturalist* 41:325–328.

Vaughn, C. C., and M. Pyron. 1995. Population ecology of the endangered Ouachita Rock Pocketbook Mussel, *Arkansia wheeleri* (Bivalvia: Unionidae), in the Kiamichi River, Oklahoma. *American Malacological Bulletin* 11:145–151.

Vaughn, C. C., and C. M. Taylor. 1999. Impoundments and the decline of freshwater mussels: A case study of an extinction gradient. *Conservation Biology* 13:912–920.

Wilhm, J., H. Namminga, and C. Ferraris. 1978. Species composition and diversity of benthic macroinvertebrates in Greasy Creek, Red Rock Creek and the Arkansas River. *American Midland Naturalist* 99:444–453.

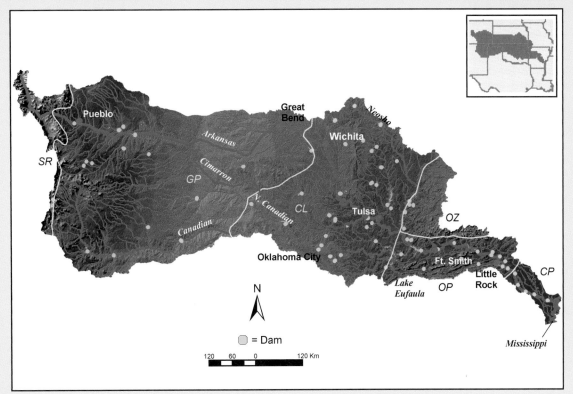

FIGURE 7.7 Map of the Arkansas River basin. Physiographic provinces are separated by yellow lines.

ARKANSAS RIVER

Relief: 4340 m

Basin area: 414,910 km^2

Mean discharge: 1004 m^3/s

River order: 7

Mean annual precipitation: 70.8 cm

Mean air temperature: 15°C

Mean water temperature: 17.9°C

Physiographic provinces: Southern Rocky Mountains (SR), Great Plains (GP), Central Lowland (CL), Ozark Plateaus (OZ), Ouachita Province (OP), Coastal Plain (CP)

Biomes: Temperate Mountain Forest, Temperate Grasslands, Temperate Deciduous Forest

Freshwater ecoregions: Southern Plains, Central Prairie, Ozark Highlands, Mississippi Embayment

Terrestrial ecoregions: 6 ecoregions (see text)

Number of fish species: 171 (141 native)

Number of endangered species: 1 fish, 1 bird

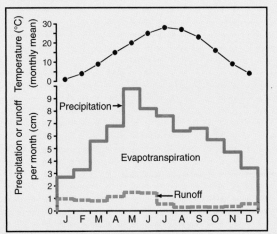

FIGURE 7.8 Mean monthly air temperature, precipitation, and runoff for the Arkansas River basin.

Major fishes: paddlefish, gars, gizzard shad, red shiner, river shiner, emerald shiner, plains minnow, smallmouth buffalo, bigmouth buffalo, river carpsucker, channel catfish, flathead catfish, plains killifish, western mosquitofish, white bass, largemouth bass, spotted bass, sunfishes, river darter

Major other aquatic vertebrates: plains leopard frog, American bullfrog, Blanchard's cricket frog, snapping turtle, spiny softshell turtle, smooth softshell turtle, yellow mud turtle, common slider, false map turtle, painted turtle, northern water snake, diamondback water snake, American white pelican, great blue heron, belted kingfisher, beaver, muskrat

Major benthic invertebrates: mayflies (*Caenis, Hexagenia, Stenonema*), caddisflies (*Cheumatopsyche, Hydropsyche*), chironomid midges (*Polypedilum, Glyptotendipes*), crayfishes (*Orconectes palmeri, Procambarus simulans*), mussels (washboard, threeridge, mapleleaf, pink papershell)

Nonnative species: Asian clam, zebra mussel, ~30 fish species (common carp, grass carp, striped bass), nutria in Arkansas

Major riparian plants: silver maple, box elder, hackberry, cottonwood, willow, cattails, American bulrush

Special features: arises as strongly flowing mountain river, almost disappears in western Kansas due to water withdrawal and evaporation; recharged near Great Bend, Kansas

Fragmentation: five major reservoirs on main stem, plus 17 locks and dams; part of the Kerr-McClellan Navigation System

Water quality: pH = 7.4, alkalinity = 52 mg/L as CaCO$_3$, NO$_3$-N = 0.25 mg/L, PO$_4$-P = 0.02 mg/L, but with wide range of values from headwaters to lower main river

Land use: western basin: mining, oil and gas production, 50% rangeland, 50% cropland; eastern basin: 50% forest, 15% cropland, 25% pasture; urban in Wichita, Tulsa, Little Rock

Population density: 14.6 people/km^2

Major information sources: Branson 1982, 1983, 1984; Cross et al. 1986; Reimer 1969; Sealander and Heidt 1990; Solley et al. 1998

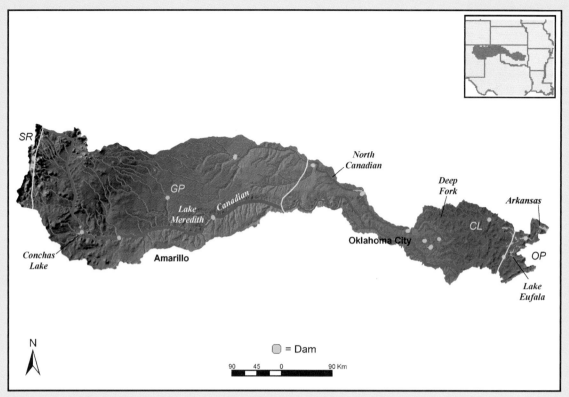

FIGURE 7.9 Map of the Canadian River basin. Physiographic provinces are separated by yellow lines.

CANADIAN RIVER

Relief: 4132 m

Basin area: 122,070 km²

Mean discharge: 174 m³/s

River order: 6

Mean annual precipitation: 52.5 cm

Mean air temperature: 15°C

Mean water temperature: 18°C

Physiographic provinces: Southern Rocky Mountains (SR), Great Plains (GP), Central Lowland (SL), Ouachita Province (OP)

Biomes: Temperate Mountain Forest, Temperate Grasslands, Temperate Deciduous Forest

Freshwater ecoregion: Southern Plains

Terrestrial ecoregions: Western Short Grasslands, Central and Southern Mixed Grasslands, Central Forest Grassland Transition Zone

Number of fish species: 63 (native)

Number of endangered species: 1 fish, 1 bird

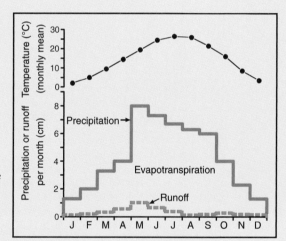

FIGURE 7.10 Mean monthly air temperature, precipitation, and runoff for the Canadian River basin.

Major fishes: gizzard shad, red shiner, Arkansas River shiner, emerald shiner, plains minnow, bluntnose minnow, fathead minnow, plains killifish, western mosquitofish, river carpsucker, channel catfish, white bass, largemouth bass, longear sunfish, green sunfish

Major other aquatic vertebrates: snapping turtle, yellow mud turtle, stinkpot turtle, smooth softshell turtle, beaver

Major benthic invertebrates: oligochaetes, midges (*Bezzia, Chironomus, Cryptochironomus*), mayflies (*Tricorythodes, Caenis*), caddisflies (*Cheumatopsyche*), crayfish (*Orconectes nais*), fingernail clams (*Sphaerium, Pisidium*), mussels (pink papershell, fragile papershell, yellow sandshell, white heelsplitter)

Nonnative species: Asian clam, Red River pupfish, inland silversides, common carp, blue tilapia, saltcedar

Major riparian plants: silver maple, box elder, American elm, hackberry, sandbar willow, ash, cottonwood, saltcedar

Special features: crosses arid grasslands in west, sometimes desiccating, mesic forest in east; shallow, shifting "sand bed" rivers create harsh environments, limiting richness and persistence of fauna

Fragmentation: four impoundments on main stem

Water quality: pH = 8.2, alkalinity = 156 mg/L as $CaCO_3$, NO_3-N = 0.46 mg/L, total phosphorus = 0.32 mg/L

Land use: 50% rangeland or pasture, 30% cropland; up to 55% forested in east

Population density: 9.1 people/km²

Major information sources: Branson 1982, 1983, 1984, Byre 2000, Cross et al. 1986, Matthews and Hill 1979, 1980, Solley et al. 1998, Reimer 1969

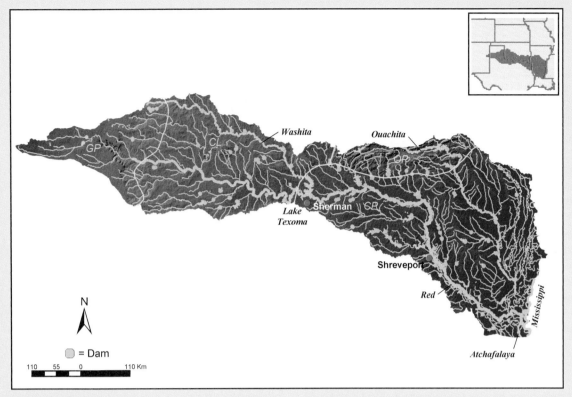

FIGURE 7.11 Map of the Red River basin. Physiographic provinces are separated by yellow lines.

RED RIVER

Relief: 1347 m
Basin area: 169,890 km²
Mean discharge: 852 m³/s
River order: 7
Mean annual precipitation: 82 cm
Mean air temperature: 18°C
Mean water temperature: 19.3°C
Physiographic provinces: Great Plains (GP), Central Lowland (CL), Ouachita Province (OP), Coastal Plain (CP)
Biomes: Temperate Grasslands, Temperate Deciduous Forest
Freshwater ecoregions: Southern Plains, Ouachita Highlands, Mississippi Embayment
Terrestrial ecoregions: Western Short Grasslands, Central and Southern Mixed Grasslands, Central Forest Grassland Transition Zone, Piney Woods Forests
Number of fish species: 171 (152 native)
Number of endangered species: 1 bird

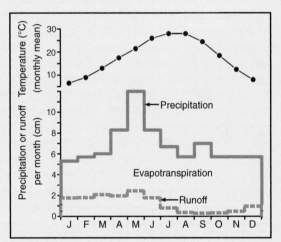

FIGURE 7.12 Mean monthly air temperature, precipitation, and runoff for the Red River basin.

Major fishes: alligator gar, longnose gar, gizzard shad, red shiner, emerald shiner, Red River shiner, chub shiner, emerald shiner, blacktail shiner, bluntnose minnow, plains minnow, blue sucker, smallmouth buffalo, river carpsucker, channel catfish, blue catfish, plains killifish, Red River pupfish, sunfishes, white bass, largemouth bass, bigscale logperch
Major other aquatic vertebrates: alligator snapping turtle, common slider, spiny softshell turtle, false map turtle, yellow mud turtle, plain-bellied water snake, cottonmouth, American alligator, great blue heron, beaver, muskrat, nutria
Major benthic invertebrates: Chironomid midges (*Glyptotendipes*, *Dicrotendipes*, *Chironomus*), mayflies (*Hexagenia*, *Caenis*, *Stenonema*), caddisflies (*Cyrnellus*, *Hydropsyche*), crayfishes (*Orconectes palmeri*, *O. nais*, *Procambarus simulans*), mussels (pink papershell, fragile papershell, bleufer)
Nonnative species: Asian clam, nutria, striped bass, walleye, threadfin shad, inland silversides, common carp, grass carp
Major riparian plants: cottonwood, willows, box elder, silver maple, slippery elm, sweetgum, post oak
Special features: spans gradient from driest to some of wettest climatic conditions in North America; high salinity in headwaters, frequently drying; Great Raft, once a logjam of gigantic proportions upstream from Shreveport
Fragmentation: one major impoundment (Lake Texoma) on main stem; four locks and dams in Louisiana
Water quality: pH = 8.0, alkalinity = 131 mg/L as CaCO₃, NO₃-N = 0.18 mg/L, PO₄-P = 0.11 mg/L, total phosphorus = 0.12 mg/L; chloride extremely high in headwaters, conductivity to 35 μS/cm, decreasing to about 2 μS/cm near Lake Texoma
Land use: 40% to 60% rangeland and 30 % cropland in west; 50% forest, 20% cropland, and 10% pasture in east
Population density: 9.1 people/km²
Major information sources: Branson 1982, 1983, 1984, Creel 1964, Cross et al. 1986, Echelle et al. 1972, Hoagland 2000, Valentine and Stansbery 1971

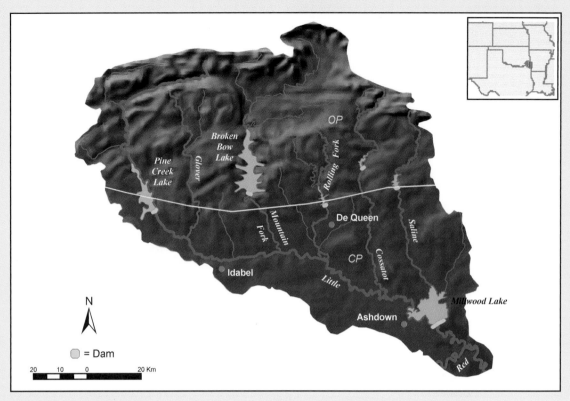

FIGURE 7.13 Map of the Little River basin. Physiographic provinces are separated by a yellow line.

LITTLE RIVER

Relief: 741 m
Basin area: 10,720 km²
Mean discharge: 183 m³/s
River order: 6
Mean annual precipitation: 123 cm
Mean air temperature: 16°C
Mean water temperature: 16.5°C
Physiographic provinces: Ouachita Province (OP), Coastal Plain (CP)
Biome: Temperate Deciduous Forest
Freshwater ecoregion: Ouachita Highlands
Terrestrial ecoregions: Ozark Mountain Forests, Central Forest Grassland Transition Zone, Piney Woods Forests
Number of fish species: 110
Number of endangered species: 2 mussels, 1 fish

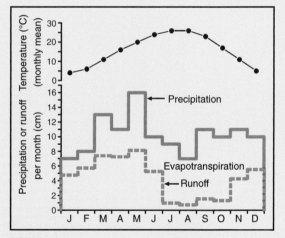

FIGURE 7.14 Mean monthly air temperature, precipitation, and runoff for the Little River basin.

Major fishes: gars, rocky shiner, blacktail shiner, central stoneroller, river redhorse, golden redhorse, blackstriped topminnow, grass pickerel, flier, bantam sunfish, pirate perch, dusky darter, crystal darter, orangethroat darter, orangebelly darter, largemouth bass, spotted bass, sunfishes, leopard darter
Major other aquatic vertebrates: snapping turtle, common slider, razor-backed musk turtle, diamondback water snake, northern water snake, cottonmouth, swamp rabbit, beaver, river otter, mink
Major benthic invertebrates: mussels (threeridge, mucket, pimpleback), limpets (*Ferrissia*), snails (*Elimia, Heliosoma*), crayfishes (*Orconectes palmeri, Procambarus acutus*), mayflies (*Stenonema, Ephemerella, Heptagenia, Isonychia*), beetles (*Stenelmis, Psephenus*), stoneflies (*Acroneuria, Neoperla*), caddisflies (*Cheumatopsyche, Helicopsyche, Chimarra*)
Nonnative species: Asian clam, brown trout, rainbow trout, common carp; grass carp and striped bass likely
Major riparian plants: river birch, sycamore, smooth alder, sugar maple, box elder, willow oak, blue beech, bald cypress
Special features: some of last well-preserved upland rivers in central United States; regional "hot spot" of biodiversity
Fragmentation: two reservoirs on main stem (Pine Creek and Millwood); four large reservoirs on main tributaries
Water quality: pH = 7.2, alkalinity = 24 mg/L as CaCO₃, NO₃-N = 0.33 mg/L, PO₄-P = 0.06 to 0.09 mg/L
Land use: 75% forest, 10% cropland, 15% pasture; no urban areas
Population density: 3.2 people/km²
Major information sources: Hoagland 2000, Hoagland et al. 1996, Jenkins 1956, Reimer 1969, Shackelford and Whitaker 1997, Solley et al. 1998, Vaughn 2000, Vaughn and Taylor 1999

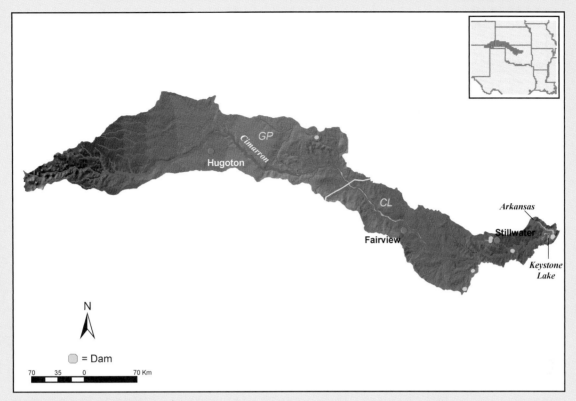

FIGURE 7.15 Map of the Cimarron River basin. Physiographic provinces are separated by a yellow line.

CIMARRON RIVER

Relief: 2036 m
Basin area: 50,540 km²
Mean discharge: 42 m³/s
River order: 4
Mean annual precipitation: 55.3 cm
Mean air temperature: 15°C
Mean water temperature: 18.4°C
Physiographic provinces: Great Plains (GP), Central Lowland (CL)
Biomes: Temperate Grasslands, Temperate Deciduous Forest
Freshwater ecoregion: Southern Plains
Terrestrial ecoregions: Western Short Grasslands, Central and Southern Mixed Grasslands, Central Forest Grassland Transition Zone
Number of fish species: 48
Number of endangered species: 1 fish, 1 bird
Major fishes: red shiner, plains minnow, plains killifish, gizzard shad, white bass, channel catfish, western mosquitofish; Arkansas River shiner now much reduced in abundance

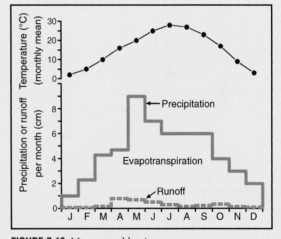

FIGURE 7.16 Mean monthly air temperature, precipitation, and runoff for the Cimarron River basin.

Major other aquatic vertebrates: snapping turtle, beaver
Major benthic invertebrates: mollusks (fingernail clam), crayfish (*Orconectes nais*), oligochaete worms (*Chaetogaster, Limnodrilus*), mayflies (*Caenis, Baetis*), caddisflies (*Cheumatopsyche, Hydropsyche*), beetles (*Stenelmis*)
Nonnative species: Asian clam, Red River shiner, striped bass, saltcedar
Major riparian plants: silver maple, box elder, ash, hackberry, cottonwood, sandbar willow, black willow, saltcedar, American elm
Special features: drains some of most arid lands of southern Great Plains; long reaches of western main stem intermittent; harsh conditions, but relatively diverse fish fauna
Fragmentation: two large reservoirs on main stem; other fragmentation by natural or human-enhanced desiccation of main-stem reaches
Water quality: pH = 8.1, alkalinity = 169 mg/L as CaCO₃, NO₃-N = 1.32 mg/L, total phosphorus = 0.84 mg/L
Land use: 50% rangeland or pasture, 35% cropland, up to 30% forest in east; no large cities
Population density: 6.7 people/km²
Major information sources: Hoagland 2000, ONHI 2001, Solley et al. 1998

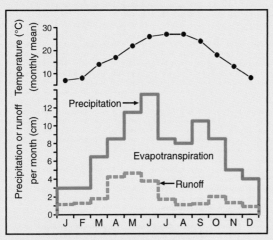

FIGURE 7.18 Mean monthly air temperature, precipitation, and runoff for the Neosho River basin.

NEOSHO (GRAND) RIVER

Relief: 325 m
Basin area: 54,550 km²
Mean discharge: 254 m³/s
River order: 6
Mean annual precipitation: 91 cm
Mean air temperature: 12°C
Mean water temperature: 15.4°C
Physiographic provinces: Central Lowland (CL), Ozark Plateaus (OZ)
Biomes: Temperate Grasslands, Temperate Deciduous Forest
Freshwater ecoregion: Central Prairie
Terrestrial ecoregion: Central Forest Grassland Transition Zone
Number of fish species: 94 native
Number of endangered species: 1 fish, 1 bird, 1 mussel
Major fishes: upstream: Topeka shiner, orangethroat darter, cardinal shiner, southern redbelly dace, endemic Neosho madtom; downstream: paddlefish, gizzard shad, numerous native minnows, smallmouth buffalo, river carpsucker, white bass, largemouth bass, sunfishes
Major other aquatic vertebrates: hellbender (threatened in Kansas), mudpuppy, snapping turtle, spiny softshell turtle, smooth softshell turtle, common slider, false map turtle, Ouachita map turtle, painted turtle, diamondback water snake
Major benthic invertebrates: 33 species of mussels (threeridge, monkeyface, Neosho mucket), caddisflies (*Hydropsyche, Potamyia*), crayfishes (*Orconectes virilis, O. neglectus*), chironomid midges (*Glyptotendipes*)
Nonnative species: Asian clam, common carp, rainbow trout
Major riparian plants: silver maple, box elder, red maple, river birch, hackberry, pecan, eastern swamp privet, ash, blackgum, sycamore, cottonwood, pin oak, American elm
Special features: drains unique uplifted region of Kansas known as "Flint Hills"; streams comprising clear water "outposts" disjunct from and containing species common to Ozark Plateaus
Fragmentation: four impoundments on main stem
Water quality: pH = 7.4, alkalinity = 52 mg/L as CaCO₃, NO₃-N = 0.25 mg/L, PO₄-P = 0.02 mg/L
Land use: upper basin: 60% rangeland and pasture, 20% crops; lower basin: 75% forest, 20% pasture, 5% crops; no large cities
Population density: 13.5 people/km²
Major information sources: Covich et al. 1978, Hoagland 2000, Obermeyer et al. 1997, Nulty 1980, Reimer 1969, Solley et al. 1998, Vaughn 1998

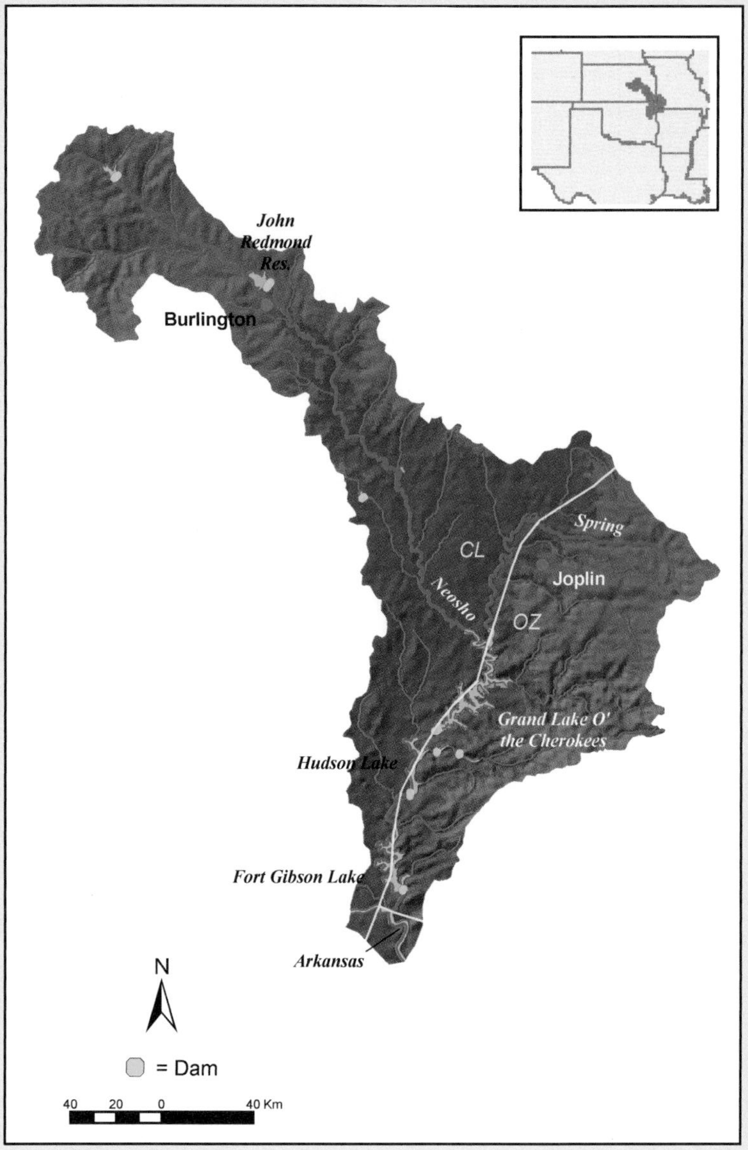

FIGURE 7.17 Map of the Neosho River basin. Physiographic provinces are separated by a yellow line.

320

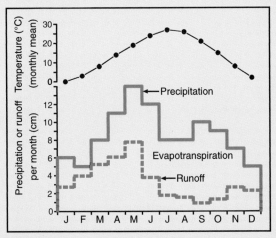

FIGURE 7.20 Mean monthly air temperature, precipitation, and runoff for the Illinois River basin.

ILLINOIS RIVER

Relief: 390 m
Basin area: 4260 km^2
Mean discharge: 54 m^3/s
River order: 5
Mean annual precipitation: 103 cm
Mean air temperature: 14°C
Mean water temperature: 16.5°C
Physiographic province: Ozark Plateaus (OZ), Ouachita Province (OP)
Biome: Temperate Deciduous Forest
Freshwater ecoregions: Ozark Highlands, Central Prairie
Terrestrial ecoregion: Ozark Mountain Forests
Number of fish species: 101
Number of endangered species: 1 bird and 1 mussel federally threatened
Major fishes: southern brook lamprey, longnose gar, central stoneroller, bigeye shiner, Ozark minnow, rosyface shiner, cardinal shiner, redspot chub, golden redhorse, slender madtom, blackspotted topminnow, brook silverside, smallmouth bass, spotted bass, longear sunfish, green sunfish, white bass, stippled darter, orangethroat darter, greenside darter, banded darter, banded sculpin
Major other aquatic vertebrates: American bullfrog, common slider, spiny softshell turtle, false map turtle, water snakes, cottonmouth
Major benthic invertebrates: 22 species of mussels (elktoe, threeridge, pigtoe), fingernail clams (*Sphaerium*, *Pisidium*), snails (*Elimia*, *Physa*), crayfish (*Orconectes meeki*), mayflies (*Baetis*, *Pseudocleon*), caddisflies (*Helicopsyche*, *Cheumatopsyche*)
Nonnative species: Asian clam, striped bass, rainbow trout
Major riparian plants: silver maple, box elder, red maple, river birch, pecan, eastern swamp privet, possum haw, sycamore, black gum, cottonwood, pin oak, American elm
Special features: state Scenic River (Oklahoma); generally a clear upland river, but water quality deterioration (municipal wastes) in headwaters (Arkansas); heavily used for recreation
Fragmentation: one major impoundment (Tenkiller); old spillway (former Lake Frances) at Arkansas–Oklahoma border impedes upstream movement of fishes
Water quality: pH = 7.4, alkalinity = 52 mg/L as CaCO$_3$, NO$_3$-N = 0.25 mg/L, PO$_4$-P = 0.02 mg/L
Land use: 70% forest, 20% pasture, 10% cropland; no large cities
Population density: 18.3 people/km^2
Major information sources: Gordon et al. 1979, Hoagland 2000, ONHI 2001, Moore and Paden 1950, Reimer 1969

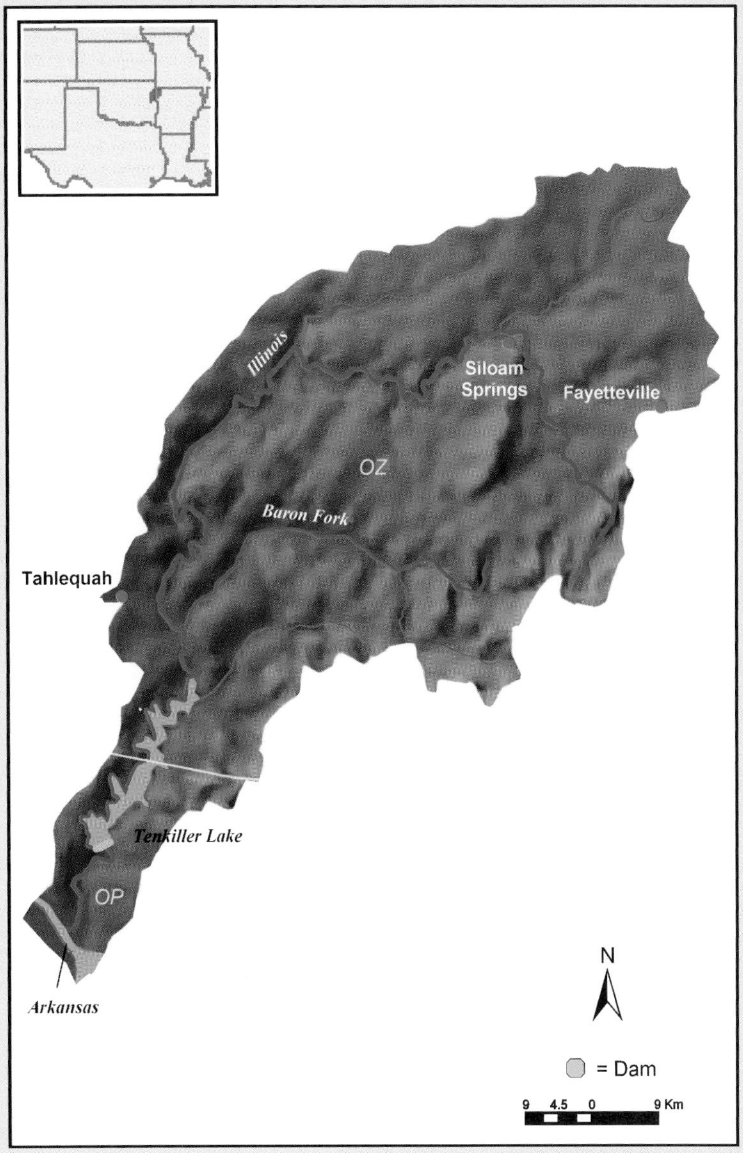

FIGURE 7.19 Map of the Illinois River basin. Physiographic provinces are separated by a yellow line.

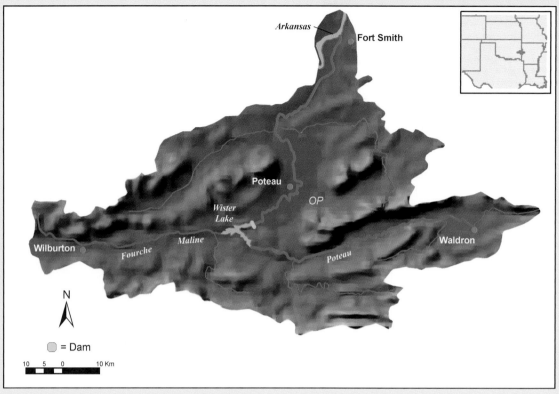

FIGURE 7.21 Map of the Poteau River basin.

POTEAU RIVER

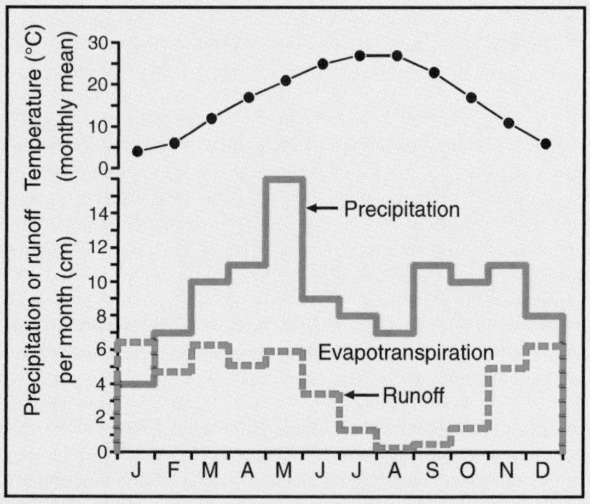

Relief: 680 m
Basin area: 4840 km^2
Mean discharge: 68 m^3/s
River order: 4
Mean annual precipitation: 112 cm
Mean air temperature: 16°C
Mean water temperature: 17°C
Physiographic province: Ouachita Province (OP)
Biome: Temperate Deciduous Forest
Freshwater ecoregions: Ouachita Highlands, Central Prairie
Terrestrial ecoregion: Ozark Mountain Forests
Number of fish species: 95
Number of endangered species: none
Major fishes: spotted gar, grass pickerel, pugnose minnow, steelcolor
shiner, redfin shiner, bluntnosed minnow, central stoneroller,
smallmouth buffalo, river carpsucker, creek chubsucker, channel
catfish, blackstriped topminnow, western mosquitofish, largemouth

FIGURE 7.22 Mean monthly air temperature,
precipitation, and runoff for the Poteau River basin.

bass, spotted bass, longear sunfish, white crappie, logperch, redfin darter, orangethroat darter
Major other aquatic vertebrates: mudpuppy, common slider, snapping turtle, beaver, river otter
Major benthic invertebrates: 32 species of mussels (threeridge, washboard, fluted shell), freshwater shrimp (*Palaemonetes kadiakensis*), mayflies (*Pseudocleon, Choroterpes*), caddisflies (*Cheumatopsyche, Hydroptila*), beetles (*Microcylloepus, Stenelmis, Dubiraphia*)
Nonnative species: Asian clam, common carp, striped bass, inland silversides
Major riparian plants: silver maple, box elder, red maple, smooth alder, blue beech, ash, black gum, cottonwood, willow oak, American elm, slippery elm
Special features: arises in steep upland slopes of Ouachita Mountain "Ridge and Valley" structural features but makes abrupt transition to low-gradient, turbid conditions before flowing into Arkansas River; commercially harvested for mussels
Fragmentation: one large impoundment (Lake Wister)
Water quality: pH = 7.1, alkalinity = 19 mg/L as CaCO$_3$, NO$_3$-N = 1.11 mg/L, total phosphorus = 0.15 mg/L
Land use: 65% forest, 20% pasture, 5% to 10% cropland; no large cities
Population density: 9.4 people/km^2
Major information sources: Cross and Moore 1952, Hoagland 2000, Lindsay et al. 1983, Solley et al. 1998

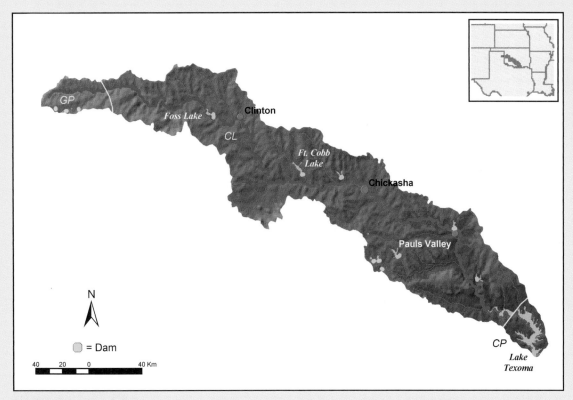

FIGURE 7.23 Map of the Washita River basin. Physiographic provinces are separated by yellow lines.

WASHITA RIVER

Relief: 714 m
Basin area: 20,230 km²
Mean discharge: 44 m³/s
River order: 4
Mean annual precipitation: 76 cm
Mean air temperature: 16°C
Mean water temperature: 18.4°C
Physiographic provinces: Great Plains (GP), Central Lowland (CL), Coastal Plain (CP)
Biomes: Temperate Grasslands, Temperate Deciduous Forest
Freshwater ecoregion: Southern Plains
Terrestrial ecoregions: Central and Southern Mixed Grasslands, Central Forest Grassland Transition Zone
Number of fish species: 51
Number of endangered species: none
Major fishes: gizzard shad, speckled chub, channel catfish, longear sunfish, green sunfish, bluegill, red shiner; carpsuckers common in lower river
Major other aquatic vertebrates: common slider, false map turtle, plain-bellied water snake, beaver
Major benthic invertebrates: At least 10 mussels (bleufer, white heelsplitter, yellow sandshell), chironomid midges (31 genera), mayflies (*Baetis*, *Choroterpes*), caddisflies (*Hydropsyche*, *Hydroptila*), beetles (*Dubiraphia*, *Heterelmis*)
Nonnative species: Asian clam, striped bass, threadfin shad, inland silversides, common carp, saltcedar
Major riparian plants: silver maple, box elder, ash, hackberry, cottonwood, bur oak, sandbar willow, black willow, saltcedar, American elm
Special features: lower main stem one of most turbid rivers in North America, extremely heavy load of silt or clay; very muddy bottoms; large snag piles common, likely to be major habitat
Fragmentation: two impoundments on main stem
Water quality: pH = 8.0, alkalinity = 170 mg/L as CaCO₃, NO₃-N = 2.53 mg/L, PO₄-P = 0.051 mg/L
Land use: 55% rangeland or pasture, 30% crops, 5% to 10% forest; no large cities
Population density: 8.7 people/km²
Major information sources: Branson 1982, 1983, 1984, Hoagland 2000, Johnson and Vaughn 1995, Magdych 1984, Morris and Madden 1978, ONHI 2001, Solley et al. 1998

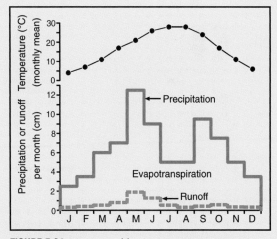

FIGURE 7.24 Mean monthly air temperature, precipitation, and runoff for the Washita River basin.

323

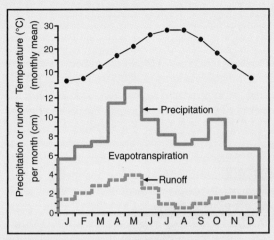

FIGURE 7.26 Mean monthly air temperature, precipitation, and runoff for the Blue River basin.

BLUE RIVER

Relief: 330 m
Basin area: 1650 km^2
Mean discharge: 9 m^3/s
River order: 4
Mean annual precipitation: 100 cm
Mean air temperature: 17°C
Mean water temperature: 17°C
Physiographic provinces: Central Lowland (CL), Coastal Plain (CP)
Biome: Temperate Deciduous Forest
Freshwater ecoregions: Ouachita Highlands, Southern Plains
Terrestrial ecoregion: Central Forest Grassland Transition Zone
Number of fish species: 85
Number of endangered species: none
Major fishes: redspot chub, central stoneroller, bigeye shiner, blacktail shiner, rocky shiner, spotted sucker, golden redhorse, largemouth bass, longear sunfish, channel darter, orangethroat darter, smallmouth buffalo, river carpsucker, blue sucker; unique forms of the orangebelly darter and striped shiner are distinct from those found elsewhere in their range
Major other aquatic vertebrates: westernmost tributary of Red River with razor-backed musk turtle, cottonmouth
Major benthic invertebrates: 24 species of mussels (threeridge, pigtoe, pistolgrip), amphipod crustaceans (*Hyalella azteca*), crayfish (*Orconectes virilis*), mayflies (*Stenonema, Tricorythodes*), beetles (*Dubiraphia, Helichus*), caddisflies (*Cheumatopsyche, Hydropsyche*)
Nonnative species: Asian clam, common carp, rainbow trout, possibly rudd
Major riparian plants: silver maple, box elder, hackberry, pecan, ash, American elm, sycamore, desert false indigo
Special features: upper portion strongly marl-depositing; marl dikes a prominent feature forming pools and small waterfalls; disjunct populations of seaside alder and redspot chub
Fragmentation: no large impoundments
Water quality: pH = 8.2, alkalinity = 186 mg/L as CaCO$_3$, NO$_3$-N = 1.65 mg/L, total phosphorus = 0.43 mg/L
Land use: 50% pasture, 20% cropland, 30% forest; no large cities
Population density: 18.9 people/km^2
Major information sources: Hoagland 2000, ONHI 2001, Reimer 1969, Vaughn 1997

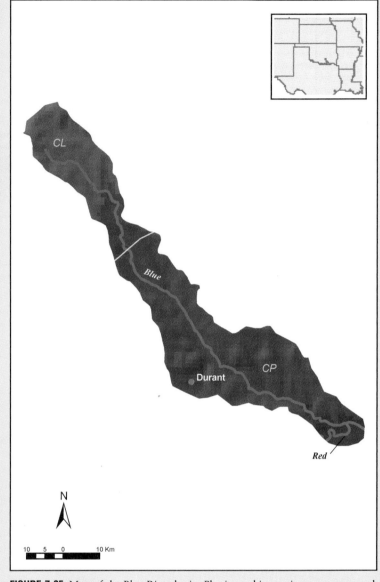

FIGURE 7.25 Map of the Blue River basin. Physiographic provinces are separated by a yellow line.

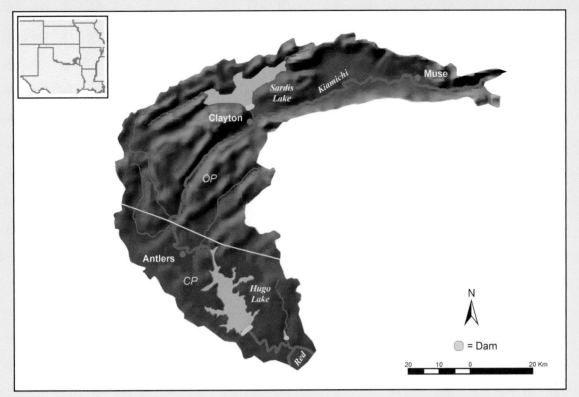

FIGURE 7.27 Map of the Kiamichi River basin. Physiographic provinces are separated by a yellow line.

KIAMICHI RIVER

Relief: 701 m
Basin area: 4650 km^2
Mean discharge: 48 m^3/s
River order: 5
Mean annual precipitation: 110 cm
Mean air temperature: 17°C
Mean water temperature: 16.7°C
Physiographic provinces: Ouachita Province (OP), Coastal Plain (CP)
Biome: Temperate Deciduous Forest
Freshwater ecoregion: Ouachita Highlands
Terrestrial ecoregions: Ozark Forests (although not in Ozark Mountains proper), Central Forest Grassland Transition Zone
Number of fish species: 86
Number of endangered species: 2 mussels
Major fishes: orangebelly darter, Johnny darter, dusky darter, central stoneroller, bigeye shiner, redfin shiner, rocky shiner, steelcolor shiner, spotted sucker, flathead catfish, smallmouth bass, spotted bass, largemouth bass, blackstriped topminnow, red shiner, gizzard shad, gars, blue sucker, river carpsucker

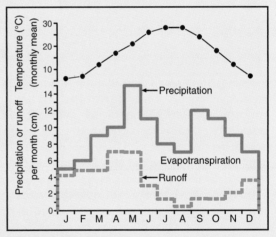

FIGURE 7.28 Mean monthly air temperature, precipitation, and runoff for the Kiamichi River basin.

Major other aquatic vertebrates: snapping turtle, false map turtle, stinkpot turtle, spiny softshell turtle, plain-bellied water snake, cottonmouth, beaver
Major benthic invertebrates: 29 mussel species (threeridge, mucket, pigtoe, pimpleback) freshwater shrimp (*Palaemonetes kadiakensis*), crayfish (*Octonectes menae*), mayflies (*Stenonema, Caenis*), caddisflies (*Oecetis, Nectopsyche*), beetles (*Stenelmis, Microcylloepus*)
Nonnative species: Asian clam, common carp, striped bass in lower main stem, threadfin shad
Major riparian plants: silver maple, box elder, red maple, smooth elder, river birch, blue beech, ash, sweetgum, swamp tupelo, black gum, sycamore, cottonwood, willow oak, American elm, slippery elm
Special features: identified by The Nature Conservancy as one of most critical watersheds in United States for protecting freshwater biodiversity; population of endangered Ouachita rock pocketbook mussel; reintroduction of river otters
Fragmentation: one major impoundment on main stem; one major impoundment on tributary
Water quality: pH = 7.3, alkalinity = 120 mg/L as CaCO$_3$, NO$_3$-N = 0.31 mg/L, total phosphorus = 0.71 mg/L
Land use: 75% forest, 15% pasture, 10% crops; no large cities
Population density: 5.6 people/km^2
Major Information sources: Echelle and Schnell 1976, Hoagland 2000, Master et al. 1998, Moulton and Stewart 1996, Pigg and Hill 1974, Vaughn and Pyron 1995, Vaughn et al. 1996

8

UPPER MISSISSIPPI
RIVER BASIN

MICHAEL D. DELONG

INTRODUCTION

UPPER MISSISSIPPI RIVER

MINNESOTA RIVER

ST. CROIX RIVER

WISCONSIN RIVER

ILLINOIS RIVER

ADDITIONAL RIVERS

ACKNOWLEDGMENTS

LITERATURE CITED

INTRODUCTION

The Upper Mississippi River basin, which represents 10% of the 3rd largest drainage basin in the world, begins as a 1st order stream draining Lake Itasca in the bog and spruce swamps of northern Minnesota and flows south to join the Ohio River as a 10th order alluvial river to form the largest river in North America (Fig. 8.2). The progression of the river from lake outlet to great river creates an impressive range of physical, chemical, and biological diversity throughout the basin. The Upper Mississippi River basin includes areas within Minnesota, South Dakota, Wisconsin, Illinois, Indiana, Iowa, and Missouri, ranging in latitude from 47°N to 37°N. Because of its generally north–south flow across the temperate zone of North America, climatic conditions vary considerably from its source to the confluence with the Ohio River. Describing the Upper Mississippi River as beginning at Lake Itasca and ending at the confluence of the Ohio River would include the Missouri River basin; however, details on the Missouri River basin are given elsewhere and all information provided for the Upper Mississippi River excludes the Missouri (see Chapter 10). The basin encompasses five terrestrial ecoregions, three biomes,

and three physiographic provinces. Despite the variability created by climate and geology, commonalities are evident among rivers at the physiographic province and terrestrial ecoregion levels.

Archaeological finds suggest that human history in the upper Mississippi River basin dates back 9000 or more years. Ceremonial and community mounds and other signs of man-made structures found throughout the basin hint at the cultural diversity present in the basin prior to European settlement. The first Europeans credited with exploring the Upper Mississippi, Louis Joliett and Father Jacques Marquette, arrived in 1673. Settlement of the basin began slowly, with a few isolated groups in Missouri and Illinois in the early eighteenth century and in the northern reaches of the basin in the early nineteenth century. Expansion of settlements came with the advent of the paddle wheeler, which opened the fertile soils of the basin to immigrants seeking to farm their own land. The river today maintains its significant role as a center of commerce for the transportation of goods by barge.

The rivers described in this chapter were selected as representative within each region and to reflect both the common threads among rivers in the Upper Mississippi River basin and their unique attributes.

◀ **FIGURE 8.1** Chippewa River, Wisconsin (PHOTO BY TIM PALMER).

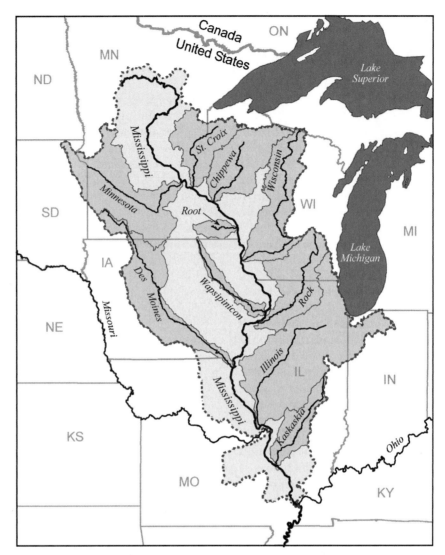

FIGURE 8.2 Rivers of the Upper Mississippi River basin covered in the chapter.

Five rivers are discussed in detail: Upper Mississippi, Minnesota, St. Croix, Wisconsin, and Illinois. Abbreviated descriptions of physical, chemical, and biological attributes are included for six additional rivers: Chippewa, Root, Wapsipinicon, Rock, Des Moines, and Kaskaskia.

Physiography and Climate

The Upper Mississippi River basin lies within three physiographic provinces: Superior Upland, Central Lowland, and Ozark Plateau. The dominant physiographic feature, representing almost 85% of the basin, is the Central Lowland physiographic province (Fig. 8.11). The features of the northern part of the province, which includes the Upper Mississippi basin,

are a result of glacial processes, as evidenced by its low altitude (<620 m asl) and limited local relief (Hunt 1967). Although the Central Lowland appears to be relatively homogeneous and creates some common physical features among rivers (i.e., low stream gradient), glacial and other geological processes have provided some measure of variation in the landscape. The Central Lowland is divided into six physiographic sections, five of which lie within the Upper Mississippi River basin.

The drainage basin begins in the Western Lakes physiographic section, characterized by deep glacial drift left by the Wisconsin glaciation, which created abundant kettle and moraine lakes as well as wetlands and bogs. The drift lies over a bed of Precambrian igneous and metamorphic rock extending from the Superior Upland physiographic province, which is replaced in the southern limit of the Western Lakes section of the Central Lowland Province by Paleozoic sandstones. The Eastern Lakes section includes eastern Wisconsin and is also covered by Wisconsin Drift, under which lies Paleozoic limestones and shales (Nielsen et al. 1984). The Driftless Area of southeastern Minnesota and central-southwestern Wisconsin lies between the Eastern and Western Lakes sections. The major glaciations passed around this area, leaving outcroppings consisting of alternating layers of limestone and sandstone rather than the low-relief terrain typical of the areas over which the glaciers moved. Actions of rivers and streams in the Driftless Area have created a rolling bluff topography, with some bluffs extending 100 to 200 m above the floor of the plain. Although not touched directly by glaciation, the area was repeatedly filled with debris from glacial outwash from the glacial River Warren, the predecessor of the Upper Mississippi.

The part of the basin that encompasses Iowa, Illinois, Missouri, and Indiana is divided into the two sections of the Central Lowland province: the Till Plain section to the east of the Upper Mississippi River proper and the Dissected Till Plain section to the west. Both are covered by highly weathered Illinoian and Kansan Drift and loess deposits. Loess in the Dissected Till Plain is commonly more than 10 m thick and gradually thins eastward into the Till Plain. Paleozoic limestones and shales lie deep beneath the overlying drift and loess. Glaciation of the Dissected Till Plain occurred earlier than in the Till Plain, which was also partly covered by the Wisconsin glaciation. Although the Till Plain still retains plains features with little relief, the area to the west of the Upper Mississippi has been dissected by streams and rivers (Hunt 1967).

The Upper Mississippi flows from the Central Lowland province into the Ozark Plateau province near St. Louis, Missouri, before emptying into the Coastal Plain province, through which the Lower Mississippi River flows (Chapter 6). Only about 5% of the basin lies within the Ozark Plateau province, which is a transition into the broader floodplain of the Lower Mississippi River below the confluence of the Upper Mississippi and Ohio rivers.

Although the main-stem Upper Mississippi begins in the Central Lowland province, the northeastern corner, or about 10%, of the basin lies within the Superior Upland physiographic province. The Superior Upland is an extension of the Precambrian Shield of Canada created by an upwarp of Precambrian rock (Hunt 1967). The result is a gently rolling landscape with limited erosion by the streams and rivers flowing through this region, which creates river systems very different from the rest of those found in the drainage. As described earlier, the geological features of the Superior Upland province influence the geology of the Western and Eastern Lakes sections of the Central Lowland physiographic province.

Climatic conditions change considerably from the northern extreme of the basin to its southern boundary at the confluence with the Ohio River. The average annual air temperature ranges from 3°C in the north to 15°C in the southern portion of the basin. Precipitation demonstrates a similar gradient, increasing from 60 cm/yr in the north to 81 cm/yr in the middle of the basin to 122 cm/yr in the south. Precipitation is usually highest from April to June in the northern basin and from April to July in the southern part of the basin. January and February are the driest months throughout the basin, but precipitation above 42°N latitude is low November through March and is primarily in the form of snow. Average runoff yield is 24% throughout the basin but varies from 5% to 40% as a function of location (Nielsen et al. 1984).

Basin Landscape and Land Use

The Upper Mississippi River basin covers three biomes: Boreal Forest, Temperate Deciduous Forest, and Temperate Grassland. The basin is further divided into five terrestrial ecoregions (Ricketts et al. 1999) that reflect attributes of the biomes and transitional gradations between biomes. The Western Great Lakes Forest ecoregion overlaps the Boreal Forest and Temperate Deciduous Forest biomes and is typified by a forest composition that includes paper birch, black spruce, jack pine, white pine, sugar maple, red maple, and balsam fir. The Upper Midwest Forest/Savanna Transition ecoregion lies to the south of the Western Great Lakes ecoregion. It is an ecotonal unit separating the Great Plains from the eastern forests (Ricketts et al. 1999). The region was heavily forested with oak, maple, and basswood, with oak savannas present throughout the forested areas.

The Upper Midwest Forest/Savanna Transition gives way to the Central Tall Grasslands to the southwest (southern Minnesota and most of Iowa) and the Central Forest/Grassland Transition Zone ecoregion to the southeast (most of Illinois and northern Missouri). The Central Forest/Grassland Transition is also an ecotonal unit, separating the eastern deciduous forests from tallgrass and mixed-grass prairies. The Central Tall Grasslands ecoregion supported the greatest diversity of tallgrasses—big bluestem, switchgrass, and Indian grass—due largely to the high levels of rainfall (100 cm/yr; Ricketts et al. 1999). The southernmost part of the basin extends into the Central U.S. Mixed Hardwood Forests ecoregion. Although primarily viewed as an oak–hickory forest, the ecoregion once retained the savanna-prairie mosaic characteristic of the Central Forest/Grassland Transition Zone and is, therefore, reminiscent of that ecoregion (Ricketts et al. 1999). Tree species diversity is low in this ecoregion, but it once possessed the richest diversity of shrub and herbaceous vegetation in North America.

The landscape of the Upper Mississippi River, although once diverse, is drastically different from the presettlement period. The Great Lakes Forest ecoregion could be considered the most pristine of the ecoregions, but only 20% of the forest remains (Ricketts et al. 1999). Intensive logging has eliminated most old-growth stands of white pine and red pine, which have been replaced by younger stands of birch and aspen or have been converted to agriculture or urban and recreational developments. The Upper Midwest Forest/Savanna has faired even less well, with less than 5% of the ecoregion considered intact. Conversion of land for crop agriculture and overgrazing has been the primary causes of habitat loss. The Central Tall Grasslands and the Central Forest/Grassland Transition Zone ecoregions are now the Corn Belt, with over 80% of the available land used for agriculture. Streams and rivers in these ecoregions have among the highest nitrate concentrations in the country and have been identified as a likely source of much of the nitrogen loading contributing to the formation of a biological "dead zone" in the Gulf of Mexico to the west of the Mississippi River alluvial fan.

The Rivers

The entire Upper Mississippi River basin represents the Mississippi freshwater ecoregion (Abell et al. 2000). If viewed from this larger scale of representing a single freshwater ecoregion and almost entirely in a single physiographic province (Central Lowland), it might be concluded that there is little variability among the rivers of the Upper Mississippi River system. Certainly, commonalities in the physical, chemical, and biological attributes of these rivers do exist, but there are also obvious differences. These differences stem from the influence of glaciation,

specifically the movement of ice sheets over the surface of the Upper Mississippi River basin on several occasions. The most logical approach to identify unique features and common threads between rivers in the Upper Mississippi River system is based on the physiographic sections through which they flow and climatic differences resulting from the 10-degree range of latitude in the basin.

One grouping of rivers includes those flowing through the Superior Upland and the Western Lakes section of the Central Lowland province (Fig. 8.2, Fig. 8.11). This includes the Mississippi headwaters as well as the St. Croix, Chippewa, and Wisconsin rivers. The upper ends of all of these systems are relatively high gradient, with beds of gravel, cobble, boulder, and bedrock. Of these, however, only the St. Croix retains these features for most of its length. The Chippewa and Wisconsin rivers decrease in stream gradient farther downstream as they flow into the Driftless Area, where substrata change from gravel to sand-dominated beds. Most of the St. Croix River is afforded the protection of being in the St. Croix National Scenic Riverway. This is in sharp contrast to the Chippewa and Wisconsin rivers, which have been heavily dammed for hydroelectric production and flood control. Only the lower portions of both rivers are free of dams. The northern rivers are influenced by snow pack, with discharge highest in the spring (April to May) as spring rains fall on the melting snow. Discharge decreases rather abruptly following peak discharge. Minimum discharges occur in the winter as the rivers and the surrounding drainage basin freeze. These systems also exhibit a hydrological feature not evident in other rivers within the Upper Mississippi River system (with the exception of the Upper Mississippi River proper). All three rivers have a slight increase in discharge in the autumn due to increased surface runoff following leaf fall and a reduction in evapotranspiration as air temperature declines. Although not as substantial as the spring peak, this rise appears to be a critical component in the population dynamics of some benthic invertebrates.

The Minnesota River is the first major tributary to join the Mississippi headwaters and flows primarily through the Western Lakes section of the Central Lowland province. Glacial processes created a channel morphology that causes the Minnesota to stand out from the other rivers of the basin. Bottom sediments are similar to those of the other northern rivers—gravel, cobble, and sand—but large quantities of silt are also present in the river channel, originating from the erosion of the large amount of loess present. The Minnesota is also hydrological similar to the northern rivers. Peak flows are in the spring, resulting from snowmelt and spring rains over the snow, and lowest flows are in the winter. There is, however, no autumn increase in discharge, making the hydrograph resemble those of the rivers in the Dissected Till Plain.

The rivers of the Driftless Area, represented by the Root River, stand out from the other rivers of the basin in that they flow through a region that escaped the direct effects of glaciation. Rivers flowing through the Driftless Area have carved out deep valleys, or coulees, as they flow to the Upper Mississippi River. Geology influences both the physical and chemical characteristics, causing high alkalinity (>200 mg/L as $CaCO_3$) and a basic pH (8 to 9). The tributaries are coldwater, owing to their groundwater sources, and the main rivers are coolwater systems. Bottom sediments are gravel, cobble, and sand, some of which includes glacial outwash. Rivers on the Minnesota side of the Driftless Area originate in the Dissected Till Plain, which explains the presence of silt deposits in and along the river channels of this region. Natural erosion from the Dissected Till Plains did occur in these systems, but agriculture has exacerbated the problem, making suspended sediment transport a critical water-quality concern. Hydrologically, rivers in the Driftless Area reach peak flows from March to April as spring rains fall on the snow pack, and reach their minimum over the winter. They are susceptible to flash flooding because of their steep-banked valleys and geology. These rivers are largely free flowing, with no major impoundments.

Downriver are the rivers flowing through the Dissected Till Plains and the Till Plains. Although these rivers share many common features, they are best viewed as two separate groups. Rivers flowing through northern Iowa (Wapsipinicon and Des Moines rivers) and northern Illinois (Rock River) begin in areas with surface geology comprised of Wisconsin Drift and thin loess. Bottom sediments in the upper reaches consist of cobble, sand, sand–gravel, and silt. As they work their way into the more weathered Illinoian and Kansan Drift, bottom substrates change to predominantly silt and sand–silt. The hydrology of the Wapsipinicon and Rock rivers more closely resembles rivers in the northern part of the basin. Annual discharge for the Des Moines, which flows into the Upper Mississippi River much farther to the south, more closely resembles the southern rivers, with a protracted peak flow period (March to July) and minimum flows in the fall (September to October) as evapotranspiration becomes the controlling factor.

The last group of rivers includes the Illinois and Kaskaskia rivers, both of which flow almost entirely through the Till Plains. Rivers in this region exhibit low gradients and bottom substrata consisting of sand, sand–gravel, and silt. Some cobble and gravel areas are found in the upper Illinois River and its tributaries to the north. Rivers in the southern portion of the basin exhibit a more protracted falling limb in their annual hydrograph, probably attributable to subsurface and groundwater inputs and because snowmelt and thawing ground are not part of their hydrological cycle. A trend for rivers of this region is to shift from peak discharges in the spring (Illinois River) to maximum flows in the winter (Kaskaskia) farther south. Minimum discharge occurs in September and October.

The biological distinctiveness of the Upper Mississippi River is rated as only bioregionally outstanding (Abell et al. 2000). This is partly due to the fact that there is only one species of fish, one species of crayfish, and one freshwater mussel that are endemic to the region. Another reason is that the Upper Mississippi, along with the Lower Mississippi, has served as a faunal refugium and has been the source for fauna now inhabiting adjacent ecoregions. The value of the Upper Mississippi River as a faunal refugium is evident by the presence of ancestral species, such as the paddlefish and alligator gar, and because the northernmost biogeographical range of many species occurs only along the Upper Mississippi River (Abell et al. 2000). The conservation status of the basin is endangered, owing to the extensive conversion of the prairie and forest landscape to agriculture and alteration of the rivers for navigation and flood control. The major environmental threats are channelization, flood-control and navigation impoundment, and the still lingering effects of point-source pollution. Introduction of nonnative species, particularly following the linkage of the Upper Mississippi to the Great Lakes, is another serious threat to the diversity and productivity of the basin.

UPPER MISSISSIPPI RIVER

The Upper Mississippi River proper refers exclusively to the main channel of the river and all habitats encompassed within the boundaries of its floodplain (see Fig. 8.11). The main-stem Upper Mississippi River begins at Lake Itasca in northern Minnesota and flows 2008 km, receiving water from tributaries

FIGURE 8.3 Upper Mississippi River above Little Falls, Minnesota (PHOTO BY TIM PALMER).

draining a 444,185 km² area before joining with the Missouri River to form a 10ᵗʰ order stream (Fig. 8.3). The waters of the Missouri increase the flow of the Upper Mississippi by one-third and add another 1,331,810 km² to the total area drained by the Upper Mississippi–Missouri river system. The river flows another 312 km, drawing water from an additional 45,325 km² (not including the Missouri), before joining the Ohio River to form the Lower Mississippi River (see Fig. 8.11). Representing over half of the total length of the Mississippi River, the Upper Mississippi drains some of the most unique landscape found in the Mississippi River basin, including boreal forest, glacial lakes, and bog and spruce swamps.

The importance of the Upper Mississippi River as a center of commerce and community extends well before the arrival of Europeans. Shards of pottery, skeletal remains, and spear tips suggest that humans were in the Upper Mississippi River Valley as early as 8000 years ago, but the best known ancient peoples date back 3000 years. Predominant among these was the Mississippian culture and their settlement, Cahokia, near St. Louis. Cahokia is believed to have been home to 5000 or more people. Both burial and ceremonial mounds are evident throughout the Upper Mississippi basin, including effigy mounds in Iowa and lizard mounds in Wisconsin. Early cultures were later replaced by the Ojibwas, Mesquakie,

Sauk, Dakota, Iowa, and Winnebago Indian nations. The first Europeans credited with seeing the Upper Mississippi River were Louis Joliett and Father Jacques Marquette, who led an expedition from Quebec in 1673 to find a route to China. They arrived at the Upper Mississippi near present-day Prairie du Chien, Wisconsin, and continued down to the Lower Mississippi River near the mouth of the Arkansas River (Waters 1977). The first steamboat reached St. Anthony Falls in St. Paul, Minnesota, in 1828, opening the basin to agriculture, timber harvest, and urbanization.

Physiography, Climate, and Land Use

The Upper Mississippi River proper flows almost entirely through the Central Lowland (CL) physiographic province, with the last 300 km flowing along the Ozark Plateau (OZ) province (see Fig 8.11). Soils in the basin and bottom sediments in the river, however, change considerably as the river flows through areas impacted differently by glaciation. These differences are evident when it is considered that the channel winds through five different terrestrial ecoregions: Western Great Lakes Forests, Upper Midwest Forest/Savanna Transition, Central Tall Grasslands, Central Forest/Grassland Transition, and Central U.S. Hardwood Forests (Ricketts et al. 1999). Many of the characteristics of these ecoregions are determined by whether they were influenced by glaciation and the timing of the last glacial event. The upland vegetation reflects the influence of glaciation on the basin as well as the transition from north temperate to south temperate climates. Upland vegetation in the upper reaches of the river basin includes balsam fir, black spruce, and white pine. Upland vegetation then shifts to oak–maple forest before transitioning into oak-savannas. The middle of the basin was dominated by grasses of the Central Tall Grasslands prior to conversion for agriculture. The lower basin returns to an oak–hickory forest as the river meets the Central U.S. Hardwood Forests as it nears the confluence with the Ohio River.

Climatic conditions change considerably from the northern end of the river to the confluence with the Ohio River. Because of this range, the values shown for temperature and precipitation in Figure 8.12 reflect averages for the length of the river. Precipitation is usually highest from April to July in the northern basin and from April to September in the southern part of the basin. January and February are the driest months throughout the basin, but precipitation above 42°N latitude is low from November through March and is primarily in the form of snow.

The human population density in the basin is 45 people/km^2. There are 18 metropolitan areas with populations >100,000 people, with three of these (Minneapolis–St. Paul, Minnesota; the Quad Cities of Iowa and Illinois; and St. Louis, Missouri) occurring on the Upper Mississippi River proper. Despite this, only 5% of the basin has been converted to urban areas. Nearly 70% of the basin has been converted to agriculture, with the most intense activities in the Corn Belt. Corn and soybeans are the most common row crops. Livestock (hogs, cattle) production is also common in this portion of the basin. The concentration of livestock in small feedlots has become a growing environmental issue on smaller tributaries. Approximately 25% of the basin is forested, with the majority of the remaining forested areas in the Mississippi headwaters, the upper ends of the Chippewa and Wisconsin rivers, and the Driftless Area.

River Geomorphology, Hydrology, and Chemistry

The entire Upper Mississippi River, including all tributaries, is within its own freshwater ecoregion, the Mississippi (Abell et al. 2000). The Upper Mississippi River flows from Lake Itasca, Minnesota, at an elevation of 440 m asl. The headwaters flow through bogs, spruce swamps, sand plains, glacial lake beds, and moraine (Fremling et al. 1989). Subsequently, it alternates between low-gradient and high-gradient reaches. At the end of the headwaters is St. Anthony Falls, formed by erosion of the glacial River Warren and now largely obscured by higher water levels created by the nearby lock and dam. St. Anthony Falls served as a barrier to fishes and was the end point for navigation on the Upper Mississippi prior to construction of the lock and dam system.

The Upper Mississippi is believed to flow along its preglacial course from Minneapolis to Davenport, Iowa (also known as the Quad Cities area; Patrick 1998). As the late Wisconsin glacier retreated 14,000 years ago, it cut off river drainage to Hudson Bay, forming glacial Lake Agassiz and glacial Lake Duluth. The overflow of these lakes was carried by the Minnesota and St. Croix rivers, respectively. The massive flows of both rivers carved deeply, as much as 90 m, into the Mississippi River valley (Fremling et al. 1989). As flow from the glaciers diminished, the river valley became partly filled with glacial outwash, leaving bottom sediments of sand and gravel. From the confluence of the Minnesota downstream to the Driftless Area, the channel is highly braided due to inputs of sand. Channel morphology is most complex near the inflow of tributaries (Nielsen et al. 1984). Braiding diminishes below the Driftless Area, where sediments transported by tributaries in Iowa, Illinois, and Missouri are dominated by silt (Patrick 1998).

Mean annual discharge near Thebes, Illinois (77 Rkm upstream of the confluence of the Ohio River) is 5923 m³/s. Once the contribution of the Missouri River is deducted, average annual discharge for the Upper Mississippi and its tributaries is 3576 m³/s. The annual discharge pattern of the river remains consistent from headwaters to the confluence with the Ohio. Discharge is highest in the spring (April and May) in response to snowmelt and spring rains (Fig. 8.12). Spring peak flows remain at flood stage or higher for 6 to 8 wk before receding. Evapotranspiration increases from late May to June, causing discharge to decline through the summer. Discharge increases slightly in September or October as the effects of evapotranspiration diminish with cooling air temperatures and increased surface runoff following leaf fall (Patrick 1998). The river maintains an annual pattern of discharge comparable to pre-navigation dam conditions, except that minimum discharge levels increased following dam construction. In addition, the peak in the annual hydrograph diminishes as the river approaches a navigation dam to the point that it is almost an inverse of the hydrograph prior to dam construction (Sparks 1995).

The Upper Mississippi is a hardwater system, with slightly alkaline pH. Alkalinity decreases from an average of 184 mg/L as CaCO₃ at Lake Itasca to 156 mg/L as CaCO₃ at St. Louis (Fremling et al. 1989). The pH remains around 8 throughout the length of the river but is more basic between St. Paul and Rock Island, Illinois, where limestone is more abundant. Hardness follows a similar pattern. Nitrate-N and total phosphorus concentrations are low in the headwaters, ranging from 0.11 to 0.33 mg/L and from 0.03 to 0.15 mg/L, respectively (Fremling et al. 1989). Nitrate-N concentrations increase in the Driftless Area, reaching concentrations of 1.58 mg/L, and are as high as 5.70 mg/L in the Corn Belt. Nitrate-N concentrations decline slightly at St. Louis (1.2 to 4 mg/L), but remain high. Total phosphorus concentrations approach 0.5 mg/L by the time the river reaches Keokuk, Iowa, and increase further by the time the river reaches St. Louis (0.18 to 0.8 mg/L). Suspended sediment concentrations are low (<20 mg/L)

from the headwaters down to the confluence with the Missouri River (Patrick 1998). Below the Missouri, suspended sediment concentrations average 340 mg/L (Fremling et al 1989). Specific conductance also increases dramatically as higher quantities of dissolved solids, particularly sulfates and chlorides, enter the Upper Mississippi via the Missouri River.

Average annual water temperature above Minneapolis–St. Paul is 10°C (range 0°C to 25°C) at Royalton, Minnesota, the northernmost gaging station with long-term physical data. Average water temperature increases to 15.6°C, with a range from 0°C to 29°C, at Clinton, Iowa. Average water temperature at Thebes, Illinois, the last gaging station with long-term data, is 16.6°C, with a range from 0°C to 31.5°C. Water temperatures reach maximum levels from late July to early August, but only remain at these levels for short periods (2 to 3 wk).

River Biodiversity and Ecology

Algae and Cyanobacteria

Benthic algae are largely limited to nearshore areas of the main channel, side channels, and backwaters. Epiphytic algae are particularly abundant, as is algal growth on snags. Diatoms are the predominant taxa among the microalgae, with *Gomphonema*, *Synedra*, *Navicula*, and *Diatoma* among the most abundant. Green algae become slightly more abundant in the periphyton during the summer months. The filamentous green algae *Cladophora* is also common and produces dense mats on rocks in wing dams and rip-rap. Diatoms are also the most abundant of the phytoplankton, in some cases representing as much as 85% of the community (Patrick 1998). The diatoms *Melosira*, *Asterionella*, *Stephanodiscus*, and *Cyclotella* are among the most abundant phytoplankton in the main river. Diatoms are the most abundant of the phytoplankton in backwaters early in the year (February to April), but diversity increases from June to August, with increased abundance of green algae and, to a lesser extent, cyanobacteria (Patrick 1998).

Plants

The greatest complexity in the riparian and floodplain vegetative communities can be found in the headwaters, where the river flows through sedge meadow, willow–alder shrub swamp, conifer swamps, lowland hardwood forest, and maple–basswood forest. Riparian and floodplain vegetation downstream of the confluence with the Minnesota River is relatively uniform, with only the relative abundance of the representative species changing. This may not have been the case historically, as studies have indicated that wide tracts of prairie and savanna may once have been an integral part of the terrestrial floodplain community (Nelson and Sparks 1997). Urban development, agriculture, and, probably most important, fire suppression have led to the replacement of prairie and savanna habitats by flood-tolerant and fire-intolerant species. Throughout the Upper Mississippi River floodplain the predominant trees are silver maple, cottonwood, and black willow. River birch is present in the floodplain in Minnesota and Wisconsin, but diminishes downstream.

The numerous backwaters and shallow border habitats along main and side channels provide ample habitat for aquatic macrophytes. Macrophyte abundance in these areas often exceeds the upper range of biomass reported in the literature (Patrick 1998), but the occurrence of macrophytes generally declines farther downriver as the abundance of backwaters and shallow margins declines in response to natural and anthropogenic changes in channel morphology. The abundance of macrophytes in navigation reaches 5 to 8 results in dense mats floating downriver and collecting in snags and on shore as macrophytes senesce from late August through October. Arrowheads are among the most abundant emergent macrophytes, typically occurring in monotypic stands. Several species of submergent vegetation are common, with coontail, waterweed, and wild celery among the most widespread. Coontail and waterweed are limited to backwaters, whereas wild celery prefers higher-current areas along the channel borders. American lotus and American waterlily are the most common floating-leaved plants and are often found in association with arrowheads. Duckweed is common in quiet habitats throughout the year, but often forms dense mats in late summer through fall that drift into the main channel to collect along the shoreline.

Invertebrates

The diversity of benthic invertebrates reflects the tremendous habitat diversity of the Upper Mississippi River. A compilation of past studies estimated that almost 430 invertebrate taxa are found in the Upper Mississippi (Patrick 1998), but even this is an underestimate, as many of the insect taxa are not identified to species. Invertebrate studies in the headwaters have been limited, but taxa found here are more representative of a small lotic system, with black flies and net-spinning caddisflies representing the most abundant groups.

Taxa composition for the river below the headwaters is typical of a large river system, with distribution influenced primarily by current velocity and substrate type. Although species represented do change longitudinally, the generic makeup of invertebrate communities remains fairly consistent. Taxa richness in soft substrates is relatively low (Seagle et al. 1982). Taxa richness and organismal abundance are much greater on hard substrates. Most rocky substrate in the river is artificial, consisting of wing dams and rip-rap, rocks placed along erosional banks to increase stability. Submerged wood, or snags, found in the nearshore areas of channels, backwaters, and floodplain lakes are the most important natural hard substrate. Taxa on snags in nearshore areas of channels possess a greater diversity of caddisflies, including filterers (*Hydropsyche, Cheumatopsyche, Potamyia*), herbivores (*Hydroptila, Nectopsyche*), and predators (*Oecetis*). Mayflies are also abundant and are represented primarily by the grazers *Stenonema, Stenacron, Caenis,* and *Heptagenia*. Beetles are represented primarily by the riffle beetles *Stenelmis, Macronychus,* and *Dubiraphia*. Common chironomid midges include *Rheotanytarsus, Dicrotendipes, Polypedilum, Ablabesmyia,* and *Cryptochironomus*. Other invertebrate groups well represented on snags in nearshore channel habitats include snails (*Physella, Gyraulus, Pleurocera, Ferrissia*), leeches (*Placobdella, Erpobdella*), stoneflies (*Acroneuria, Perlesta*), and odonates (*Enallagma, Macromia*).

Invertebrate fauna on snags in the lentic habitats (backwaters and floodplain lakes) are quite different from the channel fauna, although some overlap occurs. For example, the midges *Dicrotendipes, Polypedilum,* and *Cricotopus* are abundant but represented by different species. Other midges in the lentic habitats include *Glyptotendipes* and *Chironomus*. Caddisflies are represented by the filterer *Polycentropus* and gatherer/scrapers like *Nectopsyche, Orthotrichia,* and *Ceraclea*. Snails are well represented, including *Physella, Campeloma,* and *Helisoma*. The Odonata are abundant in lentic habitats, particularly *Enallagma, Argia,* and *Dromogomphus*.

There are 51 documented species of freshwater mussels in the Upper Mississippi, but only 44 species have been found in the last 35 years (Havlik and Sauer 2000). There are three species on the federally endangered list: winged mapleleaf, Higgin's eye, and fat pocketbook. Thirteen species are listed as endangered by one or more of the states along the river.

Vertebrates

Habitat heterogeneity also greatly influences the diversity of fishes in the Upper Mississippi River. Approximately 145 species have been identified in the Upper Mississippi River, including 44 species of Cyprinidae (carp and minnows), 19 species of Catostomidae (suckers), 11 species of Centrarchidae (sunfishes), and 10 species of Ictaluridae (Burr and Page 1986, Fremling et al. 1989). Like the invertebrates, there are considerable differences in the species and their distribution longitudinally and in response to habitat heterogeneity. There are 67 species in the headwaters, some of which did not occur above St. Anthony Falls until the falls were bypassed with the completion of a lock in 1963. Common fishes in the headwaters include goldeye, northern pike, common shiner, bluegill, and yellow perch (Fremling et al. 1989). Gizzard shad, common carp, and emerald shiners are found in abundance throughout the Upper Mississippi River in both lentic and lotic habitats.

The most abundant species inhabiting lotic habitats (including nearshore areas with slower flow) from Lake Pepin to Keokuk, Iowa, include river shiner, spotfin shiner, shorthead redhorse, channel catfish, white bass, and sauger. Fishes common to backwaters or floodplain lakes are bullhead minnow, smallmouth buffalo, bluegill, largemouth bass, and black crappie. Silver redhorse and freshwater drum are found in both lentic and lotic habitats (Patrick 1998, Burkhardt et al. 1998). Farther downriver to the confluence with the Missouri River, distinctions between lotic and lentic species diminishes, with shortnose gar, threadfin shad, river carpsucker, smallmouth buffalo, channel catfish, white bass, bluegill, freshwater drum, common carp, gizzard shad, and emerald shiners representing the most abundant fishes to occur in both lentic and lotic habitats (Burkhardt et al. 1998). Fishes present in abundance in backwaters of this region include bullhead minnow, river shiner, golden shiner, black crappie, and white crappie. The pallid sturgeon and Topeka shiner are the only federally endangered fish species on the Upper Mississippi. Each state maintains a list of locally threatened and endangered species. A total of five fishes are listed as endangered in at least one state and seven fishes are listed as threatened. Common among state lists is the skipjack herring, as the construction of the lock and dam system is believed to have cut off their migratory path.

Based on available lists, there are 11 species of salamanders, 14 species of turtles, 14 species of frogs, 6 species of snakes, and 3 species of mammals in the Upper Mississippi River. No vertebrate species are listed as federally endangered, but each state lists several reptiles and amphibians as locally threatened or endangered. Most of the aquatic vertebrates found on the Upper Mississippi primarily use side channels, backwaters, and floodplain lakes, but some species, especially among the turtles, can be found in slow-flow nearshore areas where cover (snags and rock-covered wing dams) is abundant. The most common turtles include the painted turtle, spiny and smooth softshells, and the common map turtle. The choruses of treefrogs can be heard throughout the floodplain, as can the calls of leopard frogs and bullfrogs. Some herpetofauna appear to be extending their range north using the Upper Mississippi as a corridor. For example, the range of the lesser siren extends north of Tennessee only along the Upper Mississippi (Abell et al. 2000). Beaver and muskrat are common in floodplain lakes, with the macrophyte and mud dens of muskrats looking like small towns as winter approaches.

Ecosystem Processes

As a floodplain river, the predictable flood cycle is of critical importance in shaping physical and chemical processes of the Upper Mississippi River. The annual spring flood inundates the 1 to 10 km wide floodplain, eroding and redepositing sediments and changing the physical features of the floodplain, islands, and channels (Sparks 1995). Nutrient exchanges between the river and floodplain provide benefits to both terrestrial and aquatic organisms, with the river depositing nutrient-laden sediments when the floodplain is inundated during the spring flood and aquatic autotrophs increasing rates of primary production in the shallow and rapidly warming waters. The annual flood cycle is also critical for the reproduction of fishes dependent on floodplain spawning habitats. Floodplain spawners enter the floodplain as the river overflows its banks in the spring; however, construction of levees cuts off the river from the floodplain, reducing the amount of reproductive habitat available for floodplain spawners. This is particularly true where the river flows along the heavily agricultural areas of Illinois, Iowa, and Missouri.

Quantification of organic matter and trophic processes has focused on the origins of organic matter transported in the water column. Organic matter inputs into Reach 19, near Keokuk, Iowa, have been estimated to consist of almost 86% materials transported from upstream sources, 7% from tributaries, 6% from floodplain vegetation, 0.15% from aquatic macrophytes, and 0.15% from phytoplankton (Patrick 1998). Phytoplankton contributions to organic matter inputs are highly variable along the length of the river and appear to be related to the abundance of main-channel and side-channel slackwater habitats as well as frequency of backwaters (e.g., Fremling et al. 1989). Grubaugh and Anderson (1989) noted that organic matter inputs are seasonal, with floodplain contributions highest during autumn leaf fall and the spring flood.

Missing from estimates of the amount of organic matter present has been attempts to link potential food sources to consumers. A study using stable isotope ratios of carbon and nitrogen identified a linkage between fine transported organic matter (FTOM) and invertebrate consumers (Delong et al. 2001). The association between FTOM and consumers continued through to insectivorous and piscivorous fishes. Based on stable isotope and C/N ratios it was hypothesized that FTOM consisted primarily of phytoplankton and other autotrophic organic matter produced within the river rather than organic matter from the floodplain, although the latter was more abundant. A separate study, which looked at seasonal patterns of stable isotope ratios of transported organic matter, also concluded that organic matter in transport was composed primarily of matter that was riverine in origin (Kendall et al. 2001).

Human Impacts and Special Features

The location of the Upper Mississippi River places it at the crossroads for transition from the eastern deciduous forests and the Great Plain as well as from the boreal forests of the north to the temperate forests of the south. This transition, coupled with gradations in the landscape formed by various glacial events, creates a diverse landscape that, in turn, influences the many rivers found within the basin. The river's location, however, has also influenced its use and subsequent development. As a center of commerce and civilization from prehistoric times through today, the river has been impacted by human activities intended to alter the river for navigation and to make the surrounding floodplain available for agricultural and urban development. Despite these changes, the Upper Mississippi River still retains much of its natural condition, with >80% of the floodplain still connected to the river and a hydro-

logical pattern that, at least regionally, still closely resembles natural conditions. It is apparent, however, that human activity in and along the river has adversely impacted the ecological integrity of the upper half of the mighty Mississippi.

The most obvious human impact on the Upper Mississippi River is the 26 locks and dams located between St. Paul and the confluence with the Missouri River (see Fig. 8.11). Two dams were in place prior to 1920, with the other dams built in the 1930s following authorization by the U.S. Congress of a 2.75 m navigation channel. In contrast to the large flood-control dams associated with many rivers, dams on the Upper Mississippi are low-head navigation dams, designed to maintain a 2.75 m deep navigation channel during periods of low flow but not altering peak flows. Many hydrological and physical features of the river remain despite the presence of the dams, but there are obvious impacts. Regulation of the river has decreased river–floodplain interaction on the upstream side of the dams through the inundation of the floodplain (Sparks 1995). Dams have also increased the rate of sedimentation in backwaters and side channels, decreasing the quality of these critical habitats. The primary function of navigation dams is to keep water in the channel during periods of low flow. The result has been a reduction in the amount of slackwater habitats, zones of increased primary production, in the upper portion of the navigation reaches, and the formation of a large slow-flow area as the river approaches the next dam. Dam construction is not limited to the river south of Minneapolis–St. Paul. There are 11 dams in the Mississippi headwaters, functioning for either flood control, hydroelectric production, or as navigation aids. Changes in the river for navigation did not begin with the construction of the lock and dam system. Snag removal (removing woody debris to prevent it from interfering with boat traffic) dates back to 1824. Wing dams, structures composed of rock debris and willow, were laid perpendicular to the shoreline and at the upstream ends of side channels beginning in the 1870s to divert water to the main channel and force sediment deposition to the shoreline rather than in the channel.

Agriculture is probably the single greatest water-quality concern in the river today. The effect of agriculture is evident in the nitrogen and phosphorus concentrations described previously, with the amount of both increasing downstream to the confluence with the Ohio River. Agriculture has also contributed to decreased connectivity between the river and floodplain through the construction of levees. Levees above the confluence of the Rock River are limited to urban areas, so that about 95% of the floodplain is still inundated by the river. Large protective levees become more abundant downriver, where they are built near the river channel to prevent inundation of bottomland converted to farmland.

Point-source pollution issues have been greatly diminished since the enactment of the Clean Water Act. Large zones depauperate of life were common in the river below large urban areas, but better industrial waste management (with the exception of occasional spills) and sewage treatment limit point-source impacts to highly localized areas.

Several nonnative species are present in the river, with some causing ecological concerns. Introductions have come primarily through the flooding of ponds in areas adjacent to the river (e.g., grass carp, bighead carp, silver carp) or via the Illinois River and its connection to the Great Lakes (e.g., zebra mussel). The Chicago Sanitary and Ship Canal linking the Great Lakes to the Upper Mississippi system will continue to expose the river to additional introductions. Common carp, intentionally introduced as a potential forage fish, is among the most common fish, by biomass, in the Upper Mississippi. Feeding and spawning activity by common and grass carp tears out vegetation in slackwater areas, decreasing water clarity. Bighead and silver carp, which are largely limited to the river around and below the confluence with the Illinois River, are both zooplanktivores. Capable of rapid growth, both species may outcompete larvae of native fishes that are dependent on zooplankton as a food source. Zebra mussels, present in the Upper Mississippi River since 1992, consume phytoplankton and have the potential to be a major disruption to the food web through their removal of this important resource. By attaching to hard substrate, rocks, and snags, zebra mussels may outcompete some invertebrate taxa for space, thus changing invertebrate community structure and reducing the resources available for invertivorous fishes.

Water quality has improved considerably in the Upper Mississippi River since the 1970s, but it is apparent that new concerns about water quality have arisen. Nonpoint-source pollution compounded with changes in hydrological processes and physical attributes of the river threaten the health and sustainability of this great river–floodplain system. Despite this, the Upper Mississippi River retains more of its natural attributes than perhaps any other large river–floodplain ecosystem in the eastern United States and should be considered the starting point for gaining a better understanding of the dynamics of great rivers and the role model for river–floodplain restoration.

MINNESOTA RIVER

The Minnesota River, the first major tributary of the Upper Mississippi River, increases the Mississippi's discharge by almost 50% (Fig. 8.13). Draining about one-fifth of Minnesota, plus a small part of South Dakota and northern Iowa, the Minnesota River is the first tributary to drain primarily agricultural land and the first tributary from the Central Tall Grasslands. The Upper Mississippi does not receive tributaries that flow through the Central Tall Grasslands again for 500 Rkm downstream. The Minnesota is also a river of contrasts. Although possessing some of the most scenic falls and river corridors in Minnesota, it is also heavily impacted by human activity, with the lower Minnesota receiving a U.S. Environmental Protection Agency watershed indicator index of 6—more serious problems and high vulnerability—due to the potential for future degradation of water quality.

The river played an important role in Minnesota history and in the settling of the northern plains. It was a point of navigation for goods and supplies into the Dakotas in the nineteenth century and was the route that many immigrants took in the 1850s after treaties were arranged with the upper and lower bands of the Sioux nations along the Minnesota River. The Minnesota River basin was also the site of one of the worst conflicts between Native Americans and settlers during the Sioux Uprising of 1862 (Waters 1977). Like other parts of the Upper Mississippi River basin, evidence of human settlement dates back at least 8000 years.

Physiography, Climate, and Land Use

The Minnesota River lies entirely within the Central Lowland (CL) physiographic province and almost entirely within the Central Tall Grasslands terrestrial ecoregion. A small portion near the confluence with the Upper Mississippi lies within the Upper Midwest Forest/Savanna Transition zone (Ricketts et al. 1999). The basin has been greatly influenced by glaciation, as demonstrated by the identification of 27 different geomorphic sections based on glacially formed features. Soils in the basin are primarily Wisconsin Drift overlying igneous and metamorphic rock or poorly consolidated shales and sandstones. The poor drainage of soils in some regions, especially the upper end of the river, historically created broad areas of wetlands and lakes. Characteristic of the mesic conditions of the Central Tall Grasslands, dominant vegetation in much of the basin prior to settlement were tallgrasses, primarily big blue-stem, switchgrass, and Indian grass. Vegetation in the eastern one-third of the basin consisted of oak–prairie mix and forests of oak, maple, bass-wood, ash, and elm. Only remnants of presettlement vegetation can be found today, with much of this limited to the river valley.

Average annual precipitation varies from 60 cm/yr in the west to 76 cm/yr near the mouth. Precipitation is primarily rain, occurring in the greatest amount during the summer (Fig. 8.14). Precipitation is lowest in the winter, falling as snow. The major climatic event in the basin is in the spring, when rain falls over the melting snow and thawing ground. Average snowfall in the basin is 117 cm/yr (as snow). Mean annual temperature is about 7.5°C, with an average monthly range of −11°C to 23°C (see Fig. 8.14).

About 95% of the basin is used for agriculture, with corn and soybeans as the major crops. Wheat production is also important in the basin, as is hog and cattle production. Less than 1% of the basin is urban, with Mankato and part of the Minneapo-lis–St. Paul metropolitan area as the major urban areas on the river. The remainder of the basin is an agriculture/forest mix, although a few areas of wetland, prairie, oak forest, and prairie-woodland persist, largely in the river valley and riparian zone.

River Geomorphology, Hydrology, and Chemistry

The Minnesota River has been described as a small stream in an oversized valley (Hunt 1967). The geo-logical term for this is an underfit river. This desig-nation is because the Minnesota River flows through a valley as much as 80 m deep and 8 km wide (Waters 1977). The valley was cut by the glacial River Warren during the Wisconsin Glaciation. The glacier cut off drainage of Lake Agassiz to Hudson Bay. A natural dam on the southern end of the lake ruptured and a torrent of water was released. The glacial River Warren followed a path to the southeast, then turned to the northeast, a path dictated by the southern edge of the ice sheet and moraine. The massive flows of the River Warren continued for as much as 9000 to 12,000 years before eventually subsiding after the glacier had retreated sufficiently to reopen flow from Lake Agassiz to Hudson Bay. Although the Minnesota River flows as a small river through an immense valley, the valley itself creates impressive scenery as many of the tributaries flow into the valley over falls or through gorges and rapids as they descend from the plains (Waters 1977).

The Minnesota River begins as Big Stone Lake, a channel lake formed as the sediments carried by the Whetstone River formed a fan in the channel of the Minnesota River (see Fig. 8.13). The river does not originate from mountains or plateau, but from a depression, specifically a glacial river valley where a slightly higher area within the depression serves as a watershed divide (Waters 1977). Two other channel lakes, Marsh Lake and Lac qui Parle, are in the upper Minnesota River and are formed by suspended sedi-ment deposits from the Pomme de Terre and Lac qui Parle rivers, respectively. The river, at least histori-cally, ran through and by many bogs and marshes formed by the poor drainage of soils in the basin, especially in the upper Minnesota. The river mean-ders from bluff to bluff as it moves along the valley floor (Fig. 8.4). These meanders have created many oxbow lakes. Additional floodplain lakes formed by the deposition of natural levees are also present. Most of the river flows over glacial drift, moraine, and outwashes that include sand, gravel, and some boulders. Bedrock is exposed in some areas, a remnant of the erosive force of the River Warren. Sediment composition shifts in the middle Minnesota River, near Mankato, as the river begins to flow over younger sedimentary rock, including limestones, sandstones, and dolomite (Waters 1977). The river meanders over these deposits in response to their varying thickness, often behaving like a typical low-gradient (15 cm/km) plains river.

Average annual discharge is 125 m³/s (1935 to 1999), with peak flows in April in response to snowmelt, ground thawing, and spring rains (see Fig. 8.14). Discharge declines through the rest of the year as evapotranspiration increases. Discharge reaches its lowest levels in January and February, when soils freeze deeply and lock up most of the subsurface runoff. Hydrological processes can be very erratic, with rapid rises and falls throughout the basin because the tiling of fields increases the rate of water removal from soils. There are 12 major subbasins, which can be divided into three groups. The first are the tributaries flowing from the northern part of the basin—Pomme de Terre River, Chippewa River, and Hawk Creek—which supply the Min-nesota with 10% of its total discharge (Waters 1977). The Lac qui Parle, Yellow Medicine, Redwood, and Cottonwood rivers flow across the southwest low plains of the basin. Last is the Blue Earth River, the

FIGURE 8.4   Minnesota River at flood stage, Minnesota (PHOTO BY TIM PALMER).

largest and last major tributary, joining the river as it changes to a northeasterly path. The Blue Earth River provides 25% of the Minnesota's discharge.

The Minnesota River is best described as a hardwater, turbid river. Alkalinity exceeds 200 mg/L as $CaCO_3$ but varies along the gradient of the river in response to changing geology. Before shifting its course to the northeast over sedimentary rock, the relatively high total dissolved solids concentrations are from NaCl and $CaSO_4$ (gypsum). The concentrations of sulfates, sulfides, and chlorides decline after the river passes Mankato, and bicarbonate concentrations increase (Downing et al. 1999). Other aspects of water chemistry do not change appreciably longitudinally, but do reflect the high dissolved solids and limestone/dolomite topography of the area. Conductivity is high (865 μS/cm), as is hardness (435 mg/L as $CaCO_3$), and pH is slightly basic (8.0). Nutrient concentrations are consistently high throughout the length of the river. Total phosphorus ranges from 0.04 to 0.48 mg/L and NO_3-N concentrations frequently exceed 20 mg/L (Payne 1994).

Suspended sediment concentrations also change longitudinally, but this is due to a combination of natural landscape and human activities. Suspended sediment concentrations were high prior to agriculture, particularly below tributaries such as the Blue Earth River, which drain areas with more weathered soils. Suspended sediment concentrations are lower in the upper reaches of the river (approximately 40 mg/L), but this is probably due primarily to settling in upstream reservoirs (Payne 1994). Concentrations increase continually downstream until reaching their maximum (sometimes >300 mg/L) approximately 125 Rkm from the mouth. A substantial increase occurs as the Blue Earth River, which contributes about 55% of the total suspended sediment load, empties into the Minnesota. Concentrations begin to decline as the river approaches the Minneapolis–St. Paul Metropolitan area. Much of the sediment load in the lower river enters the Upper Mississippi River, as there is very little long-term deposition in the mainstem Minnesota River (Downing et al. 1999). Average annual water temperature is 10.5°C, with a range of 0°C (January) to 29°C (July) near Mankato.

River Biodiversity and Ecology

Algae and Cyanobacteria

The diatoms *Nitzschia amphibia*, *Gomphonema parvulum*, *Navicula radiosa*, and *Achnanthes lanceolata* are the major members of the benthic algal community and are found nearly throughout the Minnesota River. Other taxa found include *Surirella*, *Cymbella*, and *Cyclotella*. Cyanobacteria and filamentous algae are less common and are typically found in slower-flowing waters. There are no known studies on the phytoplankton taxa of the Minnesota River.

Plants

The riparian zone is considerably degraded in many areas, but there are still many natural areas of scenic value (Waters 1977). When present, cottonwood, green ash, black willow, and sandbar willow are in the riparian zone. Intense agricultural activity has resulted in corn and soybeans displacing natural vegetation in many areas. Aquatic macrophytes are common in the lakes and wetlands in the upper Minnesota River. Cattails and sedge grasses are the predominant emergent plants, with submerged pondweeds being the most common submerged plants. Macrophytes appear to be limited in abundance throughout much of the main stem, owing to the high turbidity of the system, but are found in the marshes and oxbow lakes.

Invertebrates

Studies of benthic invertebrates on the Minnesota River have largely been for assessing pollution impacts and for developing an index of biotic indicators (Zischke et al. 1994). Approximately 100 invertebrate taxa have been identified and are largely representative of large-river fauna. Chironomid midges (22 genera) are the most diverse group, with *Glyptotendipes*, *Polypedilum*, and *Tanytarsus* typically having the greatest numbers of individuals. Mayflies, of which 14 genera have been identified, are represented primarily by the scrapers *Stenonema* and *Stenacron*, the filterer *Potamanthus*, and the collector-gatherer *Tricorythodes*. The filterers *Hydropsyche*, *Cheumatopsyche*, and *Cyrnellus* are the most abundant of the nine genera of caddisflies identified. Zischke et al. (1994) examined benthic invertebrate community structure from below Lac qui Parle Lake to Jordan, Minnesota (Rkm 80). They found that amphipods were the most abundant invertebrates immediately below Lac qui Parle Lake, whereas caddisflies were most abundant at all other sites, with midges as the second most abundant group. All sites received a macroinvertebrate biotic index pollution rating of poor. Substrate consisted of gravel, sand, silt, and woody debris at all locations, but no quantification of these is given.

The status of benthic invertebrates is best exemplified by freshwater mussels. Historical accounts note the presence of 40 species of freshwater mussels. Bright et al. (1990) found only 20 species living in the Minnesota River, with only 14 species abundant enough to be considered healthy populations. Many of the other species are present in low densities and appear to no longer be reproducing. Possible causes for the decline of freshwater mussels include unstable substrates, excessive siltation, and chemical pollution (Bright et al. 1990).

Vertebrates

The fish community consists of 87 species, including 29 species of Cyprinidae (minnows and carp), 14 species of Percidae (perches), 10 species of Centrarchidae (sunfishes), and 8 species of Catostomidae (suckers). The most abundant fishes from below Lac qui Parle Lake downstream to Jordan, Minnesota, are fairly consistent throughout this portion of the river: shorthead redhorse, quillback, common carp, freshwater drum, emerald shiner, spotfin shiner, sand shiner, channel catfish, and gizzard shad (Bailey et al. 1994). There is a slight increase in the number of species and an increase in fish abundance as the geology of the river valley changes from drift and igneous/metamorphic rock to limestones, sandstones, and shale, but increased abundance is primarily from greater numbers of emerald shiners and gizzard shad. Channel catfish also decline considerably in the lower river (Bailey et al. 1992c).

Although the most common fishes, the Cyprinidae, are trophic generalists, many of the other fishes are small insectivorous species. It is likely that they are focusing on benthic invertebrates, particularly the abundant caddisflies and midges. The most abundant piscivorous fish, walleye, is present in low numbers, and smallmouth bass and largemouth bass are rare (Bailey et al. 1994). Four fishes are listed as extirpated from the Minnesota, including three Catostomidae (Burr and Page 1986): skipjack herring, bigmouth buffalo, spotted sucker, and river redhorse. Paddlefish is listed as threatened by the state of Minnesota. Five other fish species designated statewide as a special concern are found in the Minnesota River.

Three species of salamanders, six species of frogs, five species of turtles, and two species of snakes are recorded in the Minnesota River valley. Herpeto-fauna in the Minnesota River are consistent with other rivers of the region, with painted turtles, common snapping turtles, northern water snakes, treefrogs, northern leopard frogs, and mudpuppies among the species found. No reports on the status of aquatic vertebrates in the Minnesota River were located. Habitat surveys indicate that snag abundance and instream cover is low (Bailey et al. 1994, Kischke et al. 1994). This, combined with drainage of wetlands, has probably been detrimental to aquatic vertebrates. Muskrat and beaver are present primarily in the upper Minnesota.

Ecosystem Processes

No comprehensive studies of ecosystem processes have been conducted in the Minnesota River. Historical accounts of the Minnesota River indicate that it was a turbid system prior to conversion of agricultural fields. Despite this, instream primary producers are abundant in some areas, even where suspended sediment concentrations are high. Phytoplankton biovolume approaches $30\,mm^3/L$ at Jordan, compared to $<1\,mm^3/L$ in the Mississippi headwaters above the Minneapolis–St. Paul metropolitan area (Stark et al. 2001). Abundance of phytoplankton could be sufficient to support the invertebrate community that appears to consist primarily of filter feeders such as caddisflies and midges (Tanytarsini). Reliance on instream primary production may also be a necessity in light of the degradation of riparian vegetation. This could result in low availability of particulate organic matter from terrestrial sources, contributing further to the environmental stresses placed on this agriculturally impacted river. High chlorophyll concentrations ($>15\,\mu g/L$) relative to the Mississippi headwaters and St. Croix River may, however, be entirely an artifact of high nutrient inputs generating low food-quality phytoplankton and benthic algae, the abundance of which may create problems stemming from high biological oxygen demand during algal decomposition (Payne 1994, Downing et al. 1999).

Human Impacts and Special Features

The Minnesota River, representing the first river draining the central grasslands to empty into the Upper Mississippi, stands out as a small river flowing through an oversized river valley. The river valley, formed by the glacial River Warren, is just one of many remnants of glacial processes that characterize this river. The valley has also been a blessing ecolog-ically, for it is within the river valley that much of the vegetation and wildlife that was once characteristic of the entire basin can still be found. Although most of the basin has been converted for agricultural use, the river valley still provides many areas of scenic beauty and recreational opportunities (Waters 1977).

Suspended sediments are the primary water-quality concern for the Minnesota River. Sediment load during summer rain events can range between 2 and 500 tons/d below Lac qui Parle Lake (465 Rkm above the mouth) to approximately 40 to 4000 tons/d at Henderson. Most sediment delivery is through tributaries, with the major contributors being tributaries draining the southern and eastern part of the basin (Watonwan, Blue Earth, and Le Sueur rivers), where soils consist of finer particle sizes and precipitation is slightly higher in summer (Payne 1994).

Agriculture has also created concern because of high nutrient and pesticide concentrations. The longitudinal pattern for NO_3-N closely reflects that of suspended sediments, with the inflow of the Watonwan, Le Sueur, and Blue Earth rivers each progressively increasing concentrations within the main-stem Minnesota River. Nitrate-N concentrations exceeded the national standard of 10 mg/L in 11% of samples collected at Jordan (Stark et al. 2001). The Minnesota basin is included in the Upper Mississippi River basin National Water Quality Assessment (NAWQA) program, along with the Upper Mississippi River down to Lake Pepin and the St. Croix River. Of the three rivers, the Minnesota has by far the greatest occurrence of pesticides, especially atrazine, deethylatrizine, metolachlor, acetoxchlor, and alachlor (Stark et al. 2001). Many of these pesticides are also found in the Prairie du Chien–Jordan aquifer, which serves as a major water source for the Minneapolis–St. Paul metropolitan area.

The major change in agricultural practices in the basin was the tiling of the land, which dates back to the 1880s. The poor drainage of the soils in the basin made it difficult to grow crops in the rich soils, and tiling systems were placed below the soil surface to rapidly move water from fields into ditches that would drain into nearby streams. Tiling also allowed for marshy areas to be drained and plowed and changed the hydrology of the basin. Prior to tiling, much of the precipitation percolated through the soils and was stored in aquifers. Delivery to surface waters has gone from a long-term process to being almost immediate. Discharge has become very erratic, reflecting shortened delivery time that creates a rapid increase in discharge followed by a rapid decline without subsurface and groundwater flow to dampen the falling limb of the hydrograph. Tiling has also increased the delivery rate of contaminants to the river.

Hydrology is further impacted by dams and water diversions. There are 63 dams on the Minnesota and its tributaries. The greatest number of dams, most of which are small diversion dams, is on the upper Minnesota and Chippewa rivers. Dams have been placed at the outlets of the three large natural lakes, Big Stone, Marsh, and Lac qui Parle, for flood control and to control lake levels. In addition, flow from the Chippewa River basin is diverted to Lac qui Parle Lake. All flow in excess of $28\,m^3/s$ is diverted during spring and summer floods and about one half of the flow is diverted when $<28\,m^3/s$ (Payne 1994). Ironically, dams on some parts of the Minnesota actually improve fish abundance immediately downstream relative to unimpounded portions of the river (Bailey et al. 1994). The dams and diversions, combined with extensive tiling, have contributed to the rapid rises and falls seen in the hydrograph of the Minnesota River.

It is clear that human activities have adversely impacted the ecology and ecosystem dynamics of the Minnesota River. Bailey et al. (1994) collected only 54 species of fishes, although 87 species are supposed to be in the Minnesota River (Burr and Page 1986). The relatively low diversity of invertebrate taxa and apparent extirpation of 17 species of freshwater mussels further suggest ecosystem degradation. Increasing nutrient delivery, while increasing phytoplankton abundance, is enhancing the production of mostly undesirable taxa. Conversely, increased rates of soil erosion and pesticide delivery from the tiling system degrade habitat quality and enhance exposure to higher concentrations of contaminants. Declines in organismal diversity, coupled with changes to the landscape, clearly demonstrate that the Minnesota River, overall, is an ecosystem in peril.

ST. CROIX RIVER

The St. Croix River was one of the original eight rivers protected under the National Wild and Scenic River Act of 1968 and is viewed as one of the best recreational rivers in the Midwest (Waters 1977). It is the second major tributary of the Upper Mississippi, joining the river 53 Rkm downstream of the Minnesota River (see Fig. 8.2). The St. Croix flows from the north, beginning in the northern highlands of Wisconsin and continuing south for 276 km through rapids, pools, and gorges before flowing into Lake St. Croix and the Upper Mississippi River (Fig. 8.15).

Prior to the presence of Europeans, the Dakota nation was the primary occupant of the basin until the 1500s. At this time, the Chippewa nation began to move from the east into the basin and the two nations battled continuously before the Chippewa eventually drove the Dakota out of the St. Croix valley. The St. Croix was important as a point of passage early in the exploration of the upper Midwest because of the proximity of its headwaters to Lake Superior. It became a route, first found by Sieur du Luth in 1680, for the fur trade, first by the French and later by the British (Waters 1977). The river later became one of the busiest logging rivers in the Midwest (Fago and Hatch 1993).

Physiography, Climate, and Land Use

Roughly two-thirds of the river flows through the Superior Upland (SU) physiographic province, in which most of the Western Great Lakes ecoregion lies (see Fig. 8.15). The lower one-third flows through the Central Lowland (CL) as the river approaches the Upper Mississippi. The Superior Upland is an extension of the Canadian Shield, composed of Precambrian igneous and metamorphic rock exposed by the movement of glaciers. The St. Croix region was on the edge of the last ice sheet, which left moraine and drift (up to 70 m deep in some areas) over much of the landscape (Hunt 1967). The materials overlaying bedrock are not extensively weathered, which, when combined with the underlying bedrock, leaves the headwaters poorly drained, with almost 40% of the headwaters area covered in lakes and marshes. Bedrock in the lower St. Croix is Cambrian sandstones and dolomite, in addition to some basaltic bedrock. The basin is divided into the upper and lower St. Croix at Taylor Falls Dam (Rkm 85), the former site of the St. Croix Falls.

The St. Croix begins in the Western Great Lakes Forest terrestrial ecoregion before entering the Upper Midwest Forest/Savanna Transition ecoregion (Ricketts et al. 1999). Vegetation present on the St. Croix today is similar to what was present prior to settlement, although it is mostly second-growth forest. The portion of the basin within the Western Great Lakes ecoregion has a northern hardwood–conifer mixed forest, with sugar maple, basswood, and yellow birch as the dominant hardwoods. White pine, balsam fir, and white spruce are the representative conifers (National Park Service 1997). Also common in the upper St. Croix are conifer swamps and bogs, with black spruce, tamarack, and white cedar. In the lower basin the forest contains red oak, basswood, black ash, green ash, and yellow birch. Oak barrens and tallgrass prairie are also present because the lower St. Croix basin includes the northern limit of the Forest/Savanna Transition zone.

Average annual air temperature is 6.3°C at St. Croix Falls, Wisconsin (Fig. 8.16). Air temperature is lowest in January (−12°C) and highest in July (22°C). Average annual precipitation is 78 cm/yr and is highest during summer months (>9 cm/mo), dropping to only 2 cm/mo in January and February (see Fig. 8.16). Snowfall is also important in the basin, with average snow accumulation of 107 cm/yr over the entire basin and snowfall approaching 130 cm/yr in the headwaters.

Land use is 27% agriculture, mostly in the lower St. Croix, 50% forest, and around 2% urban, most of which is near the confluence with the Upper Mississippi. Most of the remainder of the basin is wetland and open lake. Corn and dairy production are the predominant forms of agriculture in the basin, with a number of other activities creating a diverse, if limited, agriculture (Waters 1977). Timber harvesting was important in the basin in the nineteenth century but is limited today with the designation of the St. Croix as a protected National Wild and Scenic River. Most of the urban development is in the lower basin as the river approaches the Minneapolis–St. Paul metropolitan area.

FIGURE 8.5 St. Croix River above Grantsburg, Wisconsin (PHOTO BY TIM PALMER).

River Geomorphology, Hydrology, and Chemistry

As with the St. Croix basin, the features of the channel were shaped by glaciation. Runoff from glacial Lake Duluth drained to the south through the present St. Croix River valley. The torrent eroded through the igneous and metamorphic bedrock as it flowed south toward the glacial River Warren. As lake levels receded, the middle reaches of the glacial river dried, leaving one river flowing to the south that drained the glacial melt and precipitation falling on the side of the divide and another river flowing to the north (Bois Brule) into what would become Lake Superior (Hunt 1967). The modern St. Croix River begins in the northern highlands of Wisconsin as an outflow of Upper St. Croix Lake. The river changes considerably at the confluence with the Namekagon River, a tributary actually larger than the St. Croix when they join. The channel widens and current velocity decreases, except when rapids are encoun-

tered (Fig. 8.5). The river slows again as it approaches Taylors Falls Dam (Waters 1977). The remaining 84 km provide some of the best scenery on the river. Glacial meltwaters cut deeply into lava rock, creating deep vertical channels, palisades, and large rock formations that jut out over the river (Waters 1977). There are also many islands throughout this section of the river. The river ends in Lake St. Croix, a 26.4 km channel lake formed by the deposition of sand and silt across the mouth of the St. Croix by the Upper Mississippi (Fago and Hatch 1993). Stream gradient is 47 cm/km in the upper river but has a gradient of 105 cm/km in the 13 km Kettle River Rapids, located on the St. Croix above and below its confluence with the Kettle River (Waters 1977). Gradient averages 15 cm/km in the lower river (Fago and Hatch 1993).

There are six major tributaries in the St. Croix system, the most noteworthy of which are the Namekagon, Kettle, and Snake rivers. The Kettle and

Snake offer some of the most spectacular and treacherous rapids in the St. Croix system (see Fig. 24.1), with each having a gradient approaching 100 cm/km (Waters 1977). Average annual discharge near Taylors Falls Dam is 122 m³/s. The Apple River, the only major tributary below this site, has an annual discharge of 8.5 m³/s. Other smaller tributaries probably only add slightly to discharge before the river enters the Upper Mississippi. Discharge in the St. Croix reaches a peak in midspring (April) in conjunction with snowmelt and spring rains (see Fig. 8.16). Average discharge then decreases with decreasing snowmelt and increasing rates of evapotranspiration until reaching low flows in late summer (August). High retention of water in the numerous lakes and bogs also contributes to lower discharge, even during periods of heavy rain. Lowest annual flows occur from December through February, when parts of the river and the surrounding drainage basin are frozen.

The St. Croix River is best described as a softwater, brown-water river (Waters 1977). Staining of the water reflects the drainage of bogs and marshes in the upper reaches and from tributaries such as the Kettle River. Turbidity is low, averaging less than 5 NTU throughout much of the year. Nutrient concentrations are also low, with total nitrogen averaging <1 mg/L and total phosphorus averaging <0.09 mg/L throughout the length of the river. Low suspended sediment and nutrient concentrations of the St. Croix serve to benefit the Upper Mississippi by diluting the elevated concentrations of suspended sediments and nutrients entering the Upper Mississippi River from the Minnesota River (Stark et al. 2001). Conductivity is also low (150 μS/cm) and pH is close to neutral, averaging around 7.5. Alkalinity (59 mg/L as CaCO₃) and hardness (65 mg/L as CaCO₃) are also low and reflect the limited presence of limestone in the basin. Water chemistry remains relatively consistent along the length of the river. Average annual water temperature is 10.4°C, with a range of 0°C to 27°C. The water temperature has never exceeded 26°C in the upper St. Croix for the period of record (1976 to 2000).

River Biodiversity and Ecology

Algae and Cyanobacteria

Studies of phytoplankton have been limited to quantification of chlorophyll concentrations and have not addressed taxonomic composition. Diatoms are the predominant representatives of the benthic algal community, but no taxonomic breakdown is available.

Plants

An important feature of the riparian vegetation of the upper St. Croix are conifer swamps, with black spruce, tamarack, and white cedar as the dominant trees. Lowland hardwood forests, where basswood, black ash, green ash, red oak, and yellow birch are the dominant trees, are interspersed with the conifer swamps. Hardwood swamps composed of red maple, paper birch, and black ash are also common through the middle reaches. The floodplain forest of the lower river consists primarily of silver maple and green ash (National Park Service 1997). Aquatic macrophytes in the upper basin are associated primarily with bogs and swamps.

Invertebrates

Surveys have identified 332 species of invertebrates throughout the St. Croix, including 71 species of Diptera, 54 species of mayflies, 37 species of caddisflies, and 19 species of beetles. The invertebrate community changes longitudinally, with 218 species identified in the upper river and 167 species in the lower (Fago and Hatch 1993). Major faunal differences are in the numbers of mayflies, caddisflies, and stoneflies, with 46 species from these orders known in the lower river versus 113 species in the upper. Although many invertebrates are representative of large-river fauna, there are also species not seen in other large rivers in the Upper Mississippi system owing to the unique features of the St. Croix, particularly in the upper reaches.

Many of the abundant invertebrates in the upper St. Croix are described functionally as being collector-gatherers/grazers, with the choice of diet probably dictated by resource availability. Most abundant among these are mayflies, particularly *Baetis*, *Siphlonurus*, and *Ephemerella* (Lillie 1995). The riffle beetle *Dubiraphia* is another common genus that falls into this category. The mayflies *Brachycercus* and *Cercobrachys* are characterized as collector-gatherers, whereas the mayflies *Heptagenia* and *Stenonema* and the caddisfly *Helicopsyche* are considered almost exclusively algal scrapers. The riffle beetle *Stenelmis* is another abundant scraping invertebrate in the upper St. Croix. One of the more abundant filterers is the mayfly *Anthopotamus*. The caddisflies *Hydropsyche* and *Ceratopsyche* represent the other abundant filterers (Lillie 1995). Shredding invertebrates include the caddisflies *Nectopsyche* and *Lepidostoma*, the stonefly *Pteronarcys*, and the beetle

Peltodytes. Predaceous invertebrates are represented by 45 different species of dragonflies and damselflies (most of which occur in tributaries), seven species of perlid stoneflies, and the beetle *Gyrinus*. A breakdown of invertebrates to genus was not found for the lower St. Croix, but family lists indicate that many of the common large-river invertebrates are represented. Principal among these is the mayfly family Heptageniidae, the caddisfly family Hydropsychidae, and 24 genera of midges (Fago and Hatch 1993).

The St. Croix hosts a diverse and abundant community of 40 species of freshwater mussels. In some locations, 20 to 30 species might be found together, with many locations having densities of >20 mussels/m^2 and some areas even approaching 200 mussels/m^2 (Hornbach 2001). Mussels are abundant throughout the river, but the species composition does shift, probably due to the cutoff of fish migration by Taylors Falls Dam (Fago and Hatch 1993, Hornbach 2001). The most common mussels in the river are threeridge, deertoe, spike, and Wabash pigtoe. Two federally endangered species are present, the Higgin's eye and winged mapleleaf. There are seven mussels listed as endangered and three listed as threatened by Wisconsin. Minnesota lists three endangered and six threatened species of mussels for the St. Croix.

Vertebrates

There are 110 species of fishes recorded in the St. Croix basin, including 43 species of Cyprinidae (minnows and carp), 16 species of Percidae (perches), 12 species of Catostomidae (suckers), and 9 species of Centrarchidae (sunfishes). Seven species have not been reported since 1974 and are believed to be extirpated, but it is believed that the St. Croix was a marginal habitat for these species (Fago and Hatch 1993). Another six species are nonnative, including common carp, rainbow trout, brown trout, and lake trout. There are 22 species found in the lower St. Croix, but not the upper St. Croix, and 18 species found only in the upper river (Fago and Hatch 1993). St. Croix Falls and, later, Taylors Falls Dam are believed to have been a barrier to fish migration from lower to upper river. Of the 18 species of fishes listed by the State of Wisconsin as threatened or endangered, 8 are in the St. Croix and have healthy populations (National Park Service 1997). Of note on this list is the river redhorse, which is reported to have declined in numbers in many of the other tributaries of the Upper Mississippi River.

The majority of riverine fishes found throughout the St. Croix River (Gordan Dam to mouth) are characterized as insectivores. Examples include golden redhorse, shorthead redhorse, river redhorse, spottail shiner, mimic shiner, quillback, and Johnny darter. Common carp, an omnivore, are also found throughout the length of the river. The major carnivores, smallmouth bass and walleye, are found throughout the river, although walleye are more abundant in the upper St. Croix than the lower. Bluegill and yellow perch, both insectivores, are common in more lentic habitats throughout the river. Planktivorous gizzard shad and emerald shiner, an insectivore, are among the most abundant fishes found exclusively in the lower St. Croix. Five species of darters are also found exclusively in the lower river. Many of the fishes found only in the upper St. Croix, such as longnose dace, brassy minnow, and hornyhead chub, are also benthic insectivores. There are several specialists, including the algivorous central stoneroller, found only in the upper St. Croix. The bigmouth shiner is among the most abundant omnivores found exclusively in the upper St. Croix.

Eighteen species of amphibians and 14 reptile species are known to occur in the St. Croix. Very little is known about their numbers and current status (National Park Service 1997), except that the seven turtle species present are more abundant in the lower river. Turtles known to occur in the region include painted turtles, common map turtles, wood turtle, and spiny softshell turtles. A number of treefrogs are known to the area, including spring peepers. Tiger salamanders are also known to occur in the basin. Mammals known to be associated with the river include muskrat, beaver, mink, river otter, and moose.

Ecosystem Processes

There have been no comprehensive ecosystem studies of the St. Croix River. Water quality and the diversity of trophic specialists among the invertebrates and fishes suggest, however, that the river has a healthy and diverse trophic structure. The description of the St. Croix as a brown-water river infers that terrestrial inputs from riparian vegetation and connected swamps and bogs are important to system dynamics. Terrestrial inputs are, mostly likely, a result of direct litterfall and from transported organic matter delivered by tributaries considering that the often-incised nature of the channel valley creates a floodplain of limited size throughout much of the river. Phytoplankton abundance is low, only about 10% of that seen in the Upper Mississippi at the confluence of these two rivers, but seasonal algal blooms develop in slackwater and backwater habitats

(National Park Service 1997). Benthic algae are reported in both lentic and lotic habitats, and the presence of several algivorous invertebrates and fishes (e.g., central stoneroller) suggests it is available in quantities to support part of the food web. Abundant rock substrate and snags throughout the river create structural heterogeneity that might otherwise be lacking given the generally constrained nature of the river channel. Emergent and submergent aquatic vegetation is present in both lentic and lotic habitats and could contribute to energy flow, particularly as detritus.

Human Impacts and Special Features

The upper St. Croix from Gordon Dam (Rkm 247) downstream to Taylors Falls Dam (Rkm 85) and all of the Namekagon River were among the first eight rivers afforded protection under the Wild and Scenic Rivers Act of 1968. The lower St. Croix was added in 1972. Together, they form the St. Croix National Scenic Riverway, managed by the National Park Service. Thus, the river and a narrow terrestrial corridor on each side of the river are protected from further development. The primary human impact on the river today is through recreation, particularly camping, hiking, and canoeing. Lake St. Croix is designated as recreational, allowing for the use of motorized boats. Water quality and the physical structure of the river remain intact through the protection afforded by the scenic river designation, giving it a physical and chemical stability not seen in any of the other large tributaries of the Upper Mississippi River.

The designation of the St. Croix as a scenic river, however, does not mean that it has not been influenced by human activity. As is the case with other rivers extending into the Superior Upland, the basin was logged heavily in the nineteenth century and into the early twentieth century. White and red pines, in particular, were cut throughout most of the basin, leaving very little old-growth forest in the basin. The river served for transport of huge log rafts downstream to the many sawmills built along the lower St. Croix (Waters 1977). Natural features of the river actually aided in the delivery of logs to the mills. The many channels created by islands above Lake St. Croix were used to sort logs prior to their arrival at the mills. Timber harvesting declined into the early twentieth century, leaving most of the basin relatively undisturbed after 1920 and allowing for the regrowth of native vegetation. Agriculture is limited primarily to the lower St. Croix, and impacts appear minimal, based on low nutrient and suspended sediment concentrations throughout the length of the river. Lake St. Croix has a history of industrial development that is still evident today through the issuance of fish consumption advisories. Consumption advisories are primarily due to the presence of PCBs in lake sediments. A navigation channel is maintained in the lake, but barge traffic, which was once as high as 120 tows/yr, is minimal today, with most materials transported by rail (Fago and Hatch 1993).

In addition to the nonnative fishes already mentioned, zebra mussels and the Asiatic clam have been reported in Lake St. Croix. Asiatic clams have only been reported in proximity to outflows from a power plant and do not appear abundant. Federal and state agencies worked throughout the early 1990s through educational programs and boat inspections to prevent the migration of zebra mussels from the Upper Mississippi into the St. Croix. Zebra mussels, however, have been reported in the lower St. Croix in very low numbers since 1994.

The St. Croix River is almost an anomaly in the Upper Mississippi basin. Although, like all other rivers in the basin, it has been impacted by human activity, the St. Croix River maintains more of its natural condition than perhaps any other system in the Upper Mississippi basin. With its designation as a National Wild and Scenic River and the protection afforded by its inclusion in the National Park System, the river will continue to retain its beauty and splendor, thus remaining a jewel at the top of the Upper Mississippi River.

WISCONSIN RIVER

The Wisconsin River is the largest and longest river system in Wisconsin, draining approximately 22% of the state and totaling 692 km in length. The river begins in the northern highlands of Wisconsin, winding south into the glacial drift left behind by the Wisconsin glaciation, then turning west toward the Upper Mississippi River (Fig. 8.17). The Wisconsin River is heavily dammed for hydroelectric production and flood control. Most of the dams occur in the upper and middle river, with the headwaters and the last 133 km of the river flowing freely. It is important locally for recreation, including boating, canoeing, fishing, and hunting. Despite its size and importance within the state, however, comprehensive studies of the ecology of the river are limited, with most studies focusing on the "large river" lower Wisconsin.

Human presence in the Wisconsin River basin dates back to the end of the last ice age (9000 to 12,000 years ago), but their presence is known only from a few spear points and tools (Nichols 1984). The Woodland cultures (2100 to 400 years ago) included the mound builders found throughout the Upper Mississippi valley. Only three Native American tribes are definitively known to have been in the basin before the late seventeenth century: the Menominee, Winnebago, and Santee Sioux. Other tribes, including the Mascoutan, Potawatomi, Kickapoo, Fox, and Huron, moved into the area by the late seventeenth century as European settlement of the east pushed tribal nations to the west (Nichols 1984). The first Europeans on the Wisconsin River were Father Marquette and Louis Jolliet during their search for a water route to China. After canoeing down the Fox River, they portaged to the Wisconsin and took it to the confluence with the Upper Mississippi River near Prairie du Chien. Although they did not find their route to China, the French would later establish a lucrative fur trade with the Indian nations along the Wisconsin River. Settlers moved into the basin in greater numbers beginning in the 1830s, drawn by the lead mines of southern Wisconsin. The river served as a means of transporting people and trade goods, in addition to log rafts from the pine forests to the north.

Physiography, Climate, and Land Use

The Wisconsin River begins in the Superior Upland (SU) physiographic province, with the remainder of the basin in the Central Lowland (CL) province (see Fig. 8.17). The terminal moraine of the late Wisconsin glaciation forms an approximate boundary between the two provinces near Merrill, Wisconsin (Martin 1932). The basin within the Superior Upland is characterized by numerous lakes and muskegs (open marshes) owing to the poor drainage of the soils. The drainage basin changes markedly in the Central Lowland, where the basin alternates between glaciated areas with older, more weathered drift overlying Cambrian sandstones and the Driftless Area. The Driftless Area, a region of extensive topographic relief near the terminus of the drainage basin, was bypassed by the glaciers, leaving an area of extensive topographic relief formed by erosion from the numerous streams and small rivers flowing through the region. The only natural lakes in the Driftless Area are fluvial in origin.

The Wisconsin River begins in the Western Great Lakes Forest terrestrial ecoregion, but much of the

basin lies within the Upper Midwest Forest/Savanna Transition ecoregion (Ricketts et al. 1999). The Western Great Lakes region corresponds roughly with the Superior Upland physiographic province. The forest of the upper Wisconsin historically was a mixed conifer–hardwood forest. Conifers were represented by white pine and red pine and the hardwoods were primarily sugar maple, birch, and ash. Timber harvest, especially for pines, was intense in this area, but stands of the original vegetation are still present, largely as second growth. Aspen is also abundant in clear-cut areas. Forest vegetation in the lower Wisconsin basin is represented primarily by sugar maple, red oak, white oak, burr oak, basswood, and elm. Prairies are also present, particularly along the south-facing slopes of the bluffs of the Driftless Area. Prairie vegetation includes big bluestem, northern dropseed, Indiangrass, pasqueflower, coneflowers, columbine, and blackeyed Susan (Nichols 1984).

Average annual air temperature in the headwaters is 4°C, compared to 9°C near the mouth of the Wisconsin. Average monthly temperatures range from a maximum of 19°C and a minimum of −13°C in the north to 23.5°C and −8°C in the south (Fig. 8.18). Precipitation is similar in both regions, averaging 85 cm/yr, with precipitation heaviest (8 to 11 cm/mo) from April through September (see Fig. 8.18). Snowfall, as would be expected, does differ considerably. Average annual snowfall is 252 cm/yr in the headwaters, with snow occurring during all seasons except summer. Snowfall in the south totals only 92 cm/yr, with most falling in winter.

Land cover in the basin is 27% agriculture, 54% forest, and 3% urban. The remainder is mostly open water and wetland. Agriculture is predominantly dairy production, corn, hops, and oats. Cranberries are an important crop in the bog regions in the upper Wisconsin River basin. Wausau, located on the upper Wisconsin River, is the major metropolitan area. The area around Wisconsin Dells, Portage, and Baraboo is a major summer recreational area, with activities on and around the river.

River Geomorphology, Hydrology, and Chemistry

The Wisconsin River begins in the northern highlands of Wisconsin as an outlet to Lac Vieux Desert. As the upper Wisconsin River flows south, it traverses drift and moraine. Rapids are common throughout the glaciated areas through which the

FIGURE 8.6 Wisconsin River at Prairie du Sac, Wisconsin (Photo by Tim Palmer).

river flows because of areas with steeper than average gradient and the cobble–boulder substrate. The upper river drops from an elevation of 510 m asl to 285 m asl, a gradient of 67 cm/km. Below Merrill the middle Wisconsin River forms a broad, shallow valley across sandstones that were only lightly marked by glaciation. The river flows through this area as a low-gradient system with few tributaries. Bottom substrate consists primarily of gravel and sand, with some cobble in this region. The channel narrows to a steep-walled canyon known as the Wisconsin Dells. The river flows for 11 km by sandstone cliffs as high as 31 m. The channel widens once it has passed through the Dells. The river turns to the west as it enters the Driftless Area, just downstream from Prairie du Sac Dam, the last dam on the river (Fig. 8.6). The landscape changes from a relatively broad plain to limestone and sandstone bluffs approaching 125 m high on both sides of the river. The lower Wisconsin River is free-flowing for 133 km from the Prairie du Sac Dam to the confluence with the Upper

Mississippi. The lower river has a low gradient, 28 cm/km, as it flows through an incised river valley. Islands and sand bars are abundant, creating backwaters and side channels as the river flows over glacial outwash.

Annual discharge in the headwaters is 19 m³/s. This increases to 76 m³/s at Merrill, where the river moves from the Superior Upland to the Central Lowland. By the time the river reaches Muscoda, the last gaging station on the river (Rkm 72), annual discharge increases to 247 m³/s. Discharge begins rising in February at Muscoda as the snow melts and spring rains begin, reaching annual peak discharge in April (see Fig. 8.18). Discharge decreases to near annual minimum flows in the summer as evapotranspiration increases and runoff rates decline following development of canopy and ground cover foliage. Unlike other rivers in the northern Upper Mississippi basin, summer and winter minimum discharges are nearly identical on the lower Wisconsin. The annual peak discharge in the spring is reduced in magnitude by the retention of water in lakes and muskegs of the headwaters (Martin 1932). A sustained increase in discharge from September through November following the minimum low flow in August is probably a combination of late summer–early fall rains and reduced evapotranspiration and leaf fall.

Water chemistry from the headwaters to Merrill suggests that pH is acidic (6.6 to 6.9) and hardness is low (25 to 35 mg/L as $CaCO_3$). The headwaters receive substantial inflow from the numerous lakes and muskegs of the region, which typically contain large quantities of decomposing organic matter. The limited amount of limestone in the soils of the Superior Upland would have a low buffering capacity. These conditions are further supported by the description of the lower Wisconsin as a brown-water system containing large quantities of particulate and dissolved organic matter transported from upstream (Lillie and Hilsenhoff 1992).

Despite the brown coloration from natural and anthropogenic organic inputs, turbidity in the lower Wisconsin is low, averaging 5 NTU. Suspended sediment concentrations, averaging 21 mg/L, are low relative to other rivers flowing through the same physiographic provinces as the Wisconsin (e.g., Chippewa River, Black River). Suspended sediment concentrations generally remain <100 mg/L, largely due to sediment retention by dams farther upstream (Rose 1992). Hardness is higher than in the headwaters (100 mg/L as $CaCO_3$) and pH is more basic (7.9). Alkalinity is not as high in the lower Wisconsin as in other rivers flowing through the Driftless

Area, averaging 89 mg/L as $CaCO_3$. Concentrations of inorganic nitrogen are <0.6 mg N/L in the lower Wisconsin and dissolved phosphorus concentrations are <0.02 mg P/L (Lillie and Hilsenhoff 1992).

Average annual water temperature in the lower Wisconsin River is 11.8°C, with a range of 0°C to 28°C. The thermal regime is influenced by tributaries, groundwater inputs, and impoundments at Wisconsin Dells and Prairie du Sac (Lillie and Hilsenhoff 1992). Water temperatures in the spring and fall can fluctuate 2°C within 24 hours. Thermal gradients with a range as great as 7°C have been observed in the summer as much as 300 m downstream of coldwater tributaries (Lillie and Hilsenhoff 1992).

River Biodiversity and Ecology

Plants

The upper Wisconsin River floodplain includes hardwoods of the region, principally birch, aspen, and alder. Conifer swamps and muskegs are also a common feature of the upper river. The lower Wisconsin River has diverse floodplain forests composed of silver maple, river birch, swamp white oak, green ash, cottonwood, and black willow. Dry and wet prairies are also still found along the lower Wisconsin (Nichols 1984). These grass-dominated communities contain species such as big bluestem, Indiangrass, northern dropseed, and switchgrass. Aquatic vegetation is found primarily in sloughs, backwaters, and oxbow lakes off the main river (Nichols 1984). The predominant emergent vegetation includes cattails and arrowhead, whereas American lotus and American white waterlily are the most common floating-leaved plants. Submerged pondweed, coontail, northern watermilfoil, spatterdock, and bladderwort are common submerged plants found in deeper water.

Invertebrates

A total of 232 species of benthic invertebrates have been identified in the lower Wisconsin River (Lillie and Hilsenhoff 1992). Invertebrates are broadly represented, as indicated by the 53 species of beetles, 49 species of true flies, 42 species of mayflies, and 38 species of true bugs. Some of the most common taxa found were the true bug *Sigara*, a piercing herbivore, and *Trichocorixa*, a predator. Other common invertebrate predators were the beetles *Gyrinus* and *Dineutus*, the stonefly *Isoperla*, and the damselfly *Enallagma*. Common filter-feeders included the caddisflies *Hydropsyche*, *Cheumato-*

psyche, and *Potamyia*, the black fly *Simulium*, and the mayfly *Isonychia*. The most abundant of the invertebrates are characterized as grazers/collector-gatherers. Most of these are mayflies (*Baetis*, *Baetisca*, *Procloeon*, *Stenonema*, *Stenacron*, and *Caenis*). The most abundant shredder was the caddisfly *Nectopsyche* (Lillie and Hilsenhoff 1992).

Substrate and habitat type appear to be the greatest influence on invertebrate diversity and abundance in the lower Wisconsin (Lillie and Hilsenhoff 1992). Sand is the predominant substrate, found in 90% to 95% of the total available habitat, but only 16% of the invertebrates are found exclusively in sand. Another 30% of the fauna are found in areas with some sand. Snags, rock, and gravel beds are much less abundant but account for a substantially greater number of invertebrate species and total numbers of individuals than sand. Among habitats, invertebrates were most abundant in nearshore areas, followed by rocky runs. Backwaters were moderately productive, with 12% of the species found. The deeper main channel and sandy runs have the lowest diversity.

Thirty-six species of freshwater mussels are known for the Wisconsin River, including five listed as endangered and five listed as threatened by the state of Wisconsin. Higgin's eye, a federally endangered mussel, is in the lower Wisconsin River. Another endangered species, the winged mapleleaf, was extirpated in the Wisconsin by the 1930s. The most common mussels are threeridge, Wabash pigtoe, pocketbook, and pimpleback. Some species are known to have been extirpated from the middle Wisconsin and several others have declined due to industrial discharges into the 1980s (David Heath, personal communication). Water quality has improved, but mussels have failed to recover, partly due to the inability of downstream populations to restock upstream stocks. The dams on this section of the river are believed to prevent or limit upstream movement of fish hosts, thereby restricting mussel dispersal.

Vertebrates

Historically, 119 species of fishes were found in the Wisconsin River (Burr and Page 1986); however, more recent surveys have identified 100 species, including 30 species of Cyprinidae (carp and minnows), 16 species of Percidae (perches), 17 species of Catostomidae (suckers), 10 species of Centrarchidae (sunfishes), and 7 species of Ictaluridae (catfishes) (Fago 1992, Lyons et al. 2000). No federally endangered fishes are found in the Wisconsin

River. The state lists five species as endangered, six as threatened, and six as of special concern.

Four fishes are common throughout the length of the river: shorthead redhorse, Johnny darter, logperch, and walleye, although the status of walleye in the headwaters is uncertain (John Lyons, unpublished data). Thirty-nine species of fishes have been associated with the headwaters, with 12 species occurring either exclusively in the headwaters or occurring rarely below Otter Rapids Dam, approximately 60 Rkm upstream of Rhinelander, Wisconsin. Common species limited to the headwaters are the southern brook lamprey and mottled sculpin. Below Otter Rapids Dam, largemouth bass, smallmouth bass, bluegill, and yellow perch are common throughout the remaining length of the river.

The upper and middle Wisconsin are the most heavily impounded segments of the river and were historically most heavily impacted by pollution from pulp and paper mills and municipal sewage from small communities. The most heavily impacted section of the river is between Rhinelander (Rkm 550) and Petenwell (Rkm 275). Portions of river that still retain flow and riverine habitat are >16 km in length between dams on the upper river, but are usually <2 km long in the middle Wisconsin, where dam effects are more pronounced. Sixty species have been collected through the 352 km length of the upper and middle Wisconsin (excluding the headwaters), with 20 species considered common. In addition to broadly common species already mentioned, species common in this section include white sucker, silver redhorse, black bullhead, and black crappie.

The lower Wisconsin consists of the 133 km freeflowing portion of the river below Prairie du Sac Dam downstream to the Upper Mississippi. With 95 species known to occur in the lower Wisconsin and 40 species considered abundant, this portion of the river by far contains the greatest diversity of fishes. Common large-river fishes include paddlefish (threatened in Wisconsin), mooneye, river shiner, river carpsucker, smallmouth buffalo, gizzard shad, quillback, and freshwater drum.

Information on other aquatic vertebrates is limited to curator lists, which show that seven species of frogs, 11 species of turtles, and one species of snake are found in the Wisconsin River (G. Casper, personal communication). Turtles frequently seen in and along the river include painted turtle, common map turtle, common snapping turtle, and smooth and spiny softshells. Green frogs, mink frogs, and spring peepers are common, and the northern water snake is the lone snake found in proximity to the river. The

wood turtle and Blanding's turtle are listed as threatened in Wisconsin. Muskrat and beaver are also common throughout the Wisconsin River, particularly in quiet waters of sloughs, backwaters, and oxbow lakes.

Ecosystem Processes

Despite being the largest river in Wisconsin and one of the longest tributaries of the Upper Mississippi, there have been no ecosystem-level studies of the Wisconsin. The invertebrate and fish communities both contain representatives of a wide range of trophic guilds, including many generalists. Although filter-feeding invertebrates such as net-spinning caddisflies and black flies are abundant in the lower river, their presence only implies feeding on transported organic matter and not a specific form of organic matter. Algal grazing/scraping invertebrates are also well represented (e.g., heptageniid mayflies), implying, at least for hard substrates, that algal periphyton is available for consumption. No data are available on amounts of particulate organic matter transported in the water column, but it is likely that high concentrations of fine particulate organic matter and dissolved organic matter are transported from upstream sources. Limited data indicate that both periphyton ($14.2 \, mg/m^2$) and phytoplankton (30 to $59 \, \mu g/L$; Lillie and Hilsenhoff 1992) are present in quantities that could support consumers.

Human Impacts and Special Features

At nearly 700 km in length, the Wisconsin River is the longest tributary of the Upper Mississippi River, excluding the Missouri River. It is also the last tributary entering the Upper Mississippi that originates in the Superior Uplands. Although the river has been markedly influenced by human activities, it is still a focal point for recreation and still retains natural features in the floodplain in many areas. It also retains many of the physical and biological features of a large river in its lower reaches, thus making it an important point for comparison in understanding the effects of human activities on large-river ecosystems.

One of the first major impacts on the Wisconsin River was timber harvest. Like the other river basins of the Superior Uplands, the stands of white and red pines were cut throughout the nineteenth century. Mills were built to rough cut timber before it was floated downriver in huge log rafts to finishing mills on the Mississippi River, primarily in St. Louis (Nichols 1984). The amount of timber coming out of the Wisconsin valley was impressive. The largest logjam in Wisconsin history occurred at Grandfather Falls when 2.5×10^7 m of timber created a jam of logs over 6 m high. Low discharge periods slowed the movement of logs downstream; therefore dams with weirs wide enough to allow passage of log rafts were constructed in the upper Wisconsin so that logs could be floated over rapids (Nichols 1984). The timber harvest continued to dwindle into the 1890s, when there was not enough pine left to make it profitable.

The basin has a history of agriculture, beginning with wheat production in the 1850s. Despite this, the effects of agriculture impacts are not dealt with extensively for the Wisconsin River, largely because agriculture is viewed as having less of an impact on the river than flow regulation, municipal sewage, and industrial (pulp mill) effluent. Low nitrogen and phosphorus concentrations found throughout the river suggest that agricultural inputs are low, supporting efforts by managers to focus on other human impacts.

The pulp and paper mill industry and inputs from municipal sewage have had perhaps the greatest impact. This is particularly true for the upper and middle Wisconsin down to Rkm 275, where 16 pulp and paper mills and 14 municipal waste treatment plants were present prior to 1980. Water quality was poor throughout this part of the river, as is evident from comparison of fish and invertebrate studies immediately following enactment of the Clean Water Act to studies performed in later years. The number of fish species has increased, as has abundance over this period (John Lyons, personal communication). Benthic invertebrate diversity and abundance increased from 1972 to 1980, particularly among net-spinning caddisflies and *Hexagenia*. Improvement in water quality is also evident in the diversity of mayflies found in the Wisconsin River (Lillie and Hilsenhoff 1992). Effects of past and present pulp milling, however, are still evident. Mercury present in pulp effluent into the 1960s is still present in bottom sediments and mercury contamination of fishes (common carp and walleye) in the 1980s was similar to levels observed in the early 1970s (Rada et al. 1986).

Forty-seven storage reservoirs for flood control and 27 hydroelectric dams fragment the upper and middle Wisconsin River and its tributaries. In some areas the only riverine habitat is the tailwaters immediately below a dam. Changes in the physical structure are apparent in the diversity of fishes described earlier. Many lentic species are most abundant in the heavily impounded areas, whereas large-river species are not found in abundance until below the Prairie

du Sac Dam (marking the beginning of the free-flowing lower Wisconsin). Water levels in riverine sections fluctuate dramatically, particularly below dams where hydropower peaking is practiced. A bright spot to this is that there have been efforts to remove dams in tributaries. The last dam on the Baraboo River was removed in 2001, letting the river flow freely throughout its 185 km length for the first time in 150 years.

The Wisconsin River is a system that has been profoundly impacted by alteration of hydrological dynamics throughout much of its length as well as within many of its tributaries. In many instances within the upper and middle Wisconsin, lacustrine fauna have replaced lotic species. The Wisconsin also retains the legacy of heavy nonpoint-source pollution from paper mills and municipal sewage, which still has an effect of native fauna even though effluent releases have been drastically reduced over the last 20 years (e.g., Lyons et al. 2000). Still, the headwaters and lower river retain relatively natural conditions that have allowed for the retention of fauna that would be expected to occur within the region, thus serving to maintain some of the ecological integrity of this large river.

ILLINOIS RIVER

The Illinois River, at 439 km in length and draining 44% of the state, is the largest river in Illinois. Formed by the confluence of the Des Plaines and Kankakee rivers 77 km southwest of Chicago, then flowing to the southwest, the Illinois formerly drained 73,038 km^2 that also included part of Wisconsin and Indiana (Fig. 8.19). The drainage area increased to 75,136 km^2 when the Chicago Sanitary and Shipping Canal opened in 1900, reversing the flow of the Chicago and Calumet rivers, thus diverting water from Lake Michigan to the Illinois and pushing the boundaries of the basin even closer to Lake Michigan (Starrett 1971a). The Illinois is typically divided into two sections. The upper Illinois runs from the origin downstream to a section of the river known as the Great Bend (Rkm 338) and includes the Fox, Des Plaines, Kankakee, and Vermillion rivers. The remainder of the river is the lower Illinois and includes, in downstream order, the Mackinaw, Spoon, Sangamon, and La Moine rivers, until it finally reaches the Mississippi (Fig. 8.7).

Like other areas of the Upper Mississippi basin, humans were present in the Illinois valley as early as 8000 to 9000 years ago (Starrett 1971a). Evidence of

FIGURE 8.7 Illinois River (right) and Mississippi River (left) just upstream of their confluence west of Alton, Illinois (PHOTO BY TIM PALMER).

the presence of the Black Sand culture, a group that practiced agriculture, dates from 4500 to 2500 years ago. The Late Woodland Culture (1300 to 700 years ago), as evidenced by their ceremonial mounds, was also present. Immigration by Europeans and Americans into the Illinois country was slow, with settlements not growing appreciably until navigation of the Illinois River began in 1828. Population growth exploded in the late nineteenth century, going from 180,000 people in 1840 to 3.3 million in the Illinois River basin by 1900 as Chicago, Peoria, and other cities in the basin grew.

The natural history of the Illinois River has been studied extensively through the efforts of the Illinois Natural History Survey. Stephen A. Forbes established a biological station near Havana, Illinois, in 1894 that would become a center for the study of the river. Since then, detailed studies of the natural history of many of the biota inhabiting the river have

been conducted. The Survey has also monitored the impacts of pollution on the river and today tracks the Illinois River's recovery through the Illinois River Biological Station in Havana.

Physiography, Climate, and Land Use

The Illinois River lies within the Tilled Plains section of the Central Lowland (CL) physiographic province (see Fig. 8.19). Included within this area is the Central Forest/Grassland Transition terrestrial ecoregion (Ricketts et al. 1999). Impacted by later glacial periods (Illinoian and Wisconsin), the area has low relief and river valleys are not extensively incised. Glacial drift is more weathered than glaciated areas to the north. Loess is also found throughout the basin, with the thickness of this layer diminishing progressively east of the Upper Mississippi River (Martin 1932). Underlying the loess and drift are Paleozoic shales and limestones (Nielsen et al. 1984). Coal deposits are also present near the surface, primarily where loess deposits are thin.

The upper Illinois basin historically was tallgrass prairie with big bluestem, Indiangrass, and switchgrass as the dominant vegetation. Although prairie areas were found sporadically in the lower basin, areas along waterways and river and stream valleys were primarily oak, hickory, and maple forests. Some forest areas remain within the basin, as do protected prairies; however, much of the basin has been converted to agriculture.

Climate in the region is described as humid continental. Average annual air temperature varies from 12.4°C near the confluence with the Upper Mississippi to 10.3°C near the origin. Air temperature basinwide ranges from a daily average of −5.8°C in January to 24°C in July (Fig. 8.20). Precipitation ranges from 91.5 cm/yr to 97.9 cm/yr throughout the Illinois basin. Snow is not as significant as in the northern tributaries of the Upper Mississippi, with 65 cm/yr in the upper Illinois basin and 35.6 cm/yr to the southwest. Precipitation exceeds 6 cm/mo during every month except January and February but is highest from April through September (see Fig. 8.20).

Agriculture covers approximately 87% of the basin, with corn and soybeans the primary commercial crops. Approximately 7% is forested, mostly in and along the river and stream valleys, and 5% of the basin is urbanized. With five urban areas with populations >100,000 people (Chicago, Peoria, Bloomington–Normal, Springfield, and Decatur), the population density is 97 people/km^2.

River Geomorphology, Hydrology, and Chemistry

Like the Upper Mississippi's tributaries to the north, the channel features of the Illinois were shaped by glacial processes. The upper Illinois is geologically younger than the lower river and does not flow through a preglacial channel. About 4500 years ago the moraine holding back lakes to the north and east gave way, releasing a flood of water known as the Kankakee Torrent, carving a channel to the present lower Illinois. The upper Illinois has carved a channel as much as 2 km wide through the younger glacial drift. In some areas the river has cut down to rock (Martin 1932). Gradient in the upper Illinois is 20 cm/km. The lower river from Hennepin, Illinois, downstream to the mouth, flows through a preglacial channel previously occupied by the Upper Mississippi River. The Upper Mississippi River flowed through this channel until one of the Wisconsin glacial advances pushed the Mississippi west into its present valley (Martin 1932). Flow in the Illinois decreased following the final retreat of the Wisconsin glacier and the valley was partially filled with sand and gravel glacial outwash. The river carved a trench through the outwash 45 to 77 m deep and, in many places, 8 to 11 km wide. The gradient of the lower Illinois is extremely low, only 2 cm/km.

The Illinois River is an aggrading system, depositing silt and sand along the shoreline, building low natural levees, and creating a broad marshy plain, at least under natural conditions (Martin 1932, Sparks 1984). Substrates in the lower Illinois are primarily sand, silt, and some gravel. The river receives considerable sediment inputs from tributaries draining the Till Plains. Alluvial fans are often evident at a tributary mouth, causing the channel of the Illinois to migrate away from the fan. Peoria Lake is a channel lake created by tributary deposition in the Illinois River channel.

The Kankakee River contributes an annual discharge of 126.8 m³/s to the upper Illinois. The Des Plaines, through inputs from the Chicago Sanitary and Ship Canal, contributes an annual average of 100 m³/s. Annual discharge is 306 m³/s when the Illinois reaches Marseilles, 51 Rkm downstream. The Illinois better than doubles in size to 649 m³/s by the time it reaches Valley City, approximately 90 Rkm upstream of the confluence with the Upper Mississippi River. Discharge is measured at Valley City because the water level of the Upper Mississippi sometimes affects flow in the Illinois River downstream of Valley City (Groschen et al. 2000).

Discharge is typically lowest in September, as precipitation is low in August and evapotranspiration remains high into September (see Fig. 8.20). Discharge increases through the fall and winter as rates of evapotranspiration decrease. Discharge within a given year can be highly variable and is dependent on the release of water through the low-head dams, particularly the Peoria Lock and Dam (Rkm 254) and the La Grange Lock and Dam (Rkm 129) in the lower Illinois. The lack of coordination between the locks and dams creates pronounced fluctuations in discharge locally to the extent that hydrographs have different shapes from one gaging station to the next.

Most physicochemical characteristics are similar throughout the Illinois. Average annual water temperature is around 16°C, with a minimum of 0°C and a maximum of 32°C. Alkalinity averages near 275 mg/L as $CaCO_3$ and pH averages 7.8. Conductivity is high, averaging around 740 μS/cm, and hardness averages around 280 mg/L as $CaCO_3$ in both sections of the river. Nitrate-N and total phosphorus concentrations are high throughout the river, ranging from 0.24 to 9.3 mg/L and 0.04 to 0.86 mg/L, respectively. Suspended sediment concentrations differ substantially, averaging 82 mg/L in the upper river compared to 337 mg/L in the lower river. This is most likely a reflection of geology, with the lower river receiving waters from a basin with more weathered drift and abundant fine loess.

River Biodiversity and Ecology

Algae

There have been no recent assessments of the phytoplankton assemblage for the entire Illinois, and benthic algae have only been reviewed in the upper Illinois, where community composition suggests mesoeutrophic or eutrophic conditions (Leland and Porter 2000). Common taxa among the green algae included *Cladophora*, *Sphaerocystis*, and *Oocystis*. *Navicula*, *Achnanthes*, and *Nitzshia* were the diatoms found at the greatest number of sample sites in the upper Illinois.

Plants

Presettlement floodplain vegetation in the upper Illinois consisted of oak–hickory forest with patches of bluestem prairie. Only remnants of the prairie remain, as the fertile soils of the prairie were converted to agriculture beginning in the mid-nineteenth century (Arnold et al. 1999). Floodplain vegetation prior to settlement in the lower Illinois was a mixture of codominant species represented by hackberry, pecan, American elm, silver maple, and oaks. Prairies and savannas were also present (Nelson and Sparks 1997). Diversion through the Chicago Canal in 1900 and the opening of the lock and dam system each raised water levels, inundating wide tracts of floodplain and eliminating many species from the floodplain. Fire control and agriculture had a profound impact on grassland vegetation in both the floodplain and the river basin (Nelson and Sparks 1997). Typical floodplain vegetation today is dominated by silver maple, in addition to ash, hackberry, box elder, and willow.

Sporadic water fluctuations of up to 1 m from dam regulation limit the abundance of aquatic plants in the main channel of the Illinois River, but patches of submergent sago and horny pondweed, wild celery, and coontail can be found in stable areas of the main channel. Aquatic plants are considerably more abundant in backwaters and floodplain lakes, particularly in the upper Illinois, where 11 species of submerged plants have been observed. The most common submerged aquatic plants are coontail, curly-leaved pondweed, the nonnative Eurasian watermilfoil, and longleaf pondweed. Bulrush is among the most common of the emergent aquatic plants, and American waterlily is a common floating-leaved plant. Another floating-leaved plant, the nonnative water hyacinth, appears to be increasing in abundance in the upper Illinois (T. Cook, personal communication). Duckweed is also common to backwaters and floodplain lakes.

Invertebrates

Published accounts of benthic invertebrates in the upper Illinois River consist primarily of freshwater mussel surveys. Surveys of the Illinois from the confluence of the Des Plaines and Kankakee rivers to Rkm 372 between 1870 and 1900 found 38 species of mussels. A survey in 1912 found only two species, whereas a survey conducted in 1966 to 1969 found no living mussels (Starrett 1971b). Sphaeriidae, which are often used as a water-quality indicator in large rivers, are also absent from the area (Sparks 1984) even though water quality has improved. Fortunately, the Kankakee River contains a diverse and healthy assemblage of mussels (Page et al. 1992); therefore there is a potential for recruitment. Long-term pollution of bottom sediments, however, may limit the ability of invertebrates to recolonize the area (Sparks 1984).

Benthic invertebrates in the lower Illinois are representative of typical large-river fauna. Filter-feeding caddisflies (*Potamyia*, *Hydropsyche*, *Cyrnellus*) are present on hard substrates (Seagle et al. 1982). Oligochaetes and midges can also be found on hard substrates, although they are typically more abundant in nearshore areas. Main-channel midge (*Robackia* and *Rheosmittia*) and oligochaete (*Barbidrilus*) sand specialists can be abundant in the main channel (Dettmars et al. 2001). Sand–silt is the predominant substrate in the lower Illinois, with snags serving as the only natural hard substrate. Water-level fluctuations created by the dams, however, appear to clear woody debris or leave it above the water level. Siltation of hard substrates also reduces invertebrate abundance and diversity (Seagle et al. 1982). Collection records indicate that other large river mayflies (e.g., *Stenonema*, *Stenacron*, *Caenis*, *Hexagenia*) are found in the lower Illinois River (Burks 1953).

The fate of freshwater mussels in the lower Illinois River was no better than their counterparts in the main-stem upper Illinois. Pollution and habitat alteration through modification of flow are thought to be the primary reasons for their decline. Thirty-eight species were found in the lower river in 1870. This number dropped to 14 species by 1966 (Cummings 1991). The current outlook for freshwater mussels throughout the length of the Illinois River is poor (Cummings 1991, Page et al. 1992). Only 23 of the 47 species known to occur in the Illinois River were found in 1967. All of these species found in the 1967 survey were common to other rivers, and no state or federally endangered species were found living. The state of Illinois lists four species as threatened, eight as endangered (including two federally listed endangered species, Higgin's eye and fat pocketbook) that were historically found in the Illinois River.

Vertebrates

Fishes in the upper river were dealt a similar fate to that of mussels following the opening of the Chicago Canal. Increasing sewage waste caused oxygen levels to fall to near zero, and fishes were practically eliminated in the upper Illinois (Arnold et al. 1999). The river gradually improved to the point where nonnative common carp and goldfish were abundant and, still later, native fishes returned. Recent surveys identified 74 species of fishes in the upper Illinois, including 24 species of Cyprinidae (minnows and carp), 11 species of Catostomidae (suckers), 10 species of Centrarchidae (sunfishes), and 8 Percidae (perches). Green sunfish, bluntnose minnow, common carp, and white sucker were the most abundant fishes (Arnold et al. 1999). Although native fishes have returned to the Illinois, the low number of pollution-intolerant fishes, such as the Percidae, and the types of abundant fishes present suggest that water-quality problems still persist.

There are 131 species of fishes known for the lower Illinois, including 37 species of Cyprinidae, 17 species of Catostomidae, 15 species of Centrarchidae, and 14 species of Percidae. Large-river specialists are well represented in the lower Illinois, but there also appears to be more overlap into habitats with lower current velocities than is seen in the Upper Mississippi. Side channels, as is the case in the Upper Mississippi, harbor several species that are typically viewed as specialized for lotic or lentic habitats. White bass and channel catfish are common in all three habitats (Burkhardt et al. 1998). Gizzard shad, emerald shiners, and freshwater drum are common fish species in nearshore areas of the main channel. Another fish abundant in the main channel is the skipjack herring, which is rare or considered extirpated in the Upper Mississippi and tributaries above the Illinois River. The deep main channel is viewed as an area used by fishes only to move from one habitat to another, with no more time spent in the high current velocity than necessary. Trawl samples of the lower Illinois show that over 50% of the fishes collected in the main channel are present throughout much of the year, suggesting they are spending more time in the main flow and are feeding there (Dettmers et al. 2001). Common fishes in trawls of the main flow are freshwater drum, gizzard shad, and channel catfish. Fishes abundant in backwater habitats include bluegill, largemouth bass, common carp, smallmouth buffalo, and western mosquitofish (Burkhardt et al. 1998). Many of these fishes are also associated with nearshore areas of the side channel.

No federally endangered fishes are found in the Illinois River, but there are 11 fishes listed as threatened and 5 listed as endangered by the state of Illinois. Of these, 13 species are thought to be extirpated from the Illinois River (Page et al. 1992). The pallid shiner, river redhorse, and greater redhorse have been found in the river.

Collection lists indicate there are nine species of frogs, six species of turtles, seven species of salamanders, and two species of snakes in the upper Illinois. Amphibian and reptile diversity drops almost immediately below the confluence of the Des Plaines and Kankakee rivers and remains low into the upper sections of the lower Illinois. As few as three species of frogs, three species of turtles, and only one species of salamander are shown on county species lists from Rkm 436 downstream to the vicinity of Rkm 305. The number of species begins to increase slightly downstream, but an appreciable change is not seen until the river reaches Rkm 216. Additional species reappear downstream or appear for the first time to the point that 12 species of frogs, 11 species of turtles, 5 species of salamanders, and 4 species of snakes can be found throughout the lower Illinois River. Common among these are the common map turtle, painted turtle, smooth softshell turtle, slider, treefrogs, southern leopard frog, and northern water snake. The alligator snapping turtle is listed as state endangered and Blanding's turtle is listed as threatened.

Ecosystem Processes

As a river–floodplain ecosystem, functional processes within the Illinois River were dependent on periodic and prolonged inundation of the floodplain and the creation of aquatic linkages of the main river to backwaters and floodplain lakes. Historically, floodplain inundation and aquatic habitat connectivity would last as long as 6 months (Sparks 1995). This prolonged period would allow riverine organisms, including benthic invertebrates and zooplankton, access to high levels of primary productivity in shallow floodplain lakes and backwaters. These areas also served as spawning grounds and nurseries for several species of fishes. Inundated areas also provided food resources for migrating waterfowl, shorebirds, and bald eagles. Later in the year, after water withdrew from the floodplain, moist soil vegetation would grow in the newly deposited nutrient-rich sediments, creating resources for waterfowl and shorebirds as they migrated south in the fall. Much of this dynamic, as discussed later, has been lost due to disruption of annual hydrological patterns and the construction of levees that cut the river off from the floodplain.

Although most recent studies have dealt with assessing the current status of the ecological condition of the Illinois River and life histories of flora and fauna, some studies have examined trophic dynamics in the Illinois and provide an indication of current ecosystem dynamics. A study focusing on the main channel of the lower Illinois noted that 14 of 26 fish species collected were omnivorous and 11 species were insectivores (Dettmers et al. 2001). Only one primarily piscivorous species was found. Gut contents of channel catfish and freshwater drum consisted primarily of larval and pupal Chironomidae and Hydropsychidae. Zooplankton are abundant in both backwaters and the main channel, with main-channel densities averaging around 80 individuals/L. The abundant zooplankton provide an ample food source for larval and adult planktivorous fishes such as gizzard shad. Dettmers et al. (2001) concluded that both riverine primary production and organic matter from terrestrial sources are probably important to the food web of the lower Illinois. The most abundant benthic invertebrates, midges and oligochaetes, are viewed as primarily detritivorous in nature, feeding on detritus deposited in the bottom sediments. Filter-feeding caddisflies are considered omnivorous, consuming nearly all materials caught in their nets. It is assumed, therefore, that terrestrial organic matter derived from the floodplain is important for benthic-based trophic processes and instream production would be important to trophic groups associated with the phytoplankton–zooplankton assemblages. Phytoplankton counts averaging 21,100 cells/mL, with some samples as high as 150,000 cells/mL, indicate that phytoplankton are abundant, thus accounting, at least, for the abundance of zooplankton.

Human Impacts and Special Features

The Illinois River, prior to settlement, discharged about the same amount of water into the Upper Mississippi River as the Rock River located to the north (Richard Sparks, personal communication). Annual flow patterns, however, were much more like those of a large floodplain river because of the nature of the river valley of the lower Illinois. The river valley carved out by the ancestral Mississippi River is 8 to 11 km wide in some areas and the floodplain is very flat. The combination of a broad, flat floodplain and low slope of the river resulted in water remaining on the floodplain for much longer than would be expected for a river of this size. Prolonged inundation of the floodplain also allowed the river to scour out backwaters and floodplain lakes, where many species of fishes could be found feeding and spawning. These conditions resulted in the Illinois River having a commercial fishery that consistently ranked at or near the top nationally. This structural diversity also contributed to the presence of a diverse assemblage of fishes, including lotic and lentic specialists, and a high diversity of freshwater mussels. Much of this has changed, however, due to the changes inflicted on the system by humans.

The effects of population growth were probably the first human impacts to be felt in the Illinois River. The Chicago Sanitary and Ship Canal was opened in 1900 to divert flow from the Chicago River into the Des Plaines River to protect Chicago's water supply (Starrett 1971a). This wedge of pollution continued downstream over the years and was compounded by population growth on the Illinois River proper and a major tributary, the Sangamon River. The effects throughout the river are evident from the demise of freshwater mussels described earlier. Although the river is significantly cleaner than before through improved waste treatment practices, problems do linger, partly from unknown, and possibly irreversible, effects of earlier pollution (Sparks 1984) and because of continued population growth. There are still 196 wastewater treatment plants in the upper Illinois, including the Kankakee and Des Plaines basins (Arnold et al. 1999).

Current nutrient concerns relate more to agricultural activities. Approximately 54% of the nutrient load to the Illinois River in the 1980s was from nonpoint sources, whereas 46% came from point sources. More recent studies illustrate that nutrient inputs from agriculture are a continuing concern. All of the study sites in the lower Illinois NAWQA study were in the highest 10% nationally for NO_3-N, whereas total phosphorus concentrations ranked in the top 25% nationally at seven of eight sites on the lower Illinois River (Groschen et al. 2000). Pesticides are also a major water-quality concern. Concentrations of atrazine and cyanazine sometimes exceed U.S. Environmental Protection Agency guidelines (Groschen et al. 2000). Increased suspended sediment transport is contributing to habitat loss by filling backwaters and floodplain lakes and increasing turbidity in these highly productive habitats.

Hydrological alterations and flood control have also played a part in reducing habitat and species diversity. The opening of the Chicago Canal raised water levels throughout the length of the river, eliminating some aquatic habitats. Water levels were raised further when five navigation dams were opened in the 1930s to maintain a 2.75 m navigation channel. The effects of the dams compounded the habitat loss already begun by the opening of the Chicago Canal. Water levels fluctuate unpredictably throughout the year as a result of current management of releases through the dams. Many areas of the river are still connected to the floodplain, but levees have been built throughout the river valley and the land behind the levees has been drained for farming. A consequence of this is that the river can no longer deposit sediments across the breadth of the floodplain. Sediment deposition between the levees and in the channel has raised the level of the river to the point that, in some areas, the river is perched above the surrounding floodplain. Many efforts have been initiated to restore some of the natural qualities of the Illinois River, particularly in the floodplain. Over 20 state and federal areas have been established as fish and wildlife refuges or as management areas to improve the quality of fish and wildlife habitat. The Nature Conservancy of Illinois has purchased areas from private landowners with plans to restore connectivity between these former floodplain habitats and the river.

The opening of the Chicago Canal has made the Illinois and the rest of the Mississippi River susceptible to invasion by nonnative species established in the Great Lakes. This is the path through which zebra mussels entered the Illinois and moved throughout the entire Mississippi system. Several nonnative species have moved into the Illinois from the Mississippi River, including common carp, grass carp, bighead carp, silver carp, and the cladoceran zooplankter *Daphnia lumholtzi*.

Despite the extensive effects of human activity in the Illinois River, there are efforts to restore this river–floodplain system. The Illinois Natural History Survey, largely through the Illinois River Biological Station in Havana, continues to study the dynamics of the river. The Survey has an extensive history on the Illinois River, making it possible to compare the current status of the river to past conditions and identify positive and negative attributes. This is bolstered by the Survey's affiliation with the Long Term Resource Monitoring Program (LTRMP), a monitoring program established in the Upper Mississippi and Illinois rivers to access the health and stability of the ecosystem. There are also efforts to restore river–floodplain connectivity. This has been primarily through the efforts of The Nature Conservancy, which has purchased two areas formerly used for crop production with plans to reestablish linkages between these lands and the Illinois River. The goal is to recreate some of the structural diversity that once contributed to the tremendous productivity of this river–floodplain ecosystem.

ADDITIONAL RIVERS

The Chippewa River of Wisconsin, like the St. Croix and Wisconsin rivers, begins in the Superior Upland physiographic province and ends in the Central Lowland (Fig. 8.21, Fig. 8.1). The Chippewa River transports a tremendous volume of sand from the deposits left behind by the Wisconsin glaciation. Deposition of sand at the mouth of the Chippewa

River as it empties into the Upper Mississippi River led to the formation of Lake Pepin, which marks one of the widest areas of the entire Mississippi River. Common to many rivers in the upper Midwest, the Chippewa has been dammed extensively for hydroelectric production. Many of these facilities practice hydropower peaking, where water is held back for most of the day and is then released in large volumes. Discharge during the summer at the Dells Dam in Eau Claire is held at $14\,m^3/s$, then raised to as high as $142\,m^3/s$ in less than 2 hours when electricity is needed. Over 125 dams are identified on the Chippewa and its tributaries, although many of these are for small storage reservoirs.

The Root River is a small system with a drainage basin of only $4325\,km^2$ in southeastern Minnesota (Fig. 8.23), but it is an important representative of the rivers entering the Upper Mississippi from the Driftless Area. The Driftless Area was bypassed during the glacial events of the ice age, leaving behind a terrain highlighted by bluffs where the streams have carved out deep valleys known as coulees. The geology of this region creates numerous pockets of groundwater, and it is these groundwater seeps that feed the many streams and small rivers of the region to create a complex of coldwater and coolwater streams. The Root is also typical of other rivers flowing through this limestone-dominated terrain in that it is a hardwater system with alkalinity >200 mg/L as $CaCO_3$ and pH >8. Registered on the National River Inventory as a scenic river, the river is impacted by corn and soybean production and the presence of livestock concentrated into small feedlots.

The Wapsipinicon River, the first river after the Minnesota that flows almost entirely through the Central Tall Grasslands terrestrial ecoregion, has a narrow drainage basin that extends from its confluence with the Upper Mississippi River in central Iowa up into southern Minnesota (Fig. 8.25). One of the first of the larger streams to flow through the Corn Belt, it is distinct from other rivers draining this region, such as the Iowa–Skunk River system and the Des Moines River, in that it does not carry as high a nutrient load (NO_3-N average concentration is <5 mg/L). It also is one of the few rivers in Iowa that still possesses mature stands of riparian and floodplain forest (Fig. 8.8). The main-stem Wapsipinicon is free-flowing, with the only impoundments in the upper tributaries.

The basin of the Rock River traverses the Central Forest/Grassland and Upper Midwest Forest Savanna Transition Zone terrestrial ecoregions as it flows

FIGURE 8.8 Wapsipinicon River north of Tripoli, Iowa (Photo by S. Porter, U.S. Geological Survey).

FIGURE 8.9 Rock River north of Watertown, Dodge County, Wisconsin (Photo by S. Wade).

south from central Wisconsin (Fig. 8.9) through northern Illinois to its confluence with the Upper Mississippi (Fig. 8.27). The diversity of the terrain it covers is reflected in its biological diversity, with 115 species of fishes, ranging from coldwater to warm-water species. The Rock River has been extensively impounded, with 19 hydroelectric dams on the main stem and over 250 small dams, primarily for water storage, on its tributaries. Urbanization is a concern for the Rock River, as the metropolitan areas

FIGURE 8.10 Des Moines River near Des Moines, Iowa (PHOTO BY TIM PALMER).

of Madison, Wisconsin, and Rockford, Illinois, lie within the basin.

The Des Moines River is the largest of the rivers draining the Central Tall Grasslands terrestrial ecoregion to empty into the Upper Mississippi River (Fig. 8.29, Fig. 8.10). Characteristic of rivers draining the Corn Belt, nonpoint-source pollution is a major concern in the basin. Concentrations of NO_3-N average around 6 mg/L throughout the basin. Pesticide concentrations in fish tissues are low but are still considered a point of concern. Sediment loads are also high, with 6.3 million kg/d delivered to Red Rock Lake, one of two large reservoirs on the Des Moines River. Gas supersaturation–induced gas bubble trauma has been blamed for 16 fish kills below Red Rock Dam since 1983. Habitat loss due primarily to the impoundments has caused declines in the abundance of aquatic vertebrates, especially turtles.

The Kaskaskia River is the last major tributary to empty into the Upper Mississippi River before the confluence of the Ohio River (Fig. 8.31). Flowing south through central and southwestern Illinois, the river's basin lies entirely within the Temperate Decid-uous Forest biome. The largest contiguous tract of forest, including bottomland, remaining in Illinois lies within the Kaskaskia River basin. The forest includes pecan, sugar maple, bur oak, pin oak, and shellbark hickory. Nearly 40% of the vascular plants known in Illinois can be found in the Kaskaskia River basin. There are 107 dams in the basin, located primarily on the tributaries. Two large reservoirs are on the main-stem Kaskaskia, including Lake Shelbyville, a 4452 ha reservoir that receives heavy recreational use, particularly from Champaign, located in the northern end of the basin. There is a lock and dam near the mouth that forms a 21 km long pool. The last 58 km of the river have been channelized for navigation. Primary pollution concerns in the basin are nutrients and siltation from urban and agricultural runoff.

ACKNOWLEDGMENTS

I would like to thank the following people for their assistance and expertise: Terry Balding, Heath Benike, Michael Bohn, Gary Casper, Thad Cook, Kevin Cummings, Cal

Fremling, Marian Havlik, David Heath, Randolph Hoffman, Daryl Howell, Robert Hrabik, Dan Kelner, Maria Lemke, Paul La Liberte, David Lonzarich, Gregory Milton, Gary Montz, Neal Mundahl, Dan North, Mark Pegg, Clay Pierce, Robin Richardson, Joseph Rohrer, Steven Russell, Konrad Schmidt, Trent Thomas, and Scott Whitney. I am especially grateful to John Lyons and Dick Lillie for the wealth of information provided for the Wisconsin River and Don Fago and Byron Karns for information on the St. Croix.

LITERATURE CITED

Abell, R. A., D. M. Olson, E. Dinerstein, P. T. Hurley, J. T. Diggs, W. Eichbaum, S. Walters, W. Wettengel, T. Allnutt, C. J. Loucks, and P. Hedao. 2000. *Freshwater ecoregions of North America: A conservation assessment*. Island Press, Washington, D.C.

Arnold, T. L., D. J. Sullivan, M. A. Harris, F. A. Fitzpatrick, B. C. Scudder, P. M. Ruhl, D. W. Hanchar, and J. S. Stewart. 1999. Environmental setting of the upper Illinois River Basin and implications for water quality: U.S. Geological Survey Water-Resources Investigations Report 98-4268. Urbana, Illinois.

Bailey, P. A., J. W. Enblom, S. P. Hanson, P. A. Renard, and K. Schmidt. 1994. *A fish community analysis of the Minnesota River basin*. Minnesota Pollution Control Agency, St. Paul.

Brigham, A. R., and E. M. Sadorf. 2001. Benthic invertebrate assemblages and their relation to physical and chemical characteristics of streams in the Eastern Iowa Basins, 1996–1998. Water-Resources Investigations Report 00-4256. National Water Quality Assessment Program, U.S. Geological Survey.

Bright, R. C., C. Gatenby, D. Olson, and E. Plummer. 1990. *A survey of the mussels of the Minnesota River, 1989*. Bell Museum of Natural History, Minneapolis, Minnesota.

Burkhardt, R. W., M. Stopyro, E. Kramer, A. Bartels, M. C. Bowler, F. E. Cronin, D. W. Soergel, M. D. Peterson, D. P. Herzog, T. M. O'Hara, and K. S. Irons. 1998. 1997 annual status report: A summary of fish data in six reaches of the Upper Mississippi River System. U.S. Geological Survey, Program Report LTRMP 98-P008. LaCrosse, Wisconsin.

Burks, B. D. 1953. The mayflies, or Ephemeroptera, of Illinois. *Bulletin of the Illinois Natural History Survey* 26:1–216.

Burr, B. M., and L. M. Page. 1986. Zoogeography of the fishes of the Lower Ohio–Upper Mississippi basin. In C. H. Hocutt and E. O. Wiley (eds.). *The zoogeography of North American freshwater fishes*, pp. 290–324. John Wiley and Sons, New York.

Cummings, K. S. 1991. Freshwater mussels of the Illinois River: Past, present, and future. Proceedings of the 3rd biennial Governor's Conference on the Management of the Illinois River System, Peoria.

Delong, M. D., J. H. Thorp, K. S. Greenwood, and M. C. Miller. 2001. Responses of consumers and food resources to a high magnitude, unpredicted flood in the upper Mississippi River Basin. *Regulated Rivers: Research and Management* 17:217–234.

Dettmers, J. M., D. H. Wahl, D. A. Soluk, and S. Gutreuter 2001. Life in the fast lane: Fish and foodweb structure in the main channel of large rivers. *Journal of the North American Benthological Society* 20:255–265.

Downing et al. 1999.

Fago, D. 1992. *Distribution and relative abundance of fishes in Wisconsin*. Vol. 7: *Summary report*. Technical Bulletin 175. Wisconsin Department of Natural Resources, Madison.

Fago, D., and J. Hatch. 1993. Aquatic resources of the St. Croix basin. In L. W. Hesse, C. B. Stalnaker, N. G. Benson, and J. R. Zuboy (eds.). *Proceedings of the Symposium on Restoration Planning of the Mississippi River Ecosystem*, pp. 23–56. U.S. Department of the Interior, Washington, D.C.

Fremling, C. R., J. L. Rasmussen, R. E. Sparks, S. P. Cobb, C. F. Bryan, and T. O. Claflin. 1989. Mississippi River fisheries: A case history. In D. P. Dodge (ed.). *Proceedings of the International Large River Symposium (LARS)*. Canadian Special Publication of Fisheries and Aquatic Sciences 106, pp. 309–351.

Groschen, G. E., M. A. Harris, R. B. King, P. J. Terrio, and K. L. Warner. 2000. Water quality in the lower Illinois River basin, Illinois, 1995–1998. Circular 1209. U.S. Geological Survey, Denver, Colorado.

Grubaugh, J. W., and R. V. Anderson. 1989. Upper Mississippi River: Seasonal and floodplain forest influences on organic matter transport. *Hydrobiologia* 174:235–244.

Havlik, M. E., and J. S. Sauer. 2000. Native freshwater mussels of the Upper Mississippi River system. LTRMP Report PSR 2000-04. U.S. Geological Survey, La Crosse, Wisconsin.

Hornbach, D. J. 2001. Macrohabitat factors influencing the distribution of naiads in the St. Croix River, Minnesota and Wisconsin, USA. In G. Bauer and K. Wächtler (eds.). *Ecology and evolution of the freshwater mussels Unionoida*, pp. 213–230. Springer-Verlag, Berlin.

Hunt, C. B. 1967. *Physiography of the United States*. W. H. Freeman, San Francisco.

Keeny, D. R., and T. H. DeLuca. 1993. Des Moines River nitrate in relation to watershed agricultural practices: 1945 versus 1980s. *Journal of Environmental Quality* 22:267–272.

Kendall, C., S. R. Silva, and V. J. Kelly. 2001. Carbon and nitrogen isotopic compositions of particulate organic matter in four large river systems across the United States. *Hydrological Processes* 15:1301–1346.

Leland, H. V., and S. D. Porter. 2000. Distribution of benthic algae in the upper Illinois River basin in relation to geology and land use. *Freshwater Biology* 44:279–301.

Lillie, R. A. 1995. A survey of rare and endangered mayflies of selected rivers of Wisconsin. Research Report no. 170. Wisconsin Department of Natural Resources, Madison.

Lillie, R. A., and W. L. Hilsenhoff. 1992. A survey of the aquatic insects of the lower Wisconsin River, 1985–1986, with notes on distribution and habitat. Technical Report no. 178. Wisconsin Department of Natural Resources, Madison.

Lutz, D. S., S. J. Eggers, and R. L. Esser. 2001. Annual report, water quality studies: Red Rock and Saylorville Reservoirs, Des Moines River, Iowa. Engineering Research Institute, Annual Report, ISU-ERI-Ames-01336. Iowa State University, Ames.

Lyons, J., P. A. Cochran, and D. Fago. 2000. *Wisconsin fishes 2000: Status and distribution.* University of Wisconsin Sea Grant, Madison.

Martin, L. 1932. The physical geography of Wisconsin. Bulletin no. 36, Educational Series no. 4. Wisconsin Geological and Natural History Survey, Madison.

National Park Service. 1997. *General assessment plan, environmental assessment: Upper St. Croix and Namekagon rivers.* U.S. Department of Interior, Denver, Colorado.

Nelson, J. C., and R. E. Sparks. 1997. Forest compositional change at the confluence of the Illinois and Mississippi Rivers. *Transactions of the Illinois State Academy of Science* 91:33–46.

Nichols, S. 1984. *A voyageur's guide to the lower Wisconsin River.* Wisconsin Geological and Natural History Survey, Madison.

Nielsen, D. N., R. G. Rada, and M. M. Smart. 1984. Sediments of the Upper Mississippi River: Their sources, distribution, and characteristics. In J. G. Wiener, R. V. Anderson, and D. R. McConville (eds.). *Contaminants in the Upper Mississippi River: Proceedings of the 15th annual meeting of the Mississippi River Research Consortium, La Crosse, Wisconsin,* pp. 67–98. Butterworth, Boston.

Page, L. M., K. S. Cummings, C. A. Mayer, S. L. Post, and M. E. Retzer. 1992. *Biologically significant Illinois streams: An evaluation of the streams of the Illinois based on aquatic diversity.* Illinois Natural History Survey, Champaign.

Patrick, R. 1998. *Rivers of the United States.* Vol. 4, Part A: *The Mississippi River and tributaries north of St. Louis.* John Wiley & Sons, New York.

Payne, G. A. 1994. Sources and transport of sediment, nutrients, and oxygen-demanding substances in the Minnesota River basin, 1989–1992. Minnesota Pollution Control Agency Report, Mounds View.

Rada, R. G., J. E. Findley, and J. G. Weiner. 1986. Environmental fate of mercury discharged into the Upper

Wisconsin River. *Water, Air, and Soil Pollution* 29: 57–76.

Ricketts, T. H., E. Dinerstein, D. M. Olson, C. J. Loucks, W. Eichbaum, D. DellaSala, K. Kavanaugh, P. Hedao, P. T. Hurley, K. M. Carney, R. Abell, and S. Walters. 1999. *Terrestrial ecoregions of North America: A conservation assessment.* Island Press, Washington, D.C.

Rose, W. J. 1992. Sediment transport, particle sizes, and loads in lower reaches of the Chippewa, Black, and Wisconsin rivers in western Wisconsin. Water-Resources Investigations Report 90-4124. U.S. Geological Survey, Madison, Wisconsin.

Seagle, H. H., Jr., J. C. Hutton, and K. S. Lubinski. 1982. A comparison of benthic invertebrate community composition in the Mississippi and Illinois rivers, pool 26. *Journal of Freshwater Ecology* 1:637–650.

Sparks, R. E. 1984. The role of contaminants in the decline of the Illinois River: Implications for the Upper Mississippi. In J. G. Wiener, R. V. Anderson, and D. R. McConville (eds.). *Contaminants in the Upper Mississippi River: Proceedings of the 15th annual meeting of the Mississippi River Research Consortium,* La Crosse, Wisconsin. pp. 25–66. Butterworth, Boston.

Sparks, R. E. 1995. Need for ecosystem management of large rivers and their floodplains. *Bioscience* 45:168–182.

Stark, J. R., P. E. Hanson, R. M. Goldstein, J. D. Fallon, A. L. Fong, K. E. Lee, S. E. Kroening, and W. J. Andrews. 2001. Water quality in the Upper Mississippi River Basin, Minnesota and Wisconsin, South Dakota, Iowa, and North Dakota, 1995–1998. Summary Circular 1211. U.S. Geological Survey, Reston, Virginia.

Starrett, W. C. 1971a. Man and the Illinois River. In R. T. Oglesby, C. A. Carlson, and J. A. McCann (eds.). *River ecology and the impact of man,* pp. 131–169. Academic Press, New York.

Starrett, W. C. 1971b. A survey of the mussels (Unionacea) of the Illinois River: A polluted stream. *Illinois Natural History Survey Bulletin* 30:268–403.

Sullivan, D. J. 2000. Fish communities and their relation to environmental factors in the Eastern Iowa Basins in Iowa and Minnesota, 1996. Water-Resources Investigations Report 00-4194. National Water Quality Assessment Program, U.S. Geological Survey, Reston, Virginia.

Voss, K., and B. S. Beaster. 2001. The state of the lower Chippewa basin. Publication #WT554-00. Wisconsin Department of Natural Resources, Eau Claire.

Waters, T. F. 1977. *The streams and rivers of Minnesota.* University of Minnesota Press, Minneapolis.

Zischke, J. A., G. Erickson, D. Waller, and R. Bellig. 1994. *Analysis of benthic macroinvertebrate communities in the Minnesota River watershed.* Minnesota Pollution Control Agency, St. Paul.

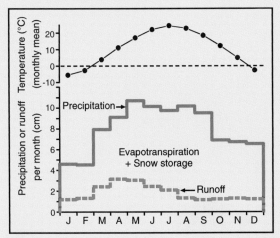

FIGURE 8.12 Mean monthly air temperature, precipitation, and runoff for the Upper Mississippi River basin.

UPPER MISSISSIPPI RIVER

Relief: 337 m
Basin area (excluding Missouri River): 489,510 km^2
Mean discharge (excluding Missouri River): 3576 m^3/s
River order: 10
Mean annual precipitation: 96 cm
Mean air temperature: 10.5°C
Mean water temperature: 14.3°C
Physiographic provinces: Central Lowland (CL),
    Superior Upland (SU), Ozark Plateau (OZ)
Biomes: Boreal Forest, Temperate Deciduous Forest,
    Temperate Grassland
Freshwater ecoregion: Mississippi
Terrestrial ecoregions: Western Great Lakes Forests,
    Upper Midwest Forest/Savanna Transition, Central
    Tall Grasslands, Central Forest/Grassland
    Transition, Central U.S. Hardwood Forests
Number of fish species: 145
Number of endangered species: 7 fishes, 3 mussels
Major fishes: carp, smallmouth buffalo, shorthead
    redhorse, river redhorse, gizzard shad, emerald
    shiner, bluntnose minnow, smallmouth bass,
largemouth bass, bluegill, walleye, sauger, channel catfish, flathead catfish, carpsucker, quillback, drum, logperch, paddlefish
Major other aquatic vertebrates: painted turtle, common snapping turtle, smooth softshell turtle, spiny softshell turtle, common
    map turtle, false map turtle, bullfrog, northern leopard frog, southern leopard frog, pickerel frog, mink frog, treefrogs,
    common mudpuppy, northern water snake, muskrat, beaver, river otter
Major benthic invertebrates: Caddisflies (*Hydropsyche, Cheumatopsyche, Potamyia, Cyrnellus, Oecetis, Hydroptila,*
    Nectopsyche), mayflies (*Hexagenia, Stenonema, Tricorythodes, Baetisca, Baetis, Caenis*), chironomid midges (*Dicrotendipes,*
    Rheotanytarsus, Cricotopus), stoneflies (*Perlesta*), damselflies (*Ischnura, Enallagma*), dragonflies (*Dromogomphus*),
    crustaceans (*Asellus, Gammarus*), snails (*Physella, Fossaria, Ferrissia*), bivalves (*Sphaerium, Pisidium*)
Nonnative species: rainbow trout, rainbow smelt, common carp, grass carp, bighead carp, silver carp, white catfish, ninespine
    stickleback, striped bass, zebra mussel, Asiatic clam, *Daphnia lumholtzi*, Eurasian watermilfoil
Major riparian plants: cottonwood, silver maple, river birch, black willow, sandbar willow, box elder, green ash
Special features: large floodplain river that still retains >80% of river–floodplain connectivity; close to natural hydrograph
Fragmentation: 11 dams in headwaters; 26 locks and low-head dams on main stem
Water quality: pH = 8, alkalinity = 170 mg/L as CaCO$_3$, NO$_3$-N = 3.2 mg/L, total phosphorus = 0.19 mg/L; nutrient inputs from
    fertilizers; sedimentation
Land use: 70% agriculture, 25% forest, 5% urban
Population density: 54 people/km^2
Major information sources: Fremling et al. 1989, Patrick 1998

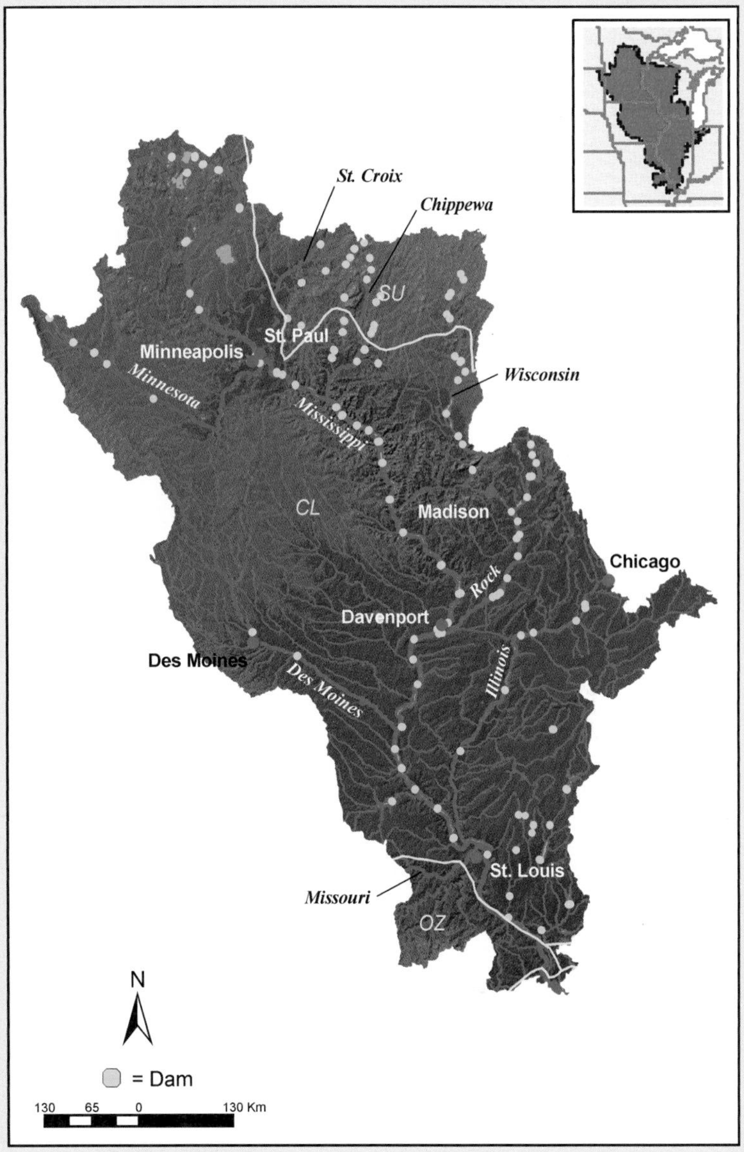

FIGURE 8.11 Map of the Upper Mississippi River basin. Physiographic provinces are separated by yellow lines.

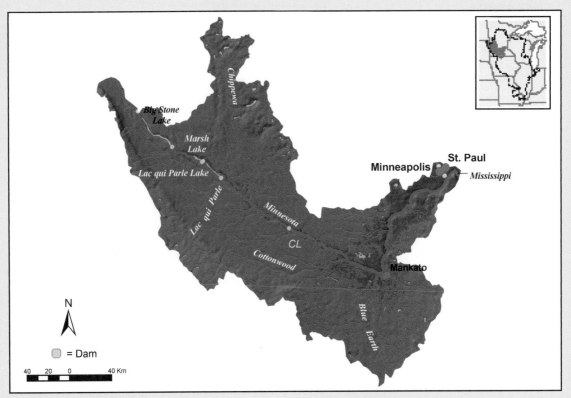

FIGURE 8.13 Map of the Minnesota River basin.

MINNESOTA RIVER

Relief: 85 m
Basin area: 27,030 km²
Mean discharge: 125 m³/s
River order: 7
Mean annual precipitation: 66 cm
Mean air temperature: 7.5°C
Mean water temperature: 10.5°C
Physiographic province: Central Lowland (CL)
Biomes: Temperate Grassland, Temperate Deciduous Forest
Freshwater ecoregion: Mississippi
Terrestrial ecoregions: Central Tall Grasslands, Upper Midwest
  Forest/Savanna Transition
Number of fish species: 87
Number of endangered species: 2 mussels (extirpated)
Major fishes: shorthead redhorse, quillback, carp, emerald shiner, spotfin shiner, sand shiner, channel catfish, freshwater drum, gizzard shad, bluntnose minnow, smallmouth buffalo, bigmouth buffalo, walleye, fathead minnow, quillback, fathead minnow

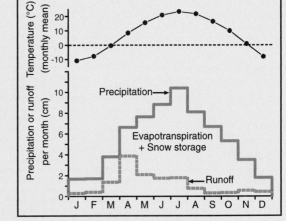

FIGURE 8.14 Mean monthly air temperature, precipitation, and runoff for the Minnesota River basin.

Major other aquatic vertebrates: northern leopard frog, mink frog, wood frog, treefrogs, mudpuppy, tiger salamander, painted turtle, common snapping turtle, false map turtle, common map turtle, northern water snake
Major benthic invertebrates: caddisflies (*Hydropsyche, Cheumatopsyche, Cyrnellus*), mayflies (*Stenonema, Stenacron, Potamanthus, Tricorythodes*), damselflies (*Enallagma, Argia*), dragonflies (*Dromogomphus*), stoneflies (*Isoperla*), chironomid midges (*Glyptotendipes, Polypedilum, Tanytarsus*), crustaceans (*Gammarus, Hyalella*), snails (*Physella*)
Nonnative species: brown trout, brook trout, goldfish, common carp, longear sunfish, Asiatic clam
Major riparian plants: cottonwood, green ash, black willow, sandbar willow
Special features: small river in a 1 to 10 km wide channel formed by glacial River Warren
Fragmentation: six dams on main stem, including outlets of three natural channel lakes
Water quality: highly impaired; high sediments, fertilizer, and pesticides from agriculture; pH = 8, alkalinity = 252 mg/L as CaCO₃, NO₃-N = 4.09 mg/L, PO₄-P = 0.08 mg/L
Land use: 95% agriculture, 1% urban, 1% forest
Population density: 6 people/km²
Major information sources: Bailey et al. 1994, Zischke et al. 1994, Downing et al. 1999

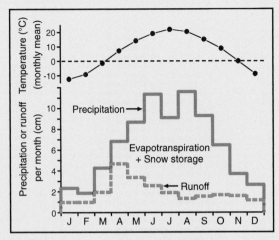

FIGURE 8.16 Mean monthly air temperature, precipitation, and runoff for the St. Croix River basin.

ST. CROIX RIVER

Relief: 319 m
Basin area: 20,018 km^2
Mean discharge: 131 m^3/s
River order: 6
Mean annual precipitation: 78 cm
Mean air temperature: 6.3°C
Mean water temperature: 10.4°C
Physiographic provinces: Superior Upland (SU), Central Lowland (CL)
Biome: Temperate Deciduous Forest
Freshwater ecoregion: Mississippi
Terrestrial ecoregions: Western Great Lakes Forest, Upper Midwest Forest/Savanna Transition
Number of fish species: 110
Number of endangered species: 2 mussels
Major fishes: emerald shiner, golden redhorse, shorthead redhorse, common carp, gizzard shad, smallmouth bass, bluegill, yellow perch, Johnny darter, spottail shiner, common shiner, spotfin shiner, bluntnose minnow, mimic shiner, brassy minnow, central stoneroller
Major other aquatic vertebrates: painted turtle, common snapping turtle, wood turtle, common map turtle, false map turtle, spiny softshell turtle, northern leopard frog, green frog, spring peeper, treefrogs, tiger salamander, muskrat, beaver, mink
Major benthic invertebrates: caddisflies (*Hydropsyche*, *Ceratopsyche*, *Nectopsyche*, *Lepidostoma*), mayflies (*Baetis*, *Stenonema*, *Stenacron*, *Heptagenia*, *Anthopotamus*, *Siphlonurus*, *Ephemerella*), stoneflies (*Pteronarcys*), beetles (*Gyrinus*, *Peltodytes*)
Nonnative species: rainbow trout, brown trout, brook trout, lake trout, rainbow smelt, common carp, zebra mussel, Asiatic clam
Major riparian plants: paper birch, slippery elm, black ash, tamarack, black spruce, white cedar, basswood, red maple, yellow birch
Special features: one of first eight rivers protected as National Wild and Scenic Rivers; spectacular gorges and rapids; most pristine river in Upper Mississippi basin
Fragmentation: two dams on main stem, but mostly free flowing; 134 small dams on tributaries
Water quality: high; pH = 7.5, alkalinity = 59 mg/L as CaCO$_3$, total nitrogen = 0.4 mg/L, total phosphorus = 0.05 mg/L
Land use: 27% agriculture, 65% forest, 2% urban
Population density: 13 people/km^2
Major information sources: Fago and Hatch 1993, National Park Service 1997

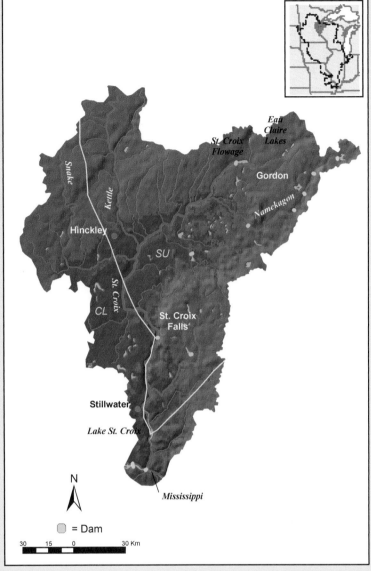

FIGURE 8.15 Map of the St. Croix River basin. Physiographic provinces are separated by a yellow line.

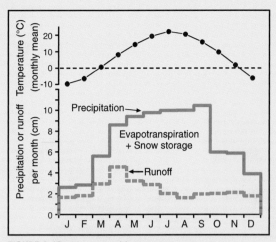

FIGURE 8.18 Mean monthly air temperature, precipitation, and runoff for the Wisconsin River basin.

WISCONSIN RIVER

Relief: 300 m
Basin area: 30,000 km²
Mean discharge: 261 m³/s
River order: 8
Mean annual precipitation: 85 cm
Mean air temperature: 7.5°C
Mean water temperature: 11.8°C
Physiographic provinces: Central Lowland (CL), Superior Upland (SU)
Biome: Temperate Deciduous Forest
Freshwater ecoregion: Mississippi
Terrestrial ecoregions: Western Great Lakes Forest, Upper Midwest Forest/Savanna Transition
Number of fish species: 119
Number of endangered species: 2 mussels
Major fishes: shorthead redhorse, Johnny darter, logperch, walleye, largemouth bass, smallmouth bass, bluegill, yellow perch, river carpsucker, smallmouth buffalo, common carp, gizzard shad, quillback, freshwater drum
Major other aquatic vertebrates: northern water snake, common snapping turtle, common musk turtle, wood turtle, painted turtle, common map turtle, false map turtle, Ouachita map turtle, smooth softshell turtle, spiny softshell turtle, green frog, mink frog, spring peeper, treefrogs, muskrat, beaver
Major benthic invertebrates: hemiptera (*Sigara*, *Trichocorixa*), caddisflies (*Cheumatopsyche*, *Hydropsyche*, *Potamyia*, *Nectopsyche*), mayflies (*Stenonema*, *Baetisca*, *Baetis*, *Caenis*, *Stenacron*, *Procloeon*, *Isonychia*), beetles (*Stenelmis*, *Macronychus*, *Peltodytes*, *Gyrinus*), stoneflies (*Isoperla*), damselflies (*Enallagma*), dragonflies (*Gomphurus*)
Nonnative species: rainbow trout, brown trout, common carp, grass pickerel
Major riparian plants: silver maple, river birch, swamp white oak, green ash, cottonwood, black willow
Special features: largest river in Wisconsin, draining 22% of state; Grandfather Falls rapids descend 28 m over 2.4 km
Fragmentation: heavily impounded in middle and upper reaches; 47 storage reservoirs and 27 hydroelectric dams on main stem and major tributaries
Water quality: point-source discharges from paper mills and municipal sewage; pH = 7.9, alkalinity = 89 mg/L as $CaCO_3$, NO_3-N = 0.52 mg/L, PO_4-P = 0.08 mg/L
Land use: 27% agriculture, 54% forest, 12% open water, 3% urban
Population density: 46 people/km²
Major information sources: Fago 1992, Lillie and Hilsenhoff 1992, Lyons et al. 2000

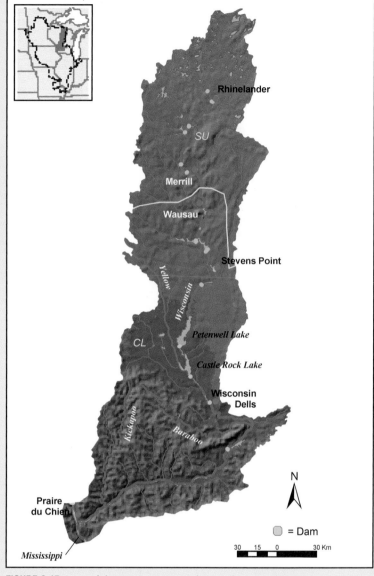

FIGURE 8.17 Map of the Wisconsin River basin. Physiographic provinces are separated by a yellow line.

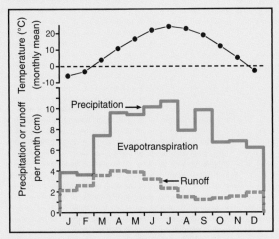

FIGURE 8.20 Mean monthly air temperature, precipitation, and runoff for the Illinois River basin.

ILLINOIS RIVER

Relief: 45 m
Basin area: 75,136 km²
Mean discharge: 649 m³/s
River order: 9
Mean annual precipitation: 92 cm
Mean air temperature: 10.4°C
Mean water temperature: 16°C
Physiographic province: Central Lowland (CL)
Biomes: Temperate Grassland, Temperate Deciduous Forest
Freshwater ecoregion: Mississippi
Terrestrial ecoregion: Central Forest/Grassland Transition
Number of fish species: 127
Number of endangered species: 2 mussels
Major fishes: gizzard shad, emerald shiner, freshwater drum, white bass, bluegill, green sunfish, largemouth bass, common carp, smallmouth buffalo, white sucker, bluntnose minnow, channel catfish, flathead catfish, bowfin, shortnose gar, grass pickerel, quillback, carpsucker
Major other aquatic vertebrates: common snapping turtle, smooth softshell turtle, common map turtle, false map turtle, slider, treefrogs, southern leopard frog, tiger salamander, northern water snake
Major benthic invertebrates: caddisflies (*Hydropsyche, Cheumatopsyche, Potamyia, Cyrnellus*), mayflies (*Stenonema, Baetis, Hexagenia, Heptagenia*), crustaceans (*Gammarus, Asellus*), bivalves (*Pisidium*), beetles (*Stenelmis*), chironomid midges (*Robackia, Rheosmittia*), flies (*Hemerodromia*), damselflies (*Argia*), oligochaete worms (*Barbidrilus*)
Nonnative species: rainbow trout, rainbow smelt, common carp, grass carp, bighead carp, silver carp, zebra mussel, Asiatic clam, round goby, *Daphnia lumholtzi*
Major riparian plants: silver maple, ash, box elder, black willow, hackberry
Special features: lower river flows through channel abandoned by Upper Mississippi; broad low-gradient floodplain results in protracted spring flood
Fragmentation: five low-head navigation dams on main stem
Water quality: pH = 7.8, alkalinity = 180 mg/L as $CaCO_3$, NO_3-N = 0.41 mg/L, PO_4-P = 0.18 mg/L; municipal sewage (upper river); nutrients, pesticides, soil erosion from agriculture throughout
Land use: 87% agriculture, 7% forest, 5% urban
Population density: 97 people/km²
Major information sources: Starrett 1971a, Page et al. 1992, Arnold et al. 1999

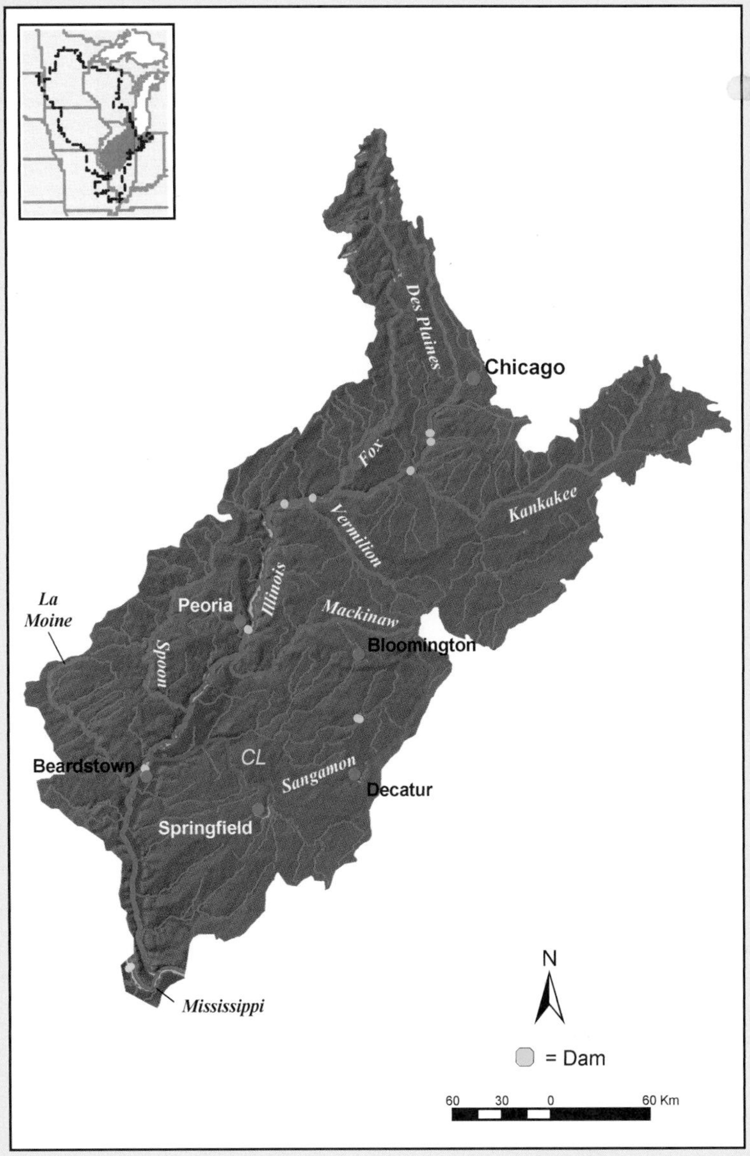

FIGURE 8.19 Map of the Illinois River basin.

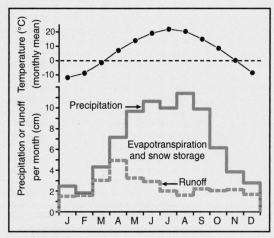

FIGURE 8.22 Mean monthly air temperature, precipitation, and runoff for the Chippewa River basin.

CHIPPEWA RIVER

Relief: 290 m
Basin area: 24,827 km²
Mean discharge: 218 m³/s
River order: 6
Mean annual precipitation: 80 cm
Mean air temperature: 6°C
Mean water temperature: 11°C
Physiographic provinces: Central Lowland (CL), Superior Upland (SU)
Biome: Temperate Deciduous Forest
Freshwater ecoregion: Mississippi
Terrestrial ecoregions: Western Great Lakes Forest, Upper Midwest Forest/Savannah Transition
Number of fish species: 110
Number of endangered species: 3 fishes, 2 mussels
Major fishes: shorthead redhorse, golden redhorse, smallmouth buffalo, carpsucker, common carp, mooneye, gizzard shad, shovelnose sturgeon, smallmouth bass, northern pike, channel catfish, walleye, sauger, muskellunge, emerald shiner, paddlefish

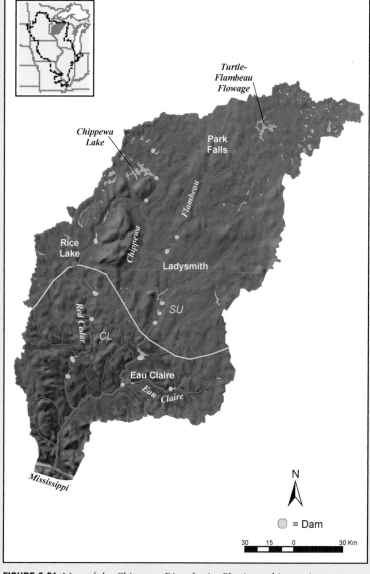

FIGURE 8.21 Map of the Chippewa River basin. Physiographic provinces are separated by a yellow line.

Major other aquatic vertebrates: muskrat, beaver, river otter, snapping turtle, common map turtle, false map turtle, spiny softshell turtle, leopard frog, green frog, treefrogs, northern water snake
Major benthic invertebrates: mayflies (*Pseudocloeon*, *Stenonema*, *Hexagenia*), alderflies (*Sialis*), dragonflies (*Dromogomphus*), stoneflies (*Allocapnia*), beetles (*Stenelmis*), caddisflies (*Cheumatopsyche*, *Hydropsyche*, *Agraylea*, *Cyrnellus*), biting midges (*Dasyhelia*), chironomid midges (*Cladotanytarsus*, *Dicrotendipes*, *Orthocladius*, *Robackia*), black flies (*Simulium*), bivalves (*Pisidium*)
Nonnative species: rainbow trout, brown trout, rainbow smelt, common carp, curly-leaved pondweed, Eurasian watermilfoil
Major riparian plants: cottonwood, silver maple, black willow, green ash, American elm, river birch, white swamp oak
Special features: sand carried by Chippewa River led to formation of Lake Pepin on Upper Mississippi River
Fragmentation: over 125 dams, including 16 impoundments >12 km²; hydrology extensively altered by hydropower
Water quality: pH = 7.6, alkalinity = 49 mg/L as $CaCO_3$, NO_3-N = 0.68 mg/L, PO_4-P = 0.05 mg/L; fertilizer, pesticide, and soil runoff from agriculture and urbanization
Land use: 36% agriculture, 54% forest, 2% urban, remainder primarily open water
Population density: 12 people/km²
Major information source: Voss and Beaster 2001

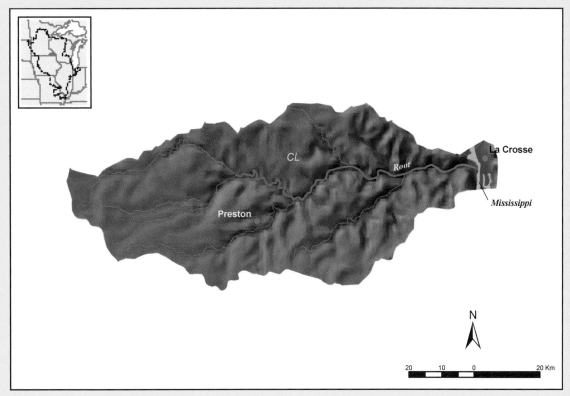

FIGURE 8.23 Map of the Root River basin.

ROOT RIVER

Relief: 170 m
Basin area: 4325 km^2
Mean discharge: 21 m^3/s
River order: 5
Mean annual precipitation: 82 cm
Mean air temperature: 6.4°C
Mean water temperature: 11°C
Physiographic province: Central Lowland (CL)
Biome: Temperate Deciduous Forest
Freshwater ecoregion: Mississippi
Terrestrial ecoregion: Upper Midwest Forest/Savanna Transition
Number of fish species: 97
Number of endangered species: unknown
Major fishes: brown trout, brook trout, slimy sculpin, white sucker, longnose dace, blacknose dace, brook stickleback, American brook lamprey, northern hog sucker, shorthead redhorse, smallmouth bass, common shiner, southern redbelly dace, longnose dace, central stoneroller, fantail darter

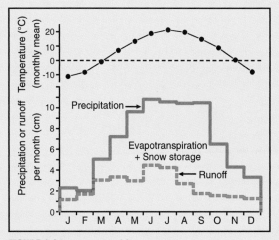

FIGURE 8.24 Mean monthly air temperature, precipitation, and runoff for the Root River basin.

Major other aquatic vertebrates: muskrat, beaver, river otter, snapping turtle, common map turtle, false map turtle, spiny softshell turtle, leopard frog, green frog, treefrogs, northern water snake
Major benthic invertebrates: caddisflies (*Hydropsyche, Cheumatopsyche, Brachycentrus*), mayflies (*Stenonema, Baetis, Isonychia*), stoneflies (*Neoperla, Pteronarcys*), beetles (*Stenelmis, Optioservus*), flies (*Atherix, Simulium, Tipula*), crustaceans (*Hyalella*), snails (*Physella*)
Nonnative species: brown trout, rainbow trout, common carp
Major riparian plants: cottonwood, silver maple, box elder, black willow, red oak, American elm
Special features: representative of the coolwater–coldwater rivers of Driftless Area; registered on the national scenic river inventory with heavy recreational use
Fragmentation: mostly free-flowing, with a few small dams in headwaters
Water quality: pH = 8.2, alkalinity = 221 mg/L as CaCO$_3$, NO$_3$-N = 3 mg/L, PO$_4$-P = 0.18 mg/L, turbidity often >50 NTU; fertilizers, pesticides, and livestock runoff (fecal coliform often >200/mL)
Land use: approximately 60% agriculture, 22% forest, 2.5% urban, 15% pasture/grassland
Population density: 9 people/km^2
Major information sources: N. Mundahl (unpublished data), K. Schmidt (unpublished data), Waters 1977

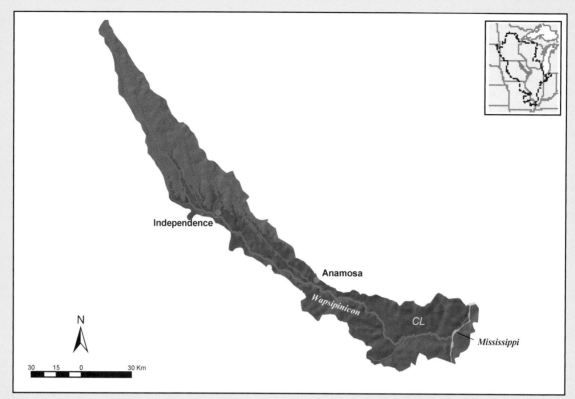

FIGURE 8.25 Map of the Wapsipinicon River basin.

WAPSIPINICON RIVER

Relief: 195 m
Basin area: 6050 km²
Mean discharge: 47 m³/s
River order: 5
Mean annual precipitation: 88 cm
Mean air temperature: 9°C
Mean water temperature: 12°C
Physiographic province: Central Lowland (CL)
Biomes: Temperate Grassland, Temperate Deciduous Forest
Freshwater ecoregion: Mississippi
Terrestrial ecoregion: Central Tall Grasslands
Number of fish species: 74
Number of endangered species: 1 fish
Major fishes: American brook lamprey, spotfin shiner, Mississippi silvery minnow, bigmouth shiner, sand shiner, bluntnose minnow, bullhead minnow, river carpsucker, shorthead redhorse, Johnny darter

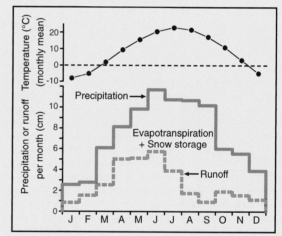

FIGURE 8.26 Mean monthly air temperature, precipitation, and runoff for the Wapsipinicon River basin.

Major other aquatic vertebrates: muskrat, beaver, river otter, snapping turtle, western painted turtle, common map turtle, false map turtle, mudpuppy, spiny softshell turtle, treefrogs, leopard frog
Major benthic invertebrates: caddisflies (*Hydropsyche, Cheumatopsyche, Ceratopsyche, Nectopsyche*), mayflies (*Tricorythodes, Heptagenia, Baetis, Caenis, Stenonema*), stoneflies (*Paragnetina, Pteronarcys*), beetles (*Macronychus, Dubiraphia*), chironomid midges (*Ablabesmyia, Cricotopus/Orthocladius*), flies (*Simulium, Atherix*), crustaceans (*Orconectes*), snails (*Physella*)
Nonnative species: common carp, rainbow trout, brown trout
Major riparian plants: silver maple, cottonwood, black willow, sandbar willow
Special features: one of few Iowa rivers that retains mature stands of riparian and floodplain vegetation; very narrow basin
Fragmentation: 21 dams, mostly small impoundments on tributaries of upper basin
Water quality: pH = 8.1, alkalinity = 137 mg/L as CaCO₃, NO₃-N = 6.2 mg/L, PO₄-P = 0.06 mg/L, Atrazine = 1.68 μg/L, Cyanazine = 0.18 μg/L, Carbofuran = 0.06 μg/L; pollution primarily from row-crop agriculture
Land use: 88% agriculture, 10% forest, 2% urban
Population density: 14 people/km²
Major information sources: Sullivan 2000, Brigham and Sadorf 2001

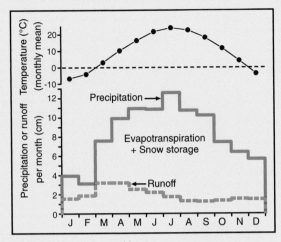

FIGURE 8.28 Mean monthly air temperature, precipitation, and runoff for the Rock River basin.

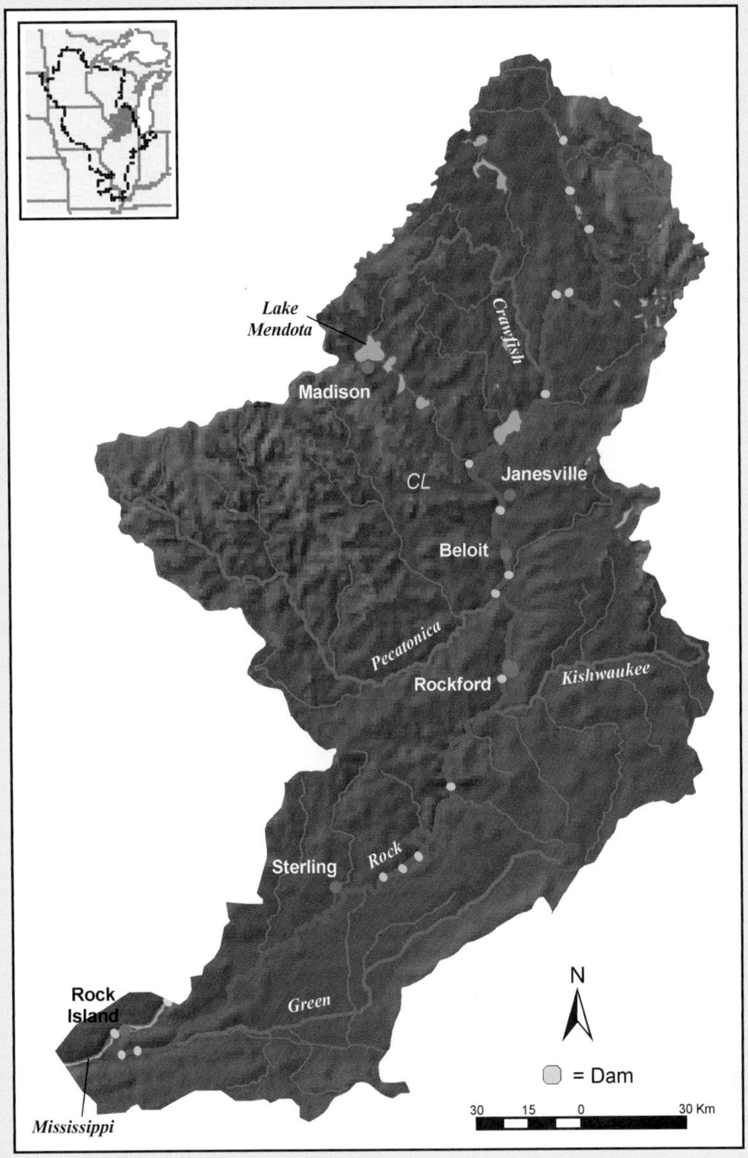

ROCK RIVER

Relief: 155 m
Basin area: 28,101 km^2
Mean discharge: 184 m^3/s
River order: 7
Mean annual precipitation: 99 cm
Mean air temperature: 10°C
Mean water temperature: 12°C
Physiographic province: Central Lowland (CL)
Biomes: Temperate Grassland, Temperate Deciduous
  Forest
Freshwater ecoregion: Mississippi
Terrestrial ecoregions: Central Forest/Grassland
  Transition, Upper Midwest Forest/Savanna
  Transition
Number of fish species: 115
Number of endangered species: 1 mussel
Major fishes: gizzard shad, carp, spotfin shiner,
  smallmouth buffalo, largemouth buffalo, channel
  catfish, white bass, smallmouth bass, walleye,
  northern pike, freshwater drum, white sucker, green
  sunfish, Johnny darter, central stoneroller
Major other aquatic vertebrates: tiger salamander, common mudpuppy, bullfrog, green frog, northern leopard frog, pickerel
  frog, treefrogs, common snapping turtle, common map turtle, painted turtle, slider, common musk turtle, northern water
  snake
Major benthic invertebrates: caddisflies (*Hydropsyche, Cheumatopsyche, Nectopsyche*), mayflies (*Stenonema, Stenacron, Baetis*),
  crustaceans (*Gammarus, Crangonyx, Hyalella, Caecidotea*), black flies (*Simulium*)
Nonnative species: rainbow trout, brown trout, lake trout, common carp, grass carp, goldfish, rusty crayfish, zebra mussel,
  Eurasian watermilfoil
Major riparian plants: cottonwood, silver maple, box elder, black willow, sandbar willow
Special features: upper river drains unique glacial formations, creating constrained channels for many tributaries
Fragmentation: 19 hydroelectric dams on main stem and 272 dams throughout the basin
Water quality: pH = 8.3, alkalinity = 220 mg/L as CaCO$_3$, NO$_3$-N = 3.2 mg/L, PO$_4$-P = 0.13 mg/L; pollution from municipal
  wastewater treatment and agriculture
Land use: 71% agriculture, 14% urban, 12% forest, 2% wetland
Population density: 52 people/km^2
Major information sources: Page et al. 1992, U.S. Army Corps of Engineers (unpublished data)

FIGURE 8.27 Map of the Rock River basin.

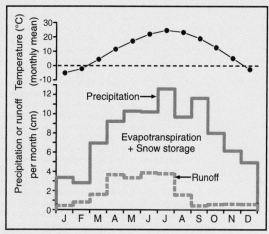

FIGURE 8.30 Mean monthly air temperature, precipitation, and runoff for the Des Moines River basin.

DES MOINES RIVER

Relief: 290 m
Basin area: 31,127 km²
Mean discharge: 182 m³/s
River order: 6
Mean annual precipitation: 96 cm
Mean air temperature: 11°C
Mean water temperature: 11.6°C
Physiographic province: Central Lowland (CL)
Biomes: Temperate Grassland, Temperate Deciduous Forest
Freshwater ecoregion: Mississippi
Terrestrial ecoregions: Central Tall Grasslands, Central Forest/Grassland Transition
Number of fish species: 84
Number of endangered species: 3 fishes
Major fishes: shorthead redhorse, common carp, smallmouth buffalo, emerald shiner, bluntnose minnow, bluegill, largemouth bass
Major other aquatic vertebrates: snapping turtle, painted turtle, spiny softshell turtle, smooth softshell turtle, red-eared slider, treefrogs, mudpuppy, leopard frog, pickerel frog, green frog, muskrat
Major benthic invertebrates: caddisflies (*Hydropsyche, Cheumatopsyche, Ceratopsyche, Nectopsyche*), mayflies (*Tricorythodes, Heptagenia, Baetis, Caenis, Stenonema*), stoneflies (*Paragnetina, Pteronarcys*), beetles (*Macronychus, Dubiraphia*), chironomid midges (*Ablabesmyia, Cricotopus/Orthocladius*), black flies (*Simulium*), snails (*Physella*)
Nonnative species: common carp, grass carp, bighead carp, striped bass
Major riparian plants: cottonwood, ash, black willow, sandbar willow
Special features: largest river in Iowa, with drainage basin covering 23% of state
Fragmentation: two major impoundments (>12 km³) and 58 small to medium-size impoundments on main stem and tributaries
Water quality: pH = 8, alkalinity = 190 mg/L as CaCO₃, NO₃-N = 5.36 mg/L, PO₄-P = 0.72 mg/L; pollution primarily from agriculture
Land use: 86% agriculture, 5% urban, 3% forest
Population density: 22 people/km²
Major information sources: Keeny and DeLuca 1993, Lutz et al. 2001

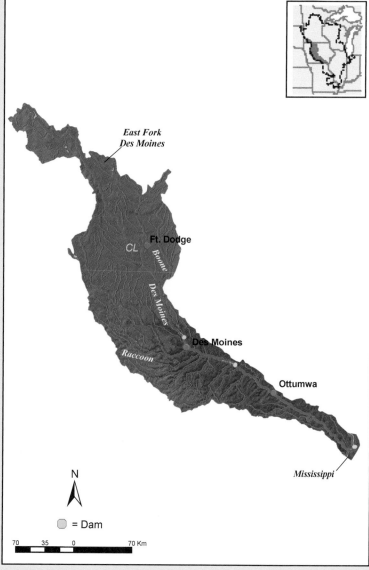

FIGURE 8.29 Map of the Des Moines River basin.

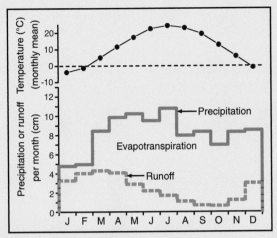

FIGURE 8.32 Mean monthly air temperature, precipitation, and runoff for the Kaskaskia River basin.

KASKASKIA RIVER

Relief: 100 m
Basin area: 15,025 km²
Mean discharge: 107 m³/s
River order: 6
Mean annual precipitation: 99 cm
Mean air temperature: 11°C
Mean water temperature: 15.2°C
Physiographic province: Central Lowland (CL)
Biome: Temperate Deciduous Forest
Freshwater ecoregion: Mississippi
Terrestrial ecoregion: Central Forest/Grassland
  Transition
Number of fish species: 112
Number of endangered species: 1 mussel
Major fishes: common carp, shorthead redhorse,
  channel catfish, freshwater drum, bluegill,
  largemouth bass, flathead catfish, white crappie,
  yellow bass, white bass, gizzard shad, sand shiner,
  bigmouth buffalo, western mosquitofish
Major other aquatic vertebrates: smooth softshell turtle,
  painted turtle, false map turtle, river cooter, smooth
  softshell turtle, slider, southern leopard frog, pickerel frog, green frog, treefrogs, tiger salamander, smallmouth salamander,
  northern water snake, diamondback water snake
Major benthic invertebrates: caddisflies (*Hydropsyche, Cheumatopsyche*), mayflies (*Stenonema, Caenis, Tricorythodes*), stoneflies
  (*Taeniopteryx*), beetles (*Stenelmis, Macronychus*), flies (*Atherix*), crustaceans (*Gammarus, Crangonyx, Hyalella,
  Procambarus, Orconectes, Lirceus, Caecidotea*)
Nonnative species: bighead carp, common carp, silver carp, goldfish, white catfish, redear sunfish, zebra mussel
Major riparian plants: silver maple, box elder, black willow, cottonwood
Special features: southernmost large tributary of upper Mississippi; largest contiguous tracts of forest, including bottomland, in
  Illinois
Fragmentation: 107 dams on the main stem and tributaries, including four impoundments >12 km²; lock and dam near mouth
  creates 21 km long pool
Water quality: primary impacts from fertilizer, pesticide and soil runoff, increased urbanization; pH = 7.5, hardness = 46 mg/L as
  CaCO₃, NO₃-N = 2.94 mg/L, PO₄-P = 0.08 mg/L
Land use: 80% agriculture, 13% forest, 4% wetland, 3% urban
Population density: 30 people/km²
Major information source: Page et al. 1992

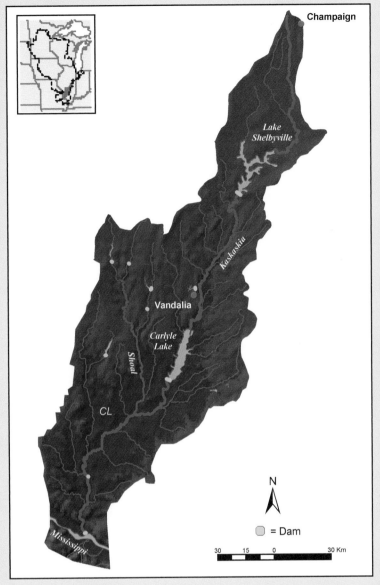

FIGURE 8.31 Map of the Kaskaskia River basin.

9

OHIO RIVER BASIN

DAVID WHITE   KARLA JOHNSTON   MICHAEL MILLER

INTRODUCTION

OHIO RIVER MAIN STEM

TENNESSEE RIVER

CUMBERLAND RIVER

WABASH RIVER

KANAWHA RIVER

ADDITIONAL RIVERS

ACKNOWLEDGMENTS

LITERATURE CITED

INTRODUCTION

The word "Ohio" comes from Iroquois, meaning "beautiful river." Early French explorers also called it "La Belle Rivière." The Ohio basin is the third largest by discharge ($8733\,m^3/s$) in the United States, accounting for more than 40% of the discharge of the Mississippi River but making up only 16% of its drainage area. The basin lies between 34°N and 41°N latitude and 77°W and 89°W longitude and drains major portions of eight states and minor parts of six additional states ($529,000\,km^2$), from New York in the northeast to Georgia and Alabama in the south to Illinois in the west (Fig. 9.2). The eastern portion of the basin has its tributary origins in the Appalachian Mountains, and the northern tributaries border on Laurentian Great Lakes drainages. Climate is continental, with abundant rainfall, cool moist winters, and warm summers. With the exception of some prairie in the north and west, the region historically was heavily forested, but today agriculture is a major feature along with many large urban areas. Patterns of glaciation, mountains, and varied geology have resulted in a highly diverse biological environment.

At the time of the first French explorers the basin was home to a wide variety of Native Americans, including Shawnee, Mosopelea, Erie–Iroquois, Cherokee, and Miami–Pottawatomie cultures. French and British traders competed with each other for the lucrative fur trade for more than a century until first the French and then the British gave up claims to territories. In the 1780s, following the American Revolutionary War, European settlers poured in from the south through Georgia, east through the Cumberland Gap, and down the Ohio River. Within 30 years they had completely displaced Native Americans and had begun to permanently alter the landscape. Development was very rapid; all the states in the Ohio basin had been admitted to the Union by 1818, and almost all of the major cities had been established.

In this chapter, we first discuss the Ohio River main stem and then four of the major rivers that reflect the basin's diversity in an upstream progression from the Tennessee River to the Cumberland River, Wabash River, and Kanawha River (see Fig. 9.2). Abbreviated descriptions of physical and biological features are provided for the Green, Kentucky, Great Miami, Scioto, Licking, Monongahela, and

◀ **FIGURE 9.1** Monongahela River north of Morgantown, West Virginia (PHOTO BY TIM PALMER).

375

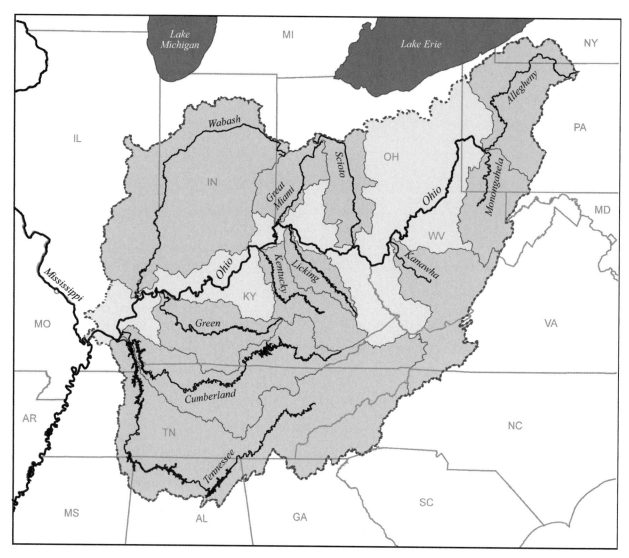

FIGURE 9.2 Rivers of the Ohio River basin covered in the chapter.

Allegheny rivers. Because of space limitations, several other large tributaries to the Ohio River are not covered here, but their major contributions to the system are recognized (e.g., the Salt River, the Little Miami River, the Big Sandy River, and the Muskingum River).

Physiography and Climate

The Ohio River basin slopes generally east to west, spanning six physiographic provinces (Hunt 1967). The eastern portions of the basin begin in the Blue Ridge, Valley and Ridge, and Appalachian Plateaus provinces of the Appalachian Highlands. The central and northern parts of the basin drain the Interior Low Plateaus province and the glaciated topography

of the Central Lowland province. At its western end the basin just touches on the Coastal Plain province.

The eastern extent of the basin is in the Blue Ridge and Valley and Ridge provinces, which run parallel to each other and contain some of the highest mountain peaks in the eastern United States. The closely spaced ridges of the Blue Ridge are made up of metamorphic Precambrian granites and gneiss along with sandstones and conglomerates. The folded mountains of the geologically younger Valley and Ridge province are primarily Paleozoic limestones, shale, and sandstones and contain heavily mined anthracite coal beds. Contiguous with the Valley and Ridge province to the west is the much larger Appalachian Plateaus province with its varied landscape of Ordovician, Silurian, and Devonian limestone, sandstone, shale, and bituminous coal formations. The western part of the Appalachian Plateaus province extends into eastern Tennessee, Kentucky, and Ohio and contains eroded low hills of Mississippian and Pennsylvanian limestone. Soils range from thin and easily eroded mountain soils to deeper podzols in the valleys. Together the three provinces contain the richest diversity of plants and animals in the United States.

The Interior Low Plateaus province slopes gently from southeast to northwest and is composed of dissected and highly eroded Ordovician limestone plateaus and domes. Notable for its high biodiversity is the Highland Rim that separates the Appalachian Plateaus province from the deeply eroded Nashville Basin in middle Tennessee. Much of the topography from northern Alabama to central Indiana is karst with extensive cave formations (e.g., Mammoth Cave, Crystal Cave). There are comparatively few small surface streams, as the landscape is dotted with sinkholes. The northwestern section contains extensive near-surface deposits of bituminous coal and has been heavily strip mined. Soils are primarily red and yellow podzols.

The northwestern part of the basin from Ohio to central Illinois lies in the Central Lowland physiographic province. The topography and surface geology were shaped and reshaped by Pleistocene glaciation, and the province was not completely ice free until 15,000 to 20,000 years ago. The landscape varies from gently rolling to extremely flat, with a mantle of glacial tills and till plains, deeply entrenched rivers, and occasional Pennsylvanian limestone outcroppings. Soils are rich neutral to basic loams, resulting in generally high-alkalinity surface waters. A few pothole lakes occur in the northern extent of the Wabash River basin, but the remainder of the province, and indeed the entire Ohio River basin, is lake free.

The climate is continental temperate with cool moist winters and warm humid summers. Monthly mean temperatures range from −7°C to 10°C in winter and 24°C to 28°C in summer. Average monthly precipitation is fairly uniform throughout the year, with significant accumulations of snow in the north and Appalachians. Annual precipitation ranges from 94 to 112 cm and is only slightly greater in higher elevations.

Basin Landscape and Land Use

The Ohio River basin covers two biomes (Temperate Grasslands and Temperate Deciduous Forest) and eight terrestrial ecoregions (Ricketts et al. 1999). The eastern part of the basin contains the Appalachian/Blue Ridge Forests, which combine the mountainous Blue Ridge and Valley and Ridge physiographic provinces of Hunt (1967). Much of this ecoregion remains forested, with evergreen spruce–fir forests dominating higher elevations and deciduous oaks and hickories lower elevations. Running from southwest to northeast and paralleling the Appalachian/Blue Ridge Forests are the mountainous Allegheny Highlands Forests and the Appalachian Mixed Mesophytic Forests ecoregions that roughly correspond with the Appalachian Plateaus physiographic province (Hunt 1967). Beech–hemlock forests that included white pine and sugar maple originally dominated the more northern Allegheny Highland Forests. The rolling hills of the Appalachian Plateaus contain extremely rich and diverse broadleaf forests of oaks, hickories, maples, and poplars in the valleys and birch–maple–hemlock forests higher up, with relict bogs.

Most of the lower part of the basin is in the Southern Great Lakes Forests and Central United States Hardwoods Forests ecoregions. Hardwood forests of oaks, maples, and hickories mark both ecoregions, differing primarily in their species composition and understories. Three additional ecoregions just touch on the Ohio Basin. Prairies of the Central Forest Grassland Transitional Zone (Temperate Grasslands biome) occur in the northwest, Mississippi Lowland Forests swamps with bald cypress and black gum are present at the confluence with the Mississippi River, and the piney Southern Mixed Forests range into northern Alabama and southern Tennessee.

The largest single land classification in the basin is forestland, which accounts for up to 80% of land use in Pennsylvania and in river basins flowing northward from Kentucky. The proportion of agricultural lands increases from 12% to 14% in the east to 64% to 73% in western Ohio, Indiana, and eastern Illinois. Although the eastern part of the basin remains extensively forested, logging and fires in the late nineteenth century greatly reduced hardwood and pine stands and increased the distribution of aspen and other softwoods. All of the forest in once glaciated parts of Ohio, Indiana, and Illinois was removed and is now in agriculture with patches of second-growth forest. Much of the forest of central and western Kentucky and Tennessee suffered a similar fate, but the soils were less fertile and more highly erosive, and more of it has reverted to second-growth forest.

The Rivers

Past the confluence of the Allegheny and Monongahela rivers the largest tributaries to the Ohio River enter around Cincinnati (Little Miami, Great Miami, Kentucky, and Licking rivers from Rkm 740 to 870) and around Paducah (Green, Wabash, Cumberland, and Tennessee rivers from Rkm 1365 to 1503). Collectively the 12 major tributaries contribute 84% of the entire basin drainage. Most of the other 57 named tributary basins are rather small, with basins <1000 km^2.

Major differences in ecology, geochemistry, and present-day land use of rivers in the Ohio basin reflect two differing geologic histories. Rivers entering the Ohio from the north and much of the Ohio main stem itself are geologically young, having been greatly affected by Pleistocene glaciation. Rivers entering from the south and east, including the Tennessee, Cumberland, Green, Kentucky, Kanawha, Monongahela, and Allegheny, are geologically much older, allowing for development of a wide array of endemic species. Indeed, the Tennessee and Cumberland have the greatest richness of aquatic invertebrates and vertebrates of any rivers in North America. Based on the differences in geomorphology and species richness, including endemics, Abell et al. (2000) divided the Ohio basin into two distinct aquatic ecoregions, the Teays–Old Ohio and the Tennessee–Cumberland.

The Teays–Old Ohio rivers can be roughly divided into three types based on slope and physiography. The first group includes the Monongahela, Allegheny, Kanawha, Licking, and Kentucky rivers that drain the Appalachian Mountains. The river slopes are high, particularly in the headwaters, and the bottoms can be rocky. The greater relief of these areas has led to extensive impoundment for hydroelectric power generation, shipping, and flood control. Alkalinity, PO$_4$-P, and NO$_3$-N are naturally low. Even though these rivers are geologically old, with the New River the oldest in the United States, there are surprisingly few endemic fishes. These rivers also have had similar human impacts from coal mining.

The second group includes the Green River (along with the Salt River), which lies entirely within the Interior Low Plateaus physiographic province. These rivers have moderately high gradients in the headwaters. Stream bottoms are sandy with chert cobble, and surface waters are generally low in alkalinity, PO$_4$-P, and NO$_3$-N. These rivers have rich and diverse fishes and benthic invertebrate faunas, and the Green River supports the highest number of endemic fishes in the ecoregion.

The Wabash, Great Miami, and Scioto rivers (along with several others) of Illinois, Indiana, and Ohio form a third group. All three begin in the glacial till plains of west-central Ohio and flow generally southward, becoming moderately to deeply entrenched in the gently sloping landscape. River bottoms are generally sandy with larger glacial till. These rivers have not been heavily dammed. Except below major cities, water-quality problems generally can be traced to land use. Alkalinity, pH, PO$_4$-P, and NO$_3$-N are naturally high compared with other rivers in the Ohio basin, and most are augmented with additional PO$_4$-P and NO$_3$-N from agricultural runoff. The rivers have high plant and animal diversity but no endemics because of the recent glaciation.

The Tennessee–Cumberland aquatic ecoregion consists of those two rivers. Both are high gradient in their Appalachian Mountain headwater tributaries and then parallel each other through a variety of lower gradient physiographic provinces and substrate types to their mouths on the Ohio River. Although not historically connected except via the Ohio River, both rivers have very high diversity and share many endemic species. Endemism of fishes and invertebrates is the highest of any aquatic ecoregion in North America. Much of the endemism occurs in the headwaters and also at the transitions between physiographic provinces. Both also share the dubious distinction of being among the most impounded of any of the major rivers in the United States. In these rivers, human impact has been much more complex, reflecting not only agricultural, industrial, and urban inputs but also acid mine drainage in the tributaries and lakelike conditions in the main-stem impoundments.

The Ohio River basin, in particular the Tennessee–Cumberland aquatic ecoregion, is considered to be globally outstanding by virtue of its high fish and invertebrate diversity (Abell et al. 2000). More species of fishes, freshwater mussels, and crayfishes occur here than in any other basin in the United States, and more are endangered than anywhere else. Although better agricultural practices have reduced runoff throughout the basin, soil erosion along with fertilizer and pesticide pollution continue to be major problems, as do impoundment, urbanization, industry, and coal mining.

OHIO RIVER MAIN STEM

The Ohio River main stem is a 9th order river that begins at the confluence of the Allegheny and

FIGURE 9.3   Ohio River at Paducah, Kentucky (PHOTO BY G. HARRIS).

Monongahela rivers in the Appalachian Mountains at Pittsburgh, Pennsylvania, and runs generally southwest for approximately 1575 km through the agricultural Midwest to the Mississippi River near Cairo, Illinois (Fig. 9.14). It is the only North American river with navigation miles numbered from the origin rather than from the mouth (Resh et al. 1976). Major tributaries include the Tennessee, Cumberland, Wabash, Green, Salt, Kentucky, Little Miami, Great Miami, Licking, Scioto, Muskingum, Kanawha, Allegheny, and Monongahela rivers. The present-day Ohio is a highly managed industrial river, with more tonnage and industrial use per kilometer

than any river of comparable length in the United States. The 20 low-water river locks and dams maintain a 3 m minimum pool depth throughout the river's length (Fig. 9.3).

Humans have lived along the river for at least the past 12,000 years. Paleo-Indian foraging cultures were present from 11,500 to 10,000 years ago, Archaic cultures followed from 10,000 to 3000 years ago, and then Woodland cultures (Adena, Hopewell) between 3000 and 1000 years ago. From 1000 years ago to the mid 1600s, extensive cities, trade, and warfare became common throughout the basin, and agriculture surpassed hunting and gathering (Lewis

1996). At the time the first French explorers visited (La Salle in 1669, Marquette and Joliet in 1674), the Ohio, Shawnee, Mosopelea, and Erie–Iroquois tribes occupied much of the main stem. The French and British vied for control of the basin and the fur trade throughout much of the seventeenth and eighteenth centuries until the Treaty of Paris (1763) ended France's claim to lands east of the Mississippi. George Rogers Clark's defeat of the British at Vincennes in 1779 transferred control of the basin to the United States. A number of battles and treaties in the interim had effectively ended most tribal rights to lands east of the Mississippi River, and the Indians moved westward (Hyde 1962, Williams 1993).

The history of the Ohio River was the history of the West in the late eighteenth and early nineteenth centuries. Europeans, led by Daniel Boone and Simon Kenton, came up the Wilderness Road through the Shenandoah Valley to settle Harrodsburg, Boonesboro, Louisville, and Limestone. The first settlers at the Falls of the Ohio (Louisville) were led by George Rogers Clark, who built blockhouses and a fort in 1778 with 20 families and 150 volunteers (Simon 1939). In the summer of 1788, 4500 people rode flatboats down the river past Marietta in the western expansion into what were once Indian lands. The first commercial packet boat passenger service between Pittsburgh and Cincinnati began in October 1793. The first steam ship, Orleans, came down the Ohio from Pittsburgh to Cincinnati and Louisville in 1811 and was stopped by the Falls of the Ohio. There were 35 steamboats on the river by 1818, and 400 by 1829 (WPA 1943).

Physiography, Climate, and Land Use

The main stem of the Ohio River flows through three physiographic provinces (see Fig. 9.14): the Appalachian Plateaus (AP), underlain by Pennsylvanian and Mississippian limestones, shale, and coal beds; the Central Lowland (CL), covered with glacial tills; and the Interior Low Plateaus (IL), underlain by Mississippian limestone sand and chert, just entering the sedimentary deposits and loess of the Coastal Plain (CP) at its juncture with the Mississippi River (Hunt 1967).

The main stem passes through four terrestrial ecoregions (Ricketts et al. 1999). It begins in the Appalachian Mixed Mesophytic Forests in the upper reaches, dominated by oaks, hickories, ashes, maples, pines, and elms. Central U.S. Hardwood Forests occur in much of the lower reaches and are primarily white oak, red oak, hickories, American elm,

tulip tree, and sweetgum. Through portions of Indiana and Ohio, the Ohio River forms the southern boundary of the Southern Great Lakes Forests, with sugar maple, American beech, basswood, oaks, elms, hickories, and ashes. At its confluence with the Mississippi River, the Ohio touches on the bald cypress–black gum swamps of the Mississippi Lowland Forests.

Precipitation for the basin is fairly evenly distributed throughout the year (Fig. 9.15) and ranges from 91.5 to 114 cm/yr (mean of 104). The basin is generally drier in the northern portions and wetter in the southern portions. The 40 to 59 days of thunderstorms per year add an element of violent weather during spring and summer. Average temperatures range from January daily lows of −6.7°C in the eastern mountains to daily highs in July of 27°C along the main stem. Snowfall ranges from 2.5 to 30 cm in the southwest to 30 to 91 cm the east, reaching >152 cm in the mountains. The average percentage of sunshine decreases from 60% to 70% in the southwest to <50% in the eastern part of the drainage (Williams 1994).

The majority of the basin's 26 million residents live in close proximity to the major rivers, including ten major cities ranging from 250,000 to 2,300,000 people. The average population density for the entire basin is approximately 49 people/km²; however, just considering counties that border the Ohio River itself the population density would be greater than 200 people/km². Basinwide, urban areas compose 4% of land use, 47% is agricultural, 47% is forested, and 1% is water. Importantly, 1% of the basin has been disturbed by mining, amounting to >2% of the land area of the Allegheny, Monongahela, Muskingum, and Big Sandy river basins.

River Geomorphology, Hydrology, and Chemistry

The upper Ohio River valley (the Teays River) was formed prior to the Illinoisan glaciation about 2 million years ago and included the Kanawha–New rivers that flowed northward through the Scioto River valley (Fig. 9.2) to a now obliterated major westerly river northwest of Springfield, Illinois, which presumably ran to the Mississippi drainage. The Illinoisan glacial advance blocked northern drainage at Hamilton, Ohio, creating 18,000 km² Lake Tight upriver of Cincinnati, which covered parts of southern Ohio, West Virginia, and Kentucky and reached an elevation of nearly 275 m asl. When the water breached divides at Cincinnati (about 400,000 years ago), it flowed southwest, forming the present channel of the Ohio River. The valley deepened by almost 46 m at Wheelersburg, Ohio, to its present level (Pohana 1992). Although not directly altering the main stem, Wisconsin glaciers changed drainage patterns in the Wabash River and other south-flowing rivers (Hough 1958).

Much of the river bottom was sand prior to impoundment and continues to be so today. Glacial gravel and cobble form 40 named islands in the river above Cincinnati. Of these, 21 constitute the Ohio River Island National Wildlife Refuge. In the lower portion of the river, there are 57 sand bar islands. Very few islands occur in the middle third of the river, but limestone outcroppings at Louisville created the Falls of the Ohio.

In 1824 the U.S. Congress allotted $75,000 for improving the Ohio River, and the U.S. Army Corps of Engineers built its first dam on the river at Henderson, Kentucky. By 1830 a set of three locks could raise small steamboats up 9 m (30 ft) over the Falls at Louisville. The first wicket lock and dam was opened in 1885 at Davis Island, and in 1929, canalization was completed, with 53 locks and a 2 m minimum pool. Modernization began in 1955 with replacement of 14 dams and reinforcement of five more, creating a 3 m deep channel at low water. Today there are 20 run-of-the-river locks and dams, with only the original wicket locks 52 and 53 remaining, which will be consolidated in a new dam by 2005 (Reid 1991).

At the convergence of the Monongahela and Allegheny rivers at Pittsburgh, elevation is 217 m asl and drops to 88 m at Rkm 1578 at Cairo, Illinois, an average slope of only 8 cm/km. The slope is higher near Pittsburgh, dropping rapidly from 50 cm/km to 10 cm/km in the first 250 km. Through the remaining distance, there is little change in slope.

Mean discharge at the Mississippi River (including the Tennessee and Cumberland rivers) is 8733 m³/s. Discharge is highest in March, during snowmelt, decreasing to lows in August and September (see Fig. 9.15). Higher evapotranspiration during summer months appears to account for much of the difference between runoff and precipitation (Patrick 1995). Floods have been a common occurrence on the Ohio River. Following the huge flood of 1937, 78 tributary flood-control dams were built, the last completed in 1990. The coordinated use of these reservoirs is alleged to take 3 m off flood crests and provide up to half the flow of the river during droughts (Reid 1991). The major flood-control reservoirs in the basin (excluding the Tennessee River) col-

lectively can hold about 26.4 km³ of water, more than six times the volume of the river (ORSANCO 1994).

Main-stem water quality has generally improved since the mid 1970s, particularly downstream from urban areas, but cities remain point sources for nutrients and industrial waste and account for much of the BOD load to the river. The main stem remains degraded by power plant thermal inputs, and industrial spills occur at least once a decade. Agriculture runoff dominates most tributary inputs, but orphaned mines still exude acid. Average pH is 7.2, increasing slightly in a downriver direction. Alkalinity averages about 70 mg/L as $CaCO_3$. Not considering reaches immediately below major cities, NO_3-N averages 1.6 mg/L and PO_4-P averages <0.2 mg/L. Both nutrients tend to increase in a downstream direction. Mean annual water temperature is 14°C, but the temperature at summer low flows can reach 35°C, aided by power plant thermal discharges.

River Biodiversity and Ecology

With the exception of the Tennessee and Cumberland rivers, the Ohio River and its tributaries compose the Teays–Old Ohio freshwater ecoregion (Abell et al. 2000). The diversity of aquatic plants and animals in this ecoregion is globally high, but the number of endemic species is comparatively low.

Algae and Cyanobacteria

Even though high turbidity and a strong current keep the Ohio River light limited throughout most of its length, there are significant numbers of phytoplankton, particularly during summer low flow. Wehr and Thorp (1997) identified 134 taxa of cyanobacteria, chlorophytes, euglenophytes, chrysophytes, pyrophytes-cryptophytes, and diatoms in a survey conducted in 1992 along a 361 km stretch from Cincinnati to Evansville, Indiana. Dominant summer cyanobacteria were *Aphanothece saxicola*, *Merismopedia punctata*, *Microcystis aeruginosa*, and *Synechococcus*. Of their 82 diatom species, the primary spring taxon was *Cyclotella*, followed by *Melosira jurgensii* and *Melosira distans* in summer. *M. aeruginosa* apparently became numerically dominant in 1997 and 1998 when zebra mussels reached peak abundance, but densities were reduced in 2000 when zebra mussels declined (M. Kannan, unpublished data).

Plants

Increased turbidity from agricultural runoff, along with impoundment, has all but eliminated submersed aquatic macrophytes. There is still extensive flood-plain and riparian forest along most of the river, except in urban areas. Trees are typical in much of the basin and include red maple, cottonwood, black willow, black gum, and American sycamore. Bald cypress occurs as far upstream as southwestern Indiana (Deam 1953). Buttonbush is common along the banks in many places. Nonnatives include purple loosestrife, brittle naiad, curly pondweed, and Eurasian watermilfoil.

Invertebrates

As with phytoplankton, the Ohio River supports a diverse zooplankton assemblage, with highest densities from May through November (Thorp et al. 1994). Rotifers dominate in all months, with *Polyarthra* most abundant, followed by copepod nauplii, the cladoceran *Bosmina longirostris*, calanoid copepods, and the rotifers *Branchionus calyciflorus* and *Keratella* spp. Density is directly related to temperature and inversely related to velocity and turbidity (Thorp et al. 1994). With the potential exception of copepod nauplii, zooplankton does not appear to affect algal densities (Wehr and Thorp 1997). The nonnative *Daphnia lumholtzi* is present, but its relation to other cladoceran species remains unknown. The effects of fishes upon the zooplankton community have been documented by Jack and Thorp (2002).

Benthic invertebrate surveys of the main stem have been made by the U.S. Environmental Protection Agency (USEPA), the Ohio River Valley Water Sanitation Commission (ORSANCO), and power plants since the early 1960s, primarily using grabs, rock baskets, or Hester-Dendy plate samplers. In the 1960s, numbers of taxa collected increased in a downstream direction, ranging from 13 to 22 at Pittsburgh to 45 to 58 near Evansville (ORSANCO 1978). Upriver sites contained primarily worms and chironomids, often dominated by pollution-tolerant *Nanocladius distinctus* and *Cricotopus intersectus*. Numbers of individuals and diversity appear to have dramatically increased since the enactment of the National Environmental Policy Act (NEPA) in 1969. Recent ORSANCO unpublished data lists 222 invertebrate taxa for the Ohio main stem, exclusive of mussels, snails, and zooplankton. Nearly 90 of the taxa are dipterans, primarily Chironomidae (*Chironomus, Coelotanypus, Dicrotendipes, Stenochironomus, Pentanura*); however, there are a variety of mayflies (*Hexagenia, Stenonema, Ephemerella, Caenis, Stenacron*), stoneflies (*Perlesta, Isoperla*), caddisflies (*Hydroptila, Cheumatopsyche, Hydropsyche, Symphitopsyche, Ceraclea, Cyrnellus, Polycentropus, Potamyia, Chimarra*), dragonflies/damselflies (*Stylurus, Argia*), and worms (*Limnodrilus, Branchiura, Nais*). Other common taxa include amphipods (*Gammarus*) and fingernail clams (*Pisidium, Sphaerium*). Although some of the increase may be due to cleaner water conditions, the greater number of taxa also may reflect more intensive sampling efforts and greater taxonomic resolution.

The Ohio River main stem is now habitat for approximately 50 species of unionid mussels, down from perhaps 80 species in pre-European times. Taylor (1989) documented 33 species from the upper Ohio River. More exhaustive recent surveys have added 9 more species, and 15 of those present from 1880 to 1920 have disappeared. Two species are extirpated from the Ohio River (leafshell and round combshell) and eight are federally listed as threatened and endangered. Common species include threehorn wartyback, Wabash pigtoe, maple leaf, white heelsplitter, three ridge, and fat mucket. The nonnative Asiatic clam and zebra mussel are common in the river today. The Asiatic clam arrived in the Ohio River by 1957 and quickly expanded into most of its tributary rivers, but it does not appear to compete with native unionids for optimum habitat.

The zebra mussel arrived in 1993, carried by barges coming down the Illinois River from the Great Lakes, and continues to move upriver as a crest of reproductive colonization. Highest densities in 2001 were near Pittsburgh (J. Hageman, unpublished data). Based on ORSANCO Hester-Dendy plate sampler studies, the periods for massive reproduction in the middle section of the river were the summers of 1997 and 1998. One-third of their samplers in 1997 had >10,000 colonists (Rkm 768 to 1460). A drought during late summer 1998 and all year in 1999 aided by once-through cooling plants created very high temperatures that may have been the proximal cause of the decline of zebra mussels in that section in subsequent years. The densities in the Ohio River have been shown to be capable of altering the phytoplankton community and rotifer densities (Jack and Thorp 2000). Predators such as common carp, channel catfish, redhorse, and smallmouth buffalo may be adapting to consume zebra mussels (Thorp et al. 1998). Zebra mussels have been responsible for the death of more than 80% of native mussel populations at some of the downriver sites in the late 1990s (J. Sickel, personal communication).

Vertebrates

Historically, the Ohio River main stem has contained 160 to 170 fish species in 25 families (Lachner

1956, Trautman 1981, Pearson and Krumholz 1984, Reash and van Hassel 1988). Constantine Samuel Rafinesque originally described many of the species in the early 1800s, including 10 that were drawn by James J. Audubon as a practical joke on Rafinesque. The joke species remained in the literature for decades (Everman 1918). Today 28 species (18%) are rare, another 21 species (13%) are of "special concern" as lost from one state or in recent decline, and only 27 species are common to the length of the Ohio River. Although species composition changes, the number of species is relatively evenly distributed from one end to the other: 122 species between Rkm 0 to 526, 132 between Rkm 527 to 1052, and 119 between Rkm 1053 to 1578. Nineteen fish species reported before 1970 have not been reported since, including lake sturgeon, Alabama shad, and the sand darter.

Fourteen species have been introduced through human activity. Rainbow trout, brown trout, and brook trout are or have been stocked. Common carp were introduced from the 1870s through the mid 1880s, when they were one of the largest fisheries in the river. Goldfish have done well, and goldfish/common carp hybrids are common. The banded killifish and mummichog were introduced as bait or forage fishes. White catfish, white bass, yellow perch, and several centrarchids are being stocked in tributary streams and even in the river. Striped bass are being released into the river as a game fish. At least five species have been added through river connections since 1984, including smelt, fathead chub, and rosefin shiner (Pearson and Pearson 1989).

Power companies and ORSANCO have documented historic trends in fish populations since 1957 (Lachner 1956, Mason et al. 1971, USEPA 1978, Pearson and Pearson 1989, EA Environmental 2001). In general, fish abundance and richness have increased since 1969. Most notable is the increase in many species in the middle and upper thirds of the river. White bass, most of the catostomids, river darter, logperch, sauger, and freshwater drum have clearly responded. Gizzard shad has once again replaced the emerald shiner as the numerically dominant species. Pollution-tolerant species have declined in the upper third of the river, namely bullheads, white sucker, and common carp. In the past 20 years, several large-river species have moved upstream: paddlefish, spotted gar, mooneye, and highfin carpsucker. Other common species throughout the main stem include skipjack herring, smallmouth buffalo, flathead catfish, sauger, largemouth bass, bluegill, silver chub, golden redhorse, and smallmouth bass.

Turtles and water snakes are plentiful along the river. Common species include the snapping turtle, stinkpot, mud turtle, common map turtle, midland painted turtle, spiny softshell, slider, Florida cooter, common water snake, and queen snake. The cottonmouth is not common but is present in the lower portion of the river. Common frogs include pickerel frog, green frog, southern leopard frog, and bullfrog (Barbour 1971). Both the bald eagle and osprey are now nesting in protected areas. A wide variety of waterfowl can be seen along the river during spring and fall migrations, and 36 species are known to nest along the river. Beaver, muskrat, and river otter are common in the basin and occasionally seen along the main stem.

Ecosystem Processes

Ecosystem processes in the Ohio River main stem have been examined in much greater detail than in any of the large tributary rivers. Primary physical controls are current velocity, light penetration, and temperature. Run-of-the-river navigation dams appear to be less important to processes than the flood storage/hydroelectric impoundments of tributaries, such as in the Tennessee and Cumberland. Summer densities of algae are high (>100,000 cells/mL) and positively correlate with temperature and negatively correlate with turbidity and velocity (Wehr and Thorp 1997). Wehr and Thorp (1997) noted that decreased velocities near dam sites did contain greater algal biomass and larger cell sizes. Algae only moderately deplete N and P, which may be amended or diluted by tributary inputs (Bukaveckas et al. 2000). Bukaveckas et al. (2000) found a summer downstream depletion of Si, which could limit phytoplankton growth. Zooplankton average >21/L from May through November, and copepod nauplii densities have been correlated with a loss of picoplankton in the pools (Wehr and Thorp 1997).

Koch (2001) estimated summer primary production at 115 and 114 g C m^{-2} yr^{-1} at the mid and lower river sections, respectively, and respiration at 780 and 562 g C m^{-2} yr^{-1}, with P/R ratios ranging from 0.15 to 0.20. Bacterial production was only 4% and 3% of the primary production. Thorp and DeLong (2002) point out that although allochthonous inputs (riparian and downstream transport) are important initial sources of organic matter and large rivers are heterotrophic on an annual basis, there is strong evidence that overall metazoan production in most large rivers ≥4th order is driven by autochthonous production that enters through algal-grazer and decomposer pathways. Their riverine productivity model (RPM) may be very important to understanding ecosystem processes throughout the basin, particularly in low-gradient tributaries.

Human Impacts and Special Features

Prior to European contact the river was clear and constantly flowing and contained lush growths of aquatic plant life and clean beds of gravel, rock, and sand (described by the Lewis and Clark expedition of 1803, Reid 1981). Despite impoundment and heavy industrial, urban, and agricultural use, the river remains remarkably biodiverse.

Today the entire river is impounded and a 3 m deep navigation channel is maintained. There are several large urban areas and numerous industrial facilities. To maintain the navigational channel, the U.S. Army Corps of Engineers dredges an average 500,000 m^3 of silt, sand, and gravel each year (1991 to 2000). Barge transportation amounts to 263 metric tons worth $45 billion, dominated by coal and aggregates. There are 67 power plants on the Ohio River and in adjacent counties. Once-through cooling for electrical power generation in summer can push thermal plumes from one power plant to the next. There are 28 major petroleum facilities, 12 grain elevators and terminals, and 29 chemical plants and terminals on the Ohio's main stem or tributary mouths.

Main-stem water quality has generally improved since the mid 1970s, particularly downstream from urban areas, but city wastewater treatment plants and combined sewer overflows remain point sources for nutrients, intestinal microbes, and pathogens. Pollution from steel mills in and around Pittsburgh and coke plants in Huntington has been reduced, but sediment contamination from industrial releases continues to be of concern, prompting fish consumption advisories. Industrial spills still occur at least once a decade. Agriculture erosion and runoff of nutrients and pesticides continue to dominate tributary inputs.

Nonnative plant invasions have not been much of a problem to date, but zebra mussels have had major effects on native species and ecosystem processes. The effects of nonnative fish species are more difficult to determine, as native fishes are still responding to water-quality mitigation.

TENNESSEE RIVER

Measured by average discharge (2000 m^3/s), the 8th order Tennessee River is the largest tributary to the Ohio River (Fig. 9.16). The Tennessee basin covers approximately 105,000 km^2 of the states of Tennessee, Kentucky, Virginia, North Carolina, Georgia, Alabama, and Mississippi. Headwaters drain portions of the southern Appalachian Mountains, flowing generally westward across four physiographic provinces. The Tennessee River itself begins at the confluence of the French Broad and Holston rivers just east of the city of Knoxville, Tennessee. From Knoxville the river runs southwest into northern Alabama before turning north through Tennessee and Kentucky and emptying into the Ohio River at Rkm 1504 just above the city of Paducah, Kentucky, a distance of about 1050 km (Fig. 9.4). Major tributaries include the French Broad, Holston, Little Tennessee, Clinch, Hiwassee, Paint Rock, Duck, and Big Sandy rivers. Additional outflow from the Tennessee occurs in the Tennessee–Tombigbee waterway, completed in 1984. The "Tenn–Tom" now directly links the Tennessee River with the Black Warrior/Tombigbee River system in Alabama. When Lake Barkley was impounded in 1966 on the Cumberland River just to the east, a shipping canal was dug between that reservoir and Kentucky Lake on the Tennessee River and water now may flow freely in either direction between the two rivers.

The history of human influence dates back at least 12,000 years to the entrance of Paleo-Indian groups that followed the warming climate. By at least 5000 years ago there was limited agriculture along the river valleys. The Woodland period from about 3100 to 1200 years ago saw the development of extensive agriculture, pottery, and trade routes. By the Mississippian period (1200 to 500 years ago) cities with temples were developed, new varieties of corn and beans were being used, and there was well-organized warfare among neighboring tribes. In 1540, Hernando de Soto was the first European explorer to see the river. Choctaw, Cherokee, Shawnee, and Chickasaw were the primary inhabitants of the Tennessee Valley in the late 1600s. European traders followed, bringing guns and disease, and the Cherokee, who became quite dependent on European goods, eventually replaced the other existing cultures. The name "Tennessee" is thought to be from the Cherokee. Its true meaning is unknown but may refer to "the place where the river bends." There were no permanent European settlements until 1769, when William Bean built a cabin on the Watauga River. Jonesborough, Tennessee, was the first chartered town in the basin (1779). By the end of the eighteenth century several more towns had been established along the Tennessee River, including Knoxville and

FIGURE 9.4 Tennessee River near Reidland, Kentucky (Photo by G. Harris).

Chattanooga. Colonization of the basin was rapid during the first half of the nineteenth century, and by the end of the nineteenth century much of the farmland was in poor condition and there was little in the way of manufacturing.

Physiography, Climate, and Land Use

The headwaters of the Tennessee basin begin primarily in the incised Blue Ridge (BL) and Valley and Ridge (VR) physiographic provinces of western Virginia, North Carolina, and eastern Tennessee (Hunt 1967) (see Fig. 9.16). The Blue Ridge geology is primarily Precambrian gneiss, granite, sandstone, conglomerates, and siltstone, whereas the Valley and Ridge is composed more of Paleozoic limestones, sandstones, and shale. The Tennessee River proper begins near Knoxville, Tennessee, and flows southward to Chattanooga, crossing into the Cumberland Plateau section of the Appalachian Plateaus (AP) province, which is underlain by Pennsylvanian and Mississippian limestones, shale, and coal beds. The river then continues southward through the

Sequatchie Valley into northern Alabama, where it turns westward across the Highland Rim of the Interior Low Plateaus (IL) province. The river then runs northward through the Western Valley, which separates the Highland Rim of the Interior Low Plateaus from the Coastal Plain (CP) physiographic province. The Interior Low Plateaus is underlain by Mississippian limestone, sand, and chert, and the Coastal Plain geology is primarily alluvium, sedimentary deposits, and loess (Hunt 1967).

The Tennessee River basin lies in four terrestrial ecoregions (Ricketts et al. 1999). Much of the headwaters are in the Appalachian/Blue Ridge Forests. Higher elevations are dominated by red spruce, balsam fir, and Frazer fir, and lower elevations are principally deciduous, including red oak, black oak, hickories, birches, and black locust. The main stem begins in Appalachian Mixed Mesophytic Forests, with American beech, sugar maple, eastern hemlock, mountain laurel, mountain maple, and rhododendrons at higher elevations and oaks, hickories, ashes, maples, pines, elms, sweetgum, black locust, and black cherry at lower elevations. Most of the lower part of the basin flows through Central United States Hardwood Forests, primarily white oak, post oak, southern red oak, hickories, American elm, tulip tree, and sweetgum, with understories of winged elm, flowering dogwood, and sassafras. A very short section of the river, primarily in northeastern Mississippi, enters the Southeastern Mixed Forests terrestrial ecoregion, which is primarily oak–hickory but also contains stands of shortleaf, loblolly, and longleaf pines.

The climate is continental temperate. Monthly mean temperatures range from about 2°C in January to about 23°C in July (Fig. 9.17). Temperatures are generally cooler in the Appalachian headwaters and warmer throughout the Interior Plateau. Winter temperatures are regularly below 0°C, particularly at night, and often exceed 38°C in July, August, and September. Annual precipitation ranges from about 80 to 130 cm (average 105 cm) and is fairly uniform throughout the year but lowest in October (see Fig. 9.17).

Human population density for the basin averages 19 people/km² and is concentrated primarily in two large cities, Knoxville at 600,000 and Chattanooga at 430,000. The Knoxville and Pigeon Forge areas have had rapid development over the past decade and are the fastest-growing areas of the basin. The upper basin contains more than 64% forest, and agriculture accounts for 27% of land use (primarily pasture). Approximately 6% is urban and the remainder is either barren, water, or in mining. There are extensive coal reserves in the Appalachian Plateau, and mining has had strong effects on water quality and biota. Approximately 55% of the lower part of the basin is forested. Row crops and pasture cover 41%, wetlands and water 3%, and urban 1% (TVA, unpublished data). Much of the lower basin has been in agriculture for the past 150 years. By 1900 agricultural land was in poor condition, but better farming practices put in place after the Great Depression have decreased erosion and increased crop yields. The primary crops are corn, soybeans, cotton, and winter wheat. Tobacco remains a cash crop throughout the basin. In the western part of the basin, pine plantations are becoming a more important crop.

River Geomorphology, Hydrology, and Chemistry

Geological evidence suggests that the upper two-thirds of the Tennessee River once drained directly into the Gulf of Mexico during the Tertiary. At some time the Tennessee merged with the ancestral Duck River, resulting in a northward flow to where it now empties into the Ohio River (Starnes and Etnier 1986). The elevation at the confluence of the French Broad and Holston is approximately 269 m asl and falls to 90 m asl at the confluence with the Ohio River. This average gradient of 17 cm/km does not mean much, however, as most of the river is now impounded, creating a stair-step gradient (Table 9.1). The present history of the Tennessee basin is directly related to the vast river modifications created by the Tennessee Valley Authority (TVA). Congress created TVA on May 18, 1933, as part of the New Deal to provide relatively inexpensive electricity to a region of the country that was agriculturally and industrially poor and had been hard hit by the Great Depression. Initially, TVA purchased a number of existing hydroelectric dams from the Tennessee Electric Power Co. The first TVA-constructed hydroelectric dam was Norris on the Clinch River in northeastern Tennessee, filled in 1937. TVA now operates 48 multipurpose dams in the basin (TVA 1980). Additional dams and reservoirs are operated by private companies and by the Army Corps of Engineers, including the lock and dam system for the Tenn–Tom Waterway located in northeast Mississippi and west-central Alabama. The Tennessee River main stem consists of nine reservoirs managed for hydroelectric power generation, flood control, and navigation, and very little would be considered free flowing except between its origin and the upper reaches of Ft. Loudoun Reservoir (see Table 9.1).

Hydrologically and limnologically, the reservoirs can be divided into three functional zones, each with its own chemical and biological characteristics: riverine, transitional, and lacustrine (Thornton et al. 1990). The locations and sizes of the three functional zones may shift depending on dam operations and rainfall patterns. Riverine zones tend to resemble the original river geomorphology and generally are within the original riverbanks. Flows, however, are quite variable and depend on release cycles from upstream dams. In the Tennessee River, riverine zones tend to have relatively high velocity (>1 m/s is common), river bottoms contain much sand and cherty gravel, and the water is usually turbid. In transitional zones, water velocity slows and larger organic and inorganic particulates drop out. Bottom substrates are often mixed sands, cherty gravel, and pockets of organic deposition. The lacustrine zones are most lakelike but retain strong longitudinal flow patterns. Lacustrine zones in the Tennessee main-

TABLE 9.1 Physical features of the Tennessee River TVA main-stem reservoirs.

Reservoir	Landscape position rank	Tailwater elevation m asl	Mean discharge m³/s	Mean retention in days	Drainage basin km²
Ft. Loudoun	1	224	396.4	10	24,734
Watts Bar	2	206	792.8	14	44,833
Chickamauga	3	193	991.1	5.6	53,846
Nickajack	4	180	1,019.4	3.1	56,643
Guntersville	5	168	1,152.5	11	63,325
Wheeler	6	154	1,401.6	7.3	63,584
Wilson	7	125	1,444.1	5.8	79,642
Pickwick Landing	8	108	1,526.2	6.4	85,003
Kentucky	9	91	1,912.0	16.7	104,117

Data from TVA 1980.

stem reservoirs occupy the entire floodplain, thus the average water depth (6 to 8 m) is quite shallow compared with many of the upstream tributary reservoirs (e.g., Norris Lake). Lacustrine zone bottom substrates are usually fine, clayey deposits of low organic content (generally <1% carbon in the main channel). All of the main-stem reservoirs have seasonally regulated pools designed to accommodate spring runoff. In Kentucky Lake, the furthest downstream in the system, winter pool (late August to March) is approximately 2 m lower than summer pool (April to early August). Tributary reservoirs such as Norris can have as much as a 20 m difference in water level between winter and summer pools (TVA 1980).

Mean discharge is 2000 m³/s. Although the system is highly regulated, long-term monthly runoff patterns are related to evapotranspiration (see Fig. 9.17). Discharge is highest in the winter months and lowest in September. Superimposed on Tennessee River discharge patterns is flood storage and release that can occur anytime during the year. Effects of a flood anywhere in the Mississippi River system can be lessened by holding back the Tennessee River until the flood crest has passed or by emptying the Tennessee reservoirs prior to a flood crest. To this end, discharge from Kentucky Lake has approached 10,000 m³/s on several occasions. Further, the discharges of the Tennessee and Cumberland rivers are often commingled through the canal linking Kentucky Lake and Lake Barkley.

Water chemistry is dependent on seasonal patterns and characteristics of water retention, dam manipulation, and landscape position (Soranno et al. 1999). NO_3-N averages 0.2 mg/L, PO_4-P averages <0.01 mg/L, pH averages 7.2, and alkalinity averages 40 mg/L as $CaCO_3$. With the exception of NO_3-N values, which decrease, most chemical concentrations increase in a downriver direction, including alkalin-

ity, Ca, SO_4, Si, and total P. Secchi depth transparency also tends to decrease as more and more fine particles are entrained in the flow. As turbidity and velocity decrease in the transitional zones, productivity may be quite high if NO_3-N and PO_4-P are present in sufficient quantities; therefore, pH can range from 7.2 to 9.5 within hours on a warm, sunny day in May. Because of the water velocity, main-stem reservoirs rarely thermally stratify, but oxygen deficits may occur in summer when discharges are low. Increased water retention coupled with constant mixing often results in the entire water mass of the Tennessee River reaching 30°C or warmer in August. With their much longer retention times, tributary impoundments such as Norris Reservoir do thermally stratify, and cool hypolimnetic releases are capable of supporting salmonid fishes.

River Biodiversity and Ecology

The Tennessee River and its sister river, the Cumberland, form the Tennessee–Cumberland freshwater ecoregion (Abell et al. 2000). The diversity of this ecoregion may be the highest of all temperate ecoregions (Starnes and Etnier 1986, Abell et al. 2000), with the greatest number of endemic species of any ecoregion in the United States. This is one of the few areas of the United States where new species of both invertebrates and fishes are still being described in tributary rivers. Although not historically connected except by the Ohio River, the two rivers share many of the same endemic species, with the Tennessee slightly more diverse.

Algae and Cyanobacteria

Because of extensive impoundment, the greatest source of autochthonous production is now phytoplankton (Yurista et al. 2001). A spring diatom peak (primarily *Melosira*) occurs in the main-stem reservoirs from late March to early April, with chlorophyll values often reaching 40μg/L. The dominant midsummer phytoplankton are green algae (*Scenedesmus, Pediastrum*), diatoms (*Asterionella, Fragilaria, Melosira, Synedra, Cymbella*), and various cyanobacteria (*Anabaena, Oscillatoria*) (Tabor 2000).

Plants

Much of the floodplain riparian vegetation has been lost or significantly altered because of impoundment and seasonally fluctuating water levels. Except in the embayments, typical upland hardwood forests grow right to the river or reservoir edges (Luken and Spaeth 2002). There is little documentation of macrophyte beds prior to impoundment, but water willow beds most likely would have been prominent in shallow river runs. Water willow appears to be resistant to reservoir level fluctuations and is one of the few native aquatic plants established in backwaters and embayments. Nonnative macrophytes have done well in embayments (parrot-feather, alligatorweed, brittle naiad, curly pondweed, Nepal grass). Eurasian watermilfoil remains common in embayments despite extensive TVA control measures in the 1980s (Kobraei and White 1996). Other common embayment taxa include red maple, buttonbush, black gum, black willow, and sweetgum. Bald cypress has been extensively planted.

Invertebrates

Phytoplankton–zooplankton interactions are an important ecological feature of the system. The spring diatom peak is quickly grazed by *Bosmina longirostris* and then by *Daphnia retrocurva*. *Leptodora kindti* is the major invertebrate predator on *Bosmina* and *D. retrocurva*, and all three cladocerans, along with a number of copepods, are a primary source of food for larval fishes (Schram and Marzolf 1993, Yurista et al. 2001). Until the 1990s, *Diaphanosoma birgei* was the primary midsummer grazing zooplankter and a major food source for larval fishes. Since then, the nonnative *Daphnia lumholtzi*, which feeds heavily on desmids, has pushed *D. birgei* populations later into the summer.

There have been no comprehensive surveys of the benthic invertebrates of the Tennessee River since its impoundment, and, with the exception of the unionids, there is not much information available on preimpoundment conditions. Most data are in reports to either the TVA or state agencies. Penning-

ton and Associates (1994) found that longitudinal benthic trends within the impoundments were consistent with riverine, transitional, and lacustrine zone reservoir bottom conditions. Riverine zones tend to be dominated by filter-feeding bivalves (Unionidae, Corbiculidae, and Sphaeriidae). Among the predominant insects are the mayflies *Stenonema* and *Tricorythodes* and the caddisfly *Cheumatopsyche*. Transitional zones have the greatest diversity and density of benthic invertebrates and are dominated by subsurface deposit-feeding worms (*Limnodrilus* spp.), with the mayflies *Stenonema, Tricorythodes*, and *Caenis*, the caddisfly *Hydroptila*, and more than 200 species of Chironomidae the primary insects.

Lacustrine zone sediments are often poor in organic matter, and deposit-feeding tubificid worms average <100/m^2 in Kentucky Lake. Benthic invertebrates are primarily sediment deposit gatherers and surface filterers. The mayfly *Hexagenia limbata* is common, as are surface-deposit-gathering Chironomidae. Including littoral areas, more than 100 Chironomidae species (e.g., *Chironomus, Parachironomus, Polypedilum, Tanytarsus, Cricotopus*) have been identified from Kentucky Lake. The most spectacular is *Chironomus major*, which has larvae up to 50mm long in the 4th instar. The majority of lacustrine zone surface filterers are unionids (24 species), fingernail clams, and the Asiatic clam. The predaceous larva of the dipteran *Chaoborus punctipennis* is abundant throughout the lacustrine zones, and the opossum shrimp is becoming abundant (Brooks et al. 1998). Sixty-five species of crayfishes (*Cambarus, Procambarus, Orconectes*) are present in the basin, but most occur only in the smaller tributaries. Forty species are endemic and have very limited distributions (Hobbs 1988).

The Tennessee River drainage has the highest diversity of freshwater mussels (Unionidae and Margaritiferidae) of any river in North America. Approximately 100 species have been recorded, of which 5 are now extinct, 22 are listed as endangered, and an additional 14 are being considered for endangered status under the Endangered Species Act (Parmalee and Bogan 1998). More than 40 species are found in the fast-flowing headwater streams of the Holston, Powell, and Clinch rivers of Virginia and North Carolina. Impoundment has greatly reduced the density and diversity of the main-stem mussel populations through habitat loss, pollution, and possibly through loss of host fish species for the parasitic glochidia stage. A number of species, however, have done very well in the reservoirs (e.g., threeridge, mapleleaf, washboard), with annual commercial mussel harvests in Kentucky Lake exceeding $1 million/yr in the mid-1990s. Asiatic clam densities have decreased slowly over the past 20 years, which may be due to an increase in predators such as carps, catfishes, and freshwater drum. The zebra mussel has localized colonies throughout the main stem as far upstream as Ft. Loudoun, and populations may be on the increase. Zebra mussel populations in the Tennessee River are not as great as in the Ohio River, making the Tennessee a potential refuge for freshwater mussels.

Vertebrates

Depending on the level of taxonomic resolution, between 225 and 240 species of fishes occur in the Tennessee River drainage (Etnier and Starnes 1993), including at least 20 nonnative forms. Percidae (primarily darters) and Cyprinidae (minnows) are the two most diverse families, followed by Ictaluridae (catfishes), Catostomidae (suckers), and Centrarchidae (sunfishes). Much of the diversity occurs in the Highland Rim interface of the Interior Low Plateaus and Valley and Ridge physiographic provinces, which is home to a large number of endemic darters. Including several species that occur in the Cumberland River, 67 species are endemic to the ecoregion, including 5 catostomids, 16 cyprinids, and 41 darters. The number of darters probably will increase as new species are described from the headwaters. Three species are listed as federally endangered: the boulder darter, the smoky madtom, and the pygmy madtom.

The fishes of the Tennessee River proper have been greatly altered by impoundment, pollution, and introductions through stocking. The principle forage fishes are now gizzard shad and threadfin shad, which are dependent on the abundant plankton in the system. Major piscivores include channel catfish, small cyprinids, largemouth bass, smallmouth bass, white bass, and striped bass. The freshwater drum feeds heavily on mollusks, particularly the Asiatic clam. Major species feeding on benthic invertebrates include the common carp, paddlefish, and 18 species of Catostomidae. Commercial species include channel catfish and paddlefish, which is sought for its caviar. Seven different nonnative salmonids have been introduced into the reservoirs and cooler tail waters of tributary dams. Rainbow trout, Ohrid trout, and brown trout have proven successful, whereas introductions of cisco, cutthroat trout, and coho and sockeye salmon have not (Etnier and Starnes 1993). Other nonnative species include

bighead carp, common carp, goldfish, grass carp, tench, golden shiner, rainbow smelt, yellow perch, and striped bass.

Reservoir backwaters and embayments provide excellent habitat for turtles, and more than a dozen species have been recorded. The common snapping turtle, stinkpot, mud turtle, common map turtle, midland painted turtle, and spiny and smooth soft-shell turtles are present almost everywhere in the basin. The alligator snapping turtle, Ouachita false map turtle, slider, and Florida cooter are present in the main-stem reservoirs. The red-eared slider occurs in the lower half of the drainage and is replaced by the Cumberland slider in the upper half. Four water snakes occur in the lower reservoirs, but only the common water snake and the queen snake are abundant throughout the drainage. The cottonmouth is present but not common in Kentucky Lake. A dozen species of frogs, including green frog, pickerel frog, and bullfrog, occur in the Tennessee basin and are common in the reservoirs (Barbour 1971).

Once rare, bald eagles and ospreys have become common sights, and both have breeding populations along the lower Tennessee. Great blue herons, gulls, terns, Canada geese, and a variety of ducks are common throughout the year. The Tennessee National Wildlife Refuge units have increased the winter abundance of waterfowl throughout the lower part of the basin. Beaver are once again common throughout the basin and considered a nuisance. River otters and muskrats also are abundant.

Ecosystem Processes

With the exception of headwater tributaries, processes occurring in the extensive series of impoundments dominate the ecosystem. Shallow shoals, large snags, and accumulations of woody debris have all but disappeared; thus, benthic processes appear to be much less important than they were prior to impoundment. Although the reservoirs are net heterotrophic and the principal organic matter supply may still be allochthonous, autochthonous primary production probably fuels most water-column processes (Thornton et al. 1990, Kimmel et al. 1990, also see Thorp and DeLong 2002). The chemistry of riverine zones is dependent on what is released from the upstream river; however, productivity is usually low because of the high turbidities. As turbidity and velocity decrease in the transitional zones, productivity may be quite high if NO_3-N and PO_4-P are present in sufficient quantities. Lacustrine zone NO_3-N, PO_4-P, and Si reach maximum water-column concentrations in March, producing chlorophyll peaks in early April (D. White, unpublished data), and productivity rates of $1400\,mg\ C\ m^{-2}\,day^{-1}$ are common (Taylor 1971). By early June NO_3-N and PO_4-P concentrations often are too low to measure. Very little primary or secondary production accumulates in bottom sediments, as they typically contain <1% organic carbon. Much of the fixed carbon probably is flushed through the dams or metabolized in the water column (Yurista et al. 2001); thus, water retention time is likely the controlling factor for productivity (Kimmel et al. 1990).

Because reservoirs create ecosystem conditions that did not exist previously in the basin, conceptually these are "new" ecosystems. Reservoir ecosystems do not reach the longitudinal and temporal equilibriums of the parent river (Thornton et al. 1990), producing conditions ripe for invasion of true nonnative plants and animals that are highly adaptable. Although most species occurred in the system prior to impoundment, the dominant species now are those adapted to the new set of environmental conditions. Understanding ecosystem processes in the Tennessee River requires a much better knowledge of the sources and fates of carbon and nutrients, as well as a knowledge of the aging processes that occur as the reservoirs fill in.

Human Impacts and Special Features

The species diversity and richness of the Tennessee River prior to European settlement was equal to any basin in North America. Despite extensive impoundment and a myriad of other human influences, a number of smaller rivers and sections of rivers have retained high biodiversity (Abell et al. 2000). The Clinch, Holston, North Fork, and Powell rivers have high diversities of fishes, freshwater mussels, invertebrates, and amphibians, although there are continued threats from strip mining and forest practices. Similar high diversities occur in portions of the Hiwassee and Little Tennessee rivers that drain parts of Georgia and North Carolina. The Duck and Buffalo rivers of central Tennessee are cool, clear streams in the Highland Rim that have retained a diverse aquatic insect fauna and macrophyte flora, including extensive beds of hornleaf riverweed. The small Obed River basin contains 73 km of river that are part of the National Wild and Scenic River system. The Tennessee River from the Kentucky Lake Dam to the Ohio River is an important site for a number of large-river freshwater mussel species because it is free flowing and has not yet been heavily colonized by the zebra mussel. Enlargement of the lock and construction of

new bridges is expected to lead to mussel habitat loss mitigation and relocation of about 70,000 individual mussels (J. Sickle, personal communication).

Many extensive impacts of the reservoirs have already been described. In addition to impoundment, the Tennessee River has been affected by siltation and turbidity, deforestation, dredging and canalization, industrial and domestic pollution and enrichment, herbicides and pesticides, acid mine drainage, diversions, thermal modifications, and nonnative species. Several hundred kilometers in the basin are considered nonsupporting of aquatic life due to industrial and domestic waste pollution. Strip mining and resulting siltation and acid mine drainage continue to be a problem in the Appalachians, particularly in the Clinch and Sequatchie rivers (Burr and Warren 1986, Etnier and Starnes 1993).

The Tenn–Tom Waterway has permitted migration of both fishes and invertebrates between two very different ecoregions, and the long-term effects remain to be seen. The Tennessee National Wildlife Refuge operates three diked and seasonally flooded waterfowl habitat units in the Kentucky Lake portion of the river. When water is drained into tributary embayments, excess nutrients result in anoxia in the bottom waters. Apatite mining that once occurred in the Duck River may still be releasing particulate phosphorus into the lower Tennessee system. As noted by Burr and Warren (1986), the specific causes for the demise of most species are often difficult to attribute to particular sources. Cumulative effects from multiple stressors may cause much longer lasting damage to a basin, and the effects on species are more difficult to detect than those from any single source alone.

CUMBERLAND RIVER

The Cumberland basin covers $46,430\,km^2$ of southeastern and western Kentucky and the north-central part of Tennessee (Fig. 9.18). The Cumberland begins at the confluence of the Poor, Martin's, and Clover forks at the town of Harlan in southeastern Kentucky. From Harlan, the river runs generally southwest into northern Tennessee through the city of Nashville, where it turns northwest until it nears the Tennessee River. The Tennessee and Cumberland rivers then parallel each other, flowing northward to the Ohio River. The Cumberland empties into the Ohio River at Rkm 1481 as a 7th order river near the town of Smithland, Kentucky, after covering a distance of about 1120 km (Fig. 9.5). In general the

FIGURE 9.5 Cumberland River near Smithland, Kentucky (PHOTO BY G. HARRIS).

upper portion of the river's landscape is low mountains with quite a bit of forest, whereas the landscape of the lower half is more rolling and agricultural. Major tributaries include the relatively pristine Rockcastle River, the Big South Fork, the Caney Fork River, the Obey River, the Harpeth River, the Red River, and the Little River. The Barkley Canal between Lake Barkley (Cumberland River) and Kentucky Lake (Tennessee River) allows water to flow freely between both rivers. Local residents quip that the Tennessee River is now the largest tributary to the Cumberland River.

Human influence began at least 12,000 years ago with the Paleo-Indian cultures that entered as the climate warmed. The limited agriculture within the central Cumberland Valley 5000 years ago grew considerably in the Woodland period from about 3100 to 1200 years ago, along with pottery and extensive trade routes. The Mississippian period (1200 to 500 years ago) saw development of cities with temples, improvements in agriculture, and warfare among neighboring tribes. It is not known who the first Europeans were to see the basin, but the original name, "Rivière des Chauouanons" (Shawnee River), is attributed to Jacques Marquette and Louis Joliet, who visited the river in 1674. The primary cultures at that time were the Choctaw, Cherokee, Shawnee, and Chickasaw. The river received its present name after William Augustus, Duke of Cumberland in 1750, by Dr. Thomas Walker, one of the first English explorers and settlers (McCague 1973). Disease and firearms brought in by traders and rapid European settlement in the late 1700s all combined to displace the native cultures. The Cumberland Valley was settled by Europeans coming from the east through the Cumberland Gap and from the south through the Tennessee River valley. Several towns were established, including what would become Nashville, the largest city on the river. By the end of the nineteenth century much of the land in the lower part of the basin had been cleared for farming and was in poor condition. Coal mining, which had begun a hundred years earlier in the Kentucky Eastern Coal Fields of the Appalachian Mountains, was greatly expanded by a number of competing railroad lines. Thus, two very different environments and cultures, mining and agriculture, were established and persist to the present (McCague 1973).

Physiography, Climate, and Land Use

The headwaters of the Cumberland begin in the Appalachian Plateaus (AP) and Valley and Ridge (VR) physiographic provinces (see Fig. 9.18). The Valley and Ridge is composed of Paleozoic limestones, sandstones, and shale, whereas low-nutrient Pennsylvanian limestone, sandstone, coal beds, and shale underlie the Appalachian Plateaus. The river then passes through the Highland Rim and Nashville Basin sections of the Interior Low Plateaus (IL) province, with Ordovician and Mississippian age limestones. The Highland Rim is rich in caves, many of which feed into the Cumberland River, raising the generally low productivity. The Nashville basin represents the remnants of the highly eroded Nashville Dome, marked by chert and soluble Ordovician limestones (Hunt 1967). The lower Cumberland River is again very cherty and underlain by Mississippian limestones.

The headwaters lie in the highly diverse Appalachian Mixed Mesophytic Forests terrestrial ecoregion, with American beech, sugar maple, eastern hemlock, mountain laurel, mountain maple, and rhododendrons at higher elevations and oaks, hickories, ashes, maples, pines, and elms at lower elevations. The lower portion is in the Central United States Hardwood Forests terrestrial ecoregion and is comprised of white oak, southern red oak, hickories, American elm, tulip tree, and sweetgum, with understories of winged elm and flowering dogwood (Ricketts et al. 1999).

The climate is continental temperate. Monthly mean temperatures range from about 2°C in January to 25°C in July (Fig. 9.19). Temperatures are generally cooler in the Appalachian Mountains and warmer throughout the Highland Rim. Temperatures in winter regularly fall below 0°C and often exceed 38°C in the summer. Annual precipitation ranges from about 90 to 175 cm (average 127 cm). It is quite evenly distributed throughout the year, with highest amounts usually in March, April, and May (11 to 13 cm) and lowest in October (about 8 cm) (see Fig. 9.19).

The basin is approximately 51% forest, agriculture accounts for about 40% of land use, about 5% is urban, about 3% is water, and the remainder is fallow or in mining (TVA, unpublished data). The principal industries in the Appalachian Plateaus region are coal mining and tourism, and both have had a decided effect on the ecosystem. The Interior

Low Plateaus areas have been in agriculture for more than 150 years. Better farming practices put in place after the Great Depression have decreased erosion and increased crop yields. Primary crops are corn, soybeans, cotton, winter wheat, and Kentucky Colonels (Hunt 1967), and tobacco remains a cash crop throughout the basin. Human population density for the basin is about 16 people/km² and is concentrated primarily within the Nashville metropolitan area (about 1,200,000 people).

River Geomorphology, Hydrology, and Chemistry

The elevation of the Cumberland at the confluence of the Poor, Martin's, and Clover forks is approximately 355 m asl, and it is 90 m asl at the mouth with an overall mean slope of 23.5 cm/km. The Cumberland's headwater streams are deeply incised but relatively low gradient, with sand and shale bottoms. Waterfalls are common, the largest of which is Cumberland Falls, which drops 20 m. Further downstream the larger tributaries cut deeply into Mississippian limestones, whereas side tributaries often are still in sandstones and cherts, giving the region a natural mosaic of substrate types. From the Highland Rim to the Ohio River, stream gradients are moderate, and where the river is free flowing, stream bottoms are generally sandy with chert cobble (Starnes and Etnier 1986, Etnier and Starnes 1993).

There are nine major impoundments in the basin, five of which are on the main stem, and there are several smaller impoundments on the tributaries. Other than for some short sections, the Cumberland River itself is no longer a free-flowing river. With the exception of the Great Falls Reservoir on the Caney Fork River (built in 1916), which is now owned and operated by TVA (TVA 1980), the reservoirs are managed by the U.S. Army Corps of Engineers for a variety of purposes, including flood control, power generation, navigation, and public water supply. The oldest reservoir is Dale Hollow Lake on the Obey River, built in 1943, and the newest is Cordell Hull Lake on the Cumberland River, built in 1973. Barges and other large traffic can lock through to Nashville, approximately 300 km upriver, and boats with less than 1 m draft can navigate as far as Burnside, Kentucky, about 825 km upriver. The largest reservoir is Lake Barkley, with the dam at Rkm 40. Cheatham Lake begins at Rkm 230, Old Hickory Lake at Rkm 346, and Cordell Hull Lake at Rkm 502. Together the four reservoirs create one nearly continuous 600 km long main-stem impoundment.

The fifth main-stem impoundment is Lake Cumberland at Rkm 738. Lake Cumberland is the 2nd largest reservoir in surface area, but unlike the other impoundments it is clear and relatively deep and is classified more as a tributary impoundment (Thornton et al. 1990).

The hydrology and chemistry of the Cumberland River are now dominated by the reservoirs, as little remains free flowing. As with the Tennessee River, the four lower main-stem reservoirs can be divided into riverine, transitional, and lacustrine functional zones, each with its distinct hydrologic, chemical, and biological characteristics (Thornton et al. 1990). Average depth of these reservoirs is shallow (<4 m), except in the old river channels. Riverine zones follow the original river geomorphology and generally occupy the original riverbanks, but the original sequences of shoals and pools no longer exist. Flow in the riverine zones depends on release cycles from the upstream dams and usually is kept at levels that provide sufficient depth for transportation. The riverine zone bottoms are primarily sand and cherty gravel. Water velocity slows and larger organic and inorganic particulates drop out in the transitional zones. Bottom substrates are often mixed sands, cherty gravel, and pockets of organic deposition. Bottom substrates in the lacustrine zones are usually composed of fine, clayey deposits of low organic content (generally <1% carbon).

The mean discharge at the mouth of the Cumberland is 862 m^3/s; however, all of the main-stem reservoirs are highly regulated to hold back floods and spring rain runoff and to provide a guaranteed year-round shipping channel depth. Runoff is highest from January through March and decreases with evapotranspiration to lows during fall months (see Fig. 9.19). In Lake Barkley, the farthest downstream in the system, winter pool (August to March) is approximately 2 m lower than summer pool (April through early August). Summer and winter pool elevation differences are much more pronounced in the tributary reservoirs and Lake Cumberland, allowing them greater storage capacity.

The chemistry of the upper portion of the Cumberland River represents a mosaic of low-nutrient, low-alkalinity tributaries flowing through shale and sandstone, which empty into a main river that has cut into more productive limestones (Etnier and Starnes 1993). Alkalinity of the upper tributaries ranges from 9 to 30 mg/L as CaCO$_3$. Alkalinity at Nashville averages 60 mg/L as CaCO$_3$ and then nearly 80 mg/L as CaCO$_3$ as it empties into the Ohio River. Calcium, SO$_4$, Si, and total P also tend to increase in a downstream direction, but their concentrations are highly dependent on annual cycles within the reservoirs. Average NO$_3$-N values again are dependent on reservoir processes but tend to decrease from Nashville downstream.

Beginning with Cordell Hull Lake, the chemistry of riverine zones reflects what is released from the upstream reservoir, but productivity is usually low because of the high turbidities. As turbidity and velocity decrease in the transitional and lacustrine zones, productivity may be quite high. Nitrate-nitrogen, PO$_4$-P, and Si reach maximum water-column concentrations in March, producing chlorophyll peaks in early April, and productivity rates of 1000 mg C m^{-2} day^{-1} are common (D. White, unpublished data). Nitrate-nitrogen and PO$_4$-P concentrations in early June are often less than 1 µg/L. Much of the fixed carbon of each reservoir probably is flushed through the dams or metabolized in the water column, as very little accumulates in the sediments. pH is generally at or above 7.2 but can increase from 7.2 to 9.5 within a matter of hours on warm, sunny days in April and May. Even in the lacustrine zones water velocity remains high, preventing thermal stratification, but oxygen deficits may occur in late summer when discharges are low.

River Biodiversity and Ecology

The Cumberland River is part of the Tennessee–Cumberland freshwater ecoregion, which contains the greatest diversity of aquatic biota in the United States (Abell et al. 2000). Both rivers have very similar fauna and share many endemic species, with the Tennessee River having only slightly greater diversity.

Algae and Cyanobacteria

Following impoundment, phytoplankton became an important part of primary production. The spring diatom peak in the reservoirs (primarily *Melosira*) occurs from late March to early April, and chlorophyll values often exceed 30 µg/L. A wide variety of diatoms (*Navicula*, *Asterionella*, *Fragelaria*, *Melosira*, *Synedra*), green algae (*Scenedesmus*, *Pediastrum*), and cyanobacteria (*Anabaena*, *Oscillatoria*) are the dominant midsummer phytoplankton.

Plants

Because of impoundment and agriculture, most of the original floodplain riparian vegetation has been lost or significantly altered along the main stem. Typical upland hardwood forests (maples, oaks, hickories, etc.) are primarily present. There are no significant macrophyte beds in the main stem, but the native water willow has become established in the backwaters and embayments, along with several nonnatives (brittle naiad, curly pondweed, alligatorweed, parrot feather). Nonnative Eurasian watermilfoil is common in reservoir embayments despite extensive control measures, and hydrilla occurs sporadically. Other native aquatic macrophytes occurring primarily in backwaters include coontails and pondweeds.

Invertebrates

Primary phytoplankton grazers are the cladoceran zooplankton *Bosmina longirostris* and *Daphnia retrocurva*. The major invertebrate predators on *B. longirostris* and *D. retrocurva* are the cladoceran *Leptodora kindti* and phantom midge (*Chaoborus punctipennis*), both of which are abundant throughout the lacustrine zones. Cladocerans along with copepods and larger rotifers (e.g., *Keratella*) are a primary source of food for larval fishes. *Diaphanosoma birgei* was the primary midsummer grazing zooplankter until the 1990s, when the nonnative *Daphnia lumholtzi* displaced it until later in summer.

There are few published surveys of the benthic invertebrates in the main stem of the Cumberland River, either before or since its impoundment, but some data are available in U.S. Army Corps and state agency reports. The longitudinal benthic trends within the impoundments appear to be consistent with riverine, transitional, and lacustrine zone bottom conditions. Filter-feeding bivalves (Unionidae, Sphaeriidae, Asiatic clams, zebra mussels) dominate the riverine zones. The mayfly *Stenonema*, the caddisflies *Cheumatopsyche* and *Ceraclea*, and a variety of Chironomidae are the dominant insects on consolidated substrates. The greatest diversity and density of benthic invertebrates occurs in transitional zones that contain both consolidated and unconsolidated substrates. Primary taxa include surface deposit-feeding tubificid worms (*Limnodrilus*, *Branchiura*), the mayflies *Stenonema*, *Tricorythodes*, and *Caenis*, the caddisflies *Hydroptila* and *Ceraclea*, and numerous Chironomidae, particularly the genera *Ablabesmyia*, *Procladius*, *Cryptochironomus*, *Polypedilum*, *Pseudochironomus*, *Tanytarsus*, and *Chironomus* (D. White, unpublished data).

Lacustrine-zone sediments have generally low organic content even in backwaters and sidearm tributaries. Subsurface deposit feeding tubificid worms (*Limnodrilus*, *Branchiura*) are present but not abundant. The mayfly *Hexagenia limbata* is a common surface deposit gatherer, as are several genera of Chironomidae. The majority of the surface filter feeders are unionids, fingernail clams, and the nonnative Asiatic clams and zebra mussels.

Next to the Tennessee, the Cumberland River drainage has the 2nd highest diversity of mussels in North America. Eighty-seven species and subspecies once occurred, of which only 55 are now presumed to exist (Parmalee and Bogan 1998). Federally listed endangered species include Cumberland bean mussel, Cumberland elktoe, little wing pearly mussel, Cumberlandian combshell, northern riffleshell, tan riffleshell, oyster mussel, purple cat's paw, pink mucket, fanshell, ring pink, orangefoot pimpleback, clubshell, and rough pigtoe. Much of the present diversity exists in the headwater streams in the Daniel Boone National Forest, notably the Rock Castle River and its tributaries (e.g., Horse Lick Creek). Logging, mining, and off-road vehicle use pose threats to mussel populations. Throughout the lower Cumberland River, impoundment has altered mussel populations through habitat loss and loss of glochidia host fish species. Lake-adapted mussels (e.g., threeridge, washboard, floater, mapleleaf, fragile papershell) have become well established in the reservoirs and are now the most common species.

More than 60 species of *Orconectes*, *Cambarus*, and *Procambarus* crayfishes exist in the Cumberland basin, of which about two-thirds are endemic to the Tennessee–Cumberland ecoregion and have very limited distributions (Hobbs 1988). Notable among these is the endangered Nashville crayfish, which is now found only in Mill and Sevenmile creeks in the metropolitan Nashville area. Despite concerns, the distribution of this species continues to decline with increased urban development.

Vertebrates

Between 172 and 186 species of fishes occur in the Cumberland River drainage, including at least 11 nonnative species (Etnier and Starnes 1993). Nine species are endemic: the blackside dace and eight darters (Burr and Warren 1986). Much of the fish diversity occurs in the transition between the Appalachian Plateaus physiographic province and the Highland Rim section of the Interior Low Plateaus physiographic province, which are home to numerous darters, many of which are endemic. Cyprinidae (44 species, primarily minnows, shiners, chubs, and daces) and Percidae (43 species, primarily darters) are the two most diverse families, followed by suckers (17 species), sunfishes (13 species), and catfishes (11 species). Several species are now thought to be extinct in the drainage, including the speckled chub, stargazing minnow, hairlip sucker, crystal darter, and longhead darter. The only native salmonid is the brook trout, but rainbow trout, brown trout, and lake trout were introduced successfully into the upstream reservoirs and some of the tributaries. An unsuccessful attempt was made to introduce cutthroat trout into Center Hill and Dale Hollow reservoirs in the mid-1950s (Etnier and Starnes 1993).

The native fish populations of the basin have been greatly altered by impoundment and introductions (Etnier and Starnes 1993). The principle reservoir forage fishes are now the planktivorous gizzard shad and threadfin shad. The planktivorous alewife was introduced into Dale Hollow Reservoir in 1976 and may be spreading throughout the system. Major piscivores now include channel catfish, gars, small cyprinids, largemouth bass, smallmouth bass, and the introduced white bass and striped bass. Species feeding primarily on aquatic invertebrates include the common carp, paddlefish, suckers, sculpins, catfishes, and a vast array of darters, sunfishes, and minnows. The freshwater drum feeds heavily on mollusks, particularly Asiatic clams, and may find an additional food source in the zebra mussel. Channel catfish and paddlefish are the primary commercial species.

Most of the Midwestern species of turtles and water snakes are found in the drainage (Barbour 1971). Much additional habitat for turtles now occurs in the reservoir backwaters and embayments. The most abundant turtles throughout are common snapping turtle, stinkpot, mud turtle, common map turtle, midland painted turtle, spiny softshell, and smooth softshell. Several more western or southern species are found in the lower portion of the river through Lake Barkley: alligator snapping turtle, Ouachita false map turtle, and slider. The red-eared slider is common in the lower half of the drainage and replaced by the Cumberland slider in the upper half. Four species of *Nerodia* water snakes are present in the lower portion of the drainage on snags and fallen trees along the riverbanks and reservoir embayments. Only the common water snake and the queen snake are widespread throughout the drainage. The cottonmouth is present in the lower part of the river, including Lake Barkley, but is not common. At least 11 species of frogs occur in the Cumberland basin, and many are common in the river and reservoirs (e.g., pickerel frog, green frog, bullfrog, leopard frog).

Bald eagles and ospreys are now a common site in the lower half of the basin. Common waterfowl include blue herons, terns, gulls, mallards, coots, and Canada geese. Loons are present in the winter and spring months. Beaver, muskrat, and river otter also are common, particularly in the tributaries and embayments.

Ecosystem Processes

The ecosystem of the main-stem Cumberland River is now almost completely dominated by impoundments. The once common shoals, large snags, and accumulations of woody debris are gone except along the shorelines and around islands, and benthic processes probably are much less important than they were prior to impoundment. Allochthonous inputs may still provide the greatest amount of organic matter, creating net heterotrophy; however, autochthonous primary production probably drives most water-column metazoan production (Thornton et al. 1990, Kimmel et al. 1990, also see Thorp and DeLong 2002). Reservoir riverine zones are dependent on upstream releases, but productivity is usually low because of the high turbidities. Turbidity and velocity decreases in the transitional and lacustrine zones promote chlorophyll peaks when NO_3-N, PO_4-P, and Si concentrations are high, particularly in March and April (D. White, unpublished data). Midsummer NO_3-N and PO_4-P concentrations often are too low to measure. Very little primary or secondary production accumulates in bottom sediments, as it most likely is flushed through the dams or metabolized in the water column. Even under low nutrient conditions, water retention time may be the overriding factor for productivity (Kimmel et al. 1990). For example, the average hydraulic retention time for Lake Barkley is about 12 days. If hydraulic retention time is increased to greater than about 20 days, plankton blooms occur (U.S. Army Corps of Engineers, unpublished data).

Many of the species that existed in the lotic conditions prior to impoundment have been displaced or are now limited to tributaries. The lack of longitudinal and temporal equilibriums in manipulated reservoirs favors plants and animals that are highly adaptable and increases the potential for invasion of nonnatives. Understanding ecosystem processes now occurring in the lower Cumberland River requires a much better knowledge of reservoir processes, including hydraulic retention, sediment biogeochemistry, and reservoir aging.

Human Impacts and Special Features

A major special feature of the Cumberland is its biodiversity, of which several pockets remain in the headwater reaches, particularly in the Daniel Boone

National Forest. Outstanding among these is the Rockcastle River, which is home to many endangered fishes and freshwater mussels and a high diversity of darters. Headwater rivers have been geologically isolated by Cumberland Falls, producing considerable endemism. More than two dozen nature preserves, some as large as 2500 ha, are providing some additional protection. Seven streams have been designated as Wild Rivers by the state of Kentucky, including portions of the Big South Fork, Martins Fork, and the Rock Castle River. Protected lands in the lower portion of the basin include the Land-Between-the-Lakes National Recreation Area (U.S. Forest Service) and the Cross Creeks National Wildlife Refuge.

The Cumberland River and its basin have two sets of impacts that are related to differences in geology and culture: pollution and dams. The lower Cumberland has been affected by municipal and industrial discharges in the Nashville metropolitan area and receives considerable agricultural runoff further downstream; for example, NO_3-N levels in the Little River (Kentucky) result in chlorophyll levels of up to $100 \mu g/L$ (D. White, unpublished data). The lower portion of the Cumberland basin also contains extensive karst flow systems that are showing signs of water-quality degradation from nutrients and pesticides. Siltation, deforestation, domestic pollution, herbicides and pesticides, diversions, and off-road vehicle use affect the upper portion of the Cumberland River basin, which contains much of the high diversity of fishes and invertebrates. Several hundred kilometers of the basin, including the main stem, have been classified as nonsupporting of aquatic life due to industrial and domestic pollution. Strip mining and resulting siltation and acid mine drainage continue to be a problem throughout the Appalachians (Burr and Warren 1986, Etnier and Starnes 1993).

As mentioned, a primary human impact on the main stem is the series of large multipurpose reservoirs that are managed for hydroelectric power generation, flood control, and navigation. The reservoirs have dramatically altered not only the river but also the surrounding landscape. Natural riverine processes have been completely altered, and the native plant, fish, and invertebrate species largely have been displaced or their functions have been supplanted by nonnative or once uncommon species.

Asiatic clam densities have generally decreased over the past 20 years, but the zebra mussel is now present and may be breeding as far upstream as Nashville. The effects of zebra mussels on the river ecology and native unionids are unknown. Identifying the effects of nonnative fishes is problematic because so much of the Cumberland has been drastically altered by impoundment.

WABASH RIVER

The name "Wabash" comes from the Miami phrase "wah-bah-shik-ki," meaning "pure white" in reference to the limestone bedrock that could be seen through the clear water in pre-European times. The Wabash is an 8th order river and the 2nd largest tributary in area to the Ohio River ($85,340 km^2$). It originates in the rolling agricultural countryside of western central Ohio that also contains the headwaters of the Maumee, Great Miami, and Scioto rivers. From there the Wabash flows west through Indiana, then south, eventually forming the border between Indiana and Illinois until it empties into the Ohio River (Fig. 9.20) at Ohio River Rkm 1365. The Tippecanoe River is the 2nd largest tributary to the Wabash in Indiana and is one of few larger rivers that is unimpounded throughout much of its length (Benke 1990, Abell et al. 2000). The major Wabash River tributaries draining Illinois are the Little Wabash and Embarras rivers. The Wabash, Tippecanoe, Little Wabash, and Embarras rivers flow through a large expanse of relatively flat agricultural landscape (Fig. 9.6). The tributary White River flows through Indiana's largest city, Indianapolis, and through the more hilly regions of southern Indiana. The Wabash and White rivers are of approximately equal discharge at their confluence downstream from Vincennes (Gammon 1991).

Archaic hunter-gatherer tribes existed in the basin from approximately 10,000 to 3000 years ago, when agriculture was introduced throughout most of eastern North America (Hyde 1962). One of the early cultures was the Adena, who built permanent settlements and the first of the burial mounds in the area. By 2200 years ago the Hopewell culture appeared and even larger settlements and more elaborate burial mounds were produced. By 1500 years ago the Hopewell culture was replaced by temple- and fortress-building Mississippian cultures. Mississippian cultures dominated through the 1500s until the first Europeans arrived in the region. In the 1600s the Iroquois, armed with European weapons, increased their raids from Canada into the Ohio Valley, eventually displacing most other Indian tribes until the late 1700s, when tribes such the Miami, Wea, Piankashaw, Potawatomi, and Kickapoo

FIGURE 9.6 Wabash River near Montezuma, Indiana (Photo by T. Harris).

returned to the Wabash Valley. The Potawatomi were defeated in 1811 by William Henry Harrison at the battle of Tippecanoe, losing their rights to land in Indiana, and eventually all tribes were moved westward as European settlement increased (Hyde 1962).

The first Europeans were Spanish explorers in the 1500s; however, it was not until the late 1600s that the French and English began to actively vie for trade and trading routes. By the early 1700s the French had built a major fort at what now is Vincennes, but in 1765 the British took control of the region. George Rogers Clark defeated the British at Vincennes in 1779, giving control of the Northwest Territory to the United States (Gammon 1991). With fertile farmland and ample water transportation routes, the European population quickly expanded to nearly 150,000 by 1820. In 1828 the decision was made to construct an extensive network of shipping canals throughout the state that would permit easy transport of goods to ready markets in the east and south. The most extensive of these was the Wabash and Erie Canal, which began at Fort Wayne, running to Lafayette and then southward to Terre Haute and

eventually ending near the town of Washington, east of Vincennes. Floods, drought, and competition from the developing railroad system continually plagued operation of the canals, and eventually they were abandoned in the late 1850s, but the building and operation of the canals had greatly increased agriculture, lumber production, quarrying, and manufacturing throughout the basin (Gammon 1991).

Physiography, Climate, and Land Use

The entire Wabash River valley lies in glacial tills of various ages and depths. The upper quarter of the Wabash is in the relatively steep Bluffton Till Plain section of the Central Lowland (CL) physiographic province. Between the towns of Logansport and Clinton (north of Terre Haute), the river flattens out, passing through the Entrenched Valley section of the Central Lowland and then through the Interior Low Plateaus (IL) province at its mouth (Hunt 1967). Glacial deposits with occasional limestone outcrops largely cover this region of the Central Lowland.

Pennsylvanian and Mississippian limestones, sand, and chert underlie the Interior Low Plateaus.

The northern and eastern portions of the basin lie in the Southern Great Lakes Forests terrestrial ecoregion, which is dominated by sugar maple, American beech, basswood, oak, elm, hickory, and ash. The middle section of the river drains a small portion of the Central Forest/Grassland Transitional Zone, with both tallgrass prairies and scattered woodlands. The lower portion lies within the Central United States Mixed Hardwood Forests ecoregion (Ricketts et al. 1999). Forests in this ecoregion are primarily white oak, red oak, hickory, American elm, tulip tree, and sweetgum, with understories of winged elm, flowering dogwood, and sassafras.

The climate is continental temperate, and the northern parts of the basin receive an average snowfall of 152 cm/yr. Average rainfall in the basin is 96 cm/yr, with the upper portion of the basin receiving about 91 cm/yr and the lower portion about 112 cm/yr. Rainfall is relatively evenly distributed throughout the year, although April through July may be slightly wetter than fall months (Fig. 9.21). January temperatures average −4°C and July temperatures average about 24°C (see Fig. 9.21). Long periods of winter temperatures below 0°C are common in the north, as are long periods greater than 35°C in the south during summer.

The mean population density as of 2000 was 62.7 people/km^2. Approximately 5% of the Wabash River basin is urban. The largest city by population in the basin is Indianapolis (731,000) on the White River portion of the Wabash drainage. Major cities through which the Wabash main stem runs include Lafayette (70,000), Terre Haute (58,000), and Vincennes (20,000). With the construction of the canals and the concurrent draining of the land, much of the basin was converted to agriculture by the mid-1800s and remains some of the richest farmland in the Midwest. Major crops are corn and soybeans; however, the basin south of Terre Haute is a prime area for peaches, melons, and vegetables. About 65.2% of the Wabash basin is in agriculture and an additional 8.2% is pasture. Only about 21.6% is forested lands, primarily second growth, and most forested lands occur in the southern portion of the basin (ORSANCO 1990).

River Geomorphology, Hydrology, and Chemistry

The Wabash flows through an ancient valley dating to the Devonian (350 million years ago) that is now filled with till left by Illinoisan and Wisconsin glaciers. The entire valley was ice covered as late as 20,000 years ago (Gammon 1991). Portions of the Wabash were once part of the Teays River system, draining west through present-day Indiana and Illinois. About 14,000 years ago the Wabash drained glacial Lake Maumee through the Maumee River but was cut off by uplifting (Hough 1958). The total length of the present main stem is 772 km. The present origin is near Fort Recovery, Ohio, at an elevation of approximately 267 m asl. Between the origin and Logansport, Indiana, the gradient is 45 cm/km. The gradient abruptly changes below Logansport and averages only 12 cm/km, entering the Ohio River at 97 m asl (Gammon 1991). Prior to the 1820s the Wabash was described as a clear river with relatively constant discharge and numerous sandbars and long cobble riffles (Gammon 1991). Although the river remains free flowing for much of its length, the long history of agriculture and canal building has changed the river to quite turbid and highly variable in discharge, and the amounts of silts and clays on the bottom have greatly increased.

Mean discharge for the entire Wabash system is 1001 m^3/s at its mouth on the Ohio. At their confluence the Wabash River contributes 341.5 m^3/s and the White River contributes 335.6 m^3/s, with most of the remainder coming from the Little Wabash River and other tributaries draining southeastern Illinois. The high amounts of evapotranspiration in late summer (see Fig. 9.21) are in part due to agricultural practices that drained marshy and swampy areas (Gammon 1991). Mean monthly discharge is nearly four times greater in March and April than in September and October (see Fig. 9.21). The mean annual discharge is approximately doubled in the wettest years compared with dry years; however, because of the incised channel the river rarely leaves its banks.

Water temperatures as measured at Lafayette, Indiana, are typically at or just above 0°C in January and reach 25°C to 30°C in July and August. Although the river was historically clear, suspended solids now average greater than 60 mg/L in the upstream reaches to greater than 150 mg/L near the mouth at New Harmony. Conductivity generally decreases in a downriver direction (565 μS/cm in upper reaches to 488 μS/cm at New Harmony), as do NO_3-N (3.57 to 2.17 mg/L), NH_3-N (0.195 to 0.145 mg/L), and SO_4-S (64.7 to 60.1 mg/L) (ORSANCO 1990). As with most large rivers, PO_4-P increases in a downriver direction from an average of 0.185 mg/L at Peru to 0.300 mg/L at New Harmony. Both phosphate and nitrate loadings were greatly reduced in the 1970s through improved sewage treatment, the ban on phosphate detergents, and better farming practices. Agriculture remains the major loading contributor (Gammon 1991). Dissolved oxygen concentrations are generally near saturation at midday but may drop to near 50% of saturation during warm summer nights. pH averages 7.2 but may range from 6.8 to 7.8 over a 24-hour period. Alkalinity ranges from 160 to 250 and averages about 200 mg/L as $CaCO_3$ (Gammon 1991).

River Biodiversity and Ecology

The Wabash lies in the Teays–Old Ohio freshwater ecoregion (Abell et al. 2000). As with other rivers in this ecoregion, the diversity of aquatic plants and animals is high. Because the entire basin was ice covered during the Illinoisan glaciation, there are no endemic species.

Algae and Cyanobacteria

Although the Wabash has notably high chlorophyll concentrations (Bukaveckas et al. 2000), the taxonomic distribution of phytoplankton is not well documented. Benthic diatoms of some of the tributaries (e.g., Embarras River) include species of *Achnanthes*, *Cocconeis*, *Gomphonema*, *Navicula*, *Nitzschia*, and *Surirella* (Vaultonburg and Pederson 1994).

Plants

Although much of the floodplain forest along the Wabash was removed in the 1800s and converted to agriculture, some forest does remain along most of the banks. Floodplain forests in the Wabash valley above Terre Haute are typical of the Great Plains and include red maple, cottonwood, black willow, black gum, and American sycamore. Below Terra Haute are the Lower Wabash valley woodland floodplains, which contain these species and the northern extensions of sugar hackberry, waterlocust, and bald cypress (Deam 1953). The contributions of the floodplain forests to the ecology of the river remain unstudied; however, it is assumed that snags and fallen trees provide significant habitat and cover along the banks and a source of organic matter. Nonnative aquatic species include purple loosestrife and Eurasian watermilfoil.

Invertebrates

As once occurred in the rivers in this part of the Midwestern United States, freshwater unionid mussels were very abundant in the Wabash prior to European settlement and about 75 species historically have occurred in the basin (Cummings and Mayer 1992). Indian tribes once used the mussels as a source of food, and early settlers used the shells as a source of lime. Until the early twentieth century the shells were made into buttons. Cummings and Mayer (1994) collected 68 species in the late 1980s, but only 51 were alive. Eleven species are either now extinct or extirpated from the river. Three species are federally listed as endangered: clubshell, fat pocketbook, and eastern fanshell. Although populations of fragile papershell, mapleleaf, and hickorynut are still common, the density and diversity of unionids have been greatly reduced through loss of habitat and water-quality changes in the river. The nonnative Asiatic clam and zebra mussel are common throughout the basin, but their effects on native species have not been examined.

Little has been published on the other benthic invertebrate groups in the Wabash. Based on unpublished drift and artificial substrate studies (D. White, unpublished data), the main stem of the Wabash contains a diverse array of aquatic insects. Bottom sands and silts contain sediment-dwelling mayflies such as *Hexagenia*, *Caenis*, *Tricorythodes*, *Anepeorus* (Wallace's mayfly), and numerous chironomid midge genera. The greatest diversity appears to be associated with logs and woody debris. The mayflies *Stenonema* and *Heptagenia* are abundant, along with the caddisflies *Cheumatopsyche*, *Hydropsyche*, *Ceraclea*, *Hydroptila*, and *Ceratopsyche*, and additional genera of Chironomidae. *Allocapnia* and other winter stoneflies are present, as well as *Acroneuria*, *Isoperla*, *Perlesta*, and *Neoperla*. The megalopterans *Corydalus* and *Sialis* are common, as are the true flies *Simulium* and *Tipula*. Common aquatic beetles include *Tropisternus*, *Berosus*, *Gyrinus*, and *Stenelmis*.

Vertebrates

In the early 1800s the Wabash was known for its abundant fish resources, particularly black bass, pickerels, sunfishes, and catfishes. Gammon (1991) recorded 76 species of fishes in the Wabash proper, which did not include several darters, small cyprinids, and other small species that are common in smaller tributaries or not often collected by electrofishing. The total number of species in the basin is probably greater than 100. The most abundant main-stem species recorded by Gammon (1991) were gizzard shad, common carp, channel catfish, longnose gar, shorthead redhorse, silver redhorse, golden redhorse, and freshwater drum. Other abundant species include steelcolor shiner, emerald shiner, and quillback, particularly in the lower half of the river. Besides common carp, nonnative species include goldfish, bighead carp, yellow perch, brook trout, and rainbow trout.

Gizzard shad along with skipjack herring are the primary main-stem planktivores. At least 17 species of minnows are present, along with 12 species of sunfishes, most of which are insectivorous. The dominant piscivores are longnose gar, shortnose gar, channel catfish, flathead catfish, white bass, largemouth bass, smallmouth bass, spotted bass, white crappie, and sauger. Bottom-feeding fishes include a diverse array of suckers (15 species), common carp, stoneroller, bullheads and catfishes, paddlefish, and shovelnose sturgeon. Paddlefish are not common but are consistently collected in the lower half of the river. Shovelnose sturgeon is fairly common in the upper half of the Wabash, and numbers may be increasing (Gammon 1991). Other species occasionally found in the river include lampreys, bowfin, mooneye, goldeye, American eel, grass pickerel, and burbot.

The amphibians and reptiles of the Wabash River are typical species of the Midwest (Barbour 1971). Snapping turtle, stinkpot, map turtle, midland painted turtle, and spiny softshell are common throughout the drainage. The Wabash below Terre Haute contains several additional turtle species with primarily southern or western distributions, including false and Ouachita subspecies of the map turtle, mud turtle, red-eared slider, Florida cooter, and smooth softshell. The alligator snapping turtle is rare but present in the extreme lower portion of the river. Four species of water snakes occur, with the common water snake the most abundant on snags and fallen trees along the riverbanks. Queen snake, red-bellied water snake, and Kirtland's water snake are also present. The cottonmouth occurs along the Ohio River and may be present in the extreme southern part of the Wabash. At least 10 species of frogs occur in the basin and many are common in the river (e.g., leopard frog, green frog, bullfrog). Several species of salamander also are found in the river and tributaries, including hellbender, red spotted newt, and mudpuppy. Bald eagle, osprey, and belted kingfisher have become common in the lower basin, along with beaver, muskrat, and river otter.

Ecosystem Processes

As with most other aspects of the Wabash River, there are few published studies on ecosystem-level processes. It can be assumed that the river is heterotrophic throughout and relies on tributary allochthonous inputs for the organic matter base. This probably is particularly true in the upper portions of the river, as extensive macrophyte beds now occur only in the relatively pristine Tippecanoe River tributary. Agricultural development, along with the creation of the canal system, removed most of the original wetlands and organic inputs that may have occurred there. A significant source of organic matter still appears to be riparian vegetation and downed trees along the main-stem banks. Phytoplankton appears to be an important source of primary production in the lower portion of the river (Lafayette southward), even though turbidity is high (Gammon 1991). Primary production most likely is fueled by nitrogen inputs from agricultural runoff, which can result in dramatic day–night swings in dissolved oxygen concentrations.

Human Impacts and Special Features

Because of the low gradient, there are relatively few dams on the main stem or major tributaries. There is a single impoundment on the Wabash itself, Huntington Lake, at Rkm 662, and from there it flows uninterrupted to the Ohio. Benke (1990, also see Abell et al. 2000) includes the Tippecanoe River in his list of major free-flowing rivers in the United States. The Tippecanoe River is 267 km long and drains 5240 km^2 of northeastern Indiana. It originates as the outflow of some 80 glacial lakes, including Tippecanoe Lake, which is the deepest lake in the state. Discharge remains fairly constant throughout the year and the water is clear except after heavy rains. Water quality has remained quite high and fishes and invertebrates are abundant and diverse despite considerable agriculture in the area. Aquatic macrophyte beds are extensive in the upper portion of the river, which has a much gentler slope than the lower half.

Human impacts throughout the basin are primarily related to agricultural activities, including soil erosion and runoff of nutrients and pesticides. Groundwater nitrate levels in heavily agricultural areas often exceed the 10 mg/L maximum contaminant level for drinking water. Pesticide levels, sewage, and industrial spills have led to occasional notable fish kills (e.g., the White River near Anderson, Indiana, in 1999) and fish consumption advisories. Urban releases (sewage, toxic organic contaminants, metals) have caused localized concerns, particularly around Indianapolis, but few basinwide problems. Several coal-fired power plants are on the Wabash and White rivers. Through the early 1970s, cooling waters were taken directly from the river and returned without regulating temperatures, and differences of more than 20°C above and below cooling water outfalls were common.

The limnology and contamination of the White River basin, which occupies 29,400 km^2 or about half the Wabash basin, have been detailed by NAWQA studies in the early 1990s (USGS, unpublished data). The primary focuses were on surface and groundwater quality, and extensive data are available through the U.S. Geological Survey on pesticide concentrations. The White River differs from the remainder of the Wabash basin in having the influences of a larger urban area (Indianapolis at 772 people/km^2), in containing a higher percentage of forested watersheds, and in having its lower to middle reaches in areas of high-relief limestone outcroppings.

KANAWHA RIVER

The name "Kanawha" comes from the Native American word for "place of white rocks," which presumably referred to the natural salt licks that occur in the basin. The Kanawha is a 6th order river and the 4th largest tributary in both area (31,690 km^2) and discharge (537 m^3/s) to the Ohio River. Most of the Kanawha River lies within the heavily forested mountains of West Virginia, with the exception of the New River, which also drains mountainous portions of western Virginia and North Carolina (Fig. 9.22). The basin generally flows in a northwesterly direction, emptying into the Ohio River at Rkm 427. The largest tributaries in order are the New, Greenbrier, Elk, and Gauley rivers. Together they compose more than 60% of the basin area and discharge (Kanawha River Basin Coordinating Committee 1971, unpublished data).

Paleo-Indian hunters were present before 11,000 years ago following retreat of the glaciers. As the large game disappeared, archaic hunter-gatherer tribes developed more permanent settlements throughout most of eastern North America (Hyde 1962). By 2200 years ago the Adena culture had appeared, with settlements, agriculture, and burial mounds. The Adena were followed by the Hopewell, but the Hopewell had all but disappeared by 1500 years ago. By the 1600s the Delaware and Shawnee were the dominant organized tribes, but in the late 1600s the Iroquois Confederacy armed with European weapons controlled most of the trade in the basin. Throughout much of the seventeenth and eighteenth centuries the French and British vied for control of the basin and

fur trade with the prevailing tribes. The Treaty of Paris (1763) ended France's claim to lands east of the Mississippi, giving control to the British. Native Americans, in particular the Shawnee, continued their fights against the British, leading King George III to prohibit settlement in Virginia west of the Allegheny Mountains. In 1768 the Indian Nations (except for the Shawnee) signed a treaty with the British relinquishing rights to land between the Ohio River and the Alleghenies. The Shawnee suffered final defeat in 1774 at the mouth of the Kanawha River. George Rogers Clark's defeat of the British at Vincennes in 1779 transferred control to the United States and effectively ended most tribal rights to lands east of the Mississippi River (Hyde 1962, Williams 1993).

The treaty of 1768 produced the first influx of settlers, who came up the Kanawha River or across the Alleghenies. The end of the Revolutionary War in 1783 saw an even greater influx of settlers (Williams 1993). Except in some of the larger river valleys, much of the rugged landscape was not conducive to agriculture, but timber and coal were plentiful. Coal mining, tourism, forestry, and chemical manufacturing remain among the top economic concerns today and have been responsible for many of the past water-quality problems.

Physiography, Climate, and Land Use

The Kanawha River and its principal tributaries drain portions of the Blue Ridge (BL), Valley and Ridge (VR), and Appalachian Plateaus (AP) physiographic provinces (Hunt 1967) (see Fig. 9.22). The geology of the Blue Ridge is largely Precambrian gneiss, granite, sandstone, conglomerates, and siltstone. The Valley and Ridge is composed more of Paleozoic limestones, sandstones, and shale, whereas the Appalachian Plateaus are primarily low-nutrient Pennsylvanian limestone, coal beds, sandstone, and shale. Landscape gradients are generally steep (often >20%) throughout the basin, limiting accumulation of soils and unconsolidated sediments and allowing high potential soil loss on disturbed sites.

The terrestrial ecoregions are the Appalachian/ Blue Ridge Forests and the Appalachian Mixed Mesophytic Forests (Ricketts et al. 1999). The higher elevations of the Appalachian/Blue Ridge Forests are dominated by red spruce, balsam fir, and Frazer fir, whereas lower elevations are principally deciduous, including red oak, black oak, hickories, birches, and black locust. American beech, sugar maple, eastern hemlock, mountain laurel, mountain maple, and rhododendrons dominate Appalachian Mixed Mesophytic Forests at higher elevations. Lower elevations are rich in a variety of oaks, hickories, ashes, maples, pines, elms, sweetgum, black locust, and black cherry (Ricketts et al. 1999).

The climate is continental temperate, and the northeastern and southern higher elevations may receive considerable amounts of snow in winter. Average precipitation in the basin is 94 cm/yr and is distributed evenly throughout the year, although May is slightly wetter than fall and winter months (Fig. 9.23). The mean basin air temperature is 13°C, with a monthly minimum average of 1.8°C in January and maximum of 23.5°C in July (see Fig. 9.23). The northeastern portion of the basin may be considerably cooler in winter, averaging only −6°C. In the summer months, temperatures greater than 35°C are common.

The mean population density as of 2000 was 29 people/km^2, with about a quarter of the population in Charleston (200,000). Other cities are small, with only five having populations as great as 10,000. The primary land use is forest, at about 70%. Because of the slope, agriculture (23% of land use) is limited to larger river valleys and some ridge tops. Agriculture is primarily row crops (soybeans, corn) or pasturelands. Approximately 3% is urban/industrial, and about 1% is water. The remaining 2% to 3% of land use is coal mining or barren from past mining activities.

River Geomorphology, Hydrology, and Chemistry

Contrary to its name, the Kanawha–New River system is often described as the oldest basin in North America (although see Chapter 4, Mobile River) and represents the upper portion and headwaters of the ancient Teays River system. The age of the basin, and in particular the New River, has been the subject of considerable debate, with estimates ranging from 30 to more than 300 million years. The Kanawha River begins at the confluence of the Gauley and New rivers, approximately 145 km upstream from the Ohio River (Fig. 9.7). About 2 km downstream from the confluence is the Great Kanawha Falls. The total relief is more than 1500 m, with most occurring above the falls; thus, slopes of all the major tributaries are quite steep. Tributary valleys are V-shaped, and some are quite deep, particularly in the New River basin. Streambeds and riverbeds are rocky, with some accumulations of sand.

Mean discharge for the Kanawha system is 537 m^3/s at its mouth on the Ohio. Of this volume, the New River contributes nearly 70%. The Elk, Gauley, and Greenbrier rivers contribute another 20%. Monthly mean runoff is greatest in March with snowmelt and lowest in September (see Fig. 9.23) when evapotranspiration is high. Even though the slope is quite steep, flooding is not common. There are four major flood-control dams, one on the Gauley River, one on the Elk River, and two on the New River. Three locks and dams on the main-stem Kanawha allow for barge traffic, hydroelectric power, and some flood control.

Water-quality conditions in the Kanawha River have greatly improved since the 1970s; however, acid mine drainage, agricultural, and industrial/urban inputs throughout the basin still cause some problems. Mean water temperature in the Kanawha main stem is 14°C, but mean temperatures in the higher-elevation tributaries may be much cooler. Winter water temperatures commonly are at or less than 1°C, whereas summer temperatures may exceed 25°C. Main-stem mean pH is 7.3, and mean alkalinity is generally low (38 mg/L as CaCO$_3$). Improvements in sewage treatment and better agricultural practices have lowered nutrient levels. Nitrate-nitrogen in the main stem averages 1.2 mg/L, and PO$_4$-P averages 0.08 mg/L.

River Biodiversity and Ecology

The Kanawha River is part of the Teays–Old Ohio freshwater ecoregion (Abell et al. 2000). Although the Kanawha–New River basin is quite old, endemism and the diversity of aquatic organisms are much less than in the contiguous Tennessee–Cumberland freshwater ecoregion. The Great Kanawha Falls provides a natural barrier to invasion by Ohio River fishes and invertebrates. Beneath the falls the river is free flowing and undisturbed for only about 8 km. This short section, however, has good water quality and supports a high diversity of fishes and freshwater mussels, including two endangered species.

Plants

With 70% of the basin in forests, snags and fallen trees provide habitat and cover along riverbanks and act as a primary source of organic matter. Typical floodplain forests include river birch, red maple, cottonwood, black willow, American sycamore, and tag and speckled alders. Aquatic macrophytes in the main stem are not particularly abundant. Tributaries such as the Greenbrier and New rivers, however, do contain periphyton and submerged and emergent macrophyte beds that have been shown to significantly contribute to the organic matter budget (Hill and Webster 1983). Typical macrophyte species are water willow, hornleaf riverweed, common cattail, elodea, and various pondweeds. Nonnative species throughout the basin include Eurasian watermilfoil, purple loosestrife, parrotfeather, alligatorweed, brittle naiad, curly pondweed, yellow iris, and watercress.

Invertebrates

Many of the >40 species of freshwater mussels occur below the Great Kanawha Falls (e.g., three-ridge, washboard, giant floater, mapleleaf, pistolgrip, and fragile papershell). Above the falls, fanshell, pink mucket, northern riffleshell, and clubshell are listed as endangered species. Some species that are common elsewhere in the Ohio River basin, such as pistolgrip and purple wartyback, are present above the falls,

but there is concern because populations are small. The same is true for the seep mudalia snail. The non-native Asiatic clam occurs above and below the falls (Rodgers et al. 1977), and zebra mussels are present below the falls and may occur in some of the tributaries. It is not known if they are affecting native species populations.

Because of the rocky bottoms and amounts of woody debris, the benthos is quite diverse above the falls and in other major tributaries. Major benthic insect taxa include the mayflies *Baetis, Stenonema, Isonychia, Baetisca, Ephemerella,* and *Leptophlebia,* the stoneflies *Taeniopteryx, Nemoura, Peltoperla,* and *Isoperla,* the hellgrammite *Corydalus,* the alderfly *Sialis,* the caddisflies *Cheumatopsyche, Hydropsyche, Neureclipsis, Macronema,* and *Brachycentrus,* black flies, chironomid midges, and the cranefly *Tipula.* Water pennies and riffle beetles are common and diverse throughout the headwaters and some of the larger streams. Common genera are *Psephenus, Stenelmis, Optioservus, Promoresia, Macronychus,*

FIGURE 9.7 Kanawha River northeast of Charleston, West Virginia (PHOTO BY J. BOYNTON).

and *Ancyronyx*. Gammon's riffle beetle (*Stenelmis gammoni*) is listed as a species of concern in the New River, as are the mayfly *Ephemerella floripara*, the stonefly *Attaneuria ruralis*, the caddisflies *Ceraclea mentiea* and *C. slossonaei*, and the dragonflies *Ophiogomphus mainensis, O. asperses, O. mainensis,* and *Stylurus scudderi* (North Carolina Division of Water Quality, unpublished reports). Several crayfishes are present, including the New River crayfish.

Vertebrates

About 126 species of fishes occur in the Kanawha River basin. Only 46 native species occur above the Great Kanawha Falls (see Fig. 9.22), of which eight are endemic. Of special concern because of limited distributions are sharpnose darter, Kanawha darter, Appalachia darter, Kanawha minnow, New River shiner, and bigmouth chub (North Carolina Division of Water Quality, unpublished data). Several species endemic downstream of the falls have been introduced above the falls, and their ranges are increasing, causing concerns about natural populations. Margined madtoms now occur in the Greenbrier River, and telescope shiner and least brook lamprey are present in the Gauley River. At least 118 large-river species occur in the Kanawha main stem and tributaries below the falls. All of these are common to the Ohio River basin and include gizzard shad, emerald shiner, common carp, channel catfish, longnose gar, golden redhorse, flathead catfish, white bass, smallmouth bass, sunfishes, spotted bass, white crappie, and sauger. A variety of nonnative fish species are present, many of which have been introduced both above and below the falls and may become problematic. In addition to common carp, they include bighead carp, goldfish, grass carp, tench, coho salmon, golden shiner, rainbow smelt, Ohrid trout, rainbow trout, brown trout, yellow perch, striped bass, bighead carp, and brook stickleback.

The amphibians and reptiles of the Kanawha River are typical species of those in the upper Ohio River basin (Barbour 1971), including snapping turtle, stinkpot, midland painted turtle, hieroglyphic river cooter, and spiny softshell. The primary water snakes are common water snake and queen snake. Common frogs in the basin are green frog, pickerel frog, and bullfrog. Numerous species of plethodontid salamander are found in the surrounding Appalachians. River salamanders include hellbender, newt, and mudpuppy. Belted kingfisher, osprey, coot, and a number of ducks are common, along with beaver, muskrat, and river otter.

Ecosystem Processes

Along with portions of the headwaters of the Cumberland and Tennessee rivers, the headwater tributaries of the New and Greenbrier rivers remain comparatively undisturbed. With the exception of the upper New River, however, little has been published on ecosystem-level processes. Hill and Webster (1983) and others have examined macrophyte production, decomposition, and subsequent contributions to the organic matter budget. Macrophytes accounted for about 13% of the total particulate organic matter input and 28% of the POM generated within their study reach. It can be assumed that macrophyte decomposition provides much more labile organic matter than does allochthonous leaf fall. Further, it provides an organic matter source that comes between summer periphyton production and winter allochthonous-based food production. The underlying geology is not particularly rich in nutrients or as a source of alkalinity; however, agricultural runoff and domestic wastes may contribute to autochthonous production. The benthos is rich in filterers, scrapers, and other deposit-feeding aquatic insects. Although the fish fauna is not diverse above the falls, the majority of the species are insectivores.

Human Impacts and Special Features

Two of the primary tributaries above the Great Kanawha Falls are worth special mention. The 266 km long Greenbrier River is listed by Benke (1990) as one of the few relatively natural and unimpounded rivers in this part of the United States. The river flows generally south-southwest in the Valley and Ridge physiographic province, emptying into the New River just downstream from the Bluestone Reservoir. Most of the basin is forested, there is very little mining, and water clarity and quality are generally very good. Benthos, fish, and aquatic macrophyte diversity are among the highest in the Kanawha Basin (NAWQA 2000b; West Virginia Department of Natural Resources, unpublished data). The New River is 510 km long, beginning in the Blue Ridge Mountains of North Carolina, where it flows northeast through Virginia and into West Virginia. This very old river is generally wide and shallow without a well-developed floodplain. The gradient is steep and averages >2 m/km. The 85 km long New River Gorge National River, located between the falls and Bluestone Reservoir (see Fig. 9.22), was created in 1978 by the National Park Service. The New River

Bridge, built across the Gorge, is the largest steel-arch bridge in the United States. Portions of the upper New River (43 km) and the Bluestone River (16 km), a tributary of the New, have been designated National Wild and Scenic Rivers.

Human impacts throughout the basin are related primarily to coal mining, chemical manufacturing, and agricultural activities (NAWQA 2000b). Mined areas greater than 10% occur in some of the Appalachian Plateau physiographic province tributary watersheds. Because much of the coal has low sulfur content, mining, including mountain-top removal, has increased since 1990, and many of the coal seams have been mined repeatedly with new technologies. The Kanawha basin provides about 7% of the coal used in the United States (NAWQA 2000b). The effects of coal mining on water quality have improved since implementation of the Surface Mining Control and Reclamation Act of 1977. Although iron and manganese concentrations in surface water have decreased and acidity is not a major problem, high levels of sulfate continue to be of concern to human health. Siltation from mining activities continues to result in the loss of mayflies (e.g., *Epeorus*), caddisflies (e.g., *Dolophilodes*, *Rhyacophila*), and riffle beetles (Elmidae).

Soil erosion and runoff of nutrients and pesticides from agriculture pose some problems throughout the entire basin but generally fewer than in most other Ohio Basin rivers (NAWQA 2000b). Bacterial concentrations often exceed human contact guidelines in the major tributaries because of poor or nonexistent treatment and allowing cattle to wander in streams. Perhaps more important are contaminants resulting in part from mining but primarily from main-stem industrial inputs in the Charleston and Elkton areas. Even though discharges have been greatly reduced since the 1970s, volatile organic compounds and heavy metals (e.g., dioxin, benzene, PCB, PAH, Ni, Zn, Pb) continue to be found in the sediments and in fish tissues (NAWQA 2000b).

There are three major locks and dams on the Kanawha between the Ohio River and the confluence with the New River, and several impoundments on the tributary rivers. The dams create barriers to movement of fish populations and alter natural runoff patterns by holding back spring rains and snowmelt. More than 50 nonnative species of crayfishes, fishes, and plants have been recorded for the Kanawha River basin (U.S. Geological Survey, unpublished reports), but their affects on native species remain largely unknown.

ADDITIONAL RIVERS

The Green River was named for Revolutionary War hero General Nathaniel Greene. The Green flows west through the Interior Low Plateaus (IL) physiographic province of central Kentucky (Fig. 9.24). The upper Green begins in the heavily agricultural rolling karst area and then flows through Mammoth Cave National Park. Most rainfall enters the system through sinkholes and is discharged from large cave springs (e.g., Mammoth Cave). Biodiversity is high (151 fish species, 71 mussel species) and there are many endemics, including fishes, mussels, and crustaceans. A portion of the Red River tributary (31 km), including that flowing through the Red River Gorge, has been designated a National Wild and Scenic River. The lower Green River flows through the western coalfield region of Kentucky (Fig. 9.8), which continues to be heavily strip mined. The lower river has seven navigation locks and dams that are used to transport coal. Another six dams create recreational and flood-control reservoirs on the major tributaries. Lock and Dam #6 in Mammoth Cave National Park is scheduled to be removed in the next few years.

The Kentucky River begins in the Appalachian Plateaus (AP) province area known as the Eastern Kentucky Coal Fields and flows northwest through the Interior Low Plateaus (Fig. 9.26). The name Kentucky comes from an Indian word meaning "at the head of the river." Acid mine drainage continues to be a serious water-quality problem throughout much of the headwaters. The middle portion of the river passes through parts of the Daniel Boone National Forest, where invertebrate and fish diversity are high but not as high as in the Green River to the west or the Cumberland River to the south. There are 14 navigation locks and dams on the main stem that are used largely for coal transportation. The lower part of the river lies in the agricultural Blue Grass region of the Interior Low Plateaus and is deeply incised into the landscape, with many steep bluffs (Fig. 9.9). The U.S. Geological Survey collected water-quality data from the Kentucky Basin from 1987 through 1990 as part of its NAWQA program, but little has been published to date.

The Great Miami River begins in the Till Plains section of the Central Lowland (CL) province and flows in a southerly direction (Fig. 9.28). Although most of the landscape is gently rolling, the river has cut deeply into the glacial till, leaving very steep-sided valley walls. Nearer the Ohio River the till is thinner

FIGURE 9.8 Green River near Rockport, Kentucky (Photo by J. Boynton).

FIGURE 9.9 Kentucky River near Tyrone, Kentucky (Photo by J. Boynton).

FIGURE 9.10 Great Miami River at Miamitown, Ohio (PHOTO BY K. WILHELM).

and the landscape much hillier (Fig. 9.10). The great Miami Buried Valley aquifer, which follows the pathway of the river from Dayton to the Ohio River, not only supplies base flow but also drinking water to more than 1.5 million people in the basin. Miami and Erie Canal was completed in the 1830s and extended 400 km, linking Toledo with Cincinnati, Ohio, using 103 lift locks. Railroads took the place of the canal in the 1850s, and today the river is not highly fragmented (one major main-stem dam). The Stillwater River has been designated an exceptional warmwater habitat.

The Licking River received its name from the numerous mineral springs that attracted buffalo, deer, and other wildlife. The Licking, as with the Kentucky River to which it runs parallel, begins in the rolling farmlands and Eastern Kentucky Coal Fields of the Appalachian Plateaus province, then passes through the biologically diverse Daniel Boone National Forest (Fig. 9.30). Most of the lower 500 km traverses the agricultural Blue Grass region of the Interior Low Plateaus as it flows northwest to the

Ohio River (Fig. 9.11). Surface coal mines and acid and brine drainage still influence biodiversity in the headwaters. Near the Ohio it receives industrial and domestic inputs from the northern Kentucky portion of the greater Cincinnati metropolitan area. Unlike other rivers flowing northward to the Ohio River, much of the main stem is free flowing. The Licking has only one major dam, which is located at Rkm 287 and forms the 61 km long Cave Run Lake. Biodiversity of fishes, mussels, and aquatic insects is high but generally less than in surrounding river basins.

The Scioto River begins in the agriculturally rich Till Plains section of the Central Lowland (CL) physiographic province and flows southeast across west-central Ohio (Fig. 9.32). The name "Scioto" comes from an Indian word meaning "hairy." The reference apparently is to the amount of hair that would be in the river when the deer were shedding. Headwater aquifers in the glacial tills are geochemically rich, and pH (7.9), alkalinity (183 mg/L as $CaCO_3$), NO_3-N (2.14 mg/L), and PO_4-P (0.33 mg/L) are all among the highest in the Ohio River basin. The main stem and

FIGURE 9.11   Licking River near Midland, Kentucky (PHOTO BY J. BOYNTON).

most of the larger tributaries are deeply incised into the till, often with steep-sided valleys. A major portion of Big and Little Darby creeks (73 km) are designated as National Wild and Scenic Rivers. Fragmentation from dams on the main stem is low, but the middle portion of the basin, including the Olentangy River, continues to show impacts from increased agriculture and urbanization (Fig. 9.12). The river from Chillicothe downstream is in the heavily forested Appalachian Plateaus (AP) province.

The Allegheny River begins in the Appalachian Mountains of north-central Pennsylvania, then flows in a broad "U" shape northwest into New York and then southwest to its confluence with the Monongahela River at Pittsburgh to form the Ohio River (Fig. 9.34). Allegheny is Indian for "fair river." The upper portion of the basin is primarily oak–hickory and beech–maple forests and quite scenic, with little urban development. About 139 km of the upper Allegheny River and 83 km of the Clarion River tributary have been designated as National Wild and Scenic Rivers. There are at least 100 undeveloped islands on the upper Allegheny main stem. The lower part of the basin is fragmented, with eight navigation locks and dams on the main stem. The last 35 km is highly urbanized, and acid mine drainage, agricultural, and industrial/urban inputs still cause water-quality problems (Fig. 9.13).

The Monongahela River derives its name from an Indian word meaning "river without islands," probably in reference to the steep banks throughout much of the basin (see Fig. 9.1). The Monongahela begins at the confluence of the Cheat and Tygart Valley rivers and flows north to Pittsburgh, meeting the Allegheny River to form the Ohio River (Fig. 9.36). The headwaters of the Cheat and Tygart Valley rivers begin in the Monongahela National Forest of southern West Virginia. Many of the tributary streams, however, are heavily impacted by coal mining and acid mine drainage. The main stem is fragmented by nine navigation locks and dams in the lower 150 km, and the entire basin is regulated for flow and navigation. Agriculture is limited in the basin, primarily in the lower half. The last 30 km flow through urban areas near Pittsburgh, with industrial and urban inputs.

FIGURE 9.12 Scioto River at Columbus, Ohio (PHOTO BY K. WILHELM).

ACKNOWLEDGMENTS

We thank Jane Benson of the Mid-American Remote Sensing Center (MARC) for calculating river basin areas, and Mechelle Woodall for assistance with gathering weather and discharge data. Funding for this project was provided in part by Murray State University's Center for Reservoir Research and by National Science Foundation grant DBI 9978797.

LITERATURE CITED

Abell, R. A., D. M. Olson, E. Dinerstein, P. T. Hurley, J. T. Diggs, W. Eichbaum, S. Walters, W. Wettengel, T. Allnutt, C. J. Loucks, and P. Hedao. 2000. *Freshwater ecoregions of North America: A conservation assessment.* Island Press, Washington, D.C.

Barbour, R. M. 1971. *Amphibians and reptiles of Kentucky.* University of Kentucky Press, Lexington.

Benke, A. C. 1990. A perspective on America's vanishing streams. *Journal of the North American Benthological Society* 9:77–88.

Brooks, C., D. Dreves, and D. White. 1998. Distribution records for *Taphromysis louisianae* with notes on ecology. *Crustaceana* 71:955–960.

Bukaveckas, P. A., E. Jourdain, R. Koch, and J. H. Thorp. 2000. Longitudinal gradients in nutrients and phytoplankton in the Ohio River. *Proceedings of the International Association for Theoretical and Applied Limnology* 27:3107–3110.

FIGURE 9.13   Allegheny River near New Kensington, Pennsylvania (PHOTO BY K. WILHELM).

Burr, B. M., and M. L. Warren Jr. 1986. *A distributional atlas of Kentucky fishes*. Kentucky Nature Preserves Commission Science and Technology Series no. 4. Frankfort, Kentucky.

Cummings, K. S., and C. A. Mayer. 1992. *Field guide to freshwater mussels of the Midwest*. Illinois Natural History Survey Manual 5. Urbana, Illinois.

Cummings, K. S., and C. A. Mayer. 1994. Historical changes in the freshwater mussel fauna (Mollusca, Unionacea) of the Wabash River Drainage. *Bulletin of the North American Benthological Society* 11:109.

Deam, C. C. 1953. *Trees of Indiana*, 3rd ed. Bookwater Company, Indianapolis, Indiana.

EA Environmental (Engineering, Science, & Technology). 2001. *2000 Ohio River Ecological Research Program.*

American Electric Power Service Corp., Cinergy, Indiana—Kentucky Electric Corporation, and Buckeye Power, Inc., Columbus, Ohio.

Etnier D. A., and W. C. Starnes. 1993. *The fishes of Tennessee*. University of Tennessee Press, Knoxville.

Evermann, B. W. 1918. The fishes of Kentucky and Tennessee: A distributional catalogue of the known species. *Bulletin of the United States Bureau of Fisheries* 25:295–368.

Gammon, J. R. 1991. The environment and fish communities of the middle Wabash River. A report for Eli Lilly and Company and PSI Energy, DePauw University, Greencastle, Indiana.

Hill, B. H., and J. R. Webster. 1983. Aquatic macrophyte contribution to the New River organic matter budget.

In T. Fontaine and S. Bartell (eds.). *Dynamics of lotic ecosystems*, pp. 273–282. Ann Arbor Science, Ann Arbor, Michigan.

Hobbs, H. H., Jr. 1988. Crayfish distributions, adaptive radiation, and evolution. In D. M. Holdich and R. S. Lowery (eds.). *Freshwater crayfish biology, management, and exploitation*, pp. 52–82. Croom Helm, London.

Hough, J. L. 1958. *Geology of the Great Lakes*. University of Illinois Press, Urbana.

Hunt, C. B. 1967. *Physiography of the United States*. W. H. Freeman, San Francisco.

Hyde, G. E. 1962. *Indians of the woodlands from prehistoric times to 1725*. University of Oklahoma Press, Norman.

Jack, J. D., and J. H. Thorp. 2000. Effects of benthic suspension feeder *Dreissena polymorpha* on zooplankton in a large river. *Freshwater Ecology* 44:569–579.

Jack, J. D., and J. H. Thorp. 2002. Impacts of fish predation on an Ohio River zooplankton community. *Journal of Plankton Research* 24:119–127.

Kanawha River Basin Coordinating Committee. 1971. Kanawha River Comprehensive Basin study Vol. IV. Department of the Army, Huntington District Corps of Engineers West Virginia.

Kimmel, B. L., O. T. Lind, and L. J. Paulson. 1990. Reservoir primary production. In K. W. Thornton, B. L. Kimmel, and F. E. Payne (eds.). *Reservoir limnology: ecological perspectives*, pp. 133–194. Wiley-Interscience, New York.

Kobraei, M. E., and D. S. White. 1996. Effects of 2,4-Dichlorophenoxyacetic acid on Kentucky algae: Simultaneous laboratory and field toxicity testings. *Archives of Environmental Contamination and Toxicology* 31:571–580.

Koch, R. W. 2001. Nutrient, light and temperature limitation of phytoplankton and bacteria in large rivers and their impoundments. Ph.D. diss., University of Louisville, Louisville, Kentucky.

Lachner, E. A. 1956. Man and the waters of the upper Ohio River. In *Special Publication #1*, pp. 64–78. Pymatuning Laboratory, Linesville, Pennsylvania.

Lewis, R. B. 1996. *Kentucky archaeology*. University of Kentucky Press, Lexington.

Luken, J. O., and J. Spaeth. 2002. Comparison of riparian forests within and beyond the boundaries of Land Between the Lakes National Recreation Area, Kentucky, USA. *Natural Areas Journal* 22:283–289.

Mason, W. T., P. A. Lewis, and J. B. Anderson. 1971. *Macroinvertebrate collections and water quality monitoring in the Ohio River Basin 1963–1967*. Environmental Protection Agency Water Quality Office, Cincinnati, Ohio.

McCague, J. 1973. *The Cumberland*. Holt, Rinehart, and Winston, New York.

National Water Quality Assessment (NAWQA). 2000a. Water quality in the Allegheny and Monongahela river basins. Circular 1202. U.S. Geological Survey, Denver, Colorado.

NAWQA. 2000b. Water quality in the Kanawha—New River basin West Virginia, Virginia, and North Carolina, 1996–1998. Circular 1204. U.S. Geological Survey, Denver, Colorado.

Ohio River Valley Water Sanitation Commission (ORSANCO). 1978. *Ohio River water quality fact book*. Ohio River Valley Water Sanitation Commission, Cincinnati.

ORSANCO. 1990. *Long-term trends assessment of fifteen water quality parameters in the Ohio River*. Toxic Substances Control Program, Ohio River Valley Water Sanitation Commission, Cincinnati.

ORSANCO. 1994. *Ohio River water quality fact book*. Ohio River Valley Water Sanitation Commission, Cincinnati.

ORSANCO. 2000. *Ohio River water quality fact book*. Ohio River Valley Water Sanitation Commission, Cincinnati.

Parmalee, P. W., and A. E. Bogan. 1998. *The freshwater mussels of Tennessee*. University of Tennessee Press, Knoxville.

Patrick, R. 1995. *Rivers of the United States*. Vol. 2: *Chemical and physical characteristics*. John Wiley, New York.

Pearson, W. D., and L. B. Krumholz. 1984. Distribution and status of Ohio River fishes. Publication no. ORNL/Sub/79-7831/1. Oak Ridge National Laboratory, Oak Ridge, Tennessee.

Pearson, W. D., and B. J. Pearson. 1989. Fishes of the Ohio River. *Ohio Journal of Science* 89:181–187.

Pennington and Associates. 1994. Survey of the macroinvertebrate fauna in Kentucky Reservoir, Tennessee, February 1988–August 1993. Tennessee Wildlife Resource Agency, Nashville.

Pohana, R. E. 1992. Pleistocene history of the Ohio River valley. In W. Haneberg, M. M. Riestenberg, R. E. Pohana, and S. C. Diekmeyer (eds.). *Guidebook no. 9*, pp. 9–13. Geological Society of America, Columbus, Ohio.

Reash, R. J., and J. H. van Hassel. 1988. Distribution of upper and middle Ohio River fishes, 1973–1985. II. Influences of zoogeographic and physiochemical tolerance factors. *Journal of Freshwater Ecology* 4:459.

Reid, R. L. 1991. *Always a river: The Ohio River and the American experience*. Indiana University Press, Bloomington.

Resh, V. H., D. S. White, D. E. Jennings, and L. A. Krumholz. 1976. Vertebral variation in the emerald shiner *Notropis atherinoides* from the Ohio River: An apparent contradiction to "Jordan's Rule." *Bulletin of the Southern California Academy of Science* 75:76–84.

Ricketts, T. H., E. Dinerstein, D. M. Olson, C. J. Loucks, W. Eichbaum, D. Dellasala, K. Kavanaugh, P. Hedao, P. T. Hurley, K. M. Carney, R. Abell, and S. Walters. 1999. *Terrestrial ecoregions of North America: A conservation assessment*. Island Press, Washington, D.C.

Rodgers, J. H. Jr., D. S. Cherry, J. R. Clark, K. L. Dickson, and J. Cairns, Jr. 1997. The invasion of asiatic clam, Corbicula manilensis, in the New River, Virginia. *The Nautilus* 91:43–46.

Schram, M. D., and G. R. Marzolf. 1993. Seasonal variation of crustacean zooplankton in Kentucky Lake, U.S.A. *Proceedings of the International Association for Theoretical and Applied Limnology* 25:1158–1161.

Simon, K. F. (ed.). 1939. *The WPA guide to Kentucky.* Works Progress Administration. University of Kentucky, Lexington.

Soranno, P. A., K. E. Webster, J. L. Riera, T. K. Kratz, J. S. Baron, P. A. Bukaveckas, G. W. Kling, D. S. White, N. Craine, R. C. Lathrop, and P. R. Leavitt. 1999. Spatial variation among lakes within landscapes: Ecological organization along lake chains. *Ecosystems* 2:395–410.

Starnes, W. C., and D. A. Etnier. 1986. Drainage evolution and fish biogeography of the Tennessee and Cumberland river drainages. In C. H. Hocutt and E. O. Wiley (eds.). *Zoogeography of North American freshwater fishes*, pp. 325–361. Wiley-Interscience, New York.

Tabor, J. 2000. Seasonal patterns of phytoplankton in Ledbetter Embayment of Kentucky Lake. Masters Thesis, Murray State University, Murray, Kentucky.

Taylor, M. P. 1971. Phytoplankton productivity response to nutrients correlated with certain environmental factors in six TVA reservoirs. In G. E. Hall (eds.). *Reservoir fisheries and limnology.* Special Publication no. 8, pp. 209–217. American Fisheries Society, Washington, D.C.

Taylor, R. W. 1989. Changes in freshwater mussel populations of the Ohio River: 1,000 BP to recent times. *Ohio Journal of Science* 89:188–191.

Tennessee Valley Authority (TVA). 1980. *TVA water control projects and other major hydro developments in the Tennessee and Cumberland valleys.* Technical Monograph no. 55. Tennessee Valley Authority, Knoxville.

Thornton, K. W., B. L. Kimmel, and F. E. Payne (eds.). 1990. *Reservoir limnology: Ecological perspectives.* Wiley-Interscience, New York.

Thorp, J. H., A. R. Black, K. H. Haag, and J. D. Wehr. 1994. Zooplankton assemblages in the Ohio River: Seasonal and navigation dam effects. *Canadian Journal of Fisheries and Aquatic Science* 51:1634–1643.

Thorp, J. H., and M. D. Delong. 2002. Dominance of autochthonous carbon in food webs of heterotrophic rivers. *Oikos* 96:543–550.

Thorp, J. H., M. D. DeLong, and A. F. Casper. 1998. *In situ* experiments on predatory regulation of a bivalve mollusk (*Dreissena polymorpha*) in the Mississippi and Ohio rivers. *Freshwater Biology* 39:649–661.

Trautman, M. B. 1981. *The fishes of Ohio*, 2nd ed. Ohio State University Press, Columbus.

U.S. Environmental Protection Agency (USEPA). 1978. *Summary of Ohio River fishery surveys, 1968–1976.* EPA 903/9-78-009. EPA Surveillance and Analysis Division, Region III, Philadelphia.

Vaultonburg, D. L., and C. L. Pederson. 1994. Spatial and temporal variation in diatom community structure in two east-central Illinois streams. Transaction of the Illinois State Academy of Science 87:9–27.

Wehr, J. D., and J. H. Thorp. 1997. Effects of navigation dams, tributaries, and littoral zones on phytoplankton communities in the Ohio River. *Canadian Journal of Fisheries and Aquatic Science* 54:378–395.

Williams, J. 1994. *The weather almanac 1995.* Vantage Books, Random House, New York.

Williams, J. A. 1993. *West Virginia: A history for beginners.* Appalachian Editions, Martinsburg, West Virginia.

Works Progress Administration (WPA). 1943. *Cincinnati: A guide to the Queen City and its neighbors.* Cincinnati Historical Society, Cincinnati, Ohio.

Yurista, P. M., G. T. Rice, and D. S. White. 2001. Long-term establishment of *Daphnia lumholtzi* in Kentucky Lake, USA. *Proceedings of the International Association for Theoretical and Applied Limnology* 27:3102–3106.

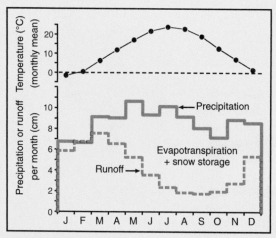

FIGURE 9.15 Mean monthly air temperature, precipitation, and runoff for the Ohio River basin.

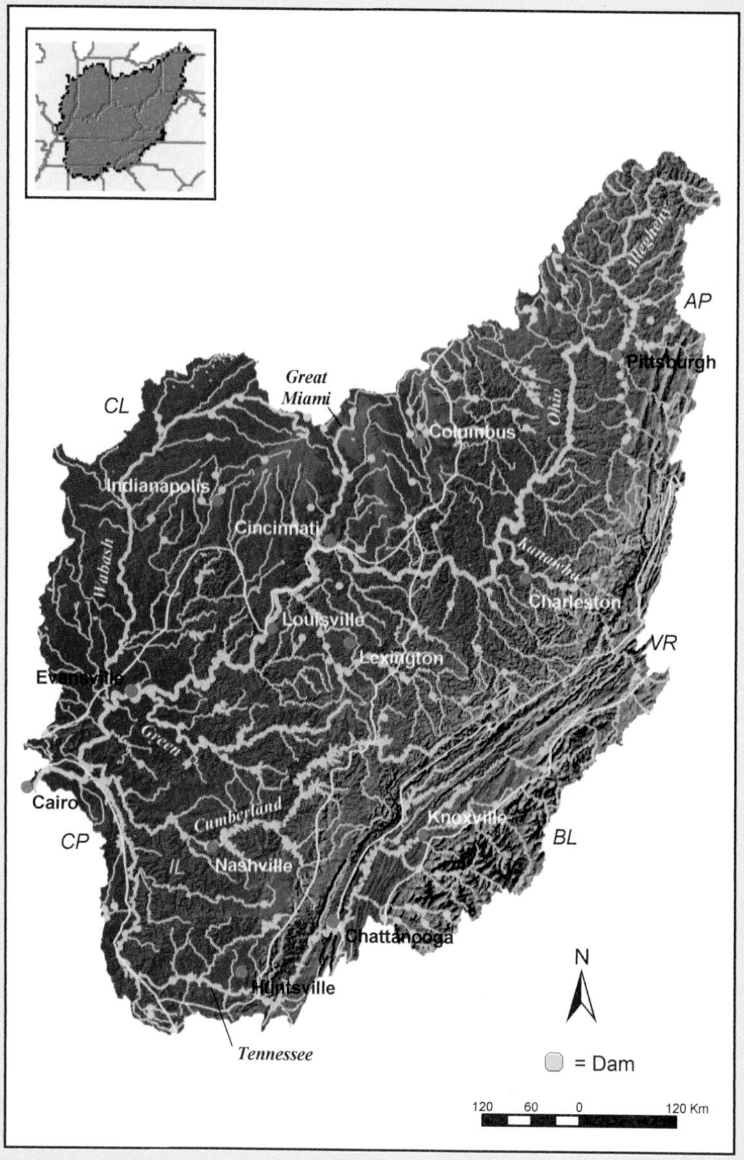

OHIO RIVER

Relief: 2300 m
Basin area: 529,000 km²
Mean discharge: 8733 m³/s
River order: 9
Mean annual precipitation: 104 cm
Mean air temperature: 12°C
Mean water temperature: 14°C
Physiographic provinces: Blue Ridge (BL), Valley and Ridge (VR), Appalachian Plateaus (AP), Central Lowland (CL), Interior Low Plateaus (IL), Coastal Plain (CP)
Biomes: Temperate Grasslands, Temperate Deciduous Forest
Freshwater ecoregions: Teays–Old Ohio, Tennessee–Cumberland
Terrestrial ecoregions: 8 ecoregions (see text)
Number of fish species: 240 to 250
Endangered species: 8 mussels
Major fishes: gizzard shad, skipjack herring, emerald shiner, white sucker, golden redhorse, black buffalo, river carpsucker, channel catfish, flathead catfish, longnose gar, shortnose gar, largemouth bass, smallmouth bass, bluegill, white crappie, freshwater drum

The following is placed beside the map:

FIGURE 9.14 Map of the Ohio River basin. Physiographic provinces are separated by yellow lines.

Major other aquatic vertebrates: snapping turtle, stinkpot, common map turtle, spiny softshell turtle, slider, common water snake, queen snake, green frog, bullfrog, pickerel frog, bald eagle, osprey, great blue heron, Canada goose, coot, mallard, beaver, muskrat, river otter
Major benthic invertebrates: bivalves (threehorn wartyback, Wabash pigtoe, white heelsplitter, fat mucket), worms (*Branchiura*), crustaceans (*Gammarus, Cambarus, Procambarus, Orconectes*), mayflies (*Hexagenia, Ephemerella, Caenis, Stenacron*), stoneflies (*Isoperla*), dragonflies (*Stylurus*), damselflies (*Argia*), caddisflies (*Hydroptila, Hydropsyche, Ceraclea, Cyrnellus, Polycentropus, Potamyia, Chimarra*), true flies (*Tipula*)
Nonnative species: *Daphnia lumholtzi*, zebra mussel, Asiatic clam, yellow perch, common carp, goldfish, rainbow smelt, fathead chub, rosefin shiner, bighead carp, grass carp, striped bass, silver carp, brown trout, brittle naiad, curly pondweed, Eurasian watermilfoil, purple loosestrife
Major riparian plants: red maple, cottonwood, black willow, sycamore, black gum, sugar hackberry, water willow, buttonbush
Special features: high biodiversity, especially fishes, freshwater mussels, crayfishes, and aquatic insects; >40% of Mississippi River discharge
Fragmentation: 20 low-water locks and dams on main stem; >700 major dams in basin
Water quality: generally improved since 1970s, agriculture runoff, power plant thermal inputs, urban/industrial discharges still cause problems; pH = 7.2, alkalinity = 70 mg/L as $CaCO_3$, NO_3-N = 1.6 mg/L, PO_4-P = <0.17 mg/L
Land use: 48% agriculture, 47% forest, 4% urban or barren, 1% water
Population density: 49 people/km²
Major information sources: ORSANCO 1994, 2000, Pearson and Pearson 1989, EA Environmental 2001, Patrick 1995

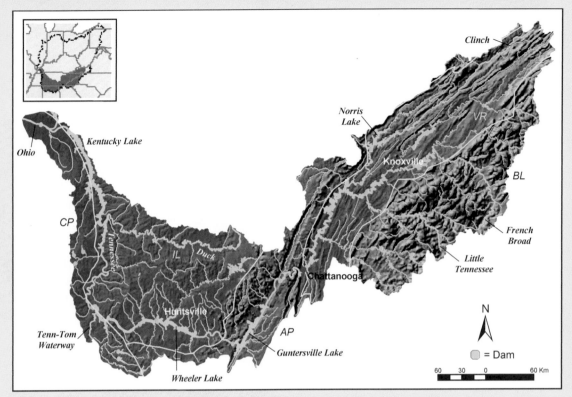

FIGURE 9.16 Map of the Tennessee River basin. Physiographic provinces are separated by yellow lines.

TENNESSEE RIVER

Relief: 1910 m
Basin area: 105,870 km²
Mean discharge: 2000 m³/s
River order: 8
Mean annual precipitation: 105 cm
Mean air temperature: 13°C
Mean water temperature: 19°C
Physiographic provinces: Valley and Ridge (VR), Blue Ridge (BL), Appalachian Plateaus (AP), Interior Low Plateaus (IL), Coastal Plain (CP)
Biome: Temperate Deciduous Forest
Freshwater ecoregion: Tennessee–Cumberland
Terrestrial ecoregions: Appalachian Mixed Mesophytic Forests, Central U.S. Hardwood Forests, Appalachian/Blue Ridge Forests, Southeastern Mixed Forests
Number of fish species: 225 to 240
Endangered species: 22 mussels, 9 fishes

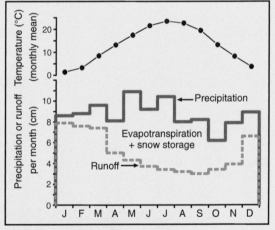

FIGURE 9.17 Mean monthly air temperature, precipitation, and runoff for the Tennessee River basin.

Major fishes: gizzard shad, threadfin shad, largemouth bass, channel catfish, white bass, striped bass, smallmouth bass, freshwater drum, paddlefish, white crappie, bluegill, flathead catfish, white sucker, spotfin shiner, striped shiner, emerald shiner, orangethroat darter, fantail darter, sauger
Major other aquatic vertebrates: snapping turtle, stinkpot, mud turtle, common map turtle, midland painted turtle, spiny softshell turtle, sliders, common water snake, queen snake, green frog, bullfrog, cottonmouth, bald eagle, osprey, great blue heron, coot, mallard, common loon, ring-billed gull, American white pelican, beaver, river otter, muskrat
Major benthic invertebrates: bivalves (threeridge, washboard, giant floater, mapleleaf, pistolgrip, fragile papershell), crustaceans (*Procambarus, Cambarus, Orconectes*), mayflies (*Hexagenia, Caenis*), stoneflies (*Acroneuria*), alderflies (*Sialis*), caddisflies (*Hydroptila, Cheumatopsyche, Ceraclea, Brachycentrus*), beetles (*Gyrinus, Tropisternus, Berosus*)
Nonnative species: *Daphnia lumholtzi*, Asiatic clam, zebra mussel, goldfish, grass carp, tench, golden shiner, rainbow smelt, Ohrid trout, rainbow trout, brown trout, yellow perch, bighead carp, brook stickleback, Eurasian watermilfoil, purple loosestrife, parrot-feather, alligatorweed, brittle naiad, curly pondweed, watercress, Nepal grass
Major riparian plants: water willow, red maple, buttonbush, cottonwood, black gum, American sycamore, black willow
Special features: most endemic fishes, mussels, and crayfishes of any river in North America; Holston, Clinch, North Fork, Duck, and Powel tributaries have retained much diversity; Obed is National Wild and Scenic River
Fragmentation: 48 multipurpose dams on main stem and major tributaries
Water quality: relatively free of major pollutants; areas of urban and industrial wastes; upper tributaries affected by acid mine drainage and logging; pH = 7.2, alkalinity = 40 mg/L as $CaCO_3$, NO_3-N = 0.2 mg/L, PO_4-P = <0.01 mg/L
Land use: 58% forest, 36% agriculture, 4% urban or barren, 2% water
Population density: 19 people/km²
Major information sources: TVA 1980, Starnes and Etnier 1986, Burr and Warren 1986, Etnier and Starnes 1993

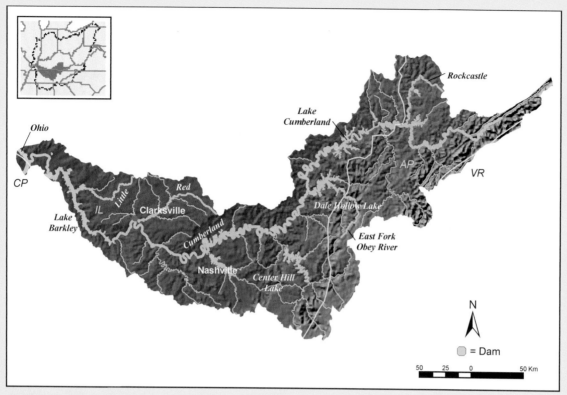

FIGURE 9.18 Map of the Cumberland River basin. Physiographic provinces are separated by yellow lines.

CUMBERLAND RIVER

Relief: 1160 m
Basin area: 46,430 km^2
Mean discharge: 862 m^3/s
River order: 7
Mean annual precipitation: 127 cm
Mean air temperature: 14°C
Mean water temperature: 16°C
Physiographic provinces: Valley and Ridge (VR), Interior Low Plateaus (IL), Appalachian Plateaus (AP)
Biome: Temperate Deciduous Forest
Freshwater ecoregion: Tennessee–Cumberland
Terrestrial ecoregions: Appalachian Mixed Mesophytic Forests, Central U.S. Hardwood Forests
Number of fish species: 172 to 186
Endangered species: 14 mussels, 1 crayfish, 1 snail, 3 fishes
Major fishes: gizzard shad, threadfin shad, channel catfish, largemouth bass, smallmouth bass, spotted bass, white bass, striped bass, common carp, white sucker, freshwater drum, shortnose gar, longnose gar, bluegill, green sunfish, white crappie, paddlefish
Major other aquatic vertebrates: common snapping turtle, stinkpot, mud turtle, common map turtle, midland painted turtle, spiny softshell turtle, common water snake, queen snake, mudpuppy, plethodontid salamanders, bullfrog, green frog, bald eagle, osprey, great blue heron, coot, mallard, ring-billed gull, beaver, muskrat, river otter
Major benthic invertebrates: bivalves (threeridge, giant floater, washboard, mapleleaf, pink heelsplitter, threehorn wartyback, fingernail clams), crustaceans (*Nashville crayfish, Orconectes, Cambarus*), mayflies (*Stenonema, Caenis, Tricorythodes, Hexagenia*), alderflies (*Sialis*), caddisflies (*Cheumatopsyche, Ceraclea, Hydroptila*), beetles (*Gyrinus, Tropisternus, Stenelmis*)
Nonnative species: Asiatic clam, zebra mussel, *Daphnia lumholtzi*, common carp, goldfish, striped bass, rainbow trout, brown trout, yellow perch, bighead carp, grass carp, tench, grass carp, golden shiner, alewife, Eurasian watermilfoil, hydrilla, brittle naiad, watercress, curly pondweed, alligatorweed, yellow iris, Uruguay seedbox, parrot-feather
Major riparian plants: water willow, buttonbush, red maple, cottonwood, American sycamore, black willow, black gum
Special features: Cumberland Falls; Rockcastle River with many endemic fishes and mussels; caves and karst flow in midregion
Fragmentation: 10 major dams and many tributary dams
Water quality: acid mine drainage in upper tributaries, high levels of nonpoint nitrogen in lower tributaries; pH = 7.2, alkalinity = 60 mg/L as CaCO$_3$, NO$_3$-N = <0.4 mg/L, PO$_4$-P = <0.02 mg/L
Land use: 51% forest, 40% agriculture, 5% urban, 3% water, 1% mining and barren
Population density: 16 people/km^2
Major information sources: McCague 1973, Starnes and Etnier 1986, Etnier and Starnes 1993, Parmalee and Bogan 1998

FIGURE 9.19 Mean monthly air temperature, precipitation, and runoff for the Cumberland River basin.

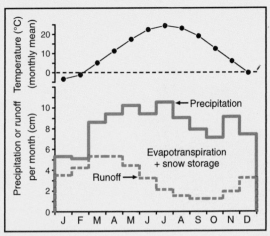

FIGURE 9.21 Mean monthly air temperature, precipitation, and runoff for the Wabash River basin.

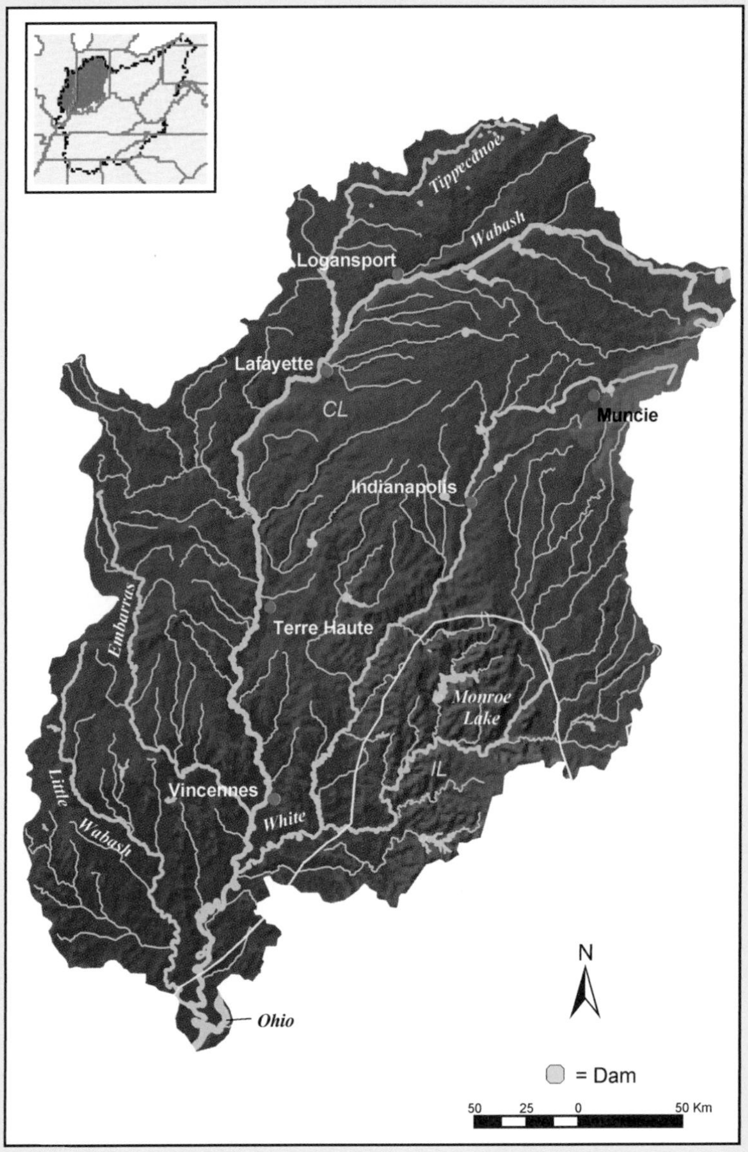

WABASH RIVER

Relief: 275 m
Basin area: 85,340 km²
Mean discharge: 1001 m³/s
River order: 7
Mean annual precipitation: 96 cm
Mean air temperature: 11°C
Mean water temperature: 15°C
Physiographic provinces: Central Lowland (CL), Interior Low Plateaus (IL)
Biomes: Temperate Grasslands, Temperate Deciduous Forest
Freshwater ecoregion: Teays–Old Ohio
Terrestrial ecoregions: Central Forest/Grassland Transitional Zone, Central U.S. Hardwood Forests, Southern Great Lakes Forests
Number of fish species: >95
Endangered species: 3 mussels
Major fishes: gizzard shad, common carp, steelcolor shiner, spotfin shiner, channel catfish, longnose gar, shortnose gar, quillback, river carpsucker, shorthead redhorse, silver redhorse, golden redhorse, freshwater drum, emerald shiner, longear sunfish, white crappie, sauger, white bass, shovelnose sturgeon
Major other aquatic vertebrates: snapping turtle, stinkpot, midland painted turtle, spiny softshell turtle, common water snake, red-bellied water snake, Kirtland's water snake, queen snake, newt, mudpuppy, leopard frog, bullfrog, green frog, bald eagle, osprey, belted kingfisher, mallard, coot, beaver, muskrat, river otter
Major benthic invertebrates: crustaceans (*Orconectes*), bivalves (fingernail clams, fragile papershell, mapleleaf, hickorynut, black sandshell, wartyback), mayflies (*Hexagenia, Caenis, Brachycerus, Tricorythodes, Stenonema, Heptagenia*), caddisflies (*Cheumatopsyche, Hydropsyche, Ceraclea*), stoneflies (*Taeniopteryx, Allocapnia, Isoperla, Perlesta*), damselflies (*Calopteryx*), beetles (*Tropisternus*), alderflies (*Sialis*)
Nonnative species: Asiatic clam, zebra mussel, common carp, goldfish, bighead carp, yellow perch, brook trout, rainbow trout, purple loosestrife, Eurasian watermilfoil
Major riparian plants: red maple, cottonwood, American sycamore, black willow, black gum, sugar hackberry, buttonbush
Special features: 267 km long Tippecanoe River remains relatively pristine, drains several northern glacial pothole lakes, and has extensive aquatic macrophytes beds
Fragmentation: one major dam at Rkm 662; free flowing from there to confluence with Ohio River
Water quality: relatively free of major pollutants, primarily agricultural runoff; pH = 7.2, alkalinity = 200 mg/L as CaCO₃, NO_3-N = 2.8 mg/L, PO_4-P = 0.24 mg/L
Land use: 65% agriculture, 8% pasture or open lands, 22% forest, 5% urban
Population density: 62.7 people/km²
Major information sources: ORSANCO 1990, Gammon 1991, Cummings and Mayer 1994

FIGURE 9.20 Map of the Wabash River basin. Physiographic provinces are separated by a yellow line.

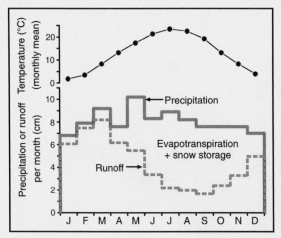

FIGURE 9.23 Mean monthly air temperature, precipitation, and runoff for the Kanawha River basin.

KANAWHA RIVER

Relief: 1545 m
Basin area: 31,690 km^2
Mean discharge: 537 m^3/s
River order: 6
Mean annual precipitation: 94 cm
Mean air temperature: 13°C
Mean water temperature: 14°C
Physiographic provinces: Blue Ridge (BL), Valley and Ridge (VR), Appalachian Plateaus (AP)
Biome: Temperate Deciduous Forest
Freshwater ecoregion: Teays–Old Ohio
Terrestrial ecoregions: Appalachian/Blue Ridge Forests, Appalachian Mixed Mesophytic Forests
Number of fish species: 126
Endangered species: 3 mussels
Major fishes: emerald shiner, spotfin shiner, white sucker, river redhorse, golden redhorse, channel catfish, flathead catfish, longnose gar, white bass, smallmouth bass, green sunfish, longear sunfish, spotted bass, white crappie, sauger, sharpnose darter, Kanawha darter, Appalachia darter, Kanawha minnow, New River shiner, bigmouth chub

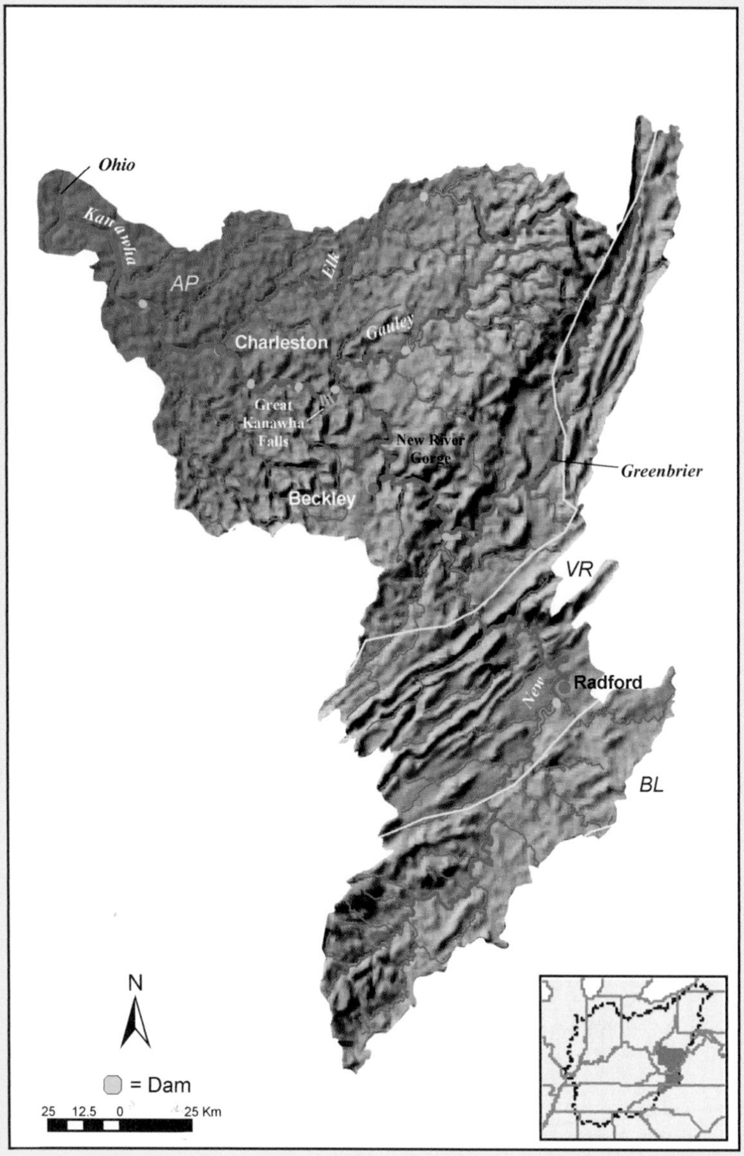

FIGURE 9.22 Map of the Kanawha River basin. Physiographic provinces are separated by yellow lines.

Major other aquatic vertebrates: snapping turtle, stinkpot, midland painted turtle, hieroglyphic river cooter, spiny softshell turtle, common water snake, queen snake, bullfrog, green frog, hellbender, newt, mudpuppy, osprey, bald eagle, mallard, coot, great blue heron, beaver, muskrat
Major benthic invertebrates: bivalves (threeridge, kidneyshell, round hickorynut, black sandshell, elephantear), crustaceans (*Orconectes*), mayflies (*Stenonema, Isonychia, Leptophlebia*), stoneflies (*Allocapnia, Taeniopteryx, Isoperla, Peltoperla*), hellgrammites (*Corydalus*), caddisflies (*Hydropsyche, Ceraclea, Hydroptila, Brachycentrus*), flies (*Tipula*), beetles (*Psephenus, Stenelmis, Optioservus, Promoresia*)
Nonnative species: Asiatic clam, zebra mussel, rusty crayfish, alewife, common carp, goldfish, golden shiner, striped bass, fathead minnow, tench, purple loosestrife, brittle naiad, curly pondweed, yellow iris
Major riparian plants: red maple, cottonwood, black willow, American sycamore, speckled alder, tulip poplar
Special features: Great Kanawha Falls limits species migration; New is oldest North American river; Greenbrier is unimpounded; New and Greenbrier have generally good water quality and significant free-flowing stretches
Fragmentation: four locks and dams on Kanawha main stem, one dam on Elk and Gurley rivers, two dams on New River
Water quality: improved since 1970s; acid mine drainage, agricultural/forest practices, and industrial/urban inputs near Charlestown still degrade water quality; pH = 7.3, alkalinity = 38 mg/L as CaCO$_3$, NO$_3$-N = 1.2 mg/L, PO$_4$-P = 0.08 mg/L
Land use: 70% forest, 23% agriculture, 3% urban/industrial, 3% mined/barren, 1% water
Population density: 29 people/km^2
Major information sources: North Carolina Division of Water Quality, unpublished data; Kanawha River Basin Coordinating Committee 1971; NAQWA 2000b

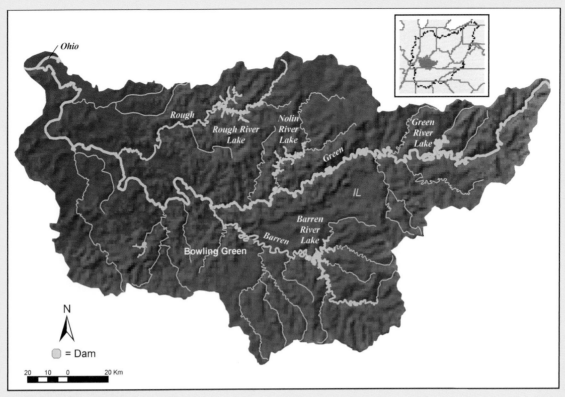

FIGURE 9.24 Map of the Green River basin.

GREEN RIVER

Relief: 385 m
Basin area: 23,850 km^2
Mean discharge: 420 m^3/s
River order: 7
Mean annual precipitation: 110 to 125 cm
Mean air temperature: 14°C
Mean water temperature: 16°C
Physiographic province: Interior Low Plateaus (IL)
Biome: Temperate Deciduous Forest
Freshwater ecoregion: Teays–Old Ohio
Terrestrial ecoregions: Appalachian Mixed Mesophytic Forests,
     Central U.S. Hardwood Forests
Number of fish species: 151
Endangered species: 1 fish, 1 cave shrimp, 9 mussels; >20 mussel
     species may be extinct

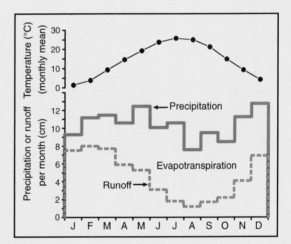

FIGURE 9.25 Mean monthly air temperature, precipitation, and runoff for the Green River basin.

Major fishes: white sucker, gizzard shad, bluegill, rock bass,
     largemouth bass, spotted bass, splendid darter, orangefin darter,
     teardrop darter, Kentucky snubnose darter, blackfin sucker, channel catfish, cavefish, spotted sunfish, flier
Major other aquatic vertebrates: banded water snake, slider, false map turtle, stinkpot, spiny softshell turtle, mudpuppy,
     hellbender, bald eagle, osprey, great blue heron, beaver, muskrat, river otter
Major benthic invertebrates: bivalves (washboard, threeridge, white heelsplitter, giant floater, mapleleaf), crustaceans
     (*Orconectes, Cambarus, Procambarus*), mayflies (*Hexagenia, Caenis, Ephemerella, Isonychia, Stenonema*), stoneflies
     (*Allocapnia, Acroneuria, Isoperla, Taeniopteryx, Neoperla*), hellgrammites (*Corydalus*), alderflies (*Sialis*), caddisflies
     (*Ceraclea, Cheumatopsyche, Hydropsyche, Hydroptila*), beetles (*Tropisternus*)
Nonnative species: Asiatic clam, zebra mussel, common carp, goldfish, striped bass, yellow perch, fathead minnow, brown trout,
     rainbow trout, yellow iris, parrot-feather, brittle naiad, watercress, Eurasian watermilfoil, purple loosestrife
Major riparian plants: red maple, buttonbush, black willow, cottonwood, sycamore, water willow, cattails
Special features: upper portion of river lies in karst topography with numerous sinkholes and caves, including Mammoth Cave
Fragmentation: 13 major dams, 7 of which have navigation locks for barge traffic
Water quality: relatively free of major pollutants; acid mine drainage occurs throughout mid-reaches; pH = 7.1, alkalinity =
     62 mg/L as CaCO$_3$, NO$_3$-N = 0.7 mg/L, PO$_4$-P = <0.1 mg/L
Land use: 55% agriculture, 39% forest, 3% urban, 2% water, and 1% barren land
Population density: 34 people/km^2
Major information sources: Kentucky Division of Water, unpublished data; U.S. Army Corps of Engineers, unpublished data;
     Burr and Warren 1986

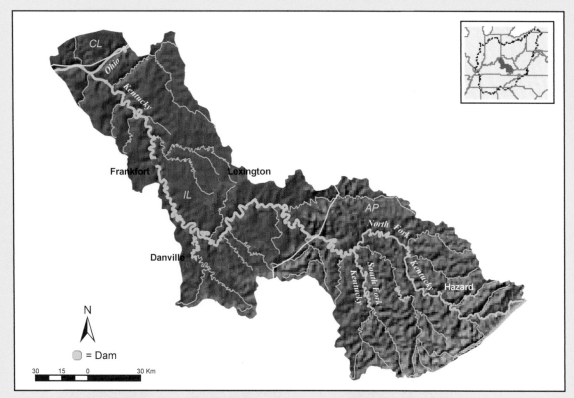

FIGURE 9.26 Map of the Kentucky River basin. Physiographic provinces are separated by yellow lines.

KENTUCKY RIVER

Relief: 840 m

Basin area: 18,025 km²

Mean discharge: 285 m³/s

River order: 6

Mean annual precipitation: 111 cm

Mean air temperature: 13°C

Mean water temperature: 15°C

Physiographic provinces: Appalachian Plateaus (AP), Interior Low Plateaus (IL)

Biome: Temperate Deciduous Forest

Freshwater ecoregion: Teays–Old Ohio

Terrestrial ecoregions: Appalachian Mixed Mesophytic Forests, Central U.S. Hardwood Forests

Number of fish species: 110 to 115

Number of endangered species: 1 fish, 1 salamander, 11 mussels

Major fishes: gizzard shad, bluegill, rock bass, largemouth bass, spotted bass, redside dace, mimic shiner, eastern sand darter, slender chub, sharpnose darter, channel catfish, spotted sucker, golden redhorse, shorthead redhorse, silver redhorse, striped shiner, longnose gar, rainbow darter, greenside darter, sauger

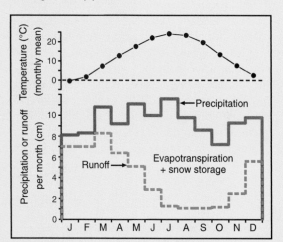

FIGURE 9.27 Mean monthly air temperature, precipitation, and runoff for the Kentucky River basin.

Major other aquatic vertebrates: banded water snake, midland painted turtle, stinkpot, spiny softshell turtle, mudpuppy, hellbender, bullfrog, green frog, bald eagle, great blue heron, beaver, muskrat

Major benthic invertebrates: bivalves (washboard, pink heelsplitter, white heelsplitter, giant floater), crustaceans (*Procambarus*, *Orconectes*), mayflies (*Ephemerella*, *Isonychia*), stoneflies (*Allocapnia*, *Acroneuria*, *Isoperla*, *Taeniopteryx*, *Perlesta*), hellgrammites (*Corydalus*), caddisflies (*Ceraclea*, *Cheumatopsyche*, *Hydropsyche*), beetles (*Stenelmis*, *Psephenus*)

Nonnative species: Asiatic clam, zebra mussel, common carp, goldfish, striped bass, yellow perch, brook stickleback, rainbow trout, brown trout, purple loosestrife, brittle naiad, curly pondweed

Major riparian plants: speckled alder, red maple, buttonbush, black willow, water willow, cottonwood, American sycamore

Special features: middle portion of river passes through Daniel Boone National Forest, where invertebrate and fish diversity are high; portions of Red tributary, including Red River Gorge, are in National Wild and Scenic River system

Fragmentation: 14 navigation dams on main stem

Water quality: relatively free of major pollutants; acid mine drainage in headwaters; pH = 7.2, alkalinity = 58 mg/L as $CaCO_3$, NO_3-N = 0.4 mg/L, PO_4-P < 0.01 mg/L

Land use: 42% agriculture, 54% forest, 4% mining, urban, and built up

Population density: 39 people/km²

Major information sources: Kentucky Division of Water, unpublished data; U.S. Army Corps of Engineers, unpublished data; Burr and Warren 1986

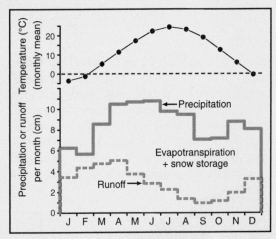

FIGURE 9.29 Mean monthly air temperature, precipitation, and runoff for the Great Miami River basin.

GREAT MIAMI RIVER

Relief: 305 m
Basin area: 13,915 km²
Mean discharge: 152 m³/s
River order: 6
Mean annual precipitation: 102 cm
Mean air temperature: 11°C
Mean water temperature: 15°C
Physiographic province: Central Lowland (CL)
Biome: Temperate Deciduous Forest
Freshwater ecoregion: Teays–Old Ohio
Terrestrial ecoregion: Southern Great Lakes Forests
Number of fish species: 120 to 125
Endangered species: 1 crayfish
Major fishes: gizzard shad, stoneroller, white sucker, hog sucker, rock bass, largemouth bass, spotted bass, smallmouth bass, mimic shiner, creek chub, sharpnose darter, channel catfish, bullhead, golden redhorse, rainbow darter, black crappie, sauger, green sunfish, spotted sucker, striped shiner, spotfin shiner, black buffalo, striped bass
Major other aquatic vertebrates: common water snake, queen snake, midland painted turtle, stinkpot, spiny softshell turtle, mudpuppy, green frog, pickerel frog, bullfrog, wood duck, great blue heron, mallard, beaver, muskrat
Major benthic invertebrates: bivalves (threeridge, washboard, pistolgrip, fragile papershell), crustaceans (*Orconectes*), mayflies (*Isonychia, Stenacron, Ephemerella, Stenonema*), stoneflies (*Allocapnia, Isoperla, Acroneuria*), true bugs (*Aquarius*), damselflies (*Argia*), alderflies (*Sialis*), caddisflies (*Hydroptila, Hydropsyche, Cheumatopsyche, Ceratopsyche, Ceraclea*), beetles (*Tropisternus, Stenelmis, Dubiraphia*), true flies (*Tipula*)
Nonnative species: Asiatic clam, zebra mussel, common carp, goldfish, striped bass, rainbow smelt, tench, brown trout, rainbow trout, purple loosestrife, Eurasian watermilfoil, European brooklime
Major riparian plants: red maple, cottonwood, sycamore, black willow, water willow
Special features: Great Miami Buried Valley aquifer extends from Dayton to Ohio River and roughly follows pathway of river; Stillwater River designated as exceptional warmwater habitat
Fragmentation: minor; Taylorsville dam north of Dayton
Water quality: water-quality problems primarily related to agricultural runoff, some urban/industrial inputs; pH = 8.1; alkalinity = 227 mg/L as $CaCO_3$, NO_3-N = 4.5 mg/L, PO_4-P = 0.27 mg/L
Land use: 80% agriculture, 13% forest, 5% urban or barren, 2% water
Population density: 134 people/km²
Major information sources: Trautman 1981, ORSANCO 1994, 2000, Ohio Environmental Protection Agency, unpublished data

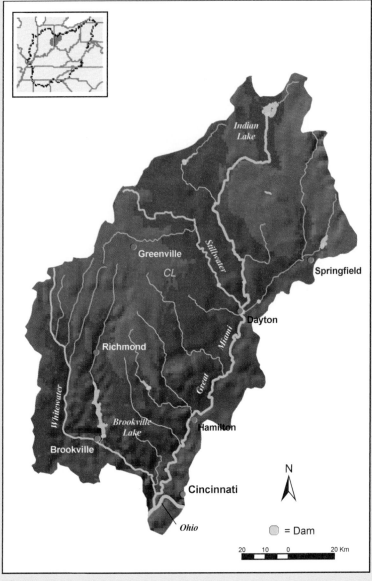

FIGURE 9.28 Map of the Great Miami River basin.

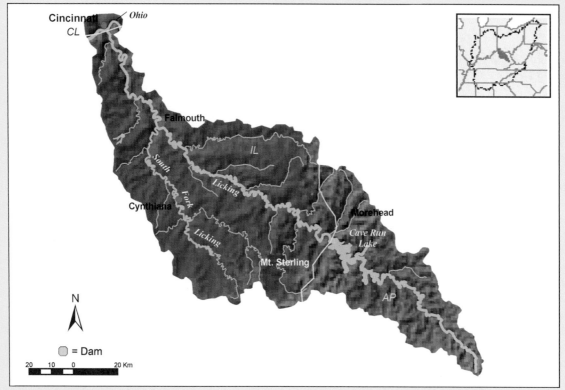

FIGURE 9.30 Map of the Licking River basin. Physiographic provinces are separated by yellow lines.

LICKING RIVER

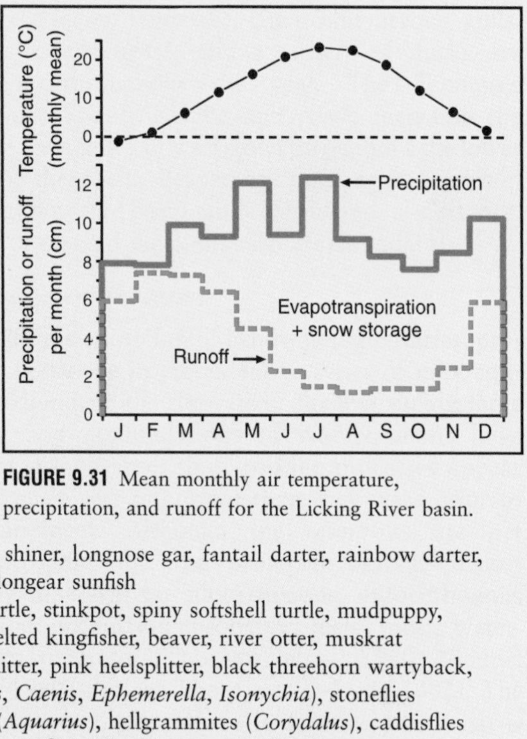

Relief: 345 m

Basin area: 9600 km²

Mean discharge: 145 m³/s

River order: 6

Mean annual precipitation: 112 cm

Mean air temperature: 12°C

Mean water temperature: 14°C

Physiographic provinces: Appalachian Plateaus (AP), Interior Low Plateaus (IL)

Biome: Temperate Deciduous Forest

Freshwater ecoregion: Teays–Old Ohio

Terrestrial ecoregions: Appalachian Mixed Mesophytic Forests, Central U.S. Hardwood Forests

Number of fish species: 110

Number of endangered species: 1 fish, 1 salamander, 11 mussels

Major fishes: rock bass, largemouth bass, spotted bass, redside dace, mimic shiner, eastern sand darter, slender chub, sharpnose darter, channel catfish, white sucker, spotted sucker, golden redhorse, striped shiner, longnose gar, fantail darter, rainbow darter, greenside darter, Johnny darter, sauger, white crappie, black crappie, longear sunfish

Major other aquatic vertebrates: banded water snake, midland painted turtle, stinkpot, spiny softshell turtle, mudpuppy, hellbender, newt, bullfrog, green frog, bald eagle, great blue heron, belted kingfisher, beaver, river otter, muskrat

Major benthic invertebrates: bivalves (threeridge, mapleleaf, white heelsplitter, pink heelsplitter, black threehorn wartyback, elephantear), crustaceans (*Procambarus*, *Orconectes*), mayflies (*Baetis*, *Caenis*, *Ephemerella*, *Isonychia*), stoneflies (*Allocapnia*, *Acroneuria*, *Isoperla*, *Taeniopteryx*, *Perlesta*), true bugs (*Aquarius*), hellgrammites (*Corydalus*), caddisflies (*Ceraclea*, *Hydropsyche*, *Hydroptila*), beetles (*Stenelmis*, *Psephenus*), true flies (*Tipula*)

Nonnative species: Asiatic clam, zebra mussel, grass carp, common carp, goldfish, rainbow trout, striped bass, brook stickleback, brittle naiad

Major riparian plants: speckled alder, red maple, buttonbush, black willow, water willow, American sycamore, cottonwood

Special features: middle portion passes through Daniel Boone National Forest, where invertebrate and fish diversity is high

Fragmentation: one major dam at Rkm 278, low-water dams on several tributaries

Water quality: relatively free of major pollutants; acid mine drainage in headwaters; pH = 7.2, alkalinity = 58 mg/L as $CaCO_3$, NO_3-N = 0.4 mg/L, PO_4-P = <0.01 mg/L

Land use: 42% agriculture, 54% forest, 4% mining, urban, and built up

Population density: 36 people/km²

Major information sources: Kentucky Division of Water, unpublished data; U.S. Army Corps of Engineers, unpublished data; Burr and Warren 1986

FIGURE 9.31 Mean monthly air temperature, precipitation, and runoff for the Licking River basin.

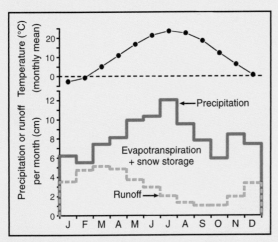

FIGURE 9.33 Mean monthly air temperature, precipitation, and runoff for the Scioto River basin.

SCIOTO RIVER

Relief: 307 m
Basin area: 16,882 km^2
Mean discharge: 189 m^3/s
River order: 6
Mean annual precipitation: 98 cm
Mean air temperature: 11°C
Mean water temperature: 15°C
Physiographic provinces: Central Lowland (CL), Appalachian Plateaus (AP), Interior Low Plateaus (IL)
Biome: Temperate Deciduous Forest
Freshwater ecoregion: Teays–Old Ohio
Terrestrial ecoregions: Southern Great Lakes Forests, Appalachian Mixed Mesophytic Forests
Number of fish species: 120 to 130
Endangered species: 2 mussels, 1 fish
Major fishes: sand shiner, silver redhorse, green sunfish, rock bass, smallmouth bass, gizzard shad, mottled sculpin, flathead catfish, rock bass, spotted bass, redside dace, mimic shiner, eastern sand darter, slender chub, sharpnose darter, spotted sucker, hog sucker, golden redhorse, spotfin shiner, striped shiner, longnose gar, rainbow darter, greenside darter, Johnny darter, sauger, black crappie
Major other aquatic vertebrates: snapping turtle, stinkpot, midland painted turtle, spiny softshell turtle, common water snake, queen snake, bullfrog, green frog, pickerel frog, bald eagle, great blue heron, osprey, beaver, river otter, common loon, mallard, coot, muskrat
Major benthic invertebrates: bivalves (threeridge, washboard, giant floater, mapleleaf, pistolgrip, fragile papershell), crustaceans (*Orconectes*), mayflies (*Acerpenna, Stenacron, Isonychia, Hexagenia*), stoneflies (*Allocapnia, Isoperla, Perlesta, Acroneuria*), alderflies (*Sialis*), caddisflies (*Hydroptila, Hydropsyche, Cheumatopsyche*), beetles (*Tropisternus, Berosus, Stenelmis, Dubiraphia, Macronychus, Optioservus*), flies (*Tipula*), black flies (*Simulium*)
Nonnative species: Asiatic clam, zebra mussel, tench, white bass, common carp, goldfish, yellow iris, purple loosestrife, Eurasian watermilfoil, brittle naiad, curly pondweed, European brooklime
Major riparian plants: red maple, cottonwood, American sycamore, black willow, black gum, buttonbush, water willow
Special features: river from Chillicothe downstream in Appalachian Plateaus is heavily forested; Big and Little Darby creeks are National Wild and Scenic Rivers
Fragmentation: one major dam on main stem, several tributary dams
Water quality: relatively free of major pollutants except below urban areas; primarily agricultural runoff, dredging, and point-source problems; pH = 7.9; alkalinity = 183 mg/L as CaCO$_3$, NO$_3$-N = 2.14 mg/L, PO$_4$-P = 0.33 mg/L
Land use: 69% agriculture, 21% forest, 9% urban, 1% water
Population density: 108 people/km^2
Major information sources: Trautman 1981, ORSANCO 1994, 2000, Ohio Environmental Protection Agency, unpublished data

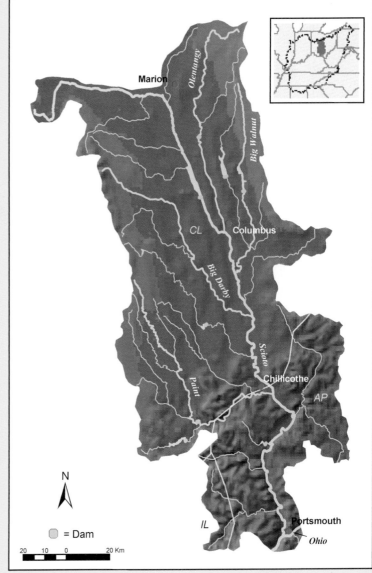

FIGURE 9.32 Map of the Scioto River basin. Physiographic provinces are separated by yellow lines.

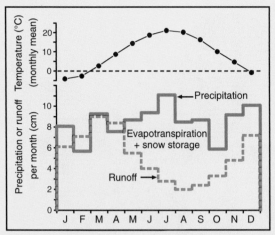

FIGURE 9.35 Mean monthly air temperature, precipitation, and runoff for the Allegheny River basin.

ALLEGHENY RIVER

Relief: 690 m
Basin area: 30,300 km^2
Mean discharge: 600 m^3/s
River order: 7
Mean annual precipitation: 104 cm
Mean air temperature: 9°C
Mean water temperature: 11°C
Physiographic province: Appalachian Plateaus (AP)
Biome: Temperate Deciduous Forest
Freshwater ecoregion: Teays–Old Ohio
Terrestrial ecoregions: Allegheny Highlands Forests, Appalachian Mixed Mesophytic Forests
Number of fish species: 114 to 120
Endangered species: 1 sedge, 2 fishes, 2 mussels
Major fishes: gizzard shad, common carp, bluegill, largemouth bass, spotted bass, channel catfish, walleye, sauger, emerald shiner, white crappie, gravel chub, blackchin shiner, river redhorse, black redhorse, longhead darter
Major other aquatic vertebrates: snapping turtle, stinkpot, midland painted turtle, spiny softshell turtle, common water snake, queen snake, bullfrog, green frog, hellbender, mudpuppy, osprey, great blue heron, mallard, beaver, muskrat
Major benthic invertebrates: bivalves (threeridge, elephantear, white heelsplitter, giant floater, black sand shell), crustaceans (*Procambarus*, *Orconectes*), mayflies (*Hexagenia*, *Ephemerella*, *Isonychia*), stoneflies (*Allocapnia*, *Isoperla*, *Acroneuria*), hellgrammites (*Corydalus*), alderflies (*Sialis*), caddisflies (*Hydroptila*, *Hydropsyche*, *Cheumatopsyche*, *Ceraclea*), beetles (*Berosus*, *Stenelmis*, *Dubiraphia*, *Optioservus*), true flies (*Tipula*)
Nonnative species: Asiatic clam, zebra mussel, common carp, goldfish, striped bass, yellow perch, brown trout, watercress, curly pondweed, brittle naiad, Eurasian watermilfoil, yellow iris, purple loosestrife
Major riparian plants: river birch, red maple, cottonwood, black willow, American sycamore, speckled alder
Special features: Upper Allegheny and Clarion rivers designated as National Wild and Scenic Rivers; upper Allegheny has >100 undeveloped islands
Fragmentation: eight navigation locks and dams on lower main stem, twelve major tributary dams
Water quality: conditions improved since 1970s; acid mine drainage, agricultural, and industrial/urban inputs cause problems in lower basin; pH = 7.0, alkalinity = 29 mg/L as CaCO$_3$, NO$_3$-N = 1.2 mg/L, PO$_4$-P = 0.07 mg/L
Land use: 64% forest, 30% agriculture, 4% urban, 1% barren/mined, 1% surface water
Population density: 39 people/km^2
Major information sources: ORSANCO 1994, 2000, NAWQA 2000a

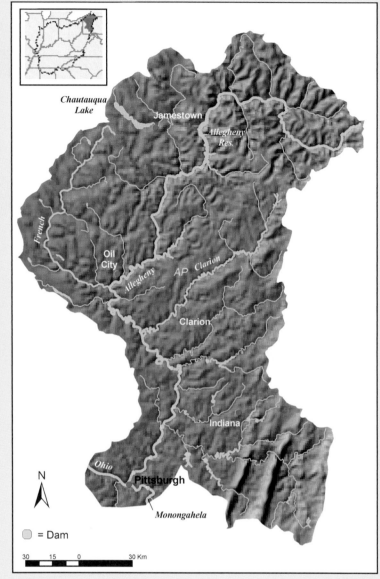

FIGURE 9.34 Map of the Allegheny River basin.

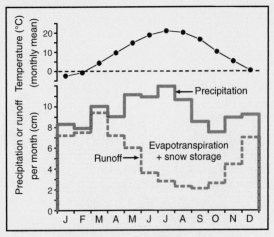

FIGURE 9.37 Mean monthly air temperature, precipitation, and runoff for the Monongahela River basin.

MONONGAHELA RIVER

Relief: 1260 m
Basin area: 19,110 km^2
Mean discharge: 377 m^3/s
River order: 7
Mean annual precipitation: 106 cm
Mean air temperature: 10°C
Mean water temperature: 14°C
Physiographic provinces: Valley and Ridge (VR), Appalachian Plateaus (AP)
Biome: Temperate Deciduous Forest
Freshwater ecoregion: Teays–Old Ohio
Terrestrial ecoregions: Appalachian/Blue Ridge Forests, Appalachian Mixed Mesophytic Forests
Number of fish species: >120
Endangered species: 0
Major fishes: gizzard shad, carp, bluegill, largemouth bass, spotted bass, channel catfish, walleye, sauger, emerald shiner, white crappie, black crappie, gravel chub, river redhorse, black redhorse, spotted sucker, flathead catfish, stoneroller, striped shiner, fantail darter, river carpsucker

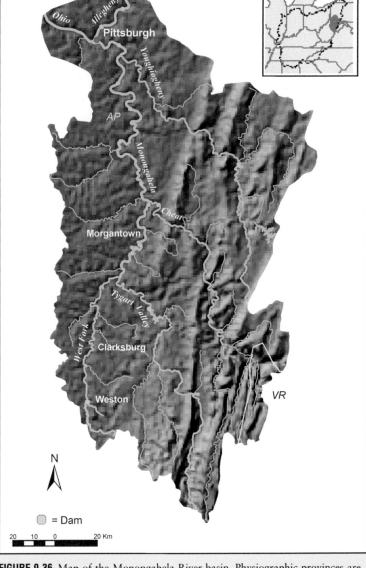

FIGURE 9.36 Map of the Monongahela River basin. Physiographic provinces are separated by a yellow line.

Major other aquatic vertebrates: snapping turtle, stinkpot, midland painted turtle, spiny softshell turtle, common water snake, bullfrog, green frog, hellbender, belted kingfisher, osprey, bald eagle, coot, common loon, beaver, river otter, muskrat
Major benthic invertebrates: bivalves (pink heelsplitter, white heelsplitter, fanshell, giant floater, fragile papershell), crustaceans (*Procambarus*, *Orconectes*), mayflies (*Hexagenia*, *Brachycentrus*, *Ephemerella*, *Isonychia*, *Stenonema*), stoneflies (*Allocapnia*, *Isoperla*, *Acroneuria*), hellgrammites (*Corydalus*), alderflies (*Sialis*), caddisflies (*Hydroptila*, *Hydropsyche*, *Cheumatopsyche*, *Ceraclea*), beetles (*Tropisternus*, *Berosus*), true flies (*Tipula*)
Nonnative species: Asiatic clam, zebra mussel, goldfish, carp, striped bass, alewife, tench, margined madtom, rainbow smelt, rainbow trout, brown trout, lake trout, yellow perch, purple loosestrife
Major riparian plants: river birch, red maple, cottonwood, black willow, American sycamore, speckled alder
Special features: headwaters of the Cheat and Tygart Valley rivers begin in the Monongahela National Forest
Fragmentation: nine major navigation locks and dams, five major tributary dams
Water quality: significant improvement since 1970s; acid mine drainage, agricultural, and industrial/urban inputs cause problems; pH = 7.1, alkalinity = 32 mg/L as CaCO$_3$, NO$_3$-N = 1.1 mg/L, PO$_4$-P = 0.05 mg/L
Land use: 65% forest, 29% agriculture, 4% urban, 1% barren/mined, 1% surface water
Population density: 65 people/km^2
Major information sources: NAWQA 2000a, West Virginia Department of Natural Resources, unpublished data; ORSANCO 2000

10

MISSOURI RIVER BASIN

DAVID L. GALAT   CHARLES R. BERRY JR.   EDWARD J. PETERS   ROBERT G. WHITE

INTRODUCTION

MISSOURI RIVER MAIN STEM

YELLOWSTONE RIVER

WHITE RIVER

PLATTE RIVER

GASCONADE RIVER

ADDITIONAL RIVERS

ACKNOWLEDGMENTS

LITERATURE CITED

INTRODUCTION

The Missouri River basin is the second largest in the United States, surpassed in area only by the Mississippi River basin of which it is a part. It drains about one-sixth of the conterminous United States (1,371,017 km^2) and about 25,100 km^2 in Canada. The Missouri basin trends in a northwest to southeast direction across the north-central United States (Fig. 10.2) and includes all or parts of 10 states, 2 Canadian provinces, and 25 Native American Tribal Reservations or Lands (74,500 km^2). Latitude of the basin ranges from 49.7°N in southwest Saskatchewan to 37.0°N in southwest Missouri.

About 20 Native American tribes belonging to four linguistic groups (Algonkian, Siouian, Caddoan, Shoshonean) lived in the Missouri River basin around 1500. After the Louisiana Purchase in 1803 put the entire Missouri River basin into federal ownership and significant Euro-American expansion into the basin began in 1848, well-documented conflicts with Native Americans ensued. Today, private interests, counties, states, or Native American tribes own about 86% of the basin.

Twelve rivers are discussed in this chapter. In addition to the Missouri main stem, we describe four tributaries in detail: the Yellowstone, White, Platte, and Gasconade rivers. These rivers extend over 56% of the Missouri River's total length and represent all three physiographic divisions in the basin. None of these tributaries have main-stem impoundments, and tributary regulation is minimal except for the North Platte and South Platte. Abbreviated physical and biological information are also provided for the Madison, Milk, Cheyenne, Niobrara, Big Sioux, Kansas, and Grand rivers. All but the Madison and Gasconade rivers rank in the top 15 of the Missouri basin by drainage basin area.

Physiography and Climate

A wide range of climatic conditions, geologic complexity, and topographic relief exist within the three physiographic divisions that contribute to the Missouri River basin. From west to east, these divisions and the number of physiographic provinces within each (in parentheses) are as follows: Rocky Mountain System (4), Interior Plains (2), and Interior Highlands (1). Basin climate is governed by interactions of four major air masses: warm, moist air originating in the Gulf of Mexico; cool, moist air from the north Pacific Ocean; cold, dry air from the north

FIGURE 10.1 Yellowstone River near north border of Yellowstone National Park (PHOTO BY C. E. CUSHING).

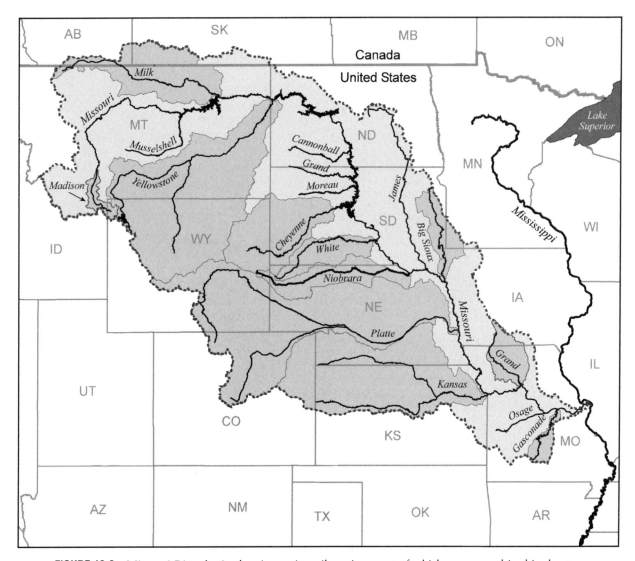

FIGURE 10.2   Missouri River basin showing major tributaries, most of which are covered in this chapter.

polar regions; and hot, dry air from plateaus in north-central Mexico. Convergence of these air masses and the midcontinent location of the basin produce extreme seasonal variability in weather. Daily fluctuations can also be sudden and severe. Winters are relatively long and cold; spring is cool, moist, and windy; summers are fair and hot; and autumn is cool, dry, and windy.

The basin is largely semiarid. About one-half receives <41 cm/yr of precipitation, with 70% of this occurring as rainfall during the growing season. Length of the freeze-free period is about 30 days at higher elevations in the Rocky Mountains, about 140 days on the Interior Plains, and about 180 days in the Interior Highlands (Missouri River Basin Commission 1977).

Rocky Mountain System

About 10.7% (~142,400 km^2) of the Missouri River's U.S. drainage basin is within this division. The Rocky Mountains are characterized by great topographic relief, with mountain summits typically >1500 m above their bases. Missouri River headwaters originate along the eastern slope of the Rocky Mountain system from Alberta south to Colorado. Climate is typical of highlands, with atmospheric conditions changing markedly over short distances in response to altitude, orientation, and slope. Annual precipitation in the mountains averages >80 cm and can exceed 127 cm, in contrast to 25 cm in some intermountain valleys. Precipitation often arrives as snowfall, and accumulations of >4 m/yr are common.

January is normally the coldest month, with mean daily minimum temperatures as low as –40°C. July is the warmest month, with daily maximum air temperatures average about 22°C.

Parts of all four physiographic provinces within the Rocky Mountain system are contained within the Missouri River basin: Northern Rocky Mountains (NR), Middle Rocky Mountains (MR), Wyoming Basin (WB), and Southern Rocky Mountains (SR) (Fig. 10.15). Geology, topography, and soils vary widely among these provinces and help define the physiochemistry of the headwater rivers to the Missouri. Elevations of the Northern Rocky Mountains range between 1200 and 3000 m asl but reach 3443 m asl at the Missouri River headwaters in the upper Madison River subbasin, Montana. Most of the Montana portion of the Northern Rockies is semiarid, receiving precipitation of <50 cm/yr, and only about one-fifth of the province drains to the east in Canada and the United States. Important tributaries to the Missouri originating in the Northern Rockies include the Milk, Marias, Big Hole, Jefferson, Madison, Gallatin, and Yellowstone rivers.

Mountains within the Middle Rocky Mountains trend in many directions and altitudes range from 1500 to about 3700 m asl. The Yellowstone subbasin is a remnant dissected volcanic plateau and thermal springs are common. Annual precipitation in the Middle Rocky Mountains ranges between 25 and 64 cm. The Bighorn and Yellowstone rivers have their origins here.

The Wyoming Basin is a series of semiarid subbasins separated by isolated low mountain ridges and knobs whose elevations range from 1800 to 2400 m asl. Extensive alluvial deposits occur in floodplains of streams and in fans at the base of mountains. Precipitation (25 to 51 cm/yr) averages lower than in the Rocky Mountain provinces and weathering is slight. The Wind, Laramie, and North Platte rivers are tributaries of the Missouri originating in the Wyoming Basin. About one-half of the Middle Rocky Mountain and Wyoming Basin drain to the Missouri River basin.

The Southern Rocky Mountains consist of a series of high-elevation ranges and intervening basins. The tallest and most rugged mountains within the division occur here, including the highest point within the Missouri River basin, Mt. Elbert, Colorado (4399 m asl), at the headwaters of the Platte River subbasin. About 50% of annual runoff from the Southern Rocky Mountains results from thunderstorms, in contrast to about 30% in the Middle Rockies and only 15% in the Northern Rockies.

Two-thirds of the Southern Rocky Mountains drain eastward into the Missouri, Arkansas, or Rio Grande basins. The North and South Platte rivers are tributaries to the Missouri River having part or all of their origins in the Southern Rocky Mountains.

Interior Plains

The largest portion of the Missouri River basin in the United States, 70.3% (~932,400 km²), is within the semiarid Great Plains (GP) province of the Interior Plains division (see Fig. 10.15). It includes parts of Alberta, Saskatchewan, Montana, North Dakota, South Dakota, Colorado, Nebraska, and Kansas. Harsh winter climate, a short growing season, and periodic severe droughts characterize the climate of much of this province. Annual precipitation averages only 36 cm and ranges from <28 to >60 cm. Minimum January air temperature averages –19°C and mean maximum July temperature is about 30°C. The Great Plains slope eastward from about 1670 m asl at their contact with the Rocky Mountain System to about 608 m asl along their eastern boundary. This province experienced at least three cycles of marine transgression and regression, resulting in deposition of extensive layers of sandstone, shale, limestone, conglomerate, and lignite. Silts, sands, and gravels deposited by eastern flowing streams from the Rocky Mountains overlie these rocks. The most northern section of the Great Plains province within the Missouri River basin is the Missouri Plateau. Surficial deposits in northern parts of this section in Alberta, Saskatchewan, Montana, North Dakota, and South Dakota are highly erodible glacial till with a gently rolling topography and belts of glacial moraines. The Missouri River and its tributaries are entrenched tens of meters into these sediments. The largest of these glacial deposits is a 745 m elevation plateaulike highland, the Missouri Coteau, located east of the Missouri River in southern Saskatchewan and northern North Dakota. Other topographic features of the Missouri Plateau are several small domed mountain groups, including the Black Hills of western South Dakota and eastern Wyoming. South of the Missouri Plateau is the High Plains section of the Great Plains province, which contains thick alluvial sediments eroded from the Rocky Mountains. Included here is the sand hills area between the Niobrara and Platte rivers.

The north portion of the lower Missouri River basin is contained within the Central Lowland province of the Interior Plains division and includes parts of North Dakota, South Dakota, Iowa, Nebraska, Kansas, and Missouri. Its western boundary with the Great Plains province approximates the

608 m asl contour and ~50 cm precipitation line. It contributes about 17.1% (~227,400 km²) to Missouri River basin's area and exhibits a relatively flat relief, with elevations ranging from 300 to 600 m asl. Most of the Central Lowland was glaciated during the Pleistocene, and glacial deposits of sands, gravels, and till dominate its contemporary surficial topography. Sections of this province within the Missouri River basin are distinguished primarily by differences in glacial histories. Portions of the Dissected Till Plains section in eastern Nebraska, southern Iowa, and northern Missouri are mantled with >9 m of highly erodible loess deposits derived from the postglacial Missouri River. About 6% of the Missouri River basin is eroded (Revenga et al. 1998), and estimated sediment yield during the 1960s was highest (>2 million kg km⁻² yr⁻¹) from the Central Lowland between the Big Sioux and Kansas rivers (Missouri Basin Inter-agency Committee 1969).

Average annual precipitation within the Central Lowland province ranges from about 40 cm in the northwest to 102 cm in the southeast. Air temperatures range from −17°C in January to 32°C in July.

Interior Highlands

The smallest area of the Missouri River basin, about 2.0% (~28,300 km²), is located in the south lower basin within the unglaciated Ozark Plateaus (OZ) province of the Interior Highlands division (see Fig. 10.15). The Ozark Plateaus are structurally and topographically a low symmetrical dome of predominately dolomites and limestones with lesser quantities of shales, cherts, and sandstones. Highest elevations within this portion of the Missouri River basin range from 444 to 532 m asl. The Ozarks have karst topography; springs and losing streams (i.e., streams where discharge decreases due to groundwater losses) are common. This province has the most mesic climate within the basin, with annual precipitation averaging >100 cm. Mean monthly air temperature ranges from about −6°C in January to about 32°C in July.

Basin Landscape and Land Use

Thirteen terrestrial ecoregions are represented in the Missouri River basin (Ricketts et al. 1999), reflecting its altitudinal, latitudinal, and climatic gradients. From west to east these ecoregions grade from Rocky Mountain coniferous forests, to shrub steppe, then short, mixed, and tall grasslands, and finally to mixed hardwood forests.

Douglas fir, Engleman spruce, sub-alpine fir, lodgepole pine, and ponderosa pine are the dominant conifers in the North Central, South Central, and Colorado Rockies Forests terrestrial ecoregions. Mountain meadows, foothill grasslands, riparian woodlands, and upper treeline–alpine tundra are other major vegetation communities within the Rockies Forests ecoregions. Northwestern Mixed Grasslands is the largest ecoregion within the Interior Plains, with grama-needlegrass and wheatgrass the principal native grass communities. Sagebrush is the dominant shrub and prickly pear is present on drier sites. Shrubby aspen, willow, cottonwood, and box elder are found on shaded valley slopes and river terraces. Typical grasses of the Northern Mixed Grasslands include grama, little bluestem, needle-and-thread grass, wheatgrass, and junegrass, whereas native grasses of the Central Tall Grasslands ecoregion are predominately big bluestem, switchgrass, and Indian grass. The Central Mixed Hardwoods ecoregion is one of the richest in North America for plants and shrubs, with over 2500 species (Ricketts et al. 1999). Widespread trees are white, red, and black oaks and bitternut and shagbark hickories. Flowering dogwood, sassafras, and hornbeams are important understory trees. Riparian areas and wetter sites favor cottonwood, American elm, sycamore, green ash, and silver maple.

The basin is 37% cropland, 30% grassland, 13% shrub, 11% forested, and 9% developed (Revenga et al. 1998). Major land-surface resources within the basin from northwest to southeast are Rocky Mountain range and forest, western range and irrigated agriculture, northern Great Plains spring wheat, western Great Plains range and irrigated agriculture, central Great Plains winter wheat and range, central feed grains and livestock, and east and central general farming and forest (Slizeski et al. 1982).

Metallic minerals (gold, silver, copper, zinc, molybdenum) are important subsurface resources in the Rocky Mountains, and nonmetallics such as flourspar, feldspar, phosphate, lime, mica, bentonite, and construction aggregate are also mined within the basin. Energy fuels are the largest and most valuable resource commodity, contributing about 55% of U.S. recoverable coal reserves and 8% of petroleum output during the early 1970s (Missouri River Basin Commission 1977).

The Rivers

Three freshwater ecoregions (Upper Missouri, Middle Missouri, and Central Prairie; Abell et al. 2000) containing 47 rivers with drainage basins >1000 km² contribute to the Missouri River. Its three

TABLE 10.1 Features of the six major earth-filled dams and reservoirs on the Missouri River. Location of dam is kilometers upstream from Missouri River mouth (see Fig. 10.15).

Feature	Fort Peck	Sakakawea	Oahe	Sharpe	Francis Case	Lewis & Clark
Name of dam	Fort Peck	Garrison	Oahe	Big Bend	Fort Randall	Gavins Point
Location of dam (Rkm)	2851	2237	1725	1588	1416	1305
Height of dam (m)	67	55	61	24	43	14
Year of dam closure	1937	1953	1958	1963	1952	1955
Total drainage area (10^3 km^2)	148.9	469.8	630.6	645.8	682.4	732.9
Length of full reservoir (km)	216	286	372	129	172	40
Surface area (km^2)	996	1538	1514	247	413	125
Gross volume (km^3)	23.1	29.3	28.5	2.3	6.7	0.6
Annual power production (MWh $\times$ 10^6)	1.2	2.5	2.9	1.1	1.8	0.7

Source: Galat and Frazier 1996, U.S. Army Corps of Engineers 2001.

largest tributaries, the Platte, Yellowstone, and Kansas (see Fig. 10.2), are ranked from 13th to 15th in descending order of drainage basin area in the United States, respectively (van der Leeden et al. 1990).

The southern limit of continental glaciation largely defines the present course of the Missouri River. Both the Upper and Middle Missouri ecoregions experienced heavy glaciation as recently as 10,000 to 15,000 years ago and harbor no known endemic fishes, mussels, crayfishes, or aquatic herpetofauna (Abell et al. 2000). Tributaries to the main stem within the Great Plains show unequal development (see Fig. 10.2). Tributaries on the west (only lightly glaciated) of the Missouri are larger and represent well-developed drainage systems dating to the early Tertiary (e.g., Knife, Cheyenne, White rivers). The Missouri River floodplain is wide where the river cuts across these preexisting alluvial valleys. Tributaries to the east (glaciated) side are small and have developed since melting of the Wisconsin ice (Johnson et al. 1976).

Thirty-seven tributaries with drainage areas >3000 km^2 enter the Missouri River, and those we discuss further (see Fig. 10.2) are italicized here. Ten tributaries have their origins within the Rocky Mountain Division, including, from upriver to downriver, the *Milk*, Marias, Sun, Jefferson, *Madison*, Gallatin, Judith, Musselshell, *Yellowstone*, and *Platte* rivers. Rivers of similar minimum basin area that originate in the Great Plains include the Dry, Redwater, Popular, Big Muddy, Little Missouri, Knife, Heart, Cannonball, Grand, Moreau, *Cheyenne*, Bad, *White*, *Niobrara*, and *Kansas*. The James, Vermillion, *Big Sioux*, Little Sioux, Nishnabotna, Platte (Mis-

souri), *Grand* (Missouri), Chariton, Lamine, and Osage rivers originate entirely or partially within the Central Lowland Province. Four rivers—the Lamine, Moreau (Missouri), Osage, and *Gasconade*—have part or all of their headwaters in the Ozark Plateaus.

Although few endemic aquatic vertebrates occur within the three Missouri River ecoregions, several fishes adapted to large, turbid, main-stem environments (i.e., "big river" fishes) are found through the main stem. These include the pallid and shovelnose sturgeons, blue sucker, and sicklefin and sturgeon chubs.

Approximately 100 multipurpose and over 1200 single-purpose reservoirs have been constructed in the basin, with a total multipurpose storage capacity of about 130.7 km^3 (U.S. Army Corps of Engineers 2001). The most prominent water development project is the six reservoirs on the main-stem Missouri (Fig. 10.15, Table 10.1). This system accounts for 69% of the basin's total multipurpose storage capacity and will be discussed further in the Missouri River section.

MISSOURI RIVER MAIN STEM

An unpredictable river in an unpredictable landscape best describes the historical Missouri River. It is the longest river in the United States (4180 km) from its headwaters in Hell Roaring Creek, Montana (its longest named source stream), and the second-longest river in North America following the Mackenzie–Slave–Peace rivers (4241 km) (van der

Leeden et al. 1990). Total length of the named Missouri River from the confluence of the Jefferson, Madison, and Gallatin rivers near Three Forks, Montana, to its junction with the Mississippi River about 24 km north of St. Louis, Missouri, is generally reported as 3768 km (Hesse et al. 1989), making it the longest named river in North America. The northernmost extent of its main channel is located near Tioga, North Dakota, at 48.17°N latitude and its southernmost extent is between Washington and Augusta, Missouri, at 38.53°N latitude (see Fig. 10.15). Elevation of the named Missouri River ranges from about 1226 m at Three Forks, Montana, to about 122 m at the mouth.

The Missouri River valley supplied abundant resources to Native Americans living along its course, including an unfailing supply of water, game, and wood. River bottom forests gave shelter from summer heat and winter wind and cold. Use of the river for travel by Native Americans was not substantial; small buffalo-skin "bull boats" were their primary craft. Major tribal groups living along the river around 1800 from its mouth northwestward included the Oto, Missouri, Omaha, Ponca, Brule, Teton Sioux, Yankton Sioux, Yanktoinai Sioux, Arikara, Hidatsa (Minitari), Mandan, Assiniboin, Atsina, and Piegan Blackfoot. Two major centers of aboriginal trade within the northern plains existed along the Missouri River: the Arikara villages near the mouth of the Grand River in South Dakota and the Mandan and Hidatsa villages near the mouth of the Knife River in North Dakota.

The Lewis and Clark expedition (1804 to 1806) gave us unprecedented detail of the unaltered Missouri's geography, natural history, and ethnography of its native peoples. Today their journals are revered as classics in the American literature of discovery and exploration (Cutright 1969). Moreover, Lewis and Clark's systematic approach to exploration of the Missouri set the pattern for future scientific expeditions throughout the Americas.

Following Lewis and Clark's "Corps of Discovery," the Missouri became the "original highway west" for Euro-American development of the United States (Thorson 1994). Keelboats were an early (~1815) mode of conveyance for explorers, trappers, immigrants, and their supplies upriver and furs and agricultural produce downriver. The glory days of Missouri River transport came during the steamboat era (1820s to 1880s), first servicing fur traders and then gold seekers and settlers.

The steamboat also played a crucial role in transformation of Native American culture. It brought the early traders with their welcome merchandise, but also their liquor and smallpox. It transported the commissioners to make treaties, and the Indian agent with the annuities those treaties guaranteed. It brought the artists George Catlin and Karl Bodmer to glorify the American frontier through portraits and scenes of their culture. It delivered the hunters who exterminated their buffalo, and finally it brought the military to pacify them.

Steamboating on the Missouri was terribly hazardous. Shifting channels, shallow sandbars, ice, countless snags, and boiler explosions sank hundreds of steamboats. Railroads began to displace river travel when rail service reached St. Joseph, Missouri, in 1859, and long-distance steamboat navigation ended by 1887.

Physiography, Climate, and Land Use

The Missouri River begins in the Northern Rocky Mountain (NR) physiographic province, then flows for most of its length through highly erodible soils of the Great Plains (GP) and Central Lowland (CL) provinces (see Fig. 10.15). It traverses six of the thirteen terrestrial ecoregions within its basin: North Central Rockies Forests, Montana Valley and Foothill Grasslands, Northwestern Mixed Grasslands, Northern Mixed Grasslands, Central Tall Grasslands, and Central Forest/Grassland Transition Zone.

Climatic ranges of the main stem reflect those described previously for the Great Plains and Central Lowlands provinces, with mean annual precipitation ranging from about 36 to 104 cm/yr. Using mean values for the entire basin, it is clear that there is a very seasonal pattern of precipitation (Fig. 10.16). Lowest precipitation occurs from January to February (~1.7 cm) and highest in June (~8 cm). January is the coldest month (−7°C to −2°C) and July the warmest (20°C to 26°C).

Land use within 5 km of the river is primarily cropland (33%) and grassland (26%), with about 17% of the area developed (Revenga et al. 1998). Basinwide land-use percentages are similar to those presented in the general introduction, except that development is higher near the river.

River Geomorphology, Hydrology, and Chemistry

The Missouri River historically was braided, shifted frequently, and contained numerous sandbars, islands, and unstable banks. Overbank floods were

(a) (b) (c) (d)

FIGURE 10.3 Illustrations of Missouri River zones discussed in text. (a) upper or least-altered zone in central Montana showing the Cow Island area (~RM 1942) where the river channel lies 150 to 300 m below the 3 to 16 km wide alluvial valley and highly eroded river bluffs (PHOTO BY C. BERRY). (b) Garrison Dam (left center), the fifth-largest earthen dam in the world, and Lake Sakakawea, nearly 300 km long in central North Dakota, is located in the interreservoir zone (PHOTO BY NASA EARTH SCIENCES; IMAGE COURTESY OF EARTH SCIENCES AND IMAGE ANALYSIS LABORATORY, NASA JOHNSON SPACE CENTER). Lake Sakakawea has the largest volume (29.3 km³) of the six U.S. Army Corps of Engineers main-stem reservoirs. (c) Middle or interreservoir zone near Bismarck, North Dakota, showing the 1.6 to 11.0 km wide alluvial valley, complex riverine and floodplain habitats, and remnant cottonwood forest (PHOTO BY NORTH DAKOTA GAME AND FISH DEPARTMENT). (d) Lower or channelized zone in Missouri, where the formerly braided channel has been engineered into a smoothly curved, self-scouring, single channel to facilitate navigation (PHOTO BY J. ROBINSON). Alluvial valley width here ranges from 2.4 to 27.4 km and averages 8.1 km.

common, turbidity high, and sediment transport enormous, giving it the nickname "Big Muddy." It is most conveniently divided into upper, middle, and lower zones based on contemporary geomorphology and hydrology (Fig. 10.3).

The upper zone is unchannelized and extends about 739 km from the origin of the named Missouri River to the first of the six major main-stem impoundments, Fort Peck Lake, Montana (see Fig. 10.15). A National Wild and Scenic River section occurs over 240 km between Fort Benton and Fort Peck Lake, representing the last largely free-flowing portion of the Missouri River that retains most of its primitive characteristics. The alluvial valley is a picturesque gorge of badlands and breaks (see Fig. 10.3a). Gradient averages 60 cm/km and ranges from <40 to >190 cm/km.

The middle or "interreservoir" zone also remains unchannelized, but it was impounded between 1937 and 1963 (see Table 10.1, Fig. 10.3b and c). It runs from the upper end of Fort Peck Lake (Rkm 3029) to Gavins Point Dam (Lewis and Clark Lake, Rkm 1305) and extends an additional 127 km to Sioux City, Iowa (Rkm 1178), where channelization begins (see Fig. 10.15). Riverbanks have been stabilized in a 44 km reach above Sioux City.

Surficial deposits within the middle zone (Great Plains province) near Bismarck, North Dakota, are a mixture of glacial till, loess, and alluvial sediments with an average depth of 30 to 35 m. Soils are predominately clays, silts, sands, and gravels. Bedrock is generally composed of shales and sandstones. These features result in highly erodible banklines and river bottoms. Slope from the top of the multiple-use pool at Fort Peck Lake to Lewis and Clark Lake is 18 cm/km. Only 547 km (3.2%) of fragmented river segments currently remain in the 1724 km between Fort Peck Lake and Gavins Point Dam (see Fig. 10.15). These remnant riverine reaches exhibit relatively sinuous to semibraided channels and retain many of the islands, backwaters, side channels, and floodplain wetlands characteristic of predam geomorphology (see Fig. 10.3c). Two sections, a 58 km reach between Fort Randall Dam and the delta of Lewis and Clark Lake and a 93 km reach below Lewis and Clark Lake, are designated as National Recreational Rivers. The latter reach is the only unchannelized segment remaining in the lower 1300 km of river; here channel widths average 720 m.

The 1178 km of the Missouri River below Sioux City to its confluence with the Mississippi River is referred to as the lower or "channelized" zone (see Fig. 10.15). This zone is now altered by channelization, bank stabilization, and floodplain levees (Fig. 10.3d). The floodplain area is about 7690 km^2 or 6.5 km^2/km of river (U.S. Fish and Wildlife Service 1980). Alluvium consists largely of clay and silt overlying sand and gravel to a depth of about 30 m. Channel width before channelization ranged from 610 to 1829 m between Rkm 1297 and Rkm 589, and channel migration was dynamic.

Gravel composes about 39% of bed materials in upper Missouri River outside bends, inside bends, and channel crossovers (where the thalweg crosses the midchannel between two bends) and decreases to 5% in the these habitats within middle and lower zones. Sand is the dominant substrate throughout the main channel. Mean percentage of sand is highest in the middle zone (86%) compared with the upper (45%) and lower (81%) zones. Percentage of silt averages <10% of bed materials in main-channel habitats but is the predominate substrate in nonconnected secondary channels and tributary confluences (Galat et al. 2001).

Spatiotemporal patterns in runoff and hydrology of the Missouri River reflect the great climatic variability among physiographic provinces within its basin. From 1967 to 1996 runoff ranged from 11.2 cm/yr for the basin above Fort Benton, Montana, decreased to 3.2 cm/yr for the area between Fort Benton and Omaha, Nebraska, and was highest in the mesic lower subbasin between Kansas City and Hermann, Missouri (29.9 cm/yr). Overall, runoff to the Missouri is relatively low, reflecting dominance of the semiarid Great Plains. For example, mean annual runoff is about four times less for the 3610 km long Missouri River at Hermann (4.6 cm/year) than for the nearby 1111 km long Upper Mississippi River at Alton, Illinois (Galat and Lipkin 2000). Such a low basin runoff reflects a combination of low precipitation and losses from evapotranspiration throughout the Great Plains, where annual evaporation from all six main-stem reservoirs is about 5% of average annual river discharge (see Fig. 10.16).

Mean annual discharge between 1929 and 1996 ranged from 204 m^3/s at Fort Benton (Rkm 3336), to 883 m^3/s at Omaha (Rkm 991), to 2256 m^3/s at Hermann (Rkm 158). Increases in flow are gradual in the semiarid Great Plains of the upper and middle Missouri River zones, but below Omaha flow increases more steeply and interannual flow variability is higher due to the more mesic climate and numerous large tributaries originating in the Central Lowland and Ozark Plateaus. Although the Missouri River from Omaha to its mouth drains only about 38% of the total basin, it contributed about 61% of the 1929 to 1996 mean annual discharge.

The seasonal increase in runoff and discharge begins in March with ice-out and runoff from prairie snowmelt. It peaks in June, corresponding to Rocky Mountain snowmelt and late spring precipitation in the lower basin, then declines in July (see Fig. 10.16). Basinwide drought in the 1930s and flooding in the 1990s resulted in 34% higher mean annual discharge and runoff after flow regulation (1967 to 1996) than before (1929 to 1948) (see Fig. 10.16). Despite this climatic effect, flow regulation now depresses the flood pulse, except in the lowermost river, and increases flows from August until navigation ceases in December. Tributary discharge ameliorates flow-reduction effects of upriver impoundments in the lowermost river, where catastrophic flooding occurred in 1993 and again in 1995. The 21,240 m³/s discharge at Boonville, Missouri (Rkm 317), on July 29, 1993 is the highest on record, and over 500 levees were overtopped or breached. Mean current velocity in the main-channel Missouri River is highest in the lower zone (1.03 m/s) and similar within the upper (0.74 m/s) and middle (0.75 m/s) zones.

Most of the Missouri is a warmwater river. Mean July to October water temperature increases about 5.5°C between upper (21.5°C) and lower (27.0°C) zones, but is affected in the middle zone by hypolimetic water releases from reservoirs. Temperature depressions from mid July to early October average 8.5°C and 6.0°C between river segments above and below Fort Peck and Garrison dams, respectively. The farthest downriver reservoir, Lewis and Clark Lake, has little effect on water temperatures because of its shallow depth and short residence time (see Table 10.1).

Turbidity was historically high over much of the Missouri River due to the erosive landscape but is now reduced by sedimentation in reservoirs. For example, turbidity is only 16 NTUs below Fort Peck Lake in the upper Missouri but increases to >150 NTUs after the confluence with the Yellowstone River. Turbidity again decreases below Lake Sakakawea and remains <20 NTU through the remainder of the interreservoir zone (Galat et al. 2001). It then increases to >200 NTUs in the channelized river of central Missouri. Areal sediment yield of the Missouri is about 159 tons $km^{-2} yr^{-1}$, second in the United States only to the Colorado River (212 tons $km^{-2} yr^{-1}$).

Throughout its length the Missouri is a hard- to very hard-water river (~140 to 250 mg/L as CaCO3 total hardness; Patrick 1998). It is alkaline (145 to 162 mg/L as CaCO3, pH 8.0 to 8.3), high in conductivity (370 to 800 μS/cm), and high in total dissolved solids (230 to 500 mg/L). Longitudinal patterns of dissolved constituents are somewhat variable, reflecting interactions between geology, physiography, reservoir evaporation, and tributary influx.

Macronutrient concentrations of the Madison and Gallatin tributaries to the upper Missouri are naturally high ($NO_3 + NO_2$-N = <0.1 to 0.33 mg/L, PO_4-P = <0.01 to 0.05 mg/L) because of hydrothermal geology in the greater Yellowstone ecosystem (Hauer et al. 1991) but decrease downstream in central Montana (Patrick 1998). Concentrations of NO_3-N + NO_2-N above Lewis and Clark Lake (Rkm 1363) range from <0.05 to 0.36 mg/L and dissolved PO_4-P from <0.01 to 0.025 mg/L (1990 to 1999). In the lower Missouri at Hermann, NO_3-N + NO_2-N ranges between 0.02 and 3.2 mg/L and dissolved PO_4-P between 0.01 and 0.23 mg/L (1980 to 1998) due to the urbanization, intensive agriculture, and livestock operations.

Concentrations of arsenic and selenium in the middle Missouri River are naturally high and may exceed state water-quality standards. Past gold mining in the Cheyenne River basin yielded high discharges of mercury, arsenic, and other contaminants to Lake Oahe and the Missouri River. Bioaccumulation of methylmercury has resulted in consumption advisories for reservoir sport fishes. Additional consumption advisories for fishes in Nebraska and Missouri are due to polychlorinated biphenyls and dieldrin.

River Biodiversity and Ecology

The Missouri is one of the few large river basins in North America that encompasses more than one freshwater ecoregion (Upper Missouri, Middle Missouri, and Central Prairie; Abell et al. 2000), reflecting its variability in physical and biological diversity. Many studies of its biology and ecology have been conducted from the upper to the lower river zones.

Algae and Cyanobacteria

High turbidity, high current velocity, and lack of adjoining lentic habitats contributed to low phytoplankton abundance prior to completion of the main-stem dams. Phytoplankton densities in the lower Missouri River before impoundment averaged 67 cells/L (Berner 1951) and diatoms (e.g., *Fragilaria*) and chlorophytes (e.g., *Pediastrum*) were dominant groups. Abundance increased markedly after impoundment, when Reetz (1982) reported phytoplankton densities in the channelized river below Lewis and Clark Lake (Rkm 1039 to 857) ranging from <1000 and >25,000 units/L, with 98% of values <10,000 units/L. Diatoms (e.g., *Asterionella*, *Stephanodiscus*) remained the dominant family, with chlorophytes (e.g., *Actinastrum*, *Ankistrodesmus*, *Scenedesmus*) becoming more abundant in summer. Cyanobacteria (e.g., *Merismopedia*) were generally <10% of total phytoplankton. Mean chlorophyll *a* concentrations ranged from 4.2 to 42 mg/m³ and showed little seasonal pattern. Increased phytoplankton densities are primarily a result of their discharge from main-stem impoundments, with numbers decreasing with increasing distance downstream from dams (Hesse et al. 1989, Patrick 1998). Primary substrates for periphyton colonization in the lower river have changed from silt, sand, and wood to rock rip-rap used for bank stabilization and channel maintenance structures. Diatoms composed 78% to 94% of periphyton taxa collected from the Missouri River above and below the Platte River, Nebraska (Farrell and Tesar 1982).

Plants

Aquatic macrophytes are nearly absent from the main-stem Missouri River due to its high turbidity, unstable substrates, and variable discharge. Perennial vegetation is sparse on sandbars and composed largely of young cottonwoods and willows. Contemporary vegetation in the Missouri River floodplain is a mosaic dominated by agriculture interspersed with remnant patches of forest, shrub/scrub, mesic prairies, and wetlands. Dominant overstory forest trees along the upper and middle Missouri River valley include plains cottonwood, green ash, box elder, and American elm. Peach-leaved willow and burr oak are subdominants (Keammerer et al. 1975, Johnson et al. 1976). Nonnative Russian olive is becoming a more common riparian plant here. Sandbar willow and black willow replace peach-leaved willow in the lower Missouri River, and hackberry, American sycamore, silver maple, red mulberry, and black walnut are added to the floodplain forest community (Bragg and Tatschl 1977). Tree species typically associate along a gradient of decreasing flood disturbance and increasing height above the channel. Willows and cottonwoods are nearest the channel, followed by sycamore, green ash, mulberry, silver maple, and box elder on intermediate sites. Burr oak, hackberry, and walnut are at higher, more protected locations.

Invertebrates

Abundance, seasonality, and taxonomic composition of contemporary Missouri River zooplankton are largely determined by discharges of reservoir assemblages. Zooplankton density and biomass generally decline with increasing distance downstream from dams. Repsys and Rogers (1982) reported 27 species of rotifers, 22 species of copepods, and 40 species of cladocerans from the main stem in central Nebraska. Dominant taxa included Rotifera (*Branchionus* spp., *Conochiloides* spp., *Conochilus* spp., *Polyarthra* spp., *Synchaeta* spp.), Copepoda (*Cyclops*

bicuspidatus, *C. vernalis*, *Diaptomus clavipes*, *D. forbesi*, *Mesocyclops edax*), and Cladocera (*Daphnia galeata*, *D. retrocurva*, and *D. leuchtenbergianum*).

Zooplankton abundance differs between unchannelized and channelized river segments and among riverine habitats. Kallemeyn and Novotny (1977) reported mean total densities of zooplankton in the unchannelized river below Gavins Point Dam (Rkm 1286 to 1253) were generally higher in the main channel (12,980 to $13,234/m^3$) and chutes (i.e., secondary channels, 8248 to $11,796/m^3$) than along the main-channel border (9326 to $9577/m^3$), whereas the main-channel border of the channelized reach had higher mean total densities ($10,041/m^3$) than did the main channel ($6084/m^3$). Recently created floodplain scours also have a rich zooplankton fauna, containing 39 species of Cladocera and 13 species of Copepoda (Havel et al. 2000).

Shifting substrates and high sediment loads in the preregulation main channel likely limited macroinvertebrates, which were most abundant on channel snags, in backwaters, and in floodplain wetlands. A rich and abundant macroinvertebrate fauna still exists in the upper Missouri River, where habitat loss is minimal and larger, more stable channel substrates occur. Hauer et al. (1991) reported mayflies (24 taxa, particularly *Baetis tricaudatus*, *Ephemerella inermis*, and *Tricorythodes minutus*) and caddisflies (13 taxa, particularly *Arctopsyche grandis*, *Hydropsyche* spp., *Cheumatopsyche* sp., *Glossosoma* sp., and *Brachycentrus* sp.) as abundant in the uppermost Missouri River from near Fort Benton to Three Forks, Montana. Densities were typically high ($>1000/m^2$) but variable among sites, being greatest below reservoirs. Dipterans such as chironomid midges and black fly larvae were also abundant, reaching densities of >3000 and $6000/m^2$, respectively. Dipterans (37%), mayflies (32%), and caddisflies (18%) were also the major macroinvertebrates downstream from Hauer et al.'s sites in the Wild and Scenic Missouri River, Montana (Berg 1981).

In the unchannelized Missouri River below Gavins Point Dam and in the channelized lower river, macroinvertebrate abundance is lowest in main channel and unvegetated secondary channels, intermediate on submerged sandbars and along sandy banks, higher along muddy banks and silty backwaters (secondary channel connected at downstream end), and highest in *Typha* marshes (Patrick 1998). Densities increase with increases in substrate stability, reduced current velocity, and increasing silt and organic matter. Highest densities and biomasses occur on hard substrates (originally snags, now largely rock dikes) rather than in shifting sand substrates.

Generalizations on relative density or biomass patterns among taxa are difficult for the lower Missouri River because of differences in collection methods and habitats sampled. Collector-gatherer taxa like the mayfly *Baetis* and collector-filterers like black flies and the caddisfly *Hydropsyche* are abundant on stable substrates in the fast-water main channel. The mayflies *Caenis* and *Hexagenia*, the caddisfly *Neureclipsis*, the dragonflies *Gomphus* and *Libellula*, the chironomids *Demicryptochironomus* and *Polypedilum*, and oligochaetes are major taxa in slow-water channel, chute, and backup marsh habitats. Densities in drift are high and major differences occur in composition and abundance between macroinvertebrates collected in benthic and drift samples. A moderately rich unionid mussel fauna of 12 species is present in the lower Missouri River despite its turbidity and unstable substrates (Hoke 1983).

Vertebrates

The Mississippi River basin, including the Missouri River, supports the richest freshwater fish fauna in North America, about 260 species (Robison 1986). There are about 183 species present in the Missouri River basin and about 136 reside in the main channel (Galat et al. in press). No fishes are endemic to the main stem and only two species (Niangua darter and bluestripe darter) are endemic to the basin. Seventy-eight percent of the river's fishes are native; two species, American eel and Alabama shad, are diadromous.

Sixty-eight percent of main-stem fishes belong to six families: Cyprinidae (46 species), Catostomidae (13), Salmonidae (12), Centrarchidae (12), and Ictaluridae (9). Species representing the archaic families Acipenseridae, Polyodontidae, Lepisosteidae, and Hiodontidae also occur in the Missouri River. About one-half of the Missouri's fishes are considered big-river species, residing primarily in the main channel.

Numerous big-river fishes share an array of eco-morphological adaptations to high turbidity, high velocity main-channel environments that make them one of the most distinctive faunas in North America. Some species are turbid-water benthic specialists, exhibiting an inferior mouth position, dorsoventral flattening of the head, streamlined or deep hump-backed body shape, sickle-shaped or enlarged pectoral fins, reduced eyes, and diverse and well-developed chemosensory organs (e.g., sturgeons, chubs, buffaloes, carpsuckers, blue suckers, catfishes,

burbot, and freshwater drum). Similar adaptations occur in fishes of the turbid Colorado River (see Chapter 11).

The pallid sturgeon is the only Missouri River fish listed as federally endangered by the U.S. Fish and Wildlife Service. Habitat and hydrologic alterations are the primary factors responsible for population declines of this obligate large-river species. Four other fishes are listed as globally vulnerable (G3; NatureServe 2002 http://www.natureserve.org/explorer/servlet/NatureServe?init=Species): lake sturgeon, Alabama shad, sicklefin chub, and sturgeon chub. Significant declines in native fishes have been reported from the middle and lower river, including paddlefish, burbot, silver chub, plains minnow, western silvery minnow, blue catfish, and sauger. Most are listed as imperiled (S1 to S3) in two or more states within the basin (NatureServe 2002 http://www.natureserve.org/explorer/servlet/NatureServe?init=Species) and all are big-river species (Cross et al. 1986). Nonnative species compose 21% of the entire river's fishes but increase to 43% in the upper river (Galat et al. 2004). Many were intentionally stocked into main-stem reservoirs for recreation or as forage for sport fishes (e.g., salmonids [10 species], centrarchids [8 species], rainbow smelt).

Approximately 78 species of reptiles and amphibians live within the Missouri River and floodplain ecosystem, although scant information is available on their distribution and ecology. Smith (1996) listed the following species richness by major group: aquatic salamanders (2) terrestrial salamanders (7), toads (6), treefrogs (2), chorus frogs (2), frogs (7), aquatic turtles (8), terrestrial turtles (2), lizards (9), aquatic snakes (5), semiaquatic snakes (6), and terrestrial snakes (22). Relative to the Mississippi River, the Missouri is characterized by western elements in the herpetofauna, with a moderately large number of snakes and lizards.

The river corridor is a migration and wintering habitat for millions of waterbirds. Most use is from spring and late summer through autumn. Shorebirds and dabbling ducks rest on islands and sandbars and forage in mudflats (shorebirds), wetlands, and grain fields (waterfowl) during migration. Seventeen species of ducks, three species of geese, and one swan species occur along the river. Common waterbirds along the lower Missouri River are American white pelican, American coot, snow goose, Canada goose, blue-winged teal, green-winged teal, northern shoveler, gadwall, mallard, wood duck, and great blue heron. Killdeer, lesser yellowlegs, pectoral, Baird's, least, spotted, and semipalmated sandpipers are common shorebird migrants. Eight species of terns and gulls occur along the lower Missouri River.

The Missouri River provides breeding habitat for interior populations of the endangered least tern, northern Great Plains populations of the threatened piping plover, and the threatened bald eagle. Presently, about 12% of the known population of interior least terns and 22% of the northern Great Plains population of piping plovers nest along the Missouri River main stem and its major tributaries (Smith 1996). Migration and wintering habitat are also used by the bald eagle and endangered whooping crane. Large numbers of bald eagles winter near open water below Missouri River main-stem dams and in the lowermost river, which seldom freezes.

Principal aquatic mammals along the Missouri River include mink, river otter, beaver, muskrat, and raccoon. Federally endangered gray and Indiana bats are reported to use lower Missouri River bluff caves for hibernation and the riparian corridor for foraging.

Ecosystem Processes

Information on ecosystem processes of the Missouri River is incomplete, but patchy data exist on production and food web structure. Seasonal mean primary production for phytoplankton has been measured only in the channelized river, Nebraska, and ranges from 7.4 mg C m^{-3}h^{-1} in winter to 203 mg C m^{-3}h^{-1} in autumn (Reetz 1982). Mean periphyton production ranged from 135 to 310 mg m^{-2}d^{-1} AFDM in the same reach (Farrell and Tesar 1982).

Light, rather than nutrients, limits algal growth in the lower Missouri River. Knowlton and Jones (2000) reported photic depths averaged 0.78 m and ranged from 0.12 to 2.4 m and mixed depth/photic depth ratios averaged 10.2 (range 1.2 to 55). Despite high turbidity, they suggested algal production was considerable in the lower Missouri River, Missouri, from 1994 to 1998. Flux of algal biomass for this area was estimated from chlorophyll concentrations as 270,000 kg/d.

High suspended sediment concentrations in the Missouri River transport the bulk of its organic carbon. Concentrations of dissolved organic carbon (DOC) and suspended sediment organic carbon (SSOC) in 1969–1970 averaged 4.6 mg/L and 20 mg/L, respectively (Malcolm and Durum 1976). Over 80% of the annual organic load is transported in the suspended fraction. Almost one-third of the SSOC load of the Mississippi River is contributed by the Missouri, which accounts for <10% of water inflow to the Mississippi. Carbon/nitrogen ratios for the Missouri are slightly higher (8.9) than Malcolm and Durum (1976) recorded from five other U.S. rivers (mean range, excluding Missouri, of 6.9 to 8.7).

Benthic insect production in the unchannelized Missouri River below Gavins Point Dam declined by 61% from 1963 to 1980 (Mestl and Hesse 1993). Secondary-channel and backwater habitats contributed 37% of this production in 1973, but their contribution declined to only 19% by 1980. Main-channel degradation draining these habitats was proposed as the cause. Insect production on woody habitats in the main channel contributed the highest amount to total production (69%, 42.4 g/m^2) in 1963 compared with mud and sand substrates and backwaters (Mestl and Hesse 1993).

No information exists on fish production, but there have been increases in the number of species collected and considerable changes in their relative abundances based on surveys made at approximately 20 yr intervals in Missouri (Pflieger and Grace 1987). Species reported to have become more abundant due to decreased turbidity are largely pelagic planktivores and sight-feeding carnivores: skipjack herring, gizzard shad, white bass, bluegill, white crappie, emerald shiner, and red shiner. Introduced fishes pose another potential threat to the native fish community structure, although their impacts have not been evaluated.

Human Impacts and Special Features

The Missouri is the longest river in North America and America's first highway to the west. Its flow patterns and braided channel geometry were once described as "uncertain as the actions of a jury or the state of a woman's mind" (*Sioux City Register*, March 28, 1868). This once highly dynamic river is today one of the most regulated rivers in the United States, containing the largest series of impoundments and >1100 km of largely flow-regulated, stable, uniform channel.

The earliest human impacts on the Missouri River were largely snag removal and deforestation. Snag removal to facilitate navigation was authorized by Congress in 1832, and between 1843 and 1846 over 60,000 channel snags were removed. A steamboat consumed >20 cords of wood per day on an upstream journey and consequently forests were nearly eliminated along the riverbanks.

Alterations of the Missouri began in earnest with passage of the Reclamation Act of 1902, "to turn the

Missouri River on itself" from Sioux City, Iowa, to St. Louis, Missouri (Ferrell 1996). A system of wooden pile dikes created a single self-scouring navigation channel (1.8 m deep, 61 m wide), and great masses of woven-willow and lumber revetments stabilized the banks of outside bends of the channel from St. Louis to Kansas City, Missouri. The first of the main-stem reservoirs, Fort Peck, was constructed in 1937 in central Montana to provide minimum flows for downriver navigation (see Fig. 10.15). The navigation channel was subsequently expanded and extended upriver to Sioux City, Iowa, with more sophisticated engineering designs replacing wooden structures. The collective effort, referred to as the Missouri River Bank Stabilization and Navigation Project, was completed by 1981.

Flooding was historically an essential process maintaining the natural character of the river–floodplain complex but an impediment to development. Construction of federal flood-control levees began in 1947 and fostered farming the floodplain. As the societal cost of flooding increased, the Pick–Sloan Plan (1944) resulted in the construction of the five remaining main-stem Missouri River dams between 1946 and 1963 (see Table 10.1). The total storage capacity of the six reservoirs is the largest of any river in the United States (90.5 km^3).

Major geomorphic changes between 1879 and 1972 occurred in the channelized river from Rkm 801 to the mouth (Funk and Robinson 1974). There was an 8% reduction in channel length, a 50% reduction in channel water surface area, a 98% reduction in island area, and an 89% reduction in the number of islands (see Fig. 10.3d).

Impoundment and flow regulation have nearly eliminated overbank flooding and sediment deposition on the floodplain in the middle and lower zones. Channel degradation and reduced rates of sandbar and island formation also occurred. For example, up to 1.8 m of bed degradation has occurred over the 32 km below Fort Randall Dam, and 3.1 m immediately below Gavins Point Dam, as substrate size has increased downstream (to gravel and cobble) since dam closure. Sediment deposition in the six main-stem reservoirs averages 113.5 km^3/yr, and large deltas have built up in the upper ends of Fort Peck (42 km long), Sakakawea (61 km), Oahe (103 km), and Lewis and Clark (23 km) lakes and at the mouths of major tributaries.

Commercial navigation on the Missouri River was projected in 1939 to transport about 10.9 million metric tons of freight per year (MMt/yr), but freight actually shipped from 1954 to 1996 averaged only 1.9 MMt/yr (U.S. Army Corps of Engineers 2001). The U.S. Army Corps of Engineers (cited in National Research Council 2002) estimated that annual net benefits of full-service commercial navigation in 1995 were <$3 million. In contrast, annual recreation benefits for 1994 from Fort Peck Lake to the mouth were estimated at $87.1 million, with about 78% generated largely from water-based recreation *within* the interreservoir zone.

Over 1600 intakes withdraw water from the Missouri River for irrigation, domestic, municipal, and industrial uses. Twenty-five thermal electric generating stations use cooling water from main-stem reservoirs and the lower river. Collectively these plants have a gross generation capacity of about 15,000 MW (U.S. Army Corps of Engineers 2001). Hydropower at the six main-stem dams adds 2435 MW of combined capacity, producing about 10.2 million MWh/yr (see Table 10.1).

Recent events have directed national attention to the Missouri River. Declines in populations of archetypical Missouri River fishes and birds resulted in multiple listings under the Endangered Species Act (ESA) and American Rivers designated the Missouri the nation's most endangered river in 1997 and again in 2001. Basinwide drought in the late 1980s and catastrophic flooding in the 1990s highlighted conflicts over water allocation. Socioeconomic values for the river and floodplain are shifting from agriculture and transportation to reservoir- and river-based recreation. There is a recognized need for more balance among all of the river's designated beneficial uses. Natural resource efforts along the river are moving away from chronicling human impacts toward designing and implementing restoration programs.

Several steps have been proposed by the National Research Council (2002) to guide Missouri River recovery within a framework of multiple uses. Recovery programs are now largely directed at acquiring floodplain lands in the lower zone and reconnecting them to the river during high flows; increasing main-channel habitat complexity, particularly the amount of shallow, low-velocity water (Fig. 10.4); and modifying reservoir water management to improve recreation and restore more natural river flows while sustaining other authorized uses. The philosophy behind these recommendations and actions is that both flow and habitat restorations are needed to benefit ecological services and broaden societal benefits of the Missouri River.

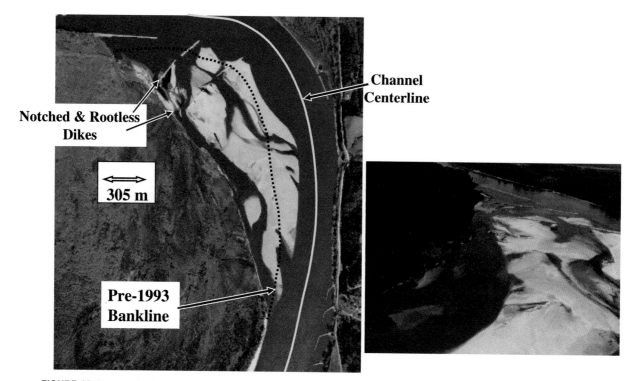

FIGURE 10.4   Notched and rootless rock dikes are used to re-create and maintain shallow-water sandbar complexes where Missouri River channel width was increased by the 1993 flood. Navigation remains along the channel centerline. Plan view (left) and oblique view (right) (From Jameson Island Unit, Big Muddy National Fish and Wildlife Refuge, Missouri, Rkm 342.7, February 2000; M. Chapman, U.S. Army Corps of Engineers, Kansas City District).

YELLOWSTONE RIVER

The Yellowstone River is the longest free-flowing river in the conterminous United States (1091 km), making it a rare model of the structure and function of large western rivers. It originates in the Absaroka Mountains in northwestern Wyoming near the southeast border of Yellowstone National Park and flows into Yellowstone Lake. From the outflow of the lake the river flows through the scenic Grand Canyon of the Yellowstone, then north and northeasterly through a drainage basin of 182,336 km² to its confluence with the Missouri River in extreme western North Dakota (Fig. 10.17). The Yellowstone has nine major tributaries, seven of which enter from the south, including the Bighorn and Powder rivers.

The origin of the name "Yellowstone," or the explorer who first called the river Roche Jaune—yellow rock—is unknown. Some Indian tribes called it Mitsiadaz, loosely translated as Yellow Rock River (Silverman and Tomlinsen 1984). The Yellowstone Valley has a rich history. Although many Native American tribes inhabited the valley for about 12,000 years, those first encountered by explorers were mainly nomadic Crow tribes. Captain William Clark led the first extensive exploration of the Yellowstone in 1806 during the return trip of the famous Lewis and Clark expedition. Clark described the vast abundance and diversity of wildlife and extensive cottonwood forests along the river. Although a variety of perturbations, including introduced species, water withdrawals, agricultural and energy developments, mining, and logging, have affected the Yellowstone, it retains much of the character it had in Clark's day (White and Bramblett 1993).

Physiography, Climate, and Land Use

The Yellowstone basin includes four physiographic provinces (Wyoming Basin [WB], Northern Rocky Mountains [NR], Middle Rocky Mountains [MR], and Great Plains [GP]) that are distinctly different in structure and stratigraphy, as well as hydrology and geomorphology (see Fig. 10.17). The landscape encompasses four terrestrial ecoregions: Wyoming Basin Shrub Steppe, South Central Rockies Forests,

Montana Valley and Foothill Grasslands, and North-western Mixed Grasslands. Vegetation varies along the river corridor, from Douglas fir in the wetter upper river, to Rocky Mountain juniper in mid-reaches, to a grassland community of blue grama or western wheatgrass in east-central Montana, to green ash and bur oak in extreme eastern Montana and western North Dakota (Silverman and Tomlinsen 1984).

Elevations of the basin range from 3660 m asl in alpine headwaters to less than 660 m asl at the mouth, resulting in large climatic differences between headwater areas and the semiarid plains. Precipitation ranges from more than 210 cm/yr in the mountains to less than 30 cm/yr in the plains. Most mountain precipitation is in the form of late-winter and early-spring snow, whereas on the plains it comes primarily as early-summer rain (Fig. 10.18). Air temperatures are extremely variable. Average air temperature varies longitudinally and through time, ranging from −7°C in January to 22°C in July. The river freezes during winter and large ice jams are common.

The Yellowstone River flows through a largely rural landscape, with only one major urban area (Billings). Human population density is 10 people/km^2; only 1% of lands adjoining the river are urbanized. Principal land uses are range (47%), forestry (28%), and agriculture (20%). Agricultural use includes dry-land farming and irrigated cropland along alluvial valleys and benches. About 30 active mines remove coal by strip mining in the Powder River Basin. Oil and gas production also occurs here as well as in the Big Horn basin.

River Geomorphology, Hydrology, and Chemistry

From its origin in northwest Wyoming, the Yellowstone flows into Yellowstone National Park and into a caldera created 600,000 years ago by the most recent major volcanic eruption that formed Yellowstone Lake. The river then flows through deep canyons (Grand Canyon of the Yellowstone, see Fig. 1.1) of erodible volcanic ash and forms large waterfalls in areas of more resistant lava rock (Silverman and Tomlinsen 1984). As the river leaves the mountains it flows over mainly Quaternary alluvial deposits, which are also the primary source of groundwater in the upper river (Fig. 10.1). Farther downstream on the plains the basin is underlain by Cretaceous and Tertiary sediments of alternating sandstone, shale, coal, and red klinker beds of the Fort Union Formation, which

contains the enormous coal reserves of Montana (Silverman and Tomlinsen 1984).

The general morphological character of the Yellowstone remains the same as described by Clark in 1806. Channel pattern varies from sinuous, to braided, to irregular meanders. The channel is often braided or split, particularly in the lower river, and long side channels are common. Islands and bars range from large stable islands with mature vegetation to unvegetated point and midchannel sand and gravel bars. Bed material is largely gravel and cobble in upstream reaches, grading to sand in the lower 50 km (Bramblett and White 2001, Koch et al. 1977). Slope ranges from 8.9 m/km in the upper river, to 1.4 m/km in the middle Yellowstone, to 46 cm/km near its mouth at the North Dakota border (Koch et al. 1977).

Characteristics of the Yellowstone River that retain any of their pristine character (hydrology, geomorphology, water quality, and biotic communities) do so because the main stem is not dammed. However, 31% of the drainage basin (mostly in the Bighorn River basin) is upstream of storage reservoirs (Koch et al. 1977). Damming of the Bighorn River, the largest tributary of the Yellowstone (see Fig. 10.17) in 1966 caused an 80% reduction in annual sediment yield (over 5 million metric tons) from that basin. Koch et al. (1977) report that the Yellowstone did not show any reduction in sediment transport at the gaging station near the mouth, indicating that it may have been out of equilibrium below the Bighorn and was degrading its bed and banks to produce extra sediment. Recent evaluation of channel changes revealed that reduction in average peak flow of 255 m^3/s and of channel maintenance flow of 57 m^3/s (5%) have resulted in channel thread stabilization, reduction in open-bar area, and vegetation encroachment downstream of the Bighorn River confluence (Womack and Associates 2001).

The undammed Yellowstone River retains a near natural hydrograph, with a small spring rise in March in downstream areas as snow melts on the plains, followed by peak flows in June from mountain runoff and precipitation (see Fig. 10.18). The difference between runoff pattern and precipitation is due to winter storage of snow in the Rocky Mountains and the subsequent spring thaw. High spring flows are important in maintaining natural channel function. Although localized flooding occurs in winter because of ice-dam formation and breakages, spring flooding is most important in determining channel form. Average annual discharge near the mouth for the 87 years of record is 362 m^3/s. The largest flow recorded was 4500 m³/s and the smallest 13.3 m³/s. The largest mean annual flow of 560 m³/s occurred in 1997. Humans have had some impact on channel morphology, largely through bank armoring, closing side channels, and clearing bank vegetation. Following two consecutive floods in 1996 and 1997, channel training (dikes and armoring) increased substantially around population centers. For example, near Billings, channel training increased from approximately 21% in 1957 to 41% in 1999, resulting in a decrease in channel length of about 5% in this area (Aquoneering and Womack and Associates 2000).

Although water quality shows a general deterioration from source to mouth, it is generally good and free from major pollution inputs. Total suspended solids, total dissolved solids, turbidity, sulfates, and water temperature increase in the lower river. Average alkalinity is 45 mg/L as $CaCO_3$, pH is 7.5, NO_3-N is 0.08 mg/L, and PO_4-P is 0.03 mg/L. Dissolved oxygen levels are usually near saturation, biological oxygen demand levels indicate no major organic pollution, and fecal coliform levels are low. Dissolved metals only rarely exceed water-use criteria (Klarich and Thomas 1977; USEPA 2001 http://www.epa.gov/storet/). Nitrogen is the major limiting nutrient in the river; sequential inputs from tributaries and wastewater discharges had relatively high levels of total soluble inorganic nitrogen, resulting in a 6- to 17-fold increase in primary production. However, neither the productivity rates nor nitrogen and phosphorus concentrations were characteristic of eutrophication; overall, the river is best characterized as mesotrophic (Klarich 1976) or oligo-mesotrophic (Peterson et al. 2001). Nuisance algae conditions have been documented in several sections of the river, possibly indicating eutrophic conditions. Levels of organochlorine insecticides in fish tissue collected throughout the river are low compared to other sites in the Rocky Mountain area and to national statistics. The concentrations of DDT and its metabolites were highest from sites in and near Yellowstone National Park, probably due to historical DDT spraying programs for spruce budworm.

River Biodiversity and Ecology

The Yellowstone River is in the Upper Missouri freshwater ecoregion. No comprehensive studies of the ecology of the Yellowstone have been completed.

Algae

Bahls (1974) identified 28 genera of algae in three samples taken from the upper, middle, and lower Yel-

lowstone River in April. *Cladophora* was the most abundant alga by volume at all sites. As a group, diatoms were the most abundant and diverse, with 75 varieties and 23 genera identified. Further analyses of samples taken in the mid 1970s at 11 sites along the river revealed well over 100 taxa of diatoms. Relative abundance of common species changed going downstream from benthic species, including nitrogen fixers in the montane zone, to a middle-river transition mix of benthic species, including some species associated with eutrophic conditions, to a zone of more planktonic species in the lower river (David Peterson, personal communication). The algal flora at the intersection of the Bighorn River is intermediate between that of a high mountain stream and a lowland plains river, coinciding with the transition between a coldwater and warmwater environment.

Plants

Composition of riparian plant communities of the Yellowstone varies longitudinally with climate, proximity, and elevation of the site to the river channel, and disturbance caused by flooding. Pioneer plant communities on disturbed sites or newly deposited bars are typically willow, followed by cottonwood. Midsere vegetation consists of mature cottonwood forests, which persist for about 100 years. Cottonwoods form extensive forests on islands and banks of the lower Yellowstone River. River-associated wetlands support many of the globally significant plants in the watershed, notably Tweedy's rush and Rolland's scirpus (Jean and Crispin 2001).

Invertebrates

The most complete survey of macroinvertebrate communities is that of Newell (1977), who sampled 20 sites from just below Yellowstone National Park to the mouth. Newell documented a rich macroinvertebrate fauna dominated by mayflies, caddisflies, and true flies; 154 taxa of benthic macroinvertebrates were collected, although more recent collections have added to Newell's list (D. Gustafson, personal communication). Densities ranged from 12,000/m² to less than 100/m². Species richness and density declined downstream. Mayfly assemblages exhibited a gradual change from a mountain fauna in the upper river (*Baetis, Ephemerella, Epeorus, Ephemera*) to a prairie fauna more adapted to the slower current velocities, warmer temperatures, and finer substrates in the lower river (*Ametropus, Lachlania, Ephoron, Caenis, Centroptilum,* and *Isonychia*). Mayfly species richness ranged from 19 species upstream to 10 species in the lower river, with a total of 37 species. Stonefly richness was highest in the upper river (21), with a total of 37 species, including *Pteronarcys, Pteronarcella, Arcynopteryx, Paraleuctra, Capnia, and Alloperla*. The number of stonefly species declined rapidly downstream, particularly in the transition zone between coldwater and warmwater communities (*Brachyptera, Acroneuria*). Caddisfly distribution was similar to that of stoneflies, with a steady decline in species richness with distance downstream. *Hydropsyche* and *Cheumatopsyche* were found at all stations and dominated the macroinvertebrate fauna in the lower 10 stations. Other caddisfly genera include *Glossosoma, Brachycentrus, Neotrichia, Oecetis*, and *Leptocella*. True flies, particularly chironomid midges, occurred throughout the river and were the most abundant and diverse of the remaining macroinvertebrate groups. Beetles were found in all samples, odonates and true bugs were found in the transition and coldwater zones, and lepidopterans were rare.

No comprehensive studies of other invertebrate groups have been made. Only two mussel species are known to occur in the Yellowstone River. One is the common fatmucket, and the other, a recent introduction, is the mapleleaf, native to the Midwestern United States. Based on size, this mussel was first established in the Tongue River and has recently spread downstream to the Yellowstone in the vicinity of the Tongue River confluence. Other recent invertebrate introductions are the New Zealand mudsnail and the rams-horn (native to the Snake River) (D. Gustafson, personal communication).

Vertebrates

The Yellowstone River supports a diverse fish community of 56 species representing 16 families; 20 species (36%) are nonnative. Except for a few sport fishes, little is known about the ecology or demographics of most species. Based on fish species distribution, the river can be divided into three zones: an upper coldwater zone extending 357 km from the headwaters to the mouth of the Boulder River, followed by a transition zone (258 km) from there to the mouth of the Bighorn River (inhabited by both coldwater and warmwater species), and a lower warmwater zone (476 km) from there to the Yellowstone mouth. Although temperature records are incomplete, maximum summer water temperatures for the three zones were 23°C, 26.5°C, and 29°C, respectively, for the period of record. Longitudinal changes in the fish community appear to correlate with water temperature. The fish community between Upper

Falls and Yellowstone Lake (part of the coldwater zone) has only two native species, Yellowstone cutthroat trout and longnose dace, and two introduced species, redside shiner and longnose sucker. Of the 14 fish species known to occur in the coldwater zone, only 7 are abundant in a portion or all of this reach (Yellowstone cutthroat trout, rainbow trout, brown trout, mountain whitefish, longnose sucker, white sucker, and mottled sculpin) (White and Bramblett 1993). In the transition zone the abundance of warmwater species rapidly increases downstream, totaling 30 species representing 7 families. Salmonids all but disappear, whereas goldeye and burbot abundance increases. New species include sauger, walleye, five minnows, river carpsucker, and three catfishes. The largest change in the fish community occurs in the warmwater zone where 49 species representing 15 families occur. Two sturgeons, one esocid, seven minnows, two suckers, one catfish, one killifish, one smelt, six sunfishes, and one drum first occur here (see White and Bramblett 1993 for complete listing). The most ubiquitous species is white sucker, classed as abundant in all zones. Goldeye, common carp, longnose dace, shorthead redhorse, burbot, longnose sucker, mountain sucker, rainbow trout, and brown trout also occur in all river zones.

The fish assemblage includes 15 game species, of which 7 are nonnative (rainbow trout, brown trout, brook trout, northern pike, smallmouth bass, largemouth bass, and walleye). The four trout species and mountain whitefish are the only game fishes in the upper river. The stream reach from the boundary of Yellowstone National Park to the mouth of the Boulder River is the longest single reach (166 km) of blue-ribbon trout stream in Montana, making up 23% of the state's 727 km of blue-ribbon waters (White and Bramblett 1993). Yellowstone cutthroat is the only trout native to the drainage. Habitat degradation in tributary streams, introduction of nonnative salmonids, and human exploitation have led to the reduction in range and abundance of Yellowstone cutthroat trout and to its designation as a species of special concern by the Montana Natural Heritage Program and the Montana Department of Fish, Wildlife, and Parks (2004). In addition to the greatly reduced native range, many of the remaining populations have been genetically contaminated by hybridization with rainbow trout. The recent discovery of nonnative lake trout in Yellowstone Lake has raised additional concern and a large effort to control this population is underway. Of the 10 sport fishes in the lower Yellowstone, six provide substantial angling opportunity (sauger, walleye, shovelnose sturgeon, paddlefish, channel catfish, and burbot). Steady declines in sauger populations resulted in its designation as a Montana species of special concern in 2001. Other state species of special concern that occur in the Yellowstone are blue sucker, paddlefish, pallid sturgeon, pearl dace, sicklefin chub, and sturgeon chub. Pallid sturgeon is also federally listed as endangered.

Riparian areas are vital habitats for many amphibians, reptiles, birds, and mammals, particularly in arid and semiarid environments. One amphibian species of concern, the northern leopard frog, is known to occur in riparian and wetland habitats in the watershed (Thompson 1982). Reptiles of concern include snapping turtle and spiny softshell turtle (both directly tied to the lower river), milk snake, and western hognose snake. A study on the lower Yellowstone showed that the riparian forest had the highest avian density and diversity of 10 habitat types evaluated (Silverman and Tomlinsen 1984). Of the 35 bird species of special concern in the watershed, 31% are associated with riparian or wetland habitat (Jean and Crispin 2001). Bald eagle and piping plover are federally listed as threatened and the interior least tern is listed as endangered. Other important birds include the great blue heron and common merganser. Most neotropical migrants rely on the riparian habitat within the watershed. Noted declines in several of these species have prompted concern for neotropical migrants as a group. Although no mammal species of special concern are dependent on the riverine habitat, many mammal species inhabit or frequent the riparian area, including river otter and beaver. Grizzly bear, a federally listed species, heavily utilize Yellowstone cutthroat trout during the spring spawning period.

Ecosystem Processes

Most ecosystem processes function today as they did historically. Fire and flooding continue to influence the landscape, providing habitats for the many species adapted to these disturbances (Jean and Crispin 2001). However, there is a general lack of information on ecosystem processes in the Yellowstone River system. Preliminary data from a U.S. Geological Survey evaluation of trophic conditions, using chemical (nutrient) and biological (algal biomass and productivity) variables, indicated oligomesotrophic conditions throughout its length (D. A. Peterson, personal communication). Nuisance algal conditions associated with eutrophy, however, have been documented in several sections of the river. Periphyton chlorophyll *a* and ash-free dry weight concentrations were largest in the middle section of the river, where *Cladophora glomerata* abundance approached nuisance proportions. Rates of stream metabolism and respiration were also largest in the middle area of the river, and productivity was associated mostly with benthic algae. Production/respiration ratios indicated autotrophic conditions in the upper river and heterotrophic conditions in the lower river. Data on functional status of macroinvertebrate communities are not yet available.

Human Impacts and Special Features

The Yellowstone River is unique among western rivers in that it has no main-stem storage or hydroelectric dams. It is among the last remaining relatively unaltered large rivers in the conterminous United States. Its upper reaches, including the spectacular Grand Canyon of the Yellowstone, are protected within Yellowstone National Park. Also, a 33km reach of Clarks Fork, a tributary arising in the Shoshone National Forest, is protected as a National Wild and Scenic River. Such free-flowing resources, important in their own right, will be key to understanding how large western rivers function, and subsequently to mitigation efforts in the Missouri River basin (White and Bramblett 1993).

Although relatively pristine compared to other large rivers in the coterminous United States, the Yellowstone is not without human impacts. Channel modification has been minimal but has accelerated in recent years, especially in the upper river. Bank armoring, closing side channels, and clearing riparian vegetation have affected channel characteristics locally. Downstream of the confluence of the Bighorn River the Yellowstone may be narrowing and deepening due to modification of sediment dynamics and flow pattern resulting from Yellowtail and Tongue river dams.

Large amounts of water are withdrawn from the river for irrigation. About 90% of all water use in the basin is for this purpose. Many tributary streams that are important as spawning habitat for migratory fish species are severely dewatered. In 1973 the Montana Water Use Act was passed, which allowed instream-flow advocates to compete with consumptive users for unreserved water. In 1978, 6.8km^3 of water at Sidney, Montana, was reserved for the Yellowstone. In addition, legislation to allow a pilot program of water leasing for instream use was passed in 1989. The first leases were on tributaries of the upper Yellowstone and the success of that program is being evaluated.

Although the Yellowstone has no dams on the main stem, there are three major cross-channel water-diversion structures in the lower river that are known to influence upstream movement of paddlefish, sturgeon, sauger, and walleye. Most other species would probably have difficulty passing upstream of these structures as well (White and Bramblett 1993). A fish passage structure was recently built at Huntley Diversion, and research on design criteria for passage of sturgeon around Intake Diversion is currently being conducted.

Nonnative species have contributed to the decline of some native fishes. Rainbow and brown trout have had a major impact on the abundance and distribution of native Yellowstone cutthroat trout in the river and its tributaries, and lake trout are thought to be having a negative impact in Yellowstone Lake. Other introduced species, such as walleye, may be influencing the abundance of native species like sauger, but little attention has been given to native–nonnative species interactions. *Myxobolus cerebral* (a myxozoan parasite), the causative agent of whirling disease, was recently detected in the upper Yellowstone Basin. This organism is thought to be responsible for the large decline in nonnative rainbow trout populations in many Montana rivers, and native Yellowstone cutthroat trout are known to be susceptible to infection. Also, the recent discovery of the New Zealand mud snail is of considerable ecological concern.

Natural values that make the Yellowstone so attractive to humans are being affected by activities such as subdivisions on agricultural lands, overuse of water, and alteration of natural habitats by noxious weeds, nonnative species, and recreation. Considerable effort to reduce habitat loss and fragmentation is occurring in the form of land trusts dedicated to open space (Jean and Crispin 2001). In 1997, through executive order, the governor of Montana established a task force to develop a set of publicly supported river-corridor management recommendations that address potential adverse cumulative effects of channel modification, floodplain development, and natural events on the human community and riparian ecosystem. A number of research projects are currently ongoing (Governor's Upper Yellowstone River Task Force 2000).

WHITE RIVER

The 816 km long White River (25,650 km^2) originates in the pine ridge region (Sioux, Dawes, Sheridan counties) of northwestern Nebraska and flows eastward through south-central South Dakota to its confluence with Lake Francis Case, a main-stem impoundment on the Missouri River (Fig. 10.19). The White River is a 6th order interjurisdictional river. It is an interstate river that also runs through the Pine Ridge and Rosebud Indian reservations, through national grasslands, forests, and wilderness areas (U.S. Forest Service), LaCreek National Wildlife Refuge (U.S. Fish and Wildlife Service) and Badlands National Park (National Park Service) (Fig. 10.5). The White River has been listed as one of 327 rivers in the United States that are critical for protecting freshwater biodiversity (Master et al. 1998).

One of the longest undammed rivers in the conterminous 48 states (Stanford and Ward 1979, Benke 1990), the White River is named for its white-gray color, derived from the heavy load of sands, clays, and volcanic ash in the middle and lower segments. Through geologic time, erosion in the White River basin has created a special landscape revealing rich fossil beds, sharp ridges and spires, and flat prairie grasslands through which winds a sinuous green riparian zone marking the course of the White River. The largest tributary, the Little White River (4100 km^2) is also unique because its origin in Nebraska's sand hills causes more stable flows, higher gradient, and cooler water temperatures than in the main stem. The Little White River is so scenic as it passes through the LaCreek National Wildlife Refuge and Rosebud Sioux Reservation that it is included in canoeing guides.

The basin has supported humans for 12,000 years, beginning with ancient mammoth hunters who were followed much later by Paleo-Indian hunter-gatherers and then by the Arikara (or Ree) and Dakota peoples in the sixteenth century. Bands of the Teton tribe of the Dakota Sioux Nation dominated the basin for about 100 years before the arrival of fur trappers, soldiers, miners, cattlemen, and homesteaders in the 1800s. The attention of the United States was focused on Dakota Territory in the late 1800s when fully one-third of the U.S. Army was in combat with the Dakota Sioux Nation. The symbolic end of the war occurred in the White River basin in 1890 with the Indian massacre at Wounded Knee, after which the Dakota people were confined to reservations that make up a large part of the basin today. During the Dakota Land Boom (1878 to 1887), pioneer settlers called the area a "sea of grass" as they established ranches, which now average from 900 ha (lower basin) to 3600 ha (upper basin) each.

FIGURE 10.5   White River, Pennington County, South Dakota, near border of Pine Ridge Indian Reservation and Badlands National Park; white mineral deposits are visible on the banks (PHOTO BY C. BERRY).

Physiography, Climate, and Land Use

The White River basin is in the unglaciated Pierre Hills and Southern Plateau regions of the Missouri Plateau section of the Great Plains (GP) physiographic province (Hunt 1974, Hogan 1995). The basin lies primarily in the Northwestern Mixed Grasslands ecoregion, with a small portion of its headwaters within the Western Short Grasslands ecoregion. Chestnut soils and steppe vegetation (short to midgrasses such as buffalo, blue, and wheat grasses) develop on a landscape that drops in a series of steps and hills from the high-plains prairie in Nebraska (1485 m asl) to the confluence with the Missouri River (402 m asl).

The Dry Continental climate features hot summers and cold winters, with great daily temperature range (average monthly range 15°C). Mean annual temperature is 8.8°C, with mean monthly variation from −8°C in January to 24°C in July and August (Fig. 10.20). Mean January air temperatures ranged from −0.6°C to −13.9°C over a seven-year period when average river ice thickness ranged from 0.1 to 1.0 m (Ferrick et al. 1995). Precipitation is very seasonal, with at least 5 cm/mo from April to August and only 1 to 2 cm/mo from November through February (see Fig. 10.20). Thunderstorms account for 85% of the precipitation.

There are no municipalities adjacent to the river and few in the basin. The largest towns are Chadron, Nebraska (about 6000), and Winner, South Dakota (about 3500). Population density varies from 0.3 to 3.0 people/km², depending on county. About 73% of the basin is in native grass or hay (e.g., alfalfa) and about 21% is in row crops, dominated by wheat with some sorghum, corn, oats, sunflowers, and soybeans. Land use in riparian zones at 11 reaches in South Dakota (Fryda 2001) was mostly pasture or prairie (66%), followed by 22% cropland and 10% wooded.

River Geomorphology, Hydrology, and Chemistry

The basin is divided into four eight-digit hydrologic units representing the upper, middle, and lower river main stem and the Little White River. The gradient is about 170 cm/km in the upper unit, 60 cm/km in the middle unit, and 50 cm/km in the lower unit. Riverine habitat data are available only for the South Dakota main stem (Fryda 2001). The river at bankfull is about 7 m wide at the Nebraska–South Dakota border and about 120 m wide near the terminus, but widths downstream from the Little White River increase from about 50 m to 110 m. The dominant riverine macrohabitat is the run (80%), followed by riffle (19%) and pool (1%). Instream cover (e.g., woody debris, undercut banks, cobble substrate) is scarce except at certain reaches near the state border. Gravel-size substrates dominate (60%) upper-river reaches and decline to about 30% in the lower White, where fines (silt, sand, clay) become increasingly prevalent and the alluvium is 7.5 to 12.5 m deep over Pierre shale (Ferrick et al. 1995, Fryda 2001). The channel is highly mobile within the floodplain, and aerial photos of one reach showed four abandoned channel segments since 1986 (Fryda 2001). Sinuosity values range between 2.1 and 2.5. The main-stem channel at 11 reaches in South Dakota was classified as entrenched (Fryda 2001).

Most stream flow occurs in response to precipitation and snowmelt during spring and early summer, with peaks in both March and May–June (about 38 m³/s) and low flows (e.g., 2 m³/s) in January (see Fig. 10.20). Runoff averages only 0.17 cm/mo due to low precipitation and high evapotranspiration. Annual discharge was about 16 m³/s (range 4.3 to 49 m³/s) between 1928 and 1997, but water yield differs among subbasins because near-surface geology differs (Ellis et al. 1971). Tributaries are unreliable sources of water year-round, but flows are more stable in the upper main stem (Nebraska) and four creeks. The river is subject to periods of no flow in its middle segment, where it is perched (Ferrick et al. 1995); however it has been perennial since 1928 in the upper and lower basin except for five years. When river flow ceases, deep pools remain wet, reflecting the water-table level. Groundwater level is below the river surface during dry seasons and rises to river level during wet seasons (Ferrick et al. 1995). In the summer months of 1999–2000, velocity was 0.2 to 0.3 m/s at 0.3 to 1.2 m³/s discharge in the upper hydrologic unit and was 0.45 to 0.55 m/s at 3.6 to 11.1 m³/s in the lower unit (Fryda 2001).

The harsh physical conditions of the White River are largely responsible for its poor water quality. As a result, the basin fails to support designated beneficial uses (i.e., limited contact recreation, warmwater semipermanent fish-life propagation) and has been listed as an impaired water body, but it is excluded from the Federal Total Maximum Daily Load program because the majority of its impairment is from natural sources (DENR 1998). The highly erosive soils in the badlands cause high suspended sediment levels and are the major source of poor water quality. Secchi depths of <5 cm are common. Upstream from the Badlands, suspended sediment concentrations average less than 250 mg/L, whereas downstream suspended sediment concentrations often exceed 5000 mg/L. Over 9.9 million metric tons of sediment is deposited annually into Lake Francis Case, where a substantial delta has formed. The delta partially blocks flow through the lake and has reduced reservoir capacity by about 10% since its creation in 1952. Elevated levels of total dissolved solids (270 to 540 mg/L), fecal coliforms, and pH (8.2 to 9.3) also contribute to nonsupport of designated uses and are highly influenced by runoff. During June, July, and August 1999–2000, dissolved oxygen ranged from 7 to 8 mg/L, morning water temperatures ranged from 18°C to 25°C, and afternoon temperatures ranged from 24°C to 34°C (Fryda 2001). Riparian shading in upstream reaches causes less diel variation in stream temperature than in downstream reaches, where summer temperatures varied about 10°C daily and reached 34°C, which is stressful to some fishes (Fryda 2001).

The Little White River is quite different than the main stem because it originates in sandhill terrain and drains mixed-grass prairie. Gradient ranges from 1.3 to 2.6 m/km, so the Little White meanders less (sinuosity 1.4 to 1.8) than the main stem. Hydrology is groundwater dominated, with high single-peak (March or April) spring flows. Average annual discharge is about 3.2 m³/s (Niehus 1999); winter discharges are about 2.6 m³/s, which accounts for most of the main-stem flow. Water quality is much better than in the main stem except in the lower Little White basin, where erosive badlands soils are intercepted and TSS concentrations become excessive (DENR 2000). Average water-quality values for selected characteristics at the Little White terminus are hardness 142 mg/L as $CaCO_3$, dissolved solids 282 mg/L, suspended sediment 1096 mg/L, NH_4-N 1.6 mg/L, and NO_3-N 1.8 mg/L (Niehus 1999).

River Biodiversity and Ecology

The White River is within the Upper Missouri freshwater ecoregion (Abell et al. 2000). As a river with harsh physical conditions in a rural environment, its biology and ecology are not well studied.

Plants

Submersed and emergent vegetation is rare in the White (Fryda 2001). Riparian vegetation includes grasses (41%), willows (36%), and other shrubs and trees (cottonwood, horsetail, green ash, wild grape, box elder, Russian olive, wild rose).

Invertebrates

Incidental observations of two mussel species (floater, fatmucket) in the upper basin were made during recent fisheries surveys, but no detailed invertebrate surveys have been done. However, a cursory survey of the adjacent river (Cheyenne) found representatives of four dipteran families (primarily chironomid midges and black flies), four families of mayflies (mostly Leptophlebiidae and Baetidae), many hydropsychid caddisflies, and a few elmid beetles and stoneflies (Hampton and Berry 1997).

Vertebrates

In spite of its harsh conditions, 41 native fish species are found in the White River, as are 8 non-natives (common carp, brown trout, rainbow trout, brook trout, black crappie, white crappie, large-mouth bass, and bluegill), which are rarely found except for common carp (Fryda 2001). Ubiquitous species found in both states in the main river and in tributaries are channel catfish, green sunfish, long-nose dace, sand shiner, and white sucker. Relative composition of the species representing >1% of recent samples (Fryda 2001) from the South Dakota portion of the main river was flathead chub (44%), channel catfish (22%), plains minnow (13%), fathead minnow (6%), sturgeon chub (4%), common carp (3%), sand shiner (2%), Western silvery minnow (2%), and goldeye and stonecat (1% each). Sturgeon chub, flathead chub, and plains minnows have declined in other parts of the Missouri River basin. There are no federally listed species, but nine species are listed as state species of concern, mostly because of the number of glacial relicts and species on the edge of their range.

Certain fish assemblages have specific geographic associations within the basin. The spring-fed stable flows in the upper basin support introduced trout and mountain sucker, a relict species. The lower main stem is used by large Missouri River species (e.g., paddlefish, flathead catfish). The Little White River has glacial relict species and other natives that are adapted to clear headwater streams (e.g., pearl dace, finescale dace, blacknose dace, blacknose shiner, northern redbelly dace, central stoneroller). A connection may have existed between the headwaters of the Little White River and the Niobrara River; hence the presence of the plains topminnow and big-mouth shiner (Mayden 1987).

Channel catfish make up 22% of the catch in hoop nets and seines (Fryda 2001). Ages ranged from 1 to 13 years, but 81% were less than 5 years old and 80% were <280 mm long. Relative weights (W_r) were 97, 79, and 85 for three length groups (<280, 281 to 410, 411 to 610 mm), which is a common pattern for regional populations (Doorenbos et al. 1999). However, growth was slow compared to other river populations in the region, especially for fish <4 years old. Channel catfish support a limited recreational fishery.

Two of South Dakota's seven turtle species (common snapping turtle, spiny softshell turtle) were collected in fish traps (Fryda 2001), and the Western painted turtle occurs in the basin. Five amphibians (e.g., northern leopard frog, bullfrog, Great Plains toad, chorus frog, Woodhouse's toad) and the tiger salamander occur in floodplain wetlands adjacent to the White River (Fischer et al. 1999).

Ecosystem Processes

Dominant factors controlling ecological processes and community structure in most of the main stem of the White River are a "harsh intermittent" or "intermittent runoff" discharge regime (Poff 1996) and unusually high turbidity. Exceptions are in the main-stem headwaters (Nebraska) and in the Little White River, where soils are less erosive and discharges more stable than in the main stem. Although no studies have been done on ecological processes, limited biological data conform to the hypothesis that abiotic controls dominate. For example, depauperate mussel fauna have been associated with high turbidity, suspended solids, and unstable substrates (Hoke 1983). The fish community, represented by channel catfish, cyprinids, and suckers, is typical of low-gradient, high-turbidity warmwater rivers (the type III river of Rabeni 1993). Species richness (n = 49) and fish feeding guilds (mostly insectivores and omnivores) are similar to communities in adjacent basins (Hampton and Berry 1997). Native fish species (e.g., sturgeon chub, flathead chub) with special adaptations to turbid conditions are present, and many fishes are classed as tolerant or moderately tolerant of environmental degradation (Barbour et al. 1999). More benign conditions in the headwaters and Little White River allow survival of coldwater species and glacial relicts.

Population data for channel catfish suggests poor habitat quality. The lack of pool habitat probably limits the density of large channel catfish, but the abundant run and riffle habitats are suitable for smaller fish and young year classes, which were more abundant than expected (Fryda 2001). The White River is probably a nursery area for channel catfish and other species (e.g., flathead catfish, paddlefish) from Lake Francis Case.

The slow growth rate of the channel catfish suggests that productivity may be limited. High turbidity, harsh flow, sand substrate, and lack of instream structure probably limit autotrophic productivity and macroinvertebrate community biomass, which makes up a large part of the diet of young channel catfish. Allochthonous inputs may also be limited. Although there is annual floodplain inundation, the frequency can be low and the duration short, depending on river segment (S. Sando, personal communication), and riparian zones and floodplains are dominated by grasses that yield small inputs of coarse organic matter and woody debris compared to forested areas. The relatively high water temperatures during summer equal or exceed the optimum for channel catfish growth (Fryda 2001).

Human Impacts and Special Features

The remote, sparsely populated White River basin is unique because of the natural beauty of hills, plateaus, tablelands and badlands topography (so named because of the lack of water and vegetation), the rugged landscape, and the hindrances to travel and agriculture. The unimpounded, free-flowing river is unique because the headwaters are relatively cool and clear compared to the warm, exceedingly turbid main stem, where harsh physical conditions challenge survival.

Human impacts have been negligible in the basin, except for poor range management in some areas. Watershed management is underway to reduce the amounts of suspended solids and fecal coliforms, especially in the middle and lower basin (DENR 2000). Livestock grazing impacts on riparian vegetation were judged low to moderate at 96% of transects measured by Fryda (2001). Trace metals were not found in sediments (Ruelle et al. 1993). Elevated levels of arsenic occur in groundwater wells in the Little White River basin but not in surface waters (USGS 1998). There are no main-stem dams or other impediments to fish migration. Stock dams (about 1120) and small reservoirs (about 20) have been constructed on tributaries, but about 6000 1st and 2nd order streams remain unimpounded. A preliminary analysis of monthly flows before and after 1950 did not indicate that main-stem discharges had changed. However, 156 irrigation appropriations in Nebraska have decreased virgin (natural) flows entering South Dakota by about 11 million m^3 (Sando 1991). Fish biomonitoring data indicate that all fish species found in historical studies (e.g., Bailey and Allum 1962) are still extant, although historical data are limited (Fryda 2001).

PLATTE RIVER

From its headwaters in the Southern Rocky Mountains and the Wyoming Basin, the Platte River sweeps easterly across the Great Plains as a river of sand festooned with interlacing ribbons of water to its confluence with the Missouri River (Fig. 10.21). For about half of its length the Platte is really two rivers: the North Platte, which primarily drains southeastern Wyoming, and the South Platte, which primarily drains north-central Colorado. Their confluence does not occur until these rivers have independently traversed their way halfway across the Great Plains to the city of North Platte, Nebraska. Within its total drainage of 230,362 km^2 the Platte River system traverses montane, foothills, and plains habitats before reaching the Missouri River just south of Omaha, Nebraska (Fig. 10.6).

Spaniards were apparently the first Europeans to see the Platte River. In 1720 an expedition under the direction of Villazur was massacred near the present site of Columbus, Nebraska. It is designated on early maps as either the Platte River or the Nebraska River, names derived from the 1739 French designation "La Riviere Plate" (Flat River), which corresponds in meaning to the Oto (*Ni brathka*), Omaha (*Ni bthaska ke*) and Pawnee (*Kits Katus*) Indian names for the stream, all of which mean "flat water" (Link 1933). Other names for the Platte include the Dakota name (*Pankeska wakpa*), which refers to its importance as a site for the trading for shells (Link 1933).

Physiography, Climate, and Land Use

The Platte basin begins in the Southern Rocky Mountains (SR) and Wyoming Basin (WB) physiographic provinces in the west, spans the entire width of the Great Plains (GP), and includes a small portion of the Central Lowland (CL) province in the east (see Fig. 10.21). The North Platte and South Platte rivers and their tributaries are the modern remnants of drainage

FIGURE 10.6   Platte River near North Bend, Nebraska (PHOTO BY E. PETERS).

systems that transported vast quantities of alluvial materials from the Rocky Mountains, uplifted approximately 60 million years ago, to the plains of what is now Nebraska during the early Cenozoic era. During the Pleistocene, wind-deposited and -rearranged sediments ranging from sand in the west to loess in the east covered much of the Platte River basin across the Great Plains (Swinehart and Diffendal 1989). For most of the past 5 million years regional uplift has resulted in channel degradation and provided abundant sediments for the Platte. Soils in the drainage are generally mollisols, with entisols appearing in areas of most recent sand deposition. In the headwaters, alfisols (boralfs) have developed in steeply sloping forested regions.

The Platte basin includes parts of five terrestrial ecoregions. The headwaters of both the North Platte and South Platte rise in the Colorado Rockies Forests ecoregion, vegetated by conifers such as Douglas fir and ponderosa pine, as well as stands of aspen and mountain meadows and foothill grasslands (Ricketts et al. 1999). The North Platte then flows through the arid Wyoming Basin Shrub Steppe ecore-

gion, where sagebrush, wheatgrasses, or fescue are common. Both the North and South Platte flow eastward across the Western Short Grasslands ecoregion, with its grama and buffalo grass. As it continues eastward the Platte crosses the Nebraska Sand Hills Mixed Grasslands ecoregion and the Central and Southern Mixed Grasslands ecoregion and terminates at the Missouri River in the Central Tall Grasslands ecoregion. There is a transition of dominant grasses capable of surviving low precipitation (sand bluestem, little bluestem, western wheatgrass) to those that require greater precipitation (big bluestem, switchgrass, and Indian grass).

The climate of the Platte River basin is generally continental, with wide variations in temperature and precipitation. Precipitation ranges from an average of about 76 cm/yr near the mouth of the Platte, to 50 cm/yr at North Platte, Nebraska, to approximately 25 cm/yr in central Wyoming, to over 100 cm/yr in mountain areas of central Colorado. However, high-intensity rainfall events and droughts produce wide annual variations around these means. At North Platte, Nebraska, average monthly rainfall totals are greatest during May, June, and July, when they average approximately 8 cm/mo, and least during December and January, when they average barely 1 cm/mo (Fig. 10.22). Average annual temperatures range from about 11°C in the eastern portion of the drainage to 9.1°C at North Platte and 3°C in the western headwater portion of the drainage. At North Platte, mean monthly air temperature is highest in July at 23.3°C and lowest in January at −5.2°C (see Fig. 10.22). The frost-free growing season ranges from less than 120 days in the west to over 160 days in the eastern portion of the basin. Another major climatic factor in the Platte River basin is the wind, which contributes to high evapotranspiration rates.

Land use in the basin is dominated by agriculture (>90%), with concentrated areas of urbanization along the river and the Front Range in Colorado. East of 100°W longitude (near North Platte, Nebraska) dry-land row crops of corn and soybeans dominate, whereas to the west wheat and livestock grazing are more important. Irrigation technologies have fostered expanded corn production to the west and increased yields throughout the basin. Forestry land use is mostly confined to montane coniferous forests of Colorado and Wyoming, but timber harvest utilizing riparian cottonwood forests along the river occurs in Nebraska.

River Geomorphology, Hydrology, and Chemistry

Both the North Platte and South Platte rivers begin in the Rocky Mountains in Colorado as incised high-gradient mountain streams. However, after they emerge from the mountains, and for their entire length across the Great Plains, they typically exhibit wide shallow braided channels with shifting sand and gravel substrates. Eastern tributaries that drain sandy and loess soils contribute finer textured materials to the Platte. Therefore, there is a general decrease in substrate particle size from west to east.

The North Platte River originates from snowmelt in northern Colorado at an elevation of approximately 3353 m asl. It flows northward into Wyoming, joins with the Sweetwater River in the Wyoming Basin province, and curves southeastward toward the panhandle of Nebraska, joining the South Platte at an elevation of 841 m asl. It traverses 1070 km, with an average slope of 2.35 m/km, and drains approximately 90,352 km^2. The South Platte River originates in central Colorado at an elevation of approximately 3810 m asl and after leaving the mountains flows

generally northeast to its confluence with the North Platte. Its average slope is 4.10 m/km over its 724 km length, and it drains approximately 62,888 km^2.

The Platte River proper begins at the confluence of the North Platte River and the South Platte River on the Great Plains near the city of North Platte, Nebraska. It flows generally eastward in a S-shaped course for 503 km, with an average slope of 1.11 m/km, until it reaches the Missouri River at an elevation of 286 m asl. The Platte and its tributaries in this 503 km reach drain an additional 77,122 km^2. Most of this area is in the Loup River, Elkhorn River, and Salt Creek drainages, which join the Platte in the lower 170 km of its length.

In pre-European settlement times, Platte River flow patterns were dominated by snowmelt in the Rockies, which produced high discharge in late spring and early summer, followed by low flows in late summer and early fall. This pattern was probably punctuated, as it is today, by spikes of flooding caused by localized heavy rain events. During times of drought, flows in the Platte River may have reduced it to a trickle or isolated pools upstream from the mouth of the Loup River (Zorich and Associates 1988), but downstream consistent flows from the groundwater-fed Loup and Elkhorn rivers likely kept the Platte flowing.

Average annual discharge near the mouth (Louisville, Nebraska) from 1953 to 2004 was 202.7 m^3/s, but diversions, dams, transbasin diversions, power developments, and groundwater withdrawals influence these flows. Mean monthly discharge from 1953 to 2000 ranged from 119.4 m^3/s (runoff 0.17 cm/mo) in August to 325.6 m^3/s (runoff 0.46 cm/mo) in March and June (see Fig. 10.22). The March high flow corresponds to snowmelt in the Great Plains, whereas June high flow results from a combination of Rocky Mountain snowmelt and runoff from rain within the basin. The low coincides with heavy irrigation demands during August.

It is difficult to find preirrigation flow data for the Platte, but flows in the North Platte River near North Platte, Nebraska, from 1896 to 1942 averaged 74.1 m^3/s and since 1943 averaged 21.9 m^3/s. The closing and filling of Lake McConaughy occurred upstream of North Platte in 1943, but after the construction of four other North Platte River dams in Wyoming. South Platte flows have been augmented since before 1900 by transmountain diversions from the Colorado River drainage, which average approximately 3.7 × 10^8 m^3/yr, making it difficult to interpret the hydrograph. A significant portion of the base flow of the Platte River comes from the groundwater-fed Loup and Elkhorn rivers, which exhibit some of the most stable flows of any rivers in the world (Bentall 1989).

Over most of its length the Platte River is generally a warm, turbid, slightly alkaline river. Temperatures range up to 36°C, suspended solids concentrations range up to 11,600 mg/L, and specific conductance readings of 3450 μS/cm have been recorded. Yu (1996) found that from 1987 to 1993, temperature and conductivity tended to be lower during years of higher discharge, whereas suspended solids concentrations tended to be higher. Median pH is 8.1, but pH values range from 6.0 to 9.0. Total alkalinity averages 153.5 mg/L as $CaCO_3$. Annual median NO_3-N concentrations range from 0.68 to 1.4 mg/L (Frenzel et al. 1998) and total PO_4-P concentrations average 0.73 mg/L.

River Biodiversity and Ecology

The Platte River basin lies within the Middle Missouri freshwater ecoregion. Changes in slope, substrate, temperature, and riparian vegetation from its headwaters to its mouth have considerable influence on the biodiversity and ecology of individual Platte River segments. The coarse substrate, steep slope, and conifer-dominated headwater reaches contrast with the fine shifting substrates, shallow slopes, and deciduous forest riparian vegetation in the lower basin.

Algae

Dominant algal genera in the Platte River include *Cyclotella* and *Fragilaria* (Bacillariophyta), *Scenedesmus* and *Dictyospharium* (Chlorophyta), and *Oscillatoria*, *Anabaena*, and *Agmenellum* (Cyanophyta).

Plants

Aquatic macrophytes are not a common feature of the Platte River, but backwater areas support stands of cattail, waterweed, and pondweed. Riparian plant species vary from east to west, but cottonwood, willows, and box elder are common along most of the Platte. The extent and density of woody riparian cover along the Platte prior to European settlement is a matter of contention among ecologists and habitat managers.

Invertebrates

Aquatic invertebrates of the Platte River in Nebraska include 18 species of unionid mollusks (Hoke 1995) and 63 taxa of insects (McBride 1995).

Common insects include mayflies (*Caenis*, *Tricorythodes*, and *Heptagenia*), stoneflies (*Isoperla*), odonates (*Argia* and *Gomphus*), true bugs (Corixidae and Gerridae), beetles (Elmidae and Dytiscidae), caddisflies (*Hydropsyche* and *Cheumatopsyche*), and chironomid midges (*Dicrotendipes*, *Cladotanytarsus*, and *Rheotanytarsus*). Most are classified in the collector-gatherer or collector-filterer functional feeding groups and occupy shoreline habitats rather than shifting sand bar habitats. Analysis of macroinvertebrate densities in the Platte River downstream from the mouth of the Loup (Peters et al. 1989) found that rock substrates supported the highest numbers of individuals per unit area ($65,245/m^2$), with most being chironomids and caddisflies. Invertebrate densities on sand, gravel, silt, and wood substrates were $8218/m^2$, $7576/m^2$, $6610/m^2$, and $6572/m^2$, respectively.

Comparisons of the current invertebrate community structure in the Platte River to the community prior to European settlement are problematic. Today, woody debris is an important substrate for aquatic invertebrates, but debate continues on whether woody plants composed a significant portion of historical Platte River vegetation. However, rock substrate in the lower Platte was virtually absent and this is apparently important to the fauna today.

Vertebrates

The fish fauna of the Platte River includes 100 species comprising 20 families, of which 76 are native to at least a portion of the basin (Peters and Schainost 2005). Flood flows with high turbidity followed by low flows with high water temperatures impose special restrictions on the biota of the Platte River. Many native main-stream species in the Platte, including red shiner, sand shiner, river shiner, bigmouth shiner, western silvery minnow, plains minnow, speckled chub, flathead chub, river carpsucker, quillback, and channel catfish, are adapted to these conditions. Pre-1900 records also indicate that several species, such as shovelnose sturgeon, sturgeon chub, and sauger, were found in the Platte River drainage as far west as Wyoming (Baxter and Stone 1995) but have been extirpated from the North Platte basin. Headwater, tributary, and spring-fed side channel reaches support species that require clear and/or coolwater conditions. Some, such as lake chub, have been extirpated from the basin, but others, such as northern redbelly dace, finescale dace, hornyhead chub, plains topminnow, and Topeka shiner, are found in isolated populations in the drainage. In general, the number of native species declines in the western portion of the basin, where 30 native species have been recorded in the North Platte basin in Wyoming and 26 species have been recorded in the South Platte in Colorado. In addition, the proportion of nonnative species increases to almost 50% in the North Platte in Wyoming and 41% in the South Platte basin in Colorado (Peters and Schainost in press). By contrast, the South Platte River tributaries are the only localities for the endangered greenback cutthroat trout, the only salmonid native to the basin (Behnke 1992). Species characteristic of the main-stem Missouri River, such as paddlefish, lake sturgeon, pallid sturgeon, shovelnose sturgeon, longnose gar, shortnose gar, and goldeye occur in the lower 100 km of the Platte River, along with big-river suckers (bigmouth buffalo, smallmouth buffalo, and blue sucker) and turbid river chubs (flathead chub, sicklefin chub, sturgeon chub, and speckled chub) (Peters and Schainost in press). Today, channel catfish are probably the most sought after sport fish of the Platte River drainage, whereas shallow-water minnows, including red shiner, sand shiner, river shiner, western silvery minnow, and plains minnow, compose the numerically most abundant species (Peters et al. 1989, Yu 1996). Side-channel and backwater habitats include both plains killifish and plains topminnow, the latter being nearly endemic to the Platte drainage (Lynch 1988).

The 24 nonnative fish species introduced to the Platte drainage include the common carp and widely stocked game fish species like walleye, largemouth bass, and bluegill. Brook trout, brown trout, and rainbow trout populations are either maintained by regular stocking or are self-sustaining in headwater tributaries, cool spring-fed reaches, and deep reservoirs. These species may compete with the native greenback cutthroat trout in Colorado. Other species, such as the western mosquitofish (Lynch 1988) and Asiatic carps, are threatening native species through competition for food resources and predation on larvae and fry.

Herpetofauna of the Platte drainage includes salamanders (2 species), frogs and toads (11), turtles (8), lizards (11), and snakes (29). The most common representatives of these groups along the Platte River are the tiger salamander, Woodhouse's toad, chorus frog, painted turtle, and spiny softshell turtle.

Records for 409 species of birds have been confirmed from the Platte River drainage, and of these at least 208 species have been confirmed nesting in the area. This list includes wood warblers (41 species), shorebirds (40), waterfowl (35), and emberizid finches (32). Probably the most spectacular avian

display along the Platte is the nearly 500,000 sandhill cranes that stage in their northward migration in the 200 km downstream from North Platte, Nebraska, each spring. Added to this are concentrations of snow geese, other waterfowl, and shorebirds that use the wet meadow complexes associated with the Platte River. In addition, beaver, muskrat, and river otter are found along the river.

Ecosystem Processes

We are aware of no studies of ecosystem processes for the Platte River, but the relative abundance of collector-filterer and collector-gatherer functional feeding groups among the invertebrates point to a mix of allochthonous and autochthonous organic inputs for the food chains (McBride 1995). The numerically most abundant fishes include plains minnow and western silvery minnow, which are primarily herbivorous, and red shiner, river shiner, and sand shiners, which are primarily insectivorous. In addition, larger fishes represented by the omnivorous carpsuckers, insectivorous shovelnose sturgeon, and piscivorous gars and large catfish are important links in the Platte's food web. The broad expanses of sandbars and shallow water in the Platte also provide energy resources critical to migratory species of waterfowl and shorebirds.

Human Impacts and Special Features

Prior to European settlement the Platte River conducted the meltwater-rich flows from the Rocky Mountains onto the drier Great Plains in a seasonal pulse that fostered a rich diversity and abundance of animal and plant life. The drainage was populated by farming cultures (Pawnee and Otoe) in the east and nomadic bison-hunting cultures (Arapaho, Cheyenne, and Dakota) in the west. The river also provided additional resources from fishes, migratory waterfowl, and freshwater clams, all of which have been found in middens along the Platte.

Initial European settlements along the Platte began by the 1850s and the first irrigation canal in Nebraska was dug in 1863. Irrigation development along the North Platte River installed diversions that totaled over 56.6 m^3/s by 1890 (Kepfield 1994). The push for more irrigation along the North Platte led to the construction of Pathfinder Dam upstream of Casper, Wyoming, in 1909. This was the first of five North Platte main-stem dams that concluded with the construction of Lake McConaughy in 1943. These reservoirs, diversions, canals, and off-channel impoundments that store water for irrigation and power generation have altered the flow regime of the North Platte River. In addition, tailrace areas below reservoirs have been changed from warmwater or coolwater to coldwater habitats that support introduced salmonids.

Extensive water diversions for mining, agricultural, and municipal uses occurred by 1858 with the discovery of gold in the South Platte basin, which apparently supported the only native trout population (greenback cutthroat). However, flows in the South Platte and several major tributaries, including the Cache la Poudre River, have been augmented by diversions from the Colorado River drainage; much of this is stored in a series of off-channel reservoirs to be used for municipal systems in Colorado. Virtually all of the flow of the South Platte River is diverted for use by the Denver metropolitan area and virtually all of the downstream flow resumes from sewage treatment facilities.

Flow is controlled and depleted by diversions downstream from the confluence of the North Platte and South Platte rivers into a series of irrigation and power generation canals that parallel the Platte. Flows from the Loup and Elkhorn basins into the lower Platte are also diminished by diversions and groundwater withdrawals in the alluvial aquifer. Instream flow water rights have been granted in the Platte River in Nebraska to protect habitat for endangered whooping crane and for fish populations, including the endangered pallid sturgeon, in downstream reaches. These flows are also important to the water supplies for the cities of Lincoln and Omaha, Nebraska.

Channelization is an important disturbance on tributaries, but bank stabilization is probably a more important perturbation on the main channels of the Platte because revetments and wing dikes also act to restrict the channel. Channels have been narrowed from 40% to 60% of their historical widths and banks that were formerly covered with mostly herbaceous vegetation are now covered with trees like eastern cottonwood, green ash, and eastern red cedar (Rothenberger 1987).

Point sources of water pollution are generally associated with municipalities; however, large cattle facilities and some industrial operations are also important. Most pollutants tend to be nitrogen compounds and oxygen-demanding wastes. Drainage from abandoned mines in headwater areas of the South Platte can raise metal concentrations to toxic levels, but they are generally present as chronic stressors to the system. Nonpoint sources of nutrients and pesticides in the lower Platte basin, especially atrazine, alachlor, and cyanazine, contribute background levels that often spike during runoff events in the spring planting season. Negative relationships between fish species tolerance and proportion of stream basin that is cropland have been shown (Frenzel et al. 1998).

Management of the Platte River ecosystem is a complex balance among competing interests for a limited and overappropriated water supply. Expansion of urban centers from the Front Range in Colorado, along the North Platte in Wyoming, and downstream to the mouth of the Platte in eastern Nebraska has exacerbated the competition for water among agricultural and environmental interests within the basin. The importance of the Platte River to migratory birds of the Central Flyway, including the endangered whooping crane and other rare and endangered species, expands concern about management of this resource to the continental and international scales.

GASCONADE RIVER

The Gasconade is a 6th order river whose 9258 km^2 basin is in the rolling uplands of south-central Missouri. It meanders northeast 436 km to join the Missouri River at Rkm 168 (Fig. 10.23). Major tributaries from south to north and their approximate proportion of the total basin are Osage Fork (14%), Roubidoux River (8%), Big Piney River (16%), and Little Piney River (8%). The basin lies about 161 km southwest of St. Louis and is roughly triangular, 80 km at its widest point and 209 km in length. It is divided into two USGS eight-digit hydrological units. Few towns are located along the river and there are no major urban centers in the basin. Large segments of the Gasconade and its tributaries are relatively undeveloped, unpolluted, and free flowing. Portions of the basin contain some of the most rugged topography and scenic areas in the region (Fig. 10.7).

Late Woodland peoples (400 to 900 AD) living in semipermanent villages were the principal prehistoric Native Americans within the basin, although the area was also under the influence of the Mississippian Mound Builder metropolis at Cahokia–St. Louis to the northeast (900 to 1200). From about 1675 until the United States gained control of "Louisiana" in 1804, the region was dominated by the powerful Osage hegemony. French trappers plied the lower reaches of the Gasconade by the 1740s. Early American settlers immigrated to the upper basin as

FIGURE 10.7 Gasconade River, Missouri (Photo by J. Rathert, Missouri Department of Conservation).

early as 1826, attracted by abundant game. Immigration increased in the 1840s, when public lands were opened for sale and further expanded with rail service in 1870. Early land-use practices of forest clearing, uncontrolled burning and livestock grazing, poor farming, and unregulated gravel mining adversely affected basin streams. Burning and grazing resulted in topsoil removal, exposing the cherty, gravelly subsoils, which eventually accumulated in streams.

Physiography, Climate, and Land Use

The Gasconade River basin lies wholly within the Springfield–Salem plateaus section of the Ozark Plateaus (OZ) physiographic province, a part of the Interior Highlands physiographic division. Its watershed is unglaciated and hillslope soils are thin and thoroughly leached. Surface geological formations are composed of Ordovician dolomites and sandstones. The main stem and tributaries cut through Gunter sandstone and Gasconade dolomite, the latter

having many springs that contribute to the river's base flow. The Gasconade formation is replaced upland from the floodplain by the Roubidoux formation, which contains sandstone and cherty dolomite. Further upland in the headwaters is a composite of well-weathered dolomite formations with numerous cracks, joints, and solution openings. The basin is characterized by a large subterranean drainage, creating numerous caves and springs.

The Gasconade basin traverses three land resource areas: Deep Loess Hills, Ozarks, and Ozark Border. The thickest soil deposits of the Deep Loess Hills are found along river bluffs, with other deposits on ridgetops and broad uplands. Most soil formations of the Ozarks and Ozark Border were formed under forest vegetation, with occasional glades. Slopes within the Ozarks contain cherty alluvium soils, whereas within the Ozark Border they contain more gravelly alluvium soils. Both areas have fragipans that can restrict plant growth. Estimated soil loss is 1.0 metric tons $ha^{-1}yr^{-1}$, and the amount reaching streams is low, about 0.29 metric tons $ha^{-1}yr^{-1}$.

Most of the Gasconade basin lies within the Central U.S. Mixed Hardwood Forests terrestrial ecoregion (Ricketts et al. 1999). Upland forests contain distinctive tree associations related to geology, soil type, temperature, and precipitation. Forests are mainly post and blackjack oak to the north, with white and black oak forest, white oak and shagbark hickory forest, and an increase in shortleaf pine and oak forest to the south.

Climate of the basin is continental, with frequent daily and seasonal changes in weather. Mean annual air temperature in the basin is 13.3°C and ranges from a mean low of –0.1°C in January to a mean high of 25.7°C in July (Fig. 10.24). Precipitation averages 107 cm/yr and monthly averages range from a low of 5.3 cm in January to a high of 12.9 cm in May (see Fig. 10.24). Snowfall is limited and seldom accumulates. The rainy season is April through June (32% of annual total) and heavy rains can cause flooding.

Logging began in the mid 1800s, and the river served to float logs to yards and railroad ties to railroad crossings. Most of the uplands of old-growth oak–hickory and short-leaved pine were harvested by 1900, particularly in the headwaters. Regeneration has been slow due to the region's steep topography and poor soils. Farming in the late 1800s included wheat and corn, and the region led the nation in apple production in the 1890s. Grain and fruit production has largely been replaced by cattle grazing and hay production. Cattle populations and grazing density have shown a general increase from the 1920s to the 1990s, averaging about 0.3/ha in 1995. Overgrazing occurs and streams and rivers in the basin are commonly used for cattle watering. Confined dairy (~2200 animal units), hog (~19,000 units), and poultry (~840,000 units) operations are also present in the basin and can be significant sources of organic pollutants. Present land use is primarily forest (55%) and grassland/pasture (35%). About 10% of the basin is in public ownership, with major portions in the Mark Twain National Forest and Ft. Leonard Wood U.S. Army Military Reservation.

Zinc, lead, and iron surface mining were important in the upper basin in the late 1800s, but little mining occurs today. Clay and limestone were also quarried, and limestone quarrying is ongoing. Ozark streams have been an important source of sand and gravel since the early 1900s. Gravel mining has been regulated since the 1990s and should be confined to gravel bars and damage to stream banks and vegetation restricted.

River Geomorphology, Hydrology, and Chemistry

Portions of the Gasconade channel meander through wooded banks or towering bluffs featuring caves and springs. Gravel bars, quiet pools, and turbulent chutes are numerous. The region's karst topography causes portions of the Gasconade and several of its tributaries to lose flow into an aquifer (i.e., segments are losing streams). About 53 km in the central portion of the main stem is the longest losing segment in the basin. The Gasconade basin is one of the most cavernous regions in the United States, with 131 named caves and 76 reported springs.

There are 9682 linear kilometers of stream (loops, braids, and disconnects excluded) within the Gasconade drainage. The main channel is highly sinuous and often entrenched in exposed limestone and dolomite formations. The most meandering section is above the confluence of the Big Piney River (see Fig. 10.23), where bluffs are 60 to 90 m tall and channel sinuosity is >30. Water depths vary from 0.3 m in the upper river to 9.1 m near the mouth, and average high-bank channel width ranges from 12.2 m to 94 m. Average gradient is 1.95 m/km for the upper Gasconade and 1.10 m/km in the lower river. Stream bottoms consist of bedrock, boulders, rubble, and gravel, with little or no sand or mud except in backwaters. Major riverine habitat groups (percentage area) for the upper Gasconade basin include pool–riffle complex (29.1%), temporary–semipermanent pool (12.7%), and gravel bar (8.3%). For the lower Gasconade basin these are gravel bar (6.6%) and temporary–semipermanent pool (4.4%) (Blanc 2001). The high percentage of gravel bars in the main-stem Gasconade and other Ozark rivers has been associated more with unstable channel conditions in headwater tributaries due to poor land-use practices than conditions within the local riparian corridor (Jacobson and Pugh 1997).

April through May is the period of highest discharge for the Gasconade River, with maximum discharge usually in April (132.5 m³/s) (see Fig. 10.24). Lowest flows occur in late summer and early fall. Most floods occur from February through June but can occur at any time of year. Runoff is rapid due to the steep valley slopes and stage increases of 1.5 to 2.4 m/d occur. Percentage of time flow exceeded a given discharge over the 75 years from 1923 to 1998 at the Jerome gage was 253 m³/s (5%), 36.1 m³/s (50%), and 12.7 m³/s (95%) (Blanc 2001). Springs make important contributions to the Gasconade River's base flow, particularly in the middle basin.

Major springs and their discharge include Bartlett Mill (1.92 m³/s), Boiling (1.84 m³/s), Roubidoux (1.65 m³/s), and Piney (1.4 m³/s). Differences in precipitation versus runoff (see Fig. 10.24) are due to three factors: (1) evapotranspiration, (2) springs contributing to river discharge somewhat independent of local precipitation, and (3) sections of the river losing discharge to the aquifer. Water temperature of Ozark springs is fairly constant at 13°C to 15°C and their discharge locally moderates temperatures of receiving streams. No dams occur on the main-stem Gasconade, whereas two 3 m high rock dams occur on its largest tributary, the Big Piney River.

Water quality of the Gasconade River is generally good, with a few problem areas. The upper Gasconade River is identified as a Category I watershed and the lower basin as Category II. Its chemistry is basic (pH 8.0), alkaline (151 mg/L as $CaCO_3$), and relatively hard (total hardness 130 to 200 mg/L as $CaCO_3$) with moderate conductivity (240 to 360 µS/cm). Turbidity is generally very low (<10 NTU) except during high water. Chemistry is fairly uniform along the river and mean ranges of annual concentrations of major dissolved inorganic constituents are calcium (24 to 39 mg/L), magnesium (14 to 25 mg/L), iron (3 to 32 mg/L), and sulfate (3.9 to 11 mg/L as SO_4). Nutrient concentrations are generally low except in reaches below municipal effluent discharge or where livestock have access to the river. Dissolved NO_2-N + NO_3-N near Jerome averages 0.37 mg/L and ranges from <0.05 to 1.20 mg/L. Dissolved PO_4-P averages 0.015 mg/L and ranges from <0.010 to 0.070 mg/L. Dissolved silica concentrations are high, ranging from 8.7 to 9.7 mg/L as SiO_2. Total phosphorus and total suspended solids concentrations are lower in Ozark streams (including the Gasconade) than in Ozark Border and Glaciated Plains streams in Missouri, whereas total nitrogen concentrations are similar. Fecal coliforms are generally below the state limit of 200 colonies/100 mL.

One of the largest oil pipeline spills in the nation occurred in December 1988, releasing >3266 m³ of crude oil into a tributary of the Gasconade River. Approximately 105 km of the lower Gasconade River were affected and only 50% of the oil from the spill was recovered.

River Biodiversity and Ecology

The Gasconade River is part of the Central Prairie freshwater ecoregion (Abell et al. 2000) and is characterized by a rich and diverse flora and fauna. Significant features within the river corridor include mesic deciduous bottomland forests, limestone and dolomite cliffs, and influent and effluent caves.

Algae

Although no studies of algae have been done in the Gasconade, periphyton from nearby Northern Ozark rivers are volumetrically dominated by Bacillariophyta (*Cocconeis*, *Cymbella*, and *Surirella*), Chlorophyta (*Cladophora*, *Cosmarium*, and *Oedigonium*), and Chrysophyta (*Bumellaria*).

Plants

Aquatic macrophytes within the channel are confined to backwater areas and include pondweeds, naiads, arrowheads, and yellow pondlily. Coldwater springs have a distinctive macrophyte community; whitewater crowfoot, watercress, water speedwell, and *Fontinalis* mosses are common taxa. Water willow is the dominant vascular plant at the water's edge on gravel bars. Black and sandbar willows invade gravel bars along the lower river where soil is mixed with gravel. At higher elevation, more stable bars support river birch, buttonbush, sycamore, silver maple, and green ash. Trees on floodplain alluvium include Shumard's oak, bur oak, and box elder, and in seldom-flooded locations, sugar maple, black walnut, and bitternut hickory. Flowering dogwood, redbud, and sassafras are common understory trees.

Invertebrates

Mussels, crayfishes, and insects compose the rich aquatic invertebrate fauna of the Gasconade River. Forty-two mussel species, composing 27 genera, have been collected. Dominant genera of mussels are *Lampsilis* (6 species), *Quadrula* (3 species), and *Fusconaia* (2 species). Asiatic clams have been introduced into the main-stem Gasconade. The plain pocketbook mussel is widely distributed. Seven mussels of state conservation concern (S1) reside in the upper basin: scale shell (also G2), elephantear, spectaclecase, elktoe, black sand shell, bullhead (also G3), and Ouachita kidneyshell. The federally endangered pink mucket is present in the main-stem Gasconade.

Five species of crayfishes are in the basin, but >99% of crayfish composition is within the genus *Orconectes*. In decreasing order of abundance the crayfishes are spothanded crayfish, golden crayfish, digger crayfish, devil crayfish, and the Salem cave crayfish. The Salem cave crayfish is listed by NatureServe (2002) as globally imperiled (G2), and another troglobite present in the Gasconade basin, the central Missouri cave amphipod, is listed as globally critically imperiled (G1). *Hyalella azteca* is a more abundant amphipod in the river.

Sampling from various locations in the basin has yielded 52 families of aquatic insects. *Acentrella*, *Baetis*, *Ephemerella*, *Stenonema*, *Tricorythodes*, and *Caenis* are major mayfly genera. Abundant odonates include *Argia*, *Erpetogomphus*, and *Gomphus*. *Chimarra*, *Cheumatopsyche*, *Hydropsyche*, *Psychomyia*, and *Helicopsyche* dominate the caddisflies. *Taeniopteryx*, *Strophopterx*, and *Neoperla* are abundant stoneflies. Beetles include *Dubiraphia*, *Psephenus*, and *Stenelmis*. *Simulium*, *Polypedilum*, *Dicrotendipes*, and *Paratanytarsus* are important Diptera. A globally imperiled (G2) stonefly, *Acroneuria ozarkensis*, is recorded only in the Gasconade River basin and a few other streams in the United States. Scrapers and collectors are the most abundant functional feeding groups.

Vertebrates

Five families dominate the rich fish fauna (105 species) of the Gasconade River. Most species occur in the Cyprinidae (29), Catostomidae and Percidae (14), Centrarchidae (12), and Ictaluridae (8). The most widely distributed cyprinids are bleeding shiner, horneyhead chub, largescale stoneroller, and central stoneroller. Longear sunfish, rock bass, bluegill, smallmouth bass, largemouth bass, and spotted bass are widely distributed centrarchids. Spotted bass are a recent addition to the Gasconade basin due to range expansion. Ten Gasconade basin fishes are of conservation concern in Missouri (S1–S3), including the southern cavefish. The anadromous Alabama shad is a candidate for federal listing, and the Gasconade is one of the few Mississippi basin rivers where it still spawns. The bluestripe darter is endemic to the Gasconade and nearby Osage River basins and is globally listed as imperiled (G2). In addition, the crystal darter is identified as globally vulnerable to extinction (G3) (http://www.natureserve.org/explorer/servlet/NatureServe?init=Species 2002).

Rock bass, smallmouth bass, largemouth bass, walleye, and channel catfish are the most sought after sport fishes. Night gigging for suckers and redhorse is also popular. Size and harvest of rock bass and smallmouth bass are controlled in several special management areas to provide a high-quality angling experience. Coldwater sections of several tributaries below springs are designated as Wild Trout Management Areas. Protection and management of self-sustaining populations of introduced rainbow trout occur here.

All major groups of amphibians and reptiles are well represented in the basin: frogs and toads (13 species), salamanders (11), turtles (12), lizards (6), and snakes (29). Stream-dwelling salamanders include the state-imperiled Ozark hellbender (S1), mudpuppy, and western lesser siren. Northern water snakes are abundant, and map turtle, red-eared slider, and western painted turtle are often seen basking on logs. Both smooth and spiny softshell turtles are observed in the Gasconade's clear waters.

Bird life along the river is outstanding, with about 290 species recorded. Wood ducks, belted kingfishers, green herons, and great blue herons are common along the river, and ospreys and bald eagles are occasionally observed during migration. Common aquatic mammals along the river corridor include river otter, beaver, mink, and muskrat.

Ecosystem Processes

There are no studies of ecosystem processes for the Gasconade River, although evidence on periphyton biomass and nutrient limitation exists from nearby rivers. Concentrations of benthic chlorophyll *a* from Northern Ozark streams (upper Moreau and Maries rivers) with similar low nitrogen and phosphorous concentrations as the Gasconade ranged from 33 to 59 mg/m^2 (Lohman et al. 1991). Northern Ozark streams appear to be primarily nitrogen limited based on nutrient enrichment studies conducted during low flow periods (Lohman et al. 1991). Low NO$_3$-N and molar TN/TP ratios (<20 to 1) support these results.

Fish growth in the Gasconade River is generally higher than average for Missouri streams (Funk 1975). Growth rates equaled or exceeded the statewide average for 9 of the 14 species examined, illustrating the river's high fish production. Growth rates of the river's two most popular sport fishes, rock bass and smallmouth bass, were excellent: 149% and 127%, respectively, of the statewide average.

Human Impacts and Special Features

The Gasconade River is one of the few free-flowing rivers in the conterminous United States (Benke 1990) and the only free-flowing major tributary to the Missouri from the Ozark Highlands. It is in a largely rural, well-forested karst basin harboring numerous coldwater springs and caves and their characteristic flora and fauna. Approximately 357 km of the Gasconade and Big Piney rivers were recommended for designation as "scenic" or "recreational" under the National Wild and Scenic Rivers Act due to their outstanding natural resource and aesthetic values; however, such designation never occurred. The river is popular for shoreline and float angling, sightseeing and nature study, boating, camping, and swimming. Days fished per total watershed area ranged from 0.038 to 0.063 between 1983 and 1991 (Blanc 2001).

Primary future threats to the high quality of this Ozark river include further expansion of cattle grazing and concentrated animal feedlots, stream bank erosion, poor gravel mining practices, expanding urban and suburban populations, and increased recreational use. Improved land-use practices, particularly excluding cattle and off-road vehicles from streams and leaving riparian buffers when logging, will help address major causes of bank erosion and nonpoint organic pollution. Agency incentives are in place to assist repair of stream bank damage, including corridor reforestation, stream bank revegetation, cedar tree revetments, and willow staking. There has been encouraging movement toward minimizing gravel mining impacts through mandated best management practices (BMPs) for commercial operations and voluntary BMPs for noncommercial uses. Enforcing existing water-quality regulations will help reduce violations and citizen activism through the Missouri Stream Team and similar programs will promote public awareness of the river's ecological values.

ADDITIONAL RIVERS

The Madison River is a coldwater system located in the Northern Rocky Mountains of Wyoming and Montana and is one of three rivers that merge to form the Missouri River (Fig. 10.25). About one-fourth of its drainage is in Yellowstone National Park, where geothermal inputs result in naturally high concentrations of arsenic. The river is relatively low gradient throughout much of its length and meanders through a broad floodplain, dominated in Montana by cattle ranches (Fig. 10.8). Human population density is low, with no major urban development, and water quality is good. Much of the river corridor in Montana is being subdivided into ranchettes. Three main-stem dams fragment the river and influence flow and water temperature. The river is very productive and internationally known for its high-quality trout fishery, dominated by nonnative rainbow trout and brown trout. Infestation of whirling disease caused major declines in the rainbow

trout population during the past decade and the recent discovery of the New Zealand mud snail is of ecological concern.

The Milk River originates in Glacier National Park, Montana, flows into Canada, and then back into Montana (Fig. 10.27). Over much of its length the meandering channel is highly braided, with unstable sand substratum (Fig. 10.9). Land use is principally range and agriculture and population density is low, with no major urban development. The river is fragmented by one major dam (Fresno), one municipal water weir, and four irrigation diversion dams, none with fish passage structures. The river above Fresno Dam is one of the few remnant, relatively intact Great Plains river ecosystems. Overall species diversity is low and nonnative fishes dominate the assemblages below Fresno Dam.

The Cheyenne River begins as two separate rivers: a north fork (Belle Fourche River), and a south fork (Little Cheyenne River) (Fig. 10.10, Fig. 10.29). With their ephemeral headwaters in the Wyoming prairie (40% of basin), both forks collect the cold water of Black Hills streams (18% of the basin) and then join to flow across a prairie landscape to Lake Oahe. The Belle Fourche captured the headwaters of the Little Missouri River and its upper Missouri River basin fishes (e.g., mountain sucker, longnose sucker). Three Bureau of Reclamation irrigation projects dominate the valley of the Belle Fourche and Little Cheyenne rivers. Contaminants from gold mining (e.g., mercury, mine tailings) have been the major pollution issue. Black Hills streams are clear and cold with stable flows, whereas the main-stem Cheyenne is a turbid, warmwater river with high spring and low winter flows. Trout in the Hills arm and walleye in the Cheyenne arm of Lake Oahe provide important sport fisheries. The main stem is used for canoeing, fishing for channel catfish, and camping on the numerous sandbars and gravel bars.

The Big Sioux River courses north to south along the borders of South Dakota, Minnesota, and Iowa (Fig. 10.11, Fig. 10.31). Much of the watershed is a fertile, rocky coteau (hill) of glacial origin that features numerous lakes and thousands of small pothole wetlands. Exposed Sioux quartzite forms the Sioux Falls (30 m drop over 0.8 km), dells, palisades and other rocky outcroppings that are unique to the region. The upper river flows over gravel and sand substrates through pastures and woodlots, whereas the lower river is sluggish and meanders through a wooded corridor. Agriculture dominates the landscape, but human density is relatively high. The river was once one of the most polluted rivers in the

FIGURE 10.8   Madison River near West Yellowstone, Wyoming (Photo by C. E. Cushing).

country, but water quality has improved since the 1960s. Nutrients, fecal coliforms, and suspended solids are still problems. Only 3% of the stream length has been altered and only 30% of the wetlands drained. Consequently, the river is relatively natural, has healthy fish populations, and supports river-oriented recreation and municipal parks.

The Niobrara River is a swift-flowing prairie river that intersects the boundaries of northern, eastern, and western forest flora. It flows in an easterly direction along most of the length of northern Nebraska before emptying into the Missouri River upstream of Lewis and Clark Lake (Fig. 10.33). It has been less affected by major habitat alterations than other eastward-flowing plains tributaries to the Missouri. A large volume of groundwater supports a consistent high-quality base flow. Ranching, supported by native grassland forage, and low human population density have resulted in good water quality. Low fragmentation by dams makes this one of the least impacted rivers in the Great Plains (Fig. 10.12).

The Kansas or Kaw River and its tributaries flow eastward, draining all of northern Kansas and southern Nebraska before joining the Missouri River at Kansas City (Fig. 10.13, Fig. 10.35). Its major tributaries, the Republican, Solomon, and Smoky Hill

FIGURE 10.9 Milk River east of Havre, Montana (PHOTO BY TIM PALMER).

rivers, arise in the high plains of eastern Colorado and southern Nebraska. The Kansas was historically a turbid, warmwater, prairie river with unstable sand substrates that was subsequently altered by water diversions, damming, and dredging. Much of the basin is in agriculture, with row crops dominating bottomlands. Pollution from pesticides, agricultural fertilizers, and municipal sewage systems has impaired water quality. Impoundments have altered habitat conditions so that mainly introduced sight-feeding fishes dominate the fauna.

The Grand River, the largest prairie river in Missouri largely unaffected by impoundments and channelization, lies within the Central Lowland glaciated prairies of northwest Missouri and southwest Iowa (Fig. 10.37). Its basin is largely rural cropland, with a declining population and no major urban areas. Although it has always been a turbid river, poor land-use practices and tributary channelization has resulted in steep eroding stream banks (Fig. 10.14). Filling of the channel with sand and silt has reduced pool habitat and coarse substrate and limits invertebrates and fishes. Sixty-one species of fishes are present, and most are habitat generalists, tolerant of tur-bidity. Angling for catfishes is a popular pastime. Water-quality standards for iron, magnesium, and fecal coliform bacteria are frequently exceeded. Most water-quality problems are associated with

FIGURE 10.10 Cheyenne River near Oral, South Dakota, showing minimal stable flow of clearer water below Angostura dam (PHOTO BY C. BERRY).

FIGURE 10.11 Big Sioux River at Sioux Falls, South Dakota. Note levees for flood control (PHOTO BY TIM PALMER).

FIGURE 10.12 Niobrara River near Merriman, Nebraska (PHOTO BY TIM PALMER).

FIGURE 10.13 Kansas River in vicinity of Lawrence, Kansas (PHOTO BY TIM PALMER).

FIGURE 10.14   Grand River at Holmes Bend access in Daviess County, Missouri (PHOTO BY G. PITCHFORD).

nonpoint-source pollutants such as soil erosion and manure runoff.

ACKNOWLEDGMENTS

We thank the following persons for information on topics included in this chapter: Sue Bruenderman (mussels), Michael Chapman (Missouri River engineering structures), David Fryda (White River), Barry Poulton and Linden Trail (insects), Matt Winston and Robert Bramblett (fishes), Julie Fleming (nonfish vertebrates), Timothy Smith (vegetation), Raymond Wood (Native Americans), Gust Annis, David Greenlee, John LaRandeau, and Scott Sowa (geography). Sandy Clark, Chad Kopplin, and Jeff Shearer helped gather and summarize data. Charles Rabeni, Nick Bezzerides, and Sandy Clark edited and reviewed portions of the chapter. The University of Nebraska, Agricultural Research Division through the School of Natural Resource Sciences provided support to E. Peters during the completion of this chapter.

This chapter is a contribution from the Missouri, Montana, and South Dakota Cooperative Fish and Wildlife Research Units (U.S. Geological Survey; Missouri Department of Conservation; Montana Department of Fish, Wildlife, and Parks; South Dakota Department of Game, Fish, and Parks; University of Missouri; Montana State University; South Dakota State University; and Wildlife Management Institute cooperating).

LITERATURE CITED

Abell, R. A., D. M. Olson, E. Dinerstein, P. T. Hurley, J. T. Diggs, W. Eichbaum, S. Walters, W. Wettengel, T. Allnutt, C. J. Loucks, and P. Hedao. 2000. *Freshwater ecoregions of North America: A conservation assessment.* Island Press, Washington, D.C.

Aquoneering, Inc., and Womack and Associates, Inc. 2000. Yellowstone River geomorphic analysis, Yellowstone County, Montana. Report to Yellowstone Conservation District, May 9, 2000.

Bahls, L. L. 1974. Microflora of the Yellowstone River: Microflora in the plankton at the confluence of the Bighorn River. Preliminary Report to Montana Department of Fish and Game, Environment and Information Division, Helena.

Bailey, R. M., and M. O. Allum. 1962. *Fishes of South Dakota.* Miscellaneous Publications No. 119, Museum of Zoology, University of Michigan, Ann Arbor.

Barbour, M. T., J. Gerritsen, B. Snyder, and J. Stribling. 1999. *Rapid bioassessment protocols for use in streams* and wadable rivers: Periphyton, benthic invertebrates and fish, 2nd ed. EPA-841-B-99-002. U.S. Environmental Protection Agency, Washington, D.C.

Baxter, G. T., and M. D. Stone. 1995. *Fishes of Wyoming*. Wyoming Game and Fish Department, Cheyenne.

Behnke, R. J. 1992. *Native trout of western North America*. Monograph 6. American Fisheries Society, Bethesda, Maryland.

Benke, A. C. 1990. A perspective on America's vanishing streams. *Journal of the North American Benthological Society* 9:77–88.

Bentall, R. 1989. Streams. In A. Bleed and C. Flowerday (eds.). *An atlas of the Sand Hills*, pp. 93–114. Resource Atlas no. 5. Conservation and Survey Division, University of Nebraska, Lincoln.

Berg, R. K. 1981. *Fish populations of the wild and scenic Missouri River, Montana*. Federal Aide to Fish and Wildlife Restoration, Project FW-3-R, Job 1-A. Montana Department of Fish, Wildlife and Parks, Great Falls.

Berner, L. M. 1951. Limnology of the lower Missouri River. *Ecology* 32:1–12.

Blanc, T. J. 2001. Gasconade River watershed: Inventory and assessment. Missouri Department of Conservation Web site: http://www.conservation.state.mo.us/fish/watershed/gascon/contents/130cotxt.htm.

Bragg, T. B., and A. K. Tatschl. 1977. Changes in floodplain vegetation and land use along the Missouri River from 1826 to 1972. *Environmental Management* 1:343–348.

Bramblett, R. G., and R. G. White. 2001. Habitat use and movements of pallid and shovelnose sturgeon in the Yellowstone and Missouri Rivers in Montana and North Dakota. *Transactions of the American Fisheries Society* 130:1006–1025.

Brooks, C. E. 1979. *The living river: A fisherman's intimate profile of the Madison River watershed, its history, ecology, lore, and angling opportunities*. Lyons Books, Garden City, New York.

Cross, F. B. 1967. *Handbook of fishes of Kansas*. Miscellaneous Publication 45. University of Kansas Museum of Natural History, Lawrence.

Cross, F. B., R. L. Mayden, and J. D. Stewart. 1986. Fishes in the western Mississippi basin (Missouri, Arkansas, and Red rivers). In C. H. Hocutt and E. O. Wiley (eds.). *The zoogeography of North American freshwater fishes*, pp. 367–412. John Wiley & Sons, New York.

Cutright, P. R. 1969. *Lewis and Clark: Pioneering naturalists*. University of Nebraska Press, Lincoln.

Department of Environment and Natural Resources (DENR). 1998. *The 1998 South Dakota 303(d) waterbody list and supporting documentation*. Department of Environment and Natural Resources, Pierre, South Dakota.

Department of Environment and Natural Resources (DENR). 2000. *South Dakota Water Quality Water Years 1995–1999*. South Dakota Department of Environment and Natural Resources, Pierre.

Dieterman, D., and C. Berry. 1998. Fish community and water quality changes in the Big Sioux River. *Prairie Naturalist* 30:199–224.

Doorenbos, R. 1998. Fishes and habitat of the Belle Fourche River, South Dakota. Master's thesis, South Dakota State University, Brookings.

Doorenbos, R., C. Berry, and G. Wickstrom. 1999. Ictalurids of South Dakota. *American Fisheries Society Symposium* 24:377–389.

Doorenbos, R., D. Dieterman, and C. Berry. 1996. Recreational use of the Big Sioux River, Iowa and South Dakota. Special Report no. 96–14. South Dakota Department of Game, Fish, and Parks, Pierre.

Ellis, M. J., J. Ficken, and D. Adoplhson. 1971. *Hydrogeology of the Rosebud Indian reservation, South Dakota*. Hydrologic Investigations Atlas HA-355. U.S. Geological Survey, Washington, D.C.

Farrell, J. R., and M. A. Tesar. 1982. Perphytic algae in the channelized Missouri River with special emphasis on apparent optimal temperatures. In L. W. Hesse, G. L. Hergenrader, H. S. Lewis, S. D. Reetz, and A. B. Schlesinger (eds.). *The middle Missouri River: A collection of papers on the biology with special reference to power station effects*, pp. 85–123. Missouri River Study Group, Norfolk, Nebraska.

Ferrell, J. 1996. *Soundings: 100 years of the Missouri River navigation project*. U.S. Army Corps of Engineers, Omaha, Nebraska.

Ferrick, M., N. Mulherin, and D. Calkins. 1995. The winter low-flow balance of the semiarid White River, Nebraska and South Dakota. Report 95-15. Cold Regions Research and Engineering Laboratory, Hanover, New Hampshire.

Fischer, T. D., D. Backlund, K. Higgins, and D. Naugle. 1999. *A field guide to South Dakota amphibians*. SDAES Bulletin 733. South Dakota State University, Brookings.

Frenzel, S. A., R. B. Swanson, T. L. Huntzinger, J. K. Stamer, P. J. Emmons, and R. B. Zelt. 1998. Water quality in the central Nebraska basins, Nebraska, 1992–95. Circular 1163. U.S. Geological Survey, Washington, D.C.

Fryda, D. D. 2001. A survey of the fishes and habitat of the White River, South Dakota. Master's thesis, South Dakota State University, Brookings.

Funk, J. L. 1975. *The fishery of the Gasconade River, Missouri, 1947–57*. Aquatic Series no. 13. Missouri Department of Conservation, Jefferson City.

Funk, J. L., and J. W. Robinson. 1974. *Changes in the channel of the lower Missouri River and effects on fish and wildlife*. Aquatic Series no. 11. Missouri Department of Conservation, Jefferson City.

Galat, D. L., and A. G. Frazier (eds.). 1996. Overview of river–floodplain ecology in the upper Mississippi River basin. Vol. 3 of J. A. Kelmelis (ed.). *Science for floodplain management into the 21st century*. U.S. Government Printing Office, Washington, D.C.

Galat, D. L., and R. Lipkin. 2000. Restoring ecological integrity of great rivers: Historical hydrographs aid in defining reference conditions for the Missouri River. *Hydrobiologia* 422/423: 29–48.

Galat, D. L., M. L. Wildhaber, and D. J. Dieterman. 2001. *Spatial patterns of physical habitat.* Vol. 2: *Population structure and habitat use of benthic fishes along the Missouri and Lower Yellowstone rivers.* U.S. Geological Survey, Cooperative Research Units, University of Missouri, Columbia. Web site: http://www.nwo.usace.army.mil/html/pd-e/benthic_fish/benthic_fish.htm

Galat, D. L., C. R. Berry, W. M. Gardner, J. C. Hendrickson, G. E. Mestl, G. J. Power, C. Stone, and M. R. Winston. In press. Spatiotemporal patterns and changes in Missouri River fishes. In J. N. Rinne, R. M. Hughes, and B. Calamusso (eds.). *Historical changes in large river fish assemblages of the Americas*, in press. American Fisheries Society Symposium 45. Bethesda, Maryland.

Governor's Upper Yellowstone River Task Force. 2000. Governor's Upper Yellowstone River Task Force annual report. Livingston, Montana.

Hampton, D., and C. Berry. 1997. Fishes of the mainstem Cheyenne River in South Dakota. *Proceedings of the South Dakota Academy of Science* 76:11–25.

Hauer, F. R., J. A. Stanford, and J. T. Gangemi. 1991. Effects of stream regulation in the upper Missouri River. Report 116-91. Flathead Lake Biological Station, University of Montana, Polson.

Havel, J. E., E. M. Eisenbacher, and A. A. Black. 2000. Diversity of crustacean zooplankton in riparian wetlands: Colonization and egg banks. *Aquatic Ecology* 34:63–76.

Hesse, L. W., J. C. Schmulbach, J. M. Carr, K. D. Keenlyne, D. G. Unkenholtz, J. W. Robinson, and G. E. Mestl. 1989. Missouri River fishery resources in relation to past, present, and future stresses. In D. P. Dodge (ed.). *Proceedings of the International Large River Symposium (LARS).* Canadian Special Publication of Fisheries and Aquatic Sciences 106:352–371.

Hogan, E. 1995. *Geography of South Dakota.* Pine Hill Press, Freeman, South Dakota.

Hoke, E. 1983. Unionid mollusks of the Missouri River on the Nebraska border. *American Malacological Bulletin* 1:71–74.

Hoke, E. 1995. A survey and analysis of the unionid mollusks of the Platte rivers of Nebraska and their minor tributaries. *Transactions of the Nebraska Academy of Sciences* 22:49–72.

Holton, G. D. 1990. *A field guide to Montana fishes.* Montana Fish, Wildlife, and Parks, Helena.

Hunt, C. B. 1974. Natural regions of the United States and Canada. W. H. Freeman and Company, San Francisco, California.

Jacobson R. B., and A. L. Pugh. 1997. Riparian vegetation controls on the spatial pattern of stream-channel instability, Little Piney Creek, Missouri. Water-Supply Paper 2494. U.S. Geological Survey, Rolla, Missouri.

Jean, C., and S. Crispin. 2001. Inventory of important biological resources in the upper Yellowstone River watershed. Report to the Environmental Protection Agency. Montana Natural Heritage Program, Helena.

Johnson, W. C., R. L. Burgess, and W. R. Keammerer. 1976. Forest overstory vegetation and environment on the Missouri River floodplain in North Dakota. *Ecological Monographs* 46:59–84.

Kallemeyn, L. W., and J. F. Novotny. 1977. *Fish and fish food organisms in various habitats of the Missouri River in South Dakota, Nebraska, and Iowa.* Publication FWS/OBS-77/25. U.S. Fish and Wildlife Service, Biological Services Program, Washington, D.C.

Keammerer, W. R., W. Carter Johnson, and R. L. Burgess. 1975. Floristic analysis of the Missouri River bottomland forests in North Dakota. *Canadian Field-Naturalist* 89:5–19.

Kepfield, S. S. 1994. El Dorado on the Platte: The development of agricultural irrigation and water law in Nebraska, 1860–1895. *Nebraska History* 75:232–243.

Klarich, D. A. 1976. Estimates of primary production and periphyton community structure in the Yellowstone River (Laurel to Huntly, Montana). Montana Water Quality Bureau, Montana Department of Health and Environmental Sciences, Helena.

Klarich, D. A., and J. Thomas. 1977. The effect of altered streamflow on water quality of the Yellowstone River basin, Montana. Technical Report 3. Montana Department of Natural Resources and Conservation, Helena.

Knowlton, M. F., and J. R. Jones. 2000. Seston, light, nutrients and chlorophyll in the lower Missouri River, 1994–1998. *Journal of Freshwater Ecology* 15:283–297.

Koch, R., R. Curry, and M. Weber. 1977. The effects of altered streamflow on the hydrology and geomorphology of the Yellowstone River basin, Montana. Technical Report 2. Montana Department of Natural Resources and Conservation, Helena.

Kuzelka, B., B. G. Volk, R. Hotchkiss, K. Hoagland, W. Hill, and H. Welch. 1993. *Environmental and natural resources of the Niobrara River Basin, Lincoln, Nebraska Water Center, University of Nebraska, Lincoln.*

Link, J. T. 1933. The origin of the place names of Nebraska. Bulletin 7, 2nd ser. Nebraska Geological Survey, University of Nebraska, Lincoln.

Lohman, K., J. R. Jones, and C. Baysinger-Daniel. 1991. Experimental evidence for nitrogen limitation in a northern Ozark stream. *Journal of the North American Benthological Society* 10:14–23.

Lynch, J. D. 1988. Introduction, establishment, and dispersal of western mosquitofish in Nebraska (Actinopterygii: Poeciliidae). *Prairie Naturalist* 20: 203–216.

Malcolm, R. L., and W. H. Durum. 1976. Organic carbon and nitrogen concentrations and annual organic carbon load of six selected rivers of the United States. Water-Supply Paper 1817-F. U.S. Geological Survey, Washington, D.C.

Master, L. L., S. R. Flack, and B. A. Stein. 1998. *Rivers of life: Critical watersheds for protecting freshwater biodiversity.* The Nature Conservancy, Arlington, Virginia.

Mayden, R. L. 1987. Faunal exchange between the Niobrara and White river systems of the North American Great Plains. *Prairie Naturalist* 19:173–176.

McBride, M. J. 1995. Benthic macroinvertebrate communities associated with forested and open riparian areas along the central Platte River. Master's thesis, University of Nebraska, Lincoln.

Mestl, G. E., and L. W. Hesse. 1993. Secondary production of aquatic insects in the channelized Missouri River, Nebraska. In L. W. Hesse, C. B. Stalnaker, N. G. Benson, and J. R. Zuboy (eds.). *Proceedings of the symposium on restoration planning for the rivers of the Mississippi River ecosystem*, pp. 341–349. Biological Report 19. National Biological Survey, Washington, D.C.

Milewski, C., C. Berry, and D. Dieterman. 2001. Use of the index of biological integrity in eastern South Dakota rivers. *Prairie Naturalist* 33:135–152.

Missouri Basin Inter-Agency Committee. 1969. *Comprehensive framework study: Missouri River basin.* Vol. 1: *Report.* U.S. Government Printing Office, Washington, D.C.

Missouri River Basin Commission. 1977. *The Missouri River basin water resources plan.* Missouri River Basin Commission, Omaha, Nebraska.

Montana Natural Heritage Program. 2004. Animal Species of Concern. Montana Natural Heritage Program Web site: http://nhp.nris.state.mt.us/animal/index.html

National Research Council. 2002. *The Missouri River ecosystem: Exploring the prospects for recovery.* Water Science and Technology Board, National Research Council, Washington, D.C.

National Research Council. 2004. *Endangered and threatened species of the Platte River.* National Academies Press, Washington, D. C.

Needham, R. G., and K. W. Gilge. 1987. *Northeast Montana fisheries study: Inventory and survey of water of the project area.* Montana Fish, Wildlife, and Parks, Helena.

Newell, R. L. 1977. Aquatic invertebrates of the Yellowstone River basin, Montana. Technical Report 5. Montana Department of Natural Resources and Conservation, Helena.

Newman, R., C. Berry, and W. Duffy. 1999. A biological assessment of four northern Black Hills streams. *Proceedings of the South Dakota Academy of Science* 78:185–197.

Niehus, C. A. 1999. Summary of water-resources data within the Little White River basin, South Dakota and Nebraska. Open-file Report 99-222. Information Services, U.S. Geological Survey, Denver, Colorado.

Patrick, R. 1998. *Rivers of the United States.* Vol. 4, part A: *The Mississippi River and tributaries north of St. Louis.* John Wiley and Sons, New York.

Peters, E. J., R. S. Holland, M. A. Callam, and D. L. Bunnell. 1989. *Platte River suitability criteria habitat utilization, preference, and suitability index criteria, for fish and aquatic invertebrates in the lower Platte River.* Nebraska Technical Series 17. Nebraska Game and Parks Commission, Lincoln.

Peters, F. J., and S. Schainost. In press. Historical changes in fish distribution and abundance in the Platte River in Nebraska. In J. N. Rinne, R. M. Hughes, and B. Calamusso (eds.). *Historical changes in large river fish assemblages of the Americas*, in press. American Fisheries Symposium 45. Bethesda, Maryland.

Peterson, D. A., S. D. Porter, and S. M. Kinsey. 2001. Algal-nutrient relations in the Yellowstone River, Montana. *Bulletin of the North American Benthological Society* 18(1):132.

Pflieger, W. L., and T. B. Grace. 1987. Changes in the fish fauna of the lower Missouri River. In W. L. Matthews and D. C. Heins (eds.). *Community and evolutionary ecology of North American stream fishes*, pp. 166–177. University of Oklahoma Press, Norman.

Pitchford, G., and H. Kerns. 2001. Grand River watershed inventory and assessment. Missouri Department of Conservation, Northwest Regional Fisheries Web site: http://www.conservation.state.mo.us/fish/watershed/grand/contents/140cotxt.htm

Poff, N. L. 1996. A hydrogeography of unregulated streams in the United States and an examination of scale-dependence in some hydrological descriptors. *Freshwater Biology* 36:71–91.

Rabini, C. F. 1993. Warmwater streams. In C. Kohler and W. Hubert (eds.). *Inland fisheries management in North America*, pp. 427–443. American Fisheries Society, Bethesda, Maryland.

Reetz, S. D. 1982. Phytoplankton studies in the Missouri River at Fort Calhoun Station and Cooper Nuclear Station. In L. W. Hesse, G. L. Hergenrader, H. S. Lewis, S. D. Reetz, and A. B. Schlesinger (eds.). *The middle Missouri River: A collection of papers on the biology with special reference to power station effects*, pp. 71–83. Missouri River Study Group, Norfolk, Nebraska.

Repsys, A. J., and G. D. Rogers. 1982. Zooplankton studies in the channelized Missouri River. In L. W. Hesse, G. L. Hergenrader, H. S. Lewis, S. D. Reetz, and A. B. Schlesinger (eds.). *The middle Missouri River: A collection of papers on the biology with special reference to power station effects*, pp. 125–145. Missouri River Study Group, Norfolk, Nebraska.

Revenga, C., S. Murray, J. Abramovitz, and A. Hammond. 1998. *Watershed of the world: Ecological value and vulnerability.* Worldwatch Institute, Washington, D.C.

Ricketts, T. H., E. Dinerstein, D. M. Olson, C. J. Loucks, W. Eichbaum, D. DellaSala, K. Kavanaugh, P. Hedao, P. T. Hurley, K. M. Carney, R. Abell, and S. Walters. 1999. *Terrestrial ecoregions of North America: A conservation assessment.* Island Press, Washington, D.C.

Robison, H. W. 1986. Zoographic implications of the Mississippi River basin. In C. H. Hocutt and E. O. Wiley (eds.). *The zoogeography of North American freshwa-*

ter fishes, pp. 267–285. John Wiley and Sons, New York.

Rothenberger, S. J. 1987. Woody plants of the lower Platte valley: An annotated checklist. *Transactions of the Nebraska Academy of Sciences* 15:53–58.

Ruelle, R., R. Koth, and C. Stone. 1993. Contaminants, fish and hydrology of the Missouri River and western tributaries, South Dakota. In L. W. Hesse, C. B. Stalnaker, N. G. Benson, and J. R. Zuboy (eds.). *Proceedings of the symposium on restoration planning for the rivers of the Mississippi River ecosystem*, pp. 449–480. Biological Report 19. National Biological Survey, Washington, D.C.

Sanders, R. M., D. G. Huggins, and F. B. Cross. 1993. The Kansas River and its biota. In L. W. Hesse, C. B. Stalnaker, N. G. Benson, and J. R. Zuboy (eds.). *Proceedings of the symposium on restoration planning for the rivers if the Mississippi River ecosystem*, pp. 295–326. Biological Report 19. National Biological Survey, Washington, D.C.

Sando, S. 1991. Estimation and characterization of the natural stream flow of the White River near the Nebraska–South Dakota state line. Water-Resources Investigations Report 91-4096. U.S. Geological Survey, Huron, South Dakota.

Silverman, A. J., and W. D. Tomlinsen. 1984. *Biohydrology of mountain fluvial systems: The Yellowstone (Part 1)*. Project G-853-02 Completion Report. U.S. Geological Survey, Reston, Virginia.

Slizeski, J. J., J. L. Anderson, and W. G. Dorough. 1982. Hydrologic setting, system operation, present and future stresses. In L. W. Hesse, G. L. Hergenrader, H. S. Lewis, S. D. Reetz, and A. B. Schlesinger (eds.). *The middle Missouri River: A collection of papers on the biology with special reference to power station effects*, pp. 15–38. Missouri River Study Group, Norfolk, Nebraska.

Smith, J. W. 1996. Wildlife use of the Missouri and Mississippi river basins: An ecological overview. In D. L. Galat and A. G. Frazier (eds.). *Overview of river–floodplain ecology in the upper Mississippi River basin*, pp. 91–111. Vol. 3 of J. A. Kelmelis (ed.). *Science for floodplain management into the 21st century*. U.S. Government Printing Office, Washington, D.C.

Stanford, J. A., and J. V. Ward. 1979. Stream regulation in North America. In J. V. Ward and J. A. Stanford (eds.). *The ecology of regulated streams*, pp. 215–236. Plenum Press, New York.

Stash, S. W. 2001. Distribution, relative abundance, and habitat associations of Milk River related to irrigation diversion dams. Master's thesis, Montana State University, Bozeman.

Swinehart, J. B., and R. F. Diffendal Jr. 1989. Geology of the pre-dune strata. In A. Bleed and C. Flowerday (eds.). *An atlas of the Sand Hills*, pp. 29–42. Conservation and Survey Division Resource Atlas no. 5. University of Nebraska, Lincoln.

Thompson, L. S. 1982. *Distribution of Montana amphibians, reptiles and mammals: Preliminary mapping by latilong*. Montana Audubon Council, Helena.

Thorson, J. E. 1994. *River of promise, river of peril: The politics of managing the Missouri River*. University of Kansas Press, Lawrence.

U.S. Army Corps of Engineers. 2001. *Missouri River master water control manual: Review and update*. Vol. 1: *Main report*. Revised Draft Environmental Impact Statement. U.S. Army Corps of Engineers, Northwest Division, Omaha, Nebraska. Web site: http://www.nwd.usace.army.mil

U.S. Fish and Wildlife Service. 1980. Missouri River stabilization and navigation project, Sioux City, Iowa, to mouth. Fish and Wildlife Coordination Act Report.

U.S. Geological Survey (USGS). 1998. *Source, occurrence, and extent of arsenic in the Grass Mountain area of the Rosebud Indian reservation, South Dakota*. U.S. Government Printing Office, Washington, D.C.

van der Leeden, F., F. L. Troise, and D. K. Todd. 1990. *The water encyclopedia*. 2nd ed. Lewis Publishers, CRC Press, Boca Raton, Florida.

Varley, J. D., and P. Schullery. 1983. *Freshwater wilderness: Yellowstone fishes and their world*. Yellowstone Library and Museum Association, Yellowstone National Park, Wyoming.

White, R. G., and R. G. Bramblett. 1993. The Yellowstone River: Its fish and fisheries. In L. W. Hesse, C. B. Stalnaker, N. G. Benson (eds.). *Proceedings of the symposium on restoration planning for the rivers of the Mississippi River ecosystem*, pp. 396–414. U.S. Department of the Interior, National Biological Survey Biological Report 19, Washington, D.C.

Womack and Associates. 2001. BNSF bank stabilization projects, lower Yellowstone River Montana: Impact assessment. Final report. Transystems Corporation, Denver, Colorado.

Yu, S. L. 1996. Factors affecting habitat use by fish species in the Platte River, Nebraska. Ph.D. diss., University of Nebraska, Lincoln.

Zorich and Associates. 1988. Assessment of the cumulative effects of major water diversion from the Platte River watershed. Final Report. City of Fremont, City of Lincoln, Omaha Metropolitan Utilities District and Lower Platte North Natural Resources District, Englewood, Colorado.

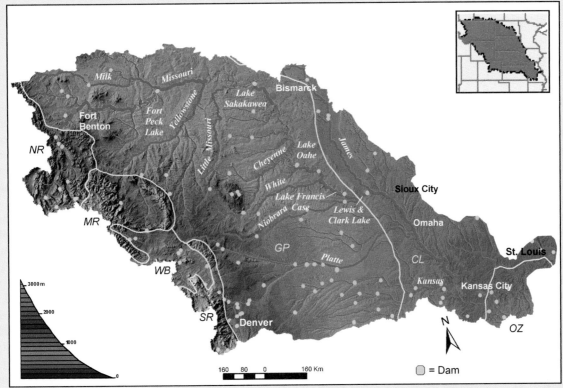

FIGURE 10.15 Map of the Missouri River basin. Physiographic provinces are separated by yellow lines.

MISSOURI RIVER

Relief: 4277 m

Basin area: 1,371,017 km²

Mean discharge: 1956 m³/s

River order: 9

Mean annual precipitation: 50.1 cm

Mean air temperature: 7.4°C

Mean water temperature: 9.3°C (Bismarck, North Dakota) 13.4°C
(Boonville, Missouri)

Physiographic provinces: 7 provinces (see text); most in Rocky
Mountains (NR, MR, SR), Great Plains (GP), Central Lowland (CL)

Biomes: Temperate Mountain Forest, Temperate Grasslands,
Temperate Deciduous Forest

Freshwater ecoregions: Upper Missouri, Middle Missouri, Central
Prairie

Terrestrial ecoregions: 13 ecoregions (see text)

Number of fish species: main stem ~136 (108 native), basin ~183 (138
native)

FIGURE 10.16 Mean monthly air temperature,
precipitation, and runoff for the Missouri River basin.

Number of endangered species: 1 insect, 1 fish, 3 birds, 2 mammals

Major fishes: shovelnose sturgeon, goldeye, gizzard shad, emerald shiner, red shiner, river shiner, flathead chub, sturgeon chub,
Hybognathus minnows, common carp, spotfin shiner, river carpsucker, shorthead redhorse, white sucker, channel catfish,
flathead catfish, white crappie, freshwater drum

Major other aquatic vertebrates: false map turtle, softshell turtles, great blue heron, wood duck, beaver

Major benthic vertebrates: mayflies (*Baetis, Heptagenia, Ephemerella, Tricorythodes, Isonychia, Hexagenia, Caenis, Stenonema*),
caddisflies (*Hydropsyche, Cheumatopsyche, Neureclipsis, Oecetis*), beetles (*Heterlimnius, Optioservus*), odonates
(*Gomphus*), crustaceans (*Isopoda*)

Nonnative species: Russian olive, reed canary grass, Johnson grass, *Daphnia lumholtzi*, Asiatic clam, 28 fishes (rainbow trout,
brown trout, cisco, common carp, goldfish, bighead carp, grass carp, rainbow smelt, white perch, striped bass), salmonid
whirling disease parasite

Major riparian plants: cottonwoods, willows, American elm, green ash, box elder, red mulberry, Virginia creeper, prairie
cordgrass, Canada wild rye, switchgrasses, giant ragweed, smartweeds

Special features: longest named river in North America (3768 km); primary water route for settlement of western United States

Fragmentation: 6 major main-stem dams; 581 total large dams

Water quality: pH = 8.0 to 8.3, alkalinity = 145 to 162 mg/L as $CaCO_3$, NO_3-N = 0.08 to 1.24 mg/L, PO_4-P = 0.007 to
0.09 mg/L

Land use: (within 5 km of river) 33% cropland, 26% grassland, 10% shrub, 6% forest, 17% developed

Population density: 8 people/km²

Major information sources: Galat and Frazier 1996, Hesse et al. 1989, Patrick 1998, Revenga et al. 1998

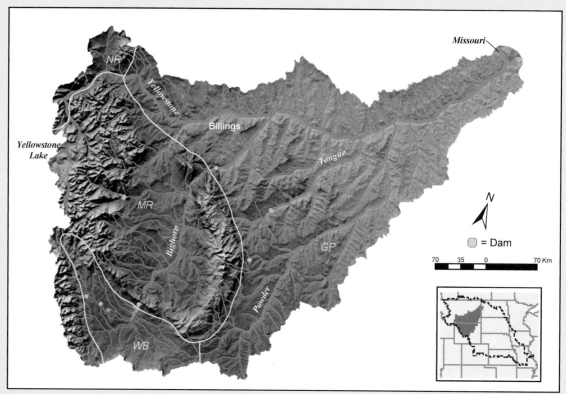

FIGURE 10.17 Map of the Yellowstone River basin. Physiographic provinces are separated by yellow lines.

YELLOWSTONE RIVER

Relief: 3050 m
Basin area: 182,336 km²
Mean discharge: 362 m³/s
River order: 8
Mean annual precipitation: 29.5 cm
Mean air temperature: 7.8°C
Mean water temperature: NA
Physiographic provinces: Northern Rocky Mountains (NR), Middle
   Rocky Mountains (MR), Wyoming Basin (WB), Great Plains (GP)
Biomes: Temperate Mountain Forest, Temperate Grasslands
Freshwater ecoregion: Upper Missouri
Terrestrial ecoregions: South Central Rockies Forests, Montana Valley
   and Foothills Grasslands, Northwestern Mixed Grasslands,
   Wyoming Basin Shrub Steppe
Number of fish species: 56 (36 native)
Number of endangered species: 1 fish

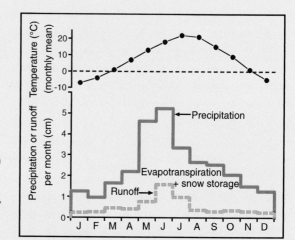

FIGURE 10.18 Mean monthly air temperature, precipitation, and runoff for the Yellowstone River basin.

Major fishes: Yellowstone cutthroat trout, rainbow trout, brown trout,
   mountain whitefish, white sucker, longnose dace, mottled sculpin, shovelnose sturgeon, paddlefish, freshwater drum, channel
   catfish, sauger, goldeye, river carpsucker, shorthead redhorse, chubs, shiners
Major other aquatic vertebrates: river otter, beaver, bald eagle, osprey, great blue heron, common merganser, spiny softshell
   turtle, snapping turtle, painted turtle
Major benthic invertebrates: mayflies (*Baetis, Ephemerella, Epeorus, Ephemera, Ametropus, Lachlania, Ephoron, Caenis,
   Centroptilum, Isonychia*), stoneflies (*Acroneuria, Pteronarcys, Pteronarcella, Arcynopteryx, Paraleuctra, Capnia, Alloperla*),
   caddisflies (*Rhyacophila, Amiocentrus, Glossosoma, Brachycentrus, Lepidostoma, Neotrichia, Oecetis, Leptocella*)
Nonnative species: black bullhead, black crappie, white crappie, brook trout, brown trout, rainbow trout, common carp, green
   sunfish, largemouth bass, northern pike, rainbow smelt, walleye, salmonid whirling disease parasite
Major riparian plants: black cottonwood, narrowleaf cottonwood, plains cottonwood, Geyer willow, wolf willow
Special features: longest free-flowing river in the conterminous United States (1091 km)
Fragmentation: no major dams
Water quality: pH = 7.5, alkalinity = 45 mg/L as $CaCO_3$, NO_3-N = 0.08 mg/L, PO_4-P = 0.3 mg/L
Land use: 47% range, 28% forest, 20% agriculture, 5% National Park, 1% urban
Population density: 10 people/km²
Major information sources: White and Bramblett 1993, Newell 1977, Varley and Schullery 1983, Silverman and Tomlinsen 1984

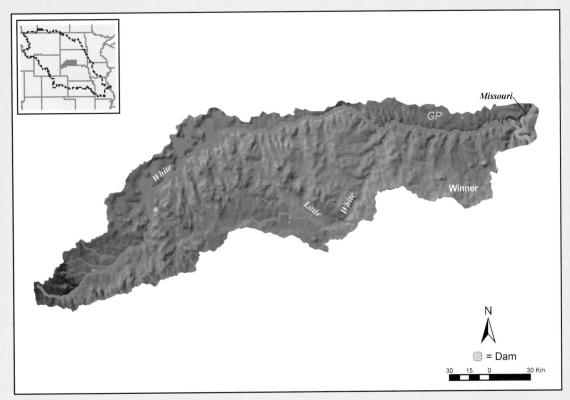

FIGURE 10.19 Map of the White River basin.

WHITE RIVER

Relief: 1112 m
Basin area: 26,418 km^2
Mean discharge: 16 m^3/s
River order: 6
Mean annual precipitation: 44.2 cm
Mean air temperature: 9.3°C
Mean water temperature: 12.7°C
Physiographic province: Great Plains (GP)
Biome: Temperate Grasslands
Freshwater ecoregion: Upper Missouri
Terrestrial ecoregions: Northwestern Mixed Grasslands, Western Short Grasslands
Number of fish species: 49 (41 native)
Number of endangered species: 1 mammal
Major fishes: flathead chub, plains minnow, fathead minnow, sturgeon chub, common carp, sand shiner, western silvery minnow, channel catfish

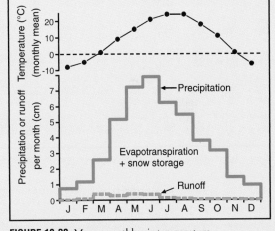

FIGURE 10.20 Mean monthly air temperature, precipitation, and runoff for the White River basin.

Major other aquatic vertebrates: northern leopard frog, bullfrog, Great Plains toad, Woodhouse's toad, chorus frog, tiger salamander, common snapping turtle
Major benthic invertebrates: NA
Nonnative species: Canada thistle, saltcedar, Russian olive, smooth brome, leafy spurge, brown trout, rainbow trout, brook trout, common carp, black crappie, white crappie, largemouth bass, bluegill
Major riparian plants: plains cottonwood, green ash, sandbar willow, buffalo grass
Special features: harsh aquatic conditions; badlands and xeric landscape; Little White River (major tributary) has more benign conditions
Fragmentation: no main-stem dams
Water quality: pH = 8.2, alkalinity = 408 mg/L as CaCO$_3$, NO$_3$-N = 1.6 mg/L, PO$_4$-P = 1.9 mg/L, total suspended solids = 4171 mg/L, fecal coliforms may exceed state standards (2000 colonies/100 mL); high suspended sediments from natural sources
Land use: 21% agriculture, 73% grassland/pasture, 2% forest/shrub, 4% sparse vegetation/badlands
Population density: 1.3 people/km^2
Major information sources: Ferrick et al. 1995, Niehus 1999, Fryda 2001

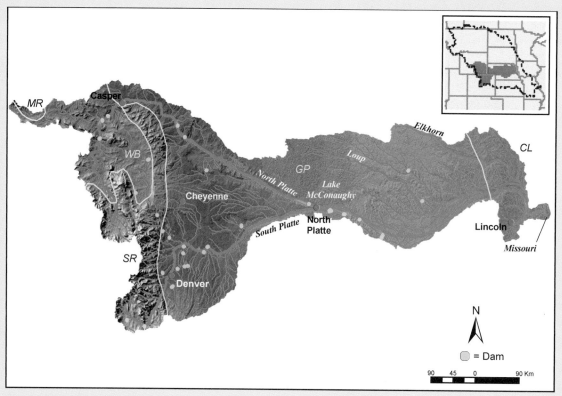

FIGURE 10.21 Map of the Platte River basin. Physiographic provinces are separated by yellow lines.

PLATTE RIVER

Relief: 3524 m

Basin area: 230,362 km²

Mean discharge: 203 m³/s

River order: 5

Mean annual precipitation: 50.2 cm

Mean air temperature: 9.1°C

Mean water temperature: 11.8°C

Physiographic provinces: Southern Rocky Mountains (SR), Wyoming Basin (WB), Great Plains (GP), Central Lowland (CL)

Biomes: Temperate Mountain Forest, Temperate Grasslands

Freshwater ecoregion: Middle Missouri

Terrestrial ecoregions: Colorado Rockies Forests, Wyoming Basin Shrub Steppe, Western Short Grasslands, Nebraska Sand Hills Mixed Grasslands, Central and Southern Mixed Grasslands, Central Tall Grasslands

Number of fish species: 100 (76 native)

Number of endangered species: 4 plants, 3 insects, 2 fishes, 1 amphibian, 6 birds, 1 mammal

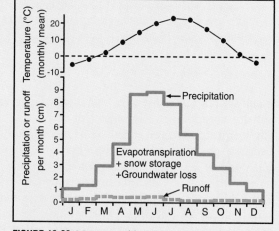

FIGURE 10.22 Mean monthly air temperature, precipitation, and runoff for the Platte River basin.

Major fishes: shovelnose sturgeon, longnose gar, flathead chub, speckled chub, sand shiner, red shiner, river shiner, western silvery minnow, plains minnow, river carpsucker, quillback, plains killifish, channel catfish, freshwater drum

Major other aquatic vertebrates: snapping turtle, spiny softshell turtle, painted turtle, migratory waterfowl, beaver, muskrat

Major benthic invertebrates: snails (*Physa*), crustaceans (*Hyalella azteca*, *Orconectes*), mayflies (*Heptagenia*, *Hexagenia*, *Caenis*, *Baetis*, *Isonychia*), odonates (*Gomphus*, *Progomphus*), stoneflies (*Acroneuria*, *Isoperla*, *Pteronarcys*), caddisflies (*Hydropsyche*, *Cheumatopsyche*), true flies (*Chaoborus*, *Simulium*, *Chernovskiia*, *Chironomus*, *Robackia*, *Saetheria*)

Nonnative species: 24 fishes (brown trout, rainbow trout, common carp, grass carp, bighead carp, western mosquitofish, striped bass, yellow perch, walleye)

Major riparian plants: eastern cottonwood, eastern red cedar, rough leaf dogwood, silver maple, green ash, sandbar willow

Special features: a wide, shallow braided river with shifting sandbars

Fragmentation: 7 main-stem dams, 20 diversions

Water quality: pH = 8.0, alkalinity = 153.5 mg/L as CaCO₃, NO₃-N = 1.35 mg/L, PO₄-P = 0.73 mg/L

Land use: >90% agriculture, 2% forest, 3% urban

Population density: 9.1 people/km²

Major information sources: National Research Council 2004, Peters and Schainost in press

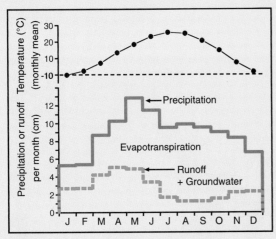

FIGURE 10.24 Mean monthly air temperature, precipitation, and runoff for the Gasconade River basin.

GASCONADE RIVER

Relief: 380 m
Basin area: 9258 km^2
Mean discharge: 87 m^3/s
River order: 6
Mean annual precipitation: 108 cm
Mean air temperature: 13.4°C
Mean water temperature: 15.2°C
Physiographic province: Ozark Plateaus (OZ)
Biome: Temperate Deciduous Forest
Freshwater ecoregion: Central Prairie
Terrestrial ecoregions: Central Forest/Grassland
   Transition Zone, Central U.S. Hardwood Forests
Number of fish species: 105 (98 native)
Number of endangered species: 1 mussel, 1 bird,
   2 mammals
Major fishes: bleeding shiner, Ozark minnow, largescale
   stoneroller, wedgespot shiner, striped shiner, bigeye
   shiner, southern redbelly dace, longear sunfish,
   smallmouth bass, rock bass, northern orangethroat
Major other aquatic vertebrates: bullfrog, green frog,
   common map turtle, red-eared slider, midland smooth softshell turtle, common snapping turtle, western painted turtle,
   northern water snake, wood duck, belted kingfisher, beaver, mink, muskrat, river otter
Major benthic invertebrates: mussels (mucket, spectaclecase, purple wartyback, Wabash pigtoe, plain pocketbook), crustaceans
   (spothanded crayfish, golden crayfish), mayflies (*Acentrella, Baetis, Isonychia, Stenonema*), odonates (*Argia, Erpetogomphus,
   Gomphus*), caddisflies (*Chimarra, Cheumatopsyche, Hydropsyche*), stoneflies (*Taeniopteryx, Strophopterx, Neoperla*),
   beetles (*Psephenus, Ectopria*)
Nonnative species: Asiatic clam, rainbow trout, common carp, goldfish
Major riparian plants: silver maple, American sycamore, green ash, river birch, Ward's willow, swamp dogwood, buttonbush,
   water willow
Special features: located in karst topography, sections lose flow to groundwater, 76 springs; largest undammed tributary to
   Missouri River draining Ozark Plateau; bluestripe and Missouri saddled darters endemic to Gasconade and adjacent
   drainages in Missouri
Fragmentation: no main-stem dams
Water quality: pH = 8.0, alkalinity = 151 mg/L as CaCO$_3$, NO$_3$-N = 0.365 mg/L, PO$_4$-P = 0.015 mg/L; 3266 m^3 oil spill in
   December 1988
Land use: 55% forest, 35% grassland/pasture, 6% cropland, 4% urban
Population density: 13.5 people/km^2
Major information source: Blanc 2001

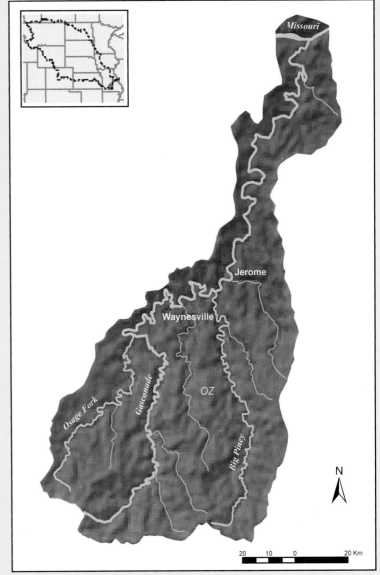

FIGURE 10.23 Map of the Gasconade River basin.

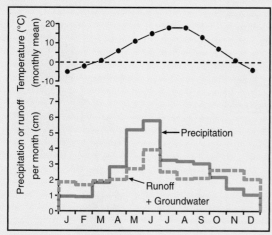

FIGURE 10.26 Mean monthly air temperature, precipitation, and runoff for the Madison River basin.

MADISON RIVER

Relief: 1959 m
Basin area: 6537 km²
Mean discharge: 50.7 m³/s
River order: 6
Mean annual precipitation: 33.7 cm
Mean air temperature: 6.4°C
Mean water temperature: NA
Physiographic province: Northern Rocky Mountains (NR), Middle Rocky Mountains (MR)
Biome: Temperate Mountain Forest
Freshwater ecoregion: Upper Missouri
Terrestrial ecoregions: South Central Rockies Forests, Montana Valley and Foothills Grasslands
Number of fish species: 17 (10 native)
Number of endangered species: 0
Major fishes: westslope cutthroat trout, rainbow trout, brown trout, mountain whitefish, mountain sucker, longnose sucker, white sucker, longnose dace, mottled sculpin
Major other aquatic vertebrates: northern leopard frog, spotted frog, osprey, bald eagle, American dipper, great blue heron, common merganser, river otter, beaver, water shrew

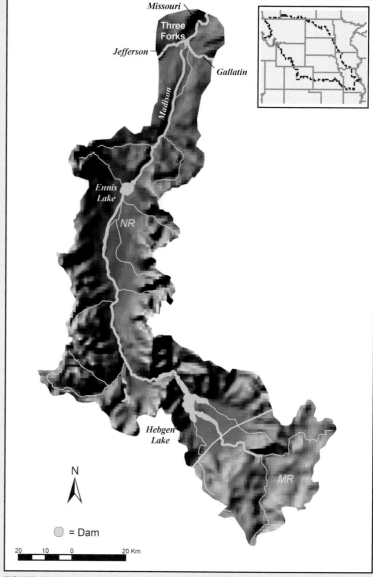

FIGURE 10.25 Map of the Madison River basin. Physiographic provinces are separated by a yellow line.

Major benthic invertebrates: mayflies (*Baetis, Epeorus, Tricorythodes, Rhithrogena, Paraleptophlebia, Ephemera*), caddisflies (*Cheumatopsyche, Hydropsyche, Brachycentrus, Micrasema, Glossosoma*), stoneflies (*Pteronarcys, Acroneuria, Claassenia, Isoperla*)
Nonnative species: rainbow trout, brown trout, brook trout, Utah chub, New Zealand mud snail, salmonid whirling disease parasite
Major riparian plants: black cottonwood, narrowleaf cottonwood, Geyer willow, wolf willow, sandbar willow
Special features: geothermal inputs, notably near headwaters in geyser basins of Yellowstone National Park
Fragmentation: Madson (Ennis), Hegben, and Quake Lake dams on main stem
Water quality: relatively free of major pollutants, naturally high concentrations of arsenic from geothermal water; pH = 8.5, alkalinity = 89 mg/L as CaCO₃, NO₃-N = 0.100 mg/L, PO₄-P = 0.01 mg/L
Land use: 26% forestry, 41% agriculture, 2% urban, 8% wilderness area, 23% National Park
Population density: 1.2 people/km²
Major information sources: Brooks 1979, Varley and Schullery 1983, Holton 1990

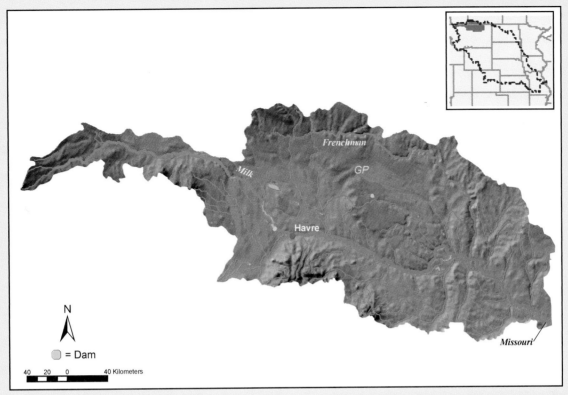

FIGURE 10.27 Map of the Milk River basin.

MILK RIVER

Relief: 1930 m
Basin area: 57,839 km²
Mean discharge: 18.9 m³/s
River order: 7
Mean annual precipitation: 28.8 cm
Mean air temperature: 6.1°C
Mean water temperature: NA
Physiographic province: Great Plains (GP)
Biomes: Temperate Mountain Forest, Temperate Grasslands
Freshwater ecoregion: Upper Missouri
Terrestrial ecoregions: Northern Mixed Grasslands, Montana Valley and Foothills Grasslands, Northwestern Mixed Grasslands
Number of fish species: 45 (30 native)
Number of endangered species: 0

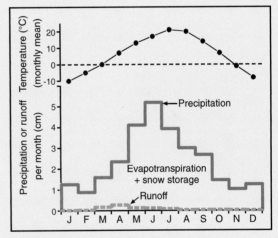

FIGURE 10.28 Mean monthly air temperature, precipitation, and runoff for the Milk River basin.

Major fishes: shovelnose sturgeon, paddlefish, freshwater drum, channel catfish, sauger, Iowa darter, river carpsucker, blue sucker, bigmouth buffalo, smallmouth buffalo, shorthead redhorse, brook stickleback, chubs, shiners
Major other aquatic vertebrates: osprey, common merganser, snapping turtle, spiny softshell turtle, painted turtle, muskrat
Major benthic invertebrates: mayflies (*Analetris, Camelobaetidius, Ametropus, Lachlania, Raptoheptagenia, Macdunnoa, Cercobrachys, Hexagenia*), odonates (*Gomphus, Ophiogomphus*), stoneflies (*Oemopteryx, Acroneuria*)
Nonnative species: common carp, spottail shiner, bluegill, smallmouth bass, white crappie, black crappie, yellow perch, walleye, black bullhead, rainbow trout, brown trout, lake whitefish, northern pike, virile crayfish
Major riparian plants: plains cottonwood, red-osier dogwood, peach-leaved willow, Rocky Mountain juniper, box elder
Special features: meandering, highly braided channel with very unstable sand bottom in upper reaches; reach above Fresno Dam one of few remnant Great Plains river ecosystems
Fragmentation: Fresno dam on main stem; one municipal water weir; four diversion dams
Water quality: pH = 7.6, alkalinity = 150 mg/L as CaCO₃, NO₃-N = 0.19 mg/L, PO₄-P = 0.02 mg/L
Land use: 55% range, 43% agriculture, 2% forestry, 1% urban
Population density: 0.8 people/km²
Major information sources: Stash 2001, Needham and Gilge 1987, D. Guftason, personal communication

475

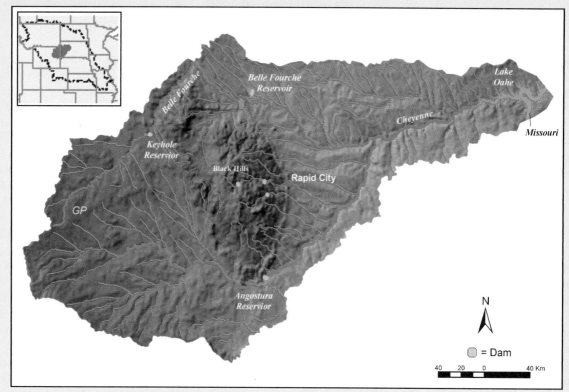

FIGURE 10.29 Map of the Cheyenne River basin.

CHEYENNE RIVER

Relief: 1410 m
Basin area: 63,455 km^2
Mean discharge: 25 m^3/s
River order: 4
Mean annual precipitation: 51.3 cm
Mean air temperature: 7.6°C
Mean water temperature: 12.7°C
Physiographic province: Great Plains (GP)
Biomes: Temperate Mountain Forest, Temperate Grasslands
Freshwater ecoregion: Upper Missouri
Terrestrial ecoregions: South Central Rockies Forests, Northwestern Mixed Grasslands
Number of fish species: 56 (37 native)
Number of endangered species: 2 birds
Major fishes: brook trout, brown trout, rainbow trout, longnose dace, white sucker, green sunfish, sand shiner, fathead minnow, common carp, flathead chub, channel catfish, plains minnow, shorthead redhorse, red shiner

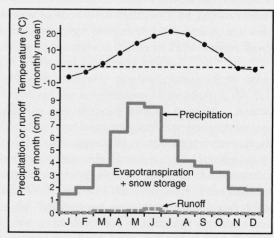

FIGURE 10.30 Mean monthly air temperature, precipitation, and runoff for the Cheyenne River basin.

Major other aquatic vertebrates: northern leopard frog, Great Plains toad, Woodhouse's toad, chorus frog, plains spadefoot toad, tiger salamander, common snapping turtle, spiny softshell turtle
Major benthic invertebrates: caddisflies (*Glossosoma, Hydropsyche, Hesperophylax*), mayflies (*Baetis*), stoneflies (*Acroneuria, Isoperla*), odonates (*Ophiogomphus, Argia*), beetles (*Narpus*), true flies (*Tipula, Atherix*), crustaceans (*Gammarus, Hyallela*), bivalves (fingernail clams, white heelsplitter)
Nonnative species: tamarisk, Russian olive, Canada thistle, leafy spurge, smooth brome, common carp, brown trout, rainbow trout, brook trout, cutthroat trout, largemouth bass, bluegill, pumpkinseed, smallmouth bass, rock bass, golden shiner
Major riparian plants: white spruce, aspen, Bebb willow, plains cottonwood, green ash, salt grass, buffalo grass, sandbar willow
Special features: Black Hills coldwater streams and hot springs; harsh conditions; section proposed for National Wild and Scenic River
Fragmentation: three major dams on main stem; Lake Oahe isolates Cheyenne from other Missouri River tributaries
Water quality: pH = 8.2, alkalinity = 236 mg/L as CaCO$_3$, NO$_3$-N = 2.0 mg/L, PO$_4$-P = 0.05 mg/L, conductivity = 2145 µS/cm, dissolved solids = 1636 mg/L, fecal coliform = 266 col/100 mL
Land use: 8% agriculture, 69% grassland/pasture, 18% forest/shrub, 5% sparse vegetation/badlands, 1% urban
Population density: 3.3 people/km^2
Major information sources: Hampton and Berry 1997, Doorenbos 1998, Newman et al. 1999

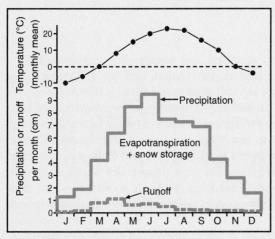

FIGURE 10.32 Mean monthly air temperature, precipitation, and runoff for the Big Sioux River basin.

BIG SIOUX RIVER

Relief: 305 m
Basin area: 23,325 km²
Mean discharge: 35.4 m³/s
River order: 5
Mean annual precipitation: 62 cm
Mean air temperature: 7.8°C
Mean water temperature: 12.0°C
Physiographic province: Central Lowland (CL)
Biome: Temperate Grasslands
Freshwater ecoregion: Middle Missouri
Terrestrial ecoregion: Central Tall Grasslands
Number of fish species: 70 (66 native)
Number of endangered species: 1 fish, 1 bird
Major fishes: northern pike, common carp, common shiner, fathead minnow, red shiner, sand shiner, river carpsucker, shorthead redhorse, white sucker, black bullhead, channel catfish, tadpole madtom, brook stickleback, green sunfish, walleye, Johnny darter, freshwater drum
Major other aquatic vertebrates: American toad, Canadian toad, chorus frog, Great Plains toad, northern leopard frog, tiger salamander, spiny softshell turtle, common snapping turtle, western painted turtle, beaver, mink
Major benthic invertebrates: caddisflies (*Cheumatopsyche, Hydroptila*), odonates (*Calopteryx, Lestes, Libellula*), mayflies (*Baetis intercalaris, Hexagenia, Caenis, Stenacron*), beetles (*Coptotomus, Laccophilus, Hydrobius*), chironomid midges (*Chironomus*), crustaceans (*Hyalella azteca*), bivalves (*Sphaerium, Pisidium, Anodonta grandis*)
Nonnative species: bighead carp, common carp, grass carp, smallmouth bass, smooth brome grass, reed canary grass, Russian olive, buckthorn, Canada thistle, bindweed
Major riparian plants: green ash, American elm, eastern cottonwood, sandbar willow, smooth brome grass, reed canary grass
Special features: relatively intact floodplain wetlands and river channel; geological features related to quartzite formations; recreational use for hunting and fishing
Fragmentation: Sioux Falls is natural barrier to fish movement; three low-head (2 to 5 m high) dams impede fish movements
Water quality: pH = 8.1, alkalinity = 260 mg/L as $CaCO_3$, NO_3-N = 4.5 mg/L, PO_4-P = 0.4 mg/L, conductivity = 975 μS/cm, turbidity = 54 NTU, fecal coliforms may be >2000 colonies/100 mL
Land use: 62% agriculture, 27% grassland/pasture, 3% forest shrub, 7% wetlands/river, 1% urban
Population density: 17.4 people/km²
Major information sources: Dieterman and Berry 1998, Doorenbos et al. 1996, Milewski et al. 2001

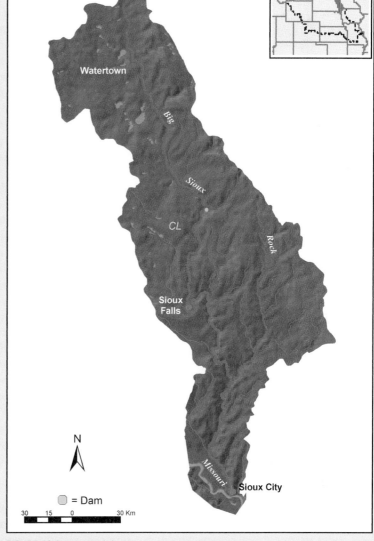

FIGURE 10.31 Map of the Big Sioux River basin.

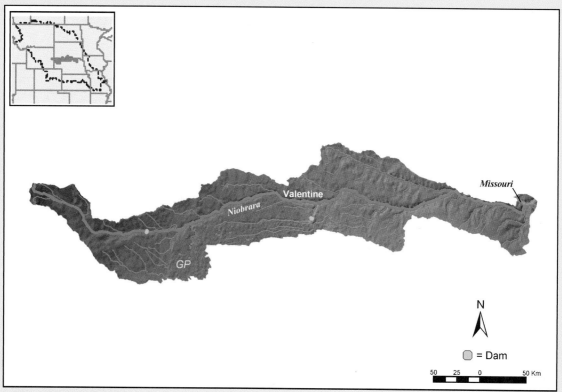

FIGURE 10.33 Map of the Niobrara River basin.

NIOBRARA RIVER

Relief: 1182 m
Basin area: 32,600 km²
Mean discharge: 49 m³/s
River order: 4
Mean annual precipitation: 47 cm
Mean air temperature: 8.3°C
Mean water temperature: 12.3°C
Physiographic province: Great Plains (GP)
Biome: Temperate Grasslands
Freshwater ecoregion: Middle Missouri
Terrestrial ecoregions: Western Short Grasslands, Nebraska Sand Hills
  Grasslands, Northwestern Mixed Grasslands
Number of fish species: 67 (43 native)
Number of endangered species: 1 plant, 1 insect, 1 fish, 2 birds,
  1 mammal
Major fishes: sand shiner, red shiner, river shiner, emerald shiner,
  bigmouth shiner, flathead chub, river carpsucker, channel catfish

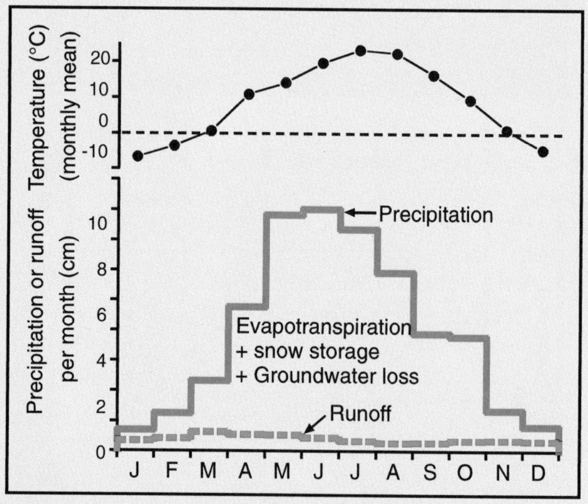

FIGURE 10.34 Mean monthly air temperature,
precipitation, and runoff for the Niobrara River basin.

Major other aquatic vertebrates: spiny softshell turtle, painted turtle, beaver, muskrat, river otter
Major benthic invertebrates: snails (*Physa*), crustaceans (*Hyalella azteca*), mayflies (*Isonychia, Caenis, Baetis, Pseudocleon,
  Callibaetis*), caddisflies (*Hydropsyche, Brachycentrus*), true bugs (*Belostoma*), chironomid midges (*Rheocricotopus*)
Nonnative species: 24 fishes (brown trout, rainbow trout, brook trout, alewife, common carp)
Major riparian plants: eastern red cedar, ponderosa pine, eastern cottonwood, box elder, paper birch, American elm
Special features: swift-flowing Great Plains prairie river with eastern, western, and northern forest species in riparian zone;
  groundwater a major source of discharge
Fragmentation: three main-stem dams
Water quality: good water quality, but high fecal coliforms; pH = 7.0, alkalinity = 23 mg/L as CaCO₃, NO₃-N = 0.40 mg/L,
  PO₄-P = 0.162 mg/L
Land use: >95% agriculture, 2% forest
Population density: 1.1 people/km²
Major information source: Kuzelka et al. 1993

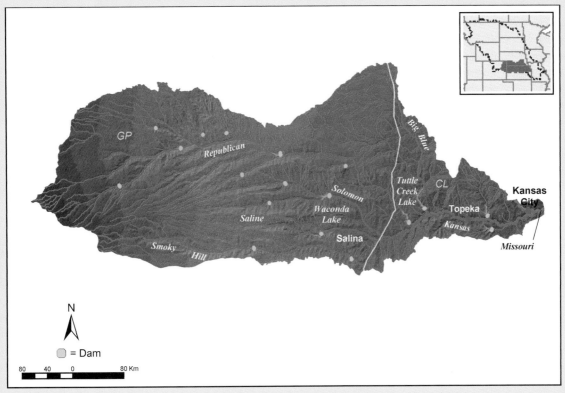

FIGURE 10.35 Map of the Kansas River basin. Physiographic provinces are separated by a yellow line.

KANSAS RIVER

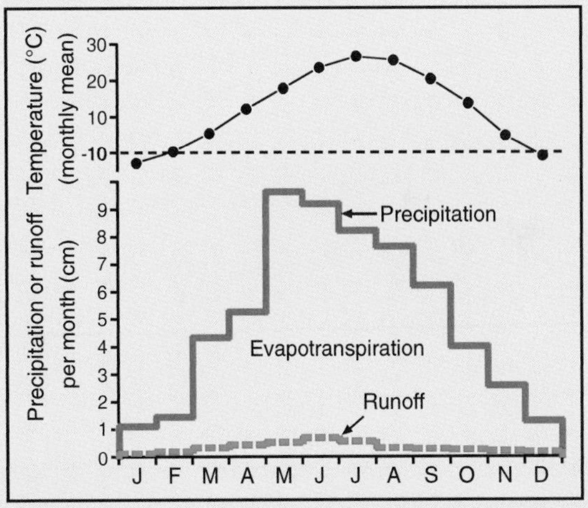

Relief: 975 m
Basin area: 159,171 km^2
Mean discharge: 214 m^3/s
River order: 7
Mean annual precipitation: 61 cm
Mean air temperature: 12.2°C
Mean water temperature: 11.6°C
Physiographic provinces: Great Plains (GP), Central Lowland (CL)
Biome: Temperate Grasslands
Freshwater ecoregion: Middle Missouri
Terrestrial ecoregions: Western Short Grasslands, Southern Mixed Grasslands, Central Tall Grasslands
Number of fish species: 99 (75 native)
Number of endangered species: 1 fish
Major fishes: shovelnose sturgeon, longnose gar, gizzard shad, creek chub, suckermouth minnow, plains minnow, sand shiner, red shiner, river carpsucker, shorthead redhorse, blue sucker, white sucker, flathead catfish, channel catfish, largemouth bass, sauger, freshwater drum

FIGURE 10.36 Mean monthly air temperature, precipitation, and runoff for the Kansas River basin.

Major other aquatic vertebrates: smooth softshell turtle, migratory waterfowl, beaver, muskrat
Major benthic invertebrates: mollusks (*Potamilus ohioensis*, *Lampsilis siliquoidea*), mayflies (*Isonychia*, *Heptagenia*, *Tricorythodes*), stoneflies (*Neoperla*, *Isoperla*), odonates (*Gomphus*, *Argia*), caddisflies (*Hydropsyche*, *Cheumatopsyche*)
Nonnative species: 24 fishes (common carp, grass carp, bighead carp, western mosquitofish, striped bass, yellow perch, walleye)
Major riparian plants: eastern red cedar, eastern cottonwood, box elder, American elm, sycamore, silver maple, willows
Special features: a large river that starts in the Great Plains and ends in the Central Lowland
Fragmentation: 18 large reservoirs, >13,000 small impoundments, dewatering by irrigation withdrawals
Water quality: reduction of turbidity from damming; pH = 8.4, alkalinity = 190 mg/L as CaCO$_3$, NO$_3$-N = 0.60 mg/L, PO$_4$-P = 0.16 mg/L
Land use: >90% agriculture, 2% forest, 3% urban
Population density: 12.7 people/km^2
Major information sources: Cross 1967, Sanders et al. 1993, Patrick 1998

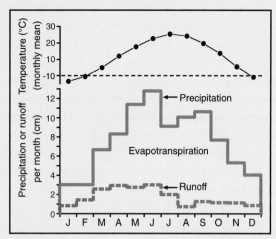

FIGURE 10.38 Mean monthly air temperature, precipitation, and runoff for the Grand River basin.

GRAND RIVER

Relief: 238 m
Basin area: 20,390 km²
Mean discharge: 117 m³/s
River order: 7
Mean annual precipitation: 92 cm
Mean air temperature: 11.8°C
Mean water temperature: 13.0°C
Physiographic province: Central Lowland (CL)
Biomes: Temperate Grasslands, Temperate Deciduous Forest
Freshwater ecoregions: Middle Missouri, Central Prairie
Terrestrial ecoregion: Central Forest/Grassland Transition Zone
Number of fish species: 61 (55 native)
Number of endangered species: 2 fishes, 1 mammal
Major fishes: shortnose gar, bigmouth shiner, red shiner, creek chub, sand shiner, central stoneroller, fathead minnow, bluntnose minnow, common carp, river carpsucker, channel catfish, flathead catfish, bluegill, green sunfish, Johnny darter

FIGURE 10.37 Map of the Grand River basin.

Major other aquatic vertebrates: bullfrog, green frog, false map turtle, red-eared slider, smooth softshell turtle, common snapping turtle, western painted turtle, northern water snake, wood duck, belted kingfisher, beaver, mink, muskrat, river otter
Major benthic invertebrates: mayflies (*Baetis, Isonychia, Stenonema, Caenis*), stoneflies (*Neoperla*), caddisflies (*Cheumatopsyche, Hydropsyche, Potamyia*), beetles (*Stenelmis*), true flies (*Hemerodromia*), mollusks (fragile papershell, pink papershell, giant floater, yellow sandshell), crustaceans (northern crayfish, papershell crayfish)
Nonnative species: common carp, goldfish, bighead carp, grass carp, rainbow smelt, white perch, striped bass, Asiatic clam, Johnson grass
Major riparian plants: cottonwood, silver maple, green ash, American sycamore, hackberry, aromatic sumac, gray dogwood, giant ragweed, stinging nettle, poison ivy, grapes
Special features: preglacial channel of ancestral Missouri River; largest prairie river in Missouri relatively unaffected by dams or channelization (mouth to Rkm 56)
Fragmentation: no main-stem dams; ~30 impoundments >20 ha on tributaries
Water quality: pH = 7.8, alkalinity = 152 mg/L as $CaCO_3$, NO_3-N = 0.74 mg/L, PO_4-P = 0.04 mg/L; standards for Fe, Mg, and fecal coliforms frequently exceeded
Land use: 57% cropland, 32% grassland/pasture, 5% forest, 6% other
Population density: 6.3 people/km²
Major information source: Pitchford and Kerns 2001

11

COLORADO RIVER BASIN

DEAN W. BLINN   N. LEROY POFF

INTRODUCTION

COLORADO RIVER

GREEN RIVER

YAMPA RIVER

LITTLE COLORADO RIVER

GILA RIVER

ADDITIONAL RIVERS

ACKNOWLEDGMENTS

LITERATURE CITED

INTRODUCTION

The Colorado River Basin lies within the Intermontane Plateaus of the American Southwest. The basin encompasses a large geographic area from 42°N to 32°N latitude ranging from the high mountains of the Rockies in Wyoming and Colorado to the Colorado River delta in Mexico (Fig. 11.2). The Colorado basin drains seven states, or nearly 8% of the United States, including parts of Colorado, Wyoming, Utah, New Mexico, Nevada, and California, and 95% of Arizona's land mass. The entire basin encompasses some 642,000 km², with the Upper Basin draining about 45% of the total area (Pope et al. 1998). The majority (75%) of the river's flow is supplied by mountain headwater streams, but most of the catchment lies in a semiarid desert. Therefore, the annual unit area discharge (29,800 m³/km²) or runoff (2.98 cm/yr) makes the Colorado basin one of the driest in the world (Stanford and Ward 1986). In spite of this water scarcity, there are heavy demands from urbanization, agriculture, and hydropower. Nearly 64% of the runoff is used for irrigation and another 32% is lost by evaporation from reservoirs (Dynesius and Nilsson 1994).

The drainage basin has been divided into upper and lower basins at the confluence of the Paria River, near Lees Ferry, Arizona (36°52′N latitude), in order to administer the allocation of Colorado River water among the western states and Mexico. Much of the Colorado that runs through the upper basin was once called the Grand River. In 1921, President Warren Harding approved the name change to the Colorado River for the entire river corridor.

A series of creeks and rivers and numerous arroyos drain the region, with water mostly flowing in a northwesterly or south-southwesterly direction. The major rivers originate in the Wind River Range of the Wyoming Rockies, the Colorado Rockies, and the Mexican Highlands of west-central New Mexico and east-central Arizona. Over 80% of the entire basin includes semiarid shrub lands and desert scrub and two major North American deserts (Sonoran and Mojave), each with sparse vegetation (Brown 1994). Consequently, rivers in the lower basin generally carry limited upland terrestrial carbon but large loads of suspended sediments, especially after storm events.

Twenty-two major rivers converge with the Colorado after it begins its descent from Rocky Mountain National Park and winds through the

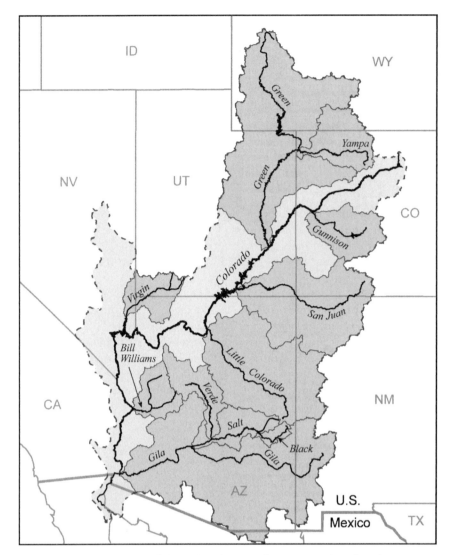

FIGURE 11.2   Rivers of the Colorado River basin covered in this chapter.

plateaus of Colorado, Utah, and Arizona, onto the deserts of southwestern Arizona, and finally into the Gulf of California, where inflows from the Río Hardy and Río Sonoita in Mexico complete the drainage. As the Colorado flows into drier and lower latitudes, significant discharges from tributaries may only occur after major summer storms. These storms play a major role in the community structure and ecological processes of these desert streams (Fisher and Grimm 1988). We will discuss the Colorado River main stem and four important tributaries: the Green, Yampa, Little Colorado, and Gila rivers. Summary paragraphs and one-page summaries are also provided for seven additional tributaries in the drainage: the Gunnison, San Juan, Virgin, Bill Williams, Verde, Black, and Salt rivers.

Physiography and Climate

Rivers in the upper basin (Upper Colorado, Green, Yampa, Gunnison, and San Juan) flow through four physiographic provinces (Middle Rocky Mountains [MR], Wyoming Basin [WB], Southern Rocky Mountains [SR], Colorado Plateaus [CO]) and four sections within these provinces (Uinta Basin, High Plateaus of Utah, Canyon Lands, and Navajo; Hunt 1974). Rivers in the lower basin (Lower Colorado, Little Colorado, Virgin, Black, Bill Williams, Verde, Salt, and Gila) pass through two provinces (Colorado Plateaus and Basin and Range [BR]) that include six sections (Mexican Highlands, Datil, High Plateaus of Utah, Navajo, Grand Canyon, and Sonoran Desert). The northernmost section of the Baja California (BC) province includes the mouth of the Colorado River (Hunt 1974).

The geology of the basin is varied, but much of it consists of uplifted, highly erodible sedimentary deposits through which rivers have carved the spectacular canyons characteristic of the region. The headwaters of the basin generally lie in crystalline, granite bedrock, whereas the larger rivers flow through expansive and highly weathered sedimentary deposits.

Soils are generally shallow and derived from shale or sandstone. These soils are relatively poor in organic matter due to low vegetative production but contain large quantities of soluble minerals. Following evaporation or the consumptive use of water by plants, these minerals accumulate on or near the surface. Invasion of nonnative plants and increased irrigation have augmented the surface deposition of these salts, which are delivered to the main channels via overland flow by numerous arroyos during storm events (Ghassemi et al. 1995).

The Intermountain Plateaus receive little precipitation compared to the uplifted mountain ranges on either side. Annual air temperatures range from subfreezing to 38°C in the upper basin and from 15°C to >45°C in the lower basin. Annual precipitation is ≤26 cm/yr in the valleys of the upper basin and from 25 to <5 cm/yr in the lower basin. In higher elevations of the upper basin an average of 2 m of snow accumulates during winter and spring, providing a surge in meltwater in late spring for much of the river system. Convective storms carry moisture from the Gulf of Mexico and the Gulf of California and cyclonic storms originating over the Pacific Ocean cause short-term storm events during the summer in the lower basin.

Basin Landscape and Land Use

The Colorado River basin encompasses a variety of landscapes, from subalpine in the Rockies to deserts in southwestern Arizona and Mexico, which are divided into nine terrestrial ecoregions (Ricketts et al. 1999). The terrestrial ecoregions in the highest elevations (>3000 m asl) include Colorado Rockies Forests, South Central Rockies Forests, Wasatch and Uinta Montane Forests, and Arizona Mountain Forests. These are primarily coniferous forests with Douglas fir, subalpine fir, and Engelmann spruce interspersed with ponderosa pine in the lower mountainous areas. The shrublands, located at intermediate elevations (1500 to 3000 m asl), include the open and arid Wyoming Basin Shrub Steppe, with sage-

brush, wheatgrass, and fescue, and the Colorado Plateau Shrublands, with sagebrush, mountain mahogany, pinyon pine, and several junipers as the dominant vegetation. The Colorado Plateau Shrublands is the largest (326,390 km^2) terrestrial ecoregion in the river basin, with approximately 90% of the ecoregion drained by the Colorado River and its tributaries (Ricketts et al. 1999). The lower elevations (>1500 m asl) of the Colorado River Basin include the Mohave, Sonoran, and Chihuahuan Desert ecoregions. The vegetation in the Mohave includes all-scale, brittlebush, creosote bush, desert holly, white burrobush, numerous species of cacti, and the endemic Joshua tree, whereas the Sonoran Desert contains creosote, white bursage, palo verde, ironwood, and an assortment of tall cacti, including the giant saguaro. The Chihuahuan is characterized by shrubs such as creosote, tarbush, mesquite, and acacia.

The riparian communities at higher elevations (>2000 m asl) in both the upper and lower basins consist of alder, dogwood, birch, elderberry, Rocky Mountain maple, and several willows. At elevations between 1800 and 2000 m asl cottonwoods and willows dominate, along with nonnative salt-cedar in many locations. In lower latitudes, Arizona sycamore and Arizona walnut can be common. As elevation decreases the width of the riparian band shrinks to only a few meters with decreasing precipitation (Brown 1994).

Common desert riparian plants include acacia, black greasewood, cat-claw, mesquite, saltbush, and seep willow. Cattails, sedges, and common reed occur in some of the backwater marshes of the lower Colorado River. These narrow bands of vegetation provide critical organic matter to desert streams otherwise surrounded by a depauperate terrestrial flora.

Several nonnative species, including salt-cedar and camelthorn, are rapidly invading the riparian communities. Salt-cedar was imported into California by farmers to control erosion, but quickly invaded the lower Gila River basin. Today salt-cedar is found all the way into the headwaters of the Gila River and is slowly moving up the entire Colorado River drainage as far as the Green River, especially in disturbed areas or flow-regulated reaches. There is concern that this nonnative plant may not perform the same ecological functions in desert stream communities as native riparian plants (Pomeroy et al. 2000).

The archaeology of the upper basin shows evidence of widespread human culture dating back 11,000 years (Smith 1974). About two millennia ago the agrarian Anasazi culture arose in the southern part of the basin; numerous cliff dwellings, including Mesa Verde National Monument in southwestern Colorado, remain as testaments to their presence. Around 800 AD the Fremont culture arose throughout most of Utah and western Colorado, but they were supplanted from the north and west by the Utes, who ultimately spread to occupy most of the upper basin province by the time of European arrival. Eventually, seven loose confederations formed in the Ute Nation. These often coincided with major river drainages. For example, the Weeminuche occupied the San Juan River valley, the Tabeguache lived in valleys along the Gunnison River, the Yampa band inhabited the Yampa River valley, and the Uintah Utes occupied the Uintah basin (Jefferson et al. 1972). The Spanish made first contact with the Utes in the 1630s.

The great influx of English-speaking settlers into the upper basin began in 1847, when the Mormons migrated to the Salt Lake valley, and about the same time the gold rush began in Colorado. Most new inhabitants settled outside the upper basin drainage divide, to the west in Salt Lake valley and to the east along the Colorado Front Range near Denver. But these human population centers quickly began exerting an influence on the rivers of the upper basin through agriculture and ranching. For example, the headwaters of the upper Colorado River have been diverted to the Missouri River drainage since 1890 via the Grand Ditch through Rocky Mountain National Park (Wohl 2001).

In the upper basin states (Wyoming, Colorado, Utah, and New Mexico), 90% of the water used is spread on land irrigated for crops, leaving 10% for urban and other uses. Of the 1.6 million acres irrigated, feed for livestock is raised on 88% of the irrigated land. The largest urban center in the drainage is Grand Junction, Colorado (population ~42,000). About 100 years ago fire spread across this region, resulting in major vegetation changes that included declines in grassy vegetation and increases in woody plants, a trend that continues today.

The lower basin states (Arizona, California, and Nevada) also have a long history of agricultural occupation. The Mogollon culture first developed an agricultural civilization in the upper Gila River region in western New Mexico as early as 2200 years ago (McNamee 1998). Evidence of their culture remains at the Gila Cliff Dwellings National Monument in New Mexico and the Casa Malpais and Raven ruins near Springerville, Arizona. Other Native American cultures later developed on the Gila River near the present site of Coolidge Dam and at the confluence of the Colorado rivers. Many of these cultures were presumably absorbed into the Hohokam culture that occupied the central region of the Gila River. These Native Americans also used the Gila and lower Colorado rivers to fish for Gila trout and the large and once abundant Colorado pikeminnow. Spanish explorers ventured into the American Southwest from Mexico in the late 1600s, where they encountered a harsh environment and a network of canals that had previously irrigated prehistoric crops along the middle Gila River (Fradkin 1981).

In the late 1800s, Mormon settlers moved into the Little Colorado River basin in eastern and central Arizona and developed extensive sheep and cattle ranches as well as crop farms where wheat, oats, and barley were grown. These regions remain active ranching communities today. Also, the San Pedro and Santa Cruz rivers, both intermittent north-flowing tributaries to the Gila River, provided southern corridors for early Europeans to explore the American Southwest and beyond (Fradkin 1981).

The arid landscape in the lower basin shows evidence of heavy use from agriculture, ranching, and mining over the last century, with dense urbanized areas in metropolitan Las Vegas, Phoenix, and Tucson. Even metropolitan areas outside the Colorado River basin, such as Denver and Los Angeles, exert an influence on the basin via transbasin diversions for municipal and agricultural water use. Heavy demands on a restricted water supply to satisfy expanding agriculture and mining needs and the rapid expansion of dense urban communities over the last century have resulted in reduced water flows, deterioration of water quality, reduced groundwater levels, and ground subsidence, especially in the Tucson area, where large quantities of groundwater are pumped for human use. Heavy grazing in certain regions has enhanced soil erosion, caused the invasion of nonnative plants, and destroyed riparian communities. Over 85% of the native riparian communities in the lower basin have been modified or lost and <2% remain natural (Brown 1994). Mine wastes are of concern along sections of the Gila River in the lower basin, and storage reservoirs and irrigation have greatly increased secondary salinization. In the lower basin, 85% of water goes to agricultural purposes, with a significant percentage going to grow feed for livestock. Of the 99 million acres in the lower basin, 82 million acres are rangeland or pasture, whereas only 500,000 acres are classified as urban.

The Rivers

The principal tributaries along the Colorado main stem in order of their occurrence include the Gunnison, Green/Yampa, and San Juan river systems in the upper basin and the Little Colorado, Virgin, Bill Williams, and Black/Salt/Verde/Gila river system in the lower basin (Fig. 11.2). Combined, the four tributaries in the upper basin deliver over 11 billion m^3 of water to the main stem annually, of which over 50% is delivered during May and June from snowmelt. In contrast, the main tributaries in the lower basin deliver <0.5 billion m^3 annually, most of which comes from the Little Colorado and Virgin rivers, often during summer monsoonal storms.

The Colorado River basin contains three freshwater ecoregions according to the classification of Abell et al. (2000). In all ecoregions, river headwaters are clear and have steep gradients and stony substrates. However, once these rivers descend onto the plateaus they become highly depositional, with a substratum of fine sediments and a high degree of embeddedness. Woody debris dams are rare in the turbid, desert rivers.

The Colorado ecoregion of southwestern Wyoming, western Colorado, eastern Utah, and northern Arizona contains several major rivers: the upper Colorado, Green, Yampa, Gunnison, San Juan, Bill Williams, and Little Colorado (Fig. 11.2). These rivers originate in the Rocky Mountains in Wyoming and Colorado and in the White Mountains in east-central Arizona. The Vegas-Virgin ecoregion of southwestern Utah and northwestern Arizona is represented by the Virgin River, which originates at the edge of the Markagunt Plateau in southwestern Utah. The Gila ecoregion of west-central New Mexico and east-central Arizona contains the Gila River and associated drainages, including the Black, Verde, and Salt rivers. The Gila River proper has headwaters in the Gila National Forest, New Mexico (Fig. 11.2). All of these rivers have similar headwater and downstream conditions as outlined for the aforementioned rivers.

The rivers in the three freshwater ecoregions of the Colorado Complex have a relatively high degree of biological distinctiveness in that all are considered continentally outstanding (Abell et al. 2000). The Colorado is the 16th largest (507,245 km^2) freshwater ecoregion in North America and has the richest native fauna of the three ecoregions, with its 29 fishes, 26 herpetofauna, and 3 mussels. The Vegas-Virgin (34,565 km^2) is the 5th smallest in North America and has a native fauna of 11 fishes, 8 herpetofauna, and 1 mussel, whereas the Gila (159,875 km^2) has a native fauna of 19 fishes, 27 herpetofauna, and 1 mussel (Abell et al. 2000).

The American Southwest has some of the highest rates of fish endemism on the continent. For example, in the Colorado River basin, 35% of all native genera and 64% of the species are endemic (Carlson and Muth 1989). Over 85% of the fish fauna in Arizona are threatened and all three ecoregions have a conservation status of either critical or endangered with a high likelihood of future threats (Abell et al. 2000). All of the endemic fish species in the Colorado and Vegas-Virgin ecoregions are considered imperiled and over 40% are imperiled in the Gila ecoregion. All three ecoregions have high vulnerability to the deterioration of water quality, with urban development, nonpoint-source pollution, groundwater pumping, mining, diversions for agriculture and ranching, and fragmentation by dams the major threats. Recently Minckley et al. (2003) proposed an extensive conservation plan for native fishes of the lower basin.

The Colorado River drainage is one of the most regulated rivers in the world, with over 40 large flow-regulation structures and countless diversions along its river corridors (Dynesius and Nilsson 1994). The large metropolitan areas of Phoenix and Tucson started to expand in the early 1900s, as did Las Vegas in the mid-1900s. With the continued growth in the American Southwest, settlers began to divert Colorado waters into the deserts, eventually leading to the construction of major dams, including Theodore Roosevelt on the Salt River in 1911, Coolidge on the Gila in 1930, and Hoover and Glen Canyon on the Colorado in 1935 and 1963, respectively. Major dams were also built in the upper basin for storage and hydropower. These include Flaming Gorge on the Green River in 1962, Navajo on the San Juan in 1963, and Blue Mesa on the Gunnison River in 1963, all of which greatly altered the hydrology, ecology, and water quality of the Colorado River (Marzolf 1991, Blinn et al. 1998, Vinson 2001). These alterations, in addition to the introduction of nonnative species, have directly contributed to the endangered or threatened status of 24 native fish species; 4 are now extinct (Minckley 1991, Minckley and Deacon 1991, Starnes 1995).

The Colorado River is managed and operated under compacts and regulatory guidelines collectively known as the "Law of the River," which apportions and regulates the management and use of the Colorado among the seven basin states and Mexico (Gleick et al. 2002). The cornerstone for the Law of the River was the Colorado River Compact of 1922, at which time President Herbert Hoover suggested the Colorado River basin be divided into an upper and lower half, each having the right to develop and use river water. Since that time there have been at least 10 additional amendments released on the management and allocation of water from the river. The Mexican Water Treaty was signed in 1944. These interstate and international issues will only escalate as water continues to become a diminishing resource in the southwestern United States. The recent applications to double the number of coal-burning power facilities within the lower basin will further impact groundwater levels and water quality. Currently, the lower basin has four such plants in operation. Future threats to water quality in the upper basin include agriculture and the possible mining of the extensive and highly saline oil shale deposits in western Colorado.

COLORADO RIVER MAIN STEM

The Colorado River originates in alpine meadows on the western slopes of the Continental Divide in Rocky Mountain National Park, Colorado, at an elevation of ~3105 m asl (Stanford and Ward 1986). The Colorado is the largest waterway in the American Southwest and has the seventh-largest drainage area and the seventh-longest river corridor in North America. The river meanders in a southwesterly direction between 105°W, 40°N and 114°W, 31°N for over 2300 km through the Great Basin and Sonoran deserts, and portions of five states (Figs. 11.2 and 11.13). On its course it winds through broad aggraded valleys and bedrock uplifts and passes through Arches and Canyonlands National Parks in Utah and Grand Canyon National Park in Arizona, and then delineates the borders of California and Arizona as it flows southward. Ultimately, what remains of its highly altered flows reach its oceanic outlet, the Gulf of California, approximately 50 km into Mexico. Nearly three-fourths of the basin lies on federal lands.

As described earlier in more detail, an ancient agricultural culture may have occupied parts of the lower basin as early as 2200 years ago (McNamee 1998), with occupation of the upper basin dating back to 11,000 years ago (Smith 1974). Native American cultures later developed on the Gila River near Coolidge Dam and at its confluence with the Colorado River. Spanish explorers ventured into the American Southwest in the late 1600s, where they found evidence of prehistoric agriculture along the middle Gila River (Fradkin 1981). Mormon settlers

migrated into the upper basin in the mid 1800s, with similar settlements along the Little Colorado River in eastern and central Arizona in the late 1800s.

Physiography, Climate, and Land Use

The Colorado River flows out of the Middle Rocky Mountains (MR), Southern Rocky Mountains (SR), and Wyoming Basin (WB) physiographic provinces, through the Colorado Plateau (CO) province for over 1500 km, into the Basin and Range (BR) province, and ultimately into the Baja California (BC) province (see Fig. 11.13) (Hunt 1974). This area corresponds to the Colorado Rockies Forests, Colorado Plateau Shrublands, Mohave Desert, and Sonoran Desert terrestrial ecoregions (Ricketts et al. 1999).

The Colorado River flows through at least seven Merriam Life-zones (Brown 1994). The river starts in the Colorado Rockies, which contain alpine fir, Douglas fir, and Engelmann spruce. The Colorado meanders through Montane Conifer Forests of ponderosa pine onto the Great Basin Conifer Woodland of pinyon-junipers; the Great Basin Grasslands, with wheatgrasses and fescues; the Desertscrub Zone of sagebrush, shadscale, and winterfat; the Mohave Desert, with all-scale, brittlebush, creosote bush, Joshua trees, white burrobush, and assorted cacti; and finally through the Sonoran Desert (<500 m asl) of creosote bush, giant saguaro, white bursage, and assorted cacti (Brown 1994, Ricketts et al. 1999).

Most of the water is delivered as snow in the upper basin, whereas summer monsoons deliver much of the precipitation in the lower basin. The lower Colorado meanders through the most arid portions of the Sonoran Desert, with summer air temperatures >45°C. Precipitation is >100 cm/yr in the headwaters of the Colorado, <15 cm/yr at the start of the lower basin (952 m asl), and about 1.5 cm/yr near the mouth of the river (Brown 1994). In the upper basin, mean monthly highs and lows in air temperature are 25.3°C and −3.6°C, respectively, compared to 33.4°C and 8.9°C for the lower basin (Fig. 11.14). Annual mean precipitation for the combined upper and lower basins averages about 16.4 cm.

The landscape through the river basin is largely influenced by agriculture, ranching, and, to a lesser degree, mining. Dense urban areas are generally lacking along the main stem except for the metropolitan area of Las Vegas, which is rapidly approaching a million residents. Other population centers along the Colorado include Grand Junction, Colorado, and Page and Lake Havasu City, Arizona, all with populations of less than 50,000 people. Even though large metropolitan areas such as Denver are outside the Colorado basin, the Colorado–Big Thompson Project in the upper basin diverts a substantial volume of water (over 370 million m³/yr) in the Colorado River headwaters to Denver and other cities in the Mississippi River drainage. Similarly, cities such as Los Angeles and Phoenix receive large quantities of river water in underground and open aqueducts from below Lake Havasu as a result of the Colorado River Compact in 1944. Water from diversions along the Colorado River irrigates over 750,000 ha of landscape and serves over 30 million people (Gleick et al. 2002). Although the basin itself has only

about 7 people/km², each drop of water in the Colorado is estimated to be used an average of 17 times.

River Geomorphology, Hydrology, and Chemistry

The Colorado River is divided into upper (Wyoming, Colorado, New Mexico, and Utah) and lower basins (Arizona, Nevada, and California) at the confluence of the Paria River near Lees Ferry, Arizona (Fig. 11.3). By the time it reaches the lower basin the river has dropped over 1750 m. The average river gradient is about 2.2 m/km in the upper basin, 0.7 m/km in the

FIGURE 11.3 Colorado River near Lees Ferry, Arizona (Photo by E. Benenati).

lower basin, and 1.8 m/km for the entire river corridor.

During its course the river flows through deep canyons comprised largely of interbedded sandstone, siltstone, and shale. The basin contains igneous and metamorphic rock types in the upper watersheds and sedimentary formations, rich in sodium chloride (halite) and calcium sulfate (gypsum) deposited in ancient marine and brackish water environments, on the plateaus (Patrick 2000). In sections of the upper basin, the river has carved deep, narrow canyons in the highly erodible sedimentary deposits at a rate of approximately 17 m/1000 yr (Stanford and Ward 1986). On the plateaus a ribbon of water winds through a wide alluvial floodplain. Through Grand Canyon, the Colorado cuts over 1100 m deep through several distinct geological strata ranging in age from 180 million (Jurassic) to nearly 2 billion years (Precambrian). The river drops over 570 m through the canyon, where whitewater rafters encounter over 75 major rapids. Much of the lower basin is ≤750 m asl and the rivers typically have low gradients, wide floodplains, elevated solutes, seasonally heavy silt loads, embedded substrates, and regulated flows.

Based on morphometric criteria, the Colorado River may be considered a 6th order stream by the time it reaches the lower basin and continues as such until it empties into the Gulf of California. Others consider the Colorado to be a 4th or 5th order stream at the head of the lower basin (Shannon, Blinn, McKinney et al. 2001). This disparity suggests that traditional classification schemes for rivers should be used with caution, especially in arid biomes. The greatest difficulty in using ordering schemes lies in properly characterizing the first several orders in the upper watersheds and ephemeral drainages (W. Osterkamp, personal communication).

Discharge in the upper basin of the Colorado River is seasonably variable (<30 to >2000 m³/s), with an annual mean stream flow of 74 m³/s at Grand Junction, Colorado, and ~210 m³/s near Cisco, Utah (USGS 2004a http://water.usgs.gov/public/nasqan). The annual virgin discharge for the Colorado River is 550 m³/s (Dynesius and Nilsson 1994). In 1905, floodwaters from the Colorado flowed into the Salton basin in southern California to form the Salton Sea.

Glen Canyon Dam (GCD) has regulated flows below Lake Powell since 1963. This 216 m high structure currently inundates over 160 km of riverine habitat through the spectacular Glen Canyon National Recreational Area. Before closure of GCD,

annual discharge through the Grand Canyon from melting mountain snowpacks ranged from ~85 m³/s during late summer, fall, and winter to ~2300 m³/s during late spring and early summer, with a flood maximum of 8500 m³/s recorded on July 7, 1884 (Pope et al. 1998). Historic shoreline deposits suggest a flood larger than 14,160 m³/s within the last 1600 years. The range of annual stage for the Colorado River in the Grand Canyon prior to GCD was an impressive 6.4 m, but after closure this was reduced to 4.1 m (Stanford and Ward 1991). Today the average depth of the river through the Grand Canyon is about 9 m, with a maximum depth of nearly 40 m near Phantom Ranch. After the closure of GCD, flows have a peak discharge of 566 m³/s and a minimum of 141 m³/s, with daily fluctuations of 170 to 226 m³/s (Shannon et al. 1996). During the spring of 1996 a controlled flood of 1275 m³/s for 4 days was conducted below GCD to determine its effect on the geomorphic and ecological resources of the Colorado through the Grand Canyon (Webb et al. 1999).

Major spring-fed tributaries in the Grand Canyon (Vasey's Paradise, Nankoweap, Bright Angel, Tapeats, Havasu, and Spring Canyon) contribute <98 million m³/yr to the Colorado, with base discharges <2 m³/s (Pope et al. 1998). The Little Colorado River is the primary tributary in the canyon, with an average annual discharge of about 8 m³/s. Short-duration flood flows (up to 700 m³/s) provide most of the annual runoff from the Little Colorado.

Discharges diminish throughout the Lower Colorado River basin due to municipal and agricultural diversions and evaporation from reservoirs. Today, the annual mean streamflow is <100 m³/s through much of the Lower Basin (USGS 2004a http://water.usgs.gov/public/nasqan). Prior to the construction of Hoover Dam (Lake Mead) in 1935 the river supplied the Colorado Delta in Mexico with 6 to 18 billion m³ of water annually, compared to today's <0.7 billion m³ (Rodriguez et al. 2001). Even though discharges from the upper basin are seasonally variable, the numerous diversions and reservoirs greatly dampen the seasonal patterns in annual runoff in the lower basin. Presently, runoff below Hoover Dam ranges from only 0.20 to 0.28 m/mo, with an annual runoff of only 2.8 m (see Fig. 11.14).

Large amounts of sediment delivered from the semiarid regions of the Colorado Plateau in southern Utah, southwestern Colorado, northwestern New Mexico, and northern Arizona are transported by the Colorado. The arid plateau represents 37% of the total drainage area, but it delivers only 15% of the total runoff and over 80% of the sediment load

(Andrews 1991). Suspended sediment loads are <5 mg/L in the headwaters of the upper Colorado River but increase to >600 mg/L by the time the river reaches the Colorado–Utah border. Prior to the closure of GCD the river transported nearly 86 million tons of sediment through the Grand Canyon each year and the Paria and Little Colorado rivers supplied only 10% of that sediment load. Today the Paria and Little Colorado rivers supply most of the sediment to the regulated canyon river (3.0 and 9.3 million tons, respectively), with the greatest deliveries from July through September, when waters rapidly turn to a thick, reddish-brown color below these two tributaries (Andrews 1991). In addition, over 65 dry washes and creeks periodically deliver sediment to the canyon river at lesser loads. Debris flows inside canyons also transport sediment to the Colorado in Grand Canyon. Presently, the desert reservoirs in the lower basin are rapidly collecting these sediments as depositional deltas in their upstream bays. In addition, the river below GCD is accumulating sediment in pools above rapids and along shoreline eddies below rapids because delivery from the Paria and Little Colorado rivers exceeds the current sediment transport capacity of the highly regulated Colorado. Highest suspended sediment concentrations (>5000 mg/L) occur during the summer, when intense local storms produce flash floods in the numerous ephemeral drainages.

There are 40 large flow-regulation structures along the Colorado, including five complete main-stem dams (Davis, Parker, Imperial, Laguna, and Morelos) across the river and two partial rock and earthen dams (Palo Verde and Headgate Rock) used for control and diversion below Hoover Dam. The 654 km corridor between Flaming Gorge Dam and the lower portions of the Green and Colorado rivers through Canyonlands National Park represents the longest unregulated portion of the Colorado River system and may approach the closest natural riverine conditions along the Colorado prior to excessive regulation (Haden et al. 2003).

The combined storage capacity for all reservoirs on the Colorado River is 17 billion m^3, equal to approximately four times the total average annual flow in the Colorado River (Andrews 1991). Most of the reservoirs on the Colorado River were completed by 1963, including the four largest main-stem reservoirs (Powell, Mead, Mojave, Havasu; see Fig. 11.13). These storage basins are used for flood control, hydropower, municipal water supplies, irrigation of crops, and recreation (Andrews 1991). The high evaporation rates on these desert reservoirs have increased salinities and adversely affected irrigated agriculture in the lower basin. Furthermore, dams below the impoundments have modified the timing and amplitude of discharge as well as water quality and water temperature throughout the basin (e.g., Rader and Ward 1988, Marzolf 1991, Blinn et al. 1998, Vinson 2001).

Dams create highly altered temperature patterns in desert streams. Air temperatures exceed 40°C in the summer in the semiarid region of Grand Canyon National Park and are below freezing in the winter, yet annual water temperatures below GCD range from 9°C to 14°C due to hypolimnial releases. Summer water temperatures approached 30°C through the Grand Canyon prior to the dam (Stanford and Ward 1986). Recently there has been discussion on increasing water temperatures below GCD by releasing water from different reservoir depths. Regulated water temperatures are 13°C to 14°C below Hoover Dam and 13°C to 26°C below Parker Dam (Lake Havasu).

Water quality changes dramatically along the 2300 km river corridor of the Colorado, especially specific conductance, in part due to high evaporation rates on desert reservoirs, where average annual air temperatures exceed 24°C in the lower basin. Conductance in the upper watershed of the Rocky Mountains is <100 µS/cm (TDS 80 mg/L). By the time the river reaches the Colorado–Utah border, conductance is an order of magnitude higher, and it is over 3000 µS/cm (TDS 1200 mg/L) at the international boundary between the United States and Mexico. Alkalinity ranges from <50 mg/L as $CaCO_3$ in the headwaters to >175 mg/L at the mouth, and pH ranges from 7.7 to 8.6 throughout the upper and lower basins. Water temperatures range from near 0°C to 15°C in the headwaters to 30°C in the lower basin, and dissolved oxygen generally exceeds 8 mg/L throughout the river corridor. The headwaters are $CaCO_3$ dominated; however, waters at the Colorado–Utah border and throughout much of the remainder of the river corridor are dominated by $CaSO_4$ due to increases in gypsum in bedrock formations. Dissolved silica concentrations average about 10.4 mg/L throughout the river corridor. Average concentrations of PO_4-P and NO_3-N are 0.002 mg/L and 0.2 mg/L in the upper basin and 0.01 mg/L and 0.3 mg/L in the lower basin, respectively (Tadayon et al. 2001).

River Biodiversity and Ecology

The main channel of the Colorado River lies within the Colorado freshwater ecoregion, although two of its tributaries (Vegas-Virgin and Gila) are sufficiently distinct to merit their own ecoregion designation (Abell et al. 2000).

Algae and Cyanobacteria

Heavy loads of sediment carried by spring floods greatly reduce the diversity and densities of algae in the river through the upper basin, with attached diatoms and filamentous green algae (*Cladophora, Microspora, Stigeoclonium*) and cyanobacteria (*Anabaena, Oscillatoria, Phormidium*) common taxa (e.g., Rader and Ward 1988). Planktonic forms are rare. Few studies were conducted on the algal flora prior to the closure of GCD (Blinn and Cole 1991). Today, attached diatoms, filamentous green algae, and cyanobacteria make up nearly 90% of the algal community in the Grand Canyon, with *Achnanthidium, Cladophora, Cymbella, Cocconeis, Diatoma, Fragilaria, Gomphonema, Nitzschia, Oscillatoria, Phormidium, Rhoicosphenia, Tolypothrix, Ulothrix,* and assorted zygnemataleans the most widespread genera in the regulated river (Czarnecki et al. 1976, Czarnecki and Blinn 1978). Many of the algal species in the Colorado in the Grand Canyon are cold-adapted forms due to the altered thermal regimes below GCD. Numerous isolated "hanging garden" springs along the canyon walls support a rich algal flora. Also, filamentous cyanobacteria interweave their mucilaginous trichomes through fine channel sediments to form extensive surface mats, called "elephant skin," in the lower reaches through Grand Canyon.

Plants

Mosses largely occupy the higher-gradient headwaters, whereas submerged macrophytes are generally found in the lower-gradient downstream reaches. As suspended sediment loads increase, aquatic macrophytes are replaced by filamentous green algae, especially *Cladophora* and zygnemataleans, and eventually by cyanobacteria. Extensive stands of an asexual horsetail population (*Equisetum ferrissi*) provide habitat for invertebrates and fishes along the shorelines of the Colorado through Grand Canyon. Also, large monospecific stands of common watermilfoil, Eurasian watermilfoil, and sago pondweed occur in the slow-moving waters of lower Colorado River. A nonnative water fern has invaded the lower Colorado River over the last several years and has spread rapidly along the Colorado River below Blythe, California, forming thick floating mats in slow-moving backwaters (C. O. Minckley, personal communication).

The riparian vegetation in the subalpine forest (>2100 m asl) includes Englemann spruce and willows, and winter deciduous shrubs such as gooseberry, raspberry, and red elderberry. In the Montane Forests (1700 to 2100 m asl), Arizona alder, Arizona sycamore, gamble oak, and narrowleaf cottonwood are common riparian constituents. Lowered streamflow has reduced the number of mixed broadleaf species to scattered individual trees, opening the canopy and reducing its desirability for wildlife. The scrubland riparian community includes catclaw, common reed, desert-broom, and mesquite. The riparian communities along the lower Colorado include common reed, Fremont cottonwood, Goodding's willow, and/or velvet mesquite. However, the continued clearing along the lower Colorado has lead to invasions by salt-cedar that have moved high into the upper drainage. Bulrushes, cattails, and common reeds border the many marshes along the lower Colorado that provide important wintering grounds and nesting sites for avifauna.

Invertebrates

The aquatic insect assemblages in the headwaters of the Colorado are quite diverse. Ward and Kondratieff (1992) reported 63 stoneflies, 57 caddisflies, and 48 mayflies in the upper reaches of the Colorado drainage. These insect categories along with true flies and beetles contribute over 97% of the total invertebrates in the headwaters. Insect diversity is greatly reduced by the time the silt-laden Colorado reaches the Colorado Plateau and even more reduced in the lower basin. Ward et al. (1986) reported the mayflies *Traverella albertana* and *Heptagenia* spp. and only a few caddisflies and stoneflies in the upper Colorado Plateau.

More recently, Haden et al. (2003) reported a more diverse fauna of 28 insect taxa approximately 20 to 110 km above Lake Powell in Canyonlands National Park, Utah. The dominant taxa included mayflies (*Baetis, Ephoron, Heptagenia, Paracloeodes, Traverella*), caddisflies (*Ceratopsyche, Cheumatopsyche, Hydropsyche, Hydroptila, Smicridea*), a hellgrammite (*Corydalus*), a damselfly (*Argia*), beetles (*Neoelmis, Microcylloepus*), and high densities of chironomid midges and black flies.

Few studies on invertebrates were conducted on the Colorado River prior to the closure of GCD (Blinn and Cole 1991). Trophic groups such as collectors and shredders are now absent in the Colorado throughout the Grand Canyon due to the extensive river alteration (Stevens, Shannon, and Blinn 1997). Filtering and piercing caddisflies (*Ceratopsyche oslari, Chimarra, Hydroptila arctia*), and three species of *Ochrotrichia* (*O. dactylophora, O. logana, O. stylata*) and dipterans, namely chironomid midges and black flies (*Simulium chromatinum*), are the primary aquatic insects in the main-stem Colorado through Grand Canyon National Park. Adjoining tributaries have a number of aquatic insect species more typical of ephemeral desert streams (Oberlin et al. 1999). Chironomid species richness increases from 11 taxa in the clear tailwaters below GCD to 24 in the highly turbid downstream waters of the Colorado through the Grand Canyon (Stevens et al. 1998). Oligochaete worms (*Limnodrilus hoffmeisteri* and *Pristina* spp.), the nonnative amphipod *Gammarus lacustris*, and numerous snail species, including a recent invasion of the New Zealand mudsnail, make up the remainder of the invertebrate fauna throughout the Grand Canyon (Spamer and Bogan 1993, Stevens, Shannon, and Blinn 1997). Oligochaetes and chironomids are the dominant benthic organisms in the lower basin. The Asiatic clam became established in the 1940s and the common American prawn, red swamp crayfish, gammarid amphipods, and snails have been introduced as fish food in various sections of the lower Colorado River (Stanford and Ward 1986). The nonnative virile crayfish occurs above Lake Powell reservoir.

Vertebrates

Historically, the Colorado River drainage had 49 native freshwater fish species, represented by seven families, plus two marine species (machete and striped mullet; Starnes 1995). Of the indigenous fish species that remain, 42 are considered endemic to the system and 40 of these are considered endangered and/or threatened. Four species (Pahranagat spinedace, Las Vegas dace, Monkey Springs chub, and the Monkey Springs pupfish) are extinct (Minckley and Deacon 1991, Starnes 1995, Minckley et al. 2002). The already extinct Monkey Springs pupfish (from museum specimens) and the headwater chub have recently been described in the lower basin (Minckley and DeMarais 2000, Minckley et al. 2002). Others suggest that only 36 or as few as 29 native fish species formerly lived in the Colorado River basin (Carlson and Muth 1989, Abell et al. 2000).

The lower basin has the greatest diversity of native fish species (35), whereas the upper basin is less diverse, with 14 indigenous species (Carlson and Muth 1989, Starnes 1995, Minckley 1991). Most of the native species are extant in the upper Colorado, but many species are rapidly declining (e.g., razor-

back suckers and bonytail chubs). Approximately 500 razorback chubs occur in the upper basin and about 4000 wild adults remain in Lake Mojave in the lower basin, where there has been no reproductive success since closure of the dam. Historically, bonytail chubs were probably one of the most abundant fishes in the Colorado River basin, but they now only occur in a few locations in Cataract Canyon and the Green River in the upper basin and in Lake Mojave in the lower basin. They have been called functionally extinct by Carlson and Muth (1989). The Colorado pikeminnow, a large (1.5 m in length and over 20 kg in weight) predatory and highly migratory fish with a life span of >40 years, was last reported in the lower basin in 1975 (Minckley 1991). Pikeminnows are relatively abundant in the upper basin, but their distribution is restricted by dams and diversions.

The humpback chub, which lives up to 35 years and was once abundant in the upper and lower basins, occurs in the Grand Canyon river corridor, especially near the confluence of the Little Colorado River and from the Paria River to Rkm 420 on the main stem (Douglas and Marsh 1996, Gorman and Stone 1999). Estimates suggest that <1500 adults remain in the canyon corridor. There are also reproducing populations of humpback chub in Black Rocks and Westwater canyons in the Colorado River, Gray and Desolation canyons in the Green River, and Yampa Canyon in the Yampa River in the upper basin. The roundtail, humpback, and bonytail chubs, Colorado pikeminnow, and razorback suckers all evolved in the large, turbulent, sediment-laden rivers of the Colorado River basin and have adapted features such as large body size, highly streamlined bodies, large predorsal humps or keels, thin caudal peduncles, tiny or absent scales, thick and leathery skins, and slow growth (Minckley 1991).

At least 72 nonnative fishes have been successfully introduced in the Colorado River system over the years (Starnes 1995). Many of these introductions, including channel and flathead catfish and red shiner, have been linked to the decline and/or displacement of native species. Reservoirs (Lake Powell and Lake Mead) contain large populations of introduced species, including black crappie, common carp, largemouth bass, and striped bass. Many of these nonnative fishes are known predators on larval and juvenile native fishes (Minckley 1991). Introductions of salmonids to establish trophy fisheries in the clear, cold tailwaters below dams have also impacted the native fish communities. Presently over 85% of the fish species through Grand Canyon National Park are nonnatives (Minckley 1991). Populations of small fish species such as the desert pupfish and the Quitobaquito pupfish occur in disjunct populations near the Salton Sea, California, on Organ Pipe National Monument, Arizona, and in hypersaline pools in the Colorado River delta. Dewatering and predation pressures from nonnative tropical fishes such as Mexican mollies and sailfin have created these disjunct populations.

Beaver occur along various reaches of the river corridor where their preferred tree species (cottonwoods and willows) occur, especially in the upper sections of the Colorado in the Grand Canyon. Recently, there have been discussions on the reintroduction of river otter into Grand Canyon National Park to help control nonnative fish species. The last sighting of a river otter in the canyon was at the mouth of Bright Angel Creek in 1958. Muskrats are common along the lower Colorado. The nonnative bullfrog is also found throughout the river corridor and the Great Basin spadefoot toad occurs along the Colorado River within the Grand Canyon.

Ecosystem Processes

Much of the work on ecosystem processes in the Colorado River has been conducted in the river corridor through Grand Canyon National Park due to concerns over the listing of several endangered fishes and the reduced transport of sediment through the canyon. The construction of GCD greatly reduced the transport of allochthonous organic matter, increased armored substrata and water clarity, dampened flood discharges, and produced an unnaturally cold tailwater environment. As a result of these alterations, the filamentous green alga *Cladophora glomerata* and especially the associated diatom epiphytes became important components in the tailwater food web (Angradi 1994, Shannon et al. 1994, Shannon, Blinn, Haden et al. 2001, Blinn et al. 1998). Although *C. glomerata* does not directly provide food to tailwater communities except as detritus, it does serve important roles as a substratum for epiphytic diatoms and refugia for invertebrates and juvenile fishes.

Upland and riparian vegetation, rather than autochthonous algal carbon, were the most available food resources throughout the Grand Canyon river corridor during the experimental "control flood" conducted in March 1996, thus providing some indication of natural processes (Shannon, Blinn, McKinney et al. 2001). More recent studies with stable isotopes revealed a food web that varied spatially downstream, where aquatic communities were continually adjusting between the effects of upstream impoundment and landscape/tributary influence (Shannon, Blinn, Haden et al. 2001).

Several investigators have examined the effects of river regulation on algal communities below Glen and Boulder dams. For example, studies have tested the role of daily fluctuations in discharge below dams on benthic communities (Angradi and Kubly 1993, Blinn et al. 1995, Peterson 1986). Others have examined the interactions of regulated flows and tributary sediment on algal community structure (Shaver et al. 1997) and the role of inflow patterns on reservoir chemistry and discharge regimes on benthic algal communities (Benenati et al. 2000).

In situ studies have shown that over 50% of the *C. glomerata* mass is lost from established algal communities along fluctuating shorelines below GCD after several weeks of repeated atmospheric exposure and resubmergence (Angradi and Kubly 1993, Blinn et al. 1995). Dislodged floating tufts of *C. glomerata* and associated invertebrates make up a large part of the downstream drift (Shannon et al. 1996, McKinney et al. 1999) and provide substantial sources of autochthonous carbon for the river food web. Regulated discharges also influenced the structure of algal communities along the river corridor (Hardwick et al. 1992). Rader and Ward (1988) reported the absence of heptageniid mayflies, reductions in stoneflies and caddisflies, and an increase in chironomids in the Colorado River below Lake Granby Reservoir in the upper basin.

The cold (11°C to 14°C) hypolimnetic releases from GCD have allowed a number of Nearctic invertebrate species to invade the river corridor through Grand Canyon National Park, many of which are dipterans (Stevens et al. 1998). In addition, the controlled flood during the spring of 1996 and recent low flows during the summer of 2000, designed to increase water temperatures for native fish spawning, are coincident with dramatic increases in the nonnative snails *Fossaria parva* and *F. obrussa* and the New Zealand mudsnail. Finally, processing times for leaf material in the river have been slowed due to the perennial cold water and the low number of leaf-shredding invertebrates (Pomeroy et al. 2000). Over 50% of the leaf mass of the native willow remained after being submerged for 142 days in the cool waters (10°C to 12°C) below GCD. Schade and Fisher (1997) reported similar losses in leaf mass for willow in <30 days in a warm, nonregulated desert stream in Arizona without leaf-shredding invertebrates.

Altered conditions below GCD not only affect structure of biotic communities, but they affect function as well. In a series of in situ experiments Shaver et al. (1997) reported nearly 12 times more energy as invertebrate mass in a clear reach (<0.005 g/L suspended sediments) below GCD (Rkm 0) than in a turbid reach below the Paria River (Rkm 5), with 300 times the elevated suspended sediment loads, probably due to elevated autochthonous production. Indeed, gross primary production estimates in the same clear-water and turbid reaches were over sixfold higher than the downstream reach, and annual secondary production estimates for the amphipod *G. lacustris* were fourfold higher.

Haden et al. (1999) reported that floating driftwood provided a suitable alternative habitat to submerged substrata for aquatic insects in the turbid Colorado River with high suspended sediment loads and channel embeddedness. Hyporheic communities at the confluences of major tributaries with the Colorado River through the Grand Canyon are poorly represented, which is likely due to the high degree of embeddedness associated with reduced main-stem sediment transport capacity (J. A. Stanford, personal communication).

The heavy stream regulation throughout the Colorado River has modified community structure in a way that reduces the food supply available to fishes all the way to the Gulf of California in Mexico. Since the completion of Hoover Dam in 1935, the Colorado River delta has experienced a near cessation in freshwater flows, which historically helped mix estuarine waters nearly 60 km into the Gulf (Carbajal et al. 1997, Gleick et al. 2002). Much of the water released into Mexico is used for agriculture and only a few canal systems are concrete lined. Therefore, much of the water is lost through seepage through old river delta sediments that are very porous.

The large reductions in freshwater flows into the Gulf have now increased salinity to >40% and dramatically changed the mixing patterns in the estuary. The greatly reduced or complete lack of freshwater flows into the Gulf has also caused concentrations of selenium in water, sediment, and fish tissue to increase (Gleick et al. 2002). These changes in freshwater influx and water quality have caused declines in the once common bivalve mollusk *Mulinia coloradoensis* (Rodriguez et al. 2001), the Gulf of California harbor porpoise, the Totoaba fish (Cisneros-Mata et al. 1995), the vaquita, and the overall shrimp industry (Gleick et al. 2002).

Human Impacts and Special Features

The Colorado River is the largest desert river in North America, with enormous canyons, spectacular desert scenery, and many endangered endemic fishes. The 360 km river corridor through Grand Canyon National Park traverses one of the most spectacular canyonlands in the world and was formerly home to some of the oldest Native American cultures in North America. Some of the exposed rock formations in the inner gorge of the canyon are nearly 2 billion years old.

The Colorado River is one of the most physically developed and controlled rivers in North America, even though it passes through some of the most arid and remote regions in North America. The high demands for water throughout the semiarid and arid regions of the Colorado River basin have created heated debates over water allotments between states and at international boundaries. Major John Wesley Powell first recognized the potential for these conflicts in his historic descent of the uncharted Colorado River in 1869.

Historically, the Colorado River drained to the Sea of Cortez (Gulf of California), forming a large estuarine delta at the mouth of the river in Mexico. Today, with the construction of numerous diversions and reservoirs, the channel is nearly dry by the time it reaches the Gulf. Floodplains were developed for agriculture and/or channelized and controlled by bank stabilization or levees. Backwater and marsh habitats are gone except for a few that are maintained by human efforts.

The disappearance of native fishes in the lower basin was primarily caused by loss and/or dramatic change in habitat following river regulation and the introduction and spread of nonnative, predaceous fishes. As a result, the rate at which native fishes are jeopardized has dramatically increased, especially in the lower basin (Minckley and Deacon 1991). Although the number of native species has declined, the total number of fish species has increased two- to threefold due to successful introductions of nonnative species.

The modified conditions below Glen Canyon Dam have resulted in increased winter and breeding waterbird populations in certain reaches of the Colorado River in the Grand Canyon (Stevens, Buck et al. 1997). Also, native species within the regulated river corridor through the Grand Canyon are now relying on nonnative species as food resources. For example, great blue heron populations have increased at Lees Ferry due to the high numbers of nonnative trout, and farther downstream, native insectivorous birds are utilizing more Nearctic species of adult aquatic dipterans (J. Shannon, personal communication). Resource managers will continue to face these and new dynamic issues in the highly altered artificial Colorado River.

Recently, three states and various stakeholders and water and power agencies in the lower basin formed a regional partnership to develop a multi-species conservation program after the Fish and Wildlife Service's designation for critical habitat for the four endangered large river fishes in the basin. The Lower Colorado River Multi-Species Conservation Program is designed to protect sensitive, threatened, and endangered species of fishes, wildlife, and their habitats.

The placement of GCD above the gateway to Grand Canyon National Park has recently caused some to advocate the decommissioning of the dam with the hope of restoring a more natural hydrologic and thermal environment in the canyon that would favor native species and historical ecological processes. Arguments against the removal of GCD include the logistical difficulties and downstream ramifications in removing a large concrete structure, and recreational boating and fishing on Lake Powell and river rafting through Grand Canyon, which are important economic factors in the region.

GREEN RIVER

The Green River originates in the Wind River Range of Wyoming at an elevation of ~4100 m asl in alpine and spruce–fir zones. As it flows southward it drains portions of Utah and Colorado and winds to the east as it encircles the eastern edge of the Uinta Mountains and ultimately joins the upper Colorado River in Canyonlands National Park at an elevation of 1200 m asl (Fig. 11.15). It is a dramatic, high desert river, as documented by John Wesley Powell on his famous descent of the Grand Canyon, which began on the Green River in Wyoming. The Green River has a drainage area of 116,200 km², and as the largest tributary of the Colorado River it contributes nearly half of the total annual flow to the Colorado. Flaming Gorge Dam was built in 1962 and it regulates 35% of the Green River drainage. Several important tributaries feed into the Green below the dam: the Yampa and White rivers draining northwestern Colorado and the Duchesne, Price, and San Rafael rivers draining eastern Utah. All these tributaries contribute water and sediment along the main course of the river. Today more than 20 dams modify the Green River and its tributaries, but a single dam

on the main stem (Flaming Gorge) in northeastern Utah has had the greatest influence on the contemporary ecology of the river.

The prehistory in the Green River basin is highlighted by the origins of the Fremont culture in Utah and western Colorado around 1200 years ago. They were supplanted from the north and west by the Utes, who eventually came to dominate the region. The Uintah Utes occupied the Uintah basin at the time of European contact. By the time of extensive western settlement in the 1850s, Shoshone had also moved into some Green River valleys. Settlement of the region occurred mostly in the 1800s, and although sparse, many travelers passed through the Green River basin. This region was a major thoroughfare, serving as part of the Mormon Trail, the Oregon Trail, and a major route to the California gold rush.

Physiography, Climate, and Land Use

The Green River basin lies in the Middle Rocky Mountain (MR), Wyoming Basin (WB), Colorado Plateaus (CO), and Southern Rocky Mountains (SR) physiographic provinces (see Fig. 11.15) and includes the South Central Rockies Forests, Wyoming Basin Shrub Steppe, Wasatch and Uinta Montane Forests, and Colorado Plateau Shrublands terrestrial ecoregions (Ricketts et al. 1999).

Although the mountainous headwaters drain alpine and spruce–fir zones, dryland vegetation dominates most of the basin, including big sage, black greasewood, pinyon pine, western red cedar, and several grasses, including Indian rice grass, needle-and-thread grass, and the invasive cheat grass.

Precipitation in the mountainous headwaters of the Green River can exceed 100 cm/yr, and most of this moisture accumulates as snow in winter (Fig. 11.16). The lower elevation portions of the basin are semiarid and generally receive less than 25 cm/yr precipitation. Peak precipitation occurs in April and May (~4 cm/mo), and maximum temperatures occur in July (~20°C), whereas minimum temperatures occur in January (~−7°C; see Fig. 11.16). Climate along the Green River is characterized by cold winters and hot, dry summers. Near Vernal, Utah, average January highs are −7°C and July highs are 32°C.

The two largest urban centers in the drainage are Rock Springs, Wyoming (population ~19,000), and Vernal, Utah (~8000). Land use in the basin is largely agricultural (80%), with about 15% forests and 5%

urban. Of the water extracted in the basin, about 90% is used for irrigated crops, with 80% of the total going for feed for livestock. About 100 years ago fires across this region caused declines in grassy vegetation and increases in woody plants, a trend that continues today. The entire basin has <0.5 people/km².

River Geomorphology, Hydrology, and Chemistry

The Green River flows through rugged terrain. There are a number of mountain streams in the basin that drop quickly in elevation as high-gradient, coarse-bedded streams and merge into larger tributaries that flow at a low to moderate gradient and eventually join the main-stem Green River. Over its 1230 km course the Green River's average gradient is 2.4 m/km. Based on morphometric criteria, the Green is a 5th order stream.

The Green River below Flaming Gorge Dam can be divided into three main reaches (Muth et al. 2000). Reach 1 is the 104 km long segment that is completely regulated above the Yampa River confluence (see Fig. 11.15). This reach is straight to meandering, with a gradient of about 2.7 m/km. It is characterized by steep-walled canyon topography, with the exception of Browns Park, a 51 km long alluvial segment (Fig. 11.4). Above Browns Park the

FIGURE 11.4 Green River in Browns Park National Wildlife Refuge, northeastern Utah (PHOTO BY D. M. MERRITT).

riverbed is heavily armored by coarse sediment and the channel width ranges from 70 to 150 m and the depth from 1 to 10 m (Vinson 2001). As the river flows through the park, the channel widens to >200 m and becomes more shallow (0.5 to 2 m), and bank erosion restores finer sediment to the bed. The 158 km long Reach 2 occurs between the confluences with the Yampa and White rivers. It is largely a meandering reach, with a milder gradient of 0.85 m/km and a more natural flow and sediment regime than Reach 1 due to the contributions from the unregulated Yampa River. Major canyons in this reach include Whirlpool Canyon and Split Mountain Canyon, and flat, meandering segments include Island Park, Rainbow Park, and the broad valley of the Unita Basin. The range of bed materials spans cobbles to sand, and vegetated and unvegetated islands are common. Reach 3 is a 394 km segment below the White River to the confluence with the Colorado River. Given the more natural sediment regime, this reach has numerous low-elevation floodplains in flat segments (the Uinta Basin and most of the last 200 km of the reach). It also has the mildest gradient of 0.58 m/km. Major canyon segments include Gray, Desolation, Labyrinth, and Stillwater canyons, where gradients approach 0.1 m/km.

The mean annual discharge for the Green River is 172 m³/s, most of which is provided by snowmelt. Therefore, natural flow is very high in late spring and early summer and diminishes rapidly in midsummer (see Fig. 11.16). Although late summer and autumn flows can increase following rain events, natural flow in late summer, autumn, and winter months is generally low.

The main-stem segment of the Green River above the junction with the Yampa River provides 37% of the total natural annual flow volume in the Green River basin. The Yampa River is the largest tributary of the Green, providing another 36% of the total annual basin yield. Other tributaries include the White River in Colorado (12%) and three Utah rivers, the Duchesne (9%), Price (2%), and San Rafael (2%).

The main stem has been greatly modified by the construction and operation of Flaming Gorge Dam in northeastern Utah, approximately 655 km above the Green River's confluence with the Colorado River. This 149 m tall dam began operation in 1962 and currently inundates about 146 km of formerly riverine habitat. It is capable of storing twice the annual inflow of the Green River (Muth et al. 2000). Most of the modifications have occurred approxi-

mately 100 km below the dam and above the Green River's confluence with the Yampa River. The dam operates to capture and store peak flows, and base flows are elevated as stored water is released. Although the annual average flow of ~60 m³/s has not changed since dam construction, the magnitude and timing of seasonal high and low flows have been greatly altered by the dam (Merritt and Cooper 2000). Since dam closure maximum daily discharges have declined from more than 300 m³/s to less than 140 m³/s, whereas minimum flows increased from less than 10 m³/s to more than 20 m³/s. However, current proposals for recovering endangered native fishes in the Green River call for restoration of higher peak flows. A smaller dam, Fontanelle, was constructed upstream of Flaming Gorge in 1964, but its influence on the main-stem river is relatively small compared to the larger downstream structure, the operation of which is central to the contemporary ecology of the main-stem Green River.

The Green River basin is largely sedimentary, and as such it yields a naturally high sediment load, most of which is contributed by the lower-elevation portions of the basin. For example, the annual suspended sediment discharge of the Green River basin (prior to regulation by large dams) was almost 25,350,000 metric tons (Iorns et al. 1965). About 13% of this total load originated in the Green River basin above the Yampa River, ~6% from the Yampa River basin, ~26% from the Green River basin between the Yampa and White rivers, and ~54% from the basin downstream of the White River. Flaming Gorge Dam captures suspended sediment in the Green River, and Andrews (1986) estimated that mean annual sediment discharge was reduced by about 50% for up to hundreds of kilometers downstream of the dam. This alteration in sediment transport has resulted in distinctive changes in channel processes far downstream from the reservoir. These include a narrowing of the channel and decreased sediment deposition on point bars (Andrews 1986).

The resulting channel incision, combined with the loss of overbanking peak flows due to reservoir operations, has contributed to an effective isolation of historic floodplains along the lower Green River. This has strongly diminished the recruitment success of native cottonwoods, which require saturated, fine-textured soils away from the active channel to establish, survive, and eventually reproduce (Cooper et al. 1999, Merritt and Cooper 2000).

Water quality in the Green River is greatly affected by regulation. Prior to construction of the Flaming Gorge Dam, the Green River was free flowing and turbid, with natural seasonal dynamics. For example, water temperatures varied from freezing to near freezing in winter to above 20°C in summer (Vanicek et al. 1970). Data collected by the U.S. Geological Survey (USGS 2004b http://water.usgs.gov/ut/nwis/qwdata) from 1990 to 1999 just below Flaming Gorge Dam shows that water temperature now ranges from 2°C to 4°C from January through March and up to 12°C to 14°C in July and August. Oxygen concentration typically exceeds 8 mg/L year-round, but occasionally drops below 7 mg/L in summer and often exceeds 11 mg/L in winter and spring. Specific conductance typically ranges from 500 to 800 µS/cm throughout the year. The pH is typically between 8.0 and 8.5 year-round. Further downstream, within 200 km of the confluence with the Colorado River, water quality is much more reflective of a large, desert river, although some residual effects of regulation by Flaming Gorge occur. Water temperature ranges from 2°C in midwinter to up to 24.5°C in late July or early August. Oxygen concentration typically drops below 7 mg/L in summer and exceeds 11 mg/L in winter. Specific conductance typically ranges from 350 to 400 µS/cm in May and June to >900 µS/cm in fall and winter. The pH is consistently between 8.0 and 8.5 year-round and alkalinity is about 165 mg/L as $CaCO_3$ (Haden et al. 2003).

River Biodiversity and Ecology

The main channel of the Green River represents the northernmost tributary within the Colorado River freshwater ecoregion (Abell et al. 2000). Biological information for the Green is sufficient to provide excellent insight into how alteration in flow, temperature, and sediment regimes by a large main-stem dam interact to influence benthic species and communities (Vinson 2001). Such documentation is relevant to assess the preimpoundment fauna for many other regulated rivers in the Colorado River basin.

Algae and Cyanobacteria

A biotic survey was conducted in the lower Green River in Dinosaur National Monument around the confluence with the Yampa River in the summer of 1962, just one month before the closure of Flaming Gorge Dam (Woodbury 1963). Attached filaments of *Cladophora*, *Stigeoclonium*, and *Vaucheria*, as well as filamentous cyanobacteria (*Anabaena*, *Nodularia*, *Oscillatoria*), were common taxa in the turbid, unregulated desert river.

Plants

The riparian vegetation along the Green River before regulation was adapted to periodic flooding associated with the snowmelt hydrograph. Annuals and scour-tolerant perennials grew along the banks, and floodplains were occupied by box elder, coyote willow, and Fremont cottonwood, among others (Holmgren 1962). Above Flaming Gorge Reservoir, spring flows are adequate to maintain some natural recruitment of these riparian plants.

The severe modification of the natural flow regime below Flaming Gorge Dam has greatly altered the riparian communities along the Green River. Conditions are no longer favorable for cottonwood recruitment, and invasive salt-cedar has become widely established along the Green; this species may even contribute to channel narrowing (Allred and Schmidt 1999). Merritt and Cooper (2000) did an extensive vegetation survey at Browns Park (1636 m asl), an unconfined reach of the Green River ~70 km downstream from Flaming Gorge Dam and above the confluence with the free-flowing Yampa River (see Fig. 11.4). On stabilized in-channel islands, wetland plants such as common three square, coyote willow, and jointed rush occur. On rarely flooded surfaces the nonnative salt-cedar dominates, often in dense stands where only shade-tolerant species such as whitetop and Wood's rose can survive in the understory. Fremont cottonwood now occurs at low densities, and the understory is dominated by native desert shrubs, such as big sage, black greasewood, and Douglas rabbitbrush, and grasses and herbs, such as alkali sacaton and salt grass, as well as an assortment of introduced species, such as beggar's tick, cheat grass, Russian thistle, and summer cypress.

Invertebrates

The macroinvertebrate fauna of the lower Green River before construction of Flaming Gorge was dominated by a diverse assemblage of warmwater species, many of which no longer occur (Woodbury 1963, Holden and Crist 1981). The extent of the change in invertebrate species has been documented by Vinson (2001), who used more than 50 years of invertebrate data to document long-term changes about 20 km below the dam. The predam fauna consisted of 27 species in 15 mayfly genera (*Ametropus, Brachycercus, Caenis, Callibaetis, Camelobaetidius, Choroterpes, Ephemera, Ephoron, Isonychia, Lachlania, Leptophlebia, Pentagenia, Pseudiron, Siphlonurus, Traverella*), 5 stonefly genera (*Alloperla, Claassenia, Isogenoides, Isoperla, Perlesta*), and

3 caddisfly genera (*Leptocerus*, *Nectopsyche*, unidentified phryganeid). The only common insect taxa following dam closure were *Baetis tricaudatus*, Chironomidae, and Simuliidae. New insect taxa found below the dam have coldwater affinities and include the mayfly *Paraleptophlebia*; the stoneflies *Arcynopteryx*, *Hesperoperla pacifica*, *Taenionema*, *Zapada*; and the caddisflies *Hesperophylax*, *Hydroptila*, *Leucotrichia*, *Oecetis*, *Psychoglypha*, and *Rhyacophila*.

In 1978, operations at Flaming Gorge were modified to increase water temperatures (by releasing water from different reservoir depths) for the rainbow trout fishery, and a dramatic shift from insects to crustaceans, particularly *Hyalella azteca*, occurred. This was facilitated by flow stabilization that allowed the proliferation of macrophytes, a good amphipod habitat. Interestingly, only a few kilometers further downstream the amphipods were replaced by insects, namely *Baetis tricaudatus*. Here, a large intermittent tributary adds warmer water and fine sediment during occasional floods to reduce growths of macrophytes and filamentous algae that favor crustaceans.

A few collections have occurred further downstream in the Green River main stem. For example, Wolz and Shiozawa (1995) characterized the invertebrate fauna of soft sediments from June to August in Ouray National Wildlife Refuge, about 350 km downstream from Flaming Gorge Dam and above the confluence with the White River. The dominant taxa in all sampled habitats (main river channel, ephemeral side channels, river backwaters, and seasonal wetlands) were 6 to 15 genera of chironomid midges, Ceratopogonidae, nematodes, and oligochaetes. Additional insects encountered in the main channel included mayflies (*Baetis*, *Tricorythodes*), a stonefly (*Isoperla*), a dragonfly (Gomphidae), and true flies (Simuliidae). In the more standing water habitats, additional insects included mayflies (*Baetis*, *Caenis*, *Callibaetis*), a damselfly (*Ischnura*), true flies (Empididae, Simuliidae), and beetles (Hydrophilidae).

Another 300 km downstream in Canyonlands National Park, Haden et al. (2003) reported a more diverse assemblage of invertebrates, which may suggest a partial recovery from the altered conditions below Flaming Gorge Dam. These included 29 species in nine mayfly taxa (*Acentrella*, *Baetis*, *Camelobaetidius*, *Ephoron*, *Heptagenia*, *Lachlania*, *Rhithrogena*, *Traverella*, *Tricorythodes*), seven stonefly genera (*Acroneuria*, *Doroneuria*, *Frisonla*, *Isogenoides*, *Isoperla*, *Oemopteryx*, *Taenionema*), seven

caddisfly genera (*Brachycentrus*, *Ceratopsyche*, *Cheumatopsyche*, *Hydropsyche*, *Hydroptila*, *Nectopsyche*, *Smicridea*), four odonate genera (*Argia*, *Erpetogomphus*, *Ophiogomphus*, *Stylurus*), the hellgrammite *Corydalus*, the beetle *Microcylloepus*, and high densities of chironomid midges and black flies. Nearly 80% of these taxa were found on floating driftwood and only 40% on cobbles (Haden et al. 1999). The signal crayfish occurs in reaches of the Green River that run through Wyoming and the virile crayfish occurs in downstream reaches in Utah (M. Vinson, personal communication).

Vertebrates

Fishes indigenous to the Green River are represented by four families, which include five cyprinid minnows (Colorado pikeminnow, humpback chub, bonytail chub, roundtail chub, speckled dace), four catostomid suckers (razorback sucker, flannelmouth sucker, bluehead sucker, mountain sucker), two salmonids (cutthroat trout, mountain whitefish), and one cottid (mottled sculpin) (Muth et al. 2000).

Several big-river fishes (bonytail chub, Colorado pikeminnow, humpback chub, razorback sucker) used to reproduce in the 100 km above the Yampa confluence, but since the closure of Flaming Gorge Dam, with its year-round release of cold water, this activity has stopped (Vanicek et al. 1970, Holden and Stalnaker 1975). This long reach is now primarily a fishery for native Colorado River cutthroat trout (which were historically uncommon in this part of the river) and nonnative trout. The river downstream of the dam does support native fishes, but they now compete with many introduced fishes.

A total of 25 nonnative fish species in nine families have been reported from reaches of the mainstem Green River between Flaming Gorge and the Colorado River confluence and from lower portions of tributaries. Of the coolwater or warmwater nonnative fishes, channel catfish, common carp, fathead minnow, sand shiner, and rainbow trout are most common.

These nonnative fishes have been implicated as contributing to reductions in the distribution and abundance of native fishes as a result of competition and predation (Carlson and Muth 1989). Behnke and Benson (1983) attributed the dominance of nonnative fishes to dramatic changes in flow regimes, water quality, and habitat characteristics. They observed that water development has converted a turbulent, highly variable river system into a relatively stable system, with flow and temperature patterns that allowed for the proliferation of nonnative fish species.

Beaver occur along various reaches of the Green River, where their preferred tree species (alders, cottonwoods, and willows) occur, as do dwarf shrews, muskrat, and river otters. Boreal western toads (Northern Rocky Mountain population) and northern leopard frogs also occur along selected reaches of the Green.

Ecosystem Processes

The presence of Flaming Gorge Dam has promoted dramatic changes in flow, sediment, and temperature regimes over most of the entire 655 km course of the Green River before its confluence with the Colorado River. These physicochemical alterations have fundamentally modified habitat availability and dynamics and organism bioenergetics, resulting in significant modifications of energy flow, food web structure, and species composition for aquatic and riparian communities. These modifications are particularly dramatic in the 100 km reach between Flaming Gorge and the Green River's confluence with the Yampa River (Muth et al. 2000).

Sediment capture by Flaming Gorge Reservoir, combined with the loss of peak flows, has resulted in channel degradation, elimination of lateral channel migration, and loss of overbank flooding along alluvial reaches below the dam, such as in Brown's Park. Consequently, invasive nonnative species (salt-cedar) now dominate these reaches due to the failure of recruitment by flood-dependent native cottonwoods (Merritt and Cooper 2000). Channel fossilization and degradation also have simplified aquatic habitat and reduced backwater habitat for fishes and invertebrates. In addition, cold temperatures have created conditions that prevent native fishes from reproducing and a shift in higher trophic levels to salmonids (a pattern seen in other regulated segments of the Colorado River). Substantially colder water temperatures, combined with flow stabilization, have also promoted large changes in the aquatic food web. Extensive macrophyte beds and filamentous green algae now flourish along some reaches of the regulated Green River, providing ideal habitat for previously rare crustaceans that now exclude aquatic insects (Vinson 2001). With tributary inputs of sediment and warmer water downstream, the Green River recovers some of its natural character as a large, turbid, high desert river driven by allochthonous energy sources and with extensive river–floodplain interactions in alluvial reaches (Wolz and

Shiozawa 1995, Muth et al. 2000, Haden et al. 2003).

Human Impacts and Special Features

The Green River above its confluence with the Colorado River and below its confluence with the unregulated Yampa River represents the longest stretch of relatively free-flowing desert river in the Colorado River basin. Many of the 26 major reservoirs (Muth et al. 2000) in the basin are in headwater reaches of the Green's tributaries. Some of these reservoirs support transbasin diversions, such as on the Duchense and Strawberry rivers in Utah. The Flaming Gorge Dam on the main stem is by far the largest dam, with a storage capacity of >4600 million m^3. The Green River currently supports two of the remaining six populations of humpback chub, the largest populations of Colorado pikeminnow, and the largest riverine population of razorback suckers (Muth et al. 2000). These species are endangered due to both river regulation and the proliferation of introduced, nonnative fish species that compete or feed upon the natives. Given the status of these native fishes, intense restoration efforts have occurred in the Green River, primarily via the operation of temperature and water releases from Flaming Gorge Dam. Starting in 1978, the dam was retrofitted with a multilevel outlet to improve downstream water temperatures, primarily to improve the nonnative rainbow trout fishery. In 1985, an interim flow agreement was established to change reservoir releases to protect endangered fish nursery habitats in the Green River downstream of the Yampa River. In 1992 the U.S. Fish and Wildlife Service issued a biological opinion concluding that a more natural hydrograph below Flaming Gorge was needed to avoid likely jeopardy of the continued existence of these rare and endangered fish species. Several years of study ensued and in 2000 a final report was issued with recommendations to the Bureau of Reclamation for managing the flow and temperature regimes below Flaming Gorge Dam for the recovery of these species (Muth et al. 2000). Chief among these recommendations is moderating flow fluctuations due to hydropower generation and largely restoring the magnitude and timing of peak and low flows. The magnitude of the annual peak flow is recommended to vary depending on annual runoff into the reservoir from snowmelt. The timing of peak releases from Flaming Gorge is recommended to coincide with peak and immediate postpeak spring flows in the Yampa River to create maximum floodplain inundation and backwater duration downstream of the confluence of these two rivers in the lower Green River. Base flows are recommended to be low and stable from August through February, with daily changes in discharge caused by hydropower operations not to exceed 3% (or a stage difference of 0.1 m) below the Yampa River confluence. Finally, release of warmer water from Flaming Gorge Dam was recommended so that native warmwater fishes would have more thermally suitable habitat.

This proposed alteration of operations at Flaming Gorge Dam is deemed critical to the continued persistence of endangered fishes in the Green River. However, it is important to note that success is largely dependent on the maintenance of the unregulated Yampa River, which not only provides critical habitat for Green River populations, but also greatly normalizes the hydrologic and thermal characteristics after 100 km of severe regulation.

YAMPA RIVER

The Yampa River is the largest tributary to the Colorado River in the upper basin that remains largely unregulated (Fig. 11.17). It originates as small mountain streams in the alpine and spruce–fir zones (~3800 m asl) in the Park Range and White River Plateau of northwestern Colorado. Its largest tributary, the Little Snake River, originates in the mountains of southern Wyoming and contributes 28% of the total drainage basin of about 24,595 km² (Tyus and Karp 1989). From its source, the Yampa flows westward for about 320 km before finding its confluence with the Green River at an elevation of 1524 m asl (Fig. 11.5).

The headwater tributaries of the Yampa River coalesce as they flow off the western slope of the Rocky Mountains. At Steamboat Springs (2060 m asl) the river has reached about 20% of its ultimate

FIGURE 11.5 Yampa River, Colorado (PHOTO BY D. J. COOPER).

size with only an 8% contributing watershed. It turns westward and begins its meandering path across the high desert toward Craig and beyond through low-gradient agricultural valleys surrounded by sagebrush highlands. It passes through the canyons of Juniper Mountain and Cross Mountain and picks up its main tributary, the Little Snake River, before entering Dinosaur National Monument and the Yampa Canyon, where it flows through dramatic and steep terrain, eventually finding its confluence with the Green River at Echo Park in extreme northwestern Colorado.

The prehistory of the Yampa River basin is not well described, but it shares affinities with the Green River basin. The Fremont culture extended into this region around 1200 years ago but was eventually supplanted from the north and west by the Utes. The Yampa band occupied the Yampa River valley until the time of European displacement.

Physiography, Climate, and Land Use

The Yampa River basin lies within the Southern Rocky Mountains (SR), Wyoming Basin (WB), and Middle Rocky Mountains (MR) physiographic provinces and drains primarily in a west-southwest direction from its headwaters to its confluence with the Green River (see Fig. 11.17). The basin includes the Colorado Rockies Forests, Wyoming Basin Shrub Steppe, and the Colorado Plateau Shrublands terrestrial ecoregions (Ricketts et al. 1999). The upper treeline-krummholz consists of bristlecone pine that is replaced by ponderosa pine and extensive aspen stands in the lower mountainous regions. The region also contains mountain meadows, foothill grasslands, and riparian woodlands. Sagebrush and various wheat and fescue grasses make up the vegetation in the sagebrush-steppe region, with pinyon pine and several junipers and a sparse understory of grama grass and sagebrush and alderleaf cercocarpus (mountain mahogany).

Climate in the Yampa Valley is characterized by long, cold winters and mild summers. Average highs along the Yampa River in December and January are as low as −8°C, and temperatures as low as −52°C have been recorded near Craig. Summers are mild, with average July highs of 29°C, although subfreezing temperatures can occur at night in the summer months. Mean monthly temperature ranges from highs in July (~19°C) to lows in January (~−9°C; Fig. 11.18). Annual mean precipitation for the drainage basin is ~43 cm/yr, with over 60 cm/yr falling in the upper elevations and less than 40 cm/yr falling in the lower elevations. Precipitation falls in the basin rather uniformly throughout the year, with much of the precipitation falling and accumulating as snow from the fall through early spring (see Fig. 11.18).

The predominant form of land use in the Yampa basin is ranching and grazing, especially along the relatively verdant floodplains. Overall, about 65% of the basin is in agriculture, with most irrigation withdrawals from the river going to support irrigated hay pastures. About 30% of the basin is forested and less than 5% is urbanized, with Craig (8700) and Steamboat Springs (6500) the largest population centers in the basin. The entire basin has <0.5 people/km².

River Geomorphology, Hydrology, and Chemistry

The Yampa drainage can be divided into three zones. The first consists of montane headwater streams characterized by cold water, steep gradients, and coarse substrata. The middle section, seasonally cold due to snowmelt runoff but warm in late summer and early fall, flows across a high plain with fine substrates and a channel width at base flow of <20 m. The lower section of the Yampa is a primarily warmwater canyon river, with high-gradient reaches of rocky runs and rapids interspersed with low-gradient reaches in incised bedrock meanders. Based on morphometric information the Yampa River is considered a 4th order stream. Over its entire length, the Yampa averages a gradient of >7 m/km. Most of this is achieved before the river reaches Steamboat Springs, after which the river descends at about 2 m/km on average, with the greatest drop occurring in the Yampa Canyon prior to the confluence with the Green River in Echo Park. In the headwaters, substrates are coarse and fine in the downstream direction, especially in the middle section of the river, where silted substrates are common. Coarser gravel and cobble characterize the Yampa Canyon.

Given its snowmelt origins and relatively high elevation, the Yampa River is perennial over its entire length. It has a typical snowmelt hydrology, characterized by low fall and winter flows and a predictable late spring pulse driven by snowmelt (see Fig. 11.18; Poff 1996). Thus, the difference between precipitation and runoff throughout the year is due to a combination of evapotranspiration and snow storage. The average annual discharge of the Yampa at its confluence with the Green is about 61 m³/s. Flow begins to rise in the Yampa River in late March due to spring runoff and can remain high through June.

The mean flow during spring runoff in the Yampa River is about 153 m³/s (Tyus and Karp 1989). During the spring runoff periods river levels may undergo large fluctuations due to rapid warming that causes flash floods. Following spring runoff, flows of the Yampa River decline toward a monthly base flow of about 14 m³/s for August through March (Tyus and Karp 1989). Late summer or fall peak flows caused by monsoonal air flow are rare in the Yampa River due to its northerly location and orographic isolation from sources of tropical moisture.

Water quality in the Yampa River is generally good. Data collected by the U.S. Geological Survey below Craig, Colorado, from 1990 to 1999 (USGS 2004b http://water.usgs.gov/co/nwis/qwdata) show that water temperature ranges from 0°C in mid-winter to up to 24°C in late July or early August. Dissolved oxygen generally exceeds 9 mg/L. Specific conductance ranges from <100 to >700 µS/cm and is generally highest in late winter and early spring. The pH ranges from about 7.5 to 9.0 and alkalinity is about 85 mg/L as CaCO₃. At this site the river ranges in size from <20 m width at low flow in late summer and fall to up to 85 m during high flow.

River Biodiversity and Ecology

The Yampa River basin, like the Green River basin, lies within the northernmost portion of the Colorado freshwater ecoregion (Abell et al. 2000). As the largest free-flowing tributary of this ecoregion, the distribution of its biota is relatively well studied along its length.

Algae and Cyanobacteria

A biotic survey was conducted in the lower Yampa River in Dinosaur National Monument around the confluence of the Green River in the summer of 1962 (Woodbury 1963). Attached filaments of *Cladophora*, *Stigeoclonium*, and *Vaucheria*, as well as filamentous cyanobacteria (*Anabaena*, *Nodularia*, *Oscillatoria*), were the common taxa in the river.

Plants

Because the Yampa River is unregulated, it retains a dynamic channel morphology that serves to maintain dynamic variation in environmental conditions for a wide variety of riparian plant species. The steep environmental gradients in soil moisture characteristic of the Yampa River floodplains maintain diverse plant communities and a largely native vegetation. Merritt and Cooper (2000) did an extensive survey

at Deerlodge Park (1697 m asl), a wide alluvial valley immediately below the confluence with the Little Snake River, where the river migrates naturally across an unconfined valley. They found a vegetation continuum along an elevational (and fluvial disturbance) gradient from the active channel to the high floodplain. Plants on active point bars include mostly annual species such as cocklebur, cudweed, mudwort, and smartweed, as well as short-lived perennials such as foxtail barley. Intermediate-elevation floodplain stands are dominated by Fremont cottonwood saplings and several species of perennial grasses and herbs, including foxtail barley, horseweed, sage, and slender wheatgrass. The highest-elevation stands have a cottonwood overstory, with an understory consisting mostly of native grasses, such as Canada bluegrass, Kentucky bluegrass, needle-and-thread grass, slender wheatgrass, and western wheatgrass.

Invertebrates

The aquatic invertebrate species and communities in the Yampa River change dramatically along the course of the river. Ames (1977) conducted a longitudinal study of the aquatic insects in riffles of the Yampa, documenting assemblages at six sites along a 177 km reach (305 m elevation drop) from the cool, high-gradient headwaters with coarse substrates to the warm, turbid waters just above the confluence with the Little Snake River. At the most upstream site, communities were composed of mayflies (45%), caddisflies (22%), beetles (13%), stoneflies (12%), and true flies (5%). Mayflies and caddisflies did not change appreciably in percentage composition downstream; however, stoneflies declined (to 4% at the last site), as did beetles (2%). True flies increased to become the second most represented group (25% at last site). Overall, a total of ten mayflies, eight stoneflies, eight caddisflies, four beetles, four true flies, and one aquatic moth taxa were collected.

The upstream site was characterized by cool-water mayflies (*Paraleptophlebia*, *Rhithrogena*), stoneflies (*Alloperla*, *Pteronarcella*, *Pteronarcys*), caddisflies (*Lepidostoma*), beetles (*Optioservus*), and true flies (*Atherix*, *Hexatoma*, black flies). The most downstream sites were characterized by mayflies with silty and/or warmwater habitats (*Choroterpes*, *Ephoron*, *Traverella*, *Tricorythodes*), stoneflies (*Capnia*), caddisflies (*Helicopsyche*, *Oecetis*, *Protoptila*), beetles (*Microcylloepsis*), true flies (chironomid midges), and aquatic moths (*Cataclysta*). Some taxa were widely distributed across the range of conditions represented, including certain mayflies

(*Baetis, Ephemerella*), stoneflies, and caddisflies (*Cheumatopsyche, Hydropsyche*). The virile crayfish occurs in certain reaches of the Yampa.

Vertebrates

The Yampa River is one of the most important rivers in the upper Colorado River basin in terms of conservation potential for native big-river fishes. In contrast to other major tributaries in the upper Colorado River basin, the Yampa River supports all its native fish fauna (including some self-sustaining populations of rare species), and because of its largely unregulated flow regime contains high-quality fish habitat. It also contributes to the maintenance and availability of useable rare fish habitat in the heavily regulated Green River below its confluence with that river because it helps normalize the Green River's flow, sediment, and thermal regimes. The distribution and abundance of fishes indigenous to the Yampa River have been studied since the early 1900s (Ellis 1914). The 12 fish species indigenous to the Yampa River are represented by four families, which include five cyprinids (Colorado pikeminnow, humpback chub, bonytail chub, roundtail chub, speckled dace), four catostomids (razorback sucker, flannelmouth sucker, bluehead sucker, mountain sucker), two salmonids (Colorado River cutthroat trout, mountain whitefish), and one cottid (mottled sculpin) (Tyus et al. 1982, Behnke and Benson 1983, Tyus and Karp 1989).

All mainstream fishes persist today despite the introduction of at least 18 nonnative fishes (Tyus et al. 1982), including channel catfish, common carp, green sunfish, northern pike, and red shiner. Colorado River cutthroat trout persist in high-elevation streams above barriers that restrict invasion by introduced salmonids, such as brook trout. Indeed, the largest population of Colorado River cutthroat trout occurs in the headwaters of the North Fork of the Little Snake River on U.S. Forest Service lands in Wyoming.

The lower Yampa is particularly important for spawning and nursery habitat for rare and endangered native cyprinids and catostomids. For example, the Colorado pikeminnow and razorback sucker depend on habitats in the Yampa River (and lower Green River) to meet certain life-history requirements. The Colorado pikeminnow now occurs only in the upper Colorado basin, and the Yampa Canyon provides important spawning habitat. Some adults make annual round-trip migrations 950km up the river to spawn here and in the nearby White River, another tributary of the Green (Irving and Modde

2000). Colorado pikeminnow migrate in response to rising waters associated with spring runoff, and they spawn after waters decline and water temperatures exceed 16°C. Eggs hatch quickly and larvae are swept downstream, where they rear in warm, backwater habitats (created by high flows) in the Green River (Muth et al. 2000).

The self-sustaining population of humpback chub in the Yampa River represents one of the few remaining extant populations of this species in the whole Colorado River basin (Muth et al. 2000). A population of up to 600 individuals has been estimated for the Yampa Canyon (compared to 500 in Cataract Canyon on the upper Colorado and 1500 for Desolation and Gray canyons on the Green; Muth et al. 2000). The species occupies a specialized niche in canyons and is a warmwater fish requiring growth temperatures in the range of 16°C to 22°C (Muth et al. 2000). The humpback chub uses shoreline eddy and run habitat in the Yampa Canyon (Rkm 19 to Rkm 64). There is some speculation that the relatively unaltered riverine conditions of the Yampa has prevented hybridization between closely related *Gila* species (humpback and bonytail), because in the regulated Green River morphologically "intermediate" forms of humpback chub are seen (Tyus and Karp 1989).

Beaver occur along selected reaches of the Yampa River where their preferred tree species (alder, cottonwoods, willows) occur, as do river otters. The boreal western toad, Great Basin spadefoot toad (Southern Rocky Mountain population), and wood frog also occur along selected reaches.

Ecosystem Processes

The largely unregulated nature of Yampa's flow regime makes this river an important reference for understanding natural biophysical processes for snowmelt-driven rivers in the upper Colorado basin (e.g., Green, White, Gunnison, upper Colorado rivers). The natural geomorphic processes operating in the Yampa promote dynamic channel migration and associated habitat diversity critical to sustaining native riparian forests and large-river fishes that continue to flourish in the lower Yampa. Further, the unaltered thermal regime provides suitable temperature conditions for native aquatic fauna.

For example, the Yampa provides important spawning habitat for the Colorado pikeminnow. Adults residing year-round in the Yampa (as far upstream as Craig) and adults from the downstream Green River migrate to spawning habitat in the Yampa Canyon. Migrations are in response to high spring flows, and fish predictably lay eggs on the descending limb of the spring hydrograph in newly scoured and silt-free bed sediments. Emerging larvae drift downstream to nursery areas, backwaters in alluvial reaches created by the high flows. Many of these nursery areas are in the Green River and have been created by the high spring flows contributed to the Green by the Yampa.

The natural flow regime on the Yampa also maintains a self-sustaining, natural riparian community. Merritt and Cooper's (2000) study showed that the vegetation community on higher floodplain surfaces in the Yampa is quite different than that in the nearby regulated Green River. The natural dynamic of channel migration during floods serves to maintain a much more diverse riparian community, including robust, multiaged stands of native cottonwoods where invasive species such as salt-cedar fail to dominate. Richter and Richter (2000) modeled the flow conditions needed to maintain cottonwood forests along the Yampa and determined the duration of flooding at or above 209 m^3/s (125% of bankfull discharge) to be important in driving the lateral channel migration that initiates ecological succession in the Yampa's riparian forest. These high flows may also help reduce the grazer-induced mortality of small cottonwood seedlings from mammals such as voles (Anderson and Cooper 2000).

Human Impacts and Special Features

The lack of dams on the Yampa River make it unique as the largest free-flowing tributary in the arid intermontane plateau of the upper Colorado basin. The relatively unmodified flow, sediment, and thermal regimes of the Yampa are recognized as critical to its natural functioning and to the restoration of big-river fishes on a regional scale. Indeed, because the Yampa helps normalize the heavily regulated Green River, recommendations for managing the flows below Flaming Gorge Dam on the Green River explicitly incorporate the natural flow regime of the Yampa (Muth et al. 2000). The Yampa retains its natural seasonal flow variability because there are only three small headwater dams within its basin and only about 10% of its annual flow is diverted for municipal or agricultural use. There is some concern that the late-summer low flows are being adversely affected by diversion for irrigation of hay meadows in the Yampa valley. The Yampa River has enjoyed special status in Colorado due to its relatively unaltered condition and critical role in sustaining federally endangered fish species. However, water demands on the western slope are growing, and in 2000 the Colorado state legislature relaxed the protected status of the Yampa, possibly opening the door to more water development in the basin.

LITTLE COLORADO RIVER

The Little Colorado River (LCR) originates in the White Mountains on Mt. Baldy in eastern Arizona in the spruce–fir zone at an elevation of ~3400 m asl (Fig. 11.19). The LCR flows mostly in a northwesterly direction through an arid region for about 550 river km and drains a basin of 69,000 km^2. Four headwater streams (West, East, and South forks and Hall Creek) converge to form the LCR near Greer, Arizona. After leaving Greer, the LCR flows through grasslands for about 150 km and continues through desert scrub for another 400 km. The surface flow of the LCR disappears into the consolidated materials of the Moenkopi sandstone below Winslow, Arizona, and reemerges in the lower 21 km, where a series of permanent springs in the Redwall and Muav Limestone discharge water. The river starts its descent through the Little Colorado River Gorge near Cameron, Arizona (Rkm 480), near the Painted Desert and finally joins the Colorado River at an elevation of 823 m asl. Over 85% of the river meanders through the Desert Grassland and Great Basin Desert Scrub (Fig. 11.6). Relatively little alpine and forest habitat occur in the basin.

Early occupants of the region were the Mogollon (MUGGY-own), mountain and desert dwellers whose homeland stretched from the Little Colorado River to Chihuahua, Mexico, and from the Pecos River in New Mexico west to the Verde River in Arizona. These early dwellers, along with the Anasazi, Sinagua, and Hohokam Indians, lived off the land as early as 2200 years ago until about 1450, as evidenced by the Casa Malpais and Raven site ruins in the Springerville/Eagar area. Spanish explorers crossed the river in the late sixteenth century and commented on the groves of cottonwoods and willows along the LCR (Colton 1937). The first permanent settlers were Mormon families from Utah in the early 1870s, who utilized LCR water for ranching.

Controversy still exists regarding the perennial nature of the middle reaches of the LCR prior to early Anglo colonization in the 1880s. Some claim the middle reaches were perennial; however, C. Hart Merriam reported the LCR was an ephemeral stream in its middle reaches during his ecological

FIGURE 11.6 Little Colorado River near Holbrook, Arizona (PHOTO BY D. BLINN).

survey of the region in 1889 (L. E. Stevens, personal communication).

Physiography, Climate, and Land Use

The Little Colorado River basin lies almost entirely in the Colorado Plateau (CO) physiographic province (see Fig. 11.19). The basin includes the Arizona Mountains Forests and the Colorado Plateau Shrublands terrestrial ecoregions (Ricketts et al. 1999). The LCR drains a number of life zones, from alpine and spruce–fir and pine forests in the White Mountains and San Francisco Peaks, through expansive pinyon juniper zones at intermediate elevations, to grasslands and desert scrub, and finally the edge of the Mohave Desert near the confluence with the Colorado River. Grazing has increased the invasion by junipers onto the grasslands, especially on rocky thin-soil habitats. Much of the desert scrub is dominated by nonpalatable perennial shrubs, including sagebrush and saltbush, and to a lesser degree by blackbrush, black greasewood, rabbitbrush, and winterfat (Brown 1994).

Mean air temperature throughout the LCR basin is 13.5°C; 8°C in the upper basin and 15°C throughout most of the lower 450 km. Mean monthly temperatures are as high as 27°C in July and fall to 5°C in December (Fig. 11.20). Extreme summer air temperatures may reach >38°C in the lower LCR. Winter extremes for the upper and lower portions of the LCR are ~−15°C and −10°C, respectively. Precipitation in the basin varies greatly with season and elevation. Winter precipitation is generally in the form of snow. Summer precipitation results from convective storms carrying moisture from the Gulf of Mexico or Gulf of California or large-scale cyclonic storms originating over the Pacific Ocean. Average precipitation in the headwaters of the LCR is over 65 cm/yr, with highs of 12 and 11 cm in July and August, respectively, during the summer monsoons (Brown 1994). Total rainfall in the remaining watershed is considerably lower, ranging from nearly 25 to 15 cm/yr, again with highs in July and August (see Fig. 11.20). Evapotranspiration rates range from 76 cm/yr in the mountains to well over 150 cm/yr in the lowest deserts.

Since the arrival of early Spanish explorers ranching has modified the LCR basin and aquatic communities along much of the river corridor, especially around Springerville/Eagar and St. Johns. Today, over 80% of the basin is federally managed (four National Forests, six National Parks and Monuments, six American Indian reservations or allotments), with <20% privately owned, and <1% agriculture and/or industry (L. E. Stevens, personal communication). Some grazing occurs on the reservations and National Forests and over 500 mines occur within the basin. Dense urban areas are generally lacking in the basin, with the small community of Greer (<1000 people) in the headwaters and Springerville/Eagar, St. Johns, Holbrook, and Winslow, each with less than 10,000 people, distributed along the river on the semiarid plateaus (see Fig. 11.19). Flagstaff, with about 53,000 people, is located on the western edge of the basin. The entire drainage basin has ~1.5 people/km^2.

River Geomorphology, Hydrology, and Chemistry

Presently, perennial flows occur only in about the upper 125 km of the LCR and in the lower 21 km before it joins the Colorado River. Snowmelt and storm events regulate the intermittent flows between the upper and lower sections of the river. The river goes underground and resurfaces at several points along the corridor, resulting in isolated sections of permanent pools and slowly flowing water. Also, Lyman Reservoir and several irrigation diversions in and around Springerville/Eagar regulate downstream flows. A dam was constructed below Zion Reservoir in 1905, but it was breached in 2000. Now water flows more freely during floods in the channel below. The LCR basin contains approximately 37,000 km of stream channel, of which only 2.5% are perennial (L. E. Stevens, personal communication).

The average gradient along the entire LCR corridor is ~4.9 m/km. A cascading falls (43 m drop), active only during spring snowmelt, is located at Grand Falls, 377 km downstream from the headwaters. This falls prevents upstream migration of fishes from the Colorado main stem, but upstream fishes may be transported to the lower reaches during heavy floods. The LCR is considered a 4th order stream when it joins the Colorado River.

The main stem in the headwaters consists of riffle/runs, with coarse materials and pools with fine sediments. The remainder of the river channel is composed primarily of fine sediments. The upper headwaters have <5% of their hard surfaces embedded by sediments compared to over 25% throughout the remainder of the river. Channel width during base flow ranges from 0.5 to 5 m in the headwaters, whereas in the central part of the basin, a floodplain up to 0.5 km wide is bisected by an intermittent, narrow ribbon of water (see Fig. 11.6). Little woody debris accumulates in the active channel due to the limited amount of woody vegetation throughout the catchment and the sporadic floods that clear and strand debris high on the floodplains.

Mean runoff for the LCR basin is <0.03 cm/mo, largely due to the high annual mean evapotranspiration and low precipitation (see Fig. 11.20). Peak runoff occurs during March and April from snowmelt. Mean discharge ranges from <0.1 to 0.8 m^3/s in the headwaters, but is more variable ("flashy") throughout the remainder of the river corridor. In the early 1900s, annual mean stream flow near St. Johns, Arizona, was about 2 m^3/s but has been reduced to <0.5 m^3/s over the past half century (Pope et al. 1998). Mean discharge for the lower river is approximately 6.5 m^3/s, but this has little meaning for a system that ranges from no flow along much of its length to large flash floods. For example, a gage at Cameron, Arizona, recorded a maximum discharge of 3400 m^3/s in September 1923. Flash discharges (>250 m^3/s) from storm events have occurred at a frequency of one every four years over the past seventy years. These flash discharges result from the sparse vegetation in the arid catchment and the high density of arroyos that deliver water to the LCR during intense monsoonal storms. During brief periods of flash floods, suspended sediment loads can exceed 100 g/L.

Three north-flowing subbasins (Silver, Chevelon, Clear creeks), all with annual mean discharges of <2.5 m^3/s, produce much of the streamflow in the middle region of the LCR. In addition, there are approximately 30 major intermittent washes along the LCR corridor that receive water from their own multidrainage networks. These intermittent drainages deliver discharges of up to 100 m^3/s into the lower 362 km of the LCR from July through September following storm events. After flash floods, water flows in the main channel for several weeks and eventually loses its ribbon of connectivity and forms standing pools. Water in the pools reaches specific conductance values >4000 μS/cm, but <1000 μS/cm during periods of flow. The headwaters have low solutes (<40 μS/cm) and are relatively clear (suspended sediments <5 mg/L).

508

The headwaters are dominated by $CaCO_3$; however, as solutes concentrate downstream, water becomes strongly dominated by NaCl. Alkalinity averages <50 mg/L as $CaCO_3$ in the headwaters and >165 mg/L in the river corridor below. Water temperatures range from near 0°C in the winter to 18°C in the summer in the headwaters and ~3°C in the winter to >26°C in the summer throughout the remainder of the river corridor. Nutrient concentrations vary along the river corridor depending on the intensity of agricultural use.

Blue Springs is the largest of the springs in the lower 21 km reach, supplying approximately 56% of the 6.3 m^3/s base flow. Discharges from Blue Springs have high specific conductance (>4500 μS/cm) and alkalinity (>400 mg/L as $CaCO_3$), high concentrations of dissolved CO_2 (>300 mg/L), and a pH of 8 (Robinson et al. 1996). As CO_2 degasses with exposure to the atmosphere and aggressive photosynthesis, carbonate precipitates to form travertine dikes that are important geomorphic fish barriers. The rate of carbonate precipitation has been estimated to be about 1×10^{-5} moles $L^{-1} s^{-1}$. The high mineral content of the water gives the lower LCR its characteristic light blue-green color during base flow. However, the traditional aqua-blue color can be quickly replaced with a chocolate brown from flash floods as far away as St. Johns some 500 river km upstream.

River Biodiversity and Ecology

The Little Colorado River lies in the east-central part of the Colorado River freshwater ecoregion, one of the driest portions of the ecoregion (Abell et al. 2000). Although there have been a few studies in the upper tributaries and lower reaches of the LCR, studies in the intermittent main stem are largely lacking. Ecologically important riparian habitats are found in the headwater reaches and along the perennial flows below Blue Springs to the confluence with the Colorado River.

Algae and Cyanobacteria

The canopied riparian reaches of the headwaters of the LCR are dominated by epilithic diatoms, with *Eunotia, Frustulia, Gomphonema, Meridion, Nedium, Pinnularia,* and *Staurosira* the dominant taxa. Diatom assemblages in the lower river corridor include more salt-tolerant taxa, such as *Campylodiscus, Mastogloia, Navicula, Nitzschia, Rhopalodia,* and *Surirella* that occupy the fine sediments. The cyanophyte *Nostoc* and the aquatic buttercup are also common in headwater reaches.

Plants

The riparian vegetation in the headwaters includes Douglas fir, Engelmann spruce, subalpine fir, thin-leaf alder, and willows, with understory shrubs of blueberry, elderberry, and hawthorn. Disturbed north-facing sites consist of lodgepole pine and/or quaking aspen. The riparian community along the remainder of the river corridor is either open or invaded by salt-cedar and desert grasses, although native willows still flourish along many of the more perennial reaches. In the open, high meadows of the headwaters and in selected lowland areas where the water table is relatively high, broadleaf cattails, common reed, and sedges border the river channel. Aquatic macrophytes are typically absent in the lower, more turbid, and intermittent reaches of the LCR.

Invertebrates

Summer macroinvertebrate assemblages in the cool headwaters near Greer are diverse, with relatively high densities. Mean densities for macroinvertebrates are >700 animals/m^2 in the headwaters, reduced to one-half that number ~50 km downstream and <150 animals/m^2 250 km downstream in the highly turbid waters. Macroinvertebrate diversity also decreases quickly downstream from the headwaters. For example, the headwaters have over 25 species of caddisflies, with the dominant taxa including *Anabolia bimaculata, Atopsyche sperryi, Brachycentrus americanus, B. occidentalis, Glossosoma ventrale, Gumaga, Helicopsyche borealis, Hesperophylax occidentalis, Hydropsyche* spp., *Lepidostoma unicolor, Limnephilus sperryi, Oecetis disjuncta, Oligophlebodes minutus, Polycentropus arizonensis,* and *Rhyacophila* (D. W. Blinn and D. E. Ruiter, unpublished data). In contrast, only three caddisfly species (*Cheumatopsyche enonis, Hydropsyche* sp., and *Limnephilus lithus*) have been reported 50 km downstream and a few microcaddisflies reported from the intermittent and highly turbid sections of the river corridor. Coolwater stoneflies (*Claassenia sabulosa, Suwallia pallidula, Sweltsa coloradoensis*) and mayflies (*Baetis, Cinygmula, Drunella grandis,* and *Epeorus*) are common in the headwaters. Stoneflies have not been reported 50 km downstream and headwater mayflies are replaced by *Baetis, Choroterpes, Leucrocuta,* and *Tricorythodes* and not reported in the intermittent section of the river corridor. The macroinvertebrate assemblages in the intermittent and highly turbid reaches are composed primarily of chironomids and oligochaetes.

Although the reach below Blue Springs is perennial and frequently clear, total annual macroinvertebrate biomass is low (<0.2 g/m^2) due to high carbonate deposition (Oberlin et al. 1999). With the exception of the Paria River, the LCR has the fewest species and lowest invertebrate biomass of any major tributary in Grand Canyon National Park. The lower reach has primarily one mayfly (*Baetis*), several caddisflies (*Hydropsyche* and microcaddisflies), an hemipteran (*Rhagovelia*), and several chironomid species (Oberlin et al. 1999).

The nonnative virile crayfish is common in the LCR between Greer and Springerville and has greatly restructured the invertebrate and plant communities, as well as fish communities, in the river. White (1995) conducted a series of in situ experiments in a headwater tributary of the LCR and found that crayfish preyed heavily on eggs of the native Little Colorado spinedace. A few specimens of the Asiatic clam have been reported in the LCR below Springerville, and there are records of the native California floater in the headwaters.

Vertebrates

The Little Colorado River basin is isolated from the Colorado River system by a series of falls, especially Grand Falls, and intermittent reaches and therefore harbors several unique fishes. There are 33 species of fishes in the LCR Basin, only 9 of which are native; 5 in the upper reaches and 4 in the lower 14 km reach. The native fish species in the upper 75 km are represented by three families, which include two cyprinids (Little Colorado spinedace and speckled dace), two catostomids (bluehead sucker and Little Colorado River sucker), and one salmonid (Apache trout). Native species have been forced from their natural habitats by competition and predation by nonnative fish species and degradation and loss of habitat (Minckley 1991, Blinn et al. 1993). For example, roundtail chub are presently found only in the Chevelon and Clear creek basins and not in the LCR (Young and Lopez 1999). In addition, the Zuni bluehead sucker historically occurred in the LCR basin above Grand Falls but is presently restricted to the upper Zuni River drainage of west-central New Mexico. Two of the native fishes, Apache trout and roundtail chub, are sport fishes with regulated takes. The Little Colorado spinedace, which occurs in disjunct populations in north-flowing tributaries to the LCR, is federally listed as threatened and endangered (Blinn et al. 1993), and most of the other native species are listed as wildlife of special concern in Arizona.

The lower 14 km of the LCR has received more attention than any other section of the river corridor due to concerns over the federally endangered humpback chub. Other native fishes in the reach include bluehead sucker, flannelmouth sucker, and speckled dace (Douglas and Marsh 1996, Gorman and Stone 1999). All four native species inhabit this lower reach, but only speckled dace are found above the Blue Springs complex and in the higher-elevation reaches.

The common nonnative fishes in the upper LCR basin include rainbow trout, brown trout, fathead minnow, green sunfish, and red shiner. Most of these fishes were introduced into the LCR basin in the early 1900s (Young and Lopez 1999). No resident fish populations reside in the intermittent sections of the river. Nonnative fishes in the lower 14 km include channel catfish, common carp, fathead minnow, and red shiner. Beaver are present in the headwaters of the LCR where their preferred tree species (alders, cottonwoods, willows) occur, but are rarely encountered in the lower region. Muskrats are also found in the headwater system.

Ecosystem Processes

Although the LCR is one of the longest intermittent streams in North America, studies on ecological processes such as primary and secondary production, autochthonous versus allochthonous carbon, and autotrophy versus heterotrophy have received virtually no attention. However, based on the limited riparian vegetation along much of the river in combination with high suspended sediment loads and unstable channel substrates, one would predict an autochthonous system dominated by cyanobacteria and diatoms, with limited external energy inputs. Aquatic macrophytes are typically lacking due to high suspended sediment loads and unstable channels.

The interrupted periods of flow, standing pools, and dry riverbed, along with the general "flashy" nature of the LCR, are instrumental in structuring the aquatic communities within the channel and along the riparian zone. Conditions along the river corridor change from torrential flows to standing or no water in a matter of days, with associated rapid swings in salinity and water temperature. This harsh, flashy environment is characterized not only by temperature, moisture, and salinity extremes, but also by high levels of ultraviolet radiation during periods of low water. Hyporheic communities are poorly represented in the LCR except in the headwaters, due to the high degree of embeddedness throughout much of the river corridor.

Although the lower 21 km of the LCR are perennial, the high concentrations of dissolved CO_2 emitted from the alkaline springs along the limestone walls, along with stochastic floods, greatly reduce overall biomass and species richness. As high concentrations of dissolved CO_2 degas from the water with exposure to the atmosphere and aggressive photosynthesis, carbonate precipitates on aquatic vegetation and invertebrates and interferes with photosynthesis, respiration, and general metabolism.

The intermittent nature of the LCR along a gradient of harsh, environmental conditions provides an open opportunity to study the strategies of organisms under a variety of extreme conditions and the role of ephemeral systems as evolutionary loci in aquatic ecosystems. It also provides an opportunity to compare genetic relationships within fragmented populations. A recent study has shown that the threatened and endangered Little Colorado spinedace population in the perennial upper reaches of the LCR is genetically different from disjunct populations in north-flowing tributaries joining the LCR that are separated by intermittent flows some 200 km downstream (Greenberg 1999). The loss of connectivity in intermittent streams such as the LCR and long dispersal distances across arid landscapes in the Southwest contribute to greater genetic variation within populations.

Human Impacts and Special Features

The LCR basin was chosen as the first watershed-based fisheries management program in Arizona, in part because it contains several indigenous fishes, four of which are listed under the Endangered Species Act (Young and Lopez 1999). The upper headwaters of the LCR were recommended as an important site for the conservation of freshwater biodiversity in North America by the World Wildlife Fund (Abell et al. 2000), and the mouth of the LCR is a major spawning reach for the displaced humpback chub (Douglas and Marsh 1996).

There is concern regarding the effects that groundwater development will have on surface-water resources. For example, the LCR basin has the dubious distinction of having three of the four coal-burning power plants in Arizona. These facilities draw heavily on groundwater and greatly influence discharges from the more than 900 isolated desert springs along the river corridor (L. E. Stevens, personal communication). The recent applications to double the number of power plants within Arizona will potentially further impact these groundwater

resources. In addition, agriculture in localized regions continues to contribute to alterations in the LCR system.

The entire Colorado River freshwater ecoregion, which includes the LCR basin, has a critical conservation status (Abell et al. 2000). Between 50% and 89% of the catchment has been altered, with a very high proportion of the original habitat fragmented and a very high occurrence of nonnative species. Further degradation will occur without aggressive conservation of these varied and dynamic arid fluvial systems.

GILA RIVER

Three major tributaries (East, West, and Middle forks) arise at ~3100 m asl from the Mogollon, Black, and Pinos Altos mountains to form the main-stem Gila (HEE-luh) River at about 1750 m asl in the Gila Wilderness Area in southwestern New Mexico (Fig. 11.7). The river flows south in New Mexico for ~190 km and then enters Arizona near Duncan at an elevation of 1325 m asl. The river goes underground at Verdin, New Mexico, and reemerges 8 km downstream near Duncan. From Duncan, the river meanders westward for about 770 km through grasslands, scrublands, the northern edge of the Chihuahuan Desert, and much of the Sonoran Desert until it reaches the Colorado near Yuma, Arizona, at an elevation of ~50 m asl (Fig. 11.21).

The Salt River is a major tributary of the Gila and drains much of the higher elevations to the northeast via the White and Black rivers and parts of the Mogollon Rim by Tonto Creek. The Verde River also flows into the Salt and delivers water from the rim country in central Arizona into the Gila River (see Fig. 11.21). Much of this water is utilized by metropolitan Phoenix.

Like the Colorado River, the Gila River basin is divided into an upper and lower basin at the confluence with the Salt River. Today, nearly all tributaries in the lower basin are intermittent. Only the upper 500 km of river corridor is perennial with either unregulated or regulated flows. The remainder of the river corridor in the lower basin has been intermittent since the closure of Coolidge Dam (San Carlos Lake) in 1928 and the expansion of agriculture and rapid growth of metropolitan Phoenix. Coolidge Dam was constructed to control the periodic heavy floods from snowmelt in the higher elevations of New Mexico.

The name "Gila" originated from the Yuma Indians, who referred to the stream flowing into the

FIGURE 11.7   Upper Gila River at the Gila Preserve, Grant County, New Mexico (Photo by D. M. Merritt).

Colorado River as "Hahquahsaael" (McNamee 1998). Spanish soldiers shortened the name to "Xila," which ultimately became Gila. The upper corridor of the Gila River in New Mexico is rich in Native American culture. The area is located in the Gila National Wilderness, lands that conservationist Aldo Leopold helped establish in 1924. The Mogollon and Anasazi cultures occupied the region as early as 2200 years ago and likely abandoned the area starting in the thirteenth century due to drying southwestern climates. Evidence of the Mogollon culture can be found at the Gila Cliff Dwellings National Monument, New Mexico. Several Native American tribes, including the Mogollon, Anasazi, San Carlos, Gila, and Yuma, have utilized the Gila River drainage for centuries.

A fur trader by the name of Ewing Young trapped along the Salt and Verde rivers and was the first Anglo American to explore the Gila to its mouth in the early 1820s (Trimble 1989). In the 1850s, army personnel explored and documented the flora and fauna along the Gila River corridor. The first important gold discovery along the Gila came ~30 km east of Yuma, Arizona, in 1858.

Prior to the arrival of Anglo Americans, the Gila River normally flowed to its mouth as a wide, shallow stream, and river float trips were occasionally conducted on the river. In fact, Lieutenant William H. Emory made the first scientific survey of the Gila River in 1870 and reported its flow as about one-half the Colorado River at their confluence (Fradkin 1981).

Physiology, Climate, and Land Use

The Gila River basin lies mostly in the Basin and Range (BR) physiographic province and flows primarily in a westerly direction from its headwaters in New Mexico to its convergence with the Colorado River near Yuma (see Fig. 11.21). The Gila begins in the Arizona Mountains Forests terrestrial ecoregion and flows through sections of the Madrean Sky Island Montane Forests and Chihuahuan Desert ecoregions and finally through much of the Sonoran Desert ecoregion.

Quaking aspens and spruce/fir occur in the upper watershed (>2400 m asl) of the Gila River. The vegetation at the intermediate elevations (2400 to ~2000 m asl) include mixed conifer and alligator juniper, pinyon pine, ponderosa pine, and scrub oak between 2000 and ~1200 m asl (Brown 1994). The desert vegetation (<1000 m asl) throughout much of the upper and lower Gila basin is dominated by creosote bush, desert saltbush, ragweed, screwbean, and assorted Cactaceae and Fouquieriaceae. The nonnative cheat grass has displaced much of the native bunch grasses of the Gila River basin.

The climate of the basin is strongly affected by convective storms that originate in the Gulf of Mexico and the Gulf of California, and cyclonic storms that originate over the Pacific Ocean deliver most of the precipitation throughout the Gila River corridor in the summer. Precipitation and temperatures are highly variable due to the variations in altitude throughout the basin. Mean monthly precipitation for the basin ranges from 0.5 cm/mo during April and May to 4.4 cm during July (see Fig. 11.22). Average rainfall in the headwaters is over 55 cm/yr, with highs of around 9 cm during the summer monsoons in July and August (Brown 1994). Total rainfall in the remaining basin is lower, averaging <15 cm/yr, with nearly 60% of the precipitation falling in the late summer. Precipitation at the mouth of the Gila River near Yuma is <2 cm/yr, one of the most arid regions in North America. Mean monthly temperatures for the basin range from 27°C in July and fall to <5°C in December and range from <−10°C to 26°C in the higher elevations and 5°C to 39°C in the lower elevations (see Fig. 11.22). Diurnal temperature variations of 17°C or more are characteristic of the arid regions and daily high temperatures near the mouth of the Gila can exceed 45°C. Annual evaporation rates range from <100 cm in the higher elevations to >170 cm in the lower elevations.

Eight of the fifteen Indian tribes in Arizona live in some part of the Gila River basin and use the land for fishing and/or agriculture. In addition, many regions in the drainage basin have been under heavy agriculture, ranching, mining, flood control, and municipal development, which have placed heavy demands on the river for well over a century (Fradkin 1981). Crops in the area include cotton, alfalfa, pecans, citrus, wheat, and winter vegetables. In the late 1590s, Spanish ranchers drove cattle and churro sheep into the Gila drainage, where the native tallgrasses were reported to stand as high as a man on horseback (McNamee 1994). As a result, a third of the Southwest's grazing land is now severely desertified: Permanent streams have been degraded, their banks collapsed by cattle, their waters contaminated by defecation, their topsoils reduced through erosion, and their soils compacted from passing cattle, and there has been an invasion of many nonnative species (McNamee 1994). Arizona's crop farming in the late 1800s essentially drained the water from the middle and lower Gila River. By 1920, over 80,000 ha along the Salt River was planted in cotton. Mines sprouted all over the Gila basin in the 1850s, and by 1870 there were ~6500 mining operations within the basin extracting a variety of minerals, including gold, silver, lead, zinc, manganese, molybdenum, and copper (McNamee 1994). Mining still occurs today in the "Copper Basin" alongside the Gila River in the vicinity of Globe and Clifton, Arizona.

Dense urban areas are generally lacking throughout most of the basin. Major communities in the upper basin include Safford and San Carlos, each with <10,000 people, and in the lower basin Yuma has a population of over 100,000 people. However, the rapid growth of the "Valley of the Sun" in metropolitan Phoenix, with over 2 million people, and metropolitan Tucson, with well over half a million people, has played a major role in the fate of the Gila River, with additional impacts yet to come. There are ~25 people/km^2 in the entire basin, but over 6500 people/km^2 are localized in the region of metropolitan Phoenix.

River Geomorphology, Hydrology, and Chemistry

The catchment basin in the perennial corridor of the river drains approximately 47,000 km^2, of which 13,274 km^2 is below Coolidge Dam (Pope et al. 1998). The entire 960 km of river has a catchment basin of about 149,832 km^2, or nearly half the area of Arizona. Less than 7% of the Gila River catchment is in New Mexico. Over 80% of the Gila River corridor meanders through desert scrublands. The Gila River is considered a 5th order stream, with an average gradient along the entire 960 km river corridor of ~3.2 m/km.

The upper 250 km of the Gila main stem consists of riffle/runs and associated pools. Much of the remainder of the river channel is composed of fine sediments and patchy hard substrates, with a high degree of channel embeddedness. Channel width in the upper 150 km of the main stem ranges from 2 to 6 m during base flow, whereas the floodplain in the lower corridor may extend to over 0.5 km wide.

The principal south-flowing tributaries into the Gila River are the San Francisco, San Carlos, and Salt/Verde rivers and the ephemeral Aqua Fria and

Hassayampa rivers; the San Pedro along with the ephemeral San Simon and Santa Cruz tributaries flow north. Prior to the early 1900s these rivers combined carried up to 4.9 billion m^3 of water annually (McNamee 1998). During the last 50 years these tributaries have delivered <20% of that amount of water to the Gila River due to upstream storage reservoirs and irrigation diversions for agriculture and mining. The San Pedro now stands as the last surviving desert river without dams and with yearly rejuvenating monsoonal flooding during August and September that contributes a major portion of the unregulated flow below Coolidge Dam.

Runoff throughout much of the Gila basin is very low due to high mean annual evapotranspiration (>150 cm) and low mean annual precipitation (24.7 cm; see Fig. 11.22). Mean monthly runoff ranges from a high of 0.2 cm in January to a low of ≤0.02 cm during June and July for the drainage basin above Coolidge Dam. These values are more representative of the basin before the extensive impact of human extractions and regulation below the dam. Runoff values near the mouth are much lower today.

Since closure of Coolidge Dam (San Carlos Reservoir), annual mean streamflow has averaged less than 10 m^3/s, or less than half that prior to the placement of the dam. There are currently many demands for water in the semiarid regions above Coolidge Dam, including diversions for irrigation (about 33,000 ha irrigated land), metallurgical treatment of ore, and municipal water supplies.

In addition to Coolidge Dam, there are several downstream earthen dams, the largest of which are Gillespie and Painted Rock. Painted Rock Reservoir (nearly 3.1 billion m^3) was built in 1959 for flood control. The reservoir is usually dry but fills about every six to eight years from floods in the upper catchment (Pope et al. 1998). Floods deliver intermittent flows of up to 1400 m^3/s above the dam, and have exceeded 170 m^3/s about every 5 years over the last 50 years. The river channel above and below the dam is dry for most of the year because of transbasin diversions, storage reservoirs, power developments, groundwater discharge, diversions for irrigation, and municipal and industrial uses.

The last time the Gila River flowed from headwaters to the confluence with the Colorado River was during the 1992–1993 El Niño storms in the mountains of New Mexico, when flows averaged 185 m^3/s near the mouth of the Gila. Annual mean streamflows have gradually declined over the past 150 years. Before settlement of the Gila River basin, maximum runoff occurred in the winter, with flows

up to 1500 m^3/s (Sykes 1937). Annual mean streamflow in the Gila was over 40 m^3/s near its mouth in the early 1900s but has averaged <6 m^3/s over the last 75 years, most of which is recharged water from irrigation and municipalities (USGS 2004a http://water.usgs.gov/public/nasqan).

The headwaters of the Gila River, as well as base flows in the perennial river corridor in the upper basin, are relatively clear (suspended sediments <5 mg/L), have low levels of solutes (≤1000 µS/cm), and are dominated by $CaCO_3$ salts (Earl and Blinn 2003). In contrast, base flows in the lower basin are frequently turbid, with flows carrying suspended sediment loads >500 mg/L during storm events, have high concentrations of solutes (≥4500 µS/cm), and are dominated by NaCl salts (Pope et al. 1998). Also, a large proportion of the base flow in the lower basin contains wastewater from irrigated lands and from the Chandler treatment plant (Pope et al. 1998). Total nitrogen concentrations in the lower basin may exceed 10 mg/L, compared to ≤0.05 mg/L in the upper basin (Pope et al. 1998). Wildfires in the upper watershed also periodically deliver fire ash to the Gila River and elevate nutrients in the region by an order of magnitude during summer monsoons (Earl and Blinn 2003).

Alkalinity is <50 mg/L as $CaCO_3$ in the upper Gila River and increases to >260 mg/L by the time the river reaches the lower basin. pH varies along the corridor but averages about 8.3. Water temperatures normally range from 28°C in the summer to <5°C in the winter in the upper basin and >28°C in the summer to about 12°C in the winter in the lower basin (Earl 1999, Pope et al. 1998).

River Biodiversity and Ecology

The Gila River lies within the Gila freshwater ecoregion, which is considered continentally outstanding in terms of its biological distinctiveness (Abell et al. 2000). It shares many species with the Colorado ecoregion with which it connects, but has several distinctive species. In spite of this distinctiveness, the Gila main stem has received the least attention, from an ecological standpoint, of any large river in southwestern United States, even though it drains nearly half of the state of Arizona.

Algae and Cyanobacteria

The algal communities in the canopied headwaters of the East and West Fork tributaries in New Mexico are quite diverse, with benthic assemblages dominated by epilithic diatoms. In a recent four-year study, Earl and Blinn (2003) reported that *Achnan-*

thidium spp., *Cocconeis placentula*, *Cymbella affinis*, *Diatoma vulgare*, *Epithemia sorex*, *Fragilaria pinnata*, *Navicula cryptocephala*, *Nitzschia frustulum*, *Staurosira construens*, and *Synedra ulna* made up over 75% of the diatom taxa in the upper watershed.

One hundred kilometers downstream (Gila Bird Area of the Gila National Forest), in a reach that traverses through an intense agricultural region, diatom taxa were dominated by *Cocconeis pediculus*, *Navicula amphibia*, *N. frustulum*, and *N. subminuscula* (Earl and Blinn 2003). Filamentous green algae, namely *Cladophora glomerata* and zygnemataleans, occurred in wider, more exposed sections of the channel.

Plants

The riparian vegetation in the headwaters (>2500 m) of the Gila River consists of Bebb willow, narrowleaf cottonwood, and thin-leaf alder; whereas the vegetation along the river channel at elevations between 1500 and 2000 m is primarily Arizona sycamore, box elder, Fremont cottonwood, Gooding willow, and velvet ash. Much of the riparian community in the Sonoran Desert (<1500 m) consists of scattered mesquite bosques, commonly supported by floodplain groundwaters. Historically, mesquite bosques were common along ephemeral and intermittent stream channels, but they have been largely replaced by salt-cedar. Salt-cedar was imported into California by farmers to control erosion, but quickly invaded the lower Gila River basin. Today salt-cedar is found all the way into the headwaters of the Gila River.

Invertebrates

Riffle habitats in the same headwater sites described for algae were typically composed of the following insect categories: 32% mayflies, 26% caddisflies, 23% true flies (Diptera), 13% stoneflies, and 5% beetles (Earl and Blinn 2003). The most abundant mayflies in riffle habitats were *Choroterpes*, *Epeorus*, *Leptohyphes*, and *Serratella*. *Cheumatopsyche* and *Hydropsyche* were common caddisfly taxa, and chironomid midges and black flies made up most of the dipteran assemblage. Insect densities were typically higher in May and June compared to late summer and fall. Riffles averaged 524 animals/m^2 compared to 1954 animals/m^2 in pool habitats.

Total insect densities were comparable at the same downstream site used for algae; however, insect diversity was reduced. Insect proportions below the agricultural reach were 44% dipterans, 34% mayflies, and 20% caddisflies in riffles (Earl and Blinn 2003). Stoneflies were absent at the lower site.

Vertebrates

Many native fishes in the Gila drainage are well adapted to the periodic flash flooding, heavy suspended sediment loads, and harsh conditions of the desert environment as described for the Little Colorado River. When stream channels stop flowing during the summer months, fishes move into standing pools that undergo dramatic diel physicochemical changes. Some fishes have been reported to survive in tiny volumes of water beneath mats of filamentous algae (Rinne and Minckley 1991).

The 14 native fishes presently found in the Gila River basin are represented by four families, including eight cyprinids (bonytail chub, Gila chub, headwater chub, roundtail chub, loach minnow, longfin dace, speckled dace, and spike dace), two catostomids (desert sucker and Sonora sucker), two salmonids (Apache trout and Gila trout), one cyprinidontoid (desert pupfish), and one poecilid (Gila topminnow).

The desert sucker, spike dace, and Gila and Apache trout occur only in the headwater streams of the Gila River, whereas the Sonora sucker and speckled dace are widespread throughout the perennial sections of the Gila. The loach minnow, desert sucker, headwater chub, and Sonora sucker occur in the middle elevations of the Aravaipa and San Carlos tributaries of the Gila. Typically, native fishes predominate above the confluence of the San Francisco River in the upper basin, whereas nonnatives are the dominant fishes below. Historically, the Gila chub occurred only in the headwaters but is now restricted to the middle section of the Gila River. All native fishes are currently listed as either federally endangered or threatened except for longfin dace, desert sucker, and Sonora sucker, which are under review for a listed status.

Bonytail chub, Colorado pikeminnow, flannelmouth sucker, and razorback sucker have all been extirpated from the Gila River main stem. The last confirmed report for Colorado pikeminnow in the Gila River was in 1950. Pikeminnow were once abundant in the Gila and an important source of food for Native Americans along the river until the late 1800s. A dramatic reduction in numbers occurred shortly after the construction of Hoover Dam on the Colorado main stem in the early 1930s. Gila trout formerly occurred in the headwaters of the Gila and Verde rivers of the Gila River drainage, but they are now restricted to a few remote headwater streams in New Mexico. Reasons for the extirpation and decline in these and other native fishes include regulated flow below dams, competition and predation by nonnative salmonids, hybridization with nonnatives, loss and degradation of habitat, and changes in water quality and quantity (Minckley and Deacon 1991, Rinne and Minckley 1991, Blinn et al. 1993).

Numerous nonnative fishes have been introduced and become established in the Gila River drainage, including black bullhead, black crappie, bluegill, brown trout, channel catfish, common carp, fathead minnow, flathead catfish, golden shiner, green sunfish, goldfish, guppy, largemouth bass, mosquitofish, Mozambique mouthbrooder, redear sunfish, red shiner, rainbow trout, Rio Grande killifish, Rio Grande sucker, sailfin molly, smallmouth bass, threadfin shad, and yellow bullhead. Yellow bass and white bass occur in reservoirs in the Gila basin.

Some of these nonnative fish species were introduced into canals and reservoirs in the lower Colorado River in the early or mid-1990s as sport or aquarium fishes and have migrated into the Gila River system. The Rio Grande sucker was recently introduced into the headwaters of the Gila River from the Rio Grande River in Texas (Rinne 1995). Sailfin mollies were introduced in 1952 and now occur throughout the lower Gila basin in canals and wastewater ponds. Mozambique mouthbreeders were introduced into the lower Colorado River in the early 1960s and presently occur in water refuges in the lower Gila River. Striped mullet moved into the lower reaches of the Gila River from the Gulf of California through the lower Colorado River and became landlocked in the brackish waters above diversion dams in the Gila River. It is now considered by some to be native to the lower Colorado River.

Beaver occur along upper sections of the Gila River where their preferred tree species (alders, cottonwoods, willows) occur, as do muskrat. The Chiricahua leopard frog and northern leopard frog have been reported from the higher elevations of the Gila River. The Ramsey Canyon leopard frog is found exclusively in the San Pedro River valley and the Rio Grande leopard frog has been inadvertently introduced into the lower Gila between Phoenix and Yuma. The lowland leopard frog has been extirpated from the lower Gila River.

Ecosystem Processes

Information on ecosystem processes is conspicuously lacking for the Gila River. However, like the LCR and other Sonoran Desert rivers, it is probable that much of the Gila River corridor is an autochtho-

nous system, depending on diatoms and cyanobacteria as organic energy sources, due to the limited woody vegetation along stream channels and the xeric landscape. Aquatic macrophytes are typically absent due to the high suspended sediment loads and unstable channels.

Although the main stem has received limited attention, two small tributaries, Aravaipa and Sycamore creeks, have been well studied. Aravaipa Creek is a tributary to the San Pedro River and Sycamore Creek is a tributary to the Verde River, each located in semiarid mountainous terrain. Studies have shown that flood-related disturbances and succession between disturbance events contribute greatly to the structure of these desert streams (Fisher and Grimm 1988). Due to intermittent flooding, succession is a continual process. Floods tend to thoroughly scour stream channels, which is followed by high primary ($3.3\,\mathrm{g}$ C $\mathrm{m^{-2}\,d^{-1}}$) and secondary ($135\,\mathrm{g}$ dry mass $\mathrm{m^{-2}\,yr^{-1}}$) production (Fisher 1995). Hydrologic linkage between surface and hyporheic subsystems increases ecosystem stability following flash floods (Valett et al. 1994).

Wildfires frequently occur in the higher elevations of the Gila Wilderness Area in New Mexico during summer electrical storms and deliver large quantities of fire ash to the Gila. Earl and Blinn (2003) reported a shift in diatom composition to smaller, more adnate forms, namely *Achnanthidium* spp. and *Cocconeis* spp., as a result of fire ash delivery. They also reported that stream drift was increased by tenfold and insect densities were reduced by nearly two orders of magnitude following the entry of coarse fire ash into streams.

Human Impacts and Special Features

The early development of the American Southwest has been largely at the expense of the Gila River basin. Early American Indians and Anglo settlers used its drainages as corridors to explore regions in the Southwest and its once perennial waters as a fishery and for agriculture. It has a rich history in American Indian culture from its headwaters to its mouth. The Gila River drainage continues to serve as an oasis for numerous desert animals that rely on its waters for food and protection. Unfortunately, few studies have been conducted on the ecological processes in this desert river, even though tributaries like the San Pedro River, which originates in Mexico, have been given special status as biodiversity reserves. Even the San Pedro is now threatened by the expansion of Fort Huachuca, Arizona.

The basin of the Gila River has been greatly changed by agriculture, ranching, mining, groundwater pumping, flood control, and municipal development for well over a century. Historically, sections of the Gila River were made up of large marshy areas and oxbow lakes that would become several kilometers wide in flood (Rinne 1994). The natural waters of the Gila no longer reach their historical mouth at the Colorado River but are contaminated by agriculture, mining, and municipal activities. By the early 1900s the surface flow was used up before the halfway mark of the river corridor at Coolidge Dam; now only flash floods and return flows from irrigation fields deliver intermittent water to the Colorado River. In the last three decades groundwater levels have dropped by nearly 25 m in parts of the lower basin (McNamee 1998).

The large zooplankter *Daphnia lumholtzi* has been introduced into a number of warmwater lakes, including Saguaro, Bartlett, Roosevelt, Canyon, and Apache reservoirs. This crustacean undergoes strong seasonal cyclomorphosis to avoid fish predation and may potentially alter the zooplankton structure of the reservoir systems (Dobberfuhl and Elser 2002).

Major cities, such as Phoenix, have been highly dependent upon the Gila River drainage for over a century and the groundwater beneath agricultural lands carries toxic agricultural pesticides. Sections of the lower Gila have been placed on the Environmental Protection Agency Superfund cleanup roster (McNamee 1998). As a result, many native fishes have disappeared from the Gila River and the cottonwoods and willows that once lined the river are generally gone. In spite of these enormous human impacts, the Gila River drainage still serves as a critical habitat for nearly all desert animals. Continued depletion of water will lead to further reduction in numbers of desert dwellers and perhaps to their extinction.

ADDITIONAL RIVERS

The Gunnison River begins as small streams in the high mountains of central Colorado (>4200 m asl) and flows to the west and north (Fig. 11.23). The winters in the upper Gunnison Basin are extremely cold (record minimum −42.7°C). As headwater streams drop rapidly through spruce and pine forests, they converge to create the East and Taylor rivers, which themselves unite at Almont to form the Gunnison River. As the river flows southwesterly toward and through Gunnison, it picks up Ohio Creek from the north, Tomichi Creek from the east, and Cebolla Creek from the south. The natural flow of the upper Gunnison is influenced by transbasin diversions to the eastern slope of the Rockies and by Taylor Reservoir on the Taylor River. The main stem below Gunnison is heavily regulated by three main-stem dams of the Aspinall Unit (Blue Mesa, Morrow Point, and Crystal). During the irrigation season, about 16% of the annual runoff is diverted via the East Portal to the southwest near Montrose in the heavily agricultural Uncompahgre River basin. Below this diversion the Gunnison flows through a spectacular chasm, the Black Canyon of Gunnison National Park. The river winds north until its confluence with the North Fork, at which time it turns west to Delta, where it is joined by the Uncompahgre River, which carries high selenium loads from irrigation return water. The Gunnison then flows to the northwest until its confluence with the upper Colorado at Grand Junction (~1600 m asl) (Fig. 11.8).

FIGURE 11.8 Gunnison River above Whitewater, Colorado (PHOTO BY TIM PALMER).

FIGURE 11.9 Virgin River above Virgin, Utah (Photo by Tim Palmer).

The San Juan is the river of the Four Corners area, originating in extreme southwestern Colorado and then arcing south and west through New Mexico before flowing briefly back through Colorado to the northwest into Utah (Fig. 11.25). Almost all of its flow is contributed by snowmelt runoff from permanent streams in southwestern Colorado. Heading in the San Juan Mountains at ~4000 m asl, the river flows southerly through Pagosa Springs and then into the Navajo Reservoir on the Colorado–New Mexico state line, where it is joined by the Piedro River. Below the reservoir the San Juan flows westward to Farmington, where it receives two major tributaries, the Animas and La Plata rivers, flowing southward out of Colorado across the Southern Ute Indian reservation. The river valleys of all of these rivers in this area are arid and extensively developed via irrigation withdrawals. The San Juan enters the Navajo Indian reservation near Shiprock, New Mexico, and then flows toward the Four Corners, where it briefly crosses Southern Ute Indian reservation lands in

Colorado before reentering Navajo lands in southeastern Utah. For the last ~150 km of its length the San Juan continues westerly through desert lands, eventually entering the impounded Colorado River at Lake Powell at ~1128 m asl (Fig. 11.1).

The Virgin River begins at about ~3000 m asl and drains the edge of the Markagunt Plateau. The Virgin flows southwesterly for 300 km along the western boundary of the Colorado Plateau through southwestern Utah, northwestern Arizona, and southeastern Nevada until it enters Lake Mead, Nevada, at about 365 m asl (Fig. 11.27). Lake Mead currently floods the lower 50 km of the Virgin. The terrain along the river corridor changes from steep-walled, mountainous canyons and sheer gradients near Zion National Park, Utah, to Colorado Plateau shrublands with broad, open canyons and low stream gradients and on to the arid Mohave Desert as it approaches Lake Mead (Fig. 11.9). A series of channel springs (Pah Tempe Springs) near La Verkin, Utah, have a major influence on downstream water quality. The springs contribute warm water (38°C to 42°C) at a rate of about 18 m³/min, and a salinity of about 9650 mg/L. Three native fishes (woundfin, Virgin spinedace, and Virgin River chub) are restricted to the Virgin River drainage and all have received some form of federal protection.

Three intermittent streams, the Big Sandy River, Santa Maria River, and Date Creek, join to form the Bill Williams River at ~295 m asl. Alamo Dam was built at the head of the Bill Williams in 1968 (Fig. 11.29). The Bill Williams flows westerly through a series of wetlands and backwaters that are periodically flooded from Alamo Dam. Annual mean discharge (4 m³/s) since the closure of the dam has been similar to before closure. A maximum flow of 1845 m³/s was recorded during August 1951 following a major summer storm. The lower 16 km of the river runs through the Bill Williams River National Wildlife Refuge before converging with the Colorado. The riparian corridor through the refuge is 10°C to 15°C cooler in the summer than nearby desert communities due to high evapotranspiration rates in the dense riparian vegetation. The refuge has the distinction of being one of a few reaches of southwestern rivers that still maintains a naturally regenerating riparian forest with most of its original communities intact. The periodic flood releases through the narrow channel from January through March are largely responsible for these conditions. The refuge serves as a sanctuary of plants and animals that were once present in the lower Colorado but are now found only on the refuge.

The Black River begins in the White Mountains in eastern Arizona at over 2700 m asl and flows onto semiarid grasslands (Fig. 11.10, Fig. 11.31). Two major headwater streams (West and East forks) flow south and eventually form the Black River below Buffalo Crossing. The headwaters contain some of the most natural riparian communities remaining in Arizona. From this point the Black flows through sections of the Bear Wallow Wilderness Area, a wild and scenic isolated mountain wilderness. The Black continues to flow southwesterly along the borders of the Fort Apache and San Carlos Indian reservations until it meets the White River at ~1300 m asl to form the Salt River. The Black has no major diversions except for three small reservoirs (Big Lake, Sierra Blanca Lake, and Crescent Lake) in the upper headwaters. The cool, clear, nearly unaltered waters in the higher elevations support one of the most diverse aquatic insect assemblages in Arizona, with over 30 species of caddisflies representing 22 genera reported at one site (Three-Forks) on the East Fork. These tribu-

FIGURE 11.10 Black River at Wildcat Bridge, Arizona (PHOTO BY D. BLINN).

taries also include some of the best sport fisheries in Arizona.

The Verde River begins at the headwaters of the intermittent Big Chino Wash, Arizona, at ~1325 m asl and flows for over 300 km through the Verde Valley (Fig. 11.33). Presently, the upper 60 km are relatively undisturbed. However, developers have proposed a recreational community near the springs that supply nearly 80% of the upper river's flow. The Verde flows easterly to Clarkdale, southeasterly to Camp Verde, and then arches sharply south through semiarid lands until it joins the Salt River near Phoenix at ~400 m asl (Fig. 11.11). Mean evaporation rates within the basin range from about 120 to 160 cm/yr. Six primary tributaries, including Sycamore Creek, Oak Creek, Wet Beaver, West Clear, Fossil Creek, and the East Verde River, deliver water to the Verde, each of which is <2.5 m³/s at its confluence. Oak Creek supports one of the most diverse aquatic insect assemblages in Arizona and Fossil Creek has active travertine spring deposits. Nonnative virile crayfish have greatly restructured the invertebrate community in the

FIGURE 11.11   Verde River below Camp Verde, Arizona (PHOTO BY TIM PALMER).

middle corridor of the Verde River. There are two main-stem dams, Horseshoe and Bartlett, ~50 km above the confluence with the Salt River that affect the hydrograph below; discharges average <17 m³/s compared to >23 m³/s prior to their completion. A 65 km section of the Verde between Camp Verde and Horseshoe Dam has been designated as a National Wild and Scenic River. The river contains one of few native fish communities in Arizona.

The Salt River forms at the junction of the White and Black rivers at ~1300 m asl and flows through Salt River Canyon in central Arizona (Fig. 11.35). The Verde River joins the Salt on the Fort

McDowell reservation and contributes an average discharge of about 17 m³/s. Air temperatures along the Salt range from 46°C in the lower valleys in the summer to −18°C during winter nights in the mountains. Annual precipitation ranges from 25 cm in the valleys to 64 cm in the mountains. Annual evapotranspiration approaches 178 cm. The first 200 km is unimpeded by dams, but the river is then fragmented by Theodore Roosevelt Dam and its lake, Horse Mesa Dam (Apache Lake), Mormon Flat Dam (Canyon Lake), and Stewart Memorial Dam (Saguaro Lake). Water stored in these reservoirs is diverted to the Granite Reef Diversion by the Salt

FIGURE 11.12 Salt River near Phoenix, Arizona (PHOTO BY C. E. CUSHING).

River Project and used by metropolitan Phoenix. Historically, the Salt ran through the present metropolitan Phoenix area and joined the Gila River ~70 km southwest of Phoenix (Fig. 11.12). Estimates show that annual mean discharge at the site of Roosevelt Dam were ~60 m³/s between 1888 and 1913. Since then annual discharge has been <25 m³/s. When the Salt finally reaches the Gila, even with contributions from the Verde, little water is left except for that rediverted from irrigation and municipal canals (USGS 2004a http://water.usgs.gov/public/nasqan). Water in sections of the Salt can reach >5000 μS/cm at base flow due to evaporite deposits on soils, high evapo-

ration rates on desert reservoirs, and recharged water from irrigation activities.

ACKNOWLEDGMENTS

We thank the following people for their contributions to the preparation of this chapter: Robert J. Hart and Cheryl K. Partin for climatological, hydrological, and water-quality information on the lower basin, and Kathleen Blair, Paul Boucher, Owen T. Gorman, Michael Lopez, Charles O. Minckley, John N. Rinne, Phil C. Rosen, Patti Spindler, Michael J. Sredl, David E. Ruiter, Mark J. Wetzel, Eric C. Dinger, and Michael Yard for general information on the

521

biota and fisheries throughout the lower basin. Mark R. Vinson provided information on rivers in the upper basin, James E. Deacon provided information on the Virgin River, and David Merritt assisted with botanical taxonomy and color photographs. Emma Benenati and David Cooper also provided color photographs, and Kirsten Rowell provided information on the Colorado River delta. Finally, we would like to give special thanks to Julian Older for providing general library research on the upper basin. Dana Franz also assisted with general library research on the upper basin. We also thank Joseph Shannon for insightful discussions of the Colorado River basin and a review of an earlier draft of the chapter, and the editors, Arthur Benke and Colbert E. Cushing, for their valuable comments on this chapter.

LITERATURE CITED

Abell, R. A., D. M. Olson, E. Dinerstein, P. T. Hurley, J. T. Diggs, W. Eichbaum, S. Walters, W. Wettengel, T. Allnutt, C. J. Loucks, and P. Hedao. 2000. *Freshwater ecoregions of North America: A conservation assessment.* Island Press, Washington, D.C.

Allred, T. M., and J. C. Schmidt. 1999. Channel narrowing by vertical accretion along the Green River near Green River, Utah. *Geological Society of America Bulletin* 111:1757–1772.

Ames, E. L. 1977. Aquatic insects of two western slope rivers, Colorado. Master's thesis, Colorado State University, Fort Collins.

Anderson, D. C., and D. J. Cooper. 2000. Plant–herbivore–hydroperiod interactions: Effects of native mammals on floodplain tree recruitment. *Ecological Applications* 10:1384–1399.

Andrews, E. D. 1986. Downstream effects of Flaming Gorge Reservoir on the Green River, Colorado and Utah. *Geological Society of America Bulletin* 97:1012–1023.

Andrews, E. D. 1991. Sediment transport in the Colorado River basin. In G. R. Marzolf (ed.). *Colorado River ecology and dam management: Proceedings of a symposium, May 24–25, 1990, Santa Fe, New Mexico,* pp. 54–74. National Academy Press, Washington, D.C.

Angradi, T. R. 1994. Trophic linkages in the lower Colorado River: Multiple stable isotope evidence. *Journal of the North American Benthological Society* 13:479–495.

Angradi, T. R., and D. M. Kubly. 1993. Effects of atmospheric exposure on chlorophyll *a*, biomass, and productivity of the epilithon of a tailwater river. *Regulated Rivers* 8:345–358.

Behnke, R. J., and D. E. Benson. 1983. Endangered and threatened fishes of the upper Colorado River basin. Bulletin 503A. Colorado State University Cooperative Extension Service, Ft. Collins.

Benenati, E. P., J. P. Shannon, D. W. Blinn, K. P. Wilson, and S. J. Hueftle. 2000. Reservoir–river linkages: Lake Powell and the Colorado River, Arizona. *Journal of the North American Benthological Society* 19:742–755.

Blinn, D. W., and G. A. Cole. 1991. Algal and invertebrate biota in the Colorado River: Comparison of pre- and post-dam conditions. In G. R. Marzolf (ed.). *Colorado River ecology and dam management: Proceedings of a symposium, May 24–25, 1990, Santa Fe, New Mexico,* pp. 102–123. National Academy Press, Washington, D.C.

Blinn, D. W., C. Runck, D. A. Clark, and J. N. Rinne. 1993. Effects of rainbow trout predation on Little Colorado spinedace. *Transactions of the American Fisheries Society* 122:139–143.

Blinn, D. W., J. P. Shannon, P. L. Benenati, and K. P. Wilson. 1998. Algal ecology in tailwater stream communities: The Colorado River below Glen Canyon Dam, Arizona. *Journal of Phycology* 34:734–740.

Blinn, D. W., J. P. Shannon, L. E. Stevens, and J. P. Carter. 1995. Consequences of fluctuating discharge for lotic communities. *Journal of the North American Benthological Society* 14:233–248.

Brown, D. (ed.). 1994. *Biotic communities: Southwestern United States and Northwestern Mexico.* University of Utah Press, Salt Lake City.

Carbajal, N., A. Sousa, and R. Durazo. 1997. A numerical model of the ex-ROFI of the Colorado River. *Journal of Marine Systems* 12:17–33.

Carlson, C. A., and R. T. Muth. 1989. The Colorado River: Lifeline of the American Southwest. In D. P. Dodge (ed.). *Proceedings of the International Large River Symposium (LARS).* Canadian Special Publication of Fisheries and Aquatic Sciences 106:220–239.

Cisneros-Mata, M. A., G. Montemayor-Lopez, and M. J. Roman-Rodriquez. 1995. Life history and conservation of *Totoaba maconaldi*. *Conservation Biology* 9:806–814.

Colton, H. L. 1937. Some notes on the original condition of the Little Colorado River: A side light on the problems of erosion. *Museum Notes, Museum of Northern Arizona* 10:17–20.

Cooper, D. J., D. M. Merritt, D. C. Andersen, and R. A. Chimner. 1999. Factors controlling the establishment of Fremont cottonwood seedlings on the upper Green River, USA. *Regulated Rivers* 15:419–440.

Cordy, G. E., J. A. Rees, R. J. Edmonds, J. B. Gebler, L. Wirt, D. J. Gellenbeck, and D. W. Anning. 1998. Water-quality assessment of the central Arizona basins, Arizona and northern Mexico: Environmental setting and overview of water quality. U.S. Geological Survey. Water Resources Investigations Report 98–4097.

Cross, J. N. 1985. Distribution of fish in the Virgin River, a tributary of the lower Colorado River. *Environmental Biology of Fishes* 12:13–21.

Czarnecki, D. B., and D. W. Blinn. 1978. Diatoms of the Colorado River in Grand Canyon National Park and vicinity. *Bibliotheca Phycologica* 38:1–182. J. Cramer, Stuttgart, Germany.

Czarnecki, D. B., D. W. Blinn, and T. Tompkins. 1976. A periphytic microflora analysis of the Colorado River and major tributaries in Grand Canyon National Park and vicinity. Publication no. 6. Colorado River Research Program, Flagstaff, Arizona.

Dobberfuhl, D. R., and J. J. Elser. 2002. Distribution and potential competitive effects of an exotic zooplankter (*Daphnia lumholtzi*) in Arizona reservoirs. *Journal of the Arizona–Nevada Academy of Science* 34:89–94.

Douglas, M. E., and P. C. Marsh. 1996. Population estimates/population movements of *Gila cypha*, an endangered cyprinid fish in the Grand Canyon region of Arizona. *Copeia* 1996:15–28.

Dynesius, M., and C. Nilsson. 1994. Fragmentation and flow regulation of river systems in the northern third of the world. *Science* 266:753–762.

Earl, S. R. 1999. Implications of forest fires, land use, and discharge on water chemistry and biota along the Gila River, New Mexico. Master thesis, Northern Arizona University, Flagstaff, Arizona.

Earl, S. R., and D. W. Blinn. 2003. Effects of wildfire ash on water chemistry and biota in southwestern USA streams. *Freshwater Biology* 48:1015–1030.

Ellis, M. A. 1914. Fishes of Colorado. *University of Colorado Study Series* 11:1–136.

Fisher, S. G. 1995. Stream ecosystems of the western United States. In C. E. Cushing, K. W. Cummins, and G. W. Minshall (eds.). *River and stream ecosystems*, pp. 61–87. Vol. 22 of *Ecosystems of the World*. Elsevier, Amsterdam.

Fisher, S. G., and N. B. Grimm. 1988. Disturbance as a determinant of structure in a Sonoran Desert ecosystem. *Verhandlungen der internationalen Vereinigung für theoretische und angewandte Limnologie* 23:1183–1189.

Fradkin, P. L. 1981. *A river no more: The Colorado River and the west.* University of California Press, Berkeley and Los Angeles.

Ghassemi, F., A. J. Jakeman, and H. A. Nix. 1995. *Salinization of land and water resources: Human causes, extent, management, and case studies.* University of New South Wales Press, Sydney, Australia.

Gido, K. B., and D. L. Propst. 1999. Habitat use and association of native and nonnative fishes in the San Juan River, New Mexico and Utah. *Copeia* 1999:321–332.

Gleick, P., W. C. G. Burns, E. L. Chalecki, M. Cohen, K. K. Cushing, A. S. Mann, R. Reyes, G. H. Wolff, and A. K. Wong. 2002. *The world's water: The biennial report on freshwater resources, 2002–2003.* Island Press, Washington, D.C.

Gorman, O. T., and D. M. Stone. 1999. Ecology of spawning humpback chub (*Gila cypha*) in the Little Colorado River near Grand Canyon, AZ. *Environmental Biology of Fishes* 55:115–133.

Greenberg, D. L. 1999. An analysis of genetic variation in Little Colorado spinedace (*Lepidomeda vittata*) using amplified fragment length polymorphic markers. Master's thesis, Northern Arizona University, Flagstaff.

Gregory, S. C., and J. E. Deacon. 1994. Human induced changes to native fishes in the Virgin River drainage. In R. A. Marston (ed.). *Effects of human induced changes on hydrologic systems*, pp. 435–444. American Water Resources Association, Technical Publication Series 94-3.

Haden, G. A., D. W. Blinn, J. P. Shannon, and K. P. Wilson. 1999. Driftwood: An alternative habitat for macroinvertebrates in a large desert river. *Hydrobiologia* 397:179–186.

Haden, G. A., J. P. Shannon, K. P. Wilson, and D. W. Blinn. 2003. Benthic community structure of the Green and Colorado Rivers through Canyonlands National Park, Utah, USA. *Southwestern Naturalist* 48:23–35.

Hardwick, G. G., D. W. Blinn, and H. D. Usher. 1992. Epiphytic diatoms on *Cladophora glomerata* in the Colorado River, Arizona: Longitudinal and vertical distribution on a regulated river. *Southwestern Naturalist* 37:148–156.

Hauer, F., J. A. Stanford, and J. V. Ward. 1989. Serial discontinuities in a Rocky Mountain River. II. Distribution and abundance of Trichoptera. *Regulated Rivers* 3:177–182.

Holden, P. B. (ed.). 1999. *Flow recommendations for the San Juan River.* San Juan River Basin Recovery Implementation Program, U.S. Fish and Wildlife Service, Albuquerque, New Mexico.

Holden, P. B., and L. W. Crist. 1981. Documentation of changes in the macroinvertebrate and fish populations in the Green River due to inlet modification of Flaming Gorge Dam. Final Report PR-16-5. BIO/WEST, Logan, Utah.

Holden, P. B., and C. B. Stalnaker. 1975. Distribution of fishes in the Dolores and Yampa River systems of the upper Colorado River basin. *Southwestern Naturalist* 19:403–412.

Holmgren, A. H. 1962. *The vascular plants of the Dinosaur National Monument.* Utah State University, Logan.

Hunt, C. B. 1974. *Natural regions of the United States and Canada.* W. H. Freeman and Company, San Francisco.

Iorns, W. V., C. H. Hembree, and G. L. Oakland. 1965. Water resources of the upper Colorado River basin. U. S. Geological Survey Professional Paper 441. U. S. Government Printing Office, Washington, D.C.

Irving, D., and T. Modde. 2000. Home-range fidelity and use of historical habitat by adult Colorado squawfish (*Ptychocheilus lucius*) in the White River, Colorado and Utah. *Western North American Naturalist* 60:16–25.

Jefferson, J., R. W. Delaney, and G. C. Thompson. 1972. *The Southern Utes: A tribal history.* Southern Ute Tribe, Ignacio, Colorado.

Marzolf, G. R. (ed.). 1991. *Colorado River ecology and dam management: Proceedings of a symposium, May*

24–25, 1990, Santa Fe, New Mexico. National Academy Press, Washington, D.C.

McAda, C. W. 2000. Flow recommendations to benefit endangered fishes in the Colorado and Gunnison River. Final Report. Fishery Investigations of the Lower Gunnison River Drainage, Colorado Division of Wildlife, Fort Collins.

McKinney, T., A. D. Ayers, and R. S. Rogers. 1999. Macroinvertebrate drift in the tailwaters of a regulated river below Glen Canyon Dam, Arizona. *Southwestern Naturalist* 44:205–242.

McMahon, J. P., T. Moody, S. I. Apfelbaum, S. M. Lehnhardt, and V. LaGesse. 2001. The potential for restoration along the Virgin River in Zion National Park. Final Report. Grand Canyon Trust, St. George, Utah.

McNamee, G. 1998. *Gila: The life and death of an American river.* University of New Mexico Press, Albuquerque.

Merritt, D. M., and D. J. Cooper. 2000. Riparian vegetation and channel change in response to river regulation: A comparative study of regulated and unregulated streams in the Green River basin, USA. *Regulated Rivers* 16:543–564.

Minckley, W. L. 1991. Native fishes of the Grand Canyon region: An obituary? In G. R. Marzolf (ed.), *Colorado River ecology and dam management: Proceedings of a symposium, May 24–25, 1990, Santa Fe, New Mexico,* pp. 123–177. National Academy Press, Washington, D.C.

Minckley, W. L., and J. E. Deacon (eds.). 1991. *Battle against extinction: Native fish management in the American West.* University of Arizona, Tucson.

Minckley, W. L., and B. DeMarais. 2000. Taxonomy of chubs (Teleostei, Cyprinidae, genus *Gila*) in the American Southwest with comments on conservation. *Copeia* 2000:251–256.

Minckley, W. L., P. C. Marsh, J. E. Deacon, T. E. Dowling, P. W. Hedrick, W. J. Matthews, and G. Mueller. 2003. A conservation plan for native fishes of the Lower Colorado River. *BioScience* 53:219–234.

Minckley, W. L., R. R. Miller, and S. M. Norris. 2002. Three new pupfish species, *Cyprinodon* (Teleostei, Cyprinodontidae) from Chihuahua, Mexico, and Arizona, USA. *Copeia* 2002: 687–705.

Muth, R. T., L. W. Crist, K. E. LaGory, J. H. Hayse, K. R. Bestgen, T. P. Ryan, J. K. Lyons, and R. A. Valdez. 2000. Flow and temperature recommendations for endangered fishes in the Green River downstream of Flaming Gorge Dam. Final Report, Project FG-3. Upper Colorado River Endangered Fish Recovery Program, Lakewood, Colorado.

Novy, J., and M. Lopez. 1991. *East fork of Black River fish management report.* Arizona Game and Fish Department, Phoenix.

Oberlin, G. E., J. P. Shannon, and D. W. Blinn. 1999. Watershed influence on the macroinvertebrate fauna of ten major tributaries of the Colorado River through Grand Canyon, Arizona. *Southwestern Naturalist* 44:17–30.

Patten, D. T. and L. E. Stevens. 2001. Restoration of the Colorado River ecosystem using planned flooding. *Ecological Applications* 11:633–634.

Patten, D. T., D. A. Harpman, M. I. Voita, and T. J. Randle. 2001. A managed flood on the Colorado River: background, objectives, design, and implementation. *Ecological Applications* 11:635–643.

Patrick, R. 2000. *Rivers of the United States.* Vol. 5, part A: *The Colorado River.* John Wiley and Sons, New York.

Peterson, C. G. 1986. Effects of discharge reduction on diatom colonization below a large hydroelectric dam. *Journal of the North American Benthological Society* 5:278–289.

Poff, N. L. 1996. A hydrogeography of unregulated streams in the United States and an examination of scale-dependence in some hydrological descriptors. *Freshwater Biology* 36:71–91.

Pomeroy, K., J. P. Shannon, and D. W. Blinn. 2000. Leaf breakdown in a regulated desert river: Colorado River, Arizona, U.S.A. *Hydrobiologia* 434:193–199.

Pope, G. L., P. D. Rigas, and C. E. Smith. 1998. Statistical summaries of streamflow data and characteristics of drainage basins for selected streamflow-gaging stations in Arizona through water year 1996. Water Resources Investigations Report 98-4225. U.S. Geological Survey, Water Resources Division, Denver, Colorado.

Rader, R. B., and J. V. Ward. 1988. Influence of regulation on environmental conditions and the macroinvertebrate community in the Upper Colorado River. *Regulated Rivers* 2:597–618.

Richter, B. D., and H. E. Richter. 2000. Prescribing flood regimes to sustain riparian ecosystems along meandering rivers. *Conservation Biology* 14:1467–1478.

Ricketts, T. H., E. Dinerstein, D. M. Olson, C. J. Loucks, W. Eichbaum, D. DellaSala, K. Kavanaugh, P. Hedao, P. T. Hurley, K. M. Carney, R. Abell, and S. Walters. 1999. *Terrestrial ecoregions of North America: A conservation assessment.* Island Press, Washington, D.C.

Rinne, J. N. 1994. Declining southwestern aquatic habitats and fishes: Are they sustainable? In W. W. Covington and L. F. DeBano (eds.). *Sustainable ecological systems: Implementing an ecological approach to land management,* pp. 256–264. General Technical Report RM-247. Rocky Mountain Forest and Range Experiment Station, Fort Collins, Colorado.

Rinne, J. N. 1995. Reproductive biology of the Rio Grande Sucker, *Catostomus plebeius* (Cypriniformes), in a montane stream, New Mexico. *Southwestern Naturalist* 40:237–241

Rinne, J. N. 1998. Fish community structure in the Verde River, Arizona, 1974–1977. Proceedings of Hydrology Section of the Arizona–Nevada Academy of Science. 28:75–80.

Rinne, J. N., J. A. Stefferud, A. Clark, and P. Sponholtz. 1998. Fish community structure in the Verde River, Arizona, 1975–1997. Hydrology and Water Resources in Arizona and the Southwest 28:75–80.

Rinne, J. N., and W. L. Minckley. 1991. Native fishes of arid lands: A dwindling resource of the desert southwest. General Technical Report RM-206. U.S. Forest Service, Mountain Forest and Range Experiment Station, Fort Collins, Colorado.

Robinson, A. T., D. M. Kubly, R. W. Clarkson, and E. D. Creef. 1996. Factors limiting the distributions of native fishes in the Little Colorado River, Grand Canyon, Arizona. *Southwestern Naturalist* 41:378–387.

Rodriguez, C. A., K. W. Flessa, and D. L. Dettman. 2001. Effects of upstream diversion of the Colorado water on the estuarine bivalve mollusc *Mulinia coloradoensis*. *Conservation Biology* 15:249–258.

Schade, J. D., and S. G. Fisher. 1997. Leaf litter in a Sonoran desert stream ecosystem. *Journal of the North American Benthological Society* 16:612–626.

Shannon, J. P., D. W. Blinn, P. L. Benenati, and K. P. Wilson. 1996. Organic drift in a regulated desert river. *Canadian Journal of Fisheries and Aquatic Sciences* 53:1360–1369.

Shannon, J. P., D. W. Blinn, G. A. Haden, E. P. Benenati, and K. P. Wilson. 2001. Food web implications of ^{13}C and ^{15}N variability over 370 km of the regulated Colorado River, USA. *Isotopes in Environmental and Health Studies* 37:179–191.

Shannon, J. P., D. W. Blinn, T. McKinney, E. P. Benenati, K. P. Wilson, and C. O'Brien. 2001. Aquatic food base response to the 1996 test flood below Glen Canyon Dam, Colorado River, Arizona. *Ecological Applications* 11:672–685.

Shannon, J. P., D. W. Blinn, and L. E. Stevens. 1994. Trophic interactions and benthic animal community structure in the Colorado River, Arizona, USA. *Freshwater Biology* 31:213–220.

Shaver, M. L., J. P. Shannon, K. P. Wilson, P. L. Benenati, and D. W. Blinn. 1997. Effects of suspended sediment and desiccation on the benthic tailwater community in the Colorado River, USA. *Hydrobiologia* 357:63–72.

Silvey, W., J. N. Rinne, and R. Sorenson. 1986. *Index to the natural drainage systems of Arizona: A computer compatible digital identification of perennial lotic waters.* Wildlife Unit Technical Series, Southwestern Region, Phoenix, Arizona.

Smith, A. M. 1974. *Ethnography of the Northern Utes.* Papers in Anthropology no. 17. Museum of New Mexico Press, Albuquerque, New Mexico.

Spamer, E. E., and A. E. Bogan. 1993. Mollusca of the Grand Canyon and vicinity, Arizona: New and revised data on diversity and distributions, with notes on Pleistocene–Holocene mollusks of the Grand Canyon. *Proceedings of the Academy of Natural Sciences of Philadelphia* 144:21–68.

Stanford, J. A., and J. V. Ward. 1986. The Colorado River system. In B. R. Davies and K. F. Walker (eds.). *The ecology of river systems*, pp. 353–373. Monographiae Biologicae Vol. 60. Dr. W. Junk, The Hague.

Stanford, J. A., and J. V. Ward. 1991. Limnology of Lake Powell and the chemistry of the Colorado River. In G. R. Marzolf (ed.). *Colorado River ecology and dam management: Proceedings of a symposium, May 24–25, 1990, Santa Fe, New Mexico*, pp. 75–101. National Academy Press, Washington, D.C.

Starnes, W. C. 1995. Colorado River basin fishes. In E. T. LaRoe, G. S. Ferris, C. E. Puckett, P. D. Doran, and M. J. Mac (eds.). *Our living resources: A report to the nation on the distribution, abundance, and health of U.S. plants, animals, and ecosystems*, pp. 149–152. U.S. Department of the Interior, National Biological Service, Washington, D.C.

Stevens, L. E., K. A. Buck, B. T. Brown, and N. C. Kline. 1997. Dam and geomorphological influences on Colorado River waterbird distribution, Grand Canyon, Arizona, USA. *Regulated Rivers* 13:151–169.

Stevens, L. E., J. P. Shannon, and D. W. Blinn. 1997. Colorado River benthic ecology in Grand Canyon, Arizona, USA: Tributary and geomorphological influences. *Regulated Rivers* 3:129–149.

Stevens, L. E., J. E. Sublette, and J. P. Shannon. 1998. Chironomidae (Diptera) of the Colorado River, Grand Canyon, Arizona, USA. II. Factors influencing distribution. *Great Basin Naturalist* 58:147–155.

Sykes, G. 1937. *The Colorado delta.* Special Publication 19. American Geological Society, New York.

Tadayon, S., N. R. Duet, G. G. Fisk, H. F. McCormack, C. K. Partin, G. L. Pope, and P. D. Rigas. 2001. *Water resources data Arizona: Water year 2000.* U.S. Department of the Interior, Tucson, Arizona.

Trimble, M. 1989. Arizona: *A calvalcade of history.* Treasure Chest, Tucson, Arizona.

Tyus, H. M., B. D. Burdick, R. A. Valdez, C. M. Haynes, T. A. Lytle, and C. R. Berry. 1982. Fishes of the upper Colorado River basin: Distribution, abundance and status. In W. H. Miller, H. M. Tyus and C. A. Carlson (eds.). *Fishes of the upper Colorado River system: Present and future*, pp. 12–70. American Fisheries Society, Bethesda, Maryland.

Tyus, H. M., and C. A. Karp. 1989. Habitat use and streamflow needs of rare and endangered fishes, Yampa River, Colorado. Biological Report 89. U.S. Fish and Wildlife Service, Washington, D.C.

Valett, H. M., S. G. Fisher, N. B. Grimm, and P. Camill. 1994. Vertical hydrologic exchange and ecological stability of a desert stream ecosystem. *Ecology* 75:548–560.

Vanicek, C. D., R. H. Kramer, and D. R. Franklin. 1970. Distribution of Green River fishes in Utah and Colorado following closure of Flaming Gorge Dam. *Southwestern Naturalist* 14:297–315.

Vinson, M. R. 2001. Long-term dynamics of an invertebrate assemblage downstream from a large dam. In D. T. Patten and L. E. Stevens (eds.). Restoration of the Colorado River ecosystem using planned flooding, pp. 711–730. *Ecological Applications* 11:633–730.

Ward, J. V., and B. C. Kondratieff. 1992. *An illustrated guide to the mountain stream insects of Colorado.* University Press of Colorado, Niwot.

Ward, J. V., H. J. Zimmermann, and L. D. Cline. 1986. Lotic zoobenthos of the Colorado system. In B. R. Davies and K. F. Walker (eds.). *The ecology of river systems,* pp. 403–423. Monographiae Biologicae Vol. 60. Dr. W. Junk, The Hague.

Webb, R. H., J. C. Schmidt, G. R. Marzolf, and R. A. Valdez (eds.). 1999. *The controlled flood in Grand Canyon.* Geophysical Monograph Series 110. American Geophysical Union, Washington, D.C.

White, J. N. 1995. Indirect effects of predation by crayfish on Little Colorado spinedace. Master thesis, Northern Arizona University, Flagstaff.

Winget, R. N., and R. W. Baumann. 1977. Virgin River, Utah–Arizona–Nevada, aquatic habitat, fisheries and macroinvertebrate studies, Section B. In A. Owen (ed.).

Impact of Warner Valley project on endangered fish of the Virgin River. Vaughn Hansen Associates, Salt Lake City.

Wohl, E. E. 2001. *Virtual rivers: Lessons from the mountain rivers of the Colorado Front Range.* Yale University Press, New Haven, Connecticut.

Wolz, E. R., and D. K. Shiozawa. 1995. Soft-sediment benthic macroinvertebrate communities of the Green River at the Ouray National Wildlife Refuge, Uintah County, Utah. *Great Basin Naturalist* 55:213–224.

Woodbury, A. M. (ed.). 1963. *Studies of the biota in Dinosaur National Monument, Utah and Colorado.* Miscellaneous Papers 1. Division of Biological Science, Institute of Environmental Biological Research, University of Utah, Salt Lake City.

Young, K. L., and P. Lopez (eds.). 1999. Integrated fisheries management plan for the Little Colorado River basin. Technical Report 146. Arizona Game and Fish Department, Phoenix.

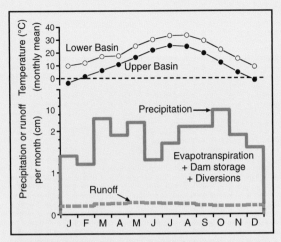

FIGURE 11.14 Mean monthly air temperature, precipitation, and runoff for the Colorado River basin.

COLORADO RIVER

Relief: ~4100 m

Basin area: 642,000 km^2

Mean discharge: 550 m^3/s (virgin) ~40 m^3/s (present)

River order: 6

Mean annual precipitation: 22.1 cm (upper basin), 10.6 cm (lower basin)

Mean air temperature: 14°C (upper basin), 21.1°C (lower basin)

Mean water temperature: 11°C (upper basin), 21°C (lower basin)

Physiographic provinces: Middle Rocky Mountains (MR), Southern Rocky Mountains (SR), Wyoming Basin (WB), Colorado Plateaus (CO), Basin and Range (BR), Baja California (BC)

Biomes: Tundra, Temperate Mountain Forest, Desert

Freshwater ecoregion: Colorado

Terrestrial ecoregions: 9 ecoregions (see text)

Number of fish species: ~75 to 85 (42 native)

Number of endangered species: 16 fishes

Major fishes: bonytail chub, brook trout, brown trout, channel catfish, common carp, fathead minnow, flannelmouth sucker, humpback chub, longfin dace, rainbow trout, razorback sucker, red shiner, roundtail chub, speckled dace

Major other aquatic vertebrates: American coot, American widgeon, Arizona toad, beaver, bullfrog, Great Basin spadefoot toad, great blue heron, mallard, muskrat, northern leopard frog, snowy egret, Sonoran mud turtle, Woodhouse toad

Major benthic invertebrates: mayflies (*Baetis*, *Drunella*, *Ephemerella*, *Epeorus*, *Heptagenia*, *Paraleptophlebia*, *Traverella*, *Tricorythodes*), stoneflies (*Capnia*, *Paraleuctra*, *Suwallia*), caddisflies (*Ceratopsyche*, *Chimarra*, *Hydropsyche*, *Hydroptila*, *Ochrotrichia*, *Oecetis*), chironomid midges (*Eukiefferiella*, *Orthocladius*), snails (*Lymnaea*, *Physa*)

Nonnative species: Asiatic clam, bullfrog, >30 fishes carp, channel catfish, rainbow trout, threadfin shad), *Daphnia lumholtzi*, *Gammarus lacustris*, New Zealand mudsnail, red swamp crayfish, saltcedar, virile crayfish, water fern

Major riparian plants: Arizona sycamore, bulrush, coyote willow, Fremont cottonwood, gamble oak, narrowleaf cottonwood, saltcedar, southern cattail, velvet mesquite

Special features: large, highly regulated desert river; drains nearly 8% of United States; most native fishes are threatened; many nonnative species; runs through Grand Canyon

Fragmentation: over 40 flow-regulation structures, 4 large main-stem reservoirs and numerous diversions; one of most regulated rivers in world

Water quality: pH = 8.3, alkalinity = 105 mg/L as CaCO$_3$, NO$_3$-N = 0.12 mg/L, PO$_4$-P = 0.04 mg/L, specific conductance = 1000 μS/cm, suspended sediments = 28 mg/L; some sections subject to mine contamination

Land use: 67% agriculture, 25% forest, 8% urban

Population density: 7 people/km^2

Major information sources: Tyus et al. 1982, Stanford and Ward 1986, Marzolf 1991, Starnes 1995, Webb et al. 1999, Patrick 2000, Patten and Stevens 2001, Patten et al. 2001

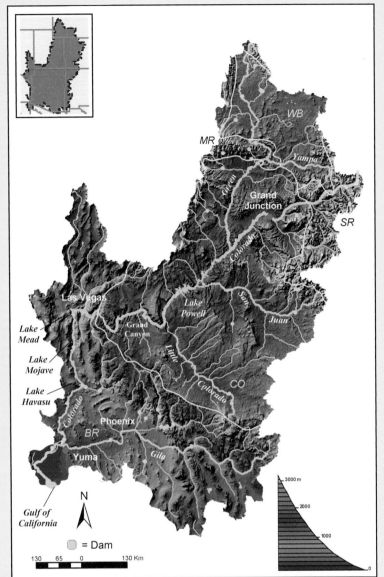

FIGURE 11.13 Map of the Colorado River basin. Physiographic provinces are separated by yellow lines.

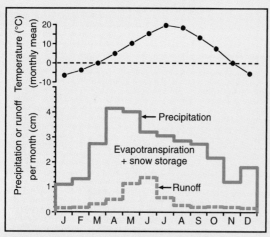

FIGURE 11.16 Mean monthly air temperature, precipitation, and runoff for the Green River basin.

GREEN RIVER

Relief: ~2950 m

Basin area: 116,200 km²

Mean discharge: 172 m³/s

River order: 5

Mean annual precipitation: 31.9 cm

Mean air temperature: 6.3°C

Mean water temperature: 2°C to 14°C below Flaming Gorge Dam, 2°C to 25°C at Rkm 189

Physiographic provinces: Middle Rocky Mountains (MR), Wyoming Basin (WB), Southern Rocky Mountains (SR), Colorado Plateau (CO)

Biomes: Temperate Mountain Forest, Desert

Freshwater ecoregion: Colorado

Terrestrial ecoregions: Wyoming Basin Shrub Steppe, South Central Rockies Forests, Wasatch and Uinta Montane Forests, Colorado Plateau Shrublands

Number of fish species: 37 (12 native)

Number of endangered species: 3 fishes

Major fishes: bonytail chub, Colorado River cutthroat trout, Colorado pikeminnow, humpback chub, razorback sucker, roundtail chub, rainbow trout

Major other aquatic vertebrates: beaver, boreal western toad, Clark's grebe, muskrat, northern leopard frog, spadefoot toad, tiger salamander, white-faced ibis, whooping crane

Major benthic invertebrates: mayflies (*Baetis, Drunella, Ephemerella, Heptagenia*), stoneflies (*Arcynopteryx, Capnia, Hesperoperla, Taenionema*), caddisflies (*Brachycentrus, Helicopsyche, Hesperophylax, Hydroptila, Leucotrichia, Oecetis, Psychoglypha, Rhyacophila*), crustaceans (*Gammarus, Hyalella,* signal crayfish, virile crayfish)

Nonnative species: channel catfish, common carp, fathead minnow, rainbow trout, saltcedar, signal crayfish, virile crayfish

Major riparian plants: alkali sacaton, big sagebush, black greasewood, Fremont cottonwood, saltcedar, saltgrass

Special features: large desert river; important habitat for endangered fishes indigenous to Colorado River system

Fragmentation: major reservoirs on main stem in upper basin (Flaming Gorge and Fontenelle); major tributaries extensively dammed or diverted, except Yampa River

Water quality: pH = 8.3, alkalinity = 165 mg/L as $CaCO_3$, NO_3-N = 0.04 mg/L, PO_4-P = <0.001 mg/L, specific conductance = 585 µS/cm, suspended sediment = 900 mg/L

Land use: 80% agriculture, 15% forest, 5% urban

Population density: <0.5 people/km²

Major information sources: Woodbury 1963, Wolz and Shiozawa 1995, Muth et al. 2000, Merritt and Cooper 2000, Vinson 2001

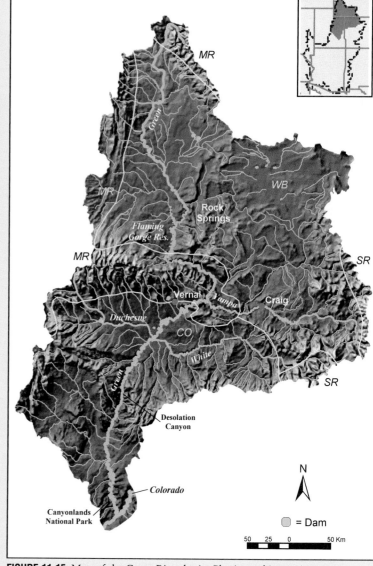

FIGURE 11.15 Map of the Green River basin. Physiographic provinces are separated by yellow lines.

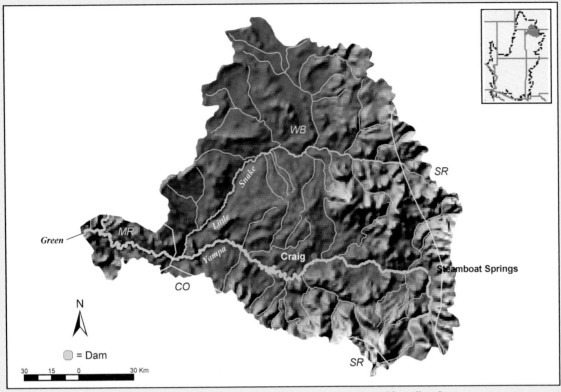

FIGURE 11.17 Map of the Yampa River basin. Physiographic provinces are separated by yellow lines.

YAMPA RIVER

Relief: ~2276 m
Basin area: 24,595 km²
Mean discharge: 64 m³/s
River order: 4
Mean annual precipitation: 42.8 cm
Mean air temperature: 5.9°C
Mean water temperature: 9°C
Physiographic provinces: Southern Rocky Mountains (SR), Middle Rocky Mountains (MR), Wyoming Basin (WB)
Biomes: Tundra, Temperate Mountain Forest, Desert
Freshwater ecoregion: Colorado
Terrestrial ecoregions: Colorado Rockies Forests, Colorado Plateau Shrublands
Number of fish species: 30 (12 native)
Number of endangered species: 4 fishes
Major fishes: bluehead sucker, bonytail chub, channel catfish, Colorado pikeminnow, Colorado River cutthroat trout, flannelmouth sucker, green sunfish, humpback chub, mountain sucker, mountain whitefish, northern pike, razorback sucker, roundtail chub, speckled dace

FIGURE 11.18 Mean monthly air temperature, precipitation, and runoff for the Yampa River basin.

Major other aquatic vertebrates: beaver, boreal western toad, Great Basin spadefoot toad, muskrat, northern leopard frog, river otter, wood frog
Major benthic invertebrates: mayflies (*Baetis, Choroterpes, Ephemerella, Heptagenia, Paraleptophlebia, Tricorythodes, Rhithrogena, Traverella*), stoneflies (*Alloperla, Isoperla, Pteronarcella, Pteronarcys*), caddisflies (*Cheumatopsyche, Helicopsyche, Hydropsyche, Lepidostoma, Oecetis, Polycentropus*), Coleoptera (*Optioservus*), crustaceans (*virile crayfish*)
Nonnative species: channel catfish, green sunfish, northern pike, saltcedar, virile crayfish
Major riparian plants: biennial sage, Canada bluegrass, coyote willow, foxtail, Fremont cottonwood, horseweed, slender wheatgrass, western wheatgrass
Special features: last remaining free-flowing river in upper Colorado Basin; montane, high plains, and canyon river with perennial flow; critical to recovery of endangered native fishes in upper basin
Fragmentation: minimal, small headwater dams and water abstraction ≤10% of annual flow
Water quality: pH = 8.4, alkalinity = 85 mg/L as CaCO₃, NO₃-N = 0.04 mg/L, PO₄-P = 0.005 mg/L, specific conductance = 500 μS/cm, suspended sediments = <500 mg/L near confluence with Green River
Land use: 65% agriculture, 30% forest, 5% urban
Population density: <0.5 people/km²
Major information sources: Woodbury 1963, Ames 1977, Tyus and Karp 1989, Muth et al. 2000, Merritt and Cooper 2000

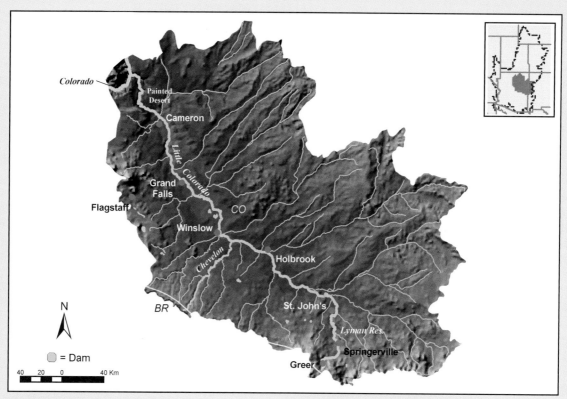

FIGURE 11.19 Map of the Little Colorado River basin. Physiographic provinces are separated by a yellow line.

LITTLE COLORADO RIVER

Relief: ~2600 m
Basin area: ~69,000 km²
Mean discharge: 6.5 m³/s
River order: 4
Mean annual precipitation: 17.2 cm
Mean air temperature: 13.5°C
Mean water temperature: 18°C
Physiographic province: Colorado Plateaus (CO)
Biomes: Temperate Mountain Forest, Desert
Freshwater ecoregion: Colorado
Terrestrial ecoregions: Arizona Mountains Forests, Colorado Plateau Shrublands
Number of fish species: 33 (9 native)
Number of endangered species: 2 fishes, 1 clam
Major fishes: Apache trout, bluehead sucker, brown trout, fathead minnow, flannelmouth sucker, green sunfish, humpback chub, Little Colorado River sucker, Little Colorado spinedace, rainbow trout, roundtail chub, speckled dace

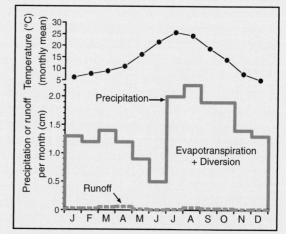

FIGURE 11.20 Mean monthly air temperature, precipitation, and runoff for the Little Colorado River basin.

Major other aquatic vertebrates: Arizona toad, beaver, bullfrog, Chiricahua leopard frog, great blue heron, belted kingfisher, mallard, muskrat, red-spotted toad, Sonoran mud turtle, striped chorus frog
Major benthic invertebrates: mayflies (*Baetis, Cinygmula, Drunella, Epeorus, Tricorythodes*), caddisflies (*Anabolia, Atopsyche, Cheumatopsyche, Glossosoma, Gumaga, Helicopsyche, Hesperophylax, Hydropsyche, Limnephilus, Oecetis, Oligophlebodes, Onocosmoecus, Polycentropus, Rhyacophila*), stoneflies (*Claassenia, Sweltsa*), crustaceans (virile crayfish) chironomid midges, black flies
Nonnative species: Asiatic clam, brown trout, bullfrog, rainbow trout, fathead minnow, green sunfish, saltcedar, virile crayfish
Major riparian plants: Bebb willow, bulrush, cattails, common reed, Goodding willow, thin-leaf alder, saltcedar
Special features: large, ephemeral desert river; river dry or pooled >60% of year; headwaters recommended for conservation by World Wildlife Fund
Fragmentation: no major dams but has Lyman Lake; Zion Reservoir used for flood control until 2000; numerous diversions
Water quality: pH = 8.3, alkalinity = 195 mg/L as CaCO₃, NO₃-N = 0.7 mg/L, PO₄-P = 0.03 mg/L, specific conductance = 2500 µS/cm (4500 µS/cm at confluence with Colorado River), suspended sediments = >500 mg/L
Land use: <1% agriculture, 15% forest, <5% urban, 55% Indian reservation, allotments, and trusts
Population density: 1.5 people/km²
Major information sources: Silvey et al. 1986, Young and Lopez 1999

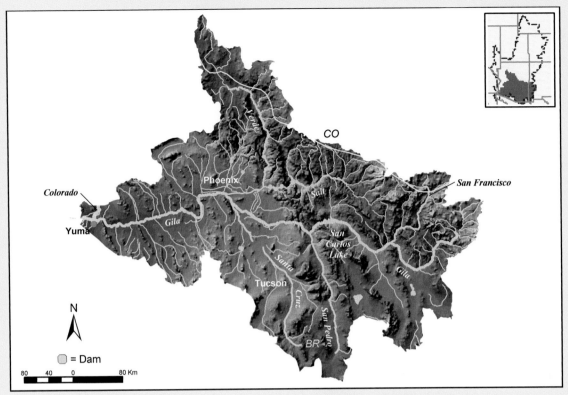

FIGURE 11.21 Map of the Gila River basin. Physiographic provinces are separated by a yellow line.

GILA RIVER

Relief: ~3050 m
Basin area: ~149,832 km²
Mean discharge: >40 m³/s (virgin), <6 m³/s present
River order: 6
Mean annual precipitation: 24.7 cm
Mean air temperature: 15.2°C
Mean water temperature: 21°C
Physiographic provinces: Basin and Range (BR), Colorado Plateaus (CO)
Biomes: Temperate Mountain Forest, Desert
Freshwater ecoregion: Gila
Terrestrial ecoregions: Arizona Mountains Forests, Madrean Sky Islands Montane Forests, Chihuahuan Desert, Sonoran Desert
Number of fish species: 36 (19 native)
Number of endangered species: 2 fishes
Major fishes: channel catfish, desert sucker, flathead catfish, loach minnow, longfin dace, roundtail chub, smallmouth bass, Sonora sucker, speckled dace, spike dace
Major other aquatic vertebrates: beaver, bullfrog, muskrat, Arizona toad, Chiricahua leopard frog, lowland leopard frog, Sonoran mud turtle, red-spotted toad, Sonoran Desert toad, Woodhouse toad
Major benthic invertebrates: mayflies (*Choroterpes, Epeorus, Ephemerella, Serratella, Thraulodes, Tricorythodes, Traverella*), caddisflies (*Cheumatopsyche, Chimarra, Helicopsyche, Hydropsyche, Polycentropus, Oecetis, Protoptila, Ochrotrichia, Smicridea, Zumatrichia*), hellgrammites (*Corydalus*), chironomid midges, black flies, crustaceans (virile crayfish)
Nonnative species: bullfrog, channel catfish, *Daphnia lumholtzi*, flathead catfish, Rio Grande leopard frog, saltcedar, smallmouth bass, virile crayfish
Major riparian plants: Arizona sycamore, Bebb willow, box elder, Fremont cottonwood, Goodding willow, green ash, mesquite, narrowleaf cottonwood, thin-leaf alder, saltcedar
Special features: several sites in upper basin recommended for conservation by the World Wildlife Fund; ≥800 bird species in Gila Cliff Valley
Fragmentation: Coolidge, Painted Rock, and Gillespie dams on main stem; numerous main-stem diversions
Water quality: pH = 8.3, alkalinity = 160 mg/L as CaCO₃, NO₃-N = 0.03 mg/L, PO₄-P = 0.1 mg/L, specific conductance = 900 μS/cm, suspended sediments = 300 mg/L
Land use: 70% agriculture, 20% forest, 10% urban; heavy agricultural, mining, and municipal use
Population density: ~25 people/km²
Major information sources: Silvey et al. 1986, Cordy et al. 1998, Earl and Blinn 2003, Abell et al. 2000

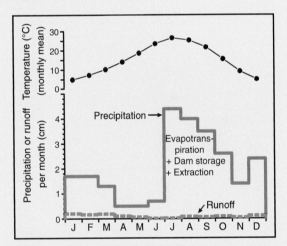

FIGURE 11.22 Mean monthly air temperature, precipitation, and runoff for the Gila River basin.

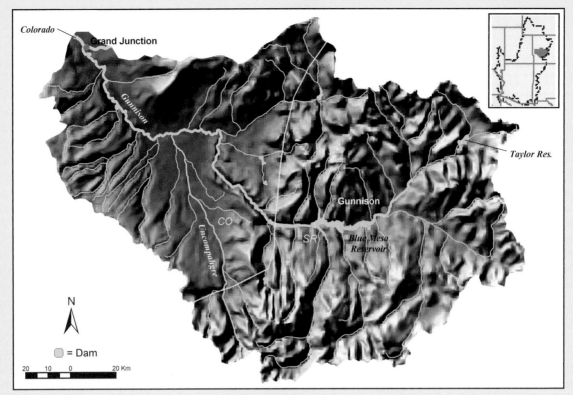

FIGURE 11.23 Map of the Gunnison River basin. Physiographic provinces are separated by a yellow line.

GUNNISON RIVER

Relief: ~2600 m

Basin area: 21,000 km²

Mean discharge: 74 m³/s

River order: 4

Mean annual precipitation: 27 cm

Mean air temperature: 9.5°C

Mean water temperature: 10°C

Physiographic provinces: Southern Rocky Mountains (SR), Colorado Plateaus (CO)

Biomes: Tundra, Temperate Mountain Forest, Desert

Freshwater ecoregion: Colorado

Terrestrial ecoregions: Colorado Rockies Forests, Colorado Plateau Shrubland

Number of fish species: 30 (11 native) and 4 hybrids

Number of endangered species: 4 fishes

Major fishes: bluehead sucker, bonytail chub, Colorado pikeminnow, Colorado River cutthroat trout, flannelmouth sucker, humpback chub, mottled sculpin, mountain whitefish, razorback sucker, roundtail chub, speckled dace

Major other aquatic vertebrates: beaver, boreal western toad, Great Basin spadefoot toad, muskrat

Major benthic invertebrates: mayflies (*Baetis*, *Heptagenia*, *Ephemerella*), stoneflies (*Isoperla*), caddisflies (*Hydropsyche*), true flies (*Atherix*), chironomid midges, black flies

Nonnative species: brown trout, common carp, rainbow trout, red shiner, sand shiner, virile crayfish, white sucker

Major riparian plants: bent grass, bluegrass, box elder, Canada bluegrass, canary grass, goldenrod, muhly, scouring rush, smooth horsetail, spikerush, woolly sedge

Special features: critical habitat for four endangered Colorado River fishes; river runs through Black Canyon of Gunnison National Park

Fragmentation: highly fragmented by headwater dams and three main-stem dams of Aspinall Unit; transbasin diversion through Gunnison Tunnel in lower river

Water quality: below the Aspinall Unit dams pH = 6.6 to 8.2, alkalinity = 100 to 170 mg/L as $CaCO_3$, specific conductance = 1200 μS/cm; historical problems with selenium pollution from agricultural runoff in lower river

Land use: 50% agriculture, 45% forest, 5% urban

Population density: 2 people/km²

Major information sources: Hauer et al. 1989, McAda 2000

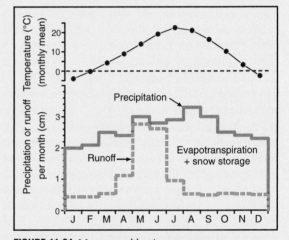

FIGURE 11.24 Mean monthly air temperature, precipitation, and runoff for the Gunnison River basin.

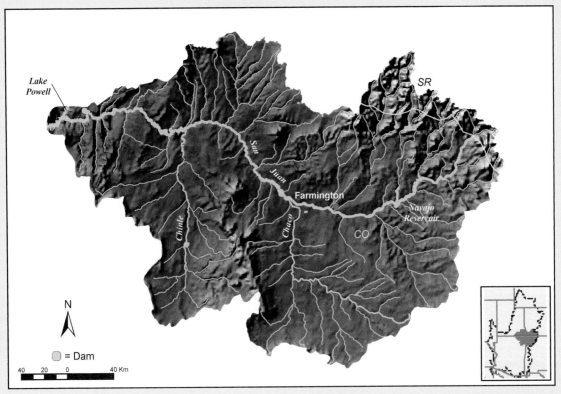

FIGURE 11.25 Map of the San Juan River basin. Physiographic provinces are separated by a yellow line.

SAN JUAN RIVER

Relief: ~2800 m

Basin area: 59,600 km²

Mean discharge: 65 m³/s

River order: 4

Mean annual precipitation: 25 cm

Mean air temperature: 11.3°C

Mean water temperature: 4°C to 18°C (Navajo Dam), 2°C to 23°C (Shiprock, New Mexico)

Physiographic provinces: Southern Rocky Mountains (SR), Colorado Plateaus (CO)

Biomes: Temperate Mountain Forest, Desert

Freshwater ecoregion: Colorado

Terrestrial ecoregions: Colorado Rockies Forest, Colorado Plateau Shrublands

Number of fish species: 26 (7 native) and 3 hybrids

Number of endangered species: 2 fishes

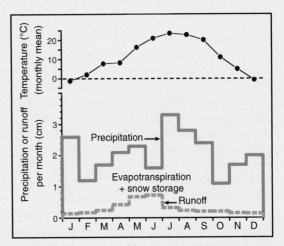

FIGURE 11.26 Mean monthly air temperature, precipitation, and runoff for the San Juan River basin.

Major fishes: bluehead sucker, channel catfish, Colorado pikeminnow, common carp, fathead minnow, flannelmouth sucker, mottled sculpin, razorback sucker, roundtail chub, speckled dace

Major other aquatic vertebrates: beaver, boreal western toad, bullfrog, muskrat, northern leopard frog, northern water shrew, plains leopard frog, tiger salamander, western chorus frog

Major benthic invertebrates: mayflies (*Acentrella, Baetis, Callibaetis, Drunella, Epeorus, Ephemera, Ephemerella, Heptagenia, Rhithrogena, Tricorythodes*), caddisflies (*Cheumatopsyche, Hydropsyche, Hydroptila, Oecetis, Ochrotrichia, Smicridea*), hellgrammites (*Corydalus*), damselflies (*Argia, Enallagma, Ischnura*), dragonflies (*Erpetogomphus*), stoneflies (*Isoperla*), bugs (*Rhagovelia*)

Nonnative species: black bullhead, channel catfish, common carp, red shiner, fathead minnow, largemouth bass, Rio Grande killifish, Russian olive, saltcedar

Major riparian plants: Fremont cottonwood, Russian olive, upland herbs and shrubs, wetland herbs, willows

Special features: large desert river; important for the recovery of two native endangered fishes; mimicking of natural flow regime implemented for restoration; runs through Four Corners

Fragmentation: Navajo Reservoir in upper basin; flows into Lake Powell at mouth

Water quality: pH = 8.2, alkalinity = 105 mg/L as CaCO₃, NO₃-N = 0.05 mg/L, PO₄-P = <0.01 mg/L, specific conductance = 550 μS/cm, suspended sediments = 3500 mg/L

Land use: 75% agriculture, 20% forest, 5% urban; irrigated agriculture, dispersed livestock grazing, petroleum extraction

Population density: 0.6 people/km²

Major information sources: Gido and Propst 1999, Holden 1999

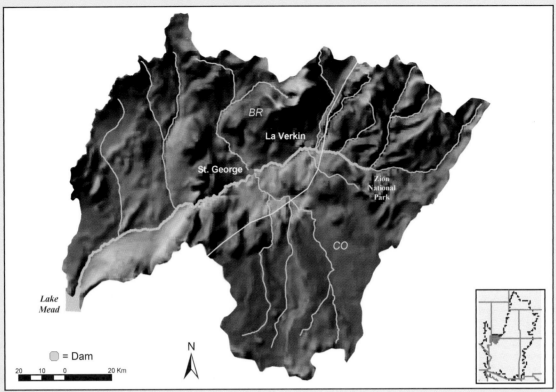

FIGURE 11.27 Map of the Virgin River basin. Physiographic provinces are separated by a yellow line.

VIRGIN RIVER

Relief: ~2635 m
Basin area: 13,200 km²
Mean discharge: 6.8 m³/s
River order: 3
Mean annual precipitation: 27 cm
Mean air temperature: 17.1°C
Mean water temperature: 17°C
Physiographic provinces: Basin and Range (BR), Colorado Plateaus (CO)
Biomes: Temperate Mountain Forest, Desert
Freshwater ecoregion: Vegas-Virgin
Terrestrial ecoregions: Colorado Plateau Shrublands, Mohave Desert
Number of fish species: 21 (11 native)
Number of endangered species: 5 fishes, 21 reptiles and amphibians
Major fishes: black bullhead, desert sucker, flannelmouth sucker, green sunfish, mosquitofish, red shiner, speckled dace, Virgin River spinedace, woundfin

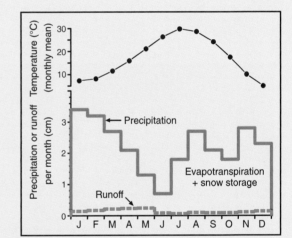

FIGURE 11.28 Mean monthly air temperature, precipitation, and runoff for the Virgin River basin.

Major other aquatic vertebrates: American white pelican, beaver, belted kingfisher, Clark's grebe, great egret, lowland leopard frog, muskrat, northern leopard frog, osprey, snowy egret
Major benthic invertebrates: mayflies (*Baetis*, *Tricorythodes*), caddisflies (*Cheumatopsyche*, *Chimarra*, *Hydropsyche*, *Nectopsyche*, *Ochrotrichia*), beetles (*Microcylloepus*), dragonflies (*Ophiogomphus*), snails (*Physella*), crustaceans (virile crayfish), bugs (*Ambrysus*), chironomid midges, biting midges, black flies
Nonnative species: black bullhead, bullfrog, green sunfish, red shiner, Russian olive, saltcedar, virile crayfish
Major riparian plants: box elder, coyote willow, Emory baccharis, Fremont cottonwood, Russian olive, saltcedar
Special features: runs through Zion National Park to Lake Mead; highest water quality through Zion; recommended for conservation by World Wildlife Fund
Fragmentation: no dams but two major main-stem diversions include Quail Lake diversion near Virgin, Utah, and the Washington Fields diversion near Saint George, Utah
Water quality: pH = 7.9, alkalinity = 240 mg/L as CaCO₃, relatively high salinity (9.6 g/L) near La Verkin from Pah Tempe Springs, NO₃-N = 0.17 mg/L, PO₄-P = 0.03 mg/L, specific conductance = 2400 µS/cm, suspended sediments = 1050 mg/L
Land use: 60% agriculture, 30% forest, 10% urban
Population density: ~7 people/km²
Major information sources: Winget and Baumann 1977, Cross 1985, Gregory and Deacon 1994, McMahon et al. 2001, Abell et al. 2000

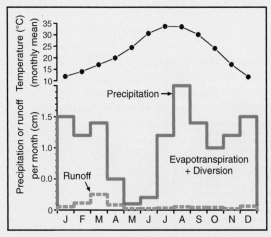

FIGURE 11.30 Mean monthly air temperature, precipitation, and runoff for the Bill Williams River basin.

BILL WILLIAMS RIVER

Relief: ~1250 m
Basin area: 13,950 km³
Mean discharge: 4.3 m³/s
River order: 3
Mean annual precipitation: 13.2 cm
Mean air temperature: 22.4°C
Mean water temperature: 20°C
Physiographic provinces: Basin and Range (BR), Colorado Plateaus (CO)
Biome: Desert
Freshwater ecoregion: Colorado
Terrestrial ecoregion: Sonoran Desert
Number of fish species: 13 (1 native)
Number of endangered species: 1 fish
Major fishes: bluegill, common carp, fathead minnow, green sunfish, largemouth bass, mosquitofish, razorback sucker, red shiner, yellow bullhead
Major other aquatic vertebrates: Arizona toad, beaver, lowland leopard frog, muskrat, river otter, red-spotted toad, Sonoran Desert toad, spiny-spotted turtle

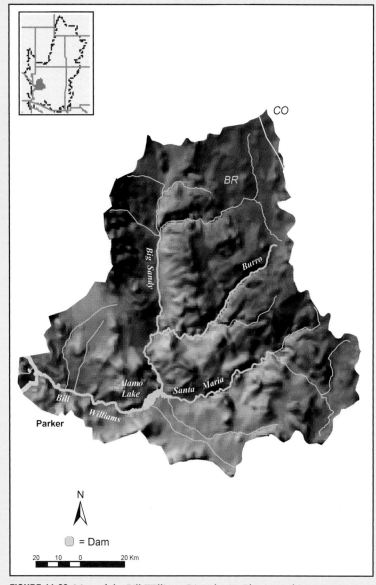

FIGURE 11.29 Map of the Bill Williams River basin. Physiographic provinces are separated by a yellow line.

Major benthic invertebrates: mayflies (*Baetis*), caddisflies (*Cheumatopsyche, Culoptila, Helicopsyche, Hydroptila, Leucotrichia, Nectopsyche, Ochrotrichia, Protoptila, Smicridea*), beetles (*Helichus, Microcylloepus, Tropisternus*), damselflies (*Argia*), bugs (*Ambrysus*), crustaceans (*Hyalella*, red swamp crayfish), snails (*Fossaria, Gyraulus, Physella*), bivalves (*Corbicula, Pisidium*)
Nonnative species: bluegill, common carp, fathead minnow, golden shiner, goldfish, green sunfish, largemouth bass, mosquitofish, redear sunfish, red shiner, red swamp crayfish, saltcedar, spiny-spotted turtle, yellow bullhead, *Daphnia lumholtzi* in Alamo Lake
Major riparian plants: broadleaf cattail, bulrush, coyote willow, Fremont willow, Goodding willow, narrowleaf cattail, saltcedar
Special features: runs through Bill Williams Wildlife Refuge; high bird density (≥335 species); dramatic lateral and vertical variations in microclimate
Fragmentation: almost entirely regulated by Alamo Dam; low to zero flows from July through October
Water quality: pH = 8.0, alkalinity = 235 mg/L as CaCO₃, NO₃-N = 0.05 mg/L, specific conductance = 1000 μS/cm, suspended sediment = <10 mg/L
Land use: 75% agriculture, 25% forest
Population density: 1.5 person/km²
Major information source: Silvey et al. 1986

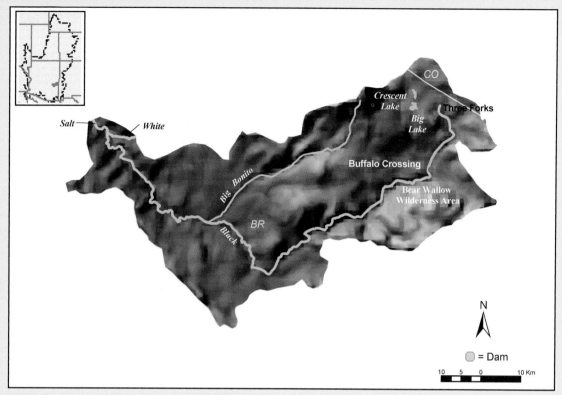

FIGURE 11.31 Map of the Black River basin. Physiographic provinces are separated by a yellow line.

BLACK RIVER

Relief: ~1400 m

Basin area: 3400 km²

Mean discharge: 12 m³/s

River order: 4

Mean annual precipitation: 53.5 cm

Mean air temperature: 14°C

Mean water temperature: 15°C

Physiographic provinces: Basin and Range (BR), Colorado Plateaus (CO)

Biomes: Temperate Mountain Forest, Desert

Freshwater ecoregion: Gila

Terrestrial ecoregion: Arizona Mountains Forests

Number of fish species: 13 (5 native)

Number of endangered species: 2 fishes, 1 frog, 1 clam, 1 snail

Major fishes: Apache trout, brown trout, channel catfish, desert sucker, fathead minnow, rainbow trout, roundtail chub, smallmouth bass, Sonora sucker, speckled dace

Major other aquatic vertebrates: beaver, belted kingfisher, great blue heron, mallard, muskrat, osprey, river otter

Major benthic invertebrates: mayflies (*Baetis, Epeorus Heptagenia, Tricorythodes, Thraulodes*), caddisflies (*Agapetus, Atopsyche, Brachycentrus, Ceratopsyche, Cheumatopsyche, Chimarra, Glossosoma, Helicopsyche, Hydropsyche, Lepidostoma, Limnephilus, Marilia, Micrasema, Phylloicus*), beetles (*Psephenus*), bugs (*Ambrysus*), chironomid midges, black flies, crustaceans (virile crayfish)

Nonnative species: bluegill, brown trout, bullfrog, channel catfish, cutthroat trout, fathead minnow, green sunfish, rainbow trout, smallmouth bass, saltcedar, virile crayfish

Major riparian plants: Bebb willow, coyote willow, Fremont cottonwood, Geyer willow, Goodding willow, narrowleaf cottonwood, thin-leaf alder, saltcedar

Special features: some of most natural riparian communities remaining in Arizona; passes through scenic Bear Wallow Wilderness Area and Fort Apache and San Carlos Indian reservations

Fragmentation: relatively unfragmented; three small reservoirs in headwaters fed by tributaries during snowmelt

Water quality: pH = 8.8, alkalinity = 150 mg/L as CaCO₃

Land use: 15% agriculture, 80% forest, 5% urban

Population density: ~1 person/km²

Major information sources: Silvey et al. 1986, Novy and Lopez 1991

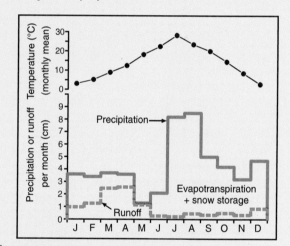

FIGURE 11.32 Mean monthly air temperature, precipitation, and runoff for the Black River basin.

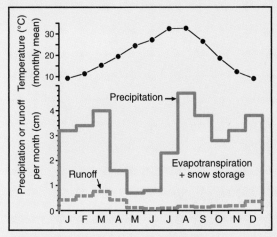

FIGURE 11.34 Mean monthly air temperature, precipitation, and runoff for the Verde River basin.

VERDE RIVER

Relief: ~925 m
Basin area: 16,190 km²
Mean discharge: 17 m³/s
River order: 4
Mean annual precipitation: 34.5 cm
Mean air temperature: 20°C
Mean water temperature: 16.5°C
Physiographic provinces: Basin and Range (BR),
    Colorado Plateaus (CO)
Biome: Desert
Freshwater ecoregion: Gila
Terrestrial ecoregion: Arizona Mountains Forests
Number of fish species: 27 (10 native)
Number of endangered species: 8 fishes
Major fishes: common carp, desert sucker, flathead
    catfish, Gila chub, green sunfish, headwater chub,
    largemouth bass, mosquitofish, red shiner,
    roundtail chub, smallmouth bass, Sonora sucker,
    yellow bullhead

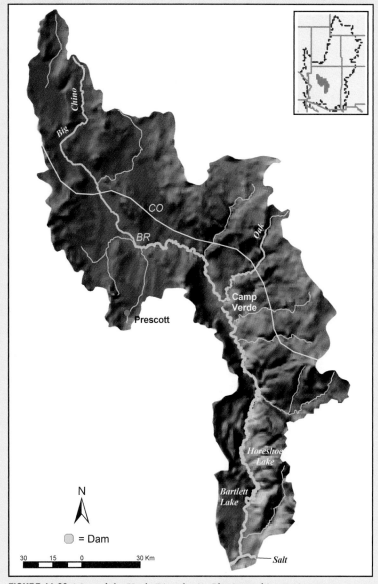

FIGURE 11.33 Map of the Verde River basin. Physiographic provinces are separated by a yellow line.

Major other aquatic vertebrates: Arizona toad, beaver,
    belted kingfisher, bullfrog, Chiricahua leopard frog,
    great blue heron, lowland leopard frog, river otter, Sonoran mud turtle, striped chorus frog, Woodhouse toad
Major benthic invertebrates: mayflies (*Baetis*, *Heptagenia*), caddisflies (*Cheumatopsyche*, *Chimarra*, *Helicopsyche*, *Heterelmis*,
    Hydropsyche, *Hydroptila*, *Ochrotrichia*, *Oecetis*, *Polycentropus*, *Protoptila*, *Smicridea*), hellgrammites (*Corydalus*),
    damselflies (*Enallagma*), bugs (*Ambrysus*), black flies, biting midges, chironomid midges, crustaceans (*Gammarus*, virile
    crayfish)
Nonnative species: Asiatic clam, bullfrog, common carp, flathead catfish, green sunfish, largemouth bass, mosquitofish, red
    shiner, smallmouth bass, saltcedar, virile crayfish, yellow bullhead
Major riparian plants: Arizona alder, Arizona sycamore, Arizona walnut, arroyo willow, box elder, cattails, common reed,
    coyote willow, Fremont willow, Goodding willow, saltcedar
Special features: one of the largest perennial rivers in Gila basin; upper section recommended for conservation by World Wildlife
    Fund; upper 72 km one of few remaining sections with roundtail chub; 65 km National Wild and Scenic River
Fragmentation: lower third regulated by dams (Horseshoe and Bartlett lakes); 7 water-diversion dams
Water quality: pH = 8.4, alkalinity = 250 mg/L as $CaCO_3$, NO_3-N = 0.03 mg/L, PO_4-P = 0.01 mg/L, specific conductance =
    550 μS/cm, suspended sediment = 60 mg/L
Land use: 70% agriculture, 25% forest, 5% urban
Population density: 3 people/km²
Major information sources: Cordy et al. 1998, Rinne et al. 1998

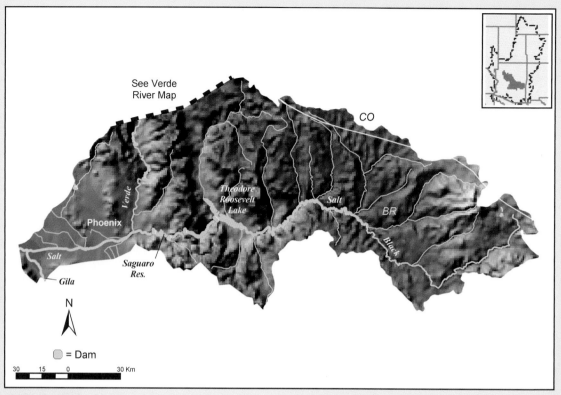

FIGURE 11.35 Map of the Salt River basin. Physiographic provinces are separated by a yellow line.

SALT RIVER

Relief: ~2540 m
Basin area: 35,480 km²
Mean discharge: < 25 m³/s
River order: 5
Mean annual precipitation: 39.6 cm
Mean air temperature: 22°C
Mean water temperature: 19°C
Physiographic provinces: Basin and Range (BR), Colorado Plateaus (CO)
Biomes: Temperate Mountain Forests, Desert
Freshwater ecoregion: Gila
Terrestrial ecoregions: Arizona Mountain Forests, Sonoran Desert
Number of fish species: 16 (9 native)
Number of endangered species: 7 fishes
Major fishes: bluegill, channel catfish, common carp, desert sucker, fathead minnow, flathead catfish, golden shiner, green sunfish, longfin dace, red shiner, Sonora sucker, speckled dace
Major other aquatic vertebrates: Arizona toad, bullfrog, Chiricahua leopard frog, northern leopard frog, lowland leopard frog, Sonoran mud turtle, virile crayfish
Major benthic invertebrates: mayflies (*Baetis*, *Callibaetis*, *Tricorythodes*), caddisflies (*Hydropsyche*, *Hydroptila*), chironomid midges
Nonnative species: bluegill, bullfrog, channel catfish, common carp, fathead minnow, flathead catfish, golden shiner, green sunfish, red shiner, Rio Grande leopard frog, saltcedar, virile crayfish
Major riparian plants: mesquite, saltcedar
Special features: lower section flows through scenic Salt River Canyon
Fragmentation: lower sections highly regulated by Roosevelt, Stuart Mountain (Saguaro Res.), Mormon Flat, and Horse Mesa dams; heavy use of diversion canals
Water quality: pH = 8.3, alkalinity = 185 mg/L as CaCO₃, NO₃-N = 0.01 mg/L, PO₄-P = 0.03 mg/L, specific conductance = 3700 µS/cm, suspended sediment = 125 mg/L
Land use: 75% agriculture, 15% forest, 5% urban
Population density: >90 people/km²
Major information sources: Silvey et al. 1986, Cordy et al. 1998

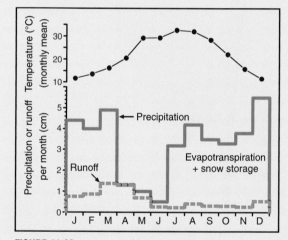

FIGURE 11.36 Mean monthly air temperature, precipitation, and runoff for the Salt River basin.

538

12

PACIFIC COAST RIVERS OF THE COTERMINOUS UNITED STATES

JAMES L. CARTER   VINCENT H. RESH

INTRODUCTION

SACRAMENTO RIVER

SAN JOAQUIN RIVER

SALINAS RIVER

KLAMATH RIVER

ROGUE RIVER

ADDITIONAL RIVERS

ACKNOWLEDGMENTS

LITERATURE CITED

INTRODUCTION

The rivers discussed in this chapter are mainly located in the southern portion of the Pacific Mountain System (Hunt 1974) and discharge into the Pacific Ocean (Fig. 12.2). Located from south of the Columbia River to southern California, these river basins occupy an area just over 10° of latitude and 10° of longitude. The rivers and the biota of this region have been influenced more than any other rivers in North America by tectonic activity that is both geologically recent and ongoing, including (1) periods of mountain building through uplift, volcanism, and accretion; (2) periods of massive erosion; (3) changes in sea level; and (4) large-scale faulting. These factors have influenced the aspect and gradient of the rivers and created a topography that affects subregional precipitation patterns, which in turn influences the hydrology, geomorphology, and consequently the present-day biology of the region's rivers. The importance of the recency of these events is shown by the uniqueness and high endemism of the regional flora and fauna (e.g., Ricketts et al. 1999, Abell et al. 2000, Moyle 2002).

From the time of the arrival of the first people to this central Pacific region of North America, which occurred approximately 11,000 years ago (Heizer and Elsasser 1980), rivers have provided humans with food, water, and transportation. Furthermore, in the northern part of the Pacific region the life cycle of the Native American tribes, their religions, and their wars focused on the rivers and, in particular, on the salmon occurring there. In contrast, the tribes in the southern portion of the region utilized a more diverse resource base; fishes, although part of their economy, were less important to them than to the tribes of the north. The diversity of freshwater and estuarine species as well as abundant terrestrial wildlife and plants, particularly acorns, supported some of the highest densities of Native Americans in North America (Heizer and Elsasser 1980).

Spanish exploration began in the early to mid-sixteenth century. Throughout the 1700s numerous

◄ FIGURE 12.1 South Fork of Eel River in Humboldt Redwoods, California (PHOTO BY TIM PALMER).

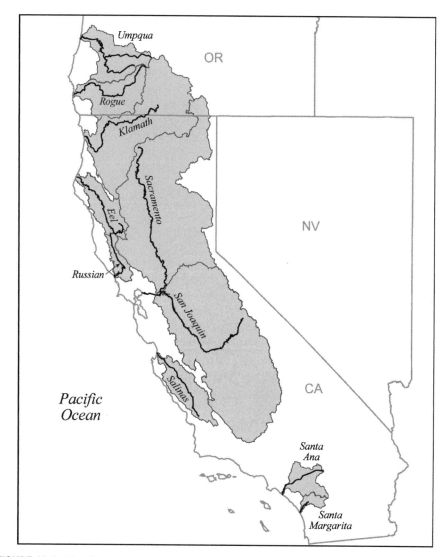

FIGURE 12.2   Pacific Coast rivers of the coterminous United States covered in this chapter.

settlements along the Pacific Coast were established around Spanish missions. However, their northern extent was limited, and when Russian fur traders arrived in California and settled at Fort Ross (100 km north of San Francisco Bay) in 1812, occupation by the Spanish was not much further north than the Bay. The small Russian settlement at Fort Ross was abandoned around 1840.

Even though Native American population densities were high relative to other areas of North America, they (as well as the early Spanish and Russian colonists) appeared to have had relatively little impact on the region. However, Native Americans did use fire as a game-management method and, at least within their area of influence, significantly modified the landscape (Keeley 2002). One of the

principal influences of the Spanish occupation was the introduction of European annual grasses, which displaced native bunchgrasses (Schoenherr 1992).

Clearly, the defining event in the human history of the Pacific region was John Marshall's discovery of gold at Sutter's Mill, California, in 1848. Immigration associated with this event significantly increased the population of the region (e.g., by 1850, 100,000 miners had arrived). The negative effects of excess sediment from hydraulic mining on stream channels and the contamination of water and biota from the mercury that was used to extract gold led to the first great assault on the region's rivers. However, today the most significant human influences on lotic systems in this region are associated with the alteration of natural flows and sediment transport (the amount and timing of) caused by impoundments and withdrawals, with the water principally (~80%) used for irrigated agriculture.

Numerous major rivers are found in the southern portion of the Pacific Mountain System. We will discuss the Sacramento, San Joaquin, Salinas, Klamath, and Rogue in detail and include abbreviated descriptions for five additional rivers: the Umpqua, Eel, Russian, Santa Ana, and Santa Margarita.

Physiography and Climate

The Pacific Mountain System includes three physiographic provinces (Hunt 1974): (1) the Cascade–Sierra Nevada Mountains province, which extends from southern California north into British Columbia; (2) the Pacific Border province, which includes those lands to the west of the Cascades and Sierra Nevada; and (3) the Lower California (Peninsular Range) province, which extends between the Salton Trough and the southern California coast. These three provinces contain the greatest range in climate, geology, geomorphology, hydrology, and biology in North America. This diversity of environmental setting has produced an enormous range in basin physiography. Importantly, the Pacific Mountain System lies along the very tectonically active Pacific Rim, and the recency of its formation and its continuing tectonic activity strongly influence both the geomorphology and biology of its rivers.

The topography of the Pacific Mountain System includes mountain ranges of five types: granitic (e.g., Sierra Nevada, Klamath); volcanic (Cascades); pre-Tertiary, complexly folded and faulted formations (Northern California Coast Ranges); Tertiary, moderately folded but much faulted (Oregon Coast Ranges, Southern California Coast Ranges); and dome (e.g., Marysville Buttes in California's Central Valley). These base materials provide for a diversity of soil types and nutrient availability for the rivers that originate in them. Differences in the geology and time of formation of these ranges also influence the geomorphology, hydrology, and biology of the streams and rivers that occur in individual regions.

The mountain ranges of the entire Pacific Mountain System have been described as being arranged like a chain (Hunt 1974). The northern link consists of the Cascade Range, the Oregon Coast Range, and the Olympic Mountains. This link is joined at the south by the Klamath Mountains, with the Puget Trough forming the "hole." The Sierra Nevada and California Coast Range form the middle link, with

the Central Valley forming the hole. The Transverse Range and the Lower California Province, with the Salton Trough as the hole, form the southernmost link.

The Cascade–Sierra Nevada Mountains province extends for 1600 km from southern California to British Columbia. The range has uplifted and, particularly in the Cascades, is topped by volcanoes and lava flows. The northern and southern extremes of the range are higher than the middle and consist of granitic rocks and metamorphosed sedimentary formations. The middle portion of the range (near the northern extent covered in this chapter) is lower and the rocks are buried by volcanic flows. The only two rivers covered in this chapter that cross the range (the Klamath and Pit) are located in this topographically depressed area. Even though geologic differences exist between the Sierra Nevada and the southern Cascade Mountains, the altitudinal zonation of flora and fauna is characteristic (although species assemblages vary) regardless of mountain range, and stream form (gradient, width, etc.) is often a function of longitudinal location, which is also highly correlated to altitude.

The Pacific Border province represents those lands west of the Cascade–Sierra Nevada ranges and north of (and including) the Transverse Range in southern California (Hunt 1974). This province includes a wide variety of landforms, such as the Great Valley of California (the Central Valley), the Coast Ranges, the Transverse Ranges, and the Klamath Mountains. This province contains some of the wettest and driest areas in the region. Land cover and land use within this province is the most varied, with the Klamath Mountains supporting vast, species-rich coniferous forests.

The Lower California (Peninsular Range) province is the smallest of the three provinces. It includes the ranges south of the Transverse Ranges, as well as the very densely populated and highly developed coastal lowland areas of southern California (Hunt 1974), and is essentially the northern extent of the Baja California peninsula.

The climate of the Pacific Mountain System is possibly the most varied of all the regions reviewed in this volume. As can be seen in the figures of this chapter, precipitation is extremely seasonal. Each figure depicts a summer dry period and a wet winter period. This annual cycle is very consistent from year to year. However, extreme interannual variation occurs and can lead to extended periods of drought and years of extremely high rainfall and runoff (e.g., El Niño events). Spatial differences in the amount and type of precipitation also occur. In general, precipitation decreases from north to south; however, spatial variation in precipitation (amount and form) is strongly influenced by topographic factors. For example, high elevations tend to receive more precipitation as snowfall and greater precipitation in general than low elevations. Low-elevation valleys and leeward slopes can be extremely dry (e.g., Tulare basin) because of rain shadow effects and adiabatic evaporation. The differences in the amount and form of precipitation greatly influence the runoff patterns and consequently the geomorphology and biology of the region's rivers.

Temperatures across the region vary considerably. Distance from the coast, altitude, and north-to-south location greatly influence temperature regimes. Coastal summer temperatures are influenced by the upwelling of deep, cold ocean waters and are generally cool. These waters also moderate winter temperatures, and freezing rarely occurs along the coast. Further inland, summer temperatures often exceed 38°C. Temperature decreases with altitude and can drop below freezing throughout the year at the highest elevations; however, in this region winter temperatures never approach the extreme cold of midcontinent.

Basin Landscape and Land Use

Rivers of this region are rarely contained within a single ecoregion, and more often traverse several ecoregions along their courses. Geology, topography, and precipitation are the dominant controls over river form and terrestrial vegetation. The number of combinations of these three factors is evident in the variety of terrestrial ecoregions found in the region. Ricketts et al. (1999) identified 10 terrestrial ecoregions in the southern portion of the Pacific Mountain System covered here. These 10 terrestrial ecoregions are grouped into three major habitat types: the Temperate Coniferous Forest, the Mediterranean Shrub and Savannah, and the Temperate Grasslands/Savannah/Shrub.

The Temperate Coniferous Forest habitat type encompasses just less than two-thirds of the total area. It includes the Central Pacific Coastal Forests, Central and Southern Cascades Forests, Eastern Cascades Forests, Klamath-Siskiyou Forests, Northern California Coastal Forests, and Sierra Nevada Forests terrestrial ecoregions. This habitat type is found in the northern and eastern portion of the region, both along the coast and above the foothills in montane areas. It includes most of the Sierra Nevada, Cascade, and Klamath mountains, and the Coast Ranges north of Monterey Bay, California. In general, this habitat contains forested areas that are dominated by Douglas fir, western hemlock, western red cedar, and ponderosa pine. In suitable coastal areas between Monterey Bay and the Oregon border, 10% of the pre-Columbian population of redwoods still exist. The remaining giant sequoia occur further inland in isolated areas in the central Sierra Nevada. Throughout these forests plant associations are greatly influenced by edaphic factors, precipitation, and elevation and form predictable sequences. Although this habitat is still extensively forested, it continues to be impacted by logging, road building, fire suppression, grazing, flow capture, and loss of riparian habitat. Many areas within this habitat type are considered unique because the level of plant and animal endemism is extremely high, particularly in the Klamath Mountains to the north and the Sierra Nevada in the south.

The Mediterranean Shrub and Savannah habitat type represents approximately one-quarter of the area and includes the California Interior Chaparral and Woodlands, California Montane Chaparral and Woodlands, and California Coastal Sage and Chaparral terrestrial ecoregions. The California Interior Chaparral and Woodlands ecoregion is the largest in this habitat type and forms a perimeter around the Central Valley of California, separating the valley floor from the more montane areas. The other two ecoregions are mainly located south of the Central Valley; however, the Santa Lucia Range, located just south of Monterey Bay, also is included in this habitat type.

Visually, Mediterranean Shrub and Savannah habitat is a mixture of open forests, low chaparral scrub, and grasslands. Forest types include oak woodland with scattered California buckeye, closed cone pine–yellow pine forests, sugar pine–white fir forests, lodgepole pine forests, and desert piñon–juniper woodlands. The shrub habitat includes lower chaparral and chamise, and upper chaparral dominated by manzanita, desert chaparral, and coastal sage shrub. This habitat type was severely impacted by the loss of native bunchgrasses through the introduction of annual European grasses during the Spanish settlement; much of the habitat in the southern portion of California continues to be lost to urban and agricultural development. Shrublands throughout California have been subject to physical and chemical removal to provide additional lands for grazing.

Last, the Temperate Grasslands/Savannah/Shrub habitat type only includes the California Central Valley Grassland terrestrial ecoregion. Although representing only about 10% of the area, it was probably the most productive habitat type in the region. This habitat type and the ecoregion are represented by California's Central Valley. The California Central Valley Grassland ecoregion was the location of extensive freshwater marshes, vernal pools, and the largest lake (Tulare Lake) west of the Mississippi River prior to agricultural development and modification. Extensive tule marshes, riparian woodlands, and native bunchgrasses existed before the 1850s. Riparian woodlands contained willows, western sycamore, box elder, Fremont cottonwood, and valley oak. Reportedly, riparian widths on the lower sections of the Sacramento and San Joaquin rivers exceeded 30 km (Ricketts et al. 1999). This entire habitat type has been extensively modified from its original condition through the development of agriculture.

The Rivers

The ecology of the rivers flowing into the Pacific Ocean is as diverse as the lands through which they flow, and this region contains one of the most diverse assemblages of watersheds on the continent. This diversity is a result of the range of climate zones (from deserts to Mediterranean areas to temperate rainforests) but especially the result of millions of years of activity along a tectonic plate boundary.

The mosaic of geologic form, the recent tectonic activity of the area, and the climatic characteristics that determine the frequency, intensity, and type of storms ultimately shape the characteristics of each of the watersheds in this region. In addition to the ecoregion classification for the terrestrial environments already described, many different classification systems exist for the drainage systems as well. Although every basin has unique characteristics, the lotic environments of this region have been classified into three freshwater ecoregions (Abell et al. 2000).

The Pacific Mid-Coastal freshwater ecoregion, with an area of 108,880 km², extends along the Pacific Coast of Oregon and California to the north shore of San Francisco Bay. It encompasses the western drainages of the Klamath and Siskiyou mountains and includes the Umpqua, Rogue, and Klamath as well as many other important rivers. This area has the highest yearly rainfalls of the region, sometimes exceeding 500 cm/yr. Although the Klamath Mountains and Trinity Alps have heavy snowfall, most precipitation in this ecoregion occurs

as rain. The rivers of this area also have the highest sediment yields, which are caused by high rainfall, high rates of uplift, unstable rock types and soil, and logging and grazing practices that promote erosion. The smaller, coastal watersheds in this region exhibit a rapid hydrographic response to rainfall events, whereas the larger, eastern rivers generally exhibit less flashy hydrographs and higher overall base flows because of precipitation storage as snow and snowmelt.

The Pacific Central Valley freshwater ecoregion has an area of 184,129 km². It lies entirely within California and encircles the Central Valley. The largest rivers are the Sacramento and San Joaquin, but it also includes the high-water-quality rivers located in the western drainages of the Sierra Nevada. This area has high precipitation, with approximately 50% occurring as snowfall. The rapid Pleistocene uplift and global cooling produced glaciers (e.g., throughout the Sierra Nevada) that altered the profiles of many rivers. In contrast to the Pacific Mid-Coastal ecoregion, sediment yields of watersheds here are lower, but land use (logging, grazing, and especially the historic effects of hydraulic mining) has increased sediment yields locally. Because of accumulations of precipitation as snowpack, lag times are long, peak runoffs are dampened, and extensive spring runoffs from snowmelt occur.

The San Francisco Bay and portions of the central coast are also part of this ecoregion but are probably best considered as a distinct hydrological area (and are separate in most water classification schemes). The northern portions of this region have higher rainfall totals over mountain ranges than in the interior valleys. The precipitation in southern regions is generally about 50% of that in northern ones, and snowmelt influences are insignificant. The tectonics associated with the San Andreas Fault system have influenced the orientation and location of the major river valleys. Sediment yields are high because of high-intensity rainfall, high rates of uplift, and unstable rocks. These small, steep watersheds have short lag times and high peak runoffs, and rivers often flood during winter storms.

The South Pacific Coastal freshwater ecoregion has an area of 170,320 km² and encompasses the coastal area that begins just south of Monterey, California, and extends to the southern tip of the Baja California peninsula in Mexico. The high frequency of low-rainfall years, interspersed with occasional high-rainfall years, results in most runoff being the consequence of intense subtropical storms. Like the Central Coast, many watershed characteristics are a product of past and present activity along the San Andreas Fault system. Although sedimentary rocks in this region are prone to landslides and produce locally high erosion rates, the high level of relief, limited vegetation cover, and wildfires are other causes of erosion. Most rivers in this region are ephemeral, with limited base flow and snowmelt, and are prone to highly variable flows and floods.

Of these different ecoregions, the coastal rivers of the San Francisco Bay area and the central and southern coastal areas are unique hydrologically when compared to most of the rivers covered in this book. The rivers in these areas are physically, chemically, and biologically shaped by sequential, predictable, seasonal periods of flooding and drying over an annual cycle. Intermittency is common in many tributaries. Correspondingly, the aquatic communities that occur in the tributaries of these rivers (and consequently in the main stems) undergo a yearly cycle whereby the communities are dominated by abiotic (i.e., environmental) factors during floods and biotic factors (e.g., predation, competition) as discharges approach base-flow conditions. However, as the dry season progresses, habitat conditions become harsher, and environmental pressures again become important regulators of stream populations and community structure (Gasith and Resh 1999).

Rivers in these areas, besides experiencing high seasonal variation (more than 80% of the rain falls in the three months of winter, often with precipitation from a few major storms responsible for most of the flooding), also experience high variation in inter-annual precipitation. Deviations in discharge of more than 30% from a multiannual average are common and lead to low constancy within a highly predictable seasonal discharge pattern. Besides the temporal differences in flow, this area is also characterized by extreme spatial differences in rainfall. From north to south, coastal to inland areas, and mountaintops to valleys, precipitation in this region varies not only in its form (rain or snow) but also its amount (e.g., from <10 cm to >500 cm).

Given the high seasonal predictability of floods and drying in streams and rivers of this region, we might expect differences in the life cycles of organisms inhabiting lotic habitats in this region compared to other areas of North America. For example, many fishes and aquatic invertebrates that occur here reproduce in spring during the declining stage of the hydrograph when temperatures are increasing. In contrast, in streams and rivers in other areas, where flooding may occur at almost any time of the year, the spawning seasons of most species are protracted

or staggered. Consequently, the composition, at least of the fish assemblage, may vary far less from year to year in streams of coastal parts of the Pacific Central Valley and South Pacific Coastal ecoregions than in many of the other North American systems (Moyle and Vondracek 1985).

Humans have significantly influenced the rivers addressed in this chapter since the mid 1800s, when the principal settlement of the region occurred. Within this relatively short period of time the natural functioning of many of the rivers has been severely impacted. Although there have been and continue to be a variety of human-induced factors contributing to these impacts, two stand out.

Hydraulic mining, which entailed directing a high-pressure jet of water onto an exposed placer deposit of gravel, sand, and/or silt and then running the sediment through sieves to remove the gold, severely impacted the region's rivers during the mid-1800s. Outlawed in California in 1884, this mining technique washed large portions of mountains into local streams, which then transported this debris into the lower reaches of the rivers. This excess sediment input exceeded the rivers' transport capacity, led to severe flooding, and permanently altered river channels in California and Oregon. Gold mining also left a legacy of mercury contamination of lotic, lentic, and estuarine systems throughout the region. Recreational mining for gold still contributes to local sediment impacts.

Flow capture and diversion have severely altered the rivers of this region. The rivers described in this southern portion of the Pacific Mountain System are extremely seasonal, with high winter–spring flows and low to intermittent summer–autumn flows. The geomorphological and biological form and function of the rivers have evolved in concert within this natural discharge regime. The extreme variability in the amount of water, spatially and temporally, has conflicted with human needs since the region was first settled. Dams have been constructed for flood control and energy production, and to support one of the most intensively agricultural and densely settled areas in North America, and the rivers of this region have been captured and diverted more than anywhere else in North America. This has led to a complex set of interrelated impacts to the channels and the biology of the rivers located in this ecologically diverse region.

On the positive side, there are thousands of water-resource professionals, scores of volunteer and citizen action groups, and many federal, state, and local programs dedicated to the maintenance and restoration of the region's rivers, particularly the restoration of threatened and endangered salmonids and other native fishes. However, many challenges exist given the limited total supply of water and an ever-increasing demand for high-quality water to meet both societal and environmental needs.

SACRAMENTO RIVER

The Sacramento River is the largest river in California (Fig. 12.12). It is 644 km long and has a basin of 72,132 km^2 that ranges in altitude from 4000 m asl in the Sierra Nevada to sea level at its mouth, which is located in the Sacramento–San Joaquin Delta. The Sacramento begins just west of Mt. Shasta, and although most of the basin is located in the northern half of California, the basin of a northern tributary, the Pit River, extends into southern Oregon. Below Shasta Dam numerous tributaries flow into the lower Sacramento from the western slopes of the northern Sierra Nevada and the eastern slopes of the Coast Range. The largest tributaries that flow from the Sierra Nevada are the Feather and American. The Feather itself has two large tributaries, the Yuba and Bear. Although the Sacramento proper does not have any reaches designated as Wild and Scenic, portions of the Feather and American Rivers have been given Wild and Scenic status.

The Sacramento basin contained a rich diversity of Native Americans (Heizer and Elsasser 1980). To the north were the Pit River tribes (Achomawi and Atsugewi), the Modoc, and the Yana and Yahi (of which the famous Ishi was a member). On the west side of the Sacramento in the northern portion of the valley lived the Wintu, in the central portion of the valley the Central Wintun, and in the southern portion of the valley the Southern Wintun (Patwin). To the east of the Sacramento River in the Feather and American river basins lived the Maidu. The Miwok inhabited the central portion of the Sacramento–San Joaquin Valley and were divided into numerous geographic groups. The Plains and Coast Miwok, the northern portion of the Yokuts and the Costanoans (Olhone) lived in and around the Sacramento–San Joaquin Delta and San Francisco Bay.

Differences in local habitat greatly influenced the lifestyles of the Native Americans inhabiting the Sacramento basin. The salmon that once seasonally filled the Sacramento and its tributaries were used whenever available. The vast wetlands of the valley and shallow northern lakes provided native fishes and mussels, seasonally available waterfowl, and abundant starchy cattail roots and tule seeds. The valley and foothills contained abundant deer, elk, pronghorn, and rabbit. Acorns from the many species of oak were a staple and were harvested during the autumn and stored throughout the winter.

Other than a few explorers and trappers, the principal settlement of the Sacramento valley by Europeans occurred in relationship to the Gold Rush during the mid-1800s. Agriculture to support the miners rapidly developed.

Physiography, Climate, and Land Use

The Sacramento River basin is part of the Basin and Range (BR), Cascade–Sierra Nevada Mountains (CS), and Pacific Border (PB) physiographic provinces (see Fig. 12.12) (Hunt 1974). Only a small portion of the Sacramento basin (the Pit River basin) is located in the Basin and Range province. The Cascade–Sierra Nevada Mountains province includes most of the high-elevation montane areas within the basin. The Southern Cascade Mountain section is principally volcanic in origin and remains volcanically active. Mount Shasta and Mount Lassen are the most prominent topographic features of this section.

The Sierra Nevada Mountains section begins just south of Lassen Peak and continues for 640 km. Plutonic rocks in the Sierra Nevada are late Jurassic to late Cretaceous, with some in the northern portion dating to the Paleozoic. The principal impression of the range is one of a massive granitic block; however, there are considerable sedimentary and volcanic rocks in the northern portion. Climatic, topographic, and precipitation patterns create a situation where several major tributaries originate in this section.

Much of the Sacramento River proper lies in the Pacific Border province. The Sacramento Valley represents the northern portion of the California Trough section. Most of the valley contains alluvial sediments from both continental and marine origins, which range in age from the Jurassic to the Holocene. In some locations the estimated depth of sediments exceeds 16 km (Domagalski et al. 1998).

The California Coast Ranges section forms the western border of the Sacramento Valley. These are northwest–southeast aligned, low-elevation mountains formed in the north by faulting and active subduction of the Pacific plate beneath the continental plate. The largest extant natural lake entirely within California, Clear Lake, is located in this section.

The Sacramento basin is a part of at least five terrestrial ecoregions (Ricketts et al. 1999). Differences in vegetation are strongly influenced by altitude, moisture, and soil type. The upper portion of the basin is part of the Eastern Cascades Forests ecoregion and includes much of the Pit River drainage. The dominant forest type is ponderosa pine, but there also are large areas of shrublands and grasslands. To the southeast lies the Klamath–Siskiyou Forests ecoregion, which is considered a global center of biodiversity (the area will be covered more thoroughly in the section on the Klamath River). The eastern part of the Sacramento drainage is in the northern portion of the Sierra Nevada Forests ecoregion, which contains one of the most diverse temperate conifer forests in the world. The coniferous forests are arranged in altitudinal belts, with lowest elevations (~1500 m asl) dominated by ponderosa pine, grading into mixed forests consisting of ponderosa and sugar pines, Douglas fir, and white fir. From 2100 to 2700 m asl forests are a mix of lodgepole, Jeffrey, and western white pines and juniper. Above 2700 m asl mountain hemlock and whitebark, foxtail, and limber pines are found. To the west, across the Sacramento Valley, is the California Interior Chaparral and Woodlands ecoregion, which is a mix of chaparral, grasslands, oak savannahs and woodlands, closed-cone pine forests, montane conifer forests, wetlands, and riparian forests. Tree species are dominated by blue oak, but many other oaks are found, including coast, canyon, valley, and interior live oaks. In many areas chaparral plants dominate the landscape and include California buckeye, manzanita, and scrub oak.

The final terrestrial ecoregion in the Sacramento River drainage is the California Central Valley Grasslands, the region in which the higher-stream-order section of the Sacramento River lies. This region once contained extensive native grasslands, oak woodlands, and dense riparian and floodplain vegetation, including expansive tule marshes. Much of these lands are now used for agriculture.

The climate in the Sacramento basin is as varied as the landforms it contains. Variability in precipitation is extremely high both intra- and interannually. Average annual precipitation for the basin is 90 cm/yr, most of which occurs from November through March (Fig. 12.13). However, the annual mean precipitation in different parts of the basin varies from <10 to >200 cm/yr. The amount and form of precipitation often is orographically determined. Precipitation increases with altitude, and the form of precipitation changes from rain at lower altitudes, to rain and occasional snow in the coastal range, to extremely high snow levels in the Sierra Nevada and southern Cascade Range. These spatial and temporal precipitation patterns significantly affect the amount and timing of river flows. Temperature also varies with altitude. Average temperature for the basin is 12.9°C, and monthly averages range from 4.5°C to 22.4°C, with subzero temperatures frequent in winter (see Fig. 12.13). Summer daytime temperatures often exceed 40°C, both in the central valley and in the foothills and mountains. Summer temperatures in the Sierra Nevada can drop to below freezing at any time of the year.

Land cover in the basin can be divided into urban, agricultural, and nonagricultural uses. Urban land cover comprises only 1.7% of the basin, with the state capital, Sacramento, the largest city upstream of the San Francisco Bay area. Nonurban land use can be partitioned into four categories: forest (49.3%), shrub (13.7%), grasslands (16.2%), and wetlands (0.8%). Agricultural land use makes up 15.1% of the basin. Approximately 2% of the basin is classified as open water (Vogelmann et al. 2001).

River Geomorphology, Hydrology, and Chemistry

The Sacramento River contains two distinct geomorphological sections: the upper Sacramento, which is upstream of Shasta Reservoir, and the lower Sacramento. The upper Sacramento is a high-gradient (10.4 m/km) montane stream that contains cobble and coarse gravels in areas that are not underlain by bedrock. The lower Sacramento, below Shasta and Keswick dams, has a moderate to high gradient (0.8 m/km) and contains a mixed substratum, including spawning gravels. Further downstream in the valley proper the Sacramento becomes low gradient (0.26 m/km), with well-established levees and much finer sediments (Fig. 12.3). The river meanders to the delta through a highly developed agricultural setting.

In general, rivers in California are characterized by extreme inter- and intraannual variations in flow. However, the Sacramento River, as most rivers in California, is highly regulated and has numerous impoundments, water withdrawals, and irrigation return flows that alter its natural discharge pattern. These controls are obvious when mean precipitation and runoff are compared through the annual cycle (see Fig. 12.13). Although much of the winter precipitation is in the form of snow, total runoff within the channel is far less than would be expected in the absence of flow capture. Peak flows in spring have all but been eliminated. In addition, summer flows are higher than would occur under a natural hydrograph.

FIGURE 12.3 Sacramento River below Chico, California (PHOTO BY TIM PALMER).

However, this winter water storage is necessary because of the uneven seasonal and interannual distribution of precipitation, flood-control, and agricultural needs during the summer growing season. All major tributaries (McCloud, Pit, Feather, Yuba, and American rivers) of the Sacramento are also impounded. Total reservoir storage within the basin far exceeds one-half of the total annual runoff of $27.6 \, km^3$ of water; consequently, over half of the annual flow out of the basin can be captured for human use. In the lower section of the river, near Sacramento, flood-diversion structures and ship channels also influence main-channel flow characteristics. In the Sacramento–San Joaquin Delta, channels are leveed and a significant portion of the flow of the Sacramento is diverted to the southern portion of the state via the California Aqueduct and Delta–Mendota Canal.

The highest recorded extreme flow in the Sacramento River, $3313 \, m^3/s$, occurred on February 19, 1986; however, an observed (but not measured) extreme flow event through the Sacramento–San Joaquin Delta occurred in 1862 and was estimated to be $120,000 \, m^3/s$ (Peterson et al. 1985). That suggests a maximum discharge of $>60,000 \, m^3/s$ or 20 times the maximum recorded flow (assuming one-half of the estimated flow originated in the Sacramento basin).

In general, the Sacramento River has high-chemical-quality water because much of the flow is derived from snowmelt-dominated tributaries draining the Sierra Nevada. In the main stem, dissolved oxygen is often near saturation and conductivity is low, $\sim 100 \, \mu S/cm^2$ (Domagalski and Dileanis 2000), as is alkalinity, $\sim 50 \, mg/L$ as $CaCO_3$. pH is just under 8. Although nutrients can be locally high, particularly in feeder streams and agricultural drains, in general NO_2-N + NO_3-N is $<0.2 \, mg/L$.

River Biodiversity and Ecology

The Sacramento River basin lies within the Pacific Central Valley freshwater ecoregion, which is considered one of the richest aquatic ecoregions west of the Rocky Mountains. However, the ecology of the system has been altered greatly by humans.

Algae

Algae found in the Sacramento basin are dominated by chrysophytes and chlorophytes. Common diatoms include *Nitzschia dissipata*, *N. inconspicua*, *Cymbella muelleri*, *Cymbella minuta*, *Fragilaria pinnata*, *F. construens*, *Cocconeis pediculus*, *Cocconeis placentula*, and *Achnanthes minutissima*. Common green algae include *Cladophora glomerata*, *Oedogonium* sp., *Hyalotheca* sp., and *Scenedesmus* sp.

Plants

Riparian vegetation throughout the basin varies according to latitude and altitude. Trees and shrubs commonly found are arroyo, black, narrowleaf, Pacific, and red willows; black and Fremont cottonwoods; California sycamore; mulefat; mountain and white alders; buttonbush; and water birch. Prior to the rivers of the Sacramento basin being dammed, floodwaters supported extensive floodplain vegetation (e.g., cattails and sedges), particularly in the delta.

Invertebrates

As with most biological aspects of the rivers of the Sacramento basin, the distributions of aquatic invertebrates are influenced by the diversity of habitats found in the basin. Upland stream invertebrate assemblages are dominated by cold stenothermic taxa that are often located in high-gradient streams containing high levels of dissolved oxygen. Aquatic insects common to these areas include the stonefly families Perlidae (e.g., *Hesperoperla pacifica*), Chloroperlidae (*Suwallia autumna*), and Capniidae (*Eucapnopsis brevicauda*); the mayfly families Baetidae (*Baetis tricaudatus, B. bicaudatus*), Heptageniidae (*Epeorus longimanus*), and Ephemerellidae (*Drunella grandis, D. doddsi*); and the caddisfly families Rhyacophilidae (*Rhyacophila*) and Hydropsychidae (*Arctopsyche grandis, A. californica, Hydropsyche*) (Carter and Fend 2001).

As one enters the foothills and lower-elevation streams the composition of the invertebrate fauna changes. Taxa more frequently found in lower-elevation streams include stoneflies in the families Perlodidae (*Isoperla*) and Nemouridae (*Malenka*), mayflies in the families Siphlonuridae (*Siphlonurus*) and Baetidae (*Baetis tricaudatus, Callibaetis*), and caddisflies in the families Hydropsychidae (*Hydropsyche californica, Cheumatopsyche mickeli*) and Hydroptilidae (*Hydroptila*).

Relatively little is known of the native molluscan fauna; however, species probably common to the basin included the California floater, western ridge mussel, and western pearl shell. Nonnative bivalves such as the Asiatic clam now dominate some freshwater streams and rivers. *Potamocorbula amurensis*, a recently introduced bivalve, now dominates the benthos in the downstream portion of the delta and in portions of San Francisco Bay and may have altered the area's food web by influencing nutrient availability to the crustacean *Neomysis mercedis*, which once was an important component in the diet of the introduced striped bass (Orsi and Mecum 1996).

Vertebrates

Moyle (2002) provides a fascinating account of the distribution and ecology of both native and introduced fishes of California. The long period of isolation of the California fauna from most of the rest of the continent (~17 to 10 million years ago), and both the long- and short-term tectonic history of the state, has greatly influenced the existence and distribution of fish species. In addition, the temporal and spatial environmental variability within the state led to a native fauna of freshwater fishes well adapted to seasonal extremes in flow and temperature.

Nonanadromous native fishes once common to the Central Valley included Sacramento perch, the only native of the Centrarchidae west of the Rocky Mountains; tule perch; numerous cyprinids, including Sacramento blackfish, hardhead, hitch, California roach, Sacramento splittail, Sacramento pikeminnow, thicktail chub (now extinct), and speckled dace; Sacramento sucker; and riffle sculpin. There also are seasonal runs of chinook salmon and steelhead trout that spawn in the cold, upland streams; however, many of these spawning areas are now inaccessible. In addition, both white and green sturgeon spawn in the Sacramento River.

The Pit River drainage contains the Goose Lake redband trout and the Goose Lake lamprey, Pit-Klamath brook lamprey, marbled and rough sculpin, tui chub, and redband trout. The McCloud River, a former tributary of the Pit, contained bull trout.

The evolution of the Clear Lake fauna occurred after the uplift of the Coast Range restricted the upstream migration of Sacramento Valley species. Consequently, numerous species or subspecies distinct from but related to Sacramento Valley fishes are only found in Clear Lake, including Clear Lake splittail, Clear Lake hitch, and Clear Lake tule perch.

There are approximately 40 nonnative fish species throughout the Sacramento basin, including the delta and San Francisco Bay. These species include numerous species of bass, sunfishes, catfishes, and trout.

Ecosystem Processes

There are no adequate accounts of the ecological functioning of the Sacramento basin prior to the massive changes caused by human modifications. We would speculate that the seasonally predictable discharge overlain by extreme variability in interannual precipitation and runoff greatly influenced the river's form and, consequently, its biota. Upland reaches were, and still are, principally in a high-elevation forested landscape dominated by conifers, with broadleaf species along its riparian corridor. The unaltered hydrograph and sediment regime in upland reaches were substantially influenced by winter precipitation and snowmelt runoff in the spring, creating very favorable conditions for salmonid spawning and rearing.

Lowland reaches most likely contained extensive riparian forests of willows, cottonwoods, and sycamores interspersed with seasonally flooded tule marshes. Floodplain vegetation must have dominated a great deal of the valley's floor, providing habitat for native lowland fishes and extremely high dissolved and particulate organic matter to the river. Given the naturally low nutrient conditions, this must have created a system very much driven by allochthonous inputs.

Currently, however, every major stream is impounded at least once along its course for the retention of water for flood control, irrigation, and/or hydroelectric power production. These impoundments have drastically changed the intra- and interannual hydrograph, often eliminating flood flows and the seasonal floodplain wetlands that once dominated the valley floor. Dams creating the impoundments have eliminated both upstream and downstream migrations of anadromous fishes and greatly altered natural sediment regimes necessary for maintaining spawning habitat and natural channel functioning.

When taken as a whole, and particularly in comparison to its historic state, the Sacramento River basin is basically a water storage and delivery system used to support agriculture in the Sacramento and San Joaquin valleys and urban centers in the Central Valley, the San Francisco Bay area, and the Los Angeles basin.

Human Impacts and Special Features

Although the Sacramento proper has been substantially modified in its lower reaches, several of its tributaries retain much of their natural features. Two of these tributaries, the Feather and American rivers, are designated as National Wild and Scenic rivers in portions of their basins. The Middle Fork of the Feather River has a 125 km portion that includes Feather Falls, the third-highest waterfall (195 m) in the United States, designated as Wild and Scenic. The North Fork of the American River has 62 km designated as Wild and the American main stem has 37 km flowing through the state capital of Sacramento designated as Recreational.

In contrast to these remaining scenic tributaries, much of the Sacramento system has been dramatically altered by dams, flow diversions, agriculture, mining, logging, and the introduction of nonnative species. California has one of the most extensive systems of water storage and conveyance in the world. Approximately 1400 dams, >8000 km of levees, and >140 aqueducts and canals exist in the state (Mount 1995). The three major purposes for dams are (1) municipal and agricultural water supply, (2) flood and debris control, and (3) hydropower production (Kondolf and Matthews 1993). Ditches and flumes built during the Gold Rush period in the mid- to late-1800s were the first major flow diversions.

However, the large-scale State Water Project (SWP) and federally supported Central Valley Project (CVP) systems were built in the twentieth century.

The CVP includes 16 reservoirs (including Lake Shasta) and 39 pumping plants and delivers approximately 8.6 km^3/yr (California State Water Project Atlas 1999). The SWP is designed to deliver 5.2 km^3/yr, with 70% going to urban users and 30% to agriculture. The extensive network of dams and water diversions are necessary to maintain the current irrigated agriculture and geographic population distribution in California because 75% of the runoff occurs from the northern portion of the state, whereas 80% of the demand for water is in the southern portion of the state (Mount 1995). However, this extensive system has numerous detrimental effects on the hydrology and ecology of rivers in the Sacramento basin. Water retention behind dams modifies normal hydrologic patterns by trapping sediment, reducing peak flood flows, and altering seasonal flows by limiting winter high flows and often increasing summer low flows. Most large dams also eliminate the transport of coarser gravels that would be suitable as salmonid spawning habitat (Kondolf and Matthews 1993). Reduction of peak flood flows in combination with levees and canalization practically eliminates any connection between the river and its floodplain. Elimination of flood flows also allows riparian encroachment to occur. Dam releases also can alter the seasonal temperature regime, whether releases are from the surface or hypolimnion. Seasonal changes in total discharge can also alter normal temperature regimes. However, the primary effect of dams on migratory salmonids is to exclude them from much of their historic spawning habitat; for example, chinook spawning habitat loss has been estimated to be >90%. In addition, many fishes are killed by unscreened water diversions throughout the state and at the massive screened diversions associated with the large-scale water diversions in the delta.

Mount (1995) suggests that agriculture has altered the rivers of California more than any other industry. Impacts include increased erosion, degradation of riparian corridors, and increased concentrations of pesticides and salts. Agriculture accounts for 80% (~38.1 km^3) of the total nonenvironmental water use in California. Alfalfa, irrigated pasture, cotton, and rice represent approximately one-third to one-half of this agricultural water use. When viewed on a county basis, the total number of cattle in the Sacramento basin is ~656,000 (California Agricultural Statistics Service 2001 http://www.nass.usda.gov/ca/coest/104lvstp.htm). During California's dry summers cattle often enter riparian corridors, degrading streamside habitat quality; the American Fisheries Society has listed grazing as the most important cause of riparian corridor degradation in western streams (Mount 1995).

Three types of mining—gold, hardrock, and gravel—have influenced and in some cases continue to influence rivers in the Sacramento basin. Hydraulic mining (outlawed in 1884) associated with gold extraction washed entire hillsides into streams and severely degraded channels and riparian corridors. In total, 42,500,000 m^3 of mining debris was washed into the Central Valley. Along with gold mining came the contamination of the tributaries of the Sacramento, and in particular the contamination of the delta and San Francisco Bay with mercury. Mercury was used to amalgamate gold in the Sierra Nevada and over time entered streams and was transported to the delta and San Francisco Bay, where it was readily transformed to the more biologically available and toxic methylmercury. Health advisories have been issued on the consumption of many non-migratory fishes found in the bay and delta, in part because of the methylmercury toxicity.

Of the 1500 abandoned mines in California, approximately 150 are discharging metal-rich waters (Mount 1995). Within the Sacramento basin, Iron Mountain Mine, an Environmental Protection Agency Super Fund Site, is the most thoroughly studied. The mining area releases approximately 544 kg of Cu and 363 kg of Zn to Keswick Reservoir located just below Shasta Dam on the Sacramento River. Elevated levels of Cd, Cu, Pb, and Zn have been detected in the caddisfly *Hydropsyche californica* up to 120 km below the reservoir (Cain et al. 2000).

Over 1 billion tons (>900 billion kg) of sand and gravel were mined in California between 1985 and 1995, and although not all was extracted from streams, this represents ~10 times the bedload of the state's rivers (Mount 1995). The effect of gravel mining is often long lasting, particularly downstream of dams. When Shasta Dam was built in 1944, 5,400,000 m^3 of gravel was removed from the Sacramento for its construction; the bed was excavated to a depth of 15 m. Because of the initial excavation and the elimination of the supply of new gravel from upstream by Shasta Dam, to date the bed of the reach downstream of the dam contains only coarse cobble and bedrock (Kondolf and Matthews 1993). Gravel mining often leads to incision of stream channels. Cache Creek, a tributary to the

Sacramento, has incised >3.7 m over a 50-year period of gravel mining (Mount 1995).

Logging also degrades rivers in the basin. California produces approximately 11,800,000 m^3 of lumber annually. Although the majority of the state's lumber is derived from the north coast forests, logging is an important industry in the Sierra Nevada and some Coastal Range areas. Some logging practices have led to the degradation of stream habitat by causing soil compaction, increased fine sediment runoff, and reduction in riparian quality.

Many nonnative invertebrates and fishes have been both intentionally and unintentionally introduced into the lakes, rivers, and bays of the Sacramento basin. One of the first (1869) recorded introductions to San Francisco Bay was the eastern oyster, which became the bay's most valuable product by the late 1890s. Two early fish introductions included the American shad (1872) and the striped bass (1879). A commercial fishery for striped bass existed between 1889 and 1935. Many fishes have been introduced to the Sacramento basin. In a recent study, of the 35 species collected throughout the lower portion of the basin only 12 were native species (May and Brown 2002, Domagalski et al. 2000).

Introductions continue to occur. For example, an introduced estuarine bivalve, *Potamocorbula amurensis*, which was first collected in 1987 (Carlton et al. 1990), has the potential to greatly alter food webs (Alpine and Cloern 1992) and possibly affect contaminant transport in the delta and San Francisco Bay. A very recent introduction to the bay/delta/river system is the Chinese mitten crab. This crab is catadromous, spawning in the bay and then migrating upstream to mature. It has the potential to cause substantial damage to natural streambanks and levees (particularly in the delta) because of its extensive burrowing habits (Rudnick et al. 2003).

SAN JOAQUIN RIVER

The San Joaquin River originates high (>4000 m asl) in the Sierra Nevada near the middle of California (Fig. 12.14). The river is 560 km long and has a drainage area of 83,409 km^2. The San Joaquin lies entirely within California and drains the southern portion of the Central (or Great) Valley. The basin can be divided into two regions, with a low topographic divide formed by the alluvial fan of the Kings River separating the southern Tulare basin from the San Joaquin River. Only during exceptionally wet periods does water flow from the Tulare basin into

FIGURE 12.4 San Joaquin River, within the Central Valley, California (Photo courtesy of Great Valley Museum, Modesto, California).

the San Joaquin (Gronberg et al. 1998). Today, almost all of the water of the San Joaquin is diverted for agricultural and municipal use (Fig. 12.4).

All large tributaries draining into the San Joaquin basin originate in the Sierra Nevada. Tributaries draining directly into the San Joaquin include the Stanislaus, Tuolumne, and Merced rivers. Tributaries draining into the Tulare basin include the Kings, Kaweah, Tule, and Kern rivers. Portions of four tributaries (Tuolumne, Merced, Kings, and Kern) have been designated as Wild and Scenic Rivers.

The total number of Native Americans (~70,000) and number of tribes (~50) attests to the diversity of habitat and productivity of the pre-Columbian San Joaquin Valley. The tribes inhabiting the basin included the Plains and Sierra Miwok in the north, the Yokuts in the south, and the Tübatulabal in the Kern River canyon (Heizer and Elsasser 1980). In the lakes area (Tulare, Kern, and Buena Vista lakes), the Southern Valley Yokuts fished and hunted waterfowl using rafts made of tule. They gathered the starchy roots of cattail and seeds from tules and

grasses and hunted pronghorn, elk, and rabbits in the foothills. Many of the villages in the valley were constructed on higher ground because of the annual spring flooding of the valley during snowmelt. The main European settlement began in the mid-1850s during the Gold Rush.

Physiography, Climate, and Land Use

The San Joaquin River basin is part of the Cascade–Sierra Mountains (CS) and Pacific Border (PB) physiographic provinces (see Fig. 12.14) (Hunt 1974). To the east and south is the Sierra Nevada Mountains section. These mountains begin in the north, just south of Lassen Peak, and continue for 640 km in a southeasterly direction. They are considered the structural backbone of California (Hunt 1974). They are composed of pre-Tertiary granitic rocks that are separated from the valley floor by Mesozoic and Paleozoic marine rocks and Mesozoic metavolcanic rocks. The southern Sierra Nevada are higher in altitude and have less volcanic overburden than the northern Sierra Nevada. The highest point in the basin, Mt. Whitney, is the highest point (4418 m asl) in the contiguous United States. The basin is bordered on the south by the Tehachapi Mountains.

The remainder of the basin is contained within the Pacific Border province. To the west of the Central Valley lies the California Coast Ranges section. These are northwest–southeast aligned, low-elevation mountains formed in the south by tectonic movement along the San Andreas Fault zone. These mountains are composed of a core of Franciscan assemblage dating from the late Jurassic to Cretaceous or Paleocene age and Mesozoic ultra-mafic rocks. Marine and continental sediments of Cretaceous to Quaternary age overlie these rocks, with some Tertiary volcanic rocks included (Gronberg et al. 1998).

The Central Valley lies in the California Trough section of the Pacific Border province. This portion of the Central Valley can be divided into two parts: a northern portion, extending from the Sacramento–San Joaquin Delta to just south of the San Joaquin River, and a southern portion, the Tulare basin, extending from south of the San Joaquin River to the Tehachapi Mountains. The Tulare basin is a hydrologically closed system. Historically, the Kings River drained into Tulare Lake, which overflowed into Buena Vista Lake to the south and, during extremely wet periods, the San Joaquin to the north (Schoenherr 1992).

The San Joaquin basin is part of three terrestrial ecoregions (Ricketts et al. 1999). The California Central Valley Grasslands ecoregion represents the southern portion of the Central Valley of California (in contrast to the northern portion, drained by the Sacramento River) and is the most anthropogenically modified from its pre-Columbian condition.

Surrounding the floor of the Central Valley at a slightly higher altitude is the California Interior Chaparral and Woodlands ecoregion. It represents foothill communities, which are dominated by grass-lands, chaparral, oak savannahs, and oak wood-lands. Numerous plants and animals are endemic to this ecoregion.

To the east and at higher altitude is the Sierra Nevada Forests ecoregion. This southern portion of the ecoregion has one of the highest concentrations of terrestrial endemic species. As in the north, plant assemblages are arranged in altitudinal zones forming a gradient from ponderosa pine forest, to mixed conifers (including the giant sequoia), to the subalpine zone of lodgepole and Jeffrey pines, moun-tain juniper, and aspen, to mountain hemlock and whitebark pine at >2700 m asl.

As throughout all of California, the basin experiences extremes in climate that are often orographically driven. In general, this southern portion of the Central Valley, including the surrounding lower ranges, is much drier than the Sacramento River drainage. There is both an east–west and a north–south gradient in precipitation. Precipitation in the coastal mountains ranges from 25 to >50 cm/yr. Precipitation in the valley, which lies in the rain shadow of the coast range, averages from 38 cm/yr in the north to only 12 cm/yr in the south. Precipitation increases as moisture-laden air rises over the Sierra Nevada and ranges from 50 cm/yr at lower altitudes to >200 cm/yr at the summit. As throughout this region, precipitation mainly occurs from October through April, with only occasional thunderstorms occurring during summer (Fig. 12.15). Average temperature for the basin is 15.7°C, but monthly averages range from 6.8°C in January and December to 25.2°C in July. Maximum daily average summer temperature in the basin, 30.8°C, occurs in July. Temperature in the basin generally decreases with increasing altitude.

Land cover in the basin can be divided into urban, agricultural, and nonurban/nonagricultural uses. Urban land cover comprises only 1.9% of the basin. The four most populous cities, Fresno, Bakersfield, Stockton, and Modesto, represent almost 50% of the total population of the basin. Nonurban/nonagricultural land cover includes forest (26.8%), shrub (13.4%), grasslands (23.0%), and wetlands (0.4%). Agricultural land use is the dominant single classification and represents 30% of land use in the basin (Vogelmann et al. 2001).

River Geomorphology, Hydrology, and Chemistry

Although the natural course of the San Joaquin River is interrupted at many locations, it originates in the high Sierra as an extremely high-gradient (22.7 m/km) stream flowing over granitic bedrock. As it transitions between the high Sierra and the foothills, it has a riparian corridor of willows and cottonwoods and flows over cobble and gravel sub-strates at a gradient of 4.8 m/km. Once in the lower valley the San Joaquin changes to a low-gradient (0.16 m/km) sand-bed river and takes a meandering course through the Central Valley to the delta (see Fig. 12.4). It once supported extensive tule marshes.

Tributary stream types in the basin include high-altitude, high-gradient streams, large tributary streams draining the Sierra Nevada, and lower-elevation and foothill intermittent streams.

It is difficult to imagine a river that is more hydrologically modified by humans than the San Joaquin. As with many other rivers in the Central Valley of California, it is primarily used to store and deliver water and remove municipal, industrial, and agricultural wastes. The basin has experienced a long history of flow capture and diversion; almost all the surface-water flow of the basin had been diverted by as early as 1910 (Williamson et al. 1989). Nevertheless, a substantial portion of the water used in the basin for irrigation is derived from local groundwater and surface water delivered by canals from the more northern, wetter Sacramento River basin.

Numerous small reservoirs located in the headwaters of the San Joaquin retain a portion of its flow; however, just before the river enters the valley Friant Dam captures the remaining flow at Millerton Lake (0.641 km^3). Most of the captured water is used for irrigation in the Central Valley. Water pumped from the Sacramento–San Joaquin Delta that flows through the Delta–Mendota canal functionally reverses the normal south to north flow of the San Joaquin River (see Fig. 12.14). Downstream of Mendota Pool, discharge is augmented from tributaries (Stanislaus, Tuolumne, and Merced rivers) entering from the Sierra Nevada.

All major tributaries to the San Joaquin are now impounded at least once along their course. Major reservoir storage is ~9.498 km^3 in the San Joaquin drainage (excluding the Tulare basin) and is only about one-half the storage capacity of the Sacramento basin. Prior to impoundments, all of the rivers flowing from the Sierra Nevada would have had spring–early summer peak flood flows fed by snowmelt runoff. These waters would inundate a great deal of the valley floor and supported lush tule marsh habitats with their associated biota. The lakes and marshes of the Tulare basin were drained in the mid 1800s, and now the four major rivers in the basin (Kings, Kaweah, Tule, and Kern rivers) have all been impounded.

The mean annual discharge to the basin from the Sierra Nevada is approximately 10.855 km^3/yr (344 m^3/s). Discharge from the Coast Ranges is only 0.114 km^3/yr (3.6 m^3/s) (Nady and Larragueta 1983), just over 1% of the total flowing from the Sierra. The potential annual discharge from the San Joaquin basin for the period from 1979 to 1992 has been estimated to be 7.524 km^3/yr (238.6 m^3/s); however,

actual discharge downstream on the San Joaquin near the town of Vernalis is only 4.564 km^3/yr (144.7 m^3/s), which is ~60% of the potential discharge from the basin (Gronberg et al. 1998). The small difference between 144.7 m^3/s and the value shown in the San Joaquin summary (132 m^3/s) is due to differences in coverage periods.

Water management to provide for irrigation during the summer growing season severely alters the natural hydrologic cycle. Normal spring peak flows are dampened by reservoir retention of winter–spring floodwaters, and normally low summer–autumn flows are augmented by reservoir releases and seasonal irrigation return flows (see Fig. 12.15). Runoff now only reaches 0.7 cm/mo in spring, which is only about 15% of the spring runoff in the Sacramento River.

The natural water chemistry of the drainage is influenced directly by precipitation patterns, bedrock geology, and soil types. Because most of the water in the basin drains from the Sierra Nevada over granitic bedrock and flows through poorly developed, low-solubility quartz and feldspar soils, the waters contain very low levels of dissolved solids. However, waters draining from the Coast Ranges, where soils are derived from marine sediments that are rich in soluble calcium, sodium, and magnesium sulfates, are much higher in dissolved solids. Extensive irrigation of these soils located on the west side of the San Joaquin Valley has led to the contamination of some areas by selenium (Ohlendorf et al. 1986).

The water chemistry of the San Joaquin is strongly influenced by agricultural tailwaters and inflows from the Sierra Nevada tributaries. From the upper portion of the basin to just below Friant Dam the San Joaquin has very low conductivity (<100 μS/cm) and hardness (<20 mg/L as CaCO$_3$) (Kratzer and Shelton 1998). However, downstream of the city of Mendota conductivity increases to >1000 μS/cm and hardness is >250 mg/L as CaCO$_3$. Dissolved oxygen is in excess of 8.0 mg/L and pH is slightly alkaline, ranging from 7.1 to 8.0. Nutrients are extremely low in Sierran waters and moderate in the downstream main stem. Median NO$_3$-N levels just exceed 2 mg/L and total phosphorus is <0.5 mg P/L. Dissolved ortho-P is normally <0.2 mg P/L (Kratzer and Shelton 1998). However, many of the agriculturally and municipally influenced drains, canals, and sloughs in the valley often have extremely high concentrations of measured constituents (conductivity <3000 μS/cm; hardness >800 mg/L as CaCO$_3$, NO$_3$-N >20 mg/L).

River Biodiversity and Ecology

The San Joaquin River lies entirely within the Pacific Central Valley freshwater ecoregion (Abell et al. 2000). This ecoregion has been isolated from the rest of North America for approximately 10 to 17 million years. This isolation and the diversity of habitats in the basin led to the evolution of a unique fish fauna with a high level of endemism.

Algae and Cyanobacteria

Leland et al. (2001) found that the diatoms *Navicula recens* and *Nitzschia inconspicua* dominated the benthos of the San Joaquin. The principal cyanobacteria was *Oscillatoria*. Diatoms common in phytoplankton included *Cyclotella meneghiniana*, *Skeletonema* cf. *potamos*, *Cyclostephanos invisitatus*, *Thalassiosira weissflogii*, *Nitzschia acicularis*, *N. palea*, and *N. reversa*. Salinity appeared to be an important factor in the distribution of algae within the river system.

Plants

Historically, the southern portion of the San Joaquin Valley supported some of the largest freshwater marshes in the west. Tulare Lake was the largest ($1800 \, km^2$) lake west of the Mississippi River in the mid 1800s (Schoenherr 1992). Marsh plants were distributed in predictable zonation patterns. Open-water plants included watercress, Pacific marsh purslane (marsh seedbox), water fern, and duckweeds. Within the water proper were rushes, bulrushes, sedges, and cattails. Higher in the marsh were black, red, and Pacific willows. Plants at the highest level included mulefat and buttonbush.

Invertebrates

Invertebrate species are generally distributed along an altitudinal gradient. In the upper reaches of the San Joaquin invertebrate faunas include upland taxa such as the mayflies *Baetis bicaudatus*, *Epeorus*, and *Drunella*; the stoneflies *Hesperoperla pacifica*, *Suwallia autumna*, and *Eucapnopsis brevicauda*; and the caddisflies *Rhyacophila*, *Arctopsyche grandis*, and *A. californica*.

Little is known of the original distribution of invertebrates in lower reaches, but they currently appear to be influenced by available habitat and water chemistry, particularly agriculturally derived salinity (Leland and Fend 1998). Abundant taxa in lower reaches include the caddisfly *Hydropsyche californica* and the Asiatic clam, along with chironomid midges and oligochaete worms. Downstream reaches of the large tributary streams often contain the mayflies *Caenis*, *Heptagenia*, *Fallceon*, and, less frequently, *Ephemerella*. Also abundant is the caddisfly *Nectopsyche* (Leland and Fend 1998). The sooty crayfish (possibly extinct) may also have been present in the basin. The most common crayfish now is the nonnative swamp crayfish (Schoenherr 1992).

Vertebrates

The Central Valley of California has been a "center of fish speciation" in California (Moyle 2002). Its freshwater dispersant fauna became isolated from the rest of the fish fauna of western North America 10 to 17 million years ago (Minckley et al. 1986). It contains 17 endemic species (40 if subspecies and distinct salmon runs and species shared with adjacent drainages are included) and 5 euryhaline marine species in the delta.

As a result of this long-term isolation, the fishes of the Central Valley appear to have limited ancestry. The Sacramento perch is the only native member of the Centrarchidae (sunfishes) west of the Rocky Mountains. The closest relatives of hardhead and Sacramento perch are only known from late Pliocene fossils found in Idaho. Tule perch and delta smelt have a marine origin. The Sacramento sucker and Sacramento pikeminnow have ancestry outside of the system. Numerous minnows (splittail, hitch, Sacramento blackfish, California roach, and thicktail chub) are also present. In addition to the strictly freshwater fauna, numerous anadromous species are present from the families Petromyzontidae, Acipenseridae, Salmonidae, and Gasterosteidae.

Altitude, stream gradient, and stream order strongly influenced the historical composition of the native fish fauna (Brown 1996). Prior to the introduction of trout in the high Sierra the only fishes found at the highest elevations were the Little Kern River golden trout, the Kern River rainbow trout, and the California golden trout. At elevations between 450 and 1000 m asl the fauna was dominated by rainbow trout and included riffle sculpin, California roach, and Sacramento pikeminnow and Sacramento suckers. Between 30 and 450 m asl the fauna was dominated by Sacramento pikeminnow and Sacramento suckers and included California roach, riffle or prickly sculpin, rainbow trout, and hardhead. At the lowest elevations were deep-bodied fishes, including thicktail chub (now extinct), Sacramento perch, hitch, Sacramento tule perch, Sacramento blackfish, Sacramento splittail, and large Sacramento suckers and pikeminnows. The principal migratory salmonids in the San Joaquin River include chinook salmon, steelhead rainbow trout, and Pacific lamprey.

Numerous fishes found in the San Joaquin basin have been listed as threatened or endangered. These include the delta smelt, Sacramento splittail, Little Kern golden trout, and Lahontan and Paiute cutthroat trout. Also listed as threatened is the Central Valley spring-run chinook salmon population.

Today, approximately 40 nonnative fishes are found in the San Joaquin basin. These nonnative species include bass, sunfishes, catfishes, perch, and trout. The severely reduced freshwater marshes associated with the San Joaquin provide important habitat for migratory waterfowl, muskrat, and beaver in the northern portions of the valley (Schoenherr 1992).

Ecosystem Processes

As with the Sacramento basin, there are no adequate accounts of ecosystem processes prior to the massive changes caused by human modifications. We would speculate, however, that the productivity of the San Joaquin basin was probably very high. For example, the seasonally predictable discharge in the San Joaquin, and particularly the streams flowing from the Sierra Nevada, provided excellent spawning and rearing habitat for salmonids.

Lowland reaches and the upland areas around Tule Lake contained extensive marshes and extensive riparian forests of willows, cottonwoods, and sycamores. Floodplain vegetation must have dominated a great deal of the valley floor, providing habitat for native lowland deep-bodied fishes, and contributing large amounts of dissolved and particulate organic matter to the river. Given the naturally low-nutrient waters flowing from the Sierra Nevada, this must have created a system very much driven by floodplain allochthonous inputs.

The productivity of the Tulare basin must have been extremely high to support the waterfowl and fishes reported there in the early to mid 1800s (Schoenherr 1992). Under natural hydrologic conditions the Tulare basin portion of the San Joaquin basin most likely overflowed into the San Joaquin via Fresno Slough whenever wet years increased lake levels; this also provided extremely high loads of suspended and dissolved organic matter. A more frequent connection between the two basins most likely influenced the hydrology, chemistry, and biology of both the northern and southern basins within the San Joaquin Valley.

Human Impacts and Special Features

Although the San Joaquin River has been greatly altered in the Central Valley, several of its scenic Sierra Nevada tributaries retain many of their natural features. The Tuolumne, Merced, Kings, and Kern rivers all have long sections listed as National Wild and Scenic Rivers. The Tuolumne, starting at its source and continuing to Don Pedro Reservoir, has approximately 134 km (76 km Wild, 37 km Scenic, and 21 km Recreational) designated. The Merced, originating in spectacular Yosemite National Park, has two main sections totaling 197 km (114 km Wild, 26 km Scenic, and 57 km Recreational). The Middle and South Forks of the Kings have a combined total of 130 km (105 km Wild, 25 km Recreational). The North Fork and South Fork of the Kern are designated for a total of 243 km (198 km Wild, 34 km Scenic, 11 km Recreational). The South Fork begins in Inyo National Forest and is golden trout habitat; the California golden trout is the state fish of California. The North Fork begins in Sequoia National Park and flows through the Golden Trout Wilderness.

The major alterations to the San Joaquin River and its tributaries are primarily associated with dams, flow diversion, agriculture, and nonnative species introductions. As in the Sacramento River basin, dams and flow diversions significantly influence the hydrology and biology of the San Joaquin basin. The San Joaquin itself and each of its major tributaries are dammed at least once, and most tributaries have multiple dams and/or diversions along their courses. For example, the Stanislaus River has over 40 dams upstream of the main dam (Kondolf and Matthews 1993).

The effects of dams are similar to those described for the Sacramento basin. These include alteration of the natural hydrograph, temperature regimes, and coarse sediment transport, all of which negatively influence native biota. For example, prior to building Friant Dam on the San Joaquin chinook salmon runs were estimated at 300,000 to 500,000 fish. After construction, salmon were separated from historic spawning habitat and the spring run was eliminated (Brown 1996). Dams on the Merced, Tuolumne, and Stanislaus rivers also significantly affected fall-run chinook and in 1990 fewer than 1000 adult salmon were observed in the drainage.

The California Aqueduct and Delta–Mendota canals serve principally to convey water from the delta into the San Joaquin and Los Angeles basins. Associated with agricultural withdrawals is the loss of fishes (larvae and adults) via unscreened diversions and large screened diversions within the delta. Under low-water conditions and during periods of high water pumping from the delta normal downstream flow can actually be reversed in some channels (Mount 1995).

Agriculture is the major user of water and the major source of pollution in the San Joaquin basin. Of the 14.9 km^3 of water used in 1990, 94.9% was used for irrigated agriculture and 1.5% was used for livestock (Kratzer and Shelton 1998). Fresno, Kern, Kings, and Tulare counties ranked in the top five nationally in nitrogen fertilizer applications, and Fresno and Kern ranked first and second for phosphorus application in 1985 (Kratzer and Shelton 1998). In the area around the lower San Joaquin, the use of nitrogen fertilizer increased by 500% and phosphorus by 285% from 1950 to 1990 (Kratzer and Shelton 1998). Even though nitrogen levels in some tributaries exceed drinking-water standards, only slight increases have been detected in the main stem (Dubrovsky et al. 1998). Livestock, also a source of nitrogen and phosphorus to the basin, have increased in recent years; for example, dairies in Tulare, Stanislaus, and Merced counties increased in number of dairy cows by 39%, 25%, and 42%, respectively, between 1991 and 2001 (California Agricultural Statistics Service 2001 http://www.nass.usda.gov/ca/coest/104lvstp.htm).

Pesticide use in the basin also is high. Pesticides are frequently detected in the San Joaquin River and at times have been found in concentrations that are toxic to aquatic life. Although many different pesticides are used in the basin, high concentrations of diazinon have been responsible for 40% of the water-quality exceedences (Dubrovsky et al. 1998).

In addition to the large quantities of irrigation water needed to farm in arid conditions, periodic flushing of excess salts is necessary in portions of the basin that have naturally saline soils. Unfortunately, these practices led to the contamination of surface waters with selenium. During the 1980s deformities found in some wildlife species in portions of the San Joaquin valley were attributed to selenium toxicity (Ohlendorf et al. 1986), which led to changes in the management of irrigation waters.

Nonnative species have considerably altered the natural environment of the San Joaquin River. One of the most visually obvious nonnative plants in the San Joaquin is water hyacinth; however, little is known about the effect it has on resident aquatic communities. Also present as nonnative species are the yellow pond lily and Brazilian waterweed.

Of 31 fish taxa collected by Brown (1998) in the lower San Joaquin River and lower portions of the large east-side tributaries draining the Sierra Nevada, only 10 were native to the basin. However, an assemblage of native species characteristic of predevelopment conditions existed in the portions of the large tributaries just below major impoundments. The more downstream reaches of the tributaries, and particularly the San Joaquin main stem, were dominated by nonnative species. Nonnative species characteristic of the main stem were fathead minnow, inland silverside, red shiner, and threadfin shad. Brown (1998) attributed the current distribution of taxa to species introductions and hydrological, chemical, and habitat modifications associated with agricultural development of the valley.

SALINAS RIVER

The Salinas River begins at 671 m asl in the portion of the Los Padres National Forest located in the La Panza Range of central California (Fig. 12.16). It flows in a northwesterly direction for 288 km and discharges into Monterey Bay on the Pacific Ocean about 18 km north of Monterey, California. The Salinas River is considered the third-longest river that flows entirely within California; it drains a basin of 10,983 km^2. The Salinas is known as the "upside-down river" (Fisher 1945, Anderson 2000) because it flows south to north and naturally flowed underground for approximately 129 km of its length within the lower Salinas Valley during the summer–autumn dry season. The Salinas Valley is often referred to as "America's Salad Bowl" (Anderson 2000) because of the high production of vegetables. The principal tributaries to the Salinas River are the Estrella, Nacimiento, San Antonio, and Arroyo Seco rivers. The Salinas River has been known by various names, including the Santa Delfina, San Antonio, and Rio de Monterey (McDonnell 1962). During the 1820s it was also believed to be the mythical river San Buenaventura that flowed from the Sierra Nevada to the Pacific Ocean.

Prior to European settlement several Native American groups inhabited the Salinas Valley area. The Salinan inhabited an area that extended from the middle of the valley to the southern portion of the basin. The Esselen lived in the Santa Lucia Mountains and along the Arroyo Seco River. The Costanoans (or Ohlone) lived from King City north to and including the San Francisco Bay area. Those Native Americans living near the coast harvested vast quantities of shellfish, as shown by many large middens containing marine invertebrate remains. As throughout most of California, acorns were a staple. Fire was regularly used to convert shrublands to more productive grasslands that better supported pronghorn, deer, and elk. Spanish settlement began in the mid to late eighteenth century and Spanish ranchos were formed at the very end of the eighteenth century (Anderson 2000).

Physiography, Climate, and Land Use

The Salinas River lies entirely within the California Coast Ranges section of the Pacific Border (PB) physiographic province (see Fig. 12.16) (Hunt 1974). The Salinas basin has a complex and rather recent geologic history, with ongoing tectonic activity. The Salinas River is a structurally controlled stream lying in a northwest–southeast synclinal trough with a long history of sediment deposition. The southern portion of the drainage, as well as the San Antonio and Nacimiento rivers, lie in a system of folds and faults, which are common throughout the coastal ranges (Norris and Webb 1990).

The Salinas Valley, which is west of the San Andreas Fault, was once located south of the Sierra Nevada (Schoenherr 1992). Movement along the fault over the last 80 million years has transported this region 600 km northward. Movement has been slow (2 cm/yr) but can also be very rapid (8 to 12 m during a single event). There are widely distributed Miocene rocks in the Coast Ranges that consist of marine sediments deposited as diatomites and silicic ash beds. Pliocene and Pleistocene alluvial deposits and lakebed deposits are also widespread in the region (Norris and Webb 1990). Serpentine, which is associated with distinctive plant assemblages because of its unique mineral content, is part of the Franciscan rocks also common to the basin.

The Salinas River basin is predominately in the California Interior Chaparral and Woodlands terrestrial ecoregion (Ricketts et al. 1999), although a small but important part of the basin in terms of water resources is in the California Montane Chaparral and Woodlands ecoregion located in the Santa Lucia Range. Both ecoregions represent Mediterranean Scrub and Savannah habitat types (Ricketts et al. 1999).

The California Interior Chaparral and Woodlands ecoregion ranges from about 90 to 910 m asl in elevation. Grasslands were originally dominated by perennial bunchgrasses but now are dominated by nonnative annual grasses. Chaparral is composed of

a mixture of herbaceous plants and shrubs and is often dominated by manzanita. At lower montane elevations foothill pine and blue oak are frequent. Although tree species are dominated by blue oak, oak diversity is high and includes coast, canyon, valley, and interior live oaks. In many areas, chaparral plants dominate the landscape and include California buckeye, manzanita, and scrub oak. In the southern Coast Ranges, north-facing slopes at higher elevations are dominated by mixed evergreen forest consisting of tanoak and Pacific madrone, with Coulter pine in the south and foothill pine in the north; both are drought-resistant yellow pines (Schoenherr 1992). The Santa Lucia fir is endemic. Serpentine soils that are formed from serpentinite (the state rock of California) are common in the area and contain unique plant assemblages with many endemics. These communities often contain leather oak, muskbrush (Jepson ceanothus), interior silktassel, milkwort, and streptanthus (Miles and Goudey 1997).

The only portion of the Salinas River basin included in the California Montane Chaparral and Woodlands terrestrial ecoregion is the portion of the Santa Lucia Mountains bordering the Salinas Valley on the southwest. This range runs parallel to the coast and is the origin of the basin's major tributaries because it receives higher levels of precipitation than the rest of the basin. Redwoods grow just west of the basin in the Big Sur area along the coast, and evergreen communities include coast live oaks, madrone, coastal sage, and chamise chaparral at low elevations. At higher elevations, tanoak and canyon live oak are frequently found. At even higher elevations the community transitions to an assemblage consisting of ponderosa, sugar, Jeffrey, and Coulter pines. Knobcone pines are often found on serpentine soils and the Santa Lucia fir is endemic to this ecoregion.

The climate of the area is mild, with relatively low precipitation. Average annual temperature is 14.4°C, with average winter temperature about 10°C and summer temperature about 18.9°C (Fig. 12.17). Most rainfall occurs from November though April. Average annual precipitation in the valley is only 36.3 cm. However, in the portion of the Ventana Wilderness on the western side of the Santa Lucia Range in the area of Big Sur, average annual precipitation is ~102 cm/yr and increases to ~127 cm/yr at higher elevations. A weighted mean annual precipitation for the Salinas basin based on precipitation isopleths is 44.5 cm/yr (http://endeavor.des.ucdavis.edu/newcara/).

FIGURE 12.5 Salinas River, California, looking upstream (south-southeast) from approximately Rkm 102. City in upper left is King City and highway on right is U.S. 101. Highest peaks are part of Santa Lucia Range (Photo by K. Ekelund, Monterey County Water Resources Agency).

Land use in the Salinas basin is dominated by agriculture. Land cover in the basin is 17.4% forest, 15.9% shrub, 49% grasslands, and <1% wetlands. Agriculture represents only 12.8% of the basin but has profound effects on the water quality of the northern portion of the basin. Urban areas represent 0.7% of the basin (Vogelmann et al. 2001).

River Geomorphology, Hydrology, and Chemistry

The Salinas River begins as a relatively high-gradient (8.8 m/km), pool–riffle stream flowing over bedrock and cobble to gravel substrates. However, once the river enters the lower portion of the Salinas Valley it becomes a low-gradient (0.78 m/km), meandering, sand-bed river flowing through an intensive agricultural setting (Fig. 12.5).

The first major tributary is the Estrella River, which enters from the southeast and flows out of the Gabilan Range. Further downstream major tribu-

taries entering from the southwest include the Nacimiento and San Antonio rivers that flow from the Santa Lucia Range. About 70% of the runoff to the Salinas River comes from the Santa Lucia Range (Irwin 1976). Both of these southern tributaries originate about 6 km from the Pacific Coast at about 915 to 1220 m asl. Further downstream the Arroyo Seco River, also originating about 6 km from the coast at approximately 1400 m asl, flows out of the Ventana Wilderness.

As throughout much of the west the influence of dams on river discharge is extreme. The two largest reservoirs, Nacimiento and San Antonio, are located on the southwest side of the basin in the Santa Lucia Range. The Santa Margarita Reservoir is located on the main stem in the southern portion of the drainage. Flow along the Salinas River is very much a function of the soil permeability, season, reservoir releases, the extent of groundwater pumping, and the river's longitudinal position in the valley. Runoff is very much lower than precipitation during winter because of flow capture and infiltration (see Fig. 12.17).

Flows in the upstream portion of the Salinas historically were at a minimum by September, and often <0.0142 m³/s. Discharge began to increase at Pozo in October as the autumn rains began. However, at Paso Robles, downstream of Santa Margarita Reservoir, discharge was most often zero from October through December (U.S. Geological Survey 2001 http://water.usgs.gov/nwis). Prior to building Nacimiento Reservoir in 1957, the Salinas River discharge at Bradley, approximately 6 km downstream of the confluence of the Nacimiento and Salinas rivers, was >2.8 m³/s between January and May and decreased to just a few tenths of m³/s during summer and early autumn. However, by 1958 the reservoir began releasing water for irrigation and aquifer recharge during the summer and discharge at the same location increased to 11 to 14 m³/s, at least an order of magnitude higher then prereservoir operation. Downstream of Bradley there is a general decrease in discharge from July through November as the Salinas River water recharges the underlying aquifers that are pumped for irrigation water. During this period discharge decreases in a downstream direction from 2.8 to 5.6 m³/s at Soledad, to 1.4 to 2.8 m³/s at Chualar, to <1.4 m³/s near Salinas. During drought periods, such as those in the late 1980s and early 1990s, the Salinas River went dry at these downstream locations.

Water chemistry of the Salinas River is strongly influenced by the intensive agricultural land use occurring in the Salinas Valley. Although the chemistry changes along the course of the river, the following estimates are for the Salinas River just upstream from Salinas. The waters are moderately hard (321 mg/L as $CaCO_3$) and somewhat alkaline (mean pH 7.8). Conductivity is very high, averaging 1003 µS/cm, as are nutrients (NO_3-N + NO_2-N averages 2.9 mg/L and PO_4-P averages 4.4 mg/L) (U.S. Geological Survey 2001 http://water.usgs.gov/nwis).

River Biodiversity and Ecology

The Salinas River is in the Pacific Central Valley freshwater ecoregion and has the major habitat type of Temperate Coastal Rivers, Lakes, and Springs (Abell et al. 2000). Research on its fishes and benthic biota has begun only recently.

Plants

Much of the native riparian habitat of the Salinas River has been compromised by channel alterations and agriculture. Large riparian plants that are native to the stream include willows (arroyo, narrowleaf, red, and Sitka), buttonbush, California sycamore, Fremont cottonwood, mulefat, and white alder.

An extremely invasive plant, the giant reed, is widely dispersed within the basin. Early Spanish settlers introduced this grass to North America in the 1820s (Douce 1993) and used it for building, animal feed, and erosion control. In the past, its cultivation, particularly for erosion control, has been encouraged. Giant reed frequently becomes established after disturbances such as floods and fires. When it displaces native riparian plants such as willows and cottonwoods, important habitat is lost for native riparian species, particularly streamside nesting birds.

Invertebrates

Macroinvertebrates in the Salinas basin tend to be low-elevation, fine-substratum, large-river taxa. Common mayflies include *Tricorythodes*, *Diphetor*, *Fallceon*, *Acentrella*, *Centroptilum*, *Choroterpes*, and *Baetis*. There are relatively few stoneflies other than fairly low abundances of *Suwallia*, *Sweltsa*, *Triznaka*, and *Malenka*. In the spring, the most abundant stonefly taxon is *Isoperla*.

Numerically dominant and widespread caddisflies are *Hydropsyche* and *Cheumatopsyche*. Two additional caddisflies found in the Salinas system are *Nectopsyche* and *Gumaga*. These two genera are often associated with finer substratum, which is common to the Salinas River. Naidid worms, chironomid midges, and the amphipod *Hyalella azteca* also make up a substantial portion of the invertebrate assemblage.

Vertebrates

Prior to modern human disturbances, the Salinas River system contained a freshwater fish fauna that was similar to the Central Valley fauna (Moyle 2002). The similarity between the Salinas and the Central Valley faunas is attributed to colonization via the Pajaro River system (located just to the north). Even though there is no present-day hydrologic connection between the Pajaro and Salinas system and the Central Valley of California, migration to the Salinas systems presumably occurred during the Pleistocene. Two routes have been hypothesized: (1) head capture of streams in the San Francisco Bay area by the Pajaro River and its tributaries and (2) historical drainage of Coyote Creek to the Monterey area (Coyote Creek presently discharges to San Francisco Bay).

Fishes present include Sacramento sucker, California roach, hitch, Sacramento blackfish, Sacramento pikeminnow, speckled dace, Sacramento perch, tule perch, and riffle sculpin. Sucker, roach, and hitch may be sufficiently different from Central Valley forms to be considered subspecies (Moyle 2002). Steelhead (rainbow) trout and coho salmon also are part of the basin's fauna. Moyle (2002) lists this drainage as the southern limit of coho salmon. Steelhead were historically found in the headwater areas of the Salinas River, in the Arroyo Seco River, and possibly in other tributaries. As in many other basins in this region, the Salinas basin contains many nonnative species. These include bass, sunfishes, catfishes, and trout, particularly in the reservoirs.

The basin is rich in other vertebrates as well, including both red- and yellow-legged frogs, Pacific giant salamander, and the Santa Cruz garter snake. As found in most waterways of the region, beaver, muskrat, and raccoons also are present in the basin.

Ecosystem Processes

There are no accounts of the ecological functioning of the Salinas River prior to human habitation. Early accounts of Spanish exploration indicate that the river appeared like a green ribbon running through the Salinas Valley, presumably because its banks were lined with willows and cottonwoods (Fisher 1945). This riparian corridor probably added needed allochthonous inputs to the river given that the autochthonous production was low because of the instability of its sandy bed. Historical accounts also noted the extreme differences in climate, and consequently, flow. Differences in wetted channel width at the same location ranged from a channel that could be stepped across to one that was >3 km wide. High frequency cycles of flooding and drought were common. Reports indicate that the valley was transformed into a great muddy lake in 1914 because of flooding. In contrast, during the period 1828 to 1830, 22 months passed without any rain (Fisher 1945). These extremes in flow and precipitation must have severely stressed in-stream and riparian communities.

Human Impacts and Special Features

Although no portions of the Salinas River proper are under consideration for Wild and Scenic status, a section of the Arroyo Seco is under consideration. This segment is considered some of the best habitat remaining in the basin for a portion of the threatened Central Coast steelhead population, which is designated as one of several ecologically significant units (ESU) that have been identified along the Pacific Coast.

Irrigation dominates the Salinas Valley. The lack of rainfall in coastal California during the summer growing season necessitates crop irrigation. Frequent dry periods made agriculture difficult for the inhabitants of the early Spanish missions and the first irrigation projects were small diversions from local creeks and springs to support the agricultural needs of these missions.

Prior to the 1860s, the dominant land use in the basin was grazing. There were an estimated 70,000 head of cattle in the basin in the mid-1800s, which is similar to the 75,000 present now (California Agricultural Statistics Service 2001 http://www.nass.usda.gov/ca/coest/104lvstp.htm). However, during the early 1860s two consecutive years of drought severely impacted the grasses in the valley and led to a decline in the valley's beef production and a breakup of the traditional ranchos. These factors led the way for the development of other forms of agriculture.

Irrigation in the valley has progressed through a series of stages. Initially, water was acquired from artesian wells, but canals were built as more and more land was developed for agriculture. In 1874, a small dam was placed on the Salinas River near the present city of Gonzales to provide irrigation water to the local ranchero. The first major irrigation canal (diverting about $28 \text{ m}^3/\text{s}$ of Salinas River water) was built between San Ardo and Bradley in 1884. There were several early attempts to pump irrigation water directly from the Salinas River. However, the river periodically went dry and floods often destroyed irrigation pumps (Anderson 2000).

Currently, the principal source of water used for both municipal purposes and irrigation is groundwater. As in the San Joaquin Valley, groundwater overdraft is an ongoing concern, particularly during periods of drought when there is decreased recharge and increased evapotranspiration. This overdraft has led to the intrusion of saline ocean waters into the aquifer.

The principal human impact on water quality and quantity of the Salinas River is agriculture. Discharge, particularly flows from the San Antonio and Nacimiento reservoirs, is managed to maximize groundwater recharge. The timing and amount of discharge deviates substantially from the precontrol period and functionally reverses normal hydrologic patterns. Increases in nutrients, particularly NO_3, are also related to agricultural practices.

There has only been minor gold and mercury mining in the Salinas basin. However, some mercury contamination exists in the basin. Petroleum fields have been developed in the southern portion of the valley near San Ardo.

It is difficult to imagine today, but prior to the mid 1800s the Salinas Valley contained pronghorn, tule elk, grizzly bears, and a Central Valley–type assemblage of native freshwater fishes. Currently, native blacktail deer, San Joaquin kit fox, coyote, mountain lion, and bobcat as well as large numbers of raptors, including bald and golden eagles, remain. However, numerous species have also been introduced to the area. For example, introduced boar, which are extremely fecund, have devastating effects on the native vegetation. In addition, as throughout most of California, there are many nonnative fishes, particularly in the reservoirs of the basin. These include largemouth bass, smallmouth bass, white bass, striped bass, black crappie, bluegill, inland silverside, threadfin shad, goldfish, green sunfish, redear sunfish, common carp, channel catfish, and white catfish. Many of these nonnative taxa are also present in the Salinas River. Bullfrogs were introduced for food around the turn of the century and have been implicated in the reduction of native frogs in the basin as well as in other areas of California and the west.

KLAMATH RIVER

The Klamath River begins in southern Oregon and flows through northwestern California to the Pacific Ocean (Fig. 12.18). Beginning at an altitude of 1265 m asl, the river travels 462 km to its mouth at Requa, California, just south of Crescent City. Its basin is approximately 40,700 km². The Klamath River accounts for about 18% of the discharge from California (Norris and Webb 1990). The Klamath River drainage is divided into the upper Klamath basin, located in southeastern Oregon and north-central California, and the lower Klamath basin, located almost entirely in California. Because the Klamath River proper begins just downstream of Upper Klamath Lake in Oregon, there are no large tributaries confluent with the Klamath in the upper basin. However, waters flowing into the Klamath River from Upper Klamath Lake originate in the Sprague, Sycan, Williamson, and Wood rivers in Oregon. Large tributaries of the Klamath in California include the Shasta, Scott, Salmon, and Trinity rivers. The Klamath basin has more river kilometers designated as National Wild and Scenic Rivers than

any other basin covered in this chapter, and much of the river flows through deep gorges. Crater Lake, which is located in the caldera of ancient Mount Mazama, is also within the upper basin. Crater Lake is the deepest (590 m) lake in the United States.

The Klamath basin has a rich and complex Native American cultural history. Environmental differences between the upper and lower basins strongly influenced the native peoples living in both regions. In the upper basin the Klamaths lived around Upper Klamath Lake and Klamath Marsh, the Modoc lived in the area bordering Oregon and California in the vicinity of the Lost River, Clear Lake Reservoir, and Goose Lake, and the Yahooskin Band lived in the area around Yamsay Mountain in the far northeastern portion of the basin. The Klamaths hunted deer and waterfowl, collected mussels, and gathered nuts, berries, and seeds. Particularly important was fishing the spring runs of Lost River and shortnose suckers in the rivers discharging into Upper Klamath and Agency Lakes (Klamath Tribes 2003 http://www.klamathtribes.org/suckers.htm). In autumn, the seeds of the "wocas" or water lily were gathered (Oregon Water Resources Department [OWRD] 1999). The abundant tules were used for constructing houses and rafts. Many Native American tribes inhabited the lower basin. These included the Tolowa, Yurok, Hupa, Karok, Shasta, Wiyot, Wailaki, Chimariko, and Wintun. These tribes were generally associated with riverine habitats and were dependent on the seasonal runs of anadromous fishes, including chinook and coho salmon, steelhead trout, and Pacific lamprey. Acorns also were a staple (Heizer and Elsasser 1980).

Similar to the Rogue basin, the first European trappers entered the Klamath basin in 1826 (Clark and Miller 1999). It was not until the 1850s that fur trappers and a few cattle ranchers began to spend extended periods of time in the basin. The permanent settlement began after Fort Klamath was established in 1863. Very rapid development in the basin followed the building of the railroad in 1909.

Physiography, Climate, and Land Use

The Klamath basin spans three physiographic provinces (Hunt 1974). The upper Klamath lies mostly within the Great Basin section of the Basin and Range (BR) province (see Fig. 12.18). The Klamath then passes through the Southern Cascade Mountains section of the Cascade–Sierra Nevada (CS) province. The lower portion of the Klamath flows through the Klamath Mountains section of the Pacific Border (PB) province. The Klamath Mountains section contains numerous spectacular ranges, including the Siskiyou, South Fork, Salmon Mountains, and Trinity Alps. These three provinces contrast sharply in geology and climate, which in turn greatly influences the river's hydrology and biology.

The upper basin lies in a fault-formed graben in the rain shadow of the Cascade Mountains, which border it on the west. Much of the upper basin was part of ancient Lake Modoc, a pluvial lake formed during the Pleistocene that covered an area of $2850\,km^2$. The remnant of Lake Modoc is now Upper Klamath Lake, which is the largest lake in Oregon. The surrounding lands greatly reflect the former presence of the ancient lake, and wetlands and marshes are characteristic of the area. Other lakes in the upper basin include Lower Klamath Lake, Agency Lake, and the formerly hydrologically closed basin containing Clear Lake Reservoir, the Lost River, and Tule Lake. Crater Lake is also located in the upper basin.

The geology of the lower portion of the Klamath drainage is very complex and at first appears similar to the northern Sierra Nevada. Much of the range was formed by a series of accretions that occurred during the Mesozoic Era (Harden 1998). Fossilized materials suggest that the origin of some of the accreted oceanic terranes were from very distant oceanic locations. The area was intruded by granitic plutons about 150 million years ago (Norris and Webb 1990). Approximately 130 million years ago the Klamath Mountains were located just north of the Sierra Nevada. Through a process known as extension, they separated from the Sierra Nevada and moved westward to their present location.

The Klamath basin is part of the Eastern Cascades Forests, the Klamath–Siskiyou Forests, and the Northern California Coastal Forests terrestrial ecoregions (Ricketts et al. 1999). The Central and Southern Cascades Forests border the upper portion of the basin on the west. All four terrestrial ecoregions are within the Temperate Coniferous Forests habitat type.

The upper Klamath basin is predominantly in the Eastern Cascades Forests terrestrial ecoregion. Natural vegetation is a mosaic of shrublands, grasslands, and coniferous forests, with ponderosa pine the predominate conifer. Numerous plant zones have been characterized based on temperature, elevation, and moisture. Western juniper and ponderosa pine occur in drier areas, with Douglas fir and grand fir found in more mesic, midslope areas.

Most of the Wild and Scenic portion of the Klamath River flows through the Klamath–Siskiyou Forests terrestrial ecoregion. This ecoregion contains one of the four richest temperate coniferous forests in the world; its 33 species of conifers represent a global diversity maximum (Ricketts et al. 1999). There are about 3500 different plant species known from the ecoregion and the area is considered one of the global centers of biodiversity.

Near the mouth of the Klamath a short portion of the river flows through the Northern California Coastal Forests terrestrial ecoregion. This region is most prominently characterized by the presence of redwoods in areas of fog and high moisture, and Douglas fir, tan oaks, and closed-cone pine forests in drier areas.

Precipitation in the upper and lower basins differs substantially; however, both basins have a very strong seasonal cycle (Fig. 12.19). The upper basin is in the rain shadow of the Cascade Range; therefore, the average annual precipitation for the basin is only about 41.9 cm. In contrast, the lower basin includes some of the wettest areas in the drainage and in California in general (in excess of 198.1 cm/yr). Average annual precipitation is 156.5 cm. Average annual temperature in the upper basin is 7.6°C, and frost can occur any day of the year. Temperatures near the coast are generally more moderate than temperatures inland, and average annual temperature in the lower basin is 12.9°C, slightly higher than in the upper drainage.

The upper portion of the basin is dominated by agriculture and grazing; however, land use in the lower portion includes forestry, mining, and grazing. When land cover for the entire basin is considered, 66.3% is classified as forest, 14.0% as shrub, 8% as grasslands, and 2.3% as wetlands. Agriculture represents only 5.9% of the basin but has a profound effect on the water quality in the basin. Urban areas represent only 0.2% of the basin (Vogelmann et al. 2001).

River Geomorphology, Hydrology, and Chemistry

The extreme differences in physiographic setting between the upper and lower portions of the Klamath basin has lead to significant differences in the geomorphology, hydrology, chemistry, and biology of the two areas (Clark 1999). The headwaters of the Klamath River include the streams, marshes, and lakes of the upper Klamath basin; however, the Klamath River itself begins just downstream of Upper Klamath Lake, Oregon (see Fig. 12.18). Although some of the streams in the upper basin that flow from the eastern side of the Cascades and other areas of high relief are high gradient, much of the area contains low-gradient streams, marshes, and lakes, the most prominent of which are Upper Klamath, Agency, Tule, and Clear lakes and the Klamath and Sycan marshes. The four principal rivers are the Sprague, Sycan, Williamson, and Wood. The Lost River is in the southeastern portion of the basin, and through natural channels and canals it connects a number of reservoirs, as well as Tule Lake, to the Klamath River. This area originally formed a hydrologically closed basin, with the Lost River flowing from Clear Lake Reservoir to Tule Lake. The water level of Upper Klamath Lake determines, in part, the water available to the Klamath River. Downstream of Iron Gate Reservoir the lower Klamath is high gradient (2.5 m/km) and flows freely to the coast for >300 km (Fig. 12.6).

There are 67 dams in the Oregon portion of the Klamath basin. Three of these dams (Link, Keno, and J. C. Boyle) are on the Klamath River proper. There are only two significant dams directly on the Klamath in California: Copco No. 1 and Iron Gate. Although there are no other dams on the lower Klamath, the Shasta River has 14 dams, the Scott has 3, the South Fork of the Trinity has 1, and the Trinity has 3 (only the largest shown on Fig. 12.18). These impoundments, along with private water withdrawals, affect discharge and temperature in the main stem and alter the quality and total amount of salmonid spawning and rearing habitat in the Klamath basin.

The Klamath has the second-highest mean discharge (501 m³/s) of any Pacific river south of the Columbia River. The highly seasonal precipitation creates a strongly seasonal discharge pattern, with peak runoff (>7 cm/mo) from January through March (see Fig. 12.19). Runoff slowly decreases to <1 cm/mo in August and September before increasing again in response to the beginning of the wet season in midautumn.

The upper Klamath basin is the most intensely managed basin in Oregon and water quality is greatly influenced by human as well as natural factors (http://www.deq.state.or.us/lab/WQM/WQI/klamath/klamath3.htm). A portion of the upper basin (Williamson subbasin) has good-to-excellent water quality; however, the other portions of the upper basin have poor water quality, with high levels of PO_4, pH, BOC, total solids, and nitrogen compounds (http://www.deq.state.or.us/lab/WQM/WQI/klamath/klamath3.htm). Water quality differs dramatically between the upper and lower Klamath basins. Near the town of Klamath, on the coast, total hardness is 70 mg/L as $CaCO_3$ and conductivity is about 170 μS/cm. The waters are alkaline, averaging approximately a pH of 8, and are well oxygenated (dissolved oxygen 10 mg/L). Nutrients are generally low, with dissolved NO_3-N + NO_2-N about 0.155 mg/L and PO_4-P averaging 0.037 mg/L.

River Biodiversity and Ecology

The Klamath basin is part of the Pacific Mid-Coastal freshwater ecoregion, which has a habitat type of Temperate Coastal Rivers, Lakes, and Springs.

Algae

Within Upper Klamath Lake, several companies commercially harvest the alga *Aphaizomenon flosaquae*, which is sold as a nutritional supplement. Unfortunately, more algae is produced by the lake than can be used, and the lake often suffers from low dissolved oxygen in the latter part of the summer at a time when water temperatures also are high, thereby exacerbating possible biological effects.

Plants

The upper Klamath basin once contained extensive tule marshes that were intensively used by the Native Americans of the area. However, approximately 75% of the marshes have been converted to agriculture and grazing lands. In many reaches, the lower Klamath flows through deeply incised canyons; however, where lower banks are present it has a well-developed riparian corridor that consists of willows (arroyo, black, narrowleaf, Pacific, red, sandbar, and Sitka), black and Fremont cottonwoods, mulefat, white alder, and water birch.

Invertebrates

Mayflies most often encountered include *Baetis tricaudatus*, *Drunella doddsi*, *Rhithrogena*, *Cinygmula*, *Epeorus*, *Ironodes*, *Paraleptophlebia*, and *Tricorythodes*. Commonly found stoneflies include *Yoraperla nigrisoma*, *Calineuria*, *Hesperoperla pacifica*, *Zapada columbiana*, *Malenka*, and *Sweltsa*. The most commonly collected caddisflies in the basin are in the genus *Hydropsyche*. Two leptocerids, *Nectopsyche gracilis* and *Oecetis disjuncta*, have been reported from the main stem. *Helicopsyche*, *Glossosoma*, *Agapetus*, *Lepidostoma*, *Micrasema*, and *Rhyacophila* caddisflies are also abundant in the basin. As in most systems, the most abundant aquatic invertebrates found in the lower Klamath basin are chironomid midges and oligochaete worms.

The three most commonly encountered beetles are the elmids *Optioservus*, *Cleptelmis*, and *Zaitzevia*; water pennies also are present, with

FIGURE 12.6   Klamath River between the towns of Seiad Valley and Happy Camp, California (PHOTO BY STEVE FEND, USGS).

Eubrianax as the most commonly encountered genus. The corixid *Sigara* is found in slower reaches and backwaters. Snails include *Juga* and the Klamath pebble snail. Mussels found in the basin include the western pearlshell and western ridge.

Vertebrates

Moyle (2002) classifies the Klamath system into three parts: (1) the upper Klamath River that lies above Klamath Falls; (2) the lower Klamath River, including the Trinity River; and (3) the Rogue River. There are 30 native fishes with 8 endemics. Faunas of the upper and lower basin are very distinct because they evolved in regions that experienced different geologic histories and because the upper and lower basins contain extremely different habitat types.

The distinctive fish fauna of the upper Klamath (15 natives) indicates that a long period of isolation existed between it and the adjacent systems. Some of the closest relatives to these fishes are found in the Great Basin. This region was connected to the Columbia drainage via the ancestral Snake River during the Eocene (~55 to 34 million years ago) and again during the Pliocene (~5 to 2 million years ago). It also was historically connected to the Pit River drainage (currently a tributary of the Sacramento River).

Just five families (Catostomidae [suckers], Cyprinidae [minnows], Cottidae [sculpins], Salmonidae [trout, salmon], and Petromyzontidae [lampreys]) are found in the upper Klamath. Suckers include shortnose sucker, Lost River sucker, and Klamath largescale sucker. Minnows include blue chub, Klamath tui chub, and speckled dace (both upper and lower Klamath). Sculpins include slender sculpin and Klamath Lake sculpin (both found only in Oregon) and marbled sculpin (both upper and lower). Salmonidae include two forms of rainbow trout (redband and coastal rainbow trout). Bull trout also are present. Four lamprey species are recognized: Miller Lake lamprey, Pit-Klamath brook lamprey (also found in the Pit River), Klamath River lamprey (confined to the Upper Klamath), and dwarf Pacific lamprey. All of these were derived from the anadromous Pacific lamprey.

The presence of Pit-Klamath brook lamprey and marbled sculpin in both the upper Klamath and the Pit River drainage indicates that these systems were connected at one time. There also is evidence that the upper Klamath was once hydrologically connected to the Great Basin (see Chapter 14) because the tui chub, speckled dace, and shortnose sucker found in the upper Klamath are related to taxa found in the Great Basin (Moyle 2002).

The lower Klamath contains 21 native species, of which 17 are saltwater dispersants. These include two anadromous lamprey species, two sturgeon species, six salmonids, two smelt, one stickleback, and two amphidromous sculpins. Freshwater dispersants include Klamath speckled dace, lower Klamath marbled sculpin, Klamath smallscale sucker, and Pacific brook lamprey.

The upper and lower basins share only two species, the Klamath speckled dace and the marbled sculpin. These two species probably dispersed during the Pleistocene when the water level in Upper Klamath Lake overflowed and eroded the divide, thereby creating a permanent connection between the upper and lower basins.

Other vertebrates found in the Klamath include the Pacific giant salamander, southern torrent salamander, and rough skinned newt. Mammals intimately associated with the lotic environment include beaver, raccoon, river otter, and mink. The wetlands and lakes of the upper Klamath basin support some of the largest populations of ducks in North America and provide habitat for other sensitive species, including yellow rails and sandhill cranes. The area also contains the densest concentration of wintering bald eagles in the world.

Ecosystem Processes

Little is known of the ecosystem processes that made up the Klamath basin. The contrasting environmental settings of the upper and lower portions of the basin surely influenced the processes within each location. The principal linkage between the two areas is the Klamath itself, but because of in-stream barriers and extreme differences in the habitats available to aquatic organisms, faunal differences between the two areas increased over time (e.g., suckers in the upper basin and anadromous salmonids in the lower basin). The upper basin still contains extremely high numbers of migratory waterfowl, and one can imagine that the numbers were even greater before 75% of the marshes were converted to agriculture (OWRD 1999).

Downstream of Upper Klamath Lake the Klamath River and its tributaries were free flowing, which allowed salmonids to populate all of the sub-basins that contained sufficient water and suitable spawning and rearing habitat. It is probable that seasonally Upper Klamath Lake provided high nutrient loads in the form of dissolved and particulate organic matter that originated from its marshes to the Lower Klamath on a seasonal basis. These nutrients likely formed the basal level of the river's trophic structure, which aided the development of a rich invertebrate fauna that ultimately supported salmonid development. Although there is still a contribution of organic matter and nutrients to the lower Klamath from the upper basin, its value is tempered by the poor water quality delivered to the lower Klamath from agricultural tailwaters originating above Upper Klamath Lake and from the Lost River subbasin.

Clearly, the upper Klamath basin contributed nutrients to the lower Klamath basin and thereby influenced salmonid populations, but prior to the relatively recent decrease in salmonid populations the lower basin had an internal source of nutrients as well. This internal source resulted from the breakdown of spent salmonid carcasses after spawning in the main stem and tributaries of the system. The decline in salmonid populations in the lower basin has had an unknown effect on the nutrient budget of the lower basin, and consequently its ability to support historic salmonid production.

Human Impacts and Special Features

The Klamath basin has >900 km of reaches designated as Wild and Scenic. The upper basin has a portion of the Sycan River and a short portion of the Klamath River (J. C. Boyce Powerhouse to the California border) designated. In the lower basin, 460 km of the main stem, as well as portions of the Salmon, New, and Trinity rivers, North and South forks, and Scott and Wooley creeks also have Wild and Scenic status.

As in most of the rest of the west, the distribution of available water between agriculture and in-stream use is problematic. Agriculture is extremely well developed in the upper basin; approximately 75% of the marshes and shallow lakes have been drained and converted for agricultural use. The largest irrigation project in the region, the Klamath Project, provides water to irrigate over 930 km^2. Crops grown include alfalfa, potatoes, onions, and sugar beets. Cattle ranching is the largest agricultural endeavor and accounts for more than one-third of the farm income in the region.

Agricultural practices in the upper basin affect the quantity and quality of water available to the lower

Klamath basin as well as affecting some resident fauna, such as the federally and state listed endangered Lost River sucker and shortnose sucker and the federally listed threatened bull trout. High nutrients and the addition of organic matter in agricultural return flows negatively affect water chemistry, which leads to the area having some of the poorest water-quality ratings in Oregon. The degree to which these factors (extensive reductions of freshwater marshes, reduced flows, degraded water chemistry) alone or in combination affect the fauna of the upper Klamath basin is poorly known. However, coincident with increased human development of the basin, reductions in native fauna, particularly the Lost River sucker and shortnose sucker, have occurred.

Extensive use of water in the upper basin and water captured by tributary dams within the lower basin (some of which is transported out of the basin; e.g., portions of the Trinity River) reduces the total amount of water available to the lower basin. This reduction in total discharge, along with changes in the amount and timing of peak and low flows, leads to changes in sediment supply and stream temperature (Kondolf and Matthews 1993). Both of these factors influence the quality of salmonid habitat. These human modifications most likely reduce available critical rearing habitat for chinook and coho salmon and steelhead. The sensitivity of salmonids to high water temperatures is well known, as is their need for high-quality spawning gravels.

Even though the Klamath basin is strongly influenced by human modifications, the lower Klamath supports one of the better anadromous fisheries in California. Unfortunately, the current runs in the Klamath are much lower than they were historically. The allocation of water in space and time is contentious at best (Levy 2003). Determining the amount of water to allocate for agriculture and for the restoration and maintenance of healthy aquatic habitats, particularly fishery resources, is dependent on extensive, high-quality data and is an ongoing challenge for resource managers here as it is throughout the west.

ROGUE RIVER

The Rogue River, located in southwestern Oregon, begins at 1600 m asl in Crater Lake National Park near the northern slope of ancient Mount Mazama (now Crater Lake), and flows westerly for approximately 340 km to the Pacific Ocean (Fig. 12.20). The

Rogue River drainage basin is about 13,400 km². The Rogue has many tributaries that flow from the high Cascades, but some of the more significant tributaries flow from the south. Bear Creek enters the Rogue after flowing past the rapidly developing towns of Ashland and Medford. The Applegate River enters the Rogue just downstream of Grants Pass. The Illinois River is the most downstream tributary and enters the Rogue near the town of Agness. The Rogue River was one of the first rivers to have reaches receive National Wild and Scenic designation in 1968, and portions of the Illinois River were designated as Wild and Scenic in 1984.

Prior to the settlement of the basin by Europeans the area was inhabited by numerous groups of Native Americans as early as 8560 years ago, although most researched archaeological sites are much younger (Douthit 1999). Within the Athabascan group, the Tal-tush-tun-tude lived on Galice Creek, a tributary to the Rogue; the Chas-ta-costa lived on the north side of the Rogue; and the Tu-tut-ni lived near the mouth. Further upstream, near Medford, lived the Shastan group, and the Ta-kel-man lived in the very upper reaches of the Rogue. All of the inhabitants of the basin were very dependent, as were most tribes of the northwest, on anadromous fishes. They also hunted deer and elk and gathered acorns and other foods (Purdom 1977).

The first Europeans in the basin were trappers and fur traders who arrived as early as 1826 (Douthit 1999). Almost immediately, hostile encounters occurred between the Europeans and the Native Americans. As a result, no settlements were established for many years. These early negative encounters led to the territory and its inhabitants being labeled "Rogue," which is the basis for the river's name. There were continuing conflicts between Europeans and Native Americans, and in the mid 1850s, during the Rogue River Indian War, several terrible massacres of whites and Native Americans occurred. When gold was discovered on Jackson Creek (a tributary to the Rogue) in 1852, settlement of the area increased substantially. From 1850 to 1890 the economy was based on gold and coal mining, agriculture, logging, shipbuilding, and commercial salmon fishing (Douthit 1999).

Physiography, Climate, and Land Use

Although originating in the Middle Cascades section of the Cascades–Sierra Nevada (CS) Mountains physiographic province, the Rogue River lies principally in the Klamath Mountain section of the Pacific Border (PB) province (see Fig. 12.20) (Hunt 1974). It flows south and southwest across the western slope of the Cascades and then sinuously through the Klamath Mountains before reaching the Pacific Ocean at Gold Beach, Oregon.

Ancient Mount Mazama was one of a series of volcanic peaks that compose the high Cascades (Hunt 1974). These peaks are geologically younger (Mio-Plio-Pleistocene) than the western Cascades and separate drier eastern Oregon from wetter western Oregon. Bear Creek, flowing from the southeast, separates the western portion of the Cascades from the Klamath Mountains. The Klamath Mountains are composed of folded and faulted pre-Tertiary strata with intruded granitic plutons (Baldwin 1981). The Rogue maintained its westward passage during the Cenozoic uplift of the region by down-cutting; consequently, some sections of the Rogue have steep, narrow gorges that contain many rapids.

Serpentinite occurs widely in the Klamath Mountains. These soils influence plant distributions, and vegetation is often sparse on serpentine soils. Granitic intrusions, occurring during the late Jurassic, brought solutions of metals containing gold, copper, and other base metals, which have been mined in the basin. Terraces of Plio-Pleistocene alluvial deposits were also the source of placer gold mined in the mid to late 1800s.

The Rogue basin contains three terrestrial ecoregions (Ricketts et al. 1999). The major habitat type in all three ecoregions is Temperate Coniferous Forests. The Central and Southern Cascades Forests ecoregion extends from mid-Washington to the Oregon–California border and includes both the western and high Cascades. The influence of historic and recent volcanism is evident throughout the region. The area is highly dissected by many perennial streams and rivers. Potential vegetation for the region includes western hemlock, Pacific silver fir, and western red cedar. Fire suppression and logging have significantly influenced natural disturbance regimes. Although the region, in general, contains moderate biological diversity, the southern portion of the region (including areas within the Rogue drainage) is identified as containing extremely high species richness within many groups of plants and animals (Ricketts et al. 1999).

The Klamath–Siskiyou Forests ecoregion contains one of the four richest temperate coniferous forests in the world (33 species). There are about 3500 different plant species known from the ecoregion, and the area is considered one of the global centers of biodiversity (Ricketts et al. 1999). This high species diversity results from the area's complex geologic and climatologic history, present-day diversity in soils and climates, and topographically complex gradients of moisture and temperature from the coast to inland areas.

The Central Pacific Coastal Forests ecoregion extends from Vancouver Island in the north to southern Oregon. The forests of this region are some of the most productive in the world. Dominant species include Douglas fir, western hemlock, and western red cedar, along with grand fir, and western white pine. Associated with these conifers are numerous ferns, lichens, mosses, and herbs. Sitka spruce and lodgepole pine also are present in the region. Even though most of the region is still forested, only about 4% remains intact. Impacts on the region have and continue to be logging, grazing, burning (or lack thereof), mining, and species introductions.

Precipitation in the basin varies greatly and depends on distance from the coast and local orographic influences. Precipitation at Gold Beach averages 198.6 cm/yr, but precipitation inland at Ashland is only 48.8 cm/yr. Most precipitation occurs during winter and spring, with less than 1 cm/mo in summer and >16 cm/mo in winter (Fig. 12.21). Average precipitation for the drainage is approximately 96.5 cm/yr. Temperatures are generally mild, with average monthly temperatures highest in July (19.7°C) and August (19.6°C) and lowest in December (4.4°C). Temperature varies with altitude; therefore, temperature estimates based on lower-elevation urban centers most likely overestimate temperature in the basin.

When land cover for the entire basin is considered, 82.7% is classified as forest, 2.9% as shrub, 6.4% as grasslands, and 0.2% as wetlands. Agriculture represents only 5.9% of the basin and urban areas represent only 0.8% (Vogelmann et al. 2001). Some of the more inland valleys are within what is termed the "Banana Belt" in Oregon and support a substantial amount of agriculture. Approximately 20% of the basin is National Forests.

River Geomorphology, Hydrology, and Chemistry

The Rogue River and its upstream tributaries begin in the high Cascades (Hunt 1974). Originating in Boundary Springs in Crater Lake National Park, the Rogue flows in a very sinuous but predominately western direction through the rugged Klamath

FIGURE 12.7 Rogue River, Oregon (Photo by Tim Palmer).

Mountains (Fig. 12.7). Numerous tributaries draining the volcanic peaks of the west slope of the Cascades, including the South Fork of the Rogue, enter as the Rogue flows predominantly southwest in the direction of Medford through older volcanic rocks. From its headwaters to Prospect the Rogue has a very high gradient (11.8 m/km), flowing over bedrock with many reaches containing substrates of cobble, boulders, and mixed gravels. Near Medford the Rogue enters a lower-gradient (1.9 m/km) section containing smaller substrates as it flows through the Rogue River Valley and Sam's Valley area, which are just north of Medford. Bear Creek enters the Rogue near this location, which is a location of intensive gravel mining. Just downstream of Grants Pass the Applegate River enters from the south and the Rogue turns northwest, beginning the portion of the lower river designated as Wild and Scenic. About 33 km downstream of Grants Pass, Grave Creek enters and the Rogue turns west. Between Galice and Agness the river twists and turns for approximately 64 km and its gradient increases to 3.4 m/km; from Agness to the coast the gradient decreases (0.7 m/km). Just below Agness the Illinois River, also designated as a Wild and Scenic River, enters from the south.

Discharge of the Rogue River near Agness (Rkm 47.8) averages approximately 170 m³/s, but the Illinois River draining a portion of the Klamath–Siskiyou ecoregion discharges into the Rogue at river km 46.6 and the average flow increases to approximately 285 m³/s. Minimum discharge in the Rogue occurs in September and in the Illinois in August (see Fig. 12.21). Maximum average monthly discharge downstream of the confluence of the Rogue and Illinois rivers occurs in December and is estimated at 650 m³/s. During the flood flows of 1996–1997 the discharge of the Rogue near Agness during December and January exceeded 1000 m³/s.

There are approximately 80 nonhydroelectric dams in the Rogue basin. Most of them are quite small and used for local irrigation projects. Four of the larger reservoirs are Lost Creek Lake, Applegate Lake, Emigrant Lake, and Fish Lake.

The waters of the Rogue are quite soft and somewhat alkaline, with an average hardness of 41 mg/L as $CaCO_3$ and a pH of 7.5. Conductivity is also low, with average conductance of approximately 100 μS/cm. Dissolved oxygen is very near saturation and averages 10.5 mg/L. Nutrients in the Rogue are also low. Nitrogen as dissolved NO_3-N + NO_2-N is approximately 0.135 mg/L and PO_4-P is 0.051 mg/L.

River Biodiversity and Ecology

The Rogue River is entirely within the Pacific Mid-Coastal freshwater ecoregion (Abell et al. 2000). The region extends along the coast of Oregon and south to San Francisco Bay. Although it runs principally along the coast, in the area of the Umpqua, Rogue, and Klamath rivers it extends inland. In general, this eastward extension is associated with increased precipitation and similarity of aquatic habitat in the western Cascades, Siskiyou, and Klamath mountains. This region is part of the southern range of coho salmon and cutthroat trout.

Plants

Riparian trees and shrubs along the Rogue include sandbar, Geyer, Pacific, yellow, and Scouler willow, white and red alder, black cottonwood, and Oregon ash. Two nonnative aquatic plants found in the basin are Brazilian elodea and curly-leaved pondweed (Michael Parker, personal communication). Once established, these plants inhibit flow and thereby lead to increased temperatures. High water temperature is considered a water-quality impairment throughout much of the basin.

Invertebrates

Little is known about the invertebrates inhabiting the Rogue River proper, although invertebrates have been collected in many tributary streams throughout the basin. The basin has a rich fauna of mayflies, stoneflies, caddisflies, and chironomid midges. Dominant taxa within the mayflies include baetids (*Baetis tricaudatus*, *Diphetor hageni*, *Acentrella turbida*), ephemerellids (*Ephemerella inermis/infrequens*, *Drunella*, *Caudatella hystrix*), heptageniids (*Rhithrogena*, *Epeorus*, *Ironodes*, *Cinygmula*), and the leptophlebiid *Paraleptophlebia*.

There also is a diverse assemblage of stoneflies. By far the numerically dominant stoneflies are the nemourid *Zapada cinctipes* and the perlid *Calineuria californica*. Other nemourids, *Z. oregonensis* group, *Z. columbiana*, and *Malenka* are also very numerous, as is the chloroperlid *Sweltsa*.

As in most of this region, the caddisfly fauna is dominated numerically by Hydropsychidae, including *Hydropsyche*, *Cheumatopsyche*, and, in lower abundance, *Arctopsyche grandis* and *Parapsyche elsis*. Several species of *Lepidostoma* are also common. Rhyacophilids are particularly diverse, with *Rhyacophila betteni* group, *R. narvae*, and *R. brunnea* group dominating. Second only to *Hydropsyche* in abundance is the brachycentrid *Micrasema*. *Glossosoma* and the limnephilids *Ecclisomyia* and *Hydatophylax hesperus* are common.

Some taxa more often found in lower latitudes also are found in the basin but in lower abundance, including the stoneflies *Sierraperla* and *Salmoperla* and the caddisflies *Tinodes* and *Gumaga*. The basin also has a rich fauna of psephenids (water pennies) that includes *Acneus*, *Eubrianax*, and *Psephenus*. Also abundant are numerous species of elmids (riffle beetles), including *Cleptelmis*, *Optioservus, and Zaitzevia* (Robert Wisseman, personal communication).

The molluscan fauna of the Rogue basin is extremely diverse, particularly with regard to Gastropoda. In a recent survey of the Klamath–Siskiyou region, Terrance Frest (personal communication) estimated that 46% of the malacofauna was undescribed. In the upper Rogue basin, numerous new species of *Fluminicola* were identified as well as a new species of *Juga*. Widespread genera of freshwater snails in the basin include *Juga*, *Fossaria*, *Lymnaea*, *Physella*, and *Gyraulus*. *Ferrissia rivularis* has only been found in the main-stem Rogue. Freshwater bivalves are less diverse and include the Oregon floater (uncommon), western ridge mussel, and western pearlshell.

Vertebrates

The Rogue River contains extremely high-quality salmonid habitat and has one of the finest salmonid fisheries in the west. However, most stocks are less abundant than they were historically and indications of their decline began appearing early in the last century. Dominating this fishery are chinook and coho salmon and steelhead and cutthroat trout. Within these species there are several spawning runs. For example, there are fall and spring runs of chinook and winter and summer runs of steelhead.

Most taxa present in the Klamath River are also found in the Rogue. However, the smallscale sucker found in the Rogue may be distinct from the Klamath smallscale sucker. Moyle (2002) also notes that the Rogue is the most southern basin containing the reticulate sculpin, which is abundant in streams further to the north. Fishes other than salmon commonly encountered include rainbow trout, speckled dace, redside shiner, coastrange sculpin, torrent sculpin, riffle sculpin, and Pacific lamprey. Other vertebrates include Pacific giant salamander, rough skinned newt, and foothill yellow-legged frog.

As throughout the west, numerous species introductions have occurred in the Rogue River. Although precisely documenting when an introduction occurs is difficult, Rivers (1963) lists approximate dates when nonnative species were first discovered. The first was in 1883, when American shad were collected near the mouth of the Rogue. Carp were thought to have entered the Rogue from a flooded farm pond in 1890. Many other warmwater fishes were intentionally introduced into the basin in the late 1920s, including white and black crappie, bluegill, largemouth bass, and catfish. Smallmouth bass also are found in the main-stem Rogue but in relatively low abundance. More recently (~1987) an illegal introduction of Umpqua pikeminnow from the more northern Umpqua basin was made into the Rogue River. When grown, these cyprinids prey on smaller fishes, such as juvenile salmonids. Their effects on salmonid stocks are now being evaluated. Also common to the basin is the nonnative bullfrog.

Ecosystem Processes

Very little is known of the original functioning of the Rogue River. Like most West Coast rivers the seasonally predictable high and low flows probably contributed to the diversity of anadromous fishes present in the system. However, unlike most of the other rivers in this chapter the Rogue's channel is rather well constrained except in the vicinity of the Rogue River Valley and Sam's Valley. Therefore, allochthonous floodplain contributions were probably less important to the functioning of the Rogue than they were to the lower Klamath from the upper Klamath basin or to the Sacramento and San Joaquin from their extensive floodplains.

Nevertheless, several factors contribute to the quality of the salmonid habitat found in the Rogue. The river has a relatively high gradient and is well constrained in many reaches. The available sediments of the area are more easily eroded, particularly in the upper reaches, than the granites of the Sierra Nevada. Near-unimpeded high winter and spring flows aid the transport of these coarse sediments within the river. The presence of these coarse sediments within a complex channel contributes to the formation of many main-stem and tributary reaches containing high-quality spawning gravels.

Extensive forests and well-developed riparian corridors provided ample shade that maintained cool water temperatures and contributed high-quality organic matter. Within these old growth forests were extremely large streamside trees that contributed to greater channel stability and complexity (unfortunately, many of the large old-growth trees have been removed). These conditions aided the development of a diverse and productive aquatic invertebrate community, which was conducive to the development of large and healthy salmonid populations.

Human Impacts and Special Features

Parts of the Rogue River basin are relatively pristine, particularly in comparison to some of the basins located in the southern portion of the Pacific Mountain System covered in this chapter. The Rogue River was one of the first rivers to have reaches receive National Wild and Scenic designation. The first such reach, designated in 1968, begins at the mouth of the Applegate River and runs downstream for 136 km. This section is considered one of the best white-water runs in North America. In 1988, a 65 km section of the river extending from near its headwaters to the town of Prospect was added. In 1984, a section of the Illinois River was designated as Wild and Scenic. The Rogue is well known for its white-water recreational use and salmon and steelhead sport fishery.

As detailed previously, European settlement of the basin was principally by gold miners in the 1850s. At that time gold was sufficiently important that in 1854 the name of the river was changed to Gold River; however, a year later it was renamed Rogue by the territorial legislature. As in California, miners first worked placer deposits, then lode deposits, which were located by following the placer deposits upstream. By 1911 most of the placer deposits had been depleted (Bredensteiner et al. 2001). Hydraulic mining began as early as 1856 and reached a maximum in 1885 (Purdom 1977). Early mining had devastating effects on the channels and the riparian habitats. Since 1960 suction dredge mining has been the preferred method of mining.

The Klamath–Siskiyou region (i.e., the Illinois River subbasin) also is rich in other mineral deposits, including platinum, nickel, zinc, copper, and cobalt (Bredensteiner et al. 2001). This portion of the basin has recently been the subject of a heated debate regarding increased mining because of the international recognition of the region's floral and faunal richness and its enormous ecological uniqueness (Ricketts et al. 1999). Gravel mining also has impacted the Rogue River. Although gravel mining occurs on the middle and lower portion of the Rogue and Applegate rivers, there is substantial mining on Bear Creek, particularly near its confluence with the Rogue.

Additional factors influencing water quality include water capture and diversion for agricultural development (pears, grapes, alfalfa, and corn), logging, and urban development. Urban development has been particularly important in the Medford and Ashland area of the Bear Creek drainage.

The Oregon Department of Environmental Quality (2004) summarized water quality in the Rogue basin from 1986 to 1995 (http://www.deq.state.or.us/lab/wqm/wqi/rogue/rogue4.htm). They found that water quality in the upper Rogue basin generally is excellent. However, water quality in the middle Rogue basin generally is of poor quality. This portion of the basin is influenced by the quality of water in the Bear Creek basin, which is affected by point and nonpoint sources, particularly the input of excessive organic matter. Water quality in the Applegate basin is considered good but impacted by irrigated agriculture and mining. Water quality of the Illinois basin is excellent and rivals the quality of the upper Rogue. The lower Rogue also is reported to have good water quality. The cause of impairment most often listed

for reaches in the Rogue basin is the exceedance of temperature criteria principally established for salmonids.

ADDITIONAL RIVERS

The Umpqua River is formed by the confluence of the North Umpqua and South Umpqua rivers (Fig. 12.22). The basin is located in the Pacific Mid-Coastal freshwater ecoregion and is part of three terrestrial ecoregions (Central Pacific Coastal Forests, Klamath–Siskiyou Forests, Central and Southern Cascades Forests). The basin has the fewest dams per area of any other large river basin in Oregon. The basin is approximate 13,000 km^2, of which half is under federal management. The almost contiguous Umpqua National Forest alone represents about 4000 km^2. The Umpqua basin, particularly the North Umpqua, is world renowned for its summer-run steelhead fishery and also its cutthroat trout, chinook, and coho salmon runs. National Wild and Scenic status was afforded to approximately 55 km of the North Umpqua in 1988. Within the basin are Douglas fir and western hemlock forests, meadows, near-vertical cliffs and spires, and many other spectacular geologic features (Fig. 12.8).

The Eel River is located in the Pacific Mid-Coastal freshwater ecoregion and is part of two terrestrial ecoregions (Klamath–Siskiyou Forests, Northern California Coastal Forests). The Eel begins in the high pine forests of Mendocino National Forest and flows in a northwesterly direction for 322 km past some of the most spectacular redwood groves in California (Fig. 12.1, Fig. 12.24). The Eel basin was once one of California's top three salmon and steelhead fisheries; water capture and diversion are suspected of contributing to its decline. The Eel is noted for producing the highest discharge (21,294 m^3/s) ever recorded in California. This remarkable event, given the small basin area, occurred during the December 1964 flood, at which time the Eel transported more sediment in one day then it had in the previous 23 years. In 1981, approximately 156 km of reaches on the main stem and tributaries were designated as National Wild and Scenic River.

The Russian River is located in the Pacific Mid-Coastal freshwater ecoregion and is part of two terrestrial ecoregions (Northern California Coastal Forests, California Interior Chaparral and Woodlands). The river begins just northwest of Clear Lake, flows southeast through a long valley for 111 km, and then turns abruptly west for 66 km, flowing

FIGURE 12.8   Umpqua River at Elkton, Oregon (PHOTO BY TIM PALMER).

partly through rugged canyons within the Coast Range before reaching the coast at the town of Jenner (Fig. 12.26). In the downstream, west-flowing section, the Russian is lined with cottages and homes (Fig. 12.9). The interbasin transfer of water from the Trinity River drainage augments flow in the Russian River. The Russian basin is an excellent example of a basin influenced by a Mediterranean-type climate regime.

The Santa Ana River, one of the largest basins in southern California, is located in the South Pacific Coastal freshwater ecoregion and is part of two terrestrial ecoregions (California Montane Chaparral and Woodlands, California Coastal Chaparral and Woodlands). The Santa Ana begins high in the San Bernardino Mountains in San Bernardino National Forest, where it is surrounded by a riparian corridor and forest of pine, fir, and oak (Fig. 12.28). However, downstream of the first 30 km the river is leveed, channelized, and functionally dewatered (Fig. 12.10). Most of its water is diverted to infiltration ponds that are used to recharge local aquifers to support the 2,000,000 to 4,500,000 people living in the basin.

The Santa Margarita River is located in the South Pacific Coastal freshwater ecoregion and the California Coastal Sage and Chaparral terrestrial

FIGURE 12.9 Russian River below Guerneville, California (PHOTO BY TIM PALMER).

FIGURE 12.10 Santa Ana River downstream of Imperial Highway (Highway 90) in Orange County, California (PHOTO BY C. BURTON, USGS).

FIGURE 12.11   Lower portion of Santa Margarita River, California, flowing through hills of coastal sage scrub approximately 15 km from Pacific Ocean (PHOTO BY CHERYL S. BREHME, USGS).

ecoregion. The Santa Margarita, one of the last free-flowing rivers in coastal southern California, begins at the confluence of Temecula and Murrieta creeks at the Temecula gorge (Fig. 12.30). It flows for 43 km to the ocean and has one of the most continuous riparian corridors in the region (Fig. 12.11). Water use upstream has decreased base flow, particularly during extended dry periods. Although many non-native species have been introduced to the system, the Santa Margarita has one of the largest remaining populations of arroyo chub. A short segment of the river is being considered for National Wild and Scenic River status.

ACKNOWLEDGMENTS

We would like to thank the many researchers who provided information for this chapter: Mary Adams (Regional Water Quality Control Board), Aaron Borisenko (Oregon Department of Environmental Quality), Kim Bredensteiner (Conservation Biology Institute), Larry Brown (USGS), Carmen Burton (USGS), Terrence Frest (Deixis Consultants), Joseph Furnish (USDA Forest Service), David Gibson (Regional Water Quality Control Board, San Diego), Rich Hafele (Oregon Department of Environmental Quality), James Harrington (California Department of Fish and Game), Jeanette Howard (U.C. Berkeley), Jon Keeley (USGS), Jonathan Lee (Jon Lee Consulting), Douglas F. Markle (Oregon State University), Raphael D. Mazor (U.C. Berkeley), Peter Moyle (U.C. Davis), Peter Ode (California Department of Fish and Game), Kelly Palacios (Conservation Biology Institute), Ron Rogers (California Department of Fish and Game), Deborah Rudnick (U.C. Berkeley), Eric Scheuering (U.C. Berkeley), Sue Vrilakas (Oregon Natural Heritage Information Center), and Robert Wisseman (Aquatic Biology Associates, Inc.). Leila Gass and JoAnn Gronberg from the USGS helped us with many of the GIS-derived data. We are particularly grateful to

Larry Brown and Steve Fend for their thorough and helpful reviews.

LITERATURE CITED

Abell, R. A., D. M. Olson, E. Dinerstein, P. T. Hurley, J. T. Diggs, W. Eichbaum, S. Walters, W. Wettengel, T. Allnutt, C. J. Loucks, and P. Hedao. 2000. *Freshwater ecoregions of North America: A conservation assessment.* Island Press, Washington, D.C.

Alpine, A. E., and J. E. Cloern. 1992. Trophic interactions and direct physical effects control phytoplankton biomass and production in an estuary. *Limnology and Oceanography* 37(5):946–955.

Anderson, B. 2000. *America's salad bowl: An agricultural history of the Salinas Valley.* Monterey County Historical Society, Salinas, California.

Anonymous. 1999. The Klamath tribes: A photoessay. pp. 11–18. In L. W. Powers (ed.), *A river never the same: A history of water in the Klamath basin 1999.* Shaw Historical Library, Oregon Institute of Technology, Klamath Falls.

Baldwin, E. M. 1981. *Geology of Oregon,* 3rd ed. Kendall/Hunt, Dubuque, Iowa.

Bredensteiner, K. C., J. R. Strittholt, and D. A. DellaSala. 2001. Comments supporting the proposed mineral withdrawal for National Forest and Bureau of Land Management lands in the Siskiyou Wild Rivers Area of Oregon and California, Conservation Biology Institute, Corvallis, Oregon.

Brown, L. R. 1996. Aquatic biology of the San Joaquin–Tulare basins, California: Analysis of available data through 1992. Water-Supply Paper 2471. U.S. Geological Survey, Reston, Virginia.

Brown, L. R. 1998. Assemblages of fishes and their associations with environmental variables, Lower San Joaquin River Drainage, California. Open-File Report 98–77. U.S. Geological Survey, Reston, Virginia.

Cain, D. J., J. L. Carter, S. V. Fend, S. N. Luoma, C. N. Alpers, and H. E. Taylor. 2000. Metal exposures in a benthic macroinvertebrate, *Hydropsyche californica,* related to mine drainage in the Sacramento River. *Canadian Journal of Fisheries and Aquatic Sciences* 57:380–390.

California State Water Project Atlas. 1999. California State Water Project, Sacramento.

Carlton, J. T., J. K. Thompson, L. E. Schemel, and F. H. Nichols. 1990. Remarkable invasion of San Francisco Bay (California, USA) by the Asian clam *Potamocorbula amurensis.* I. Introduction and dispersal. *Marine Ecology Progress Series* 66:81–94.

Carter, J. L., and S. V. Fend. 2001. Inter-annual changes in the benthic community structure of riffles and pools in reaches of contrasting gradient. *Hydrobiologia* 459:187–200.

Clark, A. H. 1999. A brief discussion of Klamath basin geology and geography related to water resources. In L. W. Powers (ed.), *A river never the same: A history of water in the Klamath basin 1999,* pp. 1–10. Shaw Historical Library, Oregon Institute of Technology, Klamath Falls.

Clark, M., and E. D. Miller. 1999. Notes on early water use in the Klamath basin. In L. W. Powers (ed.). *A river never the same: A history of water in the Klamath basin 1999,* pp. 19–41. Shaw Historical Library, Oregon Institute of Technology, Klamath Falls.

Domagalski, J. L., D. L. Knifong, D. E. MacCoy, P. D. Dileanis, B. J. Dawson, and M. S. Majewski. 1998. Water quality assessment of the Sacramento River basin, California: Environmental setting and study design. Water-Resources Investigations Report 97–4254. U.S. Geological Survey, Reston, Virginia.

Domagalski, J. L., and P. D. Dileanis. 2000. Water-quality assessment of the Sacramento River basin, California: Water quality of fixed sites, 1996–1998. Water-Resources Investigations Report 00–4247. U.S. Geological Survey, Reston, Virginia.

Domagalski, J. L., D. L. Knifong, P. D. Dileanis, L. R. Brown, J. T. May, V. Connor, and C. N. Alpers. 2000. Water quality in the Sacramento River basin, California, 1994–98. Circular 1215. U.S. Geological Survey, Reston, Virginia.

Douce, R. S. 1993. The biological pollution of *Arundo donax* in river estuaries and beaches. pp. 27–33. In N. E. Jackson, P. Frandsen, and S. Douthit (eds.). *Proceedings of the* Arundo Donax *Workshop, Nov. 1993, Ontario, California.* California Exotic Pest Plant Council, Riverside, California.

Douthit, N. 1999. *A guide to Oregon south coast history: Traveling the Jedediah Smith Trail.* Oregon State University Press, Corvallis.

Dubrovsky, N. M., C. R. Kratzer, L. R. Brown, J. M. Gronberg, and K. R. Burow. 1998. Water quality in the San Joaquin–Tulare basins, California, 1992–95. Circular 1159. U.S. Geological Survey, Reston, Virginia.

Fisher, A. B. 1945. *The Salinas: Upside-down river.* Farrar and Rinehart, New York.

Gasith, A., and V. H. Resh. 1999. Streams in Mediterranean climate regions: Abiotic influences and biotic responses to predictable seasonal events. *Annual Reviews of Ecology and Systematics* 30:51–81.

Gronberg, J. M., N. M. Dubrovsky, C. R. Kratzer, J. L. Domagalski, L. R. Brown, and K. R. Burow. 1998. Environmental settings of the San Joaquin–Tulare basins, California. Water-Resources Investigations Report 97–4205. U.S. Geological Survey, Reston, Virginia.

Harden, D. R. 1998. *California geology.* Prentice Hall, Upper Saddle River, New Jersey.

Heizer, R. F., and A. B. Elsasser. 1980. *The natural world of the California Indians.* University of California Press, Berkeley and Los Angeles.

Hunt, C. B. 1974. *Natural regions of the United States and Canada.* W. H. Freeman, San Francisco.

Irwin, G. A. 1976. Water-quality investigation, Salinas River, California. Water-Resources Investigations 76–110. U.S. Geological Survey, Reston, Virginia.

Keeley, J. E. 2002. Native American impacts on fire regimes of the California coastal ranges. *Journal of Biogeography* 29:303–320.

Kondolf, G. M., and W. V. G. Matthews. 1993. Management of coarse sediment on regulated rivers. Report no. 80. California Water Resources Center, University of California, Davis.

Kratzer, C. R., and J. L. Shelton. 1998. Water quality assessment of the San Joaquin–Tulare basins, California: Analysis of available data on nutrients and suspended sediment in surface water, 1972–1990. Professional Paper 1587. U.S. Geological Survey, Reston, Virginia.

Leland, H. V., L. R. Brown, and D. K. Mueller. 2001. Distribution of algae in the San Joaquin River, California, in relation to nutrient supply, salinity and other environmental factors. *Freshwater Biology* 46:1139–1167.

Leland, H. V., and S. V. Fend. 1998. Benthic invertebrate distributions in the San Joaquin River, California, in relation to physical and chemical factors. *Canadian Journal of Fisheries and Aquatic Sciences* 55:1051–1067.

Levy, S. 2003. Turbulence in the Klamath River basin. *BioScience* 53:315–320.

May, J. T., and L. R. Brown. 2002. Fish communities of the Sacramento River basin: Implication for conservation of native fishes in the Central Valley, California. *Environmental Biology of Fishes* 63:373–388.

McDonnell, L. R. (ed.). 1962. *Rivers of California*. Pacific Gas and Electric Company, San Francisco.

Miles, S. R., and C. B. Goudey (comp.). 1997. *Ecological subregions of California: Section and subsection descriptions*. Book no. R5-EM-TP-005. U.S. Department of Agriculture, San Francisco.

Minckley, W. L., D. A. Hendrickson, and C. E. Bond. 1986. Geography of western North American freshwater fishes: Description and relationships to intercontinental tectonism. pages In C. H. Hocutt and E. O. Wiley (eds.). *The zoogeography of North American freshwater fishes*, pp. 519–614. John Wiley and Sons, New York.

Mount, J. F. 1995. *California rivers: The conflict between fluvial process and land use*. University of California Press, Berkeley and Los Angeles.

Moyle, P. B. 2002. *Inland fishes of California: Revised and expanded*. University of California Press, Berkeley and Los Angeles.

Moyle, P. B., and B. Vondracek. 1985. Structure and persistence of the fish assemblage in a small Sierra stream. *Ecology* 66:1–13.

Nady, P., and L. L. Larragueta. 1983. Estimated average annual streamflow into the Central Valley of California. Hydrologic Investigations Atlas HA-657, scale 1:500,000. U.S. Geological Survey, Reston, Virginia.

Norris, R. M., and R. W. Webb. 1990. *Geology of California*, 2nd ed. John Wiley and Sons, New York.

Ode, P. R., A. Rehn, and J. M. Harrington. 2002. California Region Water Quality Control Board, San Diego region 2002 biological assessment report: Results of May 2001 reference site study and preliminary index of biotic integrity. Water Quality Assessment Series, California Department of Fish and Game, Rancho Cordova.

Ohlendorf, H. M., D. J. Hoffman, M. K. Saiki, and T. W. Aldrich, 1986. Embryonic mortality and abnormalities of aquatic birds: Apparent impacts by selenium from irrigation drainwater. *Science of the Total Environment* 52:49–63.

Oregon Natural Heritage Information Center. 2002. GIS and database information. Portland, Oregon.

Orsi, J. J., and W. L. Mecum. 1996. Food limitation as the probable cause of a long-term decline in the abundance of *Neomysis mercedis* the opossum shrimp in the Sacramento–San Joaquin estuary. In J. T. Hollibaugh (ed.). *San Francisco Bay: The ecosystem: Further investigations into the natural history of San Francisco Bay and Delta with reference to the influence of man*, pp. 375–401. AAAS, San Francisco.

Peterson, D. H., R. E. Smith, S. W. Hager, D. D. Harmon, R. E. Herndon, and L. E. Schemel. 1985. Interannual variability in dissolved inorganic nutrients in Northern San Francisco Bay Estuary. *Hydrobiologia* 129:37–57.

Purdom, W. B. 1977. Guide to the geology and lore of the wild reach of the Rogue River, Oregon. Bulletin no. 22. Museum of Natural History, University of Oregon, Eugene.

Oregon Water Resources Department (OWRD). 1999. Resolving the Klamath: Klamath Basin General Stream Adjudication. OWRD, Salem, Oregon.

Ricketts, T. H., E. Dinerstein, D. M. Olson, C. J. Loucks, W. Eichbaum, D. DellaSala, K. Kavanagh, P. Hedao, P. T. Hurley, K. M. Carney, R. Abell, and S. Walters. 1999. *Terrestrial ecoregions of North America: A conservation assessment*. Island Press, Washington, D.C.

Rivers, C. M. 1963. *Rogue River fisheries*. Vol. 1: *History and development of the Rogue River basin as related to its fishery prior to 1941*. Unpublished manuscript. Oregon Department of Fish and Wildlife, Portland.

Rudnick, D. A., K. Hieb, K. F. Grimmer, and V. H. Resh. 2003. Patterns and processes of biological invasion: The Chinese mitten crab in San Francisco Bay. *Basic and Applied Ecology* 4:249–262.

Schoenherr, A. A. 1992. *A natural history of California*. University of California Press, Berkeley and Los Angeles.

Vogelmann, J. E., S. M. Howard, L. Yang, C. R. Larson, B. K. Wylie, and N. Van Driel. 2001. Completion of the 1990s National Land Cover Data Set for the conterminous United States from Landsat Thematic Mapper data and ancillary data sources, *Photogrammetric Engineering & Remote Sensing* 67:650–662.

Williamson, A. K., D. E. Prudic, and L. A. Swain. 1989. Ground-water flow in the Central Valley, California. Professional Paper 1401-D. U.S. Geological Survey, Reston, Virginia.

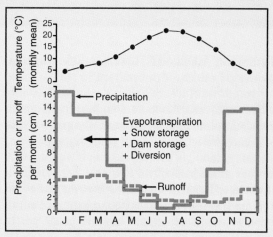

FIGURE 12.13 Mean monthly air temperature, precipitation, and runoff for the Sacramento River basin.

SACRAMENTO RIVER

Relief: 4317 m
Basin area: 72,132 km²
Mean discharge: 657 m³/s
River order: 8 (estimated)
Mean annual precipitation: 89.7 cm
Mean air temperature: 12.9°C
Mean water temperature: NA
Physiographic provinces: Pacific Border (PB), Cascades–Sierra Nevada Mountains (CS), Basin and Range (BR)
Biomes: Chaparral, Temperate Mountain Forest
Freshwater ecoregions: Pacific Mid-Coastal, Pacific Central Valley, Oregon Lakes
Terrestrial ecoregions: Eastern Cascades Forests, Klamath–Siskiyou Forests, Sierra Nevada Forests, California Central Valley Grasslands, California Interior Chaparral and Woodlands
Number of fish species: 69 (29 native)
Number of endangered species: >50 threatened and endangered
Major fishes: Pacific lamprey, river lamprey, white sturgeon, green sturgeon, Sacramento blackfish, hardhead, hitch, Sacramento pikeminnow, tui chub, Sacramento splittail, California roach, speckled dace, Sacramento sucker, delta smelt, longfin smelt, Modoc sucker, chinook salmon, rainbow trout, coho salmon, threespine stickleback, Sacramento perch, tule perch, staghorn sculpin, riffle sculpin
Major other aquatic vertebrates: California newt, Sierra newt, California red-legged frog, foothill yellow-legged frog, mountain yellow-legged frog, western leopard frog, western toad, Yosemite toad, western pond turtle, western aquatic garter snake, water shrew, mountain beaver, beaver, muskrat, raccoon, river otter, ermine (short-tailed weasel), long-tailed weasel, mink
Major benthic invertebrates: mayflies (*Siphlonurus, Baetis, Drunella, Ephemerella, Serratella, Epeorus, Rhithrogena, Cinygmula*), stoneflies (*Pteronarcys, Malenka, Zapada, Hesperoperla, Calineuria, Isoperla, Suwallia, Sweltsa, Eucapnopsis*), caddisflies (*Arctopsyche, Hydropsyche, Cheumatopsyche, Rhyacophila, Hydroptila*), bivalves (*Corbicula fluminea, Potamocorbula amurensis*), crustaceans (*Neomysis mercedis*)
Nonnative species: eastern oyster, bullfrog, >40 species of fishes (e.g., American shad, striped bass, and many warmwater fishes)
Major riparian plants: arroyo willow, black willow, narrowleaf willow, Pacific willow, red willow, black cottonwood, Fremont cottonwood, California sycamore, mulefat, mountain alder, white alder, buttonbush, water birch
Special features: largest river in California; several National Wild and Scenic reaches; one of largest salmonid runs in California; discharges through the highly productive Sacramento–San Joaquin Delta into San Francisco Bay
Fragmentation: hundreds of dams and water withdrawals within basin
Water quality: pH near 8, alkalinity = 52 mg/L as CaCO₃, NO₃-N = 0.25 mg/L, PO₄-P = 0.02 mg/L
Land use: 49.3% forest, 13.7% shrub, 16.2% grassland/herbaceous, 0.8% wetlands, 15.1% agriculture, 1.7% urban
Population density: 23.6 people/km²
Major information sources: Moyle 2002, Schoenherr 1992

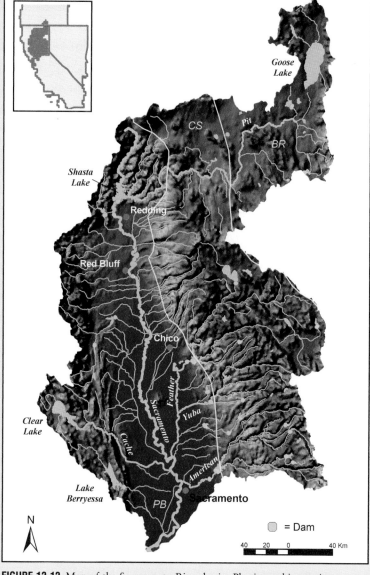

FIGURE 12.12 Map of the Sacramento River basin. Physiographic provinces are separated by yellow lines.

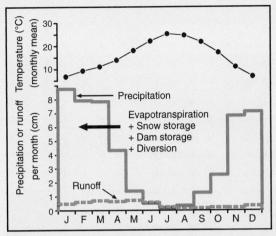

FIGURE 12.15 Mean monthly air temperature, precipitation, and runoff for the San Joaquin River basin.

SAN JOAQUIN RIVER

Relief: 4418 m
Basin area: 83,409 km²
Mean discharge: 132 m³/s
River order: 8 (estimated)
Mean annual precipitation: 48.8 cm
Mean air temperature: 15.7°C
Mean water temperature: NA
Physiographic provinces: Pacific Border (PB),
    Cascade–Sierra Nevada Mountains (CS)
Biomes: Chaparral, Temperate Mountain Forest
Freshwater ecoregion: Pacific Central Valley
Terrestrial ecoregions: California Central Valley
    Grasslands, California Interior Chaparral and
    Woodlands, Sierra Nevada Forests
Number of fish species: 63 (native and nonnative)
Number of endangered species: >50 threatened and
    endangered
Major fishes: Pacific lamprey, white sturgeon,
    Sacramento blackfish, hardhead, hitch, Sacramento
    pikeminnow, California roach, speckled dace,
    Sacramento sucker, delta smelt, longfin smelt,
    chinook salmon, rainbow trout, threespine stickleback, Sacramento perch, tule perch, staghorn sculpin, riffle sculpin
Major other aquatic vertebrates: California newt, Sierra newt, tailed frog, California red-legged frog, foothill yellow-legged frog,
    mountain yellow-legged frog, bullfrog, western leopard frog, western toad, Yosemite toad, western pond turtle, western
    aquatic garter snake, water shrew, mountain beaver, beaver, muskrat, nutria, raccoon, river otter, ermine (short-tailed
    weasel), long-tailed weasel, mink
Major benthic invertebrates: mayflies (*Baetis, Epeorus, Caenis, Heptagenia, Fallceon, Ephemerella, Drunella*), stoneflies
    (*Hesperoperla, Suwallia, Eucapnopsis*), caddisflies (*Hydropsyche, Arctopsyche, Rhyacophila, Nectopsyche*), bivalves
    (*Corbicula*)
Nonnative species: bullfrog, >40 species of fishes (e.g., most warmwater species, such as largemouth bass, smallmouth bass,
    crappie, bluegill, catfishes)
Major riparian plants: arroyo willow, black willow, narrowleaf willow, Pacific willow, red willow, black cottonwood, Fremont
    cottonwood, buttonbush, California sycamore, mulefat, white alder
Special features: begins in high Sierra Nevada; several tributaries designated National Wild and Scenic Rivers; flows through
    intensive agricultural region before discharging through highly productive Sacramento–San Joaquin Delta into San Francisco
    Bay
Fragmentation: hundreds of dams and water withdrawals within basin
Water quality: pH = 7.6, alkalinity = 93 mg/L as $CaCO_3$, dissolved NO_3-N + NO_2-N = 1.36 mg/L, dissolved PO_4-P = 0.15 mg/L
Land use: 26.8% forest, 13.4% shrub, 23.0% grassland/herbaceous, 0.4% wetlands, 30.0% agriculture, 1.9% urban
Population density: 29.3 people/km²
Major information sources: Moyle 2002, Schoenherr 1992

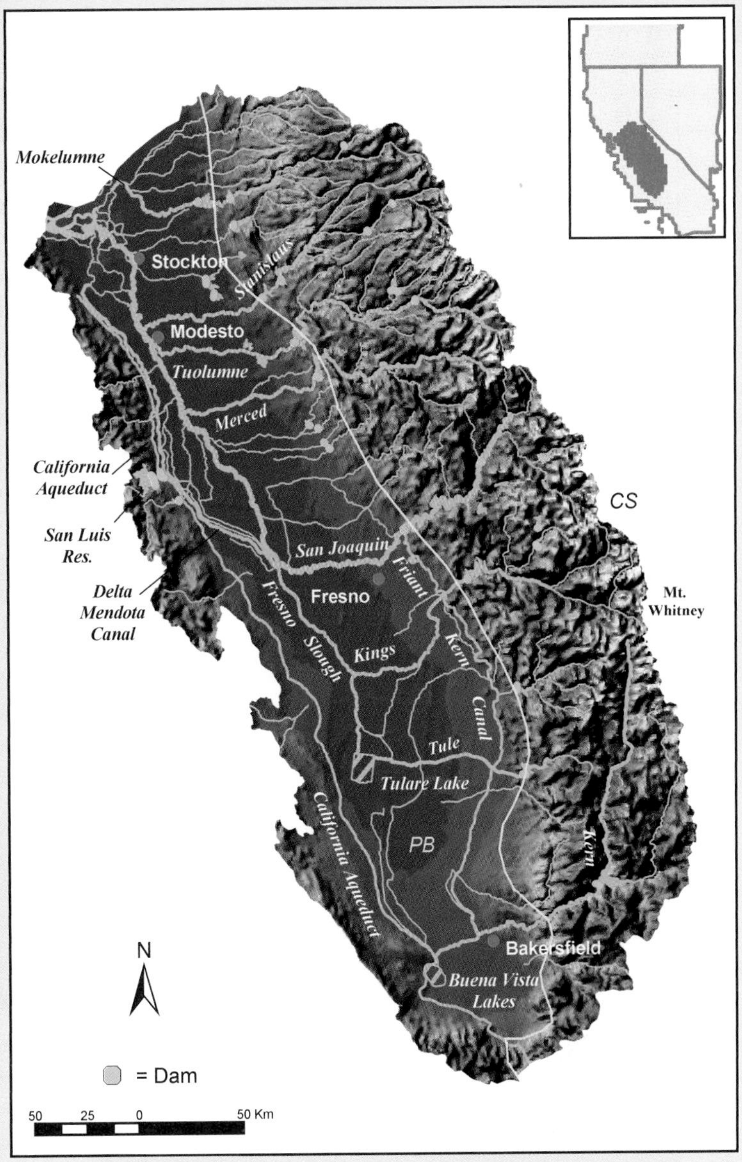

FIGURE 12.14 Map of the San Joaquin River basin. Physiographic provinces are separated by a yellow line.

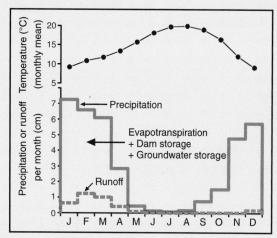

FIGURE 12.17 Mean monthly air temperature, precipitation, and runoff for the Salinas River basin.

SALINAS RIVER

Relief: 1787 m
Basin area: 10,983 km²
Mean discharge: 12.7 m³/s
River order: 6 (estimated)
Mean annual precipitation: 36.4 cm
Mean air temperature: 14.6°C
Mean water temperature: NA
Physiographic province: Pacific Border (PB)
Biomes: Chaparral, Temperate Mountain Forest
Freshwater ecoregion: Pacific Central Valley
Terrestrial ecoregions: California Interior Chaparral and Woodlands, California Montane Chaparral and Woodlands
Number of fish species: 36 (16 native)
Number of endangered species: 42 threatened and endangered
Major fishes: Pacific lamprey, Pacific brook lamprey, Sacramento blackfish, hitch, Sacramento pikeminnow, California roach, speckled dace, Sacramento sucker, coho salmon, steelhead, rainbow trout, threespine stickleback, tidewater goby, staghorn sculpin, coastrange sculpin, prickly sculpin, riffle sculpin
Major other aquatic vertebrates: California giant salamander, California newt, Pacific chorus frog, California chorus frog, California red-legged frog, foothill yellow-legged frog, western toad, arroyo toad, western pond turtle, aquatic/Santa Cruz garter snake, beaver, muskrat, raccoon, long-tailed weasel
Major benthic invertebrates: mayflies (*Acentrella, Baetis, Diphetor, Fallceon, Centroptilum, Procloeon, Serratella, Tricorythodes*), caddisflies (*Hydropsyche, Cheumatopsyche*), beetles (*Optioservus*), crustaceans (*Hyalella azteca*)
Nonnative species: bullfrog, threadfin shad, common carp, golden shiner, fathead minnow, channel catfish, white catfish, brown bullhead, black bullhead, brown trout, mosquitofish, inland silverside, white bass, black crappie, white crappie, green sunfish, bluegill, redear sunfish, largemouth bass, smallmouth bass
Major riparian plants: arroyo willow, narrowleaf willow, red willow, Sitka willow, buttonbush, California sycamore, Fremont cottonwood, mulefat, white alder
Special features: flows through the Salinas Valley, which is often referred to as "America's Salad Bowl"; under natural flow conditions it is one of longest underground rivers in North America
Fragmentation: 17 dams
Water quality: pH = 7.8, alkalinity = 258 mg/L as $CaCO_3$, dissolved NO_3-N + NO_2-N = 2.94 mg/L, dissolved PO_4-P = 1.73 mg/L
Land use: 17.4% forest, 15.9% shrub, 49.0% grasslands/herbaceous, 12.8% agriculture, 0.7% urban
Population density: 10.1 people/km²
Major information sources: Moyle 2002, Schoenherr 1992

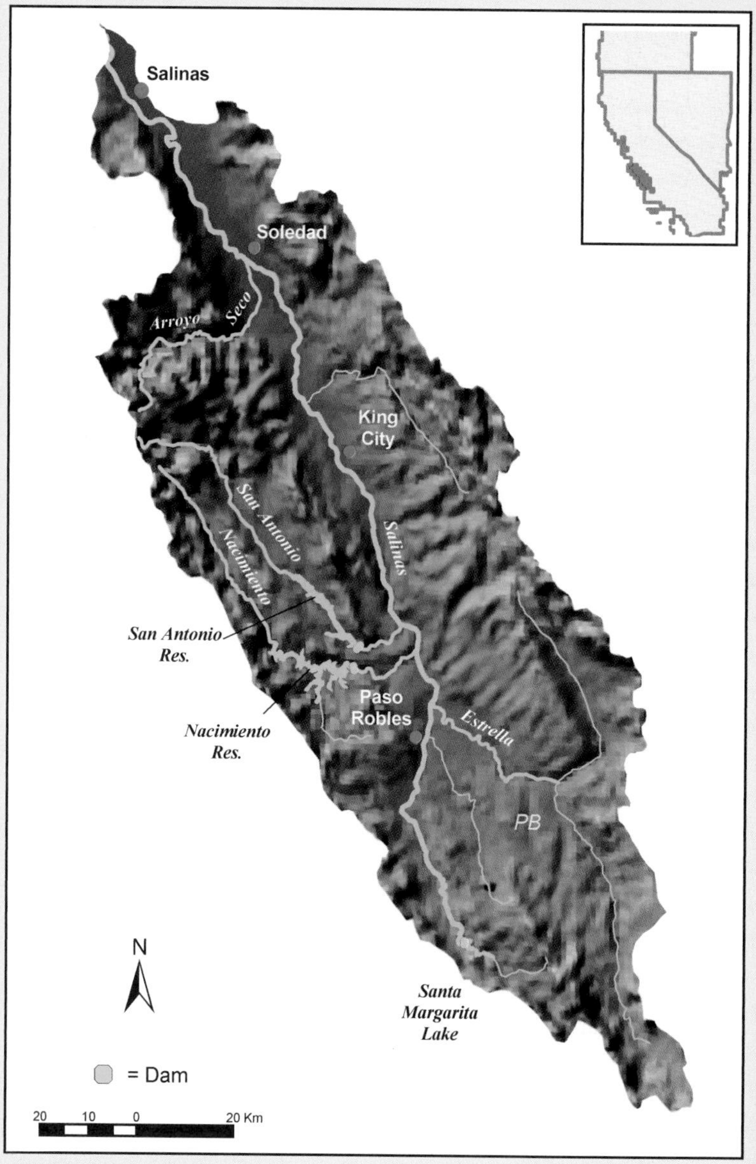

FIGURE 12.16 Map of the Salinas River basin.

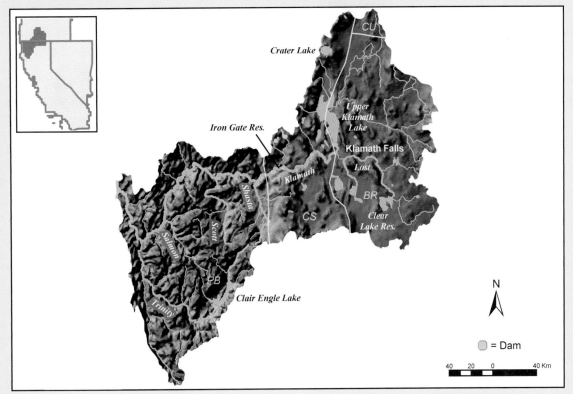

FIGURE 12.18 Map of the Klamath River basin. Physiographic provinces are separated by yellow lines.

KLAMATH RIVER

Relief: 2894 m

Basin area: 40,608 km²

Mean discharge: 501 m³/s

River order: 7 (estimated)

Mean annual precipitation: 85 cm

Mean air temperature: 10.5°C

Mean water temperature: NA

Physiographic provinces: Pacific Border (PB), Basin and Range (BR), Cascade–Sierra Nevada Mountains (CS)

Biomes: Temperate Mountain Forest, Desert

Freshwater ecoregion: Pacific Mid-Coastal

Terrestrial ecoregions: Eastern Cascades Forests, Klamath–Siskiyou Forests, Northern California Coastal Forests

Number of fish species: 48 (30 native)

Number of endangered species: 41

FIGURE 12.19 Mean monthly air temperature, precipitation, and runoff for the Klamath River basin.

Major fishes: Pacific lamprey, Klamath River lamprey, white sturgeon, green sturgeon, blue chub, tui chub, speckled dace, Lost River sucker, shortnose sucker, Klamath smallscale sucker, Klamath largescale sucker, longfin smelt, chum salmon, coho salmon, chinook salmon, rainbow trout, cutthroat trout, steelhead, staghorn sculpin, slender sculpin, coastrange sculpin

Major other aquatic vertebrates: northwestern salamander, southern torrent salamander, rough skinned newt, red-legged frog, California red-legged frog, foothill yellow-legged frog, Cascades frog, spotted frog, western leopard frog, western toad, western pond turtle, western aquatic garter snake, mountain beaver, beaver, muskrat, nutria, raccoon, river otter, mink

Major benthic invertebrates: mayflies (*Baetis, Drunella, Rhithrogena, Cinygmula, Epeorus, Ironodes, Paraleptophlebia*), stoneflies (*Yoraperla, Calineuria, Hesperoperla, Zapada, Malenka*), caddisflies (*Hydropsyche, Nectopsyche, Oecetis, Helicopsyche, Glossosoma, Lepidostoma, Agapetus*), beetles (*Optioservus, Cleptelmis, Zaitzevia*), bugs (*Sigara*), snails (*Juga*)

Nonnative species: bullfrog, American shad, goldfish, golden shiner, fathead minnow, brown bullhead, black bullhead, wakasagi, kokanee, brown trout, brook trout, brook stickleback, Sacramento perch, black crappie, white crappie, green sunfish, bluegill, pumpkinseed, largemouth bass, spotted bass, yellow perch

Major riparian plants: arroyo willow, Hooker willow, black willow, narrowleaf willow, Pacific willow, red willow, sandbar willow, Sitka willow, black cottonwood, Fremont cottonwood, mulefat, white alder, water birch

Special features: flows through rugged, species-rich Klamath Mountains; below most downstream dam flows unimpounded for >450 km to Pacific; one of best salmonid fisheries in California; >900 km designated as National Wild and Scenic River

Fragmentation: 24 dams

Water quality: pH = 8.0, alkalinity = 70 mg/L as $CaCO_3$, NO_3-N + NO_2-N = 0.155 mg/L, PO_4-P = 0.037 mg/L

Land use: 66.3% forest, 14.0% shrub, 8% grasslands, 2.3% wetlands, 5.9% agriculture, 0.2% urban

Population density: 1.9 people/km²

Major information sources: Moyle 2002, Schoenherr 1992

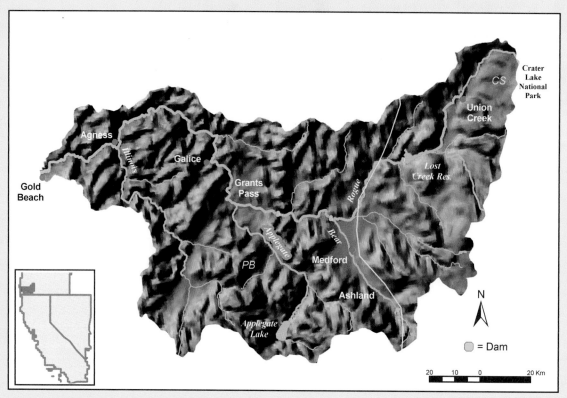

FIGURE 12.20 Map of the Rogue River basin. Physiographic provinces are separated by a yellow line.

ROGUE RIVER

Relief: 2894 m
Basin area: 13,348 km²
Mean discharge: 285 m³/s
River order: 7 (estimated)
Mean annual precipitation: 97 cm
Mean air temperature: 11.6°C
Mean water temperature: NA
Physiographic provinces: Cascade–Sierra Nevada Mountains (CS), Pacific Border (PB)
Biome: Temperate Mountain Forest
Freshwater ecoregion: Pacific Mid-Coastal
Terrestrial ecoregions: Central Pacific Coastal Forests, Klamath–Siskiyou Forests, Central and Southern Cascades Forests
Number of fish species: 23 (14 native)
Number of endangered species: 11
Major fishes: Pacific lamprey, green sturgeon, white sturgeon, coastal cutthroat, pink salmon, coho salmon, rainbow trout, Chinook salmon, speckled dace, tui chub, Klamath smallscale sucker, steelhead

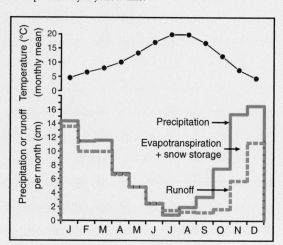

FIGURE 12.21 Mean monthly air temperature, precipitation, and runoff for the Rogue River basin.

Major other aquatic vertebrates: northwestern salamander, Pacific giant salamander, southern torrent salamander, rough skinned newt, red-legged frog, foothill yellow-legged frog, Cascades frog, spotted frog, western toad, western pond turtle, western aquatic garter snake, water shrew, mountain beaver, beaver, muskrat, raccoon, river otter, ermine, long-tailed weasel, mink
Major benthic invertebrates: mayflies (*Baetis, Diphetor, Acentrella, Ephemerella, Drunella, Rhithrogena, Epeorus, Paraleptophlebia*), stoneflies (*Zapada, Calineuria, Malenka, Sweltsa*), caddisflies (*Hydropsyche, Cheumatopsyche, Arctopsyche, Parapsyche, Lepidostoma, Rhyacophila, Micrasema, Glossosoma*), snails (*Fluminicola, Juga, Fossaria, Lymnaea, Physella, Gyraulus, Ferrissia*), bivalves (*Gonidea*, western pearlshell mussel)
Nonnative species: brook trout, common carp, golden shiner, Umpqua pikeminnow, redside shiner, brown bullhead, smallmouth bass, largemouth bass, black crappie, bullfrog
Major riparian plants: sandbar willow, Geyer willow, Pacific willow, yellow willow, Scouler willow, white alder, red alder, black cottonwood, Oregon ash
Special features: begins in high Cascade Mountains on slopes of Crater Lake National Park; one of first National Wild and Scenic Rivers; one of best salmonid fisheries in west; renowned for white-water boating
Fragmentation: ~80 dams
Water quality: pH = 7.6, alkalinity = 44 mg/L as CaCO₃, NO₃-N + NO₂-N = 0.135 mg/L, PO₄-P = 0.051 mg/L
Land use: 82.7% forest, 2.9% shrub, 6.4% grasslands, 0.2% as wetlands, 5.9% agriculture, 0.8% urban
Population density: 12.4 people/km²
Major information sources: D. F. Markle, personal communication, Oregon Natural Heritage Information Center 2002

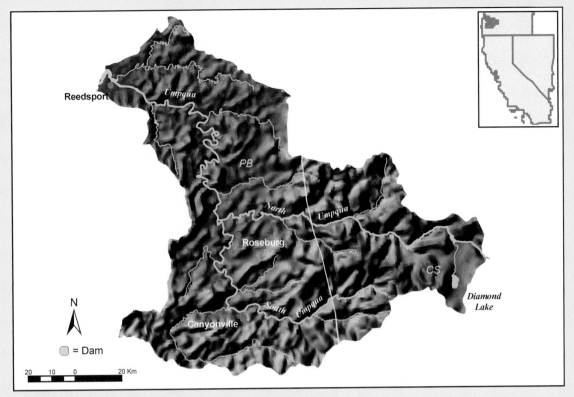

FIGURE 12.22 Map of the Umpqua River basin. Physiographic provinces are separated by a yellow line.

UMPQUA RIVER

Relief: 2799 m
Basin area: 12,133 km²
Mean discharge: 211 m³/s
River order: 6 (estimated)
Mean annual precipitation: 115 cm
Mean air temperature: 11.7°C
Mean water temperature: NA
Physiographic provinces: Cascade–Sierra Nevada Mountains (CS),
   Pacific Border (PB)
Biome: Temperate Mountain Forest
Freshwater ecoregion: Pacific Mid-Coastal
Terrestrial ecoregions: Central Pacific Coastal Forests, Klamath–
   Siskiyou Forests, Central and Southern Cascades Forests
Number of fish species: 27 (19 native)
Number of endangered species: 9

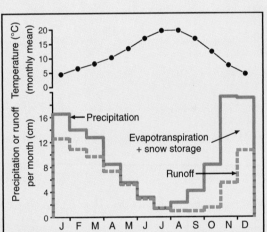

FIGURE 12.23 Mean monthly air temperature, precipitation, and runoff for the Umpqua River basin.

Major fishes: river lamprey, western brook lamprey, Pacific lamprey, coastal cutthroat, chum salmon, coho salmon, rainbow trout, chinook salmon, Umpqua chub, Umpqua pikeminnow, Umpqua dace, speckled dace, tui chub, largescale sucker, threespine stickleback, coastrange sculpin, prickly sculpin, riffle sculpin, reticulate sculpin
Major other aquatic vertebrates: northwestern salamander, Pacific giant salamander, southern torrent salamander, rough skinned newt, Pacific chorus frog, red-legged frog, foothill yellow-legged frog, Cascades frog, spotted frog, western toad, western aquatic garter snake, water shrew, mountain beaver, beaver, muskrat, nutria, raccoon, river otter, mink
Major benthic invertebrates: mayflies (*Baetis tricaudatus*, *Diphetor*, *Paraleptophlebia*, *Rhithrogena*), stoneflies (*Calineuria*, *Hesperoperla*, *Sweltsa*, *Malenka*, *Zapada*), caddisflies (*Hydropsyche*, *Glossosoma*, *Neophylax*), snails (*Juga*)
Nonnative species: American shad, redside shiner, yellow bullhead, western mosquitofish, bluegill, smallmouth bass, largemouth bass, black crappie, brook trout, bullfrog
Major riparian plants: sandbar willow, Geyer willow, Pacific willow, yellow willow, Scouler willow, white alder, red alder, black cottonwood, Oregon ash
Special features: begins in Cascades; world renowned steelhead and salmon fishery; portions have National Wild and Scenic River status
Fragmentation: 64 dams in Douglas County, Oregon; fewest dams of any large basin in Oregon
Water quality: North Umpqua very good water quality; South Umpqua lesser quality; temperatures higher than desired; pH = 7.4, alkalinity = 31 mg/L as CaCO₃, NO₃-N + NO₂-N = 0.109 mg/L, PO₄-P = 0.028 mg/L
Land use: 85.9% forest, 2% shrub, 4.9% grasslands, 0.1% wetlands, 4.6% agriculture, 0.5% urban
Population density: 5.2 people/km²
Major information sources: D. F. Markle, personal communication, Oregon Natural Heritage Information Center 2002

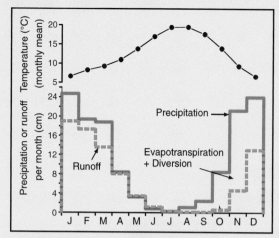

FIGURE 12.25 Mean monthly air temperature, precipitation, and runoff for the Eel River basin.

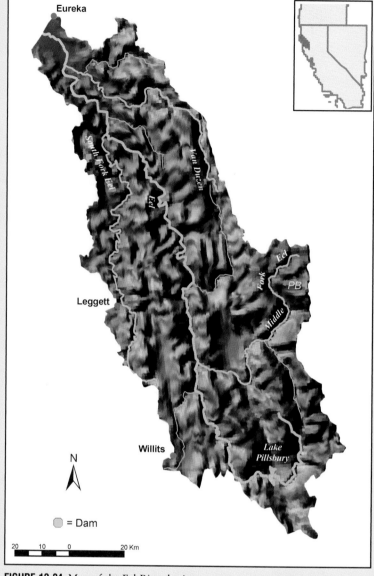

FIGURE 12.24 Map of the Eel River basin.

EEL RIVER

Relief: 2270 m
Basin area: 9456 km²
Mean discharge: 210 m³/s
River order: 6 (estimated)
Mean annual precipitation: 133 cm
Mean air temperature: 12.7°C
Mean water temperature: NA
Physiographic province: Pacific Border (PB)
Biomes: Temperate Mountain Forest, Chaparral
Freshwater ecoregion: Pacific Mid-Coastal
Terrestrial ecoregions: Klamath–Siskiyou Forests, Northern California Coastal Forests
Number of fish species: 25 (15 native)
Number of endangered species: 12
Major fishes: Pacific lamprey, river lamprey, Pacific brook lamprey, Sacramento sucker, coho salmon, chinook salmon, rainbow trout, cutthroat trout, threespine stickleback, staghorn sculpin, coastrange sculpin, prickly sculpin
Major other aquatic vertebrates: northwestern salamander, Pacific giant salamander, southern torrent salamander, rough skinned newt, California newt, red-bellied newt, red-legged frog, foothill yellow-legged frog, western toad, western pond turtle, western aquatic garter snake, aquatic/Santa Cruz garter snake, beaver, muskrat, raccoon, river otter, mink
Major benthic invertebrates: mayflies (*Baetis, Ephemerella, Epeorus, Cinygmula, Rhithrogena, Paraleptophlebia*), stoneflies (*Malenka, Calineuria, Sweltsa*), caddisflies (*Hydropsyche, Rhyacophila, Helicopsyche*), beetles (*Psephenus, Optioservus*)
Nonnative species: bullfrog, American shad, threadfin shad, golden shiner, Sacramento pikeminnow, California roach, speckled dace, fathead minnow, brown bullhead, green sunfish, bluegill
Major riparian plants: arroyo willow, Hooker willow, black willow, narrowleaf willow, Pacific willow, red willow, sandbar willow, Sitka willow, black cottonwood, Fremont cottonwood, mulefat, white alder
Special features: begins in Six Rivers and Mendocino National Forests; flows past some of the most spectacular ancient redwood groves in California; more kilometers designated as National Wild and Scenic Rivers than any basin in California
Fragmentation: 14 dams
Water quality: pH = 8.0, alkalinity = 96 mg/L as CaCO₃, NO₃-N + NO₂-N = 0.082 mg/L, PO₄-P = 0.020 mg/L
Land use: 65.1% forest, 12.2% shrub, 19.2% grassland, 1.9% agriculture, 0.2% urban
Population density: 3.3 people/km²
Major information sources: Moyle 2002, Schoenherr 1992

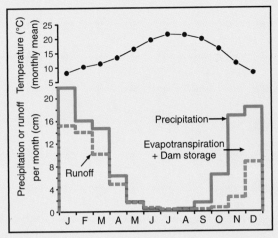

FIGURE 12.27 Mean monthly air temperature, precipitation, and runoff for the Russian River basin.

RUSSIAN RIVER

Relief: 1324 m
Basin area: 3728 km²
Mean discharge: 66 m³/s
River order: 6 (estimated)
Mean annual precipitation: 105 cm
Mean air temperature: 14.9°C
Mean water temperature: NA
Physiographic province: Pacific Border (PB)
Biomes: Chaparral, Temperate Mountain Forest
Freshwater ecoregion: Pacific Mid-Coastal
Terrestrial ecoregions: Northern California Coastal Forests, California Interior Chaparral and Woodlands
Number of fish species: 41 (20 native)
Number of endangered species: 43
Major fishes: Pacific lamprey, river lamprey, Pacific brook lamprey, hardhead, hitch, Sacramento pikeminnow, California roach, Sacramento sucker, Klamath largescale sucker, longfin smelt, coho salmon, chinook salmon, rainbow trout, threespine stickleback, tule perch, staghorn sculpin, coastrange sculpin, prickly sculpin, riffle sculpin

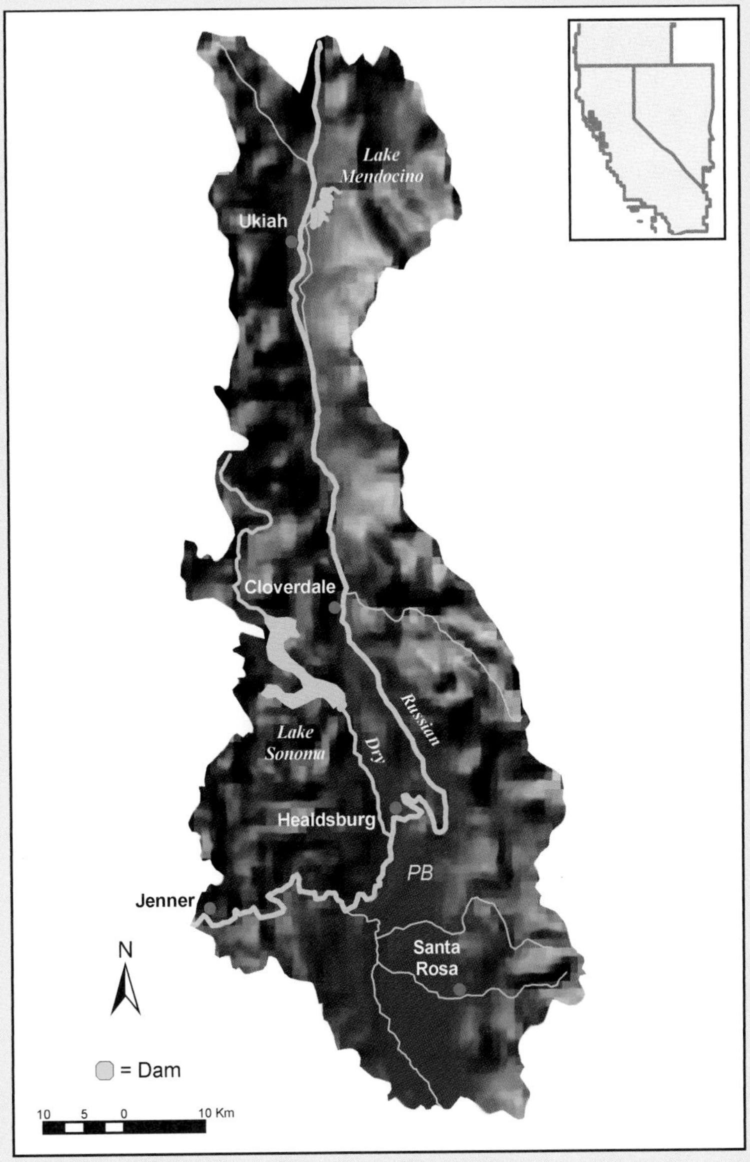

Major other aquatic vertebrates: northwestern salamander, Pacific giant salamander, California giant salamander, southern torrent salamander, rough skinned newt, California newt, sierra newt, red-bellied newt, California red-legged frog, foothill yellow-legged frog, mountain yellow-legged frog, western toad, western aquatic garter snake, aquatic/Santa Cruz garter snake, mountain beaver, beaver, muskrat, mink

Major benthic invertebrates: mayflies (*Baetis, Diphetor, Tricorythodes, Drunella, Ephemerella, Paraleptophlebia, Rhithrogena, Epeorus*), stoneflies (*Calineuria, Malenka, Sweltsa*), caddisflies (*Hydropsyche, Rhyacophila, Lepidostoma*), beetles (*Optioservus*)

Nonnative species: bullfrog, American shad, threadfin shad, common carp, goldfish, golden shiner, Sacramento blackfish, fathead minnow, brown bullhead, green sunfish, bluegill, blue catfish, channel catfish, white catfish, brown bullhead, black bullhead, brown trout, mosquitofish, inland silverside, black crappie, white crappie, redear sunfish, largemouth bass, smallmouth bass

Major riparian plants: arroyo willow, Hooker willow, narrowleaf willow, Pacific willow, red willow, sandbar willow, Sitka willow

Special features: begins in Coast Range of California; flows through a valley of mixed land use; classic Mediterranean-type river in California

Fragmentation: 62 dams

Water quality: pH = 7.8, alkalinity = 111 mg/L as $CaCO_3$, NO_3-N + NO_2-N = 0.266 mg/L, PO_4-P = 0.082 mg/L

Land use: 50.2% forest, 8.0% shrub, 23.3% grasslands/herbaceous, 0.1% wetlands, 14.3% agriculture, 2.8% urban

Population density: 62.5 people/km²

Major information sources: Moyle 2002, Schoenherr 1992

FIGURE 12.26 Map of the Russian River basin.

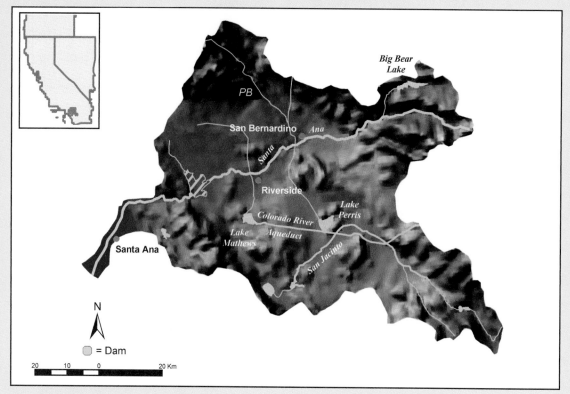

FIGURE 12.28 Map of the Santa Ana River basin.

SANTA ANA RIVER

Relief: 3506 m

Basin area: 6314 km²

Mean discharge: 1.7 m³/s

River order: 6 (estimated)

Mean annual precipitation: 34.1 cm

Mean air temperature: 17.2°C

Mean water temperature: NA

Physiographic province: Pacific Border (PB)

Biome: Chaparral

Freshwater ecoregion: South Pacific Coastal

Terrestrial ecoregions: California Montane Chaparral and Woodlands, California Coastal Sage and Chaparral

Number of fish species: 45 (9 native)

Number of endangered species: 54

Major fishes: arroyo chub, speckled dace, rainbow trout, California killifish, threespine stickleback, striped mullet, tidewater goby, staghorn sculpin, prickly sculpin

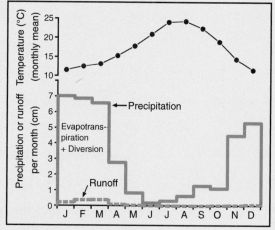

FIGURE 12.29 Mean monthly air temperature, precipitation, and runoff for the Santa Ana River basin.

Major other aquatic vertebrates: California newt, Sierra newt, California red-legged frog, mountain yellow-legged frog, western leopard frog, African clawed frog, western toad, arroyo toad, western pond turtle, western aquatic garter snake, beaver, muskrat, raccoon, river otter, long-tailed weasel

Major benthic invertebrates: mayflies (*Baetis tricaudatus, Fallceon quilleri, Caudatella, Drunella, Tricorythodes*), damselflies (*Argia vivida*), caddisflies (*Hydropsyche californica, H. occidentalis, Helicopsyche*), moths (*Petrophila*), snails (*Physella*)

Nonnative species: bullfrog, threadfin shad, common carp, goldfish, hitch, Sacramento pikeminnow, Colorado pikeminnow, red shiner, fathead minnow, catfish, brown trout, rainwater killifish, mosquitofish, sailfin molly, inland silverside, crappie, green sunfish, bluegill, pumpkinseed, redear sunfish, largemouth bass, spotted bass, smallmouth bass, bigscale logperch, redbelly tilapia, tule perch, yellowfin goby, Shimofuri goby; list includes Los Angeles basin

Major riparian plants: arroyo willow, black willow, narrowleaf willow, Pacific willow, red willow, California sycamore, Fremont cottonwood, mulefat, white alder

Special features: one of the largest river systems in southern California, but intermittent, channelized, and highly urbanized; upper portion still retains natural characteristics

Fragmentation: 52 dams

Water quality: pH = 7.8, alkalinity = 214 mg/L as $CaCO_3$, NO_3-N + NO_2-N = 5.0 mg/L, PO_4-P = 1.56 mg/L

Land use: 57% vacant land and forest, 11% agriculture, 32% urban

Population density: 334.1 people/km²

Major information sources: Moyle 2002, Schoenherr 1992

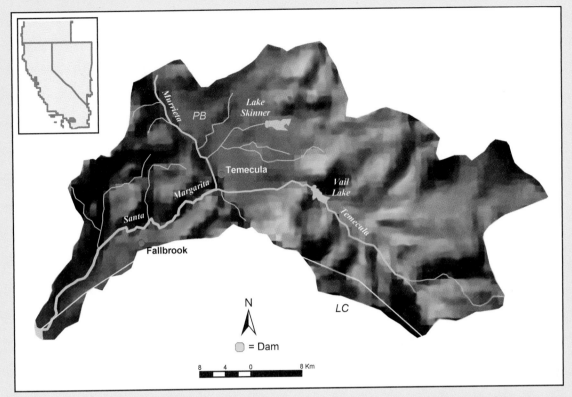

FIGURE 12.30 Map of the Santa Margarita River basin. Physiographic provinces are separated by a yellow line.

SANTA MARGARITA RIVER

Relief: 2076 m
Basin area: 1896 km²
Mean discharge: 1.2 m³/s
River order: 5 (estimated)
Mean annual precipitation: 49.5 cm
Mean air temperature: 14.6°C
Mean water temperature: N.A.
Physiographic provinces: Lower California (LC), Pacific Border (PB)
Biome: Chaparral
Freshwater ecoregion: South Pacific Coastal
Terrestrial ecoregion: California Coastal Sage and Chaparral
Number of fish species: 17 (6 native)
Number of endangered species: 52
Major fishes: arroyo chub, rainbow trout, California killifish, striped mullet, longjaw mudsucker, staghorn sculpin
Major other aquatic vertebrates: California newt, Pacific chorus frog, California chorus frog, California red-legged frog, mountain yellow-legged frog, western leopard frog, African clawed frog, western toad, arroyo toad, western pond turtle, western aquatic garter snake, beaver, muskrat, raccoon, long-tailed weasel
Major benthic invertebrates: mayflies (*Baetis, Tricorythodes, Fallceon quilleri, Centroptilum/Procloeon*), stoneflies (*Isoperla, Malenka, Zapada*), caddisflies (*Hydropsyche, Cheumatopsyche, Amiocentrus, Micrasema*), black flies (*Simulium*), crustaceans (*Hyalella azteca, Pacifasticus leniusculus*), bivalves (*Corbicula, Pisidium*), snails (*Physa/Physella, Gyraulus*)
Nonnative species: bullfrog, common carp, channel catfish, black bullhead, yellow bullhead, mosquitofish, black crappie, white crappie, green sunfish, bluegill, largemouth bass, redeye bass, yellowfin goby
Major riparian plants: arroyo, black, narrowleaf, Pacific, and red willow; California sycamore; Fremont cottonwood; mulefat; white alder
Special features: one of the last free-flowing rivers in southern California; one of the most intact and continuous riparian corridors in region
Fragmentation: 9 dams
Water quality: pH = 8.5, alkalinity = 184 mg/L as CaCO₃, NO₃-N + NO₂-N = 0.54 mg/L, PO₄-P = 0.152 mg/L
Land use: 11.2% forest, 58.5% shrub, 13.2% grasslands/herbaceous, 0.1% wetlands, 11.6% agriculture, 3.2% urban
Population density: 51.5 people/km²
Major information sources: Ode et al. 2002, Moyle 2002, Schoenherr 1992

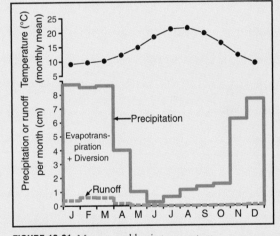

FIGURE 12.31 Mean monthly air temperature, precipitation, and runoff for the Santa Margarita River basin.

589

13

COLUMBIA RIVER BASIN

JACK A. STANFORD   F. RICHARD HAUER
STANLEY V. GREGORY   ERIC B. SNYDER

INTRODUCTION

FLATHEAD RIVER

SNAKE/SALMON RIVER

YAKIMA RIVER

WILLAMETTE RIVER

COLUMBIA RIVER MAIN STEM

ADDITIONAL RIVERS

ACKNOWLEDGMENTS

LITERATURE CITED

INTRODUCTION

The Columbia River drains 724,025 km² of the American Pacific Northwest, including parts of Washington, Oregon, Nevada, Utah, Idaho, Wyoming, Montana, and British Columbia (Fig. 13.2). It is a 9th order river based upon analysis of the stream network at 1:100,000 scale. Landscape diversity is high, with parts of 13 terrestrial (Ricketts et al. 1999) and three freshwater (Abell et al. 2000) ecoregions included in the basin. The Columbia obtains much of its runoff from distant Rocky Mountain headwaters, flows through a huge and mostly dry interior basin, gains significant runoff from the eastern slopes of the Cascade Mountains, squeezes through a narrow gap in the Cascade cordillera, picks up the Willamette River, and softly passes through cool rainforests of the low Pacific Coast mountains. The river is a great water road for salmon and people, a purveyor of energy and irrigation water that has figured decisively in the development of the nation (White 1995).

The significance of the river was not clear to Boston fur trader and whaler Robert Gray when he first viewed and recorded the mouth of the Columbia in 1792, perhaps because of the rapid dissolve of the wood-jammed river into the mists of the coastal rain forests. Nonetheless, the captain named the river after his ship, the Columbia Rediviva. Caucasian sailor-explorers had some contact and perhaps traded for furs with native people of the Columbia, but they had no idea that the river system was populated by over 100,000 Native Americans, including Nez Perce, Salish, Kootenai, Umatilla, and Yakama people with strong Pacific coastal affinity, along with several tribes such as Blackfeet and Shoshone of Great Plains origin using the headwaters.

Indeed, hunter-gatherer people have an ancient tenure in the Columbia River basin, and recent discoveries have renewed and helped reform theories regarding the peopling of the New World (Chatters 2000). The earliest evidence of human occupation corresponds to the last catastrophic floods from glacial outwash approximately 12,000 years ago (Attwater 1986). Salmon apparently were present in fairly large numbers at the Dalles on the lower main stem of the Columbia River by this time (Cressman et al. 1960, Butler 1993). Around 9000 years ago salmon were being caught at Kettle Falls on the upper Columbia (Chance and Chance 1977) and in Hells Canyon on the Snake River. Human populations during this period are characterized as "broad spectrum foraging," indicating small group size, high seasonal mobility, and low population densities (Schalk

FIGURE 13.1 Grande Ronde River just upstream from its junction with Snake River (PHOTO BY C. E. CUSHING).

591

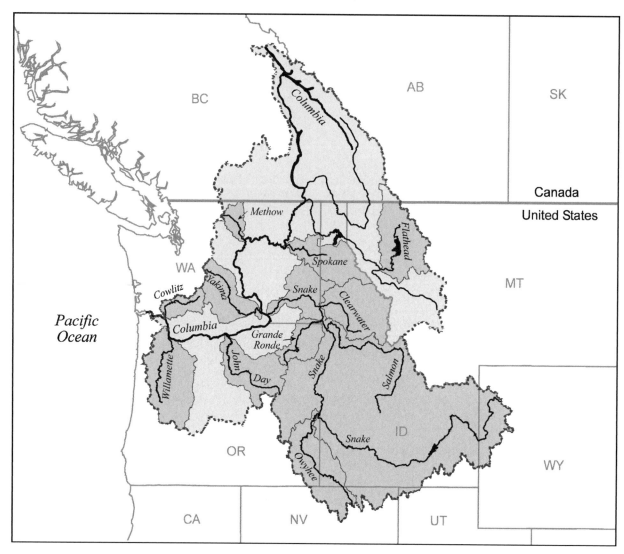

FIGURE 13.2 Rivers of the Columbia River basin covered in this chapter.

and Cleveland 1983). Archaeological data and tribal oral traditions from this early time period indicate the resources of the Columbia River and its tributary streams were important to early cultural systems. Human skeletal material from the middle Columbia indicates marine-derived resources were important dietary components (Chatters 2000). Hunter-gatherers shifted their life ways, adapting to changes in resource patterns associated with climatic shifts marking the early and middle Holocene period. From about 4000 years ago through the early historic period, salmon in association with gathered roots and hunting formed the basis of the human economy in the Columbia basin. Associated with floodplains of the major rivers and particularly at river junctures, large "winter villages complexes" emerged (Nelson

1973). These river-oriented villages formed the winter homes of peoples who gathered plants and hunted across a much larger territory (Uebelacker 1986). Variations in abundance and timing of fish and game migrations and broad spatial distribution of a large number of native plants with edible roots explain, in part, the cultural diversity that has so characterized the history of the Basin (see Walker 1998 for an exhaustive review of prehistoric and historic occupancy of the Columbia River basin).

The great expedition of Lewis and Clark (1804 to 1806) clearly revealed the physiographic breadth and ecological scope of the Columbia. The river and its expansive landscape and diverse native people were daunting, solicitous, and preeminent. Indeed, within 150 years the rapids were tamed and the cultural and ecological landscape dramatically altered. Today, the river is aptly described as the great organic machine, owing to the rich legacy of fishing, mining, ranching, farming, commerce, and industry driven by the river's copious hydropower (White 1995).

In this chapter, we describe in some detail four of the Columbia tributaries, the Flathead, Snake, Yakima, and Willamette, and the Columbia main stem, along with brief summaries for seven others: Cowlitz, Grande Ronde, John Day, Methow, Spokane, Owyhee, and Clearwater. The Flathead River characterizes the distant upper segments of the Columbia. Due to its large size, the Snake and its major tributaries, the Clearwater and the Salmon, are key to understanding the Columbia. The Yakima represents the mid-Columbia rivers. The Yakima is especially noteworthy because it offers great potential for salmon restoration owing to expansive floodplains that remain largely intact, coupled with the opportunity to return irrigation diversions to the river through augmentation from the main-stem Columbia. Finally, the Willamette stands out as a culturally important large alluvial river that remains in a normative condition in spite of increasing urbanization, intense farming, and a long legacy of interaction between the timber industry and river ecology.

Physiography and Climate

The source of the Columbia River generally is conceded to be Columbia Lake in British Columbia almost exactly 2000 km from the Pacific. However, the most distant origin is via its largest tributary, the Snake, which begins on the Yellowstone Plateau in Wyoming, 2290 km from the ocean (Patrick 1995). The river and its tributaries traverse eight physiographic provinces (Hunt 1974). The Rocky Mountain headwaters include a long stretch of the Continental Divide cordillera, from the Tetons in Wyoming to the Columbia Ice Fields in Canada, and include the following physiographic provinces: the Middle Rocky Mountains, Northern Rocky Mountains, Rocky Mountains in Canada, and Coast Mountains of British Columbia and Southeast Alaska. Sequential mountain ranges extend across the Canadian portion of the catchment from the Rockies to the Cascades and other Pacific coastal ranges. The uplifted granite batholith that forms the extensive mountain ranges of Idaho west of the Rockies, including the Bitterroot, Selway, and Sawtooth mountains, also is a very prominent headwater feature of the basin. The comparatively arid Interior Columbia basin is represented by the

Columbia–Snake River Plateaus province, which includes the expansive Snake River Plain, the Blue Mountains, and the high desert steppe of eastern Oregon. Lower elevation and relatively dry ranges (e.g., Steens, Owyhee, Albion, Portneuf mountains) of northern Nevada and Utah separate the interior from the Great Basin to the south. However, some small tributaries of the Snake River drain from the northern edge of the Basin and Range province. The broad, fertile Willamette Valley is contained between the mountains of the Cascade–Sierra Nevada province and the coastal ranges of the Pacific Border province. Prominent Cascade volcanoes, Rainier, Hood, St. Helens, and Adams, are the sentinels of the Columbia basin.

The climate of the basin is predominately Pacific maritime, although cold, dry weather associated with northern continental weather intrudes into the headwaters frequently. Thus, the temperature range in the basin is extreme. Temperatures in the Rocky Mountain areas often are well below zero for many days in midwinter. Indeed, the coldest non-Alaska temperature recorded in the United States was a bitter $-57°C$ at Rogers Pass on the Continental Divide at the headwaters of the Clark Fork of the Columbia on January 20, 1954. Summer temperatures in the mountain valleys seldom reach above 35°C, but areas in the arid interior often exceed 38°C degrees for many days in midsummer. Temperatures moderate toward the coast and the Cascade Range produces a profound rain shadow over the interior. Precipitation in the Cascades can exceed 350 cm/yr, whereas in the interior precipitation in places is 15 cm or less. Precipitation rises back to over 200 cm annually in the Rockies, although spatial variability is high owing to the complexity of the mountain and valley landscape and increasing influence of continental climate.

Basin Landscape and Land Use

Triple Divide Peak in Glacier National Park, Montana, is the common source of three of North America's great rivers: the Columbia (via the Flathead River), the Missouri (via the Milk River), and the Saskatchewan (via the Saint Mary's River). Early naturalist and glaciologist James Bird Grinnell called this mountainous area the Crown of the Continent. The glaciated landscape epitomizes the headwaters of the Columbia in the Rocky Mountains and in much of the Cascades as well. Sharp peaks tower over U-shaped valleys with large alluvial floodplains, the legacy of glacial scour and subsequent deposition of fluvial outwash deposits later reworked by the rivers.

Lobes of the Quaternary continental glaciers extended south through the Canadian ranges into Montana via the Rocky Mountain Trench, forming Flathead Lake, and into Idaho via the Purcell Trench, forming Lake Pend Oreille. Mountain glaciers draped all of the basin's ranges. Advances of the Purcell lobe, 12,000 to 15,000 years ago, dammed the Clark Fork multiple times, creating a proglacial Lake Missoula that covered over 20,000 km² to depths of several hundred meters. Cataclysmic floods burst over eastern Washington as the ice dams broke and Lake Missoula drained. Deep channels were scoured through the lava formations of eastern Washington, creating the Scablands. Water and sediments flooded into the tributary valleys of the middle and lower Columbia, modifying the river landscape (Allen et al. 1986). Glacial Lake Bonneville (now the Great Salt Lake) of the Great Basin in Utah and Nevada also spilled into the Snake in at least one cataclysmic flood, leaving many islands in the river channel as the outwash was retarded by Hells Canyon. Subsequent floods have never overtopped these distinctive features of the Snake River that now compose the Deer Flats National Wildlife Refuge (Connor 1993). Hence, the entire Columbia basin reflects the legacy of glaciation, either by ice scour or proglacial flooding or both. Glacial modification of the landscape is superimposed on the much older legacy of volcanic ash and lava deposition, and mountain uplift mediated by collision of continental plates.

Great landscape and climatic complexity is reflected by 13 terrestrial ecoregion designations in the basin (Ricketts et al. 1999). The headwaters of the North Central Rockies Forests and South Central Rockies Forests ecoregions have diverse alpine plant assemblages and conifers dominated by subalpine fir, spruce, larch, Douglas fir, and lodgepole pine. The conifers extend onto the valley floors on the wetter sites or grade into dry savannah–steppe communities dominated by bunchgrasses, bitter- and sagebrush, and ponderosa pine. The rivers cut through the mountain ranges, with precipitous canyons and expansive alluvial floodplains arrayed like beads on a string. The floodplains have distinctive cottonwood, spruce, and willow riparia. The interior Columbia basin includes the sagebrush, juniper, and ponderosa steppe of eastern Oregon and the historically expansive intermountain grasslands, the Palouse Prairie of Idaho and eastern Washington (Daubenmire 1978). These correspond to the Snake/Columbia Shrub Steppe and the Palouse Grasslands ecoregions. The Palouse Prairie today is almost completely converted to dry-land grain farms.

The Snake River plain in Idaho also is classified as grass and sagebrush steppe, but much of it has been converted to row-crop agriculture (e.g., potatoes, sugar beets, onions) by virtue of massive irrigation diversions that substantially dewater the middle reaches of the river. A robust apple, cherry, grape (wine), and other fruit industry developed from irrigation projects in the mid-Columbia region. Huge Douglas fir and Sitka spruce of the temperate rain forest zone inhabit the Eastern Cascades Forests, Central and Southern Cascades Forests, Willamette Valley Forests, and Central Pacific Coastal Forests ecoregions. Timbering, agriculture, and urban and exurban expansion has substantially cleared and otherwise altered much of this landscape, particularly in the Willamette Valley.

Placer and tunnel mining for gold, silver, copper, and other metals, especially in the hard bedrocks of the Idaho batholith, produced the first wave of immigrants as the fur trade waned. Ranches, farms, the salmon fishing/canning industry, and associated infrastructure followed. Open-pit mining and aluminum production from imported ore, along with other heavy industries, came with hydropower development. Currently, land use in the Columbia is 50% forest/alpine, 34% scrub/grassland, 15% agriculture, and 2% urban.

The Rivers

The rivers of the basin may be grouped in relation to basin physiography. The Rocky Mountain tributaries include (1) the Flathead-Clark Fork and Kootenay rivers that flow from the Continental Divide ranges of Montana and British Columbia into the upper Columbia above Grand Coulee Dam (Lake Roosevelt), (2) the Spokane–St. Joe–Coeur d'Alene that flows from the Bitteroot Range into the mid-Columbia, and (3) the upper Snake, originating on the Continental Divide in Wyoming, plus the Payette, Boise, Weiser, Salmon, and Clearwater rivers that drain the Idaho batholith and feed the Snake. A suite of mid-Columbia tributaries, the Okanagan, Methow, Wenachee, and Yakima, drain the east or leeward side of the Cascades, adding significant flow upstream of the Snake confluence. All of these rivers have some reaches of wooded floodplains between or upstream of steep and often quite long canyon sections. The Salmon River is notably constrained in canyons, including the "River-of-No-Return" stretch that stymied Lewis and Clark. The high desert steppe of the interior has a suite of similar rivers. The Owyhee River originates in northern ranges of the

Great Basin and flows north to the Snake through long canyons. The lowermost big tributary of the Snake, the Grande Ronde, drains the Blue Mountains, the largest interior range. Other high-desert tributaries flowing into the Columbia below the Snake confluence are the Umatilla, John Day, and Deschutes. The first two have long floodplain segments, whereas the Deschutes is more constrained by incision into ancient lava (basalt) flows. The Willamette is by far the largest of the lower Columbia tributaries in the temperate rain forest zone and has expansive floodplains throughout. The Cowlitz is one of several important lower-river tributaries and is noteworthy because it is recovering from severe sedimentation after the eruption of Mount St. Helens. Like the lower Snake, the main-stem Columbia River largely is incised throughout its course owing to the immense scouring of Lake Missoula and Bonneville flooding. Nonetheless, the river was predominately alluvial, with long segments of gravel–cobble floodplains, as in the Hanford Reach. But these ecologically important features are entirely inundated today. The John Day and Methow rivers are the only free-flowing rivers of any size on the mid- and lower Columbia and the Salmon is the only completely free-flowing tributary of any size in the Snake system.

A rich literature describes the ecology of the Columbia River, mainly as a result of intensive efforts in the last two decades to research and manage declining salmon runs. The Federal Northwest Power Planning Act of 1986 mandated funds for fisheries and wildlife management and restoration from hydropower revenue as mitigation for lost habitat associated with the construction and operation of the many dams and reservoirs in the basin. The Web site of the Northwest Power Planning Council (http://www.nwppc.org) and an associated database system called STREAMNET (http://205.230.28.30/) provide subbasin management plans, river and fisheries statistics, literature citations, and links to other agency and commerce information. A recent review of river ecology (Williams 2005) done in context of the council's recovery goals is a detailed and indispensable reference.

The waters of the Columbia are generally rich in ions owing to the wide array of rock and soil types in the basin and copious irrigation return. Headwaters of the granite batholith (Bitteroot Range) and the argillites (Rocky Mountains in Glacier National Park) are the most dilute. Organic matter is rapidly transported through the canyon reaches (Minshall et al. 1992), underscoring the importance of produc-

tion and retention of organic matter in the floodplain segments. Many streams and river reaches of the basin are impaired from mining (cf. Beckwith 2002) and agriculture pollution, based on listings per the Federal Clean Water Act by the basin states and ongoing superfund restorations (e.g., Clark Fork and Coeur d'Alene rivers, Hanford site). Lower reaches of some mid-Columbia tributaries (e.g., Yakima) and the lower Snake River in late summer approach temperatures that are lethal to salmonid fishes, owing to irrigation diversions that reduce flow and limit flux of water through floodplain aquifers and impoundment by hydroelectric dams.

Food webs in the free-flowing rivers are species rich and productive, most often dominated by hydropsychid caddisflies and midges, as will be described later for the Flathead and Yakima rivers. Of course, the trophic ecology of the impounded reaches is dramatically changed, as will be described later for the main-stem Columbia. Historically, the river system contained abundant beds of freshwater clams (*Margaritifera* spp.) that were intensely harvested by native people, especially during times of reduced salmon runs related to climate variation (Chatters et al. 1995). The floodplain backwaters also were filled with beds of arrowroot or wapato, with tuberous roots that were a preferred food of the Indians and also sustained the Lewis and Clark party at times.

The Columbia River system historically was a natural salmon factory. Over 200 different runs or stocks of four species of Pacific salmon spawned abundantly in the river system: chinook, sockeye, chum, and coho. Multiple runs of steelhead (anadromous rainbow or redband trout) also routinely made their way deep into the basin. Stock diversity apparently was driven by local adaptation to the widely varying riverine habitat conditions of the basin. Each stock had fidelity to particular rivers and in some cases even particular segments of rivers. The floodplains likely were especially important as rearing habitat. The total numbers are unknown but may have approached 20 million fishes per year. Fertility of the river therefore may have been substantially driven by the marine subsidy provided by the biomass of decomposing salmon that died every year after spawning (Cederholm et al. 1999; Schindler et al. 2003).

Other anadromous species of cultural importance were sturgeon that reached 3 m or more in length (500 kg) and dense runs of eulachon so rich in oil they burned like candles. Coastal cutthroat trout likely were locally abundant. In all there were 65 native species in the ichthyofauna of the river system, including three lamprey species (Ward and Ward 2004, Wydoski and Whitney 2003). Sculpin were notably diverse, with at least 13 species. The anadromous species ascended the entire main-stem Columbia but were stopped by falls on the Clark Fork below Pend Oreille Lake and by the Shoshone Falls on the Snake. Above the natural barriers the fish fauna was more limited but notably included three subspecies of cutthroat trout and bull charr. The zoogeography of Columbia River ichthyofauna clearly is related to Quaternary population radiation and subsequent isolation by the barrier falls (McPhail and Lindsey 1986), but the details are clouded by the unfortunate legacy of overfishing of many stocks and fish introductions dating to the 1880s.

At least 53 nonnative fishes have become well established in the Columbia system (Ward and Ward 2004, Wydoski and Whitney 2003). Notably robust throughout the basin are brook and lake trout, smallmouth bass, and walleye. One of the most abundant anadromous runs of the lower river is American shad. Atlantic salmon, escapees from farming operations in coastal net pens, are now routinely recorded migrating upstream, perhaps to spar with native salmon for the limited spawning habitat. Further complicating matters, the dams are migration barriers for some species but not for others. For example, formally anadromous sturgeon populations are trapped between dams on the Snake and Kootenay rivers and remaining anadromous runs cannot ascend salmon ladders at Bonneville Dam. The reservoirs promote nonnative dominance of food webs and function as predator gauntlets for outmigrating juvenile wild salmon.

The first dams of the Columbia basin were small irrigation or mining diversions; then came the big hydropower projects that eventually harnessed the river system completely. Today more than 400 dams exist in the basin, ranging from behemoth structures that store massive amounts of water to run-of-the-river structures on the main stem and small diversion dams for abstraction of irrigation water. Every major tributary of the Columbia except the Salmon is totally or partially regulated by dams and diversions to support hydropower and irrigation. The total water storage in the Columbia River system is 67.8 billion m^3, of which 51.8 billion m^3 are available for coordinated operation (FCRPS 2001). These dams have an installed electrical generation capacity of about 37,000 megawatts. The river system irrigates nearly 2 million ha of fruit, vegetables, hops, and hay and transports 1.2 million metric tons of grain and other commerce annually.

The development of the Columbia as an organic machine was coherent with the demise of the salmon runs, caused by overfishing and habitat loss associated with impoundment and water diversion for irrigation. Ladders for salmon passage were installed on the lower river dams, but migration was permanently blocked first at Grand Coulee and later at Chief Joseph dams on the Columbia and at Hells Canyon Dam on the Snake. The most productive floodplains of the river system are either flooded by hydropower dams or dewatered by irrigation diversions. Today, the only stable run of wild salmon left is the fall chinook stock that returns to the Hanford Reach. Some wild spring chinook and steelhead still return to some of the tributaries but are listed as threatened or endangered. Restoration of the fisheries so far has not been very successful, even with 98 salmon and steelhead hatcheries operating in the basin, expensive fish screens on almost all irrigation diversions, and sophisticated adult and juvenile bypass systems on the dams. Too little quality habitat remains to sustain wild runs, even though egg to adult survival of individual wild fishes is much higher than for hatchery fishes. Returns of hatchery fishes have increased in years of favorable ocean conditions, but by any measure the return on investment in fisheries restoration has been dismal indeed. Recovery, even of a few stocks of the once huge and diverse Columbia salmon fisheries, requires a new management paradigm. Specifically, management must focus on harvest control and habitat restoration, mainly through major changes in dam operations or dam removal, as opposed to hatchery operations (Williams 2005).

Abell et al. (2000) divide the Columbia River basin into three freshwater ecoregions, based primarily on fish distributions, for conservation purposes. The Columbia Glaciated ecoregion was glaciated during the Pleistocene and includes the northern part of the basin that is mountainous and heavily forested. The Columbia Unglaciated ecoregion was never glaciated and includes the lower Columbia main stem and most of the tributaries to the south (except the upper Snake) and is considered continentally outstanding in its biological distinctiveness. The Upper Snake ecoregion, in the southeastern corner of the basin upstream of Shoshone Falls, contains native fish species found nowhere else in the Columbia basin. All three ecoregions are considered endangered by Abell et al. (2000), primarily as a result of dams, diversions, introduction of nonnative fish species, and damage to riparian zones and floodplains.

FIGURE 13.3 Flathead River near Bad Rock Canyon, Montana (PHOTO BY C. E. CUSHING).

FLATHEAD RIVER

The Flathead River originates in the Waterton-Glacier International Peace Park of southern British Columbia and northern Montana and the massive mountain ranges of the Great Bear–Scapegoat–Bob Marshall Wilderness complex that straddle the Continental Divide of the Rocky Mountains (Fig. 13.15). The basin is 22,241 km², with 2300 km² in British Columbia, based on data from the Flathead Lake Biological Station derived from recent remote imagery. Headwater streams fed largely by snowmelt emerge from the talus slopes in the alpine and cascade off the flanks of the mountains, coalescing in the glaciated mountain valleys to form three forks of the Flathead River and the Swan River. These tributaries are characterized by expansive alluvial floodplains that are variously constrained within glacially sculptured mountain valleys. The floodplains occur longitudinally like beads on a string separated by startlingly steep and deep canyons (Fig. 13.3). Periodically, very deep glacial lakes, rather than floodplains, retard the river flow, sparkling like deep blue jewels in photos from spacecraft. The biggest jewel is 480 km² Flathead Lake, where all the waters of the Flathead commingle. From the big lake, the Flathead River courses through the intermountain prairie of the Mission Valley, picking up a bit more flow from the Little Bitterroot and Jocko rivers. Confluence with the Clark Fork of the Columbia is near Paradise, Montana, 285 km downstream from the source at Triple Divide Peak.

The Flathead basin is part of the original homeland of the Kootenai (Kootenay in Canada) and Salish Indians (Fahey 1974). These Native Americans hunted elk and deer in river floodplains and uplands, caught huge bull trout and abundant cutthroat trout and other fishes from the streams and lakes, and harvested edible plants, such as bitterroot and balsam root, that grew profusely on the intermountain prairies. They regularly crossed the Continental Divide via the Middle Fork of the Flathead on well-worn trails to hunt the abundant bison of the Great Plains and raid the Piegan Blackfeet Indians. The Blackfeet, just as regularly and reputedly more viciously, returned the favor.

In early March 1812, the Canadian geographer and explorer David Thompson stood on a ridge between the lower Flathead and the Clark Fork. He was the first nonnative to view and record the expansive landscape of Flathead Lake and its valley (Nisbet 1994). Canadian fur trappers and merchants working with Thompson established trading posts, first on the Clark Fork and later in the Flathead. By the 1880s the vast forests and valleys of the Flathead were penetrated by the Great Northern and other railroads and the Salish and Kootenai were confined to the Flathead Indian reservation that today encompasses the south half of Flathead Lake and lands extending south to the Clark Fork. The rails brought many white settlers into the Flathead. Soon steamboats plied Flathead Lake and the upper river, linking southern rail routes to the main High Line that followed the old Indian trail on the Middle Fork through the Rockies to Kalispell and onward west to Seattle. The Flathead reservation was opened to non-Indian settlement and the federal government built an irrigation system that insured long-term commingling of native and nonnative people.

Physiography, Climate, and Land Use

The mountain ranges of the basin all are part of the Northern Rocky Mountains (NR) and Rocky Mountains in Canada (RM) physiographic provinces (see Fig. 13.15) (Hunt 1974). In this area these mountains notably feature Precambrian Belt Series argillites (metamorphosed mud and silt stones) and dolomites uplifted 3500 m asl on the highest peaks in Glacier National Park. These ancient argillites contain stromatolites, fossil algae, among the oldest on Earth. The Lewis Overthrust is the predominant geologic feature and reflects the gradual movement of the Precambrian formations on the west side of the Continental Divide overtopping Mesozoic shales and limestones on the east (Ross and Rezak 1959).

Basin physiography is dominated by the effects of Quaternary glaciation, as mountain glaciers flowed down all of the major tributaries to confluence with the massive lobe of the continental glacier that flowed south from Canada in the Rocky Mountain Trench. The terminal moraines of this huge valley glacier lay south of Flathead Lake. It was a primary source of ice and water that contributed to the ice-dammed Lake Missoula, whose cataclysmic floods created the Scablands of Washington and flooded the Willamette, Yakima, and other tributary valleys of the lower Columbia. The mountain glaciers originating in high-elevation cirques scoured expansive U-shaped valleys, some with deep lakes.

Flathead Lake is of tectonic origin but modified by glacial scour. The outlet cuts through the terminal moraine to a bedrock sill at the town of Polson. The Mission Valley south of the lake is dotted with kettle lakes in the intermountain prairie landscape that formed as various glacial advances melted away. Today, few active glaciers are left in the basin and those have rapidly melted in the last decade (Hall and Fagre 2003).

The climate most of the time is of Pacific maritime origin, with weather fronts primarily from the southwest. However, sometimes the continental air masses from the northeast spill over the Continental Divide, creating extended very cold and dry periods. Precipitation varies with elevation: <50 cm/yr in the dry valleys to as much as 250 cm/yr in the mountains (Finklin 1986). Because of this great variation, mean monthly precipitation collected at a low elevation (Fig. 13.16) underestimates the basin average. Although precipitation is not highly seasonal, much of it accumulates in the snowpack during the winter months. Winter temperatures in the valleys average near freezing, with rare periods of −20°C and much colder in the mountains, especially toward the Continental Divide, where high winds also may occur. Summers are generally cool but with periods reaching 35°C, even in the mountains.

Valley bottom and lowland mountain forests are predominately ponderosa pine steppe or expansive lodgepole pine stands. Stands of quaking aspen exist in wetter areas or, if the site is very wet, white or water birch may by mixed into the conifer forest. Large expanses of the Mission and Kalispell valleys, located south and north of the big lake (see Fig. 13.15), historically were intermountain prairies dominated by bunchgrass (Montana Valley and Foothills Grasslands terrestrial ecoregion). The upland forests are Douglas fir and larch on the mountain flanks, with occasional stands of lodgepole pine on dry sites and white pine and western red cedar in mesic microclimates of narrow mountain valleys. Spruce are present in wet sites at all elevations. Subalpine fir and subalpine larch and stands of whitebark pine occur on the ridge tops to the timber line. This complex forest landscape, classified as the North Central Rockies Forests ecoregion, historically was maintained by succession in a mosaic of ages mediated by frequent late-summer fires. Forbe diversity is high in the extensive alpine areas.

About 80% of the basin is forested or alpine. Glacier National Park and the wilderness complex encompass 42% of the basin. The forested areas of the basin not included in the park or wilderness are intensively managed for timber products. Most of the ponderosa steppe and grasslands of the Kalispell and Mission valleys north and south of Flathead Lake were converted to agriculture by 1950. Dry-land cereal grains are very productive on thick loam soils of the upland benches. Crops irrigated from the alluvial aquifer of the Kalispell Valley and via the Flathead Irrigation Project in the Mission Valley include potatoes, mint, and alfalfa. Cattle production is important as well, from rangeland and irrigated pastures and some feedlot and dairy operations. The massive volume of Flathead Lake moderates winter and summer temperatures, especially along the east shore, allowing the production of cherries and some apples. Agriculture and rangeland is about 18% of the basin area. From the mid 1980s through the early part of the twenty-first century, farm and ranch land was gradually subsumed by exurban development associated with increasing economic emphasis on tourism and footloose business. Basin population currently is estimated at ~100,000, but almost all of these people live in the Kalispell and Mission valleys.

River Geomorphology, Hydrology, and Chemistry

The expansive floodplains of the alluvial valleys dominate the river landscapes. Notice the wide valley form of the North and South forks and the Swan River on the Flathead map (see Fig. 13.15). These valley bottoms were initially formed by glaciation and then modified by the recent legacy of flooding and plant succession. The Middle Fork is more constrained in canyon segments. An 80 km floodplain segment on the South Fork is inundated by Hungry Horse Reservoir.

The floodplains of the Flathead rivers are composed of deep deposits of fluvially sorted sand, gravel, and rock. These coarse bed sediments transmit large volumes of river water through alluvial aquifers. Groundwater and surface water exchange is a dominant process that interacts with flood-mediated scour and deposition, plant succession, and channel avulsion to produce the shifting habitat mosaic of these floodplains.

Spring snowmelt hydrographs characterize the rivers (see Fig. 13.16). Flow may change two orders of magnitude as it rises from winter minimum to the peak of spring freshet. Rainfall on snow can vastly accelerate the meltwater flow. The flood of record occurred in early June 1964 when a very wet maritime front collided with continental cold air, causing intense, sustained rainfall on the remaining snowpack at the higher elevations across the basin. The three forks of the Flathead each were carrying more than 2700 m^3/s (compared to usual base flow of 30 to 50 m^3/s and the mean annual flow of 340 m^3/s). In spite of large retention of South Fork water in Hungry Horse Reservoir, floodwaters extended 3 km beyond the active floodplain in the Kalispell Valley. Erosion and deposition radically altered the stream and river corridors. Course sediment layers in cores taken from the Flathead Lake bed suggest such floods occur on less than 50 year return intervals.

Hydrology of the basin, especially late-summer base flow, may have been significantly influenced by melting glaciers until recent decades, but this has not been quantified. Hungry Horse Dam on the South Fork and Kerr Dam at the outlet of Flathead Lake alter the natural hydrographs of the main-stem river, the lake, and the lower river, tending to smooth out the monthly hydrograph compared to free-flowing rivers (see Fig. 13.16). Wave erosion associated with Kerr Dam holding the lake some 3 m above its natural base level has reconfigured much of the shoreline, especially the Flathead River delta area at the north end of the lake.

Because much of the Flathead basin bedrocks are geologically very old and therefore well leached, the streams and lakes of Glacier National Park are characterized by very low ion concentrations (conductivity <10 μS/cm) and aquatic productivity is ultralimited by a paucity of plant-available phosphorus (usually below detection of 0.5 μg/L). Some high-elevation lakes have nearly the same chemical characteristics as precipitation and therefore are slightly acidic, with very little calcium carbonate buffer capacity (Ellis et al. 2003). Acid rain, which is not occurring, would be very harmful to these waters. The headwaters of the Middle and South forks are beyond the Lewis Overthrust and cut through massive headwall-type outcrops of Mesozoic limestone, some extending for over 50 km. Glaciers carried tills from these limestone areas downstream. Hence, lowland waters of the basin are well buffered, although nutrient concentrations remain naturally low and Flathead Lake is oligotrophic (Ellis and Stanford 2001).

River Biodiversity and Ecology

The Flathead River is part of the Columbia Glaciated freshwater ecoregion (Abell et al. 2000). River and lake ecological studies are very diverse and detailed, encompassing faunal and floral studies and ecosystem processes. A literature archive is available at the Flathead Lake Biological Station (http://www.umt.edu/flbs).

Algae, Cyanobacteria, and Protists

Owing to the shifting cobble bottoms and generally very low nutrient loads, periphyton of the rivers is mainly diatoms, dominated by *Achnanthes minutissima*, *Gomphonema olivaceoides*, and *Cymbella minuta*, indicators of well-oxygenated, clean water. *Ulothrix zonata* and *Hydrurus foetidus* are the common attached filamentous algae. In floodplain spring brooks, large patches of *Enteromorpha* sp. and *Nitzschia* spp., indicators of naturally high nitrogen concentrations, often occur at the spring heads. Ellis et al. (1998) described a rich assemblage of bacteria and protists in the alluvial aquifers. Stanford and Prescott (1988) described a rare, perhaps endemic alga (*Cladophora gyrfaconium*) that formed a furry covering in the littoral zone of a remote, high-altitude lake in Glacier National Park. The macroalgae *Chara* spp. commonly occurs in floodplain ponds and shallow lakes if alkalinity is fairly high. Heavy growths of *Cladophora glomerata* and *Vaucheria* sp. characterize artificially enriched areas of the river.

Over 300 species and varieties of diatoms and other algae have been identified in the phytoplankton of Flathead Lake. *Rhizosolenia eriensis*, *Cyclotella* spp., *Stephanodiscus* spp., *Coscinodiscus* spp., *Tabellaria fenestrata*, *Dinobryon divergens*, *D. bavaricum*, *D. cylindricum*, *Cryptomonas erosa*, *C. ovata*, *Asterionella formosa*, and *Gymnodinium* spp., in that order, dominate biomass. However, over 90% of the productivity in the water column is from phytoplankton less than 10 μm in size. The pollution alga *Anabaena flos aqua* has produced weak blooms in the lake along with the green alga *Botryocccus* sp. in some years, suggesting gradual eutrophication. This trend is corroborated by gradually increasing phytoplankton productivity measured by carbon uptake over 30 years since the mid 1970s (Stanford and Ellis 2002).

Plants

Aquatic vascular plants are not usually present in the benthos of the rivers, but many species of pondweed, hornwort, watermilfoil, waterweed, cattail, pondlily, and duckweed and a wide variety of rushes and sedges grow in the floodplain wetlands and shallow lakes. The endangered water howellia occurs in a few oxbows of the Swan River.

Gallery forests of black cottonwood and spruce intermixed with willow and alder dominate the floodplains of the river corridors. These floodplains are hot spots of plant richness and Mouw and Alaback (2003) showed 64% of the basin's vascular flora occurred on the Nyack Flood Plain of the Middle Fork.

Invertebrates

The benthic invertebrates of the Flathead rivers have been intensively studied throughout the river corridor. Over 500 species have been recorded and species are arrayed in relation to annual patterns of water temperature from the valley bottoms upstream to the alpine (e.g., Hauer et al. 2000). The fauna is dominated by insects and is notably species rich among caddisfly, stonefly, and mayfly species. Indeed, 82% of the stonefly species reported for the entire Rocky Mountains from the Yukon to New Mexico are documented in the Flathead basin. Forty-two species coexist in the Flathead River in the Kalispell Valley and emerge sequentially from January to October. Seven species are amphibitic, that is, the larvae inhabit the alluvial aquifers, coming to the surface only to emerge as flying adults that mate and lay eggs in the river. Some of the more abundant stonefly species are *Utacapnia* (four species), *Capnia confusa*, *Isocapnia missouri*, *Pteronarcella badia*, *Isoperla fulva*, *Suwallia pallidula*, and *Claassenia sabulosa*.

A rich Crustacean stygofauna also is present in the alluvial aquifers, including the amphipod *Stygobromus* spp., various Bathynellidae, and Copepoda (Stanford et al. 1994). Net-spinning caddisflies (e.g., *Arctopsyche grandis*, *Parapsyche elsis*, *Hydropsyche* [3 species], *Cheumatopsyche campyla*) often dominate benthic biomass in the rivers, but *Brachycentrus americanus*, *Dicosmoecus gilvipes*, and *Glossosoma alesence*, among over 40 other species of caddisflies, often are notably abundant. Side channels and gravel-bar ponds often have an abundance of larvae of the limnephilid *Psychoglypha subborealis* and nymphs of the mayfly *Siphlonurus* spp. Mayflies, although not often dominant in biomass, are numerically abundant and species rich in western Montana streams and lakes. Mayflies can be found in all but the most severely polluted environments. The mayfly fauna of western Montana has not been completely cataloged, but records indicate that over 110 species have been collected in Montana. Some of the most commonly occurring lotic taxa are *Ameletus* spp., *Baetis* (>4 species), *Caudatella* (2 spp.), *Cinygmula*, *Diphetor*, *Drunella* (5 spp.), *Epeorus* (4 spp.),

Ephemerella (2 spp.), *Paraleptophlebia* (2 spp.), *Rhithrogena* (3 spp.), and *Serratella*. Some common lentic genera are *Caenis*, *Callibaetis*, *Ephemera*, and *Hexagenia*. Several mayfly taxa are occasionally found in great numbers as occasional hyporheos near the river banks (up to 100 m from the river channel), including *Ameletus* (2 spp.), *Paraleptophlebia* (2 spp.), and *Siphlonurus* (2 spp.). A wide variety of chironomid midges, black flies, and other true flies, such as *Hexatoma* and *Atherix variegata*, are usually present in river benthos samples.

Mollusks are not usually present in benthos samples today. However, large freshwater clams, western pearlshell (Newell 2003, Smith 2001), were routinely collected by Salish people until after 1900, but only a few beds are known today, and all are in the lower river tributaries below Flathead Lake. The signal crayfish is native below Flathead Lake, and the virile crayfish was introduced sometime around the 1960s and is now abundant in Flathead Lake and likely moving upstream.

Vertebrates

Fishes are perhaps the least species-rich group of aquatic animals in the Flathead, with only 12 native species: bull and westslope cutthroat trout, largescale and longnose sucker, Rocky Mountain and pigmy whitefishes, peamouth minnow, longnose dace, northern pikeminnow, redside shiner, and slimy and shorthead sculpin. At least 17 nonnative species are naturally reproducing in the basin and some of these, notably lake, brook, and rainbow trout, are rapidly expanding their range upstream and attendant hybridization, competition, predation, and food web cascades are problematic for native species. Bull trout are listed as threatened and considerable concern exists for cutthroats. Native assemblages of fishes are largely intact only in one subbasin in Glacier National Park owing to no stocking and presence of a natural barrier falls that prevents immigration of nonnative fishes and other biota. Natives also are intact in the entire South Fork drainage above Hungry Horse Dam, except for some introgression of the cutthroats by a very few introduced rainbows.

The northern leopard frog historically was widely reported from the valley bottom wetlands of the Flathead, but for unknown reasons today is rarely observed in many of the previously reported habitats. The floodplains harbor western toads in large numbers; tadpoles sometimes number in the thousands within small scour pools. Mass migration of juvenile toads from the floodplains toward the uplands occurs annually. The spring brooks are pop-

ulated by the Columbia spotted frog and the tailed frog is commonly found in fishless, high-gradient headwater streams. The long-toed salamander is common in fishless lakes and streams (Maxell et al. 2003). One specimen was pumped from the Kalispell alluvial aquifer. All of these species are declining throughout their ranges, probably due to loss of floodplain and wetland habitat. In contrast, the common garter snake is abundant in all aquatic habitats. A notable nonnative species is the snapping turtle, although only a few confirmed reports exist.

Semiaquatic vertebrates living in the river floodplains or along stream corridors are river otter, mink, muskrat, beaver, bald eagle, osprey, water ouzel (American dipper), harlequin and other ducks and mergansers, and Canada geese. Elk, moose, and white-tail and mule deer require the thermal cover and browse of riparian vegetation in the floodplains during winter, and cottonwood gallery forests are primary birthing and nursery sites for elk and white-tail deer.

Ecosystem Processes

In addition to climate, including temperature as the master variable (Hall et al. 1992, Ward and Stanford 1982) and the zoogeographic setting of the basin, four primary natural processes interact to control biodiversity and bioproduction of the Flathead rivers. The first is fire from lightening strikes in late summer. Fires periodically reset successional trajectories of basin vegetation, thereby naturally altering water and nutrient flux and storage in the uplands and sometimes even within riparian forests. In 1988, 2002, and 2003, wildfires burned over 200,000 ha of the landscape in Glacier National Park and adjacent wilderness areas. These fires were predominately allowed to burn uncontrolled and reset most of the senescent stands of lodgepole pine. In places they even burned across floodplains. The result was a renewed landscape mosaic, a key natural process of Rocky Mountain ecosystems. Second, cut-and-fill alluviation mediated by flooding constantly is resetting the habitat mosaic of the floodplains. Third, flux and transformation of carbon, nitrogen, and phosphorus ultimately determines productivity. Finally, herbivory, predation, burrowing, and other disturbances by animals modulate the other three processes.

These linked processes play out on a river–lake ecosystem stage. Water and nutrient inputs control productivity of the valley bottom lakes and adfluvial (grow to maturity in the lake, reproduce in the tributaries) fishes, and other vertebrates carry lake nutrients back upstream. The lakebed sediments and the floodplains function as materials and energy sinks, or storehouses. Hyporheic exchange of river water with the alluvial aquifers of the floodplains is a primary process controlling productivity and diversity of the riverine landscape. The wide riparian corridor is a dynamic ecotone that strongly influences distribution and abundance of nearly all native biota of the basin (Stanford and Ellis 2002).

Primary production in Flathead Lake is low (average 95 g C m^{-2} yr^{-1} over 30 years), owing to the natural oligotrophic condition, but has increased 30% since 1977 (Stanford and Ellis 2002). Bioavailable phosphorus concentrations in the lake and rivers usually are not detectable (<1 µg/L) and primary production usually is colimited by paucity of labile nitrogen and phosphorus. Dissolved organic matter concentrations usually are below 1 mg/L in surface and ground waters. Benthic production measured by diel measures of community metabolism in the rivers varies but also is low (133 to 1063 mg C m^{-2} d^{-1}; Stanford and Ward 1983). More recent measures of chlorophyll in river biofilms over a wide range of habitat types was 50.6 mg/m^2 (M. Anderson, unpublished data).

Natural river food webs in the Flathead generally are based on primary production from diatom-dominated biofilms, often enhanced locally by nutrient subsidy from upwelling groundwater from the alluvial aquifers, and on inputs of leaves from the floodplain forests. The species-rich invertebrate assemblage has representatives of all the trophic guilds (e.g., scrapers, collectors, shredders, predators), which, along with inputs of terrestrial invertebrates, support juvenile fishes of all species and adults of mountain whitefish and cutthroat trout. Flathead sculpins, in spite of their small size, and the much larger pikeminnows and bull charrs are voracious predators of the other fishes. Ospreys, bald eagles, mink, and river otters are the natural top carnivores, although grizzly bears occasionally feed on spawning cutthroat trout.

Human Impacts and Special Features

The most significant aquatic features of the Flathead rivers are the big valley bottom lakes and expansive floodplains. Ecology of the rivers, and perhaps also the big lakes, is controlled or vastly influenced by floodplain processes, including natural water storage and flow regulation, nutrient transformation and storage, and species packing associated with the shifting habitat mosaic produced by interactive flooding,

cut-and-fill alluviation, and riparian plant succession. The floodplains and glacial lakes are the jewels in the Crown of the Continent landscape of Glacier National Park and the adjacent wilderness areas. Water quality throughout the basin remains near likely historic levels and Flathead Lake remains one of the cleanest big lakes in the temperate areas of the world (Stanford and Ellis 2002).

Therefore, the main environmental concern is maintaining high water quality by minimizing increasing nitrogen and phosphorus loads from human sources, including atmospheric fallout and steadily increasing urban and exurban land conversion upstream of Flathead Lake. Some years 30% or more of the bioavailable nutrients reaching Flathead Lake come as dry and wet deposition on the lake surface. Smoke and fugitive road dust are the main sources. Historically, forest fires occurred only in late summer, but slash burning in harvest zones of the basin in winter and spring and home heating coupled with increasing vehicle use fills the valleys with smoke during inversions year-round. Unit area fallout is three times higher at Flathead Lake than in the mountains. The human population of the main valleys more than doubled in the last two decades and land conversion from forest or grassland occurred commensurately. Sewage treatment plants are state-of-the-art for the urban areas owing to documentation of water-quality decline associated with urban and exurban expansion. Of course, the largest input to the lake is natural catchment runoff in the nondeveloped areas of the basin. Seventy-five percent of the total N and P load reaching Flathead Lake occurs during the May to June spring freshet, and it settles rapidly to the lake bottom in association with fine sediments. By summer labile N and P are not detectable in the photic zone of the lake because uptake by the algae has exceeded supply. Nonetheless, primary production in the lake has steadily increased over the last three decades in relation to increasing inputs from human sources, based upon the continuous measures made at the Flathead Lake Biological Station. The problem likely would be much greater if most of the basin were not congressionally designated park or wilderness areas. Indeed, 85% of the water entering Flathead Lake annually comes from the high-mountain portions of the catchment, which is mostly protected and essentially pristine (Stanford and Ellis 2002).

Some other aspects of water pollution detract from the generality of high-quality water in the Flathead. Mercury is biomagnified in the Flathead Lake food web, resulting in levels of concern in lake trout tissues (Stafford et al. 2002). Sources are undocumented but could also involve aerosols from forest fires and slash burning. Nonpoint runoff produces cyanobacteria blooms in some lakes and reservoirs in the basin (e.g., Echo and Mary Ronan lakes, Crow Reservoir) and herbicides and pesticides are of localized concern (Beckwith 2002).

Hydroelectric production at the two big dams is done in ways that produce erratic and nonseasonal flows and lake levels that substantially compromise near-shore food-web structure and productivity and wash juvenile fishes out of nursery areas in the rivers (Stanford and Hauer 1992). Hungry Horse Dam on the South Fork was retrofitted in 1998 for multiple-level releases, allowing a normative temperature pattern downstream. Kerr Dam at the lake outlet always has been surface release and the historic temperature pattern of the lower Flathead River has changed little. Hence, temperature problems often associated with large storage dams in the western United States do not exist in the Flathead today.

Without question the introduction of nonnative biota is the most pervasive and persistent human impact. The historically very abundant cutthroat trout in the valley bottom lakes were compromised by introduction of kokanee salmon, which is a more efficient feeder on zooplankton, and by genetic introgression from planted rainbows that have persistently immigrated upstream. In a misguided effort to increase kokanee production in the valley bottom lakes, managers introduced *Mysis relicta*, a small shrimp that feeds on zooplankton at night and rests on the lake bottom during the day. The *Mysis* introduction backfired because kokanee and cutthroat feed only in daylight. Rapidly expanding *Mysis* populations destroyed robust kokanee fisheries in Flathead and other big lakes in a few years as the population of shrimp exploded to more than 130/m². Lake trout and lake whitefish stay in deep water and therefore immediately benefited from the mysids (the three species are native in Canadian Shield lakes). Native bull trout have declined concurrently with lake trout expansion (Stanford and Ellis 2002).

The floodplains of the Flathead are designated by conservation groups as "critical lands" because they are the primary protectorates of the native biota and clearly function as natural water-cleansing systems. However, protection is uncertain owing to management of flood threat in ways that are charitable to urban and exurban expansion and gravel mining. Protection of the river corridor in the valleys needs to be aggressively pursued along with control of

nutrient pollution if the Flathead is to remain one of the most pristine river–lake ecosystems in the world.

SNAKE/SALMON RIVER

The Snake River basin (~281,000 km²) represents almost 40% of the entire Columbia River basin area. It extends from the small subalpine lakes, meadows, and streams of the Yellowstone Plateau in northwest Wyoming to the arid lands of south-central Washington (Fig. 13.17). The river flows westward and then northward for over 1400 km before its confluence with the main-stem Columbia River. Along its path to the Columbia the Snake changes from a mountain stream as it departs Yellowstone National Park to a large impounded river with ocean shipping passing through locks on its way to and from Lewiston, Idaho.

The upper Snake contains the picturesque scenery of the Jackson Hole Valley and southern Yellowstone National Park and produces blue-ribbon fishing for native cutthroat trout in a few places. Likewise, the Salmon River, its largest tributary (by area), is famous for its wilderness beauty and wild salmon and steelhead fisheries (Fig. 13.4). Unfortunately, the once abundant wild chinook and sockeye salmon runs have declined to endangered status in the Snake basin and the contemporary steelhead runs are predominately hatchery stocks. The loss of wild stocks is due to a combination of habitat loss, river corridor discontinuities and blockages to migration by large dams, chronic overharvest, and supplementation by hatcheries.

The Snake and its tributary rivers possess a rich natural and cultural history. In 1805, the Lewis and Clark expedition crossed the Continental Divide at Lemhi Pass and descended to the Salmon River, where they were able to confirm their presence in the Columbia basin by the salmon spawning in the river. Along the Salmon, Clearwater, and lower Snake rivers, the "Corps of Discovery" encountered several Native American peoples with predominantly salmon-based economies, notably the Nez Perce and Walla Walla tribes.

Physiography, Climate, and Land Use

The upper Snake River begins in the Middle Rocky Mountain (MR) physiographic province of western Wyoming (see Fig. 13.17). This area has some of the most spectacular scenery in North America with many mountain ranges, broad intermontane valleys,

FIGURE 13.4   Salmon River below Redfish Creek, Idaho (PHOTO BY TIM PALMER).

and scenic lakes and rivers. Gannet Peak (4207 m), located on the Continental Divide in the northern Wind River Mountain Range in the Fitzpatrick and Bridger Wilderness Areas of west-central Wyoming, is the highest peak in the Snake River basin. The Snake River enters the Columbia–Snake River Plateaus (CU) province along the easternmost segment of the Snake River Plain, which extends in a crescent across southern Idaho (see Fig. 13.17). After winding through Hells Canyon on the Idaho–Oregon border the Snake River is joined first by the Salmon River draining the rugged Northern Rocky Mountains (NR) province of central Idaho from the southeast, then by the Grande Ronde River draining the Blue Mountain region of

the Columbia-Snake River Plateau from the west, and finally by the Clearwater River flowing from the Northern Rocky Mountains just north of the Salmon subbasin (Fig. 13.5). The remaining section of the lower Snake River flows through a portion of the Columbia–Snake River Plateau in southeastern Washington before joining the Columbia River.

The headwaters of the Snake River and much of the Salmon River drain the South Central Rockies terrestrial ecoregion (Ricketts et al. 1999). In this ecoregion the dominant vegetation type is coniferous forest, and the most common trees are Douglas fir, Englemann spruce, subalpine fir, and lodgepole pine. The Clearwater River drains the southern portion of

FIGURE 13.5   Snake River below confluence with Grande Ronde River, Washington (Photo by C. E. Cushing).

the adjoining North Central Rockies Forests ecoregion. Coniferous forest also characterizes this ecoregion, which also includes hemlocks, white spruce, alpine fir, and larch. The middle Snake River primarily flows through the Snake–Columbia Shrub Steppe ecoregion of the Columbia–Snake River Plateaus. The dominant vegetation here is sagebrush, with various wheatgrasses and bunchgrasses. As the Snake flows north into northeastern Oregon it enters the Blue Mountains Forests ecoregion, a relatively arid mountainous area characterized by sagebrush and various conifers, such as pinyon–juniper, and mountain grasslands. As the river leaves the Blue Moun-

tains it enters the Palouse Grasslands ecoregion, which has become an intensive agricultural region with only patches of natural grassland vegetation.

The climate of the Snake basin is variable owing to the rain shadow of the Cascades, which keeps the interior plain quite dry but increases precipitation in the mountains. Weather is mainly Pacific maritime, but cold, dry continental air masses occasionally predominate. Overall, annual precipitation is about 36 cm, with highest amounts occurring in the winter as snow (Fig. 13.18). Temperatures in Jackson Hole, Wyoming, near the headwaters, average between −9°C to 16°C over the year, with annual precipitation averaging 41 cm. Downstream, the Snake River plain is cool and dry; temperatures average –2°C to 21°C over the year in Twin Falls, Idaho, and annual precipitation averages 25 cm. The Salmon River subbasin of mountainous central Idaho is colder and wetter. The climate of the Salmon subbasin is typified at Galena, Idaho (elevation 2228 m asl), which has an annual mean temperature of 1.6°C, with average winter lows of –17°C and highs of 24°C in summer. Annual precipitation is 62.8 cm, with nearly 70% as snowfall from November through April. Further downstream at Lewiston, Idaho, temperature averages 11.1°C, typically ranging from 1°C to 21°C over the year, with annual precipitation of 32.5 cm.

Much of the mountain country of the Snake basin is designated wilderness: Sawtooth, Selway-Bitterroot, Frank Church, Gospel Hump, Hells Canyon, Teton, and Gros Ventre. Yet many of the valleys and hill slopes of the upper Snake/Salmon basin have a history of resource extraction, particularly timber harvest and mining on both private and public lands. The Snake River plain of southern Idaho is dominated by irrigated crop farming of potatoes, sugar beets, onions, cereal grains, and alfalfa. Approximately 10% to 15% of the Snake basin is forest/alpine, 50% scrub/rangeland, 30% agriculture, 4% barren, and 1% urban.

River Geomorphology, Hydrology, and Chemistry

The main-stem Snake River may be separated into three sections defined by geomorphic, hydrologic, and chemical characteristics: the headwaters (including the Palisades), the Snake River plain segment, and the lower river segment that extends upriver through Hells Canyon.

The headwaters section runs from Yellowstone National Park to the Snake River plain and includes the Henrys Fork draining the west slope of the Teton Range in Idaho. These rivers have a spring snowmelt hydrograph and bed sediments dominated by boulders, cobble, and well-sorted gravel. Confined reaches with gradients up to 10 to 50 m/km alternate with unconfined floodplain segments where the gradient is 1 to 5 m/km. River geomorphology is influenced by large wood eroded from hillslope and floodplain forests. Water chemistry of this section is characterized by high water clarity, particularly outside of the spring snowmelt discharge pulse. Mean conductivity (268 µS/cm), pH (7.9), alkalinity (75 mg/L as $CaCO_3$), and hardness (91 mg/L as $CaCO_3$) are typical of waters draining the Yellow-

stone Plateau. There are two dams with storage reservoirs in the headwaters of the Snake River in Wyoming, the relatively small Grassy Reservoir and the larger Jackson Lake Dam at the outlet to Jackson Lake. Approximately 100 km downriver of Jackson Lake the Snake River enters eastern Idaho, where Palisades Dam impounds the river. This is the first of a series of storage reservoirs feeding an elaborate diversion and canal system that irrigates much of the Snake River plain.

The Snake River plain segment begins at the confluence of the Snake and Henrys Fork in eastern Idaho. Approximately 60 km downriver of this confluence the Snake is impounded by American Falls and then Minidoka dams, the two largest impoundments in this segment. Irrigation diversions along the Snake River plain nearly dewater the river in late summer of dry years. Flow stabilizes in the Deer Flats area (Swan Falls Dam to Hells Canyon) owing to inputs from the Thousand Springs, various small tributaries, and irrigation returns. Throughout Deer Flats the river is constrained and riparian vegetation is limited to a fringe of willows along the shoreline of the river and numerous islands formed by the proglacial Lake Bonneville flood. The snowmelt hydrograph characteristic of the headwaters apparently was substantially moderated in this segment by natural storage of runoff within the huge areas of basalt of the Snake River plain and its fringe areas. In spite of substantial irrigation influences, average conductivity (336 µS/cm), pH (8.7), and alkalinity (75 mg/L as $CaCO_3$) values are lower than values at the upstream end of the segment, apparently due to ion retention in the reservoirs and limited dissolution from the basalt bedrock.

The lower section begins at the entrance to Hells Canyon. Hells Canyon is among the deepest river gorges on earth, as the river is 1600 to 2400 m below the canyon rim. Today, this magnificent canyon is impounded (Brownlee, Oxbow, and Hells Canyon dams) except for 60 km below Hells Canyon Dam. These dams, built between 1959 and 1967, are operated by Idaho Power Company and have a power generation nameplate capacity of 585 mW, 220 mW, and 450 mW, respectively. A short, constrained section below Hells Canyon Dam to the Salmon River is unimpounded, but the rest of the river is continually impounded to its confluence with the Columbia River by four dams: Lower Granite (1984), Little Goose (1970), Lower Monumental (1969), and Ice Harbor (1961). These are "run-of-the-river" hydroelectric and navigation dams and are equipped for fish passage.

Mean discharge of the Snake River below Ice Harbor Dam just before it enters the Columbia River is $1565\,m^3/s$. The Clearwater ($433\,m^3/s$) and the Salmon ($315\,m^3/s$) rivers together contribute almost half of the total flow of the Snake. In spite of extensive regulation and diversions all along its path, the Snake still has a seasonal pattern in mean monthly discharge and runoff (see Fig. 13.18). Runoff is usually highest from April to June, largely the result of melting of the snowpack. Runoff declines in late summer as precipitation falls and continues low into the winter as precipitation is stored in the mountains as snow. The dams undoubtedly have played a major role in reducing the variation in high and low discharge since the 1960s. Runoff is a higher fraction of precipitation than one might expect in a relatively arid region, but much of the flow is due to the accumulation of snow collected in mountainous areas with greater precipitation than in the Columbia–Snake River Plateaus.

River Biodiversity and Ecology

The Snake River flows through two freshwater ecoregions (Abell et al. 2000). The Upper Snake ecoregion is primarily that portion of the Snake River upstream of 65 m high Shoshone Falls, near Twin Falls, Idaho (see Fig. 13.17). Shoshone Falls once represented the upstream range of anadromous salmon before construction of Hells Canyon Dam. The remainder of the Snake River basin, including the Salmon River, is located in the Columbia Unglaciated ecoregion. Hauer et al. (2002) provide a comprehensive review of river ecology of the headwaters segment of the Snake, particularly as it relates to the legacy of flow regulation. The ecological characteristics of the Salmon River have been summarized by Minshall et al. (1992). A great deal of river and fisheries information and data collected by state, federal, and tribal agencies are available through STREAMNET (http://205.230.28.30/) and the Columbia Basin Fish and Wildlife Authority (www.cbfwa.org).

Algae, Cyanobacteria, and Protists

Minshall et al. (1992) reported attached algae, measured as chlorophyll, ranged from 21 to $179\,mg/m^2$ longitudinally on the Salmon River. The dominant species in the colder upper reaches were *Hydrurus*, *Gomphonema*, and *Ulothrix*. The filamentous green algae *Cladophora glomerata* occurred abundantly in association with warmer, shallow water and higher nutrient concentrations. Data are lacking, but this description probably characterizes the river system in general.

Plants

In broad valleys with low gradients, fine sediments dominate streams and rivers of the Snake basin. Where the water table tends to remain relatively close to the surface throughout the summer these low-gradient floodplain reaches possess riparian vegetation composed largely of willows and sedge grasses. A good example of this can be seen along the Lemhi River in the Salmon drainage, where the stream often meanders between the stabilizing structure of mature willows. In contrast, on higher-gradient floodplains in reaches with high discharge and expansive gravel–cobble bed sediments, the water table is seasonally dynamic and the scoured broad river channel is inundated annually with the snowmelt. Here, the riparian vegetation is dominated by black cottonwood or narrow leaf cottonwood (Hauer et al. 2002). Excellent examples of this can be seen along the upper reaches of the Snake River, above and below Jackson Lake in Wyoming, along the upper Snake River in Idaho, and on the Salmon River in the floodplain reaches between Challis and Salmon, Idaho. In the many canyon reaches of the Snake and Salmon rivers only a narrow ribbon of willow and alder fringe the shoreline as a boundary between the river and Douglas fir forests in montane landscapes or Ponderosa pine along the steppe of the foothills.

Dams, diversions, and levee systems have had a profound effect on the natural vegetation along significant portions of the Snake River. For example, below Palisades Dam on the upper Snake, reduction in peak flows since the mid 1950s has reduced scour required for cottonwood regeneration. Although the historical gallery forest remains intact, a demographic shift toward old (>150 years) senescent trees has occurred (Merigliano 1996). Ute lady tresses, a white-flowered orchid, occurs in wetlands of abandoned channels in association with willows, spike-rushes, sedges, and horsetails (Moseley 1998). Even in the canyon-bound reaches, the effect of regulation is profound. For example, the island-based waterfowl nesting habitat of the Deer Flats National Wildlife Refuge are progressively connecting to the mainland by establishment of native and nonnative hydrophytes, particularly purple loosestrife, due to lack of flood scour and inundation (Johnson et al. 1995).

Invertebrates

The cold gravel and cobble-bed habitat of the unaltered reaches of the Snake system are ideal habitat for stoneflies (e.g., predatory *Claassenia sabulosa*, *Hesperoperla pacificum*, and detritivorous *Pteronarcys californica*), caddisflies (e.g., *Arctopsyche grandis*, *Hydropsyche cockerelli*, *Hydropsyche occidentalis*), and ephemerellid, heptageniid, and drunellid mayflies that occur in species-rich assemblages (e.g., Clearwater River; Munn and Brusven 1991). Minshall et al. (1992) found macroinvertebrate density along the Salmon River from headwater reaches to the large main-stem river decreased in density from 2550 to 7525/m^2 at headwater sites in the Stanley subbasin and 400 to 3700/m^2 not far from the mouth. Shredders were prevalent in the headwater reaches where riparian vegetation is abundant relative to stream size, but were rare in the large canyon-bound segments downstream. The collector-gatherers were increasingly abundant, coherent with increasing size of the river.

On the upper Snake River, particularly the Henrys Fork and the South Fork of the Snake, abundant resident fisheries apparently are maintained by relatively high but unmeasured secondary production of benthic invertebrates. These reaches are particularly well known by fly fishers for their emergence of salmon flies (*Pteronarcys californica*), ephemerellid mayflies (*Ephemerella inermis*), and several species of caddisflies (*Hydropsyche* spp.). However, production is compromised by flow regulation below the big dams (Munn and Brusven 1987, 1991). On the lower Snake River, where slackwater conditions predominate, the benthic invertebrates are dominated by chironomid midges.

Four snails (Bruneau Hot Springsnail, Utah valvata snail, Snake River physa, Bliss Rapids snail), listed as either rare or endangered, are found in nearshore habitat of the bedrock rapids of the Thousand Springs–Shoshone Falls area of the river, where flow regulation and water abstraction are problematic for their survival. Gravel bars in the Deer Flats segment are littered with native freshwater clam shells; however, status of the population is not documented (U.S. Fish and Wildlife Service, unpublished data).

Vertebrates

The Snake system once was a wild salmon factory, with robust runs of chinook, coho, chum, sockeye, and steelhead. Sockeye were able to travel over 1500 km up the Snake/Salmon to spawn and rear in small lakes of the Stanley subbasin. Today, only a few wild chinook and steelhead stocks remain and those are limited by dams to the Grande Ronde, Salmon, and Clearwater rivers. The last runs, near extinction, are propped up by hatchery operations. Recovery plans suggest the lower Snake impoundments are particularly problematic for outmigrating salmon and steelhead because the reservoirs are too warm and full of predatory fishes, such as pikeminnow and nonnative smallmouth bass. Removal of these dams may be the only salvation for Snake River anadromous fisheries (Williams 2005).

Several important resident salmonids exist in the upper subbasins above the natural barriers to anadromous runs. Yellowstone cutthroat and Snake River finespotted cutthroat trout are sympatric (Behnke 2002) and remain in relatively high populations in portions of the Snake River above Shoshone Falls in Idaho and Wyoming. Cutthroats are locally migratory and probably require large reaches of floodplain rivers to be sustained. The future of cutthroats may depend upon on naturalization of flows in the upper subbasin, especially below Jackson Lake and Palisades Reservoir (Hauer et al. 2002).

Bull trout are an integral part of the fish communities of the Snake River and its tributaries below Shoshone Falls. Many populations are isolated by the discontinuities and barriers to movement associated with dams. For example, adfluvial bull trout achieve maturity in CJ Strike Reservoir, where the Bruneau River joins the Snake, and migrate into the smaller tributaries, such as the Jarbidge River, to spawn.

Native redband trout, closely related to steelhead and possibly a subspecies, historically existed throughout the Snake basin below Shoshone Falls, but pure populations are rare owing to extensive introduction of rainbow trout stocked from various hatcheries (Behnke 2002). Brown, brook, and lake trout, smallmouth bass, and many other nonnative fishes also have been introduced throughout the basin. The reported composition of fishery uses in the upper Snake River, based on creel census, was 68% naturally spawning cutthroat, 4% hatchery cutthroat, 9% brown trout, 2% spawning rainbow, 2% hatchery rainbow, 0.5% lake trout, and 14% whitefish (Moore 1980).

Huge-bodied sturgeon historically were found in the deep channels of the Snake downstream from Shoshone Falls. Populations continue to survive below Swan Falls Dam upstream of Deer Flats, in the deep channels of the Snake below Hells Canyon Dam, and in the lower Salmon and Clearwater rivers. However, these populations are no longer migratory due to dams and their inability to ascend the fish ladders. Natural reproduction has been essentially eliminated by regulated flows; thus, they persist in the wild mainly owing to their longevity. Many individuals are well over 50 years old. The species now is commercially produced in springs and ponds in the area of Hagerman, Idaho (Williams 2005).

Montane riparian areas of the Snake and Salmon occupy a relatively small portion of the total landscape, yet they are naturally heterogeneous and function as centers of high diversity and abundance of birds. The Snake River corridor and riparian areas are well-known sites for raptor nesting. Nineteen raptorial species were documented within a study area along the Snake River in Idaho, and seven more were suspected as periodic residents (Whitfield et al. 1995). A study by Saab (1999) on the Snake River illustrated that the best predictors of high bird species richness were natural and heterogeneous landscapes, large cottonwood patches, close proximity to other cottonwood patches, and microhabitats with open canopies. The most frequent significant predictor of bird species occurrence was landscape coverage. Habitat use increased with natural upland landscapes and decreased with agriculture (Saab 1998). Both interior and edge specialists were found in linear cottonwood forests with large amounts of edge. Nest predators, brood parasites, and nonnative species correlated with human-altered landscapes.

Ecosystem Processes

Comprehensive studies of organic matter production and flux were conducted on the Salmon River (Minshall et al. 1983, Minshall et al. 1992). The riverine food web is complex and allocation of organic matter from riparian and instream sources varied longitudinally with river size and slope. Instream primary production, as opposed to inputs of riparian organic matter, was the source of the majority of organic matter production. However, the conduitlike nature of the canyon-bound river resulted in very little organic matter retention. The food web of the system was predominately driven by benthic feeding on attached algae or on ultrafine particles of organic matter transported by river flow. Inputs of leaves from riparian trees was greatest (340 to 420 g/m^2) in headwater streams and generally decreased downstream (24 to 90 g/m^2) and correlated with the increasing width of the river. However, Hauer et al. (2002) noted the likelihood of considerable variability associated with streamside vegetation, floodplain width, and other geomorphic characteristics.

Human Impacts and Special Features

The Snake River headwaters in Wyoming is the only section of this long river that is in relatively pristine condition, but it is far removed and isolated from the lower river by dams and natural barriers (e.g., Shoshone Falls). The free-flowing Salmon and Clearwater rivers (except the Clearwater's north fork), on the other hand, are premier white-water rivers, remain in relatively pristine condition, and have the highest potential for salvation of Snake River salmonid fisheries. The Salmon River is one of the longest free-flowing rivers in the lower 48 states, with a large section of the main stem and its Middle Fork designated as National Wild and Scenic Rivers. The Middle Fork of the Clearwater River, as well as two major tributaries (Lochsa and Selway rivers), are also Wild and Scenic Rivers. The Salmon drains the extremely rugged Frank Church River-of-No-Return Wilderness Area, and much of the Clearwater drainage is in the Selway-Bitterroot Wilderness Area.

No large municipalities exist in the upper Snake or Salmon river drainages, nor are there major industries with point sources of pollution. Nonetheless, isolated sites of pollution from mining and broadscale impacts from cattle grazing and irrigation withdrawal and return flows are documented. Although the relatively pristine character of the Snake and Salmon headwaters are similar in origin and character, these rivers are dramatically different with respect to the dominant human impact of river regulation by large dams. The Snake River proper is regulated by dams and diversions almost immediately after its exit from Yellowstone National Park and sequentially all the way to the Columbia River. Hydrologic modification has reduced the ecological integrity of the river in various ways, most significantly by substantial reduction in riparian and river productivity across all trophic levels. Historically the Owyhee, Boise, and Payette rivers had major runs of salmon, as did the main-stem Snake all the way to Shoshone Falls. However, this highly productive salmon fishery has been completely lost due to dam construction in Hells Canyon.

The lower Snake, Lower Granite, Little Goose, Lower Monumental, and Ice Harbor dams and their associated reservoirs cause very high mortality of outmigrating anadromous salmon smolts from the Salmon and Clearwater rivers. In an effort to reduce mortality, for two decades the U.S. Army Corps of Engineers has captured many of the fish at Lower Granite dam, put them in special barges, and transported them downstream to Bonneville Dam, where

FIGURE 13.6   Yakima River in Yakima Canyon below Ellensburg, Washington (PHOTO BY TIM PALMER).

they are released to continue on to the ocean. Unfortunately, this massively expensive operation is of little or no avail to Snake River fisheries (Williams 2005).

YAKIMA RIVER

The Yakima is a 5th order river located in south-central Washington and has a drainage basin of 15,900 km². It flows 344 km from the very wet Cascade Range southeast into the very dry interior basin of the Columbia River and is located between latitude 46.1°N to 47.4°N and longitude 119.2°W to 121.3°W (Fig. 13.19). The river has steep headwater tributaries that feed three large glacial lakes. Downstream the river flows through broad valleys separated by canyons cut through lateral basalt (lava) formations (Fig. 13.6). The valleys have expansive alluvial floodplains with cottonwood gallery forests transitioning into prairie steppe on the uplands. The lower river flows through a long canyon segment to its confluence with the Columbia at Richland, Washington.

Native Americans in tribes and bands that later composed the Yakama Nation and others lived in these valleys in large villages owing to the natural goods and services provided by the river and its floodplains. Ross (1855) described herds of horses totaling several thousand along with several hundred tepees in the Kittatas Valley (Ellensburg area). The largest Indian village, estimated at around 3000 individuals, was located near the current city of Yakima, where a large river floodplain exists. The ethnographic and archaeological evidence indicates salmon, particularly chinook because of their ability to be cached, provided a large portion of the food and trade (Uebelacker 1986).

Missionaries arrived in 1848 and the Yakama Indian reservation was created in 1855 to make farmers out of the Indians. White and Hispanic immigrants settled throughout the valleys and increasingly abstracted flow for irrigation of crops. By 1902, 121,000 acres of reservation and other lands were irrigated and diversions dewatered the river on dry years. Thus, in 1905, a huge federal water storage and irrigation canal project was initiated that today drives a robust agricultural economy. As throughout the Columbia basin, these cultural changes have resulted in water pollution and vastly reduced salmon production.

Physiography, Climate, and Land Use

The western part of the Yakima basin in the Cascade–Sierra Mountain (CS) physiographic province drains the eastern Cascade Mountains, which are uplifted Eocene sandstone, shale, and some coal layers, with pre-Miocene volcanic, intrusive, and metamorphic formations (see Fig. 13.19). This part of the basin intersects with the Walla Walla Plateau section of the Columbia–Snake River Plateaus (CU) physiographic province and is characterized by numerous Tertiary lava flows over continental formations. In the Yakima basin, the Plateau deformed into a series of southeast-trending anticlinal ridges and synclinal valleys through which the Yakima River down-cut, creating the upper Yakima Canyon, Selah and Union gaps, and lower Yakima Canyon, with the broad floodplain valleys in between (see Fig. 13.6) (Kinnison and Sceva 1963).

The climate is extremely variable, ranging from maritime in the heavy snow belt of the Cascades to arid in the interior Columbia basin. Annual precipitation ranges from 350 cm to approximately 18 cm (Rinella et al. 1992). Precipitation in the lower river basin occurs primarily in late fall and early winter (>2 cm/mo) but declines to less than 1 cm/mo during summer (Fig. 13.20). Clearly, the monthly precipitation shown in the figure underestimates precipitation for the entire basin, as 85% of the annual discharge comes from the Cascade headwaters. Mean monthly temperatures in the lower basin range from 1°C in midwinter to 24°C in July.

Elevation varies from around 2500 m asl in the Cascades to 98 m asl at the Columbia River confluence. The higher areas located in the Eastern Cascades Forests terrestrial ecoregion receive the most precipitation and are thickly covered with Douglas fir, larch, lodgepole, and ponderosa pine. The conifer zone grades into the Snake/Columbia Shrub Steppe ecoregion, which is dominated by Idaho fescue and other grasses plus sage, bitter brush, and occasional stands of ponderosa in the foothills and valley flanks with floodplain cottonwood forests along the river (Franklin and Dyrness 1973). The boundary between these two ecoregions is visible along Bristol Canyon, northwest from Ellensburg along Highway 10, where the vegetation changes from grand and Douglas fir to shrub steppe. On the east side of the Cascades the northernmost distribution of Oregon white oak is found at the confluence of the Swuak and Yakima rivers north of Ellensburg.

Within the Yakima basin about 36% is forested, 47% is shrub, and 16% has been converted to agriculture, primarily in the Walla Walla Plateau. Irrigation serves over 200,000 ha, ranking the Yakima basin among the more productive agricultural areas in the United States. Apples, cherries, peaches, grapes, and other fruits are grown along with hops, mint, asparagus, alfalfa, and timothy. Yakima County recently ranked 5th in the country for annual agricultural revenue. Viniculture is a growing industry. The foothills of the basin are range cattle country and enough timber is harvested annually from extensive federal and private holdings in the mountains to sustain several processing mills. The two main cities are Yakima and Ellensburg, but considerable exurban fragmentation of farm and ranch land is occurring in the upper basin due to proximity to the Seattle metroplex. About 500,000 people live in the basin, but only about 1% of the land is considered urban.

River Geomorphology, Hydrology, and Chemistry

The river may be divided into five geomorphic provinces: step-pool headwaters, glacial lake district, braided alluvial valleys (Cle Elum-Teanaway, Kittatas, Selah, Naches, Yakima-Union Gap, and Wapato), meandering alluvial valley (Lower Wapato), and main-stem canyons (upper and lower Yakima canyons). Riverbed sediments of the expansive floodplains are coarse sand, gravel, and cobble, except in the meander segment above the lower canyon, where sands and silts are more prevalent. Glacial ice scour advanced only as far as the upper Yakima Canyon, but outwash from Cascade glaciers and Lake Missoula flood deposits are extensive throughout the basin; for example, in the meander reach. Storage for irrigation comes from two reservoirs on the main tributary, the Naches River, and from elevation and regulation of the three glacial lakes on the upper Yakima. Together they capture approximately one-third of the annual basinwide runoff (1.23 billion m³). A key point is that storage is insufficient to prevent scouring floods from occurring on about a five-year return interval, which in part explains the presence of existing complex floodplain habitat, especially at the lower ends of the major floodplains (Snyder and Stanford 2001).

Mean discharge is quite high (102 m³/s) for the Yakima River, given its relatively small basin area (<16,000 km²). Peak flow in the lower basin historically occurred from April through June from mountain snowmelt, adding substantially to a relatively high annual runoff of about 20 cm/yr (see Fig. 13.20). In contrast, the lower basin, located in the Snake/Columbia Shrub Steppe ecoregion, receives relatively little precipitation (see Fig. 13.20) and is obviously insufficient to account for the observed runoff. Fall floods of short duration and high peak discharge generated by rain events also distinguish the discharge records (see Fig. 13.20). Flooding recharges the alluvial aquifers of the floodplains and promotes ground- and surface-water exchange and associated habitat mosaics. Much of the historic base flow during late summer and winter likely was sustained by discharge from the alluvial aquifers, which also would have cooled the river and prevented anchor ice in winter. On several of the floodplains large drainage ditches were constructed to drain aquifers to prevent "souring" of soils for irrigated crops. This, of course, destroyed the natural flow and temperature buffering of the aquifer system.

Stream water in the basin is neutral to slightly alkaline and specific conductance increases longitudinally from ca 75 µS/cm in the headwaters to greater than 300 µS/cm in the lowlands (Carter et al. 1996). Similarly, the quantity of nutrients (nitrogen and phosphorus) increases longitudinally as a response to increasing anthropogenic alteration of the watershed (urban runoff, agricultural return flows, etc.; Cuffney et al. 1997). For example, total phosphorus increases from 0.04 mg/L to >0.1 mg/L and NO_3-N + NO_2-N increases from 0.13 mg/L to >1.0 mg/L in the mountains and agricultural drains, respectively (Rinella et al. 1992). Groundwater pollution by fertilizers and pesticides contributes substantially to the river pollution load. Nutrient enrichment and biomagnification of toxic chemicals are primary problems and warnings against human consumption of fishes have been issued in recent years. However, fish kills and other extreme pollution problems are not reported and biotic productivity in the river is high (Snyder and Stanford 2001).

River Biodiversity and Ecology

The Yakima River is located in the southwestern corner of the Columbia Glaciated freshwater ecoregion (Abell et al. 2000). The Yakima River ecological literature is relatively rich, although mostly unpublished (see review by Snyder and Stanford 2001). A literature archive is available at the Flathead Lake Biological Station (http://www.umt.edu/flbs).

Algae and Cyanobacteria

Throughout its length the Yakima has a rich periphyton assemblage of diatoms, with areas of attached algae, mainly *Cladophora* spp. and *Stigeoclonium* spp. In synoptic sampling along the river continuum during the fall of 1989, Leland (1995) found a total of 132 algal taxa, 119 of which were diatoms. Dominant diatom taxa included *Achnanthes lanceolata*, *A. minutissima*, *Cocconeis placentula*, *Cymbella minuta*, *Diatoma vulgare*, *Navicula radiosa* var. *tenella*, *N. dissipata*, *N. frustulum*, *N. frustulum* var. *perpusilla*, *N. palea*, and *Synedra ulna*. The Cyanobacteria *Oscillatoria* sp. and *Schizothrix calcicola* were also quite abundant. Taxa were grouped according to physiographic changes from headwaters to mouth, including the density and composition of riparian vegetation, land use, and surficial geology.

Plants

The floodplains of the Yakima River are a dominant ecological feature of the basin that is structured in large part by the dominant woody taxa, mainly black cottonwood and narrowleaf cottonwood, along with several species of willow and alder (Franklin and Dyrness 1973, Guard 1995). Dunlap and Stettler (1996) described an abrupt change in black cottonwood phenotype in which populations in the xeric lower river had less stem volume, later spring flush, and earlier autumn budset and leaf fall relative to the more mesic upper river (upstream of the Teanaway River). They concluded that moisture availability was a key selective agent. This conclusion was supported by genetic analyses conducted by Reed (1995), who documented genetic differences between mesic and xeric stands. The abundant fringing wetlands and marshes are also loaded with aquatic plants, particularly arrowroot or wapato.

Nonnative grass species included the perennial reed canary grass, which is broadly distributed and more abundant throughout the river corridor than native annual grasses. In addition, purple loosestrife, silver maple, Russian olive, and European willows are commonly found, particularly along the lower reaches of the river. Braatne and Jamieson (2001) observed that younger age classes of native willow and cottonwood were largely absent and attribute this to alteration of the natural flow regime.

The physiography of the basin is such that extensive riparian galleries have developed on the broader floodplains, whereas the more constricted zones in the canyons are dominated by mixed stands of cottonwoods and ponderosa pine. The processes of cut-and-fill alluviation have greatly enhanced structural complexity of the broad floodplains, leading to a high abundance of off-channel habitat with stagnant water where hydrophytes are common.

Invertebrates

Over 300 species of macroinvertebrates occur basinwide, with site-specific richness ranging from 20 to 69 species and mean densities ranging from <1000 to >2100 individuals/m^2 from headwaters to lower river reaches. Taxa richness in general was negatively related to the intensity of agricultural activity and positively related to canopy closure. Species found exclusively in the headwaters included the stoneflies *Zapada columbiana* and *Z. fridida* and the caddisflies *Parapsyche elsis*, *Rhyacophila* nr *blarina* Ross, *R. vocala*, and *R. valuma*. Headwater (1009 m asl) dominants included the flatworm *Polycelis coronata* and the ice worm *Mesenchytraeus* sp., as well as the mayflies *Baetis tricaudatus*, *Rhithrogena* nr *robusta*, *Drunella doddsi*, *Cinygmula* sp., and nr *Doddsia occidentalis*. Downstream (781 m asl) the community was dominated by some of the same taxa (*Mesenchytraeus* sp., *Cinygmula* sp., and *B. tricaudatus*) in addition to the mayfly *Diphetor hageni* and the stoneflies *Zapada cinctipes* and *Sweltsa* sp. At lower elevations (mean altitude 669 m) some of the same taxa again dominated the community (*Mesenchytraeus* sp., *B. tricaudatus*, *Cinygmula* sp., *Z. cinctipes*, and *Sweltsa* sp.) in addition to an increase in abundance of *Nais behningi*, an aquatic earthworm, the mayfly *Ephemerella* nr. *infrequens*, and the dipteran *Antocha* sp. At 656 m the caddisfly *Cheumatopsyche* sp., *Aulodrilus pluriseta* (a tubificid), and the chironomid *Eukiefferiella claripennis* were most abundant, whereas most of the upstream species were no longer present. Finally, within the Wapato floodplain segment, where seasonal temperature variation and pollution concentrations were maximized, the mayflies *Baetis tricaudatus* and *Ephemerella* nr *infrequens*, the caddisfly *Hydropsyche* nr *californica*, as well as the chironomids *Cricotopus trifascia* gr. sp., *Polypedilum* nr *convictum*, *Eukiefferiella claripennis* gr., and *Thienemanniella* sp. and the planarian *Dugesia* sp. were most abundant (Carter et al. 1996, Cuffney et al. 1997).

Snyder et al. (2003) found the amphibitic stoneflies *Kathroperla perdita* and *Paraperla* spp. in the alluvial aquifers of the upper Yakima. As in the Flathead River, these genera are indicators of hydrologic connectivity between the floodplain and the river, but they were not present in the aquifers most influenced by pollution and draining. They also described the longitudinal distribution of *Paraperla* spp. and surficial groundwater crustaceans. The distribution and abundance of these hyporheic invertebrates decreased from upstream to downstream; 88% of the variation was explained by ambient concentrations of NO_3-N + NO_2-N.

The silver-bordered bog fritillary is a rare colonial butterfly that pollinates the equally rare northern bog violet (Pyle 1974, 1992). Within the Yakima basin this interesting association is found in only one place within a fringing floodplain bog.

Of the six species of freshwater mussels reported in Washington and Oregon, four have been recommended for listing under various categories of ESA (Frest and Johannes 1995). These include western pearlshell, western ridge mussel, Willamette floater, and California floater. The other two are Oregon floater and western floater. Extensive mussel beds, probably western pearlshell, were observed in a spring brook side-channel complex in the upper Yakima, apparently owing to the stable flow of the spring brook (E. Snyder, personal observation). No other beds have been reported in the basin, although they apparently were widely distributed historically.

Vertebrates

At least 50 species of fishes exist in the Yakima basin, of which 12 are nonnative and most abundant in the warmer reaches of the lower river. The more abundant resident species are (E, nonnative; N, native) rainbow (E), brook (E), and cutthroat (N) trout, mountain whitefish (N), sculpins (N), largescale (N) and bridgelip (N) suckers, northern pikeminnow (N), speckled (N) and longnose (N) dace, smallmouth bass (E), black crappie (E), channel catfish (E), and yellow perch (E). Historically, the Yakima basin sustained four anadromous (chinook, coho, steelhead, and sockeye) and five resident salmonid species (bull charr, rainbow, kokanee and cutthroat trout, and mountain whitefish), with numerous stocks or runs associated with the great habitat heterogeneity of the basin. For example, the headwater glacial lakes supported robust runs of sockeye salmon that coexisted with all of the other salmonids, except possibly coho, along with burbot and several other native fishes. Today, all of the native salmonids and some of the other native fishes have declined to a few stocks in critical condition or have been extirpated.

Chinook salmon are grouped as spring, summer, and fall stocks, corresponding to the time when the adults return from the ocean. Spring and fall chinook can be divided into three and two stocks, respectively.

Of the latter, the more abundant stock, representing approximately 70% of the total run, is derived from the Hanford Reach of the Columbia River (Busack et al. 1991). The second stock spawns in a large drainage ditch (Marion Drain) in the Wapato floodplain and is genetically similar to stocks found in the Snake and Deschutes rivers. It is believed that this population at least partially represents the original Yakima River stock, although some genetic mixing has likely occurred with hatchery fish from the Hanford Reach (Busack et al. 1991).

Coho, locally extirpated many years ago, have been reestablished in the Yakima from hatchery releases. Sockeye runs of 200,000 or more existed in the glacial lakes of the upper basin, but they also were extirpated by the dams constructed on the outlets of the glacial lakes to allow storage of irrigation water.

Steelhead were once ubiquitous throughout the basin, occurring in all reaches that supported spring chinook as well as numerous tributaries. Because of declining runs, steelhead were listed as endangered in 1997. Genetic analysis indicated that three reproductively isolated populations still exist in two tributaries on the Yakima Indian reservation and the Naches–upper Yakima (Busack et al. 1991). Resident rainbow trout from a long legacy of hatchery stocking are common and self-reproducing in the river and a popular sport fishery exists in the Kittatas Valley and Yakima Canyons.

Native bull trout likely were distributed throughout the basin, but today they exist only in isolated headwaters (Craig 1997). A few have occasionally been observed in the upper reaches of the Yakima and Naches rivers.

Amphibians found along the river include the northern leopard frog, Columbia spotted frog, tailed frog, and the western toad, all of which have experienced declines due to numerous factors, including habitat destruction, sedimentation, altered temperature and flow (Nordstrom and Milner 1997), and predation by nonnative bullfrogs. The rare Larch Mountain salamander also has been reported. Reptiles found along the river corridor include garter snakes and the Western pond turtle, which is listed by the state of Washington as endangered.

The Yakima floodplains historically contained most of the Columbia basin's wintering ducks and geese (from 250,000 to 300,000; Oliver 1983). Species include the mallard, pintail, green-winged teal, cinnamon teal, blue-winged teal, wood duck (largest population in eastern Washington), shoveler, redhead, ruddy duck, ring-necked duck, Canada goose, white-fronted goose, tundra swan, and trumpeter swan (Parker 1989). Since the 1970s numbers have declined (to between 30,000 and 40,000) as the population has shifted from the lower Yakima basin to the lower Columbia River (Lloyd et al. 1983). The change is attributed to increased surface water due to hydroelectric development on the lower Columbia River, changes in cropland patterns, and improved refuge conditions (Thompson et al. 1988). Bald eagle, osprey, and sandhill crane nest and feed in the floodplains. The floodplains and prairie also supported large herds of elk and deer.

Ecosystem Processes

Recent monitoring of shallow alluvial wells indicated that in the lateral dimension, key nutrients (various forms of nitrogen and phosphorus) and dissolved organic carbon tend to increase from the main stem to spring brooks to floodplain (Stanford et al. 2002). In addition, nutrients increased longitudinally and were quite variable through time, as predicted based on the fluctuating hydrologic regime. This pattern was mirrored by a general increase in chlorophyll-*a* concentrations from upstream to downstream. Pigment concentrations ranged from 0.5 to $27 \mu g/cm^2$, and organic content ranged from 0.4 to $6 \, mg \, AFDM/cm^2$. Although primary productivity and organic matter spiraling were not measured, the pattern and concentrations of chlorophyll suggest that allochthonous sources of energy would likely dominate in the upper reaches, with a gradual shift to autochthonous energy supply in the lower reaches. This pattern is likely enhanced by anthropogenic nutrification of the lower reaches. Additional patterns in chlorophyll *a* and AFDM suggested off-channel habitats, such as spring brook and side-channel complexes, maintained higher pigment concentrations and biomass, particularly in the upper reaches. These patterns were correlated with heavy use of spring brooks and side channels by juvenile salmonids (coho and chinook).

These fringing floodplain features likely represent areas of refuge from native and nonnative fish predators. For example, McMichael et al. (1998) found that predation rates during downstream smolt migration were particularly substantial for fall chinook. The three main predators included the indigenous northern pikeminnow and two nonnative piscivorous species, the smallmouth bass and channel catfish. The abundance of nonnative species increases downstream as a function of accumulating anthropogenic impacts, as well as proximity to a source pool (namely the Columbia River main stem).

Salmonids likely are not food limited, because estimates of macroinvertebrate biomass ranged from 6.4 to 12.6 g/m^2 in samples from the main channel in the floodplain reaches (Nightengale 1998). However, juvenile salmonids favor shallow edge habitat where benthic biomass is reduced or missing entirely due to flow fluctuations associated with irrigation transfers and diversions. Moreover, sudden changes in velocity caused by irrigation transfers may wash juvenile fishes downstream or laterally into unproductive zones (Snyder and Stanford 2001).

Human Impacts and Special Features

The expansive floodplains and headwater glacial lakes are unique features of the Yakima. The floodplains are more extensive and habitat rich than generally exists elsewhere in the Columbia basin and few large lakes exist within the natural range of anadromous fishes. In spite of considerable modification by human activities, much of the system retains habitat complexity and connectivity. This is mainly because water storage in the basin is not sufficient to prevent big floods (~25-year return events). Thus, critical habitat-forming processes, namely cut-and-fill alluviation, still occur sufficiently to create and maintain floodplain habitat. The shifting habitat mosaic crucial to salmon production is intact to some extent throughout the river corridor.

Human impacts include water abstraction, flow regulation, pollution, nonnative species, and habitat alteration, especially in the floodplain segments, all of which have compromised the ecological integrity of the Yakima River. For example, 66% of the total historical area of the five floodplains has been functionally disconnected from the river by revetments and human structural encroachment (Eitemiller et al. 2002, Snyder et al. 2003). Water abstraction for summer irrigation has left the river dry some years and substantially below preregulation norm most years in the Wapato segment downstream of the Sunnyside diversion. In reference to the Federal Clean Water Act (sec. 303d), 72 stream and river segments are listed as impaired by the Washington Department of Ecology and 83% exceed temperature standards. For example, Lilga (1998) found temperatures in the lower river from June through November (1996) that were lethal (>15.6°C) for salmon egg and fry incubation between 60% and 85% of the time. Temperatures are stressful for juveniles (>18.3°C) between 25% and 65% of the time and stressful for adults (>15.6°C) between 60% and 85% of the time. Standards set for DDT, PCB, and other toxic chemicals were exceeded in 15% of the listed (303d) reaches. In addition, high concentrations of DDT (and its breakdown products DDE and DDD) were found in fish tissue from the lower river; these concentrations are among the highest recorded in the United States (Rinella et al. 1993).

The main problem in the Yakima River, however, is artificial manipulation of base flows to facilitate irrigation. Conservation and restoration of the floodplain segments through implementation of a basinwide naturalized flow regime is a crucial next step in the recovery of anadromous salmon and steelhead. However, the only effective ways to do this are to substantially reduce water abstraction, which will compromise agriculture production, or import water from outside the basin (Stanford et al. 2002). Whited et al. (2003) demonstrated that the ecological condition of the lower river could be improved by diversion of irrigation water from the Columbia via a process called pump-exchange, allowing Yakima water to stay in channel. Similar measures could be implemented for most of the rest of the Yakima, albeit at great expense, because of the proximity and similar channel elevation of the Columbia (U.S. Bureau of Reclamation, unpublished data).

Returning the Yakima to a free-flowing condition would be a hallmark of river restoration experiments. The idea is substantially enhanced by the fact that the Yakima flows into the only free-flowing part of the Columbia main stem (Hanford reach), which also happens to sustain the most robust salmon population left in the entire interior Columbia basin. Uncertainties beyond affordability include whether such a large-scale restoration effort can forestall inexorable floodplain encroachment and at the same time overcome documented negative influences of hatchery operations, outmigration mortality in the main-stem Columbia, variation in subscribed harvest, and human-mediated variation in oceanic productivity, among other internal and external impediments (cf. Williams 2005).

WILLAMETTE RIVER

The Willamette River is a 7th order river (1:100,000 scale) in the western edge of the Columbia River basin in western Oregon (Fig. 13.21). The Willamette River runs north for more than 230 km through the heart of the Willamette Valley and enters the Columbia River just downstream of Portland, Oregon. The Willamette River basin is the 13th largest river in the contiguous 48 states and produces more water per land area than any of the larger rivers of the United States (Kammerer 1990). Its total annual discharge makes up 12% to 15% of the total flow of the Columbia River (Hulse et al. 2002). The Willamette River basin is 290 km long, 161 km wide, and 29,728 km² in area, bounded on the west by the Coast Range and on the east by the Cascade Mountain Range. This basin, centered at 45°N latitude and 123°W longitude, makes up 12% of the land area of Oregon and contains 68% of the state's population. The three largest cities in Oregon—Portland, Salem, and Eugene—are located on the banks of the Willamette River. The water quality is good in the headwater streams and moderately good in the mainstem river. Fishing is a major recreational use in the river and its tributaries, especially for Pacific salmon, steelhead, nonnative trout, and sturgeon. Wildlife is abundant along the rivers of the basin, with frequent evidence of the state's icon, the beaver.

Native Americans have occupied the Willamette Valley for more than 10,000 years. Lewis and Clark actually missed the mouth of the Willamette River on their expedition down the Columbia River and also on the return trip upriver. After a tribe on the Sandy River gave him directions, William Clark went back downstream and entered the mouth of the Willamette River in April 1806. Euro-American settlement began with fur trading in the early 1800s and continued, with the first settlement in Oregon on the floodplain of the Willamette in 1829. Agriculture, forestry, and commercial fishing became established after 1840. The Willamette Valley was a mosaic of wet and dry prairie, oak savannas maintained by Native American burning, and floodplain forests, creating many obstacles for transportation. Riverboats on the Willamette River served as the first major method for transporting people and materials along the valley floor.

Physiography, Climate, and Land Use

The Willamette River basin includes three major types of geological parent material. Within the Pacific Border (PB) physiographic province on the western side of the basin (see Fig. 13.21) are marine volcanic and sedimentary rocks in the Oregon Coast Range section, alluvial deposits in the Willamette Valley, and volcanic basalts and andesite that occupy the southern end of the Puget Trough section. Volcanic basalts and andesite also occupy the Middle Cascade Mountains section of the Cascade–Sierra Mountains (CS) physiographic province on the eastern side of the basin. These geological surfaces were formed by

the combined effects of tectonics and volcanism over the last 30 million years and by recent deposits of glacial floods between 15,000 to 13,000 years ago. Roughly 35 million years ago (mya), an ocean slab formed the floor of Willamette basin. The valley rose and became dry 20 mya. Two periods of volcanic activity formed the old Cascade Mountains approximately 20 to 30 mya and the new Cascade Mountains at higher elevation 10 to 15 mya. On the west side of the basin tectonic and volcanic activity formed the Coast Range around 15 mya.

The middle of the basin is a large flat valley whose present land surfaces were created by a series of glacial outburst floods from the Lake Missoula system between 15,500 and 13,000 years ago. The surface of the valley floor was modified by the deposits of the glacial Lake Missoula floods from Montana at the end of the last glacial period. As ice dams forming Lake Missoula were breached, walls of water poured down the Columbia River with discharges that exceeded the annual discharge of all present-day rivers of the world combined. The Willamette basin became a backwater of the Columbia River and filled to depths of more than 120 m. Alluvial deposits carried by these floods formed the modern surfaces of the Willamette Valley.

The climate of the Willamette River basin is relatively mild, influenced by the moderating effect of the Pacific Ocean to the west (Uhrich and Wentz 1999). Across the basin, average annual air temperature ranges from 4°C to 18°C depending on elevation. In the valley, mean minimum air temperature in January averages 3°C to 5°C, whereas in the summer monthly means range from 17°C to 20°C (Fig. 13.22). Precipitation averages approximately 100 cm/yr at low elevation and approaches 500 cm/yr near crests of the Cascade Mountains and Coast Range (Hulse et al. 2002). Across the Willamette basin, estimates of average precipitation and recharge (surface-water runoff plus groundwater recharge) are 130 cm/yr and 50 cm/yr, respectively (Lee and Risley 2002, Uhrich and Wentz 1999). Most of this precipitation (70% to 80%) occurs between October and March during the wet season and <5% of the precipitation occurs in July and August during the dry season (see Fig. 13.22).

The climate, geology, soils, and hydrology create three major terrestrial ecoregions within the Willamette basin: Central Pacific Coastal Forests, Willamette Valley Forests, and Central and Southern Cascade Forests (Ricketts et al. 1999). The Central Pacific Coastal Forests ecoregion on the western edge of the basin is characterized by Douglas fir, western hemlock, and western red cedar. The central Willamette Valley Forests ecoregion once was a prairie with native perennial grasses and oak savannas, groves of Douglas fir, ponderosa pine, and other tree species, and was strongly influenced by anthropogenic fire. Very little of the prairie remains, however, having been replaced by agriculture. The Central and Southern Cascade Forests ecoregion on the eastern side of the basin includes western hemlock, western red cedar, and Pacific silver fir.

The Willamette River basin currently contains a population of 1,970,000, of which 86% live in urban portions of the landscape (Hulse et al. 2002). The majority (64%) of the land base is held in private ownership. Of the 36% public lands in the Willamette River basin, the U.S. Forest Service manages 30%, the U.S. Bureau of Land Management 5%, and the state of Oregon 1%. Forestlands account for most of the land use across the basin (68%) because it is so steep and heavily forested. Agriculture makes up 19% of the total land use and almost entirely occurs in the Willamette Valley. Urban areas occupy 5% of the basin. Transportation systems place a strong imprint on the land and rivers, with more than 130,000 km of roads, resulting in overall road densities of 2 to 3 km/km^2. The human population of the Willamette basin is projected to double to 3,900,000 people by 2050, with an increase to 93% living in urban settings.

River Geomorphology, Hydrology, and Chemistry

The highest point in the Willamette Basin is Mount Jefferson at 3199 m asl in the Cascade Mountains. The Willamette River has an elevation of 137 m asl where it enters the valley floor and descends to an elevation of 3 m asl at its confluence with the Columbia River. The river is approximately 100 to 200 m wide along the main stem, with an average gradient of 50 cm/km. Substrates in the upper Willamette River are dominated by cobble and gravel, with sand and clays becoming more abundant in the lower Willamette River. There are three major reaches in the river. The lower reach is extremely low gradient and is tidally influenced by the Columbia River. Willamette Falls creates a drop of 12 m in the lower reach at Oregon City, 72 km above the mouth (Fig. 13.7). The river flows between volcanic intrusions in the valley floor between Newberg and Albany, creating a highly variable floodplain. Above Albany the gradient increases and the floodplain widens, repre-

senting a large depositional area upstream of foothill landforms extending across the valley floor. Large wood contributed from the adjacent floodplain forests or transported from upstream reaches is a common feature of the shallow habitats and islands in the upper section of the Willamette River.

Mean discharge for the Willamette River at its mouth averages 917 m^3/s, but varies seasonally from 233 m^3/s in August to over 2230 m^3/s in December, reflecting the seasonal pattern of precipitation (see Fig. 13.22). Much of the summer runoff depends on the high-elevation snowpack because more than 35% of the precipitation occurs as snow above 1200 m in elevation. As a result of the rain-dominated hydrograph at low elevations and snowmelt-dominated hydrograph at high elevations, stream flows overall are highest in winter and lowest in late summer. Twelve major subbasins deliver flow into the main-stem Willamette River from upstream to downstream: Middle Fork Willamette, Coast Fork Willamette, McKenzie, Long Tom, Mary's, Calapooia, Luckiamute, Santiam, Yamhill, Molalla/Pudding, Tualatin, and Clackamas rivers (see Fig. 13.21). The main stem of the Willamette River is not dammed, but there are 13 major flood-control reservoirs in the larger tributaries in the Coast Range and Cascade Mountains and many small dams for irrigation or power generation. Reservoirs control approximately 27% of the flow of the Willamette River (USACE 1999).

Historically, most major floods were rain-on-snow events and caused floodplain inundation for weeks or months. The largest flood on record was the 1861 flood, which inundated 130,000 ha with a discharge of 9700 m^3/s. In contrast, the public was alarmed by a recent flood in 1996 that had a peak flow of only 3300 m^3/s. Floodplains defined by the Federal Emergency Management Agency (FEMA) closely match the historical extent of floodplain inundation in the lower Willamette River, but FEMA floodplains only represent 49% of the historically inundated floodplain in the broader depositional areas of the upper Willamette River (Hulse et al. 2002). Studies of gaging stations in the main-stem Willamette River indicate the channel down-cut at a rate of 0.3 m per decade from 1930 to 1960. Possible causes for this channel degradation include dredging, gravel mining, channelization, reservoir control, and climatic decreases in peak flows (USACE 1999, Hulse et al. 2002). Limited evidence indicates this down-cutting (in conjunction with flood-control dams) has led to less frequent overbank flooding; a flood that historically (prior to 1964) had a 1-in-10-

FIGURE 13.7   Willamette River at Oregon City, Oregon. Willamette Falls is seen in background. During summer, most of the water goes through historic locks built in the 1870s rather than over the falls (PHOTO BY TIM PALMER).

year chance of occurrence now has a 1-in-100-year chance of occurrence. This trend indicates a decrease in floodplain function and a need to restore both the geomorphology and the hydrology of the Willamette River system.

The main stem of the Willamette River is a low-gradient river, which served as a major transportation corridor for early settlers in the valley. The channels of the Willamette River have been greatly simplified over the last 150 years (Hulse et al. 2002). From 1850 to 1995 total area of river channel and islands in the main-stem Willamette River decreased from 16,600 ha to 9200 ha, and the total length decreased from 571 km to 424 km, primarily through loss of side channels. Much of this loss has been caused by channelization and bank hardening. Only 27% of the length of the main-stem Willamette has revetments or rip rap on one or both banks, but more than two-thirds of the meanders in the river are hardened by revetments.

Water quality has been a major environmental concern since the late 1920s. Dissolved oxygen concentrations were less than 1 mg O_2/L at sites along the entire main stem in the 1920s and 1930s. Elimination of most point-source discharges and extensive sewage treatment along the river improved conditions but were not sufficient to improve water quality. Summer low flows in the Willamette River are now double their historical levels, increasing to roughly 140 m^3/s from historical August flows of 70 m^3/s. Although this "dilution of pollution" has reduced the risks of high temperatures and low dissolved oxygen, it has altered the natural flows of the river and raised numerous questions about cottonwood forest regeneration and fish passage. Water quality in the middle and upper Willamette River is generally ranked as good, but the lower Willamette River is classified by the Oregon Department of Environmental Quality as "poor to marginal." The Willamette River was one of the 50 sites studied as part of the National Water

Quality Assessment. Overall, nutrients and contaminants in the Willamette River were ranked as typical of concentrations found in other NAWQA sites. Nitrate-nitrogen concentrations ranged from 0.054 to 22 mg/L in 98% of the samples, and increased as area of agricultural land increased (Wentz et al. 1998). Soluble reactive phosphorus concentrations in streams ranged from 0.01 to 0.93 mg/L in 89% of the samples.

In contrast to results for nutrients, the occurrence of abnormalities in fishes and the proportions of non-native fishes were far worse than most NAWQA sites. Thirty-six toxic compounds were detected in water sampling throughout the Willamette River basin. Six pesticides, including atrazine, diuron, and metolachlor, exhibited higher concentrations in agricultural lands than in urban areas. Other pesticides, such as carbaryl, diazinon, and dichlobenil, exhibited higher concentrations in urban areas than in agricultural lands.

River Biodiversity and Ecology

The Willamette River is in the westernmost portion of the Columbia Unglaciated freshwater ecoregion, which includes most of the southern tributaries of the Columbia basin, such as the lower Snake, Salmon, and John Day rivers (Abell et al. 2000). The ecology of streams and rivers in the Willamette River basin has been reviewed and summarized in a number of recent publications (Altman et al. 1997, Uhrich and Wentz 1999, Hulse et al. 2002). The following section highlights a few of the major studies of the aquatic ecosystems in the Willamette River basin.

Algae and Cyanobacteria

Algal assemblages of the main-stem Willamette River and the thousands of miles of smaller streams in the basin have not been studied extensively. Common algae in the main-stem Willamette River include diatoms such as *Melosira*, *Stephanodiscus*, *Cymbella*, *Achnanthes*, *Nitzschia*, and *Fragilaria*, which have been noted as the dominant types of benthic and planktonic algae (Wille 1976, Rickert et al. 1977). A study of the main stem and a major tributary, the Santiam River, also observed a dominance of diatoms in the 86 species observed in the phytoplankton and benthic algae and noted no major differences in algal community structure between the main stem and the tributary (Rinella et al. 1981). A subsequent study of benthic algae in 23 sites along the main-stem Willamette River and eight major tributaries found 35 genera of algae (Gregory 1993).

More than 80% of the cell numbers were comprised of cyanobacteria, predominantly in the genera *Anabaena*, *Aphanocapsa*, and *Chroococcus*, with diatoms making up most of the remaining 20%. Benthic algal abundance, as indicated by standing stock of chlorophyll *a*, increased downstream along the main-stem Willamette River in this study.

Algal communities in smaller streams also include many genera of diatoms (e.g., *Achnanthes*, *Diatoma*, *Fragilaria*, *Gomphoneis*, *Gomphonema*, *Cymbella*, *Cocconeis*), green algae (e.g., *Zygnema*, *Spirogyra*, *Ulothrix*, *Cladophora*, *Chlamydomonas*), and cyanobacteria (e.g., *Nostoc*, *Phormidium*, *Oscillatoria*, *Schizothrix*, *Calothrix*) (Clifton 1985, Lyford and Gregory 1975). In general, heavily shaded streams in forested reaches are dominated by diatoms and cyanobacteria, but open stream reaches include greater abundance of filamentous green algae, diatoms, and yellow-green algae.

Plants

The floodplain forests of the Willamette River are dominated by extensive stands of black cottonwood, Oregon ash, bigleaf maple, white alder, and willow. In 1850, the floodplain forest averaged 1.5 km to 3 km in width and was up to 11 km wide at the confluence of the Santiam River (Hulse et al. 2002). Even from a more narrow perspective, riparian vegetation has been greatly diminished by land conversion. In a 120 m band along both banks in 1850, 89% of the area would have been covered by forests of deciduous trees, conifers, or mixed stands. By 1990 only 37% of the riparian area within 120 m was forested, 30% was agricultural fields, and 16% was urban or suburban lands. Loss of floodplain forest has been one of the most dramatic changes in the Willamette River since Euro-American settlement.

Invertebrates

The macroinvertebrate fauna of the main-stem Willamette River has received little study, with much greater attention given to headwater streams in the Coast Range and Cascade Mountains. In a small stream in the Willamette Valley, more than 325 species of macroinvertebrates were observed over a 25-year period (Anderson and Hansen 1987). The River Continuum studies in the McKenzie River basin of the Willamette observed shifts in feeding functional groups from a dominance of shredders in small headwater streams, to scrapers in midorder streams, to collectors in larger rivers (Hawkins and Sedell 1981). Abundances of caddisflies (e.g., *Hydropsyche*, *Lepidostoma*, *Heteroplectron*, *Glosso-*

soma, Dicosmoecus), stoneflies (e.g., *Taeniopteryx, Nemoura, Yoraperla*), and mayflies (e.g., *Rhithrogena, Baetis, Paraleptophlebia, Ephemerella*) have been used as indicators of stream habitat quality in small streams of the Willamette basin (Van Sickle et al. 2004). In addition to aquatic insects, small streams in lower elevations also support crayfish (*Pacifastacus*) and gastropods (*Juga plicifera, Juga silicula*). Abundance of these taxa that require higher-quality habitat was inversely related to the extent of development and agricultural conversion within the riparian areas along the streams (Van Sickle et al. 2004). The proportion of the good- to high-quality habitat for invertebrates was projected to be approximately 75% of the total stream habitat in 1850 and had decreased to approximately 40% in 1990. Many macroinvertebrates are associated with wood in streams. A study in the Willamette basin documented 37 invertebrate taxa that are closely associated with wood and 67 taxa that are facultatively associated (Dudley and Anderson 1982). In particular, a unique elmid beetle, *Lara avara*, lives primarily in wood, consuming wood along grooves in the wood surface (Steedman and Anderson 1985).

Human and natural disturbances potentially alter the abundance and composition of macroinvertebrate communities in streams. Studies of headwater streams revealed that abundance of invertebrates increased in open reaches, such as clear-cuts (Hawkins et al. 1982). Catastrophic disturbances also have been observed to cause abrupt changes in aquatic macroinvertebrates. A 5000 m³ landslide dramatically reduced the abundance of invertebrates in the 3rd order stream in the Cascade Mountains, but numbers of invertebrates had recovered to levels comparable to predisturbance conditions by the end of summer seven months later (Lamberti et al. 1991, Anderson 1992).

Aquatic macroinvertebrates in the main-stem Willamette River include many mayflies, stoneflies, and caddisflies (Richard Miller, personal communication), which are commonly used as indicators of better water quality. Study of a middle section of the Willamette River near a pulp mill found a midge, *Rheotanytarsus*, and a caddisfly, *Hydropsyche*, to be the most abundant invertebrate taxa (Richard Miller, personal communication). A survey of macroinvertebrates in the main-stem Willamette River found that the composition reflected degraded water quality increasingly downstream (Tetra Tech, Inc. 1994). Noninsect macroinvertebrates also are major components of the invertebrate fauna of the main-stem Willamette River. A study of rip rap or revetments in the upper Willamette River found that polychaete worms (*Manayunkia*) and amphipods (*Anisogammarus*) were the most numerous macroinvertebrates (Hjort et al. 1984).

Vertebrates

The fish assemblage of the Willamette River basin exhibits low diversity, with 31 native fish species (Hulse et al. 2002). However, 29 species of fishes have been introduced, mostly in the main-stem river and lower tributaries. Native fish species include salmonids (cutthroat trout, bull trout, rainbow trout, mountain whitefish, Chinook salmon, steelhead, coho salmon, chum salmon, sockeye salmon), suckers (largescale sucker, mountain sucker, bridgelip sucker), minnows (northern pikeminnow, Oregon chub, chiselmouth, peamouth, leopard dace, longnose dace, speckled dace, redside shiner), sculpins (mottled sculpin, prickly sculpin, Paiute sculpin, reticulate sculpin, riffle sculpin, shorthead sculpin, torrent sculpin), lampreys (Pacific lamprey, western brook lamprey, river lamprey), and other species (white sturgeon, sandroller, eulachon, threespine stickleback). Some of the major nonnative species include largemouth bass, smallmouth bass, bluegill, walleye, crappie, common carp, grass carp, brown bullhead, western mosquitofish, brook trout, brown trout, kokanee, and lake trout. The lower rivers contain the largest number of fish species, with more than 20 species occurring within a 1 km reach of the main-stem Willamette River (Hulse et al. 2002). Headwater streams contain far fewer numbers of species, with cutthroat trout commonly the only fish species present in very small headwater streams. Several native fish species are currently listed as threatened or endangered under the Endangered Species Act (1973) or as sensitive species under the Oregon Endangered Species Act. Federally listed fish species in the Willamette River include spring Chinook salmon, winter steelhead, chum salmon (found only at mouth of the Willamette River), and Oregon chub. Coho salmon (found in tributary rivers below Willamette Falls) is considered to be critically sensitive by the state of Oregon.

One of the first extensive surveys of fishes in the main-stem Willamette River found increasing numbers of pollution-tolerant species near the mouth in Portland (Dimick and Merryfield 1945). Two later studies concluded that numbers of pollution-intolerant species had increased since the mid-twentieth century, reflecting improved water quality in the Willamette River as a result of pollution control measures (Hughes and Gammon 1987, Tetra Tech,

Inc. 1994). A recent study found similar patterns, but noted that nonnative species increased in abundance downstream and comprised more than 70% of the number of fish species in the lower river near the mouth (Hulse et al. 2002).

Other vertebrates in the Willamette basin include 18 species of amphibians, 15 species of reptiles, 154 species of birds, and 69 species of mammals. Not all of these species are aquatic. Pacific giant salamanders are a common vertebrate in streams of the Willamette basin and may be more abundant than fishes in headwater streams (Murphy and Hall 1981). Three species of garter snakes are very common in the Willamette River and its riparian forests. Native turtles include the western pond turtle, placed on the state sensitive species list because of its low population size. One riparian-dependent bird, the yellow-billed cuckoo, has been extirpated in the Willamette basin since Euro-American settlement. This species required closed-canopy riparian forests along the Willamette River and its extirpation may be directly related to loss of floodplain forests. One of the common aquatic birds in mountain streams of the Willamette River basin is the American dipper, a small, robin-sized bird that swims underwater and feeds on aquatic insects (Parsons 1975). Harlequin ducks migrate from the Pacific Ocean to mountain streams in the spring and early summer for nesting and rearing, also feeding on aquatic insects in high-gradient streams. Osprey, bald eagle, great blue heron, and green heron are common along the main-stem Willamette River. Osprey numbers have increased over the last two decades, and researchers have hypothesized that their increase is due to the ban on DDT and learning to use power poles for nesting (Henny and Kaiser 1996). Beaver populations are generally increasing throughout the basin, but the numbers are presumed to be far lower than historical populations and their distribution has been greatly restricted. River otter and mink also are common native mammals along the rivers and streams, and nutria and muskrat have been introduced into the region.

Ecosystem Processes

One of the four major basins for research in the River Continuum Concept was the McKenzie River at the southern end of the Willamette basin (Vannote et al. 1980, Minshall et al. 1983). This pioneering research revealed that the energy base for the smallest headwater stream, Devil's Club Creek, was dominated by allochthonous inputs from the surrounding forests, and algal production increased in the larger streams (Mack Creek, Lookout Creek, and McKenzie River) as the increasing stream width created larger openings over the stream for solar radiation. Shredders dominated the smaller sites and scrapers and collectors became more dominant in larger streams. The riparian forests along these streams in the Willamette basin have been shown to contribute many ecosystem functions, including litter input, shading, large wood inputs, nutrient uptake, and bank stabilization (Gregory et al. 1991). Streams in the H. J. Andrews Experimental Forest in the McKenzie River basin were one of the first locations where the ecological role of large wood in stream ecosystems was documented (Swanson and Lienkaemper 1978, Harmon et al. 1986). These studies have led to the recognition of the ecological role of wood in river networks from the smallest headwaters to large floodplain rivers throughout the world (Gregory et al. 2003). The floodplains of the Willamette River and their forests are critical for aquatic ecosystems and their fish assemblages. Tributary junctions and multiple channel reaches of the Willamette River support more diverse riparian forest communities and richer fish assemblages (Hulse et al. 2002). Floods and other natural disturbances are important processes for creating complex habitat structure, diverse and abundant macroinvertebrate and fish communities, and mosaics of riparian plant communities (Lamberti et al. 1991, Anderson 1992, Johnson et al. 2000).

Human Impacts and Special Features

The Willamette River is a unique component of the Columbia River system, providing 12% to 15% of its total flow from only 4% of its area. The basin drains two major mountain ranges, the Coast Mountains to the west and the Cascade Mountains to the east, before flowing north through a low-gradient valley. It is an important basin for the state of Oregon, containing more than two-thirds of the state's population. It is home to several species of regional importance: spring Chinook salmon, coho salmon, winter steelhead, coastal cutthroat trout, beaver, river otter, bald eagle, and spotted owl. The headwater streams exhibit good water quality and the main-stem Willamette River continues to have water quality problems. The Willamette River lies in the heart of the valley and is a central feature in the lives of the communities within this basin.

Concerns over water quality in the Willamette River and declines in fish populations led to a public referendum and establishment of the state's Pollution

Control Authority in the late 1930s. These concerns surfaced again in the late 1960s; Governor Tom McCall gained national attention through the state's well-publicized efforts to clean up the Willamette River. Those actions decreased the impacts of raw sewage and industrial wastes and led to the formation of the Willamette Greenway. New concerns over continued loss of floodplain habitat and associated natural resources caused Governor John Kitzhaber to form a Willamette River Basin Task Force in 1998, which led to the development of the Willamette Restoration Strategy in 2001.

The first large dam was constructed in the city of Portland in 1894 to provide domestic water. Communities and agencies in the Willamette River basin have constructed 371 licensed dams with a storage capacity of 3 billion m^3 (Hulse et al. 2002). This amount of water is equivalent to filling the entire Willamette Valley floor to a depth of more than 0.3 m (1 ft). These dams have blocked passage of anadromous salmonids and altered habitats downstream. Dams are one of the major causes of the declines of salmon in the Pacific Northwest (Nehlsen et al. 1991). Reductions in peak flows and deliberate channelization of the river have greatly reduced riverine habitats. Since the mid-1800s more than $2.7 billion has been spent in direct channel alteration and operation of flood-control reservoirs in the Willamette River basin.

In addition to the creation of dams to store or divert water, water is withdrawn directly from the Willamette River and its tributaries. Most water rights (75%) for water withdrawal are recent, dating after 1960. Most streams in the Willamette basin have been fully allocated and have no remaining water to be distributed during normal years. On average, 77% of the surface waters in the Willamette River basin are withdrawn and 40% of this withdrawal is consumptive removal (the water does not return to the stream) (Hulse et al. 2002). Irrigation accounts for 49% of total withdrawal, domestic use 15%, industrial use 13%, and commercial use 20%. Most water withdrawal in the basin comes from surface waters; 88% of all withdrawals from public water-supply sources comes from surface waters. Projections of present and future water use in the Willamette basin indicate more than 130 km of streams that ran perennially in 1850 are now dry during summer low flows. That distance is projected to double to approximately 275 km by 2050 under current environmental policies (Hulse et al. 2002).

Introduction of nonnative species into the basin raises major concerns. In the upper main-stem Willamette River, only 4% of the fish species are nonnative. In the lower Willamette River near Portland, more than 70% of the fish species are nonnative. In addition, rates of abnormalities increase markedly. These nonnative species and the abnormalities in native and nonnative fishes are indications of potential stresses on the aquatic ecosystems of the Willamette River.

Loss of floodplain forest and channel complexity in the Willamette River limits the current function and structure of the ecosystem and demonstrates a trend of continued decline and impairment. As society moves closer and closer to the Willamette River and calls for greater control of flows and simplification of its channel, the condition of the Willamette River ecosystem will decline. Society demands clean water, aesthetic parks and scenic resources, and opportunities for recreation, bird watching, fishing, and hunting. At the same time, the pressures of land conversion, rural residential development, population growth, and urbanization place those values in jeopardy. The state of Oregon developed a Willamette River Conservation Strategy in 2001. The future of the Willamette River and its aquatic ecosystem will depend on the effectiveness of that strategy and its implementation.

COLUMBIA RIVER MAIN STEM

The Columbia is the fourth-largest river that flows to the sea in North America. Its main stem flows from Columbia Lake in British Columbia 2000 km to the ocean at Astoria, Oregon (Fig. 13.23). Along the way it receives water from 12 tributaries that are 6th order or larger, draining vast mountain ranges from the Rockies to the Cascades. The main stem is a canyon river, constrained by mountain flanks in the ranges (Fig. 13.8) and deeply entrenched in the basalt of the Columbia–Snake River Plateaus by Pleistocene floods of the Lake Missoula cataclysms (Allen et al., 1986). Only in the tidewater zone downstream from the confluence of the Willamette does it slow, historically allowing massive sediment deposition that created a shifting mosaic of large midchannel islands and fringing floodplain wetlands.

The Columbia, like all the great rivers of the world, is a conveyance for water, materials, and people. Although the tributaries organize the river as a complex stream network, natural (floods, sediment transport, salmon runs) and human cultural (commerce transport, hydropower, social interactions) attributes converge consequently along the main

FIGURE 13.8 Columbia River Gorge, looking upstream from Crown Point State Park. This tidewater section of river is just downstream of Bonneville Dam, the lowermost dam on the river (PHOTO BY A. C. BENKE).

stem. The noted historian Richard White, in his wonderfully insightful book (White 1995), characterized the river as an organic machine, a metaphor that uniquely captures the essence of the Columbia: a complex, natural–cultural system, constantly changing in time and space. The main-stem Columbia is a high-volume, high-gradient, and therefore powerful river, fed by snowmelt from vast headwater mountain ranges. Historically, it was the conveyance and habitat for enormous and diverse salmon and steelhead runs, equally diverse native peoples, and a wide variety of immigrants from all over the world. The native peoples lived at specific sites along the river corridor for at least 10,000 years, loosely united by their salmon culture (Cressman et al. 1960, Butler 1993). In less than 200 years the immigrants used labor and ingenuity to parlay water, salmon, and water power into capital, ultimately creating a legacy that fished, dammed, diverted, farmed, and urbanized the river into an organic machine (White 1995). Indeed, the Columbia River today is among the most developed of the world's great rivers and its salmon and native people have suffered in consequence.

623

Physiography, Climate, and Land Use

The main stem of the Columbia may be divided into three physiographic domains. The coastal rain-forest segment from Columbia Gorge (see Fig. 13.8) through the Cascade Range to the estuary at Astoria cuts through the Cascade–Sierra Nevada (CS) and Pacific Border (PB) physiographic provinces (see Fig. 13.23). The dry interior segment of the river in the Cascade rain shadow cuts through the Columbia–Snake River Plateaus (CU) province from Kettle Falls at the United States–Canada border to the Columbia Gorge. The headwaters segment from its source at Columbia Lake in British Columbia to Kettle Falls cuts through the Rocky Mountains in Canada (RM), the Coast Mountains of British Columbia and Southeast Alaska (PM), and the Northern Rocky Mountains (NR) provinces. Additional major headwater tributaries are the Kootenay and Pend Orielle (combined Flathead and Clark Fork flows) that originate in the Precambrian Belt Series formations of the Rocky Mountains and adjacent western ranges. The Okanagan, Methow, Wenachee, Yakima, and Spokane are the major mid-Columbia tributaries that drain the volcanic Cascade Range. The main stem becomes a 9th order river at its confluence with the Snake in the middle of the Columbia–Snake River Plateau province.

The highest point in the basin is Mount Rainier at 4392 m, one of several Cascades volcanoes that dominate the scenery when viewed westerly from the river in the interior. Many mountains in the headwaters rise above 3500 m, however, and frame the river valleys of the subbasins. The elevation of Columbia Lake is 810 m asl; thus, the average gradient of the main stem is about 0.41 m/km.

Out of the 13 terrestrial ecoregions found in the entire Columbia basin, the main-stem river actually passes through 10 of them. Its upper reaches in Canada and northern Washington drain from a wide diversity of coniferous forests, including those in the North Central Rockies Forests, the Okanagan Dry Forests, and the Cascade Mountains Leeward Forests ecoregions. After passing through the coniferous forests of northeastern Washington, the main stem enters the arid and much flatter Palouse Grasslands and Snake/Columbia Shrub Steppe ecoregions, much of which has been converted to agriculture. After flowing many kilometers through the shrub steppe, the river eventually crosses through the following forested ecoregions on its path to the Pacific: Eastern Cascades Forests, Central and Southern Cascades Forests, Willamette Valley Forests, Puget Lowlands Forests, and Central Pacific Coastal Forests. Most of these ecoregions are dominated by conifer forests, except in the developed lowlands and valleys.

The climate is extremely variable along the main stem. The continental air masses often dominate the Columbia Valley in British Columbia resulting in winter temperatures commonly −40°C or colder; summers are correspondingly cool. In extreme contrast, the interior is dry and hot, often exceeding 38°C. Precipitation can exceed 350 cm/yr in the Purcell Mountains fringing the river in British Columbia and in the Cascades, mostly deposited as winter snow, but snowpack along the river rarely exceeds 1 m. Along the river near Richland, Washington, in the very dry interior, precipitation is less than 20 cm/yr and snow is rare. Using monthly averages collected throughout the basin, it appears that basinwide precipitation is generally highest from November through January and lowest in July and August (Fig. 13.24). However, it should be kept in mind that this is a crude approximation of monthly averages and that they vary greatly throughout the basin.

Land use in the Columbia Valley below Columbia Lake is largely pasture-based ranching and recreational development. Downstream there are a few localized wood-products and mining operations, but the economy of the Canadian portion of the Columbia is based on hydropower and recreational fishing in the reservoirs and urban and exurban development associated with the full array of recreation opportunities in the adjacent mountains. In contrast, agriculture is a key economic driver on the U.S. portion of the river, driven by irrigation from the river. Water diversion facilitated by the Grand Coulee Dam provides irrigation water to 730,000 ha of former scrubland now converted to orchards and row crops in the area from Moses Lake to Pasco, Washington. From the Okanagan River confluence (north-central Washington) to Bonneville Dam (just east of Portland) irrigation water is pumped from the river via many different intakes to the adjacent terraces and uplands along the river. All of the pump intakes and other diversion structures have been screened at enormous expense to prevent entrainment of salmon and steelhead juveniles. The dominant crops are apples, cherries, and other fruits, but hops and poplar (pulp wood) plantations are common, all substantially subsidized by federal hydropower development that provides cheap electricity and the complex irrigation infrastructure that delivers water to the farms (White 1995).

River Geomorphology, Hydrology, and Chemistry

Throughout its 2000 km length the main-stem river is predominately constrained within canyons. In the headwaters, the river cuts through the mountain ranges of British Columbia. In the middle reaches, the river geomorphology was determined largely by glacial floods, particularly the Lake Missoula floods of 13,000 to 15,000 years ago. These floods entrenched the main-stem channel deep within the ancient lava (basalt) formations of the Columbia Plateau, laid down some 30 million years ago. The lower river cuts through the Cascades and the coastal ranges in deep gorges, finally spilling into the ocean through a small estuary. In 1811, Canadian surveyor David Thompson noted that the river was 800 to 1000 m wide at base flow but narrowed to ~50 m at the great cataracts of the Dalles des Morts (death rapids) in the lower river and Kettle Falls (Nisbet 1994). The channel was a uniform, undivided ribbon with only a few islands and rapids throughout its 2000 km length. Unpublished predam surveys of the area that was inundated by John Day Reservoir show large gravel bars very similar to the huge Vernita Bar, famous as a fall chinook spawning site historically and today, downstream of Priest Rapids Dam in the unimpounded Hanford Reach (Fig. 13.9). This 75 km river segment is the last major spawning and rearing area for salmon left on the main stem. Sand deposited high on the shoreline during floods and reworked into dunes by wind and sandstorms plagued early travelers on the interior portions of the river. Mid-channel sand bars, shoals, and a shifting array of fringing floodplain wetlands were common in the tide zone of the river from the Willamette confluence downstream. Today the river is almost totally impounded by dams (see Fig. 13.23) and the tidewater reach is routinely dredged to accommodate ship traffic to Portland, Oregon.

The main-stem Columbia has a spring snowmelt hydrograph that is considerably moderated (flattened) by winter rains in the Cascade and coastal ranges, and by the dams (see Fig. 13.24). Sixty percent of the variation in annual runoff is coherent with the El Niño circulation pattern of the Pacific Ocean. Strong El Niño years tend to be dry (Redmond and Cayan 1994). Mean discharge at the mouth is 7730 m³/s. Major floods, five times the average flow, occurred in 1894 (the largest), 1876, 1894, and 1948, with the latter completely destroying Vanport, built on the floodplain near Portland and then the 2nd largest city in Oregon with 20,000

residents. Prior to the dams, flooding occurred annually and with great power owing to gradient and canyon constraint. Thompson noted trees 1 m in diameter suspended on rock ledges 14 m above the base flow level of the river at the Dalles rapids (White 1995). Of course, today flooding has been mostly eliminated by impoundments throughout the basin.

Main-stem water-quality problems are generally related to urban and agricultural runoff, latent heat storage in the reservoirs, and the legacy of plutonium manufacture and atomic energy research at the Hanford Nuclear Reservation (originally called the Hanford Engineer Works) upstream from Richland, Washington. Concern exists for radioactive isotope and heavy metal (particularly arsenic and aluminum) pollution of groundwater discharging into the river from the reservation and contaminated sediments retained in McNary Reservoir downstream. Toxic chlordane and PCBs from urban runoff and air pollution are a concern in the Columbia Slough, a remnant of a complex of shifting floodplain wetlands encompassing the confluence with the Willamette River. Hot-water pollution from Hanford, a chronic problem during the years of reservation nuclear reactor operations, has ceased, but the impoundments increase heat retention and much of the main stem from McNary Reservoir to the estuary periodically experiences summer water temperatures above 21°C that can induce severe stress if not cause death of native salmon and steelhead adults and juveniles. Dissolved nitrogen and carbon dioxide entrainment causes gas bubble disease in fishes when very turbulent water is spilled over the dams at high flows. Changes in spillway architecture to reduce turbulence have eliminated much of this chronic problem, and intentional spill, coupled with fish guidance structures to bypass salmon juveniles migrating downstream (as an alternative to turbine entrainment), has resulted in nearly 100% dam passage survival at some of the dams. Survival associated with reservoir passage is much lower due to predation (Williams 2005). Water-quality records of the U.S. Geological Survey from 1993 to 2003 provide the following average measures at the Beaver Army Terminal downstream of Portland: turbidity 8.2 NTU, specific conductance 135.4 µS/cm, alkalinity 51.9 mg/L as $CaCO_3$, NO_3-N 0.26 mg/L, phosphorus (total) 0.06 mg/L, calcium 13.9 mg/L, magnesium 4.0 mg/L, sulfate (water, filtered) 8.9 mg/L, cadmium (water, filtered) 0.05 µg/L, lead (water, filtered) 0.55 µg/L, aluminum (water, filtered) 12.09 µg/L, arsenic 1.05 µg/L, DDE 0.005 µg/L, dieldrin 0.002 µg/L, and PCB 0.15 µg/L. Values were similar at a monitoring site

FIGURE 13.9 The Hanford Reach of the Columbia River, the last free-flowing section, upstream of Richland, Washington (PHOTO BY D. D. DAUBLE).

upstream near Hanford at Richland, Washington, for the same time period.

River Biodiversity and Ecology

The main-stem Columbia River flows through two of the three freshwater ecoregions of the basin: the Columbia Glaciated ecoregion from its headwaters to just above the Washington–Oregon border and the Colorado Unglaciated ecoregion from there to the Pacific (Abell et al. 2000). The ecology of the main stem is a reservoir story with periodic retrospectives about the historic river that is now predominately inundated and progressively invaded by nonnative biota more in tune with lacustrine than dynamic riverine conditions, such as bass and walleye. Many reservoirs fisheries studies exist, particularly focused on assessments of salmon survival during in and out migration. These were reviewed in detail by Independent Scientific Group (2000). Here we summarize the limited biotic data for the free-flowing segments, mainly the Hanford Reach.

Algae and Cyanobacteria

Algal assemblages in the Hanford Reach of the Columbia River include biofilms (periphyton) that grow profusely on the permanently wetted portion of the channel, where the substratum is rocky, and phytoplankton. Benthic algal assemblages include the filamentous green algae *Ulothrix* and *Cladophora* and a rich benthic diatom flora, including the genera *Gomphonema*, *Nitzschia*, *Cocconeis*, *Melosira*, *Synedra*, *Achnanthes*, and *Cyclotella* (Neitzel et al. 1982). Phytoplankton populations are abundant in the reach, but are mostly derived from rich populations in the upstream impoundments. Spring population peaks are dominated by *Asterionella formosa* and *Fragilaria crotonensis*; the latter species often dominates the fall population peak. Other significant genera include *Tabellaria*, *Synedra*, and *Melosira* (Cushing and Rancitelli 1972, Cushing 1967).

Plants

The riparian zone of the main stem was limited historically, simply because much of the area through which the river flows is arid and the river was so constrained by its canyons that little floodplain existed. Hence, the riparian zone was mainly black cottonwood and various willow species growing on limited expanses of exposed gravel bars and floodplain benches near the river channel. Except for limited areas of wetlands created by impoundment, the perimeters of the reservoirs are either rip rapped to prevent erosion and thus have no vegetation or are marked by a thin fringe of willows on sandy shorelines. Two notable exceptions exist. The floodplain wetlands of the unregulated river segment from Columbia Lake to Kenbasket Reservoir near Golden, British Columbia, remain intact and ecologically functional, with large, black cottonwood gallery forests and apparently rich but undocumented understory and aquatic plant communities. A fringing wetlands and island system associated with the tide zone also exists from the Willamette confluence downstream. Here cottonwood, ash, and willow stands are dispersed around sloughs and floodplain lakes that abound with aquatic and semiaquatic plants, such as scouring rushes (*Equisetum arvense*, *E. telmateia*, *E. hyemale*), wapato, cow parsnip, cattail, camas, and bracken fern. Roots (or meristem shoots of scouring rushes) of all of these plants, especially wapato and camas, were important food of Native Americans. A wide variety of native pondweeds also are present in these wetlands.

Invertebrates

Newell (2003) listed 145 macroinvertebrate taxa that have been identified from the free-flowing Hanford Reach since 1951, although far fewer are present today. Net-spinning caddisflies (*Hydropsyche*, *Cheumatopsyche*), ephemerellid (*Ephemerella*) mayflies, three genera of stoneflies (*Isogenus*, *Perlodes*, *Pteronarcys*), and chironomid midges were notably abundant. Additional caddisflies collected since 1951 include *Brachycentrus*, *Glossosoma*, *Hydroptila*, *Lepidostoma*, *Leptocella*, and *Rhyacophila*, and additional mayflies include *Ephemera*, *Ephoron*, *Hexagenia*, *Heptagenia*, *Stenonema*, and *Tricorythodes*. Other taxa found since 1951 include stoneflies (*Arcynopteryxa*), mollusks (*Pisidium*, *Fluminicola*, Asiatic clam, western pearlshell), and crustaceans (*Corophium*, *Pacifastacus*, *Gammarus*). Considering that the Hanford Reach remains similar to the historic condition of the rest of the main stem, it seems likely that the main-stem river historically throughout its length contained a rich food web that supported substantial salmon production.

Today, a rich phytoplankton community and abundant fine particulate organic matter support a dominant macroinvertebrate population of net-spinning caddisflies (*Hydropsyche* and *Cheumatopsyche*), chironomids, and other filter-feeding organisms (mollusks), although many invertebrate genera found prior to dam closures upstream and downstream from the Hanford Reach are no longer found. The lower river reservoirs also have been colonized by estuarine amphipods.

Vertebrates

Among the 65 native fishes of the Columbia basin, at least 12 anadromous species existed (Ward and Ward 2004). Most notably, chinook salmon and white sturgeon spawned on the formerly abundant gravel and cobble bars throughout the main stem. Moreover, the main stem likely was an important rearing and presmolt conditioning area for main-stem and tributary fishes, but this was not firmly documented before overfishing and dam operations began to substantially constrain natural production. Indeed, the great salmon fisheries crash of the Columbia occurred because management focused mostly on adults and hatchery operations with blind regard to habitat constraints as attempts were made to compensate for extreme reductions in population size and structure associated with overfishing of the stocks from 1880 to 1930 (Lichatowich 1999). Perhaps the largest run of anadromous fishes today is American shad, a nonnative species.

Like the anadromous species, the native resident fishes of the Columbia River have been extensively impacted by the numerous dams along its main stem and the attendant change of the environment from lotic to lentic conditions. Ward and Ward (2004) point out that although the total number of fish species prior to dam building is unknown, there are currently at least 53 native resident species in the Columbia basin. The most species-rich families are the Cyprinidae (12), Catostomidae (6), Cottidae (13), and resident Salmonidae (14 species or subspecies). Stober and Nakatani (1992) have estimated that there are 61 fish species (native and nonnative) in the main stem between Lake Bonneville and Lake Roosevelt (Grand Coulee Dam). Native coldwater species such as kokanee, mountain whitefish, and white sturgeon still provide important sport fisheries in different reaches of the river. Native nonsport species important in ecosystem food webs in the main stem include the northern pikeminnow, Pacific lamprey, peamouth, chiselmouth, largescale sucker, bridgelip sucker, and redside shiner (Stober and Nakatani 1992).

At least 53 nonnative fish species have been introduced to the Columbia basin, bringing the total number of fish species to approximately 118. Many nonnative fishes reside in the reservoirs, including popular sport fishes, such as walleye, smallmouth bass, and yellow perch (Stober and Nakatani 1992). Prominent among the nonnative species are several additional Centrarchidae (nine species) and Ictaluridae (eight species) (Stober and Nakatani 1992, Ward and Ward 2004). The combination of dams and predatory nonnative species is suspected to have caused substantial declines in native resident species, as well as the more publicized anadromous species. Li et al. (1987) noted that the proportion of nonnatives was far less in the free-flowing Hanford Reach than in the reservoirs.

The Columbia Wetlands in the headwaters is a major waterfowl production area, used by most of the western North American ducks, geese, swans, and cranes. Bald eagle and other raptors nest in the area and muskrat, raccoon, river otter, painted and pond turtle, and a wide variety of other wildlife are present. The lower river wetlands also are important waterfowl production and resting areas, as are portions of some of the reservoirs, such as the head of McNary Reservoir at the confluence with the Yakima. The riparian areas and wetlands of Columbia Slough provide cover and food for over 120 species of birds.

Ecosystem Processes

The size and complexity of the Columbia River make detailed studies of ecosystem processes difficult, particularly in the postdam era. Nevertheless, studies of periphyton and phytoplankton productivity have been undertaken in the Hanford Reach. Cushing (1967) found net production rates of periphyton biomass ranging from $0.005\,mg\,DM\,cm^{-2}\,d^{-1}$ in winter to $0.070\,mg\,DM\,cm^{-2}\,d^{-1}$ in August; mean value for the 10-month experiment was $0.029\,mg\,DM\,cm^{-2}\,d^{-1}$. Neitzel et al. (1982) report ^{14}C uptake values ranging from zero (in winter) to $0.033\,mg\,C\,L^{-1}\,hr^{-1}$ in June and September.

Although ecosystem processes of the Columbia River prior to dams can only be guessed, it is clear that the natural and cultural history of this large-river ecosystem revolved around salmon. The Columbia as a whole historically produced more chinook and coho salmon and steelhead than any other river in the world (Netboy 1980). It is unknown how many fishes were conveyed annually to spawning grounds in the main stem and the many tributaries. The Northwest Power Planning Council, in developing its salmon recovery goals for the river during the 1990s, concluded that 10 to 16 million chinook, chum, coho, sockeye, pink salmon, and steelhead annually entered the river prior to commercial fishing (National Research Council 1996). Some scientists view this as an absurdly conservative number given the size and complexity of the Columbia system. In any case, there were enough salmon to populate every major tributary from the river mouth to the Rocky Mountains some 2000 km inland. Only the upper Pend Oreille (including the Clark Fork and Flathead) and upper Snake were blocked by falls that the salmon could not ascend. Chinook dominated the run, with 8 to 10 million fish, including "June hogs" of the main-stem summer run to areas above Grand Coulee that commonly exceeded 25 kg each.

This dynamic salmon ecosystem obviously was driven by the connectivity of the main stem to the tributary network and the estuary. Water, sediment, nutrients, and organic matter transported by the river created habitat for the various life stages of the fishes. By dying in great numbers after spawning, the fishes undoubtedly enriched system productivity (Schindler et al. 2003), although the pathways for this transformation of marine-derived energy and nutrients are not clearly understood in the Columbia or any other salmon river ecosystem. In any case, many millions of juvenile salmon made their way back to the ocean from the spawning grounds annually. Chums, pinks, and sockeye migrated out with urgency in huge numbers, coherent with the speed of the main-stem flow, while the more individualistic chinook, coho, and steelhead dallied in the river and estuary of the Columbia or adjacent coastal areas, feeding and growing into robust smolts, energetically fit for ocean survival and return to the river. Hence, the salmon ecosystem includes (1) the tributary network, where complex processes that entrain and transform nutrients and organic matter from the uplands into productive salmon habitat; (2) the main-stem conduit to the estuary that likewise was characterized by complex materials flux and transformation processes; (3) the estuary and coastal zone, where the fishes acclimated to salt water on the way out and freshwater on the way in; and (4) the variable environments of the north Pacific Ocean.

Connectivity of the river channel to its riparian floodplain zone was extremely important. Bank storage of water during floods drained back into the river during lower flows, naturally cooling the river in the hot summers of the interior when the flow in the river was lowest. The riparian forest of cottonwood and willow, although limited on the main stem, nonetheless provided leaves and tree boles to the channel that were food sources and attachment sites for invertebrates and other fish forage. Backwaters created by flood scour were rich in aquatic plants and other invertebrates and provided resting and feeding areas for very young salmon.

Owing to abundant salmon, white sturgeon, freshwater clams, and aquatic plants with nutritious roots in the backwaters, many native cultures existed along the river but were focused on several rapids where the fishes could be caught with dip nets as they jumped upstream. First in sequence from the ocean was the Cascades, where Bonneville Dam exists today (hence the name of the mountain range through which the river passes in the narrow Columbia Gorge that included the rapids), then the Dalles with its Celilo Falls upstream, Priest Rapids upstream from the Yakima confluence, and finally at Kettle Falls far upstream at the United States–Canada border. Location was important because the fishes entering the river had high caloric content from fat stored for the long, tough journey inland. Caloric content declined with distance upstream. Only 50% was left by the time the fishes got to the Nez Perce people at the confluence of the Snake, and they were 75% depleted at the Kutenai tribal fishing site on the Kootenay River near Nelson, British Columbia.

The Dalles was an especially important native fishing site because the fish were still in very good shape at that point in the migration, and they could be caught in large numbers by dip netting the fish as they jumped up the cataracts. Indeed, 500 fishes per day could be taken by a single fisherman. The dry air of the site (owing to the rain shadow of the Cascade Mountains) allowed efficient drying of the fishes for long-term storage. The 600 or so native residents at Celilo Village at the Dalles swelled to over 3000 during the salmon season (Ross 1969).

For the native people of the Columbia, salmon belonged to everyone and thus were not currency. Trade involved things other than salmon, such as dentialium shells, and early explorers had difficulty "buying" salmon initially. The natural tendency of the native people was to allow most of the fishes to pass to the spawning grounds and elaborate ceremonies expressed anxiety about future returns rather than concern for maximum possible harvest. Hence, salmon and people coexisted on the Columbia for thousands of years. The immigrants of course did not see it that way and their legacy is one of total exploitation of the salmon runs and maximum development of the river for transportation, hydropower, and farming, which they accomplished in less than 200 years (White 1995).

Human Impacts and Special Features

Today, 15 huge dams impound most of the main stem and, along with the big storage reservoirs in the headwaters, totally regulate the flow of the river. The main-stem dams back water from one to the next throughout the river corridor from Bonneville, the most downstream dam, to Mica, far upstream in British Columbia. Indeed, only three flowing reaches remain: the headwaters from Columbia Lake to the first main-stem reservoir (Kenbasket), the Hanford Reach from Priest Rapids Dam to McNary Reservoir, and the tidewater segment from Bonneville Dam at the Cascades to the estuary. Rock Island Dam, a public utility project in the mid-Columbia, was the first main-stem dam, completed in 1932, and was built with fish ladders to pass migrating salmon around the dam. Then came the concrete behemoths—Bonneville and Grand Coulee—that the federal government completed in 1938 and 1941, respectively, the latter with no fish passage and thus blocking salmon from over 1500 km of prime river habitat. The 1964 Columbia River Treaty between the United States and Canada funded Mica, Keeleyside, and Duncan dams with U.S. money. The United States gained flood control and both countries gained enormous power benefits, but Canada lost its Columbia River salmon runs entirely. Revelstoke Dam completed the impoundment of the main-stem Columbia in 1984.

The main-stem dams were justified not just for production of hydropower for the cities but also to provide power to massive pumps that irrigate thousands of hectares along the river. Apple and cherry orchards, vineyards, hops plantations, and, believe it or not, water-loving poplar plantations for paper pulp line both sides of the river, abruptly changing the brown desert landscape of the uplands to verdant green. Thus, hydropower subsidizes agriculture throughout the lower main stem to Columbia Gorge. All of the pumps are screened at great expense to prevent entrainment of fishes.

The dams also vastly simplified barging commerce on the river by impounding the rapids. Most of the grain grown throughout the basin is transported from railheads at Lewiston, Idaho, to the oceanic shipping docks in Portland via barges that slide along the river like massive snails.

The reservoirs of the river provide recreation in many forms. The winds that howl through the Columbia Gorge provide some of the best board sailing in the world and sport fishing is a huge economic boon as anglers focus on abundant walleye, smallmouth bass, sturgeon, and salmon. But these fisheries are almost entirely a consequence of invasion of nonnatives up the reservoir-dominated environment of the main stem and sport fishing and commercial fishing in the Columbia is decidedly at odds with conservation of native wild stocks.

Over 200 distinct stocks (populations) of all five species of Pacific salmon plus an array of steelhead life-history types historically passed through all or part of the main stem (Nehlsen et al. 1991). Almost all of them are extirpated or extinct today. Most of the remaining stocks are listed as threatened or endangered (www.nwr.noaa.gov/1salmon/salmesa 2004). All of the listings in some way pertain to the main stem because it is the migratory pathway to the ocean from the tributaries. That some survive in this intensively regulated system is a testament to the resiliency of the beasts. For example, a few chum salmon return annually to spawn in a single spring system that emerges through the dredge spoils on the north bank of the river adjacent to the urbanized area of Vancouver, Washington, across from Portland.

Impoundment of the main stem is a major impediment to salmon recovery because of negative influences on food web structure and feeding habits of juvenile salmon. Juvenile salmon in the Hanford Reach feed on newly emerged adult midges, hydropsychid caddisflies, and shallow-water zooplankton. These insects, plus zooplankton discharged from the reservoir, are important food for salmon in the unimpounded reach below Bonneville Dam. In contrast, the river downstream from Grand Coulee is impounded by 10 run-of-the-river dams and therefore turnover of water is rapid, from days to a few weeks depending on river discharge. Thus, the reservoirs function rather like the estuary; they are shallow at the head and deeper near the dam, with shifting sandy bottoms and rapid water exchange. From McNary downstream they have been colonized by estuarine invertebrates *Corophium salmonis* and *Neomysis mercedes*. Apparently these brackish-water amphipods hitched a ride upstream in barges, perhaps those used by the U.S. Army Corps of Engineers to transport salmon as a dam mitigation effort. These two species apparently are the most abundant macroinvertebrates in the reservoirs today, although midges also are abundant. Juvenile salmon migrating through the reservoirs are able to feed on *Corophium* in the shallow-water areas, especially below Bonneville. However, *Neomysis* likely is largely unavailable as a food source because it lives on the bottom of the deeper portions of the reservoirs, rising at night to feed voraciously on zooplankton near the surface. Salmon stay near the surface and are daytime sight feeders. Thus, *Neomysis* avoid fish predation and likely reduce the zooplankton forage base for the salmon. The conclusion is that the reservoirs compromise salmon by increasing temperatures and maintaining nonriverine food webs with voracious predators including native (northern pikeminnow) and nonnative (walleye, smallmouth bass) species. Impoundment has severed the interactive pathways that link the main stem to its tributaries. Ecological connectivity to riparian and backwater environments also has been severed throughout most of the river corridor by the prevention of seasonal flooding and by electrical load variations at the dams that rapidly fluctuate base flow discharge, which periodically dries out and thus sterilizes shallow shoreline environments that are crucial rearing areas. Indeed, even in the Hanford Reach, "raisin fish" or the dried bodies of juvenile fall chinook recently emerged from the redds can be observed annually in backwater areas dried out suddenly by hydropower operations at Priest Rapids Dam. The very young fish cannot endure the power of the main-stem flow without being washed downstream prematurely and thus naturally entrain in the backwaters where they are vulnerable to flow fluctuations.

The main-stem Columbia River is a ribbon of water from the Rockies to the Pacific, slowed and harnessed by dam after dam. The cataracts at the

Cascades, the Dalles with Celilo Falls, and Kettle Falls were the historic gathering places for fishes and people, the social foci of the main river. They lie buried under slack waters that now carry trade, irrigate the interior dry lands, and drive the turbines. The sloughs of the lower river and the wetlands in the headwaters remain as special natural features and the Hanford Reach with its Vernita Bar certainly is important as a salmon refugium. The dams, especially Bonneville and Grand Coulee, are special features to many people. Certainly, the electric light show nightly during the summer produced on the face of Grand Coulee Dam is special as it chronicles the Emersonian triumph of human labor and American capitalism on the Columbia.

But the pervasive persistence of the salmon culture on the Columbia somewhat belies the dam light show. Some $500 million per year has been spent during the last two decades to enhance Columbia River salmon by federal and state management agencies and the power distributor, Bonneville Power Administration, on a wide array of mitigation actions, such as changes in dam operations and evaluation studies, habitat improvement activities, hatcheries, and a variety of oddly justified actions, such as barging of smolts and payment of bounties to fishermen to harvest native pikeminnow because they prey on salmon smolts as they pass stunned over the dams or through the turbines (http://www.efw.bpa.gov/cgi-bin/efw/E/Welcome.cgi). Throughout the basin, 98 hatcheries produce some 2 million smolts annually that spill downstream to the main stem, where they feed pikeminnow, walleye, smallmouth bass, and other predators that flourish in the reservoirs. Less than 1%, usually far less, of the hatchery fishes survive to return as adults, whereas naturally produced fishes, such as those from the Hanford Reach, fare much better, with a 10% or greater return rate. Canadian managers currently are spreading commercial fertilizer in Arrow Reservoir behind Duncan Dam to increase resident fish production, apparently gauged to replace the salmon that can no longer get to Canada. The salmon culture of the Columbia seems substantially confounded for the lack of a unifying management paradigm for recovery (Williams 2005).

Perhaps the dilemma for conservation of the main stem and the large-river portions of the major tributaries in general is best illustrated by the plight of white sturgeon, a giant fish commonly over 3 m long and weighing upward of 500 kg. They were historically numerous in the estuary and deep portions of

the river and migrated into shallow riparian areas to spawn during the spate of spring snowmelt. The ponderous sturgeon cannot ascend the fish ladders at the dams and populations are land locked. White sturgeon remain somewhat numerous only below Bonneville. The remnant upriver populations hang on because water quality in the main-stem rivers is generally good, they are bottom-feeders and therefore have reasonably good food from invertebrate-laced organic matter accumulating in the reservoirs and on river bottoms owing to regulated flow, and they live a long time, 80 years or more (Wydoski and Whitney 2003). They spawn on gravel and cobble scoured clean by high flows and some recruitment does occur because the main-stem rivers do flood some in very wet years. But juvenile sturgeon require shallow, productive habitat for rearing, underscoring the natural ecosystem linkage of the river channel to the riparian floodplains. Most floodplain and riparian habitat, of course, is either totally destroyed or nonfunctional because it rarely floods. But operation of the dams could be done in ways that seasonally flood the remaining nodes of riparian area, specifically as a conservation measure to increase spawning and rearing success (Coutant 2004). Otherwise they will gradually die out like an old-growth forest, no matter how much protection they are accorded.

The future of the Columbia as a salmon river ecosystem is bleak. Columbia salmon and white sturgeon, among other native fishes, are almost gone because it is not really a river any more. Critical ecological connectivity—main stem to tributaries and main stem to riparian floodplains—has been severed by dams and associated activities such as revetments, floodplain deforestation and urban encroachment, and proliferation of hatcheries and nonnative fishes. The dams serve their purpose very successfully and the Columbia is indeed an organic machine driving a huge hydropower, agricultural, and increasingly urban culture. The natural salmon culture of the river remains persistently valued, however. Enhancement of this traditional attribute of the Columbia requires reconnection of the severed ecological pathways (Stanford et al. 1996). In the main stem and the tributaries, additional nodes of functional riverine habitat, like the Hanford Reach, must be identified, naturalized, and reconnected laterally and longitudinally in an ecosystem context by substantially changing dam operations (e.g., permanent reservoir drawdown allowing river habitat to reemerge; implementation of seasonality of flow) and by allowing natural reproduction without harvest of returning fishes so that the Columbia can regain at least some of its former status as one of the world's great salmon rivers.

ADDITIONAL RIVERS

The Owyhee River drains a region located in the arid reaches of southwestern Idaho, southeastern Oregon, and northern Nevada (Fig. 13.25). A 5th order stream, it flows in a northward direction before joining the Snake River on the Oregon–Idaho border. The Owyhee occupies the driest subbasin in the Columbia River drainage and only 16% of the drainage system flows year-round. Wild and Scenic River designation has been accorded to segments of the main stem and some tributaries. The region is sparsely populated, with only 5% of the basin devoted to agriculture; the majority (82%) of the region consists of shrub-steppe desert. A single dam, impounding Lake Owyhee, is located near the mouth. Only 25 of the 49 fish species present are native, but none are endangered.

The Grande Ronde River flows in a northeasterly direction, draining a little over 10,000 km² of extreme northeastern Oregon and a small portion of southeastern Washington (Fig. 13.27). It is a 5th order stream draining the Blue Mountains and a small area of Palouse Grassland before entering the Snake River just north of the Oregon–Washington–Idaho juncture (Fig. 13.1). Wild and Scenic designation has been given to portions of several tributaries and the lower Grande Ronde itself. Of the 38 species of fishes found in the Grande Ronde, 23 are native and 4 are endangered or threatened (fall Chinook salmon, spring/summer Chinook salmon, steelhead trout, and bull trout). Two hydropower dams are present in the system. The higher elevations of the drainage basin are forested and the lower regions are devoted to rangeland and limited agriculture.

The 7th order Clearwater River and its major tributaries (North, Middle, and South forks) drain a large area of the lower Idaho panhandle, extending from the Montana border westward to its confluence with the Snake River at the eastern border of Washington state (Fig. 13.29). The basin is 80% forested and the river contains 30 species of fishes, 19 of which are native (2 endangered; steelhead trout and fall Chinook). The Middle Fork of the Clearwater and its two major tributaries, the Lochsa and Selway rivers, are free flowing and designated as National Wild and Scenic Rivers. The North Fork (Fig. 13.10) is blocked near its mouth by Dworshak Dam, a major hydroelectric and storage facility in the Columbia basin. Located as it is, essentially at the mouth, and with no fish passage facilities, the dam has eliminated historic runs of salmon and steelhead trout.

The 6th order Spokane River drains portions of the northern Idaho panhandle and eastern Washington (Fig. 13.31). It flows in a northwesterly direction before joining the Columbia River approximately 80 km west of Spokane. The basin is largely forested, with portions occurring in the Palouse Grassland ecoregion. One major lake, Lake Coeur d'Alene, is present and the largest tributary is the St. Joe River, one of the longest free-flowing rivers in the Columbia River basin. Twenty-four species of fishes are found in the basin, none of which are endangered. Six low-head dams for diversion and hydropower are present. Most of the basin is lightly populated, with the highest concentration living in Spokane, Washington, the largest city in the interior Columbia basin (Fig. 13.11).

The Methow River is a free-flowing 5th order stream in northern Washington, just south of the Canadian border (Fig. 13.12). It flows in a southerly direction before joining the Columbia River at Pateros (Fig. 13.33). The Methow has a relatively small basin (4831 km²) lying just east of the crest of the Cascade Mountains; small parts of the eastern basin are in the Palouse Grasslands. Of the 32 species of fishes present, 25 are native and 3 (steelhead trout, Chinook salmon, and bull trout) are endangered or threatened. The headwaters are in wilderness and the North Cascades National Park, and the Chewak River is a major tributary. Seven small irrigation-diversion dams are in the basin, which is lightly populated.

The 6th order John Day River drains a large area of north-central Oregon, flowing northwest and north before joining the Columbia River on the Oregon–Washington border (Fig. 13.35). The northern region of the drainage, prior to entering the Columbia River (Fig. 13.13), is arid, whereas the headwater regions are forested, mainly draining the Blue Mountains. The North, Middle, and South forks are the main tributaries. The region is extremely sparsely populated; 85% of the basin is scrub and rangeland. Of the 27 species of fishes present, 17 are native. The system is largely unregulated except for irrigation diversions. Three segments are designated as Wild and Scenic, and the river contains two of the last remaining intact wild populations of anadromous fishes in the Columbia River basin (spring Chinook salmon and summer steelhead trout).

FIGURE 13.10 North Fork of the Clearwater River near Kelly Creek Ranger Station, Idaho (PHOTO BY C. E. CUSHING).

FIGURE 13.11 Spokane River below sewage plant near Spokane, Washington (PHOTO BY TIM PALMER).

FIGURE 13.12 Lower Methow River, Washington (Photo by Tim Palmer).

FIGURE 13.13 Lower John Day River, looking upstream of Highway 206, Oregon (Photo by A. C. Benke).

FIGURE 13.14 Cowlitz River at Randle, Washington (Photo by Tim Palmer).

The Cowlitz River, a 5th order stream, originates in the Cascade Mountains, flowing west and then south before draining into the Columbia River in the Puget Sound Lowlands near Longview, Washington (Fig. 13.37, Fig. 13.14). The drainage basin is largely forested. The system was heavily impacted by the 1980 Mount St. Helens eruption; it was choked by sediments but is gradually recovering. Thirty-two species of fishes are present and two are endangered or threatened (steelhead trout and chum salmon). The Toutle River is a major tributary and was even more heavily impacted by the St. Helens eruption. Three hydroelectric dams are on the main stem, impounding Mayfield and Riffle lakes.

ACKNOWLEDGMENTS

Preparation of this chapter was supported by the Jessie M. Bierman Professorship at the Flathead Lake Biological Station. The authors thank Phil Matson, Marie Kohler, and Morris Uebelacker for research and editorial help.

LITERATURE CITED

Abell, R. A., D. M. Olson, E. Dinerstein, P. T. Hurley, J. T. Diggs, W. Eichbaum, S. Walters, W. Wettengel, T. Allnutt, C. J. Loucks, and P. Hedao. 2000. *Freshwater ecoregions of North America: A conservation assessment.* Island Press, Washington, D.C.

Allen, J. E., M. Burns, and S. C. Sargent. 1986. *Cataclysms on the Columbia.* Timber Press, Portland, Oregon.

Altman, B., C. M. Henson, and I. R. Waite. 1997. Summary of information on aquatic biota and their habitats in the Willamette Basin, Oregon, through 1995. Water-Resources Investigations Report 97-4023. U.S. Geological Survey, Portland, Oregon.

Anderson, N. H. 1992. Influence of disturbance on insect communities in Pacific Northwest streams. *Hydrobiologia* 248:79–92.

Anderson, N. H., and B. P. Hansen. 1987. An annotated check list of aquatic insects collected at Berry Creek, Benton County, Oregon, 1960–1984. Occasional Publication no. 2. Systematic Entomology Laboratory, Oregon State University, Corvallis.

Attwater, B. S. 1986. Pleistocene glacial-lake deposits of the Sanpoil River Valley, Northeastern Washington. U.S. Geological Survey Bulletin 1661. U.S. Government Printing Office, Washington, D.C.

Bastasch, R., A. Bilbao, and G. Sieglitz. 2002. *Draft Willamette subbasin summary.* Northwest Power Planning Council, Portland, Oregon.

Beckwith, M. A. 2002. Selected trace-element and synthetic-organic compound data for streambed sediment from the Clark Fork–Pend Oreille and Spokane River basins, Montana, Idaho, and Washington, 1998. Open-File 02-336. U.S. Geological Survey, Helena, Montana.

Behnke, R. J. 2002. *Trout and salmon of North America.* Free Press, New York.

Braatne, J. H., and B. Jamieson. 2001. *The impacts of flow regulation on riparian cottonwood forests of the Yakima River.* Bonneville Power Administration, Portland, Oregon.

Busack, C., C. Knudsen, A. Marshall, S. Phelps, and D. Seiler. 1991. Yakima hatchery experimental design. Annual Progress Report DOE/BP-00102. Bonneville Power Administration, Division of Fish and Wildlife, Portland, Oregon.

Butler, V. L. 1993. Natural versus cultural salmonid remains: Origin of the Dalles roadcut bones, Columbia River, Oregon. U.S.A. *Journal of Archaeological Science* 20(1):1–24.

Carter, J. L., S. V. Fend, and S. S. Kennelly. 1996. The relationships among three habitat scales and stream benthic invertebrate community structure. *Freshwater Biology* 35:109–124.

Cederholm, C. J., M. D. Kunze, T. Murota, and A. Sibatani. 1999. Pacific salmon carcasses: Essential contributions of nutrients and energy for aquatic and terrestrial ecosystems. *Fisheries* 24(10):6–15.

Chance, D. H., and J. V. Chance. 1977. *Kettle Falls 1976: Salvage archaeology in Lake Roosevelt.* Anthropological Research Manuscript Series 39. Laboratory of Anthropology, University of Idaho, Moscow.

Chatters, J. C. 2000. The recovery and first analysis of an early Holocene human skeleton from Kennewick, Washington. *American Antiquity* 65:291–316.

Chatters, J. C., V. L. Butler, M. J. Scott, D. M. Anderson, and N. A. Neitzel. 1995. A paleoscience approach to estimating the effects of climatic warming on salmonid fisheries of the Columbia River Basin. In R. J. Beamish (ed.). Climate change and northern fish populations. *Canadian Special Publication of Fisheries and Aquatic Sciences* 121:489–496.

Cichosz T., D. Saul, A. Davidson, W. Warren, D. Rollins, J. Willey, T. Tate, T. Papanicolaou, and S. Juul. 2001. *Draft Clearwater subbasin summary.* Northwest Power Planning Council, Portland, Oregon.

Clifton, D. G. 1985. Analysis of biological data collected in the Bull Run watershed, Portland, Oregon, 1978 to 1983. Water-Resources Investigations Report 85–4245. U.S. Geological Survey, Portland, Oregon.

Connor, J. E. 1993. *Hydrology, hydraulics and geomorphology of the Bonneville Flood.* Geological Society of America, Boulder, Colorado.

Cooper, D. J., and K. Meiring. 1989. *A handbook of wetland plants of the Rocky Mountains.* EPA Region VIII, Denver, Colorado.

Coutant, C. C. 2004. A riparian habitat hypothesis for successful reproduction of white sturgeon. *Reviews in Fisheries Sciences* 12:23–73.

Craig, S. D. 1997. Habitat conditions affecting bull trout, *Salvelinus confluentus*, spawning areas within the Yakima River basin, Washington. Master thesis, Central Washington University, Ellensburg.

Cressman, L. S., D. L. Cole, W. A. Davis, T. M. Newman, and D. J. Scheans. 1960. Cultural sequences at the Dalles, Oregon: A contribution to Pacific Northwest prehistory. *Transactions of the American Philosophical Society* 50(10), Philadelphia.

Cuffney, T. F., M. R. Meador, S. D. Porter, and M. E. Gurtz. 1997. Distribution of fish, benthic invertebrate, and algal communities in relation to physical and chemical conditions, Yakima River basin, Washington, 1990. Water-Resources Investigations Report 96-4280. U.S. Geological Survey, Raleigh, North Carolina.

Cushing, C. E. 1967. Concentration and transport of ^{32}P and ^{65}Zn by Columbia River plankton. *Limnology and Oceanography* 12:330–332.

Cushing, C. E., and L. A. Rancitelli. 1972. Trace element analysis of Columbia River water and phytoplankton. *Northwest Science* 46:115–121.

Dammers, W., P. Foster, M. Kohn, C. Morrill, J. Serl, and G. Wade. 2002. *Draft Cowlitz River subbasin summary.* Northwest Power Planning Council, Portland, Oregon.

Daubenmire, R. 1978. *Plant geography: With special reference to North America.* Academic Press, New York.

Dimick, R. E., and F. Merryfield. 1945. The fishes of the Willamette River system in relation to pollution. Engineering Experiment Station Bulletin Series no. 20. Oregon State College, Corvallis.

Dudley, T. L., and N. H. Anderson. 1982. A survey of invertebrates associated with wood debris in aquatic habitats. *Melanderia* 39:1–21.

Dunlap, J. M., and R. F. Stettler. 1996. Genetic variation and productivity of *Populus trichocarpa* and its hybrids. IX. Phenology and Melampsora rust incidence of native black cottonwood clones from four river valleys in Washington. *Forest Ecology and Management* 87:233–256.

Eddy, B., and M. C. Nowak. 2001. *Draft Grande Ronde subbasin summary.* Northwest Power Planning Council, Portland, Oregon.

Eitemiller, D. J., C. P. Arango, K. L. Clark, and M. L. Uebelacker. 2002. *The effects of anthropogenic alterations to lateral connectivity on seven select alluvial floodplains within the Yakima River Basin, Washington.* Department of Geography and Land Studies, Central Washington University, Ellensburg.

Ellis, B. K., and J. A. Stanford. 2001. Pollution of Flathead Lake by atmospheric fallout: Analyses of loads, comparison to other sites and potential sources. Open File Report 167-01. Prepared for the Flathead Basin Commission, Kalispell, Montana, by Flathead Lake Biological Station, University of Montana, Polson.

Ellis, B. K., J. A. Stanford, J. A. Craft, D. W. Chess, F. R. Hauer, and D. C. Whited. 2003. Plankton communities of alpine and subalpine lakes in Glacier National Park, Montana, U.S.A., 1984–1990. *Verhandlungen der Internationale Vereinigung für theoretische und angewandte Limnologie* 28:1542–1550.

Ellis, B. K., J. A. Stanford, and J. V. Ward. 1998. Microbial assemblages and production in alluvial aquifers of the Flathead River, Montana, USA. *Journal of the North American Benthological Society* 17:382–402.

Endangered Species Act. 1973. *Statutes at Large* 87:884.

Fahey, J. 1974. *The Flathead Indians.* University of Oklahoma Press, Norman.

Federal Columbia River Power System (FCRPS). 2001. *The Columbia River system inside story*, 2nd ed. DOE/BP-3372. Prepared for Federal Columbia River Power System, System Operation Review, by a joint project of the U.S. Bureau of Reclamation, U.S. Army Corps of Engineers, and Bonneville Power Administration, Portland, Oregon.

Finklin, A. 1986. A climatic handbook for Glacier National Park, with data for Waterton Lakes National Park. General Technical Report INT-204. USDA Forest Service Intermountain Research Station, Ogden, Utah.

Foster, J., and others. 2002. *Draft Methow basin summary.* Northwest Power Planning Council, Portland, Oregon.

Franklin, J. F., and C. T. Dyrness. 1973. Natural vegetation of Oregon and Washington. General Technical Report PNW-8. USDA Forest Service, Pacific Northwest Forest and Range Experimental Station, Portland, Oregon.

Frest, T. J., and E. J. Johannes. 1995. Interior Columbia basin mollusk species of special concern. Final Report to Interior Columbia Basin Ecosystem Management Project. Deixis Consultants, Seattle, Washington.

Gregory, S. V. 1993. Willamette River basin study—Periphyton algal dynamics. Final report prepared for Oregon Department of Environmental Quality. Department of Fish and Wildlife, Oregon State University, Corvallis.

Gregory, S. V., M. A. Meleason, and D. J. Sobota. 2003. Modeling the dynamics of wood in streams and rivers. In S. V. Gregory, K. L. Boyer, and A. M. Gurnell (eds.). *The ecology and management of wood in world rivers,* pp. 315–335. Symposium 37. American Fisheries Society, Bethesda, Maryland.

Gregory, S. V., F. J. Swanson, W. A. McKee, and K. W. Cummins. 1991. An ecosystem perspective of riparian zones. *Bioscience* 41:540–551.

Guard, B. J. 1995. *Wetland plants of Oregon and Washington.* Lone Pine, Renton, Washington.

Hall, M. H. P., and D. B. Fagre. 2003. Modeled climate-induced glacier change in Glacier National Park, 1850–2100. *BioScience* 53:131–140.

Hall, C. A. S., J. A. Stanford, and F. R. Hauer. 1992. The distribution and abundance of organisms as a consequence of energy balances along multiple environmental gradients. *Oikos* 65:377–390.

Harmon, M. E., J. F Franklin, F. J. Swanson, P. Sollins, S. V. Gregory, J. D. Lattin, N. H. Anderson, S. P. Cline, N. G. Aumen, J. R. Sedell, G. W. Lienkaemper, K. Cromack Jr., and K. W. Cummins. 1986. Ecology of coarse woody debris in temperate ecosystems. *Advances in Ecological Research* 15:133–302.

Hauer, F. R., B. J. Cook, M. S. Lorang, and J. A. Stanford. 2002. Review and synthesis of riverine databases and ecological studies in the upper Snake River, Idaho. Part A: Relationships of flow, geomorphology and river habitat interactions. Open File Report 175-02. Prepared for U.S. Department of the Interior, Bureau of Reclamation, Boise, Idaho, by Flathead Lake Biological Station, University of Montana, Polson.

Hauer, F. R., C. N. Dahm, G. A. Lamberti, and J. A. Stanford. 2003. Landscapes and ecological variability of rivers in North America: Factors affecting restoration strategies. In R. C. Wissmar and P. A. Bisson (eds.). *Strategies for restoring river ecosystems: Sources of variability and uncertainty in natural and managed systems,* pp. 81–105. American Fisheries Society, Bethesda, Maryland.

Hauer, F. R., and M. S. Lorang. 2004. River regulation, decline of ecological resources, and potential for restoration in an arid lands river in the western USA. *Aquatic Sciences* 66:388-401.

Hauer, F. R., J. A. Stanford, J. J. Giersch, and W. H. Lowe. 2000. Distribution and abundance patterns of macroinvertebrates in a mountain stream: An analysis along multiple environmental gradients. *Verhandlungen der Internationale Vereinigung für theoretische und angewandte Limnologie* 27:1485–1488.

Hawkins, C. P., M. L. Murphy, and N. H. Anderson. 1982. Effects of canopy, substrate composition, and gradient on the structure of macroinvertebrate communities in the Cascade Range streams of Oregon. *Ecology* 63:1840–1856.

Hawkins, C. P., and J. R. Sedell. 1981. Longitudinal and seasonal changes in functional organization of macroinvertebrate communities in four Oregon streams. *Ecology* 62:387–397.

Henny, C. J., and J. L. Kaiser. 1996. Osprey population increase along the Willamette River, Oregon, and the role of utility structures, 1976–1993. In D. M. Bird, D. E. Varland, and J. J. Negro (eds.). *Raptor adaptations to human influenced environments,* pp. 97–108. Academic Press, London.

Hjort, R. C., P. L. Hulett, L. D. LaBolle, and H. W. Li. 1984. Fish and invertebrates of revetments and other habitats in the Willamette River, Oregon: Vicksburg, Mississippi. Technical Report E–84–9. Prepared for U.S. Army Corps of Engineers by Waterways Experiment Station, Oregon State University, Corvallis.

Hughes, R. M., and J. R. Gammon. 1987. Longitudinal changes in fish assemblages and water quality in the Willamette River, Oregon. *Transactions of the American Fisheries Society* 116:196–209.

Hulse, D., S. V. Gregory, and J. Baker. 2002. *Willamette basin planning atlas: Trajectories of environmental and ecological change.* Oregon State University Press, Corvallis.

Hunt, C. B. 1974. *Natural regions of the United States and Canada.* W. H. Freeman, San Francisco.

Johnson, S. L., F. J. Swanson, G. E. Grant, and S. M. Wondzell. 2000. Riparian forest disturbances by a mountain flood: The influence of floated wood. *Hydrological Processes* 14:3031–3050.

Johnson, W. C., M. D. Dixon, R. Simons, S. Jenson, and K. Larson. 1995. Mapping the response of riparian vegetation to possible flow reductions in the Snake River, Idaho. *Geomorphology* 13:159–173.

Kammerer, J. C. 1990. Largest rivers in the United States. U.S. Geological Survey Open-File Report 87-242. U.S. Geological Survey, Washington, D.C.

Kinnison, H. B., and J. E. Sceva. 1963. *Effects of hydraulic and geologic factors on streamflow of the Yakima River basin, Washington.* U.S. Government Printing Office, Washington.

Knapp, S., and T. Unterwegner. 2001. *Draft John Day sub-basin summary.* Northwest Power Planning Council, Portland, Oregon.

Lamberti, G. A., S. V. Gregory, L. R. Ashkenas, R. C. Wildman, K. M. S. Moore. 1991. Stream ecosystem recovery following a catastrophic debris flow. *Canadian Journal of Fisheries and Aquatic Sciences* 48:196–208.

Lee, K. K. and J. C. Risley. 2002. Estimates of groundwater recharge, base flow, and stream reach gains and losses in the Willamette River basin, Oregon. Water-Resources Investigations Report 01–4215. U.S. Geological Survey, Portland, Oregon.

Leland, H. V. 1995. Distribution of phytobenthos in the Yakima River basin, Washington, in relation to geology, land use, and other environmental factors. *Canadian Journal of Fisheries and Aquatic Sciences* 52:1108–1129.

Li, H. W., C. B. Schreck, C. E. Bond, and E. Rexstad. 1987. Factors influencing changes in fish assemblages of Pacific Northwest streams. In W. J. Matthews and D. C. Heins (eds.), *Community and evolutionary ecology of North American stream fishes*, pp. 193–202. University of Oklahoma Press, Norman.

Lichatowich, J. 1999. *Salmon without rivers: A history of the Pacific salmon crisis.* Island Press, Washington, D.C.

Lilga, M. C. 1998. Effects of flow variation on stream temperatures in the lower Yakima River. Master thesis, Washington State University.

Lloyd, T. M., R. Denny, and G. Constantino. 1983. Wintering waterfowl redistribution plan for the Columbia basin of Oregon and Washington. Cooperative Report. U.S. Fish and Wildlife Service, Washington Department of Fish and Wildlife, Oregon Department of Fish and Wildlife.

Lyford, J. H., Jr., and S. V. Gregory. 1975. The dynamics and structure of periphyton communities in three Cascade Mountain streams. *Verhandlungen der Internationale Vereinigung für theoretische und angewandte Limnologie* 19:1610–1616.

Maxell, B. A., J. K. Werner, P. Hendricks, and D. L. Flath. 2003. Herpetology in Montana: A history, status summary, checklists, dichotomous keys, accounts for native, potentially native, and exotic species, and indexed bibliography. *Northwest Fauna* 5:1–138.

McMichael, G. A., A. L. Fritts, and J. L. Dunnigan. 1998. Lower Yakima River predatory fish census: Feasibility study 1997. In T. N. Pearsons, G. A. McMichael, K. D. Ham, E. L. Bertrand, A. L. Fritts, and C. W. Hopley (eds.). *Draft Yakima River species interaction studies*, pp. 224–229. Progress Report 1995–1997. Bonneville Power Administration, Portland, Oregon.

McPhail, J. D., and C. C. Lindsey. 1986. Zoogeography of the freshwater fishes of Cascadia (the Columbia system and rivers north to the Stikine). In C. H. Hocutt and E. O. Wiley (eds.). *Zoogeography of North American freshwater fishes*, pp. 615–637. John Wiley and Sons, New York.

Merigliano, M. F. 1996. Ecology and management of the South Fork Snake River cottonwood forest. Technical

Bulletin 96-9. School of Forestry, University of Montana, Missoula.

Minshall, G. W., R. C. Petersen, T. L. Bott, C. E. Cushing, K. W. Cummins, R. L. Vannote, and J. R. Sedell. 1992. Stream ecosystem dynamics of the Salmon River, Idaho: An 8th-order system. *Journal of the Northern American Benthological Society* 11:111–137.

Minshall, G. W., R. C. Petersen, K. W. Cummins, T. L. Bott, J. R. Sedell, C. E. Cushing, and R. L. Vannote. 1983. Interbiome comparison of stream ecosystem dynamics. *Ecological Monographs* 51:1–25.

Moore, V. 1980. River and stream investigations—South Fork Snake River fisheries investigations: Creel census and tributary inventory. Project F-73-R-2, Subproject IV, Study VIII, Jobs I and II, Annual Report. Idaho Department of Fish and Game, Boise.

Moseley, R. K. 1998. Ute ladies tresses (*Spiranthes diluvialis*) in Idaho: 1997 status report. Unpublished report on file at the Conservation Data Center, Idaho Department of Fish and Game, Boise.

Mouw, J. E. B., and P. B. Alaback. 2003. Putting floodplain hyperdiversity in a regional context: An assessment of terrestrial–floodplain connectivity in a montane environment. *Journal of Biogeography* 30:87–103.

Munn, M. D., and M. A. Brusven. 1987. Discontinuity of Trichopteran (caddisfly) communities in regulated waters of the Clearwater River, Idaho, U.S.A. *Regulated Rivers: Research and Management* 1:61–69.

Munn, M. D., and M. A. Brusven. 1991. Benthic macroinvertebrate communities in nonregulated and regulated waters of the Clearwater River, Idaho. *Regulated Rivers: Research and Management* 6:1–11.

Murphy, M. L., and J. D. Hall. 1981. Varied effects of clearcut logging on predators and their habitat in small streams of the Cascade Mountains, Oregon. *Canadian Journal of Fisheries and Aquatic Science* 38:137–145.

National Research Council. 1996. *Upstream: Salmon and society in the Pacific Northwest.* National Academy Press, Washington, D.C.

Nehlsen, W., J. E. Williams, and J. A. Lichatowich. 1991. Pacific salmon at the crossroads: Stocks at risk from California, Oregon, Idaho, and Washington. *Fisheries* 16:4–21.

Neitzel, D. A., T. L. Page, and R. W. Hanf Jr. 1982. Mid-Columbia River microflora. *Journal of Freshwater Ecology* 1:495–505.

Nelson, C. M. 1973. Prehistoric cultural change in the intermontane plateau of western North America. In C. Renfrew (ed.). *The explanation of culture change: Models in prehistory*, pp. 371–390. Proceedings of the Research Seminar in Archaeology and Related Subjects, University of Sheffield, 1971. University of Pittsburgh Press, Pittsburgh, Pennsylvania.

Netboy, A. 1980. *The Columbia River salmon and steelhead trout: Their fight for survival.* University of Washington Press, Seattle.

Newell, R. L. 2003. Aquatic macroinvertebrates. In J. R. Evans, M. P. Lih, and P. W. Dunwiddie (eds.). *Biodiversity studies of the Hanford site: Final report 2002–2003*, pp. 73–96. Nature Conservancy, Seattle, Washington.

Nightengale, T. L. 1998. A survey of the benthic macroinvertebrate fauna in the Yakima and Naches rivers. Annual Report to the Yakima Joint Board, Seattle, Washington.

Nisbet, J. 1994. *Sources of the river: Tracking David Thompson across Western North America.* Sasquatch Books, Seattle, Washington.

Nordstrom, N., and R. Milner. 1997. Columbia spotted frog. In E. M. Larsen (ed.). *Management recommendations for Washington's priority species.* Vol. 3: *Amphibians and reptiles*, pp. 4–14. Washington Department of Fish and Wildlife, Olympia.

Oliver, W. H. 1983. *Farm–wildlife history, relationships and problems on the Yakima Indian reservation.* Wildlife Resource Management, Yakama Nation, Toppenish, Washington.

Parker, R. C. 1989. *South central waterfowl management plan.* Washington Department of Fish and Wildlife, Olympia.

Parsons, D. R. 1975. Time and energy budgets of a population of dippers during winter in the Cascade Range of Oregon. Master thesis, Oregon State University, Corvallis.

Patrick, R. 1995. *Rivers of the United States.* Vol. 2: *Chemical and physical characteristics.* John Wiley and Sons, New York.

Perugini, C., D. Saul, C. Rabe, A. Davidson, W. Warren, D. Rollins, and S. Lewis. 2002. *Draft Owyhee subbasin summary.* Northwest Power Planning Council, Portland, Oregon.

Pyle, R. M. 1974. *Watching Washington butterflies: An interpretive guide to the state's 134 species, including most of the butterflies of Oregon, Idaho, and British Columbia.* Seattle Audubon Society, Seattle.

Pyle, R. M. 1992. *The Audubon Society field guide to North American butterflies.* Alfred A. Knopf, New York.

Redmond, K. T., and D. R. Cayan. 1994. El Nino/Southern oscillation and western climate variability. In *Sixth Conference on Climate Variations, January 23–28, 1994, Nashville, Tennessee*, pp. 141–145. American Meteorological Society, Boston, Massachusetts.

Reed, J. P. 1995. Factors affecting the genetic architecture of black cottonwood populations. Master thesis, University of Washington, Seattle.

Rickert, D. A., R. R. Peterson, S. W. McKenzie, W. G. Hines, and S. A. Wille. 1977. Algal conditions and the potential for future algal problems in the Willamette River, Oregon. Circular 715–G. U.S. Geological Survey, Portland, Oregon.

Ricketts, T. H., E. Dinerstein, D. M. Olson, C. L. Loucks, W. Eichbaum, D. DellaSala, K. Kavanagh, P. Hedao, P. T. Hurley, K. M. Carney, R. Abell, and S. Walters. 1999.

Terrestrial ecoregions of North America: A conservation assessment. Island Press, Washington, D.C.

Rinella, J. F., P. A. Hamilton, and S. W. McKenzie. 1993. *Persistence of the DDT pesticide in the Yakima River basin, Washington.* U.S. Geological Survey Circular 1090. U.S. Government Printing Office, Washington, D.C.

Rinella, J. F., S. W. McKenzie, and G. J. Fuhrer. 1992. Surface-water-quality assessment of the Yakima River basin, Washington: Analysis of available water-quality data through 1985 water year. Open File Report 91-453. U.S. Geological Survey, Portland, Oregon.

Rinella, F. A., S. W. McKenzie, and S. A. Wille. 1981. Dissolved-oxygen and algal conditions in selected locations of the Willamette River Basin, Oregon. Open-File Report 81–529. U.S. Geological Survey, Portland, Oregon.

Ross, A. 1855. *The fur hunters of the far West: A narrative of adventures in the Oregon and Rocky Mountains.* Smith, Elder and Co., London.

Ross, A. 1969 (original edition 1849). Adventures of the first settlers on the Oregon or Columbia River. Edited by Milo M. Quaife. Citadel Press, New York.

Ross, C. P., and R. Rezak. 1959. The rocks and fossils of Glacier National Park: The story of their origin and history. Professional Paper 294-K. U.S. Geological Survey, Washington, D.C.

Saab, V. A. 1998. Effects of recreational activity and livestock grazing on habitat use by breeding birds in cottonwood forests along the South Fork Snake River. Unpublished report on file with USDA, Forest Service, Rocky Mountain Research Station, Boise, Idaho.

Saab, V. A. 1999. Importance of spatial scale to habitat use by breeding birds in riparian forests: A hierarchical analysis. *Ecological Applications* 9:135–151.

Schalk, R. F., and G. C. Cleveland. 1983. A chronological perspective on hunter-gatherer land use strategies in the Columbia Plateau. In R. F. Schalk (ed.). Cultural resource investigations for the Lyons Ferry Fish Hatchery Project, near Lyons Ferry, Washington, pp. 11–56. Report no. 8. Laboratory of Archaeology and History, Washington State University, Pullman.

Schindler, D. E., M. D. Scheuerell, J. W. Moore, S. M. Gende, T. B. Francis, and W. J. Palen. 2003. Pacific salmon and the ecology of coastal ecosystems. *Frontiers of Ecology and the Environment* 1:31–37.

Smith, D. G. 2001. Systematics and distribution of the recent Margaritiferidae. In G. Bauer and K. Wachtler (eds.). *Ecological studies.* Vol. 145: *Ecology and evolution of the freshwater mussels Unionoida*, pp. 33–49. Springer-Verlag, Berlin.

Snyder, E. B., D. J. Eitemiller, C. P. Arango, M. L. Uebelacker, and J. A. Stanford. 2003. Floodplain hydrologic connectivity and fisheries restoration in the Yakima River, U.S.A. *Verhandlungen der Internationale Vereinigung für theoretische und angewandte Limnologie* 28:1653–1657.

Snyder, E. B., and J. A. Stanford. 2001. Review and synthesis of river ecological studies in the Yakima River, Washington, with an emphasis on flow and salmon habitat interactions. Open File Report 163-01. Prepared for U.S. Department of Interior, Bureau of Reclamation, Yakima, Washington, by Flathead Lake Biological Station, University of Montana, Polson.

Stafford, C. P., J. A. Stanford, F. R. Hauer, and E. B. Brothers. 2002. Changes in lake trout growth associated with *Mysis relicta* establishment: A retrospective analysis using otoliths. *Transactions of the American Fisheries Society* 131:994–1003.

Stanford, J. A., and B. K. Ellis. 2002. Natural and cultural influences on ecosystem processes in the Flathead River Basin (Montana, British Columbia). In J. S. Baron (ed.). *Rocky Mountain futures: An ecological perspective*, pp. 269–284. Island Press, Washington, D.C.

Stanford, J. A., and F. R. Hauer. 1992. Mitigating the impacts of stream and lake regulation in the Flathead River Catchment, Montana, USA: An ecosystem perspective. *Aquatic Conservation: Marine and Freshwater Ecosystems* 2:35–63.

Stanford, J. A., and G. W. Prescott. 1988. Limnological features of a remote alpine lake in Montana, including a new species of *Cladophora* (Chlorophyta). *Journal of the North American Benthological Society* 7:140–151.

Stanford, J. A., E. B. Snyder, M. S. Lorang, D. C. Whited, P. L. Matson, and J. L. Chaffin. 2002. The Reaches Project: Ecological and geomorphic studies supporting normative flows in the Yakima River Basin, Washington. Open File Report 170-02. Prepared for U.S. Department of the Interior, Bureau of Reclamation, Yakima, Washington, by Flathead Lake Biological Station, University of Montana, Polson.

Stanford, J. A., and J. V. Ward. 1983. Insect species diversity as a function of environmental variability and disturbance in stream systems. In J. R. Barnes and G. W. Minshall (eds.). *Stream ecology: Application and testing of general ecological theory.* pp. 265–278. Plenum Press, New York.

Stanford, J. A., J. V. Ward, and B. K. Ellis. 1994. Ecology of the alluvial aquifers of the Flathead River, Montana. In J. Gibert, D. L. Danielopol, and J. A. Stanford (eds.). *Groundwater ecology*, pp. 367–390. Academic Press, San Diego.

Stanford, J. A., J. V. Ward, W. J. Liss, C. A. Frissell, R. N. Williams, J. A. Lichatowich, and C. C. Coutant. 1996. A general protocol for restoration of regulated rivers. *Regulated Rivers* 12:391–413.

Steedman, R. J., and N. H. Anderson. 1985. Life history and ecological role of the xylophagous aquatic beetle, *Lara avara. Freshwater Biology* 15:535–546.

Stober, Q. J., and R. E. Nakatani. 1992. Water quality and biota of the Columbia River system. In C. D. Becker and D. A. Neitzel (eds.). *Water quality in North American river systems*, pp. 51–83. Battelle Press, Columbus, Ohio.

Swanson, F. J., and G. W. Lienkaemper. 1978. Physical consequences of large organic debris in Pacific Northwest streams. General Technical Report PNW-69. U.S. Department of Agriculture, Forest Service, Pacific Northwest Forest and Range Experiment Station, Portland, Oregon.

Tetra Tech, Inc. 1994. Willamette River basin water quality study: Phase II, ecological monitoring component—Benthic metric selection and data evaluation. Prepared for Oregon Department of Environmental Quality by Tetra Tech, Inc., Redmond, Washington.

Thompson, B. C., J. E. Tabor, and C. L. Turner. 1988. Diurnal behavior patterns of waterfowl wintering on the Columbia River, Oregon and Washington. In M. W. Weller (ed.). *Waterfowl in winter*, pp. 153–167. University of Minnesota Press, Minneapolis.

Uebelacker, M. L. 1986. Geographic exploration in the Southern Cascades of Eastern Washington: Changing land, people and resources. Ph.D. diss., University of Oregon, Engene.

Uhrich, M. A., and D. A. Wentz. 1999. Environmental setting of the Willamette basin, Oregon. Water-Resources Investigations Report 97-4082-A. U.S. Geological Survey, Portland, Oregon.

U.S. Army Corps of Engineers (USACE). 1999. Willamette River basin, Oregon, floodplain restoration project: Section 905(b) analysis. Willamette River floodplain restoration study, Section 905(b), Reconnaissance report, Portland, Oregon.

Vannote, R. L., G. W. Minshall, K. W. Cummins, J. R. Sedell, and C. E. Cushing. 1980. The river continuum concept. *Canadian Journal of Fisheries and Aquatic Science* 37:130–137.

Van Sickle, J., J. Baker, A. Herlihy, P. Bayley, S. Gregory, P. Haggerty, L. Ashkenas, and J. Li. 2004. Projecting the biological condition of streams under alternative scenarios of human land use. *Ecological Applications* 14:368–380.

Walker, D. E. (ed.). 1998. *Handbook of North American Indians*. Vol. 12. Smithsonian Institution Press, Washington, D.C.

Ward, J. V., and J. A. Stanford. 1982. Thermal responses in the evolutionary ecology of aquatic insects. *Annual Review of Entomology* 27:97–117.

Ward, N. E., and D. L. Ward. 2004. Resident fish in the Columbia River Basin: Restoration, enhancement, and mitigation for losses associated with hydroelectric development and operations. *Fisheries* 29(3):10–18.

Wentz, D. A., B. A. Bonn, K. D. Carpenter, S. R. Hinkle, M. L. Janet, F. A. Rinella, M. A. Uhrich, I. R. Waite, A. Laenen, and K. E. Bencala. 1998. Water quality in the Willamette Basin, Oregon, 1991–95. Circular 1161. U.S. Geological Survey, Portland, Oregon.

Whalen, J. 2000. *Draft Spokane River subbasin summary.* Northwest Power Planning Council, Portland, Oregon.

White, R. 1995. *The organic machine: The remaking of the Columbia River.* Hill and Wang, New York.

Whited, D. C., J. A. Stanford, and J. S. Kimball. 2003. Application of airborne multispectral digital imagery to characterize the riverine habitat. *Verhandlungen der Internationale Vereinigung für theoretische und angewandte Limnologie* 28:1373–1380.

Whitfield, M. B., P. Munholland, and M. E. Maj. 1995. Inventory and monitoring of bald eagles and other raptorial birds of the Snake River, Idaho. Technical Bulletin no. 95-12. Idaho Bureau of Land Management, Boise.

Wille, S. A. 1976. Influence of light on algal growth in the lower Willamette River. Master thesis, Portland State University, Portland, Oregon.

Williams, R. N. (ed.). 2005. *Return to the river: Restoring Salmon to the Columbia River*. Elsevier, Inc. San Diego, California.

Wydoski, R. S., and R. R. Whitney. 2003. *Inland fishes of Washington.*, 2nd ed. American Fisheries Society in association with University of Washington Press, Bethesda, Maryland.

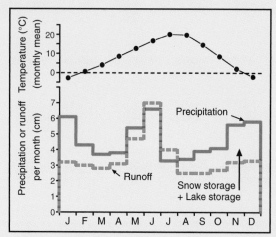

FIGURE 13.16 Mean monthly air temperature, precipitation, and runoff for the Flathead River basin.

FLATHEAD RIVER

Relief: 1676 m
Basin area: 22,241 km²
Mean discharge: 340 m³/s
River order: 6
Mean annual precipitation: 56 cm
Mean air temperature: 8.6°C
Mean water temperature: 11.1°C
Physiographic provinces: Northern Rocky Mountains (NR), Rocky Mountains in Canada (RM)
Biome: Temperate Mountain Forest
Freshwater ecoregion: Columbia Glaciated
Terrestrial ecoregions: North Central Rockies Forests, Montana Valley and Foothills Grasslands
Number of fish species: 29 (12 native)
Endangered species: 1 fish (threatened), 1 plant (threatened)
Major fishes: bull trout, westslope cutthroat trout, lake trout, mountain whitefish, lake whitefish, pigmy whitefish, sculpin, peamouth chub, longnose sucker, largescale sucker, northern pikeminnow
Major other aquatic vertebrates: spotted frog, boreal toad, painted turtle, tailed frog, three-toed salamander, beaver, river otter, mink, osprey, bald eagle, American merganser, harlequin duck, water ouzel
Major benthic invertebrates: caddisflies (*Parapsyche, Arctopsyche, Hydropsyche, Cheumatopsyche, Glossosoma, Brachycentrus, Rhyacophila*), stoneflies (*Taeniopteryx, Pteronarcys, Pteronarcella, Hesperoperla, Claassenia, Isocapnia, Paraperla*), mayflies (*Rhithrogena, Baetis, Ephemerella, Drunella*), true flies (*Chironomidae, Simuliidae, Hexatoma, Atherix*), crayfish (signal), mollusks (*Margaritinopsis*)
Nonnative species: lake trout, lake whitefish, kokanee, yellow perch, northern pike, rainbow trout, brook trout, largemouth bass, smallmouth bass, pumpkinseed, black bullhead, virile crayfish
Major riparian plants: black cottonwood, green alder, coyote willow, sandbar willow, Drummond's willow, red-osier dogwood, Englemann spruce, beaked sedge, other sedges, floodplain Drayas, spotted knapweed
Special features: National Wild and Scenic River segments; 42% of upper basin designated national park or wilderness; Flathead Indian reservation; three National Wildlife Refuges; Flathead Lake largest (surface area) lake in western United States
Fragmentation: two major dams (Hungry Horse, Kerr); Flathead Irrigation Project has 17 reservoirs
Water quality: relatively free of major pollutants; pH = 8.1, alkalinity = 86 mg/L as $CaCO_3$, NO_3-N = 0.050 mg/L, PO_4-P = <0.001 mg/L
Land use: 80% forest/alpine, 10% range, 8% agriculture, 2% urban
Population density: 2.7 people/km²
Major information sources: Stanford and Hauer 1992, Stanford and Ellis 2002, http://water.usgs.gov/nwis 2003, http://www.wcc.nrcs.usda.gov/climate/wetlands.html 2003

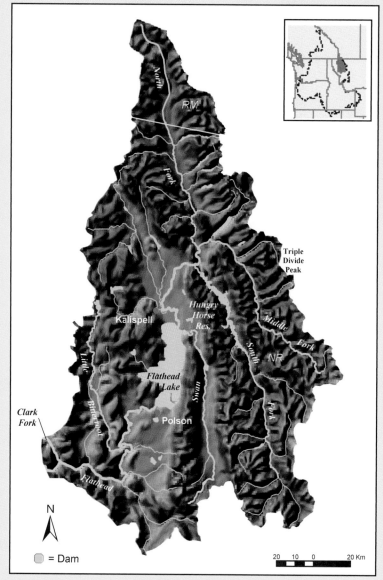

FIGURE 13.15 Map of the Flathead River basin. Physiographic provinces are separated by a yellow line.

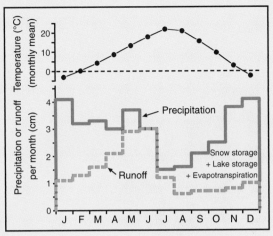

FIGURE 13.18 Mean monthly air temperature, precipitation, and runoff for the Snake River basin.

SNAKE/SALMON RIVER

Relief: 3048 m

Basin area: 281,000 km^2

Mean discharge: 1565 m^3/s

River order: 8

Mean annual precipitation: 36 cm

Mean air temperature: 9.1°C

Mean water temperature: 11.5°C

Physiographic provinces: Columbia–Snake River Plateaus (CU), Northern Rocky Mountains (NR), Middle Rocky Mountains (MR), Basin and Range (BR)

Biomes: Temperate Mountain Forest, Desert

Freshwater ecoregions: Columbia Glaciated, Columbia Unglaciated, Upper Snake

Terrestrial ecoregions: Snake/Columbia Shrub Steppe, Palouse Grasslands, Blue Mountain Forests, South Central Rockies Forests, North Central Rockies Forests

Number of fish species: 39 (19 native)

Endangered species: 5 fishes (4 threatened, 1 endangered), 6 snails (1 threatened, 5 endangered), 1 plant (threatened)

Major fishes: Yellowstone cutthroat trout, chinook salmon, steelhead, rainbow trout, bull trout, mountain whitefish, chiselmouth, carp, northern pikeminnow, longnose dace, speckled dace, Utah chub, yellow perch, black crappie, redside shiner, largescale sucker, pumpkinseed, smallmouth bass, largemouth bass, sculpin

Major other aquatic vertebrates: northern leopard frog, Columbia spotted frog, tailed frog, western painted turtle, wood duck, mallard duck, Canada goose, trumpeter swan, sandhill crane, great blue heron, white pelican, bald eagle, osprey, beaver, river otter, muskrat

Major benthic invertebrates: caddisflies (*Brachycentrus, Glossosoma, Arctopsyche, Cheumatopsyche, Hydropsyche*), mayflies (*Baetis, Drunella, Ephemerella, Rhithrogena, Tricorythodes*), stoneflies (*Sweltsa, Zapada, Claassenia, Hesperoperla, Skwala*), true flies (*Chironomidae, Simulium, Tipula*), mollusks (*Pyrgulopsis, Valvata, Physa, Taylorconcha*)

Nonnative species: smallmouth bass, walleye, largemouth bass, carp, brown bullhead, black crappie, mosquitofish, pumpkinseed, yellow perch, brown trout, brook trout, channel catfish, black bullhead, bluegill, New Zealand mudsnail

Major riparian plants: black cottonwood, red-osier dogwood, willow, alder, purple loosestrife

Special features: headwaters of Snake in Yellowstone and Grand Teton National Parks; several tributaries are National Wild and Scenic Rivers; midchannel islands of central Snake created by late Pleistocene flooding from glacial Lake Bonneville (Deer Flats National Wildlife Refuge); Salmon River's "River-of-No-Return"

Fragmentation: several major dams on main stem, including three dams in Hells Canyon

Water quality: agricultural runoff in middle reaches; pH = 7.9, alkalinity = 69 mg/L as CaCO$_3$, NO$_3$-N = 0.37 mg/L, PO$_4$-P = 0.02 to 0.06 mg/L

Land use: 10% to 15% forest/alpine, 50% scrub/rangeland, 30% agriculture, 4% barren, 1% urban

Population density: 15 people/km^2

Major information sources: Hauer et al. 2002, Hauer et al. 2003, Hauer and Lorang 2004, http://water.usgs.gov/nwis 2003, www.wcc.nrcs.usda.gov/climate/wetlands.html 2003

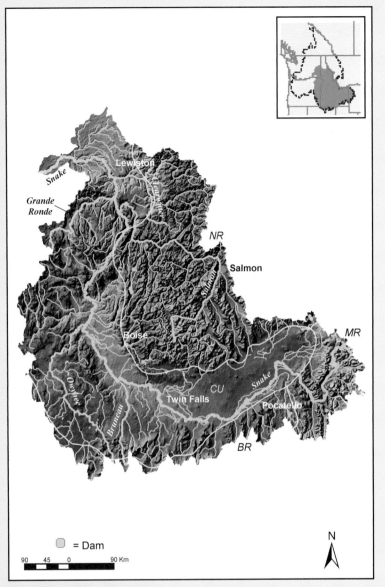

FIGURE 13.17 Map of the Snake River basin. Physiographic provinces are separated by yellow lines.

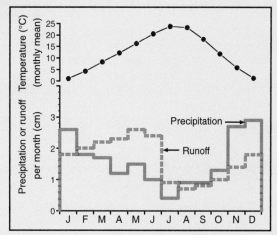

FIGURE 13.20 Mean monthly air temperature, precipitation, and runoff for the Yakima River basin.

YAKIMA RIVER

Relief: 2334 m
Basin area: 15,900 km²
Mean discharge: 102 m³/s
River order: 5
Mean annual precipitation: 19 cm
Mean air temperature: 12.3°C
Mean water temperature: 13.3°C
Physiographic provinces: Cascade–Sierra Mountains (CS), Columbia–Snake River Plateaus (CU)
Biomes: Temperate Mountain Forest, Desert
Freshwater ecoregions: Columbia Glaciated, Columbia Unglaciated
Terrestrial ecoregions: Eastern Cascades Forests, Snake/Columbia Shrub Steppe, Cascade Mountains Leeward Forests
Number of fish species: 50
Endangered species: 2 fishes (threatened)
Major fishes: chinook salmon, steelhead, rainbow trout, mountain whitefish, chiselmouth, carp, northern pikeminnow, longnose dace, speckled dace, redside shiner, bridgelip sucker, largescale sucker, pumpkinseed, smallmouth bass, largemouth bass, sculpin

Major other aquatic vertebrates: northern leopard frog, Columbia spotted frog, northwest salamander, Pacific giant salamander, western pond turtle, wood duck, mallard duck, Canada goose, tundra swan, trumpeter swan, sandhill crane, bald eagle, osprey, beaver, river otter, muskrat

Major benthic invertebrates: caddisflies (*Brachycentrus, Glossosoma, Arctopsyche, Cheumatopsyche, Hydropsyche*), mayflies (*Baetis, Drunella, Ephemerella, Rhithrogena, Tricorythodes*), stoneflies (*Sweltsa, Zapada, Claassenia, Hesperoperla, Skwala*), true flies (*Simuliidae, Chironomidae, Tipula*)

Nonnative species: smallmouth bass, walleye, largemouth bass, carp, brown bullhead, black crappie, mosquitofish, pumpkinseed, yellow perch, brown trout, brook trout, channel catfish, black bullhead, bluegill

Major riparian plants: black cottonwood, willow, alder

Special features: five expansive floodplain segments; headwaters in Alpine Lakes Wilderness; rich agriculture area, much within Yakama Indian reservation; extremely arid in lower river

Fragmentation: six headwater storage reservoirs; seven irrigation diversion dams on main stem

Water quality: 72 segments with temperatures lethal to salmonid fishes due to irrigation withdrawal; agricultural runoff in lower river; pH = 7.9, alkalinity = 103 mg/L as CaCO₃, NO₃-N = 0.79 mg/L, PO₄-P = 0.33 mg/L

Land use: 36% forested, 16% agriculture, 47% shrub, 1% urban

Population density: 31 people/km²

Major information sources: Snyder and Stanford 2001, Stanford et al. 2002, http://water.usgs.gov/nwis 2003, www.wcc.nrcs.usda.gov/climate/wetlands.html 2003

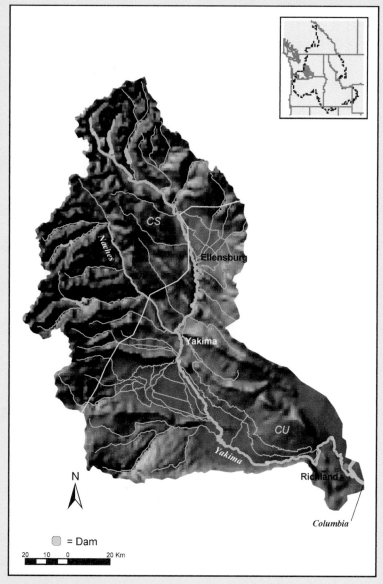

FIGURE 13.19 Map of the Yakima River basin. Physiographic provinces are separated by a yellow line.

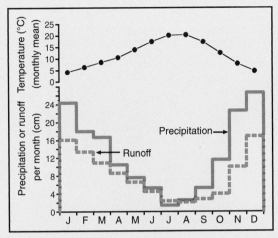

FIGURE 13.22 Mean monthly air temperature, precipitation, and runoff for the Willamette River basin.

WILLAMETTE RIVER

Relief: 3048 m
Basin area: 29,728 km^2
Mean discharge: 917 m^3/s
River order: 7
Mean annual precipitation: 153 cm
Mean air temperature: 11.9°C
Mean water temperature: 13.3°C
Physiographic provinces: Cascade–Sierra Mountains (CS), Pacific Border (PB)
Biome: Temperate Mountain Forest
Freshwater ecoregion: Columbia Unglaciated
Terrestrial ecoregions: Central Pacific Coastal Forests, Willamette Valley Forests, Central and Southern Cascade Forests
Number of fish species: 61 (~31 native)
Endangered species: 5 fishes (4 threatened, 1 endangered)
Major fishes: mountain whitefish, northern pikeminnow, largescale sucker, mountain sucker, white sturgeon, pacific lamprey, western brook lamprey, river lamprey, Oregon chub, coho salmon, sockeye salmon, rainbow trout, chiselmouth, peamouth chub, speckled dace, redside shiner, bridgelip sucker, threespine stickleback, mottled sculpin, torrent sculpin
Major other aquatic vertebrates: bullfrog, western pond turtle, spotted frog, painted turtle, clouded salamander, western toad, red-legged frog, bald eagle, great blue heron, beaver, yellow warbler
Major benthic invertebrates: crayfish (*Pacifastacus*), caddisflies (*Hydropsyche, Cheumatopsyche, Lepidostoma, Heteroplectron, Glossosoma, Dicosmoecus*), stoneflies (*Taeniopteryx, Nemoura, Yoraperla*), mayflies (*Rhithrogena, Baetis, Paraleptophlebia, Ephemerella*), true flies (*Lipsothrix, Rheotanytarsus*), mollusks (*Juga plicifera, Juga silicula*), polychaete worms (*Manayunkia*), amphipods (*Anisogammarus*)
Nonnative species: largemouth bass, smallmouth bass, bluegill, walleye, crappie, common carp, grass carp, brown bullhead, western mosquitofish, brook trout, brown trout, kokanee, lake trout
Major riparian plants: black cottonwood, bigleaf maple, Oregon ash, Douglas fir
Special features: Willamette Valley one of major agriculture areas of western U.S.; richest fish assemblage in Columbia basin
Fragmentation: 13 tributary dams regulate flow; 24 hydropower facilities
Water quality: relatively free of major pollutants; pH = 7.2, alkalinity = 24 mg/L as CaCO$_3$, NO$_3$-N = 0.7 mg/L, PO$_4$-P = 0.21 mg/L
Land use: 68% forest, 19% agriculture (one-third irrigated), 5% urban
Population density: 66 people/km^2
Major information sources: Bastasch et al. 2002, Hulse et al. 2002, http://water.usgs.gov/nwis 2003, www.wcc.nres.usda.gov/climate/wetlands.html 2003

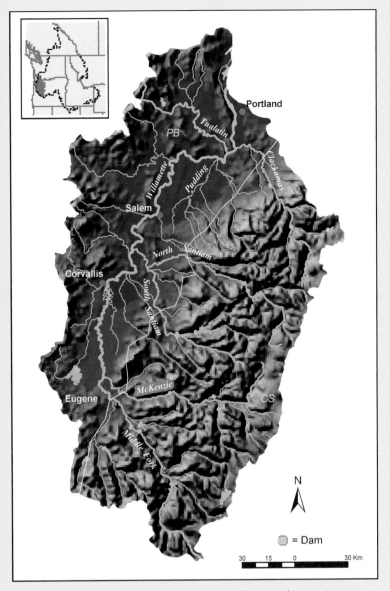

FIGURE 13.21 Map of the Willamette River basin. Physiographic provinces are separated by a yellow line.

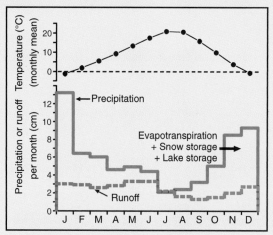

FIGURE 13.24 Mean monthly air temperature, precipitation, and runoff for the Columbia River basin.

COLUMBIA RIVER

Relief: 4392 m
Basin area: 724,025 km²
Mean discharge: 7730 m³/s
River order: 9
Mean annual precipitation: 70 cm
Mean air temperature: 9.7°C
Mean water temperature: 13.3°C
Physiographic provinces: Northern Rocky Mountains (NR), Middle Rocky Mountains (MR), Rocky Mountains in Canada (RM), Columbia–Snake River Plateaus (CU), Cascade–Sierra Mountains (CS), Pacific Border (PB), Basin and Range (BR)
Biomes: Temperate Mountain Forest, Desert
Freshwater ecoregions: Columbia Unglaciated, Columbia Glaciated, Upper Snake
Terrestrial ecoregions: 13 ecoregions (see text)
Number of fish species (basin): 103 (53 native) resident, 15 (12 native) anadromous, 4 marine
Endangered species (basin): 7 fishes (12 listings), 6 snails, 2 aquatic plants (threatened)
Major fishes (main stem): chinook salmon, steelhead, northern pikeminnow, largescale sucker, mountain whitefish, eulachon (candle fish), sculpins, speckled dace, chiselmouth, American shad, carp, smallmouth bass, walleye, channel catfish, yellow perch
Major other aquatic vertebrates (main stem): beaver, painted turtle, bullfrog, muskrat, nutria, American merganser, northwest pond turtle, northern leopard frog, chorus frog, bald eagle, Canada goose, osprey, mink, river otter
Major benthic invertebrates (main stem): caddisflies (*Glossosoma, Cheumatopsyche, Hydropsyche, Hydroptila*), stoneflies (*Arcynopteryx*), mayflies (*Baetis, Ephemerella, Ephemera, Ephoron, Heptagenia, Stenonema, Tricorythodes*), crustaceans (*Corophium, Pacifastacus, Gammarus*), mollusks (*Pisidium, Fluminicola, Corbicula*, western pearlshell)
Nonnative species (main stem): American shad, brown trout, lake trout, lake whitefish, carp, grass carp, goldfish, tench, channel catfish, black bullhead, brown bullhead, yellow bullhead, mosquitofish, largemouth bass, smallmouth bass, black crappie, white crappie, warmouth, bluegill, pumpkinseed, walleye, yellow perch, carp, northern pike
Major riparian plants (main stem): black cottonwood, Russian olive, western hemlock, water hemlock, box elder, alder, willow, red-osier dogwood, reed canary grass, cattail, bulrush, sedges, purple loosestrife
Special features: fourth-largest river flowing to ocean in North America; 62 subbasins in seven states and British Columbia; Mount Rainier is highest point in basin; Hanford Reach is major unimpounded section
Fragmentation: main-stem river almost completely impounded by large dams
Water quality: pollution includes high summer temperatures, pesticides, heavy metals, and nutrients; turbidity = 8.2 NTU, specific conductance = 135.4 µS/cm, alkalinity = 51.9 mg/L as $CaCO_3$, NO_3-N = 0.26 mg/L, phosphorus (total) = 0.06 mg/L, aluminum = 12.09 µg/L, arsenic = 1.05 µg/L, PCB = 0.15 µg/L
Land use: 50% forest/alpine, 34% scrub/grassland, 15% agriculture, 2% urban
Population density: 13.8 people/km²
Major information sources: http://water.usgs.gov/nwis 2003, www.wcc.nrcs.usda.gov/climate/wetlands.html 2003, Newell 2003, www.cbfwa.org 2003, BPA 2004, Cooper and Meiring 1989, Ward and Ward 2004, Wydoski and Whitney 2003

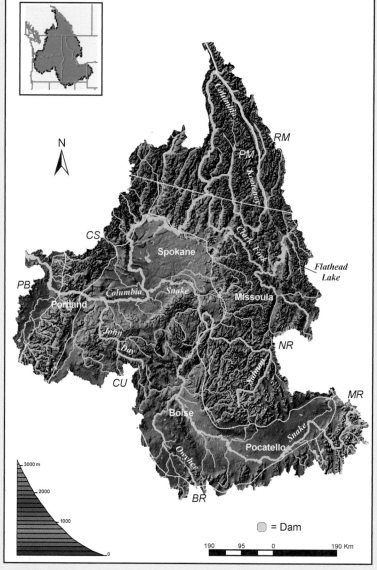

FIGURE 13.23 Map of the Columbia River basin. Physiographic provinces are separated by yellow lines.

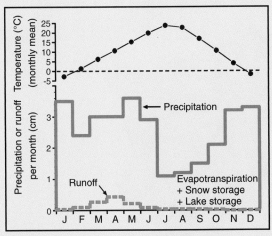

FIGURE 13.26 Mean monthly air temperature, precipitation, and runoff for the Owyhee River basin.

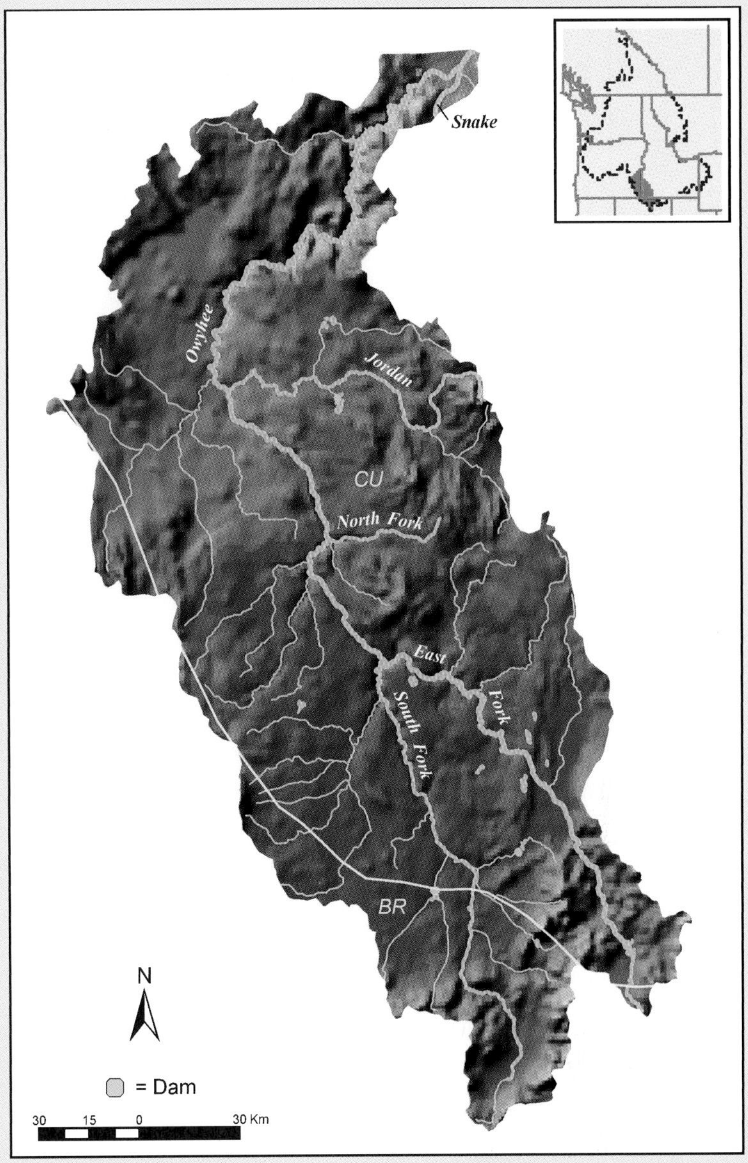

OWYHEE RIVER

Relief: 2484 m
Basin area: 28,617 km^2
Mean discharge: 12 m^3/s
River order: 5
Mean annual precipitation: 31 cm
Mean air temperature: 10.6°C
Mean water temperature: 12.3°C
Physiographic provinces: Columbia–Snake River Plateaus (CU), Basin and Range (BR)
Biome: Desert
Freshwater ecoregion: Columbia Unglaciated
Terrestrial ecoregions: Snake/Columbia Shrub Steppe, Great Basin Shrub Steppe
Number of fish species: 49 (25 native)
Endangered species: 0, but 6 extinct since settlement
Major fishes: yellow perch, white crappie, speckled dace, redside shiner, redband trout, sculpin, largescale sucker, Lahontan tui chub, largemouth bass, flathead minnow, carp, channel catfish, brown bullhead, black bullhead, brown trout, brook trout, bluegill

FIGURE 13.25 Map of the Owyhee River basin. Physiographic provinces are separated by a yellow line.

Major other aquatic vertebrates: Columbia spotted frog, Woodhouse's toad, bald eagle, white face ibis, leopard frog, bullfrog
Major benthic invertebrates: caddisflies (*Hydropsyche*, *Cheumatopsyche*), stoneflies (*Taeniopteryx*, *Pteronarcys*), mayflies (*Rhithrogena*, *Baetis*, *Ephemerella*), mollusks (*Physella*)
Nonnative species: black crappie, bluegill, brook trout, brown trout, black bullhead, brown bullhead, channel catfish, carp, flathead minnow, largemouth bass, Lahontan tui chub, oriental weatherfish, pumpkinseed, rainbow trout, smallmouth bass, tadpole madtom, Utah chub, warmouth, westslope cutthroat trout, white crappie, yellow perch
Major riparian plants: black cottonwood, juniper, red-osier dogwood, alder, willow
Special features: driest subbasin in Columbia basin; only 16% of stream network flows year-round; some segments of main stem and tributaries are National Wild and Scenic Rivers
Fragmentation: Owyhee Dam near mouth; many irrigation diversions
Water quality: Relatively free of major pollutants; pH = 8.07, alkalinity = 189 mg/L as CaCO$_3$, NO$_3$-N = 3.06 mg/L, PO$_4$-P = 0.74 mg/L
Land use: 82% shrubland, 4% forest, 9% grasslands/herbaceous; 5% agriculture
Population density: 1.3 people/km^2
Major information sources: Perugini et al. 2002, http://water.usgs.gov/nwis 2003, www.wcc.nrcs.usda.gov/climate/wetlands.html 2003

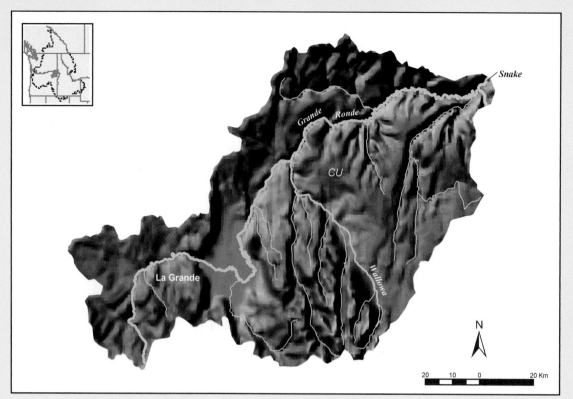

FIGURE 13.27 Map of the Grande Ronde River basin.

GRANDE RONDE RIVER

Relief: 2042 m
Basin area: 10,360 km^2
Mean discharge: 88 m^3/s
River order: 5
Mean annual precipitation: 46 cm
Mean air temperature: 7.6°C
Mean water temperature: 11.8°C
Physiographic province: Columbia–Snake River Plateaus
Biome: Temperate Mountain Forest
Freshwater ecoregion: Columbia Unglaciated
Terrestrial ecoregions: Blue Mountains Forests, Palouse Grasslands
Number of fish species: 38 (23 native)
Endangered species: 4 fishes (threatened)
Major fishes: chinook salmon, steelhead, bull trout, mountain whitefish, brook trout, sculpin, northern pikeminnow, peamouth chub, longnose dace, speckled dace, redside shiner, largescale sucker, bridgelip sucker, mountain sucker, redband trout

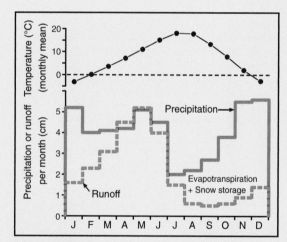

FIGURE 13.28 Mean monthly air temperature, precipitation, and runoff for the Grande Ronde River basin.

Major other aquatic vertebrates: boreal toad, painted turtle, tailed frog, chorus frog, beaver, muskrat, mink, river otter, great blue heron, wood duck
Major benthic invertebrates: caddisflies (*Hydropsyche, Cheumatopsyche, Glossosoma, Rhyacophila*), stoneflies (*Taeniopteryx, Hesperoperla*), mayflies (*Baetis, Ephemerella*)
Nonnative species: brook trout, carp, lake trout, bluegill, pumpkinseed, warmouth, yellow perch, black crappie, white crappie, largemouth bass, smallmouth bass, channel catfish, flathead catfish, brown bullhead
Major riparian plants: black cottonwood, Englemann spruce, red-osier dogwood, alder, willow, yellow star thistle, diffuse knapweed, spotted knapweed
Special features: National Wild and Scenic designation for portions of four tributaries and the lower Grande Ronde
Fragmentation: two hydropower dams
Water quality: relatively free of major pollutants; pH = 7.8, alkalinity = 86 mg/L as CaCO$_3$, NO$_3$-N = 0.133 mg/L, PO$_4$-P = 0.050 mg/L
Land use: no data, but higher elevations forested, lower elevations rangelands with limited agriculture
Population density: 6.4 people/km^2
Major information sources: Eddy and Nowak 2001, http://water.usgs.gov/nwis 2003, www.wcc.nrcs.usda.gov/climate/wetlands.html 2003

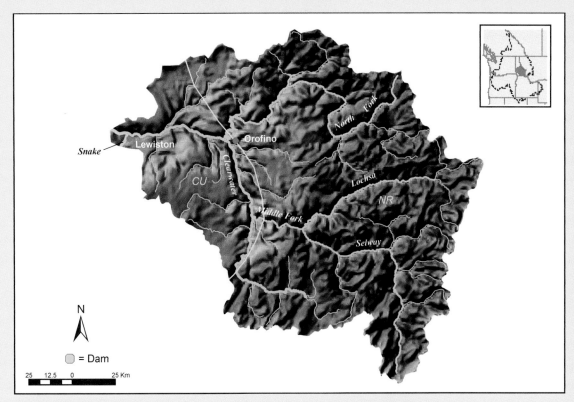

FIGURE 13.29 Map of the Clearwater River basin. Physiographic provinces are separated by a yellow line.

CLEARWATER RIVER

Relief: 2353 m

Basin area: 31,080 km²

Mean discharge: 433 m³/s

River order: 7

Mean annual precipitation: 74 cm

Mean air temperature: 11.5°C

Mean water temperature: 9.5°C

Physiographic provinces: Northern Rocky Mountains (NR), Columbia–Snake River Plateaus (CU)

Biome: Temperate Mountain Forest

Freshwater ecoregion: Columbia Unglaciated

Terrestrial ecoregions: North Central Rockies Forests, South Central Rockies Forests, Palouse Grasslands (near mouth)

Number of fish species: 30 (19 native)

Endangered species: 2 fishes (threatened)

Major fishes: steelhead, chinook salmon, westslope cutthroat trout, brook trout, mountain whitefish, northern pikeminnow, chiselmouth, peamouth chub, longnose dace, speckled dace, redside shiner, largescale sucker, sculpin

Major other aquatic vertebrates: bullfrog, northern leopard frog, painted turtle, muskrat, beaver, river otter

Major benthic invertebrates: caddisflies (*Arctopsyche, Hydropsyche, Cheumatopsyche, Glossosoma, Brachycentrus*), stoneflies (*Taeniopteryx, Pteronarcys, Hesperoperla, Claassenia*), mayflies (*Rhithrogena, Baetis, Ephemerella*), true flies (*Chironomidae, Simuliidae, Hexatoma, Atherix*)

Nonnative species: kokanee, brook trout, golden trout, arctic grayling, tiger muskie, carp, channel catfish, brown bullhead, black bullhead, smallmouth bass, largemouth bass, bluegill, pumpkinseed, black crappie

Major riparian plants: black cottonwood, red-osier dogwood, alder, willows, sedges

Special features: major tributaries, Lochsa and Selway, are free flowing; Clearwater and tributaries mostly constrained in deep canyons; headwaters mostly designated wilderness

Fragmentation: Dworshak Dam on North Fork a major hydroelectric and storage facility

Water quality: relatively free of major pollutants; pH = 7.3, alkalinity = 69 mg/L as $CaCO_3$, NO_3-N = 0.034 mg/L, PO_4-P = 0.006 mg/L

Land use: 20% agriculture, 80% forest

Population density: 1.9 people/km²

Major information sources: Cichosz et al. 2001, http://water.usgs.gov/nwis 2003, www.wcc.nrcs.usda.gov/climate/wetlands.html 2003

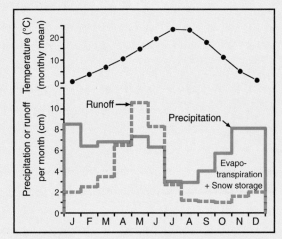

FIGURE 13.30 Mean monthly air temperature, precipitation, and runoff for the Clearwater River basin.

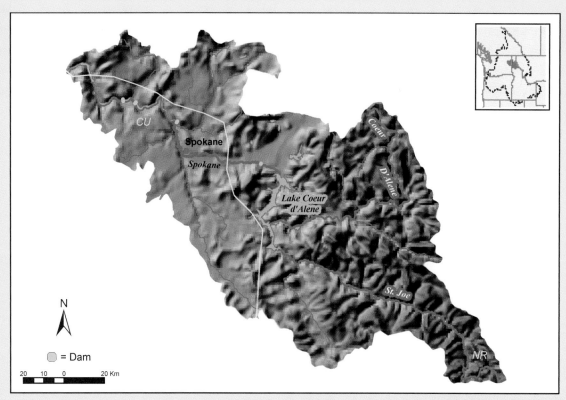

FIGURE 13.31 Map of the Spokane River basin. Physiographic provinces are separated by a yellow line.

SPOKANE RIVER

Relief: 1681 m
Basin area: 15,590 km²
Mean discharge: 225 m³/s
River order: 6
Mean annual precipitation: 63 cm
Mean air temperature: 7.5°C
Mean water temperature: 12.2°C
Physiographic provinces: Northern Rocky Mountains (NR), Columbia–Snake River Plateaus (CU)
Biomes: Temperate Mountain Forest, Desert
Freshwater ecoregion: Columbia Glaciated
Terrestrial ecoregions: North Central Rockies Forests, Palouse Grasslands, Okanagan Dry Forests
Number of fish species: 24
Endangered species: 0, but 2 fishes extinct since settlement
Major fishes: largemouth bass, yellow perch, tench, brown trout, largescale sucker, redside shiner, northern pikeminnow, chiselmouth, kokanee, westslope cutthroat trout, bull trout
Major other aquatic vertebrates: Columbia spotted frog, beaver, muskrat, white pelican, common loon, bald eagle
Major benthic invertebrates: caddisflies (*Hydropsyche*, *Cheumatopsyche*, *Glossosoma*), stoneflies (*Taeniopteryx*), mayflies (*Rhithrogena*, *Baetis*, *Ephemerella*)
Nonnative species: rainbow trout, largemouth bass, yellow perch, tench, brown trout
Major riparian plants: black cottonwood, red-osier dogwood, willow
Special features: St. Joe River, one of longest free-flowing rivers in Columbia basin; Lake Coeur d'Alene, a large glacial lake on main stem; Palouse Prairie, an expansive intermountain grassland, mostly cultivated
Fragmentation: Six low-head diversion and hydropower dams
Water quality: Upper Coeur d'Alene River polluted by mining and undergoing reclamation; eutrophication from Spokane sewage outfall; pH = 7.5, alkalinity = 47 mg/L as $CaCO_3$, NO_3-N = 0.34 mg /L, PO_4-P = 0.16 mg /L
Land use: 50% forest, 15% range, 15% cropland, 8% urban
Population density: 80 people/km²
Major information sources: Whalen 2000, http://water.usgs.gov/nwis 2003, www.wcc.nrcs.usda.gov/climate/wetlands.html 2003

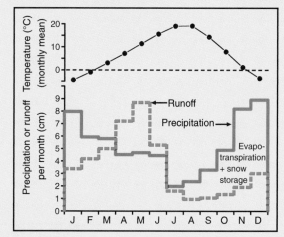

FIGURE 13.32 Mean monthly air temperature, precipitation, and runoff for the Spokane River basin.

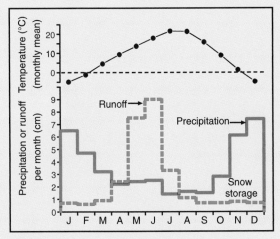

FIGURE 13.34 Mean monthly air temperature, precipitation, and runoff for the Methow River basin.

METHOW RIVER

Relief: 2347 m
Basin area: 4831 km²
Mean discharge: 45 m³/s
River order: 5
Mean annual precipitation: 34 cm
Mean air temperature: 8.6°C
Mean water temperature: 9.5°C
Physiographic province: Cascade–Sierra Mountains (CS)
Biome: Temperate Mountain Forest
Freshwater ecoregion: Columbia Glaciated
Terrestrial ecoregions: Cascade Mountains Leeward
   Forests, Palouse Grasslands
Number of fish species: 32 (25 native)
Endangered species: 3 fishes (1 threatened,
   2 endangered)
Major fishes: chinook salmon, steelhead, bull trout,
   westslope cutthroat trout, bluegill, carp, bass,
   chiselmouth, brook trout, sculpin, largescale
   sucker, redside shiner, Pacific lamprey
Major other aquatic vertebrates: Columbia spotted frog,
   bald eagle, great blue heron, wood duck, mink
Major benthic invertebrates: caddisflies (*Hydropsyche, Cheumatopsyche, Glossosoma*), stoneflies (*Claassenia, Hesperoperla,*
   Isocapnia, Taeniopteryx), mayflies (*Rhithrogena, Baetis, Ephemerella*)
Nonnative species: smallmouth bass, largemouth bass, bluegill, carp, brook trout, brown bullhead, black crappie
Major riparian plants: black cottonwood, Englemann spruce, red-osier dogwood, alder, willow
Special features: free flowing; headwaters in wilderness and North Cascades National Park; upper river with expansive
   floodplain but dry reach at low flow; lower river constrained in canyon; two of last (endangered) wild spring chinook
   salmon and steelhead runs in Columbia basin
Fragmentation: seven small irrigation-diversion dams
Water quality: relatively free of major pollutants; pH = 7.8, alkalinity = 96 mg/L as CaCO₃, NO₃-N = 0.230 mg /L, PO₄-P =
   0.020 mg/L
Land use: 86.5% forest, 9.6% range, 1.6% cropland, 2.3% other
Population density: 6.2 people/km²
Major information sources: Foster et al. 2002, http://water.usgs.gov/nwis 2003, www.wcc.nrcs.usda.gov/climate/wetlands.html
   2003

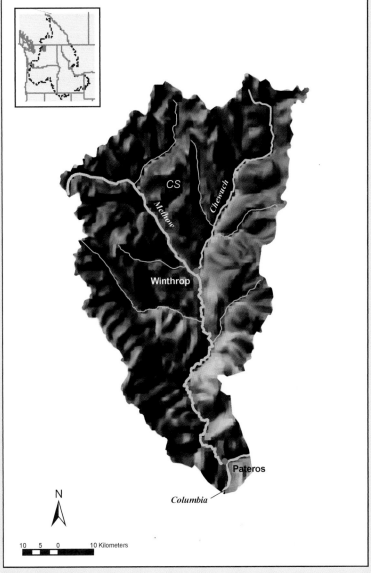

FIGURE 13.33 Map of the Methow River basin.

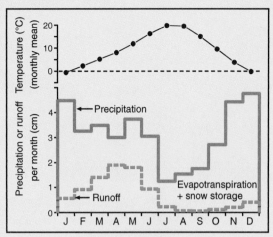

FIGURE 13.36 Mean monthly air temperature, precipitation, and runoff for the John Day River basin.

JOHN DAY RIVER

Relief: 2682 m
Basin area: 20,980 km^2
Mean discharge: 60 m^3/s
River order: 6
Mean annual precipitation: 37 cm
Mean air temperature: 9.2°C
Mean water temperature: 12.5°C
Physiographic province: Columbia–Snake River Plateaus (CU)
Biomes: Temperate Mountain Forest, Desert
Freshwater ecoregion: Columbia Unglaciated
Terrestrial ecoregions: Palouse Grasslands, Blue Mountains Forests, Snake/Columbia Shrub Steppe
Number of fish species: 27 (17 native)
Endangered species: 2 fishes (threatened)
Major fishes: chinook salmon, bull trout, westslope cutthroat, mountain whitefish, brook trout, speckled dace, longnose dace, redside shiner, chiselmouth, largescale sucker, northern pikeminnow, black bullhead, brown bullhead, lamprey, sculpin
Major other aquatic vertebrates: northern leopard frog, spotted frog, western pond turtle, bullfrog, tailed frog, painted turtle, beaver, muskrat, great blue heron, wood duck
Major benthic invertebrates: caddisflies (*Hydropsyche, Cheumatopsyche, Glossosoma, Brachycentrus, Cryptochia neosa*), stoneflies (*Taeniopteryx, Pteronarcys, Hesperoperla, Claassenia*), mayflies (*Rhithrogena, Baetis, Ephemerella*), mollusks (*Anodonta, Fluminicola, Monadenia*)
Nonnative species: brook trout, carp, black bullhead, brown bullhead, channel catfish, largemouth bass, smallmouth bass, black crappie, bluegill
Major riparian plants: black cottonwood, red-osier dogwood, alder, willow
Special features: unregulated river; three segments designated National Wild and Scenic Rivers; spring chinook salmon and summer steelhead populations are two of last remaining intact wild populations of anadromous fishes in Columbia River basin; lower river very arid landscape
Fragmentation: irrigation diversions only
Water quality: relatively free of major pollutants; pH = 8.3, alkalinity = 109 mg/L as $CaCO_3$, NO_3-N = 0.110 mg/L, PO_4-P = 0.022 mg/L
Land use: 10% forested, 85% scrub and rangeland, 5% agriculture
Population density: 0.9 people/km^2
Major information sources: Knapp and Unterwegner 2001, http://water.usgs.gov/nwis 2003, www.wcc.nrcs.usda.gov/climate/wetlands.html 2003

FIGURE 13.35 Map of the John Day River basin.

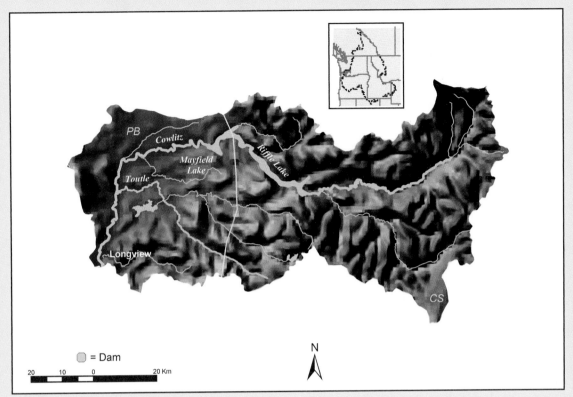

FIGURE 13.37 Map of the Cowlitz River basin. Physiographic provinces are separated by a yellow line.

COWLITZ RIVER

Relief: 4346 m

Basin area: 8870 km²

Mean discharge: 261 m³/s

River order: 5

Mean annual precipitation: 164 cm

Mean air temperature: 10.9°C

Mean water temperature: 10.4°C

Physiographic provinces: Cascade–Sierra Mountain (CS), Pacific Border (PB)

Biome: Temperate Mountain Forest

Freshwater ecoregion: Columbia Unglaciated

Terrestrial ecoregions: Central and Southern Cascades Forests, Puget Lowland Forests

Number of fish species: 32

Endangered species: 2 fishes (threatened)

Major fishes: chinook salmon, white sturgeon, green sturgeon, pacific lamprey, rainbow trout, largescale sucker, bridgelip sucker, mountain sucker, mountain whitefish, sculpin, longnose dace, speckled dace, western brook lamprey, northern pikeminnow, brook trout

Major other aquatic vertebrates: northwestern pond turtle, mink, blue heron, bald eagle

Major benthic invertebrates: caddisflies (*Hydropsyche, Cheumatopsyche, Glossosoma*), stoneflies (*Taeniopteryx, Pteronarcys, Hesperoperla, Claassenia*), mayflies (*Rhithrogena, Baetis, Ephemerella*)

Nonnative species: largemouth bass, smallmouth bass, brook trout, crappie, brown bullhead, tiger muskie

Major riparian plants: black cottonwood, red-osier dogwood, alder, willow

Special features: drains highest point in Columbia basin (Mount Rainier); river choked with sediments after 1980 Mount St. Helens volcanic eruption but gradually recovering

Fragmentation: three hydroelectric dams on main stem

Water quality: relatively free of major pollutants; pH = 7.2, alkalinity = 69 mg/L as $CaCO_3$, NO_3-N = 0.021 mg/L, PO_4-P = 0.011 mg/L

Land use: 16% agriculture, 84% forest

Population density: 8.5 people/km²

Major information sources: Dammers et al. 2002, http://water.usgs.gov/nwis 2003, www.wcc.nrcs.usda.gov/climate/wetlands.html 2003

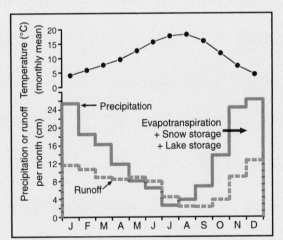

FIGURE 13.38 Mean monthly air temperature, precipitation, and runoff for the Cowlitz River basin.

653

14

GREAT BASIN RIVERS

DENNIS K. SHIOZAWA   RUSSELL B. RADER

INTRODUCTION

BEAR RIVER

SEVIER RIVER

HUMBOLDT RIVER

TRUCKEE RIVER

ADDITIONAL RIVERS

ACKNOWLEDGMENTS

LITERATURE CITED

INTRODUCTION

In October 1843, John Charles Fremont, an explorer with the U.S. Army Topographical Corps, crossed the Malheur basin of Oregon and saw the Cascade Range to the Southwest. He assumed these connected with the Sierra Nevada, which was south of his position. In his journal he named the land between these western ranges and the Rocky Mountains the Great Basin (Egan 1977). The Great Basin boundaries have been redefined many times since then but generally refer to the series of contiguous desert basins and mountain ranges that lie between the Sierra Nevada and southern Cascade Range to the west (121°W longitude), the Wasatch Range to the east (111°W longitude), the Snake River Plain of Idaho (42°N latitude), the Blue Mountains and High Lava Plains of Oregon (44°N latitude) to the north, and the Sonoran and Mohave deserts (32°N latitude) and the plateaus of southern Utah (37°N latitude) to the south (Fig. 14.2).

Great Basin rivers are small with low discharge. The Great Basin is a cold desert, below freezing in the winter and hot in the summer. Nevada and Utah, which encompass the bulk of the Great Basin, are the two driest states in the United States. Thus, rivers of the Great Basin would be considered streams else-

where, but the basin's dryness has enhanced their significance and they are as regionally important as are larger rivers in other parts of North America. Rivers and streams in the Great Basin often flow into valley playas, but the two largest subbasins within the Great Basin, the Bonneville and the Lahontan, have permanent terminal lakes.

Evidence of human habitation in the Great Basin dates to over 11,000 years ago. The earliest culture lived in caves around lakes and used a fluted point, similar to the Clovis point. This culture may have been contemporaneous with or replaced by a culture using another technology, stemmed points. The stemmed point culture utilized lake margins, but by the middle Holocene, 7500 years ago, the climate dried and many of the sites were abandoned.

Wetlands rebounded 4500 years ago. Wetlands in the Carson and Owens valleys supported high human population densities and became the center of the Numic culture (Grayson 1993, Kelly 2001). Numic tribes migrated seasonally, often wintering at sites with good pinyon nut crops and foraging opportunistically, utilizing insects, shellfish, small fishes, small mammals, cattail, and bulrush seeds (Egan 1977, Grayson 1993, Durham 1997, Kelly 2001). About 1000 years ago the Numic people are thought

◄ FIGURE 14.1 East Walker River at the Elbow, Lyon County, Nevada (PHOTO BY D. K. SHIOZAWA).

655

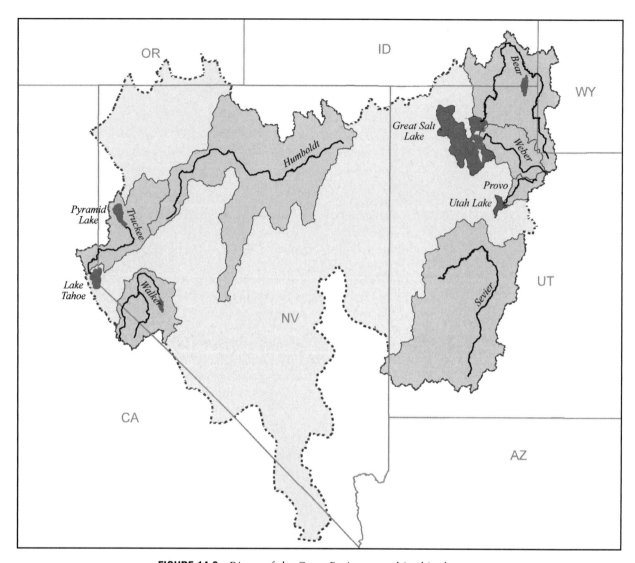

FIGURE 14.2 Rivers of the Great Basin covered in this chapter.

to have begun to expand into the northern and eastern Great Basin (Kelly 2001).

The Fremont occupied the Bonneville basin from 400 to 1350 A.D. They lived in adobe houses and constructed irrigation ditches, raising domesticated crops (Grayson 1993). They vanished about the time the Numic tribes are thought to have reached the eastern Great Basin. Numic tribes may have displaced the Fremont, but a 150-year drought beginning about 1350 (Cook et al. 2004) could be related to their disappearance (Grayson 1993).

By the 1700s the Great Basin was predominantly inhabited by Numic tribes: the Northern Paiute in the Lava and Lakes area and Lahontan basin; the Mono, the Panamint, and the Kawaiisu in the southwestern Great Basin; the Shoshone in the central Great Basin and upper Bonneville basin; and the Utes in the southern Bonneville basin. The Washoe, in the Truckee and Carson region of the Lahontan basin, are linguistically related to tribes in western California.

In the 1770s the Spanish explored the Great Basin as they established overland routes from Mexico to California (Durham 1997). Fur trappers entered the northern Great Basin in the early 1800s, but by the 1840s the fur trade was in decline and pioneers began crossing the Great Basin on their way to Oregon and California. The Oregon and California trails split in the eastern Great Basin, and the Oregon Trail continued west to Oregon, whereas the California Trail dropped south, following the Humboldt River to the Sierra Nevada.

In 1847 the Mormons settled the Bonneville basin and then established small communities throughout much of the Great Basin. Oregon became a state in 1859. The discovery of the Comstock Lode that same year, near today's Virginia City, Nevada, initiated a mining boom that led to statehood for Nevada in 1864. On May 10, 1869, the transcontinental railroad was completed at Promontory Point, Utah. It crossed the Great Basin along the Truckee, Humboldt, and Weber rivers. Utah, after multiple attempts, joined the union in 1896.

In this chapter, we will consider the Bear, Sevier, Provo, and Weber rivers from the Bonneville basin and the Humboldt, Truckee, and Walker rivers from the Lahontan basin (see Fig. 14.2). We will focus on the Bear and Sevier rivers from the Bonneville basin and the Humboldt and Truckee rivers from the Lahontan basin. Abbreviated physical and biological information is provided for the Provo, Weber, and Walker rivers. Several other Great Basin rivers not treated here have similar regional significance. For example, the Carson River of the Lahontan basin, the center of Nevada's early mining industry, is now a superfund site because of mercury contamination. The Owens River, which fed Owens Lake, California, in the Death Valley basin, now supplies water for Los Angeles, California. The lake dried and became a controversial source of wind-borne dust.

Physiography and Climate

The Great Basin is the largest section within the Basin and Range physiographic province of North America and comprises approximately 4% of the United States (Hunt 1974). The hydrographic Great Basin, which includes all contiguous drainages with no active connections to either the Pacific or Atlantic oceans, is approximately 25% larger. The Great Basin includes Death Valley in California and the Malheur basin in Oregon, almost all of Nevada, and the western half of Utah. Over 90 mountain ranges with peaks commonly reaching 2100 to 3500 m asl occur in the Great Basin. Most have a north-to-south orientation.

The Great Basin has five subdivisions: (1) the Lava and Lakes area, in the northwest corner, (2) Lahontan Basin, encompassing the west central Great Basin, (3) the central basin, consisting of numerous separate elevated basins, (4) Bonneville Basin, encompassing the eastern side, and (5) the southern area, in southern Nevada and adjacent California (Hunt 1974).

Some drainages that flow into the Great Basin originate outside of Hunt's (1974) physiographic Great Basin. Streams and rivers entering the Great Basin from the west originate in the Cascade–Sierra Mountain physiographic province. Those entering the eastern edge of the basin originate in the Middle Rocky Mountain physiographic province, and streams entering the southeastern margins originate in the High Plateau section of Hunt's (1974) Colorado Plateau physiographic province.

We consider the hydrographic Great Basin to be the relevant border in this chapter. The hydrographic eastern border, from north to south, is made up of the Tunp Range of western Wyoming, the western Uinta Mountains (3000 to 3800 m asl), and the Wasatch Plateau, Escalante Mountains, Awapa Plateau, and Paunsaugunt Plateau (2500 to 3500 m asl). The Sierra Nevada (2000 to 3800 m asl) and southern Cascade Mountains (1900 to 2900 m asl) form the western border of the Great Basin, whereas the northern border consists of a series of low-lying mountain ranges (2100 to 2700 m asl) and lava plains (1400 to 1800 m asl). Southern boundaries of the Great Basin include parts of the Sonoran and Mohave deserts (600 m asl or less) and high plateaus in Utah (2700 to 3000 m asl) and mountain ranges in Nevada and California (1300 to 2100 m asl). Although Death Valley, on the southwestern edge of the Great Basin, drops to −89 m asl, the floor of the Great Basin averages 1500 to 1700 m asl, about one km higher than the physiographic regions to the north or south (Eaton 1982).

During Pleistocene glacial intervals, approximately 80 pluvial lakes covered an estimated 112,500 km^2 of the Great Basin (Grayson 1993). Lake Lahontan, on the western edge, covered some 22,442 km^2 of the Lahontan basin in Nevada to a maximum depth of 274 m. It was fed by the Humboldt, Walker, Carson, Susan, Truckee, and Quinn rivers. Most drain the eastern slope of the Sierra Nevada Mountains. Lake Bonneville, on the eastern edge, covered 51,722 km^2 of the Bonneville basin of Utah, with depths up to 372 m (Grayson 1993). It was fed by the Bear, Sevier, and other rivers originating on the western slopes of the Wasatch and Uinta mountains. Today, Lake Lahontan remains as the Carson Sink, Humboldt Sink, Honey Lake, Walker Lake, and Pyramid Lake. The remnants of Lake Bonneville are Utah Lake, the Great Salt Lake, and an ephemeral saline playa, Lake Sevier. Lakes Lahontan and Bonneville dried about 10,000 and 11,000 years ago, respectively.

Great Basin climate changed significantly during the Holocene. In the early Holocene, 9000 years ago, lakes and wetlands were numerous in the northern Great Basin (Bradley 1999), but by the mid-Holocene (7500 to 4500 years ago) many of the wetlands and lakes had dried (Grayson 1993, Kelly 2001). From 4000 to 2000 years ago precipitation became winter dominated (Kelly 2001), producing a resurgence of lakes and wetlands. But 2000 years ago the climate again dried, culminating in severe droughts from 1130 to 1300 and 1350 to 1500 (Cook et al. 2004). After 1500 precipitation in the western Great Basin again became winter dominated, generating the climate of today.

Although changes in Holocene climate suggest a role of orbital forcing (precession of the equinox, obliquity, and eccentricity of the earth's orbit; Lunine 1999), a number of geographical factors also influence the Great Basin climate. The Great Basin is slightly north of the Northern Hemisphere arid zone (30°N latitude), which should generate grasslands rather than a cold desert. But the Sierra Nevada, rising 1200 m above the basin floor, is in the direct path of the westerly winds from the Pacific Ocean, resulting in a strong rain shadow. The rain shadow is strongest in the western Great Basin. Mean annual precipitation in Reno, Nevada (1344 m asl), at the base of the Sierra Nevada, is 19.1 cm, and Fallon, Nevada (elevation 1208 m asl), approximately 88 km east, receives 13.5 cm. Eureka, Nevada (elevation 1993 m), at the same general latitude but further east in the central basin, has a mean annual precipitation of 33.5 cm, and Ely, Nevada (elevation 1908 m asl), at the eastern edge of the central basin, receives 25.7 cm/yr. Nephi, Utah (elevation 1562 m), at the edge of the Wasatch Mountains, receives 39.5 cm/yr precipitation.

Another factor is the Arizonal monsoon. From July through September inland heating draws moisture-laden air from the Gulf of Mexico and the Gulf of California/Pacific Ocean into the southwest. This moisture is most important in the southeastern Great Basin. For example, Panguitch, Utah (37°49′N latitude), on the southeast, Beaver (38°17′N latitude) and Delta (39°21′N latitude) in central Utah, and Corinne (41°33′N latitude) in the northern Bonneville basin receive 42%, 31%, 23%, and 18% of their precipitation, respectively, as monsoon storms between July and September. Monsoonal moisture seldom reaches the far northwestern Great Basin.

Mountains influence the amount and timing of moisture availability in the river basins. Some mountain ranges are too low to capture enough water to maintain perennial streams; however, the eastern and western borders and a few large ranges in the center of the Great Basin can capture significant quantities of water. Snow is the primary source of moisture in these higher mountains ranges and is released to the rivers in the spring and early summer. The importance of winter precipitation is greatest in the western Great Basin and spring precipitation becomes progressively more important east across the basin. Forty percent of Reno, Nevada's annual precipitation falls from December to February and 24% of its moisture falls in the spring (March to May). Elko, Nevada, obtains 29% of its precipitation in the winter and 28% of its moisture in the spring. Ogden, Utah, receives 26% of its moisture in the winter and almost 33% of its annual precipitation in the spring. This difference may be related to the intensification of the Sierra Nevada rain shadow as the jet stream makes its seasonal shift to the north.

In the western and central Great Basin, daily July temperatures range between 10°C at night and 33°C during the day. In the eastern Great Basin, summer temperatures are 10°C to 13°C at night and 33°C to 35°C during the day. Mountain locations are cooler because of their higher elevation. The southern Great Basin has mean summer high and low temperatures of 39°C and 25°C, respectively. January temperatures in the western and central Great Basin valleys range from average highs of about 9°C to lows of –4°C. In the southern Bonneville basin, high-elevation valley temperatures have a daily January range from –8°C to 7°C, and northern Bonneville basin temperatures range from –3°C to 6°C, as they are moderated by the Great Salt Lake.

Basin Landscape and Land Use

Great Basin vegetation can be separated by terrestrial ecoregions (Ricketts et al. 1999). The Lava and Lakes area includes the Snake/Columbia Shrub Steppe ecoregion, which is dominated by sagebrush and perennial bunch grasses but also has juniper woodlands and higher-elevation patches of Douglas fir and quaking aspen. The central Great Basin includes both Great Basin Montane Forests and Great Basin Shrub Steppe ecoregions. The relatively stable Mohave Desert ecoregion occupies the southern Great Basin. In contrast to the majority of the Great Basin, it is a warm desert, dominated by creosote bush, Joshua tree, desert holly, and all-scale. The rivers and streams in this system are not perennial, although the Owens and Mohave rivers do flow in their headwaters. The Armagosa River is predominantly underground, and only occasionally exposed at the surface as pools and springs in the river bed.

The Lahontan basin includes the endangered Sierra Nevada Forests ecoregion along its western edge, the Great Basin Shrub Steppe ecoregion in the valleys, and the Great Basin Montane Forests ecoregion of the internal mountain ranges. The Sierra Nevada Forests include montane forests with ponderosa pine, Jeffrey pine, and lodgepole pine, subalpine forests, and pinyon–juniper woodlands. The relatively stable Great Basin Shrub Steppe ecoregion is predominantly sagebrush and shadscale (Ricketts et al. 1999). The Great Basin Montane Forests include juniper, singleleaf pinyon pine, white fir, limber pine, and bristlecone pine. White fir and, more sporadically, lodgepole pine occur on north-facing slopes and in ravines on higher south-facing slopes, limber pine at the heads of east-facing canyons, and mountain hemlock and whitebark pine at the heads of west-facing canyons.

The Bonneville basin includes three ecoregions. The northeastern corner is in the South Central Rockies Forests ecoregion, characterized by Douglas fir, subalpine fir, Engelmann spruce, lodgepole pine, and, at high elevations, whitebark pine. The Great Basin Shrub Steppe ecoregion dominates the northernmost eastern edge of the basin and the western Bonneville basin. The Wasatch and Uinta Montane Forests, an endangered ecoregion, forms the eastern border of the basin with Douglas fir, subalpine fir, ponderosa pine, and Engelmann spruce. Quaking aspen, white fir, blue spruce, and Gambel oak are also present.

The complex structure of many of the overlapping ecoregions in the Great Basin can be categorized in eight altitudinally based vegetation zones, influenced by both aspect and latitude (Thompson 1990). The first zone includes valley-floor playas and saline springs with areas of no vegetation or of saltgrass and black greasewood. The second zone is classified as a shadscale desert (Great Basin Shrub Steppe). Above that is the third zone, a lower sagebrush and grass zone (Snake/Columbia Shrub Steppe, Great Basin Shrub Steppe) that completes the basin floor and extends up the foot of the ranges. Sagebrush extends through the next three zones, but other plants define them. The fourth zone, the woodland zone (Snake/Columbia Shrub Steppe, Great Basin Montane Forests, Sierra Nevada Forests), is often called the pinyon–juniper zone because of the presence of singleleaf pinyon pine and both Utah and Western juniper. This zone starts between 1500 to 2100 m asl, varies in width from 300 to 600 m, and may also contain serviceberry and curl-leaf mountain mahogany. Singleleaf pinyon pine may be absent in

northern and northeastern parts of the Great Basin. The zone's elevation is thought to be strongly influenced by winter inversions in the valleys, which make lower elevations inhospitable to trees (Grayson 1993).

In the central and northern Great Basin the upper sagebrush-grass zone, a fifth zone, occurs above the woodland zone but not in the eastern and southeastern part of the Great Basin, where it is directly replaced by the sixth zone, the montane zone (Great Basin Montane Forests, Sierra Nevada Forests, Wasatch and Uinta Montane Forests). The montane zone, from 2100 to 2600 m asl, has white fir, Douglas fir, ponderosa pine, and quaking aspen. The seventh zone is the subalpine zone (Great Basin Montane Forests, Sierra Nevada Forests, Wasatch and Uinta Montane Forests). This zone extends to the treeline and is characterized by bristlecone pine, limber pine, Engelmann spruce, subalpine fir, and whitebark pine. The final zone, the alpine tundra, begins at about 3000 m asl in the Wasatch Range (Hayward 1945) and about 3200 m asl in the central basin.

Water was the key to the location of early settlements in the Great Basin and population centers still reflect that fact. Today the Great Basin has about 4.16 million residents. About 33% live in the southern Great Basin area, mainly metropolitan Las Vegas, where they are heavily supplemented by water from the Colorado River. The Lahontan basin contains 13.6% of the total Great Basin population and five of six people live in the Reno–Sparks metropolitan area, where water from the Sierra Nevada is available. The Bonneville basin, with 49.3% of the total population, has eight of ten of its residents living along the Wasatch Front in the greater Salt Lake metropolitan area. The Lava and Lakes area has less than 3% and the central basin just 1.4% of the Great Basin's population. These have densities of 1.0 and 5.2 people/km^2, respectively, compared to Salt Lake County's 3175/km^2 (www.census.gov 2004).

Desert covers approximately 41% of the Great Basin, and sagebrush-grass accounts for about 39% of the area. Pinyon–juniper, montane, and alpine forests make up the remaining 20% of the basin (Brussard et al. 1998). Private land ownership is focused in the valleys, especially along perennial rivers and streams, where the prime agriculture lands are found. Crops in high-elevation valleys consist mainly of hay. A wide variety of grains, onions, tomatoes, and other crops are grown in the lower-elevation valleys. Urbanization has increased in the eastern Great Basin and to a lesser extent in the western Great Basin, shifting land usage to residential dwellings, business, and industrial.

After Alaska, Nevada and Utah have the greatest total of federally owned land; 87% of Nevada and 67% of Utah. Federal ownership generally consisted of lands that had no immediate value for farming or settlement. Such lands are especially abundant in the Great Basin. Low-elevation lands are largely managed by the Bureau of Land Management (BLM), whereas the U.S. Forest Service manages the mountainous regions. Both forest and BLM lands are widely used for grazing, and timber is also harvested on forestlands. Part of the federal land is under military control: Hill Air Force Range, Desert Test Center, Wendover Range, and Dugway Proving Grounds in the Great Salt Lake Desert; Nellis Air Force Bombing and Gunnery Range and Nevada Test Site in Nevada; and China Lake Naval Weapons Center and Fort Irwin National Training Center in California. These military reserves remove large blocks of land from public access. Two national parks, Death Valley National Park, California, and Great Basin National Park, Nevada, are fully within the Great Basin. Bryce Canyon National Park of Utah straddles the divide between the Bonneville and Colorado River basins.

Great Basin native grasses were quickly eliminated by intense grazing in the mid to late 1800s. The impact was recognized in as little as 30 years (Grayson 1993). Heavy grazing fostered the spread of woody plants, which cattle did not utilize, and also favored nonnative species, one of which was cheat grass. Cheat grass was in the eastern Great Basin by the late 1800s, and by 1930 it had spread throughout the basin (Grayson 1993). Cheat grass is a fire species, drying in the early summer, after which it is easily ignited. Its seeds are somewhat resistant to burning, but sagebrush, native grasses, and other plants, if not killed, are severely stunted. The frequency of fires in the Great Basin has increased with cheat grass expansion.

The Rivers

The rivers covered in this chapter are from the two largest freshwater ecoregions within the Great Basin, which correspond to the two major physiographic subdivisions: the Lahontan and Bonneville. Thus, the Bonneville freshwater ecoregion includes the Bear River, Sevier River, Weber River, and Provo River. These rivers originate in the middle Rocky Mountain and Colorado Plateau physiographic provinces and flow west to terminal lakes in the Great Salt Lake

Desert. The Lahontan Basin freshwater ecoregion includes the Truckee River, Humboldt River, and Walker River. The Walker and Truckee rivers originate in the Cascade–Sierra Mountains Province to the west of the Great Basin and terminate in separate desert lakes of the Lahontan Basin freshwater ecoregion (Hunt 1974). The upper Humboldt River is part of the Central Basin physiographic subdivision of the Great Basin. It terminates in marshlands in Hunt's Lahontan Basin subdivision.

A number of parallels exist among these rivers, although each also has unique characters that separate it from the other systems. The Bear and Truckee rivers both head in high mountains, have lakes in the upper drainages, flow about half their length in Great Basin Desert Shrub Steppe, and terminate in large natural lakes. The Humboldt and Sevier rivers begin in drier montane habitats and flow for most of their lengths through the Great Basin desert. The Weber and Provo rivers start in high-elevation alpine habitat and flow most of their length in mesic montane or high-elevation valleys. Only a relatively short portion or these rivers flow in Desert Shrub Steppe before they terminate in the Great Salt Lake. Both the Truckee and Walker rivers start in relatively soft-water environments in the Sierra batholith, whereas the Sevier and Humboldt rivers head in limestone regions and thus begin as hard-water streams. The Bear, Weber, and Provo rivers head less than 10 km apart in Precambrian sandstones of the Uinta Mountains. They begin as soft-water streams but by their middle reaches enter limestone regions and become hard-water streams.

The freshwater ecoregions (Abell et al. 2000) categorize the aquatic fauna more accurately than do the physiographic subdivisions. Relationships with adjacent freshwater ecoregions give some insight into dispersal paths into and/or out of the Lahontan and Bonneville basins. Parts of the central Great Basin, isolated from the Lahontan drainage, contain unique endemics (e.g., the relict dace); however, the general faunal associations are closest to those in the Lahontan basin. The Death Valley freshwater ecoregion shares three of its ten native fish species with the Lahontan basin (i.e., Owens sucker, tui chub, and speckled dace). The Oregon lakes and upper Klamath have exchanged fishes with the Lahontan system (Minckley et al. 1986), sharing tui chub and Paiute sculpin. The Lahontan redside of the Lahontan basin reflects an ancient connection to either the Bonneville or Snake River basin, as do mountain whitefish. The Bonneville basin shares many of its fishes with the upper Snake River (Hubbs and Miller 1948, Martin et al. 1985, Smith et al. 2002). Spinedace, of the lower Colorado River basin, comprise the sister taxa to the leatherside chub of the Bonneville basin (Dowling et al. 2002), reflecting a previous connection with the Bonneville basin, whereas both speckled dace and roundtail chub in the upper Colorado River basin genetically show a linkage with the Bonneville basin (Shiozawa et al. 2002, McKell 2003). Corollaries are known from the distribution of mollusks (Taylor 1985).

The aquatic communities in the Lahontan and Bonneville basins and the Great Basin in general are the result of multiple invasions, speciations, and extinctions over at least the last 15 million and possibly the last 30 million years (Smith 1981, Minckley et al. 1986). The north–south orientation of the Great Basin ranges has focused the exchange of organisms along the northern and southern borders of the basin. Once aquatic organisms enter the Great Basin the harsh environment determines the final composition of the fauna and flora. The hot, dry southern Great Basin and the isolated basins of the central Great Basin provide effective barriers to even the more vagrant aquatic organisms.

BEAR RIVER

The Bear River originates on the north slope of the Uinta Mountains of Utah and flows approximately 560 km through three states (Utah, Wyoming, and Idaho; Fig. 14.9). It is a 6th order river in a drainage basin of 19,631 km². From its headwaters the river flows north to Wyoming, follows the Utah–Wyoming state line, and then enters the Bear Lake Valley of Idaho, passing the north end of Bear Lake (1805 m asl). It continues north to the lava fields near Soda Springs (Fig. 14.3), Idaho, where it turns abruptly south and ultimately reaches Bear River Bay (1280 m asl) near Brigham City, Utah, as one of three major inflows to the Great Salt Lake.

Artifacts indicative of the fluted point culture have been found at the edge of the Bear River basin north of the Great Salt Lake (Greer et al. 1981). Although no dates have been associated with them, they probably date to 11,000 years ago. The stemmed point culture (11,000 to 7500 years ago) has been well documented from caves near the wetlands around the Great Salt Lake (Grayson 1993). The lower Bear River basin wetlands were occupied through the mid-Holocene (7200 to 4500 years ago). Fremont sites (1600 to 600 years ago) are also known from the basin (Janetski 1990). By the late prehistoric

period (beginning 550 years ago) the Bear River basin was occupied by Numic tribes. One of these, the Northern Shoshone, occupied the basin when European explorers arrived, although other tribes (Bannock and Blackfeet) also used it.

In the early 1800s, beaver in the Bear River basin attracted American fur trappers operating out of St. Louis, Missouri. This concerned the British-chartered Hudson's Bay Company, which in 1825 sent Peter Skene Ogden to the Bear River with orders to exterminate the beaver so that the American trappers would leave (Durham 1997). Ogden trapped the Bear River in Cache Valley as well as the Weber River basin's Ogden (Huntsville, Utah) and Morgan valleys. His expedition withdrew after confronting American trappers. The American trappers held their rendezvous in Cache Valley (1826 and 1831) and Bear Lake (1827 and 1828; Greer et al. 1981). The beaver were gone by the early 1840s, but emigration to the Pacific had begun.

Both the Oregon and California trails followed the Bear River from near Evanston, Wyoming, to Beer Springs (Soda Springs, Idaho). There, the Oregon Trail went northwest along the Portneuf River and followed the Snake River to Oregon. The California Trail turned southwest, crossing the Malad River of the Bear River Drainage before entering western Utah. Tens of thousands of emigrants walked the upper Bear River. In 1845 alone, 3000 immigrants took the Oregon Trail to Oregon. Settlements began to appear in the Bear River basin in the 1860s, but the Northern Shoshone resisted. In January 1863 government troops stationed at Fort Douglas in Salt Lake City, Utah, and guided by Mormon scout Orrin Porter Rockwell, attacked a winter camp of Shoshone at Battle Creek on the Bear River, near Preston, Idaho. The ensuing battle and massacre opened the area to settlement (Durham 1997, Greer et al. 1981), and by 1870 towns were established throughout the basin (Haws and Hughes 1973).

Physiography, Climate, and Land Use

The Bear River basin has a total relief of 2593 m, from the 3873 m asl Yard Peak in the Uinta Mountains to the Great Salt Lake (1280 m asl). The upper basin primarily drains from the Middle Rocky Mountain (MR) physiographic province (see Fig. 14.9). After passing Soda Springs the river enters the lower basin as it turns south, mostly draining from the Basin and Range (BR) province. The basin

FIGURE 14.3   Bear River, cutting through lava flows near Grace, Idaho (PHOTO BY D. K. SHIOZAWA).

borders form a square, with a panhandle at the southeast corner (see Fig. 14.9). The eastern border is set by the Tunp Range (2300 to 2500 m asl, 110°35′W longitude) and high plains (2280 to 2400 m asl, 110°50′W longitude) of Wyoming. The northern border (42°30′N to 42°45′N latitude) is set by the low-elevation (2100 to 2500 m asl) Aspen, Chesterfield, and Portneuf ranges of southeastern Idaho. The western border of the basin (112°20′W to 112°30′W longitude) is formed by low-lying hills (2000 to 2300 m asl) north of the Great Salt Lake. The southern border (41°20′N latitude) abuts the Ogden River drainage of the Weber River basin. The

Bear River Range (2500 m to 2800 m; 111°32′W longitude) separates the upper and middle Bear River basin (averaging about 1820 m asl) from the lower, western basin (averaging 1370 m asl). The Bear River headwaters, in the panhandle, begin in the Uinta Mountains (3000 to 3873 m asl, 40°45′N latitude).

The northern edge of the Bear River basin is dominated by late Cenozoic basalts of the Snake River Plain. The eastern basin is dominated by folded Mesozoic strata, and overthrust Paleozoic limestones dominate the western basin. Block faulting has exposed the thick limestone strata in the Bear River Range (Stokes 1986). The Uinta Mountains are comprised of Proterozoic sandstones uplifted 40 to 60 million years ago.

The late Eocene upper Bear River flowed into the Green River of Wyoming (Stokes 1979). The middle Bear River developed in the Miocene along the eastern edge of the Bear River Range. The stream flowed north, merging with the Portneuf River of the Snake River system northwest of Soda Springs, Idaho. Back cutting by the middle Bear River near present-day Evanston, Wyoming, captured the upper Bear River from the Green River in the Pliocene. The upper and middle Bear River remained tributary to the Snake River until the late Pleistocene.

Late Pleistocene volcanism blocked the Bear and Portneuf rivers, forming a lake and eventually dividing it. The southern lake, fed by the Bear River, reached 1660 m asl and spilled over its lowest divide into the Bonneville basin, cutting the Oneida Narrows (Bright 1963). The Bear River capture helped Lake Bonneville reach its pluvial maximum, where it established the Bonneville bench (1552 m asl; 51,700 km^2).

About 14,500 years ago Lake Bonneville's threshold at Red Rock Pass collapsed catastrophically, resulting in the Bonneville flood (Malde 1968, Grayson 1993). The lake then stabilized, establishing the Provo bench (1444 m asl). Increasing Holocene aridity caused the lake to recede further, and the lower Bear River cut its channel through the deposits of the drying lake.

The climate is highly variable for such a relatively small basin, with high precipitation in the mountains and much lower precipitation in the valleys. The Bear River Range, the first significant mountains the westerlies encounter in the northern Bonneville basin, receives over 100 cm/yr precipitation and, around the peaks, 125 cm/yr. The upper Bear River drainage in the Uinta Mountains receives over 75 cm/yr, as do the headwaters of Smith's Fork in the Tunp Range. The Bear River Range rain shadow influences the eastern basin. Evanston, Wyoming, Woodruff, Utah, and Montpelier, Idaho, have lower average annual precipitation (28.6, 23.4, 36.0 cm/yr, respectively) than western basin cities (Preston, Idaho, 49.0 cm/yr; Logan, Utah, 42.1 cm/yr; Corinne, Utah, 43.9 cm/yr). Over the entire basin the average annual precipitation is 55.9 cm, with the peak in January and the low in July (Fig. 14.10).

The mean annual temperature of the basin is 6.5°C, but in the Uinta Mountains mean annual temperatures range from −7°C to 0°C and in the Bear River Range from 1°C to 2°C (Haws and Hughes 1973). A series of high-elevation sinks in the Bear River Range collect cold air in the winter and some-times register the lowest temperatures in the United States. One of these was −56°C, on February 1, 1985, the lowest temperature recorded in Utah. The western valley average annual low temperatures are 3°C to 13°C warmer and the average annual high temperatures 1°C to 5°C warmer than the eastern valleys. The record high at Bear River Bay, Utah, in the western basin is 40°C, whereas in the eastern basin, Evanston, Wyoming, has a record high of 36.7°C. Record lows for these two locations are −32.2°C and −38.9°C, respectively. The average basin air temperatures are lowest in January and peak in July (see Fig. 14.10).

The Bear River basin encompasses parts of four terrestrial ecoregions (Ricketts et al. 1999). The headwaters begin in alpine tundra of the Wasatch and Uinta Montane Forests ecoregion and progressively flow through subalpine fir and Engelmann spruce, followed by montane forests of lodgepole pine, white fir, blue spruce, and quaking aspen. From Wyoming to Idaho the basin valley is in the Great Basin Shrub Steppe ecoregion. Forests bordering the northeastern edge of the basin are an extension of the South Central Rockies Forests ecoregion, with lodgepole pine, blue spruce, subalpine fir, and quaking aspen. The lower basin is in the Great Basin Shrub Steppe ecoregion, but with shadscale more common. The Wasatch and Uinta Montane Forests ecoregion occurs in the mountains surrounding the lower Bear River basin, with Gambel oak, sagebrush, and bigtooth maple dominating the hillsides and box elder and narrowleaf cottonwood streamside in the canyons. Fremont cottonwood replaces narrowleaf cottonwood in the valleys.

Agricultural lands are concentrated between 1310 to 1480 m asl in the Great Basin Shrub Steppe ecoregion, yet the mean elevation of the entire basin is 1981 m asl. Only 12% to 13% of the basin is less than 1480 m asl (Haws and Hughes 1973). Approximately 2234 km^2 (11.4%) of the Bear River basin are irrigated (Utah Board of Water Resources 1992). The growing season in Corinne, Utah (1286 m asl), in the western basin is 174 days. Woodruff, Utah (1926 m asl), within the eastern basin, has a 94-day growing season. The agricultural lands in the eastern basin are mainly used for grazing and hay production. In contrast to most Great Basin drainages, the majority of the land in the Bear River basin, approximately 55%, is privately held. National forests comprise 18% of the basin and the BLM manages another 18%. State lands make up 7% and national parks and wildlife refuges 2%. Approximately 9.7% of the Bear River basin is irrigated agricultural land.

River Geomorphology, Hydrology, and Chemistry

The overall fall in the river from its headwaters to the Great Salt Lake is 2133 m (3.8 m/km). The river falls rapidly at its headwaters and in regions where interbasin captures have occurred but has relatively shallow gradients in the intervening basins. From its origin at 3413 m asl to the southernmost Utah–Wyoming border, the river falls 25.9 m/km. The next 40 km to Evanston, Wyoming, averages 7.4 m/km, and from Evanston to Woodruff Narrows Reservoir at the Wyoming–Utah border the fall averages 5.4 m/km. The next 100 km of river meanders in a relatively level 16 km wide valley (0.5 m/km) within both Utah and Wyoming. The river then enters the middle of the Bear Lake Valley (1883 m asl) just north of Bear Lake, where it remains for about 60 km before entering a region of hills and narrow canyons before emerging in Gem Valley at Soda Point Reservoir (2.0 m/km) near Soda Springs, Idaho. At Soda Springs the river turns south and enters the Oneida Narrows, flowing 19 km with an average gradient of 5.8 m/km. Once through the narrows the river enters the north end of Cache Valley, falling 1.4 m/km over a distance of 177 km to the Great Salt Lake.

The streambed substrates reflect the gradient. The headwater region has riffles with boulder, cobble, and rubble substrates, although pools have gravel and sand. The low-gradient stretches of river, such as the Bear Lake Valley, consist of sand and fine gravel bottoms. As the gradient increases in the Soda Point region, boulder and cobble again become abundant. The Oneida Narrows would have always contained large substrate sizes because of its high gradient. However, the use of the river for power generation in this region has increased the armoring of these reaches, and these substrates are now highly embedded. The lower Bear River is mainly sand with some gravel.

Of the average annual precipitation over the entire Bear River Basin (55.9 cm/yr), 11.4 cm/yr enters the runoff (Haws and Hughes 1973) and the remainder is lost to evapotranspiration or otherwise used on site (see Fig. 14.10). If the runoff were to enter the Great Salt Lake without diversion, the mean discharge of the river would be 71.2 m^3/s. However, today the average discharge at the Bear River Bird Refuge, the basin terminus, is 52.0 m^3/s, indicating an average annual consumptive use of 19.2 m^3/s. The maximum recorded annual mean discharge was 143.4 m^3/s in 1984, and the minimum was 17.3 m^3/s in 1961 (Utah Board of Water Resources 1992).

The annual hydrograph, despite water storage and irrigation diversions, peaks in May at 82 m^3/s (see Fig. 14.10), reflecting snowmelt in the headwaters. The low-discharge period is July through September (15 m^3/s), when irrigation is at its maximum. The average discharge during the remainder of the year is about 40 to 48 m^3/s (Utah Board of Water Resources 1992).

Ion concentrations are lowest during high water and highest in the fall and winter. The headwaters reflect the lower solubility of the Uinta core. Calcium, magnesium, alkalinity, and sulfate levels begin low (26 mg/L, 6.2 mg/L, 83 mg/L, and 6.9 mg/L, respectively) but increase rapidly and concentrations plateau as the river flows through the eastern basin (averaging 54.7 mg/L, 32.3 mg/L, 213.7 mg/L, and 49.7 mg/L, respectively). These four ions reach 65 mg/L, 39 mg/L, 289 mg/L, and 73 mg/L, respectively, at Corrine, Utah, near the river terminus.

Sodium and chloride ions, initially at 1.7 mg/L and 1.9 mg/L in the headwaters, reach 34.1 mg/L and 40.0 mg/L in the eastern basin, paralleling the other ions. When the river reaches Logan, Utah, sodium and chloride ions show a considerable increase (64 mg/L and 78 mg/L), and at Corrine, Utah, the concentrations have increased thirtyfold (254 mg/L and 415 mg/L) above the levels in the headwaters, reflecting leaching of the Lake Bonneville bed by irrigation. During low water the river at Corrine can have 1100 mg/L sodium and 1700 mg/L chloride.

River Biodiversity and Ecology

The Bear River is part of the Bonneville freshwater ecoregion of the Great Basin complex (Abell et al. 2000). Much of the work on the Bear River system has focused on inventories and those are usually conducted in headwater reaches.

Plants

The riparian community changes from the headwaters to the lower basin. In the upper reaches willow tends to predominate, with narrowleaf cottonwood lining the river in areas with wide floodplains. River birch and red-osier dogwood are also present, but these are more abundant in the canyons at lower elevations. The lower Bear River riparian zone includes box elder and Fremont cottonwood, willow, and wild rose.

Invertebrates

The highest diversity of invertebrates occurs in the headwaters and tributaries of the Bear River. The intense regulation of the middle reaches of the river and the soft substrate in the lower river act to reduce the number of taxa in those parts of the main stem. The dominant mayflies in the Bear River drainage are *Ameletus, Baetis, Caenis, Drunella, Ephemerella, Epeorus, Rhithrogena, Stenonema,* and *Tricorythodes. Hexagenia limbata* occurs in the Malad River and lower Bear River. Stoneflies include the genera *Capnia, Utacapnia, Sweltsa, Paraleuctra, Podmosta, Prostoia, Claassenia, Isoperla, Kogotus, Skwala, Pteronarcys,* and *Taenionema.* At least 11 families of caddisflies have been recorded, including the genera *Amiocentrus, Brachycentrus, Micrasema, Culoptila, Protoptila, Arctopsyche, Cheumatopsyche, Hydropsyche, Lepidostoma, Ceraclea, Nectopsyche, Limnephilus,* and *Oecetis.* The caddisfly *Helicopsyche* is abundant in the Oneida Narrows. Aquatic moths (*Petrophila*) and alderflies (*Sialis*) are also present. Odonates typical of the region include the dragonfly families Gomphidae (*Gomphus* and *Ophiogomphus*), Aeshnidae (*Aeshna*), and Libellulidae (*Cordulia, Somatochlora, Leucorrhinia, Sympetrum*) and the damselfly families Coenagrionidae (*Amphiagrion, Argia, Enallagma, Ischnura*) and Lestidae (*Lestes*). Two large bivalves, the Oregon floater and Nuttal's high wing floater, and the crayfish *Pacifastacus gambelii* are native to the Bear River (Chamberlin and Jones 1929, Johnson 1986).

Vertebrates

The Bear River system has 17 native fishes. Four species (Bear Lake sculpin, Bonneville cisco, Bonneville whitefish, and Bear Lake whitefish) are endemic to Bear Lake, a graben lake, which, prior to water diversion and damming for downstream power generation, was isolated from the Bear River by wetlands. The other species include two salmonids (Bonneville cutthroat trout and mountain whitefish), six cyprinids (Utah chub, redside shiner, least chub, leatherside chub, speckled dace, and longnose dace), three suckers (bluehead sucker, Utah sucker, and mountain sucker), and two sculpins (mottled sculpin and Paiute sculpin). The native Bonneville cutthroat trout, least chub, and leatherside chub are species of concern but have not yet reached threatened or endangered status.

The Bear River capture by the Bonneville basin and the subsequent connection of Lake Bonneville to the upper Snake River system allowed fishes to invade both basins (Hubbs and Miller 1948). The nature of the exchange is still not fully understood. For example, Bonneville cutthroat trout in the

Bonneville basin were thought to have entered with the Bear River (Behnke 1992), but molecular data indicate that cutthroat trout in the modern Bear River system have diverged only slightly from Yellowstone cutthroat trout in the Snake River (Loudenslager and Gall 1980, Martin et al. 1985), whereas Bonneville cutthroat trout in the main basin have a distinctly older phylogenetic position (Smith et al. 2002). The Bear River also acted as a path for dispersal of Bear Lake fishes into Lake Bonneville. Broughton et al. (2000) report Bonneville cisco, Bear Lake whitefish, and Bear Lake sculpin were in Lake Bonneville 11,200 years ago. It is unlikely that these fishes evolved in Lake Bonneville because of its ephemeral history.

Fishes have been introduced since the mid 1800s. The prominent nonnative species in the lower basin and warmwater reservoirs include common carp, fathead minnow, western mosquitofish, channel catfish, black bullhead, bluegill, green sunfish, walleye, and largemouth bass. Cooler streams and lakes have brown trout, rainbow trout, brook trout, lake trout, and Yellowstone cutthroat trout. Brook trout displace native Bonneville cutthroat trout in headwater streams, and brown trout, being more piscivorous than the native Bonneville cutthroat trout, can reduce the numbers of most native fish species in a stream. When coupled with channelization, this can lead to the extirpation of native minnows. Furthermore, rainbow trout and Yellowstone cutthroat trout will introgress with Bonneville cutthroat trout, leading to the loss of unique evolutionary lineages.

Other aquatic vertebrates include amphibians (tiger salamander, northern leopard frog, Pacific chorus frog, Columbia spotted frog, boreal toad), the common garter snake, water vole, muskrat, beaver, river otter, raccoon, and mink. Kingfisher, dipper, osprey, bald eagle, double-crested cormorant, great blue heron, green heron, snowy egret, cattle egret, and many shore birds and waterfowl occur in the basin. Several hundred species of birds utilize the Bear River Migratory Bird refuge, either for breeding grounds or during migration.

The Columbia spotted frog and boreal toad are species of concern because of declining numbers. Land-use practices have played a major role in their decline and nonnative species have had an impact as well. Cattle egret, a relatively new addition to the North American bird fauna, does not seem to be a significant problem, but raccoons have also invaded the basin. They are aggressive omnivores, impacting both terrestrial and aquatic systems.

Ecosystem Processes

Only a few studies have investigated questions of ecosystem function in Bear Lake and its tributaries (e.g., Logan River, Blacksmith Fork, Little Bear River). These included physical–chemical descriptions and energy flows through algae and microbes and the response of invertebrates to chlorophyll concentrations in leaf packs and the resultant effects on leaf processing rates in tributary systems (Giddings and Stephens 1999). McConnell and Sigler (1959) found that chlorophyll *a* standing crop and gross primary production of epilithic algae were 4 to 5 times greater in lower, regulated sections of the Logan River (1.42 g m^{-2} and 5.0 kg m^{-2} yr^{-1}) than unregulated, upper sections (0.14 g m^{-2} and 1.2 kg m^{-2} yr^{-1}). This difference was attributed to a reduction in scouring (spring bed load and winter anchor ice) and an increase in water clarity downstream from the reservoir. Midge and mayfly (*Baetis* spp., *Drunella coloradensis*, and *Cinygmula* sp.) biomass and production were on average 5.2 times and 2.0 times greater, respectively, in an open section compared to a shaded section of a Bear River tributary (Behmer and Hawkins 1986). However, black fly production was 1.7 times greater in the shaded section. The authors attributed this pattern to higher food quality (algae and detritus) in the open section, or a phototactic attraction to sunlit areas. Osborn (1981) found that macroinvertebrate production was significantly higher in four streams with high alkalinities (>150 mg/L) in the Bear River drainage (34.5 g AFDM m^{-2} yr^{-1}) compared to four streams with low alkalinities (<50 mg/L) in the Yellowstone River drainage of Wyoming (4.7 g AFDM m^{-2} yr^{-1}). Higher rates of macroinvertebrate production correlated with higher rates of primary production and detrital processing in the alkaline streams.

Human Impacts and Special Features

Following his 1843 expedition Fremont wrote of the potential of the eastern shore of the Great Salt Lake for settlement. This report and encouragement from entrepreneur Lansford Hastings influenced Brigham Young, the leader of the Mormon Church, then located in Nauvoo, Illinois (Durham 1997). In 1847 the Mormons began their emigration to the Great Salt Lake valley, arriving in late July. In the next decade, 36,600 more came to Utah (Greer et al. 1981) and they began to settle the Bear River valley as well as other regions of the Great Basin. The Mormon pioneers were the first people in North America to utilize irrigation on a wide scale (Durham 1997). In the process they established independent water companies that tightly control water distribution, even to the point of diverting the entire flow of rivers.

Today the Bear River is managed by legal agreements among water users in three states as well as with the federal government. The Bear River has the highest discharge of any Great Basin river and over 200 water companies take water from the Bear River basin (Utah Board of Water Resources 1992). The Bear River basin has 22 reservoirs with a storage capacity of over 4.9 km^3. Six of these are on the main stem (see Fig. 14.9). In spite of this, water resource studies have determined that Bear River water is underutilized and additional reservoir storage development is projected (Utah Board of Water Resources 1992). Local opposition to further damming and export of water has developed. Further diversions could significantly impact the wildlife refuges at the terminus of the river and would also amplify the existing impacts of stream-flow regulation. Those species that rely on high spring discharge to generate or maintain microhabitats will undoubtedly be negatively affected.

Reaches of the river above Bear Lake have irrigation return flow as their main water source during the summer. Throughout the basin, leaching of irrigated lands has increased total dissolved solids in the river. The greatest problems in water quality in the lower Bear River are high levels of orthophosphate, high turbidity, high fecal coliform, and increased salinity. Upstream impacts include high turbidity, high BOD, phosphorus, and sediment input (Utah Board of Water Resources 1992). Many of the problems are associated with nonpoint sources such as dairies, feedlots, fertilizer application on croplands, septic tanks in areas of high water tables, and land-use practices in the riparian zone.

The top 6.6 m of Bear Lake is used to augment and regulate downstream flows for hydroelectric power generation. To facilitate this the Bear River is diverted into the northern end of Bear Lake with locks and lift stations. Conflicts between the power company and recreational users of Bear Lake develop when, during low water years, the company lowers the lake below the 6.6 m level in order to maintain power generation. Conflicts also occur between the power company and irrigators over irrigation diversions of water claimed by the power company (Utah Board of Water Resources 1992). Daily fluctuation in water levels has enhanced bank erosion in some parts of the river and direct impacts on the Bear River below hydropower reservoirs is evident in the armoring of the channel and the development of pondweed beds.

The headwaters of the Bear River are part of a major recreation area in the Uinta Mountains. Seasonal road access to forest campgrounds and private holdings has facilitated summer use of the region. Camping opportunities range from well-developed campgrounds along highway 150, the major highway in the upper Bear River basin, to primitive camping along unimproved roads in forestlands. A number of spur roads provide trail access for backpackers hiking into the High Uinta Wilderness Area.

Bear Lake is also a major recreation destination. The clear waters and cool air temperatures at the 1805 m asl lake make it ideal for summer recreation. Boating and fishing are both popular. Although fishing in Bear Lake occurs year-round, it is best in the fall and winter months and appears to be associated mainly with spawning of resident fishes. Lake trout are most accessible during their spawning period from late October to mid November. The same is true of Bonneville whitefish, which begin spawning in November. In early to mid-January the Bonneville cisco spawn and are readily captured with dip nets along the east shore of the lake. Bear Lake whitefish spawn from late January to February. The Bear Lake cutthroat trout and the native whitefish, feeding on eggs, concentrate in these spawning grounds and are thus more easily caught by anglers. Snowmobile recreation has also become a major wintertime use of the Bear Lake area. Recreational housing development on the western side of the lake at the base of the Bear River range has helped foster that use.

The lower Bear River, because of its high discharge, played a significant role in late Pleistocene Bonneville basin and lower Snake River drainage dynamics. That has made the region geologically important. Evidence of the Bear River capture into the Bonneville basin, the scars left by an expanded Lake Bonneville, and the results of the subsequent catastrophic flood as Lake Bonneville drained to the north are classical geology textbook topics.

At the terminus of the river is the 26,300 ha Bear River Bay Migratory Bird Refuge, the first waterfowl refuge formed in the United States. This, with adjacent state (10,900 ha) and private (20,300 ha) holdings, is one of the largest freshwater wetlands in the United States (Utah Board of Water Resources 1992). State and federal agencies currently hold established water rights for the marshes. Bear River Bay is a major stopover point for migrating waterfowl in the Pacific flyway. Because of its extensive wetlands, it has become a major area for bird watching, and it is also managed for waterfowl hunting.

SEVIER RIVER

The Sevier River is a 5th order desert river in south central Utah (Fig. 14.11). Its basin is approximately 290 km long by 190 km wide (42,025 km^2). The river flows north for the first two-thirds of its length through a broad valley of high-elevation sagebrush scrub desert. It then turns abruptly southwest to its terminus in Lake Sevier, 50 km southwest of Delta, Utah. Its southerly path is in the bed of Pleistocene Lake Bonneville. Two major tributaries, the East Fork, heading near Bryce Canyon National Park, and the San Pitch River, heading near Fairview, Utah, enter the Sevier River. The Beaver River, which arises in the central basin, provides no input into the Sevier River, although prior to the early 1900s it was connected.

Prehistoric occupation of the Sevier River basin may extend back to 11,000 years ago as evidenced by fluted and stemmed points that have been found in the Beaver Bottoms and the Sevier Desert near Delta, Utah. Archaic sites (8500 to 2500 years ago) have been documented on the Wasatch Plateau, and Fremont occupation is known throughout the Sevier River basin. The Fremont (1600 to 650 years ago) in this region vanished abruptly and may have been replaced by Numic-speaking tribes (Southern Paiutes, Utes). Southern Paiutes and Utes were the inhabitants of the basin at the time of first European contact.

The first Europeans known to enter the basin were led by Franciscan padres Fray Silvestre Velez de Escalante and Fray Francisco Atanasio Dominguez. Their party left Santa Fe, New Mexico, in 1776 to find an inland route to Monterey, California. They followed the Duchesne River in eastern Utah and entered the Great Basin through Diamond Fork of the Spanish Fork River, east of Utah Lake. After resting at Utah Lake the expedition traveled south and found the Sevier River (Greer et al. 1981). They initially thought it was El Rio de San Buenaventura (the Green River), which they had crossed earlier in their expedition, but the river was too small, and they named it El Rio de Santa Isabel (Warner 1995). In 1813, traders Moricio Arce and Lagos Garcia named the Sevier River the Rio Sebero or Rio Seviro, which is thought to be the source of its present name (Utah Board of Water Resources 1999).

In 1826, Jedediah Smith, an American trapper on his way to California, opened a trail through Salina Canyon, near Salina, Utah, and followed the Sevier River upstream (south) to Clear Creek (Durham 1997), establishing what was to become part of the old Spanish Trail, a major overland route from Santa Fe, New Mexico, to Spanish southern California. In 1849, the first Mormon settlers came into the Sevier River basin, and the following year began to divert water for irrigation (Utah Board of Water Resources 1999).

Physiography, Climate, and Land Use

The Sevier River basin can be divided into four sections: the upper, middle, and lower sections of the Sevier River itself and the Beaver River subbasin. The eastern border of the upper Sevier River basin consists of the Wasatch, Fish Lake, Awapa, and Aquarius plateaus, of the high-plateau section of Hunt's (1974) Colorado Plateau (CO) physiographic province (see Fig. 14.11). The Wasatch plateau, dominated by Jurassic, Cretaceous, and early Tertiary sedimentary rocks (Hintze 1988), is over 3300 m asl and its highest peak is the 3440 m asl South Tent Mountain. The underlying strata of the Fish Lake (3000 m asl) and Awapa (2500 to 2800 m asl) plateaus are similar to the Wasatch Plateau to the north (Hintze 1988). The Fish Lake Plateau's highest peak is Mount Terrill (3520 m asl), and the plateau is capped with 100 to 300 m of mid-Tertiary lava. The Aquarius Plateau peaks at 3453 m asl and is primarily composed of basaltic flows (Stokes 1986) from the southern Great Basin hot spot (Fillmore 2000). The plateaus are in the Wasatch and Uinta Montane Forests terrestrial ecoregion, with subalpine fir, Engelmann spruce, and limber pine at high elevations, quaking aspen, blue spruce, white fir, and Douglas fir at intermediate elevations, and Gambel oak, bigtooth maple, and Utah juniper at lower elevations.

The Sevier River heads at the southernmost reaches of the Bonneville basin, adjacent to the Colorado Plateau. The western border of the Upper Sevier is produced by the Markagunt Plateau (3000 to 3446 m asl; Colorado Plateau physiographic province; southwest of Kingston, Utah) and the Tushar Mountains (Basin and Range [BR] physiographic province; west of Kingston). The Markagunt plateau, which slopes to the east, consists of basaltic dikes and lava flows with exposed Paleocene and Eocene limestones (Fillmore 2000). The Tushar Mountains, of mid-Tertiary igneous intrusives,

(Stokes 1986) include Delano Peak, the highest peak in the Sevier River basin (3710 m asl). These mountainous regions, like the eastern border, are in the Wasatch and Uinta Montane Forests ecoregion and the vegetation is similar to the eastern border, although Utah juniper tends to be more abundant. Soils are predominantly mollisols (Utah Board of Water Resources 1999).

The southern border of the upper Sevier River basin is formed by the Paunsaugunt and Markagunt plateaus. Here, late-Tertiary volcanic dikes associated with the southern Great Basin hot spot (Nelson and Tingey 1997, Fillmore 2000) intrude through uplifted plateaus of Paleocene and Oligocene lake sediments. The southern edge of the Paunsaugunt Plateau consists of these eroding limestones (Clarion Formation of Paleocene Lake Flagstaff; Hintze 1988), which form the spires of Bryce Canyon National Park. Dominant plants are ponderosa pine, Douglas fir, singleleaf pinyon pine, Utah juniper, and sagebrush, which are part of the Wasatch and Uinta Montane Forests ecoregion, plus several intervening tongues of Colorado Plateau Shrublands.

The middle Sevier River basin begins as the river leaves the Colorado Plateau physiographic province and enters the Basin and Range province, south of Kingston, Utah. The middle section ends just past the northern bend of the river at Leamington, Utah, which is northeast of Delta, Utah. Its northern border includes the San Pitch Mountains (2667 m asl) of the Wasatch Front Range, the East Tintic Mountains (2290 m asl), the Sheeprock Mountains (2500 m asl), the Simpson Mountains (2562 m asl), and the intervening alluvial and lake deposits (1600 to 2000 m asl; Great Basin Shrub Steppe ecoregion). The East Tintic, Sheeprock, and Simpson mountains are Wasatch and Uinta Montane Forests, with Utah juniper, singleleaf pinyon pine, and sagebrush at lower elevations and quaking aspen, Douglas fir, and white fir at higher elevations. The San Pitch Mountains, east of Sevier Bridge Reservoir, have similar high-elevation vegetation (also subalpine fir, Engelmann spruce, and blue spruce), and midelevation vegetation of Gambel oak, bigtooth maple, singleleaf pinyon pine, and Utah juniper. The soils are mollisols. The southern border of the middle Sevier River basin is generated by the Canyon Mountains. Valley floors are Great Basin Shrub Steppe with aridisols, and the vegetation is predominantly sagebrush and grasses.

The lower Sevier River basin, including Lake Sevier, has a western border formed by the House Range (2947 m asl) and Wah Wah Mountains (2568 m asl), both with Wasatch and Uinta Montane

Forests (sagebrush, Utah juniper, singleleaf pinyon pine, quaking aspen, ponderosa pine, Douglas fir, and western bristlecone pine). It is entirely within the Basin and Range physiographic province. The low-lands to the east of these mountains are typical Great Basin Shrub Steppe, with sagebrush, black grease-wood, and shadscale dominating the community. The Cricket (2280 m asl) and San Francisco (2944 m asl) ranges form the western interior border with the Beaver River subbasin. An ancient riverbed (72 km in length, 30 m deep, and 1200 to 6000 m wide) southwest of the Simpson Mountains is the Sevier River basin's remnant connection with the receding Lake Bonneville. The channel was active until Lake Sevier fell below 1411 m approximately 10,000 years ago (Grayson 1993).

The Beaver River subbasin is usually treated as a separate watershed because the Beaver River only flows into the Sevier River under exceptionally wet conditions. Prior to European settlement, however, it maintained a perennial connection to the Sevier River (Utah Board of Water Resources 1995). The eastern, northern, and most of the western borders of the Beaver River subbasin (see Fig. 14.11) are adjacent to the Sevier River subbasin. The southwestern border extends slightly into Nevada (Great Basin Shrub Steppe) and the southern border is formed by the Bull Valley (2230 m asl), Pine Valley (2780 m asl), and Harmony (2202 m asl) mountains west of Cedar City, Utah. These are the southerly and westerly extents of the Wasatch and Uinta Montane Forests ecoregion, with ponderosa pine, limber pine, Douglas fir, and white fir. Gambel oak and Utah juniper occur in the foothills. The entire Beaver River subbasin is within the Basin and Range physiographic province.

Climate in the Sevier River basin is influenced by mountain ranges, latitude (from 37°26′N to 39°47′N latitude), and the increasing strength of the Arizonal monsoon in the southeastern basin. The annual temperature extremes within the basin are –27°C to 39°C. The record high temperature is 43.3°C (at Delta, Utah) and the record low is –40°C (at Scipio, Utah; Utah Board of Water Resources 1999). Overall, the mean annual temperature of the basin is 9.3°C, but both elevation and latitude influence local mean annual temperatures. Bryce Canyon, Utah, in the upper Sevier River basin is 4.7°C and Delta, Utah, is 9.8°C. Precipitation is >40 cm/yr in mountain areas and one-third of that in the western basin. The basin receives about half as much precipitation as the northern Wasatch Range, and the northern Sevier River basin averages 15 cm more precipitation than the southern part (Utah Board of

Water Resources 1999). Maximum precipitation (>4 cm/mo) occurs in August in concert with the summer monsoon season (see Fig. 14.12).

The majority of the Sevier River basin is managed by the federal government. The U.S. Forest Service and Bureau of Land Management administer 23.1% and 47.9% of the land, respectively. In the early 1900s mountainous lands were added to the U.S. forest system because of severe overgrazing and resultant flooding of valley communities. National Parks and other federal landholders make up about 0.2%, and state lands 7.9%, of the total area, leaving 21.1% of the basin privately owned. Irrigated lands cover 4.5% of the basin. Dry crops are raised in just over 0.4% of the land in midelevation reaches. In contrast, 80% of the entire Sevier River basin (90% of the Beaver River and 75% of the Sevier River basins) is grazed. Cattle production is the primary agricultural activity (Utah Board of Water Resources 1995, 1999) although timber harvest occurs. Bryce Canyon National Park and Cedar Breaks National Monument are major tourist destinations. Camping, fishing, hunting, and snowmobiling are important recreational uses of forest and BLM lands. Eight state parks are located in the basin. The Beaver River subbasin supports some mining. Iron County has the only extensive source of iron ore in the Bonneville basin. The Mormons began mining it in 1852. Lead and silver were mined near Minersville, Utah, beginning in the 1870s (Greer et al. 1981). The Sevier River basin also includes the majority of the potential geothermal energy sites in the Bonneville basin.

River Geomorphology, Hydrology, and Chemistry

The Sevier River heads at 2370 m asl on the southern edge of the Markagunt Plateau in Southern Utah. The overall descent of the Sevier River from its headwaters to Lake Sevier (1369 m asl) 535 river km away is about 1.9 m/km. Asay Creek, heading at 2652 m asl, and Mammoth Creek, heading at 3200 m asl, drain the Markagunt Plateau from the west and join the headwaters of the Sevier River south of Hatch, Utah, increasing the Sevier River streamflow about threefold. Mammoth Creek falls 28.9 m/km before it meets the Sevier River. The Beaver River heads in the Tushar Mountains and falls 2243 m over 196 km (11.4 m/km) to its terminus in the Sevier Desert. The upper tributaries in the Sevier River basin are rocky, cold mountain streams with clear water and mixed riffle–pool habitat. Meadows have sand and pebble bottoms.

FIGURE 14.4 Sevier River at Big Rock Candy Mountain, Utah (downstream of Paiute Reservoir) (PHOTO BY D. K. SHIOZAWA).

Below the confluence with Mammoth Creek the Sevier River flows through Panguitch Valley and is joined by the East Fork of the Sevier River at Kingston, Utah. The East Fork originates at 2500 m asl on the Paunsaugunt Plateau and parallels the Sevier River in the next valley to the east. The Sevier River continues northward, falling 3.0 m/km (Fig. 14.4). In Circleville Canyon the gradient increases to 4.2 m/km. Below Circleville Canyon the river descends at about 2.3 m/km, but when it enters Sevier Canyon above Richfield, Utah, it falls 5.8 m/km.

From the mouth of Sevier Canyon the river descends at 1.3 m/km to the middle section of the Sevier River near Sevier Bridge Reservoir. There the river turns northwest and then southwest as it skirts the Canyon Mountains. The river falls at 2.5 m/km through this section. At Leamington, Utah, the river enters the Sevier Desert, the lower section of the Sevier River. It terminates 190 river km away in Lake Sevier, falling just 0.6 m/km in this section, a reflection of its path in the bed of ancient Lake Bonneville.

The highest tributary to the Beaver River heads at 3633 m asl near Delano Peak, 32 km upstream from the mouth of Beaver Canyon (1828 m asl), for a fall of about 56 m/km. From Beaver, Utah, to Minersville, Utah, on the edge of the Escalante Desert, the river falls 258 m for an average descent of 7.6 m/km. Once in the Escalante Desert the Beaver River flows north 130 km through the Beaver Bottoms, falling an additional 180 m (1.4 m/km) to its terminus at 1390 m asl in the Sevier Desert. The tributaries to the south and west of the Beaver River are not connected with the Beaver River, being dewatered for irrigation, evaporating, or simply entering the groundwater in the Escalante Desert.

By the time the Sevier River begins to track its main channel near Hatch, Utah, in the upper Sevier River basin it is turbid and has a sand, silt, and imbedded rubble bottom with deeply undercut banks. Even near the head of the East Fork of the Sevier River, in a region utilized for open-range grazing, the water is turbid from precipitating carbonates. Near Kingston, Utah, the Sevier River is largely sand and silt with relatively steep, raw banks from down-cutting and erosion. Where the river cuts through rock outcrops, such as Circleville Canyon and Sevier Canyon, the riverbed is partially embedded boulder and rubble. The middle and lower Sevier River bed is sand and silt with some gravel.

Desert valley streams are often influent streams feeding groundwater aquifers in their alluvial plains, and their hydrology is strongly affected by evaporative losses, irrigation diversions, and other consumptive uses. Total runoff in the Sevier River, excluding input from the Beaver River, should be capable of generating an average annual discharge of $32.2\,m^3/s$ (Utah Board of Water Resources 1999; data not available for the Beaver River), but the average annual discharge into Lake Sevier today is approximately $1.3\,m^3/s$. At Hatch, Utah, in the upper Sevier River basin, the average discharge is $3.1\,m^3/s$ and at Piute dam, after the East Fork of the Sevier joins the river, it is $5.3\,m^3/s$. Near Salina it falls to $3.9\,m^3/s$, but below the San Pitch River inflow the average annual discharge increases to $7.8\,m^3/s$ (Utah Board of Water Resources 1999). Peak runoff occurs in May (see Fig. 14.12) as high-elevation snowmelt feeds the river. Summer monsoon storms moderate the post-snowmelt decline in runoff.

The concentration of dissolved ions reflects both land use and the geology of the region. The headwaters of both the Sevier and the East Fork of the Sevier originate in soluble Paleocene and Eocene limestones. At Hatch, Utah, alkalinity is $175\,mg/L$ as $CaCO_3$ and calcium is $40\,mg/L$. These climb steadily downstream (Salina, Utah, $278\,mg/L$ and $91\,mg/L$). Other ions show a similar increase, but at Sigurd, Utah, sodium, chloride, and sulfate, which were $2\,mg/L$, $3\,mg/L$, and $4\,mg/L$, respectively, at Hatch, Utah, significantly increase ($74\,mg/L$, $81\,mg/L$, and $170\,mg/L$) as the river enters a region of exposed Jurassic marine shales (Arapien formation; Utah Board of Water Resources 1999). Irrigation return flows also increase salt concentrations. Below Delta, Utah, the total dissolved solids reach $2730\,mg/L$ ($1000\,mg/L$ chloride and $880\,mg/L$ sulfate). High NO_3-N concentrations (up to $45\,mg/L$) have been detected in shallow groundwater in the San Pitch valley, resulting from nonpoint sources, especially feedlots, and septic systems. Large poultry farms occur throughout the valley, which is the center of Utah's turkey industry.

River Biodiversity and Ecology

The Sevier and Beaver rivers are part of the Bonneville freshwater ecoregion of the Great Basin complex (Abell et al. 2000). Some descriptive biological information of aquatic plants and animals in the Sevier basin is available, but ecological studies are sparse.

Plants

The riparian vegetation of the upper Sevier River includes willow, red-osier dogwood, scattered Fremont cottonwood, willow, and river birch. Both Circleville and Sevier canyons have narrowleaf cottonwood, willow, wild rose, and river birch in the riparian zone. At higher elevations box elder, narrowleaf cottonwood, red-osier dogwood, willow, and wild rose may also line the banks, but the exact community make-up depends on the deepness of the canyon. The nonnative Russian olive and tamarisk have become invasive pests along parts of the floodplain.

Invertebrates

Mayflies occur throughout the Sevier River system. *Baetis* and *Tricorythodes* are found in the turbid upper and middle Sevier. Headwater and high-gradient streams have a greater diversity of mayflies (e.g., *Rhithrogena*, *Ephemerella*, *Drunella*) and stoneflies (*Hesperoperla*, *Isoperla*, *Nemoura*, *Pteronarcys*, and *Pteronarcella*). Caddisflies include *Brachycentrus* and *Hydropsyche*, which are found in the main rivers, as well as limnephilids and leptocerids in tributary streams. Aquatic moths and the amphipod *Hyalella* are also present. Dragonflies occur in slower waters, ponds, and wetlands throughout the entire system. These include *Gomphus*, *Ophiogomphus*, *Aeshna*, *Anax*, *Libellula*, *Erythemis*, *Plathemis*, and *Sympetrum*. Damselflies include Coenagrionidae (*Amphiagrion*, *Argia*, *Enallagma*, *Ischnura*). True bugs such as *Ambrysus* and *Lethocerus* are common and even occur in highly turbid, warm waters. Two mussels, the Oregon floater and Nuttal's high wing floater, are native to the system (Chamberlin and Jones 1929). The Asiatic clam has become established.

Vertebrates

Eight species of native fishes (Utah chub, leatherside chub, redside shiner, speckled dace, mottled sculpin, mountain sucker, Utah sucker, and Bonneville cutthroat trout) were reasonably widespread in the Sevier River basin prior to water development. Longnose dace and Paiute sculpin may also have been present, and least chub, a species of concern, occurs in the upper edge of the middle Sevier River basin. Today, Bonneville cutthroat trout are confined to headwater reaches of small streams and the leather-side chub has been greatly reduced in abundance. All other species are still in the upper Sevier River. Mountain whitefish and bluehead sucker of the northern Bonneville basin are not known to be native to the Sevier River basin, but mountain whitefish have been introduced. Other nonnative fishes include Yellowstone cutthroat trout, brook trout, brown trout, rainbow trout, fathead minnow, western mosquito-fish, channel catfish, largemouth bass, yellow perch, walleye, common carp, and bluegill. Most were introduced for recreational fisheries. Trout are in the upper Sevier and Beaver rivers and tributaries, and the spiny rayed fishes (yellow perch, walleye, etc.) are more common in low-elevation reservoirs. Nonnative fishes compete with, hybridize with, eat, or otherwise displace native fishes.

Among the nonfish vertebrates in the Sevier River basin, amphibians include tiger salamander, Woodhouse's toad, boreal toad, northern leopard, Pacific chorus frog, and Columbia spotted frog. Both the Boreal toad and the Columbia spotted frog are species of concern in Utah and are potential candidates for federal listing. Habitat destruction and introduction of nonnative fishes into previously fishless areas have impacted these species. The common garter snake is found throughout the basin. Aquatic mammals of the Sevier River basin include the water vole, found in the Wasatch Plateau, and muskrat, beaver, and mink throughout the valleys (Durrant 1952). Dippers can be found in the mountain streams and osprey and bald eagles use reservoirs and lakes in the basin. Numerous waterfowl and wading birds utilize the reservoirs along the river as well as state-owned wetlands.

Ecosystem Processes

No studies of ecological processes have been conducted in the Sevier River or any of its tributaries. Fish investigations conducted by the Utah Division of Wildlife Resources consist primarily of inventories. The U.S. Forest Service has also sampled invertebrates and conducted stream assessments, but those studies are establishing baseline data, not life histories or energetics information. This river, like other Great Basin rivers, has alpine headwaters where riparian communities are likely to function much as is seen in other high-mountain regions, with deciduous and coniferous litter input as a major food source. However, once the river enters the valley reaches, siltation, turbidity, and bank erosion become dominant factors. The riparian cover becomes dominated by grasses and low shrubs and this reduces

allochthonous input. Silt and sand reduce or eliminate interstitial habitat in the riffles. Runs and pools with sandy bottoms comprise the majority of the river bed. In these areas chironomid midge larvae, odonates, and true bugs tend to dominate. Black flies and the net-spinning caddis *Hydropsyche*, the case builder *Brachycentrus*, and the mayflies *Baetis* and *Tricorythodes* are found in the riffles and attached to any hard structures, such as sticks and branches that happen to be imbedded in the streambed. Many of these are filter feeders, but it is not known if their primary food source is derived from autochthonous sources (instream production) or allochthonous detritus, input from the herbaceous riparian and irrigation return flows. In any case, the impact of man is so great that the communities existing today are unlikely to resemble or function like those that existed 150 years ago.

Human Impacts and Special Features

The Sevier River is the major south–north flowing river in the Bonneville basin, tracking the linear mountain ranges formed with the extension of the Great Basin. Other rivers in the Bonneville basin flow through mesic mountain valleys before entering the Great Basin Shrub Steppe, usually for just short distances to their base-level lakes. But the Sevier River is in dry Great Basin Shrub Steppe for most of its length. It is also the only major river in the Great Basin with significant monsoonal influence.

About 57,000 people live in the Sevier River basin and an additional 38,000 live in the Beaver River subbasin. Water resources are fully allocated. The first reservoir in this basin, Scipio Reservoir, was built in 1860. By 1886 the first litigation over water use was filed (Utah Board of Water Resources 1999), indicative of dewatering of the river in the downstream reaches. The Sevier River basin has 65 reservoirs and lakes that are used for surface-water management, with four reservoirs on the main river. Additional sites for new reservoirs have been investigated, two on the main river. The basin is short on water relative to demand by about 15.2 million m^3 annually. On average, 1015 million m^3 of water per year are available for irrigation, yet 1114 million m^3 of water are actually diverted for agriculture use. Just over 11.5 million m^3 are diverted into the Sevier River basin from the Price and San Rafael rivers of the Green River and 166 million m^3 of groundwater are pumped from wells. About 40% of the water diversions utilize return flows from upstream irrigation (Utah Board of Water Resources 1999). No instream flow requirements exist on the main Sevier River. Entire sections of the river can be dewatered if demands are high and if downstream uses are not required. Irrigation return flows and side streams usually prevent complete drying of the Sevier River. That is not the case in the Beaver River subbasin. In the 1800s, wetlands existed in the Beaver Bottoms, and the river discharged into the Sevier River. With the construction of Minersville Reservoir in 1913 the wetlands dried, and the river no longer reaches the Sevier River.

The reduced flows associated with irrigation diversions, irrigation return flows, and increased turbidity combine with siltation to increase water temperatures in the lower basin. This restricts coldwater taxa to upstream locations and in tributaries. Many of the native fishes are tolerant of a wide range of temperatures and salinity simply because they evolved in the harshness of the Great Basin desert. However, the relegation of Bonneville cutthroat trout to the headwaters is likely partially influenced by the decline in habitat and water quality.

HUMBOLDT RIVER

The Humboldt River is a 6th order desert river in north-central Nevada. Mary's River, beginning in the Jarbidge Mountains (north of Wells, Nevada), is generally considered to be the headwaters of the Humboldt River because it is the northeastern-most tributary from the river's terminus, 559 river km away at the Humboldt Sink (Fig. 14.13). It is the only major river basin (43,597 km^2) completely contained within the state of Nevada. The primary source of water for the Humboldt River is snowmelt from the Jarbidge (3202 m asl), Ruby (3451 m asl), and Independence (2660 m asl) mountain ranges. Palisade Canyon, a narrow confined stretch of the river southwest of Elko, Nevada (see Fig. 14.13), separates the Humboldt River basin into upper and lower sections. Below Palisade Canyon the river begins to decline in size, with no significant input for its remaining length.

Evidence of prehistoric habitation comes from a stemmed point site in the lower Humboldt River basin, which was overlain by volcanic ash from Mount Mazama (6800 years ago). Stemmed points were in use from 11,000 to 7500 years ago. During the middle Holocene (7500 to 4500 years ago) aridity in the basin was accompanied by the abandonment of wetland sites, but a burial and an artifact cache from a site (Leonard Rockshelter) near Lovelock, Nevada, indicate that the basin was inhabited during

this period. As the late Holocene began, the frequency of occupied sites increased. The inclusion of sedges and cattails in the diet reflects the expansion of wetlands (Grayson 1993). Pinyon nuts also became an important component in the diets of the inhabitants of the Humboldt River basin. At the time of first contact with Europeans, Numic-speaking tribes occupied the Humboldt River basin. Whether these people were the same as those in the basin at the beginning of the late Holocene is unknown, although several have hypothesized that the early inhabitants were replaced by Numic-speaking people about 1000 years ago.

In 1828, Peter Skene Ogden of the British Hudson's Bay Company traveled into the northern Great Basin and found a river running from east to west. The next year he followed it to its terminus southwest of present-day Lovelock, Nevada. In 1831, Ogden's successor with the Hudson's Bay Company, John Work, returned to the upper Humboldt River basin with a crew of 100 men and effectively trapped out the beaver. In 1841, the first emigrant group, the Bidwell-Bartleson Company, followed Ogden's trail along the Humboldt on their way to California, and in 1845, John Fremont's party mapped the trail and it became the main route across Nevada to California. Until that time the river had many names, including the Unknown River, the Ogden River, and Mary's River (after Ogden's wife). However, Fremont renamed the river in honor of Alexander von Humboldt, the name by which it is known today (Durham 1997).

Physiography, Climate, and Land Use

The Humboldt River lies within the Lahontan basin, a part of the Great Basin section of the Basin and Range (BR) physiographic province (see Fig. 14.13). The western borders of the Humboldt River basin include the desiccated bed of Lake Lahontan, the Blackrock Desert, and the Trinity Range (118°45′W longitude). The Jarbidge, Santa Rosa, and Independence mountains form the high northern border of the basin (41°50′N latitude), and the Owyhee Bluffs and the Snake Mountains form the lower-elevation northern border. These are predominantly igneous rock. The northeastern edge of the basin is comprised of a series of low ranges (2100 to 2700 m asl) and highlands (1890 to 2050 m asl; 114°50′W longitude). The East Humboldt Range and the Ruby Mountains (3050 to 3470 m asl) form the southeastern border of the basin. About 10 smaller mountain ranges, with

FIGURE 14.5   Humboldt River downstream of Tonka siding, east of Carlin, Nevada (downstream from Elko, Nevada) (Photo by D. K. Shiozawa).

peaks between 2500 and 3000 m asl, form the southern border (38°45′N to 40°00′N latitude), with the Toiyabe Range and Shoshone Mountains bordering the southern extension of the Reese River (south of Battle Mountain, Nevada). The basin relief, from Ruby Dome Peak (33,470 m asl) to the Humboldt Sink (1186 m asl), is 2283 m.

The Humboldt River basin supports two terrestrial ecoregions. The low-altitude volcanic rock, plains, and valleys, which comprise most of the watershed, are in the Great Basin Shrub Steppe ecoregion, with sagebrush, bunch grasses, and invasive nonnative cheat grass. Shadscale and black greasewood dominate in the lower basin. Mountain ranges, most notably the Ruby, Independence, Santa Rosa, and Jarbidge mountains, are in the Great Basin Montane Forests ecoregion in their upper elevations (Ricketts et al. 1999) and are comprised of quaking aspen, juniper, singleleaf pinyon pine, limber pine, bristlecone pine, white fir, and whitebark pine.

Although located in the center of the Great Basin, climate is surprisingly variable in the Humboldt drainage. Wells and Elko, Nevada, in the upper basin have lower average annual temperatures (6.6°C and

8.2°C) than cities in the lower basin (e. g., Lovelock, Nevada, average annual temperature 11.2°C). The average basin temperature peaks at 21.7°C in July and the average low temperature is −1.7°C in January (Fig. 14.14). Precipitation is at its minimum in August, is greater in the higher elevations of the upper basin, and decreases as elevation drops and as the river moves closer to the rain shadow of the Sierra Nevada. Elko and Wells, Nevada, average over 25 cm of precipitation annually. Valleys of the lower basin only average 15 to 20 cm of precipitation per year, and their evaporation rates exceed 100 cm per year (http://water.nv.gov/Water%20planning/humboldt/hrchrono.htm 1999).

Agriculture, primarily livestock grazing, is by far the most extensive land use in both the upper and lower basins. During the 1870s and 1880s the entire basin was heavily grazed. Initially this included open-range winter grazing, but following a severe winter in 1889 to 1890, when many cattle died, ranchers began supplementing grazing with hay crops. This led to increased irrigation and thus conflicts over water. By the early 1900s grazing-induced vegetation destruction and subsequent erosion in the upper basin caused the federal government to include a number of the mountain headwater areas as part of the National Forest system. Today National Forests cover 12% of the drainage. The remaining open rangelands, which were lower-elevation federal lands, began to be managed in 1935, again in response to habitat degradation from overgrazing. These lands, 54% of the basin, are now managed by the Bureau of Land Management. Private lands compose 32% and reservations, state lands, and other federal holdings make up the remaining 2% of the land ownership in the Humboldt River basin. Essentially all federal lands are still grazed.

Mining, especially in the lower basin, has been a dominant economic activity since the 1860s. Mining, mostly for silver in the late 1800s, led to deforestation as timber was cut for ties and beams. Water was utilized for processing ore. One company even claimed all of the water in the Humboldt River for its mining operations, but subsequent vigilante action by local residents eliminated that claim by dynamiting the company's dam. In the early 1900s, gold was discovered in the Humboldt River basin, and gold mining continues today, accounting for about a quarter of the basin's employment (http://water.nv.gov/Water%20planning/humboldt/hrchrono.htm 1999).

River Geomorphology, Hydrology, and Chemistry

The upper basin above Palisade Canyon composes less than 30% of the drainage area but provides all of the water for the lower basin. The Mary's River from its headwaters to the Humboldt is 92 km in length. From its confluence with the Mary's River the Humboldt River flows 116 km past the towns of Elko and Carlin, Nevada, to Palisade Canyon (Fig. 14.5). Along this reach the Humboldt River receives water from tributaries draining the Ruby Mountains to the south and the Independence Mountains in the north.

The hydrology of the Humboldt is unusual in that discharge currently peaks near Palisade Canyon, well upstream of its terminus, at $11.4 \, m^3/s$. Streams in the lower Humboldt River basin stop flowing a considerable distance from the river. The Little Humboldt River, whose subbasin comprises over 10% of the Humboldt River basin, dries 10 km from the Humboldt River, and the Reese River, comprising almost 21% of the basin, stops flowing 16 to 32 km short of the Humboldt River. Below Rye Patch Reservoir the river discharge has fallen to $7.2 \, m^3/s$ (http://water.nv.gov/Water%20planning/walker/wrchrono.htm 1999) and it still has to flow 80 km through irrigated land before reaching the Humboldt Sink. After evaporation, irrigation, and influent losses to depleted groundwater aquifers, little water remains to enter the sink. An average discharge of $0.9 \, m^3/s$ was recorded between 1900 and 1960, but no discharge data are available for the terminus of the Humboldt River prior to water diversions. We therefore used the Humboldt River's estimated percentage of total Lahontan basin flow (Grayson 1993) to generate a prediversion discharge of $25.8 \, m^3/s$. Based on available data, prediversion runoff was still very low, peaking in June as the result of spring snowmelt from the mountains, with extremely low runoff from August through February due to evapotranspiration losses (see Fig. 14.14).

Most of the upper basin headwaters consist of mountain streams with clear water, unembedded rocky bottoms (gravel, cobbles, rubble), and mixed riffle–pool habitat. The gradient of the Mary's River is 15.1 m/km. From the confluence of Mary's River and the Humboldt River to Palisade, the river falls at 1.5 m/km. It meanders through a broad valley and habitat heterogeneity decreases, with patchy rubble and gravel substrates embedded in sand and silt.

Below the Palisade Canyon the river enters the lower basin and flow decreases as evaporative losses and agricultural diversions exceed inflows. From Palisade to the Humboldt Sink the river gradient averages about 0.7 m/km. Sand and silt become predominant. The mountains of the lower basin are far from the main channel and most of the tributaries only reach the main river during extreme flood events. Rye Patch Reservoir, the only reservoir on the main-stem Humboldt River, is located south of Winnemucca, Nevada. It was constructed in 1936 to provide irrigation water for Lovelock Valley. From Rye Patch Reservoir the river flows 70 km past a number of low-head irrigation-diversion dams to a series of canals that divert water to the Humboldt Lakes (Toulon and Humboldt lakes). Prior to European settlement the main stem of the Humboldt River annually flooded its terminus 10 km below the Humboldt Lakes, the Humboldt Sink. In 1829, Peter Skene Ogden described the sink as a 13 km long and 3.5 km wide lake, with dense vegetation (Egan 1977). Now, because of water diversions, the Humboldt River reaches the sink only during high-water years. The Humboldt Sink is about 10 km from the Carson Sink, the endpoint of the Carson River, which flows from the southwest. In extremely wet years water in the Humboldt Sink can flow into the Carson Sink, which is about 6 m lower in elevation.

Basalt, rhyolite, and other volcanic rocks dominate the Humboldt River basin, and these influence the water chemistry of the basin. Calcium concentrations are similar in the upper and lower basin. At Elko, Nevada, in the upper basin, calcium is 46 mg/L, and this changes very little in the lower basin (47 mg/L). However, sodium, chloride, and sulfate are consistently two to three times greater in the lower basin (126 mg/L, 128 mg/L, and 93 mg/L, respectively).

River Biodiversity and Ecology

The Humboldt River is part of the Lahontan freshwater ecoregion of the Great Basin complex (Abell et al. 2000). Although the Humboldt River is not well studied in terms of its biology and ecology, there is probably more known about its fishes than anything else.

Plants

The riparian community of the upper mountainous portion of the basin (Manning and Padgett 1995) is dominated by various conifers (white fir, lodgepole pine, ponderosa pine, Engelmann spruce), some

deciduous trees (river birch, narrowleaf cottonwood, quaking aspen), willow, and shrubs (e.g., red-osier dogwood, wild rose, sedges, rushes, and grasses). Forbs are common but rarely abundant. Riparian vegetation of the lower portion of the basin is relatively unstudied and poorly understood. It is less diverse than in the upper basin, and conifers are rare, whereas cottonwoods, some species of willow, sedges, and grasses are common. Two nonnative species, halogeton and cheatgrass, are often the most abundant plants (riparian and otherwise) in the lower basin. Halogeton, which favors high sodium levels in the soil, made its first appearance in North America in the Humboldt River basin in the 1930s. It produces toxic oxalates and has been responsible for thousands of livestock deaths.

Invertebrates

Although little information is available, invertebrate diversity in the upper basin is probably quite high and similar to other mountainous portions of the Great Basin. A decline in habitat heterogeneity in the lower basin is reflected in low invertebrate diversity where abundances are dominated by dipterans (true flies), some caddisflies, and beetles. Mayflies are primarily represented by *Baetis* but also include *Acentrella, Camelobaetidius, Centroptilum, Baetisca, Ephemerella, Ephemera, Hexagenia, Heptagenia, Rhithrogena, Tricorythodes, Paraleptophlebia, Traverella,* and *Ephoron*. Stoneflies include *Capnura, Taenionema, Isogenoides, Isoperla,* and *Acroneuria,* and caddisflies include *Cheumatopsyche, Hydropsyche, Micrasema, Anagapetus, Hydroptila, Brachycentrus, Chyranda, Limnephilus, Rhyacophila,* and *Nectopsyche albida* (Baumann and Kondratieff 2000). Dipterans (e.g., *Tipula,* Chironomidae, Empididae) and beetles (Hydraenidae, Elmidae, Hydrophilidae) are the most diverse groups. True bugs (*Ambrysus*) and odonates (dragonflies) are also present. The signal crayfish, native to the Humboldt River basin, is present in the Mary's River.

Vertebrates

Only seven fishes are native to the Humboldt River. These are the Lahontan cutthroat trout (threatened), Paiute sculpin, tui chub, Lahontan redside, speckled dace, Tahoe sucker, and mountain sucker. The low number of native fishes reflects the isolated nature of the basin and the associated harsh environment. At least 19 nonnative taxa occur in the basin, including common carp, goldfish, brook trout, brown trout, rainbow trout, Yellowstone cutthroat trout, black bullhead, channel catfish, walleye, bluegill, green sunfish, black crappie, Sacramento perch, largemouth bass, smallmouth bass, and western mosquitofish.

Among the other aquatic vertebrates, Northern leopard frog, boreal toad, and Columbia spotted frog occur in parts of the Humboldt River basin, as does the Great Basin spadefoot toad. Bullfrogs have been introduced into springs within the basin. The common garter snake occurs throughout the basin, feeding in the river on fishes and invertebrates. Beaver, which attracted trappers as the first Europeans in the region and were commercially trapped out by those trappers, are still present in the Humboldt River system, as are muskrat.

Ecosystem Processes

Little has been done concerning ecosystem processes, although a number of studies have been conducted on Lahontan cutthroat trout metapopulation in the Mary's River (e.g., Dunham et al. 1999). The Humboldt River heads in montane forests and high-elevation desert scrub. Many stream riparian zones in these upper reaches are lined with willows and cottonwoods, which indicate a significant role of allochthonous detritus in low-order systems. As the streams leave the mountainous regions the banks become down-cut and both silt and turbidity increases. The main Humboldt River has little riparian cover other than patches of willows, and often has steeply cut banks. In these reaches it is likely that autochthonous input is significant and perhaps dominant. In late summer attached filamentous algae can be seen in the riffles. Daily and seasonal fluctuations in temperature are more extreme in the lower basin and are important factors associated with the structuring of the aquatic community of the main river system.

Human Impacts and Special Features

The Humboldt River is a desert river that once terminated in a large wetland, the Humboldt Sink. Of the major rivers in the Great Basin, the Humboldt River is the most strongly influenced by the Great Basin climate, having its headwaters in central Great Basin mountains and the majority of its length in the Great Basin Shrub Steppe covered valleys. The Humboldt River provided a low-elevation path from eastern Nevada to the Humboldt Sink in west central Nevada, which made the overland wagon trek to California possible. Immigrants traveling the California Trail in the mid 1800s looked negatively at the Humboldt River and allowed their cattle herds to denude much of the land along the river valley.

The Humboldt River basin has a long history of agriculture (primarily grazing of cattle and sheep) and mining. Both industries have impacted the natural flow regime, riparian vegetation, rates of erosion, and water quality in the Humboldt River. The Humboldt River basin generates about 43% of the total crops and almost 45% of the livestock production in Nevada. Both operations rely heavily on the water in the Humboldt River. During the mid to late 1800s ranchers in the Humboldt River basin gained ownership of land along valley streams and springs, which in turn gave them control of many times that amount of adjacent public lands. Several ranches owned over $800\,km^2$ within the Humboldt River basin. Farming in the lower basin was impacted by upstream water diversions, and the problem was brought to a head by a drought in 1888–1889 and a severe winter in 1889–1890. Upper basin ranchers, especially in the Elko region, realized that the open range could not support their herds over the winter and they increased water diversions to raise feed. This reduced the amount of water reaching the lower basin. Because of this conflict Nevada passed and repealed several water-rights laws over the next three decades, finally settling the conflict in the early 1930s. The construction of Rye Patch Reservoir in the 1930s allowed spring runoff to be stored for later use in the lower basin. In addition, a number of ranches in the Battle Mountain area were purchased and their water rights transferred to the lower basin.

Much of the gold is extracted from open-pit mines and groundwater is pumped to prevent flooding. The state of Nevada requires that the water be returned to the groundwater or that it be utilized in lieu of existing groundwater allocations (such as irrigation of crops). If the water is discharged as surface flow, federal requirements designate the acceptable levels of contaminants in the water. A major unknown associated with mining operations is the future filling of the open pits with water once mining operations have ceased. One mine, for example, when filled will become the third-largest lake fully contained within Nevada (exceeded only by Pyramid Lake and Walker Lake). Such pit lakes may have great recreational value, but the water quality in the lakes, the impact on groundwater flow, the impact of evaporative losses, and the ultimate effect on Humboldt River discharge are unknown. These pits will eventually hold an estimated 1.9 to $2.5\,km^3$ of water (http://water. nv.gov/Water%20planning/walker/wrchrono.htm 1999).

FIGURE 14.6 Truckee River near Reno, Nevada (PHOTO BY D. K. SHIOZAWA).

TRUCKEE RIVER

The Truckee River of the Lahontan basin is a 5th order river with a drainage basin of 7925 km². It heads in the Sierra Nevada of California, near Lake Tahoe, and terminates in Pyramid Lake, Nevada (Fig. 14.15). The basin is divided into upper and lower parts. The upper Truckee River basin includes the headwaters above Lake Tahoe and the tributaries to the Truckee River above Reno, Nevada. The lower Truckee River basin begins just above Reno and extends to Pyramid Lake (Fig. 14.6). The upper Truckee River feeds Lake Tahoe from the south. The main-stem Truckee River begins as the outflow from Lake Tahoe. It flows northeast, is joined by the Little Truckee River and several other streams, and descends the Sierra Nevada to Reno, on the edge of the Nevada desert. Once past Reno the river continues east, and at Wadsworth, Nevada, within the Pyramid Lake Indian reservation, it then turns north, eventually entering Pyramid Lake. Pyramid Lake, 48 km long and 17 km wide (574 km²), has a maxi-

mum depth of 102 m. The Truckee River's total length, including the upper Truckee River, is 225 km.

The upper basin was originally occupied by the Washoe Indians, whereas the Northern Paiute occupied the lower basin around Pyramid Lake. The Pyramid Lake tribe was known as the Kuyuidokado (fish eaters) after the planktivorous cui-ui that provided a major part of their diet. The government explorer John Fremont named the lake on January 10, 1844, after its pyramid-shaped island (Egan 1977). Paiutes camping at the mouth of the Truckee River shared their catch of large trout with the expedition and Fremont named the river the Salmon Trout River (James 1978). After leaving Pyramid Lake Fremont traveled south, crossed the Carson River, and followed the Walker River into the Sierra Nevada (Egan 1977). On February 14, 1844, he and his cartographer climbed Red Lake Peak in the headwaters of the Carson River. About 30 km to the north they saw a large lake that Fremont described as being surrounded by mountains (James 1978). They were the first Europeans to see Lake Tahoe

(Egan 1977) and, unknowingly, were just a kilometer southwest of the head of the upper Truckee River.

In the late fall of that same year, the California-bound Stevens–Townsend emigrant party followed the Humboldt River to the Humboldt Sink and crossed the desert to the Salmon Trout River (the Truckee River). They followed the river upstream into the Sierra Nevada and crossed into California. They named the river the Truckee after their Paiute guide (James 1978, Durham 1997) and their route became the preferred trail to California. The next fall another pioneer party used the Stevens–Townsend trail but became stranded by snow. That group, the Donner–Reed party, resorted to cannibalism to survive and brought notoriety to the pass.

Physiography, Climate, and Land Use

The Truckee River basin extends from southwest of Lake Tahoe northeast 80 km to Pyramid Lake. The crest of the Sierra Nevada (2400 to 2700 m asl) forms the western border and the Carson Range east of Lake Tahoe forms the eastern border of the upper basin. The upper basin drainages of the Truckee River extend 30 km south and 44 km north of Lake Tahoe. The lower basin's western borders are the east side of the Carson Range. The Lake and Fox ranges, the Terraced Hills, and the Virginia Mountains form the northeastern basin border (1500 to 2300 m asl) at about 40° 25′ N latitude. The southeastern border of the lower basin is formed by the low-elevation (2100 m asl) Truckee and Virginia ranges.

The Truckee River basin includes two physiographic provinces (Hunt 1974) and two terrestrial ecoregions (Ricketts et al. 1999) that roughly correspond. The montane forests on the west shore of Lake Tahoe are typical of the Sierra Nevada Forests ecoregion but the Carson Range to the east of Lake Tahoe is dryer, and Jeffrey pine and white fir predominate. These mountainous areas are within the Cascade Sierra (CS) physiographic province (see Fig. 14.15). Patches of montane chaparral (curl-leaf mountain mahogany, green leaf manzanita, and huckleberry oak) occur between 1830 to 2900 m asl. Pinyon–juniper woodlands occur between 1520 to 2740 m asl, especially in well-drained gravely soils (Howald 2000). The lower basin is in the Great Basin Shrub Steppe ecoregion, which roughly corresponds to the Great Basin section of the Basin and Range (BR) physiographic province. Big sagebrush, rubber rabbitbrush, and antelope bitterbrush dominate the alluvial reaches. In the driest, most saline and alkaline sites the community is dominated by shadscale (Howald 2000).

Temperature and precipitation vary greatly across the basin. Only 25% of the Truckee River basin lies in California, but it captures the majority of the precipitation, mostly as snowfall. Tahoe City, California, on the west side of Lake Tahoe, receives an average of 82 cm/yr of precipitation, but Glenbrook, Nevada, just across the lake to the east, receives only 40 cm/yr, and the lower basin receives <20 cm/yr. The average precipitation for the entire basin is 43 cm/yr, with 5 to 6 cm/mo falling in winter and <1 cm/mo in July and August (Fig. 14.16). Tahoe City has a mean annual temperature of 6.4°C, similar to other parts of the upper basin. In the lower basin, Reno, Wadsworth, and Nixon, Nevada, all have mean annual temperatures near 10°C (www.ncdc.noaa.gov 2004). Although the upper basin mean annual high temperature is 4°C to 9°C cooler than the lower basin (mean July temperature 18.6°C), the mean annual low temperatures are similar throughout the entire basin, at about 0°C from December to February (see Fig. 14.16).

Lake Tahoe in the upper basin and Pyramid Lake at the basin terminus dominate the basin today. Lake Tahoe, 20 km wide and 35 km long, is a graben lake, the third-deepest lake in North America and the tenth deepest in the world, with a maximum depth of 501 m (Horne and Goldman 1994). An additional 790 m of sediments underlay the lake. Pyramid Lake is 48 km long and 17 km wide and has a surface area of 446 km^2 and a maximum depth of 102 m. Twelve major tributary streams enter Lake Tahoe, but only three minor streams and the Truckee enter Pyramid Lake. During the arid mid-Holocene (7000 to 4500 years ago), Lake Tahoe was 6 or 7 m lower than today and was not flowing into the Truckee River. Submerged trees up to 1 m in diameter, dating to about 4500 years ago, have been found in the south end of the lake. Based on archaeological sites, modern shores were established by 3500 years ago (Grayson 1993). Pyramid Lake, with a volume of 25.3 km^3, has an average evaporative rate of 120 cm/yr.

The BLM manages 42% of the Truckee River basin, the U.S. Forest Service manages 9%, and 9% is Indian reservation. State lands constitute less than 7% of the land area. Private holdings comprise 33% of the land, including urbanized areas, especially around Lake Tahoe. The Lake Tahoe region is a major recreation center. Ski resorts, vacation housing, and gambling combine to make this area a year-round recreational destination.

River Geomorphology, Hydrology, and Chemistry

The Truckee River drops 1603 m in its 225 km length, for an average gradient of 7.1 m/km. The upper Truckee River begins at 2757 m asl and descends to Lake Tahoe (1899 m asl) over a distance of 37 km for a 23.2 m/km gradient. From Lake Tahoe to Reno, Nevada, the Truckee drops 528 m in 91 km, with an average gradient of 5.8 m/km. The river enters the Great Basin desert at Reno, flows east 53 km, and then turns north for another 44 km to Pyramid Lake (1156 m asl). It descends 215 m for a gradient of about 2.2 m/km. Higher-gradient sections of the Truckee River have rubble to boulder substrates, including the Reno area. Downstream of Reno large substrates occur in confined river reaches, but siltation is noticeable in the summer.

Discharge varies seasonally (see Fig. 14.16). Average annual flow for the Truckee River at the Lake Tahoe outlet is 6.3 m³/s but in high-water years it reaches 32.6 m³/s and has been as low as 0.004 m³/s. The outlet is regulated by a dam so that water can be stored in the lake. Initially established for use in hydropower generation, it is now controlled by the Bureau of Reclamation, but flows are still regulated for downstream power-generation facilities. At the California–Nevada border the average annual flow is 21.2 m³/s, ranging from 5.2 to 69.2 m³/s. Nine streams enter the Truckee River in the upper basin. None of these exceed the Truckee River outflow from Lake Tahoe, but collectively they provide 68% of the flow at the California–Nevada border. The largest of these are Donner Creek with 12% of the discharge, Prosser Creek with 13%, and the Little Truckee River with 25%. At Reno the average annual discharge drops to 18.7 m³/s, and below Derby Dam, 64 km upstream of Pyramid Lake, the flow is reduced to 10.5 m³/s. By the time the river reaches Nixon, Nevada, near Pyramid Lake the average annual discharge is 14.2 m³/s. Irrigation return flows boost the river volume.

Although no records exist to directly give the discharge of the Truckee River into Pyramid Lake prior to water diversions, it is known that the lake fluctuated about 6 m annually and, based on its surface area at that time, that represents the evaporation of almost 586 million m³ of water annually. This suggests a prediversion average annual discharge of

18.6 m³/s from the Truckee River into Pyramid Lake. The greatest precipitation in the basin occurs in February and runoff in the Truckee River peaks in May with snowmelt (see Fig. 14.16).

The exposed Sierra batholith in the upper basin is low in nutrient content, so the water is soft. In the upper Truckee River, calcium is 4.9 mg/L, magnesium 1.2 mg/L, sodium 4.6 mg/L, chloride 4.8 mg/L, alkalinity 24.3 mg/L as $CaCO_3$, and sulfate 0.8 mg/L. By the time the river reaches Reno these values have increased to 28 mg/L, 8.1 mg/L, 37 mg/L, 31.8 mg/L, 56 mg/L as $CaCO_3$, and 6.6 mg/L, respectively. At Wadsworth, Nevada, the same six variables are 64 mg/L, 27 mg/L, 92 mg/L, 86.5 mg/L, 174 mg/L as $CaCO_3$, and 160 mg/L. Alkalinity and calcium concentrations increase seven- to thirteenfold, sodium and chloride about twentyfold, and sulfates by seventy-fivefold. These increases are much lower than in Bonneville basin rivers, where sedimentary rocks dominate. This is likely the reason Pyramid Lake has been able to maintain fishes while the Great Salt Lake has not. Nevertheless, tufa deposits in Pyramid Lake reflect the high TDS. Nitrate and orthophosphate increase downstream. In the early 1980s, these two ions were 0.02 and 0.007 mg/L, respectively, at the outflow of Lake Tahoe, and 0.02 and 0.198 mg/L at Reno. Near the river's terminus at Nixon, Nevada, NO_3-N was 0.32 mg/L and orthophosphate was 0.593 mg/L. The Reno-Sparks sewage treatment plant installed equipment to remove phosphorus in 1982 and denitrification towers in 1988. By the 1990s both NO_3-N and orthophosphate at Nixon had decreased to 0.01 mg/L.

River Biodiversity and Ecology

The Truckee River is part of the Lahontan freshwater ecoregion of the Great Basin complex (Abell et. al. 2000). Several tributaries to the Truckee River have been the sites of studies. One of the most notable was a study on the variability of invertebrates sampled in a single riffle (Needham and Usinger 1956), which has influenced the design of macrobenthos sampling since it was published. The Sagehen Creek Field Station, operated by the University of California, Berkeley, is located on a tributary to the Truckee River. This station has facilitated a number of studies that relate to the Truckee River basin, from fish life histories (Jones 1972, Erman and Hawthorn 1976), to the role of beaver in aquatic systems (e.g., Beier and Barrett 1987), to the importance of peat fens in stream energetics (Erman and Chouteau 1979).

Plants

The riparian plant community varies with elevation. In the high elevations of the Sierra Nevada, mountain alder, black cottonwood, quaking aspen, narrowleaf cottonwood, willow, and red-osier dogwood are present. At lower elevations river birch, willow, and Fremont cottonwood can form the riparian overstory (Manning and Padgett 1995, Howald 2000).

Invertebrates

Collections in the Monte L. Bean Life Science Museum at Brigham Young University include stoneflies and caddisflies, mostly from tributaries in the Truckee basin. Stoneflies are represented by *Alloperla*, *Haploperla*, *Suwallia*, *Sweltsa*, *Capnia*, *Mesocapnia*, *Utacapnia*, *Megaleuctra*, *Paraleuctra*, *Nemoura*, *Zapada*, *Doroneuria*, and *Isoperla*. Caddisflies include *Apatania*, *Hydroptila*, *Polycentropus*, *Psychomyia*, *Ochrotrichia*, *Psychoglypha*, *Wormaldia*, and *Rhyacophila*. The mayfly genera *Ephemerella*, *Heptagenia*, *Rhithrogena*, and *Baetis* are also present in the drainage (Allen and Murvosh 1991). The mayflies *Ameletus*, *Caenis*, *Leptophlebia*, *Anthopotamus*, and *Choroterpes* are in Lake Tahoe.

Lake Tahoe is possibly late Pliocene in age, and a number of unique endemic native organisms are present. The benthic invertebrates are dominated by oligochaetes, amphipods, and dipterans. The oligochaete *Rhyacodrilus* occurs from the deepest waters to 30 m depth, whereas the oligochaetes *Arcteonais* and *Uncinais* occur in water depths less than 30 m. Crustaceans include *Hyalella azteca* in the shallow waters and two endemic species of *Stygobromus* in deeper waters. Perhaps most interesting are two species of endemic flightless stoneflies. A shallow-water species, *Utacapnia tahoensis*, emerges along the lake shore, but the deep-water species, *Capnia lacustra*, spends its entire life cycle underwater at depths down to 274 m (Frantz and Cordone 1996). In 1963 the opossum shrimp was introduced to Lake Tahoe and by 1971 the major zooplankton species in the lake had crashed (Richards et al. 1975). From 1985 to 1986 an attempt to recollect the deepwater stonefly failed, a serious concern, as this species had been in densities of 38/m^2 in the 1960s (Frantz and Cordone 1996). The signal crayfish may also be native to the drainage, although populations in Lake Tahoe are thought to be introduced.

Vertebrates

The native fishes of the Truckee River include Lahontan cutthroat trout, mountain whitefish, Paiute sculpin, speckled dace, tui chub, Lahontan redside, Tahoe sucker, and mountain sucker. A ninth species, the cui-ui, listed as endangered in 1967, occurs only in Pyramid Lake and spawns in the Truckee River. Lahontan cutthroat trout were listed as threatened in 1970.

One of the most recognized human impacts on the Truckee River is the extinction of the Pyramid Lake strain of the Lahontan cutthroat trout. This piscivorous trout reached 1.2 m long. The record sport catch weighed 18.6 kg, and the record commercial catch was estimated at 27.3 kg. Although the trout survived the high salinity and temperature in Pyramid Lake, the low-elevation reaches of the Truckee River were probably too warm for good fry survival. In the late winter and early spring, the trout would swim up the Truckee from Pyramid Lake to tributaries and lakes in the Sierra Nevada to spawn (e.g., Lake Tahoe and Donner Lake). Dams blocked trout access to the upper Truckee River, and with reduced discharge due to irrigation diversions Pyramid Lake fell in elevation. A bar developed across the mouth of the river, preventing upstream fish movement after 1938 (Gerstung 1988). With that final impact, the largest cutthroat trout in North America went extinct. The cui-ui in Pyramid Lake was also unable to spawn, but its long lifespan allowed it to survive the obstruction.

A number of nonnative fishes are established in the basin (Sada 2000). These include rainbow trout, brown trout, lake trout, sockeye salmon, common carp, channel catfish, black bullhead, white bass, and Sacramento perch. The impacts of nonnative fishes are the same as elsewhere in the Great Basin. They are slowly destroying the native fish communities. Fortunately, Pyramid Lake's high salinity has prevented nonnative fishes from invading there.

The endangered mountain yellow-legged frog and foothill yellow-legged frog are part of the native Sierra Nevada fauna in the upper Truckee River basin. The Pacific tree frog and boreal toad occur in the upper basin as well. In the lower basin, Woodhouse's toad and the northern leopard frog are native. The common garter snake and the Western aquatic garter snake are both found in the Truckee basin, as are beaver and muskrat.

During wet years the Truckee River would overflow a small divide a few kilometers upstream from Pyramid Lake into the large, marshy Winnemucca Lake (see Fig. 14.15). This wetland once offered important habitat to waterfowl on their migratory flights along the Pacific Flyway. However, once water diversion into the Carson basin via Derby Dam began, Lake Winnemucca completely dried and has remained dry since 1938.

Ecosystem Processes

Prosser Creek, California, in the Truckee River drainage, is the site of Needham and Usinger's (1956) classic study of the variability among invertebrate samples taken from a single riffle. Their results have been influential in the design of quantitative stream sampling programs. Sagehen Creek, another tributary, has been a focal area for a number of studies of the Truckee River drainage. Erman and Hawthorn (1976) showed that an intermittent stream entering Sagehen Creek was important spawning habitat for rainbow trout. About 40% of the rainbow trout utilized it for reproduction, although brook trout could not because the stream was dry in the fall. The influence of such ephemeral habitats on life histories could also be reflected in the outcomes of interactions between the two species. Erman and Chouteau (1979) showed that fine particulate organic matter input from fens (small peatlands) into Sagehen Creek varied from 5 to 20 mg m^{-2}yr^{-1}. The input fluctuated daily, with the lowest occurring in the afternoon. Black flies increased in densities below the fens, although other macroinvertebrates did not. Beaver colonies in the Truckee River system above Verdi at the Nevada–Californian border were more successful in river reaches with lower gradients and abandoned sites were more common in reaches with narrow channels and steep gradients (Beier and Barrett 1987). Beaver also significantly reduced or eliminated stands of both black cottonwood and quaking aspen, which reduced the input of leaves into the river. But woody debris input was increased by the dams, and both sediments and particulate organic material were retained by the dams. These studies suggest that overall detrital dynamics in the river can be influenced by import (fens, streamside vegetation) and the presence of instream structures (beaver dams), which in turn are related to the gradient of the river.

Human Impacts and Special Features

The Truckee River begins in Lake Tahoe, one of the deepest lakes in North America, and terminates in the Nevada desert in Pyramid Lake, the largest remnant of Pleistocene Lake Lahontan. Pyramid Lake, despite being a terminal basin lake, still has a freshwater fauna, although its total dissolved solid loading is very high. Prior to watershed development spawning fishes from Pyramid Lake could swim the entire river, to Lake Tahoe as well as surrounding lakes.

The Truckee River basin is the center of western Nevada's urbanization, it provides a major route over the Sierra Nevada into California (the California Trail, the Central Pacific Railroad, Interstate 80), and it includes a major recreation area in the Lake Tahoe region. As the Truckee River basin was settled demands increased on the region's resources. In the late 1800s between 45 to 90 metric tons of trout were shipped annually from the Pyramid Lake fishery (Gerstung 1988). Trees in the watershed were cut for timbers for the mining industry, wastes from mills increased pollution, and grazing increased sediment loadings.

Like all rivers of the Great Basin, the Truckee River is a highly regulated drainage. Most of the water in the Truckee River originates from snowmelt in the Sierra Nevada. Very little input comes from the Truckee Valley in Nevada. The majority of the water-storage reservoirs are on previously free-flowing tributaries to the Truckee River, but the outflows from Lake Tahoe and most of the smaller natural lakes in the upper part of the basin (e.g., Donner Lake) are also regulated by dams. Furthermore, four hydroelectric facilities are located at or near the California–Nevada border and water diversions to these facilities dewatered reaches of the Truckee River, significantly fragmenting the system. Two small diversion dams exist on the lower Truckee within the Pyramid Lake reservation.

Seven diversions take water out of the Truckee River basin. The Tahoe basin exports approximately 0.36 m^3/s of treated wastewater to Alpine County, California, and the Carson River basin, Nevada. In the 1950s, nitrates leaching from septic systems were causing eutrophication in Lake Tahoe and by late 1960 wastewater began to be exported. Yet in 1995 the Tahoe Regional Planning Agency reported that algal primary productivity was still over 3.5 times greater than in 1968 (http://water.nv.gov/Water%20 planning/truckee/trchrono.htm 1997). In 1905 the U.S. Bureau of Reclamation began the Newlands Project, constructing Derby Dam on the Truckee River (Coleman and Johnson 1988). This dam facilitated transport of up to 50 million m^3/yr (1.6 m^3/s) of water from the Truckee to the Carson River basin, where it has been used to augment agricultural water supplies. As a result, Pyramid Lake has fallen 26 m (Coleman and Johnson 1988), losing 33% of its total volume and increasing in salinity to 5100 mg/L. Since 1970 the diversion at Derby Dam has been set at 30% of the average annual Truckee River flow to reduce impacts on the remaining Pyramid Lake fishery.

FIGURE 14.7 Provo River near Midway, Wasatch County, Utah. This is the only unchannelized section of the middle reach of the Provo River (PHOTO BY D. K. SHIOZAWA).

ADDITIONAL RIVERS

The 119 km long Provo River heads in Utah in small alpine and subalpine lakes in the Uinta Mountains, flows southwest, and terminates in Utah Lake (Fig. 14.7, Fig. 14.17). Mean annual precipitation ranges from 102 cm in the mountains (Wasatch and Uinta Montane Forests ecoregion) to 25 to 30 cm around Utah Lake (Great Basin Shrub Steppe ecoregion). Snowmelt feeds the river, resulting in peak runoff in May (Fig. 14.18). The Provo River has two major reservoirs, Jordanelle and Deer Creek, three major diversions to Heber Valley, Provo City, and Salt Lake City, plus numerous smaller diversions. The largest consumptive use of surface water is irrigation (78%). The Provo River provides one of the top trout fisheries in the western United States and minimum flows from the reservoirs protect these tailwater fisheries. The lower 7.8 km of the river is critical habitat for the endangered June sucker. The most serious threats to water quality of the Provo River are nutrient enrichment and pesticides associated with agriculture in Heber Valley, pollutants attributed to recreational activities in the reservoirs (500,000 people visit Jordanelle and Deer Creek reservoirs annually), and nonpoint-source pollutants associated with runoff from Provo City.

The Weber River is the second-largest tributary to the Great Salt Lake. Mountainous terrain, above 1515 m asl, accounts for 84% of the Weber River basin (Fig. 14.8). It flows northwest and is joined by the Ogden River, the largest of four major tributaries, just before it enters the Great Salt Lake (Fig. 14.19). Mean annual precipitation, enhanced by evaporation from the Great Salt Lake, ranges from 25 to 87 cm in the mountains and from 10 to 19 cm near the Great Salt Lake. Seasonal runoff is strongly influenced by snowmelt, which peaks in May (>5 cm/mo) and declines to ≤0.3 cm/mo from July to September (Fig. 14.20). The Weber River, approximately 200 km long, is within the Wasatch and Uinta Mountain Forests terrestrial ecoregion from its headwaters to the mouth of Weber Canyon. Upon leaving Weber Canyon it enters the Great Basin Shrub Steppe

FIGURE 14.8 Weber River, west of Coalville, Summit County, Utah (Photo by D. K. Shiozawa).

ecoregion in the bed of Lake Bonneville. It terminates in the Ogden Bay Wildlife Management Area. Mountainous areas are used for limited crop production, livestock grazing, timber harvest, and winter recreation. Mountain communities are mostly small rural towns, with the exception of Park City, which is the fastest-growing area in Utah. Reservoirs store 54% of the basin's total average annual runoff. Five interbasin transfers have been developed, two to the Provo River, one to the Bear River, and one each from the Jordan and Provo rivers back into the Weber River.

The Walker River is formed by the confluence of the East and West Walker rivers about midway in the drainage basin south of Yerington, Nevada (Fig. 14.1, Fig. 14.21). Both of these tributaries head in the Sierra Nevada of California at a maximum elevation of 3420 m asl. The upper 25% of the drainage supplies over 90% of the basin's total surface water. The highest precipitation occurs from December through March, but it is stored as snow before being released in the late spring and early summer (Fig. 14.22). Cattle grazing and agriculture are the primary land uses in the lower basin and timber

harvest and agriculture dominate the upper basin. The surface waters of the Walker River are over-appropriated, resulting in heavy groundwater pumping. Walker Lake, the terminal lake in the basin, supports a major recreational fishery for Lahontan cutthroat trout and is an important waterfowl stopover area. Since 1882 the lake has fallen 40 m in elevation and 77% in volume. TDS concentrations (primarily sodium chloride, sulfate, and bicarbonate) have increased from 2560 mg/L to 13,000 mg/L. With current water availability (3.5 m^3/s), Walker Lake is projected to stabilize at 9.7% of its 1882 volume in the year 2046. An average inflow of 4.2 m^3/s would stabilize the lake at its 1994 volume. In 1996 the conservation organization American Rivers designated the Walker River one of thirty threatened or endangered rivers in the United States.

ACKNOWLEDGMENTS

We would like to acknowledge R. W. Baumann, M. L. Bean Life Science Museum, Brigham Young University, for allowing us to examine the insect collection and for providing information, especially on the Humboldt River. Mark Vincent, Utah State University, provided species lists from the Sevier and Bear rivers. Chris Rosamond, University of Nevada, Reno, provided a list of invertebrates from the Truckee River system. Patrick Collins, Mount Nebo Scientific, Inc., kindly reviewed the plant community information. Kent Hatch and Riley Nelson of Brigham Young University provided information about Lahontan basin species. M. McKell, E. McLaughlin, K. Redlin, J. Hansen, S. Piper, B. Florence, A. Pace, A. V. Bell, and A. R. Bell aided in compiling data.

LITERATURE CITED

Abell, R., D. M. Olson, E. Dinerstein, P. T. Hurley, W. Eichbaum, S. Walters, T. Allnutt, W. Wettengel, and C. J. Loucks. 2000. *Freshwater ecoregions of North America: A conservation assessment.* Island Press, Washington, D.C.

Allen R. K., and C. M. Murvosh. 1991. A biogeographically based assessment of the potential mayfly fauna of Nevada. *Pan-Pacific Entomologist* 67:206–215.

Baumann, R. W., and B. C. Kondratieff. 2000. A confirmed record of the ephemeropteran genus *Baetisca* from west of the continental divide and an annotated list of the mayflies of the Humboldt River of Nevada. *Western North American Naturalist* 60:459–461.

Behmer, D. J., and C. P. Hawkins. 1986. Effects of overhead canopy on macroinvertebrate production in a Utah stream. *Freshwater Biology* 16:287–300.

Behnke, R. J. 1992. *Native trout of Western North America.* Monograph 6. American Fisheries Society, Bethesda, Maryland.

Beier, P., and R. Barrett. 1987. Beaver habitat use on impact in the Truckee River Basin, California. *Journal of Wildlife Management* 51:649–654.

Bradley, R. S. 1999. *Paleoclimatology: Reconstructing the climates of the Quaternary.* 2nd ed. Academic Press, San Diego, California.

Bright, R. C. 1963. Pleistocene lakes Thatcher and Bonneville, southeastern Idaho. Ph.D. diss., University of Minnesota, Minneapolis.

Broughton, J. M., D. B. Madsen, and J. Quade. 2000. Fish remains from Homestead Cave and lake levels of the past 13,000 years in the Bonneville basin. *Quaternary Research* 53:392–401.

Brussard, P. F., D. A. Charlet, and D. S. Dobkin. 1998. Regional trends of biological resources: Great Basin—Mojave Desert Region. In M. J. Mac, P. A. Opler, C. E. Puckett Haecker, and P. D. Doran (eds.). *Status and trends of the nation's biological resources.* Vol. 2, pp. 505–542. U.S. Department of the Interior, U.S. Geological Survey, Reston, Virginia

Chamberlin, R. V., and D. T. Jones. 1929. A descriptive catalog of the mollusca of Utah. *Bulletin of the University of Utah* 19(4):1–203.

Coleman, M. E., and V. K. Johnson. 1988. Summary of trout management at Pyramid Lake, Nevada, with emphasis on Lahontan cutthroat trout, 1954–1987. In R. E. Gresswell (ed.). *Status and management of interior stocks of cutthroat trout,* pp. 107–115. American Fisheries Society Symposium 4. Bethesda, Maryland.

Cook, E. R., C. Woodhouse, C. M. Eakin, D. M. Mecko, and D. W. Stahle. 2004. Long-term aridity changes in the western United States. *Science* 305:1015–1018.

Dowling, T. E., C. A. Tibbets, W. L. Minckley, and G. R. Smith. 2002. Evolutionary relationships of the Plagopterins (Teleosti: Cyprinidae) from cytochrome b sequences. *Copeia* 2002:665–678.

Dunham, J. B., M. M. Peacock, B. E. Rieman, R. E. Schroeter, and G. L. Vinyard. 1999. Conservation implications of local and geographic variability in the distribution of stream-living Lahontan cutthroat trout. *Transactions of the American Fisheries Society* 128:875–889.

Durham, M. S. 1997. *Desert between the mountains.* University of Oklahoma Press, Norman.

Durrant, S. D. 1952. *Mammals of Utah.* University of Kansas Press, Lawrence.

Eaton, G. P. 1982. The basin and range province: Origin and tectonic significance. *Annual Review of Earth and Planetary Science* 10:409–440.

Egan, F. 1977. *Fremont: Explorer for a restless nation.* University of Nevada Press, Reno.

Erman, D. C., and W. C. Chouteau. 1979. Output of fine particulate organic carbon from some Sierra Nevada fens and its importance to invertebrates. *Oikos* 32:409–415.

Erman, D. C., and V. M. Hawthorn. 1976. The quantitative importance of an intermittent stream in the spawning of rainbow trout. *Transactions of the American Fisheries Society* 105:675–681.

Fillmore, R. 2000. *The geology of the parks, monuments and wildlands of southern Utah.* University of Utah Press, Salt Lake City.

Frantz, T. C., and A. J. Cordone. 1996. Observations on the macrobenthos of Lake Tahoe, California. *California Fish and Game* 82:1–41.

Gerstung, E. R. 1988. Status, life history, and management of the Lahontan cutthroat trout. *American Fisheries Symposium* 4:93–106.

Giddings, E. M., and D. Stephens. 1999. Selected aquatic biological investigations in the Great Salt Lake basins, 1875–1998, national water-quality assessment program. Water-Resources Investigations Report 99-4132. U.S. Geological Survey, Salt Lake City, Utah.

Grayson, D. K. 1993. *The deserts past: A natural history of the Great Basin.* Smithsonian Institution Press, Washington, D.C.

Greer, D. C., K. D. Gurgel, W. L. Wahlquist, H. A. Crusty, and G. B. Peterson. 1981. *Atlas of Utah.* Brigham Young University Press, Provo, Utah.

Haws, F. W., and T. C. Hughes. 1973. Hydrologic inventory of Bear River study unit. Developed for the

Utah Division of Water Resources by Utah Water Research Laboratory, College of Engineering, Utah State University, Logan.

Hayward, L. C. 1945. Biotic communities of the southern Wasatch and Uinta mountains, Utah. *Great Basin Naturalist* 6:1–124.

Hintze, L. F. 1988. Geologic history of Utah. Brigham Young University Geological Studies Special Publication 7, Provo.

Horne, A. J., and C. R. Goldman. 1994. *Limnology.* McGraw Hill, New York.

Howald, A. 2000. Plant communities. In G. Smith (ed.). *Sierra east: Edge of the Great Basin*, pp. 94–207. California Natural History Guides. University of California Press, Berkeley and Los Angeles.

Hubbs, C. L., and R. R. Miller. 1948. The Great Basin with emphasis on glacial and post glacial times. II. The zoological evidence. *Bulletin of the University of Utah* 38(20):18–166.

Hunt, C. B. 1974. *Natural regions of the United States and Canada.* W. H. Freeman, San Francisco.

James, G. W. 1978. *Lake of the sky: Lake Tahoe in the High Sierras of California and Nevada.* Outbooks, Olympic Valley, California (facsimile of the 1915 edition).

Janetski, J. C. 1990. Wetlands in Utah Valley prehistory. In J. C. Janetski and D. B. Madsen (eds.), *Wetland adaptations in the Great Basin*, pp. 233–257. Museum of Peoples and Cultures Occasional Papers 1. Brigham Young University Press, Provo, Utah.

Johnson, J. E. 1986. Inventory of Utah crayfishes with notes on current distribution. *Great Basin Naturalist* 46:625–631.

Jones, A. C. 1972. Contributions to the life history of the Paiute sculpin in Sagehen Creek, California. *California Fish and Game* 58:285–290.

Kelly, R. L. 2001. *Prehistory of the Carson Desert and Stillwater Mountains.* University of Utah Anthropological Papers no. 123.University of Utah Press, Salt Lake City.

Loudenslager, E. J., and G. A. E. Gall. 1980. Geographic patterns of protein variation and subspeciation in cutthroat trout, *Salmo clarki. Systematic Zoology* 29:27–42.

Lunine, J. I. 1999. *Earth: Evolution of a habitable world.* Cambridge University Press, New York.

Malde, H. E. 1968. The catastrophic late Pleistocene Bonneville flood in the Snake River plain. Professional Paper 596. U.S. Geological Survey, Washington, D.C.

Manning, M. E., and W. G. Padgett. 1995. *Riparian community type classification for Humboldt and Toiyabe National Forests, Nevada and Eastern California.* Publication R4-Ecol-95-01. U.S. Forest Service Intermountain Region, Ogden, Utah.

Martin, M. A., D. K. Shiozawa, E. J. Loudenslager, and J. N. Jeensen. 1985. An electrophoretic study of cutthroat trout populations in Utah. *Great Basin Naturalist* 45:677–687.

McConnell, W. J., and W. F. Sigler. 1959. Chlorophyll and productivity in a mountain river. *Limnology and Oceanography* 4:335–351.

McKell, M. D. 2003. Phylogeography of speckled dace, *Rhinichthys osculus* (Teleostei:Cyprinidae), in the Intermountain West, USA. Master thesis, Brigham Young University, Provo, Utah.

Minckley, W. L., D. A. Hendrickson, and C. E. Bond. 1986. Geography of western North American freshwater fishes: Description and relationships to intracontinental tectonism. In C. H. Hocutt and E. O. Wiley (eds.). *The zoogeography of North American freshwater fishes*, pp. 519–613. John Wiley and Sons, New York.

Needham, P. R., and R. L. Usinger. 1956. Variability of the macrofauna of a single riffle in Prosser Creek, California, as indicated by the Surber sampler. *Hilgardia* 24:383–409.

Nelson S. T., and D. G. Tingey. 1997. Time-transgressive and extension-related basaltic volcanism in southwest Utah and vicinity. *Geological Society of America Bulletin* 109:1249–1265.

Osborn, T. C. 1981. Stream insect production as a function of alkalinity and detritus processing. Ph.D. diss., Utah State University, Logan.

Richards, R. C., C. R. Goldman, T. C. Frantz, and R. Wickwire. 1975. Where have all the Daphnia gone? The decline of a major cladoceran in Lake Tahoe, California-Nevada. *Verhandlungen der Internationalen Vereinigung für Theoretische und Angewandte Limnologie* 19:835–842.

Ricketts, T. H., E. Dinerstein, D. M. Olson, C. J. Loucks, W. Eichbaum, D. DellaSala, K. Kavanagh, P. Hedao, P. T. Hurley, K. M. Carney, R. Abell, and S. Walters. 1999. *Terrestrial ecoregions of North America: A conservation assessment.* Island Press, Washington, D.C.

Sada, D. 2000. Native fishes. In G. Smith (ed.). *Sierra east: Edge of the Great Basin*, pp. 246–264. California Natural History Guides. University of California Press, Berkeley and Los Angeles.

Shiozawa, D. K., M. McKell, B. A. Miller, and R. P. Evans. 2002. Genetic assessment of four native fishes from the Colorado River drainages in western Colorado: The result of DNA analysis. Final Report to Colorado Division of Wildlife Resources, Fort Collins.

Smith, G. (ed.). 2000. *Sierra east: Edge of the Great Basin.* California Natural History Guides. University of California Press, Berkeley and Los Angeles.

Smith, G. R. 1981. Late Cenozoic freshwater fishes of North America. *Annual Review of Ecology and Systematics* 12:163–193.

Smith, G. R., T. Dowling, K. Gobalet, T. Lugaski, D. K. Shiozawa, and R. P. Evans. 2002. Biogeography and timing of evolutionary events among Great Basin fishes. In Robert Hershler, David B. Madsen, and Donald R. Currey (eds.). *Great Basin aquatic systems history*, pp. 175–234. Smithsonian Contributions to the Earth Sciences no. 33. Smithsonian Institution Press, Washington, D.C.

Stokes, W. L. 1979. Paleohydrographic history of the Great Basin. In G. W. Newmann and H. D. Goode (eds.). *Basin and Range Symposium and Great Basin Field Conference*, pp. 345–352. Rocky Mountain Association of Geologists and Utah Geological Association, Denver, Colorado.

Stokes, W. L. 1986. *Geology of Utah*. Occasional Paper 6. Utah Museum of Natural History, University of Utah Press, Salt Lake City.

Taylor, D. W. 1985. Evolution of freshwater drainages and molluscs in western North America. In C. J. Smiley (ed.). *Late Cenozoic history of the Pacific northwest*, pp. 265–321. Pacific Division American Association for the Advancement of Science and the California Academy of Science, San Francisco, California.

Thompson, R. S. 1990. Late Quaternary vegetation and climate in the Great Basin. In J. L. Betancourt, T. R. Van Devender, and P. S. Martin (eds.). *Packrat middens: 40,000 years of biotic change*, pp. 200–239. University of Arizona Press, Tucson.

Utah Board of Water Resources. 1992. *State water plan: Bear River basin*. Utah Department of Natural Resources, Salt Lake City.

Utah Board of Water Resources. 1995. *State water plan: Cedar/Beaver basin*. Utah Department of Natural Resources, Salt Lake City.

Utah Board of Water Resources. 1999. *State water plan: Sevier River basin*. Utah Department of Natural Resources, Salt Lake City.

Warner, T. J. (ed.) and A. Chavez (trans.). 1995. *The Dominguez–Escalante journal: Their expedition through Colorado, Utah, Arizona, and New Mexico in 1776*. University of Utah Press, Salt Lake City.

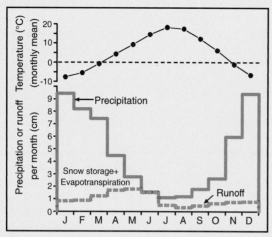

FIGURE 14.10 Mean monthly air temperature, precipitation, and runoff for the Bear River basin.

BEAR RIVER

Relief: 2593 m
Basin area: 19,631 km²
Mean discharge: 71.2 m³/s (virgin); 52.0 m³/s (present)
River order: 6
Mean annual precipitation: 55.9 cm
Mean air temperature: 6.5°C
Mean water temperature: 10.1°C
Physiographic provinces: Middle Rocky Mountains (MR), Basin and Range (BR)
Biomes: Temperate Mountain Forest, Desert
Freshwater ecoregion: Bonneville
Terrestrial ecoregions: Great Basin Shrub Steppe, Wasatch and Uinta Montane Forests, South Central Rockies Forests
Number of fish species: 33 (17 native)
Number of endangered species: 0
Major fishes: Bonneville cutthroat trout, mountain whitefish, Bonneville whitefish, Bear Lake whitefish, Bonneville cisco, speckled dace, longnose dace, redside shiner, leatherside chub, Utah chub, Utah sucker, mountain sucker, bluehead sucker, Paiute sculpin, mottled sculpin, Bear Lake sculpin

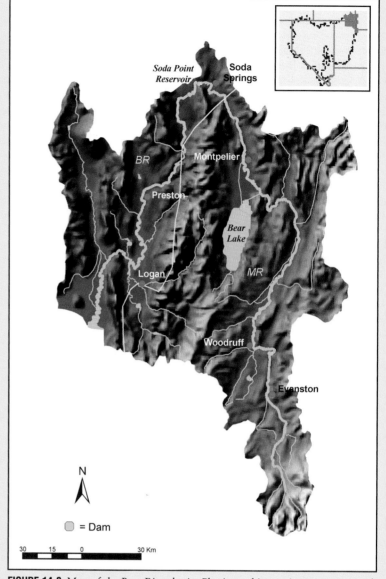

FIGURE 14.9 Map of the Bear River basin. Physiographic provinces are separated by a yellow line.

Major other aquatic vertebrates: Columbia spotted frog, boreal toad, Woodhouse's toad, northern leopard frog, tiger salamander, common garter snake, muskrat, beaver, dipper
Major benthic invertebrates: mayflies (*Baetis, Drunella, Ephemerella, Epeorus, Rhithrogena, Stenonema*), stoneflies (*Sweltsa, Claassenia, Isoperla, Zapada, Skwala*), caddisflies (*Brachycentrus, Micrasema, Helicopsyche, Arctopsyche, Cheumatopsyche, Hydropsyche, Hydroptila, Lepidostoma, Nectopsyche, Oecetis, Chyranda, Dicosmoecus, Hesperophylax, Limnephilus, Polycentropus, Rhyacophila, Oligophlebodes*)
Nonnative species: common carp, brown trout, rainbow trout, brook trout, lake trout, Yellowstone cutthroat trout, sockeye salmon, black bullhead, channel catfish, bluegill, green sunfish, yellow perch, walleye, largemouth bass, western mosquitofish, fathead minnow
Major riparian plants: Fremont cottonwood, narrowleaf cottonwood, river birch, red-osier dogwood, willow, box elder, wild rose
Special features: Bear Lake, Bear River Bay Bird Refuge, Uinta Mountains, Thatcher basin, Red Rock Pass, Oneida narrows
Fragmentation: 6 dams on main stem
Water quality: pH = 8.2, conductivity = 816 µS/cm, alkalinity = 225 mg/L as $CaCO_3$, Ca = 53 mg/L, Mg = 30 mg/L, Na = 59 mg/L, Cl = 83 mg/L, SO_4 = 53 mg/L, NO_3-N = 0.5 mg/L, PO_4-P = 0.2 mg/L
Land use: agriculture, recreation, grazing. Land ownership 55% private, 7% state land, 18% National Forest, 18% BLM, 2% refuges etc.
Population density: 9.0 people/km²
Major information sources: Haws and Hughes 1973, Utah Board of Water Resources 1992, http://waterdata.usgs.gov 2004, http://www.ncdc.noaa.gov 2004

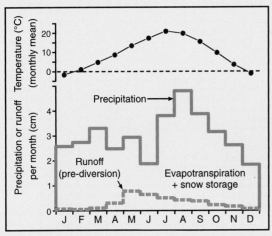

FIGURE 14.12 Mean monthly air temperature, precipitation, and runoff for the Sevier River basin.

SEVIER RIVER

Relief: 2341 m

Basin area: 42,025 km^2

Mean discharge: 32.2 m^3/s (virgin); 1.3 m^3/s (present)

River order: 5

Mean annual precipitation: 36 cm

Mean air temperature: 9.3°C

Mean water temperature: 11.3°C

Physiographic provinces: Colorado Plateaus (CO), Basin and Range (BR)

Biome: Desert

Freshwater ecoregion: Bonneville

Terrestrial ecoregions: Colorado Plateau Shrublands, Great Basin Shrub Steppe, Wasatch and Uinta Montane Forests

Number of fish species: 21 (9 native)

Number of endangered species: 0

Major fishes: Bonneville cutthroat trout, speckled dace, redside shiner, leatherside chub, least chub, Utah chub, mountain sucker, Utah sucker, mottled sculpin

Major other aquatic vertebrates: boreal toad, Woodhouse's toad, northern leopard frog, tiger salamander, common garter snake, muskrat, beaver, dipper

Major benthic invertebrates: mayflies (*Baetis, Rhithrogena*), stoneflies (*Capnia, Utacapnia, Alloperla, Claassenia, Hesperoperla, Diura, Isogenoides, Isoperla, Megarcys*), caddisflies (*Brachycentrus, Hydropsyche, Hydroptila, Ochrotrichia, Nectopsyche, Oecetis, Amphicosmoecus, Hesperophylax, Limnephilus, Onocosmoecus, Rhyacophila, Oligophlebodes*)

Nonnative species: Asiatic clam, common carp, brown trout, rainbow trout, brook trout, Yellowstone cutthroat trout, mountain whitefish, channel catfish, black bullhead, walleye, yellow perch, fathead minnow, western mosquitofish, tamarisk, Russian olive

Major riparian plants: Fremont cottonwood, narrowleaf cottonwood, river birch, red-osier dogwood, willow, box elder, wild rose

Special features: desert river through much of length; usually dries before reaching Sevier Lake, which is itself usually dry because of diversions; National Parks in high plateaus

Fragmentation: 8 dams on main stem

Water quality: pH = 8.1, conductivity = 1340 mS/cm, alkalinity = 263 mg/L as CaCO$_3$, Ca = 78 mg/L, Mg = 57 mg/L, Na = 170 mg/L, Cl = 235 mg/L, SO$_4$ = 240 mg/L, PO$_4$-P = 0.2 mg/L

Land use: 4.5% irrigated agriculture, 80% rangeland grazing; land ownership 21% private, 8% state, 23% National Forest, 48% BLM

Population density: 1.4 people/km^2

Major information sources: Utah Board of Water Resources 1999, http://waterdata.usgs.gov 2004, http://www.ncdc.noaa.gov 2004

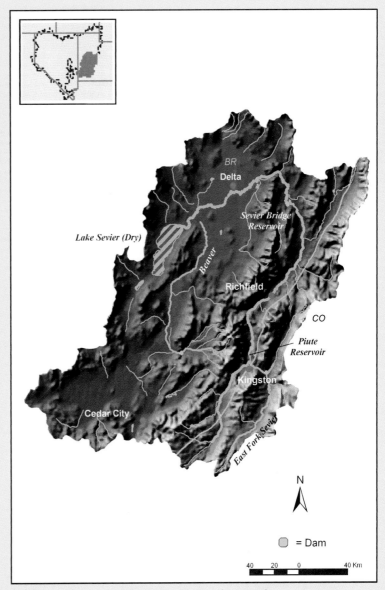

FIGURE 14.11 Map of the Sevier River basin. Physiographic provinces are separated by a yellow line.

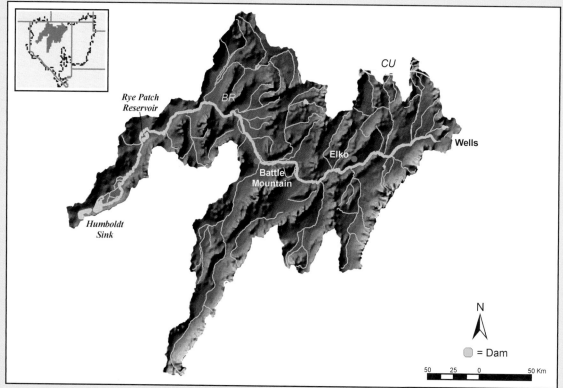

FIGURE 14.13 Map of the Humboldt River basin.

HUMBOLDT RIVER

Relief: 2036 m
Basin area: 43,597 km²
Mean discharge: 25.8 m³/s (virgin, estimated); 0.9 m³/s (present)
River order: 6
Mean annual precipitation: 22 cm
Mean air temperature: 9.4°C
Mean water temperature: 12.3°C
Physiographic province: Basin and Range (BR)
Biome: Desert
Freshwater ecoregion: Lahontan
Terrestrial ecoregions: Great Basin Montane Forests, Great Basin Shrub Steppe
Number of fish species: 23 (7 native)
Number of endangered species: 1 threatened
Major fishes: Lahontan cutthroat trout, Paiute sculpin, tui chub, Lahontan redside, speckled dace, Tahoe sucker, mountain sucker
Major other aquatic vertebrates: Columbia spotted frog, northern leopard frog, boreal toad, Great Basin spadefoot toad, common garter snake, western aquatic garter snake, muskrat, beaver
Major benthic invertebrates: mayflies (*Baetis, Acentrella, Camelobaetidius, Centroptilum, Baetisca, Ephemerella, Ephemera, Hexagenia, Heptagenia, Rhithrogena, Tricorythodes, Paraleptophlebia, Traverella, Ephoron*), stoneflies (*Isogenoides, Isoperla, Capnura, Taenionema, Acroneuria*), caddisflies (*Brachycentrus, Micrasema, Anagapetus, Cheumatopsyche, Hydropsyche, Hydroptila, Chyranda, Limnephilus, Nectopsyche albida, Rhyacophila*), crustaceans (signal crayfish)
Nonnative species: common carp, goldfish, brook trout, brown trout, rainbow trout, Yellowstone cutthroat trout, black bullhead, channel catfish, walleye, bluegill, green sunfish, black crappie, Sacramento perch, largemouth bass, bullfrog
Major riparian plants: quaking aspen, Fremont cottonwood, black cottonwood, narrowleaf cottonwood, willow, river birch
Special features: desert river traversing Great Basin east to west; longest river in Great Basin; terminates in Humboldt Sink, once a large wetland, now usually dry due to diversions; most tributaries in lower basin do not reach main stem
Fragmentation: 1 dam on main stem, numerous diversions in lower basin
Water quality: pH = 8.5, conductivity = 695 µS/cm, alkalinity = 238 mg/L as $CaCO_3$, Ca = 46 mg/L, Mg = 15 mg/L, Na = 82 mg/L, Cl = 61 mg/L, SO_4 = 72 mg/L, NO_3-N = 0.04 mg/L, PO_4-P = 0.4 mg/L
Land use: agriculture, grazing, mining; ownership 32% private, 2% state, reservations, etc., 12% National Forest, 54% BLM
Population density: 1.5 people/km²
Major information sources: Nevada Division of Water Resources 1999, Grayson 1993, http://waterdata.usgs.gov 2004, http://www.ncdc.noaa.gov 2004

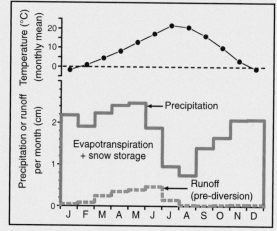

FIGURE 14.14 Mean monthly air temperature, precipitation, and runoff for the Humboldt River basin.

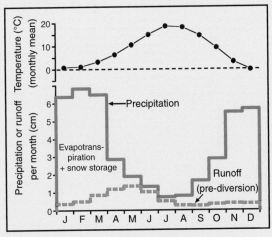

FIGURE 14.16 Mean monthly air temperature, precipitation, and runoff for the Truckee River basin.

TRUCKEE RIVER

Relief: 2159 m

Basin area: 7925 km²

Mean discharge: 18.6 m³/s (virgin, estimated); 14.2 m³/s (present)

River order: 5

Mean annual precipitation: 43 cm

Mean air temperature: 8.4°C

Mean water temperature: 11.4°C

Physiographic provinces: Cascade–Sierra Mountains (CS), Basin and Range (BR)

Biomes: Temperate Mountain Forest, Desert

Freshwater ecoregion: Lahontan

Terrestrial ecoregions: Sierra Nevada Forests, Great Basin Shrub Steppe

Number of fish species: 21 (8 native)

Number of endangered species: 1 fish, 1 frog, 1 threatened fish

Major fishes: Lahontan cutthroat trout, Paiute sculpin, tui chub, Lahontan redside, speckled dace, Tahoe sucker, mountain sucker, cui-ui

Major other aquatic vertebrates: Pacific tree frog, boreal toad, Great Basin spadefoot toad, mountain yellow-legged frog, foothill yellow-legged frog, northern leopard frog, river otter, muskrat, beaver

Major benthic invertebrates: mayflies (*Ephemerella, Drunella, Epeorus, Heptagenia, Rhithrogena, Cinygmula, Baetis*), stoneflies (*Capnia, Utacapnia, Skwala, Sweltsa, Malenka, Kogotus, Paraleuctra, Perlomyia, Prostoia, Soyedina, Zapada, Claassenia, Cultus, Isoperla*), caddisflies (*Glossosoma, Arctopsyche, Hydropsyche, Parapsyche, Lepidostoma, Limnephilus, Wormaldia, Rhyacophila*), crustaceans (signal crayfish)

Nonnative species: common carp, brook trout, brown trout, rainbow trout, Yellowstone cutthroat trout, sockeye salmon, black bullhead, channel catfish, bluegill, green sunfish, Sacramento perch, smallmouth bass, western mosquitofish, bullfrog

Major riparian plants: mountain alder, quaking aspen, black cottonwood, Fremont cottonwood, narrowleaf cottonwood, willow, red-osier dogwood, river birch

Special features: desert river flowing through Lake Tahoe and terminating in Pyramid Lake; Lake Tahoe recreation area; major transportation corridor from Nevada to California

Fragmentation: four hydroelectric diversion dams, one diversion dam in lower basin (Derby Dam), several small diversion dams on main stem; most large tributaries and lakes, including Lake Tahoe, regulated by dams

Water quality: pH = 8.0, conductivity = 226 µS/cm, alkalinity = 76 mg/L as CaCO₃, Ca = 23 mg/L, Mg = 8 mg/L, Na = 32 mg/L, Cl = 32 mg/L, SO₄ = 45 mg/L, NO₃-N = 0.04 mg/L, PO₄-P = 0.07 mg/L

Land use: recreation, agriculture, grazing, urban; land ownership 33% private, 7% state, 9% National Forest, 42% BLM, 9% reservation

Population density: 46.9 people/km²

Major information sources: Nevada Division of Water Resources 1997, Smith 2000, http://waterdata.usgs.gov 2004, http://www.ncdc.noaa.gov 2004

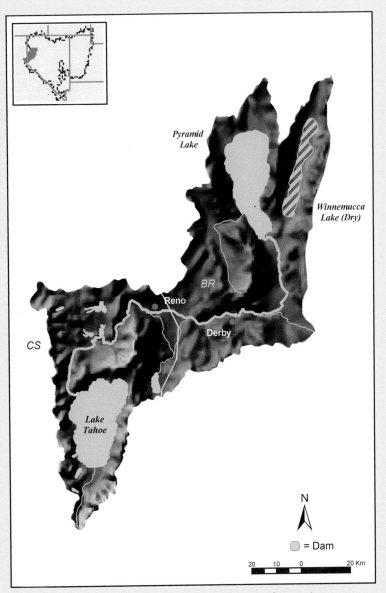

FIGURE 14.15 Map of the Truckee River basin. Physiographic provinces are separated by a yellow line.

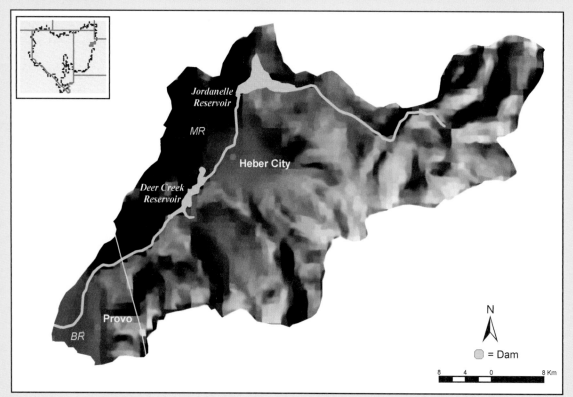

FIGURE 14.17 Map of the Provo River basin. Physiographic provinces are separated by a yellow line.

PROVO RIVER

Relief: 2294 m

Basin area: 1761 km^2

Mean discharge: 18.1 m^3/s (virgin, estimated); 9.9 m^3/s (present)

River order: 5

Mean annual precipitation: 64.2 cm

Mean air temperature: 9.3°C

Mean water temperature: 7.7°C

Physiographic provinces: Middle Rocky Mountains (MR), Basin and Range (BR)

Biomes: Temperate Mountain Forest, Desert

Freshwater ecoregion: Bonneville

Terrestrial ecoregions: Great Basin Shrub Steppe, Wasatch and Uinta Montane Forests

Number of fish species: 28 (13 native)

Number of endangered species: 1 fish

Major fishes: Bonneville cutthroat trout, mountain whitefish, redside shiner, leatherside chub, Utah chub, speckled dace, longnose dace, mountain sucker, Utah sucker, June sucker, mottled sculpin, Paiute sculpin, Utah Lake sculpin (extinct)

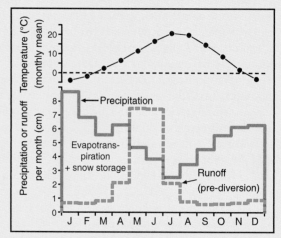

FIGURE 14.18 Mean monthly air temperature, precipitation, and runoff for the Provo River basin.

Major other aquatic vertebrates: Columbia spotted frog, northern leopard frog, Pacific chorus frog, Woodhouse's toad, boreal toad, tiger salamander, beaver, muskrat, mink, river otter, water vole, dipper

Major benthic invertebrates: mayflies (*Baetis, Paraleptophlebia, Drunella, Rhithrogena*), stoneflies (*Pteronarcys, Claassenia, Hesperoperla, Diura, Isogenoides, Isoperla, Megarcys*), caddisflies (*Brachycentrus, Hydropsyche, Hydroptila, Oecetis, Hesperophylax, Limnephilus, Onocosmoecus, Rhyacophila*), true flies (*Atherix*), crustaceans (*Gammarus, Hyalella*)

Nonnative species: virile crayfish, common carp, brown trout, rainbow trout, brook trout, Yellowstone cutthroat trout, channel catfish, black bullhead, largemouth bass, smallmouth bass, green sunfish, bluegill, black crappie, white bass, yellow perch, western mosquitofish, tamarisk, Russian olive

Major riparian plants: Fremont cottonwood, narrowleaf cottonwood, river birch, red-osier dogwood, willow, box elder

Special features: one of top trout streams in North America; middle reach undergoing restoration; recreation in Uinta Mountain

Fragmentation: two dams on main stem, numerous diversion structures

Water quality: pH = 7.8, alkalinity = 127 mg/L as CaCO$_3$, Ca = 54.1 mg/L, Mg = 11.8 mg/L, Na = 7.0 mg/L, Cl = 8.7 mg/L, SO$_4$ = 54.4 mg/L, NO$_3$-N = 0.18 mg/L, PO$_4$-P = 0.8 mg/L

Land use: urban, agriculture, grazing, recreation; land ownership 46.5% private, 7.5% state (including state parks), 45.5% National Forest, 1.0% BLM

Population density: 158.7 people/km^2

Major information sources: Utah Board of Water Resources 1999, http://waterdata.usgs.gov 2004, http://www.ncdc.noaa.gov 2004

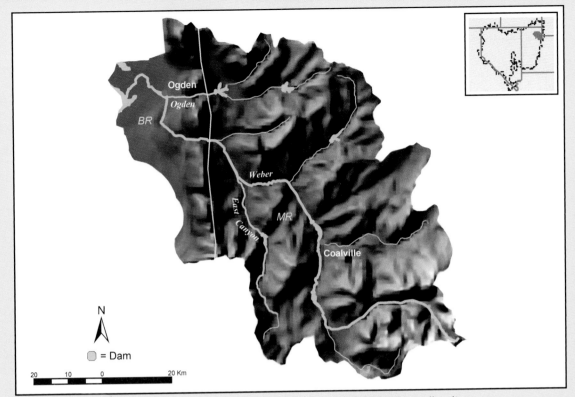

FIGURE 14.19 Map of the Weber River basin. Physiographic provinces are separated by a yellow line.

WEBER RIVER

Relief: 2358 m

Basin area: 6070 km²

Mean discharge: 38.3 m³/s (virgin, estimated); 12.2 m³/s (present)

River order: 5

Mean annual precipitation: 53.3 cm

Mean air temperature: 7.7°C

Mean water temperature: 10.0°C

Physiographic provinces: Middle Rocky Mountains (MR), Basin and Range (BR)

Biomes: Temperate Mountain Forest, Desert

Freshwater ecoregion: Bonneville

Terrestrial ecoregions: Great Basin Shrub Steppe, Wasatch and Uinta Montane Forests

Number of fish species: 26 (11 native)

Number of endangered species: 0

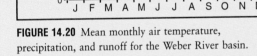

FIGURE 14.20 Mean monthly air temperature, precipitation, and runoff for the Weber River basin.

Major fishes: redside shiner, Utah chub, speckled dace, longnose dace, mottled sculpin, Paiute sculpin, Bonneville cutthroat trout, mountain whitefish, mountain sucker, Utah sucker, bluehead sucker

Major other aquatic vertebrates: Woodhouse's toad, tiger salamander, northern leopard frog, boreal toad, Pacific chorus frog, Columbia spotted frog, Great Basin spadefoot toad, water vole, muskrat, beaver, mink, river otter

Major benthic invertebrates: mayflies (*Baetis, Rhithrogena, Drunella*), stoneflies (*Capnia, Utacapnia, Alloperla, Pteronarcys, Amphinemura, Podmosta, Prostoia, Claassenia, Hesperoperla, Diura, Isogenoides, Isoperla*), caddisflies (*Brachycentrus, Hydropsyche, Hydroptila, Hesperophylax, Limnephilus, Oligophlebodes*), crustaceans (*Pacifastacus gambelii*)

Nonnative species: tamarisk, Russian olive, common carp, rainbow trout, Yellowstone cutthroat trout, brown trout, brook trout, yellow perch, smallmouth bass, largemouth bass, black bullhead, channel catfish, green sunfish, bluegill, black crappie, walleye, western mosquitofish

Major riparian plants: Fremont cottonwood, narrowleaf cottonwood, river birch, red-osier dogwood, willow, box elder

Special features: mountainous river system with most of basin privately owned; second-largest tributary of Great Salt Lake

Fragmentation: seven major dams, two on main stem

Water quality: alkalinity = 97 mg/L as $CaCO_3$, Ca = 58.9 mg/L, Mg = 17.6 mg/L, Na = 33.1 mg/L, Cl = 45.0 mg/L, SO_4 = 34.5 mg/L, PO_4-P = 0.65 mg/L

Land use: urban, agriculture, grazing, recreation; land ownership 80.9% private, 2.5% state, 15% National Forest, 1.6% BLM, Bureau of Reclamation, and Department of Defense

Population density: 80.7 people/km²

Major information sources: Utah Board of Water Resources 1999, http://waterdata.usgs.gov 2004, http://www.ncdc.noaa.gov 2004

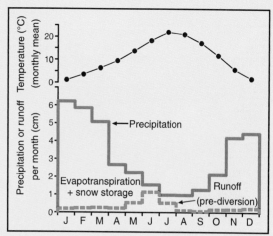

FIGURE 14.22 Mean monthly air temperature, precipitation, and runoff for the Walker River basin.

WALKER RIVER

Relief: 2576 m

Basin area: 7894 km²

Mean discharge: 9.7 m³/s (virgin); 3.5 m³/s (present)

River order: 6

Mean annual precipitation: 37.6 cm

Mean air temperature: 23.7°C

Mean water temperature: 10.8°C

Physiographic provinces: Cascade–Sierra Mountains (CS), Basin and Range (BR)

Biomes: Pinyon–Juniper, Basin Sagebrush

Freshwater ecoregion: Lahontan

Terrestrial ecoregions: Sierra Nevada Forests, Great Basin Shrub Steppe

Number of fish species: 15 (7 native)

Number of endangered species: 1 frog, 1 threatened fish

Major fishes: Lahontan redside, speckled dace, tui chub, Paiute sculpin, Lahontan cutthroat trout, mountain sucker, Tahoe sucker

Major other aquatic vertebrates: mountain yellowlegged frog, Yosemite toad, northern leopard frog, boreal toad, Great Basin spadefoot toad, beaver, muskrat, mink, common loon

Major benthic invertebrates: mayflies (*Baetis, Heptagenia, Rhithrogena, Ephemerella*), stoneflies (*Capnia, Pteronarcys, Utacapnia, Sweltsa, Isoperla, Prostoia, Paraleuctra, Claassenia, Kogotus*), caddisflies (*Brachycentrus, Micrasema, Lepidostoma, Hydropsyche, Hydroptila, Nectopsyche, Limnephilus, Rhyacophila, Glossosoma*), crustaceans (signal crayfish)

Nonnative species: Russian olive, common carp, channel catfish, black bullhead, Sacramento perch, white bass, brown trout, rainbow trout, Yellowstone cutthroat trout

Major riparian plants: narrowleaf cottonwood, Fremont cottonwood, river birch, red-osier dogwood, willow, mountain alder

Special features: best-studied basin in Lahontan basin; terminates in Walker Lake, which has major cutthroat trout fishery; lake faces destruction from increased salinity due to diversions and groundwater pumping

Fragmentation: nine major reservoirs, four on East Walker River, four on West Walker River, one on main Walker River

Water quality: pH = 8.2, alkalinity = 99.9 mg/L as $CaCO_3$, Ca = 24.8 mg/L, Mg = 6.1 mg/L, Na = 35.3 mg/L, Cl = 9.6 mg/L, SO_4 = 21.3 mg/L, NO_3-N = 0.11 mg/L, PO_4-P = 0.21 mg/L

Land use: agriculture, grazing, rangeland; land ownership 9.6% private, 0.5% state, 25.8% National Forest, 52.4% BLM, 4.1% Bureau of Reclamation and Department of Defense, 7.7% reservation

Population density: 2.3 people/km²

Major information sources: Nevada Division of Water Resources 2002, Grayson 1993, http://waterdata.usgs.gov 2004, http://www.ncdc.noaa.gov 2004

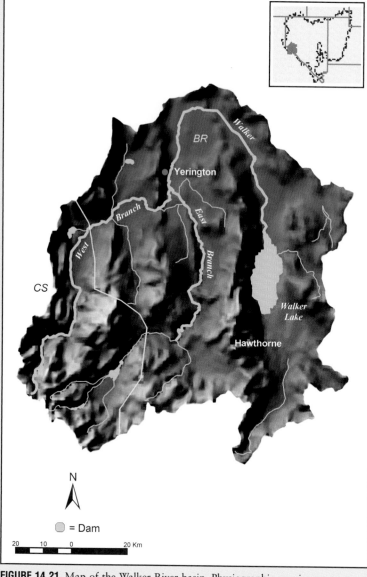

FIGURE 14.21 Map of the Walker River basin. Physiographic provinces are separated by a yellow line.

15

FRASER RIVER BASIN

TREFOR B. REYNOLDSON JOSEPH CULP
RICK LOWELL JOHN S. RICHARDSON

INTRODUCTION

FRASER RIVER MAIN STEM

THOMPSON RIVER

NECHAKO RIVER

STUART RIVER

ADDITIONAL RIVERS

ACKNOWLEDGMENTS

LITERATURE CITED

INTRODUCTION

The Fraser River, with its headwaters in the Rocky Mountains, flows across the dry Fraser Plateau through coastal mountain ranges to the Pacific Ocean (Fig. 15.2). This river has been the primary trade route through southern and central British Columbia for millennia. In recent times it has been the focus of settlement and industry and sustains Vancouver as one of North America's great cities. By any measure the Fraser is one of Canada's great rivers, with an abundance and diversity of natural resources that rivals almost any other river in the world. Of Canadian rivers it has the third-highest mean flow (3972 m³/s), it is the fifth longest (1375 km), and its drainage basin is the fifth largest (234,000 km²). The diversity of geology, climate, and the landscape is so great that it includes 11 of the 14 biogeoclimatic zones identified in British Columbia, and from its headwaters the river discharges to the Straits of Georgia in the city of Vancouver, the most densely populated region of the province.

The Fraser River was named after Simon Fraser, who while working for the North West Company was the first nonnative North American to reach the mouth of the river in 1808. The Fraser is far more than a single watercourse. Beyond the main-stem river is a vast and intricate network of tributary rivers that reach out to envelop more than a quarter of British Columbia. The entire catchment area of the Fraser River reaches as far north as Bulkley House on the Stuart River system and stretches westward from the highest summits of the Coast Mountain range to the heights of the Rocky Mountains in the east. It is the water flowing through these tributaries that links the massive Fraser basin system, providing essential nutrients, migration routes, and habitat that support the diverse ecosystems within it.

The river and its basin occupy a special historic and political place in the development of North America. For at least 10,000 years native peoples have occupied the Fraser River basin and used its resources (Kew and Griggs 1991). Within British Columbia there is a greater degree of indigenous cultural and linguistic diversity than in any other region of Canada. The Fraser basin itself was and remains inhabited by native peoples speaking six separate and distinct languages belonging to two great language families. Within the basin can be distinguished both Northwest-coast cultures on the lower part of the river and adjoining sea coast and Plateau cultures in the middle and upper parts of the basin. This remark-

FIGURE 15.1 North Fork of the Thompson River above Vavenby, British Columbia (PHOTO BY TIM PALMER).

697

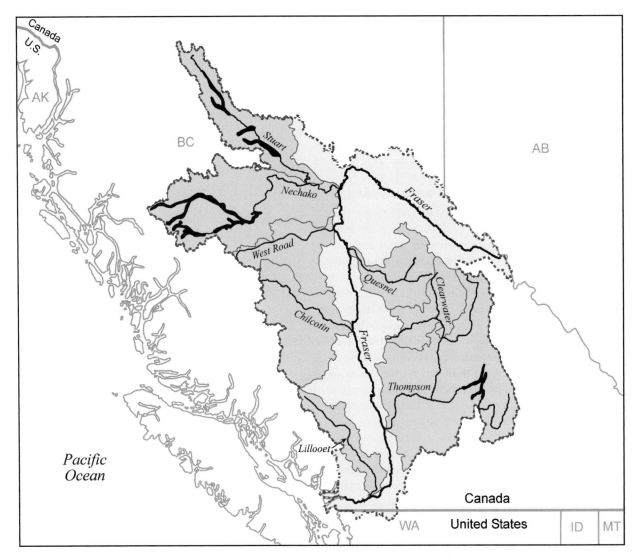

FIGURE 15.2   The Fraser River and its tributaries covered in this chapter.

able variety and diversity is consistent with the 10 millennia of human presence in the basin, and it indicates a large degree of stability. It also suggests that to a large degree, with the notable exception of occasional trade and barter or raids, each language group remained independent of its neighbors and was largely self-contained and self-sufficient within its traditional territories. Prior to contact with European explorers, an estimated 50,000 people lived in the basin (Dorcey 1991). However, these numbers were likely already half the original population after the ravages of disease introduced by earlier explorers of the continent by the time Simon Fraser canoed the river some 200 years later in 1808. Although salmon fishing and fur trading led to minor settlement and agricultural development around Fort Langley and

New Westminster before 1855, it was not until the discovery of gold on the Thompson tributary in 1857 that the consequent influx of people raised the total population to precontact levels. The gold strikes at Barkerville in the headwaters of the Willow River resulted in the development of famous land and water routes up the Fraser, including the Harrison Trail and Cariboo Road, and provided the impetus to use the river as a path through the coastal mountain barrier.

In this chapter, we describe the main-stem Fraser River as well as eight major tributaries that illustrate the wide range of geologic, climatic, and biological diversity in the basin (see Fig. 15.2). The main-stem Fraser, Thompson, Nechako, and Stuart rivers are described in some detail. Abbreviated descriptions of physical and biological characteristics are provided for the West Road, Quesnel, Chilcotin, and Lillooet and Harrison rivers on the Fraser and the Clearwater River on the Thompson.

Physiography and Climate

The basin is contained within two physiographic provinces (Fig. 15.5). A portion of the Fraser drains from the Rocky Mountains in Canada (RM) province in the east and a portion drains from the Coast Mountains of British Columbia and Southeast Alaska (PM) province. Thus, the basin is dominated by rugged, mountainous terrain that incorporates several major interior plains. The plains are more extensive in the north and extend out as intermontane valleys toward the southern half of the basin. Most of these plains and valleys are covered by glacial moraine and to some degree fluvial and lacustrine deposits, whereas the mountains consist largely of colluvium and rock outcrops. The soils of the lower valley floors to the south are often chernozems and support grasslands. These grade into arid environments in the Okanagan area toward the Canada–United States border.

The climate of the basin ranges from subarid to arid and mild in southern lower valleys to humid and cold at higher elevations in the northern reaches, reflecting the interaction of the dominant westerly circulation with the mountain ranges. Moist Pacific air and the effect of orographic rainfall control the precipitation pattern such that both rain shadows and wet belts are generated within the basin, often in close geographic proximity to each other. The rain shadow cast by the massive Coast Mountains results in some of the driest climates in Canada in the valley bottoms of the south-central part of the basin. During winter a relatively steady succession of low-pressure systems move eastward from the Pacific Ocean, bringing wet conditions associated with cyclonic, frontal, and, in the mountains, orographic uplift. Outbreaks of arctic air occasionally occur during the winter, bringing cold conditions and brisk winds, particularly in the Lower Fraser Valley. In summer the westerly circulation weakens and a persistent high-pressure area develops off the coast, resulting in a decreasing frequency of storms and frequent periods of fine weather. The Rocky Mountains also impede the westward flow of cold continental Arctic air masses. For most of the basin mean annual temperatures range between 0.5°C in the northwest (Skeena Mountains) and 7.5°C in the Okanagan area along the Canada–United States border. Mean

summer temperatures range from 11°C to 16.5°C. Mean winter temperatures range from −11°C to −1°C. The coastal portion of the basin has some of the warmest and the wettest climatic conditions in Canada. Mean annual temperatures along the coast range from 4.5°C in the north to 9°C in the Georgia-Puget basin–Lower Mainland regions. Mean summer temperatures range from 10°C in the north to 15.5°C in the south. Mean winter temperatures range from −0.5°C to 3.5°C, and relative to the rest of Canada there is little variation between the mean monthly temperatures through the year.

Basin Landscape and Land Use

The basin lies between latitudes 49°N and 56°N and longitudes 118°W and 125°W and encompasses six terrestrial ecoregions (Ricketts et al. 1999): North Central Rockies Forests, Fraser Plateau and Basin Complex, Okanagan Dry Forests, Cascade Mountains Leeward Forests, British Columbia Mainland Coastal Forests, and Puget Lowland Forests. These are some of the most diverse of all the Canadian ecoregions, ranging from alpine tundra to dense conifer forests to dry sagebrush and grasslands.

Vegetative cover is extremely diverse; alpine environments contain various herb, lichen, and shrub associations, whereas the subalpine environment has tree species such as lodgepole pine, alpine fir, and Engelmann spruce. With decreasing elevation the vegetation of the mountainous slopes and rolling plains separates into three general groups: a marginal band of forests characterized by Engelmann spruce, alpine fir, and lodgepole pine; forests characterized by ponderosa pine, interior Douglas fir, lodgepole pine, and trembling aspen in much of the southwest and central portions; and forests characterized by western hemlock, western red cedar, interior Douglas fir, and western white pine in the southeast. Shrubs found in the dry southern interior include sagebrush, rabbitbrush, and antelope bush. Most of the natural grasslands that existed in the dry south have vanished, to be replaced by urban settlement and agriculture. In the western coastal zone the temperate coastal forests are composed of mixtures of western red cedar, yellow cedar, western hemlock, Douglas fir, amabilis fir, mountain hemlock, Sitka spruce, and alder. Many of these trees reach very large dimensions and grow to great ages, and formed ancient or old-growth forests in this region, although these are now restricted to watersheds to the north of the Fraser River. Mountain hemlock is usually associated with higher elevations. Variations in altitude account for the presence of widely contrasting ecosystems within the coastal region, ranging from mild humid coastal rainforest to cool boreal and alpine conditions at higher elevations. Characteristic mammals include woodland caribou, mule, blacktail and white-tailed deer, moose, mountain goat, bighorn sheep, otter, raccoon, coyote, wolf, black and grizzly bear, hoary marmot, and Columbian ground squirrel. Typical bird species include blue grouse, Steller's jay, and black-billed magpie. In coastal areas American black oystercatcher, California and mountain quail, tufted puffin, and chestnut-backed chickadee occur, as well as pygmy owl, Steller's jay, and northwestern crow.

Following the discovery of gold in the 1850s most new immigrants arrived by sea at the mouth of the river. Prospectors and miners could travel inland to the gold fields by riding steamboats that operated as far as Yale just below the canyon and again on the river between Soda Creek and Quesnel. Following the gold rush, settlement expanded and development rapidly began to shape the basin to the resource uses and communities seen at present. Valley bottom forests were cleared for agriculture and the forest industry that is now the core of the provincial economy soon began to increase timber exports. Salted and canned salmon were also part of the initial export economy. With the completion in 1886 of the transcontinental railway that ran beside the Fraser and Thompson rivers, access was opened not only to the interior of the basin but also to the rest of Canada east of the Rockies. Growing east–west trade led to the construction of grain elevators in Vancouver to export grain from the prairies, and the completion of the Panama Canal in 1914 opened new markets in Europe for many industries and led to further expansion after World War 1. In 1894 the largest freshet on record caused major flooding in the Lower Fraser Valley, prompting construction of a dyking and drainage system that consolidated the patterns of both the river channel and delta for agricultural and urban development. At the present time greater Vancouver is the third-largest metropolitan area in Canada, with 2.1 million people at the 2001 census, representing two-thirds of the provincial population, and more than three-quarters of the basin's population is concentrated in the Lower Fraser Valley. In the interior only Kelowna (96,288), Kamloops (77,281), and Prince George (72,406) have similar populations to the medium-size lower Fraser Valley municipalities (e.g., Coquitlam 112,890 and Chilliwack 62,927). Other interior basin towns are quite small; in the middle Fraser region the major centers are Williams

Lake (11,153) and Quesnel (10,044), and in the Thompson basin Salmon Arm (15,210). The basin economy has grown with these communities but retains its resource dependence despite significant changes taking place in recent decades.

Commercial forest operations have been established in many parts of the basin, particularly in the northern interior sections. Forest productivity in some coastal areas is the highest in Canada and commercial forest operations are of major economic importance to Canada's forest industry. British Columbia accounts for 60% of Canada's lumber exports, almost half of its chemical pulp exports, and 36% of its kraft paper exports. The forest industry continues to be of major importance in the basin, with harvesting and processing in the interior increasingly significant in the past two decades. The Fraser River estuary, where the coastal forest industry collects, sorts, processes, and exports water-borne logs, is still of critical importance to the international competitiveness of the industry. The lowlands of the Fraser Valley possess the area's main expanse of highly productive agricultural soils, as well as urban lands. Mining, oil and gas production, and tourism are the other significant activities.

The mining industry, historically the second most important contributor to the gross provincial product, operates in many parts of the basin. The province produces almost 70% of Canada's coal exports and over 98% of its copper and molybdenum ore exports and potential exists for mining small high-grade ore deposits and anthracite.

Recreation and tourism has overtaken mining as the second most important provincial economic sector. The recreational freshwater and saltwater fisheries associated with the Fraser are a major component of this, along with other water-related attractions throughout the basin. Already a major world exporter of canned and frozen salmon, British Columbia has a growing aquaculture industry that is now a significant contributor to the provincial economy. In 2002 the farmgate value of the salmon, shellfish, and trout sectors combined was $304.5 million, with farmed salmon British Columbia's largest agricultural export product. The commercial fishing fleets find moorings in the estuary and depend heavily on the Fraser River salmon runs, which they share with the native and recreational fisheries.

In the eastern Rocky and Columbia mountains, national and provincial parks have been established for recreational use or as reserves for wildlife habitat. It is mainly in the valleys that areas have been improved for range or are farmed; near streams, where water is available for irrigation, row crops and hay are grown. The southern valleys are nationally important for their orchards and vineyards.

British Columbia's agriculture industry is the third-largest component of its resource-based sector. Agriculture is an important activity in the Lower Fraser Valley, where crop production utilizes some of the most productive lands in Canada, and in the interior subbasins, where ranching is of greater significance.

The Rivers

The entire Fraser basin is contained within the North Pacific Coastal freshwater ecoregion (Abell et al. 2000). This ecoregion extends from southeastern Alaska through the southwestern portion of the Yukon Territory and western and central portions of British Columbia to northwestern Washington. It is considered a cool, high-rainfall area, formerly covered by rain forests on the coast and drier forest and grasslands in the interior. The ecoregion is considered continentally outstanding with regard to biological distinctiveness.

Because the various tributaries across the basin experience a great range of topographic features and precipitation patterns, they are highly variable in their monthly stream flows. Low-elevation coastal basins are dominated by autumn and winter rainfall, with high autumn and winter flows and low summer flow when rainfall is low and evaporation losses are high. However, most of the other basin types respond to melting of winter snow, with the bulk of the runoff occurring in spring and summer and low flows occurring in winter when most of the precipitation falls as snow and is stored in the snowpack. Runoff is greater in basins dominated by mountainous terrain and/ or near the coast (Lillooet, Coquihalla, McGregor, Salmon, North Thompson, and South Thompson) than basins that dominantly drain the interior plateau (Nautley, Nicola, West Road, and Salmon near Prince George).

Although British Columbia ranks third nationally in hydroelectricity generation and second in natural gas production, only a small proportion of the basin's total energy supply is produced from tributaries of the Fraser, which remains one of the few minimally impounded great rivers of North America (with the exception of the Nechako River). However, the basin acts as a major conduit of energy for the entire province. The human population largely resides in the last few 100 km of the river, hemmed in between mountain barriers. The basin, compared to other large North American river systems, is relatively unimpacted by human activities. The two major activities in the basin are logging and mining. Effects from tree removal and pulp and paper mills have been of most concern. The impacts of pulp mill discharges are generally small, as they are located on the major rivers and the impacts from harvesting tend to be local. Mining activities do have local impacts.

FRASER RIVER MAIN STEM

The Fraser is one of four major North American river systems arising within a few kilometers of each other in the Canadian Rocky Mountains. The others include the northward-flowing Athabasca River, part of the Peace–Mackenzie system; the North Saskatchewan, part of the Nelson system; and the Columbia River. The Fraser River begins in Mount Robson Provincial Park in the Rocky Mountains on the British Columbia–Alberta border, initially flowing in a northwesterly direction (see Fig. 15.5). Just northeast of Prince George it turns sharply south until it reaches Hope (Fig. 15.3), in southern British Columbia, and then turns west before emptying into the Straits of Georgia. The Fraser River is the second-largest river in North America with no main-stem dams, after the Yukon River in Alaska (Dynesius and Nilsson 1994). The Fraser is also one of the greatest salmon rivers in the world (Northcote and Larkin 1989).

Physiography, Climate, and Land Use

The headwaters of the Fraser River are in the Rocky Mountains in Canada (RM) physiographic province of southeastern British Columbia (see Fig. 15.5). Elevations rise to over 3000 m asl along the continental divide and include the highest mountain in the Canadian Rocky Mountains, Mount Robson, at just over 3600 m asl. The river first flows northwest for approximately 250 km through the Rocky Mountain Trench and drains a relatively narrow and steep valley that is hemmed in by the Rocky Mountains (3950 m asl) to the east and the Cariboo Mountains to the west (2590 m asl). The Southern Rocky Mountain Trench is a linear, steep-walled, faulted valley about 480 km long. The valley floor is relatively level and can vary in width from less than 1 km to 20 km. The rivers that drain into and meander along the valley floor of the trench have formed large floodplains and wetlands. At Prince George is the confluence of the Nechako–Stuart systems with the Fraser River. In this

FIGURE 15.3 Fraser River at Hope, British Columbia (PHOTO BY TIM PALMER).

middle region the river has entered the Coast Mountains of British Columbia and Southeast Alaska (PM) physiographic province. Before actually reaching the Coast Mountains, however, the river flows across a high plateau with a dry continental climate bordered by the Coast Mountains (3000 m asl) to the west and the Cariboo Mountains to the east. In succession, the river flow is augmented by the West Road, Quesnel, Chilcotin, and Bridge rivers. In the lower sections of this region, the river drops rapidly and enters a very arid area. Here, the silt-loaded waters of the Fraser are joined by the clear water of the Thompson, another great river system, before entering the long narrow Fraser Canyon that cuts through the barrier

of the Coast Range. Below the canyon the river turns west into the more humid coastal region and flows through the lower Fraser Valley in a widening V bordered by the Coast Mountains to the north and the Cascades on the U.S. border to the south. Entering into the Fraser River estuary the river then divides, with 85% of its flow going down the Main Arm and the remainder down the North Arm, dividing again to form the Fraser Delta.

The Coast Mountains receive the main westerly weather systems and experience heavy precipitation during the autumn and winter. Mean annual precipitation can exceed 300 cm in some areas. At higher elevations much of the precipitation is in the form of snow and remains in storage until the spring melt. The windward slopes of the mountain ranges experience the heaviest precipitation, whereas the leeward slopes tend to exhibit a rain-shadow effect; precipitation also generally increases with elevation. The interior plateau lies in the rain shadow of the Coast Mountains and mean annual precipitation ranges from about 40 to 80 cm. In contrast to the Coast Mountains, stations in the interior plateau experience a precipitation peak in the summer, reflecting the influence of summer convective storms, and most of the winter precipitation falls as snow. The mean annual temperature for the area is approximately 3°C, with a summer mean of 12.5°C and a winter mean of −8°C. As easterly moving air masses encounter the Eastern Mountains they produce precipitation. However, having already lost moisture over the Coast Mountains the air masses are not as productive and mean annual precipitation in the region ranges from 100 to perhaps over 200 cm. As in the Coast Mountains, elevation and exposure to the prevailing winds influence precipitation. The autumn precipitation peak is not as marked as in the Coast Mountains. In the headwaters the lower-elevation valleys are marked by warm, showery summers and mild, snowy winters. Subalpine summers are cool, showery, and prone to frosts. Winters are moderately cold and snowy. The mean annual temperature for the headwater valleys is approximately 3.5°C, with a summer mean of 13°C and a winter mean of −6.5°C. The mean annual precipitation in the major valleys is 70 to 80 cm. In the subalpine zone the annual precipitation climbs to 120 cm. Approximations of mean annual precipitation and temperature are 80 cm and 6.1°C, respectively, based on nine stations throughout the basin. The high autumn–winter precipitation in the mountainous areas has the predominant influence on the basin average (Fig. 15.6).

The main stem of the Fraser River flows through five terrestrial ecoregions: North Central Rockies Forests, Fraser Plateau and Basin Complex, Cascade Mountains Leeward Forests, Mainland Coastal Forests, and Puget Lowland Forests (Ricketts et al. 1999). The upper reach of the Fraser flows through the North Central Rockies Forests ecoregion, which is predominantly composed of subalpine and alpine ecosystems and, in the major valley systems, is covered by montane forests. Montane forests are composed of western hemlock and western red cedar in the north, trending to white spruce and alpine fir forests in the south. Subalpine forests are composed of Engelmann spruce, alpine fir, and lodgepole pine.

Valley bottom vegetation ranges from bunchgrass, ponderosa pine, and Douglas fir in the south, to western red cedar and western hemlock in the central portion, to white and black spruce and lodgepole pine in the northern portions of the region. Just north of Prince George the main stem enters the Fraser Plateau and Basin Complex. In this interior plateau the forests are characterized by mixed stands of trembling aspen, paper birch, lodgepole pine, and the climax species, white and black spruce. The subalpine zone that occurs above 1200 m asl supports forests of lodgepole pine, which develop after fires, as well as Engelmann spruce and alpine fir. The plateau is a drier region where the annual precipitation range is 25 to 30 cm in the area around the junction of the Chilcotin and Fraser rivers. Here, bunchgrass-dominated grasslands occur at valley bottom elevations along the Fraser and Chilcotin rivers. Upon leaving the plateau the Fraser enters the Cascade Mountains Leeward Forests ecoregion. In this area, where the river passes through the Interior Transition Ranges between Williams Lake and Lillooet, there are ecosystems ranging from alpine at the highest elevations, to subalpine forests of Engelmann spruce, alpine fir, and lodgepole pine, to montane forests of lodgepole pine, trembling aspen, white spruce, and Douglas fir. At the lowest elevations there is a parkland of scattered ponderosa pine in a matrix of bluebunch wheat grass and sagebrush grasslands. The Fraser then passes through the British Columbia Coast Forests ecoregion. The Pacific Ranges of this ecoregion are high, irregular, steeply sloping mountains that form the main southern part of the rugged Coast Mountains. The area incorporates three main ecological zones: the coastal forest zone, which ranges from sea level to about 900 m asl, the subalpine zone, from about 900 to 1800 m asl, and the alpine zone above 1800 m asl. Vegetative cover of the low-elevation slopes includes very productive stands of western hemlock, western red cedar, and amabilis fir. Drier sites support stands of western hemlock and Douglas fir. The subalpine zone is dominated by forests of mountain hemlock and amabilis fir with some yellow cedar. Finally, the river flows the through the Puget Lowland Forests ecoregion (Lower Mainland region), where native vegetation is characterized by forests of Douglas fir with an understory of salal, Oregon grape, and moss. Mixed stands of Douglas fir and western hemlock with some dogwood and arbutus are common on drier sites. Red alder is common where sites have been disturbed. Wet sites support Douglas fir, western hemlock, and western red cedar.

In the upper reaches, as in the rest of the basin, the primary land use is forestry (97.2% for the whole basin), with water impoundment, grazing, hunting, recreation, livestock, forage-crop production, and tourism. Forestry and ranching are the main land uses in the interior plateau, along with outdoor recreation, including hunting and fishing. Passing out of the interior plateau to the interior ranges, land use reflects high recreational and wildlife values in alpine and subalpine zones, whereas forestry and agriculture tend to become more important in lower, warmer zones. Mineral exploration occurs throughout this region. There are 50,000 ha of farmland in the region, the most productive of which are the irrigated forages, mainly alfalfa, on the benches of the river, which is also an important transportation corridor. In the Pacific Ranges land uses reflect high recreational and wildlife values at upper elevations and forestry values at lower elevations. This area contains some of the most productive forestlands in Canada. Important land uses include pulp and sawlog forestry, production of hydroelectric power, water-oriented recreation, and tourism. Much of the forestland is under Tree Farm License. The Lower Mainland region is an urban and agricultural region, containing the largest population center in British Columbia, although compared to the total basin area both urban (2.2%) and agriculture (0.6%) land use is small. Intensive agriculture does occur on the valley bottoms of the Fraser River valley, where it competes with urban development. Forestry operations occur on higher slopes along the mountains. There are about 87,000 ha of highly productive farmland in the area. Coastal salt marshes are important wildlife habitat on the Fraser River delta and adjacent Boundary Bay. Urban and suburban development continues in the Vancouver area and is scattered among many communities in the Fraser River valley and Sunshine Coast.

River Geomorphology, Hydrology, and Chemistry

In the headwaters the mountain ranges are composed of Paleozoic limestones and quartzites. Glaciation has sculpted great U-shaped valleys and left valley floors filled with glaciofluvial and morainal sediments. Rock outcrops predominate at the highest elevations. Permafrost occurs in isolated patches in alpine areas. Underlain by Paleozoic and Proterozoic strata, the trench is covered with a variety of glacial deposits, including ground moraine, outwash plains

and terraces, drumlins, eskers, glacial lake terraces, and recent alluvium. The average slope of the upper river from the headwaters to Prince George is about 2.1 m/km. In the interior plateau the region is underlain by flat-lying Tertiary and volcanic bedrock that generally lies below 1000 m asl. It has a gently rolling surface covered by thick glacial drift into which the Fraser River and its major tributaries are commonly incised. The glacial deposits include moraine with well-developed drumlin features, glaciofluvial terraces, eskers, and large areas of glacial lake deposits. From Prince George to Quesnel the average slope is 1.58 m/km. Below Quesnel the river enters a broad rolling plateau, which generally lies 1150 to 1800 m asl. Surface deposits include glacial till with well-developed drumlinoid features, pitted terraces, simple and compound eskers, and areas of glacial lake (lacustrine) deposits. The average slope from Quesnel to Williams Lake is also 1.58 m/km but increases to 2.45 m/km from Williams Lake to Lillooet. The river cuts through the rugged Coast Mountains in the spectacular Fraser Canyon. Finally, the river flows through the Lower Mainland region, which is underlain by unconsolidated glaciofluvial deposits, silty alluvium, silty and clayey marine sediments, and glacial till. Below Lillooet the average slope decreases to 0.92 m/km.

The hydrology of an area reflects interactions between climatic, physiographic, geologic, and vegetative factors as well as human activity. Hydrologically the Fraser River can be divided into three zones: the Coast Mountains, the interior plateau, and the eastern mountains (Columbia and Rocky mountains) (Slaymaker 1990). Rainfall, snowmelt, and glacier melt contribute varying amounts of water to the Fraser basin. The magnitude and timing of snowmelt varies through the basin depending on the interaction of snow accumulation and melt processes. At lower elevations, especially in the Coast Mountains, midwinter melt events are common and decrease the amount of snowpack, which can be released during the spring melt. At higher elevations with cooler temperatures, midwinter melt events occur less frequently, and more of the precipitation falls as snow. The timing of the main spring thaw varies according to location and elevation. Runoff in the Fraser basin is predominantly determined by the melt of the snowpack in the headwaters and can produce great variation within and between years.

Mean annual discharge at the Mission hydrometric gaging station is 3370 m^3/s, with mean monthly flow varying from 1400 m^3/s in winter to over 8000 m^3/s in summer. With great between-year variation, however, monthly flows range from as low as 450 m³/s to as high as 20,000 m³/s. Mean runoff at Mission averages less than 2 cm/mo from January through March, but runoff from the spring snowmelt peaks in June at about 9 cm/mo (see Fig. 15.6). Mean discharge estimated at the mouth (3972 m³/s) is somewhat higher than at the Mission station. The tidal effect on water level reaches as far inland as Chilliwack, 120 km from the mouth, at low flow and 75 km upstream (Mission) during the spring freshet. The salt wedge ranges from 22 km upstream (low flow) to just to the river mouth (high flow). It is estimated that sediment is added to the river delta at a rate of 12 million m³/yr, and the Sturgeon Bank at the delta is advancing seaward at a rate of 25 m/yr.

Glacier melt is also an important source of water, mainly in the headwater basins in the Coast, Cariboo, and Rocky mountains. The melt normally begins in May in higher elevations in the Coast Mountains. The Fraser River in this region is sufficiently large that only spatially widespread and prolonged seasonal melting of snow and ice produces peak flows. In the interior plateau, the main melt begins in March (e.g., Prince George), as reflected in the hydrograph, and flow begins to increase almost a month earlier than in the eastern mountains. Because the temperature decreases both with elevation and latitude the relative importance of snow as winter precipitation and its consequent effects on runoff and hydrology is affected by these two factors. Upstream of Prince George the melt varies from March (low altitude) to May (high altitude). Glacier melt is also an important source of water in the headwater basins of the Rocky Mountains.

The most noticeable feature of the water chemistry in the Fraser River is the high sediment load, which is a natural phenomenon of the river and was noted by Simon Fraser in the spring of 1808 (Slaymaker 1991). This results from natural erosion processes as the river flows through the glacial deposits and drift material in the central plateau. However, extensive logging occurs throughout the basin, and although it does not affect sediment loads in the main stem, it may be an issue on smaller tributaries. Water chemistry along the length of the main stem in different seasons shows changes in turbidity, total phosphorus, and iron are positively related to discharge, and five variables (chloride, calcium, sodium, reactive silica, and sulphate) are negatively associated with discharge (Whitfield 1983, Whitfield and Schreier 1982). Three zones have been identified from the water chemistry: a headwaters zone that extends only as far as Red Pass, only 70 km from the

source, a midstream zone from Red Pass to between Prince George and Quesnel, and a downstream zone (Whitfield 1983). The mean ranges in these three zones, respectively, are turbidity 1.0 to 6.0 JTU, 6.0 to 10.0 JTU, and 19.0 to 25.0 JTU; specific conductance 100 to 145 µS/cm, 133 to 146 µS/cm, and 114 to 139 µS/cm; alkalinity 35 to 60 mg/L as CaCO₃, 51 to 65 mg/L as CaCO₃, and 47 to 61 mg/L as CaCO₃; hardness 50 to 73 mg/L as CaCO₃, 65 to 74 mg/L as CaCO₃, and 53 to 67 mg/L as CaCO₃; chloride 0.2 to 0.5 mg/L, 0.4 to 1 mg/L, and 1.1 to 1.5 mg/L; silica 2.6 to 3.8 mg/L, 3 to 4 mg/L, and 3.9 to 5.6 mg/L; and sodium 0.6 to 0.9 mg/L, 0.8 to 1.5 mg/L, and 2.1 to 2.7 mg/L.

River Biodiversity and Ecology

The Fraser is the largest river system within the North Pacific Coastal freshwater ecoregion (Abell et al. 2000). In spite of its size, it is probably the best studied in terms of its biodiversity and ecology, both in the main stem and in several of its tributaries.

Algae and Cyanobacteria

In the main stem above Prince George between 11 and 15 taxa have been recorded in the Fraser River, primarily diatoms, but with two species of Cyanobacteria (*Lyngbya* spp.) and one green algal species (*Scenedesmus* sp.) (Rosenberg et al. 1998). There is generally more variability in the dominant taxa upstream of Prince George, with three dominant species, the green alga *Ulothrix* sp. and the diatoms *Synedra ulna* and *Achnanthes minutissima*. Other common species are *Amphipleura pellucida*, *Fragilaria vaucheriae*, *Gomphonema olivaceum*, *Hannaea arcus*, and *Navicula* sp. There is a trend to increased species richness downstream of Prince George, with up to twice as many taxa observed (5 to 12 upstream and 15 to 23 downstream). Downstream of Prince George the diatom *Cymbella* sp. is most abundant at all but one site, and other diatoms, such as *Achnanthes minutissima*, *Fragilaria capucina*, *Gomphonema olivaceum*, *Melosira varians*, *Melosira* sp., *Navicula radiosa*, *Nitzschia palea*, and *Synedra ulna*, are also common (more than 10% of sample) (Rosenberg et al. 1998).

Plants

There is little in the way of macrophyte development in the fast-flowing Fraser River. Riparian vegetation shifts from domination by conifers such as white spruce, lodgepole pine, and Douglas fir upstream of Prince George to birch, cottonwood, various willows, and various grasses in the downstream reaches.

Invertebrates

According to Reece and Richardson (2000), little research has been done on benthic communities in streams of noncoastal regions of British Columbia. Rosenberg et al. (1998) examined changes in invertebrate communities along the length of the Fraser River from the headwaters to Agassiz, near the mouth, and identified over 50 families at 28 sampling sites. Of these, 11 families represented more than 90% of the invertebrates found. The most widely distributed and abundant were Chironomidae (midges), which were present at 97.5% of sites. Also abundant and common were stoneflies (Capniidae, Taeniopterygidae, and Perlodidae), mayflies (Heptageniidae, Ephemerellidae, and Baetidae), annelid worms (Naididae and Enchytraeidae), caddisflies (Hydropsychidae), and true flies (Empididae). There was a general trend of increasing abundance from the headwaters (300 to 400 organisms per minute of sampling) to Quesnel (1200 to 1500 organisms per minute of sampling), followed by a region of low abundance to Lillooet (<100 organisms per minute of sampling) and then a gradual increase in abundance to Agassiz, furthest downstream. Diversity (number of taxa) decreases from the headwaters to the lower reaches. The changes in abundance and diversity are partially reflected by the major families. Upstream sites are dominated by mayflies. The baetid mayflies occur primarily in the headwaters and the ephemerellid mayflies occur in the headwater and upstream sections. Chironomid midges are particularly abundant in the middle reaches. Common genera in the Fraser River are the stoneflies *Capnia* and *Isogenoides*, the mayflies *Rhithrogena*, *Ameletus*, *Ephemerella*, and *Baetis*, and two naidid oligochaetes, *Nais* and *Specaria*.

Vertebrates

Three major assemblages have been proposed (McPhail 1998) based on the distribution of native freshwater fish species and the eight marine species that regularly enter the river. First, an estuarine assemblage occupies the estuary as far as Sturgeon Bank and the river to the upper limits of tidal influence at Mission. This assemblage is comprised of six abundant salmonid species (pink, chum, coho, sockeye, and Chinook salmon and rainbow trout) as well as marine species, such as starry flounder, and many small forage fishes, such as peamouth chub, redside shiner, and sticklebacks (Richardson et al.

2000). Second, a lower-river assemblage occupies the river from Mission upstream to the Fraser Canyon and is characterized by eight native species (river lamprey, western brook lamprey, green sturgeon, longfin smelt, eulachon, coastal cutthroat trout, chum salmon, and threespine stickleback) that do not occur upstream of this reach of the Fraser. Finally, a distinctive upper-river assemblage occupies the river upstream of the Fraser Canyon, characterized by six species (chiselmouth, lake chub, white sucker, pygmy whitefish, slimy sculpin, and torrent sculpin) and one subspecies (westslope cutthroat trout) that do not occur below the canyon. Two species were formerly restricted to the upper basin (lake trout and lake whitefish) but have been transplanted to the lower basin. This upper-river assemblage has been separated by McPhail (1998) into a lower and upper community, with the geographic boundary the confluence of the Fraser and Bowron rivers. Upstream of the Bowron the community is characterized by the loss of species and is dominated by species adapted to swift glacial rivers.

The main-stem Fraser River is noteworthy for providing spawning habitat for at least three species of Pacific salmon: pink, chum, and chinook (Northcote and Burwash 1991). From 1957 to 1989 an annual average of 765,000 pink salmon spawned in the main stem in the area of Hope, representing 90% of the total pink salmon production to the lower Fraser subbasin. Chinook salmon use the headwater reaches of the main stem and make up 32% of the salmon escapement to the drainage basin.

Other aquatic vertebrates in the basin include birds and mammals, as well as several amphibian species. Aquatic mammals found in the river drainage include beaver, common muskrat, and river otter. The South Arm Marshes and Ladner Marsh in the Fraser River delta together make up internationally important stopovers for hundreds of thousands of migratory birds on the Pacific Flyway. The area provides breeding and feeding habitat for one of the largest overwintering waterfowl populations in Canada, comprising some 40 species of ducks, geese, and swans. The entire range of the coastal giant salamander in Canada (red-listed in Canada) is within the Fraser River drainage.

Ecosystem Processes

Although considerable information exists for water chemistry, fisheries, and other biological features, there have been few studies on ecological processes on the main-stem Fraser River. Calculations of the organic loading of the lower Fraser River, primarily from agricultural and municipal sources, suggest large increases in ecosystem productivity during the twentieth century (Healey and Richardson 1996). However, much more information on ecosystem processes is available on its largest tributary, the Thompson (as will be discussed later in this chapter).

Human Impacts and Special Features

The Fraser River has to be considered one of the less-impacted major river systems in North America. Along most of its length there are few urban centers. Municipal and pulp mill discharges are the major point-source discharges and seem to be generally producing mild enrichment for a short distance. The main stem provides passage to some of the world's most famous salmon fisheries. The major nonpoint-source effects are from forestry and these tend to be localized and to affect only small streams. The major effects are observed in the major urban area, Vancouver, at the mouth of the river. One of the most notable features of the Fraser is its dramatic transition as the river flows through the spectacular Fraser Canyon and Hell's Gate as it breaks through the Coastal Range. Gold prospector W. Champness described this in 1863:

> In other parts of the journey especially in the river gorges, our track conducted us along the most frightful precipices. There was no help for this, as we could select no route more passable. The river flows often through dark and awful gorges whose rocky sides tower perpendicularly from a thousand to fifteen hundred feet. By a series of zigzag paths, often but a yard in width, man and beast have to traverse these scenes of grandeur. Sad and fatal accidents often occur, and horses and their owners are dashed to pieces on the rocks below, or drowned in the deep foaming waters rushing down the narrow defiles from the vast regions of the mountain snow melting in the summer heat. (www.barkerville.ca/barkerville/documents 2004)

The major potential sources of human impact are effluents from point sources such as pulp and paper mills located at Prince George and Quesnel. Also, there are nonpoint sources of contaminants in the urbanized lower Fraser Valley and agricultural sub-basins in the lower Fraser, and exposures to effluents in the delta foreshore from large municipal wastewater treatment plants. The main-stem river upstream of the estuary and its major tributaries does not exhibit significant concentrations of contaminants at most locations, and low levels of contaminants have been recorded in wildlife (Wilson et al.

1998) and benthic invertebrates (Richardson and Levings 1996). Mesocosm experiments (Culp and Lowell 1998) have shown that pulp mill effluents stimulate growth of invertebrates at the lowest dilution levels currently observed in the Fraser River, but that an increase in effluent concentration could induce toxic effects. Analyses of fish (Raymond et al. 1998) and bird (Wilson et al. 1998) tissue show a greater induction of the detoxifying enzyme, mixed-function oxygenase, downstream of urban centers and pulp mill discharges. Furthermore, two fish species, peamouth chub and mountain whitefish, had a relatively high incidence of abnormalities throughout the basin and highly pigmented livers and kidneys and reduced gonad development at sites downstream of pulp mills. However, the ecological significance of these observations is not known. Contamination is also evident in the Fraser River estuary, where the levels of chemicals such as PAHs and some dioxin and furan congeners in sediment exceeded guidelines or draft guidelines established for the protection of aquatic life (Brewer et al. 1998). Runoff from agricultural and urban areas is now a significant source of contaminants to the basin, and the lower Fraser Valley is one of the most important agricultural regions in British Columbia. The key issues are excess nitrogen, ammonia, eutrophication, and coliform problems in small streams, as well as elevated nitrate values in groundwater and high Zn in sediment (Schreier et al. 1998). Results have shown that water-quality conditions are limiting the distribution and breeding success of native amphibians in areas of the Sumas watershed. Finally, persistent organic pollutants have been found in the rivers headwaters. The likely source of PCBs, DDT and its metabolites, and toxaphene found in fishes from headwater lakes is long-range atmospheric transport and deposition coupled with the release of historic deposits of contaminants from melting glaciers and permanent snowfields.

THOMPSON RIVER

Rising near the continental divide in cold headwater streams of the Cariboo Mountains, the Thompson River flows 300 km as the North Thompson River until it joins the flow of the South Thompson from Shuswap Lake (Fig. 15.7). The Thompson River then travels approximately 40 km through Kamloops Lake before exiting as one of the larger rocky-bottom rivers in North America. At its confluence with the Fraser River the 7th order Thompson River is more than 500 km long and drains approximately 55,000 km². As the Thompson cuts across south-central British Columbia it dissects the diverse montane cordillera, with alpine tundra and dense conifer forests in the upper basin; dry sagebrush grasslands and ponderosa pine forests cover the lower basin.

Simon Fraser named the Thompson River during his 1808 exploration of the Fraser River to honor David Thompson. Thompson explored passes west of the Saskatchewan and Athabasca rivers, as well as the Columbia River from 1806 to 1811, in search of new trade routes, but never reached as far northwest as the river system that bears his name. The Shuswap tribe of the Salish nation named the confluence of the North and South Thompson rivers near present-day Kamloops "cume-loups," apparently meaning "meeting of the waters." Between 1810 and the 1850s, fur traders and then gold seekers and farmers developed the area around the lower Thompson River as an important regional trade center. Although the river was not of great importance as a navigational route because of dangerous rapids, terraces along its banks provided natural roadways. Today the Thompson is an important transportation corridor, with highways and railways following the river from Kamloops to its confluence with the Fraser River.

Physiography, Climate, and Land Use

The Thompson River flows south to join the South Thompson River at Kamloops, British Columbia, then travels southwest to its confluence with the Fraser River. Laying between latitudes 50°N to 53°N and longitudes 122°W to 118°W, this large drainage lies within the Coast Mountains of British Columbia and Southeast Alaska (PM) physiographic province (see Fig. 15.7). Much of the basin is composed of folded sedimentary and volcanic strata, with metamorphic rocks of Paleozoic and Mesozoic age covered by a thick mantle of glacial drift. Soils across the region are quite varied, with luvisols and brunisols the most common; podzols occur in the mountain ranges of the wetter, eastern portion of the basin and chernozems form the soils of lower valley floors to the south. Plains and valleys are covered by glacial moraine and fluvial and lacustrine deposits, whereas mountains consist largely of colluvium and rock outcrops.

The Thompson basin encompasses three terrestrial ecoregions: the North Central Rockies Forests, the Okanagan Dry Forests, and the Cascade Mountains Leeward Forests (Ricketts et al. 1999). The basin is an incredibly diverse region, with habitats ranging from alpine tundra to sagebrush grassland. The diverse landscape varies with elevation from alpine environments containing herbs, lichens, and shrubs to well-treed subalpine forests. With decreasing elevation, forests are characterized by ponderosa and lodgepole pine, Douglas fir, and trembling aspen in the southwest and central portions of the basin and western hemlock, western red cedar, and Douglas fir in the southeast. In the driest areas west of the North and South Thompson confluence, grasslands of blue-bunch wheat grass, blue grass, june grass, sagebrush, rabbitbrush, and antelope bush occur.

Mean annual air temperature for the area is 9.7°C, with mean monthly temperatures ranging from −2.3°C in January to 21.4°C in July and August (Fig. 15.8). Kamloops has the highest average daytime maximum temperature of all Canadian cities at 28°C. The lower portion of the basin is one of the driest regions in Canada. Indeed, annual precipitation exhibits a strong gradient across the basin, ranging from 27 cm in the lower Thompson Valley at Kamloops to more than 100 cm in mountain headwaters. Mean monthly precipitation at Lytton ranges from 6.7 cm in December to 1.4 cm in July (see Fig. 15.8). Approximately 30% to 40% of the precipitation arrives as snowfall in winter.

Many of the natural grasslands have been replaced by urban settlement and irrigated agricultural operations. Land uses reflect high recreational and wildlife values in alpine and subalpine zones, whereas forestry and agriculture tend to become more important in lower, warmer zones. The majority of the basin is managed for sustainable forest harvesting, although river valleys generally contain sparsely populated rural districts and forage agriculture is common along the terraces of the lower Thompson River. Forest and shrubs cover approximately 75% of the basin and agricultural lands less than 3%. Mineral exploration occurs throughout the region. Population density across the basin is sparse and varies considerably, with less than 3 people/km². Approximately 70% of the population resides in Kamloops, the main urban area within the basin (estimated population of 85,000 in 1996).

River Geomorphology, Hydrology, and Chemistry

The North Thompson is a high-velocity erosional river flowing over a gravelly bottom, ranging to boulders and bedrock in some reaches (Fig. 15.1). The upper part of the North Thompson between the

Clearwater and Blue rivers is fairly steep, with an average slope of 2.90 m/km. The lower part of the North Thompson between the Clearwater and the confluence with the South Thompson has an average slope of 0.54 m/km. The North Thompson comprises 38% (20,700 km^2) of the total drainage area of the whole Thompson River basin, but contributes 55% (428 m^3/s) of the mean annual flow for the whole river (787 m^3/s). In contrast to the South Thompson, the North Thompson is a free-flowing river, without intervening lakes, and can be greatly affected by heavy rains and sudden snowmelt. Mean monthly discharge ranges from 81 m^3/s in February to 1360 m^3/s during the freshet peak in June.

The South Thompson between Shuswap Lake and the North Thompson is slow flowing. It is typified by a sand-gravel bottom and has an average slope of 0.24 m/km. The South Thompson drainage area (17,800 km^2) is 32% of the basin total for the Thompson River and contributes 37% (289 m^3/s) of the mean annual flow. The flow in the South Thompson is modulated by upstream lake storage, most notably Shuswap Lake. Mean monthly discharge ranges from 88 m^3/s in March (late winter) to 855 m^3/s in June. Thus, although only comprising 69% of the whole Thompson River drainage, the North and South Thompson subdrainages contribute 92% of the total discharge, reflecting the arid conditions in the lower Thompson. Both the North and South Thompson can be ice covered for several months of the year.

Water chemistry is quite similar in the North and South Thompson rivers. In the North Thompson, specific conductance is 91 µS/cm, alkalinity 36 mg/L as CaCO$_3$, hardness 91 mg/L as CaCO$_3$, chloride 0.3 mg/L, silica 5.4 mg/L, sodium 1.5 mg/L, and pH 7.6. In the South Thompson, specific conductance is 101 µS/cm, alkalinity 38 mg/L as CaCO$_3$, hardness 91 mg/L as CaCO$_3$, chloride 0.5 mg/L, silica 5.3 mg/L, sodium 2.0 mg/L, and pH 7.7 (Nordin and Holmes, 1992). The North Thompson, however, is more turbid (3.9 NTU), with more suspended sediments (30 mg/L) than the South Thompson (1.7 NTU, 15 mg/L). Both the North and South Thompson are fairly pristine and nutrient poor. Whereas levels of inorganic nitrogen do not appear to be limiting (~70 to 130 µg/L), phosphorus levels are quite low (soluble reactive phosphorus 0.7 to 1.1 µg/L, total dissolved phosphorus 2.7 to 3.5 µg/L), resulting in phosphorus limitation (Bothwell 1985). The 15 km length of the lower Thompson from the confluence of the North and South Thompson to Kamloops Lake is the major point source of phosphorus loading to the river

system; this is the location of the outfalls for the bleached kraft pulp mill and the city of Kamloops municipal sewage treatment plant. Increased algal growth below Kamloops Lake has been attributed to this phosphorus loading (Lowell et al. 2000).

An unusual eutrophication pattern is apparent in the lower Thompson River. Whereas the main point sources of phosphorus loading are located just upstream of Kamloops Lake and there is evidence of nutrient enhancement downstream of the lake, the lake itself remains oligotrophic. This appears to be a result of the circulation patterns in the lake (Carmack et al. 1979). Kamloops Lake is fairly long (25 km), narrow (mean width 2.1 km), and deep (mean depth 71 m). The annual mean flushing time is only 60 days, which helps to prevent nutrients from accumulating in the lake. Furthermore, periods of high river flow into the lake during the spring and summer are associated with high suspended sediment load, restricting light penetration. In combination with deep mixing in the water column this keeps algal production in the lake within the oligotrophic range. During the warmer months of the year thermally driven differences in river inflow density versus lake water density results in mixing of river water in the lake, which, together with greater river flow and dilution, leads to lesser nutrient inputs to the river downstream of the lake. During the limnological winter, however, river inflows of less-dense 0°C water tend to flow across the surface of the more-dense 4°C lake water and exit the lake with relatively little intermixing. Because this is also the period of lowest flow, dilution is minimal and greater levels of nutrients are delivered to the river downstream of the lake in the winter.

The remaining 100 km of the lower Thompson from Kamloops Lake to the Fraser River is fairly fast flowing, with a substratum consisting of boulders, cobbles, and gravel. Unlike the North and South Thompson, this section of the Thompson is usually ice free, due in part to the lake's thermal inertia. Levels of soluble reactive phosphorus (3 to 4 µg/L) are often three or more times those in the North and South Thompson (Bothwell 1985). The final section of the lower Thompson from Spences Bridge to the Fraser River passes through steep-sided gorges and is a well-known white-water rafting area. Despite the fact that precipitation is highest from November through February, discharge in the lower river peaks annually during mountain runoff between May and July and reaches the annual minimum in late winter between January and March (see Fig. 15.8). Thus, the annual discharge pattern is driven by snowmelt in spring.

River Biodiversity and Ecology

As with the rest of the Fraser system, the Thompson is within the North Pacific Coastal freshwater ecoregion (Abell et al. 2000). The river is relatively well studied in terms of its biodiversity and ecology, including studies on ecosystem processes.

Algae

The Thompson River system is naturally oligotrophic and primary producers of the river respond rapidly to nutrient additions (Bothwell 1985). Food webs in the Thompson main stem appear to be based largely on epilithic algal production (Bothwell et al. 1992, Wassenaar and Culp 1996), although the importance of allochthonous detritus probably increases toward the headwaters. Epilithic algal communities are dominated by several diatom genera in the South, North, and main-stem Thompson rivers, including *Achnanthes, Cymbella, Diatoma, Fragilaria, Gomphonema, Hannaea, Nitzschia, Synedra,* and *Tabellaria* (Federal–Provincial Thompson River Task Force 1976, Bothwell 1985). Species diversity is highest in the South Thompson River, where algal biomass is low. Algal biomass increases downstream of effluent discharges at Kamloops Lake, and the filamentous diatom *Gomphonema olivaceum* often dominates the species composition of the lower river.

Plants

Macrophytes are seldom abundant in the Thompson River. Riparian vegetation in the heavily forested North Thompson and South Thompson rivers consists of various willow species, trembling aspen, and conifers such as lodgepole pine or Douglas fir. Along the lower Thompson River the riparian vegetation is sparse and is composed mostly of grasses, shrubs (e.g., sagebrush), and the occasional balsam poplar or ponderosa pine.

Invertebrates

Early river-monitoring studies (1973 to 1975) showed that the benthic macroinvertebrate faunal composition of the North and South Thompson rivers was dominated by families of mayflies (Baetidae, Ephemerellidae, Siphlonuridae, Heptageniidae), stoneflies (Chloroperlidae, Nemouridae, Perlodidae), caddisflies (Hydropsychidae, Leptoceridae), and midges (Chironomidae) (Federal–Provincial Thompson River Task Force 1976). In contrast, taxonomic richness in the main-stem Thompson below major effluent sources was greatly reduced and comprised mostly of chironomids, oligochaetes, and nematodes

but recovered to more pollution-intolerant forms by the 1990s (Lowell and Culp 2002).

The annual freshet appears to be an important reset mechanism for the benthos, as many insects emerge prior to its onset. Insect abundance (density and biomass) is lowest in late summer following the annual freshet and early larval instars are common. Abundance of most taxa (i.e., mayflies, midges, caddisflies) increases to an annual maximum by late winter during the period of high algal biomass and low, stable discharge. Total densities during this period often exceed 20,000 individuals/m^2 and the river is dominated by taxa belonging to the collector-gatherer, filterer, scraper, and predator functional feeding groups. Insects present in late winter are generally late instars nearing emergence. Common genera in the Thompson River are the mayflies *Baetis, Ephemerella, Paraleptophlebia,* and *Rhithrogena;* the caddisflies *Arctopsyche, Brachycentrus, Cheumatopsyche, Glossosoma, Hydropsyche,* and *Hydroptila;* the stoneflies *Arcynopteryx* and *Skwala;* and the midges *Cardiocladius, Cricotopus,* and *Eukiefferiella.* Bothwell and Culp (1993) report a clear seasonal pattern of insect abundance in the Thompson River, which they summarized for two taxa, *Baetis tricaudatus* mayflies and Orthocladiinae midges.

Vertebrates

The Thompson River basin provides important spawning, rearing, and migratory habitat for several species of Pacific salmon (coho, chinook, sockeye, and pink) as well as steelhead. Of special importance is the Adams River sockeye migration, the largest in North America, and the Thompson River steelhead run (Hume 1992). At least 40 streams and rivers in the Thompson drainage provide spawning habitat for coho salmon. Over the last decade abundance of coho returning to the Thompson River watershed has declined by 90% due to a combination of factors, such as changes in land use, fishing, and climate change (Bradford and Irvine 2000). Pink salmon spawn extensively in the Thompson River below Kamloops Lake (Federal–Provincial Thompson River Task Force 1976). Recent surveys found that juvenile chinook salmon also use the main stem as winter habitat, where they feed on benthic macroinvertebrates such as caddisflies and stoneflies. In fact, electrofishing surveys revealed that the lower Thompson had the highest winter densities of juvenile sockeye in the Fraser basin. Finally, recent studies indicate that the Thompson River stocks of chinook and coho salmon form evolutionarily significant units of importance for biodiversity conservation (Small et al. 1998, Teel et al. 2000, Nelson et al. 2001).

Although Pacific salmon are the charismatic component of the basin fish fauna, the Thompson River watershed contains a total of 24 species. These include round whitefish, largescale sucker, bridgelip sucker, northern squawfish, longnose dace, and slimy sculpin. Although there is little information on most of the fish species of the Thompson other than salmon, Bothwell and Culp (1993) reviewed the seasonal cycles of growth and abundance of longnose dace, a key benthic predator in the lower Thompson River. This minnow spawns in summer with young-of-the-year (YOY) appearing in August along the river margins. YOY dace grow rapidly through the fall, attaining body lengths of 30 to 40 mm by November. Over this period YOY dace densities decline, indicating that autumn is a period of high mortality for this age class. Densities of older year classes are also highest in late summer and decrease through late winter (March and April).

Other aquatic vertebrates in the basin include birds and mammals, as well as several amphibian species. Only four species of amphibians are associated with flowing waters or ponds adjacent to streams in forested landscapes: long toed salamander, Pacific tree frog, Great Basin spadefoot, and the Columbia spotted frog (Corkran and Thoms 1996). Aquatic mammals found in the river drainage include beaver, muskrat, and river otter. The South Thompson River is an important winter habitat for waterfowl. Winter surveys of this stretch of the river have recorded 100 to 400 trumpeter swans, which represents as much as 2% of the global population. Breeding birds of the Thompson River include wood ducks, osprey, common merganser, and dipper.

Ecosystem Processes

Primary production in the Thompson River is limited by soluble reactive phosphorus (SRP) availability, as the river is extremely sensitive to nutrient dynamics (Bothwell 1985). Nutrient-saturated growth rates of benthic diatoms in the lower Thompson River occur near 1 µg SRP/L. However, areal biomass accumulated on epilithic surfaces increases with elevation of phosphorus concentration to at least 30 µg SRP/L (Bothwell 1989). Nutrient limitation varies with season in the lower Thompson River: Periphytic communities are nutrient limited in the autumn but relatively insensitive to elevation in phosphorus during the winter (Bothwell and Culp 1993, Dubé et al. 1997). Chlorophyll *a* varies about twenty-fivefold during the year in the Thompson main stem, with the lowest levels ($2 \mu g/cm^2$) occurring after the spring freshet (Bothwell and Culp 1993). Algal biomass increases from summer through late winter as a result of complex mixing patterns in Kamloops Lake (described previously) that produce conditions of nutrient limitation in the fall and eutrophication in the winter (Dubé et al. 1997). Late-winter biomass values can peak as high as $40 \mu g/cm^2$ prior to the algal mat sloughing as it becomes thick and physically unstable. Although grazers such as midges can reduce the amount of benthic algae, particularly in the autumn when water temperatures are warm, the effectiveness of insect grazing is reduced by the presence of predatory benthic fishes.

The lower Thompson River is an open-canopy river and would be predicted to have a food web based on carbon derived from algal production. However, stable isotopic analyses indicate that both the carbon and nitrogen isotopic signatures of insects collected 50 to 100 km downstream of the pulp mill contained 50% to 80% of mass derived from terrestrial sources (Wassenaar and Culp, 1996). This carbon is most likely material from pulp mill effluent and indicates that the discharge contributes directly and significantly to the lower Thompson food chain. Filter-feeding caddisflies contained the greatest percentages of terrestrial carbon. Like their insect prey, the fish isotopic signatures downstream of the mill were correlated to that of the effluent.

Human Impacts and Special Features

The Thompson River is one British Columbia's most important rivers, as it provides water to municipalities, industry, and agriculture as well as providing important habitat for commercial and recreational fisheries. The Thompson is the spawning route for millions of sockeye, chinook, coho, and pink salmon and is home to some of the largest steelhead found in North America. The North Thompson River has few sections of wild water with chinook salmon and Dolly Varden the primary targets for anglers. The South Thompson River is relatively slow moving and is the passageway for the famous Adams River run of sockeye salmon. Below Kamloops the landscape surrounding the Thompson River changes from forest to a semiarid landscape of grass and sagebrush. Many of the river terraces along this stretch of the river are used for cattle grazing and fodder crops. Both the Canadian Pacific and Canadian National transcontinental railroad lines follow the river from Kamloops to Lytton. The river reach from Kamloops

Lake to the Spences Bridge area is famous for steelhead fishing, with fish sometimes reaching 8 to 10 kg. The river section downstream of Spences Bridge to Lytton is wild and fast, traveling through steep-sides canyons and gorges. This section of the river features 18 Class III (or greater) rapids, beginning with Frog Rapid, a turbulent river section that squeezes around a house-sized rock that resembles its namesake at base flow. Thousands of people raft this section of the river each summer to experience huge standing waves, the hot climate, and desert-canyon scenery.

The primary human impact on the Thompson River is the effect of the bleached kraft pulp mill and municipal sewage effluent discharged at Kamloops. Coincident with increases in effluent loading from the pulp mill in the early 1970s were reports of excessive algal growth and changes in water quality downstream of Kamloops Lake. This raised concern about effects on Pacific salmon stocks and led to studies by the Federal–Provincial Thompson River Task Force (1976). Phosphorus was identified as the nutrient most likely responsible for the excessive increase of attached algae (later confirmed experimentally by Bothwell et al. [1992] and others). Recommendations prompted the city of Kamloops and provincial regulators to initiate tertiary treatment of the municipal effluent by alum flocculation as a routine treatment in 1977. Although phosphorus in the sewage effluent was reduced by 90%, scientists at the National Water Research Institute devised an additional "winter-holdback" strategy to further reduce phosphorus discharge to the river. This plan calls for storage of all municipal effluent during the low-flow period (December to March) in order to reduce nutrient enrichment in the lower Thompson River. Discharge commences again during the spring freshet, when high dilution and mixing within Kamloops Lake minimize nutrient impact in the river.

This strategy for reducing nutrient loading markedly lowered phosphorus discharge from the sewage plant throughout the year. Abundances of several key invertebrate taxa (mayflies [Heptageniidae, Baetidae, Ephemerellidae], stoneflies [Perlodidae], and caddisflies [Hydropsychidae]) were greater in later years during which mill output of effluent solids and phosphorus was greater. However, these effects were not pronounced enough to produce increased abundances of more pollution-tolerant taxa, such as nematodes and oligochaetes, as was observed in the 1970s prior to improved effluent treatment. Outdoor artificial stream experiments set up alongside the river demonstrate that effluent concentrations from 1% to 10% increase the growth of algae and the biomass of mayflies (*Baetis tricaudatus*) and chironomids, although some inhibitory effects on the invertebrates occurred at effluent concentrations ≥5% (Lowell et al. 1995, Dubé and Culp 1996, Culp and Lowell 1999, Lowell and Culp 1999).

NECHAKO RIVER

The Nechako River flows northeast from the Coast Mountains, and after it joins its major tributary, the Stuart River, flows east to join the Fraser River at Prince George (Fig. 15.9). This is the most regulated of the basins within the Fraser River catchment, with a 906 km^2 reservoir stored behind the rock-filled Kenney Dam. The river from the dam to the confluence with the Fraser River is now 290 km long, but prior to the dam installation was approximately 440 km in length (Fig. 15.4). The river, along with its tributaries, the Stuart and Nautley rivers, now makes up about 8% of the total flow of the Fraser River.

Native Americans have lived in the catchment for thousands of years, and include the Carrier (Dakelhne) and Wet'suwet'en peoples, part of the Athapaskan language group. Prior to European contact one of the largest settlements of native people in the basin was at the confluence of the Nechako and Stuart rivers in a place known as Chinlac, site of an intertribal massacre of Carrier people in 1745. The earliest recorded European movement into the basin is attributed to Simon Fraser, one of the early British explorers, who established trading posts and forts in 1806, one of which was Fort Fraser. The earliest farming in the area began in about 1811. The first major wave of settlement in the region occurred in 1903, when news that the Grand Trunk Pacific Railway would be built through the area, which was completed at Vanderhoof in 1914.

Physiography, Climate, and Land Use

The Nechako sits in the Coast Mountains of British Columbia and Southeast Alaska (PM) physiographic province, although most of the basin is actually eastward of the Coast Mountains (see Fig. 15.9). The headwaters of the Nechako River in the west of the basin drain a mostly igneous, granitic batholith with low solute concentrations that makes up the Coast Mountains. These headwaters are on the eastern edge of the British Columbia Mainland Coastal Forests terrestrial

FIGURE 15.4   Nechako River at Highway 16, British Columbia (PHOTO BY TIM PALMER).

ecoregion. The areas near the mountains are steep and consist of boreal, temperate forest and tundra biomes. The majority of the basin has a gently rolling topography. About 85% of the basin drains the Nechako Plateau, made up primarily of volcanic rock, with some sedimentary areas and higher solute concentrations (Hall et al. 1991). This plateau is part of the Fraser Plateau and Basin Complex terrestrial ecoregion, typified by deep, postglacial lacustrine soils.

The forests of the basin are predominantly a sub-boreal type characterized by lodgepole pine, white spruce, and aspen. At higher elevations these forests grade into subalpine zones along the leeward side of the Coast Mountains, with subalpine fir, mountain hemlock, and Engelmann spruce most common. These forests experience large-scale disturbances from fire and insect outbreaks. In recent years outbreaks of mountain pine beetle have resulted in extensive damage to forests.

The main portion of the basin is in the rain shadow of Coast Mountain range and gets only about 40 to 60 cm/yr of precipitation (Fig. 15.10).

Rainfall through the basin varies with proximity to the mountains and with elevation. Near the Coast Mountains at the west end of the catchment, average annual precipitation is 196 cm/yr, whereas in the eastern portion of the basin it averages only 20 to 40 cm/yr. Precipitation falls throughout the year without any distinct seasonal pattern, with the exception of slightly lower winter precipitation. During winter the low temperatures result in storage of precipitation as snow throughout the basin. The annual temperature pattern is characteristic of the northern continental climate, with freezing winter temperatures and mean summer temperatures of about 15°C (see Fig. 15.10), but summer temperatures can reach the mid-30s. Summer rainfall is often cyclonic in nature and results in frequent thunderstorms.

The major land use in the area is timber harvesting. Forest covers about 88% of the basin (Revenga et al. 1998). Agriculture, mostly ranching, occupies about 6% of the basin, and very little of that is irrigated farming. Forest cover and soils are similar to those in the Stuart River basin (which will be discussed later in this chapter). The largest towns in the catchment, which occupy only about 0.2% of the basin area, are Vanderhoof, Burns Lake, and Fraser Lake, with a combined population of about 7800 (2002 figures). Most water for domestic use comes from groundwater sources. There are no pulp mills in the basin.

River Geomorphology, Hydrology, and Chemistry

The river has a cobble bottom through much of its course downstream of the Kenney Dam, almost to the town of Vanderhoof. Chinook salmon spawn in the reaches upstream of the confluence with the Stuart River and can have enough influence on the bed during their digging of redds that dunes of up to 15 m in period and amplitude of 0.75 m form in large sections (Tutty 1986). Downstream of Vanderhoof until the confluence with the Fraser River the gradient is lower and the river takes on a slow, meandering form, with many backwaters (sloughs) and a main channel depth approaching 10 m.

The biggest tributary of the Nechako River is the Stuart River, which joins the Nechako downstream of Vanderhoof, 92 km before its confluence with the Fraser River at Prince George. The other major tributary of the Nechako River is the Nautley River, joining near the town of Fort Fraser. The Nechako drains a catchment area of 52,000 km². The river at the mouth is 7th order and now contributes, along with its major tributaries, the Stuart and Nautley rivers, about 8.3% of total Fraser River flow. The predam discharge from the river was approximately 434 m³/s. The 93 m high Kenney Dam was completed in 1952 to impound the Nechako River and diverts 60% to 70% of the average annual flow of the basin above the dam from the west end of the reservoir to the Aluminum Company of Canada's power plant on the Pacific coast near Kemano. The withdrawal of water is equivalent to about 3% of the total flow of the Fraser River at its mouth. The surface of the reservoir is at 853 m asl, and it has live storage of 7.1×10^9 m³. The elevation drop from the reservoir surface to the penstock of the power generating plant at 61 m asl is 792 m along a 16 km tunnel through the Coast Mountains (Mundie and Bell-Irving 1986). The power is used for aluminum smelting operations in Kitimat. One of the primary considerations in the management of the Nechako Reservoir is the provision of cool water for salmon migrations. The two salmon species of most concern are chinook salmon

that spawn in the Nechako River downstream of the dam and the Stuart River stock of sockeye salmon that migrates up the lower Nechako, as sockeye are particularly sensitive to the higher summer temperatures resulting from flow regulation (Mundie and Bell-Irving 1986).

Although only 15% of the drainage area flows from the Coast Mountains, it contributes more than 15% of the discharge of the river, given the higher rainfall amounts in the mountains. Snowmelt peaks in late spring or early summer dominate the annual hydrograph (Moore 1991). A second peak in the hydrograph is sometimes evident from autumn rains prior to freezing.

The hydrology is heavily regulated by impoundment by the Kenney Dam. There is no direct release of water from the Kenney Dam and water released from the reservoir is regulated at the Skins Lake Spillway, passing through Cheslatta Lake and River to enter the Nechako River about 9 km downstream of the dam. The hydrograph upstream of Fort Fraser has been shifted to reduce the spring and summer snowmelt peaks that characterized the natural pattern of flow (see figures in French and Chambers 1995). Nonetheless, the runoff near Prince George still peaks in June and July (about 3 cm/mo), and is ≤1 cm/mo during winter (see Fig. 15.10). In contrast, peak discharge in the Stuart River is >5 cm during July. The provincial government has forced the additional release of water when migrating fish stocks are moving up the Nechako River from Prince George, primarily the early Stuart River run, but also those of the Nadine and Stellako catchments. The high flow released during mid to late summer is to provide sufficient water depth for migration, but especially to provide water of a particular temperature to protect the sockeye salmon. There has been very careful modeling of predicted temperatures to ensure target upper temperatures of the Nechako River. The typical temperature peak is about 19°C (Russell et al. 1983), and the temperature is regulated to ensure it rarely exceeds 20°C. The low winter flows typical of rivers in the region lead to formation of frazil and anchor ice, and ice may form into the substrate (Blachut 1988).

This is a clearwater river, and total dissolved solids range from 1 to 100 mg/L. The large influence of lakes and reservoirs, as well as a large portion of the water coming from the granitic batholith of the Coast Mountains, probably contribute to the low turbidity. The water is typically low in Ca^{++} (13 to 17 mg/L) and SO_4 (3.3 to 5.3 mg/L). Total phosphorus ranges from 17 to 36 μg/L, and NO_3-N values are 216 to 295 μg/L

(Hall et al. 1991). Groundwater is the primary source of drinking water in the basin, and is mostly hard water (>150 mg/L as $CaCO_3$) with elevated Fe (>0.3 mg/L) and Mn (>0.05 mg/L) concentrations (Hall et al. 1991). Copper has sometimes been reported to be at concentrations exceeding water-quality guidelines, and the source is presumed to be natural mineralization.

The primary sources of pollution in this catchment come from domestic waste, particularly the town of Vanderhoof. Concentrations of NO_3-N typically double and orthophosphorus is four to eight times higher below the Vanderhoof sewage treatment plant than upstream (Slaney et al. 1994).

River Biodiversity and Ecology

As part of the Fraser River system the Nechako River belongs to the North Pacific Coastal freshwater ecoregion (Abell et al. 2000). There has been concerted study of the postimpoundment conditions in the river because of concern for the chinook salmon stock and the possibility of nutrient-enrichment impacts on macrophyte growth in reaches near Vanderhoof. As a major corridor for passage of salmon, especially sockeye salmon, to their spawning areas, there has been continued evaluation of those fish stocks relative to river temperatures and other changes.

Algae and Cyanobacteria

The periphyton in the Nechako River was mostly comprised of the diatoms Bacillariophyceae (~80% by area) (*Synedra, Fragilaria, Tabellaria, Hannaea*) and Chlorophyta (*Zygnema* and *Ulothrix*), with traces of Cyanobacteria (Slaney et al. 1994, Perrin and Richardson 1997). Estimates of algal biomass varied from about 3 to 20 mg chl a/m^2 in the main channel of the river (Slaney et al. 1994). Experimental additions of inorganic nutrients resulted in a large increase in the accrual of algal biomass in the main channel of the river, indicating the oligotrophic status of the upper part of the river (Slaney et al. 1994). A set of 12 wadeable streams in the Cheslaslie drainage in the central part of the basin were sampled for algae by Rosenberg et al. (1998) and they found at least 68 species of algae. The most abundant of the algal species in those smaller streams were *Achnanthes minutissima, Tabellaria flocculosa, Gomphonema olivaceum, Tabellaria fenestrata, Synedra ulna, Hannaea arcus, Epithemia turgida, Cymbella ventricosa, Cymbella turgida,* and *Melosira granulata* (in order of abundance). Estimates of algal biomass from these small streams within the basin were 0.09 mg chl a/cm^2 (Rosenberg et al. 1998).

Plants

There is a diverse macrophyte assemblage in the river consisting of several species of pondweed (e.g., sago pondweed, Richardson's pondweed), spiked watermilfoil, elodea, water buttercup (whitewater crowfoot), and coontail (French and Chambers 1996). Elodea is the most abundant of the river macrophytes and at times creates a nuisance for users of the river. Most of these vascular plants are found in the slower-velocity reaches of the river, particularly downstream of Fort Fraser, where the river gradient is lower and velocities slower. Mosses were the primary macrophyte in waters with current speeds in excess of 60 cm/s (French and Chambers 1996). The riparian forest surrounding the Nechako River includes black cottonwood, balsam poplar, aspen, Sitka alder, and willows. Thimbleberry, American fly honeysuckle, and cowparsnip are common understory plants near the riparian margins of smaller streams.

Invertebrates

In one study of macroinvertebrates in the Nechako River, estimates of benthic densities were $400/m^2$ (range 200 to 1600), with a biomass estimate of about $200 mg/m^2$ (range ~50 to 800) depending upon site and date (Russell et al. 1983). In mesocosm studies estimates from the Nechako were 14,800 (+1141 SE)/m^2 based on sorting down to 250 μm size (Perrin and Richardson 1997). Orthocladiinae chironomids were the most numerically abundant invertebrates, followed by the mayflies *Serratella tibialis* and *Baetis* spp. Some other large species included the stoneflies *Pteronarcys* sp. and *Hesperoperla pacifica* and hydropsychid caddisflies.

Vertebrates

Twenty-six species of fishes are found in the Nechako River basin. Prior to construction of the dam the canyon where the dam is located created a natural barrier to fish dispersal. The most abundant fish species in the river include cyprinids (e.g., northern pike minnow, redside shiners), largescale suckers, burbot, and slimy sculpin. There is a substantial run of chinook salmon that spawn in the upper part of the Nechako River in the section below the dam and upstream of the confluence with the Nautley River. There are also small populations of rainbow trout and Dolly Varden trout in the river. In the reservoir the most abundant of the fish species include kokanee (a landlocked form of sockeye salmon), rainbow trout, mountain whitefish, and northern pikeminnow (squawfish). Less-abundant species of fishes in the reservoir include longnose and largescale suckers, prickly and slimy sculpins, burbot, and peamouth chub.

The white sturgeon population in the Nechako and Stuart rivers is isolated from other populations in the Fraser River downstream. Sturgeon number about 600 individuals and are apparently declining in numbers (Ted Downs, personal communication), perhaps because of flow regulation and diminished habitat as a result of the Kenney Dam. Few to no sturgeon under the age of 15 years have been captured from the population during intensive surveys, suggesting serious impairment of the population. The population has been designated vulnerable to critically imperiled. Recent studies have confirmed that the Nechako River population showed genetic divergence from downstream populations in the Fraser River main stem. The low-gradient reach of the river in the 60 km upstream of the confluence with the Fraser River lacks deep pools and is considered a dispersal barrier to white sturgeon.

Two other fish species are of concern in the Nechako River basin. One is the bull trout, which is considered vulnerable throughout its British Columbia range and is on the province's blue list. The brassy minnow has a disjunct population in the basin, isolated from the remainder of the species' range.

Substantial populations of sockeye salmon migrate through the Nechako River's downstream reaches from the Fraser River to the Stuart River system each year. The management of flows and water temperatures of water released from the dam are based around the protection of the summer runs of sockeye (July and August).

The chinook salmon is the most commercially important of the Nechako fishes and escapements to the upper river have been in the range of 600 to 2000 adult spawners. Chinook salmon have juveniles of two sorts, one of which overwinters in the river as 0+ age fish and another that migrates soon after hatching. Chinook salmon numbers have apparently declined in the period following construction of the Kenney Dam. The reduced numbers have been attributed in part to changes in flow, modification of temperature patterns speeding in-gravel development, and changes in predator numbers (Bradford 1994).

There are many riparian-dependant vertebrates in the basin. Common mammals are beaver, muskrat, river otter, moose, and mink. There are many birds associated with the river, including osprey, merganser, bald eagle, and goldeneye duck.

Ecosystem Processes

The Nechako River basin is influenced by cold winter temperatures that freeze over most of the water bodies in the catchment. Late in the autumn, freezing temperatures and dark nights can result in re-radiation, causing the formation of anchor ice that can have negative impacts on benthic life, including incubating salmon eggs. The influence of the dam has changed the nature of many processes, including reducing the movement of bed sediments both through blocking transport and by reducing peak flows in the spring. Whole-river fertilization experiments in the upper Nechako River (below the dam) showed strong responses of periphyton to additions of nitrogen and phosphorus, and the periphyton growth acceleration was noted up to 50 km downstream with a lag of several months, demonstrating the slow spiraling of nutrients within this oligotrophic river (Slaney et al. 1994). However, despite the increase in periphyton production through the upper river there was no evidence of enhanced growth of fishes there (Slaney et al. 1994), suggesting other sinks for production within the river food web.

Human Impacts and Special Features

The Nechako River catchment includes several large wilderness areas, some now isolated by the waters of the reservoir. It provides some productive trout and other game fishing areas, which supply recreational opportunities to the region. The limited access to the basin has preserved most of its wildlife populations. Most of the Nechako basin is sparsely settled but has been exploited for timber production and agriculture for over a century. The major town in the catchment is Vanderhoof, along with a number of smaller towns, such as Burns Lake, Fort Fraser, and Fraser Lake. Agriculture is mostly restricted to the lower part of the basin in the plateau areas west of Prince George and includes cattle production and some cereal crops. A large open-pit molybdenum mine, the Endako Mine, is in the basin east of Fraser Lake.

There have been large changes to the hydrology as a consequence of the Kenney Dam built in the mid- to late 1950s about 280 km upstream of the confluence with the Fraser River. About 60% of the inflow to the 906 km² reservoir in most years is diverted westward directly to the turbines on the Pacific Coast through 16 km long tunnels through the Coast Mountains. Attempts to mitigate some of the effects of the dam have employed flow regulation, temperature controls (deep release), and fertilization (Perrin and Richardson 1997). A second phase to raise the dam and divert another river into the reservoir was stopped by the British Columbia government prior to development.

STUART RIVER

The Stuart River drains a 14,600 km² catchment that represents the northernmost part of the Fraser River basin. It is an unregulated 280 km long lake and river system draining southeast from high on the leeward side of the Coast Range Mountains, and joins the Nechako River just before the latter empties into the Fraser (Fig. 15.11). The northern half of the catchment is tightly confined by mountains on both sides, yielding to a relatively flat plateau south of Stuart Lake in the southern half of the basin. There are a series of large, natural lakes in the drainage, all important sockeye nursery areas. The northern half of the basin is largely forested, with limited amounts of forest harvesting. The southern half of this catchment has had more extensive forest harvesting, and a small portion is used for agriculture.

Native Americans make up the largest proportion of the residents in the basin, largely members of the Carrier–Sekani Tribal Council, and they have occupied the area for thousands of years. The first European contact is considered to have been with Simon Fraser, one of the early explorers seeking a route to the Pacific Ocean overland. He established a post called Fort St. James in 1806, and a Hudson Bay Company trading post constructed to trade with the native people. Farming in the southern portion of the basin began in the early 1800s. The town of Fort St. James remains the largest settlement in the catchment.

Physiography, Climate, and Land Use

The Stuart River basin sits in the Coast Mountains of British Columbia and Southeast Alaska (PM) physiographic province (see Fig. 15.11) (Hunt 1974). Most of the basin is on the Nechako Plateau and is primarily underlain by volcanic rock, with some clastic subbasins (Gabrielse and Yorath 1991). This region is part of a series of terranes known as the Intermontane Belt and the Omineca Belt (Gabrielse and Yorath 1991). The surface of much of the basin consists of fine lacustrine deposits from postglacial lakes and basal tills. The soil types of this area are primarily luvisols (Canadian classification system), having a large clay component in the subsoil derived from leaching above (Farley 1979).

The climate of the Stuart River basin is continental. Mean annual temperature is only about 3°C, with the lowest mean monthly temperature reaching −11.7°C in January and the highest mean monthly temperature about 15°C in July (Fig. 15.12). Sitting in the rain shadow of the Coast Mountains the basin receives about 35 to 60 cm of precipitation per year, depending on elevation and aspect (Moore 1991). Mean annual precipitation is about 49 cm and is distributed fairly evenly throughout the year, although it is lowest from February through April (see Fig. 15.12). Low winter temperatures result in fall and winter precipitation being stored as snow. Summer rainfall often occurs as late-afternoon thunderstorms in the basin.

The Stuart basin is within the Fraser Plateau and Basin Complex terrestrial ecoregion. The primary forest type is the Subboreal Spruce zone, with areas of Engelmann spruce–subalpine fir forest at higher elevations. The primary forest tree species are lodgepole pine, Engelmann spruce, hybrid Engelmann x white spruce, and subalpine fir.

Over 90% of the land area is covered by forest, with forest harvesting and farming the two biggest land uses in the basin. Only about 1% of the catchment is in agricultural use of any sort, and the amount of that that is irrigated is negligible, mostly for hay production. Urban land use accounts for <0.1% of the catchment area and the biggest town in the catchment is Fort St. James.

River Geomorphology, Hydrology, and Chemistry

The sources of the river begin in the Driftwood River between the Omineca Mountains and the Bait Range of the Skeena Mountains (elevations up to 2000 m asl). The river flows in a narrow valley between these two ranges and then through a series of large lakes: Takla Lake, Trembleur Lake, and, finally, Stuart Lake (see Fig. 15.11). Takla Lake and Trembleur Lake are connected by the Middle River, which has more characteristics of a lake than a river. Many of the tributary streams flowing into the lakes in the northern half of the catchment are relatively short (most <20 km) and steep. The basin also has a great many lakes and wetlands, which affect flow and temperature conditions in many of the tributary streams.

After leaving Stuart Lake the river flows through part of the Nechako Plateau, a relatively flat area, resulting in a lower gradient as it flows toward its confluence with the Nechako River downstream of the town of Vanderhoof.

The rivers in the northern half of the basin are cobble-bottom. The smallest tributaries have step-pool morphologies, formed by imbricated boulders as steps. Further downstream in these tributaries, where gradients drop, gravels are the appropriate size for spawning fishes, including sockeye salmon, and there are large amounts of large woody debris, about 31 pieces/100 m of stream (Fuchs et al. 2003). Sockeye spawn in the tributaries of the large lakes and estimates of bedload and suspended loads in these small tributaries suggest sockeye can be a major agent of sediment transport. These areas have sufficient groundwater inflows to prevent low winter temperatures freezing all the way into the substrate, where developing sockeye salmon eggs spend the winter.

The hydrology of the Stuart River exhibits pronounced snowmelt peaks, usually in June, July, and August, when runoff typically exceeds 4 cm/mo (see Fig. 15.12). Runoff is lowest from October through April, when much of the precipitation falls as snow and remains on the land until late spring. Smaller streams in the lower part of the valley may have their peak runoff in May and June (Moore 1991). The topographic variation from high mountains to the valley bottoms makes it difficult to characterize the whole basin, but the runoff from the Stuart River integrates the flows of all the smaller drainages. The average annual discharge is about 128 m^3/s near Fort St. James.

Water quality of the Stuart River is good, and turbidity levels are in general low, in part due to several large and deep lakes in the basin. There is no glacial source for any part of the Stuart River basin. The levels of total dissolved solids are in the range from 1 to 100 mg/L but mostly toward the lower end of the range (Hall et al. 1991). Average total phosphorus concentrations are 8 to 9 μg/L, NO_3-N 39 to 49 μg/L, Ca^{++} 14 to 15.5 mg/L, and SO_4 4.1 to 5 mg/L (Hall et al. 1991). Copper is sometimes in excess of criterion guidelines, probably from natural mineralization (Hall et al. 1991). Groundwater hardness is generally hard to very hard, usually >150 mg/L as $CaCO_3$, and often with iron and manganese exceeding drinking water standards (Fe >0.3 mg/L and Mn >0.05 mg/L) (Hall et al. 1991).

The water temperatures in the river have an average of about 6.8°C (based on a single year). The water reaches 0°C in the winter and the river and lakes are ice covered for several months each year. Summer maximum temperature is around 20°C (range over 7 years, 18.7°C to 20.5°C), which is typically attained in late July or early August (July 6 to August 16, depending on year). Studies in some of the small tributary streams (2nd to 3rd order) in the area have recorded water temperatures from 0°C to 18°C, with a daily summer average of ~13°C (MacDonald and Herunter 1998).

Stuart Lake is the most downstream lake in the river and lake system. The lake has a water residence time of about 2.3 years, despite being relatively shallow, with a mean depth of only 26 m (Macdonald et al. 1998). Sedimentation is primarily of organic materials, as the chain of upstream lakes restricts transport of heavier materials. The lake has good water quality, but cores of its sediments indicate a pulse of elevated mercury during the middle part of the twentieth century, peaking in about 1950 (now returned to background levels), as a consequence of a mercury mine in the basin at Pinci Lake that operated in the 1940s (Macdonald et al. 1998).

River Biodiversity and Ecology

The Stuart River is part of the North Pacific Coastal freshwater ecoregion, as are all Fraser River tributaries (Abell et al. 2000). Information on algae and invertebrates is primarily available from small tributaries, but more extensive information is available on anadromous fishes, particularly sockeye salmon.

Algae

Estimates of algal biomass from small streams in the Driftwood River basin at the north end of the Stuart catchment were 7.0 mg chl *a*/cm^2 (Rosenberg et al. 1998). There were a total of 38 species of algae identified in those streams. Among the most abundant of the algae species were *Achnanthes minutissima*, *Gomphonema olivaceum*, *Diatoma tenue*, *Fragilaria vaucherie*, *Synedra ulna*, and *Gomphonema parvulum* (Rosenberg et al. 1998). In some smaller streams near the Middle River, estimates of algal biomass were 0.011 μg/m^2 as chlorophyll *a* and 3.04 mg/m^2 as ash-free dry mass (Shirley Fuchs, personal communication).

Plants

Sitka alder, various willows, black cottonwood, and balsam poplar are the common trees in the riparian areas. Thimbleberry, black twinberry, and cowparsnip are common understory plants near the riparian margins of smaller streams.

Invertebrates

Macroinvertebrates have been sampled in two sets of streams, one near the Middle River and another further north in the Driftwood River sub-basin. In the Middle River area, macroinvertebrate densities were in the range of 3000 to 8100 individuals/m^2 and biomass ranged from $1.8\,g/m^3$ (under forest cover) to $3.0\,g/m^2$ (recent clearing) (Fuchs et al. 2003). Some of the common taxa reported included the mayfly *Drunella*, the caddisflies *Glossosoma* and *Rhyacophila*, the stonefly *Despaxia*, chironomid midges (Orthocladiinae and Tanytarsini), oligochaete worms, and mites (Acari) (Fuchs et al. 2003). In small streams of the Driftwood River basin, other taxa predominated, such as the mayflies *Ephemerella*, *Rhithrogena*, and *Diphetor*, the stoneflies *Capnia*, *Paraleuctra*, *Sweltsa*, and *Taenionema*, and chironomids (e.g., *Micropsectra*, *Polypedilum*, and *Rheocricotopus*) (Rosenberg et al. 1998).

Vertebrates

There are major early and late runs of sockeye to the Stuart system, distinguished as separate stocks. These stocks differ in their timing of entry to the Fraser River. Both spawn in the streams and rivers of the Stuart River system and rear in Trembleur Lake or Takla Lake. The early Stuart run reaches the spawning grounds in late July to early August, making them vulnerable to warm summer temperatures during the long migration, and making them the primary focus of temperature manipulation from the regulated Nechako River (see Nechako discussion). The early Stuart run has a four-year dominant cycle, with average numbers in peak years (e.g., 1993, 1997, 2001) of over 300,000 adults and in the three other years of the cycle adult numbers of about 60,000. The late Stuart run spawns further upstream on average, making it the longest migrating sockeye salmon population in the Fraser River basin, with a migration of about 1200 km from the ocean. The late run spawns in mid-September. The nonmigratory form of sockeye salmon known as kokanee lives its entire life in the freshwater environment but produces fertile hybrids with anadromous sockeye (Foote et al. 1989).

As with many streams of the north with continental climates, low flows in winter and freezing into the gravel is a hazard for fishes, especially for eggs deposited in the substrate to overwinter. Spawning sockeye select upwelling areas for their redds (spawning areas), which are less likely to freeze. The timing of development for fishes that lay their eggs late in the season seems to result in eggs hatching before intense freezing. Sockeye eggs hatch during the winter and there is evidence that the alevins (larvae) are capable of moving deeper within the substrate to avoid ice formation near the surface. Sockeye fry move from the spawning streams out to the lakes soon after emerging from the gravels.

The white sturgeon found in the Stuart River system are part of the Nechako River population, which are considered distinct from other Fraser River populations (see Nechako discussion). The current estimate of total population numbers, including immatures, is about 600 individuals, and the species is considered vulnerable in British Columbia. Recent evidence shows that there has been negligible recruitment in the past 15 years, for unknown reasons (Ted Downs, personal communication).

There are many other species of fishes found in the Stuart River system. Lake trout and rainbow trout are popular sport fishes in the lakes and the Middle River part of the basin. Most of the other species are not sought after by recreational or commercial fishers, so there is much less known about their populations in the Stuart River. These include mountain whitefish, largescale sucker, and a number of common cyprinid fishes, such as northern pikeminnow, peamouth chub, lake chub, and redside shiner. These species occupy the larger river and lake systems. There are few species that reside year-round in the small streams of the basin, presumably because of the risk of winter freezing.

Beaver densities are high and provide an important alteration of sediment transport processes in smaller streams. Newly created beaver dams can result in smothering of eggs and alevins of salmonids by sedimentation of fine sediments. Meanwhile, beavers create habitat diversity important to some fish and invertebrate species, and beaver ponds are regularly used as rearing habitat by salmonids. Most riparian-dependant wildlife populations in the basin, such as moose, river otter, muskrat, merganser, osprey, and bald eagle, have relatively stable numbers.

Ecosystem Processes

Although no studies of basic ecosystem processes have been done in the Stuart River, it is clear that a number of processes have large ecosystem effects. The sediments, either as bedload or suspended load, are highly mobile, with movement driven by spring freshet and by the actions of spawning salmon. One estimate has sockeye salmon contributing to 10% to 39% of the bedload movement (Scrivener and Macdonald 1998). Freezing temperatures also have several effects on these river systems, including storage of snow through the winter, ice cover of most water bodies, and early season potential for anchor ice formation deep into the substrate. This river system is relatively productive in contrast to some other drainages of the Fraser catchment, perhaps in part due to the large amounts of marine-derived nutrients delivered by returning Pacific salmon (Naiman et al. 2002).

Human Impacts and Special Features

This is a relatively pristine river system and is somewhat unusual in being a narrow lake-dominated system with short tributary streams flowing into its major lakes, Takla and Stuart. Furthermore, it is special in being the most distant tributary from the Fraser's mouth and supporting the longest migration of sockeye salmon populations in the system. The Canadian Department of Fisheries and Oceans has a large-scale experimental forestry-fisheries study underway in a set of small streams in the Stuart River catchment.

Human impacts to the river are minor. Most water withdrawal is for human use and almost all of that is from pumped groundwater. The largest town in the catchment is Fort St. James, with a population slightly greater than 2000. Forest harvesting and wood processing (milling and manufacture) are the main industrial activities in the basin, but there are no pulp mills.

ADDITIONAL RIVERS

The West Road (Blackwater) River flows eastward before joining the Fraser River north of Quesnel (Fig. 15.13). It is contained entirely within the Fraser Plateau and Basin Complex rising in the Ilgachuz Mountains, and with a slight gradient it forms numerous lakes along the way. The basin is very sparsely populated (<1 person/km^2), and its even terrain and easy canoeing are the reasons for its name. The West Road River was a major path for travel between the coast and the interior. In 1793 it was the route chosen by Sir Alexander MaKenzie, and it is now a heritage hiking trail to Bella Coola. Although it has relatively low discharge, the West Road River is one of the finest trout streams in British Columbia.

The Quesnel River begins in Quesnel Lake on the western slopes of the Cariboo Mountains and flows in a northwest direction before joining the Fraser River at Quesnel (Fig. 15.15). Historically, the Quesnel River produced the Fraser River's largest sockeye salmon runs. With the construction of the Quesnel River dam—built for placer mining in 1896–1897—and the Hell's Gate slide on the Fraser in 1913, the Horsefly and Mitchell river populations (the two most productive tributaries for sockeye in the Quesnel system) had declined to a total run of about 5000 fish and less than 1000 spawners by 1941. The Quesnel River sockeye stocks began to rebuild in the 1950s under the watchful eye of the Pacific Salmon Commission's recovery strategy, which included the construction of fish passage structures. From 1958 to 1977 the Quesnel River sockeye made a remarkable comeback. The last dominant year for Quesnel River sockeye occurred in 1993, producing a run of 12.2 million fish, with 2.5 million fish making it to the Quesnel River headwater tributaries to spawn. The Quesnel River has now surpassed the world-famous Adams River for sockeye production and is, once again, the Fraser's greatest sockeye producer.

The Chilcotin River, with headwaters in the wilderness region near the Ilgachuz Range, winds its way through mountains, open valleys, grasslands, and canyons as it flows in a generally eastward direction to the Fraser (Fig. 15.17). Major tributaries are the Chilanko and Chilco rivers. The watershed includes a wide range of habitats, from high alpine to desert canyons, and the river has major rapids, eddies, rocks, narrow spots, standing waves, and boulder gardens. The Chilcotin is rated among the best and most challenging in North America for kayaking and white-water rafting, and has some of the most spectacular scenery in the basin.

The Clearwater River begins in Wells Gray Provincial Park, a 522,000 ha wilderness area, and flows south to its confluence with the North Thompson River (Fig. 15.19). The basin includes volcanic formations, splendid alpine meadows, six major lakes, excellent fishing and rafting, superb waterfalls, and countless prominent topographical features. The area is known worldwide for spectacular mountain scenery and hiking trails. One of the most spectacular views is Helmcken Falls, which plunges 141 m down a narrow canyon. This is a clearwater river, as it flows through igneous rocks and silt is deposited in upstream ponds.

The Lillooet River arises in the Coast Mountains at the foot of Mount Dalgleish and flows in a southeasterly direction for 209 km before entering Harrison Lake (Fig. 15.21). The headwater Lillooet River system runs for about 190 km of white-water paddling and with spectacular autumn Sockeye salmon runs. Harrison Lake is famous for being the largest lake in southwestern British Columbia, with over 72 km of beautiful lake settings and hot springs. The outlet from Harrison Lake is known as the Harrison River and discharges to the Fraser River. This is a very sparsely populated catchment encompassing a wide range of habitats, from alpine to desert to coastal rain forest. High precipitation in the basin results in extremely high annual runoff (>160 cm; Fig. 15.22).

ACKNOWLEDGMENTS

The authors would like to acknowledge the Fraser River Action Plan (Environment Canada), without which much of this information would not be available, and also the Water Survey of Canada and the Meteorological Service of Canada, which provided much of the site-specific information. Finally, Taina Tuominen of Environment Canada was extremely helpful in acquiring much of data for this chapter.

LITERATURE CITED

Abell, R. A., D. M. Olson, Ricketts, E. Dinerstein, P. T. Hurley, J. T. Diggs, W. Eichbaum, S. Walters, W. Wettengel, T. Allnutt, C. J. Loucks, and P. Hedao. 2000. *Freshwater ecoregions of North America: A conservation assessment.* Island Press, Washington, D.C.

Blachut, S. P. 1988. The winter hydrologic regime of the Nechako River, British Columbia. Canadian Manuscript Report of Fisheries and Aquatic Sciences 1964.

Bothwell, M. L. 1985. Phosphorus limitation of lotic periphyton growth rates: An intersite comparison using continuous-flow troughs (Thompson River system, British Columbia). *Limnology and Oceanography* 30: 527–542.

Bothwell, M. L. 1989. Phosphorus-limited growth dynamics of lotic periphytic diatom communities: areal biomass and cellular growth rate responses. *Canadian Journal of Fisheries and Aquatic Sciences* 45:261–270.

Bothwell, M. L., and J. M. Culp. 1993. Sensitivities of the Thompson River to phosphorus: Studies on trophic dynamics. Contribution no. 93006. National Hydrology Research Institute.

Bothwell, M. L., G. Darken, R. N. Nordic, and J. M. Culp. 1992. Nutrient and grazer control of algal biomass in the Thompson River, British Columbia: A case history of water quality management. In R. R. Roberts and M. L. Bothwell (eds.). *Aquatic ecosystems in semi-arid regions: Implications for resource management.* NHRI Symposium Series 7. Environment Canada, Saskatoon, Saskatchewan.

Bradford, M. J. 1994. Trends in the abundance of Chinook salmon (*Oncorhynchus tshawytscha*) of the Nechako River, British Columbia. *Canadian Journal of Fisheries and Aquatic Sciences* 51:965–973.

Bradford, M. J., and J. R. Irvine. 2000. Land use, fishing, climate change and the decline of Thompson River, British Columbia, coho salmon. *Canadian Journal of Fisheries and Aquatic Sciences* 57:13–16.

Brewer, R., S. Sylvestre, M. Sekela, and T. Tuominen. 1998. Sediment quality in the Fraser River basin. In C. Gray and T. Tuominen (eds.). *Health of the Fraser River aquatic ecosystem: A synthesis of research conducted under the Fraser River action plan.* Vol. 1, pp. 93–108. DOE FRAP 1998-11. Environment Canada, Vancouver, British Columbia.

Carmack, E. C., C. B. J. Gray, C. H. Pharo, and R. J. Daley. 1979. Importance of lake–river interaction on seasonal patterns in the general circulation of Kamloops Lake, British Columbia. *Limnology and Oceanography* 24: 634–644.

Corkran, C. C., and C. Thoms. 1996. *Amphibians of Oregon, Washington and British Columbia.* Lone Pine, Renton, Washington.

Culp, J. M., and R. B. Lowell. 1999. Pulp mill effluent impacts on benthic invertebrate communities and selected fish species. In C. Gray and T. Tuominen (eds.). *Health of the Fraser River aquatic ecosystem: A synthesis of research conducted under the Fraser River action plan.* Vol. 2, pp. 13–34. DOE FRAP 1998-11. Environment Canada, Vancouver, British Columbia.

Dorcey, A. H. J. 1991. Water in the sustainable development of the Fraser River Basin. In H. J. Dorcey, A. H. J., and J. R. Griggs (eds.). *Water in sustainable development: Exploring our common future in the Fraser River basin.* Vol. 2, pp. 3–18. Westwater Research Centre, University of British Columbia, Vancouver.

Dubé, M. G., and J. M. Culp. 1996. Growth responses of periphyton and chironomids exposed to biologically treated bleached-kraft pulp mill effluent. *Environmental Toxicology and Chemistry* 15:2019–2027.

Dubé, M. G., J. M. Culp, and G. J. Scrimgeour. 1997. Nutrient limitation and herbivory: Processes influenced by bleached kraft pulp mill effluent. *Canadian Journal of Fisheries and Aquatic Sciences* 54:2584–2595.

Dynesius, M., and C. Nilsson. 1994. Fragmentation and flow regulation of river systems in the northern third of the world. *Science* 266:753–762.

Farley, A. L. 1979. *Atlas of British Columbia: People, environment, and resource use.* University of British Columbia Press, Vancouver.

Federal–Provincial Thompson River Task Force. 1976. Sources and effects of algal growth, colour, foaming and fish tainting in the Thompson River system. Summary Report.

Foote, C. J., C. C. Wood, and R. E. Withler. 1989. Biochemical genetic comparison of sockeye salmon and kokanee, the anadromous and nonanadromous forms of *Oncorhynchus nerka*. *Canadian Journal of Fisheries and Aquatic Sciences* 46:149–158.

French, T. D., and P. A. Chambers. 1995. Nitrogen and phosphorus in the upper Fraser River in relation to point and diffuse source loadings. DOE FRAP Report no. 1995-09. Fraser River Action Plan, Environmental Conservation Branch, Environment Canada, Vancouver, British Columbia.

French, T. D., and P. A. Chambers. 1996. Habitat partitioning in riverine macrophyte communities. *Freshwater Biology* 36:509–520.

Fuchs, S. A., S. G. Hinch, and E. Mellina. 2003. Effects of streamside logging on stream macroinvertebrate communities and habitat in the sub-boreal forests of British Columbia, Canada. *Canadian Journal of Forest Restoration* 33:1408–1415.

Gabrielse, H., and C. J. Yorath (eds.). 1991. *Geology of the Cordilleran Orogen in Canada*. Geology of Canada 4. Geological Survey of Canada, Ottawa, Ontario.

Gray, C., and T. Tuominen. 1998. *Health of the Fraser River aquatic ecosystem: A synthesis of research conducted under the Fraser River action plan*. Vol. 2. DOE FRAP 1998-11. Environment Canada, Vancouver, British Columbia.

Hall, K. J., H. Schreier, and S. J. Brown. 1991. Water quality in the Fraser River basin. In A. H. J. Dorcey and J. R. Griggs (eds.). *Water in sustainable development: Exploring our common future in the Fraser River basin*, pp. 41–75. Westwater Research Centre, University of British Columbia, Vancouver.

Healey, M. C., and J. S. Richardson. 1996. Changes in the productivity base and fish populations of the lower Fraser River associated with historical changes in human occupation. *Archiv für Hydrobiologie Supplementbände* 113:279–290.

Hume, M. 1992. *The run of the river*. New Star Books, Vancouver, British Columbia.

Hunt, C. B. 1974. *Natural regions of the United States and Canada*. W. H. Freeman, San Francisco.

Kew, J. E. M., and J. R. Griggs. 1991. Native Indians of the Fraser basin: Towards a model of sustainable resource use. In H. J. Dorcey and J. R. Griggs (eds.). *Water in sustainable development: Exploring our common future in the Fraser River basin*. Vol. 1. Westwater Research Centre, University of British Columbia, Vancouver.

Lowell, R. B., and J. M. Culp. 1999. Cumulative effects of multiple effluent and low dissolved oxygen stressors on mayflies at cold temperatures. *Canadian Journal of Fisheries and Aquatic Sciences* 56:1624–1630.

Lowell, R. B. and J. M. Culp. 2002. Implications of sampling frequency for detecting temporal patterns during environmental effects monitoring. *Water Quality Research Journal of Canada* 37:119–132.

Lowell, R. B., J. M. Culp, and M. G. Dubé. 2000. A weight-of-evidence approach for northern river risk assessment: Integrating the effects of multiple stressors. *Environmental Toxicology and Chemistry* 19:1182–1190.

Lowell, R. B., J. M. Culp, and F. J. Wrona. 1995. Stimulation of increased short-term growth and development of mayflies by pulp mill effluent. *Environmental Toxicology and Chemistry* 14:1529–1541.

MacDonald, J. S., and H. E. Herunter. 1998. The Takla fishery/forestry interaction project: Study area and project design. In M. K. Brewin and D. M. A. Monita (eds.). *Forest-fish conference: Land management practices affecting aquatic ecosystems*, pp. 203–207. Natural Resources Canada, Calgary, Manitoba.

Macdonald, R. W., D. P. Shaw, and C. Gray. 1998. Contaminants in lake sediments and fish. In C. Gray and T. Tuominen (ed.). *Health of the Fraser River aquatic ecosystem: A synthesis of research conducted under the Fraser River action plan*. Vol. 1, pp. 23–45. DOE FRAP 1998-11. Environment Canada, Vancouver, British Columbia.

McPhail, J. D. 1998. Fishes. In I. M. Smith and G. G. E. Scudder (eds.). *Assessment of species diversity in the Montane Cordillera Ecozone*. Burlington: Ecological Monitoring and assessment Network, http://www.naturewatch.ca/eman/reports/publications/99_montane/fishes/intro.html

Moore, R. D. 1991. Hydrology and water supply in the Fraser River basin. In A. H. J. Dorcey and J. R. Griggs (eds.). *Water in sustainable development: Exploring our common future in the Fraser River basin*. Vol. X, pp. 21–40. Westwater Research Centre, University of British Columbia, Vancouver.

Mundie, J. H., and R. Bell-Irving. 1986. Predictability of the consequences of the kemano hydroelectric proposal for the natural salmon populations. *Canadian Water Resources Journal* 11:14–25.

Naiman, R. J., R. E. Bilby, D. E. Schindler, and J. M. Helfield. 2002. Pacific salmon, nutrients, and the dynamics of freshwater and riparian ecosystems. *Ecosystems* 5:399–417.

Nelson, R. J., M. P. Small, T. D. Beacham, and K. J. Supernault. 2001. Population structure of Fraser River chinook salmon (*Oncorhynchus tshawytscha*): An analysis using microsatellite DNA markers. *Fishery Bulletin Seattle* 99:94–107.

Nordin, R. N., and D. W. Holmes. 1992. Thompson River water quality assessment and objectives. Technical appendix, B.C. Environment Report. British Columbia Ministry of Environment, Lands, and Parks, Victoria.

Northcote, T. G., and P. A. Larkin. 1989. The Fraser River: A major salmonine production system. In D. P. Dodge (ed.). *Proceedings of the international large river symposium (LARS)*. Canadian Special Publication of Fisheries and Aquatic Sciences 106:172–204.

Northcote, T. G., and M. D. Burwash. 1991. Fish and fish habitats of the Fraser River basin. In H. J. Dorcey and

J. R. Griggs (eds.). *Water in sustainable development: Exploring our common future in the Fraser River basin.* Vol. 2, pp. 117–141. Westwater Research Centre, University of British Columbia, Vancouver.

Perrin, C. J., and J. S. Richardson. 1997. N and P limitation of benthos abundance in the Nechako River, British Columbia. *Canadian Journal of Fisheries and Aquatic Sciences* 54:2574–2583.

Raymond, B., D. P. Shaw, and K. Kim. 1998. Fish health assessment. In C. Gray and T. Tuominen (eds.). *Health of the Fraser River aquatic ecosystem: A synthesis of research conducted under the Fraser River action plan.* Vol. 1, pp. 143–160. DOE FRAP 1998-11. Environment Canada, Vancouver, British Columbia.

Reece, P. F., and J. S. Richardson. 2000. Benthic macroinvertebrate assemblages of coastal and continental streams and large rivers of southwestern British Columbia, Canada. *Hydrobiologia* 439:77–89.

Revenga, C., S. Murray, J. Abramovitz, and A. Hammond. 1998. *Watersheds of The World: Ecological Value And Vulnerability.* World Resources Institute and Worldwatch Institute, Washington, D.C.

Richardson, J. S., and C. D. Levings. 1996. Chlorinated organic contaminants in benthic organisms of the lower Fraser River, British Columbia. *Water Quality Research Journal of Canada* 31:153–162.

Richardson, J. S., T. J. Lissimore, M. C. Healey, and T. G. Northcote. 2000. Fish communities of the lower Fraser River (Canada) and changes through time. *Environmental Biology of Fishes* 59:125–140.

Ricketts, T. H., E. Dinerstein, D. M. Olson, C. J. Loucks, W. Eichbaum, D. DellaSala, K. Kavanagh, P. Hedao, P. T. Hurley, K. M. Carney, R. Abell, and S. Walters. 1999. *Terrestrial ecoregions of North America: A conservation assessment.* Island Press, Washington, D.C.

Rosenberg, D. M., T. B. Reynoldson, and V. H. Resh. 1998. Establishing reference conditions for benthic invertebrate monitoring in the Fraser River catchment, British Columbia, Canada. Fraser River Action Plan DOE FRAP 1998-32.

Russell, L. R., K. R. Conlin, O. K. Johansen, and U. Orr. 1983. Chinook salmon studies in the Nechako River: 1980, 1981, 1982. *Canadian Manuscript Report of Fisheries and Aquatic Sciences* 1728.

Schreier, H., K. J. Hall, S. J. Brown, B. Wernick, C. Berka, W. Belzer, and K. Petit. 1998. Non-point source contamination in the urban environment of Greater Vancouver: A case study of the Brunette River watershed. In C. Gray and T. Tuominen (eds.). *Health of the Fraser River aquatic ecosystem: A synthesis of research conducted under the Fraser River action plan.* Vol. 2, pp. 109–134. DOE FRAP 1998-11. Environment Canada, Vancouver, British Columbia.

Scrivener, J. C., and J. S. Macdonald. 1998. Interrelationships of streambed gravel, bedload transport, beaver activity and spawning sockeye salmon in Stuart-Takla tributaries, British Columbia, and possible impacts from forest harvesting. In M. K. Brewin and D. M. A. Monita. Forest-fish conference: Land management practices affecting aquatic ecosystems. *Proceedings of the Forest-Fish Conference, May 1–4, 1996,* Calgary, Alberta. Nat. Resour. Can., Can. For. Serv., North. For. Cent., Edmonton, Alberta. Inf. Rep. NOR-X-356. pp. 267–282.

Slaney, P. A., B. O. Rublee, C. J. Perrin, and H. Goldberg. 1994. Debris structure placements and whole-river fertilization for salmonids in a large regulated stream in British Columbia, *Bulletin of Marine Science* 55(2–3): 1160–1180.

Slaymaker, O. 1990. Climate change and erosion processes in mountain regions of western Canada. *Mountain Research and Development* 10:171–182.

Slaymaker, O. 1991. Implications of the processes of erosion and sedimentation for sustainable development in the Fraser River basin. In A. H. J. Dorcey and J. R. Griggs (eds.). *Water in sustainable development: Exploring our common future in the Fraser River basin.* Vol. 2. Westwater Research Centre, University of British Columbia, Vancouver.

Small, M. P., T. D. Beacham, R. E. Withler, and R. J. Nelson. 1998. Discriminating coho salmon (*Oncorhynchus kisutch*) populations within the Fraser River, British Columbia, using microsatellite DNA markers. *Molecular Ecology* 7:141–155.

Teel, D. J., G. B. Milner, G. A. Winans, and W. S. Grant. 2000. Genetic population structure and origin of life history types in chinook salmon in British Columbia, Canada. *Transactions of the American Fisheries Society* 129:194–209.

Tutty, B. D. 1986. Dune formations associated with multiple redd construction by chinook salmon in the upper Nechako River, British Columbia, Canada. *Canadian Manuscript Report of Fisheries and Aquatic Sciences* 1893.

Wassenaar, L. I., and J. M. Culp. 1996. The use of stable isotopic analyses to identify pulp mill effluent signatures in riverine food webs. In M. Servos, J. Carey, K. Munkittrick, and G. Van Der Kraak (eds.). Environmental fate and effects of pulp mill effluents, pp. 413–424. St. Lucie, DelRay Beach, Florida.

Whitfield, P. H. 1983. Regionalization of water quality in the upper Fraser River basin, British Columbia. *Water Research* 17:1053–1066.

Whitfield, P. H., and H. Schreier. 1982. Hysteresis in relationships between discharge and water chemistry in the Fraser River basin, British Columbia. *Limnology and Oceanography* 26:1179–1182.

Wilson, L., M. Harris, and J. Elliott. 1998. Contaminants in wildlife indicator species from the Fraser basin. In C. Gray and T. Tuominen (eds.). *Health of the Fraser River aquatic ecosystem: A synthesis of research conducted under the Fraser River action plan.* Vol. 1, pp. 161–188. DOE FRAP 1998-11. Environment Canada, Vancouver, British Columbia.

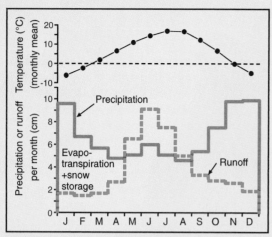

Mean monthly air temperature,
precipitation, and runoff for the Fraser River basin.

FRASER RIVER

Relief: 3954 m

Basin area: 234,000 km²

Mean discharge: 3972 m³/s

River order: 8

Mean annual precipitation: 80 cm (approx. basin mean)

Mean air temperature: 6.1°C (approx. basin mean)

Mean water temperature: NA

Physiographic Provinces: Coast Mountains of British
  Columbia and Southeast Alaska (PM), Rocky
  Mountains in Canada (RM)

Biomes: Temperate Mountain Forest, Tundra

Freshwater ecoregion: North Pacific Coastal

Terrestrial ecoregions: North Central Rockies Forests,
  Fraser Plateau and Basin Complex, Okanagan Dry
  Forests, Cascade Mountains Leeward Forests,
  British Columbia Mainland Coastal Forests,
  Puget Lowland Forests

Number of fish species: 40 freshwater (native),
  8 marine

Number of endangered species: 7 fishes

Major fishes: starry flounder, coho salmon, chinook
  salmon, sockeye salmon, pink salmon, chum salmon, rainbow trout, river lamprey, pacific lamprey, eulachon, surf smelt,
  longfin smelt, northern squawfish, peamouth chub, redside shiner, longnose dace, largescale sucker, longnose sucker, white
  sucker, bridgelip sucker, prickly sculpin

Major other aquatic vertebrates: great blue heron, bald eagle, river otter, mink

Major benthic invertebrates: mayflies (*Baetis, Ephemerella, Drunella, Rhithrogena*), stoneflies (*Capnia, Sweltsa, Taenionema,
  Zapada*), chironomid midges (*Eukiefferiella, Micropsectra, Tvetenia*), caddisflies (*Rhyacophila*)

Nonnative species: American shad, common carp, brown bullhead, goldfish, fathead minnow, Atlantic salmon, brook trout,
  yellow perch, pumpkinseed, largemouth bass, black crappie

Major riparian plants: white spruce, lodgepole pine, trembling aspen, Douglas fir, Engelmann spruce, alpine fir, common paper
  birch, black cottonwood, trembling aspen, various willows, various grasses

Special features: Canadian Heritage river; Fraser Canyon between Yale and Boston Bar

Fragmentation: none except for dam in Nechako River

Water quality: alkalinity = 47 to 61 mg/L as CaCO₃, pH = 7.8 to 7.9, NO₃-N = 0.13 to 0.21 mg/L, total phosphorus = 0.07 to
  0.11 mg/L

Land use: 2.2% urban, 0.6% agriculture, 97.2% forest and park

Population density: <10.7 people/km²

Major information sources: Dorcey 1991, Gray and Tuominen 1998, Rosenberg et al. 1998, Richardson et al. 2000, Whitfield
  1983, Northcote and Larkin 1989

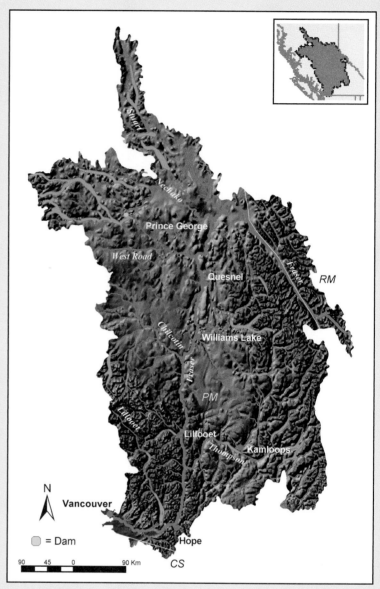

FIGURE 15.5 Map of the Fraser River basin. Physiographic provinces are
separated by a yellow line.

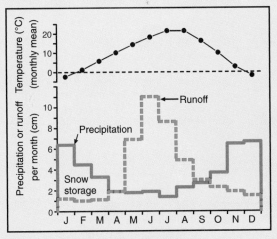

FIGURE 15.8 Mean monthly air temperature, precipitation, and runoff for the Thompson River basin.

THOMPSON RIVER

Relief: 3250 m
Basin area: 55,400 km²
Mean discharge: 787 m³/s
River order: 7
Annual mean precipitation: 43.2 cm
Mean air temperature: 9.7°C
Mean water temperature: 9.0°C
Physiographic province: Coast Mountains of British Columbia and Southeast Alaska (PM)
Biome: Temperate Mountain Forest
Freshwater ecoregion: North Pacific Coastal
Terrestrial ecoregions: North Central Rockies Forests, Okanagan Dry Forests, Cascade Mountains Leeward Forests
Number of fish species: 24
Number of endangered species: 2 fishes
Major fishes: round whitefish, coho salmon, chinook salmon, sockeye salmon, pink salmon, largescale sucker, bridgelip sucker, northern squawfish, longnose dace, slimy sculpin

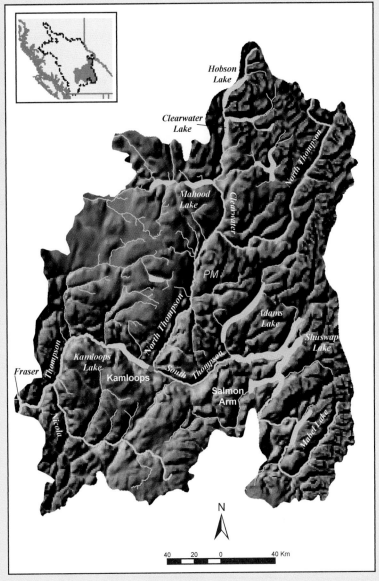

FIGURE 15.7 Map of the Thompson River basin.

Major other aquatic vertebrates: beaver, muskrat, dipper, merganser, osprey
Major benthic invertebrates: mayflies (*Baetis, Ephemerella, Paraleptophlebia, Rhithrogena*), caddisflies (*Arctopsyche, Brachycentrus, Cheumatopsyche, Glossosoma, Hydropsyche, Hydroptila*), stoneflies (*Arcynopteryx, Skwala*), chironomid midges (*Cardiocladius, Cricotopus, Eukiefferiella*)
Nonnative species: carp
Major riparian plants: common paper birch, black cottonwood, trembling aspen, various willows, various grasses
Special features: white-water rapids between Spences Bridge and the confluence with the Fraser River; Adams River run of Sockeye salmon
Fragmentation: none
Water quality: alkalinity = 38.2 mg/L as CaCO₃, pH = 7.7, NO₃-N = 0.08 mg/L, orthophosphate = 0.004 mg/L, total dissolved phosphorus = 0.005 mg/L, total phosphorus = 0.012 mg/L, total dissolved nitrogen = 0.16 mg/L
Land use: 0.3% urban, 2.4% agriculture, 5.1% alpine, 4.3% freshwater, 41.0% young forest, 25.9% old forest, 8.2% recently logged, 4.5% rangeland, 8.6% other
Population density: <3 people/km²
Major information sources: Bradford and Irvine 2000, Bothwell 1985, Culp and Lowell 1999, Dubé et al. 1997, Federal–Provincial Thompson River Task Force 1976

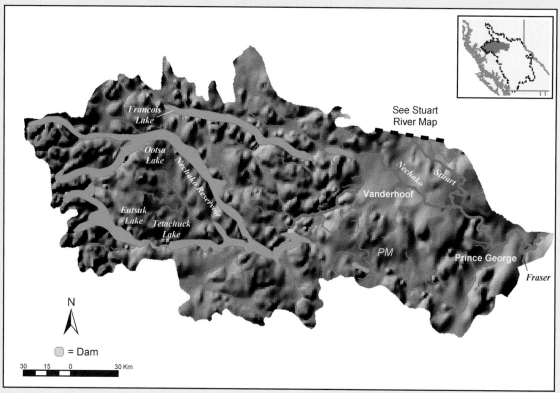

FIGURE 15.9 Map of the Nechako River basin.

NECHAKO RIVER

Relief: ~1400 m

Basin area: 42,500 km^2

Mean discharge: 434 m^3/s (virgin, estimated); 284 m^3/s (present)

River order: 7

Mean annual precipitation: 60.1 cm

Mean air temperature: 3.7°C

Mean water temperature: NA

Physiographic province: Coast Mountains of British Columbia and Southeast Alaska (PM)

Biome: Temperate Mountain Forest

Freshwater ecoregion: North Pacific Coastal

Terrestrial ecoregions: Fraser Plateau and Basin Complex, British Columbia Mainland Coastal Forests

Number of fish species: 26

Number of endangered species: 2 fishes

Major fishes: burbot, mountain whitefish, lake whitefish, lake trout, chinook salmon, sockeye salmon (kokanee), rainbow trout, prickly sculpin, largescale sucker, longnose sucker, redside shiner, peamouth chub, lake chub

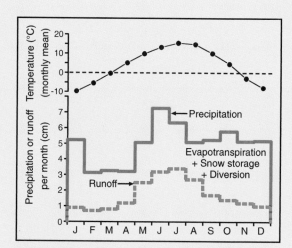

FIGURE 15.10 Mean monthly air temperature, precipitation, and runoff for the Nechako River basin.

Major other aquatic vertebrates: beaver, moose, merganser, goldeneye duck

Major benthic invertebrates: mayflies (Baetidae, *Ephemerella*, *Serratella*), stoneflies (Capniidae, Chloroperlidae, *Hesperoperla pacifica*, *Pteronarcys*), caddisflies (Hydropsychidae), chironomid midges (Tanypodinae, Tanytarsini, Orthocladiinae)

Nonnative species: goldfish, brook trout

Major riparian plants: willows, black cottonwood, balsam poplar, alder, aspen

Special features: remote basin with large areas of pristine forest in the western part of its basin near the Coast Range

Fragmentation: major dam with large influence on river regulation and diversion of water to a power generating station on the Pacific coast

Water quality: alkalinity = 40.4 mg/L as CaCO$_3$, pH = 7.6, NO$_3$-N + NO$_2$-N = 0.2 mg/L, PO$_4$-P = 0.029 mg/L

Land use: in 1996, 0.2% urban, 2.5% agriculture, 11.6% freshwater, 8.3% recently logged, 28.4% old forest, 42.4% new forest, 6.3% other

Population density: <0.1 person/km^2

Major information sources: www.statcan.ca 2002, Environment Canada 2002, Rosenberg et al. 1998, McPhail et al. 1998, Perrin and Richardson 1997

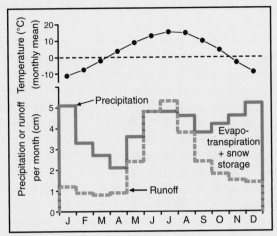

FIGURE 15.12 Mean monthly air temperature, precipitation, and runoff for the Stuart River basin.

STUART RIVER

Relief: 1097 m

Basin area: 14,600 km²

Mean discharge: 128 m³/s

River order: 6

Mean annual precipitation: 48.7 cm

Mean air temperature: 2.8°C

Mean water temperature: 6.8°C

Physiographic province: Coast Mountains of British Columbia and Southeast Alaska (PM)

Biome: Temperate Mountain Forest

Freshwater ecoregion: North Pacific Coastal

Terrestrial ecoregion: Fraser Plateau and Basin Complex

Number of fish species: 23

Number of endangered species: 2 fishes

Major fishes: mountain whitefish, lake whitefish, lake trout, sockeye salmon, rainbow trout, largescale sucker, slimy sculpin, northern pikeminnow, redside shiner, peamouth chub, lake chub

Major other aquatic vertebrates: beaver, river otter, muskrat, moose, merganser

Major benthic invertebrates: mayflies (*Diphetor hageni, Ephemerella, Leucrocuta, Rhithrogena*), stoneflies (*Capnia, Paraleuctra, Sweltsa, Taenionema*), chironomid midges (*Micropsectra, Polypedilum, Rheocricotopus [Rheocricotopus] eminellobus, Tanytarsus, Tvetenia bavarica group*)

Nonnative species: none

Major riparian plants: black cottonwood, Sitka alder, willows

Special features: lake-dominated system with short tributary streams flowing into Takla and Stuart lakes; system supports sockeye salmon populations with longest migration in Fraser River system

Fragmentation: none

Water quality: median of 10 sites, alkalinity = 38.4 mg/L as $CaCO_3$, pH = 7.7, NO_3-N + NO_2-N = 0.003 mg/L, total Kjeldahl nitrogen = 0.091 mg/L, total phosphorus = 0.004 mg/L

Land use: 0.1% urban, 0.4% agriculture, 12.3% freshwater, 8.2% recently logged, 29.8% old forest, 43.6% new forest, 5.6% other

Population density: <0.1 person/km²

Major information sources: www.statcan.ca 2002, Environment Canada 2002, Rosenberg et al. 1998, McPhail et al. 1998

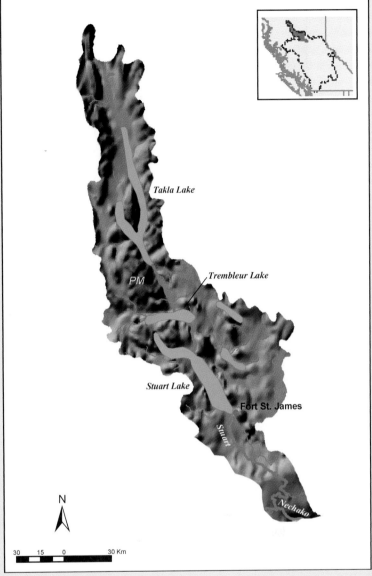

FIGURE 15.11 Map of the Stuart River basin.

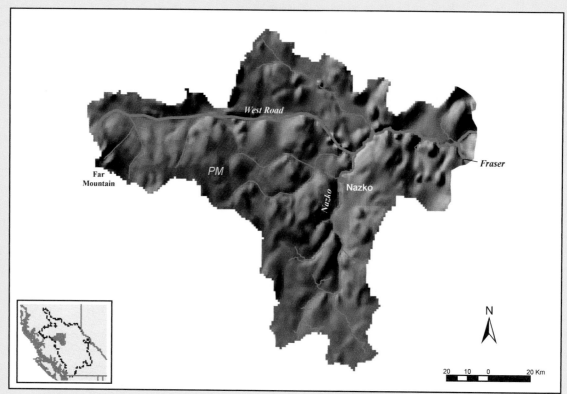

FIGURE 15.13 Map of the West Road River basin.

WEST ROAD RIVER

Relief: 2000 m
Basin area: 12,400 km²
Mean discharge: 38 m³/s
River order: 7
Mean annual precipitation: 64.1 cm
Mean air temperature: 4.7°C
Mean water temperature: NA
Physiographic province: Coast Mountains of British Columbia and Southeast Alaska (PM)
Biome: Temperate Mountain Forest
Freshwater ecoregion: North Pacific Coastal
Terrestrial ecoregion: Fraser Plateau and Basin Complex
Number of fish species: 21
Number of endangered species: none
Major fishes: burbot, mountain whitefish, bull trout, Chinook salmon, sockeye salmon, rainbow trout, largescale sucker, longnose sucker, redside shiner, northern squawfish, peamouth chub

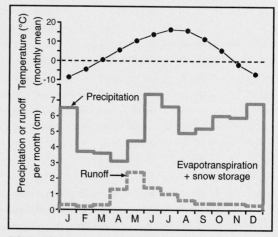

FIGURE 15.14 Mean monthly air temperature, precipitation, and runoff for the West Road River basin.

Major other aquatic vertebrates: great blue heron, bald eagle, river otter, mink
Major benthic invertebrates: mayflies (*Baetis tricaudatus, Paraleptophlebia, Leucrocuta, Serratella*), stoneflies (*Capnia, Sweltsa, Zapada cinctipes*), chironomid midges (*Micropsectra, Polypedilum, Tanytarsus, Tvetenia bavarica group, Zavrelimyia*)
Nonnative species: brook trout
Major riparian plants: willows, black cottonwood, alder, aspen
Special features: one of the finest trout streams in British Columbia; route chosen by Sir Alexander Mackenzie in 1793—now a heritage hiking trail to Bella Coola
Fragmentation: none
Water quality: median of 38 sites, alkalinity = 54.8 mg/L as $CaCO_3$, pH = 7.9, NO_3-N + NO_2-N = 0.003 mg/L, total Kjeldahl nitrogen = 0.269 mg/L, total phosphorus = 0.021 mg/L
Land use: 0.7% agriculture, 9.4% freshwater, 16.9% old forest, 10.8% recently logged, 60.3% young forest, 1.9% other
Population density: <1 person/km²
Major information sources: Environment Canada 2002, Rosenberg et al. 1998, McPhail et al. 1998

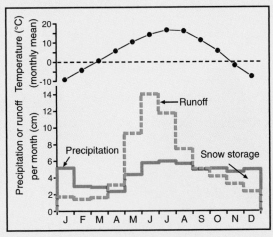

FIGURE 15.16 Mean monthly air temperature, precipitation, and runoff for the Quesnel River basin.

QUESNEL RIVER

Relief: 2500 m
Basin area: 11,500 km²
Mean discharge: 239 m³/s
River order: 6
Mean annual precipitation: 53.9 cm
Mean air temperature: 4.9°C
Mean water temperature: NA
Physiographic province: Coast Mountains of British Columbia and Southeast Alaska (PM)
Biome: Temperate Mountain Forest
Freshwater ecoregion: North Pacific Coastal
Terrestrial ecoregions: North Central Rockies Forests, Fraser Plateau and Basin Complex
Number of fish species: 23
Number of endangered species: none
Major fishes: sockeye salmon, rainbow trout, largescale sucker, longnose sucker, redside shiner, northern squawfish, peamouth chub, lake chub
Major other aquatic vertebrates: moose, beaver, merganser
Major benthic invertebrates: NA
Nonnative species: none
Major riparian plants: white spruce, lodgepole pine, trembling aspen, Douglas fir
Special features: remarkable comeback of Quesnel River sockeye; run of 12.2 million fish in 1993, with 2.5 million fish making it to Quesnel headwaters to spawn; once again the Fraser's greatest sockeye producer, surpassing world-famous Adams River for sockeye production
Fragmentation: none
Water quality: alkalinity = 53.8 mg/L as $CaCO_3$, pH = 7.9, orthophosphate = 0.003 mg/L, total phosphate = 0.3 mg/L
Land use: 0.03% urban, 1.5% agriculture, 9.8% alpine, 6.8% freshwater, 11.2% recently logged, 29.1% old forest, 34.2% young forest, 7.4% other
Population density: 2.2 people/km²
Major information sources: Environment Canada 2002, Rosenberg et al. 1999, McPhail et al. 1998

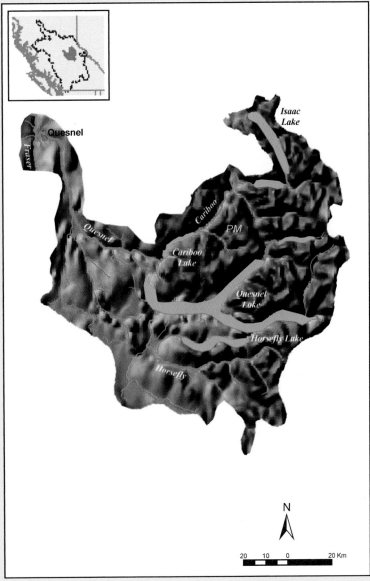

FIGURE 15.15 Map of the Quesnel River basin.

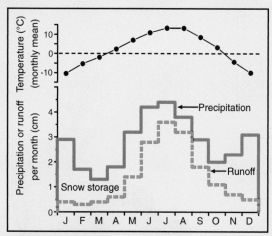

FIGURE 15.18 Mean monthly air temperature, precipitation, and runoff for the Chilcotin River basin.

CHILCOTIN RIVER

Relief: 3250 m

Basin area: 19,300 km²

Mean discharge: 102 m³/s

River order: 6

Mean annual precipitation: 33.5 cm

Mean air temperature: 2.2°C

Mean water temperature: NA

Physiographic province: Coast Mountains of British Columbia and Southeast Alaska (PM)

Biome: Temperate Mountain Forest

Freshwater ecoregion: North Pacific Coastal

Terrestrial ecoregion: Fraser Plateau and Basin Complex

Number of fish species: 16

Number of endangered species: none

Major fishes: mountain whitefish, rainbow trout, sockeye salmon, bull trout, longnose sucker, redside shiner, largescale sucker, northern squawfish, peamouth chub, lake chub

Major other aquatic vertebrates: beaver, moose, merganser

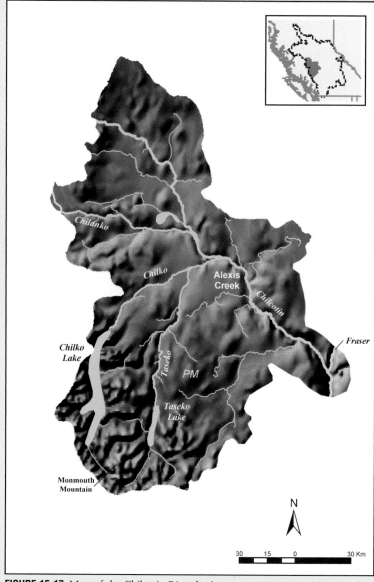

FIGURE 15.17 Map of the Chilcotin River basin.

Major benthic invertebrates: mayflies (*Baetis tricaudatus, Epeorus deceptivus, Paraleptophlebia, Rhithrogena*), stoneflies (*Capnia, Sweltsa, Taenionema, Zapada cinctipes, Zapada columbiana*), chironomid midges (*Brillia retifinis, Eukiefferiella brehmi group, Micropsectra, Orthocladius [Eudactylocladius]*)

Nonnative species: brook trout

Major riparian plants: white spruce, lodgepole pine, trembling aspen, Douglas fir, Engelmann spruce, alpine fir, bunchgrass

Special features: among best and most challenging river in North America for kayaking and white-water rafting; spectacular scenery, such as Farwell Canyon, where river cuts deeply into sandstone cliffs with native pictographs on overhang

Fragmentation: none

Water quality: alkalinity = 51 mg/L as CaCO₃, pH = 7.7, NO₃-N + NO₂-N = 0.002 mg/L, total Kjeldahl nitrogen = 0.060 mg/L, total phosphate = 0.042 mg/L

Land use: 0.05% urban, 0.53% agriculture, 0.9% alpine, 7.9% water, 7.1% recently logged, 18.3% old forest, 49.7% new forest, 1.8% rangeland, 13.7% other

Population density: <1 person/km²

Major information sources: Environment Canada 2002, Rosenberg et al. 1998, McPhail et al. 1998

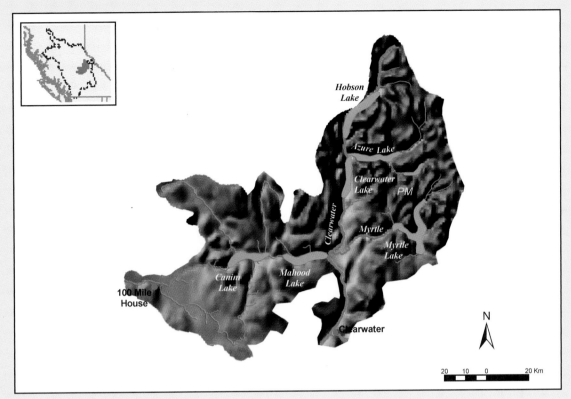

FIGURE 15.19 Map of the Clearwater River basin.

CLEARWATER RIVER

Relief: 2750 m
Basin area: 10,200 km²
Mean discharge: 223 m³/s
River order: 5
Mean annual precipitation: 45.1 cm
Mean air temperature: 6.2°C
Mean water temperature: NA
Physiographic province: Coast Mountains of British Columbia and Southeast Alaska (PM)
Biome: Temperate Mountain Forest
Freshwater ecoregion: North Pacific Coastal
Terrestrial ecoregion: North Central Rockies Forests
Number of fish species: 16
Number of endangered species: none
Major fishes: mountain whitefish, chinook salmon, sockeye salmon, coho salmon, rainbow trout, largescale sucker, longnose sucker
Major other aquatic vertebrates: beaver, moose, merganser

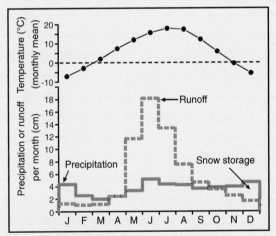

FIGURE 15.20 Mean monthly air temperature, precipitation, and runoff for the Clearwater River basin.

Major benthic invertebrates: mayflies (*Baetis tricaudatus*, *Drunella doddsi*, *Ephemerella*, *Paraleptophlebia*, *Rhithrogena*), stoneflies (*Sweltsa*, *Zapada*), chironomid midges (*Micropsectra*, *Tvetenia bavarica group*), true flies (*Chelifera*)
Nonnative species: none
Major riparian plants: white spruce, lodgepole pine, trembling aspen, Douglas fir, Engelmann spruce, alpine fir
Special features: headwaters rise in Wells Gray Provincial Park; six major lakes; numerous spectacular waterfalls; incredible mountain scenery and hiking trails; Helmcken Falls plunges 141 m down narrow canyon
Fragmentation: none
Water quality: alkalinity = 37.1 mg/L as $CaCO_3$, pH = 7.8, NO_3-N + NO_2-N = 0.017 mg/L, total Kjeldahl nitrogen = 0.081 mg/L, total phosphorus 0.012 mg/L
Land use: 0.3% urban, 0.7 % agriculture, 5.6% recently logged, 32.7% old forest, 35.4% young forest, 9.7% alpine, 7.5% water, 8.1% other
Population density: <1 person/km²
Major information sources: Environment Canada 2002, Rosenberg et al. 1998, McPhail et al. 1998

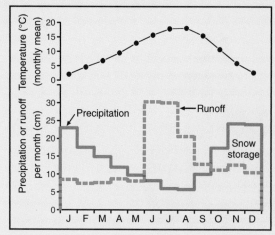

FIGURE 15.22 Mean monthly air temperature, precipitation, and runoff for the Lillooet and Harrison River basin.

LILLOOET AND HARRISON RIVERS

Relief: 2750 m
Basin area: 7870 km^2
Mean discharge: 445 m^3/s
River order: 5
Mean annual precipitation: 172.6 cm
Mean air temperature: 10.2°C
Mean water temperature: NA
Physiographic province: Coast Mountains of British Columbia and Southeast Alaska (PM)
Biome: Temperate Mountain Forest
Freshwater ecoregion: North Pacific Coastal
Terrestrial ecoregions: Cascade Mountains Leeward Forests, British Columbia Mainland Coastal Forests
Number of fish species: 20
Number of endangered species: none
Major fishes: chinook salmon, sockeye salmon, coho salmon, chum salmon, pink salmon, rainbow trout, cutthroat trout, largescale sucker, redside shiner
Major other aquatic vertebrates: beaver, moose, merganser
Major benthic invertebrates: mayflies (*Drunella doddsi*, *Rhithrogena*), stoneflies (*Megarcys*, *Taenionema*), caddisflies (*Oligophlebodes*)
Nonnative species: brown bullhead
Major riparian plants: white spruce, lodgepole pine, trembling aspen, Douglas fir, Engelmann spruce, alpine fir
Special features: headwater Lillooet River runs for almost 200 km of white-water paddling with spectacular autumn sockeye salmon runs; Lillooet empties into Harrison Lake, the largest lake in southwestern British Columbia
Fragmentation: none
Water quality: alkalinity = 19.0 mg/L as CaCO$_3$, pH = 7.1, NO$_3$-N + NO$_2$-N = 0.017 mg/L, total Kjeldahl nitrogen = 0.031 mg/L, total phosphorus = 0.015 mg/L
Land use: 0.15% urban, 0.6 % agriculture, 25.3% alpine, 4.5% freshwater, 6.1% recently logged, 31.3% old forest, 14.7% young forest, 12.3% other
Population density: <1 person/km^2
Major information sources: www.statcan.ca 2002, Environment Canada 2002, Rosenberg et al. 1998, McPhail et al. 1998

FIGURE 15.21 Map of the Lillooet and Harrison River basin.

16

PACIFIC COAST RIVERS OF CANADA AND ALASKA

JOHN S. RICHARDSON   ALEXANDER M. MILNER

INTRODUCTION

KUSKOKWIM RIVER

SUSITNA RIVER

KENAI RIVER

STIKINE RIVER

SKEENA RIVER

ADDITIONAL RIVERS

ACKNOWLEDGMENTS

LITERATURE CITED

INTRODUCTION

The Pacific coast rivers of Canada and Alaska (Fig. 16.2) cover the region extending from British Columbia's north coast (north of the Fraser River, Chapter 15) to north of the Aleutian Islands (south of the Yukon River, Chapter 17), and have a latitudinal range from 54°N to 64°N. The southernmost rivers in this region have their catchments entirely within British Columbia, and the northernmost rivers have their basins within Alaska. In 1867 the United States purchased Alaska from Russia, creating a situation whereby the northern half of British Columbia is isolated from the ocean by the Alaskan panhandle. Thus several rivers of this region have a large portion of their catchment within Canada before flowing through Alaska to the Pacific Ocean; these rivers are known as transboundary rivers.

Most of the coastal rivers of the northeastern Pacific and southern Bering Sea are characterized by catchments originating at high elevations within a series of mountain ranges. These are the classic wild salmon rivers of the West Coast and include some of the most pristine drainages in the north temperate region (Dynesius and Nilsson 1994). Each one of the

basins described here has its upper drainages in mountainous terrain, some originating from glaciers, and discharge to the ocean. Several have their highest points well above 3000 m asl within the Alaska Range (including Mount McKinley), the Wrangell Mountains, the St. Elias Mountains, and the Coast Mountains. All the basins have considerable topographic relief from highest source to mouth.

Native peoples have a close association with most of these large rivers, which provided important food resources, notably salmon. There are different views of when people arrived in western North America, but it seems most likely that human settlement dates from at least the end of the Wisconsinan glacial period (approximately 11,000 years ago) when the Bering land bridge between Asia and Alaska (Beringia) still existed (McGhee 1996). Rich aquatic resources and abundant wildlife supported subsistence living in these coastal regions of North America. Large winter villages and a dependence on salmon appear to be among the factors leading to the rich set of traditions and culture associated with coastal peoples (Muckle 1998). There are many native groups in the region, largely differentiated on the basis of language, but there are differences of opinion, even among native

◄ FIGURE 16.1 Susitna River, Alaska, showing different channels (PHOTO BY A. MILNER).

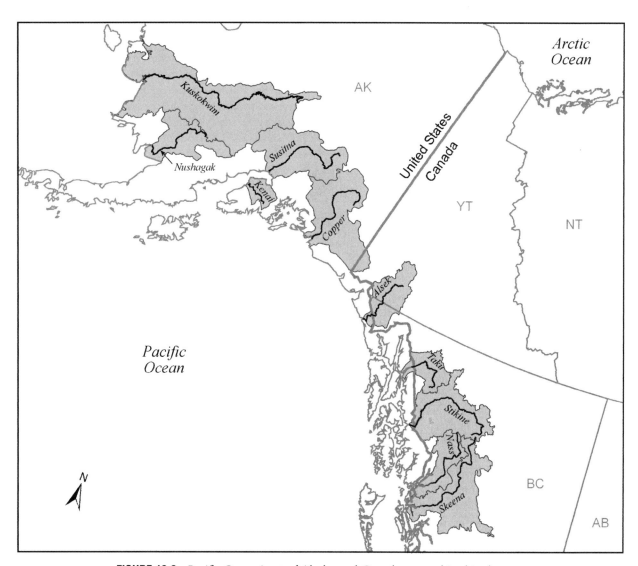

FIGURE 16.2   Pacific Coast rivers of Alaska and Canada covered in this chapter.

peoples, as to how the different bands or tribes should be grouped (Muckle 1998). Some of the native groups associated with the area covered in this chapter include the southern Eskimos, the Yuit of the Yukon–Kuskokwim delta and southwest Alaska (McGhee 1996), and the Athabascan, Haida, Tsimshian, Tlingit, Tahltan, and Nisga'a Indians, to name a few. Some "nations" have traditional areas defined in part by catchment boundaries, which partly reflects their association with the rivers. Europeans reached coastal areas of British Columbia and Alaska during the late 1700s and trading posts and fishing ports were established during the early 1800s. It was not until the gold rushes (especially the Klondike gold rush) of the mid-1800s that large numbers of European settlers moved inland. Today

the populations of much of this region (villages or smaller communities) are still well represented by aboriginal peoples with many traditional occupations, such as trapping and fishing.

We will discuss 10 major rivers from this region, several of which are among the 30 largest rivers in North America (Leopold 1994; see Fig. 16.2). The five focus rivers to be covered were selected to illustrate the diversity of drainages, although many similarities among rivers in this sparsely populated and relatively "wild" part of North America exist. The focus rivers (with their world rank in terms of average annual discharge) are Kuskokwim (40th), Susitna (60th), Kenai, Stikine (48th), and Skeena (46th). Although the Kenai River is not as large as some rivers in the region, it is very important economically, supporting significant commercial and sports fisheries. Abbreviated physical and biological descriptions of five additional rivers are also provided: Nushagak (57th), Copper (45th), Alsek, Taku, and Nass (69th).

Physiography and Climate

The region includes parts of three physiographic provinces, Yukon Basin, South Central Alaska, and Coast Mountains of British Columbia and Alaska. The latter two are provinces within the Pacific Mountain System physiographic division. These coastal areas of British Columbia and Alaska are tectonically active where the convergence of the Pacific plate and the North American plate results in extensive deformation of the crust and marked topographic relief. A series of mountain ranges run approximately parallel to the margin of the intersection of these two main plates. The Pacific Mountain System, which incorporates the Alaska Range, the Chugach Mountains, the Wrangell Mountains, the St. Elias Mountains, and the Coast Mountains extends south to Baja California in Mexico. Some of the largest mountains in North America (some >4000 m asl, including Mount McKinley, Mount Logan, and Mount St. Elias) occur in these mountain ranges. These ranges heavily influence the weather and other attributes of coastal catchments. Other legacies of the tectonic activity are the large number of earthquakes and the resulting changes to some river drainages. There are several geothermal areas where hot springs are common, particularly within the Stikine River basin (Farley 1979). There are also many ancient and recent volcanoes along the coastal mountain ranges throughout the region (recent volcanoes principally in the Aleutian Range apart from Mount Wrangell and Mount Edgecumbe near Sitka).

From the Rocky Mountains westward are land areas, or terranes, that have been added to the original North American plate during or since the Phanerozoic era. Large portions of the region are underlain by rocks of volcanic origin, intermixed with igneous intrusives and some sedimentary formations (Gabrielse and Yorath 1991). The coastal mountains are formed by an extensive intrusive granitic and crystalline gneiss batholith from late Cretaceous to early Tertiary age (Farley 1979), which is highly resistant to weathering.

Most of the region included in this chapter has been subject to glaciation over the past million years and was covered by ice during the Wisconsinan glacial period (75,000 to 10,000 years ago), with the exception of the "Beringian refuge" and some small coastal areas (Hunt 1974). The Beringian refuge includes much of the Yukon and Kuskokwim lowlands, which receive very low amounts of precipitation in contrast to coastal regions and higher elevations. Recolonization of previously glaciated regions by freshwater species following ice retreat was predominantly from the Beringian and Pacific refuges, and, for a few species, the Mississippi refuge (McPhail and Lindsay 1970, Oswood et al. 2000).

The mountains show signs of glacial scouring, with U-shaped valleys and basal till forming a thin layer over bedrock in many parts of the region. The thin ground layer means that there is relatively low groundwater storage in many areas compared to unglaciated regions. Most of these systems also respond rapidly to precipitation events because of their low storage capacity. This till contributes to high bedload transport and is actively reworked to create alluvial structures in river sections with lower gradients. Broad, braided reaches of these rivers are common where the gradient permits. There are no extensive limestone deposits, and thus no cave systems in this region.

The soils of the area are predominantly podzols. There is a broad band of ferro-humic podzols (in the Canadian series, and the equivalent United States classes are humic cryorthod and humic haplorthod) along the coastal regions of about 150 km width (Farley 1979). Further inland there is a wide expanse of humo-ferric podzols (cryorthod, haplorthod). Some parts of the Skeena River basin originate in a wide, interior, postglacial plateau characterized by gray luvisols (boralf). In the more northerly basins, cryosols (permafrost areas) and brunisols are more extensive.

Given the northern location and the extensive area of high mountains in this region, a large portion of the annual precipitation falls as snow. The wide altitudinal relief in these basins and sparse distribution of monitoring stations (mostly in valley bottoms) results in severe underestimation of catchmentwide precipitation. Climate varies throughout these large basins and simple measures cannot adequately characterize entire catchments. All of the catchments included in this chapter have some glacial influence, with the Nass and Skeena rivers to the south having the lowest contributions. The dominant weather systems along these coastal areas are Pacific westerlies that bring extensive moisture to the coast, as well as moderating temperatures, creating a relatively mild maritime climate. Precipitation is highest along the coast, and especially in mountainous areas, but diminishes rapidly inland. This is a very cold region, with mean monthly temperatures in January for the Copper and Alsek basins as low as –21°C and July mean temperatures for any of the basins not exceeding 16°C. Temperature patterns inland of the Coast Mountains and the other mountain ranges are more continental, with wider seasonal fluctuations in temperature.

Basin Landscape and Land Use

The region includes 14 terrestrial ecoregions (see river descriptions), including coastal, temperate rainforests, interior coniferous forests, taiga, tundra, and alpine tundra (Ricketts et al. 1999). Most of the dominant forest trees are conifers. The Northern Pacific Coastal Forests ecoregion encompasses a band reaching a maximum width of 200 km that borders coastal regions. The dominant coastal forest tree species include western red cedar, Sitka spruce, western hemlock, Douglas fir, and Pacific silver fir, and further north, yellow cedar (also known as Alaska cypress) and mountain hemlock. Other species found in these forests include red alder, other alder species, vine maple, and western yew. These coastal temperate forests of British Columbia and southeastern Alaska are considered a globally outstanding habitat (Ricketts et al. 1999). The treeline in Alaska is about 800 to 1000 m asl and is higher further south. Interior forests of northern British Columbia are boreal (Northern Cordillera Forests ecoregion) and dominated by white spruce, lodgepole pine, black spruce, subalpine fir, and quaking aspen. In northern British Columbia, the southern part of the Yukon Territory, and Alaska the coastal forest grades into the boreal forest or taiga (white spruce, black spruce, quaking aspen, balsam poplar, Sitka spruce, cottonwood, and paper birch), then into tundra. The tundra ecoregions (Pacific Coastal Mountain Tundra and Ice Fields, and Alaska/St. Elias Range Tundra) are characterized by dwarf shrubs, including species of the family Ericaceae, birch, willows, and alders. Additional ground cover in the tundra is provided by other ericaceous species (e.g., mountain heath) and a range of herbaceous species, such as mountain aven. Additional forest types occur in the more southerly basins, as described in this chapter.

Historically, land use in the region was subsistence hunting and fishing by indigenous peoples. Current land use throughout the region continues to predominantly involve resource extraction, namely timber harvest, mining for minerals, and fisheries. Many areas are sufficiently remote from markets,

mills, and other infrastructure, or timber is of lower quality, such that forestry is still a minor activity and there are extensive tracts of original forest. Parts of the region have fewer than 60 frost-free days per year (Farley 1979), and thus permafrost dominates here. Agriculture is thus not viable in most areas, except in the Matanuska Valley in south-central Alaska and in the southernmost part of this region in British Columbia. Trapping for furs remains an important activity for native and nonnative inhabitants.

Most of the region is sparsely populated, with the major population centers on the coast. Most of these towns and cities (e.g., Anchorage and Prince Rupert) occur at the ocean mouths of the large rivers. Most other towns and villages are small (<1000 inhabitants), often associated with forestry activities or native settlements. Even so, these villages are tied to the rivers for their fisheries value and other resources.

Sport fishing and hunting are two important economic activities in these catchments. Tourism is also a major force that continues to increase and is now the second most important activity for the Alaska economy after oil. One of the commercially dominant tourism activities is the cruise ship industry, taking passengers from Vancouver to southeast Alaska, including Juneau and Glacier Bay National Park and Preserve. Rafting, kayaking, and wildlife viewing (e.g., grizzly bear, Kermode bear [variety of black bear], and other species) are also popular activities in the area. The Alsek, Tatshenshini, and Taku rivers are popular for float trips with rafts.

The region includes several large parks and protected areas. These parks are largely wilderness areas, further evidence of the region's low level of development. Among those parks is an extensive, contiguous U.S.–Canada protected area of 13,926 km², which includes Wrangell–St. Elias National Park and Glacier Bay National Park in Alaska, Kluane National Park in Yukon Territories, and Tatshenshini–Alsek Park in British Columbia (the Kluane and Tatshenshini parks protect most of the Alsek River basin). Much of this area was designated a World Heritage Site by the United Nations in 1979. Other protected areas in the region include Denali National Park (Alaska), Spatsizi Wilderness Park (British Columbia), and the Tongass National Forest (Alaska), although some forest harvest is allowed therein.

The Rivers

Two freshwater ecoregions are represented by the rivers covered in this chapter: the North Pacific

Coastal and the Yukon (Abell et al. 2000). The Kuskokwim and Nushagak Rivers are part of the Yukon ecoregion, and the others belong to the North Pacific Coastal ecoregion. These rivers share similar properties: snowmelt- and/or glaciermelt-dominated hydrology (large storage of precipitation as ice or snow), influence of past glaciation on landforms, low conductivity, and at least some portion of each basin draining steep, mountainous areas. Although runoff is largely based on snowmelt, glaciermelt, or both, there is also a rainfall-driven signal in the hydrology as a consequence of the mild westerly flow of moist air across the Pacific falling as rain in the low-lying portions of the basins and especially coastal areas. Few meteorological stations exist on these rivers, and the ones that do are often associated with population centers in valley-bottom sites, whereas the majority of the basin areas are at elevation. Hence, precipitation records do not accurately represent inputs to each basin. Two factors that underestimate inputs are the well-known effects of orographic influence, with rainfall increasing from 50% to 400% across the elevations of 1000s of m asl in the region, and the predominance of snow as input, which is difficult to accurately measure (e.g., Jones 1997, Dingman 2002). The set of rivers here represent a latitudinal gradient in terms of forest types, from coastal temperate rainforests to tundra. That latitudinal gradient is also inversely correlated with gradients in human use.

A unique aspect of the glacial influence on some of these rivers is the periodic occurrence of outburst floods, also known by their Icelandic name, *Jökulhlaup* events. Outburst floods take place when lake discharge is restricted by glacial activity, which eventually releases suddenly through the glacier during summer. This typically occurs as a result of the ice barrier becoming raised or water tunneling through the ice as it warms. An outburst flood can raise instantaneous discharges twofold to fivefold above the already swollen late-summer flood stage (e.g., Brabets 1997). Examples include Miles Lake on the Copper River, Lake Nolake and Tulsequah Lake in the Taku River drainage (the latter being close to a proposed mine site), and the Snow River basin within the Kenai River system, which is discussed in this chapter. These outburst floods can take place in the course of a few days and the large flows play a major role in reshaping channel geomorphology, as well as disturbing floodplain vegetation and potentially damaging bridges, roads, and other structures.

All rivers sustain important spawning and rearing areas for Pacific salmon (sockeye, chinook, coho, chum, and pink) and steelhead. Rainbow trout (the

nonanadromous form of steelhead), cutthroat trout, and Dolly Varden (char) also have many resident populations that remain entirely in freshwater. Studies indicate that less than 5% of salmon stocks in southeastern Alaska are in decline, with <1% rated at moderate to high risk of extinction (Baker et al. 1996). The species in British Columbia waters are more threatened (mostly toward the south), with 1.5% of stocks listed as extinct and 7.3% of stocks classed at moderate to high risk of extinction (Slaney et al. 1996). Abundance data are not available for many salmon stocks (57% of stocks in Alaska, 43% in British Columbia), so complete assessments are not possible.

The Canadian and U.S. governments frequently debate the allocation of salmon fisheries along the southeast Alaska coast. Salmon stocks primarily spawn and rear in Canadian waters, but juvenile fishes spend time in river habitats and estuaries in Alaskan waters before migrating to sea for one to four years, depending upon species (Murphy et al. 1997). Debates persist over salmon entitlements, particularly the commercial catch in Alaskan waters of salmon migrating to natal areas in Canada. Currently the Pacific Salmon Treaty determines allocations between the two nations.

A freshwater fish fauna of 44 species in the Pacific drainages of Canada and Alaska is depauperate compared to other regions due to the legacy of glaciation (Scott and Crossman 1973). As noted earlier, all rivers covered in this chapter have large, self-sustaining populations of the seven Pacific salmon and trout species (*Oncorhynchus* spp.). Of the 44 freshwater species, 18 are common to the river drainages discussed in this chapter, or at least should be within their range. It is also worth noting that 19 of the species are salmonids (including salmon, trout, char, and whitefish).

Few aquatic species in this region of North America are considered at risk. Crayfishes are widely distributed throughout much of North America, but the coast of British Columbia supports only a single species, the signal crayfish. However, its range may not extend as far north as the Skeena River, and there are no crayfish in Alaska (Taylor et al. 1996). There are 25 species of mollusks listed for the region (Clarke 1981), many of which have widespread distributions across northern North America. Some species are found only in the northern Pacific region, such as the Alaskan pond snail, the muskeg stagnicola, the subarctic lake stagnicola, the giant northern pea clam, and the Arctic-alpine pea clam; however, the ranges of these species are not well described. One species of mussel, the Yukon floater, is endemic to the Yukon basin, including the Kuskokwim River (Clarke 1981).

A corollary of the predominance of Pacific salmon in the trophic ecology of most of these rivers and their tributaries and lake systems is the contribution of marine-derived nutrients to these mostly oligo-trophic systems following salmon spawning (e.g., Helfield and Naiman 2001, Naiman et al. 2002). Some of these systems have broad and shallow alluvial channels, with riparian vegetation set back from the channels, and thus a predominance of algal-based food webs, excepting those with high degrees of glacially influenced turbidity (Oswood et al. 2000). Tributary systems, especially toward the southern portion of the region, have a greater dependence upon allochthonous sources of energy.

The sources of water impose various characteristics on these systems. For instance, water that varies in terms of organic staining, glacially derived turbidity, or clear water, has different thermal properties. Backscattering from suspended particles was shown by Edmundson and Mazumder (2002) to result in water temperatures of 5.9°C in turbid lakes compared to 7.2°C to 7.4°C in stained or clear lakes. This kind of variation in turbidity and staining likely has impacts on river thermal properties and other characteristics, but there has been little detailed study of variation caused by these sources.

Populations of most wildlife species are not threatened in this region. Some of the species commonly associated with the majority of rivers in this region include those that depend on fishes for parts of their diet, including grizzly bear, black bear, American mink, and river otter. Many other species are common throughout much of the region and associated in some ways with the rivers, including moose, beaver (except southwest Alaska), and muskrat. Among river-associated birds, merganser, belted kingfisher, and American dipper are species common to the rivers of the region (Oswood et al. 1995).

The rivers in this chapter are unaffected by fragmentation or flow regulation (Dynesius and Nilsson 1994). Minor amounts of water extraction occur for domestic use, and some small reservoirs exist on tributaries to a couple of these rivers. Rivers of this region represent one of the few remaining places in the world to contain the possibility of an "international preservation network of representative, un-regulated and unfragmented" large-river systems (Dynesius and Nilsson 1994). Only three rivers in the northern hemisphere with discharge >350 m³/s are unregulated and drain areas other than boreal forest and taiga (i.e., the Kamchatka River of Russia and

the Stikine and Skeena rivers of British Columbia; Dynesius and Nilsson 1994), the latter two of which are discussed in this chapter.

Major current and future threats to rivers within this region are impoundments to generate hydroelectric power (especially on the Stikine River), contamination from mining effluents and sediments, timber harvest, and introduced (nonnative) species. Less than 0.1% of the length of Alaska rivers are listed as National Wild and Scenic, but at 5165 km, Alaska contains nearly one-third of the protected rivers in the United States (Karr et al. 2000). Many of these rivers have a large potential for generating electricity, and there is little protection from that fate in the future for most rivers. The rate of timber harvesting is increasing throughout the region, although some of the forests are not currently viable because of distance from mills and markets. Atlantic salmon escaped from sea-pen farming operations have been found as far north as the Bering Sea (Brodeur and Busby 1998), but potential impacts of these escaped fish on native, anadromous salmonids remain unclear. To date these threats remain relatively minor pressures on the rivers covered in this chapter.

Global climate change is predicted to significantly influence rivers within the region. Current predictions are that warmer temperatures will cause earlier snowmelt and higher peak flows from more rapid melting. Flows will be potentially lower in late summer and autumn (McCarthy et al. 2001). In rivers dominated by glacial flow, climate changes will likely cause increased glacial melt and increase the peak summer flow. Colder temperatures are likely to occur further downstream from the glacier margin (McGregor et al. 1995, Milner et al. 2001). These effects may put stress on riverine fish, as well as cause changes to river channels, estuaries, and coastal wetlands from changes in hydrographs. Climate warming could also alter turbidity and concentrations of dissolved organic carbon, which would alter stream food webs further (Milner et al. 2000, Edmundson and Mazumder 2002). Oswood et al. (1995) reported that dissolved organic carbon levels have increased and will continue to increase in permafrost-rich catchments in Alaska.

Much less is known about the biodiversity and ecology of most of these rivers than rivers in more populated and anthropogenically influenced basins of North America. Salmon production in all of these river systems supports a large industry, and thus comprehensive data on Pacific salmon numbers have been recorded. Knowledge of many aspects of the biology of some basins is sparse.

KUSKOKWIM RIVER

The Kuskokwim is a large subarctic river that drains mostly wilderness, including the north side of the Alaska Range, and then flows to the southwest into Bristol Bay in the Bering Sea (Fig. 16.9). On the basis of average annual discharge, the Kuskokwim River is ranked fortieth largest in the world (Leopold 1994) and the ninth in the United States. The Kuskokwim is the second-largest and second-longest river in Alaska (at 1130 km) after the nearby Yukon River (see Chapter 17), the latter of which has three times the average annual flow of the Kuskokwim. Large portions of the Yukon basin, including much of the Kuskokwim basin, were not glaciated during the most recent ice age, leaving an area known as Beringia, which acted as a refugium for terrestrial and aquatic organisms that later recolonized those areas that were glaciated (Oswood et al. 2000).

The area has been occupied for thousands of years by the Yuit or southern Eskimos (speakers of the Yupik language) near the coast and the Athabaskan Indians (or Athapaskan) of the Ingalik and Upper Kuskokwim tribes in the interior of the basin. The Eskimos used distinct winter villages and summer camps as appropriate to the resources available. The Athabaskan Indians are believed to have been nomadic. The area is thought to have been first visited by Europeans as part of Captain Cook's third voyage in 1778. Russian traders soon followed at the end of the eighteenth century, and trading posts were established in the early nineteenth century. Later growth of towns was a result of missionary settlements, first Russian Orthodox, then others. The territory was sold to the United States in 1867. The 1898 Yukon gold rush brought major changes associated with the large numbers of people seeking their fortune in gold in the Yukon and other large Alaskan rivers.

Physiography, Climate, and Land Use

The basin has a variety of landforms, with tributaries originating from the northern flanks of the Alaska Range at elevations exceeding 3000 m asl. It also drains the Kuskokwim Mountains to the northwest. The Kuskokwim basin is made up of the South Central Alaska (SC) and Yukon Basin (YB) physiographic provinces (see Fig. 16.9).

The Kuskokwim basin includes four terrestrial ecoregions (Ricketts et al. 1999). The Beringian Lowland Tundra and the Beringian Upland Tundra ecoregions are found along the coast near Bristol Bay, much of which is in the Kuskokwim delta. The delta is a large wetland complex formed by low topographic relief and permafrost that restricts drainage and results in extensive areas of thaw lakes (Oswood et al. 1995). Vegetation includes sedges and shrubs, as well as alder and willow trees. Much of the upper basin is within the Interior Alaska/Yukon Lowland Taiga and the Alaska/St. Elias Range Tundra terrestrial ecoregions. Vegetation includes open taiga, mountain tundra, and even areas of moderately productive closed taiga forest. The predominant tree species include black spruce, white spruce, aspen, larch, birch, and poplars (Oswood et al. 1995), giving way to alpine tundra shrubs, rocky slopes, and glaciers at higher elevations (Ricketts et al. 1999).

Climate in the basin is described from only two weather stations. At Bethel, at the head of the estuary, monthly average temperatures reach 12.9°C in July and −14.5°C in January (Fig. 16.10). The average annual temperature at Bethel, probably the warmest point in the basin, is −1.6°C. About 63% of precipitation falls from June through October, and August is the wettest month, although this may underestimate inputs from snow, which are difficult to accurately measure (see Fig. 16.10).

At least 99% of the basin is undeveloped tundra, taiga forest, and wetlands. Most of the land use in the region is subsistence fishing, hunting, and trapping. There is a minor amount of forestry in the upper part of the basin. Mining for gold (mostly placer mining) and mercury occurred in the eastern part of the river basin, although most of the mercury mining has ceased. The Kuskokwim River basin is populated by about 14,000 people distributed among 56 towns and villages. The largest center is Bethel, with a population exceeding 5200, with smaller settlements at McGrath, Crooked Creek, Aniak, and Akiak.

River Geomorphology, Hydrology, and Chemistry

The Kuskokwim flows approximately 1300 km from its source in the interior of Alaska to its mouth at Bristol Bay. Many portions of the river are wide, braided, cobble-bottom channels with low gradients. The gradient of the river up to McGrath, about 455 km upstream from Bethel, is only about 20 cm/km. Many sections flow through thick alluvial terraces. The source of this river and many of its tributaries is glaciers at high elevation on the north side of the Alaska Range. The lower portion up to Bethel is tidal or tidally influenced.

The main tributaries include North Fork, East Fork, and South Fork upstream of McGrath, as well as Holitna River, Stony River, Swift River, Kisaralik River, and Kwethluk River (all upstream of Bethel). The most downstream water-sampling station from which the most water-quality and discharge data are available is the confluence with Crooked Creek, some 350 km from the mouth (Wang 1999). Mean width at the confluence with Crooked Creek is 340 m.

The average annual discharge at Crooked Creek is 1145 m³/s. The discharge is relatively low for the size of the basin as a result of most of the basin receiving relatively low amounts of precipitation because it is in the rain shadow of the Alaska Range. Other tributaries contribute to the discharge of the river to Bristol Bay, so the river discharge is much higher than that at Crooked Creek; Dynesius and Nilsson (1994) have estimated annual discharge to be 1900 m³/s. Winter discharge is very low, as precipitation falls mostly as snow and is locked up in snowpack and glaciers throughout the basin (see Fig. 16.10). Peaks in discharge occur from snowmelt and glaciermelt and can occur from May through August, depending upon year. Peak flows of 7362 to 11,100 m³/s have been recorded in some years (record period 1952 to 2000, USGS data, Crooked Creek station [http://waterdata.usgs.gov/nwis/sw]). A very minor amount of water is withdrawn from the river for power generation and domestic consumption.

Several aspects of water quality vary throughout the Kuskokwim River basin depending upon elevation, degree of glacial influence, and underlying parent materials. The mean annual water temperature was about 5°C (based on sparse data records from 1951 to 1999 by the USGS [http://waterdata.usgs.gov/nwis/qw]). The temperature usually remains at 0°C from December through March, when most of the river is covered by ice. The river warms quickly in May and June to reach a summer maximum of up to 20°C in July (range of summer maxima from 13.6°C to 20.6°C at the Crooked Creek station). Turbidity shows a strong seasonal pattern as glacier water melts in summer. Turbidity increases from 2 NTU (monthly average) during the cold part of the year (November through April) to a peak of 124 NTU in July (long-term July average; Fig. 16.3). Suspended solids show a similar pattern, averaging 112 mg/L. Average conductivity is 175 μS/cm (maximum 317), and hardness is 86.9 mg/L as CaCO₃. Some tributaries of the Kuskokwim River drain from a naturally elevated source of mercury (Wang 1999). Nitrate-N is 210 μg/L and PO₄-P is 123 μg/L.

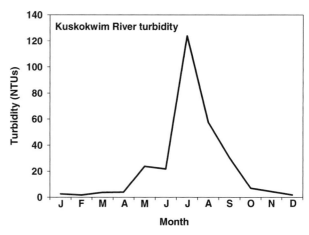

FIGURE 16.3   Annual variation in turbidity (NTUs) of the Kuskokwim River based on data from the USGS for the Crooked Creek monitoring site. This seasonal pattern of high turbidity in summer is characteristic of glacier-influenced discharge, a feature common to a large proportion of streams in this region.

River Biodiversity and Ecology

The Kuskokwim River is within the Yukon freshwater ecoregion, which includes rivers draining into the Bering and Beaufort seas (Abell et al. 2000). Besides the Kuskokwim and Yukon rivers, these are the Kuzitrin, Colville, Sagavanirktok, Kuparuk, and Nushagak, among others of northern Alaska. Unfortunately, the ecology of the Kuskokwim River has been little studied, aside from salmon escapement estimates. The larger Yukon ecoregion is considered continentally outstanding in biological distinctiveness, but much remains to be learned of this region. It includes at least two species endemic to the Beringian refuge, the Alaskan blackfish (McPhail and Lindsay 1970) and a freshwater mussel, the Yukon floater (Clarke 1981). Other species endemic to the Beringian refuge (Squanga whitefish, Bering cisco, and Angayukaksurak char) may also be found in the Kuskokwim (Abell et al. 2000), although we have found no mention of their occurrence there.

Algae and Cyanobacteria

Little is known about algae and cyanobacteria in the Kuskokwim, but there have been some studies in small tributaries of the nearby Nushagak basin to the south, which are similar to many of those in the Kuskokwim (Peterson and Foote 2000). Summer algal biomass ranged from negligible to >10 mg/m² and showed generally higher levels following salmon spawning.

Plants

Riparian vegetation includes alders and willows. Most of the riparian vegetation is set back on river terraces away from the active channel.

Invertebrates

As with algae, little is known about invertebrates in the Kuskokwim. Studies in the small Nushagak tributary streams found benthic invertebrate densities at approximately 5500/m^2(Peterson and Foote 2000).

Vertebrates

The main stem of the Kuskokwim and the north side of its basin in the Yukon plains was unglaciated during the Wisconsinan glaciation, forming a part of the Beringian refuge. The fish communities of the Kuskokwim basin differ from those of south-central and southeastern Alaska, from which it is isolated by the high peaks of the Alaska Range, but are similar to those of the nearby Yukon basin (Oswood et al. 2000; see Chapter 17).

Pacific salmon (sockeye, chum, chinook, coho, and pink) and steelhead are the most commercially valuable fishes in this river and spawn in various parts of the basin. The single biggest commercial sockeye fishery in the world is based in Bristol Bay (one-third of Alaska's total commercial salmon catch). Bristol Bay fisheries are based on fishes originating from the Kuskokwim, Nushagak, Kvichak, and other southwest Alaska rivers. Other fish species found in the river include Pacific lamprey, humpback whitefish, lake trout, Dolly Varden, longfin smelt, boreal smelt, eulachon, longnose sucker, burbot, threespine stickleback, and coastrange sculpin.

The abundant salmon support large populations of vertebrates that come to the river to feed, including grizzly and black bears, mink, and river otters. Moose and muskrat are common to the wetlands of the basin, although moose were considered rare in the early period of Russian exploration in the late eighteenth century. Many predatory birds spend summers around the river and its tributaries, including mergansers, belted kingfishers, and dippers. Large breeding populations of many wetland birds are found in the delta (e.g., emperor goose, spectacled and common eiders, other geese, ducks, and shorebirds; Sedinger 1997). This area is thus critical habitat for many breeding waterfowl and many other kinds of wildlife.

Ecosystem Processes

Relatively few ecosystem processes have been studied in the Kuskokwim River. However, because of the importance of contaminant sources in the catchment, details of sediment transport have been determined (Nelson et al. 1977, Duffy et al. 1999). The bedform of the river is highly mobile and strongly influenced by summer freshet from snowmelt and glaciermelt. Furthermore, temperatures at these latitudes have important influences on ice cover and formation of anchor ice, which have important biological consequences. As noted, in rivers in southeast Alaska, the relatively less turbid periods toward the end of summer and into autumn may be important to primary production given the lack of canopy cover along most of the river and its tributaries (Milner et al. 2001).

Human Impacts and Special Features

The Kuskokwim River, like the larger Yukon River on its northern border, is largely pristine and shares many of the unique characteristics of the Beringian refuge. The Kuskokwim remains free flowing and is largely unaltered by humans. It does, however, have some contaminant problems, and could be in danger of future development. The river has an anomalous problem with mercury, which originates from natural sources in some basin tributaries (Nelson et al. 1977). There was some limited mining for mercury, but erosion of natural sediments (partly exacerbated by mining) has contributed mercury to the river system. The concentrations in tributaries can average 570 mg/kg of channel sediments, but these values are diluted downstream to levels closer to an average of 0.5 mg/kg of sediment in the main stem (Nelson et al. 1977). Levels of mercury in fishes consumed by native peoples are elevated and frequently exceed the EPA's reference dose, and this has motivated continuing studies of the sources and fates of mercury (Duffy et al. 1999).

The Iditarod trail, famous for its annual dogsled race, traverses the river basin to provide winter access to the upper part of the catchment.

SUSITNA RIVER

The Susitna River is an unregulated river in south-central Alaska, fed by glaciers on the southern slopes of the Alaska Range (Fig. 16.11), with a drainage area of approximately 51,800 km^2, the sixth largest in Alaska (Alaska Power Authority 1985). It flows west and then southwest before emptying into the Cook Inlet to the west of Anchorage, Alaska's largest city at 260,000. Glaciers cover 11% of the basin, and 58% of the catchment is considered to be in high mountains (Kyle and Brabets 2001). The Susitna valley is a major transportation corridor for road and railway links between the cities of Anchorage and Fairbanks (Fig. 16.1). This river makes for a nice comparison to the Kenai River on the Kenai Peninsula, both of which drain through the Cook Inlet Taiga terrestrial ecoregion before emptying into Cook Inlet.

Prehistoric humans probably invaded the Susitna valley soon after the glaciers receded about 9000 years ago. Several eras of native settlements followed, with the Pacific Eskimos being displaced by the Athabaskan Dena'ina Indians of the Tanaina tribe at least several hundred years ago. Talkeetna, where the three main tributaries meet, means "river of plenty" in the Dena language, and was a major center for the native Indians living in the catchment. James Cook is reportedly the first European to explore the coastal area near the mouth of the river close to Anchorage in 1778. Russians explored and traded in the area beginning in the late eighteenth century and established trading posts before the sale of Alaska in 1867 from Russia to the United States. Discovery of gold in the Susitna River in 1886 lured nonnative prospectors far inland along the Alaskan rivers in the late 1800s, including deep into the Susitna River valley. In the early 1900s, the Alaska Railroad was built along the east side of the lower Susitna, leading to the establishment of towns such as Willow and Anchorage.

Physiography, Climate, and Land Use

The Susitna River lies within the South Central Alaska (SC) physiographic province. The drainage basin lies in a zone of discontinuous permafrost, where well-drained upland soils support white and black spruce and poorly drained valley bottoms support muskeg. The headwaters of the Susitna begin in the Copper Plateau Taiga terrestrial ecoregion to the east (see Fig. 16.11), a relatively flat area that was the site of a large Pleistocene lake (Ricketts et al. 1999). At a relatively low elevation (<1000 m asl), this plateau contains coniferous forest dominated by black spruce. The river then passes through the Alaska/St. Elias Range Tundra ecoregion, a long region of rugged mountains where elevations are often >4000 m asl. Some tributaries to the north (Chulitna River) drain from Mount McKinley at an

elevation of >6100 m asl. These mountains mostly consist of rocky slopes, ice fields, and glaciers. The Susitna (and many of its tributaries, such as the Yentna River) then enter the relatively flat, low-elevation Cook Inlet Taiga ecoregion. Here the vegetation includes various combinations of black and white spruce, Sitka spruce, quaking aspen, balsam poplar, black cottonwood, and paper birch (Ricketts et al. 1999).

The Susitna River originates in the continental climate zone and then flows south into the transitional climate zone. Due to maritime influence, temperatures are more moderate at the lower elevations of the basin (within the Cook Inlet Taiga) than in the interior mountains. In these more mild lowlands (where records are available), mean monthly air temperatures range from −12°C in January to 15°C in July (Fig. 16.12). Precipitation in the lowlands is quite seasonal, with maximum values of 12 cm/mo in August, but values of only 4 cm/mo or less from January through April (see Fig. 16.12). This precipitation pattern is an obvious underestimation of total precipitation for the basin, because precipitation is less than total runoff, an anomaly that is too large to be explained by the melting of glaciers.

The basin is almost entirely pristine wilderness, with no major land-use activities or settlements within the drainage. There are a number of remote cabins used for recreation and a number of homestead sites. There is no mining or timber harvest of note. Talkeetna is the largest settlement in the drainage at the junction of the Susitna, Talkeetna, and Chulitna rivers. It has a permanent population of <300 and serves primarily as an aircraft access point for climbers and sightseers to Denali National Park and Mount McKinley and for sports fishing.

River Geomorphology, Hydrology, and Chemistry

The 510 km long river (Alaska Power Authority 1985) originates as a number of small tributaries draining the East Fork, Susitna, West Fork, and MacLaren glaciers. This upper river flows in a braided channel across the alluvial floodplain of the Copper Plateau Taiga ecoregion for the first 80 km, with an average gradient of 2.6 m/km. Downstream from there the river flows for 115 km through a steep-walled canyon of the Alaska Range, known as Devil Canyon, with a gradient of 5.8 m/km. For the next 155 km, known as the middle river, it flows south through a well-defined and relatively stable series of channels until its confluence with the Talkeetna and Chulitna rivers (both glacier fed) about 160 km from the river mouth. This confluence near the settlement of Talkeetna is frequently termed the "three-rivers" confluence. Below Devil Canyon the gradient decreases from 2.6 m/km above Talkeetna to 1.52 m/km at Talkeetna. Downstream of this confluence the Susitna River valley broadens into a large coastal lowland, where the lower river is heavily braided with negligible riparian vegetation. As the river braids below Talkeetna the gradient further decreases to 1.05 m/km and in the last 70 km falls to 0.3 m/km (Schoch and Bredthauer 1983). The Yentna (a glacier-fed river) is the largest tributary joining the Susitna River from the west, with its confluence 45 km from the Susitna mouth, below Willow.

Six principal habitat categories were identified by the Alaska Department of Fish and Game (1984a) in the middle river section of the Susitna River based on similar morphologic, hydrologic, and hydraulic characteristics: main stem, side channel, side slough, upland slough, tributary, and tributary mouth. Some habitats may shift from one category to another depending upon main-stem discharges (Klinger and Trihey 1984); for example, turbid side channels at summer high flow to clear side sloughs at lower flows in the fall or spring. These habitats support characteristically different fishes and benthic macroinvertebrate communities.

The main water sources to the Susitna River—glacier melt, tributary inflow, surface runoff, and groundwater input—change seasonally, with approximately 25% of flow being glacial in the middle river during summer, falling to less than 3% in winter when flows are dominated by groundwater. Consequently, the Susitna River is characterized by highly turbid streamflow in the summer and low clearwater flow in the winter and early spring (see Fig. 16.12). There are scattered open areas of water where the river is not frozen over due to groundwater upwelling or high current velocity. Mean discharge just below the confluence with the Yentna River is 1427 m³/s. Peak flows typically occur in July, exceeding 1400 m³/s at Gold Creek in the middle river and averaging 5432 m³/s (1975 to 1992) toward the river mouth. The record peak of 8839 m³/s was recorded at Susitna Station in October 1986 following heavy rains. The river below the confluence with the Chulitna River regularly experiences flooding during summer storms, which add to the snowmelt and icemelt flows (Schoch and Bredthauer 1983).

Winter has a major effect on the Susitna River, with ice generation, staging, ice jamming, and break-up processes all having a major impact on aquatic habitat and biotic communities (Schoch and Bredthauer 1983). The size and configuration of existing sloughs along the river is dependent on the frequency of ice jamming in the adjacent main stem. Ice jamming creates floes that can modify the elevation of entrance berms to the sloughs or enlarge sloughs or side channels by scouring the banks and bed material. Ice processes do not appear to play such an important role in the morphology of the Susitna River below the Chulitna confluence. Frazil ice can scour surfaces and move organisms and stored organic matter downstream (Oswood 1997).

At Gold Creek in the Middle River, pH is circumneutral at 7.3, with alkalinity 50 mg/L as $CaCO_3$, NO_3-N 0.15 mg/L, and total phosphorus 0.12 mg/L. Turbidity values reach 200 NTU in the summer but fall to <1 NTU in winter.

River Biodiversity and Ecology

The Susitna River represents the upper end of the North Pacific Coastal freshwater ecoregion (Abell et al. 2000). This ecoregion stretches down the Pacific coast until it reaches (but does not include) the Columbia River. The region is considered continentally outstanding in biological distinctiveness.

Plants

Upstream of Talkeetna the riparian vegetation is relatively well established along the main channel and on island bars and includes black and white spruce, alder, and willow. Downstream of Talkeetna the river migrates markedly across a wide floodplain and riparian vegetation is limited due to the instability of the substrate.

Invertebrates

Studies of macroinvertebrates are limited to the middle river in association with a proposed hydroelectric project (as discussed in this chapter). Macroinvertebrates in the main channel during summer are low in diversity and abundance, limited principally to Diamesinae and Orthocladiinae chironomid midges. In side channel and side slough habitats, diversity increased, and in addition to chironomids, the principal taxa found were oligochaete worms, capniid stoneflies, baetid mayflies (principally *Baetis*), and heptageniid mayflies (principally *Cinygmula* and *Epeorus*) (Hansen and Richards 1985). When side channels and side sloughs were breached by main-stem water, drifting organisms increased, providing an important food source for rearing juvenile chinook salmon (principally chironomids).

Vertebrates

Pacific salmon are the most prominent of the fish species in the Susitna River and the group of fishes about which the most is known. The primary spawning habitat of chinook, coho, and pink salmon within the Susitna River drainage is in the tributaries, whereas sockeye and chum salmon spawn mainly in side slough habitats. Upwelling water is an important factor for the selection of sites by chum and sockeye salmon to enhance egg survival (Alaska Department of Fish and Game 1984b).

Depending upon the season, juvenile rearing salmon are found within all aquatic habitat types. Rearing juvenile coho salmon utilize tributary and side slough habitat, whereas juvenile sockeye salmon are found principally in side slough and upland slough habitats. Side channels are used by juvenile king (chinook) salmon, and the turbidity and invertebrate drift into these channels can provide cover and food sources. Juvenile chinook overwinter in these habitats. Some resident species, such as rainbow trout and Arctic grayling, use the main-stem channel for overwintering. Burbot are found exclusively in the main stem and side channels due to their preference for turbid water (Alaska Department of Fish and Game 1984c). Other fish species known from the river basin are Pacific lamprey, humpback whitefish, lake trout, Dolly Varden, longfin smelt, boreal smelt, eulachon, longnose sucker, threespine stickleback, and coastrange sculpin.

The majority of the upper Cook Inlet chum and coho commercial fishery originates in the Susitna River basin, and in the mid-1980s the estimated contribution of the river to the important and most valuable species, sockeye, was between 10% and 30% of the total upper Cook Inlet sockeye fishery (Alaska Department of Fish and Game 1984a).

The seasonal spawning migration of Pacific salmon lures many species of wildlife to the river, including bears (grizzly and black), bald eagles, and scavengers. Many other piscivorous species are found, including mink, river otter, osprey, merganser, and kingfisher. Dippers are also found in the basin. Moose, beaver, and muskrat are common species associated with the river margins and wetlands.

Ecosystem Processes

The ecology of the river in the main stem and side channel habitats is dominated by the hydrology and the influence of turbidity and sediment from glacial runoff. During the window of late winter–spring, when turbidities are low, algal growth and consequent macroinvertebrate production may be high. As discharges start to rapidly rise during late May and sediment concentrations increase, scouring removes the majority of algal production in these habitats and, with light penetration decreasing to <0.15 m, primary production is limited (Van Nieuwenhuyse 1984). A second algal bloom typically occurs in autumn as stream flows moderate and turbidities fall below 20 NTU. Some of this production may be lost due to the detrimental effects of ice regimes during freeze-up. In the summer, the main stem serves principally as a migratory corridor for adult salmon and outmigrating smolts.

Human Impacts and Special Features

The Susitna River is one of the most pristine basins in North America and is unique in draining some of the highest glacial mountains, including Mount McKinley and others in the Denali National Park of Alaska. However, with its mouth at Cook Inlet just west of Anchorage the river has and will probably continue to be threatened by human development.

At present there is relatively little impact from human population within the Susitna drainage, as the largest town is Talkeetna, near the three-rivers confluence, with a population of <300. Otherwise the settlement of the drainage is principally cabins and homestead sites, with a few small settlements along the Parks Highway bordering the Susitna downstream of Talkeetna. The highway does not appear to have much influence on the river. No major industries are present in the basin. In the early 1980s, however, a series of dams were proposed at Watana (294 Rkm) and Devil's Canyon (243 Rkm), which were scheduled for completion in 1994 and 2002 with a generating capacity of 1600 megawatts. Due to an economic downturn in oil prices in the 1980s the project was cancelled. It seems safe to say that the threat of damming of this pristine wilderness will always be present.

KENAI RIVER

The Kenai River is located on the northern portion of the Kenai Peninsula, just south of Anchorage. The Kenai River has a drainage area of approximately 5206 km^2 and is an unregulated river in south-central

FIGURE 16.4 Kenai River, Kenai Peninsula, Alaska, at low flow in May (Photo by A. Milner).

Alaska fed by glaciers on the western slopes of the rugged Kenai Mountain Range. The river flows westerly through two large lakes and across lowlands before emptying into the Cook Inlet at the town of Kenai (Fig. 16.13). It is one of the most productive and economically important rivers in Alaska due to its support of commercial and sports fisheries, including its world-renowned king salmon fishery (Fig. 16.4).

The Kenai Peninsula has been occupied by humans for many thousands of years. From about 3000 to 1000 years ago the area along the river was occupied by the Kachemak Riverine culture. About 1000 years ago they were replaced by the Kenaitze Indians, Athabascans of the Tanaina tribe (Boggs et al. 1997) who built settlements along the major rivers. At least 20 settlements are estimated to have existed on the Kenai Peninsula, with a maximum population in 1805 of 3000 people (Pederson 1983). The Kenai River was a major source of fishes, and summer camps were established to catch and dry salmon, and also to hunt muskrat, waterfowl, and caribou (Langdon 1987). The river and some of its

tributaries would also be used for transportation through the wetland areas of the low-lying Kenai Peninsula. Indians occupied the area when, in 1778 in search of the Northwest Passage, the English naval Captain James Cook sailed up the inlet that is now named after him. By this time Russian fur traders had been visiting the region, and a fortified Russian trading post, Fort St. Nicholas (Kenai), was established in 1791. Interactions between natives and traders were not always peaceful, and several battles between the Russians and Indians in the Kenai area resulted in many deaths. When the United States purchased Alaska in 1867 from Russia, the post was renamed Fort Kenai for the local Indians. Nonaboriginal American expansion into the catchment was spurred on by the discovery of gold near Cooper Landing in the 1880s. Kenai and the surrounding areas were opened to homesteading in 1947. Since that time the recreational value of the basin has attracted development, particularly in recent times. Most of this development has occurred along the main highway, which is in close proximity to the river for much of its length to Skilak Lake.

Physiography, Climate, and Land Use

The Kenai River basin is located in the South Central Alaska (SC) physiographic province (see Fig. 16.13). It is dominated by headwaters originating in the rugged Kenai Mountains rising up to 1500 m asl, two large lakes, and the river flowing for a great part of its length across lowlands dominated by wetlands. The drainage incorporates two terrestrial ecoregions. In the mountainous uplands to the east is the Pacific Coastal Mountain Tundra and Ice Fields. The lowlands to the west are within the Cook Inlet Taiga ecoregion. The lowland geology is dominated by unconsolidated Quaternary deposits (Brabets et al. 1999). The vegetation in the lowlands is dominated by closed spruce forest, closed mixed forest, and tall shrub.

The climate is classified as transitional and rainfall averages 48 cm in the coastal town of Kenai. The highest precipitation occurs in late summer, with a mean monthly accumulation of 8 cm in September (Fig. 16.14). Mean air temperature is only 0.9°C, with a minimum monthly mean of −11°C in January and a maximum of 12°C in July (Milner et al. 1997; see Fig. 16.14). Toward the mountains more precipitation and wider temperature fluctuations typically occur.

The land within the Kenai lowlands is predominantly state or native owned, except along the river in the lower reaches, where private ownership dominates. Above and in the environs of Skilak Lake the land is federally owned, the majority within the Kenai National Wildlife Refuge, which has 61 km of river frontage. Three main communities occur along the main river: Sterling (60 km from the mouth) at its junction with the Moose River, with a population of approximately 6000; Soldotna (40 km from the mouth), with 4000; and Kenai (at the river mouth), with 7000. The Borough of Kenai had a censused population over 51,000 in 2002. Most of the basin is uninhabited, except for numerous remote cabins and lodges along the lower river. Most of the Kenai River basin has been infested by the spruce bark beetle, and hence some salvage logging has occurred in these areas.

River Geomorphology, Hydrology, and Chemistry

Approximately 10% of the basin is glacierized (Dorava and Milner 2000). The Kenai River is similar to the other large rivers in the region in being dominated by glacial runoff, with many tributaries, both glacial and clearwater. However, none of the tributaries attain the size of those within the Susitna drainage, which accounts for the Kenai River's relatively lower discharge. Another major feature of the Kenai drainage is the presence within the catchment of two large glacial lakes, Kenai and Skilak (see Fig. 16.13), which have a significant influence on river hydrology and the associated river ecology (as discussed later in the chapter). Kenai Lake is 55.9 km², with a mean depth of 91 m and a maximum depth of 165 m, whereas Skilak Lake is 100 km², with a mean depth of 73 m and a maximum depth of 160 m. Many of the streams and rivers flowing into Kenai Lake, most notably the Snow River, are fed by lakes and glaciers. Many of the tributaries upstream of and including the Killey River (10 km downstream of the outlet of Skilak Lake) are principally glacial in origin. However, between Skilak and Kenai Lake, the Russian River is a clearwater tributary fed by lakes well known for their sport sockeye salmon fishery. A major input into Skilak Lake is the Skilak River from Skilak Glacier, a major outflow from the Harding Ice Field. Downstream of the Killey River various clearwater tributaries, fed by snowmelt and rainfall, drain the Kenai Lowlands and frequently flow through wetlands and lakes.

The series of transitions of the Kenai River from glacial mountains to the Cook Inlet result in it being divided into five distinct reaches based on geomorphology (Liepitz 1994): (1) an intertidal reach from the mouth to Rkm 16, (2) a transitional reach from Rkm 16.0 to Rkm 28.3, (3) an entrenched reach from Rkm 28.3 to Rkm 68.3 where the river is confined within the Soldotna terrace, (4) an upper reach from the glacial moraine to the outlet of Skilak Lake, and (5) the interlake reach between Kenai and Skilak lakes. The entrenched reach resulted from the river down-cutting due to sea-level decline and plays a major factor in the stability of the channel. The bed is armored in this reach, with 38% of the substrate between 6.4 and 12.7 cm and 41% between 12.7 and 25.4 cm (Scott 1982). Upstream the substrate is generally larger. In the lower reaches the substrate is markedly smaller, with 40% sand and silt (<0.6 cm) and 51% gravel (0.6 to 6.4 cm; Bendock and Bingham 1988). Because of the large lakes, the Kenai River has a lower width-to-depth ratio (57) than three other glacier-fed rivers in Alaska without lakes (ratio of 85) (Dorava and Milner 2000). These lower ratios in the Kenai River create deeper water habitats, which is important for fish production. Gradient from Kenai Lake to Skilak Lake is relatively steep at 6.8 m/km, but from Skilak Lake to Cook Inlet it is typically <3 m/km.

Continuous discharge data for the main-stem Kenai River are available from two stations, Cooper Landing located downstream of Kenai Lake (since 1948) and Soldotna, 40 km from the mouth with a drainage area of 5206 km^2 (from 1966). Minimum mean monthly discharge occurs from February to April, falling to <51 m^3/s, with maximum discharges in August with a monthly mean of 410 m^3/s (1966 to 1994 measured at Soldotna), fed predominantly by snowmelt and glaciermelt (see Fig. 16.14). The two large lakes buffer flow variations such that the river channel downstream of Skilak Lake is not braided and multichanneled, as is so typical of many glacier-fed rivers in Alaska (e.g., the Susitna River). Instead, a single meandering or sinuous channel occurs, with relatively stable banks and riparian vegetation. Nevertheless, major floods can still occur in the Kenai River, as evidenced by flows in October 1999 after a period of sustained rainfall, with discharges reaching >1000 m^3/s at Soldotna. Floods may also occur due to sudden discharges from ice-dammed lakes in the drainage, a *Jökulhlaup* event. This frequently occurs at a glacial lake within the Snow River basin (every two to four years), and on one occasion caused a flood peak of 954 m^3/s at Soldotna (Scott 1982).

The presence of the large lakes in the Kenai River system markedly reduces the average sediment load carried by this major glacier-fed river. The average load in the Lower Kenai River is 0.09 tons d^{-1} km^{-2}. This can be compared with an average of 2.16 tons d^{-1} km^{-2} for glacier-fed rivers in south-central Alaska without major lakes in their systems (Dorava and Milner 2000).

In the main-stem Kenai River, water temperature ranges from near 0°C to 13°C, with a mean of 6.4°C (Litchfield and Kyle 1992). Maximum water temperature typically occurs in July. Dissolved oxygen in Kenai River water is always close to saturation, although it can be lower in some of the shallow slow-flowing tributaries on the Kenai Lowlands. Litchfield and Kyle (1992) reported turbidities in the main stem from 2 to 18 NTU and suspended sediment concentrations between 10 and 100 mg/L. Conductivity ranges from 30 to 81 μS/cm, with pH close to neutral. Alkalinities are low, typically averaging 25 mg/L as CaCO$_3$. Studies in 1991 and 1992 indicated that nitrate and nitrite decreased downstream, with values in 1991 exceeding 230 μg/L below Skilak Lake but 150 μg/L near the mouth. Orthophosphate increased near the mouth but still remained below 10 μg/L.

River Biodiversity and Ecology

Like the Susitna River, the Kenai River is one of the most northern rivers within the North Pacific Coastal freshwater ecoregion (Abell et al. 2000). Although the fisheries of the Kenai are extremely well studied, less attention has been given to its lower trophic levels and general ecology. Nonetheless, more is known about this system than most of the northern Pacific rivers.

Plants

Due to the stability of the river channel below Skilak Lake there is extensive growth of riparian vegetation along the river, largely made up of alders and willows. Where wetland areas adjoin the river, typical wetland communities of sedges and mosses occur. Macrophytes are not extensive in the main-stem river due to low water clarity in the summer and high velocities.

Invertebrates

By attenuating peak flows, sustaining high flows throughout the summer, supplementing winter flows, settling bedload and suspending sediment, and enhancing downstream water temperature, the lakes have a major influence on the biotic productivity of the Kenai River system (Dorava and Milner 2000). The macroinvertebrate fauna of the Kenai River is dominated by chironomid midges, particularly the subfamilies Diamesinae and Orthocladiinae. Other dipterans, such as crane flies (*Dicranota*), dance flies (Empididae), and black flies, have also been reported (Burger et al. 1983, Milner and Gabrielson 1994). Densities have been found to increase downstream and range from 592 organisms/m^2 to 11,050 organisms/m^2 (Milner and Gabrielson 1994). The mayflies were dominated by *Baetis*, *Cinygmula*, *Drunella*, and *Ephemerella*; stoneflies by *Capnia*, *Plumiperla*, *Paraleuctra*, and *Isoperla*; and caddisflies by *Glossosoma* and *Brachycentrus*. Richness of nondipteran taxa was typically between 10 and 15 species. Macroinvertebrate richness is typically higher in the clear-water tributaries. However, due to the stabilizing influence of the lakes on the Kenai River, overall abundance and macroinvertebrate diversity in the main-stem river is an order of magnitude higher than in a glacial system without lakes. For example, macroinvertebrate densities in the Johnson River, a glacier-fed river without lakes on the west side of Cook Inlet, averaged only 460 organisms/m^2 (Dorava and Milner 1999).

Vertebrates

All five species of Pacific salmon (chinook, coho, sockeye, chum, and pink) use the Kenai River, which supports extensive commercial and sports fisheries. Dorava and Scott (1998) compared the salmon escapement per unit length of river, normalized for catchment area, and found the Kenai River to be 2.3 times higher than the Deshka River, a well-known clearwater salmon river in south-central Alaska. The availability of suitable overwintering habitat is critical to the survival of juvenile salmon to smolts in coastal Alaska and British Columbia (Tschaplinski and Hartman 1983). Glacial lakes provide important overwintering habitat for juvenile fishes, thereby reducing mortality and enhancing production (Reynolds 1997). A large percentage of 0+ juvenile chinook salmon were found to move over 50 km upstream to overwinter in or near the outlet of Skilak Lake (Bendock 1989).

Chinook salmon have two major runs into the Kenai River, an early run entering the Kenai River before June 30 and spawning principally in tributaries, with about 15% spawning in the main stem. Late-run chinook salmon enter the Kenai River after June 30 and spawn in the main stem, principally in the transitional and middle river reaches (Burger et al. 1983). Early chinook runs typically range between 10,000 to 27,000 fish, whereas late runs are between 39,000 and 80,000, with a mean of 58,360 (1984 to 1994) (Hammarstrom 1995). Of these, an average of 16,570 were caught in the commercial harvest in Cook Inlet and 10,800 were taken by sports fishermen in the river, giving an average escapement of 30,990 (19,580 to 48,040). Juvenile chinook salmon typically remain in fresh water for one year and rear in the margins of the main river.

Coho salmon also have two overlapping runs, with the early run beginning in late July and running until late August and the late run extending into November. Similar to chinook, early-run coho spawn primarily in tributary streams from September through early October, whereas the late-run fish spawn primarily in the main stem from October through February. Juvenile coho salmon typically spend two years in fresh water before going to the ocean as smolts, principally in low-velocity areas in the main stem or side sloughs. However, these juveniles can migrate extensively within the drainage and may emigrate from the main-stem Kenai River to overwinter in tributaries and lakes.

The species of most commercial value is the sockeye salmon. Similar to the other two species, sockeye possess two distinct stocks, an early run

entering the river in mid-May and a late run that typically peaks in mid July. The early run supports an important sports fishery at the confluence of the Russian River and the main-stem Kenai, with an average catch of 22,530 or 42% of the river return (1963 to 1993) (Nelson 1985). The Russian River escapement has usually exceeded 30,000, with an average of over 23,000 fishes (1968 to 1993). These fishes spawn in the Russian River and rear exclusively in Upper Russian Lake.

The late run of sockeye salmon enters the Kenai River in early July, and the run is usually completed by early August. The escapement goal is for between 400,000 and 700,000 spawners. The average (1968 to 1992) has been 635,000, with a minimum of 53,000 and a maximum of 1.6 million following the Exxon Valdez oil spill in 1989, when commercial fishing was closed down due to possible tainting of flesh from oil. Sockeye salmon spawn in the main stem below Skilak Lake and several tributary streams, including the Russian River and Hidden Lake. The majority of main-stem and tributary sockeye salmon rear principally in Kenai and Skilak lakes, with about 20% in Kenai Lake. Sockeye salmon juveniles typically spend one year in fresh water, but high winter mortality is typical. Tarbox and Brannian (1994) estimated about 10% survival, which is positively correlated to fall copepod density and fry condition (Schmidt et al. 1995).

Other fishes known from the Kenai River catchment include Pacific lamprey, humpback whitefish, lake trout, Dolly Varden, longfin smelt, boreal smelt, eulachon, longnose sucker, burbot, threespine stickleback, and coastrange sculpin.

Wildlife within the basin is abundant, and the availability of upstream-migrating salmon and eulachon (hoolican) provides a short-lived but critical resource to grizzly bears, black bears, bald eagles, and many scavengers. The river supports river otter, mink, merganser, kingfisher, osprey, and dipper. Beaver and muskrat are common within the catchment, and moose are abundant and associated with the river margins and wetlands.

Ecosystem Processes

Turbidity is likely to limit primary production during the high summer discharges, but when water clarity increases in late spring and early fall, windows for primary production exist when the river is not ice covered, and mats of filamentous algae have been observed. However, we have no indication of the role of primary production relative to the energetics of the main-stem Kenai River and its tributaries. Indeed, there is a lack of studies of lower trophic levels in the Kenai River and their interactions, as most of the focus has been on fisheries populations for management objectives. We have no indication of the role of benthic organic matter in the Kenai River, whether the lakes are a sink or source of organic carbon, or the role of tributaries in supplying organic carbon. Nutrient dynamics are poorly understood, particularly the role of salmon carcasses and wetland areas.

Human Impacts and Special Features

The Kenai River is unique, with its rugged undeveloped catchment, its strong influence of lakes and glaciers, and its low-elevation wetlands, all within the Kenai Peninsula. It is also the most important sport fishing river in Alaska, with many fishermen seeking out the mighty king (chinook) salmon because of the size reached in the Kenai River. The state record for a king salmon is 44 kg in the Kenai River. The river is also important in providing sockeye salmon to the commercial fishery in Cook Inlet, both by set nets along the beach and drift gill netting by boat. The intense recreational use has potentially damaged riparian habitat and spawning areas by increased bank erosion due to trampling and boat wakes, principally from fishing guides in the lower river (Dorava and Moore 1997). Development near the river, the infilling of wetlands, and construction of structures within the river have added to habitat effects (Dorava 1995). Instream structures like jetties and groins may increase velocities and sheltered areas reduce velocities, which are outside the preferable range of juvenile chinook salmon (Burger et al. 1983). High velocities may hinder juvenile salmon migration, which is essential for overwinter survival.

Commercial and recreational fisheries have the most direct impact on Kenai River salmon. Early-run chinook and sockeye are not targeted by the commercial fishery, but some late-run chinook are removed as incidental by-catch of the commercial sockeye fishery in Cook Inlet, creating conflict with the sports fishermen. Similar conflicts can occur when the sports fishery for sockeye is restricted due to lower bag limits or hours of fishing as a result of low catches within the commercial fishery. Commercial sockeye salmon harvests typically range from 500,000 to 3 million fish. Additional pressure has occurred since 1989 from a subsistence dip-net fishery at the mouth of the river that permits Alaska residents to remove 35 sockeye salmon for personal use.

STIKINE RIVER

The Stikine River is a transboundary river originating within and having most of its basin in Canada but reaching the Pacific through the Alaska panhandle (southeast Alaska). The 539 km long river initially flows west and then southwesterly from its source east of the Coast Range Mountains, and is joined by its major tributary, the Iskut River, close to its mouth (Fig. 16.15). This largely pristine system is one of only three (together with the Skeena River of British Columbia and the Kamchatka River of Russia) large (>350 m^3/s) free-flowing rivers in the Northern Hemisphere that drain a biome (Temperate Mountain Forest) other than the Boreal Forest and Tundra (Dynesius and Nilsson 1994) (Fig. 16.5). As with all the rivers covered in this chapter, the Stikine River has a large component of glacial influence. However, less of the basin is influenced by glaciers compared to the more northerly rivers.

Two native groups have occupied the river basin for more than a thousand years, the Tlingit on the coast and the Tahltan in the interior. The Tahltan claimed winter hunting grounds down the frozen Stikine River but were not great river travelers in the summer, whereas the Tlingit claimed the Stikine salmon-spawning tributaries. In the summer the Tahltan would trade products of the interior with the items the Tlinglit would bring along before they returned to the coast. The Stikine River, meaning "great river" in the native Tlingit language, was missed by Captain Vancouver but "discovered" by American fur traders. European contact was first recorded in the late 1700s, when the Russian-American Company began trading with Tlingit peoples in the delta of the Stikine River. The Russian-American Company operated there from 1799 to 1839. About 1834, the Hudson Bay Company established a trading post in the interior of the Stikine basin and eventually took over trading operations, including the coast. The Stikine provided a transportation corridor to the gold rushes of the north: the Stikine gold rush in 1862, followed by the Cassiar gold rush of 1874–1876, and finally in the Yukon in 1897. Steamboats were a common mode of transport and between 1862 and 1969, 107 commercial boats operated on the river. The head of navigation was Telegraph Creek, 250 km from the mouth, which connected to trails and later roads. The population of Telegraph Creek is 300, mostly Tahltan Indians. The Stikine was also surveyed as a potential route for a telegraph line in the 1860s. The area was also used as a staging and supply post for construction of part

FIGURE 16.5 Stikine River, British Columbia (PHOTO BY TIM PALMER).

of the Alaska Highway in 1941–1942. Today the Cassiar Highway passes through the basin, but it is still a sparsely settled and remote location. The Tahltan nation claims most of the catchment as its traditional territory and remains the primary occupants of the area, relying on the river fishery (BC-LUCO 2000).

Physiography, Climate, and Land Use

The Stikine River belongs to the Coast Mountains of British Columbia and Southeast Alaska (PM) physiographic province (see Fig. 16.15). The Stikine basin includes several forest types from the coastal temperate forest (western red cedar, Sitka spruce, western hemlock), part of the Northern Pacific Coastal Forests terrestrial ecoregion, to the eastern extent of the basin. Upstream of the coast range the river passes through a cold, wet forest type, known in British Columbia as the Interior Cedar–Hemlock zone, and is part of the Northern Cordilleran Forests ecoregion (includes western red cedar, western hemlock, "interior" spruce [*Picea engelmannii* ×

glauca], and lodgepole pine in drier sites). At higher elevations along the coast is the mountain hemlock zone, replaced away from the coast by the Engelmann spruce–subalpine fir zone. Two other forest types that cover small portions of the basin in drier, lower-elevation zones are known as the subboreal spruce (Engelmann–White spruce hybrid, subalpine fir) and the boreal white and black spruce (black spruce, white spruce, trembling aspen) zones, part of the Northern Cordillera Forests ecoregion. The mid-course of the Iskut River, the Stikine's main tributary, flows through the Northern Transitional Alpine Forests ecoregion, representing a mix of forest types from rainforest in the valley bottoms to tundra and permafrost at higher elevations. The largest portion of the basin includes alpine tundra (40%; Pacific Coastal Mountain Tundra and Ice Fields ecoregion) and spruce–willow–birch forests (28%).

The Stikine River basin had a mean annual air temperature of only 2°C (1951 to 1980) at the one valley-bottom recording station of Telegraph Creek. Mean monthly temperature was lowest (−18°C) in January and highest (16°C) in July (Fig. 16.16). Most of the basin would be cooler on average. The average annual precipitation at Telegraph Creek in the rain shadow of the coast range was about 40 cm, likely an underestimate of the average for the basin. The rate of precipitation was highest from October through December and lowest in spring, but did not vary greatly throughout the year (see Fig. 16.16).

Only about 50% of this basin is forested, with much of the remainder covered by alpine tundra and ice fields. About 73% is considered to be in wilderness or near-wilderness condition (BC-LUCO 2000). Only about 1300 people reside within the catchment, centered on the towns of Telegraph Creek and Dease Lake.

River Geomorphology, Hydrology, and Chemistry

The two main rivers of the Stikine drainage, the Iskut and Stikine, join near the coast. Both rivers are glacially influenced and have large braided channels in their lower reaches, which in the Stikine is nearly 130 km long before entering the Pacific Ocean. The primary tributaries to the Stikine River include the Tahltan, Spatsizi, Klappan, and Tuya rivers. The catchment has its origins in the extensive mountain areas, with the highest point (2938 m asl) being Amaman Mountain in the Coast Range Mountains. Extensive lava beds in the basin are evidence of the past volcanic history of the region. The river substrate includes clasts up to house size in some of the canyon reaches. One spectacular feature of the 540 km long river is the 95 km long Grand Canyon of the Stikine, with gradients between 2.5 and 7 m/km. This gorge (up to 300 m deep) separates the inland and coastal portion of the river. Below the canyon the river is navigable and has an average gradient of 1 m/km from the settlement of Telegraph Creek to the border with Alaska some 180 km downstream. From Telegraph Creek to the border the river width expands from 100 m to about 250 m across the active channel. The Iskut River has a gradient of 1.5 to 2.5 m/km upstream of the confluence with the Stikine, but in the 5 km reach downstream of its confluence with the Forrest Kerr River the river goes through a canyon section up to 120 m deep and with a 27 m/km gradient. The Stikine is considered one of only three large free-flowing rivers in the Northern Hemisphere that drain biomes (Temperate Mountain Forest) other than Boreal Forest and Tundra (Dynesius and Nilsson 1994). Nevertheless, there is continuing pressure to consider development of the enormous potential for hydroelectric power generation from the basin.

Mean discharge for the Stikine River is 1587 m^3/s. Runoff is highest during June and July (>19 cm/mo) due to the domination of the hydrograph by snowmelt and to a lesser extent glaciermelt (mostly from the Coast Range Mountains) (see Fig. 16.16). Runoff appears to be higher than precipitation, but this is caused by the underestimation of precipitation for the basin as a whole, as seen for most northern Pacific rivers with few valley-bottom weather stations. Winter precipitation is mostly stored as snow throughout the basin, and the mountains hold snow until the middle of spring before melting.

Water quality of the Stikine River has been measured at a station about 8.2 km upstream of its confluence with the Iskut River (Jang and Webber 1996). There were no significant trends in the water quality. Most parameters varied seasonally along with flow conditions. As with other glacially influenced rivers, turbidity has a considerable range, with fall and winter values of 1 to 5 NTUs and summer peaks between 50 to 350 NTUs, depending upon year. Metal concentrations were relatively high, suggesting natural sources: Copper regularly exceeded the criterion for protection of aquatic life, with concentrations frequently >10 μg/L and up to 60 μg/L; iron regularly exceeded the criterion for aquatic life of 0.3 mg/L (range 0.2 to 11 mg/L); average zinc concentrations were 15 μg/L; aluminum was ~2 mg/L; and calcium concentrations were 15 to 32 mg/L (suggesting low sensitivity to acid inputs). Alkalinity was 45 to 90 mg/L as CaCO$_3$ and hardness had a range of 50 to 110 mg/L as CaCO$_3$, with higher values during high-flow periods. Nitrate-N and NO$_2$-N both averaged 0.1 mg/L (range 0.02 to 0.35 mg/L) and total phosphorus was ~0.2 mg/L (range 0.005 to 1.0 mg/L). Conductivity ranged from 100 to 240 μS/cm (at 25°C) and pH was ~7.8 (range 7.5 to 8.3). Temperatures were not continuously measured, but summer peak temperatures were between 11°C and 12°C (Jang and Webber 1996), peaking from late June to mid July.

River Biodiversity and Ecology

Like most of the rivers in this chapter, the Stikine is located within the North Pacific Coastal freshwater ecoregion (Abell et al. 2000). In spite of the size and pristine nature of this Pacific river, relatively few biological and ecological studies have been done, particularly for algae and invertebrates.

Plants

In the lower reaches of the Stikine River, riparian vegetation includes cottonwood, alder, and willow trees. Sitka spruce and hemlock are found at somewhat higher elevations.

Invertebrates

We are not aware of any published reports on the Stikine or its tributaries. However, samples taken in the Iskut River near the Forrest Kerr River show this river to have similar assemblages to many other glacially influenced rivers. The predominant taxa were the mayfly *Baetis*, the stoneflies *Capnia*, *Doddsia*, *Suwallia*, *Podmosta*, and *Zapada*, and the chironomid genera *Diamesa*, *Cricotopus*, and *Orthocladius* (L. Fanning, unpublished data).

Vertebrates

The Pacific salmon are all large contributors to the fish assemblages of the Stikine River system, and all are commercially exploited. Sockeye and chinook salmon populations are substantial. The sockeye salmon populations are unique because they rear in the main-stem Stikine and not in lakes; this also occurs in other rivers in southeast Alaska, the Taku River (Murphy et al. 1997). Escapements of sockeye salmon to the basin are approximately 38,000 per year (range 6400 to 90,600 from 1979 to 1999). Despite the importance of anadromous fishes in this system, reliable population data are limited (available for <10% of the individual stocks). Of the stocks that are sufficiently well documented, none appear to be at risk (Slaney et al. 1996).

Sockeye salmon in this glacial system were thought to come primarily from Tahltan Lake, one of the few clearwater lakes in the basin. However, there is evidence that >30% of sockeye from the Stikine may spend their freshwater time in glacial rivers ("river type") or even go to sea immediately after fry emerge from the gravels ("sea type"). This is partly attributed to life history plasticity and the need for alternative rearing strategies in northern rivers (Wood et al. 1987). The bull trout, found in most of the Taku, Stikine, Nass, and Skeena catchments but not further north, is the only fish in the Stikine River basin on British Columbia's blue list (Williams et al. 1989). Other fish species found in the catchment are Pacific lamprey, humpback whitefish complex, lake trout, Dolly Varden, longfin smelt, boreal smelt complex, eulachon, longnose sucker, burbot, threespine stickleback, and coastrange sculpin.

Bears (grizzly and black) are drawn to the migrating adult salmon. There are many other piscivorous species, including bald eagle, river otter, mink, kingfisher, merganser, and osprey. Other species common to the river and its margins are beaver, muskrat, dippers, and moose. Wildlife within the basin is abundant and many species found here in abundance are considered threatened or endangered in other parts of their range.

Ecosystem Processes

There have been no ecosystems studies to date in the basin. Most of the work that has been done is related to hydrology for estimation of hydroelectric power potential.

Human Impacts and Special Features

As a very large river flowing through a rugged mountainous wilderness, there is very little human development in the Stikine River basin. About 17% of the basin is currently under protected status, which will likely increase to over 25% with the implementation of a proposed land-use plan. The Spatsizi Wilderness Area is a large protected area (6568 km^2). The Stikine River Recreation Area (2170 km^2) borders the upper Stikine River, primarily serving as a rafting and canoeing area. Further protected areas include Mount Edziza Provincial Park (2302 km^2) and the Stikine–LeConte Wilderness Area within the Tongass National Forest (Alaska).

Despite the extensive forest cover of the region, <1% is considered commercially viable, partly because of the remote location, which makes harvesting uneconomical. Forestry accounts for only 2% of employment within the basin (BC-LUCO 2000). Mining accounts for 5% of employment and includes the Eskay Creek mine (silver and gold), which is the world's fifth-largest producer of silver. Large high-grade coal deposits occur in the basin, notably at Mount Klappen, but these are not currently being mined (BC-LUCO 2000). Government and other services account for over 50% of employment within the basin. Guide outfitting and ecotourism are other sources of employment. Contentious and largely unresolved issues over fisheries values remain, but these are partly addressed by the Pacific Salmon Treaty and similar legislation. This currently unimpounded river has enormous hydroelectric potential and the river remains at risk of being developed for energy in several reaches, including the Grand Canyon of the Stikine and the gorge on the Iskut River.

FIGURE 16.6 Skeena River below Terrace, British Columbia (Photo by Tim Palmer).

SKEENA RIVER

The Skeena River arises in the interior plateau of British Columbia and the central mountains and flows west 579 km through the Coast Range Mountains to the Pacific (Fig. 16.17). This is the third-largest catchment within British Columbia (after the Fraser River [Chapter 15] and the Stikine River) at 42,200 km^2 (Fig. 16.6). Like the Stikine River, the Skeena is one of only three (also the Kamchatka River of Russia) large (>350 m^3/s), free-flowing rivers in the Northern Hemisphere that drain a biome (Temperate Mountain Forest) other than the Boreal Forest and Tundra (Dynesius and Nilsson 1994).

The Skeena River catchment is the southernmost among the Pacific Coast rivers covered in this chapter. Like most of the coastal rivers, the Skeena River is a large producer of Pacific salmon. This basin supports the greatest diversity of forest types within the basins covered in this chapter and the highest intensity of land use, although still relatively low.

The area of the Skeena (the "river of mist") has been settled by Native Americans for several thousand years, with the Tsimshian people near the coast and the Gitksan in the interior of the basin. These peoples developed a rich culture and permanent settlements based on the resources provided by returning salmon and eulachon, plants, abundant wildlife, and, for the Tsimshian, the sea. The British Captain Vancouver is thought to be the first European to make contact with the Tsimshian in 1793 during surveying of the Pacific Coast. The first trading post was established by the Hudson's Bay Company in 1834 at Port Simpson on the coast, and in 1879 they established a trading post near Hazelton to serve the interior of the basin using the lower Skeena River as the transportation corridor from the coast. The Grand Trunk Railway, completed to the coast in 1913, opened the basin to further development.

Physiography, Climate, and Land Use

The Skeena River sits in the Coast Mountains of British Columbia and Southeast Alaska (PM) physiographic province (see Fig. 16.17). The basin drains from many mountain ranges, including parts of the east side of the Coast Range Mountains, the Nass Ranges, the Bulkley Ranges, the Babine Range, and the Skeena Mountains. Many peaks in the catchment exceed 2000 m asl and most of the basin is mountainous.

The different forest types found in the Skeena River basin are somewhat similar to the Stikine River (previously discussed). The British Columbia Mainland Coastal Forests terrestrial ecoregion penetrates almost 180 km from the coast along the Skeena River. This lowland portion of the basin includes western hemlock, western red cedar, and amabilis fir in the low elevations, and mountain hemlock, subalpine fir, and yellow cedar in the subalpine zone (Ricketts et al. 1999). At higher elevations is the Northern Transitional Alpine Forests ecoregion, which is dominated by subalpine fir and mountain hemlock. There is also the Fraser Plateau and Basin Complex ecoregion, which is characterized by Engelmann spruce, lodgepole pine, and subalpine fir. The highest portion of the Skeena basin drains from the Central British Columbia Mountain Forests ecoregion. This region includes Engelmann spruce, alpine fir, lodgepole pine, quaking aspen, white and black spruce, and a small amount of alpine tundra (<0.3% of catchment area).

The climate of the Skeena basin is modified by the rain shadow imposed by the Coast Range, and annual precipitation in the valley bottom averages 62 cm. Precipitation is not highly seasonal in the valley, but is lowest (<4 cm/mo) from February through May and highest (up to 8 cm/mo) from October to January (Fig. 16.18). As with many of the northern basins in western North America, there are relatively few weather recording stations. Most are in the valleys, associated with airports and towns,

and do not reflect adequately the overall conditions in the majority of the basin, which is at higher elevation and receives more precipitation and is colder than the recording stations. Average air temperature for the towns in the valley is about 4°C, with a low of −11°C in January and a high of 16°C in July (see Fig. 16.18). However, there is a large amount of variation across a catchment of over 50,000 km².

Most of the basin remains in forest (>90%). There are several industries in the basin, including forestry, mining, recreation, and service providers. Forestry is a major industry within the basin, accounting for >20% of employment within the region (Tamblyn and Horn 2001). However, only 39% of the catchment is considered productive forestland, and only 6% is considered operable timber-producing area because of steep terrain and other limitations (Tamblyn and Horn 2001). The primary tree species harvested are western hemlock, western red cedar, amabilis fir, Sitka spruce, yellow cedar, and cottonwood.

The primary mineral resources are gold, silver, copper, zinc, and lead. There are at least 14 active sites for mineral exploration or extraction in the catchment. Most of these are small operations. Sport fishing, hunting, and other outdoor recreation in the basin are popular due to its wilderness setting (Tamblyn and Horn 2001). Agriculture is very scarce in the area and the potential for agriculture is low. Thus, there is no irrigation use of water from the river or its tributaries.

River Geomorphology, Hydrology, and Chemistry

The Skeena River reaches the ocean in a broad, braided alluvial channel as it passes through the Coast Mountain Range. The majority of the river's basin is inland of the Coast Range, with only minor tributaries west of the city of Terrace (see Fig. 16.17). The river gradient up to Terrace from the estuary is only 0.63 m/km and navigable (see Fig. 16.6). Two large tributaries, the Bulkley and Kispiox rivers, meet the Skeena River near the town of Hazelton. Several other large tributaries to the river include the Morice, Babine, and Kitsumkalum rivers. Upstream of Terrace the gradient of the river through the 180 km to the confluence with the Babine River is about 1.42 m/km and alternates between reaches with gravel bars and alluvial development and strongly constrained sections with steep margins, such as the reach adjacent to the Skeena Mountains. Upstream of the confluence with the Babine River the river gradient reaches 4.0 m/km until the junction with the Sustut River, and even steeper upstream of that. The river is cobble bottom throughout and gravel bars in alluvial reaches are common in many segments of the river downstream of the Babine River.

The hydrology of the river is snowmelt dominated, although coastal areas receive a large amount of rain in winter, but this includes only a small portion of the basin. The Skeena basin has the least glacier influence of any of the rivers treated in this chapter. The Skeena River system has a reservoir on one of its tributaries, with a gross capacity of about 0.2% of its flow (Dynesius and Nilsson 1994). Runoff is dominated by snowmelt and peaks in May and June (see Fig. 16.18), earlier than the other rivers in this chapter. Parts of the Skeena River basin in the Nechako Plateau have greater groundwater storage than other parts of the region and can supply up to 90 L/min through pumping.

Water chemistry has been measured by Environment Canada nearly every two weeks (1984 to 2000) at Usk near the Pacific Ocean (www.ec.gc.ca/water). Like most rivers in the region the water temperature is near 0°C for nearly six months of the year, with an approximate average annual temperature of 6°C (based on unweighted data from biweekly samples). Summer maximum temperature at that station is about 14°C to 16°C in most summers, peaking about late July to early August, but temperatures as high as 24°C and 26°C have been recorded during that interval. Turbidity does not have the regular seasonal extremes of more northern and glacial-influenced rivers, but peaks in excess of 200 NTUs have been recorded a couple of times during the period of record. Mean values for other standard water-chemistry values include alkalinity 41 mg/L as $CaCO_3$, pH 7.6, NO_3-N 0.08 mg/L, and total phosphorus 0.057 mg/L.

River Biodiversity and Ecology

The Skeena is the southernmost river from this chapter located within the North Pacific Coastal freshwater ecoregion (Abell et al. 2000). As with many of these rivers, the biological and ecological information is relatively sparse, with the exception of the salmon populations.

Plants

Riparian plants associated with the margins of the river include alders, willows, black cottonwood, and trembling aspen. Thimbleberry and black twinberry provide an important shrub layer in the understory.

Invertebrates

We are unaware of any published studies of invertebrates in the Skeena, but some unpublished data have been collected by Shauna Bennett (University of British Columbia) and Ian Sharpe (British Columbia Ministry of Water, Land, and Air Protection). The most common mayfly genera are *Baetis*, *Rhithrogena*, *Epeorus*, *Cinygmula*, *Drunella*, *Serratella*, and *Paraleptophlebia*. Caddisflies include hydropsychids and *Rhyacophila*, and stoneflies include *Sweltsa* and *Zapada*. As in most riverine systems, oligochaete worms are present, and the dipterans are well represented by chironomid midges and black flies.

Vertebrates

As with all the rivers in this chapter, this is a major salmon-producing river, with the five species of Pacific salmon (chinook, coho, sockeye, chum, and pink) and steelhead present. About half of the stocks in this river lack sufficient data to determine trends. From the 515 individual stocks (species and breeding populations) for which data are available, 2 are known to have gone extinct and 45 more are considered to be at moderate to high risk of extinction (Slaney et al. 1996). The nonanadromous form of sockeye (kokanee) is found in some lakes of the region, notably Babine Lake.

The rest of the fish community includes Pacific lamprey, humpback whitefish complex, lake trout, Dolly Varden, longfin smelt, boreal smelt complex, eulachon, longnose sucker, burbot, threespine stickleback, and coastrange sculpin. Bull trout, found in most of the Taku, Stikine, Nass, and Skeena catchments but not further north, is on British Columbia's blue list (Williams et al. 1989). Giant pygmy whitefish (*Prosopium* sp. nr. *coulteri*) is found in only two lakes in the world, one of which is in the Skeena drainage, and appears to have diverged from *P. coulteri* (Cannings and Ptolemy 1998). This species has not been formally described, but as a distinct form is listed on British Columbia's red list.

The tailed frog is found in some of the tributaries of the Skeena and Nass river systems, mostly in the Coast Mountains, and is listed as vulnerable (blue list) in British Columbia. It is the only amphibian or reptile associated with the rivers in this chapter, which are past the northern limits of most species of these groups of organisms.

Grizzly bears typically occur at higher densities in coastal areas and are thought to be secure in the catchment, with densities of approximately 0.035 per km^2 (Tamblyn and Horn 2001). Salmon form an important food resource for coastal grizzly bear (also known as brown bears). Fishes also sustain good populations of river otter, mink, bald eagle, osprey, kingfisher, and merganser. Dippers and spotted sandpiper are common in the catchment. Beaver reach high densities and are thought to still be recovering from heavy exploitation during the fur-trading days of the early nineteenth century in this part of the continent. Muskrat and moose are also common inhabitants of riparian areas of the streams and wetlands of the catchment.

Ecosystem Processes

Studies of ecosystem processes are lacking, except for assessments of salmonid production. The Skeena system is considered to be among the three greatest salmonid-producing rivers in the world, after the Fraser and Columbia rivers (Northcote and Larkin 1989). Pink salmon is the most abundant of the anadromous fishes in the Skeena River system, with escapements numbering from 0.5 to 3 million fish per year (Tamblyn and Horn 2001). Pink salmon have an odd-even year alternation of dominant runs, with odd years being higher. Adult sockeye salmon escapements to the river number up to 1.6 million fish per year, of which the Babine Lake stock makes up the largest proportion in the Skeena system. The Babine Lake sockeye run numbers from 0.5 to 1.5 million fish per year. Numbers of adult Chinook salmon range from 40,000 to 50,000 per year, and chum salmon about 10,000 to 20,000 per year (Tamblyn and Horn 2001). Numbers of most stocks appear to be improving after lower numbers during the 1980s and early 1990s. Annual harvests of salmon returning to the Skeena system have averaged 5.8 million pink salmon, ~2 million sockeye, 0.5 million chum, 300,000 coho, and about 80,000 chinook (Tamblyn and Horn 2001).

Human Impacts and Special Features

Although the Skeena River basin is the most populated basin covered in this chapter, with towns mostly based on service industries to resource extraction, population densities in the drainage are still relatively low. Much of the Skeena basin is still largely a forested wilderness area, and the main-stem river is undammed. The major population centers include Terrace, Smithers, and Houston. The city of Prince Rupert is not within the basin proper but lies at the mouth of the river where it enters the Pacific Ocean and is thought to have been the largest Native

American center north of Mexico because of its rich resources and strategic location (Tamblyn and Horn 2001). Overall population densities in the drainage are low relative to other parts of North America.

Forestry and mining occur in the basin, with a minor amount of agriculture. Forest harvesting occurs in areas around the Skeena basin and other coastal areas and generates the highest rates of timber harvest for basins within this region. A plan to divert part of the Skeena basin into the Nechako Reservoir (Fraser basin) was stopped by the government in the early 1990s. At present the Skeena River basin appears to be free of proposals for major dams.

ADDITIONAL RIVERS

The Nushagak River drains a nearly pristine area of southwest Alaska south of the Alaska Range, and near the town of Dillingham flows into Bristol Bay in the south Bering Sea (Fig. 16.19). The lower 100 km of the river has a low gradient and meanders through a large wetlands complex. There are many lakes in the catchment supporting a large sockeye salmon production, and the basin also produces many other salmon. There is less glacial influence on turbidity in this river system because of the low elevation of most of the basin and the presence of many lakes, which themselves are still turbid with glacial "flour." The basin is sparsely populated, largely by Native Americans with a fishing- and hunting-based economy. The University of Washington has maintained a research station at Woods Lakes in the basin for many years. There are two National Wild and Scenic Rivers in the basin, the Mulchatna River and the Chilikadrotna River.

The Copper River is a large braided river flowing south into the Gulf of Alaska just east of Cordova (Fig. 16.21). The river is highly influenced by glaciers from which it originates in the Alaska Range to the north and east parts of the basin. The catchment has little development, but the highways connecting Anchorage and Valdez to the interior parts of the state, including Fairbanks, follows the river north. Most study of the river has been devoted to examining the geomorphology and hydrology of the river to safeguard highway construction, especially bridges associated with the highways, as it has a highly mobile river bed and high rates of bedload movement. The Copper River is also prone to outburst floods (*Jökulhlaup* event) from the sudden release of glacier-dammed lakes, which can cause large flood peaks, potentially causing damage to developments

FIGURE 16.7 Copper River (left to right) at confluence with Chitina River (center in distance), its largest tributary, at Chitina, Alaska (PHOTO BY M. C. T. SMITH).

downstream (Brabets 1997). The river exports an average of 70 million tonnes of suspended sediment per year, as much as the Yukon River, which drains an area >10 times larger (Brabets 1997) (Fig. 16.7). As with other rivers in the region, it helps support a productive fishery for Pacific salmon.

The Alsek River and its major tributary, the Tatshenshini River, are highly braided and geomorphically active glacial rivers, parts of which are in the Canadian Heritage Rivers system. The dramatic mountain landscape of the basin shows the effects of glacial advance and retreat, including the periodic breaking of glacier ice dams, even in modern times

as glaciers have retreated during the past century (Fig. 16.8). The tributaries of the rivers arise from high in the Alaska Range in the Yukon Territories and British Columbia and flow southwest into the Gulf of Alaska (Fig. 16.23). The lower part of the Alsek River passes between high mountains on its way to the ocean. This pristine basin is almost entirely protected in the Kluane National Park in the Yukon Territories and Tatshenshini–Alsek Park in British Columbia. These parks, along with the adjacent Wrangell–St. Elias National Park and Glacier Bay National Park of Alaska, make up the world's largest protected area and have large populations of

FIGURE 16.8 Alsek River (top) at confluence with Tatshenshini River (right), its largest tributary, in British Columbia (PHOTO BY M. C. T. SMITH).

wildlife, including grizzly bear, golden eagle, moose, Dall sheep, and many others. This is another wild salmon river, and the productivity of the Pacific salmon has long been part of the subsistence basis of the native Tlingit people. The river is a world-class destination for river rafting, and recreational tourism is the main activity in the basin.

The Taku River is a highly braided, glacially influenced river flowing from glaciers in the Alaska Range southwest through British Columbia, then flowing westward to enter the Gulf of Alaska near Juneau (Fig. 16.25). This turbid river also experiences outburst floods nearly every year from Tulsequah Lake, which cause sudden increases in flow as glacier-dammed lakes empty and may cause damage downstream. The river supports substantial populations of all the Pacific salmon and steelhead. The highly active riverbed and high amounts of suspended load cause unstable channels and fine sediment deposition in lower-gradient sections but highly productive spawning habitats for salmon in other reaches. Down-

stream of the U.S.–Canada border the river becomes wide for 28 km, and the lower 30 km of the river is broad and tidally influenced, providing nursery habitat for juvenile salmonids prior to going to the ocean (Murphy et al. 1997).

The Nass River originates in the high mountains in the Coast Range Mountains and Hazelton Mountains of British Columbia and flows into the Portland Canal, a fjord of the Pacific Ocean (Fig. 16.27). The braided river has a variety of sources, some highly influenced by glaciermelt and others predominated by snowmelt. The basin is the traditional territory of the Nisga'a people. The land is mostly pristine and as with most rivers in this chapter has no regulation of flows; however, there has been logging and mining activity in the catchment. The basin shows signs of volcanic activity, including more than $15\,km^2$ of Nisga'a lava beds, formed within the past 300 years, and other earlier flows from the Tseax cone. This is another of the productive salmon rivers of the north Pacific.

ACKNOWLEDGMENTS

We are grateful to the many people who provided information in response to our inquiries, particularly Jim Edmondson from the Alaska Department of Fish and Game, and others from the U.S. Geological Survey and B.C. Ministry of Water, Land, and Air Protection. We thank Shauna Bennett, Ian Sharpe, Benoit Godin, and Len Fanning for use of unpublished invertebrate data. We greatly appreciate the considerable efforts of Art Benke and Colbert Cushing in reviewing and editing this chapter.

LITERATURE CITED

Abell, R. A., D. M. Olson, T. H. Ricketts, E. P. Dinerstein, T. Hurley, J. T. Diggs, W. Eichbaum, S. Walters, W. Wettengel, T. Allnutt, C. J. Loucks, and P. Hedao. 2000. *Freshwater ecoregions of North America: A conservation assessment.* Island Press, Washington, D.C.

Alaska Department of Fish and Game. 1984a. Adult anadromous fish investigations, May to October 1983. Susitna Hydro Aquatic Studies Report no. 1. Prepared for the Alaska Power Authority, Anchorage.

Alaska Department of Fish and Game. 1984b. An evaluation of chum and sockeye salmon spawning habitat in sloughs and side channels of the middle Susitna River. Susitna Hydro Aquatic Studies Report no. 3. Prepared for the Alaska Power Authority, Anchorage.

Alaska Department of Fish and Game. 1984c. Resident and juvenile anadromous fish investigations, May to October 1983. Susitna Hydro Aquatic Studies Report no 2. Prepared for the Alaska Power Authority, Anchorage.

Alaska Power Authority. 1985. Instream flow relationships report for the Susitna hydroelectric project. Vol. 1. Prepared for Harza-Ebasco Susitna Joint Venture by E-Woody Trihey and Associates and Woodward-Clyde Consultants.

Baker, T. T., A. C. Wertheimer, R. D. Burkett, R. Dunlap, D. M. Eggers, E. I. Fritts, A. J. Gharrett, R. A. Holmes, and R. L. Wilmot. 1996. Status of Pacific salmon and steelhead escapements in southeastern Alaska. *Fisheries* 21(10):6–18.

Bendock, T. 1989. Lakeward movements of juvenile chinook salmon and recommendations for habitat management in the Kenai River, Alaska, 1986–1988. Fishery Manuscript Series no. 7. Alaska Department of Fish and Game, Division of Sports Fish, Anchorage.

Bendock, T. N. and A. E. Bingham. 1988. Juvenile salmon seasonal abundance and habitat preference in selected regions of the Kenai River, Alaska, 1987–1988. Fishery Manuscript no. 70, Alaska Department of Fish and Game, Division of Sports Fish, Anchorage.

Boggs, K., J. C. Davis, and A. M. Milner. 1997. Aquatic and terrestrial resources of the Kenai River watershed: A synthesis of publications. Report 910/R-97-001. U.S. Environmental Protection Agency, Seattle.

Brabets, T. P. 1997. Geomorphology of the lower Copper River, Alaska. Paper 1581. U.S. Geological Survey, Denver, Colorado.

Brabets, T. P., G. L. Nelson, J. M. Dorava, and A. M. Milner. 1999. Water quality assessment of the Cook Inlet basin: Environmental setting. Water Resources Investigations Report 99-4025. U.S. Geological Survey, Denver, Colorado.

British Columbia Land Use Coordination Office (BC-LUCO). 2000. Cassiar Iskut–Stikine land and resource management plan. Land Use Coordination Office, British Columbia Ministry of Sustainable Resource Management, Victoria.

Brodeur, R. D., and M. S. Busby. 1998. Occurrence of an Atlantic salmon *Salmo salar* in the Bering Sea. *Alaska Fishery Research Bulletin* 5:64–66.

Burger, C. V., R. L. Wangaard, R. L. Wilmot, and A. M. Palmisano. 1983. Salmon investigations in the Kenai River, Alaska: 1979–1981. National Fishery Research Center Report. U.S. Fish and Wildlife Service.

Cannings, S. G., and J. Ptolemy. 1998. *Rare freshwater fish of British Columbia.* British Columbia Ministry of Environment, Lands and Parks, Victoria.

Clarke, A. H. 1981. *The freshwater molluscs of Canada.* National Museums of Canada, Ottawa.

Dingman, S. L. 2002. *Physical hydrology,* 2nd ed. Prentice-Hall, Upper Saddle River, New Jersey.

Dorava, J. M. 1995. Hydraulic characteristics near streamside structures along the Kenai River, Alaska. Water-Resources Investigations Report 95. U.S. Geological Survey, Denver, Colorado.

Dorava, J. M., and A. M. Milner. 1999. Effects of recent volcanic eruptions on aquatic habitat in the Drift River, Alaska: Implications at other Cook Inlet region volcanoes. *Environmental Management* 23:217–230.

Dorava, J. M., and A. M. Milner. 2000. Role of lake regulation on glacier-fed rivers in enhancing salmon productivity: The Cook Inlet watershed, south-central Alaska, USA. *Hydrological Processes* 14:3149–3159.

Dorava, J. M., and G. W. Moore. 1997. Effects of boat-wakes on streamside erosion, Kenai River, Alaska. Water-Resources Investigations Report 97-4015. U.S. Geological Survey, Denver, Colorado.

Dorava, J. M., and K. M. Scott. 1998. The role of glaciers and glacial deposits in the Kenai River watershed and the implications for aquatic habitat. In J. E. Gray and J. R. Ritchie (eds.). *Geological studies in Alaska by the US Geological Survey 1996.* Professional Paper 1595. U.S. Geological Survey, Denver, Colorado.

Duffy, L. K., E. Scofield, T. Rodgers, M. Patton, and R. T. Bowyer. 1999. Comparative baseline levels of mercury, Hsp 70 and Hsp 60 in subsistence fish of the Yukon–Kuskokwim delta region of Alaska. *Comparative Biochemistry and Physiology* C 124:181–186.

Dynesius, M., and C. Nilsson. 1994. Fragmentation and flow regulation of river systems in the northern third of the world. *Science* 266:753–762.

Edmundson, J. A., and A. Mazumder. 2002. Regional and hierarchical perspectives of thermal regimes in subarctic, Alaskan lakes. *Freshwater Biology* 47:1–17.

Farley, A. L. 1979. *Atlas of British Columbia.* University of British Columbia Press, Vancouver.

Gabrielse, H., and C. J. Yorath (eds.). 1991. *Geology of the Cordilleran Orogen in Canada.* Geology of Canada 4. Geological Survey of Canada, Ottawa.

Hammarstrom, S. L. 1995. Stock assessment of the return of late run chinook salmon to the Kenai River, 1994. Fishery Data Series no. 95-3. Alaska Department of Fish and Game, Division of Sports Fish, Anchorage.

Hansen, T. C., and J. C. Richards. 1985. Availability of invertebrate food sources for rearing juvenile chinook salmon in turbid Susitna River habitats. Susitna Aquatic Studies Program Report no. 6. Alaska Department of Fish and Game, Anchorage.

Helfield, J. M., and R. J. Naiman. 2001. Effects of salmon-derived nitrogen on riparian forest growth and implications for stream productivity. *Ecology* 82:2403–2409.

Hunt, C. B. 1974. *Natural regions of the United States and Canada.* W. H. Freeman, San Francisco.

Jang, L., and T. Webber. 1996. *State of water quality of Stikine River above Choquette River, 1981–1996.* British Columbia Ministry of Environment, Lands and Parks, Victoria.

Jones, J. A. A. 1997. *Global hydrology: Processes, resources and environmental management.* Longman, Singapore.

Karr, J. R., J. D. Allan, and A. C. Benke. 2000. River conservation in the United States and Canada. In P. J. Boon, B. R. Davies, and G. E. Petts (eds.), *Global perspectives on river conservation: Science, policy and practice,* pp. 3–39. Wiley, Toronto.

Klinger, S., and E. W. Trihey. 1984. Response of aquatic habitat surface areas to mainstem discharge in the Talkeetna to Devil Canyon reach of the Susitna River, Alaska. Report for Alaska Power Authority, Susitna Hydroelectric Project. E. Woody Trihey and Associates, Anchorage.

Kyle, R. E., and T. P. Brabets. 2001. Water temperature of streams in the Cook Inlet basin, Alaska, and implications of climate change. Water-Resources Investigations Report 01-4109. U.S. Department of Interior and U.S. Geological Survey, Anchorage, Alaska.

Langdon, S. J. 1987. *The native people of Alaska.* Greatland Graphics, Anchorage, Alaska.

Liepitz, G. S. 1994. An assessment of cumulative impacts of development and human uses on fish habitat in the Kenai River. Technical Report 94-6. Alaska Department of Fish and Game, Habitat and Restoration Division, Anchorage.

Leopold, L. B. 1994. *A view of the river.* Harvard University Press, Cambridge, Massachusetts.

Litchfield, V. P., and G. B. Kyle. 1992. Kenai River water quality investigation. Enhancement and Development Report no. 111. Alaska Department of Fish and Game, Division of Fisheries Rehabilitation, Anchorage.

McCarthy, J. J., O. F. Canziani, N. A. Leary, D. J. Dokken, and K. S. White (eds.). 2001. *Climate change 2001: Impacts, adaptation, and vulnerability.* Cambridge University Press, Cambridge.

McGhee, R. 1996. *Ancient people of the Arctic.* University of British Columbia Press, Vancouver.

McGregor, G. M., A. M. Gurnell, G. E. Petts, and A. M. Milner. 1995. Sensitivity of alpine stream ecosystems to climate change. *Aquatic Conservation: Marine and Freshwater Ecosystems* 5:233–247.

McPhail, J. D., and R. Carveth. 1999. *Field key to the freshwater fishes of British Columbia.* Resources Inventory Committee, British Columbia Ministry of Forests, Victoria.

McPhail, J. D., and C. C. Lindsey. 1970. *Freshwater fishes of northwestern Canada and Alaska.* Bulletin 173. Fisheries Research Board of Canada, Ottawa.

Milner, A. M., J. E. Brittain, E. Castella, and G. E. Petts. 2001. Trends of macroinvertebrate community structure in glacier-fed rivers in relation to environmental conditions: A synthesis. *Freshwater Biology* 46:1833–1847.

Milner, A. M., and E. B. Gabrielson. 1994. Bioassessment of the Kenai River, May 1993. University of Alaska Environment and Natural Resources Institute unpublished report to the Alaska Department of Fish and Game.

Milner, A. M., J. G. Irons III, and M. W. Oswood. 1997. The Alaskan landscape: An introduction for limnologists. In A. M. Milner and M. W. Oswood (eds.). *Freshwaters of Alaska: Ecological syntheses,* pp. 1–44. Springer-Verlag, New York.

Milner, A. M., E. Knudsen, C. Soiseth, A. L. Robertson, D. Schell, I. T. Phillips, and K. Magnusson. 2000. Stream community development across a 200 year gradient in Glacier Bay National Park, Alaska, USA. *Canadian Journal of Fisheries and Aquatic Science* 57:2319–2335.

Mount, J. F., P. Moyle, and S. Yarnell (eds.). 2002. *Glacial and periglacial process as hydrogeomorphic and ecological drivers in high-latitude watersheds.* University of California at Davis Geology Department.

Muckle, R. J. 1998. *The First Nations of British Columbia.* University of British Columbia Press, Vancouver.

Murphy, M. L., K. V. Koski, J. M. Lorenz, and J. F. Thedinga. 1997. Downstream migrations of juvenile Pacific salmon (*Oncorhynchus* spp.) in a glacial transboundary river. *Canadian Journal of Fisheries and Aquatic Science* 54:2837–2846.

Naiman, R. J., R. E. Bilby, D. E. Schindler, and J. M. Helfield. 2002. Pacific salmon, nutrients, and the dynamics of freshwater and riparian ecosystems. *Ecosystems* 5:399–417.

Nelson, D. C. 1985. Russian River sockeye salmon study. Project no. F-9-17, 26(G-II-C), pp. 1–59. Alaska Department of Fish and Game, Division of Sports Fish, Anchorage.

Nelson, H., B. R. Larsen, E. A. Jenne, and D. H. Sorg. 1977. Mercury dispersal from lode sources in the Kuskokwim River drainage, Alaska. *Science* 198:820–824.

Northcote, T. G., and P. A. Larkin. 1989. The Fraser River: A major salmonine production system. In D. P. Dodge (ed.). *Proceedings of the International Large River Symposium (LARS)*, Canadian Special Publication of Fisheries and Aquatic Sciences 106:172–204.

Oswood, M. W. 1997. Streams and rivers of Alaska: A high latitude perspective on running waters. In A. M. Milner and M. W. Oswood (eds.). *Freshwaters of Alaska: Ecological Syntheses*, pp. 331–356. Springer-Verlag, New York.

Oswood, M. W., J. G. Irons III, and A. M. Milner. 1995. River and stream ecosystems of Alaska. In C. E. Cushing, K. W. Cummins, and G. W. Minshall (eds.). *River and stream ecosystems*, pp. 9–32. Elsevier, New York.

Oswood, M. W., J. B. Reynolds, J. G. Irons III, and A. M. Milner. 2000. Distributions of freshwater fishes in ecoregions and hydroregions of Alaska. *Journal of the North American Benthological Society* 19:405–418.

Pederson, E. (ed.). 1983. *A larger history of the western Kenai.* Adams Press, Chicago.

Peterson, D. P., and C. J. Foote. 2000. Disturbance of small-stream habitat by spawning sockeye salmon in Alaska. *Transactions of the American Fisheries Society* 129:924–934.

Reynolds, J. B. 1997. Ecology of overwintering fishes in Alaskan freshwaters. In A. M. Milner and M. W. Oswood (eds.). *Freshwaters of Alaska: Ecological syntheses*, pp. 281–302. Springer-Verlag, New York.

Ricketts, T. H., E. Dinerstein, D. M. Olson, C. J. Loucks, W. Eichbaum, D. DellaSala, K. Kavanaugh, P. Hedao, P. T. Hurley, K. M. Carney, R. Abell, and S. Walters. 1999. *Terrestrial ecoregions of North America: A conservation assessment.* Island Press, Washington, D.C.

Schmidt, D. C., K. E. Tarbox, G. V. Kyle, and S. R. Carlson. 1995. Sockeye salmon overescapement: 1993 annual report on Kenai River and Kodiak Investigations, Alaska. Alaska Department of Fish and Game, Commercial Fish Division, Regional Information Report Series 6J95-15, Anchorage.

Schoch, C., and S. R. Bredthauer. 1983. Environmental effects of ice processes on the Susitna River. In J. W. Edrich (ed.). *Managing water resources for Alaska's development.* Report IWR-105. Institute of Water Resources, University of Alaska, Anchorage.

Scott, W. B., and E. J. Crossman. 1973. *Freshwater fishes of Canada.* Bulletin 184. Fisheries Research Board of Canada, Ottawa.

Scott, K. M. 1982. Erosion and sedimentation in the Kenai River, Alaska. Paper 1235. U.S. Geological Survey, Denver, Colorado.

Sedinger, J. S. 1997. Waterfowl and wetland ecology in Alaska. In A. M. Milner and M. W. Oswood (eds.). *Freshwaters of Alaska: Ecological syntheses*, pp. 155–178. Springer-Verlag, New York.

Slaney, T. L., K. D. Hyatt, T. G. Northcote, and R. J. Fielden. 1996. Status of anadromous salmon and trout in British Columbia and Yukon. *Fisheries* 21(10):20–35.

Tamblyn, G. C., and H. Horn. 2001. *Current conditions report: North Coast land and resource management plan.* Land Use Coordination Office, British Columbia Ministry of Water, Land and Air Protection, Victoria.

Taylor, C. A., M. L. Warren Jr., J. F. Fitzpatrick Jr., H. H. Hobbs III, R. F. Jezerinac, W. L. Pfleiger, and H. W. Robison. 1996. Conservation status of crayfishes of the United States and Canada. *Fisheries* 21(4):25–38.

Tarbox, K. E., and L. K. Brannian. 1994. An estimate of juvenile fish densities in Skilak and Kenai Lakes, Alaska through the use of dual-beam hydroacoustic techniques. Technical Fishery Report no. 94-14. Alaska Department of Fish and Game, Division of Commercial Fisheries, Anchorage.

Tschaplinski, P. J. and G. F. Hartman. 1983. Winter distribution of juvenile coho salmon (Oncorhynchus kisutch) before and after logging in Carnation Creek, British Columbia, and some implications for overwinter survival. *Canadian Journal of Fisheries and Aquatic Sciences* 40:452–461.

Van Nieuwenhuyse, E. 1984. Preliminary analysis of the relationships between turbidity and light penetration in the Susitna River, Alaska. Anchorage Technical Memorandum prepared for Harza-Ebasco Susitna Joint Venture. University of Alaska, Anchorage.

Wang, B. 1999. Spatial distribution of chemical constituents in the Kuskokwim River, Alaska. Water-Resources Investigations Report 99-4177. U.S. Geological Survey, Anchorage, Alaska.

Williams, J. E., J. E. Johnson, D. A. Hendrickson, S. Contreras-Balderas, J. D. Williams, M. Navarro-Mendoza, D. E. McAllister, and J. E. Deacon. 1989. Fishes of North America: Endangered, threatened, or of special concern: 1989. *Fisheries* (Bethesda) 14:2–20.

Wood, C. C., B. E. Riddell, and D. T. Rutherford. 1987. Alternative juvenile life histories of sockeye salmon (Oncorhynchus nerka) and their contribution to production in the Stikine River, northern British Columbia. In H. D. Smith, L. Margolis, and C. C. Wood (eds.). *Sockeye salmon* (Oncorhynchus nerka) *population biology and future management*, Canadian Special Publication of Fisheries and Aquatic Sciences 96:12–24.

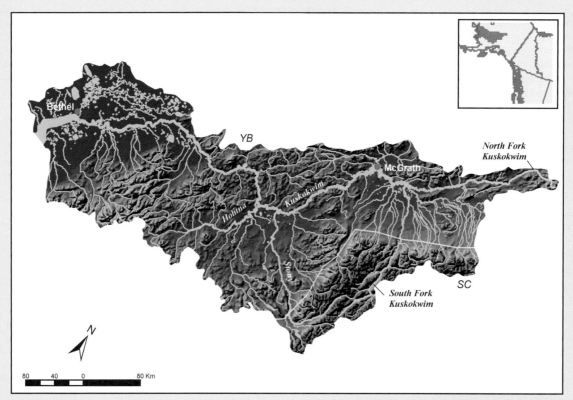

FIGURE 16.9 Map of the Kuskokwim River basin. Physiographic provinces are separated by the yellow line.

KUSKOKWIM RIVER

Relief: >3550 m
Basin area: 124,319 km²
Mean discharge: 1900 m³/s
River order: 9 (approximated)
Mean annual precipitation: 42.0 cm
Mean air temperature: ~1.6°C
Mean water temperature: ~5°C
Physiographic provinces: Yukon Basin (YB), South Central
     Alaska (SC)
Biomes: Tundra, Boreal Forest
Freshwater ecoregion: Yukon
Terrestrial ecoregions: Beringian Lowland Tundra, Beringian Upland
     Tundra, Alaska/St. Elias Range Tundra, Interior Alaska/Yukon
     Lowland Taiga
Number of fish species: 27 to 31
Number of endangered species: none

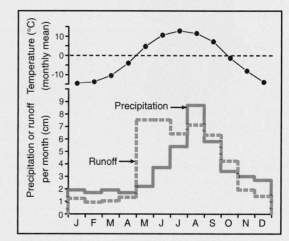

FIGURE 16.10 Mean monthly air temperature, precipitation, and runoff for the Kuskokwim River basin.

Major fishes: sockeye salmon, chinook salmon, coho salmon, chum
     salmon, pink salmon, Pacific lamprey, humpback whitefish, lake trout,
     Dolly Varden, longfin smelt, boreal smelt, eulachon, longnose sucker, burbot, threespine stickleback, coastrange sculpin
Major other aquatic vertebrates: river otter, muskrat, moose, mink, merganser, belted kingfisher, American dipper
Major benthic invertebrates: NA
Nonnative species: none
Major riparian plants: willows, alder
Special features: globally important wildlife area in Yukon–Kuskokwim delta, especially for breeding waterfowl; part of
     Beringian glacial refuge and home to several endemic species; a variety of water sources, including glacier runoff, snowmelt,
     wetlands, and forest; mostly pristine wilderness
Fragmentation: none
Water quality: generally high suspended sediment loads; nutrient concentrations low; NO_2-N + NO_3-N = 0.04 to 0.08 mg/L,
     total phosphorus = 0.01 to 0.05 mg/L; mercury elevated in much of basin from natural sources and past mining
Land use: >99% taiga, tundra, or ice fields, with <1% of the basin developed; subsistence hunting and fishing; Bethel is the
     primary settlement in basin
Population density: ~0.1 people/km²
Major information sources: Nelson et al. 1977, Wang 1999, Dynesius and Nilsson 1994

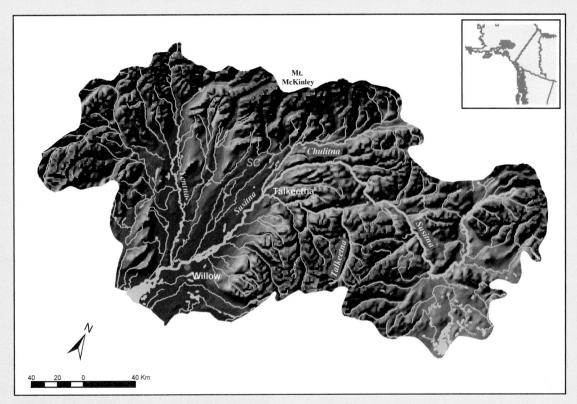

FIGURE 16.11 Map of the Susitna River basin.

SUSITNA RIVER

Relief: >4000 m
Basin area: 51,800 km²
Mean discharge: 1427 m³/s
River order: 7
Mean annual precipitation: 70.8 cm (underestimate)
Mean air temperature: 0.8°C
Mean water temperature: 4.3°C
Physiographic province: South Central Alaska (SC)
Biomes: Tundra, Boreal Forest
Freshwater ecoregion: North Pacific Coastal
Terrestrial ecoregions: Cook Inlet Taiga, Alaska/St. Elias Range
  Tundra, Copper Plateau Taiga
Number of fish species: 25 to 29
Number of endangered species: none
Major fishes: chinook salmon, coho salmon, chum salmon, pink
  salmon, sockeye salmon, Bering cisco, round whitefish, rainbow
  trout, Dolly Varden, Arctic grayling, Arctic lamprey, burbot

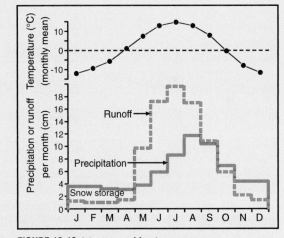

FIGURE 16.12 Mean monthly air temperature, precipitation, and runoff for the Susitna River basin.

Major other aquatic vertebrates: river otter, muskrat, moose, mink, merganser, belted kingfisher, American dipper
Major benthic invertebrates: chironomid midges, oligochaete worms, stoneflies (capniids), mayflies (*Baetis, Cinygmula*)
Nonnative species: none
Major riparian plants: willow, alder, cottonwood
Special features: major glacier-fed system that dominates landscape to the southwest of Alaska Range with many complex, off-
  channel habitats; drains from Mount McKinley, highest point in North America, and other glaciated peaks of the Alaska
  Range (U.S.)
Fragmentation: none
Water quality: no major pollutants; middle river (Gold Creek) pH = 7.3, alkalinity = 50 mg/L as CaCO₃, NO₃-N = 0.15 mg/L,
  total phosphorus = 0.12 mg/L, turbidity (summer) = 200 NTU
Land use: 95% forest and tundra, 5% glaciers; mostly wilderness
Population density: ~0.01 people/km²
Major information sources: Kyle and Brabets 2001; Alaska Department of Fish and Game 1984a, 1984b, 1984c; Alaska Power
  Authority 1985

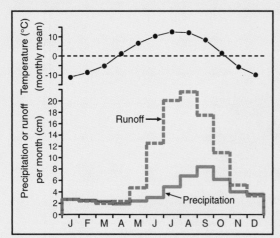

FIGURE 16.14 Mean monthly air temperature, precipitation, and runoff for the Kenai River basin.

KENAI RIVER

Relief: >1500 m
Basin area: 5206 km²
Mean discharge: 167.7 m³/s
River order: 6
Mean annual precipitation: 48.8 cm (underestimate)
Mean air temperature: 0.9°C
Mean water temperature: 6.4°C
Physiographic province: South Central Alaska (SC)
Biome: Temperate Coniferous Forest
Freshwater ecoregion: North Pacific Coastal
Terrestrial ecoregions: Northern Pacific Coastal Forests, Pacific Coastal Mountain Tundra and Ice Fields
Number of fish species: 28
Number of endangered species: none
Major fishes: Dolly Varden, Arctic grayling, chinook salmon, sockeye salmon, coho salmon, pink salmon, chum salmon, rainbow trout, steelhead, Pacific lamprey, humpback whitefish, lake trout, longfin smelt, boreal smelt, eulachon, longnose sucker, burbot, threespine stickleback, coastrange sculpin

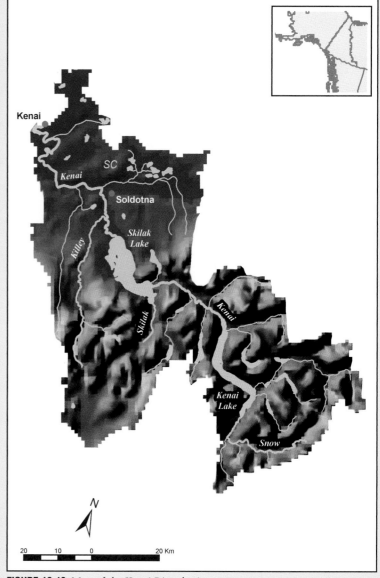

FIGURE 16.13 Map of the Kenai River basin.

Major other aquatic vertebrates: river otter, muskrat, beaver, moose, mink, merganser, belted kingfisher, American dipper
Major benthic invertebrates: chironomid midges (*Diamesa, Cricotopus/Orthocladius, Pagastia, Parakiefferiella, Eukiefferiella, Rheotanytarsus*), mayflies (*Baetis, Cinygmula, Ephemerella*), stoneflies (*Paraleuctra, Plumiperla, Isoperla*), caddisflies (*Brachycentrus, Ecclisocosmoceus, Glossosoma, Hydropsyche*)
Nonnative species: none
Major riparian plants: willow, alder
Special features: diversity of clearwater, glacier-influenced, and wetland-stained river systems; very productive salmon river; large tourism industry based on freshwater and tidal zone fishing; valuable economies through commercial and sports fisheries
Fragmentation: none
Water quality: no major pollutants except possible organic enrichment around Soldotna; pH = ~7.0, alkalinity = 25 mg/L as CaCO₃, conductivity = 30 to 81 μS/cm, NO₃-N + NO₂-N = 150 to 250 μg/L, total phosphorus = <10 μg/L, turbidity = 2 to 18 NTU
Land use: >95% forest and wetlands, with small amounts of development; primarily undeveloped with many wetland areas; recreational cabins along river with some guiding/tourist facilities, mostly on main stem
Population density: ~3.3 people/km²
Major information sources: Boggs et al. 1997, Brabets et al. 1999, Litchfield and Kyle 1992

766

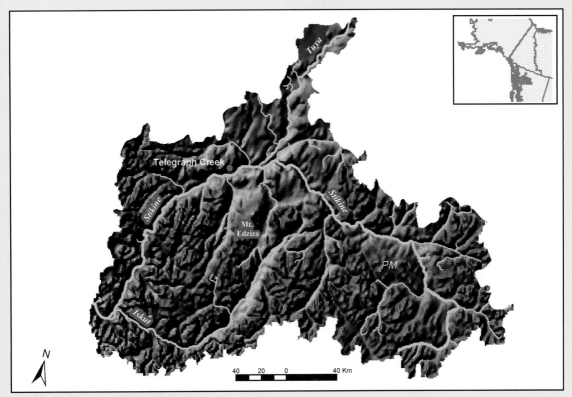

FIGURE 16.15 Map of the Stikine River basin.

STIKINE RIVER

Relief: >2900 m
Basin area: 51,592 km^2
Mean discharge: 1587 m^3/s
River order: 8
Mean annual precipitation: 37.7 cm (underestimate)
Mean air temperature: 2.0°C
Mean water temperature: 6.6°C
Physiographic province: Coast Mountains of British Columbia and Southeast Alaska (PM)
Biomes: Temperate Mountain Forest, Boreal Forest, Tundra
Freshwater ecoregion: North Pacific Coastal
Terrestrial ecoregions: Northern Cordillera Forests, Pacific Coastal Mountain Tundra and Ice Fields, Northern Pacific Coastal Forests, Northern Transitional Alpine Forests
Number of fish species: 22 to 26
Number of endangered species: 1 fish (threatened)

FIGURE 16.16 Mean monthly air temperature, precipitation, and runoff for the Stikine River basin.

Major fishes: sockeye salmon, chinook salmon, Arctic grayling, burbot, chum salmon, coho salmon, cutthroat trout, longnose sucker, mountain whitefish, rainbow trout, lake chub, pink salmon, threespine stickleback, sculpins
Major other aquatic vertebrates: river otter, muskrat, beaver, moose, mink, merganser, belted kingfisher, American dipper
Major benthic invertebrates: mayflies (*Baetis, Cinygmula, Rhithrogena, Ameletus*), stoneflies (*Suwallia, Capnia, Doddsia, Isoperla, Podmosta, Zapada*), chironomid midges (*Diamesa, Cricotopus, Orthocladius, Euryhapsis, Eukiefferiella*), Acarina
Nonnative species: none; threats of Atlantic salmon; historical records of American shad
Major riparian plants: willows, alders, cottonwood
Special features: one of the largest free-flowing rivers draining a temperate biome; largely pristine wilderness with some glacial drainage; drains Edziza Provincial Park and Spatsizi Plateau Wilderness Provincial Park
Fragmentation: none
Water quality: naturally elevated metals, copper usually = >10 µg/L (5 to 60 µg/L), iron usually = >0.3 mg/L (0.2 to 11 mg/L), zinc = ~15 µg/L; alkalinity = 45 to 90 mg/L as CaCO$_3$, pH = 7.8, NO$_3$-N = 0.1 mg/L, total phosphorus = 0.2 mg/L, turbidity (summer maximum) = 50 to 350 NTU
Land use: subsistence hunting and fishing, mining, and low levels of timber extraction; over 95% of the basin is covered by forest, alpine tundra, and ice fields
Population density: <0.025 people/km^2
Major information sources: www.dfo-mpo.gc.ca 1999, L. Fanning, unpublished data

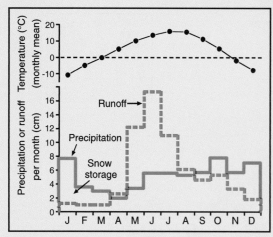

FIGURE 16.18 Mean monthly air temperature, precipitation, and runoff for the Skeena River basin.

SKEENA RIVER

Relief: 2755 m
Basin area: 54,400 km²
Mean discharge: 1760 m³/s
River order: 8
Mean annual precipitation: 62.5 cm (underestimate)
Mean air temperature: 4.3°C
Mean water temperature: ~6.0°C
Physiographic province: Coast Mountains of British Columbia and Southeast Alaska (PM)
Biomes: Temperate Mountain Forest, Boreal Forest, Tundra
Freshwater ecoregion: North Pacific Coastal
Terrestrial ecoregions: British Columbia Mainland Coastal Forests, Northern Transitional Alpine Forests, Fraser Plateau and Basin Complex, Central British Columbia Mountain Forests
Number of fish species: 33
Number of endangered species: 1 fish
Major fishes: pink salmon, sockeye salmon, chinook salmon, coho salmon, steelhead, rainbow trout, cutthroat trout, Dolly Varden, eulachon, burbot, mountain whitefish, chum salmon, prickly sculpin, Pacific lamprey, burbot, largescale sucker, peamouth chub, pygmy whitefish, white sucker, white sturgeon, northern pikeminnow
Major other aquatic vertebrates: river otter, muskrat, beaver, moose, mink, merganser, belted kingfisher, American dipper, tailed frog
Major benthic invertebrates: chironomid midges, black flies, Oligochaete worms, mayflies (*Baetis, Rhithrogena, Epeorus, Cinygmula, Drunella, Serratella, Paraleptophlebia*), stoneflies (*Sweltsa, Zapada*), caddisflies (Hydropsychidae, *Rhyacophila*)
Nonnative species: none; threats of Atlantic salmon; historical records of American shad
Major riparian plants: willows, alders, vine maple, western red cedar, western hemlock
Special features: largest free-flowing river in North America draining a temperate biome; productive salmon river
Fragmentation: very low; dam in tributary affects about 0.2% of total flow
Water quality: alkalinity = 41 mg/L as CaCO₃, pH = 7.6, NO₃-N = 0.08 mg/L, total phosphorus = 0.057 mg/L, turbidity = 16.9 NTUs (summer maximum up to 250)
Land use: >95% forest; moderate levels of timber extraction, some mining, limited agriculture
Population density: <3 people/km²
Major information sources: McPhail and Carveth 1999, Shauna Bennett and Ian Sharpe, personal communication, Dynesius and Nilsson 1994

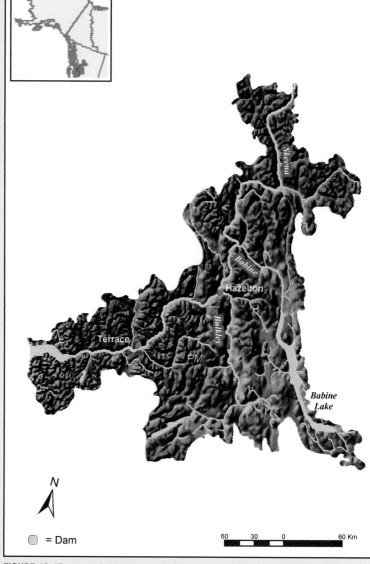

FIGURE 16.17 Map of the Skeena River basin.

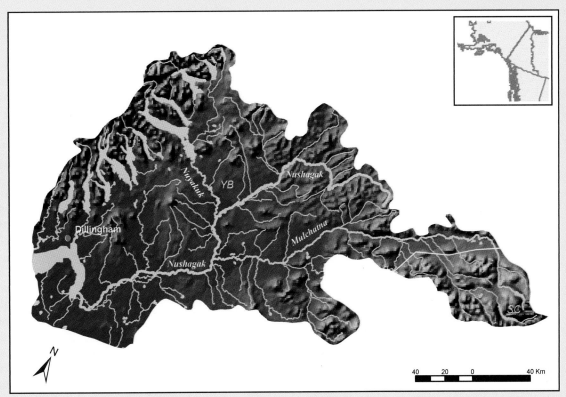

FIGURE 16.19 Map of the Nushagak River basin. Physiographic provinces are separated by the yellow line.

NUSHAGAK RIVER

Relief: ~600 m
Basin area: 34,706 km²
Mean discharge: 1000 m³/s
River order: 8
Mean annual precipitation: 66.3 cm (underestimate)
Mean air temperature: 0.9°C
Mean water temperature: NA
Physiographic provinces: Yukon Basin (YB), South Central Alaska (SC)
Biomes: Tundra, Boreal Forest, Temperate Mountain Forest
Freshwater ecoregion: Yukon
Terrestrial ecoregions: Beringia Lowland Tundra, Interior
    Alaska/Yukon Lowland Taiga, Alaska/St. Elias Range Tundra
Number of fish species: 27 to 31
Number of endangered species: none
Major fishes: sockeye salmon, chinook salmon, coho salmon, chum
    salmon, pink salmon, Alaska blackfish, Dolly Varden, northern
    pike, rainbow trout, slimy sculpin, Arctic grayling, lake trout,
    threespine stickleback, rainbow smelt, humpback whitefish, Pacific lamprey, burbot
Major other aquatic vertebrates: river otter, muskrat, moose, mink, merganser, belted kingfisher, American dipper
Major benthic invertebrates: mayflies (Ephemerellidae), chironomid midges, triclad flatworms, stoneflies (Chloroperlidae,
    Nemouridae), crustaceans (ostracods), caddisflies (Hydroptilidae, Glossosomatidae), bivalves
Nonnative species: none
Major riparian plants: grasses, willow, alder
Special features: pristine wilderness; Mulchatna and Chilikadrotna tributaries are U.S. National Wild and Scenic Rivers; large
    glacial lakes are predominant feature of basin; long-term studies of salmon and their food webs by University of Washington
    at Woods Lakes research station
Fragmentation: none
Water quality: high quality, no pollution
Land use: >99% of the basin is forest, taiga, and tundra; mostly wilderness; subsistence hunting and fishing by native
    population; Dillingham is largest settled area, with ~2500 people
Population density: <0.1 people/km²
Major information source: Peterson and Foote 2000

FIGURE 16.20 Mean monthly air temperature, precipitation, and runoff for the Nushagak River basin.

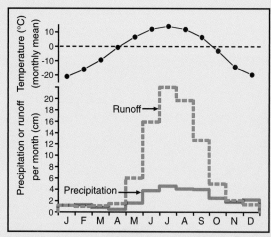

FIGURE 16.22 Mean monthly air temperature, precipitation, and runoff for the Copper River basin.

COPPER RIVER

Relief: >2500 m
Basin area: 63,196 km²
Mean discharge: 1785 m³/s
River order: ~7
Mean annual precipitation: 28.4 cm (underestimate)
Mean air temperature: −2.7°C
Mean water temperature: NA
Physiographic province: South Central Alaska (SC)
Biomes: Tundra, Boreal Forest
Freshwater ecoregion: North Pacific Coastal
Terrestrial ecoregions: Copper Plateau Taiga, Pacific
    Coastal Mountain Tundra and Ice Fields, Alaska/
    St. Elias Range Tundra
Number of fish species: 24 to 27
Number of endangered species: none
Major fishes: coho salmon, longnose sucker, arctic
    grayling, Dolly Varden, chinook salmon, sockeye
    salmon, chum salmon, round whitefish, slimy
    sculpin
Major other aquatic vertebrates: river otter, muskrat,
    beaver, moose, mink, merganser, belted kingfisher, American dipper
Major benthic invertebrates: chironomid midges (particularly *Diamesa*), black flies, craneflies, mayflies (Baetidae, Ameletidae,
    Ephemerellidae, Heptageniidae), stoneflies (Nemouridae, Chloroperlidae), caddisflies (Lepidostomatidae)
Nonnative species: none
Major riparian plants: willows, alders
Special features: Gulkana River, a tributary, is a U.S. National Wild and Scenic River; drains from Wrangell–St. Elias National
    Park (U.S.) and Kluane National Park (Canada)
Fragmentation: none
Water quality: no pollution
Land use: >99% of the basin is in forest, tundra, or ice; undeveloped
Population density: <0.05 people/km²
Major information source: Mount et al. 2002

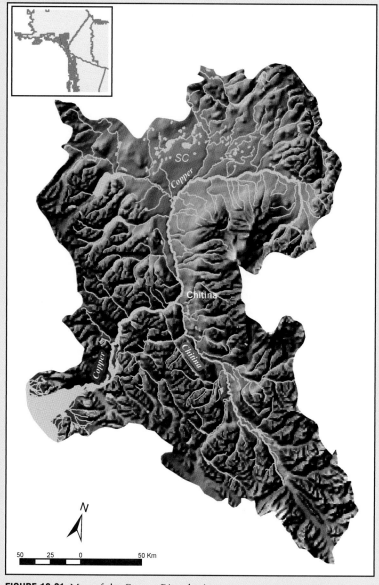

FIGURE 16.21 Map of the Copper River basin.

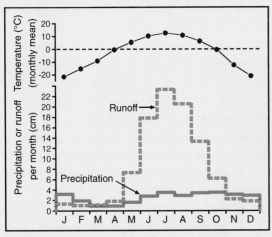

FIGURE 16.24 Mean monthly air temperature, precipitation, and runoff for the Alsek River basin.

ALSEK RIVER

Relief: >3000 m
Basin area: 28,023 km^2
Mean discharge: 862.6 m^3/s
River order: 6
Mean annual precipitation: 30.6 cm (underestimate)
Mean air temperature: −2.9°C
Mean water temperature: NA
Physiographic provinces: South Central Alaska (SC), Coast Mountains of British Columbia and Southeast Alaska (PM), Yukon Basin (YB)
Biomes: Tundra, Boreal Forest, Temperate Mountain Forest
Freshwater ecoregion: North Pacific Coastal
Terrestrial ecoregions: Pacific Coastal Mountain Tundra and Ice Fields, Alaska/St. Elias Range Tundra, Northern Cordillera Forests
Number of fish species: ≥32
Number of endangered species: none
Major fishes: Arctic grayling, chinook salmon, coho salmon, cutthroat trout, Dolly Varden, sockeye salmon, steelhead, chum salmon, burbot, northern pike, lake trout
Major other aquatic vertebrates: river otter, muskrat, beaver, moose, mink, merganser, belted kingfisher, American dipper
Major benthic invertebrates: chironomid midges (*Diamesa, Cricotopus, Rheotanytarsus, Eukiefferiella, Thienemannimyia*), mayflies (*Cinygmula, Baetis, Ephemerella, Rhithrogena, Drunella*), stoneflies (*Sweltsa, Capnia*)
Nonnative species: none; threats of Atlantic salmon; historical records of American shad
Major riparian plants: willows, alders
Special features: most of basin protected within wilderness areas and National Parks; Alsek and its major tributary, the Tatshenshini River, are part of Canadian Heritage River system and part of World Heritage Site (UNESCO); popular for river rafting
Fragmentation: one tributary has dam affecting ~2% of Alsek flow in live storage
Water quality: sparse data for Alsek River above Bates River; alkalinity = ~70 mg/L as CaCO$_3$, hardness = ~90 mg/L as CaCO$_3$, conductivity = ~180 µS/cm, NO$_3$-N = ~0.08 mg/L, total phosphorus = ~0.2 mg/L, pH = 8.2, turbidity (summer) peaks = 65 to 175 NTU
Land use: largely pristine and protected in parks and reserves, such as Kluane National Park in the Yukon Territories; minor amounts of development; >95% of the basin is covered in forest, taiga, tundra, and ice
Population density: <0.01 people/km^2
Major information sources: Benoit Godin, personal communication, Dynesius and Nilsson 1994

FIGURE 16.23 Map of the Alsek River basin. Physiographic provinces are separated by yellow lines.

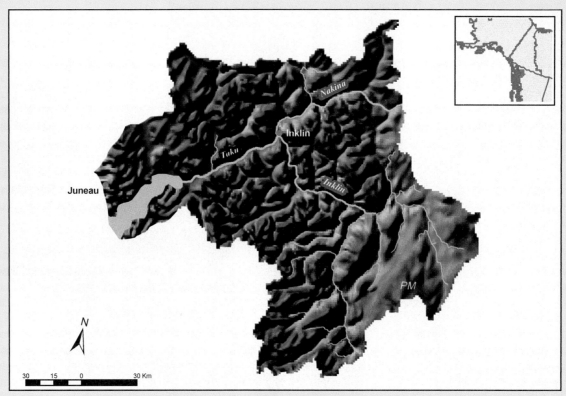

FIGURE 16.25 Map of the Taku River basin.

TAKU RIVER

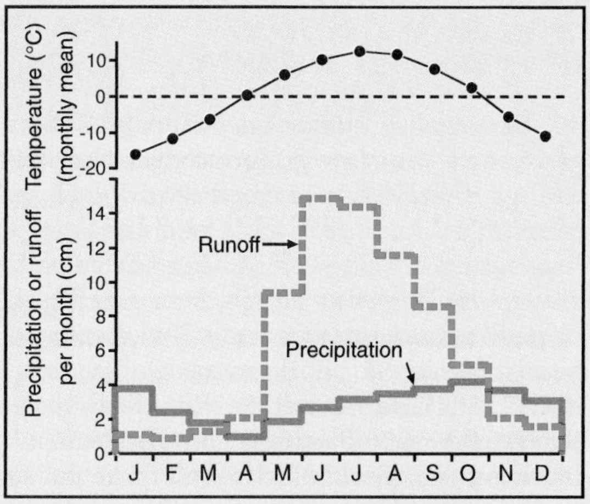

Relief: >2300 m

Basin area: 29,800 km^2

Mean discharge: 600 m^3/s

River order: 8

Mean annual precipitation: 33.8 cm (underestimate)

Mean air temperature: 0°C

Mean water temperature: NA

Physiographic province: Coast Mountains of British Columbia and Southeast Alaska (PM)

Biomes: Tundra, Boreal Forest, Temperate Mountain Forest

Freshwater ecoregion: North Pacific Coastal

Terrestrial ecoregions: Pacific Coastal Mountain Tundra and Ice Fields, Northern Cordillera Forests, Yukon Interior Dry Forests

Number of fish species: ≥32

Number of endangered species: none

Major fishes: sockeye salmon, pink salmon, coho salmon, chinook salmon, Dolly Varden, cutthroat trout, steelhead, chum salmon, eulachon, longfin smelt, Pacific lamprey, round whitefish, slimy sculpin, threespine stickleback

Major other aquatic vertebrates: river otter, muskrat, beaver, moose, mink, merganser, belted kingfisher, American dipper

Major benthic invertebrates: NA

Nonnative species: none; threats of Atlantic salmon; historical records of American shad

Major riparian plants: alders, willows

Special features: primarily wilderness basin; peak flows fill Tulsequah Lake and cause lake outburst events most years (as lake swells the water can float the glacier or melt it sufficiently to cause sudden releases); popular river for rafting; very productive salmon river

Fragmentation: none

Water quality: elevated Cu, Pb, and Zn concentrations; turbidity = 200 NTUs (summer peak); very little "clear" water (not glacier influenced) in basin

Land use: basin is covered by forest, tundra, and ice fields, with <1% of basin developed; subsistence hunting and fishing, ecotourism, forestry, some placer mining; threats from mine developments, especially proposed reactivation of Tulsequah Chief mine (copper and other metals)

Population density: <0.025 people/km^2

Major information source: McPhail and Carveth 1999

FIGURE 16.26 Mean monthly air temperature, precipitation, and runoff for the Taku River basin.

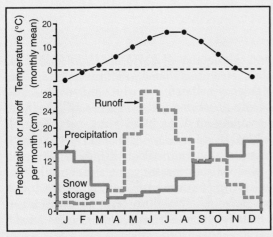

FIGURE 16.28 Mean monthly air temperature, precipitation, and runoff for the Nass River basin.

NASS RIVER

Relief: 2368 m
Basin area: 21,100 km²
Mean discharge: 892 m³/s
River order: 7
Mean annual precipitation: 129.5 cm
Mean air temperature: 6.1°C
Mean water temperature: NA
Physiographic province: Coast Mountains of British Columbia and Southeast Alaska (PM)
Biomes: Temperate Mountain Forest, Boreal Forest, Tundra
Freshwater ecoregion: North Pacific Coastal
Terrestrial ecoregions: British Columbia Mainland Coastal Forests, Northern Transitional Alpine Forests, Pacific Coastal Mountain Tundra and Ice Fields
Number of fish species: 27
Number of endangered species: none
Major fishes: chinook salmon, steelhead, chum salmon, coho salmon, sockeye salmon, pink salmon, cutthroat trout, eulachon, mountain whitefish, Dolly Varden, Pacific lamprey, threespine stickleback
Major other aquatic vertebrates: river otter, muskrat, beaver, moose, mink, merganser, belted kingfisher, American dipper
Major benthic invertebrates: NA
Nonnative species: none; threats of Atlantic salmon; historical records of American shad
Major riparian plants: willows, alders, cottonwood
Special features: largely pristine catchment draining extensive areas in lee of Coast Range Mountains; flows 380 km from most inland source to Pacific; drains large areas of high-elevation forest and tundra; productive salmon river
Fragmentation: none
Water quality: high quality
Land use: >95% forest, tundra, and a small amount of ice fields; largely wilderness; low levels of timber harvest and subsistence fishing, hunting, and trapping
Population density: ~0.14 people/km²
Major information source: McPhail and Carveth 1999

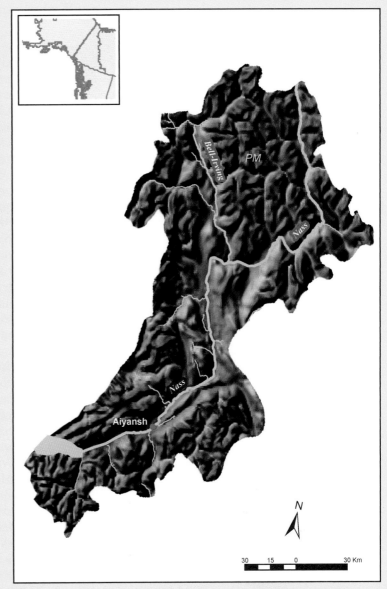

FIGURE 16.27 Map of the Nass River basin.

17

YUKON RIVER BASIN

ROBERT C. BAILEY

INTRODUCTION

YUKON RIVER MAIN STEM

TANANA RIVER

KOYUKUK RIVER

ADDITIONAL RIVERS

ACKNOWLEDGMENTS

LITERATURE CITED

INTRODUCTION

The Yukon basin is the 7th largest in North America (after the Mississippi, Missouri, St. Lawrence, Nelson, Mackenzie, and Rio Grande), with an area of $839,200\,km^2$. Approximately 61% of the basin is in Alaska and 39% is in Canada (Fig. 17.2). Most (>90%) of the Canadian portion of the basin is in the Yukon Territory, although many of the headwater lakes and streams are in northern British Columbia. The Yukon River is one of the wildest and is by far the longest free-flowing river in North America (Dynesius and Nilsson 1994), with only a single dam at its headwaters and one each on two of its major tributaries. From its headwater lakes less than 30 km from the Pacific Ocean in British Columbia to its outflow into the Bering Sea 3200 km downstream in Alaska, the vast Yukon River and its basin encompass a heterogeneous collection of often spectacular ecosystems, from ephemeral mountain streams to minideserts, from opaque glacial runoff rivers to clear boreal forest creeks, from torrential running waters to productive wetland "flats" of thousands of km^2 of small lakes and ponds. The southern origin of the Yukon River, draining the eastern side of the Coast Mountains and the west side of the Rockies in

northern British Columbia, is at a latitude of 59°N and a longitude that is about 15° west of Los Angeles (133°W). The Bering Sea outflow of the river at 63°N is 30° west (164°W) of the headwaters, and the river basin itself extends to a northern limit of about 68°N in the Porcupine River watershed.

Although sparsely populated, the Yukon River basin has been prehistorically and more recently very important to the human population of North America and the world. Most anthropologists agree that the first human colonization of North America was via a land bridge that crossed what is now Bering Strait >12,000 years ago (Yesner 2001). Subsequent to the colonization, the basin became an important resource-rich environment for the development and evolution of northern aboriginal communities. In the last 300 years the Yukon River basin has attracted fur traders and gold miners from Russia, southern North America, and around the world. More recently, and not without controversy, the basin has been part of a transportation route to the south for oil and gas rather than humans.

This chapter discusses the major features of the main-stem Yukon River, as well as two of its major tributaries, the Tanana and Koyukuk rivers (see Fig. 17.2). Abbreviated physical and biological descriptions are included for several smaller tributary rivers,

← **FIGURE 17.1** South Fork of the Koyukuk River, from the Dalton Highway along the Trans-Alaskan pipeline, Alaska (PHOTO BY D. SHIOZAWA).

775

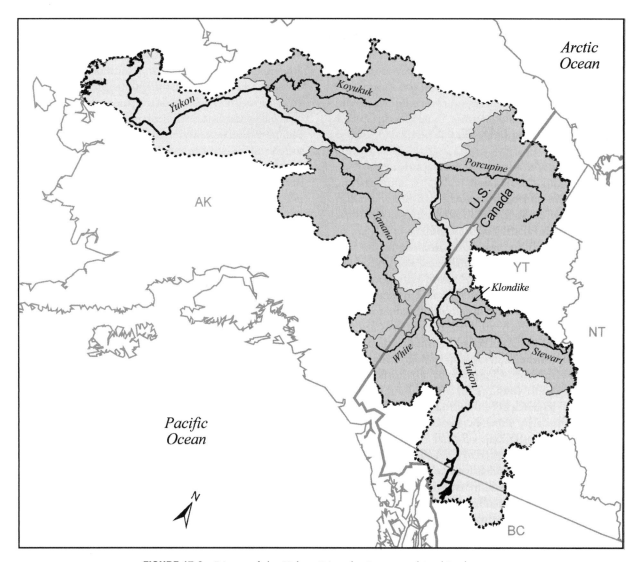

FIGURE 17.2 Rivers of the Yukon River basin covered in this chapter.

including the historically and economically important Klondike River and Stewart River, as well as the largely glacial meltwater White River and the tundra-draining Porcupine River.

Physiography and Climate

With the exception of its headwater lakes, which are in the Coast Mountains of British Columbia and Alaska (PM) physiographic province to the southeast, most of the body of the Yukon River's catchment area is in the Yukon Basin province (YB) (Fig. 17.7). The Mackenzie Mountains (MM), Brooks Range (BR), Seward Peninsula and Bering Coast Uplands (SB), and South Central Alaska (SC) physiographic provinces define the watershed in the

northeast, north, northwest, and southwest, respectively. As Hunt (1974) describes, the Yukon Basin is structurally the highest part of Alaska, exposing the oldest (Precambrian) rock in northwestern North America. Its watershed is largely defined by a south-to-north divergence of the Rocky Mountains. This divergence includes the Cassiar, Selwyn, Ogilvie, Richardson, and British mountains and Brooks Range to the east and north, and the larger Coast Mountains, including the St. Elias (with Mount Logan, Canada's highest peak at 5959 m asl) and Alaska ranges (with Mount McKinley, North America's highest peak at 6194 m asl), to the south of the basin.

Except for the outflow of the basin on the west coast of Alaska, which has a Transitional-

Continental climate, the Yukon River basin lies in the Continental climate zone (National Climate Center 1982), characterized by cool summers and very cold winters but perhaps best described as seasonally variable (Wahl et al. 1987). Temperature and particularly precipitation are affected by the mountain ranges that line the southern portion of the basin, where the moist warm air from the Pacific Ocean is often stalled. Areas directly east of the St. Elias Mountains, in the White River subbasin and headwater lakes of the Yukon River, receive as little as 20 cm/yr precipitation (Wahl et al. 1987). The famous Carcross Desert, near Bennett and Tagish lakes (part of the headwaters of the Yukon River), is a striking example of a rain shadow in the region. Although there is more precipitation in the central and lower parts of the basin, it generally totals less than 50 cm, considerably lower than the continental, temperate part of North America (75 to 100 cm) and much lower than maritime areas (>100 cm) but similar to southern rain-shadow areas, such as Alberta in Canada (40 to 50 cm).

Basin Landscape and Land Use

It is ironic that much of the lower (downstream of the Canada–U.S. border) Yukon River basin, with a climate that is cold relative to most of North America, was largely unglaciated during the various advances of the Cordilleran ice sheet in the Wisconsin glaciation of the Pleistocene (Hopkins et al. 1982). Hunt (1974) points out that both cold and precipitation are necessary for glaciation, and the relative dryness of the Yukon basin spared most of it from the ice front. Approximately 18,000 years ago sea level was low enough to expose a 1500 km wide land bridge across the Bering Sea, and populations of large herbivorous mammals, such as woolly mammoth, horse, and bison, colonized East Beringia, or what is now the Yukon River basin (Hopkins et al. 1982). The periodically emerging land bridge is considered very significant in the biogeographical history of North American fauna, whereas the Bering Strait is a key marine biogeographic linkage between the Pacific and Atlantic oceans during periods when the land bridge is submerged.

Debate about the specific nature of the ecosystems of the Yukon River basin between 8000 and 20,000 years ago has raged for many years. Divergent views on the nature of the "steppe-tundra" vegetation community, in particular its ability to support a diverse megafauna of herbivores, have been proposed and defended, often with some rancor (Yurtsev 2001). It

is clear, at any rate, that the Beringean refugium from the most recent glaciations provided an opportunity for evolutionary divergence while much of temperate North America was "shut down" by ice cover. However, climatic conditions in the basin were even harsher than at present, with the consequence that we do not see much evidence of evolutionary radiation of aquatic or terrestrial taxa during this period. The unique evolutionary opportunity was largely unrealized, such that the region has been far more important from a biogeographic than an evolutionary perspective.

The Yukon River basin encompasses several terrestrial ecoregions within the Boreal Forest and Tundra biomes, including Yukon Interior Dry Forests in the headwater area, Interior Yukon/Alaska Alpine Tundra in the middle part of the basin, Interior Alaska/Yukon Lowland Taiga in the most northerly part of the basin (including the significant and productive Yukon Flats area), and the Beringia Lowland Tundra at the deltaic outflow of the Yukon River into the Bering Sea (Ricketts et al. 1999). Ecoregions in the mountain ranges defining the basin include Ogilvie/MacKenzie Alpine Tundra, Brooks/British Range Tundra, Beringia Upland Tundra, and Alaska/St. Elias Range Tundra (Ricketts et al. 1999). Tundra shrubs and sedges occur throughout much of the Yukon basin, depending on the elevation and microhabitat. White and black spruce are common throughout the basin, and alpine fir and lodgepole pine are quite common in the southern headwaters. Alaska larch is fairly abundant in the middle reaches of the Yukon River and its tributaries in central Alaska. Lowland areas throughout the basin are characterized by several species of willows, as well as balsam poplar, quaking aspen, white birch, Alaska paper birch, and green alder. Riparian tree vegetation often includes mountain and water birch (Farrar 1995).

Not surprising for an area almost entirely above 60°N latitude, the basin has much lower species richness in both aquatic (fishes, mussels, crayfishes, and herpetofauna; Abell et al. 2000) and terrestrial (birds, mammals, butterflies, reptiles, land snails, and vascular plants; Ricketts et al. 1999) communities than similar-size areas in southern North America, although it harbors considerably more biodiversity than similar latitudes of eastern North America. There have been many hypotheses proposed to account for the often-observed pattern of decreasing diversity as one moves from the equator toward the poles, including a harsher physical-chemical environment, less predation pressure, lower primary productivity, more intrayear climatic variation, and younger evolutionary "age" in the far north than in the tropics (Begon et al. 1996). Wiggins and Parker (1997), in discussing the caddisflies in particular, note not only a decline in diversity as one moves north but a decline of lotic species with a proportional increase of lentic species, and propose that lentic species were better at crossing the Bering land bridge than lotic species. Cobb and Flanagan (1980) attribute the low diversity of mayfly fauna in northern latitudes to constrained dispersal of adults in cold weather and increased reliance on dispersal of nymphs in the aquatic environment. None of these explanations has been thoroughly refuted or supported, but it is reasonable to assume that some combination of them is responsible for the low biodiversity of the Yukon River basin. Many of the relatively small number of species that occur in the Yukon River basin have adaptations to arctic or subarctic environment (e.g., insects; Downes 1965, Danks et al. 1997).

In addition to its importance for North American biogeography in general, Beringia (Yukon Territory, Alaska, eastern Siberia) has played a key role in the establishment and dispersal of humans into and throughout North America (Clark 1991, Yesner 2001). About 25,000 years ago humans in Eurasia learned how to live in the far north. Although there is much debate among archaeologists about the specific date, humans dispersed from Eurasian into North American Beringia at least 12,000 years ago (Yesner 2001). Following the Wisconsin glaciation, some humans dispersed up the Yukon River basin, south and east into central North America. The first peoples who stayed in the Yukon basin, largely descendants of the Athapaskans, have maintained small but persistent and increasingly powerful populations. They have in recent decades achieved a measure of self-government through land-claim settlements in both Alaska and Yukon Territory. Traditional territories of aboriginal peoples within the Yukon River basin include the nations derived from eighteenth-century Tlingit, Tagish, Tutchone, Han, Tanana, Gwichin, Tanana, Tanacross, Koyukon, Kolchan, and Ingalik tribes of what is now Yukon Territory and Alaska (Clark 1991).

Secondary human colonization of northwestern North America by Russia began during the reign of Peter the Great in the early eighteenth century but was mainly focused on the coast rather than the Alaskan interior. In 1724, the tsar decided that he wanted to know whether and where Russia and America were separated, so he sent Danish captain Vitus Bering on his first voyage (Hayes 1999). Bering actually reached the strait that bears his name by 1728, but the proximity (and separation) of North America and Asia was first elucidated by Mikhail Gwosdev in 1732 (Hayes 1999). The British, Spanish, Americans, and others joining in the search for the northwest passage and exploitation of abundant wildlife and fisheries resources in and off the coast of Alaska followed throughout the late eighteenth and nineteenth centuries. In 1825, the Russians and British established 141°W longitude as the boundary between Russian and British North America, in addition to what is now known as the Alaskan panhandle (Hayes 1999). In 1867, a treaty was negotiated by Secretary of State William H. Seward by which the United States purchased Alaska for $7.2 million, or about 5 cents/ha.

There had been modest gold discoveries within the Yukon River basin when, in 1896, gold nuggets were found in Rabbit (later Bonanza) Creek, a tributary of the Klondike River just outside of Dawson City, Yukon. A gold rush began in 1897 and swelled in 1898, as miners and adventurers poured in, mainly from the United States. Placer mining continues to this day in the Klondike River, Stewart River, and adjacent basins, and to a lesser degree in parts of the headwater lakes basin. Although placer mining is usually a relatively small-scale activity, the effluent from the process can have significant impacts on downstream ecosystems (Bailey et al. 1998, Laperriere and Reynolds 1997, Pentz and Kostaschuk 1999). There are also several larger-scale mining developments (mainly zinc, silver, and copper) that operate with varying amounts of activity, depending on the metal markets (Burke 2002, Swainbank et al. 2002).

Although the first oil claims were staked in the Cook Inlet area of southern Alaska (near Anchorage) in the nineteenth century, the modern Alaskan oil and gas era began in the mid-1950s, just prior to statehood in 1958. A major oil discovery in the mid-1960s in the north slope of Alaska led to plans for a pipeline. Oil transportation became significant to the Yukon River basin with the construction of the Trans-Alaskan Pipeline, completed in 1977. The 122 cm diameter pipeline allows movement of about 25% of the annual crude oil production of the United States from Prudhoe Bay on the Arctic Ocean coast to Valdez in the Gulf of Alaska of the Pacific Ocean. In addition to the well-known Exxon Valdez accident of March 1989, the pipeline itself has been the subject of scrutiny from environmentalists. Its construction did, however, motivate some of the initial work on previously unstudied ecosystems north of the Yukon River, and was deemed a "success story" with respect to environmental impact analysis, measured effects, and remediation (Alexander and Van Cleve 1983, Maki 1992). Currently there is (controversial) interest in further oil exploration in the north slope of Alaska and adjacent Porcupine River basin in and around the Arctic National Wildlife Refuge. A natural gas pipeline that may be built close to the Alaska Highway (passing through the Tanana River, White River, and headwater lakes basins of the Yukon River basin) is now in the preconstruction phase of development ("Producers to study Alaska natural gas pipeline" 2001).

Other than mining and fossil fuel developments and transportation, direct human impacts on the Yukon River basin are modest. There are about 100,000 people in the entire Yukon River basin, mostly concentrated in the small cities of Whitehorse, Yukon (20,000), on the upper Yukon River and Fairbanks, Alaska (82,000), on the lower Tanana River near its confluence with the lower Yukon River.

The Rivers

The Yukon River main stem and most of its major tributaries (Stewart, White, Tanana, and Koyukuk) begin as high-gradient rivers draining the rugged mountain ranges that define the basin (see Fig. 17.2). Beginning upstream, the east-flowing, relatively short (265 km) White River has a broad drainage basin (50,500 km^2), with large inputs of turbid glacial meltwater from the steep east slope of the St. Elias and Wrangell Mountains, the highest mountain range in North America (maximum elevation 6200 m). The turbidity of the White River is also enhanced by erosion in its basin of the White River Ash, a 10 to 20 cm layer of volcanic ash that was deposited following the eruption of Mount Churchill (near the southern end of the Yukon–Alaska border) in about 700 A.D. (Clark 1991). The west-flowing Stewart River drains a basin of similar size (51,000 km^2) but travels 650 km from its headwaters on the western side of the Selwyn Mountains (maximum elevation ~3000 m asl) to its confluence with the Yukon just south of Dawson City at 335 m asl. The Porcupine River, although similar in length to the Stewart, has its headwaters on the more modest north slope of the Ogilvie Mountains (maximum elevation ~1800 m asl), from which it curls counterclockwise, and meets the Yukon River well downstream of Dawson City in the labyrinth of the Yukon Flats at 130 m asl. Its flatter basin has over twice the drainage basin area (118,000 km^2) of the Stewart River. Well downstream

from the Porcupine confluence, the Yukon is joined by its largest tributary, the Tanana River. The Tanana drains the north slope of the Alaska Range and flows in a northwesterly direction for about 1000 km roughly parallel to the Yukon River before joining it. The last major tributary of the Yukon is the Koyukuk River, which flows 800 km south from the Brooks Range through tundra to taiga biomes. It is a largely unstudied river at a critical interface between anadromous fish populations migrating from the Bering Sea and human development of fossil fuel transportation from the north slope of Alaska. The Yukon River culminates in a 100 km wide delta at the Bering Sea.

The entire Yukon River basin lies within the Yukon freshwater ecoregion within the Arctic Complex of the Arctic–Atlantic Bioregion (Abell et al. 2000). Rivers in the Yukon basin are not well studied but are believed to have relatively low diversity and productivity, probably due to complex interactions among a number of abiotic and biotic factors, as proposed by Oswood (1997). Schindler (1998) argued that although the lower diversity of aquatic ecosystems in northern boreal areas such as the Yukon River basin may attract less attention from those wanting to protect biodiversity (Abell et al. 2000), it also makes such ecosystems more susceptible to perturbation because of their lack of ecological redundancy. This is, of course, a restatement of the "diversity begets stability" hypothesis of ecology (MacArthur 1955), which has often been proposed and debated but rarely tested (Lehman and Tilman 2000). It is clear from modern assessments of the area at an ecoregional scale (Abell et al. 2000) that the Yukon River basin has neither the biodiversity nor the anthropogenic threats to merit the same amount of vigilance as southern areas with higher diversity and larger human populations. However, it is also clear at smaller scales, such as the Chena River, a 6th order tributary of the Tanana River near Fairbanks, Alaska, that the same kind of threats from human activity, including sewage treatment plant effluent, flood-control structures, and fish harvesting, can alter the ecosystem (Oswood et al. 1992). Whether or not these northern aquatic ecosystems are more, less, or similarly sensitive to human activity remains a testable hypothesis, but we obviously must continue to assess the effects of proposed and ongoing human stressors on northern streams.

Climate and topography of the basin are the primary determinants of the balance between glacial melt and accumulation, distribution of permafrost, and the dynamics of ice formation and breakup in rivers and streams, which are the major structuring forces in far northern aquatic ecosystems such as the Yukon River basin (Hamilton and Moore 1996; Power and Power 1995; Prowse 1994; Prowse 2001a, 2001b; Yarie et al. 1998). Permafrost (soil or rock that remains at or below 0°C over at least two consecutive winters and an intervening summer) is spatially "sporadic" or patchily distributed in the upper Yukon basin, and occurs more frequently but is still "discontinuous" in most of the lower basin. Only in the extreme downstream Yukon Delta and far northern area of the basin in the upper Porcupine River is there continuous permafrost underlying the stream network. Even in areas of continuous permafrost, large lakes and rivers are underlain by unfrozen substrata. As Prowse (1994) describes, the spatial configuration of permafrost and unfrozen zones ("taliks") in a northern stream's drainage basin greatly affects the water quality in streams and essentially controls midwinter flow. The dynamics of ice within the stream also have significant effects on the structure and function of the stream ecosystem and its valley. The hydrological year in northern streams, including floods and low flows, is much more a product of the timing of freeze-up, the effects of developing ice in the channel, and the date of breakup than seasonal patterns of precipitation within the watershed. Other physical processes governed, or at least affected, by ice dynamics include erosion and deposition of sediment and the production and transport of oxygen (Prowse 2001a). Biological processes, such as seasonal movements of fishes and invertebrates as well as the dynamic creation and destruction of habitats within the stream, are also affected by ice dynamics in northern streams (Prowse 2001b).

YUKON RIVER MAIN STEM

Although the Yukon River main stem is generally considered to be the fifth-longest river in North America (Gleick 1993), it is by far the longest wild and free-flowing river on the continent (Dynesius and Nilsson 1994). It has just one dam near its headwaters in Whitehorse, Yukon, and then flows unimpeded for >3100 km through boreal forests of the Yukon Territories and Alaska to the Yukon Delta and the Bering Sea (see Fig. 17.7). There are several major inflows in the upper basin, including the Teslin, Big Salmon, Nordenskiöld, Pelly, White, Stewart, and Klondike rivers. In the lower basin, there are only a few major inflows (albeit the largest in terms of discharge), including the Porcupine, Tanana, and Koyukuk rivers.

The earliest inhabitants of the Yukon River migrated across the Bering land bridge more than 12,000 years ago, and as the climate warmed and the Ice Age ended these Paleo-Indians continued to expand their population southward and eastward. Although doing detailed archaeological reconstruction is even more difficult in the far north than elsewhere in North America, there has been clear evidence of big-game hunting and tool use found at placer mines outside of Dawson City that has been dated to 11,350 years ago (Clark 1991). In more recent prehistory (1300 years ago to European contact), moving from upstream to downstream in the Yukon River, the Tlingit, Tagish, Tutchone, Han, Gwichin, and Koyukon tribes of the Athapaskan people established traditions and cultures that have evolved to the modern First Nations of today (Clark 1991).

Physiography, Climate, and Land Use

The main stem of the Yukon River lies primarily in the Yukon Basin (YB) physiographic province, characterized by plateaus and lowlands (see Fig. 17.7). The main stem primarily flows through two terrestrial ecoregions, the Interior Yukon/Alaska Alpine Tundra in the upper basin and the Interior Alaska/Yukon Lowland Taiga in the lower basin. White and black spruce are the most common trees of the boreal forests. Tundra vegetation at higher elevations includes shrubs and sedges.

The climate of the Yukon basin is characterized by cool summers, very cold winters, and a relatively low annual precipitation of about 33 cm. The warmest air temperatures are in July, when mean temperatures are about 12°C and rarely exceed 15°C (Fig. 17.8). January is the coldest month, with mean temperatures falling to almost −20°C and sometimes below −30°C. Daily extremes at both times of year often exceed these averages. Mayo, in the Stewart River subbasin of the Yukon, established the record high temperature for the Yukon Territory of 36°C, and the town of Snag on the White River tributary to the Yukon River, in southwestern Yukon, recorded the record low temperature in North America of −63°C. There is strong seasonal precipitation throughout much of the basin, with the highest amounts falling primarily as rain in August (see Fig. 17.8). However, as precipitation declines in the early

FIGURE 17.3 Upper Yukon River at Rink Rapids between Whitehorse and Dawson City, Yukon Territory (Photo by Tim Palmer).

autumn it turns to snow and accumulates across the landscape until late spring.

The landscape along most of the Yukon River is natural tundra and boreal forest. Land use along the Yukon River main stem is largely confined to the immediate area of Whitehorse, Yukon Territory, near the headwaters. There is a hydroelectric dam on the river there, and sewage treatment facilities associated with the human population (~20,000). There is some agricultural activity (primarily hay production and livestock grazing) in the southern, upstream part of the river, just north of Whitehorse.

River Geomorphology, Hydrology, and Chemistry

The Yukon River can be more or less naturally subdivided into two regions. Upstream of its confluence with the Porcupine River, the Yukon River has a low to moderate gradient, declining in a downstream direction. The change in elevation along the Yukon River from the headwater Little Atlin Lake to the Porcupine River, a distance of 1600 km, is about 500 m, for a mean slope of 30 cm/km. The change in elevation over the remaining 1600 km to the Bering

Sea is only 130 m, for a very low gradient of 8 cm/km.

In the northwest corner of British Columbia in Canada, just a few kilometers over mountains and glaciers from the coast of Alaska, the headwaters of the Yukon River are collected into Teslin, Atlin, Tagish, and Bennett lakes. Teslin Lake is fairly typical of the group; a large, deep, oligotrophic lake in the boreal forest surrounded by alpine landscapes, with a linear shape on a north–south axis. All of the lakes have a relatively active recreational and subsistence (First Nations) fishery for lake trout, among other species, but there is little significant industrial activity in these lake basins and very small human populations.

Atlin Lake, the largest of the series, drains west via the Atlin River into Graham Inlet of Tagish Lake, where flow is north to the inflow from the lunar landscape of Tutshi Lake and Bennett Lake. Tagish Lake, and Marsh Lake directly downstream, form the final collecting points of the Yukon River headwater basin. The outflow of Marsh Lake, just south of Whitehorse, is where the Yukon first really defines itself as a river. In the upper basin (upstream of the U.S.–Canada border), the Shakwak Trench in the west and the Tintina Trench in the east broadly define the main Yukon River channel (Hunt 1974).

From Whitehorse to Dawson City (near the Canada–U.S. border), the Yukon is a moderate-gradient large river, as it accumulates flows from several large tributaries (Fig. 17.3). The declining gradient of the Yukon becomes apparent downstream of the international border (at 141°W), west of Dawson City. This is particularly evident as it enters the large Yukon Flats area, a web of channels and ponds that is currently protected as the Yukon Flats National Wildlife Refuge Area. It is here that the Yukon River changes substantially in character relative to its energetic nature further upstream, as it notches to its northernmost point at Fort Yukon, Alaska, at the Arctic Circle. This wetland-rich, low-relief part of the Yukon River includes the inflow of the Porcupine River.

Emerging from the Yukon Flats, the river is crossed by the 122 cm diameter Trans-Alaskan Pipeline, completed in 1977. Beyond the pipeline the Yukon again becomes an inflow-structured system, flowing west and sharply south as it confronts the Nulato Hills at the base of the Seward Peninsula on the west coast of Alaska. The large Tanana River and Koyukuk River, as well as the Innoko River (draining the west slope of the Kuskokwim Mountains), meet the Yukon in this region and greatly influence

its physical, chemical, and biological nature as it moves toward the final leg of its journey.

The wetland-dominated Yukon Delta defines the final "personality" of the Yukon River as it approaches the Bering Sea. The coastal arctic tundra is the last and most extreme climate directly encountered by the river. The delta is very similar topographically to the Mississippi River Delta, although about twice as large (Hunt 1974).

For much of its length the Yukon River has cut hundreds of meters into a complex plateau, often resulting in spectacular bluffs bounding the wide valley of the river (Hunt 1974). Upstream of its confluence with the White River, just south of Dawson City, it is a clear, fast-flowing river. Its predominately sand and gravel substrate produces bars and small islands that come and go with seasons and annual variation in ice and temperature dynamics. The lower Yukon, including the Yukon Flats in Fort Yukon, Alaska, and the deltaic outflow of the river into the Bering Sea, is a relatively turbid river, with substantial inputs from melting glaciers via major inflows like the White River and the Tanana River.

Like most northern rivers, the hydrological dynamics of the Yukon River are controlled on a massive scale by a combination of seasonal and annual variations in both temperature and precipitation and their effect on the supply of glacial meltwater and the distribution of permafrost (Prowse 2001a, 2001b). Mean annual discharge is 6340 m³/s, which is the 6th highest in North America. From mid-autumn through late spring, runoff is low (<1 cm/mo), as much of the potential runoff is locked up as ice, whether in glaciers or in snow and ice on and in the river and its drainage basin (see Fig. 17.8). Spring snowmelt begins in May as mean air temperature exceeds 0°C. Runoff peaks in June (5 cm/mo) but relatively high runoff continues throughout much of the summer, as peak rainfall occurs from July through September. As would be expected in such a cold environment, evapotranspiration is relatively low and annual runoff is a high fraction of precipitation (see Fig. 17.8). The strong influence of temperature on runoff suggests that climate change would be an important factor in future changes to the river's overall function and seasonal dynamics.

It is a truism in stream ecology that in terms of water chemistry a river is "what it catches." The complexity and heterogeneity of the surficial geology of the Yukon River basin, combined with the seasonal dynamics of its water sources and flow, make it difficult to generalize about the river's water quality. In most places, the river is circumneutral pH, with moderate conductivity (~200 µS/cm) and relatively low nutrient concentrations (PO$_4$-P = 0.02 mg/L). Tidewater and its effect on water quality extend about 150 km upstream of the Bering Sea outflow.

River Biodiversity and Ecology

The Yukon River is contained within and largely defines the Yukon freshwater ecoregion, the largest freshwater ecoregion in North America (Abell et al. 2000). However, this ecoregion also comprises all but the southern portion of Alaska and includes several other rivers draining into the Bering and Beaufort seas (see Chapters 16 and 20). Although the ecological characteristics of the Yukon River are undoubtedly as vast and variable as its geography, biodiversity and ecology of the river are poorly known. Danks and Downes (1997) make a clear case that not enough is known about the various insect taxa in Yukon Territory; the same could be said for all other taxonomic groups outside of the narrow range of exploited fishes and wildlife. The cost of adequately studying such a large and remote area has limited detailed study in the past, as has the decline in the research resources dedicated to museum field expeditions and other taxonomic research. This is unfortunate, because the Yukon River, as rivers in other remote areas of the world, might be altered by climate change or other human stressors before it is possible to conduct adequate ecological assessments.

Plants

The upper southern Yukon River basin has alpine tundra vegetation surrounding the mountain tributary streams, with boreal forest, including white and black spruce, some lodgepole pine, and some balsam poplar, trembling aspen, and white birch bordering the main-stem river (Farrar 1995). In the riparian zone, mountain alder and various shrub willow species are very common, and some water birch is present (Farrar 1995). In the northern Yukon Flats area there is more characteristic taiga vegetation, including scattered black spruce with sedges and willow. This grades back into a boreal forest community that also includes Alaska larch, until at the coast, in the Yukon Delta, low-lying tundra vegetation is more common.

Invertebrates

As with other taxa, the overall diversity of stream invertebrates is lower in the Yukon River basin than in similar-size temperate and subtropical basins. This is mainly attributable to the extreme climate, as well as the isolation of appropriate habitats (Danks et al. 1997, Wiggins and Parker 1997, Cobb and Flanagan 1980). The lower amounts of allochthonous CPOM provided to streams from limited riparian production (Oswood et al. 1995), autochthonous production with a limited growing season, and often turbid glacial meltwater also contribute to the lack of productivity and diversity in these streams (Hershey et al. 1997). Oswood (1997) puts these various factors together in a conceptual model that summarizes why we see the relatively low productivity and diversity of invertebrates in far-northern rivers and streams. Although this model has not really been tested in any rigorous fashion, it integrates the most probable structuring factors and provides a research plan for future improved understanding of northern running waters. Furthermore, virtually all of the invertebrate taxa that have been found and confirmed in the basin have actually been collected from the smaller tributaries of the main stem of the Yukon River (see, e.g., taxonomic and collecting-site descriptions in Danks et al. 1997 for the insects). Until the massive gaps in ecological research begin to be filled in this area, we can only hypothesize about the species that will be found in the river itself.

Mollusks are poorly represented, with just one species (albeit endemic) of Unionidae mussel (Yukon floater), and eight Sphaeriidae clams (the Arctic fingernail clam and seven pea clam [*Pisidium*] species) (Clarke 1981). There are also thirteen snail species in the families Valvatidae (*Valvata sincera*), Lymnaeidae (mostly *Lymnaea* and *Stagnicola* species), Physidae (*Physa gyrina* and *Aplexa hypnorum*), and Planorbidae (mostly *Gyraulus* species). Many of these species are primarily lentic and so would only be found in the more quiescent pool environments of the river.

Although the insect community in the Yukon River is dominated by the Diptera (true flies) (Oswood et al. 1995), our knowledge of the two major groups of Diptera in the Yukon Basin (Chironomidae and Simuliidae) is quite incomplete. Oliver and Dillon (1997) catalogued 100 species of chironomid midges collected from the Yukon "north slope" and adjacent areas (i.e., draining into the Arctic Ocean), but there is no such comprehensive list of chironomid species for the Yukon River. A clearer picture exists for the black flies (Simuliidae). Currie (1997) noted that 76 of the estimated 265 North American black fly species occur in the Yukon (some of them not yet formally described), which is diverse relative to similar-size areas in temperate North America (e.g., Ontario has 63 species) but not in the same league as relatively small Mono County in central California, with 55 to 60 species. There are five endemic species in the Yukon River basin, including *Gymnopais dichopticus* Stone, *Gymnopais holopticus* Stone, *Gymnopais fimbriatus* Wood, *Simulium* sp. near *giganteum*, and *Simulium (Hellichiella)* sp. (Currie 1997). Many of the chironomid and black fly species show adaptations to the colder climate, including reduced antennal brushes and, in the case of the black flies, a far greater occurrence (~25%) of autogenous (female does not require a blood meal to reproduce) species than in southern North America (~10%).

The next most abundant insect taxa in Yukon streams, including the Yukon River itself, are the mayflies and stoneflies (Oswood et al. 1995). Harper and Harper (1997) noted that of the 30 taxa (mostly species) of mayflies found in the Yukon, none are narrowly endemic to the Yukon Basin (East Beringia). The mayfly fauna is poorer than in the adjacent Mackenzie Delta (54 taxa) and is dominated by (in usual order of abundance) Baetidae (primarily *Baetis spp.*), Heptageniidae (primarily *Epeorus spp.* and *Heptagenia pulla* [Clemens]), Siphlonuridae (including Ameletidae; primarily *Siphlonurus spp.*), Ephemerellidae (primarily *Drunella* spp.), and Leptophlebiidae (primarily *Leptophlebia* spp.) (Oswood 1989, Harper and Harper 1997). There are 71 species (in eight families) of stoneflies in the Yukon basin. One Chloroperlidae, *Alaskaperla obivovis* (Ricker), is an endemic (Stewart et al. 1991). Only six of the species have external gills (three Taeniopterygidae, one Perlidae, and two Pteronarycidae), and just one of these, *Hesperoperla pacifica* (Banks), is a large-bodied predator, probably reflecting the environmental challenges facing such stoneflies in the Yukon River. The most diverse and abundant families (Capniidae, Leuctridae, Nemouridae; the "winter stoneflies") share many morphological and behavioral adaptations for a cold climate. *Capnia, Eucapnopsis, Utacapnia, Zapada, Taenionema, Alloperla, Suwallia,* and *Isoperla* are among the more common genera (Stewart and Ricker 1997). In recent distributional mapping combining about 30 years of small-scale impact-assessment studies (Doug Davidge, personal communication), *Utaperla soplandora* (Ricker) has been associated with streams near significant placer gold deposits, suggesting that its common name should be the "prospector stonefly."

The caddisflies are less important in abundance than mayflies and stoneflies in the Yukon Basin (Oswood et al. 1995), but, with respect to diversity, Wiggins and Parker (1997) catalogued 145 species in the basin, with six endemics that include one lentic Leptoceridae (*Ylodes schmidi* Manuel and Nimmo), one Apataniidae (*Allomyia picoides* [Ross]), and four Limnephilidae (including lotic *Grammotaulius alascensis* Schmid and four lentic species). There is also a diverse but relatively rare freshwater beetle fauna in the basin, with representatives of the families Haliplidae, Amphizoidae, Dytiscidae (including the endemic *Oreodytes recticollis*), Gyrinidae, Hydrophilidae, and Elmidae. A few other insects (e.g., corixid hemipterans) are present in the Yukon River but much less abundant than in more southern rivers.

Vertebrates

Like other taxa, the fish species in the Yukon River have a reduced diversity (30 species) compared to more southern latitudes, with the species showing behavioral, morphological, or physiological adaptations to the cold climate (Reynolds 1997). Two fishes, the Arctic grayling (Salmonidae) and the Alaska blackfish (Umbridae), are in many ways icons of the Yukon River. The Arctic grayling, one of the most important sport fishes in the Yukon and Alaska, is found throughout the river (Scott and Crossman 1973, Morrow 1980). It spawns in early spring, congregates at the mouths of clear tributaries and often proceeds upstream in the channels cut into the ice by surface flow, and then in autumn heads quickly back downstream for winter residence (Reynolds 1997). The Alaska blackfish is found in weedy areas of the lower Yukon River below its confluence with the Porcupine River (Scott and Crossman 1973). Its modified esophagus, which enables it to "breathe air" in oxygen-depleted waters or dewatered, frozen side channels (Reynolds 1997), inspired a failed introduction to several farm ponds in southern Ontario (Scott and Crossman 1973).

There are several other ecologically and socially important salmonids in the Yukon River system. Lake trout are found primarily in the headwater lakes of the Yukon River upstream of its confluence with the Pelly River. Its distribution more or less matches the furthest advance of the Pleistocene glaciation (Lindsey 1964). Dolly Varden is more commonly lotic but only found in the upper reaches of the Yukon River, whereas chinook salmon are found from the mouth of the Yukon River to its headwaters, making full use of the Whitehorse fish ladder during their summer spawning run (Scott and Crossman 1973, Morrow 1980).

Other noteworthy fish species include the endemic Arctic lamprey, found in the river as far upstream as Dawson City; the ubiquitous, winter-spawning burbot; and the inconnu (Yukon Territory), also known as the sheefish (Alaska), which causes as much debate about its palatability as its common name. Also found in the Yukon are Bering cisco, whitefish, chum salmon, coho salmon, and northern pike. Finally, it should be noted that rainbow trout is a deliberately introduced but popular nonnative species at several sites in the Yukon River and its tributaries.

Other important vertebrates that influence the river ecosystem either directly through their consumption of prey or indirectly through habitat modification in the river system include bald eagle, beaver, muskrat, brown and black bear, and moose. The role of bears and eagles, particularly in the food web and nutrient cycling of anadromous salmon in tributaries and the main stem of the Yukon River, has been explored by Kline et al. (1997).

Ecosystem Processes

There has been surprisingly (for its size and significance) little work done to understand ecosystem processes in the Yukon River. Ecological work has largely been limited to "collecting trips" in an attempt to catalogue biological diversity of plants and insects (e.g., Danks and Downes 1997), fisheries research targeted at game fishes (e.g., Beacham et al. 1989; Bradford et al. 2001), biological monitoring with benthic invertebrate communities (e.g., Bailey et al. 1998), or analyses of contaminant flux in food webs (Kidd et al. 1995, Kidd et al. 1998). Although no substantive research on ecosystem structure and function has been done in the Yukon River, there are probably useful lessons from Oswood's (1997) model of ecosystem processes in northern rivers.

The study of contaminant flux is a good example of how applied research forces us to understand the more fundamental properties of an ecosystem. Just 35 km downstream (north) of Whitehorse the Yukon River flows into Lake Laberge (the lake made famous by Robert Service's poem about Sam McGee). Kidd et al. (1995) and Kidd et al. (1998) found elevated levels of organochlorines in lake trout, burbot, and northern pike from Lake Laberge. They speculated that this was from long-range transport of atmos-

pheric pollutants rather than upstream (Whitehorse) human activity and showed how differences among lakes (they also looked at nearby Fox Lake and Kusawa Lake) in contaminants accumulating in fishes were at least partially explained by differences in the length of the food chains.

Human Impacts and Special Features

The size and large-scale ecological heterogeneity of the wild Yukon River, together with its context in subarctic North America and the Beringean Refugium, its historical importance to humans as a colonization route from northern Asia, and its importance as a source of valued resources from the nineteenth to the twenty-first centuries, make it one of the premier North American rivers. The wild status of the river and its landscape has been recognized by the inclusion of many of its smaller tributaries in the U.S. Wild and Scenic River system (e.g., Andreafsky River, Alatna River, North Fork of Koyukuk River, Wind River, Sheenjek River, Nowitka River, Charley River, and Forty Mile River), and the Canadian Heritage River System (Thirty Mile River). Unfortunately, no reaches of the main stem are recognized by either system.

Human impacts on the Yukon River are relatively minor at this time. Several negative impacts on the river may become substantially worse in the future. These include effects of climate change, dams, activities associated with fossil fuel development, forestry, and pollution.

Climate change is probably the most significant potential human impact on the river, as the hydrological dynamics and ecosystem processes of the river are for the most part ultimately controlled by the seasonal temperature regime. Global warming is likely to be felt most strongly in such cold climates, where organisms have adapted to the environmental conditions created by more than seven months of subzero temperatures. Warmer temperatures can dramatically change the hydrological patterns controlled by autumn freeze-up and spring snowmelt.

Activities associated with exploration for and development and transportation of fossil fuels in northern Alaska also have the potential for devastating consequences. The Trans-Alaskan pipeline crosses two of the Yukon River's largest tributaries and crosses the Yukon River itself in central Alaska. The direct effects of the pipeline on the Yukon River have been modest so far and mostly associated with disruption of the river and its basin, with its sensitive taiga and alpine–arctic tundra biomes. However, a major leak from the pipeline to the main-stem river or its tributaries could have an effect similar to that of major spills from oceanic tankers.

Relatively few dams have been built in the Yukon basin, with one in its headwaters near Whitehorse and one each on tributaries of the Stewart (near Mayo, Yukon) and Tanana (near Fairbanks, Alaska) rivers. The White Horse Rapids in Miles Canyon, from whence Whitehorse got its name, have disappeared because of a hydroelectric dam in the city with a summer capacity of 40 MW. The demand for electricity is inversely related to its production; as demand for electrical heat and light increases substantially in winter, freeze-up lowers the production of the generating plant to 24 MW. There are still chinook salmon and other fishes that manage to spawn upstream of the dam, thanks to what is reputedly the longest wooden fish ladder in the world. Other urban stressors on the river as it flows through Whitehorse include channelization and stabilization

of banks, as well as effluent from two sewage treatment facilities.

TANANA RIVER

The Tanana River drains the north-facing slopes of the Alaska Range and is the most developed area within the Yukon basin (Fig. 17.9). The basin, which lies almost entirely within Alaska, is 114,000 km^2. The headwaters of the river coalesce in the Tetlin National Wildlife Refuge, rich in small lakes and ponds, near the Yukon–Alaska border at the western edge of the White River basin. The river flows generally northwest from its origin, with several significant inflows from the Alaska Range mountains to the south, including the Chisana, Nabesna, Robertson, Gerstie, Delta, Nenana, and Kantishna rivers (Fig. 17.4). Most of these derive much of their flow from glacial meltwaters, which makes them and the Tanana River main stem quite turbid, with discharge dependent more on air temperature than precipitation (Prowse 2001a). Major inflows from the north,

FIGURE 17.4 Tanana River above Tok, Alaska (PHOTO BY TIM PALMER).

including the Healey, Goodpaster, Salcha, Chena, Chatanika, and Tolovana rivers, tend to be lower gradient and clear flowing.

Like much of the Yukon River basin on the Alaska side of the present-day international border, human colonization of the Tanana River occurred more than 12,000 years ago and progressed with the receding Cordilleran ice sheet as the Ice Age ended (Clark 1991). Evidence of the microblade people of the Paleo-Arctic tradition of about 11,000 years ago has been found at several sites in the Tanana River basin. More recently, the Tanana tribe has developed a culture and tradition over the last 1200 years in the basin that survives to the present (Clark 1991).

Physiography, Climate, and Land Use

The South-Central Alaska (SC) physiographic province defines the upper watershed of the Tanana River basin in the mountainous southwest of Alaska (see Fig. 17.9; Hunt 1974). The headwaters of the Tanana River flow from the Alaska Range to the south and east and the Yukon–Tanana Upland (Intermontane Plateau) to the north. The main-stem river is in the Yukon Basin (YB) physiographic province or, more specifically, the Tanana–Kuskokwim lowland of the Intermontane Plateaus of Alaska (Hartman and Johnson 1978). Three terrestrial ecoregions occupy the Tanana basin: Interior Alaska/Yukon Lowland Taiga, Alaska/St. Elias Range Tundra, and Interior Yukon/Alaska Alpine Tundra. As throughout much of the Yukon basin, black and white spruce are common, as well as mountain alder and quaking aspen.

Climate in the Tanana River is Continental, with large diurnal and seasonal variation in temperature and relatively low precipitation and humidity. Mean monthly air temperatures at Fairbanks range from a low of −23°C in January to a high of 17°C in July (Fig. 17.10). There is about 28 cm of precipitation, with around 40 "wet days" (>2.5 mm water precipitation) a year (Hartman and Johnson 1978). Mean precipitation ranges from <1 cm/mo in March and April to a peak of about 5 cm/mo in August (see Fig. 17.10). About 30% of precipitation falls as snow, which accumulates from mid-autumn through late spring.

The city of Fairbanks on the Tanana River has been a major center of placer gold mining for the last 150 years. It has developed into a moderate-size population center (~80,000 people) with associated infrastructure and modest industrial activity. Gold mining has continued in the Tanana River basin since the middle of the nineteenth century and ebbs and flows depending on the price of gold. Some agricultural activity also occurs in the river valley near Fairbanks. Otherwise, the basin is largely unmodified wilderness.

River Geomorphology, Hydrology, and Chemistry

Like the Yukon River, the Tanana River is in a wide valley that cuts through a broad plain just north of the Alaska Range. It is bounded by the foothills of the Alaska Range to the south and high bluffs to the north. It has a shifting sand and gravel substrate with dynamic formation and destruction of point and mid-channel bars. Near Fairbanks the active floodplain can be as wide as 2 km.

The Tanana River, as the largest tributary of the Yukon, has an average annual discharge of $682\,m^3/s$ at Nenana, which drains close to 60% of the basin. Discharge at the mouth thus has been approximated at $1185\,m^3/s$. Runoff is <1 cm/mo from December through April with freeze-up and low precipitation (see Fig. 17.10). As with the Yukon, spring snowmelt begins in May and runoff peaks in July and August at more than 6 cm/mo. Thus, hydrological dynamics tend to track the freeze–thaw seasonal cycle (Prowse 2001a) more closely than seasonal patterns in precipitation. Although evapotranspiration is low, runoff appears to be higher than precipitation (see Fig. 17.10), probably due to the fact that the weather station at Fairbanks underestimates the average precipitation for the entire basin.

The river is primarily fed by glaciermelt and snowmelt from the south and groundwater from the north, so the water is moderately hard (total hardness 114 mg/L as $CaCO_3$). Turbidity is usually quite high in the main stem due to the collection of glacial meltwater, although some snowmelt- and spring-fed tributary streams (primarily from the north and northeast) run clear. Nutrient concentrations in the river are moderate to high in the lower river (e.g., PO_4-P 0.3 mg/L), partially because of human inputs at or near Fairbanks. Metal concentrations are elevated because of the nature of the surficial geology in the basin.

River Biodiversity and Ecology

Like the Yukon River, the Tanana River is part of the Yukon freshwater ecoregion (Abell et al. 2000).

Many of the aquatic organisms found in the Yukon River as a whole are found in the Tanana River. There has been perhaps more management and manipulation of the biota in this subbasin than any other in the Yukon River basin. There is reasonably high fishing and hunting activity and there have been introductions of nonnative fish species in the past (e.g., rainbow trout).

Plants

In most of the subalpine riparian areas of the Tanana there is a typical boreal forest cover of primarily white spruce, birch, aspen, and willow. At higher elevations (above the tree line), hardier dwarf willow and other shrubs, as well as sedges and lichens, are common. The successional processes of floodplain vegetation in the Tanana basin have been studied extensively as a model for the northern boreal forest (taiga) ecosystem (Kielland et al. 1997, Kielland and Bryant 1998, Mann et al. 1995). Willow and mountain alder, as well as balsam poplar and quaking aspen, are common on the floodplain.

Invertebrates

Invertebrates of the Tanana River are poorly studied, and only a couple of more general studies have included collections of invertebrates in the Tanana and its tributaries that have found the same general pattern at the order level as described for the Yukon River (e.g., Oswood et al. 1992). Like other northern rivers, the benthos is dominated by species of chironomid midges. Stoneflies, mayflies, and caddisflies are present but not nearly as abundant as the chironomids. There are no published reports of the endemic stonefly *Alaskaperla obivovis* or endemic Yukon floater mussel having been collected in the Tanana River basin. Given the ecological parallels with the Yukon basin as a whole, it is probable that many of the invertebrate species found in tributaries of the rest of the Yukon River basin would also be found in the Tanana River basin.

Vertebrates

The Alaska Department of Fish and Game cites 18 indigenous and 2 introduced (stocked) species of fishes (rainbow trout, Arctic char). Several salmonids are common in the Tanana and its tributaries, including Arctic grayling, chinook salmon, least cisco, round whitefish, humpback whitefish, and inconnu (Oswood 1997). Additional common species include slimy sculpin, longnose sucker, northern pike, burbot, and Arctic lamprey.

There is perhaps more recreational fishing pressure in the Tanana River and its tributaries than any other part of the Yukon River basin. By far the greatest sport fishery harvest is of the Arctic grayling. As much of the river is largely glacial fed, the most significant fishery pressure comes in the clearwater sloughs and small inflow streams, although there is an active winter burbot fishery.

As in the Yukon River, other important vertebrates include bald eagles and brown bears, which can both be important in their role as part of the food web that includes the near or already dead anadromous salmon after their spawning runs (Kline et al. 1997). The Tanana has very occasionally hosted beluga whales that have migrated up the Yukon River from the Bering Sea.

Ecosystem Processes

Little work has been done on ecosystem processes of the Tanana River. However, Oswood's (1997) model of ecosystem processes in northern rivers is pertinent. The Tanana River's drainage basin is a mixture of largely glacial sources to the south with nonglacial inflows to the north, so the increase in discharge with increased seasonal temperature can result in relatively clear or turbid flow depending on variability in the weather. In citing unpublished data concerning primary production in the Chena River, a major tributary of the Tanana River just upstream from Fairbanks, Oswood et al. (1992) concluded that gross primary production rivaled that in temperate streams in the summer but was virtually nil in the long winter. Because the Chena River is a subsurface and precipitation-fed river (not glacial fed), one has to assume this represents the best conditions for primary production in terms of light availability. Allochthonous inputs of energy and nutrients to rivers such as the Tanana are also likely to be modest. This, together with the extreme seasonal disturbances of freeze-up and ice-out, would suggest a system with low productivity and diversity, but this needs to be confirmed with future ecosystem-scale research and detailed determination of species distributions.

Human Impacts and Special Features

The Tanana River, as the largest tributary of the Yukon, is an unregulated and largely wild river. As a relatively intact large and wild river with easier accessibility than most Alaskan rivers (Fairbanks), it could be a prime candidate for more detailed studies of biodiversity and ecology in an unaltered river. The southern uplands of the Tanana River basin include part of the 2.1 million acre Denali Wilderness Area in the west and the 9.7 million acre Wrangell–Elias Wilderness Area in the east. The Delta River tributary is in the U.S. Wild and Scenic River system.

Although still a relatively wild river, the Tanana, of all the major and minor streams in the Yukon River basin, is exposed to the greatest magnitude and diversity of human stresses. Nonetheless, these impacts are likely to be small in comparison to most rivers at lower latitudes. Along with the city of Fairbanks, the river and its tributaries are traversed by the Trans-Alaskan pipeline, the Alaska Highway, the Richardson Highway, the Tok Cutoff (Highway), and the Taylor ("Top of the World") Highway, along with associated commercial and recreational traffic of these transportation corridors. Impacts of these energy and transportation corridors are largely unknown.

Pollution impacts on the Tanana River include sewage effluent, a tributary diversion (Fairbanks Flood Control Project Chena River; Oswood et al. 1992), high levels of placer mining activity, and other development (Swainbank et al. 2002). In addition, active recreational fisheries and other tourist activity may have negative impacts. For example, overfishing may have been responsible for declines in the Arctic grayling of the Chena River tributary (Oswood et al. 1992).

KOYUKUK RIVER

The last major tributary of the Yukon River upstream of the outflow into the Bering Sea is the Koyukuk River (Fig. 17.11). Its basin (83,500 km²) is entirely within Alaska, and it is an almost pristine river, flowing from the southern edge of the Brooks Range in northern Alaska. Its headwaters are in the 3.4 million ha Gates of the Arctic Wilderness Area (established in 1980), and the eastern edge of the basin includes part of the Arctic National Wildlife Refuge, just above the Arctic Circle. The eastern branch of the headwaters of the Koyukuk River defines the pathway of the Trans-Alaskan pipeline from Prudhoe Bay to Valdez as it proceeds south over the Brooks Range, but there have been surprisingly few ecological studies done within its basin or the river itself. It enjoys some popularity as a canoeing and outfitting destination, but for such a large river, surprisingly little is known about it.

The Koyukuk River was colonized by humans early in the west-to-east Beringean migration from

what is now Asia to what is now northwestern North America (Clark 1991). There was also a south-to-north migration by humans who had previously come to North America across the Bering land bridge and then turned north. Evidence of tools used by Paleo-Indian and Paleo-Arctic cultures has been found at several sites along the Koyukuk River (Clark 1991). About 1300 years ago this developed into the Koyukon and Gwitchin tribes that developed into the First Nations present in the Koyukuk River basin in modern times.

Physiography, Climate, and Land Use

The Koyukuk River's headwaters are in the Endicott Mountains of the Brooks Range (BR) physiographic province, whereas most of the lower two-thirds of the river's length is in the Yukon Basin (YB) province (see Fig. 17.11). Some smaller eastward-flowing streams drain from the Seward Peninsula and Bering Coast Uplands (SB) province near the mouth of the Koyukuk River. There is quite gentle relief in the lower Koyukuk (Koyukuk Flats) area, similar to the Yukon Flats and Yukon Delta regions of the main-stem Yukon River. The basin contains parts of three terrestrial ecoregions: Brooks/British Range Tundra, Interior Yukon/Alaska Alpine Tundra, and Interior Yukon/Alaska Lowland Taiga. Besides extensive areas of tundra vegetation, including shrubs and sedges, various taiga trees are found, such as white and black spruce, lodgepole pine, quaking aspen, and alpine fir. This landscape is almost entirely intact wilderness, with few human settlements.

Climate in the Koyukuk River basin is Continental, with large diurnal and seasonal variation in temperature and relatively low precipitation and humidity. There is about 31 cm/yr of precipitation, with around 40 "wet days" (>2.5 mm water precipitation) a year (Hartman and Johnson 1978). Maximum precipitation (>5 cm/mo) occurs during the summer (Fig. 17.12). Mean monthly temperatures are very low, ranging from a high of only 10°C in July and falling to below −20°C for five months.

River Geomorphology, Hydrology, and Chemistry

The Koyukuk River drains south from the Brooks Range, more modestly dissecting the same plateau cut by the Yukon and Tanana rivers. The river's headwater tributaries, including the John River, the South, Middle, and North forks, and the Alatna River, collect at the north end of the Kanuti National Wildlife Refuge in the village of Bettles, Alaska (Fig. 17.1). Major inflows as the Koyukuk River flows southwest include the Kanuti, Batzu, Hogatzu, Huslia, Dubli, Kateel, and Gisasa rivers.

The Koyukuk has a relatively high average annual discharge ($664 \, m^3/s$), but like most rivers of the far north discharge is quite variable seasonally, as the annual hydrological cycle is determined more by temperature than precipitation patterns. Because of its smaller size and more northern drainage basin, the Koyukuk shows this annual hydrological pattern much more strongly than either the Yukon or Tanana rivers (see Fig. 17.12). The ratio of maximum to minimum monthly average discharge in the Koyukuk is about $100:1$. The same ratio for the Yukon and Tanana is $10:1$ and $8:1$, respectively. This is because the precipitation becomes almost totally locked up in snow and ice from November through April, with very minimal runoff (see Fig. 17.12). Ironically, maximum flow occurs in May and June following the lowest months of precipitation, a clear indication of the influence of the spring snowmelt effect from ice and snow that accumulated during the previous winter.

The Koyukuk River does not predominately drain glaciers, so its maximum flow in May and June tends to be relatively clear. It has a pH of 7.6, moderate conductivity ($\sim$200 μS/cm) and hardness (105 mg/L as $CaCO_3$), and low nutrient concentrations (PO_4-P 0.01 mg/L).

River Biodiversity and Ecology

The Koyukuk River is part of the vast Yukon freshwater ecoregion (Abell et al. 2000) and like the rest of the Yukon River it is poorly studied.

Plants

Skirting the Arctic Circle at relatively high altitude, much of the upper Koyukuk riparian vegetation is sparse. Spruce–lichen forest cover with dwarf willow and other shrubs, as well as sedges and lichens, are common. Tundra vegetation grades into somewhat more substantial taiga and boreal forest-plant communities in the lower Koyukuk. The Koyukuk Flats are a low-gradient, poorly drained muskeg environment with seasonally abundant populations of sedges and other herbaceous species. In the higher-gradient floodplains of the Koyukuk River, mountain alder and trembling aspen are common, as well as white and black spruce.

Invertebrates

No studies of invertebrates in the Koyukuk River and its major tributaries have been published, although very general information about some Koyukuk tributaries is included in a report associated with the pipeline construction (Johnson and Rockwell 1979). This study just reiterates the common, order-level pattern for northern rivers and streams: several common true flies, probably including several chironomid midge and black fly species, and many mayfly and stonefly species present, with a somewhat lesser abundance of caddisflies at a given location. As with other, even major tributaries within the Yukon River system (and the main stem itself), there is plenty of important ecological and taxonomic research waiting to be done in this area. Some experimental oil spills were carried out to assess their effects on biota, including benthic invertebrates, but these were done in the somewhat different north-slope tundra ecosystems (Nauman and Kernodle 1975).

Vertebrates

Because the Koyukuk River is the last major inflow into the Yukon River before its outflow into the Bering Sea, there are some fish species found here that either occur rarely or not at all further upstream in the Yukon basin. Arctic lamprey are found commonly in the lower Koyukuk. Sockeye salmon ascend to the lower Koyukuk for spawning. More commonly, some of the other anadromous salmonids, such as chinook, chum, and sockeye salmon, also make it up into the Koyukuk and its tributaries.

The upper Koyukuk's major tributaries are important migration routes for caribou, and several studies were undertaken to assess the effect of pipeline construction and operation on the caribou population, which numbers about 5000 in this area (Alexander and Van Cleve 1983). Other major aquatic or semiaquatic vertebrates include brown and black bear, beaver, mink, marten, river otter, and bald eagle. Like the Tanana River, the lower Koyukuk has very occasionally been host to beluga whales.

Ecosystem Processes

Although there was some aquatic ecosystem research done in the 1970s in preparation for the construction of the Trans-Alaskan pipeline (Alexander and Van Cleve 1983, Hobbie 1997, Milner and Oswood 1997), most of this was carried out on the north slope of the Brooks Range. No detailed research of ecosystem processes has been carried out in the Koyukuk River system, and given its susceptibility to disturbance from human activities associated with fossil fuel exploration, development, and transportation, this is a glaring gap in our knowledge of North American rivers.

As with the other rivers in the Yukon River basin, Oswood's (1997) model of ecosystem processes in northern rivers seems an appropriate albeit untested hypothesis of how the Koyukuk River functions. There is much less glacial-source inflow to the Koyukuk relative to the Tanana and White rivers, so differences in light regime and water quality may make the relative role of these factors and their interaction somewhat different from the other major tributaries.

Human Impacts and Special Features

The Koyukuk River is undoubtedly one of the most pristine tributaries of an otherwise largely undeveloped Yukon wilderness. This has been recognized by the designation of four tributaries upstream of Allakaket (Alatna River, John River, Tinayguk and North Fork of Koyukuk River) as U.S. National Wild and Scenic Rivers. They all drain the Brooks Range within the pristine Gates of the Arctic National Park and Preserve. Thus, human impacts to date are small.

In the Koyukuk River basin, human activities include subsistence hunting and fishing by native communities, southern tourists participating in fly-in fishing and hunting, and canoe and raft tripping. There is modest placer mining activity in the basin, and virtually no hard-rock mining exploration or development (Laperriere and Reynolds 1997, Swainbank et al. 2002).

Significantly, the upper eastern portion of the Koyukuk River basin is traversed by the Trans-Alaskan pipeline, which always has the potential for impacts to the riverine environment. This renders both the headwaters and downstream areas susceptible to future exploration or development on the north slope of Alaska, or accidents during transportation of fossil fuel across Alaska.

ADDITIONAL RIVERS

The White River is a free-flowing high-gradient river that flows in a northerly direction before joining the Yukon River south of Dawson City (Fig. 17.13). This largely pristine river drains several glaciated areas in the relatively high St. Elias and Wrangell Mountains of the southeastern Yukon (and southwestern Alaska), as well as the Donjek River outflow of the large, submontane Kluane Lake. At the headwaters of the White River in Alaska, Mount Churchill was the source of the volcanic eruption more than 1200 years ago that is thought by some to have motivated the southern migration of the Athapaskan peoples (Moodie et al. 1992), although this is controversial (Clark 1991). Because of its large burden of suspended glacier-derived sediment and a thick layer of White River Ash that underlays much of the basin (Fig. 17.5), the White River adds considerable turbidity at its confluence with the Yukon.

The Stewart River is in the Yukon Territory, with its headwaters on the western slope of the Mackenzie Mountains (Fig. 17.15). It flows in a westerly direction before joining the Yukon River just downstream of the White River confluence. The river has no dams on the main stem, but there is a small (2 MW capacity) hydroelectric dam on the Mayo River tributary near the town of Mayo, Yukon (population 500). Streams in the Stewart River system have been actively mined in the last 100 years or so, both for gold (usually in small placer mining operations) and silver (at a now dormant hard-rock mine at Elsa, Yukon). The Nacho Nyak Dun have a settled land claim encompassing a large amount of territory in the Stewart River basin and have actively pursued collaborative research on the assessment and monitoring of the stream ecosystems in the area.

The Klondike River drains a small basin just north of the Stewart River basin (Fig. 17.17). Its confluence with the Yukon is at Dawson City, "urban service center" for the famous Klondike gold rush (Fig. 17.6). At the height of the gold rush in the late nineteenth century there were about 30,000 people in Dawson. Presently, the permanent population is only about 500, although there is a significant "tourist rush" in the summer months thanks to good road access. Although the gold rush ended long ago, there is still mining activity in streams of the Klondike basin, often involving reanalysis of previously processed "waste" rock made economically feasible by the ups and downs of gold prices. Together with the adjacent Indian River basin (inflow to the Yukon River), it is still the most significant gold mining area in the Yukon, with 133,000 ounces recovered (62% of total gold production) in 1991–1992 (Placer Mining Section 1993).

The Porcupine River is the northernmost tributary of the Yukon basin and drains the largest subbasin (118,000 km^2; Fig. 17.19). The Porcupine is a

FIGURE 17.5 White River at the Alaska Highway near Koidern, Yukon Territory (Photo by R. Bailey).

FIGURE 17.6 Klondike River inflow to the Yukon River at Dawson City, Yukon Territory (Photo by J. Schwindt).

large pristine Arctic river that has no dams. A major tributary of the Porcupine is the Sheenjek River, which drains a portion of the Arctic National Wildlife Refuge. The upper half of the Porcupine River basin is in northern Yukon Territory, where the headwater streams drain the eastern slope of the Richardson Mountains and the northern slope of the Ogilvie Mountains, from which the Porcupine River curls north to Old Crow, Yukon, and then southwest into Alaska to meet the Yukon River at Fort Yukon. Precipitation is only about half that of the Tanana basin, of similar area, and there is no glacial meltwater input into the Porcupine as there is the Tanana, so the Porcupine River has lower discharge (414 m^3/s) and runoff (Fig. 17.20). The upper Porcupine basin is primarily taiga and alpine tundra, with stunted black spruce and larch. It has large patches of continuous permafrost with turbic cryosolic soils, and it is rich in peatlands (Scudder 1997). The greatest concern regarding human impacts are associated with potential oil exploration in and around the Arctic National Wildlife Refuge and the transportation of oil and gas out of this area.

ACKNOWLEDGMENTS

Many thanks to Mark Poustie and Brendan Bailey for literature and map research, and John Schwindt and John Bailey for help with field logistics and photography in an expedition to the Yukon basin in 2002 and 2003. Jim McDonald suggested several valuable references on the human prehistory of the Yukon River basin. Art Benke and Bert Cushing provided critical and patient comments throughout work on this project.

LITERATURE CITED

Abell, R., D. Olson, E. Dinerstein, P. Hurley, J. Diggs, W. Eichbaum, S. Walters, W. Wettengel, T. Allnut, C. Loucks, and P. Hedao. 2000. *Freshwater ecoregions of North America: A conservation assessment*. Island Press, Washington, D.C.

Alexander, V., and K. Van Cleve. 1983. The Alaska pipeline: A success story. *Annual Review of Ecology and Systematics* 14:443–463.

Bailey, R. C., M. G. Kennedy, M. Z. Dervish, and R. M. Taylor. 1998. Biological assessment of freshwater ecosystems using a reference condition approach: Comparing predicted and actual benthic invertebrate communities in Yukon streams. *Freshwater Biology* 39:765–774.

Beacham, T. D., C. B. Murray, and R. E. Withler. 1989. Age, morphology, and biochemical genetic-variation of Yukon River chinook salmon. *Transactions of the American Fisheries Society* 118:46–63.

Begon, M., J. Harper, and C. R. Townsend. 1996. *Ecology: Individuals, populations and communities*. Blackwell Science, Oxford.

Bradford, M. J., J. A. Grout, and S. Moodie. 2001. Ecology of juvenile chinook salmon in a small non-natal stream of the Yukon River drainage and the role of ice conditions on their distribution and survival. *Canadian Journal of Zoology* 79:2043–2054.

Burke, M. 2002. Yukon mining and exploration overview, 2001. Exploration and Geological Services Division, Yukon Region, Indian and Northern Affairs Canada, Whitehorse, Yukon.

Clark, D. W. 1991. *Western subarctic prehistory*. Canadian Museum of Natural History, Hull, Providence of Quebec.

Clarke, A. H. 1981. *The freshwater molluscs of Canada*. National Museum of Natural Sciences/National Museums of Canada, Ottawa, Ontario.

Cobb, D. G., and J. F. Flanagan. 1980. The distribution of Ephemeroptera in northern Canada. In J. F. Flannagan and K. E. Marshall (eds.). *Advances in Ephemeroptera biology*, pp. 155–166. Plenum, New York.

Currie, D. C. 1997. Black flies (Diptera: Simuliidae) of the Yukon with reference to the black-fly fauna of northwestern North America. In H. V. Danks and J. A. Downes (eds.). *Insects of the Yukon*, pp. 513–614. Biological Survey of Canada (Terrestrial Arthropods), Ottawa.

Danks, H. V. and J. A. Downes (eds.). 1997. *Insects of the Yukon*. Biological Survey of Canada (Terrestrial Arthropods), Ottawa.

Danks, H. V., J. A. Downes, D. J. Larsen, and G. G. E. Scudder. 1997. Insects of the Yukon: Characteristics and history. In H. V. Danks and J. A. Downes (eds.). *Insects of the Yukon*, pp. 963–1013. Biological Survey of Canada (Terrestrial Arthropods), Ottawa.

Downes, J. A. 1965. Adaptations of insects in the Arctic. *Annual Review of Entomology* 10:257–274.

Dynesius, M., and C. Nilsson. 1994. Fragmentation and flow regulation of river systems in the northern third of the world. *Science* 266:753–762.

Ennis, G. L., A. Cinader, S. McIndoe, and T. Munsen. 1982. An annotated bibliography and information summary on the fisheries resources of the Yukon River basin in Canada. *Canadian Manuscript Report of Fisheries and Aquatic Sciences* 1657.

Farrar, J. L. 1995. *Trees in Canada*. Fitzhenry and Whiteside Limited, Markham, Ontario, and the Canadian Forest Service (Natural Resources Canada) Ottawa.

Geraghty, J. J. 1973. *Water atlas of the United States*. 2nd ed. Water Information Center, Port Washington, New York.

Gleick, P. H. 1993. *Water in crisis. A guide to the world's fresh water resources*. Oxford University Press, New York.

Hamilton, A., and R. Moore. 1996. Winter streamflow variability in two groundwater-fed sub-Arctic rivers, Yukon Territory, Canada. *Canadian Journal of Civil Engineering* 23:1249–1259.

Harper, P. P. and F. Harper. 1997. Mayflies (Ephemeroptera) of the Yukon. In H. V. Danks and J. A. Downes (eds.). *Insects of the Yukon*, pp. 151–167. Biological Survey of Canada (Terrestrial Arthropods), Ottawa.

Hartman, C., and P. Johnson. 1978. *Environmental atlas of Alaska*. Institute of Water Resources, University of Alaska, Fairbanks.

Hayes, D. 1999. *Historical atlas of British Columbia and the Pacific Northwest*. Cavendish Books, Vancouver.

Hershey, A. E., W. B. Bowden, L. A. Deegan, J. E. Hobbie, B. J. Peterson, G. W. Kipphut, G. W. Kling, M. A. Lock, R. W. Merritt, M. C. Miller, J. Vestal, and J. A. Schuldt. 1997. The Kuparuk River: A long-term study of biological and chemical processes in an arctic river. In A. M. Milner and M. W. Oswood (eds.). *Freshwaters of Alaska: Ecological syntheses*, pp. 107–129. Springer-Verlag, New York.

Hobbie, J. E. 1997. History of limnology in Alaska: Expeditions and major projects. In A. M. Milner and M. W. Oswood (eds.). *Freshwaters of Alaska: Ecological syntheses*, pp. 45–60. Springer-Verlag, New York.

Hopkins, D. M., J. Matthews, C. Schweger, and S. Young. 1982. *Paleoecology of Beringia*. Academic Press, New York.

Hunt, C. B. 1974. *Natural regions of the United States and Canada*. W. H. Freeman, San Francisco.

Jack, M. E., T. R. Osler, and B. E. Burns. 1983. Water quality: Yukon River basin. Water Quality Work Group Report #1. Indian and Northern Affairs Canada, Ottawa, Ontario.

Johnson, R. L., and J. Rockwell Jr. 1979. List of streams and other water bodies along the Trans-Alaska pipeline route. Pipeline Office, U.S. Department of the Interior, Anchorage, Alaska.

Kidd, K. A., D. W. Schindler, R. H. Hesslein, and D. C. G. Muir. 1998. Effects of trophic position and lipid on organochlorine concentrations in fishes from subarctic lakes in Yukon Territory. *Canadian Journal of Fisheries and Aquatic Sciences* 55:869–881.

Kidd, K. A., D. W. Schindler, D. C. G. Muir, W. L. Lockhart, and R. H. Hesslein. 1995. High concentrations of toxaphene in fishes from a sub-Arctic lake. *Science* 269:240–242.

Kielland, K., and J. P. Bryant. 1998. Moose herbivory in taiga: Effects on biogeochemistry and vegetation dynamics in primary succession. *Oikos* 82:377–383.

Kielland, K., J. P. Bryant, and R. W. Ruess. 1997. Moose herbivory and carbon turnover of early successional stands in interior Alaska. *Oikos* 80:25–30.

Kline, T. C., Jr., J. J. Goering, and R. Piorkowski. 1997. The effect of salmon caracasses on Alaskan freshwaters. In A. M. Milner and M. W. Oswood (eds.). *Fresh-*

waters of Alaska: Ecological syntheses, pp. 179–204. Springer-Verlag, New York.

Laperriere, J. D., and J. B. Reynolds. 1997. Gold placer mining and stream ecosystems of interior Alaska. In A. M. Milner and M. W. Oswood (eds.). *Freshwaters of Alaska: Ecological syntheses*, pp. 265–280. Springer-Verlag, New York.

Lehman, C. L., and D. Tilman. 2000. Biodiversity, stability, and productivity in competitive communities. *American Naturalist* 156:534–552.

Lindsey, C. 1964. Problems in zoogeography of the Lake Trout, *Salvelinus namycush*. *Journal of the Fisheries Research Board of Canada* 21:977–994.

MacArthur, R. 1955. Fluctuations of animal populations, and a measure of community stability. *Ecology* 36:533–536.

Maki, A. W. 1992. Of measured risks: The environmental impacts of the Prudhoe Bay, Alaska, oil-field. *Environmental Toxicology and Chemistry* 11:1691–1707.

Mann, D. H., C. L. Fastie, E. L. Rowland, and N. H. Bigelow. 1995. Spruce succession, disturbance, and geomorphology on the Tanana River floodplain, Alaska. *Ecoscience* 2:184–199.

Milner, A. M., and M. W. Oswood (eds.). 1997. *Freshwaters of Alaska: Ecological synthesis*. Springer-Verlag, New York.

Moodie, D. W., A. J. W. Catchpole, and K. Abel. 1992. Northern Athapaskan oral traditions and the White River volcano. *Ethnohistory* 39:148–171.

Morrow, J. E. 1980. *The freshwater fishes of Alaska*. Alaska Northwest, Anchorage.

National Climate Center. 1982. *Climate of Alaska*. NOAA Environmental Data Service, Asherville, North Carolina.

Nauman, H., and D. Kernodle. 1975. The effect of a fuel oil spill on benthic invertebrates and water quality on the Alaskan arctic slope, Happy Valley Creek near Sagwan Alaska. *J. Res. US Geological Survey* 3:495–500.

Oliver, D., and M. Dillon. 1997. Chironomids (Diptera: Chironomidae) of the Yukon Arctic North Slope and Herschel Island. In H. Danks and J. Downes (eds.). *Insects of the Yukon*, pp. 615–635. Biological Survey of Canada (Terrestrial Arthropods), Ottawa.

Oswood, M. W. 1989. Community structure of benthic invertebrates in interior Alaskan (USA) streams and rivers. *Hydrobiologia* 172:97–110.

Oswood, M. W. 1997. Streams and rivers of Alaska: A high latitude perspective on running waters. In A. M. Milner and M. W. Oswood (eds.). *Freshwaters of Alaska: Ecological syntheses*, pp. 331–356. Springer-Verlag, New York.

Oswood, M. W., J. G. Irons, and A. M. Milner. 1995. River and stream ecosystems of Alaska. In C. E. Cushing, K. W. Cummins, and G. W. Minshall (eds.). *River and stream ecosystems*, pp. 9–32. Elsevier, Amsterdam.

Oswood, M. W., J. B. Reynolds, J. D. Laperriere, R. Holmes, J. Hallberg, and J. H. Triplehorn. 1992. Water quality and ecology of the Chena River Alaska. In C. Becker and D. Neitzel (eds.). *Water quality in North American river systems*, pp. 5–27. Battelle Press, Columbus, Ohio.

Pentz, S. B. and R. A. Kostaschuk. 1999. Effect of placer mining on suspended sediment in reaches of sensitive fish habitat. *Environmental Geology* 37:78–89.

Placer Mining Section. 1993. *Yukon placer industry 1991 to 1992*. Mineral Resources Directorate, Indian and Northern Affairs Canada, Yukon.

Power, G., and M. Power. 1995. Ecotones and fluvial regimes in arctic lotic environments. *Hydrobiologia* 303:111–124.

Producers to study Alaska natural gas pipeline. 2001. *Pipeline & Gas Journal* 228:2.

Prowse, T. D. 1994. Environmental significance of ice to streamflow in cold regions. *Freshwater Biology* 32: 241–259.

Prowse, T. D. 2001a. River-ice ecology. I: Hydrologic, geomorphic, and water-quality aspects. *Journal of Cold Regions Engineering* 15:1–16.

Prowse, T. D. 2001b. River-ice ecology. II: Biological aspects. *Journal of Cold Regions Engineering* 15: 17–33.

Reynolds, J. B. 1997. Ecology of overwintering fishes in Alaskan freshwaters. In A. M. Milner and M. W. Oswood (eds.). *Freshwaters of Alaska: Ecological syntheses*, pp. 281–302. Springer-Verlag, New York.

Ricketts, T., E. Dinerstein, D. Olson, C. Loucks, W. Eichbaum, D. DellaSala, K. Kavanagh, P. Hedao, P. Hurley, K. Carney, R. Abell, and S. Walters. 1999. *Terrestrial ecoregions of North America: A conservation assessment*. Island Press, Washington, D.C.

Schindler, D. W. 1998. Sustaining aquatic ecosystems in boreal regions. *Conservation Ecology* 2(2): article 18. http://www.ecologyandsociety.org/vol2/iss2/art18.

Scott, W. and Crossman, E. J. 1973. Freshwater fishes of Canada. Fisheries Research Board of Canada.

Scudder, G. G. E. 1997. Environment of the Yukon. In H. V. Danks and J. A. Downes (eds.). *Insects of the Yukon*, pp. 13–57. Biological Survey of Canada (Terrestrial Arthropods), Ottawa.

Stewart, K. W., and W. E. Ricker. 1997. Stoneflies of the Yukon. In H. V. Danks and J. A. Downes (eds.). *Insects of the Yukon*, pp. 201–222. Biological Survey of Canada (Terrestrial Arthropods), Ottawa.

Stewart, K. W., R. E. Dewalt, and M. W. Oswood. 1991. *Alaskaperla*, a new stonefly genus (Plecoptera: Chloroperlidae) and further descriptions of related Chloroperlidae. *Annals of the Entomological Society of America* 84:239–247.

Wahl, H. E., D. B. Fraser, R. C. Harvey, and J. B. Maxwell. 1987. *Climate of Yukon*. Environment Canada (Atmospheric Environment Service) 40.

Wiggins, G. B., and B. R. Parker. 1997. Caddisflies (Trichoptera) of the Yukon. In H. Danks and J. Downes (eds.). *Insects of the Yukon*, pp. 513–614. Biological Survey of Canada (Terrestrial Arthropods), Ottawa.

Yarie, J. 1993. Effects of selected forest management practices on environmental parameters related to successional development on the Tanana River floodplain, interior Alaska. *Canadian Journal of Forest Research* 23:1001–1014.

Yarie, J., L. Viereck, K. Van Cleve, and P. Adams. 1998. Flooding and ecosystem dynamics along the Tanana River: Applying the state-factor approach to studies of ecosystem structure and function on the Tanana River floodplain. *BioScience* 48:690–695.

Yesner, D. R. 2001. Human dispersal into interior Alaska: Antecedent conditions, mode of colonization, and adaptations. *Quaternary Science Reviews* 20:315–327.

Yurtsev, B. A. 2001. The Pleistocene "Tundra-Steppe" and the productivity paradox: The landscape approach. *Quaternary Science Reviews* 20:165–174.

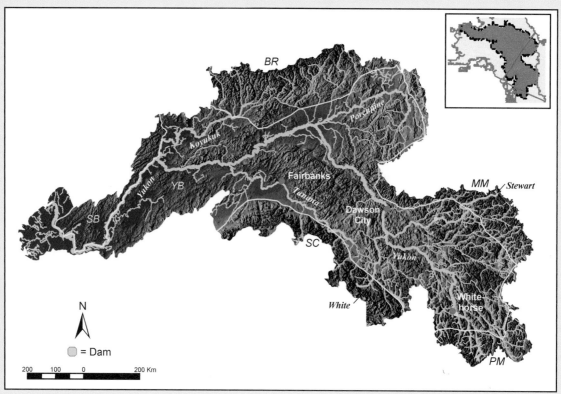

FIGURE 17.7 Map of the Yukon River basin. Physiographic provinces are separated by yellow lines.

YUKON RIVER

Relief: 6200 m
Basin area: 839,200 km²
Mean discharge: 6340 m³/s
River order: 9
Mean annual precipitation: 33 cm
Mean air temperature: −3°C
Mean water temperature: 7.7°C
Physiographic provinces: Yukon Basin (YB), Seward Peninsula and Bering Coast Uplands (SB), Brooks Range (BR), Mackenzie Mountains (MM), South-Central Alaska (SC), Coast Mountains of British Columbia and Alaska (PM)
Biomes: Boreal Forest, Tundra
Freshwater ecoregion: Yukon
Terrestrial ecoregions: Northern Cordillera Forests, Yukon Interior Dry Forests, Interior Yukon/Alaska Alpine Tundra, Ogilvie/Mackenzie Alpine Tundra, Alaska/St. Elias Range Tundra, Brooks/British Range Tundra, Interior Yukon/Alaska Lowland Taiga, Beringia Lowland Tundra
Number of fish species: 30
Number of endangered species: none

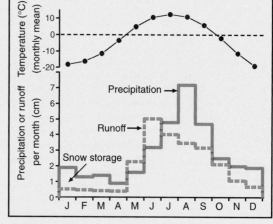

FIGURE 17.8 Mean monthly air temperature, precipitation, and runoff for the Yukon River basin.

Major fishes: Bering cisco, whitefish, chinook salmon, chum salmon, coho salmon, Arctic grayling, inconnu (sheefish), lake trout, Alaska blackfish, northern pike, burbot, Dolly Varden, Arctic lamprey
Major other aquatic vertebrates: muskrat, brown bear, black bear, mink, river otter, bald eagle
Major benthic invertebrates: chironomid midges (>100 species), black flies (*Simulium,* several endemic *Gymnopais* species), mayflies (*Baetis, Epeorus, Heptagenia pulla, Siphlonurus, Drunella, Leptophlebia*), stoneflies (*Alaskaperla obivovis* ([endemic]), *Capnia, Eucapnopsis, Utacapnia, Zapada, Taenionema, Alloperla, Suwallia, and Isoperla*), mussel (*Pyganodon beringiana* [endemic])
Nonnative species: rainbow trout, threespine stickleback
Major riparian plants: mountain alder, water birch, black spruce
Special features: largest wild river in North America
Fragmentation: free-flowing except for dam at Whitehorse
Water quality: conductivity = 199 µS/cm; pH = 7.0; total hardness = 93.3 mg/L as CaCO₃; PO₄-P = 0.02 mg/L
Land use: <0.01% urban, agricultural, industrial, >99% forest and tundra
Population density: <0.1 people/km²
Major information sources: Danks and Downes 1997, Oswood et al. 1995, Ennis et al. 1982, Farrar 1995

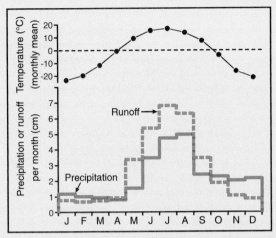

FIGURE 17.10 Mean monthly air temperature, precipitation, and runoff for the Tanana River basin.

TANANA RIVER

Relief: 6100 m
Basin area: 114,000 km^2
Mean discharge: 1185 m^3/s
River order: 8
Mean annual precipitation: 28 cm
Mean air temperature: −2.9°C
Mean water temperature: 6.7°C
Physiographic provinces: South-Central Alaska (SC), Yukon Basin (YB)
Biomes: Boreal Forest, Tundra
Freshwater ecoregion: Yukon
Terrestrial ecoregions: Interior Yukon/Alaska Alpine Tundra, Alaska/St. Elias Range Tundra, Interior Yukon/Alaska Lowland Taiga
Number of fish species: 20
Number of endangered species: none
Major fishes: Bering cisco, whitefish, chinook salmon, chum salmon, coho salmon, Arctic grayling, inconnu (sheefish), lake trout, Alaska blackfish, northern pike, burbot
Major other aquatic vertebrates: brown bear, black bear, beaver, mink, marten, river otter, bald eagle
Major benthic invertebrates: unconfirmed; see Yukon River
Nonnative species: rainbow trout, Arctic char
Major riparian plants: black spruce, mountain alder willow, quaking aspen
Special features: mixture of turbid glacial input from the south and clear subsurface inflow from the north
Fragmentation: free-flowing except for flood-control dam on Chena River
Water quality: conductivity = 243 μS/cm, total hardness = 114 mg/L as CaCO$_3$, PO$_4$-P = 0.31 mg/L
Land use: <1% urban, agricultural, industrial; subsistence and recreational fishing and hunting; >99% tundra and boreal forest
Population density: 0.26 people/km^2
Major information sources: Oswood et al. 1992, Geraghty et al. 1973, Hartman and Johnson 1978, Yarie et al. 1998

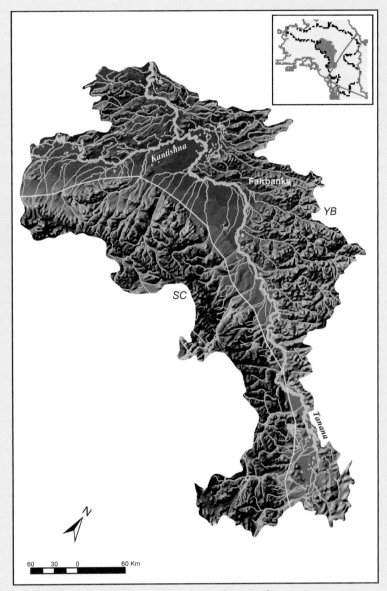

FIGURE 17.9 Map of the Tanana River basin. Physiographic provinces are separated by a yellow line.

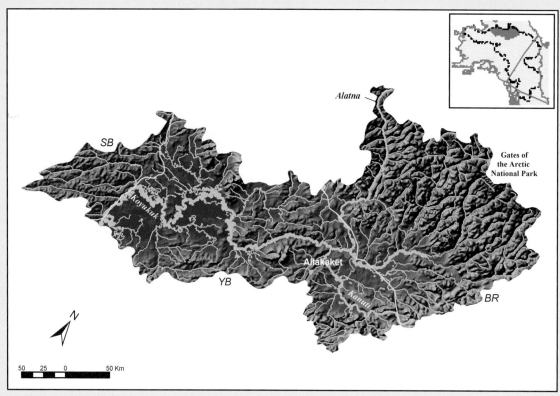

FIGURE 17.11 Map of the Koyukuk River basin. Physiographic provinces are separated by yellow lines.

KOYUKUK RIVER

Relief: 2280 m
Basin area: 83,500 km²
Mean discharge: 664 m³/s
River order: 8
Mean annual precipitation: 31 cm
Mean air temperature: −9.5°C
Mean water temperature: 8.6°C
Physiographic provinces: Brooks Range (BR), Yukon Basin (YB), Seward Peninsula and Bering Coast Uplands (SB)
Biomes: Boreal Forest, Tundra
Freshwater ecoregion: Yukon
Terrestrial ecoregions: Interior Yukon/Alaska Lowland Taiga, Brooks/British Range Tundra, Interior Yukon/Alaska Alpine Tundra
Number of fish species: 25 (unconfirmed)
Number of endangered species: none
Major fishes: unconfirmed; see Yukon River
Major other aquatic vertebrates: see Yukon River
Major benthic invertebrates: unconfirmed; see Yukon River
Nonnative species: rainbow trout
Major riparian plants: black spruce, mountain alder, balsam poplar, quaking aspen, willow
Special features: almost pristine tundra river; important spawning areas for anadromous fishes; headwaters contact Trans-Alaskan pipeline
Fragmentation: none
Water quality: conductivity = 213 µS/cm, pH = 7.6, total hardness = 105 mg/L as CaCO₃, PO₄-P = 0.01 mg/L
Land use: subsistence and recreational fishing and hunting; pipeline right of way; >99% tundra and boreal forest
Population density: <0.01 people/km²
Major information sources: Oswood et al. 1995, Geraghty et al. 1973, Hartman and Johnson 1978

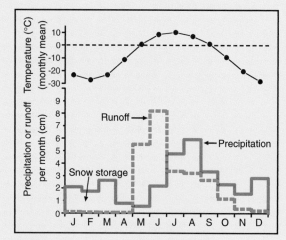

FIGURE 17.12 Mean monthly air temperature, precipitation, and runoff for the Koyukuk River basin.

798

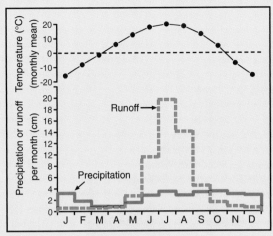

FIGURE 17.14 Mean monthly air temperature, precipitation, and runoff for the White River basin.

WHITE RIVER

Relief: 5635 m
Basin area: 50,500 km²
Mean discharge: 927 m³/s
River order: 6
Mean annual precipitation: 31 cm
Mean air temperature: 3.8°C
Mean water temperature: NA
Physiographic provinces: South-Central Alaska (SC), Yukon Basin (YB)
Biomes: Boreal Forest, Tundra
Freshwater ecoregion: Yukon (Arctic Complex)
Terrestrial ecoregions: Interior Yukon/Alaska Alpine Tundra, Alaska/St. Elias Range Tundra, Interior Yukon/Alaska Lowland Taiga
Number of fish species: 28
Number of endangered species: none
Major fishes: Bering cisco, whitefish, chinook salmon, chum salmon, coho salmon, Arctic grayling, inconnu (sheefish), lake trout, Alaska blackfish, northern pike, burbot
Major other aquatic vertebrates: see Yukon River
Major benthic invertebrates: unconfirmed; see Yukon River
Nonnative species: none
Major riparian plants: black spruce, mountain alder, quaking aspen, balsam poplar
Special features: pristine river with primarily glacial inputs; extremely turbid
Fragmentation: none
Water quality: pH = 7.8, total hardness = 122 mg/L as CaCO₃, NO₃-N = 0.3 mg/L
Land use: <0.01% urban, industrial; subsistence and recreational fishing and hunting; >99% forest and tundra
Population density: <0.01 people/km²
Major information sources: Ennis et al. 1982, Jack et al. 1983, Scudder 1997

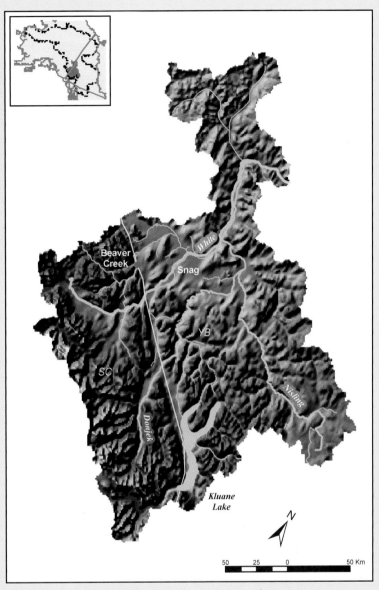

FIGURE 17.13 Map of the White River basin. Physiographic provinces are separated by a yellow line.

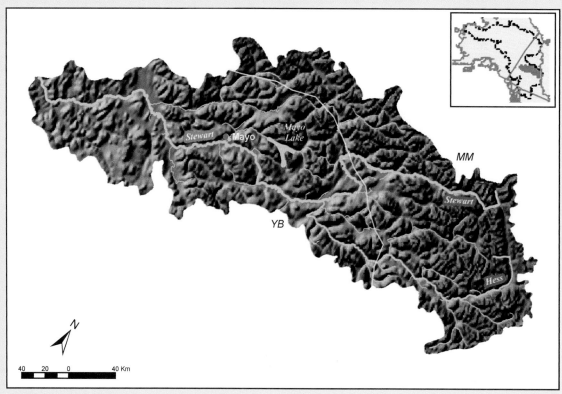

FIGURE 17.15 Map of the Stewart River basin. Physiographic provinces are separated by a yellow line.

STEWART RIVER

Relief: 2635 m
Basin area: 51,000 km²
Mean discharge: 675 m³/s
River order: 7
Mean annual precipitation: 28 cm
Mean air temperature: −2.9°C
Mean water temperature: NA
Physiographic provinces: Mackenzie Mountains (MM), Yukon Basin (YB)
Biome: Boreal Forest
Freshwater ecoregion: Yukon (Arctic Complex)
Terrestrial ecoregions: Yukon Interior Dry Forests, Interior Yukon/Alaska Alpine Tundra, Ogilvie/Mackenzie Alpine Tundra
Number of fish species: 29 (unconfirmed)
Number of endangered species: none
Major fishes: Bering cisco, whitefish, chinook salmon, chum salmon, coho salmon, Arctic grayling, inconnu (sheefish), lake trout, northern pike, burbot
Major other aquatic vertebrates: see Yukon River
Major benthic invertebrates: unconfirmed; see Yukon River
Nonnative species: rainbow trout, lake trout
Major riparian plants: see Yukon River
Special features: mostly pristine river, heavy placer activity in lower Stewart River tributaries
Fragmentation: none; one dam on tributary (Mayo River)
Water quality: pH = 7.5, total hardness = 123 mg/L as CaCO₃, NO₃-N = 0.35 mg/L, PO₄-P = 0.02 mg/L
Land use: <1% urban, agricultural, industrial; subsistence and recreational fishing and hunting; >99% forest and tundra
Population density: <1 people/km²
Major information sources: Ennis et al. 1982, Jack et al. 1983, Scudder 1997

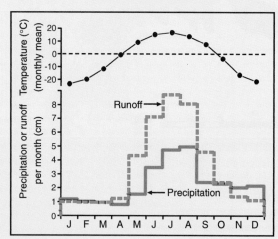

FIGURE 17.16 Mean monthly air temperature, precipitation, and runoff for the Stewart River basin.

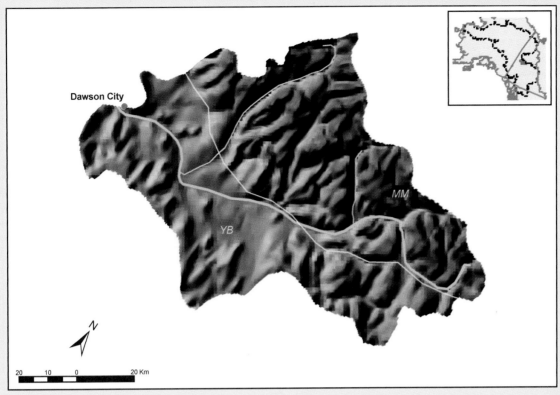

FIGURE 17.17 Map of the Klondike River basin. Physiographic provinces are separated by a yellow line.

KLONDIKE RIVER

Relief: 2200 m
Basin area: 7800 km²
Mean discharge: 63 m³/s
River order: 5
Mean annual precipitation: 32 cm
Mean air temperature: −5.2°C
Mean water temperature: NA
Physiographic provinces: Mackenzie Mountains (MM), Yukon Basin (YB)
Biomes: Boreal Forest, Tundra
Freshwater ecoregion: Yukon
Terrestrial ecoregions: Ogilvie/Mackenzie Alpine Tundra, Interior Yukon/Alaska Lowland Taiga
Number of fish species: 25
Number of endangered species: none
Major fishes: Bering cisco, whitefish, chinook salmon, chum salmon, coho salmon, Arctic grayling, inconnu (sheefish), burbot
Major other aquatic vertebrates: brown bear, black bear, beaver, mink, marten, river otter
Major benthic invertebrates: unconfirmed; see Yukon River
Nonnative species: none
Major riparian plants: black spruce, mountain alder, balsam poplar, quaking aspen
Special features: Klondike gold rush of 1898
Fragmentation: none
Water quality: pH = 7.0, total hardness = 117 mg/L as $CaCO_3$, NO_3-N = 0.15 mg/L, PO_4-P = 0.02 mg/L
Land use: <1% urban, industrial; subsistence and recreational fishing and hunting; >99% tundra and forest
Population density: 0.10 people/km²
Major information sources: Ennis et al. 1982, Jack et al. 1983, Scudder 1997

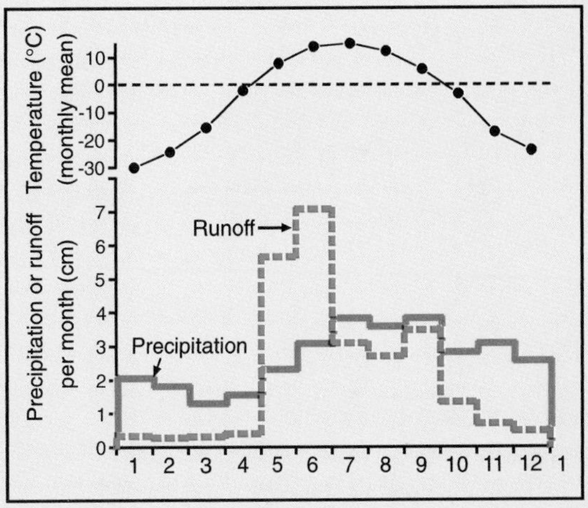

FIGURE 17.18 Mean monthly air temperature, precipitation, and runoff for the Klondike River basin.

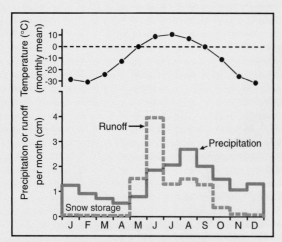

FIGURE 17.20 Mean monthly air temperature, precipitation, and runoff for the Porcupine River basin.

PORCUPINE RIVER

Relief: 1670 m
Basin area: 118,000 km²
Mean discharge: 414 m³/s
River order: 8
Mean annual precipitation: 16 cm
Mean air temperature: −11°C
Mean water temperature: 5.5°C
Physiographic provinces: Mackenzie Mountains (MM), Brooks Range (BR), Yukon Basin (YB)
Biomes: Boreal Forest, Tundra
Freshwater ecoregion: Yukon
Terrestrial ecoregions: Brooks/British Range Tundra, Interior Yukon/Alaska Lowland Taiga
Number of fish species: 29 (unconfirmed)
Number of endangered species: none
Major fishes: Bering cisco, whitefish, chinook salmon, chum salmon, coho salmon, Arctic grayling, inconnu (sheefish), lake trout, Alaska blackfish, northern pike, burbot
Major other aquatic vertebrates: brown bear, black bear, beaver, mink, marten, river otter, bald eagle
Major benthic invertebrates: unconfirmed; see Yukon River
Nonnative species: rainbow trout, threespine stickleback
Major riparian plants: black spruce, mountain alder
Special features: large pristine tundra river; northernmost basin within Yukon system
Fragmentation: none
Water quality: conductivity = 171 μS/cm, pH = 7.4, total hardness = 76 mg/L as CaCO₃
Land use: subsistence fishing and hunting; >99% tundra and forest
Population density: <0.1 people/km²
Major information sources: Oswood et al. 1995, Geraghty et al. 1973, Hartman and Johnson 1978

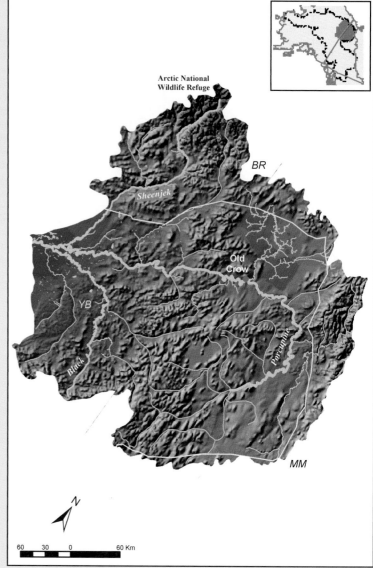

FIGURE 17.19 Map of the Porcupine River basin. Physiographic provinces are separated by yellow lines.

18

MACKENZIE RIVER BASIN

JOSEPH M. CULP   TERRY D. PROWSE   ERIC A. LUIKER

INTRODUCTION

MACKENZIE RIVER MAIN STEM

LIARD RIVER

SLAVE RIVER

PEACE RIVER

ATHABASCA RIVER

ADDITIONAL RIVERS

ACKNOWLEDGMENTS

LITERATURE CITED

INTRODUCTION

The catchment of the Mackenzie River basin encompasses an enormous geographic area extending over 15° of latitude and 37° of longitude from central Alberta to the Arctic Ocean, an area larger than central Europe (Fig. 18.2). Rosenberg and Barton (1986) and the Mackenzie River Basin Committee (1981a) provide a detailed description of the river's characteristics, noting that it ranks fourth in discharge relative to other Arctic Ocean rivers and is the 12th-largest drainage basin in the world. Canada's largest river basin composes 20% of the nation's territory and is the largest northerly flowing river in North America. The system includes a number of major rivers, including, from south to north, the Athabasca, Peace, Slave, Liard, Arctic Red, and Peel rivers, as well as the main-stem channel of the Mackenzie River. At the mouth the river empties through the Mackenzie Delta, the second-largest Arctic delta in the world, into the Beaufort Sea. In addition, the system contains two large freshwater deltas (the Peace–Athabasca and the Slave) and three major lakes (Lake Athabasca, Great Slave Lake, and Great Bear Lake). Both the Lower and Upper Mackenzie freshwater ecoregions of the Arctic Complex are contained within the watershed (Abell et al. 2000).

The first inhabitants in the Mackenzie watershed area were the ancestral American Indians, who moved to the area from Asia across the Bering Sea approximately 12,000 years ago. Immigration by the Inuit followed, and they eventually occupied arctic coastal areas, including the Mackenzie Delta and other tundra areas north of the tree line. Early contacts between the aboriginal peoples of the Mackenzie River basin and Europeans appear to have occurred in the late 1600s and early 1700s as a result of fur trading (Rosenberg and Barton 1986). Many of the early explorers worked for the North West Company or the Hudson's Bay Company, which were amalgamated in 1821 after years of competition (MRBC 1981a). Several of these explorers, including David Thompson and Simon Fraser, traveled through parts of the basin in search of new trade routes to the Pacific and the North. However, the river bears the name of Sir Alexander Mackenzie because he explored the main stem from Great Slave Lake to its Arctic outlet in 1789. The Mackenzie River main stem from Great Slave Lake to the Beaufort Sea has been a key transportation route for the settlement of the region, with wood-burning paddle wheelers operating on the river from the 1880s until the 1940s. Barge traffic remains an important method of transporting commodities and supplies to northern communities.

FIGURE 18.1 Hay River, south of the town of Hay River, Northwest Territories (PHOTO BY T. CARTER).

805

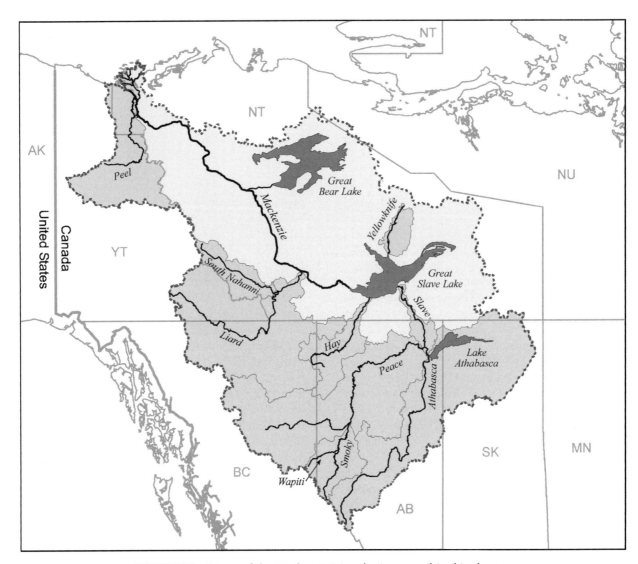

FIGURE 18.2   Rivers of the Mackenzie River basin covered in this chapter.

We discuss five major rivers of the Mackenzie River System to demonstrate the large range of natural diversity and human impacts evident within the basin. These rivers include the Athabasca, Peace, and Slave, which drain the most southern reaches of the watershed, and the Liard River and the main stem of the Mackenzie River, both of which are located north of Great Slave Lake. In addition, summary information is provided on physical, chemical, and biological characteristics and human uses of these rivers, and for five other rivers of the Mackenzie River basin (Smoky River, Hay River, Yellowknife River, South Nahanni River, and Peel River).

Physiography and Climate

During the Wisconsin time the Laurentide ice sheet covered 80% of the basin, from its eastern edge to the foothills, and western Cordillera glaciers covered many of the mountain valleys (Brunskill 1986). Glacial meltwaters formed large lakes, such as Lake McConnell, and encompassed the entire area of several present-day water bodies (Lake Athabasca, Great Slave Lake, Great Bear Lake). In addition, meltwater channels contributed to the formation of many aspects of present-day drainages. Glacial refugia extended from the South Nahanni River in the Liard basin to the Richardson Mountains in the north, and this glacial history is thought to have greatly influenced the composition of present-day aquatic fauna (McPhail and Lindsey 1970, Rosenberg and Barton 1986).

Seven physiographic provinces are included in the Mackenzie basin: Mackenzie Mountains (MM), Coast Mountains of British Columbia and Southeast Alaska (PM), Rocky Mountains in Canada (RM), Great Plains (GP), Athabasca Plains (AT), Bear–Slave–Churchill Uplands (BC), and Arctic Lowlands (AL) (Hunt 1974). The western portion of the basin is largely underlain by sedimentary bedrock and is bounded by the Western Cordillera (MM, PM, and RM provinces), which contains numerous high mountain peaks and vast plateaus separated by wide river valleys and lowlands. This region receives the greatest amount of precipitation in the basin and thus produces all of the major tributaries to the Mackenzie River main stem. Central portions of the Mackenzie River basin lie within the GP province, an area that extends from the Athabasca River in the south to the subarctic Mackenzie Delta. Terrain in this region is flat and rolling, with occasional ranges of hills and small mountains, all underlain by sedimentary materials, such as sandstones, shales, and limestones (Brunskill 1986). Much of the area contains extensive permafrost, particularly north of 60°N latitude, and this produces vast areas of poorly drained soils and wetland habitat. To the east of the GP province the basin is bounded by AT and BC provinces, both within the Canadian (Precambrian) Shield physiographic division. Low amounts of runoff originate from this region to other parts of the basin. Bedrock here is near the surface, soils are thin, permafrost is extensive, and the landscape is a mosaic of lakes and bogs. Finally, the Mackenzie Delta, which lies within the AL province, is an extensive area of poorly developed levees formed by sediments transported from the Mackenzie River over the last 13,000 years. In addition, the delta forms a complex network of lakes, ponds, and river channels. The delta is the 10th largest in the world and provides critical summer habitat for migratory birds using the Pacific and Central flyways of North America.

Two major climatic zones, the arctic and subarctic, divide the basin at tree line. Climate is strongly influenced by continental climatic conditions, with both arctic and subarctic regions having extremely cold and long winters. Mean daily temperatures are approximately −20°C to −30°C in January and 14°C to 20°C in July. Precipitation in the basin ranges between 25 and 40 cm/yr east of the Mackenzie River and 50 to 160 cm/yr in the mountainous Cordillera

of the western basin (Brunskill 1986). Depending on the location within the catchment, from 30% to 70% of precipitation falls as snow. Southern portions of the basin receive about 8 hours of sunlight in December and 17 hours in July. At the mouth of the river Inuvik has 24 hours of darkness in December and early January and 24 hours of sunlight in early July.

Basin Landscape and Land Use

Because of the enormous size of the basin (37° of longitude, 15° of latitude), the landscape of the Mackenzie River basin is quite varied from west to east and south to north. Ricketts et al. (1999) divided the basin into 15 ecoregions within four broader categories: Temperate Coniferous Forests, Temperate Grasslands/Savanna/Shrub, Boreal Forest/Taiga, and Tundra. The forested mountains in the southwest correspond to the Central British Columbia Mountain Forests, the Alberta Mountain Forests, and the Alberta/British Columbia Foothills Forests ecoregions. To the south the basin overlaps with the Canadian Aspen Forest and Parklands ecoregion, which forms a transition to the vast midcontinental prairie grasslands. Much of the central and eastern portion of the basin is covered by the Interior Alaska/Yukon Lowland Taiga, Northwest Territories Taiga, Northern Cordillera Forests, Muskwa/Slave Lake Forests, Northern Canadian Shield Taiga, Midcontinental Canadian Forests, and Midwestern Canadian Shield Forests ecoregions. In the north, the extreme climate produces tundra, including the Ogilvie/Mackenzie Alpine Tundra, Brooks/British Range Tundra, Arctic Coastal Tundra, and Low Arctic Tundra ecoregions.

Below 60°N latitude much of the landscape of the Mackenzie River basin is closed boreal forest bordered in the south by aspen parkland (Boreal Forest biome). At higher elevations to the west forests are dominated by Engelmann and white spruce, lodgepole pine, and alpine fir (mountain provinces and ecoregions). Black spruce and tamarack occur in lower, poorly drained areas, whereas balsam poplar and white spruce are common in river valleys. Northern areas of the basin consist of open boreal forest grading into arctic tundra. The lower Mackenzie valley and delta have extensive bogs and fens, with dwarf black spruce, willow, and alder. The Arctic ecoregions are characterized by low shrubs, lichens, and mosses on drier sites, and by cottongrass–sedge meadows on poorly drained sites (MRBC 1981a).

Early development in the basin centered on fur trading, with settlements generally established at trading posts on major waterways (MRBC 1981a).

Rivers provided important transportation corridors, first for freighter canoes and later for scheduled barge service. Northern residents still rely on barge shipments for many of their commodities. Much of the basin remained relatively pristine until the early 1900s, when mineral exploration began. Mining in the basin began in the 1920s and has included the extraction of many minerals (e.g., asbestos, coal, copper, gold, silver, tungsten, and uranium). Present-day development in the Mackenzie River basin is focused around oil fields at Norman Wells, massive oil sands developments in Alberta, mineral (e.g., gold, silver, uranium) and diamond mining, hydroelectric facilities on the Peace River, and large forest harvesting operations located primarily in the Peace and Athabasca river drainages. Petroleum exploration and production continues to increase through major expansion of oil sand extraction in Alberta and oil and gas production in British Columbia and Alberta. Hydroelectric development is mostly restricted to the massive W. A. C. Bennett dam on the Peace River. Although forests have only been harvested on a large scale for pulp and paper (as well as timber) since the 1950s, southern forests have already become fragmented. Most agricultural lands are located in the Athabasca and Peace basins and are locally extensive. However, most of the basin is in a relatively pristine state, with the landscape remaining primarily as forests, shrubland, and some agricultural land. Development in the basin has progressed from the southern headwaters northward; consequently, the major cities are found in Alberta (Grande Prairie, Fort McMurray) and the southern Northwest Territories (Yellowknife). The Mackenzie River basin is very sparsely populated (<1 person/km²) and less than 350,000 people resided there in 1996.

The Rivers

The Mackenzie River basin includes the Upper and Lower Mackenzie freshwater ecoregions (Abell et al. 2000). Rivers discussed in this chapter differ mostly in basin size and mean annual temperature. Because much of the basin drains sedimentary rock, water chemistry of the rivers is similar, the exception being rivers, such as the Great Bear, that flow from the granitic bedrock of the Precambrian Canadian Shield. In addition, the flora and fauna of the rivers are quite depauperate relative to more southerly rivers in North America, largely because of recent glaciation and the severe climate. All of the major rivers are wide and moderately swift flowing.

The southern headwaters of the Mackenzie River are the Athabasca and Peace rivers, which have a unique confluence at the Peace–Athabasca Delta (PAD) where the Athabasca drains into and out of Lake Athabasca. Depending on the hydrological stage of the Peace River, waters of the delta and Lake Athabasca either flow to join the Peace at its confluence with the Slave River, or flows are reversed, resulting in flooding of PAD lakes and perched basins (Prowse and Conly 2000). The river continues flowing northward as the Slave River before emptying into Slave River Delta and Great Slave Lake. Sediments from the Peace and Athabasca rivers are mostly removed by the deltas and lake; thus the main-stem Mackenzie River is very clear when it forms at the outflow of the Great Slave Lake. The first major tributary to join the Mackenzie is the warmer and much siltier Liard River. Numerous rivers flow into the Mackenzie as it travels north to the Arctic Ocean. The major northern tributaries include the Great Bear, Peel, and Arctic Red. Finally, the Mackenzie River flows through the many channels of the Mackenzie Delta before emptying into the Beaufort Sea.

The greatest change in biota occurs in the transition from the western mountainous headwaters to the lower-gradient main stems of the major tributaries. Discharge patterns in all rivers are characterized by high flows in spring or early summer due to snowmelt (and in some cases glacial melt), followed by declining flow until the following spring, except when rainfall produces secondary flood peaks during the open-water season. Rivers are ice covered generally from sometime in the fall until late winter, and river ice breakup can have significant effects on physicochemical and biological processes (Prowse and Conly 1996, Scrimgeour et al. 1994, Prowse and Culp 2003).

Major threats to the rivers are land-use changes, including agriculture, forestry, and mining, hydrologic fragmentation through the creation of reservoirs, and point-source inputs from industry and municipalities. These developments have placed important demands on rivers as the receivers of nonpoint and point effluents. For example, nutrient and contaminant loadings from five pulp mills impact the Athabasca River (Chambers, Brown et al. 2000, Chambers, Dale, and Scrimgeour 2000, Culp, Cash, and Wrona 2000, Culp, Podemski, and Cash 2000). Municipal sewage effluent impacts have accompanied resource development as cities increase in size, a trend that is particularly evident in Alberta on the Athabasca and Wapiti rivers. In addition, damming of the Peace River has caused impacts on the Peace–Athabasca Delta approximately 1200 km downstream of Williston Reservoir (Prowse and Conly 2000). Commercial fishing is important on basin lakes such as Great Slave Lake and Lake Athabasca, and aboriginal peoples of the basin continue to use wildlife resources for food and furs.

There are four tributaries of the Mackenzie River that have been designated as Canadian Heritage Rivers, including the Arctic Red, Athabasca, Bonnet Plume, and South Nahanni rivers (www.chrs.ca 2003). The Canadian Heritage Rivers System was established in 1984 by the federal, provincial, and territorial governments to conserve and protect Canada's river heritage.

MACKENZIE RIVER MAIN STEM

The Mackenzie River main stem originates at the outflow of Great Slave Lake, having drained over 970,000 km^2 of the Athabasca, Peace, and Lake Athabasca catchments. It then flows 1800 km northwest to the Beaufort Sea in the Arctic Ocean. At its mouth the Mackenzie River is part of an enormous 9th order system that is 4200 km long and has a drainage area of 1,787,000 km^2 (Fig. 18.9). The Mackenzie's most significant northern tributary, the Liard River, joins the river about 300 km downstream of Great Slave Lake at Fort Simpson (Fig. 18.3), delivering an enormous sediment load to the Mackenzie River in spring and doubling the discharge (Brunskill 1986). For many kilometers downstream the waters of the silty Liard River and the clear Mackenzie River are visually distinct (MRBC 1981a). Many other tributaries, such as the Great Bear, Arctic Red, and Peel rivers, join the Mackenzie as it flows north to the delta. Over its final 240 km the river meanders through numerous large channels in the vast Mackenzie Delta, which has an active area of 12,170 km^2 (MRBC 1981a, Brunskill 1986).

Early inhabitants of the main stem of the Mackenzie River were the Thule, ancestors of the Inuit of this region (Morrison and Germain 1995). They arose in Northwestern Alaska approximately 1100 years ago. Over the next two centuries they spread east throughout the Arctic to Northern Greenland. The Thule lived in permanent sod-houses in the winter, undergoing hunting based migrations during other times of the year. The transition from Thule to the Inuvialuit (Inuit of the Northwest Territories region) included adding fishing as a way of obtaining food, and use of kayaks for hunting beluga whales. From archaeological records there seemed to be little cultural change from 700 years ago until the arrival of the Europeans 200 years ago. At the time of European contact the Inuvialuit were composed of six distinct territorial groups, most having a main village as a focal point. One of the larger groups were the Kittegaryumuit, who resided in a village near the mouth of the Mackenzie River. During this period there were approximately 2500 Inuvialuit.

Early contact with Europeans occurred in the 1780s, when Peter Pond established a trading post at Fort Resolution at the mouth of the Slave River (Rosenberg and Barton 1986). In 1789, Sir Alexander Mackenzie explored from Great Slave Lake to the Arctic Ocean along the length of the river that bears his name. The river became an important transportation route, so much so that the Hudson's Bay Company established a regular shipping service along the river.

Physiography, Climate, and Land Use

Although the Mackenzie basin drains seven physiographic provinces, the main-stem river flows through only three provinces (see Fig. 18.9). The main stem, situated between 60°N and 69°N latitude and 115°W and 140°W longitude, begins within the Great Plains physiographic province. As it turns north above Ft. Simpson it enters the Mackenzie Mountains province, flowing in the lowlands between the Mackenzie Mountains to the west and the smaller Franklin Mountains to the east. Near the Mackenzie Ramparts the river flows into the Arctic Lowlands province. Along its path the main stem receives runoff from tributaries of the mountain provinces to the west and the Precambrian Shield (Bear–Slave–Churchill Uplands province) to the east. Much of the region is underlain by sedimentary rock, such as limestone, shale, and sandstone. Poor drainage and widespread permafrost favors the formation of cryosolic, gleysolic, and organic soils; however, brunisolic and luvisolic soils are present in uplands. Regosols have developed within morainal, alluvial, and lacustrine landscapes. Permafrost is widespread throughout the region and is continuous in northern areas, such as the Mackenzie Delta, where it can extend to a depth of 100 m (MRBC 1981a).

Although the Mackenzie basin drains from 15 terrestrial ecoregions (Ricketts et al. 1999), the main stem flows primarily through only two. The main

FIGURE 18.3 Mackenzie River (left) at its confluence with the Liard River (right), Northwest Territories, illustrating the contrast in sediment dynamics and ice breakup. The Liard River is in active breakup and laden with sediment. The Mackenzie River is relatively clear and undergoing a thermal breakup (PHOTO BY T. D. PROWSE).

stem begins in the Muskwa/Slave Lake Forests ecoregion and then borders and eventually enters the Northwest Territories Taiga ecoregion along its path to the Mackenzie River Delta. Vegetation along the Mackenzie River Valley consists mostly of stands of black spruce and jack pine, but white spruce, balsam fir, and trembling aspen occur in warmer and moister sites to the south. The northern tree line, which has a poorly defined boundary, marks the transition zone between subarctic forests to the south and the northerly arctic tundra. The tundra is characterized by low shrubs, mosses, and lichens on drier sites and

cottongrass and sedge meadows on poorly drained sites (MRBC 1981a).

Climate of the main-stem Mackenzie River is extremely harsh, with very short and cool summers followed by bitterly cold winters. At settlements north of the Arctic Circle, such as Inuvik, Northwest Territories, mean annual temperature is about −10°C, with mean temperatures in July and January of 14° and −29°C, respectively (Fig. 18.10). Temperature extremes recorded at Inuvik are 32° and −51°C. The southern part of the region is milder; for example, annual mean temperature at Yellowknife is

−5°C, with temperatures in July and January averaging 16.5°C and −28°C. Precipitation ranges between 15 and 20 cm/yr throughout the river valley, with 50% to 70% of this moisture arriving as snow. At Inuvik mean monthly precipitation is lowest in February, at 1.1 cm/month, and increases monthly to a maximum of 4.4 cm/month in August (see Fig. 18.10). Areas in the mountains to the west have much higher precipitation (up to 70 cm/yr), which is why western tributaries, such as the Liard River, are so important to Mackenzie River discharge. The river is ice covered from October to April. In fact, public ice roads that follow the Mackenzie River are vital links between the communities of Fort Good Hope and Norman Wells.

The Mackenzie River and Valley are very sparsely populated, with <1 person/km². Excluding Yellow-knife (population 17,300), which is located along the north shore of Great Slave Lake, the major communities in the basin (i.e., Fort Nelson, Inuvik, Hay River, Fort Smith, Fort Simpson) have a combined population of less than 15,000 people. Very little agriculture occurs in the region, and the primary human activities are mining, oil and gas extraction, and some forestry and tourism. In addition, hunting, trapping, and fishing are primary subsistence activities in the local economies. Land types for the main stem of the Mackenzie River include 36.5% taiga (tree crown density below 10%, may include significant shrub cover), 27.3% open land (land containing usually less than 10% of tree crown density, may contain shrubs, lichen, herbaceous vegetation cover, bare soil, rock, or small water bodies), 23.5% forest cover (trees of greater than 5 m in height with a crown density greater than 10% of the land area), 8.7% shrubland (land covered mainly by low [less than 1 m], and intermediate [less than 2 to 3 m] height woody shrubs), and 4.0% water (land covered with water in liquid form) (http://atlas.gc.ca/site/english/index.html 2003).

River Geomorphology, Hydrology, and Chemistry

Over its length from Great Slave Lake to the Mackenzie River Delta the main stem falls approximately 150 m, resulting in a comparatively low-gradient river (10 cm/km). The exceptions are fast-flowing reaches such as the Ramparts (downstream from Norman Wells and upstream from Fort Good Hope), where gradients can be 1.5 m/km (MRBC 1981a). The river channel is very straight and wide (1 to 2 km) and has seen little development since deglacia-

tion (Brunskill 1986). Rivers from the western mountains (Liard, Redstone, Arctic Red, Peel) bring silt-laden water to the west side of the Mackenzie (see Fig. 18.3), whereas eastern rivers (Great Bear) pour clear or humic-colored waters into the main stem (Brunskill 1986).

Mean monthly runoff remains below 1 cm/mo from November to April, with lowest levels of 0.5 cm/mo from February to April (see Fig. 18.10). Runoff in May and June exceeds precipitation due to the addition of snowmelt to precipitation, with a maximum of 3.1 cm/mo in June. The greatest difference between precipitation and runoff occurs in August due to evapotranspiration. Average monthly discharge of the Mackenzie River near the delta is 9020 m³/s, with peak discharge occurring in June (20,626 m³/s) and minimum values in March (3373 m³/s). Spring breakup of river ice on the upstream portions of the Mackenzie main stem occurs in late April to May and is typically triggered by flow from the Liard River (see Fig. 18.3), which can double the Mackenzie discharge in spring (Prowse 1986). Breakup advance along the main stem usually takes a month before reaching the Mackenzie Delta. More than 75% of the runoff in the Mackenzie River occurs during open water between May and October.

Water chemistry of the Mackenzie River main stem is strongly affected by Cordilleran tributary rivers that drain basins underlain by sedimentary materials (Brunskill 1986, Bodaly 1989). Thus this is an alkaline (average 93 mg/L as $CaCO_3$), hard-water system with basic pH (7.9) and moderate levels of cations such as calcium and potassium. Because much of the sediment load of the Peace and Athabasca rivers is trapped in deltas and lakes, the high levels of suspended sediments in the main stem during the spring and summer originate in western mountain drainages of the Yukon, Northwest Territories, and British Columbia. Although total phosphorus is high during the summer (0.24 mg/L), this is mostly due to particulate forms, as dissolved phosphorus concentration is very low (0.011 mg/L). Total nitrogen mean annual concentration is 0.37 mg/L. Metal concentrations in the water of the main stem are low (Environment Canada 1981).

River Biodiversity and Ecology

The main-stem Mackenzie is within the Lower Mackenzie freshwater ecoregion (Abell et al. 2000). Biological data on primary and secondary producers of the Mackenzie River system is very limited (Bodaly et al. 1989). Most available information relates to fish populations and fisheries, but even these data are scarce. Authoritative reviews on the biota of the Mackenzie River include Barton (1986), McCart (1986), Rosenberg (1986), and Mackay (1995).

Algae

Most detailed information on the aquatic flora of the Mackenzie River relates to studies of delta lakes. Diatom assemblages in these lakes are dominated by benthic microflora, namely *Navicula*, *Nitzschia*, *Coccoconeis*, and *Gomphonema* (Hay et al. 2000). In addition, Wiens et al. (1975) recorded >100 species of algae from 36 genera, with *Achnanthes*, *Gomphonema*, *Navicula*, *Nitzschia*, *Pinnularia*, and *Synedra* the most diverse algal taxa.

Plants

Macrophyte growth is higher in lakes that are not strongly influenced by river flooding. Furthermore, macrophytes can be abundant in the main river channel, particularly in the less-turbid reach of the Mackenzie between Great Slave Lake and the Liard River confluence (Barton 1986). Wiens et al. (1975) provide an extensive species list of aquatic plants in the Mackenzie basin (>50 macrophyte species), with sedges, pondweed (*Potamogeton*), and the macrophytic alga *Chara* the most diverse genera.

Invertebrates

Information on benthic invertebrate assemblages of the Mackenzie River is even more limited than that for primary producers. Published data on biota of the main channel is restricted to Barton's (1986) review. He reports accumulations of large amounts of allochthonous debris and a fauna dominated by true flies (chironomid midges, black flies, biting midges), stoneflies, mayflies, and caddisflies in baskets of stones suspended in the water column. In contrast, samples collected from river channel sediments in the Mackenzie Delta contained large fractions of oligochaete worms, midges, and snails during both the summer and winter. Mean densities in the river channels of the delta were very low and ranged from 166 to 417 individuals/m². Rosenberg and Snow (1975) report that benthic invertebrate densities in Mackenzie Delta lakes were higher than in the river channel, with standing crop reduced by up to 50% when clear lakes were flooded by sediment-laden Mackenzie River waters in the spring. Benthic macroinvertebrate densities were highest in streams with low suspended sediments (<20 mg/L; Rosenberg and Snow 1975). The limited information available

for several small tributaries of the main stem indicates that these streams have a fauna typical of fast-flowing waters (i.e., mayflies, stoneflies, caddisflies, and midges; Rosenberg and Snow 1975; Rosenberg and Wiens 1976, 1978). Wiens et al. (1975) provide a list of 381 genera and 464 species of invertebrates found in the Mackenzie basin, including detailed lists for several main-stem tributaries. Common invertebrates of the main-stem Mackenzie include several true flies (biting midges, chironomid midges, black flies), mayflies (*Ametropus*, *Baetis*, *Ephemerella*, *Heptagenia*), stoneflies (*Isoperla*), and caddisflies (*Brachycentrus*).

Vertebrates

The fish fauna of the Mackenzie River basin is composed of 52 species from 14 families. The main stem of the river has 34 of these species, with the dominant fish groups consisting of the salmonids (3 families, 13 species) and cyprinids (7 species) (McCart 1986), with salmonids the most important group for subsistence and commercial fisheries. Other families include Petromyzontidae (lamprey), Hiodontidae (mooneyes), Esocidae (pike), Osmeridae (smelts), Catostomidae (suckers), Percopsidae (trout-perches), Gadidae (codfish), Gasterosteidae (sticklebacks), Cottidae (sculpins), and Percidae (perches) (McCart 1986).

Most of the Mackenzie River was covered by ice during Wisconsin glaciation and composition of the species pool is largely explained by numerous hydrological connections during deglaciation that allowed dispersal of species from the Mississippi (Rempel and Smith 1998), Bering, and Pacific refugia (McPhail and Lindsay 1970). Much of the following discussion of fish populations and fisheries in the Mackenzie River main stem is based on the reviews of McCart (1986) and Bodaly et al. (1989).

Large-scale movements of fishes are common in the Mackenzie River because of the lack of major barriers to fish movement (Bodaly et al. 1989). Anadromous fishes undertake extensive migrations in the river, with the arctic cisco traveling from the Mackenzie Delta to spawning areas in the Liard River (McLeod and O'Neil 1983). Other major anadromous fishes include Arctic char, least cisco, lake whitefish, broad whitefish, and inconnu. McCart (1986) suggests that fishes maximize their use of habitat by migrating among widely separated spawning, feeding, and overwintering habitats. For example, the broad whitefish migrates to spawning areas in the main-stem Mackenzie (e.g., Ramparts Rapids) and overwinters in the outer Mackenzie

Delta (Bodaly et al. 1989). Juveniles are hypothesized to leave the system for the Beaufort Sea with the spring flood, only to utilize shallow lake systems along the coast as summer feeding areas. Freshwater species also undertake long migrations. For example, Arctic grayling spawn in small tributaries of the Mackenzie main stem, yet feed during the summer and overwinter in Great Bear River and Great Bear Lake. In addition, inconnu, lake whitefish, and long-nose suckers appear to migrate between the Liard and Mackenzie rivers (MacDonald 1992).

There are no known reptiles that live within the Mackenzie River main-channel watershed, and only one species of amphibian, the wood frog (www.canadianbiodiversity.mcgill.ca/english/ecozones/taigaplains/taigaplains.htm 2003). The main channel of the Mackenzie River is an important route for migrating geese and other aquatic birds. Large portions of the Western Central Flyway snow geese population use the Mackenzie River as a migration route (www.ibacanada.com 2003). Other birds identified in the main channel of the river and the delta include black brant, greater white-fronted goose, tundra swan, sandhill crane, glaucous gull, Arctic tern, dabbling duck, and shorebirds (www.ibacanada.com 2003).

During the spring the Mackenzie Delta is home to a distinct stock of beluga whales, estimated at 5000 animals, which use the delta to calve and to molt (www.greatcanadianrivers.com/rivers/mack/mack-home.html 2003). The Mackenzie Delta also provides ideal habitat for muskrat, producing a thriving population in this region, which supports an active fur-harvesting industry. Other mammals found within the riparian zone include moose, mink, and beaver.

Ecosystem Processes

In general, primary productivity in these cold and turbid waters is low, such that annual phytoplankton productivity in delta lakes is $<10\,g\ C/m^2$ and annual macrophyte production is $<30\,g\ C/m^2$ (Bodaly et al. 1989). Production of periphyton in delta lakes is approximately equal to that of the phytoplankton. Although little information is available on allochthonous inputs to the Mackenzie River, detrital sources probably are very important to river metabolism given the substantial amount of dissolved and particulate organic matter contributed during spring breakup and flooding (Prowse and Culp 2003).

The highest amounts of primary production and algal biomass occur in clear delta lakes that have low connectivity with the river. Bodaly et al. (1989) hypothesize that primary production in delta lakes and the river channel is light limited during the open-water season because of the high concentration of suspended sediments.

Fish production in the Mackenzie main stem and delta lakes is thought to be very low; however, only commercial fisheries yields are available to provide quantitative support for this observation (Mackay 1995). Low productivity can be attributed to high turbidity that limits primary production and unstable river substrate that restricts the growth of macrophytes (Bodaly et al. 1989). Recently, Fisheries and Oceans Canada began a multiyear study to collect information on fish populations (e.g., individual growth, age structure, contaminant body burden) in the lower Mackenzie to facilitate management of important traditional fisheries (Stewart et al. 1997).

Human Impacts and Special Features

The main channel of the Mackenzie River between Great Slave Lake and the Beaufort Sea is 1706 km long, is relatively straight, and has a relatively low gradient. There are only nine communities along this stretch of the river, including Fort Simpson, Norman Wells, Arctic Red River, and Inuvik. Thus most of the river is uninhabited and undisturbed, providing unique adventure tourism opportunities. Perhaps the most dramatic stretch of the river occurs upstream of Fort Good Hope at the Ramparts, a limestone-walled canyon approximately 2 km long and 400 to 1300 m wide (www.greatcanadianrivers.com/rivers/mack/mack-home.html 2003). There are three internationally recognized Important Bird Areas along the Mackenzie River, including the Lower Mackenzie River Islands near Fort Good Hope, the Middle Mackenzie Islands near Fort Norman, and the Mackenzie River Delta (www.ibacanada.com 2003). These areas provide important migration stopover and summer habitat areas for snow geese, Canada geese, and tundra swans, as well as many duck and shorebird species.

The Mackenzie River drains into Canada's largest and the world's 12th-largest delta. The Mackenzie Delta is 210 km in length, with an average width of 62 km. The delta consists of islands, tidal flats, ponds, and at least 25,000 shallow lakes. The moderating temperature effect of the delta allows for the tree line to continue further north in this area than other surrounding areas of the arctic. The Mackenzie Delta also contains 1500 pingos (cone-shaped ice hills formed in areas of permafrost), the largest concentration in the world (www.greatcanadianrivers.com/rivers/mack/mack-home.html 2003). The Kendall Island Migratory Bird Sanctuary is located on the outer margins of the delta and provides summer breeding habitat for several species of geese, swans, gulls, terns, and shorebirds.

The sparse population and remote, northern location of the Mackenzie River main stem have maintained this system in a relatively unpolluted state relative to other large-river systems in Canada. Today the river remains an important corridor for barge traffic from Hay River (Canada's most northern railhead) to Tuktoyaktuk. Barge service operates between Hay River on Great Slave Lake and the Arctic Ocean during the open-water season. In addition, the Mackenzie Valley is proposed as a primary pipeline route for transporting oil and natural gas from the north to Alberta.

Although forestry, mining, and agriculture occur in several of its major tributaries, the primary effects of humans on the system are most likely related to fisheries exploitation, hydrocarbon extraction, and inputs of atmospheric pollutants (Rosenberg 1986, Bodaly et al. 1989, Muir et al. 1990). In addition, hydroelectric and pipeline developments (Rosenberg 1986, Bodaly et al. 1989), as well as climate change (Reist 1994), are anticipated to threaten the Mackenzie River in the future.

The Mackenzie basin is an important region for energy production in Canada; thus upstream developments and oil production along the main stem near Norman Wells have the potential to produce acute and chronic toxicity effects on aquatic biota. Other than through minor oil spills, the Mackenzie biota does not appear to have been seriously threatened by hydrocarbon development. Rosenberg (1986) reviews the research on oil-spill impacts that was undertaken in the 1970s as part of the Mackenzie Valley Pipeline Study. Experiments conducted in tributary streams of the Mackenzie main stem found that exposure to Norman Wells crude led to increases in periphyton biomass (Rosenberg and Wiens 1976). Increased algal production may have been an indirect response to the depletion or loss of grazers due to toxicity effects or because of oil-stimulated microflora production. Oil pollution increased chironomid abundance but reduced the density and diversity of mayflies (Rosenberg et al. 1977). These studies have renewed relevance because governments and industry are once again undertaking environmental assessments in anticipation of an oil and gas pipeline being built along the Mackenzie Valley to

bring energy reserves from the northern frontier to southern markets.

Hydroelectric power production remains an ongoing environmental threat to the Mackenzie River (Rosenberg 1986, Bodaly et al. 1989). There are no dams on the main channel of the Mackenzie River; however, there are eight dams within the greater basin, including two in Lake Athabasca, two on the Peace River, and four on the Great Slave Lake subbasin (Rosenberg 1986). Although existing dams in the Mackenzie basin do not appear to affect the Mackenzie main stem, proposed regulation of Cordilleran flows below Great Slave Lake have the potential to severely disrupt the Mackenzie River (Bodaly et al. 1989). For example, flows from the Liard River double the discharge of the Mackenzie in summer and trigger spring river-ice breakup. If the Liard is dammed, effects could be severe in the Mackenzie Delta, where spring flooding appears to be a crucial cue for migration of coregonid fishes. In addition, the spring flood serves as an environmental reset mechanism that controls delta lake biology and river channel morphology.

Global climate change is predicted to increase mean annual temperature in the Mackenzie Basin by up to 5°C by the middle of the century. Reist (1994) reviewed the possible effects of this level of change on fish species of the Mackenzie and listed several outcomes, including changes in fish distribution, effects on fish abundance and growth, and modified fisheries yields. Coolwater species at the northern edge of their distribution, such as yellow walleye and yellow perch, may increase in abundance. Arctic species with physiological intolerance to warm temperatures may experience restricted geographical ranges, with arctic coregonides (broad whitefish, least cisco, Arctic cisco) at particular risk. If fish species are below their physiological optimum temperature, higher average environmental temperature may increase growth. Thus simple fisheries models predict higher yield with increased water temperatures. McCart (1986) presents data that corroborate this prediction for Arctic grayling in the Mackenzie basin, indicating that the length of year-old fry increases with the number of growing degree days (days above 5.5°C) between latitudes 61°N and 69°N.

An obvious but poorly studied impact of humans on the ecology of the Mackenzie River is that of the commercial and subsistence fisheries (reviewed by Bodaly et al. 1989). Fishing in the river and delta lakes commences after spring spawning and extends until freeze-up. Winter fishing occurs during the period when ice is safe for travel. Most commercial fisheries are managed by quota systems and the size of fishes caught is regulated by minimum mesh size of gill nets. Often the regulatory system for commercial operations is based on year-end questionnaires; size of the subsistence catch is more poorly documented. Bodaly et al. (1989) report that genetic stocks are not well understood; thus quotas may be inappropriate, particularly when migrating stocks are mixed, as in the Arctic Red River fishery. Despite these shortcomings, recent studies have attempted to gather the critical data necessary to facilitate fisheries management (Stewart et al. 1997).

LIARD RIVER

The Liard River arises in the Pelly Mountains in southeastern Yukon and flows through northeastern British Columbia before turning northeast to its confluence with the Mackenzie River at Fort Simpson, Northwest Territories (Fig. 18.11). The basin spans latitudes 57°N to 61°N and longitudes 121°W to 131°W and also includes parts of Alberta. With a total drainage area of 277,000 km² and a length of

1115 km, this 8th order river is the largest northern tributary of the Mackenzie River and has the seventh-largest annual river discharge in Canada. Mountains with high peaks and vast plateaus separated by large valleys and lowlands typify much of the basin. This unregulated river is quite hazardous in some stretches, particularly in the rapids between its confluence with the Trout and Toad rivers in the Grand Canyon of the Liard, a section that includes the Rapids of the Drowned and Hell Gate. Major tributaries of the Liard River include the South Nahanni and Fort Nelson rivers.

The Liard River region was occupied by aboriginals for thousands of years before the onset of European settlers. Groups included the Kaska Dene and the Slave, who were seasonal migrants in the region. The area now known as Fort Nelson was settled in 1775 by the Slave, who were driven to this area by the Cree, who had already acquired firearms, giving them a significant advantage in battle (Young 1980). Further downstream the confluence of the Liard with the Mackenzie River (known today as Fort Simpson) has for centuries been an important place to gather in the summer to celebrate after the ice breakup (www.fortsimpson.com/fshs.html 2003).

FIGURE 18.4 Liard River, Northwest Territories, illustrating ice-jam flooding, forcing backwater to overbank and develop a secondary flow channel (PHOTO BY T. D. PROWSE).

Early references to the river included the names "Courant-Fort" and "Rivière aux Liards." Liard is a French name for poplar trees that occur along the riparian zone of the river. Surveyed by Richard McConnell in 1887 and an important route that gold prospectors used to access northern Yukon in the late 1890s, the Liard River had been an important fur-trading route for most of the century (Marsh 2001a). Fort Simpson is the oldest continuous trading post in the basin, dating back to the North West Company in 1804, and is named after Hudson's Bay Company governor Sir George Simpson (Pool 2001).

Physiography, Climate, and Land Use

Much of the Liard River drainage comes from three Cordillera physiographic provinces: the Coast Mountains of British Columbia and Southeast Alaska (PM), the Rocky Mountains in Canada (RM), and the Mackenzie Mountains (MM) (Hunt 1974, Rosenberg and Barton 1986). A smaller part of the basin also drains from the Great Plains province, through which the lower main-stem river flows before joining the Mackenzie River at Fort Simpson (see Fig. 18.11). The basin was heavily glaciated and surface materials are primarily glacial drift, colluvium, and rock outcrops. Extensive permafrost and poor drainage in the basin have created favorable conditions for cryosolic, gleysolic, and organic soils, whereas brunisolic and luvisolic soils are common at warmer, lower-elevation locations. The dominant geology of the basin includes sedimentary materials, such as Carboniferous Paleozoic limestone, and Cretaceous shale and sandstone.

The Liard basin landscape is a mosaic of five terrestrial ecoregions (Ricketts et al. 1999). In the mountain provinces, the basin primarily drains through the Northern Cordillera Forests ecoregion but also drains portions of the Ogilvie/Mackenzie Alpine Tundra and the Alberta/British Columbia Foothills Forests. As the river enters the Great Plains it primarily flows through the Muskwa/Slave Lake Forests ecoregion but also the Northwest Territories Taiga. White and black spruce, lodgepole pine, aspen, and white birch dominate river valleys of the western Cordillera; alpine fir is abundant near tree-line. In the Great Plains portion of the Liard River basin, white spruce, balsam poplar, willow, and alder are common (MacDonald 1992).

The Liard River basin has a very cold and relatively dry Continental climate. Coastal mountains to the west prevent much of the moist Pacific air mass

from moving inland. Mean temperature decreases from −1°C in the southwestern basin to −4°C at Fort Simpson. At Fort Simpson average temperature in January is extremely cold (−27°C), with mean monthly values for July rising to 17°C (Fig. 18.12). Precipitation is low throughout the basin, with western portions receiving the most moisture (basin range ~35 to 50 cm). Monthly mean precipitation at Fort Simpson ranges from 1.64 cm/mo in April to 5.33 cm/mo in July (see Fig. 18.12). Approximately 60% of the precipitation falls as rain between the months of May and October.

At present the Liard River basin is relatively pristine; however, there is substantial forest harvesting and mineral resource extraction in the headwaters (MacDonald 1992). Forests of white and black spruce, lodgepole pine, and aspen are harvested, mainly in the river valleys of British Columbia. Oil and gas exploration and production in the basin occurs mostly in British Columbia in the Fort Nelson and Petitot river subbasins and has increased rapidly in the last decade. In addition, 29 mining operations, mostly for asbestos, zinc, lead, and silver, are located in the headwater subbasins of the Coal, Dease, Francis, Little Moose, Meister, Rancheria, and Toad rivers in British Columbia and the Yukon. Some placer mining activity has also occurred within the basin. There is very little commercial agriculture in the Liard River basin, and crops of hay, canola, and grains are largely used locally. Land types for the region include forest (55.4%), taiga (17.8%), open land (13.8%), shrubland (11.5%), and water (1.4%) (http://atlas.gc.ca/site/english/index.html 2003).

During World War II (1941) the American military constructed an airport in Fort Nelson as part of the Northwest Staging Route, and in 1942 the Alaska Highway was constructed, running right through town (Artibise 2001). These two construction projects greatly increased the size and importance of Fort Nelson as a transportation hub (Artibise 2001). Today the basin is sparsely populated, with <1 person/km², as only about 8500 people inhabit the basin. Besides forest harvesting and extraction of petroleum and other minerals, the main human activities are subsistence hunting and trapping, big game hunting and guiding, water-oriented recreation, and tourism.

River Geomorphology, Hydrology, and Chemistry

The Liard River drops more than 1200 m from its headwaters in the Yukon to the Mackenzie River confluence. Tributary streams in the Yukon have high slopes and contain bed materials of boulder, cobble, and gravel. Immediately above its confluence with the Black River the channel of the upper Liard River is straight, with cobble and gravel substrate and a slope of 2.1 m/km. The meandering channel of the middle river reaches, located in northern British Columbia, is often braided, with a moderate slope. However, from just upstream of the Kechika River confluence the Liard River descends about 50 m over a distance of 25 km (2 m/km slope) in a series of rapids before resuming its moderate gradient. Below Liard Hot Springs the river slope increases to 1 m/km as it cascades through dangerous rapids, such as Hell Gate Rapids. In its lower reach, the river channel is both meandering and straight, depending on the location, and has numerous midchannel bars and islands (Fig. 18.4). The slope of this section ranges broadly from 10 cm/km to 60 cm/km. Bed material also exhibits a wide range, with different reaches composed of varying proportions of sand and gravel. The final 100 km of the Liard River is characterized by a single, well-incised channel with occasional small bars and islands (BC Hydro 1985). During the annual flood peak the average river width and depth within this reach is 250 to 500 m and 3 to 8 m, respectively. Substrate along this stretch of river is predominantly coarse gravel, cobbles, boulders, and bedrock sills.

Mean monthly runoff at Fort Simpson is lowest during February, at 0.39 cm/mo, and highest in June, at 6.84 cm/mo (see Fig. 18.12). For May, June, and July runoff is greater than precipitation due to increasing precipitation and increased snowmelt as the daily mean temperature rises. Discharge in the unregulated Liard River is strongly affected by mountain runoff and reaches a monthly maximum in June of 7300 m³/s. Minimum discharge occurs in late winter in March before river ice breakup (~400 m³/s). Mean monthly discharge in the Liard River is approximately 2500 m³/s. The Liard River has a major influence on the Mackenzie River and, relative to other tributary rivers, has the most impact on temperature, sediment load, and breakup of the Mackenzie (MRBC 1981a). In most years the warmer Liard River is the trigger mechanism for river-ice breakup of the Mackenzie River (Prowse 1986, 1990).

Available water quality data for the Liard River at Fort Liard and Fort Simpson was summarized by Macdonald (1992), who found very similar trends for both sites. Given the importance of sedimentary bedrock in the basin, it is not surprising that the alkaline, hard water system has a high pH (8.2) and conductivity (287 μS/cm; Macdonald 1992, Brunskill

1986). Available measurements of plant nutrients suggest the Liard River may be phosphorus limited, as annual total dissolved phosphorus at Fort Liard was 0.009 mg/L (Macdonald 1992). Mean nonfilterable residue was quite high (144 mg/L), with maximum values ranging up to 865 mg/L. Suspended sediment concentrations can be extreme (>1000 mg/L) in spring during major runs of river ice (Prowse 1993). The Liard River is the single largest contributor to the Mackenzie River in terms of wash load, which includes fine material such as silt, clay, and fine sand (Carson et al. 1998). Throughout the basin metal and metalloid concentration is low; however, annual variability is large because of high suspended sediment loads in spring and early summer (MacDonald 1992). Nutrient and metal concentrations in the South Nahanni River are indicative of pristine conditions (Environment Canada 1991).

River Biodiversity and Ecology

River habitats of the Liard River are part of the Lower Mackenzie freshwater ecoregion (Abell et al. 2000). Relatively little is known about the ecological structure and function of the Liard River, and few articles about this ecosystem have been published. Most information on the biodiversity and ecology of the river is contained in unpublished government and consulting reports. Information on the benthic community of the Liard River is sparse and information on bacteria, periphyton, or macrophytes was not located. MacDonald (1992) summarizes much of this information for the basin and is an excellent reference source for available reports.

Plants

In general, there are over 700 species of vascular plants recorded in Nahanni National Park Reserve. Terrestrial vegetation is primarily boreal forest with lowland wetlands to alpine tundra. Riparian zones are dominated by white spruce and trembling aspens (http://parkscanada.gc.ca/pn-np/nt/nahanni 2003).

Invertebrates

A summary of benthic macroinvertebrate families recorded in the Liard River is presented by MacDonald (1992). Although no abundance data were located, longitudinal zonation of the macroinvertebrate assemblage is evident from this presence/absence information for dipterans (true flies), mayflies, stoneflies, and caddisflies. Dipteran families and subfamilies include Diamesinae, Empididae, Orthocladiinae, Simuliidae, Tanyderidae, and Tipulidae. Most mayfly (Baetidae, Ephemerellidae, Heptageniidae) and stonefly (Capniidae, Chloroperlidae, Nemouridae, Perlidae) families have been recorded throughout the river's length. However, several families have been recorded only in the Yukon headwaters of the Liard River. These include ephemerellid mayflies, Isogeninae and nemourid stoneflies, and several caddisflies (Brachycentridae, Limnephilidae, Rhyacophilidae). Other caddisflies have not been sampled in the lower Northwest Territories portions of the river. Clearly, these trends are very preliminary, as there has not been a rigorous effort to collect and describe the benthic macroinvertebrate fauna of this basin.

Vertebrates

The ecology of socially and economically important fish species is much better known than other biota of the Liard River, yet the knowledge base for even this group is minimal for this basin (MacDonald 1992). Surveys of the Liard River have recorded 34 species from 12 families, with most species belonging to the Salmonidae or Cyprinidae. Two of the important anadromous species are chum salmon and Arctic cisco. Both are fall spawners and their fry move downstream in spring after ice breakup. Arctic cisco, which are important to native peoples for human consumption and sled-dog food, appear to have rearing areas in the Mackenzie River and its delta. Several other fall-spawning species are notable, such as mountain and lake whitefish, inconnu, and bull trout. Mountain whitefish is one of the most abundant species in the river, with adults and juveniles occurring throughout the main stem and upper tributaries. Bull trout are widely distributed in the river but are rare in the lower river reaches. Portions of the Liard River in the vicinity of Nahanni National Park may have been an important glacial refugium for fish species, such as the lake whitefish (Clayton et al. 1992).

Important spring-spawning species of the Liard River include longnose sucker, Arctic grayling, goldeye, northern pike, and walleye (MacDonald 1992). Longnose suckers are ubiquitous in the Liard River and appear to be the most abundant large fish species in the basin. This species is used as food for humans and dogs. The most abundant sport-fish species in the basin is Arctic grayling, which is a resident of the main stem and spawns on gravel and cobble in small streams and tributaries. The goldeye spends much of its life cycle in the main stem and is thought to be the species most dependent on

these habitats. Several surveys on tributaries failed to collect goldeye; in contrast, they are the most abundant sport fish in the lower reaches of the main-stem Liard River. Similarly, walleye are abundant in the lower river but are rare upstream of the Fort Nelson River confluence. Other species in the river system include burbot, which spawn under the ice in late winter, and many smaller forage fishes, such as various minnows and sculpins for which no information was found.

Based on surveys conducted within the Nahanni National Park Reserve, the Liard watershed area has a total of 42 species of mammals, including many that are closely associated with the river (moose, beaver, river otter, muskrat, and mink) (http://parkscanada.gc.ca/pn-np/nt/nahanni 2003). There are 180 species of birds, including loons, grebes, ducks, and bald eagle. In addition, Yohin Lake, which is located within the National Park Reserve, supports a small nesting population of trumpeter swans (http://parkscanada.gc.ca/pn-np/nt/nahanni 2003). There are no known species of reptiles in this region; however, there are a few species of amphibians, with the most common the wood frog (http://parkscanada.gc.ca/pn-np/nt/nahanni 2003).

Ecosystem Processes

Although studies on ecosystem function of the Liard River have not been undertaken, one can speculate that the important processes follow trends of the river continuum concept. Decomposition of terrestrial material is likely an important carbon source in headwater reaches, whereas autochthonous carbon sources likely dominate downstream. River-ice breakup is an important ecological reset mechanism in spring, as this disturbance scours the riverbed and adds large amounts of terrestrial carbon to the ecosystem (see Fig. 18.3).

Human Impacts and Special Features

The Liard River is known for its wild and rough stretches of river, with its Rapids of the Drowned and Hell Gate (Artibise 2001). Another important feature in the basin is Liard River Hot Springs, which is a British Columbia Provincial Park (http://wlapwww.gov.bc.ca/bcparks/explore/parkpgs/liard_hs/nat_cul.htm 2004). Located just off the main channel near the community of Liard River, it has thermal springs that flow into an intricate system of swamps, creating a microclimate that allows a unique vegetative community to thrive there. The South Nahanni River is a tributary of the Liard and is known for its natural beauty. The approximately 100 m high Virginia Falls on the South Nahanni is 1.5 times the height of Niagara Falls and is part of Nahanni National Park, a U.N. World Heritage Site since 1978. Fort Nelson, the largest community in the watershed (approximately 6000 people), is home to the largest gas-processing plant in North America and the world's largest chopstick manufacturer.

Although development of forestry, mining, and petroleum resources in the upper Liard River is locally intensive, this basin remains relatively pristine and adverse environmental changes have not been observed (MacDonald 1992). Water-quality measurements throughout the system are generally well below Canadian Water Quality Guidelines (CCME 1999) for the protection of aquatic life. In addition, the very limited measurements of heavy metals in sediments and fishes from the main stem revealed levels similar to reference conditions in pristine areas throughout British Columbia and the Yukon. Nevertheless, areas of concern do exist, such as resource-extraction developments that have occurred upstream of Nahanni National Park. Of more serious concern, however, are the proposals for several hydroelectric developments along the main-stem Liard River. MacDonald (1992) outlines numerous adverse changes that these dams could cause to water quality and temperature regimes. In fact, he suggests that this hydroelectric development could cause damage to the fisheries resource through impacts to spawning and rearing habitats, as well as fragmentation of critical migration routes.

SLAVE RIVER

The 8th order Slave River is formed near the confluence of the Peace River and the outflow from Lake Athabasca, known as Rivière des Rochers, and flows north approximately 420 km to Great Slave Lake (Fig. 18.13). For its last 200 km the river meanders through the active portions of the massive Slave River Delta, which is 8300 km^2 in area and up to 70 km wide. It is the major southern tributary of the Mackenzie River and drains approximately 615,000 km^2. This area represents the combined discharge of the Peace and Athabasca rivers, the Peace–Athabasca Delta (PAD), and the Lake Athabasca drainage system. Surprisingly, the drainage area located between the Slave's origination at the confluence of the Peace and Rivière des Rochers rivers is only 15,100 km^2, <3% of the catchment (Alberta Environment 1987). Flow contributed to the Slave

River from the PAD originates primarily from Lake Athabasca, largely located within Saskatchewan, and the Athabasca River, fed from central portions of Alberta (Prowse and Conly 1996). Exposed bedrock occurs all along the river, with the topography of the basin consisting of rolling hills in the southeast and low scarps of limestone and small lakes in other parts of the basin.

The Slave River and Great Slave Lake are named for the Slavey peoples. The Slavey are part of the larger Dene nation, whose ancestors arrived from the Bering Strait. It is estimated that permanent residency along the Slave River occurred approximately 2500 years ago. The Slave River was a main travel route for aboriginals traveling northward. The Slavey had a seminomadic way of life due to the scattered and seasonal occurrence of the migratory animals that were their food source and harsh weather conditions (Crowe 1974). They survived primarily on fishes, moose, and caribou. The Slavey were in frequent conflict with the mountain people of the Nahanni area and today live in a number of communities of the Northwest Territories.

Early European explorations were conducted by English explorers Samuel Hearne (discovery of Great Slave Lake in 1771), Donald Alexander Smith (namesake for the town of Fort Smith), Sir Alexander Mackenzie, and David Thompson. During the 1800s the Hudson's Bay Company established a trading post along the banks of the Slave River at Fort Smith, and this became the primary trade link to the Mackenzie River.

Physiography, Climate, and Land Use

The Slave River flows northwest between 58°N and 62°N latitude and 110°W and 114°W longitude along the boundary of two physiographic provinces, the Bear–Slave–Churchill Uplands (BC) to the east and the Great Plains (GP) to the west (Rosenberg and Barton 1986) (Fig. 18.13). Precambrian granites and gneisses in the BC province, and Devonian sedimentary rocks in the Great Plains dominate the geology of the basin. Much of the northern part of the catchment drains a flat area of lacustrine deposits through which the river meanders. Unconsolidated glacial drift, postglacial alluvial sediments, and recent organic deposits cover most of the surficial geology of the Slave River basin (Alberta Environment 1987). Eolian deposits of sand dunes, formed by wind action immediately after glacial Lake McConnell drained, are common in the southwest basin. Soils of the system include brunisols, gleysols, regosols, fibrisols, and cryosols; however, only the brunisols have a well-developed soil horizon.

Landscape along the Slave River (excluding the upstream Peace and Athabasca basins) is composed of the Northern Canadian Shield Taiga and Mid-continental Canadian Forests terrestrial ecoregions (Ricketts et al. 1999). Vegetation includes closed stands of trembling aspen, balsam poplar, and jack pine, with white and black spruce and balsam fir dominating late successional stages. Some areas of semiopen grassland with clumps of willow and aspen are found in uplands. Sedge–grass meadows often occupy depressions on lacustrine and deltaic plains (Alberta Environment 1987). Fens and bogs are covered with tamarack, black spruce, and mosses, and up to 50% of the area is covered by wetlands. Discontinuous permafrost is occasionally evident in organic deposits. The basin contains Wood Buffalo National Park, which is the home of the world's largest bison herd.

This basin has a harsh Continental climate characterized by short cool summers and long cold winters. The mean annual temperature at Fort Smith, Northwest Territories, is approximately −3°C, with the monthly mean temperature rising to 16°C in July and falling to −25°C in January (Fig. 18.14). Daily maximum temperatures range between 23°C and −20°C. Annual precipitation in the basin is low, with approximately 60% of the 30 to 40 cm of moisture falling as rain between May and October (see Fig. 18.14). At Fort Smith monthly precipitation ranges from 1.35 cm/mo in April to a maximum of 5.68 cm/mo in July.

Currently the largest land-use activity in the Slave River watershed region is forestry; other activities include hunting, fishing, trapping, and some mineral exploration. Population of the region is <10,000 people (<1 person/km²) and contains the Northwest Territories towns of Fort Resolution and Fort Smith. Much of the west side of the Slave River watershed lies within Wood Bison National Park. Land types in the watershed consist of forest cover (40.3%), taiga (25.8%), shrubland (20.9%), agriculture (land covered with herbaceous, typically annual crops), which may contain a small proportion (<10%) of trees and shrubs (1.6%), open land (1.4%), and water (9.9%) (http://atlas.gc.ca/site/english/index.html 2003).

Geomorphology, Hydrology, and Chemistry

The river is dominated by regional geology but has moved laterally, depositing alluvial material and

creating floodplain areas (Prowse and Conly 1996). Numerous alluvial islands and channel-bar complexes are common along the upper reaches, where the western margin of the Canadian Shield restricts river movement. At Fitzgerald the Slave River passes over a Precambrian sill of bedrock, falling 35 m over a 30 km distance. Progradation of the Slave River Delta has been occurring for 8000 years at a rate of about 20 m/yr and has progressively filled in the south arm of glacial Great Slave Lake (Prowse and Conly 1996). Scroll bars, abandoned channels, and oxbow lakes dominate river morphology within the delta.

Flow of the Slave River comes largely from the Peace River or from Lake Athabasca via the Rivière des Rochers and nearby PAD channels, depending on the hydrological stage of the Peace River (Alberta Environment 1987). During high spring flows the Peace River can act as a hydraulic dam, reversing the flow of channels of the PAD, such as the Rivière des Rochers. Thus spring floodwaters from the Peace River can be stored in the PAD–Lake Athabasca system and released later in the season (Fig. 18.5). During the remainder of the year about 90% of the outflow from the lake drains through the Rivière des Rochers or the smaller Chenal des Qauatres Fourches and Revillon Coupe. Spring melt in the Peace River basin has a major influence on the Slave River hydrograph and floods are normally caused by Peace River flows (Prowse and Conly 1996).

At Fort Smith mean monthly runoff ranges from 0.82 cm/mo in February to 2.41 cm/mo in June (see Fig. 18.14). Runoff increases in May with increasing precipitation and the initiation of snowmelt. Evapotranspiration increases the precipitation-to-runoff ratio from June to October. Peak annual discharge occurs in June and July when, on average, about 25% of annual runoff is observed (see Fig. 18.14). In contrast, from August through March decrease in discharge in the Slave River is less than the Peace River due to input from the lake. Mean monthly discharge for the Slave River at Fitzgerald is approximately 3400 m³/s; minimum and maximum monthly discharge occur in February (2000 m³/s) and June (5600 m³/s), respectively (see Fig. 18.14). The normal date of freeze-up is mid-November; river ice tends to break up in mid-May.

The Peace and Athabasca rivers and Lake Athabasca strongly affect water quality in the Slave River. The water is slightly basic, with an annual mean pH of 7.8, and alkaline, with high concentrations of calcium bicarbonate, suspended sediments, and turbidity (Alberta Environment 1987). Conduc-

FIGURE 18.5 Peace–Athabasca Delta showing typical perched basin and adjacent channel system. Note contrast in sediment conditions between the basin and channel, indicating lack of flow connection at this water level (PHOTO BY T. D. PROWSE).

tivity (253 μS/cm) and alkalinity (82 mg/L as $CaCO_3$) are lower than the Peace River, probably as a result of significant runoff from the Canadian Shield (Brunskill 1986, see also Peace and Slave river sections; Prowse and Conly 1996). In contrast, sodium and chloride ion concentration increases in the river near the delta as a result of inflows from the surrounding karst area. Nutrient levels in the Slave River are relatively low and similar to values for the Peace River, with average total nitrogen concentrations of 0.40 mg/L and average total phosphorus concentrations of 0.21 mg/L.

River Biodiversity and Ecology

The Slave River lies within the Upper Mackenzie freshwater ecoregion (Abell et al. 2000). The biodiversity and ecology of the river are relatively poorly understood, with most of the information coming from surveys of McCarthy, Robertson et al. (1997), McCarthy, Stephens et al. (1997), and McCarthy, Williams et al. (1997) and reports summarized by Prowse and Conly (1996).

Plants

Although the deltaic flora of the Slave River basin is reasonably well described, little information is available on the primary producers of the river channel. Tallman (1996) speculates that high levels of suspended sediments and turbidity prevent macrophyte growth in the river channel. In contrast, willow, horsetail, and grasses are located along the frequently flooded riparian margins of the main stem (Prowse and Conly 1996). Shallow lakes and wetlands of the PAD have an abundance of emergent and submergent vegetation, as well as sedges and reed grasses. Much of the Slave River Delta is covered by emergent vegetation, such as horsetail. For a detailed description of vegetational succession in the PAD, see Prowse and Conly (2000).

Invertebrates

Recent information on benthic invertebrate communities of the Slave River is restricted to the baseline survey work of McCarthy, Robertson et al. (1997). They found that benthic environments in the river were composed of sand, silt, and clay, and sediment organic carbon was low, averaging only 2.5%. A total of 69 taxa were identified, the most abundant members of the benthic community being chironomid midges and oligochaete worms. Common chironomids were *Stictochironomus quagga*, *Procladius* sp., *Chironomus anthracinius*, and *Polypedilum scalaenum*. Oligochaetes were composed mostly of *Limnodrilus hoffmeisteri* and *Chaetogaster diaphanus*. Other common taxa included ostracods and biting midges. Total benthic density in the Slave River ranged between about 1250 and 1700 individuals/m^2 in 1990 and 1991, and chironomids accounted for >80% of this fauna. The primary factors correlated with benthic invertebrate distribution were water depth and organic carbon content. Furthermore, oligochaetes and ostracods were more abundant in shallow water and sediments of high organic content, and *Procladius* sp. was the only taxon exhibiting higher abundance with increasing water depth. McCarthy, Robertson et al. (1997) found that autumnal benthic communities were very similar between years. Clearly, McCarthy, Roberson et al. (1997) did not sample the complete faunal pool, as Tallman (1996) found that several other orders of benthic invertebrates comprised important proportions of fish diet in the Slave River; namely, amphipods, caddisflies, corixids, dytiscids, mayflies, and stoneflies.

Vertebrates

The fish fauna of the Slave River Basin contains 45 species, 28 of which can be found in the Slave River main stem and delta (McCart 1986). Similar to other tributaries of the Mackenzie River, most of the species are either salmonids or cyprinids. The primary information source on the Slave River fish community is Tallman (1996), who provides information on the fish fauna of the lower Slave River north of 60°N latitude. Fish species composition varies considerably among major areas in this part of the river. For example, goldeye were the most abundant fish species sampled during surveys of the main stem, whereas northern pike were common in the delta. In both the river and delta, longnose sucker and lake whitefish were present in low abundance. Most fishes in the Slave River system appear to be generalists, acting as opportunistic feeders consuming a number of different prey over the seasons (Little et al. 1998). Species such as flathead chub, goldeye, longnose sucker, and lake whitefish forage heavily on invertebrates. Top predators include the predominantly piscivorous walleye, inconnu, and large northern pike.

The Slave River is an important spawning, rearing, and feeding habitat for many fish species. Not surprisingly, spawning migrations cause seasonal changes in fish-community composition in the river. Spring spawners such as goldeye, flathead chub, longnose and white suckers, northern pike, and walleye dominate in late May shortly after river ice breakup. Fall spawners, including lake and small cisco, lake whitefish, and inconnu, become abundant between September and October. Northern pike, flathead chub, and goldeye remain relatively abundant throughout the open-water season. Radio tagging studies indicate that inconnu move from Great Slave Lake to the Rapids of the Drowned by mid-October and vacate the site before November. As ice cover forms in late fall, burbot move into the river, returning to the lake after the February spawning period (Howland et al. 2000, Tallman 1996). Based on the comparison of gillnet catches, the diversity of major fish species in the Slave River has not changed substantially since the 1970s.

The Slave River watershed contains the only known breeding habitat for the endangered whooping crane, located in the northeastern portion of Wood Buffalo National Park (www.hww.ca/hww2.asp?id-79 2003). The nesting area consists of six small areas totaling about 400 km^2. Most of these areas are close to the Sass, Klewi, and Nyarling rivers, which flow into the Slave River (www.hww.ca/hww2.asp?id-79 2003). In addition, the Slave River is home to the most northerly nesting place in North America for the American white pelican. These nests are located on islands within the main channel of the Slave River. Other aquatic birds include snow geese, Canada geese, loons, grebes, mergansers, osprey, and bald eagles (www.nwtwildlife.rwed.gov.nt.ca/monitoring/speciesmonitoring 2003).

There are six known species of amphibians within the Slave River watershed, including chorus frog, wood frog, northern leopard frog, Canadian toad, boreal toad, and long toed salamander. There is one species of reptile, the red sided garter snake. Aquatic mammals found in the Slave River include beaver, muskrat, river otter, and mink (www.nwtwildlife.rwed.gov.nt.ca/monitoring/speciesmonitoring 2003).

Ecosystem Processes

Information on ecological processes of the Slave River system is limited to recent studies investigating food web structure in the main stem or the interplay of hydrologic processes and ecosystem structure and function of the PAD. First, studies of stable isotope ratios of sulphur, carbon, and nitrogen suggest that fishes in the Slave River are influenced by the pelagic food chain of Great Slave Lake and by upstream river processes (McCarthy, Robertson et al. 1997). Upstream food sources appear to be organic matter transported from the Peace and Athabasca rivers, suggesting heterotrophic processes dominate the metabolism of the Slave River. Second, ecological processes of the PAD are dependent on the hydrology of the downstream river (Prowse and Conly 1996, 2000). When ice jams or open-water floods in the Peace and Slave rivers produce water levels that are higher than those of Lake Athabasca, the connecting channels between the rivers and lake reverse their flow, filling the wetlands and perched basins of the PAD.

Human Impacts and Special Features

The Slave River drains from Lake Athabasca and the PAD in Alberta into Great Slave Lake located in the Northwest Territories. The PAD is the largest inland freshwater delta in the world and the largest alluvial-wetland habitat in the region (see Fig. 18.5) (www.ibacanada.com 2003). Its significance cannot be overstated, as it is critical to migratory birds using the Central and Mississippi Flyways. The PAD is also important to fishes migrating between the delta lakes and the rivers, and to muskrat populations in the delta. Great Slave Lake, at 28,400 km², is the fifth-largest lake in North America and, at 618 m, the deepest (Asch 2001).

The river is the natural eastern border for Wood Bison National Park, which contains the largest free-roaming bison herd (5000 to 6000 animals) in the world (www.nwtwildlife.rwed.gov.nt.ca/monitoring/speciesmonitoring 2003). Salt Plains, located on the eastern banks of the Slave River, is an area of underground springs that transport salt from below the surface and deposit it along the flat open areas of the plain, creating unique saline microhabitats.

Historically, the Slave River has been an important transportation route to the north for the Slavey and other aboriginal groups and fur traders. Most of the river is easily boatable. However, Fort Fitzgerald and Fort Smith, the two major towns along the length of the Slave River, are located at the beginning and the end of a treacherous set of rapids spanning approximately 25 km. The towns are located in areas that have traditionally been used as portage routes. Islands located within these rapids provide the most northerly nesting habitat for the American white pelican (www.ibacanada.com 2003). Past proposals for a major hydroelectric-power dam at Pelican Rapids near Fort Smith are currently moribund.

The Slave River receives chemical compounds from several sources, including industrial, agricultural, long-range transport, and natural pathways. Concern over the threat of upstream developments in the Slave River basin, particularly the expansion of pulp mills and oil sands on the Athabasca River and damming of the Peace River, led to a multiyear monitoring program of the river in the early 1990s, which provided much of the data summarized here. Compounds in the water, such as chlorinated compounds, PAHs, and PCBs, were very low and seldom reached levels above detection limits (McCarthy, Williams et al. 1997). Furthermore, pesticides were not detected in the Slave River. In contrast, metal concentration in the Slave River often exceeded federal water-quality guidelines. Nevertheless, McCarthy, Williams et al. (1997) concluded that these relatively high levels (e.g., mercury) were most likely from natural sources rather than from upstream industrial inputs. Levels of organic and inorganic contaminants were also very low in suspended sediments. Taken together, it was concluded that contaminant levels were unlikely to have adverse effects on the river ecosystem (McCarthy, Williams et al. 1997). This conclusion was supported by bioassay experiments that demonstrated suspended sediments from the river were not toxic to *Daphnia magna* (i.e., 48 h acute toxicity) or the bacterium *Photobacterium phosphoreum* (i.e., *Microtox* bioluminescence assays). Similarly, although benthic invertebrates occur at very low densities in the Slave River and most of the organisms are small chironomid midges, McCarthy, Robertson et al. (1997) indicate that few changes in diversity or composition occurred over the period of increased input of point and nonpoint impacts upstream (e.g., industrial effluents and forest harvesting).

Tissue from several species of fishes from the Slave River (i.e., burbot, lake whitefish, longnose suckers, northern pike, walleye) had consistently low concentrations of heavy metals, chlorinated phenolics, dioxins and furans, organochlorine pesticides, PAHs, and PCBs, indicating that fishes were fit for human consumption (McCarthy, Stephans et al. 1997). Elevated levels of toxaphene were found in the livers of burbot, which are a traditional country food, prompting Health Canada to set a human consumption limit of one burbot liver per week. Because the application of toxaphene was restricted in 1970 and its use has been banned in Canada since 1983, the source of toxaphene in the Slave River is thought to be long-range transport. Finally, biochemical indicators of contaminant exposure and effects on fishes demonstrated that contaminant effects on Slave River fish populations were minimal (Williams et al. 1997, Cash et al. 2000).

The hydrological regime of the Slave River is affected by one of the world's largest reservoirs, located more than 1200 km upstream on the Peace River, its main tributary. A major impact of flow regulation has been a reduction in the frequency and magnitude of flooding of the PAD (see Fig. 18.5) that has complex hydrological interconnections with the Slave River (Prowse and Conly 1996, 2000). Flow regulation has resulted in drying of the delta since 1968 because postreservoir discharge and ice formation are now higher in the winter, thereby leading to a reduced probability of large ice-jam flooding of the delta (Prowse and Lalonde 1996). In addition, changes in atmospheric circulation patterns and associated reductions in winter snowfall have combined to decrease the amount of snow available for spring runoff. Ultimately, decreased flooding of the delta has caused various ecological responses, including loss of wetland habitat and biota of perched basins.

Recently, flow-augmentation strategies have been proposed and successfully tested with the aim of returning the flood regime of the delta to preregulation levels and providing improved remediation strategies for wetland vegetation and wildlife (Prowse et al. 2002).

PEACE RIVER

Originating within the alpine zone of the Rocky Mountains in northeastern British Columbia, the Peace River is formed at the junction of the north-flowing Parsnip and the south-flowing Finlay rivers from headwater streams, many of which are glacial in origin (Fig. 18.15). With a drainage area of $293,000 \, km^2$, the Peace River is the major tributary of the Slave River (Prowse and Conly 1996, 2000) and is regulated by a massive hydroelectric development as it feeds into the Williston Reservoir formed by the W.A.C. Bennett Dam. Regulated flow from this hydroelectric operation then passes downstream into northern Alberta and ultimately to the Peace–Athabasca Delta approximately 1200 km downstream (see Fig. 18.5). At its confluence with the PAD, one of the world's largest freshwater deltas, the Peace River is an 8th order river that is almost 2000 km in length.

Approximately 10,000 years ago the last glacial ice sheets retreated from the northern parts of the Canadian prairies. Groups of nomadic hunters probably passed through the area whenever the Rocky Mountain and Keewatin ice sheets separated briefly. There is evidence of them pausing briefly near Fort St. John, on the Peace River, around 10,000 years ago. The ice-free corridor allowed people from Asia to penetrate deep into the Americas. With the end of the Ice Age many of these groups moved back north. They followed the herds of grazing animals, which were, in turn, following the grasses northward in the warming climate. The previous introductory information and that which follows is drawn from as essay by Clare (www.calverley.ca/briefhistory.html 1998).

The arrival of the Hudson's Bay Company in Eastern Canada in 1670 eventually had a major impact in the Peace River area, as guns made their way westward as trade goods and the Cree Tribe began to push the Beaver Tribe further west. The Beaver, in turn, pushed the Sekani deep into the Rocky Mountain Trench in the mid-1700s. A truce was eventually agreed to by the Cree and the Beaver and the Peace River became the boundary between their hunting territories. Early residents of the Peace

River basin were members of the Athapaskan linguistic group.

The two rival fur-trading companies, the Hudson's Bay Company and the North West Company, pushed westward in the late 1700s. Fort Chipewayan on Lake Athabasca became the headquarters for the Northwest Company's attempts to reach the Pacific Ocean. Sir Alexander Mackenzie is thought to be the first European to explore the river during his overland search in 1793 for the Columbia River and a route to the Pacific. An era of fur trading began in the Peace with the first forts built within a few years of Mackenzie's great trip. Simon Fraser stopped over at Hudson's Hope in 1806 before he pushed south to follow the river named after him to its mouth near Vancouver. Posts such as Dunvegan, Fort Vermilion, Fort St. John, and McLeod Lake became centers of the northern fur trade. Both the Catholic and Anglican churches established missions in the Peace along with the fur traders. There was a brief flurry of gold panning in the 1870s on the Omineca and on the Peace, but it was never of much importance. Klondikers passed through the Peace in 1898 on their way to the Yukon, but this was one of the worst routes to follow to the gold fields.

Physiography, Climate, and Land Use

The Peace River basin is comprised of three major physiographic provinces (Hunt 1974) and spans latitudes 53°N to 59°N and longitudes 111°W to 127°W (see Fig. 18.15). The western Cordillera provinces include the Coast Mountains of British Columbia and Southeast Alaska (PM) and the Rocky Mountains in Canada (RM) (Rosenberg and Barton 1986, Prowse and Conly 1996). About two-thirds of the basin, however, drains the Great Plains (GP) province. Soils in the Peace basin are quite diverse and include chernozemic, brunisolic, gleysolic, luvisolic, organic, and cryosloic soils. Much of the southwestern portion of the basin, including the Rocky Mountains, is composed of sedimentary rocks with deep valleys and high plateaus covered by glacial tills. However, the Great Plains is underlain by recent metamorphic and sedimentary bedrock of Devonian and Cretaceous age. The effects of Pleistocene glaciation have modified landforms here, resulting in a surficial geology consisting of glacial drift, postglacial alluvial, and aeolian deposits (Prowse and Conly 1996).

The Peace River drains from a mosaic of terrestrial ecoregions (Ricketts et al. 1999). Ecoregions in the mountainous southwestern portion of the basin include the Central British Columbia Mountain Forests, Northern Cordillera Forests, Fraser Plateau and Basin Complex, and Alberta Mountain Forests. Within the Great Plains province the main-stem Peace River flows primarily through the Canadian Aspen Forest and Parklands ecoregion, but also drains portions of the Alberta/British Columbia Foothills Forests, Muskwa/Slave Lake Forests, and Mid-continental Canadian Forests. In the upper elevations, forests of aspen, balsam poplar, spruce, and lodgepole pine are dominant (Prowse and Conly 1996). Vegetation in most of the basin, however, consists primarily of trembling aspen and balsam poplar, and secondarily of white spruce, black spruce, lodgepole pine, and jack pine. In drier areas semiopen grasslands may also be found. Black spruce and peat lands are abundant in the Taiga regions.

The climate in the Peace River basin is relatively dry, with cool summers and cold winters. Mean annual air temperature for the area is approximately −0.9°C, with mean monthly temperatures ranging from −23°C to 17°C during the year (Fig. 18.16; data are climate normals for Fort Vermillion, Alberta, from 1908 to 1985 [www.climate.weatheroffice. ec.gc.ca/climate_normals/index_e.html 2002]). Winter temperatures can be several degrees colder in northeastern portions of the basin. Average precipitation varies from 46.8 cm/yr at Fort St. John in the upper Peace River basin to 38.1 cm/yr at Fort Vermilion in the lower river. At Fort Vermillion monthly precipitation ranges from 1.75 cm/mo to 6.44 cm/mo in July (see Fig. 18.16). Approximately 30% of this moisture arrives as snow. The main stem is ice covered from late November through early April.

The construction of the huge W.A.C. Bennett Dam at Hudson Hope and the creation of Williston Lake on the Upper Peace in the mid-1960s initiated energy extraction in the Peace River region (www.calverley.ca/Briefhistory.html 2000). Coal, oil, natural gas, and hydroelectricity provide energy sources for markets in southern Canada, the United States, and Pacific Rim countries. This region also contains an active forest industry, including pulp and paper, lumber, and particle-board mills in Chetwynd, Dawson Creek, Taylor, and Fort Nelson. In addition, there are areas of agriculture, which produce primarily oil seeds, grains, cattle, bison, and other livestock (www.calverley.ca/briefhistory.html 2000). Land type in the basin is dominated by forest cover (57.4%), followed by agriculture (16.2%), taiga (12.5%), and shrubland (11.4%) (http://atlas.gc.ca/site/english/

index.html 2003). Population in the basin is approximately 110,000 and population density is <1 person/km^2.

River Geomorphology, Hydrology, and Chemistry

The average gradient of the Peace River in the Great Plains of Alberta is only 16 cm/km (Prowse and Conly 1996). The upper Peace River downstream of W.A.C. Bennett Dam is incised 200 m into the surrounding Alberta Plateau. This reach has a primarily straight channel with occasional islands and gravel bars. At this point the river is approximately 500 m wide and 4 m deep (Prowse and Conly 1996, Kellerhals et al. 1972). The middle river reach from the confluence of the Smoky River to Fort Vermilion mostly drains the Peace River lowlands and has mean widths ranging from 500 to 650 m. Intermittent islands and bar complexes are common, and the river can be characterized as partly entrenched and confined with irregular meanders (Prowse and Conly 1996). Bed materials are primarily sand and fine gravel, with silt and erodible bedrock occurring along the banks (Kellerhals et al. 1972). The lower river reach from Fort Vermilion to Peace Point cuts through the lacustrine deposits of glacial Lake McConnell. The channel is weakly sinuous, with split channels and island complexes, and channel width ranges between 700 and 1500 m (Prowse and Conly 1996). Substrate composition within this reach is predominantly shallow sand with local areas of gravel.

Mean monthly runoff ranges from 1.02 cm/mo in February to 3.84 cm/mo in June (see Fig. 18.16). High runoff-to-precipitation ratios in May and June are due to snowmelt and increased monthly precipitation. From July through October evapotranspiration increases the difference between precipitation and runoff. Mean monthly discharge of the Peace River at Peace Point, near the mouth, is 2188 m^3/s. About two-thirds of this discharge originates in British Columbia. As in the Athabasca River, peak discharge is associated with snowmelt from mountain runoff, which occurs in late May to early June (~4300 m^3/s). Summer rainfall events can produce secondary flood peaks during most years (Prowse and Conly 1996). Minimum monthly discharge values of approximately 200 m^3/s occur during late winter (February to March) and are strongly regulated by the upstream dam (Peters and Prowse 2001). Spring flood peaks of the tributaries and the main-stem

Peace play a critical role in the type of river-ice breakup in the lower Peace, and thus the recharge of the Peace–Athabasca Delta as a result of ice-jam formation (Prowse and Conly 2000).

The enormous discharge of the Peace River results in considerable dilution of dissolved chemical constituents in the water. Shaw et al. (1990) indicate the Peace main stem has much higher dissolved oxygen concentration (always near saturation) and low values of most water-chemistry variables (e.g., anions, cations, nutrients) compared to tributary waters. For the main channel of the Peace River, water color is turbid, conductivity is 257 μS/cm (Brunskill 1986), annual average pH is 7.9, alkalinity is 96 mg/L as CaCO$_3$, total nitrogen is 0.52 mg/L, and total phosphorus is 0.16 mg/L (Shaw et al. 1990). Although tributaries can be adversely affected by anthropogenic point-source effluents (Shaw et al. 1990, Chambers, Brown et al. 2000, Chambers, Dale, and Scrimgeour 2000), synoptic surveys conducted by Alberta Environment in 1988–1989 indicate no measurable impacts of tributary discharges to the Peace main stem. Water chemistry of the main stem varies longitudinally, with upper reaches (headwaters to Smoky River confluence) characterized by low values of suspended solids (<150 mg/L) and nutrients (e.g., total phosphorus 0.1 mg/L). The lower reaches of the Peace River (Fort Vermilion to Peace Point) have higher concentrations of suspended solids and associated variables, including turbidity, BOD, total metals, and nutrients (Shaw et al. 1990). It is notable that many of the tributaries in this lower reach arise in peat lands and are brown-water streams. Shaw et al. (1990) also indicate that relative to the Athabasca River, the other major southern tributary of the Mackenzie, the Peace River tends to have lower and more constant concentrations of dissolved substances and higher amounts of particulate matter.

River Biodiversity and Ecology

The Peace River is a vast, sparsely populated basin and relatively little is known about its ecology and biodiversity. The Peace falls entirely within the Upper Mackenzie freshwater ecoregion (Abell et al. 2000). The available scientific literature largely originates from surveys by Alberta Environment or from studies that have examined the effects of river regulation or industrial impacts (Culp et al. 2000a, Gummer et al. 2000).

Algae and Cyanobacteria

Information on the microbial and algal communities of the Peace River is limited to data collected for the purpose of impact assessment (i.e., Shaw et al. 1990, Chambers, Brown et al. 2000, Chambers, Dale, and Scrimgeour 2000). Bacterial analyses on the Peace River in 1988–1989 indicate that total coliform counts tended to be highest in the summer and lowest in winter, often increasing downstream of point sources, such as effluents or tributaries that received large municipal or industrial loadings (e.g., Smoky River; Shaw et al. 1990). Although no information is available on algal community composition for the Peace River, diatom species probably dominate the potamoplankton and periphyton of this cold-water river. Potamoplankton biomass in the water column was low during an Alberta Environment synoptic survey, with mean concentrations ranging between 1 and 4 µg/L chlorophyll *a* along the entire river (Shaw et al. 1990). Tributary streams exhibited similar values, except for occasionally higher recordings associated with extreme sediment loads. Very few values of epilithic biomass are available for the main stem; however, observations in 1988 indicate that periphyton biomass during spring discharge was <25 mg/m^2 of chlorophyll *a*. Although late summer values reached levels as high as 99 mg/m^2 in river reaches upstream of the confluence with the Smoky River, chlorophyll *a* levels were largely <50 mg/m^2. Values throughout the open-water season tended to decrease along the length of the Peace River (Shaw et al. 1990). Epilithic algal biomass can be considerably higher in tributary streams receiving substantial point-source discharges. For example, algal biomass in the Wapiti River exceeded 100 mg/m^2 of chlorophyll *a* during fall 1994 (Scrimgeour and Chambers 2000, Chambers, Dale, and Scrimgeour 2000).

Plants

Riparian vegetation of the Peace River consists of poplar, spruce, alder, willow, and horsetail (Prowse and Conley 1996). Shoreline areas that are flooded frequently are vegetated by horsetail and grass species that tolerate high sediment deposition. With increasing distance from the channel margin, sediment- and flood-tolerant groups are replaced by willow and alder. Eventually, poplar and spruce occupy the drier and more stable portions of the upper riparian zone. Since the construction of the W.A.C. Bennett Dam, riparian vegetation has encroached upon the channel margins as a result of reductions in large flood peaks and bank scour (Prowse and Conly 1996).

Invertebrates

An extensive longitudinal survey of benthic invertebrate communities along the Peace River main stem recorded a diverse assemblage of macroinvertebrates consisting of 106 taxa (Shaw et al. 1990). Mean taxa richness/sample was between 23 and 28 at upstream sites, but decreased to <10 taxa/sample in the lower 600 km of the river. This longitudinal trend was also evident in benthic invertebrate densities. For example, mean densities at sites along the British Columbia border exceeded 20,000 individuals/m^2. In contrast, densities in middle reaches of the river decreased to approximately 2000 individuals/m^2 and to <500 individuals/m^2 below Fort Vermilion. Oligochaete worms, chironomid midges (Orthocladiinae, Tanypodinae, Tanytarsini), and nematodes were the dominant taxa in the Peace River, their combined relative abundance making up at least 50% of the individuals at most sampling sites. Oligochaetes (Enchytraeidae, Naididae) were the most abundant groups upstream of the Smoky River confluence. Stoneflies (*Isoperla*, *Isogenoides*, Capniidae, Taeniopterygidae) and mayflies (*Baetis*, *Ephemerella*, *Heptagenia*, *Isonychia*, *Rhithrogena*) were an important component in the middle reaches of the river. Caddisflies were never abundant, with only a few genera recorded in the Peace River.

Vertebrates

The fish community of the Peace River is depauperate, containing only 31 species belonging to 12 families (McCart 1986). Eighteen of these species are either cyprinids or salmonids, and six families are represented by only one species. Most fishes in the Peace basin originate from the Mississippi–Missouri or Columbia refugia (Prowse and Conly 1996). Knowledge of the ecology of the Peace River fish community is limited but was recently updated during the Northern River Basins Study (Boag 1993). The dominant forage fishes collected in these surveys were minnows, troutperch, and sculpins. The most abundant sport fish was goldeye, followed by walleye, burbot, northern pike, and mountain whitefish. Kokanee, rainbow trout, bull trout, and lake whitefish were seldom captured. Abundance of fishes varied longitudinally, with more fishes caught in middle and lower reaches on the main-stem Peace compared to upstream sections. However, species richness was higher in upper and middle reaches of the river. Most fishes in the main stem were concentrated in backwaters and tributary confluences, with fishes selecting areas of calm water apparently to avoid the high main-stem velocity. Boag (1993) considers the Vermilian Chutes on the Peace to be a partial barrier to upstream movement of fishes. In addition, fish surveys indicate that the Peace is used for overwintering by most of the dominant species, particularly the deep-water habitats upstream of Fort Vermilion. Goldeye and other dominant species also use the river below Peace Point for overwintering and rearing. Knowledge of the tributaries is incomplete, but spawning runs of walleye, longnose sucker, and northern pike have been observed (e.g., Wabasca River runs; Boag 1993). Furthermore, most tributaries supported populations of forage fishes, as well as juvenile longnose sucker and white sucker.

Amphibian species within the Peace River watershed include long toed salamander, western toad, spotted frog, wood frog, and striped chorus frog (PWFWCP 2000). There are only two species of reptiles: western garter snake and common garter snake. Common aquatic mammals include beaver, muskrat, river otter, and mink (PWFWCP 2000). Birds found near the river include snow goose, Canada goose, trumpeter swan, loons, grebes, merganser, osprey, and bald eagle (PWFWCP 2000).

Ecosystem Processes

Little information is available for most ecosystem processes of the Peace River. Upstream of Williston Lake attached algae are likely the most important source of autochthonous primary production. In contrast, Shaw et al. (1990) found low levels of periphytic algae (<1 mg/L chlorophyll *a*) downstream of the reservoir. In addition, this lower river reach had low values of chlorophyll *a* in the potamoplankton. The metabolism of this lower section of the river is likely based largely on decomposition of various carbon sources from upstream sources, as turbidity levels are generally high.

Human Impacts and Special Features

The Peace River is 1923 km long and one of the principal tributaries of the Mackenzie River system (Marsh 2001b). This river valley is an ancient course, which during the early Cretaceous period

(125 million years ago) was warm and lush with extensive swamps and broad bodies of fresh water, creating an ideal environment for dinosaur track formation (Currie and Sarjeant, 1979). During the construction of the W.A.C. Bennett Dam near Hudson's Hope large numbers of dinosaur fossils were unearthed (1700 tracks in >100 trackways; Currie and Sarjeant 1979). Tracks include some of the oldest records of bird footprints in the world, including a primitive shore bird, similar to killdeer, *Aquatilavipes*. Other fossils found in the area include duck-billed dinosaurs (Hadrosaurs), at least four species of carnivorous dinosaurs (including *Irenesauripus* and *Ornitholestes*), and horned dinosaurs (*Camptosaurus* and *Sauropelta*) (Currie and Sarjeant 1979). Further downstream the Peace River flows into Wood Buffalo National Park and its associated Peace–Athabasca Delta. This delta is a wetland ecosystem recognized by the Ramsar Convention on Wetlands and is listed as a U.N. World Heritage Site.

Agriculture, natural gas, forestry, and the tourism industry are the basis of the local and regional economies, which are focused in Fort St. John, the basin's largest community. The Peace valley is fertile and is the northernmost commercially important agricultural region of North America. The valley produces 85% of British Columbia's grain crops and most of the province's canola. The W.A.C. Bennett Dam located at the confluence of the Finlay and Parsnip rivers creates British Columbia's largest reservoir, Williston Lake. The Gordon M. Shrum hydroelectric power station near Hudson's Hope was built from 1968 to 1980 and at 2416 MW is the third largest in Canada.

The major impact of human activities along the Peace River main stem is regulation by the W.A.C. Bennett Dam. Filling of the Williston Reservoir between 1968 and 1972 reduced mean annual peak flows, with the magnitude of this reduction decreasing with distance downstream (Prowse and Conly 1996). Mean flood peaks at Hudson Hope near the dam outlet were reduced by about 30% relative to flows during the decade prior to regulation. Because this is a hydroelectric facility, winter flows are higher and summer flows have diminished relative to the preregulation hydrograph (Peters and Prowse 2001). For example, Prowse and Conly (1996) determined that winter flow at Peace Point, approximately 1200 km downstream of the dam, was 2.5 times greater than preregulation discharge. Although sediments appear to be transported easily through the reservoir, reductions in peak flows have reduced the capacity of the river to transport sediments below the dam. Thus old floodplains are being transformed into new low terraces and fine sediments are accumulating along channel margins and in former back channels. As a result, semiaquatic and shoreline vegetation is prograding down the banks, leading to progressively more narrow channels (Church 1995). Timescales for establishment of new forest communities are estimated in the order of centuries, but channel geometry adjustments appear to take only a few decades.

The Peace River main stem has not been affected by point-source effluents, probably because of its large volume relative to discharges from effluents and tributaries (Shaw et al. 1990). This is not the case for several of the tributary streams. For example, the Wapiti River (a tributary of the Smoky River, which has significant data available for analysis; see Additional Rivers section later in the chapter) receives point-source discharges from a pulp mill and a municipality, with the combined volume from these treated effluents adding 125 kg/d total phosphorus and 711 kg/d total nitrogen (Chambers, Dale, and Scrimgeour 2000). These effluents average approximately 1% of the river flow during the year and increase annual total phosphorus and nitrogen concentrations in the river downstream as far as the confluence with the larger Smoky River. In fact, during the low-flow period of winter (December to April) approximately 41% of the total phosphorus and 34% of the total nitrogen in the Wapiti River originates from municipal and pulp mill inputs. Above these effluent discharges the Wapiti River is oligotrophic and periphyton biomass is less than $10 \, \text{mg/m}^2$ of chlorophyll *a*. Chambers, Dale, and Scrimgeour (2000) indicate that immediately downstream of the municipal discharge algal biomass increased more than tenfold and reached a peak of approximately $140 \, \text{mg/m}^2$ of chlorophyll *a* downstream of the pulp mill discharge. The bottom-up effects of nutrient enrichment are transferred to higher trophic levels as benthic invertebrate densities increase below the pulp mill (Cash et al. 1996), and longnose sucker from this reach have greater mesenteric fat storage and condition factor (Swanson et al. 1994).

The increase in BOD, as a direct result of the effluent and indirectly from decomposing algal biomass, appears to depress dissolved concentration during the period of winter ice cover in the Wapiti River (Chambers, Brown et al. 2000). Although the concentration of dissolved oxygen in the water column remains near the current Canadian guidelines, dissolved oxygen in the substrate can be more than 3 mg/L lower than those measured in the overlying water (Lowell and Culp 1999). Thus there are concerns that winter conditions in the Wapiti River may adversely affect benthic invertebrates or egg development of fall-spawning fishes.

Environmental contamination in the Peace basin is low compared to other major river systems in Canada (Cash et al. 2000, Wrona et al. 2000). Levels of contaminants increased downstream of the pulp mill on the Wapiti River; however, improvements in the pulp mill bleaching process led to decreased levels of dioxins, furans, and chlorinated resin acids by 1996. Elevated concentrations of polychlorinated biphenyls and organochlorine pesticides were observed in burbot livers from fishes collected downstream of the pulp mill on the Wapiti River (Cash et al. 2000); however, the study concluded that the observed concentrations were not a significant risk to human health.

ATHABASCA RIVER

The Athabasca River originates in Jasper National Park in western Alberta, Canada, and is the most southerly tributary of the Mackenzie River, extending across latitudes 52°N to 58°N and longitudes 108°W to 119°W (Fig. 18.17). From its source in the Columbia Ice Field in the Rocky Mountains, this unregulated river flows northeast across boreal forests and grasslands to the Athabasca Delta and Lake Athabasca (Chambers, Dale, and Scrimgeour 2000), a natural water body covering $7940 \, \text{km}^2$. With the Peace River the Athabasca forms the north-flowing Slave River. The Athabasca is approximately 1200 km in length, drains an area of $154,880 \, \text{km}^2$, and is a 6th order river at its confluence with the lake. Resource extraction is advanced within the watershed, as open-pit coal mines operate in the upper basin, expansive oil sands developments are situated in the lower basin, and the boreal forest is harvested throughout the drainage. Because of this extensive development, the Athabasca River receives more point-source effluent than other rivers in the Mackenzie River basin (NRBS 1996).

Prior to the arrival of the white man, Sekani, Shuswap, Kootenay, Salish, Stoney, and Cree tribes hunted and fished along the Athabasca River (www.chrs.ca 2003). With the advent of the fur trade, Iroquois also inhabited this area. Early exploration of the Athabasca River occurred in 1778 when Peter Pond transported goods from the lower Saskatchewan River to the Athabasca in order to

FIGURE 18.6   Athabasca River in the Rocky Mountains of Canada in Jasper National Park, Alberta (PHOTO BY C. E. CUSHING).

establish trade with the First Nations peoples. One of the earliest European settlements in the basin occurred at Fort Chipewyan in 1788 on the southwestern shore of the Lake Athabasca. This settlement became an important trading post in the region, as fur traders extensively used the river as a navigation route. In 1811, David Thompson established a fur-trade route to the Pacific by following the upper Athabasca River through Athabasca Pass to Wood River, a Columbia River tributary. This pass was a primary route across the Rocky Mountains for several years. By the 1870s the Athabasca Landing Trail had been established, a route that used the site of the present-day town of Athabasca on the river as the loading point for barges transporting supplies downriver to northern outposts.

Physiography, Climate, and Land Use

Similar to the Peace River to the north, the Athabasca River basin flows through four major physiographic provinces (Hunt 1974). Moving from southwest to northeast, these include the Rocky Mountains in Canada (RM), Great Plains (GP), Athabasca Plain (AT), and Bear–Slave–Churchill Uplands (BC) (see Fig. 18.17). Terrestrial ecoregions follow a similar trajectory and begin with the Alberta Mountain Forests ecoregion in the southwest portion of the basin (Ricketts et al. 1999). Moving from southwest to northeast, the drainage passes through the Alberta/British Columbia Foothills Forests, Canadian Aspen Forest and Parklands, Mid-continental Canadian Forests, Midwestern Canadian Shield Forests, and Northern Canadian Shield Taiga ecoregions. Similarly, soils are diverse and include brunisolic, gleysolic, luvisolic, and organic materials. Headwaters in the Rocky Mountains are dominated by mixed forests of lodgepole pine, trembling aspen, and white spruce, with balsam fir, balsam poplar, and paper birch characteristic of the region. The lower elevations of the basin form part of the continuous midboreal mixed coniferous and deciduous forest that extends from northwestern Ontario to the foothills of the Rocky Mountains. Sites in this region that are cold and poorly drained are often covered with tamarack and black spruce. In general, conifers are more prevalent on the cooler and higher elevations of the foothills; aspen tends to be dominant in drier sites in the lower plains.

The Athabasca River basin has a relatively harsh climate, with short cool summers followed by cold winters. Temperature decreases from the southwestern to the northeastern portions of the basin, but precipitation is lowest in the southwest. Jasper, in the Rocky Mountains, has an average temperature of 3°C, with mean monthly temperatures ranging from −11°C to 15°C. Maximum readings of 22°C occur in July, whereas mean minimum temperatures of −16°C are expected in January. In contrast, Fort McMurray has a daily mean temperature of 0.2°C; the mean monthly temperature range spans 36°C (i.e., −20°C to 16°C; Fig. 18.18). Precipitation ranges from 39.4 cm/yr at Jasper to 46.5 cm/yr near Fort McMurray. Monthly precipitation at Fort MacMurray ranges from 1.6 cm in February to 7.9 cm in July (see Fig. 18.18). Approximately 70% of the moisture received within the basin falls as rain. A detailed analysis of the climate of the Athabasca basin can be found in Hudson (1997).

Although nearly 100,000 people reside in the basin, the population density is <1 person/km². Fort McMurray, with a population of 35,000 in 1996, is the largest city; other towns in the basin have 2000 to 10,000 residents. Land-use activities in the Athabasca River basin include commercial pulpwood and saw-log forestry, coal mining, oil and natural gas production, various agricultural activities, tourism, wildlife trapping, and hunting. Five pulp mills discharge to the river and major oil sands developments are ongoing near Fort McMurray. Land types within this subwatershed include forest (59.8%), taiga (17.4%), agriculture (9.8%), shrubland (6.0%), open land (4.3%), water (2.7%), and permanent ice and snow (0.1%) (http://atlas.gc.ca/site/english/index.html 2003).

River Geomorphology, Hydrology, and Chemistry

The Athabasca River descends more than 1000 m from Athabasca Falls on the upper main stem to its mouth at Lake Athabasca. River gradient exhibits a normal concave profile, with the greatest changes in average channel slope occurring in the mountainous region of Jasper National Park. In this upper reach the gradient decreases from approximately 4 m/km near Athabasca Falls to 90 cm/km at the eastern boundary of the park (Fig. 18.6). Mean channel width is 100 m in this area and mean depth is about 1.5 m (Kellerhals et al. 1972). Middle reaches of the river near the town of Athabasca have a low slope (40 cm/km), large mean channel width (200 m), and a mean depth of 2.2 m (Kellerhals et al. 1972). The river below Fort McMurray is very wide (500 m) and deep (2.4 m), with minimal slope (<20 cm/km). Substrate in the river grades from largely cobble and gravel in Jasper National Park to sand below Fort McMurray, with local areas of gravel and limestone bedrock. Multiple channels dominate the upper and lower sections of the river; however, middle reaches of the river are rarely braided (R. L. and L. Environmental, Ltd. 1994a).

Mean monthly runoff for the Athabasca River basin ranges from 0.31 cm/mo in February to 2.67 cm/mo in July (see Fig. 18.18). Mean monthly runoff pattern is similar to precipitation, with highest levels in summer (July) and lowest in winter (February). From June to September evapotranspiration results in a greater difference between precipitation and runoff. Mean monthly discharge increases from 88 m³/s near Jasper to 433 m³/s at Athabasca and 783 m³/s at the river mouth (R. L. and L. Environmental, Ltd. 1994b). River discharge is dominated by mountain runoff in the spring such that average values for July exceed 1500 m³/s at the river mouth (see Fig. 18.18). The hydrograph descends the remainder of the year, except during periods of high rainfall. Minimum discharge values of about 200 m³/s occur in February and March prior to river-ice breakup.

The Athabasca is a hardwater river with ion concentrations dominated by bicarbonate and calcium (Anderson 1989), resulting in an average conductivity of 332 μS/cm. River discharge strongly influences the concentration of the major cations (calcium, magnesium, sodium, and potassium) and anions (bicarbonate, sulphate, and chloride). The carbonate ions provide good buffering capacity, with an average river alkalinity of 117 mg/L as $CaCO_3$. Seasonal changes in concentration are greater than spatial patterns within a season, as the highest values of most chemical constituents occur during minimum winter discharge and the lowest values occur in summer during annual mountain runoff (Anderson 1989). Relative to the more northerly Peace River, the Athabasca River has a higher and more variable concentration of dissolved substances (average TDS

187 mg/L) and a lower amount of particulate matter (Shaw et al. 1990). Dissolved oxygen under ice shows significant declines over an 800 km river distance from the most upstream pulp mill to the re-aeration zone at Grand Rapids (upstream from Fort McMurray) (Chambers, Brown et al. 2000). Between 1990 and 1993, late winter values ranged from saturation in the headwaters (~12 mg/L) to approximately 8 mg/L upstream of the rapids. Average nutrient concentrations for the whole river are 0.54 mg/L for total nitrogen and 0.10 mg/L for total phosphorus. Analysis of long-term trends in total phosphorus and total nitrogen from 10 sites indicates that the concentrations of both nutrients increase along the length of the Athabasca River, with the lowest concentrations occurring in Jasper National Park (Chambers, Dale, and Scrimgeour 2000). Median total phosphorus was less than 0.01 mg/L at Jasper and increased to approximately 0.04 mg/L below Fort McMurray. Nutrient concentration is typically highest in winter and lowest during annual peak discharge in summer (Chambers 1996).

River Biodiversity and Ecology

The Athabasca, along with the Peace River basin, composes most of the Upper Mackenzie freshwater ecoregion (Abell et al. 2000). The biodiversity and ecology of the Athabasca is the best understood of any system in the Mackenzie River basin, largely as a result of the Northern River Basins Study (NRBS 1996, Culp et al. 2000b) and its successor, the Northern Ecosystems Research Initiative.

Algae and Cyanobacteria

Although algal biomass throughout the length of the river is typically low, biomass values >700 mg/m^2 chlorophyll *a* were recorded below point-source effluents such as pulp mills and municipal sewage. Algal standing crop as chlorophyll *a* in Jasper National Park, upstream of all major human impacts, was <20 mg/m^2 chlorophyll *a* and increased approximately sixteenfold immediately downstream of Jasper, the most upstream town along the river (Chambers, Dale, and Scrimgeour 2000). Information on algal community composition is limited to two surveys: a limited survey of epilithic algae near Hinton, Alberta, by Podemski (1999), and a study by McCart et al. (1977) near Fort McMurray that examined the seasonal patterns of algae colonizing glass slides. Podemski found that autumnal periphyton communities on the upper Athabasca were dominated by diatom species, with more than 40 species recorded. The most common genera were *Achnanthes*, *Diatoma*, *Fragilaria*, *Cymbella*, and *Nitzschia*. On the lower Athabasca River the temporal study of McCart et al. recorded a rich algal flora, with 191 species from December through October. Standing stocks were dominated by cyanobacteria in the lower river (e.g., *Hapalosiphon*, *Lyngbya*, *Oscillatoria*, *Phormidium*), except during the summer, when diatoms were most abundant.

Plants

Macrophytes are rare in the main channel. Barton (1986) hypothesized this was because of high concentrations of suspended solids and unstable riverbed substrate. Riparian vegetation includes peat and brown mosses, reindeer lichens, sedges, willow, and Labrador tea. Tree species include white spruce, balsam poplar, tamarack, black spruce, aspen, and willow.

Invertebrates

Information on the upper Athabasca River (above the Grand Rapids) is limited to a few surveys (Ouellett and Cash 1996). Mean macroinvertebrate densities in the spring (20,000 to 50,000 individuals/m^2) were typically lower than those observed in the fall (30,000 to 200,000 individuals/m^2) (Anderson 1989). Similarly, more taxa were found in the fall (18 to 33) than in the spring (12 to 21) at the reference site. Actual species richness is much higher than these figures imply, because taxonomic identifications were made only to the lowest practical unit, which was genus for mayflies, stoneflies, and caddisflies. During both seasons macroinvertebrate assemblages consisted primarily of chironomid midges (Chironomini, Orthocladiinae, Tanypodinae), mayflies (*Ameletus*, *Baetis*, *Cinygmula*, *Heptagenia*, *Ephemerella*, *Rhithrogena*), stoneflies (*Alloperla*, *Capnia*, *Isoperla*, *Hesperoperla*, *Pteronarcella*, *Taenionema*), and caddisflies (*Brachycentrus*, *Glossosoma*, *Hydropsyche*).

Limited information is available on the macroinvertebrate communities of the lower Athabasca River (Barton 1986). Faunal composition on bedrock in this reach resembled that of stony tributary streams, but standing stocks were very low (about 2000 individuals/m^2). Bedrock fauna included the mayfly *Ephemerella inermis* and four species of the stonefly

Isoperla. Barton (1980) reports that the benthic invertebrate fauna of unstable sand and gravel substrates consisted mostly of chironomid midges and oligochaete worms, with standing stocks of 10 to 9000 individuals/m². Dominant chironomids included *Polypedilum*, *Paracladopelma*, and *Cyphomella*. In addition, the dragonfly *Gomphus notatus* and the mayfly *Ametropus neavei* were often present in the sand and gravel substrates.

In contrast to most other smaller tributaries within the Mackenzie watershed, benthic invertebrate information has been collected on the Bigoray River, a tributary of the Pembina River, which flows into the upper Athabasca River (Mackay 1995). This slow-flowing brown-water stream is typical of muskeg areas in the region (Mackay 1995) and is dominated by chironomids (112 species), ostracods, and 66 species of beetles, mayflies, and caddisflies. (Boerger 1981, Clifford 1978). Standing crop densities range between 20,000 to 50,000 individuals/m², and average annual standing crop biomass in the river was estimated at 0.86 g DM/m² (Clifford 1978). Areas of fast-flowing habitat that occur at beaver dam outflows are thought to increase overall biodiversity in muskeg streams (Mackay 1995).

Vertebrates

The fish community of the Athabasca River contains 36 species distributed among 12 families (McCart 1986). Cyprinids, coregonids, and salmonids account for 22 of these species; five families have only one species. Postglacial colonization of the river originated mostly from source populations located in the Mississippi–Missouri and Columbia refugia. Recent surveys conducted for the Northern River Basins Study provide most of the limited information available on the Athabasca River fish community (R. L. and L. Environmental, Ltd. 1994a, 1994b). These surveys found that fish species richness was lowest in the upper Athabasca River and tributary streams and highest in the lower Athabasca. The major sport-fish species included mountain whitefish, walleye, goldeye, northern pike, burbot, and Arctic grayling. Bull trout, rainbow trout, and lake whitefish were rarely encountered. Mountain whitefish dominated the sport fishery in the upper and middle reaches of the river, whereas the lower river contained several codominant species, including goldeye, northern pike, and walleye. Longnose and white sucker were distributed throughout the main stem. Forage fishes, such as flathead chub, trout-perch, and lake chub, were the most abundant species

in the main stem. Several species spawn in the main stem, including mountain whitefish, walleye, and goldeye; however, others use the tributaries for spawning, such as the bull trout. Finally, McCart (1986) indicates that runs of lake whitefish move from Lake Athabasca in summer to spawn in rapids of the Athabasca main stem; longnose suckers exhibit a similar spawning run in late winter and spring.

The majority of the Athabasca River basin is covered by mixed boreal forest, providing habitat for a large number of wildlife species. Amphibians in the watershed are represented by five species, including boreal chorus frog, western toad, wood frog, northern leopard frog, and long toed salamander. Reptiles are represented by one species, the wandering garter snake (www.biology.ualberta.ca/courses.hp/200301. hp/t-elegans.html 2000). Fifty-five species of mammals have been identified within the Athabasca River basin; those that can be found near the river include grizzly and black bear, mink, river otter, moose, muskrat, and beaver. The basin is also home to 222 bird species (http://scienceoutreach.ab.ca/resource_resources.htm #resources_inventories 2003). There are areas that support large numbers of waterfowl, including the Peace Athabasca Delta and the Pocahontas Marsh, which has nesting habitat for over 60 species (CHRB 2003). Birds that live along the river include sandpipers, plover, sandhill crane, bald eagle, osprey, American white pelican, cormorants, grebes, herons, loons, and geese (http://scienceoutreach.ab.ca/resource_ resources.htm#resources_inventories 2003).

Ecosystem Processes

Over the river's length the accumulation of nutrients from multiple point and nonpoint sources increases algal biomass and bottom-up food web effects (Chambers 1996, Cash et al. 1996). In this ecosystem, nutrient enrichment appears to have more important ecological effects on the ecosystem than most contaminants. Nutrient enhancement is evident in periphyton biomass, with levels in the Athabasca River 2 to 50 times higher below effluent discharges, often at distances of >100 km downstream from the point source (Chambers, Dale, and Scrimgeour 2000). Nutrient limitation varied substantially along the river, primarily due to phosphorus concentration. However, periphyton growth in the middle and lower Athabasca River was also limited by nitrogen or by both nutrients, depending on the location (Scrimgeour and Chambers 2000). Effluent loadings from pulp mills or municipalities alleviated nutrient limitation for a considerable distance below an effluent discharge. Abundance of total insects and dominant taxa was higher in areas of the river containing pulp mill effluent (Anderson 1989). Differences observed between assemblages upstream and downstream of impact sites were caused by changes in relative taxonomic abundance rather than the loss or gain of taxa (Cash et al. 1996, Culp, Podemski, and Cash 2000). This increased macroinvertebrate production is transferred to fishes downstream of the effluent, as these downstream populations exhibit increased body condition, gonad size, and egg mass compared to unexposed fishes (Gibbons et al. 1998).

Human Impacts and Special Features

The Athabasca River is the longest river in Alberta (1538 km), flowing from the glaciers of Jasper National Park to Lake Athabasca in Wood Buffalo National Park (www.chrs.ca 2003). Because of its natural beauty and significance and importance for river recreation, the section of the Athabasca River within Jasper National Park is designated as a Canadian Heritage River (www.chrs.ca 2003). Key features along the river include Athabasca Falls in Jasper National Park and the Grand Rapids in the lower Athabasca. The Grand Rapids re-aerates the river with dissolved oxygen in winter and is an important spawning area for fishes, such as lake whitefish and walleye (NRBS 1996). The watershed also includes the Columbia Ice Field, the largest ice field in the Rocky Mountains. The river is a popular canoeing and rafting destination.

Concern about water quality in the Athabasca River dates back to 1957, when the first pulp mill began operations on the upper river. By 1994 five pulp mills were operational, their combined input of total phosphorus (TP) and total nitrogen (TN) loads to the river totaling approximately 331 and 1033 kg/day, respectively (Chambers, Dale, and Scrimgeour 2000). Analysis of TP and TN concentrations measured monthly at 10 sites on the Athabasca River between 1980 and 1993 demonstrated that nutrient concentrations were lowest in Jasper National Park, upstream of any effluent loadings, and increased along the length of the river (Chambers, Dale, and Scrimgeour 2000).

A particularly impaired section of the river was the reach immediately downstream of Hinton, where high concentrations of contaminants and nutrients combined to raise concern for human and ecosystem health (Culp et al. 2000a). In the middle river reaches

(Whitecourt to Athabasca), the low concentrations of DO and moderate levels of nutrients and contaminants contributed to degraded water quality. The primary concern in the lower Athabasca River (below Fort McMurray) was the effect of contaminants (e.g., mercury in older predatory fishes exceeded Health Canada guidelines) (Cash et al. 2000). In addition to nutrient enhancement, effluent from pulp mills was responsible for declines in under-ice DO concentrations of 1 to 2 mg/L (Chambers et al. 1997, Chambers, Brown et al. 2000).

Recorded levels of chlorinated organic and metal contaminants in water or sediments are low and have not exceeded Canadian health and environmental guidelines (Wrona et al. 2000). Contaminant body burdens of caddisflies and mayflies occasionally exceeded the Canadian Water Quality Guideline for polychlorinated dibenzofurans (PCDFs) (Culp, Podemski, and Cash 2000). Low levels of contaminants were also found in fishes, although levels of dioxins, furans, PCBs, and mercury in biota occasionally exceeded Canadian guidelines (Cash et al. 2000). Stable isotope analyses indicated that a primary route of exposure to pulp mill contaminants was through the food chain (Cash et al. 2000). Pulp mill effluent appeared to be the source of these contaminants and also appeared to cause the endocrine disruption measured in spoonhead sculpin downstream of the first pulp mill effluent to the river. In addition, fishes captured immediately downstream of pulp mills exhibited an increase in external abnormalities, such as tumors, lesions, and skin discoloration (Cash et al. 2000).

ADDITIONAL RIVERS

The majority of the South Nahanni River basin is within the Mackenzie Mountains of the Northwest Territories (Fig. 18.19). The river originates in the ice fields of the Selwyn mountains, flows in a southeastern direction, and enters the Great Plains just before its confluence with the Liard River, 540 km downstream (www.chrs.ca 2003). Mean monthly discharge can range from 60 to 1300 m^3/s, providing significant flow to the lower Liard River and ultimately to the main-stem Mackenzie (Fig. 18.20). The river passes through mountain valleys and has canyons with 500 to 1000 m high walls (Brunskill 1986). The river is located within the Nahanni National Park Reserve (an U.N. World Heritage Site) and is designated as a Canadian Heritage River (www.chrs.ca 2003). There are many significant

FIGURE 18.7   South Nahanni River at Last Chance Harbor above Virginia Falls (Photo by D. Bicknell).

natural features along the watershed, including Virginia Falls (almost twice the height of Niagara Falls), canyons, hot springs, and geological formations that escaped glaciation (Fig. 18.7) (www.chrs.ca 2003). This remote river can only be reached by plane or boat. There is no land development in the watershed; however, there have been mineral deposits discovered within the area. The South Nahanni is renowned for its adventure tourism, focused on river touring by canoe, kayak, or raft.

The Smoky River is located in the western central region of Alberta and has its headwaters in the Rocky Mountains (Fig. 18.21). The river flows in a northeasterly direction across the foothills of Alberta into the Great Plains, where 492 km downstream it empties into the Peace River near the town of Peace River (Chambers, Dale et al. 2000). The Smoky River is not regulated and has two main tributaries, the Wapiti River and the Little Smoky River. Mean monthly discharge varies from a low of $43\,m^3/s$ in February to a high of $1024\,m^3/s$ in June during mountain snowpack melt (Fig. 18.22). The majority of the watershed is forested, with about 22% of the land area used for agriculture (http://atlas.gc.ca/site/english/index.html 2003). The headwaters of the Smoky River are located in Willmore Wilderness Park, just north of Jasper National Park. The terrain is mountainous, creating many diverse alpine and subalpine ecoregions, with white spruce, lodgepole pine, balsam fir, and aspen poplar. Kimiwan Lake,

FIGURE 18.8 Yellowknife River near outlet of Lower Carp Lake, Northwest Territories (PHOTO BY C. SPENCE).

located at the crossing of several flyways, is an area of great bird diversity. Grande Prairie (on the Wapiti River), with a population of 40,226, began as a Hudson's Bay post in 1881. Grande Cache (upper portion of the Smoky River), with a population of about 4000, was named for the large fur cache established by the Hudson's Bay Fur Trading Company in 1821 to store large shipments of fur from British Columbia.

The Hay River originates in the foothills of British Columbia's Rocky Mountains and flows in a northeasterly direction through the Great Plains of Alberta before draining into the southwest shore of Great Slave Lake (Fig. 18.23). Mean monthly discharge ranges from a low of only $3\,\mathrm{m^3/s}$ in March during snow and ice storage in winter to $423\,\mathrm{m^3/s}$ in May during snowmelt (Fig. 18.24). The Hay River drains an area that primarily consists of black spruce boreal muskeg/forest and wetlands, resulting in water that is humic and turbid in color (Fig. 18.1). Within the Hay River basin in northwestern Alberta is a remote lowland wetland complex known as the Zama and Hay Lakes area. This area is globally significant for its huge concentrations of geese and ducks (up to 1 million birds) during spring and fall migrations

(www.ibacanada.com 2003). Approximately 50 km from the mouth of the Hay River, the Hay River Gorge and the Twin Falls provide interesting scenery for tourists. The town of Hay River (population 3800), located near the mouth, was initiated in 1868 as a fur trading post by the Hudson's Bay Company. Today it is the northernmost railroad accessible area in the Northwest Territories and an important hub for barge travel along the Mackenzie River. Areas of the Hay River basin have had considerable oil and gas exploration.

The Yellowknife River basin is completely within the Northwest Territories and drains in a southwesterly direction from Greenstocking Lake to Great Slave Lake, a distance of approximately 260 km (Fig. 18.8, Fig. 18.25). There are several significant lakes within the watershed, including Greenstocking, Reindeer, Drybones, Thistlethwaite, Duncan, and Gordon lakes. Mean monthly discharge ranges from a low of $21\,\mathrm{m^3/s}$ in April to $72\,\mathrm{m^3/s}$ in July (Fig. 18.26). The upper portion of the watershed is dominated by gently rolling hills and eskers of the barrenland, with black spruce as the dominant tree, followed by poplar and birch. Terrain near the mouth becomes more rocky, with taller wooded areas. The northern end of the basin is part of the wintering grounds for the Bathurst caribou herd, and sections of the river are known for Arctic grayling. The name Yellowknife originated from the copper-wielding Chipewyan tribe, who resided near the present location of the city of Yellowknife (the only major community in the basin, with a population of 18,000) and the capital of the Northwest Territories. The city, founded in 1934, is known for its history of gold mining (www.assembly.gov.nt.ca/VisitorInfo/NWT MapandHistory/Yellowknife.html 2003). A dam is located on the lower portion of the river to provide hydroelectricity to the city.

The Peel River borders northeastern Yukon and northwestern Northwest Territories and straddles the Arctic Circle, flowing from south to north (Fig. 18.27). This is the most northerly tributary of the Mackenzie River, with headwaters initiating in the rugged Mackenzie Mountains, flowing into the Arctic Lowlands, and finally discharging into the Mackenzie River Delta, a total distance of 644 km. Mean monthly discharge ranges from only 60 m³/s in November to 143 m³/s in May (Fig. 18.28). In the upper headwaters, the valleys lie within the tundra region, with massive scree slopes devoid of vegetation and open stands of black and white spruce along the longer slopes. In the lower sections of the watershed, white spruce is dominant, with sections of black spruce and larch in poorly drained areas (www.chrs.ca 2003). There are several major tributaries of the Peel River, including the Hart, Bonnet Plume (a Canadian Heritage River), and Snake rivers, known for their ruggedness and beauty, providing many opportunities for adventure tourism (www.chrs.ca 2003). The Peel River watershed is mostly uninhabited, with Fort McPherson (near the confluence with the Mackenzie River) the largest community in the region (population 910), established in 1840 by the Hudson's Bay Company. The community is one of three located along the Dempster Highway, the most northerly public highway in Canada, linking Dawson City with Inuvik, a total distance of 736 km.

ACKNOWLEDGMENTS

Financial support for this research was provided by the National Water Research Institute (Environment Canada) and a Natural Sciences and Engineering Research Council Discovery Grant to J. M. Culp. Technical assistance and information collection during manuscript preparation was provided by C. Casey, N. Glozier, D. Halliwell, H. Popoff and M. Meding. P. Harker (Natural Resources Canada) provided geographical support. Photographs were graciously supplied by D. Bicknell, T. Carter, C. Cushing and C. Spence. The manuscript benefited from constructive criticisms by A. Benke, C. Cushing and K. Roberts.

LITERATURE CITED

Abell, R. A., D. M. Olson, E. Dinerstein, P. T. Hurley, J. T. Diggs, W. Eichbaum, S. Walters, W. Wettengel, T. Allnut, C. J. Loucks, and P. Hedao. 2000. *Freshwater ecoregions of North America: A conservation assessment.* Island Press, Washington, D.C.

Alberta Environment. 1987. Slave River basin overview. Report for Alberta/Northwest Territories Technical Committee. Alberta Environment, Edmonton.

Anderson, A. M. 1989. *An assessment of the effects of the combined pulp mill and municipal effluents at Hinton on the water quality and zoobenthos of the Athabasca River.* Alberta Environment, Edmonton.

Artibise, A. F. 2001. Fort Nelson. In *The Canadian Encyclopedia.* 2001 World Edition. Stewart House, Etobicoke, Ontario.

Asch, M. I. 2001. Great Slave Lake. In *The Canadian Encyclopedia.* 2001 World Edition. Stewart House, Etobicoke, Ontario.

Barton, D. R. 1980. Benthic macroinvertebrate communities of the Athabasca River near Ft. Mackay, Alberta. *Hydrobiologia* 74:151–160.

Barton, D. R. 1986. Invertebrates of the Mackenzie system. In B. R. Davies and K. F. Walker (eds.). *The ecology of river systems*, pp. 473–492. Junk, Boston.

B. C. Hydro. 1985. Liard River hydroelectric development, downstream hydrology. Interim Report no. H1794. British Columbia Hydro and Power Authority, Hydroelectric Generation Projects Division, Vancouver.

Boag, T. D. 1993. A general fish and riverine habitat inventory, Peace and Slave rivers, April to June, 1992. Report no. 9. Northern River Basins Study, Edmonton, Alberta.

Bodaly, R. A., J. D. Reist, D. M. Rosenberg, P. J. McCart, and R. E. Hecky. 1989. Fish and fisheries of the Mackenzie and Churchill river basins, northern Canada. In D. P. Dodge (ed.). *Proceedings of the International Large River Symposium (LARS).* Canadian Special Publication of Fisheries and Aquatic Sciences 106, pp. 128–144.

Boerger, H. 1981. Species composition, abundance and emergence phenology of midges (Diptera, Chironomidae) in a brown-water stream of west-central Alberta, Canada. *Hydrobiologia* 80:7–30.

Brunskill, G. J. 1986. Environmental features of the Mackenzie system. In B. R. Davies and K. F. Walker (eds.). *The ecology of river systems*, pp. 435–471. Junk, Boston.

Canadian Council of Ministers of the Environment (CCME). 1999. *Canadian environmental quality guidelines.* Canadian Council of Ministers of the Environment, Winnipeg, Manitoba.

Carson, M. A., J. N. Jasper, and F. M. Conly. 1998. Magnitude and sources of sediment input to the Mackenzie Delta, Northwest Territories. *Arctic* 51:116–124.

Cash, K. J., W. N. Gibbons, K. R. Munkittrick, S. B. Brown, and J. Carey. 2000. Fish health in the Peace, Athabasca and Slave river systems. *Journal of Aquatic Ecosystem Stress and Recovery* 8:77–86.

Cash, K. J., M. S. J. Ouellett, and F. J. Wrona. 1996. An assessment of the utility of benthic macroinvertebrate and fish community structure in biomonitoring the Peace, Athabasca and Slaver river basins. Report no. 123. Northern River Basins Study, Edmonton, Alberta.

Chambers, P. A. 1996. Nutrient enrichment in the Peace, Athabasca and Slave rivers: Assessment of present conditions and future trends. Synthesis Report no. 4. Northern River Basins Study, Edmonton, Alberta.

Chambers, P. A., S. Brown, J. M. Culp, R. Lowell, and A. Pietroniro. 2000. Dissolved oxygen decline in ice-covered rivers and its effects on aquatic biota. *Journal of Aquatic Ecosystem Stress and Recovery* 8:27–38.

Chambers, P. A., A. Dale, and G. J. Scrimgeour. 2000. Nutrient enrichment from point and non-point nutrient loadings. *Journal of Aquatic Ecosystem Stress and Recovery* 8:53–66.

Chambers, P. A., G. J. Scrimgeour, and A. Pietroniro. 1997. Winter oxygen conditions in ice-covered rivers: The impact of pulp mill and municipal effluents. *Canadian Journal of Fisheries and Aquatic Sciences* 54:2796–2806.

Church, M. 1995. Geomorphic response to river flow regulation: Case studies and time-scales. *Regulated Rivers: Research and Management* 11:3–22.

Clayton, J. W., C. C. Lindsey, and R. A. Bodaly. 1992. Evolution of late whitefish (*Coregonus clupeaformis*) in North America during the Pleistocene: Evidence for a Nahanni glacial refuge race in the northern Cordillera region. *Canadian Journal of Fisheries and Aquatic Sciences* 49:760–768.

Clifford, H. F. 1978. Descriptive phenology and seasonality of a Canadian brown-water stream. *Hydrobiologia* 58: 213–231.

Crowe, K. J. 1974. A history of the original peoples of northern Canada. Mcgill-Queens Press, Montreal.

Culp, J. M., K. J. Cash, and F. J. Wrona. 2000a. Cumulative effects assessment for the Northern River Basins Study. *Journal of Aquatic Ecosystem Stress and Recovery* 8:87–94.

Culp, J. M., K. J. Cash, and F. J. Wrona. 2000b. Integrated assessment of ecosystem integrity of large northern rivers: The Northern Rivers Basins Study example. *Journal of Aquatic Ecosystem Stress and Recovery* 8:1–5.

Culp, J. M., C. L. Podemski, and K. J. Cash. 2000. Interactive effects of nutrients and contaminants from pulp mill effluents on riverine benthos. *Journal of Aquatic Ecosystem Stress and Recovery* 8:67–75.

Currie, P. J. and W. A. S. Sarjeant. 1979. Lower cretaceous dinosaur footprints from the Peace River Canyon, British Columbia, Canada. *Palaeogeography, Paleoclimatology, Palaeoecology* 28:103–115.

Environment Canada. 1981. *Water quality data, Mackenzie River Basin 1960–1979.* Inland Waters Directorate, Water Quality Branch, Ottawa.

Environment Canada. 1991. Protecting the waters of Nahanni National Park Reserve, Northwest Territories. Report no. TR91-I/NAH. Inland Waters Directorate and Canadian Parks Service, Yellowknife, Northwest Territories.

Environment Canada. 1999. *Surface Water and Sediment Data,* Hydat CD-ROM, Version 96–1.05.8. Atmospheric Environment Service.

Gibbons, W. N., K. R. Munkittrick, and W. D. Taylor. 1998. Monitoring aquatic environments receiving industrial effluents using small fish species. 1: Response of spoonhead sculpin (*Cottus ricei*) downstream of a bleached-kraft pulp mill. *Environmental Toxicology and Chemistry* 17:2227–2237.

Golder Associates. 2000. Weyerhaeuser Canada Ltd., Grand Prairie Environmental Effects Monitoring Cycle 2, Final Report. Golder Associates, Calgary, Alberta.

Gummer, W. D., K. J. Cash, F. J. Wrona, and T. D. Prowse. 2000. The Northern River Basins Study: Context and design. *Journal of Aquatic Ecosystem Stress and Recovery* 8:7–16.

Hay, M. B., N. Michelutti, and J. P. Smol. 2000. Ecological patterns of diatom assemblages from Mackenzie Delta lakes, Northwest Territories, Canada. *Canadian Journal of Botany* 78:19–33.

Howland, K. L., R. F. Tallman, and W. M. Tonn. 2000. Migration patterns of freshwater and anadromous inconnu in the Mackenzie River system. *Transactions of the American Fisheries Society* 129:41–59.

Hudson, E. 1997. The climate of the Peace, Athabasca and Slave river basins. Report no. 124. Northern River Basins Study, Edmonton, Alberta.

Hunt, C. B. 1974. *Natural regions of the United States and Canada.* W. H. Freeman, San Francisco.

Kellerhals, R. C., C. R. Neil, and D. I. Bray. 1972. *Hydraulic and geomorphic characteristics of rivers in Alberta.* Alberta Environment, Cooperative Research Program in Highway and River Engineering Technical Report, Edmonton.

Little, A. S., W. M. Tonn, R. F. Tallman, and J. D. Reist. 1998. Seasonal variation in diet and trophic relationships within the fish communities of the lower Slave River, Northwest Territories, Canada. *Environmental Biology of Fishes* 53:429–445.

Lowell, R. B., and J. M. Culp. 1999. Cumulative effects of multiple effluent and low dissolved oxygen stressors on mayflies at cold temperatures. *Canadian Journal of Fisheries and Aquatic Sciences* 56:1624–1630.

MacDonald, D. D. 1992. An assessment of ambient environmental conditions in the Liard River basin, Northwest Territories. Prepared for Water Resources Division, Indian and Northern Affairs Canada, Yellowknife, Northwest Territories.

MacDonald, D. D. 1994. A discussion paper on the development of ecosystem maintenance indicators for the transboundary river systems within the Mackenzie River basin: Slave, Liard. Prepared for Water Resources Division, Indian and Northern Affairs Canada, Yellowknife, Northwest Territories.

Mackay, R. J. 1995. River and stream ecosystems of Canada. In C. E. Cushing, K. W. Cummins, and G. W. Minshall (eds.). *River and stream ecosystems*, pp. 33–60. Elsevier, New York.

Mackenzie River Basin Committee (MRBC). 1981a. Mackenzie River Basin Study report. A report prepared for the Mackenzie River Basin Committee. Environment Canada, Inland Directorate, Regina, Saskatchewan.

Mackenzie River Basin Committee. 1981b. Water quality data, Mackenzie River basin 1960–1979. A report prepared for the Mackenzie River Basin Committee. Environment Canada, Inland Directorate, Ottawa.

Marsh, J. 2001a. Liard River. In *The Canadian Encyclopedia*. 2001 World Edition. Stewart House, Etobicoke, Ontario.

Marsh, J. 2001b. Peace River. In *The Canadian Encyclopedia*. 2001 World Edition. Stewart House, Etobicoke, Ontario.

McCart, P. J. 1986. Fish and fisheries of the Mackenzie system. In B. R. Davies and K. F. Walker (eds.). *The ecology of river systems*, pp. 493–515. Junk, Boston.

McCart, P. J., P. Tsui, W. Grant, and R. Green. 1977. Baseline studies of aquatic environments in the Athabasca River near Lease 17. Environmental Research Monograph 1977–2. Syncrude Ltd., Ft. McMurray, Alberta.

McCarthy, L. H., K. Robertson, R. H. Hesslein, and T. G. Williams. 1997. Baseline studies in the Slave River, NWT, 1990–1994. Part IV: Evaluation of benthic invertebrate populations and stable isotope analyses. *Science of the Total Environment* 197:111–125.

McCarthy, L. H., G. R. Stephens, D. M. Whittle, J. Peddle, S. Harbicht, C. LaFontaine, and D. J. Gregor. 1997. Baseline studies in the Slave River, NWT, 1990–1994. Part II: Body burden contaminants in whole fish tissue and livers. *Science of the Total Environment* 197:55–86.

McCarthy, L. H., T. G. Williams, G. R. Stephens, J. Peddle, K. Robertson, and D. J. Gregor. 1997. Baseline studies in the Slave River, NWT, 1990–1994. Part I: Evaluation of the chemical quality of water and suspended sediment from the Slave River (NWT). *Science of the Total Environment* 197:21–53.

McLeod, C. L., and J. P. O'Neil. 1983. Major range extensions of anadromous salmonids and first record of chinook salmon in the Mackenzie River drainage. *Canadian Journal of Zoology* 61:2183–2184.

McPhail, J. D., and C. C. Lindsey. 1970. *Freshwater fishes of Northwestern Canada and Alaska*. Fisheries Research Board of Canada Bulletin 173.

Morrison, D. and G. H. Germain. 1995. *Inuit: Glimpses of an Arctic Past*. Canadian Museum of Civilization, Ottawa, Ontario.

Muir, D. C. G., C. A. Ford, N. P. Grift, D. A. Netner, and W. L. Lockhart. 1990. Geographic variation of chlorinated hydrocarbons in burbot (*Lota lota*) from remote lakes and rivers in Canada. *Archives of Environmental Contamination and Toxicology* 19:530–542.

Northern River Basins Study (NRBS). 1996. *Report to the ministers*. Northern River Basins Study, Edmonton, Alberta.

Ouellett, M. S. J., and J. K. Cash. 1996. BONAR: A database for benthos of Peace, Athabasca and Slave River basins, user's guide. Report no. 143. Northern River Basins Study, Edmonton, Alberta.

Peace Williston Fish and Wildlife Compensation Program (PWFWCP). 2000. Fish and wildlife species list for the Williston and Dinosaur reservoir watersheds. BC Hydro, Prince George, British Columbia.

Peters, D. L., and T. D. Prowse. 2001. Regulation effects on the lower Peace River, Canada. *Hydrological Processes* 15:3181–3194.

Podemski, C. L. 1999. Ecological effects of a bleached kraft pulp mill effluent on benthic biota of the Athabasca River. Ph.D. diss., University of Saskatchewan.

Pool, A. 2001. Fort Simpson. In *The Canadian Encyclopedia*. 2001 World Edition. Stewart House, Etobicoke, Ontario.

Prowse, T. D. 1986. Ice jam characteristics, Liard–Mackenzie river confluence. *Canadian Journal of Civil Engineering* 13:653–665.

Prowse, T. D. 1990. Heat and mass balance of an ablating ice jam. *Canadian Journal of Civil Engineering* 17:629–635.

Prowse, T. D. 1993. Suspended sediment concentration during river ice breakup. *Canadian Journal of Civil Engineering* 20:872–875.

Prowse, T. D., and F. M. Conly. 1996. Impact of flow regulation on the aquatic ecosystem of the Peace and Slave rivers. Synthesis Report no. 1. Northern River Basins Study, Edmonton, Alberta.

Prowse, T. D., and F. M. Conly. 2000. Multiple-hydrologic stressors of a northern delta ecosystem. *Journal of Aquatic Ecosystem Stress and Recovery* 8:17–26.

Prowse, T. D., and Culp, J. M. 2003. Ice breakup: A neglected factor in river ecology. *Canadian Journal of Civil Engineering* 30:128–144.

Prowse, T. D., and V. Lalonde. 1996. Open-water and ice-jam flooding of a northern delta. *Nordic Hydrology* 27:85–100.

Prowse, T. D., D. Peters, S. Beltaos, A. Pietroniro, L. Romolo, J. Töyrä, and R. Leconte. 2002. Restoring ice-jam floodwater to a drying delta ecosystem. *Water International* 27(1):58–69.

R. L. and L. Environmental Services, Ltd. 1994a. A general fish and riverine habitat inventory, Athabasca River April to May, 1992. Report no. 32. Northern River Basins Study, Edmonton, Alberta.

R. L. and L. Environmental Services, Ltd. 1994b. A general fish and riverine habitat inventory, Athabasca River October, 1993. Report no. 40. Northern River Basins Study, Edmonton, Alberta.

Rempel, L. L., and D. G. Smith. 1998. Postglacial fish dispersal from the Mississippi refuge to the Mackenzie River basin. *Canadian Journal of Fisheries and Aquatic Sciences* 55:893–899.

Reist, J. D. 1994. An overview of the possible effects of climate change on northern freshwater and anadromous fishes. Interim Report no. 2. Mackenzie Basin Impact Study, Environment Canada, Downsview, Ontario.

Ricketts, T. H., E. Dinerstein, D. M. Olson, C. J. Loucks, W. Eichbaum, D. DellaSala, K. Kavanagh, P. Hedao, P. T. Hurley, K. M. Carney, R. Abell, and S. Walters. 1999. *Terrestrial ecoregions of North America: A conservation assessment.* Island Press, Washington, D.C.

Rosenberg, D. M. 1986. Resources and development of the Mackenzie system. In B. R. Davies and K. F. Walker (eds.). *The ecology of river systems*, pp. 517–540. Junk, Boston.

Rosenberg, D. M., A. P. Wiens, and O. A. Saether. 1977. Responses to crude oil contamination by *Cricotopus* (*Cricotopus*) *bicinctus* and *C. (C.) mackenziensis* (Diptera Chironomidae) in the Fort Simpson area, Northwest Territories. *Journal of Fisheries Research Board of Canada* 34:254–261.

Rosenberg, D. M., and D. R. Barton. 1986. The Mackenzie River system. In B. R. Davies and K. F. Walker (eds.). *The ecology of river systems*, pp. 425–433. Junk, Boston.

Rosenberg, D. M., and N. B. Snow. 1975. Ecological studies of aquatic organisms in the Mackenzie and Porcupine river drainages in relation to sedimentation. Freshwater Institute, Winnipeg, Manitoba. Technical Report (Canada Fisheries and Marine Service. Research and Development Directorate); 547.

Rosenberg, D. M., and A. P. Wiens. 1976. Community and species responses of Chironomidae (Diptera) to contamination of fresh waters by crude oil and petroleum products, with special reference to the Trail River, Northwest Territories. *Journal of the Fisheries Research Board of Canada* 33:1955–1963.

Rosenberg, D. M., and A. P. Wiens. 1978. Effects of sediment addition on macrobenthic invertebrates in a northern Canadian river. *Water Research* 12:753–763.

Scrimgeour, G. J., and P. A. Chambers. 2000. Cumulative effects of pulp mill and municipal effluents on epilithic biomass and nutrient limitation in a large northern river ecosystem. *Canadian Journal of Fisheries and Aquatic Sciences* 57:1342–1354.

Scrimgeour, G. J., T. D. Prowse, J. M. Culp, and P. A. Chambers. 1994. Ecological effects of river ice break-up: A review and perspective. *Freshwater Biology* 32:261–275.

Shaw, R. D., L. R. Noton, A. M. Anderson, and G. W. Guenther. 1990. *Water quality of the Peace River in Alberta.* Alberta Environment, Edmonton.

Statistics Canada. 2003. Canadian Statistics, Geography Web site: http://www.statcan.ca/english/Pgdb/geogra.htm

Stewart, D. B., G. Low, W. E. F. Taptuna, and A. C. Day. 1997. Biological data from experimental fisheries at special harvesting areas in the Sahtu Dene and Meti settlement area, NWT. Vol. 1: The Upper Ramparts and Little Chicago areas of the Mackenzie River. *Canadian Data Report of Fisheries and Aquatic Sciences* 1020 I–VI, 1–60.

Swanson, S. M., R. Schryer, R. Shelast, P. Kloepper-Sams, and J. W. Owens. 1994. Exposure of fish to biologically treated bleached kraft mill effluent. 3: Fish habit and population assessment. *Environmental Toxicology and Chemistry* 13:1497–1507.

Tallman, R. F. 1996. Synthesis of fish distribution, movements, critical habitat and food web for the lower Slave River north of the 60th parallel: A food chain perspective. Synthesis Report no. 13. Northern River Basins Study, Edmonton, Alberta.

Wiens, A. P., D. M. Rosenberg, and N. B. Snow. 1975. Species list of aquatic plants and animals collected from the Mackenzie and Porcupine River watersheds from 1971–1973. Freshwater Institute, Winnipeg, Manitoba. Technical Report (Canda. Fisheries and Marine Service. Research and Development Directorate); 547.

Williams, T. G., W. L. Lockhars, D. A. Metner, and S. Harbicht. 1997. Baseline studies in the Slave River, NWT, 1990–1994. Part III: MFO enzyme activity in fish. *Science of the Total Environment* 197:87–109.

Wrona, F. J., J. C. Carey, B. Brownlee, and F. E. R. McCauley. 2000. Contaminant sources, distribution and fate in the Athabasca, Peace and Slave river basins, Canada. *Journal of Aquatic Ecosystem Stress and Recovery* 8:39–51.

Young, G. 1980. *The Fort Nelson story.* Gerry Young, Fort Nelson, British Columbia.

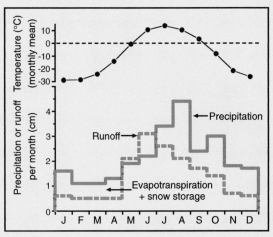

FIGURE 18.10 Mean monthly air temperature, precipitation, and runoff for the Mackenzie River basin.

MACKENZIE RIVER

Relief: 3620 m
Basin area: 1,787,000 km²
Mean discharge: 9020 m³/s
River order: 9
Mean annual precipitation: 25.8 cm
Mean air temperature: −9.5°C
Mean water temperature: 5.3°C
Physiographic provinces: Mackenzie Mountains (MM), Coast Mountains of British Columbia and Southeast Alaska (PM), Rocky Mountains in Canada (RM), Great Plains (GP), Athabasca Plains (AT), Bear–Slave–Churchill Uplands (BC), Arctic Lowlands (AL)
Biomes: Tundra, Boreal Forest, Temperate Mountain Forest
Freshwater ecoregions: Upper Mackenzie, Lower Mackenzie
Terrestrial ecoregions: 15 ecoregions (see text); main stem flows primarily through Northwest Territories Taiga, Muskwa/Slave Lake Forests
Number of fish species: 34 (main stem), 52 (basin)
Number of endangered species: none (main stem)
Major fishes: Arctic lamprey, goldeye, Arctic cisco, lake cisco, Arctic char, least cisco, lake whitefish, broad whitefish, mountain whitefish, pond smelt, rainbow smelt, lake chub, flathead chub, longnose dace, inconnu, Arctic grayling, lake trout, northern pike, longnose suckers, white sucker, troutperch, burbot, slimy sculpin, spoonhead sculpin, walleye
Major other aquatic vertebrates: wood frog, beluga whale (Mackenzie Delta), muskrat, moose, mink, beaver, river otter, snow goose, black brant, greater white-fronted goose, tundra swan, sandhill crane, glaucous gull, Arctic tern, canvasback duck
Major benthic invertebrates: true flies (Ceratopogonidae, Chironomidae, Simuliidae), mayflies (*Ametropus, Baetis, Ephemerella, Heptagenia*), stoneflies (*Isoperla*), caddisflies (*Brachycentrus*)
Nonnative species: none
Major riparian plants: horsetail, bulrush, cattail, Labrador tea, willow, sedges, balsam poplar, white spruce, black spruce, alder, birch, tamarack; reverse delta on Mackenzie supports grasses
Special features: largest river in Canada; Limestone Canyon of the Ramparts, Mackenzie Delta, Great Bear Lake, Great Slave Lake
Fragmentation: none on main stem
Water quality: pH = 7.9, conductivity = 269 μS/cm, alkalinity = 93 mg/L as CaCO₃, total N = 0.37 mg/L, total P = 0.13 mg/L
Land use (main stem): 36.5% taiga, 27.3% open land, 23.5% forest, 8.7% shrubland, 4.0% water
Population density: <1 person/km²
Major information sources: Barton 1986; Bodaly et al. 1989; Brunskill 1986; MRBC 1981a, 1981b; McCart 1986; Rosenberg and Barton 1986

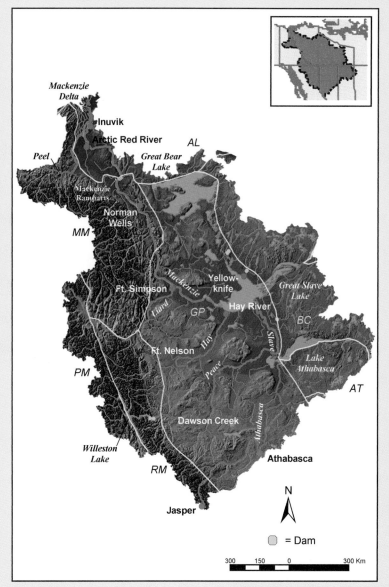

FIGURE 18.9 Map of the Mackenzie River basin. Physiographic provinces are separated by yellow lines.

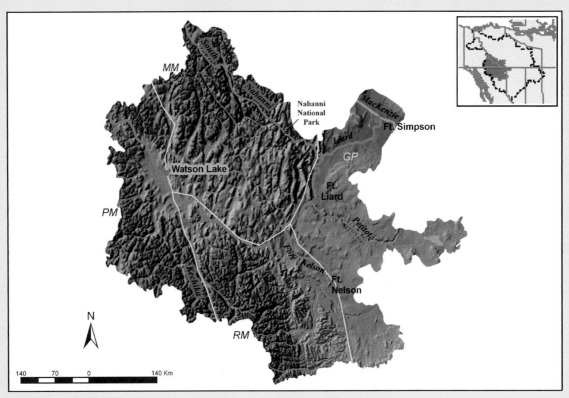

FIGURE 18.11 Map of the Liard River basin. Physiographic provinces are separated by yellow lines.

LIARD RIVER

Relief: 2573 m
Basin area: 277,000 km^2
Mean discharge: 2446 m^3/s
River order: 8
Mean annual precipitation: 36 cm
Mean air temperature: −3.7°C
Mean water temperature: 6.1°C
Physiographic provinces: Coast Mountains of British Columbia and Southeast Alaska (PM), Rocky Mountains in Canada (RM), Great Plains (GP), Mackenzie Mountains (MM)
Biomes: Boreal Forest, Temperate Mountain Forest
Freshwater ecoregion: Lower Mackenzie
Terrestrial ecoregions: Alberta/British Columbia Foothills Forests, Northern Cordillera Forests, Muskwa/Slave Lake Forests, Brooks/British Range Tundra, Arctic Coastal Tundra
Number of fish species: 34
Number of endangered species: 1 mollusk

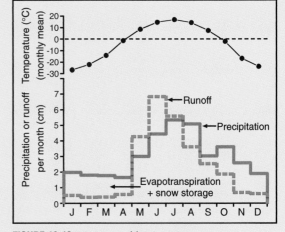

FIGURE 18.12 Mean monthly air temperature, precipitation, and runoff for the Liard River basin.

Major fishes: chum salmon, Arctic cisco, mountain whitefish, lake whitefish, inconnu, bull trout, longnose sucker, white sucker, Arctic grayling, goldeye, northern pike, walleye, burbot, lake chub, flathead chub, northern squawfish, longnose dace, troutperch, slimy sculpin, spoonhead sculpin
Major other aquatic vertebrates: wood frog, moose, beaver, river otter, muskrat, mink, trumpeter swan (nesting population at Yohin Lake), common loon, red-necked grebe, common goldeneye, bald eagle
Major benthic invertebrates: true flies (Diamesinae, Empididae, Orthocladiinae, Simuliidae, Tanyderidae, Tipulidae), mayflies (Baetidae, Ephemerellidae, Heptageniidae), stoneflies (Capniidae, Chloroperlidae, Nemouridae, Perlidae), caddisflies (Brachycentridae, Limnephilidae, Rhyacophilidae)
Nonnative species: none
Major riparian plants: black spruce, white spruce, trembling aspen, balsam poplar, aspen, lodgepole pine, willow, sedges
Special features: Grand Canyon of the Liard, Liard Hot Springs, Hell Gate Rapids, Virginia Falls, Nahanni National Park
Fragmentation: none
Water quality: l color = turbid, pH = 8.0, conductivity 287 μS/cm, alkalinity = 125 mg/L as CaCO$_3$, total N = 0.43 mg/L, total P = 0.11 mg/L
Land use: 55.4% forest, 17.8% taiga, 13.8% open land, 11.5% shrubland, 1.4% water
Population density: <1 person/km^2
Major information sources: Environment Canada 1999, 2002a, 2002b; MacDonald 1992; MRBC 1981a, 1981b; Rosenberg and Barton 1986; http://wlapwww.gov.bc.ca/bcparks/explore/parkpgs/liard_hs/nat_cul.htm 2004; www.statcan.ca

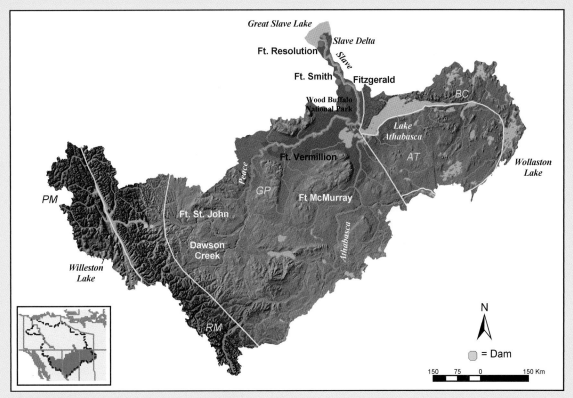

FIGURE 18.13 Map of the Slave River basin. Physiographic provinces are separated by yellow lines.

SLAVE RIVER

Relief: 3500 m
Basin area: 615,000 km²
Mean discharge: 3437 m³/s
River order: 8
Mean annual precipitation: 35.3 cm
Mean air temperature: −3.0°C
Mean water temperature: NA
Physiographic provinces: Coast Mountains of British Columbia and Southeast Alaska (PM), Rocky Mountains in Canada (RM), Great Plains (GP), Athabasca Plain (AT), Bear–Slave–Churchill Uplands (BC)
Biomes: Boreal Forest, Temperate Mountain Forest
Freshwater ecoregion: Upper Mackenzie
Terrestrial ecoregions: Northern Canadian Shield Taiga, Mid-continental Canadian Forests
Number of fish species: 28 (main stem), 45 (basin)
Number of endangered species: none

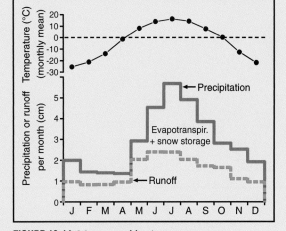

FIGURE 18.14 Mean monthly air temperature, precipitation, and runoff for the Slave River basin.

Major fishes: Arctic lamprey, goldeye, northern pike, Dolly Varden, lake cisco, lake trout, longnose sucker, white sucker, troutperch, lake whitefish, walleye, inconnu, flathead chub, spottail shiner, pearl dace, burbot, ninespine stickleback, slimy sculpin, yellow perch
Major other aquatic vertebrates: chorus frog, wood frog, northern leopard frog, beaver, muskrat, river otter, mink, whooping crane, white pelican, snow goose, Canada goose, common loon, red-necked grebe, common merganser, osprey, bald eagle
Major benthic invertebrates: chironomid midges (*Stictochironomus quagga*, *Procladius sp.*, *Chironomus anthracinius*, *Polypedilum scalaenum*), oligochaete worms (*Limnodrilus hoffmeisteri*, *Chaetogaster diaphanous*), caddisflies, true bugs (*Corixidae*), beetles (*Dytiscidae*), mayflies, stoneflies
Nonnative species: NA
Major riparian plants: white spruce, black spruce, tamarack, balsam poplar, aspen, Labrador tea, reindeer lichens, peat mosses, sago weed, pickerelweed, river horsetail, sedges, cattails, reed grasses, rushes, burrweed, willow, alder
Special features: Wood Buffalo National Park (whooping crane breeding sites), Slave River Delta, Peace–Athabasca Delta
Fragmentation: submerged weir on the Riviere des Rochers at Little Rapids and Revillion Coupe; major dam on (Peace River)
Water quality: pH = 7.8, conductivity = 253 μS/cm, alkalinity = 82 mg/L as CaCO₃, total N = 0.40 mg/L, total P = 0.21 mg/L
Land use: 40.3% forest, 25.8% taiga, 20.9% shrubland, 9.9% water, 1.6% agriculture, 1.4% open land
Population density: <1 person/km²
Major information sources: Alberta Environment 1987, MacDonald 1994, MRBC 1981b, McCarthy, Robertson et al. 1997, McCarthy, Stephens et al. 1997, McCarthy, Williams et al. 1997, Prowse and Conly 1996, Rosenberg and Barton 1986, Tallman 1996

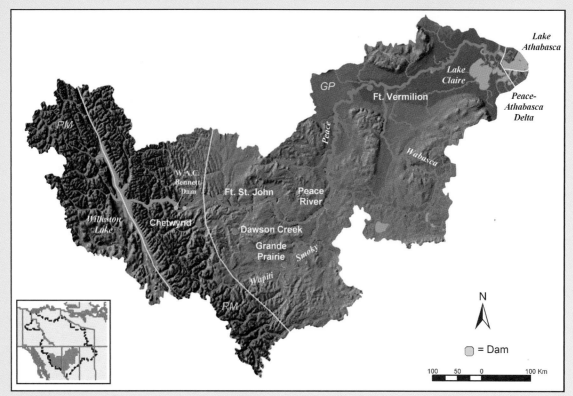

FIGURE 18.15 Map of the Peace River basin. Physiographic provinces are separated by yellow lines.

PEACE RIVER

Relief: 2130 m

Basin area: 293,000 km^2

Mean discharge: 2118 m^3/s

River order: 8

Mean annual precipitation: 31 cm

Mean air temperature: −0.9°C

Mean water temperature: NA

Physiographic provinces: Great Plains (GP), Rocky Mountains in Canada (RM), Coast Mountains of British Columbia and Southeast Alaska (PM)

Biomes: Boreal Forest, Temperate Mountain Forest

Freshwater ecoregion: Upper Mackenzie

Terrestrial ecoregions: Central British Columbia Mountain Forests, Alberta/British Columbia Foothills Forests, Canadian Aspen Forest and Parklands, Muskwa/Slave Lake Forests, Mid-continental Canadian Forests, Northern Cordillera Forests, Fraser Plateau and Basin Complex, Alberta Mountain Forests

Number of fish species: 31

Number of endangered species: none

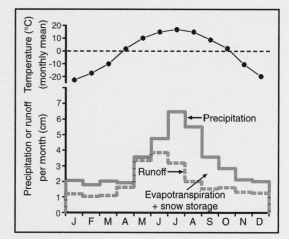

FIGURE 18.16 Mean monthly air temperature, precipitation, and runoff for the Peace River basin.

Major fishes: goldeye, walleye, burbot, northern pike, rainbow trout, Dolly Varden, lake trout, mountain whitefish, lake whitefish, lake chub, flathead chub, northern squawfish, longnose dace, trout perch, prickly sculpin, spoonhead sculpin

Major other aquatic vertebrates: long toed salamander, spotted frog, wood frog, beaver, muskrat, river otter, mink, snow goose, Canada goose, trumpeter swan, common loon, red-necked grebe, common merganser, osprey, bald eagle

Major benthic invertebrates: chironomid midges (Orthocladiinae, Tanypodinae, Tanytarsini), oligochaete worms (Enchytraeidae, Naididae), stoneflies (*Isoperla, Isogenoides*, Capniidae, Taeniopterygidae), mayflies (*Baetis, Ephemerella, Heptagenia, Isonychia, Rhithrogena*)

Nonnative species: spottail shiner, fathead minnow, westslope cutthroat trout, brook trout

Major riparian plants: black spruce, trembling aspen, balsam poplar, willow, sedges, wheat grass, sedge, horsetail, Labrador tea

Special features: Vermilion Chutes, Boyer Rapids, Peace–Athabasca Delta

Fragmentation: W.A.C. Bennett Dam at Williston Lake on main stem

Water quality: pH = 7.9, conductivity = 257 µS/cm, alkalinity = 96 mg/L as CaCO$_3$, total N = 0.52 mg/L, total P = 0.16 mg/L

Land use: 57.4% forest, 16.2% agriculture, 12.5% taiga, 11.4% shrubland, 2.1% water, 0.5% open land

Population density: <1 person/km^2

Major information sources: Boag 1993, Cash et al 2000, Culp et al. 2000a, Chambers, Brown et al. 2000, MRBC 1981b, Prowse and Conly 1996, Prowse and Conly 2000, Rosenberg and Barton 1986, Shaw et al. 1990, Wrona et al. 2000

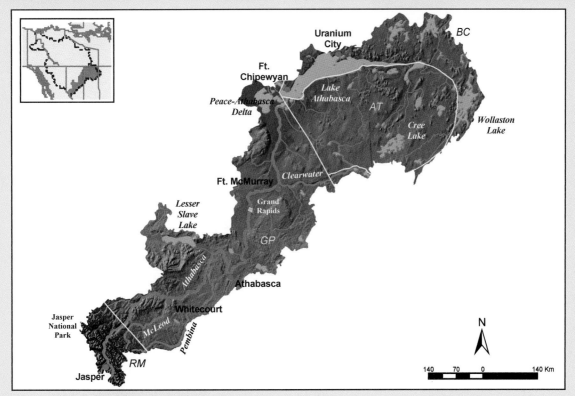

FIGURE 18.17 Map of the Athabasca River basin. Physiographic provinces are separated by yellow lines.

ATHABASCA RIVER

Relief: 3420 m
Basin area: 154,880 km²
Mean discharge: 783 m³/s
River order: 6
Mean annual precipitation: 46.5 cm
Mean air temperature: 0.2°C
Mean water temperature: 8.2°C
Physiographic provinces: Athabasca Plain (AT), Bear–Slave–Churchill Uplands (BC), Great Plains (GP), Rocky Mountains in Canada (RM)
Biomes: Boreal Forest, Temperate Mountain Forest
Freshwater ecoregion: Upper Mackenzie
Terrestrial ecoregions: Alberta/British Columbia Foothills Forests, Alberta Mountain Forests Canadian Aspen Forest and Parklands, Northern Canadian Shield Taiga, Mid-continental Canadian Forests, Midwestern Canadian Shield Forests
Number of fish species: 36
Number of endangered species: none

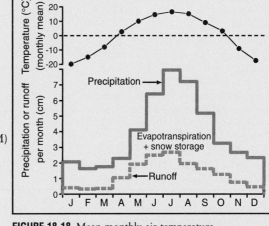

FIGURE 18.18 Mean monthly air temperature, precipitation, and runoff for the Athabasca River basin.

Major fishes: lake cisco, lake whitefish, round whitefish, mountain whitefish, rainbow trout, Dolly Varden, walleye, lake trout, goldeye, northern pike, burbot, Arctic grayling, longnose sucker, white sucker, flathead chub, longnose dace, troutperch, slimy sculpin, brook stickleback, lake chub
Major other aquatic vertebrates: boreal chorus frog, western toad, wood frog, northern leopard frog, long toed salamander, mink, river otter, moose, muskrat, beaver, greater yellowlegs (sandpiper), killdeer, sandhill crane, bald eagle, osprey, American white pelican, double-crested cormorant, red-necked grebe, great blue heron, common loon, Canada goose, tundra swan
Major benthic invertebrates: chironomid midges (Chironomini, Orthocladiinae, Tanypodinae), mayflies (*Ameletus, Baetis, Cinygmula, Heptagenia, Ephemerella, Rhithrogena*), stoneflies (*Alloperla, Capnia, Isoperla, Hesperoperla, Pteronarcella, Taenionema*), caddisflies (*Brachycentrus, Glossosoma, Hydropsyche*)
Nonnative species: brown trout, brook trout
Major riparian plants: white spruce, balsam poplar, tamarack, black spruce, aspen willow, water sedge, marsh reed grass, Labrador tea, strawberry-blite, Ross' sedge, reindeer lichen, peat moss, brown moss
Special features: Athabasca Falls, Grand Rapids, Peace–Athabasca Delta, Canadian Heritage River
Fragmentation: none
Water quality: TDS = 187 mg/L, conductivity = 332 μS/cm, pH = 7.8, alkalinity = 117 mg/L as $CaCO_3$, total N = 0.54 mg/L, total P = 0.10 mg/L
Land use: 59.8% forest, 17.4% taiga, 9.8% agriculture, 6.0% shrubland, 4.3% open land, 2.7% water
Population density: <1 person/km²
Major information sources: Cash et al. 2000; Culp et al. 2000a; Chambers, Brown et al. 2000; Chambers, Dale, and Scrimgeour 2000; Environment Canada 1999; MRBC 1981b; Rosenberg and Barton 1986; Wrona et al. 2000; www.scienceoutreach.ab.ca 2003; www.climate.weatheroffice.ec.gc.ca/rel_arch/envirodat/index_e.html 2002

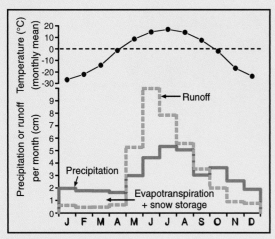

FIGURE 18.20 Mean monthly air temperature, precipitation, and runoff for the South Nahanni River basin.

SOUTH NAHANNI RIVER

Relief: 2573 m
Basin area: 33,388 km^2
Mean discharge: 404 m^3/s
River order: 6
Mean annual precipitation: 36 cm
Mean air temperature: −3.7°C
Mean water temperature: NA
Physiographic provinces: Mackenzie Mountains (MM), Great Plains (GP)
Biome: Boreal Forest
Freshwater ecoregion: Lower Mackenzie
Terrestrial ecoregions: Northwest Territories Taiga, Northern Cordillera Forests, Ogilvie/Mackenzie Alpine Tundra
Number of fish species: 13
Number of endangered species: none
Major fishes: Arctic grayling, Dolly Varden, lake trout, northern pike, lake whitefish, longnose sucker, round whitefish, inconnu
Major other aquatic vertebrates: wood frog, muskrat, beaver, mink, river otter, moose, trumpeter swan, bald eagle, golden eagle, common loon, red-necked grebe
Major benthic invertebrates: NA
Nonnative species: NA
Major riparian plants: white spruce, poplar, black spruce, sedges, wild mint, goldenrod, yellow monkey flower
Special features: Virginia Falls, Rabbitkettle Hotsprings, Sandblowouts, Nahanni National Park, Canadian Heritage River
Fragmentation: none
Water quality: pH = 8.0, conductivity = 298 µS/cm, alkalinity = 105 mg/L as CaCO$_3$, total N = 0.45 mg/L, total P = 0.11 mg/L
Land use: 44.2% open land, 27.2% taiga, 19.1% forest, 8.4% shrubland, 1.1% water
Population density: <1 person/km^2
Major information sources: Environment Canada 1999, www.nsttravel.nt.ca/html/nahriver.htm, MacDonald 1992, Rosenberg and Barton 1986, www.statcan.ca/english/pgdb/geogra.html 1996, www.unep-wcmc.org/sites/wh/nahanni.html 2002, www.climate.weatheroffice.ec.gc.ca/rel_arch/envivodat/index_e.html 2002, www.explorenwt.com/resources/northern-library/PDF/nahriver.pdf

FIGURE 18.19 Map of the South Nahanni River basin. Physiographic provinces are separated by a yellow line.

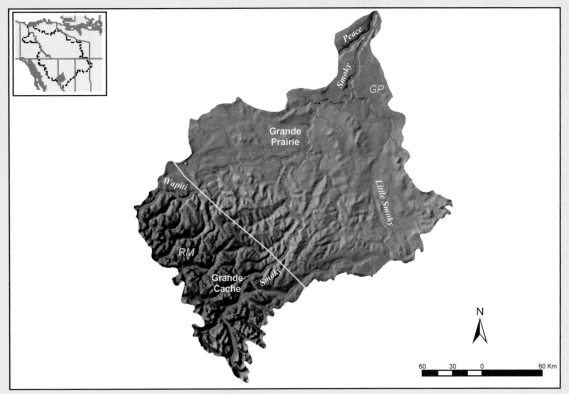

FIGURE 18.21 Map of the Smoky River basin. Physiographic provinces are separated by a yellow line.

SMOKY RIVER

Relief: 2605 m

Basin area: 49,584 km²

Mean discharge: 347 m³/s

River order: 7

Mean annual precipitation: 45 cm

Mean air temperature: 1.6°C

Mean water temperature: NA

Physiographic provinces: Rocky Mountains in Canada (RM), Great Plains (GP)

Biomes: Boreal Forest, Temperate Mountain Forest

Freshwater ecoregion: Upper Mackenzie

Terrestrial ecoregions: Alberta/British Columbia Foothills Forests, Alberta Mountain Forests, Canadian Aspen Forest and Parklands

Number of fish species: 14

Number of endangered species: none

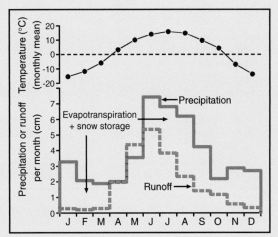

FIGURE 18.22 Mean monthly air temperature, precipitation, and runoff for the Smoky River basin.

Major fishes: longnose sucker, white sucker, walleye, northern pike, mountain whitefish, Arctic grayling, Dolly Varden, lake chub, longnose dace, redside shiner, pearl dace, burbot, slimy sculpin, spoonhead sculpin

Major other aquatic vertebrates: wood frog, moose, beaver, muskrat, mink, river otter, American white pelican, great blue heron, trumpeter swan, tundra swan, snow goose, common loon, red-necked grebe, common goldeneye

Major benthic invertebrates: chironomid midges (Chironomini, Orthocladiinae, Tanypodinae), mayflies (*Baetis*, *Cinygmula*, *Drunella*, *Rhithrogena*), stoneflies (*Alloperla*, *Capnia*, *Isoperla*, *Hesperoperla*, *Pteronarcella*, *Taenionema*), caddisflies (*Brachycentrus*, *Glossosoma*, *Hydropsyche*)

Nonnative species: NA

Major riparian plants: trembling aspen, balsam poplar, black spruce, willow, sedges, wheat grass, sedge, Labrador tea, peat moss, brown moss

Special features: Willmore Wilderness Park, dinosaur tracks (Grande Cache area)

Fragmentation: none

Water quality: pH = 8.0, alkalinity = 132 mg/L as CaCO₃, total N = 0.46 mg/L, total P = 0.13 mg/L

Land use: 63.2% forest, 22.4% agriculture, 5.1% shrubland, 4.4% open land, 4.2% taiga, 0.5% water

Population density: <1 person/km²

Major information sources: Chambers, Brown et al. 2000; Chambers, Dale, and Scrimgeour 2000; Golder Associates 2000; MRBC 1981b; Rosenberg et al. 1986; www.statcan.ca/english/pgdb/geogra.htm 1996

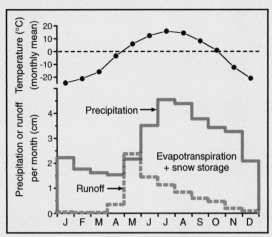

FIGURE 18.24 Mean monthly air temperature, precipitation, and runoff for the Hay River basin.

HAY RIVER

Relief: 1060 m
Basin area: 47,900 km²
Mean discharge: 113 m³/s
River order: 6
Mean annual precipitation: 34.2 cm
Mean air temperature: −3.4°C
Mean water temperature: 6.4°C
Physiographic province: Great Plains (GP)
Biome: Boreal Forest
Freshwater ecoregion: Lower Mackenzie
Terrestrial ecoregions: Alberta/British Columbia
  Foothills Forests, Muskwa/Slave Lake Forests
Number of fish species: ~18
Number of endangered species: 0
Major fishes: lake chub, flathead chub, longnose dace,
  longnose sucker, white sucker, northern pike, lake
  whitefish, mountain whitefish, Arctic grayling,
  Arctic char, lake trout, walleye, burbot, brook
  stickleback, slimy sculpin, spoonhead sculpin
Major other aquatic vertebrates: wood frog, moose,
  beaver, river otter, muskrat, mink, snow goose, Canada goose, osprey, bald eagle, Arctic tern, herring gull, northern pintail
Major benthic invertebrates: NA
Nonnative species: NA
Major riparian plants: balsam poplar, black spruce, white spruce, tamarack, dwarf birch, willows, sedges, Labrador tea, peat
  moss, brown moss
Special features: Zama and Hay Lakes Wetland area (IBA), Hay River Gorge, Twin Falls; Hay River (community) is
  northernmost railroad accessible area in Northwest Territories and important hub for barge travel along Mackenzie River
Fragmentation: none
Water quality: pH = 7.8, conductivity = 456 µS/cm, alkalinity = 151/L as mg CaCO₃, total N = 1.01 mg/L, total P = 0.09 mg/L
Land use: NA; for Great Slave subbasin, 35.2% taiga, 23.1% forest, 17.9% water, 17.3% open land, 6.2% shrubland, 0.4%
  agriculture
Population density: <1 person/km²
Major information sources: www.cd.gov.ab.ca/preserring/parks/anhic/wetlandmixedwood.asp 2001, MRBC 1981b, Rosenberg
  and Barton 1986, www.gov.ab.ca/env/parks/anhic/esa.html 1997

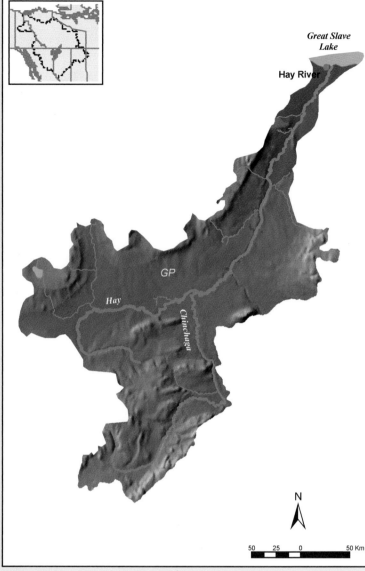

FIGURE 18.23 Map of the Hay River basin.

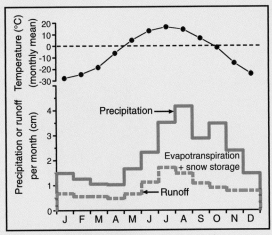

FIGURE 18.26 Mean monthly air temperature, precipitation, and runoff for the Yellowknife River basin.

YELLOWKNIFE RIVER

Relief: 706 m
Basin area: 11,300 km²
Mean discharge: 39 m³/s
River order: 5 (approximated)
Mean annual precipitation: 26.7 cm
Mean air temperature: −5.1°C
Mean water temperature: NA
Physiographic provinces: Great Plains (GP),
     Bear–Slave–Churchill Uplands (BC)
Biome: Boreal Forest
Freshwater ecoregion: Lower Mackenzie
Terrestrial ecoregion: Northern Canadian Shield Taiga
Number of fish species: NA
Number of endangered species: NA
Major fishes: northern pike, Arctic grayling, lake trout
Major other aquatic vertebrates: moose, beaver,
     muskrat, river otter, bald eagle, red-breasted
     merganser, common loon, northern pintail
Major benthic invertebrates: NA
Nonnative species: none
Major riparian plants: black spruce, birch, poplar,
     willow, alder, Labrador tea, cranberry, bunchberry, bearberry
Special features: wilderness river draining northern region of Canadian Shield; small rivers connected by many lakes; wintering
     grounds for Bathurst caribou herd (north section of basin); gold mining
Fragmentation: one dam near mouth
Water quality: pH = 7.2, alkalinity = 16 mg/L as CaCO₃, total N = 0.22 mg/L, total P = 0.01 mg/L
Land use: NA; for Great Slave subbasin, 35.2% taiga, 23.1% forest, 17.9% water, 17.3% open land, 6.2% shrubland, 0.4%
     agriculture
Population density: <1 person/m²
Major information sources: Environment Canada 1999, Rosenberg and Barton 1986, www.climate.weatheroffice.ec.gc.ca/
     rel_arch/envivodat/index_e.html 2002, www.explorenwt.com/resources/northern-library/PDF/yriver.pdf

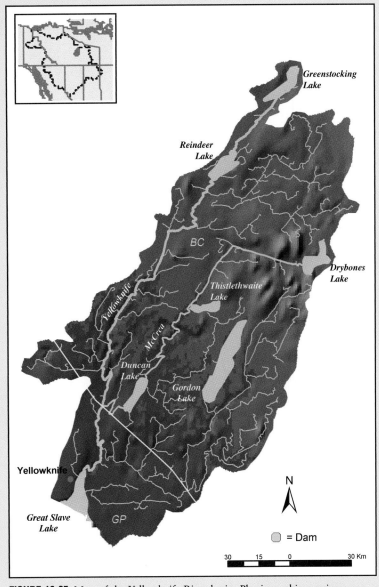

FIGURE 18.25 Map of the Yellowknife River basin. Physiographic provinces are separated by a yellow line.

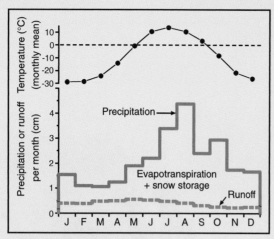

FIGURE 18.28 Mean monthly air temperature, precipitation, and runoff for the Peel River basin.

PEEL RIVER

Relief: 2650 m
Basin area: 68,134 km²
Mean discharge: 103 m³/s
River order: 6
Mean annual precipitation: 25.8 cm
Mean air temperature: −9.5°C
Mean water temperature: NA
Physiographic provinces: Arctic Lowlands (AL),
    Mackenzie Mountains (MM)
Biomes: Tundra, Boreal Forest
Freshwater ecoregion: Lower Mackenzie
Terrestrial ecoregions: Interior Alaska/Yukon Lowland
    Taiga, Northwest Territories Taiga,
    Ogilvie/Mackenzie Alpine Tundra, Brooks/British
    Range Tundra
Number of fish species: 17
Number of endangered species: none
Major fishes: northern pike, mountain whitefish,
    round whitefish, Arctic grayling, Arctic char,
    Dolly Varden, longnose sucker, cyprinids,
    ciscos, sculpins

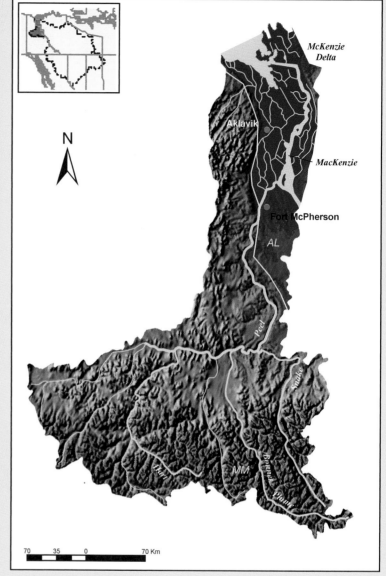

FIGURE 18.27 Map of the Peel River basin. Physiographic provinces are separated by a yellow line.

Major other aquatic vertebrates: moose, muskrat, beaver, river otter, mink, peregrine falcon, gyrfalcon, bald eagle, common
    loon, tundra swan, white-winged scoter
Major benthic invertebrates: NA
Nonnative species: NA
Major riparian plants: black spruce, white spruce, balsam poplar, larch, alder, willow
Special features: most northerly Mackenzie River tributary; several major tributaries known for their ruggedness and beauty,
    including Bonnet Plume (Canadian Heritage River), Hart, and Snake rivers; Mount MacDonald
Fragmentation: none
Water quality: pH = 7.9, conductivity = 314 μS/cm, alkalinity = 127 mg/L as CaCO₃, total N = 0.52 mg/L, total P = 0.17 mg/L
Land use: 43.7% open land, 29.1% shrubland, 18.8 % taiga, 7.5% forest, 0.9% water
Population density: <1 person/km²
Major information sources: Environment Canada 1999; www.nwttravel.nt.ca/htmlpeelriver.htm; MacDonald 1992, 1994;
    MRBC 1981b; Rosenberg and Barton 1986; www.statcan.ca/english/pgdb/geogra.htm 1996;
    www.climate.weatheroffice.ec.gc.ca/rel_arch/envivodat/index_e.html 2002; www.explorenwt.com/resources/northern-
    library/PDF/peelriver.pdf

19

NELSON AND CHURCHILL RIVER BASINS

DAVID M. ROSENBERG   PATRICIA A. CHAMBERS   JOSEPH M. CULP
WILLIAM G. FRANZIN   PATRICK A. NELSON   ALEX G. SALKI
MICHAEL P. STAINTON   R. A. BODALY   ROBERT W. NEWBURY

INTRODUCTION

SASKATCHEWAN RIVER

RED RIVER OF THE NORTH–ASSINIBOINE RIVER

WINNIPEG RIVER

NELSON RIVER MAIN STEM

ADDITIONAL RIVERS

ACKNOWLEDGMENTS

LITERATURE CITED

INTRODUCTION

Two great Canadian rivers, the Nelson and the Churchill, drain waters mainly from the interior of Canada, cut through the Canadian Shield of northern Manitoba, and empty into Hudson Bay (Newbury and Malaher 1972) (Fig. 19.2). The rivers flow through a "valley" in northern Manitoba that has been eroded by numerous glaciations along the boundary between two major geological zones, the Superior and Churchill provinces of the Canadian Shield (Newbury 1990b). This valley allows the western part of the Churchill, the Saskatchewan, the Red–Assiniboine, and the Winnipeg river systems to collect in Manitoba and flow through the Nelson and Churchill channels into Hudson Bay.

Waters of the Nelson system (latitude ~45.5°N to 57°N, longitude ~90°W to 117.5°W) begin their journey from the west on the eastern slopes of the Rocky Mountains and cross the three Canadian Prairie provinces (Alberta, Saskatchewan, and Manitoba) via the Saskatchewan River, which empties into the northern part of Lake Winnipeg. Waters from the east originate in northwestern Ontario and

flow via the Winnipeg River into the southern part of Lake Winnipeg. Waters flowing from the south drain parts of Minnesota and North Dakota via the Red River of the North and also empty into the southern part of Lake Winnipeg. The Assiniboine River flows from the Canadian west into the Red River. The Nelson River proper originates at the outflow of Lake Winnipeg and carries its continental collection of water to the sea. In total, the Nelson system covers >1,000,000 km² of the interior of North America, 892,300 km² in Canada and 180,000 km² in the United States (*National Atlas of Canada* 1985). The total distance from the mountain headwaters of the Bow River to Hudson Bay exceeds 3000 km (Lane and Sykes 1982).

The Churchill system (latitude 53°N to 59.5°N, longitude 94.5°W to 112.5°W) is much smaller than the Nelson (catchment area 281,300 km²; *National Atlas of Canada* 1985) and is contained entirely within Canada. The headwaters of the Churchill begin in Beaver Lake, northeast of Edmonton, Alberta, and, to the north, in the Wollaston Lake area of northeastern Saskatchewan. Churchill waters flow mainly through Saskatchewan and Manitoba. From

FIGURE 19.1 Bow River downstream of Lake Louise (PHOTO BY P. CHAMBERS)

© 2005, Elsevier Inc. and Her Majesty the Queen in right of Canada. All rights reserved.

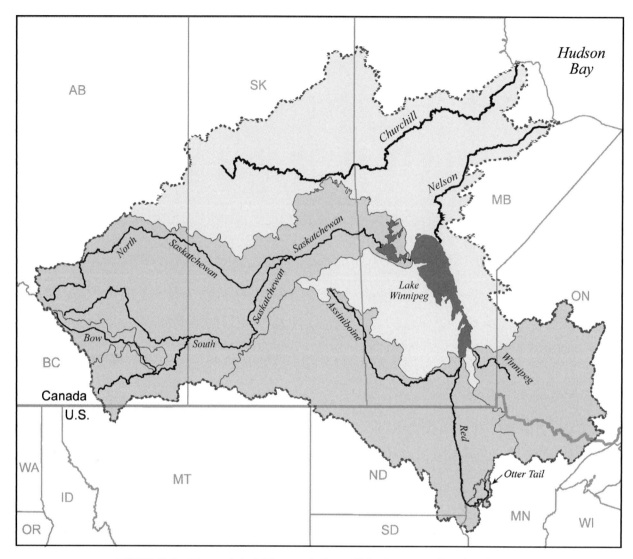

FIGURE 19.2 Rivers of the Nelson and Churchill basins covered in this chapter.

the Saskatchewan–Manitoba border the river flows northeastward through a complex chain of lakes, roughly paralleling the Nelson system to the south, and into Hudson Bay. The Churchill River is only briefly considered here because it was joined to the Nelson by a diversion in 1976.

The two river systems are separated by several hundred kilometers through most of their length, but Southern Indian Lake (Churchill catchment) and the Rat–Burntwood River (Nelson catchment) come close together in northern Manitoba. The possibility of diverting Churchill River flows into the Nelson catchment was first considered in the mid 1950s and became reality in the mid 1970s by connecting Southern Indian Lake with the Rat–Burntwood River (Hecky et al. 1984). Seventy-five percent of the

Churchill River flow was diverted into the lower Nelson River valley to augment flows for long-term hydropower development in the lower Nelson (Newbury et al. 1984), effectively joining the two rivers for their final run into Hudson Bay (Fig. 19.3, Fig. 19.4).

The Nelson and Churchill rivers have a 400 year history of European exploration and the opening of the interior of Canada to the fur trade. European explorers searching for the Northwest Passage to India had mapped the estuaries of the Seal, Churchill, Nelson, and Hayes rivers on Hudson Bay by the early 1600s (Newbury 1990b). Henry Hudson was abandoned somewhere on Hudson Bay by a mutinous crew in 1611 (Newbury 1990a). The following year Sir Thomas Button was sent to search for Hudson

FIGURE 19.3 Lower Churchill River between Southern Indian Lake and Hudson Bay before diversion into the Nelson River system (PHOTO BY A. P. WIENS).

and overwintered locked in the ice of Hudson Bay. Button built a land-based camp at the mouth of a large river, which he named "Nelson" after his sailing master, Frances Nelson of the HMS Resolution. By 1668 the French explorers Radisson and Grossilliers were trading furs at the mouth of the Nelson. The Hudson's Bay Company (still today a major Canadian retailer) was formed in 1670 and established trading depots at the mouths of all the major rivers. The Churchill River was named after John Churchill, a governor of the Hudson's Bay Company from the 1680s to the 1690s who would later become the Duke of Marlborough. In 1690, Henry Kelsey discovered that the river leading to the York Factory depot (the Hayes River) could be used to gain access to the interior of the continent. However, the actual route to the interior was unknown to Europeans until 1745, when David Thompson discovered the secret (see Newbury 1990b for the explanation). Passage from the docks of England through Hudson Bay to the western plains and mountains of North America was established, and York Factory became central to

this trading activity for the next 150 years. The "York Boat" route brought the first traders, and later settlers, to western Canada.

The Nelson River catchment is comprised of 11 major subcatchments (Lane and Sykes 1982). We will discuss four of the largest: (1) the Saskatchewan (334,100 km^2 in Canada and 1800 km^2 in the United States for a total 335,900 km^2; *National Atlas of Canada* 1985), which drains the western part of the catchment; (2) the Red–Assiniboine (138,600 km^2 in Canada and 148,900 km^2 in the United States for a total of 287,500 km^2; *National Atlas of Canada* 1985), which drains the western and southern parts of the catchment; (3) the Winnipeg (106,500 km^2 in Canada and 29,300 km^2 in the United States for a total of 135,800 km^2; *National Atlas of Canada* 1985), which drains the eastern part of the catchment; and (4) the main-stem Nelson (89,000 km^2, all in Canada; Lane and Sykes 1982). One-page summaries of physical and biological information are provided for two additional rivers in the Nelson system: the Bow and Otter Tail.

FIGURE 19.4   Lower Churchill River between Southern Indian Lake and Hudson Bay after diversion (~75% of flow) into the Nelson River system (PHOTO BY A. P. WIENS).

Physiography and Climate

The Nelson catchment occupies six physiographic provinces (Hunt 1974): Rocky Mountains in Canada, Great Plains, Central Lowland, Bear–Slave–Churchill Uplands, Superior Upland, and Hudson Bay Lowland. In the Hunt (1974) classification, the Great Plains and Central Lowland provinces combine to form the Interior Plains division, and the Bear–Slave–Churchill Uplands, Superior Upland, and Hudson Bay Lowland combine to form the Canadian Shield division. The Rocky Mountains in Canada province, hereafter referred to as the Rocky Mountains province, is part of the Rocky Mountain System division. The Churchill catchment occupies four of these physiographic provinces (Great Plains, Central Lowland, Bear–Slave–Churchill Uplands, and Hudson Bay Lowland; Hunt 1974).

The Rocky Mountains province includes the eastern slopes of several mountain ranges along the British Columbia–Alberta border and consists of a series of parallel mountain ridges and valleys aligned in a northwest–southeast direction (Lane and Sykes 1982). The mountain ranges (e.g., Rockies) rise >2000 m above the valley floors. The valleys have been shaped by glaciation and subsequent erosion. There is considerable precipitation in the mountains that drains eastward from this province. The soil on steep slopes is generally thin and coarse, allowing rapid infiltration and subsurface movement into streams.

The Interior Plains division (i.e., Central Lowland and Great Plains provinces; Hunt 1974) is a series of three steps, beginning with the Manitoba Plain in the east (Bear–Slave–Churchill Uplands westward to Manitoba Escarpment), extending in a second step to the Missouri Coteau, and then in a third step to the foothills (Lane and Sykes 1982). The Interior Plains in Manitoba and Saskatchewan are bordered on the north and east by the Bear–Slave–Churchill Uplands and Superior Upland (Canadian Shield division; Hunt 1974). The Interior Plains division has a general slope from southwest to northeast. Soils range from brown, dark brown, black, and dark grey soil groups in the southern areas to podzols in wetter regions. Wetter regions contain widespread organic soils in muskeg. There are extensive permafrost areas in the north.

The Canadian Shield division (Hunt 1974) is a major physiographic feature of the Churchill catchment in the northern part of the prairie provinces (Bear–Slave–Churchill Uplands) and in the eastern part of the Nelson catchment in Manitoba and Ontario (Superior Upland). Glacial erosion has left extremely hard, erosion-resistant rock throughout the region (Lane and Sykes 1982). The Shield is characterized by numerous lakes and marshy depressions and has a poorly organized drainage system. The many lakes and marshes result in a large surface-water resource.

The mouths of the Churchill and Nelson rivers lie in the Hudson Bay Lowland province of the Canadian Shield division of Hunt (1974). In this area, flat muskeg plains are dotted with shallow lakes, often interrupted by raised gravel beaches (Lane and Sykes 1982).

The region occupied by the Nelson and Churchill catchments generally experiences long, cold winters (Lane and Sykes 1982). A winter ridge of high pressure typically develops between low-pressure areas over the Gulf of Alaska, extending to the region south of Hudson Bay in the east and bringing clear cold weather to western Canada. The coldest month is usually January (mean daily air temperature of ~–10°C in the south to ~–27.5°C in the north). Higher mean temperatures in the southwestern part of the Nelson catchment in Alberta are caused by foehn winds ("chinooks"). These warm, dry winds provide respite from the severe cold, but they also reduce snow cover, which reduces the amount of soil moisture and can reduce the amount of spring runoff. The warmest month of the year is usually July (mean daily air temperatures from ~17.5°C in the south to ~15°C in the northern part of the two catchments). The frost-free period varies from ~120 days in the south to ~60 to 70 days in the northern and western fringes of widespread agricultural activity.

The mean annual precipitation for the Nelson and Churchill catchments varies from ~40 cm in northern areas to ~60 cm in the Rocky Mountains province (Lane and Sykes 1982). However, a large, dry area (30 to 40 cm/yr) exists in southeastern Alberta and southwestern Saskatchewan. The relative dryness of western Canada east of the foothills is caused by the rain-shadow effect of the Rocky Mountains. However, most of the precipitation received in the western parts of the Interior Plains region still originates from moist Pacific air masses. Southern air masses more strongly affect southern Manitoba. Another major source of precipitation in western Canada is convection storms on hot days in summer. These storm events are sometimes accompanied by hail, and the associated rainfall is usually intense but short.

Basin Landscape and Land Use

Both catchments occupy two major habitat types (Ricketts et al. 1999): Temperate Grasslands/Savanna/Shrub, which roughly corresponds to the Great Plains and Central Lowland, and Boreal Forest/Taiga, which roughly corresponds to the Bear–Slave–Churchill Uplands, Superior Upland, and the Hudson Bay Lowland. The Nelson catchment also occupies small parts of two other major habitat types: Temperate Coniferous Forests, which runs through the Rocky Mountains province, and Temperate Broadleaf and Mixed Forests in northwestern Minnesota and southwestern Ontario, through which part of the Winnipeg River subcatchment flows. Ricketts et al. (1999) further subdivide major habitat types into three ecoregions for the Churchill catchment and twelve ecoregions for the major subcatchments of the Nelson system.

Thirty-two percent of the Nelson catchment is forest (Revenga et al. 1998, www.waterandnature. org/eatlas/html/index.html 2004). Comparable data are not available for the Churchill catchment, although the proportion is almost certainly higher. The Boreal Forest/Taiga major habitat type extends in a broad band across the Nelson and Churchill catchments from Lake Winnipeg to north-central Alberta. The Temperate Coniferous Forests major habitat type is located along the extreme western edge of the Nelson catchment along the British Columbia–Alberta border. These forests support a number of commercial forestry operations (concentrated across the central part of Saskatchewan in the Churchill catchment and northeast and southeast of Lake Winnipeg in the Nelson catchment), provide important habitat for wildlife, and support considerable recreation activity (Lane and Sykes 1982). Forests in the foothills and mountains also influence the quality, quantity, and timing of water yields. The Canadian Aspen Forest and Parklands ecoregion (Temperate Grasslands/Savanna/Shrub major habitat type) to the south of the boreal forest is a mixture of woodland and grassland cover. The forests in this zone are generally too small to support forestry operations.

Forty-seven percent of the Nelson catchment is cropland (Revenga et al. 1998, Water and Nature Initiative 2004 http://www.waterandnature.org/eatlas/html/index.html). Comparable data are not available for the Churchill catchment, although the proportion is almost certainly lower. Only rich soils in the parklands are cultivated, but the soils in the Temperate Grasslands/Savanna/Shrub south of the parklands are very widely used (Lane and Sykes 1982). These grasslands comprise the largest single expanse of agricultural land in Canada. Uncultivated land supports large herds of beef cattle, but most of the best land is tilled. Brown, dark brown, and black soils predominate. Solonetzic soils are also present, but their cultivation is inhibited by salinity.

Nonrenewable resources include petroleum, especially in the Interior Plains division (under most of Alberta and smaller areas in Saskatchewan and Manitoba); coal, underlying some of the Rocky Mountains province and Interior Plains division; and potash, extending in a belt from the Alberta border through central Saskatchewan down to southwestern Manitoba (Lane and Sykes 1982). Most potash mines operate in Saskatchewan, where reserves are enormous. Metal mining is concentrated in the Canadian Shield division (Bear–Slave–Churchill Uplands) of northern Manitoba and northern Saskatchewan (Lane and Sykes 1982).

Water use in the Nelson and Churchill catchments can be categorized as either offstream use, which removes water from streams or lakes and returns the unused portion (e.g., irrigation, industrial, and municipal use), or instream use, which uses water directly in the stream or lake (e.g., fishing, recreation, hydropower production) (Lane and Sykes 1982). Instream water uses, dependent on reliable annual flows, are characteristic of northern parts of the study area. Offstream water uses, dependent on the total water available for use, are characteristic of southern parts of the study area. These two types of water use are discussed in more detail later in the chapter.

Subsistence hunting, fishing, and trapping, mainly by aboriginal people, are closely related to water quality and quantity (Lane and Sykes 1982). These activities are mainly pursued in more northern areas of the Nelson and Churchill catchments and often are in conflict with activities such as hydro development (e.g., Waldram 1988, Usher and Weinstein 1991). The proposed Wintego Dam on the Churchill River in northeastern Saskatchewan (Churchill River Study 1976) was successfully opposed by aboriginal peoples.

The wildlife of the Nelson and Churchill catchments, especially in the Interior Plains division, has been changed over the past 200 years by extensive hunting and trapping, and more recently by activities such as agriculture, forestry, urban and industrial expansion, mining, dam construction, and road construction (Lane and Sykes 1982). Nevertheless, wildlife is still abundant in many areas. For example, deer, elk, moose, bighorn sheep, and mountain goat can be found in the mountains and foothills. Antelope, small fur-bearers, prairie dog, rabbit, hare, waterfowl, and other birds are common in the grasslands. Deer, elk, grouse, and waterfowl are part of the aspen parkland. Deer, elk, moose, bear, and a variety of small mammals and birds live in the boreal forest. Hunting and trapping for commercial and subsistence purposes are important throughout the two catchments.

Five major urban centers (populations from ~200,000 to ~1,000,000) are located in the southern parts of the Nelson catchment: Calgary, Edmonton, Winnipeg, Saskatoon, and Regina. Each of these cities, except for Regina, is located on a major river. Moreover, there has been a recent exodus from rural areas into the cities. Populations in the northern parts of the Nelson catchment and in the Churchill catchment are small and stable.

The Rivers

Both the Nelson and Churchill catchments are part of the Hudson Bay Complex of the Arctic–Atlantic Bioregion and occupy only two major habitat types: Temperate Headwaters and Lakes and Arctic Rivers and Lakes (Abell et al. 2000). The Nelson and Churchill catchments contain four freshwater ecoregions (Abell et al. 2000). The Churchill River mostly occupies the Lower Saskatchewan ecoregion. Of the major Nelson subcatchments discussed here, the Nelson River (main stem) occupies the Lower Saskatchewan ecoregion, the Saskatchewan River mostly occupies the Upper and Lower Saskatchewan ecoregions, and the Red–Assiniboine and Winnipeg rivers mostly occupy the English–Winnipeg Lakes ecoregion. In addition, the headwaters of the Saskatchewan River emanate from the Canadian Rockies ecoregion.

From a physiographic perspective the Interior Plains physiographic division includes the Saskatchewan River (which flows through the Rocky Mountains, Great Plains, and Central Lowland provinces) and Red–Assiniboine rivers (Central Lowland province). The Canadian Shield division includes the Churchill River (Great Plains, Central Lowland, Bear–Slave–Churchill Uplands, and Hudson Bay Lowland provinces), Nelson River (Bear–Slave–Churchill Uplands and Hudson Bay Lowland provinces), and Winnipeg River (Central Lowland and Superior Upland provinces). The different physiographic regions heavily influence the hydrology, chemistry, and biology of the rivers we consider.

High precipitation and considerable relief in the Rocky Mountains province produce an annual yield of water to rivers in the foothills and the plains beyond. The Interior Plains division gets most of its streamflow from the mountains; water originating in the Interior Plains is limited because of low precipitation and high evapotranspiration (Lane and Sykes 1982). Soils in the Interior Plains have large moisture storage capacities, which can affect the timing of drainage to rivers following periods of rain. The Canadian Shield division has a high yield relative to precipitation, and the streams have a more seasonally balanced flow than those in the Rocky Mountains province because of the large natural storage capacity of lakes in the Canadian Shield.

Rivers of the Canadian Shield division, Interior Plains division, and Rocky Mountains province present three different hydrographic scenarios. Rivers in the Canadian Shield are regular in their flow characteristics, rivers of the Interior Plains tend to be erratic, and rivers of the Rocky Mountains are intermediate, although they resemble Canadian Shield rivers more than Interior Plains rivers.

Most mountain streams have good water quality, although streams draining glaciers can have high loads of suspended solids. Water quality of rivers arising on the Interior Plains is variable, but water is usually high in dissolved and suspended solids because of erosion. In the Canadian Shield division, low concentrations of dissolved and suspended solids usually result in good water quality, although high mineral and organic concentrations can occur. Dissolved and suspended solids concentrations are higher in the Hudson Bay Lowland province than in the Bear–Slave–Churchill Uplands and Superior Upland provinces because of more easily eroded surface material in the Hudson Bay Lowland.

Regional differences in water quality are also apparent, except for pH (7.5 to 8.5 throughout both catchments; Lane and Sykes 1982). Specific conductance is lowest in the Canadian Shield, followed by waters in the foothills east of the Rocky Mountains, and highest in the Interior Plains. Canadian Shield rivers flow over impermeable substrate, so their specific conductance tends to be low. Interior Plains rivers flow over highly erosive areas, so their specific conductances are higher. Rocky Mountain rivers can carry high suspended solids during high-flow periods, and their specific conductances tend to be intermediate. Turbidity values follow a similar pattern. Canadian Shield waters tend to have low turbidity values, mountain streams have intermediate values, and Interior Plains streams can have the highest values because of erosion during periods of high flow. River regulation often lowers ranges of turbidity values.

Dissolved oxygen concentrations are usually high in Rocky Mountain and Canadian Shield streams during open water and under ice. Most Interior Plains rivers do not have oxygen problems in the open-water season (unless there are upstream sewage effluents), but a number of Interior Plains rivers have oxygen depressions under ice because of a lack of aeration caused by four to five months of ice cover.

Nitrogen and phosphorus concentrations are intrinsically low in rivers traversing the Rocky Mountains province and Canadian Shield division (i.e., the headwaters of the Bow, South Saskatchewan, and North Saskatchewan rivers, and the Churchill River) because of the largely granitic bedrock and shallow soils. However, concentrations increase as the rivers cross the Interior Plains division because of inputs of nutrients from naturally fertile soils as well as from anthropogenic sources (municipal, industrial, and agricultural activities).

Water use is extensive in the Nelson and Churchill catchments. Offstream uses include irrigation, municipal centers, industrial water needs, thermal power production, and water diversions (Lane and Sykes 1982). Irrigation is the largest consumer of water in the study area, and is done mostly in southern Alberta (Bow and Oldman rivers) but also in southern Saskatchewan (South Saskatchewan River) and southern Manitoba (Red–Assiniboine rivers). Significant expansion of irrigation is generally hampered by a lack of available water. Municipal centers return most of the water they use, but they require assured supplies. All the major cities have adequate supplies, although Winnipeg and Regina may need expansions to their supplies, which come from distant sources. In addition, Lethbridge and Medicine Hat, both Alberta cities located on rivers used extensively by upstream irrigation, depend on flows regulated to meet their needs. Industrial water needs are widespread through the study area. Most mining and forestry operations are located in remote areas, but small industries and petrochemical plants are usually located near large cities. Most water needs for industrial, petrochemical, and mining uses are being met, although disposal practices must be monitored to prevent pollution of surface and groundwater systems. Thermal power production is important in

Saskatchewan and Alberta, but not in Manitoba, where hydropower production predominates (Lane and Sykes 1982). Significant expansion of thermal power generating facilities in both Saskatchewan and Alberta may be hampered by inadequate water supplies, although water diversion may be feasible for some developments in Alberta. Minor and major water diversions can be found throughout the study area and serve three purposes: (1) reducing drought vulnerability, (2) flood control, and (3) enhancing or stabilizing flows for hydropower generation. A number of water diversions are still on the drawing board or have been delayed because of environmental concerns.

Instream uses include commercial fishing, water-based recreation, subsistence natural-resource harvesting activities, and hydropower development (Lane and Sykes 1982). Commercial fishing is either a stable or declining industry in most parts of the study area. Most commercial fishing is done in Saskatchewan and, especially, in the large lakes of Manitoba. There are fears that completion of the Garrison Diversion in North Dakota, which would mix Missouri River waters with those of the Hudson Bay drainage, could introduce fish diseases, fish parasites, and undesirable fish species to Lake Winnipeg (e.g., Loch et al. 1979). Water-based recreation is an important instream use throughout the study area. For example, the sport-fishing industry in Saskatchewan is worth many times more than commercial fishing. Resource development has opened previously inaccessible areas to recreational uses. However, there is competition with other uses. For example, water-quality and low-flow conditions are a concern to recreation in parts of the Saskatchewan River subcatchment. Other threats to recreational fishing include biocides, metals (e.g., mercury), and eutrophication. Waterfowl conservation and production also conflict with other land and water uses, such as the draining of wetlands and some tillage practices. The protection of wild rivers for esthetic and recreational purposes conflicts with regulation of rivers for hydroelectric purposes. This conflict is most often felt in high-use recreational areas of mountain rivers or in wild and scenic rivers like the formerly superb lower Churchill (Newbury and Malaher 1972). However, river regulation can benefit recreation on the Interior Plains rivers by providing more reliable flows. Hydropower has been extensively developed in the study area. Most of this development has been in Manitoba, followed by Alberta and Saskatchewan. The Nelson River dissipates ~5400 MW of hydraulic power in its fall from the exit of Lake Winnipeg to Hudson Bay (Newbury et al. 1984), so development in Manitoba has focused on the lower Nelson. The three most recently constructed dams generate >3000 MW of power. The Winnipeg River is also highly regulated (560 MW). Hydropower developments in Alberta are centered on the Bow and North Saskatchewan rivers, and in Saskatchewan on the South Saskatchewan, Saskatchewan, and Churchill rivers.

Only seven species of fishes (northern pike, a subspecies of longnose dace, white sucker, troutperch, burbot, brook stickleback, and walleye) are broadly distributed throughout the Nelson and Churchill catchments, probably because of the glacial history of the area, the time since waters became available for colonization after the Wisconsinian glaciation (~12,000 to 15,000 years ago), and the physical and ecological characteristics of both the habitat and the fishes available to invade it (Crossman and McAllister 1986). Well-known species such as lake whitefish and lake trout are also broadly distributed, except for the extreme southwest or prairies. Crossman and McAllister (1986) recorded 39 nonnative species in the Hudson Bay drainage, including species native to some part of the drainage but introduced to other parts (e.g., brook trout in Alberta).

Rivers of the Nelson and Churchill catchments are not significantly biologically distinct (Abell et al. 2000). Total species richness is highest in the English–Winnipeg Lakes freshwater ecoregion (Red and Winnipeg river subcatchments) and lowest in the Lower Saskatchewan ecoregion (Nelson River subcatchment) (Abell et al. 2000). Endemic species only occur in the Canadian Rockies ecoregion. Fish species richness is highest in the English–Winnipeg Lakes ecoregion (79 species) but low in most of the rest of the area (Lower Saskatchewan, including the Hayes and Seal rivers, 38 species; Upper Saskatchewan, 41 species). Similar low species richness applies to crayfishes (1 species in each of the Lower and Upper Saskatchewan), unionid mussels (4 species in each of the Lower and Upper Saskatchewan), and herpetofauna (Lower Saskatchewan, 5 species; Upper Saskatchewan, 10 species). Very low percentages of fish or herpetofauna species in the ecoregions of the Nelson and Churchill catchments are imperiled. Taken together, biological distinctiveness is "continentally outstanding" for the English–Winnipeg Lakes, Lower Saskatchewan, and Canadian Rockies ecoregions and "nationally important" for the Upper Saskatchewan (Abell et al. 2000).

Ecoregions of the Nelson and Churchill catchments show a variety of "final conservation status"

FIGURE 19.5 Aerial view of the South Saskatchewan River as it passes through Saskatoon, Saskatchewan (Photo from Tourism Saskatchewan).

conditions (Abell et al. 2000). The Upper Saskatchewan ecoregion is "endangered." This ecoregion contains rivers in the driest part of the Interior Plains division, in southern Alberta and southwestern Saskatchewan. The English–Winnipeg Lakes and Canadian Rockies ecoregions are "vulnerable," and the Upper Saskatchewan ecoregion is "relatively stable" (Abell et al. 2000).

SASKATCHEWAN RIVER

Canada's fourth-longest river, the Saskatchewan, is formed at the confluence of the North and South Saskatchewan rivers >1200 km downstream of its glacial origins. This 8th order river is one of the most diverse rivers in North America. With its headwaters in the Rocky Mountains, the Saskatchewan River flows west to east across foothills, prairies, and boreal forest to Lake Winnipeg, via Cedar Lake (Fig. 19.11). The Saskatchewan River subcatchment has an area of ~335,900 km^2 and drains part of Montana, Alberta, Saskatchewan, and Manitoba.

Saskatchewan is derived from the Cree word *Kisiskatchewan*, meaning "swift-flowing" or "rapid." The river was an important transportation route for aboriginal peoples, and archaeologists have located >650 sites throughout the subcatchment showing human presence as long ago as 10,500 years. The earliest European reference to this river is in 1750 on a map sent to France showing the discoveries of a family of explorers known as "La Verendryes" (Douglas 1925). This map labeled the river the Poskaiao; however, by 1790 maps prepared for the Hudson's Bay Company used the present name, Saskatchewan River. It was an important route for exploration and commerce in the eighteenth century, carrying fur traders, missionaries, and settlers deep within the continent. River steamers once traveled its waters, but with the arrival of the railroad the Saskatchewan no longer had substantial navigational value.

Physiography, Climate, and Land Use

The Saskatchewan River is formed in the Great Plains (GP) and Central Lowland (CL) physiographic provinces of central Saskatchewan by the confluence of the North Saskatchewan and South Saskatchewan rivers before flowing eastward through Cedar Lake to Lake Winnipeg (see Fig. 19.11). However, the river originates in the glaciers, snowfields, and cold headwater streams of the Rocky Mountains in Canada (RM) province in Alberta before flowing across the plains. Primarily flowing eastward between latitudes 49°N to 53°N and longitudes 117°W to 100°W, the subcatchment encompasses seven terrestrial ecoregions, including Alberta Mountain Forests, Alberta/British Columbia Foothills Forests, Canadian Aspen Forest and Parklands, Montana Valley and Foothill Grasslands, Northern Mixed Grasslands, Northwestern Mixed Grasslands, and Mid-continental Canadian Forests (Ricketts et al. 1999).

Landscapes at the western edge of the subcatchment are dominated by mountains formed from Cretaceous sediments, with brunosolic, podzolic, and luvisolic soils, and forests of lodgepole pine, Engelmann spruce, alpine fir, and trembling aspen. In contrast, the northern arc of the subcatchment consists of boreal forest, with luvisolic, chernozemic, gleysolic, and mesisolic soils, and white spruce, balsam fir, black spruce, willow, and aspen. To the south the prairie has mostly chernozemic soils on glacial moraine and gently undulating lacustrine deposits underlain by Cretaceous shales and Paleozoic limestone. This area is the northern extension of open grasslands in the Great Plains of North America, once dominated by fescue, blue grama, and wheatgrasses.

Mean annual air temperature varies considerably across the subcatchment and exhibits a strong continental climate pattern (Fig. 19.12). For example, in the south (e.g., Calgary, Medicine Hat) mean annual temperature generally ranges from 3°C to 5°C, daily mean January temperatures are near −10°C, and daily mean temperatures rise to 16°C in July. In the north (e.g., Edmonton, Saskatoon, Prince Albert) mean annual temperatures are 0.5°C to 2°C and monthly means range from −20°C to 19°C. Annual precipitation is low across the subcatchment, ranging from ~32 to nearly 50 cm. Peak precipitation occurs during summer (see Fig. 19.12). Approximately 30% to 50% of this moisture is from snowfall, with snow amounts highest in mountain headwaters and lowest in the southern prairies.

Today the relatively high natural fertility and good moisture-holding capacity of the chernozemic soils in most of the subcatchment east of the mountains makes these lands highly productive for agriculture (67% of the subcatchment is cropland; Revenga et al. 1998). The relatively flat topography common to much of the aspen parkland, and particularly the prairies to the south, is conducive to highly mechanized farming and has resulted in this semiarid region becoming the most human-altered landscape in Canada. About 3 million people live in the subcatchment (P. A. Chambers, unpublished data). Most of these people live in the three largest cities (www.statcan.ca/english/Pgdb/People,Population/demo05.htm 2001): Calgary (~972,000), Edmonton (~957,000), and Saskatoon

(~231,000). Other major urban areas in the subcatchment include Lethbridge, Red Deer, Medicine Hat, and Prince Albert. Although the subcatchment contains several highly urbanized areas, particularly in Alberta, overall population density is quite low (<9.6 people/km^2; Statistics Canada 2003), with the highest densities occurring in the western portion.

River Geomorphology, Hydrology, and Chemistry

The North Saskatchewan River arises from the Saskatchewan Glacier in the Rocky Mountains (see Fig. 19.11). It exhibits a typical concave longitudinal profile, with average gradients from ~2 to 3 m/km in the mountains upstream of Rocky Mountain House to ~40 cm/km around Edmonton and <15 cm/km in Saskatchewan (Kellerhals et al. 1972). The North Saskatchewan River was originally one channel of a braided stream draining proglacial Lake Edmonton; remnants of the braided channel can still be seen in reaches upstream of Edmonton. The river channel is braided and sinuous near the glacial source and quickly changes to an entrenched and frequently confined channel in the foothills. Downstream of Edmonton the river consists of a single channel (100 to 200 m wide) that meanders through a valley that is now 2 to 4 km wide and bordered by cliffs 25 to 200 m high. Bed material in the North Saskatchewan River is predominantly gravel, becoming sandy over easily erodible shale near the Alberta–Saskatchewan border (Anderson et al. 1986).

The Bow River headwaters to the South Saskatchewan River mouth also exhibits a classic concave longitudinal profile, with average gradients from ~7 m/km in the mountains near Banff to <50 cm/km below Medicine Hat on the South Saskatchewan (Kellerhals et al. 1972). The system was a glacial spillway during the late Pleistocene and the present-day river meanders through this ancient channel. The river channel is wide and relatively shallow throughout its extent. For example, the Bow River at Calgary is ~100 m wide and 1 m deep, and the South Saskatchewan River near the Alberta border above its confluence with the Red Deer River has a mean width and depth of ~180 m and 2 m, respectively (Kellerhals et al. 1972). Well-developed riffle, run, and pool sequences are evident along the river, although riffle habitats are rare below Lake Diefenbaker (Fig. 19.5). Substrate in the upper portion of the South Saskatchewan system (e.g., Bow

River) is coarse and mostly composed of gravel, cobble, and boulders (Longmore and Stenton 1981); downstream the South Saskatchewan has a substrate of gravel, sand, and silt.

The Saskatchewan River is formed by the convergence of the North and South Saskatchewan rivers in east-central Saskatchewan. Although it historically occupied a single well-defined channel, an avulsion occurred in the 1870s just west of the Manitoba–Saskatchewan border and converted >500 km² of floodplain into a belt of anastomosing channels, wetlands, and small lakes (Smith et al. 1998). Approximately 90% to 95% of the flow is diverted through this network of channels and eventually enters Cumberland Lake, later rejoining the old channel of the Saskatchewan through three outlets from the lake (see Fig. 19.11). The area encompassing the active and abandoned channels of the Saskatchewan River, as well as an extensive wetland south of the old channel, is known as the Cumberland Marshes (sometimes called the "Saskatchewan Delta"), a ~8000 km² region comprising major waterfowl breeding grounds and muskrat trapping areas. Downstream of Cumberland Lake the river continues eastward and broadens into Cedar Lake. The Saskatchewan River continues as the outlet from Cedar Lake, from which, prior to regulation in 1965, it flowed ~40 km and dropped 36 m over a series of waterfalls into Lake Winnipeg.

Precipitation and runoff patterns appear closely related in the Saskatchewan River subcatchment (see Fig. 19.12), although individual river systems may be strongly influenced by snowmelt. Mean daily flows of the North Saskatchewan River near its mouth average 241 m³/s, with peak flows occurring in July during mountain snowpack melt and lowest flows in February (Environment Canada 2001). Most of the river is ice covered from early November to mid-April, except downstream of effluent discharges or hydroelectric dams. Regulation of the upper main stem in 1972 for hydroelectric power generation has increased mean monthly discharge in winter (November to March) by two to three times and reduced summer (June to September) discharge by 20% to 30% compared to preregulation values. Annual discharge of the Bow–South Saskatchewan system typically peaks in the spring as the mountain snowpack melts, but summer rains can occasionally produce flood peaks equal to or greater than this spring runoff (Culp and Davies 1982). Much of the system is ice covered in winter (December to March), except where ice is prevented from forming because of groundwater discharges in headwaters or downstream of large reservoirs (e.g., Gardiner Dam on the

South Saskatchewan). Preregulation discharge of the Saskatchewan River at the Pas, upstream of Cedar Lake, was 684 m³/s; postregulation discharge is 585 m³/s. Postregulation discharge at the river's inflow to Lake Winnipeg is 567 m³/s (Environment Canada 2001). Regulation of the river for hydroelectric power generation raised water levels in Cedar Lake by 3.7 m, created a 3500 km² reservoir, and reversed the natural discharge pattern so that at the inflow to Lake Winnipeg highest flows now occur in winter (January to February) and lowest flows occur in spring (April to May), in contrast to the preregulation runoff (see Fig. 19.12).

Like all rivers traversing the prairies, the North Saskatchewan carries a heavy silt load (8480 metric tons annual mean for 1962 to 1978; Environment Canada 1992). Typical of many lakes and rivers in the prairies, the water is hard (125 to 160 mg/L as $CaCO_3$), with calcium and bicarbonate the dominant ions (Saskatchewan Environment 1984, Anderson et al. 1986, Shaw et al. 1994). Total solute concentration ranges from 200 to 300 mg/L, with dilute mountain runoff determining solute concentrations along the entire length of the main stem (Taylor and Hamilton 1993). The river is intrinsically nutrient poor (14 μg/L total phosphorus [TP], 3 μg/L total dissolved phosphorus, 127 μg/L total nitrogen [TN], and 91 μg/L nitrogen as $NO_2 + NO_3$ near the headwaters); however, concentrations increase twofold to tenfold downstream of Edmonton and remain elevated along the remainder of the river. Discharge of treated wastewater from the city of Edmonton is also responsible for increased concentrations of sodium, chloride, potassium, and fecal coliform bacteria (Shaw et al. 1994). Rivers in the South Saskatchewan system have hard water because the basin bedrock is a limestone–shale complex. For example, Culp et al. (1992) reported high long-term mean values of specific conductance (357 μS/cm), total alkalinity (127 mg/L as $CaCO_3$), total dissolved solids (TDS; 200 mg/L), and hardness (163 mg/L as $CaCO_3$) near the mouth of the Bow River. All of these variables, as well as total organic carbon (6 mg/L) and turbidity (8 JTU), gradually increase downstream (Culp et al. 1992). In addition, the key limiting nutrient, phosphorus, increases downstream along the river continuum, as exemplified by patterns in the Oldman and South Saskatchewan rivers (Culp and Davies 1982, Chambers and Prepas 1994, Carr and Chambers 1998). Water quality in the Saskatchewan River upstream of Cedar Lake is fair (50 μg/L TP, 0.537 μg/L TN, 47 mg/L TDS) and generally improves downstream of the dam (Manitoba Environment 1997).

River Biodiversity and Ecology

The Saskatchewan River subcatchment is included within the Canadian Rockies and Upper and Lower Saskatchewan freshwater ecoregions (Abell et al. 2000).

Algae and Cyanobacteria

When wastewater was subjected to secondary treatment (Carr and Chambers 1999), concentrations of both planktonic and epilithic algae in the North Saskatchewan River near Edmonton were uniformly low (<1 mg/m^3 and 8.7 mg/m^2 chlorophyll *a*, respectively) upstream of Edmonton but increased substantially downstream of the municipal sewage treatment plant (to ~40 mg/m^3 and 200 mg/m^2 chlorophyll *a*, respectively). Planktonic algae were the dominant primary producers in spring and summer when the water was turbid, whereas epilithic algae dominated in the fall. Information on algal taxonomic composition was not available at this site. The epilithic algal community of the Bow and Oldman rivers is dominated by diatoms and cyanobacteria, whereas filamentous green algae tend to dominate in the South Saskatchewan downstream of the Bow–Oldman confluence (Culp and Davies 1982, Charlton et al. 1986, Culp et al. 1992). Davies et al. (1977) recorded 192 algal species in the Oldman–South Saskatchewan system. Common diatom genera included *Achnanthes*, *Cymbella*, *Fragilaria*, *Gomphonema*, *Navicula*, and *Nitzschia*. Cyanobacteria were represented by many filamentous forms, such as *Anabaena*, *Oscillatoria*, and *Phormidium*, whereas filamentous green algae (i.e., *Cladophora glomerata*, *Oedogonium*) accounted for 80% to 90% of the biomass in the lower South Saskatchewan (Green and Davies 1980). Algal biomass increases downstream in the river as a result of increased availability of plant nutrients. Seasonal trends suggest that spates, particularly during spring runoff, reduce epilithic biomass to minimum values. Peak periphyton biomass occurs during summer months, with mean summer values ranging from $<10 \mu g/cm^2$ upstream of major centers (e.g., Calgary) to $>200 \mu g/cm^2$ downstream (Culp and Davies 1982, Charlton et al. 1986). Adoption of advanced phosphorus removal by Calgary in the early 1980s resulted in declines of epilithic algal abundance downstream of the city, where TDP $<10 \mu g/L$ (Sosiak 2002). Epilithic chlorophyll *a* concentrations ranged from 104 to 1223 mg/m^2 upstream of Saskatoon compared with 49 to 2074 mg/m^2 just downstream of the city's sewage disposal (tertiary treatment) (Constable 2001). Comparatively little is known about the biodiversity and ecology of algae in the main-stem Saskatchewan River because the low population density in the system and high discharge have averted many water-quality problems.

Plants

Moisture conditions in the higher-elevation portion of the North Saskatchewan River subcatchment are similar near the river and in the upland so that upland vegetation extends close to the river bank. White spruce or aspen forests are often found on fluvial fans and terraces. At lower elevations, particularly once the river passes Edmonton, the wide river valley and its steep cliffs are colonized by balsam poplar and trembling aspen, shrubs such as alder, Saskatoon berry, chokecherry, and prickly rose, and a mixture of grasses and forbs. For the South Saskatchewan River subcatchment (Oldman and South Saskatchewan rivers), similar moisture conditions near the river and in the upland in the higher-elevation portion mean that trees and shrubs typical of the upland forest also occur close to the river margin. As elevations decline through the foothills, white spruce or aspen forests often occur on fluvial fans and terraces. Trees and shrubs in the prairies of southern Alberta and Saskatchewan are typically only found in river valleys where there is enough moisture to support riparian woodlands. In these areas, cottonwood forests occur, although human impacts (e.g., crop and livestock production, settlement, river damming and diversion) have decreased the extent of these forests. Downstream of Saskatoon the prairies give way to a boreal transition zone, where riparian vegetation is dominated by willow, aspen, and shrubs (chokecherry, Saskatoon berry), with an understory of mixed grasses and forbs.

Aquatic macrophytes (particularly pondweeds) in the North Saskatchewan River are infrequent, vary considerably in density among years and, when present, are limited to narrow bands near the river banks. Macrophytes (e.g., pondweeds) in the South Saskatchewan River system show distinct longitudinal zonation, their biomasses generally increasing sharply downstream of major municipal sewage discharges (Culp et al. 1992, Chambers and Prepas 1994, Carr and Chambers 1998, Sosiak 2002). Downstream of Calgary, plant biomass declined following phosphorus and nitrogen removal from sewage. Plant diversity in the Cumberland Marshes of the main-stem Saskatchewan River is high, with >50 emergent and submergent species recorded. Fen meadows dominated by sedges give way to cattail, bur-reed, bulrush, sweet-flag, and grant reed-grass along the margins of shallow lakes, protected bays of deeper lakes, and streams (Dirschl and Dabbs 1969, Dirschl 1972). Pondweeds, yellow pond-lily, spiked watermilfoil, hornwort, and bladderworts are abundant in the shallower lakes and along the shores of deeper lakes.

Invertebrates

Organic pollution, sedimentation, and epilithic algal growth have resulted in classic changes in the abundance and composition of benthic invertebrates in the North Saskatchewan River (Anderson et al. 1986, Shaw et al. 1994). Upstream of Edmonton the invertebrate assemblage shows little evidence of anthropogenic impacts and is characterized by a relatively low density and high diversity of organisms. Taxa of mayflies, stoneflies, and caddisflies are, together, as important numerically as chironomid midges, oligochaete worms, and nematodes. Within and immediately downstream of Edmonton total densities are higher, diversity is lower, and chironomids, oligochaetes, and nematodes comprise up to 98% of total numbers. Evidence of recovery is observed ~100 km downstream of Edmonton, but impacts persist for at least 300 km downstream. Further downstream the riverbed changes from gravel to one dominated by shifting sandbars. Benthic invertebrate abundance in these areas is low and dominated by chironomids and oligochaetes. The mayfly genus *Baetis* and the caddisfly genus *Hydropsyche* predominate where gravel patches occur (Fredeen 1983, Golder Associates 2000, J. J. Merkowsky, personal communication). The macroinvertebrate communities of the Oldman and South Saskatchewan rivers correspond broadly with the terrestrial zones of subalpine forest, fescue prairie, and mixed prairie, although zonation distinctness varies seasonally (Culp and Davies 1982). A similar pattern of zonation in stoneflies was also observed for the Bow River (Donald and Mutch 1980). In the summer and fall, upstream invertebrate communities are dominated by collector-scrapers (the mayflies *Cinygmula*, *Rhithrogena*, and *Ephemerella inermis* and the caddisfly *Glossosoma*). In contrast, invertebrates in downstream reaches of the South Saskatchewan are almost exclusively collector-gatherers (e.g., Orthocladiinae midges and the mayflies *Ephoron album* and *Ephemera simulans*) and filter-feeders (the caddisflies *Symphitopsyche* and *Cheumatopsyche* and the mayfly *Traverella albertana*). Shredders (e.g., the stoneflies *Zapada* and *Capnia*) increase in impor-

tance during late winter in the subalpine forest zone, and the fauna of the lower longitudinal zones is dominated by collector-gatherers (Orthocladiinae and Chironominae midges). These longitudinal differences in faunal composition appear to be related to downstream changes in food resources in all seasons and downstream increases in thermal degree days during the open-water period (Lehmkuhl 1972, Culp and Davies 1982). No information is available on benthic invertebrates in the main-stem Saskatchewan River.

Vertebrates

The North Saskatchewan River supports a diverse community of 36 fish species distributed throughout the system and following patterns similar to those observed in macroinvertebrate communities. Coldwater species, such as cutthroat trout, rainbow trout (nonnative), bull trout, brook trout (nonnative), brown trout (nonnative), mountain whitefish, longnose sucker, and longnose dace, dominate forested, mountainous reaches. As the river flows out of the mountains into the prairies the fish community includes warmwater species, such as northern pike, walleye, sauger, goldeye, yellow perch, quillback, and shorthead redhorse. The number of fish species in the Saskatchewan portion of the North Saskatchewan River doubled from 1957–1958 to 1985–1986, likely because of improved water quality, particularly dissolved oxygen concentrations (J. J. Merkowsky, personal communication). The relative number of goldeye and northern pike was also reported to have declined over the same period, whereas walleye and sauger populations increased because of greater abundance of spawning areas.

The pattern of fish distribution that Longmore and Stenton (1981) described for the South Saskatchewan system resembles trends observed for macroinvertebrate communities. Forested, upstream reaches are dominated by coldwater species, such as cutthroat trout, rainbow trout, bull trout, brook trout, brown trout, mountain whitefish, longnose sucker, and longnose dace. Warmwater reaches in the lower Bow, Oldman, Red Deer, and South Saskatchewan rivers are inhabited by a wide range of species, including northern pike, walleye, goldeye, yellow perch, quillback, and shorthead redhorse. The transitional zone between the foothills and prairies contains both cold- and warmwater species. In the headwaters, the nonnative rainbow trout has displaced the endemic cutthroat trout in much of the system (Longmore and Stenton 1981, Culp et al. 1992). Although angling is popular throughout the system, the best known and most heavily angled section is the 50 km reach below Calgary, where >52,000 angler-days were spent during 1985 in search of mountain whitefish, rainbow trout, and brown trout (Culp et al. 1992). The South Saskatchewan River also has an important recreational fishery for lake sturgeon.

Forty-eight species of fishes have been recorded from the Cumberland Marshes of the main-stem Saskatchewan River. The marshes are critical staging and breeding habitat for many migratory waterfowl, including large populations of ducks, geese, swans, shorebirds, grebes, and the endangered whooping crane.

Other aquatic vertebrates found in or near the rivers of the Saskatchewan subcatchment include frogs, salamanders, muskrat, beaver, river otter, mink, weasel, and water shrew.

Ecosystem Processes

Most of our knowledge of ecosystem processes comes from field studies and experiments conducted in the South Saskatchewan system. Using ^{14}C methods, Charlton et al. (1986) found that mean primary productivity of epilithic algae was <40 mg C $m^{-2} hr^{-1}$ upstream of large sewage plant effluents during the ice-free period and increased by two to five times in enriched reaches downstream; algal biomass followed a similar pattern. Aquatic macrophytes also increased in abundance downstream of sewage discharges; however, in contrast with epilithic algae the riverbed sediments were the primary source of nutrients (Chambers et al. 1989, Carr and Chambers 1998). The magnitude of the enrichment response was affected by both current velocity and sediment composition (Chambers et al. 1991, Chambers and Prepas 1994).

Analysis of invertebrate functional-feeding group composition in this system suggests that coarse particulate organic matter is important as a food base in upstream mountain and foothill river reaches (Culp and Davies 1982). Downstream invertebrate communities appear to be based upon fine particulate organic matter and epilithic algal production. Furthermore, Culp and Davies (1982) speculated that downstream reaches of the main-stem rivers are strongly affected by enrichment and autotrophic processes in summer and are heterotrophic in winter. In addition, longitudinal thermal regime is an important factor affecting macroinvertebrate distribution, particularly through the transitional zone from mountains and foothills to prairies (Donald and Anderson 1977, Culp and Davies 1982).

Human Impacts and Special Features

Much of the area in the Rocky Mountains in which the Saskatchewan River originates is a World Heritage Site and is protected by national and provincial parks. The area is a striking mountain landscape of peaks, glaciers, lakes, waterfalls, canyons, and limestone caves. The headwaters of the Saskatchewan in Banff National Park are also designated a Canadian Heritage Rivers System to recognize their outstanding nature and to ensure future management to protect their heritage value. An upper reach of the South Saskatchewan River (Empress to Lancer Ferry) and Lake Diefenbaker (a reservoir on the South Saskatchewan) have been designated globally important bird areas by Bird Life International because of their concentrations of waterfowl, some of which are considered rare species. Tobin Lake (a reservoir) and the Cumberland Marshes on the Saskatchewan main stem are also designated globally important bird areas by Bird Life International because both areas are populated by globally significant numbers of waterfowl. For example, ~2500 tundra swans, 2.3% of the total eastern tundra swan population, congregate at Tobin Lake, and 72,000 nesting ring-necked ducks (~10% of the 1970s world nesting population) use the Cumberland Marshes (www.bsc-eoc.org/iba/canmap.cfm?lang=en 2001).

Three main human actions impact the Saskatchewan River system: dams and reservoirs, municipal sewage effluents, and agricultural activities. Dams and reservoirs are present on all of the major tributaries. They serve primarily to generate hydroelectricity, store water for use during drought, and alleviate flooding during peak flows (Environment Canada 1996). Regulation of the Saskatchewan River subcatchment began in the 1890s with the initiation of irrigation projects and works to divert and deliver water to land in southern Alberta. Some of the larger irrigation works presently operating have maximum annual diversions of ~150 to 800 × 10^6 m^3; reservoir storage capacities range from ~200 to 350 × 10^6 m^3. Collectively, a maximum of 3 × 10^9 m^3 is permitted for diversion to irrigation districts in the Bow and Oldman river systems, servicing ~5000 km^2 of irrigated land, rural households, and recreational needs. The startup of irrigation diversions in the late 1890s was followed by regulation for hydroelectric power generation, first in the upper reaches of the Bow River (1911 to 1955) and then in the upper North Saskatchewan River (1965 to 1972) and Saskatchewan River (1963 to 1985). There are now eleven hydroelectric generating stations, six storage reservoirs (one on the main stem and five on tributaries), and one regulating reservoir (upper Bow River). The storage reservoirs are filled during spring and summer and then drawn down during fall and winter, storing a total of ~$700 \times 10^6\,m^3$. There are also two hydroelectric storage reservoirs located in the upper North Saskatchewan River system and another three reservoirs on the Saskatchewan River. Hydroelectric developments have significantly altered the hydrology of the Cumberland Marshes by reducing flood frequency. In recent years, developments and management activities by Ducks Unlimited have attempted to restore some of the hydrologic events that were lost as a result of altered flow regimes. In addition to dams and diversions constructed initially for irrigation or hydroelectric power generation, dam–reservoir systems on the Red Deer River, the Oldman River, and the South Saskatchewan River provide assured year-round water supply to downstream users and generate hydroelectric power, and one provides flood control (Red Deer River). The Gardiner Dam (Lake Diefenbaker) on the South Saskatchewan River has changed benthic invertebrate communities as a result of altered discharge patterns and reductions in water temperature as a result of the release of cold hypolimnetic water (Lehmkuhl 1972).

Municipal sewage discharges from major centers (i.e., Banff, Calgary, Edmonton, Medicine Hat, Red Deer, Prince Albert, Saskatoon) to rivers in the Saskatchewan River subcatchment have significant impacts on aquatic biota, largely because of effects from nutrient loading. For example, macrophyte growth in the Bow River downstream of Calgary and the South Saskatchewan River below Saskatoon is enhanced largely because of increased nutrient availability in sediments as a result of municipal wastewater discharge (Chambers et al. 1989, Chambers and Prepas 1994, Carr and Chambers 1998). Eutrophication effects are most thoroughly documented for the Bow River at Calgary and have been reviewed by Culp et al. (1992) and Sosiak (2002). The Alberta Department of Public Health noted severe impacts to the aquatic biota of the Bow River as early as 1944. Secondary sewage treatment at Calgary beginning in 1970 greatly reduced the oxygen depletion evident in the fall and winter during the late 1960s, but nutrient loadings continued to cause excessive growth of macrophytes (e.g., the sheathed pondweed) and epilithic algae (largely diatoms) from Calgary to the confluence with the Oldman River. Enrichment also decreased benthic invertebrate richness but increased densities. Although higher invertebrate density appears to have increased rainbow trout growth, low dissolved oxygen levels in summer may have caused episodic fish mortality prior to 1982 (Longmore and Stenton 1981). Advanced phosphorus removal began in late 1982, followed by nitrogen removal (1987 to 1990). These upgrades initially had little effect on nuisance macrophyte growth, likely because the bottom sediments acted as a nutrient reserve for some time after reduction in open-water nutrients (Carr and Chambers 1998), but macrophyte biomass downstream of the sewage treatment plant has decreased approximately tenfold from peak values of 1000 to $2000\,g/m^2$ dry mass in the early 1980s (Sosiak 2002).

Agricultural operations have become increasingly specialized, so that most farms are now either high-density livestock operations or intensive cash-crop farms. This development has resulted in a geographic separation between intensive livestock operations and cash-crop farms and the overapplication of manure as a fertilizer in some locales. Studies by CAESA (1998) have shown that for 27 Alberta streams (most of which were located in the Bow, Oldman, and North Saskatchewan river systems) TN and TP concentrations often exceeded interim provincial water-quality guidelines (1 mg/L TN, 0.05 mg/L TP) for the protection of aquatic life in streams in high- and moderate-intensity agricultural areas. Fecal coliform concentrations nearly always exceeded human drinking-water-quality guidelines (0/100 mL); however, pesticides were rarely detected. In addition, parts of the Cumberland Marshes have been diverted and drained for agricultural uses.

The Saskatchewan River and its tributaries do not harbor many nonnative species because of the presence of large areas of relatively undisturbed mountain and boreal coniferous forests, a continental climate, a less diverse agricultural base than in central Canada and the United States, and fewer cities and towns in which to develop local centers for the establishment of nonnative species. However, purple loosestrife is found in some wetland areas along the Saskatchewan main stem.

RED RIVER OF THE NORTH–ASSINIBOINE RIVER

The Red River of the North ("Red River") begins at the junction of the Otter Tail and Bois de Sioux rivers at Breckenridge, Minnesota, and Wahpeton, North Dakota, respectively (Fig. 19.13). The Bois de Sioux River is the southernmost tributary and originates with Lake Traverse near the junction of the state boundaries of Minnesota and North and South Dakota. The Red River subcatchment, which is near the geographic center of North America, includes portions of those three U.S. states and the southern parts of the provinces of Manitoba and Saskatchewan in Canada. It truly is an international river, originating in the United States and flowing north into Canada, but with some tributaries running across the international boundary from within both Canada and the United States. Most of the Red River main stem lies in the United States, but the Assiniboine River, a large eastward-flowing tributary that joins the Red River in Winnipeg, Manitoba, has branches that extend west into central and southern Saskatchewan (Fig. 19.15). The Souris River, a long tributary of the Assiniboine River, originates in southern Saskatchewan, flows into North Dakota, and then flows back into Canada in Manitoba. Keeping the international flavor of the subcatchment, the Pembina River flows from southwestern Manitoba into North Dakota before meeting the Red River, and the Roseau River originates in northeastern Minnesota and joins the Red River in Manitoba. The headwaters of the Red River subcatchment are intertwined with headwaters of the Missouri, Mississippi, and Winnipeg rivers, and in wet years exchanges of water may occur among these systems because of the low relief of shared swampy regions.

The Red–Assiniboine River axis had great historic importance as a trade-route junction, first among First Nations and later among First Nations and the fur traders of the North West Company and the Hudson's Bay Company. There has been a long period of contact in this area among the First Nations of the Cree in the north and northwest, the Assiniboine (a branch of the Lakota Sioux) in the southwest, the Ojibwa in the southeast, and the Saulteaux in the northeast (Taylor 2002). Representatives of all three First Nations continue to inhabit the subcatchment.

The first documented visits to the area by Europeans include that of La Verendrye in 1735, who noted that the river was called *Miskwagama Sipi* or "red water river" by local First Nations people. The North West Company referred to the Assiniboine River (named after the First Nation through whose territory it flowed) as the Upper Red River and the main-stem Red River as the Lower Red River. The Assiniboine River was originally called the Beaver River by the Assiniboines, but by about 1820 it became known by its present name. The Beaver River name remains as one of the headwater tributaries. By

1897 the Red River became known as the Red River of the North to distinguish it from several other Red Rivers in the United States (Manitoba Conservation 2000).

European settlement in the Red River subcatchment began in earnest with the arrival of settlers brought to the Red–Assiniboine junction by Lord Selkirk in 1811. All of the Red River subcatchment originally was British territory, but with the establishment of the border at the 49th Parallel in 1818 the upper part of the subcatchment became U.S. territory. Settlement of the subcatchment proceeded slowly for the first few decades of the 1800s but grew rapidly west and south with the coming of railroads from the east and south in the 1880s. American settlement of the upper part of the subcatchment began in the 1870s, somewhat later than in the Canadian portion. Until the early 1900s small river steamers were used during periods of high flow to aid commerce between Winnipeg and communities along both the Red and Assiniboine rivers.

Physiography, Climate, and Land Use

Virtually all of the present-day Red River subcatchment falls within the Central Lowland (CL) physiographic province, although some headwater tributary streams on the west side of the subcatchment originate on the margins of the Great Plains (GP) province (see Figs. 19.13 and 19.15). There are several important physiographic areas of the Central Lowland part of the Red River subcatchment that affect land use. The Red River main stem flows north along the center of former glacial Lake Agassiz, a gigantic post-Pleistocene glacial lake that deposited deep beds of lake-bottom sediments over the glacial tills of the Wisconsinan glaciation. The lake covered much of central southern Manitoba, eastern North Dakota, and northwestern Minnesota (Teller and Clayton 1983). Glacial Lake Agassiz and its deposits are an overriding feature of the Red River subcatchment, whether its former presence is reflected in the lake-bottom clay deposits, peripheral beach ridges, and associated remnant lagoons, or by outwash fans deposited into the lake by influent rivers. The Red River Valley Lake Plain dominates the main axis of the Red River valley, from the headwaters to Lake Winnipeg. The Pembina Escarpment forms the western margin of the valley and rises sharply from the plain to the Drift Prairie physiographic area. The east side of the valley is demarcated by the Lake Washed Till Plain and Moraine areas that formed at the margins of the ice sheet that once was the eastern edge of Lake Agassiz (Stoner et al. 1993). Some of these deposits are extensive, particularly in the southeast, whereas others formed low hills above the surrounding plain. The Assiniboine River system lies mainly on the Drift Prairie, north and west of the Pembina Escarpment. The soils of the region vary. Those of the Red River Valley Lake Plain tend to be heavy and wet, whereas the sandy soils along the western and eastern edges tend to be poorly developed and wet. Upland soils in the western Drift Prairie frequently are rich and light and prone to wind erosion under dry conditions. Most of the subcatchment, aside from moraine areas, is well suited to agricultural exploitation.

The Red River subcatchment includes parts of six terrestrial ecoregions (Ricketts et al. 1999). The Northern Tall Grasslands ecoregion includes the north–south axis of the Red River main stem. The Canadian Aspen Forest and Parklands and Northern Mixed Grasslands ecoregions form the western edge of the subcatchment. A small component in the northern part of the Assiniboine River system falls into the Mid-continental Canadian Forests ecoregion. Finally, the southeastern side of the subcatchment is comprised of the Western Great Lakes Forests and Upper Midwest Forest/Savanna Transition Zone ecoregions (Ricketts et al. 1999). These latter two ecoregions encompass the headwaters of the Otter Tail River in the southeast corner of the Red River subcatchment.

The climate of the Red River subcatchment is continental, characterized by relatively short, hot summers (mean air temperature ~20°C in July) and long, cold winters (mean air temperature ~−18°C in January) (see Figs. 19.14 and 19.16). Peak precipitation occurs in June and July (6 to 7 cm). The western part of the subcatchment is classified as dry subhumid, which grades to subhumid on the eastern side of the subcatchment. Weather primarily results from a predominantly western flow resulting from upper-level winds, such as the subpolar jet stream. Red River subcatchment weather changes rapidly, and temperatures may rise and fall significantly in a matter of hours because of a lack of intervening relief for more than 1000 km to the west. Precipitation arrives predominantly from the southwest, either as winter snowstorms or as thunderstorm fronts in summer. Mean annual precipitation ranges from ~43 cm in the western part of the subcatchment to ~66 cm in the east (Stoner et al. 1993).

The terrestrial ecoregions of Ricketts et al. (1999) comprise the Prairie and Boreal Plain ecozones of Environment Canada's 1996 State of the Environment Report (Environment Canada 1996), which estimated that ~97% of the Prairie ecozone and 16% of the Boreal Plain ecozone were in agricultural uses. About 70% of the Prairie ecozone was in annual cultivation, with the remainder in pasture. The agricultural portion of the Boreal Plain ecozone was mainly in pasture, with growing forestry use of the remaining 84%. In the U.S. portion of the Red River subcatchment, Stoner et al. (1993) reported that ~74% of the subcatchment was in agricultural use, of which 66% was in cropland and 8% was in pasture. The remaining 26% mainly consisted of forests (12%), water and wetlands (4%), urban use (3%), and other categories (7%). The distribution of land-use types and cropping practices are largely determined by the distribution of soil types and moisture (Stoner et al. 1993). The two largest cities in the subcatchment are Winnipeg (685,000) and Regina (~198,000) (www.statcan.ca/english/Pgdb/People,Population/demo05.htm 2001).

River Geomorphology, Hydrology, and Chemistry

The Red River flows north from the junction of the Otter Tail and Bois de Sioux Rivers ~877 km to Lake Winnipeg. The Assiniboine River flows east 1266 km from Preeceville, Saskatchewan, to join the Red at Winnipeg. The subcatchment is very flat along the Red River Valley (south to north), with a total relief of 70 m and an average slope of 8 cm/km. The Assiniboine is also considered a low-slope river (west to east), with a total relief of 350 m and an average slope of 28 cm/km (Andres and Thompson 1995). The depths of Wisconsinan tills and Lake Agassiz sediments decline northward in the Red River Valley as Lake Winnipeg is approached, and underlying sedimentary Paleozoic carbonate-dominated limestone and dolomite bedrock outcrops appear in the river bed and along channel sides of the lower river. These rocks have been eroded away from west to east nearing Lake Winnipeg, and the overlying Canadian Shield is exposed in a southeast-trending line from near the mouth of the Winnipeg River to ~100 km east of Winnipeg. Northwest of the Red River Delta at Lake Winnipeg the Paleozoic rocks are covered by shallow glacial till and Lake Agassiz deposits with frequent outcrops, whereas northeast of the Red River mouth the Canadian Shield, incompletely covered with glacial and lake-bottom sediments, bounds the east side of Lake Winnipeg. River-channel sediments are largely determined by Wisconsinan

FIGURE 19.6   Red River of the North during the 1997 "Flood of the Century" at the town of Emerson, Manitoba (Manitoba–U.S. border). North is to the left. The international boundary is the thin diagonal line just to the south of Emerson. The town stands dry because of a ring dike. The river channel is defined by the tree line (PHOTO BY B. OSWALD, MANITOBA DEPARTMENT OF WATER STEWARDSHIP, WINNIPEG).

glaciation and glacial Lake Agassiz deposits (Elson 1967). As the river meanders across the floodplain the surficial Lake Agassiz sediments are eroded down to the deeper, glacially determined sediments. Another consequence of meandering is the regular distribution of snag habitats on outside bends and point-bar formations on inside bends, both of which provide fish species ample habitat for various life-history stages while supporting an array of habitats for aquatic invertebrates. Unlike many streams in the United States, there have been no major initiatives to remove snag habitats because of the relatively low use of the river by pleasure boaters and commercial shipping, even in the major urban centers. Most of the anthropogenic alterations of the subcatchment are in the form of flood-control and irrigation reservoirs in the tributaries, as well as rather extensive drainage networks in agricultural regions that facilitate rapid runoff during heavy rain events (Goldstein

1995). The low slope of the subcatchment generates the potential for large-scale flooding, as occurred in 1997 (Fig. 19.6).

The median annual hydrograph of the Red River at Lockport, just north of Winnipeg, has a peak discharge of ~740 m^3/s during mid-April; discharge then drops off rapidly through May and normally declines steadily until ice-up (Fig. 19.14). Discharge begins to increase again in late March prior to ice breakup in early to mid-April. Mean annual discharge at Lockport is 236 m^3/s, including the flow of the Assiniboine River. The median peak flow of ~140 m^3/s on the Assiniboine River at Headingley, just west of Winnipeg, occurs in May; discharge then declines steadily through summer until March of the following year (Fig. 19.16). Mean discharge of the Assiniboine, which represents 56% of the combined catchment area, is only 47 m^3/s or ~20% of the total Red River flow. The flood-control reservoir at

Shellmouth Dam (Lake of the Prairies) on the Saskatchewan–Manitoba border normally is drawn down through the winter to make room for spring discharge, so Assiniboine River flows just downstream rise in November and frequently remain higher than summer discharges all winter. The flood-control dam at Portage la Prairie, ~100 km west of Winnipeg, shunts lower Assiniboine River flood peaks north into the south end of Lake Manitoba (out of the Assiniboine subcatchment) and maintains undersized peak discharges in the lower 160 km of the river.

Evapotranspiration tends to be high in the Red–Assiniboine catchment. Even though precipitation is highest in spring and summer, runoff during summer is very low (<0.3 cm/mo) apparently because of high evapotranspiration (see Fig. 19.14). The low combined runoff of the entire catchment is heavily influenced by the more arid Assiniboine subcatchment (see Fig. 19.16). Thus, it is clear that the spring peak in the hydrograph, when precipitation is relatively low, is caused primarily by the melting of the winter snowpack. Precipitation and river discharge follow long-term trends (~25 to 30 years) of dry and wet periods known as the prairie drought cycle, which has major influences on hydrology and land use in the whole subcatchment.

The Red River supplies an average 2,255,898 metric tons/yr of suspended sediments to Lake Winnipeg, an estimated 7.85 metric tons $km^{-2} yr^{-1}$ from the subcatchment (Brunskill et al. 1980). As a result, the waters of the Red River subcatchment are generally turbid (Secchi visibility 0.35 m in the Red River). The transported sediment load is largely suspended silts plus an undocumented amount of bedload sand. The rates of transport for major dissolved elements (10^3 mol $km^{-2} yr^{-1}$) are Ca (40.1), Mg (28.6), Na (43.0), K (5.55), SO_4 (30.2), chloride (23.1), HCO_3 (103.8), and Si (5.15), with total N and P rates of 5.18 and 0.37 (Brunskill et al. 1980). Total dissolved solids in the Red River generally are <600 mg/L, with mean values varying from ~347 mg/L near the headwaters to ~406 mg/L at the international boundary (Stoner et al. 1993). Water quality of headwater streams varies greatly with discharge and generally is high in dissolved constituents during low flow, reflecting groundwater chemistry. Headwater streams on both sides of the Red River valley typically run fairly clear at low flow but become very turbid from sheet and bank erosion during spring runoff and spates at other times of the year.

Concentrations of dissolved major ions and other variables for the Red River at Emerson, Manitoba, are available from the U.S. Geological Survey. Median values from multiple-year periods were specific conductance 690 μS/cm, pH 8.1, alkalinity 215 mg/L as $CaCO_3$, bicarbonate 255 mg HCO_3/L, Ca 64 mg/L, Mg 30 mg/L, NO_3-N 0.34 mg/L, Na 34 mg/L, chloride 35 mg/L, SO_4 94 mg/L, Al 20 μg/L, As 3 μg/L, Pb <2 μg/L, and total dissolved solids 381 mg/L. Additional values from David Donald (personal communication) were total P 0.3 mg/L, and turbidity (JTU) 125.1. Water-quality variables (means or ranges) for the Assiniboine River at Headingley, Manitoba, just upstream of the confluence with the Red River were specific conductance 865 μS/cm, pH 8.2, bicarbonate 140 to 400 mg HCO_3/L, Ca 60 to 120 mg/L, Mg 20 to 55 mg/L, total N 1.52 mg/L, Total P 0.25 mg/L, Na 50 mg/L, chloride 26 mg/L, SO_4 60 to 240 mg/L, K 5 to 10 mg/L, and total dissolved solids 552 mg/L (Gurney 1991).

Pesticides and herbicides are used regularly in croplands of the subcatchment on both sides of the international boundary. Data provided by the U.S. Geological Survey (Stoner and Lorenz 1996) indicate that pesticides occur in amounts <2% of the amounts applied and in concentrations generally far less than accepted drinking-water standards (Tornes and Brigham 1995). The amounts of agricultural herbicides applied annually are fairly constant, but higher concentrations of herbicides occur in groundwater in regions of high precipitation (Cowdery 1995).

River Biodiversity and Ecology

The whole of the Red River subcatchment falls within the English–Winnipeg Lakes freshwater ecoregion (Abell et al. 2000).

Plants

Instream macrophytes, such as bulrushes and cattails, are common only in backwater areas of the subcatchment. There are few submerged macrophytes in the main-stem larger rivers because of high turbidity, but headwater streams may have abundant submerged and emergent macrophytes.

The Red and Assiniboine rivers have narrow riparian zones comprising a gallery forest of sandbar willow associated with sand bars and point bars, peach-leaved willow, green ash, elm, cottonwood, Manitoba maple, and to a lesser extent basswood and bur oak. The riparian zone provides a thin barrier between the river channel and the floodplain, which is largely an agricultural landscape. In many areas, riverbanks are in annual crops or pasture, with little or no shrubs or trees.

Invertebrates

The main stems of the Red and Assiniboine rivers have a very diverse benthic invertebrate fauna. Mussels (Unionidae) are represented by 12 species belonging to the genera *Fusconaia, Amblema, Quadrula, Lasmigona* (2 species), *Anodontoides, Pyganodon, Strophitus, Ligumia, Lampsilis* (2 species), and *Potamilus* (Watson et al. 1998). Fingernail clams (Sphaeriidae) are also present. Inter-mediate-size faunas include the snails *Physa* sp. and *Helisoma* sp. and the northern crayfish. The mayflies include (but are not limited to) the following families (and common genera): Baetidae (*Baetis*), Heptageniidae (*Heptagenia*), Polymitarcydae (*Ephoron*), Ephemeridae (*Hexagenia, Pentagenia*), Isonychiidae (*Isonychia*), Tricorythidae (*Tricorythodes*), and Siphlonuridae. There are two abundant caddisfly families, Hydropsychidae (*Ceratopsyche, Hydropsyche*) and Brachycentridae (*Brachycentrus*). The stonefly families include Perlidae (*Acroneuria*) and Pteronarcyidae (*Pteronarcys*). Other aquatic insects include the dragonfly (*Gomphus*), elmid riffle beetles (*Narpus*), and the true fly families Tipulidae (*Tipula*), Ceratopogonidae (*Bezzia*), and Chironomidae (*Axarus*). The true bug families Belostomatidae and Corixidae also are well represented. Clam shrimps and round worms (Tylaenchidae) also are present. These groups represent primary to tertiary consumers and predators. Microbenthic invertebrates include primary consumer groups like water mites and crustaceans, fish lice (copepods, cyclopoid and calanoid), water fleas (cladocerans), and seed shrimps (ostracods).

Vertebrates

The Red and Assiniboine rivers have a very diverse fish community represented by 18 families and ~94 species. Eleven species are known from the Red River and its tributaries in the United States that are not known from the Red River north of the mouth of the Pembina River (Stewart et al. 2001). Several fish species are secondary consumers (some cyprinids, young-of-the-year fishes, quillback sucker, and bigmouth buffalo), whereas the grazers include catostomids (white sucker, silver redhorse, golden redhorse, shorthead redhorse), and cyprinids (common carp, silver chub, shiners, flathead chub). Top predators include walleye, northern pike, channel catfish, and burbot. Several nonnative fishes occur in the Red River subcatchment, including common carp, white bass, black and white crappie, and smallmouth and largemouth bass.

Amphibian vertebrates include wood frog, western chorus frog, spring peeper, northern leopard frog, gray treefrog, American toad, and isolated populations of the rare spadefoot toad. There are several salamander species in the subcatchment, including tiger salamander, blue spotted salamander, and mud puppy. Reptilian vertebrates include common snapping turtles and western painted turtles. Aquatic mammalian vertebrates include muskrat and beaver. Aquatic summer migratory birds include great blue heron, belted kingfisher, American white pelican, double-crested cormorant, and many species of ducks (e.g., nesting wood duck) and geese.

Ecosystem Processes

Ecosystem processes in the Red–Assiniboine remain largely unexplored. Although much of the landscape of the Red River is in agricultural uses, a narrow riparian gallery forest remains throughout much of the subcatchment. The gallery forest provides a rich source of allochthonous carbon to stream ecosystems. Stream hydrology of the main rivers (Otter Tail, Red Lake, Souris, Roseau, Pembina, and Assiniboine) is much altered but the rivers still support robust flora and fauna. All trophic levels are supported in the main-stem rivers and major tributaries. The lack of information on lower taxa (invertebrates, algae, etc.) makes large-scale generalizations on stream ecology impossible (Goldstein 1995). However, there is a trend of increasing fish species richness with increasing drainage area and number of ecoregions (after Stoner et al. 1993) through which a stream flows (Goldstein 1995). For example, the high biodiversity seen in the Otter Tail River (see Additional Rivers section of this chapter) is attributable partly to the fact that it flows through four ecoregions on its way to its confluence with the Red River and partly to the high abundance of lakes and the amount of forest cover in the Otter Tail system.

An ecosystem-oriented study of the entire Red River subcatchment never has been attempted, although studies on the U.S. side have begun (Stoner et al. 1993, USGS 2001 http://mn.usgs.gov/redn/cits.html). Renard et al. (1986), Peterka and Koel (1996), and Koel (1997) have reported on the distributions of fish fauna and habitats in the Minnesota and North Dakota parts of the Red River. Comprehensive data do not exist for the Canadian portion of the subcatchment, including the Assiniboine and Souris rivers, although there are unpublished studies of the Red River near Winnipeg (e.g., Clarke et al. 1980,

City of Winnipeg 2000) and of the Assiniboine River. However, both Canadian and international compilations of existing biological data are lacking.

Monitoring of water quality and quantity has been done at border stations on all the rivers crossing the international boundary because of the Boundary Waters Treaty of 1909. This reporting has been greatly enhanced by the large flood on the Red River in 1997 (see Fig. 19.6), which stimulated renewed scientific cooperation between the U.S. and Canadian federal governments and their border states and provinces. International catchment groups have been formed and seamless GIS-based maps of the subcatchment have been developed and are accessible on the internet (e.g., www.rrbdin.org).

Human Impacts and Special Features

Channelization, ditching, and water-level control dams are the major alterations that affect all the rivers. The largest dams are Shellmouth Dam (built in 1969) on the upper Assiniboine, which created Lake of the Prairies; Portage la Prairie Dam and spillway (built in 1970); and Lockport Dam on the lower Red River (built in 1922), which enhances navigation between Winnipeg and Lake Winnipeg. The major hydrologic impact on the Red River main stem is the continued operation of the Lockport Dam, which maintains water levels within the city of Winnipeg ~2 m to 3 m above normal summer levels, mainly for recreational purposes.

Operation of the Shellmouth Dam on the Assiniboine River greatly affects the hydrograph of the river. Prior to the installation of dams the lowest flows typically occurred in midwinter; lowest flows now tend to be in late autumn, prior to releasing stored water from the reservoir to make room for spring inflows. The flow regime during winter frequently is manipulated to reflect the latest predictions for spring flooding. Winter flows can be quite stable for long periods under low snowfall conditions, but if a major snowstorm brings significant precipitation to the upper part of the subcatchment flows may increase by as much as four to six times over a period of a few weeks. The effects of these flow variations on biota are unknown and virtually unstudied.

Other major dams in the subcatchment include Rafferty (built in 1991) and Alameda (built in 1994) dams on the upper Souris River in Saskatchewan, built for flood control, irrigation, and recreation. Additional low-head weirs are located at major towns in the U.S. portion of the main stem, but several were modified in the late 1990s to improve

fish and small-boat passage (Luther Aadland, personal communication). There also are two low-head weirs on the Assiniboine River in Brandon ~200 km west of Winnipeg. In addition to these main-stem structures, there are reservoir control structures on the Red Lakes, Otter Tail, and Sheyenne rivers and at the outlet of Lake Traverse (Stoner et al. 1993; see Fig. 19.13) and many water-level control weirs on tributary streams throughout the subcatchment (<350 in the U.S. portion of the subcatchment).

Some of the existing dams and weirs are barriers to fish passage, particularly at low flows. Fish distributions show, for example, that the Portage la Prairie dam on the Assiniboine is a barrier to the ongoing postglacial range expansions of recently colonizing species, such as the golden redhorse and the bigmouth buffalo, even though the dam is opened each fall.

The Manitoba government has initiated a study of instream flow needs for fish and river processes (designed according to recommendations of the Instream Flow Council 2002, www.instreamflowcouncil.org), but the analyses are not yet available. The impetus for this study was increasing demands for irrigation water for potato agriculture. A parallel study on water-quality issues in the river also is underway.

WINNIPEG RIVER

The Winnipeg River (latitude 47.39°N to 51.99°N, longitude 89.90°W to 96.37°W) drains northwestern Ontario, southeastern Manitoba, and northern Minnesota, flowing northerly and westerly before entering Lake Winnipeg at Traverse Bay (Fig. 19.17). The river has a subcatchment area of 135,800 km^2 (Ontario and Manitoba, 106,500 km^2; Minnesota, 29,300 km^2; *National Atlas of Canada* 1985) in a region straddling the weathered granite of the Canadian Shield and the sedimentary deposits of glacial Lake Agassiz. The river is intersected by numerous lakes. The 260 km long main-stem Winnipeg River collects water from its two main tributaries, the 553 km long Rainy River, beginning at Rainy Lake, and the 615 km long English River, arising in the Firesteel River at Lake Selwyn near the western shore of Lake Superior. The subcatchment has five main systems: English River–Lac Seul (35,308 km^2 or 26%), English–Wabigoon (12,222 km^2 or 9%), Rainy River–Rainy Lake (35,308 km^2 or 26%), Rainy River–Lake of the Woods (28,518 km^2 or 21%), and the Winnipeg River main stem (24,444 km^2 or 18%). About 20% of the Rainy River system is within the

United States, 66% is in Ontario, and the balance is in Manitoba. The term "Winnipeg" is variously attributed to the Cree words *Win ni pak* or *Wi nipi* for muddy or turbid waters (a plausible source for the name of Lake Winnipeg, whose waters are quite murky) or similar words *winipi* or *winnepe* for unsettled, turbulent, or confused (a plausible source for the name of the Winnipeg River).

The geological and hydrological characteristics of the Winnipeg River system made it an historical corridor for exploration and transport of European trade and culture to the heart of Canada. Although evidence of human habitation extends back 8000 to 10,000 years to the end of the last glacial retreat, it was the Woodland Amerindians, emerging around 3000 years ago, who first used the birchbark canoe to discover the network of waterways that would prove indispensable to the fur trade within the last few centuries. Jacques de Noyon was the first European to see Rainy Lake, in 1679, and Pierre de la Verendrye, who came in search of the wealth of furs and the prestige of discovery of the Great Western Sea, was probably the first white man to explore the Winnipeg River, in 1733. Passage across the subcatchment was relatively easy because much of the river's length consisted of large, often long and narrow lakes connected by sections of relatively slow-flowing river punctuated by waterfalls and rapids. By the mid 1800s these waterways were already targeted for the construction of control structures and locks to regulate water levels and facilitate transport of goods and settlers to the Red River region (Dawson 1859).

Lake Winnipeg and Lake of the Woods also played a role in the exploration and arrival of trade and European culture to the prairies. These two lakes provided a route either to the prairies and the Rocky Mountains or down the Nelson and Hayes rivers to York Factory on Hudson Bay.

Physiography, Climate, and Land Use

The Winnipeg River subcatchment drains the Superior Upland (SU) province of the Canadian Shield division and the Central Lowland (CL) province of the Interior Plains division to the south and west (Hunt 1974) (Fig. 19.17). The system encompasses two terrestrial ecoregions: the Western Great Lakes Forests to the southeast and the Midwestern Canadian Shield Forests to the northwest (Ricketts et al. 1999).

The subcatchment is underlain with crystalline, acidic, Archean bedrock of the Canadian Shield,

forming broadly sloping uplands and lowlands. Hummocky bedrock outcrops, covered with discontinuous acidic, sandy, granitic tills, are common in the northern half of the subcatchment, interspersed with undulating glaciolacustrine deposits. The Lake of the Woods area is interspersed with fluvioglacial outwash deposits, bare rock outcrops, and lacustrine deposits, forming clay plains in lowlands. The southern portion of the subcatchment ranges from bedrock outcroppings to dystic brunisols on ridged to hummocky, discontinuous, sandy morainal deposits on uplands. Lowlands are covered by rock-bound lakes, fine carbonate-rich sediments, and deep organic deposits. Dystic brunisols are dominant throughout the region, with major areas of gray luvisolic, mesisolic, gleysolic, fibrisolic, and fibrosolic organic soils. Areas of wetland are extensive throughout the region (up to 25% in the Lac Seul area), particularly in the vicinity of Lake of the Woods and the Rainy River, with treed, bowl bogs, and peat-margin swamps the predominant forms.

Presettlement vegetation in the Winnipeg River subcatchment is best known from studies in the Rainy River system, where jack pine, white pine, red pine, and hardwood–conifer forests of balsam fir, white spruce, white birch, and trembling aspen characterized the Border Lakes and Little Fork–Vermillion Uplands area of northeastern Minnesota (Heinselman 1973). Red and white pines were less common in the northern English River system. Peatlands with black spruce–sphagnum bog, and northern white cedar and black ash swamps were common in the Agassiz Lowlands (www.pca.state.mn.us 2003). Peat deposition began ~7000 years ago in the western part of the subcatchment, where deposits are now thickest, and expanded westward (Glaser 1992). The stratigraphic record indicates that the peatlands supported forests in the past, peat is becoming the dominant vegetation (Heinselman 1973), and several changes in forest vegetation corresponding to climatic changes have occurred (Heinselman 1970).

The contemporary boreal forest in northwestern Ontario is replete with common flora, including 71 species of trees and shrubs, 11 graminoids, 40 herbs, 18 bryophytes and lichens, and 16 ferns (www.rom.on.ca 2004). Dominant trees include jack pine, black spruce, white birch, and trembling aspen. Warmer portions of the subcatchment support red and eastern white pine and red and sugar maple. Cooler, wetter sites support black spruce and tamarack. Most of the Border Lakes area remains forested, with stand composition and structure similar to the original community. The Superior Upland in the southeastern portion of the Rainy River system is still predominantly forested. Many upland sites are occupied by aspen, either in relatively pure stands or mixed with balsam fir.

Climate in the basin is cold temperate Continental with modification in the vicinity of larger lakes (Brunskill and Schindler 1971). Temperatures vary slightly across the basin, with mean annual values from 0.5°C to 2°C, mean summer values from 14°C to 15.5°C, and mean winter values from −12.5°C to −14.5°C. The coldest month is usually January; the warmest month is usually July (Fig. 19.18). Annual precipitation is moderate, ranging from 50 to 70 cm, with most (80%) falling as rain. Highest precipitation is in June and July; lowest precipitation is in February (see Fig. 19.18).

The Winnipeg River subcatchment has low population density (0.6 people/km^2), thousands of lakes, and an economy based on renewable energy, forestry, mining, and recreation, but remains relatively unimpacted. Forestry, trapping, hunting, and tourism are the dominant land uses, though a significant portion of the Rainy River system is used in mixed farming or grazing. Twelve percent of the subcatchment has park designation or protected status. In Ontario, 81 provincial parks covering 3.1 million ha and the Pukaskwa National Park (188,000 ha) have been established. Several of Minnesota's most famous walleye fisheries and trout streams, as well as the Voyageurs National Park and the Boundary Waters Canoe Area Wilderness, are found in this area. Populations are concentrated in a few small towns (Kenora, Dryden, Red Lake, Sioux Lookout, Atikokan, and International Falls) devoted to forestry processing and tourism. Significant cottage development is found on Lake of the Woods and along the Winnipeg River from Lake of the Woods to Lake Winnipeg, which seasonally increases population to >1 person/km^2. Overall, 30% of the subcatchment is devoted to forestry activities, <5% is in agriculture, and <1% is urban. The remainder of the subcatchment is natural.

River Geomorphology, Hydrology, and Chemistry

The Winnipeg River subcatchment can be divided into two broad areas of geology, hydrology, and water quality: the Superior Upland part of the Canadian Shield and the bed of former glacial Lake Agassiz, which lies in the Central Lowland. The English River and upper eastern reaches of the Rainy River drain the Superior Upland, whereas the lower westerly run of the Rainy River and Lake of the Woods flow through, drain, and occupy the clay and silt deposits of glacial Lake Agassiz.

The English River drops a total of 79 m (33 cm/km), 24 m from its source to Lac Seul and a further 55 m from Lac Seul to its confluence with the Winnipeg River. This entire region is Canadian Shield, dotted with thousands of small lakes in granitic, relatively watertight basins that yield high runoff with little groundwater. Their collected waters ultimately spill through fractured or glaciated channels into Lake St. Joseph and Lac Seul, both of which are now used as reservoirs. The river channel therefore consists of several long reservoirs of low gradient connected by narrow channels of relatively steep gradient. Much of the length of the Rainy River is occupied by several long, low-gradient (3 cm/km) lake reservoirs, and the river drops only 18 m over its length. The upper, easterly reaches of the Rainy River system are, like the English River system, characterized by many small lakes in granite basins that spill through either fractured or glaciated channels. Waters from thousands of these small lakes eventually empty into Lake Namakan and Rainy Lake. Water yields are high, with little groundwater loss. The Winnipeg River connects Lake of the Woods to Lake Winnipeg. The river drops 105 m over its length in a series of waterfalls and rapids separated by long stretches of deep, turbulence-free waters (average gradient 40 cm/km), providing ideal conditions for both historical water transport and contemporary production of hydropower. Flows from the English River join the Winnipeg River at Tetu Lake, near Whitedog, just before the Manitoba border.

The Winnipeg River subcatchment has relatively high precipitation (62 cm/yr) and a relatively impermeable substrate, so subcatchment storage and water yields are high (20 cm/yr), providing for large and consistent river flows. The river has an average discharge of 850 m^3/s as it enters Lake Winnipeg. Most of the river and reservoirs of the English River are ice covered from early December to mid-April, except for raceways at control structures. Control structures at the outlet of Lake St. Joseph and Lac Seul, operated by the Lake of the Woods Control Board, have produced hydroelectric power since 1929 and have inverted the normal discharge pattern to the river downstream of Lake St. Joseph and Lac Seul from a spring/summer to a fall/winter maximum flow. Lake Namakan and Rainy Lake of the Rainy River system are both regulated for hydroelectric production by the International Rainy Lake Board of Control of the

International Joint Commission. Most of the river and reservoirs of this system are ice covered from early December to mid-April, except for raceways at control structures.

Waters of the English River system are of low ionic strength (conductivity 70 μS/cm, calcium 5 mg/L, alkalinity 15 mg/L as $CaCO_3$), except for a few lakes that are intersected by calcareous glacial deposits (Fee et al. 1989). Waters in lakes and rivers of the system tend to be nutrient poor, with most of their total nitrogen (TN 325 μg/L) and total phosphorous (TP 10 μg/L) supplied from atmospheric deposition. On the other hand, coniferous forests and wetlands are a source of relatively high levels of dissolved organic carbon (DOC 500 μmol/L). With low population densities, the English River system remains relatively unaffected by domestic waste or nonpoint-source pollutants. Resource-based industrial activities (pulp and paper and mining) are point sources of specific contaminants. Chloralkali plants associated with the pulp and paper industry were a significant source of mercury delivered to the Wabigoon River and the lower reaches of the English River in the 1960s and 1970s. Rainy River headwaters again are typical of those found on the Superior Upland: characteristically of low ionic strength (conductivity 30 μS/cm, calcium 3 mg/L, alkalinity <10 mg/L as $CaCO_3$) and nutrient poor (TN 300 μg/L, TP 7 μg/L) (M. P. Stainton, unpublished data). Waters from this part of the Canadian Shield flow westward, accumulate in Rainy Lake, and spill over Koochiching Falls off the Shield onto the flat bed of glacial Lake Agassiz sediments. Below these falls, in the western half of the system, the Rainy River flows fairly straight, with little relief, through a narrow valley carved into the old bed of glacial Lake Agassiz. Waters from the Rapid, Big Fork, and Little Fork rivers, which drain large areas of bog, swampland, and glacial clay, join the Rainy River before it empties into Lake of the Woods. These westerly waters are higher in ionic strength (conductivity >100 μS/cm), alkalinity (50 mg/L as $CaCO_3$), suspended solids (5 mg/L), and nutrients (TN 650 μg/L, TP 31 μg/L; Anderson et al. 2000). Winnipeg River water tends to be relatively dilute (conductivity 100 μS/cm) and somewhat alkaline (45 mg/L as $CaCO_3$), with levels of TN (550 μg/L) and TP (31 μg/L) that seem to have increased over the past 30 years (Jones and Armstrong 2001).

River Biodiversity and Ecology

The Winnipeg River subcatchment belongs to the English–Winnipeg Lakes freshwater ecoregion, which also includes the Red River subcatchment (Abell et al. 2000). Following retreat of the Laurentide ice sheet ~11,000 years ago, inundation of the region by proglacial Lake Agassiz initiated the gradual reestablishment of biological assemblages, particularly species originating from the southern Mississippi refugium (Bajkov 1930, Dadswell 1974, Patalas in press). Lake Agassiz eventually drained to the north, leaving a patchwork of isolated and interconnected lakes varying in size from a few hectares to several thousand km^2, each with a similar compliment of Agassiz fauna. Boreal lakes contain few species compared to temperate or tropical lakes, particularly those lakes in richer geological settings and unglaciated regions (Schindler 1990). Because of the paucity of species, boreal ecosystems have been largely ignored in studies of biodiversity, and knowledge of Winnipeg River biodiversity thus remains limited. In addition, few ecological studies of the rivers and streams in this sparsely populated region have been undertaken.

Algae and Cyanobacteria

Although substantial amounts of information are available for lake algae, and some lake algal species may survive in the potamon zone of streams, little is known of true stream phytoplankton or epilithon in the Winnipeg River subcatchment (D. Findlay and H. Kling, personal communication). The three major classes of algae of importance in streams and rivers of other regions (Cushing and Allan 2001)—diatoms (Bacillariophyceae), green algae (Chlorophyceae), and cyanobacteria (Cyanophyceae)—are likely typical of Winnipeg River subcatchment streams. Humic, tea-colored streams draining wetlands often have reduced amounts of algal material.

Plants

Stream plants are usually diverse within boreal forest areas but with less biomass than occurs in similar streams in open country (Haslam 1978). At least 38 plant varieties, including bulrushes, cattails, hornworts, reeds, watermilfoil, sedges, waterlilies, arrowheads, and wild rice, are common in wetlands and medium-size streams of the subcatchment (Eggers and Reed 1997, Newmaster et al. 1997). Beaver dams, common in the region, increase stream width and water depth and decrease flow, allowing plants such as lesser duckweed and waterlilies to grow. The full range of wetland conditions, including bogs, fens, swamps, marshes, and open-water wetlands, is described in the Northwestern Ontario

Wetland Ecosystem Classification (Harris et al. 1996).

Invertebrates

Wide-scale characterization of stream invertebrates in the Winnipeg River subcatchment has not been done, although limited studies of the impact of pulp mill effluents, logging, aerial spraying, or acidification on stream biota are available (e.g., Hall 1994). Larvae of black flies (Simuliidae), mosquitoes (Culicidae), horse and deer flies (Tabanidae), dragonflies (*Libellula*, *Aeshna*), caddisflies (*Hydropsyche*, *Cheumatopsyche*), mayflies (*Caenis*, *Hexagenia*), and sand flies (Ceratopogonidae), and various life stages of crustaceans (*Hyalella*, *Gammarus*, *Orconectes*, *Candona*, *Canthocamptus*) are common in streams and shallow ponds of the Experimental Lakes Area (ELA) of northwestern Ontario (e.g., Bilyj and Davies 1989). Mayflies, caddisflies, stoneflies, true bugs (hemipterans), and bivalve mollusks are also found in streams throughout the region. Crustacean plankton diversity and abundance in streams of the Winnipeg River subcatchment, as in most lotic environments, are low.

Vertebrates

Of the ~101 species of fishes in the Hudson Bay drainage, 69 species, representing 16 of the 19 families present, exist in the Winnipeg River subcatchment. This number is surpassed only by the Red River of the North, with 94 species. Most (58 of 69) species, including northern pike, walleye, sauger, yellow perch, lake whitefish, lake trout, brook trout, smallmouth bass, and white sucker, are dispersed throughout the subcatchment. A few species, such as longear sunfish and green sunfish, are found only in Quetico Park and the Lake of the Woods region (Meredith and Houston 1988a, 1988b). Anecdotal evidence suggests that fish populations are decreasing in both accessible and remote parts of the Winnipeg River subcatchment. For example, daily limits of lake trout in remote northwestern Ontario locations have decreased from 12 fishes to 1 to 2 during the past 30 years as access by aircraft and snow machine improved and pressure from ice fishing increased. Lake sturgeon, which were once used by aboriginal people as a staple food, are still found in the Maligne River, in Sturgeon and Russell lakes of Quetico Park, Ontario, and in the Winnipeg River (Rusak and Mosindy 1997), but many populations have now been extirpated.

Fur-bearing vertebrates, such as beaver, river otter, muskrat, and mink, are numerous enough to sustain commercial trapping operations throughout the subcatchment. About 250 species of birds have been recorded in the Quetico, Atikokan, and ELA areas of the subcatchment (www.birdsontario.org 2004, www.queticofoundation.org 2004). Primary species found on or near lakes and rivers include common loon, common merganser, great blue heron, mallard, American black duck, and belted kingfisher.

Ecosystem Processes

The generally low nutrient levels, rocky substrates, and high flows characteristic of most streams in the subcatchment do not support high primary productivity. Smaller brooks are usually devoid of plants in either summer dry or summer flooding periods (Haslam 1978). Actively functioning phytoplankton are restricted to large, slow-flowing, deep-river segments (Cushing and Allan 2001), such as impounded main-stem Winnipeg River sections, where excessive algal growth stimulated by nutrient inputs from shoreline cottage developments has recently developed. Direct assessments of in situ productivity are not available for the English or Rainy rivers. However, a maximum estimate of chlorophyll *a* concentrations of epilithon and phytoplankton in Canadian Shield streams may be taken from transparent ELA lakes, where values range from 0.05 to 0.5 mg/m^2 (Watkins et al. 2001), and 1 to 5 µg/L, respectively (Armstrong and Schindler 1971). Increased algal productivity in lower reaches of the Winnipeg River has contributed to high abundances of zooplankton in the river's inflow to Lake Winnipeg during the past two decades (Patalas and Salki 1992, A. Salki, unpublished data). Wetlands are important sources of DOC, and their capacity to sequester sulfates and nitrates in acidic precipitation partially protects downstream waters.

Human Impacts and Special Features

The Winnipeg River subcatchment simultaneously contains some of the oldest exposed rock (>3.5 billion years) and some of the youngest sedimentary and glacial deposits (<10,000 years) on earth. Low human population densities and relatively restricted regional economic development have left much of the area essentially undisturbed since the last glacial retreat, ~9000 years ago. Ancient fur trade routes, aboriginal petroforms, stands of old-growth pine, pristine lakes, and abundant wildlife are still intact, all of which attract more boating and canoeing enthusiasts than any other wilderness area in North

America. In contrast, rivers of the Winnipeg–Rainy subcatchment are among the most highly regulated waterways (for hydropower production) on the continent.

Human activity has changed the Winnipeg River ecosystem, starting with the explorations of La Verendrye, who introduced the cultivation of maize, squash, and beans on islands in Lake of the Woods to aboriginal peoples (Combet 2001). The aquatic and terrestrial components of the Superior Upland and Central Lowland provinces are changing as a result of contemporary anthropogenic and natural stresses (Urquizo et al. 1998). Human settlement, hydroelectric dams, forestry, mining, agriculture, recreational activities, and climate change are leaving measurable imprints. Even the English and Rainy rivers, which begin their journeys in ancient, near-pristine wilderness, are affected by quality-altering nutrients and contaminants as they flow westward, en route to their Winnipeg River confluence, past pulp mills, hydro dams, small urban centers, and burgeoning cottage developments. Contaminant levels in fishes within some areas of the Winnipeg River subcatchment are high enough to warrant consumption advisories in both Ontario and Minnesota fishing guides.

Reservoirs

As early as the mid 1800s Queen Victoria dispatched S. J. Dawson to survey the best routes for water transport and sites for settlement between Lake Superior and the Red River settlement at Winnipeg. Among Dawson's recommendations were the construction of several dams and locks along the Rainy River to facilitate steamboat travel (Dawson 1859). The relatively high relief that made the English River system unattractive for early exploration and transport now provides for most of the hydropower generation. The Winnipeg River has been constrained by several large dams to control water levels and provide storage to generate power for two provinces and two pulp and paper companies. Today international regulatory boards oversee water-level and flow regulation, using 34 control structures that provide water for 11 power-generating facilities with a combined capacity of >900 MW. Ontario Hydro has constructed four generating stations on the English and Winnipeg rivers with a total generating capacity of 240 MW. Boise Cascade, a paper company, owns and operates a single generating station on the Rainy River at International Falls. Manitoba Hydro operates six generating stations on the Winnipeg River main stem with a total capacity of ~560 MW, includ-

ing two plants at Point de Bois and Slave Falls, recently acquired from the city of Winnipeg. Most of the structures that regulate level and flow on the Winnipeg River were put in place without environmental impact assessments, so little is known of habitat that has been disrupted or lost. There is, however, some evidence indicating an improvement in walleye and perch habitat in the reservoirs of the Winnipeg River, making these reservoirs highly attractive to sport fishers. Lake of the Woods has been managed as a reservoir since the 1920s with the construction of the Norman Dam at Kenora, Ontario, which raised the lake 3 m above historic levels.

Radionuclides

The Whiteshell Nuclear Research Establishment (WNRE) was established on the Winnipeg River at Pinawa, Manitoba, in 1963 by Atomic Energy of Canada Limited. ^{137}Cs discovered in Lake Winnipeg sediment cores was traceable to leakage during the 1970s from the nuclear reactor, now slated for decommissioning (Lockhart et al. 2000). Levels of ^{137}Cs in Winnipeg River walleye downstream of WNRE were two times higher than in upstream fish from 1992 to 1997 (Graham et al. 1998).

Forestry and Agriculture

Alterations of stream-water quantity, quality, and habitat have occurred with the intensification of logging and forestry in northwestern Ontario (Feller and Kimmins 1984, Paterson et al. 1998, Garcia and Carignan 1999, Carignan et al. 2000, Schindler 2001). Three pulp mills, nine lumber mills, and two panel-production facilities, using 12 local tree species, are located in the subcatchment. Clearcutting has resulted in faster runoff, soil erosion, increased stream siltation, and amplified exposure to ultraviolet (UV) radiation from lost shading along waterways. The Pine Falls pulp mill on the Winnipeg River has had severe effects on the downstream benthic community (Gregory and Loch 1973). Mercury contamination from pulp and paper mills on the English–Wabigoon River system (Rudd et al. 1983, Salki et al. 1985, Parks et al. 1991) during the 1960s forced closure of fishing in 1970. Recent government regulations and technological improvements have minimized aquatic pollution from bark and fiber waste, mercury, and organochlorine compounds. In the Rainy River system in Minnesota, fishes are generally of good quality, although contaminants have been found in fishes from certain waters. Juvenile fishes downstream of pulp mill effluents accumulated polychlorinated biphenyls and chlorinated phenols (Merriman et al. 1991), and leeches and mussels (*Elliptio complanata*) contained chlorophenol congeners as far as 100 km downstream (Metcalfe and Hayton 1989). Agriculture in the Winnipeg River region is generally inhibited because of limited soil over bedrock, lack of nutrients in the soil, or the predominance of wet peatlands. Hay is produced in the Minnesota counties of the Rainy River system, but during the past two decades the extent of pastureland and cultivated cropland has declined from 115,000 to 85,000 acres and from 170,000 to 90,000 acres, respectively (www.nass.usda.gov/census/census97/volume1/mn-23/toc297.htm 1997). The structure of animal agriculture has changed dramatically in parallel fashion over the last two decades. Small- and medium-size livestock operations have been replaced by large operations at a steady rate. The total number of livestock has remained relatively unchanged, but more livestock are kept in confinement. These changes in animal agriculture have resulted in increased problems associated with the use and disposal of animal waste. However, the USDA reported that loading of nitrogen and phosphorous to soils in the Minnesota Rainy River system is not in excess. Aquatic communities in the Baudette River and Williams Creek of the Rainy River system in the United States have been impaired by lowered oxygen levels related to low stream flows and possibly increased nutrient loadings (www.pca.state.mn.us 2003).

Fires and Climate Change

Forest fires have influenced the contemporary physical, chemical, and biological properties of streams and lakes in the Winnipeg River subcatchment. Following fire, increases in sulfate and nitrate and hydrological outputs from streams draining the burned areas persist from a few to several years, depending on fire severity and weather and climate (Bayley, Schindler, Beaty et al. 1992, Bayley, Schindler, Parker et al. 1992, Schindler et al. 1992, Schindler, Bayley et al. 1996). Recent evidence of climate change in the Winnipeg River subcatchment suggests that stream, lake, and terrestrial ecosystems will come under additional stress in the future (Schindler 2001). Relatively small changes (1.6°C) in mean annual air temperatures at the ELA between 1970 and 1990 have resulted in shorter ice-cover periods for lakes and streams, reduced flows in 1st order streams, and decreased transport of nutrients and DOC to streams and lakes. Decreases of 70% to 80% in DOC in many boreal Canadian Shield lakes and streams have resulted in increases of up to 900% in UV exposure in aquatic environments (Schindler, Curtis et al. 1996). This exposure may be especially important to the communities of lake-outflow streams in the boreal forest. Subtle environmental changes that induce behavioral and distributional responses of biota may ultimately impact lower trophic levels and dependent fish communities. Impoverishment of DOC contributes to deeper UV penetration and lessening of phosphorous inputs, which lead to lower · productivity of aquatic ecosystems.

Nonnative Species

The close proximity of the Winnipeg River and the Lake Superior catchment headwaters is facilitating the dispersal of Great Lakes nonnative invertebrate and fish species into central Canada through lakes and rivers of the Winnipeg River subcatchment. The cladoceran *Eubosmina coregoni* and the rainbow smelt are two species that have recently been found in Lake of the Woods and Lake Winnipeg (Franzin et al. 1994, Salki 1996). There is increasing potential for further introductions to occur as the subcatchment is spanned from east to west by a major highway with rising traffic volumes. A second nonnative cladoceran, *Bythotrephes cederstroemi*, was recently found in Lake Saganaga in the Boundary Waters Canoe Area Wilderness along the Minnesota–Canada border (www.seagrant.mnu.edu 2004).

NELSON RIVER MAIN STEM

The main-stem Nelson River is the natural outflow of Lake Winnipeg, which is a 26,000 km^2 remnant of glacial Lake Agassiz (Newbury 1990a). Lake Winnipeg collects flows from the Saskatchewan, Red–Assiniboine, and Winnipeg river subcatchments. The river runs for 680 km before emptying into Hudson Bay. The upper Nelson (the first 350 km) flows through a complex series of lakes as a series of short cascades between bedrock-controlled lake basins down to Split Lake (Fig. 19.19). Two major dams are located on this stretch of the river. The Grass and Burntwood rivers join the Nelson from the west in the area of Split Lake. The Nelson then runs fairly straight from Split Lake to its mouth on Hudson Bay in a single channel over a series of steps that end with rapids. However, it is interrupted by three major dams before reaching Hudson Bay.

Stretches have steep banks that gradually decrease in slope as they approach the estuary of the river at Hudson Bay. The lower 150 km of river is part of the marine intrusion zone that has emerged above sea level since the last glaciation (7000 to 9000 years ago).

The Nelson River is of considerable historical importance in the development and settlement of the interior of Canada. Although European exploration began in the early seventeenth century, aboriginal peoples had already occupied the area for thousands of years, and had highly developed societies (www.mbchiefs.mb.ca/efa/history2.html). At least 39 different aboriginal groups have occupied the area of the lower Nelson and lower Churchill rivers since the glaciers retreated (G. Dickson, personal communication). North–south movements of various cultures were dynamic, often coinciding with climatic changes (e.g., cooling ~3500 years ago, warming ~3000 years ago). Early historical documents identify the Churchill River as the rough boundary between the Dene from the north and the Cree from the south, but this boundary fluctuated until ~1000 years ago. The Cree, and the Assiniboine from further south, acted as middlemen for the Hudson's Bay Company, with the Cree becoming the home guard. In fact, the help of the First Nations was essential to the success of the European colonization.

Captain Thomas Button, an Englishman, landed near the mouth of the Nelson River in 1612, two years after Europeans had begun exploring Hudson Bay (http://trulycanadian.freeservers.com/manitoba_history.htm). Exploration of the region was begun in earnest in the late seventeenth century by the Hudson's Bay Company, which established a post on the Nelson River in 1670 and on the neighboring Churchill River in 1688. A flourishing fur trade developed from these posts, although access to the interior of the continent was through the smaller, more navigable Hayes River, which runs parallel to the Nelson River.

The importance of the Nelson and Hayes rivers to the fur trade and access to the interior of Canada diminished over time as new modes of transportation developed and consumers' interests changed. The key trading posts were abandoned by the early 1900s. The York Factory post on the Nelson–Hayes estuary is now a national historic monument, and the Hayes River is currently a candidate for Canadian Heritage River status. However, the Nelson River became interesting for another reason: its hydropower potential. The Nelson River today is sparsely populated by aboriginal communities on isolated lake segments along the main channel. Its central importance to the southern population of Manitoba is its present and future hydropower generation capability.

Hydroelectric power development was the impetus for the physical, chemical, and biological characterization of the Nelson main stem. Unfortunately, the bulk of the information has been published in difficult-to-access technical reports of special study boards, the provincial and federal governments, consultants, and other sources. Few of the data appear in primary journal publications or books.

Physiography, Climate, and Land Use

The Nelson River flows in a northeasterly direction roughly along the boundary between the Bear–Slave–Churchill Uplands (BC) and Superior Upland (SU) provinces of the Canadian Shield division and, in its lower reaches, through the Hudson Bay Lowland (HB) (see Fig. 19.19). Most of the river flows through the Midwestern Canadian Shield Forests terrestrial ecoregion, and the lowest part flows through the Southern Hudson Bay Taiga ecoregion (Ricketts et al. 1999). The region has been heavily glaciated and is covered by thin (<2 m) glacial till overburden and poorly drained peat-based wetlands. Typical vegetation includes stunted subarctic black spruce, jack pine, aspen, and willows on south-facing banks. Moisture and soil conditions are the main determinants of species distributions, and forest communities can change rapidly over short distances (Province of Manitoba 1974). However, black spruce is the climax forest through much of the subcatchment. Fire and logging cause abrupt changes in the environment and have a significant effect on the vegetation of the area. For example, the pinkish hue of fireweed is characteristic of a burned-over landscape and is the first stage of vegetative succession. Jack pine usually colonizes burned-over areas and hardwood species follow logging. The area of the Nelson River upstream of Stephens Lake is characterized by rolling, undulating topography of low relief that forms part of the Bear–Slave–Churchill Uplands (Province of Manitoba 1974). The area downstream of Kettle Rapids has numerous broad depressions.

The dominant bedrock is granite, but gneissic rocks occur in wide belts. Volcanic and metamorphosed rocks have a local distribution. Glacial till, deposited by melting ice as ground moraine, forms a thin mantle, and is often composed of loamy sand and Precambrian rock fragments. Some locations are marked by a loamy clay of calcareous till that resulted from a readvance of the glaciers over lacustrine sediments. Till is locally concentrated in streamlined hills or drumlins, but the relief of the ground moraine is determined by bedrock. Esker deposits of sand and gravel were laid down by rivers flowing under glacial ice and now form meandering sandy ridges of considerable relief. Fine-textured lacustrine clay deposits from glacial Lake Agassiz cover a major portion of the subcatchment.

Permafrost is widespread in clays of the northern reaches of the subcatchment (downstream of Split Lake) and some poorly drained clays in the middle reaches (Playgreen Lake to Split Lake) but does not occur in the clays of southern reaches (outflow of Lake Winnipeg). Most mineral and organic soils in the north contain permafrost, whereas permafrost only occurs in organic soils in the south. Mineral soils are lurisolic, brunisolic, and gleysolic. Organic soils are accumulations of various kinds of peat that are saturated for most of the year.

The climate of the subcatchment is continental and characterized by short, cool summers and long, cold winters (Province of Manitoba 1974). Mean annual temperature is <0°C. Mean monthly temperatures have the greatest range between January and July (~40°C) (Fig. 19.20). Mean daily air temperatures are highest in July and range from ~17.5°C in the southern part of the subcatchment to ~12.5°C at the mouth of the river (Lane and Sykes 1982). Mean daily temperatures are lowest in January and range from ~−22.5°C in the southern part of the subcatchment to ~−27.5°C in its lowest reaches. Mean monthly precipitation is highest in July and lowest in February (see Fig. 19.20). Annual precipitation is ~50 cm (A. Warkentin, personal communication). About two-thirds of the annual precipitation falls from May to October. Usually <2.5 cm/mo of precipitation falls during the winter months.

The Nelson main stem has a very low population density (likely <0.5 people/km² compared to 5 people/km² for the entire Nelson subcatchment; Revenga et al. 1998, www.waterandnature.org/eatlas/html/index.html 2004). The largest aggregation of population in the main-stem subcatchment is Thompson (~15,000) on the Burntwood River. However, there is extensive industrial development in the subcatchment, including a large nickel mine at Thompson and three sizeable dams on the main stem. The aboriginal population is involved in natural-resource harvesting (hunting, fishing, and trapping) for domestic use and commercial markets (e.g., Usher and Weinstein 1991). Land-use proportions for the entire Nelson catchment are as follows: cropland 47%;

forest 32%; grassland, savanna and shrubland 6%, urban and industrial 7% (Revenga et al. 1998, www.waterandnature.org/eatlas/html/index.html 2004).

River Geomorphology, Hydrology, and Chemistry

The upper Nelson channel is irregular in form and winds through the rocks of the Canadian Shield division, which are overlain by a few meters of lacustrine deposits (Province of Manitoba 1974). The channel is primarily bounded by Precambrian bedrock, with a shallow, coarse till cover. The river runs primarily off the bedrock surface between rapids and has steep, rocky tributaries. Transported sediments are mainly colloidal clays originating upstream. Average slope from Lake Winnipeg to Split Lake (a 50 m drop concentrated in three major falls) is 13 cm/km. There are numerous lakes on the upper Nelson, including Playgreen, Cross, Sipiwesk, and Split. This portion of the channel has numerous sets of rapids (see Fig. 19.7). The lower Nelson lies in the Hudson Bay Lowland, where Paleozoic limestones overlain with deep marine sediments prevail. The channel is controlled by bedrock outcrops (Precambrian in the upper reaches, Silurian dolomites and limestones in the lower reaches). The riverbed becomes gravel/cobble/bedrock when it enters the Hudson Bay Lowland downstream of Kettle Rapids because it is recutting its channel in the infill valley that has emerged since glaciation. Tributaries in the lower part of the main stem are also gravel because of the infill, which accounts for their resident brook trout populations, which were exploited by natives. Average slope from Split Lake to the estuary (a 156 m drop with no major falls) is 57 cm/km, quite a bit steeper than the upper Nelson main stem and the reason the lower main stem was so attractive for hydropower production. The river transports cobbles, gravels, and sands as well as colloidal inputs from upstream. There are no major natural lakes on the Nelson below Split Lake, and the channel straightens out (Fig. 19.8).

Downstream of Split Lake the river has a series of rapids, which become long chutes over horizontally bedded limestone covering the Canadian Shield surface (Newbury 1990a). Shores and islands in reaches that end in steep rapids and chutes sometimes display well-developed trimlines caused by ice accumulation well above the maximum open-water stage. The trimline often does not extend to the upper end of ice-generating reaches because of the lack of ice accumulation before the winter ends, and trees

FIGURE 19.7 Upper Nelson River crossing the rugged Canadian Shield above the Hudson Bay Lowlands. Profile of the river is controlled by large bedrock outfalls from lakes, as shown here at Whitemud Falls, the outlet of Cross Lake. Hydro developments have replaced several of the outlet falls with dams (PHOTO BY R. NEWBURY).

extend down the valley walls to the maximum open-water levels.

Care must be taken in interpreting pre- and postregulation runoff for the Nelson main stem at Kelsey Dam above Split Lake (see Fig. 19.20). Runoff at Kelsey involves a lot of basin storage effects (lakes, reservoirs, river channels) and these effects can carry forward for months (A. Warkentin, personal communication). Although summer precipitation is greater than for other seasons, summer runoff tends to be relatively low compared to other seasons because of evapotranspiration.

Redevelopment records indicate that the mean annual natural discharge of the Nelson main stem at Hudson Bay was 2480 m³/s (Newbury et al. 1984), but the hydrology of the Nelson main stem has been greatly altered by the post 1970s effects of Lake Winnipeg regulation and the added 850 m³/s licensed diversion from the Churchill River. Peak flows in the Nelson have increased substantially, and seasonal discharge patterns have been altered. For example, with completion of the Lake Winnipeg storage dam

at Jenpeg and operation of the Churchill diversion, mean natural summer flows to Hudson Bay have increased from 2689.5 m³/s to 3250.0 m³/s, whereas mean winter flows have increased from 1885.5 m³/s to 2859.3 m³/s in recent years (FEMP 1992c). The reversal of the seasonal pattern on the lower Nelson River is obscured by the addition of Churchill River flows, but the alteration has shifted 7% of the combined annual delivery from the open-water season to the ice-covered period from November to May.

Day-to-day operation of the hydropower plants in the lower Nelson produces dramatic fluctuations in discharge. For example, at Kettle Dam (Fig. 19.9), approximately midway between Split Lake and the Nelson River estuary, most of the day-to-night decreases in hourly mean discharge were >2000 m³/s in winter and often ~3000 m³/s in summer (FEMP 1992c). Baker (1989), working in the Nelson River estuary, commented on discharge fluctuations caused by the upstream hydro dams: "Discharge is controlled . . . according to southern power demand." Mean hourly summer discharge rates (1978 to 1988

FIGURE 19.8 The lower Nelson River channel, shown here in summer, lies in a deep valley that is incised into the flat Hudson Bay lowlands. The near vertical riverbanks are permanently frozen. The south-facing banks shown here have an active thawed zone only where seasonal surface water flows over the banks, forming a series of "flat-iron" crenulations that are typical of permafrost regions (Photo by R. Newbury).

data) have ranged between 0 and 7200 m³/s (see also FEMP 1992b). Daily discharge rates varied between 340 and 6500 m³/s. Mean daily minimum and maximum discharge typically varied between 0.6 and 700 m³/s and 4.5 and 4950 m³/s, respectively. Mean discharge during weekdays normally exceeds mean discharge during weekends by several orders of magnitude. Discharge fluctuations are attenuated somewhat in the distance between the last dam on the river and the estuary (Baker 1989). The ecological effects of these dramatic fluctuations remain largely unstudied.

Efforts to determine changes in water chemistry as a result of Lake Winnipeg regulation and Churchill River diversion led to a gathering of scarce predevelopment data for the Nelson River subcatchment (e.g., Williamson and Ralley 1993). Data from the Burntwood River, a major tributary of the Nelson main stem at Thompson, are compared to three Nelson main-stem stations (Norway House, located on the East Channel of the Nelson River just below

its outflow from Lake Winnipeg; Sipiwesk Lake, still in the upper Nelson; and Split Lake, the repository of the Burntwood River in the lower Nelson). High values in the Burntwood River relative to the Nelson main stem for color (47 TCU versus 15 to 21 TCU), total suspended solids (14 mg/L versus 8 to 13 mg/L), turbidity (14 NTU versus 7 to 13 NTU), and total organic carbon (13 mg/L versus 9 to 11 mg/L) indicate a tributary that is stained by flowing through peatlands during part of its course and through erodible sediments in other parts. Lower relative values for total alkalinity (63 mg/L versus 98 to 103 mg/L as CaCO₃), conductivity (139 mg/L versus 298 to 311 mg/L), pH (7.70 versus 8.02 to 8.06), and hardness (68 mg/L versus 115 to 121 mg/L as CaCO₃) indicate more dilute waters than in the Nelson main stem. Low pH and alkalinity in the Burntwood River may be consistent with colored (humic) water. Total nitrogen (Kjeldahl) concentrations (0.81 mg/L versus 0.56 to 0.73 mg/L) and total phosphorus concentrations (29 μg/L versus 24 to 34 μg/L) were similar

FIGURE 19.9 The Kettle Rapids dam on the Lower Nelson River was the first of a series of dams that have flooded out the rapids and divided the deeply incised river channel into a series of flooded valley reservoirs. Ice damming and ice jams no longer occur (PHOTO BY R. NEWBURY).

between the Burntwood River and Nelson main-stem stations. Assessment of the changes in water quality caused by Lake Winnipeg regulation and Churchill River diversion is difficult because of a lack of adequate predevelopment data (FEMP 1992c, Williamson and Ralley 1993). However, a number of general observations have been made (FEMP 1992c): (1) natural river sites showed little change; (2) sites along the Churchill River diversion had higher postdevelopment turbidity levels, caused by bank and shoreline erosion; (3) sites along the Churchill River diversion showed significant changes, such as decreased hardness and alkalinity, which reflected differences in water quality between the diverted Churchill River water and water quality in the Rat and Burntwood rivers through which the diverted waters flow; (4) variables such as hardness and conductivity decreased at the outlet of Split Lake as a direct result of the diversion, although annual mean turbidity did not increase; (5) phosphorus increased at most sites immediately after Lake Winnipeg regulation and Churchill River diversion, probably because of increased erosion, and then tended to stabilize as a new balance was reached between discharge and shoreline erosion; and (6) many

discharge-correlated variables shifted from being negatively correlated prior to development to being positively correlated after development for some sites along the diversion route and at the outlet of Split Lake. These shifts were attributed to the increased importance of shoreline erosion.

River Biodiversity and Ecology

The entire Nelson main stem is within the Lower Saskatchewan freshwater ecoregion of Abell et al. (2000). As indicated, few data on biological characteristics for the Nelson main stem appear in primary journal publications; the bulk of information is located in difficult-to-access technical reports.

Algae and Cyanobacteria

Substantial information is available on phytoplankton in lakes and reservoirs of the Nelson main stem. However, information on algae in rivers is limited. For example, Pip (1992c) collected a filamentous alga, *Mougeotia*, in the lower Nelson River in a search for naturally occurring phenolic compounds. None were found in *Mougeotia*.

Plants

Pip and Stepaniuk (1992a) collected 28 taxa of macrophytes from a variety of habitats in the lower Nelson River. The most frequently collected taxa were sedges, narrowleaf bur-reed, water horsetail, and common mare's tail. Major riparian plants include black spruce, tamarack, willows, alders, swamp birch, paper birch, aspen, and white spruce.

Pip and Stepaniuk (1992b) reported that mean macrophyte species richness was lower in habitats downstream of hydroelectric dams on the lower Nelson River than in other stream habitats sampled. Pip (1992c) also surveyed naturally occurring phenolic compounds in 20 species of submerged, floating, and emergent macrophytes in the lower Nelson and its tributaries. Of nine phenolic compounds identified, p-hydroxybenzoic acid was commonest.

Invertebrates

Once again, substantial information is available on zooplankton and benthic invertebrates in lakes of the Nelson main-stem area. However, river information is limited. Baker (1989) conducted a baseline survey of the zoobenthos and zooplankton of the Nelson River estuary and divided the estuary into four distinct zones; the invertebrate fauna of the "riverine" (i.e., freshwater) zone is of most interest here. Chironomid midges, copepods, oligochaete worms, and snails were the most abundant taxa and had the highest species diversity. Baker did not identify the insects to lower taxonomic levels, but he provided genus- and species-level identifications for Hydrozoa (4 taxa), Ectoprocta (2 species), oligochaetes (14 taxa), crustaceans (mysids, 1 species; branchiopods, 2 genera; copepods, 10 taxa), snails (7 taxa), bivalves (2 species), and water mites (1 species). Few of the species of mollusks reported by Baker were found by Pip (1992a) in the lower Nelson or its tributaries. The snails *Gyraulus* sp., *Physa gyrina*, three-ridge valvata snail, and swamp lymnaea snail were the most frequently collected taxa (Pip 1992a). She also collected three bivalve species. Pip claimed that gastropod richness was adversely affected by hydroelectric dams on the lower Nelson.

Vertebrates

A number of lake-based fisheries surveys were done on the upper Nelson prior to Lake Winnipeg regulation and Churchill River diversion. However, Ayles (1974) surveyed the east channel of the Nelson River and reported 13 species of fishes (walleye, cisco, burbot, northern pike, yellow perch, trout perch, longnose sucker, white sucker, shorthead redhorse, mooneye, mottled sculpin, slimy sculpin, spottail shiner). The species composition resembled that of Cross and Playgreen lakes but with less productivity because of the riverine habitat. No commercial fishing existed on the East Channel. MacDonnell and Bernhardt (1992) summarized fisheries studies on the lower Nelson River over the period from 1915 to 1992. Fisheries studies in the lower Nelson began in 1915 when N. A. Comeau led the Burleigh expedition to examine the fisheries resources of Hudson and James bays. However, there was little further interest in fisheries resources until hydroelectric development started (~1974). With completion of the Kettle Generating Station and ongoing construction of the Long Spruce Generating Station in 1974, concern over impacts to the environment, and more specifically to brook trout (considered a heritage species in Manitoba), led the Manitoba government to do a number of studies on brook trout (~1974 to 1981). Construction of the Limestone Generating Station in the 1980s and plans for the Conawapa Generating Station in the early 1990s (not built yet) led to more fisheries studies by the Manitoba government and the consulting industry as part of environmental impact assessments. Studies covered the Nelson main stem from Long Spruce Generating Station to the Nelson River estuary, tributary rivers to the Nelson, and headwater lakes of the tributaries. Forty fish species were recorded from the lower Nelson main stem and 32 species were collected from its tributaries. Considerable information on distribution, abundance, life history, general ecology, and effects of impacts is available for brook trout from the area (MacDonnell and Bernhardt 1992). Lesser information is available for lake sturgeon, lake cisco, lake whitefish, white sucker, longnose sucker, and northern pike. Distribution records, but very little life-history information, are available for the other fish species collected from the lower Nelson River (MacDonnell and Bernhardt 1992).

Beaver, muskrat, mink, and lynx occur through the area (Webb 1973, Webb and Foster 1974). Baker (1989) reported large numbers of beluga whale in the riverine and nearshore estuarine zones during his spring and summer surveys. The Nelson estuary harbors most of the western Hudson Bay beluga population and may represent the largest concentration of beluga in the world (Baker 1989). Bearded seal were also commonly observed, mostly in the autumn.

Ecosystem Processes

Biological studies of the Nelson River have not progressed much past the descriptive phase, although

FIGURE 19.10 Mid-January ice damming in Wapicho Rapids on the lower Nelson River before construction of hydropower dams. Ice dams were formed over rapids in a winter-long sequence that began at Hudson Bay in early November and proceeded up the river over 250 km, ending in late March. Hydro developments have stopped the upstream accumulation by dividing the river into a series of frozen reservoirs (PHOTO BY R. NEWBURY).

information is available on primary productivity in lakes. Limited data on the production of some species of fishes in the lower Nelson main stem are summarized in MacDonnell and Bernhardt (1992). For example, production of sculpins was estimated to be 0.4 to 3.4 kg ha^{-1} yr^{-1} in lower Nelson nursery streams, and production of yearling and older brook trout in two lower Nelson tributaries were 9.0 and 1.2 kg ha^{-1} yr^{-1}, respectively (1980a as in MacDonnell and Bernhardt 1992).

Human Impacts and Special Features

The Nelson is the only northward-flowing subarctic river in North America that gathers most of its flow from a temperate southern basin. As a result, warmwater inflows cause ice to form over the whole winter season in the lower Nelson, starting in its estuary and gradually ascending the river as a series of accumulated covers between hanging ice dams in rapids sections (Fig. 19.10). Spring occurs before the whole of the lower Nelson is ice covered, so there are sections at the head of the upper reaches that are never ice covered. This natural ice regime has been drastically altered by hydropower dams on the lower Nelson. Reservoirs that freeze over in early autumn now halt the upstream progression of the ice cover, and upper reaches are ice scoured for the first time. Bank erosion and channel braiding now occur in these formerly stable reaches. Formerly safe winter ice routes have become hazardous, leading to tragic drownings as the new ice regime becomes established.

In ice-accumulation reaches of the lower Nelson, islands have a "top hat" appearance (i.e., a central core of trees surrounded by a brim of grasses), and shorelines look "trimmed" because of the natural ice regime (Newbury 1990a). The Nelson main stem also has numerous pictographic sites scattered throughout the region.

The Nelson River is a good example of single-purpose development on a massive scale (Newbury 1981). Manitoba's power planners have been aware since the early 1900s of the tremendous hydroelectric potential of the province's northern rivers, especially the Nelson (Newbury et al. 1984, www.hydro.mb.ca/export_minnesota/history.html). The first generating station, Kelsey, was built in 1961 to provide power for the International Nickel Company's (INCO) mining and smelting operation at Thompson, upstream on the Burntwood River. Federal–provincial studies were initiated in 1964 to examine the feasibility of developing further generating stations on the Nelson and Churchill rivers for markets in southern Canada and the northern United States. In 1966, nine sites on the Nelson were identified. Instead of building generating stations on the Churchill River, diversion of the Churchill River into the Nelson River was recommended to increase flows through the Nelson River dams. The dam at Kettle Rapids (1272 MW, see Fig. 19.9) was completed in 1974, and a transmission system was built on the west side of Lake Winnipeg to bring power to the south. Diversion of the Churchill into the Nelson was completed in 1976, followed by Lake Winnipeg regulation in 1977. Lake Winnipeg regulation involved installing a complex series of channels and dikes through the outlet lakes downstream of Lake Winnipeg and building the Jenpeg control dam just upstream of Cross Lake. Long Spruce Dam (980 MW) was completed in 1979 and Limestone Dam (~1300 MW) was completed in 1992. Because the Nelson–Churchill dams were primarily built for power exports, plans to build the Conawapa Dam

further downstream were shelved in 1991 as power demand slumped in the northern United States and Ontario. Although plans to build Conawapa have recently reemerged, a number of smaller power-generating sites have already been identified along the Rat–Burntwood river system, which receives water from the Churchill diversion, and one or more of these sites will likely be started within the next few years.

Diversion of the Churchill River into the Nelson system and the flooding of Southern Indian Lake was extremely contentious in Manitoba; a provincial government went down to defeat over the issue. The environmental and social impacts resulting from diversion activities were immediate and severe and still continue today. For a summary of these effects, see Bodaly et al. (1984), Hecky et al. (1984), Canada–Manitoba Agreement on the Study and Monitoring of Mercury in the Churchill River Diversion (1987), Waldram (1988), Usher and Weinstein (1991), FEMP (1992c), and Rosenberg et al. (1995). The debate over compensation for the permanent loss of resources to people living on the Nelson River, and on the advisability of developing export power, has deeply divided Manitoba aboriginal communities, politicians, academics, and resource professionals (see Tritschler Report 1979 in Waldram 1988).

Mercury contamination of fishes is a by-product of reservoir formation almost everywhere in the world (Rosenberg et al. 1995), and the Nelson River main stem is no exception (Canada–Manitoba Agreement on the Study and Monitoring of Mercury in the Churchill River Diversion 1987). Elevated mercury levels were first detected in fishes from the Churchill River diversion route in 1978 to 1979, just after diversion was completed. Levels in piscivorous fishes (walleye, northern pike) exceeded the Canadian marketing limit of $0.5\,\mu g/g$ in almost all of the lakes along the diversion route after their impoundment. Mercury levels in nonpredatory fishes will remain elevated for 10 to 20 years after flooding, and longer (20 to 30 years) in piscivorous fishes.

Other metals have also been studied in the Nelson River system. Cadmium, copper, and lead concentrations have been analyzed from biota collected in a variety of stream habitats in the lower Nelson River: sediments, algae, and macrophytes (Pip and Stepaniuk 1992a, 1992b), gastropods (Pip 1992b), and fishes (Pip and Stepaniuk 1997). Concentrations of some metals were sometimes higher downstream of hydroelectric dams in some biota, but no clear overall trends were obvious.

The only other major industrial land use in the main-stem Nelson subcatchment is a large nickel mine at Thompson. The mine poses localized problems because of leaching from mine tailings and more widespread problems from high atmospheric emissions during periods of full production (Lane and Sykes 1982).

ADDITIONAL RIVERS

The Bow River is located in southern Alberta, Canada (Fig. 19.21). Its source is the Bow Glacier in Banff National Park, one of the most famous and visited nature reserves in the world and a U.N. World Heritage Site. From its source, the Bow River flows southeast into the aqua blue waters of Lake Louise before emerging from the mountains in the Bow River Pass and entering the aspen parklands of the foothills (Fig. 19.1). The river then traverses the city of Calgary, home to ~1,000,000 people, passing through wide green valleys lined with cottonwood and spruce. The river continues across the Great Plains, where it joins with Oldman River near the Alberta–Saskatchewan border to form the South Saskatchewan River. The Bow is a major tourist attraction near its montane headwaters, sustains a world-class fly fishery for brown and rainbow trout (introduced in the 1920s and 1930s) along an 80 km reach downstream of Calgary, and is a source of water for millions of ha of prairie farmland.

The Otter Tail River is one of two tributaries that, with their confluence, begin the Red River of the North (Fig. 19.23). The other tributary is the Bois de Sioux, draining Lake Traverse. The Otter Tail is unique among the Red River tributaries in that it has a high incidence of lakes in its drainage area and it traverses three ecoregions along its length. These two aspects contribute to its singular character: generally clear waters and an exceptionally diverse biota. Several fish species known from the Otter Tail, for example, occur nowhere else in the subcatchment. The Otter Tail historically was dominated by forest cover (50%), but present forest cover has been reduced to 20%. This value is still higher than most other rivers in the Red River subcatchment. The Otter Tail has also maintained about 13% wetland cover over time and contains 14% open waters in its many lakes. Only 43% of the subcatchment is agricultural land. The lakes, wetlands, and streams of the headwaters help to maintain relatively constant streamflow, contributing to the maintenance of biodiversity. The Otter Tail is one of Minnesota's prime recreational areas, with abundant fishing, hiking, and camping opportunities. In addition, the Otter Tail has a long history of water-powered mills and hydroelectric generation with dams and is the only river in the Red River subcatchment to provide significant power generation.

ACKNOWLEDGMENTS

We thank the following people for their valuable inputs to various sections of this chapter: Saskatchewan River, C. Casey, M. Guy, and M. Meding; Red–Assiniboine River, J. Hatch, S. Nelson, K. Schmidt, and K. Stewart; Winnipeg River, B. Bilyj; Nelson River, R. Baker, G. Dickson, M. Lawrence, D. MacDonnell, E. Pip, R. Remnant, D. Riddle, A. Warkentin, and D. Williamson; and Otter Tail River, T. Waters. We also thank D. Williamson for reviewing the Nelson River section.

LITERATURE CITED

Aadland, L. P. 1993. Stream habitat types: Their fish assemblages and relationship to flow. *North American Journal of Fisheries Management* 13:790–806.

Abell, R. A., D. M. Olson, E. Dinerstein, P. T. Hurley, J. T. Diggs, W. Eichbaum, S. Walters, W. Wettengel, T. Allnutt, C. J. Loucks, and P. Hedao. 2000. *Freshwater ecoregions of North America: A conservation assessment.* Island Press, Washington, D.C.

Anderson, J., B. Paakh, S. Heiskary, and T. Heinrich. 2000. Lake of the Woods trophic status report. Minnesota Pollution Control Agency Report 39-0002. Minnesota Pollution Control Agency, St. Paul.

Anderson, R. S., A. M. Anderson, A. M. Akena, J. S. Livingstone, A. Masuda, P. A. Mitchell, T. B. Reynoldson, D. O. Trew, and M. Vukadinovic. 1986. *North Saskatchewan River: Characterization of water quality in the vicinity of Edmonton (1982–1983). Part I: Introduction, water chemistry, chlorophyll, bacteriology.* Pollution Control Division, Alberta Environment, Edmonton.

Andres, D., and J. Thompson. 1995. Summary of existing hydraulic and geomorphic data on the Assiniboine River between Winnipeg, Manitoba and Preeceville, Saskatchewan. Report no. T95-05. Trillium Engineering and Hydrographics Inc., Edmonton, Alberta.

Armstrong, F. A. J., and D. W. Schindler. 1971. Preliminary chemical characterization of waters in the Experimental Lakes Area, northwestern Ontario. *Journal of the Fisheries Research Board of Canada* 28:171–187.

Ayles, H. A. 1974. The fisheries of the East Channel of the Nelson River. Lake Winnipeg, Churchill and Nelson Rivers Study Board, Canada–Manitoba, 1971–1975. In *Fisheries and limnology.* Vol. 2, *Technical Report*

Appendix 5. Lake Winnipeg, Churchill and Nelson Rivers Study Board, Winnipeg, Manitoba.

Bajkov, A. 1930. Biological conditions of Manitoba lakes. *Contributions to Canadian Biology and Fisheries* 5:383–422.

Baker, R. F. 1989. An environmental assessment and biological investigation of the Nelson River estuary: A report prepared for Manitoba Hydro. North/South Consultants Inc., Winnipeg, Manitoba.

Bayley, S. E., D. W. Schindler, K. G. Beaty, B. R. Parker, and M. P. Stainton. 1992. Effect of multiple fires on nutrient yields from streams draining boreal forest and fen watersheds: Nitrogen and phosphorus. *Canadian Journal of Fisheries and Aquatic Sciences* 49:584–596.

Bayley, S. E., D. W. Schindler, B. R. Parker, M. P. Stainton, and K. G. Beaty. 1992. Effects of forest fire and drought on acidity of a base-poor boreal forest stream: Similarities between climatic warming and acidic precipitation. *Biogeochemistry* 17:191–204.

Bilyj, B., and I. J. Davies. 1989. Descriptions and ecological notes on seven new species of *Cladotanytarsus* (Chironomidae: Diptera) collected from an experimentally acidified lake. *Canadian Journal of Zoology* 67:948–962.

Bodaly, R. A., D. M. Rosenberg, M. N. Gaboury, R. E. Hecky, R. W. Newbury, and K. Patalas. 1984. Ecological effects of hydroelectric development in northern Manitoba, Canada: The Churchill–Nelson diversion. In P. J. Sheehan, D. R. Miller, G. C. Butler, and P. Bourdeau (eds.). *Effects of pollutants at the ecosystem level,* pp. 273–309. Scientific Committee on Problems in the Environment (SCOPE). John Wiley and Sons, New York.

Brunskill, G. J., S. E. M. Elliott, and P. Campbell. 1980. Morphometry, hydrology, and watershed data pertinent to the limnology of Lake Winnipeg. Canadian Manuscript Report of Fisheries and Aquatic Sciences no. 1556. Department of Fisheries and Oceans, Winnipeg, Manitoba.

Brunskill, G. J., and D. W. Schindler. 1971. Geography and bathymetry of selected lake basins, Experimental Lakes Area, northwestern Ontario. *Journal of the Fisheries Research Board of Canada* 28:139–155.

Canada–Alberta Environmentally Sustainable Agriculture Agreement (CAESA). 1998. *Agricultural impacts on water quality in Alberta: An initial assessment.* CAESA, Lethbridge, Alberta.

Canada–Manitoba Agreement on the Study and Monitoring of Mercury in the Churchill River Diversion. 1987. Summary report. Environmental and Workplace Safety and Health, Government of Manitoba, Winnipeg, and Environment Canada, Government of Canada, Hull, Québec.

Carignan, R., P. D'Arcy, and S. Lamontagne. 2000. Comparative impacts of fire and forest harvesting on water quality in Boreal Shield lakes. *Canadian Journal of Fisheries and Aquatic Sciences* 57:105–117.

Carr, G. M., and P. A. Chambers. 1998. Macrophyte growth and sediment phosphorus and nitrogen in a Canadian prairie river. *Freshwater Biology* 39:525–536.

Carr, G. M., and P. A. Chambers. 1999. Spatial and temporal patterns in nutrients and plant biomass in Alberta. Report for the Federal–Provincial Prairie Provinces Water Board. NWRI Contribution no. 98-225. Environment Canada, National Water Research Institute, Saskatoon, Saskatchewan.

Chambers, P. A., and E. E. Prepas. 1994. Nutrient dynamics in riverbeds: The impact of sewage effluent and aquatic macrophytes. *Water Research* 28:453–464.

Chambers, P. A., E. E. Prepas, M. L. Bothwell, and H. R. Hamilton. 1989. Roots versus shoots in nutrient uptake by aquatic macrophytes in flowing waters. *Canadian Journal of Fisheries and Aquatic Sciences* 46:435–439.

Chambers, P. A., E. E. Prepas, H. R. Hamilton, and M. L. Bothwell. 1991. Current velocity and its effect on aquatic macrophytes in flowing waters. *Ecological Applications* 1:249–257.

Charlton, S. E. D., H. R. Hamilton, and P. M. Cross. 1986. *The limnological characteristics of the Bow, Oldman and South Saskatchewan rivers (1979–1982). Part II: The primary producers.* Pollution Control Division, Alberta Environment, Edmonton.

Churchill River Study (Missinipe Probe). 1976. Summary report. Churchill River Study, Saskatoon, Saskatchewan.

City of Winnipeg. 2000. Red and Assiniboine ammonia criteria study 1998–2000. Final Report. City of Winnipeg, Winnipeg, Manitoba.

Clarke, R. M., R. W. Boychuck, and D. A. Hodgins. 1980. Fishes of the Red River at Winnipeg, Manitoba. Unpublished manuscript report. Western Region, Department of Fisheries and Oceans, Winnipeg, Manitoba.

Combet, D. 2001. *In search of the western sea: Selected journals of La Verendrye.* Great Plains Publications, Winnipeg, Manitoba.

Constable, M. 2001. Ecological survey of the South Saskatchewan River downstream of the city of Saskatoon wastewater treatment plant. EPS 5/AT/2. Environmental Protection Branch, Environment Canada, Edmonton, Alberta.

Cowdery, T. K. 1995. Similar agricultural areas, different ground-water quality: Red River of the North basin, 1993–95. Open-File Report 95-441. U.S. Geological Survey, Mounds View, Minnesota.

Cross, P. M., H. R. Hamilton, and S. E. D. Charlton. 1986. *The limnological characteristics of the Bow, Oldman and South Saskatchewan rivers (1979–1982). Part I: Nutrient and water chemistry.* Pollution Control Division, Alberta Environment, Edmonton, Alberta.

Crossman, E. J., and D. E. McAllister. 1986. Zoogeography of freshwater fishes of the Hudson Bay drainage, Ungava Bay and the Arctic Archipelago. In C. H.

Hocutt and E. O. Wiley (eds.). *The zoogeography of North American freshwater fishes*, pp. 53–104. John Wiley and Sons, New York.

Culp, J. M., and R. W. Davies. 1982. Analysis of longitudinal zonation and the river continuum concept in the Oldman–South Saskatchewan river system. *Canadian Journal of Fisheries and Aquatic Sciences* 39:1258–1266.

Culp, J. M., R. W. Davies, H. Hamilton, and A. Sosiak. 1992. Longitudinal zonation of the biota and water quality of a northern temperate river system in Alberta, Canada. In C. D. Becker and D. A. Neitzel (eds.). *Water quality of North American river systems*, pp. 29–49. Batelle Press, Columbus, Ohio.

Cushing, C. E., and J. D. Allan. 2001. *Streams: Their ecology and life.* Academic Press, San Diego.

Dadswell, M. J. 1974. *Distribution, ecology, and post-glacial dispersal of certain crustaceans and fishes in eastern North America.* National Museum of Canada Publications in Zoology no. 11. National Museum of Canada, Ottawa, Ontario.

Davies, R. W., J. M. Culp, R. Green, M. O'Connell, G. Scott, and C. Zimmermann. 1977. River classification of the South Saskatchewan River basin. Technical Report. Pollution Control Division, Alberta Environment, Edmonton.

Dawson, S. J. 1859. *Report on the exploration of the country between Lake Superior and the Red River settlement.* John Lovel Printer, Toronto, Ontario.

Dirschl, H. J. 1972. Geobotanical processes in the Saskatchewan River Delta. *Canadian Journal of Earth Sciences* 9:1529–1549.

Dirschl, H. J., and D. L. Dabbs. 1969. A contribution to the flora of the Saskatchewan River Delta. *Canadian Field-Naturalist* 83:212–228.

Donald, D. B., and R. S. Anderson. 1977. Distribution of the stoneflies (Plecoptera) of the Waterton River drainage, Alberta, Canada. *Syesis* 10:111–120.

Donald, D. B., and R. A. Mutch. 1980. The effect of hydroelectric dams and sewage on the distribution of stoneflies (Plecoptera) along the Bow River. *Quaestiones Entomologicae* 16:658–670.

Douglas, R. 1925. Saskatchewan River bore another name. *Canoma* 7:22–23.

Eggers, S. D., and D. M. Recd. 1997. Wetland plants and communities of Minnesota and Wisconsin. U.S. Army Corps of Engineers, St. Paul District, St. Paul, Minnesota.

Elson, J. A. 1967. Geology of the glacial Lake Agassiz. In W. J. Mayer-Oakes (ed.). *Life, land and water*, pp. 37–95. University of Manitoba Press, Winnipeg.

Environment Canada. 1992. *Saskatchewan sediment data.* Water Survey of Canada, Environment Canada, Ottawa, Ontario.

Environment Canada. 1996. *The state of Canada's environment: 1996.* Environment Canada, Ottawa, Ontario.

Environment Canada. 2001. *Surface water data*, Hydat CD ROM, Version 99-2.00. Water Survey of Canada, Environment Canada, Ottawa, Ontario.

Federal Ecological Monitoring Program (FEMP). 1992a. *Federal ecological monitoring program, final report.* Vol. 1. Environment Canada and Department of Fisheries and Oceans, Winnipeg, Manitoba.

Federal Ecological Monitoring Program. 1992b. *Federal ecological monitoring program, final report.* Vol. 2. Environment Canada and Department of Fisheries and Oceans, Winnipeg, Manitoba.

Federal Ecological Monitoring Program. 1992c. *Summary report.* Environment Canada and Department of Fisheries and Oceans, Winnipeg, Manitoba.

Fee, E. J., R. E. Hecky, M. P. Stainton, P. Sandberg, L. L. Hendzel, S. J. Guildford, H. J. Kling, G. K. McCullough, C. Anema, and A. G. Salki. 1989. Lake variability and climate research in northwestern Ontario: Study design and 1985–1986 data from the Red Lake district. Canadian Technical Report of Fisheries and Aquatic Sciences 1662. Central and Arctic Region, Department of Fisheries and Oceans, Winnipeg, Manitoba.

Feller, M. C., and J. P. Kimmins. 1984. Effects of clearcutting and slash burning on streamwater chemistry and watershed nutrient budgets in southwestern British Columbia. *Water Resources Research* 20:29–40.

Franzin, W. G., B. A. Barton, R. A. Remnant, D. B. Wain, and S. J. Pagel. 1994. Range extension, present and potential distribution, and possible effects of rainbow smelt in Hudson Bay drainage waters of northwestern Ontario, Manitoba and Minnesota. *North American Journal of Fisheries Management* 14:65–76.

Fredeen, F. J. H. 1983. Trends in numbers of aquatic invertebrates in a large Canadian river during four years of black fly larviciding with methoxychlor (Diptera: Simuliidae). *Quaestiones Entomologicae* 19:53–92.

Garcia, E., and R. Carignan. 1999. Impact of wildfire and clear-cutting in the boreal forest on methyl mercury in zooplankton. *Canadian Journal of Fisheries and Aquatic Sciences* 56:339–345.

Glaser, P. H. 1992. Peat landforms. In H. E. Wright, B. A. Coffin, and N. E. Aaseng (eds.). *The patterned peatlands of Minnesota*, pp. 3–14. University of Minnesota Press, Minneapolis.

Golder Associates. 2000. Environmental Effects Monitoring Cycle II: Final interpretive report for Weyerhaeuser's Prince Albert Pulp and Paper Mill. Golder Associates Ltd., Saskatoon, Saskatchewan.

Goldstein, R. M. 1995. Aquatic communities and contaminants in fish from streams of the Red River of the North basin, Minnesota and North Dakota. Water-Resources Investigations Report 95-4047. U.S. Geological Survey, Mounds View, Minnesota.

Goldstein, R. M., J. C. Stauffer, P. R. Larson, and D. L. Lorenz. 1996. Relation of physical and chemical characteristics of streams to fish communities in the Red River of the North basin, Minnesota and North Dakota, 1993–95. Water-Resources Investigations Report 96-4227. U.S. Geological Survey, Mounds View, Minnesota.

Graham, R. D., N. Soonawala, and T. A. Niemi. 1998. 1997 annual report of radiological monitoring results from the Chalk River and Whiteshell Laboratories sites. Atomic Energy Commission Limited Miscellaneous Report 362–97. Vol. 3. Environmental Monitoring. Scientific Document Distribution Office, Atomic Energy of Canada, Ltd., Chalk River, Ontario.

Green, R., and R. W. Davies. 1980. The epilithic algae of the Oldman–South Saskatchewan River system, Alberta, Canada. *Nova Hedwigia* 33:261–278.

Gregory, L. A., and J. S. Loch. 1973. Benthos studies (1971 and 1972) on the Winnipeg River in the vicinity of the Abitibi Manitoba Paper Company, Pine Falls, Manitoba. Canada Fisheries and Marine Service Technical Report 73–330. Central Region, Resource Management Branch, Department of Fisheries and Oceans, Winnipeg, Manitoba.

Gurney, S. 1991. Proposed water quality objectives through Manitoba's watershed classification process: Red and Assiniboine rivers and their tributaries within and downstream of the city of Winnipeg (including Appendix A: Water quality data summary). Technical document. Manitoba Environment, Winnipeg.

Hall, R. J. 1994. Responses of benthic communities to episodic acid disturbances in a lake outflow stream at the Experimental Lakes Area, Ontario. *Canadian Journal of Fisheries and Aquatic Sciences* 51:1877–1892.

Hamilton, H. R., and L. J. North. 1986. Bow River water quality monitoring 1970–1980. Pollution Control Division, Alberta Environment, Edmonton.

Harris, A. G., S. C. McMurray, P. W. C. Uhlig, J. K. Jeglum, R. F. Foster, and G. D. Racey. 1996. *Field guide to the wetland ecosystem classification for northwestern Ontario*. Field Guide FG-01. Ontario Ministry of Natural Resources Information Centre, Peterborough, Ontario.

Haslam, S. M. 1978. *River plants: The macrophytic vegetation of watercourses*. Cambridge University Press, Cambridge.

Hecky, R. E., R. W. Newbury, R. A. Bodaly, K. Patalas, and D. M. Rosenberg. 1984. Environmental impact prediction and assessment: The Southern Indian Lake experience. *Canadian Journal of Fisheries and Aquatic Sciences* 41:720–732.

Heinselman, M. L. 1970. Landscape evolution, peatland types and the environment in the Lake Agassiz Peatlands Natural Area, Minnesota. *Ecological Monographs* 40:235–242.

Heinselman, M. L. 1973. Fire in the virgin forests of the Boundary Waters Canoe Area, Minnesota. *Quarternary Research* 3:329–382.

Hunt, C. B. 1974. *Natural regions of the United States and Canada*. W. H. Freeman, San Francisco.

Jones, G., and N. Armstrong. 2001. Long-term trends in total nitrogen and total phosphorous concentrations in Manitoba streams. Manitoba Conservation Report no. 2001-07. Manitoba Conservation, Government of Manitoba, Winnipeg.

Kellerhals, R. C., C. R. Neil, and D. I. Bray. 1972. Hydraulic and geomorphic characteristics of rivers in Alberta. Cooperative Research Program in Highway and River Engineering Technical Report. Alberta Environment, Edmonton.

Koel, T. M. 1997. Distribution of fishes in the Red River of the North basin on multivariate environmental gradients. Ph.D. diss., North Dakota State University, Fargo.

Lane, R. K., and G. N. Sykes. 1982. *Nature's lifeline: Prairie and northern waters*. Canada West Foundation, Calgary, Alberta.

Lehmkuhl, D. M. 1972. Change in thermal regime as a cause of reduction of benthic fauna downstream of a reservoir. *Journal of the Fisheries Research Board of Canada* 29:1329–1332.

Loch, J. S., A. J. Derksen, M. E. Hora, and R. B. Oetting. 1979. Potential effects of exotic fishes on Manitoba: An impact assessment of the Garrison Diversion Unit. Fisheries and Marine Service Technical Report no. 838. Fisheries and Environment Canada, Winnipeg, Manitoba.

Lockhart, W. L., P. Wilkinson, B. N. Billeck, G. A. Stern, R. A. Danell, J. Delaronde, and D. C. G. Muir. 2000. Studies of dated sediment cores from Lake Winnipeg, 1994. In B. J. Todd, C. F. M. Lewis, D. L. Forbes, L. H. Thorleifson, and E. Neilson (eds.). 1996 Lake Winnipeg project: Cruise report and scientific results, pp. 257–267. Geological Survey of Canada Open File Report 3470. Natural Resources Canada, Ottawa, Ontario.

Longmore, L. A., and C. E. Stenton. 1981. *The fish and fisheries of the South Saskatchewan River basin*. Planning Division, Alberta Environment, Edmonton.

MacDonnell, D. S., and W. J. Bernhardt. 1992. A synthesis of fisheries studies conducted on the lower Nelson River between 1915 and 1992 and general life history descriptions of the resident fish species. Report prepared for internal use by Manitoba Hydro. North/South Consultants Inc., Winnipeg, Manitoba.

Manitoba Conservation. 2000. *Geographical names of Manitoba*. Manitoba Conservation, Winnipeg.

Manitoba Environment. 1997. *State of the environment report for Manitoba*. Manitoba Environment, Winnipeg.

McCulloch, B. R., and W. G. Franzin. 1996. A summary of fish collections from the Assiniboine River drainage with distribution maps. Canadian Technical Report of Fisheries and Aquatic Sciences 2087. Department of Fisheries and Oceans, Winnipeg, Manitoba.

Meredith, G. N., and J. J. Houston. 1988a. Status of the green sunfish, *Lepomis cyanellus*, in Canada. *Canadian Field-Naturalist* 102:270–276.

Meredith, G. N., and J. J. Houston. 1988b. Status of the longear sunfish, *Lepomis megalotis*, in Canada. *Canadian Field-Naturalist* 102:277–285.

Merriman, J. C., D. H. J. Anthony, J. A. Kraft, and R. J. Wilkinson. 1991. River water quality in the vicinity of bleached kraft mills. *Chemosphere* 23:1605–1615.

Metcalfe, J. L., and A. Hayton. 1989. Comparison of leeches and mussels as biomonitors for chlorophenol pollution. *Journal of Great Lakes Research* 15:654–668.

National Atlas of Canada. 1985. 5th ed. Energy Mines and Resources Canada, Ottawa.

Newbury, R. W. 1981. Some principles of compatible hydroelectric design. *Canadian Water Resources Journal* 6:284–294.

Newbury, R. W. 1990a. The Nelson river. In M. G. Wolman and H. C. Riggs (eds.), *The geology of North America*. Vol. 0-1, *Surface water hydrology*, pp. 287–292. Geological Society of America, Boulder, Colorado.

Newbury, R. W. 1990b. Northern waters: The discovery and development of the rivers of northern Manitoba. Presentation to People and Land in Northern Manitoba Conference, 2–4 May 1990, University of Manitoba, Winnipeg.

Newbury, R. W., and G. W. Malaher. 1972. The destruction of Manitoba's last great river. *Nature Canada* 1(4):4–13.

Newbury, R. W., G. K. McCullough, and R. E. Hecky. 1984. The Southern Indian Lake impoundment and Churchill River diversion. *Canadian Journal of Fisheries and Aquatic Sciences* 41:548–557.

Newmaster, S., A. Harris, and L. Kershaw. 1997. *Wetland plants of Ontario*. Lone Pine, Edmonton, Alberta.

Parks, J. W., C. Curry, D. Romani, and D. D. Russell. 1991. Young northern pike, yellow perch and crayfish as bioindicators in a mercury contaminated watercourse. *Environmental Monitoring and Assessment* 16:39–73.

Patalas, K. In press. Lake Winnipeg, a remnant of glacial Lake Agassiz, an efficient dispersal route of planktonic crustaceans through central and northern Canada. *Aquatic Ecosystem Health and Management.*

Patalas, K., and A. Salki. 1992. Crustacean plankton in Lake Winnipeg: Variation in space and time as a function of lake morphology, geology, and climate. *Canadian Journal of Fisheries and Aquatic Sciences* 49:1035–1059.

Paterson, A. M., B. F. Cumming, J. P. Smol, J. M. Blais, and R. L. France. 1998. Assessment of the effects of logging, forest fires and drought on lakes in northwestern Ontario: A 30-year paleolimnological perspective. *Canadian Journal of Forest Research* 28:1546–1556.

Peterka, J. J. 1978. Fishes and fisheries of the Sheyenne River, North Dakota. *Proceedings of the North Dakota Academy of Science* 32:29–44.

Peterka, J. J. 1992. Survey of fishes in six streams in northeastern North Dakota. Completion report. North Dakota Game and Fish Department, Bismarck.

Peterka, J. J., and T. M. Koel. 1996. *Distribution and dispersal of fishes in the Red River basin.* Zoology Department, North Dakota State University, Fargo.

Pip, E. 1992a. The ecology of subarctic molluscs in the lower Nelson River system, Manitoba, Canada. *Journal of Molluscan Studies* 58:121–126.

Pip, E. 1992b. Cadmium, copper and lead in gastropods of the lower Nelson River system, Manitoba, Canada. *Journal of Molluscan Studies* 58:199–205.

Pip, E. 1992c. Phenolic compounds in macrophytes from the Lower Nelson River system, Canada. *Aquatic Botany* 42:273–279.

Pip, E., and J. Stepaniuk. 1992a. Cadmium, copper and lead in sediments and aquatic macrophytes in the lower Nelson River system, Manitoba, Canada. I. Interspecific differences and macrophyte–sediment relations. *Archiv für Hydrobiologie* 124:337–353.

Pip, E., and J. Stepaniuk. 1992b. Cadmium, copper and lead in sediments and aquatic macrophytes in the lower Nelson River system, Manitoba, Canada. II. Metal concentrations in relation to hydroelectric development. *Archiv für Hydrobiologie* 124:451–458.

Pip, E., and J. Stepaniuk. 1997. Cadmium, copper and lead in fish from the lower Nelson River system in northern Manitoba. *Canadian Field-Naturalist* 111:403–406.

Province of Manitoba. 1974. Physical impact study. Lake Winnipeg, Churchill and Nelson Rivers Study Board, Canada–Manitoba, 1971–1975. In *Hydrologic, hydraulic and geomorphologic studies*. Vol. 1, *Technical Report Appendix 2*. Lake Winnipeg, Churchill and Nelson Rivers Study Board, Winnipeg, Manitoba.

Renard, P. A., S. R. Hanson, and J. W. Enblom. 1986. Biological survey of the Red River of the North. Special Publication no. 142. Division of Fish and Wildlife, Minnesota Department of Natural Resources. St. Paul.

Ricketts, T. H., E. Dinerstein, D. M. Olson, C. L. Loucks, W. Eichbaum, D. DellaSala, K. Kavanagh, P. Hedao, P. T. Hurley, K. M. Carney, R. Abell, and S. Walters. 1999. *Terrestrial ecoregions of North America: A conservation assessment*. Island Press, Washington, D.C.

Rosenberg, D. M., R. A. Bodaly, and P. J. Usher. 1995. Environmental and social impacts of large-scale hydroelectric development. *Global Environmental Change* 5:127–148.

Rudd, J. W. M., M. A. Turner, A. Furutani, A. L. Swick, and B. E. Townsend. 1983. The English–Wabigoon River system: I. A synthesis of recent research with a view towards mercury amelioration. *Canadian Journal of Fisheries and Aquatic Sciences* 40:2206–2217.

Rusak, J. M., and T. Mosindy. 1997. Seasonal movements of lake sturgeon in Lake of the Woods and the Rainy River, Ontario. *Canadian Journal of Zoology* 75:383–395.

Salki, A. G. 1996. The crustacean plankton community of Lake Winnipeg in 1929, 1969 and 1994. In B. J. Todd, C. F. M. Lewis, L. H. Thorleifson, and E. Neilson (eds.). Lake Winnipeg project: Cruise report and scientific results, pp. 319–344. Geological Survey of Canada Open File Report 3113. Natural Resources Canada, Ottawa, Ontario.

Salki, A. G., M. Turner, K. Patalas, J. Rudd, and D. Findlay. 1985. The influence of fish–zooplankton–phytoplankton interactions on the results of selenium toxicity experiments within large enclosures. *Canadian Journal of Fisheries and Aquatic Sciences* 42:1132–1143.

Saskatchewan Environment. 1984. North Saskatchewan River water quality, 1970–1982: Summary report. Water Pollution Control Branch, Saskatchewan Environment, Regina.

Schindler, D. W. 1990. Experimental perturbation of whole lakes and tests of hypotheses concerning ecosystem structure and function. *Oikos* 57:25–41.

Schindler, D. W. 2001. The cumulative effects of climate warming and other human stresses on Canadian freshwaters in the new millennium. *Canadian Journal of Fisheries and Aquatic Sciences* 58:1–12.

Schindler, D. W., S. E. Bayley, P. J. Curtis, B. R. Parker, M. P. Stainton, and C. A. Kelly. 1992. Natural and man-caused factors affecting the abundance and cycling of dissolved organic substances in precambrian shield lakes. *Hydrobiologia* 229:1–21.

Schindler, D. W., S. E. Bayley, B. R. Parker, K. G. Beaty, D. R. Cruikshank, E. J. Fee, E. U. Schindler, and M. P. Stainton. 1996. The effects of climate warming on the properties of boreal lakes and streams at the Experimental Lakes Area, northwestern Ontario. *Limnology and Oceanography* 41:1004–1017.

Schindler, D. W., P. J. Curtis, S. E. Bayley, B. R. Parker, K. G. Beaty, and M. P. Stainton. 1996. Climate-induced changes in the dissolved organic carbon budgets of boreal lakes. *Biogeochemistry* 36:9–28.

Scott, W. B., and E. J. Crossman. 1973. *Freshwater fishes of Canada*. Bulletin no. 184. Fisheries Research Board of Canada, Ottawa, Ontario.

Shaw, R. D., P. A. Mitchell, and A. M. Anderson. 1994. Water quality of the North Saskatchewan River in Alberta. Alberta Environmental Protection, Edmonton.

Smith, N. D., R. L. Slingerland, M. Pérez-Arlucea, and G. S. Morozova. 1998. The 1870s avulsion of the Saskatchewan River. *Canadian Journal of Earth Sciences* 35:453–466.

Sosiak, A. J. 1990. An evaluation of nutrients and biological conditions in the Bow River, 1986–88. Alberta Environment, Calgary.

Sosiak, A. J. 2002. Long-term response of periphyton and macrophytes to reduced municipal nutrient loading to the Bow River (Alberta, Canada). *Canadian Journal of Fisheries and Aquatic Sciences* 59:987–1001.

Statistics Canada. 2003. *Human activity and the environment: Annual statistics 2003*. Catalogue no. 16-201-XPE. Statistics Canada, Ottawa, Ontario.

Stewart, K. W., W. G. Franzin, B. R. McCulloch, and G. F. Hanke. 2001. Selected case histories of fish species invasions into the Nelson River system in Canada. In J. A. Leitch and M. J. Tenamoc (eds.). *Science and policy: Interbasin water transfer of aquatic biota*, pp. 63–81. North Dakota State University Press, Fargo.

Stoner, J. D., and D. L. Lorenz. 1996. National Water-Quality Assessment Program: Data collection in the Red River of the North basin, Minnesota, North Dakota, and South Dakota, 1992–95. Fact Sheet FS-172-95. U.S. Geological Survey, Mounds View, Minnesota.

Stoner, J. D., D. L. Lorenz, G. J. Wiche, and R. M. Goldstein. 1993. Red River of the North basin, Minnesota, North Dakota, and South Dakota. *Water Resources Bulletin* 29:575–615.

Taylor, B. R., and H. R. Hamilton. 1993. Regional and temporal patterns of total solutes in the Saskatchewan River basin. *Water Resources Bulletin* 29:221–234.

Taylor, C. F. 2002. *The American Indian*. Courage Books, Philadelphia.

Teller, J. T., and L. Clayton. 1983. *Glacial Lake Agassiz*. Geological Association of Canada Special Paper 26. University of Toronto Press, Toronto, Ontario.

Tornes, L. H., and M. E. Brigham. 1995. Pesticide amounts are small in streams in the Red River of the North basin, 1993–94. Open-File Report 95-283. U.S. Geological Survey, Mounds View, Minnesota.

Urquizo, N., T. Brydges, and H. Shear. 1998. Boreal ecozone pilot project assessment. In *Proceedings of the 4th National Science Meeting, Ecological Monitoring and Assessment Network, LaMalbaie, Quebec*. Ecological Monitoring and Assessment Network Coordinating Office, Canada Centre for Inland Waters, Environment Canada, Burlington, Ontario. Available from EMAN. Web site: http://www.eman-rese.ca

U.S. Department of Agriculture (USDA). 1997. 1997 Census of Agriculture, Volume 1: Part 23, Chapter 2, Minnesota County-Level Data Web site: www.nass.usda.gov/census/census97/volume1/mn-23/toc297.htm

Usher, P. J., and M. S. Weinstein. 1991. Towards assessing the effects of Lake Winnipeg regulation and Churchill River diversion on resource harvesting in native communities in northern Manitoba. Canadian Technical Report of Fisheries and Aquatic Sciences Report 1794. Department of Fisheries and Oceans, Winnipeg, Manitoba.

Waldram, J. B. 1988. *As long as the rivers run: Hydroelectric development and native communities in western Canada*. University of Manitoba Press, Winnipeg.

Watkins, E. M., D. W. Schindler, M. A. Turner, and D. L. Findlay. 2001. Effects of solar radiation on epilithic metabolism, nutrients and community composition in a clear-water boreal lake. *Canadian Journal of Fisheries and Aquatic Sciences* 58:2059–2070.

Watson, E. T., L. C. Graham, and W. G. Franzin. 1998. The distribution of Unionidae (Mollusca: Bivalvia) in the Assiniboine River drainage in Manitoba. Canadian Technical Report of Fisheries and Aquatic Sciences 2232. Department of Fisheries and Oceans, Winnipeg, Manitoba.

Webb, R. 1973. Wildlife resource impact assessment, Lake Winnipeg, Churchill, and Nelson rivers hydroelectric project. No. 1, Outlet lakes. Lake Winnipeg, Churchill and Nelson Study Board, Canada–Manitoba, 1971–1978. In *Wildlife studies, Technical Report Appendix* 6. Lake Winnipeg, Churchill and Nelson Rivers Study Board, Winnipeg, Manitoba.

Webb, R., and J. Foster. 1974. Wildlife resource impact assessment, Lake Winnipeg, Churchill and Nelson rivers hydroelectric project. No. 3, Lake Winnipeg, lower Churchill River, Rat-Burntwood diversion. Lake Winnipeg, Churchill and Nelson Study Board, Canada–Manitoba, 1971–1978. In *Wildlife studies, Technical Report Appendix* 6. Lake Winnipeg, Churchill and Nelson Rivers Study Board, Winnipeg, Manitoba.

Williamson, D. A., and W. E. Ralley. 1993. A summary of water chemistry changes following hydroelectric development in northern Manitoba, Canada. Water Quality Management Section Report no. 93-2. Manitoba Environment, Winnipeg.

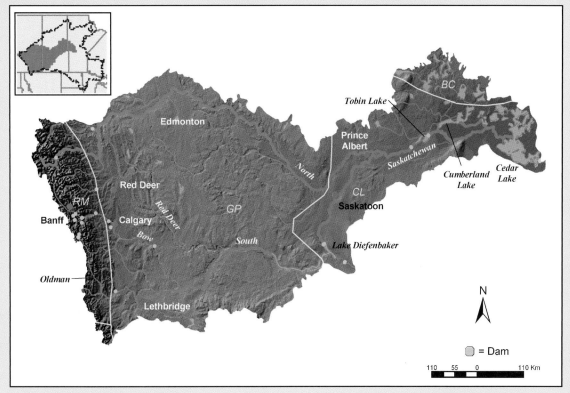

FIGURE 19.11 Map of the Saskatchewan River basin. Physiographic provinces are separated by yellow lines.

SASKATCHEWAN RIVER

Relief: 3307 m

Basin area: 335,900 km^2

Mean discharge: 567 m^3/s (postregulation)

River order: 8

Mean annual precipitation: 45.2 cm

Mean air temperature: −0.3°C

Mean water temperature: 9.7°C

Physiographic provinces: Rocky Mountains in Canada (RM), Great Plains (GP), Central Lowland (CL), Bear–Slave–Churchill Uplands (BC)

Biomes: Temperate Mountain Forest, Temperate Grasslands, Boreal Forest

Freshwater ecoregions: Canadian Rockies, Upper Saskatchewan, Lower Saskatchewan

Terrestrial ecoregions: 7 ecoregions (see text)

Number of fish species: ≥48

Number of endangered species: NA

Major fishes: cutthroat trout, rainbow trout, bull trout, brook trout, brown trout, mountain whitefish, longnose sucker, longnose dace, northern pike, walleye, goldeye, yellow perch, quillback, shorthead redhorse, lake sturgeon

Major other aquatic vertebrates: beaver, mink, white pelican, river otter, muskrat, tundra swan, ring-necked duck

Major benthic invertebrates: mayflies (*Baetisca, Baetis, Ephemera, Ephemerella, Ephoron, Heptagenia, Tricorythodes*), stoneflies (*Isoperla, Choroterpes*), caddisflies (*Brachycercus, Cheumatopsyche, Helicopsyche, Symphitopsyche, Traverella*), true flies (Chironominae, Tanypodinae, Orthocladiinae), crustaceans (*Orconectes*)

Nonnative species: brown trout, rainbow trout, brook trout, purple loosestrife, curly pondweed

Major riparian plants: red-osier dogwood, sandbar willows, poplar, water birch

Special features: originates in glaciers and snowfields of Rocky Mountains in Alberta, a World Heritage Site; headwaters of North Saskatchewan River in Banff National Park designated Canadian Heritage Rivers; designated globally important bird areas in portions of prairies and boreal forests

Fragmentation: dams throughout for hydropower and irrigation

Water quality: pH = 8.0, alkalinity = 131 mg/L as CaCO$_3$; relatively free of pollutants in mountains (NO$_3$-N = 0.075 mg/L, PO$_4$-P = 0.009 mg/L); higher nutrient concentrations below major cities and in agricultural areas

Land use: 67% cropland, 3% shrub, 7% grassland, 22% forest

Population density: 9.6 people/km^2

Major information sources: Donald and Mutch 1980, Culp and Davies 1982, Charlton et al. 1986, Cross et al. 1986, Hamilton and North 1986, Sosiak 1990, Culp et al. 1992, Chambers and Prepas 1994, Carr and Chambers 1998, Environment Canada 2001, www.climate.weatheroffice.ec.gc.ca/climate_normals/index_e.html 2002

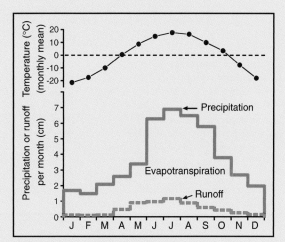

FIGURE 19.12 Mean monthly air temperature, precipitation, and runoff for the Saskatchewan River basin.

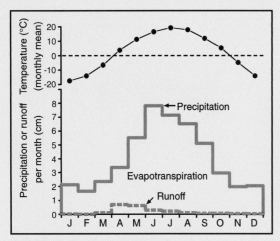

FIGURE 19.14 Mean monthly air temperature, precipitation, and runoff for the Red River of the North basin.

RED RIVER OF THE NORTH

Relief: 350 m (includes Assiniboine)
Basin area: 287,500 km² (including Assiniboine basin)
    116,600 km² (upstream of Assiniboine confluence)
Mean discharge: 236 m³/s (includes Assiniboine)
River order: 8
Mean precipitation: 48.9 cm
Mean air temperature: 2.4°C
Mean water temperature: NA
Physiographic province: Central Lowland (CL)
Biome: Temperate Grasslands
Freshwater ecoregion: English–Winnipeg Lakes
Terrestrial ecoregions: Northern Tall Grasslands,
    Canadian Aspen Forest and Parklands, Northern
    Mixed Grasslands, Western Great Lakes Forests,
    Upper Midwest Forest/Savanna Transition Zone
Number of fish species: ~94
Number of endangered species: 14 fishes
Major fishes: channel catfish, black bullhead, walleye,
    sauger, freshwater drum, common carp, white
    sucker, shorthead redhorse, goldeye, mooneye, silver
    chub, emerald shiner, black crappie

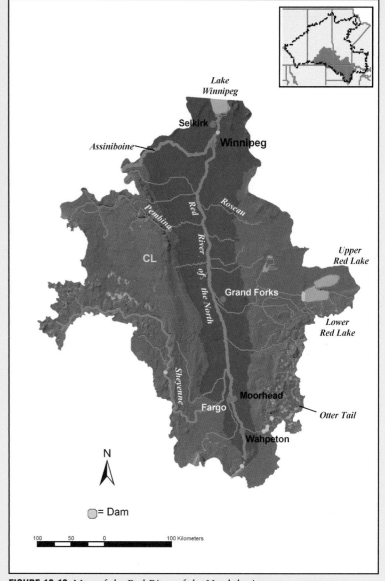

FIGURE 19.13 Map of the Red River of the North basin.

Major other aquatic vertebrates: beaver, muskrat, western painted turtle, common snapping turtle, wood frog, chorus frog, spring peeper, northern leopard frog, gray treefrog, American toad, tiger salamander, blue spotted salamander, mud puppy
Major benthic invertebrates: mayflies (*Baetis, Heptagenia, Ephoron, Hexagenia, Pentagenia, Isonychia, Tricorythodes*), caddisflies (*Ceratopsyche, Hydropsyche, Brachycentrus*), stoneflies (*Acroneuria, Pteronarcys*), true flies (*Tipula, Bezzia, Axarus*), bivalves (*Fusconaia, Amblema, Quadrula, Lasmigona, Anodontoides, Pyganodon, Strophitus, Ligumia, Lampsilis, Potamilus*)
Nonnative species: common carp, white bass, largemouth bass, smallmouth bass
Major riparian plants: cottonwood, green ash, peach-leaved willow, burr oak, basswood, elm
Special features: very low-gradient large river
Fragmentation: main stem largely continuous; all major tributaries dammed
Water quality: specific conductance = 690 μS/cm, pH = 8.1, alkalinity = 215 mg/L as $CaCO_3$, NO_3-N = 0.34 mg/L, total P = 0.3 mg/L, Na = 34 mg/L, total dissolved solids = 381 mg/L
Land use: U.S. portion, 74% agriculture (66% cropland, 8% pasture), 26% forest
Population density: 4.8 people/km² (United States); 59 people/km² (Canada)
Major information sources: Peterka 1978, 1992, Renard et al. 1986, Stoner et al. 1993, Goldstein 1995, Goldstein et al. 1996, Peterka and Koel 1996, Koel 1997

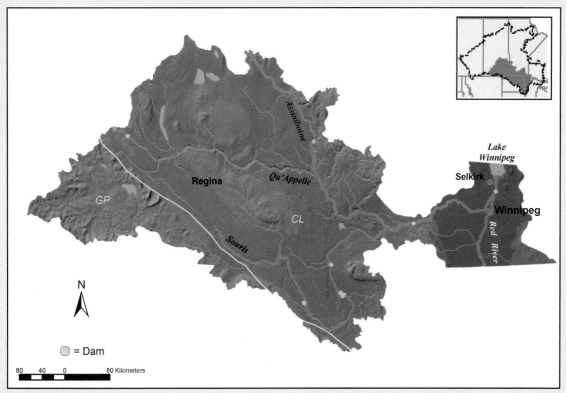

FIGURE 19.15 Map of the Assiniboine River basin. Physiographic provinces are separated by a yellow line.

ASSINIBOINE RIVER

Relief: 350 m
Basin area: 162,000 km²
Mean discharge: 47.4 m³/s
River order: ~7
Mean precipitation: 45.4 cm
Mean air temperature: 2.4°C
Mean water temperature: 9.0°C
Physiographic province: Central Lowland (CL), Great Plains (GP)
Biome: Temperate Grasslands
Freshwater ecoregion: English–Winnipeg Lakes
Terrestrial ecoregions: Mid-continental Canadian Forests, Northern
    Mixed Grasslands, Northern Tall Grasslands, Canadian Aspen
    Forests and Parklands
Number of fish species: 55
Number of endangered species: none
Major fishes: walleye, sauger, channel catfish, goldeye, mooneye,
    common carp, white sucker, silver redhorse, golden redhorse,
    shorthead redhorse, quillback
Major other aquatic vertebrates: beaver, muskrat, wood frog, western chorus frog, spring peeper, northern leopard frog, gray
    treefrog, American toad, snapping turtle
Major benthic invertebrates: bivalves (*Fusconaia, Amblema, Quadrula, Lasmigona, Anodontoides, Pyganodon, Strophitus,
    Ligumia, Lampsilis, Potamilus,* Sphaeriidae), snails (*Physa, Helisoma*), crustaceans (*Orconectes*), mayflies (*Baetis,
    Heptagenia, Ephoron, Hexagenia, Pentagenia, Isonychia, Tricorythodes*), caddisflies (*Ceratopsyche, Hydropsyche,
    Brachycentrus*), stoneflies (*Acroneuria, Pteronarcys*), dragonflies (*Gomphus*), beetles (*Narpus*), true flies (*Tipula, Bezzia*)
Nonnative species: common carp, white bass
Major riparian plants: several grasses, red-osier dogwood, peach-leaved willow, American elm, cottonwood, green ash,
    basswood, cattail
Special features: very long, relatively intact prairie river
Fragmentation: several dams on tributaries; two dams on main stem
Water quality: specific conductance = 865 µS/cm, pH = 8.2, Ca = 60 to 120 mg/L, Mg = 20 to 55 mg/L, total N = 1.52 mg/L,
    total P = 0.25 mg/L, Na = 50 mg/L, chloride = 26 mg/L, total dissolved solids = 552 mg/L
Land use: ~70% to 80% agriculture (~70% cropland, 10% to 15% pasture), remainder forested uplands
Population density: 1 person/km² upstream of city of Winnipeg
Major information sources: Andres and Thompson 1995, McCulloch and Franzin 1996, Gurney 1991,
    www.climate.weatheroffice.ec.gc.ca/climate/climate_normals/index_e.cfm 2002

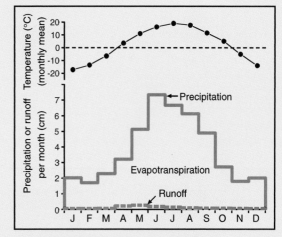

FIGURE 19.16 Mean monthly air temperature,
precipitation, and runoff for the Assiniboine River basin.

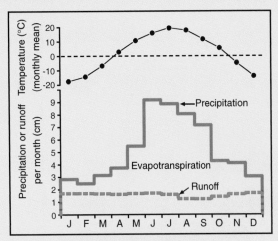

FIGURE 19.18 Mean monthly air temperature, precipitation, and runoff for the Winnipeg River basin.

WINNIPEG RIVER

Relief: 195 m
Basin area: 135,800 km²
Mean discharge: 850 m³/s
River order: NA
Mean annual precipitation: 62 cm
Mean air temperature: 2.4°C
Mean water temperature: NA
Physiographic provinces: Central Lowland (CL), Superior Upland (SU)
Biome: Boreal Forest
Freshwater ecoregion: English–Winnipeg Lakes
Terrestrial ecoregions: Western Great Lakes Forests, Midwestern Canadian Shield Forests
Number of fish species: 69
Number of endangered species: 1 snail, 2 fishes, 1 snake
Major fishes: northern pike, walleye, sauger, yellow perch, lake whitefish, lake trout, muskellunge, brook trout, smallmouth bass, white sucker, rainbow smelt
Major other aquatic vertebrates: beaver, river otter, muskrat, mink, loon, Canada goose, mallard, wood duck, great blue heron, herring gull, painted turtle
Major benthic invertebrates: oligochaete worms (*Limnodrilus, Nais*), leeches (*Erpobdella, Helobdella, Placobdella*), crustaceans (*Hyalella, Gammarus, Orconectes, Candona, Canthocamptus*), odonates (*Aeshna, Enallagma, Libellula*), mayflies (*Caenis, Hexagenia*), caddisflies (*Hydropsyche, Cheumatopsyche*), true flies (*Chaoborus, Bezzia, Procladius, Cricotopus, Psectrocladius*), bivalves (*Pisidium, Musculium*), snails (*Gyraulus, Amnicola*)
Nonnative species: rainbow smelt, *Eubosmina coregoni, Bythotrephes cederstroemi*
Major riparian plants: bunchberry, Canada mayflower, chokecherry, dogwood, fragrant bedstraw, green alder, highbush cranberry, lady fern, lowbush cranberry, mountain maple, northern bluebell, oak fern, pin cherry, pussy willow, rattlesnake plantain, slender wood grass, rose twisted stalk, sensitive fern, skunk currant, snowberry, starflower, white cedar
Special features: pristine wilderness except for dams; white-water rivers
Fragmentation: 33 major dams and control structures, many for hydropower
Water quality: no major pollutants but increased nutrients; total N = 0.30 to 1.00 mg/L, total P = 0.006 to 0.06 mg/L, alkalinity = 15 to 45 mg/L as $CaCO_3$
Land use: <5% agriculture, 30% forestry activities, <1% urban, remainder is natural
Population density: 0.6 people/km²
Major information sources: Heinselman 1970, 1973, www.rom.on.ca 2004, Schindler 1990, 2001, Scott and Crossman 1973, www.pca.state.mn.us 2003

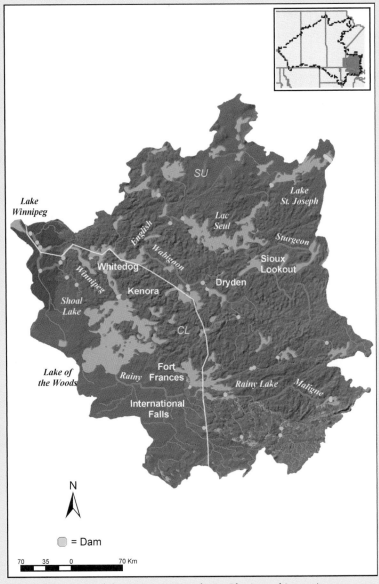

FIGURE 19.17 Map of the Winnipeg River basin. Physiographic provinces are separated by a yellow line.

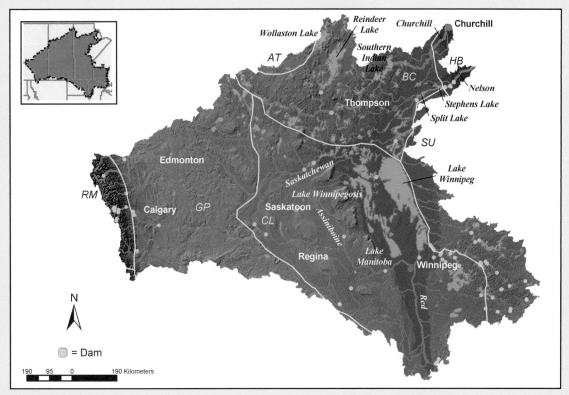

FIGURE 19.19 Map of the Nelson River basin. Physiographic provinces are separated by yellow lines.

NELSON RIVER

Relief: 218 m (main stem only) 3370 m (entire basin)

Basin area: 89,000 km² (main stem only) 1,093,442 km² (entire basin)

Mean discharge: 2480 m³/s

River order: 9

Mean annual precipitation: 52 cm

Mean air temperature: −3.4°C

Mean water temperature: NA

Physiographic provinces: Bear–Slave–Churchill Uplands (BC), Hudson Bay Lowland (HB) (main stem only); five other provinces for entire basin

Biome: Boreal Forest

Freshwater ecoregion: Lower Saskatchewan

Terrestrial ecoregions: Midwestern Canadian Shield Forests, Southern Hudson Bay Taiga (main stem only)

Number of fish species: 46 (main stem); >94 (entire basin)

Number of endangered species: NA

Major fishes: lake sturgeon, northern pike, brook trout, lake cisco, lake whitefish, longnose sucker, white sucker, burbot, walleye

Major other aquatic vertebrates: beaver, muskrat, mink, beluga whale, bearded seal

Major benthic invertebrates: chironomid midges, copepod crustaceans, oligochaete worms, snails (three-ridge valvata, swamp lymnaea, *Gyraulus*, *Physa gyrina*), bivalves

Nonnative species: NA (for main stem)

Major riparian plants: black spruce, tamarack, willow, alder, swamp birch, paper birch, trembling aspen, white spruce

Special features: on main stem, ice-caused "top hat" appearance of islands and "trimmed" shorelines in some reaches; pictographic sites throughout the region

Fragmentation: five hydropower dams on main stem; sites identified for future dams

Water quality: no major pollutants, some problems near population centers, sediment problems connected with Churchill River diversion; pH = ~8.0, alkalinity = ~100 mg/L as $CaCO_3$ (total), $NO_3 + NO_2\text{-}N$ = ~15 mg/L, total N (Kjeldahl) = ~0.70 mg/L, total P = ~30 µg/L

Land use: main stem largely natural boreal forest, localized mining, hydropower, natural resource harvesting (aboriginal population); entire basin 34% forest, 51% cropland, 5% grassland, 2% shrub

Population density: ≤0.5 people/km² (main stem); 5 people/km² (entire basin)

Major information sources: FEMP 1992a, 1992b, 1992c, Newbury 1990a, Revenga et al. 1998

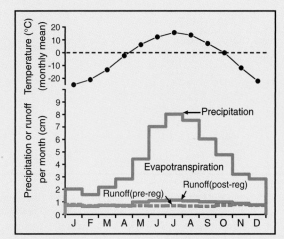

FIGURE 19.20 Mean monthly air temperature, precipitation, and runoff for the Nelson River basin.

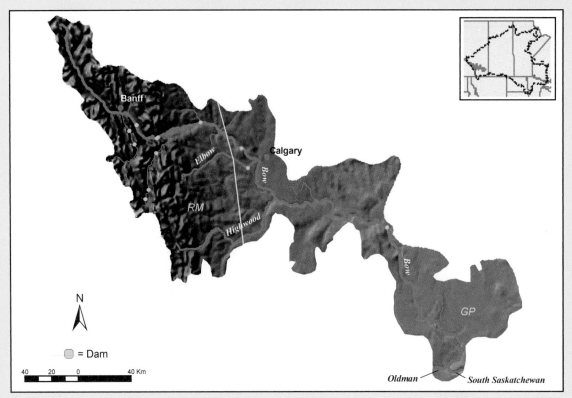

FIGURE 19.21 Map of the Bow River basin. Physiographic provinces are separated by a yellow line.

BOW RIVER

Relief: 2803 m

Basin area: 26,200 km^2

Mean discharge: 91 m^3/s

River order: 6

Mean annual precipitation: 40 cm

Mean air temperature: 3.9°C

Mean water temperature: 8.0°C

Physiographic provinces: Rocky Mountains in Canada (RM), Great Plains (GP)

Biomes: Temperate Mountain Forest, Temperate Grasslands

Freshwater ecoregions: Canadian Rockies, Upper Saskatchewan

Terrestrial ecoregions: Alberta Mountain Forests, Alberta/British Columbia Foothills Forests, Canadian Aspen Forest and Parklands, Montana Valley and Foothill Grasslands, Northern Mixed Grasslands, Northwestern Mixed Grasslands

Number of fish species: 29

Number of endangered species: NA

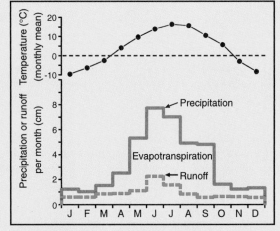

FIGURE 19.22 Mean monthly air temperature, precipitation, and runoff for the Bow River basin.

Major fishes: mountain whitefish, brook trout, brown trout, rainbow trout, bull trout, cutthroat trout, burbot, lake whitefish, northern pike, walleye, sauger, yellow perch, lake sturgeon, goldeye, mooneye, lake chub, emerald shiner, river shiner, spottail shiner, pearl dace, northern redbelly dace, quillback, longnose sucker, common white sucker, shorthead redhorse

Major other aquatic vertebrates: muskrat, beaver, mink, white pelican, Canada goose, bufflehead, common goldeneye, western grebe, gadwall, green-winged teal, horned grebe, lesser scaup, mallard, northern shoveler, pie-billed grebe, redhead

Major benthic invertebrates: mayflies (*Baetis, Cinygmula, Ephemerella, Epeorus, Heptagenia, Isonychia, Rhithrogena*), caddisflies (*Cheumatopsyche, Glossosoma, Rhyacophila, Symphitopsyche*), stoneflies (*Capnia, Isoperla, Taenionema, Zapada*), true flies (Chironominae, Orthocladiinae, *Pericoma*)

Nonnative species: brook trout, brown trout, rainbow trout

Major riparian plants: river alder, willow, poplar, water birch, red-osier dogwood, balsam poplar, sandbar willow

Special features: originates in glaciers and snowfields of Rocky Mountains in Alberta; World Heritage Site

Fragmentation: dams in mountains (for hydropower) and grasslands (for irrigation)

Water quality: pH = 8.1, alkalinity = 123 mg/L as CaCO$_3$; relatively free of pollutants in mountains (NO$_3$-N = 0.075 mg/L, soluble reactive phosphorus = 0.010 mg/L); nutrients higher below Calgary and in agricultural areas

Land use: 60% farmland (44% cropland), 2% urban, 38% forest

Population density: 42 people/km^2

Major information sources: Donald and Mutch 1980, Culp and Davies 1982, Charlton et al. 1986, Cross et al. 1986, Hamilton and North 1986, Sosiak 1990, Culp et al. 1992, Environment Canada 2001, www.climate.weatheroffice.ec.gc.ca/climate_normals/index_e.html

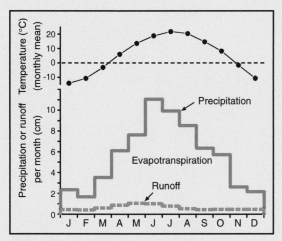

FIGURE 19.24 Mean monthly air temperature, precipitation, and runoff for the Otter Tail River basin.

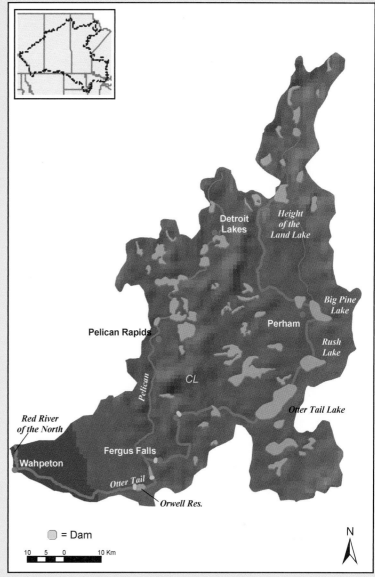

FIGURE 19.23 Map of the Otter Tail River basin.

OTTER TAIL RIVER

Relief: 190 m
Basin area: 4507 km^2
Mean discharge: 11.4 m^3/s
River order: ~5
Mean precipitation: 67 cm
Mean air temperature: 5.4°C
Mean water temperature: 14°C
Physiographic province: Central Lowland (CL)
Biome: Temperate Grasslands
Freshwater ecoregion: English–Winnipeg Lakes
Terrestrial ecoregions: Upper Midwest Forest/Savanna
    Transition Zone, Western Great Lakes Forests,
    Northern Tall Grasslands
Number of fish species: 72
Number of endangered species: 1 lichen
Major fishes: golden redhorse, northern hogsucker,
    shorthead redhorse, sand shiner, logperch, spotfin
    shiner, rock bass, black bullhead, yellow bullhead,
    brown bullhead, white sucker, pumpkinseed,
    largemouth bass, yellow perch, black crappie,
    walleye, common shiner, northern pike, channel catfish, smallmouth bass
Major other aquatic vertebrates: beaver, muskrat, snapping turtle, western painted turtle, mud puppy, tiger salamander, wood
    frog, western chorus frog, northern leopard frog, American toad, Great Plains toad, Canadian toad, Cope's gray treefrog,
    gray treefrog, mink frog, green frog
Major benthic invertebrates: mayflies (Heptageniidae, Oligoneuridae, Tricorythidae, Potamanthidae), stoneflies (Perlidae),
    caddisflies (Hydropsychidae, Leptoceridae)
Nonnative species: rainbow trout, brown trout, smallmouth bass
Major riparian plants: sugar maple, red oak, pin oak, tamarack, aspen, paper birch (east), basswood, bur oak, white oak (rare),
    American hazelnut, black willow, sandbar willow, peach-leaved willow, green ash, silver maple
Special features: high fish species richness for river size
Fragmentation: dams in upper and lower reaches
Water quality: pH = 8.2, specific conductance = 418 µS/cm, total P = 0.09 mg/L, total N = 0.22 mg/L, total dissolved solids =
    252 mg/L
Land use: 43% cropland, 20% forest, 14% water, 13% wetland
Population density: low
Major information sources: Aadland 1993, http://nd.water.usgs.gov, http://mn.water.usgs.gov/wrd/index.html,
    www.dnr.state.mn.us 2005, www.pca.state.n.us/ 2003, http://mcc.sws.uiuc.edu/ 2004, www.otpco.com 2004

20

RIVERS OF ARCTIC NORTH AMERICA

ALEXANDER M. MILNER   MARK W. OSWOOD   KELLY R. MUNKITTRICK

INTRODUCTION

NOATAK RIVER

KUPARUK RIVER

SAGAVANIRKTOK RIVER

MOOSE RIVER

ADDITIONAL RIVERS

ACKNOWLEDGMENTS

LITERATURE CITED

INTRODUCTION

This chapter covers rivers that lie within the Arctic region of North America, a vast area that encompasses northern regions of Alaska and Canada (Fig. 20.2). Some of the largest Arctic rivers (Mackenzie, Nelson–Churchill, and Yukon) are treated in other chapters. Although the Arctic is sometimes delineated by the Arctic Circle (66°32′N) (e.g., Remmert 1980), which approximates the southern boundary of the midnight sun, this definition does not accurately reflect the characteristics of the region due to the overriding influence on climate of ocean currents and land mass topography (Barry et al. 1993). Consequently, the Arctic is frequently defined as regions north of the tree line, where mean July temperature does not exceed 10°C and at least one month is <0°C. These treeless areas or tundra cover an estimated 2.8×10^6 km^2 of Alaska and Canada (Bliss and Matveyeva 1992). There is a transitional zone (ecotone) between the treeless arctic tundra and the continuous closed-canopy woodlands of the boreal forest, including the subarctic, where white and black spruce dominate, and mean monthly air temperature does not exceed 10°C for more than four months of the year and at least one month is <0°C (Remmert 1980). Thus, the coverage of the Arctic region in this chapter (see Fig. 20.2) includes Hudson Bay in Canada, which extends as far south as 51°20′, although the southerly part of the bay is more accurately considered the subarctic. The tree line also approximates the southern boundary of the zone of continuous permafrost, where the ground is permanently frozen and the surface (the active layer) thaws for only two to three months a year. As Hudson Bay is frozen for the greater part of the year, the Arctic front of cold air masses extends further south in Canada than in Alaska due to the presence of this inland sea The treeless tundra extends to 58°40′ south of Churchill.

Most archaeologists believe that the migration of humans into the New World occurred sometime between 15,000 and 50,000 years ago during the last Ice Age, when the Bering Land Bridge created a connection between Alaska and Siberia, termed Beringia (Langdon 1989). However these Paleo-Indians did not remain in Alaska but migrated southward to North, Central, and South America. When the Paleo-Eskimos (Old Eskimos) arrived across the Bering Straits from the Chukchi Peninsula of Siberia, they discovered the last major region on earth unoccupied by humans. The first discoveries of settlements date back 3000 to 4000 years, but they may have arrived earlier. By approximately 4000 years ago most areas

FIGURE 20.1 Sagavanirktok River, Alaska, as it flows into the Arctic Ocean (on horizon) (PHOTO BY C. WHITE).

903

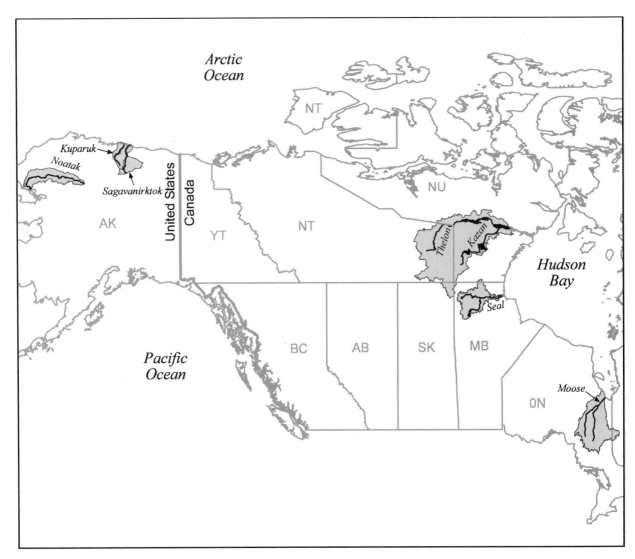

FIGURE 20.2   Rivers of Arctic North America covered in this chapter.

of Arctic Alaska, Canada, and Greenland were home to small and scattered bands of these Paleo-Eskimos (McGhee 1996). Although termed "Old Eskimos," these groups had little connection to the Eskimo groups typical of the region today. Their way of life was very different, as they lived in very marginal conditions, particularly during the winter. The Northern Eskimos, the Inuit of Alaska and Canada, arrived more recently and descended from inhabitants originally in Alaska.

In this chapter, we will focus on four rivers (three from Alaska and one from Canada) within the Arctic region. These rivers represent a spectrum from pristine unimpaired waters (the Noatak and Kuparuk rivers) to those potentially influenced by oil development (the Sagavanirktok River) and finally to a large

river system flowing into Hudson Bay that has been influenced by hydroelectric dams and other basin development, including mining and forestry (the Moose River). Additional river catchments (Seal, Thelon/Kazan), described briefly, lie within Canada; each is designated as a Canadian Heritage River.

Physiography and Climate

The Noatak, Kuparuk, and Sagavanirktok rivers have their sources within the Brooks Range (BM) physiographic province in northern Alaska (Hunt 1974). Here the mountains are lower than the Alaska Range, with elevations to 3000 m. The Noatak catchment lies entirely within this province, flowing westward to its estuary at Kotzebue Sound of the Chukchi

Sea. Most reaches of the Kuparuk and Sagavanirktok rivers flow across the Arctic Slope (AS) province before entering the Beaufort Sea of the Arctic Ocean. The Sagavanirktok River essentially divides the Arctic Slope into two distinct areas: a broader western section and a narrower eastern section. The Arctic Slope is considered a continuation of the interior plains of North America, whereas the Brooks Range is an extension of the Rocky Mountains (Wahrhaftig 1965). South of the Arctic Lowlands but north of the enormous Nelson River basin are the Thelon/Kazan and Seal rivers, both of which flow into the western side of Hudson Bay. The Thelon/Kazan River drains both the Thelon Plains and Bear River Lowland (TB) and the Bear–Slave–Churchill Uplands (BC), and the Seal River drains entirely from the Bear–Slave–Churchill Uplands. Much further to the south and east is the Moose River. The Moose and its tributaries have their headwaters in the Superior Upland (SU) and then flow northward across the Hudson Bay Lowland (HB) before flowing into Hudson Bay.

During the Wisconsin glaciation, the last of the Pleistocene stages, which ended about 10,000 years ago, half of North America was covered by major ice sheets. These ice sheets were present in the Brooks Range and the Alaska Range, but much of the Arctic Slope remained ice free.

The Arctic climate is characterized by extremely low annual precipitation, typically <15 cm/yr, which falls mostly as snow, although higher levels are typically found in the mountain ranges and the Hudson Bay region (70 to 90 cm in the Moose River drainage). The low rates of precipitation are similar to deserts, explaining why the High Arctic (northern portion of the Canadian Arctic Islands) is sometimes classed as a polar desert, with sparse vegetation. Mean monthly temperatures in the Alaskan arctic range from −29°C in January to 10°C in July (University of Alaska 1989), whereas in the Moose River drainage of Canada the range is −19°C to 17.3°C (Environment Canada 1982a). Daylight is continuous in the summer, but in the winter the sun may not appear over the horizon for four months. Cold dominates the landscape and influences both physicochemical and biotic processes in terrestrial and aquatic environments.

Basin Landscape and Land Use

Permafrost has a major influence on the landscape of the Arctic region and rivers flowing within it, resulting in dominance by thermokarst features. Per-

mafrost can be up to 600 m in thickness (Brown and Kreig 1983), with depths of thaw as little as 50 cm. Discontinuous permafrost is commonly associated with the boreal forest, particularly on south-facing slopes, where the depth of the seasonal thaw may exceed 2 m.

Regular freezing and thawing of permafrost within the active layer creates surficial thermokarst features, which may appear as characteristic polygons, circles, or stripes from the air. These features are termed "patterned ground." Another common feature is pingos, which are large ice-cored mounds up to 600 m in diameter but typically less than 20 m high (Pissart 1988). Innumerable thaw lakes, characteristic of the landscape, are created where permafrost melts and the overlying soil collapses below the water table. Over 95% of the wet coastal tundra of the Arctic is made up of thaw lakes, which range in length from several hundred meters to several kilometers (Billings and Peterson 1980).

The three Alaskan rivers encompass three terrestrial ecoregions (Ricketts et al. 1999): the Brooks/British Range Tundra, the Arctic Foothills Tundra, and the Arctic Coastal Tundra (except the Noatak River). Subalpine vegetation of the Brooks/British Range is restricted to valleys and lower slopes and is composed mainly of stunted white spruce, willow, dwarf birch, and blueberry. At higher elevations alpine tundra is made up of lichens, mountain avens, sedge, and cotton grass (Ricketts et al. 1999). The Arctic Foothills Tundra forms a transition between the flat low-lying Arctic Coastal Tundra and the steeper terrain of the Brooks/British Range Tundra. The Arctic Foothills Tundra is better drained than the Arctic Coastal Tundra; the vegetation is dominated by sedges, cotton grass, and dwarf shrubs of birch, blueberry, crowberry, and Labrador tea. The Arctic Coastal Tundra has poor drainage, so sedges and grasses predominate on the wet soils. The cover of moss is nearly continuous, although better-drained sites may support dwarf shrub communities. Along the river corridors riparian dwarf birch and willow are characteristic but rarely exceed 1 m in height (Hershey et al. 1997). In Canada, the Thelon/Kazan basin lies partly within the Low Arctic Tundra terrestrial ecoregion. This ecoregion stretches across much of northern Canada, from (west to east) the Northwest Territories, through southern Nunavut, and edging into Quebec. Nearly the entire region is underlain by permafrost. Much of the vegetation in the rolling uplands and lowlands of the Low Arctic Tundra is shrubby tundra, with some areas having trees (including white spruce, black spruce, and tamarack) and so transitional between treeless tundra to the north and taiga forest to the south. Further south, in central Canada along the western shore of Hudson Bay, is the Northern Canadian Shield Taiga ecoregion. The vegetation is transitional, with stunted black spruce and tamarack dominant. Much of the basins of the Thelon/Kazan and Seal rivers drain from this ecoregion. The Seal also traverses a short stretch through the Southern Hudson Bay Taiga ecoregion. Still further south and to the east, the Moose River drainage traverses, from its southern headwaters to its mouth at James Bay, three ecoregions: the Eastern Forest/Boreal Transition, the Central Canada Shield Forests, and the Southern Hudson Bay Taiga. The Moose River thus flows through a latitudinal and ecological transect, from forests of mixed deciduous and coniferous species and temperate climate in the south to stunted conifers and a subarctic climate in the north.

The inclement climate, with long, cold, and dark winters, limits human populations in Arctic Alaska and Canada. Barrow, the largest village in the Alaskan Arctic, has approximately 2400 residents as a mix of Eskimos and Caucasians. Barrow lies 16 km southwest of Point Barrow, the most northerly point in the United States (71°23′29″N). Point Barrow was named by Captain Beechey in 1826 for Sir John Barrow, then president of the Royal Geographical Society. Urban populations are higher in some of the more southerly parts of Arctic Canada. In the Moose River drainage, the largest towns are Kapuskasing, with over 12,000, and Timmons, exceeding 45,000.

Low population densities have limited urbanization effects on rivers of the Arctic region. However, the Arctic region is rich in natural resources, particularly oil and gas. Exploitation and export of oil on Alaska's North Slope is the main driver of the Alaskan economy, and is responsible for more than 80% of state income. Construction and operation of the Trans-Alaskan Pipeline, which connects the oilfields at Prudhoe Bay with the ice-free oil terminal in Valdez, runs along or crosses a number of rivers, particularly the Sagavanirktok River.

The Rivers

The river systems discussed in this chapter span a diagonal swath across the top of North America, from the Noatak River drainage in the northwest extreme of Alaska to the Moose River drainage discharging into the south of Hudson Bay at the southeast extreme (see Fig. 20.2). The Alaskan rivers have their origins in the Brooks Range and are fed principally by snowmelt, with some groundwater and glacial influence. Because of the low elevations in the Canadian Arctic, rivers are generally fed by lakes, of which there are many large ones, including Great Bear Lake and Great Slave Lake of the Mackenzie River system. As is typical for most of interior and northern Alaska, biological data for some of the drainages are largely derived from infrequent "expeditions" meant to provide surveys of major taxa and information on fish species of sport, subsistence, or commercial importance. Studies of ecological processes (e.g., biological production, carbon and nutrient budgets) require long-term frequent measurements. These studies have mostly been done in proximity to research stations (e.g., the Naval Arctic Research Laboratory in Barrow and the Toolik Lake Biological Station) or as "Big Science" (well-funded, multidisciplinary projects), such as the International Biological Program (IBP) and the Long Term Ecological Research (LTER) program (Hobbie 1997).

The freshwater ecoregions of the North American Arctic are primarily nested within two large complexes. The Noatak, Kuparuk, Sagavanirktok, Thelon/Kazan, and Seal systems are included in the Arctic complex. The Moose River is the only river in this chapter found in the Hudson Bay complex. The Noatak, Kuparuk, and Sagavanirktok of Alaska are found within the Yukon freshwater ecoregion even though they are not directly connected to the Yukon River to the south. The East Arctic ecoregion along the western shore of Hudson Bay includes the Thelon/Kazan and Seal rivers. Belonging to the Hudson Bay complex is the South Hudson ecoregion, which is found along the southern shore of Hudson Bay and includes the Moose River. The East Hudson ecoregion is along the east shore of the bay, but its rivers are not covered in this chapter.

The unifying theme of these river systems is not entirely an Arctic climate (as climates of these river systems range from Arctic to subarctic) but rather a general "northern syndrome." This northern syndrome derives from the ecological and historical consequences of cold: ice in rivers (surface ice and frazil ice), ice in the landscape (glaciers and permafrost), thermal limitations on carbon flow and growth rates, and the biogeographical consequences of Pleistocene ice sheets (Oswood 1997). At the maximum extent of glaciers during the Wisconsin period the Laurentide ice sheet (in the east) and the Cordilleran ice sheet (in the west) covered nearly all of present-day Canada and a strip across the northern United States (McPhail and Lindsey 1970, Pielou 1991). However, much of interior Alaska was unglaciated and connected via the Bering Land Bridge to Asia, producing a glacial refugium (Beringia). Following the retreat of continental glaciers at the Pleistocene/Holocene transition (approximately 10,000 to 12,000 years ago) the fish fauna of northern river systems derived from three centers of dispersal: Beringia; a Pacific refuge (south of the continental glaciers and west of the Rocky Mountains); and a Mississippi/Missouri refuge (northern tributaries of the Gulf of Mexico drainage; west of the Appalachian mountains and east of the Rocky mountains) (McPhail and Lindsey 1970). Consequently, the fish fauna of Alaska is dominated by fishes derived from the Beringian refuge, including endemics (e.g., the Alaskan blackfish) (Pielou 1991, Power 1997) and fishes from the Pacific refuge, many moving northward along the Pacific coast via saltwater excursions (McPhail and Lindsey 1970, Oswood et al. 2000). Conversely, inland dispersal from the Mississippi/Missouri refuge dominated postglacial colonization of northern river systems in north-central North America (McPhail and Lindsey 1970).

Streams and rivers in Arctic Alaska and Canada display unique physicochemical characteristics, due in many respects to the influence of snow and ice. Hydrological regimes are characterized by very low or no winter flows from October through early May. Subsequently, high solar radiation creates rapid snowmelt, so that 80% to 90% of the annual flow in many Arctic streams occurs within a two-week period during June (Milner et al. 1997). Low discharges then occur until convective thunderstorms increase flows in late July and early August. Where glacial influence is present, flows are more sustained through the summer months. Due to the influence of continuous permafrost, meltwaters are unable to infiltrate deep into the soil, creating a major paradox of the Arctic: the wetness of the Arctic region, with countless ponds, lakes, and muskegs, even though rainfall is more characteristic of desert environments.

In glacier-fed rivers, major diurnal fluctuations in discharge occur as air temperatures vary widely over a 24 hr period during the summer months, influencing the extent of ice melt. Water quality (e.g., sediment load) in glacier-fed systems may also vary both seasonally and diurnally. Groundwater-dominated systems display greater stability in discharge, water temperature, and sediment load and may provide refugia to biota from the harsher habitats of snowmelt and glacier-fed systems, particularly during the winter. In nonglacially influenced rivers, water temperature may increase downstream as rivers approach their estuaries and the ocean.

Taxonomic richness and growth rates of consumers in rivers typically decrease with increasing latitude in the Arctic (Oswood et al. 1995). Low annual solar radiation and low air temperature result in extensive ice formation in river basins, which has a major influence on the overwintering survival of both macroinvertebrates and fishes, particularly during ice breakup (Scrimgeour et al. 1994, Prowse 1994). Mortality at this time of year can limit ecosystem productivity, but its effect is extremely variable from year to year depending upon the severity of the winter and the ice breakup. Overwintering influences can cause macroinvertebrate taxa to be present in a river one year but then absent the next (Milner et al. 2005). The annual thermal regime also determines taxa distributions, organism adaptations, and ecosystem productivity (Ward and Stanford 1982). Many taxa typical of more temperate waters have difficulty completing their life cycles in Arctic freshwaters and thus are absent or rare (Oswood 1997). Hence, water temperature acts as a limiting factor or reach-scale filter (Poff 1997) to taxa diversity such that Diptera (true flies) dominate the macroinvertebrate fauna in Arctic rivers (Oswood 1989).

Because shallow streams and rivers generally freeze to the substrate, resident and juvenile anadromous fishes migrate in the fall to ice-free refugia for overwintering. These refugia can be lakes, ponds, spring- or groundwater-fed streams, or deep wintering pools. When overwintering, fishes undergo behavioral and physiological changes to limit energy loss and territoriality is reduced (Reynolds 1997). Wintering pools are often used by several species and life stages of fishes and high densities can lead to mortality through oxygen depletion and accumulation of metabolic wastes (Schmidt et al. 1989). Overwintering mortality is probably the most limiting factor to fish production throughout the northern region. Amphidromous fishes (e.g., Arctic char, Arctic cisco, broad whitefish, and Dolly Varden) spawn and overwinter in rivers and streams and migrate to the lower reaches or deltas of larger rivers each summer (e.g., the Sagavanirktok and Colville rivers) (Reynolds 1997). The summer is a period of intense feeding, when most of the yearly growth occurs to survive the long winter (Fechhelm et al. 1995, Fechhelm et al. 1996).

Some Arctic rivers along the northern coastlines of North America, particularly the larger drainages, provide transport pathways from the terrestrial environment to the ocean and may carry pollutants from contaminated land to the continental shelves of the northern oceans (AMAP 1997). Pollutants are transported in the water, bound to sediment, or associated with river ice. Some contaminants settle in delta and estuarine areas, whereas others reach the marine environment. Many of these contaminants originate outside the Arctic region from industrialized nations in Europe and North America.

Streams and rivers of the Arctic are potentially more likely to be influenced by climate change than many other North American basins, as climate warming is predicted to be more amplified at higher latitudes (Melack et al. 1997). In a landscape dominated by permafrost influences, enhanced melting of permafrost and associated increases in the depth of the active layer would markedly influence hydrology and watershed biogeochemistry (Oswood et al. 1992). Tundra and boreal forest soils are major stores of carbon and increases in the depth of the active layer and soil warming may enhance carbon availability and increase levels of dissolved organic carbon transported to surface waters (Oswood et al. 1992). We have already discussed the harsh temperature regimes in Arctic rivers limiting the diversity of aquatic insects and fishes, but with an amelioration of water temperature regimes from global climate change a northward extension of certain taxa may occur, increasing their geographical distribution (Poff et al. 2001). Although macroinvertebrate diversity may thus increase in Arctic rivers, certain present-day taxa well adapted to the harsher environment may be eliminated with enhanced diversity, as they are typically poor competitors (i.e., fugitive species; see Flory and Milner 1999).

Unlike most of North America, hydroelectric power generation in northern Alaska has not had a significant influence on river basins, but in Canada, particularly on a number of the rivers flowing into Hudson Bay, dams have exerted a major influence on river hydrology. Nevertheless many free-flowing rivers remain and the Arctic region provides a unique collection of pristine rivers where longitudinal changes along the continuum can be studied without anthropogenic influence.

NOATAK RIVER

The Noatak River flows primarily in a westward direction in northwest Alaska and is the only major river lying almost entirely within the Brooks Range (Fig. 20.8). In contrast, other major arctic rivers of Alaska (e.g., the Kuparuk and Sagavanirktok) flow northward from the northern foothills of the Brooks Range, draining into the Arctic Ocean. The Brooks

Range is the northern extension of the Rocky Mountains, arcing east–west across northern Alaska from the Canadian border to within 100 km of the Chukchi Sea. The Brooks Range demarcates a major biogeographic divide: arctic tundra to the north and taiga forest to the south. Elevations are higher in the central and eastern mountains (up to 2400 m asl) and lower at the western end (up to 800 m asl). The Brooks Range is comprised of several major mountain groups. The Noatak River arises at Mount Igikpak in the Schwatka Mountains, then flows westward between the Baird Mountains and the De Long Mountains to the village of Noatak (Fig. 20.3). From the village of Noatak the river swings south to flow into the Kotzebue Sound of the Chukchi Sea, north of the town of Kotzebue (Williams 1958, Gallant et al. 1995).

The Noatak River valley is a historical and current home of the northern Eskimos—the Inupiat—one of five major groups of Alaska's indigenous people (Langdon 1989). Archaeological work in northern Alaska suggests occupation as early as 11,700 years ago. Contact with European explorers began in the mid-1800s and gold-mining exploration produced the first maps of the area by the early 1900s (www.nps.gov/asko/akarc/index.htm 2004). The Inupiat people occupying the Noatak basin before contact with Western civilization belonged to the Noatagmiut (middle and upper Noatak) and Naupaktomiut (lower Noatak) (Hall 1975, www.nps. gov/asko/akarc/index.htm 2004). The "miut" suffix means "people of" (Langdon 1989). Noatak was a fishing and hunting camp in the nineteenth century (Maniilaq Association 2001). A federally sponsored mission school begun in 1908 made Noatak a regional center for school, trade, and religion (www.nps.gov/asko/akarc/index.htm 2004), and a post office was established in 1940.

Physiography, Climate, and Land Use

The Noatak River basin lies entirely within the Brooks Range (BM) physiographic province (see Fig. 20.8; Hunt 1974) and the Brooks/British Range Tundra and Arctic Foothills Tundra terrestrial ecoregions (Ricketts et al. 1999). The entire basin is underlain by permafrost, with depths likely as great as 183 to 244 m (Childers and Kernodle 1981). Although the Arctic tree line is often used as an approximate boundary between the Arctic and subarctic biomes, the basin contains biotic elements of both the Arctic and subarctic (taiga) (Young 1974, Gallant et al. 1995). In the lower Noatak drainage, there are scattered groves of

FIGURE 20.3   Noatak River, Alaska (PHOTO BY A. MILNER).

white spruce, as well as balsam poplar groves throughout the drainage. However, most (95%) of the Noatak River valley supports tundra vegetation, with taiga vegetation less than 5% (Young 1974).

The climate of the Noatak drainage changes from its lower reaches near the Bering Sea coast, where climate is moderated by ocean influences, to its upper reaches in the Brooks Range Mountains, where a more rigorous continental climate prevails (Milner et al. 1997, Green 1999). Winter in northwestern Alaska and the Noatak drainage is long and cold; summer is short and often wet (Childers and Kernodle 1981). Weather records for Noatak village

are sparse; we use weather data from Kotzebue as a surrogate for the lower Noatak valley and from Kobuk as a surrogate for the upper Noatak valley. The average temperature in July (warmest month) is 12°C in Kotzebue (29°C record high) and 14°C in Kobuk (33°C record high), providing an estimated average of 13°C for the Noatak basin (Fig. 20.9). The average temperature in February (coldest month) is –20°C in Kotzebue (–47°C record low) and –23°C in Kobuk (–56°C record low), for an estimated average of –22°C in the Noatak basin. The cooler summer/warmer winter temperatures in coastal Kotzebue compared to inland Kobuk highlight the transition from maritime influence to a cold continental climate inland from the coast. Most precipitation falls as rain in the summer, with rainfall peaking in August (see Fig. 20.9). Total precipitation averages 24.1 cm (126.5 cm of snow) in Kotzebue and 42.4 cm (137.2 cm of snow) in Kobuk. The Noatak River is north of the Arctic Circle and so is a "place of deep cold, the midnight sun, and the long polar night" (Young 1989).

Noatak village is the only current settlement in the Noatak valley. Thus, land use in the basin is essentially 100% natural tundra and taiga, and human population density is close to 0 people/km². Subsistence use of natural resources remains central to the people of Noatak, with harvest of chum salmon, whitefish, waterfowl, caribou, and moose (www.Alaska.com 2002). There is no road access to the Noatak River basin, but there are scheduled flights to Noatak village and bush plane access throughout the valley. Boats and ATVs (summer) and dog sleds and snow machines (winter) are used for local travel. For the Inupiat people of Northwest Alaska, it is "as if you're viewing the aftermath of a violent collision between the past and the present"; the challenge is to "walk in two worlds with one spirit—to hunt caribou and then sit down at a computer, or sew mukluks and later balance a checkbook" (Jans 1993).

River Geomorphology, Hydrology, and Chemistry

The Noatak River basin occupies 32,626 km², with an elevation drop of approximately 1067 m from headwaters to the mouth (Childers and Kernodle 1981); Mount Igikpak is the highest elevation (2612 m asl) in the drainage. The elevational profile shows three distinct gradients: short steep reaches of the headwater branch and mountain tributaries, a modest gradient (approximately 76 cm/km) over most of the Noatak main stem, and a low-gradient segment (about 19 cm/km) in the estuarine lower reaches (Childers and Kernodle 1981). Young (1974)—working from Smith (1913) and Wahrhaftig (1965)—designates six major regions along the longitudinal profile of the Noatak drainage. These regions can be summarized as follows, from headwaters to the mouth:

• Headwater Mountains: Tributary streams, fed by snow and mountain glaciers, enter the main stem from U-shaped (glacially carved) valleys. The main stem meanders through a 3.2 km wide valley in glacial and fluvial sediments.

• Aniuk Lowlands: The Noatak meanders through gently sloping, rolling hills; in some locations the river is incised from 15 to 61 m in the glacial and fluvial sediments. Two large tributaries enter in this province: the Aniuk and Cutler rivers.

• Cutler River Upland: Pinched between the Baird Mountains and De Long Mountains, this region contains the upper reaches of the Cutler River. The Noatak River has cut a deep (183 to 244 m) canyon through quartzite, sandstones, limestones, and igneous intrusives; elsewhere, steep bluffs of gravel limit the floodplain.

• Mission Lowland: The Noatak turns south, flowing through a broad expanse of tundra, with numerous thaw lakes and pingos in the wet tundra of the floodplain. In the north and central portions of this province, the Noatak is a wide (3.2 km) braided river, coalescing into a single channel in the southern portion.

• Igichuck Hills: The Noatak cuts through limestone and calcareous schist, creating a canyon (Lower Noatak Canyon); floodplain development is minimal.

• Coastal Lowland: In this region of low relief, the Noatak flows in a single channel through alluvial, marine, and glacial sediments. As the river nears its mouth at Kotzebue Sound it branches into distributaries and forms a small delta.

Along the main stem, from near the headwaters (Ipnelivik River) to the confluence of the Eli River (near Noatak village), the river has a pool–riffle sequence with pool lengths ranging from 305 to >1524 m, pool widths from 61 to 213 m, and pool (maximum) depths from ~0.6 m to ~2.4 m. Bed materials along this sequence range from coarse gravel to large boulders (Childers and Kernodle 1981).

Permafrost is essentially continuous in the Noatak basin. Permafrost restricts infiltration of precipitation into deeper groundwater, producing "flashy" hydrographs in response to precipitation (Childers and Kernodle 1981, MacLean 1997, Milner et al. 1997). Streams begin to freeze in October, with streamflow from most streams ceasing by December; discharge measurements show no flow from the upper basin (upstream of Noatak Canyon) by late winter. Breakup of river ice occurs in May, with snowmelt producing high streamflow in June (Childers and Kernodle 1981). Some streams and river segments are frozen to the bottom but deeper sections may have water (with little or no current) beneath deep ice cover. Occasional springs, visible in winter as open water and massive icings, have perennial flow (Childers and Kernodle 1981). The winter freezing in Alaskan Arctic running waters, with ice extending to the substrate or isolating shallow water beneath deep ice cover, forces migrations of fishes to suitable overwintering refugia (Craig 1989, Reynolds 1997) and limits the benthic macroinvertebrate fauna to taxa with behavioral (e.g., movement into deep streambed sediments) or physiological adaptations for avoiding or tolerating freezing (Oswood et al. 1991).

To our knowledge, there has only been one gaging station on the Noatak River (USGS 2003), located in the lower Noatak Canyon; winter data are sparse and unreliable (Childers and Kernodle 1981). Streamflow (see Fig. 20.9) is low in May (average flow 73 m³/s), peaks in June with snowmelt (average flow 1954 m³/s), and declines from July (820 m³/s) to August (756 m³/s) to September (493 m³/s). Freeze-up commences in October, producing much reduced flow (average 115 m³/s). Childers and Kernodle (1981) assessed streamflow and river conditions along a longitudinal profile of the Noatak River in 1978, once in late winter (April) and once in late summer (August). Winter streamflows were at or near zero in the upper basin and very low (<2.8 m³/s) in the lower basin. At low-flow conditions of late winter the lower Noatak River (near Noatak village) is tidally influenced, as indicated by high concentrations of sodium and chloride and high total dissolved solids. August streamflows along the Noatak River main stem ranged from approximately 28 m³/s in the upper basin to approximately 368 m³/s in the lower basin, near Noatak village.

Most surface waters in the Noatak basin are dominated by calcium bicarbonate, although magnesium is the dominant cation at some locations (Childers and Kernodle 1981). In late winter (April), water remaining beneath deep ice cover may have high concentrations of dissolved solids (Childers and Kernodle 1981). This cryoconcentration is due to exclusion of dissolved matter from the growing ice mass and hence concentration in the water remaining beneath the ice cover (Cheng et al. 1993). Similarly, dissolved oxygen concentrations were low in late winter at some sites in the Noatak drainage (Childers and Kernodle 1981) because of biotic respiration and isolation of water from atmospheric reaeration. High salinity, low oxygen concentrations, and ice encroachment beneath deep ice cover may make overwintering conditions inhospitable in Arctic running waters, forcing migrations of fishes to more suitable overwintering sites (Power et al. 1993, Reynolds 1997). In late summer (August), concentrations of dissolved oxygen were at or near satura-

tion throughout the basin, with water temperatures ranging from 10°C to 13°C. Along the main-stem sampling sites alkalinity ranged from 82 to 110 mg/L as $CaCO_3$ and pH from 7.8 to 8.2. The dominant cations were calcium (range 37 to 57 mg/L) and magnesium (range 8 to 16 mg/L); dissolved NO_2-N + NO_3-N ranged from 0.07 to 0.21 mg/L (Childers and Kernodle 1981).

River Biodiversity and Ecology

The Noatak River belongs in the Yukon ecoregion, the largest freshwater ecoregion in North America, incorporating most of Alaska and the Yukon Territory (Abell et al. 2000). In the remote and roadless regions of Alaska, biological data are often derived from expeditions and collecting trips. These expeditions generally provide qualitative data (e.g., taxa lists) from short-term visits (usually in summer). The isolation that makes the Noatak River such an outstanding biological reserve is mirrored in the lack of logistical support afforded by a research station or road access; the summaries here reflect the limitations of expeditionary biology.

Algae and Cyanobacteria

Data for running waters of the Noatak basin are sparse. Wiersma et al. (1986) collected epilithic periphyton samples from the Avan River, a tributary of the Noatak River. Cyanobacteria included *Coelospherium* and *Oscillatoria*. Chlorophyte algae included *Gleocystis*, *Mougeotia*, and the desmids *Cosmarium* and *Staurastrum*. Typical of northern Alaska, the taxonomic composition was dominated by diatoms: *Achnanthes*, *Amphora*, *Cymbella*, *Diatoma*, *Hannaea*, *Fragilaria*, *Melosira*, *Meridion*, *Synedra*, *Didymosphenia*, *Navicula*, *Pinnularia*, and *Nitzschia*.

Plants

Aquatic macrophytes appear to be sparse in the running waters of the Noatak Valley, confined to slow-moving streams in lowland areas. Young (1974) notes that aquatic vegetation is not well represented, even in small slow-moving streams, but that a few kinds, particularly northern bur-reed and Pallas' buttercup, are often abundant.

Invertebrates

Information on benthic macroinvertebrates is limited, but data (to order or family) are available from 13 lotic sites in August (Childers and Kernodle 1981). Nonbiting midges (Diptera: Chironomidae) were found at every site and were a substantial portion of the fauna, averaging 33% of organisms collected (numerical abundance). Craneflies (Diptera: Tipulidae) were found at most sites but in low abundance. Black flies (Diptera: Simuliidae) were found at some sites, in low abundance. Mayflies followed a similar pattern, with the families Baetidae and Heptageniidae found at nearly every site, and mayflies often composed a considerable portion (25% to 50%) of the fauna. The Brachycentridae dominated the caddisfly fauna, but even brachycentrids were only found at low abundance at fewer than half of the collecting sites. Other caddisfly families (Leptoceridae, Hydropsychidae, Molannidae, Rhyacophilidae) were found sporadically at low abundances. Stoneflies (Chloroperlidae, Nemouridae, Perlodidae) likewise occurred sporadically, nearly always at low abundance. Among the noninsect macroinvertebrates, snails were found at only a few sites. However, water mites (Acarina) were ubiquitous and generally major elements of the fauna. The presence at many collecting sites of crustaceans and insects generally associated with lentic habitats—seed shrimps (Ostracoda), water fleas (Cladocera), diving beetles (Dytiscidae), water boatmen (Corixidae)—likely indicates drift from upstream ponds, lakes, or slow-moving streams. Taxa in life history stages (at late summer) not likely to be captured by dip net sampling of benthic substrates (eggs, recently hatched larvae, and terrestrial adults) will be missed or underrepresented. Nonetheless, the general dominance of the benthic fauna by true flies (especially chironomid midges), dominance by mayflies of the families Baetidae and Heptageniidae, and sparse representation of caddisflies is consistent with northwestern Alaska in particular and Alaska in general (Oswood 1989).

Vertebrates

Information on fish fauna is also very limited, but there appear to be about 18 fish species in the running waters of the Noatak drainage (McPhail and Lindsey 1970, O'Brien and Huggins 1974, Morrow 1980, Cappiello 1995). The fish fauna of the Noatak River is dominated by the family Salmonidae, including whitefish, grayling, trout, and salmon. Apparently, all five species of Pacific salmon are present in the Noatak drainage: chinook, sockeye, coho, chum, and pink. Chum salmon are the only salmon taken in quantity from the rivers of northwestern Alaska (O'Brien and Huggins 1974). In 1982, a salmon hatchery (Sikusuliaq Springs) was constructed on the lower Noatak River, rearing chum salmon fry. Returning adults were taken in a commercial fishery in Kotzebue Sound and

for subsistence use by residents of Noatak village (Wilson and Kelly 1986). The hatchery has since closed. Arctic grayling are common throughout the drainage. Char (probably synonymous with northern Dolly Varden [Reynolds 1997]) support an important subsistence fishery, taken by seine, gill net, and hook and line from concentrations of fishes in overwintering areas (Cappiello 1995). Three (possibly five) whitefishes are found in running waters of the Noatak drainage: humpback whitefish, round whitefish, and least cisco. The Noatak drainage lies within the known ranges of two other whitefishes, broad whitefish and Bering cisco. Whitefishes are important in native subsistence fisheries on the Noatak River. A memorable essay describing subsistence gill-net fishing on the Kobuk River, just south of the Noatak River, can be found in Jans (1993).

Other families of fishes are apparently represented by one species each: northern pike, burbot (a freshwater member of the cods), Alaska blackfish (a Beringian relict), longnose sucker, slimy sculpin, and ninespine stickleback. The ecological limitations of high-latitude running waters—low food supply, severe hydrothermal regime (Oswood 1997), and barriers to dispersal—constitute a selective "filter" (Poff 1997), so that the fish fauna of Alaska in general and northern Alaska in particular is a highly selective subset of the North American fish fauna (Oswood et al. 2000). The fish fauna of the Noatak River is notably lacking fish families that are often dominant components of running waters in temperate regions (Cushing and Allan 2001): minnows (Cyprinidae); perches, walleye, and darters (Percidae); sunfishes and bass (Centrarchidae); and bullheads and catfishes (Ictaluridae).

Although the Noatak Valley is home to the full range of the Alaskan megafauna—grizzly bear, wolf, caribou, moose, wolverine—truly lotic vertebrates (other than fishes) appear to be very limited. Mammalian river specialists are limited to the river otter in the lower river. Manuwal (1974), in a short-term survey of the birds of the Noatak drainage, lists 18 species of birds with a habitat affinity for "fluvatile waters," including loons, ducks, five species of shorebirds, and gulls. Of the potential avian "river specialists" in the Noatak drainage (Schroeder 1996), belted kingfishers are common and harlequin ducks (nesting alongside streams, consuming stream macroinvertebrates and fish eggs) are rare.

Ecosystem Processes

No ecosystem studies have been conducted on the Noatak River; however, processes are likely to be similar to those described for the Kuparuk River.

Human Impacts and Special Features

The Noatak River arises in the Gates of the Arctic National Park and Preserve, then flows west through the Noatak National Preserve. All of the preserve (26,710 km^2) is designated wilderness (except for about 2833 km^2 near the village of Noatak), one of the largest wilderness areas in the United States. From its headwaters to its confluence with the Kelly River, 531 km of the Noatak River are designated Wild and Scenic, the longest continuous river segment in the U.S. National Wild and Scenic system and the largest mountain-ringed river system in natural condition in the United States. The Noatak River is an "island out of time"—kudos to Horton (1996)—a river free flowing through a landscape nearly untouched by the industrial world and inhabited by people still largely living by subsistence take of animal resources.

KUPARUK RIVER

The Kuparuk River is another pristine Arctic river that affords several useful contrasts to the other Alaskan rivers (the Noatak and Sagavanirktok) considered in this chapter. Like the Sagavanirktok, the Kuparuk river flows north from the eastern Brooks Range, arising in the northern foothills, then traverses the low-gradient, poorly drained Arctic Coastal Plain, terminating in the Beaufort Sea of the Arctic Ocean (Fig. 20.10). In contrast, the Noatak River valley is almost contained within the mountains of the western Brooks Range, flowing west to the Chukchi Sea. The Kuparuk and Sagavanirktok river drainages share a harsh continental climate and tundra vegetation, whereas the gradient of marine-influenced and continental climate of the Noatak River permits an anomalous mix of taiga (forest) and tundra vegetation. The drainage area of the Kuparuk River is approximately 8100 km^2, with a basin length of nearly 250 km (McNamara et al. 1998). Maximum elevation in the upper Kuparuk drainage is 1464 m asl. The entire region is underlain by continuous permafrost and so surface waters are isolated from deep groundwater (McNamara et al. 1998).

Of the three Arctic Alaskan rivers considered in this chapter, the Kuparuk River has received far more intensive ecological study than either the Noatak or the Sagavanirktok rivers. However, these multidisci-

plinary and long-term studies have taken place in the headwaters and at a midcontinuum reach, rather than in the lower main stem of the river. We include brief discussions of this "stream" (as opposed to "river") ecology because these studies provide insights into the ways that climate and physiography mold the biota and ecological processes of high-latitude running waters that are likely applicable across Arctic Alaska and along most of the river continuum.

Physiography, Climate, and Land Use

The Kuparuk River traverses two physiographic provinces (Hunt 1974), with headwaters in the Brooks Range (BM) and mouth in the Arctic Slope (AS) (see Fig. 20.10). Although the Kuparuk River lies entirely within the Yukon freshwater ecoregion of Abell et al. (2000), the Kuparuk drainage is north of the continental divide and so not within the drainage of the Yukon River. Near the headwaters the Kuparuk River traverses sediments eroded from the Brooks Range, with exposed sedimentary rocks (shale, sandstone, and gravel) of Cretaceous origin; nearer the coast there are younger (Tertiary) sediments (Connor and O'Haire 1988). The entire area is underlain by continuous permafrost (250 to 600 m depth), with summer thawing (active layer) typically from 25 to 40 cm but up to 100 cm deep (McNamara et al. 1998). Permafrost impedes drainage, so that in spite of low precipitation many areas of tundra are wet. The Kuparuk River crosses three terrestrial ecoregions on its path to the Beaufort Sea: Brooks/British Range Tundra, Arctic Foothills Tundra, and Arctic Coastal Tundra. This tundra landscape is dominated by herbaceous, tussock-forming vegetation, with dwarf shrubs along watercourses.

The Kuparuk River drainage lies entirely north of the Brooks Range and well above the Arctic Circle. The climate is therefore characterized by long, very cold winters and short cool summers. Average annual air temperature at the Imnavait Creek watershed in the headwaters of the Kuparuk is –7.4°C (Hinzman et al. 1996); near the mouth, on the Arctic Coastal Plain, the average annual air temperature is –12.1°C (www.wrcc.dri.edu). In the headwaters (Imnavait Creek), air temperatures (seven-day running averages) peak in late June/early July at about 15°C; winter low air temperatures range from about –15°C to –25°C, with a record low (1985–1993) of about –45°C (Hinzman et al. 1996). Near the mouth (www.wrcc.dri.edu) average monthly air temperatures (Fig. 20.11) peak in July at about 8°C and are lowest in February at about −29°C. Estimates of annual precipitation range from 35 cm in the headwaters (Hinzman et al. 1996), to 15 to 25 cm at mid-continuum (Selkregg 1977), to 9 cm near the mouth (www.wrcc.dri.edu). About two-thirds of the annual precipitation falls as rain in the summer (www.wrcc.dri.edu, Hinzman et al. 1996).

Although we have found no references to occupation of the Kuparuk River basin by indigenous peoples, the Kuparuk River is within the region of Alaska long home to the Inupiat Eskimos (Langdon 1989). There are no year-round communities in the Kuparuk River basin; the nearest habitations are the village of Nuiqsut at the mouth of the Colville River to the west of the mouth of the Kuparuk River and the oil-field developments of Prudhoe Bay and Deadhorse to the east of the Kuparuk River mouth. Thus, the basin is 100% Arctic tundra, with a human population density of zero. The remoteness of the Kuparuk River (no settlements and almost no road access) has thus far largely protected it from resource development, except for oil field developments near the mouth.

Geomorphology, Hydrology, and Chemistry

During Pleistocene glaciation the region north of the Alaska Range (including the present-day Kuparuk drainage) was ice free, except immediately north of the Brooks Range (Connor and O'Haire 1988, Gallant et al. 1995). Present-day glaciers in the Brooks Range contribute glacial meltwater to many north-flowing drainages (Oswood et al. 1992). Although adjacent watersheds have active glaciers, the Kuparuk River basin has no active glaciers, so the Kuparuk River is classified as a clear-water tundra stream in the Craig and McCart (1975) classification of Arctic Alaskan running waters. At least in the middle reaches (4th order river, near the Toolik Lake Research Station) the Kuparuk River meanders through tundra, with alternating pools and riffles and rocky substrates (Fig. 20.4).

The extreme seasonal eccentricity of climate is mirrored in the annual hydrological and thermal regimes. In the upper drainage basin, snowmelt (May to early June) is a dominant hydrological event, with spring floods flowing over anchor ice and frozen streambed sediments (Oswood et al. 1996, Hershey et al. 1997). Summer streamflows often approach zero because the catchment is sealed by permafrost and so there is no sustaining contribution from groundwater. Conversely, the underlying permafrost prevents percolation of precipitation into groundwater, so that summer rainstorms produce rapid but transient rises in streamflow. Flow decreases dramatically in September, with drying of riffles and freezing of the stream to the substrate by October except in the deepest pools (Oswood et al. 1989, Oswood et al. 1996, Hershey et al. 1997). On the lower river, near the mouth, deep winter flows (from December through April) are very low, averaging less than $0.1 \, m^3/s$ (see Fig. 20.11). Discharge increases in May (average $45 \, m^3/s$) as air temperatures rise above freezing and runoff begins. Discharge spikes in June ($305 \, m^3/s$) as snowmelt runoff peaks. From July to September average (monthly) discharge ranges from $31 \, m^3/s$ to $43 \, m^3/s$, with flows sustained by rain (or occasional snow) throughout the basin. From October to November discharge decreases precipitously as declining precipitation turns from rain to snow and the active layer of the soil refreezes, suppressing water delivery from soil to streams (see Fig. 20.11).

As noted, estimates of annual precipitation show a gradient from 9 cm near the mouth, to ~18 cm at the middle of the river continuum, to 35 cm at a 1st order tributary. Monthly precipitation data shown in Figure 20.11 are from a climate station at the river mouth. Although the seasonal pattern of precipitation shown is likely representative of the Kuparuk River basin, the amounts of monthly precipitation are almost certainly substantial underestimates of the basinwide average. If the actual, basinwide annual precipitation is close to the midrange value of 18 cm, the monthly means shown in the figure would be approximately doubled, providing a more believable hydrological balance (annual precipitation about 18 cm versus runoff about 15 cm). This suggests that even after adjusting for basinwide annual precipitation annual runoff is still a very high percentage of precipitation (>80%).

At Imnavait Creek (a headwater stream) summer water temperatures (monthly means) range from 6°C to 14°C, with extreme maximum values ranging from 19°C to 21°C (Irons and Oswood 1992). In the middle river continuum (Kuparuk River near Toolik Lake), summer water temperatures average 8°C to 10°C; during low flows, water temperatures may reach 20°C (Hershey et al. 1997). In the lower river (near the mouth), summer water temperatures (sporadic determinations; mean monthly values over 16 years) warm to 4°C in June, peak at 12°C in July, decline slightly to 10°C in August, and fall rapidly in the short autumn to 2°C in September (http://nwis.waterdata.usgs.gov/ak/nwis). In winter (October to May), water temperatures beneath the ice cover are

FIGURE 20.4 Upper Kuparuk River, Alaska. Mountains of the Brooks Range can be seen in the distance. Vegetation along the bank is primarily willow shrubs (PHOTO BY S. PARKER).

approximately 0°C; in shallow waters, surface ice can reach the streambed, with temperatures at the substrate surface considerably below freezing (Irons and Oswood 1992).

Nutrient concentrations are low in the midcontinuum Kuparuk River; typical concentrations are phosphorus (SRP) 0.4 µg/L, NH$_4$-N 18.0 µg/L, and NO$_3$-N + NO$_2$-N 5.2 µg/L (Hershey et al. 1997). Calcium values appear to increase downstream, from 0.8 to 1.4 mg/L in a headwater stream (Everett et al. 1996), to 1.7 mg/L in a midcontinuum site (Hershey et al. 1997), to 25.6 mg/L at the mouth (http://nwis. waterdata.usgs.gov/ak/nwis). Values of pH range from 5.6 to 6.3 in a peat-dominated headwater stream (Everett et al. 1996), to 7.1 for the midcontinuum Kuparuk River (Hershey et al. 1997), to 7.3 near the river mouth (http://nwis.waterdata.usgs.gov/ ak/nwis). Organic matter concentrations in stream waters of the Kuparuk drainage are dominated by leaching of dissolved organic matter from the peaty tundra, with water flowing through the soil active layer (over permafrost) en route to the stream channel. Dissolved organic carbon concentrations (averaged over the open-water season) are about 10 mg C/L in a peat-bottomed headwater stream (Oswood et al. 1996), 6.4 mg C/L in a (rocky-bottomed) midcontinuum river (Peterson et al. 1986), and 9.3 mg C/L near the mouth (http://nwis. waterdata.usgs.gov/ak/nwis). Concentrations of particulate organic carbon are at least an order of magnitude lower than DOC concentrations.

River Biodiversity and Ecology

The Kuparuk River is located within the Yukon freshwater ecoregion (Abell et al. 2000). The "Big Science" projects in the headwaters of the Kuparuk River—the R4D project (Imnavait Creek) and especially the Tundra LTER studies (Kuparuk River near the Toolik Lake Biological Station)—have produced a uniquely rich picture of running-water ecology in Arctic Alaska. Unless otherwise noted, information on the midcontinuum Kuparuk River is based on the summary by Hershey et al. (1997); biological information on Imnavait Creek, a headwaters stream, is from Bond (1988), Viavant (1989), and Oswood et al. (1996).

Algae and Cyanobacteria

Fifty-two genera of algae have been identified from Imnavait Creek. A red alga (*Batrachospermum*) is a major component of the algal community, likely dominating the algal biomass. There are 15 genera of cyanobacteria. The green algae show the highest generic diversity (23 genera), with desmids a common component of the green algae. Diatoms, represented by 9 genera, are infrequent. The algal community of Imnavait Creek seems consistent with other studies of peat-bottomed streams with acidic (pH 5 to 6), oligotrophic, and soft waters. The relative scarcity of diatoms is likely due to the very low concentrations of silica (<0.5 mg/L) in Imnavait Creek. In contrast, the epilithic algal community of the midcontinuum Kuparuk River is dominated by diatoms (187 species; 12 genera comprise up to 95% of individual cells); numerically dominant genera include *Achnanthes*, *Cymbella*, *Hannaea*, and *Diatoma*. Filamentous or macroalgae are found in patches and include the red algae *Lemonia* and *Batrachospermum*, the cyanophyte *Calothrix*, and the green algae *Stigeoclonium*, *Mougeotia*, and *Spirogyra*.

Plants

In the pools of Imnavait Creek, deeper waters have a community of common mare's-tail and northern bur-reed, with shallow pond margins characterized by yellow marsh marigold, common mare's tail, northern bur-reed, and sphagnum mosses (Walker and Walker 1996). At the midcontinuum Kuparuk River site, Agassiz's schistidium moss occurs in low abundance; in an experimentally fertilized (phosphorus) reach, hyrgrohypnum moss proliferated (Lee and Hershey 2000). Riparian plants are primarily dwarf birches and willows.

Invertebrates

As is typical for Alaskan running waters (Oswood 1989), true flies (Diptera) dominate the macroinvertebrate fauna of both the midcontinuum Kuparuk River and headwater Imnavait Creek. However, geomorphological differences between the Kuparuk River (rocky-bottomed stream with substantial flow during the open-water season) and Imnavait Creek

(largely peat-bottomed stream with pondlike pools at low flows) are reflected in the invertebrate fauna. Black flies are the numerically dominant macroinvertebrates in the Kuparuk River, with *Prosimulium perspicuum* the most abundant consumer and largest contributor to secondary production. The fauna of Imnavait Creek is completely dominated (nearly 99% of numerical abundance) by chironomid midge larvae. Midges dominate the taxonomic richness of both streams (20 genera in Imnavait Creek; 25 genera in the Kuparuk River). *Orthocladius rivulorum* (midge) larvae are a major contributor to secondary production of benthos in the Kuparuk River. Mayflies (*Baetis*, *Cinygmula*, *Ephemerella*), stoneflies (*Alaskaperla*, *Nemoura*), and caddisflies (*Brachycentrus*, *Rhyacophila*) are represented by only two or three species in each order in the Kuparuk River. However, *Baetis* mayfly nymphs and *Brachycentrus* caddisfly larvae are the dominant components of benthic secondary production (along with the dipterans *Prosimulium* and *Orthocladius*).

Vertebrates

Moulton and George (2000) summarize fish species found in the lower Kuparuk River (near the mouth of the river in the oil field region). Five fish species are listed: broad whitefish, arctic grayling, burbot, slimy sculpin, and ninespine stickleback. In the upper basin, arctic grayling are reported in both the Kuparuk River and Imnavait Creek. Arctic grayling spawn in spring immediately after breakup and feed in summer in the warm waters of tundra streams. Because tundra streams typically freeze solid in winter, arctic grayling must migrate (often long distances) to overwintering sites in deep pools, lakes, river deltas, and springs (Reynolds 1997). Also observed in the upper Kuparuk basin are slimy sculpin and lake trout, the latter found within connecting headwater lakes (Alexander Huryn, personal communication). Pink salmon have been reported in the lower Kuparuk, but are likely strays from other natal rivers rather than evidence of a sustained run (Craig and Haldorson 1986). A few other fish species are possible (but unconfirmed) inhabitants of the Kuparuk River, occurring in adjacent rivers (the Colville and Sagavanirktok rivers; Moulton and George 2000) or within the general distributional range of the species in Alaska (McPhail and Lindsey 1970, Morrow 1980). These taxa include the arctic lamprey, several whitefishes (least cisco, arctic cisco, and round whitefish), northern pike, and longnose sucker.

This area of the Arctic Coastal Plain and the adjacent coast of the Beaufort Sea are home to the "charismatic megafauna" for which Alaska is famous, including moose, muskox, and especially caribou. Large predators include grizzly bear, polar bear, arctic fox, and (rarely) wolf, but there are no aquatic or semiaquatic mammals (e.g., river otter) of which we are aware. During the brief summer, wetlands dominate the Arctic Coastal Plain, providing breeding habitat for 180 bird species (nearly all present only in the summer), with absolute numbers of birds on the North Slope approaching 10 million individuals (Gilders and Cronin 2000).

Ecosystem Processes

Organic carbon dynamics of the upper Kuparuk River (midcontinuum, 4th order site) are dominated by allochthonous inputs (Peterson et al. 1986, Hershey et al. 1997). However, inputs of leaf litter from riparian vegetation (dwarf birches and willows) are apparently negligible. Instead, allochthonous inputs derive from dissolved organic carbon leached from the tundra and from peat particles (200 to 300 g C m^{-2}yr^{-1}) eroded from the stream banks. In contrast, autochthonous inputs—net primary production of periphyton—are only about 13 g C m^{-2}yr^{-1}. Bioassay experiments have shown that phosphorus is the principle nutrient limiting biomass and production of periphyton. However, periphyton biomass is also limited by insect grazers. Secondary production of the major macroinvertebrate consumers ranges from 5.7 to 14.4 g dry mass m^{-2}yr^{-1}, with production by black flies dominating. Arctic grayling adults (5 to 12 years old, 30 to 42 cm length, 300 to 500 g weight) move into the 4th order Kuparuk River site soon after ice-out for spawning. By early July young-of-the-year emerge from stream gravels and feed for the remainder of the summer (older juveniles appear to spend several years in tributary streams before moving to the river as adults). On average, adults gain 40 g weight over the summer feeding season but growth is highly variable from year to year, with little growth or even loss of weight in poor years (low summer water flow). Gut content analyses of the four major macroinvertebrate consumers indicate that the diets of a black fly (filter-feeder) and a brachycentrid caddisfly (filter-feeder and grazer) overlap, with fine particulate organic matter dominating and smaller quantities of diatoms and animal prey; brachycentrid larvae also consume filamentous algae, presumably via grazing. A tube-building midge larva consumes only diatoms, obtained by grazing the diatom "lawn" on its silk case. Baetid mayfly nymphs consume detrital FPOM and diatoms via epilithic grazing. A long list of other taxa of midge larvae are

presumably mostly collector-gatherers, but these small-bodied midge larvae contribute very little (<3%) to overall secondary production. Food items in grayling stomachs are mostly detritus (presumably of little nutritional value); baetid mayfly nymphs, brachycentrid caddisfly larvae, and terrestrial insects comprise most of the macroinvertebrates in grayling guts.

Human Impacts and Special Features

This pristine river, although relatively small, is probably the most extensively studied of all Arctic rivers in North America. Flowing from the northern foothills of the Brooks Range across the Arctic Slope (or Coastal Plain), its seasonal amplitude of physical conditions is among the most extreme on earth. The long and extremely cold winter forces migrations of most organisms and limits biological diversity, but the long days of the short summer generate a burst of productivity that sustains resident organisms and attracts one of the great avian migrations of North America. The Kuparuk River is free flowing over its entire length but runs parallel and countercurrent to crude oil flowing south in the Trans-Alaskan Pipeline. The attraction of ancient carbon in massive oil and gas reserves has drawn the only industrial incursion into this otherwise pristine landscape.

The lower reaches and mouth of the Kuparuk River traverse the Prudhoe Bay Oil Field, the largest oil and natural gas discovery in North America (Gilders and Cronin 2000). Confusingly, the more recently developed Kuparuk Oil Field is to the west of the Kuparuk drainage. Some concerns about the effects of oil field development on the Kuparuk River and its lower tributaries center on minimizing the risks (and scale) of contamination from crude oil, fuels, and materials associated with drilling, and from nutrient additions from sewage and domestic wastes. Because most species of riverine fishes migrate seasonally between freshwater and marine systems or between overwintering and summer spawning and feeding habitats, maintenance of migration corridors is essential. Early oil field and pipeline construction activities created some road crossings with culverts inadequate for passage of migrating arctic grayling, a problem solved by replacing inadequate culverts. Gravel is mined locally for construction of drilling pads and roads, providing a stable surface and thermal insulation (preventing melting of the underlying permafrost and subsequent soil subsidence). Some resulting gravel pits have been filled with water and connected to small streams (spawning and summer feeding habitat), providing the deep water necessary for overwintering of grayling and thus developing new reproducing populations of arctic grayling (Moulton and George 2000). At present, only the lower reaches of the Kuparuk River (near the mouth) are potentially affected by industrial development; the upper Kuparuk drainage is unaffected by human activities except at infrequent river crossings by the Trans-Alaskan Pipeline and Dalton Highway. Therefore, the current and near-future impacts of resource development on the fisheries and aquatic ecosystems of the Kuparuk River seem manageable.

SAGAVANIRKTOK RIVER

The Sagavanirktok River (often termed simply the "Sag") is the second-largest river, after the Colville, on the North Slope of Alaska. Many variations of the river's name exist, and the Eskimo name, Sawanukto River, was used at one time, which meant "strong current" (Orth 1967). The river basin covers an area of 14,890 km² located in the central third of Alaska's North Slope; the river originates in the Brooks Range (between the Endicott and the Philip Smith mountains) before flowing northward approximately 270 km to enter the Beaufort Sea (Hodel 1986) (Fig. 20.12). The Trans-Alaskan Pipeline and the haul road (now the Dalton Highway) to the North Slope oilfields closely parallel the river for a large portion of its length on the Arctic Slope (Coastal Plain). Consequently, the Sagavanirktok is the most easily accessible river on the Alaskan North Slope. The Sagavanirktok is fed by Galbraith Lake, via the Atigun River, and by a number of tributaries originating in small glaciers within the Brooks Range, rising to 2400 m asl. Close to its mouth between Foggy Island and Prudhoe Bay the river forms an extensive delta approximately 20 km wide, containing numerous channels (Fig. 20.1).

Following the Paleo-Eskimos, who lived in isolated groups throughout the region, the Inupiat Eskimos constituted the native community and comprised two groups in the Sagavanirktok River area: the Tareumuit of the coastal regions and the Nunamiut further inland. The total population of these groups of Inupiat Eskimos in Alaska at the time of European contact between 1850 and 1870 was estimated to be 3350 people (Langdon 1987). Contact was nearly 100 years later here than in coastal regions of Alaska due to its greater inaccessibility, and it was not until the Yankee whalers fol-

lowed the bowhead whale through the Bering Strait that extensive contact with Europeans was made.

Physiography, Climate, and Land Use

The basin geology is characterized by sedimentary rocks, with Paleozoic rocks dominating in the mountains and Mesozoic and Cenozoic rocks in the foothills. Three glacial deposits occur in the upper reaches of the basin (Schallock and Mueller 1981). The watershed encompasses two physiographic provinces (see Fig. 20.12): the Brooks Range (BM) and the Arctic Slope (AS). The river rises in the Brooks/British Range Tundra terrestrial ecoregion, where vegetation is restricted to valleys and lower slopes and is composed mainly of stunted white spruce, willow, and dwarf birch. Wet sites contain sedges and cottongrass, whereas drier sites support mountain avens and blueberry (Ricketts et al. 1999). The river then flows through the Arctic Foothills Tundra and the Arctic Coastal Tundra ecoregions, containing herbaceous plants and dwarf shrubs.

The climate in the lower basin is typical of the Arctic Slope, with cold long winters and low precipitation levels. From approximately mid-April to the end of August the sun remains above the horizon. Mean annual precipitation is from 10 to 15 cm across the Coastal Plain, peaking in August (almost 3 cm), but is ≤0.5 cm/mo from January through May (Fig. 20.13). The coastal precipitation totals, however, greatly underestimate precipitation for the entire basin, with annual totals of 50 to 100 cm in the Brooks Range. Nearly 50% of this precipitation occurs as snowfall from September to May (Hodel 1986). Mean monthly air temperature in winter varies from −26°C to −35°C near the coast to −30°C to −40°C along the mountains, and in summer ranges from a monthly mean of 5°C to 13°C near the coast to 10°C to 18°C inland (Scott 1978; see Fig. 20.13).

Most of the drainage is unpopulated Arctic tundra, but installations related to oil development are present. Sagwon, upstream of the confluence with the Ivishak River, was an exploration facility with a large airstrip but now is usually referred to as a reference for the USGS gaging site and is close to pump station 3 on the Trans-Alaskan Pipeline.

River Geomorphology, Hydrology, and Chemistry

Approximately 45 km from the mouth the river divides into an east and a west channel (see Fig. 20.12). Boothroyd and Timson (1983) have suggested that the Sagavanirktok River is degradational (sediment eroding from streambed) for much of its length and aggradational (sediment deposited on the streambed) for 20 km toward the delta, but Hodel (1986) suggests the aggradational distance is longer. The west channel has a width up to 2.4 km and a gradient of 13 cm/km. Median grain size is 4.6 mm (Lunt et al. 2004). Significant suspended sediment loads are carried into the Beaufort Sea, estimated to reach 340 mg/L during the breakup flood in the west channel (Lunt et al. 2004).

Most of the readily accessible hydrologic information for the Sagavanirktok River derives from USGS gaging stations higher up the drainage. Although oil industry consultants monitor river flow close to the mouth, the deltaic nature of the river in these reaches makes total measurements difficult. Some preliminary data collected in 1981 and 1982 indicated in July and August the west channel carried between 68% to 70% of the flow, but at peak flow in June the channels carried similar flows (Ecological Research Associates 1982). The west channel has bankfull flows in the region of 600 m³/s.

The longest discharge record is for the period from 1983 to 1999 for the USGS station near Sagwon (close to pump station 3) at an elevation of 351 m asl and with a drainage area of 4800 km². This site does not include some of the tributaries (notably the Ivishak River), and the drainage area is only about one-third of the total drainage. Mean annual streamflow was 44 m³/s, with a maximum peak of 1200 m³/s in August 1992. Assuming the Sagwon gage to represent about one-third of the discharge, using relative basin size would give mean annual streamflow at the mouth of 132 m³/s. The river is frozen between September and May and thaws in late May to June. Peak flows at Sagwon were typically in June and early July, averaging 539 m³/s. Mean monthly flows at Sagwon ranged from 161 m³/s in June to 0.42 m³/s in March. Thus, runoff from June to August exceeded 6 cm/mo, declining to almost zero during winter (see Fig. 20.13). Interestingly, flow was zero in the river during March from 1983 to 1995 (except a negligible flow in 1986) but averaged 1.78 m³/s from 1996 to 1999. These data could possibly indicate climate change, but the length of record is relatively short and is restricted to one site. Such speculations, however, indicate the importance of time-series data to understand long-term changes, but unfortunately this gage was discontinued in September 2000. Annual runoff appears to be greater than precipitation (see Fig. 20.13) due to underestimation of precipitation for the basin as a whole.

Water chemistry in the Sagavanirktok River varies seasonally. One of the most interesting characteristics is the variation in dissolved oxygen, from a saturated 13.3 mg/L in August to a low of 1.1 mg/L in winter during April (Schallock and Mueller 1981). This oxygen depletion is more pronounced in the lower regions of the river due to low water flow and the limitation of atmospheric reaeration through ice cover. Water temperature in June varies from 3.9°C to 14°C, with typically higher water temperature reported in the foothills and decreasing toward the ocean. Schallock and Mueller (1981) attributed the decrease to the characteristic fog banks that cover the Arctic coast in the summer, reducing solar insolation even though there is 24 hr daylight. Temperatures of 0°C have been recorded in November in free-flowing water beneath ice cover. Conductivity varied from 140 to 310 µS/cm, with a mean of 203 µS/cm (1969 to 1975) for the gaging station near Sagwon. Highest turbidity values were recorded in June, reaching approximately 70 NTU, and decreased as the summer progressed, with most values in August <20 NTU. One of the major tributaries, the Ivishak River, typically has lower turbidity than the main channel of the Sagavanirktok River, upstream of the Ivishak River (Schallock and Mueller 1981). Total phosphorus values in 1969 averaged between 0.01 and 0.02 mg/L in the headwaters, increased to 0.05 mg/L near the mouth (Schallock and Mueller 1981), but were typically below detection (<0.01 mg/L) at the USGS gaging site at Sagwon. Nitrate-N typically ranged between 0.10 and 0.20 mg/L, indicating phosphorus is potentially the limiting nutrient (of N and P) to primary production. Silica is relatively high, reaching 12.5 mg/L, possibly accounting for the diverse and abundant diatom community.

River Biodiversity and Ecology

The Sagavanirktok River is located within the Yukon freshwater ecoregion (Abell et al. 2000), as are the Noatak and Kuparuk rivers. None of these rivers, however, has a direct connection with the Yukon River to the south. In contrast to the Kuparuk River to the west, biological and ecological information for the Sagavanirktok River is sparse.

Algae

An examination of diatoms collected at these sites produced a remarkable list of 261 species and species variations in 44 genera, as identified by Ruth Patrick at the Academy of Natural Sciences (Schallock and

Mueller 1981). Some of these species were locally abundant and widespread, indicating the wide diversity of diatoms at these high latitudes. The common genera were *Achnanthes, Cymbella, Eunotia, Fragilaria, Gomphonema, Navicula, Nitzschia,* and *Pinnularia.*

Invertebrates

The macroinvertebrate community is in marked contrast to the diatom diversity. Limited information is available on the benthic communities of the Sagavanirktok River, but it is clear that diversity is low. Most of the available data are from the study by Schallock and Mueller (1981), who sampled 13 sites on the main river in 1969 and 4 sites on three tributaries, the Atigin River, the Ribdon River, and the Ivishak River. Ephemeroptera, Plecoptera, Trichoptera, Diptera, and Annelida dominate the Sagavanirktok River community. The most widely distributed and abundant stoneflies were Capniidae (*Capnia*), Nemouridae (*Nemoura*), and Perlodidae (*Isogenus*). Stonefly abundance decreased downstream from the headwaters in the Brooks Range. Three genera of mayflies were dominant: *Baetis, Cinygmula,* and *Ameletus.* However, caddisflies were relatively poorly represented, with only three genera in two families (Brachycentridae and Limnephilidae) collected sporadically. Chironomidae (nonbiting midges) was typically the overall dominant family, particularly at sites near the river mouth, with sixteen genera documented. Orthocladiinae was the dominant subfamily, with nine genera identified, whereas only two Diamesinae genera were found. Although chironomid midges were abundant throughout the open-water season, black flies showed marked seasonal variations, sparse in June but widely distributed in August. Oligochaete worms were ubiquitous in the Sagavanirktok River, although abundance varied widely. Macroinvertebrate communities in the Sagavanirktok River tributaries and the Canning River showed similar patterns to the main river.

Vertebrates

Fish species known to occur within the Sagavanirktok River drainage include arctic char, arctic grayling, Dolly Varden, northern pike, broad whitefish, round whitefish, burbot, slimy sculpin, and ninespine stickleback. Some of these species are both resident and diadromous. Indeed, the current understanding of Arctic diadromous fishes derives mainly from summer fish monitoring studies within the Sagavanirktok delta (Gallaway and Fechhelm 2000). Arctic cisco and threespine stickleback are found principally within the estuary. Char and arctic grayling are important to the sports fishery.

Arctic char are also found in Galbraith Lake but are likely nonmigratory (McCart and Craig 1971). Although the lake is not land-locked, steep fast-flowing sections linking Galbraith Lake with the Sagavanirktok River are likely barriers to fish migration. A few pink and chum salmon have been reported in the drainage (Bill Wilson, personal communication).

The Sagavanirktok River is believed to support the largest Dolly Varden population on the North Slope (McCart et al. 1972). Dolly Varden typically arrive in the river mouth in June before migrating upstream to spawn in the foothill streams of the Brook Ranges in late July and August. Dolly Varden typically spawn every two to three years, with their juveniles remaining in their natal streams a similar period before their first seaward migration. Dolly Varden in this region typically overwinter in stream headwater areas near perennial springs during years in which they have been to sea (Gallaway and Fechhelm 2000). Typically the summer feeding of juvenile Dolly Varden is close to shore but long-distance movements have been reported, including documentation of traveling 1690 km to Russia (DeCiccio 1992). Fechhelm et al. (1997) found that Dolly Varden rearing juveniles grew slower in summer than juvenile whitefish and arctic cisco as a life history strategy to avoid similar habitats as the other species, thereby minimizing interspecific competition for food resources. The presence of perennial springs appears essential for Dolly Varden spawning in these arctic rivers (Fechhelm et al. 1997). The Sagavanirktok River is the only river between the Colville and the McKenzie that supports diadromous fishes apart from Dolly Varden; namely, arctic and least ciscos and broad whitefish. Arctic cisco typically leave the river by age three, possibly due to insufficient overwintering habitat for older fishes (Gallaway and Fechhelm 2000). Although broad whitefish occur in the Sagavanirktok River, the population is limited by the steeper channel gradient and the low number of accessible lakes (Moulton and George 2000). Broad whitefish also displayed wide population fluctuations over a nine-year period from 1990 to 1999 (Gallaway and Fechhelm 2000).

Lake trout are also found in Galbraith Lake. Arctic grayling are widespread throughout the drainage, although they migrate to specific areas for rearing and spawning (Moulton and George 2000). One of the features for resident fishes is the lack of flow in most of the river during late winter and consequently the need to migrate to suitable overwintering habitat to survive. These habitats include deep river pools, springs, deltas, and lakes (Craig 1989, Reynolds 1997). One problem with deep river pools is the possibility of low dissolved oxygen levels, as this habitat is a closed system in winter with negligible water flow or atmospheric reaeration.

Caribou, grizzly bears, wolves, Dall sheep, snowshoe hare, arctic foxes, squirrels, and a number of rodents are found throughout the Coastal Plain of the Sagavanirktok drainage. Over 100 bird species have been recorded along this Coastal Plain, of which only six are resident year-round: rock and willow ptarmigan, snowy owl, common raven, gyrfalcon, and dipper (Lyons and Trawicki 1994). Tundra habitat within the river delta is an important nesting and staging area for shorebirds, swans, ducks, and geese (Sedinger 1997).

Human Impacts and Special Features

The Sagavanirktok River is characterized by being unregulated and flowing across the unique Arctic tundra. However, the most notable feature of the Sagavanirktok River is that it is the most anthropogenically influenced river on the Arctic North Slope. Geological investigations in the environs of the Sagavanirktok River played an important role in the discovery of oil in this region. It was in 1963 that geologists from the Richfield Oil Company (later to become Atlantic Richfield Company [ARCO], and now Conoco Phillips) located a sand outcrop soaked in oil on the banks of the Sagavanirktok River that indicated possible oil reserves in the region, and it was "Sag River well No. 1," an exploratory well located near the mouth of the river, that eventually confirmed the considerable extent of the petroleum reservoir of the Prudhoe Bay fields (Roderick 1997).

Within the Sagavanirktok drainage developments have been linked principally to oil operations at Prudhoe Bay and the Trans-Alaskan Pipeline and Dalton Highway that runs along much of its length. The city of Deadhorse lies in close proximity to the west channel of the river delta, principally to facilitate the oil industry. One of the largest impacts has been the removal of gravel from the floodplain for road construction and oil pad building. Gravel extraction techniques have changed over time. Initially, gravel was removed by shallow scraping of riverine deposits but, due to environmental concerns, gravel extraction has shifted to deep mining, up to 20m below the surface from abandoned ox-bows or relic channels to reduce surface disturbance

(Hemming 1991). To rehabilitate gravel extraction sites within the delta, connection of both shallow and deep mined sites to the main channel has occurred and their subsequent fish use monitored. Hemming (1991) found grayling, round whitefish, and burbot used both types of extraction site, whereas ninespine stickleback and broad whitefish were found only in flooded shallow mined sites. Large numbers of broad whitefish in August indicate these sites act as important summer rearing habitat for this species. Passage facilities, like bridges or culverts, are now designed to provide both high- and low-flow access for fishes to feeding and overwintering habitats (Moulton and George 2000).

MOOSE RIVER

The Moose River basin, located in Northeastern Ontario, originates in the Superior Upland, flowing north through the Hudson Bay Lowland before emptying into James Bay near the town of Moosonee (Fig. 20.14). The drainage basin is >100,000 km^2, but the human population is <100,000, with more than 50% living in the upper (southern) reaches of the basin. The Moose River is a 7th order river but consists of five major river systems from west to east: the Missinaibi, Kapuskasing, Groundhog, Mattagami, and Abitibi (Brousseau and Goodchild 1989) (Table 20.1). The rivers are of similar size, and flow within the middle reaches is similar in gradient. All the river systems within the Moose River basin, with the exception of the Missinaibi, have undergone development to some extent, including hydroelectric dams and pulp and paper mills (Fig. 20.5). However, development is chiefly restricted to the Superior Upland region, principally since the early 1900s.

The native peoples (Cree) of the Moose River basin were nomadic and their presence dates to about

TABLE 20.1 Physical characteristics of Moose River tributaries.

	River order	Length (km)	Basin size (km^2)	High flow (m^3/s)	Average flow (m^3/s)
Missinaibi	6	430	22,530	1,740	105
Kapuskasing	5	324	8,633	963	77
Groundhog	5	363	12,518	1,844	144
Mattagami	6	491	41,672	1,230	113
Abitibi	5	285	33,987	3,210	384

Source: Modified from Table 1 in Brousseau and Goodchild (1989).

FIGURE 20.5 Harmon hydroelectric facility on the Mattagami River tributary of the Moose River, Ontario, 88 km north of Kapuskasing (Photo by W. Gibbons).

7000 years ago. The Hudson's Bay Company outpost at the river mouth dates from 1673. Upper basin mining prospecting started in the early 1900s, with settlements after 1910. The national transcontinental railway construction in the area commenced in 1905 and the Kapuskasing pulp mill dates to around 1920. During World War I a large prisoner of war camp was sited near Kapuskasing, representing the first settlement in the midbasin. Further settlements occurred after the war, with people attracted to the area by land grants.

Physiography, Climate, and Land Use

The Moose River drainage basin spans two physiographic provinces with the Canadian (Precambrian) Shield physiographic division. The upper (southern) drainage is within the Superior Upland (SU) province, whereas the lower reaches cross the Hudson Bay Lowland (HB) province (see Fig. 20.14) (Hunt 1974). The Superior Upland is underlain by the Canadian Shield and covered by glacial, glaciofluvial, and glaciolacustrine deposits. An escarpment near 50°N marks the beginning of the Hudson Bay Lowland, where sedimentary deposits of the Mesozoic and Paleozoic eras (limestone bedrock) lie beneath the northern part of the drainage basin (Brousseau and Goodchild 1989). The bedrock is covered with marine and glacial deposits, typically overlain by peat.

The drainage basin encompasses three terrestrial ecoregions: The headwaters lie in the Eastern Forest/Boreal Transition, the midreaches in the Central Canadian Shield Forests, and the lower reaches in Southern Hudson Bay Taiga (Ricketts et al. 1999). Mixed-species forests characterize the Eastern Forest/Boreal Transition, which includes white spruce, balsam fir, quaking aspen, paper birch, and yellow birch. Red, white, and jack pine favor the drier sites, and black spruce is found on wetter sites. Most of the land within these uplands is under timber-management plans. Black spruce dominates the Central Canadian Shield coniferous forests, but jack pine and some paper birch are also found due to the frequency of fire. The other species in the Eastern Forest/Boreal Transition are also present, particularly on south-facing slopes. Stunted black spruce and tamarack dominate the vegetation of the Southern Hudson Bay Taiga. Where uplifting of the land has occurred due to isostatic rebound, raised beaches are formed that support white spruce. Poorly drained areas are characterized by sedge and moss.

Mean annual temperature decreases in a south–north direction, with the mean annual temperature at Smoky Falls on the Mattagami River 1°C lower than at Timmins near the southern end of the basin. Seasonal variation in temperature of the central basin is wide, with a January mean of −19°C, approximately 36°C lower than the July mean of 17.3°C (Fig. 20.15). This wide variation is due to the continental climate of inland Canada but is less variable at the mouth due to the ameliorating effect of the inland sea of Hudson Bay. Mean annual precipitation is between 70 and 90 cm (Environment Canada 1982a, 1982b). Precipitation is highest during the summer, with peak rainfall in July and September (see Fig. 20.15). The majority of precipitation during the winter falls as snow, which contributes 30% to 40% of the annual precipitation.

The low population density in the basin (<1 person/km^2), the concentration of developments in the middle reaches, and the absence of a road transportation network throughout much of the northern portion of the basin means that development is localized. About 94% of the basin is forested, 5% is agricultural, and 1% is urban and industrial. In many parts of the basin it is still possible to have relatively pristine conditions. Past developments are primarily related to dams and reservoirs, the forest industry, and mining. Additional information on development within the Moose River basin can be found in reports associated with previous workshops related to cumulative effects within the Moose River basin (Greig et al. 1992; ESSA Technologies 1996a, 1996b).

River Geomorphology, Hydrology, and Chemistry

Within the Superior Upland there are hundreds of bedrock outcrops that create waterfalls and rapids, many of which are barriers to fish migration. In most areas these deposits are thin and heterogeneous, but in the southeastern portion of the watershed there is a large glaciolacustrine clay deposit. The river systems in the southeastern region (i.e., Abitibi and Frederickhouse rivers) flow through this clay belt and are characterized by high turbidity.

The escarpment at the beginning of the Hudson Bay Lowland is characterized by waterfalls on the Missinaibi (Thunderhouse Falls), Mattagami (Mattagami generating station complex), and Abitibi Rivers (Abitibi River Canyon). North of the escarpment the rivers have low gradients and consist of long, straight stretches with sand and gravel shoals and numerous riffle areas (Fig. 20.6). The peaking hydroelectric facilities result in a significant amount of available aquatic habitat being dewatered during peaking cycles.

Overall, river gradients in the upper basin are moderate, averaging 50 cm/km, before the streams descend over the edge of the Superior Upland. The change in elevation from the headwaters to James Bay is on the order of 325 m, and reaches of particularly high gradient occur on the Missinaibi and Mattagami rivers, where they descend from the Superior Upland (gradient 520 to 580 cm/km) to the lowlands. In several rivers the gradient has been altered due to hydroelectric facilities exploiting significant elevation drops (see Fig. 20.5).

The Moose River has an annual discharge of at least 1370 m^3/s. For the four western tributaries, the discharge of the largest of the rivers (Groundhog River) is roughly twice that of the smallest (Kapuskasing River) (see Table 20.1). Seasonal discharge patterns and patterns of year-to-year variability in discharge are similar between the rivers (Fig. 20.7). The peak river discharge occurs in late April

FIGURE 20.6   The confluence of two tributaries of the Moose River, Ontario (PHOTO BY W. GIBBONS).

and May as the snowpack melts. However, there is considerable dampening of the extreme peak and low flows in those rivers with upstream storage reservoirs, such as the Mattagami River, that are not present in the unregulated Missinaibi River. Monthly runoff calculations also illustrate the strong snowmelt effect reaching 10 cm/mo in May (see Fig. 20.15). Seasonal changes in snow storage, evapotranspiration, and possibly lake storage account for the strong seasonal pattern of runoff in spite of the highest precipitation occurring after peak runoff (see Fig. 20.15).

Water chemistry is similar in all the main tributary rivers, with conductivity ranging from 126 to 179 μS/cm and turbidity from 5 to 61 NTU (levels are typically higher downstream of the pulp mills). pH values range from 7.1 to 7.7, with relatively high levels of alkalinity (54 to 96 mg/L as $CaCO_3$) (Farwell 1999) and similar levels of dissolved oxygen among rivers. Seasonal fluctuations included lower values of alkalinity, dissolved inorganic carbon, and pH in the spring compared to the fall, whereas total suspended solids, particulate organic carbon, and nitrogen were higher in the spring (Farwell 1999). Water chemistry of the Kapuskasing and Mattagami rivers recently improved as the major pulp mills installed secondary waste treatment (Munkittrick et al. 2000). In a broader analysis of smaller tributaries in the basin,

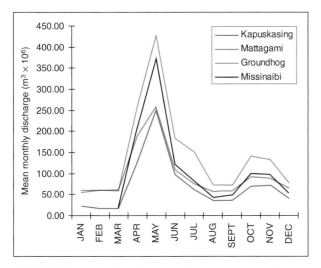

FIGURE 20.7 Monthly discharge for the western tributaries of the Moose River basin.
Source: Munkittrick et al. (2000).

Kilgour et al. (2000) found pH levels between 6.3 and 8.1, conductivity between 7 and 312 μS/cm, alkalinity between 0.3 and 165 mg/L as $CaCO_3$, and total phosphorus between 0.01 and 0.05 mg/L.

Water temperatures remain close to 0°C from November until late April in the main channel, rising rapidly to the low 20s for the summer period. Munkittrick et al. (2000) also found water temperature decreased as the main rivers flowed north. Comparisons of the rivers to the hydroelectric reservoirs demonstrated reservoirs were buffered from warmer and colder temperatures, as well as daily fluctuations.

River Biodiversity and Ecology

In contrast to the other rivers in this chapter, which are located within the Arctic complex of freshwater ecoregions, the Moose River is found within the South Hudson ecoregion of the Hudson Bay complex. The biology of the river is relatively well studied, at least for the invertebrates and fishes.

Plants

Aquatic vegetation consists of a narrow 1 m band along most of the rivers, due to the steep-sided channel morphology and the water's turbidity. Annual water-level fluctuations vary from 1 to 2 m, and significant areas of ice scour occur along river banks. Aquatic macrophytes are common in the shallow nearshore areas, and dominant plants are bulrush, eelgrass, watermilfoil, waterlily, common waterweed, sawtooth (*Najas*), crowfoot, hornwort, arrowhead, and pondweed.

Reservoirs above impoundments and shallow bays flooded by the dams on some of the tributaries have extensive beds of floating and submergent species, including yellow pond-lily, Richardson's pondweed, watermilfoil, and threadleaf crowfoot. Flooded areas also have numerous logs and stumps. Aquatic macrophytes are not obvious downstream of impoundments, where water-level fluctuations are large. Bands of woody shrubs (dogwood, alder, and willow), grasses and sedges, and emergent plants (rushes, horsetails, and arrowheads) colonize the riparian margin in upper portions of the basin.

Invertebrates

Macroinvertebrate diversity in the Moose River drainage is relatively well studied. Fiset (1995a) identified 288 macroinvertebrate taxa in a basinwide study, whereas Kilgour et al. (2000), collecting at 79 sites (1st to 6th order streams), identified 278 taxa, with chironomid midges the most dominant and diverse group (77 taxa; common subfamilies include Chironominae, Orthocladiinae, and Tanypodinae). Mayflies (19 taxa, especially the genera *Baetis*, *Caenis*, *Ephemerella*, *Hexagenia*, *Heptagenia*, and *Stenonema*), stoneflies (12 taxa, especially the families Perlodidae and Pteronarcyidae), and caddisflies (34 taxa, especially Hydropsychidae, Limnephilidae, and Polycentropodidae) were well represented, as were oligochaete worms (29 taxa, including *Nais*, *Arcteonais*), dragonflies (15 taxa, especially Aeshnidae, Corduliidae, Macromiidae, and Gomphidae), snails (15 taxa, especially Physidae and Planorbidae), beetles (13 taxa, especially in the families Dytiscidae, Elmidae, and Gyrinidae), and Tubificidae (including *Limnodrilus* and *Tubifex*). Only four species of bivalve mollusks were found (all Sphaeriidae), although Unionidae shells were also collected. Crustaceans include crayfishes in the genera *Cambarus* and *Orconectes* (Fiset 1995b). High-order sites (5th to 6th order) supported higher relative abundance of grazers and collector-gatherers. Low-order sites typically had higher relative abundance of deposit feeders, such as chironomid midges and worms.

Vertebrates

The fish community of the river basin is diverse for a northern river, with at least 14 families and 40 species occurring in the nontidal portion of the river. Typically these fishes are divided into large-bodied and small-bodied species (Fiset 1995a, Seyler 1994, Portt et al. 1999, Kilgour et al. 2000), with the families Acipenseridae, Coregonidae, Catostomidae, Centrarchidae, Cottidae, Cyprinidae, Ecoscidae,

Gadidae, Gasterosteidae, Hiodontidae, Ictaluridae, Percopsidae, Percidae, and Salmonidae represented. Percids (16 species) and cyprinids (12 species) are the dominant families. Many of the small-order streams are low gradient, with fish species such as pearl dace and stickleback (Kilgour et al. 2000). Higher-order stream sections tended to have species adapted to faster flows. Among the larger fish species the most abundant are lake sturgeon, northern pike, walleye, and white sucker. The most widely distributed small-bodied species are pearl dace, longnose dace, trout-perch, and spotfin shiner. No species in the Moose River basin are identified as endangered, rare, or threatened in Ontario.

Aquatic mammals or mammals often associated with the Moose River include muskrat, beaver, fisher, marten, mink, river otter, and moose, particularly throughout the upper and middle parts of the basin. There are reports of seal and beluga whale in the lower river in spring to feed on whitefish. The upper basin is on the North Atlantic flyway, and a wide variety of migratory waterfowl can be found, including Canada goose, lesser snow goose, white-fronted goose, Brant's goose, surf scoter, and many other sea ducks.

Human Impacts and Special Features

Although certain parts of the Moose River drainage have been significantly influenced by anthropogenic activities, particularly in the Central Canadian Shield Forests, overall the Missinaibi represents a large undeveloped river that has been designated a Heritage River and an Ontario Provincial Park. This tributary is protected from land-use activities within the park boundary that extends from 120 m to 200 m from the water's edge for much of the river's length. The dominant anthropogenic impacts within the Moose River basin are related to hydroelectric development, forestry, and mining activities, with dams having the greatest negative effects. Agricultural activity is minimal (other than silviculture) (ESSA 1998).

There are 20 hydroelectric generating facilities, with 10 operated by Ontario Hydro and 10 as nonutility generating facilities (NUGs). The existing hydroelectric facilities range from 0.15 to 285 MW in capacity (total 950 MW), and the NUGs vary from 0.5 to 50 MW in capacity (total 120 MW). The oldest facilities are on the Mattagami River (1911, 1912, 1923, and 1931), with most of the remaining facilities constructed in the 1960s. There are also 50 non-hydroelectric water-control structures within the system; 27 are flood-control or water-diversion facil-

ities and 23 are associated with work by Ducks Unlimited related to wetlands enhancement projects.

In the Moose River basin, sections of 23 forest-management units are harvested for timber. Previously, extensive log-driving operations existed on several rivers, but these ceased in the 1980s and 1990s. The major effluent discharges related to the forest industry are from three pulp mills within the basin. The pulp mills are distributed on the Kapuskasing River (380,000 tonne/yr thermo-mechanical pulp [TMP] mill located at Kapuskasing), the Mattagami River (170,000 tonne/yr bleached kraft mill located at Smooth Rock Falls), and the Abitibi River (310,000 tonne/yr mill, recently converted to a TMP process, located at Iroquois Falls). The mills underwent extensive renovations in the mid-1990s, including the installation of secondary waste treatment facilities. In addition to the three pulp mills, a variety of wood-processing facilities exist, including an oriented strand board mill in Timmins, sawmills in Hearst, Cochrane, Kirkland Lake, South Porcupine, and Timmins, and a plywood and veneer mill in Cochrane. No new development is predicted for the forest industries, other than modest increases in production associated with existing facilities (ESSA 1999).

During the 1990s there were nine operating mines, eight operating mills, and one refining facility associated with gold, copper, zinc, lead, silver, hedmonite, and talc. There were also two sites associated with peat extraction. Most of the mines are near the headwaters of the Abitibi, Frederickhouse, and Mattagami rivers. There are also eight advanced mining exploration sites in the basin (seven gold, one nickel) and over 100 additional exploration activities during 1994 alone (reviewed in ESSA 1996b). Additional potential exists within the basin for mining of phosphates, aluminum, lignite, gypsum, diamonds, gold, copper, zinc, lead, silver, nickel, cadmium, indium, kaolin, and peat. Although there are currently no aggregate facilities, potential exists within the basin. There are also extensive peat deposits, estimated at >9 billion m^3. Approximately 39% of the deposits are fuel grade and 24% have good horticultural potential (ESSA 1996a). Potential impacts associated with peat extraction include impacts on drainage and erosion.

ADDITIONAL RIVERS

The Seal River is the largest undammed river in northern Manitoba and flows in an easterly direction before emptying into Hudson Bay (Fig. 20.16). Much of the basin is drained by the South Seal and North Seal rivers, which flow into Shethanei Lake (northeast of the larger Tadoule Lake on the South Seal). The main-stem Seal thus begins at Shethanei Lake, which is ringed by sand-crowned eskers, a distinguishing feature along much of the river. Much of the basin contains subarctic boreal forest of the Precambrian Shield, before giving way to Arctic tundra as it crosses into the Hudson Bay Lowland. Like the larger Thelon/Kazan basin to its north, the Seal is home to a diverse and abundant fauna. Particularly noteworthy are its harbor seals (which give the river its name), found 200 km upstream of Hudson Bay, and a large population of beluga whales, which feed and calve in its estuary. Although the basin is entirely undeveloped wilderness and there are no dams on the main stem or tributaries, only the main-stem Seal is in the Canadian Heritage River system. The Seal is popular for white-water river trips.

The Thelon River and its largest tributary, the Kazan River, flow in an easterly and northeasterly direction, respectively, in the Nunavut Territory before emptying into Baker Lake (Fig. 20.18). The outlet from Baker Lake then flows into Chesterfield inlet at the northwestern corner of Hudson Bay. The main rivers and their tributaries have no dams, but they often widen into lakes along their course, including Aberdeen, Dubawnt, Kasba, Yathkyed, and Baker lakes (see Fig. 20.18). The basin is a boreal Arctic wilderness that supports diverse and abundant wildlife; notably caribou, muskox, grizzly bear, lynx, wolf, wolverine, peregrine falcon, tundra swan, and Canada goose. The only major human presence in the Thelon/Kazan basin is about 1000 Caribou Inuit who live in the community of Bakers Lake, near the mouth. Both the Thelon and the Kazan are in the Canadian Heritage Rivers system. The rivers are very scenic, including white water and several falls, such as the 25 m Kazan Falls.

ACKNOWLEDGMENTS

We would like to thank Bill Wilson of LGL Research in Anchorage for discussions regarding the Sagavanirktok River and Dr. Ian Lunt for access to unpublished data. We are grateful to the meticulous editing of Art Benke, who improved the final manuscript.

LITERATURE CITED

Abell, R. A., D. M. Olson, E. Dinerstein, P. T. Hurley, J. Diggs, W. Eichbaum, S. Walters, W. Wettengel, T. Allnutt, C. J. Loucks, and P. Hedao. 2000. *Freshwater ecoregions of North America: A conservation assessment.* Island Press, Washington, D.C.

Arctic and Monitoring and Assessment Programme (AMAP). 1997. *Arctic pollution issues: A state of the Arctic report.* Arctic and Monitoring and Assessment Programme, Oslo, Norway.

Barry, R. G., M. C. Serreze, J. A. Maslanik, and R. H. Preller. 1993. The Arctic Sea–Ice climate system. *Review of Geophysics* 31:397–422.

Bliss, L. C., and N. V. Matveyeva. 1992. Circumpolar arctic vegetation. In F. S. Chapin, R. L. Jefferies, J. F. Reynolds, and G. R. Shaver (eds.). *Arctic ecosystems in a changing climate: An ecophysiological perspective*, pp. 15–33. Academic Press, San Diego.

Billings, W. D., and K. M. Peterson. 1980. Vegetational change and ice-wedge polygons through the thaw-lake cycle in arctic Alaska. *Arctic and Alpine Research* 12:413–432.

Bond, B. J. 1988. Microbial ecology of an Alaskan tundra beaded stream. Master thesis, University of Alaska, Fairbanks.

Boothroyd, J. C., and B. S. Timson. 1983. The Sagavanirktok and adjacent river systems, eastern North Slope, Alaska: An analog for ancient fluvial terrain on Mars. In *Permafrost: Proceedings of the Fourth International Conference*, pp. 74–79. National Academy Press, Washington, D.C.

Brousseau, C. S., and G. A. Goodchild. 1989. Fisheries and yields in the Moose River basin, Ontario. In D. P. Dodge (ed.), Proceedings of the International Large Rivers Symposium. *Canadian Special Publication of Fisheries and Aquatic Sciences* 106:145–158.

Brown J., and R. A. Kreig. 1983. *Guidebook to permafrost and related features along the Elliott and Dalton Highways, Fox to Prudhoe Bay.* Alaska Guidebook no. 4. Fourth International Conference on Permafrost, University of Alaska, Fairbanks.

Cappiello, T. A. 1995. *Spawning run characteristics of Dolly Varden in the Kugururok River, Noatak drainage, Alaska.* University of Alaska, Fairbanks.

Cheng, H., S. Leppinen, and G. Whitley. 1993. Effects of river ice on chemical processes. In T. D. Prowse and N. C. Gridley (eds.). *Environmental aspects of river ice*, pp. 155. NHRI Science Report 5. National Hydrology Research Institute, Saskatoon, Saskatchewan.

Childers, J. M., and D. R. Kernodle. 1981. Hydrologic reconnaissance of the Noatak River basin, 1978. U.S. Geological Survey, Anchorage, Alaska.

Connor, C., and D. O'Haire. 1988. *Roadside geology of Alaska.* Mountain Press, Missoula, Montana.

Craig, P. C. 1989. An introduction to anadromous fishes in the Alaskan Arctic. In D. W. Norton (ed.). *Research advances on anadromous fish in arctic Alaska and Canada: Nine papers contributing to an ecological synthesis*, pp. 27–54. Biological Papers of the University of Alaska no. 24. Institute of Arctic Biology, Fairbanks, Alaska.

Craig, P. C., and L. Haldorson. 1986. Pacific salmon in the North American Arctic. *Arctic* 39:2–7.

Craig, P. C., and P. J. McCart. 1975. Classification of stream types in Beaufort Sea drainages between Prudhoe Bay, Alaska, and the Mackenzie Delta, N.W.T., Canada. *Arctic and Alpine Research* 7:183–198.

Cushing, C. E., and J. D. Allan. 2001. *Streams: Their ecology and life*. Academic Press, San Diego.

DeCicco, A. L. 1992. Long-distance movements of anadromous Dolly Varden between Alaska and the U.S.S.R. *Arctic* 45:120–123.

Environment Canada. 1982a. *Canadian climate normals*. Vol. 2: *Temperature, 1951–1980*. Environment Canada, Atmospheric Environment Service, Downsview, Ontario.

Environment Canada. 1982b. *Canadian climate normals*. Vol. 3: *Precipitation, 1951–1980*. Environment Canada, Atmospheric Environment Service, Downsview, Ontario.

ESSA Technologies. 1996a. Catalogue of development activities in the Moose River Basin. Prepared for the Moose River Environmental Information Partnership, Cochrane, Ontario. ESSA Technologies, Ltd., Richmond Hill, Ontario.

ESSA Technologies. 1996b. Planned and potential future development activities in the Moose River Basin: Final report. Prepared for the Moose River Environmental Information Partnership, Cochrane, Ontario. ESSA Technologies, Ltd., Richmond Hill, Ontario.

ESSA Technologies. 1998. Cumulative effects assessment in the Moose River basin: Background literature review. Prepared for the Environmental Information Partnership, South Porcupine, Ontario. ESSA Technologies, Ltd., Richmond Hill, Ontario.

ESSA Technologies. 1999. Report on the preparation of a planned and potential future development activities in the Moose River Basin. Final report prepared for the Environmental Information Partnership, South Porcupine, Ontario. ESSA Technologies, Ltd., Richmond Hill, Ontario.

Everett, K. R., D. L. Kane, and L. D. Hinzman. 1996. Surface water chemistry and hydrology of a small Arctic drainage basin. In J. F. Reynolds and J. D. Tenhunen (eds.). *Landscape function and disturbance in Arctic tundra*, pp. 185–201. Springer-Verlag, Berlin.

Farwell, A. J. 1999. Stable isotope study of riverine benthic food webs influenced by anthropogenic developments. Ph.D. diss., University of Waterloo, Ontario.

Fechhelm, R. G., J. D. Bryan, W. B. Griffths, and L. R. Martin. 1997. Summer growth patterns of northern dolly varden (*Salvelinus malma*) smolts from the Prudhoe Bay region of Alaska. *Canadian Journal of Fisheries and Aquatic Sciences* 54:1103–1110.

Fechhelm, R. G., W. B. Griffths, L. R. Martin, and B. J. Gallaway. 1996. Intra- and interseasonal changes in the relative condition and proximate body composition of arctic ciscoes from the Prudhoe Bay region of Alaska. *Transactions of the American Fisheries Society* 125:600–612.

Fechhelm, R. G., W. B. Griffths, W. J. Wilson, B. J. Gallaway, and J. D. Bryan. 1995. Intra- and interseasonal changes in the relative condition and proximate body composition of broad whitefish from the Prudhoe Bay region of Alaska. *Transactions of the American Fisheries Society* 124:508–519.

Fiset, W. 1995a. A review of aquatic invertebrate studies conducted in the Moose River basin. NEST Technical Report TR-024. Ontario Ministry of Natural Resources, Cochrane.

Fiset, W. 1995b. Taxonomic list of the aquatic invertebrates collected in the Moose River basin. NEST Technical Report TR 023. Ontario Ministry of Natural Resources, Cochrane.

Flory, E. A., and A. M. Milner. 1999. The role of competition in invertebrate community development in a recently formed stream in Glacier Bay National Park, Alaska. *Aquatic Ecology* 33:175–184.

Gallant, A. L., E. F. Binian, J. M. Omernik, and M. B. Shasby. 1995. *Ecoregions of Alaska*. U.S. Geological Survey, Washington, D.C.

Gallaway, B. J., and R. G. Fechhelm. 2000. Anadromous and amphidromous fishes. In J. C. Truett and S. R. Johnson (eds.). *The natural history of an Arctic oil field*, pp. 349–369. Academic Press, San Diego.

Gilders, M. A., and M. A. Cronin. 2000. North slope oil field development. In J. C. Truett and S. R. Johnson (eds.). *The natural history of an Arctic oil field*, pp. 15–33. Academic Press, San Diego.

Green, J. 1999. *Alaska's climates*. Williwaw, Anchorage, Alaska.

Greig, L. A., J. K. Pawley, C. H. R. Wedeles, P. Bunnell, and M. J. Rose. 1992. Hypotheses of effects of development in the Moose River basin: Workshop summary report. Prepared for the Department of Fisheries and Oceans, Burlington, Ontario. ESSA Technologies, Ltd., Richmond Hill, Ontario.

Hall, E. S. Jr. 1975. *The Eskimo storyteller: Folktales from Noatak, Alaska*. The University of Tennessee Press, Knoxville.

Hemming, C. R. 1991. Fish and habitat investigations of flooded North Slope gravel mine sites, 1990. Technical Report 91-3. Alaska Department of Fish and Game, Fairbanks.

Hershey, A. E., W. B. Bowden, L. Deegen, J. E. Hobbie, B. Peterson, G. W. Kipphut, G. W. Kling, M. A. Lock, R. W. Merritt, M. C. Miller, J. R. Vestal, and J. A. Schuldt. 1997. The Kuparuk River: A long-term study of biological and chemical processes in an arctic river. In A. M. Milner and M. W. Oswood (eds.). *Fresh waters of Alaska: Ecological syntheses*, pp. 107–129. Springer-Verlag, New York.

Hinzman, L. D., D. L. Kane, C. S. Benson, and K. R. Everett. 1996. Energy balance and hydrological processes in an Arctic watershed. In J. F. Reynolds and J. D. Tenhunen (eds.), *Landscape function and disturbance in Arctic tundra*, pp. 131–154. Springer-Verlag, Berlin.

Hobbie, J. E. 1997. History of limnology in Alaska: Expeditions and major projects. In A. M. Milner and M. W. Oswood (eds.), *Freshwaters of Alaska: Ecological syntheses*, pp. 45–60. Springer-Verlag, New York.

Hodel, K. L. 1986. The Savanirtok River, North Slope Alaska: Characterization of an Arctic stream. Open-File Report 86–267. U.S. Geological Survey, Menlo Park, California.

Horton, T. 1996. *An island out of time: A memoir of Smith Island in the Chesapeake.* W.W. Norton & Co., New York.

Hunt, C. B. 1974. *Natural regions of the United States and Canada.* W. H. Freeman, San Francisco.

Irons, J. G., III, and M. W. Oswood. 1992. Seasonal temperature patterns in an arctic and two subarctic Alaskan (U.S.A.) headwater streams. *Hydrobiologia* 237:147–157.

Jans, N. 1993. *The last light breaking.* Alaska Northwest Books, Seattle, Washington.

Kilgour, B. W., D. G. Dixon, R. C. Bailey, and T. B. Reynoldson. 2000. Development of a reference condition approach (RCA) to assess fish and benthic communities and in-stream habitat attributes in the Moose River basin. Report Submitted to Forest Ecosystem Science Co-operative, Inc., Thunder Bay, Ontario.

Langdon, S. 1989. *The native people of Alaska*, 2nd ed. Greatland Graphics, Anchorage, Alaska.

Lee, J. O., and A. E. Hershey. 2000. The effects of aquatic bryophytes and long-term fertilization on arctic stream insects. *Journal of the North American Benthological Society* 19:697–708.

LGL Ecological Research Associates. 1982. Sagavanirtok River: Hydrology and hydraulics. *Environmental Summer Studies (1982) for the Endicott Development.* Vol 11. Bryan, Texas.

Lunt, L. A., Bridge, J. S. and Tye, R. S. 2004. A quantitative, three-dimensional depositional model of gravelly braided rivers. *Sedimentology* 51:377–414.

Lyons, S. M., and J. M. Trawicki. 1994. Water resource inventory and assessment, coastal plain, Arctic National Wildlife Refuge: 1987–1992 Final Report. U.S. Fish and Wildlife Service, Water Resource Branch Anchorage, Alaska. WRB 94-3.

MacLean, R. 1997. The effect of permafrost on the biogeochemistry of two subarctic streams, Master thesis. University of Alaska, Fairbanks.

Maniilaq Association. 2001. *Noatak.* NC Studios. Kotzebue, Alaska.

Manuwal, D. A. (1974). Avifaunal investigations in the Noatak River valley. In S. B. Young (ed.). *The environment of the Noatak River basin, Alaska*, pp. 252–325. Center for Northern Studies, Wolcott, Vermont.

McCart, P., and P. C. Craig. 1971. Meristic differences between anadromous and freshwater resident arctic char (*Salvelinus alpinus*) in the Sagavanirktok River drainage, Alaska. *Journal of the Fisheries Research Board of Canada* 28:115–118.

McCart, P., P. Craig, and H. Bain. 1972. Report of fisheries investigations in the Sagavanirktok River and neighboring drainages. Unpublished Report. Alyeska Pipeline Service Company, Anchorage, Alaska.

McGhee, R. 1996. *Ancient people of the Arctic.* University of British Columbia Press, Vancouver.

McNamara, J. P., D. L. Kane, and L. D. Hnzman. 1998. An analysis of streamflow hydrology in the Kuparuk River basin, Arctic Alaska: A nested watershed approach. *Journal of Hydrology* 206:39–57.

McPhail, J. D., and C. C. Lindsey 1970. *Freshwater fishes of northwestern Canada and Alaska.* Fisheries Research Board of Canada, Ottawa, Ontario.

Melack, J., J. Dozier, C. Goldman, D. Greenland, A. M. Milner and R. J. Naiman. 1997. Effects of climate change on inland waters of the Pacific Coastal Mountains and Western Great Basin of North American. *Hydrological Processes* 11:971–992.

Milner, A. M., J. G. Irons III, and M. W. Oswood. 1997. The Alaskan landscape: An introduction for limnologists. In A. M. Milner and M. W. Oswood (eds.). *Fresh waters of Alaska: Ecological syntheses*, pp. 1–44. Springer-Verlag, New York.

Milner, A. M., S. C. Conn, J. M. Ray. 2005. Development of a long-term ecological monitoring program for Denali National Park and Preserve: Design of methods for monitoring stream communities. University of Alaska Fairbanks report to the United States Geological Survey, Fairbanks.

Morrow, J. E. 1980. *The freshwater fishes of Alaska.* Alaska Northwest, Anchorage.

Munkittrick, K. R., M. McMaster, G. Van Der Kraak, C. Portt, W. Gibbons, A. Farwell, and M. Gray. 2000. *Development of methods for effects-based cumulative effects assessment using fish populations: Moose River project.* SETAC Press, Pensacola, Florida.

Moulton, L. L., and J. C. George. 2000. Freshwater fishes in the Arctic oil-field region and coastal plain of Alaska. In J. C. Truett and S. R. Johnson (eds.). *The natural history of an Arctic oil field*, pp. 327–348. Academic Press, San Diego.

O'Brien, W. J. and D. G. Huggins. 1974. The limnology of the Noatak drainage. pp. 158–223 In S.B. Young (ed.). *The environment of the Noatak River basin, Alaska.* 584 pp. Contributions from the Center for Northern Studies no. 1. Wolcott, Vermont.

Orth, D. J. 1967. *Dictionary of the Alaska place names.* U.S. Geological Survey Professional Paper 567. U.S. Government Printing Office, Washington, D.C.

Oswood, M. W. 1989. Community structure of benthic invertebrates in interior Alaskan (USA) streams and rivers. *Hydrobiologia* 172:97–110.

Oswood, M. W. 1997. Streams and rivers of Alaska: A high latitude perspective on running waters. In A. M. Milner and M. W. Oswood (eds.). *Fresh waters of Alaska: Ecological syntheses*, pp. 331–356. Springer-Verlag, New York.

Oswood, M. W., J. G. Irons, and A. M. Milner. 1995. River and stream ecosystems of Alaska. pp. 9–32 In Cushing, C.E., Cummins, K.W., and G. W. Minshall (eds.). River and stream ecosystems. Vol. 22 *Ecosystems of the World*. Elsevier Science.

Oswood, M. W., K. M. Everett, and D. M. Schell. 1989. Some physical and chemical characteristics of an arctic beaded stream. *Holarctic Ecology* 12:290–295.

Oswood, M. W., J. G. Irons III, and D. M. Schell. 1996. Dynamics of dissolved and particulate carbon in a tundra stream in arctic Alaska. In J. F. Reynolds and J. D. Tenhunen (eds.), *Landscape function and disturbance in Arctic tundra*, pp. 275–289. Springer-Verlag, Berlin.

Oswood, M. W., L. K. Miller, and J. G. Irons III. 1991. Overwintering of freshwater benthic invertebrates. In R. E. J. Lee and D. L. Denlinger (eds.), *Insects at low temperatures*, pp. 360–375. Chapman and Hall, New York.

Oswood, M. W., A. M. Milner, and J. G. Irons III. 1992. Climate change and Alaskan rivers and streams. In P. Firth and S. Fisher (eds.), *Global warming and freshwater ecosystems*, pp. 192–210. Springer-Verlag, New York.

Oswood, M. W., J. B. Reynolds, J. G. Irons Jr., and A. M. Milner. 2000. Distributions of freshwater fishes in ecoregions and hydroregions of Alaska. *Journal of the North American Benthological Society* 19:405–418.

Peterson, B. J., J. E. Hobbie, and T. L. Corliss. 1986. Carbon flow in a tundra stream ecosystem. *Canadian Journal of Fisheries and Aquatic Sciences* 43:1259–1270.

Pielou, E. C., 1991. *After the Ice Age: The return of life to glaciated North America*. University of Chicago Press, Chicago.

Pissart, A. 1988. Pingos: An overview of the present state of knowledge. In M. J. Clark (ed.). *Advances in periglacial geomorphology*, pp 279–297. Wiley, New York.

Portt, C. B., B. W. Kilgour, and R. K. Recoskie. 1999. Development impacts in the Moose River basin: Classification, data gaps and management needs. Final report prepared for the Ontario Ministry of Natural Resources, Moose River Basin Environmental Information Partnership, Northeast Region. Ontario Ministry of Natural Resources, Cochrane.

Poff, N. L. 1997. Landscape filters and species traits: Towards mechanistic understanding and prediction in stream ecology. *Journal of the North American Benthological Society* 16:391–409.

Poff, N. L., P. L. Angermeier, S. D. Cooper, P. S. Lake, K. D. Fausch, K. O. Winemiller, L. A. K. Mertes, M. W. Oswood, J. Reynolds, and F. J. Rahel. 2001. Fish diversity in streams and rivers. pp. 315–349. In F. S. Chapin, O. E. Sala, and E. Huber-Sannwald (eds.). *Global biodiversity in a changing environment: Scenarios for the 21st century*. Springer-Verlag, New York.

Power, G. 1997. A review of fish ecology in arctic North America. *American Fisheries Society Symposium* 19:13–39.

Power, G., R. Cunjak, J. Flannagan, and C. Katopodis. 1993. Biological effects of river ice. In T. D. Prowse and N. C. Gridley (eds.). *Environmental aspects of river ice*, pp. 97–119. Ministry Supply and Services Canada, Saskatoon, Saskatchewan.

Prowse, T. D. 1994. Environmental significance of ice to streamflow in cold regions. *Freshwater Biology* 32:241–259.

Remmert, H. 1980. *Arctic Animal Ecology*. Springer-Verlag, Berlin.

Revenga, C., S. Murray, J. Abramovitz, and A. Hammond. 1998. *Watersheds of the world: Ecological value and vulnerability*. World Resources Institute, Washington, D.C.

Reynolds, J. B. 1997. Ecology of overwintering fishes in Alaskan freshwaters. In A. M. Milner and M. W. Oswood (eds.), *Fresh waters of Alaska: Ecological syntheses*, pp. 281–302. Springer-Verlag, New York.

Reynolds, J. F., and J. D. Tenhunen. 1996. *Landscape function and disturbance in Arctic tundra*. Springer-Verlag, Berlin.

Ricketts, T. H., E. Dinerstein, D. M. Olson, C. J. Loucks, W. Eichbaum, D. DellaSala, K. Kavanagh, P. Hedao, P. T. Hurley, K. M. Carney, R. Abell, and S. Walters. 1999. *Terrestrial ecoregions of North America: A conservation assessment*. Island Press, Washington, D.C.

Roderick, J. 1997. *Crude dreams: A personal history of oil and politics in Alaska*. Epicenter Press, Seattle, Washington.

Schroeder, M. 1996. *Birds of northwest Alaska areas*. National Park Service. Jamestown, North Dakota. Available at the Northern Prairie Wildlife Research Center Web site: www.npwrc.usgs.gov/resource/orthrdata/chekbird/r7/noalaska.htm.

Schallock, E. W., and E. W. Mueller. 1981. Physical, chemical and biological conditions of the Sagavanirktok River and nearby control streams, Shaviovik and Canning rivers. U.S. EPA report no. 600/3-811-034. NTIS, Springfield, Virginia.

Schmidt, D. R., W. B. Griffths, and L. R. Martin. 1989. Overwintering biology of anadromous fish in the Sagavanirktok River delta, Alaska. *Biological Papers, University of Alaska* 24:55–74.

Scott, K. M. 1978. Effects of permafrost on stream channel behavior in arctic Alaska. U.S. Geological Survey Professional Paper no. 1068, Anchorage, Alaska.

Scrimgeour, G. J., T. D. Prowse, J. M. Culp, and P. A. Chambers. 1994. Ecological effects of river ice breakup: A review and perspective. *Freshwater Biology* 32:261–275.

Sedinger, J. S. 1997. Waterfowl and wetland ecology in Alaska. In A. M. Milner and M. W. Oswood (eds.). *Fresh waters of Alaska: Ecological syntheses,* pp. 155–178. Springer-Verlag, New York.

Selkregg, L. L. 1977. Alaska Regional Profiles: Arctic Region. Arctic and Environmental Information and Data Center. Anchorage, Alaska.

Seyler, J. 1994. Biology of selected riverine fish species in the Moose River basin. IR-024. OMNR, Northeast Science and Technology, Tiimmins, Ontario.

Smith, P. S. 1913. The Noatak–Kobuk Region, Alaska. *U.S. Geological Survey Bulletin* 536, 160 pp., maps.

University of Alaska. 1989. *Alaska climate summaries,* 2nd ed. Alaska Climate Center Technical Note no. 5, Arctic Environmental Information and Data Center, University of Alaska, Anchorage.

U.S. Geological Survey (USGS). 2003. Monthly streamflow statistics for Alaska: USGS 15746000 Noatak River at Noatak AK. U.S. Geological Survey, Anchorage.

Viavant, T. R. 1989. Community structure, trophic relationships, and habitat ecology of the benthic macroinvertebrates in an Alaskan tundra beaded stream. Master thesis, University of Alaska, Fairbanks.

Wahrhaftig, C. 1965. Physiographic divisions of Alaska. U.S. Geological Survey Professional Paper, Vol. 482. 52 pp., U.S. Government Printing Office, Washington, D.C.

Walker, D. A., and M. D. Walker. 1996. *Terrain and vegetation of the Imnavait Creek watershed.* Springer-Verlag, Berlin.

Ward, J. V., and J. A. Stanford. 1982. Thermal responses in the evolutionary ecology of aquatic insects. *Annual Review of Entomology* 27:97–117.

Wiersma, G. B., C. Slaughter, J. Hilgert, A. McKee, and C. Halpern. 1986. Reconnaissance of Noatak National Preserve and Biosphere Reserve as a potential site for inclusion in the integrated global background monitoring network. U.S. Man and the Biosphere Program, Springfield, Vermont.

Williams, H. 1958. *Landscapes of Alaska: Their geologic evolution.* University of California Press, Berkeley and Los Angeles.

Wilson, W. J., and M. D. Kelly. 1986. Cultured and wild salmon populations in the Noatak River, Alaska. In D. Guthrie (ed.). *Wild trout, steelhead, and salmon in the 21st century,* pp. 93–110. Oregon State University Press, Portland.

Young, S. B. 1974a. An overview of the environment of the Noatak study area and account of research activities carried out during the summer of 1973. pp. 1–57. In S. B. Young (ed.). *The environment of the Noatak River basin, Alaska.* Contributions from the Center for Northern Studies no. 1. Wolcott, Vermont.

Young, S. B. 1974b. Vegetation of the Noatak River valley, Alaska. In S. B. Young (ed.). *The environment of the Noatak River basin, Alaska,* pp. 584. Center for Northern Studies, Wolcott, Vermont.

Young, S. B. 1989. *To the Arctic: An introduction to the far northern world.* John Wiley and Sons, Inc., New York.

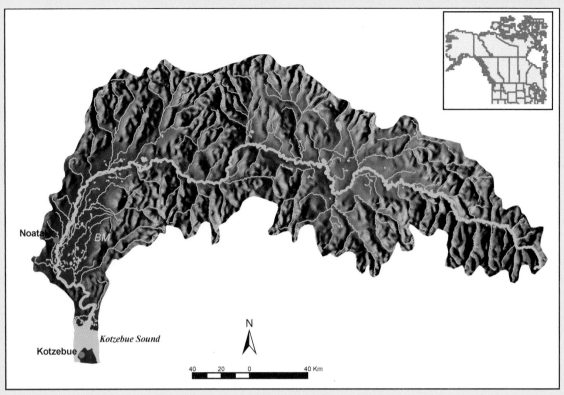

FIGURE 20.8 Map of the Noatak River basin.

NOATAK RIVER

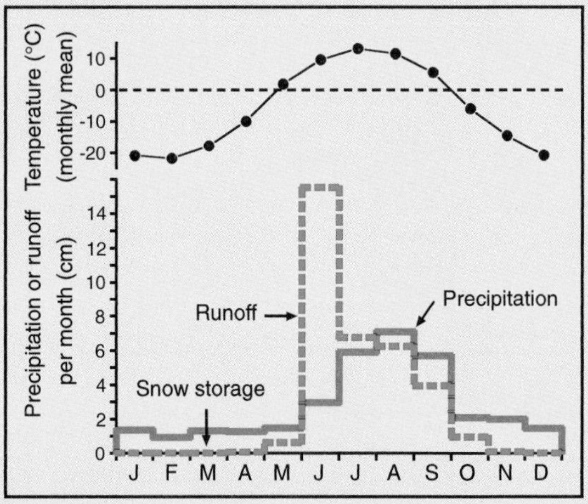

Relief: 2612 m
Basin area: 32,626 km²
Mean discharge: 469 m³/s
River order: NA
Mean annual precipitation: 33 cm
Mean air temperature: −5.8°C
Mean water temperature: NA
Physiographic province: Brooks Range (BM)
Biomes: Tundra, Boreal Forest
Freshwater ecoregion: Yukon
Terrestrial ecoregions: Brooks/British Range Tundra, Arctic Foothills
    Tundra
Number of fish species: 15 to 18
Number of endangered species: none
Major fishes: chum salmon, Arctic char, Arctic grayling, humpback
    whitefish, round whitefish, least cisco, ninespine stickleback,
    slimy sculpin
Major other aquatic vertebrates: river otter, grizzly bear, loons, ducks, belted kingfisher
Major benthic invertebrates: water mites (Acarina), mayflies (Baetidae, Heptageniidae), chironomid midges, stoneflies
    (Chloroperlidae)
Nonnative species: none
Major riparian plants: willows, cottongrass, sedges
Special features: pristine Arctic tundra river within Gates of the Arctic National Park and Preserve and Noatak National
    Preserve; longest river segment in U.S. Wild and Scenic River system; little or no winter flow under deep ice cover;
    subsistence use of riverine resources by Inupiat Eskimos
Fragmentation: none
Water quality: no significant pollution sources; alkalinity = 82 to 110 mg/L as CaCO₃, NO₂-N + NO₃-N = 0.07 to 0.2 mg/L,
    pH = 7.8 to 8.2
Land use: 100% tundra and boreal forest wilderness
Population density: 0 people/km²
Major information sources: Childers and Kernodle 1981, Young 1974

FIGURE 20.9 Mean monthly air temperature,
precipitation, and runoff for the Noatak River basin.

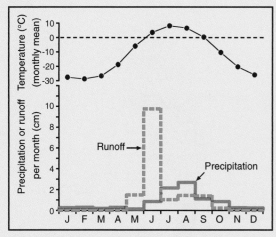

FIGURE 20.11 Mean monthly air temperature, precipitation, and runoff for the Kuparuk River basin.

KUPARUK RIVER

Relief: 1464 m
Basin area: 8107 km²
Mean discharge: 40 m³/s
River order: 5
Mean annual precipitation: 22 cm
Mean air temperature: −9.8°C
Mean water temperature: 2.6°C
Physiographic provinces: Arctic Slope (AS), Brooks Range (BM)
Biome: Tundra
Freshwater ecoregion: Yukon
Terrestrial ecoregions: Arctic Coastal Tundra, Arctic Foothills Tundra, Brooks/British Range Tundra
Number of fish species: ≥6
Number of endangered species: none
Major fishes: Arctic grayling, broad whitefish, ninespine stickleback, slimy sculpin, burbot
Major other aquatic vertebrates: loons, swans, many duck species, including spectacled eider and king eider
Major benthic invertebrates: chironomid midges (*Orthocladius*), black flies (*Prosimulium*), mayflies (*Baetis*), caddisflies (*Brachycentrus*)
Nonnative species: none
Major riparian plants: dwarf willow, birch
Special features: pristine tundra river originating in the foothills of Brooks Range and traversing Arctic Coastal Plain to Beaufort Sea
Fragmentation: none
Water quality: no pollution, except potential from oil field near mouth; phosphorus (SRP) = 0.4 µg/L, NO_3-N + NO_2-N = 5.2 µg/L, pH = 5.6 to 7.3, DOC = 6 to 10 mg/L
Land use: almost 100% tundra wilderness
Population density: 0 people/km²
Major information sources: Hershey et al. 1997, Reynolds and Tenhunen 1996

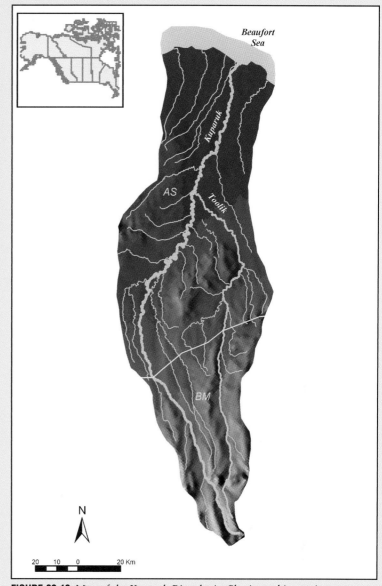

FIGURE 20.10 Map of the Kuparuk River basin. Physiographic provinces are separated by a yellow line.

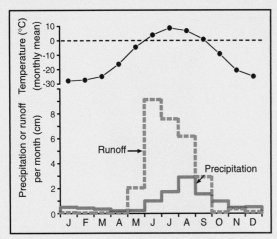

FIGURE 20.13 Mean monthly air temperature, precipitation, and runoff for the Sagavanirktok River basin.

SAGAVANIRKTOK RIVER

Relief: 2400 m
Basin area: 14,890 km^2
Mean discharge: 132 m^3/s
River order: 7
Mean annual precipitation: 10.8 cm (underestimate)
Mean air temperature: −7.7°C
Mean water temperature: NA
Physiographic provinces: Brooks Range (BM),
   Arctic Slope (AS)
Biome: Tundra
Freshwater ecoregion: Yukon
Terrestrial ecoregions: Brooks/British Range Tundra,
   Arctic Foothills Tundra, Arctic Coastal Tundra
Number of fish species: 10
Number of endangered species: none
Major fishes: Dolly Varden, Arctic grayling, Arctic
   cisco, broad whitefish, round whitefish, lake trout
Major other aquatic vertebrates: grizzly bear,
   spectacled eider, king eider, yellow-billed loon
Major benthic invertebrates: stoneflies (*Capnia*,
   Nemoura, *Isogenus*), mayflies (*Baetis*, *Cinygmula*,
   Ameletus), chironomid midges (Orthocladiinae), black flies
Nonnative species: none
Major riparian plants: dwarf willow, birch
Special features: one of major rivers on Alaskan Arctic Slope supporting large fish populations; unregulated; delta provides
   important bird and fish habitat; oil development, including Trans-Alaskan Pipeline, along most of river's length
Fragmentation: none
Water quality: no pollution, except potential near mouth from oil development; total P ≥0.05 mg/L, NO$_3$-N ≥0.2 mg/L
Land use: almost 100% tundra wilderness
Population density: 0 people/km^2
Major information sources: Schallock and Mueller 1981, Hodel 1986, Gallaway and Fechhelm 2000

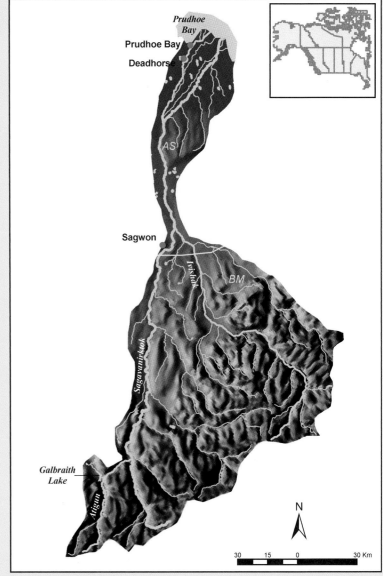

FIGURE 20.12 Map of the Sagavanirktok River basin. Physiographic provinces are separated by a yellow line.

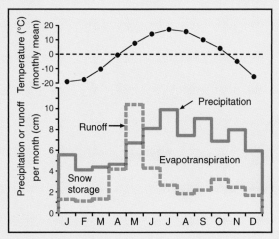

FIGURE 20.15 Mean monthly air temperature, precipitation, and runoff for the Moose River basin.

MOOSE RIVER

Relief: 325 m
Basin area: 109,000 km²
Mean discharge: 1370 m³/s
River order: 7
Mean annual precipitation: 80 cm
Mean air temperature: 0.1°C
Mean water temperature: 7.0°C
Physiographic provinces: Superior Upland (SU), Hudson Bay Lowland (HB)
Biome: Boreal Forest
Freshwater ecoregion: South Hudson
Terrestrial ecoregions: Eastern Forest/Boreal Transition, Central Canadian Shield Forests, Southern Hudson Bay Taiga
Number of fish species: 40
Number of endangered species: none
Major fishes: lake sturgeon, northern pike, walleye, white sucker, troutperch, pearl dace, longnose dace, spotfin shiner, emerald shiner
Major other aquatic vertebrates: muskrat, beaver, fisher, marten, mink, river otter, moose, lesser snow goose
Major benthic invertebrates: snails (*Valvata, Amnicola*), mayflies (*Hexagenia, Baetis, Caenis, Ephemerella, Heptagenia, Stenonema*), crustaceans (*Hyalella, Cambarus, Orconectes*), chironomid midges (*Chironomus, Paratanytarsus, Cladopelma*), stoneflies (Perlodidae, Pteronarcyidae), dragonflies (Aeshnidae, Corduliidae, Macromiidae, Gomphidae), oligochaete worms (*Dero, Pristina*)
Nonnative species: none
Major riparian plants: white birch, trembling aspen, white spruce, black spruce, jack pine, dogwood, alder, willow, grasses, sedges, rushes, horsetails, arrowheads, pondweed, coontail
Special features: >200 waterfalls and the Missinaibi pristine Heritage River
Fragmentation: more than 40 dams and water-control structures throughout basin
Water quality: in undisturbed areas, pH = 7 to 7.8, conductivity = 100 to 200 µS/cm, alkalinity = 50 to 100 mg/L as CaCO₃, turbidity = <20 NTU, total P = <0.020 mg/L
Land use: 94% forests, 1% developed, 5% open
Population density: <1 person/km²
Major information sources: Environment Canada 1982a, 1982b, Munkittrick et al. 2000, Brousseau and Goodchild 1989

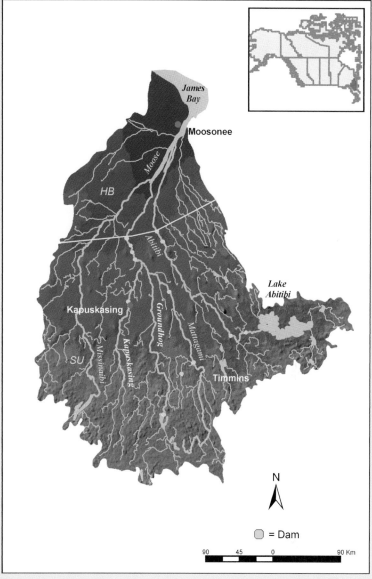

FIGURE 20.14 Map of the Moose River basin. Physiographic provinces are separated by a yellow line.

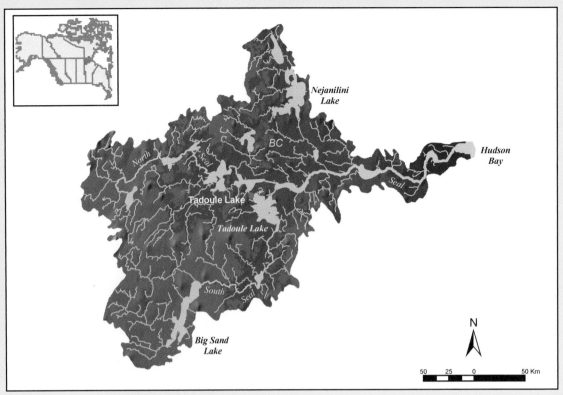

FIGURE 20.16 Map of the Seal River basin.

SEAL RIVER

Relief: NA
Basin area: 48,100 km²
Mean discharge: 346 m³/s
River order: NA
Mean annual precipitation: 41 cm
Mean air temperature: −7.1°C
Mean water temperature: NA
Physiographic province: Bear–Slave–Churchill Uplands (BC)
Biomes: Boreal Forest, Tundra
Freshwater ecoregion: East Arctic
Terrestrial ecoregions: Northern Canadian Shield Taiga, Southern
 Hudson Bay Taiga
Number of fish species: ≤18
Number of endangered species: NA
Major fishes: lake trout, northern pike, Arctic grayling
Major other aquatic vertebrates: river otter, beaver, beluga whale,
 harbor seal
Major benthic invertebrates: NA
Nonnative species: none
Major riparian plants: willow, birch
Special features: Canadian Heritage River; one of Canada's wildest and most remote rivers; upper reaches characterized by
 magnificent sand-crowned eskers
Fragmentation: none; largest remaining undammed river in Manitoba
Water quality: no pollution
Land use: 100% undeveloped tundra and montane
Population density: 0 people/km²
Major information source: Environment Canada

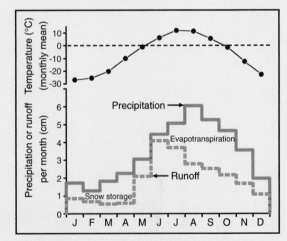

FIGURE 20.17 Mean monthly air temperature,
precipitation, and runoff for the Seal River basin.

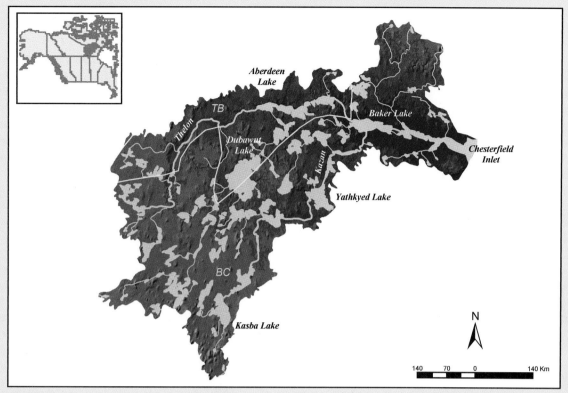

FIGURE 20.18 Map of the Thelon/Kazan River basin. Physiographic provinces are separated by a yellow line.

THELON/KAZAN RIVER

Relief: 300 to 500 m
Basin area: 239,332 km²
Mean discharge: 1380 m³/s
River order: NA
Mean annual precipitation: 17.2 cm
Mean air temperature: –8.7°C
Mean water temperature: NA
Physiographic provinces: Thelon Plains and Bear River Lowland (TB), Bear–Slave–Churchill Uplands (BC)
Biomes: Tundra, Boreal Forest
Freshwater ecoregion: East Arctic
Terrestrial ecoregions: North Canadian Shield Taiga, Low Arctic Tundra, Middle Arctic Tundra
Number of fish species: 13
Number of endangered species: none
Major fishes: lake trout, Arctic grayling, northern pike, Arctic char, humpback whitefish, round whitefish, cisco, slimy sculpin, spoonhead culpin, lake chub
Major other aquatic vertebrates: grizzly bear, bald eagle, tundra swan, Canada goose
Benthic invertebrates: NA
Nonnative species: none
Major riparian plants: willow shrub, bog birch, birch, spruce
Special features: Canadian Heritage River (both Thelon and Kazan); large remote wilderness river; Thelon Game Sanctuary; Kazan Falls
Fragmentation: NA
Water quality: no pollution
Land use: 100% wilderness (11% wetlands, 6% forest, 21% grassland, 48% barren)
Population density: <1 person/km²
Major information sources: Revenga et al. 1998, Environment Canada

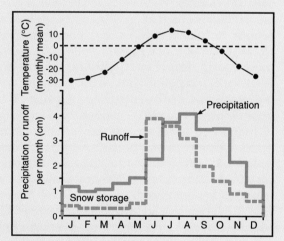

FIGURE 20.19 Mean monthly air temperature, precipitation, and runoff for the Thelon/Kazan River basin.

21

ATLANTIC COAST RIVERS OF CANADA

RICHARD A. CUNJAK   ROBERT W. NEWBURY

INTRODUCTION

EXPLOITS RIVER

MIRAMICHI RIVER

ST. JOHN RIVER

MOISIE RIVER

ADDITIONAL RIVERS

ACKNOWLEDGMENTS

LITERATURE CITED

INTRODUCTION

The rivers of eastern Canada flow into the Atlantic Ocean from Newfoundland, Labrador, Nova Scotia, Prince Edward Island (PEI), New Brunswick, and parts of eastern Quebec. Geographically, the region is bounded by the Atlantic Ocean on the south and east, 70°W longitude on the west, and 55°N latitude in the north. The northern boundary generally follows a height of land between rivers flowing north to Ungava Bay or south into the Gulf of St. Lawrence (Fig. 21.2). To the north the rivers run through near-wilderness regions of the rocky Canadian Shield. To the south many of the rivers flow through the Atlantic lowlands in wide valleys formed by large glacial meltwater rivers. Towns, industries, and farmland now occupy the valley flats and shallow plains beside the smaller present-day rivers. To the west the interior drainage of eastern Canada (the Great Lakes and various downstream tributaries) flows into the Atlantic through the St. Lawrence River, the second-largest river (by discharge) in North America (see Chapter 22).

The rivers have been used by aboriginal peoples for over 9000 years: the Inuit along the coast of Labrador, the Innu in central and southern Labrador, the Montagnais and Mi'kmaqs of the Maritimes, and the now-extinct Beothuks of insular Newfoundland. The last Beothuk died in 1829 (www.tolatsga.org/Compacts.html 2002). These rivers have long served these peoples for movement between seasonal settlements and for accessing hunting and fishing areas. Travel to eastern Canada by Europeans includes some of the earliest explorers. Basques traveled regularly to southern Labrador between 1530 and 1600, primarily in search of whales and cod; Norwegian sailors are known to have reached northwestern Newfoundland before the tenth century. Colonization of eastern Canada, primarily by French and English explorers such as Jacques Cartier and John Cabot, began in the early sixteenth century, mainly along the coast of the Bay of Fundy and along the St. Lawrence River, the route to the interior of North America.

There are 37 seaboard rivers with drainage areas greater than $100 km^2$ that lead directly into the Atlantic (National Atlas of Canada 1974). The major rivers north of the Gulf of St. Lawrence include the Manicouagan, Outardes, Moisie, and Churchill rivers in Labrador. On insular Newfoundland are the Humber, Exploits, and Gander rivers. In the Maritime Provinces the major rivers are the St. John,

FIGURE 21.1 Aerial view of the Southwest Miramichi River near Doaktown, New Brunswick. Note the wide glacial river valley and sand and gravel bed braided channel (PHOTO BY R. NEWBURY).

939

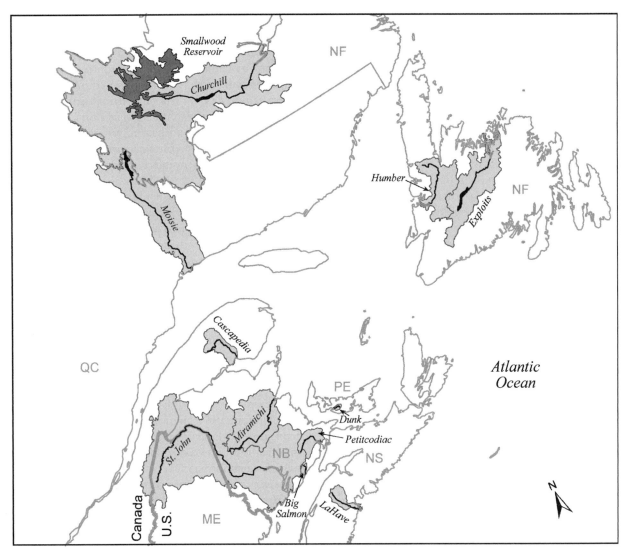

FIGURE 21.2   Atlantic Coast rivers of Canada covered in this chapter.

Restigouche, Miramichi, and Margaree rivers. In this chapter we will focus on four of these rivers: the Moisie, Exploits, St. John, and Miramichi. Together, they represent a range of human colonization and perturbation as well as wild beauty and impressive productivity. Abbreviated physical and biological descriptions are provided for the Churchill, Humber, and several smaller rivers (Cascapedia, Dunk, LaHave, Petitcodiac, and Big Salmon).

Physiography and Climate

The Atlantic Canada region includes four physiographic provinces. The northern portion of the region is located within the Labrador Highlands (LB)

province at the eastern extension of the Canadian Shield physiographic division. Here, the Torngat Mountains can be found, with peaks in excess of 1500 m. The Laurentian Highlands (LU) province occupies the southwestern corner of the region, mainly represented by the lands of eastern Quebec. The coastal lowlands of the Gulf of St. Lawrence, including the Quebec north shore, Anticosti Island, and the west coast of insular Newfoundland, are located within the St. Lawrence Lowland (SL) province. Finally, the New England/Maritime (NE) province covers most of the rest of the region's land mass. The mountain ranges are old, weathered, and of moderate height. These include the northeastern extension of the Appalachian physiographic division (with Mount Carleton, at 820 m asl, the highest point of the mainland provinces) and the Long Range Mountains (800 m asl) of western Newfoundland.

The drainage pattern is dominated by the Canadian Shield in the north and the extension of the Appalachian Mountains in the south. Drainage from the Shield, a raised arc of predominantly granitic pre-Cambrian rock that surrounds Hudson Bay, is inward to the Bay and outward into the St. Lawrence and Mackenzie river systems. In the east, the Shield drainage flows from southern Quebec and Labrador directly into the Gulf of St. Lawrence or the Atlantic. South of the Canadian Shield the Appalachian Mountain region extends for 4000 km from Alabama in the southwest to the New England/Maritime province, which includes New Brunswick, Nova Scotia, Prince Edward Island, and Newfoundland. The drainage basins are heavily glaciated and composed of a wide variety of granitic, metamorphic, and sedimentary rock. Surficial deposits vary from a thin till cover in Labrador and Newfoundland to thick riverine and tidal deposits in the southern river valleys. Sea and land elevations have changed dramatically in the last 100,000 years. The region was depressed by as much as 300 m under the weight of the Pleistocene ice sheets. About half of this depression has been recovered in the last 12,000 years. Consequently, the lower reaches of the larger rivers, such as the St. John and Miramichi, flow through broad flat valleys infilled with former marine and estuarine proglacial deposits.

The climate varies from subarctic to boreal in the north to wet maritime in the south. Cool summers and extremely cold winters are typical of the northern portion of the region, with the climate strongly influenced by proximity to the coast. Summers can be hot and humid in the interior of New Brunswick, where maximum air temperatures >30°C are common in July and August. By contrast, summer maxima of 20°C are rare along the Labrador coast. Mean annual air temperatures in the region range from −3°C in the north to 7°C in the south. Snow is the dominant form of precipitation in winter, except along the Bay of Fundy and southern Nova Scotia coasts, where <25% of the annual precipitation is from snow. By contrast, approximately 50% of annual precipitation falls as snow in Labrador. Churchill Falls (interior) has an annual accumulation of 480 cm of snow, making it one of the snowiest places in Canada. In northern portions of the region, the ground is snow covered for six to eight months of the year. The mean total annual precipitation varies from 80 cm in the north to 150 to 165 cm in the southeast (water equivalent).

Basin Landscape and Land Use

The Atlantic Canada region is distinctive for two features of its landscape: an extensive marine coastline and an extensive forest cover surrounding abundant inland waters. The land mass covers an area of 503,000 km², or about 5.5% of Canada's land mass. The major biome of the region is Boreal Forest, found throughout Labrador and the northern peninsula of insular Newfoundland and extending south into Nova Scotia, Prince Edward Island, and New Brunswick. A small portion of eastern New Brunswick and the western half of Nova Scotia along the Bay of Fundy is considered Temperate Deciduous Forest.

The major terrestrial ecoregions of the Canadian Atlantic river basins include portions of the Eastern Canadian Shield Taiga of Labrador, the Eastern Canadian Forests of eastern Quebec and Newfoundland, the New England/Acadian Forests of parts of New Brunswick and Nova Scotia, and the Gulf of St. Lawrence Lowland Forests of northern New Brunswick and Prince Edward Island (Ricketts et al. 1999). The Eastern Canadian Shield Taiga ecoregion primarily consists of stunted black spruce and tamarack. The Eastern Canadian Forests include black spruce, but also include balsam fir and white spruce resulting from the maritime influence. The New England/Acadian Forests is a transition zone between this spruce–fir forest and the deciduous forest to the south, thus including sugar maple, American beech, and yellow birch. The Gulf of St. Lawrence Lowland Forests ecoregion contains an even stronger mix of hardwoods, but also includes balsam fir, red spruce, and eastern white pine.

Population density is low (13.2 people/km²), with less than 2.3 million people living in the region (7.7% of the population of Canada). Large urban centers are scarce; instead, a pattern of isolated small towns follows the river valleys and coastlines. The economy of the region is based on the exploitation of natural resources, agriculture, and tourism. Commercial fisheries (including aquaculture) and forestry are the two major industries of the region, and recreational pursuits like fishing and hunting are practiced by a large number of residents of the region.

The commercial harvesting of trees for wood products in Atlantic Canada dates back to the 1700s, when lumber was exported to Britain, largely for use in shipbuilding. During the Napoleonic Wars (1800 to 1815) the strong demand for eastern white pine for masts and spars in the British navy fueled the forestry industry in Nova Scotia and New Brunswick

(Wright 1944). In 2000 the value of the forestry exports from Atlantic Canada was $4.7 billion (Canadian), with approximately 35,000 jobs stemming directly from the industry. Agricultural land is extensive in Prince Edward Island, parts of Nova Scotia (Annapolis Valley), and New Brunswick (mainly within the St. John River basin); elsewhere, farming is restricted to the broader valley bottomlands.

The Rivers

One natural resource is common to all four of the focus rivers and to many of the rivers of the region: the Atlantic salmon, a symbol of economic value and environmental quality. The rivers and lakes of the region have supported recreational angling since the first colonists arrived. By the nineteenth century the rivers of New Brunswick were the most prized destination for Atlantic salmon angling among Europeans and Americans. The value of recreational fisheries in the region is less impressive than that of forestry but still significant. According to 1995 statistics direct expenditures made by anglers in the Atlantic provinces amounted to approximately $100 million (Canadian), or 4% of the value of the recreational fishery for the entire country (EPAD 1997). Indeed, the state of each of the focus rivers today—what's right and what's wrong—can be assessed by understanding the state of Atlantic salmon in each of the rivers. Each of the four rivers represents a different situation in its relation to the salmon. The Miramichi River, a large unregulated river in New Brunswick, is considered the premier river for Atlantic salmon in North America and is reputed to produce more Atlantic salmon than any other river in the world, but stocks are in decline and deep concern exists for the future. The Moisie River is wild and relatively unexploited and has some of the oldest private fishing lodges of anywhere in North America; it represents the best-case scenario for a rather unique stock of large Atlantic salmon. The Exploits River is a success story for salmon enhancement, primarily because of fish stocking in reaches that were not utilized by salmon because of impassable upstream migration barriers. In the past decade 15,000 to 30,000 adult salmon have returned annually to spawn. Finally, the St. John River represents the worst case for wild Atlantic salmon survival in eastern Canada. It was once characterized as a salmon river rivaling the Miramichi in terms of fish production. Today the combination of hydroelectric dams, land-use activity that degrades habitats, pollution from

large urban centers and industries along the watercourse, and historic overexploitation have contributed to the situation where a small fraction of the historic runs of wild salmon now return to the system. Some unique stocks are now considered extinct. However, despite the gloomy scenario, the resilient nature of the river ecosystem and recent community-led proenvironment initiatives mean that such rivers can and should be viewed with hope, rather than despair, in terms of the potential for restoring ecological integrity to these once proud waterways.

In addition to the status of their salmon resources, the rivers in the region display a wide range of conditions and development. Some, like the Petitcodiac River, suffer from a causeway situated below head of tide that has changed sediment flow patterns and resulted in channels that are blocked and filled with agricultural runoff. The St. John River, an international river and the largest in the region, has suffered numerous similar insults by way of dams, urban pollution, and industrial discharges. By contrast, the Moisie, Humber, and Grand Cascapedia rivers are treasured for their pristine waters and the opportunity for realizing a wilderness experience. Some of the rivers are singularly developed for hydroelectric power, like the Churchill and Exploits rivers, whereas others, like the Dunk and LaHave rivers, provide multiple uses for agriculture, fisheries, and municipalities. The Dunk River, the smallest basin described here, is a low-gradient, short river dominated by groundwater discharge, which is characteristic of the island watercourses. A few are so scenically unique that they are found in and near National Parks, like the Big Salmon River along the Bay of Fundy and the Humber River in western Newfoundland.

Winter conditions greatly influence the hydrology, recreation, land use, and ecology of these rivers. Groundwater and the coastal maritime climate moderate the thermal and flow regimes and ice cover of the Dunk, Big Salmon, and LaHave rivers. The interior and northern rivers, like the Miramichi, Moisie, Grand Cascapedia, and Churchill, experience intact ice cover for six to eight months of the year, and most of the region's rivers experience a strong spring snowmelt influence on the hydrograph. A large dam on the Churchill has totally eliminated the annual pattern of discharge downstream.

The Canadian Atlantic rivers cover three freshwater ecoregions as defined by Abell et al. (2000). The North Atlantic–Ungava ecoregion, located in Labrador, is considered nationally important in biological distinctiveness and includes the Churchill River in this chapter. The Lower St. Lawrence ecoregion to the south is considered continentally outstanding and includes most other rivers in this chapter. As indicated, a major reason the ecoregion is considered outstanding is because of the runs of anadromous Atlantic salmon. The St. John and LaHave rivers are at the north edge of the North Atlantic ecoregion, which is considered nationally important and is also distinguished by runs of Atlantic salmon and shad.

EXPLOITS RIVER

As early as 1773 the Exploits River appeared on navigators' charts. A 6th order river, the Exploits is the largest in insular Newfoundland, flowing in a northeasterly direction for over 236 km into the Bay of Exploits on the northern edge of the island (Fig. 21.8). This 11,272 km^2 basin includes the second-largest lake in Newfoundland, Red Indian Lake (181 km^2), the last outpost of the now extinct Beothuk nation. Like most insular Newfoundland river systems, there is extensive lacustrine habitat in the catchment, in this case representing approximately 99% of the total fish-rearing habitat. The reg-

ulated flow for hydroelectric production has altered the seasonal patterns of discharge and natural ice regime in the river (Fig. 21.3). Spectacular ice jams may now occur in midwinter, instantly flooding and freezing riverside towns like Badger.

Evidence of the aboriginal Beothuk people of insular Newfoundland dates back to AD 200. Originally a coastal people during summer, hostilities with British salmon fishers and trappers who settled seasonally along the coast in Notre Dame Bay (early eighteenth century) led to the Beothuks being driven further inland, particularly in the Exploits River valley around Red Indian Lake. There they experienced a steadily deteriorating life of subsistence, disease, starvation, and finally extinction. The first record of any permanent station by white settlers was in 1778 near the mouth of the Exploits River. The fur and salmon fishing industries were closely linked during the early settlement of the area. In summer they resided on the coast for easy access to the salmon rivers, but wintered inland in the more wooded, protected areas, hunting and fur trapping. By the nineteenth century the shores of Newfoundland, including the Bay of Exploits, were overtaken by year-round European fishers, fur trappers, and loggers. In the

FIGURE 21.3 The Exploits River and hydroelectric dam at Bishop's Falls, Newfoundland. A fishway is present on the far side of the river at the base of the waterfall (Photo by C. Bourgeois).

Exploits River valley of the early twentieth century, the Anglo-Newfoundland Development Company —known by its apt acronym as the "A.N.D. Company"—controlled much of the region's industrial and commercial sector. It founded the pulp and paper industry in Grand Falls on the lower Exploits and conducted the original exploration of the Buchans Mine just north of Red Indian Lake.

Physiography, Climate, and Land Use

The drainage basin lies in the New England/Maritime (NE) physiographic province on the northeastern edge of the Appalachian Provinces physiographic division (see Fig. 21.8). The headwaters lie in the Newfoundland Highlands, whereas the main channel flows through the Notre Dame region of the Newfoundland Central Lowlands (Stockwell 1963). The basin is underlain by volcanic and metamorphic rocks associated with slate, sandstone, and conglomerates. The rocks in the upper reaches of the basin, near Red Indian Lake, are principally Ordovician in origin and volcanic in structure; there are significant granitic formations in the area that probably give the river its characteristic water quality of low dissolved solids and low conductivity. The northeasterly trending geological structures gave rise to the parallel alignment of lakes and streams that are easily identifiable in the Exploits basin. Maximum relief is 490 m asl in the headwaters of the Lloyds River in the Annieopsquotch Mountains. Unlike the other focus rivers, the Exploits basin has an abundance of lacustrine habitat (approximately $337 \times 10^6 \, m^2$; Bourgeois et al. 2001).

The Exploits basin is primarily within the Eastern Canadian Forests terrestrial ecoregion (Ricketts et al. 1999). Black spruce and balsam fir are the predominant tree species; the most common hardwood is white birch. Higher elevations are considered to be in the Newfoundland Highland Forests ecoregion, which also includes dwarf kahnia in addition to spruce–fir forest. Although most of the basin is forested, bog lands and rocky barrens also make up a significant portion of the area.

The average annual precipitation in the region is 99.1 cm (110 cm midbasin), evenly distributed throughout the year (Fig. 21.9). However, approximately half of the precipitation is stored as snowfall until the spring melt period. Mean monthly temperatures range from −9.3°C in February to 15.6°C in July (see Fig. 21.9). There are approximately 115 frost-free days.

Forestry is the main industry for this part of insular Newfoundland. A pulp and paper industry

has operated at Grand Falls since 1909. It discharges its effluent into the main stem of the Exploits River in the same general area where domestic sewage enters from the towns of Grand Falls and Windsor. In 1995, Grand Falls Pulp and Paper began aerating their effluent as part of a secondary waste treatment process; this has resulted in a reduction in suspended solids and BOD within the effluent (Bourgeois et al. 2000). Timber harvest for pulp wood is generally by clear-cutting, and has occurred over much of the Exploits basin. For example, forest harvesting operations in the area around Red Indian Lake date back to 1903 (Morry and Cole 1977). Log driving on the river and tributaries, often involving the erection and operation of small "splash-dams," was discontinued in the mid-1960s (O'Connell et al. 1983), although log driving on the main stem continued to the late 1980s and early 1990s (C. Bourgeois, personal communication).

River Geomorphology, Hydrology, and Chemistry

Several times during the Pleistocene epoch ice sheets covered the island as they did continental North America. One of the last ice caps was centered near Grand Falls. As the glaciers moved seaward they scraped most of the soil from the land surface and left a cover of gravel, sand, and till strewn with glacial erratics (boulders). The glaciers were also largely responsible for the topography of the basin, where they scooped out the bedrock to form basins for lakes and ponds and planed the hills into asymmetrical forms featuring a gentle slope on one side and a steep drop on the other. The uneven distribution of the glacial drift resulted in poorly drained regions characterized by an abundance of bogs. Average gradient for the main channel from the headwaters to the head of tide is approximately 1.6 m/km.

Flow data are recorded at a gauging station on the main stem with an area of $8640 \, km^2$ (about two-thirds of the total basin area). The average discharge at this station is $273 \, m^3/s$, varying between a maximum daily flow of $2090 \, m^3/s$ and a minimum of $36 \, m^3/s$ in 29 years of record. Approximated mean discharge at the mouth is $356 \, m^3/s$. Peak snowmelt flow occurs between April and June (see Fig. 21.9). A dam at the outlet of Red Indian Lake regulates the flow for hydroelectric production further downstream, so the seasonal pattern is not as strong as it otherwise would be.

Water-chemistry data collected from the lower reach (Grand Falls and Bishop's Falls) indicate that the Exploits River is similar to the Miramichi River (i.e., a soft-water stream that is typically low in productivity). Surveys carried out in Red Indian Lake in 1984 indicated an oligotrophic, coolwater lake (7°C to 11.5°C), with transparencies between 5.1 and 6.1 m (Hammar and Filipsson 1985). Limnological studies conducted by Morry and Cole (1977) also indicated low productivity in the lake; they suggested that this was partly due to unstable shoreline conditions resulting from severe fluctuations of water levels due to regulation at the outflow dam. Headwater tributaries are suggestive of even softer water, with average conductivities measured between 20 and 35 µS/cm (Randall et al. 1989). Typical pH measurements range between 5.8 and 6.8 but there is some indication that the waters of the Exploits may be more acidic at times, with pH values reaching 4.1 to 4.2 (Randall et al. 1989), probably during spring snowmelt conditions. Dissolved organic carbon has ranged from 9.7 to 10.5 mg/L, total N from 0.05 to 0.10 mg/L, and total P from 0.006 to 0.012 mg/L (Morry and Cole 1977, Randall et al. 1989).

River Biodiversity and Ecology

The Exploits River is found in the Lower St. Lawrence freshwater ecoregion (Abell et al. 2000). There have been surprisingly few studies of the biota of the system despite the hydroelectric development that has impacted the environment. In comparison with the other focus rivers, the Exploits River has a rather species-depauperate flora and fauna, a reflection of its island status and limited opportunity for colonization.

Algae

No information is available on the phytoplankton community or primary production in the Exploits River system. Morry and Cole (1977) characterized the community of phytoplankton in Red Indian Lake as depauperate, but provided no data.

Plants

Black spruce and balsam fir are the most common riparian trees in the basin and are the basis of the pulp and paper industry in the lower river valley; speckled alder is commonly found along the banks of tributaries. As is the case in much of Newfoundland, edible berries, such as blueberries, partridgeberries, and cloudberries, are abundant throughout the basin, including the riparian zone, except where there are

ponds and bogs. The berries provide food for numerous birds and mammals, including man. They were especially important to the Beothuk people, who historically used the berries for food as well as medicine (e.g., bunchberry). Pitcher plants are common in the acidic bogs that dot the floodplain. Aquatic macrophytes are rare in the Exploits River except for occasional mats of aquatic buttercup in the slow to moderately flowing shallows of small tributary streams, where they can be an important food source for moose.

Invertebrates

Morry and Cole (1977) provided the only data on zooplankton and macrobenthos in the Exploits River system, at Red Indian Lake; the following summary is based on their 109 plankton tows in 1974. Species diversity was low, as was the number of zooplankters and their small size, suggesting a low-productivity lake, as expected based on the water chemistry. Only 12 species of Entomostraca were identified. Of these, two cladocerans (*Bosmina longirostris* and *Holopedium gibbernum*), one calanoid copepod (*Diaptomus minutus*), and one cyclopoid copepod (*Cyclops scutifer*) were the most commonly occurring species. All are characteristic species of low-productivity waters in eastern Canada.

The benthic community of Red Indian Lake was also sampled by Morry and Cole (1977). Less than 25% of the 109 grab samples yielded any macroinvertebrates. Five taxa were present; these included leaches (Hirudinea), sphaeriid clams, chironomid midges, and two families of aquatic worms (Oligochaeta). All the species are common to Newfoundland waters and are not associated with polluted conditions. Rather, the lack of species diversity and abundance reflect the unproductive state of the waters in the system, as was the case for zooplankton. No other information is available on the macrobenthos in the Exploits River system, although insects such as mayflies, stoneflies, caddisflies, and midges are present; biting flies such as black flies and mosquitoes are particularly abundant in the summer.

Vertebrates

In addition to Atlantic salmon (both the landlocked [ouananiche] and anadromous forms), four other fish species are present in the river proper: brook trout, Arctic char, American eel, and threespine stickleback. Anadromous rainbow smelt are known to occur as far upriver as Bishop's Falls.

The largest brook trout (approximately 50 cm long) captured by Hammar and Filipsson (1985) during their fish surveys of Newfoundland lakes were from Star Lake in the upper Exploits River basin; the largest ouananiche captured, at 55.6 cm, was a specimen taken from Red Indian Lake. The sampling in this large lake also yielded Arctic char, the only species found at depths >20 m, although their results indicated a general low density of fish in Red Indian Lake. Char are rare downstream of the lake. Unfortunately, very little information exists on the status of trout and char elsewhere in the river system. Eels are most common in the lakes, being found upstream of Red Indian Lake; sticklebacks prefer slow-flowing shallow waters, often in littoral zones of ponds and lakes. All the fishes occurring in the Exploits River are indigenous to insular Newfoundland; no non-native species are known to occur.

Beaver, river otter, and muskrat are common in the Exploits River valley. They were trapped for their pelts in the early twentieth century, but this activity has largely ceased.

Ecosystem Processes

There are no known studies of ecosystem processes in the Exploits River. Given the predominance of lakes and the decomposition-resistant nature of riparian carbon sources (largely coniferous needles), it is suspected that autochthonous production may drive much of the energy cycling within the river. This is further supported by the high concentrations of filter-feeding macroinvertebrates often found at the outflows of riverine ponds and lakes (R. A. Cunjak, unpublished data).

There are no direct measurements of fish production in the lakes of the Exploits River. However, Randall et al. (1989) made estimates using values from other Newfoundland systems. Based on measurements from Western Arm Brook (northwest of the Exploits), where lacustrine habitat represents 98% of the rearing habitat (similar to the situation in the Exploits), 67% of smolts were produced in standing waters. Therefore, salmonid production in the lakes is significant, and the total production rate for the Exploits system was estimated to average 0.03 smolts/m^2, slightly less than the 0.04 smolts/m^2 estimated for the Miramichi River (Randall et al. 1989). The authors also noted that although fewer smolts are produced in the Exploits, production rate is similar to the Miramichi River because Exploits smolts are larger on average (54 g versus 25 g) and older (mean age of 3.4 versus 2.7 yr). Total fluvial production rate was estimated to be 3 to 5 g m^{-2} yr^{-1} (Randall et al. 1989). More recent data compiled by Bourgeois et al. (2000) indicate some significant

changes in smolt characteristics over the past decade. Between 1990 and 2000 smolts leaving the Exploits have been smaller (average weight 40.1 g), whereas age has remained similar (mean age 3.3 yr), suggesting that production rate may have decreased.

Human Impacts and Special Features

Much of the Exploits River lies within the central part of insular Newfoundland, where access is very limited. Hence, the river offers a true wilderness experience for those willing to venture into the remote parts of the upper reaches. In addition, Red Indian Lake offers an opportunity to view one of Newfoundland's largest natural lakes and the site where the Beothuks last lived.

The Exploits River provides a prime example of the potential success of a well-planned salmon enhancement program. Prior to 1957, <10% of the catchment area was available to diadromous fishes such as salmon due to a combination of natural and man-made barriers to fish passage. With a series of enhancement actions, including adult transfers (lower reaches), planting of swim-up fry (into the middle Exploits), and the transfer of adult salmon and fry stocking into tributaries flowing to Red Indian Lake, >90% of fish habitat within the catchment is presently available for anadromous salmon production (O'Connell and Bourgeois 1987). These actions have yielded impressive smolt–grilse return rates as high as 8% (Bourgeois et al. 1987). Since 1993 salmon production has been maintained entirely by a self-sustaining population. Part of the success can be attributed to measures outside the river system, such as the 1992 moratoria on commercial salmon fishing around insular Newfoundland, and on northern cod (thereby eliminating the by-catch of salmon in cod-fishing gear). For example, the number of returning adult salmon to Bishop's Falls averaged 8864 fish when there was a commercial fishery (1975 to 1991) compared with 20,647 after the moratorium (1992 to 2001; C. Bourgeois, personal communication). In addition, the hydroelectric industry has mitigated its environmental impacts by constructing fish-passage facilities at dams and, in 1996, by installing fish deflectors (louvers) to deter the entry of smolts and kelts into an artificial power canal associated with the dam at Grand Falls (Bourgeois et al. 2000).

Recreational fisheries target the three salmonid species in the river: Atlantic salmon, brook trout, and Arctic char. Most Atlantic salmon angling occurs in the lower reaches of the river, especially in the area between Bishop's Falls and Grand Falls. Catch and effort have both increased since the mid 1970s, presumably because of the enhancement program. To allow for a recreational fishery while the Exploits stock is still developing, the federal government (DFO) has established a management target of 13,000 spawners, a target that was reached in all years but one since its establishment (Bourgeois et al. 2000).

Hydroelectric development has resulted in significant impacts to the Exploits River system (O'Connell et al. 1983). Beginning in 1909, two hydroelectric dams were constructed on the main stem of the river, near the head of tide, at Grand Falls (45 MW capacity) and Bishop's Falls (see Fig. 21.8 and Fig. 21.3); both continue to operate today, and each site has some form of fish-passage facility. Historically, Bishop's Falls was not a barrier to fish migration, whereas Grand Falls was a complete barrier to upstream migration by diadromous fishes. Water-storage dams occur at the outlet of Red Indian Lake (constructed in 1909, and an impassable barrier), Long Lake, North Twin Lake, South Twin Lake, and Sandy Lake. Goodyear's Dam, 3 km upstream of the Grand Falls dam, was built for log driving in 1975 (two fishways present). In addition to dams, in 1968, waters from a 1068 km^2 subcatchment of the upper reaches of the Exploits River (Victoria River) were diverted south to another river system as part of the Bay d'Espoir power development (O'Connell et al. 1983). Recent concern has also been voiced over the potential impacts to fish passage at Grand Falls as a result of ongoing hydro development.

Morry and Cole (1977) suggested that the low productivity and paucity of char, trout, and ouananiche in Red Indian Lake were related to the severe water-level fluctuations, in excess of 9 m, resulting in an unstable littoral environment. More recently (~1999) a hydroelectric development was completed at the outflow from Star Lake, a subbasin draining into Red Indian Lake. The dam was built with no fish-passage facility and the subsequent inundation flooded the main spawning tributaries for a trophy-size population of resident brook trout. The project generated much controversy and concern for the unique populations of Arctic char and brook trout that inhabit the subbasin (Gibson et al. 1999).

Forestry is the main industry for this part of Newfoundland. Timber harvest is generally by clear-cutting and has occurred over much of the basin, dating back to 1903 in the area around Red Indian Lake (Morry and Cole 1977). Log driving on the main stem continued into the late 1980s and early 1990s (C. Bourgeois, personal communication). A pulp and paper industry has operated at Grand Falls since 1909. In 1995, the mill began aerating its effluent as part of a secondary waste treatment process; this has significantly reduced suspended solids and BOD (Bourgeois et al. 2000).

A base-metal mine (copper, zinc, lead) began operation in 1927 at Buchans. Prior to 1966, mine tailings were discharged directly into Buchans Brook and subsequently into Red Indian Lake; after 1966, tailings were diverted into a settling pond and the effluent treated (O'Connell et al. 1983). Today most of the mining activity in the catchment (including the Buchans mine) has ceased except for some base-metal exploration in the subbasin of Noel Paul's Brook (C. Bourgeois, personal communication).

Development is concentrated in the lower reaches of the river in communities ranging from Grand Falls/Winsor (population 17,500) to Millertown (population 750). Of the seven communities located in the drainage basin of the Exploits River, only one, Badger, treats its domestic waste prior to discharge into the river (O'Connell et al. 1983). The town of Grand Falls/Windsor recently constructed a sewage treatment plant and the town of Bishop's Falls is in the process of doing the same. No information was found detailing the possible environmental impacts of these discharges.

MIRAMICHI RIVER

The Miramichi River flows northeastward from central New Brunswick into the Gulf of St. Lawrence through the Barrier Islands at 47°N latitude (Fig. 21.10). It is the 2nd largest river system in the Maritime Provinces after the St. John River, draining an area of about 14,000 km^2 and contributing about half of the freshwater inputs to the southern portion of the Gulf of St. Lawrence. The entire river is unregulated and unobstructed and has the largest production of Atlantic salmon and other diadromous fishes in North America. The valley is lightly populated with widely separated settlements (Fig. 21.1), often located at abandoned log dumping and removal sites that were used in the nineteenth and early twentieth centuries during riverwide log drives.

For at least 10,000 years before the coming of the Europeans the area was inhabited by Mi'kmaq Indians, who called the river the Lustagoocheehk ("little goodly river"). Basque and French fishermen came to fish in the Miramichi Bay each summer from the early 1500s, but no attempt was made at settlement until French colonies were established in the seventeenth century.

Physiography, Climate, and Land Use

The Miramichi River basin lies wholly within the New England/Maritime (NE) physiographic province of the Appalachian Provinces division (see Fig. 21.10). Approximately half of the basin rises in the New England and New Brunswick Uplands. The lower river flows across the New Brunswick Lowlands. Three main tributaries, the Northwest, Little Southwest, and Southwest, rise in the predominantly granitic and volcanic highlands. In the lower reaches, the Southwest and Northwest Miramichi river branches combine to form the main channel at Beaubear's Island, several kilometers below the head of tide. The highland (western) region of the river basin is underlain by Ordovician, Silurian, and Devonian rocks, consisting primarily of granite, quartz, monzite, and granodiorite. The eastern portion of the Miramichi basin lies in the Maritime Plain, which is underlain with sandstones, conglomerates, and siltstones dating from the Pennsylvanian period and earlier (Chiasson 1995).

Although the basin is considered to be in the Boreal Forest biome, it is actually a transitional zone represented by portions of three terrestrial ecoregions: the Eastern Canadian Forests, New England/Acadian Forests, and Gulf of St. Lawrence Lowland Forests terrestrial ecoregions (Ricketts et al. 1999). The typical boreal species of balsam fir, white spruce, and black spruce are most common in the highlands and northern portions of the basin. Temperate deciduous species, such as maple, yellow birch, beech, and oak, are relatively abundant in the floodplains and ridges of the lower basin.

The climate is more continental than marine influenced due to the general movement of air masses from west to east. The mean annual air temperature is 4.3°C, with the highest mean monthly temperature of 18.8°C in July (Fig. 21.11). However, summer maximum temperatures in the interior typically reach 25°C to 30°C. Air temperatures cool rapidly in October and reach the freezing point by November. Mean temperatures from December through February are below −7°C, with the Miramichi River and estuary freezing over by mid-December on average. Spring breakup occurs between mid-March and late April (Fig. 21.4). The region has >120 frost-free days over the year. In an average year in the region, 113 cm of precipitation occurs on 160 days as rain or snow. This is distributed relatively evenly throughout the year, with the highest amount, almost 12 cm, falling in December (see Fig. 21.11).

More than 90% of the basin is forested. Since 1779, when the first contract was granted to export square timber to Britain, the harvesting of the forests has taken place throughout the Miramichi basin. This activity increased during the Napoleonic Wars (1800 to 1815) with the demand for lumber, especially white pine, for use as masts and spars for the British navy. In 1825, the Great Miramichi Fire burned much of the forest; what survived was selectively harvested in the latter half of the nineteenth century to support a thriving shipbuilding and lumber-export industry. In the twentieth century, forestry in the Miramichi changed focus to support the pulp and paper industry. Technological advancements and market demand shifted the woodland operations from selective harvest to clear-cutting.

A pattern of isolated small towns follows the river valleys, with some local farming on the broader valley bottomlands. Forestry and tourism are the major sources of income in the region and support a workforce of approximately 60,000 people. Mining for precious metals and base metals like zinc, lead, copper, and cadmium was once a major industry, particularly in the headwaters of the Northwest Miramichi River, but this activity has been discontinued in the last decade due to poor markets and depleted resources. The largest settlement is the city of Miramichi (~25,000 people) located on the upper estuary on Miramichi Bay.

River Geomorphology, Hydrology, and Chemistry

The entire Miramichi basin was covered in ice during the last Wisconsin glaciation. Ice advances from the west and north reduced the headwater bedrock contours to rounded hills. At the beginning of the deglaciation period about two-thirds of the lowland portion of the basin was submerged in a shallow marine estuary. As glacial rebound occurred large meltwater rivers extended through the lowlands to the retreating shoreline, scouring broad meandering valleys. The present-day river occupies about one-quarter of the width of the meltwater valley bottom, leaving intervals of fertile bottomland utilized by small farms and towns along the river's course. The predominantly bedrock-controlled channel changes from boulder-filled reaches in the upper valley to broad reaches of coarse gravel in the middle and

FIGURE 21.4 The Southwest Miramichi River during late winter, looking downstream from Quarryville, New Brunswick, near the head of tide. Note the 3 m high ice walls and the large ice fragments (within and outside the channel) that were formed during a recent breakup in the river (PHOTO BY R. CUNJAK).

lower valley. The river gradient changes from 5 m/km to 1 m/km between the upper and lower reaches. The mean tidal amplitude of 1 m in the Miramichi estuary also affects the water levels in the last few km of the channel.

Estimated mean annual discharge for the Miramichi is 322 m^3/s, based upon extrapolating from an upstream gaging station. The average annual runoff based on five streamflow gauging stations was estimated as 71.4 cm (Caissie and El-Jabi 1995). This represents a high percentage of the mean annual basin precipitation of 113 cm (63%). Although precipitation occurs fairly evenly throughout the year, the major portion of the runoff occurs as spring snowmelt events (see Fig. 21.11). Minimum flows in late August can threaten fish movement and survival.

Water-chemistry data collected from tributaries and main-stem sites in the Miramichi River catchment indicate a near-neutral, soft-water stream (mean conductivity range 28 to 55 µS/cm) that is typically low to moderate in productivity (see Randall et al. 1989, MREAC 1992, Komadina-Douthwright et al. 1999). Although the water in the Miramichi system is generally well buffered (mean pH 6.5 to 7.8), the melting snowpack during the spring thaw can occasionally result in localized, short-term acidic pulses. For example, the pH of water samples collected in the early spring of 1990 was occasionally <5.2 in the headwaters of the Southwest Miramichi River (MREAC 1992). Excessive loading of phosphorus or nitrogen is rare in this system, reflecting the relatively sparse population along the river's upper and middle reaches and the lack of agricultural or industrial inputs until one reaches the estuary (mean range of total N 0.06 to 0.19 mg/L, total P 0.006 to 0.016 mg/L). Below the head of tide, pulp and paper industries and urban discharges add significantly to the dissolved organic carbon, N, and P loadings. For example, average DOC values for samples collected from tributaries to Miramichi Bay are >25 mg/L (MREAC 1992).

River Biodiversity and Ecology

The Miramichi River is located within the Lower St. Lawrence freshwater ecoregion (Abell et al. 2000). Although it is one of the few unregulated rivers in this region, its biology and ecology have not been especially well studied, except for its well-known Atlantic salmon and brook trout populations.

Algae

No published information is available on the species distribution of freshwater algae of the Miramichi River. Diatom growth is pronounced on rock surfaces in the open main-stem reaches from midsummer to autumn. Filamentous algae such as *Cladophora* is locally abundant in slow-flowing reaches of the lower river, particularly in areas associated with nutrient enrichment, such as near human settlements.

Plants

A lush mixed forest occurs in the floodplain. Common tree species, such as white spruce, white pine, black spruce, yellow birch, and northern red oak, tend to grow well back from the water's edge along the principal branches of the river due to the annual scouring of ice each spring; fast-growing bushes, such as willows and sweet gale, abound in these locations. In the headwaters and tributaries, speckled alder is prolific. Macrophytes are scarce in the upper and midreaches of the river. Pondweed, cordgrass, and bur-reed are common in a wide variety of still and flowing waters, from full freshwater to the salt marshes of the estuary (where eelgrass or *Zostera* spp. are locally abundant). Aquatic mosses, particularly *Fontinalis* spp., cover the boulders and rubble in shaded headwater streams throughout the basin.

Invertebrates

A 1951 study (Bousfield 1955) found 24 zooplankton taxa in the lower Miramichi, ranging from freshwater species, such as the cladoceran *Bosmina longirostris*, to marine species, such as the copepods *Pseudocalanus elongatus* and *Tortanus discaudatus*. Locke and Courtenay (1995) carried out one of the few other studies of zooplankton (historic and recent) of the Miramichi River, also restricted to the lower river reaches and estuary, yielding 73 invertebrate taxa or more than three times the number found in 1951. Most common were copepods (e.g., *Eurytemora affinis*, *Acartia hudsonica*), the cladoceran *Bosmina* spp., barnacle nauplii, and the aquatic

stages of various insects (Plecoptera, Ephemeroptera, and Diptera). Much more information is needed on the ecology and distribution of phytoplankton and zooplankton in eastern Canadian rivers such as the Miramichi.

There has been little sampling of the benthic invertebrate fauna in the Miramichi River system. A few tributaries have been sampled but there is no published information on the benthic invertebrates of the main-stem branches: the Northwest, Southwest, and Little Southwest Miramichi rivers. From limited data it appears that the macrobenthic fauna is diverse and rich. In a 3rd order tributary of the Little Southwest Miramichi River (Catamaran Brook), Giberson and Garnett (1995) identified >95 genera of aquatic insects, with the greatest diversity found in the mayflies (25 genera) and caddisflies (19 genera); dipterans were the most common insects found in emergence traps, reaching a peak of approximately 3000/m^2 (mainly black flies) at one site in July. The most common genera included the mayflies *Baetis*, *Ephemerella*, and *Stenonema*; the stoneflies *Leuctra*, *Alloperla*, and *Pteronarcys*; the caddisflies *Hydropsyche*, *Dolophilodes*, *Glossosoma*, and *Pycnopsyche*; and the true flies *Prosimulium* and *Tipula*. Chironomid midges and elmid beetles were also abundant. Brachycentrid caddisflies are very common in boulder–rubble riffles in lower river reaches. As evidence of the poor state of knowledge of aquatic insects in Atlantic Canada, a study of the stoneflies in Catamaran Brook by Giberson and Garnett (1996) identified 31 species, of which 8 were new provincial records.

Few freshwater mollusks are known to occur, but no quantitative surveys have been carried out in the river, a reflection of the lack of targeted sampling for this environmentally sensitive group of organisms. Four species have been positively identified from the Miramichi River but more may exist (D. McAlpine and A. Martel, personal communication). These include the freshwater pearl mussel, the eastern lamp mussel, the eastern floater, and the eastern elliptio.

Vertebrates

The Miramichi River is world renowned for its sport fishing, primarily for anadromous Atlantic salmon and brook trout. Indeed, it is believed to produce more wild Atlantic salmon than any other river in North America (an estimated 200,000 adult salmon returned to the river to spawn in 1992). In addition to these salmonid species, the Miramichi River supports significant numbers of eight other diadromous fishes (sea lamprey, American eel, alewife, blueback herring, American shad, rainbow smelt, Atlantic tomcod, and striped bass). The Atlantic sturgeon, also diadromous, is occasionally found in the lower reaches of the Miramichi River. Historically (about 2500 years ago) this species was the focus for the food fishery of the Mi'kmaq people in the Miramichi region (P. Allen, personal communication). Although fish production in the fluvial freshwater portion of the river is low, the fisheries yield is high because of the high biomass produced in the marine environment (Randall et al. 1989).

Two unique characteristics of this river system, its moderate gradient (~2 m/km average slope) and the absence of barriers to fish passage (natural and man-made) on the three principal branches, make much of the system accessible to migrant species. Some, like the salmon, trout, lamprey, and eel, are capable of reaching the very headwaters in each of the three principal branches, whereas tomcod and striped bass rarely move above head of tide during their migrations.

After reaching a peak in abundance in 1992, Atlantic salmon returns to the Miramichi River have steadily declined over the past decade, presumably a consequence of poor marine survival. The decline in salmon returns has been realized despite increasing numbers of juvenile salmon in the river, the closure of commercial salmon fisheries since 1984, and a highly regulated recreational and aboriginal fishery (collectively accounting for <8% of potential egg deposition; DFO 2001a).

The Miramichi River has a diverse fish fauna for the maritime provinces. In addition to the 11 diadromous species, numerous freshwater and estuarine/marine fishes reside in the variety of habitats offered by this large river system. The brackish waters and rich food base of the Miramichi estuary make this an important rearing environment for >20 species whose larvae use this environment, including "true" estuarine fishes, like tomcod and sticklebacks, as well as marine species, such as flounder, herring, and capelin (Locke and Courtenay 1995). An extensive survey of the fish community in the Miramichi estuary was carried out by Hanson and Courtenay (1995), who found that the estuary serves as a temporally important habitat for >40 species of adult fishes. Some of these fishes use the area for brief periods en route upriver or to the sea (e.g., salmon), for spring/summer feeding (trout), or for overwintering (tomcod, smelt), whereas others used the estuary for their entire life (smooth flounder, mummichog). Recent radio-tracking data have shown that some postspawned Atlantic salmon move >50 km to over-

winter below head of tide, presumably because of the lack of ice-related disturbance relative to lotic reaches of the river (Cunjak et al. 1998, Komadina-Douthwright et al. 1997).

In the river proper, above the head of tide, the number of species in the fish community is less than in the estuary but still impressive. Randall et al. (1989) listed 21 species (representing nine families) that were found entirely in freshwater. (The only addition to this list is the recently introduced brown trout, which is found in a few isolated locations of the system.) The Cyprinidae, with nine species, form the largest group and are especially abundant in the lakes and warm, slow-flowing reaches of the main river channels. Based on electrofishing data from 27 sites sampled annually for 11 years, Randall et al. (1989) found that four species accounted for most of the fish community. These were, in order of abundance, Atlantic salmon, blacknose dace, slimy sculpin, and brook trout. Total biomass of these four dominant species was estimated to be approximately $2 g/m^2$ (of which salmon contributed 50%), thereby indicating a production rate of $2 g m^{-2} yr^{-1}$ in fluvial habitat of the Miramichi River. In the 1st order headwater streams of the Miramichi, often above barriers to anadromous fishes, fish communities are less diverse; in these situations, brook trout and occasionally slimy sculpin are the only fishes present.

Nonnative fishes in the Miramichi system include white perch, brown trout, and chain pickerel. Their distribution is limited to small sections of the basin. There is much concern about the potential impacts on native Atlantic salmon and brook trout from further introductions, especially of piscivorous species (e.g., smallmouth bass).

Commercial fisheries are confined to the estuary (mainly Miramichi Bay) and the lower reaches of the river below the head of tide. Gaspereau (a term used to collectively refer to the freshwater clupeids, blueback herring and alewife) and smelt account for >90% of the total biomass of commercially exploitable diadromous fish passing through the estuary (Chaput 1995). Average landings of these species ranged between 1900 and 2900 tons in the last 10 years (Chaput 1995, DFO 2001b). Striped bass and shad, two species at the northern limit of their distribution, are exploited at low levels, and their populations have displayed significant declines in recent years.

Beaver, mink, and river otter are the most common aquatic mammals occurring in the Miramichi River. Common merganser and double-crested cormorant are abundant in the upper and lower reaches of the river, respectively. Their piscivorous habits and large numbers have made them the target of angling groups, who blame them for a significant portion of the decline in salmonid abundance in the river, although direct evidence for the impact is lacking. Two-lined salamanders are commonly found in the headwater streams.

Ecosystem Processes

Large-scale or process-oriented research is rare in the Miramichi River system. One exception is the Catamaran Brook research project (Cunjak et al. 1990, Cunjak 1995), with its ecosystem approach to the study of population dynamics of the aquatic fauna. Invertebrate sampling in this $52\,km^2$ catchment have found a predominance of shredders in headwaters (1st and 2nd order streams) and increasing abundance of scrapers in the larger, open sections in the lower reaches of Catamaran Brook (Cunjak 1995). Doucett et al. (1996) provided supportive evidence of this trend using stable isotope analyses. They demonstrated that the food webs of headwater streams were largely dependent on allochthonous carbon compared with the larger 5th order river (Little Southwest Miramichi), where food webs are more dependent on autochthonously produced carbon (e.g., algal production). How such food webs and nutrient cycling may be impacted by large-scale human disturbances (e.g., the extensive forest harvest in the Miramichi headwaters) or point-source discharges, such as those concentrated in the estuary, require more attention. Certainly, the research on endocrine-disrupting chemicals (from forest spraying for pest control) and their potential impact on salmon survival (Fairchild et al. 1999) is suggestive of how complex such environmental assessment needs to be.

Human Impacts and Special Features

The Miramichi River, set in a broad glacial river valley, remains unregulated and unobstructed. The river produces more Atlantic salmon than any other river in North America, attracting fly fishers and river enthusiasts from around the globe. Historic fishing lodges, unique folklore, and a culture of acclaimed fishing guides and lumberjacks exists on this river, which is like no other in the Atlantic region.

Although many reaches of the river exist in near pristine condition, the basin has been subjected to several major environmental disasters, beginning with the 1825 forest fire that burned over 25% of

New Brunswick's forest and riverwide log drives in the main-stem channels (Kranck 1988). Widespread DDT spraying occurred in the 1950s as part of an annual spruce budworm control program, and heavy metal pollution was found in the Northwest Miramichi branch from poorly controlled mining activities between 1956 and 1971 (Zitko 1995).

The environmental impacts of forest harvest operations, together with log drives on large and small tributaries, were likely responsible for significant degradation of aquatic habitats but poorly documented and quantified (Cunjak 1995). Today forest managers are more considerate of environmental impacts and guidelines are more strictly enforced than in the past. Zitko (1995) summarized the three most important contemporary impacts on the flora and fauna of the Miramichi River as follows: (1) forest spraying activities to control insect pests, (2) base-metal mining activities and discharges, and (3) municipal and wood-processing discharges in the lower river and estuary.

Beginning in 1952, the Miramichi basin had the largest and longest-running spruce budworm control spray program in Canada. Early research on the effects of DDT on survival of Atlantic salmon (e.g., Elson 1974) led to banning the chemical in North America. Subsequent forest spraying (after 1963) with fenitrothion was less damaging to fish but caused increased rates of macroinvertebrate drift (Symons 1977). Some recent scientific research has linked forest spray chemicals (that resemble natural hormonal compounds) to disruption of physiological mechanisms in Atlantic salmon, which may partly explain population fluctuations (Fairchild et al. 1999). In the 1990s, virtually all spraying in Miramichi forests was with Bt, a biological control agent.

Effluents from a base-metal (copper and zinc) mine developed in the Northwest Miramichi in the 1950s resulted in mortality of resident fishes and downstream movement of large numbers of adult salmon (Elson et al. 1973), and reduction in aquatic insect abundance (Peterson 1978). Elevated concentrations of copper and zinc continue to affect the biota of this section of the Northwest Miramichi River.

The estuary is the site of greatest biological productivity and most human activity in the Miramichi basin (Chadwick 1995). The estuary receives large inputs of municipal discharges, wood-processing and wood-treatment waste from nearby industries, and heavy metals from mining discharges upriver.

Although some effluent concentrations have been reduced with new treatment facilities, stored contaminants, particularly in the sediments of the estuary, continue to be released during dredging operations and through the erosion of former dredge spoil areas (Buckley 1995, Zitko 1995, MREAC 1992).

In addition to these concerns, overexploitation of fisheries resources, such as salmon, shad, smelt, tomcod, and eels, is of concern to commercial and recreational fishers, government regulatory agencies, and citizens interested in the environmental health of this great river. Finally, conflicts over resource harvesting in the estuary, such as the Burnt Church confrontations between native and commercial lobster

fishermen, continue without a definitive resolution in sight.

ST. JOHN RIVER

The native people of the St. John River valley called it *Wolastoq*, the "Beautiful River" (Blair 2001). Its current name, the St. John River, is attributed to Samuel de Champlain, who arrived at the river mouth in 1604 on the feast day of St. John the Baptist. The 673 km long river, with a basin area of 55,110 km^2, is the longest in this region. It flows northeast through northern Maine (Fig. 21.5) and into Quebec before flowing southeast through New

FIGURE 21.5 St. John River headwaters, downstream from Moody Bridge in northern Maine, approximately 430 km from the river mouth. Note the lack of significant woody vegetation on the immediate riparian edge, a consequence of annual ice scouring (PHOTO BY M. GAUTREAU).

Brunswick and into the Bay of Fundy at 45°15′N latitude in the city of Saint John (Fig. 21.12). Indeed, between the Gulf of St. Lawrence and the Gulf of Mexico only the Susquehanna River is longer (Folster 2001). There are numerous important and substantive tributaries along the river's length. These include the Kennebecasis, Nashwaak, and Tobique rivers in New Brunswick, the Allagash and Aroostook rivers in Maine, and the Madawaska River in Quebec.

The St. John valley was first occupied by Maliseet and Mi'kmaq tribes, who raised agricultural crops on the rich valley bottomlands. French Acadians and then English Loyalists fleeing the American Revolution occupied the valley in the seventeenth and eighteenth centuries. Despite its size, its long history of navigation and colonization, its being the location of significant eighteenth-century British–American conflict (at one time home of Benedict Arnold), and its being the river on which the famous northwoods Chestnut canoe (MacGregor 1999) was perfected, the river is little known outside of New Brunswick and Maine.

Physiography, Climate, and Land Use

The St. John River basin lies in the New England/Maritime (NE) physiographic province (see Fig. 21.12). From headwaters in the Appalachian highlands to the estuary there is a vertical drop of 481 m as the river flows through three distinct physiographic zones. The upper basin flows from the river's source in Maine to Grand Falls in New Brunswick. Thirty-six percent of the basin arises in Maine in the forested hill and lake country of the "North Maine Woods." This section also includes the branch draining the Témiscouata–Madawaska River region of eastern Quebec (15% of the river basin). The upper St. John basin is sparsely populated and in relatively pristine condition. The central basin lies in the western sedimentary plain of the New Brunswick lowlands in a wide, smoothly glaciated meltwater valley (see Fig. 21.12). This reach is impounded by hydroelectric storage dams for much of its length. The lower basin from the Mactaquac Dam (west of Fredericton) to the ocean is tidal. This recently emerged coastal zone has little relief, collecting water to form three long bays, an inland delta, and several large shallow lakes. Two bedrock sills (Kennebecasis Bay, with an 11 m drop, and Reversing Falls, with a 5 m drop) limit the exchange of flows (and saline waters) between the deeper estuarine waters and the Bay of Fundy.

The St. John River basin primarily lies within the New England/Acadian Forests terrestrial ecoregion,

but also drains small portions of the Eastern Canadian Forests ecoregion and the Gulf of St. Lawrence Lowland Forests ecoregion near its mouth (Ricketts et al. 1999). Common upland vegetation includes black spruce, balsam fir, trembling aspen, and white birch. In the broad river valley, white spruce, white pine, red maple, yellow birch, beech, and red oak are common.

The climate is humid continental throughout most of the basin, with a mean annual air temperature decreasing from 5.5°C at Saint John in the estuary to 2.2°C in the northern headwaters. Mean monthly temperatures range from −9.6°C in January to 19.1°C in July (Fig. 21.13). Precipitation is evenly distributed throughout the year, varying from only 7.4 cm/mo to 11.4 cm/mo (see Fig. 21.13). Annual precipitation varies from 140 cm in the south to 90 cm in the northern headwaters, removed from maritime influence.

Settlements along the river were linked traditionally by river transportation, although this has now been interrupted by hydroelectric dams. Potatoes are the most important cash crop in the province, with approximately 22,000 ha in production, most of it in the St. John River valley principally between Grand Falls and Woodstock. Poultry and hog farms are concentrated in the upper and middle basins. Beef and dairy cattle farming are of similar importance, with dairy farms most common in the lower river valley around Sussex and the Kennebecasis subbasin. Grain crops, such as barley and oats, are of moderate importance and largely support the cattle industry. There is commercial farming of blueberries in the highlands and cranberries in the floodplains of the lower valley. Many small communities still rely on forestry, particularly in the upper basin (Maine). The most intensive harvesting for the pulp and paper industry is concentrated in the headwaters of the Tobique, upper St. John, and Nashwaak rivers. Wood-processing mills are located along the river, with the largest plants at Edmundston, Grand Falls, Nackawik, and Saint John.

River Geomorphology, Hydrology, and Chemistry

The entire St. John River basin was covered in ice during the last glaciation, reducing the headwater bedrock contours to rounded hills. The middle reach of the river follows a broad valley that was scoured by southerly trending glacial flows. As the ice melted the valley served as a major spillway from the interior. The preimpoundment river occupied a small channel within the valley. The present river widens to flood the whole valley bottom in hydroelectric reservoirs. Average main-channel gradients range from 1.3 m/km in the upper basin, to 0.5 m/km in the central basin, to 0.3 m/km in the lower basin.

Mean annual discharge for this large river is about 1110 m^3/s. The average annual runoff varies from 90 cm in the south to 64 cm in the headwaters, reflecting the spatial distribution of rainfall. A larger portion of the precipitation runs off the rugged headwater region (71%) than in the lower basin (64%). Although the precipitation is evenly distributed throughout the year, much of it is stored in the northern snowpack. The largest runoff (>15 cm/mo) occurs when the snow melts in April and May (see Fig. 21.13). The highest stages of the river occur in ice-jam-prone reaches.

Water chemistry is very much influenced by the degree of land use and the influence of industrial and urban discharges along the river corridor. Based on recent sample collections in the main river from the headwaters (Maine) to Fredericton, the following results characterize water chemistry in the St. John River during the ice-free period (R. A. Curry, unpublished data): average pH 7.7 (range 7.3 to 8.1), alkalinity 51.2 mg/L as CaCO$_3$ (25 to 102), conductivity 154 µS/cm (83 to 275), total N 0.47 mg/L (<0.3 to 1.37), total P 0.115 mg/L (0.005 to 1.33), dissolved organic carbon 7.1 mg/L (3.2 to 14.0).

River Biodiversity and Ecology

The St. John is the northernmost river within the North Atlantic freshwater ecoregion (Abell et al. 2000). In describing the fish fauna of the St. John River we have divided the basin into three sections. The upper basin includes those waters above Grand Falls, a natural barrier to upstream passage by diadromous fishes. The central basin refers to that section of the St. John River between Grand Falls and the Mactaquac Dam (west of Fredericton, and essentially head of tide). The lower basin refers to the section of the St. John River between the Mactaquac Dam and Reversing Falls (in the town of Saint John, where water flows upstream during rising tides generated from the Bay of Fundy).

Algae and Cyanobacteria

The general pattern of phytoplankton species composition and abundance is one of general similarity along the length of the river and indicative of a moderate rate of productivity. Seasonal patterns in

the abundance of phytoplankton have been well documented since the 1960s. Winter (January to April) populations of dinoflagellates, greens, cyanobacteria, and diatoms tend to be very low (<100 cells/ml) for all phytoplankters combined (Watt 1973); this is followed by a rise in abundance in May, with peaks in July and August before declining in the autumn (October). These peaks are coincident with declining water levels in the river when physical conditions (discharge) are sufficient to permit time for growth prior to being flushed out of system. The late-summer decline in phytoplankton abundance seems to be related to zooplankton grazing, as the population trends roughly parallel those of the phytoplankters (SJRBB 1974). The relative abundance of chrysophytes (mainly ochromonads) during late winter in the impoundments of Woodstock and Beechwood (middle reach) may have been the result of heterotrophic activity in these deep, low-flow water bodies (Watt 1973).

The diatom *Melosira*, typically associated with eutrophic waters (Hutchinson 1967), was common at all main-stem (middle reach) stations in the St. John River but was rare in the upper basin and headwaters (Watt 1973). Instead, the phosphate-sensitive chrysophte *Dinobryon serularia* was prominent only in the relatively unpolluted (phosphorous-limited?) headwaters (i.e., Tobique). Filamentous algae, such as *Cladophora* spp., can be locally abundant in the central and lower river basins during summer where point-source discharges of nutrients occur. Dense mats can become a nuisance, especially where low water levels are exacerbated by river-flow regulation in proximity to hydroelectric dams.

Plants

In the floodplains, speckled alder, balsam poplar, and willows are common riparian plants that grow in abundance. Ostrich ferns are especially common in the floodplains of the lower basin, where they are harvested every spring and sold as "fiddleheads." The St. John River lousewort is a large but rare parasitic member of the snapdragon family that became the first plant in Maine to be listed as endangered. The only known place in the world that this plant grows is in the floodplain of a very small section of the northern St. John River in New Brunswick and Maine (Allagash subbasin).

Invertebrates

There is very limited information concerning the abundance and diversity of invertebrates in the St. John River. Zooplankton diversity is based on sampling efforts in the impoundments coordinated by the St. John River Basin Board (SJRBB 1974). They noted that the zooplankton community was predominantly crustaceans (Copepoda and Cladocera). *Bosmina coregoni* and *Daphnia* spp. were the most common taxa. Monthly sampling in the early 1970s showed that densities were low in winter (primarily copepods), increased in May and June (mostly cladocerans), before declining in the autumn, a trend that generally matches the seasonal abundance of phytoplankton, presumably the main food source of zooplankton in the headponds.

In the upper basin, at First Green Lake, zooplankton abundance and density were the highest recorded in the study, similar to the situation for phytoplankton, suggesting that the Green River is a naturally productive subbasin of the St. John River system. Community composition in First Green Lake and Second Falls were similar to those in the impoundments, but the Glasier Lake species composition was different.

Limited sampling of the benthic invertebrate fauna has been carried out. In the early 1970s, a series of bottom "dredge" samples were taken from the six impoundments in the St. John River. Tubificid worms (oligochaetes) and chironomid midges were the most common benthic organisms collected in these samples and are generally indicative of polluted conditions. In the Grand Falls and Beechwood headponds, tubificids exceeded 100,000 worms/m^2, suggesting gross organic pollution (SJRBB 1974). The density of these benthic indicator species was much less in the upper part of the Mactaquac and Tobique impoundments.

Recent riverine samples from riffle habitats extending from the headwaters in Maine to Fredericton indicate that chironomid midges were abundant at all sites; black flies were common in summer. In the upper basin, heptageniid and baetid mayflies, chloroperlid stoneflies, and philopotamid and hydropsychid caddisflies were most commonly found. In the central basin, heptageniid and ephemerellid mayflies, perlid stoneflies, and philopotamid and hydropsychid caddisflies were common (R. A. Curry, unpublished data).

No quantitative descriptions of the molluscan fauna yet exist for the St. John River, an unfortunate situation that typifies much of Atlantic Canada's freshwaters. Recent (unpublished) underwater surveys have noted a rich community of freshwater mussels in the lower reaches of the St. John River (below Mactaquac Dam), where seven to eight species have been confirmed (D. McAlpine and A.

Martel, personal communication). These species include those found in the Miramichi River (i.e., eastern pearl mussel, eastern floater, and eastern elliptio), as well as the yellow lamp mussel, the alewife floater, and the tidewater mucket. In Belle Isle Bay, large numbers of freshwater mussels, mainly eastern elliptio, were found in densities estimated to be 10 to 60 individuals/m^2 (A. Martel, personal communication). These organisms are recognized as sensitive, long-lived indicators of environmental health. Further, they are probably important agents of cycling nutrients (e.g., C and N) in the upper layers of river sediments, especially where they occur in such high densities.

Two species of freshwater mollusks are of special environmental concern. The alewife floater has a very restricted distribution in Canada but is widespread in the St. John River, and the yellow lamp mussel is significant because it is known from only two locations in Canada: Sydney River (Cape Breton, Nova Scotia) and the St. John River (D. McAlpine, personal communication). The latter has been designated an endangered species in the United States and is presently being considered for similar designation in Canada by the Committee on the Status for Endangered Wildlife in Canada (COSEWIC).

Vertebrates

The early data (1970s) compiled for the SJRBB were biased to samples collected in the impoundments and selected lakes (e.g., Glasier Lake), mainly by gill nets and seine nets. These data make for an interesting comparison with a recent report on the fish community of the St. John River by Curry et al. (2001), who sampled 11 riverine sites between Fredericton and northern Maine using seine nets and electrofishing in 2000. Relatively little published information is available from the tributaries along the river.

In the 1970s, a total of 28 fish species were identified in the upper basin, representing 11 families; 43 fish species were found in the central basin, due primarily to the occurrence of diadromous fishes such as striped bass, American eel, shad, and gaspereau (Meth 1973). Cyprinidae (11 species) were the most common family in both sections of the river. More recently, Curry et al. (2001) found 36 fish species in the upper and central basins of the St. John River, with common shiners and white suckers the most widely distributed fishes and lake trout, Arctic char, and round whitefish noticeably absent.

Freshwater fishes are typically found in the upper two-thirds of the lower basin, whereas salinities near the river mouth are sufficient to allow at least periodic invasions by some marine species (Meth 1971). Various species, including commercially significant species, such as Atlantic salmon, American eel, alewife, and striped bass, use the estuary for some portion of their life as they stage in these environments during their spawning migrations. A fisheries survey in 1971 found 54 species of fishes in the lower basin, including obligate freshwater, marine, and diadromous species (Meth 1972).

Prior to 1890 and the construction of the Caribou Dam on the Aroostook River, virtually all of the river system downstream of Grand Falls, an area of $33,300 \, \text{km}^2$, was accessible to diadromous fishes. Historically, the St. John River has been considered the third-largest producer of Atlantic salmon in New Brunswick, behind the Miramichi and Restigouche rivers (Washburn and Gillis Associates 1996). However, given the paucity of data prior to the building of the first dams and the large drainage basin, it is possible that the St. John River was once the greatest producer of diadromous fishes in eastern Canada.

In the St. John River, restrictions on recreational fishing for Atlantic salmon started in the late 1960s with a five-year ban initiated between 1973 and 1977, followed by annual extensions and quotas in the 1980s (L. Marshall, personal communication). A complete closure on salmon angling in the river has been in place since 1998 because the number of returning adult salmon has not reached conservation targets (DFO 2001a). In 2000, the number of returning wild multi-sea-winter (MSW) salmon to Mactaquac Dam was 277 fish, the lowest count in the 31-year record and in stark contrast to 6000 to 8000 MSW salmon between 1975 and 1980 (DFO 2001a).

Of special note is the unique "Serpentine stock" of Atlantic salmon that typically entered the rivers in late May and resided in freshwater for more than a year before spawning in headwater tributaries of the St. John River in the autumn of the year following river entry. Despite the numerous main-stem dams, a remnant of this stock still exists and deserves special protection to ensure its continued existence.

Numerous introductions have also occurred in the St. John River, most notably muskellunge (Stocek et al. 1999), chain pickerel, and rainbow trout. Smallmouth bass, another nonnative species, was introduced to New Brunswick waters, including the St. John system, in the late nineteenth century; they have now increased in distribution and represent a considerable sport fishery (especially in the impoundments), contributing approximately $5 million annually to the New Brunswick economy.

In the 1970s, commercial fisheries for species such as gaspereau, American eel, and sturgeon were concentrated in the lower basin. The value of the gaspereau fishery (>$250,000) was more than four times the combined value of all other species, with the majority of fish destined for use as pet food (Meth 1972). Today very limited commercial fisheries continue. For example, there is only one commercial license to fish Atlantic sturgeon in the St. John River.

Beaver, mink, and river otter are common aquatic mammals occurring in the tributaries of the St. John River; muskrat are common on the main river. Great blue heron, common merganser, and double-crested cormorant are the most abundant aquatic birds along the river in the ice-free season.

Ecosystem Processes

Studies of ecosystem processes in the St. John River are primarily limited to estimates of primary production. The fragmented nature of the river and the numerous impoundments must certainly influence the energy cycling within the river. Estimates of total primary productivity for the impoundments and main stem of the St. John River system in the spring to autumn of 1972 were 8.5 to $25.5 \, \text{g C m}^{-2} \, \text{d}^{-1}$, moderate to low rates that are not indicative of eutrophication (Watt 1973). Watt (1973) considered the impoundments to be severely heterotrophic where primary production contributed <1% of the total carbon balance above Beechwood and about 15% in the Mactaquac headpond. In the tidal waters downstream of Fredericton, primary production in the riverine sites was lower (48 to $99 \, \text{mg C m}^{-2} \, \text{d}^{-1}$) than for the bays, like Washademoak Lake and Kennebecasis Bay (124 to $276 \, \text{mg C m}^{-2} \, \text{d}^{-1}$). As was the case with the impoundments in the middle reach of the river, production in the lower river reaches was limited by the depth of the photic zone and occasionally by low phosphorous concentrations. Rates of primary production tend to be negligible beneath snow and ice cover between January and March.

Based on phytoplankton abundance estimated in the early 1970s by Watt (1973) and the classification of lakes by Vollenweider (1970), the St. John headwaters and upper basin are oligotrophic (<$100 \, \text{mg C/m}^3$), changing to mesotrophic in the middle and lower reaches of the river (300 to $500 \, \text{mg C/m}^3$).

Human Impacts and Special Features

The St. John River, the longest river in Atlantic Canada, has a long and proud history of colonization, commercial development, forestry, agriculture, fishing, and hunting. An international river, it runs through one state (Maine) and two provinces (Quebec and New Brunswick) before discharging into the Bay of Fundy, its high tides responsible for the famous Reversing Falls at the river's mouth in Saint John. Some of the most beautiful wild scenery in maritime Canada and New England still exists within the extensive river valley. The Allagash River in northern Maine is a protected National Wild and Scenic River.

Dams for hydroelectric power generation began in the St. John River in 1890 with the construction of the Caribou Dam on the Aroostook River (Carr 2001). Today there are 11 dams in the St. John River system, including 3 main-stem dams (Mactaquac, Beechwood, and Grand Falls) with valleywide impoundments in the middle basin above tidal waters (see Fig. 21.12). Fish passage is obstructed by the dams, although trapping and trucking of some fish species, primarily Atlantic salmon, from the Mactaquac Dam to the impoundments above Beechwood and Tobique takes place. The issue of reduced biodiversity of freshwater mussels (and their specific fish hosts) is another consequence of the numerous barriers, because dams preclude movement of many migratory species of fishes that served as hosts for glochidia larvae of mollusks.

Hydroelectric facilities continue to impose significant obstacles to upstream passage and no opportunity for safe downstream passage. Stressfully high summer water temperatures, especially in the impoundments, restrict growth for many species. Altered flow regimes may affect in-river movements (e.g., migrant smolts; Carr 2001), and the abundance of piscivorous species, many of them nonnative (e.g., smallmouth bass, chain pickerel, muskellunge), provides additional constraints to juvenile salmon production in the river. Finally, there is potential for "genetic swamping" by farmed salmon escapees from nearby aquaculture operations in the Bay of Fundy and Maine (DFO 1999).

There is much concern about the impact of introduced nonnative species (e.g., muskellunge; Stocek et al. 1999). Their distribution continues to expand, as noted by the recent smallmouth bass record above Grand Falls (C. Collett, personal communication). There is also concern about possible invasion by zebra mussels from the St. Lawrence River system.

Organic loading from food and wood processing in the upper and middle basins of the river far exceeds the assimilative capacity of the river. Watt (1973) found that heavy organic carbon loading (from industrial effluents) created a zone of extreme

heterotrophy that extended from Edmundston to Beechwood in the summer. This condition is exacerbated in the slow-flowing, depositional impoundments, where less dissolved oxygen is available to meet the biological oxygen demand, resulting in low dissolved oxygen levels. In winter, with low temperatures and reduced aeration (under ice cover), the heterotrophic condition extended as far downriver as the Mactaquac Dam. In the lower river, Watt concluded from the phosphate data and estimates of primary productivity in 1972 that there was no immediate eutrophication threat to the estuary.

Contamination continues to affect the aquatic biota of the St. John River some 30 years after those initial surveys. Physiological impairment in white suckers and slimy sculpin has been found downstream of hydroelectric facilities and food-processing industries in the upper and middle basins (Munkittrick et al. 2000). Pulp mill effluents in the St. John River estuary still have sublethal (physiological) effects on estuarine fishes (e.g., mummichog) that may alter reproductive function (Dube 2000). Elevated polychlorinated biphenol (PCB) concentrations were found in five species of fishes sampled from the Aroostook River in 2001 (R. Parker, personal communication), and brook trout and Atlantic salmon egg survival seem to be negatively affected by intensive agriculture and flow regulation in small subbasins (Flanagan 2003, Munkittrick et al. 2000).

MOISIE RIVER

The Moisie River, with a basin area of 19,871 km², flows southward into the Gulf of St. Lawrence 25 km east of the town of Sept-Iles, Quebec (Fig. 21.14). The river flows for approximately 500 km, from an elevation of 533 m asl in the Laurentian Highlands in Labrador through a deep glaciated river valley to the St. Lawrence Lowlands into the Gulf of St. Lawrence (Fig. 21.6). Principal tributaries are the Nipissis, Caopacho, Pékans, and Ouapetec. The Moisie River is one of the largest rivers of the Quebec North Shore and has a worldwide reputation as a pristine wilderness river and a destination for Atlantic salmon angling and wilderness paddlers. It flows through steeply walled canyons rising 300 m above the river, and waterfalls, gorges, and big rapids attract outfitters and paddlers to what they refer to as "the Nahanni of the East" (see Chapter 18).

For at least 2000 years the river has been used by the Montagnais Indians as a traditional route into the interior of the Quebec woods (Laurentian Plateau) for hunting and trapping. The Montagnais ("mountain people"), a name given the local Innu by French explorers, refer to the river as *Mis-te-shipu*, meaning "the great river" (Weeks 1971). The abundance of salmon made it a favored site for Indian settlements because salmon was a diet staple; a major Montagnais community (Malioténam) still exists near the mouth of the river. The explorer Jacques Cartier was the first known European to visit the lower river valley and bay, in 1535. The Jesuit missionaries who established themselves in the area by 1600 first used the name "Moisie," from the French word *moisi*, meaning "damp" or "muddy." Henry Youle Hind, a professor of chemistry and geology at the University of Toronto in the middle of the nineteenth century, was the first white man to explore the Moisie to its source. Weeks (1971) provides a detailed account of the history of the Moisie River and its inhabitants.

Physiography, Climate, and Land Use

The Moisie River basin is found primarily in the Laurentian Highlands (LU) physiographic province of the Canadian Shield physiographic division (see Fig. 21.14). The Precambrian Canadian Shield bedrock is widely exposed and thinly covered throughout most of the basin (Stockwell 1963). Metamorphic rocks dominate the basin, mainly in the form of granitic gneisses and migmatite. Sedimentary rocks in the Pékans River tributary near Mont Wright contain significant iron-ore deposits (Lalonde et al. 1990). The basin lies in the Eastern Canadian Forests and Eastern Canadian Shield Taiga terrestrial ecoregions, with upland forests typical of the Boreal Forest biome. The dominant upland species are white and black spruce, white birch, trembling aspen, and balsam fir.

The climate is subalpine, with a mean annual air temperature of only 1°C (Mackay 1995). The mean monthly temperature, as recorded at nearby Sept-Iles, Quebec, ranges from a high of about 15°C in July to a low of −15°C in January (Power 1981; Fig. 21.15). Mean annual precipitation from the northern to the southern portions of the basin ranges from 113.1 cm to 87.6 cm (Lalonde et al. 1990). Precipitation is fairly uniform throughout the year, ranging from about 7 cm/mo in February to 11 cm/mo in September (see Fig. 21.15). An extensive snowpack is common in the region, with average accumulation of 42.3 cm until spring snowmelt begins in May (Power 1981).

Forests continue to cover >90% of the Moisie River basin. In the latter half of the twentieth century, land use in the basin was limited to peat extraction in the lower basin and some mining activity, principally for iron ore, in the headwaters (Nipissis and Pékans subbasins). The level of mining activity is much reduced from what it was 30 years ago. Very limited trapping for fur-bearers is practiced in the Moisie catchment. Targeted species included (in order of importance) beaver, muskrat, mink, lynx, wolf, and fox (Bertrand 1978).

River Geomorphology, Hydrology, and Chemistry

The Moisie River is a 9th order river fed by numerous headwater lakes on the Labrador plateau. The Moisie River valley lies directly south of the center of outflow of the late Wisconsin Laurentide ice sheet. The heavily glaciated valley leads from the Labrador Plateau in the uplands through a steep-walled valley leading to the St. Lawrence River (see Fig. 21.6). In the lower river, the slope is 1.6 m/km and the bottom substrate is a mix of sand and cobble (Naiman et al. 1987). The relative lack of any abrupt changes in slope in the lower reaches of the Moisie means that early travelers to the region and diadromous fishes could ascend much of the length of the river. At the mouth there is a well-defined estuary, but given the relatively steep slope the maximum extent of the salt wedge is only 2 km upstream (Black et al. 1983).

The annual hydrograph is typically dominated by the spring discharge peak, with 50% of the annual discharge occurring in May and June following snowmelt (see Fig. 21.15). The maximum daily flood recorded is 3820 m³/s. The mean annual flow is 426 m³/s, with summer low flows of <300 m³/s. The lowest flows (<130 m³/s) actually occur from January through March, when most of the precipitation accumulates as snow. Atlantic storms occasionally stall in the St. Lawrence valley, dropping enormous amounts of rainfall and causing extensive flooding. For example, in the Saguenay River basin, a similar North Shore basin located to the west of the Moisie River, over 20 cm of rain fell in 36 hours in July 1996. Widespread flooding and several dam failures occurred in all of the nearby rivers.

Maximum water temperatures in the river rarely exceed 20°C and are often much less in the tributaries, reflecting the boreal climatic zone. Ice forms on the river in November, forming a complete cover by December that persists until April or May. Mean annual water temperature is 6°C, accumulating only about 2100 degree-days (Mackay 1995).

The Moisie River is a soft-water river (alkalinity range from 8 to 18 mg/L, conductivity 10 to 77 µS/cm)

FIGURE 21.6   Katchapahum Falls on the main branch of the Moisie River in Quebec (Rkm 175). Note the steep granite cliffs, coniferous forest, and tea-colored water typical of this boreal river (PHOTO BY K. SCHIEFER).

colored by humic acids, with a stable (neutral) pH (6.3 to 7.5) and low levels of dissolved organic carbon (2.9 to 23.3 mg/L) (Environment Canada 1982, Naiman 1982, Schiefer 1982). Total nitrogen and total phosphorous were measured in the early 1980s by Naiman (1982) and Hamilton (1983), who found mean values of 134 to 392 μg/L and 4 to 22 μg/L, respectively. Although the water-chemistry data are more than 20 years old, there is little reason to expect a different situation today, as catchment activities are essentially the same as they were in the early 1980s. A 1983 study of sedimentary diatoms in the headwater lakes of the Moisie and adjacent Matamek river systems found evidence of slightly acidic soft-water conditions; pH ranged from 4.6 to 5.8 and conductivity from 9 to 20 μS/cm (Hudon et al. 1986).

River Biodiversity and Ecology

The Moisie River is located within the Lower St. Lawrence freshwater ecoregion, as are all the rivers draining the Quebec North Shore. Given the remote and largely pristine nature of the basin, it is somewhat surprising that a fair amount of biological and ecological research has been conducted in the Moisie River. These studies have focused primarily on fisheries research on Atlantic salmon and studies of ecosystem processes, the latter combined with studies of the adjacent and much smaller Matamek River (e.g., Naiman et al. 1987, Mackay 1995).

Algae and Cyanobacteria

Hamilton (1983) sampled in both the Moisie and Matamek rivers to determine factors controlling periphyton productivity and biomass. In spring, riffle periphyton growth was positively correlated with the increase in light and water temperature, and there was some suggestion that insect grazing controlled periphyton biomass. In addition to documenting drift patterns by benthic periphyton species, such as *Tetrospora cylindrica*, *Eunotia pectinalis*, *Oscillatoria minima*, and *Synedra acus*, there was evidence of algal drift by less known planktonic species, such as *S. minutum* and *Kephrion boreale*. Hudon et al. (1987) studied the physiology of the periphytic community in the Matamek River and found 64 species of diatoms belonging to 23 genera. Two filamentous taxa, *Tabellaria flocculosa* var. *flocculosa* and *Eunotia pectinalis* var. *pectinalis*, composed >70% of the total cell counts. In their 1983 sampling of the surficial sediments of 20 headwater lakes, Hudon et al. (1986) identified 158 species of diatoms representing 27 genera. Only six taxa (*Melosira distans*,

T. flocculosa, *T. fenestrata*, *E. pectinalis* var. minor, *E. pectinalis* var. *pectinalis*, and *Frustulia rhomboides*) were abundant in the sediments.

Plants

Dominant riparian plants include black spruce, white spruce, speckled alder, Labrador tea, and willows. Macrophytes (mainly pondweed) are relatively common in the lentic habitats in the lower reach of the Moisie River.

Invertebrates

Studies of invertebrates are limited in the Moisie, but Environment Canada (1982) collected invertebrate data from several sites associated with a study of environmental impacts of mining activities on a Moisie tributary (the Pékans). In general, it was found that the headwaters of the Pékans Rivers were rich in benthic fauna, with the number of benthic taxa similar between sites (30 to 37). In the vicinity of an iron-ore mining operation in the headwaters (Webb Creek), the fauna was less diverse and dominated by chironomid midges. The preponderance of oligochaete worms and the relative paucity of bivalve mollusks below the mining operation was indicative of pollution. A total of 10 bivalve species representing two genera, *Pisidium* and *Sphaerium*, were collected at 20 sampling sites in the river system (Environment Canada 1982). These were most abundant in the Pékans River rather than the main river.

For the insects, stoneflies were predominantly represented by *Paracapnia* and *Taeniopteryx* and several species of Leuctridae in the headwaters; in the Pékans River, *Paracapnia*, *Isoperla*, and species of Leuctridae and Perlidae were abundant (Environment Canada 1982). The most common mayfly genera were *Ephemerella*, *Leptophlebia*, and *Heptagenia* in the headwaters and in the Pékans River; *Ephemerella*, *Heptagenia*, and *Ameletus* (Siphlonuridae) were common in the Moisie River. Caddisfly larvae were most abundant and diverse in the Pékans and Moisie rivers, with *Lepidostoma* and *Chimarra* common; *Hydropsyche* were common in all reaches. Chironomid midges were the most abundant and widespread dipteran larvae collected.

Vertebrates

The Moisie River is most famous for its stock of Atlantic salmon. An estimated 20,000 adult salmon enter the river each year to spawn, making it the largest salmon-producing river on the Quebec North Shore (Friends of the Moisie River 1989). It is the exceptionally large individual and average sizes of the

Moisie salmon that make them distinctive. Power (1981), in his description of salmon stocks of Quebec and Newfoundland–Labrador rivers, found that grilse (adult salmon spending one winter at sea) were relatively scarce in the Moisie, and two- and three-sea-winter salmon and repeat spawners averaged 74.6, 90.7, and 103.8 cm in length, respectively.

Scientific fisheries research in the Moisie River, primarily directed at salmon, has been conducted since the 1920s, when scale samples were collected for ageing (Schiefer 1982). The most extensive research on salmon biology in the river was carried out by Schiefer (1971, 1972, 1982) and Bielak (1984). Growth rates were higher for parr in upstream sections, presumably because of low densities and lack of intraspecific competition in years immediately following installation of the fishway at Katchapahum Falls. Smolts have a mean age of 3.7 years and a fork length of 12.0 cm, relatively small for most salmon stocks of the region (Power 1981).

In addition to Atlantic salmon and brook trout (resident, or landlocked, and sea-run forms), northern pike, longnose sucker, white sucker, burbot, American eel, sea lamprey, lake chub, lake whitefish, and round whitefish are present in the river system. A full biological account of these nonsalmonid species in the Moisie River system is lacking. More species are likely to occur in the river but we are unaware of an inventory of the fish fauna of the Moisie River. However, Power et al. (1973) provided a list of species found in the nearby Matamek River that is probably similar to the Moisie. In addition to those already listed, they found Arctic char (mostly in the larger lakes); threespine, fourspine, and nine-spine sticklebacks; and rainbow smelt in freshwater habitats. Alewife, tomcod, shorthorn sculpin, and winter flounder were captured in the estuary. No nonnative fish species are known to occur in the Moisie River. Among the nonfish vertebrates, beaver are particularly abundant in the tributary streams and headwaters of the Moisie River. Mink and river otter are also common.

Ecosystem Processes

There is no evidence of eutrophication in the Moisie. Productivity is low, but few data are available on primary or secondary production. Based on research carried out largely in the adjacent Matamek River basin and a single site in the lower Moisie River, most in situ organic matter was produced by periphyton (small-order streams) or by moss and macrophytes (high-order streams), or originated from material being transported downstream;

allochthonous input from riparian vegetation was minor (Naiman 1982, 1983; Naiman et al. 1987). Gross production ranged from 0.3 to 0.8 g O_2 m^{-2} d^{-1} in small streams and from 0.8 to 1.08 g O_2 m^{-2} d^{-1} in the larger river sites (Naiman 1983). Macrophytes in the lower reach of the Moisie River (mainly pondweed) contributed 3 to 30 g/m^2 but only for the summer period before plants senesced (Naiman 1983).

Regarding fish (salmon) production, the growing season seems to be a bit shorter in the Moisie than in the Miramichi but similar to the Exploits, with spring ice-out occurring in early May and summer water temperatures in the main river rarely exceeding 21°C. Therefore, fish production rates may be similar to those in Newfoundland rivers like the Exploits (e.g., Randall et al. 1989). Morin and Naiman (1990) estimated maximum fish biomass of <9 g/m^2 and production of approximately 10 g m^{-2} yr^{-1} in 4th order streams of the nearby Matamek River.

Large-scale or process-oriented research is largely lacking for the Moisie River system. Scientific research to date has focused largely on the biology and fisheries management of the Atlantic salmon (Schiefer 1971, 1972; Bielak 1984) and on the potential impacts of mining pollution (Schiefer 1982, Environment Canada 1982). The only exception is the thorough study of suspended sediment and organic carbon transport relative to the discharge regime in the Moisie River and four other nearby boreal streams of decreasing stream order (Naiman 1982, 1983; Naiman et al. 1987; see also review by Mackay 1995). That research showed that during the spring snowmelt freshet 71% to 92% of the annual sediment load is exported from the river but only about half of the annual load of dissolved and particulate organic carbon. There was a rapid loading of carbon between 1st and 2nd order streams, followed by biological and physical processing along the river continuum. At the mouth of the Moisie River organic carbon export was reduced to a rate comparable to the 1st order streams. Primary production increased in importance in a downstream direction, with production by macrophytes particularly significant in larger rivers. Many of the results were in accordance with the predictions of the River Continuum Concept (Vannote et al. 1980). Also, Naiman (1982) concluded that in such pristine boreal rivers instream processing and retention devices (e.g., beaver ponds and large woody debris) exert considerable control over the quantity of organic matter transported; river discharge, by contrast, was more important in determining sediment concentrations. Such ecosystem-scale studies are important for understanding how organic matter is processed spatially and temporally in boreal river systems.

Hudon et al. (1996) measured carbon transport in 47 Quebec rivers, including the Moisie. They found that carbon loads were strongly related to runoff values in all rivers but that the specific dissolved organic carbon load of tundra rivers was about half that of rivers flowing through forested basins to the Gulf of St. Lawrence (e.g., Moisie).

Human Impacts and Special Features

The Moisie River is famous for its exceptional scenery and wilderness setting, and for its stock of large Atlantic salmon. The river is also popular for outdoor enthusiasts, for camping, canoeing, and kayaking. The basin continues to be relatively inaccessible, as the only road system occurs at the mouth of the river; land travel upstream is possible only by a railway that is wholly owned by the iron-ore company operating in the basin. The rail route is along the main stem of the lower Moisie for approximately 40 km before heading up the Nipissis River subbasin.

Sport fishing, primarily for Atlantic salmon, has been occurring in the Moisie River for more than 150 years. Salmon abundance was the reason for fishing settlements and commercial exploitation by French colonists and the Hudson's Bay Company. Private salmon clubs like the Moisie Salmon Club (originally the Adams Camp) have been in operation since the 1850s (Weeks 1971), with exclusive fishing rights to all but the first 15 km above the river mouth. By the mid-1800s as many as 70 salmon nets (approximately 27 km in total length) were set in the lower river and bay, where salting and icing plants were built to process the fish. By the late 1800s sport-fishing interests converged on the Moisie River, largely as a result of depleted stocks and poor fishing in the rivers of New England and parts of the Maritimes. No commercial fisheries exist today in the Moisie River or estuary. Salmon (and brook trout) continue to be exploited by the Montagnais Indians, who harvest some fishes as part of a subsistence fishery in the lower river and estuary.

Salmon enhancement has taken place in the Moisie River. Two fishways have been constructed, one on the main stem at Rkm 175 (Katchapahum Falls) between 1965 and 1975 and the other on the Nipissis River tributary at Rkm 50 (McDonald Falls) in 1973 (Bertrand 1978). The number of adult salmon enumerated passing upstream of the former barrier at Kathchapahum between 1975 and 1981 ranged from 640 to 5800 fish (Schiefer 1981).

Various mining operations have caused the only significant pollution in the Moisie River. During the 1970s mining discharge and tailings pollution caused significant environmental impacts in the upper Pékans River, particularly in the Webb Creek drainage, where elevated levels of mercury, iron, nickel, and copper were measured in sediments (Environment Canada 1982) and the red iron-rich plume was detected 300 km downstream from the source (Schiefer 1981). Benthic diversity and abundance were affected in the tributaries, as was the abundance of salmonids; the Moisie River below the confluence with the Pékans appeared not to be affected by the mining discharge to the river (Environment Canada 1982). In 1996, a significant nickel discovery was made in the Nipissis River subbasin, a tributary to the Moisie. Hundreds of claims have been staked in the area, posing a new threat to the basin ecosystem if developments take place (Native Forest Network 1997).

Although no dams currently exist in the Moisie River, Hydro-Québec proposed a hydroelectric development scheme in the 1980s that would have diverted 74% of the flow of the Pékans River to an adjacent regulated river basin, the Ste. Marguerite River (Friends of the Moisie River 1989). The diversion would have resulted in an estimated reduction of 13% of the flow of the Moisie as measured at the river mouth. Due to a combination of public opposition, unfavorable economics, and the recommendations of an expert review panel, the project was never pursued. Hydro-Québec continued to pursue development in the basin. However, in early 2003, the Quebec government assured the future protection of the Moisie River by announcing the creation of an aquatic reserve (3897 km^2) for the basin. The reserve status prohibits logging, mining, and hydro/oil development in a corridor between 6 and 30 km wide for approximately 320 km, thereby including the Pekans and Carheil subbasins.

ADDITIONAL RIVERS

The Big Salmon River is actually a very small river located in southern New Brunswick. It flows primarily in a southerly direction to empty into the Bay of Fundy (Fig. 21.16). This is a unique coastal basin with pristine old-growth forest still present in the deeply incised river valley (Fig. 21.7). Currently, there

FIGURE 21.7 Lower reach of the Big Salmon River (Rkm 2). Note boulder and rubble substrate and steep forested valley that characterizes much of the length of this river (Photo by R. Newbury).

are only small water-level control dams at the outlets of several headwater lakes (e.g., Walton Lake; see Fig. 21.16). Historically, it was one of the most productive of the Inner Fundy (IF) salmon rivers, but today so few salmon return that the stock has been listed as "at risk" (endangered). To protect this unique genetic stock of IF salmon from extinction, a gene-banking program was initiated in 1998.

The Dunk River is a very small southwest-flowing river located on the small island province of Prince Edward Island (Fig. 21.18). The Dunk is a low-gradient river with high in-stream productivity that flows into a shallow nutrient-rich estuary. It has a stable thermal regime from numerous groundwater sources. The Dunk has been very popular for sport fishing, especially for sea-run trout, but PEI rivers like the Dunk have experienced massive fish kills in recent years, presumably from pesticides entering watercourses during summer rainstorms. It has five small irrigation dams that reflect the importance of agricultural activity in most PEI river valleys, mainly potato farming.

The LaHave River flows in a southeasterly direction, primarily through a forested region in the southeastern portion of Nova Scotia (Fig. 21.20). The LaHave is fed by more than 113 lakes in the basin and has been severely impacted by acid rain deposition, reducing pH to <4.8 in some tributaries during spring snowmelt (Fig. 21.21). This has resulted in poor survival of many aquatic species. Enhancement of Atlantic salmon was initiated in 1970 with construction of a fishway at Morgan Falls (Rkm 25) on the main stem of the LaHave and subsequent stocking in upper reaches.

The Grand Cascapedia River is a southeasterly flowing river on the Gaspé Peninsula of Quebec that empties into Chaleur Bay of the Gulf of St. Lawrence (Fig. 21.22). The basin is mostly forested, with a population density of <1 person/km². There is a particularly strong spring snowmelt in this river, occurring in May (Fig. 21.23). The river is very popular for anglers and is unique for having the highest percentage of large Atlantic salmon (maiden three-sea-winter) of any North American river. A successful alliance of native (with Mi'kmaq) and nonnative partners for river management has seen salmon returns increase tenfold in the past three decades.

The Petitcodiac River flows northeast in the southeastern corner of New Brunswick before emptying into Shepody Bay in the uppermost portion of the Bay of Fundy (Fig. 21.24). Forestry and agriculture predominate in the upper basin, and much of the lower basin is urbanized near the city of Moncton (up to 69 people/km²). The Petitcodiac had a historic tidal bore twice a day, as tides from the Bay of Fundy pushed upriver beyond Moncton, but the size of the bore was much reduced after construction of a causeway in the estuary in 1968. The causeway also caused infilling with sediments and blocked migratory movements of diadromous fishes.

The Humber River is located on insular Newfoundland, just to the west of the Exploits River basin, and flows in the opposite direction, emptying into the Gulf of St. Lawrence (Fig. 21.26). The basin includes Grand Lake, the largest lake in Newfoundland, which is also used for hydropower. The seasonal hydrograph for the Humber is unusually flat (Fig. 21.27), apparently the result of flow releases from Grand Lake. The Humber is popular for sport fishing for Atlantic salmon and brook trout. The headwaters are nearly pristine, part of which are in Gros Morne National Park. Domestic sewage from Deer Lake and other towns causes pollution problems in the lower river. The basin is mostly forested, and trees are harvested throughout the basin, mainly for the pulp and paper mill in Cornerbrook.

The Churchill River (unlike the river of the same name that runs across Central Canada into Hudson Bay, see Chapter 19) flows east across Labrador (Newfoundland) to empty into Lake Melville and eventually the Labrador Sea (Fig. 21.28). The Churchill is the second-largest river in North America that eventually empties into the Atlantic Ocean (by way of the Labrador Sea, and not counting the Koksoak River, Quebec, which flows north into Ungava Bay). This large river drains the Labrador Highlands and Laurentian Highlands physiographic provinces of the Boreal Forest biome. It was mostly undeveloped boreal forest until the Churchill Falls Hydro-Electric project, which created the huge Smallwood Reservoir (see Fig. 21.28) was completed in 1974. This controversial project flooded 6700 km², including Innu hunting lands and burial sites. Much of the power generated by the project is sold by Hydro-Québec to the United States. The governments of Newfoundland and Quebec are currently developing plans for two more power stations on the Churchill River. Without regulation by the Churchill Falls dam, this northern river experienced "spring snowmelt" from May through July (Fig. 21.29). With regulation, however, the spring snowmelt has been totally eliminated, and a uniform seasonal hydrograph has been created.

ACKNOWLEDGMENTS

The authors wish to thank A. Fraser, S. McWilliam, K. Long, M. Gauthier, E. Arseneau, F. Cowie. R. A. Curry, N. Duke, J. O'Keefe, B. Dempson, J. Ritter, J. Deschenes, C. Hudon, R. MacFarlane, K. Teather, D. Guignion, M. Gauthier, D. Caissie, A. Martel, D. MacAlpine, R. Parker, D. Giberson, and C. Collett for contributing data and carrying out research for this chapter, and H. Collins, R. Jones, L. Marshall, M. O'Connell, C. Bourgeois, A. Bielak, and K. Schiefer for kindly reviewing earlier drafts of the manuscript. Funding was provided by the Canada Research Chairs program.

LITERATURE CITED

Abell, R. A., D. M. Olson, E. Dinerstein, P. T. Hurley, J. Diggs, W. Eichbaum, S. Walters, W. Wettengel, T. Allnutt, C. J. Loucks, and P. Hedao. 2000. *Freshwater ecoregions of North America: A conservation assessment.* Island Press, Washington, D.C.

Amiro, P. G. and H. Jansen. 2000. Impact of low-head hydropower generation at Morgans Falls, Lattave River, on migrating Atlantic salmon (*salme salar*). Canadian Technical Report of Fisheries and Aquatic Sciences 2323:25.

Anderson, T. C. 1985. *The rivers of Labrador.* Canadian Special Publication of Fisheries and Aquatic Sciences 81.

Bertrand, P. 1978. La Rivière Moisie. Report prepared for the Ministère de Tourisme, Chasse et Peche (MTCP), Dossier "Rivières à saumon," Direction Regionale de la Côte-Nord, Quebec City, Quebec.

Bielak, A. T. 1984. Quebec North Shore Atlantic salmon stocks. Ph.D. diss., University of Waterloo, Ontario.

Black, G. A., W. L. Montgomery, and F. G. Whoriskey. 1983. Abundance and distribution of *Salmincola edwardsii* (Copepoda) on anadromous brook trout, *Salvelinus fontinalis* (Mitchill) in the Moisie river system, Quebec. *Journal of Fish Biology* 22:567–575.

Blair, S. 2001. The Wolastoq and its people: A study in commemoration. New Brunswick Manuscripts in Archaeology 31. Archaelogical Services of New Brunswick, Fredericton.

Bourgeois, C. E., J. Murray, and V. Mercer. 2000. Status of the Exploits River stock of Atlantic salmon (*Salmo salar* L.) in 1999. Research Doc. 2000/029. Canadian Stock Assessment Secretariat, Ottawa.

Bourgeois, C. E., J. Murray, and V. Mercer. 2001. Status of the Exploits River stock of Atlantic salmon (*Salmo salar* L.) in 2000. Research Doc. 2001/026. Canadian Stock Assessment Secretariat, Ottawa.

Bourgeois, C. E., M. F. O'Connell, and D. C. Scott. 1987. Cold-branding and fin-clipping Atlantic salmon smolts on the Exploits River, Newfoundland. *North American Journal of Fish Management* 7:154–156.

Bousfield, E. L. 1955. Ecological control of the occurrence of barnacles in the Miramichi estuary. *Bulletin of the National Museum of Canada*, Biological Series 137:1–65.

Buckley, D. E. 1995. Sediment and environmental quality of the Miramichi estuary: New perspectives. In E. M. P. Chadwick (ed.), *Water, science and the public: The Miramichi ecosystem*. Canadian Special Publication of Fisheries and Aquatic Sciences 123, pp. 179–190.

Cairns, D. K. (ed.). 2002. *Effects of land use practices on fish, shellfish, and their habitats on Prince Edward Island*. Canadian Technical Report of Fisheries and Aquatic Sciences 2408.

Caissie, D. 2000a. Hydrological conditions for Atlantic salmon rivers in 1999. Canadian Stock Assessment Secretariat (CSAS) Research Document 2000/011Department of Fisheries and Oceans Science, Ottawa.

Caissie, D. 2000b. *Hydrology of the Petitcodiac River basin in New Brunswick*. Canadian Technical Report of Fisheries and Aquatic Sciences, 2301.

Caissie, D., and N. El-Jabi. 1995. Hydrology of the Miramichi River drainage basin. In E. M. P. Chadwick (ed.), *Water, science and the public: The Miramichi ecosystem*. Canadian Special Publication of Fisheries and Aquatic Sciences 123, pp. 83–93.

Carr, J. 2001. *Downstream movements of juvenile Atlantic salmon (Salmo salar) in the dam-impacted St. John River drainage*. Canadian Manuscript Report of Fisheries and Aquatic Sciences 2573.

Chadwick, E. M. P. (ed.). 1995. *Water, science, and the public: The Miramichi ecosystem*. Canadian Special Publication of Fisheries and Aquatic Sciences 123.

Chaput, G. 1995. Temporal distribution, spatial distribution, and abundance of diadromous fish in the Miramichi River watershed. In E. M. P. Chadwick (ed.). *Water, science and the public: The Miramichi ecosystem*. Canadian Special Publication of Fisheries and Aquatic Sciences 123, pp. 121–140.

Chiasson, A. G. 1995. The Miramichi Bay and estuary: An overview. In E. M. P. Chadwick (ed.). *Water, science and the public: The Miramichi ecosystem*. Canadian Special Publication of Fisheries and Aquatic Sciences 123, pp. 11–27.

Cox, P. 1899. Fresh water fishes and Batrachia of the Peninsula of Gaspe, P.Q., and their distribution in the maritime provinces of Canada. *Transactions of the Royal Society of Canada*, 2nd ser., 5(4): 140–154.

Cunjak, R. A. 1995. Addressing forestry impacts in the Catamaran Brook basin: An overview of the prelogging phase. In E. M. P. Chadwick (ed.). *Water, science and the public: The Miramichi ecosystem*. Canadian Special Publication of Fisheries and Aquatic Sciences 123, pp. 191–210.

Cunjak, R. A., D. Caissie, N. El-Jabi. 1990. *The Catamaran Brook habitat research project: Description and general design of study*. Canadian Technical Report of Fisheries and Aquatic Sciences 1751.

Cunjak, R. A., D. Caissie, N. El-Jabi, P. Hardie, J. H. Conlon, T. L. Pollock, D. J. Giberson, and S. Komadina-Douthwright. 1993. *The Catamaran Brook (New Brunswick) habitat research project: Biological, physical, and chemical conditions (1990–1992)*. Canadian Technical Report of Fisheries and Aquatic Sciences 1914.

Cunjak, R. A., T. D. Prowse, and D. L. Parrish. 1998. *Atlantic salmon (Salmo salar) in winter: The season of parr discontent?* Canadian Journal of Fisheries and Aquatic Sciences 55(Supplement 1): 161–180.

Curry, R. A., K. R. Munkittrick, and S. L. Currie. 2001. *Fish community of the St. John River from Fredericton to the Maine Northwoods, 2000*. New Brunswick Cooperative Fish and Wildlife Research Unit Series, Fredericton. Report #02-03.

DFO. 1999. Interaction between wild and farmed Atlantic salmon in the Maritime Provinces. DFO Maritimes Regional Habitat Status Report 99/1E, Department of Fisheries and Oceans, Dartmouth, Nova Scotia.

DFO. 2001a. *Atlantic salmon Maritime provinces overview for 2000*. DFO Science Stock Status Report D3-14(2001), Department of Fisheries and Oceans, Dartmouth, Nova Scotia.

DFO. 2001b. *Gaspereau, maritime provinces overview*. DFO Science Stock Status Report D3-17(2001), Department of Fisheries and Oceans, Dartmouth, Nova Scotia.

Doucett, R. R., G. Power, D. R. Barton, R. J. Drimmie, and R. A. Cunjak. 1996. Stable isotope analysis of nutrient pathways leading to Atlantic salmon. *Canadian Journal of Fisheries and Aquatic Sciences* 53:2058–2066.

Dube, M. G. 2000. Sources of contaminants in the bleached kraft pulping process and their effect on the mummichog (*Fundulus heteroclitus*). Ph.D. diss., University of New Brunswick, St. John.

Dube, M. G., R. Dostie, G. Parent, and J.-P. Jette. 1999. Diagnostic des Evenements Entrainant des Apports de Sediments dans le Reseau Hydrographique du Bassin Versant de la Branche du Lac de la Riviere Cascapedia. Report C-144. Direction de l'environnement forestier, Quebec City, Quebec.

Economic and Policy Analysis Directorate (EPAD). 1997. 1995 survey of recreational fishing in Canada. Economic and Commerical Analysis Report No. 154: 127 p.

Elson, P. F. 1974. Impact of recent economic growth and industrial development on the ecology of Northwest Miramichi Atlantic salmon (*Salmo salar*). *Journal of the Fisheries Research Board of Canada* 31:521–544.

Elson, P. F., A. L. Meister, J. W. Saunders, R. L. Saunders, and V. Zitko. 1973. Impact of chemical pollution on Atlantic salmon in North America. *International Atlantic Salmon Foundation Special Publication* 4:83–110.

Environment Canada. 1982. Impact des activites minieres sur l'ecologie des rivieres aux Pekans et Moisie. Environmental Protection Report Series no. SPE 8-RQ-82-1F. Environment Canada, Quebec region, Ouebec.

Fairchild, W. L., E. O. Swansburg, J. T. Arsenault, and S. B. Brown. 1999. Does an association between pesticide use and subsequent declines in catch of Atlantic salmon (*Salmo salar*) represent a case of endocrine disruption? *Environmental Health Perspectives* 107(5): 349–357.

Flanagan, J. J. 2003. The impacts of fine sediments and variable flow regimes on the habitat and survival of Atlantic salmon (*Salmo salar*) eggs in some New Brunswick streams. Master thesis, University of New Brunswick, Fredericton.

Folster, D. 2001. The soul of a river. In M. Bourgeois and D. Folster (eds.). *Proceedings, Third Canadian River Heritage Conference, NB Department of the Environment Fredericton*, pp. 41–43.

Friends of the Moisie River. 1989. The Moisie River and its giants: Natural resources to preserve. Unpublished report.

Giberson, D. J., and H. L. Garnett. 1995. Distribution and seasonal patterns of aquatic insects in Catamaran Brook, a tributary of the Little Southwest Miramichi River. In E. M. P. Chadwick (ed.). *Water, science and the public: The Miramichi ecosystem*. Canadian Special Publication of Fisheries and Aquatic Sciences 123, pp. 272–273.

Giberson, D. J., and H. L. Garnett. 1996. Species composition, distribution, and summer emergence phenology of stoneflies (Insecta: Plecoptera) from Catamaran Brook, New Brunswick. *Canadian Journal of Zoology* 74:1260–1267.

Gibson, R. J., J. Hammar, and G. Mitchell. 1999. The Star Lake hydroelectric project: An example of the failure of the Canadian Environmental Assessment Act. In P. M. Ryan (ed.). *Assessment and impacts of megaprojects*, pp. 147–176. Proceedings of the 38th Annual Meeting of the Canadian Society of Environmental Biologists in collaboration with the Newfoundland and Labrador Environment Network, St. John's, Newfoundland, Canada, October 1–3, 1998. Canadian Society of Environmental Biologists, Toronto.

Gray, R. W. 1986. *Biological characteristics of Atlantic salmon* (Salmo salar L.) *in the upper LaHave River basin*. Canadian Technical Report of Fisheries and Aquatic Sciences 1437.

Gray, R. W., J. D. Cameron, and E. M. J. Jefferson. 1989. *The LaHave River: Physiography and potential for Atlantic salmon production*. Canadian Technical Report of Fisheries and Aquatic Sciences 1701.

Hamilton, P. B. 1983. The role of the physical environment in the river waterbasin: Direct evidence of physically controlled periphyton productivity and biomass in the Matamek and Moisie River waterbasins, Quebec, Canada. Master thesis, University of Waterloo, Ontario.

Hammar, J., and O. Filipsson. 1985. Ecological testfishing with the Lundgren gillnets of multiple mesh size: The Drottningholm technique modified for Newfoundland Arctic char populations. Report of the Institute of Freshwater Research, Drottningholm 62.

Hanson, J. M., and S. C. Courtenay. 1995. Seasonal abundance and distribution of fishes in the Miramichi estuary In E. M. P. Chadwick (ed.). *Water, science and the public: The Miramichi ecosystem*. Canadian Special Publication of Fisheries and Aquatic Sciences 123, pp. 141–160.

Hudon, C., H. C. Duthie, and B. Paul. 1987. Physiological modifications related to density increase in periphytic assemblages. *Journal of Phycology* 23:393–399.

Hudon, C., H. C. Duthie, S. M. Smith, and S. A. Ditner. 1986. Relationships between lakewater pH and sedimentary diatoms in the Matamek watershed, northeastern Quebec, Canada. *Hydrobiologia* 140:49–65.

Hudon, C., R. Morin, J. Bunch, and R. Harland. 1996. Carbon and nutrient output from the Great Whale River (Hudson Bay) and a comparison with other rivers around Quebec. *Canadian Journal of Fisheries and Aquatic Sciences* 53:1513–1525.

Hutchinson, G. E. 1967. *A treatise on limnology*. Vol. 2: *Introduction to lake biology and the limnoplankton*. John Wiley and Sons, New York.

Jessop, B. M. 1986. *Atlantic salmon* (Salmo salar) *of the Big Salmon River, New Brunswick*. Canadian Technical Report of Fisheries and Aquatic Sciences 1415.

Johnston, C. E. 1980. Observations on the foods of brook trout, *Salvelinus fontinalis*, and rainbow trout, *Salmo gairdneri*, in the Dunk River System, Prince Edward Island. Proceedings Nova Scotia Institute of Science 30:31–40.

Komadina-Douthwright, S. M., D. Caissie, and R. A. Cunjak. 1997. *Winter movement of radio-tagged*

Atlantic salmon (Salmo salar) *kelts in relation to frazil ice in pools of the Miramichi River.* Canadian Technical Report of Fisheries and Aquatic Sciences 2161.

Komadina-Douthwright, S. M., T. Pollock, D. Caissie, R. A. Cunjak, and P. Hardie. 1999. *Water quality of Catamaran Brook and the Little Southwest Miramichi River, N.B. (1990–1996).* Canadian Data Report of Fisheries Aquatic Sciences 1051.

Krank, K. 1988. Miramichi River. In J. H. Marsh (ed.), *The Canadian encyclopedia.* 2nd ed., pp. xx–xx. Hurtig.

Lalonde, Girouard, Letendre and Associates. 1990. Riviere Moisie: Etude du milieu physique. Vol. 1(Rapport de synthese). Rapport presente a la Vice-Presidence Environnement, Hydro-Québec, Quebec City, Quebec.

Locke A. and R. Bernier. 2000. *Annotated bibliography of aquatic biology and habitat of the Petitcodiac River system, New Brunswick.* Canadian Manuscript Report of Fisheries and Aquatic Sciences 2518.

Locke, A., and S. C. Courtenay. 1995. Ichthyoplankton and invertebrate zooplankton of the Miramichi estuary: 1918–1993. In E. M. P. Chadwick (ed.). *Water, science and the public: The Miramichi ecosystem.* Canadian Special Publication of Fisheries and Aquatic Sciences 123, pp. 97–120.

Mackay, R. J. 1995. River and stream ecosystems of Canada. In C. E. Cushing, K. W. Cummins, and G. W. Minshall (eds.). *Ecosystems of the World 22,* pp. 33–60. Elsevier, Amsterdam.

MacGregor, R. 1999. *When the chestnut was in flower.* Plumsweep Press, Landsdowne, Ontario.

Meth, F. F. 1971. Fishes of the St. John estuary. Report 7a prepared for the St. John River Basin Board, Fredericton.

Meth, F. F. 1972. Ecology of the St. John River basin. V: Status of the estuary fisheries. Report prepared for the St. John River Basin Board, Fredericton.

Meth, F. F. 1973. Fishes of the upper and middle St. John River. Report No. 7c, St. John River Basin Board, Fredericton.

Martin, K. 1981. *Watershed red.* Ragweed Press, Charlottetown, Prince Edward Island.

Miramichi River Environmental Assessment Committee (MREAC). 1992. Summary: Final report, 1989–1992. Miramichi River Environmental Assessment Committee, Miramichi.

Morin, R., and R. J. Naiman. 1990. The relation of stream order to fish community dynamics in boreal forest watersheds. *Polskie Archiwum Hydrobiologii* 37:135–150.

Morry, C. J. and L. J. Cole. 1977. Limnology and fish populations of Red Indian lake, a multi-use reservoir. Environment Canada Fisheries and Marine Service, Technical Report 691.

Munkittrick, K. R., R. A. Curry, R. A. Cunjak, L. M. Hewitt, and G. Van Der Kraak. 2000. TSRI progress report (unpublished) for Project 205: Development of a cumulative effects assessment strategy for the St. John River.

Mullins, C. C., and R. R. Claytor. 1989. *Recreational Atlantic salmon catch, 1987 and 1988, and annual summaries, 1973–1988, for West Newfoundland and South Labrador, Gulf Region.* Canadian Data Report of Fisheries and Aquatic Sciences 748.

Naiman, R. J. 1982. Characteristics of sediment and organic carbon export from pristine boreal forest watersheds. *Canadian Journal of Fisheries and Aquatic Sciences* 39:1699–1718.

Naiman, R. J. 1983. The annual pattern and spatial distribution of aquatic oxygen metabolism in boreal forest watersheds. *Ecological Monographs* 53:73–94.

Naiman, R. J., J. M. Melillo, M. A. Lock, T. E. Ford, and S. R. Reice. 1987. Longitudinal patterns of ecosystem processors and community structure in a subarctic river continuum, *Ecology* 68:1139–1156.

National Atlas of Canada. 1974. 4th ed. MacMillan Company of Canada, Ltd., Toronto, Ontario, in association with the Department of Energy, Mines, and Resources and Information Canada, Ottawa.

Native Forest Network. 1997. Nickel deposit found on the North Shore: Moisie River watershed faces a new threat. *Nitassinan News,* 31 January (1997) Burlington, Vermont.

O'Connell, M. F., and C. E. Bourgeois. 1987. Atlantic salmon enhancement in the Exploits River, Newfoundland, 1957–1984. *North American Journal of Fisheries Management* 7:207–214.

O'Connell, M. F., J. P. Davis, and D. C. Scott. 1983. *Stocking of Atlantic salmon fry in the tributaries of the middle Exploits River, Newfoundland.* Canadian Technical Report of Fisheries and Aquatic Sciences 1225.

Peterson, R. H. 1978. Physical characteristics of Atlantic salmon spawning gravel in some New Brunswick streams. Canadian Fisheries Marine Service Technical Report 785.

Power, G. 1981. Stock characteristics and catches of Atlantic salmon (*Salmo salar*) in Quebec, and Newfoundland and Labrador in relation to environmental variables. *Canadian Journal of Fisheries and Aquatic Sciences* 38:1601–1611.

Power, G., G. F. Pope, and B. W. Coad. 1973. Postglacial colonization of the Matamek River, Quebec, by fishes. *Journal of the Fisheries Research Board of Canada* 30:1586–1589.

Randall, R. G., M. F. O'Connell, and E. M. P. Chadwick. 1989. Fish production in two large Atlantic coast rivers: Miramichi and Exploits. In D. P. Dodge (ed.). *Proceedings of the International Large River Symposium (LARS).* Canadian Special Publication of Fisheries and Aquatic Sciences 106, pp. 292–308.

Ricketts, T. H., E. Dinerstein, D. M. Olson, C. J. Loucks, W. Eichbaum, D. DellaSala, K. Kavanagh, P. Hedao, P. T. Hurley, K. M. Carney, R. Abell, and S. Walters. 1999. *Terrestrial ecoregions of North America: A conservation assessment.* Island Press, Washington, D.C.

St. John River Basin Board (SJRBB). 1974. Water use and aquatic ecology. Summary Report S-15. St. John River Basin Board, Fredericton.

Schiefer, K. 1971. Ecology of Atlantic salmon, with special reference to occurrence and abundance of grilse, in North Shore Gulf of St. Lawrence Rivers. Ph.D. diss., University of Waterloo, Ontario.

Schiefer, K. 1972. Salmon of the Quebec North Shore. *Atlantic Salmon Journal* 1:3–10.

Schiefer, K. 1982. A report on Atlantic salmon studies of the Moisie River, 1981. 19pp. International Atlantic Salmon Foundation, St. Andrews, New Brunswick.

Scott, W. B., and E. J. Crossman. 1964. Fishes occurring in the freshwaters of insular Newfoundland. Queens Printer, Ottawa.

Stocek, R. F., P. J. Cronin, and P. D. Seymour. 1999. The muskellunge, *Esox masquinongy*, distribution and biology of a recent addition to the ichthyofauna of New Brunswick. *Canadian Field-Naturalist* 113:230–234.

Stockwell, C. H. (ed.). 1963. *Geology and economic minerals of Canada.* Economic Geology Series no. 1. Geological Survey of Canada, Ottawa.

Symons, P. E. K. 1977. Dispersal and toxicity of the insecticide fenitrothion: Predicting hazards of forest spraying. *Residue Reviews* 38:1–36.

Vannote, R. L., G. W. Minshall, K. W. Cummins, J. R. Sedell, and C. E. Cushing. 1980. The River Continuum Concept. *Canadian Journal of Fisheries and Aquatic Sciences* 37:130–137.

Vollenweider, R. A. 1970. *Scientific fundamentals of the eutrophication of lakes and flowing waters, with particular reference to nitrogen and phosphorous as factors in eutrophication.* Organization for Economic Co-operation and Development, Paris.

Washburn and Gillis Associates, Ltd. 1992. Economic costs of soil erosion in two selected watersheds in Prince Edward Island. Report submitted to Agriculture Canada, Charlottetown, and Environment Canada, Halifax.

Washburn and Gillis Associates, Ltd. 1996. Assessment of Atlantic salmon smolt recruitment in the St. John River. Final Report submitted to SALEN, Inc., Edmundston, New Brunswick.

Watt, W. D. 1973. Aquatic ecology of the St. John River. Vol. 1: Background, design, methods, nutrients, bacteria, phytoplankton and primary production. Report 15f. St. John River Basin Board, Fredericton.

Weeks, E. 1971. *The Moisie salmon club.* Barre Publishers, Barre, Massachusetts.

Wright, E. C. 1944. *The Miramichi: A study of the New Brunswick river and of the people who settled along it.* Tribune Press, Sackville, New Brunswick.

Zitko, V. 1995. Fifty years of research on the Miramichi River. In E. M. P. Chadwick (ed.). *Water, science and the public: The Miramichi ecosystem.* Canadian Special Publication of Fisheries and Aquatic Sciences 123, pp. 29–41.

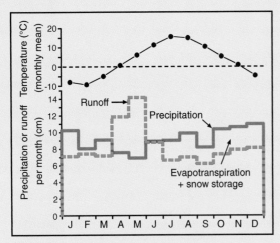

FIGURE 21.9 Mean monthly air temperature, precipitation, and runoff for the Exploits River basin.

EXPLOITS RIVER

Relief: 490 m
Basin area: 11,272 km^2
Mean discharge: 270 m^3/s
River order: 6
Mean annual precipitation: 99.1 cm
Mean air temperature: 4.4°C
Mean water temperature: NA
Physiographic province: New England/Maritime (NE)
Biome: Boreal Forest
Freshwater ecoregion: Lower St. Lawrence
Terrestrial ecoregion: Eastern Canadian Forests
Number of fish species: 6
Number of endangered species: none
Major fishes: Atlantic salmon, brook trout, threespine stickleback, rainbow smelt, American eel, Arctic char
Major other aquatic vertebrates: beaver, river otter, mink
Major benthic invertebrates: in Red Indian Lake, leeches (Hirudinea), sphaeriid clams, chironomid midges, aquatic worms (Oligochaeta); no data on river invertebrates
Nonnative species: NA
Major riparian plants: black spruce, balsam fir, speckled alder, bog laurel, Labrador tea
Special features: large island river with second largest lake in Newfoundland (Red Indian Lake); historic meeting place of Europeans and last of Beothuk peoples at Red Indian Lake; major midwinter ice jams below storage dams caused instant and catastrophic flooding in riverside towns.
Fragmentation: nine hydroelectric and water-storage dams in basin, some with no fish-passage facilities (e.g., Star Lake); water diversion from Victoria River to adjacent river basin
Water quality: very good in upper and middle reaches, but deteriorates in lower reaches; pH = 5.8 to 6.8, dissolved organic carbon = 10 mg/L, total nitrogen = 0.05 to 0.10 mg/L, total phosphorus = 0.006 to 0.012 mg/L
Land use: forestry, base-metal mining (historic) in Buchans area
Population density: 5 people/km^2
Major information sources: www.msc-smc.ec.gc.ca/climate/climate_normals/html/ 2004, www.nationalgeographic.com/ wildworld/terrestrial 2001, Morry and Cole 1977, Gibson et al. 1999, www.gov.nf.ca/env/Env/waterres/Surfacewater/WQI/ Exploits-WQI-study.pdf 2004, http://atlas.gc.ca/site/english/index.html, Scott and Crossman 1964

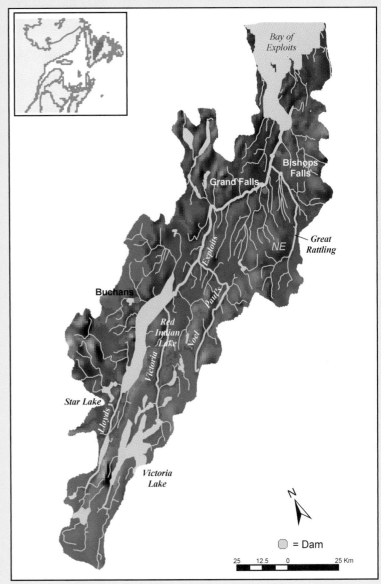

FIGURE 21.8 Map of the Exploits River basin.

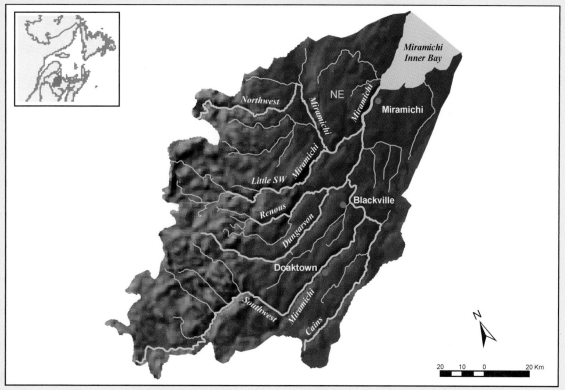

FIGURE 21.10 Map of the Miramichi River basin.

MIRAMICHI RIVER

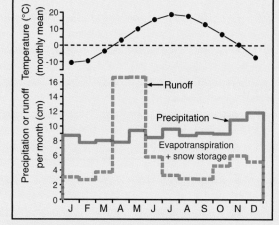

FIGURE 21.11 Mean monthly air temperature, precipitation, and runoff for the Miramichi River basin.

Relief: 764 m
Basin area: 14,000 km²
Mean discharge: 322 m³/s
River order: 8
Mean annual precipitation: 111.5 cm
Mean air temperature: 4.7°C
Mean water temperature: NA
Physiographic province: New England/Maritime (NE)
Biome: Boreal Forest
Freshwater ecoregion: Lower St. Lawrence
Terrestrial ecoregions: New England/Acadian Forests, Gulf of St. Lawrence Lowland Forests, Eastern Canadian Forests
Number of fish species: 21 freshwater, 8 diadromous
Number of endangered species: none
Major fishes: Atlantic salmon, brook trout, sea lamprey, American eel, alewife, American shad, rainbow smelt, striped bass, Atlantic sturgeon, slimy sculpin, blacknose dace, common shiner, lake chub, white sucker, sea lamprey
Major other aquatic vertebrates: beaver, river otter, mink, northern two-lined salamander, double-crested cormorant
Major benthic invertebrates: mayflies (*Baetis, Ephemerella, Stenonema*), stoneflies (*Leuctra, Alloperla, Pteronarcys*), caddisflies (*Hydropsyche, Dolophilodes, Glossosoma, Pycnopsyche*, brachycentrids), true flies (*Prosimulium, Tipula*, chironomid midges), elmid beetles, pearl mussels
Nonnative species: white perch, brown trout, chain pickerel
Major riparian plants: white spruce, white pine, black spruce, yellow birch, northern red oak, sweet gale, speckled alder
Special features: several principal branches and tributaries set in broad glacial river valley; produces more Atlantic salmon than any other river in North America; very popular for anglers and canoeists
Fragmentation: none
Water quality: headwaters and midreaches excellent; pH = 6.5 to 7.8, conductivity = 28 to 55 µS/cm; estuary with urban and industrial discharges; evidence of chemical pollutants in sediments from historic spills
Land use: forests, agriculture on valley bottomlands, historic base-metal mining, localized gravel extraction in lower river valley
Population density: 4 people/km²
Major information sources: www.msc-smc.ec.gc.ca/climate/climate_normals 2004, www.nationalgeographic.com/wildworld/terrestrial.html 2001, http://atlas.gc.ca/english/site/english/index.html, MREAC 1992, Chadwick 1995

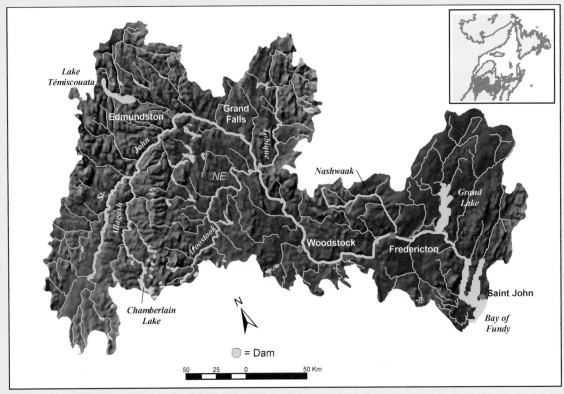

FIGURE 21.12 Map of the St. John River basin.

ST. JOHN RIVER

Relief: 820 m
Basin area: 55,110 km²
Mean discharge: 1110 m³/s
River order: 7
Mean annual precipitation: 114.3 cm
Mean air temperature: 5.3°C
Mean water temperature: NA
Physiographic province: New England/Maritime (NE)
Biome: Temperate Deciduous Forest
Freshwater ecoregion: North Atlantic
Terrestrial ecoregion: New England/Acadian Forests
Number of fish species: 36
Number of endangered species: 1 mussel, 1 plant, 1 fish (special concern)
Major fishes: common shiner, blacknose dace, brown bullhead, slimy sculpin, white sucker, Atlantic salmon, brook trout, striped bass, American eel, alewife, Atlantic sturgeon, shortnose sturgeon, sea lamprey, yellow perch

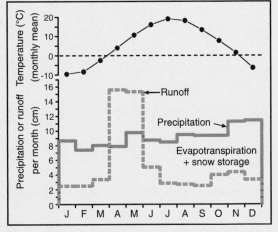

Figure 21.13 Mean monthly air temperature, precipitation, and runoff for the St. John River basin.

Major other aquatic vertebrates: muskrat, beaver, common merganser, double-crested cormorant
Major benthic invertebrates: true flies (chironomid midges, black flies), mayflies (Baetidae, Heptageniidae, Ephemerellidae), stoneflies (Chloroperlidae, Perlidae), caddisflies (Philopotamidae, Hydropsychidae), mollusks (eastern elliptio)
Nonnative species: muskellunge, chain pickerel, brown trout, rainbow trout, smallmouth bass
Major riparian plants: white spruce, white pine, balsam poplar, willows, white birch, northern red oak, speckled alder
Special features: longest river in eastern Canada; international border; "Reversing Falls" near mouth due to large tidal amplitude from Bay of Fundy; extensive marshes in lower river; >100 km under tidal influence
Fragmentation: 11 hydroelectric dams from headwaters to mouth
Water quality: effluents from pulp and paper mills, food processing plants, sewage treatment facilities; pesticide residues in sediments of intensive agriculture areas; mean alkalinity = 51.2 mg/L as CaCO₃, conductivity = 154 μS/cm, pH = 7.7, total nitrogen = 0.47 mg/L
Land use: agriculture (potatoes, beef, and dairy; cranberry farms in lower valley), forests (some harvesting)
Population density: 8 people/km²
Major information sources: www.msc-smc.ec.gc.ca/climate/climate_normals 2004, www.nationalgeographic.com/wildworld/terrestrial.html 2001, http://atlas.gc.ca/site/english/index.html, Meth 1972, Watt 1973, Washburn and Gillis Associates 1996, Curry et al. 2001

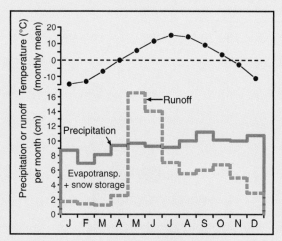

FIGURE 21.15 Mean monthly air temperature, precipitation, and runoff for the Moisie River basin.

MOISIE RIVER

Relief: 1011 m
Basin area: 19,871 km²
Mean discharge: 426 m³/s
River order: 9
Mean annual precipitation: 115.6 cm
Mean air temperature: 1.0°C
Mean water temperature: 6.0°C
Physiographic province: Laurentian Highlands (LU)
Biome: Boreal Forest
Freshwater ecoregion: Lower St. Lawrence
Terrestrial ecoregion: Eastern Canadian Forests
Number of fish species: ≥13
Number of endangered species: none
Major fishes: Atlantic salmon, brook trout, northern pike, white sucker, American eel, lake chub, lake whitefish, threespine stickleback, longnose sucker, burbot, sea lamprey
Major other aquatic vertebrates: beaver, mink, muskrat
Major benthic invertebrates: mollusks (*Pisidium*, *Sphaerium*), stoneflies (*Paracapnia*, *Isoperla*, Leuctridae, Perlidae), mayflies (*Ephemerella*, *Heptagenia*, *Ameletus*), caddisflies (*Lepidostoma*, *Chimarra*, *Hydropsyche*), chironomid midges
Nonnative species: none
Major riparian plants: black spruce, white spruce, balsam fir, white birch, speckled alder, trembling aspen, Labrador tea
Special features: exceptional scenery and wilderness setting; stock of large Atlantic salmon; traditional route to Quebec interior for trapping and hunting for Montagnais Indians; recent designation as aquatic reserve to ensure protection from development
Fragmentation: no dams, although past interest in development of hydroelectric dams and diversions
Water quality: generally excellent; historic mining discharges (1970s) in headwaters (Pékans) caused elevated base-metal concentrations in sediments; pH = 6.3 to 7.5, alkalinity = 8 to 18 mg/L as CaCO₃, dissolved organic carbon = 2.9 to 23.3 mg/L
Land use: >95% forest; some iron mining in headwaters
Population density: <1 person/km²
Major information sources: www.msc-smc.ec.gc.ca/climate/climate_normals 2004, www.nationalgeographic.com/wildworld/terrestrial.html 2001, http://atlas.gc.ca/site/english/index.html, Mackay 1995, Weeks 1971, Naiman et al. 1987, Naiman 1982, 1983

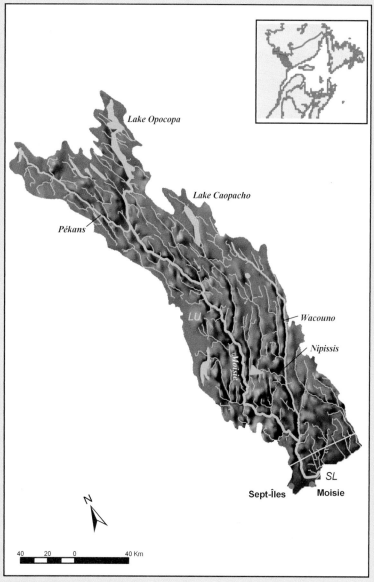

FIGURE 21.14 Map of the Moisie River basin.

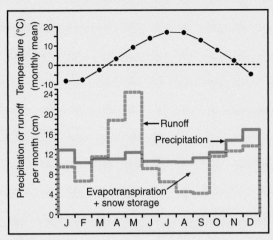

FIGURE 21.17 Mean monthly air temperature, precipitation, and runoff for the Big Salmon River basin.

BIG SALMON RIVER

Relief: 410 m
Basin area: 332 km²
Mean discharge: 11.8 m³/s
River order: 5
Mean annual precipitation: 146 cm
Mean air temperature: 5.3°C
Mean water temperature: NA
Physiographic province: New England/Maritime (NE)
Biome: Temperate Deciduous Forest
Freshwater ecoregion: North Atlantic
Terrestrial ecoregion: New England/Acadian Forests
Number of fish species: 7
Number of endangered species: 1 fish
Major fishes: Atlantic salmon, brook trout, American eel, blacknose dace
Major other aquatic vertebrates: mink, beaver
Major benthic invertebrates: mayflies (Ephemerellidae, Baetidae), stoneflies (Leuctridae), true flies (chironomid midges, black flies)
Nonnative species: rainbow trout, Arctic char (Walton Lake)
Major riparian plants: red spruce, white pine, speckled alder, white birch
Special features: unique wilderness coastal basin of Bay of Fundy with pristine, old-growth forest in deeply incised river valley; historically one of the most productive Inner Fundy salmon rivers, but today few salmon return
Fragmentation: dam near head of tide on main branch removed in 1963; currently small water-level control dams at outlets of several headwater lakes (e.g., Walton Lake)
Water quality: soft-water stream of exceptional quality and clarity; mean pH = 6.3 (range = 4.9 to 8.3), mean conductivity 20.5 to 43.2 μS/cm, mean total hardness 4.5 to 11.8 mg/L as $CaCO_3$
Land use: forest harvest in middle and upper reaches (on plateau); largely pristine, intact forest cover in lower valley; tourist center at river mouth; historic shipbuilding and logging settlement at river mouth (now abandoned)
Population density: <2 people/km²
Major information sources: Jessop 1986, R.A. Curry unpublished data, www.nationalgeographic.com/wildworld/terrestrial.html 2001, http://atlas.gc.ca/site/english/index.html, www.mscsmc.ec.gc.ca/climate/climate_normals 2004, www.cosewic.gc.ca/eng/sct0/index_e.cfm

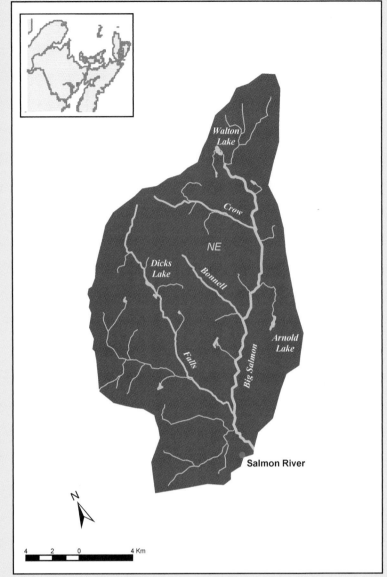

FIGURE 21.16 Map of the Big Salmon River basin.

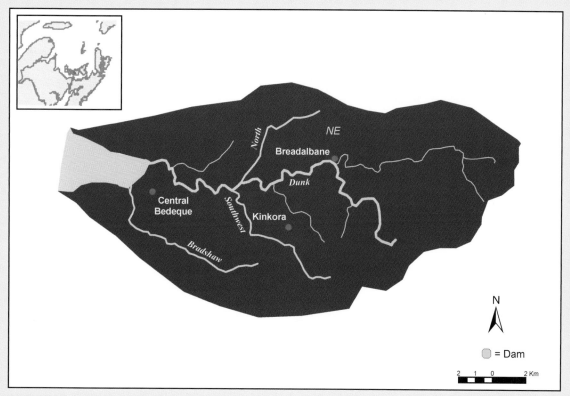

FIGURE 21.18 Map of the Dunk River basin.

DUNK RIVER

Relief: 110 m
Basin area: 217 km^2
Mean discharge: 4.9 m^3/s
River order: 4
Mean annual precipitation: 106 cm
Mean air temperature: 5.5°C
Mean water temperature: NA
Physiographic province: New England/Maritime (NE)
Biome: Boreal Forest
Freshwater ecoregion: Lower St. Lawrence
Terrestrial ecoregion: Gulf of St. Lawrence Lowland Forests
Number of fish species: 8
Number of endangered species: none
Major fishes: brook trout (resident and anadromous forms), Atlantic salmon, American eel, rainbow smelt, alewife
Major other aquatic vertebrates: beaver
Major benthic invertebrates: stoneflies (Perlidae), mayflies (*Baetis*, *Ephemerella*, Heptageniidae), caddisflies (Limnephilidae, Lepidostomatidae, *Rhyacophila*), true flies (Simuliidae, chironomid midges), Elmidae, oligochaetes
Nonnative species: rainbow trout
Major riparian plants: white spruce, red maple, white birch, yellow birch, white pine, speckled alder
Special features: short, low-gradient river with high in-stream productivity on small provincial island; shallow, nutrient-rich estuary and stable thermal regime from groundwater discharge; popular river for sport fishing, especially sea-run trout; unique in that anadromous fishes can access the source pools in headwaters
Fragmentation: five impoundments (locally called ponds)
Water quality: heavy sediment loads (mean soil loss in area estimated at 10 t ha^{-1} yr^{-1}), fertilizer, and pesticide runoff; the latter appears to be responsible for massive fish kills; pH = ~7.8, conductivity = ~210 µS/cm, turbidity = 9.7 to 700 JTU, mean total nitrogen = 3 mg/L, fecal coliforms (MPN) = 139/100 ml
Land use: intensive farming (mainly potatoes) throughout basin; limited forest harvest in headwaters
Population density: 10 to 24.9 persons/km^2
Major information sources: Martin 1981, Washburn and Gillis Associates 1992, Cairns 2002, Johnston 1980, www.nationalgeographic.com/wildworld/terrestrial.html 2001, www.msc-smc.ec.gc.ca/climate/climate_normals 2004

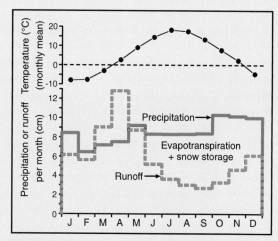

FIGURE 21.19 Mean monthly air temperature, precipitation, and runoff for the Dunk River basin.

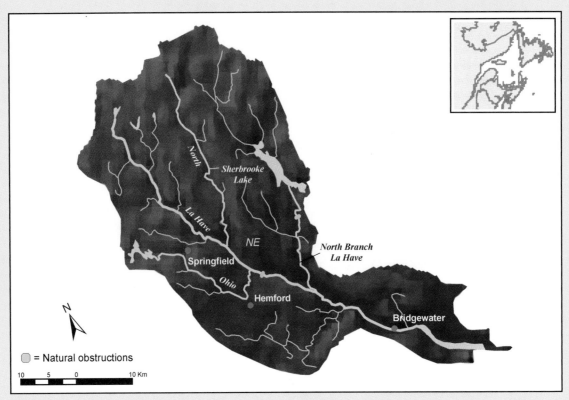

FIGURE 21.20 Map of the LaHave River basin.

LaHAVE RIVER

Relief: 195 m
Basin area: 1668 km²
Mean discharge: 45.9 m³/s
River order: 5
Mean annual precipitation: 147.3 cm
Mean air temperature: 6.7°C
Mean water temperature: NA
Physiographic province: New England/Maritime (NE)
Biome: Temperate Deciduous Forest
Freshwater ecoregion: North Atlantic
Terrestrial ecoregion: New England/Acadian Forests
Number of fish species: 7
Number of endangered species: none
Major fishes: Atlantic salmon, brook trout, alewife, American shad
Major other aquatic vertebrates: beaver, mink
Major benthic invertebrates: NA
Nonnative species: none

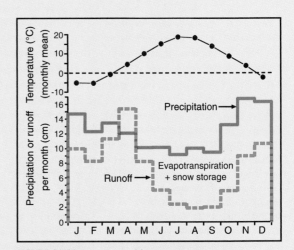

FIGURE 21.21 Mean monthly air temperature, precipitation, and runoff for the LaHave River basin.

Major riparian plants: pin cherry, white spruce, black spruce, red maple, speckled alder, white birch
Special features: river system with >113 lakes; enhancement of Atlantic salmon initiated in 1970 with construction of fishway at Morgan Falls (km 25) on main branch and subsequent annual stocking in upper reaches
Fragmentation: partial barriers at mouth of Sixty Brook and three natural obstructions on North River, many man-made obstructions removed in last 50 years and fishways constructed
Water quality: severely impacted by acid rain, reducing pH to <4.8 in some tributaries during spring snowmelt; transportation artery for oil tankers, heavy-metal pollution in Bridgewater area (estuary); very good quality of upper river but highly colored soft water; pH = 5.8, conductivity = 33.3 µS/cm, total hardness = 5 to 8 mg/L as $CaCO_3$
Land use: forested, with agricultural lands along river valley and shores of tidal zone
Population density: <10 persons/km²
Major information sources: Gray et al. 1989, Gray 1986, Caissie 2000a, Amiro and Jansen 2000, www.nationalgeographic.com/wildworld/terrestrial.html 2001, http://atlas.gc.ca/site/english/index.html, www.msc-smc.ec.gc.ca/climate/climate_normals 2004, www.cosewic.gc.ca/eng/scto/index_e.cfm

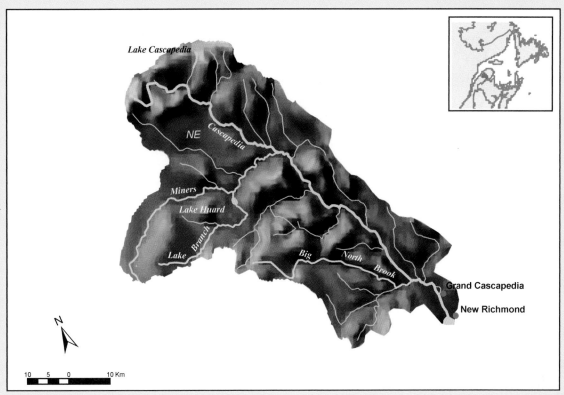

FIGURE 21.22 Map of the Grand Cascapedia River basin.

GRAND CASCAPEDIA RIVER

Relief: 490 m
Basin area: 1480 km²
Mean discharge: 36.1 m³/s
River order: 6
Mean annual precipitation: 101.9 cm
Mean air temperature: 3.7°C
Mean water temperature: NA
Physiographic province: New England/Maritime (NE)
Biome: Boreal Forest
Freshwater ecoregion: Lower St. Lawrence
Terrestrial ecoregion: Eastern Canadian Forests
Number of fish species: 11
Number of endangered species: Simuliidae, Hydropsychidae
Major fishes: Atlantic salmon, brook trout, white sucker, lake chub, slimy sculpin, common shiner, northern redbelly dace
Major other aquatic vertebrates: mink, beaver
Major benthic invertebrates: NA
Nonnative species: rainbow trout
Major riparian plants: black spruce, white pine, white birch, speckled alder
Special features: popular for anglers; unique for having highest percentage of large Atlantic salmon (maiden three-sea-winter) of all North American rivers; native/nonnative alliance has seen salmon returns increased tenfold in past 30 years
Fragmentation: no major dams
Water quality: sedimentation and turbidity from poor logging practices (roads) has affected Lake Branch; lower river and estuary repeatedly dredged and receive municipal effluents and waste from pulp and paper mill; Salmon Branch has best water quality but under threat of forestry impacts
Land use: forestry
Population density: <1 person/km²
Major information sources: Cox 1899, Dube et al. 1999, http://atlas.gc.ca/site/english/index.html, www.nationalgeographic.com/wildworld/terrestrial.html 2001, www.msc-smc.ec.gc.ca/climate/climate_normals 2004, http://magazine.andubon.org/features0111/on_the_brink.html

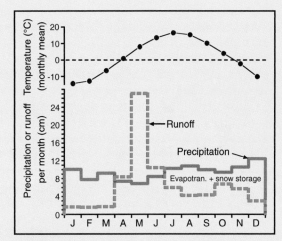

FIGURE 21.23 Mean monthly air temperature, precipitation, and runoff for the Grand Cascapedia River basin.

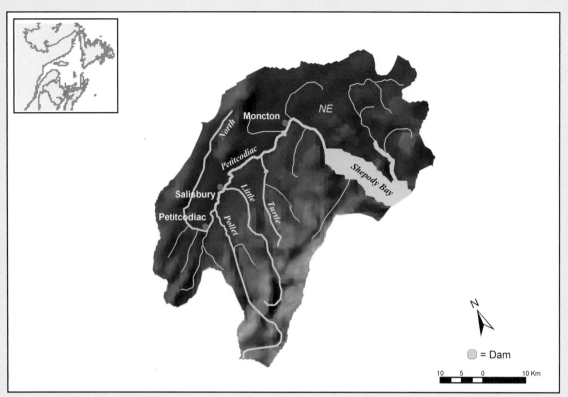

FIGURE 21.24 Map of the Petitcodiac River basin.

PETITCODIAC RIVER

Relief: 415 m
Basin area: 1360 km²
Mean discharge: 27.3 m³/s
River order: 6
Mean annual precipitation: 103 cm
Mean air temperature: 5.5°C
Mean water temperature: NA
Physiographic province: New England/Maritime (NE)
Biome: Temperate Deciduous Forest
Freshwater ecoregion: North Atlantic
Terrestrial ecoregion: New England/Acadian Forests
Number of fish species: 8
Number of endangered species: 1 fish, 1 mussel (extirpated in 1968)
Major fishes: alewife, American eel, brook trout, rainbow smelt, white sucker
Major other aquatic vertebrates: beaver, muskrat
Major benthic invertebrates: stoneflies (*Leuctra*), mayflies (*Baetis, Paraleptophlebia*), caddisflies (*Hydropsyche, Brachycentrus*), true flies (*Simulium, Tanypodinae*), Hydracarina
Nonnative species: brown bullhead, chain pickerel, smallmouth bass
Major riparian plants: willows, white spruce, white pine, speckled alder
Special features: historic tidal bore is formed twice a day as tides from Bay of Fundy push upriver beyond Moncton
Fragmentation: causeway in estuary (at Moncton) causes infilling with sediments (headpond) and blocks migratory movements of diadromous fishes; dam on Turtle Creek for municipal water supply; three abandoned dams in tributaries impede fish passage
Water quality: strong tides carry huge volumes of suspended sediment upstream twice a day; highest natural concentration of suspended sediments in North America (nickname "Chocolate River"); former Moncton landfill on riverbank suspected of leaching toxic chemicals; pH 6.4 to 7.5, conductivity 30 to 100 μs/cm
Land use: lower section urbanized; forestry and agriculture in upper basin (60% of drainage area)
Population density: 10 to 69.9 people/km²
Major information sources: Caissie 2000b, Locke and Bernier 2000, www.cosewic.gc.ca/eng/sct0/index_e.cfm, www.nationalgeographic.com/wildworld/terrestrial.html 2001, www.msc-smc.ec.gc.ca/climate/climate_normals 2004, http://atlas.gc.ca/site/english/index.html, www.petitcodiac.org/riverkeeper 2004

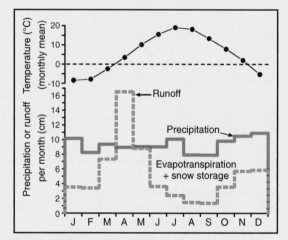

FIGURE 21.25 Mean monthly air temperature, precipitation, and runoff for the Petitcodiac River basin.

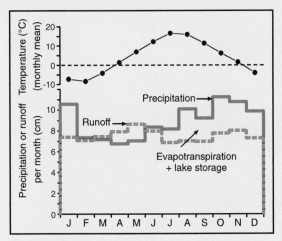

FIGURE 21.27 Mean monthly air temperature, precipitation, and runoff for the Humber River basin.

HUMBER RIVER

Relief: 700 m
Basin area: 7860 km²
Mean discharge: 246 m³/s
River order: 7
Mean annual precipitation: 118.6 cm
Mean air temperature: 5.2°C
Mean water temperature: NA
Physiographic province: New England/Maritime (NE)
Biome: Boreal Forest
Freshwater ecoregion: Lower St. Lawrence
Terrestrial ecoregions: Newfoundland Highland
    Forests, Eastern Canadian Forests
Number of fish species: 7
Number of endangered species: none
Major fishes: Atlantic salmon, brook trout, rainbow
    smelt, American eel, Arctic char, threespine
    stickleback
Major other aquatic vertebrates: mink, river otter,
    beaver

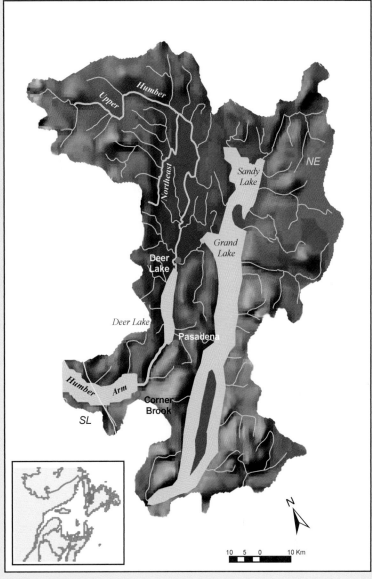

FIGURE 21.26 Map of the Humber River basin.

Major benthic invertebrates: numerous species of stoneflies, mayflies, caddisflies, black flies
Nonnative species: rainbow trout
Major riparian plants: black spruce, white birch, speckled alder, juniper, sweet gale, bog laurel
Special features: popular sport-fishing river for Atlantic salmon and brook trout; Grand Lake is largest lake in Newfoundland;
    landlocked populations of Atlantic salmon (ouananiche) and Arctic char upstream of impassable falls; parts of headwaters
    in Gros Morne National Park and provincial park
Fragmentation: natural obstruction at Main Falls (6.4 m), 113 km from river mouth; North Brook has natural obstruction 14.8
    km from mouth; Grand Lake used for hydro-power; several hydro storage dams in Grand Lake subbasin
Water quality: headwaters nearly pristine; domestic sewage from Deer Lake, Corner Brook, and other towns; pulp and paper
    effluent from Corner Brook
Land use: limited agriculture near Deer Lake; forest (harvest common throughout basin)
Population density: <10 people/km²
Major information sources: Mullins and Claytor 1989, National Atlas of Canada 1974, Scott and Crossman 1964,
    www.msc-smc.ec.gc.ca/climate/climate_normals 2004, www.nationalgeographic.com/wildworld/terrestrial.html 2001

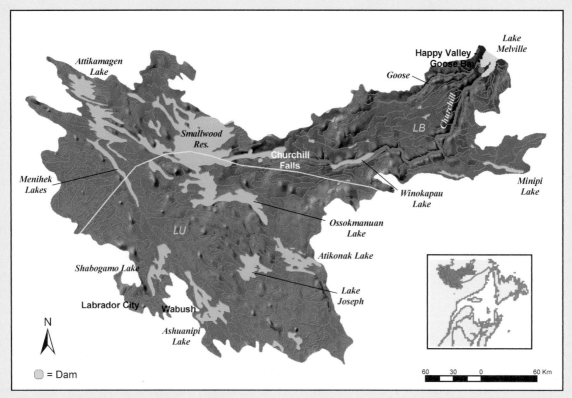

FIGURE 21.28 Map of the Churchill River basin. Physiographic provinces are separated by a yellow line.

CHURCHILL RIVER

Relief: 549 m
Basin area: 93,415 km^2
Mean discharge: 1861 m^3/s
River order: 6
Mean annual precipitation: 94.5 cm
Mean air temperature: −3.5°C
Mean water temperature: NA
Physiographic provinces: Labrador Highlands (LB), Laurentian
　　Highlands (LU)
Biome: Boreal Forest
Freshwater ecoregion: North Atlantic–Ungava
Terrestrial ecoregions: Eastern Canadian Shield Taiga, Eastern
　　Canadian Forests
Number of fish species: 20
Number of endangered species: none
Major fishes: northern pike, lake whitefish, white sucker, brook trout,
　　lake trout, Atlantic salmon (landlocked salmon only above
　　Muskrat Falls), Arctic char, lake chub, threespine stickleback, mottled sculpin, slimy sculpin
Major other aquatic vertebrates: muskrat, beaver, mink
Major benthic invertebrates: NA
Nonnative species: none
Major riparian plants: black spruce, white birch, larch, Labrador tea
Special features: second-largest river (by discharge) in Atlantic drainage; Muskrat Falls, an 8 m waterfall 40 km from mouth, is
　　complete barrier to upstream migration; original Innu name "Mishtashipu" means "big river"
Fragmentation: Churchill Falls Dam, huge hydroelectric project (5400 MW), a major energy provider for northeastern United
　　States; many dams and dykes for hydroelectric project act as unnatural barriers to fish migration
Water quality: generally very good throughout system
Land use: primarily forest; extensive area (6700 km^2) flooded by Smallwood Reservoir upstream of Churchill Falls dam
Population density: <1 person/km^2
Major information sources: Anderson 1985, http://atlas.gc.ca/site/english/index.html, National Geographic
　　2001 http://www.nationalgeographic.com/wildworld/terrestrial.html, Environment Canada 2004
　　http://www.msc-smc.ec.gc.ca/climate/climate_normals, http://www.ccge.org/ccge/english/resources/rivers

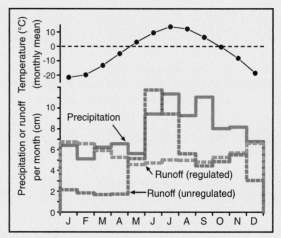

FIGURE 21.29 Mean monthly air temperature,
precipitation, and runoff for the Churchill River basin.

980

22

ST. LAWRENCE RIVER BASIN

JAMES H. THORP   GARY A. LAMBERTI   ANDREW F. CASPER

INTRODUCTION

ST. LAWRENCE RIVER MAIN STEM

OTTAWA RIVER

SAGUENAY RIVER

ST. JOSEPH RIVER

ADDITIONAL RIVERS

ACKNOWLEDGMENTS

LITERATURE CITED

INTRODUCTION

The second-largest river network in North America in annual discharge is the international St. Lawrence River–Great Lakes system. The importance of this river–lake system in times of increasing global shortage of fresh water is hard to overestimate because its basin holds about 23,000 km^3 of water (roughly 18% of the world's freshwater; Fuller et al. 1995). The river's catchment stretches from ~40°N to 50°N latitude and ~65°W to 93°W longitude (Fig. 22.2). Rivers flowing northward from there mostly enter the Labrador Sea or Hudson Bay, whereas immediately southward they pass either into the Atlantic Ocean via mid-Atlantic or northeast coastal rivers or into the Gulf of Mexico via the Mississippi River system. The St. Lawrence River–Great Lakes system forms part of the boundary between Canada and the United States, and in some places it physically divides or links various Indian nations, including the Mohawks in northern New York and southern Canada.

The ~1.6 million km^2 watershed that constitutes the St. Lawrence River–Great Lakes system is divided into three subbasins, with ~47.8% draining into the Great Lakes, 35.7% into the St. Lawrence River main stem, and 16.5% into the estuarine Gulf of St. Lawrence (St. Lawrence Centre 1996). Water in this system can travel at least 3260 km from western Lake Superior to the Cabot Strait in the estuarine Gulf of St. Lawrence. Along this lentic–lotic–estuarine pathway to the sea the river system draws sustenance from nine states (Minnesota, Wisconsin, Illinois, Indiana, Michigan, Ohio, Pennsylvania, Vermont, and New York) and at least two provinces (mostly Ontario and Quebec).

For perhaps 8500 to 9000 years the St. Lawrence River–Great Lakes system has played an important role in the lives of many nations of Native Americans, with Abnake, Algonquin, Huron, Iroquois, Montagnais, Potawatomi, and other groups thriving in this area. Although it is clearly impossible to obtain an accurate human census in this river–lake system prior to the immigration of Europeans, one estimate places the sixteenth-century population of Native Americans around the Great Lakes alone at 60,000 to 117,000 people. The earliest surviving record of the basin's exploration by Europeans dates to 1535, during a period of exploration for the fabled Northwest Passage from the Atlantic to the Pacific. In that year, the French explorer Jacques Cartier happened upon this great river and named it in honor of the coincident feast day of Saint Lawrence. In 1615, the French Explorer Samuel de Champlain or his scout Étienne Brulé were the first recorded Europeans to explore the

FIGURE 22.1 Saguenay River as it passes through the Saguenay Fjord at Cap-Eternite, Quebec (PHOTO BY M. RAUTIO).

983

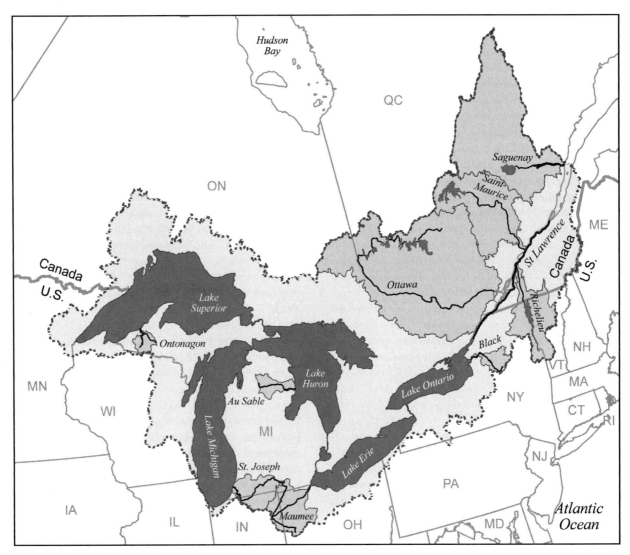

FIGURE 22.2 Rivers of the St. Lawrence River basin covered in this chapter.

Great Lakes portion of this river–lake system. The first permanent European settlement in the St. Lawrence River valley was established by Samuel de Champlain in the first decade of the seventeenth century near the present-day city of Quebec. The river changed hands in 1763 following Britain's victory in the so-called French and Indian War. The movement of mostly Euro-Americans westward into the basin was linked in the early years to the fur trade, with only isolated trading posts and forts impinging on the watershed and aquatic habitats. Environmental effects on the basin were minor until immigrants severely disturbed the watershed and streams by extensively tilling the soil, felling timber, and overharvesting populations of fishes and aquatic mammals.

Although a disproportionate amount of scientific knowledge exists about lacustrine portions of this enormous lake–river system (i.e., on the five major Great Lakes), this chapter focuses on characteristics of the main-stem St. Lawrence River and three of its major tributaries: the Ottawa, Saguenay, and St. Joseph rivers (see Fig. 22.2). Although the St. Lawrence River–Great Lakes system includes many additional tributaries worthy of discussion, abbreviated descriptions of physical and biological information are provided for six representative rivers. These are generally smaller tributaries that flow either into the Great Lakes (AuSable, Black, Maumee, and Ontonagon rivers) or the main stem (Richelieu and Saint-Maurice rivers).

Physiography and Climate

The physiographic nature of the St. Lawrence River–Great Lakes system reflects the action of numerous glacial advances and retreats over tens of thousands of years. Three major physiographic divisions and eight physiographic provinces provide water for this system (Hunt 1974): the Canadian Shield (Superior Upland, Laurentian Highlands, and Labrador Highlands provinces), Appalachian Highlands (New England/Maritime, Adirondack, St. Lawrence Lowlands, and Appalachian Plateaus provinces), and Interior Plains (Central Lowland province). The Great Lakes watershed, which is quite small compared to the volume of water retained by the lakes, is influenced primarily by the Superior Upland and Central Lowland provinces and to a lesser extent by the Appalachian Plateaus, Adirondack, and Laurentian Highlands provinces. Tributaries directly entering the freshwater and estuarine portions of the St. Lawrence River are derived mostly from the St. Lawrence Lowlands and Laurentian Highlands provinces, with additional flow from the Adirondack, New England/Maritime, and Labrador Highlands provinces.

The St. Lawrence River–Great Lakes system stretches climatically from the continental weather zone at its western edge to the oceanic zone in its eastern edge. The Great Lakes and Atlantic Ocean greatly affect precipitation patterns and dampen temperature fluctuations within this broad geographic area. This temperate region is characterized by mild, humid summers and cold snowy winters. Within the St. Lawrence–Great Lakes basin the annual precipitation (56 to 111 cm) varies considerably with a site's location upwind or downwind of a major lake, but half the precipitation entering the Great Lakes is lost to the atmosphere via transpiration and evaporation before it enters the St. Lawrence River. Summer temperatures of tributaries emptying into the Great Lakes are influenced by both cool dry air masses from the Canadian Northwest and warm moister air moving northward from the Gulf of Mexico. Around the main stem and tributaries of the St. Lawrence River these two air masses interact in the summer, with moist air flowing from the northwest Atlantic. Winter temperatures are more strongly influenced over the lakes by Arctic winds but near the sea coast by ocean winds as well. Snow averages 208 cm per year (0 to 54.6 cm/month) in the basin; this generally starts in late October, reaches a maximum in December and January, and tapers off in April. Snowfall is greatest in the lee of each of the Great Lakes.

Basin Landscape and Land Use

The two principal terrestrial biomes within the St. Lawrence River–Great Lakes system are the Temperate Deciduous Forest and the Boreal Forest, although grassland and savannah transition areas appear around major sections of Lake Michigan. Moving more-or-less westerly from the seacoast, nine terrestrial ecoregions are present in this basin: Eastern Canadian Forests, New England/Acadian Forests, Eastern Forest/Boreal Transition, Eastern Great Lakes Lowland Forests, Southern Great Lakes Forests, Western Great Lakes Forests, Central Canadian Shield Forests, Central Forest/Grassland Transition Zone, and Upper Midwest Forest/Savannah Transition Zone (Ricketts et al. 1999).

From a topographic perspective, the St. Lawrence River–Great Lakes system is relatively flat in the western half but rises to nearly 2000 m in the more mountainous eastern half of the basin. The latter includes northern portions of the Adirondack and Appalachian Mountains located southeast of the St. Lawrence River in New York and Quebec, as well as scattered mountainous areas northwest of the river in Quebec.

Nearly 60 million people live in the St. Lawrence River–Great Lakes system, with most individuals clustered within 100 km of the waterfront. The average population density in the basin is about 54 people/km^2, with a high of 97 people/km^2 in the Lake Ontario subbasin. Sixty cities in this watershed have populations of at least 100,000 people. Human densities fall off considerably in the northern Great Lakes regions and within the watershed for the main stem of the St. Lawrence River (except for cities such as Montreal, Ottawa, and Quebec).

In terms of land use and land cover, the southern portion of the Great Lakes basin is highly urbanized and industrialized, especially around Lakes Michigan, Erie, and Ontario. In contrast, northern areas of the Great Lakes and most of the watershed of the St. Lawrence River are much less developed except within ~100 km of the river. For the St. Lawrence River–Great Lakes system as a whole, the watershed is 55% forested, 20% cropland (with little or no irrigation), 22% urban, and 3% other types of land cover (Revenga et al. 1998).

The Rivers

The St. Lawrence River–Great Lakes system is nestled within the St. Lawrence Complex of the Arctic–Atlantic Bioregion, which contains five freshwater ecoregions (Abell et al. 2000). Moving westerly from the seacoast, these are the Lower St. Lawrence, Ontario, Erie, Michigan–Huron, and Superior freshwater ecoregions. The Lower St. Lawrence, which includes the main-stem river and all its tributaries below Lake Ontario, is considered continentally outstanding in terms of biological distinctiveness (Abell et al. 2000). Besides the St. Lawrence main stem, other rivers of the lower St. Lawrence ecoregion described in this chapter are the Ottawa, Saint-Maurice, Richelieu, and Saguenay. The continentally outstanding biodiversity of this river is significantly influenced by the ocean and Great Lakes as well as by the river's position on major migratory pathways for shorebirds, waterfowl, and raptors. Aside from climatic influences, the nearby Atlantic Ocean also contributes migratory marine fishes and mammals to the river, the latter especially in the lower, and to some extent upper, estuaries.

This chapter includes at least one river within each of the freshwater ecoregions that drain into the Great Lakes (see Fig. 22.2). The Ontario ecoregion is considered nationally important in biological distinctiveness (Abell et al. 2000) and includes the Black River, which drains into the eastern end of Lake Ontario from the Adirondack Mountains. The Erie ecoregion is also listed as nationally important and includes the Maumee River, which drains from northeastern Indiana and northwestern Ohio and flows into the western end of Lake Erie. The Michigan–Huron ecoregion is regarded as continentally outstanding and includes the AuSable River, which flows into Lake Huron from the west, and the St. Joseph River, which flows into southern Lake Michigan. Finally, the Superior ecoregion is listed as continentally outstanding and is represented by the Ontonagon River, which flows into Lake Superior from the upper Michigan peninsula. In addition, seven internationally designated RAMSAR wetland sites occur within the St. Lawrence–Great Lakes system.

Most of the tributary rivers of the St. Lawrence basin are small to medium size (mean discharge <200 m^3), particularly those that flow into the Great Lakes from the Central Lowland and the Superior Upland physiographic provinces. Flowing into the St. Lawrence main stem, however, are some large rivers (mean discharge >500 m^3/s), particularly the

Saguenay, Ottawa, and Saint-Maurice, which drain the Canadian Shield (see Fig. 22.2). These large rivers, as well as many smaller tributaries entering the St. Lawrence below Lake Ontario, contribute more than half the freshwater flow into the Gulf of St. Lawrence, even though they drain less than half the total basin area.

The human impacts on rivers within the St. Lawrence basin are great. Within this area are 11 large dams built for navigation, hydroelectric production, and/or flood control. Hundreds of small dams are scattered throughout the watershed, especially on Adirondack rivers, such as the Raquette River of northern New York. Pollution impacts are serious in agricultural areas of the Great Lakes freshwater ecoregions, and below cities, particularly the large cities on the St. Lawrence main stem. Recently, humans have affected the St. Lawrence and some of its tributaries through the pernicious effects of acid rain. Most rivers in the northern St. Lawrence basin originate or flow through Precambrian Shield geologic formations, which have poor buffering capacity because of generally low alkalinity and conductivity. Sensitive rivers of the St. Lawrence basin that have been studied include the Saguenay, Ottawa, and St. Maurice, and ecotoxicological effects of progressive acidification on salmon populations are well characterized in these rivers.

ST. LAWRENCE RIVER MAIN STEM

The St. Lawrence River is arguably among the world's most unique rivers, in part because about half its discharge originates from the huge Laurentian Great Lakes and their many tributaries. This river is also known as Rivière Saint-Laurent among the Francophones of Quebec and as Kaniatarowanenneh (approximately translated as "the big waterway") in the Iroquois language. It ranks among the top 16 rivers of the world in annual discharge and has the second-highest flow in North America (if the Saguenay is included). Hydrological contributions of the voluminous Great Lakes produce a river characterized by clearer water and a more stable stage level than any other large river in North America (Fig. 22.3). Indeed, the St. Lawrence transports the smallest amount of suspended sediment of the world's major rivers (Gleick 1993), a characteristic that undoubtedly affects primary productivity and food web relationships. The main portion of the river is listed here as 8th order, but this technically increases to 9th order at its confluence with the Saguenay River

FIGURE 22.3 St. Lawrence River at Thousand Islands Bridge near Ivy Lea, Ontario (PHOTO BY J. M. FARRELL, THOUSANDS ISLANDS BIOLOGICAL STATION).

at the downstream terminus of the upper estuary. However, this size designation has little utility in comparing rivers because the presence of the Great Lakes makes this river vastly larger than any other 8th or 9th order river in North America.

The main stem flows 965 km along a northeasterly path from the outlet of Lake Ontario to the mouth of the Saguenay River, which is the approximate downstream terminus of the upper estuary (Fig. 22.8). Along this pathway the river is bounded solely by Quebec, Ontario, and New York (in order of shoreline length). Another >300 km of the St. Lawrence River system constitutes the lower estuary, a section not covered in this chapter.

Although archaeological evidence indicates that Native Americans have lived in the St. Lawrence River basin for thousands of years, the earliest surviving archaeological record of construction of a longhouse dates to 1100, with maize agriculture entering this area in perhaps the fourteenth century. Many First Nations were present in this area, but the two dominant groups were tribes of the Algonquin Nation and the Confederation of the Haudenosaunee, a group now more commonly known as the Iroquois. The Haudenosaunee Confederation was established at least as early as 1570 and certainly prior to any significant contact with Europeans. By 1660 there were perhaps 25,000

Haudenosaunee living in the watershed of the main stem of the St. Lawrence River and many more elsewhere. This widely influential confederation was formed originally by five nations (Cayuga, Mohawk, Oneida, Onondaga, and Seneca) and expanded to six in 1722 with the addition of the Tuscaroras. The St. Lawrence River was vital to the culture, transportation, and livelihood of most Native Americans living near its waters, but Native Americans are thought to have had little impact on the aquatic ecosystem.

Physiography, Climate, and Land Use

The shoreline along the river's nearly 1000 km main stem and the basin's contributions to water chemistry are influenced by two major physiographic divisions: the Precambrian Shield and the Appalachian Highlands, although the river as a whole is affected by three major physiographic divisions and eight physiographic provinces because it draws water from watersheds around the Great Lakes. The huge Precambrian (or Canadian) Shield is located mostly west and north of the river, whereas the Appalachian Highlands division is present east and south of the river (see Fig. 22.8). The freshwater, nontidal main stem of the St. Lawrence is nestled primarily in the small physiographic province of the St. Lawrence Lowlands (SL) (Appalachian Highlands) and secondarily in the Laurentian Highlands (LU) (Precambrian Shield) and New England/Maritime (NE) province (Appalachian Highlands), but the river receives lesser flow through the Adirondack (AD) (Appalachian Highlands) and Labrador Highlands (LB) provinces (Precambrian Shield) (Hunt 1974). The geology of the Precambrian Shield is dominated by silicate rocks and includes metasediments (e.g., quartzite, crystalline limestone), metavolcanics, igneous rocks (e.g., granite), orthogneiss, and pegmatite. Found in the St. Lawrence Lowlands are formations of early Paleozoic rock (Cambrian and Ordovician) lying on Precambrian bedrock, with a thick horizontal deposit of sandstone, limestone, dolomite, and shale. The remaining provinces in the Appalachian Highlands are composed of sedimentary and volcanic (igneous) rock that underwent complex metamorphic deformation during the Paleozoic era. In summary, the upper catchment is dominated by silicate rocks of the Precambrian Shield, whereas downstream portions of the basin contribute large amounts of carbonates from the Paleozoic lowlands (Yang et al. 1996). Bedrock outcrops in the landscape surrounding the St. Lawrence River blend with flat plains, some rolling hills, and the ancient Adirondack Mountains southeast of the river.

Temperatures and precipitation in the valley of the St. Lawrence River are more stable than for rivers situated farther inland because of the influences of the Great Lakes and ocean. Moreover, precipitation is spread relatively evenly throughout the year (Fig. 22.9). Montreal averages 94.2 cm of precipitation per year (snow converted to rain equivalence) and has a monthly range of 6.6 cm in February to 9.1 cm in August. The historical average temperature is 6.2°C to 6.8°C, with mean monthly temperature ranging from −10°C in January to 21°C in July (see Fig. 22.9). As a result of the seasonal temperature pattern, the river is at least partially ice covered for most of the winter, especially in the fluvial lake regions. In Massena, New York (near the end of the international portion of the river), the main channel is covered with ice by late December to late January, snowfall averages 178 cm/yr, and the crop-growing season is about 139 days.

The St. Lawrence River occurs predominately in two terrestrial ecoregions. These consist of the Eastern Great Lakes Lowland Forests, through which the river cuts, and the Eastern Forest/Boreal Transition, from which the largest tributaries primarily originate. The freshwater portion of the river is also marginally affected by two other terrestrial ecoregions: the New England/Acadian Forests and the Eastern Canadian Forests. Forest communities are quite variable in the river's watershed (Ricketts et al. 1999) because this area is a geographical and ecological transition area between boreal coniferous forests and more temperate broadleaf deciduous zones.

A major portion of the Eastern Great Lakes Lowland Forests is dominated by eastern hemlock and pine, with a few species of deciduous trees: yellow birch, sugar maple, red maple, red oak, and American beech. The remainder of the ecoregion is a mosaic of deciduous stands in favorable habitats, with good soils contrasting with coniferous assemblages in less-favorable habitats with poorer soils. Drier sites feature red oak, red pine, white pine, and eastern white cedar. Some of these cedars have been aged at 700 to 800 years, making them among the oldest trees in eastern North America. Moister habitats contain red maple, elm, cottonwood, and various ashes. Wetland plant communities are abundant in the many poorly drained depressions, and small, occasionally deep lakes are present.

The Eastern Forest/Boreal Transition ecoregion is a mixed-wood forest dominated by white spruce, balsam fir, paper birch, and yellow birch (Ricketts et al. 1999). Pines are more common in drier northwestern areas. Common in southern Algonquin areas are stands of sugar maple, yellow birch, eastern hemlock, eastern white pine, and some beech. Poorly drained regions frequently have tamarack and eastern white cedar in particular.

Plant communities within the Great Lakes Lowland Forest ecoregion are highly fragmented, as no intact blocks >250 km^2 exist, and <5% of the total ecoregion is considered intact. In the surrounding Eastern Forest/Boreal Transition ecoregion, however, intact habitats remain in ~10% of the ecoregion, especially in Adirondack Park south of the river in New York.

Today millions of people in Canada and the United States live near the banks of the St. Lawrence River and benefit from its economic, municipal, and cultural attributes. For example, the 3.4 million people residing in Montreal and surrounding suburbs represent ~50% of the entire population of Quebec (www12.statcan.ca/english/census01/home/index. cfm 2001). Other large Canadian cities are Ottawa and Quebec City, at 1.06 and 0.68 million, respectively. No cities with populations of 100,000 or more occur in the United States along the river's main stem. As one moves downstream from Lake Ontario, population densities decline, especially in the United States. Agriculture becomes relatively more important in land cover, but high-density pockets of human population still exist.

Farming in the St. Lawrence Valley is still widespread in Canada but continues to decline in the United States as many dairy farmers move south, allowing former farmland to undergo reforestation. This trend has resulted in an increase in the percentage of forested lands in New York State over the last half century from ~37% to 60% (Alerich and Drake 1993). Likewise, Quebec experienced a 41% decline in the surface area devoted to agriculture from 1961 to 1991, but agriculture activities intensified on the remaining land. Beaver populations have rebounded from almost total extirpation, though trapping continues at a lower pace.

River Geomorphology, Hydrology, and Chemistry

The main stem of the St. Lawrence begins at the outlet of Lake Ontario and flows through four sections before reaching the Gulf of St. Lawrence (St. Lawrence Centre 1996): fluvial section, fluvial estuary, upper estuary (including the Saguenay River), and lower estuary (not discussed in this chapter). From Lake Ontario to the end of the lower estuary the river drops about 184 m, for an average of ~14 cm/km. The maximum topographic relief in the basin of the St. Lawrence main stem is 1270 m, but this occurs near the downstream end of the upper estuary. Many shoreline areas of the main stem feature large boulders, but the channel is composed primarily of sand and gravel except where the natural falls formerly occurred.

The 655 km long fluvial section, which extends from Lake Ontario through the international portion of the river to Cornwall-Massena in Quebec, includes uppermost braided regions, constricted channels, rapids (now mostly bypassed with navigation locks), modest floodplain areas, and four natural fluvial lakes, often widened and deepened by hydroelectric dams. Many of the latter are in the upper river in a popular tourist area called the "Thousand Island" region (see Fig. 22.3). A shipping channel with a minimum depth of 8.2 m is maintained throughout this section of the river. This channel is deeper than the 3 m minimum navigation depth maintained for many other navigable rivers in the United States because oceangoing ships rather than just barges traverse the St. Lawrence Seaway. In the fluvial section, the river varies from 1 to 2 km wide in constricted reaches to an average of 12 km wide in some fluvial lakes.

The 160 km long fluvial estuary extends from the downstream terminus of Lac Saint-Pierre to the eastern tip of the Île d'Orléans (~45 km downstream of Quebec City). Tidal forces increase gradually through this section, but the water remains fresh until the Île d'Orléans, where salinities rise to 2 PSU. The 150 km upper estuary, which runs from the eastern tip of the Île d'Orléans to the mouth of the Saguenay River, is the principal transition area between the freshwater and saltwater environments. Salinities range across this section from 2 PSU at the upstream end to 30 PSU at the confluence of the St. Lawrence and Saguenay rivers. Tidal forces are strong here (mean tidal range 4.1 m at Quebec City) and the river is wide (mean 17 km) and up to 100 m deep in places.

In contrast to the pool and riffle-run sequence in many rivers of North America, the natural fluvial lakes of the St. Lawrence River (i.e., Lake St. Lawrence and Lacs Saint-François, Saint-Louis, and Saint-Pierre) are relatively shallow (80% of the area <6 m) in comparison to the generally deeper, main riverine channels (often 10 to 12 m deep). These wide, shallow pools were present prior to construction of the hydroelectric and diversion dams, but the dams have modified the size and depth of these more lacustrine portions of the river. Because of their shallow nature, fluvial lakes support broad expanses of submerged vascular macrophytes and abundant populations of fishes, benthic invertebrates, and plankton. Fluvial lakes of the St. Lawrence should not be confused with typical lentic systems, as the water residence times in portions of fluvial lakes can be quite short (e.g., ~1.5 days in the main channel of Lac Saint-Pierre versus 2 to 3 weeks in north and south littoral areas; Jean Morin, personal communication).

Annual discharge of the St. Lawrence averages 12,101 m³/s at Quebec City (1962 to 1988) but this rises to 16,800 m³/s downstream of the confluence with the Saguenay River. The outflow of Lake Ontario averages 7,410 m³/s. Consequently, the Great Lakes contribute ~61% of the water reaching Quebec City and just under half the freshwater entering the Gulf of St. Lawrence. Most tributary inputs occur downstream of the international portion of the river. For example, at Massena, New York, and Cornwall, Ontario, which are very near the downstream end of the international section of the river, 99% of the discharge originates from Lake Ontario.

The longest-term discharge record for the St. Lawrence is at Massena–Cornwall, where peak discharge occurs from April through June in response to snowmelt, and minimum flows are in January. However, because this fluvial section of the river is so strongly influenced by discharge from Lake Ontario, it fluctuates to a very minor degree compared to other large rivers in North America. Indeed, the average maximum daily discharge from 1900 to 1989 was 11,337 m³/s at Cornwall, less than three times the minimum daily flow of 4170 m³/s over the same period. In a normal year, the minimum flow is only 19.5% less than the maximum flow in the fluvial section of the river, and the difference between average flow and peak flow is only 6.4%. This low variability is due mostly to the stabilizing influence of the high water volume held by the Great Lakes, where estimated water residence times for the five major lakes vary from 2.6 to 117 years. Runoff is relatively low and, like precipitation, is consistent among months (2.12 to 2.65 cm/mo) compared to many other rivers throughout North America (see Fig. 22.9). Farther downstream the river is more affected by seasonal flows and thus monthly variability increases. Indeed, at Quebec City, near the end of the fluvial estuary, the maximum monthly flow over the period from 1962 to 1988 (Bourgault and Koutitonsky 1999) was 62.2% higher than the minimum flow and 40.7% higher than the mean flow.

Because of the disparate nature of the sources of its water, the St. Lawrence is more poorly mixed from bank to bank than one would expect from a large river. Consequently, clear water from the Great Lakes tends to occupy the central portion of the river channel, whereas the more sediment-laden water from tributaries in Canada and the United States hugs their respective shorelines for up to 100 km. Variability in depth, current velocity, flow patterns, and water origin also influence the size of bottom sediments, causing the bottom to vary within short distances from boulders to gravel, sand, or clay.

Water chemistry in the St. Lawrence is strongly controlled by the geology of the drainage basin for each tributary, with the Great Lakes exerting the strongest influence on the river as a whole. As one moves downstream tributaries have an increasing effect on water chemistry in the main stem, and the chemical character of the water fluctuates more seasonally. Tributaries draining the Precambrian Shield are relatively pristine because of their location in areas of low concentrations of humans, industrial activity, and agriculture. They are notably low in total dissolved solids, bicarbonates, sulfates, calcium, magnesium, nitrates, and strontium but are high in silicon and total iron (Yang et al. 1996). Tributaries draining Paleozoic lowlands, with their limestone, dolostones, and evaporites, are characterized by opposite chemical attributes. They are typically in areas of high human impact. Near the terminus of the international portion of the river, historical mean water quality values are as follows: dissolved O_2 8.6 mg/L, alkalinity 90 mg/L as $CaCO_3$, hardness 130 mg/L as $CaCO_3$, pH 7.5 to 8.5, total dissolved solids 153 to 210 mg/L, conductivity 272 μS/cm, total suspended solids <5 mg/L, turbidity <52 NTU, and Secchi disk transparency 7.6 to 10.7 m (New York Power Authority 1996).

Nutrient and bacterial pollution in the St. Lawrence have been decreasing since the mid-1980s (New York Power Authority 1996). Mean nutrient concentrations now are as follows: NH_4^+-N 0.01 to 0.04 mg/L, total Kjeldahl N 0.18 mg/L, and PO_4-P 0.010 mg/L. Fecal and total coliform values were <4/100 mL and 388/100 mL of water, respectively, all much better than the New York Department of Environmental Conservation water-quality standards (200/100 mL and 2400/100 mL, respectively). Phosphorus appears to limit primary productivity in the St. Lawrence River.

River Biodiversity and Ecology

The Lower St. Lawrence freshwater ecoregion, which includes the main stem of the St. Lawrence River, is listed as continentally outstanding (Abell et al. 2000) because of its biological distinctiveness; Lac Saint-Pierre is a RAMSAR World Heritage Site based on its extreme diversity of wetland plants and habitat for fishes and waterfowl. Despite the outstanding nature of the St. Lawrence River fauna and flora, however, published studies on the ecology of the freshwater main stem seem relatively rare compared to the literature on other very large rivers like the Mississippi, and most of that is by Canadian scientists. Indeed, the St. Lawrence is not even mentioned in the index of Ruth Patrick's (1996) book on rivers of the eastern United States. Although the gray literature is abundant overall because of the activities of Environment Canada (especially in Quebec), both refereed and gray literature are minuscule in the United States. From the mostly Canadian literature (especially the excellent report by the St. Lawrence Centre [1996]), however, we can piece together a view of the ecology of the St. Lawrence River.

Algae and Cyanobacteria

The nature of phytoplankton assemblages in the St. Lawrence varies with season, tributary input, and position relative to fluvial lakes and other riverine habitats (e.g., Basu et al. 2000a, 2000b). Seasonal environmental changes appear more important to controlling phytoplankton community composition than changes in river conditions related to tributary inputs. Diatoms in particular, but also Crypto-phyceae, are abundant year-round, with Chloro-phyceae becoming important in the summer (Hudon et al. 1996). Several authors have noted a progressive downstream decrease in the total biomass and average individual size of phytoplankton along with changes in species composition that are related to distances from Lake Ontario. However, this trend can be reversed by tributaries and the presence of fluvial lakes. For example, phytoplankton assemblages below major tributaries, such as the Ottawa River, are often taxonomically richer and have a different relative species composition than those downstream of Lake Ontario, and individuals are likewise characterized by greater biomass, cell volume, and chlorophyll *a* content. Below fluvial lakes, resuspended periphyton adds to the phytoplankton assemblage.

Studies of benthic microalgae have concentrated on the fluvial section of the river and have shown that species composition and biomass vary over much smaller spatial scales than has been noted for planktonic microalgae, thus contributing to a relatively homogenous community throughout broad areas of the fluvial section (St. Lawrence Centre 1996). Periphyton biomass sampled on glass slides were higher in the St. Lawrence in clear "green" waters derived mostly from the Great Lakes than in the more turbid "brown" water entering from tributaries like the Ottawa River (Ahmad et al. 1974). In the summer, aborescent Chlorophycea (*Cladophora* sp.) are dominant at many sites, and this filamentous alga is often colonized by diatoms (e.g., *Rhoicosphenia curvata*) and Cyanophycea (e.g., *Lyngbya* sp.). Ice scouring in winter and the senescence of vascular plants (as substrates for epiphytic algae) contribute to strong seasonal fluctuations in abundance and species composition of benthic microalgae.

Plants

Species composition and abundance of aquatic vascular plants varies greatly with river section, and freshwater wetlands as a whole are high in species endemism. Within the 240 km fluvial section of the river downstream from Cornwall, Ontario, are about 30,260 ha of macrophyte beds and 12,600 ha of marshes (from Table 2.4 in St. Lawrence Centre 1996; after Gratton and Dubreuil 1990). The fluvial estuary and upper estuary contain no significant populations of submerged macrophytes but have about 5500 and 2130 ha of marshes, respectively. The dominant plants are submerged vascular macrophytes, American bulrushes, and saltmarsh cordgrass (mostly smooth cordgrass) in the fluvial section, fluvial estuary, and upper estuary, respectively. About 98% of the aquatic macrophytes in the fluvial section are submerged species, and most of the remainder have floating leaves and occur in slower waters. The dominant submerged plant is the American wild celery, which averages about 100 to 220 g/m^2 in some fluvial lakes and wetlands. Some protected habitats in the freshwater sections of the river contain abundant populations of pondweeds (clasping leaf pondweed), duckweeds (star duckweed), hornleaf riverweed, and bur-reed (branching bur-reed); mostly sedges (hairy sedge); or a combination of pond-lily, bulrushes (slender bulrush), reed manna grass, and greater yellow-cress interspersed with diamond willow and silver maple trees.

Growth of aquatic vascular plants varies considerably with river levels and season. For example, the production of plants in the fluvial Lac Saint-Pierre more than doubled in years with low summer river levels compared to years of high river stages (Hudon 1997). Moreover, ice scouring and cold temperatures significantly restrict growing seasons. Light penetration, wind–wave energy, current velocity, and physical substrate characteristics seem to control species composition and biomass (Morin et al. 2000). Water-level stabilization from damming has both favored the growth of submerged plants over emergents and roughly doubled their biomass.

Aquatic plants typically enter the riverine food web via detritivores. In the St. Lawrence River, however, muskrats and greater snow geese prey heavily on aquatic macrophytes and American bulrushes, respectively (Giroux and Bédard 1987).

Invertebrates

Benthic invertebrate assemblages within the river vary greatly with depth, substrate size, and organic content (see Table 2.14 in St. Lawrence Centre 1996, data from Ferraris 1984; see also Lemarche et al. 1982, Ricciardi et al. 1997). The latter two are influenced, in turn, by current velocity and the presence of aquatic plants. Assemblages are also affected by the source and nature of the water, such as clear Lake Ontario–derived water versus turbid tributary-derived water. As is the case for many aquatic organisms, densities and diversities of invertebrate benthos tend to increase in slackwater sites (bays, secondary and side channels, protected shorelines, etc.) and in channel macrophyte beds as long as the surface sediments are not anoxic. Rocks in stronger currents are inhabited by mayflies, chironomid midges (especially Orthocladiinae), stoneflies (*Pteronarcys*), sponges, and filter-feeding caddisflies (*Hydropsyche*). In weaker currents, rocks are colonized by amphipod crustaceans (*Gammarus*), snails (*Amnicola*), and oligochaetes. Chironomid midges (e.g., *Dicrotendipes*, *Procladius*) and oligochaete worms dominate assemblages on sediments of clay and sand, whereas some combination of snails (*Bithynia*, *Valvata*), bivalve mollusks (*Elliptio*, *Pisidium*), oligochaetes (*Peloscolex*), nematodes, and amphipods inhabit sand and gravel substrates. In aquatic vegetation within slackwater sites and fluvial lakes, one often finds snails feeding on periphyton or detritus (*Bithynia*, *Probithynella*, *Gyraulus*) and numerous bivalve mollusks (*Elliptio*, *Lampsilis*), isopod crustaceans (*Caecidotea*), amphipod crustaceans (*Hyalella*), oligochaetes, dipteran larvae, odonates (e.g., *Enallagma*), and caddisflies (*Brachycentrus*). Crayfishes (*Orconectes*, *Cambarus*) also occur in plant beds, and they are ecologically and commercially important. Recent commercial catches of 4 to 8 metric tons/yr of crayfish have been made in Lac Saint-Pierre, and the potential catch is estimated at 50 tons/yr (St. Lawrence Centre 1996). Large emergence events of mayflies (*Hexagenia*) are also known for this river, and the alderfly *Sialis* also occurs, at least in protected vegetated areas.

The St. Lawrence maintains a zooplankton community that is sustained far upstream by recruits from Lake Ontario and downstream presumably by reproduction occurring in lateral slackwater areas and in fluvial lakes of the main channel. Zooplankton biomass decreased from 40 µg/L dry mass to 16 µg/L within 90 km downstream of the outlet of Lake Ontario (Basu et al. 2000b) and continued to decline to ~10 µg/L dry mass at Montreal. Densities rise within macrophyte beds of the fluvial lakes (Basu et al. 2000a). Reproduction is minimal within the main channel but strong in slackwater areas (J. H. Thorp, unpublished data).

Zooplankton communities in the freshwater fluvial section are dominated numerically by rotifers, followed by cyclopoid copepods and small cladocera (Mills and Forney 1982, Basu and Pick 1996, Thorp and Casper 2002). Some common rotifers are *Polyarthra* (usually the overwhelming dominant), *Keratella*, *Ploesoma* (a predator), and *Synchaeta*. Fifty or so kilometers below Lake Ontario, cladocera in the river are generally small species and are dominated numerically by *Bosmina* (*Sinobosmina*) species, with smaller numbers of *Polyphemus pediculus* (a predator) and *Sida crystallina*. Chydorids, daphnids, *Polyphemus*, and other large species are more prevalent below the outlet of Lake Ontario and appear downstream in some slackwater sites. Common predaceous cyclopoid copepods are *Diacyclops thomasi* (the most abundant), *Mesocyclops edax*, and *Tropocyclops prasinus mexicanus*. *Eurytemora affinis*, an estuarine species that has recently begun colonizing large rivers, is the most common calanoid species, but *Leptodiaptomus minutus* occurs in some samples. Harpacticoid copepods are common in vegetated slackwater areas near the bottom. Some crustacean zooplankton migrate vertically, both within slackwater sites and in the main channel (J. H. Thorp and A. F. Casper, unpublished data).

Vertebrates

There are 87 freshwater and 18 migrating (diadromous) species of fishes in the St. Lawrence River (Ducharme et al. 1992), with most of the freshwater species occurring in the fluvial section. Species endemism is low, and only the pygmy smelt and the copper redhorse (considered threatened by the Canadian government) are endemic to the freshwater ecoregion. Although many fish species frequent the main channel, the warmer, more productive slackwater habitats and fluvial lakes are more commonly used by the river's ichthyofauna, including larval fishes (e.g., Werner 1977). This river is especially noteworthy as the home of the healthiest surviving stocks of Atlantic salmon in the world (albeit with declining numbers) and for supporting populations of lake and Atlantic sturgeon. Lake sturgeon can grow to a length of at least 2.4 m and live as long as 150 years (Smith 1985). Atlantic sturgeon grow even larger, with records of specimens reaching 4.3 m and 360 kg. Adult Atlantic sturgeon live in saltwater but migrate upriver several hundred kilometers to spawn and spend their first four years. Some other common species in the freshwater fluvial section of the river are smallmouth bass, largemouth bass, rock bass, brown bullhead, pumpkinseed, walleye, brook stickleback, northern pike, burbot, muskellunge, white sucker, longnose sucker, shorthead redhorse, silver redhorse, yellow perch, spottail shiner, brown trout, central mudminnow, brook char, American eel, alewife, and rainbow smelt (St. Lawrence Centre 1996).

Commercial and subsistence fisheries have declined on the river for a variety of reasons, but sport fishing still thrives. The most common commercial catches are lake sturgeon, American eel, brown bullhead, yellow perch, and sunfishes (St. Lawrence Centre 1996). Sport fishing concentrates on yellow perch, northern pike, walleye, and various bass species, with yellow perch accounting for 65% of all sporting catches in Quebec. Populations of the commercially important American eel are declining rapidly. Between 1985 and 1990 there was a ~99% drop in the number of juveniles entering Lake Ontario after passage upriver following their birth in the Sargasso Sea. Causes of this decline are probably linked to some combination of mortality from river contaminants, overharvesting of adults, loss of shallow-water habitats, and transiting the hydroelectric dams while migrating out to sea and back upriver. This is a serious situation for this East Coast species because an estimated 19% of the freshwater distribution range of the species involves the St. Lawrence River. Global factors may also be responsible, as a similar population decline has also been noted for the European eel, which also spawns in the Sargasso Sea.

Rainbow trout, common carp (which is ubiquitous along sheltered shorelines), and white perch are some of the nonnative species introduced to this river. Rainbow trout are becoming a serious competitor with Atlantic salmon. Bow fishing for carp is common, and fishing with rod and reel (often expensive rigs) is gaining in popularity with tourists from Europe.

About 115 species of waterfowl, shore birds, and raptors frequent aquatic habitats of the St. Lawrence River's main stem, with about 38 species found in the freshwater fluvial section (St. Lawrence Centre 1996). Notable species in this area are geese (snow goose, common brant, Canada goose), dabbling ducks (American black duck, mallard, northern pintail, northern shoveler, gadwall, American wigeon, wood duck, green-winged teal, blue-winged teal), diving ducks (scaup, bufflehead, goldeneye, merganser), and sea ducks (eider, oldsquaw, scoter). Also commonly seen are great blue heron, ring-billed gull, belted kingfisher, osprey, and bald eagle. There are approximately 36 active heronries in the St. Lawrence system and around 8000 or more birds in the population by the end of summer (DesGranges and Desrosiers 1995). The largest known heron colony in the world (1000 birds) was reported from the fluvial Lac Saint-Pierre. Bird populations in the St. Lawrence River continue to recover following reductions in persistent organic contaminants, habitat protection, and creation of island bird sanctuaries. Recovery of greater snow goose and common eider in particular have been linked to these efforts.

Herpetofauna are not as characteristic of the St. Lawrence River as they are in more southerly rivers in North America. Northern leopard frogs and mudpuppies are common in wetlands associated with fluvial sections of the river, and some species occasionally appear in rocky areas of the main channel. The most frequently sighted reptiles are painted and snapping turtles. Amphibians and reptile species associated with the St. Lawrence River and given priority protection status in Quebec are pickerel frogs, northern chorus frogs, brown snakes, northern water snake, map turtle, Blanding's turtle, and spiny softshell turtle (the last considered threatened in both Ontario and Quebec).

Distributions of the mostly herbivorous muskrat and the carnivorous American mink and river otter extend throughout the St. Lawrence River in alluvial wetlands and along the river banks. Muskrats in particular are quite common, and they are frequently harvested for their pelts (~27,000 were taken in the 1988–1989 season; St. Lawrence Centre 1996). Beaver can also be seen in the St. Lawrence, but their primary association derives from their residence in the river's alluvial wetlands and tributaries.

A total of about 21 species of cetaceans and pinnipeds are considered residents of or occasional visitors to the St. Lawrence from the upper estuary to the lower Gulf, but perhaps the most unusual and photogenic species to inhabit the St. Lawrence River is the rare white beluga whale. This small whale currently inhabits saline portions of the upper estuary, although it formerly moved up into lower freshwater habitats and was sighted regularly at Quebec City until the 1930s. Following a ban on hunting of this whale in 1979 the population appears to have stabilized at around 500, but it is still classified as an endangered species (Kingsley 1994). Another cetacean, the long-finned pilot whale, also inhabits the upper estuary of the St. Lawrence, occasionally appearing upstream of the confluence with the Saguenay River.

Approximately 20 species of animals in or along the St. Lawrence main stem are categorized as vulnerable, threatened, or endangered by federal, state, or provincial governments; another dozen are considered at risk. No fish species is threatened or endangered according to the U.S. federal list for New York State, but four species are on the state list: lake sturgeon (threatened), pugnose shiner (endangered), mooneye (threatened), and blackchin shiner (species of special concern). Additional species that are considered at risk are Atlantic sturgeon, American eel, American shad, northern pike, Atlantic tomcod, rainbow smelt, striped bass, anadromous brook char, and Atlantic salmon. Blanding's turtle is on the federal threatened list, along with several species of birds frequenting the shores of the St. Lawrence River, and the eastern spiny softshell turtle is threatened in Quebec.

Ecosystem Processes

The St. Lawrence is an oligotrophic river, with chlorophyll *a* rapidly declining in the main channel below Lake Ontario to ~1 µg/L and not rising to 2 to 5 µg/L until the confluence with the Ottawa River at Montreal (Basu et al. 2000b). Aside from direct effects of advective processes, low levels of primary productivity probably reflect low total phosphorus (~10 µg/L) in the upper river. However, as turbidity increases and total phosphorus levels rise to 20 to 40 µg/L below the confluence with the Ottawa River, light may be an important limiting factor. In general, however, light is probably less limiting to phytoplankton production in the St. Lawrence than in most large rivers, because this northern river is relatively sediment free compared to other large rivers of the world (Gleick 1993).

The potential role of riverine "storage zones" in nutrient spiraling and plankton productivity has recently been recognized for several American and European rivers. Compared to channel sites, slackwater habitats have faster population growth rates for phytoplankton (Reynolds and Descy 1996) and zooplankton, as demonstrated for the St. Lawrence and tributaries (Basu and Pick 1996, Basu et al. 2000b).

Food web relationships within the St. Lawrence are poorly known in general (Thorp and Casper 2002). Main-channel habitats may be controlled by advective forces, but there is some evidence that biotic interactions are important in slackwater sites (Thorp and Casper 2002, 2003). Using stable isotope data, Barth et al. (1998) concluded that in the upper St. Lawrence most suspended particulate organic carbon POC was derived from phytoplankton; this is an important food source for many zooplankton.

The zooplankton community of the St. Lawrence River is controlled by a poorly understood mixture of abiotic and biotic factors varying seasonally and among habitats (Thorp and Casper 2002). Abiotic factors regulating the potamoplankton include those influencing food abundance (e.g., intensity of photosynthetically active radiation), access to energy (e.g., hydrological mixing), mechanics of feeding (e.g., suspended sediments), downstream transport versus temporary retention (water velocity, channel configuration, impoundments, etc.), direct mortality (physical abrasion and attenuation of ultraviolet radiation), and thermal conditions. Biotic factors include competition for food, parasitism, disease, and planktivory, both by benthic and pelagic invertebrates and by most larval, some juvenile, and a few adult fishes. In an in situ experiment in the St. Lawrence on mussel predation (Thorp and Casper 2002), densities of the most abundant rotifer, *Polyarthra*, declined dramatically in enclosures with dreissenid mussels compared to controls and enclosures with unionid mussels. Rotifer densities in unionid enclosures were not different from controls. Effects on rotifers were probably from predation, as chlorophyll *a* did not vary among treatments. Densities of the dominant calanoid copepod, *Eurytemora affinis*, increased in the presence of dreissenids, probably as an indirect food web response. In another in situ experiment on effects of yellow perch predation on zooplankton of the St. Lawrence River, cladocera were the principal prey of smaller perch, followed by copepods, ostracods, and very few rotifers; larger juvenile perch fed almost exclusively on copepods and ostracods (Thorp and Casper 2003). Fish were associated directly with significant declines in densities of copepods and indirectly with increases in rotifers and cladocera.

Human Impacts and Special Features

As population densities in the St. Lawrence Valley and tributary basins rose over the last four centuries of colonization by non–Native Americans, the impacts of humans gradually increased and changed in nature. The simple removal of mammal pelts gave way to deforestation linked to rural agriculture. Pollution from urbanization and industrialization increased, and river flow was regulated nearly half a century ago to improve shipping and produce electricity. The international St. Lawrence Seaway became an important commercial link between the Great Lakes and the Atlantic Ocean, with 37 million metric tons of cargo passing the Lake Ontario–Montreal section in 1997 alone. This population density and commercial activity have come with an ecological price tag, however, both because of the construction and operation of several major hydroelectric dams, navigation locks, and diversion dams and because of pollution from industrial, municipal, and agricultural sources.

Although environmental laws now exert a stronger control over point-source and nonpoint-source pollution, regulation of the river channel continues unabated and with little significant attention to environmental consequences. Four hydroelectric dams are present on the river's main stem, and numerous channel-control structures redirect flow. Small dams on the St. Lawrence's tributaries are also quite common. In addition to these hydroelectric dams, seven navigation locks enable passage of ships through the 68 m elevational drop from Lake Ontario to the fluvial Lac Saint-Pierre, and a minimum flow channel of 8.2 m is maintained for passage of deep-draft, seagoing commercial ships. The binational International Joint Commission (IJC) was established by the Boundary Waters Treaty of 1909 for the purpose of regulating aquatic-impact structures within the Great Lakes and the St. Lawrence River and for regulating the level and flow of these boundary waters. As a result of this regulation, the annual stage variation at the mouth of Lac Saint-François has been reduced from 60 to 15 cm (St. Lawrence Centre 1996).

Major regulation of the river began in 1958 with the commissioning of the Moses-Saunders hydroelectric dam, spanning the river between Massena, New York, and Cornwall, Ontario. This project includes four dams (three of which redirect water flow) and over 17 km of dikes, which help impound the natural fluvial Lake St. Lawrence. Electricity is generated on the U.S. side of the dam by the New York Power Authority and on the Canadian side by Ontario Hydro. Together the 32 hydroelectric turbine/generator units have a capacity of 1.8 million kW. Another three hydroelectric dams are present

near Montreal. The St. Lawrence is rated as strongly fragmented because only 25% to 49% of the main-channel segments lack dams and both large and many small tributaries are impounded (Dynesius and Nilsson 1994).

Commercial traffic on the river is vital to the economy of Canada and the United States. An average of more than 10,000 trips per year are made by commercial vessels, and in 1997 alone, for example, 37 million metric tons of cargo passed through the Lake Ontario–Montreal section. An increasing number of passenger ships (including pleasure ships from Europe) traverse the St. Lawrence Seaway, and recreational boating continues to grow.

The effects of regulation of the river for power generation and navigation are poorly known because of inadequate ecosystem studies and the absence of sufficient preregulation environmental data. In the case of the Beauharnois Dam near Montreal, for example, 84% of the river's flow was diverted to pass through the dam, resulting in major hydrodynamic alterations of two nearby fluvial lakes and the river's bed with unknown biotic consequences. Nonetheless, it is clear that most river dams substantially alter the nature of a lotic ecosystem, making it more lentic and interfering with the movement of fishes and other fauna. For example, the population size of the catadromous American eel has declined precipitously in the St. Lawrence, and some environmentalists consider the major cause to be the difficulty these fishes have in passing the four hydroelectric dams to reach their oceanic breeding grounds. Several species of anadromous fishes are also hurt by the presence of dams on the main stem and tributaries. Whether the environmental price of power dams and commercial shipping is justified by their economic benefits is a political and social question outside the realm of this chapter.

Commercial fishing was a major operation in the freshwater portion of the St. Lawrence River in the 1800s, and it continues today at a lower pace. Atlantic salmon became virtually extinct by the late 1800s because of overharvesting, the presence of numerous dams on tributaries, and pollution associated with human impacts on the watershed. The large commercial sturgeon fishery also collapsed by the late 1890s from overharvesting. In 1992, 687 metric tons of multiple species of fishes were harvested in Quebec within the fluvial section of the river and another 278 tons were reported caught in the fluvial estuary (Johnson 1991, St. Lawrence Centre 1996). A much higher take is known for the saline portions of the river and the Gulf of St. Lawrence. The approximate percentages of the catch during 1992 within the fluvial Lac Saint-Pierre were 38.9% brown bullhead, 33.9% yellow perch, 11.3% lake sturgeon, 7.5% American eel, 4.1% sunfishes, and 4.3% all other species combined. Populations of bullhead and perch seem stable, but those of eel and sturgeon appear to be declining.

Most humans along the St. Lawrence live near the fluvial section of the river, especially near Montreal, where over 3 million people reside (almost 50% of Quebec's population). Two other significant urban areas are Trois-Rivières in the fluvial section and Quebec City in the fluvial estuary. Consequently, it is in those areas and downstream where most pollution occurs. Perspectives on over two decades of water quality are reviewed in Désilets and Langlois (1989). The river above Massena–Cornwall is in very good shape from a pollution perspective, and it continues to improve. Concentrations of most pollutants of potential concern are generally below levels set by New York State's Department of Environmental Conservation. Downriver, however, the situation is bleaker, and the river immediately below Massena–Cornwall was declared an EPA Area of Concern. Current or historical inputs of metals and organic compounds (PCBs, PAHs, dioxin, and various pesticides such as mirex and DDT) from point (e.g., aluminum plants and pulp and paper mills) and nonpoint sources pose significant health problems to humans and the environment. In the 1960s and 1970s, the St. Lawrence was one of the more polluted large rivers in the world, but it has now become one of the cleanest following the cleanup begun in the late 1980s.

Biotic pollution of another form has also influenced the St. Lawrence River. From the mid or late 1980s through the early 1990s, two species of bivalve mollusks, quagga and zebra mussels, spread throughout the freshwater portion of the river, reducing phytoplankton productivity and extirpating populations of native unionid mussels. These nonnative species are thought to have entered the greater St. Lawrence River ecosystem in the ballast water of ships arriving from southeastern Europe. The St. Lawrence Seaway has also been the pathway for many nonnative species to enter the Great Lakes and from there the Mississippi River drainage.

Despite imposition of dams on the main channel and many tributaries, the St. Lawrence continues to be an unusual and good-quality river. In terms of biological distinctiveness, its fauna is rated continentally outstanding (Abell et al. 2000), and Lac Saint-Pierre is a RAMSAR World Heritage Site. Aside from its uniqueness in terms of lake origin and overall size,

the St. Lawrence is less turbid than all other larger rivers of the world. On the other hand, without the navigable access to the Atlantic Ocean provided by this river, development of large cities and a strong economy along the river and around the Great Lakes would have been severely impeded over the last 200 years.

OTTAWA RIVER

The Ottawa River, or Kichesippi (meaning "Great River" in the language of the Algonquin Nation), originates in Lake Temiskaming (or Lac Temiscamingue) and flows 1271 km across the Precambrian Shield to its confluence with the St. Lawrence River near the Montreal archipelago (Fig. 22.10). The Ottawa is the largest tributary in the freshwater fluvial St. Lawrence River–Great Lakes system, with a basin of 146,334 km^2. It is roughly 65% in Quebec and 35% in Ontario, and the river forms the border between these two provinces for most of its length. Although much of the river flows through a naturally constricted channel, there are several complexes of islands and bays, such as the Petrie Island Preserve in the lower river. These environmentally sensitive areas provide a diverse and highly productive habitat for fish, invertebrate, and macrophyte assemblages.

Compared to southern tributaries of the St. Lawrence River, the Ottawa River is softer and lower in alkalinity but higher in nutrients and organic carbon. This reflects the primary origin of this northern tributary in the crystalline Precambrian Shield as well as the intensive agricultural activity in its lower basin. At its confluence with the St. Lawrence River, the more turbid waters of this northern tributary are often referred to as "brown waters" in comparison to the "green waters" derived from the Great Lakes. Despite the natural tendency of the St. Lawrence River to mix with waters of the Ottawa River, relatively distinct bodies of green and brown water can be detected at least 100 km below their confluence.

The primary Native Americans inhabiting the Ottawa River valley immediately prior to European contact were the Algonquins, who called themselves the Anishinabeg, meaning "human beings." Evidence exists of a copper-using people here 5000 years ago, but they are not thought to have been connected to the Algonquins of the historic period. The Algonquins were mostly a seasonally nomadic, hunter-gatherer nation, in part because they lived in a climatic zone that was generally unfavorable for agricultural development. Although the prehistory of the

Ottawa Valley is largely unknown, the first recorded meeting between Europeans (led by Samuel de Champlain) and the widespread Algonquin Nation occurred along the Rivière Saguenay at Tadoussac, Quebec, in the summer of 1603. The fur trade flourished thereafter; indeed, the Algonquian word *atawe*, meaning "to trade," became the European name for this river. Ancestors of a separate nation, the Ottawa Indians from the shores of Lake Huron, occasionally traveled down the Ottawa River through Algonquin lands to trade with the French, leading to some confusion in modern times about which tribes lived in the Ottawa River valley. Algonquins continued to live on the Ottawa and its tributaries in the twentieth century, with bands at Temiskaming, Mattawa, and on the Rivière Coulogne (www.civilization.ca/cmc/archeo/oracles/outaouai/30.htm).

Physiography, Climate, and Land Use

The basin, which is located along the southern edge of the Precambrian Shield–Frontenac axis, contains three physiographic provinces (see Fig. 22.10): the St. Lawrence Lowland (SL), Laurentian Highlands (LU), and Superior Upland (SU). Lowlands of the Ottawa River basin, once the floor of the ancient Champlain Sea, are bounded by two mountainous regions: the Laurentians to the west and the Algonquin dome to the east. At 968 m asl, Mount Tremblant, located just north of Montreal in the Laurentian complex, is the highest named peak in the basin and a popular skiing area. The underlying geology is dominated by a base of crystalline and crystallophyllian Precambrian rock in the north (99%) and Ordovician sedimentary rock in the south (98%). Surface soils reflect a gradation between distinctive soils in the upper and lower basin. The upper basin is characterized by well-drained organic mesisols and podzols interspersed with silt and sand deposits from the prehistoric Champlain Sea. In contrast, the lower basin features well-drained podzols interspersed with poorly drained melanic brunisols near the confluence with the St. Lawrence in the Montreal archipelago.

The climate in the Ottawa River valley is best described as humid continental. Mean annual daily temperatures are ~6°C, and the basin averages 160 to 210 growing days with temperatures of at least 5°C (Watson and MacIver 1995). Mean monthly temperatures range from −11°C in January to 21°C in July (Fig. 22.11). Annual precipitation averages 100.2 cm and is evenly distributed throughout the year (see Fig. 22.11), with rarely more than 25% falling as winter snow.

The Ottawa River basin is in the Temperate Deciduous Forest and Boreal Forest biomes, and it includes two terrestrial ecoregions: the Eastern Forest/Boreal Transition, occupying the greater part of the basin, and the Eastern Great Lakes Lowland Forests closer to the confluence with the St. Lawrence River. A white pine–aspen mix typifies northern forests, whereas more southerly regions contain mixed hardwoods (>30 species, especially sugar maple and American beech) along with white pine and hemlock. Annual productivity of these second- and third-growth stands range from ~200 to 500 g C m^{-2} yr^{-1} (Perera et al. 2000).

Most of the Ottawa River basin is forested (86%), with the remaining land cover consisting of surface waters (~10%), urban areas (2%), and agriculture (2%). Most agricultural activities are clustered in the lower portion of the basin. The largest city in this watershed is Ottawa, the nation's capital, with 1.06 million people (Canadian Census 2001).

River Geomorphology, Hydrology, and Chemistry

From its origin at Lake Temiskaming, the Ottawa River flows mostly through a naturally constricted channel interrupted in some parts by wider, shallow floodplains and island mosaics. Along this pathway the main channel slopes about 36 cm/km, whereas the maximum topographic relief in the basin is 911 m. In the upper and middle sections of the river, the constricted channel flows between artificial reservoirs. The lower section of the Ottawa River is also mainly constricted but includes some floodplain regions and areas replete with islands, especially in the section extending from its confluence with the Gatineau River downstream to the confluence with the St. Lawrence River.

The Ottawa River is the largest tributary of the fluvial section of the St. Lawrence River, with an annual discharge of 1948 m^3/s. Among its 12 major tributaries are 3 with drainage basins >5000 km^2 in Quebec: the Gatineau, Coulonge, and du Lièvre, with annual discharges of 126, 97, and 75 m^3/s, respectively. Although precipitation is fairly even throughout the year (range 7.2 to 9.8 cm/mo), spring snowmelt results in runoff peaks in April and May at three to four times (>6 cm/mo) the minimal runoff of late summer (see Fig. 22.11).

Geologic formations of the Precambrian Shield strongly influence the chemistry of this river. Mean conductivity (80 µS/cm), alkalinity (19.2 mg/L as

$CaCO_3$), hardness (29.2 mg/L as $CaCO_3$), suspended solids (6 mg/L), turbidity (4 NTU), and dissolved organic nitrogen (0.18 mg/L) are all low compared to waters of the Great Lakes and southern tributaries of the St. Lawrence. Conversely, levels of total organic phosphorus (0.053 mg/L) and organic carbon (5.6 mg/L) are higher than those other waters. Suspended detritus concentrations increase downstream, peaking in the faster-moving middle reaches. Concentrations then drop along with current velocities as the river approaches the major hydroelectric facilities at Carillon near the confluence with the St. Lawrence. Transit time from upper tributaries to Carillon is ~14 days. No consistent longitudinal patterns are evident in the mean annual levels (1979 to 1994) of NH_4-N (0.045 mg/L), NO_3-N + NO_2-N (0.17 mg/L), total nitrogen (0.41 mg/L), total phosphorus (0.029 mg/L), and chlorophyll *a* (1.9 μg/L). However, conductivity, temperature, alkalinity, total phosphorus, suspended solids, and chlorophyll *a* increase as one moves downstream, particularly below the Ottawa–Hull region, where three large tributaries join the Ottawa (Hydro-Québec 1994, Ministère de l'Environnement et de la Faune du Quebec 1996).

River Biodiversity and Ecology

The Ottawa River basin occurs in the Lower St. Lawrence freshwater ecoregion and contains a diverse flora and fauna associated with the river. However, a number of taxa are threatened, including species of fishes (American eel, American brook lamprey, silvery minnow, lake sturgeon, and river redhorse), birds (least bittern and logger-head shrike), amphibians (boreal false-cricket treefrog), reptiles (eastern spiny softshell turtle), and even some insects, such as various dragonflies (*Gomphus ventricosus, Ophiogomphus anomalus,* and *Stylurus notatus*).

Algae and Cyanobacteria

A well-developed, diatom-dominated phytoplankton-periphyton community exists in the Ottawa River. *Melosira varians* dominates an assemblage that includes at least 150 pelagic and 42 epibenthic species (De Seve and Goldstein 1981, Vis et al. 1998). Densities of periphyton are greater in the Ottawa River than in the St. Lawrence (1762 versus 691 cells/mm^2 benthic surface area). Similarly, the density of phytoplankton was also greater in the lower Ottawa than the nearby fluvial Lac Saint-Louis (>50,000 versus 25,000 cells/L; Alaerts-Smeesters and Magnin 1974).

Plants

Throughout most of its length the Ottawa River is fringed with emergent and submerged macrophyte beds. Silver maple and green ash are the principal riparian canopy trees. They are usually found with alder, frost grape, and willow as understory species. The herbaceous layer of the riparian forest along the lower Ottawa includes reed canary grass, ground nut, and the invasive purple loosestrife (Fairchild 1983, Hydro-Québec 1994). Bands of emergent sedge wetlands composed of wild rice, bur-reed, and arrowhead are common along the banks. The 43 species of emergent riverine plants are divided into groups found in the upper marsh (e.g., northern bugleweed, fox sedge, silverweed), intermediate areas (e.g., marsh speedwell, Calamus root, bulrushes), the lower marsh (e.g., Small's spikerush, bur-reed, knot-sheath sedge), and those distributed across the spectrum (e.g., bladder sedge, creeping Jenny loosestrife, red top). Shipley (1987) determined that water level significantly influenced the species composition of these emergent macrophyte assemblages. Submerged macrophyte beds are most common in depositional zones of the lower section of the Ottawa. They are largely dominated by American waterweed, wild celery, and watermilfoil. Interspersed with these dominants are coontail, waterlily, and four species of pondweed (big-leaf, ribbon-leaf, Richardson's, and spiral).

Invertebrates

Littoral benthic habitats in the Ottawa River are heterogeneous patchworks of mud, silt, and sand interspersed with gravel bars and submerged macrophyte beds, the latter supporting the greatest invertebrate diversity. As in most large rivers, overall benthic invertebrate diversity tends to correlate with increasing habitat heterogeneity, whereas total density and biomass are closely related to substrate type and current velocity. Some large-river taxa, like mussels, are apparently also limited by food availability and ionic content of the water. The abundance of benthic invertebrates and phytoplankton in the Ottawa increases progressively from the upper reaches near Lake Temiskaming to its confluence with the St. Lawrence (Ontario Water Resources Commission 1972).

In the lower reaches, more than 150 species of invertebrates can be found in vegetated and erosional zones <2 m deep at densities of up to 2000 animals/m^2. Noninsect taxa comprise much of this riverine benthic assemblage. Main-channel zones are characterized by a combination of small fingernail clams, oligochaete worms, and large unionid mussels. Overall, the oligochaete assemblage is dominated by *Stylaria* spp., *Uncinais uncinata*, *Peloscolex* sp., *Limnodrilus* sp., and *Spirosperma ferox*, and diversity, density, and biomass tend to increase with depth (Mackie 1971). Bivalve biomass is dominated by unionid mussels (eastern elliptio, spike mussel, and eastern lamp mussel), whereas fingernail clams are the most numerous. The latter consist principally of *Sphaerium* (7 species) and *Pisidium* (6 species) at 171 clams/m². There are 12 species of snails in the Montreal archipelago, the most abundant being faucet snails at 2311 snails/m², followed by mud Amnicolas, St. Lawrence pond snails, Chinese mystery snails, gyro snails, *Physella* spp., three-ridge valvata snails, and ancylid snails (sometimes called freshwater limpets) (Magnin 1970; Ontario Water Resources Commission 1972; Hamill 1975, 1979; Clair 1976; Mackie 1971). Amphipods (*Gammarus fasciatus*, *G. pseudolimaeus*, *Crangonyx pseudogracilis*, *Hyalella azteca*) are more abundant (2073 amphipods/m²) than the common isopods *Asellus communis* and *Caecidotea* sp. (953 isopods/m²). Both amphipods and isopods are more abundant in slackwaters than the main channel of the river. Four species of crayfishes have been reported.

Benthic insects comprise the remainder of the assemblage; their abundance and diversity are generally greatest in the shallow, highly heterogeneous littoral habitats containing macrophytes and cobble. Common insect taxa include dragonflies and damselflies (e.g., *Enallagma*), stoneflies (*Isoperla*), alderflies (*Sialis*), beetles (*Berosus*, *Dubiraphia*, *Haliplus*, *Microcylloepus*), black flies, 19 genera of chironomid midges, 35 species of caddisflies, and 22 species of mayflies. The chironomid assemblage includes *Procladius*, *Zavrelimyia*, *Tanypus*, *Brillia*, *Cricotopus*, *Chironomus*, *Cryptochironomus*, *Dicrotendipes*, *Einfeldia*, *Demicryptochironomus*, *Polypedilum Paracladopelma*, and *Pseudochironomus* (Magnin 1970, Fairchild 1983). Among the caddisflies, the predators *Polycentropus* and *Oecetis* were the overall dominants, but there were habitat-specific exceptions. The herbivorous caddisfly *Hydroptila* dominated weed beds, whereas the filterer *Brachycentrus* was the principal deep-water taxon. The relative abundance of each of these species was drastically lower in the faster waters of a nearby fluvial lake (Lac Saint-Louis), where *Molanna*, a smaller collector-gatherer, was dominant. Subdominant caddisfly genera include *Agraylea*, *Ceraclea*, *Cyrnellus*, *Macrostemum*, *Mystacides*, *Triaenodes*, *Hydropsyche*, *Cheumatopsyche*, *Nectopsyche*, and *Neureclipsis* (Fairchild

1983). The mayflies included *Ephemerella* spp., *Hexagenia rigida*, *H. limbata*, *Caenis simulans*, and *Stenonema bipunctatum*.

The zooplankton assemblage of the Ottawa River is particularly well-described. Forty-five species from 27 genera of cladocera are known from the lower river (Croskery 1974). *Bosmina longirostris* is an early-season dominant across all habitats, with peak abundance in June and July. Three species of *Daphnia* are subdominants that peak from late July through September. In addition to cladocera, both calanoid (*Eurytemora affinis*, *Episcura lacustris*, and four species of *Diaptomus*) and cyclopoid copepods (*Cyclops vernalis*, *Diacyclops thomasi*, *Mesocyclops edax*, *Tropocyclops parsinus mexicanus*) are common. Copepod densities are seasonally opposite of cladocera, peaking in early May, then dropping and not rising again until September. This crustacean assemblage is in turn preyed on by *Chaoborus*, *Leptodora kindtii*, and *Hydra littoralis*. As with the benthic invertebrates, pelagic abundance is inversely related to current velocity. Indicator taxa were *D. retrocurva*, *D. parvula*, and *Ceriodaphnia* spp. for low-flow conditions and *L. kindtii*, *Sida crystallina*, *Eurycercus lamellatus*, and *Camptocercus rectirostris* in moderate currents. No taxa were consistently associated with the main channel. Other widely distributed cladocera include *Diaphanosoma*, *Simocephalus*, *Eubosmina*, *Latona parviremis*, *Latonopsis occidentalis*, *Pleuroxus unicinatus*, and four species of *Chydorus*.

Vertebrates

Three major habitat assemblages of fishes exist in the lower Ottawa: (1) bays with abundant macrophyte beds, (2) the shoreline and littoral slackwaters between bays and the main channel, and (3) the channel itself. In bays, the principal species are yellow perch, brown bullhead, pumpkinseed, emerald shiner, walleye, and sauger. In main-channel habitats are channel catfish, walleye, sauger, silver redhorse, white sucker, northern pike, and mooneye. In the fluvial Lac Duchênes region, brown bullhead, emerald shiner, and yellow perch are very abundant, but carp, black crappie, and central mudminnow join the group. In addition, two regionally rare large-river species are present: lake sturgeon and river redhorse (Hydro-Québec 1994).

Density and diversity of fishes are greater in the slackwater littoral zone than in any other river habitat. More than 80% of all fish species occur here, and it is very productive habitat (e.g., black crappie occur at >80 individuals/ha or 18.4 kg/ha). Slackwater habitats range from extensive littoral macrophyte beds, sandy substrates, shoals, and sparse macrophyte beds in fluvial Lac Duchênes to narrow littoral zones with steep shores, silty substrate, and numerous submerged logs and stumps around the main channel islands of the lower Ottawa, where channel depths vary from 6 to 26 m and current velocities are 0.33 to 1.03 m/s. In slackwaters, the main secondary consumers range from insectivorous pumpkinseed to piscivorous walleye and sauger.

Seasonal changes in diet and growth rates closely follow the availability of prey items. Apex predators, such as sauger, walleye, and burbot, forage extensively on small, seasonally abundant fishes, such as silvery minnows and Iowa and Johnny darters. Walleye and sauger use slackwater zones to feed mainly on emerald shiner and small yellow perch. Intermediate predators like yellow perch, black crappie, brown bullhead, and pumpkinseed prey on amphipods, chironomids, copepods, cladocera, fish fry, and phantom midges common to the area. In addition to invertebrates, production of at least one riverine species, the brown bullhead, is supported by filamentous algae (*Spirogyra* and *Anabaena*) (Gunn et al. 1977). Fish growth in the Ottawa River reaches a maximum during the summer and in slackwater habitats (Rodgers and Qadri 1982, Osterberg 1978).

The island complexes and floodplain regions of the river are particularly favorable for birds and some herpetofauna. Two rare birds in this region, the least bittern and the sedge wren, are both periodically found in the cattail marshes adjacent to the Ottawa. Four species of herpetofauna reach the northern range limits in the Ottawa River valley and are occasionally found nearshore or in the riparian zone: spiny softshell turtle, northern water snake, common snapping turtle, and western chorus frog.

Ecosystem Processes

Productivity for most benthic invertebrates in the Ottawa peaks in mid- to late summer. Production rates for the mussels *Elliptio* and *Lampsilis* are consistently lower in the Ottawa River than in the nearby St. Lawrence (~30%), perhaps due to lower ionic content and calcium levels in the Ottawa River (Magnin and Stanczkowska 1971). Despite their small size, sphaeriids are abundant and productive (0.409 g m^{-2} yr^{-1} dry mass) across all habitats, with *Pisidium casertanum* (92 clams/m^2 and 0.26 g m^{-2} yr^{-1}) and *Musculium securis* (0.04 to 0.3 g m^{-2} yr^{-1}) dominating. As an example of the generally greater production in slackwater compared to main-channel habitats, *P. casertanum* was over twice as productive

in slackwater areas (0.39 to 0.83 g m^{-2} yr^{-1} versus 0.04 to 0.19 g m^{-2} yr^{-1}). These differences are linked primarily with current and secondarily with sediment grain size. Production of benthic crustaceans is directly correlated with abundance of macrophytes (typically common waterweed, bur-reed, and wild celery), ranging from 0.67 to 1.49 g m^{-2} yr^{-1} (Hamill 1975).

Diet studies reinforce the idea that the slackwater zones are key links between primary and secondary production in the Ottawa. *Lymnaea catascopium* and *Viviparus malleatus* ingested large amounts of detritus (~53%); however, the common, lipid-rich diatoms *Navicula*, *Fragilaria*, *Gomphonema*, *Rhoicosphenia*, and *Cocconeis* were also major items in their diets (~26%). This supports the observation that gastropods are most abundant and diverse in shallow waters due to food availability. Annual production appears to be very plastic and depends on seasonal and spatial differences in food conditions and water chemistry. *Lymnaea catascopium* produces one generation per year in the brown (soft) waters of the Ottawa compared to two in the green (hard) waters of the adjacent St. Lawrence, despite the greater food base in the Ottawa. Moreover, the distribution and abundance of gravid female isopods (*Asellus communis*) are spatially associated with abundance of diatoms. The maximum reproduction of *L. catascopium* coincides with periods of peak primary production (Magnin and Leconte 1971, Stanczkowska et al. 1972, Mackie et al. 1976, Qadri et al. 1977, Rodgers and Qadri 1982, Pinel-Alloul and Magnin 1979a, 1979b).

Production of chironomids in slackwaters (2.4 to 7.6 g m^{-2} yr^{-1}) is greater than in deeper channel sites (0.7 g m^{-2} yr^{-1}), whereas island shores are not a very productive habitat, as evidenced by the midge *Polypedilum* (peak 0.2 g m^{-2} yr^{-1} between September and April, with four cohorts a year; Clair 1976).

Human Impacts and Special Features

In spite of many dams, the Ottawa basin is largely undeveloped with most of the area forested or covered by bodies of water. Particularly noteworthy is the Petrie Island Preserve, a complex of islands and bays in the lower river below the Ottawa–Hull metropolitan area. This patchwork of habitats extends from the confluence of the Gatineau and Ottawa rivers to the latter's confluence with the St. Lawrence at Lac des Deux-Montagnes upstream of the Montreal archipelago. This stretch of small islands in the main channel is similar to both the Thousand Islands region of the upper St. Lawrence and the braided channel at the head of Lac Saint-Pierre in that it creates a mosaic of slackwaters and flooded forest habitats in close proximity to the highly advective main channel. The major difference between the Thousand Islands and the Petrie Islands area is that the latter is occasionally submerged during peak spring discharge period. The Petrie Island Preserve includes heterogeneous aquatic habitats that support very diverse fish, invertebrate, and macrophyte assemblages.

The Ottawa River is now generally considered to be in good condition following extensive environmental controls in the last 20 years. Prior to the early 1980s, the lower and middle reaches of the Ottawa River, with over 100 municipalities and 2000 farms in the drainage basin, had problems with elevated fecal coliform counts and nutrient inputs, especially downstream of metropolitan Ottawa–Hull and Montreal. Since then, more secondary treatment systems have come on-line and stricter regulations and installation of industrial wastewater treatment systems around the city of Gatineau have reduced inputs of copper, lead, and aluminum and cut emissions of persistent organics, such as PCBs. However, the greatest human impacts on the Ottawa River are probably its dams. Using criteria of Dynesius and Nilsson (1994), the river would be considered strongly fragmented because of the seven major dams on the main stem and over 300 impoundments on tributaries.

SAGUENAY RIVER

The second-largest tributary of the St. Lawrence River and the last river to enter the main stem at the downstream terminus of the upper estuary is the Saguenay River (Fig. 22.12). Flowing through a deep, rocky fjord, the Saguenay contributes over $1500\,m^3/s$ of freshwater to the St. Lawrence at a point where the larger river has a mean salinity of ~30% (Fig. 22.1). This freshwater is drawn from an $85,500\,km^2$ watershed that consists mostly of boreal forests, all within the province of Quebec. Before entering the Saguenay, however, most of the basin's tributaries first drain into Lac Saint-Jean, which in turn provides about 75% of flow at the mouth.

Native Americans of the Algonquin Nation were the primary inhabitants of this watershed prior to the exploration of the Saguenay River in 1535 by the Frenchman Jacques Cartier. In fact, the river's name is Algonquian and probably means "water flows out," possibly referring to the river's outlet from the large Lac Saint-Jean. Soon after Cartier explored the Saguenay a flood of European settlers arrived eager to exploit the vast tracks of virgin timber. Harvesting and processing this timber for lumber and paper products, along with mining and smelting of metals, have produced the major negative impacts on the river and its watershed. Agriculture and urbanization have had relatively moderate impacts because agriculture is limited by a short growing season (133 days at Chicoutimi, Quebec) and the urban population density is very low. Today the Saguenay region is the most thoroughly French-speaking region of Quebec and has become the political stronghold of the drive for secession of Quebec from the rest of Canada through support for the Parti Québécois.

Physiography, Climate, and Land Use

The Saguenay River basin is located within both the Laurentian Highlands (LU) and the Labrador Highlands (LB) physiographic provinces of the Precambrian Shield physiographic division (see Fig. 22.12). The geology of these provinces is dominated by silicate rocks and includes igneous and metamorphic (gneiss and granite) bedrock.

The local climate is cool and moist throughout much of the year. Precipitation, which averages 96.5 cm/yr, does not vary substantially among months (monthly range 7.1 to 9.6 cm; Fig. 22.13). The mean air temperature is 3.0°C, with a monthly range of −13°C in January to 17°C in July. Gardeners can count on frost from as early as late September to mid-May (at Chicoutimi). Given the cold winter air temperatures, surface water temperatures range from a low of near 0°C in December through April to a high of 17°C to 19°C in July through August. The surface of the river is frozen through much of the winter.

Balsam fir and paper birch are the dominant trees within this Boreal Forest biome and Eastern Canadian Forest terrestrial ecoregion. These forests are primarily second growth or later because of extensive timber harvesting during the nineteenth and twentieth centuries. The basin is sparsely inhabited (2.5 people/km^2) except for the urban-industrial centers of La Baie, Chicoutimi, Jonquiere, and Alma, located along the main stem of the Saguenay River below Lac Saint-Jean. There is some local agriculture, but it is restricted to the sedimentary soils immediately adjacent to this large lake. Consequently, the watershed is 90% forested, and most of the remaining basin area is composed of surface waters.

River Geomorphology, Hydrology, and Chemistry

The headwaters of the Saguenay River are often considered to be the relatively pristine Lac Saint-Jean, the fifth-largest lake in Quebec at 1053 km². About ~25% of the lake is ≤3 m deep, and the maximum depth is 63 m. Consequently, the high lotic inputs from 21 tributaries combined with a relatively shallow lake basin produce a water-renewal time of only ~2 to 3 months (Côté et al. 2002). This reduces the lentic signature of the lake's outflow to the main stem of the Saguenay. Lac Saint-Jean receives water from a heavily forested and mountainous basin of ~68,000 km². About 75% of the outflow of the lake comes from three principal rivers in this basin: the Péribonka, Ashuapmushuan, and Mistassini. From this lake, the main stem of the Saguenay River flows eastward through a heavily forested watershed of over 17,000 km² into the brackish St. Lawrence River. At least 20 small tributaries feed into the main stem of the Saguenay River from this subbasin.

The main stem of the Saguenay River is ~165 km long, ~0.6 km wide, and has a fairly steep slope (60 cm/km) as it flows through a deeply incised channel. The main stem can be divided into three sections based on depth and hydrodynamic characteristics. Two medium-size tributaries flow into the moderately deep 40 km long upper reach. The ~25 km long middle reach, with its seven tributaries, is the most urban and industrialized portion of the Saguenay. This segment is turbulent, turbid, and less than half as deep (<10 m) as the upper section. Rocky escarpments are common along the shore. In its last 100 km, the Saguenay River flows through the largest fjord in the northwest Atlantic (see Fig. 22.1) and then enters the St. Lawrence River at Tadoussac, Quebec, which is the border between the upper and lower estuaries. Along this 100 km path the river passes through cliffs towering 450 m above its deep (≤250 m) estuarine channel. Salinities in this lower river segment range from slightly brackish (5 PSU) in the top 5 m of the water column to 25 PSU below the thermohalocline. The Saguenay has a tidal amplitude >4 m at its confluence with the St. Lawrence River.

At its mouth the Saguenay contributes on average 1535 m³/s of freshwater discharge to the St. Lawrence, and roughly 75% of this water flows from Lac Saint-Jean. Currents in the Saguenay can reach 2 m/s during spring freshets, in part because of the steep riverbed and in spite of the main-stem dams. Runoff from the basin is concentrated in May and June (mean 9.35 cm/mo for those months) due to spring snowmelt, when it is more than twice as high as the average for the other ten months (see Fig. 22.13). This occurs in spite of the fact that precipitation is relatively uniform among months and peaks in July, when runoff is considerably lower (5.7 cm/mo).

The water chemistry of the Saguenay varies depending on a site's location with reference to tributaries, Lac Saint-Jean, and various municipal and industrial effluents. The main stem has an almost neutral pH (6.9), which varies only slightly among its tributaries (6.8 to 7.4). The range of conductivities among sites (20 to 120 µS/cm) is greater than for pH, and averages 34.9 µS/cm in the main stem. The mean concentrations of other basic water-quality parameters are alkalinity 7.3 mg/L as $CaCO_3$, dissolved O_2 11.3 mg/L, and dissolved organic carbon 6.2 mg/L. Nutrients and primary production vary considerably among sites and seasons, especially with reference to point-source inputs. Mean concentrations for the main stem are chlorophyll *a* 0.95 µg/L, NH_4-N 0.02 mg/L, NO_2-N + NO_3-N 0.10 mg/L, and PO_4-P 0.01 mg/L (Mousseau and Armellin 1995).

River Biodiversity and Ecology

The Saguenay River is located within the Lower St. Lawrence freshwater ecoregion, where it supports flora and fauna similar to those in other nearby tributaries of the St. Lawrence River. However, the strong salinity barrier to freshwater immigration that is present in both the lower Saguenay and the St. Lawrence River upstream of their confluence makes the freshwater communities of the Saguenay more insular and less subject to invasion by nonnative freshwater species.

Discussions of the river's flora and fauna and the functioning of this ecosystem are difficult because aquatic ecologists have rarely studied the ecology of this relatively isolated, cold-water river. In contrast, research linking industrial contamination with human health, recreational fisheries, and the marine park are more common.

Algae and Cyanobacteria

One can infer from the short length (165 km) of the Saguenay River below Lac Saint-Jean and the downstream presence of brackish habitats within the fjord that riverine microalgal communities are dominated by species commonly found in the lake. However, the relative effects of any lentic signature on the Saguenay should be tempered by the lake's short water-renewal time and inflow from >40 tributaries above and below the lake. During the summer in Lac Saint-Jean, the assemblage of phytoplankton species >20 μm is heavily dominated by the diatoms *Asterionella formosa* and *Tabellaria* (*T. fenestra* and *T. flocculosa*), and *Melosira islandica* (Thompson and Côté 1985, Côté et al. 2002). Given that diatoms tend to be relatively more important in rivers than lakes, these genera are also likely to be dominant within the main stem of the Saguenay. Indeed, the three most common species of phytoplankton in the river's brackish fjord are *A. formosa*, *M. ambigua*, *M. islandica*, and *T. fenestrata*. Total summer chlorophyll *a* concentrations and rate of primary production during 1979–1980 ranged from 4 to 79 μg/L and 1 to 88 mg C m^{-3} h^{-1}, respectively (Côté 1983). Periphyton in this section of the river includes the cyanophyte *Lyngbya nordgardii* and the diatoms *Nitzschia holsatica*, *Synedra berolineusis*, *Melosira varians*, and *Oscillatoria* sp. (Mousseau and Armellin 1995).

Plants

Aquatic plants in the Saguenay are mostly restricted to the fjord. Littoral plant assemblages in the upper 65 km of the main stem are mostly limited to sparse patches of American bulrush. The paucity of plants results both from log bashing during commercial log driving in the river and from negative effects of alternating floods and dry-downs caused by operation of power generators. Most aquatic vegetation occurs in freshwater sections of the fjord, where the emergent herbaceous assemblage is dominated by bulrush and, to a lesser extent, arrowhead, sedge, spike rush, buttercup, and silverweed (Mousseau and Armellin 1995). The lower, more brackish section of the fjord is bordered on both shores by cordgrass marshes.

Invertebrates

Studies of benthic invertebrates of the Saguenay River are sorely lacking. In the upper Saguenay River, caddisflies, chironomid midges, and aquatic earthworms are the principal invertebrate taxa. Densities are highly variable, ranging from 200 to >2000 animals/m^2 in habitats where oligochaete worms are a major component. In general, insects and isopods are numerically most abundant, followed by oligochaetes, bivalves, and gastropods (Mousseau and Armellin 1995).

Zooplankton in the Saguenay reflect in part the composition of the rotifer and microcrustacean community in Lac Saint-Jean but presumably with a slight shift to smaller species as one moves downstream. The community appears to be dominated by a few species, such as the rotifer *Polyarthra vulgaris* and the protozoan *Codonella cratera* (Côté et al. 2002). The former is also the dominant rotifer in the St. Lawrence River.

Vertebrates

From a fisheries perspective, the Saguenay can be divided into three zones: Lac Saint-Jean and its tributaries, freshwater portions of the main stem, and brackish sections of the fjord. In Lac Saint-Jean and its tributaries, species like northern pike, walleye, yellow perch, landlocked Atlantic salmon, and white sucker predominate. In the freshwater Saguenay, the assemblage is similar, with the addition of anadromous Atlantic salmon, American eel, brook (speckled) trout, and black (lake) sturgeon. Six species of migratory fishes are known from the Saguenay: rainbow smelt, Atlantic tomcod, Atlantic salmon, and threespine, brook, and ninespine sticklebacks (Mousseau and Armellin 1995).

Historical records from the commercial fishing industry indicate an overall decline in several species since the early to mid-1900s. A combination of chemical pollution, habitat degradation, commercial exploitation, and introduced species is probably responsible. Much of this loss in biodiversity has apparently been caused by acid rain and bioaccumulation of metals and organic chemicals from industrial effluents. Acid runoff from both rainfall and spring snowmelt is a regional problem with special implications for the Saguenay (Brouard et al. 1982). As continuing atmospheric inputs of acid lower the pH of these streams, aluminum, copper, iron, and manganese are liberated from the shield bedrock. Adult salmon, like most fishes, are not particularly susceptible to low pH per se, but chronic exposure can reduce reproductive capacity, depress immunological responses, and promote tumors. In contrast, eggs, smolt, and fry of salmon are especially sensitive to acid precipitation. Current models for these rivers suggest that acidification is ongoing (average pH of rain ~4.3, with high levels of both nitrates and sulfates) and coincident with elevated levels of metals (Van Coillie et al. 1982). Loss of reproductive habitat since the 1930s as a result of construction and maintenance of the locks, dams, and shipping channel has caused further population declines of resident and anadromous fishes. The catadromous American eel is an example of a species hit particularly hard by damming of their migratory pathways in the Saguenay and throughout the St. Lawrence–Great Lakes system. The dams have blocked or killed (by hydro-electric turbines) juveniles migrating upstream through the Gulf of St. Lawrence as well as sexually mature adults migrating downstream.

The low density of humans in the Saguenay River basin has allowed retention of relatively healthy populations of beaver, river otter, muskrat, and mink in the Saguenay's freshwater tributaries. Long-finned pilot whales and beautiful white beluga whales frequent the Saguenay fjord. Their populations dropped to dangerous levels in response to pollution and severe overharvesting, but this decline appears to have been halted by implementation of more rational environmental policies of provincial and federal governments.

At the confluence of the Saguenay and St. Lawrence Rivers is Batture aux Alouettes, which has been identified as a continentally and globally significant site because of its diverse waterfowl and large numbers of migratory shorebirds and raptors. Many migratory shorebirds have been recorded here in spectacularly high numbers, including red knots (1% of the North American population recorded in a single day), sanderlings (1% of the global population spotted in one day), and purple sandpipers. Diving ducks are seasonally common, including oldsquaw, common eider, and common goldeneye. Also observed here frequently are Barrow's goldeneye (a national species of concern) and harlequin duck (a Canadian endangered species). Finally, the site is considered globally significant because of fall flights of migratory raptors, including merlin, peregrine falcon, osprey, and northern goshawk.

Although a few nonnative species have been intentionally introduced (e.g., rainbow trout), the native fauna of the Saguenay is relatively unchallenged by them. Only four fishes (American eel, American shad, Atlantic tomcod, and Atlantic sturgeon) and two aquatic-related birds (great blue heron and black-crowned night heron) are included on the provincial list of endangered species in the freshwater portion of the river.

Ecosystem Processes

Because of its high discharge the river substantially influences primary productivity downriver in the brackish estuarine habitats. Although the highly stratified fjord has consistently low primary productivity due to the strong mixing action of the tides and the high flushing rate from Saguenay River discharges, transport of allochthonous carbon through the fjord and estuary has been linked to increased algal and vertebrate production in the gulf. Seasonal

dynamics are characterized by little or no spring–summer phytoplankton bloom. When annual blooms in the middle and lower Saguenay begin, they are usually closely correlated with low summer discharge and improved photic conditions. These changes shift limitations of primary production from light to nutrients. The Saguenay fjord is strongly stratified both thermally and by salinity. The top 5 m are fresh or brackish, and this thin and unstable photic zone strongly limits primary production in the fjord. Phytoplankton production is low ($25 \, \text{mg} \, \text{C} \, \text{m}^{-3} \text{hr}^{-1}$) and highly variable on a tidal regime; coefficients of variation range from 55% to 127% as a result of advective processes (Côté and Lacroix 1979a, 1979b). Consequently, historical carbon exports from the freshwater Saguenay to the estuarine food web have been mainly recalcitrant organic matter. Much of the river's suspended load is derived from weathering of local bedrock and soils, more than 90% of which accumulates in sediments of the fjord and gulf (Louchouarn et al. 1999). Since the advent of timber and paper industries, allochthonous lignin from timber paper wastes is the main form of carbon export; half of this material is quickly assimilated into marine carbon cycles (Louchouarn et al. 1999, Louchouarn and Lucotte 1998). Rich krill populations result when these exports are combined with intrusions of cold nutrient-rich upwellings in the gulf.

Human Impacts and Special Features

The Saguenay River system has several unique features, beginning with the main stem flowing out of a large lake (Lac Saint-Jean) that is fed by numerous rivers draining boreal forest. Furthermore, the large lower Saguenay flows through a spectacular fjord with towering cliffs before joining the St. Lawrence River at Batture aux Alouettes, a continentally and globally significant site for migratory birds. Among the unusual biological features of the river system are some benthic invertebrates in the fjord that are arctic relicts, two genetically distinct races of Atlantic salmon, and congregations of long-finned pilot whales and white beluga whales. Given that the human population density is so low in this basin, one might expect a relatively pristine river. Unfortunately, this is not the case, as human activities have resulted in negative impacts on the river for almost 100 years.

Although some agricultural activities occur in the Saguenay River basin, timber harvesting and paper manufacturing plants dominated the local economy until the mid-1900s, when aluminum mining and smelting and chlor-alkali refining began growth spurts. As a result, human impacts on the river have been severe and include watershed disturbance (timber harvesting, mining, and some agriculture), point- and nonpoint-source pollution, and channel regulation for hydroelectric production and navigation beginning in the late 1930s. At present there are three hydroelectric/navigation dam complexes on the main stem and more than 300 small to medium-size dams on the >40 tributaries above and below Lac Saint-Jean.

Much of the industrial pollution can be traced as far back as 1910 and has often been associated with early techniques in paper manufacturing. Persistent bioaccumulating chemicals, such as mercury, PCBs, and PAHs, represent the most severe threat to nature and humans in and around the Saguenay River. Improvements in manufacturing technology since the worst contamination of the 1950s to late 1970s, however, have led to a steady rise in water quality. Of the freshwater sport fishes in the river, only walleye continue to have mercury body burdens exceeding recommended consumption guidelines for humans.

In contrast to many rivers, eutrophication has never been a widespread problem in the Saguenay because of its fast flushing rates. Moreover, a low human population density (>2.5 people/km^2), the small number of livestock farms (~2000) relative to the size of this large basin, and secondary wastewater treatment facilities serving more than half the municipal population in the basin have contributed to low eutrophication rates. The major nutrient-related problems still existing are primarily bacterial and are related to the small percentage of the population not serviced by secondary wastewater treatment plants.

With the recent decline in the natural resource extraction economies and the creation of parks and nature preserves (the Saguenay–St. Lawrence Marine Park and the Laurentides and Saguenay National Park), recreation and tourism have become focal points for regional economic development. This shift to a tourist-based economy has helped reverse some of the negative impacts from resource extraction and manufacturing. However, other threats to the river and its tributaries exist. These include the recently proposed diversion of a significant percentage of the Manouane River's discharge (a tributary of the Peribonca River, the Saguenay's largest tributary) into the Pipmuacan Reservoir on the Betsiamites River for hydropower.

ST. JOSEPH RIVER

The St. Joseph River basin occupies portions of southwestern Michigan and north-central Indiana, with about 60% of its 12,150 km^2 basin residing in Michigan and 40% in Indiana (Fig. 22.14). The St. Joseph River is a major tributary of southern Lake Michigan, and with an average discharge of nearly 100 m^3/s it is one of the larger watersheds draining into any of the Great Lakes. The river arises near the town of Hillsdale in south-central Michigan and flows in a southwesterly direction as far as South Bend, Indiana (Fig. 22.4), bending southward into Indiana (hence the city's name) and then abruptly northward before returning to Michigan and emptying into Lake Michigan at the city of St. Joseph. Only about 20% of the ~100 km long main stem is in Indiana, a state that includes a northeastern tributary of the Maumee River that is also called the St. Joseph River. The St. Joseph River of interest here has a diverse fauna that includes nearly 100 native fish species and 23 species of mussels and clams.

The St. Joseph River has considerable historical significance (Wesley and Duffy 1999). Native Americans of the Miami, Iroquois, and Potawatomi tribes were the main inhabitants of the river basin prior to European settlement. The Miami tribe of the Algonquin Nation settled the basin around 1000 to 1200 years ago and called the river *Sauk-Wauk-Sil-Bauk*, meaning "mystery river." The Algonquin clearly harvested the abundant fishes of the St. Joseph River using traps and weirs, as shown in bones from archaeological digs. In the 1600s, the Iroquois drove most the Miami Indians from the river basin because of competition in the fur trade. Potawatomi Indians occupied the basin until the mid-1800s, when Europeans began to settle heavily in the area and clear the land for cultivation.

The French explorer LaSalle is generally credited with being the first European to float the St. Joseph River when he worked his way upstream in 1679 from Lake Michigan to what is now South Bend. From there, LaSalle portaged to the Kankakee River to begin his journey to the Mississippi River. However, another Frenchman, Medard Chauart des Groseilliers, may have canoed the river as early as 1654. Control of the river alternated between the French, British, and Indians until the Revolutionary War forged its eventual status as an American river. A French Catholic colony was established on the banks of the St. Joseph River at South Bend in the mid-1800s for the purpose of establishing the

FIGURE 22.4 St. Joseph River near South Bend, Indiana (PHOTO BY G. LAMBERTI).

University of Notre Dame. The river was valued early for its fisheries and as a navigational corridor, but its principal uses now are for hydropower, water supply, waste disposal, and recreation.

Physiography, Climate, and Land Use

The St. Joseph basin, with its generally flat, glacially molded topography, is within the Central Lowland (CL) physiographic province of the Interior Plains

(see Fig. 22.14). During the Pleistocene epoch numerous glacial lobes entered the river basin from Lakes Michigan and Erie. The advancing glaciers scoured and flattened the land surface into till plains, whereas the retreating glaciers deposited scoured materials as moraines. Well over 200 natural lakes of various sizes are found in the basin because of its glacial history, which resulted in burial of ice blocks, damming of stream channels by moraines, and irregular deposits of glacial drift that trapped water.

The geology of this basin is complex and largely reflects its glaciated history. The basin's deep basement rock consists of Precambrian igneous materials, such as granite and basalt. More shallow bedrock consists mostly of layered Paleozoic materials, including limestone, sandstone, siltstone, and shale, which were deposited by ancient inland seas. The bedrock is covered by a layer of glacial drift material, often >100 m deep. The overlying drift material is heterogeneous, ranging from sands to boulders, but mostly consists of sands and gravels. In many depressional areas, organic muds and peats have accumulated. Gravel intermixed with clay and loam tills are characteristic of the common outwash plains of the basin. Soils of the basin generally fall into one of three classes (State of Indiana 1987): (1) sandy or loamy soils developed on outwash and alluvium, (2) silt or clay developed on till, or (3) muck soils developed in depressional wetland areas.

The climate of the St. Joseph River basin is classified as temperate continental and is characterized by warm to hot summers, cold winters, and reasonably even year-round precipitation (4.8 to 10.4 cm/mo; Fig. 22.15). Mean monthly temperatures range from –4.8°C in January to 25°C in July, with an annual average of 9.7°C. Summer precipitation often occurs from thunderstorms and winter precipitation occurs as snow, which can be heavy in some years. For example, annual snowfall in South Bend averages 180 cm. Local precipitation amounts exceeding 5 cm in a day are not unusual and can result in flash flooding in tributaries of the St. Joseph River. Humidity is moderate to high, especially in the summer, when southern air masses penetrate beyond the Ohio River valley. The climate of the western portion of the basin is influenced by Lake Michigan, which produces lake-effect snow and rain but moderates the air temperature close to the lake such that a fruit-growing district, including vineyards, is possible.

From a floristic perspective, the basin is classified as a Temperate Deciduous Forest biome within the Southern Great Lakes Forest terrestrial ecoregion, although much of the landscape has been dramatically altered by humans. Row-crop agriculture is the primary land use in the basin and now occupies about 60% of the total land cover. Forests cover another 20% of the watershed, and urban/residential land use accounts for about 8% of the remaining 20%. Common forest trees are ash, elm, cottonwood, maple, oak, and poplar.

River Geomorphology, Hydrology, and Chemistry

The generally flat topography of the glaciated landscape within the St. Joseph River basin results in a small elevational change from river source to mouth (total 200 m, or about 20 cm/km in the main stem). The highest gradients (≤8 m/km) occur in the headwaters, but even these steep slopes do not match some historical gradients on the main stem that now have been inundated by impoundments, thereby also eliminating and fragmenting aquatic habitats.

The headwaters of the St. Joseph River have a narrow stream channel that is straight to meandering and has a cross-section typical of lower-gradient streams. The channel widens in the middle section with increased flow and an unconfined valley, but narrows again in the lower reaches, where it is restricted by a narrow glacial valley. Near its mouth at Lake Michigan the steam channel widens again as it flows over lake deposits. Substrate throughout the basin is generally small particles of silt, sand, and gravel, but some cobble is found in the lower section, where glacial tills predominate. The abundance of woody debris varies considerably in the tributaries related to the degree of human clearing of riparian vegetation and dredging of channels. Woody debris is relatively common along channel margins of the main-stem river, but large debris dams have been removed to accommodate navigation and recreation. Most of this clearing was accomplished by the mid-1800s, when the river was already being used as a "superhighway" to transport logs and grains by steamship and keelboat. At that time, several dams also were built on tributaries to the St. Joseph River to supply power for sawmills and grain mills.

The permeable glacial deposits of the St. Joseph basin result in fairly stable streamflows throughout the basin (mean annual discharge 96.3 m³/s) because most precipitation readily infiltrates the surficial geology and enters stream channels via groundwater (see Fig. 22.15). Exceptions occur in highly urbanized areas and intense agriculture areas because of reduced soil permeability and active drainage. Although precipitation is highest in summer, maximum runoff (>4 cm/mo) occurs in March and April as a result of spring snowmelt (see Fig. 22.15). Runoff declines to its low of about 1.6 cm/mo during August and September, apparently as a result of increased summer evapotranspiration.

Water chemistry is highly influenced by the geology of the basin, which includes extensive limestone deposits, and therefore the water naturally carries high loads of dissolved salts (State of Indiana 1987). For example, conductivity is typically 200 to 600 µS/cm and total alkalinity is 200 to 400 mg/L as $CaCO_3$. This high alkalinity effectively buffers against pH change, which ranges from 6.0 to 8.9 in the basin but is generally 7.5 to 8.0. Dissolved oxygen is quite variable, ranging from 1.2 to 10.9 mg/L in the main channel during the summer. The lower values occur in main-stem impoundments, whereas typical concentrations in the free-flowing portions of the river are 8 to 10 mg/L. Nutrient concentrations are relatively high throughout the basin because of extensive agricultural activities. Soluble reactive phosphorus typically varies from 0.02 to 0.10 mg P/L and NO_3-N ranges from 0.7 to 2.9 mg/L. However, spot measurements of nitrate in some tributaries have approached or exceeded 10 mg/L, which is the USEPA drinking water standard.

River Biodiversity and Ecology

Located within the Michigan–Huron freshwater ecoregion, the St. Joseph River and its tributaries have a relatively rich fauna and flora of aquatic species. Unfortunately, the pressures of human exploitation of the watershed have eliminated some species and continue to threaten others. Federally or state endangered fauna include river otter, fishes (greater redhorse, creek chubsucker), turtles (spotted, Blanding's, alligator snapping), snakes (eastern massasauga, copperbelly water snake), and caddisflies (*Setodes oligius*). Threatened species consist of fishes (lake sturgeon, river redhorse) and marbled salamander. Species of special concern include fishes (starhead topminnow, black buffalo, spotted gar), eastern box turtle, frogs (Blanchard's cricket, northern leopard), blue-spotted salamander, snails (pointed campeloma, swamp lymnaea), and Douglas stenelmis riffle beetle. Approximately 105 riparian plant species are endangered, threatened, or of special concern.

Aquatic macrophytes and algae have been poorly studied within the St. Joseph basin in comparison to invertebrates and fishes, and will therefore not be discussed in great detail beyond noting that wild celery and various species of pondweed are common types of aquatic macrophytes in streams of this basin. On an ecological note of interest, Horvath and Lamberti (1997) reported that wild celery was a vector for dispersal of zebra mussels into lake-outflow streams in the St. Joseph River drainage. Broken pieces of this plant from lakes served as "lifeboats" for attached zebra mussels, which could then drift many hundreds of meters in outflowing streams. Many tributary streams to the St. Joseph River exhibit interesting seasonal patterns of algal growth related to canopy cover and other factors. Shading by dense riparian canopy results in low standing crops of algae in summer, but dense algal blooms (especially *Cladophora* sp.) often completely cover the streambed by late winter or early spring in response to higher light after fall leaf abscission and perhaps elevated nutrients from the spring thaw (Lamberti and Berg 1995).

Invertebrates

The St. Joseph River has a rich array of invertebrate life, including 23 recorded species of native mussels and clams. Mussels have persisted in the river despite considerable commercial clamming pressure on some species in the early 1900s to provide shells for the button industry. It is now unlawful to harvest native mussels in Michigan without a scientific collector's permit. More recent threats to sensitive mussel species include chemical pollution and dams that halt the migration of host fishes that disperse the larval glochidia. The river basin also contains at least two species of nonnative clams: the zebra mussel and the Asiatic clam. Asiatic clams are found in sandy reaches of many streams in the basin. The spread of zebra mussels in the basin has been particularly well documented because of their recent invasion (1991). Horvath et al. (1996) correctly predicted with a "source–sink model" that lakes would have large self-sustaining populations, but streams would rely for recruitment on upstream reservoirs to sustain small populations (Horvath and Lamberti 1999). Where zebra mussels are abundant, native bivalve mollusks have largely been extirpated, because zebra mussels thoroughly encrust shells and hinder feeding and respiration.

Benthic invertebrates have been studied in only a few locations throughout the basin, and a complete species list for the basin is lacking. Some examples of important macroinvertebrates in the basin are oligochaete worms, bivalve mollusks (Corbiculidae, Dreissenidae, Sphaeriidae, Unionidae), snails (Bithyniidae, Hydrobiidae, Physidae, Pleuroceridae, Planorbidae), isopod crustaceans (*Caecidotea*), amphipod crustaceans (*Gammarus*), crayfish (*Orconectes*), beetles (*Dubiraphia*, *Macronychus*, *Optioservus*, *Stenelmis*), true flies (*Antocha*, biting midges, black flies, *Chelifera*, chironomid midges, *Hemerodromia*, *Hexatoma*, *Tipula*), mayflies

(*Baetis, Caenis, Ephemera, Hexagenia, Isonychia, Serratella, Stenacron, Stenonema, Tricorythodes*), alderflies (*Sialis*), damselflies (*Enallagma*), stoneflies (*Amphinemura, Paracapnia, Taeniopteryx*), and caddisflies (*Cheumatopsyche, Glossosoma, Hydropsyche, Hydroptila, Lepidostoma, Micrasema, Mystacides, Nectopsyche, Oecetis*).

Although information for the whole basin is sparse, concentrated invertebrate studies have been conducted in Juday Creek (South Bend, Indiana), a tributary of the St. Joseph that flows through the campus of the University of Notre Dame. This stream is of further interest because it contains naturally reproducing populations of trout related to high inputs of cool groundwater. R. A. Hellenthal and his students have studied benthic macroinvertebrates in the stream since the early 1980s (Schwenneker 1985, Berg 1989, Kohlhepp 1991, Latimore 2000), and Lamberti and Berg (1995) summarized some of these studies for the period from 1981 to 1992. Juday Creek is a 3rd order, cool-water stream whose basin experienced substantial land-use changes over that period, as woodland and some agriculture were converted to urban areas. By 1992 the watershed was dominated by agriculture (45%) and urbanization (31%), whereas only 7% of the basin remained as original woodland or prairie. A total of 119 taxa of invertebrates have been recorded from Juday Creek, and total richness did not change appreciably in the stream over the study period.

Vertebrates

Historically, the St. Joseph River contained a rich native assemblage of fishes. Wesley and Duffy (1999) document 97 native fish species in the St. Joseph River basin, although this list contains limited information about fish communities before the 1880s and prior to European settlement. Fish bones found in archaeological digs indicate that Potawatomi Indians harvested lake sturgeon, bowfin, northern pike, river redhorse, channel catfish, crappie, walleye, and freshwater drum from the river. Many additional species have been recorded in the last century, including brook trout, smallmouth bass, bluegill, white sucker, hornyhead chub, creek chub, yellow perch, logperch, pirate perch, blacknose dace, blackside darter, rainbow darter, bluntnose minnow, common shiner, common stoneroller, central mudminnow, northern hogsucker, mottled sculpin, channel catfish, and northern madtom. Many of these native species are now restricted to small portions of the basin by cumulative habitat change, nonnavigable dams, or other sources of fragmentation. Approximately 17 species of nonnative fishes occur in the basin, many of which are salmonids.

Lake sturgeon historically migrated in large numbers from Lake Michigan into the river to spawn all the way to its headwaters. Many were huge (≤130 kg and 3.5 m long) and heavily sought after by anglers in the late 1800s. A strong commercial market existed for sturgeon meat, and eggs were sold to Russia as caviar. Construction of dams in the lower river, beginning in the late 1800s, gradually eliminated sturgeon runs because they could not navigate fish ladders designed for salmonids. Spawning is now limited to the short, undammed lower river reaches.

Fishery management in Indiana and Michigan has concentrated on game fishes and stocking programs. Pacific salmon garner most attention, and steelhead, coho salmon, and chinook salmon are reared for release into the river. These fishes migrate as smolts to Lake Michigan and return on spawning runs two to five years later. Stocking efforts have created a major sport fishery in the lower portions of the river. Rainbow trout are also planted in both Michigan and Indiana as mostly a put-and-take fishery, and brown trout have been sporadically stocked in the past. Concerns exist about effects of these nonnative salmonids on native fishes, especially brook trout, which are now uncommon in the basin. Fishery managers have given more attention recently to warmwater fishes, such as walleye, northern pike, channel catfish, and smallmouth bass.

Although fishes represent the greatest diversity of vertebrates in the St. Joseph River, there are healthy populations of mammals, reptiles, amphibians, and birds. Aquatic mammals in this basin are dominated by muskrat but also include beaver, mink, and a small population of river otter. Fur-bearing mammals have declined in the last two centuries, but they played an important role in the early history of colonization by non–Native Americans and competition among Indian tribes within this basin. Amphibians and reptiles are represented by 32 known species in the basin.

Ecosystem Processes

As in many large rivers, the main stem of the St. Joseph River has been understudied in terms of ecosystem processes, such as primary and secondary production, decomposition, metabolism, and nutrient cycling. One study of note by Biddanda and Cotner (2002) demonstrated that the St. Joseph River, combined with three other rivers flowing into southern Lake Michigan, annually supplied about 5% of the carbon demand of planktonic bacteria and about 10% of the phosphorus needed by planktonic algae. Considering that these four rivers in aggregate only contribute about 1% of the volume of southern Lake Michigan on an annual basis, the river inputs are clearly important to lake productivity. On average, the St. Joseph River carried about 5 mg/L of dissolved organic carbon (mostly high quality) and 18 μg/L of total dissolved phosphorus into Lake Michigan.

In the early 1980s, secondary production of all nonchironomid benthos in Juday Creek near the University of Notre Dame campus was about 2.8 g m^{-2} yr^{-1} dry mass, but this declined 75% to about 0.7 g m^{-2} yr^{-1} by the early 1990s, apparently due to excess siltation of the stream (summarized by Lamberti and Berg 1995). Kohlhepp (1991) reported similar secondary production (to the early 1980s level) in Brandywine Creek, a Michigan tributary of the St. Joseph River that had much less sedimentation than Juday Creek. In contrast, secondary production of collector-gatherers increased by almost 300% over that period in Juday Creek, and collector-gatherer invertebrates contributed the bulk of the secondary production in both these streams. Interestingly, Berg and Hellenthal (1991) later found in Juday Creek that chironomid midges, which are often ignored in production studies, contributed much more production, about 29.7 g m^{-2} yr^{-1}, than all other benthic invertebrates combined. They termed this omission in most production estimates "the standard error of the midge." On a seasonal basis, midges dominated production in the autumn, winter, and spring (when benthic algae also were most abundant), whereas nonchironomid taxa dominated production in summer. Overall results suggest that Juday Creek historically supported a productive macroinvertebrate fauna that has gradually declined with deleterious change in land use within the watershed, a trend perhaps symptomatic of the entire basin.

The morphology and hydrology of many streams of the St. Joseph River watershed, especially the main stem, have been greatly altered by human activities. In a few remnant reaches of certain tributaries, however, reasonably natural conditions persist, including substantial amounts of large woody debris (LWD) in the channel. Ehrman and Lamberti (1992) studied LWD in Juday Creek and its influence on the retention of coarse particulate organic matter (leaves and sticks). A wooded reach of Juday Creek averaged 63 pieces of LWD and 16.3 m^3 LWD volume per 100 m of stream length, levels reasonably comparable to streams in old-growth forests of the Pacific Northwest (Harmon et al. 1986). Retention of CPOM was quite high in these reaches, with 68% of released leaves and 75% of sticks retained in 100 m of channel. Therefore, it is reasonable to infer that, at least historically, these streams were highly efficient at trapping and processing organic matter delivered by the riparian zone. In the past 100 years, however, much of this retentive capacity has been lost because of channel alteration.

Human Impacts and Special Features

The St. Joseph River is one of the larger watersheds draining into any of the Great Lakes and is of historical significance in the lives of Native Americans and settlers from other continents. Its diverse fauna includes nearly 100 native fish species and 23 species of mussels and clams, but the diversity and distribution of biota have been greatly influenced by watershed changes such as dam construction, channel modification, point-source pollution, nonpoint-source pollution, agricultural and urban land use, and introductions of nonnative species.

Point sources of pollution include 221 industrial discharges to surface waters in the basin, which are permitted to contain certain limits of parameters of concern (e.g., dissolved oxygen, metals, nutrients, organic compounds, temperature). Combined sewer overflows plague most cities and result in periodic overflows of raw sewage into rivers during heavy rain events; such outflows have sporadic but acute effects on aquatic biota.

Extensive agricultural activities and concentrated urbanization in the basin have resulted in diffuse inputs of nutrients, contaminants, and fine sediment to the river. These generally nonpoint-source inputs are probably the most important current source of water-quality impairment. Intensive agriculture of mostly row crops has required clearing of forests, draining of wetlands, and channelization of streams. Tilling of agricultural fields for drainage has altered streamflow and temperature regimes, whereas riparian clearing has resulted in bank erosion and sedimentation of many small tributaries in the basin (Wesley and Duffy 1999). Urbanization has increased direct runoff of precipitation and inputs of sediments, contaminants, and other pollutants to streams (e.g., Lamberti and Berg 1995). Wetlands have been greatly reduced in the watershed and therefore their water-filtration capacity is now limited. In many cases, agriculture has extended the stream network with the construction of ditches that drain agricultural fields and therefore carry agricultural runoff. Fine sediments and elevated nutrient levels appear to be chronic problems in the basin (Lamberti and Berg 1995, Wesley and Duffy 1999). The net cumulative effect of these practices has been reduced water quality in the larger tributaries and main stem of the St. Joseph River and chronic sedimentation problems.

Fragmentation of streams within the basin by habitat change and impoundments has had particularly severe impacts on biota. The basin contains 17 dams on the main stem and 173 on tributaries (Wesley and Duffy 1999). Dams have been classified as being for recreation (105 dams), hydroelectric power generation, irrigation, municipal water supply, or other functions. Only five dams on the lower main stem contain fish-passage structures (primarily for introduced salmonids), and these allow fishes to migrate only as far as the salmon and trout hatchery at Mishawaka, Indiana. At this point an impassable dam blocks migratory fishes from reaching the remaining 200 km of main stem and tributaries. Dams also create intermittent lentic environments where water-quality problems, such as reduced dissolved oxygen levels, often occur seasonally.

In recent years the St. Joseph River Basin Commission has focused on improving the condition of the river through proper planning and pollution reduction, and they have been helped at the local level by grassroots organizations. The latter include the Friends of the St. Joseph River Association and Friends of Juday Creek, which have often focused on habitat improvement through stream restoration. In Juday Creek, two channelized reaches were rerouted into remnant woodland, and the creek also received gravel, cobble, and large woody debris additions to enhance habitat for brown and rainbow (steelhead) trout. The restoration increased pool habitat, woody debris, flow microhabitats, and substrate diversity in the new reaches, which led initially to rapid colonization and use of the restored reaches by aquatic biota (Latimore 2000, Moerke 2000). Although length-frequency distributions of salmonids broadened and spawning activity initially increased in the restored reaches compared to channelized reaches (Moerke and Lamberti 2003), salmonid young-of-year recruitment declined over time, suggesting increased egg mortality. Through time, fine sediment deposition in restored reaches appeared to have

negated the positive effects of the habitat enhancements for fishes and macroinvertebrates (Latimore 2000). Long-term success of stream restorations in the St. Joseph River basin, as well as for most Midwestern streams, may depend on parallel attention to overall watershed conditions to control sources of degradation.

ADDITIONAL RIVERS

Flowing into Lake Superior along its south shore within the Superior freshwater ecoregion is the relatively pristine 4th order Ontonagon River (Fig. 22.16). Its 3569 km^2 watershed in the upper peninsula of Michigan is unusual for many Great Lakes tributaries in being heavily forested (>80%) and containing a National Wild and Scenic River designation. Impressive waterfalls flowing over basalt outcrops interblend with many beautiful lakes, ephemeral streams, and main-channel habitats in the Sylvania Wilderness of the Ontonagon River basin. About 50 species of fishes and many more invertebrates thrive within the Ontonagon River.

Lake Huron receives water from Michigan's 5240 km^2 eastward-flowing AuSable River basin, one-third of which is in the Huron National Forest (Fig. 22.18). The 4th order AuSable River, like the Ontonagon, is heavily forested and included within the Western Great Lakes Forests terrestrial ecoregion, but it is in the Michigan–Huron freshwater ecoregion. Nearly 40 km of the river has been designated as a National Wild and Scenic River and 560 km of the river's main stem and tributaries are included under the State Natural Rivers program. Its waters are teeming with nearly 100 species of fishes, abundant fur-bearing mammals (including river otter), and many species of other vertebrates, invertebrates, algae, and aquatic plants. Over 60 dams from 1 to 10 m high have been built on the AuSable River, with 6 hydroelectric dams on the lower river (Fig. 22.5).

Emptying into the southwestern basin of Lake Erie is the 7th order Maumee River, which draws water from a basin of 16,458 km^2 in the Erie freshwater ecoregion (Fig. 22.20). This is a river of contrasts in many ways. Clear waters of state-designated scenic river sections within some Ohio and Indiana tributaries of the Maumee support a rich fauna of fishes and invertebrates, but reaches of the lower main stem near Toledo, Ohio, were identified as an EPA Area of Concern because of high concentrations of mercury and PCBs. The diversity of fishes

FIGURE 22.5 AuSable River near mouth, Michigan (PHOTO BY TIM PALMER).

(>60 species) is lower than in the AuSable, but the Maumee attracts many fishermen because it has the largest population of migrating walleyes east of the Mississippi River. Much of the original land cover of forests in this Southern Great Lakes Forests terrestrial ecoregion have given way to agriculture (54% to 88% of the watershed) and urbanization, which have negatively affected the river ecosystem. As a result of the actions of concerned citizens and government agencies, however, the waters of the Maumee are gradually improving.

Arising from the southwestern slopes of the Adirondack Mountains of New York, the 5th order Black River flows through a 5057 km² basin before entering northeastern Lake Ontario (Fig. 22.22). Much of the upper watershed is heavily forested (75% of the entire basin is temperate deciduous forest) and quite steep in comparison to both the lower portion of its basin and the watersheds of most other Great Lakes tributaries. The upper basin is replete with many clear, cold mountain lakes that

feed the Black River within the Ontario freshwater ecoregion. Several small dams are found on the main stem and its tributaries. Small quaint towns, such as Old Forge, New York, draw many tourists eager to canoe and hike in the surrounding sugar maple and beech forests that occur within the wild natural areas in the Eastern Great Lakes Forest terrestrial ecoregion. Although its watershed is only one-third the size of the Maumee River basin, the Black River contains nearly as many fish species (52), though far fewer than are present in the AuSable River, which has a comparable basin size to the Black River but half its annual discharge.

Many tributaries flow into the main stem of the St. Lawrence River, but one of the most unusual is the Rivière Richelieu. Its basin is included with the New England/Acadian Forest terrestrial ecoregion and the Lower St. Lawrence freshwater ecoregion. Over 80% of this large watershed (23,772 km²) is spread through northeastern New York and northwestern Vermont, where many tributaries feed the

FIGURE 22.6   Rivière Richelieu at Saint-Jean-sur-Richelieu, Quebec (Photo by G. Winkler).

large natural Lake Champlain (called the 6th Great Lake by some enthusiastic proponents) (Fig. 22.24). Although the Richelieu is listed here as a 7th order river, this has little utility for comparative purposes because few other 7th order rivers have a watershed that includes such a large lake. Although the broad Lac Saint-Jean is a major feature of the upper Saguenay River watershed, this shallow lake does not exert the same influence on the outflowing river as the deep Lake Champlain. From this lake, the Rivière Richelieu flows north over Fryer Rapids, through Quebec, and into the St. Lawrence slightly downstream and across the river from Montreal (Fig. 22.6). Because much of the land use below Lake Champlain is heavily urban and agriculture, the water quality of this river has suffered.

The 43,427 km^2 watershed of the 7th order Rivière Saint-Maurice in Quebec is squeezed between the watersheds of two larger rivers, the Ottawa and Saguenay (Fig. 22.26, Fig. 22.7). Like those rivers, the Saint-Maurice partially drains a physiographic province in the Precambrian Shield, which affects the nature of its water (making it softer) and makes the watershed more susceptible to acid precipitation. A substantial portion of this watershed is included within several Canadian national (La Mauricie National Park) and provincial faunal reserves and parks. Although only a little over 30 species of fishes are recorded for this river, this figure may be an underestimate reflecting insufficient study. Seven dams are present on the main stem of this river.

ACKNOWLEDGMENTS

We appreciate the information supplied by J. David Allan, Doug Carlson, Michael Eberle, Tim Mihuc, and Rodney Renkenberger on tributaries of the Great Lakes and the St. Lawrence River. This chapter benefited greatly from information contained in the excellent Canadian government publication on the St. Lawrence River by the St. Lawrence Centre (1996).

FIGURE 22.7   Rivière Saint-Maurice at Shawinigan, Quebec (PHOTO BY G. WINKLER).

LITERATURE CITED

Abell, R. A., D. M. Olson, E. Dinerstein, P. T. Hurley, J. T. Diggs, W. Eichbaum, S. Walters, W. Wettengel, T. Allnutt, C. Loucks, and P. Hedao. 2000. *Freshwater ecoregions of North America: A conservation assessment.* Island Press, Washington, D.C.

Ahmad, A., A. Chodorowski, and P. Legendre. 1974. Studies on the effects of pollution on the diatom communities of the St. Lawrence River near Montreal. In *Proceedings of the 9th Canadian Symposium of Water Pollution Research*, pp. 135–141.

Alaerts-Smeesters, E., and E. Magnin. 1974. Étude preliminaire du phytoplancton du Lac Saint-Louis, elargissement du fleuve Saint-Laurent pres de Montréal, Québec. *Canadian Journal of Botany* 52:489–501.

Alerich, C. L., and D. A. Drake. 1993. *Forest statistics for New York: 1980 and 1993*. Resource Bulletin NE 132. USDA Forest Service Northeastern Forest Experiment Station.

Barth, J. A. C., J. Veizer, and B. Mayer. 1998. Origin of particulate organic carbon in the upper St. Lawrence: Isotopic constraints. *Earth and Planetary Science Letters* 162:111–121.

Basu, B. K., J. Kalff, and B. Pinel-Alloul. 2000a. The influence of macrophyte beds on plankton communities and their export from fluvial lakes in the St. Lawrence River. *Freshwater Biology* 45:373–382.

Basu, B. K., J. Kalff, and B. Pinel-Alloul. 2000b. Midsummer plankton development along a large temperate river: The St. Lawrence River. *Canadian Journal of Fisheries and Aquatic Sciences* 57 (Supplement 1):7–15.

Basu, B. K., and F. R. Pick. 1996. Factors regulating phytoplankton and zooplankton biomass in temperate rivers. *Limnology and Oceanography* 41:1572–1577.

Bensch, I. 1993. *Resources and trends of the Maumee River Basin, Indiana.* Indiana Department of Natural Resources.

Berg, M. B. 1989. The role of Chironomidae in stream insect secondary production. Ph.D. diss., University of Notre Dame, Indiana.

Berg, M. B., and R. A. Hellenthal. 1991. Secondary production of Chironomidae (Diptera) in a north temperate stream. *Freshwater Biology* 25:497–505.

Biddanda, B. A., and J. B. Cotner. 2002. Love handles in aquatic ecosystems: The role of dissolved organic carbon drawdown, resuspended sediments, and terrigenous inputs in the carbon balance of Lake Michigan. *Ecosystems* 5:431–445.

Bourgault, D., and V. G. Koutitonsky. 1999. Real-time monitoring of the freshwater discharge at the head of the St. Lawrence estuary. *Atmosphere–Ocean* 37:203–220.

Brouard, D., M. Lachance, G. Shooner, and R. van Coillie. 1982. Sensitivity to acidification of four salmon rivers on the north shore of the Saint Lawrence (Quebec). Canadian Technical Report of Fisheries and Aquatic Sciences 1109F.

Charlebois, P. M., and G. A. Lamberti. 1996. Invading crayfish in a Michigan stream: Direct and indirect effects on periphyton and macroinvertebrates. *Journal of the North American Benthological Society* 15:551–563.

Clair, T. A. 1976. Secondary production of the Chironomidae (Insecta: Diptera) in a section of the Ottawa River near Ottawa-Gatineau, Canada. Master thesis, Department of Biology, University of Ottawa, Ontario.

Côté, R. 1983. Aspects toxiques du cuivre sur la biomasse et la productivité du phytoplancton de la Rivière Saguenay, Quebec. *Hydrobiologia* 98:85–95.

Côté, R., D. Bussières, and P. Desgagné. 2002. Spatiotemporal distribution of phytoplankton and zooplankton in Lake Saint-Jean (Quebec), a hydroelectric reservoir. [French.] *Revue des Sciences de l'Eau* 15:597–614.

Côté, R., and G. Lacroix. 1979a. Daily variability of chlorophyll-*a* and the level of primary production in the Saguenay Fjord. *Naturaliste Canadien* 106:189–198.

Côté, R., and G. Lacroix. 1979b. Influence of high and variable freshwater run-offs on the primary production in a subarctic fjord. *Oceanologica Acta* 2:299–306.

Croskery, P. 1974. Habitat preference and frequency groupings of cladocera in a segment of the Ottawa River, near Ottawa Canada. Master thesis, Department of Biology, University of Ottawa, Ontario.

Cushing, C. E., and J. D. Allan. 2001. *Streams: Their ecology and life.* Academic Press, San Diego.

De Seve, M., and M. Goldstein. 1981. The structure and composition of epilithic diatom communities of the St. Lawrence and Ottawa rivers in the Montreal area. *Canadian Journal of Botany* 59:377–387.

DesGranges, J.-L., and A. Desrosiers. 1995. Sites de nidification et dynamique de population de Grand Héron du Saint-Laurent. Environment Canada, Canadian Wildlife Service and Ministère de l'Environnement et de la Faune, Direction de la faune et des habitats, Quebec.

Désilets, L., and C. Langlois. 1989. Variabilité spatiale et siasonnaire de la qualité de l'eau du fleuve St. Laurent. Environment Canada, Centre Saint-Laurent, Direction des Eaux Interieures, Quebec.

Ducharme, J.-L., G. Germain, and J. Talbot. 1992. Bilan de la faune 1992. Ministère du Loisir, de la Chasse et de la Pêche, Direction générale de la ressource faunique, Quebec.

Dynesius, M., and C. Nilsson. 1994. Fragmentation and flow regulation of river systems in the northern third of the world. *Science* 266:753–762.

Ehrman, T. P., and G. A. Lamberti. 1992. Hydraulic and particulate matter retention in a 3rd-order Indiana stream. *Journal of the North American Benthological Society* 11:341–349.

Environment et Faune Quebec. 1998. État de l'ecosystem aquatique du bassin versant de le Rivière Richelieu–Synthese 1998. Direction des Écosystèmes Aquatiques, le Ministère de l'Environment et de la Faune du Quebec, Envirodoq EA-13.

Fairchild, W. L. 1983. Ecology and distribution of Trichopteran larvae in the lower Ottawa River. Master thesis, Department of Entomology, McGill University, Montreal, Quebec.

Ferraris, J. 1984. Macroinvertébrés 5–Benthos et invertébrés phytophiles. Synthèse de la variabilité spatio-temporelle des macroinvertébrés benthiques et phytophiles récoltés du 7 juillet 1982 au 22 juillet 1983. Élaboration de la clé de potentiel et description des communautés associées aux habitats types. Technical Report. Ministère du Loisir, de la Chasse et de la Pêche, Service de l'aménagement, Montreal.

Fortin, G. R. A., and M. Peltier. 1995. Synthese des connaissances sur les aspects physiques et chimiques de l'eau et des sediments du Saguenay. Zones d'intervention prioritaire 22 et 23. Environment Canada, region du Quebec, Conservation de l'Environment, Centre Saint-Laurent. Rapport technique.

Fuller, K., H. Shear, and J. Wittig (eds.). 1995. *The Great Lakes: An environmental atlas and resource book*, 3rd ed. U.S. Environmental Protection Agency, Great Lakes

Environmental Program Office, Chicago; Government of Canada, Toronto, Ontario.

Giroux, J.-F., and J. Bédard. 1987. The effects of grazing by greater snow geese on the vegetation of tidal marshes in the St. Lawrence Estuary. *Journal of Applied Ecology* 24:773–788.

Gleick, P. H. 1993. *Water in crisis: A guide to the world's fresh water resources*. Oxford University Press, New York.

Gratton, L., and C. Dubreuil. 1990. Portrait de la végétation de la flore du Saint-Laurent. Ministère de l'Environnement du Quebec, Direction de la conservation et du patrimoine écologique, Quebec.

Gunn, J. M., S. U. Qadri, and D. C. Mortimer. 1977. Filamentous algae as a food source for the brown bullhead (*Ictalurus nebulosus*). *Journal of the Fisheries Research Board of Canada* 34:396–401

Hamill, S. E. 1975. Production of sphaeriid clams and amphipod crustaceans in the Ottawa River near Ottawa and Hull, Canada. Master thesis, Department of Biology, University of Ottawa, Ontario.

Hamill, S. E., S. U. Qadri, and G. L. Mackie. 1979. Production and turnover of *Pisidium casertanum* (Pelecypoda: Sphaeriidae) in the Ottawa River near Ottawa-Hull, Canada. *Hydrobiologia* 62:225–230.

Harmon, M. E., J. F. Franklin, F. J. Swanson, P. Sollins, S. V. Gregory, J. D. Lattin, N. H. Anderson, S. P. Cline, N. G. Aumen, J. R. Sedell, G. W. Lienkaemper, K. Cromack Jr., and K. W. Cummins. 1986. Ecology of coarse woody debris in temperate ecosystems. *Advances in Ecological Research* 15:133–302.

Hendrickson, G. E. 1994. *The angler's guide to twelve classic trout streams in Michigan*. University of Michigan Press, Ann Arbor.

Horvath, T. G., and G. A. Lamberti. 1997. Drifting macrophytes as a mechanism for zebra mussel (*Dreissena polymorpha*) invasion of lake-outlet streams. *American Midland Naturalist* 138:29–36.

Horvath, T. G. and G. A. Lamberti. 1999. Recruitment and growth of zebra mussels (*Dreissena polymorpha*) in a coupled lake–stream system. *Archiv fuer Hydrobiologie* 145:197–217.

Horvath, T. G., G. A. Lamberti, D. M. Lodge, and W. L. Perry. 1996. Zebra mussels in lake–stream systems: Source–sink dynamics? *Journal of the North American Benthological Society* 15:564–575.

Horvath, T. G., K. M. Martin, and G. A. Lamberti. 1999. Effect of zebra mussels, *Dreissena polymorpha*, on macroinvertebrates in a lake-outlet stream. *American Midland Naturalist* 142:340–347.

Hudon, C. 1997. Impact of water level fluctuations on St. Lawrence River aquatic vegetation. *Canadian Journal of Fisheries and Aquatic Sciences* 54:2853–2865.

Hudon, C., S. Paquet, and V. Jarry. 1996. Downstream variations of phytoplankton in the St. Lawrence River (Quebec, Canada). *Hydrobiologia* 337:11–26.

Hunt, C. B. 1974. *Natural regions of the United States and Canada*. W. H. Freeman, San Francisco.

Hydro-Québec. 1994. *Programme de stabilisation des berges Québecoises de la Rivière des Outaouais*. Rapport d'Avant-Project, Vol. 1: *Problematique et Inventaires*. Bureau d'Audiences Publiques sur l'Environment, Quebec.

Johnson, G. 1991. Statistiques de pêche commerciale de 1986–1992 (mise à jour 1992). Ministère de l'Agriculture, des Pêcheries et de l'Alimentation du Québec, Direction du développement et des activités régionales, Quebec.

Kingsley, M. C. S. 1994. Recensement, tendance et statut de la population de bélugas du Saint-Laurent en 1992. Canadian Technical Report of Fisheries and Aquatic Sciences 1938.

Kohlhepp, G. W. 1991. Life histories, secondary production rates, and trophic dynamics of five benthic invertebrates in two north temperate streams. Master thesis, University of Notre Dame, Indiana.

Laflamme, D. 1995. Qualite des eaux du bassin de la Rivière Saint Maurice, 1979–1992. Rapport QE-98, Envirodoq EN950417. Direction des ecosystemes aquatiques, environment et Faune Quebec, le Ministère de l'Environment et de la Faune du Quebec.

Lamberti, G. A., and M. B. Berg. 1995. Invertebrates and other benthic features as indicators of environmental change in Juday Creek, Indiana. *Natural Areas Journal* 15:249–258.

Latimore, J. A. 2000. Impacts of golf course construction and stream diversion on benthic invertebrates and sediment in Juday Creek, Indiana. Master thesis, University of Notre Dame, Indiana.

Lee, D. S., C. R. Gilbert, C. H. Hocutt, R. E. Jenkins, D. E. McAllister, and J. R. Stauffer Jr. 1980. *Atlas of North American freshwater fishes*. North Carolina State Museum of Natural History, Raleigh.

Lemarche, A., P. Legendre, and A. Chodorowski. 1982. Facteurs responsables de la distribution des gasteropodes dulcicoles dans le fleuve Saint-Laurent. *Hydrobiologia* 89:61–76.

Louchouarn, P., and M. Lucotte. 1998. A historical reconstruction of organic and inorganic contamination events in the Saguenay Fjord–St. Lawrence system from pre-industrial times to the present. *Science of the Total Environment* 213:139–150.

Louchouarn, P., M. Lucotte, and N. Farella. 1999. Historical and geographical variations of sources and transport of terrigenous organic matter within a large-scale coastal environment. *Organic Geochemistry* 30:675–699.

Mackie, G. 1971. Some aspects of the distribution and ecology of the macrobenthos in an industrialized portion of the Ottawa River near Ottawa and Hull, Canada. Master thesis, Department of Biology, University of Ottawa, Ontario.

Magnin, E. 1970. Faune benthique littorale du Lac St. Louis pres de Montreal (Quebec). *Annals of Hydrobiology* 1:181–195.

Magnin, E., and O. Leconte. 1971. Cycle vital d'*Asellus communis* sensu Racovitza (1920) (Crustacea, Isopoda) du Lac Saint Louis pres de Montreal. *Canadian Journal of Zoology* 49:647–655.

Magnin, E., and A. Stanczkowska. 1971. Quelqoues donnes sur la croissance, la biomasse, et la production annuelle de trois mollusques Unionidae de la region de Montreal. *Canadian Journal of Zoology* 49:491–497.

Maloney, D. C., and G. A. Lamberti. 1995. Rapid decomposition of summer-input leaves in a northern Michigan stream. *American Midland Naturalist* 133:184–195.

Mills, E. L., and J. L. Forney. 1982. Response of Lake Ontario plankton entering the international section of the St. Lawrence River. *Internationale Revue der Gesamten Hydrobiologie* 67:27–43.

Ministère de l'Environment et de la Faune du Quebec. 1996. Qualité des eaux du bassin de la Rivière des Outaouais, 1979–1994. Rapport QE-105/1, Envirodoq EN960174. Ministère de L'Environment et de la Faune, Direction des Écosystèmes Aquatiques, Quebec.

Moerke, A. H. 2000. Physical and biological responses to restoration of two Indiana streams. Master thesis, University of Notre Dame, Indiana.

Moerke, A. H., and G. A. Lamberti. 2003. Responses in fish community structure to restoration of two Indiana streams. *North American Journal of Fisheries Management* 23:748–759

Morin, J., M. Leclerc, Y. Secretan, and P. Boudreau. 2000. Integrated two-dimensional macrophytes-hydrodynamic modeling. *Journal of Hydraulic Research* 38:163–172.

Mousseau, P., and A. Armellin. 1995. Synthese des connaissances sur les communautes biologiques du Saguenay. Zones d'intervention prioritaire 22 et 23. Environment Canada, region du Quebec, Conservation de l'Environment, Centre Saint-Laurent. Rapport technique.

New York Power Authority. 1996. Initial consultation package for relicensing. St. Lawrence–F.D.R. Power Project, FERC no. 2000.

Ontario Water Resources Commission. 1972. Ottawa River basin: Water quality and its control in the Ottawa River. Vol. 2. Report to the provinces of Ontario and Quebec.

Osterberg, D. M. 1978. Food consumption feeding habits and growth of the walleye (*Stizostedion vitreum*) and sauger (*Stizostedion canadense*) in the Ottawa River near Ottawa–Hull, Canada. Ph.D. diss., Department of Biology, University of Ottawa, Ontario.

Page, L. M., and B. M. Burr. 1991. *A field guide to freshwater fishes of North America north of Mexico*. Houghton Mifflin, Boston.

Painchaud, J. 1997. La qualité de l'eau des rivières du Québec: État et tendances. Ministère de L'Environment et de la Faune, direction des écosystèmes aquatiques, Quebec.

Patrick, R. 1996. *Rivers of the United States*. Vol. 3: *The eastern and southeastern states*. Wiley, New York.

Perera, A. H., D. L. Euler, and I. D. Thompson. 2000. *Ecology of a managed terrestrial landscape: Patterns and processes of forest landscapes in Ontario*. University of British Columbia Press.

Perry, W. L., D. M. Lodge, and G. A. Lamberti. 1997. Impact of crayfish predation on exotic zebra mussels and native invertebrates in a lake-outlet stream. *Canadian Journal of Fisheries and Aquatic Sciences* 54:120–125.

Pinel-Alloul, B., and E. Magnin. 1979a. Etude de la nourriture de *Lymnaea catascopium catascopium* (Gastropoda, Lymnaeidae) dans le Lac St. Louis, Flueve Saint-Laurent, Québec. *Naturaliste Canadien* 106:277–287.

Pinel-Alloul, B., and E. Magnin. 1979b. Life cycle, growth and fecundity of five populations of *Lymnaea catascopium catascopium* (Gastropoda, Lymnaeidae) in Lake Saint-Louis, Quebec, Canada. *Malacologia* 19:87–101.

Qadri, S. U., S. E. Hamill, and T. Clair. 1977. Distribution, biomass, standing crop, and production of macrobenthos in the Ottawa River near Ottawa and Gatineau, Canada. In *Distribution and transport of pollutants in flowing water ecosystems*. Vol. 2, pp. 1–63. Ottawa River Project Final Report. National Research Council of Canada.

Revenga, C., S. Murray, J. Abramovitz, and A. Hammond. 1998. *Watersheds of the world: ecological value and vulnerability*. World Resources Institute, Washington, D.C.

Reynolds, C. S., and J.-P. Descy. 1996. The production, biomass and structure of phytoplankton in large rivers. *Archiv fuer Hydrobiologie Supplement* 113:161–187.

Ricciardi, A., F. G. Whoriskey, and J. B. Rasmussen. 1997. The role of the zebra mussel (*Dreissena polymorpha*) in structuring macroinvertebrate communities on hard substrata. *Canadian Journal of Fisheries and Aquatic Sciences* 54:2596–2608.

Richards, J. S. 1976. Changes in fish species composition in the Au Sable River, Michigan from 1920s to 1972. *Transactions of the American Fisheries Society* 105:32–40.

Ricketts, T. H., E. Dinerstein, D. M. Olson, C. J. Loucks, W. Eichbaum, D. DellaSala, K. Kavanagh, P. Hedao, P. T. Hurley, K. M. Carney, R. Abell, and S. Walters. 1999. *Terrestrial ecoregions of North America: A conservation assessment*. Island Press, Washington, D.C.

Rodgers, D. W., and S. U. Qadri. 1982. Growth and mercury accumulation in yearling yellow perch, *Perca flavescens*, in the Ottawa River, Ontario. *Environmental Biology of Fishes* 7:377–383.

Schwenneker, B. W. 1985. The contribution of allochthonous and autochthonous organic material to aquatic insect secondary production rates in a north temperate stream. Ph.D. diss., University of Notre Dame, Indiana.

Shipley, W. J. 1987. Pattern and mechanism in the emergent macrophyte communities along the Ottawa River (Canada). Ph.D. diss., Department of Biology, University of Ottawa, Ontario.

Smith, C. L. 1985. *The inland fishes of New York state.* New York State Department of Environmental Conservation, Albany.

Stanczykowska, A., M. Plinski, and E. Magnin. 1972. Étude de trois populations de *Viviparus malleatus* (Reeve) (Gastropoda, Prosobranchia) de la region de Montreal. II. Etude qualitative et quantitative de la nourriture. *Canadian Journal of Zoology* 50:1617–1624.

State of Indiana. 1987. Water resource availability in the St. Joseph River basin, Indiana. Water Resource Assessment 87-1. State of Indiana, Department of Natural Resources, Division of Water, Indianapolis.

State of Indiana. 1996. Water resource availability in the Maumee River Basin, Indiana. State of Indiana, Department of Natural Resources, Division of Water, Indianapolis.

Stelzer, R. S., and G. A. Lamberti. 1997. Condition of stream channels and fisheries in the Ontonagon River system under current flow regimes of the Bond Falls project. Report submitted to the Ottawa National Forest, USDA Forest Service.

St. Lawrence Centre. 1996. *State of the environment report on the St. Lawrence River.* Vols. 1–2: *The St. Lawrence ecosystem.* St. Lawrence Update Series. Environment Canada–Quebec Region, Environmental Conservation,

Thompson P. A., and R. Côté. 1985. Influence de la speciation du cuivre sur les populations phytoplanctoniques naturelles de la Rivière du Saguenay, Quebec, Canada. *Internationale Revue der Gesamten Hydrobiologie* 70:711–731.

Thorp, J. H., and A. F. Casper. 2002. Potential effects on zooplankton from shifts in planktivorous mussels: A field experiment in the St. Lawrence River. *Freshwater Biology* 47:107–119.

Thorp, J. H., and A. F. Casper. 2003. Importance of biotic interactions in large rivers: An experiment with planktivorous fish, dreissenid mussels, and zooplankton in the St. Lawrence. *River Research and Applications* 19:265–279.

Van Coillie, R., D. Brouard, M. Lachance, and Y. Vigneault. 1982. Physico-chemical effects of acid precipitation on four salmon rivers. *Eau Québec* 15:384–388.

Vis, C., A. Cattaneo, and C. Hudon. 1998. Periphyton in the clear and colored water masses of the St. Lawrence River (Quebec, Canada): A 20-year overview. *Journal of Great Lakes Research* 24:105–117.

Watson, B. G., and D. C. MacIver. 1995. Bioclimate mapping of Ontario. Final Unpublished Report to the Ontario Ministry of Natural Resources.

Werner, R. G. 1977. Ichthyoplankton and inshore larval fishes of the St. Lawrence River. In J. W. Gies (ed.). *Preliminary report: Biological characteristics of the St. Lawrence River*, pp. 31–60. State University of New York College of Environmental Science and Forestry, Institute of Environmental Program Affairs, Syracuse.

Wesley, J. K., and J. E. Duffy. 1999. St. Joseph River assessment. Fisheries Division Special Report 24. Michigan Department of Natural Resources, Ann Arbor.

Yang, C., K. Telmer, and K. Veizer. 1996. Chemical dynamics of the "St. Lawrence" riverine system: δD_{H2O}, $\delta^{18}O_{H2O}$, $\delta^{13}C_{DIC}$, $\delta^{34}S_{sulfate}$, and dissolved $^{87}Sr/^{86}SR$. *Geochimica et Cosmochimica Acta* 60:851–866.

Zorn, T. G. and S. P. Sendeck. 2001. AuSable River assessment. Fisheries Division Special Report 26. Michigan Department of Natural Resources, Ann Arbor.

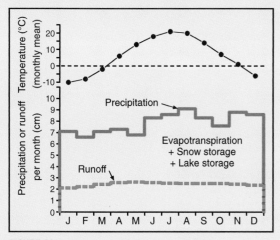

FIGURE 22.9 Mean monthly air temperature, precipitation, and runoff for the St. Lawrence River main stem.

ST. LAWRENCE RIVER MAIN STEM

Relief: 1945 m

Basin area: 574,000 km² (main stem)

Mean discharge: 12,600 m³/s (excluding Saguenay)

River order: 8 without Saguenay (9 with Saguenay)

Mean annual precipitation: 94.2 cm

Mean air temperature: 6.7°C

Mean water temperature: 9.7°C

Physiographic provinces: 4 ecoregions (see text)

Biomes: Temperate Deciduous Forest, Boreal Forest

Freshwater ecoregion: Lower St. Lawrence

Terrestrial ecoregions: Eastern Great Lakes Lowland Forests, Eastern Forest/Boreal Transition, New England/Acadian Forests, and Eastern Canadian Forests

Number of fish species: 87 freshwater and 18 diadromous species

Number of endangered species: ~20 animals and plants vulnerable, threatened, or endangered at federal, state, or provincial levels, 12 other at risk

Major fishes: lamprey, lake sturgeon, Atlantic sturgeon, gar, bowfin, American eel, alewife, gizzard shad, creek chub, fallfish, yellow perch, walleye, white sucker, silver redhorse, channel catfish, tadpole madtom, muskellunge, central mudminnow, rainbow smelt, brown trout, troutperch, banded killifish, burbot, brook silverside, mottled sculpin, sand darter, drum

Major other aquatic vertebrates: muskrat, river otter, beaver, mink, beluga whale, long-finned pilot whale, Canada goose, mallard duck, wood duck, blue-winged teal, gadwall, American wigeon, great blue heron, ring-billed gull, belted kingfisher, cormorant, tern, bald eagle, osprey, northern water snake, painted turtle, snapping turtle

Major benthic invertebrates: sponges (*Eunapius*), flatworms (*Dugesia*), bryozoans (*Plumatella*), oligochaetes (*Chaetogaster*), mollusks (*Elliptio*, *Lampsilis*, *Birgella*, *Gyraulus*), crustaceans (*Gammarus*, *Orconectes*), mayflies (*Stenonema*, *Hexagenia*), caddisflies (*Nectopsyche*), stoneflies (*Pteronarcys*), chironomid midges

Nonnative species: common carp, rainbow trout, white perch, *Bithynia tentaculata* and *Viviparus georgianus* (snails), *Pisidium amnicum* and *Sphaerium* (fingernail clams), zebra mussel, quagga mussel, *Echinogammarus ischnus* (amphipod), purple loosestrife

Major riparian plants: silver maple, red maple, black ash, green ash, black willow, American basswood, cattails, sedges, rushes, bulrushes, reed canary grass

Special features: lower river continentally outstanding; RAMSAR World Heritage Site (Lac Saint-Pierre); globally distinct site (Batture aux Alouettes) for migratory birds at confluence with Saguenay River; least turbid of world's 15 largest rivers

Fragmentation: four hydroelectric dams and seven navigation locks on main stem; many dams on tributaries

Water quality: upper river rated second-highest category by New York State; EPA Area of Concern (below Cornwall, Ontario) from PCBs and mercury; pH = 7.5 to 8.5, alkalinity = 90 mg/L as $CaCO_3$, turbidity = 2 NTU, O_2 = >8.6 mg/L, NH_4-N = 0.01 to 0.04 mg/L, PO_4-P = 0.01 mg/L

Land use: 55% forest, 20% agriculture, 22% urban, 3% other for entire St. Lawrence–Great Lakes system; percentage urban considerably lower for main-stem watershed

Population density: 10.5 people/km² (St. Lawrence subbasin) to 54 people/km² (St. Lawrence River–Great Lakes basin)

Major information sources: Painchaud 1997, St. Lawrence Centre 1996

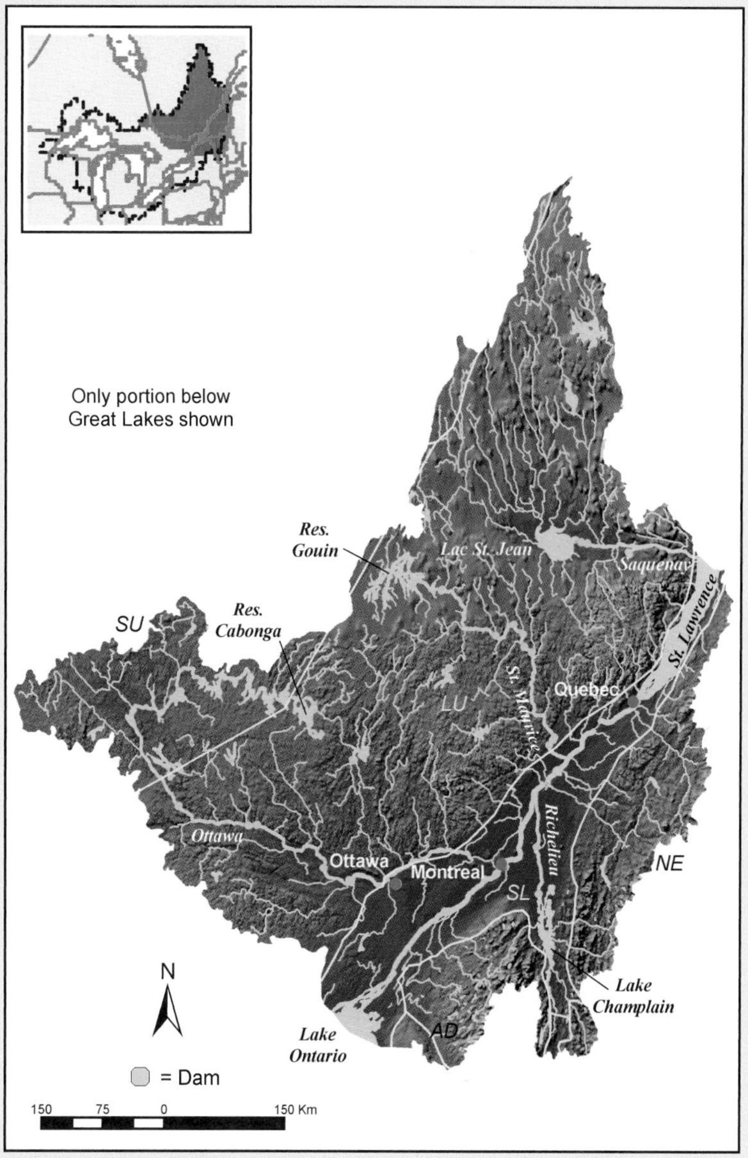

FIGURE 22.8 Map of the St. Lawrence River main stem. Physiographic provinces are separated by yellow lines.

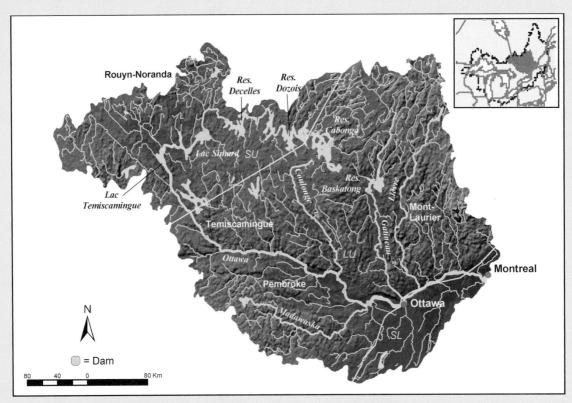

FIGURE 22.10 Map of the Ottawa River basin. Physiographic provinces are separated by yellow lines.

OTTAWA RIVER

Relief: 911 m

Basin area: 146,334 km²

Mean discharge: 1948 m³/s

River order: 8

Mean annual precipitation: 100.2 cm

Mean air temperature: 6.0°C

Mean water temperature: 9.6°C

Physiographic provinces: St. Lawrence Lowlands (SL), Laurentian Highlands (LU), Superior Upland (SU)

Biomes: Temperate Deciduous Forest, Boreal Forest

Freshwater ecoregion: Lower St. Lawrence

Terrestrial ecoregions: Eastern Forest/Boreal Transition, Eastern Great Lakes Lowland Forests

Number of fish species: 53

Endangered species: (threatened) 5 fishes, 2 birds, 1 amphibian, 1 reptile, 3 dragonflies

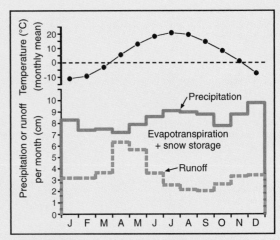

FIGURE 22.11 Mean monthly air temperature, precipitation, and runoff for the Ottawa River basin.

Major fishes: sturgeon, walleye, sauger, muskellunge, northern pike, yellow perch, crappie, lake whitefish, lake cisco, largemouth bass, smallmouth bass, channel catfish, brown bullhead, copper redhorse, suckers, cyprinids, mooneye

Major other aquatic vertebrates: muskrat, river otter, beaver, mink

Major benthic invertebrates: bivalves (*Elliptio*, fingernail clams), snails (*Bithynia*), oligochaete worms (*Stylaria*), crustaceans (*Gammarus, Caecidotea, Asellus*), 19 midge genera (*Polypedilum*), 22 mayflies (*Hexagenia, Stenonema*), damselflies (*Enallagma*), stoneflies (*Isoperla*), alderflies (*Sialis*), beetles (*Berosus*), 35 caddisflies (*Polycentropus, Brachycentrus*)

Nonnative species: brown trout, rainbow trout, zebra mussel, quagga mussel, purple loosestrife

Major riparian plants: wild rice, bur-reed, arrowhead, northern bugleweed, fox sedge, silverweed, marsh speedwell, calamus root, bulrushes, Small's spikerush, knotsheath sedge, red top, reed canary grass, silver maple, green ash, alder, willow

Special features: eight provincial faunal reserves; island–wetland complex (Petri Islands Preserve) near confluence with St. Lawrence

Fragmentation: 7 dams on main stem; >300 dams on tributaries

Water quality: good overall; pollutants from industry in Temiscamingue region and from sewage/industry near Ottawa–Hull; NH_4-N = 0.045 mg/L, NO_3-N + NO_2-N = 0.17 mg/L, total nitrogen = 0.41 mg/L, total phosphorus = 0.029 mg/L

Land use: 86% forest, 10% surface waters, 2% urban, and 2% agriculture

Population density: 2.4 people/km²

Major information source: Ministère de l'Environment et de la Faune du Québec 1996

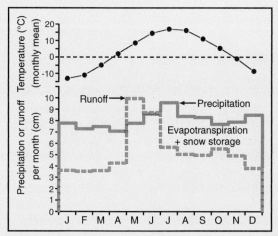

FIGURE 22.13 Mean monthly air temperature, precipitation, and runoff for the Saguenay River basin.

SAGUENAY RIVER

Relief: 1130 m
Basin area: 85,500 km²
Mean discharge: 1535 m³/s
River order: 8
Mean annual precipitation: 96.5 cm
Mean air temperature: 3.0°C
Mean water temperature: NA
Physiographic provinces: Laurentian Highlands (LU), Labrador Highlands (LB)
Biome: Boreal Forest
Freshwater ecoregion: Lower St. Lawrence
Terrestrial ecoregion: Eastern Canadian Forest
Number of fish species: 76 in river and fjord
Number of endangered species: 4 fishes, 2 birds
Major fishes: Atlantic salmon, fallfish, brown bullhead, lake whitefish, brook trout, perch, burbot, rainbow smelt, lake (black) sturgeon, American eel, emerald shiner, sauger, white sucker, longnose sucker, northern pike
Major other aquatic vertebrates: beaver, river otter, muskrat, mink, long-finned pilot whale, beluga whale, great blue heron, black-crowned night heron; many bird species in fjord, including red knot, sanderling, purple sandpiper, peregrine falcon, osprey
Major benthic invertebrates: in sandy-muddy habitats insects, oligochaetes, bivalve mollusks, gastropods, and isopods are all common, but oligochaetes dominate in gravelly areas
Nonnative species: rainbow trout, purple loosestrife and >20 other plants in riparian zone
Major riparian plants: sparse vegetation in upper third due to dams and log driving; middle third dominated by American bulrush but with ≥250 species; lower third with sparse patches of American bulrush and, where salinity intrudes, saltwater cord grass
Special features: large Lac Saint-Jean in headwaters; last 100 km flows through largest fjord in northwest Atlantic; continentally and globally distinct site (Batture aux Alouettes) for migratory birds at confluence with St. Lawrence
Fragmentation: 3 hydroelectric/navigation dams on main stem and >300 small and medium-size dams in watershed
Water quality: good in upper river; poorer in lower river from nonpoint sources, aluminum processing, pulp/paper production; conductivity = 20 to 120 mS/cm, pH = 6.8 to 7.4, alkalinity = 7.3 mg/L as $CaCO_3$, dissolved O_2 = 5.7 to 7.8 mg/L, NH_4-N = 0.02 mg/L, NO_2-N + NO_3-N = 0.2 to 6.4 mg/L, PO_4-P = 0.01 mg/L, chlorophyll a = 0.9 to 8.3 mg/L
Land use: 90% forest, 8% surface waters, 1% agriculture, and <1% urban
Population density: 2.5 people/km²
Major information sources: Fortin and Peltier 1995, Mousseau and Armellin 1995

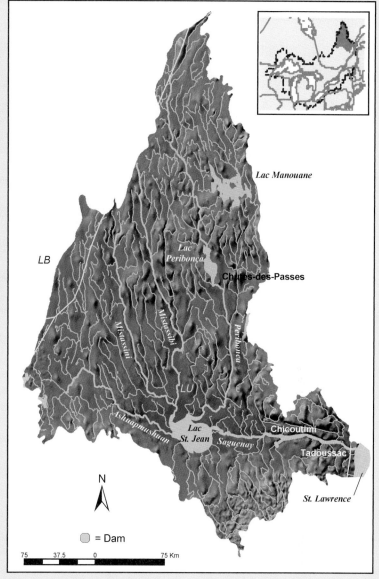

FIGURE 22.12 Map of the Saguenay River basin. Physiographic provinces are separated by yellow lines.

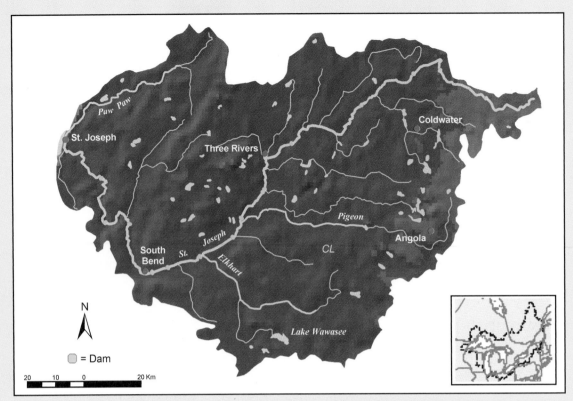

FIGURE 22.14 Map of the St. Joseph River basin.

ST. JOSEPH RIVER

Relief: 200 m
Basin area: 12,150 km^2
Mean discharge: 96.3 m^3/s
River order: 5
Mean annual precipitation: 99.4 cm
Mean air temperature: 9.7°C
Mean water temperature: 12.0°C
Physiographic province: Central Lowland (CL)
Biome: Temperate Deciduous Forest
Freshwater ecoregion: Michigan–Huron
Terrestrial ecoregion: Southern Great Lakes Forests
Number of fish species: 114
Number of endangered species: (endangered, threatened, special
   concern) 1 mammal, 7 fishes, 6 reptiles, 4 amphibians, 2 insects,
   2 snails, ~105 riparian plant species

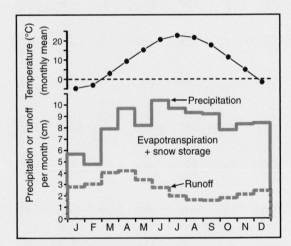

FIGURE 22.15 Mean monthly air temperature, precipitation, and runoff for the St. Joseph River basin.

Major fishes: brook trout, smallmouth bass, bluegill, walleye, white sucker, hornyhead chub, creek chub, yellow perch, logperch, pirate perch, blacknose dace, blackside darter, rainbow darter, bluntnose minnow, common shiner, common stoneroller, central mudminnow, northern hogsucker, mottled sculpin, channel catfish, northern madtom
Major other aquatic vertebrates: beaver, river otter, muskrat, mink
Major benthic invertebrates: crustaceans (*Caecidotea, Gammarus, Orconectes*), mollusks, (fingernail clams, faucet snails), beetles (*Dubiraphia, Macronychus*), mayflies (*Baetis, Hexagenia*), alderflies (*Sialis*), damselflies (*Enallagma*), stoneflies (*Amphinemura, Paracapnia, Taeniopteryx*), caddisflies (*Cheumatopsyche, Glossosoma*)
Nonnative species: brown trout, rainbow trout, chinook salmon, coho salmon, sea lamprey, common carp, goldfish, zebra mussel, Asiatic clam, purple loosestrife, Eurasian watermilfoil
Major riparian plants: ash, elm, cottonwood, maple, oak, poplar, wild celery, pondweed
Fragmentation: 17 dams on main stem, 190 dams in basin
Water quality: marginal water quality; pH = 6.0 to 8.9, alkalinity = 200 to 400 mg/L as CaCO$_3$, NO$_3$-N = 0.7 to 2.9 mg/L, PO$_4$-P = 0.02 to 0.10 mg/L
Land use: 58% agriculture, 20% forest, 8% urban, 7% open, 5% wetlands, and 2% water
Population density: 69 people/km^2
Major information sources: Horvath et al. 1999, Lamberti and Berg 1995, Lee et al. 1980, Page and Burr 1991, Perry et al. 1997, Wesley and Duffy 1999, www.census.gov, www.usgs.gov, www.ummz.lsa.umich.edu

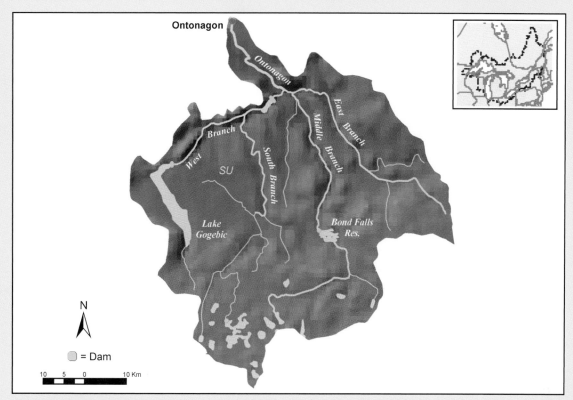

FIGURE 22.16 Map of the Ontonagon River basin.

ONTONAGON RIVER

Relief: 325 m
Basin area: 3569 km^2
Mean discharge: 39.9 m^3/s
River order: 4
Mean annual precipitation: 87.1 cm
Mean air temperature: 4.1°C
Mean water temperature: 8.3°C
Physiographic province: Superior Upland (SU)
Biome: Temperate Deciduous Forest
Freshwater ecoregion: Superior
Terrestrial ecoregion: Western Great Lakes Forests
Number of fish species: 50
Number of endangered species: (threatened or special concern)
    1 turtle, ~6 riparian plants
Major fishes: native brook trout, yellow perch, walleye, northern pike, smallmouth bass, rock bass, bluegill, pumpkinseed, black crappie, common shiner, hornyhead chub, blacknose shiner, creek chub, slimy sculpin

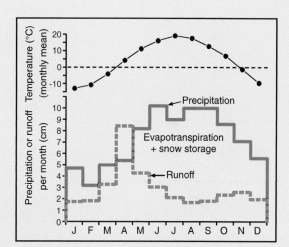

FIGURE 22.17 Mean monthly air temperature, precipitation, and runoff for the Ontonagon River basin.

Major other aquatic vertebrates: muskrat, beaver, river otter, mink
Major benthic invertebrates: crustaceans (*Gammarus, Orconectes*), beetles (*Haliplus, Stenelmis*), true flies (*Antocha, Simulium*), mayflies (*Acentrella, Baetis, Stenacron*), hellgrammites (*Nigronia*), dragonflies (*Ophiogomphus*), stoneflies (*Acroneuria*), caddisflies (*Brachycentrus, Cheumatopsyche, Glossosoma*)
Nonnative species: Eurasian ruffe, brown trout, rainbow trout, chinook salmon, coho salmon, rusty crayfish
Major riparian plants: cedar, spruce, hemlock, willow, alder
Special features: National Wild and Scenic River (several branches); Sylvania Wilderness with many high-quality lakes and ephemeral streams; impressive waterfalls on basalt outcrops
Fragmentation: four major dams direct flow to 12 MW hydroelectric facility at Victoria Lake; numerous low-head dams in basin
Water quality: generally good; improved since major logging in 1800s; most streams are lake fed and reflect lake water quality; pH = 6.7 to 8.4, alkalinity = 42 to 64 mg/L as CaCO$_3$, NO$_3$-N = 0.01 to 0.21 mg/L, PO$_4$-P = <0.01 to 0.05 mg/L
Land use: >80% forested (Ottawa National Forest), <20% urban and aquatic
Population density: 2.6 people/km^2
Major information sources: Charlebois and Lamberti 1996, Lee et al. 1980, Maloney and Lamberti 1995, Page and Burr 1991, Stelzer and Lamberti 1997

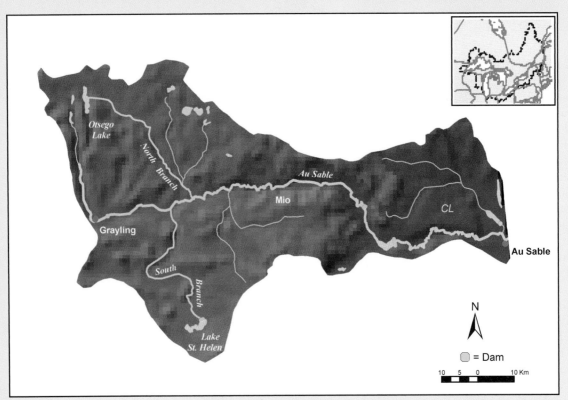

FIGURE 22.18 Map of the AuSable River basin.

AuSABLE RIVER

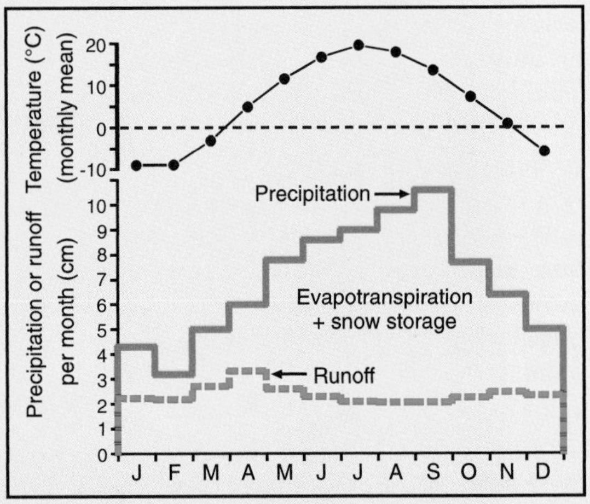

Relief: 250 m
Basin area: 5240 km^2
Mean discharge: 42.1 m^3/s
River order: 4
Mean annual precipitation: 83.3 cm
Mean air temperature: 5.5°C
Mean water temperature: 11.5°C
Physiographic province: Central Lowland (CL)
Biome: Temperate Deciduous Forest
Freshwater ecoregion: Michigan–Huron
Terrestrial ecoregion: Western Great Lakes Forests
Number of fish species: 93
Number of endangered species: (endangered, threatened, or special concern) 4 fishes, 3 reptiles, ~5 riparian plants
Major fishes: brook trout, brown trout, rainbow trout, sculpins, shiners, white sucker, other suckers, several dace, Johnny darter
Major other aquatic vertebrates: muskrat, beaver, river otter, mink
Major benthic invertebrates: crustaceans (*Gammarus*, *Caecidotea*), beetles (*Dytiscus*, *Haliplus*), dipterans (*Antocha*, *Tipula*), mayflies (*Baetis*, *Ephemerella*, *Hexagenia*, *Rhithrogena*, *Stenonema*), hellgrammites (*Nigronia*), odonates (*Calopteryx*), stoneflies (*Paragnetina*), caddisflies (*Brachycentrus*, *Helicopsyche*, *Hydropsyche*, *Rhyacophila*)
Nonnative species: ≥7 fishes, including brown trout, rainbow trout, sea lamprey
Major riparian plants: alder, ash, cedar, fir, red maple, poplar, spruce, tamarack, willow
Special features: 37 km of main stem designated National Wild and Scenic River; 560 km (157 km of main stem) in State Natural Rivers Program; about one-third of watershed in Huron National Forest
Fragmentation: 67 dams along main stem and tributaries, ranging from <1 to >10 m in height
Water quality: good due to substantial groundwater input; point sources mostly septic systems; pH = 7.4 to 8.2, alkalinity = 121 to 156 mg/L as CaCO$_3$, NO$_3$-N = <0.01 to 0.36 mg/L, PO$_4$ -P = <0.01 to 0.05 mg/L
Land use: <50% forested, >50% agricultural or urban; watershed completely logged ~1900
Population density: 17.6 people/km^2
Major information sources: Cushing and Allan 2001, Hendrickson 1994, Lee et al. 1990, Page and Burr 1991, Richards 1976, Zorn and Sendeck 2001

FIGURE 22.19 Mean monthly air temperature, precipitation, and runoff for the AuSable River basin.

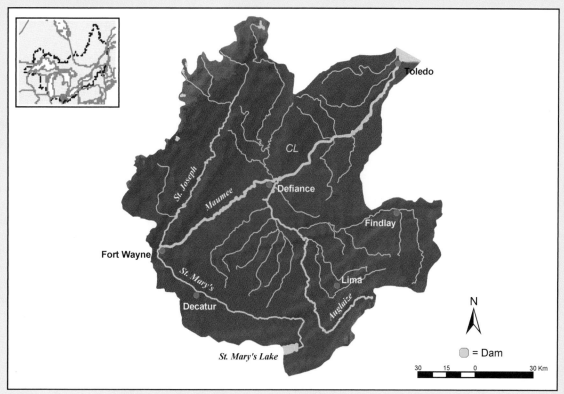

FIGURE 22.20 Map of the Maumee River basin.

MAUMEE RIVER

Relief: 73 m

Basin area: 16,458 km^2

Mean discharge: 150 m^3/s

River order: 7

Mean annual precipitation: 84 cm

Mean air temperature: 9.0°C

Mean water temperature: 12.7°C

Physiographic province: Central Lowland (CL)

Biome: Temperate Deciduous Forest

Freshwater ecoregion: Erie

Terrestrial ecoregion: Southern Great Lakes Forests

Number of fish species: >60

Number of endangered species: (listed or candidates) 4 mussels, 2 reptiles

Major fishes: topminnow, chubs, spotfin shiner, bluntnose minnow, hognose sucker, redhorse sucker, blacknose dace, mudminnow, variegated darter, green sunfish, smallmouth bass, yellow bullhead, channel catfish, madtoms, walleye, crappie, gizzard shad, longnose gar, white perch, logperch, drum

Major other aquatic vertebrates: muskrat, beaver

Major benthic invertebrates: mussels (deertoe, mapleleaf), big water crayfish, White River crayfish, beetles (*Stenelmis*), true flies (*Rheotanytarsus*, *Simulium*), mayflies (*Baetis*), moths (*Petrophila*), caddisflies (*Cheumatopsyche*)

Nonnative species: 11 fishes (goldfish, carp, sea lamprey, white perch, round goby), rusty crayfish, quagga mussel, zebra mussel, purple loosestrife

Major riparian plants: sycamores, black locust, beech, sugar maple

Special features: largest drainage area of any Great Lakes river; largest population of migrating walleye east of Mississippi River; 112 km as Ohio state scenic river; Lower Cedar Creek (Indiana) listed as outstanding segment

Fragmentation: two dams on main stem and three relatively large dams on tributaries

Water quality: lower segment EPA Area of Concern; PCBs and mercury limit fish consumption; turbidity = 54.3 NTU, pH = 7.9, hardness = 267.9 mg/L as CaCO$_3$, NH$_4$-N = 0.09 mg/L, NO$_2$-N = 0.04 mg/L, NO$_3$-N = 5.5 mg/L, PO$_4$-P = 0.05 mg/L

Land use: Lucas County (near mouth) 54% agriculture, 21% forest, 18% urban, 4% wetland; remainder of basin 79% to 88% agriculture, 6% to 10% forest and wetlands, 6% to 11% urban

Population density: 21.2 people/km^2

Major information sources: Bensch 1993, State of Indiana 1996

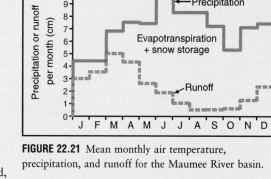

FIGURE 22.21 Mean monthly air temperature, precipitation, and runoff for the Maumee River basin.

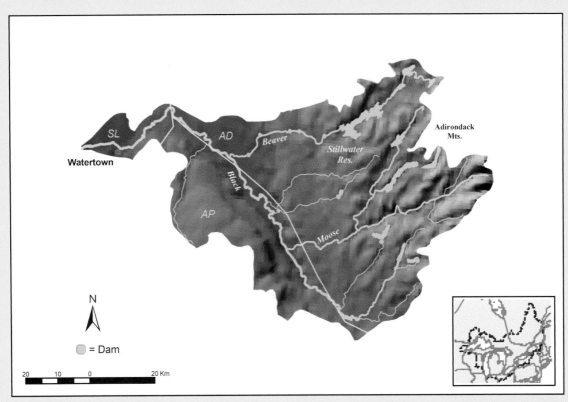

FIGURE 22.22 Map of the Black River basin. Physiographic provinces are separated by a yellow line.

BLACK RIVER

Relief: 510 m
Basin area: 5057 km²
Mean discharge: 118 m³/s
River order: 5
Mean annual precipitation: 97 cm
Mean air temperature: 6.8°C
Mean water temperature: 10.0°C
Physiographic provinces: Appalachian Plateaus (AP), Adirondack (AD)
Biome: Temperate Deciduous Forest
Freshwater ecoregion: Ontario
Terrestrial ecoregion: Eastern Great Lakes Lowland Forests
Number of fish species: ≥52
Number of endangered species: no federal; New York State
    (threatened and endangered) 2 fishes, 3 birds, 1 turtle, 1 mayfly,
    1 dragonfly
Major fishes: rock bass, smallmouth bass, brown bullhead,
    fantailed darter, tessellated darter, fallfish, northern pike,
    margined madtom, cutlips minnow, central mudminnow, yellow perch, chain pickerel, pumpkinseed, common shiner, golden
    shiner, northern hognosed sucker, white sucker, brown trout, brook trout, walleye
Major other aquatic vertebrates: muskrat, beaver
Major benthic invertebrates: mussels (*Elliptio, Anodonta, Alasidonta, Lasmigona, Strophitus, Musculium, Pisidium, Sphaerium*),
    snails (*Amnicola, Bithynia, Valvata, Viviparus, Campeloma, Lymnaea, Ferrissia, Physella, Heliosoma, Gyraulus*), caddisflies
    (*Hydropsyche, Neureclipsis*), mayflies (*Baetis, Stenonema*), true flies (*Simulium*)
Nonnative species: purple loosestrife, faucet snail, sea lamprey
Major riparian plants: sugar maple and beech, with interspersed elms and birches
Special features: one of few rivers entering Great Lakes that originates in mountains (Adirondacks); consequently, it has a steep
    slope (2.3 m/km) for a short river
Fragmentation: 23 small hydroelectric dams on main stem and tributaries
Water quality: good, but vulnerable to degradation; turbidity = 2.46 NTU, dissolved O_2 = 11.19 mg/L, pH = 7.6, hardness =
    38.8 mg/L as $CaCO_3$, NO_2-N + NO_3-N = 0.45 mg/L, NH_4-N = 0.05 mg/L, PO_4-P = 0.01 mg/L
Land use: 75% forested, 25% agriculture and urban
Population density: 16.8 people/km²

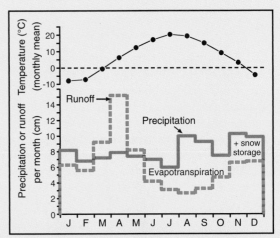

FIGURE 22.23 Mean monthly air temperature,
precipitation, and runoff for the Black River basin.

1026

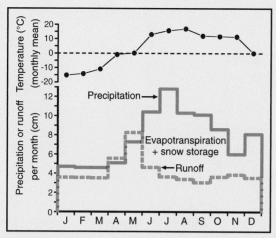

FIGURE 22.25 Mean monthly air temperature, precipitation, and runoff for the Richelieu River basin.

RIVIÈRE RICHELIEU

Relief: 1614 m
Basin area: 23,772 km²
Mean discharge: 341 m³/s
River order: 7
Mean annual precipitation: 115.2 cm
Mean air temperature: 4.1°C
Mean water temperature: NA
Physiographic provinces: St. Lawrence Lowlands (SL), Adirondack (AD), New England (NE), Valley and Ridge (VR)
Biomes: Temperate Deciduous Forest, Boreal Forest
Freshwater ecoregion: Lower St. Lawrence
Terrestrial ecoregions: New England/Acadian Forests, Eastern Great Lakes Lowland Forests, Eastern Forest/Boreal Transition
Number of fish species: 48
Number of endangered species: 0
Major fishes: black fin shiner, largemouth bass, bowfin, banded killifish, emerald shiner, spottail shiner, river redhorse, mimic shiner, American eel, fathead minnow, fallfish, greater redhorse, yellow perch, rock bass, pumpkinseed, blunt nose minnow, brown bullhead, black crappie, white sucker
Major other aquatic vertebrates: muskrat, beaver
Major benthic invertebrates: crustaceans (*Caecidotea*), chironomid midges, caddisflies (Hydropsychidae, Hydroptilidae, Leptoceridae, Limnephilidae, Polycentropidae), mayflies (Baetidae, Caenidae), beetles (Elmidae, Hydrophilidae), damselflies (Coenagrionidae)
Nonnative species: sea lamprey, quagga mussel, zebra mussel, Asiatic clam, water chestnut
Major riparian plants: primarily beech and maple, with spruce and fir at higher elevations
Special features: outlet for Lake Champlain; river falls ~20 m at Fryer Rapids near Chambly, Quebec
Fragmentation: dam and reservoir below Fryer Rapids
Water quality: pollutants from >400 industries and >2700 farms; alkalinity = 52.4 mg/L as CaCO₃, hardness 67.3 mg/L as CaCO₃, conductivity = 184 μS/cm, pH = 7.7, turbidity = 10.4 NTU, dissolved O₂ = 10.65 mg/L, total nitrogen = 0.68 mg/L, total phosphorus = 0.06 mg/L
Land use: 49% forest, 43% agriculture, 7% other uses
Population density: 21 people/km²
Major information source: Environment et Faune Quebec 1998

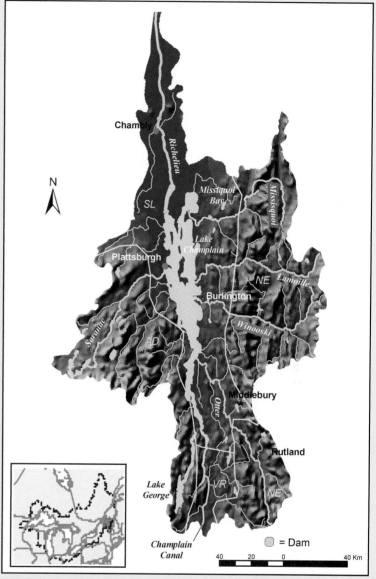

FIGURE 22.24 Map of the Richelieu River basin. Physiographic provinces are separated by yellow lines.

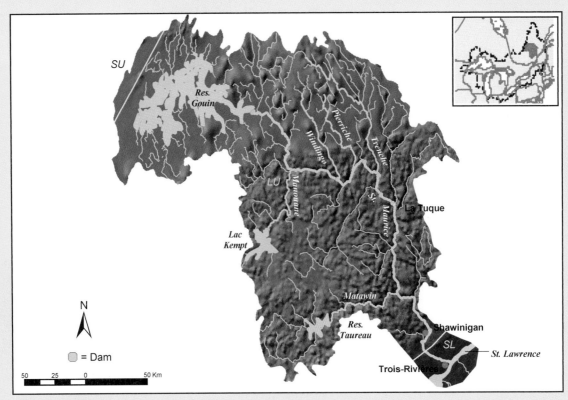

FIGURE 22.26 Map of the Saint-Maurice River basin. Physiographic provinces are separated by yellow lines.

RIVIÈRE SAINT-MAURICE

Relief: 548 m
Basin area: 43,427 km^2
Mean discharge: 670 m^3/s
River order: 7
Mean annual precipitation: 93 cm
Mean air temperature: 9.7°C
Mean water temperature: 10.0°C
Physiographic provinces: Laurentian Highlands (LU), St. Lawrence Lowland (SL)
Biome: Boreal Forest
Freshwater ecoregion: Lower St. Lawrence
Terrestrial ecoregions: Eastern Forest/Boreal Transition, Eastern Great Lakes Lowland Forests
Number of fish species: ≥30
Endangered species: 4 fishes, 1 turtle
Major fishes: northern pike, lake trout, lake whitefish, fallfish, pearl dace, creek chub, fathead minnow, longnose dace, shiners, brown bullhead, troutperch, rock bass, pumpkinseed, smallmouth bass, yellow perch, walleye, logperch, white sucker, longnose sucker, brook trout
Major other aquatic vertebrates: muskrat, mink, beaver, common loon, bufflehead, black scoter
Major benthic invertebrates: NA
Nonnative species: brook trout, pearl dace, purple loosestrife
Special features: La Mauricie National Park; Mastigouche and several other provincial faunal reserves/parks in the basin
Fragmentation: seven main-stem dams
Water quality: good overall; municipal effluents at Shawinigan and Trois-Rivières; pH = 6.6, conductivity = 29.2 mS/cm, alkalinity = 4.0 mg/L as CaCO$_3$, O$_2$ = 11.0 mg/L, NH$_4$-N = 0.03 mg/L, NO$_2$-N + NO$_3$-N = 0.07 mg/L, PO$_4$-P = 0.012 mg/L, total P = 0.031 mg/L
Land use: 85% forest, 10% surface waters, 0.2% agriculture
Population density: 2.3 people/km^2
Major information source: Laflamme 1995

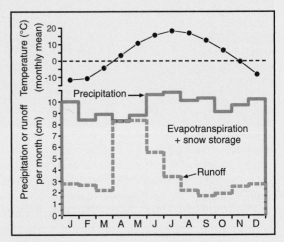

FIGURE 22.27 Mean monthly air temperature, precipitation, and runoff for the Saint-Maurice River basin.

23

RIVERS OF MEXICO

PAUL F. HUDSON   DEAN A. HENDRICKSON   ARTHUR C. BENKE
ALEJANDRO VARELA-ROMERO   ROCIO RODILES-HERNÁNDEZ
WENDELL L. MINCKLEY

INTRODUCTION
RÍO PÁNUCO
RÍOS USUMACINTA-GRIJALVA
RÍO CANDELARIA (YUCATÁN)
RÍO YAQUI

RÍO CONCHOS
ADDITIONAL RIVERS
ACKNOWLEDGMENTS
LITERATURE CITED

INTRODUCTION

Mexico, with an area of 1.97 million km², has approximately 150 large rivers and appears to have an abundance of water resources. However, the distribution of water resources is far from homogeneous, with northern Mexico extremely dry and southern Mexico among the wettest areas of North America. Located between 15°N and 33°N latitudes, Mexico is the warmest part of North America (Fig. 23.2) but has tremendous variety in climate topography. The combination of its mountainous topography producing strong orographic influences on precipitation and the fact that it straddles the temperate–tropical divide produces a diversity of runoff patterns and river environments.

With 105 million people (www.prb.org), or an average population density of 53 people/km², tremendous pressure is placed on Mexico's rivers for hydropower, irrigation, waste disposal, and domestic and industrial consumption. This is especially true in the central portion of the country, where nearly 25% of the total population lives in approximately 1% of the country's total surface area in the states of Mexico and Distrito Federal (www.citypopulation.de/Mexico.html). With a population growth rate of about 2.1% per year (www.prb.org 2004), pressure to use water from Mexico's rivers is rapidly increasing. Unfortunately, the fact that the hydrology, geomorphology, biodiversity, and ecology of Mexico's rivers are poorly studied is a serious shortcoming to understanding the environmental impacts of current and future water developments.

The history of human impacts on Mexico's rivers includes many groups of prehistoric inhabitants. Various accounts suggest the earliest inhabitants probably arrived more than 20,000 years ago and several major Mesoamerican culture regions developed large populations with sophisticated and complex cultures. Following the Olmec civilization that developed along the southeastern Mexican Gulf Coastal Plain around 4000 years ago, other major culture regions (Maya, Toltec, Huastec, and Aztec) developed within Mexico during the Classic period

FIGURE 23.1 Río Candelaria at bend in river, with surrounding floodplain and karstic hills in state of Campeche. Further investigations of basin management, pre-Hispanic floodplain manipulation, and canalization are discussed in Siemens and Soler-Graham (2003) (PHOTO BY A. H. SIEMENS).

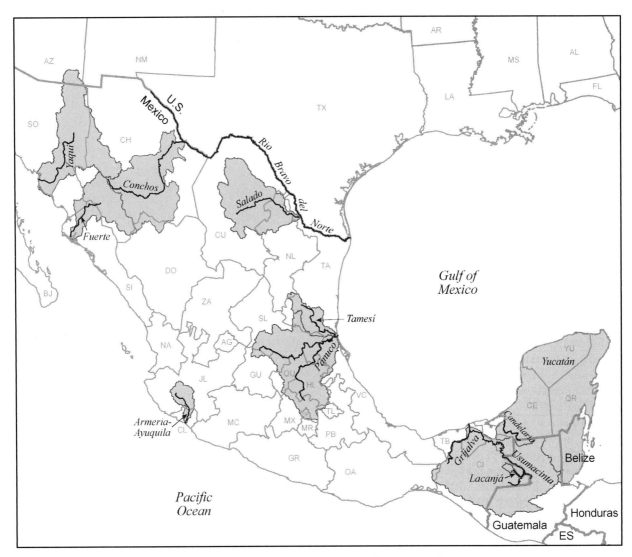

FIGURE 23.2 The rivers of Mexico described in this chapter.

(150 to 650 A.D.) (Coe and Koontz 2002). These complex civilizations were heavily dependent on Mexico's rivers and water resources until the Spanish conquest. It is now well accepted that some prehistoric societies had impacts on watershed processes of various basins. For example, O'Hara et al. (1993) found that in the Mexican Volcanic Belt, conversion of upland forests to traditional slash and burn agricultural land use by prehistoric civilizations resulted in accelerated soil erosion and associated sedimentation, with consequential impacts to downstream aquatic environments, and that the sediment continues to be stored within the basin. Beach (1998) found that slash and burn agriculture and resultant soil erosion in the Maya heartland (Yucatán) produced

distinctive soil horizons. Along the Mexican Gulf Coastal Plain, prehistoric Maya and Toltec practiced floodplain irrigation and wetland manipulation for intensifying agricultural production (Siemens 1980, 1983; Sluyter 1994; Whitmore and Turner 2002). The Olmec were primarily located within the states of Veracruz and Tabasco, from the Tuxla Mountains to the mouths of the Usumacinta, Grijalva, Coatzacoalcos, and Papaloapan rivers, and developed the first conduit drainage system in the Americas. Far to the north, other complex cultures developed, albeit to a somewhat lesser extent and in somewhat later periods. Doolittle (1985, 1988a, 1998b, 2000) provides information on the early development of agriculture, irrigation, and related erosion and other hydrologic control systems in northwestern Mexico (Sonora and Chihuahua) in areas where the Papago, Pima, Yaqui, and Tarahumara (Rarámuri) peoples, among others, still reside.

The Mayan civilization flourished from about 600 to 900 A.D., reaching a population of approximately 5 million around 700. They inhabited the Yucatán peninsula as well as the states of Chiapas and Tabasco. Covering most of the Usumacinta River basin, the Maya used the river and its tributaries extensively for travel and trade. To the north, the Huasteca region, located along the east-central Gulf Coastal Plain states of Veracruz, Tamaulipas, and San Luis Potosí, is the northernmost major culture region in Mesoamerica, reaching its apex in the Post-Classic period, around 1000. West of the Mayan region, the Aztecs dominated Central Mexico from about 900 to 1521, from the state of Veracruz on the Gulf of Mexico to the west coast states of Guerrero and Oaxaca. The Aztec population had reached an estimated 25 million when conquered by the Spaniard Hernán Cortés in 1521. Cortés's conquest resulted in Spanish rule for about 300 years until Mexican independence was declared in 1810. Thus, like many other regions of North America, prehistoric humans throughout Mexico depended on rivers for travel and food, but the complex societies and higher densities within Mexico likely impacted watershed processes and riverine resources more than in the United States and Canada (O'Hara et al. 1993, Doolittle 2000, Whitmore and Turner 2002).

In this chapter the physical and biological features of five major Mexican rivers are described in detail (see Fig. 23.2). The Río Pánuco drains arid and high-rainfall mountainous areas in east-central Mexico before flowing into the Gulf of Mexico. The combined Usumacinta–Grijalva rivers drain lush tropical rain forest before flowing together into the southern Gulf of Mexico. The Río Candelaria drains the western karstic Yucatán Peninsula. The Yaqui of northwestern Mexico is a desert river draining into the Gulf of California. The Río Conchos is an important tributary of the Río Bravo del Norte (known in the United States as the Río Grande, covered in Chapter 5). Five additional rivers described in one-page summaries include the Chihuahuan Desert's Río Salado, another tributary of the Río Bravo del Norte; the Río Tamesí, which joins the Río Pánuco near its mouth; the Río Fuerte, which flows through some of the continent's largest canyons in the Sierra Madre Occidental to the Gulf of California south of the Yaqui and Mayo rivers; the Ayuquila–Armería river system, which empties into the Pacific Ocean; and the

Río Lacanjá, a small mountainous tributary of the Usumacinta.

Physiography and Climate

Exclusive of Baja California, which has no major rivers and is not further considered, Mexico's physiography, and thus its rivers, is dominated by its mountain ranges. Most rivers drain eastward to the Gulf of Mexico (Atlantic) and Carribean or westward to the Pacific (including the Gulf of California). Two major mountain systems, the Sierra Madre Oriental and Sierra Madre Occidental, serve as the major drainage divides for the Pacific and Atlantic. However, between them lies the extensive Mexican Altiplano (or Mesa del Norte), which in some instances is characterized by having large areas with rivers draining to closed basins that prehistorically sometimes held large inland lakes, such as the Lagunas Palomas and Mayrán (Smith and Miller 1986), which undoubtedly had some degree of past interconnections. The Mexican Altiplano is bounded to the south by the Trans-Mexican Volcanic Belt, and further south is the Sierra Madre del Sur.

The physiography of Mexico's Pacific slope includes the Buried Ranges province in northwestern Mexico (Arbingast et al. 1975). This arid and hilly region, which includes part of the Sonoran Desert to the north, slopes gradually to the sea as it parallels the mountainous Sierra Madre Occidental province along a northwest–southeast shoreline. Inland, the Sierra Madre Occidental is the major physiographic province of Mexico, at 1200 km long, 200 km wide, and averaging 2000 m asl. It includes some of the world's largest canyons, such as the "Grand Canyon of Mexico" or Barranca del Cobre (Copper Canyon) of the Río Fuerte basin. These mountains parallel and form the western edge of the Basin and Range province that extends from the southwestern United States into north central Mexico and includes the Chihuahuan Desert and Desert Grasslands (northern portion of the Altiplano). Southward, across the Tropic of Cancer in the southern portion of the Altiplano, the Basin and Range rises and the Sierra Madre Occidental declines to form the northern and western edges of a vast tropical–subtropical plateau that makes up the forests and steppes of the Central Mesa province (also called the Central Plateau) and is the southern portion of the Altiplano. Further south, the Central Mesa ends at the Trans-Mexican Volcanic Belt (also called the Neovolcanic Plateau province), a chain of west-to-east-trending volcanic mountains 900 km long but only 100 km wide. These mountains have numerous active volcanoes, including Mount Orizaba (5747 m asl), the third-highest peak in North America. Significantly, much of the drainage within this extensive volcanic system is interior draining, feeding many large freshwater lakes. Although the rivers, such as the Río Lerma, draining the Mexico City metropolitan area and flowing into these lakes are not further considered, it should be noted that these lakes are distinctive within North America and include many endemic, and often now endangered, aquatic species. To the south of the Volcanic Belt lies the Sierra Madre del Sur province, which extends to Mexico's southern border.

The physiography and drainage patterns of Mexico's Atlantic slope are controlled by three mountain systems: the Sierra Madre Oriental, the Trans-Mexican Volcanic belt, and the Sierra de San Cristobal. The largest, the Sierra Madre Oriental, is a major physiographic province aligned north–south along Mexico's Gulf Coast. These mountains consist predominately of folded Cretaceous limestone, with several ridges exceeding 3500 m asl. Between the Sierra Madre Oriental and the Gulf of Mexico lies the Mexican Gulf Coastal Plain province, a continuation of the U.S. Coastal Plain with seaward-dipping Tertiary and Quaternary strata of marine and fluvial origin (Grubb and Carillo 1988) that extends to the Yucatán peninsula. Compared to the U.S. Coastal Plain, Mexico's Coastal Plain has greater relief because of the structural control imposed by the adjacent mountain systems, and it is much narrower, being widest (150 km) near the Río Bravo (Río Grande) valley and narrowing to the south to pinch out in central Veracruz as the Volcanic Belt intersects the southern edge of the Sierra Madre Oriental (de Cserna 1989). Isolated outcrops of Tertiary and Quaternary volcanics disrupt Coastal Plain drainage patterns and localized topography in southern Tamaulipas and northern Veracruz. Just north of Ciudad Veracruz a small but topographically significant volcanic range, Sierra Punta del Morro, appears to have been a biogeographical barrier to temperate and tropical fauna (Contreras-Balderas et al. 1996, Hulsey et al. 2004) as it extends to the Gulf of Mexico. South of it the Gulf Coastal Plain broadens as it arcs around southern Veracruz and Tabasco, fronting the Bahia de Campeche (part of the Gulf of Mexico). Here the Gulf Coastal Plain borders the enormous carbonate platform of the Yucatán Peninsula. In comparison to the mountainous and Gulf Coastal Plain environments of most of the rest of the country, this region represents a significant change in geomorphology and hydrology. The Yucatán is a classic example of a karstic landscape, with low relief characterized by predominantly subsurface drainage that has produced numerous caves, cenotes, springs, and solution depressions, and thus few large surface rivers. To the west and south the Yucatán Peninsula borders the Sierra de San Cristobal, a small mountain system of folded Cretaceous limestone (West and Augelli 1989) that is part of a larger system (Chiapas–Guatemala Highlands province) that extends into Guatemala and Belize, but which is distinct from the granitic Sierra de Chiapas of southern Mexico.

Approximately half of Mexico is south of the Tropic of Cancer (23.5°N) and within a tropical climatic regime. The significance of this is that the hydrological mechanisms of Mexico are distinct from those found in the temperate and northern latitudes of North America, and in particular are associated with the strengthening of easterly trade winds in summer months rather than westerly migrating midlatitude cyclones during winter months. Located at the interface between the midlatitudes and the tropics, the climate of Mexico becomes increasingly warm and humid toward the south. Mean annual temperatures exceed 20°C in low elevations of both the east and west coasts, and in the far south, mean annual lowland temperatures exceed 25°C. Temperatures are considerably cooler in mountainous areas, with mean annual values of 10°C to 15°C in the Sierra Madre Occidental and the Volcanic Belt. Seasonal differences in temperature are greater in the north and at high elevations; for example, January mean daily temperatures in the northern montane Sonoran and Chihuahuan deserts are about 5°C, whereas mean daily summer temperatures exceed 30°C. In southern Chiapas, however, mean January temperatures are about 25°C, with mean summer temperatures only a few degrees higher.

Precipitation is highly seasonal throughout Mexico, occurring mainly from summer through early fall with the strengthening of easterly trade winds associated with a northerly shift in the Intertropical Convergence Zone (ITCZ), and the occurrence of tropical cyclones between August and October (Metcalfe 1987). Mountain systems exert a significant orographic influence on precipitation, focusing Gulf moisture along the eastern mountain flanks and Coastal Plain and resulting in a steep west–east, low–high precipitation gradient. Mean annual precipitation increases toward the south. The driest regions are in the northern interior, where mean annual precipitation is less than 50 cm in the Sonoran and Chihuahua deserts. East of the

Chihuahuan Desert near the coast, the lower Río Bravo receives 80 cm, and the greatest rainfall exceeds 300 cm in portions of southeastern Veracruz (INEGI 1981c). Midlatitude cyclones (*nortes*) represent only a minor source of precipitation during winter months, occasionally penetrating as far south as Tabasco and Yucatán (West et al. 1969, Metcalfe 1987). Galindo (1995) suggests that ENSO (El Niño) has an influence on seasonal precipitation patterns in eastern Mexico. Specifically, the warm phase of ENSO is associated with increased winter precipitation along eastern Mexico due to a strengthened meridianal circulation that results in a higher frequency of midlatitude cyclones penetrating into Mexico (Diaz and Kiladis 1992). In contrast, La Niña years appear related to an increase in summer precipitation because of a higher frequency of tropical cyclones (Jauregui 1995). Thus, La Niña has more significant effects on streamflow and sediment transport than El Niño (Hudson 2003b).

Basin Landscape and Land Use

The variability of the Mexican climate allows this part of North America to support diverse vegetation, including desert, tropical, subtropical, temperate, and montane communities. More than 40 terrestrial ecoregions are recognized based on the recent analysis by the Comisión Nacional para el Conocimiento y Uso de la Biodiversidad (CONABIO) (Ricketts et al. 1999, National Geographic, www. nationalgeographic.com/wildworld/terrestrial.html). Among the largest ecoregions are the Sonoran Desert (northwest) and Chihuahuan Desert (north-central), both characterized by creosote bush, tarbush, mesquites, acacias, yuccas, and diverse cacti. Surrounded by these deserts to the east and west, the diverse Sierra Madre Occidental Pine–Oak Forest ecoregion includes 27 species of conifers and 21 species of oaks. Pine–oak forest ecoregions also characterize other mountainous areas, which roughly correspond to physiographic provinces, such as the Sierra Madre Oriental, Trans-Mexican Volcanic Belt, and Sierra Madre del Sur. In the northeast, along the lower Río Bravo del Norte, is the Tamaulipan mezquital ecoregion, with acacia, desert hackberry, javelina bush, cenizo, common bee-brush or white brush, Texas prickly pear, and tasajillo or desert Christmas cactus. South of the Tamaulipan mezquital is the Veracruz Moist Forests, the northernmost "tropical rain forests," characterized by many broad-leafed species and extremely high diversity of terrestrial plants and animals. Further south and east are

the Petén–Veracruz Moist Forests, a lowland tropical forest dominated by Mayan breadnut, sapodilla, rosadillo, and gumbo limbo. Various types of dry forest ecoregions are found primarily in western Mexico. In the middle of the continent are arid high plateaus, such as the Meseta Central Mattoral and the Central Mexican Mattoral.

According to a 1993 analysis, land use in Mexico is 12% arable land, 1% permanent crops, 39% permanent pasture, 26% forest, and 22% other (probably mostly arid lands) (www.new-agri.co.uk/02–3/countryp.html). The major agricultural products are corn, wheat, soybeans, rice, beans, cotton, coffee, fruit, tomatoes, beef, poultry, dairy products, and wood products. Approximately half of the forested area is the coniferous and broad-leaved forests of the more mountainous areas that account for 90% of Mexico's forest production (www.worldforestry.org/wfi/WF-mexic.htm). The other half of Mexico's forests is the tropical and subtropical forests of southern Mexico that account for only 10% of forest production.

The Rivers

Abell et al. (2000) defined 25 highly diverse freshwater ecoregions on the basis of the distribution and characteristics of Mexican river basins. The rivers selected for discussion in this chapter cover only a fraction of these ecoregions but were selected to illustrate the diversity of the country's rivers for which there is a reasonable amount of information. Although the majority of Mexico's 150 rivers drain into the Pacific Ocean, the majority of the streamflow is discharged from the Gulf Coastal Plain river systems, and the latter receive more attention in this chapter.

The rivers described for northern Mexico (Yaqui, Fuerte, Conchos, and Salado) are arid systems. The Conchos and Salado rivers are both tributaries of the Río Bravo del Norte (Río Grande). The Conchos (Río Conchos ecoregion), the primary drainage of the north-central state of Chihuahua, has historically provided the majority of the flow in the Río Bravo del Norte after most of the latter's water is extracted for agriculture and domestic uses in New Mexico. The Salado (Río Salado ecoregion) drains eastward across the states of Coahuila, Nuevo León, and Tamaulipas before emptying into Falcon Reservoir on the Río Bravo del Norte, northeast of the large industrial city of Monterrey. The Yaqui (Sonoran ecoregion) and Fuerte (Sinaloan Coastal ecoregion) rivers both drain into the Gulf of California (Pacific drainage) after arising in the mountains of the Sierra Madre Occidental and flowing through the Sonoran Desert. The Yaqui drains the states of Sonora and Chihuahua, as well as the extreme southeast corner of Arizona, and the Fuerte drains a large part of Chihuahua and small areas of both Durango and Sonora, with its lower reaches passing through Sinaloa. Further south, the Armería–Ayuquila river system (Manantlán–Ameca ecoregion) flows directly to the Pacific Ocean from the states of Jalisco and Colima, south of the large city of Guadalajara.

Most rivers draining into the Gulf of Mexico are from more mesic basins, with five of these, the Grijalva, Usumacinta, Pánuco, Papaloapán, and Coatzacoalcos, accounting for over 50% of Mexico's average annual flow discharged into ocean basins. Here we describe the Usumacinta–Grijalva and Pánuco in some detail. The Río Pánuco is the only major Mexican river located outside of southeastern Mexico. It and its lowermost tributary, the Tamesí, drain the Tamaulipas–Veracruz ecoregion and join near the port city of Tampico. Their headwaters begin in the arid Mexican Altiplano and the mesic Sierra Madre Oriental and flow across the Coastal Plain, draining parts of the states of Tamaulipas, San Luis Potosí, Hidalgo, Querétaro, Mexico, Guanajuato, and Veracruz. The largest river system of Mexico, the Usumacinta–Grijalva (Grijalva–Usumacinta ecoregion), drains into the southern Gulf of Mexico near Villahermosa, Tabasco, from tropical rain forests of Chiapas and Tabasco, although headwaters of both rivers are in Guatemala. The much smaller Candelaria (also Grijalva–Usumacinta ecoregion) to its northeast is one of the larger rivers draining the highly karstic Yucatán Peninsula.

In general, Mexico's major rivers are highly exploited. Northern and central Mexico, which have over 45% of the land area and roughly 60% of Mexico's population, have fewer than 10% of the country's water resources (http://worldfacts.us/Mexico-geography.htm). Since arid and semiarid regions are highly sensitive to hydrological change, it is not surprising that easily exploitable water sources have already been developed (Tortajada and Biswas 1997) and rivers in these more northerly regions have become highly degraded. Many dams have been built, primarily for crop irrigation in otherwise desert environments (Contreras-Balderas and Lozano 1994). Water management that favors irrigation, such as in the Ríos Conchos, Yaqui, and Fuerte, often results in a complete elimination of flow, ignoring aquatic biological and ecological concerns. Water pollution from discharge of domestic wastes, high salinity, and nutrients from irrigation returns, mining, and industrial wastes is widespread throughout Mexico. Such problems are particularly acute in the Pánuco and Conchos, as well as many other rivers not described in this book, such as the Lerma, San Juan, and Balsas. The most pristine waters, containing the greatest biodiversity and the most natural ecosystem functioning, are usually found in the more mountainous regions and in the tropical rain forest of southern Mexico. Even the tropical rain forest rivers are facing extensive exploitation, with plans for multiple hydropower dams that would flood large natural ecosystems as well as important archaeological sites. In particular, biosphere reserve sites in the Usumacinta basin are in considerable peril, both from impoundment and forest exploitation. Thus, although many of Mexico's rivers are already under considerable stress, the situation appears to be getting substantially worse, with an increasing population and associated land-use change and generally limited resources available for conservation of natural resources.

Although Mexico's rivers are under great stress and are poorly studied, the evidence suggests they contain unique collections of aquatic communities (Minckley et al. 1986, Smith and Miller 1986, Miller and Smith 1986, Obregón-Barboza et al. 1994, Abell et al. 2000, Arriaga-Cabrera et al. 2000). The fishes are probably the best known group, and it is clear that there are many species in Mexico found nowhere else in North America. Nelson et al. (2004) list 1277 North American freshwater fishes, with 521 occurring in Mexico (348 endemic), whereas 912 are found in the United States (544 endemic) and 212 in Canada (8 endemic). Mexican endemics thus comprise nearly 28% of the North American fish fauna, and its fauna is more highly endemic (67%) than are those of either the United States (60%) or Canada (4%) (Miller et al. in press, www.mongabay.com/fish/data/Mexico.htm). If these numbers are scaled to account for the differences in area among these countries, the number of Mexican fish species per km^2 is found to exceed that of the United States by a factor of 2.7 and that of Canada by a factor of 12.4. Numbers of endemic species per unit area in Mexico exceed that of the United States by a factor of 3 and that of Canada by a factor of 220. Such statements about the diversity and uniqueness of Mexico's fish fauna may also be true of its significantly less studied freshwater invertebrates. For example, among the aquatic insects, odonates (dragonflies and damselflies) show similar diversity patterns, with 326 species recorded from the United States, 160 from

Canada (Bick and Mauffray 2004, www.afn.org/~iori), and 342 from Mexico (Paulson and Gonzalez Soriano 2004, www.afn.org/~iori); the hellgrammite genus *Corydalus* has only two species found north of Mexico, but five within Mexico (Contreras-Ramos 1998).

RÍO PÁNUCO

The Pánuco is the second largest (98,227 km²) of the Mexican watersheds draining to the Gulf of Mexico (Hudson 2000). Three major subbasins supply the majority of the system's runoff and sediment: the Moctezuma, Tamuin, and Tamesí (Fig. 23.11). Because the Río Tamesí (19,127 km²) joins the Pánuco near its mouth, it is a distinct system and is considered separately. The Pánuco basin, without the Tamesí subbasin, drains 79,100 km². The Río Pánuco is formed by the confluence of the Ríos Tamuin (33,260 km²) and Moctezuma (42,726 km²) after they cross the Sierra Madre Oriental. The Río Pánuco then meanders 185 km before being joined by the southeasterly flowing Río Tamesí (see Fig. 23.23) at Tampico, Tamaulipas, several kilometers before discharging into the Gulf of Mexico. The small Río Topila (3114 km²) flows north within the Coastal Plain, joining the Río Pánuco 25 km upstream of Tampico.

The major prehistoric culture region associated with the Pánuco system is the Huasteca, located in east-central Mexico along the Gulf Coastal Plain and eastern ranges of the Sierra Madre Oriental (Hudson 2004). Hunters and gatherers occupied the region since at least the mid-Holocene, but there is no evidence that these primitive peoples significantly disturbed the Pánuco system. The Teenek are the major Indian group associated with the Huasteca, but that group is not considered to be homogeneous. The Teenek migrated north along the Coastal Plain before the Maya civilization reached its zenith. Indeed, although the Teenek speak a primitive Mayan language, their material culture is distinct (Ekholm 1944). Although the Huastec civilization did not construct extensive irrigation projects, they settled along the floodplains of the Gulf Coastal Plain to take advantage of the resources, such as shellfish, and to develop extensive floodplain agriculture. Thus, the Huastec would have greatly modified the riparian environments of the lower Pánuco system; however, rather than constructing large population centers, such as the Aztec in the Mexican Altiplano, the Huastec settled in small dispersed villages with the

Pánuco valley serving as the major axis within the Huastec culture region (Hudson 2004). The Huastecans were in decline by the time of Spanish arrival, and were quickly subjugated by the conquistadors, with many fleeing to the adjacent Sierra Madre Orientals, where subsistence agriculture continues to be practiced. The Spanish also contributed to extensive land-use change within the lower Pánuco system, which they found particularly useful for cattle and horse raising, and this area became the focal point of cattle raising in the Americas.

Physiography, Climate, and Land Use

The Pánuco basin drains four major physiographic provinces: the Central Mesa (CM), the north–south Sierra Madre Oriental (SO), the east–west Trans-Mexican Neovolcanic Belt (or Neovolcanic Plateau, NP), and the Mexican Gulf Coastal Plain (CP) (see Fig. 23.11). The regional physiography of the basin has a strong influence on local climate within the basin, especially spatial variability in precipitation.

The basin's precipitation exhibits a steep west–east gradient, becoming increasingly humid toward the Gulf of Mexico (INEGI 1984a, 1984b). Major mechanisms responsible for generating precipitation include trade winds and tropical cyclones during summer months and early fall. Midlatitude cyclones (*nortes*) deliver brief episodes of rainfall during winter months, but collectively are not as significant as summer precipitation mechanisms. Indeed, precipitation for this portion of Mexico averages >12 cm/mo from June through October (Fig. 23.12). Likewise, temperature is highly variable, depending on altitude, but an approximate monthly range for the entire basin is only from about 15°C in January to 24°C in June.

The eastern Central Mesa (or Central Plateau) is semiarid to arid, with average annual precipitation ranging from 30 to 40 cm and occurring mainly from May through October (INEGI 1984a, 1984b). Temperatures on the plateau are mild, but killing frosts are not uncommon during the winter along higher ridges and slopes when *nortes* sweep southward from the United States (West and Augelli 1989). The elevation of the Central Mesa increases south toward Mexico City and the increase in volcanics creates a rugged landscape with greater relief than seen in the northern Altiplano. Near Guanajuato elevations reach 3000 m asl. The major terrestrial ecoregions in this area are the Mexican Central Mattoral and the Meseta Central Mattoral (Ricketts et al. 1999). Vegetation varies within this region; it is largely limited by the lack of precipitation and consists mainly of desert scrublands, cacti, and short grasslands, with low forests of pine and oak in higher elevations (INEGI 1984c).

The central portion of the Pánuco basin drains the mountains of the Sierra Madre Oriental physiographic province. Within its folded western escarpments, headwater tributaries converge to form the Río Tampaón (draining from the west) and the Río Moctezuma (draining from the southwest), resulting in deeply incised valleys. The mountains here are mainly comprised of Cretaceous limestone, folded into a series of parallel ridges and valleys. In addition to having the greatest relief, this mountainous portion of the basin has the greatest climatic diversity. Average annual precipitation increases from 35 cm in the western mountains to 240 cm along the easternmost escarpments of the Sierra Madre Oriental (INEGI 1981b). Temperatures are moderate at higher altitudes, although light snows are common during winter months along higher ridges, which may exceed 3500 m asl. Desert scrub predominates along the western ranges of the Sierra Madre Oriental. Toward the interior of the mountains to the east and at higher elevations a high diversity of pines and oaks are present within the Sierra Madre Oriental Pine–Oak Forests terrestrial ecoregion (Ricketts et al. 1999, INEGI 1984c), although there may be dramatic differences in vegetation along a single ridge because of east- and west-facing slopes. With increasing precipitation to the east one finds the Veracruz Moist Forests ecoregion, the northern extent of tropical broadleaf evergreen forests (West and Augelli 1989). Traditional farming practices occur throughout the mountains, even on steep slopes, whereas mechanized agriculture and ranching are practiced in the larger river valleys of the Coastal Plain.

The Veracruz Moist Forests ecoregion continues across the Sierra del Abra, the easternmost ridge of the Sierra Madre Orientals, into the Mexican Gulf Coastal Plain physiographic province. The Coastal Plain extends another 90 km to the Gulf Coast at Tampico. Due to significant structural controls imposed by the adjacent Sierra Madre Orientals, the Mexican Gulf Coastal Plain is more complex than the U.S. Gulf Coastal Plain. A south-plunging anticline results in the Río Tamesí and Río Topila being diverted toward the east (Trager 1926, Muir 1936) to eventually join the Río Pánuco just upstream of Tampico. Most of the Tertiary deposits consist of weakly consolidated shale, with thin beds of friable sandstone that are highly erodible and contribute to a more diverse Gulf Coastal Plain landscape than is

seen in the U.S. Gulf Coastal Plain. Just south of Tampico, Tertiary deposits extend to the coast and reach a height of 90 m (INEGI 1984d, 1984e).

Land use and land cover exhibit considerable spatial variability across the Pánuco basin due to the west–east precipitation gradient and the regional environmental history. There is very little large-scale agriculture practiced within the arid western portions of the Sierra Madre Oriental and Altiplano, where small-scale subsistence agriculture confined to the river valleys predominates. One exception is the large-scale citrus production around the city of Río Verde in the central Sierra Madre Oriental. Toward the eastern and increasingly humid ranges of the Sierra Madre Oriental, extensive slash and burn (swidden) agriculture is practiced in river valleys and on toe and steep mountain slopes. Common crops include corn, beans, bananas, coffee, and citrus (Alcorn 1981). Although the forest has the appearance of being natural, in many instances it is secondary or tertiary growth and is heavily managed. In addition to corn and beans grown on small plots, some coffee and citrus are grown within the understory of the forest canopy. The natural vegetation of the Gulf Coastal Plain originally included thornbrush savanna on terraces and lush tropical forests within the river valleys; however, satellite imagery reveals very little forest remaining in the Gulf Coastal Plain (Crews-Meyer et al. 2004). Most of the Gulf Coastal Plain is now used for cattle ranching and farming, a land-use legacy that began in the early 1500s and continues today. Within the river valleys of the western portions of the Coastal Plain, sugar cane and citrus (mixed with some banana and papaya) are the major agricultural products, and this agriculture spread down valley during the twentieth century to displace some former cattle ranching. In part, displacement of ranching in favor of farming within these valleys is due to the Mexican government and the World Bank initiating an enormous irrigation project, designed as one of the largest irrigation systems in Latin America. Due to poor planning, however, the project is largely inoperable and was never completed (Aguilar-Robledo 1999, Hudson et al. in press).

River Geomorphology, Hydrology, and Chemistry

The Pánuco drainage system undergoes tremendous changes in river morphology and hydrology from its headwaters to the coast (Fig. 23.3). From the Central Mesa through the mountains the tributary rivers have

FIGURE 23.3 Lower Río Pánuco at average flow, 40 km upstream of Tampico. Note the small boat for scale (Photo by P. H. Hudson).

wide and shallow channels adjusted for the transportation of bed load. Within the mountains, rivers have narrow valleys and channels tend to braid with laterally active channel margins. The floodplain consists of narrow ribbons of deposits comprised of coarse sediments, and lack the complexity found in the lower portions of the basin. Because of the high-energy setting within the mountains, floodplain deposits are likely rapidly reworked during extended periods of episodic flooding.

Larger rivers within the Central Mesa have flashy discharge regimes characterized by great differences in base flow and storm flow. Smaller streams are intermittent or ephemeral, transporting water seasonally or only after precipitation events. The Río Tula, the main river in the upper Moctezuma basin, flows north from the edge of Mexico City (see Fig. 23.11). The Río Santa Maria and Río Verde are the major headwater streams for the Río Tampaón (Tamuin) basin and form to the east of the capital

city of San Luis Potosí. Here the river drains mainly Tertiary and Quaternary lacustrine and alluvial deposits of the Central Mesa about 2000 m asl, with isolated Tertiary volcanics having peaks of about 2500 m asl.

Within the mountains, river gradients are highly variable but can exceed 5 m/km in some locations. Spectacular waterfalls occur where tributaries join incised valleys. For example, the Cascada de Tamul is a 102 m waterfall at the confluence of the Ríos Gallinas and Santa Maria, forming the Río Tampaón. The Tampaón and Moctezuma, the two largest tributaries of the Pánuco, exit the Sierra Madre Orientals as formidable rivers. Along the eastern Sierra Madre Oriental, before the Tampaón exits the mountains, it encounters numerous karstic features, including sinkholes, natural bridges, and waterfalls. Surface drainage is often disrupted by solution cavities, manifested in dry valleys and disappearing streams. This water eventually returns as enormous springs along

FIGURE 23.4 Lower Río Tampaón at low stage. The Río Valles enters from the right. Further downstream the Tampaón becomes the Tamuin, and after joining the Moctezuma it becomes the Pánuco. Note the small boat for scale (PHOTO BY P. H. HUDSON).

the lower flanks of the eastern Sierra Madre Oriental. For example, the Río Coy forms from one of the largest springs in the world (Fish 1977) and joins the Río Tampaón upstream of Ciudad Tamuin to become the Río Tamuin (Fig. 23.4).

There are fewer karstic features within the lower Moctezuma basin. Tertiary shale units in the Gulf Coastal Plain produce much larger sediment loads. The Río Tempoal, receiving runoff and sediment from this lithology, enters the Moctezuma in the Coastal Plain at El Higo, resulting in an immediate increase in the sediment load of the Moctezuma and producing café-colored waters that contrast sharply with the clearer spring-fed streamflow of the Tamuin. The Moctezuma transports an annual sediment load of 4623×10^3 tons, whereas the Tamuin transports 2030×10^3 tons annually. Much of the suspended

sediment is transported in early summer and is exhausted before the arrival of larger flood events in September (Hudson 2003a). The two rivers meander through wide alluvial valleys before joining to form the Río Pánuco in the western Gulf Coastal Plain, 185 km upstream of the Gulf of Mexico. Valley gradients are predictably low, decreasing from the western edge of the Gulf Coastal Plain to the Gulf of Mexico. Between Tamazunchale and El Higo, for example, the Río Moctezuma has a gradient of 1.2 m/km, whereas downstream of Ciudad Pánuco the gradient is 0.04 m/km. This extremely low gradient of the lower Pánuco valley enables saltwater to intrude as far upstream as Ciudad Pánuco during low flows and results in a daily tidal exchange between the large lagoons downstream of Ciudad Pánuco (Hudson 2002).

Upon entering the Gulf Coastal Plain, the Moctezuma and Tamuin valleys widen considerably, and the rivers develop meanders. Holocene floodplain deposits increase in complexity and include deposits representative of lateral and vertical floodplain construction, such as point bar, natural levee, backswamp, crevasse splay, and infilled paleochannels. In several reaches the river is in contact with Tertiary outcrop, the contrasting resistance of which disrupts the planform geometry from a symmetrical meandering pattern (Hudson and Heitmuller 2003). The sinuosity (ratio of channel length to valley length) of the Río Pánuco averages 1.85 over its 185 km channel but shows considerable variability where the river is reworking more resistant Tertiary deposits (Hudson 2002). The lower 30 km of the Río Pánuco is incised into Holocene deltaic deposits with sinuosity of 1.15, essentially a straight channel. The fine-grained clayey bank sediments are sufficiently cohesive to resist erosion, and the lack of sandy bed material does not permit point bar construction. Moreover, at the lower limits of the delta, shortly before discharging into the Gulf of Mexico, the Río Pánuco is incised into a ridge of Tertiary sandstone that prevents lateral channel migration (INEGI 1984e).

Lakes and wetlands are largely absent in the mountains and Central Mesa but become more common in the lower reaches of the basin, where they are related to floodplain geomorphology (Hudson 2002). Abandoned channel courses and meander neck cutoffs serve as oxbow lakes and arcuate swamps, depending on the degree of infilling. Such lakes and wetlands are common in the lower Moctezuma and Tamuin valleys and upper Pánuco valleys. Backswamps, evidence of long-term floodplain stability, are common where valley width is sufficient. In the lower Pánuco and lower Tamesí valleys, large lakes and wetlands form between meander belts and resistant Tertiary deposits that serve as topographic drainage barriers. Although these features were naturally created, the river water is used by the city of Tampico and so is heavily regulated by a system of weirs and control structures. The lagoons and wetlands in the lower Tamesí are tidal and important to the regional fishery, and provide vital habitat for an array of North American migratory waterfowl.

Average discharge of the Río Pánuco at Pánuco, upstream of the Tamesí and Topila, is 473 m³/s, and the annual discharge regime strongly reflects the regional precipitation pattern for eastern Mexico (Hudson 2000). Discharge and runoff are uniformly low from December through early May but increase rapidly with the onset of summer trade winds in late May (see Fig. 23.12). Peak discharge events usually occur in September, often associated with an increase in tropical cyclone activity (Hudson and Colditz 2003).

In spite of the strong seasonal pattern of runoff, only the largest floods generated by tropical cyclones inundate the entire valley. Mechanisms for inundating the floodplain include a rise in the water table of the alluvial aquifer; conduits, such as crevasse channels and paleochannels connected to the active channel; and local precipitation (Hudson and Colditz 2003). The timing of flood events is similar for the Río Tamuin and Río Moctezuma, although the more humid Moctezuma basin supplies the majority of flow. In addition, the hydrograph for the Tamuin is less flashy than the Moctezuma, probably because more of its discharge derives from groundwater and spring-fed sources (Hudson 2003b). In comparison to the Usumacinta–Grijalva system to the south, the Pánuco is not as flood prone because the floodplain is approximately 10 m above average stage. For example, at Ciudad Pánuco, ~30 km west of Tampico, the floodplain is elevated above the water surface by 12 m where the valley crosses the axis of an anticline. Indeed, the abundance of prehistoric sherds along the channel banks attests to this site long being considered less likely to flood (Ekholm 1944, Hudson 2004).

Although information on water chemistry of the Pánuco system is not easily accessible, it is clear that the river and its tributaries are seriously threatened by pollution from agrochemicals, municipal discharge, and salinization (Abell et al. 2000).

River Biodiversity and Ecology

The Río Pánuco is the major river within the Tamaulipas–Veracruz freshwater ecoregion (Abell et al. 2000), where CONABIO has identified several priority sites for conservation. Although information is available on fishes of the river, other aspects of biodiversity and ecology are less well known.

Plants

Information on aquatic plants and major riparian vegetation is not easily accessible for the Pánuco system. However, nonnative hydrilla is found in some of the oxbows and sloughs of the major valleys in the Gulf Coastal Plain.

Invertebrates

Information on invertebrates also is sparse, although some information is available on mollusks, crustaceans, and hellgrammites. Crustaceans found in the Pánuco system include the freshwater shrimp *Palaemonetes mexicanus*, as well as the cave-dwelling shrimps *Troglomexicanus perezfafantae*, *T. huastecae*, and *T. tamaulipenses*. The crayfishes include *Procambarus (Ortmannicus) ortmanii*, *Procambarus (Ortmannicus) acutus cuevachicae*, and *Procambarus (Scapullicambarus) strenghi*, and mollusks include *Hydrobia tampicoensis*, *Littoridina crosseana*, and *Lithasiiopsis hinkleyi* (A. Contreras-Ramos, personal communication). Among the aquatic insects are the hellgrammites *Chloronia mexicana* and *Corydalus magnus* (Contreras-Ramos 1998). Abell et al. (2000) state that the Tamaulipas–Veracruz freshwater ecoregion (of which the Pánuco is a major part) has 17 endemic species of crayfish, although the specific species are not mentioned.

Vertebrates

The Río Pánuco (not including the Río Tamesí) harbors at least 80 described native fish species, with more than one-third of them endemic (Miller and Smith 1986, Rauchenberger et al. 1990). Much of the basin is remote and underexplored by ichthyologists, and many more species (notably cichlids, catfishes, and goodeids) surely will be discovered and described in the near future. The much smaller and more thoroughly sampled Río Tamesí has at least 93 fish species. Notable Pánuco endemics among the cichlids include the Media Luna cichlid, blackcheek cichlid, chairel cichlid, and slender cichlid. There are also several endemic goodeids, including bluetail splitfin, dusky goodea, relict splitfin, and jeweled splitfin, all but the last endemic. Among pupfishes, the endemic, endangered, monotypic genus *Cualac* is noteworthy. Swordtails (genus *Xiphophorus* of the family Poeciliidae) include many endemic species and are popular in the aquarium trade and important as research organisms (Ryan and Rosenthal 2001). They are diverse and well studied in this basin, including sheepshead swordtail, short-sword platyfish, delicate swordtail, highland swordtail, Moctezuma swordtail, barred swordtail, mountain swordtail, Pánuco swordtail, pygmy swordtail, and variable platyfish (Rauchenberger et al. 1990). The minnow family, nearing its southern distributional limit, also is represented by numerous endemics, especially in the genus *Dionda* (Pánuco minnow, bicolor minnow, chubsucker minnow, lantern minnow, flatjaw minnow, blackstripe minnow), which includes several sympatric species pairs

(Mayden et al. 1992). An extremely rare and endangered blind cave catfish, the phantom blindcat, recently discovered in the Río Guayalejo in southernmost Tamaulipas (Walsh and Gilbert 1995, Hendrickson et al. 2001) presents an interesting evolutionary enigma (Willcox et al. 2004). Many caves of this region harbor the blind form of Mexican tetra, which may well be one of the most studied nongame fishes of North American (e.g., Mitchell et al. 1977, Langecker et al. 1995, Borowsky 1996, Jeffrey 2001, Dowling et al. 2002). Several more fascinating cave fishes are likely to be discovered in this large karstic river basin. Larger fishes include the endemic large native fleshylip buffalo and the endemic Río Verde catfish.

Abell et al. (2000) mention that nonnative tilapias (*Oreochormis* spp.) are a problem in this basin, having irreversible negative impacts on the native fauna as seen throughout much of Mexico, and we concur (D. A. Hendrickson, personal observations). Fish culturists have, unfortunately, also introduced herbivorous cyprinids (grass and silver carps), channel and blue catfish, largemouth bass, and other centrarchids, to mention only some nonnatives (Garcia de León et al. in press). These will undoubtedly impact native faunas via hybridization (catfishes especially) and competition or habitat alteration.

A high diversity of other aquatic vertebrates is present in the Pánuco basin, but there are no basinwide compilations of aquatic herpetofauna, birds, or mammals. Abell et al. (2000), however, state that the Tamaulipas–Veracruz freshwater ecoregion contains at least 16 endemic species of aquatic herpetofauna.

Ecosystem Processes

Although there have been no ecosystem studies done in the Río Pánuco, this system offers an interesting contrast with the River Continuum Concept (Vannote et al. 1980). Instead of headwater streams beginning in high-altitude forests, they begin in the arid Central Mesa, then flow through pine–oak mountain forests and finally through tropical rain forests at lower elevations. The upper arid streams, both spring fed and not, appear to be supported by autochthonous production (with several herbivorous fishes, such as *Dionda* species, and abundant grazing snails in some streams). There is, however, likely to be a major shift to increasing allochthonous inputs as the major tributaries flow first through the mountains and then into the Gulf Coastal Plain as turbidities increase.

Human Impacts and Special Features

Before colonization by major civilizations, the Río Pánuco system must have exhibited some of the greatest physical and ecological diversity of the region. Draining from arid plateaus in its headwaters, its tributaries cut through the rugged pine–oak forests of the Sierra Madre Oriental, entered the Veracruz Moist Forests in the eastern ranges, and then spilled into the low-gradient tropical forests of the Coastal Plain. Among the scenic features are spectacular waterfalls, natural bridges, tremendous sinkholes, enormous springs, and some of the world's most extensive and deep cave systems, dropping from high pine and cloud forests precipitously to foothills. The transition in physical features along this river continuum was undoubtedly reflected in major transitions of its biological communities. Although many of the river's scenic physical features still exist, it is likely that remnants of its natural aquatic communities will be found primarily in tributaries that have been spared the exploitation and pollution found throughout much of the system.

Because of the physical diversity and history of land use, the Pánuco system has been impacted by humans in many ways. It is difficult to establish causal relationships on the current status of the system. However, it is worthwhile to provide a brief characterization of these different activities.

Upper arid portions of the basin are impacted by the legacy of mining. In smaller valleys, check dams (*trincheras*) are emplaced. These small structures result in infilling of sediment and increase the arable land area. Perhaps the most significant human impacts in the upper reaches involve pollution. Although Mexico City is not within the Pánuco basin, a network of canals delivers raw sewage from that metropolitan area to the Río Tula, a tributary of the Río Moctezuma. An even greater threat to the system, however, may be that headwaters of the basin drain the mining district of Mexico. Old colonial cities such as San Luis Potosí, San Miguel de Allende, Guanajuato, and Queretero were established centuries ago by Spain and extensive silver, copper, heavy-metal, and semiprecious stone mining operations remain. Unfortunately, mismanagement of these operations has resulted in extensive environmental damage that is contaminating many of the regional drainage systems, and these contaminants are likely adsorbed to fluvial sediments transported by the Pánuco system. Another form of mining, groundwater pumping for agriculture, is having profound negative impacts on some important endemic fish habitats of the arid altiplano that are drained by this river system.

Although dams are not as extensive as in some other river basins of Mexico, they interact negatively with other human impacts, and future dams are in the planning or construction phases. In the mountainous middle reaches there is a large dam on the Río Moctezuma, upstream of the major inputs of streamflow and sediment on the arid western side of the Sierra Madre Oriental. A new hydroelectric dam is under construction in the lower Río Tamuin (at Ciudad Tamuin), approximately 75 km upstream of the confluence with the Moctezuma. How this will influence river morphology and ecology remains to be seen, although at a minimum it will contribute to the hydrologic fragmentation and reduce downstream sediment loads. In the lower reaches of the basin there are few engineering modifications influencing the drainage basin hydrology, although artificial flood-control levees were constructed along selected reaches of the lower Pánuco after the devastating flood of 1955. Here the river is a freely meandering channel, as there have been no artificial cutoffs or bank protection works constructed to reduce lateral migration. Although the prehistoric legacy did not leave a conspicuous imprint, agriculture during the twentieth century extensively altered composition of the riparian corridor. Because of the significant fluctuation in the water table of the alluvial aquifer, most agriculture requires irrigation. In some instances the oxbow lakes are used for irrigation, but primarily the water is locally pumped directly from the channel. Chemical fertilizers and pesticides used for sugarcane farming throughout the lower Pánuco basin are of concern and have resulted in extensive pollution of the lower Tampaón (Tamuin) system, particularly downstream of Valles. More significant impacts, however, are likely to occur due to contamination of river sediments associated with the extensive petroleum activities along the lower river. Petroleum was discovered in the lower Pánuco basin in 1904, and this is the oldest region of petroleum development in Mexico.

RÍOS USUMACINTA–GRIJALVA

Draining 112,550 km² of southern Mexico, the Usumacinta–Grijalva drainage system is the largest in Mexico. The basin has two distinct regions, the mountainous uplands and the Coastal Plain in

FIGURE 23.5 Upper portion of Río Usumacinta (Río Lacantún) (Photo by H. Bahena).

Tabasco, both receiving ample amounts of rainfall (Fig. 23.13). The main rivers within this system are the Usumacinta and Grijalva, both with headwaters in Guatemala (35% of entire basin) but with 45% of their basin in Chiapas, Mexico's southernmost state. The Usumacinta drains western Chiapas, serves as the international border between Mexico and Guatemala, and is joined by the San Pedro downstream of Tenosique (Fig. 23.5). The Grijalva begins as the Río San Miguel in Guatemala becomes the Grijalva in Mexico and is joined by several large tributaries, including the Río Suchiapa, which drains the Sierra Madre de Chiapas of western Chiapas. As

it flows toward the Coastal Plain the Grijalva also receives drainage from the eastern margins of the states of Oaxaca and Veracruz. The Grijalva and Usumacinta join, but only partially, near Frontera, Tabasco, about 15 km above their mouth in the Gulf of Mexico (Rodiles-Hernández 2004).

The basin has a rich legacy of prehistoric human history, and prior to the Spanish arrival supported a dense population. Uplands in Chiapas were probably densely populated by the Maya during that culture's Classic period from 150 to 650 A.D. (West et al. 1969), and the area underwent land-use change as a consequence of slash and burn agriculture. The lower reaches of the basin were colonized by several different Indian groups and served as an important trading link between the Aztec to the west and the Maya in the Yucatán. Indeed, lowland Tabasco long formed the western edge of Mayan civilization during the Classic period, and the eastern fringe of the Olmec heartland, the most ancient of the great Mexican culture groups (Coe and Koontz 2002). Most prehistoric settlement in lower portions of the basin occurred along broad natural levees of mainstem channels and distributaries, with higher grounds of the levees providing safety from frequent lowlands flooding. Prehistoric peoples residing in lowlands practiced slash and burn agriculture of maize, beans, squash, and various types of tubers. The wet climate enabled cacao cultivation and that product was used extensively throughout the region for trading (West et al. 1969). Although there are few reliable population estimates from the time of Spanish conquest, archaeological evidence suggests that indigenous populations had already declined but were decimated by the Spanish after contact and did not recover until the mid-nineteenth century (West et al. 1969).

Physiography, Climate, and Land Use

Three physiographic provinces are represented in the Usumacinta–Grijalva basin (see Fig. 23.13). The largest is the Chiapas–Guatemala Highlands (CG), with its mountainous uplands, primarily in Chiapas. The other major province is the Mexican Gulf Coastal Plain (CP) to the north, mostly in Tabasco. A small part of the basin (upper Río San Pedro drainage) is in the Yucatán (YU) province to the east.

Fundamental differences in lithology of the headwaters of the Usumacinta and Grijalva result in differences in surface erosion and drainage patterns. The Usumacinta drains the extensive folded block of Cretaceous limestone of the Sierra de San Cristobal, which approaches 3000 m asl. This range is part of a more extensive mountain system that reaches 3500 m asl in northern Guatemala, where it is known as the Sierra Alto Cuchumatanes (West and Augelli 1989). The folded terrain of the upper Usumacinta produces a trellis drainage pattern. As is common in limestone regions with abundant precipitation, there is considerable subsurface dissolution that creates numerous karst landforms, particularly in the eastern portions of the basin. Common topographic features include cenotes (sinkholes), disappearing streams, and major spring systems. In contrast to the limestone of the Sierra de San Cristobal, the upper Grijalva basin drains the granitic Sierra Madre de Chiapas, part of a much larger mountain system that extends throughout Central America. Along its eastern fringes the lithology includes deeply weathered Tertiary shale and sandstone that produce much higher rates of surface erosion than are seen in the Usumacinta system.

In contrast to the Pánuco system, which has its headwaters in the arid and semiarid Central Mesa, the upper reaches of the Usumacinta–Grijalva system receive abundant precipitation (West et al. 1969). Annual precipitation ranges from more than 400 cm along the eastern side of the Sierra de San Cristobal to 80 cm in the Sierra Madre de Chiapas highlands of the western Grijalva basin (INEGI 1981c). Most of the basin is influenced by a tropical monsoon precipitation regime set up by seasonal northeasterly trade winds and tropical cyclones that persist from early summer through early fall, although less seasonality characterizes the swampy lowlands near Villahermosa, Tabasco. The highlands also are tropical, but drier in winter and less humid. Temperatures are hot in the lowlands but decrease with altitude. At lower elevations average daily high temperatures range from 29.4°C to 32.2°C. At 2000 m asl temperatures are moderate for most of the year, with average daily highs from 23.9°C to 26.7°C, but killing frosts occur several times per year (West and Augelli 1989). There can be wide variation in temperature between mountains and lowlands, but mean monthly temperatures from four climate stations throughout the basin (Villahermosa, Ville Flores, Pichucalco, Huehuetenango) give an annual mean temperature of 23°C. When viewed on a basinwide scale, mean monthly temperatures fall to only about 20°C in January and are consistently about 25°C from April through August (Fig. 23.14). On the other hand, mean monthly precipitation for the basin is strongly seasonal, being less than 8 cm/mo from February through April but exceeding 20 cm/mo from June through October (see Fig. 23.14). Annual precipitation for the entire basin is estimated to be 199 cm based on the four climate stations.

Much of the basin was originally covered in tropical rain forest, part of a larger zone of tropical broadleaf evergreen forest that extends from the southeastern portions of the Moctezuma (Pánuco) basin throughout eastern Central America to represent Mexico's most extensive forests. Large portions of this forest have been removed for coffee plantations and cattle ranching, whereas smaller plots are used for slash and burn (swidden) agriculture. What may appear to be pristine forest is now more likely secondary or tertiary growth (West and Augelli

1989). In the cooler environment above 3000 m asl, large tracts of evergreen coniferous trees are common, as well as deciduous broadleaf vegetation more commonly associated with higher latitudes. The upper Grijalva basin drains the Chiapas Depression Dry Forests terrestrial ecoregion, consisting largely of pine–oak forests (Ricketts et al. 1999, www.nationalgeographic.com/wildworld/terrestrial.html). A larger high-elevation ecoregion to the north, west, and east of the dry forests is the Central American Pine–Oak Forests, including a high diversity of endemic plant species. To the north of these pine–oak forests is the Chiapas Montane Forests ecoregion, a relatively narrow strip of extremely high precipitation on the steep northeastern slopes of the Chiapas highlands.

The Tabascan lowlands include lush tropical rain forests and extensive wetlands, but much of the tropical rain forest in the Usumacinta–Grijalva basin is part of the larger Petén–Veracruz Moist Forests ecoregion, which extends to the Pánuco basin to the northwest. Wetlands of this basin include saltwater and freshwater marshes, many within the Pantanos de Centla ecoregion, a seasonally flooded moist forest with associated bogs and swamps. Closer to the mouths of both rivers is the Usumacinta Mangroves ecoregion, a complex system of marshes and bogs, considered to be one of the most important wetlands in Mexico, with mangroves that reach 30 m in height.

Revenga et al. (1998) reported that land cover in the Usumacinta portion of the two-river drainage is 59% forest, 30% cropland, 7% grassland, and 3% developed, with a mean population density of 25 people/km^2. These same authors also reported a 37% loss of original forest. Historically, cattle ranching and tropical agriculture (cacao, bananas, sugar cane, henequen) were the most important industries in the lowlands of both rivers, but they are now second to the petroleum industry, the major economic force within Tabasco. As with the lowlands, subsistence agriculture is still practiced by small communities in the mountainous portion of the basin, although coffee production is also important.

River Geomorphology, Hydrology, and Chemistry

The Usumacinta and Grijalva rivers are Mexico's longest rivers, with a combined total of 1521 km of main-stream channel in Mexico (www.cna.gob.mx 2004). Although the Río Usumacinta remains unimpounded, the Río Grijalva has two large dams (Presa Nezahualcóyotl or Malpaso and Presa de la Angostura or Belisario Domínguez) constructed in the 1960s and 1970s, and so now serves as a major source for hydroelectric power generation (see Fig. 23.13). Both rivers flow through spectacular gorges in the upper reaches of the basin but become increasingly sinuous as the valleys widen in the Coastal Plain. Coastal Plain geology is characterized by a sequence of seaward-dipping Pleistocene terraces and, closer to the coast, Holocene deltaic deposits that onlap Pleistocene terraces. It is here that the rivers form a complex system of interconnected channels (West et al. 1969), with multiple channel bifurcations that result in no single channel carrying the entire discharge of the basin. As several channels may transport streamflow only during flood events, the result is a mosaic of wetlands interconnected by arcuate swamps and sloughs. Saltwater intrusion during the dry season creates favorable habitat for extensive mangrove swamps. Other types of wetlands include freshwater backswamps, estuarine, and abandoned channels and sloughs.

In the Tabascan lowlands, geomorphology and flood regime of the rivers are intricately related. Floods are of long duration, with the Grijalva the most prone to flooding. Flooding is important because it transports sediment overbank, forming broad flanking natural levees that are an essential component of the landscape. Coarse sediments (silty sand) and higher slope of the natural levees permits rapid drainage. For this reason natural levees do not remain inundated long after peak flood events, and they have therefore been favored sites for human settlement over the past few millennia and continue to be important for agriculture. Levees, however, also increase the severity and duration of flood events by preventing floodwaters from draining back into the river, and because of the extensive size of the delta plain, flooding can occur from local precipitation collecting in the basins.

Before joining upstream of Frontera, both the Grijalva and Usumacinta have split much of their flows into distributaries. The Grijalva, also known as the Río Mezcalapa in the lower mountains, splits into several channels after it enters the Coastal Plain. One of the channels is the Río Samaria, which becomes the Río Cañas before flowing into an extensive marsh and wetland system, never reentering the main-stem channel. Another distributary, the Río Carrizal, transports approximately one-third of the Mezcalapa discharge and subsequently becomes the Grijalva upstream of Villahermosa (West et al. 1969). To make the system even more complicated, two smaller

basins, the Chilapa (7000 km^2) and Sierra (5180 km^2) systems, flow into the Grijalva shortly before it joins the Usumacinta (West et al. 1969). Similarly, the Usumacinta has at least two major distributaries (Río Palizada, Río San Pedro) that split from the main channel before it joins the Grijalva. Thus, although the Usumacinta and Grijalva rivers partially join to form a single channel near the coast, much of the discharge from these two basins never flows within a single channel, finding multiple pathways to the Gulf of Mexico.

The discharge and runoff of the entire Usumacinta–Grijalva system is thus complicated and somewhat difficult to quantify. Daily discharge of the Río Usumacinta before it splits into distributaries was measured at the Boca del Cerro gauging station (near Tenosique) from 1949 to 1983 (UNESCO website by I. A. Shiklomanov, http://webworld.unesco.org/water/ihp/db/shiklomanov/index.shtml), and average discharge over that period was estimated to be 1857 m^3/s. In contrast to the long-term record of the Usumacinta, only a single estimate of 821 m^3/s is available from 1960 for the Grijalva (Río Mezcalapa) before it branches into distributaries within the Coastal Plain (West et al. 1969). Thus, an approximate total for the combined drainage is 2678 m^3/s, even though the partially combined rivers never carry nearly this much water in a single channel on average. A long-term estimate (most years from 1947 to 1981) of 532 m^3/s exists for the Río Samaria, a distributary of the Mezcalapa that never joins the Usumacinta. Thus, only about one-third of the Grijalva discharge (Río Carrizal) joins with only a portion of the Usumacinta discharge north of Villahermosa before they form a single channel to the Gulf of Mexico.

Estimates of monthly runoff are presented only for the single site on the Usumacinta at Boca del Cerro (UNESCO, http://webworld.unesco.org/water/ihp/db/shiklomanov/index.shtml), which represents approximately 43% of the entire two-river basin (see Fig. 23.14). For this portion of the basin, annual runoff is extremely high at 123 cm, and monthly runoff exceeds 10 cm/mo from July through December (see Fig. 23.14). Although Fig. 23.14 suggests that runoff is 62% of precipitation (estimated from four weather stations as 199 cm), this percentage is unrealistically high. Runoff for the entire Usumacinta–Grijalva basin is only about 74 cm because the Grijalva basin has less precipitation and contributes much less discharge than the Usumacinta. Thus, discharge to the Gulf of Mexico is roughly 38% of precipitation, rather than 62%. This is still a relatively

high-percentage runoff for a basin with high temperatures and high evapotranspiration.

The Río Grijalva has a much larger sediment load than the Usumacinta due to high uplands erosion rates. The Río Usumacinta transports an annual load of 6257 × 10^3 tons of sediment, whereas the Río Grijalva transported 24,134 × 10^3 tons annually (before the dams and reservoirs were constructed; West et al. 1969).

Little information was found on water chemistry for the Grijalva–Usumacinta or its tributaries. Water quality is assumed to be good in the upstream reaches where there is little development and extensive vegetation, but it likely deteriorates with industrial development in the lower reaches.

River Biodiversity and Ecology

According to Abell et al. (2000), the Usumacinta–Grijalva basin is located within the Grijalva–Usumacinta freshwater ecoregion, which also encompasses the southern portion of the Yucatán Peninsula (including the Río Candelaria). The region is poorly studied in terms of its biodiversity and ecology, although some information exists on its fishes and aquatic insects.

Plants

Riparian trees and brush of wetland forests include *Andira galeottiana, Pachira acuatica, Bravaisia integerrima, Bravaisia tubiflora,* bloodwoodtree, gregorywood, *Paquira aquatica,* willow, and mimosa. Important plants of the mangrove flooded zones are button mangrove, black mangrove, white mangrove, and American (or red) mangrove (Breedlove 1981, Ocaña and Lot 1996). Emergent aquatic plants include bent alligator-flag, common cattail, southern cattail, and common reed, and American eelgrass is an important submerged species (Lot and Novelo 1988). Floating aquatic plants form dense covers in places: water snowflake in clear waters, whereas dotleaf waterlily and nonnative water hyacinth are common in stagnant water and disturbed areas of the lower parts of the basin.

Invertebrates

Information on the invertebrates of this system is sparse. Among mollusks there are applesnail (Mexican), minute hydrobe, *Aroapyrgus clenchi, A. pasionensis, Cochliopina infundibulum, Pachychilus chrysalis,* and *P. pilsbryi* (A. Contreras-Ramos, personal communication).

Reports on aquatic insects are also rare, but Bueno-Soria et al. (in press) describe the fauna in major distributaries of the lower Grijalva system and at other sites within this region. Taxa reported here are all from the Grijalva system: the Río Carrizal, Río Samaria, and Río Mezcalapa–Grijalva. Although these lists probably only include a fraction of the taxa in this region, they give some idea of the diversity of insect communities in these rivers. The fauna includes six families of mayflies, five families of odonates (dragonflies and damselflies), twelve families of aquatic bugs, nine families of aquatic beetles, and five families of aquatic flies, but only three families of caddisflies. The hellgrammite *Corydalus luteus* is also present in the Grijalva–Usumacinta basin (A. Contreras-Ramos, personal communication).

Among the mayflies from Bueno-Soria et al. (in press) were Baetidae (*Baetis, Camelobaetidius*), Ephemeridae (*Hexagenia*), Heptageniidae (*Heptagenia, Stenonema*), Leptophlebiidae (*Leptophlebia, Traverella*), Polymitarcyidae (*Campsurus*), and Tricorythidae (*Leptohyphes, Tricorythodes*). Odonates included Gomphidae (*Archaeogomphus, Phyllocycla, Progomphus*), Libellulidae (*Libellula, Miathyrria marcella, Pachydiplax, Tauriphila*), Protoneuridae (*Protoneura, Neoneura*), Calopterygidae (*Hetaerina*), and Coenagrionidae (*Argia, Argiallagma* [= *Nehalennia*], *Heteragrion, Zonagrion*). The bugs included Belostomatidae (*Belostoma*), Corixidae (*Tenagobia*), Gelastocoridae (*Nertha*), Gerridae (*Rheumatobates, Trepobates*), Hydrometridae (*Hydrometra*), Pleidae (*Paraplea*), Veliidae (*Microvelia, Platyvelia, Rhagovelia*), Macroveliidae (*Macrovelia*), Mesoveliidae (*Mesovelia*), Gerridae (*Metrobates, Neogerris*), Naucoridae (*Ambrysus, Pelocoris*), Nepidae (*Ranatra*), and Notonectidae (*Martarega*). Caddisflies included only Hydropsychidae (*Smicridea*), Hydroptilidae (*Neotrichia, Ochrotrichia*), and Leptoceridae (*Nectopsyche*), but the hydropsychids were found at most sites. Bueno-Soria et al. found more genera of aquatic beetles than any other order, including Hydrophilidae (*Anacaena, Berosus*), Dytiscidae (*Brachyvatus, Laccophilus, Pachydrus, Thermonectus*), Hydrophilidae (*Derallus, Enochrus, Helochares optiata, Paracymus, Tropisternus ovalis*), Haliplidae (*Peltodytes*), Hydraenidae, Gyrinidae (*Gyretes boucardi*), Hydrochidae (*Hydrochus*), Limnichidae, Noteridae (*Hydrocanthus, Suphisellus*), and Scirtidae (*Ora*). Chironomidae were the most widely distributed dipterans, but they were not identified beyond family by Bueno-Soria et al. (in press). In addition to the taxa collected in the river segments mentioned already,

Bueno-Soria et al. also provide more extensive lists of the aquatic insects of Tabasco.

Vertebrates

Miller (1986) mentions a total of 115 fish species known from the Grijalva–Usumacinta system in Mexico (Minckley et al. 2005), but a recent compilation for the entire state of Chiapas documents the presence of 111 species. These are from at least 52 genera in 29 families, with 76 (74%) species native freshwater (primary and secondary), 18 (17%) marine, and the remainder marine forms now isolated in freshwater. Four families contain 58% (68) of the species: 33 cichlids (30%), 22 poeciliids (16%), 9 characids (8%) and 4 profundulids (3.6%) (Rodiles-Hernández 2004). These same families provide most of the large number of endemic species (60 to 70). Noteworthy species are from the Characidae (longjaw tetra), Profundulidae (headwater killifish), Poeciliidae (widemouth gambusia, Chiapas swordtail, sulphur molly, upper Grijalva livebearer), and Cichlidae (white cichlid, Angostura cichlid, Montechristo cichlid, Usumacinta cichlid, freckled cichlid, Teapa cichlid). Among the euryhaline species are threadfin shad, longfin gizzard shad, Maya sea catfish, freshwater toadfish, Gulf silverside, Maya needlefish, Mexican halfbeak, Mexican mojarra, and freshwater drum. In addition, there are 11 endangered species: Pénjamo tetra, Lacandon sea catfish, pale catfish, Olmec blind catfish, Chiapas killifish, Palenque priapella, Yucatán molly, Chiapas cichlid, tailbar cichlid, Petén cichlid, and Chiapa de Corzo cichlid.

Nonnative species from four families have been introduced primarily for aquaculture: Cyprinidae (common carp and grass carp), Salmonidae (rainbow trout), Centrarchidae (largemouth bass), and Cichlidae (blue tilapia, redbelly tilapia, Nile tilapia, Mozambique tilapia, and jaguar guapote) (Rodiles-Hernández 2004). Some species are fished commercially, such as tropical gar, common snook, blue catfish, white mullet, giant cichlid, and tilapias.

Though the diverse fish fauna of the Usumacinta–Grijalva has been relatively well studied, recent discovery in the Río Lacantún (upper Usumacinta) of a new species in an entirely new catfish family illustrates how little we know of this portion of North America (Rodiles-Hernández et al. 2000, Rodiles-Hernández et al. 2004). This large species (up to 500 mm standard length) is relatively common, with its description based on over 30 specimens, some of which were obtained from local residents who include it in their diets. Its evolutionary origins remain enigmatic, with extensive and detailed analyses of morphology and DNA sequence data failing to reveal any obvious relationships to any of the world's other catfish families. Obviously, if something so unusual, so large, and so obvious can remain unknown to science for so long in this remote basin, many more discoveries can be anticipated.

Amphibians and reptiles associated with riparian habitats of the Usumacinta–Grijalva system include river crocodile, swamp crocodile, common snapping turtle, tortugas blanca, and tortugas casquito. According to Abell et al. (2000), there are 82 species of native aquatic herpetofauna in the Grijalva–Usumacinta freshwater ecoregion of which 12 are endemic. Mammals include neotropical river otter, West Indian manatee, water opossum, tepezcuintle, greater bulldog bat, and Baird's tapir (March et al. 1996, Rodiles-Hernández et al. 2002).

Ecosystem Processes

There are no known studies of ecosystem processes in the Usumacinta–Grijalva system. Although much of the region is forested, ecosystem processes are likely to be quite variable, particularly between the high-gradient uplands and the influence of flooding in the low-gradient Tabascan lowlands.

Human Impacts and Special Features

The Usumacinta–Grijalva system is not only Mexico's largest river system but probably retains more invaluable natural features than any other large system, as well as having great archeological value as the center of the ancient Mayan culture. It contains part of the largest remaining tropical rain forest north of the Amazon and, in recognition of its importance, several parks and biological reserves have been established in the basin, including the Pantanos de Centla Biosphere Reserve (a RAMSAR wetland), Laguna del Tigre National Park in Guatemala (also a RAMSAR wetland), Selva Maya (Maya Forest), Montes Azules Biosphere Reserve, Lacantun Biosphere Reserve, and Maya Biosphere Reserve in the Petén of Guatamala (the region's largest "protected area").

Although the Usumacinta and Grijalva rivers include some of the most remote areas of Mexico, the drainage systems are impacted by anthropogenic activities. Deforestation is occurring in upper portions of the drainage basin, resulting in loss of Chiapas's once extensive tropical forests. The Usumacinta continues to be a free-flowing river, but the Grijalva has two large hydroelectric dams that dampen the streamflow regime and remove large amounts of sediment from the system. Unfortunately there have not been any detailed studies to examine the influence of these activities on the lower reaches of the watershed, particularly the hydrology and riparian ecology. Studies on this subject in other regions typically report a loss of plant and aquatic diversity associated with a change in the flood regime and modification to the stream channel morphology. In the lower reaches of the basin, where the Usumacinta and Grijalva interconnect through multiple bifurcating distributaries, the extensive wetland complex is being impacted by engineering activities. Several larger channels have been straightened to reduce flooding, and dikes have been constructed along the channel banks to maintain a navigable depth for transport of raw materials and agricultural products. However, the petroleum industry has probably created the most substantial damage in the lowlands. Oil pipelines and canals have drained many wetlands, and dredge spoil has created topographic barriers that alter wetland hydrology. In many cases the reduction of water and sediment to floodplain and deltaic wetlands results in dramatic changes to the ecology and hydrology of these sensitive ecosystems.

A major concern for the undammed Usumacinta River is that construction of a series of dams on the main stem appears imminent. Since at least 1987 construction of a large dam at Boca del Cerro, about 9 km south of Tenosique, has been publicly discussed. In 1987 and 1992, proposals to build a dam at Boca del Cerro met with stiff opposition and were cancelled. In 2002, however, evidence surfaced that a new plan was being developed by Mexico's Federal Commission of Electricity (CFE) for a hydroelectric dam at this site. It appears an agreement was reached between the Mexican and Guatemalan governments that calls for a dam at Boca del Cerro and a series of five upstream dams; however, little information has been released. This plan is on the watch lists of many environmental and archeological organizations, and major international protests are anticipated. Such a dam or series of dams on the Usumacinta would have enormous ecological, archaeological, and sociological impacts, potentially flooding thousands of square kilometers of tropical rain forest rich in unexplored biodiversity, as well as ancient Mayan ruins and artifacts. Upstream and downstream impacts on riparian ecology could be enormous.

RÍO CANDELARIA (YUCATÁN)

The Río Candelaria drains 10,755 km² of the western Yucatán (Gunn et al. 1995) and is one of the few significant rivers within the Yucatán Peninsula, an enormous karstic region of southeastern Mexico (Fig. 23.15). The river flows over 250 km through dense jungle and wetlands to the coastal zone, where it empties into the Laguna de Terminos, a large brackish lagoon in western Campeche (see Fig. 23.1). Most of the basin is in Mexico, although 50 km of channel extends into the Petén of northern Guatemala to account for 1158 km² in that country (Gunn et al. 1995). The Candelaria is a large river for the Yucatán and should be considered atypical of that region; most of the Yucatán lacks significant rivers, as its hydrology is controlled primarily by interior-draining cenotes aligned along large faults (INEGI 1981a).

The ancient Maya occupied the Yucatán landscape from 3000 to 1100 years ago, and at the height of their civilization had a large population within the Candelaria region. Ancient Mayan ruins located throughout the region have become popular destinations for international tourists. Although the Maya located throughout the Candelaria basin, the major ancient Mayan population centers were located in the upper basin, in the Petén of Guatemala and at Calakmul within the Maya Biosphere Reserve. Because the basin drains the old Mayan heartland, it is not surprising that surficial landscape processes have been anthropogenically modified for several thousand years (Pope and Dahlin 1989, Beach 1998). Landscape modification to exploit water resources was necessary because the permeable and fractured limestone results in the water table being as much as 100 m beneath the surface (Lesser and Weidie 1988).

The large Mayan population was decreasing by the post-Classic period (900 to 1500 A.D.), before the Spanish conquest. The Spanish did not conquer the Yucatán until a couple of decades after conquering the Aztecs in central Mexico, and the population quickly plummeted, but by the mid-nineteenth century was recovering. Because of the absence of significant mineral wealth, this region remained relatively undeveloped until the twentieth century, and in comparison to central Mexico remains largely indigenous. There was even a movement for independence after Mexico became independent from Spain. The Yucatán did not have extensive agriculture; however, henequen, a fibrous product from the agave plant used for manufacturing rope, has long been produced. Although its impact on hydrology would be difficult to speculate, the conversion of dense forest to henequen represents a significant land-use change for portions of lowland Yucatán, principally in the northwestern Yucatán state of Campeche.

Physiography, Climate, and Land Use

The Río Candelaria flows through two major physiographic provinces, the Yucatán (YU) and the Mexican Gulf Coastal Plain (CP) (Grubb and Carillo 1988) (see Fig. 23.15). More specifically the upper portion of the Candelaria basin is within the Southern Hilly Karst Plain of the Yucatán, where most of the surface drainage of the Yucatán is found (Lesser and Weidie 1988).

Few rivers drain the Yucatán because, as is common in karst settings, high infiltration and features such as swallets and cenotes prevent development of large surface drainage systems. Smaller rivers are often not connected to a surface drainage network, but instead are often diverted into solution depressions. In these settings the concept of a surface drainage divide may be inadequate because of the numerous subsurface passages that transport water across surface drainage boundaries. The northern Yucatán has a large concentration of cenotes, but almost no surface water features (Lesser and Weidie 1988). In comparison, the Candelaria region is referred to as the "Lake District" by Gunn et al. (1995) and has a higher degree of surface drainage features, such as poljes, large solution depressions that form lakes or wetlands. In this portion of the Yucatán small river channels connect these features and comprise the Candelaria drainage network.

Three rivers, the Caribe, Esperanza, and upper Candelaria, comprise the Candelaria headwaters, all funneling water from irregular networks of interconnected wetlands and small rivers toward the northwest, where they converge to form the main channel of the Candelaria on the Coastal Plain. These headwaters all drain the swampy and marshy Southern Hilly Karst Plain physiographic region. The Río Caribe begins in the east, near the border between the states of Campeche and Quintana Roo, west of the Maya Biosphere Reserve. Like the rest of this headwater region, the lithology here consists of horizontal bedded Eocene and Miocene carbonate deposits, including limestone, dolomitic limestone, and dolomite (Lesser and Weidie 1988). Here the El Tigre River flows through a poorly drained landscape with extensive wetlands. East of the swampy lowlands that make up the poorly defined drainage divide are the headwaters for the Río Hondo, the largest Mexican river draining into the Caribbean that serves as the international border with Belize. The Xpujil Hills, with a maximum elevation of 375 m asl, appear to represent the drainage divide. In general these conical karst hills are ~100 m above the swampy plain (INEGI 1998b, 1998c) and are a testament to the intensive chemical weathering that has shaped this landscape. The Río Esperanza drains the southeastern portion of the basin, extending to near the Guatemala–Mexico border (INEGI 1998b, 1998c). Finally, the uppermost reaches of the Río Candelaria extend into the Petén of Guatemala, where the drainage divide between it and Río San Pedro (which ultimately flows into the Río Usumacinta) lies in a swampy wetland and is difficult to delineate. These three major headwater rivers of the Candelaria flow northwesterly, converging near the boundary between the Southern Hilly Karst Plain and Mexican Gulf Coastal Plain, which consists of Quaternary sediments of fluvial and marine origin. From here the river flows generally north, although several resistant outcrops disrupt this pattern in the vicinity of the small town of Candelaria.

The major soil type in the Southern Hilly Karst Plain portions of the basin is rendzina (INEGI 1998a), thin clayey soils rich in organics (humus) and calcium carbonate, reflecting the limestone parent material. Caliche horizons are common and are an indication that the intensive chemical weathering that occurs during the rainy season is followed by rapid evaporation in the dry season. Like the uplands, soils within the Gulf Coastal Plain are very clayey, but these vertisols and gleysols are much deeper. Slickensides, shiny pressure surfaces on soil peds, are an indication of substantial subsurface soil churning. Slickensides are common in Coastal Plain soils and also in the deeper upland soils at the base of hills or solution cavities (INEGI 1998a, Beach 1998).

The climatic regime for the Candelaria watershed is tropical monsoon, with distinctive wet and dry seasons (West and Augelli 1989). Average annual precipitation from 1951 to 1980 was 150 cm at the village of Candelaria, and this does not likely vary too much throughout the basin because of the absence of significant topography (INEGI 1981a). The majority of precipitation falls between June and October with the onset of summer trade winds and tropical cyclones (Fig. 23.16). Average temperature over the same 29-year period was 24.6°C (Gunn et al. 1995), and monthly mean temperatures range from 21°C in December to 28°C in May.

The Río Candelaria primarily drains from two terrestrial ecoregions in its upper reaches: the Yucatán Moist Forests and the Petén–Veracruz Moist Forests (Ricketts et al. 1999). Here the climate supports growth of a dense tropical broadleaf rainforest, part of a much larger forest that extends into Belize and Guatemala and throughout southeastern Mexico. Wetland vegetation in the uplands occurs within solution depressions, varying broadly in composition between perennially and seasonally inundated basins. As the river approaches the coast it then passes through the Pantanos de Centla and Usumacinta Mangroves ecoregions. The Pantanos de Centla is seasonally flooded moist forest with associated wetlands. In the lower reaches these freshwater wetlands merge with saltwater wetlands along the coast that are dominated by mangroves.

Land use in the upper portions of the basin is limited to small-scale traditional farming, with maize most important. Logging and ranching are more important in the lower reaches of the basin, creating a mosaic of land-cover types as viewed from satellite imagery. There is little industrial development in this region.

River Geomorphology, Hydrology, and Chemistry

The channel morphology of the Río Candelaria basin reflects regional hydraulic and sedimentary controls. The course of these rivers is controlled by bedrock fractures and solution cavities. Because of an absence of sand, the bank material is comprised of resistant cohesive clayey sediments that result in a narrow and deep channel. The Río Caribe, for example, has an average depth of 3.8 m along its lower 40 km. Combined with the low energy, this setting does not permit the development of a meandering pattern. The river becomes much larger in the Coastal Plain, where it receives drainage from the Ríos Caribe and Esperanza (INEGI 1998b, 1998c). River depths in the lower Río Candelaria vary from 7.2 to 11.3 m (Gunn et al. 1995).

Karst river floodplains differ from rivers transporting clastic sediments. Because of greater channel stability, a less variable flow regime, and smaller sediment loads, these rivers tend to develop thick backswamp deposits but have smaller natural levees (Pope and Dahlin 1989). After the headwaters form the main channel, the river flows as a single channel for 90 km within a defined alluvial valley. Several kilometers downstream of the small town of Candelaria the river becomes an anastomosing channel, flowing within a network of channels for 35 km (INEGI 1998b, 1998c). This probably reflects an increase in valley width coincident with a reduction in valley gradient (e.g., Nanson and Croke 1992). A transition from anastomosing to meandering occurs toward the coast, and the river flows as a sinuous channel for 25 km until it reaches brackish coastal wetlands and discharges into the Laguna de Terminos. Before entering the Laguna de Terminos, the Candelaria and Río Mamantel, a smaller river to the north, enter into the much smaller but well-defined Laguna Panlau.

The monthly streamflow pattern for the Río Candelaria reflects the seasonal precipitation regime but also regional hydrogeologic controls. Although most tropical wet and dry rivers encounter great variability in streamflow, the Río Candelaria's discharge regime is dampened by the substantial amount of base flow supplied by groundwater, and by reduction in runoff due to the wetlands in the uplands (Gunn et al. 1995). Average daily discharge over the 29-year period, from 1958 to 1990, was $46.2\,m^3/s$. The highest ($105.5\,m^3/s$) and lowest ($18.5\,m^3/s$) average daily discharges occur in October and April, respectively. Extremes ranged from a low of $8.7\,m^3/s$ in April 1975 to $309.2\,m^3/s$ in October 1963 (Gunn et al. 1995). Large events are generally associated with intense rainfall events from tropical cyclones. In spite of the high precipitation for the basin, runoff is much lower than that calculated for the Usumacinta–Grijalva just to its southwest (compare Figs. 23.14 and 23.16). The highest monthly discharge of the Candelaria in October results in a runoff of only 2.8 cm/mo compared to a runoff of 20 cm/mo in the Usumacinta. Such large differences appear to be due to groundwater losses in the Candelaria that eventually return much of its water to the sea through submarine springs.

Surface-water features within the Candelaria basin are distinct from the Pánuco and the Usumacinta–Grijalva systems. In contrast to these two large systems, the majority of wetlands in the Candelaria occur in the upper reaches of the watershed within the swampy karstic plain that Pope and Dahlin (1989) refer to as the Yucatán Lake District. These features are poljes, which are large, irregular-shaped karstic solution depressions that lack the cylindrical morphology of cenotes. They are classified as either perennial or seasonal wetlands and further classified as forest, thicket, or herbaceous, which in part depends on the amount of sediment filling them. Collectively these features disrupt traditional surface flow paths and dampen the streamflow regime of the Río Candelaria; however, individually they may undergo great seasonal variation in hydrology. Wetlands below ~100 m asl tend to be perennial because they are close to the regional water table, and

those above ~100 m are typically seasonal, representing perched aquifers supplied by local precipitation and runoff. Evaporation, seepage, and channel outflow remove water from these systems, and there is tremendous variability in their ability to hold water. An important characteristic of these features is the degree to which they have been infilled by surficial sediments. Where infilled with clay, these features may hold water for much of the year, and if supplied by a channel may even flood adjacent areas during the rainy season. The rapid change in water results in considerable shrinking and swelling of wetland soils. The clay expands during the rainy season and contracts and cracks during the dry season, thus limiting vegetation to those plants that can adapt to such a harsh environment.

Although water-chemistry data specific to the Río Candelaria were unavailable (Arriaga-Cabrera et al. 2000), the karstic geology of the region should result in high values for alkalinity/hardness and pH values greater than 7.

River Biodiversity and Ecology

The Río Candelaria is classified by Abell et al. (2000) in the Grijalva–Usumacinta freshwater ecoregion rather than the Yucatán ecoregion. They point out that the freshwater biota is largely unexplored in the Yucatán, and this is certainly true for the Río Candelaria.

Plants

Wetlands in the lower alluvial valley are located mainly in backswamp environments. The composition of vegetation on natural levees is similar to upland forests because the coarser sediments and slight elevation of these deposits provide better drainage. Toward the coast, freshwater marsh merges with brackish swamps, much of which are predominantly mangrove, but support a variety of salt-tolerant species (Pope and Dahlin 1989).

Invertebrates

Information on invertebrates in the Candelaria is sparse, although the narrowmouth hydrobe (a snail), the hellgrammite *Corydalus bidenticulatus*, the stonefly *Anacroneuria*, and the true bug *Abedus* have been found (A. Contreras-Ramos, personal communication).

Vertebrates

Ayala-Perez et al. (1998) provide the only published summary of the fish fauna of the Río

Candelaria, but their work was based entirely on trawling samples in the shallow (3 m) 14 km² terminal estuary (1 to 22 ppt salinity) of this system, Laguna Panlau, which is fed by both the Río Candelaria and the Río Mamantél. The economically important fish faunas of many similar saline lagoons associated with the Laguna Terminos are relatively well studied, and their study was typical of that larger community. All 50 species were from 24 primary marine families, with the exception of two cichlids (Mayan cichlid and one unidentified species) and the threadfin shad. Dark sea catfish, bay anchovy, rhombic mojarra, checkered puffer, silver perch, ground croaker, sand seatrout, and spotted seatrout were the dominant species. We found no published reports on the fish fauna of the upstream freshwater system; however, a search of museum collections produced a list from a single collection (University of Michigan) from a tributary of the Río Candelaria at a highway bridge 40 km southeast of Candelaria in 1982. Not surprisingly, the fauna of this system is closely related to that of the Río Grijalva. Only one of the 17 species collected is not also recorded from the Grijalva and only one of the species collected here was also taken in the estuary by Ayala-Perez et al. (1998). Nine species of cichlids dominated the 1982 collection, comprising numerically about 75% of the 222 specimens: firemouth cichlid, blackgullet cichlid, yellow cichlid, yellowjacket, Mayan cichlid, yellowbelly cichlid, Montechristo cichlid, redhead cichlid, and giant cichlid. Cyprinodontiformes included stippled gambusia, Champoton gambusia, shortfin molly, picotee livebearer, and pike killifish. The two characids were banded tetra and Maya tetra, and a single catfish specimen was pale catfish.

Ecosystem Processes

As pointed out by Arriaga-Cabrera et al. (2000), very little or almost nothing is known about most aspects of this river system. No ecosystem studies have been done on the Candelaria but it would appear to present an interesting contrast in its hydrological and organic matter budget to most rivers.

Human Impacts and Special Features

Occupation of the Candelaria basin by the ancient Maya resulted in significant anthropogenic landscape disturbance much earlier than in most other regions of Mexico. Until the last few decades, however, this area had received relatively little human disturbance for several hundred years. Thus, much of the basin retains features of a relatively natural landscape.

Although it is part of the vast karstic Yucatán, which loses most of its water to groundwater seepage, the Candelaria is unique in retaining sufficient surface-water flow to represent one of the largest Yucatán rivers. Although its biodiversity and ecological characteristics are poorly known, this unusual hydrological system, with anastomosing channels and wetlands in its headwaters, its meandering channel in the Coastal Plain, and its coastal mangroves, is an unusual river ecosystem.

The Maya modified their landscape in two major ways: forest clearing associated with slash and burn agriculture and manipulation of the surface hydrology for agriculture, which included canals and terraces. The immediate impact of Mayan land-use practices was to increase soil erosion and lake–wetland sedimentation. This occurred on even moderate slopes, and several scholars have identified a distinctive layer of fine sediment associated with wetland infilling called the "Maya clay" (Beach 1998). Detailed soils and paleoecological analysis has identified a human imprint on the soils, including an increase in the amount of ash and pollen types that differ from natural pollen assemblages. Large-scale irrigation systems were constructed for agriculture. Even in the humid Yucatán, high infiltration associated with limestone greatly limits the availability of surface water. Within the lower reaches of the Candelaria, large canals were constructed perpendicular to the main river channel and were often connected to uplands or to a matrix of smaller canals within backswamps (Fig. 23.1). Canals were also built within perennial wetlands, although there is debate as to whether seasonal wetlands were utilized for agriculture. Pope and Dahlin (1989) argue that the rapid fluctuation of these perched aquifers greatly limited their usage for agriculture. Ayala-Pérez et al. (1998) mention that the Maya built an elevated road across the shallow estuary they studied that is still exposed during the lowest water levels.

After the collapse of the Mayan empire around 1100 years ago the landscape went fallow and forests returned. There is considerable debate regarding the degree to which this landscape is "natural." It is now widely recognized that the Maya cleared enormous tracts of forest during their 3000-year occupation of this landscape and much of the existing forest does not predate the Maya collapse. Until the late 1960s there was little development in this region other than traditional slash and burn agriculture and nondestructive extractive forestry. Since then, renewed deforestation has occurred as a result of large-scale logging operations and rapid population growth. In the uplands,

FIGURE 23.6 Upper Río Yaqui (Photo by W. E. Doolittle).

forest clearing has mainly been for small-scale slash and burn (swidden) agriculture. Although traditional in form, the pace of milpa clearing is more rapid than in prehistoric times, which may result in this being much more disruptive in comparison to prehistoric swidden practice. Coupled with this has been forest clearing for ranching, particularly in the Coastal Plain, which can be seen as permanent forest removal. Although it is difficult to quantify, Gunn et al. (1995) speculate that these actions are likely increasing runoff and storm flow through the main channel of the Candelaria. However, the coastal area is now within the Lagunas Terminos Protected Area. There is very little industry or commercial activities in the lower reaches of the river. The only significant urban area on the river is the city of Candelaria, located 57 km above the river mouth.

RÍO YAQUI

The Río Yaqui is a relatively large 6ᵗʰ order river basin in northwestern Mexico that drains an area

(73,000 km²) between 34°N and 32°N latitude in the Mexican states of Sonora and Chihuahua and the extreme southeastern corner of Arizona (Figs. 23.6 and 23.17). Based on literature and extensive field-work during the late 1970s, Hendrickson et al. (1981) provide a useful general overview of physical aspects of the basin and an account of its biota, with primary focus on its fishes. That account, as much as possible, will be updated and expanded upon here.

The main stems of two major subbasins, the southern Papigochic–Sirupa–Aros and the northern Bavispe, drain first generally north to northwest in upper reaches of the Sierra Madre Occidental in western Chihuahua. The Bavispe makes a broad turn just south of the international border, adding the San Bernardino system flowing from Arizona to flow south to southwest. The Papigochic–Sirupa–Aros system makes several similar nearly 180-degree turns as it wanders through valleys and intervening deep canyons to eventually join the Bavispe and form the main-stream Río Yaqui. Although the higher elevations and intermediate canyon reaches have high gradients, downstream the low-gradient reaches meander through a desertic forest to cross the Coastal Plain of Sonora to the Los Algodones Estuary about 24 km southwest of Ciudad Obregón.

Human impacts on the Río Yaqui date to pre-Columbian times, when the basin was occupied by the prehistoric Cahita peoples, believed to have given rise to present indigenous groups of the region, such as the Rarámuri (Tarahumara), Yaqui, Mayo, and Pima. Apache bands later roamed higher elevations of the basin, and Pueblo cultures were also present, mostly in Chihuahua. European influence began with the establishment of Spanish missions in lower elevations beginning in the early sixteenth and seventeenth centuries, with relatively little incursion into the Indian-occupied areas upstream but strong influence on the tribes of the more accessible lower reaches of the drainage.

Physiography, Climate, and Land Use

Most of the upper Yaqui basin lies in the Sierra Madre Occidental (SC) physiographic province, where it flows through rocky, complex, and often deep scenic canyons (see Fig. 23.17). A small part of the upper basin in Arizona lies in the Basin and Range (BR) province and a small part of the Buried Ranges (BU) province. As the river drains from the mountains it flows across lower elevations of the Buried Ranges province along the coast of the Gulf of California.

The Papigochic and Bavispe drainages in the eastern portion of the basin originate on the high rolling plains to the east of the crest of the Sierra Madre, and at least parts of these systems were captured from former drainages into closed basins to the east by headwater erosion of west-draining streams (Hendrickson et al. 1981, Minckley et al. 1986). These systems then drop, sometimes quite precipitously, from conifer-dominated forests and high-elevation grasslands (especially in the far southeastern reaches of the basin) to descend through the highly diverse Sierra Madre Occidental Pine–Oak Forests terrestrial ecoregion (Ricketts et al. 1999, www.nationalgeographic.com/wildworld/terrestrial.html). The northernmost portion of the drainage, including in southeastern Arizona, drains from the arid Chihuahuan Desert ecoregion, typified by creosote bush, yucca, and various cactus species, as well as grasslands. The lower Yaqui flows primarily through the Sonoran–Sinaloan Transition Subtropical Dry Forests ecoregion, a transition between the Sonoran Desert to the northwest and the Sinaloan Dry Forests to the southeast. Thus, it includes a mix

of desert vegetation, such as various cacti, and seasonally deciduous trees of the Sinaloan Forest. The very lowest portion of the Yaqui basin north of the westward-trending river (see Fig. 23.17) is part of the Sonoran Desert ecoregion.

The climate of the Yaqui basin ranges from temperate in the mountains to very dry in the desert (INEGI 2000a, 2000b). Mean monthly air temperatures and precipitation were estimated from seven weather stations throughout the basin. Monthly temperatures ranged from 11°C in January to 26°C in July (Fig. 23.18), with an annual mean of 18.4°C. Daily nighttime temperatures, however, commonly fall well below 0°C in winter at high elevations, and daytime temperatures in low and middle elevations commonly exceed 40°C in summer. Throughout the basin most of the precipitation falls (as rain) in July and August (over 11 cm/mo), comprising about half the annual amount of 48 cm (see Fig. 23.18). Precipitation is highest in the higher eastern mountains, where substantial accumulations of snow are common, and least in the lower mountains and foothills of the western side of the basin.

Human population density in the basin averages only 7 people/km², with Ciudad Obregón, Sonora, the largest city at about 345,222 people (INEGI 2000a, 2000b). In spite of this low density, agriculture and logging are pervasive. Currently, agriculture is fairly diverse but only occupies 3% of the basin, with the largest and most productive portion on the lowermost Coastal Plain. Wheat, soybeans, cotton, garbanzo, corn, cattle, poultry, pork, and even shrimp farms are found, and apples and other orchard crops are important in the upper Papigochic. Forestlands represent 75% of the basin, and intensive logging of pine forests is obvious over much of the Sierra Madre in both Sonora and Chihuahua. Recent years have seen the logging industry starting to exploit the diverse oak forests of middle elevations. Some of the few inaccessible and sometimes small but significant headwaters areas that remain in a more or less natural state are being considered for designation as natural protected areas.

River Geomorphology, Hydrology, and Chemistry

The river can be separated into four subbasins with somewhat varied physical, hydrological, and flow characteristics (INEGI 2000a, 2000b). In the southeast, the Papigochic, Sirupa, and Aros rivers drain parts of both Sonora and Chihuahua, whereas to the north the Bavispe system drains large areas of the same two states and a small part of Arizona. The third-largest subbasin, the lower Río Yaqui, harbors the system's reservoirs and most of its irrigated agriculture. The small Moctezuma–Nacozari subbasin lies on the west margin of the basin.

The name Río Yaqui is first applied starting at about 600 m asl for the reach extending downstream from the confluence of the Aros and Bavispe rivers. Here the natural uncontrolled discharge of the Aros merges with the regulated flows of the Bavispe to form a river with a mean width of 60 m. The Río Yaqui here is characterized by a well-developed channel with alternating riffles, runs, and pools, with a diversity of depth and velocity regimes, thus providing a wide variety of habitats for riverine biota. Substrates range from boulders, cobble, and well-sorted gravels with little embedding in higher reaches to high concentrations of fine benthic organic sediments with woody debris, sand, and finer particles.

The discharge of the Yaqui is now regulated by three large reservoirs: La Angostura (Lázaro Cárdenas), Plutarco Elías Calles (El Novillo), and Alvaro Obregón (El Oviachic). Two of these, Plutarco Elías Calles and Alvaro Obregón, are on the main-stream Yaqui, and La Angostura is farther upstream on the Bavispe (see Fig. 23.17). The habitat produced by these dams is intermediate between a free-flowing river and a more typical deep and established lake. At least below La Angostura, discharge occasionally is dropped to zero for days at a time. Sediments in the reservoirs are primarily the soils of the ancient canyons and bedrock and boulders that are overlain or embedded with sand and silt.

Mean annual discharge from 1976 through 1979 was 78.5 m³/s, although much of this flow is now diverted before reaching the Gulf of California. Monthly discharge and runoff are now relatively evenly distributed throughout the year by the dams (see Fig. 23.18). Highest flows occur in July and August, when precipitation exceeds 10 cm/mo.

The lower section of the lower basin includes that portion of river from just below El Oviachic Reservoir to the river's mouth, a distance of about 100 km. Concrete channels divert most of the flow here and become the most characteristic habitat, and only a small portion of the river remains to flow through the natural channel. The natural habitats here are now low velocity with an abundance of organic matter, as the remaining flows and irrigation returns course through meanders within a narrow zone of the broad natural channel through the Coastal Plain. Controlled releases resulting in removal of peak flood discharges through this reach have allowed encroachment of riparian vegetation that constrains the now small river to a narrow channel through dense vegetation. Tides are present only in the Algodones estuary, the natural mouth of the main course of the Río Yaqui, though a significant portion of the river's discharge now finds its way to sea via other, man-made routes.

Except for localized point industrial and logging contamination and somewhat more diffuse agricultural runoff, water quality is generally good along most of the upper and middle river. Numerous small sawmills in the Sierra Madre continue to discharge sawdust to stream channels, sometimes causing extensive impacts, and roads associated with logging, as well as general logging practices, have sometimes greatly increased erosion. Intensive agriculture in the upper Papigochic drainage has obvious impacts on streams of that area. Nonetheless, somewhat lower on these rivers, reservoirs support fisheries, mostly of nonnative species, that are exploited both by local residents and foreign and domestic sport fishermen.

River Biodiversity and Ecology

The Río Yaqui is located within the Sonoran freshwater ecoregion (Abell et al. 2000), which includes rivers to the west and south of it that also drain through the Sonora Desert into the Gulf of California: the Río Concepción, Río Sonora, and Río Mayo.

Algae and Cyanobacteria

Little information exists on the periphyton in the river and its wetlands, although species richness likely is quite high given the variety of habitats.

Plants

Aquatic macrophytes occur throughout the river system, but only 11 species have been collected from the river and its floodplain, with pondweed, buttercup, and *Nasturtium* the most common. Hendrickson et al. (1981) mention specific occurrences of, but did not collect, voucher specimens.

Wetlands in the lower basin are characterized by a variety of emergent and floating-leaved species, depending primarily on salinity. Typical species in freshwater areas include cattail and a few small ciénegas (desert marshlands), especially in the desert grassland areas of southern Arizona and nearby northeastern Sonora and northwestern Chihuahua (Hendrickson and Minckley 1985), where some have been given government protection (e.g., USFWS 1994). Halophytes occur in the extensive estuarine marshes and also in patches in otherwise freshwater areas where salt springs emerge. Black mangrove, white mangrove, and red mangrove dominate in salt marshes; iodinebush, saltbrush, seepweed, and Florida mayten dominate in inland areas with increased salinity. Water hyacinth (nonnative) has been recorded, but unlike what has happened in some basins further south, this problematic weed has not yet established extensively (Arriaga-Cabrera et al. 2000).

Riparian corridors in headwaters and middle elevations consist primarily of Arizona sycamore, Arizona alder, willows (Goodding, Bonpland), green ash, and Fremont cottonwood. Expansive mesquite forests occupy most terraces along larger middle- to lower-elevation rivers through the Sonoran Desert, with Fremont cottonwood, willows, common reed, salt-cedar (nonnative), and hackberry associations along the river's bank (Minckley and Brown 1994).

Invertebrates

Information on invertebrate communities in the Yaqui is sparse, but at least collections of some taxa have been made (A. Contreras-Ramos, personal communication). Among the aquatic insects collected in the system are hellgrammites (*Corydalus bidenticulatus* and *Corydalus texanus*), mayflies (*Siphlonurus occidentalis*, *Nixe salvini*, *Acentrella insignificans*), and stoneflies (*Capnia decepta*, *Mesocapnia frisoni*, and *Anacroneuria wipikupa*) (Bauman and Kondratieff 1996, Contreras-Ramos 1998, A. Contreras-Ramos, personal communication). The snail *Fossaria (Bakerilymnaea) bulimoides* has also been found, and isolated populations of the large riverine bivalve *Anodonta* sp. occur in at least the Río Sirupa in canyon reaches in Chihuahua (D. A. Hendrickson, unpublished data). Estuarine crustaceans include Cauque River prawn, blue shrimp, and white shrimp.

Vertebrates

The fish assemblage of the Yaqui is diverse, particularly because its habitats range from 2500 m asl to sea level. A few sources suggest there are at least 107 fish species, including native fishes and marine species sometimes found in freshwater or in the mouth of the river (Hendrickson et al. 1981, Minckley et al. 1986, Campoy-Favela et al. 1989, Calderón-Aguilera and Campoy-Favela 1993, Castro-Aguirre et al. 1999). There are at least seven species of Cyprinidae (chubs, shiners, and minnows), four species of Catostomidae (suckers), and two species of Poeciliidae (topmin-

nows). Endemic species are few, but include Yaqui chub (DeMarais 1991, DeMarais and Minckley 1993), Leopold sucker (Siebert and Minckley 1986), and probably Yaqui trout (undescribed, but see Hendrickson et al. 2002, Behnke and Tomelleri 2002), and Yaqui topminnow (Quattro et al. 1996). Most of the Cyprinidae and many medium-size species (Catostomidae and others) prey on aquatic and terrestrial insects. Bottom-feeding fishes include Yaqui sucker and Leopold sucker, as well as the primarily algal-grazing Río Grande sucker and Mexican stoneroller. Only the native roundtail chub and Yaqui catfish, in adult form, feed almost exclusively on fishes. Topminnows are the most abundant surface feeders from the basin in middle to low elevations, preying on aquatic algae and small insects. An interesting all-female form of topminnow (headwater livebearer) (Vrijenhoek 1993, Quattro et al. 1992) occurs in middle to lower elevations of the basin, representing the northernmost of a complex of many independently evolved asexual (clonal) lineages that exist as sexual parasites of sexual forms that gave rise to them via hybridization.

Euryhaline species are an important component of the lower basin (56 species) and comprise about of half the Yaqui fish fauna. The most common and abundant are striped mullet, machete, striped mojarra, and Heller's anchovy. Among the natives with conservation interest is Pacific gizzard shad, for which the lack of information impedes an appropriate conservation plan (Varela-Romero 1989).

The Río Yaqui has at least 17 nonnative fishes that have been introduced over the past century (Hendrickson et al. 1981, Hendrickson 1983, Campoy-Favela et al. 1989, Varela-Romero 1989). Among the most abundant are channel catfish, river carpsucker, largemouth bass, common carp, Mozambique tilapia, rainbow trout, and green sunfish. Western mosquitofish is an important nonnative that threatens the ecologically similar native topminnow. Nonnative piscivores feed primarily on smaller forage fishes, such as native cyprinids, catostomids, and poeciliids. Concern exists about the highly piscivorous flathead catfish, which has been reported in the Yaqui (Leibfried 1991), but no vouchers or subsequent specimens have been recorded. Blue catfish have long been established in the Bavispe and surely impact native faunas there. Success of these nonnatives appears directly related to the existence of reservoirs, which create less diverse habitats and function as centers for their dispersal. Largemouth bass and channel catfish now occur far above reservoirs in both the Bavispe and Papigochic–Aros–Sirupa subbasins. Genetic interactions with introduced species also endanger native stocks. Pure stocks of Yaqui catfish are essentially nonexistent because of hybridization with channel catfish.

A number of Río Yaqui fishes are considered endangered or threatened by the Mexican federal government (SEMARNAT 2002, Varela-Romero 1995). Yaqui chub is the only species considered endangered, whereas longfin dace, ornate minnow, Yaqui shiner, Cahita sucker, Río Grande sucker, and Yaqui topminnow are considered threatened and Yaqui sucker, Leopold sucker, and Yaqui catfish are categorized as under special protection. Yaqui topminnow is formally a subspecies of Gila topminnow, but some populations in the Yaqui appear to be sufficiently different to be elevated to the species level (Quattro et al. 1996). Recovery efforts for endangered Yaqui fishes are coordinated with efforts in the U.S. portion of the basin, centered at the Dexter National Fish Hatchery, New Mexico (Johnson and Jensen 1991). The final destiny of these native fishes is the San Bernardino National Wildlife Refuge (SBNWR), Arizona, and potentially one day former native habitats in Mexico (USFWS 1994).

Amphibians and reptiles directly associated with the Yaqui also are diverse. Fourteen species of frogs and toads, one salamander, five snakes, and three turtles are found either in the river main stream or the floodplain much of the year (SIUE 1992, Flores-Villela 1993, Stebbins 1985). Among the more representative amphibians are Tarahumara salamander, Tarahumara frog, and Chiricahua leopard frog, all provided legal protection by the Mexican government. The introduced American bullfrog is found in natural habitats of the northernmost Bavispe and in the lower basin, and culture of the species for commercial purposes has begun. Unfortunately, it has been amply demonstrated to have severe impacts on native amphibians and snakes, particularly those of the genus *Thamnophis*, which are diverse (six species) and common in the Yaqui. Sonoran mud turtle, common slider, and Western box turtle are all common. A major mammal associated with the Río Yaqui is the neotropical river otter, which is occasionally observed in the channelized portion of the river between the major dams (Gallo-Reynoso 1997) and also far above the reservoirs.

Ecosystem Processes

No broad-scale studies are available on ecosystem processes, though much effort has focused on sustainability of agriculture systems in the lowermost river (http://yaquivalley.stanford.edu).

Human Impacts and Special Features

The Río Yaqui is one of most important rivers of northwestern Mexico, not only because of its interesting biodiversity (Arriaga-Cabrera et al. 2000) but also because of the economic importance of the agriculture its water supports. It is, however, heavily impacted by humans. Natural discharge and inundation patterns occur only in the headwaters, upstream from its three large main-stem dams. Nonetheless, despite more than a century of exploitation and extensive, highly modified landscapes (Forbes and Hass 2000), many of the river's tributaries and their riparian zones have remained relatively intact throughout substantial portions of the basin.

Impacts are greatest in the lower basin, which continues to absorb a rapid increase in population and has been highly developed for intensive irrigated agriculture, as have the coastal plains of rivers further south in Sinaloa, such as the Ríos Mayo, Fuerte, and Sinaloa. As with those areas, human communities on the lower Río Yaqui floodplain are now experiencing problems with high pesticide levels (e.g., Arreola-Lizárraga 1995, Guillette et al. 1998) and water-supply limitations following recent record droughts. Water has recently had to be pumped over reservoir spillways to distribution canals that supply apparently unsustainable levels of agriculture, and local economies are crashing (Dean 2004).

Aquaculture is now rapidly developing in the region as a possible alternative. Despite questions regarding its sustainability, the intertidal areas of the upper estuary are being converted to extensive shrimp farms in response to government and private incentives. These activities stand to increase problems related to nonnative species brought in with aquaculture development. Still, despite its complex problems related to extensive agricultural and forestry development, the Río Yaqui basin retains long reaches of free-flowing natural desert mountain rivers, and its biota is at least more intact than those of similar desert rivers further north, such as the Gila of Arizona and New Mexico. This basin should therefore remain a high priority for management and conservation (Arriaga-Cabrera et al. 2000).

RÍO CONCHOS

The Conchos basin begins high in the Sierra Madre Occidental of northwestern Mexico along the North American continental divide. Most of the basin lies

FIGURE 23.7 Upper incised Río Conchos (PHOTO BY W. E. DOOLITTLE).

within the state of Chihuahua, but the system also drains a portion of northern Durango (Fig. 23.7, 23.19). From the pine-forested semiarid flanks of the rugged Sierra Tarahumara, a range within the Sierra Madre Occidental, the primary tributaries of the Conchos basin (Ríos Chuviscar, San Pedro, Toronto, Parral, and Florido) flow northeasterly before converging in the arid Chihuahuan Desert. Below the confluence with the Río Chuviscar the Río Conchos

flows northerly to the Río Bravo del Norte (Río Grande), joining that river at Ojinaga, Chihuahua, across from Presidio, Texas. The drainage area of the Río Conchos is estimated at 68,386 km^2, which accounts for 14% of the Río Bravo's drainage; however, because the Río Bravo now has very little flow upstream of this confluence, the Conchos contributes a large proportion to total discharge below the confluence. It is thus mostly the Conchos that provides the essential discharge through Big Bend National Park, and downstream of the park the water is heavily used for floodplain irrigation and municipal purposes by the growing populations along the Texas–Mexico border. Indeed, there is considerable debate surrounding the issue of water resources along the Río Grande, and the contribution of the Conchos basin has become a controversial topic between Mexico and the United States. A recent comprehensive overview of the Conchos basin (Kelly 2001) summarizes the water plan developed by Mexico's National Water Commission (Comisión Nacional del Agua, or CNA) in 1997.

Northern Mexico contained significant numbers of indigenous peoples. However, unlike Mesoamerican Mexico south of the Tropic of Cancer, with perhaps some exceptions (e.g., Hard and Roney 1998, Schaafsma and Riley 1999, Whalen and Minnis 2001) and at a somewhat smaller scale, it did not contain large culture regions (Coe and Koontz 2002). In addition, indigenous peoples within the area did not have irrigation projects on the scale of those of more southern populations. The Sierra Tarahumara are home to the Rarámuri, one of the largest indigenous groups in Mexico. After the conquest of Mexico the Rarámuri people migrated to this portion of Mexico to escape enslavement in Spanish silver mines. The Rarámuri, also known as the Tarahumara, did modify hydrological and sedimentological processes in a couple of ways. On hill slopes they constructed terraces to reduce runoff and soil erosion and to allow for more intensive agriculture. Along smaller streams, *trincheras* (wooden and stone structures oriented across stream channels) were constructed to reduce runoff and promote accumulation of fertile sediments for agricultural purposes (Doolittle 1985, 2000). The Tarahumara peoples have been isolated for much of history, and it is not until recently that they started to be significantly impacted by modern society. The principal disturbance to the Tarahumara people today involves logging of the extensive pine forests of the Sierra Madre Occidental.

Physiography, Climate, and Land Use

The Conchos basin drains the Sierra Madre Occidental (SC) and Basin and Range (BR) physiographic provinces (see Fig. 23.19). Its headwaters are primarily within the Sierra Madre Occidental, Mexico's most extensive mountain system. The mountains are aligned north–south and extend from the U.S. border to central Mexico. Comprised primarily of Tertiary volcanic rocks, andesite, and rhyolite, some mountains within the Conchos basin contain several ridges at least 3500 m asl, such as the headwaters of the Río Florido, a tributary draining the Sierra Tarahumara in northern Durango. Runoff from some of these peaks also enters Pacific drainages, such as the Ríos Mayo, Fuerte, Sinaloa, and Culiacán, with which the Conchos headwaters closely interdigitate. The volcanic rocks overlay older Cretaceous sedimentary rocks of marine origin that contain extensive ore deposits from igneous intrusions. These are frequently exposed where rivers have incised deep canyons, and have long been exploited for mining.

The lower portions of the basin, greater than half the total, drain the southern extension of the Basin and Range province (Hunt 1974). The major tributaries exit the mountains and form the Río Conchos within the western fringe of the Basin and Range. Chihuahua's Basin and Range topography is characterized by small northwest-trending mountains, although the ridge and valley topography is not as symmetrical as that found in the United States. This series of fault block ridges primarily consists of uplifted Tertiary igneous or Mesozoic sedimentary rocks, the elevations of which are generally around 1750 m asl, although several ridges within the lower portions of the basin exceed 2000 m asl. (INEGI 1998e). The low-lying portions include large elongated valleys infilled with Quaternary alluvial deposits. Along the flanks of the ridges the valleys have extensive *bajadas* (coalescing alluvial fans). The Basin and Range also includes (outside of, but adjoining, the Conchos basin) large internally draining playas infilled with Quaternary lacustrine deposits. These internally draining basins are fed by ephemeral streams that have undergone repeated shifts in channel patterns, complicating delineation of these drainage systems.

The terrestrial ecoregions of the basin correspond closely with the physiographic provinces (INEGI 1981d, Ricketts et al. 1999). The Sierra Madre Occidental Pine–Oak Forests ecoregion within the Sierra Madre Occidental province contains a high diversity of both pine and oak species. The Chihuahuan Desert ecoregion of the Conchos basin is largely within the Basin and Range province. This dry desert ecoregion has only xeric plants, such as creosote bush, tarbush, viscid acacia, yucca, and cacti.

The dominant precipitation mechanism within the Conchos basin is the North American monsoon, or Mexican monsoon (Douglas et al. 1993), characterized by an abrupt increase in precipitation over the basin's headwaters with more than 50% falling from mid-June to mid-September. This occurs as the subtropical high pressure cell migrates north, diverting warm moist air from the Pacific over southwestern North America. Rather than producing large frontal storms, the moist air interacts with the region's topography to create orographic rainfall with convective thunderstorms. Individual rainfall events thus tend to be of short duration but intense. Less important precipitation mechanisms include easterly migrating tropical cyclones from the Pacific and less frequently occurring westerly migrating tropical cyclones from the Gulf of Mexico. However, this mechanism is also likely influenced by synaptic scale circulation, particularly the position of the jet stream and the ridge of high pressure. The upper elevations of the headwaters (generally 2000 m asl) also experience light snowfall during winter months.

Although the basin has a generally warm and dry climate, there is substantial spatial variation in temperature and precipitation. Annual precipitation at some points in the Sierra Madre Occidental can exceed 100 cm but is as low as 20 cm in some parts of the Chihuahuan Desert. Temperature varies greatly with altitude and season, with summer daytime temperatures in the desert often exceeding 40°C and winter temperatures in the mountains falling below −10°C. Mean monthly precipitation and temperature for the basin as a whole was estimated using data from weather stations at Chihuahua (at the edge of mountains and desert) and Guanacevi (in the mountains of the Río Florido basin at 2200 m asl). Precipitation is highly seasonal, being less than 1 cm/mo from February through April and increasing to more than 10 cm/mo in July and August (Fig. 23.20). Mean monthly temperatures for the basin are close to 10°C in midwinter but rise to 23°C in June. Mean annual precipitation was estimated as 48 cm (also see www.sequia.edu.mx) and mean annual temperature was 18°C.

The economy, and thus land use, in the basin are primarily agriculture, mining, and forestry. The large amount of agriculture is made possible by large irrigation districts in the lower basin in the areas of Hidalgo de Parral, Camargo-Jiménez, Delicias, and the lower Conchos. Irrigation has made a variety of crops possible, including maize, winter wheat, alfalfa, cotton, and pecans (Kelly 2001). The population in the basin was estimated at 1.32 million in 2000 (about 17 individials/km^2) and is projected to increase to 1.77 million in 2020 (Kelly 2001). The bulk of this population is concentrated in the largest cities, such as Chihuahua (677,852), Hidalgo

de Parral (103,185), Delicias (99,137), Camargo (39,189), and Jiménez (32,966).

River Geomorphology, Hydrology, and Chemistry

The geomorphology of the Conchos system reflects significant spatial variability, which would be expected of an arid basin having close to 3000 m of relief. The upper headwaters in the Sierra Taramuhara are deeply incised within the volcanic strata (see Fig. 23.7). Here the valleys are narrow, and the rivers are primarily bedrock controlled. However, in the larger valleys in the mountains the floodplains widen and the channel is primarily alluvial. In these segments the water may be diverted for agricultural irrigation, such as in the Río Balleza and the upper Río Conchos (INEGI 1998d). The river valley near Delicias, where the Río San Pedro exits the mountains, is heavily utilized for grazing and extensive agriculture. Here the valley is 2.5 km wide with an extensive system of irrigation canals, such as Canal Principal Numero Cinco, which flows from the Francisco I. Madero reservoir (INEGI 1976a, 1976b). Overgrazing and associated runoff is becoming a significant issue with respect to water quality in these areas, and recharge of the alluvial aquifer is being affected, with consequences for sustaining biologically critical low flows. The downstream reaches of the Conchos are significantly influenced by structural controls as the river flows through the Basin and Range province. Here the river channel alternates between meandering and braided patterns. In some reaches the floodplain averages 2 km wide, whereas in other reaches it in incised into the uplifted ridges of the Basin and Range province and lacks an alluvial channel (INEGI 1976a, 1976b, 1976c, 1978). Just upstream of the river mouth, at Ojinago, the valley widens to 3 km and the meandering pattern increases in sinuosity. A low-flow structure creates a small reservoir, which is primarily used for diverting water into an irrigation system (INEGI 1998a).

A 49-year discharge record (1955 to 2003) is available for the Río Conchos from the International Boundary and Water Commission (www.ibwc.state.gov). Mean annual discharge from 1955 to 1994 was 20.5 m^3/s, varying from 7.5 to 37.1 m^3/s (Fig. 23.8). From 1995 to 2003, however, discharge dropped to an average of only 3.8 m^3/s, or only 19% of the long-term mean. Although it is commonly cited that the Río Conchos provides much of the flow of the Río Bravo del Norte (Río Grande) above the U.S. Big

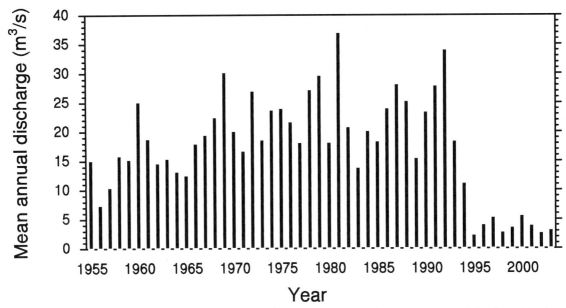

FIGURE 23.8   Long-term mean annual discharge (m³/s) on the Río Conchos at Ojinaga, Chihuahua, Mexico.

Bend National Park, flows since 1995 have dramatically reduced this contribution. Part of this flow reduction has apparently been due to a drought in the region that has reduced runoff; however, the decrease also appears due to increased diversions for agricultural irrigation from dams along the Conchos.

Water in the Río Conchos is highly regulated by a series of dams. Of the seven major reservoirs in the system, La Boquilla (also called Lago Toronto), located on the main stem of the Conchos near Camargo, is the largest (see Fig. 23.19). Other major reservoirs are the San Gabriel on the Río Florido, the F. Madero on the San Pedro, and the Luis L. León on the lower Conchos. All of the major reservoirs are used for irrigation supply, and La Boquilla is used for hydropower. The León Reservoir is of particular significance because it is the last to discharge water (or not) before the Conchos enters the Río Bravo. Not only have the mean discharge levels at the mouth of the Conchos been greatly reduced since 1995, but for extended periods water releases have been reduced to zero by dams at several points along the lower river (see 2003 reports of the Center for Space Research at the University of Texas at Austin, www.csr.utexas.edu). For example, the León Reservoir has withheld water for several months at a time, and the only flow into the Río Bravos has been from precipitation downstream of the dam.

Because the major dams on the Conchos were built before even the earliest available discharge records (1955), it is difficult to know to what extent they have influenced the natural pattern of flow, but comparison of mean monthly precipitation to mean monthly runoff provides some insight. The long-term average runoff, even from 1955 to 1994, was low, as would be expected in an arid landscape (see Fig. 23.20). Note that runoff in Figure 23.20 is multiplied by 10 to facilitate visualization of the monthly pattern. Total annual runoff from 1955 to 1994 was only 0.8 cm, or about 2% of total precipitation. Of particular interest is that the mean runoff pattern during this time is considerably flatter than might be expected in a region with highly seasonal precipitation (see Fig. 23.20). This suggests strong regulation over this 40-year period. From 1995 to 2003, however, there is not only strong regulation, but annual runoff has greatly decreased (0.16 cm or 0.3% of precipitation). This reduction is particularly obvious during the dry winter months; mean discharge and runoff from 1995 to 2003 decreased to less than 10% of prior runoff.

Although water-chemistry and water-quality information have not been readily available for the Conchos, a one-time study by Gutiérrez and Borrego (1999) provides some valuable information. Various water-quality variables were measured at 105 sampling locations, primarily in the lower river in the vicinity of Camargo, Delicias, and upstream of Ojinaga. The natural chemistry of the lower river is affected by its limestone geology, with a pH exceeding 8 and a Ca concentration greater than 100 mg/L. However, the river is so strongly polluted that it is difficult to know the baseline chemistry for most other variables. Kelly (2001) points out that there are only five water-quality monitoring stations within the basin but indicates that the most serious problem is fecal coliform contamination, primarily from Jiménez and Camargo. She also mentions that the Río Florido is contaminated with high levels of oil and grease from a chemical plant in Camargo. Because much of the river's water is used for irrigation, particularly in the Delicias area, agricultural return flows are highly contaminated with high nutrients and increased salinity. Gutiérrez and Borrego (1999) provided more specific data and indicated a great increase in total dissolved solids from 60 mg/L in the headwaters to >700 mg/L in the lower river. They showed that waters returning to the river from irrigation returns can be almost 2000 mg/L. Nutrients in the lower river were very high, with mean values of NO_3-N ranging from 2.3 to 4.6 mg/L in three different river segments. Phosphates ranged from 0.15 to 0.53 mg/L. Upstream of the irrigation districts Na usually was measured at less than 20 mg/L, whereas in the irrigated area near Delicias Na approached 300 mg/L at several sampling sites.

River Biodiversity and Ecology

The Río Conchos has its own freshwater ecoregion of the same name as part of the Río Grande complex of ecoregions (Abell et al. 2000). The river is of particular interest biologically because of substantial endemism in its fishes and herpetofauna.

Invertebrates

Information on invertebrates of the Río Conchos is sparse. One would expect the fauna of the lower Conchos to be similar to that of the Río Bravo del Norte (Río Grande), but information for the Río Grande is also limited (see Chapter 5).

Vertebrates

It is surprising that a basin as important as the Conchos is so poorly collected for fishes. As pointed out by Hendrickson et al. (2002), "The basin remains almost completely uncollected above 1700 m elevation," yet a huge area of the basin drains surfaces above that elevation to more than 3000 m asl, and lower reaches of the basin clearly still require much more extensive faunal inventories than have been realized to date. In spite of the paucity of collections, several recently published accounts of the river's fish

fauna are available, although none provide a comprehensive, detailed overview, and there is considerable disagreement about even basic information. Miller (1978) provides the first account known to us that attempted to comprehensively inventory the basin's fauna, listing 31 species. That work was updated in Smith and Miller (1986), which lists 35 species, with 7 endemic and 1 nonnative. Abell et al. (2000), in their continental-scale biodiversity analyses, stated there were 47 native fish species of which 12 were endemic. However, because they included neither a complete species list nor provided specific citations from which their data were compiled, it is impossible to determine the validity of their numbers. Arriaga-Cabrera et al. (2000) listed over 40 species for the upper Río Conchos and Río Florido, but sources for the lists are not cited and the lists contain many errors. Edwards et al. (2002a, 2002b) reported a total of 44 species from their own collections in the 1990s, but their relatively extensive collections were limited to lower and middle elevations and are not comprehensive for the entire drainage.

For purposes of this chapter, therefore, a list of species was compiled by combining data from Smith and Miller (1986), Edwards et al. (2002a, 2002b), Lozano-Vilano (2002), and records from an assortment of available ichthyology collection databases (including UMMZ, TNHC, CAS, FLMNH, INHS, ASU, and FMNH, museum codes as in Leviton et al. 1985, Leviton and Gibbs 1988). Based on that compilation and analysis, 53 fishes are recognized with apparently valid occurrence records in the basin. Of those, 38 are native (or very probably so; some doubt remains regarding a few because early collection records are inadequate to exclude the possibility that they were present before introductions started), 7 are considered endemic, and 8 are clearly introduced.

Documented Conchos basin endemics are Conchos shiner, bigscale pupfish, bighead pupfish, Salvador's pupfish, Conchos darter, yellowfin gambusia, and crescent gambusia. Species clearly introduced include goldfish, common carp, warmouth, inland silverside, white bass, rainbow trout, and perhaps plains killifish, though the latter's occurrence in rivers further south and ability to disperse through highly saline environments may cast some doubt on this conclusion.

In addition to the endemics, the remaining native fauna not surprisingly shares many species with the Río Bravo, a river that has also seen major changes in its fish fauna due to human impacts (Contreras-Balderas et al. 2003). A few marine derivatives at least formerly ascended the Río Bravo to occupy the

Conchos, including American eel and freshwater drum. Tropical freshwater components include Mexican tetra and Río Grande (or Texas) cichlid. A diversity of small minnows abound, including Mexican stoneroller, ornate minnow, red shiner, roundnose minnow, and several other shiners, including Texas shiner, Tamaulipas shiner, Chihuahua shiner, and Río Grande shiner, and longnose dace, as well as at least one species of the western genus *Gila* (*Gila pulchra* or a close relative of it). Catfishes are relatively diverse, including, in the genus *Ictalurus*, headwater, blue, and channel catfishes and an undescribed form. No trout are included as native, though Hendrickson et al. (2002) point out several lines of evidence that indicate that a still undocumented trout may live in remote high elevations of this basin. Conchos pupfish and blotched gambusia are fairly wide-ranging cyprinodontiforms. Ranges of both extend into Texas, but the latter has not been collected in that state for about a decade at least (Hubbs et al. 1991). Suckers include Yaqui sucker and Río Grande sucker, as well as Mexican redhorse. Only one gar, longnose gar, is known. Although some debate may exist regarding whether they are native or not, we consider green sunfish, bluegill, longear sunfish, and largemouth bass as native until such a hypothesis can be rejected with more conclusive data; all are native to nearby streams in both Texas and/or to the east and south in Mexico. Early collections were inadequate to rule out their presence long before introductions started.

Miller (1961, 1978), Contreras-Balderas (1978), and Contreras-Balderas et al. (2003) have all discussed the dramatic aquatic faunal changes taking place in northern Mexico, including the Conchos basin. By the time of their publications in 1978, Miller and Contreras-Balderas were separately able to provide specific examples of dramatic changes during the first three-quarters of the twentieth century in three rivers of the Conchos basin at major population centers (Chihuahua, Camargo, and Jimenez). By that time the faunas at these sites had been reduced to anywhere from 5% to 66% of their former diversity.

Abell et al. (2000) estimated that there are 46 native species of aquatic herpetofauna in the Conchos, of which 12 are endemic, but provided no information on the actual species.

Ecosystem Processes

No ecosystem studies have been done on the Río Conchos. It is difficult to speculate what preimpact processes might have been like in this system, other than being similar to that found in other desert systems studied in Arizona, New Mexico, and Texas. At present, however, any ecosystem processes have been radically altered with the heavy pollution loads, dams, and dewatering of entire river sections for months at a time.

Human Impacts and Special Features

The Conchos basin overall is heavily impacted by humans, but as is the case in many places, fish faunas and other components of aquatic communities remain most intact in headwater areas, above the majority of such impacts. These less-impacted areas are in mountainous portions of the Conchos basin, which have extensive tracts of forest. Chihuahua has more forests than any other state in Mexico, and these are primarily the pine–oak forests of the Sierra Tarahumara. These forests are essential to both the terrestrial and aquatic ecology of the region and serve as a key migratory route for North American birds. Some tributaries of the Balleza drainage, in particular, despite extensive logging activities that characterize all parts of the basin at higher elevations, appear to still harbor relatively intact high- and middle-elevation aquatic communities in some areas. Lower reaches, especially below major agricultural and/or urban areas and below reservoirs, as mentioned earlier, retain relatively little of their native diversity as a result of contamination, dewatering, and otherwise altered natural regimes. The basin, however, still retains much of its overall scenic beauty. Certainly its lower canyon, just upstream of its mouth at Ojinaga, is spectacular. The Tarahumara highlands are also a tourist attraction often visited by those en route to, or coming from, the Río Fuerte's Barranca de Cobre.

There are several major ways in which the Conchos basin has been negatively impacted by humans, including logging, dams, surface and aquifer withdrawals, irrigated agriculture, overgrazing, and pollution from various sources (Kelly 2001). The major impact in the headwaters is logging, which is primarily located within the mountainous headwaters of the basin. Recent decades have witnessed a dramatic increase in logging within this region, particularly by large international logging companies (Guerrero et al. 2001), and these semiarid mountainous regions are particularly sensitive to deforestation, which frequently results in soil erosion that has downstream consequences. Moreover, the increased runoff means that less water is available to infiltrate into the soil, and consequently groundwater recharge that supplies streams with base flow between rainfall events is reduced. A reduction in base flow has consequences for aquatic species, and in 14 locations in Chihuahua between 1901 and 1975 it is estimated that over 41% of the fish species disappeared.

As indicated earlier, the combination of dams, water withdrawals, and pollution from municipalities, agriculture, and industry has had a profound effect on the Río Conchos, particularly in the lower main stem. The dominant water use (93%) is for irrigation, with domestic withdrawals accounting for 6.4% (Kelly 2001). Contaminants are often very high and flows have sometimes been reduced to zero for extended periods, all of which is devastating to aquatic life. Here, as in many arid streams that have been dammed, there have not only been significant reductions in total flow and flow variability but a narrowing of the stream channel as vegetation, particularly nonnative salt-cedar, has encroached upon the river banks. Much of the natural grasslands have been transformed by either irrigated crops or overgrazing and the introduction of nonnative species of grass (Guerrero et al. 2001).

Water management of the Río Conchos basin has been a major point of contention between Mexico and the United States. According to the 1944 U.S.–Mexico Water Treaty, Mexico is supposed to provide a certain minimum amount of water to the United States (Kelly 2001, Center for Space Research 2003). Until the early 1990s Mexico met its treaty obligations; however, since 1992 it has failed to do so, arguing that obligations cannot be met due to persistent and extraordinary drought (Center for Space Research 2003). Simultaneously, the amount of water used for irrigation has increased. Recent heavy rainfall in 2004, however, has filled Conchos reservoirs, and Mexico has agreed to release water to the Río Grande.

ADDITIONAL RIVERS

The Río Fuerte flows in a generally southwesterly direction within the southern portion of the state of Chihuahua to cross the coastal state of Sinaloa (Fig. 23.21). Its major tributaries (Urique, San Miguel, Verde) arise in the high elevations of the Sierra Madre Occidental of Chihuahua and Durango. Of particular interest are portions of middle reaches of the basin in scenic Barranca del Cobre (Copper Canyon) Park, home to many Tarahumara (see section on the Río Conchos). Coming out of these canyons, the main-

stem river flows primarily across the Piedmont hills and deltas of the Buried Ranges physiographic province to empty into the Gulf of California, but its flow is highly regulated by dams (e.g., Miguel Hidalgo) at the lower elevations. This is a relatively arid landscape, particularly in the lower elevations near the coast, which have been converted to irrigated agricultural areas. The reservoirs have been stocked with many species of nonnative game fishes.

The Río Tamesí flows in a southeasterly direction within the northeastern state of Tamaulipas, draining an area of 19,127 km^2 (see Fig. 23.23). It joins the larger Río Pánuco near Tampico, 10 km upstream of the Gulf of Mexico. The basin's population is rather diffuse and generally low, although Ciudad Victoria (population about 260,000) is located just outside the northeastern boundary. The Tamesí headwaters arise high in the semiarid ranges of the northern Sierra Madre Oriental, west of Ciudad Victoria, and in the Sierra Tamaulipas, a small mountain range located between the Gulf of Mexico and the Sierra Madre Oriental. Lithology of the headwater areas consists predominantly of Cretaceous limestone and is considered to be heavily karstic. Much of the streamflow for the Tamesí originates as springs at the base of the mountains and enters the middle Tamesí within the Coastal Plain. Forty kilometers from its confluence with the Río Pánuco, the Río Tamesí crosses the Tamaulipas Arch, a south-plunging anticline and an extension of the Sierra Tamaulipas. The valley has an appreciably low gradient downstream of the arch, 0.49 m/km, with an extensive system of floodplain lakes. These lakes are utilized for drinking water by Tampico but, like the lagoons and lakes in the lower Pánuco, represent an important ecological component of the Coastal Plain (Fig. 23.9).

The Río Salado flows in a generally easterly direction through the northeastern states of Coahuila and Nuevo León before emptying into Falcon Reservoir on the Río Bravo del Norte in Tamaulipas (Fig. 23.25). The basin drains an arid region, spanning from the headwaters of Río Sabinas, a major tributary that begins high in the northern Sierra Madre Oriental, and flows into the Coastal Plain. A large dam (V. Carranza) just below the confluence of these rivers results in a high degree of regulation. Water from the reservoir irrigates extensive agricultural areas, particularly for cotton. Much of the main stem does not flow for extended periods. However, because of the continued international concern with irrigation and the declining condition of the Río Bravo/Grande, the Río Salado is an important tributary. Average daily discharge near its mouth in Falcon

FIGURE 23.9 Wetlands and lagoons of the lower Río Tamesí at Tampico (PHOTO BY P. H. HUDSON).

Reservoir (near Las Tortilleras, Tamaulipas) from September 1953 to June 2004 was 10.1 m³/s (www.ibwc.state.gov/wad/histflo1.htm). The Salado watershed is increasingly threatened because of a growing population and associated land-use change in northern Mexico, which places greater stress on water supplies such as springs and streams. For example, Contreras-Balderas and Lozano (1994) note that in arid Southwestern Nuevo León reduced spring flow has eliminated several endemic species of crayfish and snails.

The Río Armería flows in a southerly direction in west-central Mexico, beginning in the state of Jalisco and passing through the small state of Colima before entering the Pacific Ocean (Fig. 23.27). The Pacific coast of Mexico is along an active plate margin and therefore has a narrow Coastal Plain. The Río Armería drains both the Neovolcanic Plateau and

Sierra Madre del Sur, changing its name from Río Ayuquila to Río Armería as it approaches the coast. A 71 km section of river comprises the northeastern boundary of Sierra de Manantlán Biosphere Reserve and this basin also receives runoff from the 4240 m asl volcano in the Volcán Nevado de Colima National Park. Although 60% of the basin is in forests, its population density is relatively high at 56 people/km². Much of the river water is diverted for irrigation in a series of agricultural valleys, with 30% of the basin devoted to agriculture (largely sugarcane).

The Río Lacanjá is a small tributary of the Río Usumacinta located in southern Mexico in the Chiapas–Guatamala Highlands physiographic province (Fig. 23.29). This basin has very high annual precipitation (>200 cm) and lies within the Petén–Veracruz Moist Forests terrestrial ecoregion. The river is the boundary between two of the most

FIGURE 23.10 Río Lacanjá, a tributary of the Río Usumacinta (PHOTO BY H. BAHENA).

important biosphere reserves in Mexico, Montes Azules and Lacantún, and 80% of the basin is within reserves. The river remains in a relatively natural condition, with fast-flowing runs, a waterfall with numerous runs and rapids, floodplain clearwater lakes, a floodplain backwater, and a riparian wetland (Fig. 23.10).

ACKNOWLEDGMENTS

We acknowledge and thank Norman Mercado-Silva (University of Wisconsin and Universidad de Guadalajara, Mexico), John Lyons (State of Wisconsin Department of Natural Resources), and Luis Manuel Martínez-Rivera and Luis Ignacio Íñiguez-Dávalos (Universidad de Guadalajara, Mexico) for contributing much of the information for the one-page summary of the Armería–Ayuquila River. We also greatly appreciate the information on aquatic invertebrates provided by Atilano Contreras-Ramos for several of the rivers. Joachin Bueno-Soria provided an unpublished manuscript on the invertebrates of the Grijalva system. José Campoy-Favela helped with the Río Yaqui account. We are most grateful for the contribution of photos by A. H. Siemens and W. E. Doolittle.

LITERATURE CITED

Abell, R. A., D. M. Olson, E. Dinerstein, P. T. Hurley, J. T. Diggs, W. Eichbaum, S. Walters, W. Wettengel, T.

Allnutt, C. J. Loucks, and P. Hedao. 2000. *Freshwater ecoregions of North America: A conservation assessment.* Island Press, Washington, D.C.

Aguilar-Robledo, M. 1999. Land use, land tenure, and environmental change in the jurisdiction of Santiago de Los Valles de Oxitipa, Eastern New Spain, sixteenth to eighteenth century. Ph.D. diss., Department of Geography, University of Texas at Austin.

Alcorn, J. B. 1981. Huastec non-crop resource management: Implications for prehistoric rainforest management. *Human Ecology* 9:395–417.

Arbingast, S. A., C. P. Blair, J. R. Buchanan, C. C. Gill, R. K. Holz, C. A. Marin, R. H. Ryan, M. E. Bonine, and J. P. Weiler. 1975. *Atlas of Mexico.* University of Texas at Austin Bureau of Business Research, Austin.

Arreola-Lizárraga, J. A. 1995. *Diagnosis ecológica de bahía de Lobos, Sonora, Mexico.* Tesis de Maestría IPN-CICIMAR, La Paz, Baja, California Sur.

Arriaga-Cabrera, L., V. Aguilar-Sierra, and J. Alcocer-Durand. 2000. *Aguas Continentales y Diversidad Biológica en Mexico.* D. F. Comisión Nacional para el Conocimiento y Uso de la Biodiversidad (CONABIO). Mexico City.

Ayala-Perez, L. A., O. A. Avilés-Alatriste, and J. L. Rojas-Galaviz. 1998. Fish community structure in the Candelaria-Penlau system, Campeche, Mexico. *Revista de Biología Tropical* 46(3):763–774.

Baumann, R. W., and B. C. Kondratieff. 1996. Plecoptera. In J. Llorente, A. N. García, and E. González (eds.). *Biodiversidad, Taxonomía y Biogeografía de artrópodos de Mexico: Hacia una Síntesis de su Conocimiento,* pp. 169–174. Universidad Nacional Autónoma de Mexico, Mexico City.

Beach, T. 1998. Soil catenas, tropical deforestation, and ancient and contemporary soil erosion in the Peten, Guatemala. *Physical Geography* 19:378–405.

Behnke, R. J., and J. R. Tomelleri. 2002. *Trout and salmon of North America.* Free Press, New York.

Borowsky, R. 1996. The sierra de El Abra of northeastern Mexico: Blind fish in the world's largest cave system. *Tropical Fish Hobbyist* 44(7):178–188.

Breedlove, D. E. 1981. Introduction to the flora of Chiapas: Part 1. In D. E. Breedlove (ed.). *Flora of Chiapas,* pp. 1–33. California Academy of Sciences, San Francisco.

Bueno-Soria, J., S. Santiago-Fragoso, and R. Barba-Alvarez. In press. Insectos acuáticos. In J. Bueno-Soria, F. Alvares-Nogueda, and S. Santiago-Fragoso (eds.). *Biodiversity of Tabasco, Mexico.* Instituto de Biologia, UNAM and CONABIO. Mexico City.

Campoy-Favela, J., A. Varela-Romero, and L. Juárez-Romero. 1989. Observaciones sobre la ictiofauna nativa de la cuenca del Río Yaqui, Sonora, Mexico. *Ecológica* 1(1): 1–13.

Calderón-Aguielra, L. E., and J. R. Campoy-Favela. 1993. Bahía de Las Guásimas, Estero Los Algodones y Bahía de Lobos, Sonora. In S. I. Salazar-Vallejo and N. E.

González (eds.). *Biodiversidad Marina y Costera de Mexico,* pp. 411–419. Comisión Nacional Para la Biodiversidad (CONABIO) y Centro de Investigación de Quintana Roo, Mexico City.

Castro-Aguirre, J. L., H. Espinosa Pérez, and J. J. Schmitter-Soto. 1999. Ictiofauna estuarino-lagunar y vicaria de Mexico. Editorial Limusa, Mexico City.

Center for Space Research. 2003. *An update of surface water availability in the Río Grande basin of Mexico.* University of Texas at Austin Center for Space Research, Austin.

Coe, M. D., and R. Koontz. 2002. *Mexico: From the Olmecs to the Aztecs,* 5th ed. Thames and Hudson, New York.

Comisión Nacional del Agua (CNA). 2001. Estudio sobre Disponibilidad y Balance Hidráulico Actualizado de Aguas Superficiales de la Región Hidrológica no. 16. C. Río Armería, internal document by Altiplano de Ingeniería S.A de C.V.

Contreras-Balderas, S. 1978. Speciation aspects and man-made community composition changes in Chihuahuan Desert fishes. In R. H. Wauer and D. H. Riskind (eds.). *Transactions of the symposium on the biological resources of the Chihuahuan Desert region, United States and Mexico,* pp. 405–431. National Park Service Proceedings 3. U.S. Department of the Interior, Washington, D.C.

Contreras-Balderas, S., R. J. Edwards, Mad. L. Lozano-Vilano, and M. E. Garcia-Ramirez. 2003. Fish biodiversity changes in the Lower Río Grande/Río Bravo, 1953–1996. *Reviews in Fish Biology and Fisheries* 12(2–3):219–240.

Contreras-Balderas, S., and M. L. Lozano-Vilano. 1994. Water, endangered fishes, and development perspectives in arid lands of Mexico. *Conservation Biology* 8:379–387.

Contreras-Balderas, S., H. Obregón, and M. L. Lozano-Vilano. 1996. Punta del Morro, an interesting barrier for distributional patterns of continental fishes in north and central Veracruz, Mexico. *Acta Biológica Venezolana* 16:37–42.

Contreras-Ramos, A. 1998. *Systematics of the dobsonfly genus Corydalus (Megaloptera: Corydalus).* Thomas Say Publications in Entomology, Lanham, Maryland.

Crews-Meyer, K. A., P. F. Hudson, and R. Colditz. 2004. Landscape complexity and remote classification in eastern Mexico: Applications of Landsat 7 ETM data. *Geocarto International* 19:45–56.

Dean, A. 2004. Drought enters ninth year in birthplace of the green revolution: Crisis for farmers in the Yaqui Valley. *Encina Columns,* Spring 2004. Biannual newsletter of the Stanford Institute for International Studies, Stanford University.

de Cserna, Z. 1989. An outline of the geology of Mexico. In A. W. Bally and A. R. Palmer (eds.). *The geology of North America: An overview,* pp. 233–264. Geological Society of America, Boulder, Colorado.

DeMarais. B. D. 1991. *Gila eremica*, a new cyprinid fish from Northwestern Sonora, Mexico. *Copeia* 1991: 179–189.

DeMarais, B. D., and W. L. Minckley. 1993. Genetics and morphology of Yaqui chub *Gila purpurea*, an endangered cyprinid fish subject to recovery efforts. *Biological Conservation* 66:195–206.

Diaz, H. F., and G. N. Kiladis. 1992. Atmospheric teleconnections associated with the extreme phase of the Southern Oscillation. In H. F. Diaz and V. Markgraf (eds.). *El Nino: Historical and paleoclimatic aspects of the Southern Oscillation*, pp. 7–28. Cambridge University Press, Cambridge.

Doolittle, W. E. 1985. The use of check dams for protecting downstream agricultural lands in the prehistoric Southwest: A contextual analysis. *Journal of Anthropological Research* 41:279–305.

Doolittle, W. E. 1988a. Intermittent use and agricultural change on marginal lands: The case of smallholders in eastern Sonora, Mexico. *Geografiska Annaler* 70B: 255–266.

Doolittle, W. E. 1988b. Pre-Hispanic occupance in the Valley of Sonora, Mexico: Archaeological confirmation of early Spanish reports. Anthropological Papers of the University of Arizona. University of Arizona Press, Tucson, Arizona.

Doolittle, W. E. 2000. *Cultivated landscapes of native North America*. Oxford University Press, Oxford.

Douglas, M. W., R. Maddox, K. Howard, and S. Reyes. 1993. The Mexican monsoon. *Journal of Climate* 6:1665–1667.

Dowling, T. E., D. P. Martasian, and W. R. Jeffery. 2002. Evidence for multiple genetic forms with similar eyeless phenotypes in the blind cavefish, *Astyanax mexicanus*. *Molecular Biology and Evolution* 19:446–455.

DRSBM–IMECBIO. 2000. Propuesta para que se Considere la Cuenca del Río Ayuquila–Armería–Manantlán (Región Hidrológica XVI, Armería–Coahuayana) Como Región Hidrológica Prioritaria de la Comisión Nacional para el Conocimiento y Uso de la Biodiversidad (CONABIO). Prepared by the Dirección de la Reserva de la Biosfera Sierra de Manantlán–Comisión Nacional de Areas Naturales Protegidas–SEMARNAP and the Instituto Manantlán de Ecología y Conservación de la Biodiversidad. Departamento de Ecología y Recursos Naturales, Universidad de Guadalajara.

Edwards, R. J., G. P. Garrett, and E. Marsh-Matthews. 2002a. Conservation and status of the fish communities inhabiting the Río Conchos basin and middle Río Grande, Mexico and U.S.A. *Reviews in Fish Biology and Fisheries* 12:119–132.

Edwards, R. J., G. P. Garrett, and E. Marsh-Matthews. 2002b. An ecological analysis of fish communities inhabiting the Río Conchos basin. In M. d. L. Lozano-Vilano (ed.). *Libro Jubilar en Honor al Dr. Salvador Contreras Balderas*, pp. 43–62. Universidad Autónoma de Nuevo León, Facultad de Ciencias Biológicas, Monterrey, Nuevo León, Mexico.

Ekholm, G. F. 1944. *Excavations at Tampico and Pánuco in the Huasteca, Mexico*. Anthropological Papers of the American Museum of Natural History, V. XXVII, Part V: 508, New York.

Fish, J. E. 1977. Karst hydrogeology and geomorphology of the Sierra de el Abra and the Valles–San Luis Potosí region, Mexico. Ph.D. diss., McMaster University, Hamilton, Ontario.

Flores-Villela, O. 1993. Herpetofauna Mexicana: Annoted list of the species of amphibians and reptiles of Mexico, recent taxonomic changes, and new species. Carnegie Museum of Natural History Special Publication no. 17. Pittsburgh.

Forbes, W., and T. S. Hass. 2000. Leopold's legacy in the Río Ganlán: Revisiting an altered Mexican wilderness. *Wild Earth* 2000 (Spring): 61–67.

Galindo, I. 1995. La oscilación del sur, El Nino: el caso de Mexico. In E. Florescano and S. Swan (eds.). *Breve Historia de la Sequia en Mexico*, pp. 133–165. Universidad Veracruzana, Xalapa, Ver, Mexico.

Gallo-Reynoso, J. P. 1997. Situación y distribución de las nutrias en Mexico, con énfasis en *Lutra longicaudis annectens* Major, 1897. *Revista Mexicana de Mastozoología* 2(1): 10–32.

Garcia de León, F. J., D. Gutiérrez Tirado, D. A. Hendrickson, and H. Espinosa-Pérez. In press. Fishes of the continental waters of Tamaulipas: Diversity and conservation status. In J.-L. E. Cartron, G. Ceballos, and R. S. Felger (eds.). *Biodiversity, ecosystems, and conservation in Northern Mexico*. Oxford University Press, New York.

Grubb, H. F., and J. J. R. Carillo. 1988. Region 23, Gulf of Mexico Coastal Plain. In W. Back, J. S. Roshein, and P. R. Seaber (eds.). *Hydrogeology: The geology of North America*, pp. 219–228. Vol. O-2 Geological Society of America, Boulder, Colorado.

Guerra, L. V. 1952. Ichthyological survey of the Río Salado, Mexico. Master thesis, University of Texas, Austin.

Guerrero, M. T., F. de Villa, M. Kelly, C. Reed, and B. Vegter. 2001. *The forest industry in the Sierra Madre of Chihuahua: Economic, ecological and social impacts post NAFTA*. Texas Center for Policy Studies, Austin.

Guillette, E. A., M. M. Meza, M. G. Alquilar, A. D. Soto, and I. E. Garcia. 1998. An anthropological approach to the evaluation of preschool children exposed to pesticides in Mexico. *Environmental Health Perspectives* 106:347–353.

Gunn, J., J. F. William, and R. R. Hubert. 1995. A landscape analysis of the Candelaria watershed in Mexico: Insights into paleoclimates affecting upland horticulture in the southern Yucatán Peninsula semi-karst. *Geoarchaeology* 10:3–42.

Gutiérrez, M., and P. Borrego. 1999. Water quality assessment of the Río Conchos, Chihuahua, Mexico. *Environmental Internacional* 25:573–583.

Hard, R. J., and J. R. Roney. 1998. A massive terraced village complex in Chihuahua, Mexico, 3000 years before present. *Science* 279:1661–1664.

Hendrickson, D. A. 1983. New distribution records for native and exotic fishes in Pacific drainages of northern Mexico (in English and Spanish). *Journal of Arizona–Nevada Academy of Sciences* 18(2):33–38.

Hendrickson, D. A., H. Espinosa-Pérez, L. T. Findley, W. Forbes, J. R. Tomelleri, R. L. Mayden, J. L. Nielsen, B. Jensen, G. Ruiz-Campos, A. Varela-Romero, A. M. Van Der Heiden, F. Camarena, and F. J. García de León. 2002. Mexican native trouts: A review of their history and current systematic and conservation status. *Reviews in Fish Biology and Fisheries* 12:273–316.

Hendrickson, D. A., and J. K. Krejca. 2000. Subterranean freshwater biodiversity in northeastern Mexico and Texas. In R. A. Abell, D. M. Olson, E. Dinerstein, P. T. Hurley, J. T. Diggs, W. Eichbaum, S. Walters, W. Wettengel, T. Allnutt, C. J. Loucks, and P. Hedao (eds.). *Freshwater ecoregions of North America: A conservation assessment*, pp. 41–43. Island Press, Washington, D.C.

Hendrickson, D. A., J. K. Krejca, and J. M. Rodriguez. 2001. Mexican blindcats, genus *Prietella* (Ictaluridae): Review and status based on recent explorations. *Environmental Biology of Fishes* 62:315–337.

Hendrickson, D. A., J. C. Marks, A. B. Moline, E. Dinger, and A. E. Cohen. In press. Combining ecological research and conservation: A case study in Cuatro Ciénegas, Mexico. In L. Stevens and V. J. Meretsky (eds.). *Every last drop: Ecology and conservation of North American desert springs*. University of Arizona Press, Tucson.

Hendrickson, D. A., and W. L. Minckley. 1985. Ciénegas: Vanishing aquatic climax communities of the American Southwest. *Desert Plants* 6:131–175.

Hendrickson, D. A., W. L. Minckley, R. R. Miller, D. J. Siebert, and P. H. Minckley. 1981. Fishes of the Río Yaqui basin, Mexico and United States. *Journal of Arizona–Nevada Academy of Sciences* 15(3): 65–106.

Hendrickson, D. A., and A. Varela-Romero. 2002. Fishes of the Río Fuerte, Sonora, Sinaloa and Chihuahua, Mexico. In M. d. L. Lozano-Vilano (ed.). *Libro Jubilar en Honor al Dr. Salvador Contreras Balderas*, pp. 171–195. Universidad Autónoma de Nuevo León, Facultad de Ciencias Biológicas, Monterrey, Nuevo León, Mexico.

Hubbs, C., R. J. Edwards, and G. P. Garrett. 1991. An annotated checklist of the freshwater fishes of Texas. *Texas Journal of Science* Suppl. no. 43(4):1–56.

Hudson, P. F. 2000. Discharge, sediment, and channel characteristics of the Río Pánuco, Mexico. *Yearbook, Conference of Latin Americanist Geographers* 26:61–70.

Hudson, P. F. 2002. Floodplain styles of the lower Panuco basin, Mexico. *Journal of Latin American Geography* 1:58–68.

Hudson, P. F. 2003a. Event sequence and sediment exhaustion in the Lower Pánuco basin, eastern Mexico. *Catena* 52:57–76.

Hudson, P. F. 2003b. The influence of the El Nino Southern Oscillation on sediment yield in the Lower Pánuco basin, Mexico. *Geografiska Annaler–A* 85(3–4): 263–275.

Hudson, P. F. 2004. The geomorphic context of prehistoric *Huastec* floodplain environments: Pánuco basin, Mexico. *Journal of Archaeological Science* 31:653–668.

Hudson, P. F., and R. Colditz. 2003. Flood delineation in a large and complex alluvial valley: The lower Pánuco basin, Mexico. *Journal of Hydrology* 280:229–245.

Hudson P. F., R. Colditz, and M. Aguilar-Robledo. In press. Spatial relations between land use/land cover and floodplain environments in a large alluvial valley, lower Pánuco basin, Mexico. *Environmental Management*.

Hudson, P. F., and F. T. Heitmuller. 2003. Local- and watershed-scale controls on the spatial variability of natural levee deposits in a large fine-grained floodplain: Lower Panuco basin, Mexico. *Geomorphology* 56:255–269.

Hulsey, C. D., F. J. García de León, Y. Sánchez Johnson, D. A. Hendrickson, and T. J. Near. 2004. Temporal diversification of mesoamerican cichlid fishes across a major biogeographic boundary. *Molecular Phylogeny and Evolution* 31:754–764.

Hunt, C. B. 1974. Natural regions of the United States and Canada. W. H. Freeman and Company, San Francisco.

INE-SEMARNAP. 2000. Programa de Manejo de la Reserva de la Biosfera Montes Azules, Mexico City.

Instituto Nacional de Estadística Geografía e Informática (INEGI). 1976a. *Carta Topográfica*, 1:50,000, H13-C78. Mexico City.

Instituto Nacional de Estadística Geografía e Informática (INEGI). 1976b. *Carta Topográfica*, 1:50,000, H13-C88. Mexico City.

Instituto Nacional de Estadística Geografía e Informática (INEGI). 1976c. *Carta Topográfica*, 1:50,000, H13-D21. Mexico City.

Instituto Nacional de Estadística Geografía e Informática (INEGI). 1978. *Carta Topográfica*, 1:50,000, H13-D22. Mexico City.

Instituto Nacional de Estadística Geografía e Informática (INEGI). 1981a. *Carta Hidrologica Aguas Superficiales*, 1:1,000,000, Merida. Mexico City.

Instituto Nacional de Estadística Geografía e Informática (INEGI). 1981b. *Carta Hidrologica Aguas Superficiales*, 1:1,000,000, Mexico. Mexico City.

Instituto Nacional de Estadística Geografía e Informática (INEGI). 1981c. *Carta Hidrologica Aguas Superficiales*, 1:1,000,000, Villahermosa. Mexico City.

Instituto Nacional de Estadística Geografía e Informática (INEGI). 1984a. *Carta de Efectos Climaticos,*

Regionales Noviembre–Abril and *Regionales Mayo–Octubrei*, data record from 1921–1980, 1:250,000, Ciudad Mante, F14-5. Mexico City.

Instituto Nacional de Estadística Geografía e Informática (INEGI). 1984b. *Carta de Efectos Climaticos, Regionales Noviembre–Abril* and *Regionales Mayo–Octubre*, data record from 1921–1980, 1:250,000, Ciudad valles, F14-8. Mexico City.

Instituto Nacional de Estadística Geografía e Informática (INEGI). 1984c. *Carta uso del Suelo y Vegetacion*, data record in 1981, 1:250,000, Ciudad Valles, F14-8.

Instituto Nacional de Estadística Geografía e Informática (INEGI). 1984d. *Carta Geologica*, 1:250,000, Ciudad Mante, F14-5. Mexico City.

Instituto Nacional de Estadística Geografía e Informática (INEGI). 1984e. *Carta Geologica*, 1:250,000, Ciudad Valles, F14-8. Mexico City.

Instituto Nacional de Estadística Geografía e Informática (INEGI). 1998a. *Carta Edafologica*, 1:1,000,000, Merida. Mexico City.

Instituto Nacional de Estadística Geografía e Informática (INEGI). 1998b. *Carta Topografica*, 1:250,000, Ciudad del Carmen, E15-6. Mexico City.

Instituto Nacional de Estadística Geografía e Informática (INEGI). 1998c. *Carta Topografica*, 1:250,000, Tenosique. E15-9. Mexico City.

Instituto Nacional de Estadística Geografía e Informática (INEGI). 1998d. *Carta Topográfica*, 1:250,000, San Juanito, G13-1. Mexico City.

Instituto Nacional de Estadística Geografía e Informática (INEGI). 1998e. *Carta Topográfica*, 1:250,000, Ojinaga, H13-8. Mexico City.

Instituto Nacional de Estadística Geografía e Informática (INEGI). 2000a. Síntesis de información Geográfica del Estado de Chihuahua. Instituto Nacional de Estadística, Geografía e Informática, Mexico D.F.

Instituto Nacional de Estadística Geografía e Informática (INEGI). 2000b. Síntesis de información Geográfica del Estado de Sonora. Instituto Nacional de Estadística, Geografía e Informática, Mexico D.F.

Jauregui, E. 1995. Rainfall fluctuations and tropical storm activity in Mexico. *Erdkunde* 49:39–48.

Jeffery, W. R. 2001. Cavefish as a model system in evolutionary developmental biology. *Developmental Biology* 231(1):1–12.

Johnson, J. J., and B. L. Jensen. 1991. Hatcheries for endangered freshwater fishes. In W. L. Minckley and J. E. Deacon (eds.). *Battle against extinction*, pp. 199–217. University of Arizona Press, Tucson.

Kelly, M. E. 2001. *The Río Conchos: A preliminary overview*. Texas Center for Policy Studies, Austin.

Langecker, T. G., B. Neumann, C. Hausberg, and J. Parzefall. 1995. Evolution of the optical releasers for aggressive behavior in cave-dwelling *Astyanax fasciatus* (Teleostei, Characidae). *Behavioural Processes* 34(2):161–168.

Leibfried, W. 1991. A recent survey of fishes from the interior Río Yaqui drainage, with a record of flathead catfish *Pylodictis olivaris* from the Río Aros. In E. P. Pister (ed.). *Proceedings XX Desert Fishes Council Symposia, Death Valley, California*. Desert Fishes Council, Bishop.

Lesser, J. M., and A. E. Weidie. 1988. Region 25, Yucatán Peninsula. In W. Back, J. S. Roshein, and P. R. Seaber (eds.). *Hydrogeology: The geology of North America*, pp. 237–241. Vol. O-2. Geological Society of America, Boulder, Colorado.

Leviton, A. E., R. H. J. Gibbs, E. Heal, and C. E. Dawson. 1985. Standards in herpetology and ichthyology. Part 1: Standard symbolic codes and institutional resource collections in herpetology and ichthyology. *Copeia* 1985:802–832.

Leviton, A. E., and R. H. J. Gibbs. 1988. Standards in herpetology and ichthyology. Standard symbolic codes for institutional resource collections in herpetology and ichthyology. Supplement no. 1: Additions and corrections. *Copeia* 1988:280–282.

Lot, A., and A. Novelo. 1988. El Pantano de Tabasco y Campeche: la reserva más importante de plantas acuáticas de Mesoamérica. In *Memorias del simposio internacional sobre ecología y conservación del delta de los ríos Usumacinta y Grijalva*, pp. 537–547. INIREB y Gobierno del Estado de Tabasco, Mexico City.

Lozano-Vilano, M. d. L. 2002. *Cyprinodon salvadori*, new species from the upper Río Conchos, Chihuahua, Mexico, with a revised key to the C. *eximius* complex (Pisces, Teleostei: Cyprinodontidae). In M. d. L. Lozano-Vilano (ed.). *Libro Jubilar en Honor al Dr. Salvador Contreras Balderas*, pp. 15–22. Universidad Autónoma de Nuevo León, Facultad de Ciencias Biológicas, Monterrey, Nuevo León, Mexico.

Lyons, J., G. González-Hernández, E. Soto-Galera, and M. Guzmán-Arroyo. 1998. Decline of freshwater fishes and fisheries in selected drainages of west-central Mexico. *Fisheries* 23(4):10–18.

Lyons, J., S. Navarro-Pérez, P. Cochran, E. Santana, and M. Guzmán-Arroyo. 1995. Index of biotic integrity based on fish asssemblages for the conservation of streams and rivers in West Central Mexico. *Conservation Biology* 9:569–584.

March, I., E. Naranjo, and R. Rodiles-Hernández. 1996. Diagnóstico para la conservación y manejo de la fauna silvestre en la Selva Lacandona, Chiapas. Informe Final ECOSUR, San Cristóbal de las Casas.

Martínez, L. M., A. Carrranza, and M. García. 1999. Aquatic ecosystem pollution of the Ayuquila River, Sierra de Manantlán Biosphere Reserve, Mexico. In M. Munawar, S. Lawrence, I. F. Munawar, and D. Malley (eds.). *Aquatic ecosystems of Mexico: Status and scope*, pp. 1–17. Ecovision World Monograph Series. Blackhuys, Leiden, The Netherlands.

Mayden, R. L., R. H. Matson, and D. M. Hillis. 1992. Speciation in the North American genus *Dionda* (Teleostei: Cypriniformes). In R. L. Mayden (ed.). *Systematics, historical ecology, and North American freshwater fishes*, pp. 710–746. Stanford University Press, Palo Alto, California.

Metcalfe, S. E. 1987. Historical data and climatic change in Mexico: A review. *The Geographical Journal* 153: 211–222.

Miller, R. R. 1961. Man and the changing fish fauna of the American Southwest. *Papers of the Michigan Academy of Science, Arts, and Letters* 46:365–404.

Miller R. R. 1978. Composition and derivation of the native fish fauna of the Chihuahuan Desert region. In R. H. Wauer and D. H. Riskind (eds.). *Transactions of the symposium on biological resources of the Chihuahuan desert region, U.S. and Mexico*, pp. 365–381. National Park Service Proceedings 3. U.S. Department of the Interior, Washington, D.C.

Miller, R. R. 1986. Composition and derivation of the freshwater fish fauna of Mexico. *Anales de la Escuela Nacional de Ciencias Biológicas* 30:121–153.

Miller, R. R., W. L. Minckley, and S. M. Norris. In press. *Freshwater fishes of Mexico*. University of Chicago Press, Chicago.

Miller, R. R., and M. L. Smith. 1986. Origin and geography of the fishes of central Mexico. In C. H. Hocutt and E. O. Wiley (eds.). *The zoogeography of North American freshwater fishes*, pp. 487–517. John Wiley and Sons, New York.

Minckley, W. L. 1969. Environments of the Bolsón of Cuatro Ciénegas, Coahuila, Mexico. *Science Series, University of Texas, El Paso, Texas* 2:1–65.

Minckley, W. L., and D. E. Brown. 1982. Wetlands. In D. E. Brown (ed.). Biotic communities of the American Southwest, United States and Mexico. Special issue, *Desert Plants* 4:222–341.

Minckley, W. L., D. A. Hendrickson, and C. E. Bond. 1986. Geography of western North American freshwater fishes: Description and relations to intracontinental tectonism. In C. H. Hocutt and E. O. Wiley (eds.). *Zoogeography of freshwater fishes of North America*, pp. 519–614. Wiley Interscience, New York.

Minckley, W. L., R. R. Miller, C. D. Barbour, J. J. Schmitter-Soto, and S. M. Norris. In press. Historical ichthyogeography. In R. R. Miller, W. L. Minckley, and S. M. Norris (eds.). *Freshwater fishes of Mexico*. University of Chicago Press, Chicago.

Mitchell, R. W., W. H. Russell, and W. R. Elliott. 1977. *Mexican eyeless characin fishes of the genus Astyanax: Environment, distribution and evolution*. Texas Tech University Press, Lubbock.

Morales-Román, M., and R. Rodiles-Hernández. 2000. Implicaciones de Ctenopharyngodon idella en la comunidad de peces del río Lacanjá, Chiapas. *Hidrobiológica* 10(1):13–24.

Muir, J. M. 1936. *Geology of the Tampico Region, Mexico*. American Association of Petroleum Geologists, Tulsa, Oklahoma.

Nanson, G. C., and J. C. Croke. 1992. A genetic classification of floodplains. *Geomorphology* 4:459–486.

Nelson, J. S., E. J. Crossman, H. Espinosa-Pérez, L. T. Findley, C. R. Gilbert, R. N. Lea, and J. D. Williams. 2004. *Common and scientific names of fishes from the United States, Canada and Mexico*. Special Publication 29. American Fisheries Society, Bethesda, Maryland.

Obregón-Barboza, H., S. Contreras-Balderas, and M. L. Lozano-Vilano. 1994. The fishes of northern and central Veracruz, Mexico. *Hydrobiologia* 286(2):79–95.

Ocaña, D., and A. Lot. 1996. Estudio de la vegetación acuática vascular del sistema fluvio-lagunar-deltaico del Río Palizada, en Campeche, Mexico. *Anales del Instituto de Biologia, Universidad Nacional Autonoma de Mexico*, Series Botánica, 67:303–327.

O'Hara, S. L., F. A. Street-Perrott, and T. P. Burt. 1993. Accelerated soil erosion around a Mexican highland lake caused by prehispanic agriculture. *Nature* 362:48–51.

Pope, K. O., and B. H. Dahlin. 1989. Ancient Maya wetland agriculture: New insights from ecological and remote sensing research. *Journal of Field Archaeology* 16:87–106.

Quattro, J. M., J. C. Avise, and R. C. Vrijenhoek. 1992. An ancient clonal lineage in the fish genus *Poeciliopsis* (Atheriniformes: Poeciliidae). *Proceedings of the Academy of Natural Sciences of Philadephia* 89:348–352.

Quattro, J. M., P. L. Leberg, M. E. Douglas, and R. C. Vrijenhoek. 1996. Molecular evidence for a unique evolutionary lineage of endangered Sonoran desert fish (genus *Poeciliopsis*). *Conservation Biology* 10(1):128–135.

Rauchenberger, M., K. D. Kallman, and D. C. Morizot. 1990. Monophyly and geography of the Pánuco Basin swordtails (genus *Xiphophorus*) with description of four new species. *American Museum Novitates* 2975:1–41.

Revenga, C., S. Murray, J. Abramovitz, and A. Hammond. 1998. *Watersheds of the world: Ecological value and vulnerability*. World Resources Institute, Washington, D.C.

Ricketts, T. H., E. Dinerstein, D. M. Olson, C. J. Loucks, W. Eichbaum, D. DellaSala, K. Kavanaugh, P. Hedao, P. T. Hurley, K. M. Carney, R. Abell, and S. Walters. 1999. *Terrestrial ecoregions of North America: A conservation assessment*. Island Press, Washington, D.C.

Rodiles-Hernández, R. 2004. Diversidad de peces continentales en Chiapas. In M. González-Espinosa, N. Ramírez-Marcial, and L. Ruíz-Montoya (eds.). *Diversidad Biológica de Chiapas*, pp. 141–160. Plaza y Valdés, ECOSUR, COCYTECH, Distrito Federal, Mexico.

Rodiles-Hernández R., J. Cruz-Morales, and S. Domínguez. 2002. El sistema lagunar de playas de Catzajá, Chiapas, Mexico. In E. G. de la Lanza and J. L. García-Calderón (eds.). *Lagos y presas de Mexico*, pp. 327–337. Mexico City.

Rodiles-Hernández, R., E. Díaz-Pardo, and J. Lyons. 1999. Patterns in the species diversity and composition of the fish community of the Lacanjá River, Chiapas, Mexico. *Journal of Freshwater Ecology* 14:455–468.

Rodiles-Hernández, R., S. Domínguez Cisneros, and E. Velásquez Velásquez. 1996. Diversidad íctica del Río Lacanjá, Selva Lacandona, Chiapas, Mexico. *Zoología Informa* 34:3–18.

Rodiles-Hernández, R., D. Hendrickson, J. Lundberg, and J. Alves. 2000. A new siluriform family from southern Mexico. In *80th Annual Meeting American Society's of Ichthyologist and Herpetologists, La Paz, Baja California, Mexico.*

Rodiles-Hernández, R., J. G. Lundberg, and D. A. Hendrickson. 2004. Diagnosis of the "Chiapas catfish": An apparently ancient siluriform lineage from Mesoamerica. In *84th Annual Meeting American Society of Ichthyologist and Herpetologists*. ASIH, Norman, Oklahoma.

Ryan, M. J., and G. G. Rosenthal. 2001. Variation and selection in swordtails. In L. A. Dugatkin (ed.). *Model systems in behavioral ecology: Integrating conceptual, theoretical, and empirical approaches*, pp. 133–148. Princeton University Press, Princeton, New Jersey.

Santana-Michel, F. J., H. Luis-Guzmán, R. Enrique-Sánchez, and G. Ramón-Cuevas. 2000. Flora y vegetación del Río Ayuquila. In *Memorias del VIII Simposio Interno sobre Inventarios, Manejo de Recursos Naturales y Desarrollo Comunitario*. Universidad de Guadalajara, Instituto Manantlán de Ecología y Conservación de la Biodiversidad, Department of Natural Resources, Centro Universitario de la Costa Sur.

Schaafsma, C. F., and C. L. Riley. 1999. The Casas Grandes world. University of Utah Press, Salt Lake City.

Secretaría de Infraestructura Urbana y Ecología (SIUE). 1992. Fauna Sonorense. Gobierno del Estado de Sonora, Secretaría de Infraestructura Urbana y Ecología, Programa Ambiental Estatal-PROAMBIENTE, Hermosillo, Sonora.

Secretaría del Medio Ambiente y Recursos Naturales (SEMARNAT). 2002. NORMA Oficial Mexicana NOM-059-ECOL-2001, Protección ambiental-Especies nativas de Mexico de flora y fauna silvestres-Categorías de riesgo y especificaciones para su inclusión, exclusión o cambio-Lista de especies en riesgo. *Diario Oficial De La Federación, Mexico* Miércoles 6 de marzo de 2002(Segunda Sección):1–85.

Siebert, D. J., and W. L. Minckley. 1986. Two new catostomid fishes (Cypriniformes) from the northern Sierra Madre Occidental of Mexico. *American Museum Novitates* 2849:1–17

Siemens, A. H. 1980. Wetland agriculture in pre-hispanic Mesoamerica. *Geographical Review* 70:166–181.

Siemens, A. H. 1983. Oriented raised fields in Central Veracruz. *American Antiquity* 48:85–102.

Siemens, A. H., and J. A. Soler-Graham. 2003. Manejo prehispanico del Río Candelaria, Campeche. *Arqueologia Mexicana* 10(59):64–69.

Sluyter, A. 1994. Intensive wetland agriculture in Mesoamerica: Space, time, and form. *Annals of the Association of American Geographers* 84:557–584.

Smith M. L., and R. R. Miller. 1986. The evolution of the Río Grande basin as inferred from its fish fauna. In C. H. Hocutt and E. O. Wiley (eds.). *The zoogeography of North American freshwater fishes*, pp. 457–485. John Wiley and Sons, New York.

Stebbins, R. C. 1985. *A field guide to the western reptiles and amphibians*. 2nd ed. National Audubon Society, London.

Tortajada, C., and A. K. Biswas. 1997. Environmental management of water resources in Mexico. Water International 22(3):172-178.

Trager, E. A. 1926. The geologic history of the Pánuco River Valley and its relation to the origin and accumulation of oil in Mexico. *American Association of Petroleum Geologists Annual Meeting*: 667–696.

U.S. Fish and Wildlife Service (USFWS). 1994. *Yaqui fishes recovery plan*. U.S. Fish and Wildlife Service, Albuquerque, New Mexico.

Vannote, R. L., G. W. Minshall, K. W. Cummins, J. R. Sedell, and C. E. Cushing. 1980. The River Continuum Concept. *Canadian Journal of Fisheries and Aquatic Sciences* 37:130–137.

Varela-Romero, A. 1989. *Dorosoma petenense* (Günther), un nuevo registro para la Cuenca del Río Yaqui, Sonora, Mexico (Pisces:Clupeidae). *Ecológica* 1(1):23–25.

Varela-Romero, A. 1995. Perspectivas de recuperación y cultivo de peces nativos en el Noroeste de Mexico. *Publicaciones CICTUS* 3:1–6.

Vázquez Sánchez, M. A., and M. A. Ramos Olmos. 1992. *Reserva de la Biosfera Montes Azules, Selva Lacandona: Investigación para su Conservación*. ECOSFERA, Mexico City.

Vrijenhoek, R. C. 1984. The evolution of clonal diversity in *Poeciliopsis*. In B. J. Turner (ed.). *Evolutionary genetics of fishes*, pp. 399–429. Plenum Press, New York.

Vrijenhoek, R. C. 1993. The origin and evolution of clones versus the maintenance of sex in *Poeciliopsis*. *Journal of Heredity* 84:388–395.

Walsh, S. J., and C. R. Gilbert. 1995. New species of troglobitic catfish of the genus *Prietella* (Siluriformes: Ictaluridae) from northeastern Mexico. *Copeia* 1995:850–861.

West, R. C., and J. P. Augelli. 1989. *Middle America, its lands and peoples*. Prentice-Hall, Englewood Cliffs, New Jersey.

West, R. C., N. P. Psuty, and B. G. Thom. 1969. *The Tabasco Lowlands of Southeastern Mexico*. Technical Report no. 70., Coastal Studies Institute, Louisiana State University, Baton Rouge.

Whalen, M. E., and P. E. Minnis. 2001. *Casas Grandes and its hinterlands: Prehistoric regional organization in Northwest Mexico*. University of Arizona Press, Tucson.

Whitmore, T. M., and B. L. Turner II. 2002. *Cultivated landscapes of Middle America on the eve of conquest*. Oxford University Press, Oxford.

Wilcox, T. P., F. J. García de León, D. A. Hendrickson, and D. M. Hillis. 2004. Convergence among cave catfishes: Long-branch attraction and a Bayesian relative rates test. *Molecular Phylogenetics and Evolution* 31: 1101–1113.

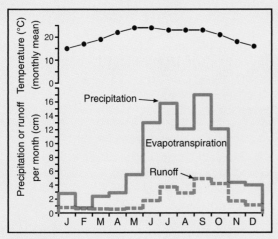

FIGURE 23.12 Mean monthly air temperature, precipitation, and runoff for the Río Pánuco basin.

RÍO PÁNUCO

Relief: 3800 m

Basin area: 98,227 km² (with Tamesí) 79,100 km² (without Tamesí)

Mean discharge: 537.4 m³/s (with Tamesí) 472.8 m³/s (without Tamesí)

River order: NA

Mean annual precipitation: 96 cm

Mean air temperature: 20°C

Mean water temperature: NA

Physiographic provinces: Neovolcanic Plateau (NP), Sierra Madre Oriental (SO), Mexican Gulf Coastal Plain (CP), Central Mesa (CP)

Biomes: Tropical Savanna, Mexican Montane Forest, Desert

Freshwater ecoregion: Tamaulipas–Veracruz

Terrestrial ecoregions: 6 ecoregions (see text)

Number of fish species: >88 (>80 native) (not including Río Tamesí)

Number of endangered species: 7 fishes

Major fishes: Media Luna cichlid, blackcheek cichlid, Chairel cichlid, slender cichlid, bluetail splitfin, dusky goodea, relict splitfin, jeweled splitfin, Media Luna pupfish, sheepshead swordtail, short-sword platyfish, delicate swordtail, highland swordtail, Moctezuma swordtail, Pánuco swordtail, variable platyfish, spottail chub, Pánuco minnow, bicolor minnow, chubsucker minnow, flatjaw minnow, Río Verde catfish, phantom blindcat, fleshylip buffalo, Mexican tetra

Major other aquatic vertebrates: NA

Major benthic invertebrates: mollusks (*Hydrobia tampicoensis*, *Littoridina crosseana*, *Lithasiiopsis hinkleyi*), crustaceans (*Palaemonetes mexicanus*, *Procambarus* [Ortmannicus] *ortmanii*, *P.* [Ortmannicus] *acutus cuevachicae*, *P.* [*Scapullicambarus*] *strenghi*, *Troglomexicanus perezfafantae*, *T. huastecae*, *T. tamaulipenses*), hellgrammites (*Chloronia mexicana*, *Corydalus magnus*)

Nonnative species: tilapias, grass carp, silver carp, channel catfish, blue catfish, largemouth bass, water hyacinth

Major riparian plants: NA, but see Río Tamesí

Special features: Cascada Tamul, 102 m high waterfall (at confluence of Río Gallinas and Río Santa Maria); natural springs; caves and sinkholes are common, some perhaps among world's deepest, many with abundant aquatic habitat; some of highest areas harbor cloud forests (El Cielo Biosphere Reserve)

Fragmentation: dams on Río Moctezuma, Río Tula, Río Topila; dam under construction on lower Río Tamuin

Water quality: highly variable, but poor in many places

Land use: sugarcane farming, citrus, and cattle ranching in Coastal Plain; traditional slash and burn, citrus, and coffee in mountains; petroleum extraction in lower basin

Population density: low

Major information sources: Contreras-Ramos 1998, www.weatherbase.com, http://webworld.unesco.org/water/ihp/db/shiklomanov/index.shtml, Hudson 2003a, Hudson 2003b, Hudson and Heitmuller 2003, Hudson and Colditz 2003, Hudson et al. in press

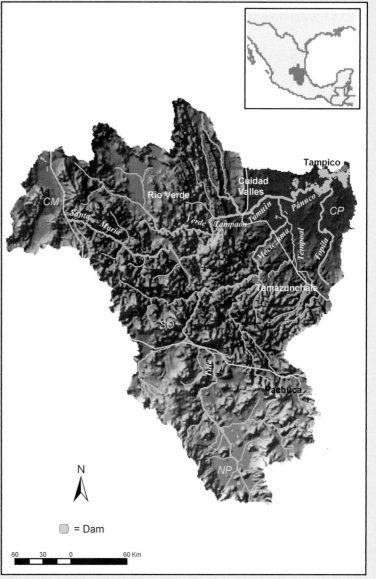

FIGURE 23.11 Map of the Río Pánuco basin. Physiographic provinces are separated by yellow lines.

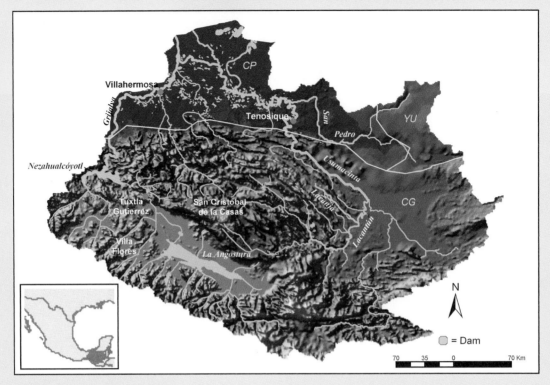

FIGURE 23.13 Map of the Ríos Usumacinta–Grijalva basin. Physiographic provinces are separated by yellow lines.

RÍOS USUMACINTA–GRIJALVA

Relief: 3800 m

Basin area: 112,550 km²

Mean discharge: 2678 m³/s

River order: NA

Mean annual precipitation: 199 cm

Mean air temperature: 23°C

Mean water temperature: NA

Physiographic provinces: Chiapas–Guatemala Highlands (CG), Mexican Gulf Coastal Plain (CP), Yucatan (YU)

Biomes: Tropical Rain Forest, Tropical Savanna

Freshwater ecoregion: Grijalva–Usumacinta

Terrestrial ecoregions: Peten–Veracruz Moist Forests, Chiapas Depression Dry Forests, Pantanos de Centla, Central American Pine–Oak Forests, Chiapas Montane Forests, Usumacinta Mangroves

Number of fish species: >112 (103 native)

Number of endangered species: 11 fishes

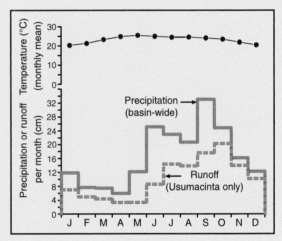

FIGURE 23.14 Mean monthly air temperature, precipitation, and runoff for the Ríos Usumacinta–Grijalva basin.

Major fishes: Pénjamo tetra, headwater killifish, Chiapas killifish, widemouth gambusia, Chiapas swordtail, sulphur molly, upper Grijalva livebearer, white cichlid, Angostura cichlid, tailbar cichlid, Petén cichlid, Montechristo cichlid, Usumacinta cichlid, Chiapa de Corzo cichlid, freckled cichlid, Teapa cichlid, longfin gizzard shad, Lacandon sea catfish, Maya sea catfish, Maya needlefish, Mexican halfbeak, Mexican mojarra

Major other aquatic invertebrates: river crocodile, swamp crocodile, common snapping turtle, tortugas blanca, tortugas casquito, neotropical river otter, West Indian manatee, water opossum, tepezcuintle

Major benthic invertebrates: mollusks (*Pomacea, Aroapyrgus, Cochliopina, Hydrobia, Pachychilus*), hellgrammites (*Corydalus luteus*), caddisflies (*Smicridea, Nectopsyche, Neotrichia*), mayflies (*Camelobaetidius, Leptophlebia, Traverella, Campsurus*), odonates (*Archaeogomphus, Phyllocycla, Protoneura, Neoneura, Heteragrion*), beetles (*Anacaena, Brachyvatus, Laccophilus*)

Nonnative species: common carp, grass carp, rainbow trout, largemouth bass, blue tilapia, redbelly tilapia, Nile tilapia, Mozambique tilapia, jaguar guapote, water hyacinth

Major riparian plants: *Andira galeottiana, Pachira acuatica, Bravaisia integerrima*, bloodwoodtree, gregorywood, *Paquira aquatica*, willow, button mangrove, black mangrove, white mangrove, American mangrove, cattails, common reed

Special features: Pantanos de Centla Biosphere Reserve; Laguna del Tigre National Park in Guatemala; Selva Maya, possibly largest remaining tropical forest in North/Central America; center of ancient Mayan culture; Montes Azules Biosphere Reserve; Lacantún Biosphere Reserve

Fragmentation: several large dams on Río Grijalva; new dams proposed on Usumacinta

Water quality: NA

Land use: 59% forest, 31% cropland, 8% grassland/savanna/shrubland, 3% urban

Population density: 28 people/km²

Major information sources: Revenga et al. 1998, A. Contreras-Ramos, personal communication, www.weatherbase.com, Rodiles-Hernández 2004, Bueno-Soria et al. in press, March et al. 1996, http://webworld.unesco.org/water/ihp/db/shiklomanov/index.shtml

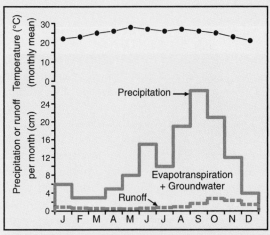

FIGURE 23.16 Mean monthly air temperature, precipitation, and runoff for the Río Candelaria basin.

RÍO CANDELARIA (YUCATÁN)

Relief: 375 m
Basin area: 10,755 km²
Mean discharge: 46 m³/s
River order: NA
Mean annual precipitation: 150 cm
Mean air temperature: 24.6°C
Mean water temperature: NA
Physiographic provinces: Yucatan (YU), Mexican Gulf Coastal Plain (CP)
Biomes: Tropical Rain Forest, Tropical Savanna
Freshwater ecoregion: Grijalva–Usumacinta
Terrestrial ecoregions: Yucatan Moist Forests, Usumacinta Mangroves, Peten–Veracruz Moist Forests, Pantanos de Centla
Number of fish species: >65 (>17 freshwater, 48 estuarine)
Number of endangered species: none
Major fishes: firemouth cichlid, blackgullet cichlid, yellow cichlid, yellowjacket, Mayan cichlid, yellowbelly cichlid, Montechristo cichlid, redhead cichlid, giant cichlid, stippled gambusia, Champoton gambusia, shortfin molly, picotee livebearer, pike killifish, banded tetra, Maya tetra, pale catfish, threadfin shad, bay anchovy, silver perch
Major other aquatic vertebrates: swamp crocodile, neotropical river otter
Major benthic invertebrates: mollusks (narrowmouth hydrobe), hellgrammites (*Corydalus bidenticulatus*), stoneflies (*Anacroneuria*), true bugs (*Abedus*)
Nonnative species: none documented
Major riparian plants: cattails, breadnut tree, bay palmetto, gumbo limbo, gregorywood, mangroves
Special features: one of the few rivers flowing through the highly karstic region of the Yucatan Peninsula; large water losses to groundwater; where ancient Mayan civilization developed
Fragmentation: no dams or reservoirs; prehistoric canal systems
Water quality: NA, but perceived to be good
Land use: slash and burn in uplands; cattle ranching in coastal plain
Population density: low, but has increased rapidly since 1960s
Major information sources: www.weatherbase.com, Gunn et al. 1995, Ayala-Perez et al. 1998

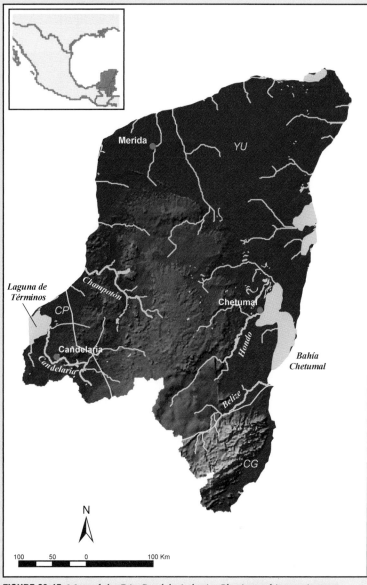

FIGURE 23.15 Map of the Río Candelaria basin. Physiographic provinces are separated by yellow lines.

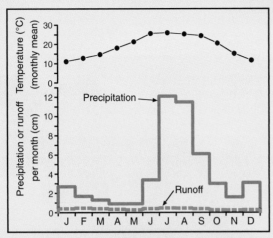

FIGURE 23.18 Mean monthly air temperature, precipitation, and runoff for the Río Yaqui basin.

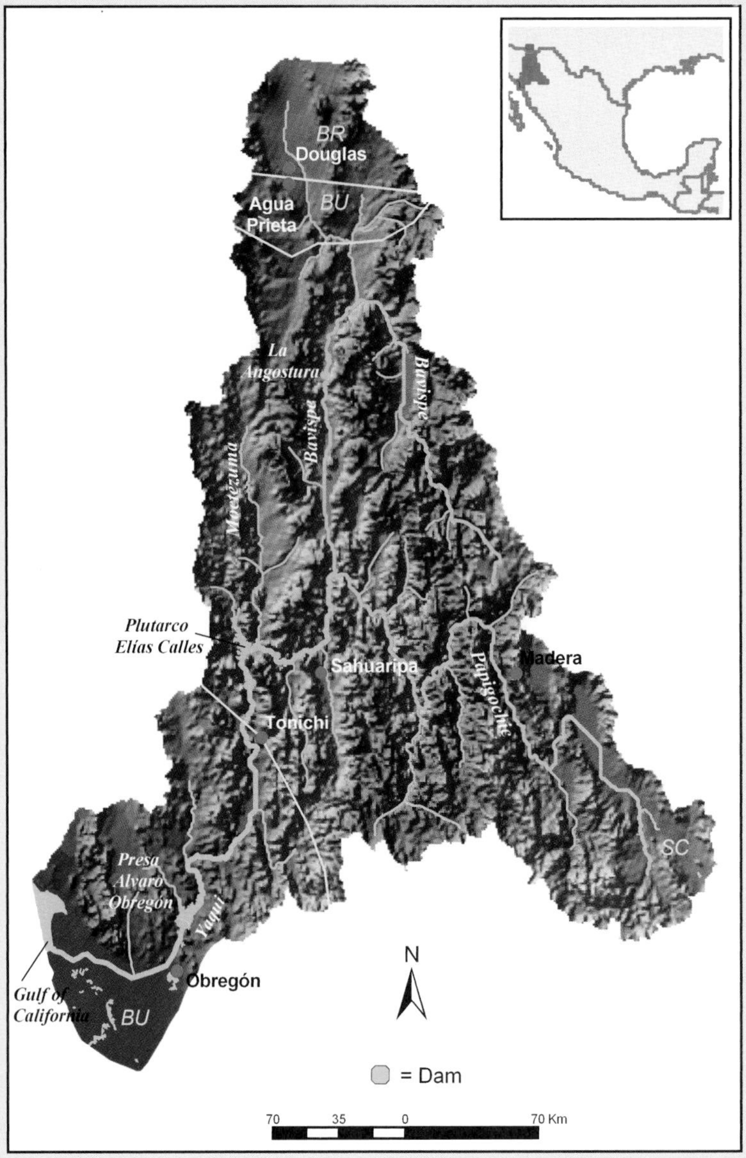

RÍO YAQUI

Relief: 2520 m
Basin area: 73,000 km^2
Mean discharge: 78.5 m^3/s
River order: 6
Mean annual precipitation: 48 cm
Mean air temperature: 18.4°C
Mean water temperature: 18.0°C
Physiographic provinces: Sierra Madre Occidental (SO), Basin and Range (BR), Buried Ranges (BU)
Biomes: Mexican Montane Forest, Desert
Freshwater ecoregion: Sonoran
Terrestrial ecoregions: Sonoran–Sinaloan Transitional Subtropical Dry Forests, Sierra Madre Occidental Pine–Oak Forests, Sonoran Desert, Chihuahuan Desert, Sinaloan Dry Forests
Number of fish species: 107
Number of endangered species: 7 fishes (threatened)
Major fishes: Yaqui trout, Yaqui sucker, Yaqui chub, roundtail chub, Mexican stoneroller, longfin dace, Yaqui catfish, Yaqui topminnow, Leopold sucker, Cahita sucker, Pacific gizzard shad, striped mullet, striped mojarra, machete, Heller's anchovy
Major other aquatic vertebrates: water snakes (*Thamnophis* spp.), Sonora mud turtle, Tarahumara frog, Tarahumara salamander, Chiricahua leopard frog, neotropical river otter
Major benthic invertebrates: crustaceans (Cauque River prawn, blue shrimp, white shrimp), mollusks (*Fossaria* [*Bakerilymnaea*] *bulimoides*), hellgrammites (*Corydalus bidenticulatus*, *Corydalus texanus*), mayflies (*Siphlonurus*, *Nixe*, *Acentrella*), stoneflies (*Capnia decepta*, *Mesocapnia frisoni*, *Anacroneuria wipikupa*)
Nonnative species: channel catfish, blue catfish, black bullhead, largemouth bass, rainbow trout, common carp, river carpsucker, green sunfish, bluegill, western mosquitofish, American bullfrog
Major riparian plants: Arizona sycamore, Arizona alder, Goodding willow, Bonpland willow, green ash, Fremont cottonwood, common reed, Chinese saltcedar, velvet mesquite
Special features: basin shared between Arizona and México (Sonora and Chihuahua); spectacular canyons (barrancas); high fish diversity for arid system, but low endemism; protected areas are Tutuaca, Papigochic, and Sierra de Ajos Bavispe
Fragmentation: three major dams on main stem; flow regulation over 50% of the basin
Water quality: NA
Land use: 62% forest, 2% cropland, 33% grassland/savanna/shrubland, 3% urban
Population density: 7 people/km^2
Major information sources: Arriaga-Cabrera et al. 2000, Hendrickson et al. 1981, Baumann and Kondratieff 1996, Watersheds of the World CD 2003, A. Contreras-Ramos, personal communication, http://webworld.unesco.org/water/ihp/db/shiklomanov/index.shtml

FIGURE 23.17 Map of the Río Yaqui basin. Physiographic provinces are separated by yellow lines.

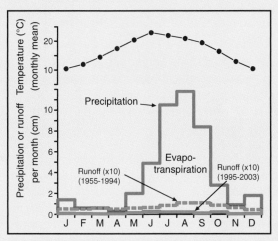

FIGURE 23.20 Mean monthly air temperature, precipitation, and runoff for the Río Conchos basin.

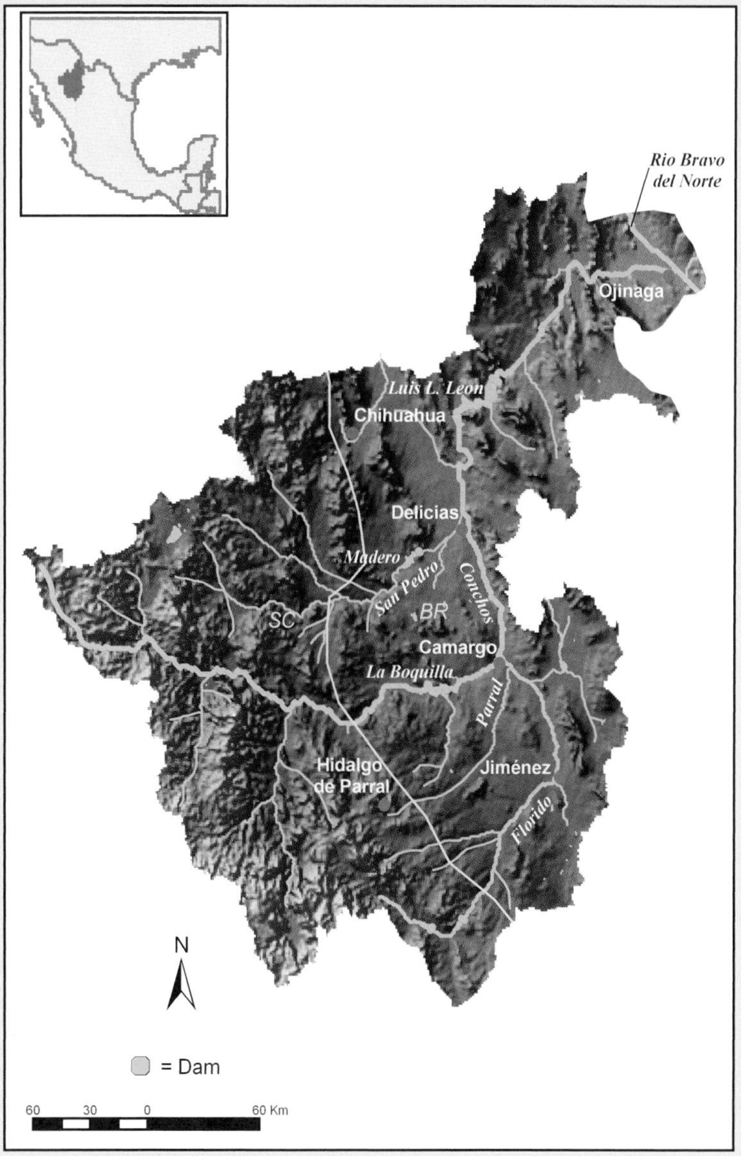

RÍO CONCHOS

Relief: 2700 m
Basin area: 68,386 km²
Mean discharge: 20.5 m³/s
River order: NA
Mean annual precipitation: 48 cm
Mean air temperature: 18°C
Mean water temperature: NA
Physiographic provinces: Sierra Madre Occidental (SC), Basin and Range (BR)
Biomes: Mexican Montane Forest, Desert
Freshwater ecoregion: Río Conchos
Terrestrial ecoregions: Sierra Madre Occidental Pine–Oak Forests, Chihuahuan Desert
Number of fish species: 53 (>38 native)
Number of endangered species: 5 fishes
Major fishes: Conchos shiner, bigscale pupfish, bighead pupfish, Salvador's pupfish, Conchos darter, yellowfin gambusia, crescent gambusia, longnose gar, largemouth bass, Mexican tetra, Río Grande cichlid, Mexican stoneroller, ornate minnow, red shiner, roundnose minnow, Tamaulipas shiner, longnose dace, Conchos chub, headwater catfish, blue catfish, channel catfish
Major other aquatic vertebrates: water snakes (*Thamnophis* spp.)
Major benthic invertebrates: virile crayfish, others NA
Nonnative species: goldfish, common carp, warmouth, inland silverside, white bass, rainbow trout
Major riparian plants: NA
Special features: major inflow for the Río Bravo del Norte upstream of Big Bend National Park; considerable endemism of fish species and herpetofauna; specialized fauna in spring and cave habitats
Fragmentation: several dams, Presa Boquilla is largest; flow is reduced to zero for months at a time
Water quality: pH = 7.7 to 8.4, total dissolved solids = 156 to 722 mg/L, NO_3-N = 2.3 to 3.8 mg/L, PO_4-P = 0.15 to 0.53 mg/L, Na = 23 to 168 mg/L, Ca = 8 to 190 mg/L; degraded in some areas by sewage, nutrients, pesticides, industrial wastes
Land use: NA
Population density: >16 people/km²
Major information sources: Kelly 2001, Abell et al. 2000, Arriaga-Cabrera et al. 2000, Gutiérrez and Borrego 1999, www.ibwc.state.gov, SEMARNAT 2002, Edwards et al. 2002a, 2002b

FIGURE 23.19 Map of the Río Conchos basin. Physiographic provinces are separated by yellow line.

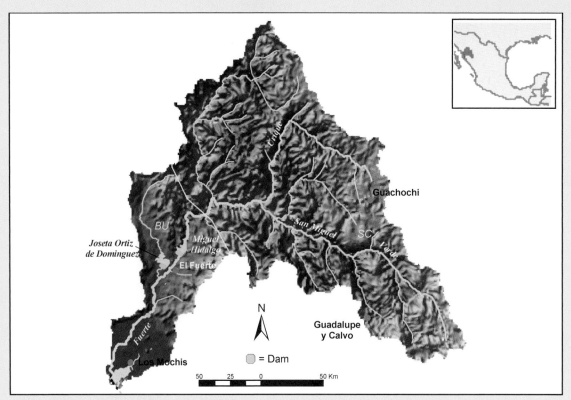

FIGURE 23.21 Map of the Río Fuerte basin. Physiographic provinces are separated by yellow line.

RÍO FUERTE

Relief: 3300 m

Basin area: 34,247 km²

Mean discharge: 31 m³/s

River order: NA

Mean annual precipitation: 78 cm

Mean air temperature: 24°C

Mean water temperature: NA

Physiographic provinces: Buried Ranges (BU), Sierra Madre Occidental (SC)

Biomes: Desert, Mexican Montane Forest

Freshwater ecoregion: Sinaloan Coastal

Terrestrial ecoregions: Sinaloan Dry Forests, Sonoran–Sinaloan Transition Subtropical Dry Forests, Sierra Madre Occidental Pine–Oak Forests

Number of fish species: 51

Number of endangered species: 6 fishes (threatened), 3 fishes (special protection)

Major fishes: Mexican stoneroller, ornate shiner, Mexican golden trout, diverse unisexual clones of topminnow, roundtail chub, Yaqui sucker, Yaqui catfish, mountain clingfish

Major other aquatic vertebrates: neotropical river otter

Major benthic invertebrates: crustaceans (*Macrobrachium*), others NA

Nonnative species: largemouth bass (important reservoir sport fishery), bluegill, green sunfish, rainbow trout, channel catfish, blue catfish, common carp, tilapia

Major riparian plants: none

Special features: major tributaries descend through Barranca del Cobre (Copper Canyon) Park; remote historic settlements; indigenous cultures (Rarámuri or Tarahumara) occupy much of the basin

Fragmentation: large dams in lower river

Water quality: NA

Land use: logging in higher-elevation pine forests; nonmechanized, dry agriculture (corn, beans, grazing) practiced by Rarámuri throughout headwaters and some canyon bottoms; extensive irrigated mechanized agriculture in Coastal Plain

Population density: NA

Major information sources: Arriaga-Cabrera 2000, www.weatherbase.com, http://webworld.unesco.org/water/ihp/db/shiklomanov/index.shtml, Hendrickson et al. 2002, SEMARNAT 2002, Hendrickson and Varela 2002, Vrijenhoek 1984

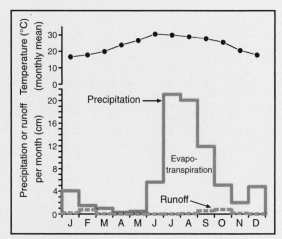

FIGURE 23.22 Mean monthly air temperature, precipitation, and runoff for the Río Fuerte basin.

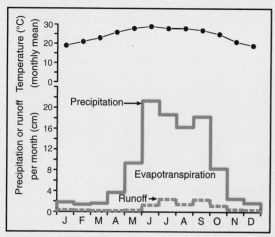

FIGURE 23.24 Mean monthly air temperature, precipitation, and runoff for the Río Tamesí basin.

RÍO TAMESÍ

Relief: 3353 m
Basin area: 19,127 km²
Mean discharge: 64.6 m³/s
River order: NA
Mean annual precipitation: 105 cm
Mean air temperature: 24°C
Mean water temperature: NA
Physiographic provinces: Sierra Madre Oriental (SO),
   Mexican Gulf Coastal Plain (CP)
Biomes: Desert/Semidesert, Tropical Savanna
Freshwater ecoregion: Tamaulipas–Veracruz
Terrestrial ecoregions: Sierra Madre Oriental
   Oak-Pine Forests, Veracruz Moist Forests,
   Alvarado Mangroves
Number of fish species: 93
Number of endangered species: 4 fishes
Major fishes: sailfin molly, Amazon molly, shortfin
   molly, Mexican tetra, alligator gar, variable
   platyfish, smallmouth buffalo, channel catfish,
   rainwater killifish, Río Grande cichlid, sheepshead
   minnow, striped mullet, hardhead catfish, phantom
   blindcat, chairel cichlid, slender cichlid, blackcheek
   cichlid, lantern minnow, Gulf gambusia, golden gambusia, spinycheek sleeper

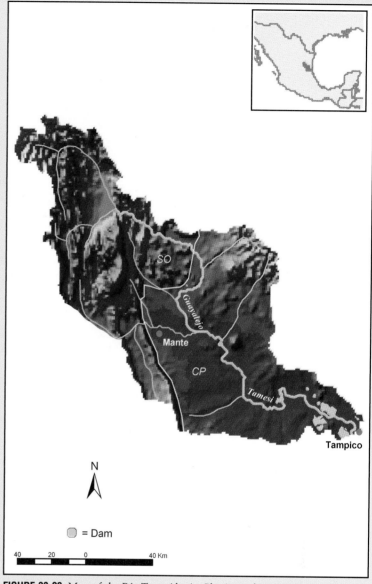

FIGURE 23.23 Map of the Río Tamesí basin. Physiographic provinces are separated by yellow line.

Major other aquatic vertebrates: NA
Major benthic invertebrates: mollusks (*Lithasiiopsis crassa, Pachychilus corpulentus, Pachychilus vallesensis*), crustaceans
   (*Palaemonetes hobbsi, P. kadiakensis, P. mexicanus, Procambarus [Ortmannicus] acutus cuevachicae, Procambarus
   [Ortmannicus] acutus*), hellgrammites (*Chloronia mexicana, Corydalus luteus*), beetles (*Tropisternus*), caddisflies
   (*Leptonema*), mayflies (*Baetis*), caddisflies (*Nectopsyche*)
Nonnative species: common water hyacinth, hydrilla, cinnamon river shrimp, grass carp, silver carp, largemouth bass,
   blue tilapia
Major riparian plants: breadnut tree, gumbo limbo, cattails, mangroves
Special features: headwaters include large El Cielo Biosphere Preserve (near Ciudad Victoria); large springs at edge of karstic
   Sierra Guatemala feed river via caves originating at high elevations in same range with numerous aquatic cave organisms;
   extensive freshwater lagoon on lower 40 km of river valley
Fragmentation: main-stem dam east of Mante; several small dams
Water quality: poor in several areas
Land use: sugarcane and citrus farming and cattle ranching in Coastal Plain; traditional slash and burn in mountains; petroleum
   extraction in lower basin
Population density: NA
Major information sources: Arriaga-Cabrera et al. 2000, Contreras-Ramos 1998, www.weatherbase.com, Garcia de León et al.
   in press, Hendrickson and Krejca 2000, Hudson 2002

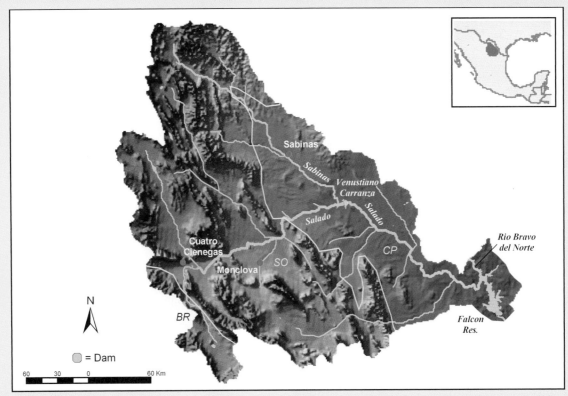

FIGURE 23.25 Map of the Río Salado basin. Physiographic provinces are separated by yellow lines.

RÍO SALADO

Relief: 2560 m
Basin area: 60,000 km^2
Mean discharge: 10 m^3/s
River order: NA
Mean annual precipitation: 31 cm
Mean air temperature: 21°C
Mean water temperature: NA
Physiographic provinces: Coastal Plain (CP), Sierra Madre Oriental
   (SO)
Biome: Desert
Freshwater ecoregion: Río Salado
Terrestrial ecoregions: Tamaulipan Mezquital, Sierra Madre Oriental
   Pine–Oak Forests, Chihuahuan Desert, Tamaulipan Matorral
Number of fish species: 52 (including Cuatro Ciénegas)
Number of endangered species: none
Major fishes: Salado shiner, tufa darter, Salado darter, Cuatro
   Ciénegas gambusia, robust gambusia, marbled swordtail,
   Cuatro Ciénegas platyfish, bolson pupfish, Cuatro Ciénegas pupfish, Mexican red shiner, Mexican tetra, largemouth bass,
   longear sunfish, Cuatro Ciénegas cichlid, Río Grande cichlid, roundnose minnow, Tamaulipas shiner, Devils River minnow,
   blue catfish, headwater catfish, channel catfish, flathead catfish, gray redhorse
Major other aquatic vertebrates: Cuatro Ciénegas box turtle, Cuatro Ciénegas red-eared slider, Cuatro Ciénegas softshell,
   diamondback water snake
Major benthic invertebrates: crustaceans (red swamp crayfish, *Hyallela*, *Paleomonetes suttkusi*), mollusks (redrim melania)
Nonnative species: warmouth, blue tilapia, spotted jewelfish, common carp, threadfin shad, red swamp crayfish, Asiatic clam,
   water hyacinth
Major riparian plants: mesquite, cottonwood, willows, common reed, giant reed, saltcedar, athel
Special features: major headwaters in Cuatro Ciénegas Protected Area for Fauna and Flora, a small desert valley with hundreds
   of large geothermal springs feeding marshes and rivers that harbor a diverse and highly endemic fauna and flora
Fragmentation: one major dam (V. Carranza) on main stem; most of main stem usually dry
Water quality: pollution from agriculture and industry; point and nonpoint sources
Land use: NA
Population density: NA
Major information sources: www.weatherbase.com, Abell et al. 2000, Arriaga-Cabrera 2000, Hendrickson et al. in press,
   www.ibwc.state.gov, Guerra 1952, Minckley 1969

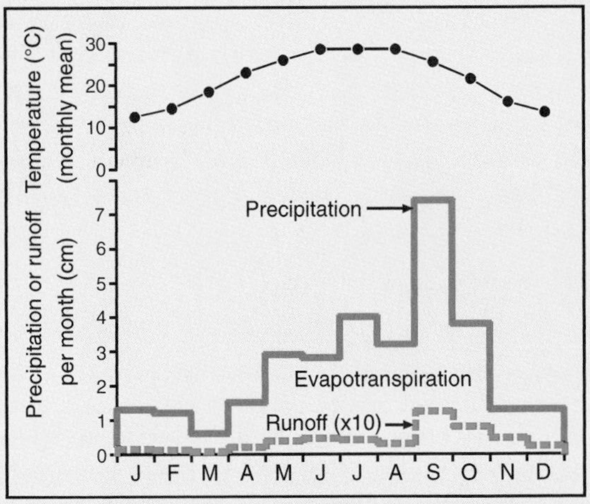

FIGURE 23.26 Mean monthly air temperature,
precipitation, and runoff for the Río Salado basin.

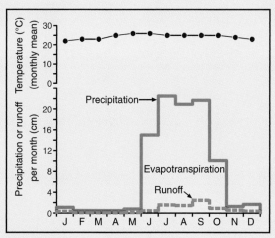

FIGURE 23.28 Mean monthly air temperature, precipitation, and runoff for the Río Armería-Ayuquila basin.

RÍO ARMERÍA-AYUQUILA

Relief: 4240 m
Basin area: 9803 km²
Mean discharge: 30.4 m³/s
River order: 6
Mean annual precipitation: 104 cm
Mean air temperature: 22.1°C
Mean water temperature: 21.8°C
Physiographic provinces: Neovolcanic Plateau (NP), Sierra Madre del Sur (SD)
Biomes: Desert, Mexican Montane Forest
Freshwater ecoregion: Manantlán–Ameca
Terrestrial ecoregion: Trans-Mexican Volcanic Belt, Oak–Pine Forests, Jalisco Dry Forests
Number of fish species: 38 (32 native), 3 extirpated
Number of endangered species: 3 fishes, 1 mammal
Major fishes: banded tetra, riffle chub, Mexican redhorse, Lerma catfish, bandfin splitfin, goldbreast splitfin, Pacific molly, Lerma livebearer, golden livebearer, Michoacán livebearer, redside cichlid, mountain mullet, mountain clingfish, Pacific sleeper, finescale sleeper, multispotted goby
Major other aquatic vertebrates: giant toad
Major benthic invertebrates: crustaceans (*Macrobrachium occidentale*, *M. americanum*, *Pseudothelphusa dilatata*, *Atya ortomannoides*, *A. margaritacea*), stoneflies (*Anacroneuria*), mayflies (*Baetis*, *Baetodes*, *Callibaetis*, *Farrodes*, *Thraulodes*, *Leptohyphes*), caddisflies (*Ceratopsyche*, *Cheumatopsyche*, *Leptonema*, *Smicridea*), mollusks (*Littoridina orcutti*), hellgrammites (*Corydalus bidenticulatus*)
Nonnative species: common carp, green swordtail, bluegill, largemouth bass, blue tilapia, redbreast tilapia
Major riparian plants: Humboldt's willow, West Indian marsh grass, common water hyacinth, fragrant flatsedge
Special features: one of the 15 largest rivers of the Mexican Pacific Slope; 71 km section of river comprises northeastern boundary of Sierra de Manantlán Biosphere Reserve; drains 4240 m volcano (Nevado de Colima; designated National Park)
Fragmentation: 40 dams in system with a total capacity of 731 million m³; series of agricultural valleys where flow is diverted for irrigation
Water quality: relatively good in isolated sections; highly impacted by organic pollutants near cities and agricultural valleys; dissolved oxygen = 1.0 to 4.3 mg/L, pH = 7.4, alkalinity = 123 mg/L as $CaCO_3$, NO_3-N = 1.91 mg/L, PO_4-P = 0.79 mg/L
Land use: 30% agriculture, 60% forestry, 10% urban
Population density: 55.9 people/km²
Major information sources: Martínez et al. 1999, Lyons, Navarro et al. 1995, Lyons, González et al. 1998, CNA 2001, DRSBM–IMECBIO 2000, Contreras-Ramos 1998, Santana-Michel et al. 2000, http://webworld.unesco.org/water/ihp/db/shiklomanov/index.shtml, www.weatherbase.com, additional data provided by Norman Mercado-Silva, John Lyons, Luis Manuel Martínez-Rivera, and Luis Ignacio Íñiguez-Dávalos

N

☐ = Dam

20 10 0 20 Km

FIGURE 23.27 Map of the Río Armería-Ayuquila basin. Physiographic provinces are separated by yellow line.

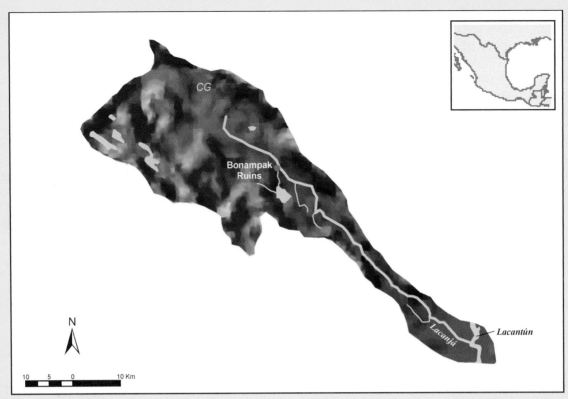

FIGURE 23.29 Map of the Río Lacanjá basin.

RÍO LACANJÁ

Relief: 350 m
Basin area: 800 km^2
Mean discharge: NA
River order: 3
Mean annual precipitation: 223 cm
Mean air temperature: 25°C
Mean water temperature: 27°C
Physiographic province: Chiapas–Guatemala Highlands (CG)
Biome: Tropical Rain Forest
Freshwater ecoregion: Grijalva–Usumacinta
Terrestrial ecoregion: Peten–Veracruz Moist Forests
Number of fish species: 44
Number of endangered species: 2 fishes, 2 reptiles, 2 mammals
Major fishes: Usumacinta cichlid, white cichlid, Petén cichlid, arroyo cichlid, freckled cichlid, bluemouth cichlid, pantano cichlid, Palenque cichlid, undescribed catfish, Lacandon sea catfish, Maya sea catfish, Mexican freshwater toadfish, white mullet, Mexican mojarra, giant cichlid, tropical gar, macabi tetra
Major other aquatic vertebrates: river crocodile, neotropical river otter
Major benthic invertebrates: crustaceans (*Potamocarcinum chajulensis, Odontothelphusa palenquensis, O. lacandona*), hellgrammites (*Platyneuromus honduranus*)
Nonnative species: grass carp, several species of tilapia
Major riparian plants: willow, cottonwood, sycamore, cypress, *Pachira acuatica, Bravaisia integerrima*, southern cattail, common reed, *Thalassia, Guadua spinosa*, waterlily
Special features: fast-flowing runs, waterfalls with numerous runs and rapids, floodplain clear-water lakes, floodplain backwater, riparian wetland; boundary between two of most important biosphere reserves in México, Montes Azules and Lacantún
Fragmentation: no dams; natural fragmentation by waterfall of 15 m
Water quality: relatively free of pollutants; pH = 7.4, conductivity = 668 μS/cm; TDS = 335.5 mg/L, dissolved oxygen = 7.18 mg/L, transparency = 3.3 m
Land use: 15% agriculture, 5% rural community, 80% ecological reserves
Population density: <1 person/km^2
Major information sources: Vázquez Sánchez and Ramos Olmos 1992, Rodiles-Hernández et al. 1996, Rodiles-Hernández et al. 1999, Morales-Román and Rodiles-Hernández 2000, INE-SEMARNAP 2000

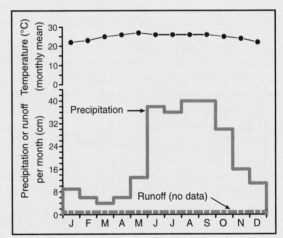

FIGURE 23.30 Mean monthly air temperature, precipitation, and runoff for the Río Lacanjá basin.

24

OVERVIEW AND PROSPECTS

J. DAVID ALLAN   ARTHUR C. BENKE

INTRODUCTION

THE VARIETY OF RIVERS

FEW RIVERS ARE PRISTINE

NORTH AMERICA'S RIVERS IN THE
TWENTY-FIRST CENTURY

LITERATURE CITED

INTRODUCTION

This concluding chapter provides an overview of information in the previous chapters and addresses some of the major challenges facing rivers in the twenty-first century. It is clear from previous chapters that the rivers of North America exhibit an almost bewildering variety of natural features and degrees of human impact. Our initial purpose, therefore, is to convey a sense of their variety and status. They differ greatly in their natural features, depending on physical, climatic, and biological factors. Few rivers can be called pristine, and the extent and type of human influence adds additional layers of complexity and variation according to how humans have used these rivers for water supply, power, navigation, waste disposal, and other purposes. Thus, the observed variability among rivers is ultimately a combination of natural variation and changes brought about by human activities. Both of these types of variation differ greatly across the continent, and this is reflected throughout the chapters.

A second purpose of this chapter is to examine the major challenges today and into the future facing North America's rivers. How will the diverse pressures from human society further alter North America's rivers? What might we anticipate about the future challenges that rivers will experience? What are the opportunities to make the twenty-first century a time of repair and restoration of the rivers of North America?

THE VARIETY OF RIVERS

The twenty-two chapters describing individual river basins or geographical regions are rich with detailed information about their regions and include one-page summaries for 218 rivers, representing most of the largest rivers on the continent. Although the majority of rivers are relatively large within their respective regions, river size exhibits wide variation across and within chapters. In addition, river basins vary greatly in their mean temperatures and precipitation, the fractions of precipitation that flow into rivers, the diversity of landscape types drained by rivers, and their natural biological diversity.

Variation in Physical Characteristics

The major rivers of North America, the focus of this book, typically are large rivers, whether assessed by river order, drainage area, or discharge. Of the five to twelve rivers described for each basin or region, some individual rivers are substantially smaller, of order as low as 3 (e.g., the Virgin and Bill Williams in the Colorado basin) or 4 (the Octonagon and AuSable of the St. Lawrence basin); their inclusion reflects their occurrence in very arid regions or their regional significance. In contrast, the lower main stems of the largest rivers are of order 9 (Mackenzie, Ohio, Missouri, St. Lawrence, Yukon, Nelson) or 10 (Mississippi). In rare instances, rivers of substantially smaller basin size (the Moisie in Atlantic Canada) have been considered as order 9. In spite of this wide range, the median value for river order in this book is generally 6 or 7, which usually represents a medium to large river.

Another way of looking at the variety of river sizes is that whereas the median drainage area of individual basins is approximately 25,000 km^2, basin areas range from as small as 217 km^2 (the Dunk River on Prince Edward Island) to 3,270,000 km^2 (the entire Mississippi basin, about 42% of the land area of the 48 coterminous states). A drainage area over 100,000 km^2 provides an arbitrary criterion for "very large"; after excluding the lower main stems of the largest river systems, some 33 individual rivers meet this criterion. However, because many basins >100,000 km^2 are in arid regions, they do not necessarily have the highest discharges.

Because of their spatial extent, most river systems encompass considerable physical heterogeneity, and this is especially true of the largest river basins. The number of physiographic provinces through which rivers run influences potential diversity in geology, gradient, channel morphology, and habitat types. The number of ecoregions not only reflects this geological diversity, but climate and vegetation as well. Thus, plant communities in particular should influence the amount and type of organic matter inputs from a basin's smallest tributaries to its main stem (Vannote et al. 1980, Webster and Meyer 1997), and climate influences precipitation and the fraction that becomes runoff.

A typical river drains two ecoregions and two physiographic provinces (median values[1]), but the largest river basins encompass much more heterogeneity. For example, the St. Lawrence drains eight physiographic provinces and nine ecoregions; these numbers are four and seven for the Saskatchewan, seven and thirteen for the Missouri, and six and eight for the Ohio. In some cases, relatively small basins can have high heterogeneity (e.g., five ecoregions and five physiographic provinces for the Potomac). Basin

---

[1] Most values reported are medians rather than averages because of occasional extreme values, and the subset of rivers included within a basin is not truly a random sample. Hence, all comparisons reported here should be interpreted as broadly indicative, but values given are approximate.

relief (from highest peak to river mouth) also varies within and among rivers. The median value for basin relief for individual rivers is approximately 1300 m, but vertical relief in some "flatland" rivers is minimal. For example, basin relief is <120 m for the Maumee River of Ohio, the Illinois and Minnesota rivers of the Upper Mississippi, and the St. Johns and Satilla rivers of the southeastern Coastal Plain. In contrast, median vertical relief for Pacific Coast rivers of Canada and Alaska is 2628 m, despite a relatively modest average basin area of about 46,000 km^2, and median relief for Mexican rivers is 3000 m, with an average basin area of about 49,000 km^2. Given such wide variation in river basin relief, the variety of habitats and ecological conditions within a river can be expected to vary accordingly.

The rivers of North America differ greatly in discharge and runoff. Of the 218 rivers described, two-thirds have a mean annual discharge exceeding 100 m^3/s, 15% exceed 1000 m^3/s, and two exceed 10,000 m^3/s. Mean annual discharge in m^3/s is an indication of how much water is exported by a river basin. It is higher for rivers with large drainage areas and wet climates (Table 24.1). The lower Mississippi (which receives inflows from the Upper Mississippi, the Missouri, and especially the Ohio) ranks 9th in the world (Leopold 1994). A number of northern river systems, including the St. Lawrence, Mackenzie, Columbia, Yukon, and Fraser, discharge very large quantities of water. River basins in arid climates have markedly lower mean annual discharges. In arid regions, the majority of rivers have discharges below 100 m^3/s. For an individual example, consider the Brazos River on the Gulf Coast of the southwestern United States and the Tennessee River in the Ohio basin. Both have slightly over 100,000 km^2 in drainage area, but annual discharge of the Tennessee is ten times higher than the more arid Brazos.

Discharge (Q) is influenced by basin area, precipitation (PPT), and the amount of PPT that becomes evapotranspiration (ET), as can be observed by comparing rivers and regions (see Table 24.1). Table 24.1 includes the largest river for each region or basin (left side), as well as the median values for the remaining rivers (right side). Annual runoff (RO, cm/yr) is another way to represent water yield (1 cm/yr = 100 m^3 ha^{-1} yr^{-1}) or the amount of water that annually runs off a unit area of basin. We estimated annual runoff directly from monthly runoff data provided by chapter authors (see figures in one-page summaries of each chapter). Median annual runoff for rivers of each region is shown in Table 24.1. It reveals the low water yields of such arid-land basins

TABLE 24.1 Some vital statistics of the rivers of North America based on data presented in the 22 chapter summaries. The name, basin area, and discharge are given for the largest single river of each basin or region, which is the main stem when all rivers drain into a single basin. Median values are given for major attributes of the rivers of each region, but main-stem rivers were excluded so that median values are not influenced by the much higher main-stem values.

Major Basin or Region	Largest River			Median Values for Rivers of Region					
	Name	Discharge (m³/s)	Basin area (km²)	Discharge (m³/s)	Basin area (km²)	Basin relief (m)	Median annual runoff (cm)	Runoff/ precip (%)	No. of rivers
Lower Mississippi	Lower Mississippi	18,400	3,270,000	98	7,773	452	44	35	9
St. Lawrence	St. Lawrence	12,600	1,600,000	150	16,458	510	42	40	9
Mackenzie	Mackenzie	9,020	1,743,058	404	68,134	2,605	18	45	9
Ohio	Ohio	8,733	529,000	420	30,300	690	54	45	11
Columbia	Columbia	7,730	724,025	225	22,667	2,353	28	58	11
Yukon	Yukon	6,340	839,200	670	67,250	2,458	29	99	6
Fraser	Fraser	3,972	234,000	231	13,500	2,625	36	77	8
Upper Mississippi	Upper Mississippi	3,576	489,510	157	25,929	163	27	30	10
Mexico	Usumacinta/ Grijalva	2,678	112,150	65	47,124	3,000	10	9	10
Nelson Churchill	Nelson/Churchill	2,480	1,093,442	164	148,900	350	7	13	6
Missouri	Missouri	1,956	1,371,017	51	32,600	1,182	6	13	11
Gulf Coast SE States	Mobile	1,914	111,369	289	20,400	220	54	37	11
Pacific Canada	Kuskokwim	1,900	124,319	1,214	43,149	2,562	90	100	10
Atlantic Canada	Churchill	1,861	93,415	246	7,860	490	73	67	11
Arctic	Thelon/Kazan	1,380	239,332	408	40,363	1,464	26	86	6
Atlantic Coast NE States	Susquehanna	1,153	71,432	361	25,707	1,358	58	55	10
Southern Plains	Arkansas	1,004	414,910	68	20,230	714	23	23	11
Pacific Coast States	Sacramento	657	72,132	171	11,158	2,847	42	72	10
Colorado	Colorado	550	642,000	17	24,595	2,600	4	15	11
Atlantic Coast SE States	Santee	434	39,500	227	25,326	372	31	31	11
Gulf Coast SW States	Brazos	249	115,566	79	46,540	720	4	5	9
Great Basin	Bear	71	19,631	26	7,925	2,341	7	13	7

as the Missouri (6 cm/yr), the Colorado (4 cm/yr), the Great Basin (7 cm/yr), the Gulf Coast of the southwestern United States (4 cm/yr), and the Nelson–Churchill (7 cm/yr). Runoff values also identify rivers with high yields, typically draining areas of high precipitation and low evapotranspiration. The highest runoff regions are the Pacific Coast rivers of Canada and Alaska (90 cm/yr), and the Atlantic Coast rivers of Canada (73 cm/yr). At least one river of Mexico (Usumacinta–Grijalva), however, has an annual runoff of at least 70 cm/yr in spite of high evapo-

transpiration, having the highest annual precipitation of any basin in North America. Several rivers of the Pacific United States have high runoff, including the Rogue and Eel, but the median for rivers of that region is lower due to the inclusion of some very arid river basins in southern California.

Another advantage of converting discharge to runoff is that it can be compared directly to precipitation (both expressed as cm/yr) to provide a rough estimate of evapotranspiration, assuming no losses or gains to groundwater or interbasin transfers. For all rivers in this volume, the median annual precipitation and runoff were 84 and 31 cm, respectively, indicating that roughly 37% of precipitation becomes runoff, and the remainder is lost to evapotranspiration (or possibly groundwater). Our estimate of median precipitation is higher than values of 67 cm (Hornberger et al. 1998) and 76 cm (Shiklomanov 1993) reported for the continent as a whole (see Chapter 1). This might suggest that the river data from this book are biased toward regions where rivers are concentrated (i.e., in higher-precipitation areas). On the other hand, the median runoff of 31 cm is similar to reported continent-wide values of 29 cm (Hornberger et al. 1998) and 34 cm (Shiklomanov 1993).

Table 24.1 gives median estimates for the individual (excluding main stem) rivers of runoff as a percentage of precipitation, where low values usually indicate high ET. The percentage of precipitation that becomes runoff varies greatly among rivers and regions and is strongly dependent on both precipitation and temperature. Not surprisingly, northern rivers with their cold climates and low plant production have very high RO/PPT ratios, approaching 100% (values for Pacific Canada likely are inflated because precipitation data are available mainly for the lower basins whereas runoff reflects higher precipitation in the upper basins). In comparison, median annual precipitation for the rivers of Mexico included in this volume is approximately 100 cm, and runoff only 10 cm, indicating that many of these low-latitude basins are relatively wet but experience high evapotranspiration (although at least one, the Candelaria, loses much of its water through groundwater seepage to the sea).

The effect of temperature on percentage runoff is particularly clear from examination of rivers draining into the Atlantic Ocean, where mean annual precipitation for 32 river basins (Chapters 2, 3, and 21) falls within the relatively narrow range of 92 to 147 cm. In these eastern rivers, mean air temperature explains 72% of the variation in percentage runoff (Fig. 24.1). In the southeast, where mean air temperatures range from 15°C to 20°C, percentage runoff is near 30%. But in eastern Canada, where mean air temperatures are <6°C, percentage runoff is typically >60%. The influence of low precipitation on percentage runoff in regions of relatively high temperature can be illustrated for the Gulf Coast rivers of the southwestern United States. The westernmost Pecos and Rio Grande have <30 cm precipitation and ≤1% runoff, whereas the easternmost rivers (Sabine and Neches) have >125 cm of precipitation and 16%

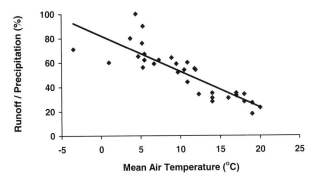

FIGURE 24.1 Annual runoff as a percentage of precipitation versus mean annual air temperature for rivers draining into the Atlantic Ocean (from Chapters 2, 3, and 21). $R^2 = 0.72$. This graph excludes the St. Lawrence River, which drains a much larger area, including the Great Lakes and their tributaries.

to 20% runoff. Of course, these estimates can be strongly influenced by human withdrawals, which also reduce runoff, and it is often impossible to separate this from natural evapotranspiration losses.

In addition to the annual water balance described here, seasonal patterns of PPT versus ET and the extent of intra- and interannual variability in both are primary determinants of a river's flow regime (Poff et al. 1997; Chapter 1). Mean and seasonal temperatures also are important, as they affect how much precipitation falls as snow and may be stored until spring thaws result in a rise in runoff. The extent of agriculture, types of crops, and reliance on irrigation drawn from surface waters (deep groundwater would not be part of the normal water budget) will further affect flows, principally during the growing season, and human withdrawals for other purposes, including municipal use, may have noticeable impacts.

In Chapter 1, several examples were given to illustrate the major factors affecting seasonal patterns of PPT and RO (see Fig. 1.4). Similar graphs of monthly PPT and RO presented throughout this book (with ET inferred from the difference) provide multiple examples that clearly distinguish the seasonal flow signatures of different regions. For example, seasonal runoff patterns of the southeastern Atlantic and Eastern Gulf drainages appear largely influenced by seasonal patterns of evapotranspiration. The northeastern Atlantic United States, the upper Mississippi, and the Ohio basin also are influenced by seasonal patterns of evapotranspiration, but in addition, they are affected to varying degrees by spring snowmelt, depending on their latitudes and monthly air temperatures. On the other hand, runoff patterns of northern rivers and those draining large western mountains are dominated by spring snowmelt. Along the Pacific Coast of the coterminous United States, however, the pattern of runoff is most obviously influenced by the distinct winter precipitation. In contrast, the runoff in most rivers of Mexico follows the strong pattern of summer precipitation.

Variation in Biological and Ecological Characteristics

The foregoing comparisons emphasize the variety of rivers in physical terms and landscape context, but what about biological and ecological variation among rivers? Chapter accounts amply document substantial diversity in the number of fish species among rivers and regions. Rivers of the Ohio basin have a reported median number of 120 fish species per individual river basin. Similarly, the median number of fish species in the southeastern Atlantic, the Gulf Coast of the southeastern United States, the upper Mississippi, and the lower Mississippi all exceed 100 per river. The Tennessee River and Mobile River each have a staggering >225 fish species. These Midwestern and southern basins support far more fish species per river basin than rivers of the west or far north. For example, the median number of native fish species reported by individual river basin was 23 for rivers of the Columbia basin, 18 for rivers of the Pacific United States, and 10 for rivers of the Colorado basin (although the entire basin of the Columbia and Colorado includes considerably more native fishes than the median for individual rivers). The Nelson–Churchill system of southern and central Canada supports a median of 62 fish species (the extensive lake habitat of the Nelson–Churchill likely contributes to this species count), and the rivers of the Yukon basin support 27 species. Physiographic and habitat variation, glacial history, and dispersal opportunities are some of the factors that underlie this enormous zoogeographic variation (Hocutt and Wiley 1986). In general, the number of fish species increases with area of drainage basin, is markedly higher in the east than the west, and is reduced in far northern rivers relative to the eastern United States. Mexican rivers have somewhat fewer species per river than the eastern United States, but they are not as well sampled. The fraction of endemic species is very high in Mexican rivers, however, and when scaled to account for differences in area by country rather than river basin, the number of Mexican fish species per km² exceeds both the United States and Canada (Chapter 23).

Biogeographic information for other taxa is scant. There was insufficient information available on a river-specific basis to describe patterns for any taxa other than fish. Molluscan diversity (approximately >300 species in North America; Master et al. 1998, Abell et al. 2000) is well known to be highest in the southeastern United States, which is globally rich in freshwater mollusks, less diverse in the west, and depauperate in the far north. This statement also applies to decapod crustaceans (nearly 300 species in North America). However, much more work is needed before species diversity trends can be identified for the abundant aquatic insects of rivers, which, based on a few well-studied systems, likely include several hundred species per river. Many studies of aquatic insects use genus-level taxonomy, species distributions are largely unknown, and large rivers are understudied. Certain genera, including *Baetis* and *Stenonema* (mayflies), *Hydropsyche* (caddisflies), *Simulium* (black flies), and *Polypedilum* (nonbiting midges) are widely reported in rivers throughout this book, indicating that species in these genera (and others) may often be important components of the majority of river communities.

Few rivers of order 6 and higher have received intensive investigation of ecosystem processes, and even description of the biota often is restricted largely to surveys of fishes, owing to the tendency of running-water ecologists to focus on smaller streams that are easier to study. In addition, our interpretation of current knowledge is complicated by the variable extent of anthropogenic disturbance. Relative to their presettlement state, many rivers have elevated turbidity and nutrients, reduced connectivity both laterally and longitudinally, and altered water budgets and water-residence times, and receive a complex brew of industrial, agricultural, and pharmaceutical chemicals.

Based on studies of ecosystem metabolism from a modest number of North American rivers, low ratios of gross primary production to respiration (P/R) appear to be common in large rivers. Among the best examples of low P/R ratios are careful studies in the Hudson, Ogeechee, and Ohio rivers (Chapters 2, 3, and 9, respectively). Of these three, the Ogeechee is arguably the least altered because it is unregulated by dams along its entire length and has a large intact floodplain. Given the tea-colored waters and extensive floodplain of this low-gradient blackwater river, which receives much allochthonous organic matter from the floodplain swamp and has relatively low primary production in the water column, the observed dominance of heterotrophic processes is expected. The carbon budget of the Hudson (Table 2.1), although also providing strong evidence of heterotrophy, nonetheless suggests a substantial role for autochthonous production, which perhaps was even greater prior to a number of human impacts.

Primary production within large rivers is expected to be limited by some combination of environmental factors, including light, nutrients, and downstream export of algal cells (Allan 1995). Light limitation is frequently reported. In the lower Missouri River (Chapter 10), reported photic depths averaged 0.78 m and mixed depth/photic depth ratios averaged 10.2. Because river water columns are well mixed and most of the water column experiences light levels too low to support algal growth, light limitation can be significant. Summer primary production by river phytoplankton for the middle and lower Ohio river sections also showed evidence of light limitation, although in this system large impoundments have notable effects by lengthening water residence time, and so phytoplankton production in river sections slowed by impoundments likely is higher than would otherwise be expected. The well-studied Hudson River (Chapter 2) illustrates several ways in which human actions may influence observed primary production. High turbidity due to silt resulting from intensive land use and reduced water residence times during summer due to human alteration of hydrology both reduce algal production. The introduced zebra mussel has had profound impacts, including through its suspension-feeding, and likely has contributed to a reduction in phytoplankton and an increase in macrophytes.

Although nutrients may seldom be limiting in the rivers just mentioned, northern rivers frequently have low background nutrient levels and therefore respond strongly to nutrient enrichment. The Wapiti River (Chapter 18), a tributary of the Smokey and Peace rivers, has low background nutrients but receives high point-source inputs of N and P from a pulp mill and a municipality. Periphyton biomass below inputs increased more than tenfold, resulting in higher macroinvertebrate densities and increased fish condition in a bottom-up trophic cascade.

Despite the dominance of ecosystem metabolism by heterotrophy, presumably reflecting the extent of inputs of allochthonous carbon, autochthonous primary production can be important to large-river food webs (Chapter 1; Thorpe and DeLong 2002). Some studies suggest that high system respiration likely is due to the metabolism of allochthonous carbon by microbes, and little of this carbon is thought to reach the metazoan food chain. Stable isotope and other evidence indicates that most metazoan production is derived from autochthonous production, often phytoplankton. Thus, the river is heterotrophic overall, with a P/R ratio well below 1, but the metazoan food webs may be primarily autotrophic. Reports for the Ohio River, where much of the work leading to this hypothesis was done, and for the Hudson River provide supporting evidence.

Snags and large wood play an important role in the ecology of large rivers, creating stable substrate for producers and invertebrates and cover for fishes (Benke et al. 1985). Snags and backwater areas support the most diverse and productive assemblages of invertebrates, as the fine substrate of main channels commonly is too unstable for invertebrates. Very high secondary production of invertebrates occurs on snags, and, exported as drift, contributes to fish biomass and production. Many large rivers had an abundance of snags in their presettlement condition, as exemplified by descriptions of the Cape Fear River in the early 1700s (Chapter 3). Clearing of snags, which was common in the 1800s and early 1900s, improved river navigation and provided access for steamboats but likely had substantial effects on ecosystem processes. Blockage of the main channel of the Red River in Louisiana with driftwood was so extensive that a major effort by the U.S. government, beginning in the 1870s, was undertaken to open the river to boat traffic (Chapter 7). Detailed studies of snag habitat in the Satilla and Ogeechee rivers (Chapter 3) reveal what has been minimized or lost in many other rivers. The main channel of the unchannelized Missouri River provides another example, where woody habitats were shown to contribute over two-thirds of total insect production compared with mud and sand substrates and backwaters (Chapter 10).

In summary, ecosystem processes have been studied for only a modest number of North American rivers, which strongly argues for the importance of further study. The lack of information on ecosystem-level processes for many of the large rivers of North America represents not only a significant frontier for further investigations but also a serious shortage of knowledge on which to base management decisions. Because these large rivers differ in their natural settings and extent of anthropogenic disturbance, the comparisons and generalizations offered here should be viewed with considerable caution, and much remains to be learned.

FEW RIVERS ARE PRISTINE

Many rivers of North America have been affected by human actions since at least European settlement and some were influenced much earlier, as described for ancient cultures in Mexico (Chapter 23). Over time the variety and magnitude of impacts has grown dramatically, although we should recognize that there have also been successes in river management. At the beginning of the twenty-first century we cannot accurately describe the condition of North America's rivers, because presently we lack effective systems of national assessment and large rivers are understudied. We do know that many of our rivers suffer from pollution, habitat degradation, fragmentation by dams, colonization by nonnative species, and more, and that climate change, water withdrawals, the continued spread of nonnative species, and the ongoing expansion of human activities will pose greater threats in the future.

Chemical contamination and water quality in rivers continue to be of major importance, reflecting the public's legitimate concern for drinking water and public health, as well as the ecological consequences of polluted waters. Although the past 30 or more years of regulatory activities focused on improvements in water quality have achieved notable successes, many rivers struggle to overcome a legacy of severe pollution, particularly from industrial pollution on the lower portions of rivers. This has been especially true in high-population areas of the Atlantic Coast rivers of the northeastern United States (Chapter 2). Water quality of rivers such as the Delaware, Hudson, and Connecticut have improved, but these rivers and others still have sediments heavily contaminated with heavy metals, PCBs, and other chemicals.

The Delaware River in the vicinity of Philadelphia reminds us that water-quality deterioration has a long history (Chapter 2). Significant pollution was reported as early as 1799; low values of dissolved oxygen, presumably due to the decomposition of organic waste, were reported in 1915; and dock workers along the river during World War II reportedly were nauseated by the stench. In the era after World War II, new, potentially highly harmful chemicals were added to the brew being discharged into rivers. The Connecticut River contains a substantial burden of contaminants, including trace metals (chromium, copper, lead, mercury, nickel, and zinc) and organic compounds (chlordane, DDT, polychlorinated biphenyls, and various polycyclic aromatic hydrocarbons). Fish and shellfish advisories have been issued in Massachusetts and Connecticut because of high levels of polychlorinated biphenyls. PCB contamination in the middle Hudson likewise is responsible for restrictions on commercial and sport fisheries and an ongoing battle over the methods and financial responsibilities for their removal (Chapter 2).

Other rivers have been greatly affected by the timber industry due to pulp mill wastes and associated factors. The Saguenay, a very large tributary of the St. Lawrence (Chapter 21), has seen considerable pollution from the timber industry and industrial pollution associated with paper manufacturing (mercury, PCBs, and PAHs) despite its relative remoteness, low population density, and sparse agricultural activity in its watershed.

Consolidated animal feeding operations on an unprecedented scale add a new dimension to chemical contamination of large rivers. The numerous industrial hog and poultry operations located in and near the floodplain of the Cape Fear River (Chapter 3), a region prone to hurricanes and extreme flooding, illustrate the risk that river ecosystems will receive high volumes of organic wastes.

Nutrient enrichment is widespread in most major river systems that drain landscapes of significant human population and disturbed land use. Heavily agricultural rivers such as the Platte (>90% agriculture, Chapter 10), the Great Miami (80% agriculture, Chapter 9), and the Minnesota (>95% agriculture, Chapter 8) typically have high concentrations of nutrients (phosphates and nitrates) and pesticides. For example, the Minnesota River, as well as other rivers of the Upper Mississippi basin, typically has concentrations of NO_3-N and PO_3-P approaching or exceeding 3 mg/L and 0.1 mg/L, respectively, each of which is substantially higher than natural background levels (Chapter 1). Although the impact of elevated nitrogen levels on primary production and the resultant hypoxia within the Gulf of Mexico is now well established (Chapter 6), the potential effect of nutrient enrichment on large-river food webs is not well studied and may be masked by other factors, including turbidity and altered water residence times. However, in some northern rivers that have experienced little disturbance and have very low nutrient concentrations the effects of elevated nutrients can be very pronounced, as in the case of the Wapiti River.

Dams and impoundments, levees, and channelization have altered the physical dimensions and dynamics of many kilometers of large rivers. Even before the twentieth-century era of large-scale river

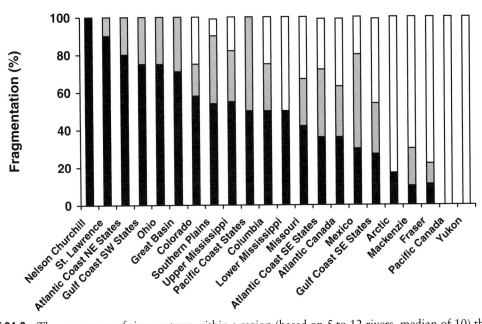

FIGURE 24.2   The percentage of river systems within a region (based on 5 to 12 rivers, median of 10) that were recorded as strongly fragmented, with multiple or large dams along their main stems (black bar); moderately fragmented (gray bar); or essentially unfragmented (unshaded bar).

engineering, North America's major rivers were altered for transport. An extensive canal system was dug in the seventeenth century to provide access and transport of goods into the Delaware valley (Chapter 2). Broad floodplains of backwaters and wetlands were transformed by the simultaneous activities of narrowing and deepening the main channel, snag removal, and drainage of adjacent bottomlands for farming. The considerable loss of channel complexity and lateral connectivity in the upper freshwater tidal Hudson, where deepening of the channel and filling of wetlands and backwaters resulted in a narrow and simplified river system, is illustrated in Figure 2.11. In the mid 1800s, less than half a century after Lewis and Clark ascended the Missouri, similar channel modification processes began, followed in the early 1900s by the construction of a system of wooden pile dikes to create a single, self-scouring navigation channel. Then came flood-control levees by the mid-twentieth century, and a series of major dams. As Chapter 10 reports, flooding historically was essential in maintaining the natural character of the river–floodplain complex. Subsequent to this chain of channel-modifying events the channelized section of the Missouri River underwent major geomorphic changes, including an 8% reduction in channel length, a 50% reduction in channel water surface area, a 98% reduction in

island area, and an 89% reduction in the number of islands (see Fig. 10.2).

Using information on size, number, and distribution of dams in main-stem rivers we have attempted to place rivers into one of three categories: rivers with few or no dams along main stems or on tributaries; rivers moderately fragmented, with at least one sizable main-stem dam; and rivers strongly fragmented, with multiple or large dams along the main stem. Figure 24.2 shows the percentage of river systems within a basin or region that were recorded in each category. Over half of the river basins or regions of North America and 10 of 13 basins or regions primarily located in the United States had at least 50% of their main rivers scored as strongly fragmented. Of the eight primarily Canadian basins and regions, the Nelson–Churchill and the St. Lawrence are highly impacted by dams, as are rivers of the Atlantic Canada region. The remaining Canadian basins and regions experience low dam fragmentation overall, although this conceals some specific rivers highly influenced by dams, including the Peace River in the Mackenzie basin and the Nechako River in the Fraser basin. Although the most fragmented Mexican rivers are found in the arid north, less fragmented rivers in the tropical south, particularly the large and free-flowing Usumacinta, are endangered due to proposed hydropower projects.

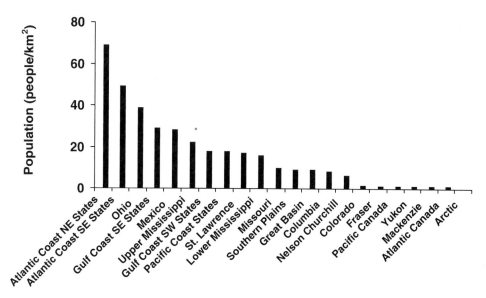

FIGURE 24.3 Median values of human population density (people/km^2) for regions and river basins. In some instances within-basin variation varied widely; for instance, the Fraser basin has a density of 11.7/km^2 for the Fraser itself, compared to <1/km^2 in most of its tributary river basins. Densities for individual Atlantic Coast rivers of Canada show a similarly wide range of values.

The extent of dams in North America has been well-documented by previous publications (Dynesius and Nilsson 1994, Graf 1999), and there is a large literature on their consequences (Ward and Stanford 1979, Hart and Poff 2002, Stanley and Doyle 2003). The imposition of lentic conditions can eliminate benthic species, including many freshwater mussels, and generally promotes a shift to plankton-based food webs. With nine main-stem reservoirs, very little of the Tennessee River main stem remains free flowing (Chapter 9), the density and diversity of the main-stem mussel populations have declined dramatically, and gizzard shad and threadfin shad, which are dependent on the abundant plankton in the system, have become the principle forage fishes. The influence of dams on any given river can vary greatly, but alterations to the river's natural flow regime are widespread. In some cases, regulation by a dam largely flattens the monthly hydrograph (e.g., Churchill Falls dam on the Churchill River, Labrador; Hoover Dam on the Colorado River). In many other cases, hydropower facilities cause enormous hourly and daily fluctuations in discharge and water height that are hidden by plots of mean monthly discharge or runoff values. Perhaps the most extreme example is on the Nelson River (Chapter 19), where dams can cause hourly summer discharge to range between 0 and 7200 m^3/s. In other cases, dams and diversions can reduce downstream flows to zero for very long periods (e.g., Rio Grande, Chapter 5; Gila River, Chapter 11; Rio Conchos, Chapter 23). By reducing the magnitude and frequency of flood events, a large reservoir on the Peace River, a major tributary of the Slave River, is causing the Slave River delta to shrink (Chapter 18). Loss of wetland habitat in the Slave River Delta and in many such wetlands throughout North America has devastating impacts on birds and other wildlife.

Human population density can be a useful proxy measure of human impacts, including nutrient loading (Cole et al. 1993). It may also serve as a very rough proxy for land-use change and urbanization, although the current phenomena of urban sprawl and depopulation of agricultural areas complicate this relationship significantly (Meyer and Turner 1994). Although human density estimates reported in this book may be approximate, they nonetheless reflect very substantial regional differences. River basins in the Atlantic states of the northeastern United States support much higher population densities than any others in North America (Fig. 24.3). Within the coterminous 48 states the Missouri and Colorado basins and river basins of the southern plains are least densely settled. The St. Lawrence and Nelson–Churchill basins support human populations comparable to less dense U.S. basins, whereas more northern river basins have comparatively very low populations.

The extent of urban and agricultural land within river basins provides another indication of human impact. It is difficult to assess the influence of between-basin differences in the extent of urban land, because urban land use commonly is a low percentage of total area yet exerts a disproportionately large influence both proximately and over distance (Paul and Meyer 2001). The river basins with the greatest percentage of urban land include Atlantic Coast rivers of the northeastern United States, Gulf Coast rivers of the southwestern United States, the St. Lawrence, and the Colorado. Agricultural land use varied considerably among basins, from near zero in some Canadian basins to 66% of the Upper Mississippi basin. Six major river basin of the United States had over 40% of their areas in agriculture: the Lower Mississippi, Upper Mississippi, Southern Plains, Ohio, Missouri, and Colorado.

The combined data on dams, population, and land use argue that all river basins of the coterminous United States experience considerable human disturbance. However, all or nearly all of these basins contain tributaries and river segments of very high quality that are suitable for designation under federal and state river protection programs. Indeed, these data, and the insightful analyses of individual chapters, clearly demonstrate the urgency and importance of protecting least-disturbed river segments whenever possible. Major river basins of the far north include many that appear little changed according to the measures reported here (which do not, however, include mineral extraction, timber harvest, or road and recreational development, which are significant threats to boreal regions; Schindler 1998), offering the potential for protection of even larger and more intact watersheds and river systems.

Nonnative species pose another, less-visible threat to river ecosystems, and in a number of instances the native fauna is largely displaced, at least in terms of biomass, energy flow, and community structure. This is especially so in western rivers, where the number of nonnative species can equal or exceed the total species richness of native fishes (Fig. 24.4). Commonly in such situations, native species persist at reduced abundances and in isolated locales, and often warrant endangered species status. From the viewpoint of species richness, the fish assemblage may appear little changed in the number of native species but be greatly augmented with nonnative species. From the perspective of the biological community, however, biomass and energy flow may be totally dominated by the nonnative component (e.g., a survey of the Colorado River in Canyonlands

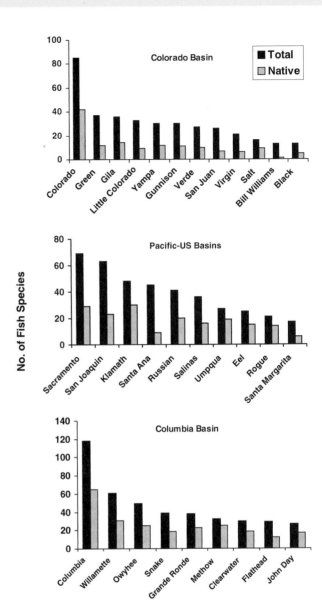

FIGURE 24.4 Native fish species and total fish species in three river basins or regions of the western United States (from Chapters 11, 12 and 13). Note that the limited native fish fauna has been substantially augmented by nonnative species.

National Park, Utah, by Valdez and Williams [1986], reported 83% percent of individuals to be nonnative), and so ecologically the system is even more altered than might be inferred from consideration of a species list.

Based on information presented in river summaries, the median number of native fish species for the 11 river basins of the Colorado system exclusive of the main stem is 10 (Chapter 11), and a median of 4 species are listed as endangered. Today the median number of all fish species for these 11 basins is 29, and so the addition of nonnatives has roughly tripled the species count, whereas much of the native fauna is in need of protected status. The 10 river basins of the Pacific Coast rivers of the United States contain 18 native species and a total of 39 fish species (median values). Within the Columbia system, based on eight rivers (excluding the main stem), chapter data indicate a median number of 23 native fishes and a total of 38 species.

The influence of nonnative fishes may be less dramatic elsewhere compared to the western United States, perhaps because high native species richness of some regions has helped to limit invasions and perhaps in far north river systems because fewer nonnative species are preadapted to these environments, and there have been fewer invasion opportunities. Nonetheless, the influence of nonnative fishes has been profound in many other rivers. For example, in several rivers of the northeastern (Chapter 2; Connecticut, Hudson, Delaware, and Susquehanna rivers) and southeastern United States (Chapter 3; James River), approximately one-third of all species are nonnatives, raising the total number of species to >100 in many cases. Nonnatives sometimes dominate fish collections in these rivers (e.g., Chapter 3; Cape Fear River, North Carolina), with unknown impacts on ecosystem function.

Finally, a changing climate driven by greenhouse gasses is likely to have profound impacts on freshwater ecosystems, altering hydrologic cycles, flow regimes, and riparian vegetation and likely facilitating range expansions and the proliferation of nonnative species. These climate-related effects are as yet poorly understood and so pose one of the most important emerging threats to rivers in coming decades.

NORTH AMERICA'S RIVERS IN THE TWENTY-FIRST CENTURY

During the twenty-first century the rivers of North America will experience growing human pressures. Water use will increase due to the growth of affluence and population, and in many areas surface-water supplies are nearly fully appropriated (NRC 2001). Reliance on groundwater will increase, with uncertain but serious consequences for river flows (Postel and Richter 2003). Some areas of the western United States are expected to exhaust their groundwater sources during this century. Surely some will call for transbasin water diversions on a scale

presently unknown in North America, potentially altering river flows and facilitating the spread of nonnative species. Conflicts between human and environmental needs for water are certain to increase. Meanwhile, all of the known threats remain, although some, such as point-source pollution, are diminished in many areas, whereas climate change adds a new challenge, likely to interact with other threats in complex ways. Opportunities exist to halt the decline in the condition of the rivers of North America and even to improve their status. However, those undertakings call for better knowledge to guide management and restoration and a stronger resolve that consumption of water resources cannot be dictated by uncontrolled growth.

Climate Change

Rivers and other freshwater ecosystems will be affected by projected climate change in multiple direct and indirect ways, of which only the most straightforward can be identified with high confidence (Firth and Fisher 1992, Meyer et al. 1999, Allan et al. 2005). A warmer climate will result in greater evaporation from water surfaces and greater transpiration by plants. However, whether rainfall will increase or decrease in a particular region is uncertain, and it also is difficult to predict whether the change in PPT or ET will be greater. Thus, increased precipitation might be accompanied by even greater evapotranspiration, leading to reduced runoff. General circulation models (GCMs) are not yet able to reliably predict how precipitation and water supplies will change at the local or regional levels (NAST 2000), and there is still a great deal of uncertainty in climate-change forecasts (Forest et al. 2002, Elzen and Schaeffer 2002). For example, Frederick and Gleick (1999) examined runoff for 18 water-resource regions of the United States using two contrasting GCMs and found that predictions often were in disagreement. The two models predicted the same direction of change in runoff in only 9 of the 18 regions, and where the direction was similar often the magnitude was not. Perhaps the greatest single challenge in evaluating aquatic ecosystem response to future climate change is the considerable uncertainty regarding the local and regional responses of the hydrologic cycle.

Despite this uncertainty, the potential for climate change to alter total runoff and the seasonality of flow regimes is considerable. Some seasonal shifts are certain to take place, because warming winter temperatures will cause some areas to experience more winter rain and episodes of rain on snow, transforming predictable spring melt runoff into highly variable winter runoff. Rivers of the far north may be most vulnerable because they will experience disproportionately greater warming, along with associated dislocations of the hydrologic cycle (Poff et al. 2001); for example, the Mackenzie basin may warm by as much as 5°C by the middle of the century (Chapter 18).

In addition to affecting the hydrologic cycle, warmer temperatures are expected to result in higher ecosystem metabolism and productivity. The biota of rivers are dominated by cold-blooded organisms, and these ectotherms generally increase their metabolism with each degree increase in temperature until very near their upper temperature tolerances. Rates generally increase by a factor of 2 to 4 with each 10°C increase in water temperature, up to about 30°C (Regier et al. 1990). In a review of over 1000 estimates of macroinvertebrate production, Benke (1993) estimated a 3% to 30% increase in biomass turnover rates for each 1°C increase in temperature. Thus, although there may be complex and unpredictable changes in species composition, an overall increase in system productivity is a potential response to climate warming.

Poleward range shifts by those taxa able to disperse are highly probable under warmer climate scenarios. Sweeney et al. (1992) estimated that a 4°C warming would result in a 640 km northward latitudinal shift in thermal regimes for macroinvertebrates, and several authors have estimated comparable distances for the range displacement of particular species of fishes. The opportunity to disperse, and the presence of corridors versus barriers, can be highly variable among river basins, however. For example, fishes in the southern Great Plains and the desert Southwest cannot move northward because those streams and rivers tend to run west and east. Because summer water temperatures now approach the upper limit for a number of fish species, just a few degrees of warming poses serious risk of extinction for native fishes in these regions (Chapter 7; Covich et al. 1997, Matthews and Zimmerman 1990). Fishes of the far north may be especially vulnerable, as warming will favor species that presently are only marginally successful. Arctic coregonids (broad whitefish, least cisco, Arctic cisco) may be particularly at risk. Temperature also sets the northern range limit for harmful nonnative species such as the zebra mussel (Strayer 1991), and so a northward range expansion seems highly probable. Given the well-established negative impacts of nonnative species on freshwater ecosystems (Allan and Flecker 1993), native biodiversity may be adversely affected by such range shifts.

Rivers draining forested landscapes and with forested riparian and headwater zones derive much of their energy as organic matter inputs from the terrestrial environment (Chapter 1). Climate change is likely to bring about shifts in terrestrial vegetation and changes in leaf chemistry and affect the processing of detritus and functioning of the microbial-shredder food web linkage in complex ways. Altered carbon/nitrogen ratios of the leaves likely will reduce palatability, temperature changes will affect leaf-processing rates, and floods may export leaf matter before it can be processed (Rier and Tuchman 2002, Allan et al. 2005). These interactions are complex and potentially offsetting, making the overall impact of climate on this important energy supply difficult to predict.

Other Threats

Although the potential impact of a changing climate deservedly is the focus of much current concern, more familiar threats may prove to be of equal or greater importance. Pollution from point and nonpoint sources continues, and legacies of past contamination are numerous. Rivers continue to be fragmented by dams, which modify flow, alter water budgets, disrupt migrating species, and interfere with ecosystem connectivity and function. Levees and other flood-control measures remain largely in place. The spread of nonnative species is virtually a certainty. Although heartening examples of improvements can be found and some of the most harmful pollution practices have largely been halted, many rivers experience a mix of stressors that include legacy effects, long-standing threats, and newly emerging challenges.

Water withdrawals and transfers comprise another potential threat to rivers but are associated with so much uncertainty they are difficult to assess. Due to the combined influence of population and economic growth, freshwater demand is expected to grow significantly during the twenty-first century (NRC 2001). On the other hand, past projections of demand have invariably proven to be overestimates due to unanticipated advances in efficiency. Nonetheless, because surface waters are largely appropriated in many regions, it seems highly likely that groundwater withdrawals will increase, with uncertain effects on hydrologic budgets and river ecosystems. Some areas of the Great Plains and the southwestern United States that now rely heavily on nonrenewable groundwater will almost assuredly desire to import water via canals and interbasin transfers.

Examples of existing out-of-basin transfers are described in several chapters. The Sacramento River in northern California (Chapter 12) has approximately 50% of its flow diverted via the California Aqueduct and Delta–Mendota Canal to meet urban and agricultural water demands in the southern part of the state. The Colorado River experiences transfers from its headwaters via the Colorado–Big Thompson project to provide water to Denver and other cities of the Mississippi drainage, whereas a diversion from the Lower Colorado below Lake Havasu to Phoenix and California sends almost 40% of the virgin river flow out of basin (Chapter 11). Water is siphoned off from three reservoirs in the Delaware River headwaters as part of the water supply of New York City and discharged into the Hudson estuary (Chapter 2). A massive diversion of water occurs from the Churchill River in northern Manitoba in order to increase hydroelectric generating capacity through dams in the nearby Nelson River. Some 75% (or >700 m^3/s) of the flow of the Churchill River was diverted into the Nelson system during the 1970s, causing severe ecological and social impacts on both the lower Churchill and the lower Nelson, where hydroelectric releases also have extreme daily fluctuations (Chapter 19). Large diversions are a common practice in Canada, where Dynesius and Nilsson (1994) report total out-of-basin transfers of 4400 m^3/s. Based on various news media accounts, interbasin water transfers have been considered from the Great Lakes to surrounding areas, between the Upper Missouri and southern tributaries of the Hudson Bay drainage via the proposed Garrison River diversion, and from the Pacific Northwest to California. Although no transfer of such magnitude has yet taken place, pressure may intensify as water shortages become more severe.

Restoration

We should aspire to make the twenty-first century a period of restoration and repair of damaged ecosystems, including rivers, for many reasons. First, we are increasingly aware of the extent of environmental degradation and have the capacity to make improvements. Second, society increasingly values biological diversity, the variability of nature, and the psychological and social benefits we derive from natural surroundings. Third, all of us benefit from a number of ecosystem goods and services, including clean water, recreation, and harvestable fishes. Healthy rivers are

of direct value. Sometimes restoration will be costly, such as in the greater Florida Everglades ecosystem, where large sums were spent to channelize the Kissimmee River and even larger sums are currently being spent to restore it. This is a call to learn from past mistakes. In other instances restoration will be cost effective, as with the purchase of riparian land to preserve water quality in the Catskill and Delaware watersheds supplying water to New York City (www.nyc.gov/html/dep/html/fadplan.html 2002), a far cheaper solution than the alternative, building treatment plants to remove excess sediments. This is a call to recognize the efficiency of sound environmental policy.

Restoration has been defined as "returning a system to a close approximation of its condition prior to disturbance, with both the structure and function of the system recreated" (NRC 1992). However, it may not be possible to know the predisturbance condition, and it may not be practical to achieve it. Various authors and scientific working groups have wrestled with the definition of restoration and the setting of goals and targets (Bradshaw 1988, Lake 2001, www.ser.org/content/ecological_restoration_primer.asp 2004). In our opinion, although we should make every effort to be aware of the historical state of the system, restoring systems to presettlement condition will very rarely be practical. In practice, restoration today seems to encompass everything we might do to repair or improve an ecosystem. Using reference sites and historical information when available, and incorporating an understanding of the dynamic processes that govern rivers and their biota are key ingredients of thoughtful restoration practice. Recognizing the importance of stakeholders, it may be sensible to consider a range of options, with public input determining the desired outcome (Hobbs and Harris 2001).

A great deal of river restoration presently takes place under the auspices of federal, state, and local governments, as well as by private citizens. Unfortunately, we do not have very good information concerning the extent and types of restoration practices or their effectiveness, and little or no accountability for the large sums being spent (Bernhardt et al. 2003). In fairness, river restoration is a relatively new area, and recent program initiatives such as CALFED (a partnership of California and federal governments, which is investing substantially in San Francisco Bay–area restoration) are requiring external scientific review, explicit hypotheses, and other key elements that should help to ensure sound science and the application of adaptive management.

The type of river restoration depends greatly on location, prioritization of threats, and how the river is valued. In addition, there may be several ways to address a particular problem. For example, in agricultural regions and wherever bank erosion is serious, maintenance of a vegetated buffer zone along river margins, which may be accompanied by planting trees, is common practice. Gabions (wire baskets of rocks) and rip-rap (individual large stones or concrete blocks, even old automobiles, referred to as "Detroit rip-rap") are less attractive means to stabilize banks. In some cases, recognizing that river channels naturally migrate, it may be appropriate to do nothing at all. Clearly, identifying the problem, understanding the fundamental processes at work, and clarifying the desired outcome are all important in arriving at the optimal solution (Palmer et al. 2005).

Flow restoration is increasingly becoming an objective of water managers, particularly those responsible for dam releases and water withdrawals. How much water a river needs is being reevaluated in light of the scientific consensus that rivers (and most ecosystems) are naturally variable and depend upon that natural variability for full geomorphic, hydrologic, and ecological function (Richter et al. 1997, Poff et al. 1997). River channels are shaped largely by the interplay among slope, sediment supply, and water supply and maintained mainly by bankfull flood events; if human actions change one of these variables, adjustments will occur in one of the others. For example, regulation of the Colorado River below Glen Canyon Dam after 1963 restricted the supply of sediments, and in response the river cut downward, lowering its slope. How much water is needed for fish populations may be difficult to determine, or perhaps society has not yet agreed to include a margin of error in its estimates. For example, in an extreme conflict between human use (for agriculture) and ecosystem use (for fishes), water-supply managers for the Klamath lake and river system experienced massive protests from farmers whose crops were threatened, and then only months later oversaw the largest die-off of Pacific salmon yet recorded, in a case marked by conflicting reports of investigating committees (Service 2003). This may indicate a need for more and better science to guide decisions and more documentation of successes and failures in order to build the case-based knowledge necessary for management of conflicts with such potentially high costs on both sides (Poff et al. 2003).

Few examples of river restoration are reported in the chapters of this book, and this likely can be taken as rough evidence that few well-documented cases can be found. This may reflect the fact that river restoration is a new agenda. In addition, restoration of large rivers is expensive and often encounters real conflicts of interest and financial constraints. In several instances, including the Columbia, Missouri, and Willamette, recovery and restoration plans date from 2000 or later and are not yet implemented. Recovery plans for anadromous fishes, including salmon and shad, typically are based on hatcheries and fish passage around dams and occasionally on dam breaching, and are perhaps most frequently cited. Examples include largely successful efforts to restore shad to the Connecticut River and largely unsuccessful efforts to restore salmon runs on the Columbia River.

Flow restoration through modifications of dam operations is at least not uncommon as a restoration tactic, although costs of lost hydropower and other commercial benefits pose a constant challenge to such efforts. In some instances, including sections of the Missouri, the Flaming Gorge Dam on the Green River of the Colorado drainage, and the Yakima River, which enters the only free-flowing section of the Columbia and is where most natural salmon reproduction occurs, biological opinion is clear, but action has yet to be taken.

Recovery proposals recommended by the National Research Council show that current science can guide actions to improve ecological services, as well as societal benefits, through a combination of acquiring floodplain lands, increasing main-channel habitat complexity, and modifying reservoir water management to restore more natural river flows (National Research Council 2002) (Chapter 10). However, it is unfortunately worth noting that efforts to resolve environmental and human uses of the Missouri River have been in political and legal gridlock for years and continue to be in controversy as of this writing.

If there is a lesson to be drawn from these few restoration examples from large rivers, it may be that the effort is at an early stage, and the extent of societal valuation of river ecosystems for their natural values and their ecosystem services remains to be determined.

In conclusion, whether North America's rivers will be in better condition at the end of the twenty-first century than at the beginning is uncertain. But they can be. Many rivers have improved due to a reduction of careless resource extraction followed by sufficient time for some natural recuperation. Additional improvement has resulted from reductions in point-source pollution and other targeted actions, including habitat improvement, dam removals, and active river restoration. Moreover, rivers have significant restorative capacity: The flow of water and movement of sediments can cleanse pollutants, and periodic floods can restore dynamism to river channels and allow natural processes to dominate. Some recovery has occurred passively and some due to sound management.

But there is discouraging news as well. At present, some good restoration plans are struggling to be implemented due to conflicting pressures and financial constraints. Water demand will increase with the growth of population and affluence, climate warming and the spread of nonnative species may pose even greater threats over the present century, and all of the familiar threats associated with human activities remain with us. We find ourselves between, on the one hand, improved knowledge, stakeholder support, and the capacity to manage and restore, and on the other hand, an array of familiar and unfamiliar threats. Although improvements in the science and practice of river restoration clearly have much to contribute, they matter little unless major strides are made to grow awareness, political will, and the policies and institutions necessary to advocate sound river management.

In this light, the growing influence of citizen groups and nongovernmental organizations concerned with river health is of great importance. By joining the dialogue on the future of our rivers and contributing scientific expertise where it is appropriate, whether through giving public lectures, serving on panels, sharing scientific knowledge with river advocates, or in any other way, scientists as individuals can help to see that knowledge is used. It will be rare (and probably should be) for decisions affecting river health to be made by a few specialists. In the long run, involvement of all sectors of society likely holds the greatest promise for the future condition of the river ecosystems so ably discussed in this volume.

LITERATURE CITED

Abell, R. A., D. M. Olson, E. Dinerstein, P. T. Hurley, J. T. Diggs, W. Eichbaum, S. Walters, W. Wettengel, T. Allnutt, C. J. Loucks, and P. Hedao. 2000. *Freshwater ecoregions of North America: A conservation assessment*. Island Press, Washington, D.C.

Allan, J. D. 1995. *Stream ecology: Structure and function of running waters*. Kluwer Academic, The Netherlands.

Allan, J. D., and A. S. Flecker. 1993. Biodiversity conservation in running waters. *BioScience* 43:32–43.

Allan, J. D., M. A. Palmer, and N. L. Poff. 2005. Freshwater ecology. In T. E. Lovejoy and L. Hannah (eds.). *Climate change and biodiversity*, pp. 272–288. Yale University Press, New Haven, Connecticut.

Benke, A. C. 1993. Concepts and patterns of invertebrate production in running waters. *Verhandlungen der Internationalen Vereinigung für Theoretische und Angewandte Limnologie* 25:15–38.

Benke, A. C., R. L. Henry III, D. M. Gillespie, and R. J. Hunter. 1985. Importance of snag habitat for animal production in southeastern streams. *Fisheries* 10:8–13.

Bradshaw, A. D. 1988. Restoration: An acid test for ecology. In W. R. Jordan III, M. E. Gilpin, and J. D. Aber (eds.). *Restoration ecology: A synthetic approach to ecological research*, pp. 23–29. Cambridge University Press, Cambridge.

Cole, J. J., B. L. Peierls, and N. F. Caraco. 1993. Nitrogen loading of rivers as a human-driven process. In M. J. McDonnell and S. T. A. Pickett (eds.). *Humans as components of ecosystems: The ecology of subtle human effects and populated areas*, pp. 141–157. Springer-Verlag, Berlin.

Covich, A. P., S. C. Fritz, P. J. Lamb, R. D. Marzolf, W. J. Matthews, K. A. Poiani, E. E. Prepas, M. B. Richman, and T. C. Winter. 1997. Potential effects of climate change on aquatic ecosystems of the Great Plains of North America. *Hydrological Process* 11:993–1021.

Dynesius, M., and C. Nilsson. 1994. Fragmentation and flow regulation of river systems in the northern third of the world. *Science* 266:753–762.

Elzen, D. M., and M. Schaeffer. 2002. Responsibility for past and future global warming: Uncertainties in attributing anthropogenic climate change. *Climatic Change* 54 (1–2): 29–73.

Firth, P., and S. G. Fisher (eds.). 1992. *Global climate change and freshwater ecosystems*. Springer-Verlag, New York.

Forest, C. E., P. H. Stone, A. P. Sokolov, M. R. Allen, and M. D. Webster. 2002. Quantifying uncertainties in climate system properties with the use of recent climate observations. *Science* 295:113–117.

Frederick, K. D., and P. H. Gleick. 1999. *Water and global climate change: Potential impacts on U.S. water resources*. Pew Center on Global Climate Change, Arlington, Virginia.

Graf, W. L. 1999. Dam nation: A geographical census of American dams and their large-scale hydrologic impacts. *Water Resources Research* 35:1305–1311.

Hart, D. D., and N. L. Poff. 2002. A special section on dam removal and river restoration. *BioScience* 52:653–655.

Hobbs, R. J., and J. A. Harris. 2001. Restoration ecology: Repairing the earth's ecosystems in the new millennium. *Restoration Ecology* 9:239–246.

Hocutt, C. H., and E. O. Wiley (eds.). 1986. *Zoogeography of North American freshwater fishes.* Wiley Interscience, New York.

Hornberger, G. M., J. P. Raffensperger, P. L. Wiberg, and K. N. Eshleman. 1998. *Elements of physical hydrology.* Johns Hopkins University Press, Baltimore.

Lake, P. S. 2001. On the maturing of restoration: Linking ecological research and restoration. *Ecological Management & Restoration* 2:110–115.

Leopold, L. B. 1994. *A view of the river.* Harvard University Press, Cambridge.

Master, L. L., S. R. Flack, and B. A. Stein (eds.). 1998. *Rivers of life: Critical watersheds for protecting freshwater biodiversity.* The Nature Conservancy, Arlington, Virginia.

Matthews, W. J., and E. G. Zimmerman. 1990. Potential effects of global warming on native fishes of the southern Great Plains and the Southwest. *Fisheries* 15:26–32.

Meyer, J. L., M. J. Sale, P. J. Mulholland, and N. L. Poff. 1999. Impacts of climate change on aquatic ecosystem functioning and health. *Journal of the American Water Resources Association* 35:1373–1386.

Meyer, W. B., and B. L. Turner (eds.). 1994. *Changes in land use and land cover: A global perspective.* Cambridge University Press, New York.

National Assessment Synthesis Team (NAST). 2000. *Climate change impacts on the United States: The potential consequences of climate variability and change.* U.S. Global Change Research Program, Washington, D.C.

National Research Council (NRC). 1992. *Restoration of aquatic systems: Science, technology, and public policy.* National Academy Press, Washington, D.C.

National Research Council (NRC). 2001. *Envisioning the agenda for water resources research in the twenty-first century.* National Academy Press, Washington, D.C.

National Research Council. 2002. The Missouri River ecosystem: Exploring the prospects for recovery. Water Science and Technology Board, National Research Council, Washington, D. C.

Palmer, M. A., D. D. Hart, J. D. Allan, E. Bernhardt, and the National Riverine Restoration Science Synthesis Working Group. 2003. Bridging engineering, ecological and geomorphic science to enhance river restoration: Local and national efforts. World Water and Environmental Resources Congress 2003 and Related Symposia. Proceedings of the World Water and Environmental Resources Congress, Philadelphia, Pennsylvania, June 24–26, 2003.

Palmer, M. A., E. S. Bernhardt, J. D. Allan, and the National River Restoration Working Group. 2005. Standards for ecologically successful river restoration. *Journal of Applied Ecology* 42:208–217.

Paul, M. J., and J. L. Meyer. 2001. Streams in the urban landscape. *Annual Reviews of Ecology & Systematics* 32:333–366.

Poff, N. L., J. D. Allan, M. B. Bain, J. R. Karr, K. L. Prestegaard, B. D. Richter, R. E. Sparks, and J. C. Stromberg. 1997. The natural flow regime: A paradigm for river conservation and restoration. *BioScience* 47:769–784.

Poff, N. L., J. D. Allan, D. D. Hart, M. A. Palmer, B. D. Richter, J. L. Meyer, and J. A. Stanford. 2003. River flows and water wars: Emerging science for environmental decision-making. *Frontiers in Ecology and the Environment* 1:298–306.

Poff, N. L., P. A. Angermeier, S. D. Cooper, P. S. Lake, K. D. Fausch, K. O. Winemiller, L. A. K. Mertes, M. W. Oswood, J. Reynolds, and F. J. Rahel. 2001. Fish diversity in streams and rivers. In F. S. Chapin III, O. E. Sala, and E. Huber-Sannwald (eds.). *Global biodiversity in a changing environment: Scenarios for the 21st Century*, pp. 315–350. Springer-Verlag, New York.

Postel, S., and B. Richter. 2003. *Rivers for life: Managing water for people and nature.* Island Press, Washington, D.C.

Regier, H. A., J. A. Holmes, and D. Pauly. 1990. Influence of temperature change on aquatic ecosystems: An interpretation of empirical data. *Transactions of the American Fisheries Society* 119:374–389.

Richter B. D., J. V. Baumgartner, R. Wigington, and D. P. Braun. 1997. How much water does a river need? *Freshwater Biology* 37:231–249.

Rier, S. T., and N. C. Tuchman. 2002. Elevated-CO_2-induced changes in the chemistry of quaking aspen (*Populus tremuloides* Michaux) leaf litter: Subsequent mass loss and microbial response in a stream ecosystem. *Journal of the North American Benthological Society* 21:16–27.

Schindler, D. W. 1998. A dim future for boreal waters and landscapes. *BioScience* 48:157–164.

Service, R. F. 2003. NRC backs ecosystem-wide changes to save Klamath fish. *Science* 302:765.

Shiklomanov, I. A. 1993. World water resources. In P. H. Gleick (ed.), *Water in crisis: A guide to the world's fresh water resources*, pp. 13–24. Oxford University Press, New York.

Stanley, E. H., and M. W. Doyle. 2003. Trading off: The ecological effects of dam removal. *Frontiers in Ecology and the Environment* 1:15–22.

Strayer, D. L. 1991. Projected distribution of the zebra mussel, *Dreissena polymorpha*, in North America. *Canadian Journal of Fisheries and Aquatic Sciences* 48:1389–1395.

Sweeney, B. W., J. K. Jackson, J. D. Newbold, and D. H. Funk. 1992. Climate change and the life histories and biogeography of aquatic insects in eastern North America. In P. Firth and S. G. Fisher (eds.). *Global climate change and freshwater ecosystems*, pp. 143–176. Springer-Verlag, New York.

Thorpe, J. H., and M. D. Delong. 2002. Dominance of authochthonous autotrophic carbon in heterotrophic rivers. *Oikos* 96:543–550.

Vannote, R. L., G. W. Minshall, K. W. Cummins, J. R. Sedell, and C. E. Cushing. 1980. The River Continuum Concept. *Canadian Journal of Fisheries and Aquatic Sciences* 37:130–137.

Valdez, R. A., and R. D. Williams. 1986. *Cataract Canyon fish study*. Final Report for U.S. Bureau of Reclamation Contract 6-CS-40-03980. U.S. Bureau of Reclamation, Salt Lake City, Utah.

Ward, J. V., and J. A. Stanford (eds.). 1979. *The ecology of regulated rivers*. Plenum, New York.

Webster, J. R., and J. L. Meyer. 1997. Organic matter budgets for streams: A synthesis. *Journal of the North American Benthological Society* 16:141–161.

APPENDIX

COMMON AND SCIENTIFIC NAMES FOR PLANTS, VERTEBRATES, AND SELECTED INVERTEBRATES

This appendix, which is provided to enable readers to locate scientific names from the common names provided in the text, was assembled from various sources. Initially, authors of the various chapters were asked to submit a list of common and scientific names for species mentioned in their chapters. Because there were differences among the lists provided by the authors, as might be expected given the constant revision of taxonomy, we have modified the final list to provide consistency. We have primarily used the U.S. Department of Agriculture's taxonomic Web site (www.itis.usda.gov) as a basis to resolve most differences; however, it is possible that some "invalid" common names still appear in the text. Some authors found that the ITIS list was not totally up-to-date, and other sources have sometimes been used. This is particularly true for fish common names, which have sometimes been updated based on Nelson et al. (2004). Many fishes from Mexico are not found in the ITIS, and Nelson et al. (2004) was used almost exclusively for those species.

PLANTS

Common names		Scientific names	Common names		Scientific names
acacia	cat-claw	*Acacia greggei*	athel		*Tamarix aphylla*
alder		*Alnus spp.*	aven	mountain	*Dryas drummondii*
	Arizona	*Alnus oblongifolia*	baccharis		*Baccharis spp.*
	brook-side (or hazel)	*Alnus serrulata*		Emory	*Baccharis emoryi*
	green	*Alnus crispa*	barley	foxtail	*Hordeum jubatum*
	mountain	*Alnus incana*	basswood	American	*Tilia americana*
	red	*Alnus rubra*	bearberry		*Arctostaphylus uva-ursi*
	seaside	*Alnus maritima*			
	Sitka	*Alnus viridis* ssp. *sinuata*	bedstraw	fragrant	*Galium triflorum*
			beech	American	*Fagus grandifolia*
	speckled	*Alnus rugosa*		blue	*Carpinus caroliniana*
	thin-leaf	*Alnus tenuifolia*			
	white	*Alnus rhombifolia*	beggar's tick		*Bidens spp.*
alligator-flag	bent	*Thalia geniculata*	berry	Saskatoon	*Amelanchier alnifolia*
alligatorweed		*Alternanthera philoxeroides*	bindweed	hedge false	*Calystegia sepium*
all-scale		*Atriplex polycarpa*	birch		*Betula spp.*
				gray	*Betula populifolia*
ammania	purple	*Ammannia coccinea*		dwarf	*Betula nana*
				paper, white	*Betula papyrifera*
anacua		*Ehretia anacua*		river	*Betula nigra*
arbutus		*Epigaea repens*		swamp	*Betula glandulosum*
arrow-arum		*Peltandra virginica*			
				sweet	*Betula lenta*
arrowhead		*Sagittaria spp.*		yellow	*Betula alleghaniensis*
	awl-leaf	*Sagittaria subulata*			
			bitterbrush	antelope	*Purshia tridentata*
	broadleaf	*Sagittaria latifolia v. latifolia*	blackberry		*Rubus spp.*
			blackbrush		*Coleogyne ramosissima*
	grassy	*Sagittaria graminea*			
			Blackeyed Susan		*Rudbeckia hirta*
	sessile fruit	*Sagittaria rigida*	blackgum		*Nyssa sylvatica*
arrowroot, wapato		*Sagittaria latifolia*	bladderwort	common	*Utricularia macrorhiza*
ash		*Fraxinus spp.*	bloodwoodtree		*Haematoxylum campechianum*
	black	*Fraxinus nigra*			
	Carolina	*Fraxinus caroliniana*	bluebell	northern	*Mertensia paniculata*
	green	*Fraxinus pennsylvanica*	blueberry		*Vaccinium spp.*
			bluegrass		*Poa palustris*
	Oregon	*Fraxinus latifolia*		Canada	*Poa compressa*
	pumpkin	*Fraxinus profunda*		Kentucky	*Poa pratensis*
			box elder (boxelder), Manitoba maple		*Acer negundo*
	velvet	*Fraxinus velvutina*			
	water	*Fraxinus caroliniana*	breadnut tree		*Brosimum alicastrum*
	white	*Fraxinus americana*	brittlebush		*Encelia spp.*
aspen	quaking	*Populus tremuloides*	brooklime (speedwell)	European	*Veronica beccabunga*

Common names		Scientific names	Common names		Scientific names
broomsedge		*Andropogon* spp.		eastern red	*Juniperus virginiana*
buckeye	California	*Aesculus californica*		eastern white, northern white	*Thuja occidentalis*
buckthorn		*Rhamnus cathartica*		salt, tamarisk	*Taramix ramosissima*
buffelgrass		*Pennisetum ciliare*		western red	*Thuja plicata*
bugleweed	northern	*Lycopus uniflorus*		yellow, Alaskan, cypress	*Chamaecyparis nootkatensis*
bulrush		*Scirpus* spp.	celery	wild	*Apium graeolens*
	American	*Schoenoplectus americanus*	chamise		*Adenostoma* spp.
	gum	*Sideroxylon lanuginosum*	cherry	black	*Prunus serotina*
	water	*Scirpus subterminalis*		pin	*Prunus pensylvanica*
bunchberry	Canadian	*Cornus canadensis*	chestnut	American	*Castanea dentata*
				water	*Trapa natans*
burrobush	white	*Hymenoclea salsola*	chokecherry		*Prunus virginiana*
burrweed		*Soliva* spp.	chuparosa		*Justica californica*
bursage	white	*Ambrosia dermosa*	cloudberry		*Rubus chamaemorus*
bush	button	*Cephalanthus occidentalis*	cocklebur		*Xanthium strumarium*
buttercup	aquatic	*Ranunculus aquatilis*	columbine		*Aquilegia* spp.
			common three-square		*Schoenoplectus pungens*
	Pallas'	*Ranunculus pallasi*	coneflower		*Rudbeckia* spp.
butternut		*Juglans cinerea*	coontail		*Ceratophyllum demersum*
buttonbush		*Cephalanthus occidentalis*	corahroot	early	*Corallomiza trifida*
cactus	prickly pear	*Opuntia humifusa*	cordgrass		*Spartina* spp.
				salt marsh	*Spartina alterniflora*
calamus root		*Acorus calamus*		salt meadow	*Spartina patens*
camelthorn		*Alhagi maurorum*		sand	*Spartina bakeri*
camas		*Camassia* spp.		smooth	*Spartina alterniflora*
cane	giant	*Arundinaria gigantea*			
	sugar	*Saccharum officinarum*	corn, maize		*Zea mays*
			cotton		*Gossypium hirsutum*
cardinal flower		*Lobelia cardinalis*			
catalpa		*Catalpa bignonioides*	cottongrass		*Eriophorum* spp.
			cottonwood		*Populus* spp.
cattail		*Typha* spp.		black	*Populus balsamifera* ssp. *trichocarpa*
	broadleaf, common	*Typha latifolia*			
	narrowleaf	*Typha angustifolia*		eastern	*Populus dentata*
				Fremont	*Populus fremontii*
	southern	*Typha domingensis*		narrowleaf	*Populus angustifolia*
ceanothus	Jepson (muskbrush)	*Ceanothus jepsonii*		plains	*Populus deltoides* ssp. *monilifera*
cedar	Atlantic white	*Chamaecyparis thyoides*		swamp	*Populus heterophylla*
	Chinese salt	*Tamarix chinensis*	cowparsnip		*Heracleum sphondylium*

Common names		Scientific names	Common names		Scientific names
cranberry		*Vaccinium macrocarpon*		American	*Ulmus americana*
	highbush	*Viburnum oppulus*		cedar	*Ulmus crassifolia*
				Siberian	*Ulmus pumila*
	lowbush	*Viburnum edule*		slippery	*Ulmus rubra*
creeper	Virginia	*Parthenocissus quinquefolia*		water	*Panera aquatica*
				winged	*Ulmus alata*
creosote bush		*Larrea tridentata*	elodea		*Elodea* spp.
crowberry		*Empertrum nigrum*		Brazilian	*Egeria densa*
			fanwort	Carolina	*Cabomba caroliniana*
crowfoot	eastern whitewater	*Ranunculus longirostris*	fern	beach	*Phegopteris* spp.
	threadleaf	*Ranunculus trichophyllus*		bracken	*Pteridium aquilinum*
	whitewater	*Ranunculus aquatilis*		lady	*Athyrium filix-femina*
cucumbervine (or fivelobe cucumber)		*Cayaponia quinqueloba*		oak	*Gymnocarpium dryopteris*
cudweed		*Gnaphalium uliginosum*		ostrich	*Matteuccia struthiopteris*
				sensitive	*Onociea sensibilis*
currant	skunk	*Ribes gladusolum*		water	*Salvinia molesta*
cypress	bald	*Taxodium distichum*	fescue		*Festuca* spp.
				Idaho	*Festuca idahoensis*
	pond	*Taxodium ascendens*	fir	alpine	*Abies lasiocarpa*
desert-broom		*Baccharis sarothroides*		balsam	*Abies balsamea*
				California red	*Abies magnifica*
ditchmoss		*Elodea nuttallii*		cascade	*Abies amabilis*
dogwood		*Cornus sericea*		Douglas	*Pseudotsuga menziesii*
	flowering	*Cornus florida*		Fraser	*Abies fraseri*
	gray	*Cornus racemosa*		grand	*Abies grandis*
	red-osier	*Cornus sericea* ssp. *sericea*		Pacific silver	*Abies amabilis*
				Santa Lucia	*Abies bracteata*
	rough leaf	*Cornus drummondii*		sub-alpine	*Abies lasiocarpa*
	silky	*Cornus amomum*		white	*Abies concolor*
	stiff	*Cornus foemina*	fireweed		*Epilobium angustifolium*
dropseed	Prairie (or northern)	*Sporobolus heterolepis*	flatsedge	fragrant (or rusty)	*Cyperus odoratus*
ducklettuce		*Ottelia alismoides*	floating heart	yellow	*Nymphoides peltata*
duckweed		*Lemna* spp.	fontinalis moss		*Fontinalis* spp.
	dotted	*Landoltia (or Spirodela) punctata*	goldenrod	(western)	*Euthamia occidentalis*
					Solidago spp.
	lesser	*Lemna minor*	gooseberry		*Ribes* spp.
eelgrass		*Zostera marina*	grape		*Vitis* spp.
	American	*Vallisneria americana*		frost	*Vitis riparia*
				Oregon	*Berberis aquifolium*
elderberry		*Sambucus nigra cerulea*	grass	big bluestem	*Andropogon gerardii*
	red	*Sambucus racemosa*		blue	*Poa* spp.
elm		*Ulmus* spp.		bluebunch wheat	*Pseudoroegneria spicata*

Common names		Scientific names	Common names		Scientific names
	blue grama	*Bouteloua gracilis*	gum	black swamp	*Nyssa sylvatica*
	buffalo	*Buchloe dactyloides*		(swamp tupelo)	*Nyssa bifolora*
	canary	*Phalaris arundinacea*	gumbo limbo		*Bursera simaruba*
	cheat	*Bromus tectorum*	hackberry		*Celtis* spp.
	cotton	*Eriophorum angustifolium*		desert	*Celtis pallida*
	Dallas	*Paspalum dilatatum*		northern or common	*Celtis occidentalis*
	drooping woodreed	*Cinna latifolia*		sugar	*Celtis laevigata*
	grama	*Bouteloua* spp., *Nassella* spp.		western	*Celtis reticulata*
	guinea	*Urochloa maxima*	halogeton		*Halogeton glomeratus*
	Indian rice	*Achnatherum hymenoides*	haw		*Viburnum* spp.
	Johnson	*Sorghum halepense*	hawthorn		*Crataegus erythropoda*
	june	*Koelaria cristata*	hazelnut		*Corylus americana*
	little bluestem	*Schyzachrium scoparium*	hemlock	eastern	*Tsuga canadensis*
	needle-and-thread	*Hesperostipa comata*		mountain	*Tsuga mertensiana*
	Nepal	*Microstegium vimineum*		western	*Tsuga heterophylla*
	panic	*Panicucum* spp.	hickory		*Carya* spp.
	prairie cord	*Spartina pectinata*		bitternut	*Carya cordiformis*
	red top bent	*Agrostis stolonifera*		black	*Carya texana*
	reed	*Calamagrostis* spp.		mockernut	*Carya tomentosa*
	reed canary	*Phalaris arundinacea*		pignut	*Carya glabra*
	reed manna	*Glyceria grandis*		shagbark	*Carya ovata*
	salt-	*Distichlis spicata*		water	*Carya aquatica*
	slender wheat	*Elymus trachycaulus*	holly	American	*Ilex opaca*
	slender wood	*Cinna latifolia*		desert	*Atriplex hymenelytra*
	smooth brome	*Bromus inermis*	honeysuckle		*Lonicera* spp.
	spike	*Distichlis spicata*		American fly	*Lonicera canadensis*
	switch	*Panicum virgatum*		Japanese	*Lonicera japonica*
	torpedo	*Panicum repens*		northern bush	*Diervilla lonicera*
	tussock grass	*Nassella* spp.		twinberry	*Lonicera involucrata*
	western wheat	*Pascopyrum smithii*	hophornbeam	eastern	*Ostrya virginiana*
	West Indian marsh	*Hymenachne amplexicaulis*	hornbeam		*Carpinus* spp.
	wheat	*Agropyron* spp.		American	*Carpinus caroliniana*
grassleaf mudplantain		*Heteranthera dubia*	hornwort		*Ceratophyllum* spp.
greasewood	black	*Sarcobatus vermiculatus*		common	*Ceratophyllum demersum*
greenbrier		*Smilax* spp.	horsefly's eye		*Dopatrium junceum*
gregorywood		*Bucida buceras*	horsetail		*Equisetum ferrissi*
groundnut		*Apios americana*		river	*Equisetum fluviatile*
				smooth	*Equisetum laevigatum*

Common names		Scientific names	Common names		Scientific names
horseweed		*Conyza canadensis*	lousewort	Furbish's	*Pedicularis furbishiae*
huisache		*Amblyotepis setigera*	lupine	Kincaid's	*Lupinus oreganos var. kincaidii*
hyacinth	common water	*Eichhornia crassipes*	madrone	Pacific	*Arbutus menziesii*
hydrilla (waterthyme)		*Hydrilla verticillata*	magnolia		*Magnolia* spp.
hygrophila	East Indian	*Hygrophila polysperma*		southern	*Magnolia grandiflora*
Indiangrass		*Sorghastrum nutans*		umbrella	*Magnolia tripetala*
indigo	false	*Amorpha fruticosa*	maidencane		*Panicum hemitomon*
iodinebush		*Allenrolfea occidentalis*	mangrove	black	*Avicennia germinans*
iris	yellow	*Iris pseudacorus*		button	*Conocarpus erectus*
ironwood		*Olneya tesota*		red	*Rhizophora mangle*
ivy	poison	*Toxicodendron radicans*		white (or American)	*Laguncularia racemosa*
jewelweed		*Impatiens pallida.*	manzanita		*Arctostaphylos* spp.
Joshua tree		*Yucca brevifolia*			
juniper		*Juniperus* spp.		green leaf	*Arctostaphylos patula*
	alligator	*Juniperus deppeana*	maple		*Acer* spp.
	Rocky Mountain	*Juniperus scopulorum*		ash-leaf	*Acer negundo*
	Utah	*Juniperus osteosperma*		bigtooth	*Acer grandidentatum*
	western	*Juniperus occidentalis*		Drummond	*Acer rubrum* var. *drummondii*
knotweed	Japanese	*Polygonum cuspidatum*		mountain	*Acer spicatum*
kudzu		*Pueraria lobata*		red	*Acer rubrum*
Labrador tea		*Ledum groenlandicum*		Rocky Mountain	*Acer glabrum*
				silver	*Acer saccharinum*
larch		*Larix* spp.		sugar	*Acer saccharum*
	subalpine	*Larix lyallii*		vine	*Acer circinatum*
	western	*Larix occidentalis*	mare's tail	common	*Hippuris vulgaris*
laurel	bog	*Kalmia polifolia*	marigold	yellow march	*Caltha palustris*
	mountain	*Kalmia latifolia*	marshweed		*Limnophila x. ludoviciana*
lichen	reindeer	*Cladinia arbuscula*	mayflower	Canada	*Maianthemum canadense*
lizard's tail		*Saururus cernuus*	mayten	Florida	*Maytenus phyllanthoides*
locust	black	*Robinia pseudoacacia*	melon		*Cucumis* spp.
	honey	*Gleditsia triacanthos*	mesquite		*Prosopis* spp.
	water	*Gleditsia aquatica*		velvet	*Prosopis velutina*
loosestrife	creeping jenny	*Lysimachia nummularia*	mile-a-minute weed		*Polygonum perfoliatum*
	purple	*Lythrum salicaria*	milkwort		*Polygala californica*
lotus	American	*Nelumbo lutea*			
	sacred	*Nelumbo nucifera*	milkwort jewelflower		*Streptanthus polygaloides*

Common names		Scientific names	Common names		Scientific names
mimosa		*Albizia* (or *Mimosa*) *julibrissin*		scrub	*Quercus turbinella*
mitrewort	naked	*Mitella nuda*		Shumard's	*Quercus shumardii*
monkey flower	yellow	*Mimulus guttatus*		southern red	*Quercus falcata*
mosquitofern	Pacific	*Azolla filiculoides*		swamp	*Quercus lyrata*
mountain mahogany		*Cercocarpus montanus*		swamp chestnut	*Quercus michauxii*
	curl-leaf	*Cercocarpus ledifolius*		swamp white	*Quercus bicolor*
	little leaf	*Cercocarpus intricatus*		tan	*Lithocarpus densiflorus*
mudplaintain, water stargrass	grassleaf	*Heteranthera dubia*		turkey	*Quercus laevis*
				water	*Quercus nigra*
mudwort		*Limosella aquatica*		white	*Quercus alba*
				willow	*Quercus phellos*
muhly		*Muhlenbergia racemosa*	orchid	lady's tresses	*Spiranthes* spp.
mulberry	white	*Morus alba*	osage-orange		*Maclura pomifera*
	red	*Morus rubra*	palmetto		*Serenoa* spp., *Sabal* spp.
mule's fat (mulefat)		*Baccharis glutinosa*	palo verde	blue	*Parkinsonia* (or *Cercidium*) *florida*
mustard	tumble	*Sisymbrium altissimum*	parrot-feather		*Myriophyllum aquaticum*
naiad		*Najas* spp.	parsnip	cow	*Heracleum maximum*
	brittle	*Najas minor*	pasqueflower		*Pulsatilla* spp.
oak		*Quercus* spp.	pawpaw		*Asimina triloba*
	black	*Quercus velutina*	pecan		*Carya illinoinensis*
	blackjack	*Quercus marilandica*	pennywort	water	*Hydrocotyle* spp.
	blue	*Quercus douglasii*	peppergrass	whitetop	*Cardaria latifolia*
	bur	*Quercus macrocarpa*	persimmon		*Diospyros virginiana*
	California white	*Quercus lobata*		Texas	*Diospyros texana*
	canyon live	*Quercus chrysolepis*	pickerelweed		*Pontederia cordata*
	chestnut	*Quercus prinus*	pine		*Pinus* spp.
	coast	*Quercus parvula*		bristlecone	*Pinus longaeva*
	coastal sage scrub	*Quercus dumosa*		California foothill	*Pinus sabiniana*
	Gamble	*Quercus gambelii*		Coulter	*Pinus coulteri*
	huckleberry	*Quercus vaccinfolia*		eastern white	*Pinus strobus*
				foxtail	*Pinus balfouriana*
	interior live	*Quercus wislizeni*		jack	*Pinus banksiana*
	laurel	*Quercus laurifolia*		Jeffrey	*Pinus jeffreyi*
	leather	*Quercus durata*		knobcone	*Pinus attenuata*
	live	*Quercus virginiana*		limber	*Pinus flexilis*
				loblolly	*Pinus taeda*
	northern red	*Quercus rubra*		lodgepole	*Pinus contorta*
	Nuttall's	*Quercus texana*		longleaf	*Pinus palustris*
	Oregon white	*Quercus garryana*		Parry (or nut)	*Pinus quadrifolia*
	overcup	*Quercus lyrata*		pinyon	*Pinus edulis*
	pin	*Quercus palustris*		ponderosa	*Pinus ponderosa*
	post	*Quercus stellata*		red	*Pinus resinosa*
				shortleaf	*Pinus echinata*

Common names		Scientific names
	singleleaf pinyon	*Pinus monophylla*
	slash	*Pinus elliottii*
	spruce	*Pinus glabra*
	sugar	*Pinus lambertiana*
	western white	*Pinus monticola*
	white, whitebark	*Pinus albicaulis*
pitcher plant		*Sarracenia purpurea*
planertree, water elm		*Planera aquatica*
plantain	rattlesnake	*Goodyera repens*
poison ivy		*Toxicodendron radicans*
pond-lily		*Nuphar luteum variegatum*
	yellow	*Nuphar lutea*
	yellow (or spatterdock)	*Nuphar lutea* ssp. *polysepala*
pondweed		*Potamogeton* spp.
	big-leaf	*Potamogeton amplifolius*
	curly-leaved or curly	*Potamogeton crispus*
	floating-leaf	*Potamogeton natans*
	Illinois	*Potamogeton illinoensis*
	leafy	*Potamogeton foliosus*
	ribbon-leaf	*Potamogeton epihydrus*
	Richardson's	*Potamogeton richardsonii*
	sago	*Stuckenia pectinatus*
	sheathed	*Stuckenia vaginatus*
	spiral	*Potamogeton spirillus*
poolmat	horned	*Zannichellia palustris*
popcornflower	rough	*Plagiobothrys hirtus*
poplar	balsam	*Populus balsamifera*
	yellow (or tulip) (or tuliptree)	*Liriodendron tulipifera*
possumhaw		*Ilex decidua*
primrose	water	*Ludwigia peploides*
privet		*Ligustrum* spp.
	eastern swamp	*Forestiera acuminata*
rabbitbrush	Douglas	*Chrysothamnus viscidiflorus*

Common names		Scientific names
	rubber	*Chrysothamnus nauseosus*
ragweed		*Ambrosia* spp.
	giant	*Ambrosia trifida*
raspberry		*Rubus* spp.
rattlesnake root		*Prenanthes* spp.
redbud		*Cercis canadensis*
red top		*Agrostis stolnifera*
redwood		*Sequoia sempervirens*
reed	bur- (or burreed)	*Sparganium* spp.
	broadfruit bur-reed	*Sparganium eurycarpum*
	common	*Phragmites australis*
	giant	*Arundo donax*
reedgrass	bluejoint	*Calamagrostis canadensis*
retama		*Poitea paucifolia*
rice		*Oryza saliva*
	wild (or northern)	*Zizania palustris*
riverweed		*Podostemum* spp.
	hornleaf	*Podostemum ceratophyllum*
Rolland's scirpus		*Scirpus rollandii*
rose	multiflora	*Rosa multiflora*
	prickly	*Rosa acicularis*
	wild	*Rosa arkansa*
	Wood's	*Rosa woodsii*
rose-mallow	swamp	*Hibiscus palustris*
rush		*Juncus* spp.
	flowering	*Butomus umbellatus*
	jointed	*Juncus nodosus*
	needle	*Eleocharis acicularis*
	needlegrass	*Juncus roemerianus*
	scouring	*Equisetum* spp.
	Small's spike	*Eleocharis smallii*
	Tweedy's	*Juncus tweedyi*
Russian olive		*Elaeagnus angustifolia*
Russian thistle		*Salsola australis*
sacaton	alkali	*Sporobolus airoides*
sagebrush		*Artemisia* spp.
	big	*Artemesia tridentata*
	coastal	*Artemisia californica*
sagewort	biennial	*Artemisia biennis*
saguaro		*Carnegia gigantea*
salal		*Gaultheria shallon*

Common names		Scientific names	Common names		Scientific names
saltbush		*Atriplex* spp.	spicebush		*Lindera benzoin*
	desert	*Atriplex polycarpa*	spiderlily		*Hymenocallis* spp.
	fourwing	*Atriplex canescens*		shoals (or Cahaba lily)	*Hymenocallis coronaria*
salt-cedar (or saltcedar)		*Tamarix ramosissima*	spikerush	common	*Eleocharis palustris*
salvinia	common	*Salvinia minima*	spongeplant (or frogbite)	American	*Limnobium spongia*
	giant	*Salvinia auriculata*	springtape		*Sagittaria kurziana*
sassafras		*Sassafras albidum*	spruce		*Picea* spp.
sawgrass		*Cladium* spp.		black	*Picea mariana*
	swamp	*Cladium mariscus*		Colorado blue	*Picea pungens*
screwbean		*Prosopis pubescens*		Englemann	*Picea engelmannii*
				red	*Picea rubens*
sea oats		*Uniola paniculata*		white	*Picea glauca*
sedge		*Carex* spp.		Sitka	*Picea sitchensis*
	beaked	*Carex rostrata*	spurge	leafy	*Euphorbia esula*
	bladder	*Carex vesicaria*	spurge (or euphorbia)		*Chamaesyce glyptosperma*
	fox	*Carex vulpinoidea*	squash		*Cucurbita* spp.
	knotsheath	*Carex retrorsa*	stalk	rose twisted	*Streptopus roseus*
	Ross'	*Carex rossii*	starflower		*Trientalis borealis*
	water	*Carex aquatilis*	stargrass	water	*Zosterella dubia*
	woolly	*Carex laevivaginata*	starwort	pond water	*Callitriche stagnalis*
seedbox	Uruguay	*Ludwigia hexapetala*		twoheaded water	*Callitriche heterophylla*
	marsh	*Ludwigia palustris*	stinging nettle		*Urtica droica*
seepweed		*Sueda sp.*	stonewort (Chara)		*Chara* spp.
sequoia	giant	*Sequoiadendron giganteum*	strawberry	woodland	*Fragaria vesca*
serviceberry		*Amelanchier* spp.	strawberry-blite		*Chenopodium capitatum*
shadscale		*Atriplex confertiflora*	sugarberry		*Celtis laevigata*
silktassel	chaparral	*Garrya condonii*		Texas	*Celtis laevigata* var. *texana*
silverweed		*Potentilla a. anserina*	sumac		*Rhus* spp.
smartweed		*Polygonum* (or *Persicaria*) spp.		aromatic	*Rhus aromatica*
			summer cypress		*Kochia scoparia*
	water	*Polygonum amphibium* var. *stipulaceum*	sweetbay		*Magnolia virginiana*
			sweet-flag		*Acorus calanus*
snowberry		*Gaultheria hispidula*	sweet gale		*Myrica gale*
			sweetgum		*Liquidambar styraciflua*
snowflake	water	*Nymphoides indica*	sweetspire		*Itea* spp.
soybean		*Glycine max*	sycamore	American	*Platanus occidentalis*
speedweed	marsh	*Veronica scutellata*		Arizona	*Platanus wrightii*
				California, western	*Platanus racemosa*
sphagnum		*Sphagnum squarrosum*	tamarack		*Larix laricina*
	Lindberg's	*Sphagnum lindbergii*	taro	wild (elephant ears)	*Colocasia esculenta*

Common names		Scientific names	Common names		Scientific names
thimbleberry		*Rubus parviflorus*		southern	*Najas guadalupensis*
thistle	Canada	*Cirsium arvense*	water plantain		*Alisma plantago-aquatica*
three square		*Scirpus pungens*			
tree-of-heaven		*Ailanthus altissima*	water shield (or watershield)		*Brasenia schreberi*
tuliptree (yellow poplar)		*Liriodendron tulipifera*	water spangles		*Salvinia minima*
			water speedwell		*Veronica anagallis*
tupelo	Ogeechee	*Nyssa ogeche*	waterweed		*Elodea* spp.
	swamp	*Nyssa biflora*		Brazilian (or common)	*Egeria densa*
	water	*Nyssa aquatica*			
Ute lady tresses		*Spiranthes diluvialis*		Canada	*Elodea canadensis*
			widgeongrass		*Ruppia maritima*
violet		*Viola* spp.	wildrice	southern	*Zizania aquatica*
	northern bog	*Viola nephrophylla*		Texas	*Zizania texana*
			wildrye	Canada	*Elymus canadensis*
Virginia creeper		*Parthenocissus quinquefolia*	willow		*Salix* spp.
walnut		*Juglans* spp.		American water	*Justicia americana*
	Arizona	*Juglans major*		arroyo	*Salix lasiolepis*
	black	*Juglans nigra*		Bebb	*Salix bebbiana*
wapato (or arrowhead)		*Sagittaria latifolia/cuneata*		black	*Salix nigra*
				Bonpland	*Salix bonplandiana*
watercelery		*Vallisneria canadensis*		coastal plain	*Salix caroliniana*
				coyote, sandbar	*Salix exigua*
water chestnut		*Trapa natans*		diamond	*Salix eriocephala*
waterclover	European	*Mariselia quadrifolia*		Drummond's	*Salix drummondiana*
watercress		*Rorippa nasturtium-aquaticum*		dusky	*Salix melanopsis*
				Geyer	*Salix geyeriana*
				Goodding	*Salix gooddingii*
water howellia		*Howellia aquatilis*		Hooker	*Salix hookerana*
				Humboldt's	*Salix humboldtiana*
water lettuce		*Pistia stratiotes*			
waterlily		*Nymphaea* spp.		hybid crack	*Salix x. rubens*
	American white	*Nymphaea odorata*		Idaho (wolf)	*Salix wolfii*
	dotleaf	*Nymphaea ampla*		narrowleaf	*Salix exigua*
	yellow	*Nymphaea mexicana*		northwest sandbar	*Salix sessilifolia*
				Pacific	*Salix lasiandra.*
waterlocust		*Gleditsia aquatica*		peach-leaved	*Salix amygdaloides*
watermilfoil				pussy	*Salix discolor*
	(or American or common)	*Myriophylllum sibiricum (= exalbescens)*		red	*Salix laevigata*
				sandbar	*Salix interior*
				Scouler	*Salix scouleriana*
	Brazilian	*Myriophyllum aquaticum*		seep	*Baccharis salicifolia*
	Eurasian (or spiked)	*Myriophyllum spicatum*		Sitka	*Salix sitchensis*
				Ward's	*Salix caroliniana var. wardii*
	loose	*Myriophyllum laxum*		water	*Justicia* spp.
waternymph	common	*Najas guadalupensis*		white	*Salix alba*
				yellow	*Salix lutea*
	slender	*Najas flexilis*	winterberry	common	*Ilex verticillata*

Common names		Scientific names	Common names		Scientific names
winterfat		*Krascheninnikovia lanata*	woodnettle	Canadian	*Laportea canadensis*
witchhazel	American	*Hamamelis virginiana*	yam		*Ipomoea batatas*
			yellow-cress	greater	*Rorippa amphibia*
	Ozark	*Hamamelis vernalis*	yew	western	*Taxus brevifolia*

INVERTEBRATES

Common names		Scientific names	Common names	Scientific names
crab	blue	*Callinectes sapidus*	Arctic fingernail clam	*Sphaerium nitidum*
	Chinese mitten	*Eriocheir sinensis*	Arctic pond snail	*Stagnicola arctica*
	Harris (or white fingered)	*Rhithropanopeus harrisii*	Arkansas brokenray mussel	*Lampsilis reeveiana brevicula*
crayfish	Appalachian brook	*Cambarus bartoni*		
	big water	*Cambarus robustus*	Arkansas fatmucket mussel	*Lampsilis powellii*
	Cajun dwarf	*Cambarellus shufeldtii*	ash snail	*Gyraulus parvus*
			Asiatic (or Asian) clam	*Corbicula fluminea*
	devil	*Cambarus diogenes*	Atlantic pigtoe mussel	*Fusconaia masoni*
	digger	*Fallicambarus fodiens*	Atlantic rangia mussel	*Rangia cuneata*
			Baltic macoma mussel	*Macoma balthica*
	golden	*Orconectes luteus*	Banbury Springs limpet	*Lanx* spp.
	longpincered	*Orconectes longidigitus*	bankclimber mussel	*Plectomerus dombeyanus*
	Nashville	*Orconectes shoupi*	black sand shell mussel	*Ligumia recta*
			bleufer mussel	*Potamilus purpuratus*
	northern	*Orconectes virilis*		
	papershell	*Orconectes immunis*	bleedingtooth mussel	*Venustaconcha pleasii*
	red swamp	*Procambarus clarkii*	Bliss Rapids snail	*Taylorconcha serpenticola*
	rusty	*Orconectes rusticus*	brook floater mussel	*Alasmidonta undulata*
	Salem cave	*Cambarus hubrichti*	Bruneau Hot Springsnail	*Pyrgulopsis bruneauensis*
	signal	*Pacifastacus leniusculus*	bullhead mussel	*Plethobascus cyphus*
	sooty	*Pacifastacus nigrescens*	butterfly mussel	*Ellipsaria lineolata*
	spinycheek	*Orconectes limosus*	California floater mussel	*Anodonta californiensis*
	spothanded	*Orconectes punctimanus*	Chinese mystery snail	*Viviparus malleatus*
	virile	*Orconectes virilis*	clubshell mussel	*Pleurobema clava*
	White River	*Procambarus acutus*	Coosa moccasinshell mussel	*Medionidus parvulus*
mollusk	ancylid snail	*Ferrissia* spp.	creeper mussel	*Strophitus undulates*
	Applesnail (Mexico)	*Pomacea flagellate*	creeping ancylid snail	*Ferrissia rivularis*

Common names	Scientific names	Common names	Scientific names
Cumberland bean mussel	*Villosa trabalis*	hickorynut mussel	*Obovaria olivaria*
		Higgin's eye mussel	*Lampsilis higginsi*
Cumberland elktoe mussel	*Alasmidonta atropurpurea*	hotwater physa snail	*Physella wrightii*
		Idaho Springsnail	*Fontelicella* (or *Pyrgulopsis*) *idahoensis*
Cumberlandian combshell mussel	*Epioblasma brevidens*		
cylindrical lioplax	*Lioplax cyclostomaformis*	James spiny mussel	*Pleurobema collina*
		kidneyshell mussel	*Ptychobranchus fasciolaris*
dark falsemussel	*Mytilopsis leucophoeta*		
dark pigtoe mussel	*Pleurobema furvum*	Lake Winnipeg snail	*Physella n.sp.*
		leafshell mussel	*Epioblasma flexuosa*
deertoe mussel	*Truncilla truncata*		
dwarf wedgemussel	*Alasmidonta heterodon*	lilliput mussel	*Toxolasma parvus*
		Lilljeborg peaclam	*Pisidium lilljeborgi*
eastern elliptio mussel	*Elliptio complanata*	little spectaclecase mussel	*Villosa lienosa*
eastern fan shell mussel	*Cyprogenia stegaria*	little wing pearly mussel	*Pegias fibula*
		Louisiana fatmucket mussel	*Lampsilis hydiana*
eastern floater mussel	*Pyganodon cataracta*		
		mapleleaf mussel	*Quadrula quadrula*
eastern lamp mussel	*Lampsilis radiata*	marsh snail	*Stagnicola elodes*
eastern oyster	*Crassostrea virginica*	minute hydrobe	*Hydrobia totteni*
		monkeyface mussel	*Quadrula metanevra*
eastern pearl mussel	*Margaritifera margaritifera*		
		mossy valvata snail	*Valvata sincera*
eastern pond mussel	*Ligumia nasuta*	mucket mussel	*Actinonaias ligamentina*
ebonyshell mussel	*Fusconaia ebena*		
elephantear mussel	*Elliptio crassidens*	mud amnicola snail	*Amnicola mimosa*
elktoe mussel	*Alasmidonta marginata*		
		mud bithynia, faucet snail	*Bithynia tentaculata*
fanshell mussel	*Cyprogenia stegaria*	narrowmouth hydrobe snail	*Texadina sphinctostoma*
fat pocketbook mussel	*Potamilus capax*	Neosho mucket mussel	*Lampsilis rafinesqueana*
fatmucket mussel	*Lampsilis siliquoidea*		
		New Zealand mudsnail	*Potamopyrgus antipodarum*
fawnsfoot mussel	*Truncilla donaciformis*		
fine-lined pocketbook mussel	*Lampsilis altilis*	northern riffleshell mussel	*Epioblasma tortulosa rangiana*
fingernail clam	*Sphaerium* spp.	Nuttal's high wing floater mussel	*Anodonta nuttalliana*
flat floater mussel	*Anodonta suborbiculata*		
		orangefoot pimpleback mussel	*Plethobasus cooperianus*
floater mussel	*Anodonta grandis*		
flutedshell mussel	*Lasmigona costata*	orange-nacre mucket mussel	*Lampsilis perovalis*
fragile papershell mussel	*Leptodea fragilis*		
frigid snail	*Lymnaea atkaensis*	Oregon floater mussel	*Anodonta oregonensis*
giant floater mussel	*Pyganodon grandis*		
giant northern peaclam	*Pisidium idahoense*	Ouachita creekshell mussel	*Villosa arkansasensis*
green floater mussel	*Lasmigona subviridis*	Ouachita kidneyshell mussel	*Ptychobranchus occidentalis*
gyro flexed snail	*Gyraulus deflectus*	Ouachita rock-pocketbook mussel	*Arkansia wheeleri*
gyro snail	*Gyraulus* spp.		

Common names	Scientific names	Common names	Scientific names
ovate clubshell mussel	*Pleurobema perovatum*	round hickorynut mussel	*Obovaria subrotunda*
oyster mussel	*Epioblasma capsaeformis*	round pigtoe mussel	*Pleurobema sintoxia*
Ozark pigtoe mussel	*Fusconaia ozarkensis*	round rocksnail	*Leptoxis ampla*
paper pondshell mussel	*Utterbackia imbecillis*	rusty peaclam	*Pisidium ferrugineum*
peaclam	*Pisidium* spp.	salamander mussel	*Simpsonaias ambigua*
pimpleback mussel	*Quadrula pustulosa*	scaleshell mussel	*Leptodea leptodon*
pink heelsplitter mussel	*Potamilus alatus*	Scioto pigtoe mussel	*Pleurobema bournianum*
pink mucket mussel	*Lampsilis abrupta*	seep mudalia snail	*Leptoxis dilatata*
pink papershell mussel	*Potamilus ohioensis*	sheepnose mussel	*Plethobasus cyphus*
		Snake River physa	*Physa natricina*
pistolgrip mussel	*Tritogonia verrucosa*	softshell (clam)	*Mya arenaria*
		southern acornshell mussel	*Epioblasma othcaloogensis*
plain pocketbook mussel	*Lampsilis cardium*	southern clubshell mussel	*Pleurobema decisum*
pocketbook mussel	*Lampsilis ovata*		
pointed campeloma snail	*Campeloma decisum*	southern fatmucket mussel	*Lampsilis straminea claibornensis*
pondhorn mussel	*Uniomerus tetralasmus*	southern mapleleaf mussel	*Quadrula apiculata*
prairie fossaria snail	*Fossaria (Bakerilymnaea) bulimoides*	southern pigtoe mussel	*Pleurobema georgianum*
purple cat's paw mussel	*Epioblasma obliquata*	southern pocketbook mussel	*Lampsilis omata*
purple lilliput mussel	*Toxolasma lividus*	southern rainbow mussel	*Villosa vibex*
purple wartyback mussel	*Cyclonaias tuberculata*	spectaclecase mussel	*Cumberlandia monodonta*
pyramid pigtoe mussel	*Pleurobema rubrum*	spike (or ladyfinger) mussel	*Elliptio dilatata*
quagga mussel	*Dreissena bugensis*		
rabbitsfoot mussel	*Quadrula cylindrica*	squawfoot mussel	*Strophilus undulatus*
rainbow	*Villosa iris*		
ramshorn marsh snail	*Planorbella trivolis*	St. Lawrence pond snail	*Lymnaea catascopium; Stagnicola emarginata*
rams-horn mussel	*Vorticifex* spp.		
rayed bean mussel	*Villosa fabalis*		
red-rimmed melania	*Melanoides tuberculatus*	swamp lymnaea snail	*Lymnaea stagnalis*
ridgebeak peaclam	*Pisidium compressum*	tan riffleshell mussel	*Epioblasma florentina walkeri*
ring pink mussel	*Obovaria retusa*		
rock fossaria snail	*Fossaria modicella*	three-forks springsnail	*Pyrgulopsis triviali*
rock-pocketbook mussel	*Arcidens confragosus*	threehorn wartyback mussel	*Obliquaria reflexa*
rough fatmucket mussel	*Lampsilis straminea*	threeridge mussel	*Amblema plicata*
		three-ridge valvata snail	*Valvata tricarinata*
rough pigtoe mussel	*Pleurobema plenum*	tidewater mucket mussel	*Leptodea ochracea*
round combshell mussel	*Epioblasma personata*	triangle floater mussel	*Alasmidonta undulata*

Common names		Scientific names	Common names		Scientific names
	triangular kidneyshell mussel	*Ptychobranchus greeni*		yellow lamp mussel	*Lampsilis cariosa*
	triangular mussel	*Pisidium variabile*		yellow lance mussel	*Elliptio lanceolata*
	ubiquitous peaclam	*Pisidium casertanum*		yellow sandshell mussel	*Lampsilis teres*
				Yukon floater mussel	*Anodonta beringiana*
	upland combshell mussel	*Epioblasma metastriata*		zebra mussel	*Dreissena polymorpha*
	Utah valvata snail	*Valvata utahensis*	shrimp	blue	*Litopenaeus stylirostris*
	variable spike mussel	*Elliptio icterina*			
	Wabash pigtoe mussel	*Fusconaia flava*		Cauque River prawn	*Macrobrachium* (or *Palaemon*) *americanus*
	wartyback mussel	*Quadrula nodulata*			
	washboard mussel	*Megalonaias nervosa*		California freshwater	*Syncaris pacifica*
	western fanshell mussel	*Cyprogenia aberti*		Cinnamon River	*Macrobrachium acanthurus*
	western floater mussel	*Anodonta kennerlyi*		eastern grass, riverine grass	*Palaemonetes paludosus*
	western pearlshell mussel	*Margaritifera (Margaritinopsis) falcata*		Kentucky cave	*Palaemonias ganteri*
	western ridge mussel	*Gonidea angulata*		Mississippi grass	*Palaemonetes kadiakensis*
	white heelsplitter mussel	*Lasmigona complanata*		Pasadena freshwater	*Synaris pasadenae*
	winged floater	*Anodonta nuttalliana*		white	*Litopenaeus setiferus*
	winged mapleleaf mussel	*Quadrula fragosa*			

FISHES

Common names		Scientific names	Common names		Scientific names
alewife		*Alosa pseudoharengus*		shadow	*Ambloplites ariommus*
anchovy	bay	*Anchoa mitchili*		shoal	*Micropterus cataractae*
	Heller's (or Gulf)	*Anchoa helleri*			
bass	Florida largemouth	*Micropterus salmoides floridanus*		smallmouth	*Micropterus dolomieu*
				spotted	*Micropterus punctulatus*
	Guadalupe	*Micropterus treculii*			
	largemouth	*Micropterus salmoides*		striped	*Morone saxatilis*
				Suwannee	*Micropterus notius*
	Ozark	*Ambloplites rupestris*		white	*Morone chrysops*
				yellow	*Morone mississippiensis*
	palmetto	*Morone chrysops* x *saxatilis*			
			blackfish	Alaska	*Dallia pectoralis*
	redeye	*Micropterus coosae*		Sacramento	*Orthodon microlepidotus*
	Roanoke	*Ambloplites cavifrons*			
			blindcat	phantom	*Prietella lundbergi*
	rock	*Ambloplites rupestris*		toothless	*Trogloglanis pattersoni*

Fishes

Common names		Scientific names	Common names		Scientific names
	widemouth	*Satan eurystomus*	char	Angayukaksurak	*Salvelinus anaktuvukensis*
bluefish		*Pomatomus saltatrix*	chiselmouth		*Acrocheilus alutaceus*
bluegill		*Lepomis macrochirus*	chub	bigeye	*Hybopsis amblops*
bowfin		*Amia calva*		bigmouth	*Nocomis platyrhynchus*
buffalo	bigmouth	*Ictiobus cyprinellus*		blue	*Gila coerulea*
	black	*Ictiobus niger*		bluehead	*Nocomis leptocephalus*
	fleshylip	*Ictiobus labiosus*		bonytail	*Gila elegans*
	smallmouth	*Ictiobus bubalus*		bull	*Nocomis raneyi*
	southern	*Ictiobus meridionalis*		Chihuahua	*Gila nigrescens*
bullhead		*Ameiurus* spp.		Conchos	*Gila pulchra*
	black	*Ameiurus melas*		creek	*Semotilus atromaculatus*
	brown	*Ameiurus nebulosus*		fathead	*Hybopsis gracilis*
	snail	*Ameiurus brunneus*		flathead	*Platygobio gracilis*
	spotted	*Ameiurus serracanthus*		Gila	*Gila intermedia*
	yellow	*Ameiurus natalis*		gravel	*Erimystax x-punctatus*
burbot		*Lota lota*		headwater	*Gila nigra*
carp	bighead	*Hypophthalmichthys nobilis*		hornyhead	*Nocomis biguttatus*
	common	*Cyprinus carpio*		humpback	*Gila cypha*
	grass	*Ctenopharyngiodon idella*		Klamath tui	*Siphateles bicolor bicolor*
	silver	*Hypophthalmichthys molitrix*		lake	*Couesius plumbeus*
carpsucker	highfin	*Carpiodes velifer*		least	*Iotichthys phlegethontis*
	river	*Carpiodes carpio*		leatherside	*Gila copei*
catfish	blue	*Ictalurus furcatus*		Monkey Springs	*Gila arcuatus*
	channel	*Ictalurus punctatus*		Oregon	*Hybopsis crameri*
	Chihuahua	*Ictalurus* spp.		peamouth	*Mylocheilus caurinus*
	dark sea	*Chathorops melanopus*		peppered	*Macrhybopsis tetranema*
	flathead	*Pylodictis olivaris*		Prairie	*Macrhybopsis australis*
	hardhead	*Ariopsis felix*		redspot	*Nocomis asper*
	headwater	*Ictalurus lupus*		riffle	*Algansea aphanea*
	Lacandon sea	*Potamarius nelsoni*		Rio Grande	*Gila pandora*
	Lerma	*Ictalurus dugesii*		roundtail	*Gila robusta*
	Maya sea	*Ariopsis assimilis*		shoal	*Macrhybopsis hyostoma*
	Olmec blind	*Rhamdia macuspanensis*		sicklefin	*Macrhybopsis meeki*
	pale	*Rhamdia guatemalensis*		silver	*Macrhybopsis storeriana*
	Rio Verde	*Ictalurus mexicanus*		slender	*Erimystax cahni*
	vermiculated highfin	*Pterogoplichthys disjunctivus*		speckled	*Macrhybopsis aestivalis*
	walking	*Clarias batrachus*		sturgeon	*Macrhybopsis geldia*
	white	*Ameiurus catus*		thicktail	*Gila crassicauda*
	Yaqui	*Ictalurus pricei*		tui	*Gila bicolor*
cavefish	northern	*Amblyopsis spelaea*		Umpqua	*Oregonichthys kalawatseti*
	southern	*Typhlichthys subterraneus*			

Common names		Scientific names	Common names		Scientific names
	Utah	*Gila atraria*		Arctic	*Coregonus autumnalis*
	Yaqui	*Gila purpurea*		Bering	*Coregonus laurettae*
chubsucker	creek	*Erimyzon oblongus*		Bonneville	*Prosopium gemmifer*
	lake	*Erimyzon sucetta*		least	*Coregonus sardinella*
cichlid	Angostura	*Cichlasoma breidohri*		shortjaw	*Coregonus zenithicus*
	arroyo	*Cichlasoma irregulare*	clingfish	mountain	*Gobiesox fluviatilis*
	blackcheek	*Cichlasoma labridens*	codfish		Gadidae
	blackgullet	*Cichlasoma pasionis*	crappie	black	*Pomoxis nigromaculatus*
	bluemouth	*Cichlasoma nourissati*		white	*Pomoxis annularis*
	chairel	*Cichlasoma pantostictum*	croker	ground	*Bairdiella ronchus*
	Chiapa de Corzo	*Cichlasoma grammodes*	cui-ui		*Chasmistes cujus*
	Chiapas	*Cichlasoma socolofi*	dace	blacknose	*Rhinichthys atratulus*
	Cuatro Ciénegas	*Cichlasoma minckleyi*		blackside	*Phoxinus cumberlandensis*
	firemouth	*Cichlasoma meeki*		finescale	*Phoxinus neogaeus*
	freckled	*Cichlasoma lentiginosum*		Klamath speckled	*Rhinichthys osculus klamathensis*
	giant	*Petenia splendida*		Las Vegas	*Rhinichthys deaconi*
	Mayan	*Cichlasoma urophthalmus*		leopard	*Rhinichthys falcatus*
	Media Luna	*Cichlasoma bartoni*		longfin	*Agosia chrysogaster*
	Montecristo	*Cichlasoma heterospilum*		longnose	*Rhinichthys cataractae*
	Palenque	*Cichlasoma rheophilus*		northern redbelly	*Phoxinus eos*
	pantano	*Cichlasoma pearsei*		pearl	*Margariscus margarita*
	Petén	*Cichlasoma intermedium*		redside	*Clinostomus elongatus*
	Pozolera (or white)	*Cichlasoma argenteum*		relict	*Relictus solitarius*
	redhead	*Cichlasoma synspilum*		southern redbelly	*Phoxinus erythrogaster*
	redside	*Cichlasoma istlanum*		speckled	*Rhinichthys osculus*
	Rio Grande	*Cichlasoma cyanoguttatum*		spike	*Meda fulgida*
	slender	*Cichlasoma steindachneri*		Umpqua	*Rhinichthys evermanni*
	tailbar	*Cichlasoma hartwegi*	darter	alonga	*Etheostoma nigrum*
	Teapa	*Cichlasoma gibbiceps*		Appalachia	*Percina gymnocephala*
	Usumacinta	*Cichlasoma ufermanni*		Arkansas	*Etheostoma cragini*
				Arkansas saddled	*Etheostoma euzomum*
	yellow	*Cichlasoma helleri*		banded	*Etheostoma zonale*
	yellowbelly	*Cichlasoma salvini*		blackbanded	*Percina nigrofasciata*
cisco		*Coregonus artedi*		blackside	*Percina maculata*
				bluehead	*Notropis hubbsi*
				bluestripe	*Percina cymatotaenia*
				bluntnose	*Etheostoma chlorosomum*
				boulder	*Etheostoma wapiti*

Common names	Scientific names
brighteye	*Etheostoma lynceum*
channel	*Percina copelandi*
coal	*Percina brevicauda*
Conchos	*Etheostoma australe*
crystal	*Crystallaria asprella*
cypress	*Etheostoma proeliare*
dusky	*Percina sciera*
eastern sand	*Ammocrypta pellucida*
fantail	*Etheostoma flabellare*
fountain	*Etheostoma fonticola*
freckled	*Percina lenticula*
gilt	*Percina evides*
goldline	*Percina aurolineata*
goldstripe	*Etheostoma parvipinne*
greenbreast	*Etheostoma jordani*
greenside	*Etheostoma blennioides*
greenthroat	*Etheostoma lepidum*
harlequin	*Etheostoma histrio*
Iowa	*Etheostoma exile*
Johnny	*Etheostoma nigrum*
Kanawha	*Etheostoma kanawhae*
Kentucky snubnose	*Etheostoma rafenesquei*
leopard	*Percina pantherina*
longhead	*Percina macrocephala*
longnose	*Percina nasuta*
mud	*Etheostoma asprigene*
naked sand	*Ammocrypta beani*
orangebelly	*Etheostoma radiosum*
orangefin	*Etheostoma bellum*
orangethroat	*Etheostoma spectabile*
paleback	*Etheostoma pallididorsum*
pearl	*Percina aurora*
rainbow	*Etheostoma caeruleum*
redfin	*Etheostoma whipplei*
Rio Grande	*Etheostoma grahami*
river	*Percina shumardi*
rock	*Etheostoma rupestre*
saddleback	*Percina vigil*
Salado	*Etheostoma segrex*

Common names		Scientific names
sand		*Ammocrypta asperella*
scaly sand		*Ammocrypta vivax*
sharpnose		*Percina oxyrhynchus*
shield		*Percina peltata*
slough		*Etheostoma gracile*
Southern sand		*Ammocrypta meridiana*
speckled		*Etheostoma stigmaeum*
splendid		*Etheostoma barrenense*
stargazing		*Percina uranidea*
stippled		*Etheostoma punctulatum*
swamp		*Etheostoma fusiforme*
teardrop		*Etheostoma barbouri*
tessellated		*Etheostoma olmstedi*
tufa		*Etheostoma lugoi*
western sand		*Ammocrypta clara*
yoke		*Etheostoma juliae*
Dolly Varden (char)		*Salvelinus malma*
drum		*Umbrina cirrosa*
	freshwater	*Aplodinotus grunniens*
eel	American	*Anguilla rostrata*
	Asian swamp	*Monopterus albus*
eulachon		*Thaleichthys pacificus*
fallfish		*Semotilus corporalis*
flier		*Centrarchus macropterus*
flounder		Bothidae, Pleuronectidae
	starry	*Platichthys stellatus*
	summer	*Paralichthys dentatus*
gambusia	Amistad	*Gambusia amistadensis*
	Big Bend	*Gambusia gaigei*
	blotched	*Gambusia senilis*
	Champoton	*Carlhubbsia kidderi*
	crescent	*Gambusia hurtadoi*
	Cuatro Ciénegas	*Gambusia longispinis*
	golden	*Gambusia aurata*
	Gulf	*Gambusia vittata*
	Pecos	*Gambusia nobilis*
	robust	*Gambusia marshi*

Common names		Scientific names	Common names		Scientific names
	San Marcos	*Gambusia georgei*		headwater	*Profundulus candalarius*
	stippled	*Gambusia sexradiata*		pike	*Belonesox belizanus*
	Tex-Mex	*Gambusia speciosa*		plains	*Fundulus zebrinus*
	widemouth	*Gambusia eurystoma*		rainwater	*Lucania parva*
	yellowfin	*Gambusia alvarezi*	lamprey		Petromyzontidae
gar	alligator	*Atractosteus spatula*		American brook	*Lampetra appendix*
	Florida	*Lepisosteus platyrhincus*		artic	*Lampetra japonica*
				Kern brook	*Lampetra hubbsi*
	longnose	*Lepisosteus osseus*		Klamath	*Lampetra similis*
	shortnose	*Lepisosteus platostomus*		least brook	*Lampetra aepyptera*
				Miller Lake	*Lampetra minima*
	spotted	*Lepisosteus oculatus*		Pacific	*Lampetra tridentata*
	tropical	*Atractosteus tropicus*		Pit-Klamath brook	*Lampetra lethophaga*
goby	blackfin	*Gobionellus atripinnis*		river	*Lampetra ayresi*
				sea	*Petromyzon marinus*
	multispotted	*Sicydium multipunctatum*		southern brook	*Ichthyomyzon gagei*
	river	*Awaous tajasica*		western brook (or Pacific brook)	*Lampetra richardsoni*
	round	*Neogobius melanostomus*	livebearer	golden	*Poeciliopsis baenschi*
	yellow river	*Awaous banana*		headwater	*Poeciliopsis monacha*
goldeye		*Hiodon alosoides*			
goldfish		*Carassius auratus*		Lerma	*Poeciliopsis infans*
goodea	dusky	*Goodea gracilis*		Michoacán	*Poeciliopsis scarlii*
grayling	Arctic	*Thymallus arcticus*		picotee	*Phallichthys fairweatheri*
guapote	jaguar	*Cichlosoma managuense*	logperch		*Percina caprodes*
guppy		*Poecilia reticulate*		Upper Grijalva	*Poeciliopsis hnilickai*
halfbeak	Mexican	*Hyporhamphus mexicanus*		bigscale	*Percina macrolepida*
hardhead		*Mylopharodon conocephalus*		Gulf	*Percina suttkusi*
				Mobile	*Percina kathae*
herring	blueback	*Alosa aestivalis*		Texas	*Percina carbonaria*
	skipjack (or blue or river)	*Alosa chrysochloris*	machete		*Elops affinis*
			madtom	caddo	*Noturus taylori*
hitch		*Lavinia exilicauda*		Coosa	*Noturus sp. cf. munitus*
	Clear Lake	*Lavinia exilicauda chi*		frecklebelly	*Noturus munitus*
				freckled	*Noturus nocturnes*
hogchoker		*Trinectes maculatus*		margined	*Noturus insignis*
Hog sucker (or hogsucker)	northern	*Hypentellum nigricans*		mountain	*Noturus eleutherus*
				Neosho	*Noturus placidus*
inconnu		*Stenodus leucichthys*		northern	*Noturus stigmosus*
				Ouachita	*Noturus lachneri*
jumprock	greater	*Scartomyzon (Moxostoma) lachneri*		Ozark	*Noturus albater*
				pygmy	*Noturus stanauli*
				Scioto	*Noturus trautmani*
				slender	*Noturus exilis*
killifish	banded	*Fundulus diaphanous*		smoky	*Noturus baileyi*
	Chiapas	*Profundulus hildebrandi*		speckled	*Noturus leptacanthus*
				tadpole	*Noturus gyrinus*
			menhaden		*Brevoortia tyrannus*

Common names		Scientific names	Common names		Scientific names
minnow	bicolor	*Dionda dichroma*		western silvery	*Hybognathus argyritis*
	blackstripe	*Dionda rasconis*	mojarra	black axillary	*Eugerres axillaris*
	bluntnose	*Pimephales notatus*		Mexican	*Eugerres mexicanus*
	brassy	*Hybognathus hankinsoni*		rhombic	*Diapterus rhombeus*
	bullhead	*Pimephales vigilax*	molly	Amazon	*Poecilia formosa*
	central stoneroller	*Campostoma anomalum*		Pacific	*Poecilia butleri*
	chiselmouth	*Acrocheilus alutaceus*		sailfin	*Poecilia latipinna*
	chubsucker	*Dionda erimyzonops*		shortfin	*Poecilia mexicana*
	cutlips	*Exoglossum maxillingua*		sulfur	*Poecilia sulphuraria*
	cypress	*Hybognathus hayi*		Yucatan	*Poecilia velifera*
	Devils River	*Dionda diaboli*	mooneye		*Hiodon tergisus*
	eastern silvery	*Hybognathus regius*	mosquitofish		*Gambusia affinis*
	fathead	*Pimephales promelas*		eastern	*Gambusia holbrooki*
	flatjaw	*Dionda mandibularis*		western	*Gambusia affinis*
	Kanawha	*Phenacobius teretulus*	mouthbrooder	Mozambique	*Tilapia mossambica*
	Lake Eustis	*Cyprinodon hubbsi*	mudminnow	central	*Umbra limi*
	lantern	*Dionda ipni*	mullet	mountain	*Agonostomus monticola*
	loach	*Rhinichthys cobitis*		striped	*Mugil cephalus*
	manantial roundnose	*Dionda argentosa*		white	*Mugil curema*
	Mississippi silvery	*Hybognathus nuchalis*	mummichog		*Fundulus heteroclitus*
	Nueces roundnose	*Dionda serena*	muskellunge		*Esox masquinongy*
	ornate (or ornate shiner)	*Cyprinella* (or *Codoma*) *ornata*	muskie	tiger (hybrid of muskellunge and northern pike)	*Esox masquinongy x Esox lucius*
	Ozark	*Notropis nubilus*	needlefish	Atlantic	*Strongylura marina*
	Pánuco	*Dionda catostomops*		Maya	*Strongylura hubbsi*
	peamouth	*Mylocheilus caurinus*	paddlefish		*Polyodon spathula*
	plains	*Hybognathus placitus*	perch	Clear Lake tule	*Hysterocarpus traski lagunae*
	pugnose	*Opsopoedus emilae*		pirate	*Aphredoderus sayanus*
	riffle	*Phenacobius catostomus*		Sacramento	*Archoplites interruptus*
	Rio Grande silvery	*Hybognathus amarus*		Sacramento tule	*Hysterocarpus traski traski*
	roundnose	*Dionda episcopa*		silver	*Bairdiella chrysoura*
	sheepshead	*Cyprinodon variegatus*		tule	*Hysterocarpus traski*
	silverjaw	*Ericymba buccata*		white	*Morone americana*
	stargazing	*Phenacobius uranops*		yellow	*Perca flavescens*
	suckermouth	*Phenacobius mirabilis*	pickerel	chain	*Esox niger*
	top	Fundulidae		grass	*Esox americanus*
				redfin	*Esox americanus americanus*
			pike	northern	*Esox lucius*
			pikeminnow	Colorado	*Ptychocheilus lucius*
				northern	*Ptychocheilus oregonensis*

Common names		Scientific names	Common names		Scientific name
	Sacramento	*Ptychocheilus grandis*		greater	*Moxostoma valenciennesi*
	Umpqua	*Ptychocheilus umpquae*		Mexican (or west Mexican)	*Moxostoma austrinum*
pipefish	opossum	*Microphis brachyurus*		river	*Moxostoma carinatum*
pirapitinga		*Piaractus brachypomus*		robust	*Moxostoma robustum*
platyfish	Cuatro Ciénegas	*Xiphophorus gordoni*		shorthead	*Moxostoma macrolepidotum*
	short-sword	*Xiphophorus continens*		silver	*Moxostoma anisurum*
	variable	*Xiphophorus variatus*	roach	California	*Hesperoleucas symmetricus*
priapella	Palenque	*Priapella compressa*	rudd	European	*Scardinius erythrophthalmus*
puffer	checkered	*Sphoeroides testudineus*	salmon	Atlantic	*Salmo salar*
pumpkinseed		*Lepomis gibbosus*		chinook	*Oncorhynchus tshawytscha*
pupfish	bighead	*Cyprinodon pachycephalus*		chum	*Oncorhynchus keta*
	bigscale	*Cyprinodon macrolepsis*		coho	*Oncorhynchus kisutch*
	bolson	*Cyprinodon atrorus*		pink	*Oncorhynchus gorbuscha*
	Comanche Springs	*Cyprinodon elegans*		sockeye (or kokanee)	*Oncorhynchus nerka*
	Conchos	*Cyprinodon eximius*	sandroller		*Percopsis transmontana*
	Cuatro Ciénegas	*Cyprinodon bifasciatus*	sauger		*Stizostedion canadense*
	desert	*Cyprinodon macularius*	sculpin		*Cottus* spp.
	Leon Springs	*Cyprinodon bovinus*		banded	*Cottus carolinae*
	Media Luna	*Cualac tessellatus*		Bear Lake	*Cottus extensus*
	Monkey Springs	*Cyprinodon arcuatus*		coastrange	*Cottus aleuticus*
	Pecos	*Cyprinodon pecosensis*		deepwater	*Myoxocephalus thompsoni*
	Quitobaquito	*Cyprinodon eremus*		Klamath Lake	*Cottus princeps*
	Red River	*Cyprinodon rubrofluviatilis*		lower Klamath River marbled	*Cottus klamathensis polyporus*
	Salvador's (or Bocochi)	*Cyprinodon salvadori*		marbled	*Cottus klamathensis*
	whitefin	*Cyprinodon albivelis*		mottled	*Cottus bairdi*
quillback		*Carpiodes cyprinus*		Paiute	*Cottus beldingi*
redhorse		*Moxostoma* spp.		prickly	*Cottus asper*
	black	*Moxostoma duquesnei*		reticulate	*Cottus perplexus*
	blacktail	*Moxostoma poecilurum*		riffle	*Cottus gulosus*
	golden	*Moxostoma erythrurum*		rough	*Cottus asperrimus*
	gray	*Scartomyzon congestus*		shorthead	*Cottus confusus*
	grayfin	*Moxostoma* sp.		slender	*Cottus tenuis*
				slimy	*Cottus cognatus*
				spoonhead	*Cottus ricei*
				staghorn	*Leptocottus armatus*
				torrent	*Cottus rhotheus*
			seatrout	sand	*Cynoscion arenarius*

Common names		Scientific names	Common names		Scientific names
	spotted	*Cynoscion nebulosus*		peppered	*Notropis perpallidus*
				phantom	*Notropis orca*
shad		*Dorosoma* spp.		plateau	*Cyprinella lepida*
	Alabama	*Alosa alabamae*		pretty	*Lythrurus bellus*
	American	*Alosa sapidissima*		proserpine	*Cyprinella proserpina*
	gizzard	*Dorosoma cepedianum*		rainbow	*Notropis chrosomus*
	hickory	*Alosa mediocris*		red	*Cyprinella lutrensis*
	longfin gizzard	*Dorosoma anale*		redfin	*Lythrurus umbratilis*
	Pacific gizzard	*Dorosoma smithi*		Red River	*Notropis bairdi*
	threadfin	*Dorosoma petenense*		redside	*Richardsonius balteatus*
shiner	Alabama	*Cyprinella calistia*			
	Arkansas River	*Notropis girardi*		ribbon	*Lythrurus fumeus*
	bigeye	*Notropis boops*		Rio Grande	*Notropis jemezanus*
	bigmouth	*Notropis dorsalis*		river	*Notropis blennius*
	blacknose	*Notropis heterolepis*		rocky	*Notropis suttkusi*
	blackspot	*Notropis atrocaudalis*		rosefin	*Lythrurus ardens*
				rosyface	*Notropis rubellus*
	blacktail	*Cyprinella venusta*		Sabine	*Notropis sabinae*
	bleeding	*Luxilus zonatus*		Salado	*Notropis saladonis*
	blue	*Cyprinella caerulea*		sand	*Notropis stramineus*
	bluntface	*Notropis camurus*		satinfin	*Cyprinella analostana*
	bluntnose	*Notropis simus*			
	Cahaba	*Notropis cahabae*		sharpnose	*Notropis oxyrhynchus*
	cardinal	*Luxilus cardinalis*			
	Chihuahua	*Notropis chihuahua*		silverband	*Notropis shumardi*
	chub	*Notropis potteri*		silverside	*Notropis candidus*
	common	*Luxilus cornutus*		silverstripe	*Notropus stilbius*
	Conchos	*Cyprinella panarcys*		skygazer	*Notropis uranoscopus*
	Coosa	*Notropis xaenocephalus*			
				smalleye	*Notropis buccula*
	duskystripe	*Luxilus pilsbryi*		spotfin	*Notropis spilopterus*
	emerald	*Notropis atherinoides*		spottail	*Notropis hudsonius*
				steelcolor	*Cyprinella whipplei*
	fluvial	*Notropis edwardraneyi*		striped	*Luxilus chrysocephalus*
	ghost	*Notropis buchanani*		swallowtail	*Notropis procne*
				taillight	*Notropus maculatus*
	golden	*Notemigonus crysoleucas*		Tamaulipas	*Notropis braytoni*
				telescope	*Notropis telescopus*
	Lahontan redside	*Richardsonius egregious*		Texas	*Notropis amabilis*
				Topeka	*Notropis longa*
	Maravillas red	*Cyprinella lutrensis blairi*		wedgespot	*Notropis greenei*
				weed	*Notropis texanus*
	Mexican red	*Cyprinella rutila*		whitefin	*Notropis niveus*
	mimic	*Notropis volucellus*		Yaqui (or	*Cyprinella formosa*
	New River	*Notropis scabricips*		beautiful)	
	orangefin	*Notropis ammophilus*	silverside	brook	*Labidesthes sicculus*
	ornate (or ornate minnow)	*Cyprinella (or Codoma) ornata*		Gulf	*Atherinella alvarezi*
				inland	*Menidia beryllina*
	Ouachita	*Lythrurus snelsoni*	sleeper	bigmouth	*Gobiomorus dormitor*
	pallid	*Hybopsis amnis*			
	Pecos bluntnose	*Notropis simus pecosensis*		finescale	*Gobiomorus polylepis*

Common names		Scientific names	Common names		Scientific names
	Pacific	*Gobiomorus maculatus*		Gulf	*Ancipenser oxyrinchus desotoi*
	spinycheek	*Eleotris pisonsis*		lake (or black)	*Acipenser fulvescen*
smelt	boreal	*Osmerus mordax*		pallid	*Scaphirhynchus albus*
	delta	*Hypomesus transpacificus*		shortnose	*Acipenser brevirostrum*
	eulachon	*Thaleichthys pacificus*		shovelnose	*Scaphirhynchus platorynchus*
	longfin	*Spirinchus thaleichthys*		white	*Acipenser transmontanus*
	pond	*Hypomesus olidus*	sucker	blackfin	*Moxostoma atripinne*
	rainbow	*Osmerus mordax*		blue	*Cycleptus elongates*
snook	(common)	*Centropomus undecimalis*		bluehead	*Catostomus discobolus*
	fat (small scale)	*Centropomus parallelus*		bridgelip	*Catostomus columbianus*
spinedace	Little Colorado	*Lepidomeda vittata*		Cahita	*Catostomus cahita*
	Pahranagat	*Lepidomeda altivelis*		desert	*Pantosteus clarki*
	virgin	*Lepidomeda mollispinis*		flannelmouth	*Catostomus latipinnis*
splitfin	bandfin	*Allodontichthys zonistius*		hairlip	*Lagochila lacera*
	bluetail	*Ataeniobius toweri*		hog	*Hypentelium nigricans*
	goldbreast	*Ilyodon furcidens*		Klamath	*Catostomus snyderi*
	jeweled	*Xenotoca variata*		largescale	
	relict	*Xenoophorus captivus*		Klamath	*Catostomus rimiculus*
splittail	Clear Lake	*Pogonichthys ciscoides*		smallscale	
				largescale	*Catostomus macrocheilus*
	Sacramento	*Pogonichthys macrolepidotus*		Leopold	*Catostomus leopoldi*
steelhead	(anadromous form of rainbow trout)	*Oncorhynchus mykiss*		Little Colorado River	*Catostomus latipinnis*
				longnose	*Catostomus catostomus*
stickleback	brook	*Culaea inconstans*		Lost River	*Deltistes luxatus*
	fourspine	*Apeltes quadracus*		Modoc	*Catostomus microps*
	ninespine	*Pungitius pungitius*		mountain	*Catostomus platyrhynchus*
	threespine	*Gasterosteus aculeatus*		Owens	*Catostomus fumeiventris*
stonecat		*Noturus flavus*		quillback	*Carpiodes cyprinus*
stoneroller	central	*Campostoma anomalum*		razorback	*Xyrauchen texanus*
	largescale	*Campostoma oligolepis*		redhorse	*Moxostoma robustum*
	Mexican	*Campostoma ornatum*		Rio Grande	*Catostomus plebeius*
sturgeon	Alabama	*Scaphirhynchus suttkusi*		Sacramento	*Catostomus occidentalis*
	Atlantic	*Acipenser oxyrinchus*		shortnose	*Chasmistes brevirostris*
	green	*Acipenser medirostris*		Sonora	*Catostomus insignis*

Common names		Scientific names	Common names		Scientific names
	southeastern blue	*Cycleptus meridionalis*		pigmy	*Xiphophorus pygmaeus*
	spotted	*Minytrema melanops*		sheepshead	*Xiphophorus birchmanni*
	Tahoe	*Catostomus tahoensis*	tench		*Tinca tinca*
	Utah	*Catostomus ardens*	tetra	banded	*Astyanax* cf. *aeneus*
	white	*Catostomus commersoni*		longjaw	*Bramocharax bronsfordii*
	Yaqui	*Catostomus bernardini*		macabi	*Brycon guatemalensis*
	Zuni bluehead	*Catostomus discobolus yarrowi*		Maya	*Hyphessobrycon compressus*
sunfish	banded	*Enneacanthus obesus*		Mexican	*Astyanax mexicanus*
				Pénjamo	*Astyanax armandoi*
			tilapia		*Oreochromis* (or *Tilapia*) spp.
	bantam	*Lepomis symmetricus*		blue	*Oreochromis aureus*
	blackbanded	*Enneacanthus chaetodon*		Mozambique	*Oreochromis mossambicus*
	bluegill	*Lepomis macrochirus*		Nile	*Oreochromis niloricus*
	bluespotted	*Enneacanthus gloriosus*		redbelly	*Tilapia zilli*
	dollar	*Lepomis marginatus*		redbreast	*Tilapia rendalli*
	green	*Lepomis cyanellus*	toadfish	Mexican freshwater	*Batrachoides goldmani*
	longear	*Lepomis megalotis*	tomcod	Atlantic	*Microgadus tomcod*
	mud	*Acantharchus pomotis*	topminnow		*Fundulus* spp.
				blackspotted	*Fundulus olivaceus*
	orangespotted	*Lepomis humilis*		blackstriped	*Fundulus notatus*
	pumpkinseed	*Lepomis gibbosus*		Blair's starheaded	*Fundulus blairae*
	pygmy (banded)	*Elassoma zonatum*		Culiche (or blackstripe livebearer)	*Poeciliopsis prolifica*
	redbreast	*Lepomis auritus*			
	redear	*Lepomis microlophus*		Gila	*Poeciliopsis occidentalis*
	redspotted	*Lepomis miniatus*		golden	*Fundulus chrysotus*
	spotted	*Lepomis punctatus*		plains	*Fundulus sciadicus*
swordtail	barred	*Xiphophorus multilineatus*		Yaqui	*Poecilliopsis occidentalis sonoriensis*
	Chiapas	*Xiphophorus alvarezi*	Totoaba fish		*Totoaba macdonaldi*
	delicate	*Xiphophorus cortezi*	trout	Apache	*Oncorhynchus apache*
	green	*Xiphophorus helleri*			
	highland	*Xiphophorus malinche*		Bonneville cutthroat	*Oncorhynchus clarki utah*
	marbled	*Xiphophorus meyeri*		brook	*Salvelinus fontinalis*
	Moctezuma	*Xiphophorus moctezumae*		brown	*Salmo trutta*
	mountain	*Xiphophorus nezahualcoyotl*		bull	*Salvelinus confluentus*
	Pánuco	*Xiphophorus nigrensis*		California golden	*Oncorhynchus mykiss aquabonita*

Common names		Scientific names	Common names		Scientific names
	coastal cutthroat	*Oncorhynchus clarki clarki*		Yellowstone cutthroat	*Oncorhynchus clarki bouvieri*
	coastal rainbow	*Oncorhynchus mykiss irideus*	troutperch		*Percopsis omiscomaycus*
	Colorado River cutthroat	*Oncorhynchus clarki pleuriticus*	wakasagi		*Hypomesus nipponensis*
	cutthroat	*Oncorhynchus clarki*	walleye		*Stizostedion vitreum*
			warmouth		*Lepomis gulosus*
	Gila	*Oncorhynchus gilae*	weakfish		*Cynoscion regalis*
	greenback cutthroat	*Oncorhynchus clarki stomias*	weatherfish	oriental	*Misgurnus anguillicaudatus*
	Kern River rainbow	*Oncorhynchus mykiss gilberti*	whitefish	Atlantic	*Coregonus huntsmani*
	Lahontan cutthroat	*Oncorhynchus clarki henshawi*		Bear Lake	*Prosopium abyssicola*
	lake	*Salvelinus namaycush*		Bonneville	*Prosopium spilonotus*
	Little Kern River golden	*Oncorhynchus whitei*		broad	*Coregonus nasus*
	Mexican golden	*Oncorhynchus chrysogaster*		humpback	*Coregonus pidschian*
	Ohrid	*Salmo letnica*		lake	*Coregonus clupeaformis*
	rainbow	*Oncorhynchus mykiss*		mountain	*Prosopium williamsoni*
	redband	*Oncorhynchus mykiss gibbsi*		pygmy	*Prosopium coulteri*
	Rio Grande cutthroat	*Oncorhynchus clarkii virginalis*		round	*Prosopium cylindraceum*
				Squanga	*Coregonus* sp.
	steelhead (anadromous form of rainbow trout)	*Oncorhynchus mykiss*	wiper		*Morone chrysops x saxatilis*
			woundfin		*Plagopterus argentissimus*
	westslope cutthroat	*Oncorhynchus clarki lewisi*	yellowjacket		*Cichlasoma friedrichsthali*
	Yaqui	*Oncorhynchus* sp.			

HERPETOFAUNA (AMPHIBIANS AND REPTILES)

Common names		Scientific names	Common names		Scientific names
alligator	American	*Alligator mississippiensis*		Cascades	*Rana cascadae*
				Chiricahua leopard	*Rana chiricahuensis*
amphiuma	two-toed	*Amphiuma means*		Columbia spotted	*Rana luteiventris*
bullfrog	American	*Rana catesbeiana*		Cope's gray treefrog	*Hyla chrysoscelis*
crocodile	river	*Crocodylus acutus*		cricket	*Acris crepitans*
	swamp	*Crocodylus moreletii*		foothill yellow-legged	*Rana boylei*
frog	Blanchard's cricket	*Acris crepitans blanchardi*		green	*Rana clamitans,*
				greenhouse	*Eleutherodactylus planirostris*
	boreal chorus	*Pseudacris triseriata maculata*		green treefrog	*Hyla cinerea*
	Brimley's chorus	*Pseudoacris brimleyi*		gray treefrog	*Hyla versicolor*

Common names		Scientific names	Common names		Scientific names
	lowland leopard	*Rana yavapaiensis*		Pacific giant	*Dicamptodon* spp.
	mink	*Rana septentrionalis*		San Marcos	*Eurycea nana*
	mountain yellow-legged	*Rana muscosa*		southern torrent	*Rhyacotriton variegates*
	northern leopard	*Rana pipiens*		Tarahumara	*Ambystoma rosaceum*
	Pacific chorus, Pacific treefrog	*Pseudacris regilla*		Texas blind	*Eurycea rathbuni*
	pickerel	*Rana palustris*		tiger	*Ambystoma tigrinium*
	pig	*Rana grylio*			
	plains leopard	*Rana blairi*		western red-backed	*Plethodon vehiculum*
	Ramsey Canyon leopard	*Rana subaquavocalis*	siren	greater	*Siren acertian*
	red-legged	*Rana aurora draytonii*		lesser	*Siren intermedia*
				western lesser	*Siren intermedia nettingi*
	Rio Grande leopard	*Rana berlandieri*	snake	blotched water	*Nerodia erythrogaster transversa*
	river	*Rana heckscheri*			
	southern leopard	*Rana sphenocephala utricularia*		broad-banded	*Nerodia fasciata confluens*
	spotted	*Rana pretiosa*		brown water	*Nerodia taxispilota*
	spotted chorus	*Pseudacris clarkii*		common garter	*Thamnophis sirtalis*
	spring peeper	*Pseudacris crucifer*			
	Strecker's chorus	*Pseudacris streckeri streckeri*		Concho water	*Nerodia paucimaculata*
	striped or western chorus	*Pseudacris triseriata*		copperbelly water	*Nerodia erythrogaster neglecta*
	tailed	*Ascaphus truei*			
	Tarahumara	*Rana tarahumarae*		copperhead	*Agkistrodon contortrix*
	wood	*Rana sylvatica*			
hellbender		*Cryptobranchus alleganiensis*		cottonmouth	*Agkistrodon piscivorous*
	Ozark	*Cryptobranchus alleganiensis bishopi*		diamondback water	*Nerodia rhombifer*
				eastern massasauga	*Sistrurus catenatus catenatus*
mudpuppy	(common)	*Necturus maculosus*			
newt		*Notophthalmus* spp.		Florida cottonmouth	*Agkistrodon piscivorus conanti*
	red spotted	*Notophthalmus v. viridescens*			
	rough skinned	*Taricha granulosa*		Florida green water	*Nerodia floridana*
salamander	Barton Springs	*Eurycea sosorum*		glossy crayfish	*Regina rigida*
	black	*Aneides flavipunctatus*		Graham's crayfish	*Regina grahamii*
				green water	*Nerodia cyclopion*
	blue-spotted	*Ambystoma laterale*		Gulf saltmarsh	*Nerodia clarkii clarkii*
	clouded	*Aneides ferreus*			
	Dunn's	*Plethodon dunni*		Harter's water	*Nerodia harteri*
	dusky	*Desmognathus* spp.		Kirtland's water	*Nerodia kirtlandi*
	long toed	*Ambystoma macrodactylum*		midland water	*Nerodia sipedon pleuralis*
	many-lined	*Stereochilus marginatus*		milk	*Lampropeltis triangulum*
	marbled	*Ambystoma opacum*		northern water	*Nerodia sipedon*
	mole	*Ambystoma talpoideum*		North Florida swamp	*Seminatrix pygaea pygaea*
	northern two-lined	*Eurycea bislineata*		plain-bellied water	*Nerodia erythrogaster*
	northwestern	*Ambystoma gracile*			

Common names		Scientific names	Common names	Scientific names
	queen	*Nerodia septemvittata*	Cuatro Ciénegas red-eared slider	*Trachemys taylori*
	rainbow	*Farancia erytrogramma*	Cuatro Ciénegas softshell	*Apalone ater*
	red sided garter	*Thamnophis sirtalis parietalis*	Cumberland slider	*Trachemys scripta troosti*
	red-bellied water	*Nerodia erythrogaster erythrogaster*	eastern box	*Terrapene carolina carolina*
			eastern chicken	*Deirochelys reticularia reticularia*
	southern water	*Nerodia fasciata fasciata*		
	wandering garter	*Thamnophis elegans vagrans*	eastern mud	*Kinosternon subrubrum subrubrum*
	water	*Nerodia* spp.		
	western aquatic garter	*Thamnophis couchi*	eastern painted	*Chrysemys picta*
			Escambia map	*Graptemys ernsti*
	western garter	*Thamnophis radix*	false map	*Graptemys pseudographica*
	western hognose	*Heterodon nasicus*		
	western ribbon	*Thamnophis proximus*	flattened musk	*Sternotherus depressus*
			Florida cooter	*Pseudemys floridana*
	yellow-bellied water	*Nerodia erythrogaster flavigaster*	Florida redbelly	*Pseudemys nelsoni*
			Florida snapping	*Chelydra serpentina osceola*
toad	American	*Bufo americanus*	Florida softshell	*Apalone ferox*
	Arizona	*Bufo microscaphus*	Gulf Coast smooth softshell	*Apalone mutica calvata*
	boreal western	*Bufo boreas*		
	Canadian	*Bufo hemiophrys*	Gulf Coast spiny softshell	*Apalone spinifera aspera*
	giant	*Bufo marinus*		
	Great Basin spadefoot	*Spea intermontana*	Kemp's ridley	*Lepidochelys kempii*
	Great Plains	*Bufo cognatus*	loggerhead musk	*Sternotherus minor*
	Gulf Coast	*Bufo valliceps*	midland painted	*Chrysemys picta marginata*
	plains spadefoot	*Spea bombifrons*		
	red-spotted	*Bufo punctatus*	midland smooth softshell	*Aplone mutica mutica*
	Sonoran desert	*Bufo alvarius*		
	western	*Bufo boreas*	Mississippi map	*Graptemys kohnii*
	Woodhouse's (or western Woodhouse's)	*Bufo woodhousei*	Mississippi mud	*Kinosternon subrubrum hippocrepis*
turtle	Alabama map	*Graptemys pulchra*	Missouri River cooter	*Pseudemys concinna metteri*
	Alabama redbelly	*Pseudemys alabamensis*		
	alligator snapping	*Macroclemys temmincki*	mud	*Kinosternon subrubrum*
	Barbor's map	*Graptemys barbouri*	musk	*Sternotherus odoratus*
	Blanding's	*Emydoidea blandingii*	northern black-knobbed (or sawback) map	*Graptemys nigrinoda nigrinoda*
	box	*Terrapene* spp.		
	Cagle's map	*Graptemys caglei*	Ouachita map	*Graptemys ouachitensis*
	common map	*Graptemys geographica*		
	common slider	*Trachemys scripta*	painted	*Chrysemys picta*
	Cuatro Ciénegas box	*Terrapene coahuila*	pallid spiny softshell	*Apalone spinifera pallida*

Common names		Scientific names	Common names		Scientific names
	Pascagoula map	*Graptemys gibbonsi*		striped mud	*Kinosternon baurii*
	razor-backed musk	*Sternotherus carinatus*		stripeneck musk	*Sternotherus minor peltifer*
	red-bellied	*Pseudemys rubriventris*		Suwannee cooter	*Pseudemys concinna suwanniensis*
	red-eared slider	*Trachemys scripta elegans*		Texas cooter	*Pseudemys texana*
	ringed map	*Graptemys oculifera*		Texas map	*Graptemys versa*
	river cooter	*Pseudemys concinna*		tortugas blanca	*Dermatemis maui*
	Sabine map	*Graptemys ouachitensis sabinensis*		tortugas casquito	*Kinosterma spp.*
				western (or ornate) box	*Terrapene ornata*
	slider	*Trachemys* spp., *Chrysemys* spp.		western painted	*Chrysemys picta bellii*
	smooth softshell	*Apalone mutica*		western pond	*Clemmys marmorata*
	snapping	*Chelydra serpentina*		western spiny softshell	*Apalone spinifera hartwegi*
	softshell	*Apalone* spp.		wood	*Clemmys insculpta*
	Sonoran mud	*Kinosternon sonoriense*		yellow mud	*Kinosternon flavescens*
	southern black-knobbed map	*Graptemys nigrinoda delticola*		yellowbelly slider	*Trachemys scripta scripta*
	southern painted	*Chrysemys picta dorsalis*		yellow-blotched map	*Graptemys flavimaculata*
	spiny softshell	*Apalone spinifera*	waterdog	Alabama	*Necturus alabamensis*
	spiny-spotted	*Trionyx spiniferous*		dwarf	*Necturus punctatus*
	spotted	*Clemmys guttata*		Gulf Coast	*Necturus beyeri*
	stinkpot	*Sternotherus odoratus*		red river	*Necturus maculosus louisianensis*

BIRDS

Common names		Scientific names	Common names		Scientific names
anhinga		*Anhinga anhinga*		Barrow's goldeneye	*Bucephala islandica*
bittern	least	*Ixobrychus exilis*		black	*Anas rubripes*
brant	black	*Branta bernicla*		black scoter	*Melanitta nigra*
chickadee	chestnut-backed	*Poecile rufescens*		blue-winged teal	*Anas discors*
coot	American	*Fulica americana*		bufflehead	*Bucephala albeola*
cormorant		Phalacrocoracidae		canvasback	*Aythya valisineria*
	double-crested	*Phalacrocorax auritus*		cinnamon teal	*Anas cyanoptera*
	great	*Phalacrocorax carbo*		common goldeneye	*Bucephala clangula*
crane	sandhill	*Grus canadensis*		gadwall	*Anas strepera*
	whooping	*Grus americana*		greater scaup	*Aythya marila*
crow	northwestern	*Corvus caurinus*		green-winged teal	*Anas crecca*
dipper	American	*Cinclus mexicanus*		harlequin	*Histrionicus histrionicus*
duck	American wigeon	*Anas americana*		lesser scaup	*Aythya affinis*

Common and Scientific Names for Plants, Vertebrates, and Selected Invertebrates

Common names		Scientific names	Common names		Scientific names
	mallard	*Anas platyrhynchos*	heron	black-crowned	*Nycticorax*
	masked	*Oxyura dominica*		night	*nycticorax*
	northern shoveler	*Anas clypeata*		great blue	*Ardea herodias*
	oldsquaw	*Clangula hyemalis*		green	*Butorides virescens*
	pintail (northern)	*Anas acuta*		greenback	*Butorides striatus*
	redhead	*Aythya americana*		little blue	*Egretta caerulea*
	ring-necked	*Aythya collaris*		tricolored	*Egretta tricolor*
	ruddy	*Oxyura jamaicensis*	ibis	white	*Eudocimus albus*
	scaup	*Aythya* spp.		white-faced	*Plegadis chihi*
	surf scoter	*Melanitta*	jay	Steller's	*Cyanocitta stelleri*
		perspicillata	killdeer		*Charadrius vociferus*
	white-winged scoter	*Melanitta fusca*	kingfisher	belted	*Ceryle alcyon*
	wood	*Aix sponsa*		green	*Chloroceryle*
eagle	bald	*Haliaeetus*			*americana*
		leucocephalus	kite	Everglades snail	*Rostrhamus*
	golden	*Aquila chrysaetos*			*sociabilis*
egret	cattle	*Bubulucus ibis*			*plumbeus*
	great	*Casmerodius albus*		Mississippi	*Ictinia*
	reddish	*Egretta rufescens*			*mississippiensis*
	snowy	*Leucophroyx thula*		swallow-tailed	*Elandoides*
eider	common	*Somateria*			*forficatus*
		mollissima	limpkin		*Aramus guarauna*
	king	*Somateria*	loon	common	*Gavia immer*
		spectabilis		yellow-billed	*Gavia adamsii*
	spectacled	*Somateria fischeri*	magpie	black-billed	*Pica pica*
falcon	peregrine	*Falco peregrinus*	merganser	common	*Mergus merganser*
gallinule	purple	*Porphyrula*		red-breasted	*Mergus serrator*
		martinica	merlin		*Falco columbarius*
goose	Canada	*Branta canadensis*	osprey		*Pandion haliaetus*
	greater white-fronted	*Anser albifrons*	owl	northern spotted	*Strix occidentals caurina*
	snow	*Chen caerulescens*		pygmy	*Glaucidium gnoma*
goshawk	northern	*Accipiter gentiles*		snowy	*Nyctea scandiaca*
grebe	Clark's	*Aechmophorus clarkii*	oystercatcher	American black	*Haematopus bachmani*
	horned	*Podiceps auritus*	pelican	American white	*Pelecanus erythrorhynchos*
	pie-billed	*Podilymbus podiceps*		brown	*Pelecanus occidentalis*
	red-necked	*Podiceps grisegena*	plocwea		*Charadriidae*
	western	*Aechmophorus occidentalis*	plover	piping	*Charadrius melodus*
grouse	blue	*Dendragapus obscurus*	ptarmigan	rock	*Lagopus mutus*
				willow	*Lagopus lagopus*
	ruffed	*Bonasa umbellus*	puffin	tufted	*Fratercula cirrhata*
	sharp-tailed	*Tympanuchus phasianellus*	quail	California	*Calipepla californica*
	spruce	*Falcipennis canadensis*		mountain	*Oreortyx pictus*
gull		*Larus* spp.	rail	yellow	*Coturnicops noveboracensis*
	glaucous	*Larus hyperboreus*	raven	common	*Corvus corax*
			red knot		*Calidris canutus*
	ring-billed	*Larus delawarensis*	sanderling		*Calidris alba*
			sandpiper	Baird's	*Calidris bairdii*
gyrfalcon		*Falco rusticolus*		least	*Calidris minutilla*
hawk	red-tail	*Buteo jamaicensis*		pectoral	*Calidris melanotos*

Common names		Scientific names	Common names		Scientific names
	purple	*Calidris maritima*		least	*Sterna antillarum*
	semipalmated	*Calidris pusilla*	turkey	wild	*Meleagris*
	spotted	*Actitis macularia*			*gallopavo*
shrike	logger-head	*Lanius ludovicianus*	warbler	golden-cheeked	*Dendroica*
snipe	common	*Gallinago gallinago*			*chrysoparia*
stork	wood	*Mycteria canadensis*		northern parula	*Parula americana*
swallow	bank	*Riparia riparia*		prothonotary	*Protonotaria citrea*
swan	mute	*Cygnus olor*	waterthrush	Louisiana	*Seiurus motacilla*
	trumpeter	*Cygnus buccinator*	wren	sedge	*Cistothorus platensis*
	tundra	*Cygnus columbianus*	yellowlegs	lesser	*Tringa flavipes*
tern		*Sterna* spp.	yellowthroat	common	*Geothlypis trichas*

MAMMALS

Common names		Scientific names	Common names		Scientific names
antelope		*Antilocapra americana*	elk		*Cervus elaphus*
			ermine		*Mustela ermina*
armadillo	nine-banded	*Dasypus novemcinctus*	fox	Arctic	*Alopex lagopus*
bat	gray	*Motis grisescens*		gray	*Urocyon cinereoargenteus*
	greater bulldog	*Noctilio leporinus*		San Joaquin kit	*Vuleps macrotis mutica*
	hoary	*Lasiurus cinereus*	goat	rocky mountain	*Oreamnos americanus*
	Indiana	*Myotis sodalis*			
	red	*Lasiurus borealis*	hare	snowshoe	*Lepus americanus*
	Seminole	*Lasiurus seminolus*	lemming	brown	*Lemmus sibiricus*
	southeastern myotis	*Myotis austroriparius*	lion	mountain	*Puma concolor*
bear	black	*Ursus americanus*	lynx	Canada	*Lynx canadensis*
	brown (or grizzly)	*Ursus arctos*	manatee	West Indian	*Trichechus manatus*
beaver		*Castor canadensis*	marten		*Martes americana*
	mountain	*Aplodontia rufa*	mink	(American)	*Mustela vison*
bison		*Bos bison*	marmot	hoary	*Marmota caligata*
boar		*Sus scrofa*	moose		*Alces alces*
bobcat		*Lynx rufus*	muskrat		*Ondatra zibethicus*
caribou		*Rangifer tarandus*	muskox		*Ovibos moschatus*
	woodland	*Rangifer tarandus caribou*	nutria		*Myocastor coypus*
			opossum	water	*Chironectes minimus*
cougar		*Puma concolor*			
coyote		*Canis latrans*	otter	neotropical (long-tailed) river	*Lontra longicaudis*
deer	blacktail	*Odocoileus hemionus columbianus*		(northern) river	*Lontra canadensis*
				sea	*Enhydra lutris kenyoni*
	Columbian white-tailed	*Odocoileus virginianus leucurus*	panther	Florida	*Puma concolor coryi*
	mule	*Odocoileus hemionus*	porpoise	Gulf of California harbor	*Phocoena sinus*
				harbor	*Phocoena phocoena*
	white-tailed	*Odocoileus virginianus*	rabbit	eastern cottontail	*Sylvilagus floridanus*
				swamp	*Sylvilagus aquaticus*
dolphin		Delphinidae	raccoon		*Procyon lotor*

Common names		Scientific names	Common names		Scientific names
seal	bearded	*Erignathus barbatus*	tapir	Baird's	*Tapirus bairdii*
	harbor	*Phoca vitulina*	tepezcuintle		*Agouti paca*
sea-lion	Steller	*Eumetopias jubatus*	vole	water	*Microtus richardsoni*
sheep		*Ovis* spp.			
	bighorn	*Ovis canadensis*	weasel	long-tailed	*Mustela frenata*
	Dall	*Ovis dalli*	whale	beluga, white beluga	*Delphinapterus leucas*
shrew	dwarf	*Sorex nanus*			
	northern water (or water)	*Sorex palustris*		bowhead	*Balaena mysticetus*
				long-finned pilot	*Globicephala melaena*
squirrel	Columbian ground	*Spermophilus columbianus*	wolf	gray	*Canis lupus*
	fox	*Sciurus niger*		red	*Canis rufus*
	gray	*Sciurus carolinensis*	wolverine		*Gulo gulo*

LITERATURE CITED

Nelson, J. S., E. J. Crossman, H. Espinosa-Pérez, L. T. Findley, C. R. Gilbert, R. N. Lea, and J. D. Williams. 2004. *Common and scientific names of fishes from the United States, Canada, and México*. Special Publication 29. American Fisheries Society, Bethesda, Maryland.

GLOSSARY

aggradation: the process by which bottom sediments accumulate due to deposition of suspended sediments.

alkalinity: a measure of the buffering capacity of water or its capacity to neutralize an acid solution; commonly presented as milligrams of calcium carbonate per liter (mg/L as $CaCO_3$)

allochthonous: organic matter that originates from the terrestrial environment and eventually enters a stream or river; e.g., tree leaves, twigs, branches.

aquifer: a geological structure (rock) containing significant quantities of groundwater.

ash free dry mass (AFDM): a measure of biomass; the dry mass of organisms after the ash (mineral) content has been subtracted.

assemblage: a group of similar organisms in a particular ecological community; e.g., the fish assemblage of the Ogeechee River.

autochthonous: organic matter in a stream or river that originates from primary production within the stream; e.g., from algae and macrophytes.

autotrophic: a term describing those organisms that produce organic matter by the process of photosynthesis; also used to describe streams or rivers capable of producing sufficient oxygen within the water to meet the needs of community respiration (see *heterotrophic* and *P/R ratio*).

base flow: the flow in a stream or river that comes from groundwater seepage.

basin: see *drainage basin*.

benthic: refers to the habitat found on the bottom of a body of water or the organisms (see *benthos*) that live on the bottom. The benthic habitat typically consists or inorganic and organic matter such as sand, rocks, mud, leaves, and pieces of wood.

benthos: organisms that live in the benthic habitat; bottom-dwelling organisms; e.g., mayflies, mussels, etc.

biodiversity: term used to describe genetic, species, and ecosystem diversity within a given environment or region; when referring to species, high biodiversity means many different species of plants, animals, or microbes and low biodiversity means few species.

biofilm (organic microlayer, "slime"): the layer of organic material that develops on surfaces found in streams or rivers. It is usually composed of a variety of algae, fungi, bacteria, other microorganisms, and detritus within a polysaccharide matrix.

biomass: the weight or mass of living things; units may be in grams per square meter (g/m^2) of bottom area. Other units used may be dry mass (DM), ash-free dry mass (AFDM), carbon (C), or in the case of algae, chlorophyll.

biome: a broad global subdivision of the earth based on terrestrial plant communities; e.g., temperate deciduous forest.

carnivores: organisms (usually animals) that eat other animals (see *predators*).

catchment: see *drainage basin*.

coarse particulate organic matter (CPOM): organic detritus particles within a river or stream that are greater than 1mm in diameter; e.g., leaves, parts of leaves, twigs, etc.

community: all the organisms within an ecosystem or specified area, such as a stream reach; often used interchangeably with "assemblage," but the latter's use implies a subset of a complete community; e.g., invertebrate community or invertebrate assemblage. Either term implies an interconnectedness of species.

conductivity: see *specific conductance*.

decomposition: in the ecological sense, refers to the process by which organic matter, such as leaves, is broken down by physical and microbial action.

degradation: when referring to bed sediments, the process by which a streambed or riverbed loses sediments due to scouring, suspension, and transport.

detritus: nonliving organic matter of either autochthonous or allochthonous origin found on the bottom or in suspension in rivers and streams (usually refers to FPOM and CPOM).

detritivore: an animal that consumes detritus and its associated microbes (e.g., a stonefly shredder)

discharge: the volume of water flowing downstream at a specific site on a river; usually measured as cubic meters per second (m³/s) or cubic feet per second (cfs).

dissolved organic matter (DOM): organic matter, usually complex hydrocarbons, occurring in a dissolved state; for practical purposes, includes all material less than 0.45 μm in size.

drainage basin: an area of land drained by a particular stream or river; also called watershed (primarily in North America) and catchment (primarily in Europe).

drift: organisms that are passively carried downstream by the current; typically refers to benthic animals that are dislodged from or leave their habitats and eventually find new habitats or are eaten by predators.

dry mass (DM): a measure of biomass; the mass of organisms after all the water has been removed by drying.

ecoregion (freshwater): a relatively large area that contains a geographically distinct assemblage of freshwater organisms, particularly fishes.

ecoregion (terrestrial): a large landscape division defined by a geographically distinct assemblage of natural species (generally by plant species) along with climate, geology, and terrain; smaller than a biome.

ecosystem: a community of organisms and their nonliving environment interacting as a unit. Boundaries of ecosystems can be somewhat arbitrary; e.g., we can narrowly define the river ecosystem or more broadly define the river–riparian ecosystem or the ecosystem of the drainage basin.

epilimnion: the upper zone of a thermally stratified lake in which the water is warmer than in the lower zone (hypolimnion).

eutrophic: the condition of a body of water (usually a lake) in which there are high concentrations of nutrients and high primary production. Many reservoirs on rivers become culturally eutrophic because of sewage outfalls from cities and non-point-source pollution from agriculture.

evapotranspiration: the combined processes of evaporation of water and transpiration by plants, which together describe the loss of water to the atmosphere from an ecosystem.

fauna: the animals from a particular habitat, ecosystem, or region.

filtering collectors: a functional feeding group that feeds by filtering small particles of organic matter (FPOM, seston) from the water column using a variety of nets, leg hairs, modified mouth parts, and other structures.

fine particulate organic matter (FPOM): organic detritus particles within a river or stream that are greater than 0.45 μm but less than 1 mm in diameter.

floodplain: a low-lying area along a river corridor that is periodically inundated during high discharge as the river overflows its banks.

flora: the plants from a particular habitat, ecosystem, or region.

functional feeding groups: groups of aquatic invertebrates defined by how they collect their food (shredders, scrapers or grazers, filtering collectors, gathering collectors, predators). May also be applied to other animal groups, such as fishes, and equivalent to the general ecological term "guild."

gaging station: a site on a stream or river where hydrological data (river height, river discharge) are obtained.

gathering collectors: a functional feeding group that feeds on organic matter (FPOM) that has settled on the riverbed and its various substrates.

geomorphology: an area of science concerned with the shape of the physical environment and the processes that created it; with respect to rivers, it includes form of the riverbed and floodplain, substrate types and their distribution, and gradient (slope).

gradient: the slope of a river as it flows in a downstream direction; can be measured as meters of decline per kilometer of length (m/km) or as a percentage. A 1% slope equals 10 m/km.

grazers: see *scrapers*.

groundwater: water located in the saturated zone below the surface of the land.

headwaters: the upper reaches of a drainage basin; e.g., 1st order streams.

herbivores: animals that consume living algal or plant material (in contrast to consuming dead plant material); e.g., a caddisfly grazer.

herpetofauna: a collective term referring to the amphibians and reptiles present in a habitat or region.

heterotrophic: a term describing those organisms that obtain their nourishment by feeding on other organisms (in contrast to autotrophic); also used to describe rivers or streams that cannot produce sufficient oxygen within the water to satisfy respiration demands of consumers and are thus dependent upon terrestrial or upstream sources of energy to supplement the autotrophic contributions (see *autotrophic* and *P/R ratio*).

hydrograph: the continuous record of streamflow, or discharge, through time.

hypolimion: the lower zone of a thermally stratified lake in which the water is colder than the upper zone (epilimnion). In an eutrophic lake, the oxygen concentration will be lower (possibly anoxic) than the epilimnion.

hyporheic zone: the zone of water in the interstitial spaces between bed sediments that is below the open-flowing water zone of a river or stream; it may extend laterally beyond the wetted perimeter of the stream or river.

inundation: becoming flooded by water, as in the inundation of a river floodplain.

Jackson turbidity units (JTU): a measure of the turbidity (cloudiness) of water. Turbidity is caused by the suspension of fine particles. Provides similar numbers as nephelometric turbidity units.

lacustrine: relating to lakes or having the characteristics of a lake; e.g., the river behind the dam is lacustrine.

landscape: in the general ecological sense, refers to a large area of land composed of many clusters of interacting ecosystems.

lentic: refers to nonflowing water; ponds and lakes are lentic systems.

levee: a man-made or natural earthen barrier along the edge of a river that prevents overflow of the river into adjacent land.

lotic: refers to running water; streams and rivers are lotic systems.

macroinvertebrates: invertebrate animals that are retained by a 0.5 mm sieve; e.g., mayflies, stoneflies, snails; in practice, however, somewhat smaller animals are often included.

macrophyte: usually refers to aquatic flowering plants entirely or partially submerged in the water; some definitions include macro-algae and mosses.

main stem: the primary downstream segment of a river, as contrasted to its tributaries.

meiofauna: the assemblage of extremely small organisms (ciliates, protists, microcrustaceans, early growth stages of macroinvertebrates); may be found in any habitat, but generally tend to be the predominant form in the interstices of the substrate, often to considerable depth.

microbes: primarily refers to microorganisms, such as bacteria and fungi.

mosses: small, nonflowering plants usually found in colder, well-oxygenated reaches of streams or rivers.

nephelometric turbidity units (NTU): a measure of the turbidity (cloudiness) of water. Turbidity is caused by the suspension of fine particles. Provides similar numbers as Jackson turbidity units.

nonpoint-source pollution: pollution that runs off a wide area of land (opposite of *point-source pollution*), such as from agricultural fields.

omnivores: animals that consume both plant and animal matter. A more general definition involves feeding on more than one trophic level.

P/R ratio: a value found by dividing the net production of oxygen (P) within the water column (from photosynthesis) by the 24 hr respiration (R) of a reach of stream. A value equal to or exceeding unity indicates an autotrophic reach and a value less than unity indicates a heterotrophic reach.

periphyton: algae, cyanobacteria, and photosynthetic protists associated with any stream substrate or surface; comprises part of the biofilm.

pH: a measure of the relative acidity or alkalinity of water. A value of 7 is considered neutral; values less than 7 are increasingly acidic and values greater than 7 are increasingly basic.

physiographic provinces: subdivisions of the continent based on topographic features, rock type, and geological structure and history; e.g., Coastal Plain, Piedmont Plateau.

phytoplankton: free-floating algae, cyanobacteria, and photosynthetic protists suspended in the water column that are capable of carrying out all life processes while suspended.

piscivores: animals (often referring to fishes) that eat fishes.

plankton: minute organisms (animals, algae, bacteria, and protists) suspended in the water column that are capable of carrying out all life processes while suspended.

point-source pollution: pollution that originates from a single source (pipe), such as from a sewage treatment plant or an industry.

predators: a functional feeding group that consumes other animals (see *carnivores*).

primary production: a measure of organic matter produced by autotrophic organisms (e.g., plants and algae) through photosynthesis over some period of time; units often reported as milligrams of carbon per square meter per day ($mg\ C\ m^{-2} d^{-1}$) or per year. Units other than carbon (AFDM, dissolved oxygen) are also used.

productivity: usually refers to the level of primary production of algae or plants or of microbes; e.g., productivity is low because of high turbidity or sedimentation (see *primary production* and *secondary production*).

refractory: refers to chemicals or organic matter that do not degrade easily and provide little available carbon to consumers.

riparian: refers to the region or corridor immediately bordering a river or stream; also refers to organisms living within this corridor; e.g. riparian vegetation.

river kilometers (Rkm): the distance in kilometers along the course of a river, including all the bends. Rkm are typically numbered from the river mouth (e.g., Rkm 120 is a location 120 km from the mouth); may also be used to indicate a distance along the river (e.g., the distance from point A to point B is 500 Rkm).

runoff: the fraction of precipitation that appears in the river. As used in this book it is a measure of the amount (height) or water that drains from a river basin over a unit of time (e.g., cm/mo or cm/yr); estimated by dividing discharge by drainage basin area.

scrapers (or grazers): a functional feeding group that scrapes food, such as algae, from a substrate, such as rocks.

secondary production: organic matter produced by heterotrophic organisms; often refers to production of animals but also applies to microbes. Often presented as mg dry mass $m^{-2} yr^{-1}$.

seston: organic matter (mainly FPOM and plankton) suspended in transport in a river; a major food source for filter-feeding animals.

shredders: a functional feeding group that consumes vascular plant material; usually refers to consumption of CPOM.

specific conductance: conductivity is a measure of the ability of water to conduct electricity, measured in microSiemens per cm (μS/cm), and thus is a measure of the concentration of ionic (dissolved) constituents. Because conductivity changes with water temperature, specific conductance is used as a measure that normalizes conductivity to 25°C. Specific conductance is sometimes used as an indication of total dissolved solids (TDS) and salinity.

stream order: a numbering system of streams based on the number of tributaries and how they are joined. The smallest-order stream is 1st order and the largest-order streams in North American are 9th or 10th order.

substrate (or substratum): generally refers to the composition of rock minerals in the bed of rivers. They are typically separated by increasing size into silt, sand, gravel, cobble, and boulder. Substrate can also apply to other surfaces that might be colonized by microorganisms or small animals, such as wood or macrophytes. For microbes, substrate can refer to an energy (carbon) source.

taxon (taxa is plural): refers to a taxonomic group that has been scientifically designated. Often refers to the species or genus level.

transpiration: the process by which plants release water vapor into the air.

turbidity: see *Jackson turbidity units* or *nephelometric turbidity units*.

watershed: see *drainage basin*.

zooplankton: minute animals suspended in the water column; usually microcrustaceans (cladocerans and copepods), rotifers, and other similar organisms that are capable of carrying out all life processes while suspended.

INDEX OF RIVERS

A

Abitibi River, 921, 923, 926, 935
Alabaha River, 122
Alabama River, 126, 130–136, 146, 168
Alapaha River, 152–155, 172
Alatna River, 785, 789, 791, 798
Allagash River, 954, 958, 972
Allegheny River, 376, 378–381, 408, 413, 423, 424
Alsek River, 736–738, 771
Altamaha River, 73, 74, 76, 97, 107, 121, 152
American River, 547, 549, 551, 580
  North Fork, 551
Andreafsky River, 785
Androscoggin River, 22, 54, 67, 68
Angelina River, 228
Animas River, 518
Aniuk River, 910
Apalachicola River, 125–129, 135, 140–147, 154, 170
Applegate River, 569, 570, 572, 573, 584
Appomattox River, 112
Aqua Fria River, 513
Aravaipa Creek, 515, 516
Arctic Red River, 805, 808, 809, 811
Arkansas River, 2, 231, 233–235, 237–239, 242, 245, 246, 273, 283–298, 303, 308, 309, 315, 316, 333, 1089
  South Fork, 291
Armagosa River, 658
Armería, Río, 1032, 1033, 1036, 1066, 1083
Aroostook River, 954, 957, 958, 972
Aros, Río, 1054, 1055, 1057
Arroyo Seco River, 558, 560, 562, 582
Asay Creek of the Sevier River, 669
Ashuapmushuan River, 1002, 1021
Assiniboine River, 853–855, 858, 859, 867–873, 879, 896, 897, 899
Atchafalaya River, 233, 234, 236, 239, 240, 241, 244, 264, 265, 272, 277, 299, 317
Athabasca River, 701, 708, 805, 806, 808, 809, 811, 819, 820, 823, 826, 829–834, 842, 844, 846
Atigun River, 918, 920, 934
Atlin River, 782
Auglaize River, 1025
AuSable River, 984, 986, 1011, 1087
  North Branch, 1024
  South Branch, 1024
Avan River, 911
Ayuquila, Río, 1032, 1033, 1036, 1066, 1083
Ayutla, Río, 1083

B

Babine River, 756, 757, 768
Bad River, 431
Balleza, Río, 1061
Balsas, Río, 1036
Baraboo River, 352, 366
Barkley Canal, 391
Barren River, 418
Battle Creek of the Bear River, 661
Batzu River, 790
Baudette River, 879
Bavispe, Río, 1054, 1055, 1057, 1078
Bayou Bartholomew, 279
Bayou De View, 265, 278
Bear Creek of the Buffalo River (AR), 256, 274
Bear Creek of the Rogue River (OR), 569, 570, 573, 584
Bear River (CA), 547

Bear River (UT, ID, WY), 656, 657, 660–667, 684, 688, 1089
  Blacksmith Fork, 666
  Smith's Fork, 663
Beaver (North Canadian) River (OK), 284, 287, 294
Beaver River (SK), 868
Beaver River (UT), 667, 668, 671, 672, 689
Belle Fourche River, 459, 476
Bell–Irving River, 773
Betsiamites River, 1005
Big Black River, 233–235, 239, 240, 242, 256–259, 272, 275
Big Blue River, 479
Big Bonito Creek, 536
Big Chino Wash, 519, 537
Big Darby Creek, 408, 422
Big Escambia Creek, 174
Big Fork River, 876
Big Hole River, 429
Bighorn River, 429, 440–444, 470
Big Muddy River, 432
Bigoray River, 833
Big Piney Creek, 255
Big Piney River, 454, 456, 458, 473
Big Salmon River (NB), 940, 942, 963, 964, 974
Big Salmon River (YT), 780
Big Sandy River (WV, KY), 376, 380, 384
Big Sandy River (AZ), 519, 535
Big Sioux River, 427, 428, 430, 431, 459, 462, 477
Big Walnut Creek, 422
Bill Williams River, 484, 487, 519, 535, 1087
Black Creek, 126, 159, 176
Black River (AK), 802
Black River (AR, MO), 246, 265, 273
Black River (AZ), 484, 487, 511, 519, 520, 536, 538
  East Fork, 519
  West Fork, 519
Black River (NC), 83, 85, 86
Black River (NY), 984, 986, 1012, 1026
Black River (WI), 349
Black River (YT), 817
Black Warrior River, 126, 129, 130, 132–135, 139, 168, 384
  Locust Fork, 139
Blackwater River (FL), 129
Blue Earth River, 339, 340, 342, 364
Blue River (BC), 709
Blue River (OK), 284, 285, 288, 289, 301, 309, 310, 324
Blue Springs, 509, 510
Bluestone River, 405
Boggy River, 301
Bogue Chitto Creek, 147, 171
Bois Brule River, 344
Bois de Sioux River, 867, 869, 888
Boise River, 594, 608
Bonnet Plume River, 809, 837, 851
Boone River, 372
Boulder River, 443
Bow River, 854, 855, 859, 860, 862–888, 895, 900
Bowron River, 706
Brandywine Creek, 1010
Bravo del Norte, Río (see Rio Grande), 184, 185, 1032, 1034, 1035, 1059, 1061–1063, 1065, 1082
Brazos River, 181–185, 203–209, 223, 1088, 1089
  Clear Fork, 203, 205, 223
  Double Mountain Fork, 203, 223

  Salt Fork, 203, 223
Bridge River, 702
Brier Creek, 304
Bright Angel Creek, 490, 494
Broad River (GA), 92
Broad River (SC), 120
Bruneau River, 608, 643
Buchans Brook, 947
Bucks Creek, 140
Buffalo (National) River (AR), 233, 235, 246, 247, 250, 251–256, 259, 272–274
Buffalo River (TN), 390
Bulkley River, 756, 768
Burntwood River, 854, 879, 881, 883, 884, 887
Burro Creek, 535
Buttahatchee River, 177

C

Cacapon River, 71
Cache Creek, 552, 580
Cache la Poudre River, 454
Cache River, 233–235, 265, 272, 273, 278
Cahaba River, 126, 129–132, 134–140, 168, 169
Cains River, 971
Calapooia River, 617
Calumet River, 352
Canadian River, 283–290, 292, 294–299, 315, 316
Cañas, Río, 1046
Candelaria, Río, 1031, 1032, 1033, 1036, 1050–1054, 1077, 1090
Caney Fork River, 391, 392
Canning River, 920
Cannonball River, 428, 431
Canoochee River, 93, 94, 115
Caopacho River, 958
Cape Fear River, 74, 76, 83–88, 113, 1092, 1093, 1097
Caribe, Río, 1050, 1051
Cariboo River, 730
Caribou River, 847
Carrizal, Río, 1046, 1048
Carson River, 657, 676
Catamaran Brook, 950, 952
Catawba River, 120
Cebolla Creek, 517
Chaco River, 533
Chama River, 188
Champlain Canal, 37
Chariton River, 431
Charley River, 785
Chatanika River, 787
Chattahoochee River, 126, 128, 129, 135, 140–147, 170
Cheat River, 408, 424
Chemung River, 66
Chena River, 779, 787, 788, 797
Cheslatta River, 714
Chevelon Creek, 508, 510, 530
Chewak River, 632
Chewuch River, 651
Cheyenne River, 427, 428, 431, 435, 459, 462, 469, 476
Chicago River, 352, 357
Chickasawhatchee Creek, 140
Chickasawhay River, 161, 176
Chicopee River, 32, 63
Chikaskia River, 287
Chilanko River, 720, 731
Chilcotin River, 698, 699, 702, 703, 720, 725, 731
Chilco River, 720, 731
Chilikadrotna River, 758

Chinchaga River, 849
Chinle Wash, 533
Chipola River, 143, 170
Chippewa River (of Minnesota River basin) (MN), 339, 343, 364
Chippewa River (WI), 328, 330, 333, 349, 358, 363, 368
Chisana River, 786
Chitina River, 759, 770
Choctawhatchee River, 125, 126, 128, 129, 157, 158, 173
Choptank River, 22
Chulitna River, 744, 745, 765
Churchill River
    (NF, Labrador), 2, 22, 939, 940, 942, 965, 980, 1089, 1095
    (SK, MB), 853–860, 880, 882–885, 887, 899, 903, 1089, 1091, 1094, 1095, 1099
Chuviscar, Río, 1059
Cimarron River, 283–285, 287–291, 293, 303, 308, 315, 319
Clackamas River, 617, 645
Clarion River, 408, 423
Clark Fork, 594–598, 624, 628, 642, 646
Clarks Fork, 444
Clear Creek of the Little Colorado River, 508, 510
Clearwater River (AB, SK), 846
Clearwater River (BC), 698, 699, 709, 720, 726, 732
Clearwater River (ID), 591, 593, 594, 603, 604, 608, 609, 632, 643, 649
    Middle Fork, 609, 632, 649
    North Fork, 609, 632, 633, 649
    South Fork, 632
Clinch River, 384, 386, 388, 390, 414
Coal River, 816
Coatzacoalcos, Río, 1032, 1036
Coldwater River, 260, 276
Colorado River (CA, NV, UT, WY, CO, AZ, NM, Mexico), 2, 16, 199, 435, 483–538, 659, 1089, 1091, 1095, 1096, 1099, 1100
Colorado River (TX, NM), 181, 183–185, 198–203, 222
Columbia River, 1, 2, 16, 591–652, 708, 758, 830, 1088, 1089, 1091, 1096, 1100
Colville River, 742, 907, 914, 916, 918, 920
Comal River, 192, 195, 197
Concepcíon, Río, 1056
Concho River (TX), 199, 201, 222
Conchos, Río (Mexico), 182–185, 188, 190, 192, 220, 1032, 1033, 1035, 1036, 1058–1065, 1079, 1095
Concord River, 69
Conecuh River, 126, 157, 174
Conestoga Creek, 49
Congaree River, 105, 107, 120
Connecticut River, 21, 30–35, 63, 1093, 1097, 1100
Contoocook River, 69
Cooper River, 76, 107, 120
Coosa River, 128, 130–135, 146, 168
Copper River, 2, 736, 737, 739, 758, 759, 770
Coquihalla River, 701
Cossatot River, 305, 306, 318
Cottonwood River, 339, 364
Couer d'Alene River, 594, 595, 650
Coulogne River, 997, 1020
Cowlitz River, 591, 593, 595, 635, 653
Cowpasture River, 77
Coy, Río, 1040
Coyote Creek, 561
Crawfish River, 371
Crooked Creek (AK), 742
Crooked Creek (AR), 255
Cumberland River, 375, 376, 378, 379, 381, 387, 390–396, 404, 405, 413, 415
    Big South Fork, 391, 396

Clover Fork, 390, 392
    Martin's Fork, 390, 392, 396
    Poor Fork, 390, 392
Current River, 231, 233, 235, 246, 247, 250, 265, 272, 273, 281
    Jack's Fork, 250, 265, 281
Cutler River, 910
Cypress Bayou, 301, 304

D

Dan River, 118
Date Creek, 519
Dead River, 67
Dease River, 816
Deep Fork River, 287, 294, 316
Deep River, 84, 86, 113
Deerfield River, 32, 63
Delaware River, 21–24, 44–49, 57, 65, 1093, 1097, 1099
    East Branch, 45, 47, 48
    West Branch, 45, 48
Delta River, 786
Deschutes River, 595, 614
Deshka River, 750
Des Moines River, 328, 331, 358, 360, 363, 372
    East Fork, 372
Des Plaines River, 352, 354–357, 367
Devil's Club Creek, 621
Devils River, 188
Donjek River, 791, 799
Donner Creek of the Truckee River, 680
Driftwood River, 718
Dry River, 431
Dubli River, 790
Duchesne River, 496, 498, 501, 528, 667
Duck River, 384, 386, 390, 414
du Liévre River, 997
Dungarvon River, 971
Dunk River, 1, 940, 942, 964, 975, 1088

E

East Canyon Creek, 694
East Pearl River, 148, 151
East River, 517
East Verde River, 519
East Walker River, 684, 694
Eau Claire River, 368
Econlochhatchee River, 101
Edisto River, 76
Eel River, 542, 543, 573, 586, 1089
    Middle Fork, 586
    South Fork, 586
Elbow River, 900
Eli River, 910
Elk River, 401, 403, 417
Elkhart River, 1022
Elkhorn River, 451, 452, 454, 472
El Tigre, Río, 1050
Embarras River, 396, 399, 416
English River, 873–878, 898
Erie Canal, 397, 407
Escambia River, 126, 128, 157, 159, 174
Escatawpa River, 129, 159, 176
Esperanza, Río, 1050, 1051
Estrella River, 558, 560, 582
Exploits River, 939, 940, 942, 943–947, 970

F

Farmington River, 32, 63
Feather River, 547, 549, 551, 580
    Middle Fork, 551
Finlay River, 824
Firesteel River, 873
Flambeau River, 368
Flathead River, 591, 593–595, 597–603, 613, 624, 628, 642
    Middle Fork, 597, 599, 600, 642
    North Fork, 599, 642

South Fork, 599, 601, 603, 642
Flint River, 126, 135, 140–147, 158, 160, 170, 175
Florido, Río, 1059–1063, 1079
Forrest Kerr River, 753, 754
Fort Nelson River, 815, 816, 843
Forty Mile River, 785
Fossil Creek, 519
Fourche Maline River, 322
Fox River, 348, 352, 367
Francis River, 816
Fraser River, 2, 697–733, 758, 1088, 1089, 1094
Frederickhouse River, 923, 926
French Broad River, 384, 386, 414
Frenchman River, 475
French River, 423
Frio River, 226
Fuerte, Río, 1032, 1033, 1035, 1036, 1058, 1060, 1064, 1065, 1080

G

Galice Creek of the Rogue River, 569
Gallatin River, 429, 431, 432, 435, 474
Gallinas, Río, 1039
Gander River, 939
Gasconade River, 427, 428, 431, 454–459, 473
    Osage Fork, 454, 473
Gatineau River, 997, 1000, 1020
Gauley River, 401–404, 417
Gerstie River, 786
Gila River, 484–488, 511–517, 521, 527, 531, 1058, 1095
    East Fork, 511, 514
    Middle Fork, 511
    West Fork, 511, 514
Gisasa River, 790
Glover River, 305, 306, 308, 318
Goodpaster River, 787
Grand Cascopedia River, 940, 942, 965, 977
Grand (Neosho) River, 284, 286, 287
Grand River (MO), 427, 428, 461, 464, 480
    Locust Fork, 480
Grand River (SD), 428, 431, 432
Grande Ronde River, 591, 593, 595, 604, 608, 632, 643, 648
Grass River, 879
Grave Creek of the Rogue River, 570
Great Bear River, 808, 809, 811, 813
Great Miami River, 375, 376, 378, 379, 396, 405, 407, 413, 420, 1093
Great Pee Dee River, 74, 76, 105, 106, 119
Great Rattling River, 970
Greenbrier River, 401, 403, 404, 417
Green River (IL), 371
Green River (KY), 375, 376, 378, 379, 405, 406, 413, 418
Green River (ME), 956
Green River (UT, CO, WY), 484, 485, 487, 491, 493, 496–506, 527, 528, 663, 667, 672, 1100
Grijalva, Río, 1032, 1033, 1036, 1041, 1043, 1052, 1076, 1089
Groundhog River, 921, 923, 925, 935
Guadalupe River, 182–185, 192–198, 221
    North Fork, 192
    South Fork, 192
Guayalejo, Río, 1042, 1081
Gulkana River, 770
Gunnison River, 484, 486, 487, 505, 517, 532
    North Fork, 517

H

Hall Creek, 506
Harpeth River, 391
Harrison River, 699, 720, 733
Hart River, 837, 851
Hassayampa River, 514

Hawk Creek, 339
Haw River, 84, 86, 88, 113
Hayes River, 855, 860, 874, 880
Hay River, 806, 836, 842, 849
Healey River, 787
Heart River, 431
Hell Roaring Creek, 431
Hess River, 800
Highwood River, 900
Hiwassee River, 384, 390
Hogatzu River, 790
Holitna River, 742, 764
Holston River, 384, 386, 388, 390
Homochitto River, 235
Hondo, Río, 1050, 1077
Horsefly River, 720, 730
Horse Lick Creek, 394
Housatonic River, 22
Hudson River, 21, 22, 24, 25, 35–43, 57, 64, 1091–1093, 1097, 1099
Humber River, 939, 940, 942, 965, 979
Humboldt River, 656, 657, 660, 673–677, 690
Hurricane Creek, 122
Huslia River, 790

I

Ichawaynochaway Creek, 140, 175
Illinois Bayou, 253, 255
Illinois River (AR, OK), 284, 285, 288–290, 309, 321
    Baron Fork, 308, 310
Illinois River (IL), 328, 331, 338, 363, 367, 382, 1088
Illinois River (OR), 569, 571, 573, 584
Imnavait Creek, 913–916
Indian River (NY), 42
Indian River (YT), 791
Inklin River, 772
Innoko River, 782
Iowa–Skunk River, 358
Ipnelivik River, 910
Iskut River, 752–754, 767
Ivishak River, 918–920, 934

J

Jackson Creek of the Rogue River, 569
Jackson River, 77
James River (SD), 431, 469
James River (VA), 57, 73, 74, 76–83, 112, 1097
Jarbridge River, 608
Jefferson River, 429, 431, 432, 474
Jocko River, 597
John Day River, 591, 593, 595, 619, 632, 634, 652
    Middle Fork, 632, 652
    North Fork, 632, 652
    South Fork, 632, 652
John River, 789, 791
Johnson River, 749
Jordan River (ID, OR), 647
Jordan River (UT), 684
Juday Creek, 1009–1011
Judith River, 431
Juniata River, 50, 66

K

Kanawha River, 375, 376, 378–380, 401–405, 413, 417
Kankakee River, 352, 354–356, 367, 1005
Kansas (Kaw) River, 283, 427, 428, 430, 431, 460, 463, 469, 479
Kantishna River, 786, 797
Kanuti River, 790, 798
Kapuskasing River, 921, 923–926, 935
Kaskaskia River, 328, 331, 360, 373
Kateel River, 790

Kaweah River, 553, 555
Kazan River, 904–906, 927, 937, 1089
Kechika River, 817, 843
Kelly River, 913
Kenai River, 736, 739, 744, 746–751, 766
Kennebec River, 21, 22, 26, 54, 55, 67, 68
Kennebecasis River, 954
Kentucky River, 375, 376, 378, 379, 405–407, 419
    North Fork, 419
    South Fork, 419
Kern River, 553, 555, 557, 581
    North Fork, 557
    South Fork, 557
Kettle River, 344, 345, 365, 1087
Kiamichi River, 284, 285, 287–289, 301, 302, 310, 325
Kickapoo River, 366
Killey River, 748, 766
Kinchafoonee Creek, 140, 175
Kings River (AR), 235
Kings River (CA), 553–555, 557, 581
    Middle Fork, 557
    South Fork, 557
Kisaralik River, 742
Kishwaukee River, 371
Kispiox River, 756
Kissimmee River, 77, 1099
Kitsumkalum River, 756
Klamath River, 542, 543, 545, 548, 563–568, 571, 572, 583, 1100
Klappan River, 753
Klewi River, 823
Klondike River, 776, 778, 780, 791, 792, 801
Knife River, 431, 432
Kobuk River, 912
Koksoak River, 2, 965
Kootenay River, 594, 596, 624, 629, 646
Koyukuk River, 775, 776, 779, 780, 782, 789–791, 796, 798
    North Fork, 785, 789, 791
    Middle Fork, 789
    South Fork, 789
Kuparuk River, 742, 904–906, 908, 913–919, 933
Kuskokwim River, 2, 736, 739, 741–743, 1089
    East Fork, 742
    North Fork, 742,
    South Fork, 742, 764
Kuzitrin River, 742
Kvichak River, 743
Kwethluk River, 742

L

Lacanjá, Río, 1032, 1033, 1066, 1067, 1076, 1084
Lacantun, Río, 1044, 1048, 1076, 1084
Lac qui Parle River, 339, 364
La Grande Rivière, 2
LaHave River, 940, 942, 943, 964, 976
Lamine River, 431
Lamington River, 70
Lamoille River, 1027
La Moine River, 352, 367
La Plata River, 518
Laramie River, 429
Lavaca River, 182
Leaf River, 161, 176
Lehigh River, 65
Lemhi River, 607
Lerma, Río, 1034, 1036
Le Sueur River, 342
Licking River, 375, 376, 378, 379, 407, 408, 421
    South Fork, 421
Lillooet River, 698, 699, 701, 720, 725, 733
Little Androscoggin River, 68
Little Bear River, 666

Little Bitterroot River, 597, 642
Little Cahaba River, 139
Little Cheyenne River, 459, 476
Little Colorado River, 186, 484, 486–488, 490, 491, 493, 515, 527, 530
    East Fork, 506
    South Fork, 506
    West Fork, 506
Little Darby Creek, 408
Little Fork River, 876
Little Miami River, 376, 378, 379
Little Missouri River, 431, 469
Little Moose River, 816
Little Pee Dee River, 119
Little Piney River, 454
Little Red River, 247
Little River (AL), 134
Little River (AR, OK), 284, 285, 288, 289, 301, 302, 304–308, 318
    Rolling Fork, 305, 306, 318
Little River (KY, TN), 391, 396, 415
Little River (of Canadian River basin) (OK), 296
Little River (TX), 223
Little Satilla River, 122
Little Sioux River, 431
Little Smoky River, 835, 848
Little Snake River, 502–505, 529
Little Southwest Miramichi River, 950, 952, 971
Little Tallahatchie River, 259, 260, 261
Little Tennessee River, 384, 390, 414
Little Truckee River, 680
Little Wabash River, 396, 398, 416
Little Washita River, 303
Little White River, 445, 447–449, 471
Llano River, 199, 201, 222
Lloyds River, 944, 970
Lochsa River, 609, 632, 649
Logan River, 666
Long Tom River, 617
Lookout Creek, 621
Lost River, 563–565, 583
Loup River, 451, 452, 454, 472
Lower Mississippi River, 231–281, 329, 332, 333, 1089, 1091, 1096
Luckiamute River, 617
Lumber River, 105, 119
Lynches River, 119

M

Mack Creek, 621
Mackenzie River, 1, 2, 192, 431, 701, 775, 805–851, 903, 906, 920, 941, 1088, 1089, 1094, 1098
Mackinaw River, 352, 367
Madawaska River, 954, 1020
Madison River, 7, 427, 428, 429, 431, 432, 435, 459, 460, 474
Malad River, 661
Maligne River, 877, 898
Mamantel, Río, 1052, 1053
Mammoth Creek, 669, 670
Manicouagan River, 939
Manouane River, 1005, 1028
Margaree River, 940
Marias River, 429, 431
Mary's River (NV), 675
Mary's River (OR), 617
Matamek River, 961, 962
Matawin River, 1028
Mattagami River, 921–926, 935
Mattaponi River, 104, 117
Maumee River, 396, 984, 986, 1005, 1011, 1012, 1025, 1088
Mayo, Río, 1056, 1058, 1060
Mayo River, 791, 800
McCleod River, 846
McCloud River, 549, 550
McGregor River, 701
McKenzie River, 617, 619, 621, 645

Mehoopany Creek, 54
Meister River, 816
Merced River, 553, 555, 557, 581
Merrimack River, 21, 22, 69
Meshoppen Creek, 51, 52
Methow River, 591, 593–595, 624, 632, 634, 651
Mezcalapa, Río (Grijalva), 1046, 1048
Middle River, 717
Milk River, 427–429, 431, 459, 461, 469, 475, 593
Mill Creek, 394
Millstone River, 70
Minnesota River, 328, 330, 331, 333, 335, 338–343, 345, 363, 364, 1088, 1093
Miramichi River, 940–942, 945–953, 957, 971
Missinaibi River, 921, 923, 925, 926, 935
Missisquoi River, 1027
Mississippi River, 1, 2, 16, 126, 147, 160, 192, 231–281, 290, 293, 299, 315, 327–773, 376, 381, 427, 428, 434, 437, 489, 775, 868, 983, 991, 996, 1005, 1088, 1091, 1096
Missouri River, 1, 2, 6, 192, 231, 237, 285, 327, 332, 334, 336, 337, 351, 363, 427–480, 486, 593, 775, 860, 868, 1088, 1089, 1092, 1094–1096, 1099, 1100
Mistassibi River, 1021
Mistassini River, 1002, 1021
Mitchell River, 720
Mobile River, 2, 6, 125–135, 140, 168, 1089, 1091
Moctezuma, Río (of Río Pánuco basin), 1037–1041, 1043, 1045, 1075
Moctezuma, Río (of Río Yaqui basin), 1056, 1078
Mohave River, 658
Mohawk River, 37, 64
Moisie River, 9, 10, 939, 940, 942, 973, 1088
Molalla River, 617
Monongahela River, 375, 376, 378–381, 408, 423, 424
    West Fork, 424
Moose River (AK), 748
Moose River (ON), 904, 905, 906, 921–926, 935
Mora River, 296
Moreau River, 428, 431
Morice River, 756
Mountain Fork River, 287, 305, 306, 308, 318
Muckalee Creek, 140
Mulberry River, 253, 255
Mulchatna River, 758, 769
Mullica River, 22
Murrieta Creek, 576, 589
Musconetcong River, 65
Muskingum River, 376, 379, 380
Musselshell River, 428, 431

N

Nabesna River, 786
Naches River, 611, 614, 644
Nacimiento River, 558–561, 582
Nadine River, 714
Nahua River, 69
Nakina River, 772
Namekagon River, 344, 347, 365
Nanawaya Creek, 147
Nashwaak River, 954
Nass River, 736, 737, 754, 757, 760, 773
Nautley River, 701, 712, 714, 715
Navasota River, 203, 208
Nazco River, 729
Nechako River, 698, 701, 712–719, 725, 727, 1094
Neches River, 182–185, 209, 215, 224, 228, 1090
Nelson River, 1, 2, 16, 775, 853–901, 903, 1088, 1089, 1091, 1094, 1095, 1099

Nenana River, 786
Neosho (Grand) River, 284, 285, 288–290, 308, 315, 320
Neuse River, 76
Neversink River, 45
New Hope River, 84
New River (CA), 568
New River (NC, VA, WV), 380, 401–405, 417
Nicola River, 701
Ninnescah River, 291
Niobrara River, 427–429, 431, 448, 460, 463, 469, 478
Nipissis River, 958, 963, 973
Nishnabotna River, 431
Nisling River, 799
Noatak River, 904, 905, 906, 908–913, 932
Noel Paul's Brook, 947, 970
Nolin River, 418
Nordenskiold River, 780
North Anna River, 104, 117
North Canadian River, 287, 288, 294, 315, 316
Northeast Cape Fear River, 83, 85, 113
Northeast Humber River, 979
North Fork River, 390
North Platte River, 427, 429, 449, 450, 451, 453, 454, 472
North Santiam River, 645
North Saskatchewan River, 701, 708, 854, 859, 860, 862–867, 895
North Seal River, 927, 936
North Thompson River, 701, 707–711, 720, 726
North Umpqua River, 573, 585
Northwest Miramichi River, 948, 950, 952, 971
Nowitka River, 785
Noxubee River, 177
Nueces River, 182, 183, 185, 213–215, 226
Nushagak River, 736, 737, 739, 742, 743, 758, 769
Nuyakuk River, 769
Nyarling River, 823

O

Oak Creek, 537
Obed River, 390
Obey River, 391, 392, 415
Ochlockonee River, 156
Ocmulgee River, 107, 121
Oconee River, 107, 121
Octonagon River, 1087
Ogden River, 662, 683, 693
Ogeechee River, 7, 10, 74, 76, 77, 93–99, 115, 1091, 1092
Ohio Creek, 517, 519
Ohio River, 1, 2, 83, 231, 237, 238, 245, 327, 329, 332, 334, 337, 360, 375–424, 1088–1092, 1096
Ohoopee River, 121
Okanagan River, 594, 624
Oklawaha River, 100, 101, 116
Oldman River, 859, 864–888, 895, 900
Old River, 240, 264, 277
Olentangy River, 408, 422
Ontonagon River, 984, 986, 1011, 1023
    East Branch, 1023
    Middle Branch, 1023
    South Branch, 1023
    West Branch, 1023
Osage River, 428, 431, 458
Ottawa River, 2, 984, 986, 991, 994, 996–1001, 1013, 1019, 1020
Otter Creek, 1027
Otter Tail River, 854, 855, 867, 869, 872, 873, 888, 896, 900
Ouachita River, 233, 235, 240, 265, 272, 279, 280, 317
Ouapetec River, 958
Outardes River, 939
Owens River, 657, 658

Owyhee River, 591, 593, 594, 609, 632, 643, 646, 647
    East Fork, 647
    North Fork, 647
    South Fork, 647

P

Paint River, 422
Paint Rock River, 384
Pajaro River, 561
Pamunkey River, 104, 117
Pánuco, Río, 1032, 1033, 1036, 1037–1043, 1045, 1065, 1075
Papaloapan, Río, 1032, 1036
Papigochic, Río, 1054, 1055, 1057, 1078
Paria River, 483, 489, 491, 495, 510
Parral, Río, 1059, 1079
Parsnip River, 824
Pascagoula River, 126, 129, 147, 149, 150, 159, 161, 176
Patsaliga Creek, 174
Patuxent River, 22
Pawnee River, 291
Paw Paw River, 1022
Payette River, 594, 609
Peace River, 431, 701, 805, 806, 808, 809, 811, 814, 819–821, 823–829, 832, 835, 842, 844, 845, 848, 1092, 1094, 1095
Pea River, 172
Pearl River, 125, 126, 128, 147–152, 171
Pecatonica River, 371
Pecos River, 181–183, 185, 186, 188, 189, 213, 215, 220, 225, 506, 1090
Pedernales River, 199
Peel River, 805, 806, 808, 809, 811, 837, 842, 851
Pékans River, 958, 959, 961, 963, 973
Pelican River, 901
Pelly River, 780, 784
Pembina River (AB), 833, 846
Pembina River (MB, ND), 868, 872, 896
Pemigewasset River, 69
Pend Orielle River, 624, 628
Penobscot River, 21, 22, 24, 25–30
    East Branch, 26
    West Branch, 26, 29
Péribonka River, 1002, 1005, 1021
Petitcodiac River, 940, 942, 965, 978
Petitot River, 816, 843
Piedro River, 518
Pierriche River, 1028
Pigeon River, 1022
Piscataquis River, 62
Pit River, 547–550, 567, 580
Platte River, 427–429, 431, 449–454, 469, 472, 1093
Pomme de Terre River, 339
Poplar River, 431
Porcupine River, 9, 10, 775, 776, 779–782, 791, 793, 796, 802
Portneuf River, 661, 663
Poteau River, 284, 285, 287–289, 309, 322
Potomac River, 21–24, 55–57, 1088
    North Branch, 71
    South Branch, 71
Powder River, 440, 441, 470
Powell River, 388, 390
Price River, 496, 498, 672
Prosser Creek, 680, 682
Provo River, 656, 657, 660, 683, 684, 692
Pudding River, 617, 645
Puerco River, 188
Purgatoire River, 291

Q

Qu'Appelle River, 897
Quesnel River, 698, 699, 702, 720, 730
Quinn River, 657

R

Rabbit (Bonanza) Creek, 778
Raccoon River, 372
Rainey River, 873, 874–878, 898
Rancheria River, 816
Rapid River, 876
Rappahannock River, 21
Raquette River, 986
Raritan River, 22, 55, 56, 70
    North Branch, 70
    South Branch, 70
Rat River, 854, 887
Red Cedar River, 368
Red Deer River, 862, 865, 867, 895
Red Lake River, 872, 873
Red River (KY), 405
Red River (OK, LA, TX, AR), 231, 235–237,
    240, 272, 277, 283–289, 294, 296, 298,
    299–304, 306, 309, 310, 317, 318, 325,
    1092
    North Fork, 299, 301
    Prairie Dog Town Fork, 299, 300, 301
    Salt Fork, 287, 290, 301
Red River (TN), 391, 415
Red River of the North, 853–855, 858, 859,
    867–874, 876, 877, 879, 888, 896, 897,
    899
Redstone River, 811
Redwater River, 431
Redwood River, 339
Reese River, 674, 675
Renous River, 971
Republican River, 460, 479
Restigouche River, 940, 957
Revuelto Creek, 296
Ribdon River, 920
Richelieu, Rivière, 984, 986, 1012, 1013,
    1019, 1027
Rio Grande, 2, 16, 181–192, 213, 220, 225,
    516, 775, 1090, 1095
Rivière des Rochers, 819, 820
Roanoke River, 74, 76, 77, 104, 106, 118
Robertson River, 786
Rockcastle (Rock Castle) River, 391, 394,
    396, 415
Rock Creek, 652
Rock River (IL, WI), 331, 338, 357–359, 363,
    371
Rock River (MN, IA), 477
Rocky River, 86
Rogue River, 542, 543, 545, 563, 567–573,
    584, 1089
    South Fork, 570
Root River, 328, 331, 358, 369
Roseau River, 868, 872, 896
Roubidoux River, 454
Rough River, 418
Russian River (CA), 542, 543, 573–575, 587
Russian River (AK), 748, 750

S

Sabinas, Río, 1065, 1082
Sabine River, 182, 183, 185, 208–213, 215,
    224, 228, 1090
    Caddo Fork, 208
    Cowleech Fork, 208, 210
    Lake Fork Creek, 208, 224
    South Fork, 208, 210
Sacandaga River, 37, 42
Saco River, 22
Sacramento River, 542, 543, 545, 547–552,
    555, 557, 567, 572, 580, 1089, 1099
Sagavanirktok River, 742, 904–908, 913, 916,
    918–921, 934
Sagehen Creek, 682
Saguenay, Rivière, 2, 983, 984, 986, 987, 989,
    994, 997, 1001–1005, 1013, 1019, 1021,
    1093
St. Croix River (ME, NB), 22

St. Croix River (MN, WI), 328, 330, 333,
    342, 343–347, 358, 363, 365, 1087
St. Francis River, 234, 235
St. Joe River, 594, 632, 650
St. John River, 26, 939–943, 947, 953–958,
    972
St. Johns River, 7, 74, 76, 99–104, 116, 1088
St. Joseph River (MI, IN), 984, 986,
    1005–1011, 1022
St. Joseph River (of Maumee River basin)
    (OH, IN), 1005, 1025
St. Lawrence River, 1, 2, 22, 192, 775, 939,
    941, 959, 983–1028, 1087–1089,
    1093–1096
Ste. Marguerite, Rivière, 963
St. Marys River (GA, FL), 76, 157
St. Mary's River (OH, IN), 1025
Saint-Maurice, Rivière, 984, 986, 1013, 1014,
    1019, 1028
Salado, Río, 182, 183, 185, 188, 220, 1032,
    1033, 1035, 1065, 1082
Salcha River, 787
Salinas River, 542, 543, 558–563, 582
Saline River (of Ouachita River basin) (AR),
    233, 235, 265, 272, 279, 280
    Alum Fork, 265
    Middle Fork, 265
    North Fork, 265
    South Fork, 265
Saline River (of Little River basin) (AR), 305,
    306, 318
Saline River (KS), 479
Salmon River (BC), 701
Salmon River (CA), 563, 568, 583
    North Fork, 568
    South Fork, 568
Salmon River (ID), 591, 593–595, 603–610,
    619, 643, 646
    Middle Fork, 609
Salt Creek, 451
Salt River (AZ), 484, 487, 511–513, 519–521,
    531, 537, 538
Salt River (KY), 376, 378, 379
Saluda River, 120
Samaria, Río, 1046, 1048
San Antonio River (CA), 558–560, 582
San Antonio River (TX), 182–185, 192–198,
    221
San Carlos River, 513, 515
Sandy River (ME), 67
Sandy River (OR), 616
San Francisco River, 513, 515, 531
Sangamon River, 352, 357, 367
San Jacinto River, 182, 184, 588
San Joaquin River, 542, 543, 545, 552–558,
    572, 581
San Juan, Río, 185, 188, 189, 1036
San Juan River, 484, 486, 487, 518, 527, 533
San Marcos River, 184, 192, 195–198, 221
San Miguel, Río, 1044, 1080
San Pedro, Río (Conchos basin) (Mexico),
    1059, 1061, 1062, 1079
San Pedro, Río (Usumacinta basin) (Mexico),
    1044, 1045, 1051, 1076
San Pedro River (AZ), 486, 514, 516, 531
San Pitch River, 667, 671
San Rafael River, 496, 498, 672
San Saba River, 222
San Simon River, 514
Santa Ana River, 542, 543, 574, 575, 588
Santa Cruz River, 486, 514, 531
Santa Fe River (FL), 152, 154, 155, 172
Santa Margarita River, 542, 543, 574, 576,
    589
Santa Maria, Río (Mexico), 1039, 1075
Santa Maria River (AZ), 519, 535
Santee River, 74, 76, 77, 105, 107, 120, 1089
Santiam River, 617, 619
Saranac River, 1027
Saskatchewan River, 593, 853–855, 858,
    860–867, 879, 895, 899, 1088

Sass River, 823
Satilla River, 9, 74, 76, 77, 107, 1088, 1092
Savannah River, 9, 10, 74, 76, 77, 88–93, 96,
    97, 114
Schuylkill River, 48, 65
Scioto River, 375, 376, 378–380, 396, 407,
    422
Scott Creek, 563, 565, 568, 583
Seal River, 860, 904, 905, 906, 926, 927, 936
Sebasticook River, 67
Selway River, 609, 632, 649
Seneca River, 89
Sepulga River, 174
Sequatchie River, 390
Sevenmile Creek, 394
Sevier River, 656, 657, 660, 667–673, 689
    East Fork, 667, 670, 671, 689
Shades Creek, 140
Shasta River, 563, 565, 583
Sheenjek River, 785, 793, 802
Shenandoah River, 71
Sheyenne River, 873, 896
Shoal River, 373
Silver Creek, 508
Silver Springs, 156
Sinaloa, Río, 1058, 1060
Sipsey River, 126, 129, 130, 134, 160, 162,
    177, 178
Sirupa, Río, 1054, 1055, 1057
Skeena River, 2, 736, 737, 739, 740, 751,
    754–758, 768
Skilak River, 766
Skuna River, 259
Slave River, 2, 431, 805, 806, 808, 809,
    819–824, 842, 844, 1095
Smoky Hill River, 560, 479
Smoky River, 806, 826, 827, 829, 835, 836,
    845, 848, 1092
Snake River (MN), 344, 345, 365
Snake River (WY, OR, ID, WA), 591,
    593–596, 603–610, 614, 619, 624, 628,
    632, 643, 646, 647, 660, 661, 663, 665
    Henrys Fork, 606, 608
    South Fork, 608
Snake River (YT), 837, 851
Snow River, 739, 748, 749, 766
Solomon River, 460, 479
Sonora, Río, 1056
Souris River, 868, 872, 873, 897
South Anna River, 117
South Canadian River, 283, 294, 295,
    298
South Nahanni River, 806, 809, 815, 818,
    834, 835, 843, 847
South Platte River, 427, 429, 449–451, 453,
    454, 472
South River, 83, 86, 113
South Santiam River, 645
South Saskatchewan River, 854, 859–888,
    895, 900
South Seal River, 927, 936
South Thompson River, 701, 707, 709–711,
    726
South Umpqua River, 573, 585
Southwest Miramichi River, 949, 950, 971
Spanish Fork River, Diamond Fork, 667
Spatsizi River, 753
Spokane River, 591, 593, 594, 624, 632, 633,
    650
Spoon River, 352, 367
Sprague River, 563, 565
Spring River, 320
Stanislaus River, 553, 555, 581
Stellako River, 714
Stewart River, 776, 778–780, 786, 791, 796,
    800
Stikine River, 2, 736, 737, 740, 751–757, 767
Stillwater River, 407, 420
Stony River, 742, 764
Strawberry River, 501
Strong River, 147, 171
Stuart River, 697, 698, 701, 712–714,
    716–720, 727, 728

Suchiapa, Río, 1044
Sulfur River, 301
Sunflower River, 259, 260, 276
Sun River, 431
Susan River, 657
Susitna River, 2, 736, 743–746, 749, 765
  East Fork, 744
  West Fork, 744
Susquehanna River, 21–24, 49–54, 57, 66,
    954, 1089, 1097
  North Branch, 50–52
  West Branch, 50, 53, 66
Sustut River, 757
Suwannee River, 126, 128, 129, 143,
    152–157, 172
Swan River, 597, 599, 600, 642
Sweetwater River, 451
Swift River, 742
Swuak River, 611
Sycamore Creek, 516, 519
Sycan River, 563, 565, 568
Sydney River, 956

T

Tahltan River, 753
Taku River, 736–739, 754, 757, 760, 772
Talkeetna River, 744, 745, 765
Tallahaga Creek, 147
Tallahatchie River, 259, 260, 276
Tallapoosa River, 130, 131, 135, 146, 168
Tamesí, Río, 1032, 1033, 1036–1038, 1041,
    1042, 1065, 1066, 1081
Tampaón, Río, 1038, 1040, 1043, 1075
Tamuin, Río, 1037, 1039–1041, 1043, 1075
Tanana River, 775, 776, 779, 780, 782,
    786–791, 793, 796, 797
Tar River, 76
Taseko River, 731
Tatshenshini River, 738, 759, 760, 771
Taylor River, 517
Teanaway River, 612
Telos Canal, 26
Temecula Creek, 576, 589
Témiscouata–Madwaska River, 954
Tempoal, Río, 1040, 1075
Tennessee River, 2, 128–130, 133, 159, 375,
    376, 378, 379, 381, 384–390, 404, 413,
    414, 1088, 1091, 1095
Tensas River, 234, 242, 265
Tensaw River Delta, 125, 131
Teslin River, 780
Thelon River, 904–906, 927, 937, 1089
Thirty Mile River, 785
Thompson River (BC), 698, 700, 702,
    707–712, 725, 726
Thompson River (MO, IA), 480
Tinayguk River, 791
Tippecanoe River, 396, 400, 416
Toad River, 815, 816
Tobique River, 954, 972
Tolovana River, 787
Tombigbee River, 126, 130, 131, 133, 159,
    160, 168, 177, 178, 384
Tomichi Creek, 517
Tongue River, 443, 470
Tonto Creek, 511
Toolik River, 933
Topila, Río, 1037, 1038, 1041, 1075
Toronto, Río, 1059
Toutle River, 635, 653
Trenche River, 1028
Trinity River (CA), 563, 565, 567, 568, 574,
    583

South Fork, 565
Trinity River (TX), 182, 183, 184, 185, 209,
    214, 215, 227, 304
  Clear Fork, 215
  West Fork, 215
Trout River, 815
Truckee River, 656, 657, 660, 678–682, 691
Tualatin River, 617, 645
Tugaloo River, 89
Tula, Río, 1039, 1043, 1075
Tule River, 553, 555, 581
Tuolumne River, 553, 555, 557, 581
Tuya River, 753, 767
Tygart Valley River, 408, 424

U

Umatilla River, 595
Umpqua River, 9, 10, 542, 543, 545, 571,
    573, 574, 585
Uncompahgre River, 517, 532
Upper Mississippi River, 2, 6, 231, 233, 237,
    239, 327–373, 1089–1091, 1093,
    1096
Urique, Río, 1080
Usumacinta, Río, 2, 1032, 1033, 1036, 1041,
    1043–1049, 1052, 1076, 1089, 1094

V

Valles, Río, 1040
Van Duzen River, 586
Verde, Río (of Río Fuerte basin) (Mexico),
    1080
Verde, Río (of Río Pánuco basin) (Mexico),
    1039, 1075
Verde River, 484, 487, 506, 511–513, 515,
    516, 519–521, 531, 537, 538
Verdigris River, 287, 288, 290, 291
Vermillion River (IL), 352, 367
Vermillion River (SD), 431
Victoria River, 947, 970
Virgin River, 9, 10, 484, 487, 534, 1087

W

Wabasca River, 828, 845
Wabash Canal, 397
Wabash River, 375, 376, 378, 379, 381,
    396–401, 413, 416
Wabigoon River, 878, 898
Wacouno River, 973
Walker River, 656, 657, 660, 684, 694
Wallawa River, 648
Walnut River, 291
Wapiti River, 806, 808, 827, 829, 835, 836,
    845, 848, 1092, 1093
Wapsipinicon River, 328, 358, 359, 370
War Eagle Creek, 255
Washita River, 283–285, 288, 289, 301, 303,
    304, 309, 310, 317, 323
Watauga River, 384
Wateree River, 105
Watonwan River, 342
Webb Creek, 963
Weber River, 656, 657, 660–662, 683, 684,
    693
Weiser River, 594
Wekiva River, 101
Wenatchee River, 594, 624
West Clear Creek, 519
Western Arm Brook, 946

Westfield River, 32, 63
West Pearl River, 148
West Road (Blackwater) River, 698, 699, 701,
    702, 720, 725, 729
West Walker River, 684, 694
Wet Beaver Creek, 519
Whetstone River, 339
White Clay Creek, 49
White River (AK, YT), 776, 779, 780, 782,
    791, 792, 796, 799
White River (AZ), 511, 519, 520, 536
White River (CO, UT), 496, 498, 505, 528
White River (IN), 396, 398, 400, 401, 416
White River (MO, AR), 233–235, 239, 242,
    245–251, 253–255, 272–274, 278,
    290
  North Fork, 247
White River (SD), 427, 428, 431, 445–449,
    469, 471
White River (TX), 203
White River (VT), 32, 63
Whitewater River, 420
Wichita River, 301
Willamette River, 591, 593, 615–622, 625,
    627, 645, 1100
  Coast Fork, 617
  Middle Fork, 617, 645
Williams Creek, 879
Williamson River, 563, 565
Willow River, 698
Windingo River, 1028
Wind River (AK), 785
Wind River (WY), 429
Winnipeg River, 853–855, 858–860, 868, 869,
    873–879, 898
Winooski River, 1027
Wisconsin River, 328, 330, 333, 347–352,
    358, 363, 366
Withlacoochee River, 152–154, 157, 172
Wood River (BC), 830
Wood River (OR), 563, 565
Wooley Creek, 568

Y

Yakima River, 591, 593–595, 610–615, 624,
    629, 644, 1100
Yalobusha River, 259, 260, 261, 276
Yamhill River, 617
Yampa River, 484, 486, 487, 493, 496,
    498–506, 527–529
Yaqui, Río, 1032, 1033, 1035, 1036,
    1054–1058, 1078
Yazoo River, 233–235, 239, 259–264, 272,
    276
Yellowknife River, 806, 836, 837, 850
Yellow Medicine River, 339
Yellow River (WI), 366
Yellowstone River, 427–429, 431, 435,
    440–445, 469, 470, 666
Yentna River, 744, 745, 765
Yockanookanay River, 147
Yocona River, 259, 260
York River, 74, 76, 104, 117
Youghiogheny River, 424
Yuba River, 547, 549, 580
Yukon River, 1, 2, 6, 192, 701, 741–743, 759,
    775–802, 903, 1088, 1089, 1091

Z

Zuni River, 510